2001

The ARRL Handbook

For Radio Amateurs

Editor
Chuck Hutchinson, K8CH

Contributing Editors
Joel Kleinman, N1BKE
R. Dean Straw, N6BV
Larry Wolfgang, WR1B

Technical Consultants
Ed Hare, W1RFI
Zack Lau, W1VT

Production
Michelle Bloom, WB1ENT
Sue Fagan, Cover
Jodi Morin, KA1JPA
David Pingree, N1NAS
Michael Daniels
Paul Lappen
Joe Shea

Proofreader
Jayne Pratt-Lovelace

Additional Contributors to the 2001 Edition:
Dennis Bodson, W4PWF
Joe Carcia, NJ1Q
Steven Karty, N5SK
Tom O'Hara, W6ORG
William Sabin, WØIYH
Doug Smith, KF6DX
Ken Stuart, W3VVN
Bob Wolbert, K6XX

The Cover

Top: The antenna relay box for NJ1Q's Automatic Antenna Switch is perched on a W1AW tower leg. The inset photo shows the control unit. Read all about it in Chapter 22. *Photos courtesy Joe Bottiglieri, AA1GW.*

Bottom left: If you're looking for power-handling capacity, you've found the right tuner. N6BV's design works with both balanced and unbalanced antenna systems. See Chapter 22.

Bottom right: WA2EBY's all-band linear amplifier uses low-cost MOSFETs. Details are in Chapter 17.

Seventy-Eighth Edition

Published by:
ARRL—the national association for Amateur Radio
Newington, CT 06111 USA

Contents

Chapter tabs

1 11 21
2 12 22
3 13 23
4 14 24
5 15 25
6 16 26
7 17 27
8 18 28
9 19 29
10 20 30

Foreword

This 2001 edition of *The ARRL Handbook* continues the tradition begun by ARRL staff in 1926 of publishing essential and useful technical information for radio amateurs. This volume represents the work of many people, some of whom are listed on the title page. The authors and editors have put together informative theory discussions as well as practical projects. Further, they have updated the References chapter to make the book even more valuable to all hams (regardless of their areas of interest), technicians and engineers.

This year's *Handbook* includes a new chapter on DSP by *QEX* editor Doug Smith, KF6DX. Doug explains how this technology works, what it will do and where you might find it used in Amateur Radio equipment.

Bob Wolbert, K6XX, wrote a new section on computer hardware for Chapter 7. With computers being used increasingly in our installations, most of us should find this material very helpful. In addition, you'll find pin connections for most common computer connectors in the References chapter.

W1AW Manager Joe Carcia, NJ1Q, presents an automatic remote antenna switch in Chapter 22. The automatic function works for Yaesu or ICOM radios, and manual control will work with any equipment you might have.

In addition, there have been many updates throughout this book. In fact, most chapters have been changed to provide you with the most up-to-date reference material available.

This is the 78th edition of *The ARRL Handbook*, and we'd like to think it's the best ever. You can help us make future editions even more useful by providing your comments and suggestions on the Feedback Form at the back of the book.

David Sumner, K1ZZ
Executive Vice President
Newington, Connecticut
September 2000

The Amateur's Code

The Radio Amateur is:

CONSIDERATE...never knowingly operates in such a way as to lessen the pleasure of others.

LOYAL...offers loyalty, encouragement and support to other amateurs, local clubs, and the American Radio Relay League, through which Amateur Radio in the United States is represented nationally and internationally.

PROGRESSIVE...with knowledge abreast of science, a well-built and efficient station and operation above reproach.

FRIENDLY...slow and patient operating when requested; friendly advice and counsel to the beginner; kindly assistance, cooperation and consideration for the interests of others. These are the hallmarks of the amateur spirit.

BALANCED...radio is an avocation, never interfering with duties owed to family, job, school or community.

PATRIOTIC...station and skill always ready for service to country and community.

—The original Amateur's Code was written by Paul M. Segal, W9EEA, in 1928.

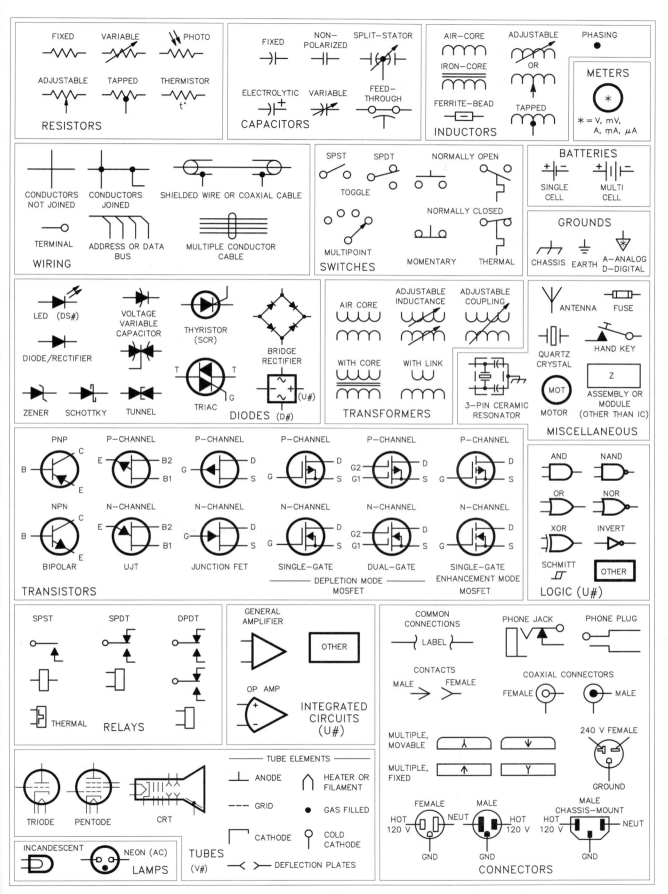

The ARRL—At Your Service

ARRL Headquarters is open from 8 AM to 5 PM Eastern Time, Monday through Friday, except holidays. Call **toll free** to join the ARRL or order ARRL products: **1-888-277-5289** (US), M-F only, 8 AM to 8 PM Eastern Time.

If you have a question, try one of these Headquarters departments . . .

	Telephone	Electronic Mail
Joining ARRL	860-594-0338	circulation@arrl.org
QST **Delivery**	860-594-0338	circulation@arrl.org
Publication Orders	860-594-0355	pubsales@arrl.org
Regulatory Info	860-594-0236	reginfo@arrl.org
Exams	860-594-0300	vec@arrl.org
Educational Materials	860-594-0301	ead@arrl.org
Contests	860-594-0232	n1nd@arrl.org
Technical Questions	860-594-0214	tis@arrl.org
Awards	860-594-0288	awards@arrl.org
DXCC/VUCC	860-594-0234	dxcc@arrl.org
Advertising	860-594-0207	ads@arrl.org
Media Relations	860-594-0328	newsmedia@arrl.org
QSL Service	860-594-0274	buro@arrl.org
Scholarships	860-594-0230	foundation@arrl.org
Emergency Comm	860-594-0265	wv1x@arrl.org
Clubs	860-594-0267	clubs@arrl.org
Hamfests	860-594-0262	hamfests@arrl.org

You can send e-mail to any ARRL Headquarters employee if you know his or her name or call sign. The second half of every Headquarters e-mail address is **@arrl.org**. To create the first half, simply use the person's call sign. If you don't know their call sign, use the first letter of their first name, followed by their complete last name. For example, to send a message to John Hennessee, N1KB, Regulatory Information Specialist, you could address it to **jhennessee@arrl.org** or **N1KB@arrl.org**.

If all else fails, send e-mail to **hq@arrl.org** and it will be routed to the right people or departments.

Technical Information Server

If you have Internet e-mail capability, you can tap into the ARRL Technical Information Server, otherwise known as the *Info Server*. To have user instructions and a handy index sent to you automatically, simply address an e-mail message to: **info@arrl.org**

Subject: **Info Request**
In the body of your message enter:
HELP
SEND INDEX
QUIT

ARRL ON THE WORLD WIDE WEB

You'll also find *ARRLWeb* at: **http://www.arrl.org/**
At the ARRL Web page you'll find the latest W1AW bulletins, a hamfest calendar, exam schedules, an on-line ARRL Publications Catalog and much more. We're always adding new features to *ARRLWeb*, so check it often!

Members-Only Web Site

As an ARRL member you enjoy exclusive access to our Members-Only Web site. Just point your browser to **http://www.arrl.org/members/** and you'll open the door to benefits that you won't find anywhere else.
• Our on-line Web magazine, the *ARRLWeb Extra* with colorful news and features you *won't* see in *QST*.
• *QST* Product Review Archive. Get copies of *QST* product reviews from 1980 to the present.
• *QST/QEX* searchable index (find that article you were looking for!)
• Previews of contest results and product reviews. See them here before they appear in *QST!*
• Access to your information in the ARRL membership database. Enter corrections or updates on line!

Stopping by for a visit?

We offer tours of Headquarters and W1AW at 9, 10 and 11 AM, and at 1, 2 and 3 PM, Monday to Friday (except holidays). Special tour times may be arranged in advance. Bring your license and you can operate W1AW anytime between 10 AM and noon, and 1 to 3:45 PM!

Would you like to write for *QST*?

We're always looking for new material of interest to hams. Send a self-addressed, stamped envelope (2 units of postage) and ask for a copy of the *Author's Guide*. (It's also available via the ARRL Info Server, and via *ARRLWeb* at **http://www.arrl.org/qst/aguide/**)

Press Releases and New Products/Books

Send your press releases and new book announcements to the attention of the *QST* Editor (e-mail **qst@arrl.org**). New product announcements should be sent to the Product Review Editor (e-mail **reviews@arrl.org**).

Strays and Up Front

Send your Strays and Up Front materials to the *QST* Features Editor (e-mail **upfront@arrl.org**). Be sure to include your name, address and daytime telephone number.

ARRL Audio News

The best way to keep up with fast-moving events in the ham community is to listen to the ARRL Audio News. It's as close as your telephone at 860-594-0384, or on the Web at **http://www.arrl.org/arrlletter/audio/**

Interested in Becoming a Ham?

Just pick up the telephone and call toll free 1-800-326-3942, or send e-mail to **newham@arrl.org**. We'll provide helpful advice on obtaining your Amateur Radio license, and we'll be happy to send you our informative Prospective Ham Package.

Handbook Software

The 1996 Edition was the first to include bundled software. The software proved to be a valuable addition for many readers. On the other hand, bundling the disk with the printed book increased its cost. A readership survey indicated clearly that keeping the cost of the book as low as possible was more important than the bundled disk.

For this reason, we will again be making the software for the 2001 Edition available separately.

DOWNLOADING HANDBOOK SOFTWARE

You can download *Handbook* software from the Product Notes section of *ARRLWeb*: **http://www.arrl.org/notes/1867**. Simply click on **Download hbk2001.exe**.

A description of the software appears on the last page of this book.

What is Amateur Radio?

1

People who pursue the hobby of using a personal radio station to communicate, purely for noncommercial purposes, with other radio hobbyists call it *ham radio* or *Amateur Radio*. They call themselves Amateur Radio operators, ham radio operators or just plain "hams."

You already know a little about the hobby—hams communicate with other hams, around the block, on a distant continent—or from an orbiting space station! Some talk via computers, others prefer to use regular voice communications and still others enjoy using one of the oldest forms of radio communication—Morse code. Some hams help save people's lives by handling emergency communications following a natural disaster or other emergency. Some become close friends with the people they talk to on the other side of the globe—then make it a point to meet one or more of them in person. Some can take a bag full of electrical parts and turn it into a station accessory that improves their station's reception of distant radio signals.

This chapter, by Rosalie White, K1STO, covers the basics—what hams do, and how they do it.

A
Hams are always willing to help others who are excited about becoming ham radio operators. You'll find more information in this chapter about how to locate ham radio operators in your local area.

B
Computers are an integral part of ham radio, and the Internet plays a large role. Rich Roznoy, K1OF and Ed Ashway, K3EIN, use their PCs to hold a video conference. Audio transmission was by way of their VHF transmitters. *(Photo courtesy of Rich Roznoy, K1OF)*

C
Back to the past. Al Brogdon, W1AB, operates a reconstruction of pioneer station W1BCG on the 75th anniversary of the first two-way short-wave message across the Atlantic. *(Photo courtesy of Paul Danzer, N1II)*

D
Ham radio, sun and fun. Peter Venlet, N8YEL, enjoys hilltop operating. Lightweight, portable rigs and small batteries give you many opportunities to pick your operating spot.

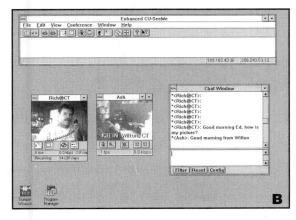

HOBBY OF DIVERSITIES

You can't imagine all of the unusual, interesting things you can do as an Amateur Radio operator. What types of people will you meet as a ham? If you walk down a city street, you'll pass men and women, girls and boys, and people of all ages, ethnic backgrounds and physical abilities. They're office workers and students, nurses and mail carriers, engineers and truck drivers, housewives and bankers. Any of them might be a ham you will meet tonight on your radio.

If you drive your car on the interstate this weekend, you'll see people on their way to a state park, a Scout camp, a convention, an airport or a computer show. The young couple going to the park to hike for the day have their hand-held ham radio transceivers in their backpacks. When they stop on a scenic hilltop for a rest, they'll pull out their radios and see how far away they can communicate with the radio's 3 watts of power. And, the radios will be handy just in case they break down on the road or lose the hiking trail.

The father and son on their way to Scout camp will soon be canoeing with their Scout troop. After setting up camp, they'll get out a portable radio, throw a wire antenna over a branch, and get on the air. Aside from the enjoyment of talking with other hams from their campsite, their radios give them the security of having reliable communications with the outside world, in case of emergency.

The family driving to the ham radio convention will spend the day talking with their ham friends, including two they've never met but know quite well from talking to them on the air every week. They will also look at new and used radio equipment, listen to a speaker talk about the latest ways computers can be used to operate on the Amateur Radio bands, and enjoy a banquet talk by a NASA astronaut who is also a ham radio operator.

The couple on the way to the airport to take a pleasure flight in their small plane have packed their hand-held radios in their flight bags. Once they're airborne, they'll contact hams on the ground all along their flight path. Up at 5,000 feet, they can receive and transmit over much greater distances than they can from the ground. Those they contact will enjoy the novelty of talking to hams in a plane. The radios are an ideal means of backup communications, too.

The two friends on their way to the computer show are discussing the best interface to use between their computers and their radios for a mode of operating called packet radio. They're looking forward to seeing a number of their ham friends who are also into computers.

What other exciting things can you look forward to on the ham bands? You might catch yourself excitedly calling (along with 50 other hams) a Russian cosmonaut in space or a sailor on the Coast Guard's tall ship *Eagle.* You could be linked via packet radio with an Alaskan sled-dog driver, a rock star, a US legislator, a major league baseball player, a ham operating the Amateur Radio station aboard the ocean liner *Queen Mary,* an active-duty soldier, a king—or someone who is building the same power supply that you are from a design in this *ARRL Handbook.*

On the other hand, a relaxing evening at home could find you in a friendly radio conversation with a ham in Frankfort, Kentucky, or Frankfurt, Germany. Unlike any other hobby, Amateur Radio knows no country boundaries and brings the world together as good friends.

Although talking with astronauts isn't exactly an everyday ham radio occurrence, more and more hams are doing just that, as many NASA astronauts are ham radio operators.

THE TECHNICIAN LICENSE: THE SHORTEST PATH TO AMATEUR RADIO FUN!

When you're ready to start, the Technician license presents an excellent way for beginners to start enjoying the fun and excitement of Amateur Radio. The only requirement for this license is that you pass a single 35-question written exam. The exam covers FCC rules and regulations that govern the airways, courteous operating procedures and techniques, and some basic electronics. There is no Morse code requirement for a Technician license.

Technician frequency privileges begin at 50 MHz and extend through the very high frequency (VHF) and ultra high frequency (UHF) ranges, and into the microwave region. All these frequency bands

A ham's operating area is called *the shack.* **It may be a corner of a room, a basement area or in this case part of a former battleship. The Mobile (AL) Amateur Radio Club installed a temporary station, W4IAX, on this imposing structure.**

| **All Types of Physical Abilities** | People who don't get around as much as they'd like to, find the world of Amateur Radio a rewarding place to |

make friends — around the block or around the globe. Many hams with and without certain physical abilities belong to the HANDI-HAM System, an international organization of radio amateurs who bring ham radio to all individuals. HANDI-HAM members live in every state and many countries around the world, and are ready to help in whatever manner they can. The HANDI-HAM System provides study materials and aids for persons with physical disabilities. Local HANDI-HAMs will assist you with studies at home. Once you receive your license, the Courage HANDI-HAM System may lend you basic radio equipment to get you started on the air.

give Technicians plenty of room to explore. They aren't restricted to only certain operating modes, either. They can use any communications methods allowed to hams.

Most new hams will operate first on the popular 2-meter band. With plenty of *repeaters* across the country, the FM voice signals from their low-power hand-held and mobile radios reach many other hams. Technicians also communicate through *satellites* and *packet radio* networks. They use *single-sideband (SSB)* voice and *Morse code (CW)*.

Technician licensees can gain operating privileges on the amateur high frequency (HF) bands by passing a 5-wpm Morse code test. Passing another 35-question written exam completes the upgrade to a General class license. Generals enjoy worldwide communications using SSB and CW as well as *slow-scan television (SSTV)* and a variety of *digital communications* modes.

One more written exam, this one containing 50 questions, will take you to the top of the Amateur Radio license ladder—the Amateur Extra class license. Amateur Extra class licensees enjoy full amateur privileges on all bands. The exam may be challenging, but many hams find it to be well worth the effort!

You can never tell what you get into. The Redondo Beach (CA) Amateur Radio Club wanted to put one end of an antenna near the top of this 100-foot high tree. No one volunteered to climb, so the local fire department was talked into taking part in the antenna raising party.

WHAT'S IN A CALL SIGN?

When you earn your Amateur Radio license, you receive a unique *call sign*. Many hams are known by their call signs (and not necessarily by their names!). All US hams get a call sign, a set of letters and numbers, assigned to them by the Federal Communications Commission (FCC). No one else "owns" your call sign—it's unique. Your Amateur Radio license with your unique call sign gives you permission to operate your Amateur Radio station on the air. US call signs begin with W, K, N or A, with some combination of letters and numbers that follow. In addition, the number in the middle of the call sign indicates the location within the country. In the US, for example, call signs that have a 9 in them indicate that the ham lived in the Midwest when the license was issued, and call signs with the number 1 indicate New England. You can tell what country issued a ham's call sign by the *prefix*—the letters before the number.

HAM RADIO ACTION

Amateur Radio and public service go together. On a warm early-summer weekend, hams can be found directing radio communications in the aftermath of a train derailment—a simulated one, that is—to help prepare for a real emergency. Others provide radio communications for a walk-a-thon. Still others hone their communications skills by setting up a station

Danny, KD4HQV, likes to operate using Morse code (CW). Here he is using a homemade set of code paddles, while his father drives. *(Photo courtesy of AC4HF)*

outdoors, away from electrical power. This largest public-service-related ham activity is called *Field Day*.

Biking and Cruising

Hams even operate their radios while riding on bicycle treks. They easily carry their lightweight hand-held radios in their packs, and can pull them out quickly, if needed. Or if they're really serious they pull along a small trailer with a sleeping bag, food and water supplies and a ham radio transceiver that they can set up in the evenings.

Ham radio manufacturers have begun offering hams adventures on ships that sail in the picturesque Caribbean. They set up radio stations for you to operate at various exotic ports on these expeditions. You get to operate the radios, relax luxuriously on the open seas, meet new friends from all over the globe and see new places. Although not all trips to exotic locales to operate ham radio stations, which hams call *DXpeditions*, are this glamorous, many hams enjoy activating a rare country to provide contacts for their fellow hams around the world.

Nets: Scheduled Get-Togethers

If you'd like to find other hams with vocational or avocational interests like yours (such as chess, gardening, rock climbing, railroads, computer programming or teaching), you'll soon learn about *nets*. A net forms when hams with similar interests get together on the air on a regular schedule. You can find your special interest—from the Armenian Amateur Radio and Traffic Net to the Zone 9 ARES Net—listed in *The ARRL Net Directory* published annually.

Awards and Contests: Competitive Fun

If you're competitive by nature, you'll want to explore ham radio awards and contests. These activities recognize your ability to contact other hams under published guidelines. In the ARRL DX Contest, for example, you'll try to contact as many DX (foreign) stations as possible over a weekend. Experienced hams with top-notch stations easily contact more than 100 different countries during a single DX Contest weekend!

Awards you can earn include *Worked All States*, earned by communicating with a ham in every US state, *Worked All VE*, earned by contacting hams in every Canadian province and the *DX Century Club*, for working stations in 100 or more different countries.

In an outdoor orienteering competition, "fox-hunters" (also called "bunny-hunters") track and locate hidden transmitters

This is a fully equipped station. John, WA2WVR, likes to operate all modes—code, voice and digital data. His shack sports several Morse code keys, a boom microphone and a computer.

by car or on foot. This activity also has its serious side: Skills learned in tracking the "fox" come in handy when there's a suspected pirate (unlicensed station) in the neighborhood.

QRP: Talk to the World with 5 Watts of Power

For a real challenge, try operating *QRP* — using low-power. Some enjoy operating with only 1 watt, or even less. It's certainly a challenge, but with decent antennas and skillful operating, QRP enthusiasts can be heard around the world. One of the best reasons for operating QRP is that equipment is lightweight, inexpensive and easy to build. Some hams use nothing but "home-brew" equipment.

Unusual Modes

If computers are your favorite aspect of today's technology, you'll soon discover that you can connect your computer and ham radio equipment and operate on such digital modes as *packet radio* and *PSK31*. With packet, you can leave messages that other packet enthusiasts will pick up and answer later. One popular packet activity is the *DX PacketCluster*, which allows hams to get real-time information about where and when rare foreign stations can be found on the bands. Another is satellites: A series of packet radio satellites provides long-distance computer-based communications with low power. PSK31 is a popular digital mode used on the amateur high-frequency (HF) bands.

Computers also can help you practice taking ham radio license examinations or improve your Morse code abilities. ARRL, and others, offer software especially designed to help you pass ham radio exams. As you become more experienced, you'll discover software for any number of ham radio applications, from keeping track of

---

Support for Expeditions	The first ham radio operator to cooperate in an expedition was League member

Don Mix, 1TS. With his radio equipment in tow, he accompanied Donald B. MacMillan to the Arctic on the schooner *Bowdoin* in 1923. In subsequent years, hams assisted with perhaps 200 other voyages and expeditions.

In the fall of 1999, Amateur Radio was part of the Border-to-Border Expedition Society's trek. Participants retraced the routes of the hearty Stampeders who searched the Klondike Gold Fields in the 1890s. Ham radio operators and others journeyed across Alaska and the Northwest Territories in Land Rover vehicles; the hams kept daily radio schedules, sharing their real-time adventures with school students and other hams. Chief radio operator, Jim Wilmerding, W2EMT, used the special events radio call sign K2A (Klondike to Alaska) just for this Trans-American trek.

The gang's all here. Who said ham radio is a solo activity? Standing are KC9XT, N9LVL, KB9ATR, WZ9M, N9IOX and N9LBT. Seated are Chris Kratzer, W9XD and KB9GRP. (Photo courtesy of Mike McCauley, KB9GNU)

have found some ingenious ways of extending the distance and improving the quality of the signals they transmit. High-frequency (HF) radio waves can be refracted or bent by a layer of the atmosphere called the ionosphere. In this way, signals are returned to Earth, often after several "hops." This *ionospheric propagation* of radio signals allows worldwide communication on the HF bands. Hams have also learned how to bounce signals off the moon, airplanes and even meteor trails! Repeaters, located on hilltops or on tall buildings, strengthen signals and transmit them much farther than would be possible without repeaters.

Phyllisan West, KA4FZI, has been teaching her middle-school students all about ham radio for years. That's Matt Clark, KC4QXJ, on the straight key.

the stations you've contacted during a contest to designing the best antenna for your location. Hams also use software to download pictures transmitted by weather satellites.

Enhancing Radio Signals

Radio signals normally travel in straight lines, which limits their range. But hams

The art of *homebrewing*, or building your own, is alive and well. This small receiver contains two transistors and one integrated circuit. It can be assembled without any special tools in one or two evenings. The thrill of listening to a station on a radio you built yourself lasts a lot longer than that! See Chapter 17 for a selection of receivers and transceivers you can build yourself.

Helping out in Emergencies

When commercial communications services are disrupted by power failures or damage that accompanies natural disasters such as earthquakes, floods and hurricanes, Amateur Radio operators are often first at the scene. Battery-powered equipment allows hams to provide essential communications even when power is knocked out. If need be, hams can make and install antennas "on the spot" from whatever materials and supports they find available.

Working with emergency personnel such as police and fire departments, the Red Cross and medical personnel, ham volunteers provide any communications necessary. Hams can handle communications between agencies whose normal radios are incompatible with one another, for example. The ability of radio amateurs to help the public in emergencies is one of the reasons Amateur Radio has survived and prospered since the early days of the 20th century.

Community Events

To keep their emergency-preparedness skills honed, and to help their community, hams enjoy assisting with communications to aid the public at any number of events and activities. Hams volunteer to provide communications for walk-a-thons, bike races, parades and other community events. In fact, it's rare to see a large community event that doesn't make use of public-spirited ham radio operators.

Build It Yourself

Another favorite activity hams enjoy is building their own radio equipment. Hams proudly stay at the forefront of technology, continually being challenged to keep up with advances that could be applied to

Hams get involved. A hiking accident in the Sierra Mountains of California led to a successful rescue effort through the efforts of KF6BEC, KE6YXE, KB6BSS, KD6ZOD and the local Sheriff's Department.

OSCARs: The Ham Satellites

You can experience the thrill of hearing your own signal returned from space by an orbiting "repeater in the sky" — a ham radio satellite. Or marvel at the clear pictures of the Earth you receive from the camera aboard another ham satellite. Hams regularly use Amateur Radio satellites, called OSCARs (for *Orbiting Satellites Carrying Amateur Radio*). VHF and UHF signals from a ham radio transceiver normally don't travel much beyond the horizon. But if you route your signal through an orbiting satellite, you can make global radio contacts on VHF and UHF.

In 1990, a series of small Amateur Radio satellites, called Microsats, were launched. One, called WEBERSAT, transmits image data that you can process into pictures using your personal computer and special software. Another is a packet radio satellite that allows messages from Earth to be stored and forwarded back down to Earth when the spacecraft is within range of the designated station.

Hams are experimenting with low-orbit global communication satellites that can store their real-time radio messages for delivery at a later time. This would allow hams to communicate with other hams in developing countries that can't afford expensive channels on geostationary satellites. Hams from around the globe are working together to design, build and launch an exciting new satellite that is the most ambitious amateur satellite project ever conceived. It will use the latest electronics technology to provide thousands of hams with reliable and pleasurable worldwide communications on several radio frequencies up through the microwave bands.

Does all of this sound futuristic or beyond your skills? It shouldn't. All it takes is a Technician license to enjoy this exciting ham radio technology.

The St Xavier High School Amateur Radio Club assembled for this photo just before a big weekend contest. Their enthusiastic efforts led them to a high score for their type of station.

If you had a radio contact with the crew of the space shuttle *Columbia*, mission STS-83 (renamed STS-94), you could have requested this picture QSL card. Crewmembers included three ham operators: Commander Jim Halsell, KC5RNL; Mission Specialist Janice Voss, KC5BTK; and Mission Specialist Don Thomas, KC5FVF. Many astronauts are ham radio operators who look forward to talking with hams back on Earth. They will be even more interested in doing so once they become crewmembers with long-duration stints aboard the International Space Station.

Most communication with space, both manned shuttles and unmanned satellites, uses VHF, UHF or microwave frequencies. This homebrewed antenna was designed to be used for a new satellite called *Phase 3D*.

the hobby. Many have an incessant curiosity and an eagerness to try new techniques. They also are constantly driven to find ways to allow the radio frequency bands to support more users, since some portions of certain bands are very popular and can be crowded.

The projects you'll find in this book provide a wide variety of equipment and accessories that make ham radio more convenient and enjoyable. Many manufacturers provide parts kits and etched circuit boards to make building even easier.

Hams in Space

In 1983, the first ham/astronaut made

history by communicating with ground-based hams from the space shuttle *Columbia*. On that mission, Payload Specialist Owen Garriott, whose Amateur Radio call sign is W5LFL, took along a hand-held amateur transceiver and placed a specially designed antenna in an orbiter window. It was the first time ham radio operators throughout the world were to experience the thrill of working an astronaut aboard an orbiting spacecraft. In 1985, Mission Specialist Tony England, WØORE, transmitted slow-scan television (SSTV) via Amateur Radio while orbiting the Earth from the shuttle *Challenger*. He named the payload *SAREX*, for Shuttle *Amateur Radio EXperiment*. This name has since been changed to *Space Amateur Radio EXperiment* because astronaut hams will be even more active from the International Space Station when it is in use.

It wasn't long before NASA was routinely scheduling SAREX missions. In 1991, each of the five members of a shuttle crew had earned an Amateur Radio license. NASA promotes ham radio activity aboard shuttle spacecraft because of its proven public relations and educational value. It's also a reliable means of backup communication. During a recent year, five shuttle crews requested that NASA include a

This is the ham radio equivalent of walking and chewing gum at the same time! Russell, KC5RWR, operates a *packet radio* station using the computer keyboard, while talking to another station on the hand-held radio. This station, KC5NTO, was operated by a group of Explorer Scouts.

SAREX payload on their flights. As for the future, hams are making plans for a ham radio station aboard a manned space station.

Plans have been under way for quite some time to have Amateur Radio on the International Space Station, in a program known as ARISS. While crew members are on board the spacecraft, they will have some leisure time. The ham astronauts will use some of that time to talk to hams, their families and school students. The space station requires long-duration stints in space, unlike space shuttle missions, so many astronauts have elected to earn their Amateur Radio licenses now.

GETTING STARTED

Now that you have an idea of what hams do, you're probably asking, "Okay, how do I get started?" The first step is to earn a license.

Most people start with the Technician license, which requires that you pass a 35-question written exam. The exams are given by local ham Volunteer Examiners (VEs). Many clubs sponsor exam sessions on a regular basis, so you shouldn't have to travel far to take your exam. They are often given on weekends or evenings. The exam questions are taken from a large pool of questions for each license class. The complete question pools for each license class are published in study guides and can even be found on the Internet. (See the URL for *ARRLWeb* in the Resources section at the end of this chapter.)

Study Guides

You can prepare for the exams on your own or in a class. Help is available at every step. The ARRL publishes complete study materials for all classes of Amateur Radio licenses. Contact ARRL's New Ham Desk (the address and phone number are at the end of this chapter) for a free package of information. It contains everything you need to get started: a list of nearby Amateur Radio clubs, Amateur Radio instructors who have registered with ARRL, and local volunteer examiners. The package also includes a description of the frequencies hams can use, and the most popular operating activities. It also has information about the latest versions of ARRL study guides.

ARRL's book for beginners, *Now You're Talking!*, includes the complete, up-to-date question pool with the correct answers, as well as clear explanations. You'll also find tips for how to choose your equipment and put together your ham radio station, how to build and install simple, inexpensive antennas and much more.

Now You're Talking! assumes no prior electronics background. Children as young as five years old have passed ham radio exams!

If you already have some electronics background or just want brief explanations to help you understand the correct answers, then *ARRL's Tech Q & A* may be just what you need for your Technician exam preparation. Every question in the

Today's nonsmokers have made the ashtray superfluous. N1II took his out and mounted a small transceiver directly in the dashboard. This mobile station takes up no room in the passenger area. *(Photo courtesy of Paul Danzer, N1II)*

Technician question pool is included, to help you prepare for that exam. When you are ready to upgrade to a higher class license, then *The ARRL General Class License Manual* and *The ARRL Extra Class License Manual* will guide your study efforts. ARRL's *Your Introduction to Morse Code* will teach you the Morse code and prepare you for the 5-wpm exam.

Elmers

Many ham radio operators learn how to get on the air through the buddy system. An experienced ham, called an "Elmer," teaches one or more newcomers about Amateur Radio on a one-to-one basis. Elmering became a ham tradition many years ago. It was first documented in ARRL's monthly magazine, *QST*, in a story that was meant to be a public thank you from an appreciative

There were no computers, satellites or TV when Clarice, W7FTX, was first licensed in 1935. One thing has not changed—the friendship of her fellow hams.

student whose mentor's name was Elmer. Elmers are there for you as you study for your exam, buy your first radio and set up your station. Many watch with pride as their newly licensed friends make their first on-the-air contacts.

Putting Together a Station

As with any other hobby, you can have fun with ham radio no matter how small (or large) your budget. You can start with a hand-held transceiver that fits in your pocket or purse, and take it along when you hike, canoe or aviate. Or you can fill your "radio shack" with the latest and fanciest radios technology offers and money can buy, and talk to people in all corners of the world. You can build a simple, inexpensive wire antenna to string between two trees in your backyard, or install giant towers, with arrays of phased beam antennas on top.

Accessories and equipment for your ham radio station come in all price ranges. *QST* contains display advertisements for new ham gear plus columns of classified ads for previously owned items. ARRL Field & Educational Services can provide you with information about where to find ham radio equipment.

Used Versus New

Hams are continually upgrading their stations, so you can always find a ready supply of good previously owned Amateur Radio gear. You can find new hand-held transceivers and used HF radios for less than $300. Many hams start with a radio that costs between $300 and $600. Antennas and other gear can add appreciably to the cost, but less-expensive alternatives, such as putting together your own antenna or low-power transceiver, are available.

HAMS AS WORLD CITIZENS

When you become an Amateur Radio operator, you become a "world citizen"—you join a group of people who have earned the privilege of talking to other hams around the corner or around the world. Hams have a long tradition of spreading international goodwill. One way hams do this is to assist with getting needed medical advice or medicine to developing countries. Another is by learning about the lives and cultures of those they contact. On the other hand, it's a good idea to avoid sensitive political or ethical issues.

Although English is the standard language on the ham bands, English-speakers will make a good impression on hams in foreign countries if they can speak a few words of the other person's language—even if it's as simple as *danke* or *sayonara*.

International Amateur Radio

Hams in other countries have formed national organizations, just as US hams organized the ARRL—the national association for Amateur Radio. These sister societies work together to have a united voice in international radio affairs, such as when governments get together to decide how radio frequencies will be divided among its various users. The International Amateur Radio Union (IARU), composed of about 150 national Amateur Radio societies, works to advance the cause of Amateur Radio at the international level. ARRL works closely with other IARU societies to help protect the amateur frequencies.

THE ADMINISTRATORS: ITU AND FCC

The laws of physics allow for a limited spectrum of radio frequencies. These radio frequencies must be shared by many competing radio services: broadcasters, land mobile, aeronautical and marine, to name a few.

The International Telecommunication Union (ITU), an agency of the United Nations, allocates these frequencies among the many services that use them. With its long tradition of public service and techno-

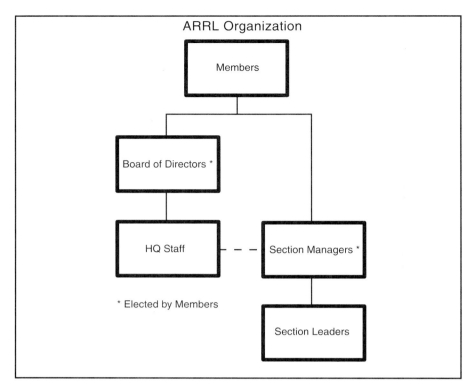

Fig 1 — Full members of the ARRL control the organization through an elected Board of Directors. Volunteers (all positions except for HQ staff) accomplish much of the work.

W1AW, the station operated by the ARRL in Newington, Connecticut, is known around the world as the home of ham radio. W1AW memorializes Hiram Percy Maxim, one of the founders of the ARRL. Visitors are welcome and you might even be able to operate. *(Photo courtesy of W2ABE)*

Many hams enjoy talking to far away stations, and gather a good knowledge of geography in the process. Here Bill, KB7JAH, and William, KB7JAG, chat with a ham in Kiribati. Don't know where Kiribati is? (Hint: Parts of the island are the first in the world to greet each new day.)

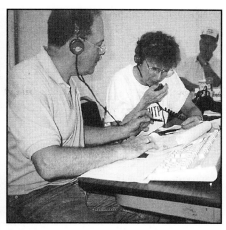

All-ham families are not unusual. Madison, AB5TV, was the first in his family to become a ham. His wife, Millie, KC5UTP, was not far behind. Here Madison proudly watches Millie make her first-ever contact.

This ham family extends across four generations. Only Andy, K4RA, and great-granddaughter Lisa, KB3BPU, are shown. Not in the picture are Lisa's grandfather Andy, Jr., KE3GY, and her father, Joe, NV3T.

logical savvy, ham radio enjoys the use of many different frequency bands.

In the US, a government agency, the FCC, regulates the radio services, including Amateur Radio. The section of the FCC Rules that deals with Amateur Radio is Part 97. Hams are expected to know the important sections of Part 97, as serious violations (such as causing malicious interference or operating without the appropriate license) can lead to fines and even imprisonment! Aside from writing and enforcing the rules governing Amateur Radio, the FCC also assigns call signs and issues licenses to those who have earned them.

THE ARRL

Since it was founded, in 1914, the ARRL — the national association for Amateur Radio has grown and evolved along with Amateur Radio. The ARRL Headquarters building and Maxim Memorial Station, W1AW, are in Newington, Connecticut, near Hartford. Through its network of dedicated volunteers and a professional staff, the ARRL promotes the advancement of the amateur service in the US and around the world.

The ARRL operates as a nonprofit, educational and scientific organization dedicated to the promotion and protection of the many privileges that ham radio operators enjoy today. Of, by and for the radio amateur, ARRL numbers within its ranks the vast majority of active amateurs in North America. Around 170,000 licensed

ham radio operators and unlicensed persons with an interest in ham radio are members. Licensed hams join as Full Members, while unlicensed persons become Associate Members who have all membership privileges except for voting in ARRL elections. Anyone who has an interest in Amateur Radio has a place in the ARRL.

The ARRL volunteer corps is called the Field Organization. Working at the state and local level, these volunteers carry out the work of ARRL to further Amateur Radio. They organize emergency communications in times of disaster and work with agencies such as the Red Cross, the National Weather Service and Civil Air Patrol. Other field volunteers keep state and local government officials abreast of the good that ham volunteers are doing at the state and local level.

Membership Services

When you join ARRL, you add your voice to those who are most involved with ham radio. The most prominent benefit of ARRL membership is *QST*, the premiere Amateur Radio magazine. *QST* has Amateur Radio news you'll want to know and need to hear. You'll also find a wide range of articles, columns and features—projects to build, announcements of upcoming ham radio activities, reviews of the latest equipment, reports on the role hams are playing in emergencies, and much more.

But being an ARRL member is far more than a subscription to *QST*. The

ARRL represents your interests to the FCC and Congress, sponsors contests and other operating events, and offers membership services at a personal level. These include:

- the QSL bureau (which lets you exchange postcards with hams in foreign countries as a confirmation of your contacts with them)
- the volunteer examiner program
- the Technical Information Service (which provides answers to your questions about any technical subject in Amateur Radio)
- low-cost equipment insurance and much more.

School Teachers and Volunteer Instructors

ARRL Field & Educational Services (F&ES) provides teachers with materials for using Amateur Radio in their schools. Thousands of teachers have found that Amateur Radio is an ideal way to provide hands-on, intercurricular learning, while enticing students to become interested in science and technology. F&ES also has materials, including newsletters and instructor guides, to help hams who wish to teach Amateur Radio licensing classes.

WELCOME!

For answers to any questions you may have about Amateur Radio, write or call ARRL Headquarters. See Resources, below, for contact information.

Glossary

Note: Words in **boldface italics** have separate entries in the Glossary.

Amateur Radio—A radiocommunication service for the purpose of self-training, intercommunication and technical investigations carried out by amateurs, that is, duly authorized persons interested in radio technique solely with a personal aim and without pecuniary interest. (*Pecuniary* means payment of any type, whether money or other goods.) Also called *ham radio*.

Amateur Radio operator — A person holding a license to operate an amateur station.

Amateur Radio station — A station licensed in the amateur service, including necessary equipment, used for amateur communication.

Amateur service — Another name for *Amateur Radio*; one of the radiocommunication services regulated in the US by the Federal Communications Commission (*FCC*).

Amateur television (ATV) — A mode of operation that Amateur Radio operators can use to exchange pictures from their radio stations.

AMSAT — An abbreviation for the *Radio Amateur Satellite Corporation.*

ARRL — The membership organization for Amateur Radio operators in the US.

Band — A range of frequencies. Hams are authorized to transmit on several different bands.

Beam antenna — A type of ham radio antenna that can be pointed in different directions.

Bunny hunt — Another name for *fox hunt.*

Call sign — A series of unique letters and numbers assigned to a person who has earned an Amateur Radio license.

Contact — A two-way communication between Amateur Radio operators.

Contest — An Amateur Radio activity in which hams and their stations compete against others to try to contact the most stations within the designated time period.

Courage HANDI-HAM System — A membership organization for Amateur Radio enthusiasts with various physical abilities.

CW — Abbreviation for *continuous wave;* another name for *Morse code* telegraphy.

Digital communications — Computer-based communications modes, such as *packet radio.*

Digital signal processing (DSP) — A recently developed technology that allows software to replace electronic circuitry.

Dipole antenna — A popular type of wire antenna often used on the high-frequency amateur bands.

DX — A ham radio abbreviation for *distance* or *foreign countries.*

DXCC — A popular ARRL award earned for contacting Amateur Radio operators in 100 different countries.

DX PacketCluster — A method of informing hams, via their computers, about the activities of stations operating from unusual locations.

DXpedition — A trip to an unusual location, such as an uninhabited island, where hams operate for a designated period of time. DXpeditions provide sought-after contacts for hams who are anxious to contact that rare location.

Elmer — A traditional term for someone who enjoys helping newcomers get started in ham radio on a one-to-one basis.

Emergency communications — Amateur Radio communications that take place during a situation where there is danger to lives or property.

Fast-scan television (FSTV) — A mode of operation that Amateur Radio operators can use to exchange live TV images from their stations.

Federal Communications Commission (FCC) — The government agency that regulates Amateur Radio in the US.

Field & Educational Services (F&ES) — Staff at ARRL Headquarters that helps newcomers get started in ham radio and provides materials to hams who want to help newcomers.

Field Day — A popular Amateur Radio activity during which hams set up radio stations outdoors and away from electrical service to simulate emergency conditions.

Field Organization — A cadre of ARRL volunteers who perform various services for the Amateur Radio community at the local level.

FM (frequency modulation) — An operating *mode* commonly used on ham radio *repeaters*.

Fox hunt — A competitive ham radio activity in which ham radio operators track down a transmitted signal. Also called *bunny hunt*.

Ham band — A range of frequencies on which amateur communications are authorized.

Ham radio — Another name for *Amateur Radio*.

Ham radio operator — A person holding a written authorization to operate an amateur station. An *Amateur Radio operator*.

High frequencies (HF) — The radio frequencies from 3 to 30 MHz.

International Amateur Radio Union (IARU) — The international organization made up of national Amateur Radio organizations such as the ARRL.

International Telecommunication Union (ITU) — An agency of the United Nations that allocates the radio spectrum among the various radio services.

Microsat — A series of small Amateur Radio satellites.

Mode — A type of ham radio communication; examples are *frequency modulation (FM), slow-scan television (SSTV)* and *packet radio.*

Morse code — A communications mode transmitted by on/off keying of a radio-frequency signal. Hams use the *international Morse code*, which differs from American (telegraph) Morse.

Net — An on-the-air meeting of Amateur Radio operators at a particular time, day and radio frequency.

OSCAR — An acronym for *Orbiting Satellite Carrying Amateur Radio*, a series of Amateur Radio satellites designed and built by the international ham radio community. Also see *AMSAT*.

Packet radio — A computer-to-computer communications mode in which information is broken into short bursts. The bursts (packets) also contain addressing and error-detection information.

Payload — A package taken onboard a space flight, such as the *Space Amateur Radio EXperiment (SAREX)*, which allows astronaut/hams to communicate with other hams from space.

Public service — Activities involving Amateur Radio that hams perform to benefit their communities.

QRP — An abbreviation for low power.

QSL bureau — A system of forwarding *QSL cards* to and from ham radio operators.

QSL cards — Postcards that serve as a confirmation of communication between two hams.

QST — The premiere Amateur Radio monthly magazine, published by the *ARRL*. QST means "calling all radio amateurs."

Radio Amateur Satellite Corporation (AMSAT) — An international membership organization that designs, builds and promotes the use of Amateur Radio satellites.

Radio frequencies (RF) — The range of frequencies that can travel through space in the form of electromagnetic radiation.

Radio shack — The room where Amateur

Radio operators keep their station.

Radiotelegraphy — See *Morse code*.

Receiver — A device that converts radio signals into a form that can be heard.

Repeater — An amateur station, usually located on a mountaintop, hilltop or tall building, that receives a signal and re-transmits it for greater range.

SAREX — An abbreviation for *Space Amateur Radio EXperiment.*

Shortwave listener (SWL) — A person who enjoys listening to radio broadcasts or Amateur Radio conversations.

Single sideband (SSB) — A common *mode* of voice operation on the amateur bands.

Slow-scan television (SSTV) — A *mode* of operation in which Amateur Radio op-erators exchange still pictures from their radio stations.

Space *Amateur Radio EX*periment — A payload of Amateur Radio equipment flown in space and operated by astro-nauts who are licensed Amateur Radio operators.

Technical Information Service — A ser-vice of the *ARRL* that helps hams solve technical problems.

Transceiver — A radio transmitter and receiver combined in one unit.

Transmitter — A device that produces radio-frequency signals.

Ultra-high frequencies (UHF) — The radio frequencies from 300 to 3000 MHz.

Very-high frequencies (VHF) — The radio frequencies from 30 to 300 MHz.

Volunteer Examiners (VEs) — Amateur Radio operators who give Amateur Radio licensing examinations.

Wavelength — A means of designating a frequency *band,* such as the 80-meter band.

Worked All States (WAS)—An *ARRL* award that is earned when an Amateur Radio operator talks to a ham in each of the 50 states in the US.

Worked All VE (WAVE)—An award that is earned when an Amateur Radio opera-tor talks to a ham in each of the Canadian provinces.

Work — To contact another ham.

RESOURCES

ARRL — the national association for Amateur Radio
225 Main St
Newington, CT 06111-1494
860-594-0200
Fax: 860-594-0259
e-mail: hq@arrl.org
Prospective hams call 1-800-32 NEW HAM (1-800-326-3942)
ARRLWeb: **http://www.arrl.org**

Membership organization of US ham radio operators and those interested in ham radio. Publishes study guides for all Amateur Radio license classes, a monthly journal, *QST*, and many books on Amateur Radio and electronics.

AMSAT NA (The Radio Amateur Satellite Corporation, Inc)
PO Box 27
Washington, DC 20044
301-589-6062

Membership organization for those interested in Amateur Radio satellites. Publishes *The AMSAT Journal*, monthly.

Courage HANDI-HAM System
3915 Golden Valley Rd
Golden Valley, MN 55422
763-520-0511

Provides assistance to persons with disabilities who want to earn a ham radio license or set up a station.

Now You're Talking! All You Need for Your First Amateur Radio License, Fourth Edition (Newington, CT: ARRL)

Complete introduction to Amateur Radio, including the exam question pool, complete explanations of the subjects covered on the exams. Tips on buying equipment, setting up a station and more.

The ARRL's Tech Q & A, Second Edition (Newington, CT: ARRL)

All the questions on the Technician exam, with the correct answers highlighted and explained in plain English. With many helpful diagrams.

Your Introduction to Morse Code (Newington, CT: ARRL)

A set of audio CDs (or cassette tapes) that make learning Morse code fun. Teaches all letters, numbers, and other required characters, and provides practice text.

Morse Tutor Gold

Software for IBM PCs and compatibles that teaches the code and provides plenty of practice at user-selected speeds from 1 to 100 words per minute.

Contents

Activities
2

One of the best things about this hobby we call Amateur Radio is its *flexibility*. In other words, Amateur Radio can be whatever *you* want it to be. Whether you are looking for relaxation, excitement, or a way to stretch your mental (and physical) horizons, Amateur Radio can provide it. This chapter was written by Larry Kollar, KC4WZK. Let's take a brief tour through the following topic areas:

Awards—the individual and competitive pursuits that make up the tradition we call "paper chasing."

Contests—the challenge of on-the-air competition.

Nets—both traffic nets, where amateurs pass messages on behalf of hams and nonhams, and the casual nets, where groups of people with common interests often meet on the air to swap equipment, anecdotes and information.

Ragchewing—meeting new friends on the air.

Amateur Radio Education—Educating current and future hams brings in new blood (and revitalizes old blood!); educating our neighbors about ham radio is good for public relations and awareness.

ARRL Field Organization—Amateur Radio in general, and the ARRL in particular, depend on the volunteer spirit. As part of the Field Organization, you can exercise your administrative, speaking and diplomatic skills in service of the amateur community.

Emergency Communications—When disaster strikes, hams often have the only reliable means to communicate with the outside world. Practice and preparation are key to fulfilling this mission.

DF (Direction Finding)—If you've ever wanted to know where a transmitter (hidden or otherwise) is located, you'll find DFing is an enjoyable and useful skill.

Satellite Operation—You may be surprised to learn that hams have their own communications satellites! Satellite operation can be great fun and a technical challenge for those who want to operate on the "final frontier."

Repeaters—Using and operating repeaters is one of the most popular activities for both new and old hams.

Image Communications—Although it's fun to talk to other amateurs, it's even more fun to *see* them.

Digital Communications—Use your computer to communicate with stations around your town or around the world.

VHF, UHF and Microwave Weak-Signal Operating—Explore the challenging, quirky and surprising world above 50 MHz.

EME (Earth-Moon-Earth), Meteor Scatter and Aurora—Making contacts by bouncing your signals off the moon, the fiery trails of meteors and auroras.

AWARDS

Winning awards, or "paper chasing,"

One of the most prized awards in Amateur Radio: the DX Century Club.

is a time-honored amateur tradition. For those who enjoy individual pursuits or friendly competition, the ARRL and other organizations offer awards ranging from the coveted to the humorous.

DX Awards

The two most popular DX awards are DXCC (DX Century Club), sponsored by the ARRL and WAC (Worked All Continents), sponsored by the International Amateur Radio Union (IARU). The WAC award is quite simple: all you have to do is work one station on each of six continents. The DXCC is more challenging: you must work at least one station in each of 100 countries! QSL cards from each continent or country are required as proof of contact.

How-to's of DXCC—Direct QSLs and DX Bureaus

Since DX stations are often inundated with QSL cards (and QSL requests) from US hams, it is financially impossible for most of them to pay for the return postage. Hams have hit upon several ways to lighten the load on popular DX stations.

The fastest, but most expensive, way to get QSL cards is the *direct* approach. You send your QSL card, with one or two International Reply Coupons—IRCs—(or one or two dollars) and a self-addressed airmail envelope to the DX station. International Reply Coupons are available from your local post office and can be used nearly anywhere in the world for return postage. Some DX hams prefer that you send one or two "green stamps" (dollar bills) because they can be used to defray posting, printing and other expenses. However, it is illegal in some countries to possess foreign currency. If you're not sure, ask the DX station or check DX bulletins available on packet radio and BBSs.

Many DX hams have recruited *QSL managers*, hams who handle the QSL chores of one or more DX stations. QSL managers are convenient for everyone. The DX station need only send batches of blank cards and a copy of the logs; hams wanting that station's card need only send a First Class stamp for US return postage and can expect a prompt reply. (In the case of QSL managers located outside the United States, you must still send IRCs [or dollars] and a self-addressed return envelope.)

The easiest (and slowest) way to send and receive large batches of QSL cards is through the incoming and outgoing QSL bureaus. The outgoing bureau is available to ARRL members. The incoming bureaus are available to all amateurs. Bureau instructions and addresses are printed periodically in *QST;* they appear in the *ARRL Operating Manual,* and they are available from ARRL Headquarters for an SASE.

DXpeditions

What does the avid DXer who has worked them all (or almost all of them) do for an encore? Answer: *become* the DX! DXpeditions journey to countries with few or no hams, often making thousands of contacts in the space of a few days.

In 1991, Albania opened its borders and legalized Amateur Radio for the first time in many years. To train the first new generation of Albanian hams and to relieve the pileups that were sure to happen, a contingent of European and American hams organized a DXpedition to Albania. The DXpedition made over 10,000 contacts and changed Albania from one of the rarest and most-desired countries to an "easy one."

DX Nets

The beginning DXer can get a good jump on DXCC by frequenting DX nets. On DX nets, a net control station keeps track of which DX stations have checked into the net. He or she then allows a small group of operators (usually 10) to check in and work one of the DX stations. This permits weaker stations to be heard instead of being buried in a pileup. Since the net control station does not tolerate net members making contacts out-of-turn, beginning operators have a better chance of snagging a new country. Nets and frequencies on which they operate vary. For the latest information on DX nets, check with local DXers and DX bulletins.

Efficient DX Operation

The best DXers will tell you the best equipment you have is "the equipment be-

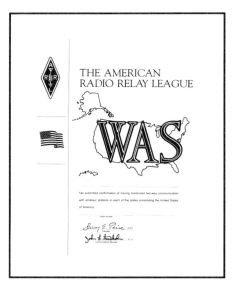

Work one station in each of the 50 states and you're eligible for the ARRL's Worked All States (WAS) award.

tween your ears." Good operators can make contacts with modest power. The details of efficient DX operating cannot be covered in such a brief space. The ARRL sells two publications that are excellent references for new DX enthusiasts: *The Complete DXer* and *The DXCC Companion.*

WAS (Worked All States)

The WAS certificate is awarded to amateurs who have QSL cards from at least one operator in each of the 50 United States. Chasing WAS is often a casual affair, although there are also nets dedicated to operators who are looking for particular states.

Endorsements

The initial DXCC or WAS award does not mean the end. There are over 300 DXCC countries. As you reach certain levels in your country count, you qualify for endorsements. Endorsements arrive in the form of stickers that you attach to your DXCC certificate.

Both WAS and DXCC offer endorsements for single-band or single-mode operation. For example, if you work all 50 United States on the 15-m band, your certificate has an endorsement for 15 m. The most difficult endorsement is the 5-band (5B) endorsement. Rare indeed is the operator who can display a 5BDXCC certificate!

Other Awards

The ARRL and other organizations offer a variety of awards for both serious and fun achievements. You can qualify for some awards, like the ARRL's RCC

(Rag Chewers' Club, for long contacts), on your very first contact! Other awards are sponsored by local clubs.

The ARRL Friendship Award is available to any ARRL member who can prove contact with 26 stations whose call signs end with each of the 26 letters of the alphabet. (For example, N1MZ*A,* KØOR*B,* W3AB*C...*K1Z*Z.)* Any frequency or mode qualifies. Most consider the Friendship Award to be the next step up from the Rag Chewers' Club. It's more difficult to earn, but just as much fun.

One of the most coveted awards is the A-1 Operator Club. Qualifying for this award is as simple as cultivating spotless operating habits and always operating by the Amateur's Code (found near the front of this *Handbook*). To receive this award, you must be recommended by two A-1 members.

CONTESTS

Some people enjoy the thrill of competition, and Amateur Radio provides challenges at all levels in the form of operating contests. Besides the competitive outlet, contests have provided many hams with a means to hone their operating skills under less-than-optimum conditions. On the VHF and higher bands, contests are one way to stimulate activity on little-used segments of the amateur spectrum.

This section briefly discusses a few ARRL-sponsored contests. The Contest Corral section of *QST* provides up-to-date information on these and other contests. The ARRL also publishes the *National Contest Journal* (*NCJ*), which is good reading for any serious (or semi-serious) contester.

Field Day

Every year on the fourth full weekend in

Elaine Larson, KD6DUT, takes a turn at logging as Fred Martin, KI6YN, works the paddles during the Conejo Valley Amateur Radio Club's Field Day operation.

June, thousands of hams take to the hills, forests, campsites and parking lots to participate in Field Day. The object of Field Day is not only to make contacts, but to make contacts under conditions that simulate the aftermath of a disaster. Most stations are set up outdoors and use emergency power sources.

Many clubs and individuals have built elaborate Field Day equipment, and that is all to the best—if a real disaster were to strike, those stations could be set up quickly, wherever needed, and need not depend on potentially unreliable commercial power!

Other Contests

Other popular contests include:

QSO Parties. These are fairly relaxed contests—good for beginners. There are many state QSO parties, and others for special interests, such as the QRP ARCI Spring QSO Party.

Sweepstakes. This is a high-energy contest that brings thousands of operators out of the woodwork each year.

Various DX contests. DX contests offer good opportunities for amateurs to pursue their DXCC award contacts. A good operator can work over 100 countries in a weekend!

VHF, UHF and microwave contests. These contests are designed to stimulate activity on the weak-signal portions of our highest-frequency bands. The ARRL VHF/UHF contests are held during the spring and fall. There is also a contest for 10-GHz operators, and another one for EME (moonbounce) enthusiasts.

Each issue of *QST* lists the contests to be held during the next two months.

NETS

A net is simply a group of hams who meet on a particular frequency at a particular time. Nets come in three classes: public service, traffic and special interest.

Public Service/Traffic Nets

Public service and traffic nets are part of a tradition that dates back almost to the dawn of Amateur Radio. The ARRL, in fact, was formed to coordinate and promote the formation of traffic nets. In those early days, nets were needed to communicate over distances longer than a few miles. (Thus the word "Relay" in "American Radio Relay League.")

Public service and traffic nets benefit hams and nonhams alike. Any noncommercial message—birthday and holiday greetings, personal information or a friendly hello—may be sent anywhere in the US and to foreign countries that have third-party agreements with the United

---

Keeping a Logbook

At one time, keeping a log of your contacts was an FCC requirement. The FCC has dropped this requirement in recent years, but many amateurs, both new and old, still keep logs.

Why Keep a Log?

If keeping a log is optional, why do it? Some of the more important reasons for keeping a log include:

Legal protection—If you can show a complete log of your activity, it can help you deal with interference complaints. Good recordkeeping can help you protect yourself if you are ever accused of intentional interference, or have a problem with unauthorized use of your call sign.

Awards tracking—A log helps you keep track of contacts required for DXCC, WAS, or other awards. Keeping a log lets you quickly see how well you are progressing toward your goal.

An operating diary—A logbook is a good place for recording general information about your station. You may be able to tell just how well that new antenna is working compared to the old one by comparing recent QSOs with older contacts. The logbook is also a logical place to record new acquisitions (complete with serial numbers in case your gear is ever stolen). You can also record other events, such as the names and calls of visiting operators, license upgrades, or contests, in your log.

Paper and Computer Logs

Many hams, even those with computers, choose to keep their logs on paper. Paper logs still offer several advantages (such as flexibility) and do not require power. Paper logs also survive hard-drive crashes!

Preprinted logsheets are available, or you can create your own. Computers with word processing and publishing software let you create customized logsheets in no time.

On the other hand, computer logs offer many advantages to the serious contester or DXer. For example, the computer can search a log and instantly tell you whether you need a particular station for DXCC. Contesters use computer logs in place of *dupe sheets* to weed out duplicate contacts before they happen, saving valuable time. Computer logs can also tell you at a glance how far along you are toward certain awards.

Computer logging programs are available from commercial vendors. Some programs may be available as shareware (you can download it from a BBS and pay for the program if you like the way it works). If you can program your computer, you can also create your own custom logging program—and then give it to your friends or even sell it!

States. Many missionaries in South America, for example, keep in touch with stateside families and sponsors via Amateur Radio.

The ARRL National Traffic System (NTS) oversees many of the existing traffic nets. Most nets are local or regional. They use many modes, from slow-speed CW nets in the Novice HF bands, to FM repeater nets on 2 m.

Since the amateur packet-radio network now covers much of the US and the world, many messages travel over packet links. Amateurs use the packet radio network not only for personal or third-party traffic, but for lively conferences, discussions and for trading equipment.

HF and Repeater Nets

HF nets usually cover a region, although some span the entire country. This has obvious advantages for amateurs sending traffic over long distances. Repeater nets usually cover only a local area, but some linked repeater nets can cover several states.

Both types of nets work together to speed traffic to its destination. For example, think of the HF nets as a "trunk" or highway that carries traffic quickly and reliably toward its *approximate* destination. From there, the local and regional nets take over and pass the traffic directly to the city or town. Finally, a local amateur delivers the message to the recipient.

Routine traffic handling keeps the National Traffic System (NTS) prepared for emergencies. In the wake of Hurricane Andrew in 1992, hams carried thousands of messages in and out of the stricken south Florida region. The work that hams do during crisis situations ensures good relations with neighbors and local governments.

Other Nets

Many nets exist for hams with common

R. C. C.

This is to Certify that

is a member-station of the

Rag Chewers' Club

and is entitled to all the privileges, prerogatives, rights, favors, glory, rank, fame, notoriety, popularity and honor of membership in that worthy organization.

◆

In accepting this certificate the member agrees to abide by the published rules of the organization.

◆

QST

The Old Sock
Chief Rag Chewer

The Rag Chewers' Club award is one of the easiest to earn. You probably qualify already!

interests inside and outside of Amateur Radio. Some examples include computers, owners of Collins radio equipment, religious groups and scattered friends and families. Most nets meet on the 80- and 20-m phone bands, where propagation is fairly predictable and there are no short-wave broadcast stations to dodge.

RAGCHEWING

Ragchewing is the fine art of the long contact. Old friends often get together on the air to catch up on current events. Family members use ham radio to keep in touch. And, of course, new acquaintances get to know each other!

In many cases, friends scattered across the country get together to create ragchewing nets. These nets are very informal and may not make much sense to the outsider listening in. The "serious" ragchewer's shack decoration is not complete without the Rag Chewers' Club (RCC) certificate. The RCC, the Friendship Award and other certificates are discussed earlier in this chapter.

AMATEUR RADIO EDUCATION

Elmering (helping new and prospective operators) is a traditional amateur activity. Much of an amateurs' educational efforts go toward licensing (original and upgrading), but there are other opportunities for education, including public relations.

Dave Hanson, KBØEVM (r), is congratulated by Volunteer Examiner (VE) Len Buonaiuto, KE2LE, after passing his exam. (*photo courtesy of KE2LE*)

Dry run just before the shuttle pass. Keilah Meuser is practicing with others looking on.

License Classes

Anyone can set up license classes. Many Amateur Radio clubs hold periodic classes, usually for the Novice and Technician elements with CW practice sessions. The ARRL supports Registered Amateur Radio Instructors, but registration is not necessary to conduct a class.

If you are looking for a class to attend, and do not have an "Elmer" to answer your questions, write the ARRL Educational Activities Department for a list and schedule of classes in your area. If you want to become an instructor, you can request the same list of classes from the Educational Activities Department—most classes will welcome another helping hand.

Volunteer Examiners (VEs)

To become a VE, you must hold a General or higher amateur license and be certified by one of the VE Coordinators (VECs). The ARRL supports the largest VE program in the nation; other organizations run VE programs on a national or regional basis. General and Advanced licensees on a VE team must be supervised by at least one Extra Class licensee.

School Presentations

Amateur Radio complements any school program. Schoolchildren suddenly find that Amateur Radio gives them a chance to apply their studies immediately. The math and science used in Amateur Radio applies equally to the classroom. Even geography takes on a new meaning when a student works a new country!

Unfortunately, many schools do not have an active Amateur Radio presence—and that is why local volunteers are important. An HF or satellite station, or even a 2-m hand-held transceiver tuned to the local repeater, can prove an exciting and educational experience for both the volunteer and the students.

Thanks to NASA's SAREX (*Shuttle Amateur Radio Experiment*) program, amateurs all over the nation have put school-children in direct contact with shuttle astronauts. Who knows how many future scientists received their inspiration while sitting behind an amateur's microphone?

ARRL FIELD ORGANIZATION

ARRL members elect the Board of Directors and the Section Managers. Each Section Manager appoints volunteers to posts that promote Amateur Radio within that Section. (The United States is divided into 15 ARRL *Divisions*. These Divisions are further broken down into 69 *Sections*.) A few of the posts include:

Assistant Section Managers—ASMs are appointed as necessary by the SM to assist the SM in responding to membership needs within the Section.

Official Observers (OO) / Amateur Auxiliary—Official Observers are authorized by the FCC to monitor the amateur bands for rules discrepancies or viola-tions. The Amateur Auxiliary is administered by Section Managers and OO Coordinators, with support from ARRL Headquarters.

Technical Coordinators (TC) and *Technical Specialists (TS)*—Technical Coordinators and Technical Specialists assist hams with technical questions and interference problems. They also represent the ARRL at technical symposiums, serve on cable TV advisory committees and advise municipal governments on technical matters.

EMERGENCY COMMUNICATIONS

The FCC Rules list emergency communications as one of the purposes of the Amateur Radio Service—and in reality, the ability to provide emergency commu-

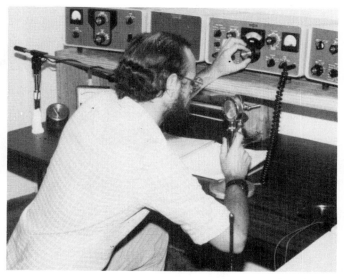

Chuck, NI5I (left), and Rick, WB5TJV, were two of the many hams who provided communications in the wake of a devastating earthquake in Mexico City. (*photos courtesy of WB5TJV*)

nications justifies Amateur Radio's existence. The FCC has recognized Amateur Radio as being among the most reliable means of medium- and long-distance communication in disaster areas.

Amateur Radio operators have a long tradition of operating from backup power sources. Through events such as Field Day, hams have cultivated the ability to set up communication posts wherever they are needed. Moreover, Amateur Radio can provide computer networks (with over-the-air links where needed) and provide other services such as video (ATV) and store-and-forward satellite links that no other service can deploy on a wide scale. One can argue, therefore, that widespread technology makes Amateur Radio even more crucial in a disaster situation.

If you are interested in participating in this important public service, you should contact your local EC (Emergency Coordinator). Plan to participate in preparedness nets and a yearly SET (Simulated Emergency Test).

ARES AND RACES

The Amateur Radio Emergency Service (ARES) and the Radio Amateur Civil Emergency Service (RACES) are the umbrella organizations of Amateur Radio emergency communications. The ARES is sponsored by ARRL (although ARRL membership is not required for ARES participation) and handles many different kinds of public-service activities. On the other hand, RACES is administered by the Federal Emergency Management Agency (FEMA) and operates only for civil preparedness and in times of civil emergency. RACES is activated at the request of a state or federal official.

Amateurs serious about emergency communication should carry dual RACES/ARES membership. RACES rules now make it possible for ARES and RACES to use the same frequencies, so that an ARES group also enrolled in RACES can work in either organization as required by the situation.

MILITARY AFFILIATE RADIO SERVICE (MARS)

MARS is administered by the US armed forces, and exists for the purpose of transmitting communications between those serving in the armed forces and their families. This service has existed in one form or another since 1925.

There are three branches of MARS: Army MARS, Navy/Marine Corps MARS and Air Force MARS. Each branch has its own requirements for membership, although all three branches require members to hold a valid US Amateur Radio license and to be 18 years of age or older (amateurs from 14 to 18 years of age may join with the signature of a parent or legal guardian).[1]

MARS operation takes place on frequencies adjoining the amateur bands and usually consists of nets. Nets are usually scheduled to handle traffic or to handle administrative tasks. Various MARS branches may also maintain repeaters or packet systems.

MARS demonstrated its importance dur-

Dave Pingree, N1NAS, hunts down a transmitter on 2-m FM. (*photo by Kirk Kleinschmidt, NT0Z*)

ing the 1991 Desert Storm conflict, when MARS members handled thousands of messages between the forces on the front lines and their friends and families at home. While MARS usually handles routine traffic, the organization is set up to handle official and emergency traffic if needed.

DIRECTION FINDING (DF)

If you've ever wanted to learn a skill that's both fun *and* useful, then you'll enjoy direction finding, or DFing. DFing is the art of locating a signal or noise source by tracking it with portable receivers and

Dick Esneault, W4IJC, a member of the original Project OSCAR team, looks over a model of the Phase 3D satellite. The body of the actual satellite will be well over 7 feet wide. This advanced Amateur Radio satellite is scheduled for launch in 2000. (*photo courtesy AMSAT-NA*)

directional antennas. Direction finding is not only fun, it has a practical side as well. Hams have been instrumental in hunting down signals from aircraft ELTs (emergency locator transmitters), saving lives and property in the process.

We will just scratch the surface of DF activities in this section. There is much more in the **Repeaters, Satellites, EME and DF** chapter.

Fox Hunting

Fox hunting, also called *T-hunting* or sometimes *bunny hunting*, is ham radio's answer to hide-and-seek. One player is designated the fox; he or she hides a transmitter and the other player attempts to find it. Rules change from place to place, but the fox must generally locate the transmitter within certain boundaries and transmit at specific intervals.

Fox hunts vary around the world. American fox hunts often employ teams of fox hunters cruising in their cars over a wide area. European and other fox hunters employ a smaller area and conduct fox hunts on foot. *Radiosport* competitions are usually European style.

Locating Interference

Imagine trying to check into your favorite repeater or HF net one day, only to find reception totally destroyed by noise or a rogue signal. If you can track down the interference, then you can figure out how to eliminate it.

Assembling a 2.4-GHz Mode-S downlink dish is easier than you think. Ed Krome, KA9LNV, put together this portable home-brew dish in less than 15 minutes. It's made of wood dowels, wire mesh and Dacron string. The helical feed and reflector plate are mounted at the focal point. (*photo by WB8IMY*)

Finding interference sources, accidental or otherwise, has both direct and indirect benefits. Touch lamps are a notorious noise source, especially on 80 m. If you can find one, the owner is legally obligated to eliminate the interference. Even better, if you can show your neighbors that something other than your station is interfering with their TV reception, you might gain an ally next time you petition the local government to let you have a higher tower!

SATELLITE OPERATION

Amateur Radio has maintained a presence in space since 1961, with the launch of OSCAR 1 (OSCAR is an acronym for *Orbiting Satellite Carrying Amateur Radio*). Since then, amateurs have launched over two dozen satellites, with over a dozen still in orbit today.

Amateurs have pioneered several developments in the satellite industry, including low-orbit communication "birds" and *PACSATs*—orbiting packet bulletin board systems.

What Does It Take?

When someone mentions satellite operation, many people conjure up an image of large dishes and incredibly complex equipment. Actually, you can probably work several OSCARs with the equipment you have in your shack right now!

The entire collection of OSCARs—and their operating modes—can be broken down into three basic types:

Voice/CW (Analog)

Analog satellites range from the low-orbit RS (Radio Sputnik) birds built and launched from Russia, to the high-orbit Phase 3 satellites, AO-10 and the soon-to-be-launched Phase 3D. Operating on analog satellites is much like operating on HF—you'll find lots of SSB and CW contacts, with some RTTY and even SSTV signals thrown in.

Packet (Digital)

Most of the digital satellites are orbiting packet mailboxes with some extra features. Many digital satellites carry one or more video cameras. These cameras snap pictures of Earth and space and make them available for downloading. Because most digital satellites are in low orbits, some clever software has been designed to allow ground stations to download images (or other data files) by monitoring a few orbital passes.

Several digital satellites carry an experiment called RUDAK, a versatile system that allows experimentation with packet, analog and crossband FM modes. The Fujisats from Japan also carry sophisticated systems that allow these birds to switch from analog to digital operation.

SAREX and *MIR*

The American and Russian space programs both recognize the value of Amateur Radio in space. The space shuttles often carry a mission called SAREX (*Shuttle Amateur Radio Experiment*) that allows hams to make packet or voice contacts with the astronauts onboard.

SAREX gives many schoolchildren the opportunity to talk to the astronauts and ask questions about their work.

The Russian *MIR* space station carries a permanent packet and FM station on board. *MIR* can occasionally be heard with a very strong signal, even on 2-m hand-held radios, and some hams have made contacts with mobile rigs.

Both of these orbiting amateur stations have proven their worth time and again. Their educational value is immense, and in case of normal communication failure Amateur Radio equipment provides a ready backup.

REPEATERS

Many amateurs make their first contacts on repeaters. Repeaters carry the vast majority of VHF/UHF traffic, making local mobile communication possible for many hams.

Hams in different regions have different opinions on repeater usage. In some areas, hams use repeaters only for brief contacts, while those in other areas encourage socializing and ragchewing. All repeater users give priority to mobile emergency communications.

The best way to learn the customs of a particular repeater is to listen for a while before transmitting. This avoids the misunderstandings and embarrassment that can occur when a newcomer jumps in. For example, in some repeater systems it is assumed that the word "break" indicates an urgent or emergency situation. Other systems recognize "break" as a simple request to join or interrupt a conversation in progress. Neither usage is more "correct," but you can imagine what might happen to

a traveling ham who was unaware of the local customs!

Most repeaters are *open*, meaning that any amateur may use the repeater. Other repeaters are *closed*, meaning that usage is restricted to members. Many repeaters have an *autopatch* capability that allows amateurs to make telephone calls. However, most autopatches are closed, even on otherwise open repeaters. The *ARRL Repeater Directory* shows repeater locations, frequencies, capabilities and whether the repeater is open or closed.

Most repeaters are maintained by clubs and other local organizations. If you use a particular repeater frequently, you should join and support the repeater organization. Some hams set up their own repeaters as a service to the community.

IMAGE COMMUNICATIONS

Several communications modes allow amateurs to exchange still or moving images over the air. Advances in technology in the last few years have brought the price of image transmission equipment within reach of the average ham's budget. This has caused a surge of interest in image communication.

ATV

Amateur TV is full-motion video over the air. (It is sometimes referred to as *fast scan*, or *FSTV*.) ATV signals use the same format as broadcast (and cable) TV. Watching an ATV transmission is the same as watching your own television. With ATV, however, you can turn a small space in your home into your own television studio. Amateur communication

takes on an exciting, new dimension when you can actually *see* the person you're communicating with!

The costs of ATV equipment have declined steadily over the years. The popularity of the camcorder has also played a significant role. (The family camcorder can do double duty as a station camera!) It is now possible to assemble a versatile station for well under $1000. Amateur groups in many areas have set up ATV repeaters, allowing lower-powered stations to communicate over a fairly wide area. If you're fortunate enough to live within range of an ATV repeater, you won't need complicated antenna arrays or high power.

If you can erect high-gain directional antennas for your ATV station, you can try your hand at DXing. When the bands are open, it's not uncommon to enjoy conversations with stations several hundred miles away. In addition to your directional antennas, you must run moderate power levels to work ATV DX. Most DXers use at least 50 W or more.

Since this is a wide-bandwidth mode, operation is limited to the UHF bands (70 cm and higher). The *ARRL Repeater Directory* and the *ARRL Operating Manual* list band plans. The *Repeater Directory* includes lists of ATV repeaters. The **Modulation Sources** chapter provides details on setting up an ATV station with dedicated or converted video gear.

SSTV

SSTV, or slow-scan TV, is a narrow-bandwidth image mode. Instead of full-motion video at roughly 24 frames per second, SSTV pictures are transmitted at 8, 16 or 32 seconds per frame. In the be-

Give him the specifications and Sam, K6LVM, can show you the radiation pattern of your antenna—via ATV! (*photo by Tom O'Hara, W6ORG*)

An SSTV image as seen on a standard TV set using a digital scan converter.

ginning, SSTV was strictly a black-and-white mode. The influx of computers (and digital interfaces) have spawned color SSTV modes. Since SSTV is a narrowband mode, it is popular on HF. Some experimenters run an SSTV net on OSCAR 13 as well.

An SSTV signal is generated by breaking an image into individual *pixels*, or dots. Each color or shade is represented by a different audio tone. This tone is fed into the audio input of an SSB transmitter, converting the tones into RF. On the receive end, the audio tones are regenerated and fed into a dedicated SSTV converter or into a simple computer interface to regenerate the picture. For more information about SSTV, see the **Modulation Sources** chapter.

Fax

Fax, or *facsimile* transmission, is one of the original image communication modes. Fax was once unavailable to amateurs due to FCC regulations, but is now a legal communication mode on most HF and higher bands.

Amateur Radio fax works much like old analog fax systems: an image is scanned from paper and converted into a series of tones representing white or black portions of a page. Amateurs are working on standards for the use of digital fax machines over radio as well.

Uses for amateur fax are as limitless as your imagination. Suppose you were having trouble with the design of your new home-brewed widget. You could fax a copy of the schematic to a sympathetic ham, who could mark in some changes and fax it back to you. And how about faxing QSL cards? No hunting for stamps or waiting for the mail to arrive!

DIGITAL COMMUNICATIONS

Digital communications predate the personal computer by many years. In fact, some amateurs consider CW to be a digital mode in which the amateur's mind handles the encoding and decoding of information. For the purposes of this *Handbook*, however, we consider digital modes to be those traditionally encoded and decoded by mechanical or electronic means. Common digital modes in use today include RTTY, AMTOR, PACTOR, G-TOR and packet radio.

As personal computers continue decreasing in price and increasing in power, many have found their way into ham shacks. Amateurs' computers perform tasks as mundane as keeping station logs and as exciting as controlling

QSLing

A QSL card (or just "QSL") is an Amateur Radio tradition. QSL cards are nearly as old as Amateur Radio itself, and the practice has spread so that SWLs can get cards from commercial shortwave and AM broadcast stations.

Most amateurs have printed QSL cards. QSL card printers usually have several standard layouts that you can choose from. Some offer customized designs at extra cost. If you are just starting out, or anticipate changing your call sign (just think, you could get a call like "KC4WZK"), you may want to purchase a pack of "generic" QSL cards available from many ham stores and mail-order outlets.

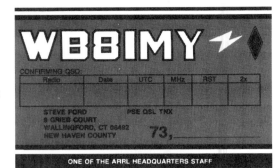

WB8IMY's QSL is a properly formatted card. Notice how all of the information is on one side of the card.

Filling Out Your Cards

QSL cards must have certain information for them to be usable for award qualification. At a minimum, the card must have:
❏ Your call sign, street address, city, state or province and country.
This information should be preprinted on one side of your QSL card.
❏ The call of the station worked.
❏ The date and time (in UTC) of the contact.
❏ The signal report.
❏ The band and mode used for the contact.
Awards for VHF and UHF operations may also require the grid locator (or "grid square") in which your station is located. If you have no plans to operate VHF and UHF, you can omit the grid square (you can always write it in later if required).

Many hams provide additional information on their QSL cards such as the equipment and antennas used during the contact, power levels, former calls and friendly comments.

Sending and Receiving Domestic QSLs

Most QSL cards can be sent as post cards within the United States, usually saving some postage costs. Back when postage was cheap, you could send out 100 post cards for a few dollars and domestic stations would send QSLs as a matter of course. Nowadays, if you really need a particular QSL, it is best to send a self-addressed stamped envelope along with your card.

QSLing for DX stations is somewhat more involved and is discussed elsewhere in this chapter.

a worldwide data network!

Packet

Packet radio is one of the fastest-growing modes of operation within the amateur community. Packet radio's strongest suits include networking and unattended operation. Do you need to give some information to an absent friend? Send an electronic mail message (or *e-mail* in networking parlance). Is your friend out of range of your 2-meter packet radio? Send your message through the packet network.

In packet radio, transmitted data is broken into "packets" of data by a *TNC* (terminal node controller). Before sending these packets over the air, the TNC calculates each packet's checksum and makes sure the frequency is clear. On the receive end, a TNC checks packets for accuracy and requests retransmission of bad packets to ensure error-free communication.

Packet radio works best on frequencies that are relatively uncrowded. On busy frequencies (or LANs), it is possible for two stations to begin transmitting at once,

Dave Patterson, WB8ISZ, checks into his local packet bulletin board as his cat Sam looks on. *(photo by WB8IMY)*

John Shew, N4QQ, (at a portable station set up by KG5OG) makes an EME CW contact with VE3ONT. *(photo by WB8IMY)*

garbling both packets (this is called a *collision*). Another common problem is the *hidden transmitter*, which happens when one of two stations (that are out of range of each other) is in contact with a third station within range of *both* (see **Fig 2.1**). Collisions can easily occur at the third station since neither of the other two stations can hear each other and thus may transmit simultaneously.

Thousands of packet radio stations have formed a worldwide network, one that parallels (and overlaps in some places) the massive Internet. Services available on the packet radio network include global e-mail, callbook servers, "white pages" servers (that provide network addresses of amateurs on the network) and gateways to (and from) the Internet. Other services, available in various locations, include libraries of program and text files, databases of equipment modifications, gateways to packet frequencies on HF and

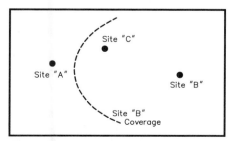

Fig 2.1—The "hidden transmitter" problem. Site B has established contact with site C, but cannot hear site A. However, site C can hear *both stations*. If site B transmits while site A is transmitting, or vice versa, a packet collision occurs.

packet satellites.

Most amateur networkers use VHF or UHF radios to access the packet network. Many amateurs, however, can be found using packet on HF and the packet satellites. Both HF and satellites see a great deal of internode network traffic as well.

AMTOR/PACTOR/G-TOR

AMTOR (*Am*ateur *T*eleprinting *O*ver *R*adio) is an "RF-hardened" digital mode. It is based on the SITOR mode (used in marine communications) and is an error-correcting mode.

A transmitting station using AMTOR sends three characters then waits for a response from the receiving station. (This produces the "cricket" sound unique to AMTOR.) The receiving station sends an ACK (send the next three characters) or a NAK (repeat the last three characters). This exchange of ACKs and NAKs ensures that only error-free text arrives at the receiving station.

Actually, there is much more than this to AMTOR; there are several submodes for calling CQ, or for reliable sending to multiple receivers.

PACTOR is a packet-like mode based on AMTOR, but with slightly longer packets and AX.25 compatibility. Unlike standard packet radio, PACTOR does not allow frequency sharing. PACTOR is much faster than AMTOR while still retaining AMTOR's ability to communi-cate through moderate noise or interference without errors. In addition, PACTOR uses the complete ASCII character set (with upper- and lower-case letters) and easily handle binary data transfers. The latest version of PACTOR is PACTOR II.

G-TOR uses several compression, checking and correction techniques along with automatic repeat requests. These techniques speed data transfer on the HF bands over that possible with AMTOR or PACTOR. The "G" in G-TOR stands for "Golay," a kind of forward-error-correction coding.

RTTY

RTTY is the original data communication mode, and it remains in use today. While RTTY does not support the features of the newer data modes, such as frequency sharing or error correction, RTTY is better suited for "roundtable" QSOs with several stations.

RTTY was originally designed for use with mechanical teleprinters, predating personal computers by several decades. Amateurs first put RTTY on the air using surplus teletypewriters (TTYs) and home-brewed vacuum-tube-based interfaces. Today, of course, RTTY uses computers or dedicated controllers, many of which also support other digital modes such as CW, PACTOR, AMTOR and packet.

CLOVER

CLOVER is a relatively new digital communications mode. It utilizes a four-tone modulation system and digital signal processing (DSP) to pass data on the HF bands at a rate much faster than AMTOR or even PACTOR. In addition, CLOVER's signal bandwidth is relatively narrow (500 Hz at −50 dB). When two CLOVER stations are linked, they share information concerning signal conditions and power output levels. As a result, CLOVER has the remarkable ability to

Where's the Action?

Hams enthusiastically adopted (and adapted) personal computers very early on. Many of the modes described in this chapter—RTTY, FAX, SSTV and others—no longer require the use of specific hardware terminals. Instead, combined with a small interface unit, they consist of software and home PCs. Now hams can operate many modes with just the press of a few keys on the PC's keyboard.

Often, the interface consists of just one IC, one or two transistors, and a handful of resistors, capacitors and diodes. It is connected between the serial or parallel port of the PC and the microphone and speaker jacks of a rig. A little software tweaking (usually just by selecting menu items) and you have instant FAX, SSTV or whatever the mode selected.

Then came the Internet. Hams quickly seized upon it to exchange information, make EME skeds, and distribute newsletters and software. It wasn't long before a few tied their rigs to their computers, and with the *chat channels* of the Internet, allowed hams from around the world to use their rigs.

A popular new activity, combining repeaters and the Internet, uses software called *IPhone*. All it takes is a PC with the *IPhone* software, a VOX interface and a VHF or UHF rig set to the local repeater (see **Fig A**). Hams from anywhere in the world can use their computers to talk through their local repeater. When two or

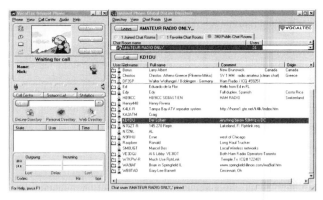

Fig A—Log in on *IPhone*, and this is what you might see. The call letters ending with -R are repeaters. *GIF file courtesy of Del (Anything Below 50 MHz is DC) Schier, KD1DU*

more repeaters are linked you can be driving down the street, using your 2-meter repeater, in contact with another ham across the country using her 440 repeater (**Fig B**). Without some other means of positive verification of identities, a control operator must be present for each repeater, to ensure the repeaters are used properly.—*N1II*

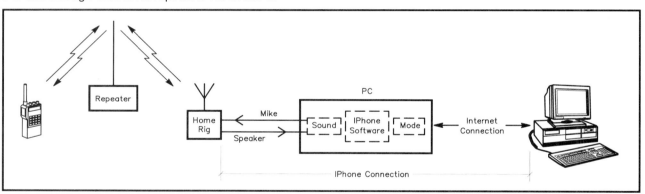

Fig B—End-to-end, this is how it works. If the PC on the right were also connected to a VHF or UHF rig, the range of two HTs would go from local to gobal!

adjust output power *automatically* to maintain a stable communication pathway. For example, if the signal begins to deteriorate at the receiving station, the transmitting station is "aware" of this fact and increases power until conditions improve. Under excellent conditions, power is adjusted *down*—sometimes to as little as a few watts!

The complete CLOVER system is contained on a card that plugs into the expansion slot of an IBM-PC or compatible computer (80286 microprocessor or better). An SSB transceiver is also required. The transceiver must be very stable because CLOVER cannot tolerate more

than approximately 15 Hz of drift after the link is established. As this *Handbook* went to press, CLOVER was primarily used as a means to move high-volume digital traffic (messages and files) on the HF bands.

MICROWAVE AND VHF/UHF WEAK-SIGNAL OPERATING

Hams use many modes and techniques to extend the range of line-of-sight signals. Those who explore the potential of VHF/UHF communications are often

known as *weak-signal* operators. Weak-signal enthusiasts probe the limits of propagation. Their goal is to discover just how far they can communicate.

They use directional antennas (beams or parabolic dishes) and very sensitive receivers. In some instances, they employ considerable output power, too. As a result of their efforts, distance records are broken almost yearly! On 2 m, for example, conversations between stations hundreds and even thousands of miles apart are not uncommon. The distances decrease as frequencies increase, but communications have spanned several hundred miles even at microwave frequencies.

EME, Meteor Scatter and Aurora

EME (Earth-Moon-Earth) communication, also known as "moonbounce," continues to fascinate many amateurs. The concept is simple: use the moon as a passive reflector for VHF and UHF signals. With a total path length of about 500,000 miles, EME is the ultimate DX.

Amateur involvement in moonbounce grew out of experiments by the military after World War II. While the first amateur signals reflected from the moon were received in 1953, it took until 1960 for the first two-way amateur EME contacts to take place. Using surplus parabolic dish antennas and high-power klystron amplifiers, the Eimac Radio Club, W6HB, and the Rhododendron Swamp VHF Society, W1BU, achieved the first EME QSO in July 1960 on 1296 MHz. Since then, EME activity has proliferated onto most VHF and higher amateur bands.

Advances in low-noise semiconductors and Yagi arrays in the 1970s and 1980s have put EME within the grasp of most serious VHF and UHF operators. Further advances in technology will bring forth sophisticated receivers with digital signal processing (DSP) that may make EME affordable to most amateurs.

EME activity is primarily a CW mode. However, improvements in station equipment now allow the best-equipped stations to make SSB contacts under the right conditions. Regardless of the transmission mode, successful EME operating requires:

- Power output as close to the legal limit as possible.

- A good-sized antenna array. Arrays of 8, 16, or more Yagis are common on the VHF frequencies, while large parabolic dish antennas are common on UHF and microwave frequencies.
- Accurate azimuth and elevation.
- Minimal transmission line losses.
- The best possible receiving equipment, generally a receiver with a low system noise figure and a low-noise preamplifier mounted at the antenna.

The ARRL sponsors EME contests to stimulate activity. Given the marginal nature of most EME contacts, EME contests designate a "liaison frequency" on HF where EME participants can schedule contacts. Contest weekends give smaller stations the opportunity to make many contacts with stations of all sizes. See the **Repeaters, Satellites, EME and DF** chapter for more about EME.

Meteor Scatter

As a meteor enters the Earth's atmosphere, it vaporizes into an ionized trail of matter. Such trails are often strong enough to reflect VHF radio signals for several seconds. During meteor showers, the ionized region becomes large enough(and lasts long enough) to sustain short QSOs.

Amateurs experimenting with meteor scatter propagation use high power (100 W or more) and beam antennas with an elevation rotor (to point the beam upward at the incoming meteors). Most contacts are made using CW, as voice modes experience distortion and fading. Re-

flected CW signals often have a rough note.

The ARRL UHF/Microwave Experimenter's Handbook contains detailed information about the techniques and equipment used for meteor scatter.

Auroral Propagation

During intense solar storms, the Earth's magnetic field around the poles can become heavily charged with ions. In higher latitudes, this often produces a spectacular phenomenon called the aurora borealis (or northern lights) in the Northern Hemisphere and the aurora australis (or southern lights) in the Southern Hemisphere. The ionization is often intense enough to reflect VHF radio signals. Many amateurs experiment with aurora contacts on 10 and 2 m. Aurora contacts are often possible even when the aurora is not visible.

Equipment used to make aurora contacts is similar to that used for meteor-scatter contacts: high power, directional antennas and CW. Antenna pointing is less critical, however, since the antenna need only be aimed at the aurora curtain. Reflected CW signals often have a rough buzzsaw-like note and can also be Doppler-shifted.

The ARRL UHF/Microwave Experimenter's Handbook contains detailed information about the techniques and equipment used for auroral propagation.

[1]You can find more information online about the three branches of the Military Affiliate Radio Service at their respective Web sites. See the **References** chapter Address List.

Contents

Modes
3

United States Amateur Radio operators are encouraged by the Federal Communications Commission (FCC) to explore, investigate and experiment, with wide latitude. The government wants hams to try new things, test new ideas and push the capabilities of two-way radio-frequency (RF) communications. In this chapter, written by Brian Battles, WS1O, you'll learn basic information about the most popular amateur operating modes. You'll see what each one is used for and its advantages and disadvantages, and you'll get a general idea of the resources needed to try each one. Details of how radio signals are created and how those signals are made to carry information is explained in later chapters of this book. This is an introduction; the other chapters of this *Handbook*, the *ARRL Operating Manual*, *QST*, *QEX* and a whole Amateur Radio publishing industry cover the continuously evolving realm of Amateur Radio modes.

WHAT IS A MODE?

In Amateur Radio, a mode is considered to be any modulation technique that permits two-way communication with another station using an identical or compatible system to receive and demodulate it. If that sounds confusing, just think of a simple analogy, such as the way we use money in modern society: A certain amount of American dollars can be used to purchase an item that may also be purchased with an equivalent amount in Japanese yen. The dollar and the yen, however, are not directly compatible—you can't simply exchange them one for one. Yet they are both forms of payment for goods and services.

This is typical of most ham radio modes. You can talk over an FM repeater, but a person with a CW receiver who tunes to your frequency won't hear a thing you say. FM and CW aren't compatible. Most Amateur Radio operating modes are incompatible with each other, and each has advantages and disadvantages that make it worthwhile to select the best mode for the intended communication. Because its information is transmitted by varying the carrier frequency substantially, an FM signal uses a relatively large amount of RF bandwidth. On a relatively wide band such as 2 m (144-148 MHz, 4 MHz wide), many hams use narrow-bandwidth FM (NBFM). NBFM requires about 15-20 kHz of spectrum for each conversation (hams call them "QSOs"). Several such FM signals would easily use up all of a narrow band such as 30 m (10.1-10.15 MHz, 0.05 MHz wide), and few conversations could take place at the same time. Other modes, such as single-sideband suppressed-carrier

Rick Castaldo, KD1BR, keeps in touch with his friends using 80-meter SSB.

amplitude modulation (SSBSCAM, or SSB) and continuous wave (CW), require much less bandwidth. Here's the trade-off: while FM has enhanced audio fidelity and relatively little background noise, SSB and CW permit many more simultaneous QSOs in any given range of frequencies.

The radio amateur has a variety of frequencies to select from when attempting to establish two-way communications. At the low end are the *medium frequencies* (MF), specifically the 160-m band (1.8-2.0 MHz). Then there are the *high frequency* (HF) or "shortwave" bands (80-10 m, or 3.5-29.7 MHz). The next steps up are the *very high frequency* (VHF) bands (6-1.25 m, or 50-225 MHz). Above these are the *ultra high frequency* (UHF) bands (70-23 cm, or 420-2450 MHz). Above all these are the *super high frequency* (SHF), *extremely high frequency* (EHF) and other segments, more simply referred to as *microwaves* (2900-250,000 MHz). The band you choose affects your signal's range (see the chapter on **Propagation**). But the mode you select is also important, for many reasons.

The key to successful Amateur Radio communications is *signal-to-noise ratio (S/N)*. This means you can communicate with anyone if your signals are loud enough to be heard through any noise present. Sometimes the noise wins; there's always *some* noise present on any radio frequency. Similarly, there's *some* propagation available to any desired location. The propagation path may not be sufficient to help conduct a weak signal halfway around the world, but theoretically, if you transmitted enough power, if there were absolutely no noise, you could contact almost anyone in the world at any time propagation conditions permitted your signal to reach its destination. The challenge in radio communications is to work within the practical constraints of your license privileges, budget, physical capabilities and atmospheric conditions to maximize your chances of being heard. If the S/N at the receiving end is sufficient, you will be heard. If the noise level is too great, you won't get anywhere.

Every portion of the radio-frequency spectrum exhibits its own unique characteristics of propagation and noise. For example, the MF or lower-frequency HF amateur bands are in a range that is particularly subject to noise. Frequencies below 5 MHz exhibit ever-present random atmospheric noise from storms, and the background of "hash" generated by man-made electrical gadgets, appliances and machinery is often very loud. If you live in an urban environment, it can be difficult to hear anyone most of the time, especially in the summer, when thunderstorms rage throughout your hemisphere. On the other hand, radio amateurs fortunate enough to live in rural settings far from factories, congested housing and other sources of noise can enjoy good reception on 160 m, especially in the winter, when thunderstorm activity is lowest.

In addition to considering the noise and propagation conditions on the various bands, an Amateur Radio operator must intelligently select the *mode* of transmission most likely to succeed. The factors of concern are (1) What modes can your transmitter produce? (2) What mode is the intended recipient using? (3) What mode will provide the receiving station with the best S/N?

The first is easiest to determine—just look in the manual or at your rig's front panel controls. The second may be more of a mystery. Unless you have some means of predicting the mode the other station is using, you must make an educated guess. The third requires an understanding of what each mode is best suited to accomplish and how to take advantage of its characteristics.

Let's examine factors (2) and (3). To make communications convenient, hams have established standard operating techniques that make it easier to choose a mode. For example, most amateur voice communications on the 160, 80 and 40-m bands (1.8-7.3 MHz) are conducted via lower sideband (LSB), while upper sideband (USB) is the normal mode on 20, 17, 15, 12 and 10 m (14-28 MHz). On 2 m (144-148 MHz) and 1.25 m (222-225 MHz), the predominant mode is frequency modulation (FM) for voice and data com-

munications. Above 420 MHz there is a mixture of FM and SSB voice communications, fast-scan amateur television (ATV), CW and experimental modes.

There are exceptions to these "rules." Data communications, such as radioteletype (RTTY) and packet, are operated almost entirely on LSB at HF; DXing (long-distance communication) on VHF and UHF is mainly conducted using USB and Morse code; 10-m repeaters use FM. Aside from such exceptions, however, you can normally assume that stations operating voice on 20 m are likely to be using USB, a 2-m repeater is almost certainly using FM, and a packet station on 30 m will be running LSB. You should set your transceiver accordingly.

Once you know what modes are used by most operators on a given frequency, what frequency and mode should you select to have the best chance of establishing contact? You won't be heard unless your signal significantly exceeds the noise at the receiving end.

There are five main categories of communications in Amateur Radio: Radiotelegraph (Morse code), radiotelephone (voice), radioteletype (Baudot and ASCII RTTY and AMTOR), digital (packet, PACTOR, CLOVER, G-TOR) and image (SSTV, FSTV and fax). Let's look at the main Amateur Radio operating modes and compare their uses and characteristics.

CW

CW *(continuous wave)* is the oldest mode of ham transmission in use today. It

CW is the perfect mode for weak-signal work. John Shew, N4QQ, uses CW to operate EME (moonbounce) during the 1993 AMSAT Space Symposium.

consists of a plain, unmodulated RF signal (or "carrier") which is transmitted by the closure of a manual key or an electronic keyer circuit (sometimes called "on-off keying"). CW conveys intelligence through the International Morse code. (This is a variation on the American Morse code used by commercial telegraphers on wired lines in the late 19th century.)

AM

Amplitude modulation was the earliest technique used to transmit the human voice over radio. It was the dominant 'phone mode until the late 1940s and 1950s, when single sideband (SSB) came along. AM uses a full carrier with two modulated sidebands, and takes up a fair amount of bandwidth. In practice, two SSB signals or eight CW signals could be

Sounding like a broadcaster from radio's Golden Age, Paul Courson, WA3VJB, operates classic AM equipment from his well-equipped shack in West Friendship, Maryland. The setup doesn't just look like a broadcast station, it's used in Paul's work in commercial radio production. *(photo courtesy of WA3VJB)*

Morse Code: A Language, Not a Mode

Digital signals are those that can be represented accurately in terms of integer numbers. Morse code has been called the most basic form of digital radio communication. It involves nothing more than three elements: a short tone, a long tone and a space. In digital perspective, these could be represented by the numerals 0, 1 and 2. If all we can do is send separate elements, we are limited to three possible pieces of information. (In practice, there would actually be only two definite expressions, the 0 and 1, or short tone and long tone. A space by itself would be the same as no signal at all.)

Mixing and matching the tones and spaces created a system to denote letters of the alphabet, the digits 0-9 and a few punctuation marks. As a result, no individual character requires more than six elements to be unique and clear.

Language expressed in nothing but long and short tones (dahs and dits) has been in use since before Samuel F. B. Morse devised the American Morse system of telegraphy. In prehistoric times, it was discovered that a prearranged method of sending signals could use a small set of such symbols, yet be a potentially powerful means of conveying complex information. The arrangement of the signals (long/short, loud/soft, bright/dim, single/double, on/off) could be used to signify much more than just two alphabetic or numeric symbols. Morse and his colleagues developed a standard code that took advantage of the frequency with which certain English letters are used in language, and made efficient use of the way our ears and minds work. Learning Morse code is as easy as learning about 40 words in a foreign language. With practice, a "speaker" of Morse code can decipher a string of long and short tones and spaces at almost half the speed of everyday spoken English conversation.

Now that we have a language, we need a means to transmit and receive it at two separate locations. Wired telegraphy was the earliest means, but the invention of wireless radio transmissions less than 50 years later made it a simple matter of turning the transmitter on and off in a particular sequence to produce a corresponding sound at the receiving end. The first equipment used to create such radio signals were spark-gap transmitters. Spark transmitters sent an electromagnetically noisy pulse to a resonant antenna system where the pulse essentially reverberated until its energy was either radiated or dissipated in the resistance of the circuit. During the reverberation, the amplitude gradually decreased, so the result is a damped wave—a wave with changing strength. The part that was radiated became an electromagnetic field. The field was strong enough for a sensitive, specially designed antenna and receiver to respond to the electromagnetic bursts by rendering an audible sound in a headset or other transducer. As experiments continued, it wasn't long before the vacuum tube was perfected. By using a tube in a tuned electrical circuit, a transmitter could generate an almost-pure, constant unmodulated radio signal on a specific frequency. The signal generated by the tube was smoother and occupied much less space on the electromagnetic spectrum.

So how did the term *continuous wave* (*CW*) come about? Many beginning hams mistakenly think that it somehow means the transmitter is emitting a constant RF signal. But how could this be? Wouldn't this mean that each press of a code key actually breaks the circuit, rather than closes it? Of course, that's backwards. A code key closes the circuit when it's pressed (or squeezed, in the case of a paddle). No, the transmitted carrier is called *continuous* because it's not damped as were the spark signals; it's continuous in the sense that the RF sent when the key is pressed is transmitted as a single radio-frequency wave of continuous strength. CW is the simplest mode to implement in a radio. You can learn more about the principles of CW in the **Modulation Sources** chapter.

---

transmitted within the bandwidth used by one AM signal. Because it's less spectrum efficient, it has been reduced to a "curiosity" on the air. There are many antique wireless enthusiasts who derive great joy from the warm, rich sound of a strong AM signal, especially when generated by a properly adjusted and maintained vintage transmitter. AM operators use specific frequencies on almost every amateur band, and most polite hams respect the rights of AMers to operate, just as AM fans avoid transmitting on portions of the bands generally used for SSB, CW, amateur television and digital modes.

ANGLE MODULATION: FM AND PM

FM is by far the most popular mode of Amateur Radio communication. The majority of newcomers to the hobby use FM transceivers to operate on VHF and UHF frequencies; the most popular activity is

FM offers clear, reliable voice communication. This is especially critical during public-service activities. Using FM portable transceivers and directional antennas, hams operate from the Command HQ tent at the Big Sur International Marathon.

Table 3-1
Equipment Requirements for Amateur Radio Operating Modes

Mode	Typical activity	Radio(s)	Antenna(s)	Other
CW, HF	Ragchewing, contesting, DXing, traffic handling	CW transmitter/receiver, Transceiver, single or multiband	Ranges from simple fixed wire antennas to tower-mounted rotatable, multi-element Yagis and phased arrays	Antenna tuner (optional) Tower (optional) Rotator (for Yagis) Amplifier (optional) CW key, paddles or keyer
CW, VHF/UHF	Ragchewing, contesting, DXing, satellites	All-mode transceiver	Omnidirectional vertical; long, extended Yagi ("boomer"), quagi or helix (depending on the application)	Tower or roof-mount for antenna desirable, Rotator (for Yagis) Key or paddle and keyer
SSB, HF	Ragchewing, contesting, DXing, traffic handling	SSB transceiver	Ranges from simple wire antennas up to tower-mounted, rotatable, multi-element Yagis and phased arrays	Antenna tuner (optional) Tower (optional) Rotator (for beams) Amplifier (optional)
SSB, VHF/UHF	Ragchewing, contesting, DXing satellites	All-mode transceiver	Omnidirectional vertical, long, extended Yagi ("boomer"), quagi or helix (depending on the application)	Tower or roof-mount for antenna desirable since DX is mostly line of sight Rotator (for Yagis)
AM, HF	Ragchewing	AM or multimode transceiver	Fixed wire or beam antennas	Antenna tuner (optional) Tower (optional) Amplifier (optional) Microphone
FM, HF	Ragchewing, repeaters	Multimode transceiver	Omnidirectional vertical or beam	Antenna tuner (optional) Tower (optional) Amplifier (optional) Microphone
FM, VHF/UHF	Ragchewing, repeaters, contesting	FM or multimode transceiver	Omnidirectional vertical or Yagi	Tower (optional) Amplifier (optional)
Digital	Ragchewing, message handling, data networking, satellites	HF or VHF/UHF transceiver	Omnidirectional or directional	Amplifier (optional) TNC or multimode processor (optional) Computer or terminal
Image	Ragchewing	HF SSB transceiver or UHF video transceiver	Omnidirectional or directional	Amplifier (optional) TV camera or image scanner Computer (for SSTV and fax) Demodulator (for SSTV and fax)

voice operation via 2-m FM repeaters. In fact, many amateurs who get their start on the air using 2-m FM and repeaters are unfortunately never exposed to other kinds of operation; they miss the pleasure of using other modes and bands. Considering that it's mainly limited to local contacts, the 2-m band offers reliable communication quality over a reasonably long range. Most VHF and UHF repeaters use FM, a great deal of direct, nonrepeater (simplex) traffic is conducted using FM, and there's an FM segment on the 10-m band (29.2-29.8). Above 30 MHz, virtually all amateur packet radio is conducted

using FM transmissions. Repeater operation typically permits hams to use hand-held and mobile transceivers to make clear contacts over ranges of 100 miles or more. FM is a quiet mode, because the technique of angle modulation greatly minimizes the effects of static and noise. The trade-off is that a rather strong signal is needed at the receiver to produce the "quieting" effect that distinguishes FM communications.

Frequency modulation and its sibling, phase modulation, is accurately known as *angle modulation* because the frequency or the phase of a transmitted signal's carrier

can be shifted to provide an "FM" signal. Amateurs use a form of FM called narrowband FM, which is about 3-20 kHz wide. This is just enough to afford decent voice communication. Commercial FM broadcast stations (88-108 MHz in the US) use wideband FM with 75-kHz deviation and 200-kHz channel spacing, which permits transmitted audio frequencies in the range of 50-12,000 Hz.

SSB

Suppressed-carrier single sideband, what we call "single sideband" is an AM signal from which the carrier and one sideband have been removed. The receiving station picks up the SSB signal, adds a carrier, and converts the RF signal back into voice.

Operators who enjoy making long-distance contacts on the VHF and UHF bands find that SSB is a better choice than FM. Even though there are many more VHF/UHF transceivers on the market that offer only FM, SSB is useful because it's much easier to copy a weak sideband signal than a weak FM transmission. The signal strength needed to quiet an FM receiver is relatively high. With low signal-to-noise ratios, it's nearly impossible to make out the other operator's voice. In conditions where weak signals are common, SSB can provide intelligible audio through a background of considerable noise.

DIGITAL MODES

Technically, any means of communication based on a simple "black-and-white" value can be considered a digital mode, because it can be expressed with simple whole numbers. A CW transmission can be viewed as a mixture of "on/off" signals, to which we could assign the values 1 and 0. Conversely, a series of 1s and 0s, in the proper order, could signify a Morse code message. In fact, many devices use this technique to encode and decode Morse code communication automatically.

In addition to simply turning a signal on and off, there are other ways to designate those digital 1s and 0s, or changes in *state*. Frequency shift keying (FSK) uses the principle of switching between two frequencies, which are used to designate the *mark* and *space* (on and off state, or digital 1 or 0). FSK is achieved by using a control circuit to switch a transmitted carrier up and down to the mark and space frequencies. We can achieve the same result by modulating an SSB transceiver with high- and low-pitched audio signals producing the functional equivalent of true FSK.

The next step toward smooth, effective Amateur Radio communications makes use of machines, specifically, electronic digital computer processing. Fortunately, the FCC permits certain kinds of amateur digital communications because they help provide error-free communications even in difficult conditions.

RTTY

Before there was radio, there was *teletype* (abbreviated *TTY*), a system of sending printed text by typing on a terminal that was connected by wires to a similar machine. The military began connecting mechanical teletype machines to HF radios during World War II. This was the birth of *radioteletype*, or *RTTY*. After some experimentation with simple on-off keying, the designers switched to FSK, which proved effective.

In the days before personal computers, enterprising hams obtained inexpensive surplus TTY machines and modified them to provide output signals that could be fed into an SSB transceiver's microphone input. The resulting signals fit within the bandwidth permitted for voice transmissions. A typical modern RTTY installation consists of three parts: a computer, a communications processor and an FSK or SSB transceiver.

In basic radioteletype (RTTY, pronounced **RIT-ee**), the operator types a continuous string of characters on the computer. A terminal program sends the characters through the serial port to the communications processor, which translates the characters into the appropriate mark/space signals as either control signals for an FSK transceiver (or audio tones for the audio input of an SSB transceiver). At the receiving station, the transceiver audio is translated by the communications processor into characters, which are sent to the computer screen via the serial port and terminal program. Amateur Radio operators use several kinds of RTTY.

Baudot (named after French engineer Émile Baudot [1845-1903], pronounced **baw-DOE**), is a method of exchanging alphanumeric characters over wires and radio links. Baudot is sometimes referred to as International Teletype Alphabet 2 (ITA2). It produces letters, numbers and a limited number of punctuation symbols with a five-bit code. Because five bits per character permit a limited number (32) of symbols, the alphabet is sent in capital (upper-case) letters. Baudot RTTY is the most widely used form of amateur RTTY on the HF bands.

ASCII (an acronym for the *American Standard Code for Information Interchange*, and pronounced **ASK-ee**) is a form of RTTY that transmits seven bits per symbol. ASCII is the same code used on most modern computers. It contains the entire upper- and lower-case alphabet, punctuation and some special symbols.

AMTOR

In a standard RTTY transmission, only one station transmits, while the other receives. Characters may be printed incorrectly if the S/N is inadequate. Atmospheric static, fading or noise bursts can garble incoming characters. A partial solution to this challenge is a form of RTTY called *AMTOR* (*Am*ateur *T*eleprinting *O*ver *R*adio), which uses a computer processor to maintain a virtually error-free communications link. Instead of one station transmitting while the other passively receives, both stations maintain a link by exchanging transmissions. The sending station sends short bursts of data and the receiving station sends shorter acknowledgment (ACK) bursts between them. Characters are sent in groups of three. The receiving station checks each character, looking for a 4:3 bit ratio. If the ratio is correct for each of the three characters, they are displayed on the receiving terminal and the receiving station sends an ACK, telling the transmitting station to continue. If errors exist, the receiving station sends a negative acknowledgment (NAK) signal, which commands the transmitting station to resend the incorrectly received group.

Because AMTOR runs a continuous validity check on the characters exchanged, it's more reliable than "plain" RTTY, which can produce lots of incorrect text. AMTOR is well suited to traffic handling and passing messages when accuracy is worth a slight trade-off in speed. A disadvantage of AMTOR is that both stations must "hear" each other well, or the communication is reduced to a lengthy exchange of poorly received character groups and NAKs.

Packet

Packet radio is an error-free mode that uses the complete ASCII character set and supports the transfer of binary data. It is called "packet" because data is not sent as a single, continuous string of characters, but rather transmitted in small bursts, or *packets*. Each packet contains not only the data to be transferred, but also *overhead* information used to route the packets and reassemble them into their original continuous whole. The overhead includes data that identifies the sending and receiving stations, enabling the packet transmission to reach the proper destination. It also provides the FCC-required station identification of the sender.

When two packet stations are exchanging data, they are said to be *connected*. One station sends a packet and then waits a specified amount of time for a reply. The receiving station checks the packet for errors and sends an ACK if the packet is error-free. If not, the receiving station does nothing. When the waiting time expires, the originating station retransmits the packet on the assumption that it did not arrive error-free. Packet stations communicate directly in many cases. If the path is too long to support a direct connection, however, packet relays known as *nodes* or *digipeaters* are used.

Packet radio stations are able to communicate with each other because they conform to a standard format, or protocol, described as AX.25. This is a specification derived from a similar protocol, X.25, used by commercial packet networks. AX.25 version 2.0 was approved by the ARRL Board of Directors in 1984; it is the general amateur packet radio protocol currently authorized by the FCC.

Packet data rates are limited to a maximum of 300 bit/s within the Amateur Radio bands below 28 MHz. Even at this relatively slow rate, noise and interference makes efficient packet communication difficult on the HF bands. Unless conditions are good at both ends of the path, packets must be repeated many times before they arrive error-free.

Above 28 MHz, data rates are not so limited. Most VHF packet users operate their systems at 1200 bit/s using 2-m FM. However, networks have been established that use much higher data rates—9600 bit/s and beyond—to link nodes and packet bulletin board systems over a wide area. The most popular networking system is NET/ROM, but there are several contenders such as TexNet, ROSE and TCP/IP. Through these systems the packet-radio network has become accessible to amateurs throughout the nation and the world.

APRS

APRS (Automatic Position Reporting System) uses the unconnected packet radio mode to graphically indicate the position of moving and stationary objects on maps displayed on a computer monitor. Unconnected packets are used to permit all stations to receive each transmitted APRS packet on a one-to-all basis rather than the one-to-one basis required by connected packets.

Virtually all VHF APRS activity occurs on 2 meters, specifically on 144.39 MHz, which is recognized as *the* APRS operat-

ing channel in the United States and Canada. Like most other 2-meter packet operations, APRS operates at 1200 bit/s.

The standard configuration for packet radio hardware (radio-to-TNC-to-computer) also applies to APRS until you add a GPS receiver to the mix. You don't need a GPS receiver for a stationary APRS installation (nor do you need a computer for a mobile or tracker APRS installation). In these cases, an extra port or special cable is not necessary. It is necessary, however, when you desire both a computer an a GPS receiver in the same installation.

One way of accomplishing this is by using a TNC or computer that has an extra serial port for a GPS receiver connection. Alternatively, you can use a hardware single port switch (HSP) cable to connect a TNC and GPS receiver to the same serial port of your computer. The HSP cable is available from a number of sources including TNC manufacturers Kantronics, MFJ and PacComm.

Whichever GPS connection you use, make sure that you configure the APRS software so it is aware that a GPS receiver is part of the hardware configuration and how the GPS receiver connection is accomplished.

For additional information see the book, *APRS Tracks, Maps and Mobiles* by Stan Horzepa, WA1LOU, published by ARRL.

PACTOR

Two German hams, Hans-Peter Helfert, DL6MAA, and Ulrich Strate, DF4KV, worked together to find a solution to the problems of HF data communication. The result of their effort is a blend of the best parts of packet radio and AMTOR: *PACTOR*, and now *PACTOR II*.

Helfert and Strate liked AMTOR because it's a simple system that works well with marginal signal-to-noise levels. However, they disliked its inadequate

error-correction capabilities, its slow effective maximum data rate (less than 35 bauds) and its use of the five-bit Baudot code character set (all upper-case letters). To make up for the deficiencies of AMTOR, Helfert and Strate devised a new system based on AMTOR that adopted some features of packet radio.

Advantages of PACTOR include:

- An error-correction algorithm called *Memory ARQ*, a method for reconstructing an original block of data by adding together the broken pieces of that block as it's repeated until the block is whole.
- Data-compression techniques (Huffman coding) that can increase the data transfer rate by up to 400% over uncompressed data.
- Compatibility with ASCII and binary data transfers.
- Automatic adjustment of its data rate to compensate for changes in radio conditions.
- Mark or space polarity is inconsequential because it is frequency-shift independent.
- Tolerates interference well, while maintaining the communication link.

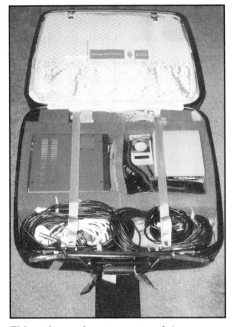

This suitcase houses a complete portable PACTOR station: transceiver, PACTOR controller, wire antenna and cables, all in one convenient package. Its owner, Joe Mehaffey, K4IHP, of Atlanta, Georgia, connects it to his laptop computer to operate mobile or from almost anywhere he pleases. He even used it to maintain a regular schedule with ham family members and neighbors while traveling around the Australian countryside. *(photo by K4IHP)*

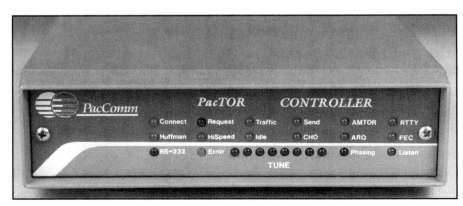

PACTOR is becoming the digital HF mode of choice for many hams. This dedicated PACTOR controller by PacComm also supports Baudot, AMTOR and CW. Many amateurs who once used AMTOR and packet on the bands below 30 MHz have switched to PACTOR for its more robust capabilities.

- It uses unique addresses (the complete call sign of a station is its PACTOR address).
- Fast, reliable changes of transmission direction and end of transmission confirmation at both ends of a connection.

Like packet and AMTOR, PACTOR is a two-way affair: A transmitting station sends data and a receiving station sends back electronic acknowledgment of each burst of characters. Unlike packet or AMTOR, however, PACTOR dynamically adapts to conditions. Rather than relying on each transmission to provide a solid block of clear characters, PACTOR can accept a series of imperfect or incomplete data segments and "intelligently" attempt to reassemble them into a solid group. In this way, the number of transmissions is reduced because the receiving station may be able to "make out" enough detail from two or three successive bursts to provide an errorless segment of data.

G-TOR

G-TOR was developed by Kantronics with twin goals: to provide greater throughput on HF channels than AMTOR and PACTOR and to be compatible with existing multimode TNCs. To increase throughput, G-TOR uses Huffman encoding to compress data, Golay forward error correction, a cyclic redundancy check to detect errors, data interleaving and automatic repeat requests to replace data that cannot be corrected. Golay encoding and interleaving work together to provide forward error correction that is effective even when long bursts of bits are corrupted. Headquarters operators have seen G-TOR provide twice the throughput of PACTOR and four times that of AMTOR under difficult conditions. More tests are needed, however. We simply don't have enough data yet to know for sure which technique is best.

CLOVER

No matter which digital mode you use—packet, AMTOR, PACTOR or RTTY—data communications on the low bands can be a struggle because of the nature of HF. When HF conditions are perfect, almost anything works, but how often are HF conditions "perfect"? Under the less-than-perfect conditions typically encountered, digital modes begin to fail. It used to be that as conditions deteriorated, all that was left was Morse code. But now there's *CLOVER*. It's a rather complex system invented for the purpose of relaying files and text on HF bands at higher speeds than packet, with faster throughput and versatile self-adjusting parameters. Developed by Ray Petit, W7GHM, and HAL Communications Corp president Bill Henry, K9GWT, CLOVER is named for the cloverleaf waveform it produces when displayed on a monitor scope. CLOVER offers improved data communication in the HF spectrum by using a high-speed, bandwidth-efficient modem and an error-correction protocol designed to counteract changing propagation conditions.

HAL Communications implemented CLOVER in a hardware/software system called PCI-4000/PC-CLOVER, for IBM-compatible computers (with 80286 CPUs or better). The user plugs a PCI-4000 board into one PC expansion slot and installs the *PC-CLOVER* software on the computer's disk drive, connects the CLOVER card to an HF transceiver, and is ready to go on the air.

On-air experience and tests indicate that CLOVER-II can reliably pass error-corrected data on HF at rates of 10-70 bytes/sec. (80 to 560 bits/sec.) Under average conditions, throughput is 20 to 40 bytes/sec. (160 to 320 bits/sec.)—2 to 10 times faster than AMTOR, Pactor, or AX.25 HF packet. While CLOVER-II with its 500-Hz wide spectra is preferred for Amateur Radio use, CLOVER-2000, the commercial version, has four times greater throughput— up to 2000 bits/sec. This waveform uses 8 tones spaced 250 Hz apart, a symbol rate of 62.5 baud, and has a characteristic spectrum from 500 to 2500 Hz. CLOVER-2000 is directly compatible with all HF SSB equipment.

The CLOVER system includes an "AUTO-ARQ" mode that provides a three-pronged attack against the problems caused by HF data signal distortion. This forward-error-correction will correct as many as 31 bytes of erroneous data for every 188 bytes of transmitted data, without requiring repeat transmissions. (AMTOR, PACTOR and packet correct errors only by retransmitting data.) When erroneous data exceeds 31 of 188 transmitted bytes, only the damaged data blocks are repeated. (AMTOR and PACTOR repeat all the data of a transmitted pulse, even for one character error.)

The CLOVER modem samples the characteristics of each received block of data (the signal-to-noise ratio, frequency offset and phase dispersion) to determine the

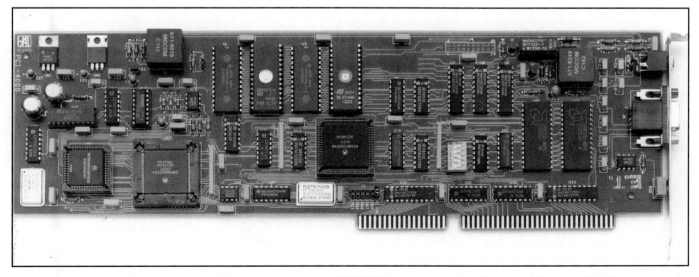

Bill Henry, K9GWT, and Ray Petit, W7GHM, are the fathers of CLOVER, an extraordinarily powerful mode of digital communication. It's an adaptive, error-correcting mode that can pass information over long-distance radio links under severe conditions of poor propagation. This is the original CLOVER adapter, the HAL PCI 4000, a one-slot add-on board for IBM-compatible personal computers.

current operating conditions. With this information, CLOVER optimizes the other station's transmitting parameters to match the measured conditions. (AMTOR and packet have no adaptive capabilities, and PACTOR uses an adaptive algorithm.)

AUTO-ARQ is available in three flavors, or "bias" settings, for three types of band conditions. For average HF conditions, the "normal" bias setting offers a good balance between error correction, data rate and throughput. For extreme HF conditions, the "robust" or "fast" bias settings are available. The robust bias setting provides the greatest degree of error correction, but to achieve this, throughput is decreased. Robust bias is recommended for fixed-frequency operation (below 7 MHz) where maintaining a connection over an unstable path is more important than the amount of data throughput. For ideal conditions (stable paths at frequencies near the MUF), the fast bias setting provides reduced error correction, greater data rates and maximum data throughput.

HAL Communications has four DSP modems that support CLOVER emissions. The P38 and DXP-38 are low cost and designed specifically for Amateur Radio use. These modems use the TMS320C25 DSP engine and a 68EC000 control processor. The PCI-4000 and DSP-4100 use the DSP-56001/2 IC and are intended primarily for commercial applications. The P38 and PCI-400 are PC ISA bus compatible plug-in circuit boards. Firmware for the on-board processors is loaded via the PC ISA Bus. The DXP-38 and DSP-4100 are cabinet modems that operate via a serial I/O connection. Firmware for these modems is stored in on-board Flash memory. Firmware updates are free and may be uploaded from the HAL web page as required. The PCI-4000/2K and DSP-4100/2K are special versions of the commercial modems that include the CLOVER-2000 waveforms. The same command, control and status report language is used in all four modems. Full details for third-party software authors are provided in a series of HAL Engineering documents, available from the HAL web page (**www.halcomm.com**).

There are many advantages and unique features in a CLOVER system. For example, a CLOVER ARQ link is always bidirectional. Like AX.25 packet radio, CLOVER does not require special "OVER" commands to be sent to change channel direction. Information may flow in either direction at any time. Further, adaptive modulation control is independent for each direction. Strong noise at one site in an ARQ link may reduce data flow to that receiver but will not impede data

flow in the other direction. CLOVER sends 8 bit data at all times. Special data parsing and reconstruction algorithms are not required. Any 8-bit file in the PC can be sent error free via CLOVER, be it an executable program, a binary data file, image file, or digital audio file.

PSK31

Despite its limited character set and lack of error correction, Baudot RTTY remains popular for conversational QSOs, round-tables and nets. Baudot RTTY was designed to work with equipment that does not meet today's standards for stability and selectivity. In the beginning, decoding and printing of Baudot signals was done in mechanical machines. Why do hams continue to use this second-only-to-CW oldest digital mode in the computer age?

There's a big difference between digital mode users who are interested in moving blocks of data from one point to another and those who only want to make two-way conversational contacts. If you're sending data over a network, you'll need speed—lots of it on the node-to-node connections. For a two-way contact you only need enough speed to keep up with your typing. Transferred data needs to be error-free, and if that adds some delays, no problem. By contrast, two-way contacts should have minimal system-imposed delays so that data exchange seems *conversational* to the users.

The error-correcting digital modes aren't designed for the "typical" ham QSO. They're great for transferring data, but not so great for conversations—especially when more than two stations are involved.

Peter Martinez, G3PLX, who brought us AMTOR, has developed a new mode that is designed for today's radio amateur using current ham gear to make QSOs with other amateurs. This mode is called PSK31. The PSK derives from the fact that

Hams can say "I'll fax you" with ease, by using a home computer with the right hardware and software. Ralph Taggart, WB8DQT, of Mason, Michigan, transmitted this picture of his daughter Jennifer with a system that doubles as a slow-scan TV adapter. *(photo by WB8DQT)*

signals use phase-shift keying. The 31 comes from the 31.25 baud rate.

In addition to your transceiver and antenna, you only need a computer with a Windows operating system and a 16-bit Soundblaster card (or compatible) to receive and transmit PSK31. A May 1999 *QST* article by Steve Ford, WB8IMY, explains in easy-to-understand language what you need to know to get in on the fun.

Additional information and software is available for free download over the Web. In addition to Windows software, you'll find versions for DOS, Linux and Macintosh. You'll also find software that does not rely on a particular operating system. You can even join an e-mail reflector to keep up with the latest developments. Point your Web browser to: **http://aintel.bi.ehu.es/psk31. html** for information and links to downloads. You may want to use a Web search engine to find other pages.

PSK31 is still in its infancy, and support for it is still under development. For that reason, it makes sense to watch the pages of *QST*, and to follow developments by visiting key Web pages.

IMAGE—FSTV, SSTV AND FAX

Amateur television (ATV) comes in two main categories: fast-scan TV (FSTV) and slow-scan TV (SSTV). ATV allows hams to send television pictures over the airwaves. SSTV originated earlier; it is a technique used to send color or black-and-white still pictures over HF, using bandwidth comparable to that used by SSB voice. Because of the large amount of information required to transmit a video image, the bandwidth restrictions for HF operation limit the mode to still pictures. FSTV, on the other hand, is used on the UHF bands, where the FCC permits hams to use much broader bandwidths. Therefore, a UHF FSTV picture can be a real-time moving picture.

Many amateur TV experimenters use home camcorders to supply the input to their FSTV transmitters. Watching a received FSTV signal looks almost the same as a commercial TV broadcast. In fact, some FSTV enthusiasts outfit their stations with "professional" video and processing equipment, such as cameras, monitors, switchers, special-effects devices and computers to control and manipulate images. There are numerous FSTV repeaters around the US.

Fax is an additional form of image communications. Fax is an abbreviation for *facsimile*, a form of transmission that sends a visual "photocopy" of a two-dimensional image, such as a piece of paper, a photograph or a diagram. The first fax machines were mechanical devices with paper wrapped around a rotating drum, or scanned by a moving light source. The light and dark spots on the paper (that is blank space and typewritten characters) were converted to electrical impulses. These impulses were sent along wires to a receiving device that converted the impulses back into an image on paper.

Modern amateur faxes are sent using PCs and special software. The images may be almost any computer file format. Images are often created by using a device called a scanner to convert a printed image into a digital format. The software converts the image into audio tones that are sent via the transmitter for decoding at the receiving end. The receiving station can display the faxed image on a computer monitor screen or printer. Fax images can also be "captured" as they are received and saved to a disk file for later access (for example viewing, printing, resending, modification). When a fax transmission begins, the receiving station must be tuned in and listening so that it can receive and decode the synchronization data sent at the beginning of the fax. Without the sync data, the receiving station can't properly interpret the image. This presents a challenge: In voice, Morse code, RTTY, packet or ATV, the operator may identify his or her station using that mode. Faxing is different: The FCC doesn't accept a faxed station identification as a fulfillment of the legal requirement to identify at least once every 10 minutes during a transmission, and interrupting a fax transmission to send a Morse code or voice identification causes the transmitting and receiving stations to lose synchronization. Because faxes can take a long time to transmit, new operators must learn what size file can be faxed in the span of 10 minutes before the transmission must be interrupted to send the required station identification.

SPREAD SPECTRUM

Spread spectrum remains largely an experimental mode. It's limited to certain frequencies and methods of implementation, and requires equipment and designs that are on the edge of radio technology. Advanced amateur and commercial development, however, may lead to wider use of this technique of conserving limited radio spectrum. You can find more information in Chapter 12.

Glossary

ACK—An abbreviation for "acknowledgment." AMTOR and PACTOR stations exchange ACKs to verify that information has been received without errors.

ACSSB—Amplitude Compandored Single Sideband. A narrow-bandwidth, low-noise AM mode designed to compete with narrow-bandwidth FM in the Land Mobile Radio Service.

AM—Amplitude modulation. Full AM transmissions use a full carrier with two modulated sidebands, however, SSB and ACSSB are both AM modes.

AMTOR—Amateur Teleprinting Over Radio. A popular method of digital communication on the HF bands.

APRS—Automatic Position Reporting System. A method that uses the unconnected packet radio mode to graphically indicate the position of moving and stationary objects on maps displayed on a computer monitor.

ASCII—American Standard Code for Information Interchange. A standard method of encoding data so that it can be understood by many computers.

AX.25—The Amateur Radio version of the CCITT X.25 packet protocol (x.25 is used for computer communications over telephone lines).

Bit—A binary digit, 0 or 1, mark or space.

Connect—To establish a data communications link between two packet stations.

CW—Continuous wave. A transmission consisting of an unmodulated carrier.

Digipeater—Digital repeater. A device that receives, temporarily stores, and then retransmits packet radio transmission directed specifically to it.

DXing—Operation to contact far-distant stations (foreign countries on HF, beyond the radio horizon on VHF and higher bands).

FM—Frequency modulation. A form of modulation where the RF carrier shifts frequency according to the amplitude of the modulating audio signal.

FSK—Frequency shift keying. Modulating a transmitter by using data signals to shift the carrier frequency. Commonly used for digital transmissions.

NAK—An abbreviation for "non-acknowledgment." AMTOR and PACTOR stations exchange NAKs to request retransmission of data (due to errors).

Node—A junction point in a packet network where data is relayed to other destinations. A node can support more than one user at a time and operate on several different frequencies simultaneously.

RTTY—Radioteletype. A method of sending text information using shifting MARK/SPACE signals or audio tones.

SSB—A form of amplitude modulation (AM) in which the carrier and one sideband are removed.

TNC (terminal node controller)—Software or hardware that processes packets.

Contents

Mathematics for Amateur Radio

4

S ooner or later, most hams will find they need to make some sort of measurement or perform a calculation as part of their hobby. This is true whether they are calculating an antenna length or designing a new piece of station equipment. When they do, they will be using mathematics. The math skills required for most electronics calculations can be developed and used by just about anyone.

This chapter, written by Larry Wolfgang, WR1B, provides a brief review of the most important math concepts needed for electronics and Amateur-Radio-related use. It will serve as a refresher for those hams who may have been familiar with the topics, but who have long since forgotten how to apply them. The examples will also help those who have no prior math background to work through

many of the calculations associated with this *Handbook*. Those readers who would like a more detailed explanation should turn to the Math Unit of ARRL's *Understanding Basic Electronics*.

Software to perform calculations is discussed in ARRL's *Personal Computers in the Ham Shack*, and new packages are reviewed from time to time in *QST*.

Mathematical Terms and Symbols

Mathematics uses letters, symbols and odd-looking characters to represent various quantities in a kind of short-hand notation we call equations. To those unfamiliar with the language of mathematics, these strange names and symbols can be very confusing. Once you have learned some basic terms and understand what the symbols represent, the elegance of an equation can begin to come through. In this section we will introduce some of the most common mathematical terms and symbols.

DEFINITIONS OF MATHEMATICAL TERMS

Algebra—The branch of mathematics that uses letter symbols to represent various quantities, and which establishes rules for manipulating these expressions. Much of the discussion in this chapter involves the rules of algebra.

Binary number system—A number system that uses only two symbols, 0 and 1. The binary system is very useful in digi-

tal electronics, because most digital electronics circuits only have to measure two voltage or current conditions: *on* or *off*. Most of these circuits represent the *on* condition as a 1 and the *off* condition as a 0. (See also *Decimal number system*, *Hexadecimal number system* and *Octal number system*.)

Cross multiplication—The most common equation-solving technique used with proportions. This involves moving terms diagonally across the equal sign. We can use the letters a, b, c and d to represent four terms of a proportion:

$$\frac{a}{b} = \frac{c}{d}$$

Then by cross multiplication, we can also write the following equivalent expressions:

$$a\,d = b\,c \text{ and } \frac{a}{c} = \frac{b}{d}$$

Cube—Multiplying a number or quantity by itself three times. Cubing a quantity

means it is raised to the third power, or has an *exponent* of 3. ($2^3 = 2 \times 2 \times 2 = 8$)

Cube root—That value which, when multiplied by itself three times, gives the value whose cube root you want to find. ($\sqrt[3]{8} = 8^{1/3} = 2$ and $\sqrt[3]{-8} = -8^{1/3} = -2$) Odd powered roots have only one possible value.

Decimal number system—A number system that uses ten symbols, 0 through 9, to count, measure and calculate. The most common number system. (See also *Binary number system*, *Hexadecimal number system* and *Octal number system*.)

Equation—A statement of mathematical balance. All that appears on one side of the equal sign (=) is equivalent to any expression on the other side. The two sides usually don't appear identical (2 = 2), but the expression on one side represents the expression on the other side ($x = 2$). The *x* here represents a *variable*, or unknown quantity.

Exponent—A value following a number, raised above the line of the number, or written as a **superscript**, to show the number is to be multiplied by itself. (10^3 indicates that 10 is to be multiplied by itself three times — $10 \times 10 \times 10$). The rules of working with exponents are covered later in this chapter.

Formula—Another name for an **equation**, especially when it represents a procedure used to calculate some quantity. ($E = I R$ is a formula that tells us to multiply current times resistance to find voltage.)

Hexadecimal number system—A number system that has 16 characters; labeled 0 through 9, A, B, C, D, E and F. The hexadecimal system (often abbreviated *hex*) is convenient for use with digital computers because hexadecimal digits can be coded as groups of four binary digits. In this case, 0001 represents hex 1, 1000 represents hex 8, 1010 represents hex A and 1111 represents hex F. (See also **Binary number system**, **Decimal number system** and **Octal number system**.)

Infinity—The term used to describe the mathematical concept of having no boundaries. There is no "largest number" or "smallest number," because you can always add 1 to obtain a larger number, or further divide to obtain a smaller one.

Integers—The "counting numbers," such as 1, 2, 3, 4, 5. Integers also include negative values. The number line of **Fig 4.1** is helpful to picture positive and negative integers. (See also **Real numbers**.)

Octal number system—A number system that uses eight characters, 0 through 7. This system is often used with digital computers, because groups of three binary digits can be coded to represent an octal digit. For example, 001 represents octal 1, 010 is the same as octal 2 and 111 represents octal 7. (See also **Binary number system**, **Decimal number system** and **Hexadecimal number system**.)

Power of 10—The exponent used with 10 when a number is written in exponential or scientific notation. The exponent tells how many places and in what direction the decimal point is moved.

Proportion—Two ratios that are equal to each other (or both equal to the same quantity). Proportions are a powerful mathematical tool because you can often write an equation to calculate some unknown quantity based on your knowledge of another ratio. Proportions are useful because when you know three of the four quantities, it is a simple matter to find the fourth. Later in this chapter we show you how to use proportions to convert between US Customary and metric system measurements.

Radical sign— $\sqrt{}$. A symbol written with a number or mathematical expression under the line, to represent a square root, such as $\sqrt{4} = 2$. If there is a superscript number in front of the radical sign, then it represents the root indicated.

$$\sqrt[3]{8} = 2$$

Ratio—A fraction, with one quantity divided by another. The value of π is the

---

Table 4.1
The Greek Alphabet and Common Electronics Quantities

Greek letter	Pronunciation	Upper Case	Common Use	Lower Case	Common Use
Alpha	'al-fə	A	Angle of a triangle	α	Transistor common-base current gain
Beta	'bā t-ə	B	Angle of a triangle	β	Transistor common-emitter current gain
Gamma	'gam-ə	Γ	Transmission line voltage reflection coefficient	γ	Phase
Delta	'del-tə	Δ	Change in quantity	δ	
Epsilon	'ep-sə-län	E		ε	Dielectric constant, permittivity
Zeta	'zāt-ə	Z		ζ	
Eta	'āt-ə	H		η	
Theta	'thāt-ə	Θ	Angles	θ	Angles
Iota	i-'ot-ə	I		ι	
Kappa	'kap-ə	K		κ	
Lambda	'lam-də	Λ		λ	Wavelength
Mu	myü	M		μ	Metric prefix for 10^{-6}, permeability
Nu	nü	N		ν	
Xi	ksi	Ξ		ξ	
Omicron	'äm-ə-krän	O		o	
Pi	pī	Π		π	3.14159 (ratio of circumference to diameter of a circle)
Rho	rō	P		ρ	Transmission line reflection coefficient, resistivity
Sigma	'sig-mə	Σ	Summation of a series	σ	
Tau	tau	T		τ	Time constant, LC circuits
Upsilon	'yüp-sə-län	Y		υ	
Phi	fi	Φ	Angles	ϕ	Angles
Chi	ki	X		χ	
Psi	si	Ψ		ψ	
Omega	o-'meg-ə	Ω	Ohm, resistance, normalized frequency	ω	Frequency in radians per second ($2\pi f$), angular velocity

ratio of the circumference of a circle (C) to the diameter of the circle (d), for example.

$$\left(\pi = \frac{C}{d} \right)$$

Voltage standing-wave ratio (VSWR or SWR) is the ratio of maximum voltage on a feed line to the minimum voltage on the feed line. Written as a fraction, we use this ratio to form an equation that shows one way to calculate SWR:

$$SWR = \frac{V_{max}}{V_{min}}$$

Real numbers—All possible numbers, including all the fractions between integers. (Fractions can be written as a ratio of two numbers, or as a decimal value that is the result of the division. 4.5 and $4\frac{1}{2}$ represent the same real number. The decimal value is often only an approximation, however, such as 6.333 and $6\frac{1}{3}$.)

Reciprocal—A quantity divided into 1 (often written as $1/x$). Reciprocals are so important that the quantity is often given a name of its own. For example, in electronics, the reciprocal of resistance is called conductance. Using letter symbols to represent the quantities (R is resistance and G is conductance).

$$\left(\frac{1}{R} = G \text{ and } \frac{1}{G} = R \right)$$

Root (of a number)—A value which, when multiplied by itself the specified number of times, gives the value whose root you want to find. Most common in electronics is the **square root** of a number, and occasionally the **cube root**. Roots may be written with a **radical** sign ($\sqrt{\ }$) or as a fractional **exponent**. ($\sqrt{4} = 4^{1/2} = 2$)

Square—Multiplying a number or quantity by itself. Squaring a quantity means it is raised to the second power, or has an **exponent** of 2. ($2^2 = 2 \times 2 = 4$)

Square root—That value which, when multiplied by itself, gives the value whose square root you want to find. Actually, there are two values for a square root. ($\sqrt{4} = 4^{1/2} = +2$ and -2) Even-powered roots have both the positive and negative values possible.

Subscript—A number or expression following a variable, written slightly lower than the line of the variable; R_1, R_2, E_3 and E_4 are examples of quantities with subscripts to distinguish similar, but different quantities.

Superscript—A number or expression written following a number or variable, written slightly higher than the line of the number or expression; 5^2, x^3, $(25 + I)^2$ and E^2 are examples of expressions with superscripts.

Variable—An expression that can take on different values. Variables are sometimes given **subscripts**.

GREEK ALPHABET

Upper- and lower-case characters of the Greek alphabet are often used to represent various measurements and constant values. Few English-speaking people are familiar with Greek, so some of these characters can look pretty strange. **Table 4.1** shows the upper and lower case Greek alphabet, the character pronunciations, and the electrical and electronics quantities some of these characters often represent.

TABLE OF MATHEMATICAL SYMBOLS

In addition to Greek characters and other letter symbols, there are many special math symbols used when we write equations. **Table 4.2** shows many of these common math symbols.

Table 4.2

Some Common Mathematical Symbols

Symbol	Meaning
+	Addition, plus
−	Subtraction, minus
±	Plus or minus
×, •, *	Multiplication, multiply by
÷, /	Division, divide by
=	Equal to
≠	Not equal to
≈	Approximately equal to
~	Similar, equivalent
<	Less than
≤	Less than or equal to
>	Greater than
≥	Greater than or equal to
:	Ratio of, is to
∝	Proportional, varies directly as
∴	Therefore
°	Degree
∠	Angle
∟	Right angle
⊥	Perpendicular to
‖	Parallel to
≡	Identical to
∞	Infinity
√	Radical, square root (Also written with a superscript before the symbol to express other roots, such as $\sqrt[3]{\ }$ to represent a cube root.)
∫	Integration
Σ	Summation

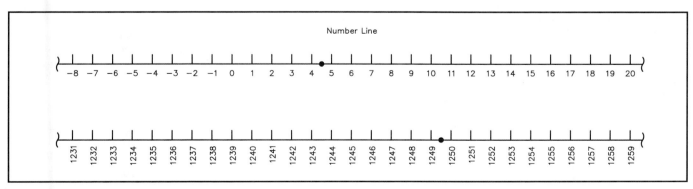

Fig 4.1 — The number line gives us a way to represent all numbers, both positive and negative, and is useful for remembering arithmetic operations.

Significant Figures and Decimal Places

Any measurement is only as good as the measuring instrument used, and as reliable as the person making the measurement. The *accuracy* of a measurement refers to how close the value is to an accepted value or standard. You might use a dip oscillator to measure the resonant frequency of a radio circuit. The measured value will probably not be very accurate, since dip oscillators are usually designed only to measure a general range. If you want to ensure that your transmitter is operating inside the amateur band edge, you might try a frequency counter or crystal calibrator, for example. So the measuring instrument plays the greatest role in determining the accuracy of a measurement. (This assumes the person taking the measurement understands how to use the instrument to take full advantage of it. An operator who does not know how to use or read the instrument properly will not obtain accurate measurements!)

Precision refers to the repeatability of a measurement. You might take five frequency readings using the dip meter mentioned above, with all five readings being 7.14 MHz. This set of measurements would have good *precision*, but because the instrument is not designed for high *accuracy*, you can't be sure of the actual frequency. Other factors might affect the precision of a measurement, such as the operator's skill at adjusting the measuring instrument, errors in reading the scale and the operation of the instrument itself.

The value you can read from the scale of a measuring instrument helps determine its precision. If you are using a ruler marked off only to the nearest quarter inch, you may be able to estimate measurements to an eighth inch, but you certainly can't read that scale to the nearest thirty-second of an inch! It will be difficult to measure two objects that differ in length by a sixteenth of an inch with this ruler. Precision also indicates the *resolution* of a measurement, or how small a change can really be detected.

The *significant figures* of a measurement represent all the digits that you can read directly from the scale, plus one digit that is estimated. **Fig 4.2** shows a voltmeter scale marked from 0 to 10 V, with lines indicating every 0.2 V. You can see that the needle indicates a value between 4.6 and 4.8 V, and perhaps you can even tell if the needle is more or less than half way between the marks. You know the reading is a little less than 4.7 V, but you really can't be sure how much less. You can estimate that the reading is 4.68 perhaps, but the 8 can't be read directly from the scale. Someone else might look at the same reading on the same meter and estimate the value at 4.67 V or even 4.69 V. None of these readings is more correct than any other, because each represents an *estimate* of the value of the last digit. This reading has three significant figures.

Any calculations made involving this measurement are limited by the accuracy of the reading. It would be completely unreasonable to say that 4.68 V produces a current of 17.01818 mA when connected to a 275-Ω resistor, even though your calculator shows all these digits.

The rules of significant figures tell us how many digits to include in any calculation based on a measured quantity. They also help us predict the accuracy of a calculation based on real component tolerances. These rules allow us to specify the accuracy of the calculated value, as it relates to the measurements on which the calculation is based. There are six rules to tell you how to count and write significant figures in a measured or calculated quantity.

1. All nonzero digits are significant: 275.4 mA has four significant digits.
2. All zeros between nonzero digits are significant: 25.004 m has five significant digits.
3. Zeros to the right of a nonzero digit, but to the left of an understood decimal point are not significant unless they are specifically indicated to be significant. You can indicate such zeros to be significant by drawing a bar over the rightmost significant zero: 21100000 hertz has three significant figures; 21100̄000 hertz has five significant figures.
4. Zeros to the right of a nonzero digit, but to the left of an expressed decimal point are significant: 21100. kHz has five significant figures.
5. Zeros to the right of a decimal point but to the left of all nonzero digits are not significant: 0.001702 A has four significant figures. (There is a zero before the decimal point to indicate that no digits to the left of the decimal point were dropped. Also notice that the zero between the 7 and the 2 is significant — remember rule 2.)
6. All zeros to the right of a decimal point and following a nonzero digit are significant: 2.00 V has three significant figures.

There are a few rules for determining the number of significant figures that result when you use measured values in a calculation. These rules are important to ensure that you don't imply a result to a greater precision than the measurements would allow.

Notice we said *measured* values here. Never use the number of digits in the value of a physical constant to limit the number of significant figures in a calculation. For example, the constant 2 or the value of π won't limit the number of significant fig-

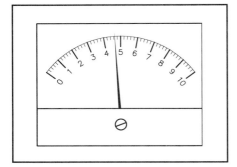

Fig 4.2 — This voltmeter scale reads from 0 to 10 V, with marks every 0.2 V. The meter is reading a value greater than 4.6 but less than 4.7 V. With care you may be able to estimate the reading as 4.68 V, but the digit 8 really only represents a guess. That digit is uncertain because you cannot read it directly from the scale.

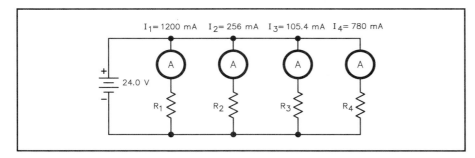

Fig 4.3 — This parallel circuit has four branches connected to the battery. Ammeters measure the current through the branches, with the readings indicated. This circuit illustrates measurements with two, three and four significant figures (I_1 and I_4, I_2, I_3 respectively).

ures in a calculation of reactance. ($X_L = 2 \pi f L$) Likewise, values of trigonometric functions and logarithm values aren't usually limited by the number of significant figures in the number used to find the function value.

When adding or subtracting measurements, remember that the rightmost significant figure represents an uncertain value. The rightmost significant figure in a sum or difference calculation occurs in the leftmost place that an uncertain value occurs in any of the measured quantities. The following example illustrates this rule.

Fig 4.3 shows a parallel circuit with four branches. Ammeters measure the current through each branch. The four current measurements are 1200 mA, 256 mA, 105.4 mA and 780 mA. What is the total current supplied to this circuit? The first measurement has only two significant figures, and the hundreds column represents an uncertain value. The second measurement has three significant figures and the units column is uncertain. The third current has four significant figures, with the tenths column being uncertain. The last current value has two significant figures, and the tens column is uncertain. **Fig 4.4** shows how to determine the last significant place in the answer. Round off the result of this addition to the hundreds place, so this circuit has a total current of 2300 mA. (The rules for rounding numbers are covered later in this section.)

When you multiply or divide measured quantities, the answer cannot have more significant figures than the least precise factor. As an example, use Ohm's Law to calculate the resistor values in Fig 4.3. **Fig 4.5** shows how to determine the significant figures for these calculations.

ROUNDING VALUES

After you determine which digit is the last significant figure in a calculation, you will have to round off the arithmetic answer. Four rules govern how to round off values properly.

1. If the first digit to be dropped is 4 or less, the preceding digit is not changed: 456351 rounded to three significant figures becomes 456000 (with no decimal point at the end).
2. If the first digit to be dropped is 6 or more, the preceding digit is increased by 1: 456351 rounded to two significant figures becomes 460000 (with no decimal point at the end).
3. If the digits to be dropped are 5 followed by digits other than zeros, the preceding digit is increased by 1: 456351 rounded to four significant figures becomes 456400 (with no decimal point at the end).
4. If the digits to be dropped are 5 followed by zeros (the digit to be dropped is exactly 5) the preceding digit is *not* changed if it is even; it is raised by 1 if it is odd: 456350 rounded to four significant figures becomes 456400 but 456450 rounded to four significant figures also becomes 456400 (with no decimal point at the end). (Another way to think of this rounding rule is that when the digit to be dropped is exactly 5, we round to the *even* value.)

```
 ↓ ──────────── Left most uncertain value
                in measurements
 1200   mA
  256   mA
  105.4 mA
+ 780   mA
 ─────────
 ↓ ──────────── Right most significant
 2341.4         figure in answer
 2300 ← Total current value rounded
        off to show the hundreds column
        as the last significant figure
```

Fig 4.4 — This calculation shows the proper use of the rule for addition and subtraction using significant figures. The current measurements from Fig 4.3 are added to calculate the total circuit current. The resulting value is then rounded off to make the hundreds column the last significant figure.

Three Significant Figures

$$R_1 = \frac{E}{I_1} = \frac{24.0\ \text{V}}{1200\ \text{mA}} = \frac{24.0\ \text{V}}{1.2\ \text{A}} = 20.\ \Omega$$

Two Significant Figures — Two Significant Figures

Three Significant Figures

$$R_2 = \frac{E}{I_2} = \frac{24.0\ \text{V}}{256\ \text{mA}} = \frac{24.0\ \text{V}}{0.256\ \text{A}} = 93.75\ \Omega = 93.8\ \Omega$$

Three Significant Figures — Three Significant Figures

Three Significant Figures

$$R_3 = \frac{E}{I_3} = \frac{24.0\ \text{V}}{105.4\ \text{mA}} = \frac{24.0\ \text{V}}{0.1054\ \text{A}} = 227.70398\ \Omega = 228\ \Omega$$

Four Significant Figures — Three Significant Figures

Three Significant Figures

$$R_4 = \frac{E}{I_3} = \frac{24.0\ \text{V}}{780\ \text{mA}} = \frac{24.0\ \text{V}}{0.78\ \text{A}} = 30.76923\ \Omega = 31\ \Omega$$

Two Significant Figures — Two Significant Figures

Fig 4.5 — These calculations show the proper use of the rule for multiplication and division using significant figures. The battery voltage and current measurements from Fig 4.3 are used to calculate the four resistor values. The resulting values are rounded off to show the proper number of significant figures.

Laws of Exponents

Exponents tell how many times a number or quantity is to be multiplied by itself. Equations often involve terms that include exponents. Two special cases with exponents are worth mention before we cover the rules for mathematical operations with exponents:

$$a^1 = a \text{ and } a^0 = 1$$

Any value raised to a power of 1 gives the value itself, and any value raised to the zero power is 1. We can also use numbers to give a few examples:

$$10^1 = 10, \ 5^1 = 5 \text{ and } 3^1 = 3$$

$$10^0 = 1, \ 5^0 = 1 \text{ and } 3^0 = 1$$

When you know the basic rules of algebra involving exponents, you will be able to manipulate the terms in an equation. There are only a few rules to remember, and they involve multiplication and division of numbers with exponents.

1. If you are adding, subtracting, multiplying or dividing two numbers involving exponents, calculate the values indicated by the exponents first, then perform the indicated operation on the numbers:
$a^x \times b^y$ can't be simplified unless you know the values of the variables.
$$2^3 \times 4^2 = 8 \times 16 = 128$$

2. If the multiplication involves a variable raised to different exponents, you can add the exponents:
$$a^x \cdot a^y = a^{x+y}$$
$$2^3 \cdot 2^4 = 2^{3+4} = 2^7 = 8 \cdot 16 = 128$$
Notice the "multiplication dot" used in this example. It is also common practice to omit any symbol when multiplication of variables is intended, and there is no chance of confusion, such as would occur if you were writing two numbers:
We can write this as $a^x a^y = a^{x+y}$ but we would not write 8 16 to indicate 8 · 16.

3. For division of a variable with exponents, subtract the denominator (bottom of the fraction) exponent from the numerator exponent.
$$\frac{a^x}{a^y} = a^{x-y}$$
(This is only true if $a \neq 0$)
$$\frac{2^4}{2^2} = 2^{4-2} = 2^2 = 4$$
As another example, the denominator exponent can be larger than the numerator exponent, resulting in a calculation with a negative exponent:

$$\frac{2^2}{2^4} = 2^{2-4} = 2^{-2} = 0.25$$

A negative exponent indicates you are to find a **reciprocal** of the quantity. We could also write the example above as:

$$\frac{1}{2^{4-2}} = \frac{1}{2^2} = \frac{1}{4} = 0.25$$

From the examples shown here, you should notice a related rule of exponents. Any factor with an exponent can be moved between the numerator and denominator of a fraction simply by changing the sign of the exponent. You will probably want to use a calculator to raise numbers to various powers. This will be much easier than doing the repeated multiplications by hand.

4. To raise a number with an exponent to some power, multiply the exponents.
$$(a^x)^y = a^{x \, y}$$
$$(2^3)^2 = 2^{3 \times 2} = 2^6 = 64$$

5. The product of two variables raised to a power is the same as raising each variable to the power and then finding the product.
$$(a \, b)^m = a^m \, b^m$$
$$(2 \times 4)^3 = 2^3 \times 4^3 = 8 \times 64 = 512$$

6. The ratio of two variables raised to a power is the same as the ratio of each variable raised to the power.
$$\left(\frac{a}{b}\right)^m = \frac{a^m}{b^m}$$
(This is only true if $b \neq 0$)
$$\left(\frac{4}{2}\right)^3 = \frac{4^3}{2^3} = \frac{64}{8} = 8$$

A quantity or expression can also have an exponent, indicating the entire quantity is to be multiplied by itself: $(3x + 12)^2$ means the quantity inside the parentheses is to be multiplied by itself. Calculating *squares* and *cubes*, *square roots* and *cube roots*, are common mathematical operations.

EXPONENTIAL AND SCIENTIFIC NOTATION

Electronics measurements and calculations often involve numbers that are very large or very small. We can represent the metric prefixes as multiples of 10. It is also convenient to use multiples of 10 to represent very large and very small numbers. Any number expressed as some multiple of 10 is written in *exponential notation* (sometimes called *engineering notation*) because the 10 is written with an *exponent*. This exponent, often called a power

of 10, represents how many times the number is multiplied by 10 to write it in *expanded notation*, or the "normal" format.

We can write 250000 in exponential notation as 25×10^4. All we had to do here was replace the four zeros with "$\times 10^4$." As another example, we can write 0.000025 as 25×10^{-6}. In this case we would have to divide 25 by 10 six times. A negative exponent means divide. If you move the decimal point to the right when you write the number in exponential notation, then use a negative exponent.

A number expressed with a single digit to the left of the decimal point and a power of 10 is written in *scientific notation*. This is just a particular form of engineering notation.

We could write the speed of light as 300000000 meters per second, for example, but it is more convenient to write this number as 3×10^8 meters per second. Notice that this number indicates one significant figure. If you wanted to indicate three significant figures, for example, you would write it as 3.00×10^8 meters per second.

You may see several other forms of exponential notation. Sometimes an E is written in place of the 10. ($3.56E6 = 3.56 \times 10^6$) Other times a P is used to represent a positive power of 10 while an N represents a negative power of 10. ($3.56P6 = 3.56 \times 10^6$ and $2.44N3 = 2.44 \times 10^{-3}$)

To write any number in scientific notation, first move the decimal point so there is one nonzero digit to the left of the decimal point. Then count how many places left or right you moved the decimal point. You will use this number as the exponent in the "$\times 10$" factor. If you moved the decimal to the left, use a positive exponent and if you moved the decimal to the right, use a negative exponent.

$$
\begin{array}{cc}
\begin{array}{r}
25.40 \times 10^3 \\
6.15 \times 10^3 \\
+ \ 0.05 \times 10^3 \\
\hline
31.60 \times 10^3 \\
\text{(A)}
\end{array}
&
\begin{array}{r}
25.40 \times 10^{-3} \\
- \ 6.15 \times 10^{-3} \\
\hline
19.25 \times 10^{-3} \\
\text{(B)}
\end{array}
\end{array}
$$

Fig 4.6 — Examples of addition and subtraction with numbers written in exponential notation. Be sure all the numbers have the same power of 10, and then write the numbers so the decimal points align. Add or subtract the number part, and use the common power of 10 with the answer.

Arithmetic Operations with Scientific Notation

One advantage of writing very large and very small numbers in exponential notation is that you don't have to keep track of so many zeros during arithmetic operations. When you work with numbers written in exponential or scientific notation you must remember a few rules, however.

To **add** or **subtract**, be certain to express all the numbers with the same power of 10. Write the numbers in a column so the decimal points align. Then add or subtract the "plain number" part as you normally would. The power of 10 for your answer is the same power in which all the numbers are expressed. **Fig 4.6** shows a sample addition and a sample subtraction using exponential notation.

To **multiply** numbers using exponential notation, first multiply the "plain number" part. Next, add the exponents for the powers of 10. Your answer is the plain-

$$\begin{array}{r} 25.40 \times 10^3 \\ \times\ 6.15 \times 10^3 \\ \hline 12700 \\ 2540 \\ 15240 \\ \hline 156.2100 \times 10^6 \\ 156 \times 10^6 \end{array}$$

Fig 4.7 — This example shows how to multiply two numbers using exponential notation. First multiply the number parts, then add the exponents for the powers of 10.

number answer times a power of 10 equal to the sum of the exponents. **Fig 4.7** shows a sample multiplication using exponential notation.

The rule for division using exponential notation is similar to the multiplication rule. To **divide** numbers written in exponential notation, first divide the "plain

$$\frac{25.40 \times 10^3}{6.15 \times 10^3} = \frac{25.40 \times 10^3 \times 10^{-3}}{6.15}$$

Fig 4.8 — This example shows how to divide two numbers using exponential notation. First divide the number parts, then subtract the denominator (the bottom part of the fraction) exponent from the numerator (the top part of the fraction) exponent. Notice in this example we moved the denominator power of 10 into the numerator and changed the sign of the exponent.

number" parts. Then subtract the denominator power from the numerator power. (The denominator is the bottom part of a fraction and the numerator is the top part.) **Fig 4.8** shows a sample division. Notice that we moved the denominator power of 10 into the numerator and changed the sign of the exponent.

Equations

Much of algebra involves manipulating equations. We know the quantities on each side of the equal sign are equivalent. Usually the goal is to find the value of some unknown quantity. We do this by isolating that unknown quantity on one side of the equal sign, and then evaluating the expression on the other side.

Here is the most important rule to remember when you try to *solve* an equation for the unknown quantity: **Be neat!** Write each step clearly. In a jumbled mess of numbers and symbols, you will soon be hopelessly lost.

The second-most-important rule is just as significant: Anything you do to one side of the equation you must also do to the other side. A few examples will illustrate some of the most common procedures.

$$2x + 4 = 8$$

We can simplify this equation by subtracting 4 from both sides of the equation:

$$(2x + 4) - 4 = 8 - 4$$

$$2x = 4$$

Now we can complete the solution by dividing both sides of the equation by 2. (This is the same as multiplying both sides by the reciprocal of the term associated with the unknown, x.)

$$\frac{1}{2} \bullet 2x = \frac{1}{2} \bullet 4$$

$$x = 2$$

Other techniques that can be used to manipulate equations were described earlier in this chapter. See *cross multiplication*, *reciprocal* and the discussion of the Laws of exponents for some important equation-solving principles.

In electronics, we often find problems in which there is more than one unknown quantity. In such cases, try to write a series of equations with the unknowns. If there are two unknown quantities, then find two equations involving those quantities. If there are three unknown quantities, find three equations, and so on. Such *systems* of *simultaneous* equations can help solve some challenging problems.

Fig 4.9 shows the schematic diagram of a simple electronics circuit. We would like to know the power dissipated in the resistor. (Power is equal to current times voltage.) In this example, we only know voltage, however. So we have two unknown quantities; current and power. Since we also know resistance, we can write a second equation from Ohm's Law, to calculate current by dividing the voltage by the resistance:

$$P = IE$$

$$I = \frac{E}{R}$$

From these two equations, we can substitute the expression for current from the second equation for current in the first equation:

$$P = \frac{E}{R} E$$

You probably recognize that E times E can be written as E^2, so we simplify this equation as:

$$P = \frac{E^2}{R}$$

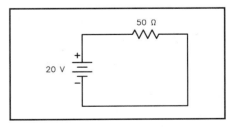

Fig 4.9 — This circuit includes a 20-V battery and a 50-Ω resistor. The text explains how we can calculate the power dissipated in the resistor.

Now it is a simple matter to fill in the known quantities of voltage and resistance to calculate power:

$$P = \frac{(20\text{ V})^2}{50\,\Omega} = \frac{400\text{ V}^2}{50\,\Omega} = 8\text{ W}$$

This example illustrates several important techniques. First, we used *substitution* to solve this problem. We substituted one expression for an equal quantity. We also used *literal equations* to solve the problem. This means we used letter symbols to represent the quantities until the last step. We could have put numbers in the equations right at the beginning, but it is often easier (and there are fewer opportunities to copy a number incorrectly) to use letter symbols. Finally, we used *dimensional analysis* with the calculation. That means we included the units associated with each measurement, and performed all the algebra operations on the units as well as the numbers. You can see this because we have volts squared in the numerator. You should check **Table 4.5** to see that volts squared divided by ohms is equivalent to watts.

Dimensional analysis is a very helpful mathematical tool if you take advantage of it. You can often use this method to help you remember the proper equation for a calculation. For example when you know that the unit of a watt can be expressed as an amp times a volt (see Table 4.5) you can write an equation that gives power as current times voltage. You can also use dimensional analysis to write a power equation involving current and resistance or voltage and resistance. Try writing

these equations with the help of Table 4.5.

Linear equations involve only unknown terms with exponents no larger than 1. For any value of one variable (x) there is a corresponding value for the second variable (y). A graph of such an equation will often help you visualize the relationship between variables. For example, if x represents the current through a circuit, y might represent the voltage across a resistor. A general expression of a linear equation is:

$$y = \text{m}x + \text{b}$$

where m represents the *slope* of the line (the change in y divided by the corresponding change in the x variable) often written as

$$\frac{\Delta y}{\Delta x}$$

and b represents the *y intercept*, or the point where the line crosses the vertical axis when $x = 0$.

An equation that involves a variable term with an exponent of 2 (a squared term) is called a *quadratic equation*. The general form of a quadratic equation is:

$$\text{a}x^2 + \text{b}x + \text{c} = 0$$

where a, b and c are constant terms, or values for a particular equation, and x is the variable quantity.

Quadratic equations always have two solutions, which means there are two values for x that satisfy the equation. Perhaps the most straightforward way to solve a quadratic equation for the unknown quantity is to use the *quadratic formula*. This

formula can be used to solve any quadratic equation.

$$x = \frac{-\text{b} \pm \sqrt{\text{b}^2 - 4\,\text{a}\,\text{c}}}{2\,\text{a}}$$

Notice the plus or minus symbol in front of the radical sign. This tells us that one solution requires that we add the resulting term to $-$b and the other solution requires that we subtract the resulting term from $-$b. This comes about because when you square a negative number, you get a positive result. There are always two solutions to a square root. A simple example will illustrate the use of the quadratic formula.

$$2x^2 + 4x - 6 = 0$$

$$x = \frac{-\text{b} \pm \sqrt{\text{b}^2 - 4\,\text{a}\,\text{c}}}{2\,\text{a}}$$

$$x = \frac{-4 \pm \sqrt{4^2 - (4 \times 2 \times (-6))}}{2 \times 2}$$

$$x = \frac{-4 \pm \sqrt{16 - (-48)}}{4}$$

$$x = \frac{-4 \pm \sqrt{64}}{4} = \frac{-4 \pm 8}{4}$$

$$x = \frac{-4 + 8}{4} = \frac{4}{4} = 1$$

$$x = \frac{-4 - 8}{4} = \frac{-12}{4} = -3$$

Measurement Units and Constants

Nearly every time we use a number, it represents some measured physical quantity. We might use a measuring tape to find the correct length of wire for an antenna, or an ohmmeter to measure the value of a resistor.

Each measurement includes a number representing its size and a unit that allows it to be compared with other measurements of a similar type. One dipole may be 126 ft, 6 inches long and another dipole may be 65 ft, 11 inches long, for example. The units used with any measurement represent standards that are generally accepted, so meaningful comparisons can be made. These comparisons are only meaningful if the measurements use the same units. It is difficult to compare a dipole

that is 65 ft 11 inches long with one specified as a 40-m dipole, because the units are not the same.

US CUSTOMARY SYSTEM

Most US residents are familiar with units like the inch, foot, quart, gallon, ounce and pound. These units represent standard measurement values used in the *US Customary* measuring system. **Table 4.3** lists some common US Customary units, and some not-so-common ones. This table shows the relationships between various *linear*, *area*, *liquid volume*, *dry volume* and *weight* measurements.

The primary disadvantage of the US Customary measuring system is that there is no logical relationship between various-

sized units of a similar type. Most electronics measurements are made in the internationally accepted *metric* system, for this reason.

METRIC SYSTEM

In the metric system, measuring units are always a multiple of 10 times larger or smaller than other units of the same type. Metric-system measurements are always based on a measurement unit and a set of prefixes to describe the larger and smaller variations of that unit. For example, a millimeter is ten times smaller than a centimeter, a meter is a hundred times larger than a centimeter and a kilometer is a thousand times larger than a meter. (In nearly every country of the world except the US,

Table 4.3
US Customary Units

Linear Units
12 inches (in) = 1 foot (ft)
36 inches = 3 feet = 1 yard (yd)
1 rod = 5½ yards = 16½ feet
1 statute mile = 1760 yards = 5280 feet
1 nautical mile = 6076.11549 feet

Area
1 ft^2 = 144 in^2
1 yd^2 = 9 ft^2 = 1296 in^2
1 rod^2 = 30¼ yd^2
1 acre = 4840 yd^2 = 43,560 ft^2
1 acre = 160 rod^2
1 mile^2 = 640 acres

Volume
1 ft^3 = 1728 in^3
1 yd^3 = 27 ft^3

Liquid Volume Measure
1 fluid ounce (fl oz) = 8 fluidrams = 1.804 in^3
1 pint (pt) = 16 fl oz
1 quart (qt) = 2 pt = 32 fl oz = 57¾ in^3
1 gallon (gal) = 4 qt = 231 in^3
1 barrel = 31½ gal

Dry Volume Measure
1 quart (qt) = 2 pints (pt) = 67.2 in^3
1 peck = 8 qt
1 bushel = 4 pecks = 2150.42 in^3

Avoirdupois Weight
1 dram (dr) = 27.343 grains (gr) or (gr a)
1 ounce (oz) = 437.5 gr
1 pound (lb) = 16 oz = 7000 gr
1 short ton = 2000 lb, 1 long ton = 2240 lb

Troy Weight
1 grain troy (gr t) = 1 grain avoirdupois
1 pennyweight (dwt) or (pwt) = 24 gr t
1 ounce troy (oz t) = 480 grains
1 lb t = 12 oz t = 5760 grains

Apothecaries' Weight
1 grain apothecaries' (gr ap) = 1 gr t = 1 gr a
1 dram ap (dr ap) = 60 gr
1 oz ap = 1 oz t = 8 dr ap = 480 gr
1 lb ap = 1 lb t = 12 oz ap = 5760 gr

Table 4.4
Metric Prefixes

Prefix	Symbol			Multiplication Factor
exa	E	10^{18}	=	1,000,000,000,000,000,000
peta	P	10^{15}	=	1,000,000,000,000,000
tera	T	10^{12}	=	1,000,000,000,000
giga	G	10^{9}	=	1,000,000,000
mega	M	10^{6}	=	1,000,000
kilo	k	10^{3}	=	1,000
hecto	h	10^{2}	=	100
deca	da	10^{1}	=	10
(unit)		10^{0}	=	1
deci	d	10^{-1}	=	0.1
centi	c	10^{-2}	=	0.01
milli	m	10^{-3}	=	0.001
micro	μ	10^{-6}	=	0.000001
nano	n	10^{-9}	=	0.000000001
pico	p	10^{-12}	=	0.000000000001
femto	f	10^{-15}	=	0.000000000000001
atto	a	10^{-18}	=	0.000000000000000001

this unit of distance measurement is spelled *metre*, which helps distinguish the distance unit from an electrical measuring instrument, also called a meter.) **Table 4.4** shows the common metric prefixes, their abbreviations and the multiplication factor associated with each one.

SI

The metric units make up an internationally recognized measuring system used by most scientists throughout the world. We call this the International System of Units, abbreviated SI (for the French, Système International d'Unités).

By the late 1700s, scientists were developing this measuring system based on multiples of ten. The original intent was to develop a measuring system based on measuring units that could be reproduced as needed. The meter was first defined as one ten millionth of the distance between the equator and the north pole, as measured along the longitude line running through Paris, France. A kilogram was originally defined as the mass of 1 liter (spelled *litre* throughout the rest of the world) of water at 4°C. As measuring instruments improve, the definitions are revised to reflect the greater measuring accuracy. For example, in 1960 the definition of a meter was revised to be a multiple of the wavelength of a particular orange-red light wave.

The metric system is based on the definitions of certain fundamental units, with all other units being based on those units. These fundamental, or defined units represent length (meter), mass — you might think of this as somewhat equivalent to weight — (gram), time (second), thermodynamic temperature (kelvin or degree celsius), luminous intensity of light (candela), the amount of substance — a measure of the number of atoms or molecules — (mole) and electric current (ampere).

All other units represent combinations of these fundamental units. For example, the unit of power (watt) is a measure of the energy required to move a one kilogram object a vertical distance of one meter in one second. **Table 4.5** lists some common units and their expression in terms of the base SI units.

Suppose you measure the frequency of a radio wave as 3825000 hertz. If we move the decimal point six places to the left, we would write this frequency as 3.825×10^6

hertz. Looking at Table 4.4, you can replace the "$\times 10^6$" part with the prefix mega. So we can write this frequency as 3.825 MHz (using the abbreviations M for mega and Hz for hertz). Similarly, you can use other metric-system prefixes to replace powers of 10 in large and small numbers.

As another example, suppose you find a capacitor marked with a value of 25 microfarads or 25 μF. From Table 4.4, you can find the multiplication factor of 10^{-6} for the prefix micro. That means you can write this capacitor value as 25×10^{-6} farads, or 0.000025 F.

CONVERTING BETWEEN US CUSTOMARY AND METRIC SYSTEMS

Sometimes it is convenient to convert between these two common measuring systems. You may know an antenna length in meters, but want to use your tape measure marked in feet and inches to cut the antenna, for example. **Table 4.6** lists most of the conversion factors you will ever need.

Using Proportions to Solve Conversion Problems

To solve US Customary and metric conversions we will use proportions. This method also illustrates some other very useful mathematical tools. The advantage of using proportions to solve conversions is that you never have to figure out if you must multiply or divide. The proportion shows you what to do!

To set up a conversion between metric and US Customary units, just make a ratio of the conversion factors and an-

Table 4.5

SI Fundamental Units

Quantity	Unit Name	Symbol	In Terms of Other Units	In terms of SI Base Units
Distance, length	meter	m		m
Mass	kilogram	kg		kg
Time	second	s		s
Thermodynamic temperature	kelvin	K		K
Luminous intensity	candela	cd		cd
Amount of substance	mole	mol		mol
Electric current	ampere	A		A

SI Derived Units

Quantity	Unit Name	Symbol	In Terms of Other Units	In terms of SI Base Units
Force, pressure	newton	N		$\dfrac{\text{m kg}}{\text{s}^2}$
Energy, work	joule	J	$N\,m,\ \Omega A^2 s$	$\dfrac{\text{m}^2\,\text{kg}}{\text{s}^2}$
Frequency	hertz	Hz		$\dfrac{\text{cycles}}{\text{s}}$
Power	watt	W	$\dfrac{V^2}{\Omega},\ A\,V$	$\dfrac{\text{m}^2\,\text{kg}}{\text{s}^3}$
Electric charge, quantity of electricity	coulomb	C		$s\,A$
Electromotive force, voltage	volt	V	$A\,\Omega,\ \dfrac{W}{A}$	$\dfrac{\text{m}^2\,\text{kg}}{\text{s}^3\,\text{A}}$
Electric resistance	ohm	Ω	$\dfrac{V}{A}$	$\dfrac{\text{m}^2\,\text{kg}}{\text{s}^3\,\text{A}^2}$
Electric conductance	siemens	S	$\dfrac{A}{V}$	$\dfrac{\text{s}^3\,\text{A}^2}{\text{m}^2\,\text{kg}}$
Capacitance	farad	F	$\dfrac{C}{V}$	$\dfrac{\text{s}^4\text{A}^2}{\text{m}^2\,\text{kg}}$
Inductance	henry	H	$\dfrac{V\,s}{A}$	$\dfrac{\text{m}^2\,\text{kg}}{\text{s}^2\,\text{A}^2}$

Table 4.6

Metric Conversion Factor		US Customary Conversion Factor
(Length)		
25.4	mm	1 inch
2.54	cm	1 inch
30.48	cm	1 foot
0.3048	m	1 foot
0.9144	m	1 yard
1.609	km	1 mile
1.852	km	1 nautical mile
(Area)		
645.16	mm^2	1 inch^2
6.4516	cm^2	1 in^2
929.03	cm^2	1 ft^2
0.0929	m^2	1 ft^2
8361.3	cm^2	1 yd^2
0.83613	m^2	1 yd^2
4047	m^2	1 acre
2.59	km^2	1 mi^2
(Mass)		(Avoirdupois Weight)
0.0648	grams	1 grains
28.349	g	1 oz
453.59	g	1 lb
0.45359	kg	1 lb
0.907	tonne	1 short ton
1.016	tonne	1 long ton
(Volume)		
16387.064	mm^3	1 in^3
16.387	cm^3	1 in^3
0.028316	m^3	1 ft^3
0.764555	m^3	1 yd^3
16.387	ml	1 in^3
29.57	ml	1 fl oz
473	ml	1 pint
946.333	ml	1 quart
28.32	l	1 ft^3
0.9463	l	1 quart
3.785	l	1 gallon
1.101	l	1 dry quart
8.809	l	1 peck
35.238	l	1 bushel
(Mass)		(Troy Weight)
31.103	g	1 oz t
373.248	g	1 lb t
(Mass)		(Apothecaries' Weight)
3.387	g	1 dr ap
31.103	g	1 oz ap
373.248	g	1 lb ap

other of the measurements. Set the two ratios equal to each other and solve the proportion for the unknown measurement.

$$\frac{\text{metric conversion}}{\text{US conversion}} = \frac{\text{metric measurement}}{\text{US measurement}}$$

Suppose we know that a dipole antenna is 126 ft 6 inches long. (For simplicity, change that to 126.5 ft.) How long is this antenna in meters? Look down the *Metric Unit* column of Table 4.6 until you find an "m" for "meters." Then go across to the *US Unit* column to find "foot." Notice there are two conversion factors for meters; one for "foot" and another for "yard." When you've located the proper conversion factor, you will see there is 0.3048 meter in 1 ft. (A meter is a little more than 3 ft.) We know the US measurement in this case, and want to find the metric measurement.

$$\frac{0.3048\ \text{m}}{1\ \text{ft}} = \frac{\text{metric measurement}}{126.5\ \text{ft}}$$

In this example we will ***cross multiply*** the US measurement term to leave the unknown measurement by itself on one side of the equal sign. When you cross multiply, just take any part of the proportion diagonally across the equal sign.

$$\frac{0.3048\ \text{m} \times 126.5\ \text{ft}}{1\ \text{ft}} = \text{metric measurement}$$

$$= \frac{38.56\ \text{m ft}}{1\ \text{ft}} = 38.56\ \text{m}$$

Notice that we include the appropriate units with the metric and US conversion factors. After we cross multiply, the units of feet that go with the dipole length cancel with the units of feet that go with the US conversion factor, leaving only units of meters in our answer. *Dimensional analysis* ensures we have solved the proportion properly.

Trigonometry

Trigonometry refers to the mathematics of angles, especially as they relate to triangles. When two lines meet or cross, they form angles. The point where the lines meet or cross is called the *vertex* of the angle. We usually measure an angle as an arc of a circle across the smallest opening between the lines. We can describe an angle as falling in one of three categories. **Fig 4.10** shows a *right angle*, an *acute angle* and an *obtuse angle*.

The figure shows the angles in terms of a degree measurement. A degree is $1/360^{th}$ of a circle, or $1/360^{th}$ of a complete revolution. We can also measure angles in radians. A radian is an angle measure obtained by taking the length of the radius of a circle and laying that length along the circumference of that circle. See **Fig 4.11**. There are 2π radians in one circle, or $360°$. Two useful conversion relationships are:

$1° = 1.745 \times 10^{-2}$ radians, and:
1 radian = $57.296°$

When three lines cross in such a manner that they form three angles, these lines form a *triangle*. We normally identify a triangle by the largest angle that it includes. This means there are three types of triangles. **Fig 4.12** shows examples of the three types of triangles. In electronics, we will use right triangles in a variety of calculations.

If you add the three angles in a triangle, you will always get a total of $180°$. For a right triangle, then, the sum of the other two angles must be $90°$.

The triangles shown in Fig 4.12 have their angles labeled with upper-case letters and their sides labeled with corresponding lower-case letters. Notice that the side opposite each angle uses the lower-case letter of its opposite angle. With a right triangle, the side opposite the right angle has a special name. It is called the *hypotenuse* of the right triangle. In Figure 4-12B, side a is *opposite* angle A, side b is *opposite* angle B and the hypotenuse (side c) is opposite the right angle (angle C). We can also say that side a is *adjacent* to angle B and side b is *adjacent* to angle A. (The hypotenuse is also adjacent to both angles A and B, but since the hypotenuse is otherwise uniquely identified we don't use this name for the hypotenuse.) The Greek letters theta (θ) and phi (ϕ) are often used to represent angles. Sometimes you will also see the Greek letters alpha (α), beta (β) and gamma (γ) used to represent the angles in a triangle.

WORKING WITH RIGHT TRIANGLES

Trigonometry defines relationships between the lengths of the sides of a right triangle and its angles. With these relationships and any combination of three sides or angles we can calculate any of the quantities we don't know. For example, if you know two sides and one angle, you can calculate the third side and the other two angles. While there are six functions defined for any right triangle, you can perform any required calculations if you know three of those functions.

The three functions we will use are the sine, cosine and tangent. Each function is defined in terms of an angle and two sides of the triangle.

$$\text{sine } \theta = \frac{\text{side opposite}}{\text{hypotenuse}}$$

$$\text{cosine } \theta = \frac{\text{side adjacent}}{\text{hypotenuse}}$$

$$\text{tangent } \theta = \frac{\text{side opposite}}{\text{side adjacent}}$$

Fig 4.12B shows the definitions of the three important trigonometry functions associated with angles A and B.

These functions are usually abbreviated as *sin*, *cos* and *tan*. Each function represents a ratio of two sides of the triangle, and this ratio is the same for any given angle, no matter how large or small the triangle: For example, **Fig 4.13** shows two right triangles that each include a $30°$ angle. The sine of the $30°$ angle is 0.5 no matter which triangle we are working with. Likewise, each of these triangles also includes a $60°$ angle, and the sine of the $60°$ angle is always 0.866.

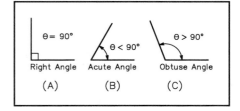

Fig 4.10 — Three types of angles. A shows a right angle, B shows an acute angle and C shows an obtuse angle.

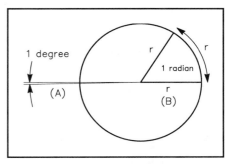

Fig 4.11 — Part A illustrates the measure of 1° as part of a circle. Part B shows the measure of 1 radian as part of a circle.

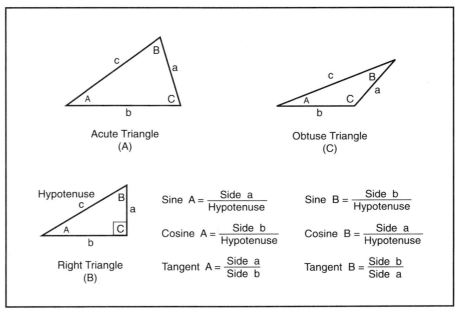

Fig 4.12 — Part A shows an acute triangle, Part B shows a right triangle and Part C shows an obtuse triangle.

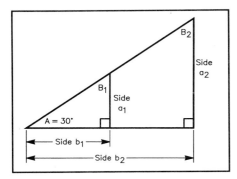

Fig 4.13 — This drawing shows two right triangles that each include a 30° angle and a 60° angle. Notice that values of the trigonometry functions don't depend on how long the sides of the triangles are; the same values of sin, cos and tan apply to each of these triangles.

Most scientific calculators include keys to find these function values. It is important to know if the calculator will understand the angle you enter as measured in degrees or radians. Most calculators work with angles measured in degrees, but some use radians. Some calculators will also use radians if you enter the proper keystrokes before starting. Computers usually work with angles measured in radians.

Suppose you know the ratio of sides, and want to know what angle is associated with that value. This is a question of finding the *inverse function*. Suppose you know that the side opposite an angle divided by the hypotenuse equals 0.5. What is the angle? Since opposite over hypotenuse is the definition of sine, we want to find the inverse sine, or arcsine (often abbreviated *arcsin*). In this example, the answer is 30°. Likewise, we can also find the arccosine (*arccos*) and arctangent (*arctan*) of an angle. You will also see these inverse functions written as $\sin^{-1}$, $\cos^{-1}$ and $\tan^{-1}$. Here the -1 exponent is simply a short-hand notation to indicate the inverse function. Do not try to follow the rules of significant figures when you find the value of a trigonometry function or its inverse function. Do follow the rules when you calculate the sides and angles of a triangle, however. For example, if you know an angle measurement to three or four significant figures, express the other angles you calculate with the same number of significant figures. If you know the length of a side to four significant figures, express the calculated sides to four significant figures.

The sine and cosine functions are used in many ways in electronics. You will often see graphs of these functions used

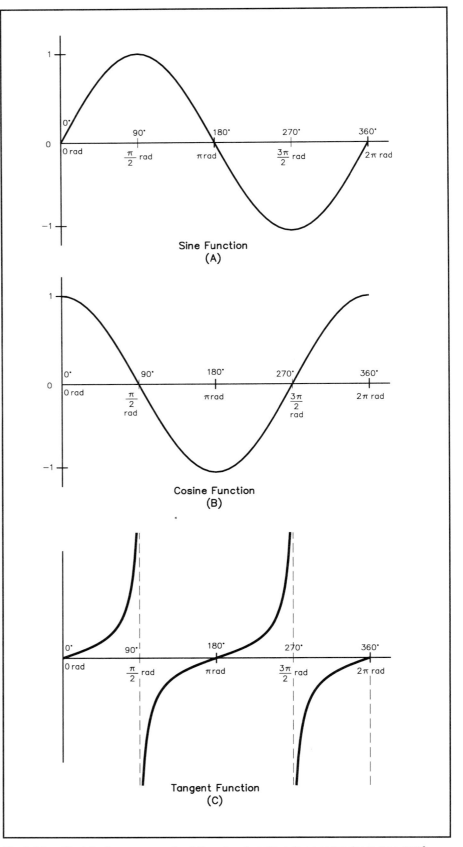

Fig 4.14 — Part A shows a graph of the sine function for angles from 0 to 360°. Part B is a graph of the cosine function and Part C is a graph of the tangent function for angles from 0 to 360°. (Note that the horizontal axis also indicates angles in radians.)

to represent the waveform of an alternating current signal. **Fig 4.14** shows graphs of the sine, cosine and tangent functions for angles from 0 to 360°. (The horizontal axis on the graph is also marked in radians, from 0 to 2π radians.)

In addition to the three trigonometry functions described here, there is one other very important relationship for working with right triangles. This principle was discovered by a Greek mathematician, Pythagoras. The *Pythagorean Theorem* states that the square of the hypotenuse is equal to the sum of the squares of the other two sides. Written as an equation, this is:

$$c^2 = a^2 + b^2$$

We can take the square root of both sides to solve this equation for the hypotenuse.

$$c = \sqrt{a^2 + b^2}$$

We can also solve the original equation for either of the other sides and then take the square root.

$$a^2 = c^2 - b^2$$

$$a = \sqrt{c^2 - b^2}$$

There is no single correct procedure for calculating the parts of a right triangle. Select a "trig" function or the Pythagorean Theorem depending on what parts of the triangle you know. **Fig 4.15** shows a right triangle. The drawing shows the two sides, and you have to calculate the hypotenuse and the two angles. We could use the Pythagorean Theorem to calculate the hypotenuse, but let's use the tangent function to find angle A first.

$$\tan A = \frac{\text{side opposite}}{\text{side adjacent}}$$

$$\tan A = \frac{25.0}{43.3} = 0.577$$

Next we must find the arctangent of this ratio:

$$A = 30.0°$$

Since the two acute angles of a right

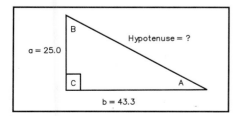

Fig 4.15 — Find the hypotenuse and the two acute angles of this right triangle. You can use any of the "trig" functions and the Pythagorean Theorem.

triangle must add up to 90°, we can see that angle B must be 60.0°. Then we can use the sine, cosine or Pythagorean Theorem to calculate the hypotenuse. Let's use the sine function for this example. (You can use the cosine to verify that it gives the same answer.)

$$\sin A = \frac{\text{side opposite}}{\text{hypotenuse}}$$

Cross multiply to solve this literal equation for the hypotenuse, and calculate the value.

$$\text{hypotenuse} = \frac{\text{side opposite}}{\sin A}$$

$$\text{hypotenuse} = \frac{25.0}{\sin 30.0°} = \frac{25.0}{0.500} = 50.0$$

As an example of using the Pythagorean Theorem, we will use that to solve for hypotenuse of this triangle also.

$$c = \sqrt{a^2 + b^2}$$

$$c = \sqrt{25.0^2 + 43.3^2} = \sqrt{625 + 1870}$$

$$c = \sqrt{2500} = 50.0$$

Notice that we have followed the rules for significant figures through these examples. In this last step, we had to round off the 43.3^2 term. Then the addition term under the radical was limited to the hundreds place, or three significant figures.

WORKING WITH ACUTE AND OBTUSE TRIANGLES

All the trigonometry functions and

techniques described in the last section apply *only* to *right triangles*. You may occasionally need to work with an acute or obtuse triangle. In this case it will be handy to remember the *Law of Sines* and the *Law of Cosines*. The Law of Sines tells us that the length of any side is proportional to the sine of the opposite angle:

$$\frac{a}{\sin A} = \frac{b}{\sin B} = \frac{c}{\sin C}$$

If you know one of the angles, the side opposite that angle, and one other side or angle, you can use this relationship to calculate the fourth side or angle. The Law of Sines is a simple proportion, and can be solved for the unknown quantity using cross multiplication. See **Fig 4.16**.

The Law of Cosines equation will remind you a bit of the Pythagorean Theorem. You can find any side of the triangle if you know the other two sides and the angle opposite the unknown side. (This is just the opposite of the Law of Sines, where you must know one angle and its opposite side.)

$$a^2 = b^2 + c^2 - 2bc\cos(A)$$

We could write similar equations solved for b^2 and c^2, but that isn't necessary, since it really doesn't matter which side is labeled a, b or c as long as each angle and its side opposite use the same letter. We can take the square root of both sides of this equation to solve for a:

$$a = \sqrt{b^2 + c^2 - 2bc\cos(A)}$$

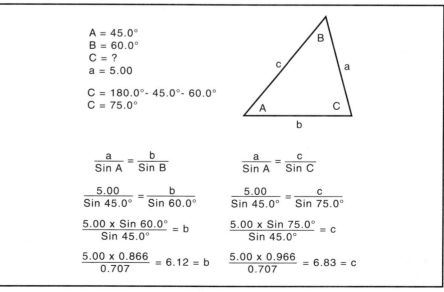

Fig 4.16 — The Law of Sines is used to calculate sides b and c of this acute triangle.

Coordinate Systems

A coordinate system helps us draw graphs to represent quantities and equations. A coordinate system provides a scale with a set of numbers to represent the location of a point on a surface. There are several such coordinate systems used in electronics. In this section we will briefly discuss three coordinate systems: the *rectangular*, or *cartesian coordinate system*, the *polar coordinate system* and the *spherical coordinate system*.

RECTANGULAR COORDINATES

Fig 4.17 shows a portion of a *rectangular*, or *cartesian coordinate system*. It is simply a pair of number lines that cross at a 90° angle. The scale on the number lines is chosen to suit the particular needs of any given situation. The graduations on one scale can be larger than the other, one or both lines can be far from 0, with an arbitrary crossing point to show the region of interest. This coordinate system represents a plane, two-dimensional surface. You have probably used graph paper drawn as a rectangular coordinate system.

The horizontal line, or *axis*, is often labeled X. This usually represents the independent, or controlled variable when an equation is being graphed.

The vertical line is often labeled Y. This usually represents the dependent variable (the value depends on the conditions set for the controlled variable).

Any point on a rectangular coordinate system can be specified by a pair of numbers, such as (−2, 5) or (5, 3). These numbers represent the distance along the X axis and the distance along the Y axis to reach the point.

POLAR COORDINATES

We specify the distance to a point with measurements along the X and Y axes with a rectangular coordinate system. Sometimes it is more convenient to specify the shortest distance from the center or *origin* to the point. In that case we can use a *polar coordinate system*. **Fig 4.18** shows an example of this system. The lines help mark the center of the system and provide a reference by dividing the circle into four equal parts, but they are not really necessary. Again we specify the location of any point on the surface with a pair of numbers, but this time the numbers represent the direct distance from the origin to the point, and an angle. The angle is usually measured counterclockwise from the line extending to the right side. The distance represents the *magnitude* or length of the value, measured as a straight line (the shortest distance) from the center to the point.

The circles represent increasing distances from the origin. You can choose any convenient scale for the radius of these circles. You don't always need the complete circles. Often you will only need one quarter or one half of the circle. You can buy graph paper marked off with a polar coordinate system, although you seldom need such graph paper. You can even use rectangular coordinate graph paper or plain paper, with a drawing compass, ruler and protractor if you do want a scale drawing.

Many times you will have to convert between rectangular and polar coordinate systems. You will find the trigonometry functions especially helpful at such times. **Fig 4.19** shows a right triangle drawn to illustrate such a conversion. The sides of

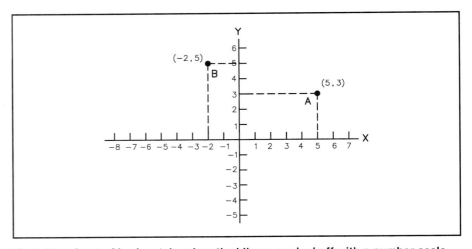

Fig 4.17 — A set of horizontal and vertical lines, marked off with a number scale, forms a rectangular coordinate system. We usually label the horizontal line, or axis, the X axis, and the vertical line the Y axis. Any point on the surface can be identified with a pair of numbers, representing the distance along the X axis and the distance along the Y axis to reach the point. When the point is identified with a pair of numbers, the convention is to list the X value first, then the Y value, as (X, Y).

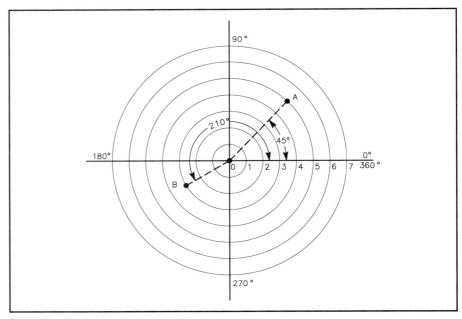

Fig 4.18 — This drawing shows a polar coordinate system. Any point on this surface is identified with a pair of numbers representing the distance from the origin directly to the point, and an angle or direction.

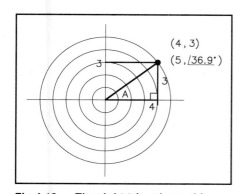

Fig 4.19 — The right triangle on this graph shows how we can specify the same point in rectangular coordinates and in polar coordinates. The rectangular coordinates (4, 3) and the polar coordinates (5, ∠36.9°) both represent the same point on this graph. You can use the trigonometry functions discussed earlier to convert between these two systems.

the triangle represent the X (4) and Y (3) values of a rectangular coordinate system. The hypotenuse of the right triangle represents the distance between the origin and the end point on a polar coordinate system. Angle A represents the polar-coordinate angle.

You should be able to use the various "trig" functions and the Pythagorean Theorem to calculate the hypotenuse (5) and angle A (36.9°) for this problem. If you knew the hypotenuse and angle A you should also be able to calculate the other two sides of the triangle, to convert from polar to rectangular coordinates.

SPHERICAL COORDINATES

Both of the coordinate systems described above represent a two-dimensional surface. This is fine for most electronics problems, but occasionally it is helpful to have a three-dimensional coor-

dinate system. It is possible to add a third axis to the rectangular coordinates, which forms a 90° angle with the other two. (Look at the corner of a room, with the two lines along the floor and walls representing the X and Y axes. Then the corner between the two walls represents the Z axis.) A set of three numbers (X, Y, Z) will represent any point in the three-dimensional space this system represents.

It is also possible to rotate the circles of the polar coordinate system to create a sphere. The resulting *spherical coordinate system* gives us another way to represent a point in three dimensions. In this system we use the radius, or distance from the center to the point, and two angles — one representing an angle measured "horizontally" and the other representing an angle measured "vertically." (Think about our Earth, and the way we draw lines of longitude and latitude.)

Complex Algebra

We most often use the rectangular and polar coordinate systems for electronics problems involving resistance, reactance and impedance (and conductance, susceptance and admittance, their reciprocals). When we work with these quantities we draw the resistance or conductance along the X axis and the reactance or susceptance along the Y axis. The hypotenuse represents impedance or admittance.

We must use some special mathematical techniques when working with these quantities because we must always distinguish between them. It is most convenient to use the algebra of what mathematicians call *imaginary numbers*, although there is nothing imaginary about these electronics quantities. Mathematicians use these techniques when they work with quantities that involve the square root of minus 1, written as $\sqrt{-1}$. It is impossible to find a number that when multiplied by itself gives −1, so this quantity is *imaginary*, yet it does show up in some mathematical procedures. Mathematicians represent this quantity with a lower-case italic *i*. Quantities that include real and imaginary parts are called *complex numbers*.

In electronics, we use a lower-case italic *j* to represent numbers on the reactance or susceptance line, or Y axis of a graph. The algebra of complex numbers provides a way to add, subtract, multiply and divide quantities that include both resistive and

reactive components. You can best think of the *j* as an *operator* that produces a 90° rotation from the resistance line. An operator is just a mathematical procedure applied to a quantity. An exponent is an operator that tells you how many times to multiply a quantity times itself and the radical sign ($\sqrt{\ }$) is an operator that tells you to take the square root.

When you see a reactance expressed as *j*250. Ω, place this quantity along the Y axis on your graph. A reactance expressed as −*j*300. Ω tells you to rotate 90° in the clockwise direction instead of the normal counterclockwise direction.

Inductive reactance is specified with a +*j* for series circuits, because the voltage across an inductor *leads* the current through it. Since voltage and current are in phase in a resistor, the voltage across the inductor leads the voltage across the resistor by 90°. (For parallel circuits, the voltage across the resistor and inductor is the same, so they are in phase. In that case the current through the inductor *lags* the current through the resistor, so the current associated with the inductive reactance gets a −*j* operator.)

Capacitive reactance is just the opposite. It is specified with a −*j* operator for series circuits because the voltage across a capacitor lags the current, so the current across the capacitor *lags* the voltage across the resistor — or is 90° behind. (For parallel circuits, the same voltage is applied

to the resistor and capacitor, so the capacitor current *leads* the resistor current. The current associated with the capacitive reactance gets a +*j* operator for parallel circuits.)

A handy memory device for these relationships is the saying, "ELI the ICE man." The E represents voltage, I represents current, L is inductance and C is capacitance.

Impedance is a combination of resistance and reactance. When we specify a series-circuit impedance as 50 + *j*200 Ω, you know this represents a circuit with a 50-Ω resistance in series with a 200-Ω inductive reactance. Likewise, an impedance of 50 − *j*200 Ω represents a circuit with a 50-Ω resistance in series with a 200-Ω capacitive reactance.

Both of these impedances can be expressed in polar-coordinate form. Plot the values on a graph and calculate the hypotenuse of the right triangle and angle A as shown on **Fig 4.20**. Then you can write these impedances as

206 Ω ∠76° and 206 Ω ∠−76°.

RULES FOR WORKING WITH COMPLEX NUMBERS

Addition and subtraction of complex numbers are best done using rectangular-coordinate form. When you add complex numbers written in rectangular notation you add the parts along the X axis, and you add the parts along the Y axis. The result gives a new set of X, Y coordinates, representing the addition of the two complex values. To subtract complex numbers you subtract one X part from the other, and subtract the corresponding Y parts. The result is a new set of X, Y coordinates, representing the subtraction of the two values.

For example, what is the total impedance of a circuit that has an impedance of $30 + j150 \, \Omega$ is series with an impedance of $40 - j100 \, \Omega$? We can write this addition as:

$$30 + j \, 150 \, \Omega$$
$$\underline{+40 - j \, 100 \, \Omega}$$
$$70 + j \, \, 50 \, \Omega$$

If you have to add or subtract impedances given in polar notation (a magnitude or length and an angle), first convert these values to rectangular-coordinate form. Use the trigonometry functions and Pythagorean Theorem described earlier in this chapter. If you need the answer specified in polar-coordinate form you can convert back to that notation after performing the addition.

Multiplication and division of complex numbers is best done in polar-coordinate form. When you multiply complex numbers in polar-coordinate form, you multiply the magnitudes and add the angles. When you divide complex numbers in polar notation you divide the magnitudes and subtract the angles.

Suppose you want to find the impedance of a circuit that has a resistor in parallel with a capacitor. When you apply 10.0 V to the circuit, you measure 0.250 A

of current. You measure the phase angle between the current and the voltage, and find the current leads the voltage by 30.0°. We can calculate the impedance of this circuit (represented by a capital Z) using Ohm's Law.

$$Z = \frac{E}{I}$$

$$Z = \frac{10.0 \, \text{V} \; \angle 0.0°}{0.250 \, \text{A} \; \angle 30.0°}$$

The components are in parallel, so the same voltage is applied to both the resistor and capacitor. Use the voltage as the phase reference for parallel-circuit calculations, so the voltage has a 0° phase angle. To perform this division, first divide the voltage magnitude by the current magnitude. Then subtract the denominator phase angle from the numerator phase angle.

$$\frac{10.0 \, \text{V}}{0.250 \, \text{A}} = 40.0 \, \Omega$$

and

$$0.0° - (30.0°) = -30.0°$$

These two values specify the impedance of the circuit in this example. We can put them together and write the circuit impedance in polar-coordinate form as:

$$40.0 \, \Omega \; \angle -30.0°$$

The negative phase angle tells us there is a capacitive reactance as part of the impedance.

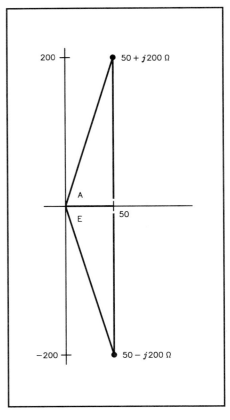

Fig 4.20 — The two triangles shown on this graph represent the impedances of two circuits. Triangle A represents a 50-Ω resistance in series with a 200-Ω inductive reactance ($50 + j200 \, \Omega$). Triangle B represents a 50-Ω resistance in series with a 200-Ω capacitive reactance ($50 - j200 \, \Omega$).

Logarithms

A logarithm is an exponent. *Common logarithms* use the number 10 as their *base*. You have some experience with "powers of 10" from writing numbers in exponential or scientific notation.

A common log, as it is usually called, is the exponent or power to which you must raise 10 to get a certain number. In the examples above, we raised 10 to the third power to get 1000. The log of 1000, then, is 3. The log of 1000000 is 6. In general, we define a common logarithm with two equations. If:

$N = 10^x$, then:

$\log (N) = x$

Sometimes you will see this written as $\log_{10} (N) = x$. This is simply to ensure that you know the base of the logarithm is 10.

Finding the log of a multiple of 10 is easy, as these examples show. You may wonder to what power you can raise 10 to get a number like 2. That is a good question, and the answer is 0.301. Logs are usually decimal fractions rather than whole numbers. Logs for numbers smaller than 10 are less than 1; logs for numbers larger than 10 are greater than 1. From the definition of a log, we can write the expression:

$2 = 10^{0.301}$

The easiest way to find any logarithm is with your calculator. Simply enter the

number whose log you want to find, and then push the button labeled "log." It is easy to find that log (5) = 0.699, for example.

It is interesting to note that log (1) = 0, because anything (including 10) raised to the zero power is 1. The log of 0 is undefined, because there is no power to which you can raise 10 and get 0.

The inverse log is called the *antilog* (often written $\log^{-1}$). When we know the log and want to find the original number, we want the antilog. To find an antilog, simply raise 10 to the given power. Your calculator probably has a button labeled "10^x" or something similar. What is the antilog of 1.845? $10^{1.845} = 70$. Don't try to follow the rules for significant figures when finding logs or antilogs. Do follow the rules with the values you calculate from logs and antilogs, however.

The second base that is frequently used for logarithms is a number usually represented by e. (Sometimes the Greek letter epsilon (ε) is used to represent e although this is an incorrect representation.) This number is approximately 2.71828. This is not an exact value, because the decimal fraction doesn't end with this last 8. This value is rounded off, but there is no exact value for e because you can never find the *last* digit. Mathematicians call such numbers with no exact value, *irrational numbers*. The number represented by e appears in several electronics calculations, and is called the *natural number*, because it appears as a constant of nature. You will use e to calculate the voltage on a capacitor as it charges or discharges, for example.

Logarithms that use e for their base are called *natural logarithms*, or *Naperian logarithms*. This can be written as $\log_e$, but to more easily distinguish it from common logs, we usually abbreviate it ln. We define natural logs the same way we define common logs. If:

$M = e^y$, then
$\log_e (M) = \ln (M) = y$

The easiest way to find a natural log is with a scientific calculator. Enter the number whose ln you want to know, then press the "ln" button on the calculator. For example, $\ln (2) = 0.693$ and $\ln (20) = 2.996$. As you might expect, $\ln (e) = 1$, $\ln (1) = 0$ and $\ln (0)$ is undefined.

Inverse natural logs, or antilogs are also easy with a calculator. Just raise e to that power: $e^{2.996} = 20$.

Computers often work only with natural logarithms. Converting between common logarithms and natural logarithms is easy, however. If you want to find a common log, and know the natural log value, divide that by the natural log of 10.

$\log(x) = \ln(x) / \ln(10) = \ln(x) / 2.3025851$

$\log(x) = 0.4342945 \ln(x)$

If you know the common log, and want to find the natural log, divide that value by the common log of e.

$\ln(x) = \log(x) / \log(e) = \log(x) / 0.4342945$

$\ln(x) = 2.3025851 \log(x)$

DECIBELS

The bel (abbreviated B) is named after Alexander Graham Bell, who did much pioneering work with sound and the way our ears respond to sound. Our ears respond to sounds ranging from an intensity less than 10^{-16} W/cm^2 to intensities larger than 10^{-4} W/cm^2 (where we begin to experience pain). This is a range of more than 10^{12} times from the softest to the loudest sounds. Logarithms provide a convenient way to represent these values, because they compress this scale into a range of 12, rather than a range of a billion.

A bel is defined as the logarithm of a power ratio. It gives us a way to compare power levels with each other and with some reference power.

$$\text{bel} = \log \left(\frac{P_1}{P_0} \right)$$

where P_0 is the reference power, or the power you want to use for comparison and P_1 is the power you are comparing to the reference level.

While the bel was first defined in terms of sound power, to describe sound intensities, in electronics we often use it to compare electrical power levels. The decibel is one-tenth of a bel, and is abbreviated dB.

It takes 10 decibels to make 1 bel, so we can write an equation to find dB directly:

$$\text{dB} = 10 \log \left(\frac{P_1}{P_0} \right)$$

How many decibels does the power increase if an amplifier takes a 1-W signal and boosts it to 50 W? Let P_0 be the 1-W signal in this example, since that is the starting point for the comparison.

$$\text{dB} = 10 \log \left(\frac{50 \text{ W}}{1 \text{ W}} \right) = 10 \log (50)$$

$$\text{dB} = 10 (1.699) = 16.99 \text{ dB}$$

The amplifier in this example has a gain of nearly 17 dB.

Sometimes when we are comparing signal levels in an electronic circuit, we know the voltage or current of the signal, but not the power. Of course we can always calculate the power, as long as we know the impedance of the circuit. We can take a shortcut to comparing the signal levels in decibels, however, *as long as the impedance is the same* in both circuits, or as long as the impedance of the circuit doesn't change when we change the voltage or current. Remember from Ohm's Law and the power equation that $P = E^2 / R$ and $P = I^2 \times R$. So we can use E^2 or I^2 in place of power in the decibel equation, *as long as the impedance is the same* in both cases.

$$\text{dB} = 10 \log \left(\frac{E_1^2}{E_0^2} \right)$$

$$\text{dB} = 20 \log \left(\frac{E_1}{E_0} \right)$$

and

$$\text{dB} = 10 \log \left(\frac{I_1^2}{I_0^2} \right)$$

$$\text{dB} = 20 \log \left(\frac{I_1}{I_0} \right)$$

Here we have also illustrated another important property of logarithms. If the quantity inside the log expression has an exponent, you can move the exponent outside the log. In this case, we move the 2 from the squared terms out front, and multiply it times the 10 already there.

Sometimes there is confusion about whether the decibel was calculated using power, voltage or current. Since the current and voltage equations use 20 instead of 10 times the log term, some hams believe the "voltage" or "current" decibel is different than one calculated using power. This is not true, however. There is only one decibel definition, and that is ten times the log of a power ratio.

There are several power ratios that you should learn to recognize and remember the decibel values that go with them. These are the decibel values for a doubling of the power and for halving the power. Let's look at the effect of doubling the power first. It doesn't matter if we are going from 1 W to 2, 50 to 100 or 500 to 1000 W. In each case the new power is twice the starting power. To find the decibel increase multiply 10 times the log of 2:

$\text{dB} = 10 \log (2)$

$\text{dB} = 10 \times 0.301 = 3.01$

Anytime you double the power, it represents approximately a 3-dB increase in power.

What is the decibel change when you cut the power in half? Again, it doesn't matter if you are going from 1000 W to 500, 100 to 50 or 2 W to 1 W; the power ratio is still 0.5.

$\text{dB} = 10 \log (0.5)$

$\text{dB} = 10 \times -0.301 = -3.01$

A negative value indicates a decrease in power. Anytime you cut the power in half there is about a 3-dB decrease in power.

Table 4.7 shows the relationship between several common decibel values and the power change associated with those values. The current and voltage changes are also included, but these are only valid if the impedance is the same for both values.

Suppose you double the power, and then double it again? The final power is four times the starting power, so you can calculate the decibel increase using the equation given. You can also calculate the total power change "by inspection" because you know each time you double the power there is a 3-dB increase. In this example you have a 3-dB increase, plus a second 3-dB increase. If you add these two decibel values, you have a 6-dB total increase. If you double the power again, you have a 9-dB total increase. Doubling the power a fourth time gives a 12-dB total increase.

The same relationship is true of power decreases. Each time you cut the power in half you have a 3-dB decrease. Cutting the power in half and then in half again is a 6-dB decrease, and so on.

The addition and subtraction of decibel values is very important in electronics. Amplification factors, gains and losses of antennas, antenna feed lines and all kinds of circuits can simply be added when they are expressed in decibels.

It is often convenient to compare a certain power level with some standard reference. For example, suppose you measured the signal coming into a receiver from an antenna and found the power to be 2×10^{-13} mW. As this signal goes through the receiver it increases and decreases in strength until it finally produces some sound in the receiver speaker or head-phones. It is convenient to describe these signal levels in terms of decibels. A common reference power is 1 mW. The decibel value of a signal compared to 1 mW is specified as "dBm" to mean decibels compared to 1 mW. In our example, the signal strength at the receiver input is:

$$dBm = 10 \log \left(\frac{2 \times 10^{-13} \text{ mW}}{1 \text{ mW}} \right)$$

$$dBm = 10 \log \left(2 \times 10^{-13} \right)$$

$$= 10 \times -12.7 = -127 \, dBm$$

There are many other reference powers used, depending upon the circuits and power levels. If you use 1 W as the reference power, then you would specify dBW. Antenna power gains are often specified in relation to a dipole (dBd) or an isotropic radiator (dBi). Anytime you see another letter following the dB, you will know some reference power is being specified.

Table 4.7
Some Common Decibel Values and Power-Ratio Equivalents

dB	P_2/P_1	V_2/V_1 or I_2/I_1
−20	10^{-2}	0.1000
−10	0.1000	0.3162
(−6.0206)	(0.2500)	(0.5000)
−6	0.2512	0.5012
(−3.0103)	(0.5000)	(0.7071)
−3	0.5012	0.7079
−1	0.7943	0.8913
0	1.000	1.000
1	1.259	1.122
3	1.995	1.413
(3.0103)	(2.0000)	(1.4142)
6	3.981	1.995
(6.0206)	(4.0000)	(2.0000)
10	10.00	3.162
20	10^2	10.00

Integration and Differentiation

You don't have to be familiar with calculus to understand modern electronics. Sometimes it is helpful to be familiar with some calculus *terminology* to understand how a circuit works or what its function is, however.

INTEGRATION

When you read that a certain op-amp circuit is designed as an "integrator" it will be easier to understand what the circuit does if you know a simple definition of *integration*. Integration is the process of calculating the area under a curve plotted on a graph.

Area always implies certain boundaries, and you want to find how much space there is inside the boundaries. A square may be the simplest surface for which to find the area. If you know the length of a side you simply square that length to find the area. If you know the length and width of a rectangle you multiply these values to calculate area.

Fig 4.21 shows a graph of a square-wave signal. If you want to *integrate* this signal, you have to find the *area* of the pulse. The scales on this graph represent voltage (on the Y axis) and time (on the X axis) so this isn't area in the most common sense, but we can perform a similar calculation. As you can imagine, if the pulse has a larger amplitude or a longer duration it will have a larger area.

Fig 4.22 illustrates a more difficult signal to integrate. Calculus methods can calculate this area from the equation that represents the curve, but we can make a reasonable approximation by drawing a series of rectangles and adding their areas. Integration is normally done over some range of values, such as x_1 and x_2 as shown on this graph. Part B shows that we can draw a series of rectangles so the *midpoint*

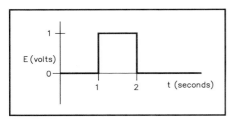

Fig 4.21 — This graph represents one cycle of a square wave. The text explains how to *integrate* this signal waveform, or find the area under the pulse.

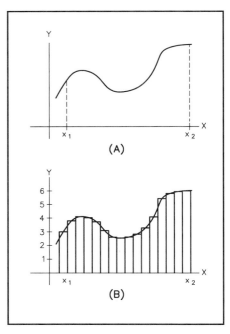

(A)

(B)

Fig 4.22 — This graph shows an irregular curve. If we want to know the area under the curve between x_1 and x_2, we can draw a series of rectangles and add their areas, as shown in Part B.

of the side of each rectangle crosses the curve. Part of the rectangle corner lies above the curve, but there is a nearly equal space below the curve that is not included. If we draw more rectangles, with smaller widths, the approximation becomes better. The concept of integration is that you can make the interval smaller and smaller until it is no longer an approximation, but an exact value.

Fig 4.23 compares a series of square-wave pulses fed into an op-amp integrator and the output waveform from the integrator. In this example the integrator changes a square-wave signal into a triangle-wave signal. The integrated signal increases while the input pulse is positive, then decreases while the signal is negative.

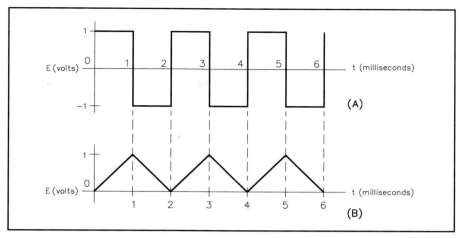

Fig 4.23 — Part A shows a series of square-wave pulses and Part B shows the output from an op-amp integrator with the square-wave input.

DIFFERENTIATION

Differentiation is another calculus procedure that may be helpful. Integration and differentiation are opposite procedures. If you integrate a function and then differentiate the result, you get the original function back again. Likewise, if you differentiate a function and then integrate the result, you get the original function back.

While integration represents a summation of *area* values over some range, differentiation represents the *slope* of a line or curve at some specific point. The slope of a straight line is equal to the change in value along the x axis divided by the corresponding change in value along the y axis.

Look at the triangle waveform of Fig 4.23 B. While the voltage is increasing, this line has a constant slope, m, such that it satisfies the equation $y = mx + b$. Since the differentiation process represents the slope of the line, the *derivative* is a constant. When the waveform begins to decrease the slope suddenly changes to a new value, which is negative this time. The derivative is again a constant value, this

time with a negative sign. If the graph in Fig 4.23B represents a signal waveform that is fed into a differentiator circuit, the waveform at A represents the output signal waveform!

We approximated the integration process for a curved-line graph by adding the areas of many small rectangles drawn to divide a curve into small segments. Similarly, we can approximate the differentiation process of a curved-line graph by finding the slope of a straight line drawn *tangent* to the curve. A tangent line touches the curve at a single point. The simplest way to show a tangent line is with a circle, as shown in **Fig 4.24**. The tangent line is perpendicular to (forms a 90° angle with) a radius line.

We can approximate the derivative of a curved-line graph at any point by drawing a tangent line at that point and calculating the slope of the line. We can even find the general trend of the derivative function by drawing a series of tangent lines at points along the line. By calculating the slope of each of those lines you can get some idea

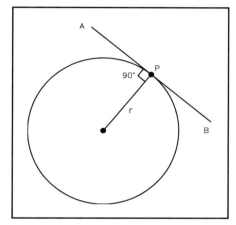

Fig 4.24 — This diagram illustrates the concept of a tangent line. Line AB is tangent to the circle at point P, and is perpendicular to the radius line, r.

of how it is changing. By selecting points closer and closer together you will find a better and better approximation to the derivative.

Contents

DC Theory and Resistive Components

5

Glossary

Alternating current — A flow of charged particles through a conductor, first in one direction, then in the other direction.

Ampere — A measure of flow of charged particles per unit time. One ampere represents one coulomb of charge flowing past a point in one second.

Atom — The smallest particle of matter that makes up an element. Consists of protons and neutrons in the central area called the nucleus, with electrons surrounding this central region.

Coulomb — A unit of measure of a quantity of electrically charged particles. One coulomb is equal to 6.25×10^{18} electrons.

Direct current — A flow of charged particles through a conductor in one direction only.

EMF — Electromotive Force is the term used to define the force of attraction between two points of different charge potential. Also called voltage.

Energy — Capability of doing work. It is usually measured in electrical terms as the number of watts of power consumed during a specific period of time, such as watt-seconds or kilowatt-hours..

Joule — Measure of a quantity of energy. One joule is defined as one newton (a measure of force) acting over a distance of one meter.

Ohm — Unit of resistance. One ohm is defined as the resistance that will allow one ampere of current when one volt of EMF is impressed across the resistance.

Power — Power is the rate at which work is done. One watt of power is equal to one volt of EMF, causing a current of one ampere.

Volt — A measure of electromotive force.

Introduction

This chapter was written by Roger Taylor, K9ALD.

The atom is the primary building block of the universe. The main parts of the atom include protons, electrons and neutrons. Protons have a positive electrical charge, electrons a negative charge and neutrons have no electrical charge. All atoms are electrically neutral, so they have the same number of electrons as protons. If an atom loses electrons, so it has more protons than electrons, it has a net positive charge. If an atom gains electrons, so it has more electrons than protons, it has a negative charge. Particles with a positive or negative charge are called ions. Free electrons are also called ions, because they have a negative charge.

When there are a surplus number of positive ions in one location and a surplus number of negative ions (or electrons) in another location, there is an attractive force between the two collections of particles. That force tries to pull the collections together. This attraction is called electromotive force, or EMF.

If there is no path (conductor) to allow electric charge to flow between the two locations, the charges cannot move together and neutralize one another. If a conductor is provided, then electric current (usually electrons) will flow through the conductor.

Electrons move from the negative to the positive side of the voltage, or EMF source. *Conventional current* has the opposite direction, from positive to negative. This comes from an arbitrary decision made by Benjamin Franklin in the 18th century. The conventional current direction is important in establishing the proper polarity sign for many electronics calculations. Conventional current is used in much of the technical literature. The arrows in semiconductor schematic symbols point in the direction of conventional current, for example.

To measure the quantities of charge, current and force, certain definitions have been adopted. Charge is measured in *coulombs*. One coulomb is equal to 6.25×10^{18}

electrons (or protons). Charge flow is measured in *amperes*. One ampere represents one coulomb of charge flowing past a point in one second. Electromotive force is measured in *volts*. One volt is defined as the potential force (electrical) between two points for which one ampere of current will do one *joule* (measure of energy) of work flowing from one point to another. (A joule of work per second represents a power of one watt. See the **Mathematics for Amateur Radio** chapter for more information about these unit definitions.)

Voltage can be generated in a variety of ways. Chemicals with certain characteristics can be combined to form a battery. Mechanical motion such as friction (static electricity, lightning) and rotating conductors in a magnetic field (generators) can also produce voltage.

Any conductor between points at different voltages will allow current to pass between the points. No conductor is perfect or lossless, however, at least not at normal temperatures. Charged particles such as electrons resist being moved and it requires energy to move them. The amount of resistance to current is measured in *ohms*.

OHM'S LAW

One ohm is defined as the amount of resistance that allows one ampere of current to flow between two points that have a potential difference of one volt. Thus, we get Ohm's Law, which is:

$$R = \frac{E}{I} \qquad (1)$$

where:
R = resistance in ohms,
E = potential or EMF in volts and
I = current in amperes.

Transposing the equation gives the other common expressions of Ohm's Law as:

$$E = I \times R \qquad (2)$$

and

$$I = \frac{E}{R} \qquad (3)$$

All three forms of the equation are used often in radio work. You must remember

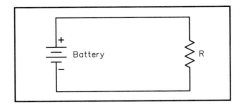

Fig 5.1 — A simple circuit consisting of a battery and a resistor.

that the quantities are in volts, ohms and amperes; other units cannot be used in the equations without first being converted. For example, if the current is in milliamperes you must first change it to the equivalent fraction of an ampere before substituting the value into the equations.

The following examples illustrate the use of Ohm's Law. The current through a 20000-Ω resistance is 150 mA. See **Fig 5.1**. What is the voltage? To find voltage, use equation 2 ($E = I \times R$). Convert the current from milliamperes to amperes. Divide by 1000 mA / A (or multiply by 10^{-3} A / mA) to make this conversion. If you are uncertain how to do these conversions, see the **Mathematics for Amateur Radio** chapter. (Notice the conversion factor of 1000 does not limit the number of significant figures in the calculated answer.)

$$I = \frac{150 \text{ mA}}{1000 \; \frac{\text{mA}}{\text{A}}} = 0.150 \text{ A}$$

Then:

$$E = 0.150 \, A \times 20000 \, \Omega = 3000 \text{ V}$$

If you are unfamiliar with the use of significant figures and rounding off calculated values, see the **Mathematics for Amateur Radio** chapter.

When 150 V is applied to a circuit, the current is measured at 2.5 A. What is the resistance of the circuit? In this case R is the unknown, so we will use equation 1:

$$R = \frac{E}{I} = \frac{150 \text{ V}}{2.5 \text{ A}} = 60. \, \Omega$$

No conversion was necessary because the voltage and current were given in volts and amperes.

How much current will flow if 250 V is applied to a 5000-Ω resistor? Since I is unknown,

$$I = \frac{E}{R} = \frac{250 \text{ V}}{5000 \, \Omega} = 0.05 \text{ A}$$

It is more convenient to express the current in mA, and 0.05 A × 1000 mA / A = 50 mA.

RESISTANCE AND CONDUCTANCE

Suppose we have two conductors of the same size and shape, but of different materials. The amount of current that will flow when a given EMF is applied will vary with the resistance of the material. The lower the resistance, the greater the current for a given EMF. The *resistivity* of a material is the resistance, in ohms, of a cube of the material measuring one centimeter on each edge. One of the best conductors is copper, and in making resistance calculations it is

Table 5.1
Relative Resistivity of Metals

Material	Resistivity Compared to Copper
Aluminum (pure)	1.60
Brass	3.7-4.90
Cadmium	4.40
Chromium	1.80
Copper (hard-drawn)	1.03
Copper (annealed)	1.00
Gold	1.40
Iron (pure)	5.68
Lead	12.80
Nickel	5.10
Phosphor bronze	2.8-5.40
Silver	0.94
Steel	7.6-12.70
Tin	6.70
Zinc	3.40

frequently convenient to compare the resistance of the material under consideration with that of a copper conductor of the same size and shape. **Table 5.1** gives the ratio of the resistivity of various conductors to the resistivity of copper.

The longer the physical path, the higher the resistance of that conductor. For direct current and low-frequency alternating currents (up to a few thousand hertz) the resistance is inversely proportional to the cross-sectional area of the path the current must travel; that is, given two conductors of the same material and having the same length, but differing in cross-sectional area, the one with the larger area will have the lower resistance.

RESISTANCE OF WIRES

The problem of determining the resistance of a round wire of given diameter and length—or its converse, finding a suitable size and length of wire to provide a desired amount of resistance—can easily be solved with the help of the copper wire table given in the **Component Data** chapter. This table gives the resistance, in ohms per 1000 ft, of each standard wire size. For example, suppose you need a resistance of 3.5 Ω, and some #28 wire is on hand. The wire table in the **Component Data** chapter shows that #28 wire has a resistance of 66.17 Ω / 1000 ft. Since the desired resistance is 3.5 Ω, the required wire length is:

$$\text{Length} = \frac{R_{DESIRED}}{\frac{R_{WIRE}}{1000 \text{ ft}}} = \frac{3.5 \, \Omega}{\frac{66.17 \, \Omega}{1000 \text{ ft}}}$$

$$= \frac{3.5 \, \Omega \times 1000 \text{ ft}}{66.17 \, \Omega} = 53 \text{ ft} \qquad (4)$$

As another example, suppose that the resistance of wire in a circuit must not exceed 0.05 Ω and that the length of wire required for making the connections totals 14 ft. Then:

$$\frac{R_{WIRE}}{1000 \text{ ft}} < \frac{R_{MAXIMUM}}{Length} = \frac{0.05 \, \Omega}{14.0 \text{ ft}} \qquad (5)$$

$$= 3.57 \times 10^{-3} \frac{\Omega}{\text{ft}} \times \frac{1000 \text{ ft}}{1000 \text{ ft}}$$

$$\frac{R_{WIRE}}{1000 \text{ ft}} < \frac{3.57 \, \Omega}{1000 \text{ ft}}$$

Find the value of R_{WIRE} / 1000 ft that is less than the calculated value. The wire table shows that #15 is the smallest size having a resistance less than this value. (The resistance of #15 wire is given as 3.1810 Ω / 1000 ft.) Select any wire size larger than this for the connections in your circuit, to ensure that the total wire resistance will be less than 0.05 Ω.

When the wire in question is not made of copper, the resistance values in the wire table should be multiplied by the ratios shown in Table 5.1 to obtain the resulting resistance. If the wire in the first example were made from nickel instead of copper, the length required for 3.5 Ω would be:

$$Length = \frac{R_{DESIRED}}{\dfrac{R_{WIRE}}{1000 \text{ ft}}} \qquad (6)$$

$$= \frac{3.5 \, \Omega}{\dfrac{66.17 \, \Omega}{1000 \text{ ft}} \times 5.1}$$

$$= \frac{3.5 \, \Omega \times 1000 \text{ ft}}{66.17 \, \Omega \times 5.1}$$

$$Length = \frac{3500 \text{ ft}}{337.5} = 10.37 \text{ ft}$$

TEMPERATURE EFFECTS

The resistance of a conductor changes with its temperature. The resistance of practically every metallic conductor increases with increasing temperature. Carbon, however, acts in the opposite way; its resistance decreases when its temperature rises. It is seldom necessary to consider temperature in making resistance calculations for amateur work. The temperature effect is important when it is necessary to maintain a constant resistance under all conditions, however. Special materials that have little or no change in resistance over a wide temperature range are used in that case.

RESISTORS

A package of material exhibiting a certain amount of resistance, made up into a single unit is called a resistor. Different resistors having the same resistance value may be considerably different in physical size and construction (see **Fig 5.2**). Current through a resistance causes the conductor to become heated; the higher the resistance and the larger the current, the greater the amount of heat developed. Resistors intended for carrying large currents must be physically large so the heat can be radiated quickly to the surrounding air. If the resistor does not dissipate the heat quickly, it may get hot enough to melt or burn.

The amount of heat a resistor can safely dissipate depends on the material, surface area and design. Typical carbon resistors used in amateur electronics ($1/8$ to 2-W re-sistors) depend primarily on the surface area of the case, with some heat also being carried off through the connecting leads. Wirewound resistors are usually used for higher power levels. Some have finned cases for better convection cooling and/or metal cases for better conductive cooling.

In some circuits, the resistor value may be critical. In this case, precision resistors are used. These are typically wirewound, or carbon-film devices whose values are carefully controlled during manufacture. In addition, special material or construction techniques may be used to provide temperature compensation, so the value does not change (or changes in a precise manner) as the resistor temperature changes. There is more information about the electrical characteristics of real resistors in the **Real-World Component Characteristics** chapter.

CONDUCTANCE

The reciprocal of resistance (1/R) is *conductance*. It is usually represented by the symbol G. A circuit having high conductance has low resistance, and vice versa. In radio work, the term is used chiefly in connection with electron-tube and field-effect transistor characteristics. The unit of conductance is the siemens, abbreviated S. A resistance of 1 Ω has a conductance of 1 S, a resistance of 1000 Ω has a conductance of 0.001 S, and so on. A unit frequently used in connection with electron devices is the μS or one millionth of a siemens. It is the conductance of a 1-MΩ resistance.

Series and Parallel Resistances

Very few actual electric circuits are as simple as Fig 5.1. Commonly, resistances are found connected in a variety of ways. The two fundamental methods of connecting resistances are shown in **Fig 5.3**. In part A, the current flows from the source of EMF (in the direction shown by the arrow) down through the first resistance, R1, then through the second, R2 and then back to the source. These resistors are connected in series. The current everywhere in the circuit has the same value.

In part B, the current flows to the common connection point at the top of the two resistors and then divides, one part of it flowing through R1 and the other through R2. At the lower connection point these two currents again combine; the total is the same as the current into the upper common connection. In this case, the two resistors are connected in parallel.

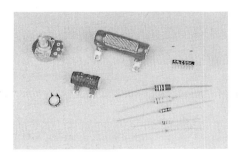

Fig 5.2 — **Examples of various resistors. In the right foreground are $1/4$-, $1/2$- and 1-W composition resistors. The two larger cylindrical components at the center are wire-wound power resistors. A surface-mount, or chip resistor is shown at the top right. The remaining two parts are variable resistors; a PC-board-mount device at the lower left and a panel-mount unit at the upper left.**

RESISTORS IN PARALLEL

In a circuit with resistances in parallel, the total resistance is less than that of the lowest resistance value present. This is because the total current is always greater than the current in any individual resistor. The formula for finding the total resistance of resistances in parallel is

$$R = \cfrac{1}{\cfrac{1}{R1} + \cfrac{1}{R2} + \cfrac{1}{R3} + \cfrac{1}{R4} + \cdots} \qquad (7)$$

where the dots indicate that any number of resistors can be combined by the same method. For only two resistances in paral-

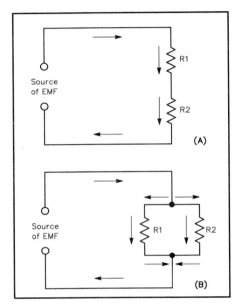

Fig 5.3 — Resistors connected in series at A, and in parallel at B.

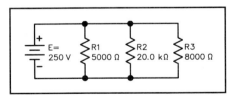

Fig 5.4 — An example of resistors in parallel. See text for calculations.

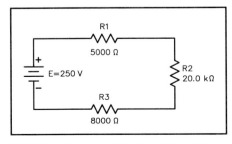

Fig 5.5 — An example of resistors in series. See text for calculations.

lel (a very common case) the formula becomes:

$$R = \frac{R1 \times R2}{R1 + R2} \qquad (8)$$

Example: If a 500-Ω resistor is connected in parallel with one of 1200 Ω, what is the total resistance?

$$R = \frac{R1 \times R2}{R1 + R2} = \frac{500\,\Omega \times 1200\,\Omega}{500\,\Omega + 1200\,\Omega}$$

$$R = \frac{600000\,\Omega^2}{1700\,\Omega} = 353\,\Omega$$

KIRCHHOFF'S FIRST LAW (KIRCHHOFF'S CURRENT LAW)

Suppose three resistors (5.00 kΩ, 20.0 kΩ and 8.00 kΩ) are connected in parallel as shown in **Fig 5.4**. The same EMF, 250 V, is applied to all three resistors. The current in each can be found from Ohm's Law, as shown below. The current through R1 is I1, I2 is the current through R2 and I3 is the current through R3.

For convenience, we can use resistance in kΩ, which gives current in milliamperes.

$$I1 = \frac{E}{R1} = \frac{250\,V}{5.00\,k\Omega} = 50.0\,mA$$

$$I2 = \frac{E}{R2} = \frac{250\,V}{20.0\,k\Omega} = 12.5\,mA$$

$$I3 = \frac{E}{R3} = \frac{250\,V}{8.00\,k\Omega} = 31.2\,mA$$

Notice that the branch currents are inversely proportional to the resistances. The 20000-Ω resistor has a value four times larger than the 5000-Ω resistor, and has a current one quarter as large. If a resistor has a value twice as large as another, it will have half as much current through it when they are connected in parallel.

The total circuit current is:

$$I_{TOTAL} = I1 + I2 + I3 \qquad (9)$$
$$I_{TOTAL} = 50.0\,mA + 12.5\,mA + 31.2\,mA$$
$$I_{TOTAL} = 93.7\,mA$$

This example illustrates Kirchhoff's Current Law: The current flowing into a node or branching point is equal to the sum of the individual currents leaving the node or branching point. The total resistance of the circuit is therefore:

$$R = \frac{E}{I} = \frac{250\,V}{93.7\,mA} = 2.67\,k\Omega$$

You can verify this calculation by combining the three resistor values in parallel, using equation 7.

RESISTORS IN SERIES

When a circuit has a number of resis-

tances connected in series, the total resistance of the circuit is the sum of the individual resistances. If these are numbered R1, R2, R3 and so on, then:

$$R_{TOTAL} = R1 + R2 + R3 + R4 + \ldots \quad (10)$$

where the dots indicate that as many resistors as necessary may be added.

Example: Suppose that three resistors are connected to a source of EMF as shown in **Fig 5.5**. The EMF is 250 V, R1 is 5.00 kΩ, R2 is 20.0 kΩ and R3 is 8.00 kΩ. The total resistance is then

$$R_{TOTAL} = R1 + R2 + R3$$

$$R = 5.00\,k\Omega + 20.0\,k\Omega + 8.00\,k\Omega$$

$$R = 33.0\,k\Omega.$$

The current in the circuit is then

$$I = \frac{E}{R} = \frac{250\,V}{33.0\,k\Omega} = 7.58\,mA$$

(We need not carry calculations beyond three significant figures; often, two will suffice because the accuracy of measurements is seldom better than a few percent.)

KIRCHHOFF'S SECOND LAW (KIRCHHOFF'S VOLTAGE LAW)

Ohm's Law applies in any portion of a circuit as well as to the circuit as a whole. Although the current is the same in all three of the resistances in the example of Fig 5.5, the total voltage divides between them. The voltage appearing across each resistor (the voltage drop) can be found from Ohm's Law.

Example: If the voltage across R1 is called E1, that across R2 is called E2 and that across R3 is called E3, then

$$E1 = I\,R1 = 0.00758\,A \times 5000\,\Omega = 37.9\,V$$

$$E2 = I\,R2 = 0.00758\,A \times 20000\,\Omega = 152\,V$$

$$E3 = I\,R3 = 0.00758\,A \times 8000\,\Omega = 60.6\,V$$

Notice here that the voltage drop across each resistor is directly proportional to the resistance. The 20000-Ω resistor value is four times larger than the 5000-Ω resistor, and the voltage drop across the 20000-Ω resistor is four times larger. A resistor that has a value twice as large as another will have twice the voltage drop across it when they are connected in series.

Kirchhoff's Voltage Law accurately describes the situation in the circuit: The sum of the voltages in a closed current loop is zero. The resistors are power sinks, while the battery is a power source. It is common to assign a + sign to power sources and a – sign to power sinks. This means the voltages across the resistors have the opposite sign from the battery voltage. Adding all the voltages yields zero. In the case of a single voltage source, algebraic manipulation implies that the sum of the individual voltage drops in the circuit must be equal to the applied voltage.

$$E_{TOTAL} = E1 + E2 + E3 \qquad (11)$$

$$E_{TOTAL} = 37.9\ V + 152\ V + 60.6\ V$$

$$E_{TOTAL} = 250\ V$$

(Remember the significant figures rule for addition.)

In problems such as this, when the current is small enough to be expressed in milliamperes, considerable time and trouble can be saved if the resistance is expressed in kilohms rather than in ohms. When the resistance in kilohms is substituted directly in Ohm's Law, the current will be milliamperes, if the EMF is in volts.

RESISTORS IN SERIES-PARALLEL

A circuit may have resistances both in parallel and in series, as shown in **Fig 5.6A**. The method for analyzing such a circuit is as follows: Consider R2 and R3 to be the equivalent of a single resistor, R_{EQ} whose value is equal to R2 and R3 in parallel.

$$R_{EQ} = \frac{R2 \times R3}{R2 + R3} = \frac{20000\ \Omega \times 8000\ \Omega}{20000\ \Omega + 8000\ \Omega}$$

$$= \frac{1.60 \times 10^8\ \Omega^2}{28000\ \Omega}$$

$$R_{EQ} = 5710\ \Omega = 5.71\ k\Omega$$

This resistance in series with R1 forms a simple series circuit, as shown in Fig 5.6B. The total resistance in the circuit is:

$$R_{TOTAL} = R1 + R_{EQ} = 5.00\ k\Omega + 5.71\ k\Omega$$

$$R_{TOTAL} = 10.71\ k\Omega$$

The current is:

$$I = \frac{E}{R} = \frac{250\ V}{10.71\ k\Omega} = 23.3\ mA$$

The voltage drops across R1 and R_{EQ} are:

$$E1 = I \times R1 = 23.3\ mA \times 5.00\ k\Omega = 117\ V$$

$$E2 = I \times R_{EQ} = 23.3\ mA \times 5.71\ k\Omega$$
$$= 133\ V$$

with sufficient accuracy. These two voltage drops total 250 V, as described by Kirchhoff's Current Law. E2 appears across both R2 and R3 so,

$$I2 = \frac{E2}{R2} = \frac{133\ V}{20.0\ k\Omega} = 6.65\ mA$$

$$I3 = \frac{E3}{R3} = \frac{133\ V}{8.00\ k\Omega} = 16.6\ mA$$

where:

I2 = current through R2 and
I3 = current through R3.

The sum of I2 and I3 is equal to 23.3 mA, conforming to Kirchhoff's Voltage Law.

THEVENIN'S THEOREM

Thevenin's Theorem is a useful tool for simplifying electrical networks. Thevenin's Theorem states that any two-terminal network of resistors and voltage or current sources can be replaced by a single voltage source and a series resistor. Such a transformation can simplify the calculation of current through a parallel branch. Thevenin's Theorem can be readily applied to the circuit of Fig 5.6A, to find the current through R3.

In this example, R1 and R2 form a voltage divider circuit, with R3 as the load (**Fig 5.7A**). The current drawn by the load (R3) is simply the voltage across R3, divided by its resistance. Unfortunately, the value of R2 affects the voltage across R3, just as the presence of R3 affects the potential appearing across R2. Some means of separating the two is needed; hence the Thevenin-equivalent circuit.

The voltage of the Thevenin-equivalent battery is the open-circuit voltage, measured when there is no current from either terminal A or B. Without a load connected between A and B, the total current through the circuit is (from Ohm's Law):

$$I = \frac{E}{R1 + R2} \qquad (12)$$

and the voltage between terminals A and B (E_{AB}) is:

$$E_{AB} = I \times R2 \qquad (13)$$

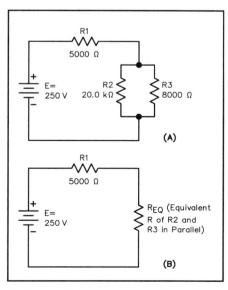

Fig 5.6 — At A, an example of resistors in series-parallel. The equivalent circuit is shown at B. See text for calculations.

By substituting the first equation into the second, we can find a simplified expression for E_{AB}:

$$E_{AB} = \frac{R2}{R1 + R2} \times E \qquad (14)$$

Using the values in our example, this becomes:

$$E_{AB} = \frac{20.0\ k\Omega}{25.0\ k\Omega} \times 250\ V = 200\ V$$

when nothing is connected to terminals A or B. With no current drawn, E is equal to E_{AB}.

The Thevenin-equivalent resistance is the total resistance between terminals A and B. The ideal voltage source, by defini-

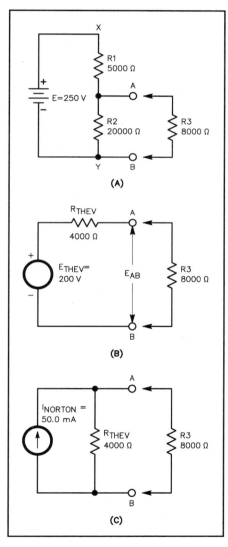

Fig 5.7 — Equivalent circuits for the circuit shown in Fig 5.6. A shows the load resistor (R3) looking into the circuit. B shows the Thevenin-equivalent circuit, with a resistor and a voltage source in series. C shows the Norton-equivalent circuit, with a resistor and current source in parallel.

tion, has zero internal resistance. Assuming the battery to be a close approximation of an ideal source, put a short between points X and Y in the circuit of Fig 5.7A. R1 and R2 are then effectively placed in parallel, as viewed from terminals A and B. The Thevenin-equivalent resistance is then:

$$R_{THEV} = \frac{R1 \times R2}{R1 + R2} \qquad (15)$$

$$R_{THEV} = \frac{5000 \ \Omega \times 20000 \ \Omega}{5000 \ \Omega + 20000 \ \Omega}$$

$$R_{THEV} = \frac{1.00 \times 10^8 \ \Omega^2}{25000 \ \Omega} = 4000 \ \Omega$$

This gives the Thevenin-equivalent circuit as shown in Fig 5.7B. The circuits of Figs. 5.7A and 5.7B are equivalent as far as R3 is concerned.

Once R3 is connected to terminals A and B, there will be current through R_{THEV}, causing a voltage drop across R_{THEV} and reducing E_{AB}. The current through R3 is equal to

$$I3 = \frac{E_{THEV}}{R_{TOTAL}} = \frac{E_{THEV}}{R_{THEV} + R3} \qquad (16)$$

Substituting the values from our example:

$$I3 = \frac{200 \ V}{4000 \ \Omega + 8000 \ \Omega} = 16.7 \ mA$$

This agrees with the value calculated earlier.

NORTON'S THEOREM

Norton's Theorem is another tool for analyzing electrical networks. Norton's Theorem states that any two-terminal network of resistors and current or voltage sources can be replaced by a single current source and a parallel resistor. Norton's Theorem is to current sources what Thevenin's Theorem is to voltage sources. In fact, the Thevenin resistance calculated previously is also used as the Norton equivalent resistance.

The circuit just analyzed by means of Thevenin's Theorem can be analyzed just as easily by Norton's Theorem. The equivalent Norton circuit is shown in Fig 5.7C. The current I_{SC} of the equivalent current source is the short-circuit current through terminals A and B. In the case of the voltage divider shown in Fig 5.7A, the short-circuit current is:

$$I_{SC} = \frac{E}{R1} \qquad (17)$$

Substituting the values from our example, we have:

$$I_{SC} = \frac{E}{R1} = \frac{250 \ V}{5000 \ \Omega} = 50.0 \ mA$$

The resulting Norton-equivalent circuit consists of a 50.0-mA current source placed in parallel with a $4000 - \Omega$ resistor. When R3 is connected to terminals A and B, one-third of the supply current flows through R3 and the remainder through R_{THEV}. This gives a current through R3 of 16.7 mA, again agreeing with previous conclusions.

A Norton-equivalent circuit can be transformed into a Thevenin-equivalent circuit and vice versa. The equivalent resistor stays the same in both cases; it is placed in series with the voltage source in the case of a Thevenin-equivalent circuit and in parallel with the current source in the case of a Norton-equivalent circuit. The voltage for a Thevenin-equivalent source is equal to the no-load voltage appearing across the resistor in the Norton-equivalent circuit. The current for a Norton-equivalent source is equal to the short-circuit current provided by the Thevenin source.

Power and Energy

Regardless of how voltage is generated, energy must be supplied if current is drawn from the voltage source. The energy supplied may be in the form of chemical energy or mechanical energy. This energy is measured in joules. One joule is defined from classical physics as the amount of energy or work done when a force of one newton (a measure of force) is applied to an object that is moved one meter in the direction of the force.

Power is another important concept. In the USA, power is often measured in horsepower in mechanical systems. We use the metric power unit of watts in electrical systems, however. In metric countries, mechanical power is usually expressed in watts also. One watt is defined as the use (or generation) of one joule of energy per second. One watt is also defined as one volt of potential pushing one ampere of current through a resistance. Thus,

$$P = I \times E \qquad (18)$$

where:

P = power in watts
I = current in amperes
E = EMF in volts.

When current flows through a resistance, the electrical energy is turned into heat. Common fractional and multiple units for power are the milliwatt (one thousandth of a watt) and the kilowatt (1000 W).

Example: The plate voltage on a transmitting vacuum tube is 2000 V and the plate current is 350 mA. (The current must be changed to amperes before substitution in the formula, and so is 0.350 A.) Then:

$$P = I \times E = 2000 \ V \times 0.350 \ A = 700 \ W$$

By substituting the Ohm's Law equivalent for E and I, the following formulas are obtained for power:

$$P = \frac{E^2}{R} \qquad (19)$$

and

$$P = I^2 \times R \qquad (20)$$

These formulas are useful in power calculations when the resistance and either the current or voltage (but not both) are known.

Example: How much power will be converted to heat in a 4000-Ω resistor if the potential applied to it is 200 V? From equation 19,

$$P = \frac{E^2}{R} = \frac{(200 \ V)^2}{4000 \ \Omega}$$

$$= \frac{40000 \ V^2}{4000 \ \Omega} = 10.0 \ W$$

As another example, suppose a current of 20 mA flows through a 300-Ω resistor. Then:

$$P = I^2 \times R = 0.020^2 \ A^2 \times 300 \ \Omega$$
$$P = 0.00040 \ A^2 \times 300 \ \Omega$$
$$P = 0.12 \ W$$

Note that the current was changed from milliamperes to amperes before substitution in the formula.

Electrical power in a resistance is turned into heat. The greater the power, the more rapidly the heat is generated. Resistors for radio work are made in many sizes, the smallest being rated to dissipate (or carry safely) about $1/16$ W. The largest resistors commonly used in amateur equipment will dissipate about 100 W. Large resistors such as those used in dummy-load antennas, are often cooled with oil to increase their power-handling capability.

If you want to express power in horsepower instead of watts, the following relationship holds:

$$1 \text{ horsepower} = 746 \text{ W} \qquad (21)$$

This formula assumes lossless transformation; practical efficiency is taken up shortly. This formula is especially useful if you are working with a system that converts electrical energy into mechanical energy, and vice versa, since mechanical power is often expressed in horsepower, in the US.

GENERALIZED DEFINITION OF RESISTANCE

Electrical energy is not always turned into heat. The energy used in running a motor, for example, is converted to mechanical motion. The energy supplied to a radio transmitter is largely converted into radio waves. Energy applied to a loudspeaker is changed into sound waves. In each case, the energy is converted to other forms and can be completely accounted for. None of the energy *just disappears*! This is a statement of the Law of Conservation of Energy. When a device converts energy from one form to another, we often say it *dissipates* the energy, or power. (Power is energy divided by time.) Of course the device doesn't really "use up" the energy, or make it disappear, it just converts it to another form. Proper operation of electrical devices often requires that the power must be supplied at a specific ratio of voltage to current. These features are characteristics of resistance, so it can be said that any device that "dissipates power" has a definite value of resistance.

This concept of resistance as something that absorbs power at a definite voltage-to-current ratio is very useful; it permits substituting a simple resistance for the load or power-consuming part of the device receiving power, often with considerable simplification of calculations. Of course, every electrical device has some resistance of its own in the more narrow sense, so a part of the energy supplied to it is converted to heat in that resistance even though the major part of the energy may be converted to another form.

EFFICIENCY

In devices such as motors and vacuum tubes, the objective is to convert the supplied energy (or power) into some form other than heat. Therefore, power converted to heat is considered to be a loss, because it is not useful power. The efficiency of a device is the useful power output (in its converted form) divided by the power input to the device. In a vacuum-tube transmitter, for example, the objective is to convert power from a dc source into ac power at some radio frequency. The ratio of the RF power output to the dc input is the efficiency of the tube. That is:

$$\text{Eff} = \frac{P_O}{P_I} \qquad (22)$$

where:
Eff = efficiency (as a decimal)
P_O = power output (W)
P_I = power input (W).

Example: If the dc input to the tube is 100 W, and the RF power output is 60 W, the efficiency is:

$$\text{Eff} = \frac{P_O}{P_I} = \frac{60 \text{ W}}{100 \text{ W}} = 0.6$$

Efficiency is usually expressed as a percentage — that is, it tells what percent of the input power will be available as useful output. To calculate percent efficiency, just multiply the value from equation 22 by 100%. The efficiency in the example above is 60%.

Suppose a mobile transmitter has an RF power output of 100. W with 52% efficiency at 13.8 V. The vehicle's alternator system charges the battery at a 5.0-A rate at this voltage. Assuming an alternator efficiency of 68%, how much horsepower must the engine produce to operate the transmitter and charge the battery? Solution: To charge the battery, the alternator must produce 13.8 V × 5.0 A = 69 W. The transmitter dc input power is 100. W / 0.52 = 190 W. Therefore, the total electrical power required from the alternator is 190 + 69 = 260 W. The engine load then is:

$$P_I = \frac{P_O}{\text{Eff}} = \frac{260 \text{ W}}{0.68} = 380 \text{ W}$$

We can convert this to horsepower using the formula given earlier to convert between horsepower and watts:

$$380 \text{ W} \times \frac{1 \text{ horsepower}}{746 \text{ W}} = 0.51 \text{ horsepower}$$

ENERGY

When you buy electricity from a power company, you pay for electrical energy, not power. What you pay for is the *work* that electricity does for you, not the rate at which that work is done. Work is equal to power multiplied by time. The common unit for measuring electrical energy is the watt-hour, which means that a power of 1 W has been used for one hour. That is:

$$\text{W hr} = P \, T$$

where:
W hr = energy in watt-hours
P = power in watts
T = time in hours.

Actually, the watt-hour is a fairly small energy unit, so the power company bills you for kilowatt-hours of energy used. Another energy unit that is sometimes useful is the watt-second (joule).

Energy units are seldom used in amateur practice, but it is obvious that a small amount of power used for a long time can eventually result in a power bill that is just as large as if a large amount of power had been used for a very short time.

One practical application of energy units is to estimate how long a radio (such as a hand-held unit) will operate from a certain battery. For example, suppose a fully charged battery stores 900 mA hr of energy, and a radio draws 30 mA on receive. You might guess that the radio will receive 30 hrs with this battery, assuming 100% efficiency. You shouldn't expect to get the full 900 mA hr out of the battery, and you will probably spend some of the time transmitting, which will also reduce the time the battery will last. The **Real-World Component Characteristics** and **Power Supplies** chapters include additional information about batteries and their charge/discharge cycles.

Circuits and Components

SERIES AND PARALLEL CIRCUITS

Passive components (resistors for dc circuits) can be used to make voltage and current dividers and limiters to obtain a desired value. For instance, in **Fig 5.8A**, two resistors are connected in series to provide a voltage divider. As long as the device connected at point A has a much higher resistance than the resistors in the divider, the voltage will be approximately the ratio of the resistances. Thus, if E = 10 V, R1 = 5 Ω and R2 = 5 Ω, the voltage at point A will be 5 V measured on a high-impedance voltmeter. A good rule of thumb is that the load at point A should be at least ten times the value of the highest resistor in the divider to get reasonably close to the voltage you want. As the load resistance gets closer to the value of the divider, the current drawn by the load affects the division and causes changes from the desired value. If you need precise voltage division from fixed resistors and know the value of the load resistance, you can use Kirchhoff's Laws and Thevenin's Theorem (explained earlier) to calculate exact values.

Similarly, resistors can be used, as shown in Fig 5.8B, to make current dividers. Suppose you had two LEDs (light emitting diodes) and wanted one to glow twice as brightly as the other. You could use one resistor with twice the value of the other for the dimmer LED. Thus, approximately two-thirds of the current would flow through one LED and one-third through the other (neglecting any effect of the 0.7-V drop across the diode).

Resistors can also be used to limit the current through a device from a fixed voltage source. A typical example is shown in Fig 5.8C. Here a high-voltage source feeds a battery in a battery charger. This is typical of nickel cadmium chargers. The high resistor value limits the current that can possibly flow through the battery to a value that is low enough so it will not damage the battery.

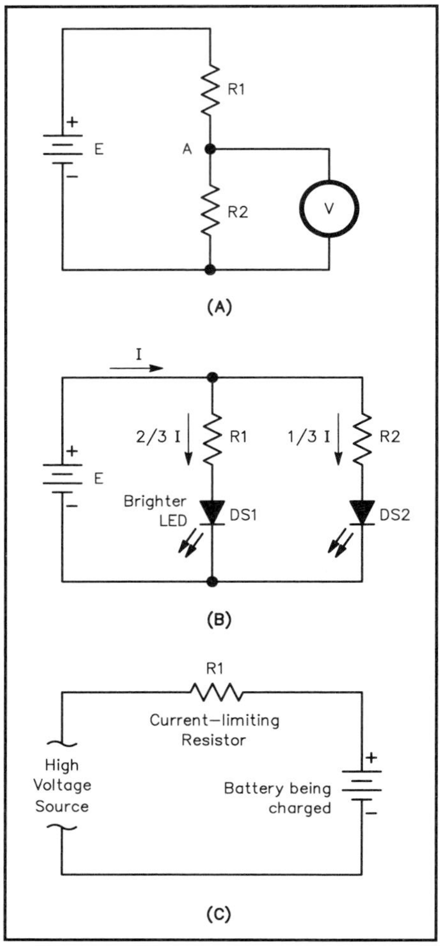

SWITCHES

Switches are used to start or stop a signal (current) flowing in a particular circuit. Most switches are mechanical devices, although the same effect may be achieved with solid-state devices. Relays are switches that are controlled by another electrical signal rather than manual or mechanical means.

Switches come in many different forms and a wide variety of ratings. The most important ratings are the voltage and current handling capabilities. The voltage rating usually includes both the breakdown rating and the interrupt rating. Normally, the interrupt rating is the lower value, and therefore the one given on (for) the switch. The current rating includes both the current carrying capacity and the interrupt capability.

Most power switches are rated for alternating current use. Because ac voltage goes through zero with each cycle, switches can successfully interrupt much more alternating current than direct current without arcing. A switch that has a 10-A ac current rating may arc and damage the contacts if used to turn off more than an ampere or two of dc.

Switches are normally designated by the number of *poles* (circuits controlled) and *positions* (circuit path choices). The simplest switch is the on-off switch, which is a single-pole, single-throw (SPST) switch as shown in **Fig 5.9A**. The off position does not direct the current to another circuit. The next step would be to change the current path to another path, and would be a single-pole, double-throw (SPDT) switch as shown in Fig 5.9B. Adding an off position would give a single-pole, double-throw, center-off switch as shown in Fig 5.9C.

Several such switches can be "ganged" to the same mechanical activator to provide double pole, triple pole or even more, separate control paths all activated at once. Switches can be activated in a variety of ways. The most common methods include lever, push button and rotary switches. Samples of these are shown in **Fig 5.10**. Most switches stay in the position set, but some are spring loaded so they only stay in the desired position while held there. These are called momentary switches.

Switches typically found in the home are usually rated for 125 V ac and 15 to 20 A. Switches in cars are usually rated for 12 V dc and several amperes. The breakdown voltage rating of a switch, which is usually higher than the interrupt rating, primarily depends on the insulating material surrounding the contacts and the separation between the contacts. Plastic or phenolic material normally provides both structural support and insulation. Ceramic material may be used to provide better insulation, particularly in rotary (wafer) switches.

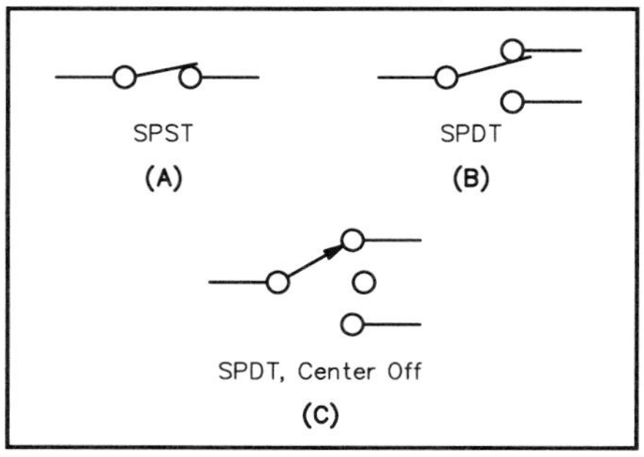

Fig 5.8 — This circuit shows a resistive voltage divider.

Fig 5.9 — Schematic diagrams of various types of switches. A is an SPST, B is an SPDT, C is an SPDT switch with a center-off position.

Fig 5.10 — This photo shows examples of various styles of switches.

The current carrying capacity of the switch depends on the contact material and size and on the pressure between the contacts. It is primarily determined from the allowable contact temperature rise. On larger ac switches, or most dc switches, the interrupt capability may be lower than the current carrying value.

Rotary/wafer switches can provide very complex switching patterns. Several poles (separate circuits) can be included on each wafer. Many wafers may be stacked on the same shaft. Not only may many different circuits be controlled at once, but by wiring different poles/positions on different wafers together, a high degree of circuit switching logic can be developed. Such switches can select different paths as they are turned, and can also "short" together successive contacts to connect numbers of components or paths. They can also be designed to either break one contact before making another, or to short two contacts together before disconnecting the first one (make before break) to eliminate arcing or perform certain logic functions.

In choosing a switch for a particular task, consideration should be given to function, voltage and current ratings, ease of use, availability and cost. If a switch is to be operated frequently, a slightly higher cost for a better-quality switch is usually less costly over the long run. If signal noise or contact corrosion is a potential problem, (usually in low-current signal applications) it is best to get gold plated contacts. Gold does not oxidize or corrode, thus providing surer contact, which can be particularly important at very low signal levels. Gold plating will not hold up under high-current-interrupt applications, however.

FUSES

Fuses self-destruct to protect circuit wiring or equipment. The fuse element that melts is a carefully shaped piece of soft metal, usually mounted in a cartridge of some kind. The element is designed to safely carry a given amount of current and to melt at a current value that is a certain percentage over the rated value. The melting value depends on the type of material, the shape of the element and the heat dissipation capability of the cartridge and holder, among other factors. Some fuses (Slo-blo) are designed to carry an overload for a short period of time. They typically are used in motor starting and power-supply circuits that have a large inrush current when first started. Other fuses are designed to blow very quickly to protect delicate instruments and solid-state circuits. A replacement fuse should have the same current rating and the same characteristics as the fuse it replaces. **Fig 5.11** shows a variety of fuse types and sizes.

The most important fuse rating is the nominal current rating that it will safely carry. Next most important are the timing characteristics, or how quickly it opens under a given current overload. A fuse also has a voltage rating, both a value in volts and whether it is expected to be used in ac or dc circuits. While you should never substitute a fuse with a higher current rating than the one it replaces, you can use a fuse with a higher voltage rating. There is no danger in replacing a 12-V, 2-A fuse with a 250-V, 2-A unit.

Fuses fail for several reasons. The most obvious reason is that a problem develops in the circuit, which causes too much current to flow. In this case, the circuit problem needs to be fixed. A fuse may just fail eventually, particularly when cycled on and off near its current rating. A kind of metal fatigue sets in, and eventually the fuse goes. A fuse can also blow because of a momentary power surge, or even turning something on and off several times quickly when there is a large inrush current. In these cases it is only necessary to replace the fuse with the same type and value. Never substitute a fuse with a larger current rating. You may cause permanent damage (maybe even a fire) to the wiring or circuit elements if/when there is an internal problem in the equipment.

RELAYS

Relays are switches that are driven by an electrical signal, usually through a magnetic coil. An armature that moves when current is applied pushes the switch contacts together, or pulls them apart. Many such contacts can be connected to the same armature, allowing many circuits to be controlled by a single signal. Usually, relays have only two positions (opening some contacts and closing others) although there are special cases.

Like switches, relays have specific voltage and current ratings for the contacts. These may be far different from the voltage and current of the coil that drives the relay. That means a small signal voltage might control very large values of voltage and/or current. Relay contacts (and housings) may be designed for ac, dc or RF signals. The control voltages are usually 12 V dc or 125 V ac for most amateur applications, but the coils may be designed to be "current sensing" and operate when the current through the coil exceeds a specific value. **Fig 5.12** shows some typical relays found in amateur equipment. Relays with 24- and 28-V coils are also common.

Coaxial relays are specially designed to handle RF signals and to maintain a characteristic impedance to match certain values of coaxial-cable impedance. They

Fig 5.11 — This photo shows examples of various styles of fuses.

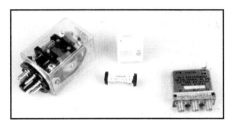

Fig 5.12 — This photo shows examples of various styles of relays.

typically are used to switch an antenna between a receiver and transmitter or between a linear amplifier and a transceiver. **Fig 5.13** shows how several relays may be controlled by a microphone button to switch various functions in an amateur station. This simple system does not include control circuitry to provide proper sequencing of the relays.

POTENTIOMETERS

Potentiometer is a big name for a variable resistor. They are commonly used as volume controls on radios, televisions and stereos. A typical potentiometer is a circular pattern of resistive material, usually a carbon compound, that has a wiper on a shaft moving across the material. For higher power applications, the resistive material may be wire, wound around a core. As the wiper moves along the material, more resistance is introduced between the wiper and one of the fixed contacts on the material. A potentiometer may be used primarily to control current, voltage or resistance in a circuit. **Fig 5.14** shows several circuits to demonstrate various uses. **Fig 5.15** shows several different types of potentiometers.

Typical specifications for a potentiometer include maximum resistance, power dissipation, voltage and current ratings, number of turns (or degrees) the shaft can rotate, type and size of shaft, mounting arrangements and resistance "taper."

Not all potentiometers have a *linear* taper. That is, the resistance may not be

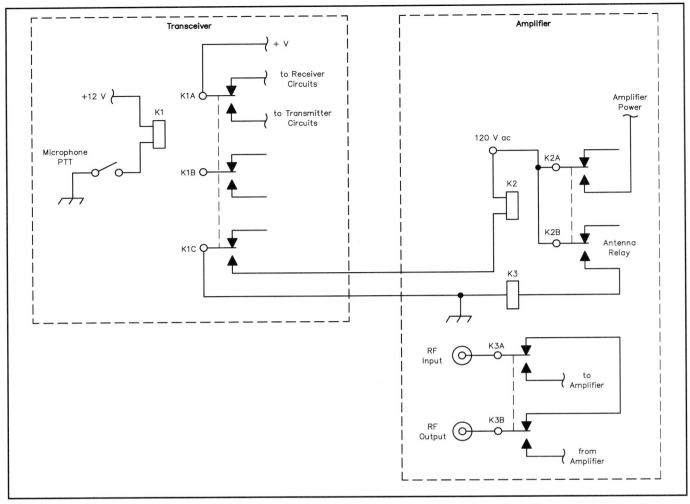

Fig 5.13 — A simple station control circuit. This example does not include control circuitry to provide proper sequencing of the relays.

the same for a given number of degrees of shaft rotation along different portions of the resistive material. A typical use of a potentiometer with a nonlinear taper is as a volume control. Since the human ear has a logarithmic response to sound, a volume control may actually change the volume (resistance) much more near one end of the potentiometer than the other (for a given amount of rotation) so that the "perceived" change in volume is about the same for a similar change in the control. This is commonly called an "audio taper" as the change in resistance per degree of rotation attempts to match the response of the human ear. The taper can be designed to match almost any desired control function for a given application. Linear and audio tapers are the most common.

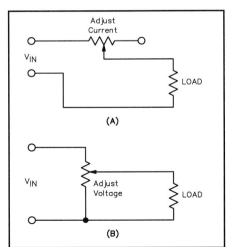

Fig 5.14 — Various uses of potentiometers.

Fig 5.15 — This photo shows examples of different styles of potentiometers.

Contents

AC Theory and Reactive Components

6

Glossary

Admittance (Y) — The reciprocal of impedance, measured in siemens (S).

Capacitance (C) — The ability to store electrical energy in an electrostatic field, measured in farads (F). A device with capacitance is a capacitor.

Conductance (G) — The reciprocal of resistance, measured in siemens (S).

Current (I) — The rate of electron flow through a conductor, measured in amperes (A).

Flux density (B) — The number of magnetic-force lines per unit area, measured in gauss.

Frequency (f) — The rate of change of an ac voltage or current, measured in cycles per second, or hertz (Hz).

Impedance (Z) — The complex combination of resistance and reactance, measured in ohms (Ω).

Inductance (L) — The ability to store electrical energy in a magnetic field, measured in henrys (H). A device, such as a coil, with inductance is an inductor.

Peak (voltage or current) — The maximum value relative to zero that an ac voltage or current attains during any cycle.

Peak-to-peak (voltage or current) — The value of the total swing of an ac voltage or current from its peak negative value to its peak positive value, ordinarily twice the value of the peak voltage or current.

Period (T) — The duration of one ac voltage or current cycle, measured in seconds (s).

Permeability (μ) — The ratio of the magnetic flux density of an iron, ferrite, or similar core in an electromagnet compared to the magnetic flux density of an air core, when the current through the electromagnet is held constant.

Power (P) — The rate of electrical-energy use, measured in watts (W).

Q (quality factor) — The ratio of energy stored in a reactive component (capacitor or inductor) to the energy dissipated, equal to the reactance divided by the resistance.

Reactance (X) — Opposition to alternating current by storage in an electrical field (by a capacitor) or in a magnetic field (by an inductor), measured in ohms (Ω).

Resistance (R) — Opposition to current by conversion into other forms of energy, such as heat, measured in ohms (Ω).

Resonance — Ordinarily, the condition in an ac circuit containing both capacitive and inductive reactance in which the reactances are equal.

RMS (voltage or current) — Literally, "root mean square;" the square root of the average of the squares of the instantaneous values for one cycle of a waveform. A dc voltage or current that will produce the same heating effect as the waveform. For a sine wave, the RMS value is equal to 0.707 times the peak value of ac voltage or current.

Susceptance (B) — The reciprocal of reactance, measured in siemens (S).

Time constant (τ) — The time required for the voltage in an RC circuit or the current in an RL circuit to rise from zero to approximately 63.2% of its maximum value or to fall from its maximum value 63.2% toward zero.

Toroid — Literally, any donut-shaped solid; most commonly referring to ferrite or powdered-iron cores supporting inductors and transformers.

Transducer — Any device that converts one form of energy to another; for example an antenna, which converts electrical energy to electromagnetic energy or a speaker, which converts electrical energy to sonic energy.

Transformer — A device consisting of at least two coupled inductors capable of transferring energy through mutual inductance.

Voltage (E) — Electromotive force or electrical pressure, measured in volts (V).

Alternating Current, Frequency and Wavelength

AC IN CIRCUITS

A circuit is a complete conductive route for electrons to follow from a source, through a load and back to the source. If the source permits the electrons to flow in only one direction, the current is *dc* or *direct current*. If the source permits the current periodically to change direction, the current is *ac* or *alternating current*. **Fig 6.1** illustrates the two types of circuits. Drawing A shows the source as a battery, a typical dc source. Drawing B shows a more abstract source symbol to indicate ac. In an ac circuit, not only does the current change direction periodically; the voltage also periodically reverses. The rate of reversal may range from a few times per second to many billions per second.

Graphs of current or voltage, such as Fig 6.1, begin with a horizontal axis that represents time. The vertical axis represents

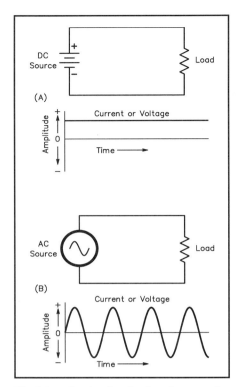

Fig 6.1 — Basic circuits for direct and alternating currents. With each circuit is a graph of the current, constant for the dc circuit, but periodically changing direction in the ac circuit.

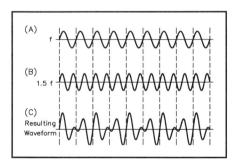

Fig 6.3 — Two ac waveforms of similar frequencies (f1 = 1.5 f2) and amplitudes form a composite wave. Note the points where the positive peaks of the two waves combine to create high composite peaks: this is the phenomenon of beats. The beat note frequency is 1.5f − f = 0.5f and is visible in the drawing.

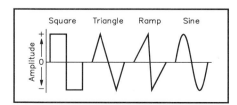

Fig 6.5 — Some common ac waveforms: square, triangle, ramp and sine.

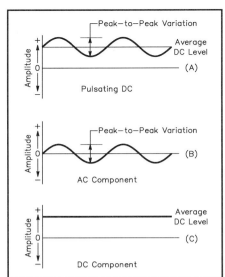

Fig 6.2 — A pulsating dc current (A) and its resolution into an ac component (B) and a dc component (C).

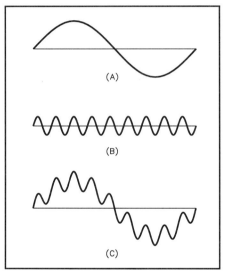

Fig 6.4 — Two ac waveforms of widely different frequencies and amplitudes form a composite wave in which one wave appears to ride upon the other.

the amplitude of the current or the voltage, whichever is graphed. Distance above the zero line means a greater positive amplitude; distance below the zero line means a greater negative amplitude. Positive and negative simply designate the opposing directions in which current may flow in an alternating current circuit or the opposing directions of force of an ac voltage.

If the current and voltage never change direction, then from one perspective, we have a dc circuit, even if the level of dc constantly changes. **Fig 6.2** shows a current that is always positive with respect to

0. It varies periodically in amplitude, however. Whatever the shape of the variations, the current can be called *pulsating dc*. If the current periodically reaches 0, it can be called *intermittent dc*. From another perspective, we may look at intermittent and pulsating dc as a combination of an ac and a dc current. Special circuits can separate the two currents into ac and dc components for separate analysis or use. There are also circuits that combine ac and dc currents and voltages for many purposes.

We can combine ac and dc voltages and currents. Different ac voltages and currents also form combinations. Such combinations will result in complex waveforms. A *waveform* is the pattern of amplitudes reached by the voltage or current as measured over time. **Fig 6.3** shows two ac waveforms fairly close in frequency, and their resultant combination. **Fig 6.4** shows two ac waveforms dissimilar in both frequency and wavelength, along with the resultant combined waveform. Note the similarities (and the differences) between the resultant waveform in Fig 6.4 and the combined ac-dc waveform in Fig 6.2.

Alternating currents may take on many useful wave shapes. **Fig 6.5** shows a few that are commonly used in practical circuits and in test equipment. The square wave is vital to digital electronics. The triangular and ramp waves — sometimes called "sawtooth" waves — are especially useful in timing circuits. The sine wave is both mathematically and practically the foundation of all other forms of ac; the other forms can usually be reduced to (and even constructed from) a particular collection of sine waves.

There are numerous ways to generate alternating currents: with an ac power generator (an *alternator*), with a transducer (for example, a microphone) or with an electronic circuit (for example, an RF oscillator). The basis of the sine wave is circular motion, which underlies the most usual methods of generating alternating current. The circular motion of the ac generator may be physical or mechanical, as in an alternator. Currents in the resonant circuit of an oscillator may also produce sine waves without mechanical motion.

Fig 6.6 demonstrates the relationship of the current (and voltage) amplitude to relative positions of a circular rotation through one complete revolution of 360°. Note that the current is zero at point 1. It rises to its maximum value at a point 90° from point 1, which is point 3. At a point 180° from point 1, which is point 4, the current level falls back to zero. Then the current begins to rise again. The direction of the current after point 4 and prior to its return to point 1, however,

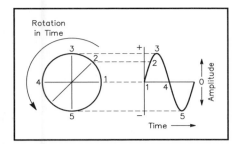

Fig 6.6 — The relationship of circular motion and the resultant graph of ac current or voltage. The curve is sinusoidal, a sine wave.

is opposite the direction of current from point 1 to point 4. Point 2 illustrates one of the innumerable intermediate values of current throughout the cycle.

Tracing the rise and fall of current over a linear time line produces the curve accompanying the circle in Fig 6.6. The curve is *sinusoidal* or a *sine wave*. The amplitude of the current varies as the sine of the angle made by the circular movement with respect to the zero point. The sine of 90° is 1, and 90° is also the point of maximum current (along with 270°). The sine of 45° (point 2) is 0.707, and the value of current at the 45° point of rotation is 0.707 times the maximum current. Similar considerations apply to the variation of ac voltage over time.

FREQUENCY AND PERIOD

With a continuously rotating generator, alternating current will pass through many equal cycles over time. Select an arbitrary point on any one cycle and use it as a marker. For this example, the positive peak will work as an unambiguous marker. The number of times per second that the current (or voltage) reaches this positive peak in any one second is called the *frequency* of the ac. In other words, frequency expresses the *rate* at which current (or voltage) cycles occur. The unit of frequency is *cycles per second*, or *hertz*— abbreviated Hz (after the 19th century radio-phenomena pioneer, Heinrich Hertz).

The length of any cycle in units of time is the *period* of the cycle, as measured from and to equivalent points on succeeding cycles. Mathematically, the period is simply the inverse of the frequency. That is,

$$\text{Frequency (f) in Hz} = \frac{1}{\text{Period (T) in seconds}} \quad (1)$$

and

$$\text{Period (T) in seconds} = \frac{1}{\text{Frequency (f) in Hz}} \quad (2)$$

Example: What is the period of a 400-hertz ac current?

$$T = \frac{1}{f} = \frac{1}{400 \text{ Hz}} = 0.00250 \text{ s} = 2.5 \text{ ms}$$

The frequency of alternating currents used in Amateur Radio circuits varies from a few hertz, or cycles per second, to thousands of millions of hertz. Likewise, the period of alternating currents amateurs use ranges from significant fractions of a second down to nanoseconds or smaller. In order to express units of frequency, time and almost everything else in electronics compactly, electronics uses a standard system of prefixes. In magnitudes of 1000 or 10^3, frequency is measurable in hertz, in kilohertz (1000 hertz or kHz), in megahertz (1 million hertz or MHz), gigahertz (1 billion hertz or GHz) and even in terahertz (1 trillion hertz or THz). For units smaller than one, as in the measurement of period, the basic unit seconds can become milliseconds (1 thousandth of a second or ms), microseconds (1 millionth of a second or μs), nanoseconds (1 billionth of a second or ns) and picoseconds (1 trillionth of a second or ps). See the **Mathematics for Amateur Radio** chapter for a complete list of prefixes and their relationship to basic units.

The uses of ac in Amateur Radio circuits are many and varied. Most can be cataloged by reference to ac frequency ranges used in circuits. For example, ac power used in the home, office and factory is ordinarily 60 Hz in the United States and Canada. In Great Britain and much of Europe, ac power is 50 Hz. For special purposes, ac power has been generated up to about 400 Hz.

Sonic and ultrasonic applications of ac run from about 20 Hz up to several MHz. Audio work makes use of the lower end of the sonic spectrum, with communications audio focusing on the range from about 300 to 3000 Hz. High-fidelity audio uses ac circuits capable of handling 20 Hz to at least 20 kHz. Ultrasonics — used in medicine and industry — makes use of ac circuits above 20 kHz.

Amateur Radio circuits include both power- and sonic-frequency-range circuits. Radio communication and other electronics work, however, require ac circuits capable of operation with frequencies up to the gigahertz range. Some of the applications include signal sources for transmitters (and for circuits inside receivers); industrial induction heating; diathermy; microwaves for cooking, radar and communication; remote control of appliances, lighting, model planes and boats and other equipment; and radio direction finding and guidance.

AC IN CIRCUITS AND TRANSDUCED ENERGY

Alternating currents are often loosely classified as audio frequency (AF) and radio frequency (RF). Although these designations are handy, they actually represent something other than the electrical energy of ac circuits: They designate special forms of energy that we find useful.

Audio or *sonic* energy is the energy imparted by the mechanical movement of a medium, which can be air, metal, water or even the human body. Sound that humans can hear normally requires the movement of air between 20 Hz and 20 kHz, although the human ear loses its ability to detect the extremes of this range as we age. Some animals, such as elephants, can apparently detect air vibrations well below 20 Hz, while others, such as dogs and cats, can detect air vibrations well above 20 kHz.

Electrical circuits do not directly produce air vibrations. Sound production requires a *transducer*, a device to transform one form of energy into another form of energy; in this case electrical energy into sonic energy. The speaker and the microphone are the most common audio transducers. There are numerous ultrasonic transducers for various applications.

Likewise, converting electrical energy into radio signals also requires a transducer, usually called an *antenna*. In contrast to RF alternating currents in circuits, RF *energy* is a form of electromagnetic energy. The frequencies of electromagnetic energy run from 3 kHz to above 10^{12} GHz. They include radio, infrared, visible light, ultraviolet and a number of energy forms of greatest interest to physicists and astronomers. **Table 6.1** provides a brief glimpse at the total spectrum of electromagnetic energy.

Table 6.1
Key Regions of the Electromagnetic Energy Spectrum

Region Name	Frequency Range		
Radio frequencies	3.0×10^3 Hz	to	3.0×10^{11} Hz
Infrared	3.0×10^{11} Hz	to	4.3×10^{14} Hz
Visible light	4.3×10^{14} Hz	to	1.0×10^{15} Hz
Ultraviolet	1.0×10^{15} Hz	to	6.0×10^{16} Hz
X-rays	6.0×10^{16} Hz	to	3.0×10^{19} Hz
Gamma rays	3.0×10^{19} Hz	to	5.0×10^{20} Hz
Cosmic rays	5.0×10^{20} Hz	to	8.0×10^{21} Hz

Table 6.2
Classification of the Radio Frequency Spectrum

Abbreviation	Classification	Frequency Range			
VLF	Very low frequencies	3	to	30	kHz
LF	Low frequencies	30	to	300	kHz
MF	Medium frequencies	300	to	3000	kHz
HF	High frequencies	3	to	30	MHz
VHF	Very high frequencies	30	to	300	MHz
UHF	Ultrahigh frequencies	300	to	3000	MHz
SHF	Superhigh frequencies	3	to	30	GHz
EHF	Extremely high frequencies	30	to	300	GHz

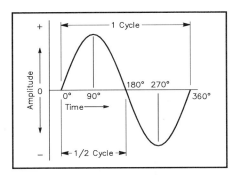

Fig 6.7 — An ac cycle is divided into 360° that are used as a measure of time or phase.

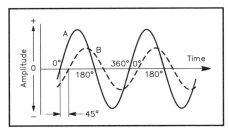

Fig 6.8 — When two waves of the same frequency start their cycles at slightly different times, the time difference or phase difference is measured in degrees. In this drawing, wave B starts 45° (one-eighth cycle) later than wave A, and so lags 45° behind A.

All electromagnetic energy has one thing in common: it travels, or *propagates,* at the speed of light. This speed is approximately 300000000 (or 3.00×10^8) meters per second in a vacuum. Electromagnetic-energy waves have a length uniquely associated with each possible frequency. The wavelength (λ) is simply the speed of propagation divided by the frequency (f) in hertz.

$$f (Hz) = \frac{3.00 \times 10^8 \left(\frac{m}{s} \right)}{\lambda (m)} \qquad (3)$$

and

$$\lambda (m) = \frac{3.00 \times 10^8 \left(\frac{m}{s} \right)}{f (Hz)} \qquad (4)$$

Example: What is the frequency of an 80.0-m RF wave?

$$f (Hz) = \frac{3.00 \times 10^8 \left(\frac{m}{s} \right)}{\lambda (m)}$$

$$= \frac{3.00 \times 10^8 \left(\frac{m}{s} \right)}{80.0 \ m}$$

$$f (Hz) = 3.75 \times 10^6 \ Hz$$

We could use a similar equation to calculate the wavelength of a sound wave in air, but we would have to use the speed of sound instead of the speed of light in the numerator of the equation. The speed of propagation of the mechanical movement of air that we call sound varies considerably with air temperature and altitude. The speed of sound at sea level is about 331 m/s at 0°C and 344 m/s at 20°C.

To calculate the frequency of an electromagnetic wave directly in kilohertz, change the speed constant to 300000 (3.00×10^5) km/s.

$$f (kHz) = \frac{3.00 \times 10^5 \left(\frac{km}{s} \right)}{\lambda (m)} \qquad (5)$$

and

$$\lambda (m) = \frac{3.00 \times 10^5 \left(\frac{km}{s} \right)}{f (kHz)} \qquad (6)$$

For frequencies in megahertz, use:

$$f (MHz) = \frac{300 \left(\frac{Mm}{s} \right)}{\lambda (m)} \qquad (7)$$

and

$$\lambda (m) = \frac{300 \left(\frac{Mm}{s} \right)}{f (MHz)} \qquad (8)$$

You would normally just drop the units that go with the speed of light constant, to make the equation look simpler.

Example: What is the wavelength of an RF wave whose frequency is 4.0 MHz?

$$\lambda (m) = \frac{300}{f (MHz)} = \frac{300}{4.0 \ MHz} = 75 \ m$$

At higher frequencies, circuit elements act like transducers. This property can be put to use, but it can also cause problems for some circuit operations. Therefore, wavelength calculations are of some importance in designing ac circuits for those frequencies.

Within the part of the electromagnetic-energy spectrum of most interest to radio applications, frequencies have been classified into groups and given names. **Table 6.2** provides a reference list of these classifications. To a significant degree, the frequencies within each group exhibit similar properties. For example, HF or high frequencies, from 3 to 30 MHz, all exhibit *skip* or ionospheric refraction that permits regular long-range radio communications. This property also applies occasionally both to MF (medium frequencies) and to VHF (very high frequencies).

Despite the close relationship between RF electromagnetic energy and RF ac circuits, it remains important to distinguish the two. To the ac circuit producing or amplifying a 15-kHz alternating current, the ultimate transformation and use of the electrical energy may make no difference to the circuit's operation. By choosing the right transducer, one can produce either an audio tone or a radio signal — or both. Such was the accidental fate of many horizontal oscillators and amplifiers in early television sets; they found ways to vibrate parts audibly and to radiate electromagnetic energy.

PHASE

When tracing a sine-wave curve of an ac voltage or current, the horizontal axis represents time. We call this the *time domain* of the sine wave. Events to the right take place later; events to the left occur earlier. Although time is measurable in parts of a second, it is more convenient to treat each cycle as a complete time unit that we divide into 360°. The conventional starting point for counting degrees is the zero point as the voltage or current begins the positive half cycle. The essential elements of an ac cycle appear in **Fig 6.7**.

The advantage of treating the ac cycle in this way is that many calculations and measurements can be taken and recorded in a manner that is independent of frequency. The positive peak voltage or current occurs at 90° along the cycle. Relative to the starting point, 90° is the *phase* of the ac at that point. Thus, a complete description of an ac voltage or current involves reference to three properties: frequency, amplitude and phase.

Phase relationships also permit the

comparison of two ac voltages or currents at the same frequency, as **Fig 6.8** demonstrates. Since B crosses the zero point in the positive direction after A has already done so, there is a *phase difference* between the two waves. In the example, B *lags* A by 45°, or A *leads* B by 45°. If A and B occur in the same circuit, their composite waveform will also be a sine wave at an intermediate phase angle relative to each. Adding any number of sine waves of the same frequency always results in a sine wave at that frequency.

Fig 6.8 might equally apply to a voltage and a current measured in the same ac circuit. Either A or B might represent the voltage; that is, in some instances voltage will lead the current and in others voltage will lag the current.

Two important special cases appear in **Fig 6.9**. In Part A, line B lags 90° behind line A. Its cycle begins exactly one quarter cycle later than the A cycle. When one wave is passing through zero, the other just reaches its maximum value.

In Part B, lines A and B are 180° *out of phase*. In this case, it does not matter which one is considered to lead or lag. Line B is always positive while line A is negative, and vice versa. If the two waveforms are of two voltages or two currents in the same circuit and if they have the same amplitude, they will cancel each other completely.

MEASURING AC VOLTAGE, CURRENT AND POWER

Measuring the voltage or current in a dc circuit is straightforward, as **Fig 6.10A** demonstrates. Since the current flows in only one direction, for a resistive load, the voltage and current have constant values until the circuit components change.

Fig 6.10B illustrates a perplexing problem encountered when measuring voltages and currents in ac circuits. The current and voltage continuously change direction and value. Which values are meaningful? In fact, several values of constant sine-wave voltage and current in ac circuits are important to differing applications and concerns.

Instantaneous Voltage and Current

Fig 6.11 shows a sine wave of some arbitrary frequency and amplitude with respect to either voltage or current. The instantaneous voltage (or current) at point A on the curve is a function of three factors: the maximum value of voltage (or current) along the curve (point B), the frequency of the wave,

and the time elapsed in seconds or fractions of a second. Thus,

$$E_{inst} = E_{max} \sin (2\pi f t) \theta \qquad (9)$$

Considering just one sine wave, independent of frequency, the instantaneous value of voltage (or current) becomes

$$E_{inst} = E_{max} \sin \theta \qquad (10)$$

where θ is the angle in degrees through which the voltage has moved over time after the beginning of the cycle.

Example: What is the instantaneous value of voltage at point D in Fig 6.11, if the maximum voltage value is 120. V and the angular travel is 60.0°?

$$E_{inst} = 120. \text{ V} \times \sin 60.0° = 120. \times 0.866$$
$$= 104 \text{ V}$$

Peak and Peak-to-Peak Voltage and Current

The most important instantaneous voltages and currents are the maximum or peak values reached on each positive and

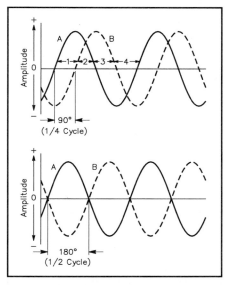

Fig 6.9 — Two important special cases of phase difference: In the upper drawing, the phase difference between A and B is 90°; in the lower drawing, the phase difference is 180°.

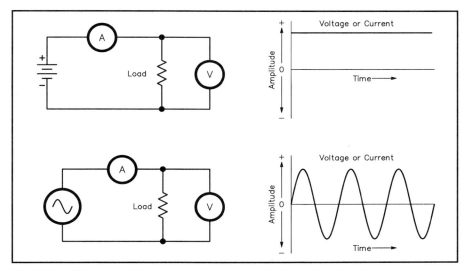

Fig 6.10 — Voltage and current measurements in dc and ac circuits.

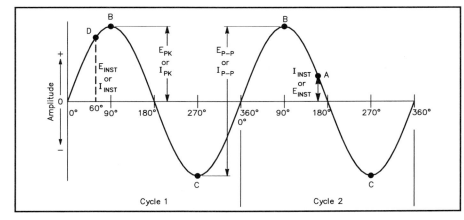

Fig 6.11 — Two cycles of a sine wave to illustrate instantaneous, peak, and peak-to-peak ac voltage and current values.

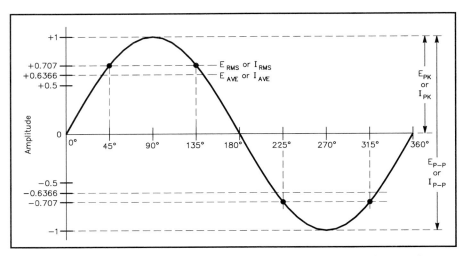

Fig 6.12 — The relationships between RMS, average, peak, and peak-to-peak values of ac voltage and current.

Table 6.3

Conversion Factors for AC Voltage or Current

From	To	Multiply By	
Peak	Peak-to-Peak	2	
Peak-to-Peak	Peak	0.5	
Peak	RMS	$1/\sqrt{2}$ or	0.707
RMS	Peak	$\sqrt{2}$ or	1.414
Peak-to-Peak	RMS	$1/(2 \times \sqrt{2})$ or	0.35355
RMS	Peak-to-Peak	$2 \times \sqrt{2}$ or	2.828
Peak	Average	$2/\pi$ or	0.6366
Average	Peak	$\pi/2$ or	1.5708
RMS	Average	$(2 \times \sqrt{2})/\pi$ or	0.90
Average	RMS	$\pi/(2 \times \sqrt{2})$ or	1.11

Note: These conversion factors apply only to continuous pure sine waves.

negative half cycle of the sine wave. In Fig 6.11, points B and C represent the positive and negative peaks of voltage or current. Peak (pk) values are especially important with respect to component ratings, which the voltage or current in a circuit must not exceed without danger of component failure.

The peak power in an ac circuit is simply the product of the peak voltage and the peak current, or

$$P_{pk} = E_{pk} \times I_{pk} \qquad (11)$$

The span from points B to C in Fig 6.11 represents the largest voltage or current swing of the sine wave. Designated the *peak-to-peak* (P-P) voltage (or current), this span is equal to twice the peak value of the voltage (or current). Thus,

$$E_{P-P} = 2E_{pk} \qquad (12)$$

Amplifying devices often specify their input limits in terms of peak-to-peak volt-ages. Operational amplifiers, which have almost unlimited gain potential, often require input-level limiting to prevent the output signals from distorting if they exceed the peak-to-peak output rating of the devices.

RMS Voltages and Currents

The *root mean square* or *RMS* values of voltage and current are the most common values encountered in electronics. Sometimes called the *effective* values of ac voltage and current, they are based upon equating the values of ac and dc power required to heat a resistive element to exactly the same degree. The peak ac power required for this condition is twice the dc power needed. Therefore, the average ac power equivalent to a corresponding average dc power is half the peak ac power.

$$P_{ave} = \frac{P_{pk}}{2} \qquad (13)$$

Since a circuit with a constant resistance is linear — that is, raising or lowering the voltage will raise or lower the current proportionally — the voltage and current values needed to arrive at average ac power are related to their peak values by the $\sqrt{2}$.

$$E_{RMS} = \frac{E_{pk}}{\sqrt{2}} = \frac{E_{pk}}{1.414} = E_{pk} \times 0.707 \quad (14)$$

$$I_{RMS} = \frac{I_{pk}}{\sqrt{2}} = \frac{I_{pk}}{1.414} = I_{pk} \times 0.707 \quad (15)$$

In the time domain of a sine wave, the RMS values of voltage and current occur at the 45°, 135°, 225° and 315° points along the cycle shown in **Fig 6.12**. (The sine of 45° is approximately 0.707.) The absolute instantaneous value of voltage or current is greater than the RMS value for half the cycle and less than the RMS value for half the cycle.

The RMS values of voltage and current get their name from the means used to derive their value relative to peak voltage and current. Square the individual values of all the instantaneous values of voltage or current in a single cycle of ac. Take the average of these squares and then find the square root of the average. This *root mean square* procedure produces the RMS value of voltage or current.

If the RMS voltage is the peak voltage divided by the $\sqrt{2}$, then the peak voltage must be the RMS voltage multiplied by the $\sqrt{2}$, or

$$E_{pk} = E_{RMS} \times 1.414 \qquad (16)$$

$$I_{pk} = I_{RMS} \times 1.414 \qquad (17)$$

Since circuit specifications will most commonly list only RMS voltage and current values, these relationships are important in finding the peak voltages or currents that will stress components.

Example: What is the peak voltage on a capacitor if the RMS voltage of a sinusoidal waveform signal across it is 300. V ac?

$$E_{pk} = 300 \text{ V} \times 1.414 = 424 \text{ V}$$

The capacitor must be able to withstand this higher voltage, plus a safety margin. The capacitor must also be rated for ac use. A capacitor rated for 1 kV dc may explode if used in this application. In power supplies that convert ac to dc and use capacitive input filters, the output voltage will approach the peak value of the ac voltage rather than the RMS value.

Example: What is the peak voltage and the peak-to-peak voltage at the usual household ac outlet, if the RMS voltage is 120. V?

$$E_{pk} = 120 \text{ V} \times 1.414 = 170 \text{ V}$$

$$E_{p-p} = 2 \times 170 \text{ V} = 340 \text{ V}.$$

Unless otherwise specified, unlabeled ac voltage and current values found in most electronics literature are normally RMS values.

Average Values of Voltage and Current

Certain kinds of circuits respond to the *average* value of an ac waveform. Among these circuits are electrodynamic meter movements and power supplies that convert ac to dc and use heavily inductive ("choke") input filters, both of which use the pulsating dc output of a full-wave rectifier. The average value of each ac half cycle is the *mean* of all the instantaneous values in that half cycle. Related to the peak values of voltage and current, average values are $2/\pi$ (or 0.6366) times the peak value.

$$E_{ave} = 0.6366 \, E_{pk} \qquad (18)$$

$$I_{ave} = 0.6366 \, I_{pk} \qquad (19)$$

For convenience, **Table 6.3** summarizes the relationships between all of the common ac values. All of these relationships apply only to pure sine waves.

Complex Waves and Peak-Envelope Values

Complex waves, as shown earlier in Fig 6.4, differ from pure sine waves. The amplitude of the peak voltage may vary significantly from one cycle to the next. Therefore, other amplitude measures are required, especially for accurate measurement of voltage and power with single sideband (SSB) waveforms. **Fig 6.13** illustrates a multitone composite waveform with an RF ac waveform as the basis.

The RF ac waveform has a frequency many times that of the audio-frequency ac waveform with which it is usually combined in SSB operations. Therefore, the resultant waveform appears as an amplitude envelope superimposed upon the RF waveform. The *peak envelope voltage* (PEV), then, is the maximum or peak value of voltage achieved.

Peak envelope voltage permits the calculation of *peak envelope power* (PEP). The Federal Communications Commission

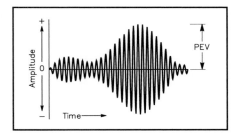

Fig 6.13 — The peak envelope voltage (PEV) for a composite waveform.

(FCC) uses the concept of peak envelope power to set the maximum power standards for amateur transmitters. PEP is the *average* power supplied to the antenna transmission line by a transmitter during one RF cycle at the crest of the modulation envelope, taken under normal operating conditions. Since calculation of PEP requires the average power of the cycle, multiply the PEV by 0.707 to obtain the RMS value. Then calculate power by using the square of the voltage divided by the load resistance.

$$PEP = \frac{(PEV \times 0.707)^2}{R} \qquad (20)$$

Capacitance and Capacitors

Without the ability to store electrical energy, radio would not be possible. One may build and hold an electrical charge in an *electrostatic field*. This phenomenon is called *capacitance*, and the devices that exhibit capacitance are called *capacitors*. See Chapter 10 for more information on practical capacitor applications and problems. **Fig 6.14** shows several schematic symbols for capacitors. Part A shows a fixed capacitor; one that has a single value of capacitance. Part B shows variable capacitors; these are adjustable over a range of values. Ordinarily, the straight line in each symbol connects to a positive voltage, while the curved line goes to a negative voltage or to ground. Some capacitor designs require rigorous adherence to polarity markings; other designs are symmetrical and nonpolarized.

CHARGE AND ELECTROSTATIC ENERGY STORAGE

Suppose two flat metal plates are placed close to each other (but not touching) and are connected to a battery through a switch, as illustrated in **Fig 6.15A**. At the instant the switch is closed, electrons are attracted from the up-

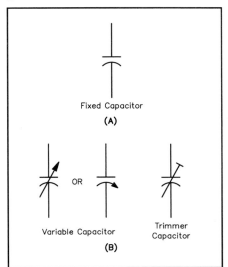

Fig 6.14 — Schematic symbol for a fixed capacitor is shown at A. The symbols for a variable capacitor are shown at B.

per plate to the positive terminal of the battery, and the same number are repelled into the lower plate from the negative battery terminal.

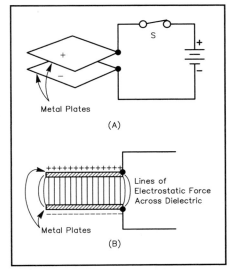

Fig 6.15 — A simple capacitor showing the basic charging arrangement at A, and the retention of the charge due to the electrostatic field at B.

Enough electrons move into one plate and out of the other to make the voltage between the plates the same as the battery voltage.

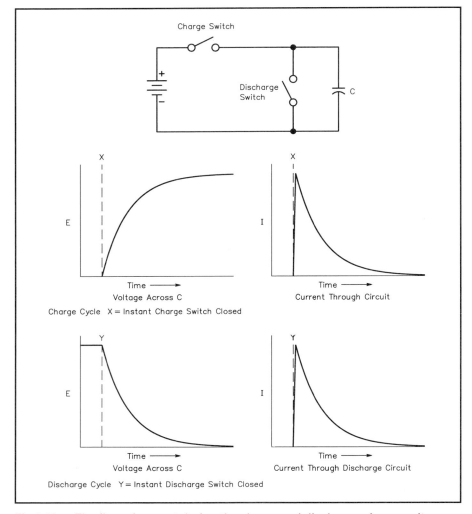

Fig 6.16 — The flow of current during the charge and discharge of a capacitor. The charge graphs assume that the charge switch is closed and the discharge switch is open. The discharge graphs assume just the opposite.

charges flowing into one plate causes a current to flow out of the other plate during one half of the cycle, resulting in a negative charge on that plate. The reverse occurs during the second half of the cycle.

The charge or quantity of electricity that can be held on the capacitor plates is proportional to the applied voltage and to the capacitance of the capacitor:

$$Q = CE \qquad (21)$$

where:
    Q = charge in coulombs,
    C = capacitance in farads, and
    E = electrical potential in volts.

The energy stored in a capacitor is also a function of electrical potential and capacitance:

$$W = \frac{E^2 C}{2} \qquad (22)$$

where:
    W = energy in joules (watt-seconds),
    E = electrical potential in volts (some texts use V instead of E), and
    C = capacitance in farads.

The numerator of this expression can be derived easily from the definitions for charge, capacitance, current, power and energy. The denominator is not so obvious, however. It arises because the voltage across a capacitor is not constant, but is a function of time. The average voltage over the time interval determines the energy stored. The time dependence of the capacitor voltage is a very useful property; see the section on time constants.

UNITS OF CAPACITANCE AND CAPACITOR CONSTRUCTION

A capacitor consists, fundamentally, of two plates separated by an insulator or *dielectric*. The *larger* the plate area and the *smaller* the spacing between the plates, the *greater* the capacitance. The capacitance also depends on the kind of insulating material between the plates: it is smallest with air insulation or a vacuum. Substituting other insulating materials for air may greatly increase the capacitance.

The ratio of the capacitance with a material other than a vacuum or air between the plates to the capacitance of the same capacitor with air insulation is called the dielectric constant, or K, of that particular insulating material. The dielectric constants of a number of materials commonly used as dielectrics in capacitors are given in **Table 6.4**. For example, if a sheet of polystyrene is substituted for air between the plates of a capacitor, the capacitance will be 2.6 times greater.

The basic unit of capacitance, the abil-

If the switch is opened after the plates have been charged in this way, the top plate is left with a deficiency of electrons and the bottom plate with an excess. Since there is no current path between the two, the plates remain charged despite the fact that the battery no longer is connected. The charge remains due to the electrostatic field between the plates. The large number of opposite charges exert an attractive force across the small distance between plates, as illustrated in Fig 6.15B.

If a wire is touched between the two plates (short-circuiting them), the excess electrons on the bottom plate flow through the wire to the upper plate, restoring electrical neutrality. The plates are discharged.

These two plates represent an electrical capacitor, a device possessing the property of storing electrical energy in the electric field between its plates. During the time the electrons are moving — that is, while the capacitor is being charged or

discharged — a current flows in the circuit even though the circuit apparently is broken by the gap between the capacitor plates. The current flows only during the time of charge and discharge, however, and this time is usually very short. There can be no continuous flow of direct current through a capacitor.

Fig 6.16 demonstrates the voltage and current in the circuit, first, at the moment the switch is closed to charge the capacitor and, second, at the moment the shorting switch is closed to discharge the unit. Note that the periods of charge and discharge are very short, but that they are not zero. This finite charging and discharging time can be lengthened and will prove useful later in timing circuits.

Although dc cannot pass through a capacitor, alternating current can. As fast as one plate is charged positively by the positive excursion of the alternating current, the other plate is being charged negatively. Positive

ity to store electrical energy in an electrostatic field, is the farad. This unit is generally too large for practical radio work, however. Capacitance is usually measured in microfarads (abbreviated µF), nanofarads (abbreviated nF) or picofarads (pF). The microfarad is one millionth of a farad (10^{-6} F), the nanofarad is one thousandth of a microfarad (10^{-9} F) and the picofarad is one millionth of a microfarad (10^{-12} F).

In practice, capacitors often have more than two plates, the alternate plates being connected to form two sets, as shown in **Fig 6.17**. This practice makes it possible to obtain a fairly large capacitance in a small space, since several plates of smaller individual area can be stacked to form the equivalent of a single large plate of the same total area. Also, all plates except the two on the ends are exposed to plates of the other group on both sides, and so are twice as effective in increasing the capacitance.

The formula for calculating capacitance from these physical properties is:

$$C = \frac{0.2248 \, K \, A \, (n - 1)}{d} \quad (23)$$

where:
C = capacitance in pF,
K = dielectric constant of material between plates,
A = area of one side of one plate in square inches,
d = separation of plate surfaces in inches, and
n = number of plates.

If the area (A) is in square centimeters and the separation (d) is in centimeters, then the formula for capacitance becomes

$$C = \frac{0.0885 \, K \, A \, (n - 1)}{d} \quad (24)$$

If the plates in one group do not have the same area as the plates in the other, use the area of the smaller plates.

Example: What is the capacitance of 2 copper plates, each 1.50 square inches in area, separated by a distance of 0.00500 inches, if the dielectric is air?

$$C = \frac{0.2248 \, K \, A \, (n - 1)}{d}$$
$$= \frac{0.2248 \times 1 \times 1.50 \, (2 - 1)}{0.00500}$$
$$C = 67.4 \text{ pF}$$

KINDS OF CAPACITORS AND THEIR USES

The capacitors used in radio work differ considerably in physical size, construction and capacitance. Representative kinds are shown in **Fig 6.18**. In variable capacitors,

Table 6.4

Relative Dielectric Constants of Common Capacitor Dielectric Materials

Material	Dielectric Constant (k)	(O)rganic or (I)norganic
Vacuum	1 (by definition)	I
Air	1.0006	I
Ruby mica	6.5 - 8.7	I
Glass (flint)	10	I
Barium titanate (class I)	5 - 450	I
Barium titanate (class II)	200 - 12000	I
Kraft paper	≈ 2.6	O
Mineral Oil	≈ 2.23	O
Castor Oil	≈ 4.7	O
Halowax	≈ 5.2	O
Chlorinated diphenyl	≈ 5.3	O
Polyisobutylene	≈ 2.2	O
Polytetrafluoroethylene	≈ 2.1	O
Polyethylene terephthalate	≈ 3	O
Polystyrene	≈ 2.6	O
Polycarbonate	≈ 3.1	O
Aluminum oxide	≈ 8.4	I
Tantalum pentoxide	≈ 28	I
Niobium oxide	≈ 40	I
Titanium dioxide	≈ 80	I

(Adapted from: Charles A. Harper, *Handbook of Components for Electronics*, p 8-7.)

which are almost always constructed with air for the dielectric, one set of plates is made movable with respect to the other set so the capacitance can be varied. Fixed capacitors — those having a single, nonadjustable value of capacitance — can also be made with metal plates and with air as the dielectric.

Fixed capacitors are usually constructed from plates of metal foil with a thin solid or liquid dielectric sandwiched between, so a relatively large capacitance can be obtained in a small unit. The solid dielectrics commonly used are mica, paper and special ceramics. An example of a liquid dielectric is mineral oil. Electrolytic capacitors use aluminum-foil plates with a semiliquid conducting chemical compound between them. The actual dielectric is a very thin film of insulating material that forms on one set of plates through electrochemical action when a dc voltage is applied to the capacitor. The capacitance obtained with a given plate area in an electrolytic capacitor is very large compared to capacitors having other dielectrics, because the film is so thin — much less than any thickness practical with a solid dielectric.

The use of electrolytic and oil-filled capacitors is confined to power-supply filtering and audio-bypass applications because their dielectrics have high losses at higher frequencies. Mica and ceramic capacitors are used throughout the frequency range from audio to several hundred megahertz.

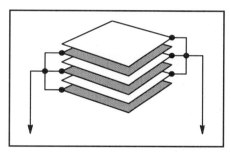

Fig 6.17 — A multiple-plate capacitor. Alternate plates are connected to each other.

New dielectric materials appear from time to time and represent improvements in capacitor performance. Silvered-mica capacitors, formed by spraying thin coats of silver on each side of the mica insulating sheet, improved the stability of mica capacitors in circuits sensitive to temperature changes. Polystyrene and other synthetic dielectrics, along with tantalum electrolytics, have permitted the size of capacitors to shrink per unit of capacitance.

VOLTAGE RATINGS AND BREAKDOWN

When high voltage is applied to the plates of a capacitor, considerable force is exerted on the electrons and nuclei of the dielectric. The dielectric is an insulator; its electrons do not become detached from atoms the way they do in conductors. If the force is great enough,

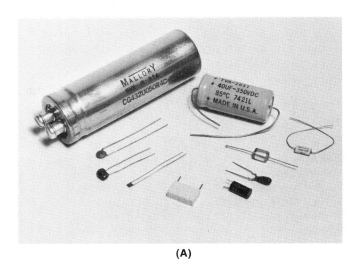

(A)

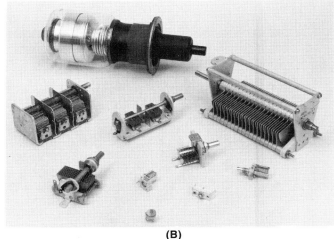

(B)

Fig 6.18 — Fixed-value capacitors are shown at A. A large computer-grade unit is at the upper left. The 40-µF unit is an electrolytic capacitor. The smaller pieces are silvered-mica, disc-ceramic, tantalum, polystyrene and ceramic-chip capacitors. The small black cylindrical unit is a PC-board-mount electrolytic. Variable capacitors are shown at B. A vacuum variable is at the upper left. The units with visible plates are air-variable capacitors. Some tiny variable capacitors use a thin piece of mica as a dielectric.

however, the dielectric will break down. Failed dielectrics usually puncture and offer a low-resistance current path between the two plates.

The *breakdown voltage* a dielectric can withstand depends on the chemical composition and thickness of the dielectric. Breakdown voltage is not directly proportional to the thickness; doubling the thickness does not quite double the breakdown voltage. Gas dielectrics also break down, as evidenced by a spark or arc between the plates. Spark voltages are generally given with the units *kilovolts per centimeter*. For air, the spark voltage or V_s may range from more than 120 kV/cm for gaps as narrow as 0.006 cm down to 28 kV/cm for gaps as wide as 10 cm. In addition, a large number of variables enter into the actual breakdown voltage in a real situation. Among the variables are the electrode shape, the gap distance, the air pressure or density, the voltage, impurities in the air (or any other dielectric material) and the nature of the external circuit (with air, for instance, the humidity affecting conduction on the surface of the capacitor plate).

Dielectric breakdown occurs at a lower voltage between pointed or sharp-edged surfaces than between rounded and polished surfaces. Consequently, the breakdown voltage between metal plates of any given spacing in air can be increased by buffing the edges of the plates. With most gas dielectrics such as air, once the voltage is removed, the arc ceases and the capacitor is ready for use again. If the plates are damaged so they are no longer smooth and polished,

they may have to be polished or the capacitor replaced. In contrast, solid dielectrics are permanently damaged by dielectric breakdown, and often will totally short out and melt or explode.

A thick dielectric must be used to withstand high voltages. Since the capacitance is inversely proportional to dielectric thickness (plate spacing) for a given plate area, a high-voltage capacitor must have more plate area than a low-voltage one of the same capacitance. High-voltage, high-capacitance capacitors are therefore physically large.

Dielectric strength is specified in terms of a dielectric withstanding voltage (DWV), given in volts per mil (0.001 inch) at a specified temperature. Taking into account the design temperature range of a capacitor and a safety margin, manufacturers specify *dc working voltage* (dcwv) to express the maximum safe limits of dc voltage across a capacitor to prevent dielectric breakdown.

It is not safe to connect capacitors across an ac power line unless they are rated for such use. Capacitors with dc ratings may short the line. Several manufacturers make capacitors specifically rated for use across the ac power line.

For use with other ac signals, the peak value of ac voltage should not exceed the dc working voltage, unless otherwise specified in component ratings. In other words, the RMS value of ac should be 0.707 times the dcwv value or lower. With many types of capacitors, further derating is required as the operating frequency increases. An additional safety margin is good practice.

Any two surfaces having different electrical potentials, and which are close enough to exhibit a significant electrostatic field, constitute a capacitor. The arrangement of circuit components and leads sometimes results in the creation of unintended capacitors. This is called *stray capacitance*: It often results in the passage of signals in ways that disrupt the normal operation of a circuit. Good design minimizes stray capacitance.

Stray capacitance may have a greater affect in a high-impedance circuit because the capacitive reactance may be a greater percentage of the circuit impedance. Also, because stray capacitance often appears in parallel with the circuit, the stray capacitor may bypass more of the desired signal at higher frequencies. Stray capacitance can often adversely affect sensitive circuits.

For further information of the physical and electrical characteristics of various types of capacitors in actual use, see the **Real-World Component Characteristics** chapter.

CAPACITORS IN SERIES AND PARALLEL

When a number of capacitors are connected in parallel, as in **Fig 6.19A**, the total capacitance of the group is equal to the sum of the individual capacitances:

$$C_{total} = C1 + C2 + C3 + C4 + \ldots + C_n \tag{25}$$

When two or more capacitors are connected in series, as in Fig 6.19B, the total capacitance is less than that of the smallest capacitor in the group. The rule for finding the capacitance of a number of series-connected capacitors is the same as that for finding the resistance of a number of parallel-connected resistors.

$$C_{total} = \cfrac{1}{\cfrac{1}{C1} + \cfrac{1}{C2} + \cfrac{1}{C3} + \ldots + \cfrac{1}{C_n}} \quad (26)$$

For only two capacitors in series, the formula becomes:

$$C_{total} = \frac{C1 \times C2}{C1 + C2} \quad (27)$$

The same units must be used throughout; that is, all capacitances must be expressed in either µF, nF or pF. Different units cannot be used in the same equation.

Capacitors are usually connected in parallel to obtain a larger total capacitance than is available in one unit. The largest voltage that can be applied safely to a parallel-connected group of capacitors is the voltage that can be applied safely to the one having the lowest voltage rating.

When capacitors are connected in series, the applied voltage is divided between them according to Kirchhoff's Voltage Law: The situation is much the same as when resistors are in series and there is a voltage drop across each. The voltage that appears across each series-connected capacitor is inversely proportional to its capacitance, as compared with the capacitance of the whole group. (This assumes ideal capacitors.)

Example: Three capacitors having capacitances of 1, 2 and 4 µF, respectively, are connected in series as shown in **Fig 6.20**. The voltage across the entire series is 2000 V. What is the total capacitance? (Since this is a calculation using theoretical values to illustrate a technique, we will not follow the rules of significant figures for the calculations.)

$$C_{total} = \cfrac{1}{\cfrac{1}{C1} + \cfrac{1}{C2} + \cfrac{1}{C3}}$$

$$= \cfrac{1}{\cfrac{1}{1\ \mu F} + \cfrac{1}{2\ \mu F} + \cfrac{1}{4\ \mu F}}$$

$$C_{total} = \cfrac{1}{\cfrac{7}{4\ \mu F}} = \frac{4\ \mu F}{7} = 0.5714\ \mu F$$

The voltage across each capacitor is proportional to the total capacitance divided by the capacitance of the capacitor in question. So the voltage across C1 is:

$$E1 = \frac{0.5714\ \mu F}{1\ \mu F} \times 2000\ V = 1143\ V$$

Similarly, the voltages across C2 and C3 are:

$$E2 = \frac{0.5714\ \mu F}{2\ \mu F} \times 2000\ V = 571\ V$$

and

$$E3 = \frac{0.5714\ \mu F}{4\ \mu F} \times 2000\ V = 286\ V$$

The sum of these three voltages equals 2000 V, the applied voltage.

Capacitors may be connected in series to enable the group to withstand a larger voltage than any individual capacitor is rated to withstand. The trade-off is a decrease in the total capacitance. As shown by the previous example, the applied voltage does not divide equally between the capacitors except when all the capacitances are precisely the same. Use care to ensure that the voltage rating of any capacitor in the group is not exceeded. If you use capacitors in series to withstand a higher voltage, you should also connect an "equalizing resistor" across each capacitor. Use resistors with about 100 Ω per volt of supply voltage, and be sure they have sufficient power-handling capability for the circuit. With real capacitors, the leakage resistance of the capacitors may have more effect on the voltage division than does the capacitance. A capacitor with a high parallel resistance will have the highest voltage across it. Adding equalizing resistors reduces this effect.

RC TIME CONSTANT

Connecting a dc voltage source directly to the terminals of a capacitor charges the capacitor to the full source voltage almost instantaneously. Any resistance added to the circuit as in **Fig 6.21A** limits the current, lengthening the time required for the voltage between the capacitor plates to build up to the source-voltage value. During this charging period, the current flowing from the source into the capacitor gradually decreases from its initial value. The increasing voltage stored in the capacitor's electric field offers increasing opposition to the steady source voltage.

While it is being charged, the voltage between the capacitor terminals is an exponential function of time, and is given by:

$$V(t) = E \left(1 - e^{-\frac{t}{RC}} \right) \quad (28)$$

where:

V(t) = capacitor voltage in volts at time t;

E = potential of charging source in volts;

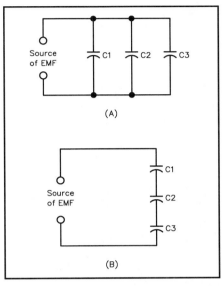

Fig 6.19 — Capacitors in parallel are shown at A, and in series at B.

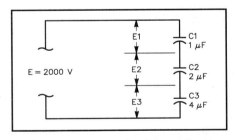

Fig 6.20 — An example of capacitors connected in series. The text shows how to find the voltage drops, E1 through E3.

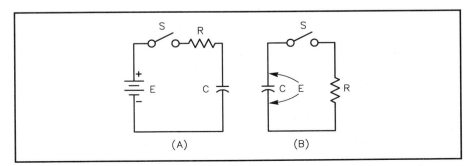

Fig 6.21 — An illustration of the time constant in an RC circuit.

t = time in seconds after initiation of charging current;

e = natural logarithmic base = 2.718;

R = circuit resistance in ohms; and

C = capacitance in farads.

Theoretically, the charging process is never really finished, but eventually the charging current drops to an unmeasurable value. For many purposes, it is convenient to let t = RC. Under this condition, the above equation becomes:

$$V(RC) = E(1 - e^{-1}) \approx 0.632\ E \qquad (29)$$

The product of R in ohms times C in farads is called the *time constant* of the circuit and is the time in seconds required to charge the capacitor to 63.2% of the supply voltage. (The lower-case Greek letter tau [τ] is often used to represent the time constant in electronics circuits.) After two time constants (t = 2τ) the capacitor charges another 63.2% of the difference between the capacitor voltage at one time constant and the supply voltage, for a total charge of 86.5%. After three time constants the capacitor reaches 95% of the supply voltage, and so on, as illus-

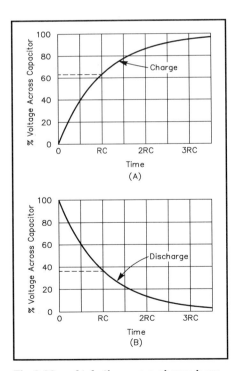

Fig 6.22 — At A, the curve shows how the voltage across a capacitor rises, with time, when charged through a resistor. The curve at B shows the way in which the voltage decreases across a capacitor when discharging through the same resistance. For practical purposes, a capacitor may be considered charged or discharged after 5 RC periods.

trated in the curve of **Fig 6.22A**. After 5 RC time periods, a capacitor is considered fully charged, having reached 99.24% of the source voltage.

If a charged capacitor is discharged through a resistor, as indicated in Fig 6.21B, the same time constant applies for the decay of the capacitor voltage. A direct short circuit applied between the capacitor terminals would discharge the capacitor almost instantly. The resistor, R, limits the current, so the capacitor voltage decreases only as rapidly as the capacitor can discharge itself through R. A capacitor discharging through a resistance exhibits the same time-constant characteristics (calculated in the same way as above) as a charging capacitor. The voltage, as a function of time while the capacitor is being discharged, is given by:

$$V(t) = E\left(e^{-\dfrac{t}{RC}}\right) \qquad (30)$$

where t = time in seconds after initiation of discharge.

Again, by letting t = RC, the time constant of a discharging capacitor represents a decrease in the voltage across the capacitor of about 63.2%. After 5 time-constant periods, the capacitor is considered fully discharged, since the voltage has dropped to less than 1% of the full-charge voltage.

Time constant calculations have many uses in radio work. The following examples are all derived from practical-circuit applications.

Example 1: A 100-μF capacitor in a high-voltage power supply is shunted by a 100-kΩ resistor. What is the minimum time before the capacitor may be considered fully discharged? Since full discharge is approximately 5 RC periods,

t = 5 × RC = 5 × 100 × 10³ Ω × 100 × 10⁻⁶ F = 50000 × 10⁻³ seconds

t = 50.0 s

(Look at the table of metric-system units in the **Mathematics for Amateur Radio** chapter to prove that *ohms* times *farads* gives units of *seconds*.)

Note: Although waiting almost a minute for the capacitor to discharge seems safe in this high-voltage circuit, never rely solely on capacitor-discharging resistors (often called *bleeder resistors*). Be certain the power source is removed and the capacitors are totally discharged before touching any circuit components.

Example 2: Smooth CW keying with-

out clicks requires approximately 5 ms (0.005 s) of delay in both the make and break edges of the waveform, relative to full charging and discharging of a capacitor in the circuit. What typical values might a builder choose for an RC delay circuit in a keyed voltage line? Since full charge and discharge require 5 RC periods,

$$RC = \frac{t}{5} = \frac{0.005\ s}{5} = 0.001\ s$$

Any combination of resistor and capacitor whose values, multiplied together, equaled 0.001 would do the job. A typical capacitor might be 0.05 μF. In that case, the necessary resistor would be:

$$R = \frac{0.001\ s}{0.05 \times 10^{-6}\ F}$$

= 0.02 × 10⁶ Ω = 20000 Ω or 20 kΩ.

In practice, a builder would likely either experiment with values or use a variable resistor. The final value would be selected after monitoring the waveform on an oscilloscope.

Example 3: Many modern integrated circuit (IC) devices use RC circuits to control their timing. To match their internal circuitry, they may use a specified threshold voltage as the trigger level. For example, a certain IC uses a trigger level of 0.667 of the supply voltage. What value of capacitor and resistor would be required for a 4.5-second timing period?

First we will solve equation 28 for the time constant, RC. The threshold voltage is 0.667 times the supply voltage, so we use this value for V(t).

$$V(t) = E\left(1 - e^{-\dfrac{t}{RC}}\right)$$

$$0.667\ E = E\left(1 - e^{-\dfrac{t}{RC}}\right)$$

$$e^{-\dfrac{t}{RC}} = 1 - 0.667$$

$$\ln\left(e^{-\dfrac{t}{RC}}\right) = \ln(0.333)$$

$$-\frac{t}{RC} = -1.10$$

We want to find a capacitor and resistor combination that will produce a 4.5 s timing period, so we substitute that value for t.

$$RC = \frac{4.5\ s}{1.10} = 4.1\ s$$

If we select a value of 10. µF, we can solve for R.

$$R = \frac{4.1\,s}{10. \times 10^{-6}\,F} = 0.41 \times 10^6\,\Omega = 410\,k\Omega$$

A 1% tolerance resistor and capacitor will give good precision. You could also use a variable resistor and an accurate method to measure the time to set the circuit to a 4.5 s period.

As the examples suggest, RC circuits have numerous applications in electronics. The number of applications is growing steadily, especially with the introduction of integrated circuits controlled by part or all of a capacitor charge or discharge cycle.

ALTERNATING CURRENT IN CAPACITANCE

Everything said about capacitance and capacitors in a dc circuit applies to capacitance in an ac circuit with one major exception. Whereas a capacitor in a dc circuit will appear as an open circuit except for the brief charge and discharge periods, the same capacitor in an ac circuit will both pass and limit current. A capacitor in an ac circuit does not handle electrical energy like a resistor, however. Instead of converting the energy to heat and dissipating it, capacitors store electrical energy and return it to the circuit.

In **Fig 6.23** a sine-wave ac voltage having a maximum value of 100 is applied to a capacitor. In the period OA, the applied voltage increases from 0 to 38; at the end of this period the capacitor is charged to that voltage. In interval AB the voltage increases to 71; that is, 33 V additional. During this interval a smaller quantity of charge has been added than in OA, because the voltage rise during interval AB is smaller. Consequently the average current during interval AB is smaller than during OA. In the third interval, BC, the voltage rises from 71 to 92, an increase of 21 V. This is less than the voltage increase during AB, so the quantity of electricity added is less; in other words, the average current during interval BC is still smaller. In the fourth interval, CD, the voltage increases only 8 V; the charge added is smaller than in any preceding interval and therefore the current also is smaller.

By dividing the first quarter cycle into a very large number of intervals, it could be shown that the current charging the capacitor has the shape of a sine wave, just as the applied voltage does. The current is largest at the beginning of the cycle and becomes zero at the maximum value of the voltage, so there is a phase difference of 90° between the voltage

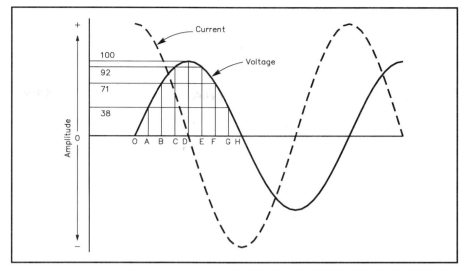

Fig 6.23 — **Voltage and current phase relationships when an alternating current is applied to a capacitor.**

and the current. During the first quarter cycle the current is flowing in the normal direction through the circuit, since the capacitor is being charged. Hence the current is positive, as indicated by the dashed line in Fig 6.23.

In the second quarter cycle — that is, in the time from D to H — the voltage applied to the capacitor decreases. During this time the capacitor loses its charge. Applying the same reasoning, it is evident that the current is small in interval DE and continues to increase during each succeeding interval. The current is flowing against the applied voltage, however, because the capacitor is discharging into the circuit. The current flows in the negative direction during this quarter cycle.

The third and fourth quarter cycles repeat the events of the first and second, respectively, with this difference: the polarity of the applied voltage has reversed, and the current changes to correspond. In other words, an alternating current flows in the circuit because of the alternate charging and discharging of the capacitance. As shown in Fig 6.23, the current starts its cycle 90° before the voltage, so the current in a capacitor *leads* the applied voltage by 90°. You might find it helpful to remember the word "ICE" as a mnemonic because the current (I) in a capacitor (C) comes before voltage (E). We can also turn this statement around, to say the voltage in a capacitor *lags* the current by 90°.

CAPACITIVE REACTANCE

The quantity of electric charge that can be placed on a capacitor is proportional to the applied voltage and the

capacitance. This amount of charge moves back and forth in the circuit once each cycle, and so the rate of movement of charge (the current) is proportional to voltage, capacitance and frequency. When the effects of capacitance and frequency are considered together, they form a quantity that plays a part similar to that of resistance in Ohm's Law. This quantity is called *reactance*. The unit for reactance is the ohm, just as in the case of resistance. The formula for calculating the reactance of a capacitor at a given frequency is:

$$X_C = \frac{1}{2\,\pi\,f\,C} \qquad (31)$$

where:

X_C = capacitive reactance in ohms,
f = frequency in hertz,
C = capacitance in farads
π = 3.1416

Note: In many references and texts, the symbol ω is used to represent $2\,\pi\,f$. In such references, equation 31 would read

$$X_C = \frac{1}{\omega\,C}$$

Although the unit of reactance is the ohm, there is no power dissipated in reactance. The energy stored in the capacitor during one portion of the cycle is simply returned to the circuit in the next.

The fundamental units for frequency and capacitance (hertz and farads) are too cumbersome for practical use in radio circuits. If the capacitance is specified in microfarads (µF) and the frequency is in megahertz (MHz), however, the reactance calculated from the previous formula retains the unit ohms.

Example: What is the reactance of a capacitor of 470. pF (0.000470 μF) at a frequency of 7.15 MHz?

$$X_C = \frac{1}{2 \pi f C}$$

$$= \frac{1}{2 \pi \times 7.15 \text{ MHz} \times 0.000470 \text{ μF}}$$

$$= \frac{1 \, \Omega}{0.0211} = 47.4 \, \Omega$$

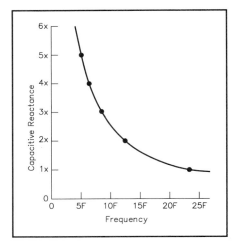

Fig 6.24 — A graph showing the general relationship of reactance to frequency for a fixed value of capacitance.

Example: What is the reactance of the same capacitor, 470. pF (0.000470 μF), at a frequency of 14.30 MHz?

$$X_C = \frac{1}{2 \pi f C}$$

$$= \frac{1}{2 \pi \times 14.30 \text{ MHz} \times 0.000470 \text{ μF}}$$

$$= \frac{1 \, \Omega}{0.0422} = 23.7 \, \Omega$$

The rate of change of voltage in a sine wave increases directly with the frequency. Therefore, the current into the capacitor also increases directly with frequency. Since, for a given voltage, an increase in current is equivalent to a decrease in reactance, the reactance of any capacitor decreases proportionally as the frequency increases. **Fig 6.24** traces the decrease in reactance of an arbitrary-value capacitor with respect to increasing frequency. The only limitation on the application of the graph is the physical make-up of the capacitor, which may favor low-frequency uses or high-frequency applications.

Among other things, reactance is a measure of the ability of a capacitor to limit the flow of ac in a circuit. For some purposes. it is important to know the ability of a capacitor to pass current. This ability is called *susceptance*, and it corresponds to conductance in resistive circuit elements. In an ideal capacitor with no resistive losses — that is, no energy lost as heat — susceptance is simply the reciprocal of reactance. Hence,

$$B = \frac{1}{X_C} \qquad (32)$$

where:
X_C is the reactance, and
B is the susceptance.

The unit of susceptance (and conductance and admittance) is the *siemens* (abbreviated S). In literature only a few years old, the term *mho* is also sometimes given as the unit of susceptance (as well as of conductance and admittance). The role of reactance and susceptance in current and other Ohm's Law calculations will appear in a later section of this chapter.

Inductance and Inductors

A second way to store electrical energy is in a *magnetic field*. This phenomenon is called *inductance*, and the devices that exhibit inductance are called *inductors*. Inductance depends upon some basic underlying magnetic properties. See Chapter 10 for more information on practical inductor applications and problems.

MAGNETISM

Magnetic Fields, Flux and Flux Density

Magnetic fields are closed fields that surround a magnet, as illustrated in **Fig 6.25**. The field consists of lines of magnetic force or *flux*. It exhibits polarity, which is conventionally indicated as north-seeking and south-seeking poles, or *north* and *south poles* for short. Magnetic flux is measured in the SI unit of the weber, which is a volt second (Wb = V s). In the *centimeter gram second* (*cgs*) metric system units, we measure magnetic flux in maxwells (1 Mx = 10^{-8} Wb).

The field intensity, known as the *flux*

density, decreases with the square of the distance from the source. Flux density (B) is represented in gauss (G), where one gauss is equivalent to one line of force per square centimeter of area across the field (G = Mx / cm²). The gauss is a *cgs* unit. In SI units, flux density is represented by the tesla (T), which is one weber per square meter (T = Wb / m²).

Magnetic fields exist around two types of materials. First certain ferromagnetic materials contain molecules aligned so as to produce a magnetic field. Lodestone, Alnico and other materials with high *retentivity* form *permanent magnets* because they retain their magnetic properties for long periods. Other materials, such as soft iron, yield temporary magnets that lose their magnetic properties rapidly.

The second type of magnetic material is an electrical conductor with a current through it. As shown in **Fig 6.26**, moving electrons are surrounded by a closed magnetic field lying in a plane perpendicular to their motion. The needle of a compass placed near a wire carrying direct current

will be deflected by the magnetic field around the wire. This phenomenon is one aspect of a two-way relationship: a moving magnetic field whose lines cut across a wire will induce an electical current in the wire, and the electrical current will produce a magnetic field.

If the wire is coiled into a solenoid, the

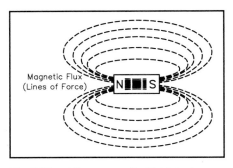

Fig 6.25 — The magnetic field and poles of a permanent magnet. The magnetic field direction is from the North to the South Pole.

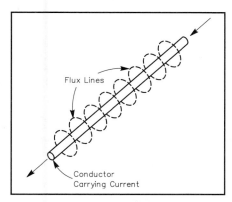

Fig 6.26 — The magnetic field around a conductor carrying an electrical current. If the thumb of your right hand points in the direction of the conventional current (plus to minus), your fingers curl in the direction of the magnetic field around the wire.

duce a magnetic field.

If the wire is coiled into a solenoid, the magnetic field greatly intensifies as the individual flux lines add together. **Fig 6.27** illustrates the principle by showing a coil section. Note that the resulting *electromagnet* has magnetic properties identical in principle to those of a permanent magnet, including poles and lines of force or flux. The strength of the magnetic field depends on several factors: the number of turns of the coil, the magnetic properties of the materials surrounding the coil (both inside and out), the length of the magnetic path and the amplitude of the current.

The magnetizing or *magnetomotive force* that produces a flux or total magnetic field is measured in gilberts (Gb). The force in gilberts equals $0.4\,\pi$ (approximately 1.257) times the number of turns in the coil times the current in amperes. (The SI unit of magnetomotive force is the ampere turn, abbreviated A, just like the ampere.) The magnetic field strength, H, measured in oersteds (Oe) produced by any particular magnetomotive force (measured in gilberts) is given by:

$$H = \frac{0.4\,\pi\,N\,I}{\ell} \qquad (33)$$

where:
H = magnetic field strength in oersteds,
N = number of turns,
I = dc current in amperes,
π = 3.1416, and
ℓ = mean magnetic path length in centimeters.

The gilbert and oersted are *cgs* units. These are given here because most amateur calculations will use these units. You may also see the preferred SI units in some literature. The SI unit of magnetic field strength is the ampere (turn) per meter.

A force is required to produce a given magnetic field strength. This implies that there is a resistance, called *reluctance*, to be overcome.

Core Properties: Permeability, Saturation, Reluctance, Hysteresis

The nature of the material within the coil of an electromagnet, where the lines of force are most concentrated, has the greatest effect upon the magnetic field established by the coil. All materials are compared to air. The ratio of flux density produced by a given material compared to the flux density produce by an air core is the *permeability* of the material. Suppose the coil in **Fig 6.28** is wound on an iron core having a cross-sectional area of 2 square inches. When a certain current is sent through the coil, it is found that there are 80000 lines of force in the core. Since the area is 2 square inches, the magnetic flux density is 40000 lines per square inch. Now suppose that the iron core is removed and the same current is maintained in the coil. Also suppose the flux density without the iron core is found to be 50 lines per square inch. The ratio of these flux densities, iron core to air, is 40000 / 50 or 800, the core's permeability.

Permeabilities as high as 10^6 have been attained. The three most common types of materials used in magnetic cores are these:

A. stacks of laminated steel sheets (for power and audio applications);
B. various ferrite compounds (for cores shaped as rods, toroids, beads and numerous other forms); and
C. powdered iron (shaped as slugs, toroids and other forms for RF inductors).

Brass has a permeability less than 1. A brass core inserted into a coil will decrease the inductance compared to an air core.

The permeability of silicon-steel power-transformer cores approaches 5000 in high-quality units. Powdered-iron cores used in RF tuned circuits range in permeability from 3 to about 35, while ferrites of nickel-zinc and manganese-zinc range from 20 to 15000. **Table 6.5** lists some common magnetic

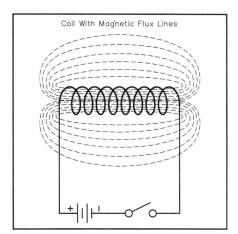

Fig 6.27 — Cross section of an inductor showing its flux lines and overall magnetic field.

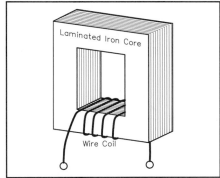

Fig 6.28 — A coil of wire wound around a laminated iron core.

Table 6.5

Properties of Some High-Permeability Materials

Material	Approximate Percent Composition					Maximum Permeability
	Fe	Ni	Co	Mo	Other	
Iron	99.91	—	—	—	—	5000
Purified Iron	99.95	—	—	—	—	180000
4% silicon-iron	96	—	—	—	4 Si	7000
45 Permalloy	54.7	45	—	—	0.3 Mn	25000
Hipernik	50	50	—	—	—	70000
78 Permalloy	21.2	78.5	—	—	0.3 Mn	100000
4-79 Permalloy	16.7	79	—	—	0.3 Mn	100000
Supermalloy	15.7	79	—	5	0.3 Mn	800000
Permendur	49.7	—	50	—	0.3 Mn	5000
2V Permendur	49	—	49	—	2 V	4500
Hiperco	64	—	34	—	2 Cr	10000
2-81 Permalloy*	17	81	—	2	—	130
Carbonyl iron*	99.9	—	—	—	—	132
Ferroxcube III**	($MnFe_2O_4 + ZnFe_2O_4$)					1500

Note: all materials in sheet form except * (insulated powder) and ** (sintered powder).
(Reference: L. Ridenour, ed., *Modern Physics for the Engineer*, p 119.)

quency sensitive, exhibiting excessive losses outside the frequency band of intended use.

As a measure of the ease with which a magnetic field may be established in a material as compared with air, permeability (μ) corresponds roughly to electrical conductivity. Permeability is given as:

$$\mu = \frac{B}{H} \qquad (34)$$

where:

B is the flux density in gauss, and

H is the magnetomotive force in oersteds.

Unlike electrical conductivity, which is independent of other electrical parameters, the permeability of a magnetic material varies with the flux density. At low flux densities (or with an air core), increasing the current through the coil will cause a proportionate increase in flux. But at very high flux densities, increasing the current beyond a certain point may cause no appreciable change in the flux. At this point, the core is said to be *saturated*. Saturation causes a rapid decrease in permeability, because it decreases the ratio of flux lines to those obtainable with the same current using an air core. **Fig 6.29** displays a typical permeability curve, showing the region of saturation. The saturation point varies with the make-up of different magnetic materials. Air and other nonmagnetic materials do not saturate and have a permeability of one. *Reluctance*, which is the reciprocal of permeability and corresponds roughly to resistance in an electrical circuit, is also one for air and other nonmagnetic cores.

The retentivity of magnetic core materials creates another potential set of losses caused by *hysteresis*. **Fig 6.30** illustrates the change of flux density (B) with a changing magnetizing force (H). From starting point A, with no residual flux, the flux reaches point B at the maximum mag-netizing force. As the force decreases, so too does the flux, but it does not reach zero simultaneously with the force at point D. As the force continues in the opposite direction, it brings the flux density to point C. As the force decreases to zero, the flux once more lags behind. In effect, a reverse force is necessary to overcome the residual magnetism retained by the core material, a *coercive force*. The result is a power loss to the magnetic circuit, which appears as heat in the core material. Air cores are immune to hysteresis effects and losses.

INDUCTANCE AND DIRECT CURRENT

In an electrical circuit, any element having a magnetic field is called an *inductor*. **Fig 6.31** shows schematic-diagram symbols and photographs of a few representative inductors: an air-core inductor, a slug-tuned variable inductor with a nonmagnetic core and an inductor with a magnetic (iron) core.

The transfer of energy to the magnetic field of an inductor represents work performed by the source of the voltage. Power is required for doing work, and since power is equal to current multiplied by voltage, there must be a voltage drop in the circuit while energy is being stored in the field. This voltage drop, exclusive of any voltage drop caused by resistance in the circuit, is the result of an opposing voltage induced in the circuit while the field is building up to its final value. Once the field becomes constant, the *induced voltage* or back-voltage disappears, because no further energy is being stored. The induced voltage opposes the voltage of the source and tends to prevent the current from rising rapidly when the circuit is closed. **Fig 6.32A** illustrates the situation of energizing an inductor or magnetic circuit, showing the relative amplitudes of induced voltage and the delayed rise in current to its full value.

The amplitude of the induced voltage is proportional to the rate at which the current changes (and consequently, the rate at which the magnetic field changes) and to a constant associated with the circuit itself: the *inductance* (or *self-inductance*) of the circuit. Inductance depends on the physical configuration of the inductor. Coiling a conductor increases its inductance. In effect, the growing (or shrinking) magnetic field of each turn produces magnetic lines of force that — in their expansion (or contraction) — cut across the other turns of the coil, inducing a voltage in every other turn. The mutuality of the effect multiplies the ability of the coiled conductor to store electrical energy.

A coil of many turns will have more

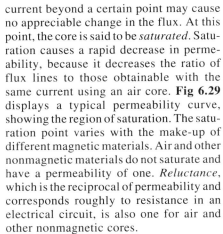

Fig 6.29 — A typical permeability curve for a magnetic core, showing the point where saturation begins.

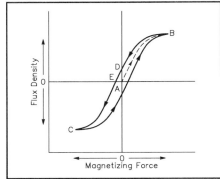

Fig 6.30 — A typical hysteresis curve for a magnetic core, showing the additional energy needed to overcome residual flux.

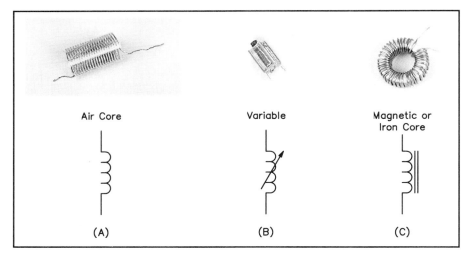

Fig 6.31 — Schematic symbols for representative inductors, including (from left to right) an air-core inductor, a variable inductor with a nonmagnetic slug, and an inductor with a magnetic core.

inductance than one of few turns, if both coils are otherwise physically similar. Furthermore, if an inductor is placed around a magnetic core, its inductance will increase in proportion to the permeability of that core, if the circuit current is below the point at which the core saturates.

The polarity of an induced voltage is always such as to oppose any change in the circuit current. This means that when the current in the circuit is increasing, work is being done against the induced voltage by storing energy in the magnetic field. Likewise, if the current in the circuit tends to decrease, the stored energy of the field returns to the circuit, and adds to the energy being supplied by the voltage source. Inductors try to maintain a constant current through the circuit. This phenomenon tends to keep the current flowing even though the applied voltage may be decreasing or be removed entirely. Fig 6.32B illustrates the decreasing but continuing flow of current caused by the induced voltage after the source voltage is removed from the circuit.

The energy stored in the magnetic field of an inductor is given by the formula:

$$W = \frac{I^2 L}{2} \qquad (35)$$

where:

W = energy in joules,
I = current in amperes, and
L = inductance in henrys.

This formula corresponds to the energy-storage formula for capacitors: energy storage is a function of current squared over time. As with capacitors, the time dependence of inductor current is a significant property; see the section on time constants.

The basic unit of inductance is the *henry* (abbreviated H), which equals an induced voltage of one volt when the inducing current is varying at a rate of one ampere per second. In various aspects of radio work, inductors may take values ranging from a fraction of a nanohenry (nH) through millihenrys (mH) up to about 20 H.

MUTUAL INDUCTANCE AND MAGNETIC COUPLING

Mutual Inductance

When two coils are arranged with their axes on the same line, as shown in **Fig 6.33**, current sent through coil 1 creates a magnetic field that cuts coil 2. Consequently, a voltage will be induced in coil 2 whenever the field strength of coil 1 is changing. This induced voltage is similar to the voltage of self-induction, but since it appears in the second coil because of

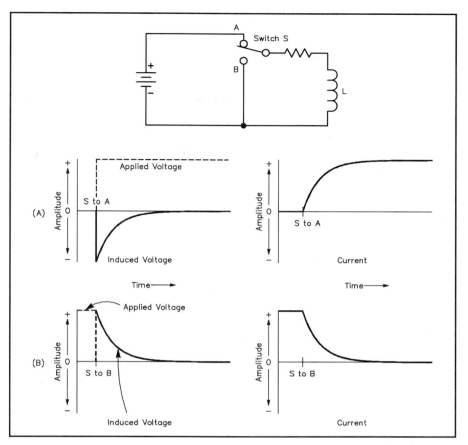

Fig 6.32 — Inductive circuit showing and graphing the generation of induced voltage and the rise of current in an inductor at A, and the decay of current as power is removed and the coil shorted at B.

current flowing in the first, it is a mutual effect and results from the *mutual inductance* between the two coils.

When all the flux set up by one coil cuts all the turns of the other coil, the mutual inductance has its maximum possible value. If only a small part of the flux set up by one coil cuts the turns of the other, the mutual inductance is relatively small. Two coils having mutual inductance are said to be *coupled*.

The ratio of actual mutual inductance to the maximum possible value that could theoretically be obtained with two given coils is called the *coefficient of coupling* between the coils. It is frequently expressed as a percentage. Coils that have nearly the maximum possible mutual inductance (coefficient = 1 or 100%) are said to be closely, or tightly, coupled. If the mutual inductance is relatively small the coils are said to be loosely coupled. The degree of coupling depends upon the physical spacing between the coils and how they are placed with respect to each other. Maximum coupling exists when they have a common axis and are as close together as possible (for example, one

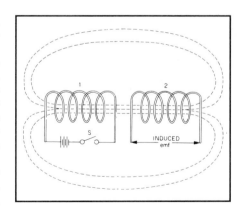

Fig 6.33 — Mutual inductance: When S is closed, current flows through coil number 1, setting up a magnetic field that induces a voltage in the turns of coil number 2.

wound over the other). The coupling is least when the coils are far apart or are placed so their axes are at right angles.

The maximum possible coefficient of coupling is closely approached when the two coils are wound on a closed iron core. The coefficient with air-core coils may run as high as 0.6 or 0.7 if one coil is wound over the other, but will be much less if the two coils are separated. Although unity coupling is suggested by Fig 6.33, such coupling is possible only when the coils are wound on a closed magnetic core.

Unwanted Couplings: Spikes, Lightning and Other Pulses

Every conductor passing current has a magnetic field associated with it — and therefore inductance — even though the conductor is not formed into a coil. The inductance of a short length of straight wire is small, but it may not be negligible. If the current through it changes rapidly, the induced voltage may be appreciable. This is the case in even a few inches of wire with an alternating current having a frequency on the order of 100 MHz or higher. At much lower frequencies or at dc, the inductance of the same wire might be ignored because the induced voltage would seemingly be negligible.

There are many phenomena, however, both natural and man-made, which create sufficiently strong magnetic fields to induce voltages into straight wires. Many of them are brief but intense pulses of energy that act like the turning on of the switch in a circuit containing self-inductance. Because the fields created grow to very high levels rapidly, they cut across wires leading into and out of — and wires wholly within — electronic equipment, inducing unwanted voltages by mutual coupling.

Short-duration, high-level voltage spikes occur on ac and dc power lines. Because the field intensity is great, these spikes may induce voltages upon conducting elements in sensitive circuits, disrupting them and even injuring components. Lightning in the vicinity of the equipment can induce voltages on power lines and other conductive paths (even ground conductors) that lead to the equipment location. Lightning that seems a safe distance away can induce

large spikes on power lines that ultimately lead to the equipment. Closer at hand, heavy equipment with electrical motors can induce significant spikes into power lines within the equipment location. Even though the power lines are straight, the powerful magnetic field of a spike source can induce damaging voltages on equipment left "plugged in" during electrical storms or during the operation of heavy equipment that inadequately filters its spikes.

Parallel-wire cables linking elements of electronic equipment consist of long wires in close proximity to each other. Signal pulses can couple both magnetically and capacitively from one wire to another. Since the magnetic field of a changing current decreases as the square of distance, separating the signal-carrying lines diminishes inductive coupling. Placing a grounded wire between signal-carrying lines reduces capacitive coupling. Unless they are well-shielded and filtered, however, the lines are still susceptible to the inductive coupling of pulses from other sources.

INDUCTORS IN RADIO WORK

Various facets of radio work make use of inductors ranging from the tiny up to the massive. Small values of inductance, as illustrated by **Fig 6.34A**, serve mostly in RF circuits. They may be self-supporting air-core or air-wound coils or the winding may be supported by nonmagnetic strips or a form. Phenolic, certain plastics and ceramics are the most common coil forms for air-core inductors. These inductors range in value from a few hundred μH for medium- and high-frequency circuits down to tenths of a μH for VHF and UHF work. The smallest values of inductance in radio work result from component leads. For VHF work and higher frequencies, component lead length is often critical. Circuits may fail to operate properly because leads are a little too short or too long.

It is possible to make these solenoid coils variable by inserting a slug in the center of the coil. (Slug-tuned coils normally have a ceramic, plastic or phenolic insulating form between the conductive slug and the coil winding.) If the slug material is magnetic, such as powdered iron, the inductance increases as the slug is centered along the length of the coil. If the slug is brass or some other conductive but nonmagnetic material, centering the slug will reduce the coil's inductance. This effect stems from the fact that brass has low electrical resistance and acts as an effective short-circuited one-turn secondary for the coil. (See more on transformer effects later in this chapter.)

An alternative to air-core inductors for RF work are toroidal coils wound on cores composed of powdered iron mixed with a binder to hold the material together. The availability of many types and sizes of powdered-iron cores has made these inductors popular for low-power fixed-value service. The toroidal shape concentrates the inductor's field tightly about the coil, eliminating the need in many cases for other forms of shielding to limit the interaction of the inductor's magnetic field with the fields of other inductors.

Fig 6.34B shows samples of inductors in the millihenry range. Among these inductors are multisection RF chokes designed to keep RF currents from passing beyond them to other parts of circuits. Low-frequency radio work may also use inductors in this range of values, sometimes wound with *litz* wire. Litz wire is a special version of stranded wire, with each strand insulated from the others. For audio filters, toroidal coils with values below 100 mH are useful. Resembling powdered-iron-core RF toroids, these coils are wound on ferrite or molybdenum-permalloy cores having much higher permeabilities.

Audio and power-supply inductors appear in Fig 6.34C. Lower values of these iron-core coils, in the range of a few henrys, are useful as audio-frequency chokes. Larger values up to about 20 H may be found in power supplies, as choke filters, to suppress 120-Hz ripple. Although some of these inductors are open frame, most have iron covers to confine the powerful magnetic fields they produce.

INDUCTANCES IN SERIES AND PARALLEL

When two or more inductors are connected in series (**Fig 6.35A**), the total inductance is equal to the sum of the individual inductances, provided that the coils

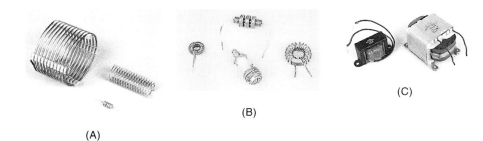

(A)

(B)

(C)

Fig 6.34 — Part A shows small-value air-wound inductors, B shows some inductors with values in the range of a few millihenrys and C shows large inductors as might be used in audio circuits or as power-supply chokes.

are sufficiently separated so that coils are not in the magnetic field of one another. That is:

$$L_{total} = L1 + L2 + L3 \ldots + L_n \quad (36)$$

If inductors are connected in parallel (Fig 6.35B), and if the coils are separated sufficiently, the total inductance is given by:

$$L_{total} = \cfrac{1}{\cfrac{1}{L1} + \cfrac{1}{L2} + \cfrac{1}{L3} + \ldots + \cfrac{1}{L_n}}$$

(37)

For only two inductors in parallel, the formula becomes:

$$L_{total} = \frac{L1 \times L2}{L1 + L2} \quad (38)$$

Thus, the rules for combining inductances in series and parallel are the same as those for resistances, assuming that the coils are far enough apart so that each is unaffected by another's magnetic field. When this is not so, the formulas given above will not yield correct results.

RL TIME CONSTANT

A comparable situation to an RC circuit exists when resistance and inductance are connected in series. In **Fig 6.36**, first consider L to have no resistance and also consider that R is zero. Closing S1 sends a current through the circuit. The instantaneous transition from no current to a finite value, however small, represents a rapid change in current, and a reverse voltage is developed by the self-inductance of L. The value of reverse voltage is almost equal and opposite to the applied voltage. The resulting initial current is very small.

The reverse voltage depends on the change in the value of the current and would cease to offer opposition if the current did not continue to increase. With no resistance in the circuit (which, by Ohm's Law, would lead to an infinitely large current), the current would increase forever, always growing just fast enough to keep the self-induced voltage equal to the applied voltage.

When resistance in the circuit limits the current, Ohm's Law defines the value that the current can reach. The reverse voltage generated in L must only equal the difference between E and the drop across R, because the difference is the voltage actually applied to L. This difference becomes smaller as the current approaches the final Ohm's Law value. Theoretically, the reverse voltage never quite disappears, and so the current never quite reaches the Ohm's Law value. In practical terms, the differences become unmeasurable after a time.

The current at any time after the switch in Fig 6.36 has been closed, can be found from:

$$I(t) = \frac{E\left(1 - e^{\frac{-tR}{L}}\right)}{R} \quad (39)$$

where:

$I(t)$ = current in amperes at time t,
E = power supply potential in volts,
t = time in seconds after initiation of current,
e = natural logarithmic base = 2.718,
R = circuit resistance in ohms, and
L = inductance in henrys.

The time in seconds required for the current to build up to 63.2% of the maximum value is called the time constant, and is equal to L / R, where L is in henrys and R is in ohms. After each time interval equal to this constant, the circuit conducts an additional 63.2% of the remaining current. This behavior is graphed in Fig 6.36. As is the case with capacitors, after 5 time constants the current is considered to have reached its maximum value. As with capacitors, we often use the lower-case Greek tau (τ) to represent the time constant.

Example: If a circuit has an inductor of 5.0 mH in series with a resistor of 10. Ω, how long will it take for the current in the circuit to reach full value after power is applied? Since achieving maximum current takes approximately five time constants,

$t = 5 L / R = (5 \times 5.0 \times 10^{-3} \text{ H}) / 10. \Omega$
$= 2.5 \times 10^{-3}$ seconds or 2.5 ms

(Look at the table of metric-system units in the **Mathematics for Amateur Radio** chapter to prove that *henrys* divided by *ohms* gives units of *seconds*.)

Note that if the inductance is increased to 5.0 H, the required time increases by a factor of 1000 to 2.5 seconds. Since the circuit resistance didn't change, the final current is the same for both cases in this example. Increasing inductance increases the time required to reach full current.

Zero resistance would prevent the circuit from ever achieving full current. All inductive circuits have some resistance, however, if only the resistance of the wire making up the inductor.

An inductor cannot be discharged in the simple circuit of Fig 6.36 because the magnetic field collapses as soon as the current ceases. Opening S1 does not leave the inductor charged in the way that a capacitor would remain charged. The energy stored in the magnetic field returns instantly to the circuit when S1 is opened. The rapid collapse of the field causes a very large voltage to be induced in the coil. Usually the induced

voltage is many times larger than the applied voltage, because the induced voltage is proportional to the rate at which the field changes. The common result of opening the switch in such a circuit is that a spark or arc forms at the switch contacts during the instant the switch opens. When the inductance is large and the current in the circuit is high, large amounts of energy are released in a very short time. It is not at all unusual for the switch contacts to burn or melt under such circumstances. The spark or arc at the opened switch can be reduced or suppressed by connecting a suitable capacitor and resistor in series across the contacts. Such an RC combination is called a *snubber network*.

Transistor switches connected to and controlling coils, such as relay solenoids,

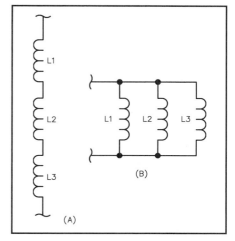

Fig 6.35 — Part A shows inductances in series, and Part B shows inductances in parallel.

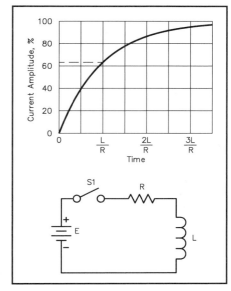

Fig 6.36 — Time constant of an RL circuit being energized.

also require protection. In most cases, a small power diode connected in reverse across the relay coil will prevent field-collapse currents from harming the transistor.

If the excitation is removed without breaking the circuit, as theoretically diagrammed in **Fig 6.37**, the current will decay according to the formula:

$$I(t) = \left(\frac{E}{R}\right)\left(e^{\frac{-tR}{L}}\right) \qquad (40)$$

where t = time in seconds after removal of the source voltage.

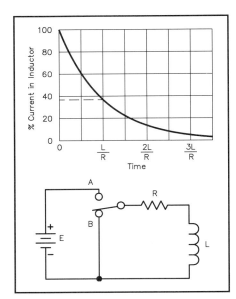

Fig 6.37 — Time constant of an RL circuit being deenergized. This is a theoretical model only, since a mechanical switch cannot change state instantaneously.

After one time constant the current will lose 63.2% of its steady-state value. (It will decay to 36.8% of the steady-state value.) The graph in Fig 6.37 shows the current-decay waveform to be identical to the voltage-discharge waveform of a capacitor. Be careful about applying the terms *charge* and *discharge* to an inductive circuit, however. These terms refer to energy storage in an electric field. An inductor stores energy in a magnetic field.

ALTERNATING CURRENT IN INDUCTORS

When an alternating voltage is applied to an ideal inductance (one with no resistance — all practical inductors have some resistance), the current is 90° out of phase with the applied voltage. In this case the current *lags* 90° behind the voltage; the opposite of the capacitor current-voltage relationship, as shown in **Fig 6.38**. (Here again, we can also say the voltage across an inductor *leads* the current by 90°.)

If you have difficulty remembering the phase relationships between voltage and current with inductors and capacitors, you may find it helpful to think of the mnemonic, "ELI the ICE man." This little phrase will remind you that voltage leads current through an inductor, because the E comes before the I, with an L between them, as you read from left to right. (The letter L represents inductance.) It will also help you remember the capacitor conditions because I comes before E with a C between them.

Interpreting Fig 6.38 begins with understanding that the primary cause for current lag in an inductor is the reverse voltage generated in the inductance. The amplitude of the reverse voltage is proportional

to the rate at which the current changes. In time segment OA, when the applied voltage is at its positive maximum, the reverse or induced voltage is also maximum, allowing the least current to flow. The rate at which the current is changing is the highest, a 38% change in the time period OA. In the segment AB, the current changes by only 33%, yielding a reduced level of induced voltage, which is in step with the decrease in the applied voltage. The process continues in time segments BC and CD, the latter producing only an 8% rise in current as the applied and induced voltage approach zero.

In segment DE, the applied voltage changes direction. The induced voltage also changes direction, which returns current to the circuit from storage in the magnetic field. The direction of this current is now opposite to the applied voltage, which sustains the current in the positive direction. As the applied voltage continues to increase negatively, the current — although positive — decreases in value, reaching zero as the applied voltage reaches its negative maximum. The negative half-cycle continues just as did the positive half-cycle.

Compare Fig 6.38 with Fig 6.23. Whereas in a pure capacitive circuit, the current *leads* the voltage by 90°, in a pure inductive circuit, the current *lags* the voltage by 90°. These phenomena are especially important in circuits that combine inductors and capacitors.

INDUCTIVE REACTANCE

The amplitude of alternating current in an inductor is inversely proportional to the applied frequency. Since the reverse voltage is directly proportional to inductance for a given rate of current change, the current is inversely proportional to inductance for a given applied voltage and frequency.

The combined effect of inductance and frequency is called inductive reactance, which — like capacitive reactance — is expressed in ohms. The formula for inductive reactance is:

$$X_L = 2\pi f L \qquad (41)$$

where:
  X_L = inductive reactance,
  f = frequency in hertz,
  L = inductance in henrys, and
  π = 3.1416.
(If $\omega = 2\pi f$, then $X_L = \omega L$.)

Example: What is the reactance of a coil having an inductance of 8.00 H at a frequency of 120. Hz?

$$X_L = 2\pi f L$$
$$= 6.2832 \times 120.\text{ Hz} \times 8.00\text{ H}$$
$$= 6030\ \Omega$$

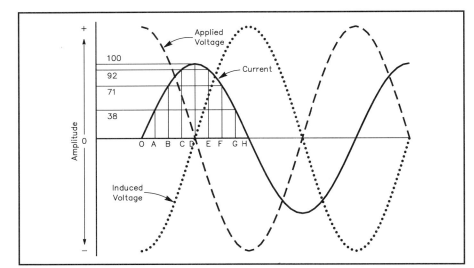

Fig 6.38 — Phase relationships between voltage and current when an alternating current is applied to an inductance.

In RF circuits the inductance values are usually small and the frequencies are large. When the inductance is expressed in millihenrys and the frequency in kilohertz, the conversion factors for the two units cancel, and the formula for reactance may be used without first converting to fundamental units. Similarly, no conversion is necessary if the inductance is expressed in microhenrys and the frequency in megahertz.

Example: What is the reactance of a 15.0-microhenry coil at a frequency of 14.0 MHz?

$$X_L = 2 \pi f L$$
$$= 6.2832 \times 14.0 \text{ MHz} \times 15.0 \text{ }\mu\text{H}$$
$$= 1320 \text{ }\Omega$$

The resistance of the wire used to wind the coil has no effect on the reactance, but simply acts. as a separate resistor connected in series with the coil.

Example: What is the reactance of the same coil at a frequency of 7.0 MHz?

$$X_L = 2 \pi f L$$
$$= 6.2832 \times 7.0 \text{ MHz} \times 15.0 \text{ }\mu\text{H}$$
$$= 660 \text{ }\Omega$$

Comparing the two examples suggests correctly that inductive reactance varies directly with frequency. The rate of change of the current varies directly with the frequency, and this rate of change also determines the amplitude of the induced or reverse voltage. Hence, the opposition to the flow of current increases proportionally to frequency. This opposition is called *inductive reactance*. The direct relationship between frequency and reactance in inductors, combined with the inverse relationship between reactance and frequency in the case of capacitors, will be of fundamental importance in creating resonant circuits.

As a measure of the ability of an inductor to limit the flow of ac in a circuit, inductive reactance is similar to capacitive reactance in having a corresponding *susceptance*, or ability to pass ac current in a circuit. In an ideal inductor with no resistive losses — that is, no energy lost as heat — susceptance is simply the reciprocal of reactance.

$$B = \frac{1}{X_L} \quad (42)$$

where:
 X_L = reactance, and
 B = susceptance.

The unit of susceptance for both inductors and capacitors is the *siemens*, abbreviated S.

Quality Factor, or Q of Components

Components that store energy, like capacitors and inductors, may be compared in terms of quality or Q. The Q of any such component is the ratio of its ability to store energy to the sum total of all energy losses within the component. In practical terms, this ratio reduces to the formula:

$$Q = \frac{X}{R} \quad (43)$$

where:
 Q = figure of merit or quality (no units),
 X = X_L (inductive reactance) for inductors and X_C (capacitive reactance) for capacitors (in ohms), and
 R = the sum of all resistances associated with the energy losses in the component (in ohms).
The Q of capacitors is ordinarily high.

Good quality ceramic capacitors and mica capacitors may have Q values of 1200 or more. Small ceramic trimmer capacitors may have Q values too small to ignore in some applications. Microwave capacitors can have poor Q values; 10 or less at 10 GHz and higher frequencies.

Inductors are subject to many types of electrical energy losses, however: wire resistance, core losses and skin effect. All electrical conductors have some resistance through which electrical energy is lost as heat. Moreover, inductor wire must be sized to handle the anticipated current through the coil. Wire conductors suffer additional ac losses because alternating current tends to flow on the conductor surface. As the frequency increases, the current is confined to a thinner layer of the conductor surface. This property is called *skin effect*. If the inductor's core is a conductive material, such as iron, ferrite, or brass, the core will introduce additional losses of energy. The specific details of these losses are discussed in connection with each type of core material.

The sum of all core losses may be depicted by showing a resistor in series with the inductor (as in Figs 6.36 and 6.37), although there is no separate component represented by the symbol. As a result of inherent energy losses, inductor Q rarely, if ever, approaches capacitor Q in a circuit where both components work together. Although many circuits call for the highest Q inductor obtainable, other circuits may call for a specific Q, even a very low one.

Calculating Practical Inductors

Although builders and experimenters rarely construct their own capacitors, inductor fabrication is common. In fact, it is often necessary, since commercially available units may be unavailable or expensive. Even if available, they may consist of coil stock to be trimmed to the required value. Core materials and wire for winding both solenoid and toroidal inductors are readily available. The following information includes fundamental formulas and design examples for calculating practical inductors, along with additional data on the theoretical limits in the use of some materials.

AIR-CORE INDUCTORS

Many circuits require air-core inductors using just one layer of wire. The approximate inductance of a single-layer air-core coil may be calculated from the simplified formula:

$$L\,(\mu H) = \frac{d^2\,n^2}{18\,d + 40\,\ell} \qquad (44)$$

where:
L = inductance in microhenrys,
d = coil diameter in inches (from wire center to wire center),
ℓ = coil length in inches, and
n = number of turns.

The notation is explained in **Fig 6.39**. This formula is a close approximation for coils having a length equal to or greater than 0.4 d. (Note: Inductance varies as the square of the turns. If the number of turns is doubled, the inductance is quadrupled. This relationship is inherent in the equation, but is often overlooked. For example, if you want to double the inductance, put on additional turns equal to 1.4 times the original number of turns, or 40% more turns.)

Example: What is the inductance of a coil if the coil has 48 turns wound at 32 turns per inch and a diameter of $^3/_4$ inch? In this case, d = 0.75, $\ell = {}^{48}/_{32} = 1.5$ and n = 48.

$$L = \frac{0.75^2 \times 48^2}{(18 \times 0.75) + (40 \times 1.5)}$$

$$= \frac{1300}{74} = 18\,\mu H$$

To calculate the number of turns of a single-layer coil for a required value of inductance, the formula becomes:

$$n = \frac{\sqrt{L\,(18\,d + 40\,\ell)}}{d} \qquad (45)$$

Example: Suppose an inductance of 10.0 µH is required. The form on which the coil is to be wound has a diameter of one inch and is long enough to accommodate a coil of $1^1/_4$ inches. Then d = 1.00 inch, ℓ = 1.25 inches and L = 10.0. Substituting:

$$n = \frac{\sqrt{10.0\,\big[(18 \times 1.00) + (40 \times 1.25)\big]}}{1}$$

$$= \sqrt{680.} = 26.1\ \text{turns}$$

A 26-turn coil would be close enough in practical work. Since the coil will be 1.25 inches long, the number of turns per inch will be 26.1 / 1.25 = 20.9. Consulting the wire table in the **References** chapter, we find that #17 enameled wire (or anything smaller) can be used. The proper inductance is obtained by winding the required number of turns on the form and then adjusting the spacing between the turns to make a uniformly spaced coil 1.25 inches long.

Most inductance formulas lose accuracy when applied to small coils (such as are used in VHF work and in low-pass filters built for reducing harmonic interference to televisions) because the conductor thickness is no longer negligible in comparison with the size of the coil. **Fig 6.40** shows the measured inductance of VHF coils and may be used as a basis for circuit design. Two curves are given; curve A is for coils wound to an inside diameter of $^1/_2$ inch; curve B is for coils of $^3/_4$-inch inside diameter. In both curves, the wire size is #12, and the winding pitch is eight turns to the inch ($^1/_8$ inch center-to-center turn spacing). The inductance values given include leads $^1/_2$-inch long.

Machine-wound coils with the preset di-

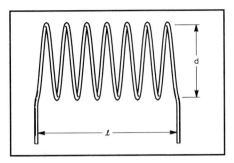

Fig 6.39 — Coil dimensions used in the inductance formula for air-core inductors.

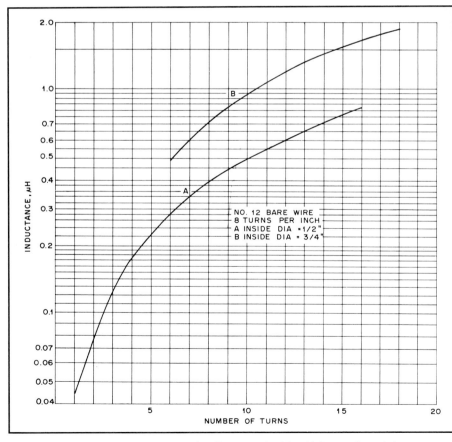

NO. 12 BARE WIRE
8 TURNS PER INCH
A INSIDE DIA =1/2"
B INSIDE DIA = 3/4"

Fig 6.40 — Measured inductance of coils wound with #12 bare wire, eight turns to the inch. The values include half-inch leads.

ameters and turns per inch are available in many radio stores, under the trade names of B&W Miniductor, Airdux and Polycoil. The **References** chapter provides information on using such coil stock to simplify the process of designing high-quality inductors for most HF applications. Forming a wire into a solenoid increases its inductance, and also introduces distributed capacitance. Since each turn is at a slightly different ac potential, each pair of turns effectively forms a parasitic capacitor. See the **Real-World Components** chapter for information on the effects of these complications to the "ideal" inductors under discussion in this chapter. Moreover, the Q of air-core inductors is, in part, a function of the coil shape, specifically its ratio of length to diameter. Q tends to be highest when these dimensions are nearly equal. With wire properly sized to the current carried by the coil, and with high-caliber construction, air-core inductors can achieve Qs above 200. Air-core inductors with Qs as high as 400 are possible.

STRAIGHT-WIRE INDUCTANCE

At low frequencies the inductance of a straight, round, nonmagnetic wire in free space is given by:

$$L = 0.00508\, b \left\{ \left[\ln\left(\frac{2b}{a}\right)\right] - 0.75 \right\} \quad (46)$$

where:
L = inductance in μH,
a = wire radius in inches,
b = wire length in inches, and
ln = natural logarithm = 2.303 × common logarithm (base 10).

If the dimensions are expressed in millimeters instead of inches, the equation may still be used, except replace the 0.00508 value with 0.0002.

Skin effect reduces the inductance at VHF and above. As the frequency approaches infinity, the 0.75 constant within the brackets approaches unity. As a practical matter, skin effect will not reduce the inductance by more than a few percent.

Example: What is the inductance of a wire that is 0.1575 inch in diameter and 3.9370 inches long? For the calculations, a = 0.0787 inch (radius) and b = 3.9370 inch.

$$L = 0.00508\, b \left\{ \left[\ln\left(\frac{2b}{a}\right)\right] - 0.75 \right\}$$

$$= 0.00508\,(3.9370) \times$$

$$\left\{ \left[\ln\left(\frac{2 \times 3.9370}{0.0787}\right)\right] - 0.75 \right\}$$

$$L = 0.0200\left[\ln(100.) - 0.75\right]$$

$$= 0.0200(4.60 - 0.75)$$

$$= 0.0200 \times 3.85 = 0.077\,\mu H$$

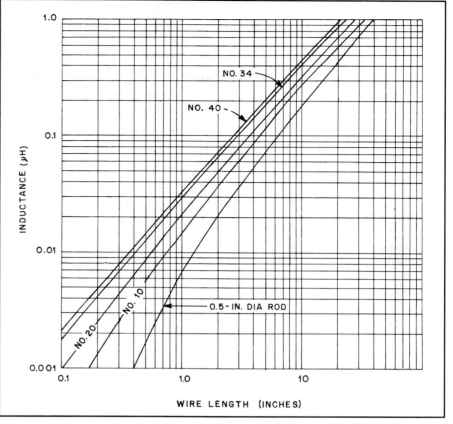

Fig 6.41 — Inductance of various conductor sizes as straight wires.

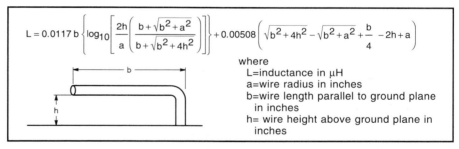

$$L = 0.0117\, b \left\{ \log_{10}\left[\frac{2h}{a}\left(\frac{b + \sqrt{b^2 + a^2}}{b + \sqrt{b^2 + 4h^2}}\right)\right] \right\} + 0.00508 \left(\sqrt{b^2 + 4h^2} - \sqrt{b^2 + a^2} + \frac{b}{4} - 2h + a \right)$$

where
L = inductance in μH
a = wire radius in inches
b = wire length parallel to ground plane in inches
h = wire height above ground plane in inches

Fig 6.42 — Equation for determining the inductance of a wire parallel to a ground plane, with one end grounded. If the dimensions are in millimeters, the numerical coefficients become 0.0004605 for the first term and 0.0002 for the second term.

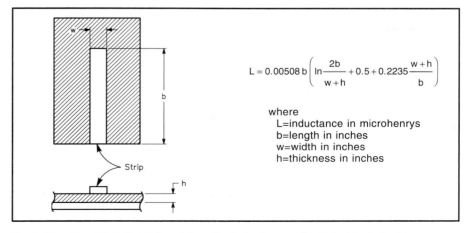

$$L = 0.00508\, b \left(\ln\frac{2b}{w+h} + 0.5 + 0.2235\frac{w+h}{b} \right)$$

where
L = inductance in microhenrys
b = length in inches
w = width in inches
h = thickness in inches

Fig 6.43 — Equation for determining the inductance of a flat strip inductor.

Fig 6.41 is a graph of the inductance for wires of various radii as a function of length.

A VHF or UHF tank circuit can be fabricated from a wire parallel to a ground plane, with one end grounded. A formula for the inductance of such an arrangement is given in **Fig 6.42**.

Example: What is the inductance of a wire 3.9370 inches long and 0.0787 inch in radius, suspended 1.5748 inch above a ground plane? (The inductance is measured between the free end and the ground plane, and the formula includes the inductance of the 1.5748-inch grounding link.) To demonstrate the use of the formula in Fig 6.42, begin by evaluating these quantities:

$$b + \sqrt{b^2 + a^2}$$
$$= 3.9370 + \sqrt{3.9370^2 + 0.0787^2}$$
$$= 3.9370 + 3.94 = 7.88$$

$$b + \sqrt{b^2 + 4\left(h^2\right)}$$
$$= 3.9370 + \sqrt{3.9370^2 + 4\left(1.5748^2\right)}$$
$$= 3.9370 + \sqrt{15.500 + 4\,(2.4800)}$$
$$= 3.9370 + \sqrt{15.500 + 9.9200}$$
$$= 3.9370 + 5.0418 = 8.9788$$

$$\frac{2\,h}{a} = \frac{2 \times 1.5748}{0.0787} = 40.0$$

$$\frac{b}{4} = \frac{3.9370}{4} = 0.98425$$

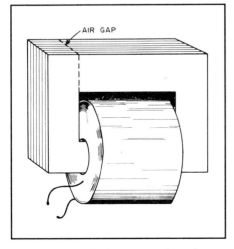

Fig 6.44 — Typical construction of an iron-core inductor. The small air gap prevents magnetic saturation of the iron and thus maintains the inductance at high currents.

Substituting these values into the formula yields:

$$L = 0.0117 \times 3.9370 \left\{ \log_{10} \left[40.0 \times \left(\frac{7.88}{8.9788} \right) \right] \right\}$$
$$+ 0.00508 \times$$
$$(5.0418 - 3.94 + 0.98425 - 3.1496 + 0.0787)$$
$$L = 0.0662\ \mu H$$

Another conductor configuration that is frequenly used is a flat strip over a ground plane. This arrangement has a lower skin-effect loss at high frequencies than round wire because it has a higher surface-area to volume ratio. The inductance of such a strip can be found from the formula in **Fig 6.43**. For a large collection of formulas useful in constructing air-core inductors of many configurations, see the "Circuit Elements" section in Terman's *Radio Engineers' Handbook* or the "Transmission Media" chapter of the *ARRL UHF/Microwave Experimenter's Manual.*

IRON-CORE INDUCTORS

If the permeability of an iron core in an inductor is 800, then the inductance of any given air-wound coil is increased 800 times by inserting the iron core. The inductance will be proportional to the magnetic flux through the coil, other things being equal. The inductance of an iron-core inductor is highly dependent on the current flowing in the coil, in contrast to an air-core coil, where the inductance is independent of current because air does not saturate.

Iron-core coils such as the one sketched in **Fig 6.44** are used chiefly in power-supply equipment. They usually have direct current flowing through the winding, and any variation in inductance with current is usually undesirable. Inductance variations may be overcome by keeping the flux density below the saturation point of the iron. Opening the core so there is a small air gap, indicated by the dashed lines in Fig 6.44, will achieve this goal. The reluctance or magnetic resistance introduced by such a gap is very large compared with that of the iron, even though the gap is only a small fraction of an inch. Therefore, the gap — rather than the iron — controls the flux density. Air gaps in iron cores reduce the inductance, but they hold the value practically constant regardless of the current magnitude.

When alternating current flows through a coil wound on an iron core, a voltage is induced. Since iron is a conductor, a current also flows in the core. Such currents are called *eddy currents.* Eddy currents represent lost power because they flow through the resistance of the iron and gen-

erate heat. Losses caused by eddy currents can be reduced by laminating the core (cutting the core into thin strips). These strips or laminations are then insulated from each other by painting them with some insulating material such as varnish or shellac. These losses add to hysteresis losses, which are also significant in iron-core inductors.

Eddy-current and hysteresis losses in iron increase rapidly as the frequency of the alternating current increases. For this reason, ordinary iron cores can be used only at power-line and audio frequencies — up to approximately 15000 Hz. Even then, a very good grade of iron or steel is necessary for the core to perform well at the higher audio frequencies. Laminated iron cores become completely useless at radio frequencies.

SLUG-TUNED INDUCTORS

For RF work, the losses in iron cores can be reduced to a more useful level by grinding the iron into a powder and then mixing it with a "binder" of insulating material in such a way that the individual iron particles are insulated from each other. Using this approach, cores can be made that function satisfactorily even into the VHF range.

Because a large part of the magnetic path is through a nonmagnetic material (the "binder"), the permeability of the iron is low compared with the values obtained at power-line frequencies. The core is usually shaped in the form of a slug or cylinder for fit inside the insulating form on which the coil is wound. Despite the fact that the major portion of the magnetic path for the flux is in air, the slug is quite effective in increasing the coil inductance. By pushing (or screwing) the slug in and out of the coil, the inductance can be varied over a considerable range. See *The ARRL Electronics Data Book* for information on a wide variety of representative slug-tuned coils available commercially.

POWDERED-IRON TOROIDAL INDUCTORS

For fixed-value inductors intended for use at HF and VHF, the powdered-iron toroidal core has become almost the standard core and material in low power circuits. **Fig 6.45** shows the general outlines of a toroidal coil on a magnetic core. Manufacturers offer a wide variety of core materials, or mixes, to provide units that will perform over a desired frequency range with a reasonable permeability. Initial permeabilities for powdered-iron cores fall in the range of 3 to 35 for various mixes. In addition, core sizes are available in the range of 0.125-inch outside diam-

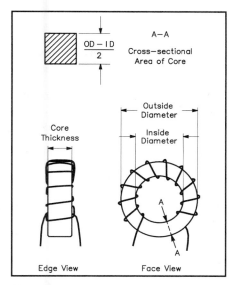

Fig 6.45 — A typical toroidal inductor wound on a powdered-iron or ferrite core. Some key physical dimensions are noted. Equally important are the core material, its permeability, its intended range of operating frequencies, and its A_L value. This is an 11-turn toroid.

eter (OD) up to 1.06-inch OD, with larger sizes to 5-inch OD available in certain mixes. The range of sizes permits the builder to construct single-layer inductors for almost any value using wire sized to meet the circuit current demands. While powdered-iron toroids are often painted various colors, you must know the manufacturer to identify the mix. There seems to be no set standard between manufacturers. Iron-powder toroids usually have rounded edges.

The use of powdered iron in a binder reduces core losses usually associated with iron, while the permeability of the core permits a reduction in the wire length and associated resistance in forming a coil of a given inductance. Therefore, powdered-iron-core toroidal inductors can achieve Qs well above 100, often approaching or exceeding 200 within the frequency range specified for a given core. Moreover, these coils are considered self-shielding since most of the flux lines are within the core, a fact that simplifies circuit design and construction.

Each powdered-iron core has a value of A_L determined and published by the core manufacturer. For powdered-iron cores, A_L represents the *inductance index*, that is, the inductance in μH per 100 turns of wire on the core, arranged in a single layer. The builder must select a core size capable of holding the calculated number of turns, of the required wire size, for the desired inductance. Otherwise, the coil calcula-

tion is straightforward. To calculate the inductance of a powdered-iron toroidal coil, when the number of turns and the core material are known, use the formula:

$$L = \frac{A_L \times N^2}{10000} \qquad (47)$$

where:
- L = the inductance in μH,
- A_L = the inductance index in μH per 100 turns, and
- N = the number of turns.

Example: What is the inductance of a 60-turn coil on a core with an A_L of 55? This A_L value was selected from manufacturer's information about a 0.8-inch OD core with an initial permeability of 10. This particular core is intended for use in the range of 2 to 30 MHz. See the **Component Data** chapter for more detailed data on the range of available cores.

$$L = \frac{A_L \times N^2}{10000} = \frac{55 \times 60^2}{10000}$$

$$= \frac{198000}{10000} = 19.8 \ \mu H$$

To calculate the number of turns needed for a particular inductance, use the formula:

$$N = 100 \sqrt{\frac{L}{A_L}} \qquad (48)$$

Example: How many turns are needed for a 12.0-μH coil if the A_L for the selected core is 49?

$$N = 100 \sqrt{\frac{L}{A_L}} = 100 \sqrt{\frac{12.0}{49}}$$

$$= 100 \sqrt{0.245} = 100 \times 0.495 = 49.5 \text{ turns}$$

If the value is critical, experimenting with 49-turn and 50-turn coils is in order, especially since core characteristics may vary slightly from batch to batch. Count turns by each pass of the wire through the center of the core. (A straight wire through a toroidal core amounts to a one-turn coil.) Fine adjustment of the inductance may be possible by spreading or squeezing inductor turns.

The power-handling ability of toroidal cores depends on many variables, which include the cross-sectional area through the core, the core material, the numbers of turns in the coil, the applied voltage and the operating frequency. Although powdered-iron cores can withstand dc flux densities up to 5000 gauss without saturating, ac flux densities from sine waves above certain limits can overheat cores. Manufacturers provide guideline limits for ac flux densities to avoid overheating. The limits range from 150 gauss at 1 MHz to 30 gauss at 28 MHz,

although the curve is not linear. To calculate the maximum anticipated flux density for a particular coil, use the formula:

$$B_{max} = \frac{E_{RMS} \times 10^8}{4.44 \times A_e \times N \times f} \qquad (49)$$

where:
- B_{max} = the maximum flux density in gauss,
- E_{RMS} = the voltage across the coil,
- A_e = the cross-sectional area of the core in square centimeters,
- N = the number of turns in the coil, and
- f = the operating frequency in Hz.

Example: What is the maximum ac flux density for a coil of 15 turns if the frequency is 7.0 MHz, the RMS voltage is 25 V and the cross-sectional area of the core is 0.133 cm²?

$$B_{max} = \frac{E_{RMS} \times 10^8}{4.44 \times A_e \times N \times f}$$

$$= \frac{25 \times 10^8}{4.44 \times 0.133 \times 15 \times 7.0 \times 10^6}$$

$$= \frac{25 \times 10^8}{62 \times 10^6} = 40. \text{ gauss}$$

Since the recommended limit for cores operated at 7 MHz is 57 gauss, this coil is well within guidelines.

FERRITE TOROIDAL INDUCTORS

Although nearly identical in general appearance to powdered-iron cores, ferrite cores differ in a number of important characteristics. They are often unpainted, unlike powdered-iron toroids. Ferrite toroids often have sharp edges, while powdered-iron toroids usually have rounded edges. Composed of nickel-zinc ferrites for lower permeability ranges and of manganese-zinc ferrites for higher permeabilities, these cores span the permeability range from 20 to above 10000. Nickel-zinc cores with permeabilities from 20 to 800 are useful in high-Q applications, but function more commonly in amateur applications as RF chokes. They are also useful in wide-band transformers (discussed later in this chapter).

Because of their higher permeabilities, the formulas for calculating inductance and turns require slight modification. Manufacturers list ferrite A_L values in mH per 1000 turns. Thus, to calculate inductance, the formula is

$$L = \frac{A_L \times N^2}{1000000} \qquad (50)$$

where:
- L = the inductance in mH,
- A_L = the inductance index in mH per 1000 turns, and
- N = the number of turns.

Example: What is the inductance of a 60-turn coil on a core with an A_L of 523? (See the **Component Data** chapter for more detailed data on the range of available cores.)

$$L = \frac{A_L \times N^2}{1000000} = \frac{523 \times 60^2}{1000000}$$

$$= \frac{1.88 \times 10^6}{1 \times 10^6} = 1.88 \text{ mH}$$

To calculate the number of turns needed for a particular inductance, use the formula:

$$N = 1000 \sqrt{\frac{L}{A_L}} \qquad (51)$$

Example: How many turns are needed for a 1.2-mH coil if the A_L for the selected core is 150?

$$N = 1000 \sqrt{\frac{L}{A_L}} = 1000 \sqrt{\frac{1.2}{150}}$$

$$= 1000 \sqrt{0.008} = 1000 \times 0.089 = 89 \text{ turns}$$

For inductors carrying both dc and ac currents, the upper saturation limit for most ferrites is a flux density of 2000 gauss, with power calculations identical to those used for powdered-iron cores. For detailed information on available cores and their characteristics, see *Iron-Powder and Ferrite Coil Forms*, a combination catalog and information book from Amidon Associates, Inc. (See the Address List in the **References** chapter for information about contacting Amidon.)

Ohm's Law for Reactance

Only ac circuits containing capacitance or inductance (or both) have reactance. Despite the fact that the voltage in such circuits is 90° out of phase with the current, circuit reactance does limit current in a manner that corresponds to resistance. Therefore, the Ohm's Law equations relating voltage, current and resistance apply to purely reactive circuits:

$$E = I X \qquad (52)$$

$$I = \frac{E}{X} \qquad (53)$$

$$X = \frac{E}{I} \qquad (54)$$

where:
E = ac voltage in RMS,
I = ac current in amperes, and
X = inductive or capacitive reactance.

Example: What is the voltage across a capacitor of 200. pF at 7.15 MHz, if the current through the capacitor is 50. mA?

Since the reactance of the capacitor is a function of both frequency and capacitance, first calculate the reactance:

$$X_C = \frac{1}{2 \pi f C}$$

$$= \frac{1}{2 \times 3.1416 \times 7.15 \times 10^6 \text{ Hz} \times 200. \times 10^{-12} \text{ F}}$$

$$= \frac{10^6 \ \Omega}{8980} = 111 \ \Omega$$

Next, use Ohm's Law:

$$E = I \times X_C = 0.050 \text{ A} \times 111 \ \Omega = 5.6 \text{ V}$$

Example: What is the current through an 8.00-H inductor at 120. Hz, if 420. V is applied?
$$X_L = 2 \pi f L = 2 \times 3.1416 \times 120. \text{ Hz} \times 8.00 \text{ H} = 6030 \ \Omega$$

$$I = \frac{E}{X_L} = \frac{420. \text{ V}}{6030 \ \Omega} = 0.0697 \text{ A} = 69.7 \text{ mA}$$

Fig 6.46 charts the reactances of capacitors from 1 pF to 100 μF, and the reactances of inductors from 0.1 μH to 10 H, for frequencies between 100 Hz and 100 MHz. Approximate values of reactance can be read or interpolated from the chart. The formulas will produce more exact values, however.

Although both inductive and capacitive reactance limit current, the two types of reactance differ. With capacitive reactance, the current *leads* the voltage by 90°, whereas with inductive reactance, the current *lags* the voltage by 90°. The convention for charting the two types of reactance appears in **Fig 6.47**. On this graph, inductive reactance is plotted along the +90° vertical line, while capacitive reactance is plotted along the −90° vertical line. This convention of assigning a positive value to inductive reactance and a negative value to capacitive reactance results from the mathematics involved in impedance calculations.

REACTANCES IN SERIES AND PARALLEL

If a circuit contains two reactances of the same type, whether in series or in parallel, the resultant reactance can be determined by applying the same rules as for resistances in series and in parallel. Series reactance is given by the formula

$$X_{total} = X1 + X2 + X3 + \ldots + X_n \qquad (55)$$

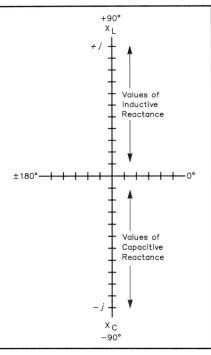

Fig 6.47 — The conventional method of plotting reactances on the vertical axis of a graph, using the upward or "plus" direction for inductive reactance and the downward or "minus" direction for capacitive reactance. The horizontal axis will be used for resistance in later examples.

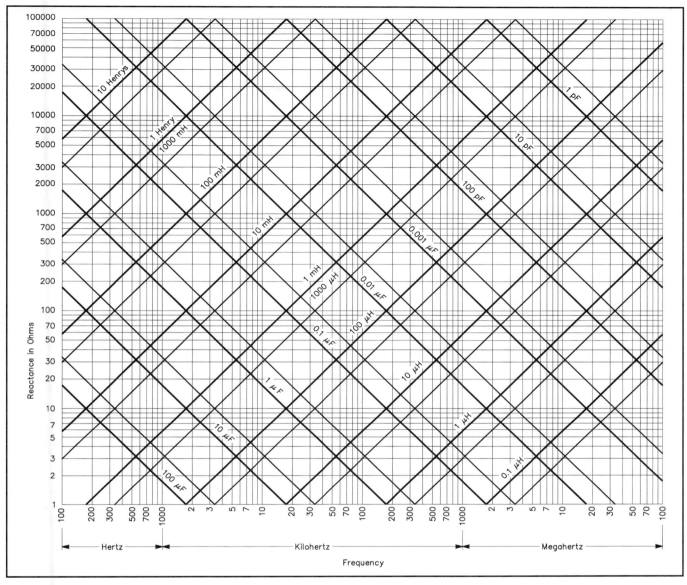

Fig 6.46 — Inductive and capacitive reactance vs frequency. Heavy lines represent multiples of 10, intermediate lines multiples of 5. For example, the light line between 10 μH and 100 μH represents 50 μH; the light line between 0.1 μF and 1 μF represents 0.5 μF, and so on. Other values can be extrapolated from the chart. For example, the reactance of 10 H at 60 Hz can be found by taking the reactance of 10 H at 600 Hz and dividing by 10 for the 10 times decrease in frequency.

Example: Two noninteracting inductances are in series. Each has a value of 4.0 μH, and the operating frequency is 3.8 MHz. What is the resulting reactance?

The reactance of each inductor is:

$$X_L = 2 \pi f L = 2 \times 3.1416 \times 3.8 \times 10^6 \text{ Hz} \times 4 \times 10^{-6} \text{ H} = 96 \ \Omega$$

$$X_{total} = X1 + X2 = 96 \ \Omega + 96 \ \Omega = 192 \ \Omega$$

We might also calculate the total reactance by first adding the inductances:

$$L_{total} = L1 + L2 = 4.0 \ \mu H + 4.0 \ \mu H$$
$$L_{total} = 8.0 \ \mu H$$

$$X_{total} = 2 \pi f L = 2 \times 3.1416 \times 3.8 \times 10^6 \text{ Hz} \times 8.0 \times 10^{-6} \text{ H}$$

$$X_{total} = 191 \ \Omega$$

(The fact that the last digit differs by one illustrates the uncertainty of the calculation caused by the uncertainty of the measured values in the problem, and differences caused by rounding off the calculated values. This also shows why it is important to follow the rules for significant figures discussed in the **Mathematics for Amateur Radio** chapter.)

Example: Two noninteracting capacitors are in series. One has a value of 10.0 pF, the other of 20.0 pF. What is the resulting reactance in a circuit operating at 28.0 MHz?

$$X_{C1} = \frac{1}{2 \pi f C}$$

$$= \frac{1}{2 \times 3.1416 \times 28.0 \times 10^6 \text{ Hz} \times 10.0 \times 10^{-12} \text{ F}}$$

$$= \frac{10^6 \ \Omega}{1760} = 568 \ \Omega$$

$$X_{C2} = \frac{1}{2 \pi f C}$$

$$= \frac{1}{2 \times 3.1416 \times 28.0 \times 10^6 \text{ Hz} \times 20.0 \times 10^{-12} \text{ F}}$$

$$= \frac{10^6 \ \Omega}{3520} = 284 \ \Omega$$

$$X_{total} = X_{C1} + X_{C2} = 568 \ \Omega + 284 \ \Omega = 852 \ \Omega$$

Alternatively, for series capacitors, the total capacitance is 6.67×10^{-12} F or 6.67 pF. Then:

$$X_{total} = \frac{1}{2 \pi f C}$$

$$= \frac{1}{2 \times 3.1416 \times 28.0 \times 10^6 \ Hz \times 6.67 \times 10^{-12} \ F}$$

$$= \frac{10^6 \ \Omega}{1170} = 855 \ \Omega$$

(Within the uncertainty of the measured values and the rounding of values in the calculations, this is the same result as we obtained with the first method.)

This example serves to remind us that *series capacitance* is not calculated in the manner used by other series resistance and inductance, but *series capacitive reactance* does follow the simple addition formula.

For reactances of the same type in parallel, the general formula is:

$$X_{total} = \frac{1}{\frac{1}{X1} + \frac{1}{X2} + \frac{1}{X3} + \ldots + \frac{1}{X_n}}$$

(56)

or, for exactly two reactances in parallel

$$X_{total} = \frac{X1 \times X2}{X1 + X2}$$

(57)

Example: Place the capacitors in the last example (10.0 pF and 20.0 pF) in parallel in the 28.0 MHz circuit. What is the resultant reactance?

$$X_{total} = \frac{X1 \times X2}{X1 + X2}$$

$$= \frac{568 \ \Omega \times 284 \ \Omega}{568 \ \Omega + 284 \ \Omega} = 189 \ \Omega$$

Alternatively, two capacitors in parallel add their capacitances.

$$C_{total} = C_1 + C_2 = 10.0 \ pF + 20.0 \ pF$$
$$= 30.0 \ pF$$

$$X_C = \frac{1}{2 \pi f C}$$

$$= \frac{1}{2 \times 3.1416 \times 28.0 \times 10^6 \ Hz \times 30.0 \times 10^{-12} \ F}$$

$$= \frac{10^6 \ \Omega}{5280} = 189 \ \Omega$$

Example: Place the series inductors above (4.0 μH each) in parallel in a 3.8-MHz circuit. What is the resultant reactance?

$$X_{total} = \frac{X_{L1} \times X_{L2}}{X_{L1} + X_{L2}}$$

$$= \frac{96 \ \Omega \times 96 \ \Omega}{96 \ \Omega + 96 \ \Omega} = 48 \ \Omega$$

Of course, equal reactances (or resistances) in parallel yield a reactance that is the value of one of them divided by the number (n) of equal reactances, or:

$$X_{total} = \frac{X}{n} = \frac{96 \ \Omega}{2} = 48 \ \Omega$$

All of these calculations apply only to reactances of the same type; that is, all capacitive or all inductive. Mixing types of reactances requires a different approach.

UNLIKE REACTANCES IN SERIES

When combining unlike reactances — that is, combinations of inductive and capacitive reactance — in series, it is necessary to take into account that the voltage-to-current phase relationships differ for the different types of reactance. **Fig 6.48** shows a series circuit with both types of reactance. Since the reactances are in series, the current must be the same in both. The voltage across each circuit element differs in phase, however. The voltage E_L *leads* the current by 90°, and the voltage E_C *lags* the current by 90°. Therefore, E_L and E_C have opposite polarities and cancel each other in whole or in part. The dotted line in Fig 6.48 approximates the resulting voltage E, which is the *difference* between E_L and E_C.

Since, for a constant current, the reactance is directly proportional to the voltage, the net reactance must be the difference between the inductive and the capacitive reactances, or:

$$X_{total} = X_L - X_C$$

(58)

For this and subsequent calculations in which there is a mixture of inductive and capacitive reactance, use the absolute value of each reactance. The convention of recording inductive reactances as positive and capacitive reactances as negative is built into the mathematical operators in the formulas.

Example: Using Fig 6.48 as a visual aid, let $X_C = 20.0 \ \Omega$ and $X_L = 80.0 \ \Omega$. What is the resulting reactance?

$$X_{total} = X_L - X_C = 80.0 \ \Omega - 20.0 \ \Omega$$
$$= +60.0 \ \Omega$$

Since the result is a positive value, reactance is inductive. Had the result been a negative number, the reactance would have been capacitive.

When reactance types are mixed in a series circuit, the resulting reactance is always smaller than the larger of the two reactances. Likewise, the resulting voltage across the series combination of reactances is always smaller than the larger of the two voltages across individual reactances.

Every series circuit of mixed reactance types with more than two circuit elements can be reduced to the type of circuit covered here. If the circuit has more than one capacitor or more than one inductor in the overall series string, first use the formulas given earlier to determine the total series inductance alone and the total series capacitance alone (or their respective reactances).

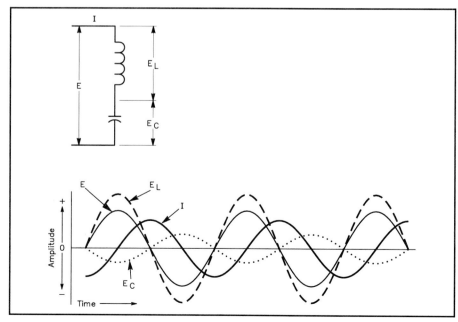

Fig 6.48 — A series circuit containing both inductive and capacitive components, together with representative voltage and current relationships.

Then combine the resulting single capacitive reactance and single inductive reactance as shown in this section.

UNLIKE REACTANCES IN PARALLEL

The situation of parallel reactances of mixed type appears in **Fig 6.49**. Since the elements are in parallel, the voltage is common to both reactive components. The current through the capacitor, I_C, *leads* the voltage by 90°, and the current through the inductor, I_L, *lags* the voltage by 90°. The two currents are 180° out of phase and thus cancel each other in whole or in part. The total current is the difference between the individual currents, as indicated by the solid, heavy line in Fig 6.49.

Since reactance is the ratio of voltage to current, the total reactance in the circuit is:

$$X_{total} = \frac{E}{I_L - I_C} \qquad (59)$$

In the drawing, I_C is larger than I_L, and the resulting differential current retains the phase of I_C. Therefore, the overall reactance, X_{total}, is capacitive in this case. The total reactance of the circuit will be larger than the smaller of the individual reactances, because the total current is smaller than the larger of the two individual currents.

In parallel circuits of this type, reactance and current are inversely proportional to each other for a constant voltage. Therefore, to calculate the total reactance directly from the individual reactances, use the formula:

$$X_{total} = \frac{-X_L \times X_C}{X_L - X_C} \qquad (60)$$

As with the series formula for mixed reactances, use the absolute values of the reactances, since the minus signs in the formula take into account the convention of treating capacitive reactances as negative numbers. If the solution yields a negative number, the resulting reactance is capacitive, and if the solution is positive, then the reactance is inductive.

Example: Using Fig 6.49 as a visual aid, place a capacitive reactance of 10.0 Ω in parallel with an inductive reactance of 40.0 Ω. What is the resulting reactance?

$$\begin{aligned} X_{total} &= \frac{-X_L \times X_C}{X_L - X_C} \\ &= \frac{-40.0\,\Omega \times 10.0\,\Omega}{40.0\,\Omega - 10.0\,\Omega} \\ &= \frac{-400.\,\Omega}{30.0} = -13.3\,\Omega \end{aligned}$$

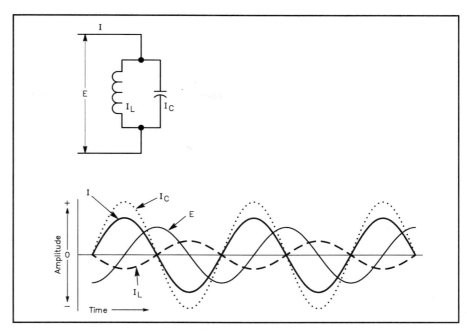

Fig 6.49 — A parallel circuit containing both inductive and capacitive components, together with representative voltage and current relationships.

The reactance is capacitive, as indicated by the negative solution. Moreover, the resultant reactance is always smaller than the larger of the two individual reactances.

As with the case of series reactances, if each leg of a parallel circuit contains more than one reactance, first simplify each leg to a single reactance. If the reactances are of the same type in each leg, the series reactance formulas for reactances of the same type will apply. If the reactances are of different types, then use the formulas shown above for mixed series reactances to simplify the leg to a single value and type of reactance.

APPROACHING RESONANCE

When two unlike reactances have the same numerical value, any series or parallel circuit in which they occur is said to be *resonant*. For any given inductance or capacitance, it is theoretically possible to find a value of the opposite reactance type to produce a resonant circuit for any desired frequency.

When a series circuit like the one shown in Fig 6.48 is resonant, the voltage E_C and E_L are equal and cancel; their sum is zero. Since the reactance of the circuit is proportional to the sum of these voltages, the total reactance also goes to zero. Theoretically, the current, as shown in **Fig 6.50**, can rise without limit. In fact, it is limited only by power losses in the components and other resistances that would be in a real circuit

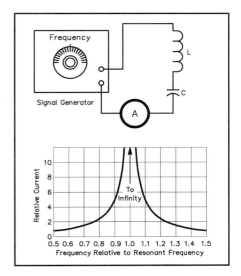

Fig 6.50 — The relative generator current with a fixed voltage in a series circuit containing inductive and capacitive reactances as the frequency approaches and departs from resonance.

of this type. As the frequency of operation moves slightly off resonance, the reactance climbs rapidly and then begins to level off. Similarly, the current drops rapidly off resonance and then levels.

In a parallel-resonant circuit of the type in Fig 6.49, the current I_L and I_C are equal and cancel to zero. Since the reactance is inversely proportional to the current, as the current approaches

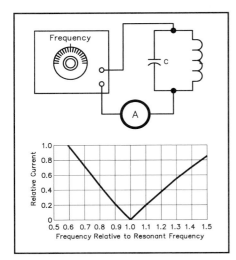

Fig 6.51 — The relative generator current with a fixed voltage in a parallel circuit containing inductive and capacitive reactances as the frequency approaches and departs from resonance. (The circulating current through the parallel inductor and capacitor is a maximum at resonance.)

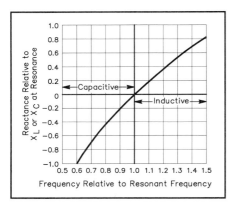

Fig 6.52 — The transition from capacitive to inductive reactance in a series-resonant circuit as the frequency passes resonance.

zero, the reactance rises without limit. As with series circuits, component power losses and other resistances in the circuit limit the current drop to some point above zero. **Fig 6.51** shows the theoretical current curve near and at resonance for a purely reactive parallel-resonant circuit. Note that in both Fig 6.50 and Fig 6.51, the departure of current from the resonance value is close to, but not quite, symmetrical above and below the resonant frequency.

Example: What is the reactance of a series L-C circuit consisting of a 56.04-pF capacitor and an 8.967-μH inductor at 7.00, 7.10 and 7.20 MHz? Using the formulas from earlier in this chapter, we calculate a table of values:

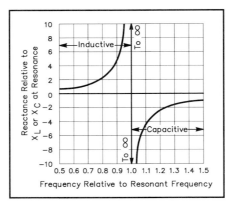

Fig 6.53 — The transition from inductive to capacitive reactance in a parallel-resonant circuit as the frequency passes resonance.

Frequency (MHz)	X_L (Ω)	X_C (Ω)	X_{total} (Ω)
7.000	394.4	405.7	−11.3
7.100	400.0	400.0	0
7.200	405.7	394.4	11.3

The exercise shows the manner in which the reactance rises rapidly as the frequency moves above and below resonance. Note that in a series-resonant circuit, the reactance at frequencies below resonance is capacitive, and above resonance, it is inductive. **Fig 6.52** displays this fact graphically. In a parallel-resonant circuit, where the reactance increases without limit at resonance, the opposite condition exists: above resonance, the reactance is capacitive and below resonance it is inductive, as shown in **Fig 6.53**. Of course, all graphs and calculations in this section are theoretical and presume a purely reactive circuit. Real circuits are never purely reactive; they contain some resistance that modifies their performance considerably. Real resonant circuits will be discussed later in this chapter.

REACTIVE POWER

Although purely reactive circuits, whether simple or complex, show a measurable ac voltage and current, we cannot simply multiply the two together to arrive at power. Power is the rate at which energy is consumed by a circuit, and purely reactive circuits do not consume power. The charge placed on a capacitor during part of an ac cycle is returned to the circuit during the next part of a cycle. Likewise, the energy stored in the magnetic field of an inductor returns to the circuit as the field collapses later in the ac cycle. A reactive circuit simply cycles and recycles energy into and out of the reactive components. If a purely reactive circuit were possible in reality, it would consume no

power at all.

In reactive circuits, circulation of energy accounts for seemingly odd phenomena. For example, in a series circuit with capacitance and inductance, the voltages across the components may exceed the supply voltage. That condition can exist because, while energy is being stored by the inductor, the capacitor is returning energy to the circuit from its previously charged state, and vice versa. In a parallel circuit with inductive and capacitive branches, the current circulating through the components may exceed the current drawn from the source. Again, the phenomenon occurs because the inductor's collapsing field supplies current to the capacitor, and the discharging capacitor provides current to the inductor.

To distinguish between the nondissipated power in a purely reactive circuit and the dissipated power of a resistive circuit, the unit of reactive power is called the *volt-ampere reactive*, or VAR. The term watt is not used; sometimes reactive power is called wattless power. Formulas similar to those for resistive power are used to calculate VAR:

$$VAR = I \, E \qquad (61)$$

$$VAR = I^2 \, X \qquad (62)$$

$$VAR = \frac{E^2}{X} \qquad (63)$$

These formulas have only limited use in radio work.

REACTANCE AND COMPLEX WAVEFORMS

All of the formulas and relationships shown in this section apply to alternating current in the form of regular sine waves. Complex wave shapes complicate the reactive situation considerably. A complex or nonsinusoidal wave can be resolved into a fundamental frequency and a series of harmonic frequencies whose amplitudes depend on the original wave shape. When such a complex wave — or collection of sine waves — is applied to a reactive circuit, the current through the circuit will not have the same wave shape as the applied voltage. The difference results because the reactance of an inductor and capacitor depend in part on the applied frequency.

For the second-harmonic component of the complex wave, the reactance of the inductor is twice and the reactance of the capacitor is half their respective values at the fundamental frequency. A third-harmonic component produces inductive reactances that are triple and capacitive reactances that are one-third those at the fundamental frequency. Thus, the overall circuit reactance is different for each harmonic component.

The frequency sensitivity of a reactive circuit to various components of a complex wave shape creates both difficulties and opportunities. On the one hand, calculating the circuit reactance in the presence of highly variable as well as complex waveforms, such as speech, is difficult at best. On the other hand, the frequency sensitivity of reactive components and circuits lays the

foundation for filtering, that is, for separating signals of different frequencies and passing them into different circuits. For example, suppose a coil is in the series path of a signal and a capacitor is connected from the signal line to ground, as represented in **Fig 6.54**. The reactance of the coil to the second harmonic of the signal will be twice that at the fundamental frequency and oppose more effectively the flow of harmonic current. Likewise, the reactance of the capacitor to the harmonic will be half that to the fundamental, allowing the harmonic an easier current path away from the signal line toward ground. See

the **Filters** chapter for detailed information on filter theory and construction.

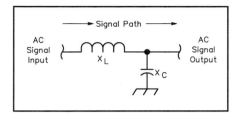

Fig 6.54 — A signal path with a series inductor and a shunt capacitor. The circuit presents different reactances to an ac signal and to its harmonics.

Impedance

When a circuit contains both resistance and reactance, the combined opposition to current is called *impedance*. Symbolized by the letter Z, impedance is a more general term than either resistance or reactance. Frequently, the term is used even for circuits containing only resistance or reactance. Qualifications such as "resistive impedance" are sometimes added to indicate that a circuit has only resistance, however.

The reactance and resistance comprising an impedance may be connected either in series or in parallel, as shown in **Fig 6.55**. In these circuits, the reactance is shown as a box to indicate that it may be either inductive or capacitive. In the series circuit at A, the current is the same in both elements, with (generally) different voltages appearing across the resistance and reactance. In the parallel circuit at B, the same voltage is applied to both elements, but different currents may flow in the two branches.

In a resistance, the current is in phase with the applied voltage, while in a reactance it is 90° out of phase with the voltage. Thus, the phase relationship between current and voltage in the circuit as a whole may be anything between zero and 90°, depending on the relative amounts of resistance and reactance.

As shown in Fig 6.47 in the preceding section, reactance is graphed on the vertical (Y) axis to record the phase difference between the voltage and the current. **Fig 6.56** adds resistance to the graph. Since the voltage is in phase with the current, resistance is recorded on the horizontal axis, using the positive or right side of the scale.

CALCULATING Z FROM R AND X IN SERIES CIRCUITS

Impedance is the complex combination

of resistance and reactance. Since there is a 90° phase difference between resistance and reactance (whether inductive or capacitive), simply adding the two values will not yield what actually happens in a circuit. Therefore, expressions like "Z = R ± X" can be misleading, because they suggest simple addition. As a result, impedance is often expressed "Z = R ± jX."

In pure mathematics, "*i*" indicates an imaginary number. Because i represents current in electronics, we use the letter "*j*" for the same mathematical operator, although there is nothing imaginary about what it represents in electronics. With re-

spect to resistance and reactance, the letter *j* is normally assigned to those figures on the vertical scale, 90° out of phase with the horizontal scale. The actual function of *j* is to indicate that calculating impedance from resistance and reactance requires *vector addition*. In vector addition, the result of combining two values at a 90° phase difference results in a new quantity for the combination, and also in a new combined phase angle relative to the base line.

Consider **Fig 6.57**, a series circuit consisting of an inductive reactance and a resistance. As given, the inductive reactance is 100 Ω and the resistance is 50 Ω. Using

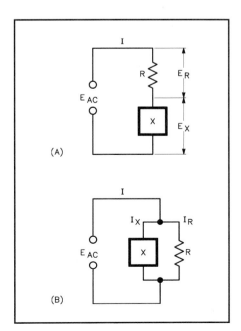

Fig 6.55 — Series and parallel circuits containing resistance and reactance.

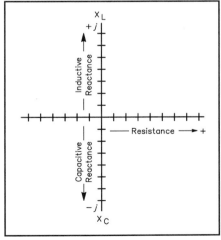

Fig 6.56 — The conventional method of charting impedances on a graph, using the vertical axis for reactance (the upward or "plus" direction for inductive reactance and the downward or "minus" direction for capacitive reactance), and using the horizontal axis for resistance.

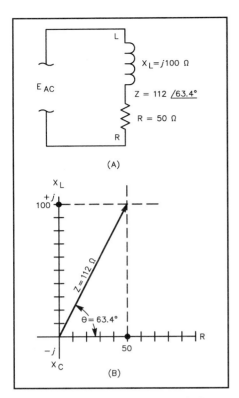

Fig 6.57 — A series circuit consisting of an inductive reactance of 100 Ω and a resistance of 50 Ω. At B, the graph plots the resistance, reactance, and impedance.

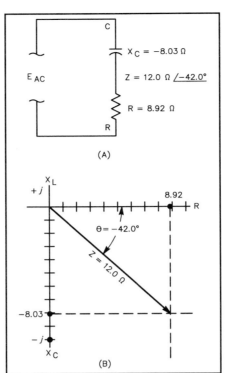

Fig 6.58 — A series circuit consisting of a capacitive reactance and a resistance: the impedance is given as 12.0 Ω ∠−42.0°. At B, the graph plots the resistance, reactance, and impedance.

rectangular coordinates, the impedance becomes

$$Z = R + jX \qquad (64)$$

where:
 Z = the impedance in ohms,
 R = the resistance in ohms, and
 X = the reactance in ohms.

In the present example,

$$Z = 50 + j100 \ \Omega.$$

As the graph shows, the combined opposition to current (or impedance) is represented by a line triangulating the two given values. The graph will provide an estimate of the value. A more exact way to calculate the resultant impedance involves the formula for right triangles, where the square of the hypotenuse equals the sum of the squares of the two sides. Since impedance is the hypotenuse:

$$Z = \sqrt{R^2 + X^2} \qquad (65)$$

In this example:

$$Z = \sqrt{(50 \ \Omega)^2 + (100 \ \Omega)^2}$$
$$= \sqrt{2500 \ W^2 + 10000 \ W^2}$$
$$= \sqrt{12500 \ W^2} = 112 \ W$$

The impedance that results from combining 50. Ω of resistance with 100. Ω of inductive reactance is 112 Ω. The phase angle of the resultant is neither 0° nor +90°. Instead, it lies somewhere between the two. Let θ be the angle between the horizontal axis and the line representing the impedance. From trigonometry, the tangent of the angle is the side-opposite the angle divided by the side adjacent to the angle, or

$$\tan \theta = \frac{X}{R} \qquad (66)$$

where:
 X = the reactance, and
 R = the resistance.

Find the angle by taking the inverse tangent, or arctan:

$$\theta = \arctan \frac{X}{R} \qquad (67)$$

In the example shown in Fig 6.57,

$$\theta = \arctan \frac{100 \ \Omega}{50 \ \Omega} = \arctan 2.0 = 63.4°$$

Combining the resultant impedance with the angle provides the impedance in *polar coordinate* form:

$$Z \angle \theta \qquad (68)$$

Using the information just calculated, the impedance is:

$$Z = 112 \ \Omega \ \angle 63.4°$$

The expressions $R \pm jX$ and $Z \angle \theta$ both provide the same information, but in two different forms. The procedure just given permits conversion from rectangular coordinates into polar coordinates. The reverse procedure is also important. **Fig 6.58** shows an impedance composed of a capacitive reactance and a resistance. Since capacitive reactance appears as a negative value, the impedance will be at a negative phase angle, in this case, 12.0 Ω at a phase angle of −42.0° or $Z = 12.0 \ \Omega \ \angle −42.0°$.

Think of the impedance as forming a triangle with the values of X and R from the rectangular coordinates. The reactance axis forms the side opposite the angle θ.

$$\sin \theta = \frac{\text{side opposite}}{\text{hypotenuse}} = \frac{X}{Z} \qquad (69)$$

Solving this equation for reactance, we have:

$$X = Z \times \sin \theta \ \text{(ohms)} \qquad (70)$$

Likewise, the resistance forms the side adjacent to the angle.

$$\cos \theta = \frac{\text{side adjacent}}{\text{hypotenuse}} = \frac{R}{Z}$$

Solving for resistance, we have:

$$R = Z \times \cos \theta \ \text{(ohms)} \qquad (71)$$

Then from our example:

$$X = 12.0 \ \Omega \times \sin (−42.0°)$$
$$= 12.0 \ \Omega \times −0.669 = −8.03 \ \Omega$$

$$R = 12.0 \ \Omega \times \cos (−42.0°)$$
$$= 12.0 \ \Omega \times 0.743 = 8.92 \ \Omega$$

Since X is a negative value, it plots on the lower vertical axis, as shown in Fig 6.58, indicating capacitive reactance. In rectangular form, $Z = 8.92 \ \Omega − j8.03 \ \Omega$.

In performing impedance and related calculations with complex circuits, rectangular coordinates are most useful when formulas require the addition or subtraction of values. Polar notation is most useful for multiplying and dividing complex numbers. The **Mathematics for Amateur Radio** chapter has information about performing addition, subtraction, multiplication and division with complex numbers.

All of the examples shown so far in this section have presumed values of reactance that contribute to the circuit impedance. Reactance is a function of frequency, however, and many impedance calculations may begin with a value of capacitance or inductance and an operating frequency. In terms of these values, the series impedance formula (Eq 65) becomes two formulas:

$$Z = \sqrt{R^2 + (2 \pi f L)^2} \qquad (72)$$

$$Z = \sqrt{R^2 + \left(\frac{1}{2 \pi f C}\right)^2} \qquad (73)$$

Example: What is the impedance of a circuit like Fig 6.57 with a resistance of 100 Ω and a 7.00-μH inductor operating at a frequency of 7.00 MHz? Using equation 72,

$$Z = \sqrt{R^2 + (2 \pi f L)^2}$$

$$= \sqrt{(100 \ \Omega)^2 + (2 \pi \times 7.00 \times 10^{-6} \ H \times 7.00 \times 10^6 \ Hz)^2}$$

$$Z = \sqrt{10,000 \ \Omega^2 + (308 \ \Omega)^2}$$

$$= \sqrt{10,000 \ \Omega^2 + 94,900 \ \Omega^2}$$

$$= \sqrt{104900 \ \Omega^2} = 323.9 \ \Omega$$

Since 308 Ω is the value of inductive reactance of the 7.00-μH coil at 7.00 MHz, the phase angle calculation proceeds as given in the earlier example (equation 67):

$$\theta = \arctan \left(\frac{X}{R}\right) = \arctan \left(\frac{308 \ \Omega}{100. \ \Omega}\right)$$

$$= \arctan (3.08) = 72.0°$$

Since the reactance is inductive, the phase angle is positive.

CALCULATING Z FROM R AND X IN PARALLEL CIRCUITS

In a parallel circuit containing reactance and resistance, such as shown in **Fig 6.59**, calculation of the resultant impedance from the values of R and X does not proceed by direct triangulation. The general formula for such parallel circuits is:

$$Z = \frac{RX}{\sqrt{R^2 + X^2}} \qquad (74)$$

where the formula uses the absolute (unsigned) reactance value. The phase angle for the parallel circuit is given by:

$$\theta = \arctan \left(\frac{R}{X}\right) \qquad (75)$$

If the parallel reactance is capacitive, then θ is a negative angle, and if the parallel reactance is inductive, then θ is a positive angle.

Example: An inductor with a reactance of 30.0 Ω is in parallel with as resistor of 40.0 Ω. What is the resulting impedance and phase angle?

$$Z = \frac{RX}{\sqrt{R^2 + X^2}} = \frac{30.0 \ \Omega \times 40.0 \ \Omega}{\sqrt{(30.0 \ \Omega)^2 + (40.0 \ \Omega)^2}}$$

$$= \frac{1200 \ \Omega^2}{\sqrt{900 \ \Omega^2 + 1600 \ \Omega^2}} = \frac{1200 \ \Omega^2}{\sqrt{2500 \ \Omega^2}}$$

$$= \frac{1200 \ \Omega^2}{50.0 \ \Omega}$$

$$Z = 24.0 \ \Omega$$

$$\theta = \arctan \left(\frac{R}{X}\right) = \arctan \left(\frac{40.0 \ \Omega}{30.0 \ \Omega}\right)$$

$$= \arctan (1.33) = 53.1°$$

Since the parallel reactance is inductive, the resultant angle is positive.

Example: A capacitor with a reactance of 16.0 Ω is in parallel with a resistor of 12.0 Ω. What is the resulting impedance and phase angle?

$$Z = \frac{RX}{\sqrt{R^2 + X^2}} = \frac{16.0 \ \Omega \times 12.0 \ \Omega}{\sqrt{(16.0 \ \Omega)^2 + (12.0 \ \Omega)^2}}$$

$$= \frac{192 \ \Omega^2}{\sqrt{256 \ \Omega^2 + 144 \ \Omega^2}} = \frac{192 \ \Omega^2}{\sqrt{400 \ \Omega^2}}$$

$$Z = \frac{192 \ \Omega^2}{20.0 \ \Omega} = 9.60 \ \Omega$$

$$\theta = \arctan \left(\frac{R}{X}\right) = \arctan \left(\frac{12.0 \ \Omega}{16.0 \ \Omega}\right)$$

$$\theta = \arctan (0.750) = -36.9°$$

Because the parallel reactance is capacitive, the resultant phase angle is negative.

ADMITTANCE

Just as the inverse of resistance is conductance (G) and the inverse of reactance is susceptance (B), so too impedance has an inverse: admittance (Y), measured in siemens (S). Thus,

$$Y = \frac{1}{Z} \qquad (76)$$

Since resistance, reactance and impedance are inversely proportional to the current (Z = E / I), conductance, susceptance and admittance are directly proportional to current. That is,

$$Y = \frac{I}{E} \qquad (77)$$

One handy use for admittance is in simplifying parallel circuit impedance calculations. A parallel combination of reactance and resistance reduces to a vector addition of susceptance and conductance, if admittance is the desired outcome. In other words, for parallel circuits:

$$Y = \sqrt{G^2 + B^2} \qquad (78)$$

where:
  Y = admittance,

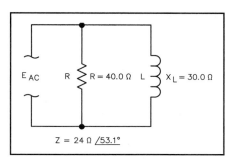

Fig 6.59 — A parallel circuit containing an inductive reactance of 30.0 Ω and a resistor of 40.0 Ω. No graph is given, since parallel impedances do not triangulate in the simple way of series impedances.

  G = conductance or 1 / R, and
  B = susceptance or 1 / X.

Example: An inductor with a reactance of 30.0 Ω is in parallel with a resistor of 40.0 Ω. What is the resulting impedance and phase angle? The susceptance is 1 / 30.0 Ω = 0.0333 S and the conductance is 1 / 40.0 Ω = 0.0250 S.

$$Y = \sqrt{(0.0333 \ S)^2 + (0.0250 \ S)^2}$$

$$= \sqrt{0.00173 \ S^2} = 0.0417 \ S$$

$$Z = \frac{1}{Y} = \frac{1}{0.0417 \ S} = 24.0 \ \Omega$$

The phase angle in terms of conductance and susceptance is:

$$\theta = \arctan \left(\frac{B}{G}\right) \qquad (79)$$

In this example,

$$\theta = \arctan \left(\frac{0.0333 \ S}{0.0250 \ S}\right) = \arctan (1.33) = 53.1°$$

Again, since the reactive component is inductive, the phase angle is positive. For a capacitively reactive parallel circuit, the phase angle would have been negative. Compare these results with the direct calculation earlier in the section.

Conversion from resistance, reactance and impedance to conductance, susceptance and admittance is perhaps most useful in complex-parallel-circuit calculations. Many advanced facets of active-circuit analysis will demand familiarity both with the concepts and with the calculation strategies introduced here, however.

More than Two Elements in Series or Parallel

When a circuit contains several resistances or several reactances in series, simplify the circuit before attempting to calculate the impedance. Resistances in series add, just as in a purely resistive circuit. Series reactances of the same kind — that is, all capacitive or all inductive — also add, just as in a purely reactive circuit. The goal is to produce a single value of resistance and a single value of reactance for the impedance calculation.

Fig 6.60 illustrates a more difficult case in which a circuit contains two different reactive elements in series, along with a further series resistance. The series combination of X_C and X_L reduce to a single value using the same rules of combination discussed in the section on purely reactive components. As Fig 6.60B demonstrates,

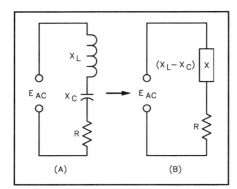

Fig 6.60 — A series impedance containing mixed capacitive and inductive reactances can be reduced to a single reactance plus resistance by combining the reactances algebraically.

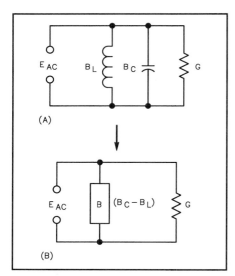

Fig 6.61 — A parallel impedance containing mixed capacitive and inductive reactances can be reduced to a single reactance plus resistance using formulas shown earlier in the chapter. By converting reactances to susceptances, as shown in A, you can combine the susceptances algebraically into a single susceptance, as shown in B.

the resultant reactance is the difference between the two series reactances.

For parallel circuits with multiple resistances or multiple reactances of the same type, use the rules of parallel combination to reduce the resistive and reactive components to single elements. Where two or more reactive components of different types appear in the same circuit, they can be combined using formulas shown earlier for pure reactances. As **Fig 6.61** suggests, however, they can also be combined as susceptances. Parallel susceptances of different types add, with attention to their differing signs. The resulting single susceptance can then be combined with the conductance to arrive at the overall circuit admittance. The inverse of the admittance is the final circuit impedance.

Equivalent Series and Parallel Circuits

The two circuits shown in Fig 6.55 are equivalent if the same current flows when a given voltage of the same frequency is applied, and if the phase angle between voltage and current is the same in both cases. It is possible, in fact, to transform any given series circuit into an equivalent parallel circuit, and vice versa.

A series RX circuit can be converted into its parallel equivalent by means of the formulas:

$$R_P = \frac{R_S^2 + X_S^2}{R_S} \qquad (80)$$

$$X_P = \frac{R_S^2 + X_S^2}{X_S} \qquad (81)$$

where the subscripts P and S represent the parallel- and series-equivalent values, respectively. If the parallel values are known, the equivalent series circuit can be found from:

$$R_S = \frac{R_P X_P^2}{R_P^2 + X_P^2} \qquad (82)$$

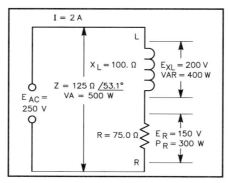

Fig 6.62 — A series circuit consisting of an inductive reactance of 100 Ω and a resistance of 75.0 Ω. Also shown is the applied voltage, voltage drops across the circuit elements, and the current.

and

$$X_S = \frac{R_P^2 X_P}{R_P^2 + X_P^2} \qquad (83)$$

Example: Let the series circuit in Fig 6.55 have a series reactance of –50.0 Ω (indicating a capacitive reactance) and a resistance of 50.0 Ω. What are the values of the equivalent parallel circuit?

$$R_P = \frac{R_S^2 + X_S^2}{R_S} = \frac{(50.0\ \Omega)^2 + (-50.0\ \Omega)^2}{50.0\ \Omega}$$

$$= \frac{2500\ \Omega^2 + 2500\ \Omega^2}{50.0\ \Omega} = \frac{5000\ \Omega^2}{50.0\ \Omega} = 100\ \Omega$$

$$X_P = \frac{R_S^2 + X_S^2}{X_S} = \frac{(50.0\ \Omega)^2 + (-50.0\ \Omega)^2}{-50.0\ \Omega}$$

$$= \frac{2500\ \Omega^2 + 2500\ \Omega^2}{-50.0\ \Omega} = \frac{5000\ \Omega^2}{-50.0\ \Omega} = -100\ \Omega$$

The parallel circuit in Fig 6.55 calls for a capacitive reactance of 100 Ω and a resistance of 100 Ω to be equivalent to the series circuit.

OHM'S LAW FOR IMPEDANCE

Ohm's Law applies to circuits containing impedance just as readily as to circuits having resistance or reactance only. The formulas are:

$$E = I Z \qquad (84)$$

$$I = \frac{E}{Z} \qquad (85)$$

$$Z = \frac{E}{I} \qquad (86)$$

where:
E = voltage in volts,
I = current in amperes, and
Z = impedance in ohms.

Fig 6.62 shows a simple circuit consisting of a resistance of 75.0 Ω and a reactance of 100 Ω. in series. From the series-impedance formula previously given, the impedance is

$$Z = \sqrt{R^2 + X_L^2} = \sqrt{(75.0\ \Omega)^2 + (100\ \Omega)^2}$$

$$= \sqrt{5630\ \Omega^2 + 10000\ \Omega^2} = \sqrt{15600\ \Omega^2}$$

$$= 125\ \Omega$$

If the applied voltage is 250 V, then

$$I = \frac{E}{Z} = \frac{250\ V}{125\ \Omega} = 2.00\ A$$

This current flows through both the resistance and reactance, so the voltage drops are:

$$E_R = I\ R = 2.00\ A \times 75.0\ \Omega = 150\ V$$

$E_{XL} = I\,X_L = 2.00\ \text{A} \times 100\ \Omega = 200\ \text{V}$

The simple arithmetical sum of these two drops, 350 V, is greater than the applied voltage because the two voltages are 90° out of phase. Their actual resultant, when phase is taken into account, is:

$$E = \sqrt{(150\ \text{V})^2 + (200\ \text{V})^2}$$
$$= \sqrt{22500\ \text{V}^2 + 40000\ \text{V}^2} = \sqrt{62500\ \text{V}^2}$$
$$= 250\ \text{V}$$

POWER FACTOR

In the circuit of Fig 6.62, an applied voltage of 250 V results in a current of 2.00 A, giving an apparent power of 250 V × 2.00 A = 500 W. Only the resistance actually consumes power, however. The power in the resistance is:

$P = I^2\,R = (2.00\ \text{A})^2 \times 75.0\ \text{V} = 300\ \text{W}$

The ratio of the consumed power to the apparent power is called the power factor of the circuit.

$$PF = \frac{P_{consumed}}{P_{apparent}} = \frac{R}{Z} \tag{87}$$

In this example the power factor would be 300 W / 500 W = 0.600. Power factor is frequently expressed as a percentage; in this case, 60%. An equivalent definition of power factor is:

$$PF = \cos\theta$$

Where θ is the phase angle. Since the phase angle equals:

$$\theta = \arctan\left(\frac{X}{R}\right) = \arctan\left(\frac{100\ \Omega}{75.0\ \Omega}\right)$$
$$= \arctan(1.33) = 53.1°$$

Then the power factor is:

$$PF = \cos 53.1° = 0.600$$

as the earlier calculation confirms.

Real, or dissipated, power is measured in watts. Apparent power, to distinguish it from real power, is measured in volt-amperes (VA). It is simply the product of the voltage across and the current through an overall impedance. It has no direct relationship to the power actually dissipated unless the power factor of the circuit is known. The power factor of a purely resistive circuit is 100% or 1, while the power factor of a pure reactance is zero. In this illustration, the reactive power is:

$$VAR = I^2\,X_L = (2.00\ \text{A})^2 \times 100\ \Omega$$
$$= 400\ \text{VA}$$

Since power factor is always rendered as a positive number, the value must be followed by the words "leading" or "lagging" to identify the phase of the voltage with respect to the current. Specifying the numerical power factor is not always sufficient. For example, many dc-to-ac power inverters can safely operate loads having a large net reactance of one sign but only a small reactance of the opposite sign. Hence, the final calculation of the power factor in this example yields the value 0.600, leading.

Resonant Circuits

A circuit containing both an inductor and a capacitor — and therefore, both inductive and capacitive reactance — is often called a *tuned circuit*. There is a particular frequency at which the inductive and capacitive reactances are the same, that is, $X_L = X_C$. For most purposes, this is the *resonant frequency* of the circuit. (Special considerations apply to parallel circuits; they will emerge in the section devoted to such circuits.) At the resonant frequency — or at resonance, for short:

$$X_L = 2\,\pi\,f\,L = X_C = \frac{1}{2\,\pi\,f\,C}$$

By solving for f, we can find the resonant frequency of any combination of inductor and capacitor from the formula:

$$f = \frac{1}{2\,\pi\,\sqrt{L\,C}} \tag{88}$$

where:
f = frequency in hertz (Hz),
L = inductance in henrys (H),
C = capacitance in farads (F), and
$\pi = 3.1416$.

For most high-frequency (HF) radio work, smaller units of inductance and capacitance and larger units of frequency are more convenient. The basic formula becomes:

$$f = \frac{10^3}{2\,\pi\,\sqrt{L\,C}} \tag{89}$$

where:
f = frequency in megahertz (MHz),
L = inductance in microhenrys (μH),
C = capacitance in picofarads (pF), and
$\pi = 3.1416$.

Example: What is the resonant frequency of a circuit containing an inductor of 5.0 μH and a capacitor of 35 pF?

$$f = \frac{10^3}{2\,\pi\,\sqrt{L\,C}} = \frac{10^3}{6.2832 \times \sqrt{5.0 \times 35}}$$
$$= \frac{10^3}{83} = 12\ \text{MHz}$$

To find the matching component (inductor or capacitor) when the frequency and one component is known (capacitor or inductor) for general HF work, use the formula:

$$f^2 = \frac{1}{4\,\pi^2\,L\,C} \tag{90}$$

where F, L and C are in basic units. For HF work in terms of MHz, μH and pF, the basic relationship rearranges to these handy formulas:

$$L = \frac{25330}{f^2\,C} \tag{91}$$

$$C = \frac{25330}{f^2\,L} \tag{92}$$

where:
f = frequency in MHz,
L = inductance in μH, and
C = capacitance in pF.

Example: What value of capacitance is needed to create a resonant circuit at 21.1 MHz, if the inductor is 2.00 μH?

$$C = \frac{25330}{f^2\,L} = \frac{25330}{(21.1^2 \times 2.00)}$$
$$= \frac{25330}{890.} = 28.5\ \text{pF}$$

For most radio work, these formulas will permit calculations of frequency and component values well within the limits of component tolerances. Resonant circuits have other properties of importance, in addition to the resonant frequency, however. These include impedance, voltage drop across components in series-resonant circuits, circulating current in parallel-resonant circuits, and bandwidth. These properties determine such factors as the selectivity of a tuned circuit and the component ratings for circuits handling considerable power. Although the basic determination of the tuned-circuit resonant frequency ignored any resistance in the circuit, that resistance will play a vital role in the circuit's other characteristics.

SERIES-RESONANT CIRCUITS

Fig 6.63 presents a basic schematic diagram of a *series-resonant circuit*. Although most schematic diagrams of radio circuits would show only the inductor and the capacitor, resistance is always present in such circuits. The most notable resistance is associated with losses in the inductor at HF; resistive losses in the capacitor are low enough at those frequencies to be ignored. The current meter shown in the circuit is a reminder that in series circuits, the same current flows through all elements.

At resonance, the reactance of the capacitor cancels the reactance of the inductor. The voltage and current are in phase with each other, and the impedance of the circuit is determined solely by the resistance. The actual current through the circuit at resonance, and for frequencies near resonance, is determined by the formula:

$$I = \frac{E}{Z} = \frac{E}{\sqrt{R^2 + \left[2\pi f L - \frac{1}{(2\pi f C)}\right]^2}} \quad (93)$$

where all values are in basic units.

At resonance, the reactive factor in the formula is zero. As the frequency is shifted above or below the resonant frequency without altering component values, however, the reactive factor becomes significant, and the value of the current becomes smaller than at resonance. At frequencies far from resonance, the reactive components become dominant, and the resistance no longer significantly affects the current amplitude.

The exact curve created by recording the current as the frequency changes depends on the ratio of reactance to resistance. When the reactance of either the coil or capacitor is of the same order of magnitude as the resistance, the current decreases rather slowly as the frequency is moved in either direction away from resonance. Such a curve is said to be *broad*. Conversely, when the reactance is considerably larger than the resistance, the current decreases rapidly as the frequency moves away from resonance, and the circuit is said to be *sharp*. A sharp circuit will respond a great deal more readily to the resonant frequency than to frequencies quite close to resonance; a broad circuit will respond almost equally well to a group or band of frequencies centered around the resonant frequency.

Both types of resonance curves are useful. A sharp circuit gives good selectivity — the ability to respond strongly (in terms of current amplitude) at one desired frequency and to discriminate against others. A broad circuit is used when the apparatus must give

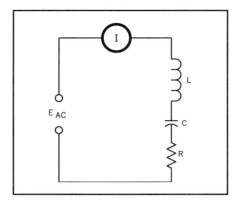

Fig 6.63 — A series circuit containing L, C, and R is resonant at the applied frequency when the reactance of C is equal to the reactance of L. The I in the circle is the schematic symbol for an ammeter.

about the same response over a band of frequencies, rather than at a single frequency alone.

Fig 6.64 presents a family of curves, showing the decrease in current as the frequency deviates from resonance. In each case, the reactance is assumed to be 1000 Ω. The maximum current, shown as a relative value on the graph, occurs with the lowest resistance, while the lowest peak current occurs with the highest resistance. Equally important, the rate at which the current decreases from its maximum value also changes with the ratio of reactance to resistance. It decreases most rapidly when the ratio is high and most slowly when the ratio is low.

Q

As noted in earlier sections of this chapter, the ratio of reactance or stored energy to resistance or consumed energy is Q. Since both terms of the ratio are measured in ohms, Q has no units and is variously known as the *quality factor*, the *figure of merit* or the *multiplying factor*. Since the resistive losses of the coil dominate the energy consumption in HF series-resonant circuits, the inductor Q largely determines the resonant-circuit Q. Since this value of Q is independent of any external load to which the circuit might transfer power, it is called the *unloaded Q* or Q_U of the circuit.

Example: What is the unloaded Q of a series-resonant circuit with a loss resistance of 5 Ω and inductive and capacitive components having a reactance of 500 Ω each? With a reactance of 50 Ω each?

$$Q_{U1} = \frac{X1}{R} = \frac{500\ \Omega}{5\ \Omega} = 100$$

$$Q_{U2} = \frac{X2}{R} = \frac{50\ \Omega}{5\ \Omega} = 10$$

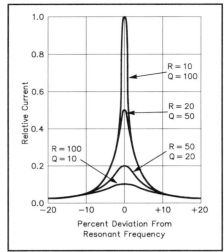

Fig 6.64 — Current in series-resonant circuits with various values of series resistance and Q. The current values are relative to an arbitrary maximum of 1.0. The reactance for all curves is 1000 Ω. Note that the current is hardly affected by the resistance in the circuit at frequencies more than 10% away from the resonant frequency.

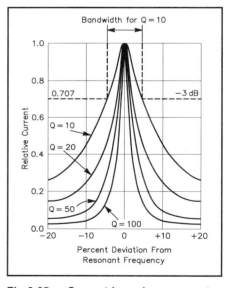

Fig 6.65 — Current in series-resonant circuits having different values of Q_U. The current at resonance is set at the same level for all curves in order to show the rate of change of decrease in current for each value of Q_U. The half-power points are shown to indicate relative bandwidth of the response for each curve. The bandwidth is indicated for a circuit with a Q_U of 10.

Bandwidth

Fig 6.65 is an alternative way of drawing the family of curves that relate current to frequency for a series-resonant circuit. By assuming that the peak current of each curve

Table 6.6
The Selectivity of Resonant Circuits

Approximate percentage of current at resonance[1] or of impedance at resonance[2]	Bandwidth (between half-power or –3 dB points on response curve)	Series circuit current phase angle (degrees)
95	f / 3Q	18.5
90	f / 2Q	26.5
70.7	f / Q	45
44.7	2f / Q	63.5
24.2	4f / Q	76
12.4	8f / Q	83

[1]For a series resonant circuit
[2]For a parallel resonant circuit

is the same, the rate of change of current for various values of Q_U and the associated ratios of reactance to resistance are more easily compared. From the curves, it is evident that the lower Q_U circuits pass frequencies over a greater *bandwidth* of frequencies than the circuits with a higher Q_U. For the purpose of comparing tuned circuits, bandwidth is often defined as the frequency spread between the two frequencies at which the current amplitude decreases to 0.707 (or $1/\sqrt{2}$) times the maximum value. Since the power consumed by the resistance, R, is proportional to the square of the current, the power at these points is half the maximum power at resonance, assuming that R is constant for the calculations. The half-power, or –3 dB, points are marked on Fig 6.65.

For Q values of 10 or greater, the curves shown in Fig 6.65 are approximately symmetrical. On this assumption, bandwidth (BW) can be easily calculated:

$$BW = \frac{f}{Q_U} \qquad (94)$$

where BW and f are in the same units, that is, in Hz, kHz or MHz.

Example: What is the bandwidth of a series-resonant circuit operating at 14 MHz with a Q_U of 100?

$$BW = \frac{f}{Q_U} = \frac{14 \text{ MHz}}{100} = 0.14 \text{ MHz} = 140 \text{ kHz}$$

The relationship between Q_U, f and BW provides a means of determining the value of circuit Q when inductor losses may be difficult to measure. By constructing the series-resonant circuit and measuring the current as the frequency varies above and below resonance, the half-power points can be determined. Then:

$$Q_U = \frac{f}{BW} \qquad (95)$$

Example: What is the Q_U of a series-resonant circuit operating at 3.75 MHz, if

the bandwidth is 375 kHz?

$$Q_U = \frac{f}{BW} = \frac{3.75 \text{ MHz}}{0.375 \text{ MHz}} = 10.0$$

Table 6.6 provides some simple formulas for estimating the maximum current and phase angle for various bandwidths, if both f and Q_U are known.

Voltage Drop Across Components

The voltage drop across the coil and across the capacitor in a series-resonant circuit are each proportional to the reactance of the component for a given current (since E = I X). These voltages may be many times the source voltage for a high-Q circuit. In fact, at resonance, the voltage drop is:

$$E_X = Q_U E \qquad (96)$$

where:
E_X = the voltage across the reactive component,
Q_U = the circuit unloaded Q, and
E = the source voltage.

(Note that the voltage drop across the inductor is the vector sum of the voltages across the resistance and the reactance; however, for Qs greater than 10, the error created by using equation 96 is not ordinarily significant.) Since the calculated value of E_X is the RMS voltage, the peak voltage will be higher by a factor of 1.414. Antenna couplers and other high-Q circuits handling significant power may experience arcing from high values of E_X, even though the source voltage to the circuit is well within component ratings.

Capacitor Losses

Although capacitor energy losses tend to be insignificant compared to inductor losses up to about 30 MHz, the losses may affect circuit Q in the VHF range. Leakage resistance, principally in the solid dielectric that forms the insulating support for the capacitor plates, is not exactly like the wire resis-

tance losses in a coil. Instead of forming a series resistance, capacitor leakage usually forms a parallel resistance with the capacitive reactance. If the leakage resistance of a capacitor is significant enough to affect the Q of a series-resonant circuit, the parallel resistance must be converted to an equivalent series resistance before adding it to the inductor's resistance.

$$R_S = \frac{X_C^2}{R_P} = \frac{1}{R_P \times (2 \pi f C)^2} \qquad (97)$$

Example: A 10.0 pF capacitor has a leakage resistance of 10000 Ω at 50.0 MHz. What is the equivalent series resistance?

$$R_S = \frac{1}{R_P \times (2 \pi f C)^2}$$

$$= \frac{1}{1.00 \times 10^4 \times (6.283 \times 50.0 \times 10^6 \times 10.0 \times 10^{-12})^2}$$

$$R_S = \frac{1}{1.00 \times 10^4 \times 9.87 \times 10^{-6}}$$

$$= \frac{1}{0.0987} = 10.1 \text{ Ω}$$

In calculating the impedance, current and bandwidth for a series-resonant circuit in which this capacitor might be used, the series-equivalent resistance of the unit is added to the loss resistance of the coil. Since inductor losses tend to increase with frequency because of skin effect, the combined losses in the capacitor and the inductor can seriously reduce circuit Q, without special component- and circuit-construction techniques.

PARALLEL-RESONANT CIRCUITS

Although series-resonant circuits are common, the vast majority of resonant circuits used in radio work are *parallel-resonant circuits*. **Fig 6.66** represents a typical HF parallel-resonant circuit. As is

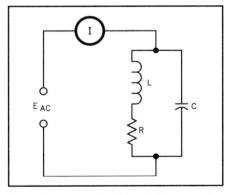

Fig 6.66 — A typical parallel-resonant circuit, with the resistance shown in series with the inductive leg of the circuit. Below a Q_U of 10, resonance definitions may lead to three separate frequencies which converge at higher Q_U levels. See text.

the case for series-resonant circuits, the inductor is the chief source of resistive losses, and these losses appear in series with the coil. Because current through parallel-resonant circuits is lowest at resonance, and impedance is highest, they are sometimes called *antiresonant* circuits. Likewise, the names *acceptor* and *rejector* are occasionally applied to series- and parallel-resonant circuits, respectively.

Because the conditions in the two legs of the parallel circuit in Fig 6.66 are not the same — the resistance is in only one of the legs — all of the conditions by which series resonance is determined do not occur simultaneously in a parallel-resonant circuit. **Fig 6.67** graphically illustrates the situation by showing the currents through the two components. When the inductive and capacitive reactances are identical, the condition defined for series resonance is met as shown in line (a). The impedance of the inductive leg is composed of both X_L and R, which yields an impedance that is greater than X_C and that is not 180° out of phase with X_C. The resultant current is greater than its minimum possible value and not in phase with the voltage.

By altering the value of the inductor slightly (and holding the Q constant), a new frequency can be obtained at which the current reaches its minimum. When parallel circuits are tuned using a current meter as an indicator, this point (b) is ordinarily used as an indication of resonance. The current "dip" indicates a condition of maximum impedance and is sometimes called the *antiresonant* point or *maximum impedance resonance* to distinguish it from the condition where $X_C = X_L$. Maximum impedance is achieved by vector addition of X_C, X_L and R, however, and the result is a current somewhat out of phase with the voltage.

Point (c) on the curve represents the *unity-power-factor* resonant point. Adjusting the inductor value and hence its reactance (while holding Q constant) produces a new resonant frequency at which the resultant current is in phase with the voltage. The inductor's new value of reactance is the value required for a parallel-equivalent inductor and its parallel-equivalent resistor (calculated according to the formulas in the last section) to just cancel the capacitive reactance. The value of the parallel-equivalent inductor is always smaller than the actual inductor in series with the resistor and has a proportionally smaller reactance. (The parallel-equivalent resistor, conversely, will always be larger than the coil-loss resistor shown in series with the inductor.) The result is a resonant frequency slightly different from the one for minimum current

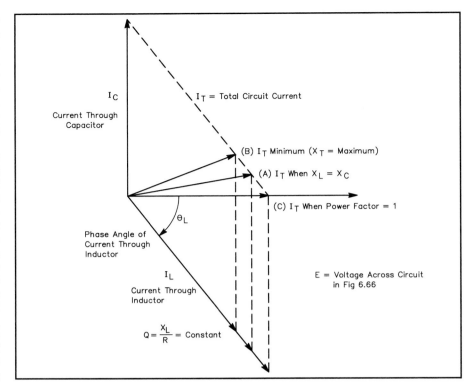

Fig 6.67 — Resonant conditions for a low-Q_U parallel circuit. Resonance may be defined as (a) $X_L = X_C$, (b) minimum current flow and maximum impedance or (c) voltage and current in phase with each other. With the circuit of Fig 6.66 and a Q_U of less than 10, these three definitions may represent three distinct frequencies.

and the one for $X_L = X_C$.

The points shown in the graph in Fig 6.67 represent only one of many possible situations, and the relative positions of the three resonant points do not hold for all possible cases. Moreover, specific circuit designs can draw some of the resonant points together, for example, compensating for the resistance of the coil by retuning the capacitor. The differences among these resonances are significant for circuit Qs below 10, where the inductor's series resistance is a significant percentage of the reactance. Above a Q of 10, the three points converge to within a percent of the frequency and can be ignored for practical calculations. Tuning for minimum current will not introduce a sufficiently large phase angle between voltage and current to create circuit difficulties.

Parallel Circuits of Moderate to High Q

The resonant frequencies defined above converge in parallel-resonant circuits with Qs higher than about 10. Therefore, a single set of formulas will sufficiently approximate circuit performance for accurate predictions. Indeed, above a Q of 10, the performance of a parallel circuit appears in many ways to be simply the inverse of the performance of a series-

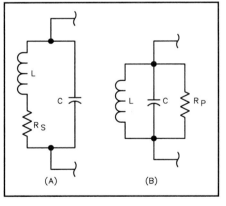

Fig 6.68 — Series and parallel equivalents when both circuits are resonant. The series resistance, R_S in A, is replaced by the parallel resistance, R_P in B, and vice versa. $R_P = X_L^2 / R_S$.

resonant circuit using the same components.

Accurate analysis of a parallel-resonant circuit requires the substitution of a parallel-equivalent resistor for the actual inductor-loss series resistor, as shown in **Fig 6.68**. Sometimes called the *dynamic resistance* of the parallel-resonant circuit, the parallel-equivalent resistor value will increase with circuit Q, that is, as the series resistance value decreases. To calculate the approximate parallel-equivalent resistance, use the formula:

$$R_P = \frac{X_L^2}{R_S} = \frac{(2 \pi f L)^2}{R_S} = Q_U X_L \qquad (98)$$

Example: What is the parallel-equivalent resistance for a coil with an inductive reactance of 350 Ω and a series resistance of 5.0 Ω at resonance?

$$R_P = \frac{X_L^2}{R_S} = \frac{(350 \, \Omega)^2}{5.0 \, \Omega}$$

$$= \frac{122{,}500 \, \Omega^2}{5.0 \, \Omega} = 24{,}500 \, \Omega$$

Since the coil Q_U remains the inductor's reactance divided by its series resistance, the coil Q_U is 70. Multiplying Q_U by the reactance also provides the approximate parallel-equivalent resistance of the coil series resistance.

At resonance, where $X_L = X_C$, R_P defines the impedance of the parallel-resonant circuit. The reactances just equal each other, leaving the voltage and current in phase with each other. In other words, the circuit shows only the parallel resistance. Therefore, equation 98 can be rewritten as:

$$Z = \frac{X_L^2}{R_S} = \frac{(2 \pi f L)^2}{R_S} = Q_U X_L \qquad (99)$$

In this example, the circuit impedance at resonance is 24,500 Ω.

At frequencies below resonance the current through the inductor is larger than that through the capacitor, because the reactance of the coil is smaller and that of the capacitor is larger than at resonance. There is only partial cancellation of the two reactive currents, and the line current therefore is larger than the current taken by the resistance alone. At frequencies above resonance the situation is reversed and more current flows through the capacitor than through the inductor, so the line current again increases. The current at resonance, being determined wholly by R_P, will be small if R_P is large, and large if R_P is small. **Fig 6.69** illustrates the relative current flows through a parallel-tuned circuit as the frequency is moved from below resonance to above resonance. The base line represents the minimum current level for the particular circuit. The actual current at any frequency off resonance is simply the vector sum of the currents through the parallel equivalent resistance and through the reactive components.

To obtain the impedance of a parallel-tuned circuit either at or off the resonant frequency, apply the general formula:

$$Z = \frac{Z_C \, Z_L}{Z_S} \qquad (100)$$

where:

Z = overall circuit impedance

Z_C = impedance of the capacitive leg (usually, the reactance of the capacitor),

Z_L = impedance of the inductive leg (the vector sum of the coil's reactance and resistance), and

Z_S = series impedance of the capacitor-inductor combination as derived from the denominator of equation 93.

After using vector calculations to obtain Z_L and Z_S, converting all the values to polar form — as described earlier in this chapter — will ease the final calculation. Of course, each impedance may be derived from the resistance and the application of the basic reactance formulas on the values of the inductor and capacitor at the frequency of interest.

Since the current rises off resonance, the parallel-resonant-circuit impedance must fall. It also becomes complex, resulting in an ever greater phase difference between the voltage and the current. The rate at which the impedance falls is a function of Q_U. **Fig 6.70** presents a family of curves showing the impedance drop from resonance for circuit Qs ranging from 10 to 100. The curve family for parallel-circuit

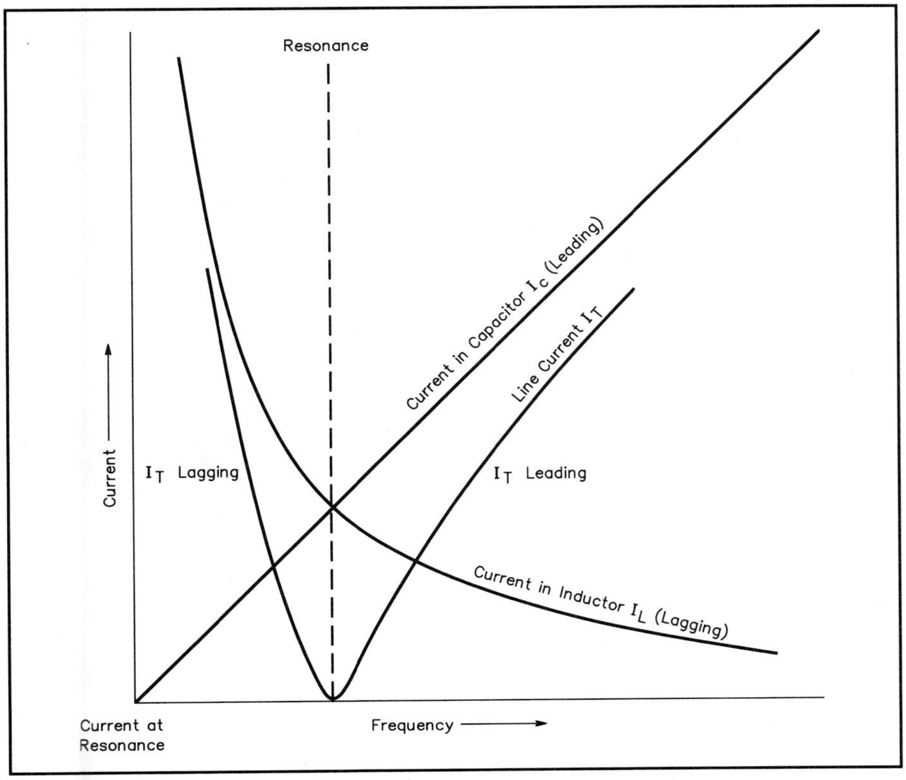

Fig 6.69 — The currents in a parallel-resonant circuit as the frequency moves through resonance. Below resonance, the current lags the voltage; above resonance the current leads the voltage. The base line represents the current level at resonance, which depends on the impedance of the circuit at that frequency.

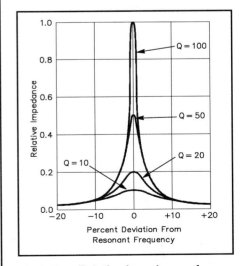

Fig 6.70 — Relative impedance of parallel-resonant circuits with different values of Q_U. The curves are similar to the series-resonant circuit current level curves of Fig 6.64. The effect of Q_U on impedance is most pronounced within 10% of the resonance frequency.

impedance is essentially the same as the curve family for series-circuit current.

As with series tuned circuits, the higher the Q of a parallel-tuned circuit, the sharper the response peak. Likewise, the lower the Q, the wider the band of frequencies to which the circuit responds. Using the half-power (−3 dB) points as a comparative measure of circuit performance, equations 94 and 95 apply equally to parallel-tuned circuits. That is, $BW = f / Q_U$ and $Q_U = f / BW$, where the resonant frequency and the bandwidth are in the same units. As a handy reminder, **Table 6.7** summarizes the performance of parallel-resonant circuits at high and low Qs and above and below resonant frequency.

It is possible to use either series or parallel-resonant circuits do the same work in many circuits, thus giving the designer considerable flexibility. **Fig 6.71** illustrates this general principle by showing a series-resonant circuit in the signal path and a parallel-resonant circuit shunted from the signal path to ground. Assume both circuits are resonant at the same frequency, f, and have the same Q. The series tuned circuit at A has its lowest impedance at f, permitting the maximum possible current to flow along the signal path. At all other frequencies, the impedance is greater and the current at those frequencies is less. The circuit passes the desired signal and tends to impede signals at undesired frequencies. The parallel circuit at B provides the highest impedance at resonance, f, making the signal path the lowest impedance path for the signal. At frequencies off resonance, the parallel-resonant circuit presents a lower impedance, thus presenting signals with a path to ground and away from the signal path. In theory, the effects will be the same relative to a signal current on the signal path. In actual circuit design exercises, of course, many other variables will enter the design picture to make one circuit preferable to the other.

Circulating Current

In a parallel-resonant circuit, the source voltage is the same for all the circuit elements. The current in each element, however, is a function of the element's reactance. **Fig 6.72** redraws the parallel-tuned circuit to indicate the line current and the current circulating between the coil and the capacitor. The current drawn from the source may be low, because the overall circuit impedance is high. The current through the individual elements may be high, how-

Table 6.7
The Performance of Parallel-Resonant Circuits

A. High and Low Q Circuits (in relative terms)

Characteristic	High Q Circuit	Low Q Circuit
Selectivity	high	low
Bandwidth	narrow	wide
Impedance	high	low
Line current	low	high
Circulating current	high	low

B. Off-Resonance Performance for Constant Values of Inductance and Capacitance

Characteristic	Above Resonance	Below Resonance
Inductive reactance	increases	decreases
Capacitive reactance	decreases	increases
Circuit resistance	unchanged*	unchanged*
Circuit impedance	decreases	decreases
Line current	increases	increases
Circulating current	decreases	decreases
Circuit behavior	capacitive	inductive

*This is true for frequencies near resonance. At distant frequencies, skin effect may alter the resistive losses of the inductor.

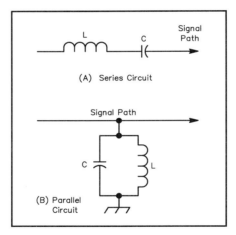

Fig 6.71 — Series- and parallel-resonant circuits configured to perform the same theoretical task: passing signals in a narrow band of frequencies along the signal path. A real design example would consider many other factors.

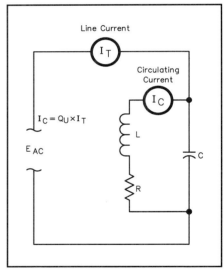

Fig 6.72 — A parallel-resonant circuit redrawn to illustrate both the line current and the circulating current.

ever, because there is little resistive loss as the current circulates through the inductor and capacitor. For parallel-resonant circuits with an unloaded Q of 10 or greater, this *circulating current* is approximately:

$$I_C = Q_U I_T \qquad (101)$$

where:

I_C = circulating current in A, mA or A,
Q_U = unloaded circuit Q, and
I_T = line current in the same units as I_C.

Example: A parallel-resonant circuit permits an ac or RF line current of 30 mA and has a Q of 100. What is the circulating current through the elements?

$$I_X = Q_U I = 100 \times 30 \text{ mA} = 3000 \text{ mA}$$
$$= 3 \text{ A}$$

Circulating currents in high-Q parallel-tuned circuits can reach a level that causes component heating and power loss. Therefore, components should be rated for the anticipated circulating currents, and not just the line current.

The Q of Loaded Circuits

In many resonant-circuit applications, the only power lost is that dissipated in the resistance of the circuit itself. At frequencies below 30 MHz, most of this resistance is in the coil. Within limits, increasing the number of turns in the coil increases the reactance faster than it raises the resistance, so coils for circuits in which the Q must be high are made with relatively large inductances for the frequency.

When the circuit delivers energy to a load (as in the case of the resonant circuits used in transmitters), the energy consumed in the circuit itself is usually negligible compared with that consumed by the load. The equivalent of such a circuit is shown in **Fig 6.73**, where the parallel resistor, R_L, represents the load to which power is delivered. If the power dissipated in the load is at least 10 times as great as the power lost in the inductor and capacitor, the parallel impedance of the resonant circuit itself will be so high compared with the resistance of the load that for all practical purposes the impedance of the combined circuit is equal to the load imped-

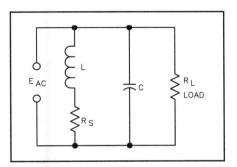

Fig 6.73 — A loaded parallel-resonant circuit, showing both the inductor-loss resistance and the load, R_L. If smaller than the inductor resistance, R_L will control the loaded Q of the circuit (Q_L).

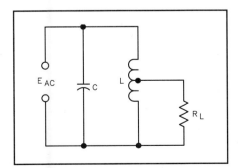

Fig 6.74 — A parallel-resonant circuit with a tapped coil to effect an impedance match. Although the impedance presented by the entire circuit is very high, the impedance "seen" by the load, R_L, is lower.

ance. Under these conditions, the load resistance replaces the circuit impedance in calculating Q. The Q of a parallel-resonant circuit loaded by a resistive impedance is:

$$Q_L = \frac{R_L}{X} \qquad (102)$$

where:
Q_L = circuit loaded Q,
R_L = parallel load resistance in ohms, and
X = reactance in ohms of either the inductor or the capacitor.

Example: A resistive load of 3000 Ω is connected across a resonant circuit in which the inductive and capacitive reactances are each 250 Ω. What is the circuit Q?

$$Q_L = \frac{R_L}{X} = \frac{3000 \ \Omega}{250 \ \Omega} = 12$$

The effective Q of a circuit loaded by a parallel resistance increases when the reactances are decreased. A circuit loaded with a relatively low resistance (a few thousand ohms) must have low-reactance elements (large capacitance and small inductance) to have reasonably high Q. Many power-handling circuits, such as the output networks of transmitters, are designed by first choosing a loaded Q for the circuit and then determining component values. See the **Amplifiers** chapter for more details.

Parallel load resistors are sometimes added to parallel-resonant circuits to lower the circuit Q and increase the circuit bandwidth. By using a high-Q circuit and adding a parallel resistor, designers can tailor the circuit response to their needs. Since the parallel resistor consumes power, such techniques ordinarily apply to receiver and similar low-power circuits, however.

Example: Specifications call for a parallel-resonant circuit with a bandwidth of 400. kHz at 14.0 MHz. The circuit at hand has a Q_U of 70.0 and its components have reactances of 350 Ω each. What is the parallel load resistor that will increase the bandwidth to the specified value? The bandwidth of the existing circuit is:

$$BW = \frac{f}{Q_U} = \frac{14.0 \ \text{MHz}}{70.0} = 0.200 \ \text{MHz} = 200 \ \text{kHz}$$

The desired bandwidth, 400 kHz, requires a circuit with a Q of:

$$Q = \frac{f}{BW} = \frac{14.0 \ \text{MHz}}{0.400 \ \text{MHz}} = 35.0$$

Since the desired Q is half the original value, halving the resonant impedance or parallel-resistance value of the circuit is

in order. The present impedance of the circuit is:

$$Z = Q_U \ X_L = 70.0 \times 350 \ \Omega = 24500 \ \Omega$$

The desired impedance is:

$$Z = Q_U \ X_L = 35.0 \times 350 \ \Omega = 12{,}250 \ \Omega$$
$$= 12.25 \ \text{k}\Omega$$

or half the present impedance.

A parallel resistor of 24500 Ω , or the nearest lower value (to guarantee sufficient bandwidth), will produce the required reduction in Q and bandwidth increase. Although this example simplifies the situation encountered in real design cases by ignoring such factors as the shape of the band-pass curve, it illustrates the interaction of the ingredients that determine the performance of parallel-resonant circuits.

Impedance Transformation

An important application of the parallel-resonant circuit is as an impedance matching device in the output circuit of an RF power amplifier. There is an optimum value of load resistance for each type of tube or transistor and each set of required operating conditions. The resistance of the load to which the active device delivers power may be considerably lower than the value required for proper device operation, or the load impedance may be considerably higher than the amplifier output impedance.

To transform the actual load resistance to the desired value, the load may be tapped across part of the coil, as shown in **Fig 6.74**. This is equivalent to connecting a higher value of load resistance across the whole circuit, and is similar in principle to impedance transformation with an iron-core transformer (described in the next section of this chapter). In high-frequency resonant circuits, the impedance ratio does not vary exactly as the square of the turns ratio, because all the magnetic flux lines do not cut every turn of the coil. A desired impedance ratio usually must be obtained by experimental adjustment.

When the load resistance has a very low value (say below 100 Ω) it may be connected in series in the resonant circuit (R_S in Fig 6.68A, for example), in which case it is transformed to an equivalent parallel impedance as previously described. If the Q is at least 10, the equivalent parallel impedance is:

$$Z_R = \frac{X^2}{R_L} \qquad (103)$$

where:
Z_R = resistive parallel impedance at resonance,

X = reactance (in ohms) of either the coil or the capacitor, and

R_L = load resistance inserted in series.

If the Q is lower than 10, the reactance will have to be adjusted somewhat — for the reasons given in the discussion of low-Q circuits — to obtain a resistive impedance of the desired value.

Networks like the one in Fig 6.74 have some serious disadvantages for some applications. For instance, the common connection between the input and the output provides no dc isolation. Also, the common ground is sometimes troublesome with regard to ground-loop currents. Consequently, a network with only mutual magnetic coupling is often preferable. With the advent of ferrites, constructing impedance transformers that are both broadband and permit operation well up into the VHF portion of the spectrum has become relatively easy. The basic principles of broadband impedance transformers appear in the following section.

Transformers

When the ac source current flows through every turn of an inductor, the generation of a counter-voltage and the storage of energy during each half cycle is said to be by virtue of *self-inductance*. If another inductor — not connected to the source of the original current — is positioned so the expanding and contracting magnetic field of the first inductor cuts across its turns, a current will be induced into the second coil. A load such as a resistor may be connected across the second coil to consume the energy transferred magnetically from the first inductor. This phenomenon is called *mutual inductance*.

Two inductors positioned so that the magnetic field of one (the *primary* inductor) induces a current in the other (the *secondary* inductor) are *coupled*. **Fig 6.75** illustrates a pair of coupled inductors, showing an ac energy source connected to one and a load connected to the other. If the coils are wound tightly on an iron core so that nearly all the lines of force or magnetic flux from the first coil link with the turns of the second coil, the pair is said to be tightly coupled. Coils with air cores separated by a distance would be loosely coupled. The signal source for the primary inductor may be household ac power lines, audio or other waveforms at lower frequencies, or RF currents. The load may be a device needing power, a speaker converting electrical energy into sonic energy, an antenna using RF energy for communications or a particular circuit set up to process a signal from a preceding circuit. The uses of magnetically-coupled energy in electronics are innumerable.

Mutual inductance (M) between coils is measured in henrys. Two coils have a mutual inductance of 1 H under the following conditions: as the primary inductor current changes at a rate of 1 A/s, the voltage across the secondary inductor is 1 V. The level of mutual inductance varies with many factors: the size and shape of the inductors, their relative positions and distance from each other,

and the permeability of the inductor core material and of the space between them.

If the self-inductance values of two coils are known, it is possible to derive the mutual inductance by way of a simple experiment schematically represented in **Fig 6.76**. Without altering the physical setting or position of two coils, measure the inductance of the series-connected coils with their windings complementing each other and again with their windings opposing each other. Since, for the two coils, $L_C = L1 + L2 + 2M$, in the complementary case, and $L_O = L1 + L2 - 2M$ for the opposing case,

$$M = \frac{L_C - L_O}{4} \qquad (104)$$

The ratio of magnetic flux set up by the secondary coil to the flux set up by the primary coil is a measure of the extent to which two coils are coupled, compared to the maximum possible coupling between them. This ratio is the *coefficient of coupling* (k) and is always less than 1. If k were to equal 1, the two coils would have the maximum possible mutual coupling. Thus:

$$M = k\sqrt{L1\ L2} \qquad (105)$$

where:

M = mutual inductance in henrys,

L1 and L2 = individual coupled inductors, each in henrys, and

k = the coefficient of coupling.

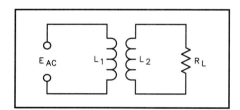

Fig 6.75 — A basic transformer: two inductors — one connected to an ac energy source, the other to a load — with coupled magnetic fields.

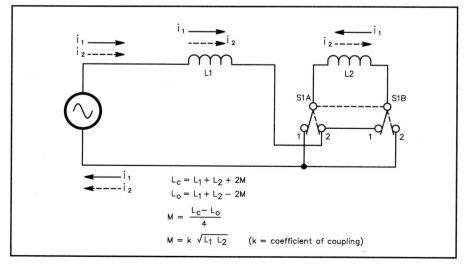

Fig 6.76 — An experimental setup for determining mutual inductance. Measure the inductance with the switch in each position and use the formula in the text to determine the mutual inductance.

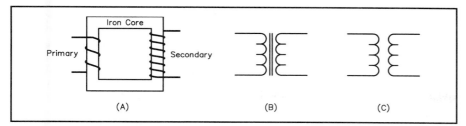

Fig 6.77 — A transformer. A is a pictorial diagram. Power is transferred from the primary coil to the secondary by means of the magnetic field. B is a schematic diagram of an iron-core transformer, and C is an air-core transformer.

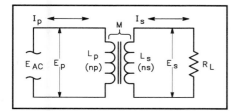

Fig 6.78 — The conditions for transformer action: two coils that exhibit mutual inductance, an ac power source, and a load. The magnetic field set up by the energy in the primary circuit transfers energy to the secondary for use by the load, resulting in a secondary voltage and current.

Using the experiment above, it is possible to solve equation 105 for k with reasonable accuracy.

Any two coils having mutual inductance comprise a *transformer* having a *primary winding* or inductor and a *secondary winding* or inductor. **Fig 6.77** provides a pictorial representation of a typical iron-core transformer, along with the schematic symbols for both iron-core and air-core transformers. Conventionally, the term *transformer* is most commonly applied to coupled inductors having a magnetic core material, while coupled air-wound inductors are not called by that name. They are still transformers, however.

We normally think of transformers as ac devices, since mutual inductance only occurs when magnetic fields are expanding or contracting. A transformer connected to a dc source will exhibit mutual inductance only at the instants of closing and opening the primary circuit, or on the rising and falling edges of dc pulses, because only then does the primary winding have a changing field. The principle uses of transformers are three: to physically isolate the primary circuit from the secondary circuit, to transform voltages and currents from

one level to another, and to transform circuit impedances from one level to another. These functions are not mutually exclusive and have many variations.

IRON-CORE TRANSFORMERS

The primary and secondary coils of a transformer may be wound on a core of magnetic material. The permeability of the magnetic material increases the inductance of the coils so a relatively small number of turns may be used to induce a given voltage value with a small current. A closed core having a continuous magnetic path, such as that shown in Fig 6.77, also tends to ensure that practically all of the field set up by the current in the primary coil will cut the turns of the secondary coil. For power transformers and impedance-matching transformers used in audio work, cores of iron strips are most common and generally very efficient.

The following principles presume a coefficient of coupling (k) of 1, that is, a perfect transformer. The value, k = 1, indicates that all the turns of both coils link with all the magnetic flux lines, so that the voltage induced per turn is the same with both coils. This condition makes the induced voltage independent of the inductance of the primary and secondary inductors. Iron-core transformers for low frequencies most closely approach this ideal condition. **Fig 6.78** illustrates the conditions for transformer action.

Voltage Ratio

For a given varying magnetic field, the voltage induced in a coil within the field is proportional to the number of turns in the coil. When the two coils of a transformer are in the same field (which is the case when both are wound on the same closed core), it follows that the induced voltages will be proportional to the number of turns in each coil. In the primary, the induced voltage practically equals, and opposes, the applied voltage, as described earlier. Hence:

$$E_S = E_P \left(\frac{N_S}{N_P} \right) \qquad (106)$$

where:
 E_S = secondary voltage,
 E_P = primary applied voltage,
 N_S = number of turns on secondary, and
 N_P = number of turns on primary.

Example: A transformer has a primary of 400 turns and a secondary of 2800 turns, and a voltage of 120 V is applied to the primary. What voltage appears across the secondary winding?

$$E_S = 120 \text{ V} \left(\frac{2800}{400} \right) = 120 \text{ V} \times 7 = 840 \text{ V}$$

(Notice that the number of turns is taken as a known value rather than a measured quantity, so they do not limit the significant figures in the calculation.) Also, if 840 V is applied to the 2800-turn winding (which then becomes the primary), the output voltage from the 400-turn winding will be 120 V.

Either winding of a transformer can be used as the primary, provided the winding has enough turns (enough inductance) to induce a voltage equal to the applied voltage without requiring an excessive current. The windings must also have insulation with a voltage rating sufficient for the voltage present.

Current or Ampere-Turns Ratio

The current in the primary when no current is taken from the secondary is called the *magnetizing current* of the transformer. An ideal transformer, with no internal losses, would consume no power, since the current through the primary inductor would be 90° out of phase with the voltage. In any properly designed transformer, the power consumed by the transformer when the secondary is open (not delivering power) is only the amount necessary to overcome the losses in the iron core and in the resistance of the wire with which the primary is wound.

When power is taken from the secondary winding by a load, the secondary current sets up a magnetic field that opposes the field set up by the primary current. For the induced voltage in the primary to equal the applied voltage, the original field must be maintained. The primary must draw enough additional current to set up a field exactly equal and opposite to the field set up by the secondary current.

In practical transformer calculations it may be assumed that the entire primary current is caused by the secondary load. This is justifiable because the magnetizing current should be very small in comparison with the primary load current at rated power output.

If the magnetic fields set up by the primary and secondary currents are to be equal, the primary current multiplied by the primary turns must equal the secondary current multiplied by the secondary turns.

$$I_P = I_S \left(\frac{N_S}{N_P} \right) \qquad (107)$$

where:

I_P = primary current,
I_S = secondary current,
N_P = number of turns on primary, and
N_S = number of turns on secondary.

Example: Suppose the secondary of the transformer in the previous example is delivering a current of 0.20 A to a load. What will be the primary current?

$$I_P = 0.20 \text{ A} \times \left(\frac{2800}{400} \right) = 0.20 \text{ A} \times 7 = 1.4 \text{ A}$$

Although the secondary voltage is higher than the primary voltage, the secondary current is lower than the primary current, and by the same ratio. The secondary current in an ideal transformer is 180° out of phase with the primary current, since the field in the secondary just offsets the field in the primary. The phase relationship between the currents in the windings holds true no matter what the phase difference between the current and the voltage of the secondary. In fact, the phase difference, if any, between voltage and current in the secondary winding will be reflected back to the primary as an identical phase difference.

Power Ratio

A transformer cannot create power; it can only transfer it and change the voltage level. Hence, the power taken from the secondary cannot exceed that taken by the primary from the applied voltage source. There is always some power loss in the resistance of the coils and in the iron core, so in all practical cases the power taken from the source will exceed that taken from the secondary.

$$P_O = n \, P_I \qquad (108)$$

where:

P_O = power output from secondary,
P_I = power input to primary, and
n = efficiency factor.

The efficiency, n, is always less than 1. It is usually expressed as a percentage: if n is 0.65, for instance, the efficiency is 65%.

Example: A transformer has an efficiency of 85.0% at its full-load output of 150 W. What is the power input to the primary at full secondary load?

$$P_I = \frac{P_O}{n} = \frac{150 \text{ W}}{0.850} = 176 \text{ W}$$

A transformer is usually designed to have the highest efficiency at the power output for which it is rated. The efficiency decreases with either lower or higher outputs. On the other hand, the losses in the transformer are relatively small at low output but increase as more power is taken. The amount of power that the transformer can handle is determined by its own losses, because these losses heat the wire and core. There is a limit to the temperature rise that can be tolerated, because too high a temperature can either melt the wire or cause the insulation to break down. A transformer can be operated at reduced output, even though the efficiency is low, because the actual loss will be low under such conditions. The full-load efficiency of small power transformers such as are used in radio receivers and transmitters usually lies between about 60 and 90%, depending on the size and design.

IMPEDANCE RATIO

In an ideal transformer — one without losses or leakage reactance — the following relationship is true:

$$Z_P = Z_S \left(\frac{N_P}{N_S} \right)^2 \qquad (109)$$

where:

Z_P = impedance looking into the primary terminals from the power source,
Z_S = impedance of load connected to secondary, and
N_P, N_S = turns ratio, primary to secondary.

A load of any given impedance connected to the transformer secondary will be transformed to a different value looking into the primary from the power source. The impedance transformation is proportional to the square of the primary-to-secondary turns ratio.

Example: A transformer has a primary-to-secondary turns ratio of 0.6 (the primary has six-tenths as many turns as the secondary) and a load of 3000 Ω is connected to the secondary. What is the impedance at the primary of the transformer?

$$Z_P = 3000 \text{ Ω} \times (0.6)^2 = 3000 \text{ Ω} \times 0.36$$
$$Z_P = 1080 \text{ Ω}$$

By choosing the proper turns ratio, the impedance of a fixed load can be transformed to any desired value, within practical limits. If transformer losses can be neglected, the transformed (reflected) impedance has the same phase angle as the actual load impedance. Thus, if the load is a pure resistance, the load presented by the primary to the power source will also be a pure resistance. If the load impedance is complex, that is, if the load current and voltage are out of phase with each other, then the primary voltage and current will show the same phase angle.

Many devices or circuits require a specific value of load resistance (or impedance) for optimum operation. The impedance of the actual load that is to dissipate the power may differ widely from the impedance of the source device or circuit, so a transformer is used to change the actual load into an impedance of the desired value. This is called impedance matching.

$$\frac{N_P}{N_S} = \sqrt{\frac{Z_P}{Z_S}} \qquad (110)$$

where:

N_P / N_S = required turns ratio, primary to secondary,
Z_P = primary impedance required, and
Z_S = impedance of load connected to secondary.

Example: A transistor audio amplifier requires a load of 150 Ω for optimum performance, and is to be connected to a loudspeaker having an impedance of 4.0 Ω. What is the turns ratio, primary to secondary, required in the coupling transformer?

$$\frac{N_P}{N_S} = \sqrt{\frac{Z_P}{Z_S}} = \frac{N_P}{N_S} \sqrt{\frac{150 \text{ Ω}}{4.0 \text{ Ω}}} = \sqrt{38} = 6.2$$

The primary therefore must have 6.2 times as many turns as the secondary.

These relationships may be used in practical work even though they are based on an ideal transformer. Aside from the normal design requirements of reasonably low internal losses and low leakage reactance, the only other requirement is that the primary have enough inductance to operate with low magnetizing current at the voltage applied to the primary.

The primary terminal impedance of an iron-core transformer is determined wholly by the load connected to the secondary and by the turns ratio. If the characteristics of the transformer have an appreciable effect on the impedance presented to the power source, the transformer is either poorly designed or is not suited to the voltage and frequency at which it is being used. Most transformers will operate quite well at voltages from slightly above to well below the design figure.

Transformer Losses

In practice, none of the formulas given so far provides truly exact results, al-

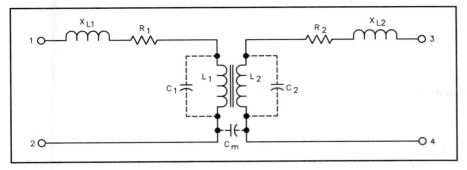

Fig 6.79 — A transformer as a network of resistances, inductances and capacitances. Only L1 and L2 contribute to the transfer of energy.

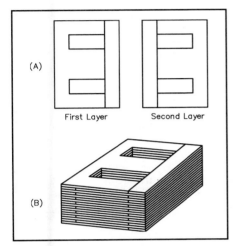

Fig 6.80 — A typical transformer iron core. The E and I pieces alternate direction in successive layers to improve the magnetic path while attenuating eddy currents in the core.

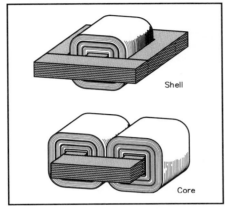

Fig 6.81 — Two common transformer constructions: shell and core.

though they afford reasonable approximations. Transformers in reality are not simply two coupled inductors, but a network of resistances and reactances, most of which appear in **Fig 6.79**. Since only the terminals numbered 1 through 4 are accessible to the user, transformer ratings and specifications take into account the additional losses created by these complexities.

In a practical transformer not all of the magnetic flux is common to both windings, although in well designed transformers the amount of flux that cuts one coil and not the other is only a small percentage of the total flux. This *leakage flux* causes a voltage of self-induction. Consequently, there are small amounts of leakage inductance associated with both windings of the transformer. Leakage inductance acts in exactly the same way as an equivalent amount of ordinary inductance inserted in series with the circuit. It has, therefore, a certain reactance, depending on

the amount of leakage inductance and the frequency. This reactance is called *leakage reactance*, shown as X_{L1} and X_{L2} in Fig 6.79.

Current flowing through the leakage reactance causes a voltage drop. This voltage drop increases with increasing current; hence, it increases as more power is taken from the secondary. Thus, the greater the secondary current, the smaller the secondary terminal voltage becomes. The resistances of the transformer windings, R1 and R2, also cause voltage drops when there is current. Although these voltage drops are not in phase with those caused by leakage reactance, together they result in a lower secondary voltage under load than is indicated by the transformer turns ratio.

Thus, the voltage regulation in a real transformer is not perfect. At power frequencies (60 Hz), the voltage at the secondary, with a reasonably well-designed transformer, should not drop more than about 10% from open-circuit conditions to full load. The voltage drop may be considerably more than this in a transformer operating at voice and music frequencies, because the leakage reactance increases directly with the frequency.

In addition to wire resistances and

leakage reactances, certain stray capacitances occur in transformers. An electric field exists between any two points having a different voltage. When current flows through a coil, each turn has a slightly different voltage than its adjacent turns, creating a capacitance between turns. This *distributed capacitance* appears in Fig 6.79 as C1 and C2. Another capacitance, C_M, appears between the two windings for the same reason. Moreover, transformer windings can exhibit capacitance relative to nearby metal, for example, the chassis, the shield and even the core.

Although these stray capacitances are of little concern with power and audio transformers, they become important as the frequency increases. In transformers for RF use, the stray capacitance can resonate with either the leakage reactance or, at lower frequencies, with the winding reactances, L1 or L2, especially under very light or zero loads. In the frequency region around resonance, transformers no longer exhibit the properties formulated above or the impedance properties to be described below.

Iron-core transformers also experience losses within the core itself. *Hysteresis losses* include the energy required to overcome the retentivity of the core's magnetic material. Circulating currents through the core's resistance are *eddy currents*, which form part of the total core losses. These losses, which add to the required magnetizing current, are equivalent to adding a resistance in parallel with L1 in Fig 6.79.

Core Construction

Audio and power transformers usually employ one or another grade of silicon steel as the core material. With permeabilities of 5000 or greater, these cores saturate at flux densities approaching 10^5 lines per square inch of cross section. The cores consist of thin insulated laminations to break up potential eddy current paths.

Each core layer consists of an E and an I piece butted together, as represented in **Fig 6.80**. The butt point leaves a small gap. Since the pieces in adjacent layers have a continuous magnetic path, however, the flux density per unit of applied magnetic force is increased and flux leakage reduced.

Two core shapes are in common use, as shown in **Fig 6.81**. In the shell type, both windings are placed on the inner leg, while in the core type the primary and secondary windings may be placed on separate legs, if desired. This is sometimes done when it is necessary to minimize capacitive effects between the primary and secondary, or when one of the windings must operate at very high voltage.

The number of turns required in the primary for a given applied voltage is determined by the size, shape and type of core material used, as well as the frequency. The number of turns required is inversely proportional to the cross-sectional area of the core. As a rough indication, windings of small power transformers frequently have about six to eight turns per volt on a core of l-square-inch cross section and have a magnetic path 10 or 12 inches in length. A longer path or smaller cross section requires more turns per volt, and vice versa.

In most transformers the coils are wound in layers, with a thin sheet of treated-paper insulation between each layer. Thicker insulation is used between adjacent coils and between the first coil and the core.

Shielding

Because magnetic lines of force are continuous and closed upon themselves, shielding requires a path for the lines of force of the leakage flux. The high-permeability of iron cores tends to concentrate the field, but additional shielding is often needed. As depicted in **Fig 6.82**, enclosing the transformer in a good magnetic material can restrict virtually all of the magnetic field in the outer case. The nonmagnetic material between the case and the core creates a region of high reluctance, attenuating the field before it reaches the case.

AUTOTRANSFORMERS

The transformer principle can be used with only one winding instead of two, as shown in **Fig 6.83A**. The principles that relate voltage, current and impedance to the turns ratio also apply equally well. A one-winding transformer is called an *autotransformer*. The current in the common section (A) of the winding is the difference between the line (primary) and the load (secondary) currents, since these currents are out of phase. Hence, if the line and load currents are nearly equal, the common section of the winding may be wound with comparatively small wire. The line and load currents will be equal only when the primary (line) and secondary (load) voltages are not very different.

Autotransformers are used chiefly for boosting or reducing the power-line voltage by relatively small amounts. Fig 6.83B illustrates the principle schematically with a switched, stepped autotransformer. Continuously variable autotransformers are commercially available under a variety of trade names;

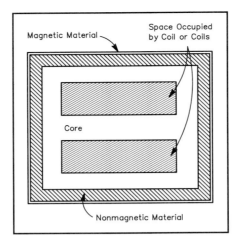

Fig 6.82 — A shielded transformer: the core plus an outer shield of magnetic material contain nearly all of the magnetic field.

Variac and Powerstat are typical examples.

Technically, tapped air-core inductors, such as the one in the network in Fig 6.74 at the close of the discussion of resonant circuits, are also autotransformers. The voltage from the tap to the bottom of the coil is less than the voltage across the entire coil. Likewise, the impedance of the tapped part of the winding is less than the impedance of the entire winding. Because leakage reactances are great and the coefficient of coupling is quite low, the relationships true of a perfect transformer grow quite unreliable in predicting the exact values. For this reason, tapped inductors are rarely referred to as transformers. The stepped-down situation in Fig 6.74 is better approximated — at or close to resonance — by the formula

$$R_P = \frac{R_L \, X_{COM}^2}{X_L} \qquad (111)$$

where:

R_P = tuned-circuit parallel-resonant impedance,

R_L = load resistance tapped across part of the coil,

X_{COM} = reactance of the portion of the coil common to both the resonant circuit and the load tap, and

X_L = reactance of the entire coil.

The result is approximate and applies only to circuits with a Q of 10 or greater.

AIR-CORE RF TRANSFORMERS

Air-core transformers often function as mutually coupled inductors for RF applications. They consist of a primary winding and a secondary winding in close prox-

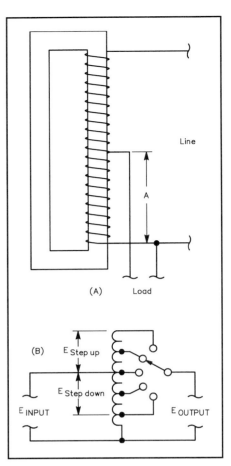

Fig 6.83 — The autotransformer is based on the transformer, but uses only one winding. The pictorial diagram at A shows the typical construction of an autotransformer. The schematic diagram at B demonstrates the use of an autotransformer to step up or step down ac voltage, usually to compensate for excessive or deficient line voltage.

imity. Leakage reactances are ordinarily high, however, and the coefficient of coupling between the primary and secondary windings is low. Consequently, unlike transformers having a magnetic core, the turns ratio does not have as much significance. Instead, the voltage induced in the secondary depends on the mutual inductance.

Nonresonant RF Transformers

In a very basic transformer circuit operating at radio frequencies, such as in **Fig 6.84A**, the source voltage is applied to L1. R_S is the series resistance inherent in the source. By virtue of the mutual inductance, M, a voltage is induced in L2. A current flows in the secondary circuit through the reactance of L2 and the load resistance of R_L. Let X_{L2} be the reactance of L2 independent of L1, that is, indepen-

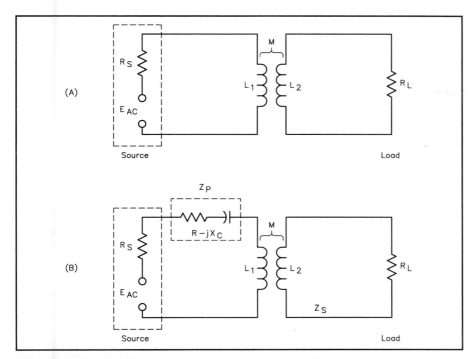

Fig 6.84 — The coupling of a complex impedance back into the primary circuit of a transformer composed of nonresonant air-core inductors.

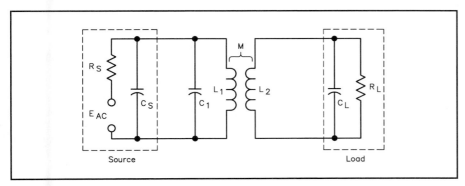

Fig 6.85 — An air-core transformer circuit consisting of a resonant primary circuit and an untuned secondary. R_S and C_S are functions of the source, while R_L and C_L are functions of the load circuit.

dent of the effects of mutual inductance. The impedance of the secondary circuit is then:

$$Z_S = \sqrt{R_L^2 + X_{L2}^2} \qquad (112)$$

where:

   Z_S = the impedance of the secondary circuit in ohms,

   R_L = the load resistance in ohms, and

   X_{L2} = the reactance of the secondary inductance in ohms.

The effect of Z_S upon the primary circuit is the same as a coupled impedance in series with L1. Fig 6.84B displays the coupled impedance (Z_P) in a dashed enclosure to indicate that it is not a new physical component. It has the same absolute value of phase

angle as in the secondary impedance, but the sign of the reactance is reversed; it appears as a capacitive reactance. The value of Z_P is:

$$Z_P = \frac{(2 \pi f M)^2}{Z_S} \qquad (113)$$

where:

   Z_P = the impedance introduced into the primary,

   Z_S = the impedance of the secondary circuit in ohms, and

   $2 \pi f M$ = the mutual reactance between the reactances of the primary and secondary coils (also designated as X_M).

Resonant RF Transformers

The use of at least one resonant circuit in place of a pair of simple reactances

eliminates the reactance from the transformed impedance in the primary. For loaded or operating Qs of at least 10, the resistances of individual components is negligible. **Fig 6.85** represents just one of many configurations in which at least one of the inductors is in a resonant circuit. The reactance coupled into the primary circuit is cancelled if the circuit is tuned to resonance while the load is connected. If the reactance of the load capacitance, C_L is at least 10 times any stray capacitance in the circuit, as is the case for low impedance loads, the value of resistance coupled to the primary is

$$R1 = \frac{X_M^2 \, R_L}{X2^2 + R_L^2} \qquad (114)$$

where:

   $R1$ = series resistance coupled into the primary circuit,

   X_M = mutual reactance,

   R_L = load resistance, and

   $X2$ = reactance of the secondary inductance.

The parallel impedance of the resonant circuit is just R1 transformed from a series to a parallel value by the usual formula, $R_P = X^2 / R1$.

The higher the loaded or operating Q of the circuit, the smaller the mutual inductance required for the same power transfer. If both the primary and secondary circuits consist of resonant circuits, they can be more loosely coupled than with a single tuned circuit for the same power transfer. At the usual loaded Q of 10 or greater, these circuits are quite selective, and consequently narrowband.

Although coupling networks have to a large measure replaced RF transformer coupling that uses air-core transformers, these circuits are still useful in antenna tuning units and other circuits. For RF work, powdered-iron toroidal cores have generally replaced air-core inductors for almost all applications except where the circuit handles very high power or the coil must be very temperature stable. Slug-tuned solenoid coils for low-power circuits offer the ability to tune the circuit precisely to resonance. For either type of core, reasonably accurate calculation of impedance transformation is possible. It is often easier to experiment to find the correct values for maximum power transfer, however. For further information on coupled circuits, see the section on Matching Networks in the **Transceivers** chapter.

BROADBAND FERRITE RF TRANSFORMERS

The design concepts and general theory

of ideal transformers presented earlier in this chapter apply also to transformers wound on ferromagnetic-core materials (ferrite and powdered iron). As is the case with stacked cores made of laminations in the classic I and E shapes, the core material has a specific permeability factor that determines the inductance of the windings versus the number of wire turns used.

Toroidal cores are useful from a few hundred hertz well into the UHF spectrum. The principal advantage of this type of core is the self-shielding characteristic. Another feature is the compactness of a transformer or inductor. Therefore, toroidal-core transformers are excellent for use not only in dc-to-dc converters, where tape-wound steel cores are employed, but at frequencies up to at least 1000 MHz with the selection of the proper core material for the range of operating frequencies. Toroidal cores are available from microminiature sizes up to several inches in diameter. The latter can be used, as one example, to build a 20-kW balun for use in antenna systems.

One of the most common ferromagnetic transformers used in Amateur Radio work is the *conventional broadband transformer*. Broadband transformers with losses of less than 1 dB are employed in circuits that must have a uniform response over a substantial frequency range, such as a 2- to 30-MHz broadband amplifier. In applications of this sort, the reactance of the windings should be at least four times the impedance that the winding is designed to look into at the lowest design frequency.

Example: What should be the winding reactances of a transformer that has a 300-Ω primary and a 50-Ω secondary load? Relative to the 50-Ω secondary load:

$$X_S = 4 \, Z_S = 4 \times 50 \, \Omega = 200 \, \Omega.$$

The primary winding reactance (X_P) is:

$$X_P = 4 \, Z_P = 4 \times 300 \, \Omega = 1200 \, \Omega.$$

The core-material permeability plays a vital role in designing a good broadband transformer. The effective permeability of the core must be high enough to provide ample winding reactance at the low end of the operating range. As the operating frequency is increased, the effects of the core tend to disappear until there are scarcely any core effects at the upper limit of the operating range. The limiting factors for high frequency response are distributed capacity and leakage inductance due to uncoupled flux. A high-permeability core minimizes the number of turns needed for a given reactance and therefore also mini-

mizes the distributed capacitance at high frequencies.

Ferrite cores with a permeability of 850 are common choices for transformers used between 2 and 30 MHz. Lower frequency ranges, for example, 1 kHz to 1 MHz, may require cores with permeabilities up to 2000. Permeabilities from 40 to 125 are useful for VHF transformers. Conventional broadband transformers require resistive loads. Loads with reactive components should use appropriate networks to cancel the reactance.

Conventional transformers are wound in the same manner as a power transformer. Each winding is made from a separate length of wire, with one winding placed over the previous one with suitable insulation between. Unlike some transmission-line transformer designs, conventional broadband transformers provide dc isolation between the primary and secondary circuits. The high voltages encountered in high-impedance-ratio step-up transformers may require that the core be wrapped with glass electrical tape before adding the windings (as an additional protection from arcing and voltage breakdown), especially with ferrite cores that tend to have rougher edges. In addition, high voltage applications should also use wire with high-voltage insulation and a high temperature rating.

Fig 6.86 illustrates one method of transformer construction using a single toroid as the core. The primary of a step-down impedance transformer is wound to occupy the entire core, with the secondary wound over the primary. The first step in planning the winding is to select a core of the desired permeability. Convert the required reactances determined earlier into inductance values for the lowest frequency of use. To find the number of turns for each winding, use the A_L value for the selected core and equation 51 from the section on ferrite toroidal inductors earlier in this chapter. Be certain the core can handle the power by calculating the maximum flux using equation 49, given earlier in the chapter, and comparing the result with the manufacturer's guidelines.

Example: Design a small broadband transformer having an impedance ratio of 16:1 for a frequency range of 2.0 to 20.0 MHz to match the output of a small-signal stage (impedance $\approx$ 500 Ω) to the input (impedance $\approx$ 32 Ω) of an amplifier.

1. Since the impedance of the smallest winding should be at least 4 times the lower impedance to be matched at the lowest frequency, $X_S = 4 \times 32 \, \Omega = 128 \, \Omega$.

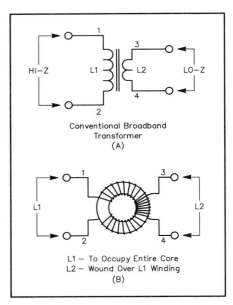

Fig 6.86 — Schematic and pictorial representation of a conventional broadband transformer wound on a ferrite toroidal core. The secondary winding (L2) is wound over the primary winding (L1).

2. The inductance of the secondary winding should be $L_S = X_S \, / \, 2 \, \pi \, f = 128 \, / \, (6.2832 \times 2.0 \times 10^6 \, \text{Hz}) = 0.010 \, \text{mH}$.

3. Select a suitable core. For this low-power application, a $^3/_8$-inch ferrite core with a permeability of 850 is suitable. The core has an A_L value of 420. Calculate the number of turns for the secondary.

$$N_S = 1000 \sqrt{\frac{L}{A_L}} = 1000 \sqrt{\frac{0.010}{420}}$$

$$= 1000 \times 0.0049 = 4.9 \text{ turns}$$

4. A 5-turn secondary winding should suffice. The primary winding derives from the impedance ratio:

$$NP = N_S \sqrt{\frac{Z_P}{Z_S}} = 5 \sqrt{\frac{16}{1}} = 5 \times 4 = 20 \text{ turns}$$

This low power application will not approach the maximum flux density limits for the core, and #28 enamel wire should both fit the core and handle the currents involved.

A second style of broadband transformer construction appears in **Fig 6.87**. The key elements in this transformer are the stacks of ferrite cores aligned with tubes soldered to pc-board end plates. This style of transformer is suited to high power applications, for example, at the input and output ports of transistor RF power amplifiers. Low-power versions of this transformer can be wound on "binocular" cores having pairs of parallel holes through them.

For further information on conventional transformer matching using ferromagnetic materials, see the Matching Networks section in the **Amplifiers** chapter. Refer to the **Component Data** chapter for more detailed information on available ferrite cores. A standard reference on conventional broadband transformers using ferromagnetic materials is *Ferromagnetic Core Design and Applications Handbook* by Doug DeMaw, W1FB, published by Prentice Hall.

TRANSMISSION-LINE TRANSFORMERS

Conventional transformers use flux linkages to deliver energy to the output circuit. *Transmission line transformers* use transmission line modes of energy transfer between the input and the output terminals of the devices. Although toroidal versions of these transformers physically resemble toroidal conventional broadband transformers, the principles of operations differ significantly. Stray inductances and interwinding capacitances form part of the characteristic impedance of the transmission line, largely eliminating resonances that limit high frequency response. The limiting factors for transmission line transformers include line length, deviations in the constructed line from the design value of characteristic impedance, and parasitic capacitances and inductances that are independent of the characteristic impedance of the line.

The losses in conventional transformers depend on current and include wire, eddy-current and hysteresis losses. In contrast, transmission line transformers exhibit voltage-dependent losses, which make higher impedances and higher VSWR values limiting factors in design. Within design limits, the cancellation of flux in the cores of transmission line transformers permits very high efficiencies across their passbands. Losses may be lower than 0.1 dB with the proper core choice.

Transmission-line transformers can be configured for several modes of operation, but the chief amateur use is in *baluns* (balanced-to-*un*balanced transformers) and in *ununs* (*un*balanced-to-*un*balanced transformers). The basic principle behind a balun appears in **Fig 6.88**, a representation of the classic Guanella 1:1 balun. The input and output impedances are the same, but the output is balanced about a real or virtual center point (terminal 5). If the characteristic impedance of the transmission line forming the inductors with numbered terminals equals the load impedance, then E2 will equal E1. With respect to terminal 5, the voltage at terminal 4 is E1 / 2, while the voltage at terminal 2 is −E1 / 2, resulting in a balanced output.

The small losses in properly designed baluns of this order stem from the potential gradient that exists along the length of transmission line forming the transformer. The value of this potential is −E1 / 2, and it forms a dielectric loss that can't be eliminated. Although the loss is very small in well-constructed 1:1 baluns at low impedances, the losses climb as impedances climb (as in 4:1 baluns) and as the VSWR climbs. Both conditions yield higher voltage gradients.

The inductors in the transmission-line transformer are equivalent to — and may be — coiled transmission line with a characteristic impedance equal to the load. They form a choke isolating the input from the output and attenuating undesirable currents, such as antenna current, from the remainder of the transmission line to the energy source. The result is a *current* or *choke* balun. Such baluns may take many forms: coiled transmission line, ferrite beads placed over a length of transmission line, windings on linear ferrite cores or windings on ferrite toroids.

Reconfiguring the windings of Fig 6.88 can alter the transformer operation. For example, if terminal 2 is connected to terminal 3, a positive potential gradient appears across the lengths of line, resulting in a terminal 4 potential of 2 E1 with respect to ground. If the load is disconnected from terminal 2 and reconnected to ground, 2 E1 appears across the load — instead of ±E1 / 2. The product of this experiment is a 4:1 impedance ratio, forming an unun. The bootstrapping effect of the new connection is applicable to many other design configurations involving multiple windings to achieve custom impedance ratios from 1:1 up to 9:1.

Balun and unun construction for the impedances of most concern to amateurs requires careful selection of the feed line used to wind the balun. Building transmission line transformers on ferrite toroids may require careful attention to wire size and spacing to approximate a 50-Ω line. Wrapping wire with polyimide tape (one or two coatings, depending upon the wire

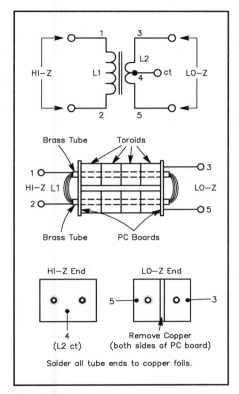

Fig 6.87 — Schematic and pictorial representation of a "binocular" style of conventional broadband transformer. This style is used frequently at the input and output ports of transistor RF amplifiers. It consists of two rows of high-permeability toroidal cores, with the winding passed through the center holes of the resulting stacks.

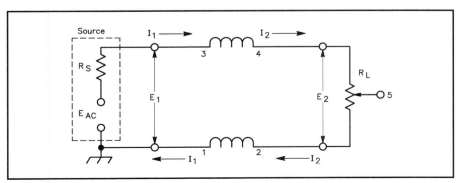

Fig 6.88 — Schematic representation of the basic Guanella "choke" balun or 1:1 transmission line transformer. The inductors are a length of two-wire transmission line. R$_S$ is the source impedance and R$_L$ is the load impedance.

size) and then glass taping the wires together periodically produces a reasonable 50-transmission line. Ferrite cores in the permeability range of 125 to 250 are generally optimal for transformer windings, with 1.25-inch cores suitable to 300-W power levels and 2.4-inch cores usable to the 5 kW level. Special designs may alter the power-handling capabilities of the core sizes. For the 1:1 balun shown in Fig 6.88, 10 bifilar turns (#16 wire for the smaller core and #12 wire for the larger, both Thermaleze wire) yields a transformer operable from 160 to 10 m.

Transmission-line transformers have their most obvious application to antennas, since they isolate the antenna currents from the feed line, especially where a coaxial feed line is not exactly perpendicular to the antenna. The balun prevents antenna currents from flowing on the outer surface of the coax shielding, back to the transmitting equipment. Such currents would distort the antenna radiation pattern. Appropriately designed baluns can also transform impedance values at the same time. For example, one might use a 4:1 balun to match a 12.5-Ω Yagi antenna impedance to a 50-Ω feed line. A 4:1 balun might also be used to match a 75-Ω TV antenna to 300-Ω feed line.

Interstage coupling within solid-state transmitters represents another potential for transmission-line transformers. Broadband coupling between low-impedance, but mismatched stages can benefit from the low losses of transmission-line transformers. Depending upon the losses that can be tolerated and the bandwidth needed, it is often a matter of designer choice between a transmission-line transformer and a conventional broadband transformer as the coupling device.

For further information on transmission-line transformers and their applications, see the **Amplifiers** chapter. Another reference on the subject is *Transmission Line Transformers*, by Jerry Sevick, W2FMI, published by Noble Publishing (see the Address List in Chapter 30 for contact information).

Contents

Digital Signal Theory and Components

7

Digital Fundamentals

Digital signal theory is an important aspect of Amateur Radio. With a knowledge of digital theory, there are many new worlds for the radio amateur to explore. Applications of digital signal theory include digital communications, code conversion, signal processing, station control, frequency synthesis, amateur satellite telemetry, message handling, word processing and other information handling operations.

This chapter, written by Christine Montgomery, KGØGN, presents digital-theory fundamentals and some applications of that theory in Amateur Radio. The fundamentals introduce digital mathematics, including number systems, logic devices and simple digital circuits. Next, the implementation of these simple circuits is explored in integrated circuits, their families and interfacing. Integrated circuits continue with memory chips and microprocessors, culminating in a synthesis of these components in the modern digital computer. Where possible, this chapter mentions Amateur Radio applications associated with the technologies being discussed, as well as pointers to other chapters that discuss such applications in greater depth.

DIGITAL VS ANALOG

An essential first step in understanding digital theory is to understand the difference between a *digital* and an *analog* signal. An analog value, a real number, has no end; for example, the number 1/3 is 0.333... where the 3 can be repeated forever, or 3/4 equals 0.7500... with infinite repeated 0s. A digital approximation of an analog number breaks the real number line into discrete steps, for example the integers. This process of approximating a value with discrete steps either truncates or rounds an analog value to some number of decimal places. For example, rounding 1/3 to an integer gives 0 and rounding 3/4 gives 1.

For a simple physical example, look at your wristwatch. A watch with a face — with the hands of the watch rotating in a continuous, smooth motion — is an analog display. Here, the displayed time has a *continuous* range of values, such as from 12:00 exactly to 12:00 and 1/3 second or any values in between. In contrast, a watch with a digital display is limited to *discrete* states. Here the displayed time jumps from 12:00 and 0 seconds to 12:00 and 1 second, without showing the time in between. (A watch with a second hand that jerks from one second to another could also fit the digital analogy.)

In the digital watch example, time is represented by ten distinct states (0, 1, 2, 3, 4, 5, 6, 7, 8 and 9). Digital electronic signals, however, will usually be much more limited in the number of states al-

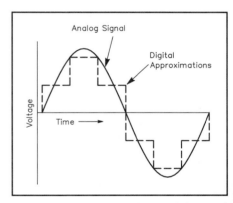

Fig 7.1 — An analog signal and its digital approximation. Note that the analog waveform has continuously varying voltage while the digital waveform is composed of discrete steps.

lowed. *Binary*, the most common system, has only two states: 0 and 1. Ternary (3 states), quaternary (4 states) and other digital systems also exist. **Fig 7.1** illustrates the contrast of an analog signal (in this case a sine wave) and its digital approximation.

While the focus in this chapter will be on digital theory, many circuits and systems involve *both* digital and analog components. Often, a designer may choose between using digital technology, analog technology or a combination.

Number Systems

In order to understand digital electronics, you must first understand the digital numbering system. Any number system has two distinct characteristics: a set of *symbols* (digits or numerals) and a *base* or radix. A *number* is a collection of these digits, where the left-most digit is the *most significant digit (MSD)* and the right-most digit is the *least significant digit (LSD)*. The value of this number is a weighted sum of its digits. The *weights* are determined by the system's base and the digit's position relative to the decimal point.

While these definitions may seem strange with all the technical terms, they will be more familiar when seen in a decimal system example. This is the "traditional" number system we are all familiar with.

DECIMAL

The decimal system is a base-10 system, with ten symbols: {0, 1, 2, 3, 4, 5, 6, 7, 8, 9}. In the decimal number, 548.21, the digits are 5, 4, 8, 2 and 1, where 5 is the most significant digit since it is positioned to the far left and 1 is the least significant digit since it is positioned to the far right. The value of this number is a weighted sum of its digits, as shown in **Table 7.1**.

The weight of a position is the system's base raised to a power, 10^P with the power determined by the position relative to the decimal. For example, digit 8, immediately to the left of the decimal, is at position 0; therefore, its weight factor is $10^0 = 1$. Similarly, digit 5 is 2 positions to the left of the decimal and has a weight factor $10^2 = 100$. The value of the number is the sum of each digit times its weight.

BINARY

Binary is a *base-2* number system that is limited to two symbols: {0, 1}. The weight factors are now powers of 2, like 2^0, 2^1 and 2^2. For example, the decimal number, 163 and its equivalent binary number, 10100011, are shown in **Table 7.2**.

The digits of a binary number are now *bits* (short for binary digit). The MSD is the *most significant bit (MSB)* and the LSD is the *least significant bit (LSB)*. Four bits make a *nibble* and two nibbles, or eight bits, make a *byte*. A *word* can consist of two or four bytes, and two words (a most significant word, *MSW*, and a least significant word, *LSW*) is sometimes called a *longword*. These groupings are useful when converting to hexadecimal notation, which is explained later.

OCTAL

Octal is a *base-8* number system, using the symbols {0,1,2,3,4,5,6,7}. The weight factors are now powers of 8, such as 8^0, 8^1 and 8^2. For example, the decimal number 163 is equivalent to octal 243.

Since $2^3 = 8$, it is easy to switch between binary and octal just by viewing the binary number in groups of 3. (Add a leading 0 on the left most group, if the number of digits doesn't divide evenly into groups of three.)

HEXADECIMAL

The hexadecimal, or hex, *base-16* number system uses both numbers and characters in its set of sixteen symbols: {0,1,2,3,4,5,6,7,8,9,A,B,C,D,E,F}. Here, the letters A to F have the decimal equivalents of 10 to 15 respectively: A=10, B=11, C=12, D=13, E=14 and F=15. Again, the weights are powers of the base, such as 16^0, 16^1 and 16^2. The decimal number 163 is equivalent to hex A3.

CONVERSION TECHNIQUES

An easy way to convert a number from decimal to another number system is to do repeated division, recording the remainders in a tower just to the right. The converted number, then, is the remainders, reading up the tower. This technique is illustrated in **Table 7.3** for hexadecimal, octal and binary conversions of the decimal number 163.

For example, to convert decimal 163 to hex, repeated divisions by 16 are performed. The first division gives 163 / 16 = 10 remainder 3. The remainder 3 is written in a column to the right. The second division gives 10 / 16 = 0 remainder 10. Since 10 decimal = A hex, A is written in the remainder column to the right. This division gave a divisor of 0 so the process is complete. Reading up the remainders column, the result is A3. The most common mistake in this technique is to forget that the Most Significant Digit ends up at the bottom.

Using the repeated division for binary is rather cumbersome, since the tower quickly grows large. Combining this technique with the grouping technique discussed earlier should make conversions fairly easy. Simply perform the tower di-

---

Table 7.1
Decimal Numbers

Digit   Weight   Position

Example: $5(10^2)$

548.21	=	$5(10^2)$	+	$4(10^1)$	+	$8(10^0)$	+	$2(10^{-1})$	+	$1(10^{-2})$
	=	5(100)	+	4(10)	+	8(1)	+	2(0.1)	+	1(0.01)
	=	500	+	40	+	8	+	0.2	+	0.01
	=	5		4		8	•	2		1
		MSD						decimal		LSD

Table 7.2
Decimal and Binary Number Equivalents

163	=	128	+	0	+	32	+	0	+	0	+	0	+	2	+	1	decimal
	=	1(128)	+	0(64)	+	1(32)	+	0(16)	+	0(8)	+	0(4)	+	1(2)	+	1(1)	
	=	$1(2^7)$	+	$0(2^6)$	+	$1(2^5)$	+	$0(2^4)$	+	$0(2^3)$	+	$0(2^2)$	+	$1(2^1)$	+	$1(2^0)$	
10100011	=	1		0		1		0		0		0		1		1	binary
		MSB												LSB			

Nibble — Nibble — Byte

---

vision on one of the larger numbers, such as hexadecimal 16, then use grouping to put the result into binary form.

Another technique that should be briefly mentioned can be even easier: get a calculator with a binary and/or hex mode option. One warning for this technique: this chapter doesn't discuss negative binary numbers. If your calculator does not give you the answer you expected, it may have interpreted the number as negative. This would happen when the number's binary form has a 1 in its MSB, such as the highest (leftmost) bit for the binary mode's default size. To avoid learning about negative binary numbers, always use a leading 0 when you enter a number in binary or hex into your calculator.

BINARY CODED DECIMAL (BCD)

Scientists have experimented with many devices out of a desire for fast computations. The first generation computers were born when J. Vincent Atanasoff decided to use binary numbers instead of decimal to do his computations. A binary number system representation is the most appropriate form for internal computations since there is a direct mathematical relationship for every bit in the number. To interface with a user — who usually wants to see I/O in terms of decimal numbers — other codes are more useful. The *Binary Coded Decimal (BCD)* system is the simplest and most widely used form for inputs and outputs of user-oriented digital systems.

In the Binary Coded Decimal (BCD) system, each decimal digit is expressed as a corresponding 4-bit binary number. In other words, the decimal digits 0 to 9 are encoded as the bit strings 0000 to 1001. To

Table 7.3
Number System Conversions

Hex	Remainder	Octal	Remainder	Binary	Remainder	
16 |163		8 |163		2 |163		
|10	3	|20	3	|81	1	LSB
|0	A	|2	4	|40	1	
		|0	2	|20	0	
				|10	0	
				|5	0	
				|2	1	
				|1	0	
				|0	1	MSB
A3 hex		243 octal		1010 0011 binary		

Table 7.4
Binary Coded Decimal Number Conversion

	0 0 0 1	0 1 1 0	0 0 1 1	BCD
=	$1(2^0)$	$1(2^2) + 1(2^1)$	$1(2^1) + 1(2^0)$	
=	(1)	(4 + 2)	(2 + 1)	
163 =	1	6	3	decimal

make the number easier to read, a space is left between each 4-bit group. For example, the decimal number 163 is equivalent to the BCD number 0001 0110 0011, as shown in **Table 7.4**.

A generic code could use any n-bit string to represent a piece of information. BCD uses 4 bits because that is the minimum needed to represent a 9. All four bits are always written; even a decimal 0 is written as 0000 in BCD.

The important difference between BCD and the previous number systems is that, starting with decimal 10, BCD loses the standard mathematical relationship of a weighted sum. Instead of using the 4-bit code strings 1010 to 1111 for decimal 10 to 15, BCD uses 0001 0000 to 0001 0101. There are other n-bit decimal codes in use and, even for specifically 4 bits, there are millions of combinations to represent the decimal digits 0-9. BCD is the simplest way to convert between decimal and a binary code; thus it is the ideal form for I/O interfacing. The binary number system, since it maintains the mathematical relationship between bits, is the ideal form for the computer's internal computations.

Physical Representation Of Binary States

STATE LEVELS

Most digital systems use the binary number system because many simple physical systems are most easily described by two state levels (0 and 1). For example, the two states may represent "on" and "off," a punched hole or the absence of a hole in paper tape or a card, or a "mark" and "space" in a communications transmission. In electronic systems, state levels are physically represented by voltages. A typical choice is

state 0 = 0 V

state 1 = 5 V

Since it is unrealistic to obtain these exact voltage values, a more practical choice is a range of values, such as

state 0 = 0.0 to 0.4 V

state 1 = 2.4 to 5.0 V

Fig 7.2 illustrates this representation of states by voltage levels. The undefined region between the two binary states is also known as the *transition region* or *noise margin*.

Transition Time

A change in state between binary 0 and binary 1 does not occur instantly. There is a *transition time* between states. This transition time is a result of the time it takes to

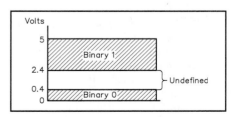

Fig 7.2 — Representation of binary states 1 and 0 by a selected range of voltage levels.

charge or discharge the stray capacitance in wires and other components because voltage cannot change instantaneously across a capacitor. (Stray inductance in the wires also has an effect because the current through an inductor can't change instantaneously.) The transition from a 0 to a 1 state is called the *rise time*. Similarly, the transition from a 1 to a 0 state is called the *fall time*. Note that these times need not be the same. **Fig 7.3A** shows an ideal signal, or *pulse*, with zero-time switching. Fig 7.3B shows a typical pulse, as it changes between states in a smooth curve.

Rise and fall times vary with the logic family used and the location in a circuit. Typical values of transition time are in the microsecond to nanosecond range. In a circuit, distributed inductances and capacitances in wires or PC-board traces may cause rise and fall times to increase as the pulse moves away from the source.

Propagation Delay

Rise and fall times only describe a relationship within a pulse. For a circuit, a pulse input into the circuit must propagate through the circuit; in other words it must pass through each component in the circuit until eventually it arrives at the circuit output. The time delay between providing an input to a circuit and to seeing a response at the output is the *propagation delay*, and is illustrated by **Fig 7.4**.

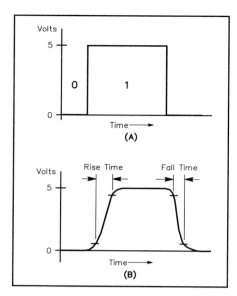

Fig 7.3 — (A) An ideal digital pulse and (B) a typical actual pulse, showing the gradual transition between states.

For modern switching logic, typical propagation delay values are in the 1 to 15 nanosecond range. (It is useful to remember that the propagation delay along a wire or printed-circuit-board trace is about 1.0

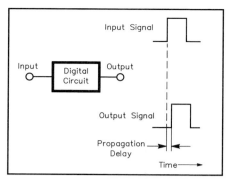

Fig 7.4 — Propagation delay in a digital circuit.

to 1.5 ns per inch.) Propagation delay is the result of cumulative transition times as well as transistor switching delays, reactive element charging times and the time for signals to travel through wires. In complex circuits, different propagation delays through different paths can cause problems when pulses must arrive somewhere at exactly the same time. Solutions to this timing problem include adding a buffer amplifier to the circuit and synchronization of circuits. These are discussed in later sections.

Combinational Logic

Having defined a way to use voltage levels to physically represent digital numbers, we can apply digital signal theory to design useful circuits. Digital circuits combine binary inputs to produce a desired binary output or combination of outputs. This simple combination of 0s and 1s can become very powerful, implementing everything from simple switches to powerful computers.

A digital circuit falls into one of two types: combinational logic or sequential logic. In a *combinational logic* circuit, the output depends only on the *present inputs*. (If we ignore propagation delay.) In contrast, in a *sequential logic* circuit, the output depends on the present inputs, the *previous sequence of inputs* and often a clock signal. John F. Wakerly, on page 147 of *Digital Design Principles and Practices*, described this difference as follows: "The rotary channel selector knob on an inexpensive TV is like a combinational circuit — its 'output' selects a channel based only on the current position of the knob ('input'). In contrast, the channel selector controlled by the up [+] and down [−]

pushbuttons on a fancy TV or VCR is a sequential circuit — the channel selection depends on the past sequence of up/down pushes... as far back as when you first powered-up the device."

The next section discusses combinational logic circuits. Later, we will build sequential logic circuits from the basics established here.

BOOLEAN ALGEBRA AND THE BASIC LOGICAL OPERATORS

Combinational circuits are composed of logic gates, which perform binary operations. Logic gates manipulate binary numbers, so you need an understanding of the algebra of binary numbers to understand how logic gates operate. *Boolean algebra* is the mathematical system to describe and design binary digital circuits. It is named after George Boole, the mathematician who developed the system. Standard algebra has a set of basic operations: addition, subtraction, multiplication and division. Similarly, Boolean algebra has a set of basic operations, called *logical operations*: NOT, AND and OR.

The function of these operators can be described by either (A) a Boolean equation or (B) a truth table. A Boolean *equation* describes an operator's function by representing the inputs and the operations performed on them. An equation is of the form "B = A," while an *expression* is of the form "A." In an assignment equation, the inputs and operations appear on the right and the result, or output, is assigned to the variable on the left.

A *truth table* describes an operator's function by listing all possible inputs and the corresponding outputs. Truth tables are sometimes written with Ts and Fs (for true and false) or with their respective equivalents, 1s and 0s. In company databooks (catalogs of logic devices a company manufactures), truth tables are usually written with Hs and Ls (for high and low). In the figures, 1 will mean high and 0 will mean low. This representation is called positive logic. The meaning of different logic types and why they are useful is discussed in a later section.

Each Boolean operator also has two circuit symbols associated with it. The traditional symbol — used by ARRL and other US publications — appears on top in each of the figures; for example, the triangle and bubble for the NOT function in **Fig 7.5**. In the traditional symbols, a small circle, or *bubble*, always represents "NOT." (This *bubble* is called a state indicator.) Appearing just below the traditional symbol is the newer ANSI/IEEE Standard symbol. This symbol is always a square box with notations inside it. In these newer symbols, a small flag represents "NOT." The new notation is an attempt to replace the detailed logic drawing of a complex function with a simpler block symbol.

Figs 7.5, **7.6** and **7.7** show the truth tables, Boolean algebra equations and circuit symbols for the three basic Boolean operations: NOT, AND and OR. All combinational logic functions, no matter how complex, can be described in terms of these three operators.

The NOT operation is also called *inversion*, *negation* or *complement*. The circuit that implements this function is called an *inverter* or *inverting buffer*. The most common notation for NOT is a bar over a variable or expression. For example, NOT A is denoted $\overline{A}$. This is read as either "Not A" or as "A bar." A less common notation is to denote Not A by A′, which is read as "A prime."

While the inverting buffer and the noninverting buffer covered later have only one input and output, many combinational logic elements can have multiple inputs. When a combinational logic element has two or more inputs and one output, it is called a *gate*. (The term "gate" has many different but specific technical uses. For a clarification of the many definitions of gate, see the section on Synchronicity and Control Signals, later in this chapter.) For simplicity, the figures and truth tables for multiple-input elements will show the operations for only two inputs, the minimum number.

The output of an AND function is 1 only if *all* of the inputs are 1. Therefore, if *any* of the inputs are 0, then the output is 0. The notation for an AND is either a dot (•) between the inputs, as in C = A•B, or nothing between the inputs, as in C = AB. Read these equations as "C equals A AND B."

The OR gate detects if one or more inputs are 1. In other words, if *any* of the inputs are 1, then the output of the OR gate is 1. Since this includes the case where more than one input may be 1, the OR operation is also known as an INCLUSIVE OR. The OR operation detects if *at least one* input is 1. Only if all the inputs are 0, then the output is 0. The notation for an OR is a plus sign (+) between the inputs, as in C = A + B. Read this equation as "C equals A OR B."

Other Common Gates

More complex logical functions are derived from combinations of the basic logical operators. These operations — NAND, NOR, XOR, XNOR and the noninverter — are illustrated in **Figs 7.8** through **7.12** respectively. As before, each is described by a truth table, Boolean algebra equation and circuit symbols. Also as before, except for the noninverter, each could have more inputs than the two illustrated.

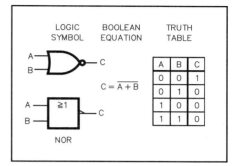

Fig 7.9 — Two-input NOR gate.

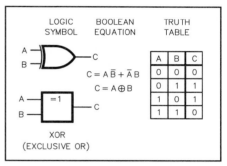

Fig 7.10 — Two-input XOR gate.

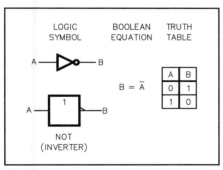

Fig 7.5 — Inverter.

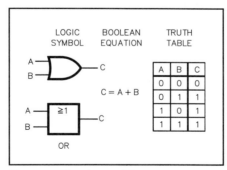

Fig 7.7 — Two-input OR gate.

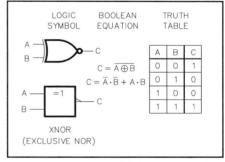

Fig 7.11 — Two-input XNOR gate.

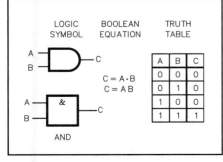

Fig 7.6 — Two-input AND gate.

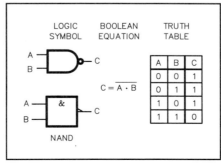

Fig 7.8 — Two-input NAND gate.

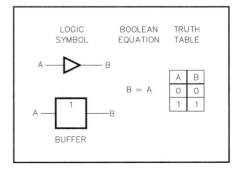

Fig 7.12 — Noninverting buffer.

The NAND gate (short for NOT AND) is equivalent to an AND gate followed by a NOT gate. Thus, its output is the complement of the AND output: The output is a 0 only if all the inputs are 1. If any of the inputs is 0, then the output is a 1.

The NOR gate (short for NOT OR) is equivalent to an OR gate followed by a NOT gate. Thus, its output is the complement of the OR output: If any of the inputs are 1, then the output is a 0. Only if all the inputs are 0, then the output is a 1.

The operations so far enable a designer to determine two general cases: (1) if *all* inputs have a desired state or (2) if *at least one* input has a desired state. The XOR and XNOR gates enable a designer to determine if *one and only one* input of a desired state is present.

The XOR gate (read as EXCLUSIVE OR) has an output of 1 if one and only one of the inputs is a 1 state. The output is 0 otherwise. The symbol for XOR is ⊕. This is easy to remember if you think of the "+" OR symbol enclosed in an "O" for *only one*.

The XOR gate is also known as a "half adder," because in binary arithmetic it does everything but the "carry" operation. The following examples show the possible binary additions for a two-input XOR.

```
0   0   1   1
0   1   0   1
0   1   1   0
```

The XNOR gate (read as EXCLUSIVE NOR) is the complement of the XOR gate. The output is 0 if one and only one of the inputs is a 1. The output is 1 either if all inputs are 0 or more than one input is 1.

Noninverter

A *noninverter*, also known as a *buffer*, *amplifier* or *driver*, at first glance does not seem to do anything. It simply receives an input and produces the same output. In reality, it is changing other properties of the signal in a useful fashion, such as amplifying the current level. The practical uses of a noninverter include (A) providing sufficient current to drive a number of gates, (B) interfacing between two logic families, (C) obtaining a desired pulse rise time and (D) providing a slight delay to make pulses arrive at the proper time.

BOOLEAN THEOREMS

The analysis of a circuit starts with a logic diagram and then derives a circuit description. In digital circuits, this description is in the form of a truth table or logical equation. The *synthesis*, or design, of a circuit goes in the reverse: starting with an informal description, determining an equation or truth table and then expanding the truth table to components that will implement the desired response. In both of these processes, we need to either simplify or expand a complex logical equation.

To manipulate an equation, we use mathematical *theorems*. Theorems are statements that have been proven to be true. The theorems of Boolean algebra are very similar to those of standard algebra, such as commutativity and associativity. Proofs of the Boolean algebra theorems can be found in an introductory digital design textbook.

BASIC THEOREMS

Table 7.5 lists the theorems for a single variable and **Table 7.6** lists the theorems for two or more variables. These tables illustrate the *principle of duality* exhibited by the Boolean theorems: Each theorem has a dual in which, after swapping all

Table 7.5
Boolean Algebra Single Variable Theorems

Identities:	$A \cdot 1 = A$	$A + 0 = A$
Null elements:	$A \cdot 0 = 0$	$A + 1 = 1$
Idempotence:	$A \cdot A = A$	$A + A = A$
Complements:	$A \cdot \overline{A} = 0$	$A + \overline{A} = 1$
Involution:	$(\overline{A}) = A$	

Table 7.6
Boolean Algebra Multivariable Theorems

Commutativity:	$A \cdot B = B \cdot A$
	$A + B = B + A$
Associativity:	$(A \cdot B) \cdot C = A \cdot (B \cdot C)$
	$(A + B) + C = A + (B + C)$
Distributivity:	$(A + B) \cdot (A + C) = A + B \cdot C$
	$A \cdot B + A \cdot C = A \cdot (B + C)$
Covering:	$A \cdot (A + B) = A$
	$A + A \cdot B = A$
Combining:	$(A + B) \cdot (A + \overline{B}) = A$
	$A \cdot B + A \cdot \overline{B} = A$
Consensus:	$A \cdot B + \overline{A} \cdot C + B \cdot C = A \cdot B + \overline{A} \cdot C$
	$(A + B) \cdot (\overline{A} + C) \cdot (B + C) = (A + B) \cdot (\overline{A} + C)$

Table 7.7
DeMorgan's Theorem

(A) $\overline{A \cdot B} = \overline{A} + \overline{B}$
(B) $\overline{A + B} = \overline{A} \cdot \overline{B}$
(C)

(1)	(2)	(3)	(4)	(5)	(6)	(7)	(8)	(9)	(10)
A	B	$\overline{A}$	$\overline{B}$	$A \cdot B$	$\overline{A \cdot B}$	$A + B$	$\overline{A + B}$	$\overline{A} \cdot \overline{B}$	$\overline{A} + \overline{B}$
0	0	1	1	0	1	0	1	1	1
0	1	1	0	0	1	1	0	0	1
1	0	0	1	0	1	1	0	0	1
1	1	0	0	1	0	1	0	0	0

(A) and (B) are statements of DeMorgan's Theorem. The truth table at (C) is proof of these statements: (A) is proven by the equivalence of columns 6 and 10 and (B) by columns 8 and 9.

ANDs with ORs and all 1s with 0s, the statement is still true.

The tables also illustrate the *precedence* of the Boolean operations: the order in which operations are performed when not specified by parenthesis. From highest to lowest, the precedence is NOT, AND then OR. For example, the distributive law includes the expression "A + B•C." This is equivalent to "A + (B•C)." The parenthesis around (B•C) can be left out since an AND operation has higher priority than an OR operation. Precedence for Boolean algebra is similar to the convention of standard algebra: raising to a power, then multiplication, then addition.

DeMorgan's Theorem

One of the most useful theorems in Boolean algebra is DeMorgan's Theorem: $\overline{A \cdot B} = \overline{A} + \overline{B}$ and its dual $\overline{A + B} = \overline{A} \cdot \overline{B}$. The truth table in **Table 7.7** proves these statements. DeMorgan's Theorem provides a way to simplify the complement of a large expression. It also enables a designer to interchange a number of equivalent gates, as shown by **Fig 7.13**.

The equivalent gates show that the duality principle works with symbols the same as it does for Boolean equations: just swap ANDs with ORs and switch the bubbles. For example, the NAND gate — an AND gate followed by an inverter bubble — becomes an OR gate preceded by two inverter bubbles. DeMorgan's Theorem is important because it means any logical function can be implemented using either inverters and AND gates or inverters and OR gates. Also, the ability to change placement of the bubbles using DeMorgan's theorem is useful in dealing with mixed logic, to be discussed next.

POSITIVE AND NEGATIVE LOGIC

The truth tables shown in the figures in this chapter are drawn for positive logic. In *positive logic*, or *high true*, a higher voltage means true (logic 1) while a lower voltage means false (logic 0). This is also referred to as *active high*: a signal performs a named action or denotes a condi-

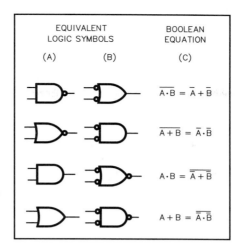

Fig 7.13 — Equivalent gates from DeMorgan's Theorem: Each gate in column A is equivalent to the opposite gate in column B. The Boolean equations in column C formally state the equivalences.

tion when it is "high" or 1. In *negative logic*, or *low true*, a lower voltage means true (1) and a higher voltage means false (0). An *active low* signal performs an action or denotes a condition when it is "low" or 0.

In both logic types, true = 1 and false = 0; but whether true means high or low differs. Company databooks are drawn for general truth tables: an "H" for high and an "L" for low. (Some tables also have an "X" for a "don't care" state.) The function of the table can differ depending on whether it is interpreted for positive logic or negative logic. **Fig 7.14** shows how a general truth table differs when interpreted for different logic types. The same truth table gives two equivalent gates: positive logic gives the function of a NAND gate while negative logic gives the function of a NOR gate.

Note that these gates correspond to the equivalent gates from DeMorgan's theorem. A bubble on an input or output terminal indicates an active low device. The

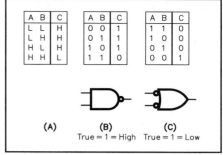

Fig 7.14 — (A) A general truth table, (B) a truth table and NAND symbol for positive logic and (C) a truth table and NOR symbol for negative logic.

absence of bubbles indicates an active high device.

Like the bubbles, signal names can be used to indicate logic states. These names can aid the understanding of a circuit by indicating control of an action (GO, /ENABLE) or detection of a condition (READY, /ERROR). The action or condition occurs when the signal is in its active state. When a signal is in its active state, it is called *asserted*; a signal not in its active state is called *negated* or *deasserted*. A prefix can easily indicate a signal's active state: active low signals are preceded by a "/," like /READY, while active high signals have no prefix. Standard practice is that the signal name and input pin match (have the same active level). For example, an input with a bubble (active low) may be called /READY while an input with no bubble (active high) is called READY. Output signal names should always match the device output pin.

In this chapter, positive logic is used unless indicated otherwise. Although using mixed logic can be confusing, it does have some advantages. Mixed logic combined with DeMorgan's Theorem can promote more effective use of available gates. Also, well-chosen signal names and placement of bubbles can promote more understandable logic diagrams.

Sequential Logic

The previous section discussed combinational logic, whose outputs depend only on the present inputs. In contrast, in *sequential logic* circuits, the new output depends not only on the present inputs but also on the present outputs. The present outputs depended on the previous inputs and outputs and those earlier outputs depended on even earlier inputs and outputs and so on. Thus, the present outputs depend on the previous *sequence of inputs* and the system has *memory*. Having the outputs become part of the new inputs is known as *feedback*.

This section first introduces a number of terms necessary to understand sequential logic: types of synchronicity, types of control signals and ways to illustrate circuit function. Numerous sequential logic circuits are then introduced. These circuits provide an overview of the basic sequential circuits that are commercially available. Depending on your approach to learning, you may choose to either (1) read the material in the order presented, definitions then examples, or (2) start with the example circuits, which begin with the flip-flop, referring back to the definitions as needed.

SYNCHRONICITY AND CONTROL SIGNALS

When a combinational circuit is given a set of inputs, the outputs take on the expected values after a propagation delay during which the inputs travel through the circuit to the output. In a sequential circuit, however, the travel through the circuit is more complicated. After application of the first inputs and one propagation delay, the outputs take on the resulting state; but then the outputs start trickling back through and, after a second propagation delay, new outputs appear. The same happens after a third propagation delay. With propagation delays in the nanosecond range, this cycle around the circuit is rapidly and continually generating new outputs. A user needs to know when the outputs are valid.

There are two types of sequential circuits: synchronous circuits and asynchronous circuits, which are analyzed differently for valid outputs. In *asynchronous* operation, the outputs respond to the inputs immediately after the propagation delay. To work properly, this type of circuit must eventually reach a *stable* state: the inputs and the fed back outputs result in the new outputs staying the same. When the nonfeedback inputs are changed, the feedback cycle needs to eventually reach a new stable state.

In *synchronous* operation, the outputs change state only at specific times. These times are determined by the presence of a particular input signal: a clock, toggle, latch or enable. Synchronicity is important because it ensures proper timing: all the inputs are present where needed when the control signal causes a change of state.

Some authors vary the meanings slightly for the different control signals. The following is a brief illustration of common uses, as well as showing uses for noun, verb and adjective. *Enabling* a circuit generally means the control signal goes to its asserted level, allowing the circuit to change state. *Latch* implies memory: (noun) a circuit that stores a bit of information or (verb) to hold at the same output state. *Gate* has many meanings, some unrelated to synchronous control: (A) a signal used to trigger the passage of other signals through a circuit (for example, "A gate circuit passes a signal only when a gating pulse is present."), (B) any logic circuit with two or more inputs and one output (used earlier in this chapter) or (C) one of the electrodes of an FET (as described in the **Analog Signals and Components** chapter). To *toggle* means a signal changes state, from 1 to 0 or vice versa. A *clock* signal is one that toggles at a regular rate.

Clock control is the most common method, so it has some additional terms, illustrated by **Fig 7.15**. The *clock period* is the time between successive transitions in the same direction; the *clock frequency* is the reciprocal of the period. A *pulse* or *clock tick* is the first edge in a clock period, or sometimes the period itself or the first half of the period. The *duty cycle* is the percentage of time that the clock signal is at its asserted level.

The reaction of a synchronous circuit to its control signal is *static* or *dynamic*. Static, *gated* or *level-triggered* control allows the circuit to change state whenever the control signal is at its active or asserted level. Dynamic, or *edge-triggered*, control allows the circuit to change state only when the control signal *changes* from unasserted to asserted. By convention, a control signal is active high if state changes occur when the signal is high or at the rising edge and active low in the opposite case. Thus, for positive logic, the convention is enable = 1 or enable goes from 0 to 1. This transition from 0 to 1 is called *positive edge-triggered* and is indicated by a small triangle inside the circuit box. A circuit responding to the opposite transition, from 1 to 0, is called *negative edge-triggered*, indicated by a bubble with the

triangle. Whether a circuit is level-triggered or edge-triggered can affect its output, as shown by **Fig 7.16**. Input D includes a very brief pulse, called a *glitch*, which may be caused by noise. The differing results at the output illustrate how noise can cause errors.

ILLUSTRATING CIRCUIT FUNCTION

Since the action of sequential circuits is more complex, many ways have been developed to examine circuit function. **Fig 7.17** shows an example for each type of table or diagram. Each type has advantages and disadvantages.

State Transition Tables

Describing a sequential circuit with the

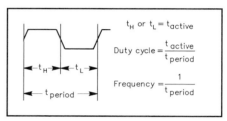

Fig 7.15 — Clock signal terms. The duty cycle would be t_H / t_{PERIOD} for an active high signal and t_L / t_{PERIOD} for an active low signal.

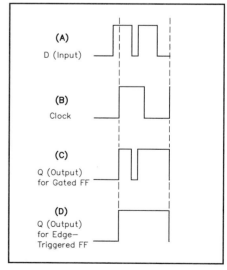

Fig 7.16 — Level-triggered vs edge-triggered for a D flip-flop: (A) input D, (B) clock input, (C) output Q for level-triggered: circuit responds whenever clock is 1. (D) output Q for edge-triggered: circuit responds only at rising edge of clock. Notice that the short negative pulse on the input D is not reproduced by the edge-triggered flip-flop.

conventional truth table would require an infinite number of previous events as the possible inputs. Instead, the sequential logic form of the truth table is called a state transition table. A *state transition* (or excitation) *table* lists all combinations of present inputs and feedback outputs — the *current state*, represented by small q, and the resulting outputs, both feedback and nonfeedback and the *next state*, represented by capital Q. The state transition table is the most common way to describe sequential circuit function, since it lists all possibilities.

Characteristic Equation

An equivalent equation for a state transition table or excitation table can be written in terms of the *state variables*, a symbol for each input and output. Although the number of state variables can be very large for a complex system, there is always a limited number of inputs and outputs; thus, the combinations of inputs and outputs is also limited. A circuit with n binary state variables has 2^n possible states. Since the possible states, 2^n, are always finite, sequential circuits are also known as *finite-state machines*. Each of the sequential logic circuits in this section includes a state table to illustrate the circuit functions. As before, these tables are for positive logic, so 1s and 0s are already substituted for Highs and Lows.

Excitation Table

An *excitation table* is derived from the truth table. Its usefulness is to show, for each possible output Q_n what inputs are needed to obtain a desired output Q_{n+1}. For some outputs, one or more of the input variables may not have an effect. In this case, the input variable corresponds to a *don't care* state, represented by an "X" or dash.

Timing Diagram

When a clock is the controlling signal, a timing diagram can describe the circuit operation. A *timing diagram* draws a circuit's signals as a function of time. This form emphasizes the cause-and-effect delays between critical signals. It is especially useful in detecting errors, as will be shown in the flip-flop implementations to be presented. There are software packages available to *simulate* a circuit. The simulation produces the timing diagram for a given set of inputs, allowing the designer to examine the timing diagram for expected results.

State Diagram

A *state diagram* depicts output changes in terms of a flow chart. Each possible

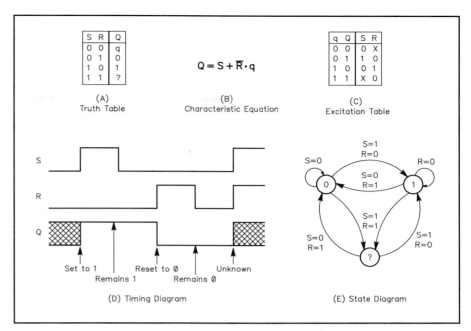

Fig 7.17 — Sequential circuit function for a clocked S-R flip-flop: (A) state transition table or truth table, (B) characteristic equation, including its derivation from the state table, (C) excitation table, (D) partial timing diagram, (E) state diagram.

state is listed inside a circle with arrows between the circles to indicate possible next states. State diagrams are especially useful for studying a sequence of inputs and the corresponding result.

FLIP-FLOPS

Flip-flops are the basic building blocks of sequential circuits. A *flip-flop* is a device with two stable states: the *set* state (1) or the *reset* state (0). (The reset state is also called the *cleared* state.) The flip-flop can be placed in one or the other of the two states by applying the appropriate input. (Since a common use of flip-flops is to store one bit of information, some use the term *latch* interchangeably with flip-flop. A set of latches, or flip-flops holding an n-bit number is called a *register*.) While gates have special symbols, the schematic symbol for most components is a rectangular box with the circuit name or abbreviation, the signal names and assertion bubbles. For flip-flops, the circuit name is usually omitted since the signal names are enough to indicate a flip-flop and its type. The four basic types of flip-flops are the S-R, D, T and J-K. The first section examines the S-R flip-flop for each of the various control methods. The next section introduces each of the other basic flip-flops and their uses.

S-R Flip-Flop

The S-R flip-flop is one of the simplest circuits for storing a bit of information. It

has two inputs, represented by S (set) and R (reset). These inputs, naturally, cause the two possible output states: if S = 1 and R = 0, then output Q is set to 1; if S = 0 and R = 1, then output Q is reset to 0; if both inputs are 0, then the output remains unchanged; and if both inputs are 1, then the output cannot be determined. The S-R flip-flop can illustrate each of the types of control signals: unclocked (asynchronous, no control signal), clocked or gated, master-slave and edge-triggered.

Unclocked/Sequential

The unclocked S-R flip-flop, shown in **Fig 7.18**, is an asynchronous device; its outputs change immediately to reflect changes on its inputs. The circuit consists of two NOR gates. The sequential nature of the circuit is a result of the output of each NOR gate being fed back as an input to the opposite gate. The state transition table shows the expected set/reset pattern of inputs to outputs. The table shows an unpredictable result for inputs S = 1 and R = 1. In actual circuits, the results vary and are usually either $Q = \overline{Q} = 1$ or $Q = \overline{Q} = 0$. While $Q = \overline{Q}$ is a logical impossibility, real flip-flops may present this output. The designer should avoid the R = S = 1 input and make no assumptions about the resulting output. The flip-flop is not predictable if both inputs go to 0 at exactly the same time.

Fig 7.18C shows an alternate implementation of the S-R flip-flop, with two NAND

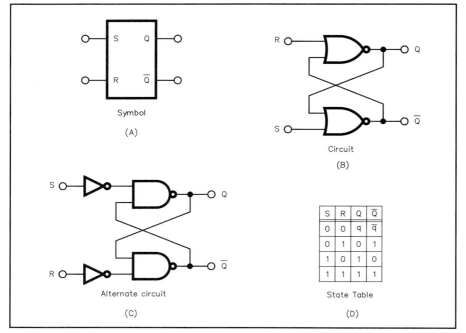

Fig 7.18 — Unclocked S-R Flip-Flop. (A) schematic symbol. (B) circuit diagram. (C) alternate circuit. (D) state table.

S	R	Q	Q̄
0	0	q	q̄
0	1	0	1
1	0	1	0
1	1	1	1

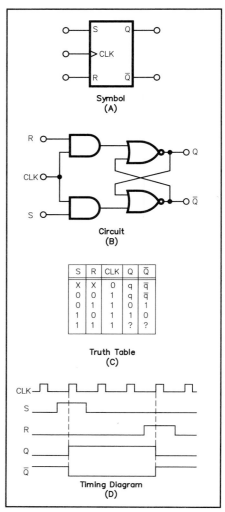

Fig 7.19 — Clocked S-R Flip-Flop. (A) schematic symbol. (B) circuit. (C) truth table. (D) timing diagram.

S	R	CLK	Q	Q̄
X	X	0	q	q̄
0	0	1	q	q̄
0	1	1	0	1
1	0	1	1	0
1	1	1	?	?

gates and two inverters. Since a NAND gate can become an inverter by having its two inputs receive the same signal, the S-R flip-flop can be implemented with four NAND gates. This alternate version is important because a 4-NAND gate chip is one of the most readily available commercial integrated circuits; thus, the 4-NAND gate S-R flip-flop can be implemented on a single IC.

Gated or Level-Triggered

The gated S-R flip-flop, or gated latch, has a controlling input in addition to its S and R inputs. The inputs S and R produce the same results as those on an unclocked S-R flip-flop, but a change in output will only occur when the control input is high. A gated S-R flip-flop is illustrated in **Fig 7.19** along with a timing diagram for a clock input. This flip-flop is also called the R-S-T flip-flop, where "T," for toggle, is the clock input. Although not often used, the R-S-T flip-flop is important because it illustrates a step between the R-S flip-flop and the J-K flip-flop.

A problem with the level-triggered flip-flop is that the Q output can change more than once while the clock is asserted. We would prefer the output to change only once per clock period for easier timing design. A second problem can occur when flip-flops are connected in series and triggered by the same clock pulse or, similarly, when a flip-flop is in series with itself, using its own output as an input. Since the series-connected flip-flop feeds back to itself, its output will be changing at about the same time as it receives new input. This can result in an erroneous output.

Master/Slave Flip-Flop

A solution to the problems of the level-triggered method is a circuit that samples and stores its inputs before changing its outputs. Such a circuit is built by placing two flip-flops in series; both flip-flops are triggered by a common clock but an inverter on the second flip-flop's clock causes it to be asserted only when the first flip-flop is not asserted. The action for a given clock pulse is as follows: The first, or *master*, flip-flop is active when the clock is high, sampling and storing the inputs. The second, or *slave*, flip-flop gets its input from the master and acts when the clock is low. Hence, when the clock is 1, the input is sampled; then when the clock becomes 0, the output is generated. A master/slave flip-flop is built with either two S-R flip-flops, as shown by **Fig 7.20**, or with two J-K flip-flops. Note that a bubble appears on the schematic symbol's clock input, reminding us that the output appears when the clock is asserted low. This is conventional for TTL-style J-K flip-flops, but it can be different for CMOS devices.

The master/slave method isolates output changes from input changes, eliminating the problem of series-fed circuits. It also ensures only one new output per clock period, since the slave flip-flop responds to only the single sampled input. A prob-

lem can still occur, however, because the master flip-flop can change more than once while it is asserted; thus, there is the potential for the master to sample at the wrong time. There is also the potential that either flip-flop can be affected by noise.

Edge-Triggered Flip-Flop

The edge-triggered flip-flop solves the problem of noise. An example of noise is the glitch shown earlier in Fig 7.16. The different outputs for the level and edge-triggered methods in this figure show how a glitch can cause an output error. Edge-triggering avoids the problem of noise by minimizing the time during which a circuit responds to its inputs: the chance of a glitch occurring during the nanosecond transition of a clock pulse is remote. A side benefit of edge-triggering is that only one new output is produced per clock period. Edge-triggering is denoted by a small rising-edge or falling-edge symbol

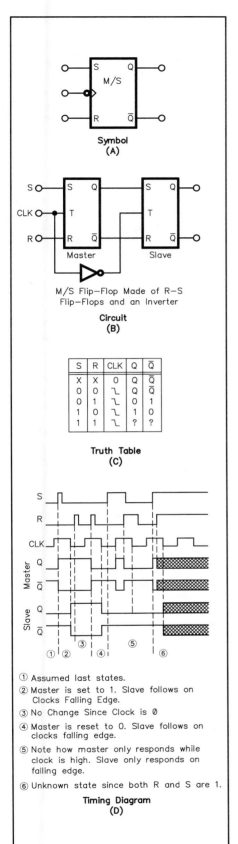

Fig 7.20 — Master/Slave S-R Flip-Flop. (A) schematic symbol. (B) circuit. (C) truth table. (D) timing diagram.

Symbol
(A)

M/S Flip–Flop Made of R–S Flip–Flops and an Inverter

Circuit
(B)

Master   Slave

S	R	CLK	Q	Q̄
X	X	0	Q	Q̄
0	0	⌐	Q	Q̄
0	1	⌐	0	1
1	0	⌐	1	0
1	1	⌐	?	?

Truth Table
(C)

① Assumed last states.
② Master is set to 1. Slave follows on Clocks Falling Edge.
③ No Change Since Clock is 0
④ Master is reset to 0. Slave follows on clocks falling edge.
⑤ Note how master only responds while clock is high. Slave only responds on falling edge.
⑥ Unknown state since both R and S are 1.

Timing Diagram
(D)

Table 7.8

Summary of Standard Flip–Flops

q = current state   Q = next state   X = don't care

Flip–Flop Type	Symbol	Truth Table	Characteristic Equation	Excitation Table
SR		S R CLK Q X X 0 q 0 0 1 q 0 1 1 0 1 0 1 1 1 1 1 ?	$Q = \left(S + \bar{R}\cdot q\right)\cdot CLK$ with $S \cdot R = 0$	q Q S R CLK 0 0 0 X 1 0 1 1 0 1 1 0 0 1 1 1 1 X 0 1
D		D CLK Q X 0 q 0 1 0 1 1 1	$Q = D \cdot CLK$	q Q D CLK 0 0 X 0 0 1 1 1 1 0 0 1 1 1 X 0
T		T CLK Q X 0 q 0 1 q 1 1 q̄	$Q = (T \oplus q)\cdot CLK$	q Q T CLK 0 0 0 X 0 1 1 1 1 0 1 1 1 1 0 X
JK		J K CLK Q X X 0 q 0 0 1 q 0 1 1 0 1 0 1 1 1 1 1 q̄	$Q = \left(J \cdot \bar{q} + \bar{K}\cdot q\right)\cdot CLK$	q Q J K CLK 0 0 0 X 1 0 1 1 X 1 1 0 X 1 1 1 1 X 0 1
JK		J K CLK Q 0 0 ⌐ 0 0 1 ⌐ 0 1 0 ⌐ 1 1 1 ⌐ q̄	$Q = \left(J \cdot \bar{q} + \bar{K}\cdot q\right)\cdot CLK$	q Q J K CLK 0 0 0 X ⌐ 0 1 1 X ⌐ 1 0 X 1 ⌐ 1 1 X 0 ⌐

such as ⌐ or ⌐; sometimes an arrow is included such as ⌐ or ⌐. This symbol appears in the circuit's truth table and can also appear, instead of the clock triangle, inside the schematic symbol.

Other Flip-flops

Table 7.8 provides a summary of the four basic flip-flops: the S-R (Set-Reset), D (Data or Delay), T (Toggle) and J-K. Each is briefly explained below, including its particular applications. The internal circuitry of each of these flip-flops is similar to the components and complexity of the S-R flip-flop. Readers may be interested in trying to design their own circuit implementation for a flip-flop type and control method; however, in practical use, a commercially available integrated circuit chip would probably be used. Company databooks include the individual circuit implementation for each IC. Digital design textbooks will also show sample circuit implementations for each of the flip-flops.

D Flip-Flop

In a D (data) flip-flop, the *data* input is transferred to the outputs when the flip-flop is enabled: The logic level at input D is transferred to Q when the clock is positive; the Q output retains this logic level until the next positive clock pulse (see **Fig 7.21**). The flip-flop is also called a *delay* flip-flop because, once enabled, it passes D after a propagation delay. A D flip-flop is useful to store one bit of information. A collection of D flip-flops forms a register.

Toggle Flip-Flop

In a T flip-flop, the output *toggles* (changes state) with each positive clock pulse. The T flip-flop is also called a

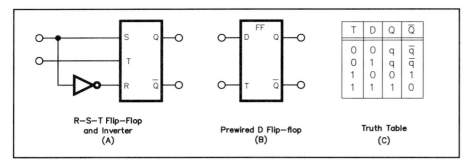

Fig 7.21 — (A) The D flip-flop. When T = 0, Q and $\overline{Q}$ states don't change. When T = 1, the output states change to reflect the D input. (C) A truth table for the D flip-flop.

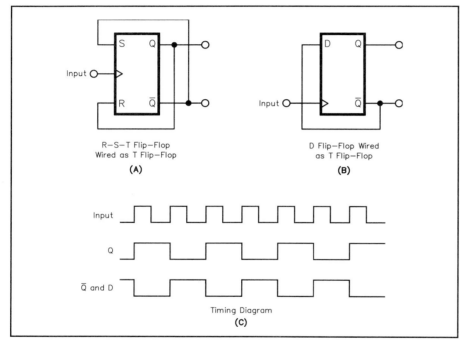

Fig 7.22 — (A) A clocked S-R-T flip-flop wired as a T flip-flop. (B) A D flip-flop wired as a T flip-flop. (C) Timing diagram. Notice that the output frequency is half the input frequency.

complementing flip-flop. **Fig 7.22** shows how a T flip-flop can be created from either an S-R or D flip-flop. The timing diagram in Fig 7.22 shows an important result of the T-flip-flop: the output frequency is one half of the input frequency. Thus, a T flip-flop is a 2:1 (also called *modulo-2* or *radix-2*) frequency divider. Two T flip-flops connected in series form a 4:1 divider and so on.

J-K Flip-Flop

It's somewhat ironic that the most readily available flip-flop, the J-K flip-flop, is discussed last and so briefly. The discussion is short because the J-K flip-flop acts the same as the S-R flip-flop (where J = S and K = R) with only one difference: The S-R flip-flop had the disadvantage of invalid results for the inputs 1,1. For the J-K flip-flop, simultaneous 1,1 inputs cause Q to change state after the clock transition.

Summary

Only the D and J-K flip-flops are generally available as commercial integrated circuit chips. Since memory and temporary storage are so often desirable, the D flip-flop is manufactured as the simplest way to provide memory. When more functionality is needed, the J-K flip-flop is available. The J-K flip-flop can substitute for an S-R flip-flop and a T flip-flop can be created from either the D or J-K flip-flop.

COUNTERS

Groups of flip-flops can be combined to make counters. Toggle flip-flops are the most common for implementing a counter. Intuitively, a counter is a circuit that starts at state 0 and sequences up through states 1, 2, 3, to m, where m is the maximum number of states available. From state m, the next state will return the counter to 0. This describes the most common counter: the *n-bit binary counter*, with n outputs corresponding to $2^n = m$ states. Such a counter can be made from n flip-flops, as shown in **Fig 7.23**. This figure shows implementations for each of the types of synchronicity. Both circuits pass the data count from stage to stage. In the asynchronous counter, Fig 7.23A, the clock is also passed from stage to stage and the circuit is called *ripple* or *ripple-carry*. In the synchronous counter, Fig 7.23B, each stage is controlled by a common clock signal.

There are numerous variations on this first example of a counter. Most counters have the ability to *clear* the count to 0. Some counters can also *preset* to a desired count. The clear and preset control inputs are often asynchronous — they change the output state without being clocked. Counters may either count up (increment) or down (decrement). *Up/down* counters can be controlled to count in either direction. Counters can have sequences other than the standard numbers, for example a BCD counter.

Counters are also not restricted to changing state on every clock cycle. An n-bit counter that changes state only after m clock pulses is called a *divider* or *divide-by-m* counter. There are still $2^n = m$ states; however, the output after p clock pulses is now p / m. Combining different divide-by-m counters can result in almost any desired count. For example, a base 12 counter can be made from a divide-by-2 and a divide-by-6 counter; a base 10 (decade) counter consists of a divide-by-2 and a BCD divide-by-5 counter.

The outputs of these counters are binary. To produce output in decimal form, the output of a counter would be provided to a binary-to-decimal decoder chip and/or an LED display.

REGISTERS

Groups of flip-flops can be combined to make registers, usually implemented with D flip-flops. A *register* stores n bits of information, delivering that information in response to a clock pulse. Registers usually have asynchronous *set* to 1 and *clear* to 0 capabilities.

Storage Register

A storage register simply stores temporary information, for example incoming information or intermediate results. The size is related to the basic size of information handled by a computer: 8 flip-flops for an 8-bit or *byte register* or 16 bits for a *word register*. **Fig 7.24** shows a typical circuit and schematic symbols for an 8-bit storage register. In (C), although the bits are passed on 8 separate lines (from 8 flip-flops), a slash and number, "/8," is used to simplify the symbol. Storage registers are important to computer architecture; this topic is discussed in depth later in the chapter.

Shift Register

Shift registers also store information and provide it in response to a clock signal, but they handle their information differently: When a clock pulse occurs, instead of each flip-flop passing its result to the output, the flip-flops pass their data to each other, up and down the row. For example, in up mode, each flip-flop receives the output of the preceding flip-flop. A data bit starting in flip-flop D0 in a left shifter would move to D1, then D2 and so on until it is shifted out of the register. If a 0 was input to the least significant bit, D0, on each clock pulse then, when the last data bit has been shifted out, the register contains all 0s.

Shift registers can be left shifters, right shifters or controlled to shift in either direction. The most general form, a *universal shift register*, has two control inputs for four states: Hold, Shift right, Shift left and Load. Most also have asynchronous inputs for preset, clear and parallel load. The primary use of shift registers is to convert parallel information to serial or vice versa. This is useful in interfacing between devices, and is discussed in detail in the Digital Interfacing section.

Additional uses for a shift register are to (1) delay or synchronize data, (2) multiply or divide a number by a factor 2^n or (3) provide random data. Data can be delayed simply by taking advantage of the Hold feature of the register control inputs. Multiplication and division with shift registers is best explained by example: Suppose a 4-bit shift register currently has the value 1000 = 8. A right shift results in the new parallel output 0100 = 4 = 8 / 2. A second right shift results in 0010 = 2 = (8 / 2) / 2. Together the 2 right shifts performed a division by 2^2. In general, shifting right n times is equivalent to dividing by 2^n. Similarly, shifting left multiplies by 2^n. This can be useful to compiler writers to make a computer program run faster. Random data is provided via a ring

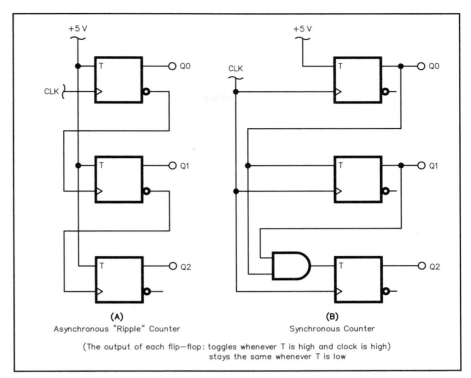

(A) Asynchronous "Ripple" Counter   **(B)** Synchronous Counter

(The output of each flip-flop: toggles whenever T is high and clock is high)
stays the same whenever T is low)

Fig 7.23 — Three-bit binary counter: (A) asynchronous or ripple counter, (B) synchronous counter.

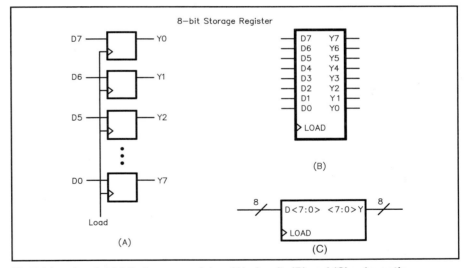

Fig 7.24 — An eight-bit storage register: (A) circuit, (B) and (C) schematic symbols.

counter. A *ring counter* is a shift register with its output fed back to its input. At each clock pulse, the register is shifted up or down and some of the flip-flops feedback to other flip-flops, generating a random binary number. Shift registers with several feedback paths can be used as a *pseudorandom number generator*, where the sequence of bits output by the genera-

tor meets one or more mathematical criteria for randomness.

MULTIVIBRATORS

A multivibrator is widely used as a switch and comes in three basic forms: bistable, monostable and astable. It is broadly defined as a closed-loop, regenerative circuit that alternates between two

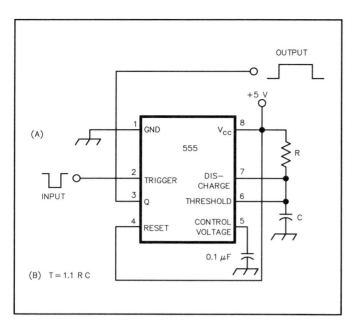

Fig 7.25 — (A) A 555 timer connected as a monostable multivibrator. (B) The equation to calculate values for R (in ohms) and C (in farads), where T is the pulse duration (in seconds).

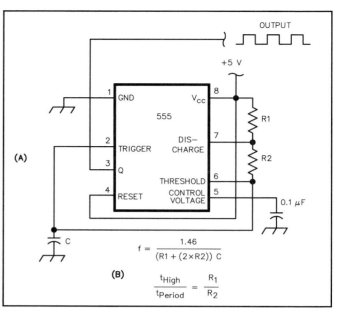

Fig 7.26 — (A) A 555 timer connected as an astable multivibrator. (B) The equations to calculate values for R1, R2 (in ohms) and C (in farads), where f is the clock frequency (in Hertz).

stable or quasi-stable states. The flip-flop is a *bistable multivibrator*: both of its two states are stable; it can be triggered from one stable state to the other by an external signal. To create *quasi-stable* or *unstable* states, energy-storing devices (capacitors) are added in the feedback loops of the multivibrator; the instability is a result of the exponential decay of the stored energy. A *monostable multivibrator* is the result of adding one energy-storing element into a feedback loop. An *astable multivibrator* is the result of adding two energy-storing elements, one in each feedback loop.

Monostable Multivibrator

A *monostable* or *one-shot* multivibrator has one energy-storing element in its feedback paths, resulting in one stable and one quasi-stable state. It can be switched, or *triggered*, to its quasi-stable state; then returns to the stable state after a time delay. Thus, the one-shot multivibrator puts out a pulse of some duration, T. (Note that T is not the period, but the duration of the quasi-stable state.) Triggering during the stable state results in the pulse, as expected. Triggering during the unstable state has two possibilities: A *nonretriggerable* multivibrator is not affected. A *retriggerable* multivibrator will start counting its pulse duration from the most recent trigger pulse. Both types of one-shots are common.

Fig 7.25 shows a 555 timer IC connected as a one-shot multivibrator. The one-shot is activated by a negative-going pulse be-

tween the trigger input and ground. The trigger pulse causes the output (Q) to go positive and capacitor C to charge through resistor R. When the voltage across C reaches two-thirds of V_{CC}, the capacitor is quickly discharged to ground and the output returns to 0. The output remains at logic 1 for a time determined by T = 1.1 RC, where R is the resistance in ohms and C is the capacitance in farads.

Astable Multivibrator

An *astable* or *free-running* multivibrator has two energy-storing elements in its feedback paths, resulting in two quasi-stable states. It continuously switches between these two states without external excitation. Thus, the astable multivibrator puts out a sequence of pulses. By properly selecting circuit components, these pulses can be of a desired frequency and width.

Fig 7.26 shows a 555 timer IC connected as an astable multivibrator. The capacitor C charges to two-thirds V_{CC} through R1 and R2 and discharges to one-third V_{CC} through R2. The ratio R1 : R2 sets the asserted high duty cycle of the pulse: t_{HIGH} / t_{PERIOD}. The output frequency is determined by:

$$f = 1.46 / (R1 + 2\,R2)\,C$$

where:
  R1 and R2 are in ohms,
  C in farads and
  f in hertz.

It may be difficult to produce a 50%

duty cycle, due to manufacturing tolerance for the resistors R1 and R2. One way to ensure a 50% duty cycle is to run the astable multivibrator at 2f and then divide by 2 with a toggle flip-flop.

Applications

An astable multivibrator is useful in generating clock pulses. When triggered by a clock pulse, the one-shot multivibrator acts to lengthen or "stretch" the pulse, which is useful to delay digital events. Either of these pulse signals, when input to the bistable multivibrator (flip-flop), can be the control of a sequential circuit. The three types of multivibrators can ensure synchronicity, so that a sequential circuit will execute correctly.

SUMMARY

Digital logic plays an increasingly important role in Amateur Radio. Most of this logic is binary and can be described and designed using Boolean algebra. Using the NOT, AND and OR gates of combinational logic, designers can build sequential logic circuits that have memory and feedback. The simplest sequential logic circuit is called a flip-flop. By using control inputs, a flip-flop can latch a data value, retaining one bit of information and acting as memory. Combinations of flip-flops can form useful circuits such as counters, storage registers and shift registers. The primary method of controlling sequential circuits is via a clock pulse, which can be created with a multivibrator.

Digital Integrated Circuits

Integrated circuits (ICs) are the cornerstone of digital logic devices. Modern technology has enabled electronics to become miniature in size and less expensive. Today's complex digital equipment would be impossible with vacuum tubes or even with discrete transistors.

An IC is a miniature electronic module of components and conductors manufactured as a single unit. All you see is a ceramic or black plastic package and the silver-colored pins sticking out. Inside the package is a piece of material, usually silicon, created (fabricated) in such a way that it conducts an electric current to perform logic functions, such as a gate, flip-flop or decoder.

As each generation of ICs surpassed the previous one, they became classified according to the number of gates on a single chip. These classifications are roughly defined as:

Small-scale integration (SSI):
10 or fewer gates on a chip.
Medium-scale integration (MSI):
10-100 gates.
Large-scale integration (LSI):
100-1000 gates.
Very-large-scale integration (VLSI):
1000 or more gates.

This chapter will primarily deal with SSI ICs, the basic digital building blocks. Microprocessors, memory chips and programmable logic devices are discussed later in the Computer Hardware section.

The previous section discussed the design of a digital circuit. To build that circuit, the designer must choose between IC chips available in various logic families. Each family and subfamily has its own desirable characteristics. This section reviews the primary IC logic families of interest to radio amateurs. The designer may also be challenged to interface between different logic families or between a logic device and peripheral device. The former is discussed at the end of this section; the latter with Computer Hardware, later in the chapter.

COMPARING LOGIC FAMILIES

When selecting devices for a circuit, a designer is faced with choosing between many families and subfamilies of logic ICs. Which subfamily is right for the application at hand is among several desirable characteristics: logic speed, power consumption, fan-out, noise immunity and cost.

Speed

Logic device families operate at widely varying clock speeds. Standard transistor-transistor logic (TTL) devices can only operate up to a few MHz while some emitter-coupled logic (ECL) ICs can operate at several GHz. Gate propagation delay determines the maximum clock speed at which an IC can operate; the clock period must be long enough for all signals within the IC to propagate to their destinations. ICs with capacitively coupled inputs have minimum, as well as maximum, clock rates. While the initial reaction may be to use the fastest available ICs, the designer must usually choose between a trade-off of high speed and low power consumption.

Power Consumption

In some applications, power consumption by logic gates is a critical design consideration. This power consumption can be divided into two parts: Dynamic power is the power consumed when a gate changes state. Static power is the power consumed when a gate is holding a state, either high or low. Each of these has different power requirements. Calculating the total required power can be a complex task when several gates and diverse functions are involved. Nominal power requirements, however, can be used to compare logic subfamilies.

Fan-out

Gate impedance is another parameter to consider. To deliver high current without dropping considerable voltage, an ideal gate would have low output impedance. To draw minimal current, an ideal gate would have infinite input impedance. Such a gate does not exist. The designer must compromise on input and output impedances.

A gate output can supply only a limited amount of current. Therefore, a single output can only drive a limited number of inputs. The measure of driving ability is called fan-out, expressed as the number of inputs (of the same subfamily) that can be driven by a single output. If a logic family that is otherwise desirable does not have sufficient fan-out, consider using noninverting buffers to increase fan-out, as shown by **Fig 7.27**.

Noise Immunity

The noise margin was illustrated in Fig 7.2. The choice of voltage levels for the binary states determines the noise margin. If the gap is too small, a spurious signal can too easily produce the wrong state. Too large a gap, however, produces longer, slower transitions and thus decreased switching speeds.

Circuit impedance also plays a part in noise immunity, particularly if the noise is from external sources such as radio transmitters. At low impedances, more energy is needed to change a given voltage level than at higher impedances.

Other Considerations

The parameters above are the basic considerations to influence the selection of a logic family for a specific application. These considerations have complex interactions that come into play in demanding low-current, high-speed or high-complexity circuits. Numerous other parameters can also be examined. These are provided in the electrical specification of a device. These are given on a data sheet and usually include four sections: (1) absolute maximum ratings specify worst-case conditions, including safe storage temperatures, (2) recommended operating conditions specify power-supply, input voltage, dc output loading and temperatures for normal operation, (3) electrical characteristics specify other dc voltages and currents observed at the inputs and outputs and (4) switching characteristics specify propagation delays for "typical" operation.

The list of parameters can seem overwhelming to the novice; but with experience, the important information will be more easily spotted. If you are designing a circuit, always consult the data sheet for specific information on the device you are considering, because these parameters vary not only between the logic families but also vary between the manufacturers and with changing technologies. Each manufacturer has data books available listing their devices and the corresponding data sheets.

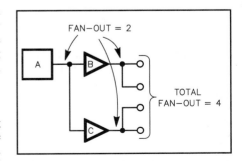

Fig 7.27 — Nonverting buffers used to increase fan-out: Gate A (fan-out = 2) is connected to two buffers, B and C, each with a fan-out of 2. Result is a total fan-out of 4.

BIPOLAR LOGIC FAMILIES

Two broad categories of digital logic ICs are *bipolar* and *metal-oxide semiconductor* (MOS). Numerous manufacturing techniques have been developed to fabricate each type. Each surviving, commercially available family has its particular advantages and disadvantages and has found its own special niche in the market.

Bipolar semiconductor ICs usually employ NPN junction transistors. (Bipolar ICs are possible using PNP transistors, but NPN transistors make faster circuits.) While early bipolar logic was faster and had higher power consumption than MOS logic, these distinctions have blurred as manufacturing technology has developed. There are several families of bipolar logic devices and within some of these families there are subfamilies. The most-used digital logic family is Transistor-Transistor Logic (TTL). Another bipolar logic family, Emitter Coupled Logic (ECL), has exceptionally high speed but high power consumption.

Transistor-Transistor Logic (TTL)

The TTL family has seen widespread acceptance because it is fast and has good noise immunity. It is by far the most commonly used logic family. TTL levels were shown earlier in Fig 7.2: An input voltage between 0.0-0.4 V will represent LOW and an input voltage between 2.4-5.0 V will represent HIGH.

TTL Subfamilies

The original standard TTL is infrequently used today. In the standard TTL circuit, the transistors saturate, reducing the operating speed. TTL variations cure this by clamping the transistors with Schottky diodes to prevent saturation, or by using a dopant in the chip fabrication to reduce transistor recovery time. Schottky-clamped TTL is the faster of these two manufacturing processes.

TTL IC identification numbers begin with either 54 or 74. The 54 prefix denotes a military temperature range of −55 to 125°C, while 74 indicates a commercial temperature range of 0 to 70°C. The next letters, in the middle of the TTL device number, indicate the TTL subfamily. Following the subfamily designation is a 2, 3 or 4-digit device-identification number. For example, a 7400 is a standard-TTL NAND gate and a 74LS00 is a low-power Schottky NAND gate. (The NAND gate is the workhorse TTL chip. Recall, from Fig 7.18, the alternative implementation of the S-R flip-flop.) The following TTL subfamilies are available:

	74xx	standard TTL
H	74Hxx	High-speed
L	74Lxx	Low-power
S	74Sxx	Schottky
LS	74LSxx	Low-power Schottky
AS	74ASxx	Advanced Schottky
ALS	74ALSxx	Advanced Low-power Schottky

Each subfamily is a compromise between speed and power consumption. Because the speed-power product is approximately constant, less power consumption results in less speed and vice versa. For the amateur, an additional consideration to the speed-versus-power trade-off is the cost trade-off. The advanced Schottky devices offer both increased speed and reduced power consumption but at a higher cost.

In addition to the above power/speed/cost trade-offs, each TTL subfamily has particular characteristics that can make it suitable or unsuitable for a specific design. **Table 7.9** shows some of these parameters. The actual parameter values may vary slightly from manufacturer to manufacturer so always consult the manufacturers' data books for complete information.

TTL Circuits

Fig 7.28A shows the schematic representation of a TTL hex inverter. A 7404 chip contains four of these inverters. When the input is low, Q1 is ON, conducting current from base to emitter through the input lead and into ground. Thus a low TTL input device must be prepared to sink current from the input. Since Q1 is saturated, Q2 is OFF because there is not enough voltage at its base. Similarly, Q4 is also OFF. With Q2 and Q4 OFF, Q3 will be ON and pull the output high, about one volt below V_{CC}. When the input is high, an unusual situation occurs: Q1 is operating in the inverse mode, with current flowing from base to collector. This current causes Q2 to be ON, which causes Q4 to be ON. With Q2 and Q4 ON, there is not enough current left for Q3, so Q3 is OFF. Q4 is pulling the output low.

By replacing Q1 with a multiple-emitter transistor, as is done with the two-input Q5 in Fig 7.28B, the inverter circuit becomes a NAND gate. Commercially available TTL NAND gates have as many as 13 inputs, the limiting factor being the number of input pins on the standard 16-pin chip. The operation of this multiple-input NAND circuit is the same as described for the inverter, the difference being that any one of the emitter inputs being low will conduct current through the emitter, leading to the conditions described above to produce a high at the output. Similarly, all inputs must be high to produce the low output.

In the TTL circuit of Fig 7.28A, transistors Q3 and Q4 are arranged in a *totem-pole* configuration. This configuration gives the output circuit a low source impedance, allowing the gate to source (supply) or sink substantial output current. The 130-Ω resistor between the collector of Q3 and +V_{CC} limits the current through Q3.

When a TTL gate changes state, the amount of current that it draws changes rapidly. These changes in current, called switching transients, appear on the power supply line and can cause false triggering of other devices. For this reason, the power bus should be adequately decoupled. For proper decoupling, connect a 0.01 to 0.1 µF capacitor from V_{CC} to ground near each device to minimize the transient currents caused by device switching and magnetic coupling. These capacitors must be low-inductance, high-frequency RF capacitors

Table 7.9

TTL and CMOS Subfamily Performance Characteristics

TTL Family	Propagation Delay (ns)	Per Gate Power Consumption (mW)	Speed Power Product (pico-joules)
Standard	9	10	90
L	33	1	33
H	6	22	132
S	3	20	60
LS	9	2	18
AS	1.6	20	32
ALS	5	1.3	6.5

CMOS Family Operating with 4.5 <V_{CC} <5.5 V		f=100 kHz	f=1 MHz	f=10 MHz	f=100 kHz	f=1 MHz	f=10 MHz
HC	18	0.0625	0.6025	6.0025	1.1	10.8	108
HCT	18	0.0625	0.6025	6.0025	1.1	10.8	108
AC	5.25	0.080	0.755	7.505	0.4	3.9	39
ACT	4.75	0.080	0.755	7.505	0.4	3.6	36

(disk-ceramic capacitors are preferred). In addition, a large-value (50 to 100 µF) capacitor should be connected from V_{CC} to ground somewhere on the board to accommodate the continually changing I_{CC} requirements of the total V_{CC} bus line. These are generally low-inductance tantalum capacitors rather than rolled-foil mylar or aluminum-electrolytic capacitors.

Darlington and Open-Collector Outputs

Fig 7.28C and D show variations from the totem-pole configuration. They are the Darlington transistor pair and the open-collector configuration respectively.

The Darlington pair configuration replaces the single transistor Q4 with two transistors, Q4 and Q5. The effect is to provide more current-sourcing capability in the high state. This has two benefits: (1) the rise time is decreased and (2) the fanout is increased.

Transistor(s) on the output in both the totem-pole and Darlington configurations provide active pull-up. Omitting the transistor(s) and providing an external resistor for passive pull-up gives the open-collector configuration. This configuration, unfortunately, results in slower rise time, since a relatively large external resistor must be used. The technique has some very useful applications, however: driving other devices, performing wired logic, busing and interfacing between logic devices.

Devices that need other than a 5-V supply can be driven with the open-collector output by substituting the device for the external resistor. Example devices include light-emitting diodes (LEDs), relays and solenoids. Inductive devices like relay coils and solenoids need a "flyback" protection diode across the coil. You must pay attention to the current ratings of open-collector outputs in such applications. You may need a switching transistor to drive some relays or other high-current loads.

Open-collector outputs can perform wired logic, rather than gated IC logic, by wire-ANDing the outputs. This can save the designer an AND gate, potentially sim-

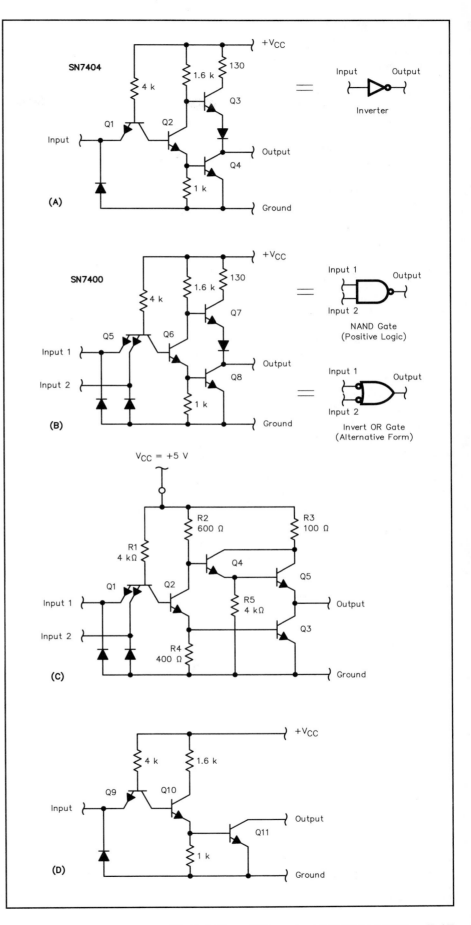

Fig 7.28 — Example TTL circuits and their equivalent logic symbols: (A) an inverter and (B) a NAND gate, both with totem-pole outputs. (C) A NAND gate with a Darlington output. (D) A NAND gate with an open-collector output. (Indicated resistor values are typical. Identification of transistors is for text reference only; these are not discrete components but parts of the silicon die.)

plifying the design. Wire-ANDed outputs are several open-collector outputs connected to a single external pull-up resistor. The overall output, then, will only be high when all pull-down transistors are OFF (all connected outputs are high), effectively performing an AND of the connected outputs. If any of the connected outputs are low, the output after the external resistor will be low. **Fig 7.29** illustrates the wire-ANDing of open-collector outputs.

The wire-ANDed concept can be applied to several devices sharing a common bus. At any time, all but one device has a high-impedance (off) output. The remaining device, enabled with control circuitry, drives the bus output.

Open-collector outputs are also useful for interfacing TTL gates to gates from other logic families. TTL outputs have a minimum high level of 2.4 V and a maximum low level of 0.4 V. When driving nonTTL circuits, a pull-up resistor (typically 2.2 kΩ) connected to the positive supply can raise the high level to 5 V. If a higher output voltage is needed, a pull-up resistor on an open-collector output can be connected to a positive supply greater than 5 V, so long as the chip output voltage and current maximums are not exceeded.

Three-State Outputs

While open-collector outputs can perform bus sharing, a more popular method is three-state output, or tristate, devices. The three states are low, high and high impedance, also called Hi-Z or *floating*. An output in the high-impedance state behaves as if it is disconnected from the circuit, except for possibly a small leakage current. Three-state devices have an additional disable input. When enable is low, the device provides high and low outputs just as it would normally; when enable is high the device goes into its high-impedance state.

A bus is a common set of wires, usually used for data transfer. A three-state bus has several three-state outputs wired together. With control circuitry, all devices on the bus but one have outputs in the high-impedance state. The remaining device is enabled, driving the bus with high and low outputs. Care should be taken to ensure only one of the output devices can be enabled at any time, since simultaneously connected high and low outputs may result in an incorrect logic voltage. (The condition when more than one driver is enabled at the same time is called *bus contention*.) Also, the large current drain from V_{CC} to ground through the high driver to the low driver can potentially damage the circuit or produce noise pulses that can affect overall system behavior.

Unused TTL Inputs

A design may result in the need for an n-input gate when only an n + m input gate is available. In this case, the recommended solution for extraneous inputs is to give the extra inputs a constant value that won't affect the output. A low input is easily provided by connecting the input to ground. A high input can be provided with either an inverter whose input is ground or with a pull-up resistor. The pull-up resistor is preferred rather than a direct connection to power because the resistor limits the current, thus protecting the circuit from transient voltages. Usually, a 1-kΩ to 5-kΩ resistor is used; a single 1-kΩ resistor can handle up to 10 inputs.

It's important to properly handle all inputs. Design analysis would show that an unconnected, or floating, TTL input is usually high but can easily be changed low by only a small amount of capacitively coupled noise.

Emitter-Coupled Logic (ECL)

ECL, also called current-mode logic (CML), is the fastest commercially available logic family, with some devices operating at frequencies higher than 1.2 GHz. The fast speed is a result of reducing the propagation delay by keeping the transistors from saturating. ECL devices operate with their transistors in the active region. The voltage swing is small, less than a volt; and the circuit internally switches between two possible paths depending on the output state. The two-path arrangement provides a significant feature of ECL: complementary output states are always available.

Naturally, high speed comes at some cost, in this case high power consumption. Heat sinking is sometimes necessary because of the great deal of power being dissipated. Because of its poor speed-power product and also because it is not directly compatible with TTL and CMOS, ECL is less popular than TTL. ECL devices are most likely to be found where performance is more important than cost, including UHF frequency counters, UHF frequency synthesizers and high-speed mainframe computers.

ECL Subfamilies

There are several ECL subfamilies, to balance the trade-off of high speed versus low power dissipation. The subfamilies differ mostly in resistance values and the presence or absence of input and output pull-down resistors.

The most popular subfamily is the 10K series, with five-digit part numbers of the form "10xxx." This family's design started one of ECL's most familiar characteristics: operation with $V_{CC} = 0$ V (ground) and V_{EE} at a negative voltage. This feature provides immunity to power supply noise, since noise on V_{EE} is rejected by the circuit's differential amplifier. The design voltage, $V_{EE} = -5.2$ V for the 10K subfamily, provides the best noise immunity; but other voltages can be used. Typically, a high logic state corresponds to -0.9 V and a low is −1.75 V.

ECL Circuits

ECL gets its name from the emitter-coupled pair of transistors in the circuit,

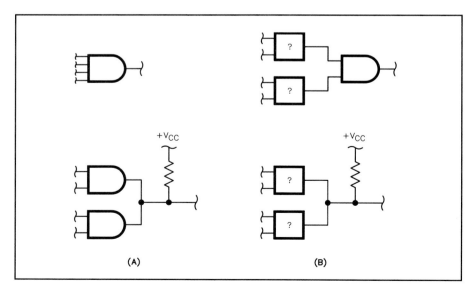

Fig 7.29 — The outputs of two open-collector-output AND gates are shorted together (wire ANDed) to produce an output the same as would be obtained from a 4-input AND gate.

connected as a differential amplifier. For example, in **Fig 7.30**, either Q1 or Q2 together with Q3 form a differential amplifier. This arrangement produces the complementary outputs available from each ECL circuit. The circuit in Fig 7.30 provides both an OR output and a NOR output. When an input is high, its transistor (Q1 or Q2) is ON but not in saturation; and Q3 is OFF. Q6 is then OFF so its emitter output is low, while Q5 is ON and its output high. Similarly, when both inputs are low, Q1 and Q2 are OFF so the NOR output from Q6 is high; and Q3 is ON, so the OR output from Q5 is low. Q4, D1, D2 and associated circuitry form a bias generator. The reference voltage at the base of Q3 determines the input switching threshold.

METAL-OXIDE SEMICONDUCTOR (MOS) LOGIC FAMILIES

While bipolar devices use junction transistors, MOS devices use field effect transistors (FETs). MOS is characterized by simple device structure, small size (high density) and ease of fabrication. MOS circuits use the NOR gate as the workhorse chip rather than the NAND. MOS families are used extensively in digital watches, calculators and VLSI circuits such as microprocessors and memories.

P-Channel MOS (PMOS)

The first MOS devices to be fabricated were PMOS, conducting electrical current by the flow of positive charges (holes). PMOS power consumption is much lower than that of bipolar logic, but its operating speed is lower. The only extensive use of PMOS is in calculators and watches, where low speed is acceptable and low power consumption and low cost are desirable.

N-Channel MOS (NMOS)

With improved fabrication technology, NMOS became feasible and provided improved performance and TTL compatibility. The speed of NMOS is at least twice that of PMOS, since electrons rather than holes carry the current. NMOS also has greater gain than PMOS and supports greater packaging density through the use of smaller transistors.

Complementary MOS (CMOS)

CMOS combines both P-channel and N-channel devices on the same substrate to achieve high noise immunity and low power consumption: less than 1 mW per gate and negligible power during standby. This accounts for the widespread use of CMOS in battery-operated equipment. The high impedance of CMOS gates

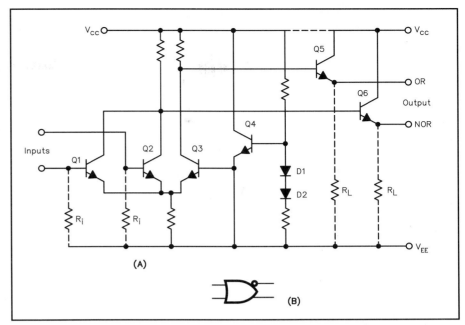

Fig 7.30 — (A) Circuit topology of the ECL family. (B) The modified logic symbol to indicates the availability of the complementary output.

makes them susceptible to electromagnetic interference, however, particularly if long traces are involved. Consider a trace $1/4$-wavelength long between input and output. The output is a low-impedance point so the trace is effectively grounded at this point. You can get high RF potentials $1/4$-wavelength away, which disturbs circuit operation.

A notable feature of CMOS devices is that the logic levels swing to within a few millivolts of the supply voltages. The input switching threshold is approximately one half the supply voltage ($V_{DD} - V_{SS}$). This characteristic contributes to high noise immunity on the input signal or power supply lines. CMOS input-current drive requirements are minuscule, so the fan-out is great, at least in low-speed systems. (For high-speed systems, the input capacitance increases the dynamic power dissipation and limits the fan-out.)

CMOS Subfamilies

There are a number of CMOS subfamilies available. Like TTL, the original CMOS has largely been replaced by later subfamilies using improved technologies. This original family, called the 4000-series, has numbers beginning with 40 or 45 followed by two or three numbers to indicate the specific device. 4000B is second generation CMOS. When introduced, this family offered low power consumption but was fairly slow and not easy to interface with TTL.

Later CMOS subfamilies provided improved performance and TTL compatibility. For simplicity, the later subfamilies were given numbers similar to the TTL numbering system, with the same leading numbers, 54 or 74, followed by 1 to 3 letters indicating the subfamily and as many as 5 numbers indicating the specific device. The subfamily letters usually include a "C" to distinguish them as CMOS.

The following CMOS device families are available:

4000	4071B	standard CMOS
C	74Cxx	CMOS versions of TTL

Devices in this subfamily are pin and functional equivalents of many of the most popular parts in the 7400 TTL family. It may be possible to replace all TTL ICs in a particular circuit with 74C-series CMOS, but this family should not be mixed with TTL in a circuit without careful design considerations. Devices in the C series are typically 50% faster than the 4000 series.

HC   74HCxx High-speed CMOS

Devices in this subfamily have speed and drive capabilities similar to Low-power Schottky (LS) TTL but with better noise immunity and greatly reduced power consumption. High-speed refers to faster than the previous CMOS family, the 4000-series.

HCT 74HCTxx   High-Speed CMOS, TTL compatible

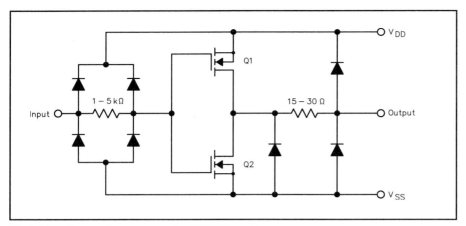

Fig 7.31 — Internal structure of a CMOS inverter.

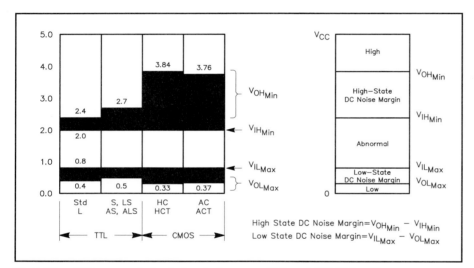

High State DC Noise Margin=$V_{OH_{Min}}$ − $V_{IH_{Min}}$
Low State DC Noise Margin=$V_{IL_{Max}}$ − $V_{OL_{Max}}$

Fig 7.32 — Differences in logic levels for some TTL and CMOS families.

Devices in this subfamily were designed to interface TTL to CMOS systems. The HCT inputs recognize TTL levels, while the outputs are CMOS compatible.

AC 74ACxxxxx Advanced CMOS

Devices in this family have reduced propagation delays, increased drive capabilities and can operate at higher speeds than standard CMOS. They are comparable to Advanced Low-power Schottky (ALS) TTL devices.

ACT 74ACTxxxxx Advanced CMOS, TTL compatible

This subfamily combines the improved performance of the AC series with TTL-compatible inputs.

As with TTL, each CMOS subfamily has characteristics that make it suitable or unsuitable for a particular design. You should consult the manufacturer's data

books for complete information on each subfamily you are considering.

CMOS Circuits

A simplified diagram of a CMOS logic inverter is shown in **Fig 7.31**. When the input is low, the resistance of Q2 is low so a high current flows from V_{CC}; since Q1's resistance is high, the high current flows to the output. When the input is high, the opposite occurs: Q2's resistance is low, Q1's is high and the output is low. The diodes are to protect the circuit against static charges.

Special Considerations

Some of the diodes in the input- and output-protection circuits are an inherent part of the manufacturing process. Even with the protection circuits, however, CMOS ICs are susceptible to damage from static charges. To protect against damage from static, the pins should not be inserted

in styrofoam as is sometimes done with other components. Instead, a spongy conductive material is available for this purpose. Before removing a CMOS IC from its protective material, make certain that your body is grounded. Touching nearly any large metal object before handling the ICs is probably adequate to drain any static charge off your body. Some people prefer to touch a grounded metal object or to use a conductive bracelet connected to the ground terminal of a three-wire ac outlet through a 10-MΩ resistor. Since wall outlets aren't always wired properly, you should measure the voltage between the ground terminal and any metal objects you might touch. Connecting yourself to ground through a 1 MΩ to 10 MΩ resistor will limit any current that might flow through your body.

All CMOS inputs should be tied to an input signal; a positive supply voltage or ground if a constant input is desired. Undetermined CMOS inputs, even on unused gates, may cause gate outputs to oscillate. Oscillating gates draw high current, overheat and self destruct.

The low power consumption of CMOS ICs made them attractive for satellite applications, but standard CMOS devices proved to be sensitive to low levels of radiation — cosmic rays, gamma rays and X rays. Later, radiation-hardened CMOS ICs, able to tolerate 10^6 rads, made them suitable for space applications. (A rad is a unit of measurement for absorbed doses of ionizing radiation, equivalent to 10^{-2} joules per kilogram.)

SUMMARY

There are many types of logic ICs, each with its own advantages and disadvantages. If you want low power consumption, you should probably use CMOS. If you want ultra-high-speed logic, you will have to use ECL. Whatever the application, consult up-to-date literature when designing logic circuits. IC databooks and applications notes are usually available from IC manufacturers and distributors.

INTERFACING LOGIC FAMILIES

Each semiconductor logic family has its own advantages in particular applications. For example, the highest frequency stages in a UHF counter or a frequency synthesizer would use ECL. After the frequency has been divided down to less than 25 MHz, the speed of ECL is unnecessary; and its expense and power dissipation are unjustified. TTL or CMOS are better choices at lower frequencies.

When a design mixes ICs from different logic families, the designer must account for the differing voltage and current requirements each logic family recognizes. The designer must ensure the appropriate interface between the point at which one logic family ends and another begins. A knowledge of the specific input/output (I/O) characteristics of each device is necessary, and a knowledge of the general internal structure is desirable, to ensure reliable digital interfaces. Typical internal structures have been illustrated for each common logic family. **Fig 7.32** illustrates the logic level changes for different TTL and CMOS families; databooks should be consulted for manufacturer's specifications.

Often more than one conversion scheme is possible, depending on whether the designer wishes to optimize power consumption or speed. Usually one quality must be traded off for the other. The following section discusses some specific logic conversions. Where an electrical connection between two logic systems isn't possible, an optoisolator can sometimes be used.

TTL Driving CMOS

TTL and low-power TTL can drive 74C series CMOS directly over the commercial temperature range without an external pull-up resistor. However, they cannot drive 4000-series CMOS directly; and for HC-series devices, a pull-up resistor is recommended. The pull-up resistor, connected between the output of the TTL gate and V_{CC} as shown in **Fig 7.33A**, ensures proper operation and enough noise margin by making the high output equal to V_{DD}. Since the low output voltage will also be affected, the resistor value must be chosen with both desired high and low voltage ranges in mind. Resistors values in the range 1.5 kΩ to 4.7 kΩ should be suitable for all TTL families under worst conditions. A larger resistance reduces the maximum possible speed of the CMOS gate; a lower resistance generates a more favorable RC product but at the expense of increased power dissipation.

HCT-series and ACT-series CMOS devices were specifically designed to interface nonCMOS devices to a CMOS system. An HCT device acts as a simple buffer between the nonCMOS (usually TTL) and CMOS device and may be combined with a logic function if a suitable HCT device is available.

When the CMOS device is operating from a power supply other than +5 V, the TTL interface is more complex. One fairly simple technique uses a TTL open-collector output connected to the CMOS input, with a pull-up resistor from the CMOS input to the CMOS power supply. Another method, shown in Fig 7.33B, is a common-base level shifter. The level shifter trans-

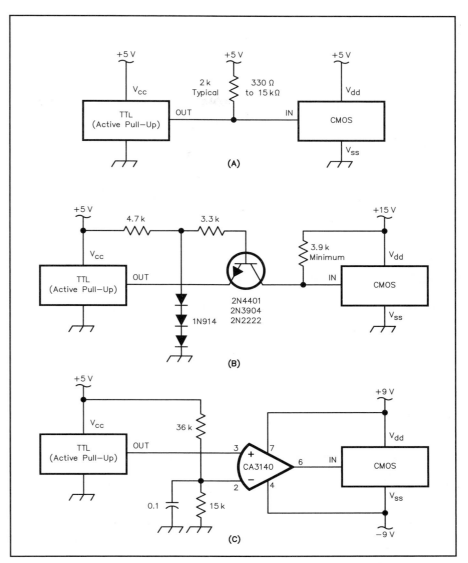

Fig 7.33 — TTL to CMOS interface circuits: (A) pull-up resistor, (B) common-base level shifter and (C) op amp configured as a comparator.

lates a TTL output signal to a +15 V CMOS signal while preserving the full noise immunity of both gates. An excellent converter from TTL to CMOS using dual power supplies is to configure an operational amplifier as a comparator, as shown in Fig 7.33C. An FET op amp is shown because its output voltage can usually swing closer to the rails (+ and − supply voltages) than a bipolar unit.

CMOS Driving TTL

Certain CMOS devices can drive TTL loads directly. The output voltages of CMOS are compatible with the input requirements of TTL, but the input-current requirement of TTL limits the number of TTL loads that a CMOS device can drive from a single output (the fan-out).

Interfacing CMOS to TTL is a bit more complicated when the CMOS is operating at a voltage other than +5 V. One technique is shown in **Fig 7.34A**. The diode blocks the high voltage from the CMOS gate when it is in the high output state. A germanium diode is used because its lower forward-voltage drop provides higher noise immunity for the TTL device in the low state. The 68-kΩ resistor pulls the input high when the diode is back biased.

There are two CMOS devices specifically designed to interface CMOS to TTL when TTL is using a lower supply voltage. The CD4050 is a noninverting buffer that allows its input high voltage to exceed the supply voltage. This capability allows the CD4050 to be connected directly between the CMOS and TTL devices, as shown in Fig 7.34B. The CD4049 is an inverting buffer that has the same capabilities as the CD4050.

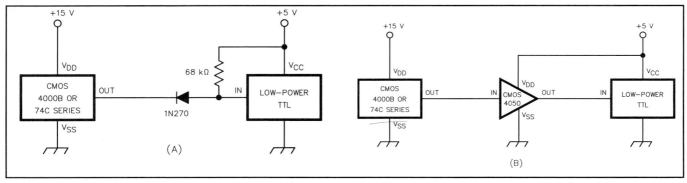

Fig 7.34—CMOS to TTL interface circuits: (A) blocking diode used when different supply voltages are used. The diode is not necessary if both devices operate with a +5 V supply. (B) CMOS noninverting buffer IC.

Computer Hardware

So far, this chapter has discussed digital logic, the implementation of that logic with integrated circuits, interfacing IC logic families and the use of memory to store information used by the ICs. The synthesis of all this technology is the microcomputer—combining a microprocessor IC, memory, peripheral devices, and user interface into the modern personal computer. A computer has both physical components, hardware and a collection of programs, software, to tell it what to do. This section (by Bob Wolbert, K6XX) will focus on the physical components of the computer: its internal physical components, their interaction, and peripheral I/O devices that communicate with other systems and the operator. While the basic theory has changed little in the past couple of decades, implementations have vastly improved. Unlike most other sections of the *Handbook*, this portion is destined for rapid obsolescence due to the extremely fast progression of the computer industry. While the underlying concepts will remain valid and useful, specific discussions of processor type and speed, memory size, hard drive capacity, etc, will appear "quaint" alarmingly quickly. The author has attempted to minimize such content, both for your benefit and to reduce his future embarrassment!

WHAT IS A COMPUTER?

The strictest definition of the term "computer" includes special purpose digital systems optimized for a particular task. For example, a modern synthesized transceiver, with its memory, I/O, serial control, DSP, etc. meets the definition of a computing device. Many of the concepts discussed in this section apply equally well to your HF rig as well as to your PC, however, our definition of a computer will be restricted

to a general purpose machine whose task is quickly and easily changed by loading or changing software: If its task cannot be readily modified—to compute a spreadsheet or compose e-mail, for example—we will exclude that system from our discussion. The personal computer (PC) will be emphasized due to its ubiquitous nature.

The three major divisions of a PC are its hardware, its software, and its firmware. See **Fig 7.35**.

The hardware includes the *central processing unit* (CPU) and input/output (I/O) devices. Software refers to the programs that are loaded into the computer to configure it for the task at hand. Firmware, also called microcode or BIOS (Basic Input/Output System), is a hybrid of both hardware and software that is used to perform specific tasks; the microcode is the basis of the microprocessor's command set that tells it how to fetch data and add numbers, for example; the BIOS is firmware generally used to start-up (boot) the system.

COMPUTER ARCHITECTURE

Unlike many present textbooks, where computer architecture is narrowly defined as only including those attributes of the system that interest programmers, our discussion deals with the structural organization and hardware design of the digital computer system. While all modern computer systems consist of three basic sections: the CPU, memory, and peripherals for interfacing with the operator and the real world—the architecture of a computer is the arrangement of these internal subsystems:. The CPU, called a *microprocessor* in personal computers, is an IC consisting of three major parts: a control unit, an arithmetic logic unit (ALU) and temporary storage registers. A *bus* — a set of wires carrying address, data and control

information — interconnects all of the subsystems. Virtually all computers are designed based on the basic "Von Neumann" architecture shown in **Fig 7.36**

The microprocessor, memory chips and other circuitry are all part of the system's hardware, the physical components of a system. The computer case, the nuts and bolts and physical parts are other parts of the hardware. A computer also includes software, a collection of programs or sequence of instructions to perform a specified task. The design of computers is so complex, however, that it is nearly impossible to design an original architecture without any bugs. Thus many designers use microprocessors that include *microcode* or microinstructions: instructions in the control unit of a microprocessor. This hybrid between hardware and software is called firmware. Firmware also includes software stored in ROM or EPROM rather than being stored on magnetic disk or tape.

Computer designers make decisions on hardware, software and firmware based on

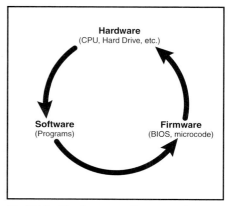

Fig 7.35—Hardware, software and firmware comprise a computer,

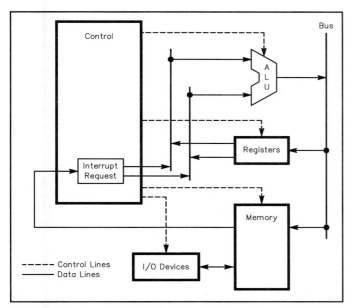

Fig 7.36—Example of a basic computer architecture.

cost versus performance. Today's computer market includes a wide range of systems, from high-performance super-computers costing millions of dollars, to the personal microcomputer, with prices in the high hundreds to a few thousands new and ranging from free on up for older used models.

THE CENTRAL PROCESSING UNIT

The central processing unit is usually a single microprocessor chip, although its subsystems can be on more than one chip. The CPU at least includes a control unit, timing circuitry, an arithmetic logic unit (ALU) and registers for temporary storage. Modern microprocessors have tens of millions of transistors and are designed in modules, as shown in the sidebar, "Microprocessor Layout."

Control Unit

The control unit directs the operation of the computer, managing the interaction between subunits. It takes instructions from the memory and executes them, performing tasks such as accessing data in memory, calling on the ALU or performing I/O. Control is one of the most difficult parts to design; thus it is the most likely source of bugs in designing an original architecture.

Microprocessors consist of both hardwired control and

Microprocessor Layout and Block Description

Block   Function

PIC   Programmable Interrupt Controller.
EBL   External front side Bus Logic.
CLK   Clock and control circuitry.
BBL   Back-side Bus interface Logic.
DTLB Data Translation Look-aside Buffer translates linear addresses to physical address required for virtual memory operation.
PMH   Page Miss Handler.
PFU   Packed Floating point arithmetic Unit.
FEU   Floating point Execution Unit.
L2   Level 2 Cache.
DCU   Data Cache Unit. Contains level one data cache along with associated fill and write back buffering.
MOB   Memory Order Buffer acts as a separate schedule and dispatch engine for data loads and stores.
IEU   Integer Execution Unit for ALU functionality of scalar intege instructions.
MIU   Memory Interface Unit. This is responsible for data conversion and formatting of floating point data types.
RAT   Register Alias Table.
SIMD SIMD integer execution unit.
BTB   Branch Target Buffer, responsible for dynamic branch prediction based on the history of past decisions paths.
ALLOC Allocator. Allocation of various resources such as ROB, MOB, and RS entries is performed here prior to micro-instruction dispatch by the RS.
RS   Reservation Station. Micro-instructions and source data are held here for scheduling and dispatch to the execution ports.
BAC   Branch Address Calculator.
TAP   Testability Access Port for production testing and debugging.
IFU   Instruction Fetch Unit.
ID   Instruction Decoder.

ROB   Re-Order Buffer. This supports a physical register file that holds temporary write-back results that can complete out of order.
MS   Micro-instruction Sequencer, which holds the micro code and sequencer.

Condensed Die Photo Detail

Left region: Functionality in this area includes bus interface circuitry and a level 2 cache. Top center region: Functionality in this region is split into assorted functions including data cache access, and allocation. Top right region: This primarily consists of the execution data path for the Pentium III processor. Bottom right region: Instruction decode, scheduling, dispatch, and retirement functionality is contained within this region.

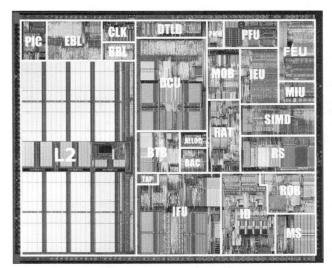

Pentium III Processor die photo (*courtesy of Intel Corporation*).

microprogrammed control. In both cases, the designer determines a sequence of states through which the computer cycles, each with inputs to examine and outputs to activate other CPU subsystems (including activating itself, indicating which state to do next). For example, the sequence usually starts with "Fetch the next instruction from memory," with control outputs to activate memory for a read, a program counter to send the address to be fetched and an instruction register to receive the memory contents. Hardwired control is completely via circuitry, usually with a programmed logic array. Microprogrammed control uses a microprocessor with a modifiable control memory, containing microcode or microinstructions. An advantage of microprogrammed control is flexibility: the code can be changed without changing the hardware, making it easier to correct design errors. **Fig 7.37** shows examples of both types of control.

RISC or CISC?

There are two fundamental schools of thought regarding microprocessor design. One, referred to as *complex* instruction set computer (CISC), believes that adding more built-in functions to the microprocessor leads to better programs and more cost-efficient overall systems. The other, the *reduced* instruction set computer (RISC) philosophy, recognizes that only about 20% of the instructions built into standard CISC processors are commonly used; if this subset of instructions were streamlined in the processor, faster overall system operation and a smaller, less expensive microprocessor would result. RISC architectures value speed and processor die size (chip area) over programming ease and main memory conservation. The choice is partially an economic one: slow, expensive memory (relative to logic) favors CISC; more balanced speeds and costs favor RISC. Another factor, reduced dependence on direct assembly language programming and improved compiler technology, reduces the attractiveness of extra instructions the compiler cannot easily use. At present, the two schools are merging, with the most popular processors designed around a RISC core but with microcode providing a multitude of other built-in instructions.

Timing

Usually, an oscillator controlled by a quartz crystal generates the microcomputer's clock signal. The output of this clock goes to the microprocessor and to other ICs. The clock synchronizes the microcomputer subunits. For example, each of the microinstructions is designed to take only one clock cycle to execute, so any components triggered by a microinstruction's control outputs should finish their actions by the end of the clock cycle. The exception to this is memory, which may take multiple clock cycles to finish, so the control unit repeats in its same state until memory says it's done. Since the clock rate effectively controls the rate at which instructions are executed, the clock frequency is one way to measure the speed of a computer. Clock frequency, however, cannot be the only criteria considered because the actions performed during a clock cycle vary for different designs, particularly processors with superscalar or pipelined designs capable of computing multiple items simultaneously.

Arithmetic Logic Unit

The *arithmetic logic unit* (ALU) performs logical operations such as AND, OR and SHIFT and two number arithmetic operations such as addition, subtraction, multiplication and division. The ALU depends on the control unit to tell it which operation to perform and also to trigger other devices (memory, registers and I/O) to supply its input data and to send out its results to the appropriate place.

The ALU often only performs simple operations. Complex operations, such as multiplication, division and operations involving decimal numbers, are performed by dedicated hardware, called floating-point processors, or *coprocessors*.

Registers

Microprocessor chips have some internal memory locations that are used by the control unit and ALU. Be-cause they are inside the microprocessor IC, these registers can be accessed more quickly than main memory locations. Special purpose registers or *dedicated registers* are purely internal, have predefined uses and cannot be directly accessed by programs. *General purpose registers* hold data and addresses in use by programs and can be directly accessed, although usually only by assembly level programs.

The dedicated registers include the instruction register, program counter, effective address register and status register. The first step to execute an instruction is to fetch it from memory and put it in the *instruction register* (IR). The *program counter* (PC) is then incremented to contain the address of the next instruction to

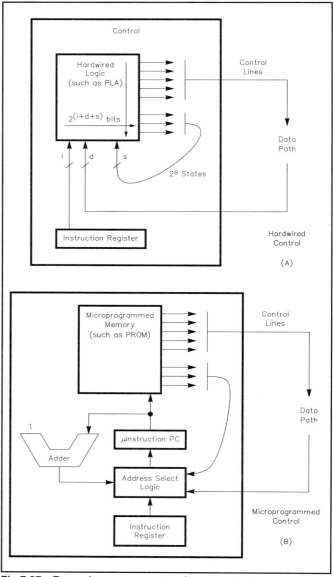

Fig 7.37—Example arrangements of a control unit and related components: (A) hardwired control and (B) microprogrammed control.

be fetched. An instruction may change the program counter as a result of a conditional branch (if-then), loop, subroutine call or other nonlinear execution. If data from memory is needed by an instruction, the address of the data is calculated and fetched with the *effective address register* (EAR). The *status register* (SR) keeps track of various conditions in the computer. For example, it tells the control unit when the keyboard has been typed on so the control unit knows to get input. It also notices if something goes wrong during an instruction execution, for example an attempted divide by 0, and tells the control unit to halt the program or fix the error. Certain bits in the status register are known as the *condition codes*, flags set by each instruction. These flags tell information about the result of the latest instruction — such as if the result was negative or positive or zero and if an arithmetic overflow or a carry error occurred. The flags can then be used by a conditional branch to decide if that branch should be taken or not.

MEMORY

Computers and other digital circuits rely on stored information, either data to be acted upon or instructions to direct circuit actions. This information is stored in memory devices, in binary form. Computers use four main types of memory, as shown in **Fig 7.38**. This section concentrates on solid state memory, and first discusses how to access an individual item in memory and then compares different memory types, which can vary how quickly and easily an item is accessed.

Accessing a Memory Item

Memory devices consist of a large number of memory cells each capable of remembering one bit of binary information. The information in memory is stored in digital form with collections of bits, called words, representing numbers and symbols. The most common symbol set is the American National Standard Code for Information Interchange (ASCII). Words in memory, just like the letters in this sentence, are stored one after the other. They are accessed by their location or address. The number of bits in each word, equal to the number of memory cells per memory location, is constant within a memory device but can vary for different devices. Common memory devices have word sizes of 8, 16 and 32 bits.

Addresses and Chip Size

An *address* is the identifier, or name, given to a particular location in memory. Since this address is expressed as a binary number, the number of unique addresses

available in a particular memory chip is determined by the number of bits to express the address. For example, a memory chip with 8 bit addresses has $2^8 = 256$ memory locations. These locations are accessed as the addresses 00000000 through 11111111, 0 through 255 decimal or 00 through FF hex. (For ease of notation, programmers and circuit designers use hexadecimal (base 16) notation to avoid long strings of 1s and 0s.) The memory chip size can be expressed as M × N, where M is the number of unique addresses, or memory locations and N is the word size, or number of bits per memory location. Memory chips come in a variety of sizes and can be arranged, together with control circuitry and decoders, to meet a designer's needs.

Basic Structure

Memory chips, no matter how large or small, have several things in common. Each chip has address, data and control lines, as shown by the example chip in **Fig 7.39**. A memory chip must have enough address lines to uniquely address each of its words and as many data lines as there are bits per word. For example, the 256 × 1 memory in Fig 7.37 has 8 address lines and 1 data line.

The control lines for a memory chip can vary. Fig 7.37 shows a simple example: two control lines, a $R/\overline{W}$ (read/write) and CS (chip select). In this case, data lines transfer both inputs (when writing) and outputs (when reading) so the $R/\overline{W}$ control line is needed to put the memory chip in read mode or write mode. The chip select, CS, control line tells the chip whether it is in use. When the chip is selected, it is "on," acting upon the address, data and $R/\overline{W}$ information presented to it. When the chip is not selected, the data line enters a high-impedance state so that it does not affect, and is not affected by, devices or circuits attached to it.

Reading and Writing

To write (store data in) or read (retrieve data from) a memory device, it is necessary to gain access to specific memory cells. A small 256 × 1 memory chip is used as an example. Later, this example will be

expanded to a larger computer memory system.

If we want to write a 1 to the 11th word of the 256 × 1 memory (such as memory location 10 decimal or 00001010 binary), we must execute the following steps:

(1) Place the correct address (00001010) on the address lines.

(2) Place the data to be written (1) on the data line.

(3) Set the $R/\overline{W}$ control line to write (low, 0).

(4) Set the CS control line to select (high, 1). (Many memory devices use an active low chip select, $\overline{CS}$.)

This writes the data on the data line (1) to the address on the address lines (00001010).

The steps to read the contents of the 11th word are similar except that the R/W control line is set to read (high, 1).

Timing

Subtle timing requirements must also be incorporated into the above steps. While writing, the address and data information must be present for a minimum setup time before and hold time after, the CS and $R/\overline{W}$ signals have been activated. This is to avoid spurious signals spraying all over the memory array. While reading, address line changes are not harmful, but the output data is only valid a minimum access time after the last address input is stable. Manufacturers' data sheets and application notes provide the timing specifications for the particular IC you are using.

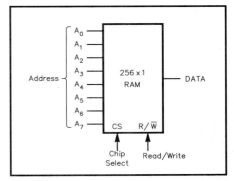

Fig 7.39—Example of a 256 × 1 memory chip.

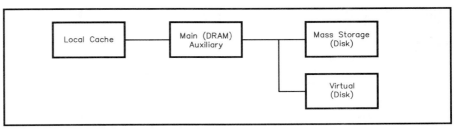

Fig 7.38—Computer memory types.

Larger Words

The one-bit-wide memory described above provides a good introduction, but usually we want a wider memory. One way to get wider memory is to use several 1-bit-wide memory chips, as shown in **Fig 7.40**. The address and control lines go to each chip, and data from each chip is used as a single bit in the large word. It is easy to see that when reading from address 0A (hex), the data lines D0 through D3 contain the data from address 0A of chips U0 through U3.

An address placed on the shared address lines (called an address bus) now specifies an entire word of data. Notice that one line of the address bus connects to the CS pin of each memory chip. This line is labeled ME, or *memory enable*, and sets all four ICs to read or write data at the same time.

If all four memory chips were put in a single package, they would make a 256 × 4 IC. This IC would look much like the chip in Fig 7.39, except that it would have 4 data lines.

More Address Space

For larger memory systems, the same principles as shown in Fig 7.40 can be applied. **Fig 7.41** shows a 1024 × 8 (1 kilobyte) memory built with four 256 × 8 memory chips. A kilobyte, or kbyte is usually abbreviated as K. Notice this is not quite the same as the metric prefix kilo, because it represents 1024, rather than 1000.

Ten address lines are needed to address 1024 locations ($2^{10} = 1024$). Eight of the 10 address lines, A0 to A7, are used as a normal address bus for chips 0 through 3. The remaining 2 address lines, A8 and A9, are run through a 2-to-4 line decoder to choose between the 4 memory chips. When employed in this manner, the 2-to-4 line decoder is called an address decoder.

To assert the CS input for one of the memory chips, ME must be 1 and the correct output of the 2-to-4 line decoder must also be 1. When an address is placed on A0 through A9, a single memory chip is selected by ME, A8 and A9. The other 8 address lines address a single word from that chip. The three chips that are not selected enter a high-impedance state and do not affect the data lines. This example shows that, using the proper memory chips and address decoding, any size memory with any word length can be built. Digital design invariably uses a modular approach, combining and cascading electronic building blocks until the desired level of capacity or complexity is achieved.

Alternate Structures

Fig 7.42 shows how the same chip can be accessed in different ways by using two decoders, a row decoder and a column decoder. The same 256 × 256 memory array can be treated as a 64 K × 1 array, a 256 × 256 array or other possibilities. In fact, most larger memory chips are made as square arrays: 32 × 32 (1024 bytes or 1 K), 64 × 64 (4096 bytes or 4 K), 256 × 256 (65536 bytes or 64 K), 1024 × 1024 (1 M), 2048 × 2048 (4 M), 4096 × 4096 (16 M) and so on. (Here the M represents a megabyte, which is 1,048,576 bytes.) The square array makes the chips more cost effective to manufacture (easier quality control and less waste) and easier to incorporate into a printed-circuit-board circuit layout. Notice that each M × N is a power

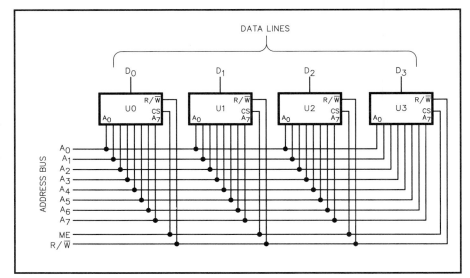

Fig 7.40—A 256 × 4 memory built with four 256 × 1 memory chips.

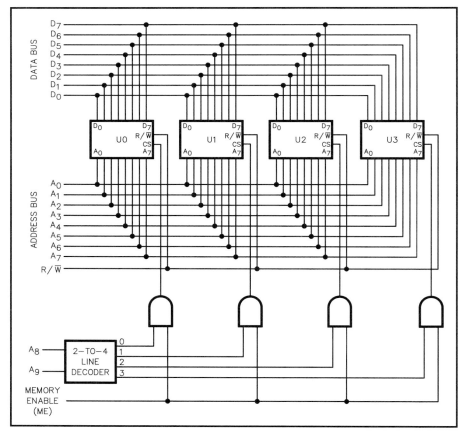

Fig 7.41—A 1024 × 8 memory built with four 256 × 8 memory chips and appropriate control circuitry.

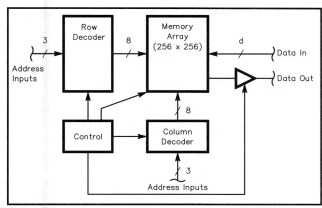

Fig 7.42—Row and column decoders allow a memory array to be accessed in a variety of formats.

of 2. So while we refer to the chips by shorter names like 1 Mbyte, the actual number of memory cells is larger than 1,000,000. The product, M × N, only refers to the number of memory cells in the chip; and designers are free to choose the word size appropriate to their needs. In fact, they may access one location as an 8-bit word, another as 16 bits, and yet another as 32-bits. In using memory, the controller chips and circuitry to access the memory can be just as important as the memory itself.

Memory Types

The concepts described above are applied to several types of random-access, semiconductor memory. Semiconductor memories are categorized by the ease and speed with which they can be accessed and their ability to "remember" in the absence of power.

SAM versus RAM

One way to categorize memory is by what memory cells can be accessed at a given instant. *Sequential-access memory* (SAM) must be accessed by stepping past each memory location until the desired location is reached. Magnetic tapes implement SAM; to reach information in the middle of the tape, the tape head must pass over all of the information on the beginning of the tape. Two special types of SAM are the queue and the push-down stack. In a *queue*, also called a first-in, first-out (FIFO) memory, locations must be read in the order that they were written. The queue is a first-come, first-served device, like a line at a ticket window. The push-down stack is also called last-in, first-out (LIFO) memory. In LIFO memory, the location written most recently is the next location read. LIFO can be visualized as a stack, always adding to and removing from the top of the stack. *Random-access memory* (RAM) allows any memory cell to be ac-

cessed at any instant, with no time wasted stepping past the beginning parts of the data. Random-access memory is like a bookcase; any book can be pulled out at any time.

It is usually faster to access a desired word in RAM than in SAM. Also, all words in RAM have the same access time, while each word in a SAM has a different access time based on its position. Generally, the semiconductor memory devices internal to computers are random-access memories. Magnetic devices, such as tapes and disks, have at least some sequential access characteristics. We will leave tapes and disks for a later section and concentrate here on random-access, solid-state memories.

Random Access Memory

Most RAM chips are *volatile*, meaning that stored information is lost if power is removed. RAM is either static or dynamic. *Dynamic RAM* (DRAM) stores a bit of information as the presence or absence of charge. This charge, since it is stored in a capacitor, slowly leaks away and must be refreshed periodically. Memory refresh typically occurs every few milliseconds and is usually performed by a dynamic RAM controller chip. *Static RAM* (SRAM) stores a bit of information in a flip-flop. Since the bit will retain its value until either power is removed or another bit replaces it, refresh is not necessary.

Both types of RAM have their advantages and disadvantages. The advantage of DRAM is increased density and ease of manufacture, making them significantly less expensive. SRAMs, however, have much faster access times. Most general purpose computers use DRAMs, since large memory size and low cost are the major objectives. Where the amount of memory required doesn't justify the use of DRAM, and the faster access time is important, SRAMs are common, for example, in embedded systems (telephones, toasters), battery powered devices, and for cache memories. Cost, power consumption and access time, provided in manufacturers' data sheets, are factors to consider in selecting the best RAM for a given application.

Read-Only Memory

Read-only memory (ROM) is nonvolatile; its contents are not lost when power is

removed from the memory. Despite its name, all ROMs can be written or programmed at least once. The earliest ROM designs were "written" by clipping a diode between the memory bit and power supply wherever a 0 was desired. Modern MOS ROMs use a transistor instead of a diode. Mask ROMs are programmed by having ones and zeros etched into their semiconductors at manufacturing time, according to a pattern of connections and nonconnections provided in a mask. Since the programming of a mask ROM must be done by the manufacturer, adding expense and time delays, this type of ROM is primarily used only in high volume applications.

For low-volume applications, the programmable ROM (PROM) is the most effective choice since the data can be written after manufacture. A PROM is manufactured with all its diodes or transistors connected. A PROM programmer device then burns away undesired connections. This type of PROM can be written only once.

Two types of PROMs that can be erased and reprogrammed are EPROMs and EEPROMs. The transistors in UV erasable PROMs (EPROMs) have a floating gate surrounded by an insulating material. When programming with a bit value, a high voltage creates a negative charge on the floating gate. Exposure to ultraviolet light erases the negative charge. Similarly, electrically erasable PROMs (EEPROMs) erase their floating-gate values by applying a voltage of the opposite polarity.

Besides being nonvolatile, PROMs are also distinguished from RAMs by their read and write times. RAM Read and write times are nearly equal, in the nanosecond range. Naturally, since PROMs are only written to infrequently, they can have slow write times (in the millisecond range). Their read times, however, are near those of RAM. Two factors make it hard to write to PROMs: (1) PROMs must be erased before they can be reprogrammed and (2) PROMs often require a programming voltage higher than their operating voltage.

ROMs are practical only for storing data or programs that do not change frequently and must survive when power is removed from the memory. The BIOS program that start a computer when it is first switched on or the memory that holds the call sign in a repeater IDer are prime candidates for ROM.

Nonvolatile RAM

For some situations, the ideal memory would be as nonvolatile as ROM but as easy to write to as RAM. The primary example is data that must not be allowed to perish despite a power failure. Low-power RAMs can be used in such applications if they are supplied with NiCd or lithium cells for backup power. A more elegant and durable solution is nonvolatile RAM (NVRAM), which includes both RAM and ROM. The standard volatile RAM, called shadow RAM, is backed up by nonvolatile EEPROM. When the RECALL control is asserted, such as when power is first applied, the contents of the ROM are copied into the RAM. During normal operation, the system reads and writes to the RAM. When the STORE control is triggered, such as by a power failure or before turning off the system, the entire contents of the RAM are copied into the ROM for nonvolatile storage. In the event of primary power failure, to successfully save the RAM data, some power must be maintained until the memory store is complete, generally about 20 ms.

Cache versus Main Memory

Memory is in high demand for many applications. To balance the trade-off of speed versus cost, most computers use a larger, slower, but cheaper main memory in conjunction with a smaller, faster, but more expensive cache memory. As you run a computer program, it accesses memory frequently. When it needs an item, a piece of data or the next part of the program to execute, it first looks in the cache. If the item is not found in the cache, it is copied to the cache from the main memory. As you run a computer program, it often repeats certain parts of the program and repeatedly uses pieces of data. Since this information has been copied to the high-speed cache, your computer game or other application can run faster. Information used less often or not being used at all (programs not currently being run) can stay in the slower main memory.

A "cache" is a place to store treasure; the treasure, the information you are using frequently, can be accessed quickly because it is in the high-speed cache. The use of cache versus main memory is managed by a computer's CPU so it is transparent to the user. The improvement in program execution time is similar to accessing a floppy disk versus the computer's internal memory.

I/O TRANSFERS

No computer will perform useful work without some means of communicating with the real world. Its input and output system allow the computer to react to and affect the outside world. The ability to interact with their environment is a primary reason why computers are so useful and cost-effective. Often, I/O is provided by a user, and a great deal of effort goes towards making computers user-friendly. Alongside the drive for user-friendly computers is the drive for automation. Data is acquired and operations are performed automatically, such as the packet bulletin board automatically forwarding a message. This section discusses the relationship of I/O to the internal operation of the computer: how the computer knows when and what I/O has been provided. The next section, on peripherals, discusses the range of devices that provide this information.

Program-Controlled I/O

Program-controlled I/O is a method of transferring data between devices that employs the CPU as part of the data path. A helpful analogy might be made by thinking of a cook who returns to the oven every few minutes to see if a meal is ready. Under program-controlled I/O, or polling, input and output events are initiated by the program currently running on the microcomputer. The program polls the I/O device, constantly checking if it is ready to accept or deliver data. When the I/O device indicates that it is ready, then the instruction that actually sends or receives the data is executed.

An advantage of program-controlled I/O is its simplicity. Program-controlled I/O is easily written and debugged. A disadvantage is wasted time. The program must spend its time checking the status of the I/O device rather than doing other useful things. If the program must have the input data before continuing, then no time is wasted; but if it could have been performing other tasks, then polling can be expensive and wasteful. Packet radio provides a familiar example of polling: the TNC repeatedly sends a packet until a confirmation message has been received from the BBS. In the PC, AT Attachment (ATA) disk transfers are implemented using PIO.

Interrupt-Driven I/O

Interrupt-driven I/O avoids wasting time in a polling loop. The cook, rather than constantly checking the oven, goes off to other work until the timer rings. This efficiency is especially important on multi-user systems, where one program may be waiting for I/O while another program is executing.

The alarm of the timer is called an *interrupt*, a temporary break in the normal execution of a program. The act of taking the food out of the oven is coded in an *interrupt service routine*. An interrupt service routine (ISR) is any code that performs the appropriate actions in response to a certain interrupt. Each interrupt has a number, and the location of each ISR is listed in a table next to its number. From the machine's perspective, the process is as follows: one of the bits in the microprocessor's status register is called the interrupt request indicator. When this bit becomes a 1, an interrupt has occurred. Circuitry indicates the number of the device requesting the interrupt. The machine temporarily suspends whatever it was working on and looks at its table of service routines. From the table, the machine finds the location of the appropriate ISR and automatically jumps to that code and begins executing it. When finished with the ISR, the machine automatically returns to whatever it was doing before the interrupt.

The "getchar" subroutine below shows how an interrupt service routine for keyboard input might look in assembly language.

```
getchar:
MOVE RCVDATA,R7; Move the
    data from the receiver to the
    temporary storage register
RTE; Return to normal execution
```

There are two key differences from the previous example: (1) No polling loop is involved. The READY bit, instead of being polled, triggers the interrupt request bit. (2) Leaving the main program and returning to it are done automatically by the machine instead of with a subroutine call inside the code.

The advantage of interrupt-driven I/O is that no time is wasted in a polling loop. This is especially advantageous in a multi-user environment where processing time must be juggled between the user demands. The disadvantage of interrupts is that program flow can become very confusing; for example, what happens if an interrupt service routine gets interrupted? This is usually handled by assigning priorities to each possible interrupt and, when inside an ISR, ignoring other interrupts of lesser or equal priority.

A familiar example of an interrupt occur each time you click a mouse button or press a key on the keyboard. Examples of a timer interrupt are the sending every ten minutes of a repeater's ID; the time-out timer of a voice repeater, and the time-keeping interrupt generated by the computer's real time clock every few milliseconds.

When the mouse button is pushed, an interrupt is generated that calls an ISR which captures and buffers the (x,y) location of the mouse along with the button status. This information is presented to the program the next time it reads mouse status.

Memory-Mapped I/O

In *memory-mapped I/O*, addresses that are treated like RAM by the microprocessor are actually I/O devices. Thus, a command that would usually be used to read or write to a memory location might actually result in an I/O operation. Since memory mapping is an addressing technique, it can be used with either interrupt-driven or polled I/O.

Direct Memory Access

Direct Memory Access (DMA) enables data to be transferred directly between memory and an I/O device without involving the CPU, as contrasted with PIO. The advantages of DMA are to provide high-speed transfer of data, such as from a peripheral disk drive or communications device, while the CPU is performing internal tasks. The data transfer operation is managed by a DMA controller, either a separate chip or internal to the microprocessors. The following illustrates some of the steps involved in the I/O transfer:

1. An I/O device requests DMA operation.
2. The DMA controller requests the bus from the CPU.
3. The CPU acknowledges the request and releases the bus.
4. The DMA controller tells the I/O device to send its information.

The DMA technique is used by I/O processors. In large computer systems, these auxiliary processors perform most of the I/O functions, thus freeing the CPU for other tasks. CD-ROMs use DMA in PCs.

BUS STRUCTURE: LOCAL BUSSES

Tying the blocks together is the data bus, the main information corridor inside the computer. The bus carries signals to and from various components, such as the CPU, the keyboard, mass storage, and communications ports. The first IBM PC and compatibles used the 62-pin, 8-bit, 8-MHz ISA bus, which was revised to the AT-bus, a 98-pin, 16-bit wide 8MHz bus. Other pre-PCI bus architectures include Apple's NuBus, the Extended Industry Standard Architecture (EISA), and the VESA Local Bus.

The slow ISA bus was a bottleneck to system performance, so a separate bus was implemented between the microprocessor and main memory. This bus was called a *local bus*, as it was local to the CPU and memory only. Eventually, graphics and hard disk drive speeds increased to the point where they could ride on the local bus as well, without impacting memory access performance. Present computer systems are built around the 188-pin Peripheral Components Interface (PCI) bus, a 32- or 64-bit wide system running at 33 or 66 MHz.

PERIPHERALS

Peripherals work with the CPU and memory to provide additional capabilities. One of the most common examples is communication with a user via input de-

vices and output devices. Peripherals may be divided into three groups: bidirectional (input/output) devices, input devices, and output devices. Bidirectional devices allow data storage and communications with the outside world. Input devices provide the computer both data to work on and programs to tell it what to do. Output devices present the results of computer operations to the user or another system and may even control an external system. Both input and output combine to provide user friendly interaction. Most of these devices have adapted to certain standards and use readily available connectors and cables, enabling easy incorporation into a system. A knowledge of how external memory devices work is more useful and will be discussed in more detail.

Bidirectional Input/Output Devices

Mass storage and communications devices provide data input as well as output. Perhaps the most important peripheral in the computer system is the mass storage unit, such as the *disk drive*. Another important I/O unit is the *modem* (a contraction of MODulator/DEModulator), which allows easy communication between computers across standard telephone lines A third is the *local area network* (LAN) card which provides high speed communications between nearby computers.

Hard Disk Drives

An electromechanical hybrid, the hard disk drive provides the largest capacity at

the lowest cost per byte of any random-access storage media. Hard drives are an essential part of present computer systems. Their key features include:
• Low cost per byte of storage.
• Large capacity available.
• Random access to data.
• Non-volatile magnetic storage.

Hard drives consist of three main units, the *head/disk assembly* (HDA), the *read/write channel*, and the *controller*. The HDA comprises the mechanical portion of the assembly, with one or more aluminum disks mounted on a spindle which is rotated by a brushless dc motor, generally between 3600 and 7200 rpm. Read/write heads are mounted on an actuator arm that sweeps across the disk surfaces. The read/write amplifier is affixed to the actuator by a flexible cable to provide the lowest possible noise pickup. Read/write heads do not touch the disk surface; instead, air flowing over the rapidly spinning disk cause them to fly slightly above the disk. "Slightly" is no exaggeration—typical flying heights of drives made in the year 2000 are approximately two millionths of an inch (about 500 nanometers). Due to these extremely tight tolerances, the entire HDA is enclosed in a sealed aluminum casting that prevents contamination. As shown in **Fig 7.43**, debris such as a hair or a smoke particle tower over the flying heads.

The read channel takes the tiny electrical signal from the read heads and determines the time location of the serial data pulses. Pulse detection is accomplished by

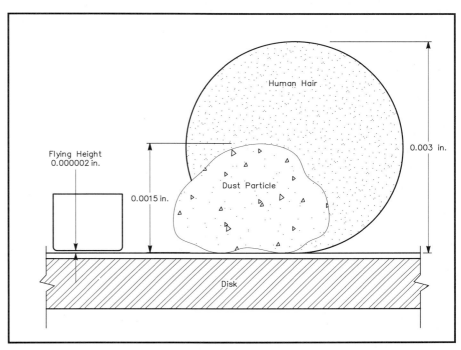

Fig 7.43—Hard disk drive head flying height compared to common debris size.

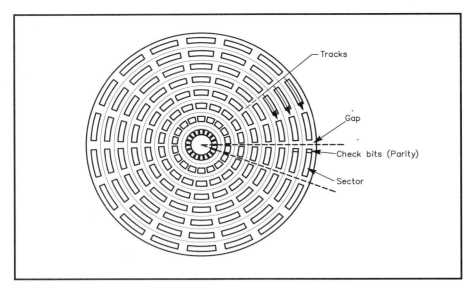

Fig 7.44—Disk sectors and tracks.

either analog peak detection or by digital signal processing (DSP). After qualification, the now logic-level pulses enter the data separator, where the embedded clock information is stripped away.

The drive controller performs data caching and communication with the host bus. Common hard drive busses include the EIDE (Extended Integrated Device Electronics, the (similar) ATA (AT-Attachment), and SCSI (small computer systems interface).

Data is stored on both surfaces of each disk in concentric arcs called sectors, as shown in **Fig 7.44**. All sectors on one surface a given distance from the disk edge constitute a track; a cylinder is the collection of all similar tracks on all surfaces of the drive. When installing a hard drive in a PC, the motherboard must be informed of the number of disk heads and sectors available. Often, the number entered is not the actual number of physical heads and sectors, as intelligent controllers map data requests to the proper physical location without burdening the CPU with the fine details.

Important hard disk drive parameters include bus type (for compatibility with your system), storage capacity, average and worst-case seek time, and size of on-board cache. Seek times are dominated by mechanical considerations, such as the time the read head arms require to settle from track to track or from one edge to the other, and by the rotational speed of the disk. Better overall performance is expected with larger on-board cache memory. The best drives have high capacity, fast seek times, and large caches.

Floppy Disk Drive

The most common storage peripheral in

the PC is the 3.5-in. 1.44 MB floppy disk drive. The floppy drive operates similarly to its higher speed/higher capacity brother, but its media is removable and it does not offer much on-board intelligence. Floppy disks enclose the magnetic-media platter in a protective casing, as shown in **Fig 7.45**, so the disk can be carried around. The floppy disk can be inserted into a disk drive and the read/write head automatically extended; when done, the read/write head is automatically retracted before the disk is ejected from the drive.

When configuring standard PCs, each floppy drive should be jumpered as drive 2. The first drive, drive A, attaches to the *end* of the standard "twisted" control cable; if only one drive is used, leave the middle connector free. The twist in the cable causes the floppy controller to recognize the end drive as drive 1, which saves time for professional computer builders since they need not deal with address jumpers on each drive. For the rest of us, it causes confusion if this feature is unknown!

Other Magnetic Disk Drives

Disk drives featuring moderate storage capacity and performance with removable media are available, such as the Iomega Zip and Jazz, and the Imation SuperDrive. These drives use the same basic techniques as the hard drives, but with somewhat looser mechanical tolerances that allow an unsealed disk chamber and more flexible media. These drives are available with EIDE, SCSI, parallel port, USBus, and PC Card interfaces for internal or external use.

Optical Storage

The audio compact disc evolved into the

CD-ROM offering over 600 MB of storage on a very low cost single-sided platter. Initially a read-only media, writeable and re-writable discs are used as removable mass storage. While media cost is low, the low performance and restricted number of write sessions, even on the re-writable discs, prevent these technologies from competing with the magnetic hard drive except in archival or other applications where removability is important. Another transplant from the consumer entertainment industry, the DVD is another important optical storage format. Similar in appearance to CD-ROM discs, DVDs offer up to 8.5GB of storage capability per disc.

Both the CD-ROM and the DVD use a laser and an optical system to detect the surface deformities that represent data. Unlike the hard disk and floppy drives which employ concentric tracks, the optical drives use a spiral pattern that works out from the center. Also, more elaborate data handling is necessary since the raw data error rate is significantly higher than that from magnetic drives.

All drives and disks eventually fail, and the data on the disk can be lost. Therefore, it is prudent to make backup copies of your disks, stored in a clean, dry, cool place.

Tape

Tape is one of the more inexpensive options for auxiliary memory. Tape access time is slow, since the data must be accessed sequentially, so tape is primarily used for backup copies of a system's hard drive. Tape is available in cassette form (common sizes are comparable to the cassettes for a portable tape player and VCR tapes) and on digital audio tape (DAT). A single 4-mm-wide DAT cartridge, which fits in the palm of your hand, can hold over 2 gigabytes (GB) of data (1 GB = 1000 MB).

Modems

Nearly all computers assembled today include a modem for connecting to the Internet via standard telephone lines. Beside connecting to an ISP or online service, this peripheral may call another modem-equipped PC or send standard facsimiles. The so-called 56-k modem uses V.90 protocol with its sophisticated DSP techniques providing echo cancellation and dynamic line equalization. This protocol uses the telephone line to gain every possible bit per second of data transfer rate. While the data rate never reaches 56 kbps, if the modem is less than four miles from the telephone central office, expect speeds of 40 kbps to 53 kbps. Achieving the 56 kbps rate would necessitate increasing transmission power above

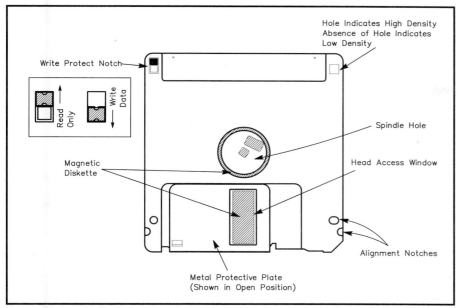

Fig 7.45—Standard 3.5-in. floppy disk.

the −9 dBm limit and could cause excessive crosstalk between lines. The V.90 specification has an unusual characteristic in that the bit rate is non-symmetrical. The modem originating a call is limited to approximately 33 kbps while the answering modem runs up to 53 kbps. This compromise was made because most modem traffic is between a user and an ISP, where the user downloads much more data from the World Wide Web than he uploads.

Modems are available in three configurations, internal, external, and PC Card. Internal modems plug directly into the computer motherboard, external devices attach to a serial port, and PC-Card (often called PCMCIA) modems plug into a PC-Card slot. The internal version is the least expensive and most common. PC Card modems are popular with notebook computers due to their small size. External modems offer the advantage of providing immediate visual feedback of all data transfer activity. Additionally, external modems provide another level of surge protection to the computer system; if a destructive surge travels through the phone line it *might* be stopped outside the PC cabinet by the external modem.

Input Devices

The keyboard is probably the most familiar input device. A keyboard simply makes and breaks electrical contacts. The open or closed contacts are usually sensed by a microprocessor built into the circuit board under the keys. This microprocessor decodes the key closures and sends the appropriate ASCII code to the main computer unit. Keyboards will generate the entire 128-character ASCII set and often, with CONTROL and ALT (Alternate) keys, the 256-character extended ASCII set.

The mouse is a close second in familiarity to the keyboard. This pointing device controls the position of a cursor on the screen, and switches on the mouse make and break connections (clicking) to select and activate items (icons) on the screen. Touchpads, trackballs, and pen input on a sensitive screen are variations of the mouse and may offer a more natural, human-friendly interface to the computer. Voice recognition systems promise even easier data entry.

Digital cameras and streaming video cameras allow quick transfer of images into the computer realm.

Image scanners digitize photographs and older printed pages, allowing reuse of material without laboriously recreating the work. Scanners are available in four major configurations: handheld, sheet feed, flatbed, and drum. The handheld scanner is low cost and very portable. The sheet feed scanner is also small and is easier to use, since pages are better aligned. Flatbed scanners are economical and moderately high resolution means of entering data from book or other bound sources. Drum scanners provide the best resolution and color reproduction, but their high cost relegates them to professional graphics shops.

Output Devices

The most familiar output device is the computer screen, or monitor. The next most common output device is the printer, to produce paper hardcopy. Sound cards provide high fidelity stereo audio.

Monitors and most printers share a common display technique: images, such as characters and graphics, are formed by tiny dots, called *pixels* (picture elements). On screens, these are dots of light turned on and off. In printers, they are dots of ink or electrostatic toner imposed onto the paper. For color displays, pixels in red, green and blue (RGB) are spaced closely together and appear as colors to the human eye.

Video Displays

Video monitors are usually specialized high resolution cathode-ray tube (CRT) displays, except in notebook computers which use screens fabricated with liquid-crystal displays (LCDs). Most monitors employ *raster scanning* techniques to turn on the screen pixels, similar to that used by standard broadcast television receivers. The electron beam paints the screen one row of pixels at a time, from left to right and top to bottom. Then, a vertical retrace brings the beam back to the top of the screen to begin again. Raster scanning signals every pixel on or off for each screen pass.

Printers

Printers suitable for hamshack use generally fall into two categories, inkjet and laser. The inkjet printer uses a controlled spray of liquid ink to produce images. Near photographic-quality full-color prints are possible when the proper paper is employed.

Laser printers produce exceptionally crisp text and graphics in black or a few colors at relatively high speed and low cost per page. While color laser printers do not (yet) produce the lifelike quality images of inkjets, they are not as fussy about the quality of the paper used, and its powdered *toner*, or "ink," does not dry up when stored in the printer over time as does the inkjet pigment.

Sound Cards

Your PC will produce and record full CD-quality audio when a suitable sound card and speaker system is deployed. Microphone and auxiliary inputs and line level outputs on the sound card let the PC serve as a contest voice keyer. When the proper software is used, it can also serve as a RTTY, CW, Pactor, PSK31, etc, terminal.

COMMUNICATIONS: INTERNAL AND EXTERNAL INTERFACING

Designing an interface, or simply using an existing interface, to connect two devices involves a number of issues. For example, digital interfacing can be categorized as parallel or serial, internal or external and asynchronous or synchronous. Additional issues are the data rate, error detection methods and the signaling format or standards. The format can be especially important since many standards and conventions have developed that should be taken into consideration. This section focuses on some basic concepts of digital communications for interfacing between devices.

Parallel Versus Serial Signaling

To communicate a word to you across the room, you could hold up flash cards displaying the letters of the word. If you hold up four flash cards, each with a letter on it, all at once, then you are transmitting in parallel. If instead, you hold up each of the flashcards only one at a time, then you are transmitting in serial. *Parallel* means all the bits in a group are handled exactly at the same time. *Serial* means each of the bits is sent in turn over a single channel or wire, according to an agreed sequence. **Fig 7.46** gives a graphic illustration of parallel and serial signaling.

Both parallel and serial signaling are appropriate for certain circumstances. Parallel signaling is faster, since all bits are transmitted simultaneously; but each bit needs its own conductor, which can be expensive. Parallel signaling is more likely to be used for internal communications. For spanning longer distances, such as to an external device, serial signaling is more appropriate. Each bit is sent in turn, so communication is slower; but it is also less expensive, since fewer channels are needed between the devices.

Most amateur digital communications use serial transmission to minimize cost and complexity. The number of channels needed for signaling also depends on the operational mode: one channel per bit for simplex (one-way, from sender to receiver only) and for half-duplex (two-way communication, but only one person can talk at a time) but two channels per bit for full-duplex (simultaneous communications in both directions).

Parallel I/O Interfacing

Fig 7.47 shows an example of a parallel input/output chip. Typically, they have eight data lines and one or more handshaking lines. *Handshaking* involves a number of functions to coordinate the data transfer. For example, the READY line indicates that data is available on all 8 data lines. If only the READY line is used, however, the receiver may not be able to keep up with the data. Thus, the STROBE line is added so the receiver can watch to ensure the transmitter is ready for the next character.

On standard PCs, a parallel port is available using a 25-pin DB-25 connector. This port was originally intended for use with a printer. Several versions of the parallel port exist, and late-model PCs feature a high speed, bidirectional parallel port that, in addition to high speed printing, may also interface with mass storage devices, scanners, and other I/O devices. The most com-

mon parallel ports are:
- Printer Mode—The most basic, output mode only port.
- Standard & Bidirectional (SPP)—The low-speed bidirectional port
- Enhanced Parallel Port (EPP)—Uses local hardware handshaking and strobing to accomplish 500 KB/s to 2 MB/s transfer rates.
- Extended Capabilities Port (ECP)—Similar to EPP, except negotiates a reverse channel with the external peripheral and requires that peripheral controls handshaking. It is optimized for the Windows operating system and uses DMA channels, a FIFO buffer, and real time data compression of up to 64:1.

Most PCs offer a choice of port protocol in the BIOS setup. Unless you have a reason not to, select the "ECP and EPP 1.9 Mode" for maximum flexibility and performance.

Serial I/O Interfacing

Serial input/output interfacing is more complex than parallel, since the data must be transmitted based on an agreed sequence. For example, transmitting the 8 bits (b7, b6, . . . b0) of a word includes specifying whether the least significant bit, b0, or the most significant bit, b7, is sent first. Fortunately, a number of standards have developed to define the agreed sequence, or encoding scheme.

Conversions

Within computers and other digital circuits, data is usually operated on, stored and transmitted in parallel. For communicating with an external device, data must usually be converted from parallel to serial format and vice versa. This conversion is usually handled by shift registers.

Shift registers can be left shifters, right shifters or controlled to shift in either direction. The most general form, a universal shift register, has two control inputs for four states: Hold, Shift right, Shift left and Load. Most also have asynchronous inputs for preset, clear and parallel load.

A register with parallel input and shift left serial output will be described, as seen in Fig. 7.24. (A serial input/parallel output register would work in the opposite fashion.) Since the register receives information in parallel, the n-bit register has n inputs, one to each flip-flop. A parallel load control input is asserted to pass the initial value. The register sends out information in serial fashion so there is only one output line. Since this example shifts left, the output comes from the left-most register, the most significant bit. On each clock pulse, one bit is output and the other flip-flops cycle their value up to the next flip-flop. A 0 is usually input to the least-significant bit so 0s will cycle up to fill the register. After

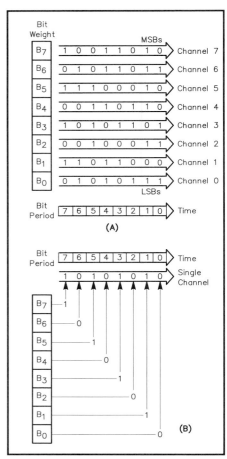

Fig 7.46—Parallel (A) and serial (B) signaling. Parallel signaling in this example uses 8 channels and is capable of transferring 8 bits per bit period. Serial transfer only uses 1 channel and can send only 1 bit per bit period.

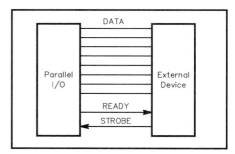

Fig 7.47—Parallel interface with READY and STROBE handshaking lines.

n clock pulses, all data bits have been shifted out and the register has a value of 0.

Asynchronous versus Synchronous Communication

To correctly receive data, the receiving interface must know when data bits will occur; it must be synchronized with the sender. In *asynchronous* communication, the receiver synchronizes on each incoming character. Each character includes start and stop bits to indicate the beginning and end of that character. In *synchronous* communication, data is sent in long blocks, without start and stop bits or gaps between characters.

Asynchronous Communication

In asynchronous communication, each transmitted character begins with a start bit and ends with a stop bit, as shown in **Fig 7.48A**. The start bit (usually a zero) tells the receiver to begin receiving a character. The stop bit (usually a one) signals the end of a character. Between characters, the transmitting circuit sends the stop bit state (steady one or zero).

Since the receiver is always told when a character begins and ends, characters can be sent at irregular intervals. This is especially advantageous for typed input, since the person typing is usually slower than the data communications equipment and will usually work at an uneven pace. Another advantage of asynchronous data is that it does not need complex circuits to keep it synchronized. Since the receiver is newly synchronized at the beginning of each character, the characters need not be sent in a steady stream and no stringent demands are made on the person or process generating the characters.

A disadvantage of asynchronous communications is the inclusion of the start and stop bits, which are not useful data. If you are transmitting 8 data bits, 1 start bit and 1 stop bit, then 20% (2 of 10 bits) is overhead.

Synchronous Communication

In synchronous communications, data is sent in blocks, usually longer than a single character, as shown in Fig 7.48B. At the beginning of each block, the sender transmits a special sequence of bits that the receiver uses for initial synchronization. After becoming synchronized at the beginning of a block, the receiver must stay synchronized throughout the block. The sender and receiver may be using slightly different clock frequencies, so it is usually not adequate for them to merely be synchronized initially. There are several ways for the receiver to stay synchronized. The transmitter may send the clock signal on a separate channel, but this is wasteful. The modulation technique used on the communications channel may convey clock information or the clock may be implicit within the data. See **Fig 7.49**.

One disadvantage of synchronous transmission is that the data must be sent as a continuous stream; characters must be placed in a buffer until there are enough to make a block. Also, while errors in asynchronous signaling usually only affect one character (the receiver can resynchronize at the beginning of the next character), error recovery on synchronous channels may be a longer process involving several lost characters or an entire lost block.

The major advantage of synchronous signaling is that it does not impose overhead (the start and stop bits) on each character. This is an important consideration during large data transfers.

Data Rate

There are a number of limitations on how fast data can be transferred: (1) The sending equipment has an upper limit on how fast it can produce a continuous stream of data. (2) The receiving equipment has an upper limit on how fast it can accept and process data. (3) The signaling channel itself has a speed limit, often based on how fast data can be sent without errors. (4) Finally, standards and the need for compatibility with other equipment may have a strong influence on the data rate.

Two ways to express data transmission rates are *baud* and *bits per second (bps)*. These two terms are not interchangeable: Baud describes the signaling, or symbol, rate — a measure of how fast individual signal elements *could* be transmitted through a communications system. Specifically, the baud is defined as the reciprocal of the shortest element (in seconds) in the data encoding scheme. For example, in a system where the shortest element is 1-ms long, the maximum signaling rate would be 1000 elements per second. (Note that, since baud is measured in elements per second, the term "baud rate" is incorrect since baud is already a measure of speed, or rate.) Continuous transmission is not required, because signaling speed is based only on the shortest signaling element.

Signaling rate in baud says nothing about actual information transfer rate. The maximum information transfer rate is defined as the number of equivalent binary digits transferred per second; this is measured in bits per second.

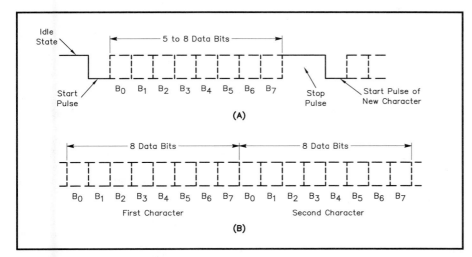

Fig 7.48—Serial data transmission format. In asynchronous signaling at A, a start pulse of one bit period is followed by the data bits and a stop pulse of at least one bit period. In synchronous signaling at B, the data bits are sent continuously without start or stop pulses.

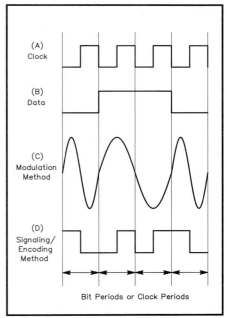

Fig 7.49—Recovering the clock (A) when the data (B) is transmitted allows a receiver to maintain synchronization during synchronous communication. The modulation method shown at C results in a transition of the received carrier at the beginning and end of each clock period. The encoding method shown at D results in a data transition in the middle of each clock period. Either of these methods provides enough information for clock recovery.

When binary data encoding is employed, each signaling element represents one bit. Complications arise when more sophisticated data encoding schemes are used. In a quadrature phase shift keying (QPSK) system, a phase transition of 90° represents a level shift. There are four possible states in a QPSK system; thus, two binary digits are required to represent the four possible states. If 1000 elements per second are transmitted in a quadriphase system where each element is represented by two bits, then the actual information rate is 2000 bps.

This scheme can be extended. It is possible to transmit three bits at a time using eight different phase angles (bps = 3 × baud). In addition, each angle can have more than one amplitude. A 9600 bps modem uses 12 phase angles, 4 of which have two amplitude values. This yields 16 distinct states, each represented by four binary digits. Using this technique, the information transfer rate is four times the signaling speed. This is what makes it possible to transfer data over a phone line at a rate that produces an unacceptable bandwidth using simpler binary encoding. This also makes it possible to transfer data at 2400 bps on 10 m, where FCC regulations allow only 1200-baud signals.

When are transmission speed in bauds and information rate in bps equal? Three conditions must be met: (1) binary encoding must be used, (2) all elements used to encode characters must be equal in width and (3) synchronous transmission at a constant rate must be employed. In all other cases, the two terms are not equivalent.

Within a given piece of equipment, it is desirable to use the highest possible data rate. When external devices are interfaced, it is normal practice to select the highest standard signaling rate at which both the sending and receiving equipment can operate.

Error Detection

Since data transfers are subject to errors, data transmission should include some method of detecting and correcting errors. Numerous techniques are available, each used depending on the specific circumstances, such as what types of errors are likely to be encountered. Some error detection techniques are discussed in the Modulation Sources chapter. One of the simplest and most common techniques, parity check, is discussed here.

Parity Check

Parity check provides adequate error detection for some data transfers. This method transmits a parity bit along with the data bits. In systems using odd parity, the parity bit is selected such that the number of 1 bits in the transmitted character (data bits plus parity bit) is odd. In even parity systems, the parity bit is chosen to give the character

an even number of ones. For example, if the data 1101001 is to be transmitted, there are 4 (an even number) ones in the data. Thus, the parity bit should be set to 1 for odd parity (to give a total of 5 ones) or should be 0 for even parity (to maintain the even number, 4). When a character is received, the receiver checks parity by counting the ones in the character. If the parity is correct, the data is assumed to be correct. If the parity is wrong, an error has been detected.

Parity checking only detects a small fraction of possible errors. This can be intuitively understood by noting that a randomly chosen word has a 50% chance of having even parity and a 50% chance of having odd parity. Fortunately, on relatively error-free channels, single-bit errors are the most common and parity checking will always detect a single bit in error. Parity checking is a simple error detection strategy. Because it is easy to implement, it is frequently used.

Standard Interface Busses

Signaling Levels

Inside equipment and for short runs of wire between equipment, the normal practice is to use neutral keying; that is, simply to key a voltage such as + 5 V on and off. In neutral keying, the off condition is considered to be 0 V. Over longer runs of wire, the line is viewed as a transmission line, with distributed inductance and capacitance. It takes longer to make the transition from 0 to 1 or vice versa because of the additional inductance and capacitance. This decreases the maximum speed at which data can be transferred on the wire and may also cause the 1s and 0s to be different lengths, called bias distortion. Also, longer lines are more likely to pick up noise, which can make it difficult for the receiver to decide exactly when the transition takes place. Because of these problems, bipolar keying is used on longer lines. Bipolar keying uses one polarity (for example +) for a logical 1 and the other (– in this example) for a 0. This means that the decision threshold at the receiver is 0 V. Any positive voltage is taken as a 1 and any negative voltage as a 0.

EIA-RS-232

The most common serial bus protocol, EIA-RS-232, addresses this issue (however, a Mark "1" is a negative voltage and a Space "0" is positive). Generally called RS-232, this protocol defines connectors and voltages between data terminal equipment (DTE) such as a PC, and data communications equipment (DCE), such as a modem or TNC. The connector is the DB-25, or the presently more popular DB-9 version. Signaling voltages are defined between + 3V and + 25V for logic "0" and between – 3V and – 25V for logic "1." Al-

though the top data rate addressed in the specification is only 20 kbps, speeds of up to 115 kbps are commonly used. Communications distances of hundreds of meters are possible at reasonable data rates.

Since neutral keying is usually used inside equipment and bipolar keying for lines leaving the equipment, signals must be converted between bipolar and neutral. Discrete level shifters or op amp circuits may perform this task, or low cost specialized IC line drivers and receivers are available.

RS-422

RS-422 is a serial protocol similar to RS-232, but employing fully differential data lines. Differential data offers the important advantage that common grounds between remote units are not necessary, and an important cause of ground loops (and their associated problems) is eliminated. Available on many Apple Macintosh computers, RS-422 systems may connect to standard RS-232 modems and TNCs by building a cable that makes the following translations:

RS-422 DTE	RS-232 DCE
RXD–	RXD
TXD–	TXD
RXD+	GND
TXD+	No Connection
GPi	CD

IrDA (Infrared Data Access)

Another high speed serial protocol is the IrDA, which is a simple, short range wireless system using infrared LEDs and detectors. Data rates up to 3 MB/sec are possible between compatible units.

Universal Serial Bus (USBus)

Not to be confused with upper sideband, USBus is a computer standard for an up-to-12 Mbps intelligent serial data transfer protocol. In addition to its higher speed than RS-232, USBus offers reasonable power availability to its loads, or *functions*. Under certain circumstances, up to 127 hubs and functions may connect to a single computer. USBus requires that each function have on-board intelligence and that it negotiate with the host for power and bandwidth allocation, and has the major advantage of *hot-pluggablity*—the PC need not reboot when new functions are added. The USBus connectors use four-conductor cable, with two bidirectional, differential data lines, power, and ground. Approximately 5 V at 100 mA is allowed per function, with up to 500 mA available if the host system has the capability. This means that relatively sophisticated devices, such as modems, small video cameras, or hand-held scanners may operate from the bus without additional power supplies. Preventing power back-flowing up from function to host is accomplished by configuring the connector shapes such that the host has a rectangular connector while that of functions are nearly square.

IEEE-1394 (FireWire)

A very high speed serial protocol, IEEE-1394 (christened "FireWire" by its creator, Apple Computer), is capable of up to 400 Mbps of sustained transfer. It is ideal for high bandwidth systems, such as live video, external hard drives, or high-speed DVD player/recorders. Up to 63 devices may daisy-chain together at once via a standard six-wire cable. Unlike USBus, 1394 is peer-to-peer, meaning any device may initiate a data transfer—the PC is not a requirement. Similar to USB, IEEE-1394 is hot-pluggable and provides power on the cable, but the voltage may vary from 7 V to almost 40 V, and may be sourced by any device. Allowable current drain per device may reach 1 A.

PC Card (PCMCIA)

The PC Card Standard is a collection of specifications for miniature plug-in peripherals. The most common is colloquially referred to as PCMCIA cards—68-pin devices the size of thick credit cards that contain a modem, LAN, GPS receiver, USBus port, FireWire port, high resolution video, an extra serial or parallel port, or memory storage expansion. The standard PC Card allows up to 5 V at 1 A peak current, or 3.3 V at 1 A, depending upon configuration; other voltages may be used if available from the host. Other portions of the specification define memory storage-only cards in even smaller footprints.

Small Computer System Interface (SCSI)

SCSI interfaces provide for up to 80 MB/s transmission rates with up to 15 devices. Used mostly with disk drives, the "skuzzy" bus also supports a very wide variety of high speed peripherals. The Apple Macintosh, with the exception of the iMac, uses SCSI for both internal and external connectivity, and expansion cards are available for other PCs. A wide variety of bus widths, speeds, and connectors exist, so incompatibility presents some problems. SCSI devices are either 8-bit (also called "narrow SCSI") or 16-bit ("wide SCSI"). SCSI devices may be "slow" (SCSI II), or "fast" (SCSI III, or "Ultra SCSI"). Generally, the faster devices and controllers will automatically reduce speeds to accommodate slower peripherals, and when all goes well, allow mixed speeds on a single SCSI device chain.

10Base2, 10BaseT, 10Base4, 100BaseT

Common office/home networks use 10BaseN protocol. 10Base2 is generally recommended for amateur radio installations, since it uses shielded cable (RG-58, renamed "thin coax" in this application), Additionally, no separate hub is needed as the connected computers work on a peer-to-peer basis. A drawback of 10Base2 is its maximum data rate is limited to 10 Mbps. The other protocols use one or more hubs, RJ-45 connectors, and unshielded Category 5 cable. 100BaseT systems are rated to 100 Mbps.

POWER SYSTEMS AND ATX

When the initial personal computers were designed, the most common logic family was TTL and CMOS that interfaced with TTL levels at 5 V. Disk drive and fan motors preferred +12 V. RS-232 demanded a higher voltage bipolar supply, so –12 V was added to complement the 12 V already used. The analog portion of the early modems required –5 V. Thus, the initial "silver box" PC supply provided +5 V at high current, +12 V at moderate current, and –5 V and –12 V at low current. Advances in semiconductor technology allowed shrinking transistor geometries; the smaller transistors were faster, but had lower breakdown voltages, so a new logic voltage of 3.3 V was introduced. Initially, computer manufacturers responded by placing IC regulators on the motherboard to power the 3.3-V circuits, but eventually the current demanded by these circuits exceeded that of the traditional +5V components and a new physical standard, called ATX, was introduced.

The ATX standard defines a layout physically different from older "AT-type" computers; the computer case, motherboard mounting holes, expansion slot location, and the power supply and its connector are all changed (ATX computers are recommended for the hamshack due to their better RFI control, resulting from careful mechanical design of connectors and consideration of card slot case penetration). ATX power supplies produce +3.3 V, +5 V, +12 V, –5 V and –12 V. They also provide an output voltage, even when the power supply is otherwise off, allowing *sleep mode* operation. Sleep mode retains the computer RAM contents and configuration so a reboot is not necessary each time the computer is used, and is especially critical to extending battery life in notebook PCs.

As semiconductor technology continued improving, the 3.3 V source became too high for the fine geometry microprocessors and an even lower voltage was required. At present, there is no standard for the next lower voltage, but devices are available that need anywhere between 1 V to 2.8 V for the microprocessor core (densest portion). Motherboard manufacturers have addressed this issue by again using on-board voltage regulators to drop an existing ATX standard voltage to the value needed by the CPU. Since there is no standard—in fact, the exact voltage preferred by a given processor family decreases as the manufacturing process evolves—motherboards provide means of selecting the matching voltage. Sometimes this process is automatic, as the on-board power supply communicates with the microprocessor before initializing and rises to the proper voltage, but other times, jumpers must be manually positioned *before* initial power is applied. Using a higher than recommended supply voltage causes excessive operating temperature and stresses the gates of the CMOS transistors, leading to reduced reliability and early (sometimes immediate) circuit death.

Power Quality

As operating voltages drop, power quality—the measure of voltage accuracy and transient response to changing load currents—becomes simultaneously more critical and more difficult. A 500 mV spike represents a 10% error in a 5-V supply—but the same 500-mV spike applied to a 2-V processor represents an overvoltage of 25%, grossly exceeding the maximum rated supply voltage for that controller. Further, if the spike is of the opposite polarity, it seriously reduces the noise margin of the logic-high levels, possibly corrupting data.

Providing clean power becomes more difficult when the effects of sleep mode are considered. Microprocessors reduce their power consumption when idle by slowing down internal clocks and other techniques, but when called back to duty, their response occurs in nanoseconds. The result is a huge change in supply current, from nearly zero to maximum current flow in those few nanoseconds. Large voltage spikes may result from this fast rise-time current step working against the inductance of the PC board power supply traces. Careful power supply design, especially during layout, and judicious use of low ESR, and low-inductance bypass capacitors mitigate the transients and keep the system reliable.

STANDARD COMPUTER CONNECTIONS

See the **References** chapter for details on computer connector pinouts. You'll also find details on cables, such as a null modem cable.

Contents

Analog Signals and Components
8

Glossary

Active Region — The region in the characteristic curve of an analog device in which the signal is amplified linearly.

Amplification — The process of increasing the size of a signal. Also called gain.

Analog signal — A signal, usually electrical, that can have any amplitude (voltage or current) value and exists at any point in time.

Anode — The element of an analog device that accepts electrons.

Base — The middle layer of a bipolar transistor, often the input.

Biasing — The addition of a dc voltage or current to a signal at the input of an analog device, which changes the signal's position on the characteristic curve.

Bipolar Transistor — An analog device made by sandwiching a layer of doped semiconductor between two layers of the opposite type: PNP or NPN.

Buffer — An analog stage that prevents loading of one analog stage by another.

Cascade — Placing one analog stage after another to combine their effects on the signal.

Cathode — The element of an analog device that emits electrons.

Characteristic Curve — A plot of the relative responses of two or three analog-device parameters, usually output with respect to input.

Clamping — A nonlinearity in amplification where the signal can be made no larger.

Collector — One of the outer layers of a bipolar transistor, often the output.

Compensation — The process of counteracting the effects of signals that are inadvertently fed back from the output to the input of an analog system. The process increases stability and prevents oscillation.

Cutoff Region — The region in the characteristic curve of an analog device in which there is no current through the device. Also called the OFF region.

Diode — A two-element vacuum tube or semiconductor with only a cathode and an anode (or plate).

Drain — The connection at one end of a field-effect-transistor channel, often the output.

Electron — A subatomic particle that has a negative charge and is the basis of electrical current.

Emitter — One of the outer layers of a bipolar transistor, often the reference.

Field-Effect Transistor (FET) — An analog device with a semiconductor channel whose width can be modified by an electric field. Also called a unipolar transistor.

Gain — see **Amplification**.

Gain-Bandwidth Product — The interrelationship between amplification and frequency that defines the limits of the ability of a device to act as a linear amplifier. In many amplifiers, gain times bandwidth is approximately constant.

Gate — The connection at the control point of a field-effect transistor, often the input.

Grid — The vacuum-tube element that controls the electron flow from cathode to plate. Additional grids in some tubes perform other control functions to improve performance.

Hole — A positively charged "particle" that results when an electron is removed from an atom in a semiconductor crystal structure.

Integrated Circuit (IC) — A semiconductor device in which many components, such as diodes, bipolar transistors, field-effect transistors, resistors and capacitors are fabricated to make an entire circuit.

Junction FET (JFET) — A field-effect transistor that forms its electric field across a PN junction.

Linearity — The property found in nature and most analog electrical circuits that governs the processing and combination of signals by treating all signal levels the same way.

Load Line — A line drawn through a family of characteristic curves that shows the operating points of an analog device for a given output load impedance.

Loading — The condition that occurs when a cascaded analog stage modifies the operation of the previous stage.

Metal-Oxide Semiconductor (MOSFET) — A field-effect transistor that forms its electric field through an insulating oxide layer.

N-Type Impurity — A doping atom with an excess of electrons that is added to semiconductor material to give it a net negative charge.

Noise — Any unwanted signal.

Noise Figure (NF) — A measure of the noise added to a signal by an analog processing stage.

Operational Amplifier (op amp) — An integrated circuit that contains a symmetrical circuit of transistors and resistors with highly improved characteristics over other forms of analog amplifiers.

Oscillator — An unstable analog system, which causes the output signal to vary spontaneously.

P-Type Impurity — A doping atom with an excess of holes that is added to semiconductor material to give it a net positive charge.

Peak Inverse Voltage (PIV) — The highest voltage that can be tolerated by a reverse biased PN junction before current is conducted.

Pentode — A five element vacuum tube with a cathode, a control grid, a screen grid, a suppressor grid, and a plate.

Plate — See anode, usually used with vacuum tubes.

PN Junction — The region that occurs when P-type semiconductor material is placed in contact with N-type semiconductor material.

Saturation Region — The region in the characteristic curve of an analog device in which the output signal can be made no larger. See **Clamping**.

Semiconductor — An elemental material whose current conductance can be controlled.

Signal-To-Noise Ratio (SNR) — The ratio of the strength of the desired signal to that of the unwanted signal (noise).

Slew Rate — The maximum rate at which a signal may change levels and still be accurately amplified in a particular device.

Source — The connection at one end of the channel of a field-effect transistor, often the reference.

Superposition — The natural process of adding two or more signals together and having each signal retain its unique identity.

Tetrode — A four-element vacuum tube with a cathode, a control grid, a screen grid, and a plate.

Triode — A three element vacuum tube with a cathode, a grid, and a plate.

Unipolar Transistor — see **Field-Effect Transistor (FET)**.

Zener Diode — A PN-junction diode with a controlled peak inverse voltage so that it will start conducting current at a preset reverse voltage.

Introduction

This chapter, written by Greg Lapin, N9GL, treats analog signal processing in two major parts. Analog signals behave in certain well defined ways regardless of the specific hardware used to implement the processing. Signal processing involves various electronic stages to perform functions such as amplifying, filtering, modulation and demodulation. A piece of electronic equipment, such as a radio, cascades a number of these circuits. How these stages interact with each other and how they affect the signal individually and in tandem is the subject of the first part of this chapter.

Implementing analog signal processing functions involves several types of active components. An active electronic component is one that requires a power source to function, and is distinguished in this way from passive components (such as resistors, capacitors and inductors) that are described in the **DC Theory and Resistive Components** chapter and the **AC Theory and Reactive Components** chapter. The second part of this chapter describes the various technologies that implement active devices. Vacuum tubes, bipolar semiconductors, field-effect semiconductors and integrated semiconductor circuitry comprise a wide spectrum of active devices used in analog signal processing. Several different devices can perform the same function. The second part of the chapter describes the physical basis of each device. Understanding the specific characteristics of each device allows you to make educated decisions about which device would be best for a particular purpose when designing analog circuitry, or understanding why an existing circuit was designed in a particular way.

Analog Signal Processing

LINEARITY

The term, *analog signal*, refers to the continuously variable voltage of which all radio and audio signals are made. Some signals are man-made and others occur naturally. In nature, analog signals behave according to laws that make radio communication possible. These same laws can be put to use in electronic instruments to allow us to manipulate signals in a variety of ways.

The premier properties of signals in nature are *superposition* and *scaling*. Superposition is the property by which signals combine. If two signals are placed together, whether in a circuit, in a piece of wire, or even in air, they become one combined signal that is the sum of the individual signals. This is to say that at any one point in time, the voltage of the combined signal is the sum of the voltages of the two original signals at the same time. In a linear system any number of signals will add in this way to give a single combined signal.

One of the more important features of superposition, for the purposes of signal processing, is that signals that have been combined can be separated into their original components. This is what allows signals that have been contaminated with noise to be separated from the noise, for example.

Amplification and attenuation scale signals to be larger and smaller, respectively. The operation of scaling is the same as multiplying the signal at each point in time by a constant value; if the constant is greater than one then the signal is amplified, if less than one then the signal is attenuated.

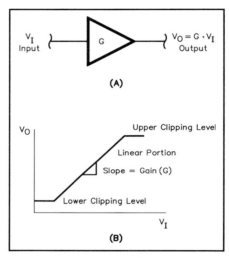

Fig 8.1 — Generic amplifier. (A) Symbol. For the linear amplifier, gain is the constant value, G, and the output voltage is equal to the input voltage times G; (B) Transfer function, input voltage along the x-axis is converted to the output voltage along the y-axis. The linear portion of the response is where the plot is diagonal; its slope is equal to the gain, G. Above and below this range are the clipping limits, where the response is not linear and the output signal is clipped.

Linear Operations

Any operation that modifies a signal and obeys the rules of superposition and scaling is a *linear operation*. The most basic linear operation occurs in an amplifier, a circuit that increases the amplitude of a signal. Schematically, a generic amplifier is signified by a triangular symbol, its input along the left face and its output at the point on the right (see **Fig 8.1**). The linear amplifier multiplies every value of a signal by a constant value. Amplifier gain is often expressed as a multiplication factor (x 5, for example).

$$\text{Gain} = \frac{V_o}{V_i} \qquad (1)$$

where V_o is the output voltage from an amplifier when an input voltage, V_i, is applied.

Ideal linear amplifiers have the same gain for all parts of a signal. Thus, a gain of 10 changes 10 V to 100 V, 1 V to 10 V and –1 V to –10 V. Amplifiers are limited by their dynamic range and frequency response, however. An amplifier can only produce output levels that are within the range of its power supply. The power-supply voltages are also called the *rails* of an amplifier. As the amplified output approaches one of the rails, the output will not go beyond a given voltage that is near the rail. The output is limited at the *clipping level* of an amplifier. When an amplifier tries to amplify a signal to be larger than this value, the output remains at this level; this is called output *clipping*. Clipping is a nonlinear effect; an amplifier is considered linear only between its clipping levels. See Fig 8.1.

Another limitation of an amplifier is its frequency response. Signals within a range of frequencies are amplified consistently but outside that range the amplification changes. At higher frequencies an amplifier acts as a low-pass filter, decreasing amplification with increasing frequency. For lower frequencies, amplifiers are of two kinds: dc and ac coupled. A dc coupled amplifier equally amplifies signals with frequencies down to dc. An ac coupled amplifier acts as a high-pass filter, decreasing amplification as the frequency decreases toward dc.

The combination of gain and frequency limitations is often expressed as a *gain-bandwidth product*. At high gains many amplifiers work properly only over a small range of frequencies. In many amplifiers, gain times bandwidth is approximately constant. As gain increases, bandwidth decreases, and vice versa. Another similar descriptor is called *slew rate*. This term describes the maximum rate at which a signal can change levels and still be accurately amplified in a particular device. There is a direct correlation between the signal-level rate of change and the frequency content of that signal.

Feedback and Oscillation

The stability of an amplifier refers to its ability to provide gain to a signal without tending to oscillate. For example, an amplifier just on the verge of oscillating is not generally considered to be "stable." If the output of an amplifier is fed back to the input, the feedback can affect the amplifier stability. If the amplified output is added to the input, the output of the sum will be larger. This larger output, in turn, is also fed back. As this process continues, the amplifier output will continue to rise until the amplifier cannot go any higher (clamps). Such *positive feedback* increases the amplifier gain, and is called *regeneration*.

Most practical amplifiers have intrinsic feedback that is unavoidable. To improve the stability of an amplifier, *negative feedback* can be added to counteract any unwanted positive feedback. Negative feedback is often combined with a phase-shift *compensation* network to improve the amplifier stability.

The design of feedback networks depends on the desired result. For amplifiers, which should not oscillate, the feedback network is customized to give the desired frequency response without loss of stability. For oscillators, the feedback network is designed to create a steady oscillation at the desired frequency.

Filtering

A filter is a common linear stage in radio equipment. Filters are characterized by their ability to selectively attenuate certain frequencies (stop band) while passing or amplifying others (pass band). Passive filters are described in the **Filters and Projects** chapter. Filters can also be designed using active devices. All practical amplifiers are low-pass filters or band-pass filters, because the gain decreases as the frequency increases beyond their gain-bandwidth products.

Summing Amplifiers

In a linear system, nature does most of the work for us when it comes to adding signals; placing two signals together naturally causes them to add. When processing signals, we would like to control the summing operation so the signals do not distort. If two signals come from separate stages and they are connected, the stages may interact, causing both stages to distort their signals. Summing amplifiers generally use a resistor in series with each stage, so the resistors connect to the common input of the following stage. **Fig 8.2** illustrates the resistors connecting to a summing amplifier. Ideally, any time we wanted to combine signals (for example, combining an audio signal with a PL tone in a 2 m FM transmitter prior to modulating the RF signal) we could use a summing amplifier.

Buffering

It is often necessary to isolate the stages of an analog circuit. This isolation reduces the loading, coupling and feedback between stages. An intervening stage, called a *buffer*, is often used for this purpose. A buffer is a linear circuit that is a type of amplifier. It is often necessary to change the characteristic impedance of a circuit between stages. Buffers can have high values of amplification but this is unusual. A buffer performs impedance transformations most efficiently when it has a low or unity gain. **Fig 8.3** shows common forms of buffers with low-impedance outputs:

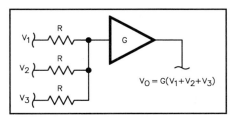

Fig 8.2 — Summing amplifier. The output voltage is equal to the sum of the input voltages times the amplifier gain, G. As long as the resistance values, R, are equal and the amplifier input impedance is much higher, the actual value of R does not affect the output signal.

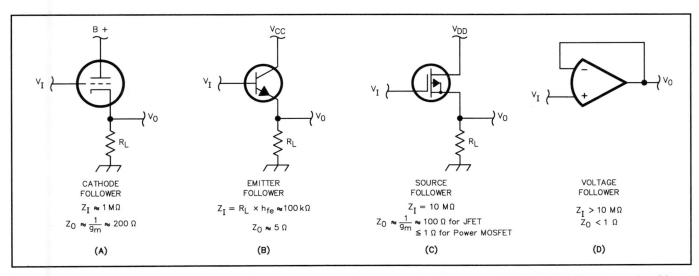

Fig 8.3 — Common buffer stages and some typical input (Z$_I$) and output (Z$_O$) impedances. (A) Cathode follower, made with triode tube; (B) Emitter follower, made with NPN bipolar transistor; (C) Source follower, made with FET; and (D) Voltage follower, made with operational amplifier. All of these buffers are terminated with a load resistance, R$_L$, and have an output voltage that is approximately equal to the input voltage (gain ≈ 1).

the cathode follower using a triode tube, the emitter follower using a bipolar transistor, the source follower using a field-effect transistor and the voltage follower, using an operational amplifier.

In some circuits, notably power amplifiers, the desired goal is to deliver a maximum amount of power to the output device (such as a speaker or an antenna). Matching the amplifier output impedance to the output-device impedance provides maximum power transfer. A buffer amplifier may be just the circuit for this type of application. Such amplifier circuits must be carefully designed to avoid distortion.

Amplitude Modulation/ Demodulation

Voice signals are transmitted over the air by amplitude modulating them on higher frequency carrier signals (see the **Mixers** chapter). The process of amplitude modulation can be mathematically described as the multiplication (product) of the voice signal and the carrier signal. Multiplication is a linear process since amplitude modulating the sum of two audio signals produces a signal that is identical to the sum of amplitude modulating each audio signal individually. When two equal-strength SSB signals are transmitted on the same frequency, the observer hears both of the voices simultaneously. Another aspect of the linear behavior of amplitude modulation is that amplitude-modulated signals can be demodulated to be exactly in their original form. Amplitude demodulation is the converse of amplitude modulation, and is represented as a division operation.

In the linear model of amplitude modulation, the signal to be modulated (such as the audio signal in an AM transmitter) is shifted in frequency by multiplying it with the carrier. The modulated waveform is considered to be a linear function of the signal. The carrier is considered to be part of a time-varying linear system and not a second signal.

A curious trait of amplitude modulation is that it can be performed nonlinearly. Each nonlinear form of amplitude modulation generates the desired linear product term in addition to other unwanted terms that must be removed. Accurate analog multipliers and dividers are difficult and expensive to fabricate. Two common nonlinear amplitude modulating schemes are much simpler to implement but have disadvantages as well.

Power-law modulators generate many frequencies in addition to the desired ones. These unwanted frequencies, often called *intermodulation products*, steal energy from the desired *first order product*. The

unwanted signals must be filtered out. The inefficiency of this process makes this type of modulator good only for low-level modulation, with additional amplification required for the modulated signal. A *square-law modulator* can be implemented with a single FET, biased in its saturation region, as the only active component.

Switching modulators are more efficient and provide high-level modulation. A single active device acts as a switch to turn the signal on and off at the carrier frequency. Both the signal and the carrier must be amplified to relatively high levels prior to this form of modulation. The modulated carrier must be filtered by a tank circuit to remove unwanted frequency components generated by the switching artifacts.

Nonlinear demodulation of an amplitude-modulated signal can be realized with a single diode. The diode rectifies the signal (a nonlinear process) and then the nonlinear products are filtered out before the desired signal is recovered.

NONLINEAR OPERATORS

All signal processing doesn't have to be linear. Any time that we treat various signal levels differently, the operation is called *nonlinear*. This is not to say that all signals must be treated the same for a circuit to be linear. High frequency signals are attenuated in a low-pass filter while low frequency signals are not, yet the filter can be linear. The distinction is that all voltages of the high-frequency signal are attenuated by the same amount, thus satisfying one of the linearity conditions. What if we do not want to treat all voltage levels the same way? This is commonly desired in analog signal processing for clipping, rectification, compression, modulation and switching.

Clipping and Rectification

Clipping is the process of limiting the range of signal voltages passing through a circuit (in other words, *clipping* those voltages outside the desired range off of the signals). There are a number of reasons why we would like to do this. Clipping generally refers to the process of limiting the positive and negative peaks of a signal. We might use this technique to avoid overdriving an amplifier, for example. Another type of clipping results in rectification. The rectifier clips off all voltages of one polarity (positive or negative) and allows only the other polarity through, thus changing ac to pulsating dc (see the **Power Supplies and Projects** chapter). Another use of clipping is when only one signal polarity is allowed to drive an

amplifier input; a clipping stage precedes the amplifier to ensure this.

Logarithmic Amplification

It is sometimes desirable to amplify a signal logarithmically, which means amplifying low levels more than high levels. This type of amplification is often called *signal compression*. Speech compression is sometimes used in audio amplifiers that feed modulators. The voice signal is compressed into a small range of amplitudes, allowing more voice energy to be transmitted without over modulation (see the **Modulation Sources** chapter).

ANALOG BUILDING BLOCKS

Many types of electronic equipment are developed by combining basic analog signal processing circuits or "building blocks." This section describes several of these building blocks and how they are combined to perform complex functions. Although not all basic electronic functions are discussed here, the characteristics of combining them can be applied generally.

An analog building block can contain any number of discrete components. Since our main concern is the effect that circuitry has on a signal, we often describe the building block by its actions rather than its specific components. For this reason, an analog building block is often referred to as a *two-port network* or a *black box*. Two basic properties of analog networks are of principal concern: the effect that the network has on an analog signal and the interaction that the network has with the circuitry surrounding it. The two network ports are the input and output connections. The signal is fed into the input port, is modified inside the network and then exits from the output port.

An analog network modifies a signal in a specific way that can be described mathematically. The output is related to the input by a *transfer function*. The mathematical operation that combines a signal with a transfer function is pictured symbolically in **Fig 8.4**. The output signal, w(t), has a value that changes with time. The output signal is created by the action of an analog transfer function, h(t), on the input signal, g(t).

While it is not necessary to understand transfer functions mathematically to work

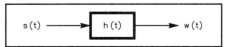

Fig 8.4 — Linear function block. The output signal, w(t) is produced by the action of the transfer function, h(t) on the input signal s(t).

with analog circuits, it is useful to realize that they describe how a signal interacts with other signals in an electronic system. In general, the output signal of an analog system depends not only on the input signal at the same time, but also on past values of the input signal. This is a very important concept and is the basis of such essential functions as analog filtering.

Cascading Stages

If an analog circuit can be described with a transfer function, a combination of analog circuits can also be described similarly. This description of the combined circuits depends upon the relationship between the transfer functions of the parts and that of the combined circuits. In many cases this relationship allows us to predict the behavior of large and complex circuits from what we know about the parts that make them up. This aids in the design and analysis of analog circuits.

When two analog circuits are cascaded (the output signal of one stage becomes the input signal to the next stage) their transfer functions are combined. The mechanism of the combination depends on the interaction between the stages. The ideal case is when there is no interaction between stages. In other words, the action of the first stage is unchanged, regardless of whether or not the second stage follows it. Just as the signal entering the first stage is modified by the action of the first transfer function, the ideal cascading of analog circuits results in changes produced only by the individual transfer functions. For any number of stages that are cascaded, the combination of their transfer functions results in a new transfer function. The signal that enters the circuit is changed by the composite transfer function, to produce the signal that exits the cascaded circuits.

Cascaded Buffers

Buffer stages that are made with single active devices can be more effective if cascaded. Two types of such buffers are in common use. The *Darlington pair* is a cascade of two common-collector transistors as shown in **Fig 8.5**. (The various amplifier configurations will be described later in this chapter.) The input impedance of the Darlington pair is equal to the load impedance times the current gain, h_{FE}. The current gain of the Darlington pair is the product of the current gains for the two transistors.

$$Z_I = Z_{LOAD} \times h_{FE1} \times h_{FE2} \qquad (2)$$

For example, if a typical bipolar transistor has $h_{FE} = 100$ and a circuit has a $Z_{LOAD} = 15$ kΩ, a pair of these transistors

in the Darlington-pair configuration would have:

$$Z_I = 15 \text{ k}\Omega \times 100 \times 100 = 150 \text{ M}\Omega.$$

The shunt capacitance at the input of real transistors can lower the actual impedance as the frequency increases.

A common-emitter amplifier followed by a common-base amplifier is called a *cascode buffer* (see **Fig 8.6**). Cascodes are also made with FETs by following a common-source amplifier by a common-gate configuration. The input impedance and current gain of the cascode are approximately the same as those of the first stage. The output impedance is much higher than that of a single stage. Cascode amplifiers have excellent input/output isolation (very low unwanted feedback) and this can provide high gain with good stability. An example of a cascode buffer made with bipolar transistors has moderate input impedance, $Z_I = 1$ kΩ, high current gain, $h_{FE} = 50$ and high output impedance, $Z_O = 1$ MΩ. There is very little reverse internal feedback in the cascode design, making it very stable, and the amplifier design has little effect on external tuning components. Cascode circuits are often used in tuned amplifier designs for these reasons.

Interstage Loading and Impedance Matching

If the transfer function of a stage changes when it is cascaded with another stage, we say that the second stage has *loaded* the first stage. This often occurs when an appreciable amount of current passes from one stage to the next.

Every two-port network can be further defined by its input and output impedance. The input impedance is the opposition to current, as a function of frequency, that is seen when looking into the input port of the network. Likewise, the output impedance is similarly defined when looking back into a network through its output port. Interstage loading is related to the relative output impedance of a stage and the input impedance of the stage that is cascaded after it.

In some applications the goal is to transfer a maximum amount of power. In an RF amplifier, the impedance at the input of the transmission line feeding an antenna is transformed by means of a matching network to produce the resistance the amplifier needs in order to efficiently produce RF power.

In contrast, it is the goal of most analog signal processing circuitry to modify a signal rather than to deliver large amounts of energy. Thus, an impedance-matched condition may not be what is desired. Instead,

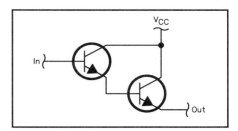

Fig 8.5 — Darlington pair made with two emitter followers. Input impedance, Z_I, is far higher than for a single transistor and output impedance, Z_O, is nearly the same as for a single transistor. DC biasing has been omitted for simplicity.

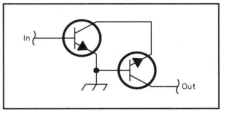

Fig 8.6 — Cascode pair made with two NPN bipolar transistors has a medium input impedance and high output impedance. DC biasing has been omitted for simplicity.

current between stages can be minimized by having mismatched impedances. Ideally, if the output impedance of a network approaches zero ohms and the input impedance of the following stage is very high, very little current will pass between the stages, and interstage loading will be negligible.

Noise

Generally we are only interested in specific man-made signals. Nature allows many signals to combine, however, so the desired signal becomes combined with many other unwanted signals, both man-made and naturally occurring. The broadest definition of noise is any signal that is not the one in which we are interested. One of the goals of signal processing is to separate desired signals from noise.

One form of noise that occurs naturally and must be dealt with in low-level processing circuits is called *thermal noise*, or *Johnson noise*. Thermal noise is produced by random motion of free electrons in conductors and semiconductors. This motion increases as temperature increases, hence the name. This kind of noise is present at all frequencies and is proportional to temperature. Naturally occurring noise can be reduced either by decreasing the bandwidth or by reducing the temperature in the system. Thermal noise voltage and current vary with the circuit impedance, according to Ohm's Law. Low-noise-amplifier-design techniques are based on these relationships (see the **Amplifiers** chapter).

Analog signal processing stages are characterized in part by the noise they add to a signal. A distinction is made between enhancing existing noise (such as amplifying it) and adding new noise. The noise added by analog signal processing is commonly quantified by the *noise factor, f*. Noise factor is the ratio of the total output noise power (thermal noise plus noise added by the stage) to the input noise power when the termination is at the standard temperature of 290 K (17°C). When the noise factor is expressed in dB, we often call it *noise figure, NF*. NF is calculated as:

$$NF = 10 \log \frac{P_{NO}}{A P_{NTH}} \qquad (3)$$

where:

P_{NO} = total noise output power,
A = amplification gain, and
P_{NTH} = input thermal noise power.

The noise factor can also be calculated as the difference between the input and output signal-to-noise ratios (SNR), with SNR expressed in dB.

In a system of many cascaded signal processing stages, each stage affects the noise of the system. The noise factor of the first stage dominates the noise factor of the entire system. Designers try to optimize system noise factor by using a first stage with a minimum possible noise factor and maximum possible gain. A circuit that overloads is often as useless as one that generates too much noise. See the **Transceivers** chapter for more information about circuit noise.

Analog Devices

There are several different kinds of components that can be used to build circuits for analog signal processing. The same processing can be performed with vacuum tubes, bipolar semiconductors, field-effect semiconductors or integrated circuitry, each with its own advantages and disadvantages.

TERMINOLOGY

A similar terminology is used for most active electronic devices. The letter V stands for voltages and I for currents. Voltages generally have two subscripts indicating the terminals the voltage is measured between (V_{BE} is the voltage between the base and the emitter of a bipolar transistor). Currents have a single subscript indicating the terminal that the current flows into (I_P is the current into the plate of a vacuum tube). If the current flows out of the device, it is generally indicated with a negative sign. Power supply voltages have two subscripts that are the same, indicating the terminal to which the voltage is applied (V_{DD} is the power supply voltage applied to the drain of a field-effect transistor). A transfer characteristic is a ratio of an output parameter to an input parameter, such as output current divided by input current. Transfer characteristics are represented with letters, such as h, s, y or z. Resistance is designated with the letter r, and impedance with the letter Z. For example, r_{DS} is resistance between drain and source of an FET and Z_i is input impedance. In some designators, values differ for dc and ac signals. This is indicated by using capital letters in the subscripts for dc and lower-case subscripts for ac. For example, the common-emitter dc current gain for a bipolar transistor is designated as h_{FE}, and h_{fe} is the ac current

gain. Qualifiers are sometimes added to the subscripts to indicate certain operating modes of the device. SS for saturation, BR for breakdown, ON and OFF are all commonly used.

The abbreviations for tubes existed before these standards were adopted so some tube-performance descriptors are different. For example, B+ is usually used for the plate bias voltage. Since integrated circuits are collections of semiconductor components, the abbreviations for the type of semiconductor used also apply to the integrated circuit. V_{CC} is a power supply voltage for an integrated circuit made with bipolar transistor technology.

Amplifier Types

Amplifier configurations are described by the *common* part of the device. The word "common" is used to describe the connection of a lead directly to a reference. The most common reference is ground, but positive and negative power sources are also valid references. The type of reference used depends on the type of device (vacuum tube, transistor [NPN or PNP], FET [P-channel or N-channel]), which lead is common and the range of signal levels. Once a common lead is chosen, the other two leads are used for signal input and output. Based on the biasing conditions, there is only one way to select these leads. Thus, there are three possible amplifier configurations for each type of three-lead device.

The operation of an amplifier is specified by its gain. A gain in this sense is defined as the change (Δ) in the output parameter divided by the corresponding change in the input parameter. If a particular device measures its input and output as currents, the gain is called a current gain.

If the input and output are voltages, the amplifier is defined by its voltage gain. If the input is a voltage and the output is a current, the ratio is called the *transconductance*.

Characteristic Curves

Analog devices are described most completely with their *characteristic curves*. Almost all devices that we deal with are nonlinear over a wide range of operating parameters. We are often interested in using a device only in the region that approximates a linear response. The characteristic curve is a plot of the interrelationships between two or three variables. The vertical (y) axis parameter is the output, or result of the device being operated with an input parameter on the horizontal (x) axis. Often the output is the result of two input values. The first input parameter is represented along the x axis and the second input parameter by several curves, each for a different value. For example, a vacuum tube characteristic curve may have the plate current along the y axis, the grid voltage along the x axis and several curves, each representing a different value of the plate bias voltage (see **Fig 8.7**).

The parameters plotted in the characteristic curve depend on how the device will be used. The common amplifier configuration defines the input and output leads, and their relationship is diagrammed by the curves. Device parameters are usually derived from the characteristic curve. To calculate a gain, the operating region of the curve is specified, usually a straight portion of the curve if linear operation is desired. Two points along that portion of the curve are selected, each defined by its location along the x and y axes. If the two points are defined by (x_1, y_1) and (x_2, y_2),

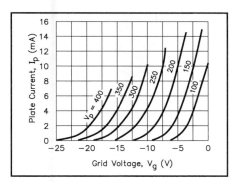

Fig 8.7 — Tube characteristic curve. Input signal is the grid voltage, V_g, along the x-axis and the output signal is the plate current, I_p, along the y-axis. Different curves are plotted for various values of plate bias voltage, V_p (also called B+).

the slope, m, of the curve, which can be a gain, a resistance or a conductance, is calculated as:

$$m = \frac{\Delta y}{\Delta x} = \frac{y_1 - y_2}{x_1 - x_2} \qquad (4)$$

A characteristic curve that plots device output voltage and current along the x and y axes permits the inclusion of an additional curve. The *load line* is a straight line with a slope that is equal to the load impedance. The intersections between the load line and the characteristic curves indicate the operating points for that circuit. Load lines are only applicable to output characteristic plots; they cannot be used with input or transfer (input versus output) characteristic curves.

BIASING

The operation of an analog signal processing device is greatly affected by which portion of the characteristic curve is used to do the processing. As an example, consider the vacuum tube characteristic curves in **Fig 8.8** and **Fig 8.9**.

The relationship between the input and the output of a tube amplifier is illustrated in Fig 8.8. The input signal (a sine wave in this example) is plotted in the vertical direction and below the graph. For a grid bias level of –5 V, the sine wave causes the grid voltage, V_g, to deviate between –3 and –7 V. These values correspond to a range of plate currents, I_p, between 1.4 and 2.6 mA. With a plate bias of 200 V and a load resistance, R_p, of 50 kΩ, the corresponding change in plate voltage, V_p, is between 70 and 130 V. Thus, this triode amplifier configuration changes a range of 4 V at the input to 60 V at the output. Also there is a change of output-signal voltage polarity; this amplifier both amplifies the signal magnitude 15 times and

shifts the phase of the signal by 180°.

In the previous example the signal was biased so that it fell on a linear (straight) portion of the characteristic curve. If a different bias voltage is selected so that the signal does not fall on a linear portion of the curve, the output signal will be a distorted version of the input signal. This is illustrated in Fig 8.9. The input signal is amplified within a curved region of the characteristic curve. The positive part of the signal is amplified more than the negative part of the signal. Proper biasing is crucial to ensure amplifier linearity.

Input biasing serves to modify the relative level (dc offset) of the input signal so that it falls on the desired portion of the characteristic curve. Devices that perform signal processing (vacuum tubes, diodes, bipolar transistors, field-effect transistors and operational amplifiers) usually require appropriate input signal biasing.

Manufacturers' Data Sheets

Manufacturer's data sheets list device characteristics, along with the specifics of the part type (polarity, semiconductor type), identification of the pins, and the typical use (such as small signal, RF, switching or power amplifier). The pin identification is important because, although common package pinouts are normally used, there are exceptions. Manufacturers may differ slightly in the values reported, but certain basic parameters are listed. Different batches of the same devices are rarely identical, so manufacturers specify the guaranteed limits for the parameters of their device. There are usually three columns of values listed in the data sheet. For each parameter, the columns may list the guaranteed minimum value, the guaranteed maximum value and/or the typical value.

Another section of the data sheet lists ABSOLUTE MAXIMUM RATINGS, beyond which device damage may result. For example, the parameters listed in the ABSOLUTE MAXIMUM RATINGS section for a solid-state device are typically voltages, continuous currents, total device power dissipation (P_D) and operating- and storage-temperature ranges.

Rather than plotting the characteristic curves for each device, the manufacturer often selects key operating parameters that describe the device operation for the configurations and parameter ranges that are most commonly used. For example, a bipolar transistor data sheet might include an OPERATING PARAMETERS section. Parameters are listed in an OFF CHARACTERISTICS subsection and an ON CHARACTERISTICS subsection that describe the conduction properties of the device for dc

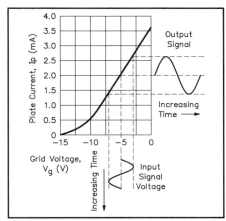

Fig 8.8 — Determination of output signal (to the right of the plot) for a given input signal (below the plot, turned on its side) with a tube characteristic curve plotted for a given plate bias. Note that the grid bias voltage, –5 V, causes the entire range of the input signal to be mapped onto the linear (diagonal straight line) portion of the characteristic curve. The output signal has the same shape as the input signal except that it is larger in amplitude.

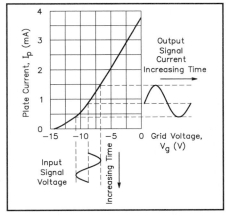

Fig 8.9 — Same characteristic curve and input signal as in Fig 8.8 except the grid bias voltage is now about –8.75 V. The input signal falls on the curved (non-linear) portion of the plot and causes distortion in the output signal. Note how the upper portion of the output sine wave was amplified more than the lower portion.

voltages. The SMALL-SIGNAL CHARACTERISTICS section often contains the guaranteed minimum Gain-Bandwidth Product (f_T), the guaranteed maximum output capacitance, the guaranteed maximum input capacitance and the guaranteed range of the transfer parameters applicable to a given device. Finally, the SWITCHING CHARACTERISTICS section lists absolute maximum ratings for Delay Time (t_d),

Rise Time (t_r), Storage Time (t_s) and Fall Time (t_f). Other types of devices list characteristics important to operation of that specific device.

When selecting equivalent parts for replacement of specified devices, the data sheet provides the necessary information to tell if a given part will perform the functions of another. Lists of equivalencies generally only specify devices that have nearly identical parameters. There are usually a large number of additional devices that can be chosen as replacements. Knowledge of the circuit requirements adds even more to the list of possible replacements. The device parameters should be compared individually to make sure that the replacement part meets or exceeds the parameter values of the original part required by the circuit. Be aware that in some applications a far superior part may fail as a replacement, however. A transistor with too much gain could easily oscillate if there were insufficient negative feedback to ensure stability.

VACUUM TUBES

Current is generally described as the flow of electrons through a conductor, such as metal. The vacuum tube controls the flow of electrons in a vacuum, which is analogous to a faucet that adjusts the flow of a fluid. The British commonly refer to vacuum tubes as *valves*. Although the physics of the operation of vacuum tubes varies greatly from that of semiconductors, there are many similarities in the way that they behave in analog circuits.

Thermionic Theory

Metals are elements that are characterized by their large number of free electrons. Individual atoms do not hold onto all of their electrons very tightly, and it is relatively easy to dislodge them. This property makes metals good conductors of electricity. Under electrical pressure (voltage), electrons collide with metal atoms, dislodging an equal number of free electrons from the metal. These collide with adjoining metal atoms to continue the process, resulting in a flow of electrons.

It is also possible to cause the free electrons to be emitted into space if enough energy is added to them. Heat is one way of adding energy to metal atoms, and the resulting flow of electrons into space is called *thermionic emission*. It is important to remember that the metal atoms don't permanently lose electrons; the emitted electrons are replaced by others that come from an electrical connection to the heated metal. Thus, an electron that flows into the

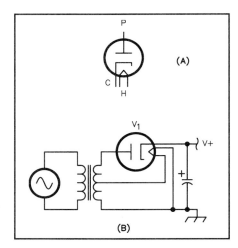

Fig 8.10 — Vacuum tube diode. (A) Schematic symbol detailing heater (H), cathode (C) and plate (P). (B) Power supply circuit using diode as a half wave rectifier.

heated metal collides with and is captured by a metal atom, knocking loose a highly energized electron that is emitted into space.

In a vacuum, there are no other atoms with which the emitted electron can collide, so it follows a straight path until it collides with another atom. A *vacuum tube* has nearly all of the air evacuated from it, so the emitted electrons proceed unhindered to another piece of metal, where they continue to move as part of the electrical current.

Components of a Vacuum Tube

A basic vacuum tube contains at least two parts; a *cathode* and a *plate*. The electrons are emitted from the *cathode*. The cathode can either be heated directly by passing a large dc current through it, or it can be located adjacent to a heating element. Although ac currents can also be used to directly heat cathodes, if any of the ac voltage mixes with the signal, ac hum will be introduced into the output. If the ac heater supply voltage can be obtained from a center tapped transformer, and the center tap is connected to the signal ground, hum can be minimized. Cathodes are made of substances that have the highest emission of electrons for the lowest temperatures and voltages. Tungsten, thoriated tungsten and oxide-coated metals are commonly used.

Every vacuum tube needs a receptor for the emitted electrons. After moving though the vacuum, the electrons are absorbed by the *plate*. Since the plate receives electrons, it is also called the

anode. Each electron has a negative charge, so a positively biased plate will attract the emitted electrons to it, and a current will result. For every electron that is accepted by the plate, another electron flows into the cathode; the plate and cathode currents must be the same. As the plate voltage is increased, there is a larger electrical field attracting electrons, causing more of them to be emitted from the cathode. This increases the current through the tube. This relationship continues until a limit is reached where further increases to the electrical field do not cause any more electrons to be emitted. This is the *saturation point* of the vacuum tube.

A vacuum tube that contains only a cathode and a plate is called a *diode tube* (di- for two components). See **Fig 8.10**. The diode tube is similar to a semiconductor diode since it allows current to pass in only one direction; it is used as a rectifier. When the plate voltage becomes negative, the electrical field that is set up repels electrons, preventing them from being emitted from the cathode.

To amplify signals, a vacuum tube must also contain a control *grid*. This name comes from its physical construction. The grid is a mesh of wires located between the cathode and the plate. Electrons from the cathode pass between the grid wires on their way to the plate. The electrical field that is set up by the voltage on these wires affects the electron flow from cathode to plate. A negative grid voltage sets up an electrical field that repels electrons, decreasing emission from the cathode because of the higher energy needed for the electrons to escape from their atoms into the vacuum. A positive grid voltage will have the opposite effect. Since the plate voltage is always positive, however, grid voltages are usually negative. The more negative the grid, the less effective the electrical field from the plate will be at attracting electrons from the cathode.

Vacuum tubes containing a cathode, a grid and a plate are called *triode* tubes (tri- for three components). See **Fig 8.11**. They are generally used as amplifiers, particularly at frequencies in the HF range and below. Characteristic curves for triodes normally relate grid bias voltage and plate bias voltage to plate current for the triode (Fig 8.7). There are three descriptors of a tube's performance that can be derived from the characteristic curves. The *plate resistance*, r_p, describes the resistance to the flow of electrons from cathode to plate. The r_p is calculated by selecting a vertical line in the characteristic curve and dividing the change in plate-to-cathode voltage (ΔV_p) of two of the

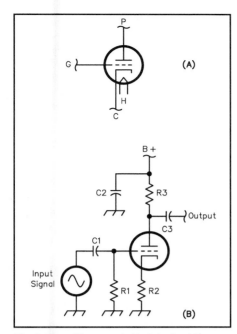

Fig 8.11 — Vacuum tube triode. (A) Schematic symbol detailing heater (H), cathode (C), grid (G) and plate (P). (B) Audio amplifier circuit using a triode. C1 and C3 are dc blocking capacitors for the input and output signals to isolate the grid and plate bias voltages. C2 is a bypass filter capacitor to decrease noise in the plate bias voltage, B+. R1 is the grid bias resistor, R2 is the cathode bias resistor and R3 is the plate bias resistor. Note that although the cathode and grid bias voltages are positive with respect to ground, they are still negative with respect to the plate.

lines by the corresponding change in plate current (ΔI_p).

$$r_p = \frac{\Delta V_p}{\Delta I_p} \qquad (5)$$

The ratio of change in plate voltage (ΔV_p) to the change in grid-to-cathode voltage (ΔV_g) for a given plate current is the *amplification factor* (μ). Amplification factor is calculated by selecting a horizontal line in the characteristic curve and dividing the difference in plate voltage of two of the lines by the difference in grid voltages that corresponds to the same points.

$$\mu = \frac{\Delta V_p}{\Delta V_g} \qquad (6)$$

Triode amplification factors range from 10 to about 100.

The plate current flows to the plate bias supply, so the output from a triode amplifier is often expressed as the voltage that is developed as this current passes through a load resistor. The value of the load resistance affects the tube amplification, as

illustrated by the dynamic characteristic curves in **Fig 8.12**, so the tube μ does not fully describe its action as an amplifier. *Grid-plate transconductance* (g_m) takes into account the change of amplification due to load resistance. The slope of the lines in the characteristic curve represents g_m. (Since the various lines are nearly parallel in the linear operating region, they have about the same slope.)

$$g_m = \frac{\Delta I_p}{\Delta V_g} \qquad (7)$$

This ratio represents a conductance, which is measured in siemens. Triodes have g_m values that range from about 1000 to several thousand microsiemens, the higher values indicating greater possible amplification.

The input impedance of a vacuum tube amplifier is directly related to the grid current. Grid current varies with grid voltage, increasing as the voltage becomes more positive. The normal operation uses a negative grid-bias voltage, and the input impedance can be in the megohm range for very negative grid bias values. This is limited by the desired operating point on the characteristic curve, however, as illustrated in Figs 8.8 and 8.9. The output impedance of the amplifier is a function of the plate resistance, r_p, in parallel with the output capacitance. Typical output impedance is on the order of hundreds of ohms.

The physical configuration of the components within the vacuum tube appear as conductors that are separated by an insulator (in this case, the vacuum). This description is very similar to that of a capacitor. The capacitance between the cathode and grid, between the grid and plate, and between the cathode and plate can be large enough to affect the operation of the amplifier at high frequencies. These capacitances, which are usually on the order of a few picofarads, can limit the frequency response of a vacuum tube amplifier and can also provide signal feedback paths that may lead to unwanted oscillation. Neutralizing circuits are sometimes used to counteract the effects of internal capacitances and to prevent oscillations.

The grid-to-plate capacitance is the chief source of unwanted signal feedback. A special form of vacuum tube has been developed to deal with the grid-to-plate capacitance. A second grid, called a *screen grid*, is inserted between the original grid (now called a *control grid*) and the plate. The additional tube component leads to the name for this new tube — *tetrode* (tetra- for four components). See **Fig 8.13**. The screen grid reduces the capacitance between the control grid and the plate, but it also reduces the electrical field from

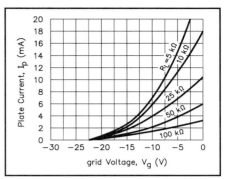

Fig 8.12 — Vacuum tube dynamic characteristic curve. This corresponds to the V_p = 300 line in Fig 8.7 with different values of load resistance. This shows how the tube will behave when cascaded to circuits with different input impedances.

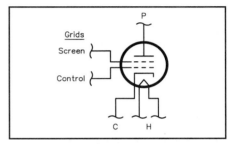

Fig 8.13 — Vacuum tube tetrode. Schematic symbol detailing heater (H), cathode (C), the two grids: control and screen and plate (P).

the plate that attracts electrons from the cathode. Like the control grid, the screen grid is made of a wire mesh and electrons pass through the spaces between the wires to get to the plate. The bias of the screen grid is positive with respect to the cathode, in order to enhance the attraction of electrons from the cathode. The electrons accelerate toward the screen grid and most of them pass through the spaces and continue to accelerate until they reach the plate. The presence of the screen grid adversely affects the overall efficiency of the tube, since some of the electrons strike the grid wires. A bypass capacitor with a low reactance at the frequency being amplified by the vacuum tube is generally connected between the screen grid and the cathode.

A special form of tetrode concentrates the electrons flowing between the cathode and the plate into a tight beam. The decreased electron-beam area increases the efficiency of the tube. *Beam tetrodes* permit higher plate currents with lower plate voltages and large power outputs with smaller grid driving power. RF power amplifiers are usually made with this type of vacuum tube.

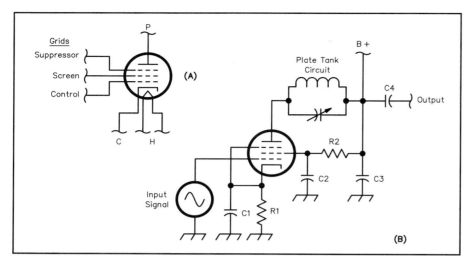

Fig 8.14 — Vacuum tube pentode. (A) Schematic symbol detailing heater (H), cathode (C), the three grids: control, screen and suppressor, and plate (P). (B) RF amplifier circuit using a pentode. C1, C2 and C3 are bypass (filter) capacitors and C4 is a dc blocking capacitor to isolate the plate bias voltage from the output signal. R1 is the cathode bias resistor and R2 is the screen voltage dropping resistor. The plate tank circuit is tuned to the desired frequency bandpass. As is common, the heater circuit is not shown.

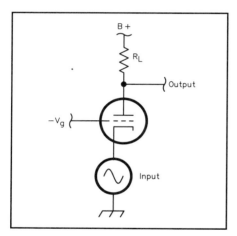

Fig 8.15 — Grounded grid amplifier schematic. The input signal is connected to the cathode, the grid is biased to the appropriate operating point by a dc bias voltage, $-V_G$, and the output voltage is obtained by the voltage drop through R_L that is developed by the plate current, I_P.

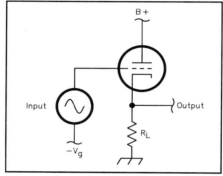

Fig 8.16 — Cathode follower schematic. The input signal is biased by $-V_G$ and fed into the grid. The plate bias, B+ is fed directly into the plate terminal. The output is derived by the cathode current (which is equal to the plate current, I_P) dropping the voltage through the load resistor, R_L.

Another unwanted effect in vacuum tubes is the emission of electrons from the plate. The electrons flowing within the tube have so much energy that they are capable of dislodging electrons from the metal atoms in the plate. These *secondary emission* electrons are repelled back to the plate by the negative bias of the grid in a triode and are of no concern. In the tetrode, the screen grid is positively biased and attracts the secondary emission electrons, causing a reverse current from the plate to the screen grid.

A third grid, called the *suppressor grid*, can be added between the screen grid and the plate. This overcomes the effects of secondary emission in tetrodes. A vacuum tube with three grids is called a *pentode* (penta- for five components). See **Fig 8.14**. The suppressor grid is negatively biased with respect to the screen grid and the plate. In some tube designs it is internally connected to the cathode. The suppressor grid repels the secondary emission electrons back to the plate.

As the number of grids is increased between the cathode and the plate, the effect of the electrical field from the positive plate voltage at the cathode is decreased. This limits the number of electrons that can be emitted from the cathode and the

characteristic curves tend to flatten out as the grid bias becomes less negative. This flattening is another nonlinearity of the tube as an amplifier, since the response saturates at a given plate current and will go no higher. Tube saturation can be used advantageously in some circuits if a constant current source is desired, since the current does not change within the saturation region regardless of changes in plate voltage.

Types of Vacuum Tube Amplifiers

The descriptions of vacuum tube amplifiers up to this point have been for only one configuration, the common cathode, where the cathode is connected to the signal reference point, the grid is the input and the plate is the output. Although this is the most common configuration of the vacuum tube as an amplifier, other configurations exist. If the signal is introduced into the cathode and the grid is at a reference level (still negatively biased but with no ac component), with the output at the plate, the amplifier is called a *grounded-grid* (**Fig 8.15**). This amplifier is characterized by a very low input impedance, on the order of a few hundred ohms, and a low output impedance, that is mainly determined by the plate resistance of the tube.

The third configuration is called the *cathode follower* (**Fig 8.16**). The plate is the common element, the grid is the input and the cathode is the output. This type of amplifier is often used as a buffer stage due to its high input impedance, similar to that of the common cathode amplifier, and its very low output impedance. The output impedance (Z_o) can be calculated from the tube characteristics as:

$$Z_o = \frac{r_p}{1 + \mu} \qquad (8)$$

where:
r_p = tube plate resistance
μ = tube amplification factor.

For a close approximation, we can simplify this equation as:

$$Z_o \approx \frac{r_p}{\mu} = \frac{1}{g_m}$$

Other Types of Tubes

Vacuum tube identifiers do not generally indicate what type of tube the device is. The format is typically a number, one or two letters and a number (such as 6AU6 or 12AT7). The first number in the identifier indicates the heater voltage (usually either 6 or 12 V). The last number often indicates the number of elements, including the heater. Some tubes also have an additional letter following the identifier (usually A or B) that indicates a revision of the tube design that represents an improvement in its operating parameters. There are also tubes that do not follow this naming convention, many of which are power amplifiers or military-type tubes (such as 6146 and 811).

To reduce stray reactances, some tubes do not have the plate connection in the tube base, where all the other connections are located. Rather, a connection is made at the top of the tube though a metallic cap. This requires an additional connector for the plate circuitry.

Tubes may share components in a single envelope to reduce size and incidental power requirements. A very common example of this is the dual triode tube (such as 12AT7 or 12AU7) that contains a single heater circuit and two complete triode tubes in the same device. Other configurations of multiple devices contained in a single vacuum tube also exist. The 6GW8 and 6EA8 tubes each contain both a triode and a pentode. The 6BN8 contains three distinct devices, one triode and two diodes.

Most common vacuum tubes are encased in glass. It is also possible to encase them in metal or ceramic materials to attain higher tube power and smaller size. Since heat dissipation from the plate is one of the major limiting factors for vacuum tube power amplifiers, the alternate materials remove heat more efficiently. These tubes can be cooled by convection, with the casing connected to a large heat sink, or with water flowing past the tube for hydraulic cooling.

A variation of the vacuum tube that is widely used in oscilloscopes and television monitors is the *cathode ray tube (CRT)*, diagrammed in **Fig 8.17**. The CRT has a cathode and grid much like a triode tube. The plate, usually referred to as the *anode* in this device, is designed to accelerate the electrons to very high velocities, with anode voltages that can be as high as tens of thousands of volts. The anode of the CRT differs from the plates of other vacuum tubes, since it is designed as a set of plates that are parallel to the electron beam. The anode voltage accelerates the electrons but does not absorb them. The electron beam passes by the anode and continues to the face of the tube. The cathode, grid and anode are all located in the neck of the CRT and are collectively referred to as the *electron gun*.

The electron beam is deflected from its path by either magnetic deflectors that surround the yoke of the tube or by electrostatic deflection plates that are built into the tube neck just beyond the electron gun.

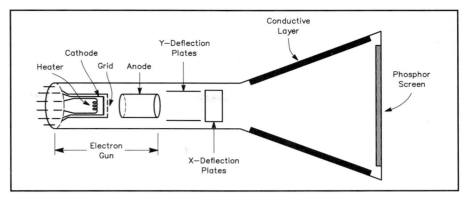

Fig 8.17 — Cross section of CRT. The electron gun generates a stream of electrons and is made up of a heater, cathode, grid and anode (plate). The electron beam passed by two pairs of deflection plates that deviate the path of the beam in the vertical (y) direction and then the horizontal (x) direction. The deflected electron beam strikes a phosphor screen and causes it to glow at that spot. Any electrons that bounce off the screen are absorbed by the conductive layer along the sides of the tube, preventing spurious luminescence

A CRT typically has two sets of deflectors: vertical and horizontal. When a potential is applied to a set of deflectors, the passing electron beam is bent, altering its path. In an oscilloscope, the time base typically drives the horizontal deflectors and the input signal drives the vertical deflectors, although in many oscilloscopes it is possible to connect another input signal to the horizontal deflectors to obtain an X-Y, or vector, display. In televisions and some computer monitors, the deflectors are typically driven by a raster generator. The horizontal deflectors are driven by a sawtooth pattern that causes the beam to move repeatedly from left to right and then retrace quickly to the left. The vertical deflectors are driven by a slower sawtooth pattern that causes the beam to move repeatedly from top to bottom and then retrace quickly to the top. The relative timing of the two sawtooth patterns is such that the beam scans from left to right, retraces to the left and then begins the next horizontal trace just below the previous one.

Beyond the deflectors, the CRT flares out. The front face is coated with a phosphorescent material that glows when struck by the electron beam. To prevent spurious phosphorescence, a conductive layer along the sides of the tube absorbs any electrons that reflect off the glass.

Vector displays have better resolution than raster scanning. The trace lines are clearer, which is the reason oscilloscope displays use this technique. It is faster to fill the screen using raster scanning, however. This is why TVs use raster scanning.

Some CRT tubes are designed with multiple electron beams. The beams are sometimes generated by different electron guns that are placed next to each other in the neck of the tube. They can also be gen-erated by splitting the output of a single electron gun into two or more beams. Very high quality oscilloscopes use two electron beams to trace two input channels rather than the more common method of alternating a single beam between the two inputs. Color television tubes use three electron beams for the three primary colors (red, green and blue). Each beam is focused on only one of these colored phosphors, which are interleaved on the face of the tube. A metal shadow mask keeps the colors separate as the beams scan across the tube.

A variation of the CRT is the *vidicon tube*. The vidicon is used in many video cameras and operates in a similar fashion to the CRT. The vidicon absorbs light from the surroundings, which charges the plate at the location of the light. This charge causes the cathode-to-plate current to increase when the raster scan points the electron beam at that location. The current increase is converted to a voltage that is proportional to the amount of light absorbed. This results in an electrical signal that represents the pattern of a visual image.

Standard vacuum tubes work well for frequencies up to hundreds of megahertz. At frequencies higher than this, the amount of time that it takes for the electrons to move between the cathode and the plate becomes a limiting factor. There are several special tubes designed to work at microwave frequencies. The *klystron* tube uses the principle of velocity modulation of the electrons to avoid transit time limitations. The beam of electrons travels down a metal drift tube that has interaction gaps along its sides. RF voltages are applied to the gaps and the electric fields that they generate accelerate or decelerate the passing electrons. The relative positions of the electrons shift due to their changing velocities causing the electron density of the beam to vary. The modulation of the electron density is used to perform amplification or oscillation. Klystron tubes tend to be relatively large, with lengths ranging from 10 cm to 2 m and weights ranging from as little as 150 g to over 100 kg. Unfortunately, klystrons have relatively narrow bandwidths, and are not retunble by amateurs for operation on different frequencies.

The *magnetron* tube is an efficient oscillator for microwave frequencies. Magnetrons are most commonly found in microwave ovens and high powered radar equipment. The anode of a magnetron is made up of a number of coupled resonant cavities that surround the cathode. The magnetic field causes the electrons to rotate around the cathode and the energy that they give off as they approach the anode adds to the RF electric field. The RF power is obtained from the anode through a vacuum window. Magnetrons are self oscillating with the frequency determined by the construction of their anodes, however they can be tuned by coupling either inductance or capacitance to the resonant anode. The range of frequencies depends on how fast the tuning must be accomplished. The tube may be tuned slowly over a range of approximately 10% of the center frequency. If faster tuning is necessary, such as is required for frequency modulation, the range decreases to about 5%.

A third type of tube capable of operating in the microwave range is the *traveling wave tube*. For wide band amplifiers in the microwave range this is the tube of choice. Either permanent magnets or electromagnets are used to focus the beam of electrons that emerges from an electron gun similar to the one described for the CRT tube. The electron beam passes through a helical *slow-wave structure*, in which electrons are accelerated or decelerated, providing density modulation due to the applied RF signal, similar to that in the klystron. The modulated electron beam induces voltages in the helix that provides an amplified tube output whose gain is proportional to the length of the slow-wave structure. After the RF energy is extracted from the electron beam by the helix, the electrons are collected and recycled to the cathode. Traveling wave tubes can often be operated outside their designed frequencies by carefully optimizing the beam voltage.

PHYSICAL ELECTRONICS OF SEMICONDUCTORS

Every atom of matter consists of, among other things, an equal number of protons and electrons. These two subatomic particles must match in number to neutralize the electric charge: one positive charge for a proton and one negative charge for an electron.

Electrons orbit the nucleus, which contains the protons, at different energy levels. The binding of the electrons to the nucleus determines how an atom will behave electrically. Loosely bound electrons are easily liberated from their nuclei; atoms with this property are called *conductors*. In contrast, tightly bound electrons require considerable energy to be dislodged from their atoms; these atoms are called *insulators*. In between these two extremes is a class of elements called *semiconductors*, or partial conductors. As energy is added to a semiconductor atom, electrons are more easily freed. This property leads to many potential applications for this type of material.

In a conductor, such as a metal, the outer, or *valence*, electrons of each atom are shared with the adjacent atoms so there are many electrons that can move about freely between atoms. The moving free electrons are the constituents of electrical current. In a good conductor, the concentration of these free electrons is very high, on the order of 10^{22} electrons / cm^3. In an insulator, nearly all the electrons are tightly held by their atoms; the concentration of free electrons is very small, on the order of 10 electrons / cm^3.

Semiconductor atoms (germanium — Ge and silicon — Si) share their valence electrons in a chemical bond that holds adjacent atoms together. The electrons are not free to leave their atom in order to move into the sphere of the adjacent atom, as in a conductor. They can be shared by the adjacent atom, however. The sharing of electrons means that the adjacent atoms are attracted to each other, forming a bond that gives the semiconductor its physical structure.

When energy is added to a semiconductor lattice, generally in the form of heat, some electrons are liberated from their bonds and move freely throughout the structure. The bond that loses an electron is then unbalanced and the space that the electron came from is referred to as a *hole*. Electrons from adjacent bonds can leave their positions and fill the holes, thus creating new holes in the adjacent bonds. Two opposite movements can be said to occur: negatively charged electrons move from bond to bond in one direction and positively charged holes move from bond to bond in the opposite direction. Both of these movements represent forms of electrical current, but this is very different from the current in a conductor. While the conductor has *free electrons* that flow re-

gardless of the crystalline structure, the current in a semiconductor is constrained to move only along the crystalline lattice between adjacent bonds.

Crystals formed from pure semiconductor atoms (Ge or Si) are called *intrinsic* semiconductors. In these materials the number of free electrons is equal to the number of holes. Each atom has four valence electrons that form bonds with adjacent atoms. Impurities can be added to the semiconductor material to enhance the formation of electrons or holes. These are *extrinsic* semiconductors. There are two types of impurities that can be added: one kind with five valence electrons *donates* free electrons to the crystalline structure; this is called an *N-type* impurity, for the negative charge that it adds. Some examples are antimony (Sb), phosphorus (P) and arsenic (As). N-type extrinsic semiconductors have more electrons and fewer holes than intrinsic semiconductors. Impurities with three valence electrons accept free electrons from the lattice, adding holes to the overall structure. These are called P-type impurities, for the net positive charge; some examples are boron (B), gallium (Ga) and indium (In).

Intrinsic semiconductor material can be formed by combining equal amounts of N-type and P-type impurity materials. Some examples of this include gallium-arsenide (GaAs), gallium-phosphate (GaP) and indium-phosphide (InP). To make an N-type compound semiconductor, a slightly higher amount of N-type material is used in the mixture. A P-type compound semiconductor has a little more P-type material in the mixture.

The conductivity of an extrinsic semiconductor depends on the charge density (in other words, the concentration of free electrons in N-type, and holes in P-type, semiconductor material). As the energy in the semiconductor increases, the charge density also increases. This is the basis of how all semiconductor devices operate: the major difference is the way in which the energy level is increased. Variations are: The *transistor*, where conductivity is altered by injecting current into the device via a wire; the *thermistor*, where the level of heat in the device is detected by its conductivity, and the *photoconductor*, where light energy that is absorbed by the semiconductor material increases the conductivity.

The PN Semiconductor Junction

If a piece of N-type semiconductor material is placed against a piece of P-type semiconductor material, the loction at which they join is called a *PN semiconductor junction*. The junction has characteristics that make it possible to develop diodes and transistors. The action of the junction is best described by a diode operating as a rectifier. Initially, when the two types of semiconductor material are placed in contact, each type of material will have only its majority carriers: P-type will have only holes and N-type will have only free electrons. The net positive charge of the P-type material attracts free electrons from across the junction and the opposite is true in the N-type material. These attractions lead to diffusion of some of the majority carriers across the junction, which neutralize the carriers immediately on the other side. The region close to the junction is then *depleted* of carriers, and, as such, is named the *depletion region* (or the *space-charge region* or the *transition region*). The width of the depletion region is very small, on the order of 0.5 µm.

If the N-type material is placed at a more negative voltage than the P-type material, current will pass through the junction because electrons are attracted from the lower potential to the higher potential and holes are attracted in the opposite direction. When the polarity is reversed, current does not flow because the electrons that are trying to enter the N-type material are repelled, as are the holes trying to enter the P-type material. This unidirectional current is what allows a semiconductor diode to act as rectifier.

Diodes are commonly made of silicon or germanium. Although they act similarly, they have slightly different characteristics. The *junction threshold voltage*, or *junction barrier voltage*, is the forward bias voltage at which current begins to pass through the device. This voltage is different for the two kinds of diodes. In the diode response curve of **Fig 8.18**, this value corresponds to the voltage at which the positive portion of the curve begins to rise sharply from the x axis. Most silicon diodes have a junction threshold voltage of about 0.7 V, while the value for germanium diodes typically is 0.3 V. The reverse biased leakage current is much lower for silicon diodes than for germanium diodes. The forward resistance of a diode is typically very low and varies with the amount of forward current.

Multiple Junctions

A bipolar transistor is formed when two PN junctions are placed next to each other. If N-type material is surrounded by P-type material, the result is a PNP transistor. Alternatively, if P-type material is in the middle of two layers of N-type material, the NPN transistor is formed (**Fig 8.19**).

Physically, we can think of the transistor as two PN junctions back-to-back, such as two diodes connected at their *anodes*

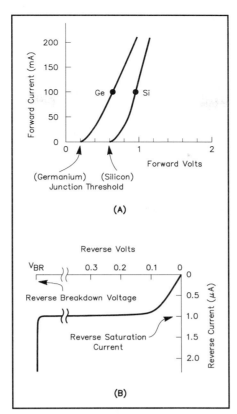

Fig 8.18 — Semiconductor diode (PN junction) response curve. (A) Forward biased (anode voltage higher than cathode) response for Germanium — Ge and Silicon — Si devices. Each curve breaks away from the x-axis at its junction threshold voltage. The slope of each curve is its forward resistance. (B) Reverse biased response. Very small reverse current increases until it reaches the reverse saturation current (I_0). The reverse current increases suddenly and drastically when the reverse voltage reaches the reverse breakdown voltage, V_{BR}.

(the positive terminal) for an NPN transistor or two diodes connected at their *cathodes* (the negative terminal) for a PNP transistor. The connection point is the base of the transistor. (You can't actually *make* a transistor this way.) A transistor conducts when the base-emitter junction is forward biased and the base-collector is reverse biased. Under these conditions, the emitter region emits majority carriers into the base region, where they are minority carriers because the materials of the emitter and base regions have opposite polarity. The excess minority carriers in the base are attracted across the base-collector junction, where they are collected and are once again considered majority carriers. The flow of majority carriers from emitter to collector can be modified by the application of a bias current to the base terminal. If the bias cur-

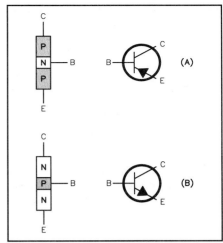

Fig 8.19 — Bipolar transistors. (A) A layer of N-type semiconductor sandwiched between two layers of P-type semiconductor makes a PNP device, the schematic symbol has three leads: collector (C), base (B) and emitter (E), with the arrow pointing in toward the base. (B) A layer of P-type semiconductor sandwiched between two layers of N-type semiconductor makes an NPN device, the schematic symbol has three leads: collector (C), base (B) and emitter (E), with the arrow pointing out away from the base.

rent has the same polarity as the base material (for example holes flowing into a P-type base) the emitter-collector current increases. A transistor allows a small base current to control a much larger collector current.

As in a semiconductor diode, the forward biased base-emitter junction has a threshold voltage (V_{BE}) that must be exceeded before the emitter current increases.

PNPN Diode

If four alternate layers of P-type and N-type material are placed together, a PNPN (usually pronounced like *pinpin*) diode with three junctions is obtained (see **Fig 8.20**). This device, when the anode is at a higher potential than the cathode, has its first and third junctions forward biased and its center junction reverse biased. In this state, there is little current, just as in the reverse biased diode. As the forward bias voltage is increased, the current through the device increases slowly until the *breakover (or firing) voltage*, V_{BO}, is reached and the flow of current abruptly increases. The PNPN diode is often considered to be a switch that is off below V_{BO} and on above it.

Bilateral Diode Switch

A semiconductor device similar to two

PNPN diodes facing in opposite directions and attached in parallel is the *bilateral diode switch* or *diac*. This device has the characteristic curve of the PNPN diode for both positive and negative bias voltages. Its construction, schematic symbol and characteristic curve are shown in **Fig 8.21**.

Silicon Controlled Rectifier

Another device with four alternate layers of P-type and N-type semiconductor is the *silicon controlled rectifier (SCR)*, or *thyristor*. In addition to the connections to the outer two layers, two other terminals can be brought out for the inner two layers. The connection to the P-type material near the cathode is called the *cathode gate* and the N-type material near the anode is called the *anode gate*. In nearly all commercially available SCRs, only the cathode gate is connected (**Fig 8.22**).

Like the PNPN diode switch, the SCR is used to abruptly start conducting when the voltage exceeds a given level. By biasing the gate terminal appropriately, the breakover voltage can be adjusted. The SCR is highly efficient and is used in power control applications. SCRs are available that can handle currents of greater than 100 A and voltage differentials of greater than 1000 V, yet can be switched with gate currents of less than 50 mA.

Triac

A five layered semiconductor whose operation is similar to a bidirectional SCR is the *triac* (**Fig 8.23**). This is also similar to a bidirectional diode switch with a bias control gate. The gate terminal of the triac can control both positive and negative breakover voltages and the devices can pass both polarities of voltage.

SCRs and triacs are often used to modify ac power sources. A sine wave with a given RMS value can be switched on and off at preset points during the cycle to decrease the RMS voltage. When conduction is delayed until after the peak (as **Fig 8.24** shows) the peak-to-peak voltage is re-duced. If conduction starts before the peak, the RMS voltage is reduced, but the peak-to-peak value remains the same. This method is used to operate light dimmers and 240 V ac to 120 V ac converters. The

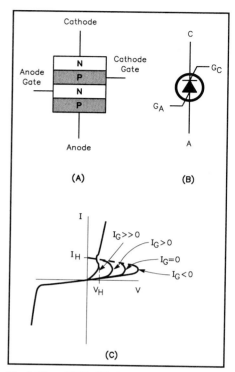

Fig 8.22 — SCR. (A) Alternating layers of P-type and N-type semiconductor. This is similar to a PNPN diode with gate terminals attached to the interior layers. (B) Schematic symbol with anode (A), cathode (C), anode gate (G_A) and cathode gate (G_C). Many devices are constructed without G_A. (C) Voltage-current response curve with different responses for various gate currents. $I_G = 0$ has the same response as the PNPN diode.

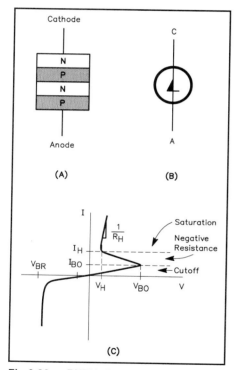

Fig 8.20 — PNPN diode. (A) Alternating layers of P-type and N-type semi-conductor. (B) Schematic symbol with cathode (C) and anode (A) leads. (C) Voltage-current response curve. Reverse biased response is the same as normal PN junction diodes. Forward biased response acts as a hysteresis switch. Resistance is very high until the bias voltage reaches V_{BO} and exceeds the cutoff current, I_{B0}. The device exhibits a negative resistance with the current increases as the bias voltage decreases until a voltage of V_H and saturation current of I_H is reached. After this the resistance is very low, with large increases in current for small voltage increases.

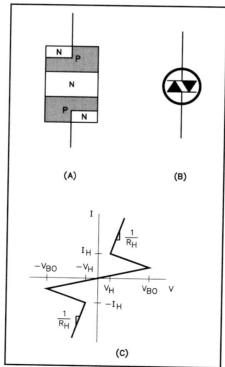

Fig 8.21 — Bilateral switch. (A) Alter-nating layers of P-type and N-type semiconductor. (B) Schematic symbol. (C) Voltage-current response curve. The right-hand side of the curve is identical to the PNPN diode response in Fig 8.20. The device responds identically for both forward and reverse bias so the left-hand side of the curve is symmetrical to the right-hand side.

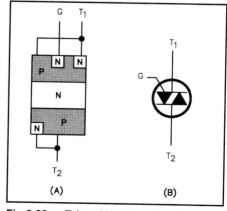

Fig 8.23 — Triac. (A) Alternating layers of P-type and N-type semiconductor. This behaves as two SCR devices facing in opposite directions with the anode of one connected to the cathode of the other and the cathode gates connected together. (B) Schematic symbol.

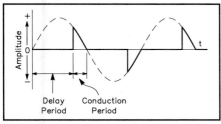

Fig 8.24 — Triac operation on sine wave. The dashed line is the original sine wave and the solid line is the portion that conducts through the triac. The relative delay and conduction period times are controlled by the amount or timing of gate current, I_G. The response of an SCR is the same as this for positive voltages (above the x-axis) and with no conduction for negative voltages.

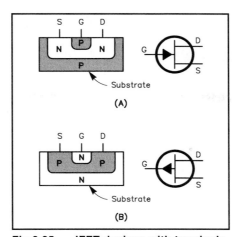

Fig 8.25 — JFET devices with terminals labeled: source (S), gate (G) and drain (D). A) Pictorial of N-type channel embedded in P-type substrate and schematic symbol. B) P-channel embedded in N-type substrate and schematic symbol.

sharp switching transients created when these devices switch are common sources of RF interference. SCRs are used as "crowbars" in power supply circuits, to short the output to ground and blow a fuse when an overvoltage condition exists.

FIELD-EFFECT TRANSISTORS

The *field-effect transistor (FET)* controls the current between two points but does so differently than the bipolar transistor. The FET operates by the effects of an electric field on the flow of electrons through a single type of semiconductor material. This is why the FET is sometimes called a *unipolar* transistor. Also, unlike bipolar semiconductors that can be arranged in many configurations to provide diodes, transistors, photoelectric devices, temperature sensitive devices and so on, the field effect is usually only used to make transistors, although FETs are also available as special-purpose diodes, for use as constant current sources.

Current moves within the FET in a channel, from the source connection to the drain connection. A gate terminal generates an electric field that controls the current (see **Fig 8.25**). The channel is made of either N-type or P-type semiconductor material; an FET is specified as either an N-channel or P-channel device. Majority carriers flow from source to drain. In N-channel devices, electrons flow so the drain potential must be higher than that of the source ($V_{DS} > 0$). In P-channel devices, the flow of holes requires that $V_{DS} < 0$. The polarity of the electric field that controls current in the channel is determined by the majority carriers of the channel, ordinarily positive for P-channel FETs and negative for N-channel FETs.

Variations of FET technology are based on different ways of generating the elec-

tric field. In all of these, however, electrons at the gate are used only for their charge in order to create an electric field around the channel, and there is a minimal flow of electrons through the gate. This leads to a very high dc input resistance in devices that use FETs for their input circuitry. There may be quite a bit of capacitance between the gate and the other FET terminals, however. The input impedance may be quite low at RF.

The current through an FET only has to pass through a single type of semiconductor material. There is very little resistance in the absence of an electric field (no bias voltage). The drain-source resistance ($r_{DS\ ON}$) is between a few hundred ohms to less than an ohm. The output impedance of devices made with FETs is generally quite low. If a gate bias voltage is added to operate the transistor near cut off, the circuit output impedance may be much higher.

FET devices are constructed on a *substrate* of doped semiconductor material. The channel is formed within the substrate and has the opposite polarity (a P-channel FET has N-type substrate). Most FETs are constructed with silicon. In order to achieve a higher gain-bandwidth product, other materials have been used. Gallium Arsenide (GaAs) has electron mobility and drift velocities that are far higher than the standard doped silicon. Amplifiers designed with *GaAs FET* devices have much higher frequency response and lower noise factor at VHF and UHF than those made with standard FETs.

JFET

There are two basic types of FET. In the

junction FET (JFET), the gate material is made of the opposite polarity semiconductor to the channel material (for a P-channel FET the gate is made of N-type semiconductor material). The gate-channel junction is similar to a diode's PN junction. As with the diode, current is high if the junction is forward biased and is extremely small when the junction is reverse biased. The latter case is the way that JFETs are used, since any current in the gate is undesirable. The magnitude of the reverse bias at the junction is proportional to the size of the electric field that "pinches" the channel. Thus, the current in the channel is reduced for higher reverse gate bias voltage.

Because the gate-channel junction in a JFET is similar to a bipolar junction diode, this junction must never be forward biased; otherwise large currents will pass through the gate and into the channel. For an N-channel JFET, the gate must always be at a lower potential than the source ($V_{GS} < 0$). The channel is as fully open as it can get when the gate and source voltages are equal ($V_{GS} = 0$). The prohibited condition is when $V_{GS} > 0$. For P-channel JFETs these conditions are reversed (in normal operation V_{GS} 0 and the prohibited condition is when $V_{GS} < 0$).

MOSFET

Placing an insulating layer between the gate and the channel allows for a wider range of control (gate) voltages and further decreases the gate current (and thus increases the device input resistance). The insulator is typically made of an oxide (such as silicon dioxide, SiO_2). This type of device is called a *metal-oxide-semiconductor FET (MOSFET)* or *insulated-gate FET (IGFET)*. The substrate is often connected to the source internally. The insulated gate is on the opposite side of the channel from the substrate (see **Fig 8.26**). The bias voltage on the gate terminal either attracts or repels the majority carriers of the substrate across the PN junction with the channel. This narrows (depletes) or widens (enhances) the channel, respectively, as V_{GS} changes polarity. For N-channel MOSFETs, positive gate voltages with respect to the substrate and the source ($V_{GS} > 0$) repel holes from the channel into the substrate, thereby widening the channel and decreasing channel resistance. Conversely, $V_{GS} < 0$ causes holes to be attracted from the substrate, narrowing the channel and increasing the channel resistance. Once again, the polarities discussed in this example are reversed for P-channel devices. The common abbreviation for an N-channel MOSFET is *NMOS*, and for a P-channel MOSFET, *PMOS*.

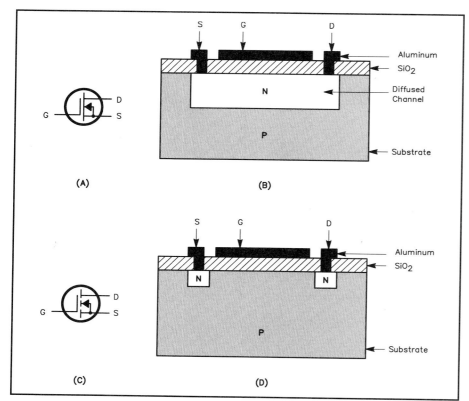

Fig 8.26 — MOSFET devices with terminals labeled: source (S), gate (G) and drain (D). N-channel devices are pictured. P-channel devices have the arrows reversed in the schematic symbols and the opposite type semiconductor material for each of the layers. (A) N-channel depletion mode device schematic symbol and pictorial of P-type substrate, diffused N-type channel, SiO₂ insulating layer and aluminum gate region and source and drain connections. The substrate is connected to the source internally. A negative gate potential narrows the channel. (B) N-channel enhancement mode device schematic and pictorial of P-type substrate, N-type source and drain wells, SiO₂ insulating layer and aluminum gate region and source and drain connections. Positive gate potential forms a channel between the two N-type wells.

Because of the insulating layer next to the gate, input resistance of a MOSFET is usually greater than 10^{12} Ω (a million megohms). Since MOSFETs can both deplete the channel, like the JFET, and also enhance it, the construction of MOSFET devices differs based on the channel size in the resting state, $V_{GS} = 0$. A *depletion mode* device (also called a *normally on MOSFET*) has a channel in resting state that gets smaller as a reverse bias is applied; this device conducts current with no bias applied (see Fig 8.26 A and B). An *enhancement mode* device (also called a *normally off MOSFET*) is built without a channel and does not conduct current when $V_{GS} = 0$; increasing forward bias forms a channel that conducts current (see Fig 8.26 C and D).

Semiconductor Temperature Effects

The number of excess holes and electrons is increased as the temperature of a semiconductor increases. Since the conductivity of a semiconductor is related to the number of excess carriers, this also increases with temperature. With respect to resistance, semiconductors have a negative temperature coefficient. The resistance of silicon *decreases* by about 8% / °C and by about 6% / °C for germanium. Semiconductor temperature properties are the opposite of most metals, which *increase* their resistance by about 0.4% / °C. These opposing temperature characteristics permit the design of circuits with opposite temperature coefficients that cancel each other out, making a temperature insensitive circuit. Left by itself, the semiconductor can experience an effect called *thermal runaway* as the current causes an increase in temperature. The increased temperature decreases resistance and may lead to a further increase in current (depending on the circuit) that leads to an additional temperature increase. This sequence of events can continue until the semiconductor destroys itself.

Semiconductor Failure

There are several common failure modes for semiconductors that are related to heat. The semiconductor material is connected to the outside world through metallic leads. The point at which the metal and the semiconductor are connected is one common place for the semiconductor device to fail. As the device heats up and cools down, the materials expand and contract. The rate of expansion and contraction of semiconductor material is different from that of metal. Over many cycles of heating and cooling the bond between the semiconductor and the metal can break. Some experts have suggested that the lifetime of semiconductor equipment can be extended by leaving the devices on all the time. While this would decrease the type of failure just described, inadequate cooling can lead to another type of semiconductor failure.

Impurities are introduced into intrinsic semiconductors by diffusion, the same physical property that lets you smell cookies baking from several rooms away. Smells diffuse through air much faster than molecules diffuse through solids. Once the impurities diffuse into the semiconductor, they tend to stay in place. Rates of diffusion are proportional to temperature, and semiconductors are doped with impurities at high temperature to save time. Once the doped semiconductor material is cooled, the rate of diffusion of the impurities is so low that they are essentially immobile for many years to come.

A common failure mode of semiconductors is due to the heat generated during semiconductor use. If the temperatures at the junctions rise to high enough levels for long enough periods of time, the impurities start to diffuse across the PN junctions. When enough of these atoms get across the junction, it stops functioning properly and the semiconductor device fails.

Thermistors

The effect of temperature on current in semiconductors is put to use in a controlled fashion in a *thermistor*. The temperature coefficients of silicon and germanium are highly dependent on the amount of doping. For stability, thermistors are made of oxides such as nickel oxide (NiO), dimanganese trioxide (Mn_2O_3) or dicobalt trioxide (Co_2O_3). If the doping concentration of a semiconductor is high enough, it will start to take on some of the properties of a metal and the temperature coefficient becomes positive. A device made from this type of material is sometimes called a *sensistor*.

Practical Semiconductors

SEMICONDUCTOR DIODES

Although many types of semiconductor diodes are available, there are not many differences between them. The diode is made of a single PN junction that affects current differently depending on its direction. This leads to a large number of applications in electronic circuitry.

The diode symbol is shown in **Fig 8.27**. Current passes most easily from anode to cathode, in the direction of the arrow. This is often referred to as the *forward* direction and the opposite is the *reverse* direction. Remember that *current* refers to the flow of electricity from higher to lower potentials and is in the opposite direction to the flow of electrons (current moves from anode to cathode and electrons flow from cathode to anode, as based on the definitions of the words, *anode* and *cathode*). The anode of a semiconductor junction diode is made of P-type material and the cathode is made of N-type material, as indicated in Fig 8.27. Most diodes are marked with a band on the cathode end (Fig 8.27). The ideal diode would have zero resistance in the forward direction and infinite resistance in the reverse direction. This is not the case for actual devices, which behave as shown in the plot of a diode response in Fig 8.18. Note that the scales of the two parts of the graph are drastically different. The inverse of the slope of the line (the change in voltage between two points on a straight portion of the line divided by the corresponding change in current) on the upper right is the resistance of the diode in the forward direction. The range of voltages is small and the range of currents is large since the forward resistance is very small (in this example, about 2 Ω). The lower left portion of the curve illustrates a much higher resistance that increases from tens of kilohms to thousands of megohms as the reverse voltage gets larger, and then decreases to near zero (a nearly vertical line) very suddenly at the peak inverse voltage (PIV = 100 V in this example).

There are five major characteristics that distinguish standard junction diodes from one another: the PIV, the current or power handling capacity, the response speed, reverse leakage current and the junction barrier voltage. Each of these characteristics can be manipulated during manufacture to produce special purpose diodes.

The most common application of a diode is to perform rectification; that is, allowing positive voltages to pass and stopping negative voltages. Rectification is used in power supplies that convert ac to dc and in amplitude demodulation. The most important diode parameters to consider for power rectification are the PIV and current ratings. The peak negative voltages that are stopped by the diode must be smaller in magnitude than the PIV and the peak current through the diode when it is forward biased must be less than the maximum amount for which the device was designed. Exceeding the current rating in a diode will cause excessive heating (based on $P = I \times V_F$) that leads to PN junction failure as described earlier.

Fast Diodes

The speed of a diode affects the frequencies that it can act on. The diode response in Fig 8.18 is a steady state response, showing how that diode will act at dc. As the frequency increases, the diode may not be able to keep up with the changing polarity of the signal and its response will not be as expected. Diode speed mainly depends on charge storage in the depletion region. Under reverse bias, excess charges move away from the junction, forming a larger space-charge region that is the equivalent of a dielectric. The diode thus exhibits capacitance, which is inversely proportional to the width of the dielectric and directly proportional to the cross-sectional surface area of the junction.

One way to decrease charge storage time in the depletion region is to form a metal-semiconductor junction. This can be accomplished with a point-contact diode, where a thin piece of aluminum wire, often called a *whisker*, is placed in contact with one face of a piece of lightly doped N-type material. In fact, the original diodes used for detecting radio signals ("cat's whisker diodes") were made this way. A more recent improvement to this technology, the *hot-carrier diode*, is like a point-contact diode with more ideal characteristics attained by using more efficient metals, such as platinum and gold, that act to lower forward resistance and increase PIV. This type of contact is known as a *Schottky barrier*, and diodes made this way are called *Schottky diodes*.

The PIN diode, shown in Fig 8.27C is a *slow response* diode that is capable of passing microwave signals when it is forward biased. This device is constructed with a layer of intrinsic (undoped) semiconductor placed between very highly doped P-type and N-type material (called P+-type and N+-type material to indicate the high level of doping), creating a PIN junction. These devices provide very effective switches for RF signals and are often used in TR switches in transceivers. PIN diodes have longer than normal carrier lifetimes,

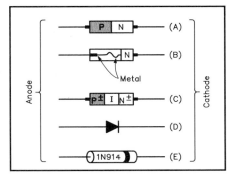

Fig 8.27 — Practical semiconductor diodes. All devices are aligned with anode on the left and cathode on the right. (A) Standard PN junction diode. (B) Point-contact or "cat's whisker" diode. (C) PIN diode formed with heavily doped P-type (P+), undoped (intrinsic) and heavily doped N-type (N+) semiconductor material. (D) Diode schematic symbol. (E) Diode package with marking stripe on the cathode end.

resulting in a slow switching process that causes them to act more like resistors than diodes at high radio frequencies.

Varactors

If the PN junction capacitance is controlled rather than reduced, a diode can be made to act as a variable capacitor. As the reverse bias voltage on a diode increases, the width of the junction increases, which decreases its capacitance. A *varactor* is a diode whose junction is specially formulated to have a relatively large range of capacitance values for a modest range of reverse bias voltages (**Fig 8.28**). Although special forms of varactors are available from manufacturers, other types of diodes may be used as inexpensive varactor diodes, but the relationship between reverse voltage and capacitance is not always reliable. When designing with varactor diodes, the reverse bias voltage must be absolutely free of noise since any variations in the bias voltage will cause changes in capacitance. Unwanted frequency shifts or instability will result if the reverse bias voltage is noisy. It is possible to frequency modulate a signal by adding the audio signal to the reverse bias on a varactor diode used in the carrier oscillator.

Zener Diodes

When the PIV of a reverse biased diode is exceeded, the diode begins to conduct current as it does when it is forward biased. This current does not destroy the diode if it is limited to less than the device's maximum allowable value. When the PIV is controlled during manufacture to be at desired levels, the device is called a *Zener diode*. Zener diodes (named after the American physicist Clarence Zener) provide accurate voltage references and are often used for this purpose in power supply regulators.

When the reverse breakdown voltage is exceeded, the reverse voltage drop across the Zener diode remains constant. With an appropriate current limiting resistor in series with it, the Zener diode provides an accurate voltage reference (**Fig 8.29**).

Zener diodes are rated by their reverse breakdown voltage and their power handling capacity. The power is a product of the current passing through the reverse biased Zener diode "in breakdown" (that is, in the breakdown mode of operation)

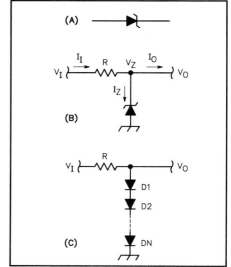

Fig 8.29 — Zener diode. (A) Schematic symbol. (B) Basic voltage regulating circuit. V_Z is the Zener reverse breakdown voltage. The Zener diode draws more current until $V_I - I_IR = V_Z$. The circuit design should select R so that when the maximum current is drawn, $R < (V_I - V_Z) / I_O$. The diode should be capable of passing the same current when there is no output current drawn. (C) For small voltages, several forward biased diodes can be used in place of Zener diodes. Each diode will drop the voltage by about 0.7 V for silicon or 0.3 V for germanium.

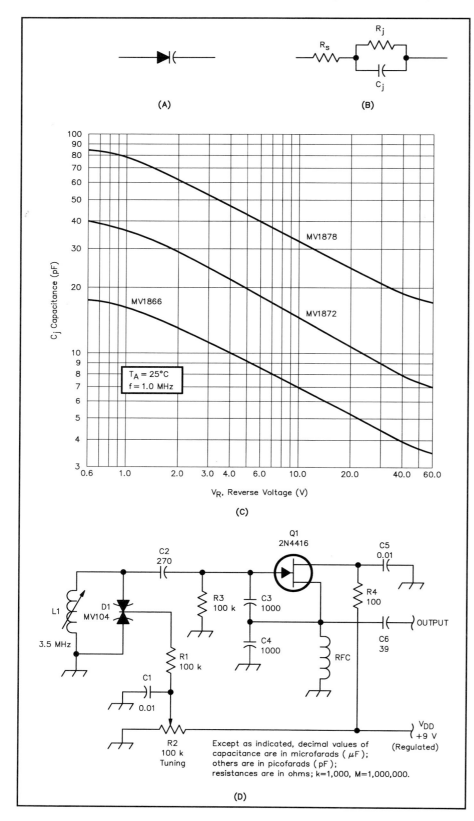

Fig 8.28 — Varactor diode. (A) Schematic symbol. (B) Equivalent circuit of the reverse biased varactor diode. R_S is the junction resistance, R_J is the leakage resistance and C_J is the junction capacitance, which is a function of the magnitude of the reverse bias voltage. (C) Plot of junction capacitance, C_J, as a function of reverse voltage, V_R, for three different varactor devices. Both axes are plotted on a logarithmic scale. (D) Oscillator circuit with varactor tuning. D1-L1 is a tuned circuit with a dual varactor diode that is controlled by the voltage from potentiometer R2. C1 is a filter capacitor to insure that the varactor bias voltage is clean dc. C2 and C6 are dc blocking capacitors. Q1 is an N-channel JFET in common drain configuration with feedback to the gate through C3. R3 is the gate bias resistor. R4 is the drain voltage resistor with filter capacitor C5.

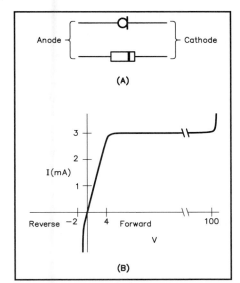

Fig 8.30 — Current regulator diode.
(A) Schematic symbol and package with
line marking cathode end. (B) Diode
characteristic curve (1N5283 device).
When forward bias voltage exceeds
about 4 V the current passing through
the device is held constant regardless of
the voltage across the device.

Fig 8.31 — Diode circuits. (A) Half wave
rectifier circuit. Only when the ac voltage is
positive does current pass through the
diode. Current flows only during half of the
cycle. (B) Full-wave center-tapped rectifier
circuit. Center-tap on the transformer
secondary is grounded and the two ends of
the secondary are 180° out of phase. During
the first half of the cycle the upper diode
conducts and during the second half of the
cycle the lower diode conducts. There is
conduction during the full cycle with only
positive voltages appearing at the output.
(C) Full-wave bridge rectifier circuit. In each
half of the cycle two diodes conduct
capacity. (D) Polarity protection for external
power connection. J1 is the connector that
power is applied to. If polarity is correct,
the diode will conduct and if reversed the
diode will block current, protecting the
circuit that is being powered. (E) Overvoltage protection circuit. If excessive
voltage is applied to J1, D1 will conduct
current until fuse, F1, is blown. (F) Bipolar
voltage clipping circuit. In the positive
portion of the cycle, D2 is forward biased
but no current is shunted to ground
because D1 is reverse biased. D1 starts to
conduct when the voltage exceeds the
Zener breakdown voltage and the positive
peak is clipped. When the negative portion
of the cycle is reached, D1 is forward
biased but no current is shunted to ground
because D2 is reverse biased. When the
voltage exceeds the Zener breakdown
voltage of D2, it also begins to conduct and
the negative peak is clipped. (G) Diode
switch. The signal is ac coupled to the
diode by C1 at the input and C2 at the
output. R2 provides a reference for the bias
voltage. When switch S1 is in the ON
position, a positive dc voltage is added to
the signal so it is forward biased and is
passed through the diode. When S1 is in
the OFF position, the negative dc voltage
added to the signal reverse biases the
diode and the signal does not get through.

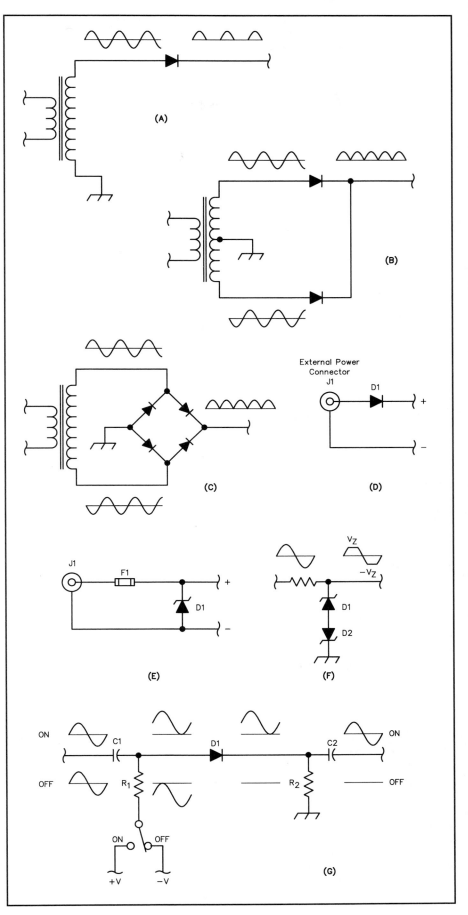

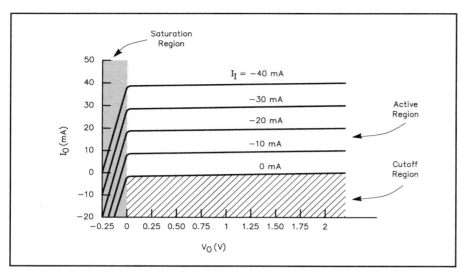

Fig 8.32 — Transistor response curve. The x-axis is the output voltage and the y-axis is the output current. Different curves are plotted for various values of input current. The three regions of the transistor are its cutoff region, where no current flows in any terminal, its active region, where the output current is nearly independent of the output voltage and there is a linear relationship between the input current and the output current, and the saturation region, where the output current has large changes for small changes in output voltage.

and the breakdown voltage. Since the same current must always pass though the resistor to drop the source voltage down to the reference voltage, with that current divided between the Zener diode and the load, this type of power source is very wasteful of current. The Zener diode does make an excellent and efficient voltage reference in a larger voltage regulating circuit where the load current is provided from another device whose voltage is set by the reference. (See the **Power Supplies and Projects** chapter for more information about using Zener diodes as voltage regulators.) The major sources of error in Zener-diode derived voltages are the variation with load current and the variation due to heat. Temperature compensated Zener diodes are available with temperature coefficients as low as 0.0005 % / °C. If this is unacceptable, voltage reference integrated circuits based on Zener diodes have been developed that include additional circuitry to counteract temperature effects.

Constant Current Diodes

A form of diode, called a *field-effect regulator diode*, provides a constant current over a wide range of forward biased voltages. The schematic symbol and characteristic curve for this type of device are shown in **Fig 8.30**. Constant current diodes are very useful in any application where a constant current is desired. Some part numbers are 1N5283 through 1N5314.

Common Diode Applications

Standard semiconductor diodes have many uses in analog circuitry. Several examples of diode circuits are shown in **Fig 8.31**. Rectification has already been described. There are three basic forms of rectification using semiconductor diodes: half wave (1 diode), full-wave center-tapped (2 diodes) and full-wave bridge (4 diodes). These are more fully described in the **Power Supplies and Projects** chapter.

Diodes are commonly used to protect circuits. In battery powered devices a forward biased series diode is often used to protect the circuitry from the user inadvertently inserting the batteries backwards. Likewise, when a circuit is powered from an external dc source, a diode is often placed in series with the power connector in the device to prevent incorrectly wired power supplies from destroying the equipment. Diodes are commonly used to protect analog meters from both reverse voltage and over voltage conditions that would destroy the delicate needle movement.

Zener diodes are sometimes used to protect low-current (a few amps) circuits from over-voltage conditions. A reverse biased Zener diode connected between the positive power lead and ground will conduct excessive current if its breakdown voltage is exceeded. Used in conjunction with a fuse in series with the power lead, the Zener diode will cause the fuse to blow when an over-voltage condition exists.

Very high, short-duration voltage spikes can destroy certain semiconductors, particularly MOS devices. Standard Zener diodes can't handle the high pulse powers found in these voltage spikes. Special Zener diodes are designed for this purpose, such as the *mosorb*. (General Semiconductor Industries, Inc calls these devices *TransZorbs*.) A reverse biased TransZorb with a low-value series resistor can decrease the voltage reaching the sensitive device. Since the polarity of the spike can be positive, negative, or both, over voltage transient suppressor circuits can be designed with two devices wired back-to-back. They protect a circuit over a range of voltages rather than just suppressing positive peaks.

Diodes can be used to clip signals, similar to rectification. If the signal is appropriately biased it can be clipped at any level. Two Zener diodes placed back-to-back can be used to clip both the positive and negative peaks of a signal. Such an arrangement is used to convert a sine wave to an approximate square wave.

Care must be taken when using Zener diodes to process signals. The Zener diode is a relatively noisy device and can add excessive noise to the signals if it operates in breakdown. The Zener diode is often specified for use in circuits that intentionally generate noise, such as the noise bridge (see the **Test Procedures and Projects** chapter.) The reverse biased Zener diode in breakdown generates wide band (nearly white) noise levels as high as $2000 \, \mu V / \sqrt{Hz}$. (The noise voltage is determined by multiplying this value by the square root of the circuit bandwidth in Hz.)

Diodes are used as switches for ac coupled signals when a dc bias voltage can be added to the signal to permit or inhibit the signal from passing through the diode. In this case the bias voltage must be added to the ac signal and be of sufficient magnitude so that the entire envelope of the ac signal is above or below the junction barrier voltage, with respect to the cathode, to pass through the diode or inhibit the signal. Special forms of diodes, such as the PIN diode described earlier, which are capable of passing higher frequencies, are used to switch RF signals.

BIPOLAR TRANSISTORS

The bipolar transistor is a *current-controlled device*. The current between the emitter and the collector is governed by the current that enters the base. The convention when discussing transistor operation is that the three currents into the device are positive (I_c into the collector, I_b into the base and I_e into the emitter).

Kirchhoff's current law applies to transistors just as it does to passive electrical networks: the total current entering the device must be zero. Thus, the relationship between the currents into a transistor can be generalized as

$$0 = I_c + I_b + I_e \qquad (9)$$

which can be rearranged as necessary. For example, if we are interested in the emitter current,

$$I_e = -(I_c + I_b) \qquad (10)$$

The back-to-back diode model is appropriate for visualization of transistor construction. In actual transistors, however, the relative sizes of the collector,

base and emitter regions differ. A common transistor configuration that spans a distance of 3 mm between the collector and emitter contacts typically has a base region that is only 25 μm across.

Current conduction between collector and emitter is described by regions in the common-base response curves of the transistor device (see **Fig 8.32**). The transistor is in its *active region* when the base-collector junction is reverse biased and the base-emitter junction is forward biased. The slope of the output current (I_O) versus the output voltage (V_O) is virtually flat, indicating that the output current is nearly independent of the output voltage. The slight slope that does exist is due to base-

width modulation (known as the "Early effect"). Under these conditions, there is a linear relationship between the input current (I_I) and I_O. When both the junctions in the transistor are forward biased, the transistor is said to be in its *saturation region*. In this region, V_O is nearly zero and large changes in I_O occur for very small changes in V_O. The *cutoff region* occurs when both junctions in the transistor are reverse biased. Under this condition, there is very little current in the output, only the nanoamperes or microamperes that result from the very small leakage across the input-to-output junction. These descriptions of junction conditions are the basis for the use of

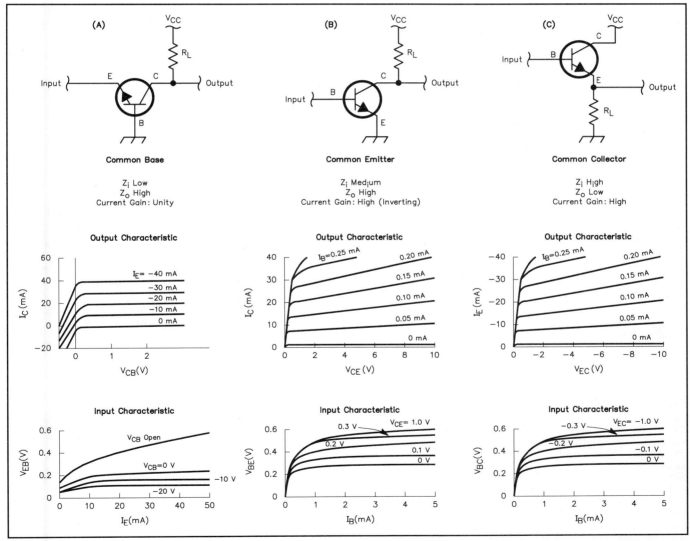

Fig 8.33 — The three configurations of transistor amplifiers. Each has a table of its relative impedance and current gain. The output characteristic curve is plotted for each, with the output voltage along the x-axis, the output current along the y-axis and various curves plotted for different values of input current. The input characteristic curve is plotted for each configuration with input current along the x-axis, input voltage along the y-axis and various curves plotted for different values of output voltage. (A) Common base configuration with input terminal at the emitter and output terminal at the collector. (B) Common emitter configuration with input terminal at the base and output terminal at the collector. (C) Common collector with input terminal at the base and output terminal at the emitter.

transistors. Various configurations of the transistor in circuitry make use of the properties of the junctions to serve different purposes in analog signal processing.

In the common base configuration, where the input is at the emitter and the output is at the collector, the current gain is defined as

$$\alpha = -\frac{\Delta I_C}{\Delta I_E} \approx 1 \tag{11}$$

In the common emitter configuration, with the input at the base and the output at the collector, the current gain is

$$\beta = \frac{\Delta I_C}{\Delta I_B} \tag{12}$$

and the relationship between α and β is defined as

$$\alpha = \frac{\beta}{1 + \beta} \tag{13}$$

Since the common-emitter configuration is the most used transistor-amplifier configuration, another designation for β is often used: h_{FE}, the forward dc current gain. (The "h" refers to "h parameters," a set of parameters for describing a two-port network.) The symbol, h_{fe}, is used for the forward current gain of ac signals. Other transistor transfer function relationships that are measured are h_{ie}, the input impedance, h_{oe}, the output admittance (reciprocal of impedance) and h_{re}, the voltage feedback ratio.

The behavior of a transistor can be defined in many ways, depending on which type of amplifier it is wired to be. A complete description of a transistor must include characteristic curves for each configuration. Typically, two sets of characteristic curves are presented: one describing the input behavior and the other describing the output behavior in each amplifier configuration. Different transistor

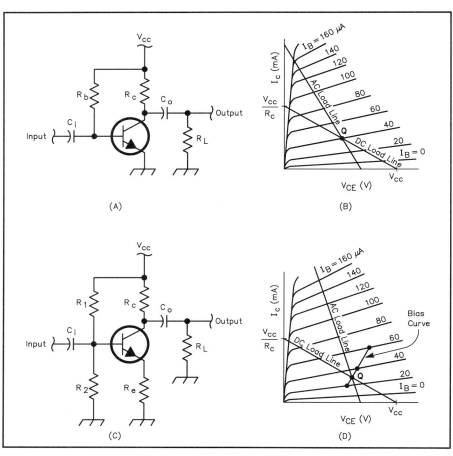

Fig 8.34 — Transistor biasing circuits. (A) Fixed bias. Input signal is ac coupled through C_i. The output has a voltage that is equal to $V_{CC} - I_C \times R_C$. This signal is ac coupled to the load, R_L, through C_o. For dc signals, the entire output voltage is based on the value of R_C. For ac signals, the output voltage is based on the value of R_C in parallel with R_L. (B) Characteristic curve for the transistor amplifier pictured in (A). The slope of the dc load line is equal to $-1/R_C$. For ac signals, the slope of the ac load line is equal to $-1/(R_C \| R_L)$. The quiescent operating point, Q, is based on the base bias current with no input signal applied and where this characteristic line crosses the dc load line. The ac load line must also pass through point Q. (C) Self-bias. Similar to fixed bias circuit with the base bias resistor split into two: R1 connected to V_{CC} and R2 connected to ground. Also an emitter bias resistor, R_E, is included to compensate for changing device characteristics. (D) This is similar to the characteristic curve plotted in (B) but with an additional "bias curve" that shows how the base bias current varies as the device characteristics change with temperature. The operating point, Q, moves along this line and the load lines continue to intersect it as it changes.

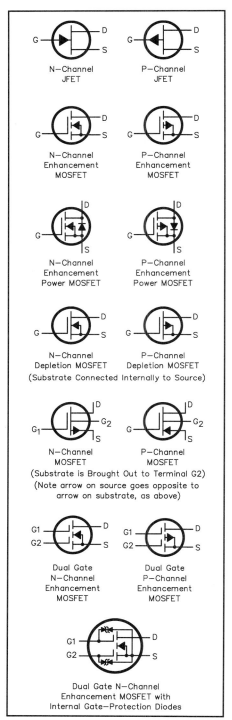

Fig 8.35 — FET schematic symbols.

amplifier configurations have different gains, input and output impedances. At low frequencies, where parasitic capacitances aren't a factor, the common emitter configuration has a high current gain (about − 50, with the negative sign indicating a 180° phase shift), medium to high input impedance (about 50 kΩ) and a medium to low output impedance (about 1 kΩ). The common collector has a high current gain (about 50), a high input impedance (about 150 kΩ) and a low output impedance (about 80 Ω). The common base amplifier has a low current gain (about 1), a low input impedance (about 25 Ω) and a very high output impedance (about 2 MΩ). Depending on the intended use of the transistor amplifier in an analog circuit, one configuration will be more appropriate than others. Once the common lead of the transistor amplifier configuration is chosen, the input and output impedance are functions of the device bias levels and circuit loading (**Fig 8.33**). The actual input and output impedances of a transistor amplifier are highly dependent on the input, biasing and load resistors that are used in the circuit.

A typical general-purpose bipolar-transistor data sheet lists important device specifications. Parameters listed in the ABSOLUTE MAXIMUM RATINGS section are the three junction voltages (V_{CEO}, V_{CBO} and V_{EBO}), the continuous collector current (I_C), the total device power dissipation (P_D) and the operating and storage temperature range. In the OPERATING PARAMETERS section, the three guaranteed minimum junction breakdown voltages are listed — $V_{(BR)CEO}$, $V_{(BR)CBO}$ and $V_{(BR)EBO}$ — along with the two guaranteed maximum collector cutoff currents — I_{CEO} and I_{CBO} — under OFF CHARACTERISTICS. Under ON CHARACTERISTICS are the guaranteed minimum dc current gain (h_{FE}), guaranteed maximum collector-emitter saturation voltage — $V_{CE(sat)}$ — and the guaranteed maximum base-emitter on voltage — $V_{BE(on)}$. The next section is SMALL-SIGNAL CHARACTERISTICS, where the guaranteed minimum current gain-bandwidth product — f_T, the guaranteed maximum output capacitance — C_{obo}, the guaranteed maximum input capacitance — C_{ibo}, the guaranteed range of input impedance — h_{ie}, the small-signal current gain — h_{fe}, the guaranteed maximum voltage feedback ratio — h_{re} and output admittance — h_{oe} are listed. Finally, the SWITCHING CHARACTERISTICS section lists absolute maximum ratings for delay time — t_d, rise time — t_r, storage time — t_s and fall time — t_f.

Transistor Biasing

Biasing in a transistor adds or subtracts

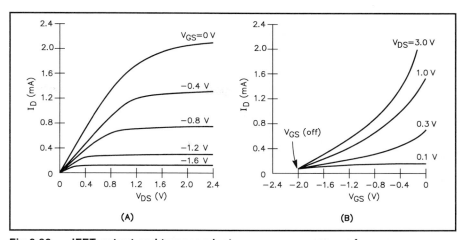

Fig 8.36 — JFET output and transconductance response curves for common source amplifier configuration. (A) Output voltage (V_{DS}) on the x-axis versus output current (I_D) on the y-axis, with different curves plotted for various values of input voltage (V_{GS}). (B) Transconductance curve has the same three variables rearranged, V_{GS} on the x-axis, I_D on the y-axis and curves plotted for different values of V_{DS}.

a fixed amount of current from the signal at the input port. This differs from vacuum tube, FET and operational amplifier biasing where a bias *voltage* is added to the input signal. Fixed bias is the simplest form, as shown in **Fig 8.34A**. The operating point is determined by the intersection between the characteristic curves, the load line and the quiescent current bias line (Fig 8.34B). The problem with fixing the bias current is that if the transistor parameters drift due to heat, the operating point will change. The operating point can be stabilized by self biasing, also called emitter biasing, as pictured in Fig 8.34C. If I_C increases due to temperature changes, the current in R_E increases. The larger current through R_E increases the voltage drop across that resistor, causing a decrease in the base current, I_B. This, in turn, leads to a decreasing I_C, minimizing its variation due to heat. The operating point for this type of biasing is plotted in Fig 8.34D.

FIELD-EFFECT TRANSISTORS

FET devices are more closely related to vacuum tubes than are bipolar transistors. Both the vacuum tube and the FET are controlled by the voltage level of the input rather than the input current, as in the bipolar transistor. FETs have three basic terminals, the gate, the source and the drain. These are related to both vacuum tube and bipolar transistor terminals: the gate to the grid and the base, the source to the cathode and the emitter, and the drain to the plate and the collector. Different forms of FET devices are pictured in **Fig 8.35**.

The characteristic curves for FETs are similar to those of vacuum tubes. The two most useful relationships are called the

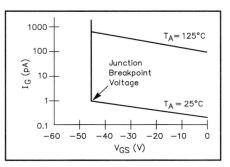

Fig 8.37 — JFET input leakage curves for common source amplifier configuration. Input voltage (V_{GS}) on the x-axis versus input current (I_G) on the y-axis, with two curves plotted for different operating temperatures, 25°C and 125°C. Input current increases greatly when the gate voltage exceeds the junction breakpoint voltage.

transconductance and output curves (**Fig 8.36**). Transconductance curves give the drain current, I_D, due to different gate-source voltage differences, V_{GS}, for various drain-source voltages, V_{DS}. The same parameters are interrelated in a different way in the output curve. For different values of V_{GS}, I_D is plotted against V_{DS}. In both of these representations, the device output is the drain current and these curves describe the FET in the common-source configuration. The action of the FET channel is so nearly ideal that, as long as the JFET gate does not become forward biased, the drain and source currents are virtually identical. For JFETs the gate leakage current, I_G, is a function of V_{GS} and this is often expressed with an input curve (**Fig 8.37**). The point at which there is a great increase in I_G is called the

Transistor Amplifier Design—a Practical Approach

The design of a transistorized amplifier is a straight-forward process. Just as you don't need a degree in mechanical engineering to drive an automobile, neither do you need detailed knowledge of semiconductor physics in order to design a transistor amplifier with predictable and repeatable properties.

This sidebar will describe how to design a small-signal "Class A" transistor amplifier, following procedures detailed in one of the best books on the subject— *Solid State Design for the Radio Amateur*, by Wes Hayward, W7ZOI, and Doug DeMaw, W1FB. For many years, both hams and professional engineers have used this classic ARRL book to design untold numbers of working amplifiers.

How Much Gain?

One of the simple, yet profound, observations made in *Solid State Design for the Radio Amateur* is that a designer should *not* attempt to extract every last bit of gain from a single amplifier stage. Trying to do so virtually guarantees that the circuit will be "touchy"—it may end up being more oscillator than amplifier! While engineers might debate the exact number, modern semiconductors circuits are inexpensive enough that you should try for no more than 25 dB of gain in a single stage.

For example, if you are designing a high-gain amplifier system to follow a direct-conversion receiver mixer, you will need a total of about 100 dB of audio amplification. We would recommend a conservative approach where you use four stages, each with 25 dB of gain. But you might risk oscillation and instability by using only two stages, with 50 dB gain each. The component cost will not be greatly different between these approaches, but the headaches and lack of reproducibility of the "simpler" two-stage design will very likely far outweigh any small cost advantages!

Biasing the Transistor Amplifier

The first step in amplifier design is to *bias* the transistor properly. A small-signal linear amplifier is biased properly when there is current at all times. Once you have biased the stage, you can then use several simple rules of thumb to determine all the major properties of the resulting amplifier.

Solid State Design for the Radio Amateur introduces several elegant transistor models. We won't get into that much detail here, except to say that the most fundamental property of a transistor is this: When there is current in the base-emitter junction, a larger current will flow in the collector-emitter junction. When the base-emitter junction is thus *forward* biased, the voltage across the base and emitter leads of a silicon transistor will be relatively constant, at 0.7 V. For most modern transistors, the dc current in the collector-emitter junction will be at least 50 to 100 times greater than the base-emitter current. This dc current gain is called the transistor's *Beta* (β).

See **Fig A**, which shows a simple capacitively coupled low-frequency amplifier suitable for use at 1 MHz. Resistors R1 and R2 form a voltage divider feeding the base of the transistor. The amount of current in the resistive voltage divider is purposely made large enough so the base current is small in comparison, thus creating a "stiff" voltage supply for the base. As stated above, the voltage at the emitter will be 0.7 V less than the base voltage for this NPN transistor. The emitter voltage V_E appears across the series combination of R4 and R5. Note that R5 is bypassed by capacitor C4 for ac current.

By Ohm's Law, the emitter current is equal to the emitter voltage V_E divided by the sum of R4 plus R5.

Now, the emitter current is made up of both the base-emitter and the collector-emitter current, but since the base current is much smaller than the collector current, the amount of collector current is essentially equal to the emitter current, at V_E / (R4 + R5).

Our design process starts by specifying the amount of current we want to flow in the collector, with the dc collector voltage equal to half the supply voltage. For good bias stability with temperature variation, the total emitter resistor should be at least 100 Ω for a small-signal amplifier. Let's choose a collector current of 5 mA, and use a total emitter resistance of 200 Ω, with R4 = R5 = 100 Ω each. The voltage across 200 Ω for 5 mA of current is 1.0 V. This means that the voltage at the base must be 1.0 V + 0.7 V = 1.7 V, provided by the voltage divider R1 and R2.

The dc base current requirements for a collector current of 5 mA is approximately 5 mA / 50 = 0.1 mA if the transistor's dc Beta is at least 50, a safe assumption for modern transistors. To provide a "stiff" base voltage, we want the current through the voltage divider to be about five to ten times greater than the base current. For convenience then, we choose the current through R1 to be 1 mA. This is a convenient current value, because the math is simplified—we don't have to worry about decimal points for current or resistance: 1 mA × 1.8 kΩ = 1.8 V. This is very close to the 1.7 V we are seeking. We thus choose a standard value of 1.8 kΩ for R2. The voltage drop across R1 is 12 V – 1.8 V = 10.2 V. With 1 mA in R1, the necessary value is 10.2 kΩ, and we choose the closest standard value, 10 kΩ.

Let's now look at what is happening in the collector part of the circuit. The collector resistor R3 is 1 kΩ, and the 5 mA of collector current creates a 5 V drop across R3. This means that the collector dc voltage must be 12 V – 5 V = 7 V. The dc power dissipated in the transistor will be essentially all in the collector-emitter junction, and will be the collector-emitter voltage (7 V – 1 V = 6 V) times the collector current of 5 mA = 0.030 W, or 30 mW. This dissipation is well within the 0.5 W rating typical of small-signal transistors.

Now, let's calculate more accurately the result from using standard values for R1 and R2. The actual base voltage will be 12 V × [1.8 kΩ / (1.8 kΩ + 10 kΩ)] = 1.83 V,

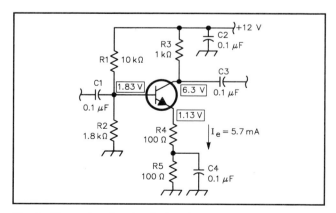

Fig A—Example of a simple low-frequency capacitively coupled transistorized small-signal amplifier. The voltages shown are the preliminary values desired for a collector current of 5 mA. The ac voltage gain is the ratio of the collector load resistor, R3, divided by the unbypassed portion of the emitter resistor, R4.

rather than 1.7 V. The resulting emitter voltage is 1.83 V − 0.7 V = 1.13 V, resulting in 1.13 V / 200 Ω = 5.7 mA of collector current, rather than our desired 5 mA. We are close enough—we have finished designing the bias circuitry!

Performance: Voltage Gain

Now we can analyze how our little amplifier will work. The use of the unbypassed emitter resistor R4 results in *emitter degeneration*—a fancy word describing a form of negative feedback. The bottom line for us is that we can use several handy rules of thumb. The first is for the ac voltage gain of an amplifier: A_V = R3 / R4, where A_V is shorthand for *voltage gain*. The ac voltage gain of such an amplifier is simply the ratio of the collector load resistor and the unbypassed emitter resistor. In this case, the gain is 1000 / 100 = 10, which is 20 dB of voltage gain. This expression for gain is true virtually without regard for the exact kind of transistor used in the circuit, provided that we design for moderate gain in a single stage, as we have done.

Performance: Input Resistance

Another useful rule of thumb stemming from use of an unbypassed emitter resistor is the expression for the ac input resistance: R_{in} = Beta × R4. If the ac Beta at low frequencies is about 50, then the input resistance of the transistor is 50 × 100 Ω = 5000 Ω. The actual input resistance includes the shunt resistance of voltage divider R2 and R1, about 1.5 kΩ. Thus the biasing resistive voltage divider essentially sets the input resistance of the amplifier.

Performance: Overload

We can accurately predict how this amplifier will perform. If we were to supply a peak positive 1 V signal to the base, the voltage at the collector will try to fall by the voltage gain of 10. However, since the dc voltage at the collector is only 7 V, it is clear that the collector voltage cannot fall 10 V. In theory, the collector voltage could fall as low as the 1.13 V dc level at the emitter. This amplifier will "run out of voltage" at a negative collector voltage swing of about 6.3 V − 1.13 V = 5.17 V, when the input voltage is 5.17 divided by the gain of 10 = 0.517 V.

When a negative-going ac voltage is supplied to the base, the collector current falls, and the collector voltage will rise by the voltage gain of 10. The maximum amount of voltage possible is the 12 V supply voltage, where the transistor is cut off with no collector current. The maximum positive collector swing is from the standing collector dc voltage to the supply voltage: 12 V − 6.3 V = 5.7 V positive swing. This occurs with a peak negative input voltage of 5.7 V / 10 = 0.57 V. Our amplifier will overload rather symmetrically on both negative and positive peaks. This is no accident—we biased it to have a collector voltage halfway

between ground and the supply voltage.

When the amplifier "runs out of output voltage" in either direction, another useful rule of thumb is that this is the *1 dB compression point*. This is where the amplifier just begins to depart from linearity, where it can no longer provide any more output for further input. For our amplifier, this is with a peak-to-peak output swing of approximately 5.1 V × 2 = 10.2 V, or 3.6 V rms. The output power developed in output resistor R3 is $(3.6)^2$ / 1000 = 0.013 W = 13 mW, which is +11.1 dBm (referenced to 1 mW on 50 Ω).

At the 1 dB compression point, the third-order *IMD* (intermodulation distortion) will be roughly 25 dB below the level of each tone. **Fig B** shows a graph of output versus input levels for both the desired signal and for third-order IMD products. The rule of thumb for IMD is that if the input level is decreased by 10 dB, the IMD will decrease by 30 dB. Thus, if input is restricted to be 10 dB below the 1 dB compression point, the IMD will be 25 dB + 30 dB = 55 dB below each output tone.

With very simple math we have thus designed and characterized a simple amplifier. This amplifier will be stable for both dc and ac under almost any thermal and environmental conditions conceivable. That wasn't too difficult, was it?—*R. Dean Straw, N6BV, ARRL Senior Assistant Technical Editor*

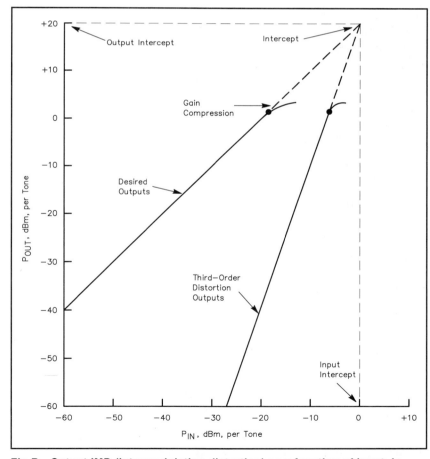

Fig B—Output IMD (intermodulation distortion) as a function of input. In the region below the 1 dB compression point, a decrease in input level of 10 dB results in a drop of IMD products by 30 dB below the level of each output tone in a two-tone signal.

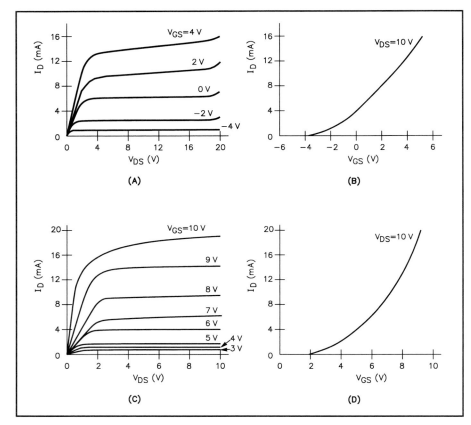

Fig 8.38 — MOSFET output [(A) and (C)] and transconductance [(B) and (D)] response curves. Plots (A) and (B) are for an N-channel depletion mode device. Note that V_{GS} varies from negative to positive values. Plots (C) and (D) are for an N-channel enhancement mode device. V_{GS} has only positive values.

junction break point voltage. The insulated gates in MOSFET devices do away with any appreciable gate leakage current. MOSFETs do not need input and reverse transconductance curves. Their output curves (**Fig 8.38**) are similar to those of the JFET.

The parameters used to describe a FET's performance are also similar to those of vacuum tubes. The dc channel resistance, r_{DS}, is specified in data sheets to be less than a maximum value when the device is biased on ($r_{DS(on)}$). For ac signals, $r_{ds(on)}$ is not necessarily the same as $r_{DS(on)}$, but it is not very different as long as the frequency is not so high that capacitive reactance becomes significant. The common source forward transconductance, g_{fs}, is obtained as the slope of one of the lines in the forward transconductance curve,

$$g_{fs} = \frac{\Delta I_D}{\Delta V_{GS}} \qquad (14)$$

When gate voltage is maximum ($V_{GS} = 0$ for a JFET) $r_{DS(on)}$ is minimum. This describes the effectiveness of the device as an analog switch.

A typical FET data sheet gives

ABSOLUTE MAXIMUM RATINGS for V_{DS}, V_{DG}, V_{GS} and I_D, along with the usual device dissipation (P_D) and storage temperature range. The OFF CHARACTERISTICS listed are the gate-source breakdown voltage, $V_{GS(BR)}$, the reverse gate current, I_{GSS} and the gate-source cutoff voltage, $V_{GS(OFF)}$. The ON CHARACTERISTIC is the zero-gate-voltage drain current (I_{DSS}). The SMALL SIGNAL CHARACTERISTICS include the forward transfer admittance, y_{fs}, the output admittance, y_{os}, the static drain-source on resistance, $r_{ds(on)}$ and various capacitances such as input capacitance, C_{iss}, reverse transfer capacitance, C_{rss}, the drain-substrate capacitance, $C_{d(sub)}$. FUNCTIONAL CHARACTERISTICS include the noise figure, NF and the common source power gain G_{ps}.

The relatively flat regions in the MOSFET output curves are often used to provide a constant current source. As is plotted in these curves, the drain current, I_D, changes very little as the drain-source voltage, V_{DS}, varies in this portion of the curve. Thus, for a fixed gate-source voltage, V_{GS}, the drain current can be considered to be constant over a wide

range of drain-source voltages.

Multiple gate MOSFETs are also available (MFE130, MPF201, MPF211, MPF521). Due to the insulating layer, the two gates are isolated from each other and allow two signals to control the channel simultaneously with virtually no loading of one signal by the other. A common application of this type of device is an automatic gain control (AGC) amplifier. The signal is applied to one gate and a rectified, low-pass filtered form of the output (the AGC voltage) is fed back to the other gate. Another common application is for mixers.

FET Biasing

There are two ways to bias an FET, with and without feedback. Source self biasing for an N-channel JFET is pictured in **Fig 8.39A**. In this common-drain amplifier circuit, bias level is determined by the current through R_S, since I_G is very small and there is essentially no voltage drop across R_G. The characteristic curve for this configuration is plotted in Fig 8.39B. The operating points of the amplifier are where the load line intersects the curves. An example of feedback biasing is shown in Fig 8.39C. R_1 is generally much larger than R_S and the load line is determined by the sum of these resistors, as shown in Fig 8.39D. Feedback biasing increases the input impedance of the amplifier, but is rarely required, since input resistance (R_G) can be made very large.

MOSFET Gate Protection

The MOSFET is constructed with a very thin layer of SiO_2 for the gate insulator. This layer is extremely thin in order to improve the gain of the device but this makes it susceptible to damage from high voltage levels. If enough charge accumulates on the gate terminal, it can punch through the gate insulator and destroy it. The insulation of the gate terminal is so good that virtually none of this potential is eased by leakage of the charge into the device. While this condition makes for nearly ideal input impedance (approaching infinity), it puts the device at risk of destruction from even such seemingly innocuous electrical sources as static electricity in the air.

Some MOSFET devices contain an internal Zener diode with its cathode connected to the gate and its anode to the substrate. If the voltage at the gate rises to a damaging level the Zener junction breaks down and bleeds the excess charges off to the substrate. When voltages are within normal operating limits the Zener has little effect on the signal at the gate, although it may decrease the input impedance of the

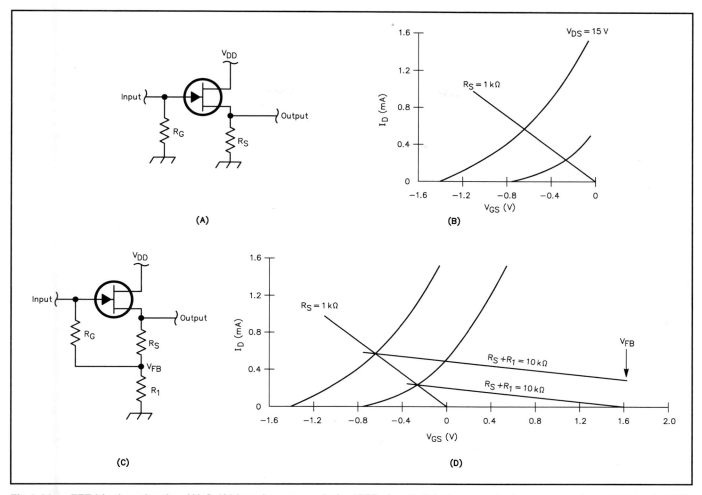

Fig 8.39 — FET biasing circuits. (A) Self biased common drain JFET circuit. (B) Transconductance curve for self biased JFET in (A). Gate bias is determined by current through R_S. Load line has a slope of $-1 / R_S$ and gate bias voltage can vary between where the load line crosses the characteristic curves. (C) Feedback bias common drain JFET circuit.

MOSFET. This solution will not work for all MOSFETs. The Zener diode must always be reverse biased to be effective. In the enhancement mode MOSFET, $V_{GS} > 0$ for all valid uses of the part. In depletion mode devices V_{GS} can be both positive and negative; when negative, a protection Zener diode would be forward biased and the MOSFET would not work properly. In some depletion mode MOSFET devices, back-to-back Zener diodes are used to protect the gate.

MOSFET devices are at greatest risk of damage from static electricity when they are out of circuit. Even though static electricity is capable of delivering little current, it can generate thousands of volts. When storing MOSFETs, the leads should be placed into conductive foam. When working with MOSFETs, it is a good idea to minimize static by wearing a grounded wrist strap and working on a grounded table. A humidifier may help to decrease the static electricity in the air. Before inserting a MOSFET into a circuit board it helps to first touch the device leads with your hand and then touch the circuit board. This serves to equalize the excess charge so that when the device is inserted into the circuit board little charge will flow into the gate terminal.

OPTICAL SEMICONDUCTORS

In addition to electrical energy and heat energy, light energy also affects the behavior of semiconductor materials. If a device is made to allow light to fall on the surface of the semiconductor material, the light energy will break covalent bonds and increase the number of electron-hole pairs, decreasing the resistance of the material.

Photoconductors

In commercial *photoconductors* (also called *photoresistors*) the resistance can change by as much as several kilohms for a light intensity change of 100 ft-candles. The most common material used in photoconductors is cadmium sulfide (CdS), with a resistance range of more than 2 MΩ in total darkness to less than 10 Ω in bright light. Other materials used in photoconductors respond best at specific colors. Lead sulfide (PbS) is most sensitive to infrared light and selenium (Se) works best in the blue end of the visible spectrum.

A similar effect is used in some diodes and transistors so that their operation can be controlled by light instead of electrical current biasing. These devices are called *photodiodes* and *phototransistors*. The flow of minority carriers across the reverse biased PN junction is increased by light falling on the doped semiconductor material. In the dark, the junction acts the same as any reverse biased PN junction, with a very low current (on the order of 10 μA)

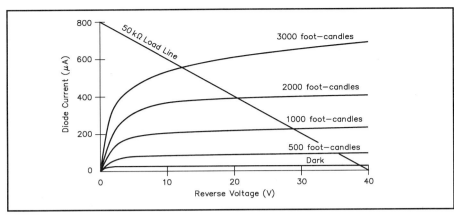

Fig 8.40 — Photodiode response curve. Reverse voltage is plotted on the x-axis and current through diode is plotted on the y-axis. Various response lines are plotted for different illumination. Except for the zero illumination line, the response does not pass through the origin since there is current generated at the PN junction by the light energy. A load line is shown for a 50 kΩ resistor in series with the photodiode.

that is nearly independent of reverse voltage. The presence of light not only increases the current but also provides a resistance-like relationship (reverse current increases as reverse voltage increases). See **Fig 8.40** for the characteristic response of a photodiode. Even with no reverse voltage applied, the presence of light causes a small reverse current, as indicated by the points at which the lines in Fig 8.40 intersect the left side of the graph. Photoconductors and photodiodes are generally used to produce light-related analog signals that require further processing. The phototransistor can often be used to serve both purposes, acting as an amplifier whose gain varies with the amount of light present. It is also more sensitive to light than the other devices. Phototransistors have lots of gain, but photodiodes normally have less noise, so they make sensitive detectors.

Photovoltaic Effect

When illuminated, the reverse biased photodiode has a reverse current due to excess minority carriers. As the reverse voltage is reduced, the potential barrier to the forward flow of majority carriers is also reduced. Since light energy leads to the generation of both majority and minority carriers, when the resistance to the flow of majority carriers is decreased these carriers form a forward current. The voltage at which the forward current equals the reverse current is called the *photovoltaic potential* of the junction. If the illuminated PN junction is not connected to a load, a voltage equal to the photovoltaic potential can be measured across it. Devices that use light from the sun to produce electricity in this way are

called *solar cells* or *solar batteries*. Common operating characteristics of silicon photovoltaic cells are an open circuit voltage of about 0.6 V and a conversion efficiency of about 10 to 15%.

Light Emitting Diodes

In the photodiode, energy from light falling on the semiconductor material is absorbed to make additional electron-hole pairs. When the electrons and holes recombine, the same amount of energy is given off. In normal diodes the energy from recombination of carriers is given off as heat. In certain forms of semiconductor material, the recombination energy is given off as light with a mechanism called *electroluminescence*. Unlike the incandescent light bulb, electroluminescence is a cold light source that typically operates with low voltages and currents (such as 1.5 V and 10 mA). Devices made for this purpose are called *light emitting diodes (LEDs)*. They have the advantages of low power requirements, fast switching times (on the order of 10 ns) and narrow spectra (relatively pure color). The LED emits light when it is forward biased and excess carriers are present. As the carriers recombine, light is produced with a color that depends on the properties of the semiconductor material used. Gallium arsenide (GaAs) generates light in the infrared region, gallium phosphide (GaP) gives off red light when doped with oxygen or green light when doped with nitrogen. Orange light is attained with a mixture of GaAs and GaP (GaAsP). Silicon doped with carbon gives off yellow light but does not produce much illumination. Other colors are also possible with different types and concentrations of dopants but usually have

lower illumination efficiencies.

The LED is very simple to use. It is connected across a voltage source with a series resistor that limits the current to the desired level for the amount of light to be generated. The cathode lead is connected to the lower potential, and is usually specially marked (flattening of the lead near the package, a dot of paint next to the lead, and a flat portion of the round device located next to the lead are all common methods).

Optoisolators

An interesting combination of optoelectronic components proves very useful in many analog signal processing applications. An *optoisolator* consists of an LED optically coupled to a phototransistor, usually in an enclosed package. The optoisolator, as its name suggests, isolates different circuits from each other. Typically, isolation resistance is on the order of 10^{11} Ω and isolation capacitance is less than 1 pF. Maximum voltage isolation varies from 1,000 to 10,000 V ac. The most common optoisolators are available in 6 pin DIP packages.

Optoisolators are used for voltage level shifting and signal isolation. The isolation has two purposes: to protect circuitry from excessive voltage spikes and to isolate noisy circuitry from noise sensitive circuitry. A disadvantage of an optoisolator is that it adds a finite amount of noise and is not appropriate for use in many applications with low level signals. Optoisolators also cannot transfer signals with high power levels. The power rating of the LED in a 4N25 device is 120 mW. Optoisolators have a limited frequency response due to the high capacitance of the LED. A typical bandwidth for the 4N25 series is 300 kHz.

As an example of voltage level shifting, the input to an optoisolator can be derived from a tube amplifier that has a signal varying between 0 and 150 V by using a series current limiting resistor. In order to drive a semiconductor circuit that operates in the −1 to 0 V range, the output of the optoisolator can be biased to operate in that range. This conversion of voltage levels, without a common ground connection between the circuits, is not easily performed in any other way.

A 1000 V spike that is high enough to destroy a semiconductor circuit will only saturate the LED in the optoisolator and will not propagate to the next stage. The worst that will happen is the LED will be destroyed, but very often it is capable of surviving even very high voltage spikes.

Optoisolators are also useful for isolating different ground systems. The input and output signals are totally isolated from each other, even with respect to the references for each signal. A common application for optoisolation is when a computer is used to control radio equipment. The computer signal, and even its ground reference, typically contains considerable wide band noise due to the digital circuitry. The best way to keep this noise out of the radio is to isolate both the signal and its reference; this is easily done with an optoisolator.

The design of circuits with optoisolators is not different from the design of circuits with LEDs and with transistors. The LED is forward biased and usually driven with a series current limiting resistor whose value is set so that the forward current will be less than the maximum value for the device (such as 60 mA in a 4N25). Signals must be appropriately dc shifted so that the LED is always forward biased. The phototransistor typically has all three leads available for connection. The base lead is used for biasing, since the signal is usually derived from the optics, and the collector and emitter leads are used as they would be in any transistor amplifier circuit.

Fiber Optics

An interesting variation of the optoisolator is the *fiber optic* connection. Like the optoisolator, the signal is introduced to an LED device that modulates light. The signal is recovered by a photodetecting device (photoresistor, photodiode, or phototransistor). Instead of locating the input and output devices next to each other, the light is transmitted in a fiber optic cable, an extruded glass fiber that efficiently carries light over long distances and around fairly sharp bends. The fiber optic cable isolates the two circuits and provides an interesting transmission line. Fiber optics generally have far less loss than coaxial cable transmission lines. They do not leak RF energy, nor do they pick up electrical noise. Special forms of LEDs and phototransistors are available with the appropriate optical couplers for connecting to fiber optic cables. These devices are typically designed for higher frequency operation with bandwidths in the tens and hundreds of megahertz.

LINEAR INTEGRATED CIRCUITS

If you look into a transistor, the actual size of the semiconductor is quite small compared to the size of the packaging. For most semiconductors, the packaging takes considerably more space than the actual semiconductor device. Thus, an obvious way to reduce the physical size of circuitry is to combine more of the circuit inside a single package.

Hybrid Integrated Circuits

It is easy to imagine placing several small semiconductor chips in the same package. This is known as *hybrid circuitry*, a technology in which several semiconductor chips are placed in the same package and miniature wires are connected between them to make complete circuits.

Hybrid circuits miniaturize analog electronic circuits by replacing much of the packaging that is inherent in discrete electronics. The term *discrete* refers to the use of individual components to make a circuit, each in its own package. One application that still exists for hybrid circuitry is microwave amplifiers. The components of the amplifier are placed in a standard TO-39 package that is only 1 cm in diameter. The small dimensions of these circuits permit operation at VHF. For example, the Motorola MWA5157 can provide over 23 dB of gain at 1 GHz.

Both discrete and hybrid circuitry require that connections be made between the leads of the components. This takes space, is relatively expensive to construct and is the source of most failures in electronic circuitry. If multiple components could be placed on a single piece of semiconductor with the connections between them as part of the semiconductor chip, these three disadvantages would be overcome.

Monolithic Integrated Circuits

In order to build entire circuits on a single piece of semiconductor, it must be possible to fabricate other devices, such as resistors and capacitors, as well as transistors and diodes. The entire circuit is combined into a single unit, or chip, that is called a *monolithic integrated circuit*.

An integrated circuit (IC) is fabricated in layers. An example of a semiconductor circuit schematic and its implementation in an IC is pictured in **Fig 8.41**. The base layer of the circuit, the *substrate*, is made of P-type semiconductor material. Although less common, the polarity of the substrate can also be N-type material. Since the mobility of electrons is about three times higher than that of holes, bipolar transistors made with N-type collectors and FETs made with N-type channels are capable of higher speeds and power handling. Thus, P-type substrates are far more common. For devices with N-type substrates, all polarities in the ensuing discussion would be reversed. Other substrates have been used, one of the most successful of which is the silicon-on-sapphire (SOS) construction that has been used to increase the bandwidth of integrated circuitry. Its relatively high manufacturing cost has impeded its use, however.

On top of the P-type substrate is a thin layer of N-type material in which the active and passive components are built. Impurities are diffused into this layer to form the appropriate component at each location. To prevent random diffusion of impurities into the N-layer, its upper surface must be protected. This is done by covering the N-layer with a layer of silicon dioxide (SiO_2). Wherever diffusion of impurities is desired, the SiO_2 is etched away. The precision of placing the components on the semiconductor material depends mainly on the fineness of the etching. The fourth layer of an IC is made of metal (usually aluminum) and is used to make the interconnections between the components.

Different components are made in a single piece of semiconductor material by first diffusing a high concentration of acceptor impurities into the layer of N-type material. This process creates P-type semiconductor — often referred to as P^+-type semiconductor because of its high concentration of acceptor atoms—that isolates regions of N-type material. Each of these regions is then further processed to form single components. A component is produced by the diffusion of a lesser concentration of acceptor atoms into the middle of each isolation region. This results in an N-type *isolation well* that contains P-type material, is surrounded on its sides by P^+-type material and has P-type material (substrate) below it. The cross sectional view in Fig 8.41B illustrates the various layers. Contacts to the metal layer are often made by diffusing high concentrations of donor atoms into small regions of the N-type well and the P-type material in the well. The material in these small regions is N^+-type and facilitates electron flow between the metal contact and the semiconductor. In some configurations, it is necessary to connect the metal directly to the P-type material in the well.

An isolation well can be made into a resistor by making two contacts into the P-type semiconductor in the well. Resistance is inversely proportional to the cross-sectional area of the well. An alternate type of resistor that can be integrated in a semiconductor circuit is a *thin film resistor*, where a metallic film is deposited on the SiO_2 layer, masked on its upper surface by more SiO_2 and then etched to make the desired geometry, thus adjusting the resistance.

There are two ways to form capacitors in a semiconductor. One is to make use of

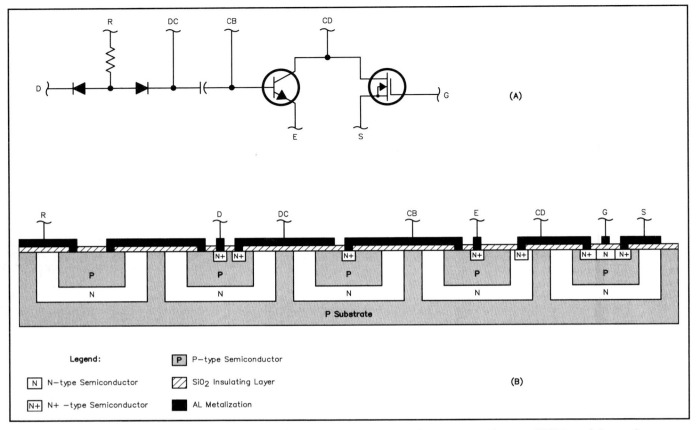

Fig 8.41 — Integrated circuit layout. (A) Circuit containing two diodes, a resistor, a capacitor, an NPN transistor and an N-channel MOSFET. Labeled leads are D for diode, R for resistor, DC for diode-capacitor, E for emitter, S for source, CD for collector-drain and G for gate. **(B)** Integrated circuit that is identical to circuit in (A). Same leads are labeled for comparison. Circuit is built on a P-type semiconductor substrate with N-type wells diffused into it. An insulating layer of SiO_2 is above the semiconductor and is etched away where aluminum metal contacts are made with the semiconductor. Most metal-to-semiconductor contacts are made with heavily doped N-type material (N+-type semiconductor).

the PN junction between the N-type well and the P-type material that fills it. Much like a varactor diode, when this junction is reverse biased a capacitance results. Since a bias voltage is required, this type of capacitor is polarized, like an electrolytic capacitor. Nonpolarized capacitors can also be formed in an integrated circuit by using thin film technology. In this case, a very high concentration of donor ions is diffused into the well, creating an N+-type region. A thin metallic film is deposited over the SiO_2 layer covering the well and the capacitance is created between the metallic film and the well. The value of the capacitance is adjusted by varying the thickness of the SiO_2 layer and the cross-sectional size of the well. This type of thin film capacitor is also known as a metal oxide semiconductor (MOS) capacitor.

Unlike resistors and capacitors, it is very difficult to create inductors in integrated circuits. Generally, RF circuits that need inductance require external inductors to be connected to the IC. In some cases, particularly at lower frequencies, the behavior of an inductor can be mimicked by an amplifier circuit. In many cases the appropriate design of IC amplifiers can obviate the need for external inductors.

Transistors are created in integrated circuitry in much the same way that they are fabricated in their discrete forms. The NPN transistor is the easiest to make since the wall of the well, made of N-type semiconductor, forms the collector, the P-type material in the well forms the base and a small region of N+-type material formed in the center of the well becomes the emitter. A PNP transistor is made by diffusing donor ions into the P-type semiconductor in the well to make a pattern with P-type material in the center (emitter) surrounded by a ring of N-type material that connects all the way down to the well material (base), and this is surrounded by another ring of P-type material (collector). This configuration results in a large base width separating the emitter and collector, causing these devices to have much lower current gain than the NPN form. This is one reason why integrated circuitry is designed to use many more NPN tran-

sistors than PNP transistors.

The simplest form of diode is generated by connecting to an N+-type connection point in the well for the cathode and to the P-type well material for the anode. Diodes are often converted from NPN transistor configurations. Integrated circuit diodes made this way can either short the collector to the base or leave the collector unconnected. The base contact is the anode and the emitter contact is the cathode.

FETs can also be fabricated in IC form. Due to its many functional advantages, the MOSFET is the most common form used for digital ICs. MOSFETs are made in a semiconductor chip much the same way as MOS capacitors, described earlier. In addition to the signal processing advantages offered by MOSFETs over other transistors, the MOSFET device can be fabricated in 5% of the physical space required for bipolar transistors. MOSFET ICs can contain 20 times more circuitry than bipolar ICs with the same chip size. Just as discrete MOSFETs are at risk of gate destruction, IC chips made with MOSFET devices have a similar risk. They should be treated with the same care to protect them from static electricity as discrete MOSFETs. Integrated circuits need not be made exclusively with MOSFETs or bipolar transistors. It is common to find IC chips designed with both technologies, taking advantage of the strengths of each.

Complementary Metal Oxide Semiconductors

Power dissipation in a circuit can be reduced to very small levels (on the order of a few nW) by using the MOSFET devices in complementary pairs (CMOS). Each amplifier is constructed of a series circuit of MOSFET devices, as in **Fig 8.42**. The gates are tied together for the input signal, as are the drains for the output signal. In saturation and cutoff, only one of the devices conducts. The current drawn by the circuit under no load is equal to the OFF leakage current of either device and the voltage drop across the pair is equal to V_{DD}, so the steady state power used by the circuit is always equal to $V_{DD} \times I_{D(off)}$. For ac signals, power consumption is proportional to frequency.

CMOS circuitry could be built with discrete components, however the number of extra parts and the need for the complementary components to be matched has made it an unusual design technique. Although CMOS is most commonly used in digital integrated circuitry, its low power consumption has been put to advantage by several manufacturers of analog ICs.

Integrated Circuit Advantages

There are many advantages of monolithic integrated circuitry over similar circuitry implemented with discrete components. The integration of the interconnections is one that has already been mentioned. This procedure alone serves to greatly decrease the physical size of the circuit and to improve its reliability. In fact, in one study performed on failures of electronic circuitry, it was found that the failure rate is not necessarily related to the complexity of the circuit, as had been previously thought, but is more closely a function of the number of interconnections between packages. Thus, the more circuitry that can be integrated onto a single piece of semiconductor material, the more reliable the circuit should be.

The amount of circuitry that can be placed onto a single semiconductor chip is a function of two factors: the size of the chip and how closely the various components are spaced. A revolution in IC manufacture occurred when semiconductor material was created in the laboratory rather than found in nature. The man-made semiconductor wafers are more pure and

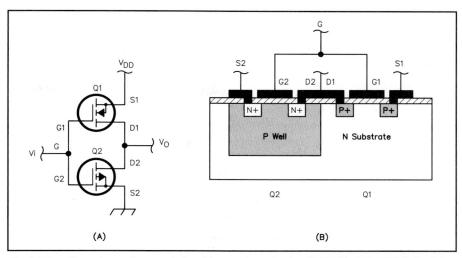

Fig 8.42 — Complementary metal oxide semiconductor (CMOS). (A) CMOS device is made from a pair of enhancement mode MOS transistors, the upper is an N-channel device and the lower is a P-channel device. When one transistor is biased on, the other is biased off so there is minimal current from V_{DD} to ground. (B) Implementation of a CMOS pair as an integrated circuit.

allow for larger wafer sizes. This, along with the steady improvement of the etching resolution on the chips, has caused an exponential increase over the past two decades in the amount of circuitry that can be placed in a single IC package. Currently, it is not unusual to find chips with more than one million transistors on them.

Decreased circuit size and improved reliability are only two of the advantages of monolithic integrated circuitry. The uncertainty of the exact behavior of the integrated components is the same as it is for discrete components, as discussed earlier. The relative properties of the devices on a single chip are very predictable, however. Since adjacent components on a semiconductor chip are made simultaneously (the entire N-type layer is grown at once, a single diffusion pass isolates all the wells and another pass fills them), the characteristics of identically formed components on a single chip of silicon should be identical. Even if the exact characteristics of the components are unknown, very often in analog circuit design the major concern is how components interact. For instance, push-pull amplifiers require perfectly matched transistors, and the gain of many amplifier configurations is governed by the ratio between two resistors and not their absolute values of resistance.

Integrated circuits often have an advantage over discrete circuits in their temperature behavior. The variation of performance of the components on an integrated circuit due to heat is no better than that of discrete components. While a discrete circuit may be exposed to a wide range of temperature changes, the entire semiconductor chip generally changes temperature

by the same amount; there are fewer "hot spots" and "cold spots." Thus, integrated circuits can be designed to better compensate for temperature changes.

A designer of analog devices implemented with integrated circuitry has more freedom to include additional components that could improve the stability and performance of the implementation. The inclusion of components that could cause a prohibitive increase in the size, cost or complexity of a discrete circuit would have very little effect on any of these factors in an integrated circuit.

Once an integrated circuit is designed and laid out, the cost of making copies of it is very small, often only pennies per chip. Integrated circuitry is responsible for the incredible increase in performance with a corresponding decrease in price of electronics over the last 20 years. While this trend is most obvious in digital computers, analog circuitry has also benefited from this technology.

The advent of integrated circuitry has also improved the design of high frequency circuitry. One problem in the design and layout of RF equipment is the radiation and reception of spurious signals. As frequencies increase and wavelengths approach the dimensions of the wires in a circuit board, the interconnections act as efficient antennas. The dimensions of the circuitry within an IC are orders of magnitude smaller than in discrete circuitry, thus greatly decreasing this problem and permitting the processing of much higher frequencies with fewer problems of interstage interference. Another related advantage of the smaller interconnections in an IC is the lower inherent inductance of the wires, and lower stray capacitance between components and traces.

Integrated Circuit Disadvantages

Despite the many advantages of integrated circuitry, disadvantages also exist. ICs have not replaced discrete components, even tubes, in some applications. There are some tasks that ICs cannot perform, even though the list of these continues to decrease over time as IC technology improves.

Although the high concentration of components on an IC chip is considered to be an advantage of that technology, it also leads to a major limitation. Heat generated in the individual components on the IC chip is often difficult to dissipate. Since there are so many heat generating components so close together, the heat can build up and destroy the circuitry. It is this limitation that currently causes many power amplifiers to be designed with discrete components.

Integrated circuits, despite their short interconnection lengths and lower stray inductance, do not have as high a frequency response as similar circuits built with appropriate discrete components. (There are exceptions to this generalization, of course. Monolithic microwave integrated circuits—MMICs—are available for operation on frequencies up through 10 GHz.) The physical architecture of an integrated circuit is the cause of this limitation. Since the substrate and the walls of the isolation wells are made of opposite types of semiconductor material, the PN junction between them must be reverse biased to prevent current from passing into the substrate. Like any other reverse biased PN junction, a capacitance is created at the junction and this limits the frequency response of the devices on the IC. This situation has improved over the years as isolation wells have gotten smaller, thus decreasing the capacitance between the well and the substrate, and techniques have been developed to decrease the PN junction capacitance at the substrate. One such technique has been to create an N^+-type layer between the well and the substrate, which decreases the capacitance of the PN junction as seen by the well. As an example, in the 1970s the LM324 operational amplifier IC package was developed by National Semiconductor and claimed a gain-bandwidth product of 1 MHz. In the 1990s the HFA1102 operational amplifier IC, developed by Harris Semiconductor, was introduced with a gain-bandwidth product of 600 MHz.

A major impediment to the introduction of new integrated circuits, particularly with special applications, is the very high cost of development of new designs. The masking cost alone for a designed and tested integrated circuit can exceed $100000. Adding the design, layout and debugging costs motivates IC manufacturers to produce devices that will be widely used so that they can recoup the development costs by volume of sales. While a particular application would benefit from customization of circuitry on an IC, the popularity of that application may not be wide enough to compel an IC manufacturer to develop that design. A designer who wishes to use IC chips must often settle for circuits that do not behave exactly as desired for the specific application. This trade-off between the advantages afforded by the use of integrated circuitry and the loss of performance if the available IC products do not exactly meet the desired specifications must be considered by equipment designers. It often leads to the use of discrete circuitry in sensitive applications. Once again, the improvements afforded by technology have mitigated this problem somewhat. The design and layout of ICs has been made more affordable by computer-based aids. Interaction between the computer aided design (CAD) software and modern chip masking hardware has also decreased the masking costs. As these development costs decrease, we are seeing an increase in the number of specialty chips that are being marketed and also of small companies that are created to fill the needs of the niche markets.

Common Types of Linear Integrated Circuits

The three main advantages of designing a circuit into an IC are to take advantage of the matched characteristics of like components, to make highly complex circuitry more economical, and to miniaturize the circuit. As a particular technology becomes popular, a rash of integrated circuitry is developed to service that technology. A recent example is the cellular telephone industry. Cellular phones have become so pervasive that IC manufacturers have developed a large number of devices targeted toward this technology. Space limitations prohibit a comprehensive listing of all analog special function ICs but a sampling of those that are more useful in the radio field is presented.

Component Arrays

The most basic form of linear integrated circuit is the component array. The most common of these are the resistor, diode and transistor arrays. Though capacitor arrays are also possible, they are used less often. Component arrays usually provide space saving but this is not the major advantage of these devices. They are the least densely packed of the integrated circuits, limited mainly by the number of off-chip connections needed. While it may be possible to place over a million transistors on a single semiconductor chip, individual access to these would require a total of three million pins and this is beyond the limits of practicability. More commonly, resistor and diode arrays contain from five to 16 individual devices and transistor arrays contain from three to six individual transistors. The advantage of these arrays is the very close matching of component values within the array. In a circuit that needs matched components, the component array is often a good method of obtaining this feature. The components within an array can be internally combined for special functions, such as termination resistors, diode bridges and Darlington pair transistors. A nearly infinite number of possibilities exists for these combinations of components and many of these are available in arrays.

Multivibrators

A *multivibrator* is a circuit that oscillates, usually with a square wave output in the audio frequency range. The frequency of oscillation is accurately controlled with the addition of appropriate values of external resistance and capacitance. The most common multivibrator in use today is the 555 (NE555 by Signetics [now Philips] or LM555 by National Semiconductor). This very simple eight-pin DIP device has a frequency range from less than one hertz to several hundred kilohertz. Such a device can also be used in *monostable* operation, where an input pulse generates an output pulse of a different duration, or in *astable* operation, where the device freely oscillates. Some other applications of a multivibrator are as a frequency divider, a delay line, a pulse width modulator and a pulse position modulator.

Operational Amplifiers

An *operational amplifier*, or *op amp*, is one of the most useful linear devices that has been developed in integrated circuitry. While it is possible to build an op amp with discrete components, the symmetry of this circuit requires a close match of many components and is more effective, and much easier, to implement in integrated circuitry. **Fig 8.43** shows a basic op-amp circuit. The op amp approaches a perfect analog circuit building block.

Ideally, an op amp has an infinite input impedance (Z_i), a zero output impedance

(Z_o) and an open loop voltage gain (A_v) of infinity. Obviously, practical op amps do not meet these specifications, but they do come closer than most other types of amplifiers. An older op amp that is based on bipolar transistor technology, the LM324, has the following characteristics: guaranteed minimum CMRR of 65 dB, guaranteed minimum A_v of 25000, an input bias current (related to Z_i) guaranteed to be below 250 nA (2.5×10^{-7} A), output current capability (which determines Z_o)

guaranteed to be above 10 mA and a gain-bandwidth product of 1 MHz. The TL084, which is a pin compatible replacement for the LM324 but is made with both JFET and bipolar transistors, has a guaranteed minimum CMRR of 80 dB, an input bias current guaranteed to be below 200 pA (2.0×10^{-10} A, almost 1000 times smaller than the LM324) and a gain-bandwidth product of 3 MHz. Philips has recently introduced the LMC6001 op amp with an input bias current of 25 fA (2.5×10^{-14} A,

almost 10000 times smaller than the TL084). This is equivalent to 156 electrons entering the device every millisecond and corresponds to nearly infinite input impedance. Op amps can be customized to perform a large variety of functions by the addition of external components.

The typical op amp has three signal terminals (see **Fig 8.44**). There are two input terminals, the noninverting terminal marked with a + sign and the inverting terminal marked with a – sign. The output of the amplifier has a single terminal and all signal levels within the op amp float, which means they are not tied to a specific reference. Rather, the reference of the input signals becomes the reference for the output signal. In many circuits this reference level is ground. Older operational amplifiers have an additional two connections for *compensation*. To keep the amplifier from going into oscillation at very high gains (increase its stability) it is often necessary to place a capacitor across the compensation terminals. This also decreases the frequency response of the op amp. Most modern op amps are internally compensated and do not have separate pins to add compensation capacitance. Additional compensation can be attained by connecting a capacitor between the op amp output and the inverting input.

One of the major advantages of using an op amp is its very high common mode rejection ratio (CMRR). Since there are two input terminals to an op amp, anything that is common to both terminals will be subtracted from the signal during amplification. The CMRR is a measure of the effectiveness of this removal. High CMRR results from the symmetry between the circuit halves. The rejection of power-supply noise is also an important parameter of an op amp. This is attained similarly, since

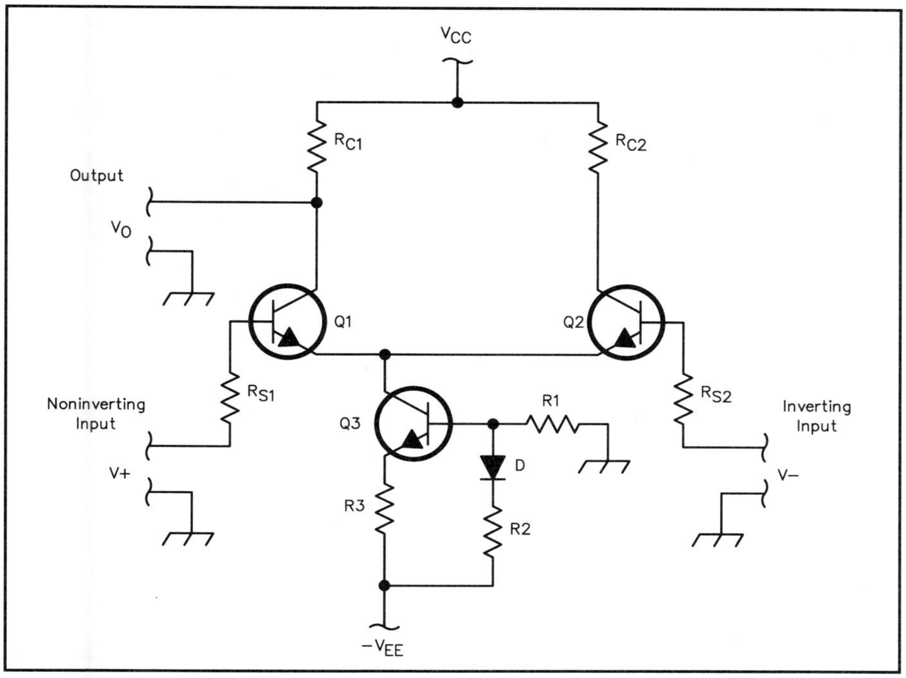

Fig 8.43 — Schematic of the components that make up an operational amplifier. Q1 and Q2 are matched emitter-coupled amplifiers. Q3 provides a constant current source. The symmetry of this device makes the matching of the components critical to its operation. This is why this circuit is usually implemented only in integrated circuitry. This simple op amp design has a large dc offset voltage at the output. Most practical designs include a level-shifting circuit, so the output voltage can exist near ground potential.

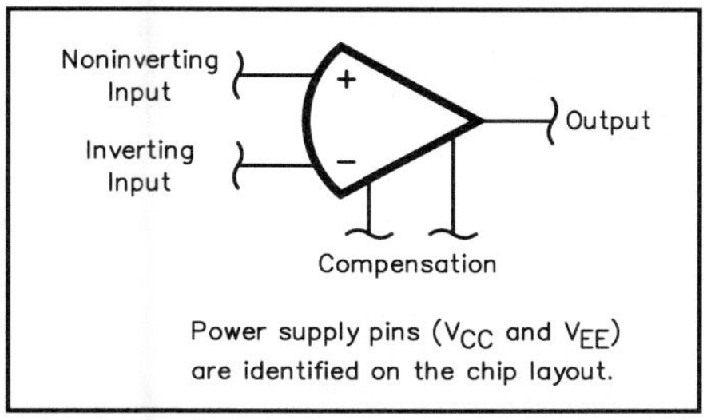

Fig 8.44 — Operational amplifier schematic symbol. The terminal marked with a + sign is the noninverting input. The terminal marked with a – sign is the inverting input. The output is to the right. On some op amps, external compensation is needed and leads are provided, pictured here below the device. Usually, the power supply leads are not shown on the op amp itself but are specified in the data sheet.

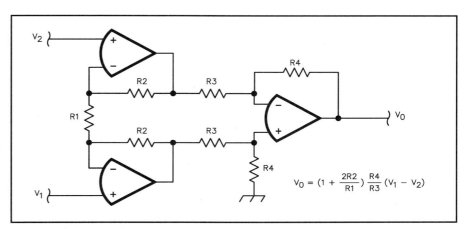

$$V_0 = \left(1 + \frac{2R2}{R1}\right)\frac{R4}{R3}(V_1 - V_2)$$

Fig 8.45 — Operational amplifiers arranged as an instrumentation amplifier. The balanced and cascaded series of op amps work together to perform differential amplification with good common-mode rejection and very high input impedance (no load resistor required) on both the inverting (V_1) and noninverting (V_2) inputs.

the power supply is connected equally to both symmetrical halves of the op amp circuit. Thus, the power supply rejection ratio (PSRR) is similar to the CMRR and is often specified on the device data sheets.

Just as the symmetry of the transistors making up an op amp leads to a device with high values of Z_i, A_v and CMRR and a low value of Z_o, a symmetric combination of op amps is used to further improve these parameters. This circuit, shown in **Fig 8.45**, is called an instrumentation amplifier.

The op amp is capable of amplifying signals to levels limited mainly by the power supplies. Two power supplies are required, thus defining the range of signal voltages that can be processed. In most op amps the signal levels that can be handled are less than the power supply limits (rails), usually one or two diode drops (0.7 V or 1.4 V) away from each rail. Thus, if an op amp has 15 V connected as its upper rail (usually denoted V^+) and ground connected as its lower rail (V^-), input signals can be amplified to be as high as 13.6 V and as low as 1.4 V in most amplifiers. Any values that would be amplified beyond those limits are clamped (output voltages that should be 1.4 V or less appear as 1.4 V and those that should be 13.6 V or more appear as 13.6 V). This clamping action is illustrated in Fig 8.1. Recently, op amps have been developed to handle signals all the way out to the power supply rails (for example, the MAX406, from Maxim Integrated Products).

If a signal is connected to the input terminals of an op amp, it will be amplified as much as the device is able (up to A_v), and

will probably grow so large that it clamps, as described above. Even if such large gains are desired, A_v varies from one device to the next and cannot be guaranteed. In most applications the op amp gain is limited to a more reasonable value and this is usually realized by providing a negative feedback path from the output terminal to the inverting input terminal. The *closed loop gain* of an op amp depends solely on the values of the passive components used to form the loop (usually resistors and, for frequency-selective circuits, capacitors). Some examples of different circuit configurations that manipulate the loop gain follow.

The op amp is often used as either an inverting or a noninverting amplifier. Accurate amplification can be achieved with just two resistors: the feedback resistor, R_f, and the input resistor, R_i (see **Fig 8.46**). If connected in the noninverting configuration, the input signal is connected to the noninverting terminal. The feedback resistor is connected between the output and the inverting terminal. The inverting terminal is connected to R_i, which is connected to ground. The gain of this configuration is:

$$\frac{V_o}{V_n} = \left(1 + \frac{R_f}{R_i}\right) \tag{15}$$

where:
V_o is the output voltage, and
V_n is the input voltage to the noninverting terminal.

In the inverting configuration, the input signal (V_i) is connected through R_i to the inverting terminal. The feedback resistor is again connected between the inverting terminal and the output. The noninverting terminal can be connected to ground or to a dc offset voltage. The gain of this circuit is:

$$\frac{V_o}{V_i} = -\frac{R_f}{R_i} \tag{16}$$

where V_i represents the voltage input to R_i.

The negative sign in equation 16 indicates that the signal is inverted. For ac signals, inversion represents a 180° phase shift. The gain of the noninverting op amp can vary from a minimum of × 1 to the maximum of which the device is capable. The gain of the inverting op amp configuration can vary from a minimum of × 0 (gains from × 0 to × 1 attenuate the signal while gains of × 1 and higher amplify the signal) to the maximum that the device is capable of, as indicated by A_v for dc signals, or the gain-bandwidth product for ac signals. Both parameters are usually specified in the manufacturer's data sheet.

A voltage follower is a type of op amp that is commonly used as a buffer stage. The voltage follower has the input connected directly to the noninverting terminal and the output connected directly to the inverting terminal (**Fig 8.47**). This configuration has unity gain and provides the maximum possible input impedance and the minimum possible output impedance of which the device is capable.

A *differential amplifier* is a special application of an operational amplifier (see **Fig 8.48**). It amplifies the difference between two analog signals and is very useful to cancel noise under certain conditions. For instance, if an analog signal and a reference signal travel over the same cable they may pick up noise, and it is likely that both signals will have the same amount of noise. When the differential amplifier subtracts them, the signal will be unchanged but the noise will be completely removed, within the limits of the CMRR. The equation for differential amplifier operation is

$$V_o = \frac{R_f}{R_i} \left[\frac{1}{\frac{R_n}{R_g} + 1} \left(\frac{R_i}{R_f} + 1 \right) V_n - V_i \right] \tag{17}$$

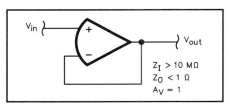

Fig 8.47 — **Voltage follower. This operational amplifier circuit makes a nearly ideal buffer with a voltage gain of about one, extremely high input impedance and extremely low output impedance.**

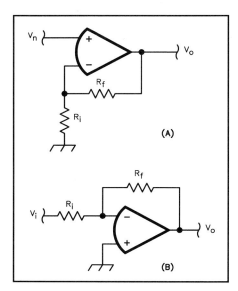

Fig 8.46 — **Operational amplifier circuits. (A) Noninverting configuration. (B) Inverting configuration.**

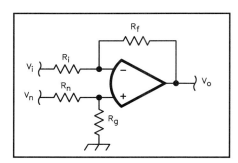

Fig 8.48 — **Differential amplifier. This operational amplifier circuit amplifies the difference between the two input signals.**

which, if the ratios $\dfrac{R_i}{R_f}$ and $\dfrac{R_n}{R_g}$ are equal, simplifies to

$$V_o = \frac{R_f}{R_i}\left(V_n - V_i\right) \tag{18}$$

Note that the differential amplifier response is identical to the inverting op amp response (equation 16) if the voltage source to the noninverting terminal is equal to zero. If the voltage source to the inverting terminal (V_i) is set to zero, the analysis is a little more complicated but it is possible to derive the noninverting op amp response (equation 15) from the differential amplifier response by taking into account the influence of R_n and R_g.

DC offset is an important consideration in op amps for two reasons. Actual op amps have a slight mismatch between the inverting and noninverting terminals that can become a substantial dc offset in the output, depending on the amplifier gain. The op amp output must not be too close to the clamping limits or distortion will occur. Introduction of a small dc correction voltage to the noninverting terminal is sometimes used to apply an offset voltage that counteracts the internal mismatch and centers the signal in the rail-to-rail range.

The high input impedance of an op amp makes it ideal for use as a *summing amplifier*. In either the inverting or noninverting configuration, the single input signal can be replaced by multiple input signals that are connected together through series resistors, as shown in **Fig 8.49**. For the inverting summing amplifier, the gain of each input signal can be calculated individually using equation 16 and, because of the superposition property, the output becomes the sum of each input signal multiplied by its gain. In the noninverting configuration, the output is the gain times the weighted sum of the m different input signals:

$$V_n = V_{n1}\frac{R_{p1}}{R_1 + R_{p1}} + V_{n2}\frac{R_{p2}}{R_2 + R_{p2}}$$

$$+ \ldots + V_{nm}\frac{R_{pm}}{R_m + R_{pm}} \tag{19}$$

where R_{pm} is the parallel resistance of all m resistors excluding R_m. For example, with three signals being summed, R_{p1} is the parallel combination of R_2 and R_3.

Other combinations of summing and difference amplification can be realized with a single op amp. The analyses of such circuits use the standard op amp equations coupled with the principle of superposition.

A *voltage comparator* is another special form of an operational amplifier. It takes in two analog signals and provides a binary output that is true if the voltage of one signal is bigger than that of the other, and false if not. A standard operational amplifier can be made to act as a comparator by connecting the two voltages to the noninverting and inverting inputs with no input or feedback resistors. If the voltage of the noninverting input is higher than that of the inverting input, the output voltage will be clamped to the positive clamping limit. If the inverting input is at a higher potential than the noninverting input, the output voltage will be clamped to the negative clamping limit (although this is not necessarily a negative voltage, depending on the value of the lower rail). Some applications of a voltage comparator are a zero crossing detector, a signal squarer (which turns other cyclical wave forms into square waves) and a peak detector.

Charge Coupled Devices

As the speed of integrated circuitry increases, it becomes possible to process some of the signals digitally while other processing occurs in analog form, all of this on the same IC chip. Such a chip is often called a *mixed modality* or *hybrid* chip (not to be confused with the hybrid circuitry discussed earlier). An example of this is the *charge coupled device (CCD)*. Pure digital analysis of signals requires digitization in two domains, namely the time sampling of a signal into individual packets and the amplitude sampling of each time packet into digital levels. CCDs perform time sampling but the time packets remain in analog form; they can take on any voltage value rather than a fixed number of discrete values. The CCD is often used to produce a delay filter. While most analog filters introduce some phase shift or delay into the signal, the relationship between the phase shift and the frequency is not always linear; different frequencies are delayed by different amounts of time. The goal of an ideal delay filter is to delay all parts of the signal by the same time. The CCD is used to realize this by sampling the signal, shifting the time packets through a series of capacitors and then reconstructing the continuous signal at the other end. The rate of shifting the time packets and the number of stages determines the amount of the delay. When originally introduced in the late 1970s, CCDs were described as bucket brigade devices (after the old fire fighting technique), where the buckets filled with signal packets are passed along the line until they are dumped at the end and recombined into an analog

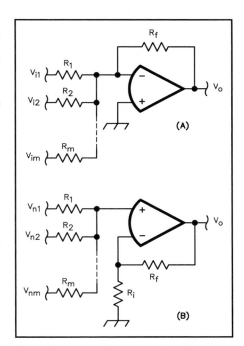

Fig 8.49 — Summing operational amplifier circuits. (A) Inverting configuration. (B) Noninverting configuration.

signal. These devices are simply constructed in an IC where each bucket is a MOS capacitor that is surrounded by two MOSFETs. When the transistors are biased to conduct, the charge moves from one bucket to the next and, while biased off, the charges are held in their capacitors. Very accurate filters, called *switched capacitor filters*, can be made with CCDs (see the **Filters** chapter).

A special form of CCD has also become quite popular in recent years, replacing the vidicon in modern camera circuitry. A two dimensional array of CCD elements has been developed with light sensitive semiconductor material; the charge that enters the capacitors is proportional to the amount of light incident on that location of the chip. The charges are held in their array of capacitors until shifted out, one horizontal line at a time, in a raster format. The CCD array mimics the operation of the vidicon camera and has many advantages. CCD response linearity across the field is superior to that of the vidicon. Very bright light at one location saturates the CCD elements only at that location rather than the blooming effect in vidicons where bright light spreads radially from the original location. CCD imaging elements do not suffer from image retention, which is another disadvantage of vidicon tubes.

Balanced Mixers

The *balanced mixer* is a device with

many applications in modern radio transceivers (see the **Mixers, Modulation and Demodulation** chapter). Audio signals can be modulated onto a carrier or demodulated from the carrier with a balanced mixer. RF signals can be downconverted to intermediate frequency (IF) or IF can be upconverted to RF with a balanced mixer. This device is made with a bridge of four matched Schottky diodes and the necessary transformers packaged in a small metal, plastic or ceramic container. The consequence of unmatched diodes is poor isolation between the local oscillator (LO) and the two signals. IC mixers often use a "Gilbert cell" to provide LO isolation as high as −30 dB at 500 MHz. The isolation improves with decreasing frequency.

Receiver Subsystems

High performance ICs have been designed that make up complete receivers with the addition of only a few external components. Two examples that are very similar are the Motorola MC3363 and the Philips NE627. Both of these chips have all the active RF stages necessary for a double conversion FM receiver. The MC3363 has an internal local oscillator (LO) with varactor diodes that can generate frequencies up to 200 MHz,

although the rest of the circuit is capable of operating at frequencies up to 450 MHz with an external oscillator. The RF amplifier has a low noise factor and gives this chip a $0.3\ \mu V$ sensitivity. The intermediate frequency stages contain limiter amplifiers and quadrature detection. The necessary circuitry to implement receiver squelch and zero crossing detection of FSK modulation is also present. The circuit also contains received signal strength ("S-meter") circuitry (RSSI). The input and output of each stage are also brought out of the chip for versatility. The audio signal out of this chip must be appropriately amplified to drive a low-impedance speaker. This chip can be driven with a dc power source from 2 to 7 V and it draws only 3 mA with a 2 V supply.

The Philips NE627 is a newer chip than the MC3363 and has better performance characteristics even though it has essentially the same architecture. Its LO can generate frequencies up to 150 MHz and external oscillator frequencies up to 1 GHz can be used. The chip has a 4.6 dB noise figure and $0.22\ \mu V$ sensitivity. The circuit can be powered with a dc voltage between 4.5 and 8 V and it draws between 5.1 and 6.7 mA. This chip is also ESD hardened so it resists damage from electrostatic dis-

charges, such as from nearby lightning strikes.

The various stages in the receiver subsystem ICs are made available by connections on the package. There are two reasons that this is done. Filtering that is added between stages can be performed more effectively with inductors and crystal or ceramic filters, which are difficult to fabricate in integrated circuitry, so the output of one stage can be filtered externally before being fed to the next stage. It also adds to the versatility of the device. Filter frequencies can be customized for different intermediate frequencies. Stages can be used individually as well, so these devices can be made to perform direct conversion or single conversion reception or other forms of demodulation instead of FM.

Older integrated circuits that are subsets of the receiver subsystems are popular. The NE602 contains one double balanced mixer and a local oscillator, along with voltage regulation and buffering (**Fig 8.50**). It contains almost everything required to construct a direct conversion receiver. Its small size, an 8 pin DIP, makes it more desirable for this purpose than using part of an MC3363, which is in a 24 pin DIP and is more expensive. The NE604 contains the IF amplifiers and quadrature detector that, together with two NE602s and an RF amplifier, could almost duplicate the functions of the MC3363 or the NE367.

Transmitter Subsystems

Single chips are available to implement FM transmitters. One implementation is the Motorola MC2831A. This chip contains a mike preamplifier with limiting, a tone generator for CTCSS or AFSK, and a frequency modulator. It has an internal voltage controlled oscillator that can be controlled with a crystal or an LC circuit. This chip also contains circuitry to check the power supply voltage and produce a warning if it falls too low. Together with an FM receiver IC, an entire transceiver can be fabricated with very few parts.

Monolithic Microwave Integrated Circuit

A class of bipolar IC that is capable of higher frequency responses is the *monolithic microwave integrated circuit (MMIC)*. There is no formal definition of when an IC amplifier becomes an MMIC and, as the performance of IC devices improves, particularly MOS based devices, the distinction is becoming blurred. MMIC devices typically have predefined operating characteristics and require few external components. An example of an

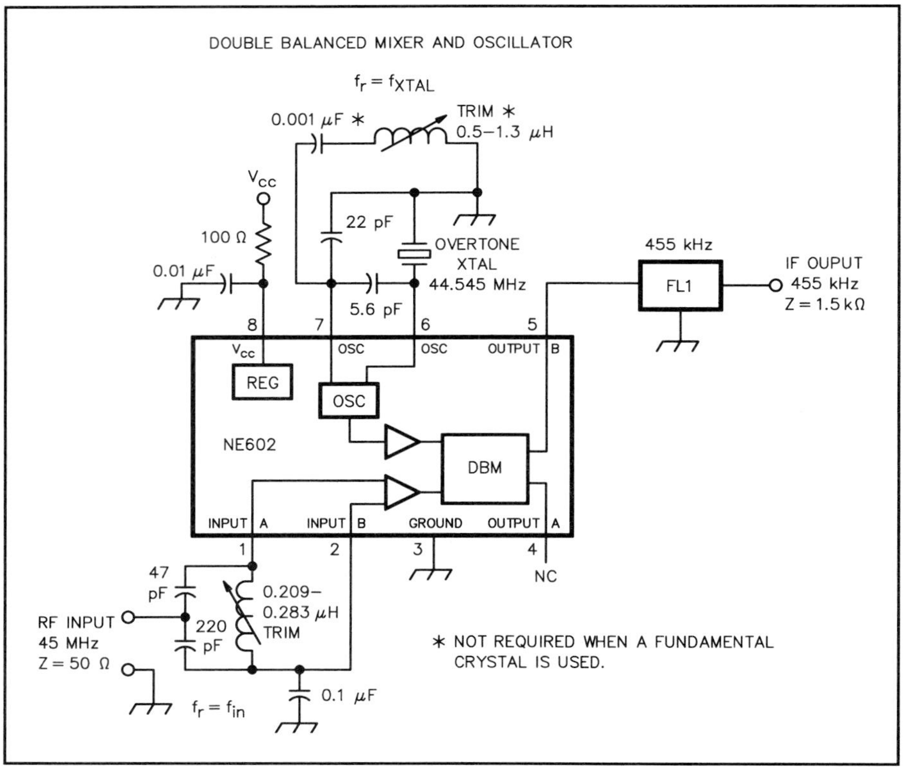

Fig 8.50 — The NE602 functional block diagram in circuit. This device contains a doubly balanced mixer, a local oscillator, buffers and a voltage regulator. This application uses the NE602 to convert an RF signal in a receiver to IF.

MMIC is a fixed gain amplifier, the MSA0204 (**Fig 8.51**), which can deliver 12 dB of gain up to 1 GHz. More modern MMIC devices are being developed with bandwidths in the tens of GHz.

Comparison of Analog Signal Processing Components

Analog signal processing deals with changing a signal to a desired form. Vacuum tubes, bipolar transistors, field-effect transistors and integrated circuitry perform similar functions, each with specific advantages and disadvantages. These are summarized here.

Of the four component types, vacuum tubes are physically the largest and require the most operating power. They have more limited life spans, usually because the heater filament burns out just as a light bulb does. Regardless of its use, a vacuum tube always generates heat. Miniaturization is difficult with vacuum tubes both because of their size and because of the need for air space around them for cooling. Vacuum tubes do have advantages, however. They are electrically robust. You need not be as concerned about static charges destroying vacuum tubes. A transmitter with vacuum tube finals usually has a variable matching network built in, and can be loaded into a higher SWR than one with semiconductor finals. Tubes are generally able to withstand the high voltages generated by reflections under high SWR conditions. They are not as easily damaged by short-term overloads or the electromagnetic pulses generated by lightning. The relatively high plate voltages mean that the plate current is lower for a given power output; thus power supplies do not need as high a current handling capability. Vacuum tubes are capable of considerable heat dissipation and many high power applications still use them. Special forms of vacuum tubes are also still used. Most video displays use CRTs, and microwave transmitting tubes are still common.

Bipolar transistors have many advantages over vacuum tubes. When treated properly they can have virtually unlimited life spans. They are relatively small and, if they do not handle high currents, do not generate much heat, improving miniaturization. They make excellent high-frequency amplifiers. Compared to MOSFET devices they are less susceptible to damage from electrostatic discharge. RF amplifiers designed with bipolar transistors in their finals generally include circuitry to protect the transistors from the high voltages generated by reflections under

high SWR conditions. Lightning strikes in the area (not direct hits) have been known to destroy all kinds of semiconductors, including bipolar transistors. Semiconductors have replaced almost all small-signal applications of tubes.

There are many performance advantages to FET devices, particularly MOSFETs. The extremely low gate currents allow the design of analog stages with nearly infinite input resistance. Signal distortion due to loading is minimized in this way. As these characteristics are improved by technology, we are seeing an increase in FET design at the expense of bipolar transistors.

The current trend in electronics is portability. Transceivers are decreasing in size and in their power requirements. Integrated circuitry has played a large part in this trend. Extremely large circuits have been designed with microscopic proportions. It is more feasible to use MOSFETs within an IC chip than as discrete components since the devices at risk are usually those that are connected to the outside world. It is not necessary to use electrostatic discharge protection circuitry on the gate of every MOSFET in an IC; only the ones that connect to the pins on the chip need this protection. This arrangement both improves the performance of the internal MOSFETs and decreases the circuit size even further. Semiconductors are slowly replacing the last tube applications. CCD chips have been so successful in video cameras that it is difficult to find an application for vidicon tubes. The liquid crystal displays (LCDs) in laptop computers have given considerable competition

to the CRT tube.

An important consideration in the use of analog components is the future availability of parts. At an ever increasing rate, as new components are developed to replace older technology, the older components are discontinued by the manufacturers and become unavailable for future use. This tends to be a fairly long term process but it is not unusual for a manufacturer to stop offering a component when demand for it falls. This has become evident with vacuum tubes, which are becoming more difficult to find and more expensive as fewer manufacturers produce them.

The major disadvantages of IC technology have been power handling capability, frequency response and noncustomized circuitry. These characteristics have improved at an amazing pace over recent years; it is a process that feeds itself. As ICs are improved they are used to make more powerful tools (such as computers and electronic test equipment) that are used in the design of further IC improvements. Entire transceivers are designed with just a few IC chips and the appropriate transistors for power amplification. The quiescent current draw of these devices has been reduced to the microampere level so they can operate effectively from small battery packs. The improved noise performance of circuitry has also decreased the need for high transmitter power, further decreasing the current requirements for these devices. If this trend continues, we should eventually see a near total switch to IC components with few discrete semiconductors and no vacuum tubes.

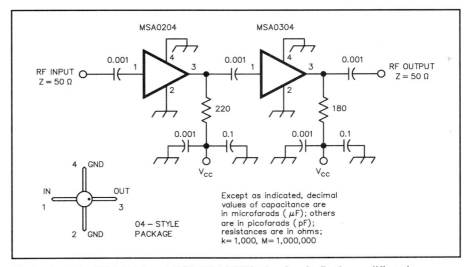

Fig 8.51 — The MSA0204 and MSA0304 MMICs in circuit. Both amplifiers have both input and output impedance of 50 Ω and a bandwidth of more than 2.5 GHz.

Contents

Safety
9

This chapter, written by James N. Woods, KC7FG, will focus on how to avoid potential hazards as we explore Amateur Radio and its many facets. We need to learn as much as possible about what could *go wrong* so we can avoid factors that might result in accidents. Amateur Radio activities are not inherently hazardous, but like many things in modern life, it pays to be informed. Stated another way, while we long to be creative and innovative, there is still the need to act responsibly. Safety begins with our attitude. Make it a habit to plan work carefully. Don't be the one to say, "I didn't think it could happen to me."

Having a good attitude about safety is not enough, however. We must be knowledgeable about common safety guidelines and follow them faithfully. Safety guidelines cannot possibly cover all situations, but if we approach each task with a measure of common sense, we should be able to work safely.

This chapter will address some of the most popular ham radio activities: building and erecting antennas, constructing radio equipment, and the testing and troubleshooting of our radios. Safety associated with emergency disaster operations are covered best by the agencies and organizations affected.

Although the RF, ac and dc voltages in most amateur stations pose a potentially grave threat to life and limb, common sense and knowledge of safety practices will help you avoid accidents. Building and operating an Amateur Radio station can be, and is for almost all amateurs, a perfectly safe pastime. Carelessness can lead to severe injury, or even death, however. The ideas presented here are only guidelines; it would be impossible to cover all safety precautions. *Remember: There is no substitute for common sense.*

Fires in well-designed electronic equipment are not common but are known to occur. Proper use of a suitable fire extinguisher can make the difference between a small fire with limited damage and loss of an entire home. Make sure you know the limitations of your extinguisher and the importance of reporting the fire to your local fire department immediately.

Several types of extinguishers are suitable for electrical fires. The multipurpose dry chemical or "ABC" type units are relatively inexpensive and contain a solid powder that is nonconductive. Avoid buying the smallest size; a 5-pound capacity will meet most requirements in the home. ABC extinguishers are also the best choice for kitchen fires (the most common location of home fires). One disadvantage of this type is the residue left behind that might cause corrosion in electrical connectors. Another type of fire extinguisher suitable for energized electrical equipment is the carbon dioxide unit. CO_2 extinguishers require the user to be much closer to the fire, are heavy and difficult to handle, and are relatively expensive. For obvious reasons, water extinguishers are not suitable for fires in or near electronic equipment.

Involve your family in Amateur Radio. Having other people close by is always beneficial in the event that you need immediate assistance. Take the valuable step of showing family members how to turn off the electrical power to your equipment safely. Additionally, cardiopulmonary resuscitation (CPR) training can save lives in the event of electrical shock. Classes are offered in most communities. Take the time to plan with your family members exactly what action should be taken in the event of an emergency, such as electrical shock, equipment fire or power outage.

Antenna and Tower Safety

Many amateurs enjoy building and installing their antennas and consider this one of the most enjoyable aspects of their hobby. Since antennas are generally outdoors, they are affected by such potentially hazardous weather as wind, ice and lightning. Learning about the potential hazards of towers and antennas and how to do antenna work safely will pay dividends.

ARRL TA Paul Krugh, N2NS, reminds us to remember that putting up a tower has a set of responsibilities associated with it.

Any heavy, large and permanent structure that fails or collapses can potentially hurt or even kill somebody. The complete installation *must* comply with all applicable structural and building codes. Professional engineers design towers to withstand code loadings—that is, dead weight, wind and ice loadings that are applicable to the environment at your particular location. The latest revision of the EIA-222 standard is the document from which professional engineers work to ensure that their tower designs are structurally safe.

To ensure structural safety and integrity, you must demonstrate that your tower has been designed by a qualified engineer to withstand EIA-222 loadings at your specific geographic area. Further, the tower, foundation, guys and anchors must be installed (and maintained) according to any drawings, instructions and specifications supplied by the professional engineer. Remember: A properly designed, installed and maintained tower should be as safe as a building or a bridge!

It is not feasible to discuss each type of antenna and tower in detail, so this section will include only highlights. For a full understanding of the specific hardware you will be working with, consult the manufacturer or supplier. You should discuss your antenna plans with a qualified engineer. The ARRL Volunteer Consulting Engineer program can steer you to a knowledgeable engineer.

In addition, your town or city will probably require that you obtain a building permit to erect a tower or antenna. This is their way to help ensure that the installation follows good practices and that the installation is safe. Wise amateurs realize that an independent review of drawings and site inspections are beneficial and can result in fewer problems in the future.

Towers must have a properly engineered support, both for the tower sections themselves as well as guy wire attachments. Sometimes towers are braced to buildings for added support. The Antenna Supports chapter of *The ARRL Antenna Book* covers this subject in greater detail. Towers are available commercially in both guyed and self-supporting styles, and constructed of both steel and aluminum materials. Masts may be wood or metal. One popular and inexpensive mast used to support small antennas is the tubular mast often sold for TV antenna use. These come in telescoping sections, in heights from 20 to 50 ft.

Aluminum extension ladders are sometimes used for temporary antenna supports, such as at Field Day sites. One problem with this approach is the difficulty in holding down the bottom section while "walking up" the ladder. Do *not* try to erect this type of support alone.

Trees are sometimes pressed into service for holding one end of a wire antenna. When using slingshots or arrows to string up the antenna, be sure no one is in range before you launch.

FACTORS TO CONSIDER WHEN SELECTING A TOWER

• Towers have design load limitations. Make very sure the tower you consider has the capacity to safely handle the antenna(s) you intend to install in the kind of environment that is applicable to your QTH.

• The antenna must be located in such a position that *it cannot possibly tangle with power lines, both during normal operation or if the structure should fall.*
• Sufficient yard space must be available to position a guyed tower properly. A rule of thumb is that the guy anchors should be between 60% and 80% of the tower height in distance from the base of the tower.
• Provisions must be made to keep children from climbing the support.
• Always write to the manufacturer of the tower before purchasing and ask for installation specifications, including guying data.
• Soil conditions at the tower site should be investigated. The footings need to be designed around actual soil conditions, particularly on a rocky site.

TOWER TIPS

• Beware of used towers. Have them professionally inspected and contact the manufacturer for installation criteria.
• Always follow manufacturer's instructions, using only parts that are designed for the model you have.
• Never rush into projects. Consult the most experienced amateurs in your community for assistance, especially if you are new to tower installation.
• Check with your local building officials.
• Liability may be increased with a tower installation. Check with your insurer to ensure your coverage is adequate.
• Consider your neighbors and any hazards your antennas may present to them.
• Don't let your installation become an "attractive nuisance." Take steps to install positive barriers so your tower cannot easily be climbed by others, particularly adventurous children.
• Use only the highest quality materials in your system.
• Make sure you have all the tools needed before starting. Some specialized tools (such as a gin pole) may be required.
• Never erect an antenna, tower or rotor during an electrical storm or rainstorm, or when lightning is a possibility.
• The assembly crew as well as those climbing the tower during erection must wear hard hats and use appropriate personal protective equipment including gloves, boots, climbing belt or harness.

Don't forget that lifelines are needed when the belt is unattached from the tower while moving.
• Be careful not to over-stress the tower when it is being assembled. The tower manufacturer can offer suggestions that will avoid jeopardizing the tower.
• Assign someone in the erection crew to monitor the use of safety equipment.
• After the tower is installed, keep the installation safe. Inspection and maintenance recommended by the tower's manufacturer should be carefully followed.
• If making attachments to houses or installations on roofs, have a qualified person determine that the method is adequate and the loading conditions are satisfactory.
• Avoid metal ladders if there are any utility lines in the vicinity. Assume that any line is energized—including cable television and telephone lines.

POWER LINES

Hundreds of people have been killed or seriously injured when attempting to install or dismantle antennas. In virtually all cases, the victim was aware of the hazards, including electrocution, but did not take the necessary steps to eliminate the risks. Never install antennas, towers and masts near power lines. How far away is considered safe? Towers and masts should be installed twice the height of the installation away from power lines. Every electrical wire must be considered dangerous. If the installation should contact power lines, you or those around you could be killed! If you have any questions about power lines, contact your electrical utility, city inspector or a qualified professional.

If, for some reason your tower starts to fall, get away from it immediately. If it touches energized lines it may be a lethal hazard if you are in contact with the antenna. If a coworker becomes energized, do not touch the person. Instead, use an insulated wooden pole to knock the energized conductor away from them. Don't become a victim yourself! If the person is not breathing, immediately start CPR and call for emergency assistance.

Further information about tower safety appears in *The ARRL Antenna Book*.

Electrical Wiring Around the Shack

The standard power available from commercial mains in the United States for residential service is 120/240-V ac. The "primary" voltages that feed transformers in our neighborhoods may range from 2000 to about 10,000 V. Generally, the responsibility for maintaining the power distribution system belongs to a utility company, electric cooperative or city. The "ownership" of conductors usually transfers from the electric utility supplier to the homeowner where the power connects to the meter or weatherhead. If you are unsure where the division of responsibility falls in your community, a call to your electrical utility will provide the answer. **Fig 9.1** shows the typical division of responsibility between the utility company and the homeowner.

There are two facets to success with electrical power: safety and performance. Since we are not professionals, we need to pursue safety first and consult professionals for alternative solutions if performance is unacceptable.

Station Concerns

The primary electrical power supplied to your radio equipment should be controlled by one master switch so that it is easy to kill the power in an emergency. One convenient means is a switched outlet strip, as used for computer equipment. The strip should be listed by a nationally recognized testing laboratory such as Underwriters Lab and incorporate a circuit breaker. See "What Does UL Listing Mean?" and "How Safe are Outlet Strips?" for warnings about poor quality products. It is poor practice to "daisy-chain" several power strips. If you need more outlets than are available on a strip, have additional convenience outlets installed.

Before adding equipment to your home,

be sure that it does not overload the circuit. National and local codes set permissible branch capacities according to a rather complex process. Here's a safe rule of thumb: consider adding a new circuit if the total load is more than 80% of the circuit breaker or fuse rating. (This assumes that the fuse or breaker is correct. If you have any doubts, have an electrician check it.)

Do It Yourself Wiring?

Amateurs sometimes "rewire" parts of their homes to accommodate their hobby. Most local codes *do* allow for modification of wiring (by building owners), so long as the electrical codes are met. Generally, the building owner must obtain an electrical permit before beginning work. Some jobs may require drawings of planned work. Often the permit fee pays for an inspector to review the work. Considering the risk of injury or fire if critical mistakes are left uncorrected, a permit and

inspection are well worth the effort. *Don't take chances*—seek assistance from the building officials or an experienced electrician if you have *any* questions or doubts about proper wiring techniques.

Ordinary 120-V circuits are the most common source of fatal electrical accidents. Never use bare wire for exposed circuits or open-chassis construction with exposed connections! Remember that high-current, low-voltage power sources can be just as dangerous as high-voltage sources.

Never work on electrical wiring with the conductors energized! Switch off the circuit breaker or remove the fuse and take

What Does UL Listing Mean?

CAUTION: Listing *does not* mean what most consumers expect it to mean! More often than not the listing *does not* relate to the performance of the listed product. The listing simply indicates that a sample of the device meets certain manufacturers' construction criteria. Similar devices from the same or different manufacturers may differ significantly in overall construction and performance even though all are investigated and listed against the same UL product category.

Fig 9.2—If the switch box feeding power to your shack is equipped with a lock-out hole, use it. With a lock through the hole on the box, the power cannot be accidentally turned back on. (*Photos courtesy of American ED-CO, at top, and Osborn Mfg Corp, at bottom*)

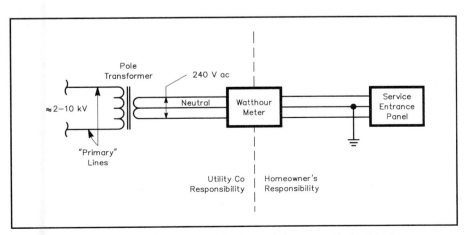

Fig 9.1—Typical division of responsibility for maintenance of electrical power conductors and equipment. The meter is supplied by the utility company.

positive steps to ensure that others do not restore the power while you are working. (**Fig 9.2** illustrates two ways of ensuring that power will be off until you want it turned on.) Check the circuit with an ac voltmeter to be sure that it is "dead" *each time you begin work.* Before restoring power, check your work with an ohm meter: There should be good continuity between the neutral conductor (white wire, "silver" screw) and the grounding conductor (green or bare wire, green screw). An ohmmeter should indicate a closed circuit between the conductors.

There should be no continuity between the hot conductor (black wire, "brass" screw) and the grounding conductor or the neutral conductor. An ohmmeter should indicate an *open* circuit between the hot wire and either of the other two conductors.

A commercially available plug-in tester is the best way to test regular three-wire receptacles.

How Safe are Outlet Strips?

CAUTION: The switch in outlet strips is generally *not* rated for repetitive *load break* duty. Early failure and fire hazard may result from using these devices to switch loads. Misapplications are common (another bit of bad technique that has evolved from the use of personal computers), and manufacturers are all too willing to accommodate the market with marginal products that are "cheap."

Nonindicating and poorly designed surge protection also add to the safety hazard of using power strips. Marginally rated MOVs often fail in a manner that could cause a fire hazard, especially in outlet strips that have nonmetallic enclosures.

A lockable disconnect switch or circuit breaker, as shown in Fig 9.2, is a better and safer station master switch.

NATIONAL ELECTRICAL CODE

Fortunately, much has been learned about how to harness electrical energy safely. This collective experience has been codified into the *National Electrical Code*, or *NEC*. The *Code* details safety requirements for many kinds of electrical installations. Compliance with the *NEC* provides an installation that is *essentially* free from hazard, but not necessarily efficient, convenient or adequate for good service (paraphrased from NEC Article 90-1a and b). For example, the *NEC* requirements discussed here are *not* adequate for lightning protection and high transient voltage events. Look at "Lightning/Transient Protection" for more information. While the *NEC* is national in nature and sees wide application, it is not universal.

Local building authorities set the codes for their area of jurisdiction. They often incorporate the *NEC* in some form, while considering local issues. For example, Washington state specifically exempts telephone, telegraph, radio and television wires and equipment from conformance to electrical codes, rules and regulations. However, some local jurisdictions (city, county and so on) do impose a higher level of installation criteria, including some of the requirements exempted by the state.

Code interpretation is a complex subject, and untrained individuals should steer clear of the *NEC* itself. The *NEC* is not written to be understood by do-it-yourselfers. Therefore, the best sources of information about code compliance and acceptable practices are local building officials, engineers and practicing electricians. With that said, let's look at a few *NEC* requirements for radio installations.

Antenna conductors—Transmitting antennas using hard-drawn copper wire: #14 for unsupported spans less than 150 ft, and #10 for longer spans. Copper-clad steel, bronze or other high-strength conductors must be #14 for spans less than 150 ft and #12 for longer spans. Open-wire transmission line conductors must be at least as large as those specified for antennas.

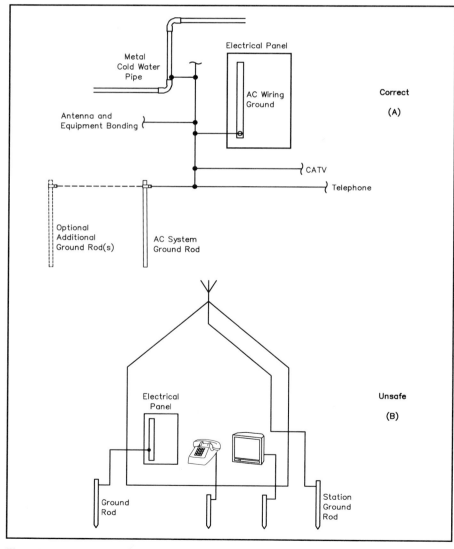

Fig 9.3—At A, proper bonding of all grounds to electrical service panel. Installation shown at B is unsafe—the separate grounds are not bonded. This could result in a serious accident or electrical fire.

Lead-ins—There are several *NEC* requirements for antenna lead-in conductors. For transmitting stations, their size must be equal to or greater than that of the antenna. Lead-ins attached to buildings must be firmly mounted at least 3 inches clear of the surface of the building on non-absorbent insulators. Lead-in conductors must enter through rigid, noncombustible, nonabsorbent insulating tubes or bushings, through an opening provided for the purpose that provides a clearance of at least 2 inches; or through a drilled window pane. All lead-in conductors to transmitting equipment must be arranged so that accidental contact is difficult.

Lightning arrestors—Transmitting stations are required to have a means of draining static charges from the antenna system. An antenna discharge unit (lightning arrestor) must be installed on each lead-in conductor that is not protected by a permanently and effectively grounded metallic shield, unless the antenna itself is permanently and effectively grounded. (The code exception for shielded lead-ins does *not* apply to coax, but to shields such as thinwall conduit. Coaxial braid is neither "adequate" nor "effectively grounded" for lightning protection purposes.) An acceptable alternative to lightning arrestor installation is a switch (capable of withstanding many kilovolts) that connects the lead-in to ground when the transmitter is not in use.

Ground Conductors

Grounding conductors may be made from copper, aluminum, copper-clad steel, bronze or similar erosion-resistant materials. Insulation is not required. *[Lightning and high-voltage transient events may require much larger conductors.—Ed.]* The "protective grounding conductor" (main conductor running to the ground rod) must be as large as the antenna lead-in, but not smaller than #10. The "operating grounding conductor" (to bond equipment chassis together) must be at least #14. There is a "unified" grounding electrode requirement—it is necessary to bond *all* ground rods to the electric service entrance ground. All utilities, antennas and any separate grounding rods used must be bonded together. **Fig 9.3** shows correct (A) and incorrect (B) ways to bond ground rods. **Fig 9.4** demonstrates the importance of correctly bonding ground rods. (Note: The *NEC* requirements do not address effective RF grounds. See the **EMI** chapter of this book for information about RF grounding practices.)

Additionally, the *Code* covers some information on safety inside the station. All conductors inside the building must be at least 4 inches away from conductors of any lighting or signaling circuit except when they are separated from other conductors by conduit or insulator. Transmitters must be enclosed in metal cabinets, and the cabinets must be grounded. All metal handles and controls accessible by the operator must be grounded. Access doors must be fitted with interlocks that will automatically disconnect all voltages above 350 when the door is opened.

Ground-Fault Circuit Interrupters

GFCIs are devices that can be used with common 120-V circuits to reduce the chance of electrocution when the path of current flow leaves the branch circuit (say, through a person's body to another branch or ground). The *NEC* requires GFCI outlets in all wet or potentially wet locations, such as: bathrooms, kitchens, any outdoor outlet with ground-level access, garages and unfinished basements. Any area with bare concrete floors or concrete/masonry walls should be GFCI equipped. GFCIs are available as portable units, duplex outlets and as individual circuit breakers. Some early units may have been sensitive to RF radiation but this problem appears to have been solved. Ham radio shacks in potentially wet areas (basements, out buildings) should be GFCI equipped.

LIGHTNING/TRANSIENT PROTECTION

Nearly everyone recognizes the need to protect themselves from lightning. From miles away, the sight and sound of lightning boldly illustrates its destructive potential. Many people don't realize that destructive transients from lightning and other events can reach electronic equipment from many sources, such as outside antennas, power, telephone and cable TV

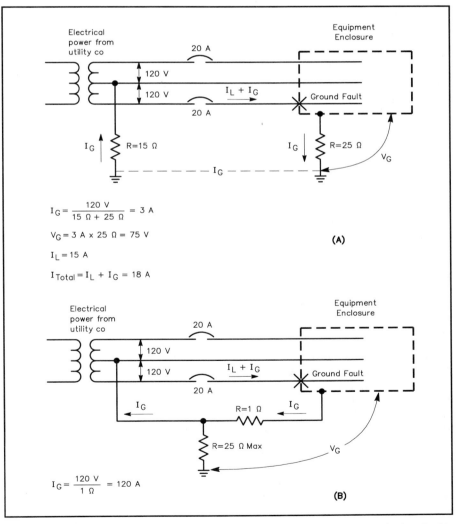

$$I_G = \frac{120 \text{ V}}{15\ \Omega + 25\ \Omega} = 3 \text{ A}$$

$$V_G = 3 \text{ A} \times 25\ \Omega = 75 \text{ V}$$

$$I_L = 15 \text{ A}$$

$$I_{Total} = I_L + I_G = 18 \text{ A}$$

(A)

$$I_G = \frac{120 \text{ V}}{1\ \Omega} = 120 \text{ A}$$

(B)

Fig 9.4—These drawings show the importance of properly bonded ground rods. In the system shown in A, the 20-A breaker will not trip. In the system in B, the 20-A circuit breaker trips instantly. There is an equipment internal short to ground—the ground rod is properly bonded back to the power system ground.

installations. Many hams don't realize that the standard protection scheme of several decades, a ground rod and simple "lightning arrestor" is *not* adequate.

Lightning and transient high-voltage protection follows a familiar communications scenario: identify the unwanted signal, isolate it and dissipate it. The difference here is that the unwanted signal is many megavolts at possibly 200,000 A. What can we do?

Hams *cannot* expect to design or install effective lightning protection systems, but reasonably complete protection from lightning is available in systems designed by lightning protection professionals. Hams *can* easily follow some general guidelines that will protect their stations against high-voltage events that are induced by nearby lightning strikes or that arrive via utility lines. Let's talk about where to find professionals first, then consider construction guidelines.

Professional Help

Start with your local government. Find out what building codes apply in your area and have someone explain the regulations about antenna installation and safety. For more help, look in your telephone yellow pages for professional engineers, lightning protection suppliers and contractors.

Companies that sell lightning-protection products may offer considerable help to apply their products to specific installations. One such source is PolyPhaser Corporation. Look under "Ground References," later in this chapter, for a partial list of PolyPhaser's publications.

Construction Guidelines

Ground rods—Ground rods should be either solid copper, copper-clad steel, hot-dipped galvanized steel or stainless steel. They should be at least 8 ft long by $^1/_2$ inch in diameter ($^5/_8$ inch diameter for iron or steel).

Bonding Conductors—Copper strapping (or *flashing*) comes in a number of sizes; use 1$^1/_2$ inches wide and 0.051 inches thick as a *minimum* for ground connections. Copper strap is a better lightning and RF ground than wire because straps have less inductance than wires. On the other hand, straps are more expensive than wire and more difficult to find.

Use bare copper for buried ground wires. (There are some exceptions; seek an expert's advice if your soil is corrosive.) Exposed runs above ground that are subject to physical damage may require additional protection (a conduit) to meet code requirements. Wire size depends on the application, but never use anything smaller than #6 AWG for bonding conductors. Local lightning-protection experts or building inspectors can recommend sizes for each application.

Tower and Antennas

Because a tower is usually the highest metal object on the property, it is the most likely strike target. Proper tower grounding is essential to lightning protection. The goal is to establish short multiple paths to the Earth so that the strike energy is divided and dissipated.

Connect each tower leg and each fan of metal guy wires to a separate ground rod. Space rods at least 6 ft apart. Bond the leg ground rods together with a #6 AWG or larger copper bonding conductor (form a ring around the tower base, see **Fig 9.5**). Connect a continuous bonding conductor between the tower ring ground and the entrance panel. Make all connections with fittings approved for grounding applications. ***Do not use solder for these connections.*** Solder will be destroyed in the heat of a lightning strike.

Unless the tower is also a shunt-fed antenna, use grounded metal guys. For

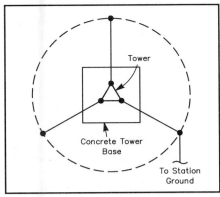

Fig 9.5—Schematic of a properly grounded tower. A bonding conductor connects each tower leg to a ground rod and a buried (1 ft deep) bare, tinned copper ring (dashed line), which is also connected to the station ground and then to the ac safety ground. Locate ground rods on the ring, as close as possible to their respective tower legs. All connectors should be compatible with the tower and conductor materials to prevent corrosion. See text for conductor sizes and details of lightning and voltage transient protection.

crank-up or telescoping towers, connect the sections with strap jumpers. Because galvanized steel (which has a zinc coating) reacts with copper when combined with moisture, use stainless steel hardware between the galvanized metal and the copper grounding materials.

To prevent strike energy from entering a shack via the feed line, ground the feed line *outside* the home. Ground the coax shield *to the tower* at the antenna and the base to keep the tower and line at the same potential. Several companies offer grounding blocks that make this job easy.

All grounding media at the home must be bonded together. This includes lightning-protection conductors, electrical service, telephone, antenna system grounds and underground metal pipes. Any ground rods used for lightning protection or entrance-panel grounding must be spaced at least 6 ft from each other and the electrical service or other utility grounds and then bonded to the ac system ground as required by the *NEC*.

A Radio Entrance Panel

We want to control the flow of the energy in a strike. Eliminate any possible paths for surges to enter the building. This involves routing the feed lines, rotator control cables, and so on at least 6 ft away from other nearby grounded metal objects.

Every conductor that enters the structure should have its own surge suppressor.

Antenna system control lines at the Radio Entrance Panel, other services where they connect to the ac system ground. They are available from a number of manufacturers, including ICE and PolyPhaser.

Both balanced line and coax arrestors should be mounted to a secure ground connection on the *outside* of the building. The easiest way to do this is to install a large metal enclosure as a bulkhead and ground block. This bulkhead serves as the last line of lightning defense, so it's critical that it be installed properly. You can home-brew a bulkhead panel from 1/8-inch copper sheet, bent into a box shape. Position the bulkhead on the building exterior, 4 to 6 inches (minimum) away from nearby combustible materials. Install a separate ground rod for this panel and connect it to the bulkhead with a short, direct connection. Bond this ground rod to the rest of the ground system. Mount all protective devices, switches and relay disconnects on the outside face wall of the bulkhead.

Lightning Arrestors

Feed line lightning arrestors are available for both coax cable and balanced line. Most of the balanced line arrestors use a simple spark gap arrangement, but a balanced line *impulse* suppresser is available from Industrial Communication Engineers, Ltd (ICE, see Address List in **References**).

Coaxial Cable Arrestors—DC blocking arrestors have a fixed frequency range. They present a high-impedance to lightning (less than 1 MHz) while offering a low impedance to RF.

DC continuity arrestors (gas tubes and spark gaps) can be used over a wider frequency range than those that block dc. Where the coax carries supply voltages to remote devices (such as a mast-mounted preamp or remote coax switch), dc-continuous arrestors *must* be used.

GROUNDS

As hams we are concerned with three kinds of ground, which are easily confused because we call each of them "ground." The first is the power line ground, which is required by building codes to ensure the safety of life and property surrounding electrical systems. The *NEC* requires that all grounds be *bonded* together; this is a very important safety feature as well as an *NEC* requirement. Ground systems to prevent shock hazards are generally referred to as the *dc ground* by amateurs, although *safety ground* is a more appropriate term.

The previous section discussed some of the features of a lightning protection grounding system. Additional information on lightning, surge and EMI grounding can be found in *The ARRL Antenna Book*. The *National Electrical Code* requires lightning protection ground rods to be separate from the power line safety grounding electrodes. As discussed later, however, all grounding systems must eventually be bonded together.

An effective safety ground system is necessary for every amateur station. It provides a common reference potential for all parts of the ac system and reduces the possibility of electrical shock by ensuring that all exposed conductors remain at that (low) potential. Three-wire electrical systems effectively ground our equipment for dc and low frequencies. Unfortunately, an effective ground conductor at 60 Hz (5,000,000 m wavelength) may be an excellent antenna for a 20 m signal.

When stray RF causes interference or other problems, we need another kind of ground—a low-impedance path for RF to reach the earth or some other "ground" that dissipates, rather than radiates, the RF energy. Let's call this an *RF ground*.

In most stations, dc ground and RF ground are provided by the same system. If you install ground rods, however, bond them to each other and to the safety ground at the electrical service entrance. In older houses, water lines are sometimes used for the service entrance panel ground. It is a good idea to check that the pipes are electrically continuous from the panel to earth. (Consider that Teflon tape is often used to seal pipe joints in modern repairs.)

For decades, amateurs have been advised to bond all equipment cabinets to an RF ground located near the station. That's a good idea, but it's not easily achieved. "Near" in this use is 10 ft or less for HF operation, even less for higher frequencies. At some stations, it is very difficult to produce an effective RF ground. When levels of unwanted RF are low, an RF ground may not be needed. (See the **EMI** chapter for more about RF grounds and interference.) Some think that RF grounds should be isolated from the safety ground system—*that is not true! All grounds, including safety, RF, lightning protection and commercial communications, must be bonded together in order to protect life and property.*

The first step in building an RF ground system is to bond together the chassis of all equipment in your station. Choose conductors large enough to provide a low-impedance path. The *NEC* requires that grounding conductors be as large as the largest conductor in the primary power circuit (#14 for a 15-A circuit, #12 for 20 A). Copper strap, sold as "flashing copper," is excellent for this application. Coax braid is a popular choice; but it is not a

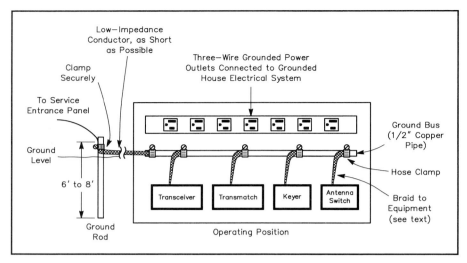

Fig 9.6—An effective station ground bonds the chassis of all equipment together with low-impedance conductors and ties into a good earth ground. Note that the ground bus is in turn bonded to the service entrance panel. This connection should be made by a licensed electrician with #6 AWG (minimum size) copper wire.

good ground conductor unless tinned, and then it's no longer very flexible. Rather use commercially made copper braid ground strap that is tinned and ampacity rated—wider straps make better RF grounds. Avoid solid conductors; they tend to break.

Grounding straps can be run from equipment chassis to equipment chassis, but a more convenient approach is illustrated in **Fig 9.6.** In this installation, a ¹/₂-inch-diameter copper water pipe runs the entire length of the operating bench. A wide copper ground braid runs from each piece of equipment to a stainless-steel clamp on the pipe.

After the equipment is bonded to a common ground bus, the ground bus must be wired to a good earth ground. This run should be made with a heavy conductor (copper braid is a good choice again) and should be as short and direct as possible. The earth ground usually takes one of two forms.

In most cases, the best approach is to drive one or more ground rods into the earth at the point where the conductor from the station ground bus leaves the house. The best ground rods to use are those available from an electrical supply house. These rods are generally 8 ft long and made from steel with a heavy copper plating. Do not depend on shorter, thinly plated rods sold by some home electronics suppliers, as they can quickly rust and soon become worthless.

Once the ground rod is installed, clamp the conductor from the station ground bus to it with a clamp that can be tightened securely and will not rust. Copper-plated

clamps made specially for this purpose (and matching the rods) are available from electrical supply houses. Multiple ground rods reduce the electrical resistance and improve the effectiveness of the ground system.

Building cold water supply systems were used as station grounds in years past. Connection was made via a low-impedance conductor from the station ground bus to a convenient cold water pipe, preferably somewhere near the point where the main water supply enters the house. (Hot water lines are unsuitable for grounding conductors.) Increased used of plastic plumbing both inside and outside houses is reducing the availability of this option. If you do use the cold water line, ensure that it has a good electrical connection to the earth and attach it *outside* the structure to reduce EMI. As with ground rods, ensure that the water line is also bonded to the service entrance panel.

For some installations, especially those located above the first floor, a conventional ground system such as that just described will make a fine dc ground but will not provide the necessary low-impedance path to ground for RF. The length of the conductor between the ground bus and the ultimate ground point becomes a problem. For example, the ground wire may be about ¹/₄ wavelength (or an odd multiple of ¹/₄ wavelength) long on some amateur band. A ¹/₄-wavelength wire acts as an impedance inverter from one end to the other. Since the grounded end is at a very low impedance, the equipment end will be at a high impedance. The likely result is RF hot spots around the station

while the transmitter is in operation. In this case, this ground system may be worse (from an RF viewpoint) than no ground at all.

Ground References

Contact information appears in the **References** chapter Address List.

Federal Information Processing Standards (FIPS) publication 94: *Guideline on Electrical Power for ADP Installations.* FIPS are available from the National Technical Information Service.

IAEI: *Soares' Book on Grounding,* available from International Association of Electrical Inspectors (IAEI).

IEEE Std 1100: *Powering and Grounding Sensitive Electronics Equipment.*

Polyphaser: *The Grounds for Lightning and EMP Protection.* PolyPhaser's quarterly newsletter, *Striking News,* contains articles on Amateur Radio station lightning protection in the February and May 1994 issues. Complimentary copies of these issues are available from PolyPhaser.

STATION POWER

Amateur Radio stations generally require a 120-V ac power source. (In residential systems voltages from 110 V through 125 V are considered equivalent, as are those from 220 V through 250 V.) 120-V ac is converted to the proper ac or dc levels required for the station equipment. Power supplies should accommodate the measured voltage range at each station. (The measured voltage usually varies by hour, day, season and location.) Power supply theory is covered in the **Power Supplies** chapter. If your station is located in a room with electrical outlets, you're in luck. If your station is located in the basement, an attic or other area without a convenient 120-V source, you may need to have a new line run to your operating position.

Stations with high-power amplifiers should have a 240-V ac power source in addition to the 120-V supply. Some amplifiers may be powered from 120-V, but they require current levels that may exceed the limits of standard house wiring. For safety, and for the best possible voltage regulation in the equipment, it is advisable to install a separate 240 or 120-V line with an appropriate current rating if you use an amplifier.

The usual line running to baseboard outlets is rated at 15 A, although 20-A outlets may be installed in newer houses. This may or may not be enough current to power your station. To determine how much current your station requires, check the ratings for each piece of gear. Usually, the manufacturer will specify the required current at 120 V; if the power consumption is rated in watts, divide that rating by 120 V to get amperes. If the total current required for your station is near 12 ($0.8 \times 15 = 12$ A), you need to install another circuit. Keep in mind that other rooms may be powered from the same branch of the electrical system, so the power consumption of any equipment connected to other outlets on the branch must be taken into account. Whenever possible, power your station from a separate, heavy-duty line run directly to the distribution panel through a disconnect switch or circuit breaker that can be locked in the off position.

If you decide to install a separate heavy-duty 120-V line or a 240-V line, consult the power company for local requirements. In some areas, this work must be performed by a licensed electrician. Others may require a special building permit. Even if you are allowed to do the work yourself, it might need inspection by a licensed electrician. Go through the system and get the necessary permits and inspections! Faulty wiring can destroy your possessions and take away your loved ones. Many fire insurance policies are void if there is unapproved wiring in the structure.

If you decide to do the job yourself, work closely with local building officials. Most home-improvement centers sell books to guide do-it-yourself wiring projects. If you have any doubts about doing the work yourself, get a licensed electrician to do the installation.

Three-Wire 120-V Power Cords

Most metal-cased electrical tools and appliances are equipped with three-conductor power cords. Two of the conductors carry power to the device, while the third conductor is connected to the case or frame.

When both plug and receptacle are properly wired, the three-contact polarized plug connects the equipment to the system ground. This grounds the chassis or frame of the appliance and prevents the possibility of electrical shock to the user. Most commercially manufactured test equipment and ac-operated amateur equipment are supplied with these three-wire cords. Unfortunately, the ground wire is sometimes improperly installed. Before connecting any new equipment, check for continuity from case to ground pin with an ohmmeter. If there is no continuity, have the equipment repaired before use. Use such equipment only with properly installed three-wire outlets. If your house does not have such outlets, consult with an electrician or local building officials to learn about safe alternatives.

Equipment with plastic cases is often "double insulated" and fed with a two-wire cord. Such equipment is safe because both conductors are completely insulated from the user. Nonetheless, there is still a hazard if, say, a double insulated drill were used to drill an improperly grounded case of a transmitter that was still plugged in. Remember, all insulation is prey to age, damage and wear that may erode its safety value.

Safe Homebrewing

Since Amateur Radio began, building equipment in home workshops has been a major part of an amateur's activity. In fact, in the early days, building equipment with your hands was the *only* option available. While times and interests change, home construction of radio equipment and related accessories remains very popular and enjoyable. Building your own gear need not be hazardous if you become familiar with the hazards, learn how to perform the necessary functions and follow some basic safe practices including the ones listed below.

Selecting tools and equipment. Selection of quality tools appropriate for the job is a good place to start. Sometimes a low-priced tool may appear too good to pass by, but give some thought to its longevity, the availability of replacement parts, whether the store will support you if needed and whether the design is one that will offer you the protection you need. Power-operated mechanical tools such as saws and drills should be listed by a nationally recognized testing laboratory.

Read instructions carefully...and follow them. The manufacturers of tools are the most knowledgeable about how to use their products safely. Tap their knowledge by carefully reading all operating instructions and warnings. Avoiding injuries with power tools requires safe tool design as well as proper operation by the user. Keep the instructions in a place where you can refer to them in the future.

Keep your tools in good condition. Always take care of your investment. Store tools in a way to prevent damage or use by untrained persons (young children, for example). Keep the cutting edges of saws, chisels and drill bits sharp. Protect metal surfaces from corrosion. Frequently inspect the cords and plugs of electrical equipment and make any necessary repairs. If you find that your power cord is becoming frayed, do not delay its repair. Often the best solution is to buy a replacement cord with a molded connector already attached.

Protect yourself. Use of drills, saws, grinders and other wood- or metal-working equipment can release small fragments that could cause serious eye damage. Always wear safety glasses or goggles when doing work that might present a flying object hazard. If you use hammers, wire-cutters, chisels and other hand tools, you will also need the protection that safety eyewear offers. Dress appropriately—loose clothing (or even hair) can be caught in exposed rotating equipment such as drill presses.

Take your time. If you hurry, not only will you make more mistakes and possibly spoil the appearance of your new equipment, you won't have time to think things through. Always plan ahead. Do not work with shop tools if you can't concentrate on what you are doing.

Know what to do in an emergency. Despite your best efforts to be careful, accidents may still occur from time to time. Ensure that everyone in your household knows basic first aid procedures and understands how to summon help in an emergency. They should also know where to find and how to safely shut down electrical power in your shack and shop. Keep your shop neat and orderly, with everything in its place. Do not store an excessive amount of flammable materials. Keep clutter off the floor so no one will trip or lose their footing. Exemplary housekeeping is contagious—set a good example for everyone!

Soldering. Soldering requires a certain degree of practice and, of course, the right tools. What potential hazards are involved?

- Since the solder used for virtually all electronic components is a lead-tin alloy, the first thing in most people's mind is lead, a well-known health hazard. There are two primary ways lead might enter our bodies when soldering: we could breathe lead fumes into our lungs or we could ingest (swallow) lead or lead-contaminated food. Inhalation of lead fumes is extremely unlikely because the temperatures ordinarily used in electronic soldering are far below those needed to vaporize lead. But since lead is soft and we may tend to handle it with our fingers, contaminating our food is a real possibility. For this reason, wash your hands carefully after any soldering (or touching of solder connections).

- Generally, solder used for electronic components contains a flux, often a rosin material. When heated the flux flows freely and emits a vapor in the form of a light gray smoke-like plume. This flux vapor, which often contains aldehydes, is a strong irritant and can cause potentially serious problems to persons who may have respiratory sensitivity conditions including those who suffer from asthma. In most cases it is relatively easy to use a small fan to move the flux vapor away from your eyes and face. Open a window, if there is one, to provide additional air exchange. In extreme cases use an organic vapor cartridge respirator.

- Although it is fairly obvious, be careful when soldering not to burn yourself. A soldering iron stand is helpful.

- Solvents are often used to remove excess flux after the parts have cooled to room temperature. Minimize skin contact with solvents by wearing molded gloves that are resistant to the solvent.

RF Burns!

There's a lot of talk about hazards of RF radiation, but most people don't think about RF burns. Happily, most ham shacks offer little exposure to RF current. Transmitters are enclosed, coaxial cable is the most common feed line, and antennas are located well out of reach.

Some people have experienced a mild tingling on their lips while operating with a metal microphone—a gentle reminder of "RF in the shack." When first licensed in 1963, I learned a stronger lesson. Lightbulbs were often used as dummy loads then: they give a nice visual indication of output power, but provide a poor load for the transmitter (not 50 Ω). Also, you can work a lot of people on such a "dummy" antenna. (Don't try this with a modern solid-state transmitter; the mismatch could be fatal to the radio!)

While tuning my Viking Adventurer one day, I bumped the lit bulb and it fell off the table. I prevented a broken lightbulb by catching it—with my finger across the cable ends that were soldered to the bulb. 50-W of RF went through my finger tip and cauterized a path about 3/16× 1/8 inch. It was an extremely painful burn; I would rather have broken the bulb. To avoid RF burns, insulate or enclose any exposed RF conductors and keep your antennas out of reach. Ground mounted vertical antennas that carry more than a few watts should be enclosed by an insulator such as a PVC pipe slipped over the radiator or an 8-ft-high fence around the antenna base.
—*Bob Schetgen, KU7G, ARRL Handbook Editor*

RF Radiation and Electromagnetic Field Safety

Amateur Radio is basically a safe activity. In recent years, however, there has been considerable discussion and concern about the possible hazards of electromagnetic radiation (EMR), including both RF energy and power-frequency (50-60 Hz) electromagnetic (EM) fields. FCC regulations set limits on the maximum permissible exposure (MPE) allowed from the operation of radio transmitters. These regulations do not take the place of RF-safety practices, however. This section deals with the topic of RF safety.

This section was prepared by members of the ARRL RF Safety Committee and coordinated by Dr. Robert E. Gold, WBØKIZ. It summarizes what is now known and offers safety precautions based on the research to date.

All life on Earth has adapted to survive in an environment of weak, natural, low-frequency electromagnetic fields (in addition to the Earth's static geomagnetic field). Natural low-frequency EM fields come from two main sources: the sun, and thunderstorm activity. But in the last 100 years, man-made fields at much higher intensities and with a very different spectral distribution have altered this natural EM background in ways that are not yet fully understood. Researchers continue to look at the effects of RF exposure over a wide range of frequencies and levels.

Both RF and 60-Hz fields are classified as *nonionizing radiation,* because the frequency is too low for there to be enough photon energy to ionize atoms. (*Ionizing radiation,* such as X-rays, gamma rays and even some ultraviolet radiation has enough energy to knock electrons loose from their atoms. When this happens, positive and negative ions are formed.) Still, at sufficiently high power densities, EMR poses certain health hazards. It has been known since the early days of radio that RF energy can cause injuries by heating body tissue. (Anyone who has ever touched an improperly grounded radio chassis or energized antenna and received an *RF burn* will agree that this type of injury can be quite painful.) In extreme cases, RF-induced heating in the eye can result in cataract formation, and can even cause blindness. Excessive RF heating of the reproductive organs can cause sterility. Other health problems also can result from RF heating. These heat-related health hazards are called *thermal effects.* A microwave oven is a positive application of this thermal effect.

There also have been observations of changes in physiological function in the presence of RF energy levels that are too low to cause heating. These functions return to normal when the field is removed. Although research is ongoing, no harmful health consequences have been linked to these changes.

In addition to the ongoing research, much else has been done to address this issue. For example, FCC regulations set limits on exposure from radio transmitters. The Institute of Electrical and Electronics Engineers, the American National Standards Institute and the National Council for Radiation Protection and Measurement, among others, have recommended voluntary guidelines to limit human exposure to RF energy. The ARRL has established the RF Safety Committee, consisting of concerned medical doctors and scientists, serving voluntarily to monitor scientific research in the fields and to recommend safe practices for radio amateurs.

THERMAL EFFECTS OF RF ENERGY

Body tissues that are subjected to *very high* levels of RF energy may suffer serious heat damage. These effects depend on the frequency of the energy, the power density of the RF field that strikes the body and factors such as the polarization of the wave.

At frequencies near the body's natural resonant frequency, RF energy is absorbed more efficiently, and an increase in heating occurs. In adults, this frequency usually is about 35 MHz if the person is grounded, and about 70 MHz if insulated from the ground. Individual body parts may be resonant at different frequencies. The adult head, for example, is resonant around 400 MHz, while a baby's smaller head resonates near 700 MHz. Body size thus determines the frequency at which most RF energy is absorbed. As the frequency is moved farther from resonance, less RF heating generally occurs. *Specific absorption rate (SAR)* is a term that describes the rate at which RF energy is absorbed in tissue.

Maximum permissible exposure (MPE) limits are based on whole-body SAR values, with additional safety factors included as part of the standards and regulations. This helps explain why these safe exposure limits vary with frequency. The MPE limits define the maximum electric and magnetic field strengths or the plane-wave equivalent power densities associated with these fields, that a person may be

exposed to without harmful effect—and with an acceptable safety factor. The regulations assume that a person exposed to a specified (safe) MPE level also will experience a safe SAR.

Nevertheless, thermal effects of RF energy should not be a major concern for most radio amateurs, because of the power levels we normally use and the intermittent nature of most amateur transmissions. Amateurs spend more time listening than transmitting, and many amateur transmissions such as CW and SSB use low-duty-cycle modes. (With FM or RTTY, though, the RF is present continuously at its maximum level during each transmission.) In any event, it is rare for radio amateurs to be subjected to RF fields strong enough to produce thermal effects, unless they are close to an energized antenna or un- shielded power amplifier. Specific suggestions for avoiding excessive exposure are offered later in this chapter.

ATHERMAL EFFECTS OF EMR

Research about possible health effects resulting from exposure to the lower level energy fields, the athermal effects, has been of two basic types: epidemiological research and laboratory research.

Scientists conduct laboratory research into biological mechanisms by which EMR may affect animals including humans. Epidemiologists look at the health patterns of large groups of people using statistical methods. These epidemiological studies have been inconclusive. By their basic design, these studies do not demonstrate cause and effect, nor do they postulate mechanisms of disease. Instead, epidemiologists look for associations between an environmental factor and an observed pattern of illness. For example, in the earliest research on malaria, epidemiologists observed the association between populations with high prevalence of the disease and the proximity of mosquito infested swamplands. It was left to the biological and medical scientists to isolate the organism causing malaria in the blood of those with the disease, and identify the same organisms in the mosquito population.

In the case of athermal effects, some studies have identified a weak association between exposure to EMF at home or at work and various malignant conditions including leukemia and brain cancer. A larger number of equally well designed and performed studies, however, have found no association. A risk ratio of between 1.5 and 2.0 has been observed in positive studies (the number of observed cases of malignancy being 1.5 to 2.0 times the "expected" number in the population). Epidemiologists generally regard a risk ratio of 4.0 or greater to be indicative of a strong association between the cause and effect under study. For example, men who smoke one pack of cigarettes per day increase their risk for lung cancer tenfold compared to nonsmokers, and two packs per day increases the risk to more than 25 times the nonsmokers' risk.

Epidemiological research by itself is rarely conclusive, however. Epidemiology only identifies health patterns in groups—it does not ordinarily determine their cause. And there are often confounding factors: Most of us are exposed to many different environmental hazards that may affect our health in various ways. Moreover, not all studies of persons likely to be exposed to high levels of EMR have yielded the same results.

There also has been considerable laboratory research about the biological effects of EMR in recent years. For example, some separate studies have indicated that even fairly low levels of EMR might alter the human body's circadian rhythms, affect the manner in which T lymphocytes function in the immune system and alter the nature of the electrical and chemical signals communicated through the cell membrane and between cells, among other things. Although these studies are intriguing, they do not demonstrate any effect of these low-level fields on the overall organism.

Much of this research has focused on low-frequency magnetic fields, or on RF fields that are keyed, pulsed or modulated at a low audio frequency (often below 100 Hz). Several studies suggested that humans and animals can adapt to the presence of a steady RF carrier more readily than to an intermittent, keyed or modulated energy source.

The results of studies in this area, plus speculations concerning the effect of various types of modulation, were and have remained somewhat controversial. None of the research to date has demonstrated that low-level EMR causes adverse health effects.

Given the fact that there is a great deal of ongoing research to examine the health consequences of exposure to EMF, the American Physical Society (a national group of highly respected scientists) issued a statement in May 1995 based on its review of available data pertaining to the possible connections of cancer to 60-Hz EMF exposure. This report is exhaustive and should be reviewed by anyone with a serious interest in the field. Among its general conclusions were the following:

1. The scientific literature and the reports of reviews by other panels show no consistent, significant link between cancer and power line fields.

2. No plausible biophysical mechanisms for the systematic initiation or promotion of cancer by these extremely weak 60-Hz fields has been identified.

3. While it is impossible to prove that no deleterious health effects occur from exposure to any environmental factor, it is necessary to demonstrate a consistent, significant, and causal relationship before one can conclude that such effects do occur.

In a report dated October 31, 1996, a committee of the National Research Council of the National Academy of Sciences has concluded that no clear, convincing evidence exists to show that residential exposures to electric and magnetic fields (EMFs) are a threat to human health.

A National Cancer Institute epidemiological study of residential exposure to magnetic fields and acute lymphoblastic leukemia in children was published in the *New England Journal of Medicine* in July 1997. The exhaustive, seven-year study concludes that if there is any link at all, it is far too weak to be concerned about.

Readers may want to follow this topic as further studies are reported. Amateurs should be aware that exposure to RF and ELF (60 Hz) electromagnetic fields at all power levels and frequencies has not been fully studied under all circumstances. "Prudent avoidance" of any avoidable EMR is always a good idea. Prudent avoidance doesn't mean that amateurs should be fearful of using their equipment. Most amateur operations are well within the MPE limits. If any risk does exist, it will almost surely fall well down on the list of causes that may be harmful to your health (on the other end of the list from your automobile). It does mean, however, that hams should be aware of the potential for exposure from their stations, and take whatever reasonable steps they can take to minimize their own exposure and the exposure of those around them.

Safe Exposure Levels

How much EM energy is safe? Scientists and regulators have devoted a great deal of effort to deciding upon safe RF-exposure limits. This is a very complex problem, involving difficult public health and economic considerations. The recommended safe levels have been revised downward several times over the years — and not all scientific bodies agree on this question even today. An Institute of Electrical and Electronics Engineers (IEEE) standard for recommended EM exposure limits was published in 1991 (see Bibliography). It replaced a 1982 American National Standards Institute (ANSI) standard. In the new standard, most of the permitted exposure levels were revised downward (made more stringent), to better reflect the current research. The new IEEE standard was adopted by ANSI in 1992.

The IEEE standard recommends frequency-dependent and time-dependent maximum permissible exposure levels. Unlike earlier versions of the standard, the 1991 standard recommends different RF exposure limits in *controlled environments* (that is, where energy levels can be accurately determined and everyone on the premises is aware of the presence of EM fields) and in *uncontrolled environments* (where energy levels are not known or where people may not be aware of the presence of EM fields). FCC regulations also include controlled/occupational and uncontrolled/general population exposure environments.

The graph in **Fig 9.8** depicts the 1991 IEEE standard. It is necessarily a complex graph, because the standards differ not only for controlled and uncontrolled environments but also for electric (E) fields and magnetic (H) fields. Basically, the lowest E-field exposure limits occur at frequencies between 30 and 300 MHz. The lowest H-field exposure levels occur at 100-300 MHz. The ANSI standard sets the maximum E-field limits between 30 and 300 MHz at a power density of 1 mW/cm^2 (61.4 V/m) in controlled environments— but at one-fifth that level (0.2 mW/cm^2 or 27.5 V/m) in uncontrolled environments. The H-field limit drops to 1 mW/cm^2 (0.163 A/m) at 100-300 MHz in controlled environments and 0.2 mW/cm^2 (0.0728 A/m) in uncontrolled environments. Higher power densities are permitted at frequencies below 30 MHz (below 100 MHz for H fields) and above 300 MHz, based on the concept that the body will not be resonant at those frequencies and will therefore absorb less energy.

In general, the 1991 IEEE standard requires averaging the power level over time periods ranging from 6 to 30 minutes for power-density calculations, depending on the frequency and other variables. The ANSI exposure limits for uncontrolled environments are lower than those for controlled environments, but to compensate for that the standard allows exposure levels in those environments to be averaged over much longer time periods (generally 30 minutes). This long averaging

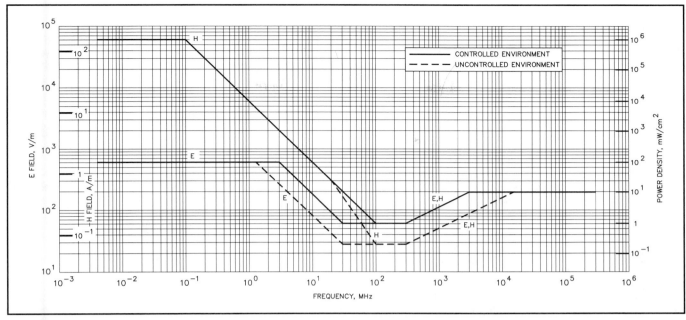

Fig 9.8—1991 RF protection guidelines for body exposure of humans. It is known officially as the "IEEE Standard for Safety Levels with Respect to Human Exposure to Radio Frequency Electromagnetic Fields, 3 kHz to 300 GHz."

time means that an intermittently operating RF source (such as an Amateur Radio transmitter) will show a much lower power density than a continuous-duty station—for a given power level and antenna configuration.

Time averaging is based on the concept that the human body can withstand a greater rate of body heating (and thus, a higher level of RF energy) for a short time than for a longer period. Time averaging may not be appropriate, however, when considering nonthermal effects of RF energy.

The IEEE standard excludes any transmitter with an output below 7 W because such low-power transmitters would not be able to produce significant whole-body heating. (Recent studies show that handheld transceivers often produce power densities in excess of the IEEE standard within the head.)

There is disagreement within the scientific community about these RF exposure guidelines. The IEEE standard is still intended primarily to deal with thermal effects, not exposure to energy at lower levels. A small but significant number of researchers now believe athermal effects also should be taken into consideration. Several European countries and localities in the United States have adopted stricter standards than the recently updated IEEE standard.

Another national body in the United States, the National Council for Radiation Protection and Measurement (NCRP), also has adopted recommended exposure

guidelines. NCRP urges a limit of 0.2 mW/cm^2 for nonoccupational exposure in the 30-300 MHz range. The NCRP guideline differs from IEEE in two notable ways: It takes into account the effects of modulation on an RF carrier, and it does not exempt transmitters with outputs below 7 W.

The FCC MPE regulations are based on parts of the 1992 IEEE/ANSI standard and recommendations of the National Council for Radiation Protection and Measurement (NCRP). The MPE limits under the regulations are slightly different than the IEEE/ANSI limits. Note that the MPE levels apply to the FCC rules put into effect for radio amateurs on January 1, 1998.

These MPE requirements do not reflect and include all the assumptions and exclusions of the IEEE/ANSI standard.

Cardiac Pacemakers and RF Safety

It is a widely held belief that cardiac pacemakers may be adversely affected in their function by exposure to electromagnetic fields. Amateurs with pacemakers may ask whether their operating might endanger themselves or visitors to their shacks who have a pacemaker. Because of this, and similar concerns regarding other sources of electromagnetic fields, pacemaker manufacturers apply design meth-

---

Table 9.1

Typical 60-Hz Magnetic Fields Near Amateur Radio Equipment and AC-Powered Household Appliances

Values are in milligauss.

Item	Field	Distance
Electric blanket	30-90	Surface
Microwave oven	10-100	Surface
	1-10	12"
IBM personal	5-10	Atop monitor
computer	0-1	15" from screen
Electric drill	500-2000	At handle
Hair dryer	200-2000	At handle
HF transceiver	10-100	Atop cabinet
	1-5	15" from front
1-kW RF amplifier	80-1000	Atop cabinet
	1-25	15" from front

(Source: measurements made by members of the ARRL RF Safety Committee)

FCC RF-Exposure Regulations

FCC regulations control the amount of RF exposure that can result from your station's operation (§§97.13, 97.503, 1.1307 (b)(c)(d), 1.1310 and 2.1093). The regulations set limits on the maximum permissible exposure (MPE) allowed from operation of transmitters in all radio services. They also require that certain types of stations be evaluated to determine if they are in compliance with the MPEs specified in the rules. The FCC has also required that five questions on RF environmental safety practices be added to Novice, Technician and General license examinations.

These rules went into effect on January 1, 1998 for new stations or stations that file a Form 605 application with the FCC. Other existing stations have until September 1, 2000 to be in compliance with the rules.

The Rules

Maximum Permissible Exposure (MPE)

All radio stations regulated by the FCC must comply with the requirements for MPEs, even QRP stations running only a few watts or less. The MPEs vary with frequency, as shown in **Table A**. MPE limits are specified in maximum electric and magnetic fields for frequencies below 30 MHz, in power density for frequencies above 300 MHz and all three ways for

frequencies from 30 to 300 MHz. For compliance purposes, all of these limits must be considered *separately*. If any one is exceeded, the station is not in compliance.

The regulations control human exposure to RF fields, not the strength of RF fields. There is no limit to how strong a field can be as long as no one is being exposed to it, although FCC regulations require that amateurs use the minimum necessary power at all times (§97.311 [a]).

Environments

The FCC has defined two exposure environments — *controlled* and *uncontrolled*. A controlled environment is one in which the people who are being exposed are aware of that exposure and can take steps to minimize that exposure, if appropriate. In an uncontrolled environment, the people being exposed are not normally aware of the exposure. The uncontrolled environment limits are more stringent than the controlled environment limits.

Although the controlled environment is usually intended as an occupational environment, the FCC has determined that it generally applies to amateur operators

Table A—(From §1.1310) Limits for Maximum Permissible Exposure (MPE)

(A) Limits for Occupational/Controlled Exposure

Frequency Range (MHz)	Electric Field Strength (V/m)	Magnetic Field Strength (A/m)	Power Density (mW/cm^2)	Averaging Time (minutes)
0.3-3.0	614	1.63	(100)*	6
3.0-30	1842/f	4.89/f	(900/f^2)*	6
30-300	61.4	0.163	1.0	6
300-1500	—	—	f/300	6
1500-100,000	—	—	5	6

f = frequency in MHz
* = Plane-wave equivalent power density (see Note 1).

(B) Limits for General Population/Uncontrolled Exposure

Frequency Range (MHz)	Electric Field Strength (V/m)	Magnetic Field Strength (A/m)	Power Density (mW/cm^2)	Averaging Time (minutes)
0.3-1.34	614	1.63	(100)*	30
1.34-30	824/f	2.19/f	(180/f^2)*	30
30-300	27.5	0.073	0.2	30
300-1500	—	—	f/1500	30
1500-100,000	—	—	1.0	30

f = frequency in MHz
* = Plane-wave equivalent power density (see Note 1).

Note 1: This means the equivalent far-field strength that would have the E or H-field component calculated or measured. It does not apply well in the near field of an antenna. The equivalent far-field power density can be found in the near or far field regions from the relationships: $P_d = |E_{total}|^2 / 3770$ mW/cm^2 or from $P_d = |H_{total}|^2 \times 37.7$ mW/cm^2.

ods that for the most part shield the pacemaker circuitry from even relatively high EM field strengths.

It is recommended that any amateur who has a pacemaker, or is being considered for one, discuss this matter with his or her physician. The physician will probably put the amateur into contact with the technical representative of the pacemaker manufacturer. These representatives are generally excellent resources, and may have data from laboratory or "in the field" studies with specific model pacemakers.

One study examined the function of a modern (dual chamber) pacemaker in and around an Amateur Radio station. The pacemaker generator has circuits that receive and process electrical signals produced by the heart, and also generate electrical signals that stimulate (pace) the heart. In one series of experiments, the pacemaker was connected to a heart simulator. The system was placed on top of the cabinet of a 1-kW HF linear amplifier during SSB and CW operation. In another test, the system was placed in close prox-

imity to several 1 to 5-W 2-meter handheld transceivers. The test pacemaker was connected to the heart simulator in a third test, and then placed on the ground 9 meters below and 5 meters in front of a three-element Yagi HF antenna. No interference with pacemaker function was observed in these experiments.

Although the possibility of interference cannot be entirely ruled out by these few observations, these tests represent more severe exposure to EM fields than would ordinarily be encountered by an amateur

Table B—Power Thresholds for Routine Evaluation of Amateur Radio Stations

Wavelength Band	Evaluation Required if Power* (watts) Exceeds:
MF	
160 m	500
HF	
80 m	500
75 m	500
40 m	500
30 m	425
20 m	225
17 m	125
15 m	100
12 m	75
10 m	50
VHF (all bands)	50
UHF	
70 cm	70
33 cm	150
23 cm	200
13 cm	250
SHF (all bands)	250
EHF (all bands)	250
Repeater stations (all bands)	*non-building-mounted antennas*: height above ground level to lowest point of antenna < 10 m *and* power > 500 W ERP *building-mounted antennas*: power > 500 W ERP

*Transmitter power = Peak-envelope power input to antenna. For repeater stations *only,* power exclusion based on ERP (effective radiated power).

and members of their immediate households. In most cases, controlled-environment limits can be applied to your home and property to which you can control physical access. The uncontrolled environment is intended for areas that are accessible by the general public, such as your neighbors' properties.

The MPE levels are based on average exposure. An averaging time of 6 minutes is used for controlled exposure; an averaging period of 30 minutes is used for uncontrolled exposure.

Station Evaluations

The FCC requires that certain amateur stations be evaluated for compliance with the MPEs. Although an amateur can have someone else do the evaluation, it is not difficult for hams to evaluate their own stations. The ARRL book *RF Exposure and You* contains extensive

information about the regulations and a large chapter of tables that show compliance distances for specific antennas and power levels. Generally, hams will use these tables to evaluate their stations. Some of these tables have been included in the FCC's information — *OET Bulletin 65* and its *Supplement B*. If hams choose, however, they can do more extensive calculations, use a computer to model their antenna and exposure, or make actual measurements.

Categorical Exemptions

Some types of amateur stations do not need to be evaluated, but these stations must still comply with the MPE limits. The station licensee remains responsible for ensuring that the station meets these requirements.

The FCC has exempted these stations from the evaluation requirement because their output power, operating mode and frequency are such that they are presumed to be in compliance with the rules.

Stations using power equal to or less than the levels in **Table B** do not have to be evaluated. For the 100-W HF ham station, for example, an evaluation would be required *only* on 12 and 10 meters.

Hand-held radios and vehicle-mounted mobile radios that operate using a push-to-talk (PTT) button are also categorically exempt from performing the routine evaluation. Repeater stations that use less than 500 W ERP or those with antennas not mounted on buildings, if the antenna is at least 10 meters off the ground, also do not need to be evaluated.

Correcting Problems

Most hams are already in compliance with the MPE requirements. Some amateurs, especially those using indoor antennas or high-power, high-duty-cycle modes such as a RTTY bulletin station and specialized stations for moonbounce operations and the like may need to make adjustments to their station or operation to be in compliance.

The FCC permits amateurs considerable flexibility in complying with these regulations. As an example, hams can adjust their operating frequency, mode or power to comply with the MPE limits. They can also adjust their operating habits or control the direction their antenna is pointing.

More Information

This discussion offers only an overview of this topic; additional information can be found in *RF Exposure and You* and on *ARRLWeb* at **http://www.arrl.org/news/rfsafety/**. *ARRLWeb* has links to the FCC Web site, with *OET Bulletin 65* and *Supplement B* and links to software that hams can use to evaluate their stations.

—with an average amount of common sense. Of course prudence dictates that amateurs with pacemakers, who use hand-held VHF transceivers, keep the antenna as far as possible from the site of the implanted pacemaker generator. They also should use the lowest transmitter output required for adequate communication. For high power HF transmission, the antenna should be as far as possible from the operating position, and all equipment should be properly grounded.

LOW-FREQUENCY FIELDS

Although the FCC doesn't regulate 60-Hz fields, some recent concern about EMR has focused on low-frequency energy rather than RF. Amateur Radio equipment can be a significant source of low-frequency magnetic fields, although there are many other sources of this kind of energy in the typical home. Magnetic fields can be measured relatively accurately with inexpensive 60-Hz meters that are made by several manufacturers.

Table 9.1 shows typical magnetic field

intensities of Amateur Radio equipment and various household items. Because these fields dissipate rapidly with distance, "prudent avoidance" would mean staying perhaps 12 to 18 inches away from most Amateur Radio equipment (and 24 inches from power supplies with 1-kW RF amplifiers).

DETERMINING RF POWER DENSITY

Unfortunately, determining the power density of the RF fields generated by an

Table 9.2
Typical RF Field Strengths Near Amateur Radio Antennas
A sampling of values as measured by the Federal Communications Commission and Environmental Protection Agency, 1990

Antenna Type	Freq (MHz)	Power (W)	E Field (V/m)	Location
Dipole in attic	14.15	100	7-100	In home
Discone in attic	146.5	250	10-27	In home
Half sloper	21.5	1000	50	1 m from base
Dipole at 7-13 ft	7.14	120	8-150	1-2 m from earth
Vertical	3.8	800	180	0.5 m from base
5-element Yagi at 60 ft	21.2	1000	10-20	In shack
			14	12 m from base
3-element Yagi at 25 ft	28.5	425	8-12	12 m from base
Inverted V at 22-46 ft	7.23	1400	5-27	Below antenna
Vertical on roof	14.11	140	6-9	In house
			35-100	At antenna tuner
Whip on auto roof	146.5	100	22-75	2 m antenna
			15-30	In vehicle
			90	Rear seat
5-element Yagi at 20 ft	50.1	500	37-50	10 m antenna

amateur station is not as simple as measuring low-frequency magnetic fields. Although sophisticated instruments can be used to measure RF power densities quite accurately, they are costly and require frequent recalibration. Most amateurs don't have access to such equipment, and the inexpensive field-strength meters that we do have are not suitable for measuring RF power density.

Table 9.2 shows a sampling of measurements made at Amateur Radio stations by the Federal Communications Commission and the Environmental Protection Agency in 1990. As this table indicates, a good antenna well removed from inhabited areas poses no hazard under any of the IEEE/ANSI guidelines. However, the FCC/EPA survey also indicates that amateurs must be careful about using indoor or attic-mounted antennas, mobile antennas, low directional arrays or any other antenna that is close to inhabited areas, especially when moderate to high power is used.

Ideally, before using any antenna that is in close proximity to an inhabited area, you should measure the RF power density. If that is not feasible, the next best option is make the installation as safe as possible by observing the safety suggestions listed in **Table 9.3**.

It also is possible, of course, to calculate the probable power density near an antenna using simple equations. Such calculations have many pitfalls. For one, most of the situations where the power density would be high enough to be of concern are in the near field. In the near field, ground interactions and other variables produce power densities that cannot be determined by simple arithmetic. In the far field, conditions become easier to predict with simple calculations.

The boundary between the near field and the far field depends on the wavelength of the transmitted signal and the physical size and configuration of the antenna. The boundary between the near field and the far field of an antenna can be as much as several wavelengths from the antenna.

Computer antenna-modeling programs are another approach you can use. *MINI-NEC* or other codes derived from *NEC* (Numerical Electromagnetics Code) are suitable for estimating RF magnetic and electric fields around amateur antenna systems.

These models have limitations. Ground interactions must be considered in estimating near-field power densities, and the "correct ground" must be modeled. Computer modeling is generally not sophisticated enough to predict "hot spots" in the near field—places where the field intensity may be far higher than would be expected, due to reflections from nearby objects. In addition, "nearby objects" often change or vary with weather or the season, so the model so laboriously crafted may not be representative of the actual situation, by the time it is running on the computer.

Intensely elevated but localized fields often can be detected by professional measuring instruments. These "hot spots" are often found near wiring in the shack, and metal objects such as antenna masts or equipment cabinets. But even with the best instrumentation, these measurements also may be misleading in the near field.

One need not make precise measurements or model the exact antenna system, however, to develop some idea of the relative fields around an antenna. Computer

Table 9.3
RF Awareness Guidelines

These guidelines were developed by the ARRL RF Safety Committee, based on the FCC/EPA measurements of Table 9.2 and other data.

- Although antennas on towers (well away from people) pose no exposure problem, make certain that the RF radiation is confined to the antennas' radiating elements themselves. Provide a single, good station ground (earth), and eliminate radiation from transmission lines. Use good coaxial cable or other feed line properly. Avoid serious imbalance in your antenna system and feed line. For high-powered installations, avoid end-fed antennas that come directly into the transmitter area near the operator.

- No person should ever be near any transmitting antenna while it is in use. This is especially true for mobile or ground-mounted vertical antennas. Avoid transmitting with more than 25 W in a VHF mobile installation unless it is possible to first measure the RF fields inside the vehicle. At the 1-kW level, both HF and VHF directional antennas should be at least 35 ft above inhabited areas. Avoid using indoor and attic-mounted antennas if at all possible. If open-wire feeders are used, ensure that it is not possible for people (or animals) to come into accidental contact with the feed line.

- Don't operate high-power amplifiers with the covers removed, especially at VHF/UHF.

- In the UHF/SHF region, never look into the open end of an activated length of waveguide or microwave feed-horn antenna or point it toward anyone. (If you do, you may be exposing your eyes to more than the maximum permissible exposure level of RF radiation.) Never point a high-gain, narrow-bandwidth antenna (a paraboloid, for instance) toward people. Use caution in aiming an EME (moonbounce) array toward the horizon; EME arrays may deliver an effective radiated power of 250,000 W or more.

- With hand-held transceivers, keep the antenna away from your head and use the lowest power possible to maintain communications. Use a separate microphone and hold the rig as far away from you as possible. This will reduce your exposure to the RF energy.

- Don't work on antennas that have RF power applied.

- Don't stand or sit close to a power supply or linear amplifier when the ac power is turned on. Stay at least 24 inches away from power transformers, electrical fans and other sources of high-level 60-Hz magnetic fields.

modeling using close approximations of the geometry and power input of the antenna will generally suffice. Those who are familiar with *MININEC* can estimate their power densities by computer modeling, and those who have access to professional power-density meters can make useful measurements.

While our primary concern is ordinarily the intensity of the signal radiated by an antenna, we also should remember that there are other potential energy sources to be considered. You also can be exposed to RF radiation directly from a power amplifier if it is operated without proper shielding. Transmission lines also may radiate a significant amount of energy under some conditions. Poor microwave waveguide joints or improperly assembled connectors are another source of incidental radiation.

FURTHER RF EXPOSURE SUGGESTIONS

Potential exposure situations should be taken seriously. Based on the FCC/EPA measurements and other data, the "RF awareness" guidelines of Table 9.3 were developed by the ARRL RF Safety Committee. A longer version of these guidelines, along with a complete list of references, appeared in a *QST* article by Ivan Shulman, MD, WC2S ("Is Amateur Radio Hazardous to Our Health?" *QST*, Oct 1989, pp 31-34). For more information or background, see the list of RF Safety References in the next section.

In addition, the ARRL has published a book, *RF Exposure and You*, that is helping hams comply with the FCC's RF-exposure regulations. The ARRL also maintains an RF-exposure news page on its Web site. See **http://www.arrl.org/news/rfsafety**. This site contains reprints of selected *QST* articles on RF exposure and links to the FCC and other useful sites.

Other Hazards in the Ham Shack

CHEMICALS

We can't seem to live without the use of chemicals, even in the electronics age. A number of substances are used everyday by amateurs without causing ill effects. A sensible approach is to become knowledgeable of the hazards associated with the chemicals we use in our shack and then treat them with respect.

A few key suggestions:

- Read the information that accompanies the chemical and follow the manufacturer's recommended safety practices. If you would like more information than is printed on the label, ask for a material safety data sheet.
- Store chemicals properly away from sunlight and sources of heat. Provide security so they won't fall off the shelf. Secure them so that children and untrained persons will not gain access.
- Always keep containers labeled so there is no confusion about the contents. Use the container in which the chemical was purchased.
- Handle chemicals carefully to avoid spills.
- Clean up any spills or leaks promptly but don't overexpose yourself in the process. Dispose of chemicals in accordance with local and/or state regulations—*not* in the sink or storm sewer! Your city waste plant operator or fire department can explain disposal procedures in your community. The best solution is to use all of the chemical if at all possible. Buy only the amount you will need.
- Always use recommended personal protective equipment (such as gloves, face shield, splash goggles and aprons).
- If corrosives (acids or caustics) are splashed on you *immediately* rinse with cold water for a minimum of 15 minutes to flush the skin thoroughly. If splashed in the eyes, direct a gentle stream of cold water into the eyes for at least 15 minutes. Gently lift the eyelids so trapped liquids can be flushed completely. Start flushing before removing contaminated clothing. Seek professional medical assistance. It is unwise to work alone since people splashed with chemicals need the calm influence of another person.
- Food and chemicals don't mix. Keep food, drinks and cigarettes *away* from areas where chemicals are used and don't bring your chemicals to places where you eat.

Table 9.4 summarizes the uses and hazards of chemicals used in the ham shack. It includes preventive measures that can minimize risk.

ERGONOMICS

Ergonomics is a term that loosely means "fitting the work to the person." If tools and equipment are designed about what people can accommodate, the results will be much more satisfactory. For example, in the 1930s research was done in telephone equipment manufacturing plants because use of long-nosed pliers for wiring switchboards required considerable force at the end of the hand's range of motion. A simple tool redesign resolved this issue. Considerable attention has been focused on ergonomics in recent years because we have come to realize that long periods of time spent in unnatural positions can lead to repetitive-motion illness. Much of this attention has been focused on people whose job tasks have required them to operate video display terminals (VDTs). While most Amateur Radio operators do not devote as much time to their hobby as they might in a full-time job, it does make sense to consider comfort and flexibility when choosing furniture and arranging it in the shack or workshop. Adjustable height chairs are available with air cylinders to serve as a shock absorber. Footrests might come in handy if the chair is so high that your feet cannot support your lower leg weight. The height of tables and keyboards often is not adjustable.

Placement of VDT screens should take into consideration the reflected light coming from windows. It is always wise to build into your sitting sessions time to walk around and stimulate blood circulation. Your muscles are less likely to stiffen, while the flexibility in your joints can be enhanced by moving around.

Selection of hand tools is another area where there are choices to make that may affect how comfortable you will be while working in your shack. Look for screwdrivers with pliable grips. Take into account how heavy things are before picking them up—your back will thank you.

ENERGIZED CIRCUITS

Working with energized circuits can be very hazardous since our senses cannot directly detect dangerous voltages. The first thing we should ask ourselves when faced with troubleshooting, aligning or other "live" procedures is, "Is there a way to reduce the hazard of electrical shock?" Here are some ways of doing just that.

1. If at all possible, troubleshoot with an ohmmeter. With a reliable schematic diagram and careful consideration of how various circuit conditions may reflect resistance readings, it will often be unneces-

sary to do live testing.

2. Keep a fair distance from energized circuits. What is considered "good practice" in terms of distance? The *National Electrical Code* specifies minimum working space about electric equipment in Sections 110-16 and 110-34, depending on the voltage level. The principle here is that a person doing live work needs adequate space so they are not forced to be dangerously close to energized equipment.

3. If you need to measure the voltage of a circuit, install the voltmeter with the power safely off, back up, and only then energize the circuit. Remove the power before disconnecting the meter.

4. If you are building equipment that has hinged or easily removable covers that could expose someone to an energized circuit, install interlock switches that safely remove power in the event that the enclosure was opened with the power still on. Interlock switches are generally not used if tools are required to open the enclosure.

5. Never assume that a circuit is at zero potential even if the power is switched off and the power cable disconnected. Capacitors can retain a charge for a considerable period of time. Bleeder resistors should be installed, but don't assume they have bled off the voltage. Instead, after power is removed and disconnected use a "shorting stick" to ground all exposed conductors and ensure that voltage is not present.

Avoid using screwdrivers, as this brings the amateur too close to the circuit and could ruin the screwdriver's blade.

6. If you must hold a probe to take a measurement, always keep one hand in your pocket. As mentioned in the sidebar on the effects of high voltages, the worst path current could take through your body is from hand to hand since the flow would pass through the chest cavity.

7. Make sure someone is in the room with you and that they know how to remove the power safely. If they grab you with the power still on they will be shocked as well.

8. Test equipment probes and their leads must be in very good condition and rated

Table 9.4

Properties and Hazards of Chemicals often used in the Shack or Workshop

Generic Chemical Name	Purpose or Use	Hazards	Ways to Minimize Risks
Lead-tin solder	Bonding electrical components	•Lead exposure (mostly from hand contact) •Flux exposure (inhalation)	•Always wash hands after soldering or touching solder. •Use good ventilation.
Isopropyl alcohol	Flux remover	•Dermatitis (skin rash)	•Wear molded gloves suitable for solvents.
		•Vapor inhalation	•Use good ventilation and avoid aerosol generation.
		•Fire hazard	•Use good ventilation, limit use to small amounts, keep ignition sources away, dispose of rags only in tightly sealed metal cans.
Freons	Circuit cooling and general solvent	•Vapor inhalation •Dermatitis	•Use adequate ventilation. •Wear molded gloves suitable for solvents.
Phenols	Enameled wire stripper	•Strong skin corrosive	•Avoid skin contact; wear suitable molded gloves.
Beryllium oxide	Ceramic insulator which can conduct heat well	•Toxic when in fine dust form and inhaled	•Avoid grinding, sawing or reducing to dust form.
Beryllium metal	Lightweight metal, often alloyed with copper.	•Same as beryllium oxide	•Avoid grinding, sawing, welding, or reducing to dust. Contact supplier for special procedures.
Various paints	Finishing	•Exposures to solvents	•Adequate ventilation; use respirator when spraying.
		•Exposures to sensitizers (especially urethane paint)	•Adequate ventilation and use respirator. Contact supplier for more info.
		•Exposure to toxic metals (lead, cadmium, chrome, and so on) in pigments	•Adequate ventilation and use respirator. Contact supplier for more info.
		•Fire hazard (especially when spray painting)	•Adequate ventilation; control of residues; eliminate ignition sources.
Ferric chloride	Printed circuit board etchant	•Skin and eye contact	•Use suitable containers; wear splash goggles and molded gloves suitable for acids.
Ammonium persulphate and mercuric chloride	Printed circuit board etchants	•Skin and eye contact	•Use suitable containers; wear splash goggles and molded gloves suitable for acids.
Epoxy resins	General purpose cement or paint	•Dermatitis and possible sensitizer	•Avoid skin contact. Mix only amount needed.
Sulfuric acid	Electrolyte in lead-acid batteries	•Strong corrosive when on skin or eyes. •Will release hydrogen when charging (fire, explosion hazard).	•Always wear splash goggles and molded plastic gloves (PVC) when handling. Keep ignition sources away from battery when charging. Provide adequate ventilation.

for the conditions they will encounter.

9. Be wary of the hazards of "floating" (ungrounded) test equipment. A number of options are available to avoid this hazard. Contact your test equipment manufacturer for suggested procedures.

10. Ground-fault circuit interrupters can offer additional protection for stray currents that flow through the ground on 120-V circuits. Know their limitations. They cannot offer protection for the plate supply voltages in linear amplifiers, for example.

11. Older radio equipment containing ac/dc power supplies have their own hazards. If working on these live, use an isolation transformer, as the chassis may be connected directly to the hot or neutral power conductor.

12. Be aware of electrolytic capacitors that might fail if used outside their intended applications.

13. Replace fuses only with those having proper ratings.

SUMMARY

The ideas presented in this chapter are intended to reinforce the concept that ham radio, like many other activities in modern life, does have certain risks. But by understanding the hazards and how to deal effectively with them, the risk can be minimized. Common-sense measures can go a long way to help us prevent accidents. Traditionally, amateurs are inventors and experimenting is a major part of our nature. But reckless chance-taking is never wise, especially when our health and well-being is involved. A healthy attitude toward doing things the right way will help us meet our goals and expectations.

BIBLIOGRAPHY

Source material and more extended discussion of topics covered in this chapter can be found in the references given below.

High-Voltage Hazards

What happens when someone receives an electrical shock?

Electrocutions (fatal electric shocks) usually are caused by the heart ceasing to beat in its normal rhythm. This condition, called ventricular fibrillation, causes the heart muscles to quiver and stop working in a coordinated pattern, in turn preventing the heart from pumping blood.

The current flow that results in ventricular fibrillation varies between individuals but may be in the range of 100 mA to 500 mA. At higher current levels the heart may have less tendency to fibrillate but serious damage would be expected. Studies have shown 60-Hz alternating current to be more hazardous than dc currents. Emphasis is placed on application of cardiopulmonary resuscitation (CPR), as this technique can provide mechanical flow of some blood until paramedics can "restart" the heart's normal beating pattern. Defibrillators actually apply a carefully controlled dc voltage to "shock" the heart back into a normal heartbeat. It doesn't always work but it's the best procedure available.

What are the most important factors associated with severe shocks?

You may have heard that the current that flows through the body is the most important factor, and this is generally true. The path that current takes through the body affects the outcome to a large degree. While simple application of Ohm's Law tells us that the higher the voltage applied with a fixed resistance, the greater the current that will flow. Most electrical shocks involve skin contact. Skin, with its layer of dead cells and often fatty tissues, is a fair insulator. Nonetheless, as voltage increases the skin will reach a point where it breaks down. Then the lowered resistance of deeper tissues allows a greater current to flow. This is why electrical codes refer to the term "high voltage" as a voltage above 600 V.

How little a voltage can be lethal?

This depends entirely on the resistance of the two contact points in the circuit, the internal resistance of the body, and the path the current travels through the body. Historically, reports of fatal shocks suggest that as little as 24 V *could* be fatal under extremely adverse conditions. To add some perspective, one standard used to prevent serious electrical shock in hospital operating rooms limits leakage flow from electronic instruments to only *50 µA* due to the use of electrical devices and related conductors inside the patient's body.

Lightning Protection Code, NFPA 780, National Fire Protection Association, Quincy, MA, 1992.

National Electrical Code, NFPA 70, National Fire Protection Association, Quincy, MA, 1993. *National Electrical Code* and *NEC* are registered trademarks of the National Fire Protection Association, Inc, Quincy, MA 02269.

R. P. Haviland, "Amateur Use of Telescoping Masts," *QST*, May 1994, pp 41-45.

For more information about soldering hazards, symptoms, and protection, see "Making Soldering Safer," by Bryan P. Bergeron, MD, NU1N (Mar 1991 *QST*, pp 28-30) and "More on Safer Soldering," by Gary E. Myers, K9CZB (Aug 1991 *QST*, p 42).

Contents

Real-World
Component Characteristics
10

When is an inductor not an inductor? When it's a capacitor! This statement may seem odd, but it suggests the main message of this chapter. In the earlier chapters about **DC** and **AC Theory**, the basic components of electronic circuits were introduced. You saw that each has its own unique function to perform. For example, a capacitor stores energy in an electric field, a diode rectifies current and a battery provides voltage. All of these unique and different functions are necessary in order to build large circuits that perform useful tasks. The first part of this chapter, up to Low-Frequency Transistor Models, was written by Leonard Kay, K1NU.

As you may know from experience, these component pictures are *ideal*. That is, they are perfect mathematical pictures. An ideal component (or *element*) by definition behaves exactly like the mathematical equations that describe it, and *only* in that fashion. For example, an ideal capacitor passes a current that is equal to the capacitance C times the rate of change of the voltage across it. Period, end of sentence.

We call any other exhibited behavior either *nonideal*, *nonlinear* or *parasitic*. Nonideal behavior is a general term that covers any deviation from the theoretical picture. Ideal circuit elements are often linear: The graphs of their current versus voltage characteristics are straight lines when plotted on a suitable (Cartesian, rectangular) set of axes. We therefore call deviation from this behavior nonlinear: For example, as current through a resistor exceeds its power rating, the resistor heats up and its resistance changes. As a result, a graph of current versus voltage is no longer a straight line.

When a component begins to exhibit properties of a different component, as when a capacitor allows a dc current to pass through it, we call this behavior *parasitic*. Nonlinear and parasitic behavior are both examples of nonideal behavior.

Much to the bane of experimenters and design engineers, ideal components exist only in electronics textbooks and computer programs. Real components, the ones we use, only approximate ideal components (albeit very closely in most cases). *Real* diodes store minuscule energy in electric fields (junction capacitance) and magnetic fields (lead inductance); *real* capacitors conduct some dc (modeled by a parallel resistance), and real battery voltage is not *perfectly* constant (it may even decrease nonlinearly during discharge).

Knowing to what extent and under what conditions real components cease to behave like their ideal counterparts, and what can be done to account for these behaviors, is the subject of component or circuit *modeling*. In this chapter, we will explore how and why the real components behave differently from ideal components, how we can account those differences when analyzing circuits, how to select components to minimize, or exploit, nonideal behaviors and give a brief introduction to computer-aided circuit modeling.

Much of this chapter may seem intuitive. For example, you probably know that #20 hookup wire works just fine for wiring many experimental circuits. When testing a circuit after construction, you probably don't even think about the voltage drops across those pieces of wire (you inherently assume that they have zero resistance). But, connect that same #20 wire directly across a car battery—with no *other* series resistance—and suddenly, the resistance of the wire *does* matter, and it changes—as the wire heats, melts and breaks—from a very small value to infinity!

LUMPED VS DISTRIBUTED ELEMENTS

Most electronic circuits that we use everyday are inherently and mathematically considered to be composed of *lumped elements*. That is, we assume each component acts at a single point in space, and the wires that connect these lumped elements are assumed to be *perfect conductors* (with zero resistance and insignificant length). This concept is illustrated in **Fig 10.1**. These assumptions are perfectly reasonable for many applications, but they have limits. Lumped element models break down when:

- Circuit impedance is so low that the small, but non-zero, resistance in the wires is important. (A significant portion of the circuit power may be lost to heat in the conductors.)
- Operating frequency (f_0) is high enough that the length of the connecting wires is a significant fraction (> 0.1) of the wave-

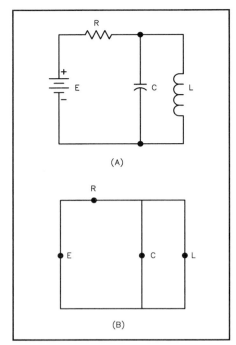

Fig 10.1—The lumped element concept. Ideally, the circuit at A is assumed to be as shown at B, where the components are isolated points connected by perfect conductors. Many components exhibit nonideal behavior when these assumptions no longer hold.

length. (Radiation from the conductors may be significant.)

• When transmission lines are used as conductors. (Their characteristic impedance is usually significant, and impedances connected to them are transformed as a function of the line length. See the **Transmission Lines** chapter for more information.)

Effects such as these are called *distributed*, and we talk of *distributed* elements or effects to contrast them to lumped elements.

To illustrate the differences between lumped and distributed elements, consider the two resistors in **Fig 10.2**, which are both 12 inches long. The resistor at A is a uniform rod of carbon—a battery anode, for example. The second "resistor" B is made of two 6-inch pieces of silver rod (or other highly conductive material), with a small resistor soldered between them. Now imagine connecting the two probes of an ohmmeter to each of the two resistors, as in the figure. Starting with the probes at the far ends, as we slide the probes toward the center, the carbon rod will display a constantly decreasing resistance on the ohmmeter. This represents a *distributed* resistance. On the other hand, the ohmmeter connected to the other

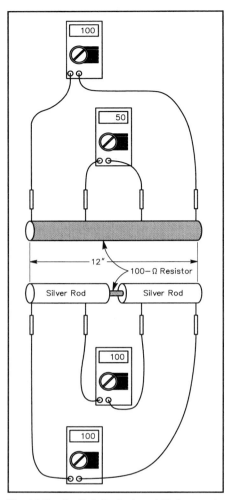

Fig 10.2—A, distributed and B, lumped resistances. See text for discussion.

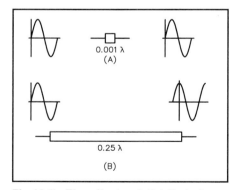

Fig 10.3—The effects of distributed resistance on the phase of a sinusoidal current. There is no phase delay between ends of a lumped element.

12-inch "resistor" will display a constant resistance as long as one probe remains on each side of the small resistance. (Oh yes, as long as we neglect the resistance of the silver rods!) This represents a *lumped* resistance connected by perfect conductors.

Lumped elements also have the very

desirable property that they introduce no phase shift resulting from propagation delay through the element. (Although combinations of lumped elements can produce phase shifts by virtue of their R, L and C properties.) Consider a lumped element that is carrying a sinusoidal current, as in **Fig 10.3A**. Since the element has negligible length, there is no phase difference in the current between the two sides of the element—*no matter how high the frequency*—precisely *because* the element length is negligible. If the physical length of the element were long, say 0.25 λ as shown in Fig 10.3B, the current phase would *not* be the same from end to end. In this instance, the current is delayed 90 electrical degrees. The amount of phase difference depends on the circuit's electrical length.

Because the relationship between the physical size of a circuit and the wavelength of an ac current present in the circuit will vary as the frequency of the ac signal varies, the ideas of lumped and distributed effects actually occupy two ends of a spectrum. At HF (30 MHz and below), where λ ≥ 10 m, the lumped element concept is almost always valid. In the UHF region and above, where λ ≤ 1 m and physical component size can represent a significant fraction of a wavelength, everything shows distributed effects to one degree or another. From roughly 30 to 300 MHz, problems are usually examined on a case-by-case basis.

Of course, if we could make resistors, capacitors, inductors and so on, very small, we could treat them as lumped elements at much higher frequencies. Thanks to the advances constantly being made in microelectronic circuit fabrication, this is in fact possible. Commercial monolithic microwave integrated circuits (MMICs) in the early 1990s often use lumped elements that are valid up to 50 GHz. Since this frequency represents a wavelength of roughly 5 mm, this implies a component size of less than 0.5 mm.

LOW-FREQUENCY COMPONENT MODELS

Every circuit element behaves nonideally in some respect and under some conditions. It helps to know the most common types of nonideal behavior so we can design and build circuits that will perform as intended under expected ranges of operating conditions. In the sections below, we will discuss the basic components in order of increasing common nonideal behavior. Please note that much of this section applies only through HF; the peculiarities of VHF frequencies and above are discussed later in the chapter.

First, remember that some of the common limitations associated with real components are manufacturing concerns: tolerance and standard values. Tolerance is the measure of how much the actual value of a component may differ from its labeled value; it is usually expressed in percent. By convention, a *"higher"* (actually *closer*) tolerance component indicates a *lower* (lesser) percentage deviation and will usually cost more. For example, a 1-kΩ resistor with a 5% tolerance has an actual resistance of anywhere from 950 to 1050 Ω. When designing circuits be careful to include the effects of tolerance. More specific examples will be discussed below.

Keep standard values in mind because not every conceivable component value is available. For instance, if circuit calculations yield a 4,932-Ω resistor for a demanding application, there may be trouble. The nearest commonly available values are 4700 Ω and 5600 Ω, and they'll be rated at 5% tolerance!

These two constraints prevent us from building circuits that do precisely what we wish. In fact, the measured performance of any circuit is the summation of the tolerances and temperature characteristics of all the components in both the circuit under test and the test equipment itself. As a result, most circuits are designed to operate within tolerances such as "up to a certain power" or "within this frequency range." Some circuits have one or more adjustable components that can compensate for variations in others. These limitations are then further complicated by other problems.

RESISTORS

Resistors are made in several different ways: carbon composition, carbon film, metal film, and wire wound. Carbon composition resistors are simply small cylinders of carbon mixed with various binding agents. Carbon is technically a semiconductor and can be *doped* with various impurities to produce any desired resistance. Most common everyday $^1/_2$- and $^1/_4$-W resistors are of this sort. They are moderately stable from 0 to 60 °C (their resistance increases above and below this temperature range). They are not inductive, but they are relatively noisy, and have relatively wide tolerances.

The other resistors exploit the fact that resistance is proportional to the length of the resistor and inversely proportional to its cross-sectional area:

Wire-wound resistors are made from wire, which is cut to the proper length and wound on a coil form (usually ceramic).

They are capable of handling high power; their values are very stable, and they are manufactured to close tolerances.

Metal-film resistors are made by depositing a thin film of aluminum, tungsten or other metal on an insulating substrate. Their resistances are controlled by careful adjustments of the width, length and depth of the film. As a result, they have very close tolerances. They are used extensively in surface-mount technology. As might be expected, their power handling capability is somewhat limited. They also produce very little electrical noise.

Carbon film resistors use a film of doped carbon instead of metal. They are not quite as stable as other film resistors and have wider tolerances than metal-film resistors, but they are still as good as (or better than) composition resistors.

Resistors behave much like their ideal through AF; lead inductance becomes a problem only at higher frequencies. The major departure from ideal behavior is their temperature coefficient (TC). The resistivity of most materials changes with temperature, and typical TC values for resistor materials are given in **Table 10.1**. TC values are usually expressed in parts-per-million (PPM) for each degree (centigrade) change from some nominal temperature, usually room temperature (77 °F/ 27 °C). A positive TC indicates an increase in resistance with increasing temperature while a negative TC indicates a decreasing resistance. For example, if a 1000-Ω resistor with a TC of +300 PPM/°C is heated to 50 °C, the *change* in resistance is 300(50 − 27) = 6900 PPM, yielding a new resistance of

$$1000 \left(1 + \frac{6900}{1000000} \right) = 1006.9 \ \Omega$$

Carbon-film resistors are unique among the major resistor families because they alone have a negative temperature coefficient. They are often used to "offset" the thermal effects of the other components (see Thermal Effects, below).

If the temperature increase is small (less than 30-40 °C), the resistance change with temperature is nondestructive—the resistor will return to normal when the temperature returns to its nominal value. Resistors that get too hot to touch, however, may be permanently damaged even if they appear normal. For this reason, be conservative when specifying power ratings for resistors. It's common to specify a resistor rated at 200% to 400% of the expected dissipation.

Wire-wound resistors are essentially inductors used as resistors. Their use is therefore limited to dc or low-frequency ac applications where their reactance is

Table 10.1
Temperature Coefficients for Various Resistor Compositions
1 PPM = 1 part per million = 0.0001%

Type	TC (PPM/°C)
Wire wound	±(30 - 50)
Metal Film	±(100 - 200)
Carbon Film	+350 to −800
Carbon composition	±800

negligible. Remember that this inductance will also affect switching transient waveforms, even at dc, because the component will act as an RL circuit.

As a rough example, consider a 1-Ω, 5-W wire-wound resistor that is formed from #24 wire on a 0.5-inch diam form. What is the approximate associated inductance? First, we calculate the length of wire using the wire tables:

$$L = \frac{R}{\Omega/ft \text{ for } \#24 \text{ wire}}$$

$$= \frac{1}{25.7 \ \Omega / 1000 \text{ ft}} \approx 39 \text{ ft}$$

This yields a total of (39 × 12 inches) / (0.5 π inch/turn) ≈ 298 turns, which further yields a coil length (for #24 wire close-wound at 46.9 turns per inch) of 6.3 inches, assuming a single-layer winding. Then, from the inductance formula for air coils in the **AC Theory** chapter, calculate

$$L = \frac{(0.5)^2 \times 298^2}{18 \times 0.5 + 40 \times 6.3} = 85 \ \mu H$$

Real wire-wound resistors have multiple windings layered over each other to minimize both size and parasitic inductance (by winding each layer in opposite directions, much of the inductance is canceled). If we assume a five-layer winding, the length is reduced to 1.8 inches and the inductance to approximately 17 μH. If we want the inductive reactance to stay below 10% of the resistor value, then this resistor cannot be used above f = 0.1 / (2 π 17 μH) = 937 Hz, or roughly 1 kHz.

Another exception to the rule that resistors are, in general, fairly ideal has to do with skin effect at RF. This will be discussed later in the chapter. **Fig 10.4** shows some more accurate circuit models for resistors at low frequencies. For a treatment of pure resistance theory, look at the **DC Theory** chapter.

VOLTAGE AND CURRENT SOURCES

An ideal voltage source maintains a constant voltage across its terminals no

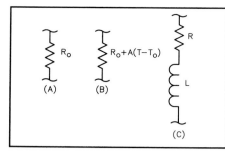

Fig 10.4—Circuit models for resistors. A is the ideal element. At B is the simple temperature-varying model for noninductive resistors. The wire-wound model with associated inductance is shown at C. For UHF or microwave designs, the model at C could be used with L representing lead inductance.

matter how much current is drawn. Consequently, it is capable of providing infinite power, for an infinite period of time. Similarly, an ideal current source provides a constant current through its terminals no matter what voltage appears across it. It, too, can deliver an infinite amount of power for an infinite time.

Internal Resistance

As you may have learned through experience, *real* voltage and current sources—batteries and power supplies—do not meet these expectations. All real power sources have a finite *internal resistance* associated with them that limits the maximum power they can deliver.

We can model a real dc voltage source as a Thevenin-equivalent circuit of an *ideal* source in series with a resistance R_{thev} that is equal to the source's internal resistance. Similarly, we can model a real dc current source as a Norton-equivalent circuit: an ideal current source in *parallel* with R_{thev}. These two circuits shown in **Fig 10.5** are interchangeable through the relation

$$V_{oc} = I_{sc} \times R_{thev} \qquad (1)$$

Using these more realistic models, the maximum current that a real voltage source can deliver is seen to be I_{sc} and the maximum voltage is V_{oc}.

We can model sinusoidal voltage or current sources in much the same way, keeping in mind that the internal *impedance*, Z_{thev}, for such a source may not be purely resistive, but may have a reactive component that varies with frequency.

Battery Capacity and Discharge Curves

Batteries have a finite energy capacity as well as an internal resistance. Because of the physical size of most batteries, a convenient unit for measuring capacity is the *milliampere hour* (mAh) or, for larger

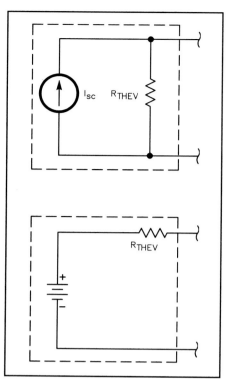

Fig 10.5—Norton and Thevenin equivalent circuits for real voltage and current sources.

units, the *ampere hour* (Ah). Note that these are units of charge (coulombs per second × seconds) and thus they measure the total net charge that the battery holds on its plates when full.

A capacity of 1 mAh indicates that a battery, when fully charged, holds sufficient electrochemical energy to supply a steady current of 1 mA for 1 hour. If a battery discharges at a constant rate, you could estimate the useful time of a battery charge by the simple formula

$$\text{Time (hr)} = \frac{\text{capacity (mAh)}}{\text{current (mA)}} \qquad (2)$$

In practice, however, the usable capacity of a rechargeable battery depends on the discharge *rate*, decreasing slightly as the current increases. For example, a 500 mAh nickel-cadmium (NiCd) battery may supply a current of 50 mA for almost 10 hours while maintaining a current of 500 mA for only 45 minutes.

The finite capacity of a battery does not *in itself* change the circuit model for a real source as shown in **Fig 10.5**. However, the chemistry of electrolytic cells, from which batteries are made, does introduce a minor, but significant, change. The voltage of a discharging battery does not stay constant, but slowly drops as time goes on. **Fig 10.6** illustrates the *discharge curve* for a typical NiCd rechargeable battery. For a given battery design, such curves are

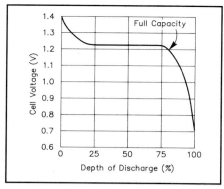

Fig 10.6—Discharge curve (terminal voltage vs percent capacity remaining) for a typical NiCd rechargeable battery. Note the dramatic voltage drop near the end.

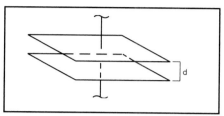

Fig 10.7—The ideal parallel-plate capacitor.

fairly reproducible and often used by battery-charging circuits to sense when a battery is fully charged or discharged.

CAPACITORS

The ideal capacitor is a pair of infinitely large parallel metal plates separated by an insulating or *dielectric* layer, ideally a vacuum (see **Fig 10.7**). (For a discussion of pure capacitance see the **AC Theory** chapter.) For this case, the capacitance is given by

$$C = \frac{A \varepsilon_r \varepsilon_0}{d} \qquad (3)$$

where
   C = capacitance, in farads
   A = area of plates, cm
   d = spacing of the plates, cm
   ε_r = dielectric constant of the insulating material
   ε_0 = permittivity of free space, 8.85×10^{-14} F/cm.

If the plates are not infinite, the actual capacitance is somewhat higher due to *end effect*. This is the same phenomenon that causes a dipole to resonate at a lower frequency if you place large insulators (which add capacitance) on the ends.

If we built such a capacitor, we would find that (neglecting the effects of quantum mechanics) it would be a perfect open circuit at dc, and it could be operated at whatever voltage we desire without breakdown.

Leakage Conductance and Breakdown

If we use anything other than a vacuum for the insulating layer, even air, we introduce two problems. Because there are atoms between the plates, the capacitor will now be able to conduct a dc current. The magnitude of this *leakage current* will depend on the insulator quality, and the current is usually very small. Leakage current can be modeled by a resistance R_L in parallel with the capacitance (in the ideal case, this resistance is infinite).

In addition, when a high enough voltage is applied to the capacitor, the atoms of the dielectric will ionize due to the extremely high electric field and cause a large dc current to flow. This is *dielectric breakdown,* and it is destructive to the capacitor if the dielectric is ruined. To avoid dielectric breakdown, a capacitor has a *working voltage* rating, which represents the maximum voltage that can be permitted to develop across it.

Dielectrics

The leakage conductance and breakdown voltage characteristics of a capacitor are strongly dependent on the composition and quality of the dielectric. Various materials are used for different reasons such as availability, cost, and desired capacitance range. In rough order of "best" to "worst" they are:

Vacuum Both fixed and variable vacuum capacitors are available. They are rated by their maximum working voltages (3 to 60 kV) and currents. Losses are specified as negligible for most applications.

Air An *air-spaced* capacitor provides the best commonly available approximation to the ideal picture. Since $\varepsilon_r = 1$ for air, air-dielectric capacitors are large when compared to those of the same value using other dielectrics. Their capacitance is very stable over a wide temperature range, leakage losses are low, and therefore a high Q can be obtained. They also can withstand high voltages. For these reasons (and ease of construction) most variable capacitors in tuning circuits are air-spaced.

Plastic film Capacitors with plastic film (polystyrene, polyethylene or Mylar) dielectrics are more expensive than paper capacitors, but have much lower leakage rates (even at high temperatures) and low TCs. Capacitance values are more stable than those of paper capacitors. In other respects, they have much the same characteristics as paper capacitors. Plastic-film variable capacitors are available.

Mica The capacitance of mica capacitors is very stable with respect to time, temperature and electrical stress. Leakage and losses are very low. Values range from 1 pF to 0.1 µF, with tolerances from 1 to 20%. High working voltages are possible, but they must be derated severely as operating frequency increases.

Silver mica capacitors are made by depositing a thin layer of silver on the mica dielectric. This makes the value even more stable, but it presents the possibility of silver migration through the dielectric. The migration problem worsens with increased dc voltage, temperature and humidity. Avoid using silver-mica capacitors under such conditions.

Ceramic There are two kinds of ceramic capacitors. Those with a low dielectric constant are relatively large, but very stable and nearly as good as mica capacitors at HF. High dielectric constant ceramic capacitors are physically small for their capacitance, but their value is not as stable. Their dielectric properties vary with temperature, applied voltage and operating frequency. They also exhibit piezoelectric behavior. Use them only in coupling and bypass roles. Tolerances are usually +100% and −20%. Ceramic capacitors are available in a wide range of values: 10 pF to 1 µF. Some variable units are available.

Electrolytic These capacitors have the space between their foil plates filled with a chemical paste. When voltage is applied, a chemical reaction forms a layer of insulating material on the foil.

Electrolytic capacitors are popular because they provide high capacitance values in small packages at a reasonable cost. Leakage is high, as is inductance, and they are polarized—there is a definite positive and negative plate, due to the chemical reaction that provides the dielectric. Internal inductance restricts aluminum-foil electrolytics to low-frequency applications. They are available with values from 1 to 500,000 µF.

Tantalum electrolytic capacitors perform better than aluminum units but their cost is higher. They are smaller, lighter and more stable, with less leakage and inductance than their aluminum counterparts. Reformation problems are less frequent, but working voltages are not as high as with aluminum units.

Electrolytics should not be used if the dc potential is well below the capacitor working voltage.

Paper Paper capacitors are inexpensive; capacitances from 500 pF to 50 µF are available. High working voltages are possible, but paper-dielectric capacitors have high leakage rates and tolerances are no better than 10 to 20%. Paper-dielectric capacitors are not polarized; however, the body of the capacitor is usually marked with a color band at one end. The band indicates the terminal that is connected to the outermost plate of the capacitor. This terminal should be connected to the side of the circuit at the lower potential as a safety precaution.

Loss Angle

For ac signals (even at low frequencies), capacitors exhibit an additional parasitic resistance that is due to the electromagnetic properties of dielectric materials. This resistance is often quantified in catalogs as *loss angle*, θ, because it represents the angle, in the complex impedance plane, between $Z_C = R + jX_C$ and X_C. This angle is usually quite small, and would be zero for an ideal capacitor.

Loss angle is normally specified as $\tan \theta$ at a certain frequency, which is simply the ratio R/X_C. The loss angle of a given capacitor is relatively constant over frequency, which means the *effective series resistance* or $ESR = (\tan \theta) / (2 \pi f C)$ goes down as frequency goes up. This resistance is placed in *series* with the capacitor because it came (mathematically) from the equation for Z_C above. It can always be converted into a parallel resistance if desired. To summarize, **Figs 10.8** and **10.9** show reasonable models for the capacitor that are good up to VHF.

Temperature Coefficients and Tolerances

As with resistors, capacitor values vary in production, and most capacitors have a tolerance rating either printed on them or listed on a data sheet. Capacitance varies with temperature, and this is important to consider when constructing a circuit that will carry high power levels or operate in a hot environment. Also, as just described, not all capacitors are available in all ranges of values due to inherent differences in material properties. Typical values, temperature coefficients and leakage conduc-

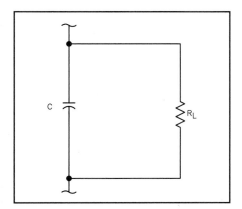

Fig 10.8—A simple capacitor model for frequencies well below self-resonance.

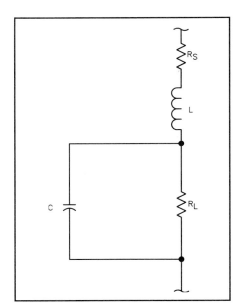

Fig 10.9—A capacitor model for VHF and above including series resistance and distributed inductance.

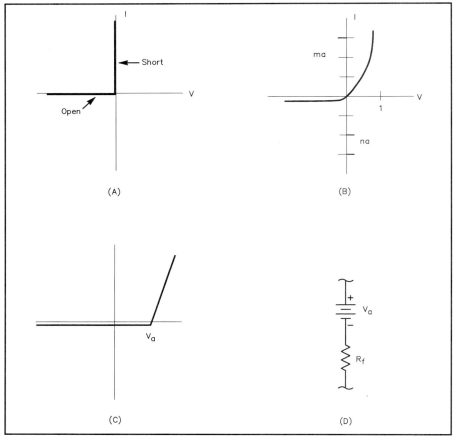

Fig 10.10—Circuit models for rectifying switches (diodes). A: I-V curve of the ideal rectifier. B: I-V curve of a typical semiconductor diode. Note the different scales for + and − current. C shows a simplified diode I-V curve for dc-circuit calculations. D is an equivalent circuit for C.

Table 10.2

Typical Temperature Coefficients and Leakage Conductances for Various Capacitor Constructions

Type	TC @ 20°C (PPM/°C)	DC Leakage Conductance (Ω)
Ceramic Disc	±300(NP0)	> 10 M
	+150/−1500(GP)	> 10 M
Mica	−20 to +100	> 100,000 M
Polyester	±500	> 10 M
Tantalum Electrolytic	±1500	> 10 MΩ
Small Al Electrolytic(≈ 100 μF)	−20,000	500 k - 1 M
Large Al Electrolytic(≈ 10 mF)	−100,000	10 k
Vacuum (glass)	+100	≈ ∞
Vacuum (ceramic)	+50	≈ ∞

tances for several capacitor types are given in **Table 10.2.**

DIODES

An ideal diode acts as a rectifying switch—it is a short circuit when forward biased and an open circuit when reverse biased. Many circuit components can be completely described in terms of current-voltage (or *I-V*) characteristics, and to discuss the diode, this approach is especially helpful. **Fig 10.10A** shows the I-V curve for an ideal rectifier.

In contrast, the I-V curve for a semiconductor diode junction is given by the following equation (slightly simplified).

$$I = I_s \left(e^{\frac{v}{V_t}} \right) \qquad (4)$$

where

I = diode current
V = diode voltage
I_s = reverse-bias saturation current
V_t = kT/q, the thermal equivalent of voltage (about 25 mV at room temperature).

This curve is shown in Fig 10.10B. The obvious differences between Fig 10.10A and B are that the semiconductor diode has a finite *turn-on* voltage—it requires a small but nonzero forward bias voltage before it begins conducting. Furthermore, once conducting, the diode voltage continues to increase very slowly with increasing current, unlike a true short circuit. Finally, when the applied voltage is negative, the current is not exactly zero

but very small (microamperes).

For bias (dc) circuit calculations, a useful model for the diode that takes these two effects into account is shown by the artificial I-V curve in Fig 10.10C. The small reverse bias current I_s is assumed to be completely negligible.

When converted into an equivalent circuit, the model in Fig 10.10C yields the picture in Fig 10.10D. The ideal voltage source V_a represents the turn-on voltage

and R_f represents the effective resistance caused by the small increase in diode voltage as the diode current increases. The turn-on voltage is material-dependent: approximately 0.3 V for germanium diodes and 0.7 for silicon. R_f is typically on the order of 10 Ω, but it can vary according to the specific component. R_f can often be completely neglected in comparison to the other resistances in the circuit. This very common simplification leaves only a pure voltage drop for the diode model.

Temperature Bias Dependence

The reverse saturation current I_S is not constant but is itself a complicated function of temperature. For silicon diodes (and transistors) near room temperature, I_S increases by a factor of 2 every 4.8 °C. This means that for every 4.8 °C rise in temperature, either the diode current doubles (if the voltage across it is constant), or if the current is held constant by other resistances in the circuit, the diode voltage will *decrease* by $V_t \times \ln 2 = 18$ mV. For germanium, the current doubles every 8 °C and for gallium arsenide (GaAs), 3.7 °C. This dependence is highly reproducible and may actually be exploited to produce temperature-measuring circuits.

While the change resulting from a rise of several degrees may be tolerable in a circuit design, that from 20 or 30 degrees may not. Therefore it's a good idea with diodes, just as with other components, to specify power ratings conservatively (2 to 4 times over) to prevent self-heating.

While component derating does reduce self-heating effects, circuits must be designed for the expected operating environment. For example, mobile radios may face temperatures from –20° to +140°F (–29° to 60°C).

Junction Capacitance

Immediately surrounding a PN junction is a *depletion layer*. This is an electrically charged region consisting primarily of ionized atoms with relatively few electrons and holes (see **Fig 10.11**). Outside the depletion layer are the remainder of the P and N regions, which do not contribute to diode operation but behave primarily as parasitic series resistances.

We can treat the depletion layer as a tiny capacitor consisting of two parallel plates. As the reverse bias applied to a diode changes, the width of the depletion layer, and therefore the capacitance, also changes. There is an additional *diffusion capacitance* that appears under forward bias due to electron and hole storage in the bulk regions, but we will not discuss this here because diodes are usually reverse biased when their capacitance is exploited. The diode junction capacitance under a reverse bias of V volts is given by

$$C_j = C_{j0} / \sqrt{V_{on} - V} \qquad (5)$$

where C_{j0} = measured capacitance with zero applied voltage.

Note that the quantity under the radical is a large *positive* quantity for reverse bias. As seen from the equation, for large reverse biases C_j is inversely proportional to the square root of the voltage.

Junction capacitances are small, on the order of pF. They become important, however, for diode circuits at RF, as they can affect the resonant frequency. In fact, *varactors* are diodes used for just this purpose.

Reverse Breakdown and Zener Diodes

If a sufficiently large reverse bias is applied to a real diode, the internal electric field becomes so strong that electrons and holes are ripped from their atoms, and a large reverse current begins to flow. This

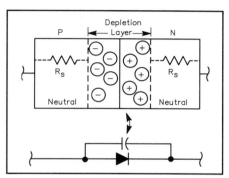

Fig 10.11—A more detailed picture of the PN junction, showing depletion layer, bulk regions and junction capacitance.

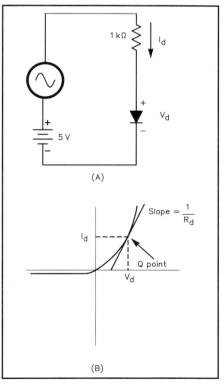

Fig 10.12—A simple resistor-diode circuit used to illustrate dynamic resistance. The ac input voltage "sees" a diode resistance whose value is the slope of the line at the Q-point, shown in B.

is called *breakdown* or *avalanche*. Unless the current is so large that the diode fails from overheating, breakdown is not destructive and the diode will again behave normally when the bias is removed.

Once a diode breaks down, the voltage across it remains relatively constant regardless of the level of breakdown current. *Zener* diodes are diodes that are specially made to operate in this region with a very constant voltage, called the *Zener voltage*. They are used primarily as voltage regulators. When operating in this region, they can be modeled as a simple voltage source. Zener diodes are rated by their Zener voltage and power dissipation.

A Diode AC Model

Fig 10.12A shows a simple resistor-diode circuit to which is applied a dc bias voltage plus an ac signal. Assuming that the voltage drop across the diode is 0.6 V and R_f is negligible, we can calculate the bias current to be $I \approx (5 - 0.6) / 1$ kΩ = 4.4 mA. This point is marked on the diodes I-V curve in Fig 10.12B. If we draw a line tangent to this point, as shown, the slope of this line represents the *dynamic* resistance R_d of the diode seen by a small ac signal, which at room temperature can be approximated by

$$R_d = \frac{25}{I} \Omega \qquad (6)$$

where I is the diode current in mA. Note that this resistance changes with bias current and should not be confused with the dc forward resistance in the previous section, which has a similar value but represents a different concept. **Fig 10.13** shows a low-frequency ac model for the diode, including the dynamic resistance and junction capacitance.

Switching Time

If you change the polarity of a signal applied to the ideal switch whose I-V

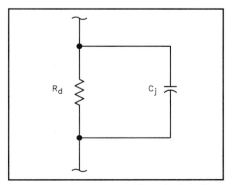

Fig 10.13—An ac model for diodes. R_d is the dynamic resistance and C_j is the junction capacitance.

curve appears in Fig 10.10A, the switch turns on or off instantaneously. A real diode cannot do this, as a finite amount of time is required to move electrons and holes in or out of the diode as it changes states (effectively, the diode capacitances must be charged or discharged). As a result, diodes have a maximum useful frequency when used in switching applications. The operation of diode switching circuits can often be modeled by the picture in **Fig 10.14**. The approximate switching time (in seconds) for this circuit is given by

$$t_s = \tau_p \left(\frac{\left(\dfrac{V_1}{R_1}\right)}{\left(\dfrac{V_2}{R_2}\right)} \right) = \tau_p \left(\frac{I_1}{I_2} \right) \qquad (7)$$

where τ_p is the minority carrier lifetime of the diode (a material constant determined during manufacture, on the order of 1 ms). I_1 and I_2 are currents that flow during the switching process. The minimum time in which a diode can switch from one state to the other and back again is therefore 2 t_s, and thus the maximum usable switching frequency is f_{sw} (Hz) = $^1/_2$ t_s. It is usually a good idea to stay below this by a factor of two. Diode data sheets usually give typical switching times and show the circuit used to measure them.

Note that f_{sw} depends on the forward and reverse currents, determined by I_1 and I_2 (or equivalently V_1, V_2, R_1, and R_2). Within a reasonable range, the switching time can be reduced by manipulating these currents. Of course, the maximum power that other circuit elements can handle places an upper limit on switching currents.

Schottky Diodes

Schottky diodes are made from metal-semiconductor junctions rather than PN junctions. They store less charge internally and as a result, have shorter switching times and junction capacitances than standard PN junction diodes.

Their turn-on voltage is also less, typically 0.3 to 0.4 V. In most other respects they behave similarly to PN diodes.

INDUCTORS

Inductors are the problem children of the component world. (Fundamental inductance is discussed in the **AC Theory** chapter.) Besides being difficult to fabricate on integrated circuits, they are perhaps the most nonideal of real-world components. While the leakage conductance of a capacitor is usually negligible, the series resistance of an inductor often is not. This is basically because an inductor is made of a long piece of relatively thin wire wound into a small coil. As an example, consider a typical air-core tuning coil from a component catalog, with L = 33 µH and a minimum Q of 30 measured at 2.5 MHz. This would indicate a series resistance of R_s = 2 π f L / Q = 17 Ω. This R_s could significantly alter the resonant frequency of a circuit. For frequencies up to HF this is the only significant nonlinearity, and a low-frequency circuit model for the inductor is shown in **Fig 10.15**. Many of the problems associated with inductors are actually due to core materials, and not the coil itself.

Magnetic Materials

Many discrete inductors and transformers are wound on a core of iron or other magnetic material. As discussed in the **AC Theory** chapter, the inductance value of a coil is proportional to the density of magnetic field lines that pass through it. A piece of magnetic material placed inside a coil will concentrate the field lines inside itself. This permits a higher number of field lines to exist inside a coil of a given cross-sectional area, thus allowing a much higher inductance than what would be possible with an air core. This is especially important for transformers that operate at low frequencies (such as 60 Hz) to ensure the reactance of the windings is high.

Core Saturation

Magnetic (more precisely, ferromagnetic) materials exhibit two kinds of nonlinear behavior that are important to circuit design. The first is the phenomenon of *saturation*. A magnetic core increases the magnetic flux density of a coil because the current passing through the coil forces the atoms of the iron (or other material) to line up, just like many small compass needles, and the magnetic field that results from the atomic alignment is *much* larger than that produced by the current with no core. As coil current increases, more and more atoms line up. At some high current, all of the atoms will be aligned and the core is *saturated*. Any further increase in current can't increase the core alignment any further. Of course, added current generates its own magnetic flux, but it is very small compared to the magnetic field contributed by the aligned atoms in the magnetic core.

The important concept in terms of circuit design is that as long as the coil current remains below saturation, the inductance of the coil is essentially constant. **Fig 10.17** shows graphs of magnetic flux linkage (N φ) and inductance (L) vs current (i) for a typical iron-core inductor. These quantities are related by the equation

$$N \; \phi = L \; i \qquad (8)$$

where
  N = number of turns,
  φ = flux density
  L = inductance
  i = current.

In the lower graph, a line drawn from any point on the curve to the (0,0) point

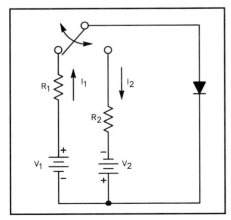

Fig 10.14—Circuit used for computation of diode switching time.

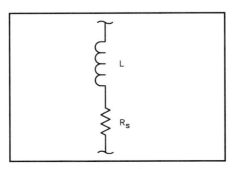

Fig 10.15—Low- to mid-frequency inductor equivalent circuit showing series resistance R_s.

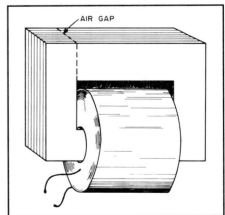

Fig 10.16—Typical construction of a magnetic-core inductor. The air gap greatly reduces core saturation at the expense of some inductance. The insulating laminations between the core layers help to minimize eddy currents.

will show the effective inductance, $L = N\phi/i$, at that current. These results are plotted on the upper graph.

Note that below saturation, the inductance is constant because both $N\phi$ and i are increasing at a steady rate. Once the saturation current is reached, the inductance decreases because $N\phi$ does not increase anymore (except for the tiny additional magnetic field the current itself provides). This may render some coils useless at VHF and higher frequencies. Air-coil inductors do not suffer from saturation.

One common method of increasing the saturation current level is to cut a small air gap in the core (see **Fig 10.16**). This gap forces the flux lines to travel through air for a short distance. Since the saturation flux linkage of the core is unchanged, this method works by requiring a higher current to achieve saturation. The price that is paid is a reduced inductance below saturation. Curves B in Fig 10.17 show the result of an air gap added to that inductor.

Manufacturer's data sheets for magnetic cores usually specify the saturation flux density. Saturation flux density, ϕ, in gauss can be calculated for ac and dc currents from the following equations:

$$\phi_{ac} = \frac{3.49\,V}{fNA} \qquad (9)$$

$$\phi_{dc} = \frac{NIA_L}{10\,A} \qquad (10)$$

where

  V = RMS ac voltage
  f = frequency, in MHz
  N = number of turns
  A = equivalent area of the magnetic path in square inches (from the data sheet)
  I = dc current, in A
  A_L = inductance index (also from the data sheet).

Hysteresis

Consider **Fig 10.18**. If the current passed through a magnetic-core inductor is increased from zero (point a) to near the point of saturation (point b) and then decreased back to zero, we find that a magnetic field remains, because some of the core atoms retain their alignment. If we then increase the current in the opposite direction and again return to zero (through points c, d and e), the curve does *not* retrace itself. This is the property of *hysteresis*.

If a circuit carries a large ac current (that is, equal to or larger than saturation), the path shown in Fig 10.18 (from b to e and back again) will be retraced many times each second. Since the curve is nonlinear, this will introduce distortion in the result-

ing waveform. Where linear circuit operation is crucial, it is important to restrict the operation of magnetic-core inductors well below saturation.

Eddy Currents

The changing magnetic field produced by an ac current generates a "back voltage" in the core as well as the coil itself. Since magnetic core material is usually conductive, this voltage causes a current to flow in the core. This eddy current serves no useful purpose and represents power lost to heat. Eddy currents can be substantially reduced by laminating the core—slicing the core into thin sheets and placing a suitable insulating material

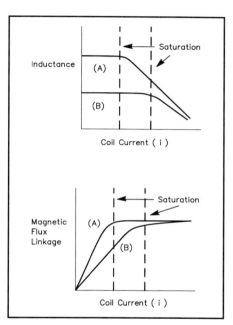

Fig 10.17—Magnetic flux linkage and inductance plotted versus coil current for (A) a typical iron-core inductor. As the flux linkage $N\phi$ in the coil saturates, the inductance begins to decrease since inductance = flux linkage / current. The curves marked B show the effect of adding an air gap to the core. The current handling capability has increased but at the expense of reduced inductance.

(such as varnish) between them (see Fig 10.16).

Transformers

In an ideal transformer (described in the **AC Theory** chapter), *all* the power supplied to its primary terminals is available at the secondary terminals. An air-core transformer provides a very good approximation to the ideal case, the major loss being the series resistances of the windings. As you might guess, there are several ways that a magnetic-core transformer can lose power. For example, useless heat can be generated through hysteresis and eddy currents; power may be lost to harmonic generation through nonlinear saturation effects.

Another source of loss in transformers (or actually, any pair of mutual inductances) is *leakage reactance*. While the main purpose of a magnetic core is to concentrate the magnetic flux entirely within itself, in a real-life device there is always some small amount of flux that does not pass through both windings. This leakage reactance can be pictured as a small amount of self-inductance appearing on each winding. **Fig 10.19** shows a fairly complete circuit model for a transformer at low to medium frequencies.

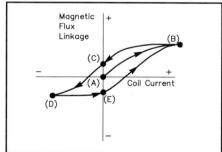

Fig 10.18—Hysteresis loop. A large current (larger than saturation current) passed through a magnetic-core coil will cause some permanent magnetization of the core. This results in a different path of flux linkage vs current to be traced as the current decreases.

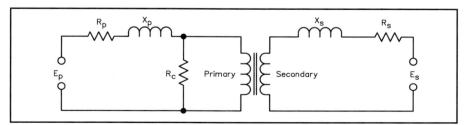

Fig 10.19—An equivalent circuit for a transformer at low to medium frequencies. R_p and R_S represent the series resistances of the windings, X_p and X_S are the leakage inductances, and R_c represents the dissipative losses in the core due to eddy currents and so on.

COMPONENTS AT RF

The models described in the previous section are good for dc, and ac up through AF and low RF. At HF and above (where we do much of our circuit design) several other considerations become very important, in some cases dominant, in our component models. To understand what happens to circuits at RF (a good cutoff is 30 MHz and above) we turn to a brief discussion of some electromagnetic and microwave theory concepts.

Parasitic Inductance

Maxwell's equations—the basic laws of electromagnetics that govern the propagation of electromagnetic waves and the operation of all electronic components—tell us that any wire carrying a current that changes with time (one example is a sine wave) develops a changing magnetic field around it. This changing magnetic field in turn induces an opposing voltage, or back EMF, on the wire. The back EMF is proportional to how fast the current changes (see **Fig 10.20**).

We exploit this phenomenon when we make an inductor. The reason we typically form inductors in the shape of a coil is to concentrate the magnetic field lines and thereby maximize the inductance for a given physical size. However, *all* wires carrying varying currents have these inductive properties. This includes the wires we use to connect our circuits, and even the *leads* of capacitors, resistors and so on.

The inductance of a straight, round, nonmagnetic wire in free space is given by:

$$L = 0.00508b \left[\ln \left(\frac{2b}{a} \right) - 0.75 \right] \quad (11)$$

where

L = inductance, in µH
a = wire radius, in inches
b = wire length, in inches
ln = natural logarithm ($2.303 \times \log_{10}$).

Skin effect (see below) changes this formula slightly at VHF and above. As the frequency approaches infinity, the constant 0.75 in the above equation approaches 1. This effect usually presents no more than a few percent change.

As an example, let's find the inductance of a #18 wire (diam = 0.0403 inch) that is 4 inches long (a typical wire in a circuit).

Then a = 0.0201 and b = 4:

$$L = 0.00508(4) \left[\ln \left(\frac{8}{0.0201} \right) - 0.75 \right]$$

$$= 0.0203 [5.98 - 0.75] = 0.106 \, \mu H$$

It is obvious that this *parasitic* inductance is usually very small. It becomes important *only at very high frequencies*; at AF or LF, the parasitic inductive reactance is practically zero. To use this example, the reactance of a 0.106 µH inductor even at 10 MHz is only 6.6 Ω. **Fig 10.21** shows a graph of the inductance for wires of various gauges (radii) as a function of length.

Any circuit component that has wires attached to it, or is fabricated from wire, will have a parasitic inductance associated with it. We can treat this parasitic inductance in component models by adding an inductor of appropriate value in *series* with the component (since the wire lengths are in series with the element). This (among other reasons) is why minimizing lead lengths and interconnecting wires becomes very important when designing circuits for VHF and above.

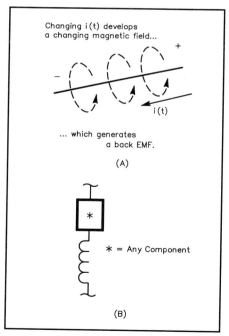

Fig 10.20—Inductive consequences of Maxwell's equations. At A, any wire carrying a changing current develops a voltage difference along it. This can be mathematically described as an effective inductance. B adds parasitic inductance to a generic component model.

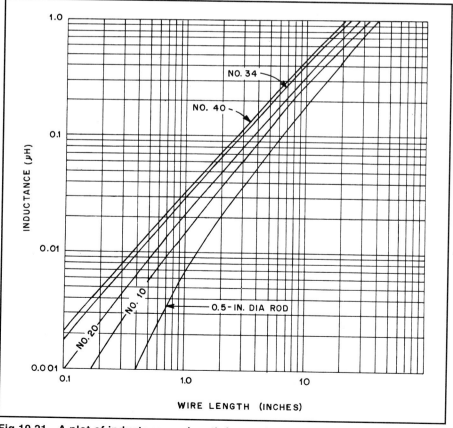

Fig 10.21—A plot of inductance vs length for straight conductors in several wire sizes.

Parasitic Capacitance

Maxwell's equations also tell us that if the voltage between any two points changes with time, a *displacement* current is generated between these points. See **Fig 10.22.** This displacement current results from the propagation, at the speed of light, of the electromagnetic field between the two points and is not to be confused with conduction current, which is caused by the movement of electrons. This displacement current is directly proportional to how fast the voltage is changing.

A capacitor takes advantage of this consequence of the laws of electromagnetics. When a capacitor is connected to an ac voltage source, a steady ac current can flow because taken together, conduction current and displacement current "complete the loop" from the positive source terminal, across the plates of the capacitor, and back to the negative terminal.

In general, parasitic capacitance shows up *wherever* the voltage between two points is changing with time, because the laws of electromagnetics require a displacement current to flow. Since this phenomenon represents an *additional* current path from one point in space to another, we can add this parasitic capacitance to our component models by adding a capacitor of appropriate value in *parallel* with the component. These parasitic capacitances are typically on the order of nF to

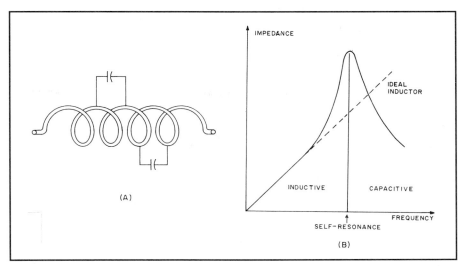

Fig 10.23—Coils exhibit distributed capacitance as explained in the text. The graph at B shows how distributed capacitance resonates with the inductance. Below resonance, the reactance is inductive, but it decreases as frequency increases. Above resonance, the reactance is capacitive and increases with frequency.

pF, so that below VHF they can be treated as open circuits (infinite resistances) and thus neglected.

Consider the inductor in **Fig 10.23**. If this coil has n turns, then the ac voltage between identical points of two neighboring turns is 1/n times the ac voltage across the entire coil. When this voltage changes due to an ac current passing through the coil, the effect is that of many small capacitors acting in parallel with the inductance of the coil. Thus, in addition to the capacitance resulting from the leads, inductors have higher parasitic capacitance due to their physical shape.

Package Capacitance

Another source of capacitance, also in the nF to pF range and therefore important only at VHF and above, is the packaging of the component itself. For example, a power transistor packaged in a TO-220 case (see **Fig 10.24**), often has either the emitter or collector connected to the metal tab itself. This introduces an extra *interelectrode capacitance* across the junctions.

The copper traces on a PC board also present capacitance to the circuit it holds. Double-sided PC boards have a certain capacitance per square inch. It is possible to create capacitors by leaving unetched areas of copper on both sides of the board. The capacitance is not well controlled on inexpensive low-frequency boards, however. For this reason, the copper on one side of a double-sided board should be completely removed under frequency-determining circuits such as VFOs. Board capacitance is *exploited* to make micro-

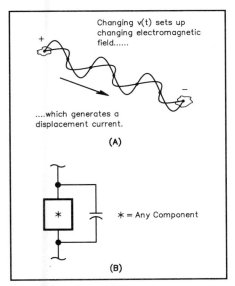

Fig 10.22—Capacitive consequences of Maxwell's equations. A: Any changing voltage between two points, for example along a bent wire, generates a *displacement* current running between them. This can be mathematically described as an effective capacitance. B adds parasitic capacitance to a generic component model.

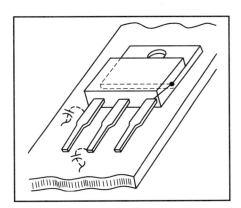

Fig 10.24—Unexpected stray capacitance. The mounting tab of TO-220 transistors is often connected to one of the device leads. Because one lead is connected to the chassis, small capacitances from the other leads to the chassis appear as additional package capacitance at the device. Similar capacitance can appear at any device with a conductive package.

wave transmission lines (microstrip lines). The capacitance of boards for microwave use is better controlled than that of less expensive board material.

Stray capacitance (a general term used for any "extra" capacitance that exists due to physical construction) appears in any circuit where two metal surfaces exist at different voltages. Such effects can be modeled as an extra capacitor in parallel with the given points in the circuit. A rough value can be obtained with the parallel-plate formula given above.

Thus, similar to inductance, *any* circuit

component that has wires attached to it, or is fabricated from wire, or is near or attached to metal, will have a parasitic capacitance associated with it, which again, becomes important only at RF.

A GENERAL MODEL

The parasitic problems due to component leads, packaging, leakage and so on are relatively common to all components. When working at frequencies where many or all of the parasitics become important, a complex but completely general model such as that in **Fig 10.25** can be used for just about any component, with the actual component placed in the box marked "*". Parasitic capacitance C_p and leakage conductance G_L appear in parallel across the device, while series resistance R_s and parasitic inductance L_p appear in series with it. Package capacitance C_{pkg} appears as an additional capacitance in parallel across the whole device. This maze of effects may seem overwhelming, but remember that it is very seldom necessary to consider all parasitics. In the Computer-Aided-Design section of this chapter, we will use the power of the computer to examine the combined effects of these multiple parasitics on circuit performance.

Self-Resonance

Because of the effects just discussed, a capacitor or inductor—all by itself—exhibits the properties of a resonant RLC circuit as we increase the applied fre-

quency. **Fig 10.26** illustrates RF models for the capacitor and inductor, which are based on the general model in Fig 10.25, leaving out the packaging capacitance. Note the slight difference in configuration; the pairs C_p, R_p and L_s, R_s are in series in the capacitor but in parallel in the inductor. This is because in the inductor, C_p and R_p are the parasitics, while in the capacitor, L_s and R_s are the added effects.

At some sufficiently high frequency, both inductors and capacitors become *self-resonant*. Just like a tuned circuit, above that frequency the capacitor will appear inductive, and the inductor will appear capacitive.

For an example, let's calculate the approximate self-resonant frequency of a 470-pF capacitor whose leads are made from #20 wire (diam 0.032 inch), with a total length of 1 inch. From the formula above, we calculate the approximate parasitic inductance

$$L(\mu H) = 0.00508(1)\left[\ln\left(\frac{2(1)}{(0.032/2)}\right) - 0.75\right]$$

$$= 0.021 \mu H$$

and then the self-resonant frequency is roughly

$$f = \frac{1}{2\pi\sqrt{LC}} = 50.6\,\text{MHz}$$

The purpose of making these calculations is to give you a rough feel for actual component values. They could be used as a rough design guideline, but should not be used quantitatively. Other factors such as lead orientation, shielding and so on, can alter the parasitic effects to a large extent. Large-value capacitors tend to have higher parasitic inductances (and therefore a lower self-resonant frequency) than small-value ones.

Self-resonance becomes critically im-

portant at VHF and UHF because the self-resonant frequency of many common components is at or below the frequency where the component will be used. In this case, either special techniques can be used to construct components to operate at these frequencies, by reducing the parasitic effects, or else the idea of lumped elements must be abandoned altogether in favor of microwave techniques such as striplines and waveguides.

Skin Effect

The resistance of a conductor to ac is different than its value for dc. A consequence of Maxwell's equations is that thick, near-perfect conductors (such as metals) conduct ac only to a certain depth that is proportional to the wavelength of the signal. This decreases the effective cross-section of the conductor at high frequencies and thus increases its resistance.

This resistance increase, called *skin effect*, is insignificant at low (audio) frequencies, but beginning around 1 MHz (depending on the size of the conductor) it is so pronounced that practically all the current flows in a very thin layer near the conductor's surface. For this reason, at RF a hollow tube and a solid tube of the same diameter and made of the same metal will have the same resistance. The depth of this skin layer decreases by a factor of 10 for every 100× increase in frequency. Consequently, the RF resistance is often much higher than the dc resistance. Also, a thin highly conductive layer, such as silver plating, can lower resistance for UHF or microwaves, but does little to improve HF conductivity.

A rough estimate of the cutoff frequency where a nonferrous wire will begin to show skin effect can be calculated from

$$f = \frac{124}{d^2} \tag{12}$$

where

f = frequency, in MHz
d = diam, in mils (a mil is 0.001 inch).

Above this frequency, increase the resistance of the wire by 10× for every 2 decades of frequency (roughly 3.2× for every decade). For example, say we wish to find the RF resistance of a 2-inch length of #18 copper wire at 100 MHz. From the wire tables, we see that this wire has a dc resistance of (2 in.) (6.386 Ω/1000 ft) = 1.06 milliohms. From the above formula, the cutoff frequency is found to be 124 / 40.3^2 = 76 kHz. Since 100 MHz is roughly three decades above this (100 kHz to 100 MHz), the RF resistance will be approximately (1.06 mΩ) (10 × 3.2) = 34 mΩ. Again, values calculated in this

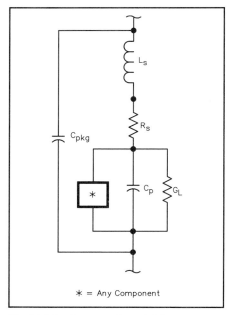

Fig 10.25—A general model for electrical components at VHF frequencies and above. The box marked "*" represents the component itself. See text for discussion.

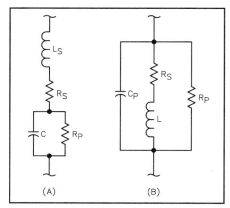

Fig 10.26—Capacitor (A) and inductor (B) models for RF frequencies.

manner are approximate and should be used qualitatively—that is, when you want an answer to a question such as, "Can I neglect the RF resistance of this length of connecting wire at 100 MHz?"

For additional information, see *Reference Data for Engineers*, Howard W. Sams & Co, Indianapolis, IN 46268. Chapter 6 contains a discussion and several design charts.

Effects on Q

Recall from the **AC Theory** chapter that circuit Q, a useful figure of merit for tuned RLC circuits, can be defined in several ways:

$$Q = \frac{X_L \text{ or } X_C \text{ (at resonance)}}{R} \qquad (13)$$

$$= \frac{\text{energy stored per cycle}}{\text{energy dissipated per cycle}}$$

Q is also related to the bandwidth of a tuned circuit's response by

$$Q = \frac{f_0}{BW_{3dB}} \qquad (14)$$

Parasitic inductance, capacitance and resistance can significantly alter the performance and characteristics of a tuned circuit if the design frequency is anywhere near the self-resonant frequencies of the components.

As an example, consider the resonant circuit of **Fig 10.27A**, which could represent the input tank circuit of an oscillator. Neglecting any parasitics, $f_0 = 1 / (2 \pi (LC)^{0.5}) = 10.06$ MHz. As in many real cases, assume the resistance arises entirely from the inductor series resistance. The data sheet for the inductor specified a minimum Q of 30, so assuming Q = 30 yields an R value of $X_L / Q = 2 \pi (10.06$ MHz) (5 μH) / 30 = 10.5 Ω.

Next, let's include the parasitic inductance of the capacitor (Fig 10.27B). A reasonable assumption is that this capacitor has the same physical size as the example from the Parasitic Inductance discussion above, for which we calculated $L_s = 0.106 \mu$H. This would give the capacitor a self-resonant frequency of 434 MHz—well above our area of interest. However, the added parasitic inductance does account for an extra 0.106/5.00 = 2% inductance. Since this circuit is no longer strictly series or parallel, we must convert it to an equivalent form before calculating the new f_0.

An easier and faster way is to *simulate* the altered circuit by computer. This analysis was performed on a desktop computer using *SPICE*, a standard circuit simulation program; for more details, see the Computer-Aided Design section below. The voltage response of the circuit

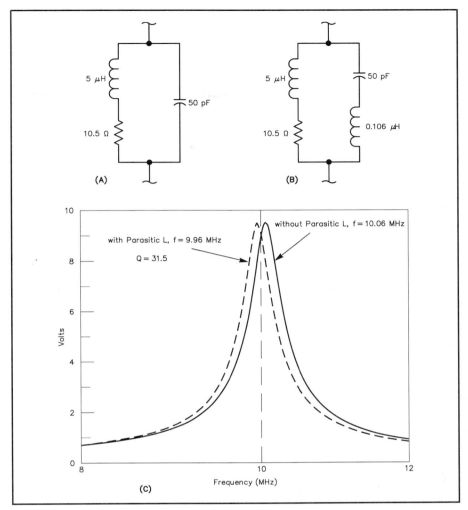

Fig 10.27—A is a tank circuit, neglecting parasitics. B same circuit including L_p on capacitor. C frequency response curves for A and B. The solid line represents the unaltered circuit (also see Fig 10.29) while the dashed line shows the effects of adding parasitic inductance.

(given an input current of 1 mA) was calculated as a function of frequency for both cases, with and without parasitics. The results are shown in the plot in Fig 10.27C, where we can see that the parasitic circuit has an f_0 of 9.96 MHz (a shift of 1%) and a Q (measured from the −3 dB points) of 31.5. For comparison, the simulation of the unaltered circuit does in fact show $f_0 = 10.06$ MHz and Q = 30.

Inductor Coupling

Mutual inductance will also have an effect on the resonant frequency and Q of the involved circuits. For this reason, inductors in frequency-critical circuits should always be shielded, either by constructing compartments for each circuit block or through the use of "can"-mounted coils. Another helpful technique is to mount nearby coils with their axes perpendicular as in **Fig 10.28**. This will minimize coupling.

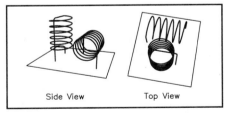

Fig 10.28—Unshielded coils in close proximity should be mounted perpendicular to each other to minimize coupling.

As an example, assume we build an oscillator circuit that has both input and output filters similar to the resonant circuit in Fig 10.27A. If we are careful to keep the two coils in these circuits uncoupled, the frequency response of either of the two circuits is that of the solid line in **Fig 10.29**, reproduced from Fig 10.27C.

If the two coils are coupled either

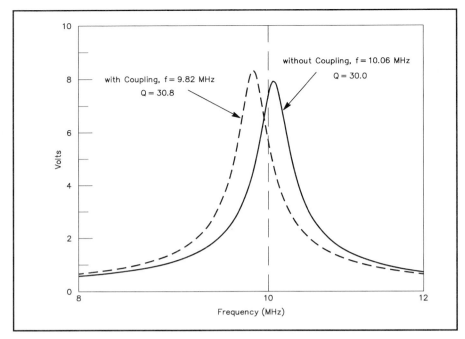

Table 10.3

Dielectric Constants and Breakdown Voltages

Material	Dielectric Constant*	Puncture Voltage**
Aisimag 196	5.7	240
Bakelite	4.4-5.4	240
Bakelite, mica filled	4.7	325-375
Cellulose acetate	3.3-3.9	250-600
Fiber	5-7.5	150-180
Formica	4.6-4.9	450
Glass, window	7.6-8	200-250
Glass, Pyrex	4.8	335
Mica, ruby	5.4	3800-5600
Mycalex	7.4	250
Paper, Royalgrey	3.0	200
Plexiglas	2.8	990
Polyethylene	2.3	1200
Polystyrene	2.6	500-700
Porcelain	5.1-5.9	40-100
Quartz, fused	3.8	1000
Steatite, low loss	5.8	150-315
Teflon	2.1	1000-2000

* At 1 MHz
** In volts per mil (0.001 inch)

Fig 10.29—Result of light coupling (k=0.05) between two identical circuits of Fig 10.27A on their frequency responses.

through careless placement or improper shielding, the resonant frequency and Q will be affected. The dashed line in Fig 10.29 shows the frequency response that results from a coupling coefficient of k = 0.05, a reasonable value for air-wound inductors mounted perpendicularly in close proximity on a circuit chassis. Note the resonant frequency shifted from 10.06 to 9.82 MHz, or 2.4%. The Q has gone up slightly from 30.0 to 30.8 as a result of the slightly higher inductive reactance at the resonant frequency.

To summarize, even small parasitics can significantly affect frequency responses of RF circuits. Either take steps to minimize or eliminate them, or use simple circuit theory to predict and anticipate changes.

Dielectric Breakdown and Arcing

Anyone who has ever watched a capacitor burn out, or heard the hiss of an arc across the inductor of an antenna tuner, while loading the 2-m vertical on 160-m, or touched a doorknob on a cold winter day has seen the effects of dielectric breakdown. When the dielectric is gaseous, especially air, we often call this *arcing*.

In the ideal world, we could take any two conductors and put as large a voltage as we want across them, no matter how close together they are. In the real world, there is a voltage limit (*dielectric strength*, measured in kV/cm and determined by the insulator between the two conductors) above which the insulator will break down.

Because they are charged particles, the electrons in the atoms of a dielectric material feel an attractive force when placed in an electric field. If the field is sufficiently strong, the force will strip the electron from the atom. This electron is available to conduct current, and furthermore, it is traveling at an extremely high velocity. It is very likely that this electron will hit another atom, and free another electron. Before long, there are many stripped electrons producing a large current. When this happens, we say the dielectric has suffered *breakdown*.

If the dielectric is liquid or gas, it can heal when the applied voltage is removed. A solid dielectric, however, cannot repair itself. A good example of this is a CMOS integrated circuit. When exposed to the very high voltages associated with static electricity, the electric field across the very thin gate oxide layer exceeds the dielectric strength of silicon dioxide, and the device is permanently damaged.

Capacitors are, by nature, perhaps the component most often associated with dielectric failure. To prevent damage, the working voltage of a capacitor—and there are separate dc and ac ratings—should ideally be 2 or 3 times the expected maximum voltage in the circuit.

Arcing is most often seen in RF circuits where the voltages are normally high, but it is possible anywhere two components at significantly different voltage levels are closely spaced.

The breakdown voltage of a dielectric layer depends on its composition and thickness (see **Table 10.3**). The variation with thickness is not linear; doubling the thickness does not quite double the breakdown voltage. Breakdown voltage is also a function of geometry: Because of electromagnetic considerations, the breakdown voltage between two conductors separated by a fixed distance is less if the surfaces are pointed or sharp-edged than if they are smooth or rounded. Therefore, a simple way to help prevent breakdown in many projects is to file and smooth the edges of conductors.

Radiative Losses and Coupling

Another consequence of Maxwell's equations states that any conductor placed in an electromagnetic field will have a current induced in it. We put this principle to good use when we make an antenna. The unwelcome side of this law of nature is the phrase "any conductor"; even conductors we don't intend as antennas will act this way.

Fortunately, the *efficiency* of such "antennas" varies with conductor length. They will be of importance only if their length is a significant fraction of a wavelength. When we make an antenna, we usually choose a length on the order of λ/2. Therefore, when we *don't* want an antenna, we should be sure that the conductor length is *much less* than λ/2, no more than 0.1 λ. This will ensure a very low-efficiency antenna. This is why 60-Hz power lines do not lose a significant fraction of the power they carry over long distances—at 60 Hz, 0.1 λ is about 300 miles!

In addition, we can use shielded cables. Such cables do allow some penetration of EM fields if the shield is not solid, but even 95% coverage is usually sufficient, especially if some sort of RF choke is used to reduce shield current.

Radiative losses and coupling can also be reduced by using twisted pairs of conductors—the fields tend to cancel. In some applications, such as audio cables, this may work better than shielding.

This argument also applies to large components, and remember that a component or long wire can *radiate* RF, as well as receive them. Critical stages such as tuned circuits should be placed in shielded compartments where possible. See the **EMI** and **Transmission Lines** chapters for more information.

REMEDIES FOR PARASITICS

The most common effect (always the most annoying) of parasitics is to influence the resonant frequency of a tuned circuit. This shift could cause an oscillator to fail, or more commonly, to cause a stable circuit to oscillate. It can also degrade filter performance (more on this later) and basically causing any number of frequency-related problems.

We can often reduce parasitic effects in discrete-component circuits by simply exploiting the models in Fig 10.26. Since parasitic inductance and loss resistance appear in series with a capacitor, we can reduce both by using several smaller capacitances in parallel, rather than of one large one.

An example is shown in the circuit block in **Fig 10.30**, which is representative of the input tank circuit used in many HF VFOs. C_{main}, C_{trim}, C1 and C3 act with L to set the oscillator frequency. Therefore, temperature effects are critical in these components. By using several capacitors in parallel, the RF current (and resultant heat) is reduced in each component. Parallel combinations are used at the feedback capacitors for the same reason.

Another example of this technique is the common capacitor bypass arrangement shown in **Fig 10.31**. C1 provides a bypass path at audio frequencies, but it has a low self-resonant frequency due to its large capacitance. How low? Even if we assume the 0.02-µH value above, f_{SC} = 355 kHz. Adding C2 (with a smaller capacitance but much higher self-resonant frequency) in parallel provides bypass at high frequencies where C1 appears inductive.

Construction Techniques

These concepts also help explain why so many different components exist with similar values. As an example, assume you're working on a project that requires you to wind a 5 µH inductor. Looking at the coil inductance formula in the **AC Theory** chapter, it comes to mind that many combinations of length and diameter could yield the desired inductance. If you happen to have both 0.5 and 1-inch coil forms, why should you select one over the other? To eliminate some other variables, let's make both coils 1 inch long, close-wound, and give them 1-inch leads on each end.

Let's calculate the number of turns required for each. On a 0.5-inch-diameter form:

$$n = \frac{\sqrt{5\left[(18 \times 0.5) + (40 \times 1)\right]}}{0.5}$$

$$= 31.3 \text{ turns}$$

This means coil 1 will be made from #20 wire (29.9 turns per inch). Coil 2, on the 1-inch form, yields

$$n = \frac{\sqrt{5\left[(18 \times 1) + (40 \times 1)\right]}}{1}$$

$$= 17.0 \text{ turns}$$

which requires #15 wire in order to be close-wound.

What are the series resistances associated with each? For coil 1, the total wire length is 2 inches + (31.3 × π × 0.5) = 51 inches, which at 10.1 Ω/1000 ft gives R_s = 0.043 Ω at dc. Coil 2 has a total wire

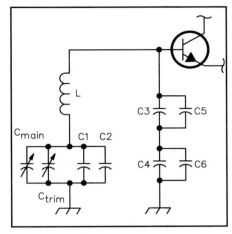

Fig 10.30—A tank circuit of the type commonly used in VFOs. Several capacitors are used in parallel to distribute the RF current, which reduces temperature effects.

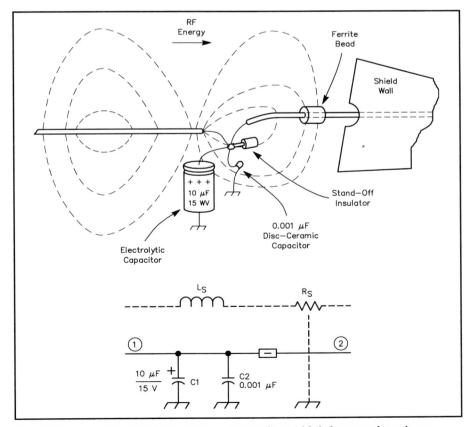

Fig 10.31—A typical method to provide bypassing at high frequencies when a large capacitor with a low-self-resonant frequency is required.

length of 2 inches + (17.0 × π × 1) = 55 inches, which at 3.18 Ω/1000 ft gives a dc resistance of $R_s = 0.015$ Ω, or about ⅓ that of coil 1. Furthermore, at RF, coil 1 will begin to suffer from skin effect at a frequency about 3 times lower than coil 2 because of its smaller conductor diameter. Therefore, if Q were the sole consideration, it would be better to use the larger diameter coil.

Q is not the only concern, however. Such coils are often placed in shielded enclosures. Rule of thumb says the enclosure should be at least one coil diameter from the coil on all sides. That is, 3×3×2 inches for the large coil and 1.5×1.5×1.5 for the small coil, a volume difference of over 500%.

THERMAL CONSIDERATIONS

Any real energized circuit consumes electric power because any real circuit contains components that convert electricity into other forms of energy.[1] This *dissipated power* appears in many forms. For example, a loudspeaker converts electrical energy into the motion of air molecules we call sound. An antenna (or a light bulb) converts electricity into electromagnetic radiation. Charging a battery converts electrical energy into chemical energy (which is then converted back to electrical energy upon discharge). But the most common transformation by far is the conversion, through some form of *resistance*, of electricity into heat.

Sometimes the power lost to heat serves a useful purpose—toasters and hair dryers come to mind. But most of the time, this heat represents a power loss that is to be minimized wherever possible or at least taken into account. Since all real circuits contain resistance, even those circuits (such as a loudspeaker) whose primary purpose is to convert electricity to some *other* form of energy also convert some part of their input power to heat. Often, such losses are negligible, but sometimes they are not.

If unintended heat generation becomes significant, the involved components will get warm. Problems arise when the temperature increase affects circuit operation by either

• causing the component to fail, by explosion, melting, or other catastrophic event, or, more subtly,

• causing a slight change in the properties of the component, such as through a temperature coefficient (TC).

In the first case, we can design conservatively, ensuring that components are rated to safely handle two, three or more times the maximum power we expect them to dissipate. In the second case, we can specify components with low TCs, or we can design the circuit to minimize the effect of any one component. Occasionally we even exploit temperature effects (for example, using a resistor, capacitor or diode as a temperature sensor). Let's look more closely at the two main categories of thermal effects.

HEAT DISSIPATION

Not surprisingly, heat dissipation (more correctly, the efficient removal of generated heat) becomes important in medium- to high-power circuits: power supplies, transmitting circuits and so on. While these are not the only examples where elevated temperatures and related failures are of concern, the techniques we will discuss here are applicable to all circuits.

Thermal Resistance

The transfer of heat energy, and thus the change in temperature, between two ends of a block of material is governed by the following heat flow equation (see **Fig 10.32**):

$$P = \frac{kA}{L} \Delta T = \frac{\Delta T}{\theta} \qquad (15)$$

where
   P = power (in the form of heat) conducted between the two points
   k = *thermal conductivity*, measured in W/(m °C), of the material between the two points, which may be steel, silicon, copper, PC board material and so on

L = length of the block
A = area of the block
ΔT = *change* in temperature between the two points;
$\theta = \dfrac{L}{kA}$ is often called the *thermal resistance* and has units of °C/W.

Thermal conductivities of various common materials at room temperature are given in **Table 10.4**.

A very useful property of the above equation is that it is *exactly* analogous to Ohm's Law, and therefore the same principles and methods apply to heat flow problems as circuit problems. The following correspondences hold:

• Thermal conductivity W/(m °C) <—> Electrical conductivity (S/m).
• Thermal resistance (°C/W) <—> Electrical resistance (Ω).
• Thermal current (heat flow) (W) <—> Electrical current (A).
• Thermal potential (T) <—> Electrical potential (V).
• Heat source <—> Current source.

For example, calculate the temperature of a 2-inch (0.05 m) long piece of #12 copper wire at the end that is being heated by a 25 W (input power) soldering iron, and whose other end is clamped to a large metal vise (assumed to be an infinite heat sink), if the ambient temperature is 25 °C (77 °F).

First, calculate the thermal resistance of the copper wire (diameter of #12 wire is 2.052 mm, cross-sectional area is 3.31 × 10^{-6} m²)

$$\theta = \frac{L}{kA} = \frac{(0.05 \text{ m})}{(390 \text{ W}/(\text{m} °\text{C}))(3.31 \times 10^{-6} \text{ m}^2)}$$

$$= 38.7 °\text{C}/\text{W}$$

Then, rearranging the heat flow equation above yields (after assuming the heat

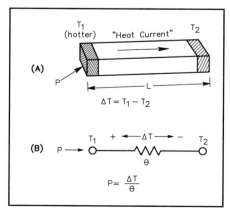

Fig 10.32—Physical and "circuit" models for the heat-flow equation.

Table 10.4

Thermal Conductivities of Various Materials

Gases at 0 °C, Others at 25 °C; from *Physics*, by Halliday and Resnick, 3rd Ed.

Material	$\left(\text{k in units of } \dfrac{W}{m \, °C} \right)$
Aluminum	200
Brass	110
Copper	390
Lead	35
Silver	410
Steel	46
Silicon	150
Air	0.024
Glass	0.8
Wood	0.08

---

[1]Superconducting circuits are the one exception; but they are outside the scope of this book.

energy actually transferred to the wire is around 10 W)

$$\Delta T = P\theta = (10 \text{ W}) (38.7°\text{C} / \text{W}) = 387°\text{C}$$

So the wire temperature at the hot end is 25 C + ΔT = 412 °C (or 774 °F). If this sounds a little high, remember that this is for the steady state condition, where you've been holding the iron to the wire for a long time.

From this example, you can see that things can get very hot even with application of moderate power levels. In the case of a soldering iron, that's good, but in the case of a 25-W power transistor in a transmitter output stage, it's bad. For this reason, circuits that generate sufficient heat to alter, not necessarily damage, the components must employ some method of cooling, either active or passive. Passive methods include heat sinks or careful component layout for good ventilation. Active methods include forced air (fans) or some sort of liquid cooling (in some high-power transmitters).

Heat Sink Design and Use

The purpose of a heat sink is to provide a high-power component with a large surface area through which to dissipate heat. To use the models above, it provides a low thermal-resistance path to a cooler temperature, thus allowing the hot component to conduct a large "thermal current" away from itself.

Power supplies probably represent one of the most common high-power circuits amateurs are likely to encounter. Everyone has certainly noticed that power supplies get warm or even hot if not ventilated properly. Performing the thermal design for a properly cooled power supply is a very well-defined process and is a good illustration of heat-flow concepts.

A 28-V, 10-A power supply will be used for this design example. This material was originally prepared by ARRL Technical Advisor Dick Jansson, WD4FAB, and the steps described below were actually followed during the design of that supply.

An outline of the design procedure shows the logic applied:

1. Determine the expected power dissipation (Pin).
2. Identify the requirements for the dissipating elements (maximum component temperature).
3. Estimate heat-sink requirements.
4. Rework the electronic device (if necessary) to meet the thermal requirements.
5. Select the heat exchanger (from heat sink data sheets).

The first step is to estimate the filtered, unregulated supply voltage under full

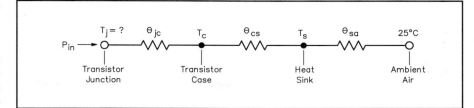

Fig 10.33—Resistive model of thermal conduction in a power transistor and associated heat sink. See text for calculations.

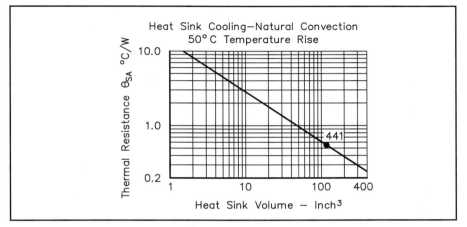

Fig 10.34—Thermal resistance vs heat-sink volume for natural convection cooling and 50°C temperature rise. The graph is based on engineering data from EG&G Wakefield.

load. Since the transformer secondary output is 32 V ac (RMS) and feeds a full-wave bridge rectifier, let's estimate 40 V as the filtered dc output at a 10-A load.

The next step is to determine our critical components and estimate their power dissipations. In a regulated power supply, the pass transistors are responsible for nearly all the power lost to heat. Under full load, and allowing for some small voltage drops in the power-transistor emitter circuitry, the output of the series pass transistors is about 29 V for a delivered 28 V under a 10-A load. With an unregulated input voltage of 40 V, the total energy heat dissipated in the pass transistors is (40 V – 29 V) × 10 A = 110 W. The heat sink for this power supply must be able to handle that amount of dissipation and still keep the transistor junctions below the specified safe operating temperature limits. It is a good rule of thumb to select a transistor that has a maximum power dissipation of twice the desired output power.

Now, consider the ratings of the pass transistors to be used. This supply calls for 2N3055s as pass transistors. The data sheet shows that a 2N3055 is rated for 15-A service and 115-W dissipation. But the design uses *four* in parallel. Why? Here we must look past the big, bold type at the top of the data sheet to such subtle characteristics as the junction-to-case thermal resistance, θ_{jc}, and the maximum allowable junction temperature, T_j.

The 2N3055 data sheet shows θ_{jc} = 1.52 °C/W, and a maximum allowable case (and junction) temperature of 220 °C. While it seems that one 2N3055 could barely, on paper at least, handle the electrical requirements, at what temperature would it operate?

To answer that, we must model the entire "thermal circuit" of operation, starting with the transistor junction on one end and ending at some point with the ambient air. A reasonable model is shown in **Fig 10.33**. The ambient air is considered here as an infinite heat sink; that is, its temperature is assumed to be a constant 25 °C (77 °F). θ_{jc} is the thermal resistance from the transistor junction to its case. θ_{cs} is the resistance of the mounting interface between the transistor case and the heat sink. θ_{sa} is the thermal resistance between the heat sink and the ambient air. In this "circuit," the generation of heat (the "thermal current source") occurs in the transistor at P_{in}.

Proper mounting of most TO-3 package power transistors such as the 2N3055 requires that they have an electrical insulator between the transistor case and the heat sink. However, this electrical insulator must at the same time exhibit a low thermal resistance. To achieve a quality mounting, use thin polyimid or mica formed washers and a suitable thermal compound to exclude air from the interstitial space. "Thermal greases" are commonly available for this function. Any silicone grease may be used, but filled silicone oils made specifically for this purpose are better.

Using such techniques, a conservatively high value for θ_{cs} is 0.50 °C/W. Lower values are possible, but the techniques needed to achieve them are expensive and not generally available to the average amateur. Furthermore, this value of θ_{cs} is already much lower than θ_{jc}, which cannot be lowered without going to a somewhat more exotic pass transistor.

Finally, we need an estimate of θ_{sa}. **Fig 10.34** shows the relationship of heat-sink volume to thermal resistance for natural-convection cooling. This relationship presumes the use of suitably spaced fins (0.35 inch or greater) and provides a "rough order-of-magnitude" value for sizing a heat sink. For a first calculation, let's assume a heat sink of roughly 6×4×2 inch (48 cubic inches). From Fig 10.34, this yields a θ_{sa} of about 1 °C/W.

Returning to Fig 10.33, we can now calculate the approximate temperature increase of a single 2N3055:

$$\delta T = P\, \theta_{total} = (110\text{ W})\,(1.52\text{ °C/W} + 0.5\text{ °C/W} + 1.0\text{ °C/W}) = 332\text{ °C}$$

Given the ambient temperature of 25 °C, this puts the junction temperature T_j of the 2N3055 at 25 + 332 = 357 °C! This is

clearly too high, so let's work backward from the air end and calculate just how many transistors we need to handle the heat.

First, putting more 2N3055s in parallel means that we will have the thermal model illustrated in **Fig 10.35**, with several identical θ_{jc} and θ_{cs} in parallel, all funneled through the same θ_{as} (we have one heat sink).

Keeping in mind the physical size of the project, we could comfortably fit a heat sink of approximately 120 cubic inches (6×5×4 inches), well within the range of commercially available heat sinks. Furthermore, this application can use a heat sink where only "wire access" to the transistor connections is required. This allows the selection of a more efficient design. In contrast, RF designs require the transistor mounting surface to be completely exposed so that the PC board can be mounted close to the transistors to minimize parasitics. Looking at Fig 10.34, we see that a 120-cubic-inch heat sink yields a θ_{sa} of 0.55 °C/W. This means that the temperature of the heat sink when dissipating 110 W will be 25 °C + (110 W) (0.55 °C/W) = 85.5 °C.

Industrial experience has shown that silicon transistors suffer substantial failure when junctions are operated at highly elevated temperatures. Most commercial and military specifications will usually not permit design junction temperatures to exceed 125 °C. To arrive at a safe figure for our maximum allowed T_j, we must consider the intended use of the power supply. If we are using it in a 100% duty-cycle transmitting application such as RTTY or FM, the circuit will be dissipating 110 W continuously. For a lighter duty-cycle load such as CW or SSB, the "key-down" temperature can be slightly higher as long as the average is less than 125 °C. In this intermittent type of service, a good conservative figure to use is $T_j = 150$ °C.

Given this scenario, the temperature rise across each transistor can be 150 – 85.5 = 64.5 °C. Now, referencing Fig 10.35, remembering the total θ for each 2N3055 is 1.52 + 0.5 = 2.02 °C/W, we can calculate the maximum power each 2N3055 can safely dissipate:

$$P = \frac{\delta T}{\theta} = \frac{64.5\text{ °C}}{2.02\text{ °C/W}} = 31.9\text{ W}$$

Thus, for 110 W full load, we need four 2N3055s to meet the thermal requirements of the design. Now comes the big question: What is the "right" heat sink to use? We have already established its requirements: it must be capable of dissipating 110 W, and have a θ_{sa} of 0.55 °C/W (see above).

A quick consultation with several manufacturer's catalogs reveals that Wakefield Engineering Co model nos. 441 and 435 heat sinks meet the needs of this application.[2] A Thermalloy model no. 6441 is suitable as well. Data published in the catalogs of these manufacturers show that in natural-convection service, the expected temperature rise for 100 W dissipation would be just under 60 °C, an almost perfect fit for this application. Moreover, the no. 441 heat sink can easily mount four TO-3-style 2N3055 transistors. See **Fig 10.36**. Remember: heat sinks should be mounted with the fins and transistor mounting area vertical to promote convection cooling.

The design procedure just described is applicable to any circuit where heat buildup is a potential problem. By using the thermal-resistance model, we can easily calculate whether or not an external means of cooling is necessary, and if so, how to choose it. Aside from heat sinks, forced air cooling (fans) is another common method. In commercial transceivers, heat sinks with forced-air cooling are common.

Transistor Derating

Maximum ratings for power transistors are usually based on a case temperature of 25 °C. These ratings will decrease with increasing operating temperature. Manufacturer's data sheets usually specify a *derating* figure or curve that indicates how the maximum ratings change per degree rise in temperature. If such information is not available (or even if it is!), it is a good

[2]There are numerous manufacturers of excellent heat sinks. References to any one manufacturer are not intended to exclude the products of any others, nor to indicate any particular predisposition to one manufacturer. The catalogs and products referred to here are from: EG&G Wakefield Engineering (see the Address List in the **References** chapter).

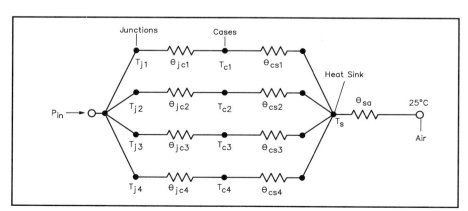

Fig 10.35—Thermal model for multiple power transistors mounted on a common heat sink.

rule of thumb to select a power transistor with a maximum power dissipation of at least twice the desired output power.

Rectifiers

Diodes are physically quite small, and they operate at high current densities. As a result their heat-handling capabilities are somewhat limited. Normally, this is not a problem in high-voltage, low-current supplies. The use of high-current (2 A or greater) rectifiers at or near their maximum ratings, however, requires some form of heat sinking. Frequently, mounting the rectifier on the main chassis (di-

(A)

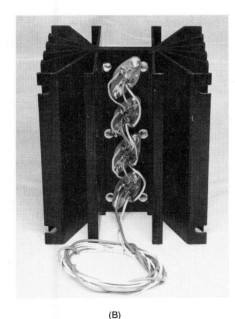

(B)

Fig 10.36—An EG&G Wakefield 441 heat sink with four 2N3055 transistors mounted.

rectly, or with thin mica insulating washers) will suffice. If the diode is insulated from the chassis, thin layers of silicone grease should be used to ensure good heat conduction. Large, high-current rectifiers often require special heat sinks to maintain a safe operating temperature. Forced-air cooling is sometimes used as a further aid.

Forced-Air Cooling

In Amateur Radio today, forced-air cooling is most commonly found in vacuum-tube circuits or in power supplies built in small enclosures, such as those in solid-state transceivers or computers. Fans or blowers are commonly specified in cubic feet per minute (CFM). While the nomenclature and specifications differ from those used for heat sinks, the idea remains the same: to offer a low thermal resistance between the inside of the enclosure and the (ambient) exterior.

For forced air cooling, we basically use the "one resistor" thermal model of Fig 10.32. The important quantity to be determined is heat generation, P_{in}. For a power supply, this can be easily estimated as the difference between the input power, measured at the transformer primary, and the output power at full load. For variable-voltage supplies, the worst-case output condition is minimum voltage with maximum current.

A discussion of forced-air tube cooling appears in the **Amplifiers** chapter.

TEMPERATURE STABILITY

Aside from catastrophic failure, temperature changes may also adversely affect circuits if the TCs of one or more components is too large. If the resultant change is not too critical, adequate temperature stability can often be achieved simply by using higher-precision components with low TCs (such as NP0/C0G capacitors or metal-film resistors). For applications where this is impractical or impossible (such as many solid-state circuits), we can minimize temperature sensitivity by *compensation* or *matching*—using temperature coefficients to our advantage.

Compensation is accomplished in one of two ways. If we wish to keep a certain circuit quantity constant, we can interconnect pairs of components that have equal but opposite TCs. For example, a resistor with a negative TC can be placed in series with a positive TC resistor to keep the total resistance constant. Conversely, if the important point is to keep the *difference* between two quantities constant, we can use components with the *same* TC so that the pair "tracks." That is, they both change by the same amount with temperature.

An example of this is a Zener reference circuit. Since the I-V equation for a diode is strongly affected by operating temperature, circuits that use diodes or transistors to generate stable reference voltages must use some form of temperature compensation. Since, for a constant current, a reverse-biased PN junction has a negative voltage TC while a forward-biased junction has a positive voltage TC, a good way to temperature-compensate a Zener reference diode is to place one or more forward-biased diodes in series with it.

RF Heating

RF current often causes component heating problems where the same level of dc current may not. An example is the tank circuit of an RF oscillator. If several small capacitors are connected in parallel to achieve a desired capacitance, skin effect will be reduced and the total surface area available for heat dissipation will be increased, thus significantly reducing the RF heating effects as compared to a single large capacitor. This technique can be applied to any similar situation; the general idea is to divide the heating among as many components as possible.

CAD TOOLS FOR CIRCUIT DESIGN

Hams today enjoy the easy availability of tremendous computing power to aid them in their hobby. A commercial application that computers perform very well is circuit simulation. Today anyone with access to a computer software bulletin board is likely to find shareware or other inexpensive software to perform such analysis. Indeed, ARRL offers *ARRL Radio Designer,* which provides excellent RF design capabilities at a reasonable price. *Radio Designer* is discussed in October 1994 *QST* and a subsequent *QST* column, "Exploring RF."

The advantages of computer circuit simulation over pencil-and-paper calculation are many. For example, while the analysis of a small circuit, say 5 to 10 components, may be recomputed without too much effort if the design changes, the same is not true for a 50 to 100-component circuit. Indeed, the larger circuit may require too many simplifying assumptions to calculate by hand at all. It is much faster to watch and adjust the response of a circuit on a computer screen than to breadboard the circuit in search of the same information.

All circuit simulators, including *SPICE,* work from a *netlist* that is simply a component-by-component list of the circuit that tells the program how components are interconnected. The program then uses standard techniques of numerical analysis and matrix mathematics to calculate the analysis you wish to perform. Common analyses include dc and ac bias and operating point calculations, frequency response curves, transient analysis (looking at waveforms in the circuit over a specified period of time), Fourier transforms, transfer functions, two-port parameter calculations, and pole-zero analysis. Many packages also perform sensitivity analyses: calculating what happens to the voltages and currents as you sweep the value(s) of an individual component (or group of components), and statistical analyses: determining the variational "window" of the system response that would result if all component values randomly varied by a given tolerance (also known as *Monte Carlo* analysis).

Beware!

Before we turn to a brief introduction to circuit simulation for the amateur, a *caveat* to the reader is definitely in order: However fast and powerful the computer may seem, keep in mind that it is only another tool in your workshop, just like your 'scope or soldering iron. Circuit simulations are only meaningful if you can interpret the results correctly, and a good initial circuit design based on real experience and common sense is mandatory. *SPICE* and other programs have no problem with a bench-top power supply delivering 10,000 V to a 5-Ω resistor because you made a mistake when specifying the circuit. Software also won't remind you that the resistor better be rated for 20 megawatts! Remember, the real power behind any simulation software is *your mind*.

SPICE

The electronics industry has been the main consumer of circuit simulation packages ever since the development of *SPICE* (or *Simulation Program with Integrated Circuit Emphasis*) on mainframe systems in the mid 1970s. While many companies today produce other simulation software, *SPICE* remains the *de facto* industry standard. *SPICE* itself now exists in many "flavors" and specially tailored versions for desktop computers are sold by several companies, including *PSPICE* by Microsim Corp and *HSPICE* by Intusoft. In fact, an "evaluation" copy of *PSPICE* has been placed in the public domain by Microsim and is probably the easiest and most inexpensive way for amateurs to begin circuit simulation. It may be obtained (with a companion reference book) from technical bookstores, and also from many sites on the Internet. See Chapter 30, **References**, for contact information for Microsim and Intusoft.

A Basic Circuit

Consider the simple RC low-pass filter in **Fig 10.37A**. To simulate this circuit, we must first precisely describe it. Convention sets ground as *node* 0, and we label all other nodes (a node is a place where two or more components meet) with numbers, as shown in the figure. We give each element a unique name, and then prepare a *netlist*, listing each element with the numbers of its "+" and "−" nodes (*SPICE* assumes the ground node to be labeled 0), and its value, as shown at B in the figure.

The question we wish to answer by simulation is the following: what is the frequency response of this filter? Of course, for this simple problem we can calculate the answer by hand. The cutoff frequency of a simple RC filter is given by:

$$f_{co} = \frac{1}{2\pi RC} \qquad (17)$$

where
  f_{co} = cutoff frequency (−3 dB)
  R = resistance, in ohms
  C = capacitance, in farads.

In this case f_{co} is 1.59 kHz. Beyond this, with one storage element (the capacitor) we expect the output to decrease by 20 dB for each decade of frequency.

Fig 10.38 shows the graph of the ratio of output to input voltages V_{out}/V_{in}, in dB for that frequency range as calculated by *SPICE*. Note the 3-dB point is indeed 1.59 kHz, and above that frequency the response drops by 20-dB/decade.

At this point we could use this circuit to ask some "What if?" questions, but for a circuit this simple we could answer those questions with pencil and paper. Below is a more extensive design to show the power of simulation, where "What if?" questions and the interaction of numerous components are much more easily and clearly examined with a computer.

Case Study No. 1: Transmitter Low-Pass Filter Design

Our first design illustration will be a low-pass transmitter filter for 10 m and down that provides good attenuation at TV channel 2 and above. Such a filter is a real example of the effects discussed in this chapter and their magnitude. The chosen design specifications are as follows:

Insertion Loss: ≤0.5 dB for f
  < 29.7 MHz
Attenuation: ≥40 dB for f >
  54 MHz (TV channel 2)

We will design the filter using the design tables found in the **References** chapter, and then use circuit simulation to correct, tweak and verify our design. Our objective is to include all the parasitic effects we have been discussing, and examine their effect and possible degradation of the filter transfer function. We will examine:

1. Attenuation due to series resistance in inductors.
2. Effects due to component tolerance.
3. Self-resonance effects of the capacitors.

From common sense, more components in the filter mean greater total possible parasitic effects, and greater total possible variation from the desired response curve. Therefore, we will choose the filter with the least number of components that can meet the specified rolloff (40 dB from 29.7 to 54 MHz). The filter tables give parameters for the three common filter classes—

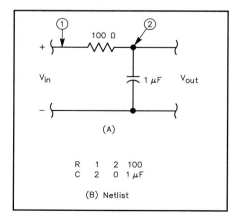

Fig 10.37—A, a simple RC low-pass filter.

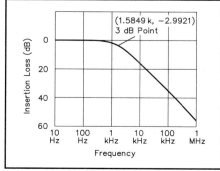

Fig 10.38—Low-pass filter output simulation.

Butterworth, Chebyshev, and Elliptic. Normalizing the filter specifications gives a ratio of $f_{40dB} / f_{co} = 54 / 29.7 = 1.82$, or equivalently, a rolloff of $40 / \log10(1.82) = 154$ dB/decade. Scanning the filter tables, we see that the minimum filter configurations that meet our needs are a 10-element Butterworth, a 7-element Chebyshev (at various ripple factors), and a 5-branch (7-element) Elliptic. The Butterworth will require 10 elements even though at first glance it seems an 8-element, at 160 dB/decade, should fit the bill. The reason for this is that the Butterworth filter has a 3-dB loss at cutoff. Obtaining < 0.5 dB in the passband requires pushing the cutoff frequency to around 33 MHz, which then requires 187 dB/decade, 10 elements.

We therefore eliminate the Butterworth. The Chebyshev and Elliptic designs tie for the fewest elements (seven). The capacitive input/output Chebyshev contains 3 Ls and 4 Cs, the inductive version contains 4 Ls and 3 Cs, and the Elliptic has 2 Ls and 5 Cs. Given these three choices, we chose the 7-element capacitive input/output Chebyshev, even though the Elliptic may perform better, because its ideal rolloff curve is simpler than the Elliptic and the effects we are trying to show will be more easily discernible.

Scanning the appropriate filter design table, (**Table 30.22**), it seems filter 52 will fit the bill if we scale it up by a factor of 10, using the procedure outlined in the **Filters** chapter. This yields a cutoff frequency $f_{co} = 32.6$ MHz and a 40-dB frequency of 51.9 MHz—plenty of room, or so it would seem. This design, which we'll call Filter A, is shown in **Fig 10.39A**.

The calculated frequency response from *SPICE* for Filter A is shown by the line labeled "Ideal" in **Fig 10.40**. (NOTE: The *SPICE* input files that were used for generating the figures are grouped at the end of the discussion). The large figure shows the complete response while the inset shows a magnified picture around the cutoff frequency. At this stage the design seems to be performing nicely. Let's build it, okay? Not so fast. At this point, the inductors and capacitors are still perfectly ideal.

We should now approximate real effects as best we can. We know the inductors have series resistance—but how much? An approximate value (within a factor of 2) should be good enough. Since this filter should be able to handle 100 W of power, we should wind our coils ourselves using fairly large wire with an air core; this will also reduce the series resistance. Using the inductance formula from the **AC Theory** chapter, a 1-inch 0.5 µH coil with a di-

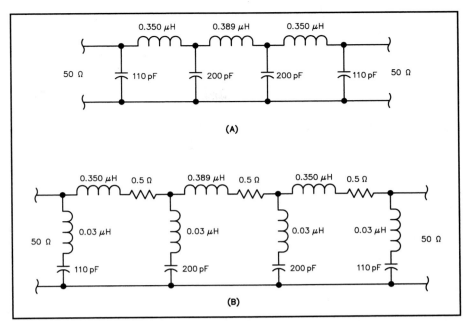

Fig 10.39—Filter A, the initial design for the low-pass TVI filter. This is filter 52 from Table 30.22 in the References chapter, scaled up 10×. Values are in µH and pF. A, ideal case. B, nominal case including parasitics.

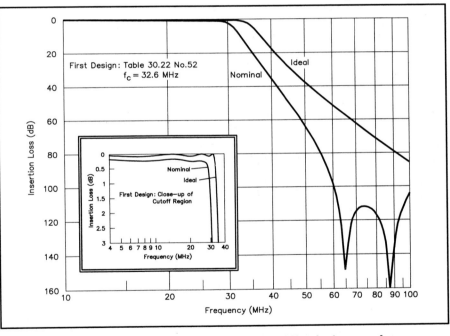

Fig 10.40—Response curves for Filter A for Ideal and Nominal cases. A more accurate model would show a smaller difference between the Ideal and Nominal.

ameter of 0.5 inch, wound from #14 wire with 0.5-inch leads on each end, takes about 7 inches of wire, which for #14 translates to a resistance of about 0.01 Ω. This is *a dc value*! Counting skin effect increases this to about 0.3 Ω. We'll round this up to 0.5 Ω to be conservative.

The capacitors also will have some parasitic inductance. Again using the value computed above for the 470-pF ca-

pacitor, we'll round this up as well to 0.03 µH. We'll ignore the parasitic capacitance of the inductors just to keep it from getting too complicated; it's probably the smallest of all these effects anyway.

With these additions, the electrical circuit for Filter A now looks like Fig 10.39B. This is now approaching the limits of hand calculations. *SPICE* has no problem, however, and the response is shown by the

"Nominal" curve in Fig 10.40, again with the cutoff region magnified in the inset. Note that the notches produced by the capacitor self-resonances actually *help* the stopband attenuation. *However*, the series resistances have now increased the insertion loss at 29.7 MHz to slightly more than 1 dB! Our filter already shows signs of failure, but we have one more step to perform before reevaluating our design.

The curves in Fig 10.40 assume that the values of our components are *exactly* what we say they are, but the real capacitors we buy will have a tolerance rating, probably in the range of ±10%. Granted, if we have the means to measure capacitors, we could hand-pick components with exactly the required values from a bag of capacitors, but our goal is to not to have to do that. We are hand-winding our inductors, and we will assume a tolerance of ±20%.

We can ask *SPICE* to perform a "Monte Carlo" or statistical analysis to answer the question, "How much can the filter response change from the nominal response if each component value actually falls randomly within a certain tolerance range?" Such an analysis is conducted by recalculating the filter response n times, each time selecting a random value for each component from the specified range. To illustrate, say we performed this analysis on the simple low-pass filter mentioned earlier, where R = 100 Ω, ±5%, and C = 1 μF, ±10%. On each run of the analysis, *SPICE* would randomly choose a resistor value between 95 and 105 Ω and a capacitor value between 0.9 and 1.1 μF, and compute the new filter response.

Obviously, the more "trials" we perform in our statistical analysis, the more confidently we can say we will see the total possible variation. For our filter design, we performed 80 trials. The results are shown in **Fig 10.41**. We now see a response "band" instead of a single response. While we can easily guarantee a cutoff of 40 dB at 54 MHz, the best we can do at 29.7 MHz is about 8 dB! In fact, from the magnified graph we see that we can guarantee 0.5 dB only up to about 17 MHz or so. Accounting for tolerances has produced the worst effect!

Let's tweak the design to compensate. The outer tolerance of our response band for 40 dB is about 43 MHz. Let's scale the filter *up* to place this point at 54 MHz. Hopefully by so doing, we will move the response band near the cutoff region inside our specs. This is a scaling of 54/43 ≈ 1.25×, placing the new cutoff at 32.6 × 1.25 = 40.7 MHz. Scaling the components appropriately, we arrive at the new circuit of **Fig 10.42**, which we'll call Filter B.

The new "Ideal" and "Nominal" graphs are shown in **Fig 10.43**. The same parasitic values for Filter A were used for Filter B. Statistical analysis results for Filter B, where we assumed the same tolerances as for Filter A, are shown in **Fig 10.44**. As anticipated, the response window at 40 dB falls snugly inside 54 MHz. Now, the f_{co} specification is still inside the response band, but the majority of curves do fall above it.

Given these results, at this point we could proceed in two directions. If we insist on the more robust design that has the tolerances factored in, we would decide that we do in fact need a steeper rolloff and would repeat this analysis for a 9-element Chebyshev filter, pushing the cutoff frequency out a little bit further. If we have the luxury of hand-picking precise component values, we could

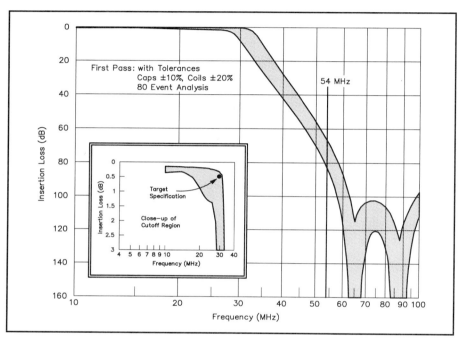

Fig 10.41—Statistical analysis results for Filter A.

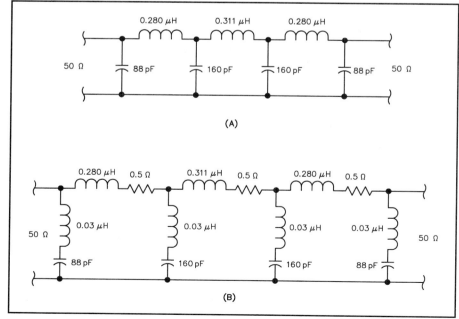

Fig 10.42—Filter B circuit and component values.

stop here, since the nominal response curve is more than adequate. The main purpose of this example was to illustrate the power of simulation when the circuit gets complicated, and we wish to ask "What if?" questions.

Case Study No. 2: Amplifier Distortion

Another application where computer circuit simulation does a much faster, easier, and more accurate job than pencil and paper is Fourier analysis. Recall that any periodic signal can be broken down into a sum of sine waves of different amplitudes whose frequencies are multiples of the fundamental frequency of the sig-nal. Performing such an analysis allows us to identify how much signal power is contained in the fundamental frequency and its harmonics. A Fourier *transform* is the representation of a signal by its frequency components, and resembles what you would see if you fed the signal into a spectrum analyzer.

Consider the amplifier circuit in **Fig 10.45**, which is a simple common-emitter audio amplifier. The frequency re-sponse of its voltage gain, calculated from *SPICE*, is shown in **Fig 10.46**. Pay particular attention to the voltage gain at 1 kHz, which is 148, or 43.4 dB. If we place a small ac signal on the input as shown in Fig 10.45, the output will be a faithfully amplified sine wave. How-ever, the amplifier is powered from a 9-V supply, so when the output reaches this level, the waveform will begin to show signs of clipping. For a

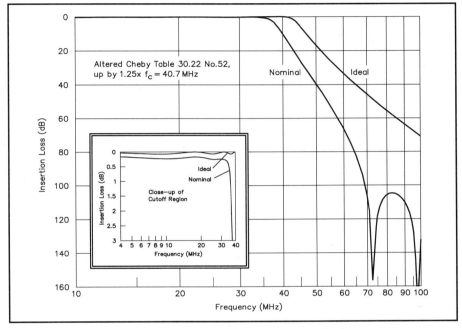

Fig 10.43—Response curves for Filter B for Ideal and Nominal cases.

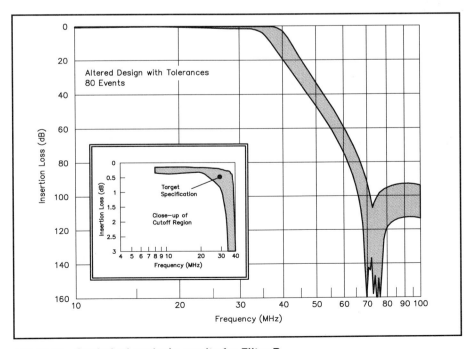

Fig 10.44—Statistical analysis results for Filter B.

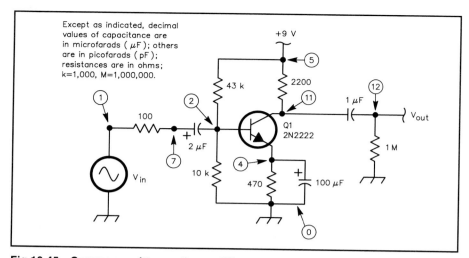

Fig 10.45—Common-emitter audio amplifier used in the text.

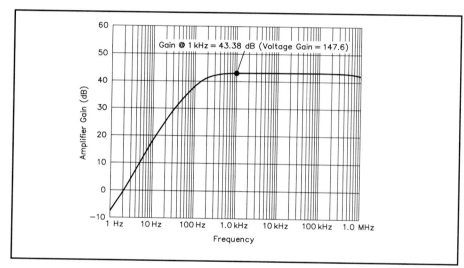

Fig 10.46—Frequency response of the voltage gain of the amplifier in Fig 10.45. The gain at 1 kHz is approximately 148 or 43 dB.

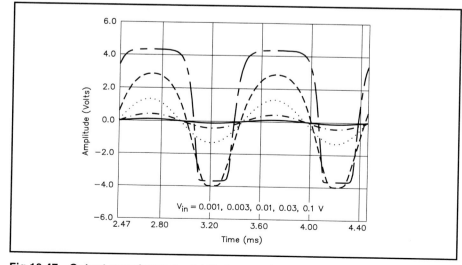

Fig 10.47—Output waveforms for 1-kHz inputs of different magnitudes. Note the clipping that begins to appear at approximately $V_{in} = 0.03$ V.

1-kHz signal, this will occur at roughly $V_{in} = 4.5/148 = 0.03$ V amplitude.

Total Harmonic Distortion is a common figure of merit for audio amplifiers, and is defined as

$$THD = \frac{P_H}{P_F} \times 100 \qquad (18)$$

where

  THD = total harmonic distortion, in percent

  P_H = total power in all harmonics above the fundamental

  P_F = power in fundamental.

SPICE readily calculates the Fourier content of a signal and also THD. Assume that we will tolerate 5% THD for our design. What is the maximum swing we can allow the input signal?

Fig 10.47 shows the output waveforms that *SPICE* calculates for our circuit with input signals of 0.001, 0.003, 0.01, 0.03 and 0.1-V amplitude. Note that, as expected, the last two input voltages show definite signs of clipping. What is not so evident is that the smaller inputs have some distortion as well, due to the small but finite nonlinearity of the amplifier even at those voltages.

Fig 10.48 shows the relative amplitudes of the harmonic content (Fourier transform) of the Fig 10.47 waveforms, all normalized to fundamental = 100%. You can see how the higher harmonics grow in strength as the signal increases, especially when the clipping starts. The THD for each case is also shown, and from this simulation we see we must limit our input to somewhere between 0.003 and 0.01 V. We would pin this down more closely by running another simulation around this "window."

A caution is in order here. Circuit models have limits just as do circuits. The models apply only over a limited dynamic range. That's why we have both small- and large-signal models. One challenge of CAD is determining the limits of your models. There is more information about this in the sections on modeling transistors.

Conclusion

Start using your computer for more challenging tasks than QSL bookkeeping! Just like any other piece of software, you'll find yourself using *SPICE* to do things you never thought of trying before. It can even do digital circuits and transmission lines (no antennas, though). Remember that a computer is only as smart as the person using it! Have fun!

References

J. Carr, *Secrets of RF Circuit Design*, Tab Books, 1991.

D. DeMaw, *Practical RF Design Manual*, Prentice-Hall, 1982.

R. Dorf, Ed., *The Electrical Engineering Handbook*, CRC Press and IEEE Press, 1993.

W. Hayward, *Introduction to RF Design*, ARRL, 1994.

P. Tuinenga, *SPICE: A Guide to Circuit Simulation and Analysis using PSPICE*, by Prentice-Hall, ISBN 0-13-747270-6. This is a wonderful *PSPICE* reference manual that comes with a copy of the software. It is available at most university or technical bookstores.

A. Vladimirescu, *The SPICE Book*, Wiley & Sons, 1994.

D. Pederson and K. Mayaram, *Analog Integrated Circuits for Communication: Principles, Simulation and Design*, 1991, Kluwer Academic Publishers. Pederson is one of the inventors of *SPICE*. This book is about *SPICE* simulation.

J. White, Thermal Design of Transistor Circuits," April 1972 *QST*, pp 30-34.

APPENDIX: *SPICE* Input Files
TVI Filter (Case Study 1)
TVI filter 1
Table 30.22 no. 52 Chebyshev

```
*
*      Define    elements    as
subcircuits.
*
..SUBCKT CA 11 13
C 11 12 CMOD 110PF
R 11 12 1GOMEGA
LX 12 13 LSELF 1UH
..ENDS CA
..SUBCKT CB 11 13
C 11 12 CMOD 200PF
R  1 12 1GOMEGA
LX 12 13 LSELF 1UH
..ENDS CB
..SUBCKT LA 11 12
L 11 13 LMOD 0.350UH
R 13 12 RSERIES 1
..ENDS LA
..SUBCKT LB 11 12
L 11 13 LMOD 0.389UH
R 13 12 RSERIES 1
..ENDS LB
*
*   Model statements for toler-
ance simulation
*    and parasitics
*
MODEL CMOD CAP (C=1 DEV 10%)
MODEL LMOD IND (L=1 DEV 20%)
MODEL LSELF IND (L=0.03)
MODEL RSERIES RES (R= 0.5)
*
*  Assemble  filter  as  larger
subcircuit.
*
```

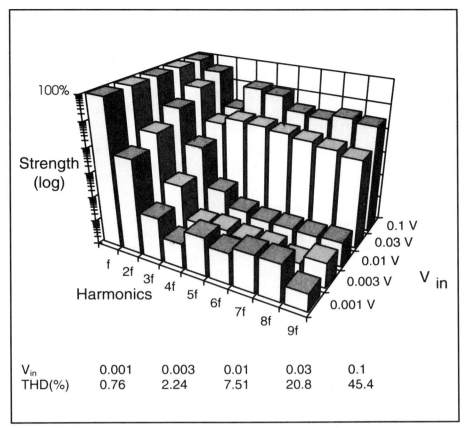

V_{in}	0.001	0.003	0.01	0.03	0.1
THD(%)	0.76	2.24	7.51	20.8	45.4

Fig 10.48—Fourier (harmonic) decomposition and Total Harmonic Distortion values for the waveforms in Fig 10.47.

```
..SUBCKT FILTER 1 4
XC1 1 0 CA
XL1 1 2 LA
XC2 2 0 CB
XL2 2 3 LB
XC3 3 0 CB
XL3 3 4 LA
XC4 4 0 CA
..ENDS FILTER
*
*  Place filter on "test bench"
*   with 50-Ωsource and load
resistances
*
VIN1 51 0   ac 2
RIN1 51 52 50
XF1  52 53 FILTER
ROUT1 53 0   50
*
*  Sweep
*
..ac DEC 80 10MEGHZ 100MEGHZ
..MC 80 ac VDB(53) FALL_EDGE -46
LIST OUTPUT(ALL)
..PROBE VDB(53)
..END
```

CE Audio Amp (Case Study 2)
```
Amplifier with distortion test
*
```

```
VSIG   1 0   ac 0.01 SIN (0
0.001 1KHZ)
RIN    1 7   100
CIN    7 2   2UF
RB1    5 2   43K
RB2    2 0   10K
Q1     11 2 4  Q2N2222A
RC     5 11   2200
RE     4 0   470
CE     4 0   100UF
COUT   11 12   1UF
ROUT   12 0    1MEG
VBIAS  5 0   DC 9
*
..MODEL Q2N2222A NPN(Is=14.34f
Xti=3 Eg=1.11 Vaf=74.03 Bf=255.9
Ne=1.307
+Ise=14.34f  Ikf=.2847  Xtb=1.5
Br=6.092 Nc=2 Isc=0 Ikr=0 Rc=1
+Cjc=7.306p Mjc=.3416 Vjc=.75
Fc=.5 Cje=22.01p Mje=.377 Vje=.75
+Tr=46.91n  Tf=411.1p  Itf=.6
Vtf=1.7 Xtf=3 Rb=10)
*National    pid=19   case=TO18
*
..AC DEC 10 1HZ 1MEGHZ
..TRAN 1MS 10MS 2MS 20US
..FOUR 1KHZ V(12)
..PROBE v(12) v(1)
..END
```

LOW-FREQUENCY TRANSISTOR MODELS

THE FUNDAMENTAL EQUATIONS

Design models are based on the physics of the components we use. The complexity of models can vary widely though, and many times we can use very simple models to achieve our goals. Increasingly complex models are developed and used only when demanded by the circuit application.

In this discussion, we will focus on simple models for bipolar transistors (BJTs) and FETs. These models are reasonably accurate at low frequencies, and they are of some use at RF. For more sophisticated RF models, look to professional RF-design literature. This discussion is adapted from Wes Hayward's *Introduction to RF Design*. Derivations of the material shown here appear in that book, an excellent text for the beginning RF designer.

This discussion is centered on NPN BJTs and N-channel JFETs. The material here applies to PNPs and P-channel FETs when you simply change the bias polarities.

First, consider the bipolar transistor as a current controlled device. When the base current controls the collector current, this equation defines the transistor operation

$$I_c = \beta I_b \tag{19}$$

where β = common-emitter current gain.

When we consider a transistor as a voltage-controlled device, this equation describes it (emitter current, I_e, in terms of base-emitter voltage, V_{be}):

$$I_e = I_{es} [\exp(qV/kT) - 1]$$
$$\approx I_{es} \exp(qV/kT) \tag{20}$$

where
$V = V_{be}$
q = electronic charge
k = Boltzmann's constant
T = temperature in kelvins (K)
I_b = emitter saturation current, typically 1×10^{-13} A.

Both equations approximate models of more complex behavior. Equation 20 is a simplification of the first Ebers-Moll model (Ref 1). More sophisticated models for BJTs are described by Getreu in Ref 2. Equations 19 and 20 apply to both transistor dc biasing and signal design.

The operation of an N-channel JFET can be characterized by

$$I_D = I_{DSS} (1 - V_{sg}/V_p)^2 \tag{21}$$

where
I_{DSS} = drain saturation current
V_{sg} = the source-gate voltage
V_p = the pinch-off voltage.

This equation applies only so long as V_{sg} is between 0 and V_p. JFETs are seldom used with the gate-to-channel diode forward biased. Drain current, I_D, is 0 when V_{sg} exceeds V_p. This equation applies to both biasing and signal design.

Now that we have some basic equations, let's go on to some other areas:
- Small-signal amplifier design (and application limits).
- Large-signal amplifier design (distortion from nonlinearity).

BIPOLAR TRANSISTORS (SMALL SIGNALS)

Transistors are usually driven by both biasing and signal voltages. Small-signal models treat only the signal components. We will consider bias and nonlinear signal effects later.

A Basic Common-Emitter Model

Fig 10.49 shows a BJT amplifier. The circuit is adequately described by equation 19. A mathematical analysis of the control and output currents and voltages yields the small-signal common-emitter amplifier model shown in **Fig 10.50**. This is the most common of all transistor small-signal models, a controlled current source with emitter resistance.

In order to use this model, however, we must have a value for r_e. That's no trouble, however, $r_e = kT/qI_0$, or $r_e = 26/I_e$, where I_e is the dc bias current in milliamperes. This value applies at a typical ambient temperature of 300 K.

The device output resistance is infinite because it is a pure current source, which is a good approximation for most silicon transistors at low frequencies. In use, the collector lead would feed a load resistor, R_L. For this model, that resistance must be small enough so that the collector bias voltage is positive for the chosen bias current.

Gain vs Frequency

Fig 10.50 is a low frequency approximation. As signal frequency increases, however, current gain appears to decrease. The low-frequency current gain is β_0. β_0 is constant through the audio spectrum, but it eventually decreases, and at some high frequency it will drop by a factor of 2 for each doubling of signal frequency. A transistor's frequency vs current gain relationship is specified by its *gain-bandwidth product*, or F_T. F_T is the frequency at which the current gain is 1. Common transistors for lower RF applications might have $\beta_0 = 100$ and $F_T = 500$ MHz. The frequency where current gain is $\beta_0 \sqrt{2}$ is called F_β and related to F_T by $F_\beta = F_T / \beta_0$.

The frequency dependence of current gain is modeled by adding a capacitor across the base resistor of Fig 10.50A; **Fig 10.51**, the *hybrid-pi* model results. The

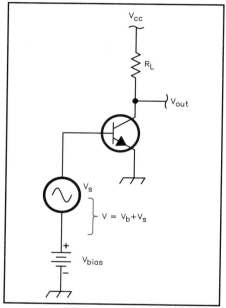

Fig 10.49—Bipolar transistor with voltage bias and input signal.

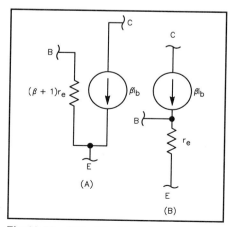

Fig 10.50—Simplified low-frequency model for the bipolar transistor, a "beta generator with emitter resistance." $r_e = 26 / I_e$(mA dc).

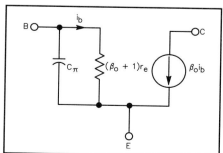

Fig 10.51—The hybrid-pi model for the bipolar transistor.

capacitor reactance should equal the low-frequency input resistance, $(\beta + 1)r_e$, at F_β. This simulates a frequency-dependent current gain.

Three Simple Models

Even though transistor gain varies with frequency, the simple model is still useful under certain conditions. Calculations show that the simple model is valid, with $\beta = F_T / F$, for frequencies well above F_β. The approximation worsens, however, as the operating frequency (f_0) approaches F_T. **Fig 10.52** shows small-signal models for the three common amplifier configurations: common emitter (ce), common base (cb) and common collector (cc).

The common-collector amplifier, unlike the common-emitter or common-base configurations, has a finite output resistance. This resistance is calculated by short circuiting the input voltage source, V_s, and "driving" the output port with either a voltage or current source. The result is the equation for R_{out} that appears in the figure.

The common-collector example shows characteristics that are more typical of practical RF amplifiers than the idealized ce and cb amplifiers. Specifically, the input resistance is a function of both the device and the termination at the output. The output resistance is critically dependent upon the input driving source resistance.

These examples have used the simplest of models, the controlled current genera-

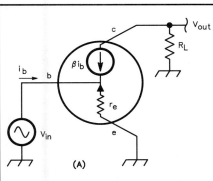

$$R_{in} = r_e (\beta + 1)$$

$$I_b = \frac{V_{in}}{(\beta + 1) r_e}$$

$$I_c = \frac{b V_{in}}{r_e (\beta + 1)} \approx g_m V_{in}$$

where $g_m = \dfrac{1}{r_e}$

and the approximation applies

when $\beta \rangle 20$ at the operating frequency.

$$V_{out} = -I_c R_L$$

$$G_V = \frac{-R_L}{r_e}$$

A simple model for an NPN common-emitter amplifier and the pertinent design equations. The minus signs indicate a 180° phase difference between input and output. The approximate voltage gain requires that β is high. Amplifier gain may be controlled by changing r_e, that is: by choosing operating current or adding resistance to the emitter circuit. Adding resistance to the emitter circuit is a form feed-back called *emitter degeneration*.

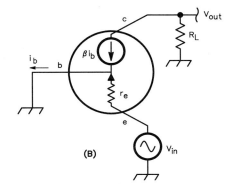

A small-signal common-base (cb) amplifier model. The background equation at the base is identical with that for the ce amplifier, but the direction of the base and collector currents is different. The calculations yield these design equations:

$$R_{in} = R_e = \frac{1}{g_m}$$

$$G_I = \frac{\beta}{\beta + 1} = \alpha$$

α **is very close to 1, because β is usually much greater than 1. The cb amplifier, when driven from a voltage source, has a voltage gain of $G_V = \alpha R_L$ (This gain is noninverting.)**

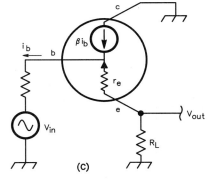

The cc amplifier, often called an emitter follower. The driving source for this example contains a resistance, R_s. Noting the direction of the assumed currents shown, the nodal equations yield:

$$V_b = \frac{V_s (\beta + 1)(r_e + R_L)}{R_s (\beta + 1)(r_e + R_L)}$$

$$V_{out} = \frac{V_b R_L}{r_e + R_L}$$

The output voltage, V_{out}, is related to the base voltage through voltage divider action.

The emitter current is related directly to the base voltage by

$$I_e = \frac{V_b}{(r_e + R_L)}$$

$$I_b = \frac{I_e}{(\beta + 1)}$$

$$R_{in} = (\beta + 1)(r_e + R_L)$$

$$R_{out} = \frac{R_s}{(\beta + 1)} + r_e$$

$$G_V = \frac{R_L}{R_L + r_e}$$

which is approximately 1 when R_L is much larger than r_e.

Fig 10.52—Application of small-signal models for analysis of (A) the ce amplifier, (B) the cb and (C) the cc bipolar transistor amplifiers.

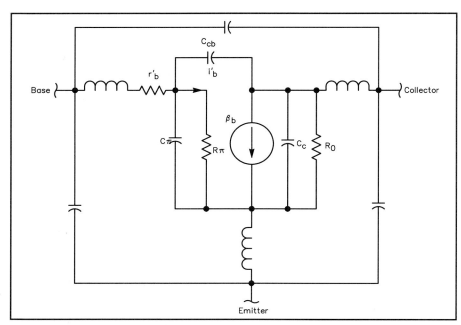

Fig 10.53—A more refined small-signal model for the bipolar transistor. Suitable for many applications near the transistor F_T.

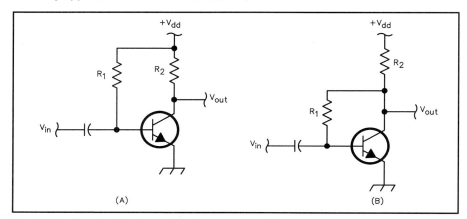

Fig 10.54—Simple biasing methods for ce amplifier. The scheme at (A) suffers if β is not well known. Negative feedback is used in (B).

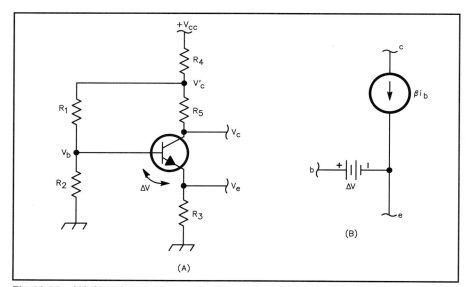

Fig 10.55—(A) Circuit used for evaluation of transistor biasing. (B) The model used for bias calculations.

tor with an emitter resistance, r_e. Better models are necessary to design at high frequencies. The simplified hybrid-pi of Fig 10.51 is often suitable.

Small-Signal Design at RF

Fig 10.53 shows a better small-signal model for RF design that expands on the hybrid-pi. Consider the physical aspects of a real transistor: There is some capacitance across each of the PN junctions (C_{cb} and C_π) a capacitance from collector to emitter (C_c). There are also capacitances between the device leads (C_e, C_b and C_c). There is a resistance in each current path, emitter to base and collector. From emitter to base, there is r_π from the hybrid-pi model and $r'b$, the "base spreading" resistance. From emitter to collector is R_O, the output resistance. The leads from the die to the circuit present three inductances.

Manual circuit analysis with this model doesn't look like much fun. It's best tackled with the aid of a computer and specialized software. Other methods are presented in Hayward's *Introduction to RF Design*.

Don't be intimidated by the complexity of the model, however. Surprisingly accurate results may be obtained, even at RF, from the simple models. Simple models also give a better "feel" for device characteristics that might be obscured by the mathematics of a more rigorous treatment. Use the simplest model that describes the important features of the device and circuit at hand.

Biasing Bipolar Transistors

Proper biasing of the bipolar transistor is more complicated than it might appear. The Ebers-Moll equation would suggest that a common-emitter amplifier could be built as shown in Fig 10.49, grounding the emitter and biasing the base with a constant voltage source. Further examination shows that this presents many problems. The worst is that constant-voltage bias ultimately leads to thermal *runaway*. Constant-voltage biasing applied to the base is almost never used.

Constant base-current biasing is shown in **Fig 10.54A**. This works reasonably well if the current gain is known, which is rarely true. A transistor with a typical β of 100 might actually have values ranging from 50 to 250. A slightly improved method is shown in Fig 10.54B, where the bias is derived from the collector. As current increases collector voltage decreases, as does the bias current flowing through R_1. This ensures operation in the transistor active region.

The most common biasing method is shown in **Fig 10.55A**. The device model used, shown in Fig 10.55B, is based on the Ebers-Moll equation, which shows that virtually no transistor current flows until the base-emitter voltage reaches about 0.6. Then, current increases dramatically with small additional voltage change. The transistor is thus modeled as a current controlled generator with a battery in series with the base. The battery voltage is ΔV.

The circuit is analyzed with nodal equations. The collector resistance, R_5, is initially assumed to be zero. The analysis results in three equations for V_b, V_c' and I_b:

$$V_b = \frac{V_{cc}R2R3 + \Delta VR2\left(\dfrac{R4+R1}{\beta+1}\right)}{R3R4 + R2R4 + R2R3 + R1R3 + \dfrac{R1R2}{\beta+1}}$$

(22)

$$V_c' = \frac{R1V_{cc}R4V_b + \beta R1R4\left(\dfrac{\Delta V - V_b}{R3(\beta+1)}\right)}{R1 + R4}$$

(23)

$$I_b = \frac{V_b - \Delta V}{R3(\beta+1)}$$

(24)

The emitter current is then $I_b(\beta + 1)$. Once the circuit has been analyzed, R5 may be taken into account. The final collector voltage is

$$V_c = V_c' - \beta I_b R5$$

(25)

The solution is valid so long as V_c exceeds V_b.

Analysis of these equations with a computer or hand-held programmable calculator shows that I_e is not a strong function of the transistor parameters, ΔV and β. In practice, the base biasing resistors, R1 and R2, should be chosen to draw a current that is much larger than I_b (to eliminate effects of β variation). V_b should be much larger than ΔV to reduce the effects of variations in ΔV.

Three additional biasing schemes are presented in **Fig 10.56**. All provide bias that is stable regardless of device parameter variations. A and B require a negative power supply. The circuit of Fig 10.56C uses a second, PNP, transistor for bias control. The PNP transistor may be replaced with an op amp if desired. All three circuits have the transistor emitter grounded directly. This is often of great importance in microwave amplifiers. These circuits may be analyzed using the simple model of Fig 10.55.

The biasing equations presented may be solved for the resistors in terms of desired

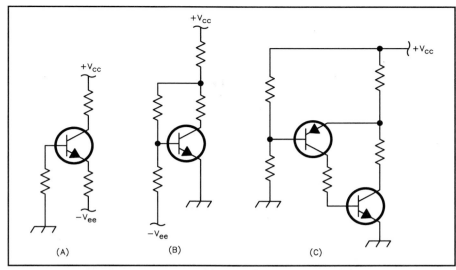

Fig 10.56—Alternative biasing methods. (A) and (B) use dual power supplies, (B) and (C) allow the emitter to be at ground while still providing temperature-stable operation.

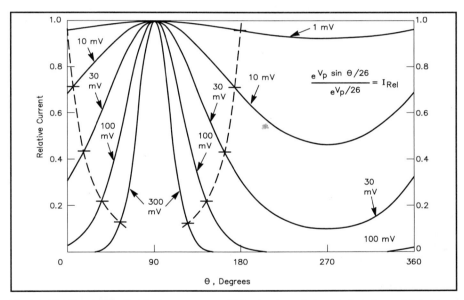

Fig 10.57—Normalized relative current of bipolar transistor under sinusoidal drive at the base.

operating conditions and device parameters. It is generally sufficient, however, to repetitively analyze the circuit, using standard resistor values.

The small-signal transconductance of a common-emitter amplifier was found in the previous section. If biased for constant current, the small-signal voltage gain will vary inversely with temperature. Gain may be stabilized against temperature variations with a biasing scheme that causes the bias current to vary in *proportion to absolute temperature*. Such methods, termed PTAT methods, are often used in modern integrated circuits and are finding increased application in circuits built

from discrete components.

Large-Signal Operation

The models presented in previous sections have dealt with small signals applied to a bipolar transistor. While small-signal design is exceedingly powerful, it is not sufficient for many designs. Large signals must also be processed with transistors. Two significant questions must be considered with regard to transistor modeling. First, what is a reasonable limit to accurate application of small-signal methods? Second, what are the consequences of exceeding these limits?

The same analysis of the Ebers-Moll

model yields an equation for collector current. The mathematics show that current will vary in a complicated way, for the sinusoidal signal voltage is embedded within an exponential function. Nonetheless, the output is a sinusoidal current if the signal voltage is sufficiently low.

The current of the equation may be studied by normalizing the current to its peak value. The result is relative current, I_r, which is plotted in **Fig 10.57** for V_p values of 1, 10, 30, 100, and 300 mV. The 1-mV case is very sinusoidal. Similarly, the 10-mV curve is generally sinusoidal with only minor distortions. The higher amplitude cases show increasing distortion.

Constant base-voltage biasing is unusual. More often, a transistor is biased to produce nearly constant emitter current. When such an amplifier is driven by a large input signal, the average bias voltage will adjust itself until the time average of the nonlinear current equals the previous constant bias current. Hence, it is vital to consider the average relative current of the waveforms of Fig 10.57. This is evaluated through calculus.

The average relative currents for the cases analyzed occur at the intersection of the curves with the dotted lines of Fig 10.57. For example, the dotted curve intersects the V_p = 300-mV waveform at an average relative current of 0.12. If an amplifier was biased to a constant current of 1 mA, but was driven with a 300-mV signal, the positive peak current would reach a value greater than the average by a factor of 1/(0.12). The average current would remain at 1 mA, but the positive peak would be 8 mA. The transistor would not conduct for most of the cycle.

The curves have presented data based upon the simplest of large-signal models, the Ebers-Moll equation. Still, the simple model has yielded considerable information. The analysis suggests that a reasonable upper limit for accurate small-signal analysis is a peak base signal of about 10 mV. The effect of emitter degeneration is also evident. Assume a transistor is biased for $r_e = 5 \ \Omega$ and an external emitter resistor of 10 Ω is used. Only the r_e portion of the 15Ω total is nonlinear. Hence, this amplifier would tolerate a 30-mV signal while still being well described with a small-signal analysis.

FETS

An often used device in RF applications is the field-effect transistor (FET). There are many kinds: JFETs, MOSFETs and so on. Here we will discuss JFETs, with the understanding that other FETs are similar.

We viewed the bipolar transistor as controlled by either voltage or current. The

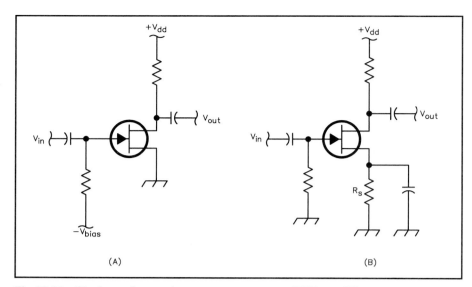

Fig 10.58—Biasing schemes for a common-source JFET amplifier. $-V_{bias}$ is normally adjusted to suit each device; there is a significant spread over a product run. Also note that some FETs can exhibit thermal runaway in some current/temperature ranges.

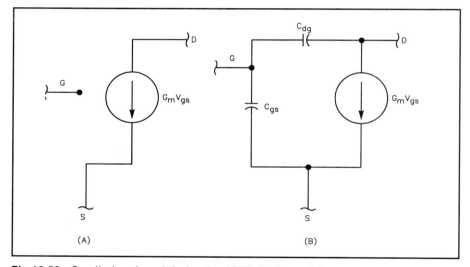

Fig 10.59—Small-signal models for the JFET. (A) is useful at low frequency, (B) is a modification to approximate high-frequency behavior.

JFET, however, is purely a voltage controlled element, at least at low frequencies. The input gate is usually a reverse biased diode junction with virtually no current flow. The drain current is related to the source-gate voltage by:

$$I_D = I_{DSS} \left(1 - \frac{V_{sg}}{V_p}\right)^2 \qquad 0 \leq V_{sg} \leq V_p$$

$$I_D = 0 \qquad\qquad V_{sg} > V_p \qquad (26)$$

where I_{DSS} is the drain saturation current and V_p is the pinch-off voltage. Operation is not defined when V_{sg} is less than zero

because the gate diode is then forward biased. Equation 26 is a reasonable approximation as long as the drain bias voltage exceeds the magnitude of the pinch-off voltage.

Biasing FETs

Two virtually identical amplifiers using N-channel JFETs are shown in **Fig 10.58**. The two circuits illustrate the two popular methods for biasing the JFET. Fixed gate-voltage bias, Fig 10.58A, is feasible for JFETs because of their favorable temperature characteristics. As the temperature of the usual FET increases, current decreases, avoiding the thermal-runaway problem of bipolar transistors.

A known source resistor, R_s in Fig 10.58B, will lead to a known source voltage. This is obtained from a solution of equation 26:

$$V_{sg} = \frac{\left[\dfrac{1}{R_s I_{DSS}} + \dfrac{2}{V_p}\right] - \left[\left(\dfrac{1}{R_s I_{DSS}} + \dfrac{2}{V_p}\right)^2 - \left(\dfrac{2}{V_p}\right)^2\right]^{0.5}}{\dfrac{2}{V_p^2}}$$

(27)

The drain current is then obtained by direct substitution.

Alternatively, a desired drain current less than I_{DSS} may be achieved with a proper choice of source resistor

$$R_s = \frac{V_p\left(1 - \dfrac{I_D}{I_{DSS}}\right)^{0.5}}{I_D}$$

(28)

The small-signal transconductance of the JFET is obtained by differentiating equation 26

$$g_m = \frac{dI_D}{dV_{sg}} = \frac{-2I_{DSS}}{V_p}\left(1 - \frac{V_{sg}}{V_p}\right)$$ (29)

The minus sign indicates that the equation describes a common-gate configuration. The amplifiers of Fig 10.58 are both common-source types and are described by equation 29 except that g_m is now positive. Small-signal models for the JFET are shown in **Fig 10.59**. The simple model is that inferred from the equations while the model of Fig 10.59B contains capacitive elements that are effective in describing high-frequency behavior. Like the bipolar transistor, the JFET model will grow in complexity as more sophisticated applications are encountered.

Large-Signal Operation

Large-signal JFET operation is examined by normalizing the previous equation to $V_p = 1$ and $I_{DSS} = 1$ and injecting a sinusoidal signal. The circuit is shown in **Fig 10.60**. Also shown in the figure are examples for a variety of bias and sinusoid amplitude conditions. The main feature is the asymmetry of the curves. The positive portions of the oscillations are farther from the mean than are the negative excursions. This is especially dramatic when the bias, v_0, is large, which places the quiescent point close to pinch-off. With such bias and high-amplitude drive, conduction occurs only over a small fraction of the total input waveform period.

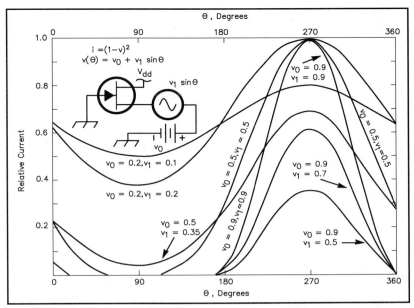

Fig 10.60—Relative normalized drain current for a JFET with constant voltage bias and sinusoidal signals. Relatively "clean" waveforms exist for low signals while large input amplitudes cause severe distortion.

The average current for these operating conditions can be determined by calculus. The average current values obtained may be further normalized by dividing by the corresponding dc bias current, $I_0 = (1 - v_0)^2$. The results are shown in **Fig 10.61**. The curves show that the average current increases as the amplitude of the drive increases. This, again, is most pronounced when the FET is biased close to pinch-off.

Although practical for the JFET, constant-voltage operation in the previous curves is not common. Instead, a resistive bias is usually employed, Fig 10.58B. With this form of bias, the increased current from high signal drive will cause the voltage drop across the bias resistor to increase. This will then move the quiescent operating level closer to pinch-off, accompanied by a reduced small-signal transconductance. This behavior is vital in describing the limiting found in FET oscillators.

The limits on small-signal operation are not as well defined for a FET as they were for the bipolar transistor. Generally, a maximum voltage of 50 to 100 mV is allowed at the input (normalized to a 1-V pinch-off) without severe distortion. The voltages are much higher than they were for the bipolar transistor. However, the input resistance of the usual common source amplifier is so high and the corresponding transconductance low enough that the available gain is no greater than could be obtained with a bipolar transis-

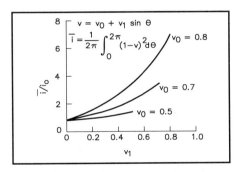

Fig 10.61—Change in average current of a JFET with increasing input signals. The average current with no input signals is I_0, while v_1 is the normalized drive amplitude, and v_0 is the bias voltage.

tor. The distortion is generally less with FETs, owing to the lack of high-order curvature in the defining equations.

Many of the standard circuits used with bipolar transistors are also practical with FETs. Noting that the transconductance of a bipolar transistor is $g_m = I_e$ (dc mA) / 26, the previous equations may be applied directly. The "emitter current" is chosen to correspond with the FET transconductance. A very large value is used for current gain. The same calculator or computer program is then used directly. In practice, much higher terminating impedances are needed to obtain transducer gain values similar to those of bipolar transistors.

References

1. Ebers, J., and Moll, J., "Large-Signal Behavior of Junction Transistors," *Proceedings of the IRE*, 42, pp 1761-1772, December 1954.
2. Getreu, I., *Modeling the Bipolar Transistor*, Elsevier, New York, 1979. Also available from Tektronix, Inc, Beaverton, Oregon, in paperback form. Must be ordered as Part Number 062-2841-00.
3. Searle, C., Boothroyd, A., Angelo, E. Jr., Gray, P., and Pederson, D., *Elementary Circuit Properties of Transistors*, Semiconductor Electronics and Education Committee, Vol 3 John Wiley & Sons, New York, 1964.
4. Clarke, K. and Hess, D., *Communication Circuits: Analysis and Design*, Addison-Wesley, Reading, Massachusetts, 1971.
5. Middlebrook, R., *Design-Oriented Circuit Analysis and Measurement Techniques*. Copyrighted notes for a short course presented by Dr Middlebrook in 1978.

Suggested Additional Readings

Alley, C. and Atwood, K., *Electronic Engineering*, John Wiley & Sons, New York, 1962.

Terman, F., *Electronic and Radio Engineering*, McGraw-Hill, New York, 1955.

Gilbert, B., "A New Wide-Band Amplifier Technique," *IEEE Journal of Solid-State Circuits*, SC-3, 4, pp 353-365, December 1968.

Egenstafer, F., "Design Curves Simplify Amplifier Analysis," *Electronics*, pp 62-67, August 2, 1971.

Contents

Power Supplies and Projects

11

Glossary

Bipolar Transistor — A term used to denote the common two junction transistor types (NPN, PNP) as opposed to the field effect families of devices (JFET, MOSFET and so on.).

Bleeder — A resistive load across the output or filter of a power supply, intended to quickly discharge stored energy once the supply is turned off.

C-Rate — The charging rate for a battery, expressed as a ratio of the battery's ampere-hour rating.

Circular Mils — A convenient way of expressing the cross-sectional area of a round conductor. The area of the conductor in circular mils is found by squaring its diameter in mils (thousandths of an inch), rather than squaring its radius and multiplying by pi. For example, the diameter of 10-guage wire is 101.9 mils (0.1019 inch). Its cross-sectional area is 10380 CM, or 0.008155 square inches.

Core Saturation (Magnetic) — That condition whereby the magnetic flux in a transformer or inductor core is more than the core can handle. If the flux is forced beyond this point, the permeability of the core will decrease, and it will approach the permeability of air.

Crowbar — A last-ditch protection circuit included in many power supplies to protect the load equipment against failure of the regulator in the supply. The crowbar senses an overvoltage condition on the supply's output and fires a shorting device (usually an SCR) to directly short-circuit the supply's output and protect the load. This causes very high currents in the power supply, which blow the supply's input-line fuse.

Darlington Transistor — A package of two transistors in one case, with the collectors tied together, and the emitter of one transistor connected to the base of the other. The effective current gain of the pair is approximately the product of the individual gains of the two devices.

DC-DC Converter — A circuit for changing the voltage of a dc source to ac, transforming it to another level, and then rectifying the output to produce direct current.

Fast Recovery Rectifier — A specially doped rectifier diode designed to minimize the time necessary to halt conduction when the diode is switched from a forward-biased state to a reverse-biased state.

Foldback Current Limiting — A special type of current limiting used in linear power supplies, which reduces the current through the supply's regulator to a low value under short circuited load conditions in order to protect the series pass transistor from excessive power dissipation and possible destruction.

Ground Fault (Circuit) Interrupter (GFI or GFCI) — A safety device installed between the household power mains and equipment where there is a danger of personnel touching an earth ground while operating the equipment. The GFI senses any current flowing directly to ground and immediately switches off all power to the equipment to minimize electrical shock. GFIs are now standard equipment in bathroom and outdoor receptacles.

Input-Output Differential — The voltage drop appearing across the series pass transistor in a linear voltage regulator. This term is usually stated as a minimum value, which is that voltage necessary to allow the regulator to function and conduct current. A typical figure for this drop in most three-terminal regulator ICs is about 2.5 V. In other words, a regulator that is to provide 12.5 V dc will need a source voltage of at least 15.0 V at all times to maintain regulation.

Inverter — A circuit for producing ac power from a dc source.

Peak Inverse Voltage — The maximum reverse-biased voltage which a semiconductor is rated to handle safely. Exceeding the peak inverse rating can result in junction breakdown and device destruction.

Power Conditioner — Another term for a power supply.

Regulator — A device (such as a Zener diode) or circuitry in a power supply for maintaining a constant output voltage over a range of load currents and input voltages.

Resonant Converter — A form of dc-dc converter characterized by the series pass switch turning on into an effective series-resonant load. This allows a zero current condition at turn-on and turn-off. The resonant converter normally operates at frequencies between 100 kHz and 500 kHz and is very compact in size for its power handling ability.

Ripple — The residual ac left after rectification, filtration and regulation of the input power.

RMS — *R*oot of the *M*ean of the *S*quares. Refers to the effective value of an alternating voltage or current, corresponding to the dc voltage or current that would cause the same heating effect.

Second Breakdown — A runaway failure

condition in a transistor, occurring at higher collector-emitter voltages, where hot spots occur due to (and promoting) localization of the collector current at that region of the chip.

Series Pass Transistor, or Pass Transistor — The transistor(s) that controls the passage of power between the unregulated dc source and the load in a regulator. In a linear regulator, the series pass transistor acts as a controlled resistor to drop the voltage to that needed by the load. In a switch-mode regulator, the series pass transistor switches between its ON and OFF states.

SOAR (Safe Operating ARea) — The range of permissible collector current and collector-emitter voltage combinations where a transistor may be safely operated without danger of device failure.

Spike — An extremely short perturbation on a power line, usually lasting less than a few microseconds.

Surge — A moderate-duration perturbation on a power line, usually lasting for hundreds of milliseconds to several seconds.

Transient — A short perturbation on a power line, usually lasting for microseconds to tens of milliseconds.

Varistor — A surge suppression device used to absorb transients and spikes occurring on the power lines, thereby protecting electronic equipment plugged into that line. Frequently, the term MOV (*M*etal *O*xide *V*aristor) is used instead.

Volt-Amperes — The product obtained by multiplying the current times the voltage in an ac circuit without regard for the phase angle between the two. This is also known as the *apparent power* delivered to the load as opposed to the actual or *real power* absorbed by the load, expressed in watts.

Voltage Multiplier — A type of rectifier circuit that is arranged so as to charge a capacitor or capacitors on one half-cycle of the ac input voltage waveform, and then to connect these capacitors in series with the rectified line or other charged capacitors on the alternate half-cycle. The voltage doubler and tripler are commonly used forms of the voltage multiplier.

Alternating-Current Power

The text for the theory portion of this chapter was written by Ken Stuart, W3VVN.

In most residences, three wires are brought in from the outside electrical-service mains to the house distribution panel. In this three-wire system, one wire is neutral and should be at earth ground. The voltage between the other two wires is 60 Hz alternating current with a potential difference of approximately 240 V RMS. Half of this voltage appears between each of these wires and the neutral, as indicated in **Fig 11.1A**. In systems of this type, the 120 V household loads are divided at the breaker panel as evenly as possible between the two sides of the power mains. Heavy appliances such as electric stoves, water heaters, central air conditioners and so forth, are designed for 240 V operation and are connected across the two ungrounded wires.

Both ungrounded wires should be fused. A fuse or switch should never be used in the neutral wire, however. Opening the neutral wire does not disconnect the equipment from an active or "hot" line, creating a potential shock hazard between that line and earth ground.

Another word of caution should be given at this point. Since one side of the ac line is grounded to earth, all communications equipment should be reliably connected to the ac-line ground through a heavy ground braid or bus wire of #14 or heavier-gauge wire. This wire must be a separate conductor. You must not use the power-wiring neutral conductor for this safety ground. (A properly wired 120-V outlet with a ground terminal uses one wire for the ac *hot* connection, one wire for the ac *neutral* connection and a third wire for the safety *ground* connection.) This not only places the chassis of the equipment at earth ground for minimal RF energy on the chassis, but also provides a measure of safety for the operator in the event of accidental short or leakage of one side of the ac line to the chassis.

Remember, the antenna system is almost always bypassed to the chassis via an RF choke or tuning circuit, which could make the antenna electrically "live" with respect to the earth ground and create a potentially lethal shock hazard. A Ground Fault Circuit Interrupter (GFCI or GFI) is also desirable for safety reasons, and should be a part of the shack's electrical power wiring.

FUSES AND CIRCUIT BREAKERS

All transformer primary circuits should be fused properly, and multiple secondary outputs should also be individually fused. To determine the approximate current rating of the fuse or circuit breaker to be used, multiply each current being drawn by the load or appliance, in amperes, by the voltage at which the current is being drawn. In the case of linear regulated power supplies, this voltage has to be the voltage appearing at the output of the rectifiers before being applied to the regulator stage. Include the current taken by bleeder resistors and voltage dividers. Also include filament power if the transformer is supplying filaments. The National Electrical Code also specifies maximum fuse ratings based on the wire sizes used in the transformer and connections.

After multiplying the various voltages and currents, add the individual products. This will be the total power drawn from the line by the supply. Then divide this power by the line voltage and add 10 or 20%. Use a fuse or circuit breaker with the nearest larger current rating. Remember that the charging of filter capacitors can take large surges of current when the supply is turned on. If turn on is a problem, use slow-blow fuses, which allow for high initial surge currents.

For low-power semiconductor circuits, use fast-blow fuses. As the name implies, such fuses open very quickly once the current exceeds the fuse rating by more than 10%.

ELECTRICAL POWER CONDITIONING

We often use the term "power supply" to denote a piece of equipment that will process the electrical power from a source, such as the ac power mains, by manipulating it so the device output will be acceptable to other equipment that we want to power. The common form of the power supply is the familiar direct-current power supply, which will power a transmitter, receiver or other of a wide variety of electronic devices.

In the strictest terms, however, the power supply is not actually a source, or "supply" of power, but is actually a processor of already existing energy. Therefore, the old term of "power supply" is becoming obsolete, and a new term has arisen to refer to the technology of the processing of electrical power: "power conditioning." By contrast, the term "power supply" is now used to refer to devices for chemical to electrical energy conversion (batteries) or mechanical to electrical conversion (generators). Other varieties include thermoelectric generators (TEGs) and radioactive thermoelectric generators (RTGs).

In this chapter, we shall examine the traditional forms of power conditioning, which consists of the following component parts in various combinations: transformer, rectifier, filter and regulator. We will also look briefly at true power supplies as we examine battery technology and emergency power generation.

POWER TRANSFORMERS

Numerous factors are considered in order to match a transformer to its intended use. Some of these parameters are listed below:

1. Output voltage and current (volt-ampere rating)
2. Power source voltage and frequency
3. Ambient temperature
4. Duty cycle and temperature rise of the transformer at rated load
5. Mechanical shape and mounting

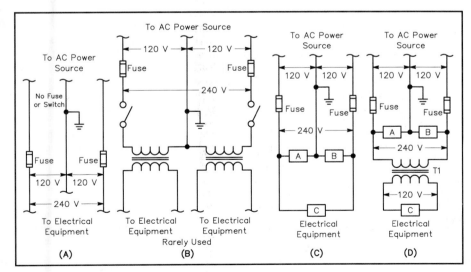

Fig 11.1 — Three-wire power-line circuits. At A, normal three-wire-line termination. No fuse should be used in the grounded (neutral) line. The ground symbol is the power company's ground, not yours! Do *not* connect *anything*, including the equipment chassis, to the power neutral wire. At B, the "hot" lines each have a switch, but a switch in the neutral line would not remove voltage from either side of the line, and so should never be used. At C, connections for both 120 and 240-V transformers. At D, operating a 120-V plate transformer from the 240-V line to avoid light blinking. T1 is a 2:1 step-down transformer.

Volt-Ampere Rating

In alternating-current equipment, the term "volt-ampere" is often used rather than the term "watt." This is because ac components must handle reactive power as well as real power. If this is confusing, consider a capacitor connected directly across the secondary of a transformer. The capacitor appears as a reactance that permits current to flow, just as if the load were a resistor. The current is at a 90° phase angle, however. If we assume a perfect capacitor, there will be no heating of the capacitor, so no real power (watts) will be delivered by the transformer. The transformer must still be capable of supplying the voltage, and be able to handle the current required by the reactive load. The current in the transformer windings will heat the windings as a result of the I^2R losses in the winding resistances. The product of the voltage and current is referred to as "volt-amperes", since "watts" is reserved for the real, or dissipated, power in the load. The volt-ampere rating will always be equal to, or greater than, the power actually being drawn by the load.

The number of volt-amperes (VA) delivered by a transformer depends not only upon the dc load requirements, but also upon the type of dc output filter used (capacitor or choke input), and the type of rectifier used (full-wave center tap or full-wave bridge). With a capacitive-input filter, the heating effect in the secondary is higher because of the high peak-to-average current ratio. The volt-amperes handled by the transformer may be several times the power delivered to the load. The primary winding volt-amperes will be somewhat higher because of transformer losses.

Source Voltage and Frequency

A transformer operates by producing a magnetic field in its core and windings. The intensity of this field varies directly with the instantaneous voltage applied to the transformer primary winding. These variations, coupled to the secondary windings, produce the desired output voltage. Since the transformer appears to the source as an inductance in parallel with the (equivalent) load, the primary will appear as a short circuit if dc is applied to it. The unloaded inductance of the primary must be high enough so as not to draw an excess amount of input current at the design line frequency (normally 60 Hz). This is achieved by providing sufficient turns on the primary and enough magnetic core material so that the core does not saturate during each half-cycle.

The magnetic field strength produced in the core is usually referred to as the *flux density*. It is set to some percentage of the maximum flux density that the core can stand without saturating, since at saturation the core becomes ineffective and causes the inductance of the primary to plummet to a very low level and input current to rise rapidly. This causes high primary currents and ex-

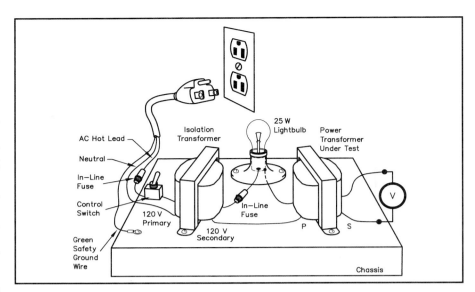

Fig 11.2 — Use a test fixture like this to test unknown transformers. Don't omit the isolation transformer, and be sure to insulate all connections before you plug into the ac mains.

treme heating in the primary windings. For this reason, transformers and other electromagnetic equipment designed for 60-Hz systems must not be used on 50-Hz power systems unless specifically designed to handle the lower frequency.

How to Evaluate an Unmarked Power Transformer

Many hams who regularly visit hamfests eventually end up with a junk box filled with used and unmarked transformers. After years of use, transformer labels or markings on the coil wrappings may come off or be obscured. There is a good possibility that the transformer is still useable. The problem is to determine what voltages and currents the transformer can supply. First consider the possibility that you may have an audio transformer or other impedance-matching device rather than a power transformer. If you aren't sure, don't connect it to ac power!

If the transformer has color-coded leads, you are in luck. There is a standard for transformer lead color-coding, as is given in the **Component Data** chapter. Where two colors are listed, the first one is the main color of the insulation; the second is the color of the stripe.

Check the transformer windings with an ohmmeter to determine that there are no shorted (or open) windings. The primary winding usually has a resistance higher than a filament winding and lower than a high-voltage winding.

A convenient way to test the transformer is to rig a pair of test leads to an electrical plug with a 25-W household light bulb in series to limit current to safe (for the transformer) levels. See **Fig 11.2**. Use an isolation transformer, and be sure to insulate all connections before you plug into the ac mains.

Switch off the power while making or changing any connections. Connect the test leads to each winding separately. BE CAREFUL! YOU ARE DEALING WITH HAZARDOUS VOLTAGES! The filament/heater windings will cause the bulb to light to full brilliance. The high-voltage winding will cause the bulb to be extremely dim or to show no light at all, and the primary winding will probably cause a small glow.

When you are connected to what you think is the primary winding, measure the voltages at the low-voltage windings with an ac voltmeter. If you find voltages close to 6 V ac and 5 V ac, you know that you have found the primary. Label the primary and low voltage windings.

Even with the light bulb, a transformer can be damaged by connecting mains power to a low-voltage or filament winding. In such a case the insulation could break down in a high-voltage winding.

Connect the voltmeter to the high-voltage windings. Remember that the old TV transformers will typically put out as much as 800 V or so across the winding, so make sure that your meter can withstand these potentials without damage. Divide 6.3 by the voltage you measured across the 6.3-V winding in this test setup. This gives a multiplier that you can use to determine the actual no-load voltage rating of the high-voltage secondary. Simply multiply the ac voltage measured across the winding by the multiplier.

The current rating of the windings can be determined by loading each winding with the primary connected directly (no bulb) to the ac line. Using power resistors, increase loading on each winding until its voltage drops by about 10% from the no-load figure. The current drawn by the resistors is the approximate winding load-current rating.

Rectifier Types

VACUUM TUBE

Once the mainstay of the rectifier field, the vacuum-tube rectifier has largely been supplanted by the silicon diode, but it may be found in vintage receivers still in use. Vacuum-tube rectifiers were characterized by high forward voltage drops and inherently poor regulation, but they were immune to ac line transients that can destroy other rectifier types.

MERCURY VAPOR

The mercury-vapor rectifier was an improvement over the vacuum tube rectifier in that the electron stream from cathode to plate would ionize the vaporized mercury in the tube and greatly reduce the forward voltage drop. Since ionized mercury is a much better conductor of current than a vacuum, these tubes can carry relatively high currents. As a result, they were popular in transmitters and RF power amplifiers.

Mercury rectifiers had to be treated with special care, however. When power was initially applied, the tube filament had to be turned on first to vaporize condensed mercury before the high-voltage ac could be applied to the plate. This could take from one to two minutes. Also, if the tube was handled or the equipment transported, filament power would have to be applied for about a half hour to vaporize any mercury droplets that might have been shaken onto tube insulating surfaces. Mercury vapor rectifiers have mostly been replaced by silicon diodes.

SELENIUM

The selenium rectifier was the first of the solid-state rectifiers to find its way into commercial electronic equipment. Offering a relatively low forward voltage drop, selenium rectifiers found their way into the plate supplies of test equipment and accessories, which needed only a few tens of milliamperes of current at about a hundred volts, such as grid-dip meters, VTVMs and so forth.

Selenium rectifiers had a relatively low reverse resistance and were therefore inefficient. Voltage breakdown per rectifying junction was only about 20 V.

GERMANIUM

Germanium diodes were the first of the solid-state semiconductor rectifiers. They have an extremely low forward voltage drop. Germanium diodes are relatively temperature sensitive, however. They can be easily destroyed by overheating during soldering, for instance. Also, they have some degree of back resistance, which varies with temperature.

Germanium diodes are used for special applications where the very low forward drop is needed, such as signal diodes used for detectors and ring modulators.

SILICON

Silicon diodes are the main choice today for virtually all rectifier applications. They are characterized by extremely high reverse resistance, forward drops of usually a volt or less and operation at high temperatures.

FAST RECOVERY

DC-DC converters regularly operate at 25 kHz and higher frequencies. Switch-mode regulators also operate in these same frequency ranges. When the switching transistors in these devices switch, voltage transitions take place within time periods usually much less than one microsecond, and the new FET switching transistors cause transitions that are often less than 100 ns.

When the transitions in these circuits occur, the previously conducting diodes see a reversal of current direction. This change tends to reverse bias those diodes, and thereby put them into an open-circuit condition. Unfortunately, solid-state rectifiers cannot be made to cease conduction instantaneously. As a result, when the opposing diodes in a bridge rectifier or full-wave rectifier become conductive at the time the converter switches states, the diodes being turned off will actually conduct in the reverse direction for a brief time, and effectively short circuit the converter for several microseconds. This puts excessive strain on the switching transistors and creates high current spikes, leading to electromagnetic interference. As the switching frequency of the converter or regulator increases, more of these transitions happen each second, and more power is lost due to this diode cross-conduction.

Semiconductor manufacturers have recognized this as a problem for some time. Many companies have product lines of specially doped diodes designed to minimize this storage time. These diodes are called fast-recovery rectifiers and are commonly used in high-frequency dc-dc converters and regulators. Diodes are available that can recover in about 50 ns and less, as compared to standard-recovery diodes that can take several microseconds to cease conduction in the reverse direction.

Amateurs building their own switching power supplies and dc-dc converters will find greatly improved performance with the use of these diodes in their output rectifiers. Fast-recovery rectifiers are not needed for 60 Hz rectification because the source voltage is a sine wave (no fast transitions) and the input frequency is too slow for transitions to be of significance.

Rectifier Circuits

HALF-WAVE RECTIFIER

Fig 11.5 shows a simple half-wave rectifier circuit. A rectifier (in this case a semiconductor diode) conducts current in one direction but not the other. During one half of the ac cycle, the rectifier conducts and there is current through the rectifier to the load (indicated by the solid line in Fig 11.5B). During the other half cycle, the rectifier is reverse biased and there is no current (indicated by the broken line in Fig 11.5B) to the load. As shown, the output is in the form of pulsed dc, and current always flows in the same direction. A filter can be used to smooth out these variations and provide a higher average dc voltage from the circuit. This idea will be covered in the section on filters.

The average output voltage — the voltage read by a dc voltmeter—with this circuit (no filter connected) is $0.45 \times E_{RMS}$ of the ac voltage delivered by the transformer secondary. Because the frequency of the pulses is low (one pulse per cycle), considerable filtering is required to provide adequately smooth dc output. For this reason the circuit is usually limited to applications where the required current is small, as in a transmitter bias supply.

The peak inverse voltage (PIV), the

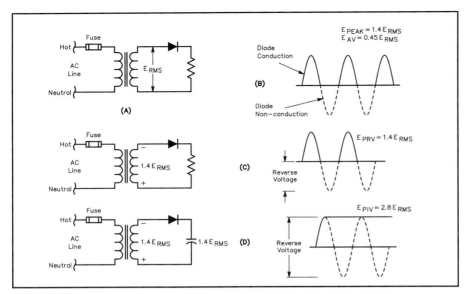

Fig 11.5 — Half-wave rectifier circuit. A illustrates the basic circuit, and B displays the diode conduction and nonconduction periods. The peak-inverse voltage impressed across the diode is shown at C and D, with a simple resistor load at C and a capacitor load at D. E_{PIV} is 1.4 E_{RMS} for the resistor load and 2.8 E_{RMS} for the capacitor load.

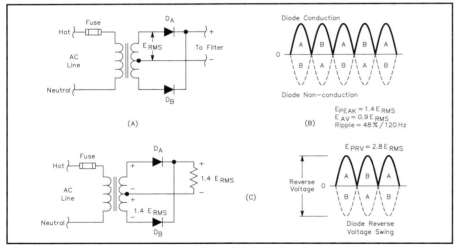

Fig 11.6 — Full-wave center-tap rectifier circuit. A illustrates the basic circuit. Diode conduction is shown at B with diodes A and B alternately conducting. The peak-inverse voltage for each diode is 2.8 E_{RMS} as depicted at C.

voltage that the rectifier must withstand when it isn't conducting, varies with the load. With a resistive load, it is the peak ac voltage (1.4 × E_{RMS}); with a capacitor filter and a load drawing little or no current, it can rise to 2.8 × E_{RMS}. The reason for this is shown in parts C and D of Fig 11.5. With a resistive load as shown at C, the voltage applied to the diode is that voltage on the lower side of the zero-axis line, or 1.4 × E_{RMS}. A capacitor connected to the circuit (shown at D) will store the peak positive voltage when the diode conducts on the positive pulse. If the circuit is not supplying any current, the voltage

across the capacitor will remain at that same level. The peak inverse voltage impressed across the diode is now the sum of the voltage stored in the capacitor plus the peak negative swing of voltage from the transformer secondary. In this case the PIV is 2.8 × E_{RMS}.

FULL-WAVE CENTER-TAP RECTIFIER

A commonly used rectifier circuit is shown in **Fig 11.6**. Essentially an arrangement in which the outputs of two half-wave rectifiers are combined, it makes use of both halves of the ac cycle. A trans-

former with a center-tapped secondary is required with the circuit.

The average output voltage is 0.9 × E_{RMS} of half the transformer secondary; this is the maximum that can be obtained with a suitable choke-input filter. The peak output voltage is 1.4 × E_{RMS} of half the transformer secondary; this is the maximum voltage that can be obtained from a capacitor-input filter.

As can be seen in Fig 11.6C, the PIV impressed on each diode is independent of the type of load at the output. This is because the peak inverse voltage condition occurs when diode A conducts and diode B does not conduct. The positive and negative voltage peaks occur at precisely the same time, a condition different from that in the half-wave circuit. As the cathodes of diodes A and B reach a positive peak (1.4 E_{RMS}), the anode of diode B is at a negative peak, also 1.4 E_{RMS}, but in the opposite direction. The total peak inverse voltage is therefore 2.8 E_{RMS}.

Fig 11.6B shows that the frequency of the output pulses is twice that of the half-wave rectifier. Comparatively less filtering is required. Since the rectifiers work alternately, each handles half of the load current. The current rating of each rectifier need be only half the total current drawn from the supply.

FULL-WAVE BRIDGE RECTIFIER

Another commonly used rectifier circuit is illustrated in **Fig 11.7**. In this arrangement, two rectifiers operate in series on each half of the cycle, one rectifier being in the lead to the load, the other being the return lead. As shown in Fig 11.7A and B, when the top lead of the transformer secondary is positive with respect to the bottom lead, diodes A and C will conduct while diodes B and D are reverse biased. On the next half cycle, when the top lead of the transformer is negative with respect to the bottom, diodes B and D will conduct while diodes A and C are reverse biased.

The output wave shape is the same as that from the simple full-wave center-tap rectifier circuit. The average dc output voltage into a resistive load or choke-input filter is 0.9 times the RMS voltage delivered by the transformer secondary; with a capacitor filter and a light load, the maximum output voltage is 1.4 times the secondary RMS voltage.

Fig 11.7C shows the inverse voltage to be 1.4 E_{RMS} for each diode. When an alternate pair of diodes (such as D_A and D_C) is conducting, the other diodes are essentially connected in parallel in a reverse-biased direction. The reverse stress is then 1.4 E_{RMS}. Each pair of diodes conducts on alternate half cycles, with the full load current through each diode during its conducting half cycle. Since each diode is not conducting during the other half cycle the average current is one half the total load current drawn from the supply.

PROS AND CONS OF THE RECTIFIER CIRCUITS

Comparing the full-wave center-tap rectifier circuit and the full-wave bridge-rectifier circuit, we can see that both circuits have almost the same rectifier requirement, since the center tap has half the number of rectifiers as the bridge. These rectifiers have twice the inverse voltage rating requirement of the bridge diodes, however. The diode current ratings are identical for the two circuits. The bridge makes better use of the transformer's secondary than the center-tap rectifier, since the transformer's full winding supplies power during both half cycles, while each half of the center-tap circuit's secondary provides power only during its positive half-cycle. This is usually referred to as the *transformer utilization factor*, which is unity for the bridge configuration and 0.5 for the full-wave, center-tapped circuit.

The bridge rectifier often takes second place to the full-wave center tap rectifier in high-current low-voltage applications. This is because the two forward-conducting series-diode voltage drops in the bridge introduce a volt or more of additional loss, and thus more heat to be dissipated, than does the single diode drop of the full-wave rectifier.

The half-wave configuration is rarely used in 60 Hz rectification for other than bias supplies. It does see considerable use, however, in high-frequency switching power supplies in what are called *forward converter* and *flyback converter* topologies.

VOLTAGE MULTIPLIERS

Other rectification circuits of interest are the so-called *voltage multipliers*. These circuits function by the process of charging one or more capacitors on one half cycle of the ac waveform, and then connecting that capacitor or capacitors in series with the opposite polarity of the ac waveform on the alternate half cycle. With full-wave multipliers, this charging occurs during both half-cycles.

Voltage multipliers, particularly doublers, find considerable use in high-voltage supplies. When a doubler is employed, the secondary winding of the power transformer need only be half the voltage that would be required for a bridge rectifier. This reduces voltage stress in the windings and decreases the chance of corona in the

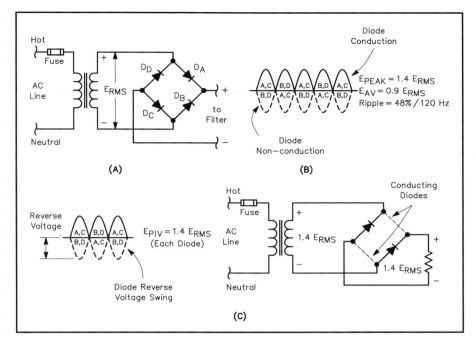

Fig 11.7 — Full-wave bridge rectifier circuit. The basic circuit is illustrated at A. Diode conduction and nonconduction times are shown at B. Diodes A and C conduct on one half of the input cycle, while diodes B and D conduct on the other. C displays the peak inverse voltage for one half cycle. Since this circuit reverse-biases two diodes essentially in parallel, 1.4 E_{RMS} is applied across each diode.

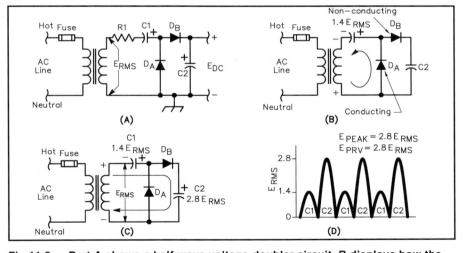

Fig 11.8 — Part A shows a half-wave voltage-doubler circuit. B displays how the first half cycle of input voltage charges C1. During the next half cycle (shown at C), capacitor C2 charges with the transformer secondary voltage plus that voltage stored in C1 from the previous half cycle. The arrows in parts B and C indicate the conventional current. D illustrates the levels to which each capacitor charges over several cycles.

windings, prolonging the life of the transformer. This is not without cost, however, because the transformer-secondary current rating has to be correspondingly doubled.

Half-Wave Doubler

Fig 11.8 shows the circuit of a half-wave voltage doubler. Parts B, C and D illustrate the circuit operation. For clarity, assume the

transformer voltage polarity at the moment the circuit is activated is that shown at B. During the first negative half cycle, D_A conducts (D_B is in a nonconductive state), charging C1 to the peak rectified voltage (1.4 E_{RMS}). C1 is charged with the polarity shown at B. During the positive half cycle of the secondary voltage, D_A is cut off and D_B conducts, charging capacitor C2. The

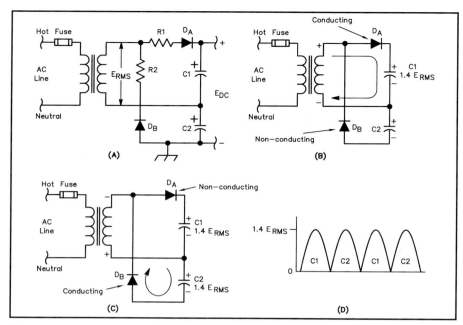

Fig 11.9 — Part A shows a full-wave voltage-doubler circuit. One half cycle is shown at B and the next half cycle is shown at C. Each capacitor receives a charge during every input-voltage cycle. D illustrates how each capacitor is charged alternately.

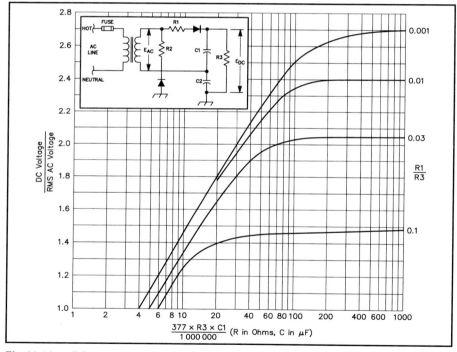

Fig 11.10 — DC output voltages from a full-wave voltage doubler circuit as a function of the filter capacitances and load resistance. For the ratio R1 / R3 and for the R3 × C1 product, resistance is in ohms and capacitance is in microfarads. Equal resistance values for R1 and R2, and equal capacitance values for C1 and C2 are assumed. These curves are adapted from those published by Otto H. Schade in "Analysis of Rectifier Operation," *Proceedings of the I. R. E.*, July 1943.

amount of voltage delivered to C2 is the sum of the transformer peak secondary voltage plus the voltage stored in C1 (1.4 E_{RMS}). On the next negative half cycle, D_B is noncon-ducting and C2 will discharge into the load. If no load is connected across C2, the capacitors will remain charged — C1 to 1.4 E_{RMS} and C2 to 2.8 E_{RMS}. When a load is connected to the circuit output, the voltage across C2 drops during the negative half cycle and is recharged up to 2.8 E_{RMS} during the positive half cycle.

The output waveform across C2 resembles that of a half-wave rectifier circuit because C2 is pulsed once every cycle. Fig 11.8D illustrates the levels to which the two capacitors are charged throughout the cycle. In actual operation, the capacitors will not discharge all the way to zero as shown.

Full-Wave Doubler

Fig 11.9 shows the circuit of a full-wave voltage doubler. The circuit operation can best be understood by following Parts B, C and D. During the positive half cycle of the transformer secondary voltage, as shown at B, D_A conducts charging capacitor C1 to 1.4 E_{RMS}. D_B is not conducting at this time.

During the negative half cycle, as shown at C, D_B conducts, charging capacitor C2 to 1.4 E_{RMS}, while D_A is nonconducting. The output voltage is the sum of the two capacitor voltages, which will be 2.8 E_{RMS} under no-load conditions. Fig 11.9D illustrates that each capacitor alternately receives a charge once per cycle. The effective filter capacitance is that of C1 and C2 in series, which is less than the capacitance of either C1 or C2 alone.

Resistors R1 and R2 in Fig 11.9A are used to limit the surge current through the rectifiers. Their values are based on the transformer voltage and the rectifier surge-current rating, since at the instant the power supply is turned on, the filter capacitors look like a short-circuited load. Provided the limiting resistors can withstand the surge current, their current-handling capacity is based on the maximum load current from the supply. Output voltages approaching twice the peak voltage of the transformer can be obtained with the voltage doubling circuit shown in Fig 11.9. **Fig 11.10** shows how the voltage depends upon the ratio of the series resistance to the load resistance, and the load resistance times the filter capacitance. The peak inverse voltage across each diode is 2.8 E_{RMS}.

Tripler and Quadrupler

Fig 11.11A shows a voltage-tripling circuit. On one half of the ac cycle, C1 and C3 are charged to the source voltage through D1, D2 and D3. On the opposite half of the cycle, D2 conducts and C2 is charged to twice the source voltage, because it sees the transformer plus the charge in C1 as its source (D1 is cut off during this half cycle.) At the same time, D3 conducts, and with the transformer and the charge in C2 as the source, C3 is charged to three times the transformer voltage.

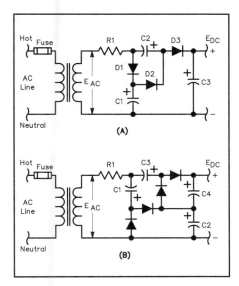

Fig 11.11 — Voltage-multiplying circuits with one side of the transformer secondary used as a common connection. A shows a voltage tripler and B shows a voltage quadrupler. Capacitances are typically 20 to 50 μF, depending on the output current demand. Capacitor dc ratings are related to E_{PEAK} (1.4 E_{RMS}):

C1 — Greater than E_{PEAK}
C2 — Greater than 2 E_{PEAK}
C3 — Greater than 3 E_{PEAK}
C4 — Greater than 2 E_{PEAK}

The voltage-quadrupling circuit of Fig 11.11B works in similar fashion. In either of the circuits of Fig 11.11, the output voltage will approach an exact multiple of the peak ac voltage when the output current drain is low and the capacitance values are high.

RECTIFIER RATINGS VERSUS OPERATING STRESS

Power supplies designed for amateur equipment use silicon rectifiers almost exclusively. These rectifiers are available in a wide range of voltage and current ratings. In peak inverse voltage (PIV) ratings of 600 or less, silicon rectifiers carry current ratings as high as 400 A. At 1000 PIV, the current ratings may be several amperes. It is possible to stack several units in series for higher voltages. Stacks are available commercially that will handle peak inverse voltages up to 10 kV at a load current of 1 A or more.

RECTIFIER STRINGS OR STACKS

Diodes in Series

When the PIV rating of a single diode is not sufficient for the application, similar diodes may be used in series. (Two 500 PIV diodes in series will withstand 1000 PIV and so on.) There used to be a general recommendation to place a resistor across each diode in the string to equalize the PIV drops.

With modern diodes, this practice is no longer necessary.

Modern silicon rectifier diodes are constructed to have an avalanche characteristic. Simply put, this means that the diffusion process is controlled so the diode will exhibit a Zener characteristic in the reverse biased direction before destructive breakdown of the junction can occur. This provides a measure of safety for diodes in series. A diode will go into Zener conduction before it self destructs. If other diodes in the chain have not reached their avalanche voltages, the current through the avalanched diode will be limited to the leakage current in the other diodes. This should normally be very low. For this reason, shunting resistors are generally not needed across diodes in series rectifier strings. In fact, shunt resistors can actually create problems because they can produce a low-impedance source of damaging current to any diode that may have reached avalanche potential.

Diodes in Parallel

Diodes can be placed in parallel to increase current-handling capability. Equalizing resistors should be added as shown in **Fig 11.12**. Without the resistors, one diode may take most of the current. The resistors should be selected to have several tenths of a volt drop at the expected peak current.

RECTIFIER PROTECTION

The important specifications of a silicon diode are:

1. PIV, the peak inverse voltage.
2. I_0, the average dc current rating.
3. I_{REP} — the peak repetitive forward current.
4. I_{SURGE}, a nonrepetitive peak half sine wave of 8.3 ms duration (one-half cycle of 60-Hz line frequency).
5. Switching speed.
6. Power dissipation and thermal resistance.

The first two specifications appear in most catalogs. I_{REP} and I_{SURGE} often are not specified in catalogs, but they are very important. Because the rectifier never allows current to flow more than half the time, when it does conduct it has to pass at least twice the average direct current. With a capacitor-input filter, the rectifier conducts much less than half the time, so that when it does conduct, it may pass as much as 10 to 20 times the average dc current, under certain conditions. This is shown in **Fig 11.13**. Part A shows a simple half-wave rectifier with a resistive load. The waveform to the right of the drawing shows the output voltage along with the diode current. Parts B and C show conditions for circuits with "low" capacitance and "high" capacitance to filter the output.

After the capacitor is charged to the peak-rectified voltage, a period of diode nonconduction elapses while the output voltage dis-

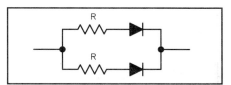

Fig 11.12 — Diodes can be connected in parallel to increase the current-handling capability of the circuit. Each diode should have a series current-equalizing resistor, with a value selected to provide a few tenths of a volt drop at the expected current.

charges through the load. As the voltage begins to rise on the next positive pulse, a point is reached where the rectified voltage equals the stored voltage in the capacitor. As the voltage rises beyond that point, the diode begins to supply current. The diode will continue to conduct until the waveform reaches the crest, as shown. Since the diode has only that short time in which to charge the capacitor with enough energy to provide power to the load for the non-conducting balance of the cycle, the current will be high. The larger the capacitor for a given load, the shorter the diode conduction time and the higher the peak repetitive current (I_{REP}).

Current Inrush

When the supply is first turned on, the discharged input capacitor looks like a dead short, and the rectifier passes a very heavy current. This current transient is called I_{SURGE}. The maximum surge current rating for a diode is usually specified for a duration of one-half cycle (at 60 Hz), or about 8.3 ms. Some form of surge protection is usually necessary to protect the diodes until the input capacitor becomes nearly charged, unless the diodes used have a very high surge-current rating (several hundred amperes). If a manufacturer's data sheet is not available, an educated guess about a diode's capability can be made by using these rules of thumb for silicon diodes commonly used in amateur power supplies:

Rule 1. The maximum I_{REP} rating can be assumed to be approximately four times the maximum I_0 rating.

Rule 2. The maximum I_{SURGE} rating can be assumed to be approximately 12 times the maximum I_0 rating. (This figure should provide a reasonable safety factor. Silicon rectifiers with 750-mA dc ratings, for example, seldom have 1-cycle surge ratings of less than 15 A; some are rated up to 35 A or more.) From this you can see that the rectifier should be selected on the basis of I_{SURGE} and not on I_0 ratings.

Although you can sometimes rely on the dc resistance of the transformer secondary to provide ample surge-current limiting, this is

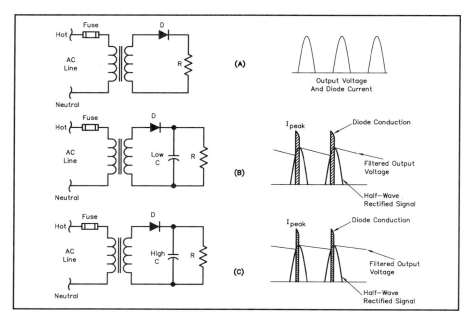

Fig 11.13 — The circuit shown at A is a simple half-wave rectifier with a resistive load. The waveform shown to the right is the output voltage and current. B illustrates how the diode current is modified by the addition of a capacitor filter. The diode conducts only when the rectified voltage is greater than the voltage stored in the capacitor. Since this time is usually only a short portion of a cycle, the peak current will be quite high. C shows an even higher peak current. This is caused by the larger capacitor, which effectively shortens the diode conduction period.

seldom true in high-voltage power supplies. Series resistors are often installed between the secondary and the rectifier strings or in the transformer's primary circuit, but these can be a deterrent to good voltage regulation.

Voltage Spikes

Vacuum-tube rectifiers had little problem with voltage spikes on the incoming power lines — the possibility of an internal arc was of little consequence, since the heat produced was of very short duration and had little effect on the massive plate and cathode structures. Unfortunately, such is not the case with silicon diodes.

Silicon diodes, because of their forward voltage drop of about one volt, create very little heat with high forward currents and therefore have tiny junction areas. Conduction in the reverse direction, however, can cause junction temperatures to rise extremely rapidly with the resultant melting of the silicon and migration of the dopants into the rectifying junction. Destruction of the semiconductor junction is the end result.

To protect semiconductor rectifiers, special surge-absorption devices are available for connection across the incoming ac bus or transformer secondary. These devices operate in a fashion similar to a Zener diode, by conducting heavily when a specific voltage level is reached. Unlike Zener diodes, however, they have the ability to absorb very

high transient energy levels without damage. With the clamping level set well above the normal operating voltage range for the rectifiers, these devices normally appear as open circuits and have no effect on the power-supply circuits. When a voltage transient occurs, however, these protection devices clamp the spike and thereby prevent destruction of the rectifiers.

Transient protectors are available in three basic varieties:
1. Silicon Zener diodes — large junction Zeners specifically made for this purpose and available as single junction for dc (unipolar) and back-to-back junctions for ac (bipolar). These silicon protectors are available under the trade name of TransZorb from General Semiconductor Corporation and are also made by other manufacturers. They have the best transient suppressing characteristics of the three varieties mentioned here, but are expensive and have the least energy absorbing capability per dollar of the group.
2. Varistors — made of a composition metal-oxide material that breaks down at a certain voltage. Metal-oxide varistors, also known as MOVs, are cheap and easily obtained, but have a higher internal resistance, which allows a greater increase of clamped voltage than the Zener variety. Varistors can also degrade with successive transients within their rated power handling limits (this is not usually a prob-

lem in the ham shack where transients are few and replacement of the varistor is easily accomplished).

Varistors usually become short circuited when they fail. Large energy dissipation can result in device explosion. Therefore, it is a good idea to include a fuse that limits the short-circuit current through the varistor, and to protect people and circuitry from debris.
3. Gas tube — similar in construction to the familiar neon bulb, but designed to limit conducting voltage rise under high transient currents. Gas tubes can usually withstand the highest transient energy levels of the group. Gas tubes suffer from an ionization time problem, however. A high voltage across the tube will not immediately cause conduction. The time required for the gas to ionize and clamp the spike is inversely proportional to the level of applied voltage in excess of the device ionization voltage. As a result, the gas tube will let a little of the transient through to the equipment before it activates.

In installations where reliable equipment operation is critical, the local power is poor and transients are a major problem, the usual practice is to use a combination of protectors. Such systems consist of a varistor or Zener protector, combined with a gas-tube device. Operationally, the solid-state device clamps the surge immediately, with the beefy gas tube firing shortly thereafter to take most of the surge from the solid-state device.

Heat

The junction of a diode is quite small; hence it must operate at a high current density. The heat-handling capability is, therefore, quite small. Normally, this is not a prime consideration in high-voltage, low-current supplies. Use of high-current rectifiers at or near their maximum ratings (usually 2-A or larger stud-mount rectifiers) requires some form of heat sinking. Frequently, mounting the rectifier on the main chassis — directly, or with thin mica insulating washers — will suffice. If insulated from the chassis, a thin layer of silicone grease should be used between the diode and the insulator, and between the insulator and the chassis, to assure good heat conduction. Large, high-current rectifiers often require special heat sinks to maintain a safe operating temperature. Forced-air cooling is sometimes used as a further aid. Safe case temperatures are usually given in the manufacturer's data sheets and should be observed if the maximum capabilities of the diode are to be realized. See the thermal design section in the **Real World Components** chapter for more information.

Filtration

The pulsating dc waves from the rectifiers are not sufficiently constant in amplitude to prevent hum corresponding to the pulsations. Filters are required between the rectifier and the load to smooth out the pulsations into an essentially constant dc voltage. The design of the filter depends to a large extent on the dc voltage output, the voltage regulation of the power supply and the maximum load current rating of the rectifier. Power-supply filters are low-pass devices using series inductors and shunt capacitors.

LOAD RESISTANCE

In discussing the performance of power-supply filters, it is sometimes convenient to express the load connected to the output terminals of the supply in terms of resistance. The load resistance is equal to the output voltage divided by the total current drawn, including the current drawn by the bleeder resistor.

VOLTAGE REGULATION

The output voltage of a power supply always decreases as more current is drawn, not only because of increased voltage drops in the transformer and filter chokes, but also because the output voltage at light loads tends to soar to the peak value of the transformer voltage as a result of charging the first capacitor. Proper filter design can eliminate the soaring effect. The change in output voltage with load is called voltage regulation, and is expressed as a percentage.

$$\text{Percent Regulation} = \frac{(E1 - E2)}{E2} \times 100\% \quad (1)$$

where:
E1 = the no-load voltage
E2 = the full-load voltage.

A steady load, such as that represented by a receiver, speech amplifier or unkeyed stages of a transmitter, does not require good (low) regulation as long as the proper voltage is obtained under load conditions. The filter capacitors must have a voltage rating safe for the highest value to which the voltage will soar when the external load is removed.

A power supply will show more (higher) regulation with long-term changes in load resistance than with short temporary changes. The regulation with long-term changes is often called the static regulation, to distinguish it from the dynamic regulation (short temporary load changes). A load that varies at a syllabic or keyed rate, as represented by some audio and RF amplifiers, usually requires good dynamic regulation (15% or less) if distortion products are to be held to a low level. The dynamic regulation of a power supply can be improved by increasing the value of the output capacitor.

When essentially constant voltage regardless of current variation is required (for stabilizing an oscillator, for example), special voltage regulating circuits described later in this chapter are used.

BLEEDER RESISTOR

A bleeder resistor is a resistance connected across the output terminals of the power supply. Its functions are to discharge the filter capacitors as a safety measure when the power is turned off and to improve voltage regulation by providing a minimum load resistance. When voltage regulation is not of importance, the resistance may be as high as 100 Ω per volt of power supply output voltage. The resistance value to be used for voltage-regulating purposes is discussed in later sections. From the consideration of safety, the power rating of the resistor should be as conservative as possible, since a burned-out bleeder resistor is dangerous!

RIPPLE FREQUENCY AND VOLTAGE

Pulsations at the output of the rectifier can be considered to be the result of an alternating current superimposed on a steady direct current. From this viewpoint, the filter may be considered to consist of shunt capacitors that short circuit the ac component while not interfering with the flow of the dc component. Series chokes will readily pass dc but will impede the flow of the ac component.

The alternating component is called ripple. The effectiveness of the filter can be expressed in terms of percent ripple, which is the ratio of the RMS value of the ripple to the dc value in terms of percentage.

$$\text{Percent Ripple (RMS)} = \frac{E1}{E2} \times 100\% \quad (2)$$
where:
E1 = the RMS value of ripple voltage
E2 = the steady dc voltage.

Any frequency multiplier or amplifier supply in a CW transmitter should have less than 5% ripple. A linear amplifier can tolerate about 3% ripple on the plate voltage. Bias supplies for linear amplifiers should have less than 1% ripple. VFOs, speech amplifiers and receivers may require a ripple no greater than to 0.01%.

Ripple frequency refers to the frequency of the pulsations in the rectifier output waveform — the number of pulsations per second. The ripple frequency of half-wave rectifiers is the same as the line-supply frequency — 60 Hz with a 60-Hz supply. Since the output pulses are doubled with a full-wave rectifier, the ripple frequency is doubled — to 120 Hz with a 60-Hz supply.

The amount of filtering (values of inductance and capacitance) required to give adequate smoothing depends on the ripple frequency. More filtering is required as the ripple frequency is reduced.

CAPACITOR-INPUT FILTERS

Capacitor-input filter systems are shown in **Fig 11.14**. Disregarding voltage

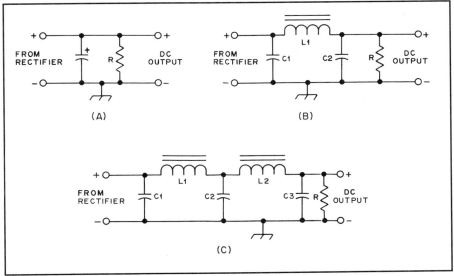

Fig 11.14 — Capacitor-input filter circuits. At A is a simple capacitor filter. B and C are single- and double-section filters, respectively.

drops in the chokes, all have the same characteristics except with respect to ripple. Better ripple reduction will be obtained when LC sections are added as shown in Fig 14B and C.

Input Versus Output Voltage

The average output voltage of a capacitor-input filter is generally poorly regulated with load-current variations. This is because the rectifier diodes conduct for only a small portion of the ac cycle to charge the filter capacitor to the peak value of the ac waveform. When the instantaneous voltage of the ac passes its peak, the diode ceases to conduct. This forces the capacitor to support the load current until the ac voltage on the opposing diode in the bridge or full wave rectifier is high enough to pick up the load and recharge the capacitor. For this reason, the diode currents are usually quite high.

Since the cyclic peak voltage of the capacitor-filter output is determined by the peak of the input ac waveform, the minimum voltage and, therefore, the ripple amplitude, is determined by the amount of voltage discharge, or "droop," occurring in the capacitor while it is discharging and supporting the load. Obviously, the higher the load current, the proportionately greater the discharge, and therefore the lower the average output voltage will be.

Although not exactly accurate, an easy way to determine the peak-to-peak ripple for a certain capacitor and load is to assume a constant load current. We can calculate the droop in the capacitor by using the relationship:

$$C \times E = I \times t \qquad (3)$$

where:
C = the capacitance in microfarads,
E = the voltage droop, or peak-to-peak ripple voltage,
I = the load current in milliamperes and
t = the time between half-cycles of the rectified waveform, in milliseconds. For 60 Hz full-wave rectifiers, t is about 7.5 ms.

As an example, let's assume that we need to determine the peak to peak ripple voltage at the dc output of a full-wave rectifier/filter combination that produces 13.8 V dc and supplies a transceiver drawing 2.0 A. The filter capacitor in the power supply is 5000 µF. Using the above relationship:

$$C \times E = I \times t$$

$$5000 \ \mu F \times E = 2000 \ mA \times 7.5 \ ms$$

$$E = \frac{2000 \ mA \ \times \ 7.5 \ ms}{5000 \ \mu F} = 3 \ V \ P-P$$

Obviously, this is too much ripple. A capacitor value of about 20000 µF would be better suited for this application. If a linear regulator is used after this rectifier/filter combination, however, and the source voltage raised to produce a dc voltage of about 20 V, the 5000-µF capacitor with its 3 V peak-to-peak ripple would work well, since the regulator would remove the ripple content before the output power was applied to the transceiver.

CHOKE-INPUT FILTERS

Choke-input filters have become less popular than they once were, because of the high surge current capability of silicon rectifiers. Choke-input filters provide the benefits of greatly improved output voltage stability over varying loads and low peak-current surges in the rectifiers. On the negative side, however, the choke is bulky and heavy, and the output voltage is lower than that of a capacitor-input filter.

As long as the inductance of the choke is large enough to maintain a continuous current over the complete cycle of the input ac waveform, the filter output voltage will be the average value of the rectified output. The average dc value of a full-wave rectified sine wave is 0.637 times its peak voltage. Since the RMS value is 0.707 times the peak, the output of the choke input filter will be (0.637 / 0.707), or 0.90 times the RMS ac voltage. For light loads, however, there may not be enough energy stored in the choke during the input waveform crest to allow continuous current over the full cycle. When this happens, the filter output voltage will rise as the filter assumes more and more of the characteristics of a capacitor-input filter.

Choke-input filters see extensive use in the energy-storage networks of switch-mode regulators.

Regulation

The output of a rectifier/filter system may be usable for some electronic equipment, but for today's transceivers and accessories, further measures may be necessary to provide power sufficiently clean and stable for their needs. Voltage regulators are often used to provide this additional level of conditioning.

Rectifier/filter circuits by themselves are unable to protect the equipment from the problems associated with input-power-line fluctuations, load-current variations and residual ripple voltages. Regulators can eliminate these problems, but not without costs in circuit complexity and power-conversion efficiency.

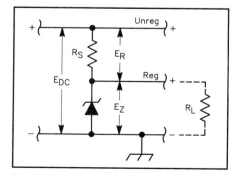

Fig 11.15 — Zener-diode voltage regulation. The voltage from a negative supply may be regulated by reversing the power-supply connections and the diode polarity.

ZENER DIODES

A Zener diode (named after American physicist Dr. Clarence Zener) can be used to maintain the voltage applied to a circuit at a practically constant value, regardless of the voltage regulation of the power supply or variations in load current. The typical circuit is shown in **Fig 11.15**. Note that the cathode side of the diode is connected to the positive side of the supply. The electrical characteristics of a Zener diode under conditions of forward and reverse voltage are given in the **Analog Signals and Components** chapter.

Zener diodes are available in a wide variety of voltages and power ratings. The voltages range from less than two to a few hundred, while the power ratings (power the diode can dissipate) run from less than 0.25 W to 50 W. The ability of the Zener diode to stabilize a voltage depends on the diode's conducting impedance. This can be as low as 1 Ω or less in a low-voltage, high-power diode or as high as 1000 Ω in a high-voltage, low-power diode.

LINEAR REGULATORS

Linear regulators come in two varieties: series and *shunt*. The shunt regulator is simply an electronic (also called "active") version of the Zener diode. For the most part, the active shunt regulator is rarely used since the series regulator is a superior choice for most applications. **Fig 11.16B** shows a shunt regulator.

The series regulator consists of a stable voltage reference, which is usually established by a Zener diode, a transistor in series with the power source and the load (called a *series pass transistor*), and an error amplifier. In critical applications a temperature-compensated reference diode would be used instead of the Zener diode. (See Fig 11.16A.)

The output voltage is sampled by the error amplifier, which compares the output (usually scaled down by a voltage divider) to the reference. If the scaled-down output voltage becomes higher than the reference voltage, the error amplifier reduces the drive current to the pass transistor, thereby allowing the output voltage to drop slightly. Conversely, if the load pulls the output voltage below the desired value, the amplifier drives the pass transistor into increased conduction.

The "stiffness" or tightness of regulation of a linear regulator depends on the gain of the error amplifier and the ratio of the output scaling resistors. In any regulator, the output is cleanest and regulation stiffest at the point where the sampling network or error amplifier is connected. If heavy load current is drawn through long leads, the voltage drop can degrade the regulation at the load. To combat this effect, the feedback connection to the error amplifier can be made directly to the load. This technique, called *remote sensing*, moves the point of best regulation to the load by bringing the connecting loads inside the feedback loop. This is shown in Fig 11.16C.

Input Versus Output Voltage

In a series regulator, the pass-transistor power dissipation is directly proportional to the load current and input/output voltage differential. The series pass element can be located in either leg of the supply. Either NPN or PNP devices can be used, depending on the ground polarity of the unregulated input.

The differential between the input and output voltages is a design tradeoff. If the input voltage from the rectifiers and filter is only slightly higher than the required output voltage, there will be minimal voltage drop across the series pass transistor resulting in minimal thermal dissipation and high power-supply efficiency. The supply will have less capability to provide regulated power in the event of power line brownout and other reduced line voltage conditions, however. Conversely, a higher input voltage will provide operation over a wider range of input voltage, but at the

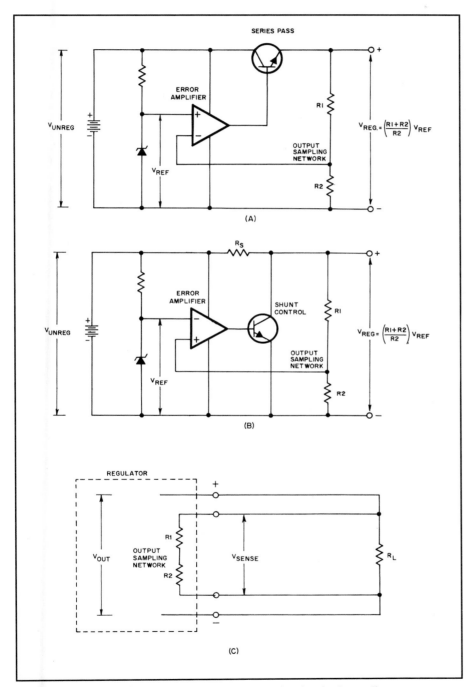

Fig 11.16 — Linear electronic voltage-regulator circuits. In these diagrams, batteries represent the unregulated input-voltage source. A transformer, rectifier and filter would serve this function in most applications. Part A shows a series regulator and Part B shows a shunt regulator. Part C shows how remote sensing overcomes poor load regulation caused by the I R drop in the connecting wires by bringing them inside the feedback loop.

expense of increased heat dissipation.

Pass Transistors

Darlington Pairs

A simple Zener diode reference or IC op-amp error amplifier may not be able to source enough current to a pass transistor that must conduct heavy load current. The Darlington configuration of **Fig 11.17A** multiplies the pass transistor beta, thereby extending the control range of the error amplifier. If the Darlington arrangement is implemented with discrete transistors, resistors across the base-emitter junctions may be necessary to prevent collector-to-base leakage currents in Q1 from being amplified and turning on the transistor pair. These resistors are contained in the envelope of a monolithic Darlington device.

When a single pass transistor is not available to handle the current required from a regulator, the current-handling capability may be increased by connecting two or more pass transistors in parallel. The circuit of Fig 11.17B shows the method of connecting these pass transistors. The resistances in the emitter leads of each transistor are necessary to equalize the currents.

Transistor Ratings

When bipolar (NPN, PNP) power transistors are used in applications in which they are called upon to handle power on a continuous basis, rather than switching, there are four parameters that must be examined to see if any maximum limits are being exceeded. Operation of the transistor outside these limits can easily result in the failure of the device. Unfortunately, not many hams (nor, sometimes, equipment manufacturers) are aware of all these parameters. Yet, for a transistor to provide reliable operation, the circuit designer must be sure not to allow his power supply or amplifier to cause overstress.

The four limits are maximum collector current (I_C), maximum collector-emitter voltage (V_{CEO}), maximum power and second breakdown (I_{SB}). All four of these parameters are graphically shown on the transistor's data sheet on what is known as a Safe Operating ARea (SOAR) graph. (See **Fig 11.18**.) The first three of these limits are usually also listed prominently with the other device information, but it is often the fourth parameter that is responsible for the "sudden death" of the power transistor after an extended operating period.

The maximum current limit of the tran-

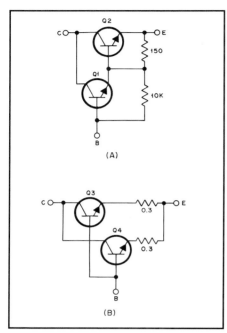

Fig 11.17 — At A, a Darlington-connected transistor pair for use as the pass element in a series-regulating circuit. At B, the method of connecting two or more transistors in parallel for high current output. Resistances are in ohms. The circuit at A may be used for load currents from 100 mA to 5 A, and the one at B may be used for currents from 6 A to 10 A.

Q1 — Motorola MJE 340 or equiv.
Q2 - Q4 — Power transistor such as 2N3055 or 2N3772.

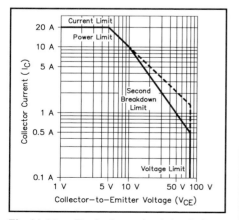

Fig 11.18 — Typical graph of the Safe Operating ARea (SOAR) of a transistor. See text for details. Safe operating conditions for specific devices may be quite different from those shown here.

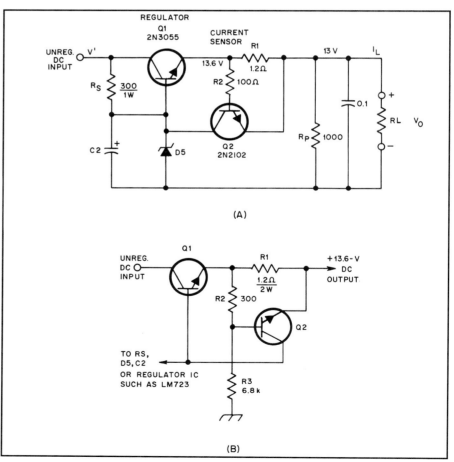

Fig 11.19 — Overload protection for a regulated supply can be implemented by addition of a current-overload-protective circuit, as shown at A. At B, the circuit has been modified to employ current-foldback limiting.

sistor (I_C MAX) is usually the current limit for fusing of the bond wire connected to the emitter, rather than anything pertaining to the transistor chip itself. When this limit is exceeded, the bond wire can melt and open circuit the emitter. On the operating curve, this limit is shown as a horizontal line extending out from the Y-axis (zero volts between collector and emitter) and ending at the voltage point where the constant power limit begins.

The maximum collector-emitter voltage limit of the transistor (V_{CE} MAX) is the point at which the transistor can no longer stand off the voltage between collector and emitter.

With increasing collector-emitter voltage drop at maximum collector current, a point is reached where the power in the transistor will cause the junction temperature to rise to a level where the device leakage current rapidly increases and begins to dominate. In this region, the product of the voltage drop and the current would be constant and represent the maximum power (P_t) rating for the transistor; that is, as the voltage drop continues to increase, the collector current must decrease to maintain the power dissipation at a constant value.

With most of the higher voltage rated transistors, a point is reached on the constant power portion of the curve whereby, with further increased voltage drop, the maximum power rating is *not* constant, but decreases as the collector to emitter voltage increases. This decrease in power handling capability continues until the maximum voltage limit is reached.

This special region is known as the forward bias second breakdown (FBSB) area. Reduction in the transistor's power handling capability is caused by localized heating in certain small areas of the transistor junction ("hot spots"), rather than a uniform distribution of power dissipation over the entire surface of the device.

The region of operating conditions contained within these curves is called the Safe Operating ARea, or SOAR. If the transistor is always operated within these limits, it should provide reliable and continuous service for a long time.

MOSFET Transistors

The bipolar junction transistor (BJT) is rapidly being replaced by the MOSFET transistor in new power supply designs due to the latter's ease of drive. Just as the BJT comes in both NPN and PNP varieties, the MOSFET is available in N-channel and P-channel types, with the N-channel being the more popular of the two. The N-channel MOSFET is equivalent to the NPN bipolar, and the P-channel is equivalent to the PNP.

There are some considerations that should be observed when using a MOSFET as a linear regulator series pass transistor. Several volts of gate drive are needed in order to start conduction of the device, as opposed to less than one volt for the BJT. MOSFETs are inherently very-high-frequency devices, and will readily oscillate with stray circuit capacitances. In order to prevent oscillation in the transistor and surrounding circuits, it is common practice to insert a small resistor of about 100 ohms directly in series with the gate of the series pass transistor to reduce the gate circuit Q.

Overcurrent Protection

Damage to a pass transistor can occur when the load current exceeds the safe amount. **Fig 11.19A** illustrates a simple current-limiter circuit that will protect Q1. All of the load current is routed through R1. A voltage difference will exist across R1; the value will depend on the exact load current at a given time. When the load current exceeds a predetermined safe value, the voltage drop across R1 will forward-bias Q2 and cause it to conduct. Because Q2 is a silicon transistor, the voltage drop across R1 must exceed 0.6 V to turn Q2 on. This being the case, R1 is chosen for a value that provides a drop of 0.6 V when the maximum safe load current is drawn. In this instance, the drop will be 0.6 V when I_L reaches 0.5 A. R2 protects the base-emitter junction of Q2 from current spikes, or from destruction in the event Q1 fails under short-circuit conditions.

When Q2 turns on, some of the current through R_S flows through Q2, thereby depriving Q1 of some of its base current. This action, depending upon the amount of Q1 base current at a precise moment, cuts off Q1 conduction to some degree, thus limiting the current through it.

Foldback Current Limiting

Under short-circuit conditions, a constant-current type current limiter must still withstand the full source voltage and limited short circuit current simultaneously, which can impose a very high power dissipation or second breakdown stress on the series pass transistor. For example, a 12-V regulator with current limiting set for 10 A and having a source of 16 V will have a dissipation of 40 W [(16 V – 12 V) × 10 A] at the point of current limiting (knee). But its dissipation will rise to 160 W under short-circuit conditions (16 V × 10 amps).

A modification of the limiter circuit can cause the regulated output current to decrease with decreasing load resistance after the overcurrent knee. With the output shorted, the output current is only a fraction of the knee current value, which protects the series pass transistor from exces-

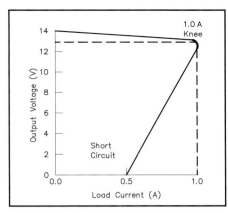

Fig 11.20 — The 1-A regulator shown in Fig 11.19B will fold back to 0.5 A under short-circuit conditions. See text.

sive dissipation and possible failure. Using the previous example of the 12-V, 10-A regulator, if the short-circuit current is designed to be 3 A (the knee is still 10 A), the transistor dissipation with a short circuit will be only 16 V × 3 A = 48 W.

Fig 11.19B shows how the current limiter example given in the previous section would be modified to incorporate foldback limiting. The divider string formed by R2 and R3 provides a negative bias to the base of Q2, which prevents Q2 from turning on until this bias is overcome by the drop in R1 caused by load current. Since this hold-off bias decreases as the output voltage drops, Q2 becomes more sensitive to current through R1 with decreasing output voltage. See **Fig 11.20**.

The circuit is designed by first calculating the value of R1 for short-circuit current. For example, if 0.5 A is chosen, the value for R1 is simply 0.6 V / 0.5 A = 1.2 Ω (with the output shorted, the amount of hold-off bias supplied by R2 and R3 is very small and can be neglected). The knee current is then chosen. For this example, the selected value will be 1.0 A. The divider string is then proportioned to provide a base voltage at the knee that is just sufficient to turn on Q2 (a value of 13.6 V for 13.0 V output). With 1.0 A flowing through R1, the voltage across the divider will be 14.2 V. The voltage dropped by R2 must then be 14.2 V –13.6 V, or 0.6 V. Choosing a divider current of 2 mA, the value of R2 is then 0.6 V / 0.002 A = 300 Ω. R3 is calculated to be 13.6 V / 0.002 A = 6800 Ω.

"Crowbar" Circuits

Electronic components *do* fail from time to time. In a regulated power supply, the only component standing between an elevated dc source voltage and your rig is one transistor, or a group of transistors wired in parallel. If the transistor, or one of the transistors in the group, happens to

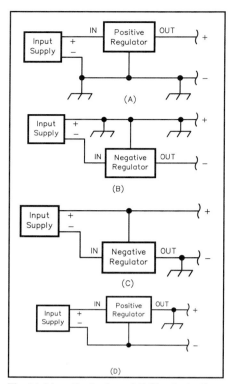

Fig 11.21 — Parts A and B illustrate the conventional manner in which three-terminal regulators are used. Parts C and D show how one polarity regulator can be used to regulate the opposite-polarity voltage.

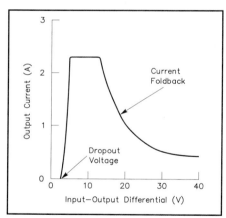

Fig 11.22 — Effects of input-output differential voltage on three-terminal regulator current.

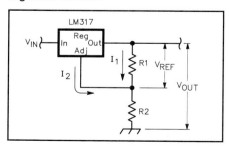

Fig 11.23 — By varying the ratio of R2 to R1 in this simple LM317 schematic diagram, a wide range of output voltages is possible. See text for details.

short internally, your rig could suffer lots of damage.

To safeguard the rig or other load equipment against possible overvoltage, some power-supply manufacturers include a circuit known as a crowbar. This circuit usually consists of a silicon-controlled rectifier (SCR) connected directly across the output of the power supply, with a voltage-sensing trigger circuit tied to its gate. In the event the output voltage exceeds the trigger set point, the SCR will fire, and the output is short circuited. The resulting high current in the power supply (shorted output in series with a series pass transistor failed short) will blow the power supply's line fuses. This is both a protection for the supply as well as an indicator that something has malfunctioned internally. For these reasons, never replace blown fuses with ones that have a higher current rating.

IC VOLTAGE REGULATORS

The modern trend in regulators is toward the use of three-terminal devices commonly referred to as *three-terminal regulators*. Inside each regulator is a voltage reference, a high-gain error amplifier, temperature-compensated voltage sensing resistors and a pass element. Many currently available units have thermal shut-down, overvoltage protection and current foldback, making them virtually destruction-proof.

Three-terminal regulators (a connection for unregulated dc input, regulated dc output and ground) are available in a wide range of voltage and current ratings. It is easy to see why regulators of this sort are so popular when you consider the low price and the number of individual components they can replace. The regulators are available in several different package styles, depending on current ratings. Low-current (100 mA) devices frequently use the plastic TO-92 and DIP-style cases. TO-220 packages are popular in the 1.5-A range, and TO-3 cases house the larger 3-A and 5-A devices.

Three-terminal regulators are available as positive or negative types. In most cases, a positive regulator is used to regulate a positive voltage and a negative regulator a negative voltage. Depending on the system ground requirements, however, each regulator type may be used to regulate the "opposite" voltage.

Fig 11.21A and B illustrate how the regulators are used in the conventional mode. Several regulators can be used with a common-input supply to deliver several voltages with a common ground. Negative regulators may be used in the same manner. If no other common supplies operate off the input supply to the regulator, the circuits of Fig 11.21C and D may be used to regulate posi-

tive voltages with a negative regulator and vice versa. In these configurations the input supply is floated; neither side of the input is tied to the system ground.

Manufacturers have adopted a system of family numbers to classify three-terminal regulators in terms of supply polarity, output current and regulated voltage. For example, National uses the number LM7805C to describe a positive 5-V, 1.5-A regulator; the comparable unit from Texas Instruments is a UA7805KC. LM7812C describes a 12-V regulator of similar characteristics. LM7905C denotes a negative 5-V, 1.5-A device. There are many such families with widely varied ratings available from manufacturers. Fixed-voltage regulators are available with output ratings in most common values between 5 and 28 V. Other families include devices that can be adjusted from 1.25 to 50 V.

Regulator Specifications

When choosing a three-terminal regulator for a given application, the most important specifications to consider are device output voltage, output current, input-to-output differential voltage, line regulation, load regulation and power dissipation. Output voltage and current requirements are determined by the load with which the supply will ultimately be used.

Input-to-output differential voltage is one of the most important three-terminal regulator specifications to consider when designing a supply. The differential value (the difference between the voltage applied to the input terminal and the voltage on the output terminal) must be within a specified range. The minimum differential value, usually about 2.5 V, is called the dropout voltage. If the differential value is less than the dropout voltage, no regulation will take place. At the other end of the scale, maximum input-output differential voltage is generally about 40 V. If this differential value is exceeded, device failure may occur.

Increases in either output current or differential voltage produce proportional increases in device power consumption. By employing a safety feature called current foldback, some manufacturers ensure that maximum dissipation will never be exceeded in normal operation. **Fig 11.22** shows the relationship between output current, input-output differential and current limiting for a three-terminal regulator nominally rated for 1.5-A output current. Maximum output current is available with differential voltages ranging from about 2.5 V (dropout voltage) to 12 V. Above 12 V, the output current decreases, limiting the device dissipation to a safe value. If the output terminals are accidentally short circuited, the input-output differential will rise, causing current foldback, and thus preventing the power-supply components from being over stressed. This protective feature makes three-terminal regulators particularly attractive in simple power supplies.

When designing a power supply around a particular three-terminal regulator, input-output voltage characteristics of the regulator should play a major role in selecting the transformer-secondary and filter-capacitor component values. The unregulated voltage applied to the input of the three-terminal device should be higher than the dropout voltage, yet low enough that the regulator does not go into current limiting caused by an excessive differential voltage. If, for example, the regulated output voltage of the device shown in Fig 11.22 was 12, then unregulated input voltages of between 14.5 and 24 would be acceptable if maximum output current is desired.

In use, all but the lowest current regulators generally require an adequate external heat sink because they may be called on to dissipate a fair amount of power. Also, because the regulator chip contains a high-gain error amplifier, bypassing of the input and output leads is essential for stable operation.

Most manufacturers recommend bypassing the input and output directly at the leads where they protrude through the heat sink. Solid tantalum capacitors are usually recommended because of their good high-frequency capabilities.

External capacitors used with IC regulators may discharge through the IC junctions under certain circuit conditions, and high current discharges can harm ICs. Look at the regulator data sheet to see whether protection diodes are needed, what diodes to use and how to place them in any particular application.

In addition to fixed-output-voltage ICs, high-current, adjustable voltage regulators are available. These ICs require little

more than an external potentiometer for an adjustable output range from 5 to 24 V at up to 5 A. The unit price on these items is only a few dollars, making them ideal for test-bench power supplies. A very popular low current, adjustable output voltage three terminal regulator, the LM317, is shown in **Fig 11.23**. It develops a steady 1.25-V reference, V_{REF}, between the output and adjustment terminals. By installing R1 between these terminals, a constant current, I1, is developed, governed by the equation:

$$I1 = \frac{V_{REF}}{R1} \qquad (4)$$

Both I1 and a 100-μA error current, I2, flow through R2, resulting in output voltage V_O. V_O can be calculated using the equation:

$$V_O = V_{REF}\left(1 + \frac{R2}{R1}\right) + I2 \times R2 \qquad (5)$$

Any voltage between 1.2 and 37 V may be obtained with a 40-V input by changing the ratio of R2 to R1. At lower output voltages, however, the available current will be limited by the power dissipation of the regulator.

Fig 11.24 shows one of many flexible applications for the LM317. By adding only one resistor with the regulator, the voltage regulator can be changed into a constant-current source capable of charging NiCd batteries, for example. Design equations are given in the figure. The same precautions should be taken with adjustable regulators as with the fixed voltage units. Proper heat sinking and lead bypassing are essential for proper circuit operation.

Increasing Regulator Output Current

When the maximum output current from an IC voltage regulator is insufficient to operate the load, discrete power transistors may be connected to increase the

current capability. **Fig 11.25** shows two methods for boosting the output current of a positive regulator, although the same techniques can be applied to negative regulators.

In A, an NPN transistor is connected as an emitter follower, multiplying the output current capacity by the transistor beta. The shortcoming of this approach is that the base-emitter junction is not inside the feed-back loop. The result is that the output voltage is reduced by the base-emitter drop, and the load regulation is degraded by variations in this drop.

The circuit at B has a PNP transistor "wrapped around" the regulator. The regulator draws current through the base-emitter junction, causing the transistor to conduct. The IC output voltage is unchanged by the transistor because the collector is connected directly to the IC output (sense point). Any increase in output voltage is detected by the IC regulator, which shuts off its internal pass transistor, and this stops the boost-transistor base current.

HIGH-PRECISION SERIES REGULATOR LOOP DESIGN

This regulator-loop-design material was contributed by William E. Sabin, WØIYH. (A similar discussion appears in May 1991 *QEX*, pp 3-9.) The discussion here concerns some of the important factors in the design and testing of a series regulator feedback loop. These techniques were used to design "A Series Regulated 4.5- to 25-V 2.5-A Power Supply," which appears later in this chapter. The values and measurements discussed here are from that circuit. **Fig 11.26** is a simplified version of that supply, which is adequate for this discussion.

The series regulator is a good example of a feedback control system. Open-loop gain and bandwidth, the phase and gain margins and the transient response are important factors. The goal of the design is to maximize the regulator closed-loop

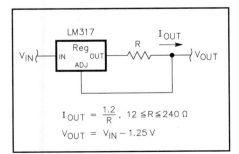

Fig 11.24 — The basic LM317 voltage regulator is converted into a constant-current source by adding only one resistor.

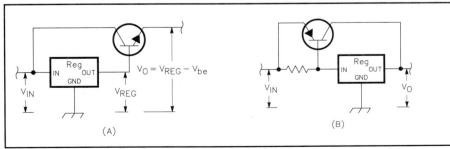

Fig 11.25 — Two methods for boosting the output-current capacity of an IC voltage regulator. Part A shows an NPN emitter follower and B shows a PNP "wrap-around" configuration. Operation of these circuits is explained in the text.

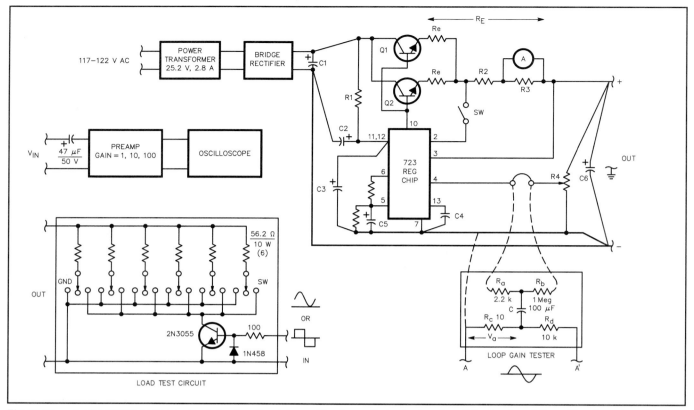

Fig 11.26 — A simplified voltage-regulator feedback loop. The Loop Gain Tester, Load Tester, Load Test Circuit and oscilloscope circuits are used to test open-loop response and gain. The heavy lines indicate critical low-impedance circuit paths.

performance. One approach is to use a high value of open-loop gain and establish the open-loop frequency response in two ways: (1) an RC low-pass filter consisting of C6, R2, R3, $\frac{1}{2}$ R_e and the out-put resistances of Q1 and Q2; and (2) a single small capacitor (C4) at the regulator IC. Note that the voltage drop across R2 and R3 is applied to pins 2 and 3 of the LM723 regulator IC, which are used for current limiting. R2 sets the current limit value, and R2's value affects the RC filter.

Test Circuits

Fig 11.26 shows three test circuits. One is an adjustable Load Test Circuit that can be modulated linearly (almost) by: (1) a sine or triangle wave from a function generator with a dc-offset adjustment (so that the waveform always has positive polarity), or (2) by a bidirectional square wave. This circuit is used to test the loop response to various load fluctuations. It has proved to be very informative, as discussed later.

The second test circuit (Loop Gain Tester) is inserted into the regulator loop so that a test signal can be injected into the loop, in order to measure the open-loop gain and frequency response. Notice that the loop is closed through the feedback path of R4, R_b and R_a at dc and very low frequencies (less than 0.5 Hz), and the dc output voltage is

pretty well regulated (which is essential to loop testing). By observing the magnitude (and rate of change) of the frequency response, it is possible to deduce information about phase shift. With this information available, the gain and phase margins (and therefore the regulation, stability, transient response and output impedance) of the closed-loop regulator can be estimated.

The third test circuit is a two-stage op amp preamplifier and oscilloscope. It is used to measure very small signals in the 0.1-Hz to 400-kHz range.

Open-Loop Tests

The test signal applied to points A and A' is reduced 60 dB by a voltage divider: R_d and R_c. (The voltage division converts convenient input voltages, which we can measure with ordinary equipment, to microvolt levels in the loop.) Capacitor C couples V_a, the voltage across R_c, to the '723 through R_a. R_a is roughly the resistance that the '723 sees in normal operation. The test signal is amplified by at most 74 dB on its way (clockwise around the loop and through the regulator) to the right-hand end of R_b. It is then attenuated 100 dB by R_b and R_c. This means that the "leak-through" back to the '723 input is much smaller than the V_a that we started with, if the frequency is 2 Hz or greater. At

dc (and very low frequencies) the regulator functions somewhat normally. Above 2 Hz, then, the magnitude of the open-loop gain at the test frequency is very nearly the ratio $|V_{out}| / |V_a|$.

The first benefit of this test circuit was that it isolated an instability in the '723. The oscillation, at several hundred kHz, was cured by adding C4 (33 pF) and C3 (100 µF / 50 V with very short leads). Normally, one would suspect an oscillation to involve the overall loop, but this was not the case. This kind of instability is common in feedback control systems: Everything appears to function (usually not to full specification), but an embedded element is not stable.

Open-Loop Frequency Response

Refer to Fig 11.26. The open-loop gain is the product of three factors. The test signal is:

1. Voltage amplified by the regulator IC (about 74 dB for the '723) on its way (clockwise) to the emitters of Q1 and Q2.
2. Low-pass filtered by C6 and R_E (the combination of $\frac{1}{2}$ R_e + R2 + R3).
3. Divided by potentiometer R4.

For the 2.5-A supply, the greatest open-loop gain values are 59 dB at 25 V output and 74 dB at 4.5 V.

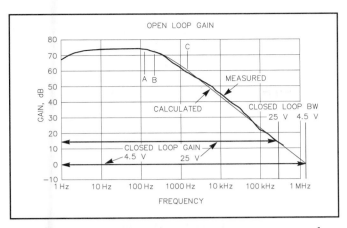

Fig 11.27 — The open-loop frequency response curves of the voltage-regulator feedback loop. Point A is 120 Hz; point B is 280 Hz; point C is 1.2 kHz (above 1.2 kHz, C6 is no longer effective).

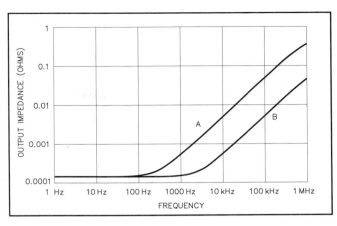

Fig 11.28 — Output impedance magnitude for the voltage-regulator feedback loop.

Fig 11.27 shows the open-loop frequency response to the top of R4 when R_E is set for a 2.5-A current limit. At very low frequencies, the drop-off is due to the gradual closure of the feedback loop, as mentioned above. At higher frequencies, the roll off results from the combined effects of C4 and C6; it occurs at a 6-dB-per-octave rate (within the errors of instrumentation). The cutoff frequency is about 280 Hz, which is:

$$f_{co} = \frac{1}{2 \times \pi \times R_E \times C6} \qquad (6)$$

where:

f_{co} = cutoff frequency in Hz
R_E = 0.57 Ω
C6 = 1000 µF

For comparison, a reference curve (6 dB per octave at the high and low ends) is superimposed. At about 1.2 kHz or so, the reactance of C6 is roughly equal to its equivalent series resistance (ESR), which is about 0.13 Ω for a small 1000-µF aluminum-electrolytic capacitor. Beyond this frequency the impedance of C6 does not diminish, and C4 takes over (thereby maintaining the 6-dB-per-octave roll-off rate). Careful measurements and computer simulations of the regulator loop verified that C4 and C6 do, in fact, collaborate pretty well in this manner.

This characteristic is the desired effect. At 120 Hz (the major ripple frequency) the loop gain is maximum, so the regulator loop works hard to suppress output ripple. At higher frequencies, the roll-off rate implies a loop phase shift in the neighborhood of 90°, which assures closed-loop stability and good transient response. Closed-loop transient response tests (using a square-wave signal injected into the

Load Test Circuit) verify the absence of ringing or large overshoot.

When R_E is increased to limit at smaller currents, the cutoff frequency decreases. In the example supply, the 0.5-A and 0.1-A range cutoff frequencies are 58 Hz and 12 Hz, respectively, and the roll-off rate remains 6 dB per octave as before. Hence, the loop gain at 120 Hz is reduced. The ripple voltage across C1, however, is also greatly reduced at lighter load currents. In the end, output ripple remains very low.

Closed-Loop Response

The closed-loop gain of the regulator is:

$$G_{CL} = 20 \log \left(\frac{V_{out}}{V_{ref}} \right) \qquad (7)$$

where:

G_{CL} = closed-loop gain, in dB
V_{out} = output voltage
V_{ref} = reference voltage.

Here, V_{out} is 4.5 V, minimum (25.0 V, maximum) and V_{ref} is 4.5 V. Fig 11.27 shows the locations of the minimum and maximum gain values and also the corresponding closed-loop bandwidths. By locating the 280-Hz cutoff frequency fairly close to the 120-Hz ripple frequency, the closed-loop bandwidth is minimized, which is desirable in a voltage regulator.

Another important regulator parameter is closed-loop output impedance. **Fig 11.28** shows a computer simulation of this parameter. Mathematical analysis and actual measurements using the Load Test Circuit with a sine-wave test signal corroborate the simulations quite well. Two results are shown. For curve A, the

C6 component (of Fig 11.26) is removed from the circuit, and C4 is increased so that the 280-Hz cutoff frequency is maintained, as we discussed before. At low frequencies the output impedance should be R_E (0.57 Ω) divided by the open-loop voltage gain (5000 maximum), about 0.11 mΩ. Above 280 Hz, though, the output impedance increases rapidly because the open-loop gain is decreasing. It will eventually reach the value of R_E.

For curve B, the original values of C6 and C4 are used, and the output impedance remains low up to about 1.2 kHz. It then increases, but remains much lower than curve A. This happens because C4 is smaller than in curve A and C4 mainly determines the frequency characteristic. In other words, the impedance of C6 (its reactance plus its ESR) is in parallel with the relatively small output impedance of a high-gain feedback amplifier. Therefore, C4 is much less influential in determining the power-supply output impedance. This situation gradually changes as frequency increases.

This discussion shows that the output-impedance characteristic of the power supply is reduced at frequencies that are significant in certain applications. Furthermore, it can be reduced by feedback to levels that are impossible with practical capacitors. To take advantage of this lower impedance, the regulator must be located extremely close to the load (remote sensing is another possibility).

When extremely tight regulation and low output impedance are important, the leads to the load must be very short, heavy straps. Multiple loads should be connected in parallel directly at the binding posts: A "daisy-chain" connection scheme does not assure equal regulation for each load; it counteracts precision regulation.

High-Voltage Techniques

The construction of high-voltage supplies poses special considerations in addition to the normal design and construction practices used for lower-voltage supplies. In general, the constructor needs to remember that physical spacing between leads, connections, parts and the chassis must be sufficient to prevent arcing. Also, the series connection of components such as capacitor and resistor strings needs to be done with consideration for voltage stresses in the components.

CAPACITORS

Capacitors will usually need to be connected in series strings to form an equivalent capacitor with the capability to withstand the applied voltage. When this is done, equal-value bypassing resistors need to be connected across each capacitor in the string in order to distribute the voltage equally across the capacitors. The *equalizing resistors* should have a value low enough to equalize differences in capacitor leakage resistance between the capacitors, while high enough not to dissipate excessive power. Also, capacitor bodies need to be insulated from the chassis and from each other by mounting them on insulating panels, thereby preventing arcing to the chassis or other capacitors in the string.

For high voltages, oil-filled paper-dielectric capacitors are superior to electrolytics because they have lower internal impedance at high frequencies, lower leakage resistance and are available with higher working voltages. These capacitors are available in values of several microfarads and have working voltage ratings of thousands of volts. Avoid older oil-filled capacitors. They may contain polychlorinated biphenyls (PCBs), a known cancer-causing agent. Newer capacitors have eliminated PCBs and have a notice on the case to that effect.

BLEEDER RESISTORS

Bleeder resistors should be given careful consideration. These resistors provide protection against shock when the power supply is turned off and dangerous wiring is exposed. A general rule is that the bleeder should be designed to reduce the output voltage to 30 V or less within 2 seconds of turning off the power supply. Take care to ensure that the maximum voltage rating of the resistor is not exceeded. The bleeder will consist of several resistors in series. One additional recommendation is that two separate bleeder strings be used, to provide safety in the event one of the strings fails.

METERING TECHNIQUES

Special considerations should be observed for metering of high-voltage supplies, such as the plate supplies for linear amplifiers. This is to provide safety to both personnel and also to the meters themselves.

To monitor the current, it is customary to place the ammeter in the supply return (ground) line. This ensures that both meter terminals are close to ground potential, as compared to the hazard created by placing the meter in the positive output line—in which case the voltage on each meter terminal would be near the full high-voltage potential. Also, there is the strong possibility that an arc could occur between the wiring and coils inside the meter and the chassis of the amplifier or power supply itself. This hazardous potential cannot exist with the meter in the negative leg. Another good safety practice is to place a low-voltage Zener diode across the terminals of the ammeter. This will bypass the meter in the event of an internal open circuit in the meter.

For metering of high voltage, the builder should remember that resistors to be used in multiplier strings have voltage breakdown ratings. Usually, several resistors need to be used in series to reduce voltage stress across each resistor. A basic rule of thumb is that resistors should be limited to a maximum of 200 V, unless rated otherwise. Therefore, for a 2000-V power supply, the voltmeter would have a string of 10 resistors connected in series to distribute the voltage equally.

Batteries and Charging

The availability of solid-state equipment makes it practical to use battery power under portable or emergency conditions. Hand-held transceivers and instruments are obvious applications, but even fairly powerful transceivers (100 W or so output) may be practical users of battery power (for example, emergency power for the home station for ARES operation).

Lower-power equipment can be powered from two types of batteries. The "primary" battery is intended for one-time use and is then discarded; the "storage" (or "secondary") battery may be recharged many times.

A battery is a group of chemical cells, usually series-connected to give some desired multiple of the cell voltage. Each assortment of chemicals used in the cell gives a particular nominal voltage. This must be taken into account to make up a particular battery voltage. For example, four 1.5-V carbon-zinc cells make a 6-V battery and six 2-V lead-acid cells make a 12-V battery. The **DC Theory and Resistive Components** chapter has more information about energy storage in batteries. In addition, the **Real** World Component Characteristics chapter has information about battery capacity and charge/discharge rates.

PRIMARY BATTERIES

One of the most common primary-cell types is the alkaline cell, in which chemical oxidation occurs during discharge. When there is no current, the oxidation essentially stops until current is required. A slight amount of chemical action does continue, however, so stored batteries eventually will degrade to the point where the battery will no longer supply the desired current. The time taken for degradation without battery use is called *shelf life*.

The alkaline battery has a nominal voltage of 1.5 V. Larger cells are capable of producing more milliampere hours and less voltage drop than smaller cells. Heavy-duty and industrial batteries usually have a longer shelf life.

Lithium primary batteries have a nominal voltage of about 3 V per cell and by far the best capacity, discharge, shelf-life and temperature characteristics. Their disadvantages are high cost and the fact that they cannot be readily replaced by other types in an emergency.

The lithium-thionyl-chloride battery is a primary cell, and should not be recharged under any circumstances. The charging process vents hydrogen, and a catastrophic hydrogen explosion can result. Even accidental charging caused by wiring errors or a short circuit should be avoided.

Silver oxide (1.5 V) and mercury (1.4 V) batteries are very good where nearly constant voltage is desired at low currents for long periods. Their main use (in subminiature versions) is in hearing aids, though they may be found in other mass-produced devices such as household smoke alarms.

SECONDARY OR RECHARGEABLE BATTERIES

Many of the chemical reactions in primary batteries are theoretically reversible if current is passed through the battery in the reverse direction.

Primary batteries should not be recharged for two reasons: It may be dangerous because of heat generated within sealed cells, and even in cases where there may be some success, both the charge and life are limited. One type of alkaline battery is rechargeable, and is so marked.

Nickel Cadmium

The most common type of small rechargeable battery is the nickel-cadmium (NiCd), with a nominal voltage of 1.2 V per cell. Carefully used, these are capable of 500 or more charge and discharge cycles. For best life, the NiCd battery must not be fully discharged. Where there is more than one cell in the battery, the most-discharged cell may suffer polarity reversal, resulting in a short circuit, or seal rupture. All storage batteries have discharge limits, and NiCd types should not be discharged to less than 1.0 V per cell.

Nickel cadmium cells are not limited to "D" cells and smaller sizes. They also are available in larger varieties ranging to mammoth 1000 Ah units having carrying handles on the sides and caps on the top for adding water, similar to lead-acid types. These large cells are sold to the aircraft industry for jet-engine starting, and to the railroads for starting locomotive diesel engines. They also are used extensively for uninterruptible power supplies. Although expensive, they have very long life. Surplus cells are often available through surplus electronics dealers, and these cells often have close to their full rated capacity.

Advantages for the ham in these vented-cell batteries lie in the availability of high discharge current to the point of full discharge. Also, cell reversal is not the problem that it is in the sealed cell, since water lost through gas evolution can easily be replaced. Simply remove the cap and add distilled water. By the way, tap water should never be added to either nickel cadmium or lead-acid cells, since dissolved minerals in the water can hasten self discharge and interfere with the electrochemical process.

Lead Acid

The most widely used high-capacity rechargeable battery is the lead-acid type. In automotive service, the battery is usually expected to discharge partially at a very high rate, and then to be recharged promptly while the alternator is also carrying the electrical load. If the conventional auto battery is allowed to discharge fully from its nominal 2 V per cell to 1.75 V per cell, fewer than 50 charge and discharge cycles may be expected, with reduced storage capacity.

The most attractive battery for extended high-power electronic applications is the so-called "deep-cycle" battery, which is intended for such use as powering electric fishing motors and the accessories in recreational vehicles. Size 24 and 27 batteries furnish a nominal 12 V and are about the size of small and medium automotive batteries. These batteries may furnish between 1000 and 1200 W-hr per charge at room temperature. When properly cared for, they may be expected to last more than 200 cycles. They often have lifting handles and screw terminals, as well as the conventional truncated-cone automotive terminals. They may also be fitted with accessories, such as plastic carrying cases, with or without built-in chargers.

Lead-acid batteries are also available with gelled electrolyte. Commonly called *gel cells*, these may be mounted in any position if sealed, but some vented types are position sensitive.

Lead-acid batteries with liquid electrolyte usually fall into one of three classes — conventional, with filling holes and vents to permit the addition of distilled water lost from evaporation or during high-rate charge or discharge; maintenance-free, from which gas may escape but water cannot be added; and sealed. Generally, the deep-cycle batteries have filling holes and vents.

Nickel Metal Hydride

This battery type is quite similar to the NiCd, but the Cadmium electrode is replaced by one made from a porous metal alloy that traps hydrogen; therefore the name of metal hydride. Many of the basic characteristics of these cells are similar to NiCds. For example, the voltage is very nearly the same, they can be slow-charged from a constant current source, and they can safely be deep cycled. There are also some important differences: The most attractive feature is a much higher capacity for the same cell size — often nearly twice as much as the NiCd types! The typical size AA NiMH cell has a capacity between 1000 and 1300 mAh, compared to the 600 to 830 mAh for the same size NiCd. Another advantage of these cells is a complete freedom from memory effect. We can also find comfort in the fact that NiMH cells do not contain any dangerous substances, while both NiCd and lead-acid cells do contain quantities of toxic heavy metals.

The internal resistance of NiMH cells is somewhat higher than that of NiCd cells, resulting in reduced performance at very high discharge current. This can cause slightly reduced power output from an HT powered by a NiMH pack, but the effect is barely noticeable, and the higher capacity and resulting longer run time far outweigh this. At least one manufacturer warns that the self-discharge of NiMH cells is higher than for NiCd, but again, in practice this can hardly be noticed. The fast-charge process is different for NiMH batteries. A fast charger designed for NiCd will not correctly charge NiMH batteries. But many commercial fast chargers are designed for both types of batteries. NiMH batteries outperform NiCd batteries whenever high capacity is desired, while NiCd batteries still have advantages when delivering very high peak currents.

At the time of this writing, many cell phones and portable computers use NiMH batteries, and several manufacturers offer NiMH packs for Amateur Radio applications. Standard-sized NiMH cells are widely available from the major electronic parts suppliers.

Lithium-Ion cells

The lithium-ion cell is another possible alternative to NiCd cells. It features, for the same energy storage, about one third the weight and one half the volume of a NiCd. It also has a lower self-discharge rate. Typically, at room temperature, a NiCd cell will lose from 0.5 to 2% of its charge per day. The Lithium-ion cell will lose less than 0.5% per day and even this loss rate decreases after about 10% of the charge has been lost. At higher temperatures the difference is even greater. The result is that Lithium-ion cells are a much better choice for standby operation where frequent recharge is not available.

One major difference between NiCd and Li-ion cells is the cell voltage. The nominal voltage for a NiCd cell is about 1.2 V. For the Li-ion cell it is 3.6 V with a maximum cell charging voltage of 4 V. You cannot substitute Li-ion cells directly for NiCd cells. You will need one Li-ion cell for three NiCd cells. Chargers intended for NiCd batteries must not be used with Li-ion batteries, and vice versa.

PROS AND CONS

Chemical Hazards of Each Battery Type

In addition to the precautions given above, the following precautions are recommended. (Always follow the manufacturer's advice.)

Gas escaping from storage batteries may be explosive. Keep flames or lighted tobacco products away.

Dry-charged storage batteries should be given electrolyte and allowed to soak for at least half an hour. They should then be charged at about a 15 A rate for 15 minutes or so. The capacity of the battery will build up slightly for the first few cycles of charge and discharge, and then have fairly constant capacity for many cycles. Slow capacity decrease may then be noticed.

No battery should be subjected to unnecessary heat, vibration or physical shock. The battery should be kept clean. Frequent inspection for leaks is a good idea. Electrolyte that has leaked or sprayed

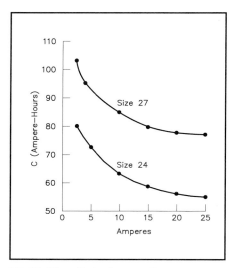

Fig 11.29 — Output capacity as a function of discharge rate for two sizes of lead-acid batteries.

from the battery should be cleaned from all surfaces. The electrolyte is chemically active and electrically conductive, and may ruin electrical equipment. Acid may be neutralized with sodium bicarbonate (baking soda), and alkalis may be neutralized with a weak acid such as vinegar. Both neutralizers will dissolve in water, and should be quickly washed off. Do not let any of the neutralizer enter the battery.

Keep a record of the battery use, and include the last output voltage and (for lead-acid storage batteries) the hydrometer reading. This allows prediction of useful charge remaining, and the recharging or procuring of extra batteries, thus minimizing failure of battery power during an excursion or emergency.

Internal Resistance

Cell internal resistance is very important to handheld-transceiver users. This is because the internal resistance is in series with the battery's output and therefore reduces the available battery voltage at the high discharge currents demanded by the transmitter. The result is reduced transmitter output power and power wasted in the cell itself by internal heating. Because of cell-construction techniques and battery chemistry, certain types of cells typically have lower internal resistance than others.

The NiCd cell is the undisputed king of cell types for high discharge current capability. Also, the NiCd maintains this low internal resistance throughout its discharge curve, because the specific gravity of its potassium hydroxide electrolyte does not change.

Next in line is probably the alkaline primary cell. When these cells are used with handheld transceivers, it is not un-

common to have lower output power, and often to have the low battery indicator come on, even with fresh cells.

The lead-acid cell, which is finding popularity in belt-hung battery packs, is pretty close to the alkaline cell for internal resistance, but this is only at full charge. Unlike the NiCd, the electrolyte in the lead-acid cell enters into the chemical reaction. During discharge, the specific gravity of the electrolyte gradually drops as it approaches water, and the conductivity decreases. Therefore, as the lead-acid cell approaches a discharged state, the internal resistance increases. For the belt pack, larger cells are used (approximately 2 Ah) and the internal resistance is consequently reduced.

The worst cell of all is the common carbon-zinc flashlight cell. With the transmit current demand levels of handheld radios, these cells are pretty much useless.

BATTERY CAPACITY

The common rating of battery capacity is ampere hours (Ah), the product of current drain and time. The symbol "C" is commonly used; C/10, for example, would be the current available for 10 hours continuously. The value of C changes with the discharge rate and might be 110 at 2 A but only 80 at 20 A. **Fig 11.29** gives capacity-to-discharge rates for two standard-size lead-acid batteries. Capacity may vary from 35 mAh for some of the small hearing-aid batteries to more than 100 Ah for a size 28 deep-cycle storage battery.

Sealed primary cells usually benefit from intermittent (rather than continuous) use. The resting period allows completion of chemical reactions needed to dispose of by-products of the discharge.

The output voltage of all batteries drops as they discharge. "Discharged" condition for a 12-V lead-acid battery, for instance, should not be less than 10.5 volts. It is also good to keep a running record of hydrometer readings, but the conventional readings of 1.265 charged and 1.100 discharged apply only to a long, low-rate discharge. Heavy loads may discharge the battery with little reduction in the hydrometer reading.

Batteries that become cold have less of their charge available, and some attempt to keep a battery warm before use is worthwhile. A battery may lose 70% or more of its capacity at cold extremes, but it will recover with warmth. All batteries have some tendency to freeze, but those with full charges are less susceptible. A fully charged lead-acid battery is safe to −30°F (−26°C) or colder. Storage batteries may be warmed somewhat by charging. Blowtorches or other flame should never be used to heat any type of battery.

A practical discharge limit occurs when

the load will no longer operate satisfactorily on the lower output voltage near the "discharged" point. Much gear intended for "mobile" use may be designed for an average of 13.6 V and a peak of perhaps 15 V, but will not operate well below 12 V. For full use of battery charge, the gear should operate well (if not at full power) on as little as 10.5 V with a nominal 12 to 13.6-V rating.

Somewhat the same condition may be seen in the replacement of carbon-zinc cells by NiCd storage cells. Eight carbon-zinc cells will give 12 V, while 10 of the same size NiCd cells are required for the same voltage. If a 10-cell battery holder is used, the equipment should be designed for 15 V in case the carbon-zinc units are plugged in.

Discharge Planning

Transceivers usually drain a battery at two or three rates: one for receiving, one for transmit standby and one key-down or average voice transmit. Considering just the first and last of these (assuming the transmit standby is equal to receive), average two-way communication would require the low rate $3/4$ of the time and the high rate $1/4$ of the time. The ratio may vary somewhat with voice. The user may calculate the percentage of battery charge used in an hour by the combination (sum) of rates. If, for example, 20% of the battery capacity is used in an hour, the battery will provide five hours of communications per charge. In most actual traffic and DX-chasing situations the time spent listening should be much greater than that spent transmitting.

Charging/Discharging Requirements

The rated full charge of a battery, C, is expressed in ampere-hours. No battery is perfect, so more charge than this must be offered to the battery for a full-charge. If, for instance, the charge rate is 0.1 C (the 10-hour rate), 12 or more hours may be needed for the charge.

Basically NiCd batteries differ from the lead-acid types in the methods of charging. It is important to note these differences, since improper charging can drastically shorten the life of a battery. NiCd cells have a flat voltage-versus-charge characteristic until full charge is reached; at this point the charge voltage rises abruptly. With further charging, the electrolyte begins to break down and oxygen gas is generated at the positive (nickel) electrode and hydrogen at the negative (cadmium) electrode.

Since the cell should be made capable of accepting an overcharge, battery manufacturers typically prevent the generation of hydrogen by increasing the capacity of the cadmium electrode. This allows the oxygen formed at the positive electrode to reach the metallic cadmium of the negative electrode and reoxidize it. During overcharge, therefore, the cell is in equilibrium. The positive electrode is fully charged and the negative electrode less than fully charged, so oxygen evolution and recombination "wastes" the charging power being supplied

In order to ensure that all cells in a NiCd battery reach a fully charged condition, NiCd batteries should be charged by a constant current at about a 0.1-C current level. This level is about 50 mA for the AA-size cells used in most hand-held radios. This is the optimum rate for most NiCds since 0.1 C is high enough to provide a full charge, yet it is low enough to prevent over-charge damage and provide good charge efficiency.

Although fast-charge-rate (3 to 5 hours typically) chargers are available for hand-held transceivers, they should be used with care. The current delivered by these units is capable of causing the generation of large quantities of oxygen in a fully charged cell. If the generation rate is greater than the oxygen recombination rate, pressure will build

in the cell, forcing the vent to open and the oxygen to escape. This can eventually cause drying of the electrolyte, and then cell failure. The cell temperature can also rise, which can shorten cell life. To prevent overcharge from occurring, fast-rate chargers should have automatic charge-limiting circuitry that will switch or taper the charging current to a safe rate as the battery reaches a fully-charged state.

Gelled-electrolyte lead-acid batteries provide 2.4 V/cell when fully charged. Damage results, however, if they are overcharged. (Avoid constant-current or trickle charging unless battery voltage is monitored and charging is terminated when a full charge is reached.) Voltage-limited charging is best for these batteries. A proper charger maintains a safe charge-current level until 2.3 V/cell is reached (13.8 V for a 12-V battery). Then, the charge current is tapered off until 2.4 V/cell is reached. Once charged, the battery may be safely maintained at the "float" level, 2.3 V/cell. Thus, a 12-V gel-cell battery can be "floated" across a regulated 13.8-V system as a battery backup in the event of power failure.

Deep-cycle lead-acid cells are best

charged at a slow rate, while automotive and some NiCd types may safely be given quick charges. This depends on the amount of heat generated within each cell, and cell venting to prevent pressure build-up. Some batteries have built-in temperature sensing, used to stop or reduce charging before the heat rise becomes a danger. Quick and fast charges do not usually allow gas recombination, so some of the battery water will escape in the form of gas. If the water level falls below a certain point, acid hydrometer readings are no longer reliable. If the water level falls to plate level, permanent battery damage may result.

Overcharging NiCds in moderation causes little loss of battery life. Continuous overcharge, however, may generate a voltage depression when the cells are later discharged. For best results, charging of NiCd cells should be terminated after 15 hours at the slow rate. Better yet, circuitry may be included in the charger to stop charging, or reduce the current to about 0.02 C when the 1.43-V-per-cell terminal voltage is reached. For lead-acid batteries, a timer may be used to run the charger to make up for the recorded discharge, plus perhaps 20%. Some chargers will switch over automatically to an acceptable standby charge.

Emergency Operations

CARE AND FEEDING OF GENERATORS

For long term emergency operation, a generator is a must, as anyone who has operated field day can attest to. The generator will provide power as long as the fuel supply holds out. Proper care is necessary to keep the generator operating reliably, however.

When the generator runs out of fuel, the operator may be tempted to rush over with the gasoline can and begin refueling. This is very hazardous, since the engine's manifold and muffler are at temperatures that can ignite spilled gasoline. The operator should wait a few minutes to allow hot surfaces to cool sufficiently to ensure safety. For these periods when the generator is shut down,

plan on having battery power available to support station operation until the generator can be brought back on line.

Check the level of the engine's lubricating oil from time to time. If the oil sump becomes empty, the engine can seize, putting the station out of operation and necessitating costly engine repairs.

Remember that the engine will produce carbon monoxide gas while it is running. The generator should never be run indoors, and should be placed away from open windows and doors to keep exhaust fumes from coming inside.

INVERTERS

For battery-powered operation of alternating-current loads, inverters are avail-

able. An inverter is a dc-to-ac converter, switching at 60 Hz to provide 120 V ac to the loads.

Inverters come in varying degrees of sophistication. The simplest, as mentioned above, produces a square-wave output. This is no problem for lighting and other loads that don't care about the input waveform. Lots of equipment using motors will work poorly or not at all when supplied with square wave power. Therefore, many higher-power inverters use waveform shaping to approximate a sine-wave output. The simplest of these methods is a resonant inductor and capacitor filter. Higher-power units employ pulse-width modulation of the converter switches to create a sinusoidal output waveform.

Power-Supply Projects

Construction of a power supply can be one of the most rewarding projects undertaken by a radio amateur. Whether it's a charger for the NiCds in a VHF hand-held transceiver, a low-voltage, high-current monster for a new 100-W solid-state transceiver, or a high-voltage supply for a new linear amplifier, a power supply is basic to all of the radio equipment we operate and

enjoy. Final testing and adjustment of most power-supply projects requires only a voltmeter, and perhaps an oscilloscope — tools commonly available to most amateurs.

General construction techniques that may be helpful in building the projects in this chapter are outlined in the **Circuit Construction** chapter. Earlier chapters in this *Handbook* contain basic information about the

components that make up power supplies.

Safety must always be carefully considered during design and construction of any power supply. Power supplies contain potentially lethal voltages, and care must be taken to guard against accidental exposure. For example, electrical tape, insulated tubing (spaghetti) or heat-shrink tubing is recommended for covering exposed wires, components leads, component solder terminals and tie-down points. Whenever possible, connectors used to mate the power supply to the outside world should be of an insulated type designed to prevent accidental contact.

Connectors and wire should be checked for voltage and current ratings. Always use wire with an insulation rating higher than the working voltages in the power supply. Special high-voltage wire is available for use in B+ supplies. The **Component Data** chapter contains a table showing the current-carrying capability of various wire sizes. Scrimping on wire and connectors to save money could result in flashover, meltdown or fire.

All fuses and switches should be placed in the hot leg(s) only. The neutral leg should not be interrupted. Use of a three-wire (grounded) power connection will greatly reduce the chance of accidental shock. The proper wiring color code for 120-V circuits is: black — hot; white — neutral; and green — ground. For 240-V circuits, the second hot lead generally uses a red wire.

Fig 11.30 — CEE-22 connectors are available with built-in line filters and fuse holders.

Power-supply Primary Circuit Connector Standard

The International Commission on Rules for the Approval of Electrical Equipment (CEE) standard for power-supply primary-circuit connectors for use with detachable cable assemblies is the CEE-22 (see **Fig 11.30**). The CEE-22 has been recognized by the ARRL and standards agencies of many countries. Rated for up to 250 V, 6 A at 65°C, the CEE-22 is the most commonly used three-wire (grounded),

chassis-mount primary circuit connector for electronic equipment in North America and Europe. It is often used in Japan and Australia as well.

When building a power supply requiring 6 A or less for the primary supply, a builder would do well to consider using a CEE-22 connector and an appropriate cable assembly, rather than a permanently installed line cord. Use of a detachable line cord makes replacement easy in case of damage. CEE-22 compatible cable assemblies are available with a wide variety of power plugs including most types used overseas.

Some manufacturers even supply the CEE-22 connector with a built-in line filter. These connector/filter combinations are especially useful in supplies that are operated in RF fields. They are also useful in digital equipment to minimize conducted interference to the power lines.

CEE-22 connectors are available in many styles for chassis or PC-board mounting. Some have screw terminals; others have solder terminals. Some styles even contain built-in fuse holders.

A SERIES-REGULATED 4.5- TO 25-V, 2.5-A POWER SUPPLY

For home-laboratory requirements, a series-regulator supply is simpler and less expensive than a switching power supply. Series-regulated supplies are also free of electrical switching noise, which is a problem with some switching supplies. During tests of sensitive low-level circuitry, the

power supply output should be pure dc; it should not contribute to problems in the circuit under test.

The power supply in **Fig 11.31** was designed and built by William E. Sabin, WØIYH. See "High-Precision Series Regulator Loop Design," earlier in this

chapter, for a discussion of the design, analysis and tests of the feedback control loop used in this power supply.

Features

This supply was designed to meet the following objectives:

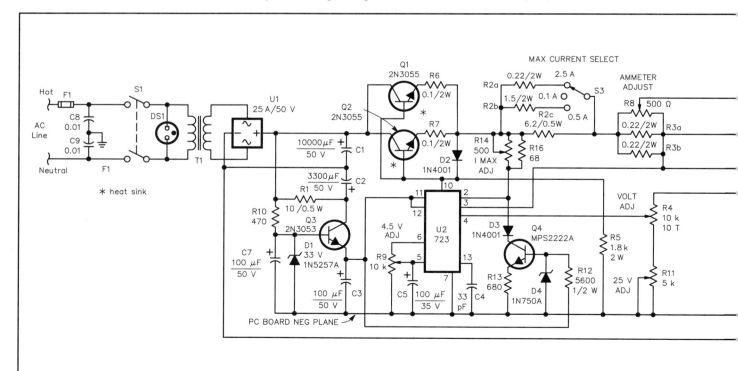

- Continuously variable output voltage from 4.5 to 25.0 V.
- Tight load voltage regulation, better than 0.03%, for load currents from 0 to 2.0 A, 0.1% to 2.5 A; 0.01% into a 2.0-A load for ac-line voltages from 117 to 122 V.
- Excellent response to load fluctuations and transients (low output impedance).
- Very low ac ripple, less than 2 µV RMS with a 2.0-A load.
- Very low random noise, less than 2 µV RMS from 0.1 Hz to 500 kHz.
- Use an off-the-shelf transformer and other easily obtainable parts.
- Load currents up to 2.5 A, continuous duty.
- Switch selectable current limiting, at 0.1, 0.5 or 2.5 A, to protect delicate circuits under test.

The Circuit

Fig 11.32 is a schematic of the power-supply circuit. An LM723 regulator was selected because it's simple, and its reference voltage is available (pin 6) for filtering. Reference noise (typical of zener diodes) is reduced to a very low level via C5, as suggested in the data sheet for the '723. The current-limiting circuitry is also accessible (at pins 2 and 3). It senses the voltage drop across R2 and R3.

The combination of R2 and R3 sets the current limit at 0.62 / (R2 + R3) A. R3 (R3a and R3b together) forms an 0.11-Ω, 4-W resistor, which acts as a shunt for the digital meter that measures load current. R8 provides meter adjustment without affecting R3 significantly. S3 selects from three values of R2 to set current limits at approximately 0.1, 0.5 or 2.5 A.

C1 has a large value to reduce output ripple voltage. Smaller values would increase ripple and require a higher transformer voltage to prevent regulator drop out. Other voltage drops between the Q1/Q2 emitters and the output terminal are minimized to assure that a standard 25.2-V transformer can do the job. The R1-C2 combination reduces ac ripple at the '723 by a factor of 25; this eliminates the need for an extremely large value of C1.

The circuitry of Q3, R10, C7 and D1 prevents the voltage on

Fig 11.31 — An exterior view of the 2.5-A power supply. The DMM is mounted on the control panel as the output meter.

pins 11 and 12 of U2 from exceeding the 40-V maximum rating of the '723 (especially with light loads and high line voltage). C7 eliminates a very small ac ripple at the dc output. As load current increases, the voltage at C1 decreases and Q3 is saturated.

When R4 is adjusted to reduce output voltage, U2 and Q1/Q2 are switched off until C6 discharges to the lower voltage. This caused the emitter-base junctions of Q1 and Q2 to break down (at about 2.0 V) so that C6 could discharge through R5. D2 provides an alternate path and prevents this breakdown. R5 provides a minimum load for U2.

The foldback circuit is interesting: Q4, D4 and R13 provide a constant current through, and therefore a constant voltage drop across, R16. This voltage makes the current limiting (pins 2 and 3 of U2) work properly over the entire 4.5- to 25.0-V range. As the current limiting action pulls the voltage at the top of R16

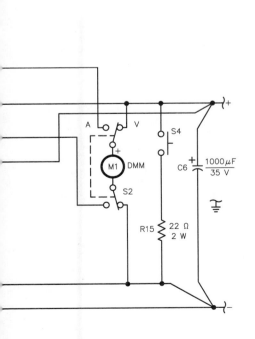

Fig 11.32 — Schematic of the 4.5 to 25-V, 2.5-A regulated dc power supply. Heavy lines indicate critical low-impedance conductors; use short conductors with large surface areas. "RS" signifies a Radio Shack part number; "CDE" is Cornell Dubilier Electric.

C1 — 10000 µF, 50 V (CDE 10000-50-AC, or equiv).
C2 — 3300 µF, 50 V (CDE 3300-50-M, or equiv).
C3, C7 — 100 µF, 50 V (RS 272-1044, or equiv).
C4 — 33 pF, 50 V.
C5 — 100 µF, 35 V (RS 272-1028, or equiv).
C6 — 1000 µF, 35 V (RS 272-1032, or equiv).
C8, C9 — 0.01 µF, ac-rated capacitor.
D1 — 1N5257A 33-V Zener diode.
D2, D3 — 1N4001 (RS 276-1101, or equiv).
D4 — 1N750A.
DS1 — Neon lamp, 120-V ac (RS 272-704, or equiv).
F1 — 1.5 A slow-blow (RS 270-1284, or equiv).
M1 — DMM (Heath SM-2300-A, or equiv).
P1 — Three-wire power cord and plug.
Q1, Q2 — 2N3055 (RS 276-2041, or equiv).
Q3 — 2N3053 (RS 276-2030, or equiv).
Q4 — MPS2222A (RS 276-2009, or equiv).
R1 — 10 Ω, 0.5 W.
R2a — 0.22 Ω, 2 W.
R2b — 1.5 Ω, 2 W.
R2c — 6.2 Ω, 5%, 0.5 W.

R3 — 0.15 Ω, 2 W.
R4 — 10 kΩ, 10-turn potentiometer (Bourns 3540S, or equiv).
R5 — 1.8 kΩ, 2 W.
R6, R7 — 0.1 Ω, 2 W.
R8, R14 — 500 Ω potentiometer (RS 271-226, or equiv).
R9 — 10 kΩ potentiometer.
R10 — 470 Ω, 0.25 W.
R11 — 5 kΩ potentiometer.
R12 — 5.6 kΩ, 0.5 W.
R13 — 680 Ω, 0.25 W.
R16 — 68 Ω 0.25 W.
S1, S2 — DPDT (RS 275-652, or equiv).
S3 — SPDT (center off) (RS 275-654, or equiv).
S4 — Normally open, momentary-contact, push-button switch (RS 275-1547, or equiv).
T1 — 120-V primary, 25.2-V, 2.8-A secondary (Stancor P-8388, or equiv).
U1 — Rectifier bridge, 25 A, 50 V (RS 276-1185, or equiv).
U2 — LM723 regulator (RS 27-1740, or equiv).
Heat sink — Wakefield 403A, or equiv (two required).

below about 4.0 V, D3 quickly stops conducting, the voltage across R16 approaches zero and the load current is limited at about 1.9 A. This limits Q1/Q2 and T1 dissipation and provides short-circuit protection. This is a regenerative positive feedback process. R14 sets the current at which foldback begins, 2.6 A. There is further discussion of foldback circuitry earlier in this chapter.

An inexpensive Heath SM-2300-A auto ranging DMM (mounted on the front panel) displays the supply output levels. The dedicated DMM can display very small output changes. S2 selects either voltage or current (divided by ten) for display. (The DMM is left in its voltage range for both measurements.) The voltage across R3 is 0.11 times the load current. R8 provides the required ammeter adjustment without significantly affecting R3. The meter shows 0.2 for 2 A. Ordinary wire-wound resistors are adequate for R3 because they do not heat significantly at 2.5 A.

R9 sets the reference voltage on pin 5 of U2 at 4.5 V to establish the minimum output voltage. R11 sets the 25.0-V upper limit. A three-wire line cord assures that the supply chassis is always tied to the ac-line ground, for safety reasons. The dc output is not referenced to chassis ground, and performance is independent of the ac-ground connection. If the load current is very small, it may take a long time for C6 to discharge when the power is switched off; a press of S4 quickly discharges the capacitors (through R15) after turnoff. The mechanical construction emphasizes heat removal, so a cooling fan is not needed.

Construction

Several areas are critical to good performance. Pay particular attention to these:

- Wire C1 with short, heavy leads to present a minimum impedance to ac ripple. Also, connect C1 directly to the negative binding post with a heavy lead. This provides a low inductance path for load-current fluctuations and keeps them off the PC-board ground plane.
- Connect C2 directly to the negative lead of C1.
- Connect C6 directly across the binding posts.
- Connect R4 to R11 with a low-impedance lead and connect R11 directly to the PC board ground plane.
- Connect the regulator PC board ground plane to the negative binding post at a single point.
- Use heavy-duty binding posts to reduce a small but significant voltage drop (from the rear to the front) at the front panel.

Fig 11.33 shows the general construction. (Fig 11.35 also shows the front-panel layout.) The cabinet and chassis surface are 1/16-in aluminum plates connected by aluminum angle stock, drilled and tapped for #6-32 screws. Ventilation screens at the top and rear provide an excellent chimney effect. The chassis plate is tightly joined to the side plates for good heat transfer. Construction of the cabinet shown is somewhat labor intensive; it involves a lot of metal work. Any enclosure of similar size with good ventilation is suitable. For example, a 7×11×2-inch chassis with a bottom cover and rubber feet, a 7 1/2×6-inch front panel and a U-shaped perforated metal cover would serve well.

Q1 and Q2 each have a Wakefield 403A heat sink. The sinks were selected according to the thermal design discussion in the **Real-World Component Characteristics** chapter. The worst-case 2N3055 junc-

tion temperature was calculated to be 145°C, based on a measured case temperature of 95°C and dissipation of 32 W (each) for Q1 and Q2. This temperature is a little higher than recommended, but it's acceptable for intermittent lab use.

The DMM is epoxied to a narrow aluminum strip, which is then screwed to the front panel. This allows easy removal to replace the DMM battery.

Fig 11.33B shows the supply underside. The PC board is mounted on standoff insulators and positioned so that the positive and negative outputs are close to the binding posts.

Fig 11.34 is a parts-placement diagram for the regulator board. A full-size etching pattern is available from ARRL — see Chapter 30, **References**, for ordering information. All components and wiring are on the etched side of the board (with the help of a few jumper wires). The other side is entirely ground plane. The components are surface mounted by bending a section of each lead and soldering it to the surface of the appropriate copper pad. The socket for U2 is mounted by bending the pins out 90° and soldering pins 11 and 12 to the pad for C3. Some leads pass through holes (marked with cross hairs in Fig 11.34) in the PC board and solder to the ground plane. The emitter and base leads of Q1 and Q2 are isolated from ground; use insulated-wire to pass from the PC-board pads through holes in the PC board and chassis to the transistors.

Fig 11.35 shows the placement of major parts in front, top and side views. Silastic and pieces of Kraft paper cover any exposed 120-V wiring.

C1, C2 and R1 are mounted on a 1/8-in thick plastic board as shown in Fig 11.35. In the supply shown, it's a piece of 1/8-in PC board with wide traces. A piece of angle

(A)   (B)

Fig 11.33 — A shows a top view of the open power-supply chassis. The 1/8-in plate holding C1 is visible at the left edge of the heat sinks. B shows a bottom view. Notice that the components are mounted on the etched side of the board. The circuit pattern is simple enough for freehand layout. The front panel is at right in both views.

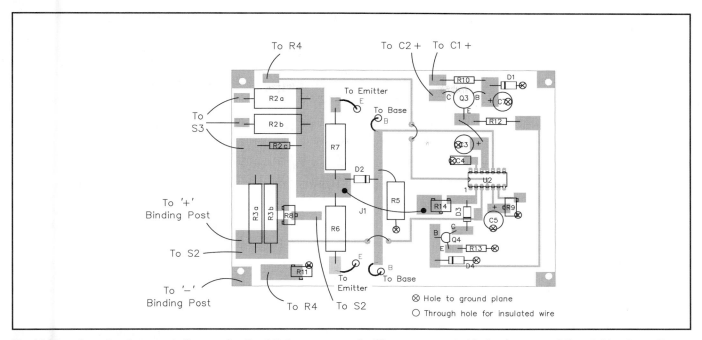

Fig 11.34 — A parts-placement diagram for the 2.5-A power supply. The component side is shown, and the etching is on the component side. Cross hairs indicate holes through the board. Component leads that terminate at cross hairs are grounded on the far side of the board. The emitter and base leads of Q1 and Q2 do not connect to the ground plane. Use insulated wire to pass through the PC board and holes in the chassis to the transistors on the heat sinks. Wire Jumper J1 is soldered to the pad surfaces without drilling holes. If you do drill holes for the jumper be sure to remove the ground plane around the holes on the non-component side. A full-size etching pattern is included in the References chapter.

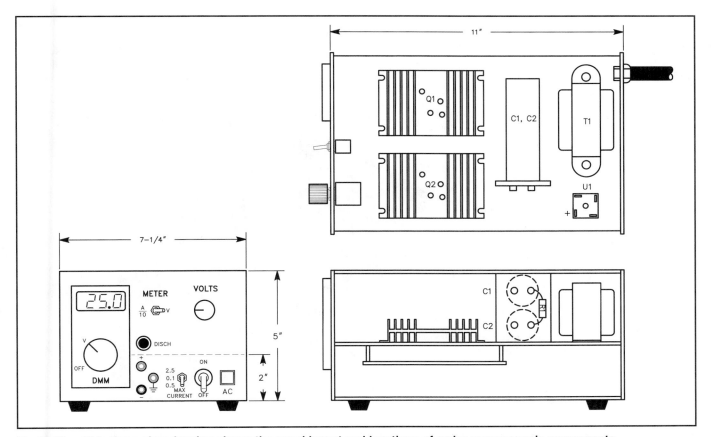

Fig 11.35 — This three-view drawing shows the panel layout and locations of major power-supply components.

stock mounts the board to the chassis.

The main voltage control, R4, is a 10-turn potentiometer, for ease of adjustment. C8 and C9 are ceramic capacitors rated for ac-line use. Once the supply is complete, double check all wiring and connections.

Adjustments

Several adjustments are needed: the dedicated AMMETER ADJUST (R8), the foldback limit (R14, I MAX ADJ), the lower (R9, 4.5 VOLT ADJ) and upper (R11, 25 V ADJ) voltage limits. First, obtain an accurate external ammeter and a test load that can draw at least 3 A at 12 V or so (automobile sealed beam headlights work well). Series connect the load and external meter to the supply binding posts. Set S3 to the 2.5-A range; set R4 for minimum output; set S2 to A and switch on the power. Advance R4 until the ammeter reads 2.0 A, and adjust R8 until the panel meter also reads 0.200. Next, increase R4 until the current slightly exceeds 2.6 A, and adjust R14 so that current begins to decrease (foldback). Switch off the supply and remove the external meter and load.

Adjust the voltage limits with no load connected and S2 set to V. Switch on the supply, set R4 for minimum voltage and adjust R9 until the meter reads 4.5. Increase R4 to maximum voltage and adjust R11 so that the meter reads 25. The supply is ready for use.

Performance and Use

The example supply met the design goals: When the line voltage was varied from 117 to 122 V (with a 25-V dc 2.0-A load), the output voltage varied less than 0.01%. When the dc load was varied from 0 to 2.0 A, the output voltage changed less than 0.03%.

When extremely tight regulation and low output impedance are important, the leads to the load must be very short, heavy straps. Multiple loads should be parallel-connected directly at the binding posts: A "Daisy chain" connection scheme does not assure equal regulation for each load; it counteracts the precision regulation of this supply.

A 13.8-V, 40-A SWITCHING POWER SUPPLY

Switching power supplies ("switchers," as they are often called) offer very attractive features—small size, low weight, high efficiency and low heat dissipation. Although some early switchers produced objectionable amounts of RF noise, nowadays you can build very quiet switchers using proper design techniques and careful EMI filtering. This power supply produces 13.8 V, regulated to better than 1%, at a continuous load current up to 40 A and with an efficiency of 88%. No minimum load is required and the ripple on the output is about 20 mV.

The supply produces no detectable RF noise at any frequency higher than the main switching frequency of 50 kHz. The author, Manfred Mornhinweg, XQ2FOD, checked this with a wire looped around his supply, tuning his TS-450 from 30 kHz to 40 MHz. The completed supply weighs only 2.8 kg (6.2 pounds)! **Fig 11.36** shows the unit built by HQ staffer Larry Wolfgang, WR1B, for the ARRL Lab.

LINEAR VERSUS SWITCHING SUPPLIES

A typical linear regulated power supply is simple and uses few parts—but several of these parts are big, heavy and expensive. The efficiency is usually only around 50%, producing *lots* of heat that must be removed by a big heat sink and often fans.

In this switching power supply, the line voltage is directly rectified and filtered at 300 V dc, which feeds a power oscillator operating at 25 kHz. This relatively high frequency allows the use of a small, lightweight and low-cost transformer. The output is then rectified and filtered. The control circuit steers the power oscillator so that it delivers just the right amount of energy needed; little energy is wasted.

While MOSFETs can switch faster, bipolar switching devices have lower conduction losses. Since very fast switching was undesirable because of RF noise, the author used bipolar transistors. These tend to be *too* slow, however, if the driving current is heavier than necessary. If the transistors must switch at varying current levels, the drive to them must also be varied. This is called *proportional driving* and is used in this project.

The switching topology used is called a *half-bridge forward* converter design (also known as a *single-ended pushpull* converter— *Ed.*). The converter is controlled using pulse-width modulation, using the generic 3524 IC.

CIRCUIT DESCRIPTION

Refer to the schematic diagram in **Fig 11.37**. Line voltage enters through P1, a connector that includes EMI filtering. It then goes through fuse F1, a 2-pole power switch and an additional common-mode noise filter (C1, L1, C2). Two NTC (negative temperature coefficient) resistors limit the inrush current. Each exhibits a resistance of about 2.5 Ω when cold and then loses most of its resistance as it heats up. A rectifier delivers the power to C3A and C3B, big electrolytic capacitors working at the 300-V dc level. The power oscillator is formed by Q1, Q2, the components near them, and the feedback and control trans-

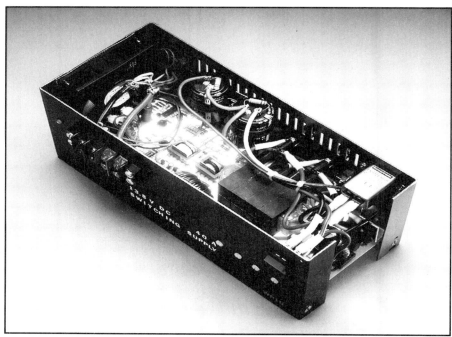

Fig 11.36—Photo of the completed 13.8 V, 40 A switching supply built by WR1B.

former T3. T2 and associated components act as a primary-current sensor.

T1 is the power transformer, delivering a 20-V square wave to the Schottky rectifiers (D6 through D9). A toroidal inductor L2 and six low equivalent-series-resistance (ESR) electrolytic capacitors form the main filter, while L3 and C23 and C24 are there for additional ripple reduction. The 13.8 V is delivered to the output through a string of ferrite beads with RF decoupling capacitors mounted directly on the output terminals.

The control circuit IC U1 is powered from an auxiliary rectifier D17. U1 senses the output voltage and the current level and controls the power oscillator through Q3 and Q4. C37, C35 and R23 are used to implement a full PID (*proportional-integral-derivative*) response in the control loop.

A quad operational amplifier U2 controls the cooling fan according to the average current level and also drives the voltage indicating tricolor LED, which glows green if the voltage is okay, orange if the voltage is too low and red if it is too high.

MORE DESIGN DETAILS

When the unit is powered up, the operating voltage builds up on C3A and C3B, and R2 and R6 bias the two power transistors Q1 and Q2 into their active zones. They start conducting a few mA, but for only a short time, because the positive feedback introduced by T3 quickly throws the system out of balance. One of the two transistors receives an increased base current from T3, while the other one sees its base drive reduced. It takes just a fraction of a microsecond for one of the transistors to become saturated and the other cut off. Which transistor will start first is unpredictable, but for this analysis let's suppose it is Q1. Because the control circuit is not yet powered, Q3 and Q4 are off at startup.

T1 sees about 150 V ac across its primary, producing about 20 V ac on the secondary. Schottky rectifiers D6 through D9 rectify this, so L2 sees 20 V across it. The current in L2 will start rising and this is reflected back to the primary side of T1. The primary current passes through the 1-turn winding of T3, forcing one-eighth as much current to flow into the base of Q1, the transistor assumed to be conducting at this moment. After some time, the ferrite core of T3 will saturate, causing the base drive of Q1 to decrease sharply. Q1 will stop and Q2 will start conducting. Now the flux in T3's core decreases, crosses zero and increases in the other direction until it saturates the core again, shutting Q2 off and turning Q1 back on. Meanwhile, the current in L2 continues to

build up and the filter capacitors C17 through C22 are charged.

For safe startup, it is essential that T3 saturates completely before T1 starts to do so. If this were not the case, the transistors would have to switch under a very high and potentially destructive current. The power supply will oscillate freely for only a few cycles, because D17 is already charging C32 and C33, powering up the control circuit so that it takes over the control of the power oscillator. Note that

---

Fig 11.37—Schematic diagram for the 13.8-V, 40-A switching power supply.

C1, 2—0.1 µF, 250 V ac polypropylene, Digi-Key P4610ND.
C3A, C3B—1500 µF, 200 V electrolytic.
C4 to C11—0.47 µF, 400 V polypropylene, Digi-Key P3496ND.
C12, C13—1 µF, 50 V ceramic multilayer.
C14—0.0033 µF, 1.6 kV polypropylene.
C15, C16—0.01 µF, 250 V ac polypropylene.
C17 to C22—1000 µF, 25 V low-impedance (low-ESR) electrolytic.
C23, C24—2200 µF, 16 V low-impedance (low-ESR) electrolytic.
C25 to C30—0.01 µF, 50 V ceramic.
C31—0.47 µF, 50 V ceramic multilayer.
C35—0.033 µF, 50 V polyester.
C36—0.0047 µF, 50 V polyester.
C37—0.33 µF, 50 V polyester or ceramic multilayer.
D1—Rectifier bridge, 1 kV, 12 A, GBPC1210 or similar.
D2, D4, D17—Ultrafast diode, 1 kV, 1 A. UF4007 or similar. Lower voltage (down to 100 V) is acceptable. The ARRL Lab used UF1007 diodes, Digi-Key UF1007DICT-ND.
D3, D5—Ultrafast diode 1 kV, 3 A. UF5408 or similar, from Techsonic.
D6 to D9—Dual Schottky diode, 100 V, 30 A total. PBYR30100CT or similar. Single diode would also be suitable. ARRL Lab used International Rectifier 30CPQ100, Digi-Key 30CPQ100-ND.
D10 to D16, D18—1N4148 switching diode.
F1, F2—Fuse, 10 A for 120-V ac operation; 5 A for 240-V ac operation.
FB1, FB2—Amidon FB-73-801 ferrite bead, slipped over wire. Available from Bytemark.
FB3 to FB14—Amidon FB-73-2401 ferrite beads, slipped six each over the two 13.8 V dc output cables. Available from Bytemark.
L1—Common mode choke, approximately 2 mH each winding, 6 A. Author used junk box specimen. We used a Magnatek CMT908-V1 choke (Digi-Key part 10543-ND) in the supply built in the ARRL Lab.
L2—20 µH, 60 A choke. 16 turns on Amidon T-200-26 toroid, wound with ten #16 enameled wires in parallel.
L3—5 µH (uncritical), 60 A choke. 10 turns on ferrite solenoid, 10 mm diameter, 50 mm long. Wound with two #12 wires in parallel. Amidon #33-050-200 used in ARRL Lab.
LED1—Dual LED, green-red, common cathode, Digi-Key LU204615-ND (pin 1 is red; pin 3 is green).
M1—12 V, 5 W brushless dc fan, approximately 120 × 120 × 25 mm, Digi-Key P9753-ND is 120 × 120 × 38 mm and 5.5 W.
NTC1, NTC2—Inrush current limiter, 2.5 Ω cold resistance, Digi-Key KC003L-ND.
P1—Male ac connector with integrated EMI filter, 250 V ac, 10 A, Newark 97F8256.
Q1, Q2—High voltage switching transistor, BUH1215 or similar. Motorola MJW16010 was used in ARRL Lab; Newark 08TMJW16012.
Q3, Q4—BC639-16 transistor, available from Newark. Must resist 100 V and 0.5 A.
Q5—BD683 Darlington transistor, from Techsonic. The back of the transistor should be facing the outside of the board.
R1, R5—10 Ω, 5 W, low inductance preferred. For the supply built in the ARRL Lab, we used three 30 Ω, 2 W film resistors wired in parallel.
R9—47 Ω, 5 W, low inductance preferred. For the supply built in the ARRL Lab, we used three 150 Ω, 2 W film resistors wired in parallel.
R10, R11—1.8 Ω, 2 W, low inductance preferred.
S1—2-pole power switch, 250 V ac, 10 A.
S2—120/240 V ac power selector slide switch, 250 V ac, 10 A. A locking tab made of aluminum locks the switch in either 240 or 120-V position.
T1—Primary 15 turns, secondary 2+2 turns. Wound with copper foil and mylar sheet. Uses four Amidon EA-77-625 ferrite E-cores (8 halves). Equivalents include Thompson GER42x21x15A, Phillips 768E608, TDK EE42/42/15.
T2—Secondary is 100+100 turns #36 enamel wire. Primary is one turn #14 plastic insulated cable, wound on secondary. Wound on Amidon EE24-25-B bobbin. Uses an Amidon EA-77-250 core. Equivalents are Thompson GER25x10x6, Phillips 812E25Q, TDK EE25/19.
T3—Control winding is 26+26 turns #28 enamel wire. Base windings are 8 turns #20 each. Collector winding is one turn #14 plastic insulated wire. Bobbin and core same as T2.
U1—Pulse-width modulator IC, LM3524, SG3524, UC3524 or similar.
U2—Quad single-supply operational amplifier, LM324 or similar.
U3—5 V voltage reference, LM336Z-5.0 or similar.
VR1 to VR3—1 kΩ PCB mounted trimpot, Digi-Key #3309P-102-ND.
Cabinet—Hammond Manufacturing, PN 1426Y-B, 12 × 6 × 5.5 inches, and internal case mounting rails, Hammond Manufacturing 1448R12, used in ARRL Lab.

[schematic drawing on next page]

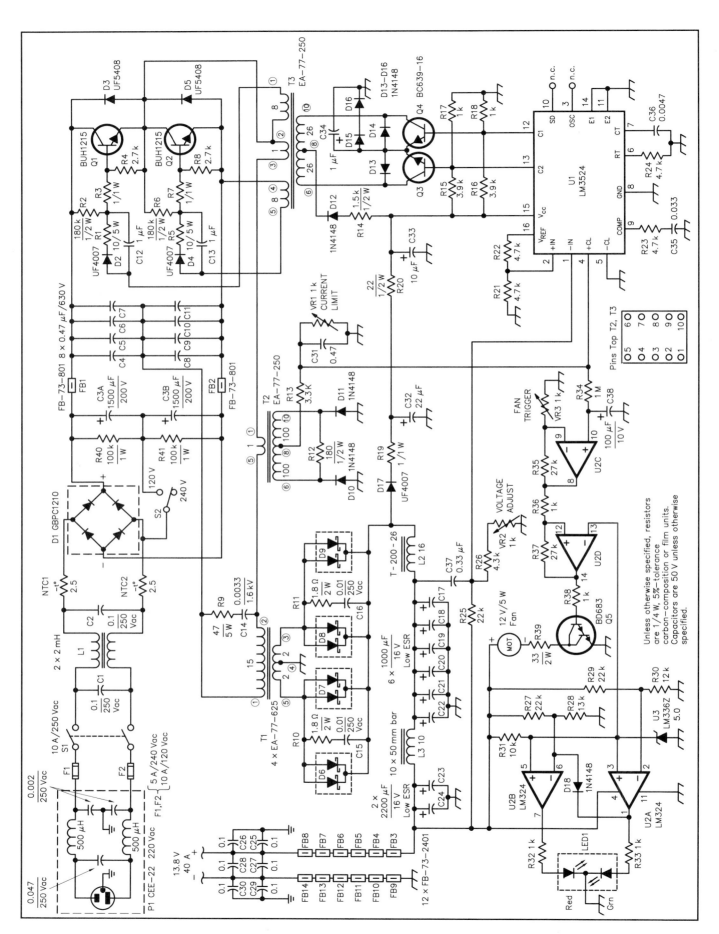

the self-oscillation frequency must be lower than the operating frequency for the feedback loop to be able to control things properly.

Q3 and Q4, together with D13 and D14, can place a short on T3's control winding. This holds the voltage across that transformer close to zero, regardless of any current that may be flowing in the windings. When U1 wishes to switch Q1 on, it simply switches pin 12 to ground, switching off Q4 and ending the short circuit on T3. Through R14 and D12, about 15 mA flow into the control winding center tap, returning to ground via Q3. This puts about 50 mA into the base of Q1, which quickly switches on. Now the heavy collector current (up to 8 A at full load) adds up to the total current flowing in T3 and puts enough drive into Q1 to keep it saturated at that heavy current. Note that by this method the strong drive current for the power transistors comes from the collector current through T3 so the control circuit does not have to provide any substantial driving power.

If U1 now determines that Q1 has been conducting long enough, it simply switches off pin 12. Q4 starts conducting again, shorting out T3. The current in T3 is dumped into Q4, which may have to take up to 300 mA. The voltage on T3 falls and Q1 switches off. Some time later, U1 grounds pin 13, starting the conduction cycle for Q2.

U1 uses two input signals to decide what to do with its outputs. One is a sample of the output voltage, taken through R25 and nearby components, while the other is a current sample taken through the primary of T2. This current transformer produces 200 times less current from its secondary than what goes through its one-turn primary. At full load, about 40 mA goes into R12, producing a maximum voltage drop of about 7 V. This is rectified and half of it is taken at the center tap, divided down by R13 and VR1 and smoothed by C31. When VR1 is properly adjusted, there will be 200 mV at pin 4 of U1 with the power supply running at full load.

A second amplifier inside U1 is used for current limiting. Its inputs are at pins 4 and 5. This amplifier is ground-referenced and has an internal offset of 200 mV. The amplifier will pull down the main error amplifier's output if the difference between pin 4 and pin 5 reaches 200 mV.

U1 also contains an internal oscillator, whose frequency is set by R24 and C36 to approximately 50 kHz. The sawtooth output of this oscillator is connected to an internal comparator, which has its other input internally connected to the output of the error amplifier. The output of the comparator is a square wave whose duty cycle depends on the dc voltage at the output of the error amplifier.

During operation at medium to high loads, the duty cycle is about 70%. At the cathodes of the Schottky rectifiers you will see a square wave that stays at about 20 V for some 14 μs, and then goes slightly below ground level for 6 μs. L2, which has its output end at a constant 13.8 V, will therefore see about 6 V for 14μs, followed by −14 V for the rest of the time. Given its inductance of about 20 μH, the current in L2 will increase by about 4 A during each conduction cycle and decrease by that same amount during rest time. As long as the current drawn from the power supply is more than 2 A, the current in L2 will never cease completely. For example, if the current is 20 A on average, the current in L2 will vary between about 18 and 22 A. As the ripple current stays basically constant while operating at up to the maximum current of the power supply, filter capacitors C17 to C22 are never exposed to more than about 1.5 A RMS total ripple current, assuring that they have a long lifetime. This is an advantage over some other types of switching power supplies, where the ripple current is much higher, forcing the designer to use more expensive capacitors or to accept reduced lifetime in these components.

If the load is less than about 2 A, the current flow in L2 is no longer continuous. The duty cycle of the power transistors starts to drop, until at zero load the duty cycle almost becomes zero too.

C37 serves several purposes. For higher frequencies it couples the first filter stage (L2 and C17 through C22) to the error amplifier, while for lower frequencies (and at dc) the output of the supply is sampled. This is necessary because each filter stage introduces 180° of phase shift at the higher frequencies. After two stages the phase shift goes through a full 360°, making it impossible to stabilize the control loop without additional circuitry. But for dc, sampling the output is desirable to compensate for the voltage drop in L3. C37 gives the error amplifier a nice PID response, together with R23 and C35. This affords the best possible transient behavior with unconditional stability. In addition, C37 provides some measure of soft starting, so the voltage does not overshoot too much when first switching on the power supply.

R34 and C38 average out the current level over a period of about 2 minutes. U2C amplifies the resulting voltage by an amount that can be adjusted. U2D acts as a Schmidt trigger to switch the fan cleanly on and off when the current average crosses the trigger level set by VR3. R39 limits the speed of the fan to a rather low value that is more than enough to keep the power supply cool. At this low speed the fan produces almost no noise and it will probably last longer than its owner.

Snubbers and EMI Filters

No transformer is perfect. Each winding has some inductance that is not magnetically coupled to the others. There is also the magnetizing current, which can be a considerable part of the total current in small transformers. At the end of a conduction cycle, a strong current flows in T1. After switching the power transistors off, some means must be provided to discharge the energy stored in the magnetic field of the core and in the leakage inductances. D3 and D5 are included for this purpose. They recover most of this energy and dump it back into C3. Another portion flows through the Schottky diodes into L2, but this cannot be more than the current flowing in L2 at the moment of switchoff.

A problem arises if the magnetizing current is bigger than the actual load current, a situation that can occur during startup. Also it must be taken into account that diodes, even fast ones, take some time to switch, and the transformer cannot wait to start dumping its energy. So some absorbing RC networks have to be included. These are commonly called *snubbers*. R9 and C14 form the primary snubber, absorbing energy during the switching of D3, D5, Q1 and Q2. On the secondary side of T1, R10, C15, R11 and C16 protect the Schottky rectifiers from inductive spikes.

Some RF noise is generated and it must be cleaned up. Between C3 and the power oscillator, two Type-73 ferrite beads FB1 and FB2 perform a critical noise-absorbing task. On the output side, L2 already absorbs most of the noise. It is wound on a high-permeability iron-powder toroid that is very lossy in the HF range. The main filter capacitors have low equivalent-series-resistances for good filtering.

L3 is another noise absorber. To minimize capacitive coupling, a ferrite solenoid was used instead of a toroid so that the input windings are well separated from the output ones. The ferrite used starts absorbing at HF, so this coil not only blocks but also absorbs RF energy. Finally, the output leads are passed through a dozen 73-material ferrite beads. The filtering is completed by bypass capacitors on the output leads to the cabinet. Note that the ground on the printed circuit board is floating to reduce stray HF currents on the enclosure.

Parts Substitution

Don't be afraid to substitute parts when you can't find the exact one specified. Here is some information for hard-to-find parts:

D1: Any rectifier bridge that can handle 8 A at 240 V ac (or 12 A at 120 V ac), with enough headroom for spikes, will do the job. Try to find one that fits the PCB or modify the board accordingly. You may also use single diodes, but mount them close to the board to get suitable heatsinking through their terminals.

D2, D4, D17: Any ultrafast diode rated for at least 100 V and 1 A is suitable. The author used the UF4007, which is an ultrafast equivalent to the 1N4007 (1 kV, 1 A). *Do not use 1N4007 diodes!* They are not fast enough for this job. You need a switching speed in the 50-ns class.

D3, D5: You can use any ultrafast diode rated for 600 V, 3 A or higher. The UF5408 is rated at 1 kV, somewhat of an overkill here. Again, *do not* use the low speed 1N5408.

D6, D7, D8, D9: PBYR30100CT dual Schottky diodes were used. A good replacement is any single or dual Schottky rectifier rated at least 100 V and 30 A total current, that comes in a TO-218 or similar package. If you use single diodes, you may have to bend the pins to fit the board properly. These 100-V Schottky diodes have been widely available only for a few years, although they are becoming more common.

Q1, Q2: BUH1215 transistors were used, which can work at a higher voltage than actually necessary in this circuit. If you need to replace them, look for any NPN power switching transistors that have a V_{CEO} of at least 400 V, I_C of at least 15 A, an h_{FE} of at least 12 at 8 A, and come in a TO-218 or similar package. The power transistors *must* maintain their beta up to at least 8 A, otherwise they will cut short the conduction cycles when the load increases. Motorola MJW16010 transistors are a suitable alternative.

Some power switching transistors have a reverse protection diode and a base-to-emitter resistor built in. Beware of these! The resistor would not allow this power supply to start. If in doubt, take a multimeter and measure the resistance between base and emitter. If you get the same low resistance (typically 50 Ω) in *both* senses, the transistor is unsuitable for this project. If you get a diode behavior, the transistor is okay.

Q3, Q4: Instead of the BC639-16 you can use any small TO-92 cased NPN transistor that has a V_{CEO} rating of 100 V and an I_C of 1 A. Be careful with the pinout, because not all TO-92 transistors use the same pinout. You may have to bend the leads to fit the printed circuit board.

Q5: Instead of the BD683 you can use any small NPN Darlington transistor that has an I_C of at least 1 A.

U3: If you have trouble finding the LM336Z-5.0 voltage reference, you have several options. You may use a reference at another voltage (2.5 V is typical) and modify the values of R27 to R30 accordingly. Or you may replace U3 with a 3-terminal regulator like the 7805, modifying the circuit as necessary. Finally, you could completely eliminate U3 and R31 and use the 5V reference provided by U1 at pin 16. In this case you would lose the voltage indicator's independence from U1. Note that using a simple zener diode instead of U3 is not suitable because zeners are not stable enough for this application.

If you cannot find low-ESR electrolytic capacitors, simply use normal capacitors. The circuit is designed to place a low ripple current on these capacitors, so standard components can be used. The noise at the output will be slightly higher, however.

Running on 240/120 V ac

The author lives in a country where the mains supply is 220 V at 50 Hz. This supply will accept input voltages between about 95 to 250 V ac, using S2 to switch from 240 to 120 V ac operation. For 120-V ac operation the fuse F1 should be rated at 10 A, 5 A for 240 V ac operation.

THE PCB

The exact size of the pc board is 120×272 mm (4.72×10.71 inches). It must be made from good quality, single-sided glass epoxy board—don't try to use a cheaper grade of board. The heavy components would stress it too much and the copper adhesion is not good enough for the heavy soldering required. A circuit board is available from FAR Circuits. See the **References** chapter.

BUILDING THE MAGNETIC COMPONENTS

The biggest challenge for most home builders will be the magnetic components. To keep things simple, Amidon cores were used. The only exceptions are L1 and L3, which were made from materials found in the author's junk box. Both of these inductors are not critical, and suitable Amidon part numbers are included in the parts list.

T1, the main power transformer, is the heart of this circuit. T1 was built using a tape-winding technique, stacking four pairs of ferrite E cores to obtain the necessary magnetic capabilities. Comments from WR1B as he constructed the transformers and inductors are included in the construction details below.

Making T1

Because four cores are stacked there is no factory-made bobbin available for this transformer, so the author made a paper bobbin. He wound the transformer using 0.1-mm thick copper strips interleaved with mylar sheets, because a thick wire needed for the heavy current would be impossible to bend around the sharp corners of the bobbin. Instead of using a lot of thin wires in parallel, it is better to use copper strips. The whole assembly is sealed in epoxy resin, with the magnetic cores glued in place with epoxy too.

Cut a piece of hardwood to serve as a form when making the bobbin. As the center legs of the four stacked cores measure 62×12 mm (2.44×0.47 inches), the wood block must be 63 mm (2.48 inches) wide and 12.5 mm (0.49 inches) thick, to allow for some play. The length of the block should be around 100 mm (4 inches). The height of the bobbin will be 28 mm (1.10 inches), so make your block long enough to hold it with the bobbin in place with room for holding onto it. The author used a belt sander to trim his wood block to the exact dimensions. Try to be precise—if the bobbin is too big you will waste valuable winding space, running the risk of not being able to fit the windings. If the bobbin comes out too small your finished winding assembly may not fit the ferrite cores, making it unusable.

Now wrap the wood block with one layer of plastic film, such as "Saran Wrap" used in the kitchen to preserve food. This material allows you to remove the bobbin from the wood block easily. Cut a strip of strong packing paper, 28 mm (1.10 inches) wide and about 1 m (39.4 inches) long. A brown-paper grocery bag is a good source of suitable paper. Mix some 5-minute epoxy glue (the author used the type sold in airplane modeling shops, which comes in good sized bottles) and apply a layer of epoxy to the paper strip. Now wind 6 layers of the paper strip very tightly around the plastic-wrapped wood block. Wrap another sheet of Saran Wrap around your work and press it between two wooden blocks held together with strong rubber bands or wood clamps so the long sides of the bobbin are flat and smooth against the wood. Now place the bobbin assembly in an oven for about 15 minutes at 50°C (122°F). The epoxy sets much more quickly and becomes somewhat stronger at that temperature.

[Comments from Larry Wolfgang, WR1B: The paper I used for my T1 bobbin was cut from a 36-inch-wide length of "kraft paper." This had been used to wrap some paper my wife had purchased at an art-supply store. It was about as heavy as the paper used for grocery bags. I used 30-minute epoxy for this step, providing a bit more "working time" than 5-minute epoxy allows. It takes *lots* of epoxy, because so much soaks into the paper. My epoxy was the kind with the double plunger, and equal amounts come out of both tubes as you push in the plunger. Wear rubber or plastic gloves to protect your hands. I squeezed out an amount that made a puddle of resin and a puddle of hardener each about $1^1/_2$ inches across and $^1/_8$ inch or so deep. This was not enough, and I had to mix more. I used a spring clamp to hold the paper to my workbench and then held the paper in one hand while spreading epoxy with a heavy toothpick. I coated the entire length and then wrapped my plastic-covered wooden block. My electronic-controlled gas oven only allows me to set the temperature as low as 170°F, so I had to watch the temperature and shut the oven off as the temp rose to about 150°F, then let it cool down. I ran it twice this way to "cure" the 30-minute epoxy I used for the bobbin.—WR1B]

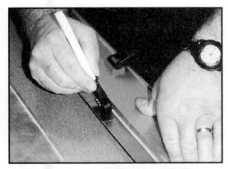

Fig 11.38—Larry Wolfgang, WR1B, using a 4-foot straightedge designed as a guide for hand-held circular saws to clamp the copper foil tape to a board on a tabletop. After carefully measuring to ensure a uniform 22 mm width, he cut the foil tape using a Fiskars rotary cutter. Be careful to keep the cutter wheel against the straightedge for the entire length. Move the tape in 4-foot intervals to cut the entire length. (Photo by Dan Wolfgang.)

Now you will need some 0.1 mm (0.004 inches = 4 mils) thick copper tape, and some mylar sheet of a similar thickness. Cut the copper in strips 22 mm (0.87 inches) wide, and the mylar in strips 28 mm (1.10 inches) wide. (The Wireman has suitable copper foil available.) If you can make long strips, say 2 m (6.56 feet), this is an advantage. Otherwise, you will have to solder individual copper strips together. In total, you will need about 7 m (23 feet) of copper tape and slightly less mylar tape. [I made 7 meters of "double-thickness" tape, using two 3-mil thick, sticky-backed copper tapes that we had in the ARRL Lab. After making the 15-turn winding, I cut the leftovers in four equal lengths to make the "four-layer tape" used in the secondary. There was less than a foot of left-over tape after the transformer was completed. The mylar tape I used was made by 3M and was 2-mil thick and 1-inch wide with adhesive backing. This thickness is sufficient for the voltages involved, provided that care is taken so that the mylar isn't punctured by accident. If like the author, you cut strips from a sheet of copper, you should file down the edges to remove burrs. See Fig 11.38.—WR1B]

Once the epoxy has had ample time to harden and has cooled, remove the rubber bands, the outer wood blocks, and the outer plastic wrapping (don't worry if it doesn't come off completely). Do not remove the plastic wrapping that separates the bobbin from the wood. You now have your wrapped wooden core and the epoxy-paper bobbin on it.

Take a 60 mm (2.36 inch) length of #12 bare copper wire. Wrap the end of one of your copper strips around the wire, so that the wire protrudes out from one side of the copper loop. Use a big soldering iron to flow some solder into the junction. Try to avoid getting solder on the outside, be-

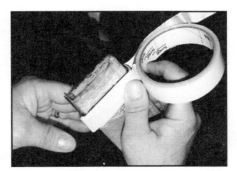

Fig 11.39—Winding the foil tape tightly on the epoxy-coated paper bobbin on the wooden block. The mylar tape is unrolled and positioned over the foil layer as you wind. (Photo by Dan Wolfgang.)

cause this could later puncture the mylar insulation. [I scraped the adhesive off the back of the sticky-backed tape where I soldered the wire. Otherwise, the solder won't stick to the back of the copper, and the layers may not have good conductivity between them.—WR1B]

Now place the copper wire on one of the narrow sides of the bobbin, so that the copper strip is centered on the width of the bobbin, leaving 3 mm (0.12 inches) room on each side. Stick the start of the copper strip to the bobbin with some thin adhesive tape. See Fig 11.39.

Position the start of a mylar strip so that it covers all the copper and is centered on the bobbin, and then tape it in place. Now wind 15 turns of this copper-mylar sandwich, as tightly as you can, keeping the mylar aligned with the bobbin sides and the copper nicely centered. Don't lose your grip, or the whole thing will spring apart! If your copper strip is not long enough, fix everything with strong rubber bands or a clamp and solder another copper strip to the end of the first one, allowing 2 mm of overlap. Before doing this, cut the first copper sheet so that the joint will be on one of the narrow sides of the bobbin, because here you have space, while the wide sides will have to fit inside the ferrite core's window. If the Mylar strip runs out, just use adhesive tape to add another strip. Make the overlap 5 mm to avoid risk of creepage between the sheets and also try to locate the joint on one of the narrow sides of the bobbin. See Fig 11.40.

When the 15 turns are complete, cut the copper strip so that the second terminal will be on the same narrow side of the bobbin as the first terminal. Solder the second terminal (another 60 mm piece of bare copper wire) to the strip, position it and wind three or four layers of Mylar to make the insulation safe between the primary and secondary. [I started my primary winding with the bulge of the wire on the corner, so that I was immediately winding along the wide side. When I finished the 15 turns, I positioned the end wire so it is on the narrow side, just beyond the corner of the long side. This way, the two bulges meet at the middle, but don't cross each other.—WR1B]

If you think this is a messy business, you are right. But it's fun too! The secondary is just a little bit messier: It is wound with a five-layer sandwich—four layers of copper and the Mylar topping layer. But it's only four turns total, so take a deep breath and do it. Solder the four copper strips together around a piece of #12 copper wire. Don't be overly worried if the outcome is not very clean; the author's was quite a mess too, yet it worked well on the

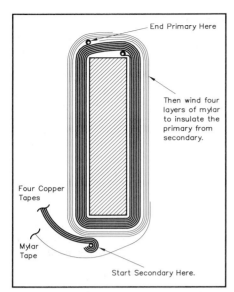

Fig 11.40—Primary 15 turns on bobbin, with start of 4-turn, center-tapped secondary winding.

first try. Just be sure you don't create sharp edges or pointed solder mounds, because these may damage the insulation. See **Fig 11.40** for details.

Now position the start of your secondary conductor so the terminal wire will come out on the same side as those of the primary, but on the other narrow side of the coil assembly. The goal is to end up with a transformer with its primary leads on one extreme and the secondary on the other, and that will also fit the printed circuit board nicely. Wind two turns, solder the center tap wire between the four copper strips, wind the other two turns, solder the last terminal wire and then wind a finishing layer of mylar and fix it in place with adhesive tape. This finishes the worst part of making T1.

What you have now is a springy, messy coil assembly that will fall apart if you let it go. You have to seal it, but this is easy to do. Temporarily hold things together with some stout rubber bands. Wrap your two wooden blocks, the same you used to press together the bobbin, in plastic film. Place them against the sides of the coil assembly, and apply hard pressure, using a clamp or a lot of rubber bands, so that the long sides of the coil straighten out completely, and any slack is displaced to the narrow sides. Now mix a fair quantity of epoxy glue, place the coil assembly so that the pins face up, and let the epoxy run into the coil. Continue supplying epoxy until it starts to set. If it drips out from the other side, no problem. (Just don't do this work over your best rug!) When the epoxy doesn't flow any longer, turn over the coil assembly, mix a new batch of epoxy and

fill the other side completely, forming a smooth surface. As the lower side is now sealed, the epoxy will not flow out there. When this epoxy has set, turn the assembly over again, mix some more epoxy and apply it to form a smooth surface there. The idea is to replace all the air between copper and mylar sheets by epoxy, and especially to fill the room left by the copper strip, which is narrower than the mylar. This filling is necessary both for mechanical and for electric safety reasons. See **Fig 11.41**. [My wooden "screw clamps" worked well for applying strong even pressure to the sides. I don't think rubber bands would apply enough pressure to minimize the air space inside the transformer.—*WR1B*]

Now place the assembly in the oven again. Let the epoxy harden completely, then remove the coil from the oven, remove the clamp, rubber bands, wooden blocks, wooden core and all remains of plastic film. You will be surprised how your messy and springy assembly changed into a very robust, hard, strong and nice coil. Now test-fit the ferrite cores. See **Fig 11.42**. See if they can be installed easily, so that each pair of facing E-cores comes together in intimate contact, without pressing on the winding. If everything is right, the winding should have some play room in the assembled core. But it is easy to get too much epoxy on the coil. If this happens to you, just take a file and work the epoxy down so that it doesn't disturb the ferrite. The ferrite core *must* close properly; otherwise you will later burn out the power transistors.

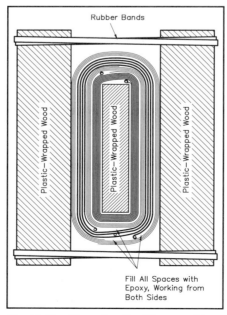

Fig 11.41—Clamping the T1 assembly and filling with epoxy.

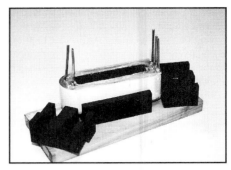

Fig 11.42—Photo showing how the core halves must fit into the completed transformer after it is removed from the wooden block. You will have to file off the rough edges of epoxy to allow the cores to meet properly. The top E-cores have not been inserted into the bobbin yet. (Photo by Larry Wolfgang.)

When the sides fit, prepare some more epoxy, apply a very thin layer to all contact faces of the ferrite cores and mount them onto the coil assembly. You can hold them in place with adhesive tape until the epoxy sets. Again, use the oven to speed up the hardening. The last thing you have to do is bend the copper wires into the proper shape to fit the printed circuit board holes. Be sure that on the secondary winding the center tap is actually in the center position. The polarity of the other pins doesn't matter. This completes the manufacture of T1. All the other transformers and coils are just child's play after making T1!

Making T2

The current sense transformer T2 has a lot of turns but they needn't be wound nicely side-by-side. You can use a winding machine with a turns counter or you can just wind T2 by hand. Get some #36 or other thin enameled wire, solder the end to one of the outer pins of the EE24-25-B bobbin, and wind 100 turns. Don't worry if your winding is criss-crossed and ugly, and don't feel guilty if you lose count and wind a few turns more or less. As long as you don't overdo it, it will just affect the position of VR1 when you adjust the completed power supply later. Solder the wire to the center pin on the same side, then wind another 100 turns in the same sense. Solder to the other outer pin on the same bobbin side and apply one or two layers of mylar, just to protect the thin wire.

Now take a piece of #14 plastic insulated wire, wind one single turn over the mylar and solder the two ends to the two outer pins of the other side of the bobbin. It doesn't matter which end goes to which side. Install the EA77-250 core with a small amount of epoxy cement and T2 is finished. [I used AWG #14 house wire here. The insulation made it a bit tight for the core, but it fit.—WR1B.]

Making T3

T3 is made using the same type of bobbin and core as T2. Wind 26 turns of #28 enameled wire. The 26 turns should fit nicely in a single layer. Study the schematic diagram to see how the windings connect to the bobbin pins. Bring the wire back to the starting side over the last half turn, for connection to the center-tap pin. Wind one layer of mylar sheet, then put on the next 26 turns. Again, bring the wire back to the starting side over the last half turn, for connection to the bobbin pin.

Wind 3 layers of mylar tape, to insulate the primary and secondary properly. Wind 8 turns of #20 wire, and solder the ends to the bobbin pins. Look at the printed circuit board drawing to see which wire is soldered to which pin. Wind a single layer of mylar, then wind the other 8-turn winding over the first one. This will leave a space at one side of the bobbin big enough to take the single turn of #14 plastic insulated wire. This completes the assembly. See **Fig 11.43** for a cross-sectional view of the windings. Now glue the core in place with epoxy cement and T3 is finished.

Making L2

L2 is wound on an Amidon T-200-26 iron powder toroid core. As it is too difficult to bend thick wire through a toroid, and tape winding it is not practical either, the author chose to make this coil with 10 pieces of #16 enameled wire in parallel.

Cut the wires to about 1.5 m (59 inches) in length and lightly twist them together. Then insert the bundle into the core, and starting from the middle of the wire bundle, wind 8 turns, using half of the core's circumference. Now wind another 7 turns, starting from the middle toward the other end of the wire bundle. The 16th turn is the one you made when you inserted the wire bundle into the core to start.

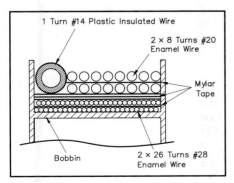

Fig 11.43—Cross-sectional view of T3 (not to scale), showing distribution of windings.

Making L3

To make L3 you must first find a suitable rod. I used a part of an old ferrite antenna rod about 10 mm in diameter (0.39 inches) diameter and 50 mm long (1.97 inches). (An Amidon number 33-050-200 rod is just the right size.) Wind 10 bifilar turns of #12 enameled wire. This wire is quite stiff, but it is still no problem to handle. You should wind the coil on a 12 mm (15/32 inches) drill bit, allow it to spring open and place it on the ferrite core. Otherwise you could crack the ferrite trying to wind directly on it. A tapered "drift punch" helped open the turns just enough to fit the core. Fix the core to the winding with some epoxy. Bend the wires so that all four of them point down with the core pointing straight up. That's the position L3 is mounted on the PCB.

PUTTING IT TOGETHER

Install and solder all parts except for Q1, Q2, and D6 to D9. Before installing D1, fashion a simple heatsink from a 30 × 80 mm (1.18 × 3.15 inches) piece of 1 mm (0.039 inches) thick aluminum sheet, bent into U shape. Drill a hole and screw the rectifier bridge onto the heatsink together with a lock washer. Then solder D1 to the board.

The author made his own enclosure, using two 3-mm (0.12 inches) aluminum plates, measuring 300 × 120 mm (11.81 × 4.72 inches) for the front and rear walls. They are screwed to the fan, the PCB and to a 120-mm (4.72-inch) long spreader tube of 6-mm (0.24-inch) diameter, so that these parts become integral to the structure. The connections between the PCB, aluminum plates and fan were made with small pieces of 10 × 10 mm (0.39 × 0.39 inches) aluminum angle stock. The assembly is surprisingly rigid.

The top and bottom covers were made from 1 mm (0.04 inch) aluminum sheet and measure 126 × 300 mm (4.96 × 11.81 inches). The bottom cover has a hole for the PCB's center mount. The side covers were cut from wire mesh to allow unrestricted airflow, and measure 122 × 126 mm (4.80 × 4.96 inches). The panels are held together with 10 × 10 mm (0.39 × 0.39 inches) aluminum angle stock, running along all edges and held with small sheet metal screws. These covers are not installed until the power supply is complete, tested and adjusted.

All the panels were painted flat black on the outside, which looks nice together with the anodized aluminum angle stock. The edges and insides were kept free of paint, in order to get proper electrical contact between the panels for good shielding.

The version made by WR1B used a Hammond Manufacturing ventilated, low-profile instrument case, catalog number 1426Y-B. This is a rugged case that also looks very nice. Larry mounted the circuit board inside the case using a pair of steel mounting rails, also from Hammond, catalog number 1448R12.

The components external to the PCB (P1, SW1, C3, the LED and the output screw terminal block) are mounted to the front and rear panels. Q1 and Q2 are mounted to the rear panel, using M3 nylon screws and 3 mm (0.12-inch) thick ceramic insulators. These thick insulators were used not only for safety reasons but also because they reduce the capacitive coupling of the transistors to the enclosure. Do not use metal screws with plastic washers, because this approach does not give enough safety margin to operate at the input line voltage. [The author's junkbox ceramic insulators proved difficult to duplicate for the supply we built in the ARRL Lab. Equivalent new parts would have nearly doubled the cost of the supply! Instead, for good heat-transfer properties, we used thin rubber insulators manufactured by Wakefield Engineering as PN 175-6-250-P, available from Newark Electronics as PN 46F7884. Individual aluminum spacers milled from aluminum blocks were used between Q1, Q2 and the Schottky diodes and the metal chassis. Care must be taken to make sure the surfaces of the spacers are parallel and free of burrs to ensure low thermal resistance.]

The Schottky diodes are mounted using the same kind of insulators and screws, but there is a heat spreader made from 6-mm (0.24-inch) aluminum plate between those insulators and the case. All surfaces requiring thermal contact are covered with heat-transfer compound before assembly. When installing the diodes and transistors, first do all the mechanical assembly and then solder the pins. Otherwise you could stress them too much while fastening the screws.

All wire connections are made next and the output filter is assembled by sliding the ferrite beads over the output cables and soldering the bypass capacitors C25 through C30. Be sure to use a nice thick wire for the output. A 40 A continuous-duty current is no joke.

The tracks on the PCB cannot be trusted to carry 40 A without some help. Use a big soldering iron (100 to 150 W) to solder lengths of #12 bare copper wire cut and bent to fit the shape of all the high-current paths. To prevent any failures due to vibration from the fan, place some drops of hot-melt glue anywhere a wire is connected to the board. Hot-melt glue is also excellent for fixing anything that would otherwise rattle, like ferrite beads.

Fig 11.44—Photo of top of PCB mounted in cabinet.

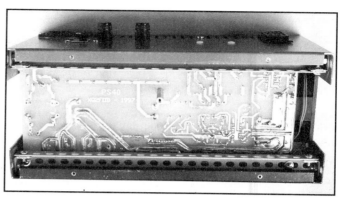

Fig 11.45—Photo of bottom of PCB mounted in cabinet.

Testing and Adjusting

Make sure you do a thorough visual check. Set the three potentiometers to mid position. Check that there is no continuity between the ac input and ground, between the ac input and the dc output or between the dc output and chassis ground.

Connect a variable voltage supply (you need 12 to 15 V for the tests) to the output leads, without plugging the switcher into the ac line. You should see the LED light up. Change the voltage fed into your project to see how the LED changes color. If you have a dual-channel oscilloscope, connect its two channels to the base-emitter junctions of the power transistors. [Since you are not connected to the ac power line, you will not be grounding it through the oscilloscope's ground leads connected to the emitter leads.—Ed.] With the external voltage at about 12 V, you should see small pulses. As you increase the voltage suddenly the pulses will disappear. If you want, you can preadjust VR2 by setting your lab power supply to exactly 13.8 V and then setting VR2 to where the pulses just disappear.

Now it's time to start up the switcher. Remove your lab supply and the oscillo-scope leads and connect the supply to the ac line in series with a 60-W light bulb. This will avoid most or all damage if something is really wrong. Connect a voltmeter to the output and switch on your supply. If everything is right, the bulb will light up, then slowly dim while the power supply starts up and delivers about 13.8 V.

Now, connect a load of about 2 A to the output—a car brake-light bulb makes a good load. At 2 A output, the bulb in the ac line will probably glow, with 13.8 V dc at the output. If everything is okay so far, now comes the big moment. Remove the series bulb from the ac circuit. Startup of the supply should be fast and you can now connect a heavier load to it. With a load of 2 to 10 A connected (the value is uncritical, given the good regulation of this supply), adjust VR2 so that you have exactly 13.8 V at the output.

Next adjust the current shutdown point. For this you need a load that can handle 40 A. You could make one by connecting a lot of car headlamps in parallel or you could use some resistance wire to build a big power resistor. The author made a 13.8 V, 550 W heater for his supply. Connect the load and adjust VR1 so that the output voltage is just at the limit of shutting down.

The last adjustment is for the fan trigger point. Connect a 65-W car headlamp or similar load that consumes about 5 A. Let the supply run for several minutes, then move VR3 to the point where the fan switches on. Now check out the trigger function by changing the load several times between about 2 and 10 A. The fan should switch off and on between 30 to 60 seconds after each load change. You may have to readjust VR3 until you get the fan to switch on at no more than 7 A continuous load and switch off at about 4 A.

And If It Doesn't Work?

If you used substitute parts for the magnetic cores and made a bad choice, the results could be dramatic. If either T1 or L2 saturates, the power transistors could burn out before the fuse has a chance to open. The protective light bulb in the ac line will avoid damage in this case, so by all means use that bulb for initial testing!

Another possible error is reversing the phase of a winding in T3. If you get one of the 8-turn windings reversed, the results will be explosive unless you have the light bulb in series. If you reverse the 1-turn winding, the power supply will simply not start.

28-V, HIGH-CURRENT POWER SUPPLY

Many modern high-power transistors used in RF power amplifiers require 28-V dc collector supplies, rather than the traditional 12-V supply. By going to 28 V (or even 50 V), designers significantly reduce the current required for an amplifier in the 100-W or higher output class. The power supply shown in **Fig 11.49** through Fig 11.53 is conservatively rated for 28 V at 10 A (enough for a 150-W output amplifier) — continuous duty! It was designed with simplicity and readily-available components in mind. Mark Wilson, AA2Z, built this project in the ARRL lab.

Circuit Details

The schematic diagram of the 28-V supply is shown in **Fig 11.50**. T1 was designed by Avatar Magnetics specifically for this project. The primary requires 120-V ac, but a dual-primary (120/240 V) version is available. The secondary is rated for 32 V at 15 A, continuous duty. The primary is bypassed by two 0.01-μF capacitors and protected from line transients by an MOV.

U1 is a 25-A bridge module available from Radio Shack or a number of other suppliers. It requires a heat sink in this application. Filter capacitor C1 is a

Fig 11.49 — The front panel of the 28-V power supply sports only a power switch, pilot lamp and binding posts for the voltage output. There is room for a voltmeter, should another builder desire one.

computer-grade 22000-µF electrolytic. Bleeder resistor R1 is included for safety because of the high value of C1; bleeder current is about 12 mA.

There is a tradeoff between the transformer secondary voltage and the filter-capacitor value. To maintain regulation, the minimum supply voltage to the regulator circuitry must remain above approximately 31 V. Ripple voltage must be taken into account. If the voltage on the bus drops below 31 V in ripple valleys, regulation may be lost.

In this supply, the transformer secondary voltage was chosen to allow use of a commonly available filter value. The builder found that 50-V electrolytic ca-

pacitors of up to about 25000 µF were common and the prices reasonable; few dealers stocked capacitors above that value, and the prices increased dramatically. If you have a larger filter capacitor, you can use a transformer with a lower secondary voltage; similarly, if you have a transformer in the 28- to 35-V range, you can calculate the size of the filter capacitor required. Equation 3, earlier in this chapter, shows how to calculate ripple for different filter-capacitor and load-current values.

The regulator circuitry takes advantage of commonly available parts. The heart of the circuit is U3, a 723 voltage regulator IC. The values of R8, R9 and R10 were

chosen to allow the output voltage to be varied from 20 to 30 V. The 723 has a maximum input voltage rating of 40 V, somewhat lower than the filtered bus voltage. U2 is an adjustable 3-terminal regulator; it is set to provide approximately 35 V to power U3. U3 drives the base of Q1, which in turn drives pass transistors Q2-Q5. This arrangement was selected to take advantage of common components. At first glance, the number of pass transistors seems high for a 10-A supply. Input voltage is high enough that the pass transistors must dissipate about 120 W (worst case), so thermal considerations dictate the use of four transistors. See the **Real World Component Characteristics** chapter for

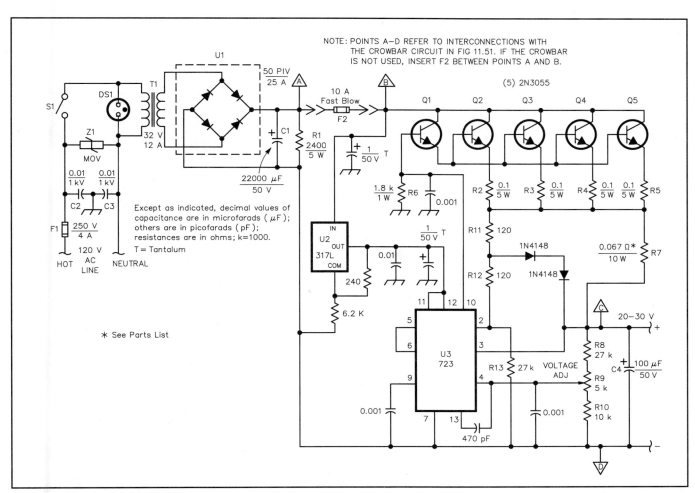

Fig 11.50 — Schematic diagram of the 28-V, high-current power supply. Resistors are 1/4-W, 5% types unless otherwise noted. Capacitors are disc ceramic unless noted; capacitors marked with polarity are electrolytic. Parts numbers given in parentheses that are preceded by the letters RS are Radio Shack catalog numbers.

C1 — Electrolytic capacitor, 22000 µF, 50 V (Mallory CGS223U050X4C or equiv., available from Mouser Electronics).
C2, C3 — AC-rated bypass capacitors.
C4 — Electrolytic capacitor, 100 µF, 50 V.
DS1 — Pilot lamp, 120-V ac (RS 272-705).
Q1-Q5 — NPN power transistor, 2N3055 or equiv. (RS 276-2041).

R2-R5 — Power resistor, 0.1 Ω, 5 W (or greater), 5% tolerance.
R7 — Power resistor, 0.067 Ω, 10 W (or greater), made from three 0.2-Ω, 5-W resistors in parallel.
T1 — Power transformer. Primary, 120-V ac; secondary, 32 V, 15 A. (Avatar Magnetics AV-430 or equiv. Dual primary version is part #AV-431. Available from Avatar Magnetics.)

U1 — Bridge rectifier, 50 PIV, 25 A (RS 276-1185).
U2 — Three-terminal adjustable voltage regulator, 100 mA (LM-317L or equiv.). See text.
U3 — 723-type adjustable voltage regulator IC, 14-pin DIP package (LM-723, MC1723, etc. RS 276-1740).
Z1 — 130-V MOV (RS 276-570).

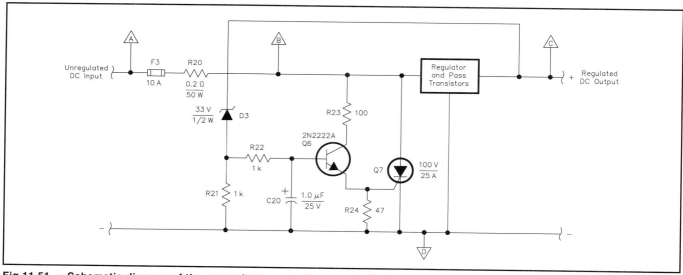

Fig 11.51 — Schematic diagram of the overvoltage protection circuit. Resistors are ¼-W, 5% carbon types unless noted.
D3 — 33 V, ½ W Zener (NTE 5036A or equiv.).
Q6 — NPN Transistor (2N2222A or equiv.).
Q7 — 100 V, 25A SCR (NTE 5522 or equiv.).

a complete discussion of thermal design. If you use a transformer with a significantly different secondary potential, refer to the thermal-design tutorial to verify the size heat sink required for safe operation.

R9 is used to adjust supply output voltage. Since this supply was designed primarily for 28-V applications, R9 is a "set and forget" control mounted internally. A 25-turn potentiometer is used here to allow precise voltage adjustment. Another builder may wish to mount this control, and perhaps a voltmeter, on the front panel to easily vary the output voltage.

The 723 features current foldback if the load draws excessive current. Foldback current, set by R7, is approximately 14 A, so F2 should blow if a problem occurs. The output terminals, however, may be shorted indefinitely without damage to any power-supply components.

If the regulator circuitry should fail, or if a pass transistor should short, the unregulated supply voltage will appear at the output terminals. Most 28-V RF transistors would fail with 40-plus volts on the collector, so a prospective builder might wish to incorporate the overvoltage protection circuit shown in **Fig 11.51** in the power supply. This circuit is optional. It connects across the output terminals and may be added or deleted with no effect on the rest of the supply. If you choose to use the "crowbar," make the interconnections as shown. Note that R20 and F3 of Fig 11.51 are added between points A and B of Fig 11.50. If the crowbar is not used, connect F2 between points A and B of Fig 11.50.

The crowbar circuit functions as

Fig 11.52 — Interior of the 28-V, high-current power supply. The cooling fan is necessary only if the pass transistors and heat sink are mounted inside the cabinet. See text.

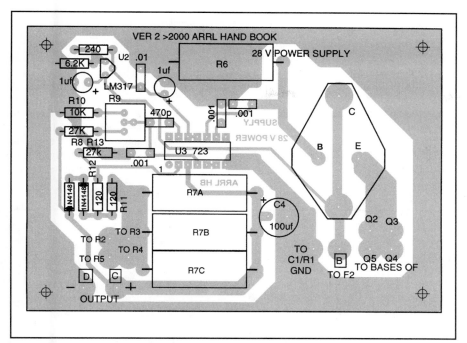

Fig 11.53 — Parts placement diagram for the 28-V power supply. A full-size etching pattern is in Chapter 30, References.

follows: The Zener holdoff diode (D3) blocks the positive regulated voltage from appearing at the base of Q6 until its avalanche voltage is exceeded. In the case of the device selected, this voltage level is 33 V, which provides for small overshoots that might occur with sudden removal of the output load (switching off a load, for instance).

In the event the output voltage exceeds 33 V, D3 will conduct, and forward bias Q6 through R22 and C20, which eliminate short duration transients and noise. When Q6 is biased on, trigger current flows through R23 and Q6 into the gate of SCR Q7, turning it on and shorting the raw dc source, forcing F3 to blow. Since some SCRs have a tendency at high temperature to turn themselves on, resistor R24 shunts any internal leakage current to ground.

Construction

Fig 11.52 shows the interior of the 28-V supply. It is built in a Hammond 1401K enclosure. All parts mount inside the box. The regulator components are mounted on a small PC board attached to the rear of the front panel. See **Fig 11.53**. Most of the parts were purchased at local electronics stores or from suppliers listed in the **References** chapter. Many parts, such as the heat sink, pass transistors, 0.1-Ω power resistors and filter capacitor

can be obtained from scrap computer power supplies found at flea markets.

Q2-Q5 are mounted on a Wakefield model 441K heat sink. The transistors are mounted to the heat sink with insulating washers and thermal heat-sink compound to aid heat transfer. Radio Shack TO-3 sockets make electrical connections easier. The heat-sink surface under the transistors must be absolutely smooth. Carefully deburr all holes after drilling and lightly sand the edges with fine emery cloth.

A five-inch fan circulates air past the heat sink inside the cabinet. Forced-air cooling is necessary only because the heat sink is mounted inside the cabinet. If the heat sink was mounted on the rear panel with the fins vertical, natural convection would provide adequate cooling and no fan would be required.

U1 is mounted to the inside of the rear panel with heat-sink compound. Its heat sink is bolted to the outside of the rear panel to take advantage of convection cooling.

U2 may prove difficult to find. The 317L is a 100-mA version of the popular 317-series 1.5-A adjustable regulator. The 317L is packaged in a TO-92 case, while the normal 317 is usually packaged in a larger TO-220 case. Many of the suppliers listed in the **References** chapter sell them, and RCA SK7644 or Sylvania ECG1900

direct replacements are available from many local electronics shops. If you can't find a 317L, you can use a regular 317 (available from Radio Shack, among others).

R7 is made from two 0.1-Ω, 5-W resistors connected in parallel. These resistors get warm under sustained operation, so they are mounted approximately $1/16$ inch above the circuit board to allow air to circulate and to prevent the PC board from becoming discolored. Similarly, R6 gets warm to the touch, so it is mounted away from the board to allow air to circulate. Q1 becomes slightly warm during sustained operation, so it is mounted to a small TO-3 PC board heat sink.

Not obvious from the photograph is the use of a single-point ground to avoid ground-loop problems. The PC-board ground connection and the minus lead of the supply are tied directly to the minus terminal of C1, rather than to a chassis ground.

The crowbar circuit is mounted on a small heat sink near the output terminals. Q7 is a stud-mount SCR and is insulated from the heat sink. The other components are mounted on a small circuit board attached to the heat sink with angle brackets.

Although the output current is not extremely high, #14 or #12 wire should be used for all high-current runs, including the wiring between C1 and the collectors of Q2-Q5; between R2-R5 and R7; between F2 and the positive output terminal; and between C1 and the negative output terminal. Similar wire should be used between the output terminals and the load.

Testing

First, connect T1, U1 and C1 and verify that the no-load voltage is approximately 44 V dc. Then, connect unregulated voltage to the PC board and pass transistors. Leave the gate lead of Q6 disconnected from pin 8 of U4 at this time. You should be able to adjust the output voltage between approximately 20 and 30 V. Set the output to 28 V.

Next, short the output terminals to verify that the current foldback is working. Voltage should return to 28 when the shorting wire is disconnected. This completes testing and setup.

The supply shown in the photographs dropped approximately 0.1 V between no load and a 12-A resistive load. During testing in the ARRL lab, this supply was run for four hours continuously with a 12-A resistive load on several occasions, without any difficulty.

A 3200-V POWER SUPPLY

This high-voltage power supply was described by Russ Miller, N7ART, in December 1994 *QST* for use with a 2-m, 1-kW amplifier using the 3CX1200Z7 triode. The supply is rated at 3200 V, at a continuous output current of 1.2 A, and an intermittent peak current of 2 A. The power supply also has a 12 V, 1 A output, used to power the control circuitry in the N7ART 2-meter amplifier. See the **RF Power Amplifiers** chapter for details.

A simple, effective step/start circuit is incorporated to ensure that the rectifier diodes are not over stressed by the heavy surge current into the output capacitor when the power supply is first turned on. A 5 kV meter is used to monitor high-voltage output.

Fig 11.54 is a schematic diagram of the supply. A power supply for a high-power linear amplifier should operate from a 240-V circuit, for best line regulation. A special, hydraulic/magnetic circuit breaker doubles as the main power switch. Don't substitute a regular switch and fuses for this breaker; fuses won't operate quickly enough to protect the amplifier in case of an operating abnormality. The

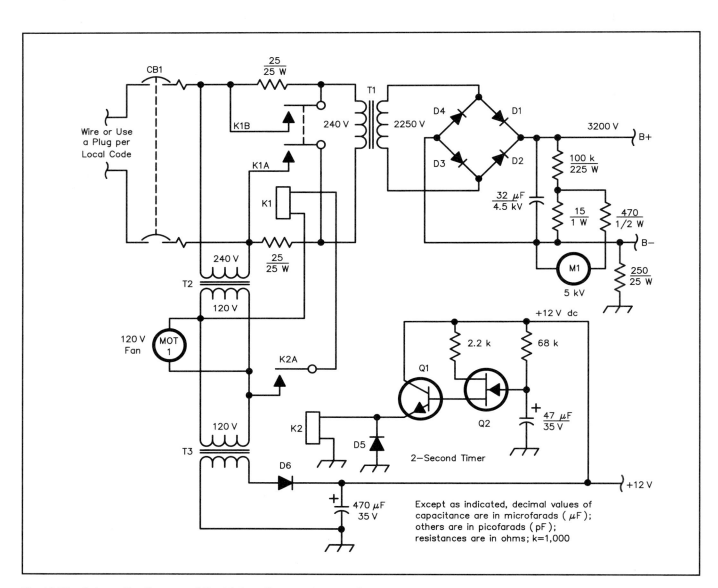

Fig 11.54—Schematic diagram of the high-voltage power supply.

D1-D4—Strings of 5 each, 1000-PIV, 3-A diodes, 1N5408 or equiv. A commercial diode block assembly is also suitable: K2AW HV14-1, from K2AW's Silicon Alley.
K1—DPST relay, 120-V ac coil, 240-V-ac, 20-A contacts (Midland Ross 187-321200 or equiv).
K2—SPDT miniature relay, 12-V dc coil (Radio Shack 275-248 or equiv).

M1—High-voltage meter, 5 kV dc full scale. Select a series resistor to give a 5 kV full scale reading with your specific meter.
MOT1—Cooling fan, Torin TA-300 or equiv.
Q1—2N2222A or equiv.
Q2—MPF102 or equiv.
CB1—20-A hydraulic/magnetic circuit breaker (Potter and Brumfield

W68X2Q12-20 or equiv).
T1—High-voltage power transformer, 240-V primary, 2250-V, 1.2-A secondary (Avatar AV-538 or equiv).
T2—Stepdown transformer, Jameco 112125, 240-V to 120-V, 100 VA.
T3—Power transformer, Jameco 104379, 120-V primary; 16.4 V, 1-A secondary (half used).

bleeder resistor dissipates about 100 W, so a small fan is included to remove the excess heat.

Power Supply Construction

The power supply can be built into a 17×13×10-inch cabinet. The power transformer is quite heavy, so use 1/8-inch aluminum for the cabinet bottom, and reinforce it with aluminum angle for extra strength. The diode bridge consists of four legs, each containing five diodes.

Power Supply Operation

When the front-panel breaker is turned on, two 25-Ω resistors in the primary circuit limit inrush current as the filter capacitor charges. After two seconds, step-start relay K1 activates, shorting both resistors and allowing full line voltage to be applied to the transformer.

As with all high-voltage power supplies, be extremely careful! Before opening the cabinet, remove the ac-line plug from its receptacle, and confirm that the filter capacitor is discharged before working on the supply.

Contents

Modulation Sources (What and How We Communicate)

12

"**M**odulation Sources"—what does that mean? An engineer might simply call this chapter *Baseband*. Engineers use that term to distinguish information before it's used to modulate a carrier. So, this chapter covers the various kinds of information we impress on RF (audio, video, digital, remote control) before that information has been moved to some intermediate frequency (IF) or the desired RF.

Baseband carries the connotation "at or near dc" because the final RF is usually much higher than the baseband frequency, yet that is misleading. For example, a baseband ATV signal usually extends up to 5 MHz, which is hardly dc. Nonetheless, compared to the operating frequencies (52 to 806 MHz for broadcast, 420 MHz and higher for amateur) it *is* practically dc.

Here we will discuss characteristics of the information (such as bandwidth), how we prepare it for transmission, optimize the transfer and process it after reception.

Nearly all of the processing discussed in this chapter can be implemented with the emerging digital-signal-processing (DSP) technology. For example, the CLOVER-II system described later in this chapter uses DSP to vary the modulation scheme as required by propagation conditions. Look to the **Digital Signal Processing** chapter for further information about DSP techniques.

BANDWIDTH

Whenever information is added to a carrier (we say the carrier is *modulated)*, sidebands are produced. Sidebands are the frequency bands on both sides of a carrier resulting from the baseband signal varying some characteristic of the carrier. The modulation process creates two sidebands: the upper sideband (USB) and the lower sideband (LSB). The width of each sideband is generally equal to the highest frequency component in the baseband signal. In some modulation systems, the width of the sidebands may greatly exceed the highest baseband frequency component.

The USB and LSB are mirror images of each other and carry indentical information. Some modulation systems transmit only one sideband and partially or completely suppress the other in order to conserve bandwidth.

According to FCC Rules, *occupied bandwidth* is:

The frequency bandwidth such that, below its lower and above its upper frequency limits, the mean powers radiated are each equal to 0.5 percent (–23 dB) of the total mean power radiated by a given emission.

In some cases a different relative power level may be specified; for example, –26 dB (0.25%) is used to define bandwidth in §97.3(a)(8) of the FCC rules.

Occupied bandwidth is not always easy for amateurs to determine. It can be measured on a spectrum analyzer, which is not available to most amateurs. Occupied bandwidth can also be calculated, but the calculations require an understanding of the mathematics of information theory and are not covered in this book.

The FCC has defined *necessary bandwidth* as:

For a given class of emission, the minimum value of the occupied bandwidth sufficient to ensure the transmission of information at the rate and with the quality required for the system employed, under specified conditions.

Voice Modes

AMPLITUDE MODULATION (AM)

AM voice was the second mode used in Amateur Radio (after Morse code). AM techniques are the basis for several other modes, such as single-sideband voice and AFSK. This material was supplied by Jeff Bauer, WA1MBK.

AM is a mixing process. When RF and AF signals are combined in a standard AM modulator, four output signals are generated: the original RF signal (carrier), the original AF signal, and two sidebands, whose frequencies are the sum and difference of the original RF and AF signals, and whose amplitudes are proportional to that of the original AF signal. The sum component is called the upper sideband (USB). It is erect: A frequency increase of the modulating AF causes a frequency increase in the RF output. The difference component is called the lower sideband (LSB), which is inverted: A frequency increase in the modulating AF produces a decrease in the output frequency. The amplitude and frequency of the carrier are unchanged by the modulation process, and the original AF signal is rejected by the RF output network. The RF envelope (sum of sidebands and carrier), as viewed on an oscilloscope, has the shape of the modulating waveform.

Fig 12.1B shows the envelope of an RF signal that is 20% modulated by an AF sine wave. The envelope varies in amplitude because it is the vector sum of the carrier and the sidebands. A spectrum analyzer or selective receiver will show the carrier to be constant. The spectral photograph also shows that the bandwidth of an AM signal is twice the highest frequency component of the modulating wave.

An AM signal cannot be frequency multiplied without special processing because the phase/frequency relationship of the modulating-waveform components would be severely distorted. For this reason, once an AM signal has been generated, its frequency can be changed only by heterodyning.

All of the information in an AM signal is contained in the sidebands, but two-thirds of the RF power is in the carrier. If the carrier is suppressed in the transmitter and reinserted (in the proper phase) in the receiver, significant advantages accrue. When the reinserted carrier is strong compared to the incoming double-sideband signal (DSB), exalted carrier reception is achieved, and distortion from selective fading is reduced greatly. A refinement called synchronous detection uses a PLL to reject interference. Suppressing the carrier also eliminates the heterodyne interference common with adjacent AM signals. More important, eliminating the carrier increases overall transmitter efficiency. Transmitter power requirements are reduced by 66%, and the remaining 34% has a light duty cycle.

Mathematics of AM

AM can happen at low levels (as in a driver or predriver stages of a transmitter) or high levels, in the final output stage. The numbers are consistent for both methods.

For example, to 100% modulate a 10 W RF carrier, 5 W of clean AF is required from the modulator. Good engineering calls for a 25% overdesign; thus 6.25 W of AF allows plenty of system headroom. The circuitry can then "loaf along" at the 100% level.

Overmodulation

Overmodulation occurs when more audio is impressed on a carrier than is needed for 100% modulation. It is also known as *flattopping*. Overmodulation causes distortion of the information conveyed and produces *splatter* (spurious emissions) on adjacent frequencies. Splatter interferes with others sharing our already-crowded bands: Prevent overmodulation at all times.

Years ago amateurs used ingenious ways to detect and prevent or control overmodulation. Nowadays with solid-state, large-scale integration (LSI) chips and microprocessor control, bullet-proof overmodulation prevention can be designed into transmitters. The most familiar method is called ALC, for automatic level control.

Although modern use of full-carrier AM on the amateur bands is very limited, there is a core group of AM aficionados experimenting with pulse duration modulation (PDM). This system was pioneered in AM broadcast transmitters.

This form of high-level modulation differs from conventional AM in that the PDM modulator operates in switching mode, with audio information contained during on-pulses. The audio amplitude is therefore determined by the duty cycle of the modulator or switching tube. Those interested in further reading on the topic should look in William Orr's (W6SAI) *Radio Handbook* (published by Howard W. Sams and Co) and the *AM Press/Exchange* (for contact information, see the Address List in the **References** chapter).

Balanced Modulators

The carrier can be suppressed or nearly eliminated by using a balanced modulator or an extremely sharp filter. Contemporary amateur transmitters often use both methods.

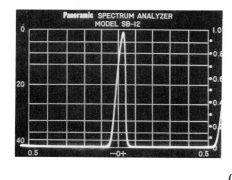
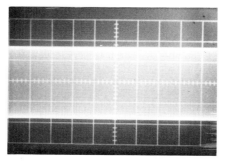

(A)

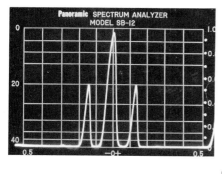
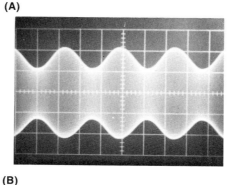

(B)

Fig 12.1—Electronic displays of AM signals in the frequency and time domains. A shows an unmodulated carrier or single-tone SSB signal. B shows a full-carrier AM signal modulated 20% with a sine wave.

Table 12.1
Guidelines for Amateur SSB Signal Quality

Parameter	Suggested Standard
Carrier suppression	At least 40-dB below PEP
Opposite-sideband suppression	At least 40-dB below PEP
Hum and noise	At least 40-dB below PEP
Third-order intermodulation distortion	At least 30-dB below PEP
Higher-order intermodulation distortion	At least 35-dB below PEP
Long-term frequency stability	At most 100-Hz drift per hour
Short-term frequency stability	At most 10-Hz P-P deviation in a 2-kHz bandwidth

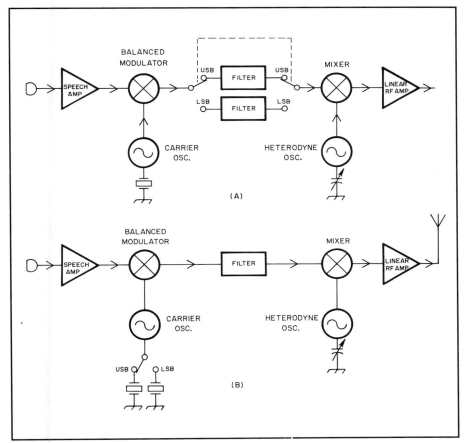

Fig 12.2—Block diagrams of filter-method SSB generators. They differ in the manner that the upper and lower sideband are selected.

The basic principle of any balanced modulator is to introduce the carrier in such a way that only the sidebands will appear in the output. The balanced-modulator circuit chosen by a builder depends on constructional considerations, cost and the active devices to be employed.

In any balanced-modulator circuit, there is (theoretically) no output when no audio signal is applied. When audio is applied, the balance is upset, and one branch conducts more than the other. Since any modulation process is the same as "mixing," sum and difference frequencies (sidebands) are generated. The modulator is not balanced for the sidebands, and they appear in the output.

SINGLE-SIDEBAND (SSB)

A further improvement in communications effectiveness can be obtained by transmitting only one of the sidebands. When the proper receiver bandwidth is used, a single-sideband (SSB) signal will show an effective gain of up to 9 dB over an AM signal of the same peak power. Because the redundant information is eliminated, the required bandwidth of an SSB signal is half that of a comparable AM or DSB emission. Unlike DSB, the phase of the local carrier generated in the receiver is unimportant. **Table 12.1** shows the qualities of a good SSB signal.

SSB Generation: The Filter Method

If the DSB signal from the balanced modulator is applied to a narrow band-pass filter, one of the sidebands can be greatly attenuated. Because a filter cannot have infinitely steep skirts, the response of the filter must begin to roll off within about 300 Hz of the phantom carrier to obtain adequate suppression of the unwanted sideband. This effect limits the ability to transmit bass frequencies, but those frequencies have little value in voice communications. The filter rolloff can be used to obtain an additional 20 dB of carrier suppression. The bandwidth of an SSB filter is selected for the specific application. For voice communications, typical values are 1.8 to 3.0 kHz.

Fig 12.2 illustrates two variations of the filter method of SSB generation. In A, the heterodyne oscillator is represented as a simple VFO, but may be a premixing system or synthesizer. The scheme at B is perhaps less expensive than that of A, but the heterodyne-oscillator frequency must be shifted when changing sidebands in order to maintain dial calibration.

The ultimate sense (erect or inverted) of the final output signal is influenced as much by the relationship of the heterodyne oscillator frequency to the fixed SSB frequency as by the filter or carrier frequency selection. The heterodyne-oscillator frequency must be chosen to allow the best image rejection. This consideration requires that the heterodyne-oscillator frequency be above the fixed SSB frequency on some bands and below it on others. To reduce circuit complexity, early amateur filter-method SSB transmitters used a 9-MHz IF and did not include a sideband switch. The result was that the output was LSB on 160, 75 and 40 m, and USB on the higher bands. This convention persists today, despite the flexibility of most modern amateur SSB equipment. Appropriate filtering methods and filters for SSB generation are discussed in the **Filters** chapter.

SSB Generation: The Phasing Method

Fig 12.3 shows another method to obtain an SSB signal. The audio and carrier signals are each split into equal components with a 90° phase difference (called *quadrature*) and applied to balanced modulators. When the DSB outputs of the modulators are combined, one sideband is reinforced and the other is canceled. The figure shows sideband selection by means of transposing the audio leads, but the

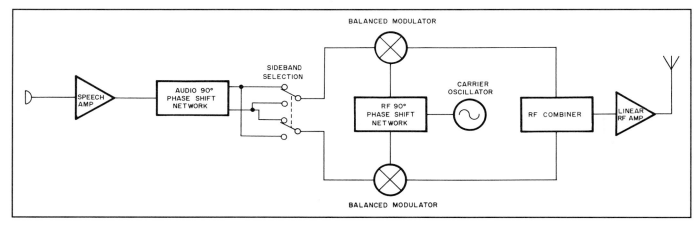

Fig 12.3—Block diagram of a phasing SSB generator.

same result can be achieved by switching the carrier leads.

The phase shift and amplitude balance of the two channels must be very accurate if the unwanted sideband is to be adequately attenuated. **Table 12.2** shows the required phase accuracy of one channel (AF or RF) for various levels of opposite sideband suppression. The numbers given assume perfect amplitude balance and phase accuracy in the other channel.

The table shows that a phase accuracy of $\pm 1°$ is required to satisfy the criteria tabulated at the beginning of this chapter. It is difficult to achieve this level of accuracy over the entire speech band. Note, however, that speech has a complex spectrum with a large gap in the octave from 700 to 1400 Hz. The phase-accuracy tolerance can be loosened to $\pm 2°$ if the peak deviations can be made to occur within that spectral gap.

The major advantage of the phasing system is that the SSB signal can be generated at the operating frequency without the need for heterodyning. Phasing can be used to good advantage even in fixed-frequency systems. A loose-tolerance ($\pm 4°$)

phasing exciter followed by a simple two-pole crystal filter can generate a high-quality signal at very low cost.

Audio Phasing Networks

It would be difficult to design a two-port network having a quadrature (90°) phase

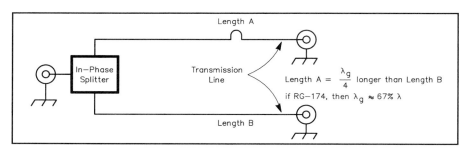

Fig 12.4—A simple RF phase shifter using transmission lines. It is practical at VHF and UHF. Examples: If L1 and L2 are made from RG-174, L1 is 69.1 inches longer than L2 at 28.5 MHz. 13.7 inches longer at 144.2 MHz and 8.86 inches longer at 222.1 MHz.

Table 12.2
Unwanted Sideband Suppression as a Function of Phase Error

Phase Error (deg.)	Suppression (dB)
0.125	59.25
0.25	53.24
0.5	47.16
1.0	41.11
2.0	35.01
3.0	31.42
4.0	28.85
5.0	26.85
10.0	20.50
15.0	16.69
20.0	13.93
30.0	9.98
45.0	6.0

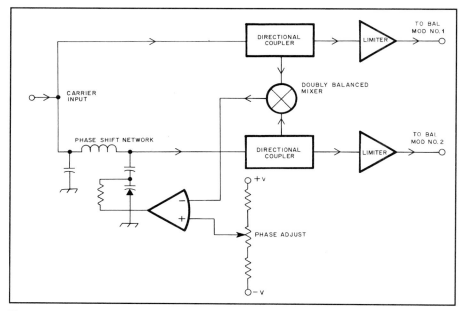

Fig 12.5—A block diagram of a PLL phase-shifting system that can maintain quadrature (90° phase difference) over a wide frequency range. The doubly balanced mixer is used as a phase detector.

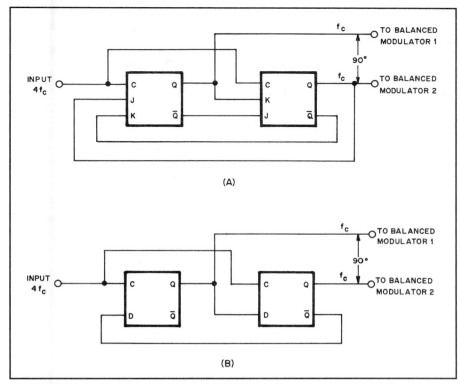

Fig 12.6—Digital RF phase-shift networks. Circuit A uses JK flip-flops; B uses D flip-flops. The carrier frequency must be quadrupled before processing.

relationship between input and output with constant-amplitude response over a decade of bandwidth. A practical approach, pioneered by Robert Dome, W2WAM, is to use two networks having a differential phase shift of 90°. This differential can be closely maintained in a simple circuit if precision components are used.

Numerous circuits have been developed to synthesize the required 90° phase shift electronically. Active-filter techniques are used in many of these systems; use precision components for good results.

RF Phasing Networks

If the SSB signal is to be generated at a fixed frequency, the RF phasing problem is trivial; any method that produces the proper phase shift can be used. If the signal is produced at the operating frequency, problems similar to those in the audio networks must be overcome.

A differential RF phase shifter is shown in **Fig 12.4**. The amplitudes of the quadrature signals won't be equal over an entire phone band, but this is of little consequence as long as the signals are strong enough to saturate the modulators.

Where percentage bandwidths are

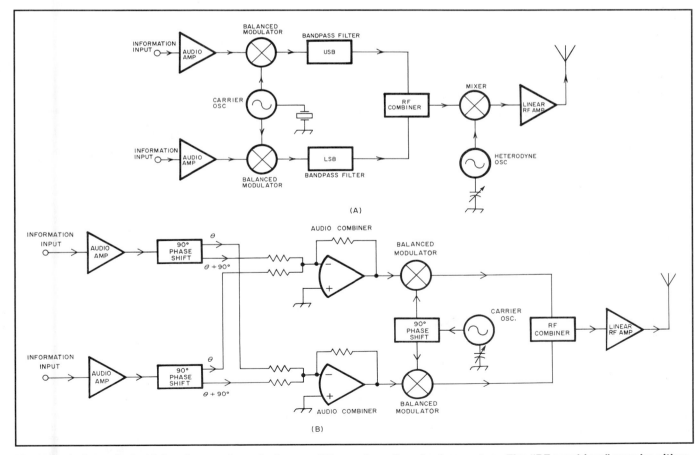

Fig 12.7—Independent-sideband generators. A shows a filter system, B a phasing system. The "RF combiner" may be either a hybrid combiner or a summing amplifier.

small, such as in the 144.1- to 145-MHz range, the RF phase shift can be obtained conveniently with transmission-line methods. If one balanced-modulator feed line is made an electrical quarter wavelength longer than the other, the two signals will be 90° out of phase. (It is important that the cables be properly terminated.)

One method for obtaining a 90° phase shift over a wide bandwidth is to generate the quadrature signals at a fixed frequency and heterodyne them individually to any desired operating frequency. Quadrature hybrids having multioctave bandwidths are manufactured commercially, by Mini-Circuits Labs and others.

Another practical approach uses two VFOs in a master-slave PLL system. Many phase detectors lock the two signals in phase quadrature. A doubly balanced mixer also has this property. One usually thinks of a PLL as having a VCO locked to a reference signal, but a phase differential can be controlled independently of the oscillator. The circuit in **Fig 12.5** illustrates this principle. Two digital phase shifters are sketched in **Fig 12.6**. If ECL ICs are used, this system can work over the entire HF spectrum.

Independent-Sideband (ISB)

If two SSB exciters, one USB and the other LSB, share a common carrier oscillator, two channels of information can be simultaneously transmitted from one antenna. Methods for ISB generation in filter and phasing transmitters are shown in **Fig 12.7**.

The most obvious amateur application for ISB is the transmission of SSTV with simultaneous audio commentary. On the VHF bands, other combinations are possible, such as voice and code or SSTV and RTTY.

Amplitude Compandored Single Sideband (ACSSB)

When SSB was tried in the Land Mobile Service, several problems arose. One was that users (who are not trained operators) couldn't master the control known as CLARIFIER to land-mobile users or receiver incremental tuning, RIT, to amateurs. In addition, users were annoyed by SSB's fading and noise performance, compared to that of FM.

So, to get the spectrum savings of SSB over FM, Land-Mobile Service engineers came up with a form of SSB that satisfied the users accustomed to FM. At present, there is almost no amateur use of this modulation.

FREQUENCY MODULATION (FM)

When the frequency of the carrier is varied in accordance with the variations in a modulating signal, the result is frequency modulation (FM). Varying the phase of the carrier current is called phase modulation (PM). Frequency and phase modulation are not independent, since the frequency cannot be varied without also varying the phase, and vice versa. This section was written by Dean Straw, N6BV.

The primary advantage of FM is its ability to produce a high signal-to-noise ratio when receiving a signal of only moderate strength. This has made FM popular for mobile communications services and high-quality broadcasting. However, because of the wide bandwidth required and the distortion suffered in skywave propagation, the use of FM has generally been limited to frequencies higher than 29 MHz.

When compared to AM or SSB, FM has some impressive advantages for VHF operation. In an FM transmitter, modulation takes place in a low-level stage. Amplifiers following the modulator can be operated Class C for best efficiency, since operation need not be linear. The frequency tolerances needed for channelized FM operation are much less severe than for SSB, helping to keep cost down.

The effectiveness of FM and PM for communication purposes depends almost entirely on the methods used for receiving. If the FM receiver responds to frequency and phase changes but is insensitive to amplitude changes, it can discriminate against many forms of noise.

Fig 12.8 is a representation of frequency modulation. When an audio modulating signal is applied, the carrier frequency is increased during one half cycle of the modulating signal and decreased during the half cycle of opposite polarity. In this figure RF cycles occupy less time (higher frequency) when the modulating signal is positive, and more time (lower frequency) when the modulating signal is negative. The change in the carrier frequency is called *frequency deviation* and is proportional to the instantaneous amplitude of the modulating signal. The deviation is small when the instantaneous amplitude of the modulating signal is small and is greatest when the modulating signal reaches its peak, either positive or negative.

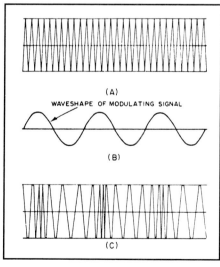

(A)

WAVESHAPE OF MODULATING SIGNAL

(B)

(C)

Fig 12.8—Graphical representation of frequency modulation. In the unmodulated carrier (A) each RF cycle occupies the same amount of time. When the modulating signal (B) is applied, the radio frequency is increased and decreased according to the amplitude and polarity of the modulating signal (C).

Phase Modulation (PM)

If the phase of the current in a circuit shifts there is an instantaneous frequency change during the time the phase is shifting. The amount of frequency change is directly proportional to how rapidly the phase is shifting and to the total amount of the phase shift. The rapidity of the phase shift is directly proportional to the frequency of the modulating signal. In a properly operating PM system, the amount of phase shift is proportional to the instantaneous amplitude of the modulating signal. Phase modulators have a built-in *pre-emphasis*, where deviation increases with modulating frequency.

FM AND PM SIDEBANDS

The sidebands generated by FM and PM occur at integral multiples of the modulating frequency on either side of the carrier. This is in contrast to AM, where a modulating frequency will produce a single set of sidebands on either side of the carrier frequency. An FM or PM signal therefore inherently occupies a wider channel than AM. The number of additional sidebands that occur in FM and PM depend on the relationship between the modulating frequency and the frequency deviation. The ratio between the frequency deviation, in hertz, and the modulating frequency, also in hertz, is called the *modulation index*.

$$\chi = \frac{D}{m} = \phi \qquad (1)$$

where

χ = modulation index
D = peak deviation ($1/2$ difference between maximum and minimum frequency)
m = modulation frequency in hertz
ϕ = phase deviation in radians (a radian = $180°/\pi$ or approximately $57.3°$).

For example, the maximum frequency deviation in an FM transmitter is 3000 Hz either side of the carrier frequency. The modulation index when the modulation frequency is 1000 Hz is 3000/1000 = 3.0. At the same deviation with 3000 Hz modulation, the index would be 1; at 100 Hz it would be 30 and so on.

Given a constant input level to the modulator, in PM the modulation index is constant regardless of the modulating frequency. In FM it varies with the modulating frequency, as shown above. In an FM system the ratio of the maximum carrier-frequency deviation to the highest modulating frequency used is called the *deviation ratio*. Thus

$$\text{deviation ratio} = \frac{D}{M} \qquad (2)$$

where

D = peak deviation

M = maximum modulation frequency in hertz.

The deviation ratio used above 29 MHz for narrow-band FM is 5000 Hz (maximum deviation) divided by 3000 Hz (maximum modulating frequency) or 1.67.

Fig 12.9 shows how the amplitudes of the carrier and the various sidebands vary with the modulation index for single-tone modulation. The first pair of sidebands are displaced from the carrier by an amount equal to the modulating frequency, the second by twice the modulating frequency, and so on. For example, if the modulating frequency is 2000 Hz and the carrier frequency is 29,500 kHz, the first sideband pair is at 29,498 and 29,502 kHz, the second pair is at 29,496 and 29,504 kHz, the third at 29,494 and 29,506 kHz, and so on. The amplitudes of these sidebands depend on the modulation index, not on the frequency deviation.

The carrier strength varies with the modulation index—at a modulation index of 2.405, the carrier disappears entirely. As the index is raised further, the carrier level becomes negative, since its phase is reversed compared to the phase without modulation. In FM and PM the energy going into the sidebands is taken from the carrier—the total power remains the same regardless of the modulation index. Since there is no change in amplitude with modulation, an FM or PM signal can be amplified without distortion by an ordinary Class-C amplifier, either as a straight-through amplifier or frequency-multiplier stage.

If the modulated signal is passed through one or more frequency multipliers, the modulation index is multiplied by the same factor as the carrier frequency. For example, if modulation is applied on 3.5 MHz and the final output is on 28 MHz, the total frequency multiplication is eight times. If the frequency deviation is 500 Hz at 3.5 MHz, it will be 4000 Hz at 28 MHz. Frequency multiplication offers a means for obtaining practically any desired amount of frequency deviation, whether or not the modulator itself is capable of giving that much deviation without distortion.

If the modulation index (with single-tone modulation) does not exceed 0.6 or 0.7, the most important extra sideband, the second, will be at least 20 dB below the unmodulated carrier level. This represents

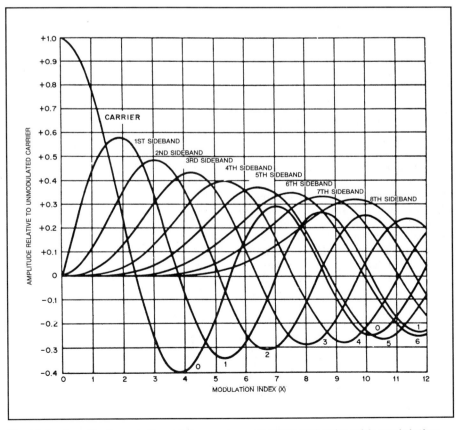

Fig 12.9—Amplitude variation of the carrier and sideband pairs with modulation index. This is a graphical representation of mathematical functions developed by F. W. Bessel. Note that the carrier completely disappears at modulation indexes of 2.405, 5.52 and 8.654.

an effective channel width about equal to an AM signal. The energy in speech is distributed among many audio frequencies. On average, the modulation index for any one frequency component is smaller than that for a single audio tone having the same peak amplitude. Thus, the effective modulation index for speech can be somewhat higher while retaining the same average bandwidth. The rule-of-thumb for determination of bandwidth requirements for an FM system using *narrow-band* (5 kHz deviation) modulation is

$$B_n = 2(M+D) \qquad (3)$$

where

 B_n = necessary bandwidth in hertz
 M = maximum modulation frequency in hertz
 D = peak deviation in hertz.

For narrow-band FM, the bandwidth equals $2 \times (3000 + 5000) = 16000$ Hz. Additional bandwidth may be needed to compensate for cumulative errors in the transmitter and receiver frequencies.

FM vs Phase Modulation (PM)

FM cannot be applied to an amplifier stage, but phase modulation (PM) can. PM is therefore readily adaptable to transmitters employing oscillators of high stability, such as the crystal-controlled oscillators. The amount of phase shift that can be obtained with good linearity yields a maximum practicable modulation index of about 0.5. Because phase shift is proportional to the modulating frequency, this index can be used only at the highest frequency present in the modulating signal, assuming that all frequencies will at one time or another have equal amplitudes.

The frequency response of the speech-amplifier system above 3000 Hz must be sharply attenuated to prevent splatter on adjacent channels. Due to its inherent preemphasis, PM received on an FM receiver sounds "tinny." The audio must be processed for PM to have the same modulation-index characteristic as an FM signal. The speech-amplifier frequency-response curve is thus shaped so the output voltage is inversely proportional to frequency over most of the voice range. When this is done the maximum modulation index can only be used below a relatively low audio frequency, perhaps 300 to 400 Hz in voice transmission, and must decrease in proportion to an increase in frequency. The net result is that the maximum linear frequency deviation is only one or two hundred hertz. In order to increase the deviation up to narrowband level, we must typically multiply the frequency by eight or more.

GENERATING FM

Direct FM

A simple circuit for producing *direct FM* in amateur transmitters is the *reactance modulator*. An active device is connected to the RF tank circuit of an oscillator to act as a variable inductance or capacitance. **Fig 12.10A** is a representative circuit using a MOSFET. This modulator acts as though an inductance were connected across the tank. The frequency increases in proportion to the amplitude of the current in this modulator. If the modulated oscillator is free running, it must usually be operated on a relatively low frequency to maintain good carrier stability. Fig 12.10B shows how a varactor may be used to FM a crystal oscillator directly. The supply voltage for either modulator and oscillator should be regulated to reduce residual FM. The oscillator frequency is multiplied up to the final output frequency.

In many modern frequency-synthesized transceivers, a VCO used in one of the phase-locked loops (PLL) is often frequency modulated directly. A PLL consists of a phase detector, a filter, a dc amplifier and a voltage-controlled oscillator (VCO). See **Fig 12.11**. The VCO runs at a frequency close to that desired when the loop is in lock. The phase detector produces an error voltage if any frequency difference exists between the VCO divided by the variable divider N and the reference signal. The error voltage is applied to the VCO to keep it locked on the carrier frequency when there is no modulation present. The loop bandwidth of the PLL is made narrow enough so that the audio can change the VCO frequency, while the PLL still keeps the unmodulated carrier frequency on-channel.

Indirect FM

The same type of reactance-modulator circuit used to vary the tuning of an oscil-

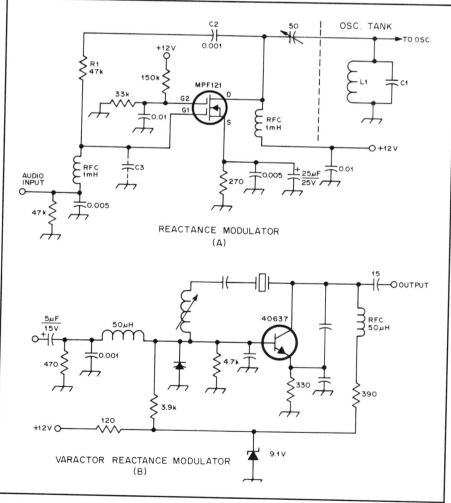

Fig 12.10—At A, reactance modulator using a high-transconductance MOSFET. At B, reactance modulator using a varactor diode.

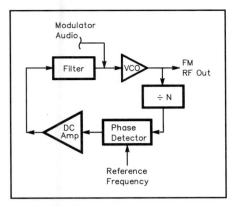

Fig 12.11—Simple phase-locked loop (PLL) where VCO is FMed directly. The loop filter is designed to be narrow enough so that the loop will lock onto the desired channel frequency, while audio frequencies will modulate the VCO outside the loop bandwidth.

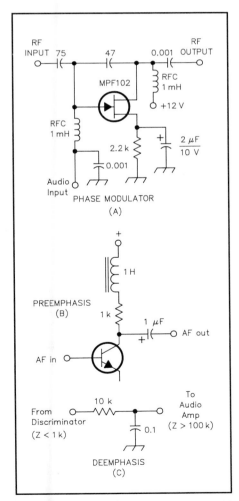

Fig 12.12— At A, a phase-shifter type of phase modulator. At B, preemphasis and at C, deemphasis circuits.

lator tank in direct FM can be used to vary the tuning of an amplifier tank. See **Fig 12.12A**. This varies the phase of the tank current to create phase modulation. When audio shaping is used in the speech amplifier, an FM-compatible signal will be generated by the phase modulator. The phase shift that occurs when a circuit is detuned from resonance depends on the amount of detuning and the Q of the circuit. The higher the Q, the smaller the amount of detuning needed to secure a given number of degrees of phase shift. Since reactance modulation of an amplifier stage results in simultaneous amplitude modulation, this must be eliminated using succeeding Class-C limiting stages.

Speech Processing for FM

Several forms of speech processing produce worthwhile improvements in FM system performance. The peak amplitude of the audio signal applied to an FM or PM modulator should be limited so that transmitter cannot be driven into overdeviation. Peak limiting is often maintained using a simple audio clipper between the speech amplifier and modulator. An audio low-pass filter with a cut-off frequency between 2.5 and 3 kHz eliminates harmonics produced by the clipper. Since excessive clipping can cause severe distortion of a voice signal, a more effective audio processor consists of a compressor followed by a clipper and low-pass filter.

An audio shaping network called *preemphasis* is added to an FM transmitter to attenuate the lower audio frequencies, spreading out the energy evenly in the audio band. Preemphasis applied to an FM transmitter gives the deviation characteristic of PM. The reverse process, called *deemphasis*, is used at the receiver to restore the audio to its original relative proportions. See Fig 12.12B and C.

RECEPTION OF FM SIGNALS

A block diagram of an FM receiver is shown in **Fig 12.13B**. The FM receiver employs a wide-bandwidth filter and an FM detector, and has one or more limiter stages between the IF amplifier and the FM detector. The limiter and discriminator stages in an FM set can eliminate a good deal of impulse noise, except noise that manages to acquire a frequency-modulation characteristic.

FM receivers exhibit a characteristic known as the *capture effect* when QRM is present. The loudest signal received, even if it is only two or three times stronger than other stations on the same frequency, will be the only transmission demodulated.

Limiters

The circuit in the FM receiver that has the task of chopping noise and amplitude modulation from an incoming signal is the *limiter*. Most types of FM detectors respond to both frequency and amplitude variations of the signal. Thus, the limiter stages preceding the detector are included so only the desired frequency modulation

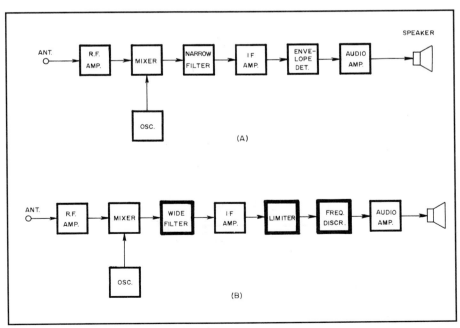

Fig 12.13—At A, block diagram of an AM receiver. At B, an FM receiver. Dark borders outline the sections that are different in the FM set.

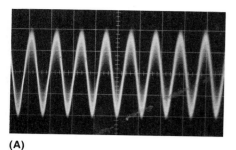

(A)

(B)

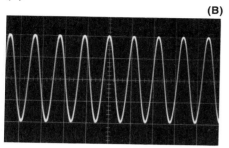

Fig 12.14—At A, input wave form to a limiter stage shows AM and noise. At B, the same signal, after passing through two limiter stages, is devoid of AM components.

will be demodulated. This action can be seen in **Fig 12.14**.

For an amplifier to act as a limiter, the applied voltages are chosen so that the stage overloads predictably, even with a small amount of signal input. Limiting action in an FM receiver should start with an RF input of 0.2 µV or less, so a large amount of gain is required between the antenna terminal and the limiter stages. ICs offer simplification of the IF system, as they pack a lot of gain into a single package.

When sufficient signal arrives at the receiver to start limiting action, the set *quiets*—that is, the background noise disappears. The sensitivity of an FM receiver is rated in terms of the amount of input signal required to produce a given amount of quieting, usually 20 dB. Modern receivers achieve 20 dB quieting with 0.15 to 0.5 µV of input signal.

Fig 12.15A shows a two-stage limiter using discrete transistors. The base bias on either transistor may be varied to provide limiting at a desired level. The input-signal voltage required to start limiting

action is called the *limiting knee*. This refers to the point at which collector current ceases to rise with increased input signal. Modern ICs have limiting knees of 100 mV for the circuit shown in Fig 12.15B, using the RCA CA3028A or Motorola MC1550G, or 200 mV for the MC1590G

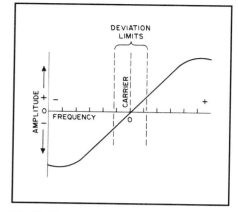

Fig 12.16—The characteristic of an FM discriminator.

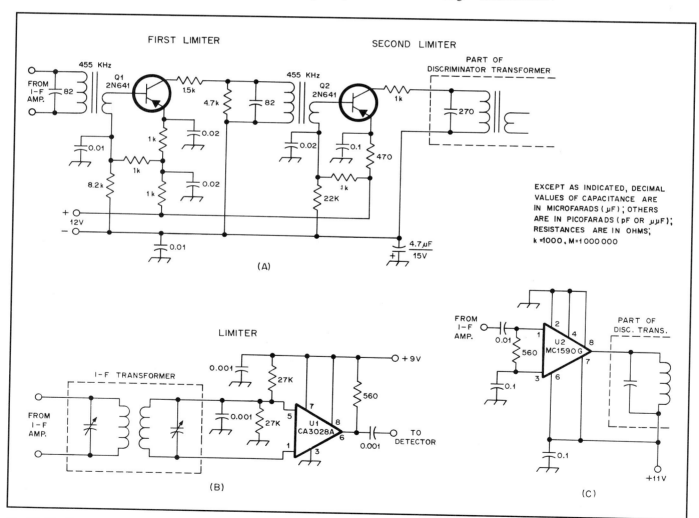

Fig 12.15—Typical limiter circuits using (A) transistors, (B) a differential IC, (C) a high-gain linear IC.

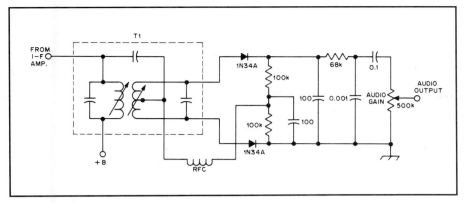

Fig 12.17—Typical frequency-discriminator circuit used for FM detection. T1 is a Miller 12-C45.

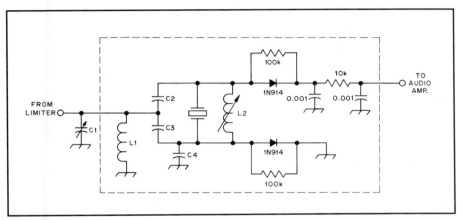

Fig 12.18—A crystal discriminator, C1 and L1 are resonant at the IF. C2 is equal in value to C3. C4 corrects any circuit imbalance so equal amounts of signal are fed to the detector diodes.

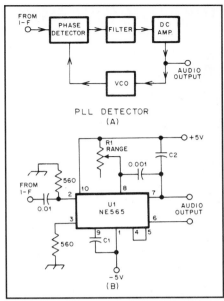

Fig 12.19—At A, block diagram of a PLL demodulator. At B, complete PLL circuit.

of Fig 12.15C. Because high-gain ICs contain many active stages a single IC can provide superior limiting performance compared to most discrete designs.

Detectors

The first FM detector to gain popularity was the *frequency discriminator*. The characteristic of such a detector is shown in **Fig 12.16**. When the FM signal has no modulation, and the carrier is at point 0, the detector has no output. When audio input to the FM transmitter swings the signal higher in frequency, the rectified output increases in the positive direction. When the frequency swings lower, the output amplitude increases in the negative direction. Over a range where the discriminator is linear (shown as the straight portion of the line), the conversion of FM to AM will also be linear. A practical discriminator circuit is shown in **Fig 12.17**.

Other Detector Designs

The difficulties often encountered in building and aligning LC discriminators have inspired research that has resulted in a number of adjustment-free FM detector designs. The crystal discriminator utilizes a quartz resonator, shunted by an inductor, in place of the tuned-circuit secondary used in a discriminator transformer. A typical circuit is shown in **Fig 12.18**.

The PLL

The *phase-locked loop* (PLL) has made a significant impact on transmitter and receiver design, both for frequency generation and for modulation/demodulation. It can act as an FM detector in a process similar to that used for direct-frequency modulation in a transmitter PLL. As the VCO tracks the frequency of an incoming signal, the voltage at the phase detector output becomes demodulated audio. See **Fig 12.19**.

Text (Digital) Modes

MORSE TELEGRAPHY (CW)

Telegraphy by on-off keying (OOK, or amplitude-shift keying ASK) of a carrier is the oldest radio modulation system. It is also known as CW (for continuous wave). While CW is used by amateurs and other communicators to mean OOK telegraphy by Morse code, parts of the electronics industry use CW to signify an unmodulated carrier.

This discussion centers on aural reception of CW, but computers are used to send and receive CW as well. A table of characters and their Morse equivalents appears in the **References** chapter.

TRANSMITTING

WPM vs Bauds

The speed of Morse telegraphy is usually expressed in WPM, rather than bauds, which are the common measure in other digital modes. The following formulas relate WPM to bauds:

$$WPM = 2.4 \times dot/s \qquad (4)$$
$$WPM = 1.2 \times B \qquad (5)$$

where

WPM = telegraph speed in words per minute

2.4 = a constant calculated by comparing dots per second with plain language Morse code sending the word "PARIS"

1.2 = a constant calculated by comparing the signaling rate in bauds with plain-language Morse code sending the word "PARIS"

B = telegraph speed in bauds.

Thus a keying speed of 25 dot/s or 50 bauds is equal to 60 WPM.

Rise Time vs Bandwidth

Keying a carrier on and off produces double (upper and lower) sidebands corresponding to the period of the keying pulse. A string of dits at 50 baud will have sidebands at multiples of 25 Hz above and below the carrier. The rise time of the pulses affects the distribution of power among the sidebands. As rise time increases or pulse rate decreases, the bandwidth of the signal decreases. In addition, the rise time affects how our ears hear the signal.

League publications have long promoted 5-ms rise and fall times for CW keying envelopes (see **Fig 12.20**). This shape is based on an assumed necessary bandwidth of 150 Hz for a 60-WPM (50-

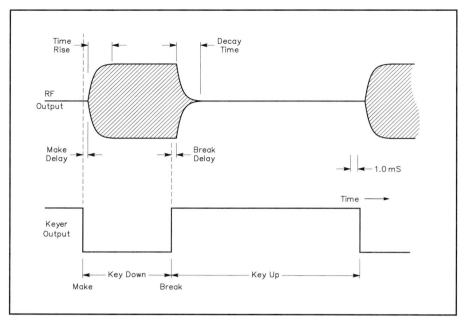

Fig 12.20—Optimum CW keying waveforms.

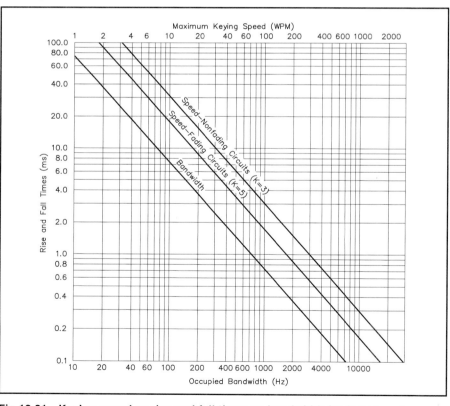

Fig 12.21—Keying speed vs rise and fall times vs bandwidth for fading and nonfading communications circuits. For example, to optimize transmitter timing for 25 WPM on a nonfading circuit, draw a vertical line from the WPM axis to the K = 3 line. From there draw a horizontal line to the rise/fall time axis (approximately 15 ms). Draw a vertical line from where the horizontal line crosses the bandwidth line and see that the bandwidth will be about 60 Hz.

Table 12.3
Baudot Signaling Rates and Speeds

Signaling Rate (bauds)	Data Pulse (ms)	Stop Pulse (ms)	Speed (WPM)	Common Name
45.45	22.0	22.0	65.00	Western Union
	22.0	31.0	61.33	"60 speed"
	22.0	33.0	60.61	45 bauds
50.00	20.0	30.0	66.67	European; 50 bauds
56.92	17.57	25.00	76.68	"75 speed"
	17.57	26.36	75.89	57 bauds
74.20	13.47	19.18	100.00	"100 speed"
	13.47	20.21	98.98	74 bauds
100.0	10.00	15.00	133.33	100 bauds

baud) pulse train and an equation relating necessary CW bandwidth to keying speed in Appendix 6 of the CCIR Radio Regulations and §2.202 of the FCC rules. The relationship is shown in **Fig 12.21.**

K is part of the bandwidth equation. Low K values produce softer keying, while high values sound harder. The CCIR and FCC recommend K = 3 for nonfading circuits and K = 5 for fading circuits with aural reception. K = 3, is the minimum used for comfortable aural reception, but K = 1 is useful for machine recognition.

Given a 150-Hz bandwidth, how fast can we communicate over a fading path? With an occupied bandwidth of 150 Hz and K = 5, Fig 12.21 yields 36 WPM. Therefore, 5-ms rise and fall times are suitable for up to 36 WPM on fading circuits and 60 WPM on nonfading circuits.

RECEIVING

For aural reception, a Morse-code OOK RF signal is not completely demodulated to its original dc pulse, because only thumping would be heard. Instead, the signal is moved (by mixing) down to AF, usually near 700 Hz.

Proper reception of a Morse-code transmission requires that the receiver bandwidth be at least that of the necessary bandwidth plus any frequency error. Thus, if you have 150-Hz receiver bandwidth, it would be necessary for you to carefully tune your receiver to receive a 150-Hz-bandwidth transmission. In practice, it is common to use 500 or 250-Hz IF filters.

Many operators find that it is easier to distinguish between multiple signals as the frequency of the desired signal is lowered to 500 Hz or less. Some modern transceivers provide a CW OFFSET adjustment to accommodate this preference. The same result can be achieved by adjusting the RIT control, although the audio from incoming signals will no longer match the sidetone with this technique.

Those who desire a narrower bandwidth often use audio filters, either an op-amp audio-peak filter or a low-pass switched-capacitor. Such filters may be part of the radio or added as accessories. Look in the **Filters** chapter for projects.

BAUDOT (ITA2) RADIOTELETYPE

The Baudot Code: ITA2

One of the first data communications codes to receive widespread use had five bits (traditionally called "levels") to present the alphabet, numerals, symbols and machine functions. In the US, we use International Telegraph Alphabet No. 2 (ITA2), commonly called *Baudot*, as specified in FCC §97.309(a)(1). The code is defined in the ITA2 Codes table in the **References** chapter. In Great Britain, the almost-identical code is called *Murray* code. There are many variations in five-bit coded character sets, principally to accommodate foreign-language alphabets.

Five-bit codes can directly encode only $2^5 = 32$ different symbols. This is insufficient to encode 26 letters, 10 numerals and punctuation. This problem can be solved by using one or more of the codes to select from multiple code-translation tables. ITA2 uses a LTRS code to select a table of upper-case letters and a FIGS code to select a table of numbers, punctuation and special symbols. Certain symbols, such as carriage return, occur in both tables. Unassigned ITA2 FIGS codes may be used for the remote control of receiving printers. This scheme can be expanded, as shown by the ASCII-over-AMTOR discussion latter in this chapter.

FCC rules provide that ITA2 transmissions must be sent using start-stop pulses as illustrated in **Fig 12.22.** The bits in the figure are arranged as they would appear on an oscilloscope.

Speeds and Signaling Rates

The signaling speeds for all forms of RTTY are those used by the old TTYs: 60, 67, 75 or 100 WPM. **Table 12.3** relates speeds, signaling rates and pulse times. In practice, the real speeds do not exactly match their names. The names have been rounded through years of common usage. The Signaling Rates table in the **References** chapter lists names, signaling rates and data patterns for common RTTY speeds.

There's a problem specifying signaling speed of RTTY because the length of the start and stop pulses vary from that of the data bits. The answer is to base the signaling speed on the shortest pulses used. The *baud* is a unit of signaling speed equal to one pulse (event) per second. The signaling rate, in bauds, is the reciprocal of the shortest pulse length. For example, the "Western Union," "60 speed" and "45 bauds" speeds all signal at 1/0.022 = 45.45 bauds.

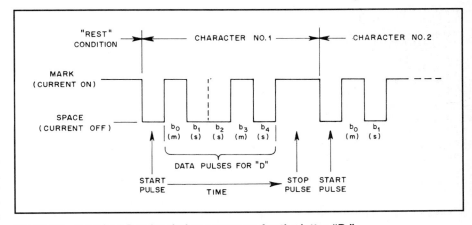

Fig 12.22—A typical Baudot timing sequence for the letter "D."

Transmitter Keying

When TTYs and TUs (terminal units) roamed the airwaves, frequency-shift keying (FSK) was the order of the day. DC signals from the TU controlled some form of reactance (usually a capacitor or varactor) in a transmitter oscillator stage that shifted the transmitter frequency. Such direct FSK is still an option with some new radios.

AFSK

Multimode communications processors (MCPs), however, generally connect to the radio AF input and output, often through the speaker and microphone connectors, sometimes through auxiliary connectors. They simply feed AF tones to the microphone input of an SSB transmitter or transceiver. This is called AFSK for "audio frequency-shift keying." When it is properly designed and adjusted, this method of modulation cannot be distinguished from FSK on the air.

When using AFSK, make certain that audio distortion, carrier and unwanted sidebands do not cause interference. Particularly when using the low tones discussed later, the harmonic distortion of the tones should be kept to a few percent. Most modern AFSK generators are of the continuous-phase (CPFSK) type. Older types of noncoherent-FSK (NCFSK) generators had no provisions for phase continuity and produced sharp switching transients. The noise from phase discontinuity caused interference several kilohertz around the RTTY signal.

Also remember that equipment is withstanding a 100% duty cycle for the duration of a transmission. For safe operation, it is often necessary to reduce the transmitter power output (25 to 50% of normal) from that safe for CW operation.

What are Low Tones?

US amateurs customarily use the same modems (2125 Hz mark, 2295 Hz space) for both VHF AFSK and HF via an SSB transmitter. Because of past problems (when 850-Hz shift was used), some amateurs use "low tones" (1275 Hz mark, 1445 Hz space). Both high and low tones can be used interchangeably on the HF bands because only the *amount* of shift is important. The frequency difference is unnoticed on the air because each operator tunes for best results. On VHF AFSK, however, the high and low tone pairs are not compatible.

Transmit Frequency

It is normal to use the lower sideband mode for RTTY on SSB radio equipment.

In order to tune to an exact RTTY frequency, remember that most SSB radio equipment displays the frequency of its (suppressed) carrier, not the frequency of the mark signal. Review your MCP's manual to determine the tones used and calculate an appropriate display frequency. For example, to operate on 14,083 kHz with a 2125-Hz AFSK mark frequency, the SSB radio display (suppressed-carrier) frequency should be 14,083 kHz + 2.125 kHz = 14,085.125 kHz.

Receiving Baudot

Surplus Baudot-encoded teletypewriters (TTY, sometimes called the "green keys") were the mainstay of amateur RTTY from 1946 through around 1977. There are still some mechanical-TTY aficionados, but most operators use computer-based terminals.

Some of the first popular home computers (VIC-20, Commodore 64, Apple II) were adapted to read signals from "terminal units" or "TUs" required by TTYs. TUs translated receiver AF output into 20-mA current-loop signals to drive a polar relay in a TTY. An interface would translate the current-loop signals (or sometimes the receiver AF) to levels appropriate for the computer. Software, unique to each computer, would then decode the stream of marks and spaces into text. This technology was convoluted in that it required many different interfaces and software packages to suit the computers in use. Thankfully, it was soon replaced by multimode communications processors.

MCPs accept AF signals from a radio and translate them into common ASCII text or graphics file formats (see **Fig 12.23**). Because the basic interface is via ASCII,

MCPs are compatible with virtually any PC running a simple terminal program. There may be compatibility problems with graphics formats, but those are fairly well standardized. Many MCPs handle CW, RTTY, ASCII, AMTOR, packet, fax and SSTV—multimode indeed!

AFSK Demodulators

An AFSK demodulator takes the shifting tones from the audio output of a receiver and produces TTY keying pulses. FM is a common AFSK demodulation method. The signal is first band pass filtered to remove out-of-band interference and noise. It is then limited to remove amplitude variations. The signal is demodulated in a discriminator or a PLL. The detector output low pass filtered to remove noise at frequencies above the keying rate. The result is fed to a circuit that determines whether it is a mark or a space.

AM (limiterless) detectors, when properly designed, permit continuous copy even when the mark or space frequency fades out completely. At 170-Hz shift, however, the mark and space frequencies tend to fade at the same time. For this reason, FM and AM demodulators are comparable at 170-Hz shift.

At wider shifts (say 425 Hz and above), the independently fading mark and space can be used to achieve an in-band frequency-diversity effect if the demodulator is capable of processing it. To conserve spectrum, it is generally desirable to stay with 170-Hz shift for 45-baud Baudot and forego the possible in-band frequency-diversity gain. Keep the in-band frequency-diversity gain in mind, however, for higher signaling rates that would justify greater shift.

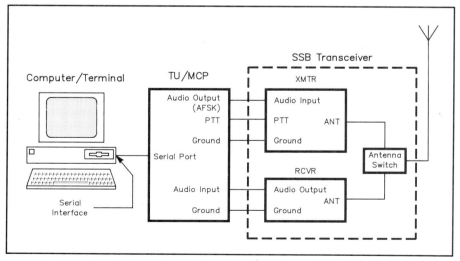

Fig 12.23—A typical MCP station. MCPs can do all available data modes as well as SSTV and fax.

Diversity Reception

Another type of diversity can be achieved by using two antennas, two receivers and a dual demodulator. This setup is not as far fetched as it may sound; some amateurs are using it with excellent results. One of the antennas would be the normal station antenna for that band. The second antenna could be either another antenna of the same polarization located at least 3/8-wavelength away, or an antenna of the opposite polarization located at the first antenna or anywhere nearby. A problem is to get both receivers on the same frequency without carefully tuning each one. Some RTTY diversity enthusiasts have located slaved receivers on the surplus market. ICOM produced the IC-7072 Transceiver Unit, which slaves an IC-720(A) transceiver to an IC-R70 receiver. Other methods could include a computer controlling two receivers so that both would track.

Two demodulators are needed for this type of diversity. Also, some type of diversity combiner or selector is needed. Many commercial or military RTTY demodulators are equipped for diversity reception.

The payoff for using diversity is a worthwhile improvement in copy. Depending on fading conditions, adding diversity may be equivalent to raising transmitter power sevenfold (8 dB).

Baudot RTTY Bibliography

Contact information for suppliers named here appears in the **References** chapter Address List.

Ford, *Your RTTY/AMTOR Companion* (Newington, CT: ARRL, 1993).

DATACOM, British Amateur Radio Teleprinter Group (BARTG).

Henry, "Getting Started in Digital Communications," Part 3, *QST*, May 1992.

Hobbs, Yeomanson and Gee, *Teleprinter Handbook*, Radio Society of Great Britain.

Ingram, *RTTY Today*, Universal Electronics, Inc.

Kretzman, *The New RTTY Handbook*, Cowan Publishing Corp.

Nagle, "Diversity Reception: an Answer to High Frequency Signal Fading," *Ham Radio*, Nov 1979, pp 48-55.

RTTY Journal.

Schwartz, "An RTTY Primer," *CQ* magazine, Aug 1977, Nov 1977, Feb 1978, May 1978 and Aug 1978.

Tucker, *RTTY from A to Z*, Cowan Publishing Corp.

GLOSSARY OF DIGITAL COMMUNICATIONS TERMINOLOGY

ACK—Acknowledgment, the control signal sent to indicate the correct receipt of a transmission block.

Address—A character or group of characters that identifies a source or destination.

AFSK—Audio frequency-shift keying.

ALOHA—A channel-access technique wherein each packet-radio station transmits without first checking to see if the channel is free; named after early packet-radio experiments at the University of Hawaii.

AMICON—AMSAT International Computer Network—Packet-radio operation on SSC L1 of AMSAT-OSCAR 10 to provide networking of ground stations acting as gateways to terrestrial packet-radio networks.

AMRAD—Amateur Radio Research and Development Corporation, a nonprofit organization involved in packet-radio development.

AMTOR—Amateur teleprinting over – radio, an amateur radioteletype transmission technique employing error correction as specified in several CCIR documents 476-2 through 476-4 and 625. CCIR Rec. 476-3 is reprinted in the *Proceedings* of the Third ARRL Amateur Radio Computer Networking Conference, available from ARRL Hq.

ANSI—American National Standards Institute

Answer—The station intended to receive a call. In modem usage, the called station or modem tones associated therewith.

ARQ—Automatic repeat request, an error-sending station, after transmitting a data block, awaits a reply (ACK or NAK) to determine whether to repeat the last block or proceed to the next.

ASCII—American National Standard Code for Information Interchange, a code consisting of seven information bits.

AX.25—Amateur packet-radio link-layer protocol. Copies of protocol specification are available from ARRL Hq.

Backwave—An unwanted signal emitted between the pulses of an on/off-keyed signal.

Balanced—A relationship in which two stations communicate with one another as equals; that is, neither is a primary (master) or secondary (slave).

Baud—A unit of signaling speed equal to the number of discrete conditions or events per second. (If the duration of a pulse is 20 ms, the signaling rate is 50 bauds or the reciprocal of 0.02, abbreviated Bd).

Baudot code—A coded character set in which five bits represent one character. Used in the US to refer to ITA2.

Bell 103—A 300-baud full-duplex modem using 200-Hz-shift FSK of tones centered at 1170 and 2125 Hz.

Bell 202—A 1200-baud modem standard with 1200-Hz mark, 2200-Hz space, used for VHF FM packet radio.

BER—Bit error rate.

BERT—Bit-error-rate test.

Bit stuffing—Insertion and deletion of 0s in a frame to preclude accidental occurrences of flags other than at the beginning and end of frames.

Bit—Binary digit, a single symbol, in binary terms either a one or zero.

BLER—Block error rate.

BLERT—Block-error-rate test.

Break-in—The ability to hear between elements or words of a keyed signal.

Byte—A group of bits, usually eight.

Carrier detect (CD)—Formally, received line signal detector, a physical-level interface signal that indicates that the receiver section of the modem is receiving tones from the distant modem.

CCIR Rec 476-4—The CCIR Recommendation used as the basis of AMTOR and incorporated by reference into the FCC Rules.

CCIR—International Radio Consultative Committee, an International Telecommunication Union (ITU) agency.

CCITT—International Telegraph and Telephone Consultative Committee, an ITU agency. CCIR and CCITT recommendations are available from the UN Bookstore.

Chirp—Incidental frequency modulation of a carrier as a result of oscillator instability during keying.

Collision—A condition that occurs when two or more transmissions occur at the same time and cause interference to the intended receivers.

Connection—A logical communication channel established between peer levels of two packet-radio stations.

Contention—A condition on a communications channel that occurs when two or more stations try to transmit at the same time.

Control field—An 8-bit pattern in an HDLC frame containing commands or responses, and sequence numbers.

CRC—Cyclic redundancy check, a mathematical operation. The result of the CRC is sent with a transmission block. The receiving station uses the received CRC to check transmitted data integrity.

CSMA—Carrier sense multiple access, a channel access arbitration scheme in which packet-radio stations listen on a channel for the presence of a carrier before transmitting a frame.

CTS—clear to send, a physical-level interface circuit generated by the DCE that, when on, indicates the DCE is ready to receive transmitted data (abbreviated CTS).

Cut numbers—In Morse code, shortening of codes sent for numerals.

DARPA—Defense Advanced Research Projects Agency; formerly ARPA, sponsors of ARPANET.

Data set—Modem.

Datagram—A mode of packet networking in which each packet contains complete addressing and control information. (compare virtual circuit).

DCE—Data circuit-terminating equipment, the equipment (for example, a modem) that provides communication between the DTE and the line radio equipment.

Destination—In packet radio, the station that is the intended receiver of the frame sent over a radio link either directly or via a repeater.

Digipeater—A link-level gateway station capable of repeating frames. The term "bridge" is used in industry.

Domain—In packet radio, the combination of a frequency and a geographical service area.

DTE—Data terminal equipment, for example a VDU or teleprinter.

DXE—In AX.25, Data switching equipment, a peer (neither master nor slave) station in balanced mode at the link layer.

EASTNET—A series of digipeaters along the US East Coast.

EIA—Electronic Industries Association.

EIA-232-C—An EIA standard physical-level interface between DTE (terminal) and DCE (modem), using 25-pin connectors.

Envelope-delay distortion—In a complex waveform, unequal propagation delay for different frequency components.

Equalization—Correction for amplitude-frequency and/or phase-frequency distortion.

Eye pattern—An oscilloscope display in the shape of one or more eyes for observing the shape of a serial digital stream and any impairments.

FADCA—Florida Amateur Digital Communications Association.

FCS—Frame check sequence. (See CRC.)

FEC—Forward error correction, an error-control technique in which the transmitted data is sufficiently redundant to permit the receiving station to correct some errors.

Field—In packet radio, at the link layer, a subdivision of a frame, consisting of one or more octets.

Flag—In packet switching, a link-level octet (01111110) used to initiate and terminate a frame.

Frame—In packet radio, a transmission block consisting of opening flag, address, control, information, frame-check-sequence and ending flag fields.

FSK—Frequency-shift keying.

Gateway—In packet radio, an interchange point.

HDLC—High-level data link control procedures as specified in ISO 3309.

Host—As used in packet radio, a computer with applications programs accessible by remote stations.

IA5—International Alphabet No. 5, a 7-bit coded character set, CCITT version of ASCII.

Information field—Any sequence of bits containing the intelligence to be conveyed.

ISI—Intersymbol interference; slurring of one symbol into the next as a result of multipath propagation.

ISO—International Organization for Standardization.

ITA2—International Telegraph Alphabet No. 2, a CCITT 5-bit coded character set commonly called the Baudot or Murray code.

Jitter—Unwanted variations in amplitude or phase in a digital signal.

Key clicks—Unwanted transients beyond the necessary bandwidth of a keyed radio signal.

LAP—Link access procedure, CCITT X.25 unbalanced-mode communications.

LAPB—Link access procedure, balanced, CCITT X.25 balanced-mode communications.

Layer—In communications protocols, one of the strata or levels in a reference model.

Level 1—Physical layer of the OSI reference model.

Level 2—Link layer of the OSI reference model.

Level 3—Network layer of the OSI reference model.

Level 4—Transport layer of the OSI reference model.

Level 5—Session layer of the OSI reference model.

Level 6—Presentation layer of the OSI reference model.

Level 7—Application layer of the OSI reference model.

Loopback—A test performed by connecting the output of a modulator to the input of a demodulator.

LSB—Least-significant bit.

Mode A—In AMTOR, an automatic repeat request (ARQ) transmission method.

Mode B—In AMTOR, a forward error correction (FEC) transmission method.

Modem—Modulator-demodulator, a device that connects between a data terminal and communication line (or radio). Also called data set.

MSB—Most-significant bit.

MSK—Frequency-shift keying where the shift in Hz is equal to half the signaling rate in bits per second.

NAK—Negative acknowledge (opposite of ACK).

NAPLPS—ANSI X3.110-1983 Videotex/Teletext Presentation Level protocol syntax.

NBDP—Narrow-band direct-printing telegraphy.

NEPRA—New England Packet Radio Association.

Node—A point within a network, usually where two or more links come together, performing switching, routine and concentrating functions.

NRZI—Nonreturn to zero. A binary baseband code in which output transitions result from data 0s but not from 1s. Formal designation is NRZ-S (non-return-to-zero—space).

Null modem—A device to interconnect two devices both wired as DCEs or DTEs; in EIA RS-232-C interfacing, back-to-back DB25 connectors with pin-for-pin connections except that Received Data (pin 3) on one connector is wired to Transmitted Data (pin 3) on the other.

Octet—A group of eight bits.

OOK—On-off keying.

Originate—The station initiating a call. In modem usage, the calling station or modem tones associated therewith.

OSI-RM—Open Systems Interconnection Reference Model specified in ISO 7498 and CCITT Rec X.200.

Packet radio—A digital communications technique involving radio transmission of short bursts (frames) of data containing addressing, control and error-checking information in each transmission.

PACSAT—AMSAT packet-radio satellite with store-and-forward capability.

PAD—Packet assembler/disassembler, a

device that assembles and disassembles packets (frames). It is connected between a data terminal (or computer) and a modem in a packet-radio station (see also TNC).

Parity check—Addition of noninformation bits to data, making the number of ones in a group of bits always either even or odd.

PID—Protocol identifier. Used in AX.25 to specify the network-layer protocol used.

PPRS—Pacific Packet Radio Society.

Primary—The master station in a master-slave relationship; the master maintains control and is able to perform actions that the slave cannot. (Compare secondary.)

Protocol—A formal set of rules and procedures for the exchange of information within a network.

PSK—Phase-shift keying.

RAM—Random access memory.

Router—A network packet switch. In packet radio, a network-level relay station capable of routing packets.

RS-232-C—See EIA-232-C.

RTS—Request to send, physical-level signal used to control the direction of data transmission of the local DCE.

RTTY—Radioteletype.

RxD—Received data, physical-level signals generated by the DCE are sent to the DTE on this circuit.

Secondary—The slave in a master-slave relationship. Compare primary.

SOFTNET—An experimental packet-radio network at the University of Linkoping, Sweden.

Source—In packet radio, the station transmitting the frame over a direct radio link or via a repeater.

SOUTHNET—A series of digipeaters along the US Southeast Coast.

SSID—Secondary station identifier. In AX.25 link-layer protocol, a multipurpose octet to identify several packet-radio stations operating under the same call sign.

TAPR—Tucson Amateur Packet Radio Corporation, a nonprofit organization involved in packet-radio development.

Teleport—A radio station that acts as a relay between terrestrial radio stations and a communications satellite.

TNC—Terminal node controller, a device that assembles and disassembles packets (frames); sometimes called a PAD.

TR switch—Transmit-receive switch to allow automatic selection between receive and transmitter for one antenna.

TTY—Teletypewriter.

TU—Terminal unit, a radioteletype modem or demodulator.

Turnaround time—The time required to reverse the direction of a half-duplex circuit, required by propagation, modem reversal and transmit-receive switching time of transceiver.

TxD—Transmitted data, physical-level data signals transferred on a circuit from the DTE to the DCE.

UI—Unnumbered information frame.

V.24—A CCITT standard defining physical-level interface circuits between a DTE (terminal) and DCE (modem), equivalent to EIA RS-232-C.

V.28—A CCITT standard defining electrical characteristics for V.24 interface.

VADCG—Vancouver Amateur Digital Communications Group.

VDT—Video-display terminal.

VDU—Video display unit, a device used to display data, usually provided with a keyboard for data entry.

Videotex—A presentation-layer protocol for two-way transmission of graphics.

Virtual circuit—A mode of packet networking in which a logical connection that emulates a point-to-point circuit is established (compare Datagram).

WESTNET—A series of digipeaters along the US West Coast.

Window—In packet radio at the link layer, the range of frame numbers within the control field used to set the maximum number of frames that the sender may transmit before it receives an acknowledgment from the receiver.

X.25—CCITT packet-switching protocol.

ASCII

The American National Standard Code for Information Interchange (ASCII) is a coded character set used for information-processing systems, communications systems and related equipment. Current FCC regulations provide that amateur use of ASCII shall conform to ASCII as defined in ANSI Standard X3.4-1977. Its international counterparts are ISO 646-1983 and International Alphabet No. 5 (IA5) as specified in CCITT Rec V.3.

ASCII uses 7 bits to represent letters, figures, symbols and control characters. Unlike ITA2 (Baudot), ASCII has both upper- and lower-case letters. A table of ASCII characters is presented as "ASCII Character Set" in the **References** chapter.

In the international counterpart code, £ replaces #, and the international currency sign ¤ may replace $ by agreement of the sender and recipient. Without such agreement, neither £, ¤ nor $ represent the currency of any particular country.

Parity

While not strictly a part of the ASCII standard, an eighth bit (P) may be added for parity checking. FCC rules permit optional use of the parity bit. The applicable US and international standards (ANSI X3.16-1976; CCITT Rec V.4) recommend an even parity sense for asynchronous and odd parity sense for synchronous data communications. The standards, however, generally are not observed by hams.

Code Extensions

By sacrificing parity, the eighth bit can be used to extend the ASCII 128-character code to 256 characters. Work is underway to produce an international standard that includes characters for all written languages.

ASCII Serial Transmission

Serial transmission standards for ASCII (ANSI X3.15 and X3.16; CCITT Rec V.4 and X.4) specify that the bit sequence shall be least-significant bit (LSB) first to most-significant bit (MSB), that is b0 through b6 (plus the parity bit, P, if used).

Serial transmission may be either synchronous or asynchronous. In synchronous transmissions, only the information bits (and optional parity bit) are sent, as shown in **Fig 12.24A**.

Asynchronous serial transmission adds a start pulse and a stop pulse to each character. The start pulse length equals that of an information pulse. The stop pulse may be one or two bits long. There is some variation, but one stop bit is the convention, except for 110-baud transmissions with mechanical teletypewriters.

ASCII-over-AMTOR

The superior weak signal performance of AMTOR, compared to RTTY and HF packet radio, has made it a popular mode for HF data networks. AMTOR BBS systems are popular for passing long-haul traffic. Traffic from VHF and UHF packet networks is converted into AMTOR (and more recently PACTOR and CLOVER) by specially equipped HF BBS stations. This system combines the best attributes of several different data transfer modes: The convenience and short-range high data rate of VHF/UHF packet is combined with the high reliability of AMTOR/PACTOR/CLOVER for long-range HF data transfer.

There was a problem with AMTOR relays, however. The CCIR-476 and CCIR-625 AMTOR symbol set has no lower-case letters and lacks many punctuation symbols common in VHF/UHF packet radio. Therefore, messages routed via AMTOR can differ from the original in format and appearance. Differences in the header text of AMTOR vs packet messages can be particularly troublesome to automated data-transfer systems.

In late 1991, G3PLX (Peter Martinez) and W5SMM (Vic Poor) devised an extended AMTOR character set that contains all of the printable ASCII symbols (ASCII control characters are not supported). Using this scheme, AMTOR-delivered messages are indistinguishable from those delivered via ASCII-based modes (such as packet radio). This "ASCII-over-AMTOR" system uses the generally unused "blank" character code ("00000" in Baudot, "1101010" in AMTOR) to toggle between the standard AMTOR character set and the new "Blank Code Extension" character set, which includes lower-case letters and ASCII punctuation symbols.

When two ASCII-over-AMTOR equipped stations first link, both controllers are set to the standard CCIR-476/

625 character set; upper-case letters are sent and the FIGS code (AMTOR "0110110") switches between letters and numbers. When the first "blank" character is sent, both stations switch to the new character set. Any following AMTOR letter codes are assumed to be lower-case letters and FIGS codes are translated into the new punctuation symbol set. A second instance of the "blank" code switches both stations back to the standard AMTOR character set. The expanded ASCII-over-AMTOR character set is shown in the Table.

The ASCII-over-AMTOR extended symbol set is supported by most commercially available AMTOR controllers and popular BBS software, such as APLINK and AMTOR MBO. The symbol set is backward compatible with stations that do not have the extended capability. A station that is not equipped with ASCII-over-AMTOR will notice very few differences when receiving these signals except that all letters will appear to be upper-case and the standard punctuation symbols will be printed.

The ASCII-over-AMTOR extension is remarkably efficient. If no nonBaudot characters are sent, there is no additional overhead to the transmission. Even if the extended set is sent, far fewer bits are transmitted than if ASCII were transmitted.

This technique, however, requires an error-correcting code such as AMTOR. The concept would not work with standard Baudot RTTY because a noise "hit" on a Blank character would result in printing from the wrong symbol set. The AMTOR error-correcting code is not infallible, but on-the-air use of ASCII-over-AMTOR has demonstrated that case-errors are very rare occurances. The system works well and is in daily use by AMTOR BBS stations throughout the world.

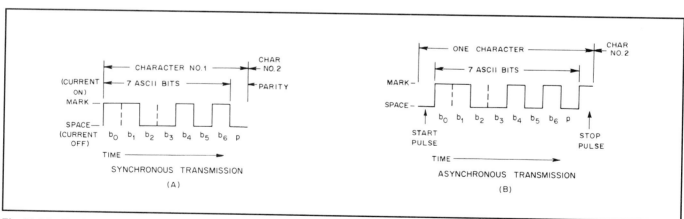

Fig 12.24—Typical serial synchronous and asynchronous timing for the ASCII character S.

ASCII Data Rates

Data-communication signaling rates depend largely on the medium and the state of the art when the equipment was selected. Numerous national and international standards that recommend different data rates, are listed in **Table 12.4**. The most-used rates tend to progress in 2:1 steps from 300 to 9600 bits/s and in 8 kbits/s increments from 16 kbits/s upward (see **Table 12.5**). For Amateur Radio, serial ASCII transmissions data rates of 75, 110, 150, 300, 600, 1200, 2400, 4800, 9600, 16000, 19200 and 56000 bits/s are suggested.

Bauds vs Bits Per Second

The "baud" is a unit of signaling speed equal to one discrete condition or event per second. In single-channel transmis-

ASCII-over-AMTOR
Blank-Code Extension Symbol Set

Bit Code	Standard CCIR-476 Ltrs	Figs	Blank Code Extension Ltrs	Figs
1000111	A	-	a	_
1110010	B	?	b	:
0011101	C	:	c	;
1010011	D	WRU	d	WRU
1010110	E	3	e	
0011011	F	%	f	'
0110101	G	@	g	}
1101001	H	#	h	{
1001101	I	8	i	
0010111	J	*	j	
0011110	K	(	k	[
1100101	L	)	l	]
0111001	M	.	m	>
1011001	N	,	n	<
1110001	O	9	o	~
0101101	P	0	p	
0101110	Q	1	q	!
1010101	R	4	r	$
1001011	S	'	s	"
1110100	T	5	t	
1001110	U	7	u	&
0111100	V	=	v	|
0100111	W	2	w	
0111010	X	/	x	\
0101011	Y	6	y	^
1100011	Z	+	z	
1111000	CR	CR	CR	CR
1101100	LF	LF	LF	LF
1011010	LTRS	LTRS	LTRS	LTRS
0110110	FIGS	FIGS	FIGS	FIGS
1011100	SP	SP	SP	SP
1101010	BLNK	BLNK	BLNK	BLNK

CR = carriage return
LF = line feed
LTRS = shift to letter characters
FIGS = shift to figure characters
SP = space
BLNK = toggle between CCIR-476 and Blank Code Extension sets.

Notes:
1. The logic state "1" represents the Mark or "Z" condition, the higher radiated radio frequency.
2. Certain FIGS-case symbols follow CCIR-476 and common European usage, differing from the "US TTY" symbols shown in the ITA2 Codes table in the **References** chapter. These differences are necessary to assure international compatibility.
3. The signal "BELL" is not supported because it is generally a nuisance to operation of otherwise silent automated message relay stations. If BELL is required, use FIGS-J in the Blank Extension set.

ted ASCII in the Amateur Radio Service. US amateurs have been slow to abandon Baudot in favor of asynchronous serial ASCII.

One cause for resistance is the reasoning that asynchronous ASCII has two (or three with a parity bit added) more bits than asynchronous Baudot and is usually sent at higher speeds. Thus, it is felt that the greater data rates and increased bandwidth needed for ASCII would make its reliability less than that of Baudot. This is true as far as it goes, but does not exhaust the theoretical possibilities, which will be discussed below.

On the practical side, some amateurs tried ASCII on the air and experienced poor results. In some cases, this can be traced to the use of modems that were optimized for 45-baud operation. At 110 or 300 bauds, the 45-baud mark and space filters are too narrow.

On the HF bands, speeds above 50 or 75 bauds are subject to intersymbol interference (ISI, slurring one pulse into the next) from multipath propagation. Multiple paths can be avoided by operating at the maximum usable frequency (MUF), where there is only one ray path. The amount of multipath delay varies according to operating frequency with respect to the MUF and path distance. Paths in the 600- to 5000-mile range are generally less subject to multipath than shorter or longer ones. Paths of 250 miles or less are difficult from a multipath standpoint. As a result, successful operation at the higher ASCII speeds depends on using the highest frequency possible as well as having suitable modems at both ends of the circuit.

Returning to the theoretical comparison of Baudot and ASCII, recall that the FCC requires asynchronous (start-stop) transmission of Baudot. This means that the five information pulses must be sent with a start pulse and a stop pulse, usually of 1.42 times the length of the information pulse. Thus, an asynchronous Baudot transmitted character requires 7.42 units. In contrast, 7 bits of ASCII plus a parity bit, a start and a two-unit stop pulse has 11 units.

However, it is possible to send only the 7 ASCII information bits synchronously (without start and stop pulses), making the number of units that must be transmitted (7 vs 7.42) slightly smaller for ASCII than for Baudot. Or, it is possible to synchronously transmit 8 bits (7 ASCII bits plus a parity bit) and take advantage of the error-detection capability of parity. Also, there is nothing to prevent ASCII from being sent at a lower speed such as 50 or 75 bauds, to make it as immune to multipath as is 45- or 50-baud Baudot RTTY. So it is

sion, such as the FCC prescribes for Baudot transmissions, the signaling rate in bauds equals the data rate in bits per second. However, the FCC does not limit ASCII to single-channel transmission. Some digital modulation systems have more than two (mark and space) states. In *dibit* (pronounced die-bit) modulation, two ASCII bits are sampled at a time. The four possible states for a dibit are 00, 01, 10 and 11. In four-phase modulation, each state is assigned an individual phase of 0°, 90°, 180° and 270° respectively. For dibit

phase modulation, the signaling speed in bauds is half the information-transfer rate in bits/s. As the FCC specifies the digital sending speed in bauds, amateurs may transmit ASCII at higher information rates by using digital modulation systems that encode more bits per signaling element. This technology is open for exploration by Amateur Radio experimenter. One such example is Clover II.

Amateur ASCII RTTY Operations

On April 17, 1980 the FCC first permit-

Table 12.4

Data Transmission Signaling-Rate Standards

Standard	Signaling Rates (bit/s)	Tolerance
CCIT		
V.5	600, 1200, 2400, 4800	±0.01%
V.6	Preferred: 600, 1200, 2400, 3600, 4800, 7200, 9600 Supplementary: 1800, 3000, 4200, 5400, 6000, 6600 7800, 8400, 9000, 10200, 10800	±0.01%
V.21	110, 150, 300 (where possible)	≤200 bit/s ≤300 bit/s
V.23	600 1200 75 (backward channel)	≤600 bit/s ≤1200 bit/s ≤75 bits
V.34	28800, 26400, 24000, 21600, 19200, 16800 or 14400	
V.35	Preferred: 48000	
V.36	Recommended for international use: 48000 Certain applications: 56000, 64000, 72000	
X.3	Packet assembly/disassembly speeds: 50, 75, 100, 134.5, 150, 200, 300, 600, 1200, 1200/75, 1800, 2400, 4800, 9600, 19200, 48000, 56000, 64000	
ANSI		
X3.1	Serial: 75, 150, 300, 600, 1200, 2400, 4800, 7200, 9600 Parallel: 75, 150, 300, 600, 900, 1200	
X3.36	Above 9600 bit/s, signaling rates shall be in integral multiples of 8000 bit/s. Selected standard rates: 16000, 56000, 1344000 and 1544000 Recognized for international use: 48000	
EIA		
RS-269-B	(Same as ANSI X3.1)	
FED STD		
-1001	(Same as ANSI X3.36) For foreign communications: 64000	
-1041	2400, 4800, 9600	

Table 12.5

ASCII Asynchronous Signaling Rates

Bits per Second	Data Pulse (ms)	Stop Pulse (ms)	CPS	WPM
110	9.091	18.182	10.0	100
150	6.667	6.667	15.0	150
300	3.333	3.333	30.0	300
600	1.667	1.667	60.0	600
1200	0.8333	0.8333	120	1200
2400	0.4167	0.4167	240	2400
4800	0.2083	0.2083	480	4800
9600	0.1041	0.1041	960	9600
19200	0.0520	0.0520	1920	19200

CPS = characters per second

$$= \frac{1}{\text{START} + 7 \text{ (DATA)} + \text{PARITY} + \text{STOP}}$$

WPM = words per minute $= \frac{\text{CPS}}{6} \times 60$

= number of 5-letter-plus-space groups per minute

easy to see that ASCII can be as reliable as Baudot RTTY, if care is used in system design.

While 45- or 50-baud RTTY circuits can provide reliable communications, this range of signaling speeds does not make full use of the HF medium. Speeds ranging from 75 to 1200 bauds can be achieved on HF with error-detection and error-correction techniques similar to those used in AMTOR. Reliable transmission at higher speeds can be accomplished by means of more sophisticated modes, which are described later in this chapter.

ASCII Bibliography

ANSI X3.4-1977, "Code for Information Interchange," American National Standards Institute.

ANSI X3.15-1976, "Bit sequencing of the American National Standard Code for Information Interchange in Serial-by-Bit Data Transmission."

ANSI X3.16-1976, "Character Structure and Character Parity Sense for Serial-by-Bit Data Communication Information Interchange."

ANSI X3.25-1976, "Character Structure and Character Parity Sense for Parallel-by-Bit Communication in American National Standard Code for Information Interchange."

Bemer, "Inside ASCII," *Interface Age*, May, June and July 1978.

CCITT V.3, "International Alphabet No. 5," International Telegraph and Telephone Consultative Committee, CCITT volumes with recommendations prefixed with the letters V and X are available from United Nations Bookstore.

CCITT V.4, "General Structure of Signals of International Alphabet No. 5 Code for Data Transmission over the Public Telephone Network."

ISO 646-1973 (E), "7-bit Coded Character Set for Information Processing Interchange," International Organization for Standardization, available from ANSI.

Mackenzie, *Coded Character Sets, History and Development*, Addison-Wesley Publishing Co, 1980.

AMTOR

RTTY circuits are plagued with problems of fading and noise unless something is done to mitigate these effects. Frequency, polarization and space diversity are methods of providing two or more simultaneous versions of the transmission to compare at the receiving station. Another method of getting more than one opportunity to see a given transmission is time diversity. The same signal sent at different times will experience different fading and noise conditions. Time diversity is the basis of AMTOR or Amateur Teleprinting Over Radio.

AMTOR always uses two forms of time diversity in either Mode A (ARQ, automatic repeat request) or Mode B (FEC, forward error correction). In Mode A, a repeat is sent only when requested by the receiving station. In Mode B, each character is sent twice. In both Mode A or Mode B, the second type of time diversity is supplied by the redundancy of the code itself.

Since 1983, AMTOR has been part of the US Amateur Radio rules. The rules recognize several documents that define AMTOR, from 476-2 (1978) to CCIR Rec 476-4 and Rec 625 (1986). Anyone interested in the design aspects of AMTOR should refer to these recommendations. You may obtain a complete reprint of Rec

Table 12.6

CCIR Rec 625 Service Information Signals[1]

Mode A (ARQ)	Bit No. 6543210	Mode B (FEC)
Control signal 1 (CS1)	1100101	
Control signal 2 (CS2)	1101010	
Control signal 3 (CS3)	1011001	
Control signal 4 (CS4)	0110101	
Control signal 5 (CS5)	1101001	
Idle signal β	0110011	Idle signal β
Idle signal α	0001111	Phasing signal 1, idle signal α
Signal repetition (RQ)	1100110	Phasing signal 2

[1]1 represents the mark condition (shown as B in CCIR recommendations), which is the higher emitted radio frequency for FSK, the lower audio frequency for AFSK.
0 represents the space condition (shown as Y in CCIR recommendations). Bits are numbered 0 (LSB) through 6 (MSB). The order of bit transmission is LSB first, MSB last.

476-3 as part of the *Proceedings* of the Third ARRL Amateur Radio Computer Networking Conference, available from ARRL Hq.

Overview

AMTOR is based on SITOR, a system devised in the Maritime Mobile Service as a means of improving communications between RTTYs using the ITA2 (Baudot) code. The system converts the 5-bit code to a 7-bit code for transmission such that there are 4 mark and 3 space bits in every character (see the ITA2 and AMTOR Codes table in the **References** chapter).

The constant mark/space ratio limits the number of usable combinations to 35. ITA2 takes up 32 of the combinations; the 3 remaining are service information signals—α, β and RQ in **Table 12.6**. The table also shows several other service signals that are borrowed from the 32 combinations that equate to ITA2. They are not confused with the message characters because they are sent only by the receiving station.

Mode B (FEC)

When transmitting to no particular station (for example calling CQ, net operation or bulletin transmissions) there is no (one) receiving station to request repeats. Even if one station were selected, its ability to receive properly may not be representative of others desiring to copy the signal.

Mode B uses a simple forward-error-control (FEC) technique: it sends each character twice. Burst errors are virtually eliminated by delaying the repetition for a period thought to exceed the duration of most noise bursts. In AMTOR, groups of five characters are sent (DX) and then repeated (RX). At 70 ms per character, there is 280 ms between the first and second transmissions of a character.

The receiving station tests for the constant 4/3 mark/space ratio and prints only unmutilated DX or RX characters. If both are mutilated, an error symbol or space prints.

The Information Sending Station (ISS) transmitter must be capable of 100% duty-cycle operation for Mode B. Thus, it may be necessary to reduce power level to 25% to 50% of full rating.

Mode A (ARQ)

This synchronous system, transmits blocks of three characters from the Information Sending Station (ISS) to the Information Receiving Station (IRS). After each block, the IRS either acknowledges correct receipt (based on the 4/3 mark/space ratio), or requests a repeat. This cycle repeats as shown in **Fig 12.25**.

The station that initiates the ARQ protocol is known as the Master Station (MS).

The MS first sends the selective call of the called station in blocks of three characters, listening between blocks. Four-letter AMTOR calls are normally derived from the first character and the last three letters of the station call sign. For example, W1AW's AMTOR call would be WWAW. The Slave Station (SS) recognizes its selective call and answers that it is ready. The MS now becomes the ISS and will send traffic as soon as the IRS says it is ready.

When an ISS is done sending, it can enable the other station to become the ISS by sending the three-character sequence FIGS Z B. A station ends the contact by sending an "end of communication signal," three Idle Signal Alphas.

On the air, AMTOR Mode A signals have a characteristic "chirp-chirp" sound. Because of the 210/240-ms on/off timing, Mode A can be used with some transmitters at full power levels.

The W1AW AMTOR Mode B transmission follows the Baudot and ASCII bulletins. A W1AW schedule appears in the **References** chapter.

AMTOR Bibliography

S. Ford, *Your RTTY/AMTOR Companion* (Newington, CT: ARRL, 1993).
CCIR, Recommendation 476-3, "Direct-Printing Telegraph Equipment in the Maritime Mobile Service." Reprint available from ARRL Hq as part of the *Proceedings* of the Third ARRL Amateur Radio Computer Networking Conference.
DATACOM, British Amateur Radio Teleprinter Group (BARTG).
Henry, "Getting Started in Digital Communications," Part 4, *QST*, June 1992.

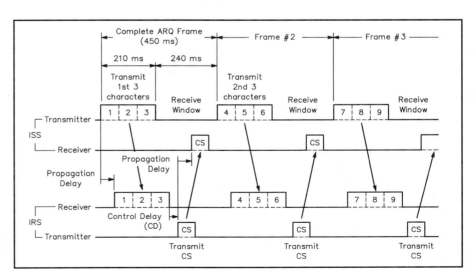

Fig 12.25—Typical AMTOR timing. Dark arrows indicate the signal path from the ISS to the IRS and vice versa. Note the propagation delays; they determine the minimum and maximum communications distances.

Martinez, "AMTOR, An Improved RTTY System Using a Microprocessor," *Radio Communication*, RSGB, Aug 1979.

Martinez, "AMTOR, The Easy Way," *Radio Communication*, RSGB, June/July 1980.

Martinez, "AMTOR—a Progress Report," *Radio Communication*, RSGB, Sep 1981, p 813.

Meyn, "Operating with AMTOR," Technical Correspondence, *QST*, July 1983, pp 40-41.

Newland, "An Introduction to AMTOR," *QST*, July 1983.

Newland, "A User's Guide to AMTOR Operation," *QST*, Oct 1985.

PACKET RADIO

Data communications is telecommunications between computers. *Packet switching* is a form of data communications that transfers data by subdividing it into "packets," and *packet radio* is packet switching using the medium of radio. This description was written by Steve Ford, WB8IMY.

Packet radio has its roots in the Hawaiian Islands, where the University of Hawaii began using the mode in 1970 to transfer data to its remote sites dispersed throughout the islands. Amateur packet radio began in Canada after the Canadian Department of Communications permitted amateurs to use the mode in 1978. (The FCC permitted amateur packet radio in the US in 1980.)

In the first half of the 1980s, packet radio was the habitat of experimenters and those few communicators who did not mind communicating with a limited number of potential fellow packet communicators. In the second half of the decade, packet radio "took off" as the experimenters built a network that increased the potential number of packet stations that could intercommunicate and thus attracted tens of thousands of communicators who wanted to take advantage of this potential. Today, packet radio is one of the most popular modes of Amateur Radio communications, because it is very effective.

It provides error-free data transfer. The receiving station receives information exactly as the transmitting station sends it, so you do not waste time deciphering communication errors caused by interference or changes in propagation.

It uses time efficiently, since packet bulletin-board systems (PBBSs) permit packet operators to store information for later retrieval by other amateurs.

It uses the radio spectrum efficiently, since one radio channel may be used for multiple communications simultaneously or one radio channel may be used to interconnect a number of packet stations to form a "cluster" that provides for the distribution of information to all of the clustered stations. The popular *DX PacketClusters* are typical examples (see **Fig 12.26**).

Each local channel may be connected to other local channels to form a network that affords interstate and international data communications. This network can be used by interlinked packet bulletin-board systems to transfer information, messages and third-party traffic via HF, VHF, UHF and satellite links.

It uses other stations efficiently, since any packet-radio station can use one or more other packet-radio stations to relay data to its intended destination.

It uses current station transmitting and receiving equipment efficiently, since the same equipment used for voice communications may be used for packet communications. The outlay for the additional equipment necessary to make your voice station a packet-radio station may be as little as $100. It also allows you to use that same equipment as an alternative to costly landline data communications links for transferring data between computers.

The TNC

The terminal node controller—or *TNC*—is at the heart of every packet station. A TNC is actually a computer unto itself. It contains the AX.25 packet protocol firmware along with other enhancements depending on the manufacturer. The TNC communicates with you through your computer or data terminal. It also allows you to communicate with other hams by feeding packet data to your transceiver.

The TNCs accepts data from a computer or data terminal and assembles it into packets (see **Fig 12.27**). In addition, it translates the digital packet data into audio tones that can be fed to a transceiver. The TNC also functions as a receiving device, translating the audio tones into digital data a computer or terminal can understand. The part of the TNC that performs this tone-translating function is known as a *modem* (see **Fig 12.28**) .

If you're saying to yourself, "These TNCs sound a lot like telephone modems," you're pretty close to the truth! The first

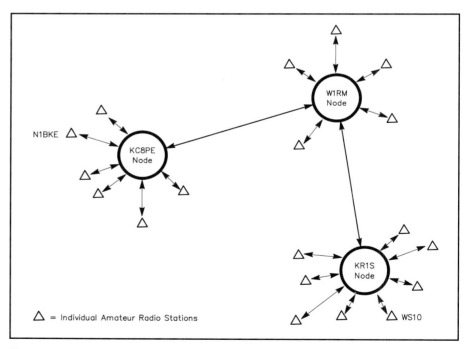

Fig 12.26—*DX PacketClusters* are networks comprised of individual nodes and stations with an interest in DXing and contesting. In this example, N1BKE is connected to the KC8PE node. If he finds a DX station on the air, he'll post a notice—otherwise known as a *spot*—which the KC8PE node distributes to all its local stations. In addition, KC8PE passes the information along to the W1RM node. W1RM distributes the information and then passes it to the KR1S node, which does the same. Eventually, WS1O—who is connected to the KR1S node—sees the spot on his screen. Depending on the size of the network, WS1O will receive the information within minutes after it was posted by N1BKE.

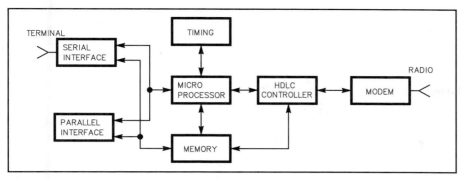

Fig 12.27—The functional block diagram of a typical TNC.

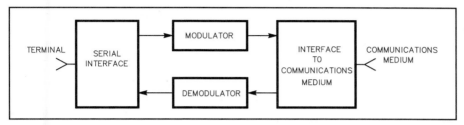

Fig 12.28—A block diagram of a typical modem.

TNCs were based on telephone modem designs. If you're familiar with so-called *smart* modems, you'd find that TNCs are very similar.

You have plenty of TNCs to choose from. The amount of money you'll spend depends directly on what you want to accomplish. Most TNCs are designed to operate at 300 and 1200 bit/s, or 1200 bit/s exclusively (see **Fig 12.29**). There are also TNCs dedicated to 1200 and 9600 bit/s operation, or 9600 bit/s exclusively. Many of these TNCs include convenient features such as personal packet mailboxes where friends can leave messages when you're not at home. Some TNCs also include the ability to easily disconnect the existing modem and substitute another. This feature is very important if you wish to experiment at different data rates. For example, a 1200 bit/s TNC with a *modem disconnect header* can be converted to a 9600 bit/s TNC by disconnecting the 1200 bit/s modem and adding a 9600 bit/s modem.

If you're willing to spend more money, you can buy a complete *multimode communications processor*, or *MCP*. These devices not only offer packet, they also provide the capability to operate RTTY, CW, AMTOR, PACTOR, FAX and other modes. In other words, an MCP gives you just about every digital mode in one box.

TNC Emulation and Internal TNCs

TNC-emulation systems exist for IBM PCs and compatibles. One is known as *BayCom,* which uses the PC to emulate the functions of a TNC/terminal while a small external modem handles the interfacing. BayCom packages are available in kit form for roughly half the price of a basic TNC.

PC owners also have the option of buying full-featured TNCs that mount *inside* their computers. TNC *cards* are available on the market. They are complete TNCs that plug into card slots inside the computer cabinet. No TNC-to-computer cables are necessary. Connectors are provided for cables that attach to your transceiver. In many cases, specialized software is also provided for efficient operation.

Transceiver Requirements

Packet activity on the HF bands typically takes place at 300 bit/s using common SSB transceivers. The transmit audio is fed from the TNC to the microphone jack or auxiliary audio input. Receive audio is obtained from the external speaker jack or auxiliary audio output. Tuning is critical for proper reception; a visual tuning indicator—available on some TNCs and all MCPs—is recommended.

These simple connections also work for 1200 bit/s packet, which is common on the VHF bands (2 m in particular). Almost any FM transceiver can be made to work with 1200 bit/s packet by connecting the transmit audio to the microphone jack and taking the receive audio from the external speaker (or earphone) jack.

At data rates beyond 1200 bit/s, transceiver requirements become more rigid. At 9600 bit/s (the most popular data rate above 1200 bit/s), the transmit audio must be injected at the modulator stage of the FM transceiver. Receive audio must be tapped at the discriminator. Most 9600 bit/s operators use modified Amateur Radio transceivers or commercial radios. The Motorola *Mitrek* transceiver is a popular choice.

In the mid '90s amateur transceiver manufacturers began incorporating data ports on some FM voice rigs. The new "data-ready" radios are not without problems, however. Their IF filter and discriminator characteristics leave little room for error. If you're off frequency by a small amount, you may not be able to pass data. In addition, the ceramic discriminator coils used in some transceivers have poor group delay, making it impossible to tune them for wider bandwidths. With this in mind, some amateurs prefer to make the leap to 9600 bit/s and beyond using *dedicated* amateur data radios such as those manufactured by Tekk and Kantronics

Fig 12.29—Four popular 1200 bit/s packet TNCs: (clockwise, from bottom left) the MFJ-1270C, AEA PK-88, Kantronics KPC-3 and the DRSI DPK-2.

(see Address List in **References** chapter), among others.

Regardless of the transceiver used, setting the proper deviation level is extremely critical. At 9600 bit/s, for example, optimum performance occurs when the maximum deviation is maintained at 3 kHz. Deviation adjustments involve monitoring the transmitted signal with a deviation meter or service monitor. The output level of the TNC is adjusted until the proper deviation is achieved.

Packet Networking

Digipeaters

A digipeater is a packet-radio station capable of recognizing and selectively repeating packet frames. An equivalent term used in industry is *bridge*. Virtually any TNC can be used as a single-port digipeater, because the digipeater function is included in the AX.25 Level 2 protocol firmware. Although the use of digipeaters is waning today as network nodes take their place, the digipeater function is handy when you need a relay and no node is available, or for on-the-air testing.

NET/ROM

Ron Raikes, WA8DED, and Mike Busch, W6IXU, developed new firmware for the TNC 2 (and TNC-2 clones) that supports Levels 3 and 4, the Network and Transport layers of the packet-radio network. NET/ROM replaces the TNC-2 EPROM (that contains the TAPR TNC-2 firmware) and converts the TNC into a

network node controller (NNC) for use at wide- and medium-coverage digipeater sites. Since it is so easy to convert an off-the-shelf TNC into an NNC via the NET/ROM route, NET/ROM has become the most popular network implementation in the packet-radio world and has been installed at most dedicated digipeater stations, thus propelling the standard AX.25 digipeater into packet-radio history.

The NET/ROM network user no longer has to be concerned with the digipeater path required to get from one point to another. All you need to know is the local node of the station you wish to contact. NET/ROM knows what path is required, and if one path is not working or breaks down for some reason, NET/ROM will switch to an alternative path, if one exists. You can be assured that NET/ROM is on top of things, because each NET/ROM node automatically updates its node list periodically, and whenever a new node comes on the air, the other NET/ROM nodes become aware of the new node's existence. In addition to automatic route updating, routing information may also be updated manually by means of a terminal keyboard or remotely using a packet-radio connection.

Once you are connected to another station via the NET/ROM network, most of your packets get through because node-to-node packet acknowledgment is used rather than end-to-end acknowledgment. Besides offering node-to-node acknowledgment, NET/ROM also allows you to build cross-frequency or cross-band

multiport nodes. This is done by installing NET/ROM in two TNCs and connecting their serial ports together. In addition to providing these sophisticated NNC functions, NET/ROM also provides the standard AX.25 digipeater function.

ROSE

Several years ago, the Radio Amateur Telecommunications Society (RATS) developed a networking protocol known as *RATS Open System Environment*, or *ROSE*. Like networks based on *NET/ROM* nodes, the objective of ROSE is to let the network do the work when you're trying to connect to another station.

Using a ROSE network is similar to using the telephone. ROSE nodes are frequently referred to as *switches*, and each switch has its own address based on the telephone area code and the first 3 digits of the local exchange. A ROSE switch in one area of Connecticut, for example, may have an address of 203555. 203 is the area code and 555 is the local telephone exchange. The ROSE network uses this addressing system to create reliable routes for packets (see **Fig 12.30**).

Unless you wish to set up a ROSE switch of your own, you won't need special equipment or software to use the network. You can access a ROSE network today if a switch is available in your area. All you need to know is the call sign of your local switch and the ROSE address of the switch nearest to any stations to want to contact.

ROSE networks are appearing in many areas of the country. They are especially popular in the southeast and midAtlantic states. ROSE addresses and system maps are available from RATS (see **References** chapter Address List). Send a business-sized SASE with your request.

TexNet

TexNet is a high speed, centralized packet networking system developed by the Texas Packet Radio Society (TPRS). Designed for local and regional use, TexNet provides AX.25-compatible access on the 2-m band at 1200 bit/s. This allows packeteers to use TexNet without investing in additional equipment or software. The node-to-node backbones operate in the 70-cm band with data moving through the network at 9600 bit/s. Telephone links are also used to bridge some gaps in the system.

The network offers a number of services to its users. Two conference levels are available by simply connecting to the proper node according to its SSID. By connecting to W5YR-2, for example, you'll join the first conference level. Con-

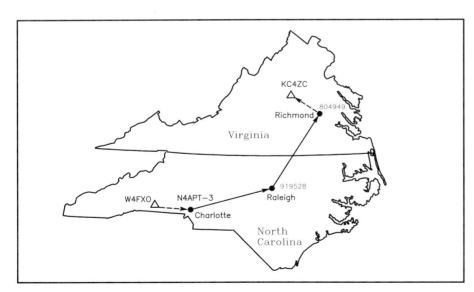

Fig 12.30—In this hypothetical example, W4FXO, near Charlotte, North Carolina, uses the ROSE network to establish a connection to KC4ZC northwest of Richmond, Virginia. All that W4FXO has to do is issue a connect request that includes his local ROSE switch (N4APT-3) and the ROSE address of the switch nearest KC4ZC (804949). When the request is sent, the network takes over. In this example, the connection to KC4ZC is established by using a ROSE switch in Raleigh.

necting to W5YR-3 places you in the second level. When you connect to a conference, you can chat with anyone else on the network in roundtable fashion.

Every TexNet network is served by a single PBBS. By using only one PBBS, the network isn't bogged down with constant mail forwarding. Even if you're some distance from the PBBS, with the speed and efficiency of TexNet you'll hardly notice the delay.

TCP/IP

If you're an active packeteer, sooner or later someone will bring up the subject of TCP/IP—Transmission Control Protocol/Internet Protocol. Of all the packet networking alternatives discussed so far, TCP/IP is the most popular. In fact, many packeteers believe that TCP/IP may someday become the standard for amateur packet radio.

Despite its name, TCP/IP is more than two protocols; it's actually a set of several protocols. Together they provide a high level of flexible, "intelligent" packet networking. At the time of this writing, TCP/IP networks are local and regional in nature. For long-distance mail handling, TCP/IP still relies on traditional AX.25 NET/ROM networks. Even so, TCP/IP enthusiasts see a future when the entire nation, and perhaps the world, will be linked by high-speed TCP/IP systems using terrestrial microwave and satellites.

Maintaining a packet connection on a NET/ROM network can be a difficult proposition—especially if the station is distant. You can only hope that all the nodes in the path are able to relay the packets back and forth. If the one of the nodes becomes unusually busy, your link to the other station could collapse. Even when the path is maintained, your packets are in direct competition with all the other packets on the network. With randomly calculated transmission delays, collisions are inevitable. As a result, the network bogs down, slowing data throughput for everyone.

TCP/IP has a unique solution for busy networks. Rather than transmitting packets at randomly determined intervals, TCP/IP stations automatically *adapt* to network delays as they occur. As network throughput slows down, active TCP/IP stations sense the change and lengthen their transmission delays accordingly. As the network speeds up, the TCP/IP stations shorten their delays to match the pace. This kind of intelligent network sharing virtually guarantees that all packets will reach their destinations with the greatest efficiency the network can provide.

With TCP/IP's adaptive networking scheme, you can chat using the *telnet* protocol with a ham in a distant city and rest assured that you're not overburdening the system. Your packets simply join the constantly moving "freeway" of data. They might slow down in heavy traffic, but they *will* reach their destination eventually. (This adaptive system is used for *all* TCP/IP packets, no matter what they contain.)

TCP/IP excels when it comes to transferring files from one station to another. By using the TCP/IP *file transfer protocol* (ftp), you can connect to another station and transfer computer files—including software. As you can probably guess, transferring large files can take time. With TCP/IP, however, you can still send and receive mail (using the *SMTP* protocol) or talk to another ham *while* the transfer is taking place.

When you attempt to contact another station using TCP/IP, all network routing is performed automatically according to the TCP/IP address of the station you're trying to reach. In fact, TCP/IP networks are transparent to the average user.

On conventional *NET/ROM* networks, access to backbone links is restricted. This isn't true on TCP/IP. Not only are you allowed to use the backbones, you're actually *encouraged* to do so. If you have the necessary equipment to communicate at the proper frequencies and data rates, you can tap into the high-speed TCP/IP backbones directly. By doing so, you'll be able to handle data at much higher rates. This benefits you and everyone else on the network.

To operate TCP/IP, all you need is a computer (it must be a computer, not a terminal), a 2-m FM transceiver and a TNC with *KISS* capability. As you might guess, the heart of your TCP/IP setup is software. The TCP/IP software set was written by Phil Karn, KA9Q, and is called *NOSNET* or just *NOS*.

There are dozens of *NOS* derivatives available today. All are based on the original *NOSNET*. The programs are available primarily for IBM-PCs and compatibles and Macintoshes. You can obtain NOS software from on-line sources such as the CompuServe *HAMNET* forum libraries, Internet ftp sites, Amateur Radio-oriented BBSs and elsewhere. *NOS* takes care of all TCP/IP functions, using your "KISSable" TNC to communicate with the outside world. The only other item you need is your own IP address. Individual IP Address Coordinators assign addresses to new TCP/IP users.

PACTOR

PACTOR (PT) is an HF radio transmission system developed by German amateurs Hans-Peter Helfert, DL6MAA, and Ulrich Strate, DF4KV. It combines the best of AMTOR and packet to make a system that is superior to both. This description was adapted from PACTOR specifications by the *Handbook* Editor. PACTOR is much faster than AMTOR, yet improves on AMTOR's error-correction scheme. It performs well under both weak-signal and high-noise conditions. PACTOR/AMTOR BBS stations operating in the US and other countries are used by amateurs all over the world. The BBSs respond automatically to both PACTOR and AMTOR calls. PACTOR carries binary data, so it can transfer binary files, ASCII and other symbol sets.

Packet-radio style CRCs (two per packet, 16 bits each) and "ARQ Memory" enable the PT system to reconstruct defective packets by overlaying good and damaged data from different transmissions, which reduces repeats and transmission time. PT's overhead is much less than that of AMTOR. PACTOR uses complete call signs for addressing. The mark/space convention is unnecessary and frequency-shift independent.

Transmission Formats
Information Blocks

All packets have the basic structure shown in **Fig 12.31**, and their timing is as shown in **Table 12.7**:

Header: contains a fixed bit pattern to simplify repeat requests, synchronization and monitoring. The header is also important for the Memory ARQ function. In each packet carrying new information the bit pattern is inverted.

Data: any binary information. The format is specified in the status word. Current choices are 8-bit ASCII or 7-bit ASCII (with Huffman encoding). Characters are not broken across packets. ASCII RS (hex 1E) is used as an IDLE character in both formats.

Status word: see **Table 12.8**

CRC: The CRC is calculated according to the CCITT standard, for the data, status and CRC.

Acknowledgment Signals

The PACTOR acknowledgment signals are similar to those used in AMTOR, except for CS4 (see **Table 12.9**). Each of the signals is 12 bits long. The characters differ in pairs in 8 bits (Hamming offset) so

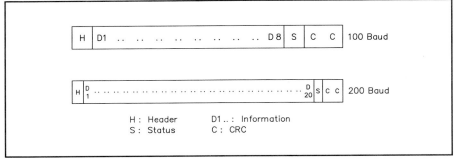

H : Header   D1.. : Information
S : Status   C : CRC

Fig 12.31—PACTOR data packet format.

Table 12.7
PACTOR Timing

Object	Length (seconds)
Packet	0.96 (200 bd: 192 bits; 100 bd: 96 bits)
CS receive time	0.29
Control signals	0.12 (12 bits at 10 ms each)
Propagation delay	0.17
Cycle	1.25

Table 12.8
PACTOR Status Word

Bit	Meaning
0	Packet count (LSB)
1	Packet count (MSB)
2	Data format (LSB)
3	Data format (MSB)
4	Not defined
5	Not defined
6	Break-in request
7	QRT request

Data Format Bits

Format	bit 3	bit 2
ASCII 8 bit	0	0
Huffman code	0	1
Not defined	1	0
Not defined	1	1

Bits 0 and 1 are used as a packet count; successive packets with the same value are identified by the receiver as repeat packets. A modulus-4 count helps with unrecognized control signals, which are unlikely in practice.

Table 12.9
PACTOR Control Signals

Code	Chars (hex)	Function
CS1	4D5	Normal acknowledge
CS2	AB2	Normal acknowledge
CS3	34B	Break-in (forms header of first packet from RX to TX)
CS4	D2C	Speed change request

All control signals are sent only from RX to TX.

that the chance of confusion is reduced. (One of the most common causes of errors in AMTOR is the small CS Hamming offset of 4 bits.)

If the CS is not correctly received, the TX reacts by repeating the last packet. The request status can be uniquely recognized by the 2-bit packet number so that wasteful transmissions of pure RQ blocks are unnecessary.

Timing

The receiver pause between two blocks is 0.29 s. After deducting the CS lengths, 0.17 s remain (just as in AMTOR) for switching and propagation delays so that there is adequate reserve for DX operation.

Contact Flow

Listening

In the listen mode, the receiver scans any received packets for a CRC match. This method uses a lot of computer processing resources, but it's flexible.

Table 12.10
PACTOR Initial Contact
Master Initiating Contact

Size (bytes)	1	8	6
Content	/Header/	SLAVECAL/	SLAVECAL/
Speed (bauds)	100	100	200

Slave Response

The receiving station detects a call, determines mark/space polarity, decodes 100-bd and 200-bd call signs. It uses the two call signs to determine if it is being called and the quality of the communication path. The possible responses are:

First call sign does not match slave's (Master not calling this slave)	none
Only first call sign matches slave's (Master calling this slave, poor communications)	CS1
First and second call signs both match the slaves (good circuit, request speed change to 200 bd)	CS4

CQ

A station seeking contacts transmits CQ packets in a FEC mode, without pauses for acknowledgment between packets. The transmit time length number of repetitions and speed are the transmit operator's choice. (This mode is also suitable for bulletins and other group traffic.) Once a listening station has copied the call, the listener assumes the TX station role and initiates a contact. Thus, the station sending CQ initially takes the RX station role. The contact begins as shown in **Table 12.10**

Speed Changes

With good conditions, PT's normal signaling rate is 200 baud (for a 600-Hz bandwidth), but the system automatically changes from 200 to 100 baud and back, as conditions demand. In addition, Huffman coding can further increase the throughput by a factor of 1.7. There is no loss of synchronization speed changes; only one packet is repeated.

When the RX receives a bad 200-baud packet, it can acknowledge with CS4. TX immediately assembles the previous packet in 100-baud format and sends it. Thus, one packet is repeated in a change from 200 to 100 baud.

The RX can acknowledge a good 100-baud packet with CS4. TX immediately switches to 200 baud and sends the next packet. There is no packet repeat in an upward speed change.

Change of Direction

The RX station can become the TX station by sending a special change-over packet in response to a valid packet. RX sends CS3 as the first section of the changeover packet. This immediately changes the TX station to RX mode to read the data in that packet and responds with CS1 and CS3 (acknowledge) or CS2 (reject).

End of Contact

PACTOR provides a sure end-of-contact procedure. TX initiates the end of contact by sending a special packet with the QRT bit set in the status word and the call of the RX station in byte-reverse order at 100 baud. The RX station responds with a final CS.

PACTOR II

· This new protocol is a significant improvement over PACTOR; yet it is fully compatible with the older mode. Invented in Germany, PACTOR uses 16PSK to transfer up to 800 bits/s at a 100-baud rate. This keeps the bandwidth less than 500 Hz. Users believe that PACTOR II is faster and more robust than CLOVER.

PACTOR II uses a DSP with Nyquist waveforms, Huffman *and* Markov compression, and powerful Viterbi decoding to increase transfer rate and sensitivity into the noise level. The effective transfer rate of text is over 1200 bits/s. Features of PACTOR II include:

- Frequency agility—It can automatically adjust or lock two signals together over a ±100-Hz window.

- Powerful data reconstruction based upon computer power—with over 2 MB of available memory.
- Cross correlation—applies analog Memory ARQ to acknowledgment frames and headers.
- Soft decision making—Uses artificial intelligence (AI) as well as digital information received to determine frame validity.
- Extended data block length—When transferring large files under good conditions, the data length is doubled to increase the transfer rate.
- Automatic recognition of PACTOR I, PACTOR II and so on, with automatic mode switching.

- Intermodulation products are canceled by the coding system.
- Two long-path modes extend frame timing for long-path terrestrial and satellite propagation paths.

This is a fast, robust mode—possibly the most powerful in the ham bands. It has excellent coding gain as well. It can also communicate with all earlier PACTOR I systems. Like packet and AMTOR stations, PACTOR II stations acknowledge each received data block. Unlike those modes, PACTOR II employs computer logic as well as received data to reassemble defective data blocks into good frames. This reduces the number of transmissions and increases the throughput of the data.

G-TOR

This brief description has been adapted from "A Hybrid ARQ Protocol for Narrow Bandwidth HF Data Communication" by Glenn Prescott, WBØSKX, Phil Anderson, WØXI, Mike Huslig, KBØNYK, and Karl Medcalf, WK5M (May 1994 *QEX*, pp 12-19).

G-TOR is short for Golay-TOR, an innovation of Kantronics, Inc. It's a new HF digital-communication mode for the Amateur Service. G-TOR was inspired by HF Automatic Link Establishment (ALE) concepts and is structured to be compatible with ALE systems when they become available.

The purpose of the G-TOR protocol is to provide an improved digital radio communication capability for the HF bands. The key features of G-TOR are:

- Standard FSK tone pairs (mark and space)
- Link-quality-based signaling rate: 300, 200 or 100 baud
- 2.4-s transmission cycle
- Low overhead within data frames
- Huffman data compression—two types, on demand
- Embedded run-length data compression
- Golay forward-error-correction coding
- Full-frame data interleaving
- CRC error detection with hybrid ARQ

- Error-tolerant "Fuzzy" acknowledgments

The primary benefit of these innovations is increased throughput—that is, more bits communicated in less time. This is achieved because the advanced processing features of G-TOR provide increased resistance to interference and noise and greatly reduce multipath-induced data errors.

The G-TOR protocol is straightforward and relatively easy to implement on existing multimode TNCs.

Propagation Problems

The miserable propagation conditions characteristic of the HF bands make effective data communication a nightmare. Received signals are often weak and subject to multipath fading; ever-present interference can impair reception. With digital communication, the human brain cannot help interpret the signal. Therefore, we need to incorporate great ingenuity into the receiving system. G-TOR uses modern communication signal processing to help us transmit error-free data via the inherently poor HF communication medium.

Worldwide HF communication may experience interference, multipath fading, random and burst noise. For data communication over the HF bands, three factors dominate: available bandwidth, signaling rate and the dynamic time behavior of the channel.

... and Answers

Transmission bandwidths of 500 Hz or less minimize the effects of multipath propagation and man-made interference. G-TOR transmits at 300 baud or less, with maximum separation of 200 Hz, for a band-width just slightly greater than 500 Hz.

The FCC does not currently permit symbol rates greater than 300 symbols per second (baud) on most HF bands. This is a reasonable limit because multipath propagation can become a serious problem with faster rates.

The HF channel has a characteristic dynamic time behavior: Conditions can change significantly in a few seconds. This indicates an optimum data-transmission length (usually 1 s or less). G-TOR transmissions are nearly 2 s long because the signal-processing techniques can overcome some propagation change.

The G-TOR Protocol

Since one of the objectives of this protocol is ease of implementation in existing

TNCs, the modulation format consists of standard tone pairs (FSK), operating at 300, 200 or 100 baud, depending upon channel conditions. (G-TOR initiates contacts and sends ACKs only at 100 baud.) FSK was chosen for economy and simplicity, but primarily because many hams already have FSK equipment.

The G-TOR waveform consists of two phase-continuous tones (BFSK) spaced 200 Hz apart (mark = 1600 Hz, space = 1800 Hz); however, the system can still operate at the familiar 170-Hz shift (mark = 2125 Hz, space = 2295 Hz), or with any other convenient tone pairs. The optimum spacing for 300-baud transmission is 300 Hz, but we trade some performance for a narrower bandwidth.

Each transmission consists of a synchronous ARQ 1.92-s frame and a 0.48-s interval for propagation and ACK transmissions (2.4 s cycle). All advanced protocol features are implemented in the signal-processing software.

Synchronous operation increases the system throughput during multipath fading and keeps overhead to a minimum. Synchronization is performed using the received data and precise timing.

Frame Structures

Data Frames—The basic G-TOR frame structure (see **Fig 12.32**) uses multiple 24-bit (triple-byte) words for compatibility with the Golay encoder. Data frames are composed of 72 (300 baud), 48 (200 baud) or 24 (100 baud) data bytes, depending upon channel conditions.

A single byte before the CRC carries command and status information:
    status bits 7 and 6: Command
        00 - data
        01 - turnaround request
        10 - disconnect
        11 - connect
    status bits 5 and 4: Unused
        00 - reserved
    status bits 3 and 2: Compression
        00 - none
        01 - Huffman (A)
        10 - Huffman (B)
        11 - reserved
    status bits 1 and 0: Frame no. ID
The error-detection code transmitted with each frame is a 2-byte cyclic redundancy check (CRC) code—the same used in AX.25. A CRC calculation determines if error correction is needed, and another tests the result.

The connect and disconnect frames are essentially identical in structure to the data frame and contain the call signs of both stations.

ACK Frames—G-TOR ACK frames are not interleaved and do not contain error-correction (parity) bits. There are five different ACK frames:
- Frame received correctly (send next data frame)
- Frame error detected (please repeat)
- Speed-up
- Speed-down
- Changeover

The ACK codes are composed of multiple cyclic shifts of a single 15-bit pseudorandom noise (PN) sequence (plus an extra 0 bit to fill 16 bits). PN sequences have powerful properties that facilitate identification of the appropriate ACK code, even in the presence of noise and interference. We refer to this concept as a "fuzzy" ACK, in that it tolerates 3 bit errors within a received ACK frame.

Change-over frames are essentially data frames in which the first 16 bits of data is the ACK changeover PN code.

Data Compression

Data compression is used to remove redundancy from source data. Therefore, fewer bits are needed to convey any given message. This increases data throughput and decreases transmission time—valuable features for HF. G-TOR uses run-length coding and two types of Huffman coding during normal text transmissions. Run-length coding is used when more than two repetitions of an 8-bit character are sent. It provides an especially large savings in total transmission time when repeated characters are being transferred.

The Huffman code works best when the statistics of the data are known. G-TOR applies Huffman A coding with the upper- and lower-case character set, and Huffman B coding with upper-case-only text. Either type of Huffman code reduces the average number of bits sent per character. In some situations, however, there is no benefit from Huffman coding. The encoding process is then disabled. This decision is made on a frame-by-frame basis by the information-sending station.

Golay Coding

The real power of G-TOR resides in the properties of the (24,12) extended Golay error-correcting code, which permits correction of up to three random errors in three received bytes. The (24,12) extended Golay code is a half-rate error-correcting code: Each 12 data bits are translated into an additional 12 parity bits (24 bits total). Further, the code can be implemented to produce separate input-data and parity-bit frames.

The extended Golay code is used for G-TOR because the encoder and decoder are simple to implement in software. Also, Golay code has mathematical properties that make it an ideal choice for short-cycle synchronous communication:
- The rare property of self-duality makes the code "invertible"; that is, the original data can be recovered by simply recoding the parity bits.
- Because of the linear block code structure of the Golay code, the encoder and decoder can be implemented using a simple table look-up procedure. An alternative decoder implementation uses the well-known Kasami decoding algorithm, which requires far less memory than the look-up table.

Error-correction coding inserts some redundancy into each (triple-byte) word so that errors occurring in the receiving process can be corrected. However, most error-correcting codes are effective at correcting only random errors. Burst errors from lightning or interference exceed the capabilities of most error-correcting codes.

Interleaving

The conventional solution is called "interleaving." Interleaving (the very last operation performed before transmission and first performed upon reception) rearranges the bit order to randomize the effects of long error bursts.

The interleaving process reads 12-bit

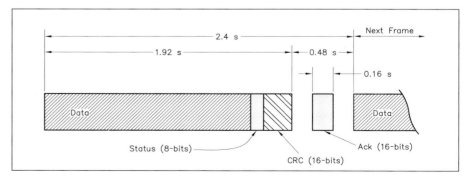

Fig 12.32—G-TOR ARQ system timing and frame structure before interleaving. The data portion may be 69 (300 baud), 45 (200 baud) or 21 (100 baud) bytes depending on the channel quality.

words into registers by columns and reads 48-bit words out by rows; see **Fig 12.33**. The deinterleaver simply performs the inverse, reading the received data bits into the registers by row and extracting the original data sequence by reading the columns. If a long burst of errors occurs—say, 12 bits in length—the errors will be distributed into 48 separate 12-bit words before error correction is applied, thus effectively nullifying the long burst. Both data and parity frames are completely interleaved.

Hybrid ARQ

G-TOR combines error detection and forward error correction with ARQ. Hybrid-ARQ uses a CRC to check for errors in every frame. Only when errors are found; does G-TOR use forward error correction (a relatively slow process) to recover the data.

The half-rate invertible Golay code provides an interesting dimension to the hybrid-ARQ procedure. With separate data and parity frames, both of which can supply the complete data, G-TOR frames alternate between data and parity frames.

When the receiver detects an error and requests a retransmission, the sending station sends the complementary portion of the frame (data or parity).

When the complementary frame arrives, it is processed and checked for errors. If it

checks, the data is accepted and a new frame is requested. If it fails the CRC check, the two frames are combined, corrected and checked.

Using this scheme, two transmissions provide three independent chances to correct any errors. If this process still fails, a retransmission is requested.

G-TOR Performance

Initial testing with G-TOR was conducted during January 1994, between Lawrence, Kansas, and Laguna Niguel, California. During these tests, TRACE was set ON at each station, enabling the raw data display of frames received with and without the aid of forward error correction and interleaving. The results were somewhat surprising. While PACTOR often dropped in transmission speed from 200 to 100 bauds, G-TOR nearly always operated at 300 bauds. Enough frames were corrected to keep the system running at maximum speed, regardless of man-made interference and mild multipath conditions. Transfer duration for the entire test files varied from 12 to 27 minutes for PACTOR, but only 5.5 to 7.5 minutes for all but one G-TOR transfer. G-TOR simply maintained its highest pace better than PACTOR, resulting in a substantial increase in average throughput.

On-air tests have shown G-TOR to have

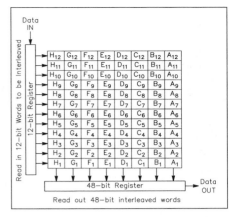

Fig 12.33—Interleaving the bits to be transmitted.

the ability to "hang in there" when channel conditions get tough. The time required to send a given binary file tends to be much less for G-TOR than for PACTOR.

This protocol should continue to be valuable when DSP-based TNCs become widely available. G-TOR has the essential characteristics to be a useful protocol for years to come.

See "A Comparison of HF Digital Protocols" in Jul 1996 *QST* for an overview of performance tradeoffs between the numerous competing protocols available.

CLOVER-II

The desire to send data via HF radio at high data rates and the problems encountered when using AX.25 packet radio on HF radio led Ray Petit, W7GHM, to develop a unique modulation waveform and data transfer protocol that is now called "CLOVER-II." Bill Henry, K9GWT, supplied this description of the Clover-II system. CLOVER modulation is characterized by the following key parameters:

- Very low base symbol rate: 31.25 symbols/second (all modes).
- Time-sequence of amplitude-shaped pulses to provide a very narrow frequency spectra. Occupied bandwidth = 500 Hz at 50 dB below peak output level.
- Differential modulation between pulses.
- Multilevel modulation.

The low base symbol rate is very resistant to multipath distortion because the time between modulation transitions is much longer than even the worst-case time-smearing

caused by summing of multipath signals. By using a time-sequence of tone pulses, Dolph-Chebychev "windowing" of the modulating signal and differential modulation, the total occupied bandwidth of a CLOVER-II signal is held to 500 Hz.

The CLOVER Waveform

Multilevel tone, phase and amplitude modulation give CLOVER a large selection of data modes that may be used (see **Table 12.11**). The adaptive ARQ mode of CLOVER senses current ionosphere conditions and automatically adjusts the modulation mode to produce maximum data throughput. When using the "Fast" bias setting, ARQ throughput automatically varies from 11.6 bytes/s (1.7 times AMTOR) to 70 bytes/s (10.5 times AMTOR).

The CLOVER-II waveform uses four tone pulses that are spaced in frequency by 125 Hz. The time and frequency domain

characteristics of CLOVER modulation are shown in **Figs 12.34, 12.35** and **12.36**. The time-domain shape of each tone pulse is intentionally shaped to produce a very compact frequency spectra. The four tone pulses are spaced in time and then combined to produce the composite output shown. Unlike other modulation schemes, the CLOVER modulation spectra is the same for all modulation modes.

Modulation

Data is modulated on a CLOVER-II signal by varying the phase and/or amplitude of the tone pulses. Further, all data modulation is differential on the same tone pulse; data is represented by the phase (or amplitude) difference from one pulse to the next. For example, when binary phase modulation is used, a data change from "0" to "1" may be represented by a change in the phase of tone pulse 1 by 180° between the first and second occurrence of

Table 12.11
CLOVER-II Modulation Modes

As presently implemented, CLOVER-II supports a total of 7 different modulation formats: 5 using PSM and 2 using a combination of PSM and ASM (Amplitude Shift Modulation).

Name	Description	In-Block Data Rate
16P4A	16 PSM, 4-ASM	750 bps
16PSM	16 PSM	500 bps
8P2A	8 PSM, 2-ASM	500 bps
8PSM	8 PSM	375 bps
QPSM	4 PSM	250 bps
BPSM	Binary PSM	125 bps
2DPSM	2-Channel Diversity BPSM	62.5 bps

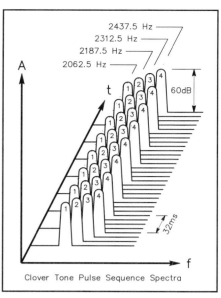

Fig 12.35—A frequency-domain plot of a CLOVER-II waveform.

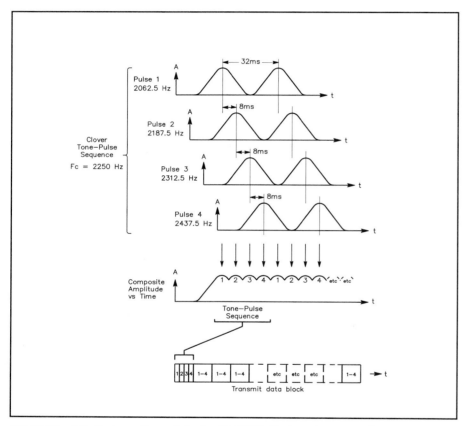

Fig 12.34—Amplitude vs time plots for CLOVER-II's four-tone waveform.

Table 12.12
Data Bytes Transmitted Per Block

Block Size	Reed-Solomon Encoder Efficiency			
	60%	75%	90%	100%
17	8	10	12	14
51	28	36	42	48
85	48	60	74	82
255	150	188	226	252

Table 12.13
Correctable Byte Errors Per Block

Block Size	Reed-Solomon Encoder Efficiency			
	60%	75%	90%	100%
17	1	1	0	0
51	9	5	2	0
85	16	10	3	0
255	50	31	12	0

that pulse. Further, the phase state is changed only while the pulse amplitude is zero. Therefore, the wide frequency spectra normally associated with PSK of a continuous carrier is avoided. This is true for all CLOVER-II modulation formats. The term "phase-shift modulation" (PSM) is used when describing CLOVER modes to emphasize this distinction.

Coder Efficiency Choices

CLOVER-II has four "coder efficiency" options: 60%, 75%, 90% and 100% ("efficiency" being the approximate ratio of real data bytes to total bytes sent). "60% efficiency" corrects the most errors but has

the lowest net data throughput. "100% efficiency" turns the encoder off and has the highest throughput but fixes no errors. There is therefore a tradeoff between raw data throughput vs the number of errors that can be corrected without resorting to retransmission of the entire data block.

Note that while the "In Block Data Rate" numbers listed in the table go as high as 750 bps, overhead reduces the net throughput or overall efficiency of a CLOVER transmission. The FEC coder efficiency setting and protocol requirements of FEC and ARQ modes add overhead and reduce the net efficiency.

Table 12.12 and **Table 12.13** detail the

relationships between block size, coder efficiency, data bytes per block and correctable byte errors per block.

CLOVER FEC

All modes of CLOVER-II use Reed-Solomon forward error correction (FEC) data encoding which allows the receiving station to correct errors without requiring a repeat transmission. This is a very powerful error correction technique that is not available in other common HF data modes such as AX.25 packet radio or AMTOR ARQ mode.

CLOVER ARQ

Reed-Solomon data coding is the primary means by which errors are corrected in CLOVER "FEC" mode (also called

Multilevel Digital Modulation Waveforms

Digital waveforms discussed so far have all used either on/off keying (OOK, that is Morse code, or CW) or frequency-shift keying (FSK, RTTY, AMTOR and packet radio). Both OOK and FSK are "simple" digital modulation waveforms; they have only two binary states that are represented by two radio-frequency states. In Morse code, the states are key-down = logical "1" and key-up = logical "0." In RTTY, AMTOR, PACTOR and packet radio, one frequency is "1" state, another is the "0" state.

More efficient use may be made of the spectrum by using multilevel modulation, in which one change in the transmitted signal may represent two or more bits of data. A simple example of multilevel modulation is quadrature phase-shift keying, known as QPSK. The simplest QPSK signal transmits a continuous carrier at a single frequency. Digital information is modulated on this carrier by changing the phase shift in 90° increments. Since there are four possible 90°-increment states (0°, 90°, 180° and 270°), four different modulation states may be signaled. Put another way, each phase state may be used to represent two bits of binary data. Examples of four common PSK modes are shown in **Fig A**.

Note that the phase of the transmitter carrier may be changed from any given state to any other state. Thus when using QPSK, if two bits of data change from "00" to "11," only one change to the transmitter carrier phase is required—from 0° to 270°. This observation illustrates the very important difference between modulation symbol rate (bauds) and data throughput rate (in bits-per-second, bits/s). In QPSK, bauds = 0.5 × bits/s (100 baud = 200 bits/s). This concept can be extended to 8PSK which has 8 phase states that represent 3 bits of data (bits/s = 3 × bauds). Carried further, each phase state in 16PSK modulation represents 4 bits of binary data and the throughput is 4 times the base symbol rate (bit/s = 4 × bauds).

Higher-level phase shift modulation schemes have been used (32PSK and 64PSK for example), but these systems require much more complex demodulator design. In particular, demodulator sensitivity to noise and distortion increases greatly as the number of possible phase states is increased. Consider the relatively simple QPSK example. The design-center phase states of 0°, 90°, 180°, and 270° represent the four possible modulation conditions. Ionosphere propagation, multipath signal reflections, and transmission distortion all conspire to insert phase "jitter" or uncertainty in the received signal. In QPSK, signals with a phase shift between 45° and 135° can be assumed to represent the 90° state, 135° to 225° for the 180° state and so on. The margin for error or "phase margin" for QPSK is ±45°. A similar calculation for 16PSK shows that its phase margin is just ±12.25°. If we consider use of a 10.000 MHz carrier with 16PSK, the period of the carrier sine wave is 0.100 microsecond and the allowable phase jitter corresponds to a time uncertainty of only ±0.003403 ms, or ±3.403 ns. Obviously, very stable phase references must be used in a 16PSK system and it does not take very much distortion or noise to make correct data detection impossible. However, such systems are commonly used in telephone-line modems.

Telephone modems carry the multilevel concept one step further and use amplitude-level modulation (amplitude-shift keying, ASK) in addition to PSK modulation. If two-level ASK is used with 16PSK, a total of 32 states may be sent. Similarly, use of 4ASK and 16PSK gives 64 unique states for each modulation change. This is commonly called "QAM" for Quadrature Amplitude Modulation and is the modulation used by most 9600-baud telephone modems. Each modulation change can represent the state of 6 bits of binary data; the data

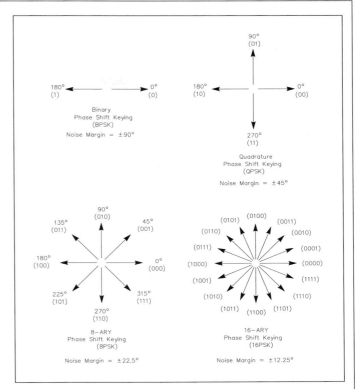

Fig A

throughput is 6 times the base modulation symbol rate (bits/s = 6 × bauds). As noted above, complex multilevel modulation schemes require very complex and expensive demodulators that are very susceptible to noise and distortion. Fortunately, modern telephone lines are relatively noise-free and stable. By use of error correction and line distortion equalization, high "speed" data transmission via telephone line is now in common use.

Unfortunately, these same techniques cannot be directly applied to radio data transmission, particularly to HF signals. Long-range HF signals are propagated via the ionosphere, which is not stable or well defined from instant to instant. Ionosphere reflection height and signal attenuation varies widely with time of day, geographic location, and solar activity. Moreover, noise levels on HF vary considerably with location as well as time of day.

With multipath propagation, multiple copies of an original signal are summed at the receiving antenna. Since each signal travels via a different path, the propagation delays are different. Multiple signals therefore arrive at the receiver at slightly different times and the "mark-to-space" transition time is different for each signal. This causes "smearing" of the exact transition times. Multipath distortion occurs commonly on HF when both single-hop and multiple-hop signals arrive at the receiving antenna with similar strengths. Multipath distortion is also common on VHF and UHF signals in highly populated areas where large buildings provide reflecting surfaces.

The HF environment is therefore complicated and hostile to data transmission. Modulation techniques that work well on stable and predictable telephone lines may also be usable for VHF and UHF radio systems, but they may seldom be directly applied to HF data radio systems. Further, data format protocols that were devised for the stable phone-line environment are generally not optimum for use on HF data radio. For example, both the FSK modulation and the protocol used for AX.25 packet radio lead to serious problems when used on HF signals.

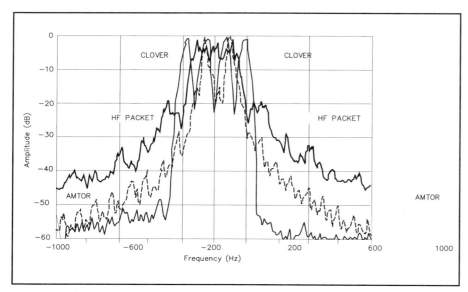

Fig 12.36—Spectra plots of AMTOR, HF packet-radio and CLOVER-II signals.

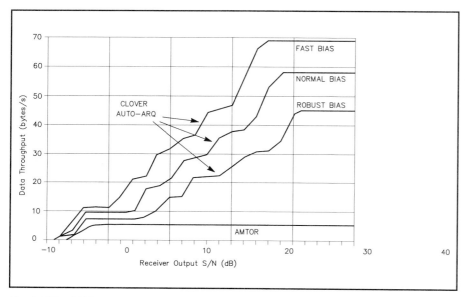

Fig 12.37—ARQ-mode data throughput vs receiver S/N ratio for AMTOR and three different CLOVER-II configurations.

II signal energy is concentrated within ±250 Hz of the center frequency. Therefore, CLOVER-II signals can be spaced as closely as 500 Hz from any data-mode signal with very little cochannel interference. Tests show that "cross-talk" between two 500-Hz spaced CLOVER-II signals is less than 50 dB. This is much better than the common spacing of AMTOR (1000 Hz) or HF packet signals (2000 Hz).

Fig 12.37 shows throughput vs S/N for AMTOR and various modes of CLOVER-II. For all values of S/N and all modes of CLOVER, the data throughput obtainable using CLOVER-II is higher than that achievable when using AMTOR. In addition, CLOVER may be used to send full 8-bit computer data whereas AMTOR is restricted to either the Baudot RTTY characters set (CCIR-476/625) or the printable subset of ASCII (ASCII-over-AMTOR).

RTTY has better automatic receive decoding performance than Morse code and is relatively inexpensive, but offers no automatic error correction. AMTOR includes error correction, has good performance under weak signal conditions and is relatively inexpensive. However, its maximum data throughput rate is low and it cannot support transmission of 8-bit data files.

AX.25 packet radio is inexpensive but its performance on HF is typically very poor. This is due both to the popular choice of modulation (200 Hz shift, 300 baud FSK) and the AX.25 protocol which was not designed to handle the burst-type errors that are common to HF propagation. The MIL-188/110A (now proposed Federal Standard pFS-1052) "Serial, Single-Tone" waveform works well on HF and can pass error-corrected 8-bit data with a throughput of up to 2400 baud. However, modems for this mode are presently very expensive, the occupied bandwidth of 3000 Hz is very wide, and ARQ or adaptive ARQ modes are still under development.

In comparison, CLOVER-II modems are moderately expensive but will adaptively match existing signal conditions and provide high data throughput rates when conditions permit. CLOVER-II will pass full 8-bit data and a CLOVER signal is the most bandwidth efficient of all modes considered.

How do They Compare?

An extensive comparison of digital modes was written by Tim Riley; Dennis Bodson, W4PWF; Stephen Rieman and Teresa Sparkman. See "A Comparison of HF Digital Protocols," *QST*, July 1996, page 35.

"broadcast mode"). In ARQ mode, CLOVER-II employs a three-step strategy to combat errors. First, channel parameters are measured and the modulation format is adjusted to minimize errors and maximize data throughput. This is called the "Adaptive ARQ Mode" of CLOVER-II. Second, Reed-Solomon encoding is used to correct a limited number of byte errors per transmitted block. Finally, only those data blocks in which errors exceed the capacity of the Reed-Solomon decoder are repeated (selective block repeat). Unlike AX.25 packet radio, CLOVER-II does not repeat blocks which have been received correctly.

With seven different modulation formats, four data block lengths (17, 51, 85 or 255 bytes) and four Reed-Solomon coder efficiencies (60%, 75%, 90% and 100%), there are 112 (7×4×4) different waveform modes that could be used to send data via CLOVER. Once all of the determining factors are considered, however, there are 8 different waveform combinations which are actually used for FEC and/or ARQ modes.

CLOVER vs AMTOR vs Packet

Fig 12.36 shows the modulator output spectra of CLOVER-II, AMTOR and HF packet radio. Nearly all of the CLOVER-

CLOVER-2000

CLOVER-2000 is a faster version of CLOVER (about four times faster) that uses eight tone pulses, each of which is 250-Hz wide, spaced at 250-Hz centers, contained within the 2-kHz bandwidth between 500 and 2500 Hz. The eight tone pulses are sequential, with only one tone being present at any instant and each tone lasting 2 ms. Each frame consists of eight tone pulses lasting a total of 16 ms, so the base modulation rate of a CLOVER-2000 signal is always 62.5 symbols per second (regardless of the type of modulation being used). CLOVER-2000's maximum raw data rate is 3000 bits per second. Allowing for overhead, CLOVER-2000 can deliver error-corrected data over a standard HF SSB radio channel at up to 1994 bits per second, or 249 charcaters (8-bit bytes) per second. These are the uncompressed data rates; the maximum throughput is typically doubled for plain text if compression is used. The effective data throughput rate of CLOVER-2000 can be even higher when binary file transfer mode is used with data compression.

The binary file transfer protocol used by HAL Communications operates with a terminal program explained in the HAL E2004 engineering document listed under references. Data compression algorithms tend to be context sensitive—compression that works well for one mode (e.g. text), may not work well for other data forms (graphics, etc.). The HAL terminal program uses the PK-WARE compression algorithm which has proved to be a good general-purpose compressor for most computer files and programs. Other algorithms may be much more efficient for some data formats, particularly for compression of graphic image files and digitized voice data. The HAL Communications CLOVER-2000 modems can be operated with other data compression algorithms in the users' computers.

CLOVER-2000 is similar to the previous version of CLOVER, including the transmission protocols and Reed-Solomon error detection and correction algorithm. The original descriptions of the CLOVER Control Block (CCB) and Error Correction Block (ECB) still apply for CLOVER-2000, except for the higher data rates inherent to CLOVER-2000. Just like CLOVER, all data sent via CLOVER-2000 is encoded as 8-bit data bytes and the error-correction coding and modulation formatting processes are transparent to the data stream—every bit of source data is delivered to the receiving terminal without modification. Control characters and special "escape sequences" are not re-quired or used by CLOVER-2000. Compressed or encrypted data may therefore be sent without the need to insert (and filter) additional control characters and without concern for data integrity. Five different types of modulation may be used in the ARQ mode—BPSM (Binary Phase Shift Modulation), QPSM (Quadrature PSM), 8PSM (8-level PSM), 8P2A (8PSM + 2-level Amplitude-Shift Modulation), and 16P4A (16 PSM plus 4 ASM).

The same five types of modulation used in ARQ mode are also available in Broadcast (FEC) mode, with the addition of 2-Channel Diversity BPSM (2DPSM). Each CCB is sent using 2DPSM modulation, 17-byte block size, and 60% bias. The maximum ARQ data throughput varies from 336 bits per second for BPSM to 1992 bits per second for 16P4A modulation. BPSM is most useful for weak and badly distorted data signals while the highest format (16P4A) needs extremely good channels, with high SNRs and almost no multipath.

Most ARQ protocols designed for use with HF radio systems can send data in only one direction at a time. For example, when using CCIR-476/625 (SITOR) or PACTOR, one station sends all of its data, ending the transmission with an "OVER" command. The second station may then send its information. Because CLOVER-2000 does not need an "OVER" command, data may flow in either direction at any time. The CLOVER ARQ time frame automatically adjusts to match the data volume to be sent in either or both directions. When first linked, both sides of the ARQ link exchange information using six bytes of the CCB. When one station has a large volume of data buffered and ready to send, ARQ mode automatically shifts to an expanded time frame during which one or more 255 byte data blocks are sent. If the second station also has a large volume of data buffered and ready to send, its half of the ARQ frame is also expanded. Either or both stations will shift back to CCB level when all buffered data has been sent. This feature provides the benefit of full-duplex data transfer but requires use of only simplex frequencies and half-duplex radio equipment. This two-way feature of CLOVER can also provide a back-channel order-wire capability. Communications may be maintained in this "chat" mode at 55 words per minute, which is more than adequate for real-time keyboard-to-keyboard communications.

Two different CLOVER-2000 modems are available from HAL Communications, the PCI-4000/2K and the DSP-4100/2K. The PCI-4000/2K is for use inside dedicated desk-top personal computers. The PCI-4000/2K may be installed in any IBM-compatible personal computer that uses an 80386 or faster microprocessor (386, 486, Pentium, etc.) and supports the ISA PC plug-in card bus. The DSP-4100/2K is for connection to a laptop or non-IBM PC, since it is a stand-alone DSP modem that may be used with any computer or data terminal having an RS-232 port.

PSK31

Peter Martinez, G3PLX, who was instrumental in bringing us AMTOR, developed PSK31 for real time keyboard-to-keyboard QSOs. This section was adapted from an article in *Radcom*, Jan 1999. The name derives from the modulation type (phase shift keying) and the data rate, which is actually 31.25 bauds. PSK31 is a robust mode for HF communications that features the 128 ASCII (Internet) characters and the full 256 ANSI character set. This mode works well for two-way QSOs and for nets. Time will tell if PSK31 will replace Baudot RTTY on the amateur HF bands.

Morse code uses a single carrier frequency keyed on and off as dits and dahs to form characters. RTTY code shifts between two frequencies one for *mark* (1) the other for *space* (0). Sequences of marks and spaces comprise the various characters.

Martinez devised a new variable-length code for PSK31 that combines the best of Morse and RTTY. He calls it *Varicode* because a varying number of bits are used for each character (see **Fig A**). Much like the Morse code, the more commonly used letters have shorter codes.

As with RTTY, there is a need to signal the gaps between characters. The Varicode does this by using "00" to represent a gap. The Varicode is structured so that two zeros never appear together in any of the combinations of 1s and 0s that make up the characters. In on-the-air tests, Martinez has verified that the unique "00" sequence works significantly better than RTTY's stop code for keeping the receiver synchronized.

With Varicode, a typing speed of about 50 words per minute requires a 32 bit/s transmission rate. Martinez chose 31.25 bit/s because it can be easily derived from the 8-kHz sample rate used in many DSP systems.

The shifting carrier phase generates sidebands 31.25 Hz from the carrier. These are used to synchronize the receiver with the transmitter. The required bandwidth is less than that for the FSK signal of 100 baud Baudot RTTY, as shown in **Fig B**.

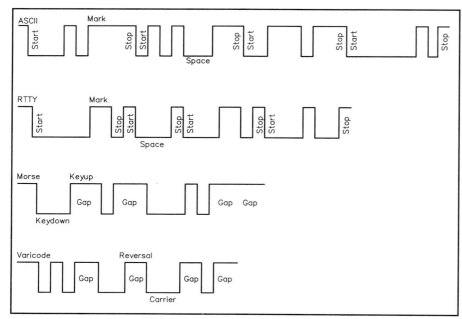

Fig A—Codes for the word "ten" in ASCII, Baudot, Morse and Varicode.

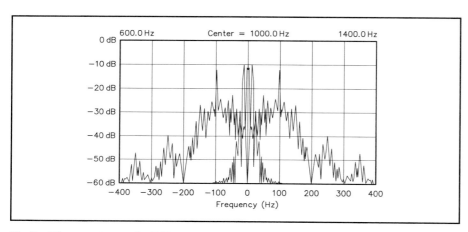

Fig B—The spectrum of a PSK31 signal compared to that of a 100 baud, 200-Hz-shift FSK signal.

Image Modes

FACSIMILE

This section, by Dennis Bodson, W4PWF, and Steven Karty, N5SK, covers several facsimile systems in most common Amateur Radio use today. For further information on the area of facsimile, its history, and the development of related standards associated with this mode, refer to *FAX: Facsimile Technology and Systems*.[1] The subject of Weather fax, while of interest to many amateurs, is not a primary activity of the Amateur Radio Service. Information on this subject is contained in the *Weather Satellite Handbook*.[2]

FACSIMILE OVERVIEW

Facsimile (fax) is a method for transmitting very high resolution still pictures using voice bandwidth radio circuits. The narrow bandwidth of the fax signal, equivalent to SSTV, provides the potential for worldwide communications on the HF bands. Fax is the oldest of the image-transmitting technologies and has been the primary method of transmitting newspaper photos and weather charts. Fax is also used to transmit high-resolution cloud images from both polar-orbit and geostationary satellites. Many of these images are retransmitted using fax on the HF bands.

The resolution of typical fax images greatly exceeds what can be obtained using SSTV or even conventional television (typical images will be made up of 800 to 1600 scanning lines). This high resolution is achieved by slowing down the rate at which the lines are transmitted, resulting in image transmission times of 4 to 10 minutes. Prior to the advent of digital technology, the only practical way to display such images was to print each line directly to paper as it arrived. The mechanical systems for accomplishing this are known as facsimile *recorders* and are based on either photographic media (a modulated light source exposing film or paper) or various types of direct printing technologies including electrostatic and electrolytic papers.

Modern desktop computers have virtually eliminated bulky fax recorders from most amateur installations. Now the incoming image can be stored in computer memory and viewed on a standard TV monitor or a high-resolution computer

ERROR CORRECTION

Martinez has added error correction to PSK31 by using QPSK (quatenary phase shift keying) and a *convolutional encoder* to generate one of four different phase shifts that corresond to patterns of five successive data bits. At the receiving end, a Viterbi decoder is used to correct errors. There are 32 possible sequences for five bits. The Viterbi decoder tracks these possibilities while discarding the least likely and retaining the most likely sequences. Retained sequences are given a score that is based on the running total. The most accurate sequence is reported, and thus errors are corrected.

Operating PSK31 in the QPSK mode should result in 100% copy under most conditions, but at a price. Tuning is twice as critical as it is with BPSK. An accuracy of less than 4 Hz is required for the Viterbi decoder to function properly.

GETTING STARTED

In addition to a transceiver and antenna, you only need a computer with a Windows operating system and a 16-bit sound card to receive and transmit PSK31. Additional information and software is available for free download over the Web. Use a search engine to find PSK31 information and links to downloads.

[1]McConnell, Ken, Bodson, Dennis, and Urban, Steve, *FAX: Facsimile Technology and Systems*, 3rd Ed., Artech House, 1999,
[2]Taggart, R.E., *Weather Satellite Handbook*, 5th Ed. (Newington: ARRL, 1994).

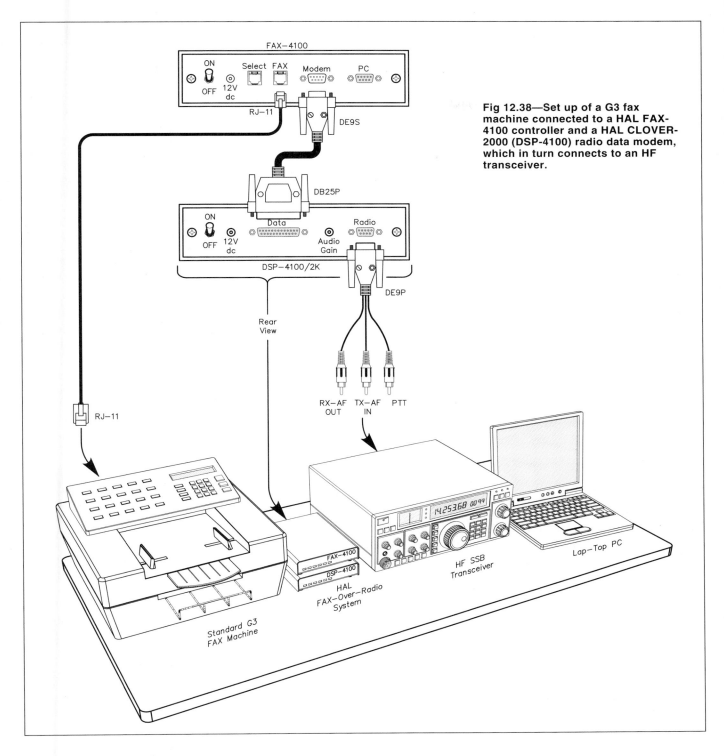

Fig 12.38—Set up of a G3 fax machine connected to a HAL FAX-4100 controller and a HAL CLOVER-2000 (DSP-4100) radio data modem, which in turn connects to an HF transceiver.

graphics display. The use of a color display system makes it entirely practical to transmit color fax images when band conditions permit. The same computer-based system that handles fax images is often capable of SSTV operation as well, blurring what was once a clear distinction between the two modes. The advent of the personal computer has provided amateurs with a wide range of options within a single imaging installation. SSTV images of low or moderate resolution can be transmitted when crowded band conditions favor short frame transmission times. When band conditions are stable and interference levels are low, the ability to transmit very high resolution fax images is just a few keystrokes away!

HARDWARE AND SOFTWARE

Electromechanical fax equipment has been replaced by personal computer hardware and software. The computer allows reception and transmission of various line per minute rates and indices of cooperation by simply pressing a key or by pointing and clicking a mouse. Many fax programs are available as either commercial software or shareware. Usually, the shareware packages (and often trial versions of the commercial packages) are available by downloading from the Internet.

A good starting point is the ARRL

software repositories. To get to them, set your browser to the *ARRL Web* and go to the FTP (files) link in the site index. You can use any commercial search site to look for "fax" AND "software." Examples of several fax programs are as follows:

JVFAX is a very popular fax program. It is DOS-based program with a large number of options for installation. It can receive and transmit several fax formats, black-and-white and color. Your computer's serial port, connected to a very simple interface, provides the connection to your transceiver.

The *FAX 480* software program can also be used with fax as well as SSTV. For more information on this program and others including website addresses, see the July 1998 *QST* article "FAX 480 and SSTV Interfaces and Software" page 32. A copy for downloading of the free software program *vester_n.zip* for *FAX 480* can be found online at the Oakland University FTP site (see the **References** chapter). This program also uses a simple interface almost identical to that for *JVFAX*.

Weatherman is a DOS-based program, using a SoundBlaster (or compatible) card as the interface. The program is shareware and provides receive-only capability. A single, shielded wire from your receiver audio output to the computer audio input is the only connection needed.

WXSat operates under *Windows 3.X*. While specifically set up to decode and store weather-satellite APT pictures, it can also be used for HF-fax reception.

Both *Weatherman* and *WXSat* are samples of what you can find during a search on the Internet. Often, programs are offered and then either withdrawn or improved over the versions previously distributed—to get the latest and greatest you have to periodically search and see what comes up. If you use an online service such as CompuServe or AOL, they are another source of fax software. Check their ham forums or sections for listings.

Many commercial multimode controllers either contain software to receive and transmit fax, or are compatible with PC-hosted software. Available controller sup-pliers include MFJ, Timewave, and Kantronics; additional software may be required for the Kam Plus. Check the advertising pages of *QST* for the latest units available.

One well-known fax page on the Internet, complete with downloadable software, is posted and maintained by Marius Rensen; it contains listings of commercial fax transmissions for you to test your software or just SWL for interest. See the **References** chapter for the URL. Before using a program taken from any Internet source, check other sources for newer versions. It is not uncommon to have older versions posted on one place and newer versions in another. It is a good idea to virus check the software before and after unzipping.

Image transmission using voice bandwidth is a trade-off between resolution and time. In the section on slow-scan television, standards are described that permit 240-line black-and-white images to be transmitted in about 36 seconds while color images of similar resolution require anywhere from 72 to 188 seconds, depending on the color format. In terms of resolution, 240-line SSTV images are roughly equivalent to what you would obtain with a standard broadcast TV signal recorded on a home VCR. This is more than adequate for routine video communication, but there are many situations that demand images of higher resolution.

HAL Communications Corporation has developed an interesting system which enables a standard fax machine (Group 3 or G3) to send commercial fax images over HF radio. HAL Communications accomplishes this with just two small ancillary devices, which connect between a standard fax machine and an ordinary HF radio transceiver (see **Fig 12.38**). This method is frequently referred to as "G3 fax over radio." Any G3 fax machine can be connected to the HAL FAX-4100 controller with just a standard RJ-11 modular connector. The FAX-4100 controller connects directly to the HAL CLOVER-2000 (DSP-4100) radio data modem, which in turn connects to the HF transceiver. This entire setup is duplicated at the opposite end of the link.

A "call" is initiated from the fax machine keypad just as if the fax machine were connected to a phone line. The FAX-4100 controller includes a built-in 9600-baud G3 modem which emulates the telephone system: The controller at the initiating end answers the ring from the originating fax machine, establishes the HF radio link (based on the "phone number"), and handshakes with the controller at the other end to start the receiving fax machine. Fax image data then passes from the fax machine into the controller's memory at the originating end. The controller also establishes a data link between the CLOVER-2000 modems at both ends, then passes the fax data through them and the controller at the receiving end, and finally into the receiving G3 fax machine. HAL has automated the HF radio operating procedures. To the user, sending a fax over HF radio is a simple three-step process:

1. Lay the page(s) on the fax machine.
2. Enter the ID number of the other station.
3. Push GO on the fax machine.

Housekeeping control functions and indications are also automated, feeding messages back to the fax machine whenever possible (link failed, other station not available, etc.). A full page can be sent in 2 to 6 minutes, depending upon ionospheric conditions and density of the page to be transmitted. The entire link set up and maintenance procedure is transparent to the fax operator, who need not know nor care that an HF radio system is part of the fax link. It all works just like a standard fax telephone transmission. An additional piece of equipment is available from HAL to enable the same fax machine to be shared between HF radio and conventional telephone lines. The HAL LI-4100 Line Interface is a "smart switch" that can be connected between the fax machine, the FAX-4100 controller, and up to two telephone lines.

SLOW-SCAN TELEVISION (SSTV)

An ancient Chinese proverb states: "A picture is worth a thousand words." It's still true today. Sight is our highest bandwidth sense and the primary source of information about the world around us. What would you think about a TV news program without pictures about the stories? Would you enjoy reading the comics if there were no drawings with the text?

Do you close your eyes when talking to someone in person? Many hams feel the same way about conversing with Amateur Radio: sending images is a wonderful way to enhance communication. This material was written by John Langner, WB2OSZ.

For decades only a dedicated few kept SSTV alive. The little commercial equipment was very expensive and home brew-ing was much too complicated for most people. Early attempts at computer-based systems were rather crude and frustrating to use.

The situation has changed dramatically in recent years. There is now a wide variety of commercial products and home-brew projects to fit every budget, and SSTV activity is experiencing rapid growth. There is even software that uses the popular Sound Blaster computer sound card for SSTV.

The early SSTV 8-second transmission standard is illustrated in **Fig 12.49**. Audio tones in the 1500 to 2300-Hz range represent black, white, and shades of gray. A short 1200-Hz burst separates the scan lines, and a longer 1200-Hz tone signals the beginning of a new picture.

Color SSTV Evolution

The early experimenters weren't content with only black and white (B&W) images and soon devised a clever way to send color pictures with B&W equipment. The transmitting station sends the same image three times, one each with red, green and blue filters in front of the TV camera lens. The receiving operator took three long-exposure photographs of the screen, placing red, green and blue filters in front of the film camera's lens at the appropriate times. This was known as the "frame sequential" method.

In the 1970s, it became feasible to save these three images in solid-state memory and simultaneously display them on an ordinary color TV. But, the frame-sequential method had some drawbacks. As the first frame was received you'd see a red and black image. During the second frame, green and yellow would appear. Blue, white, and other colors wouldn't show up until the final frame. Any noise (QRM or QRN) could ruin the image registration (the overlay of the frames) and spoil the picture.

The next step forward was the "line

sequential" method. Each line is scanned 3 times: once each for the red, green, and blue picture components. Pictures could be seen in full color as they were received and registration problems were reduced. The Wraase SC-1 modes are examples of early line-sequential color transmission. They have a horizontal sync pulse for each of the color component scans. The major weakness here is that if the receiving end gets out of step, it won't know which scan represents which color.

Rather than sending color images with

the usual RGB (red, green, blue) components, Robot Research used luminance and chrominance signals for their 1200C modes. The first half or two thirds of each scan line contains the luminance information which is a weighted average of the R, G and B components. The remainder of each line contains the chrominance signals with the color information. Existing B&W equipment could display the B&W-compatible image on the first part of each scan line and the rest would go off the edge of the screen. This compatibility was very

Table 12.15
SSTV Transmission Characteristics

Mode	Designator	Color Type	Scan Time (sec)	Scan Lines	Notes
AVT	24	RGB	24	120	D
	90	RGB	90	240	D
	94	RGB	94	200	D
	188	RGB	188	400	D
	125	BW	125	400	D
Martin	M1	RGB	114	240	B
	M2	RGB	58	240	B
	M3	RGB	57	120	C
	M4	RGB	29	120	C
HQ	HQ1	YC	90	240	G
	HQ2	YC	112	240	G
Pasokon TV	P3	RGB	203	16+480	
	P5	RGB	305	16+480	
	P7	RGB	406	16+480	
Robot	8	BW	8	120	A,E
	12	BW	12	120	E
	24	BW	24	240	E
	36	BW	36	240	E
	12	YC	12	120	
	24	YC	24	120	
	36	YC	36	240	
	72	YC	72	240	
Scottie	S1	RGB	110	240	B
	S2	RGB	71	240	B
	S3	RGB	55	120	C
	S4	RGB	36	120	C
	DX	RGB	269	240	B
Wraase SC-1	24	RGB	24	120	C
	48	RGB	48	240	B
	96	RGB	96	240	B
Wraase SC-2	30	RGB	30	128	
	60	RGB	60	256	
	120	RGB	120	256	
	180	RGB	180	256	
Pro-Skan	J120	RGB	120	240	
WinPixPro	GVA 125	BW	125	480	
	GVA 125	RGB	125	240	
	GVA 250	RGB	250	480	
JV Fax	JV Fax Color	RGB	variable	variable	F
FAX480	Fax 480	BW	138	480	
	Truscan	BW	128	480	H
	Colorfax	RGB	384	480	I

Notes

RGB—Red, green and blue components sent separately.

YC—Sent as Luminance (Y) and Chrominance (R-Y and B-Y).

BW—Black and white.

A—Similar to original 8-second black & white standard.

B—Top 16 lines are gray scale. 240 usable lines.

C—Top 8 lines are gray scale. 120 usable lines.

D—AVT modes have a 5-second digital header and no horizontal sync.

E—Robot 1200C doesn't really have B&W mode but it can send red, green or blue memory

separately. Traditionally, just the green component is sent for a rough approximation of a b&w image.

F—JV Fax Color mode allows the user to set the number of lines sent, the maximum horizontal resolution is slightly less than 640 pixels. This produces a slow but very high resolution picture. SVGA graphics are required.

G—Available only on Martin 4.6 chipset in Robot 1200C.

H—Vester version of FAX480 (with VIS instead of start signal and phasing lines).

I—Trucolor version of Vester Truscan.

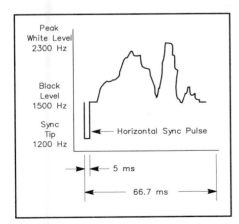

Fig 12.49—Early SSTV operators developed a basic 8-second black and white transmission format. The sync pulses are often called "blacker than black." A complete picture would have 120 lines (8 seconds at 15 ms per line). Horizontal sync pulses occur at the beginning of every line; a 30 ms vertical sync pulse precedes each frame.

SSTV Glossary

ATV—Amateur Television. Sending pictures by Amateur Radio. You'd expect this abbreviation to apply equally to fast-scan television (FSTV), slow-scan television (SSTV) and facsimile (fax), but it's generally applied only to FSTV.

AVT—Amiga Video Transceiver.
1) Interface and software for use with an Amiga computer, developed by Ben Blish-Williams, AA7AS, and manufactured by Advanced Electronic Applications (AEA);
2) a family of transmission modes first introduced with the AVT product.

Back porch—The blank part of a scan line immediately following the horizontal sync pulse.

Chrominance—The color component of a video signal. NTSC and PAL transmit color images as a black-and-white compatible luminance signal along with a color subcarrier. The subcarrier phase represents the hue and the subcarrier's amplitude is the saturation. Robot color modes transmit pixel values as luminance (Y) and chrominance (R-Y [red minus luminance] and B-Y [blue minus luminance]) rather than RGB (red, green, blue).

Demodulator—For SSTV, a device that extracts image and sync information from an audio signal.

Field—Collection of top to bottom scan lines. When interlaced, a field does not contain adjacent scan lines and there is more than one field per frame.

Frame—One complete scanned image. The Robot 36-second color mode has 240 lines per frame. NTSC has 525 lines per frame with about 483 usable after subtracting vertical sync and a few lines at the top containing various information.

Frame Sequential—A method of color SSTV transmission which sent complete, sequential frames of red, then green and blue. Now obsolete.

Front porch—The blank part of a scan line just before the horizontal sync.

FSTV—Fast-Scan TV. Same as common, full-color, motion commercial broadcast TV.

Interlace—Scan line ordering other than the usual sequential top to bottom. For example, NTSC sends a field with just the even lines in $1/60$ second, then a field with just the odd lines in $1/60$ second. This results in a complete frame 30 times a second. AVT "QRM" mode is the only SSTV mode that uses interlacing.

Line Sequential—A method of color SSTV transmission that sends red, green, and blue information for *each sequential scan line*. This approach allows full-color images to be viewed during reception.

Luminance—The brightness component of a video signal. Usually computed as Y (the luminance signal) = 0.59 G (green) + 0.30 R (red) + 0.11 B (blue).

Martin—A family of amateur SSTV transmission modes developed by Martin Emmerson, G3OQD, in England.

NTSC—National Television System Committee. Television standard used in North America and Japan.

PAL—Phase alteration line. Television standard used in Germany and many other parts of Europe.

Pixel—Picture element. The dots that make up images on a computer's monitor.

P7 monitor—SSTV display using a CRT having a very-long-persistence phosphor.

RGB—Red, Green, Blue. One of the models used to represent colors. Due to the characteristics of the human eye, most colors can be simulated by various blends of red, green, and blue light.

Robot—(1) Abbreviation for Robot 1200C scan converter; (2) a family of SSTV transmission modes introduced with the 1200C.

Scan converter—A device that converts one TV standard to another. For example, the Robot 1200C converts SSTV to and from FSTV.

Scottie—A family of amateur SSTV transmission modes developed by Eddie Murphy, GM3SBC, in Scotland.

SECAM—Sequential color and memory. Television standard used in France and the Commonwealth of Independent States.

SSTV—Slow Scan Television. Sending still images by means of audio tones on the MF/HF bands using transmission times of a few seconds to a few minutes.

Sync—That part of a TV signal that indicates the beginning of a frame (vertical sync) or the beginning of a scan line (horizontal sync).

VIS—Vertical Interval Signaling. Digital encoding of the transmission mode in the vertical sync portion of an SSTV image. This allows the receiver of a picture to automatically select the proper mode. This was introduced as part of the Robot modes and is now used by all SSTV software designers.

Wraase—A family of amateur SSTV transmission modes first introduced with the Wraase SC-1 scan converter developed by Volker Wraase, DL2RZ, of Wraase Electronik, Germany.

---

beneficial when most people still had only B&W equipment.

The luminance-chrominance encoding made more efficient use of the transmission time. A 120-line color image could be sent in 12 s, rather than the usual 24 s. Our eyes are more sensitive to details in changes of brightness than color, so the time could be used more efficiently by devoting more time to luminance than chrominance. The NTSC and PAL broadcast standards also take advantage of this

vision characteristic and use less bandwidth for the color part of the signal.

The 1200C introduced another innovation: it encodes the transmission mode in the vertical sync signal. By using narrow FSK encoding around the sync frequency, compatibility was maintained. This new signal just looked like an extra-long vertical sync to older equipment. (See the sidebar "Examining Robot's Vertical-Interval-Signaling (VIS) Code" for more details.)

The luminance-chrominance encoding

offers some benefits but image quality suffers. It is acceptable for most natural images but looks bad for sharp, high-contrast edges, which are more and more common as images are altered via computer graphics. As a result, all newer modes have returned to RGB encoding.

The Martin and Scottie modes are essentially the same except for the timings. They have a single horizontal sync pulse for each set of RGB scans. Therefore, the receiving end can easily get back in step if

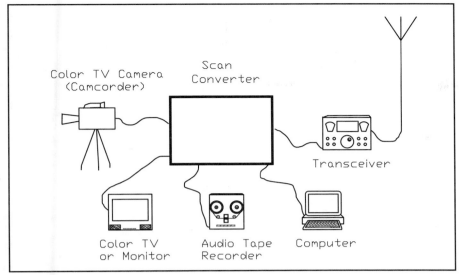

Fig 12.50—Diagram of an SSTV station based on a scan converter.

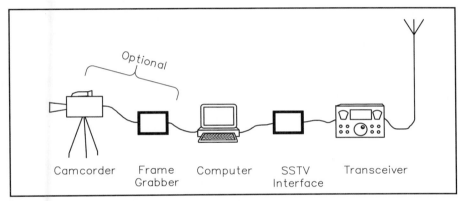

Fig 12.51—A modern, PC-based SSTV station.

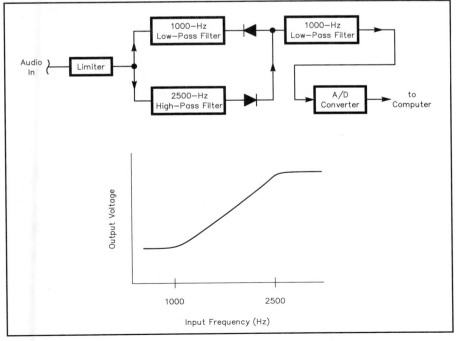

Fig 12.52—Block diagram of an analog SSTV demodulator.

Examining Robot's Vertical-Interval-Signaling (VIS) Code

The original 8-second black-and-white SSTV-image standard used a 30-millisecond, 1200-Hz pulse to signal the beginning of a new frame. In the Robot 1200C, Robot Research increased the vertical sync period by a factor of 10, encoded 8 bits of digital data into it and called it *vertical-interval signaling* (VIS). VIS is composed of a start bit, 7 data bits, an even parity bit, and a stop bit, each 30 milliseconds long. (See Fig A).

Since then, inventors of new SSTV modes (Martin, Scottie, AVT, etc) have adopted Robot's scheme and assigned codes to their particular mode that are unused by the Robot modes. So, each of the SSTV transmission modes has a unique VIS code. This allows new equipment to automatically select any of the new SSTV modes while maintaining compatibility with the older equipment.—WB2OSZ

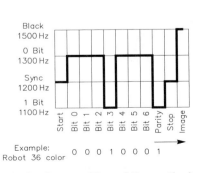

Example: Robot 36 color   0 0 0 1 0 0 0 1

Fig A—Composition of the vertical interval signaling (VIS) code.

synchronization is temporarily lost. Although they have horizontal sync, some implementations ignore them on receive. Instead, they rely on very accurate time bases at the transmitting and receiving stations to keep in step. The advantage of this "synchronous" strategy is that missing or corrupted sync pulses won't disturb the received image. The disadvantage is that even slight timing inaccuracies produce slanted pictures.

In the late 1980s, yet another incompatible mode was introduced. The AVT mode is different from all the rest in that it has *no horizontal sync*. It relies on very accurate oscillators at the sending and receiving stations to maintain synchronization. If the beginning-of-frame sync is missed, it's all over. There is no way to determine where a scan line begins. However, it's much harder to miss the 5-s header than the 300-ms VIS code. Redundant information is encoded 32 times and a more powerful error-detection scheme is used. It's only necessary to receive a small part of the AVT header in order to achieve synchronization. After this, noise can wipe out parts of the image, but image alignment and colors remain correct. **Table 12.15** lists characteristics of common modes.

Scan Converters

A scan converter is a device that converts signals from one TV standard to another. In this particular case we are interested in converting between SSTV, which can be sent through audio channels, and fast scan (broadcast or ATV), so we can use ordinary camcorders and color televisions to generate and display pictures. From about 1985 to 1992, the Robot 1200C was king.

Fig 12.50 shows a typical SSTV station built around a scan converter such as the Robot 1200C or a SUPERSCAN 2001. The scan converter has circuitry to accept a TV signal from a camera and store it in memory. It also generates a display signal for an ordinary television set. The interface to the radio is simply audio in, audio out and a push-to-talk (PTT) line. In the early days, pictures were stored on audio tape, but now computers store them on disks. Once a picture is in a computer, it can be enhanced with paint programs.

This is the easiest approach. Just plug in the cables, turn on the power and it works. Many people prefer special dedicated hardware, but most of the recent growth of SSTV has been from these lower cost PC-based systems.

SSTV with a Computer

There were many attempts to use early home computers for SSTV. Those efforts were hampered by very small computer memories, poor graphics capabilities and poor software development tools.

Surprisingly, little was available for the ubiquitous IBM PC until around 1992, when several systems appeared in quick succession. By this time, all new computers had a VGA display, which is required for this application. Most new SSTV stations look like **Fig 12.51**. Some sort of interface is used to get audio in and out of the computer. These can be external interfaces connected to a serial or printer port, an internal card specifically for SSTV or even a peripheral audio card. IBM-type PC compatible computers with VGA video display monitors can also be used with their existing SoundBlaster-compatible sound boards for the interfaces, if software such as WinSkan and WinPix Pro are used. Most of the work is done in software. System updates are performed by reading a floppy disk instead of changing EPROMs or other components. Most of these software programs are based on the work of Ben Vester, K3BC. These computer programs include Vester Truscan, Pasokon TV Lite, ProScan, JVFAX, and HamComm; they all use a simple "clipper" hardware interface, which can easily be built with less than $15 worth of RadioShack parts, because the computer program does all of the processing work previously done by more expensive hardware. See the July 1998 issue of *QST* for "FAX 480 and SSTV Interfaces and Software" on page 32. The URL for downloading Vester's software is **ftp://oak. oakland.edu/pub/hamradio/arrl/bbs/ programs.**

How It Works

Transmitting SSTV images with a computer is quite simple. All you need to do is generate fairly accurate tones and change them at the proper pixel rate. Tones in the range of 1500 to 2300 Hz correspond to the pixel intensities, and most modes use 1200-Hz sync pulses. A very low-cost system could even use the computer's built-in tone generator for transmitting, but the tones must be pure with little distortion in order to produce an acceptable RF signal via AFSK (see "AFSK" under Baudot section of this chapter).

SSTV reception is a little more difficult. First you must somehow measure the frequency of the incoming tone. You can't simply count the number of cycles in a second, or even 0.01 second, because the frequency is changing thousands of times each second. **Fig 12.52** illustrates one way of rapidly measuring the incoming tone's frequency. Two filters are designed to have maximum outputs a little beyond the ends of the frequency range of interest.

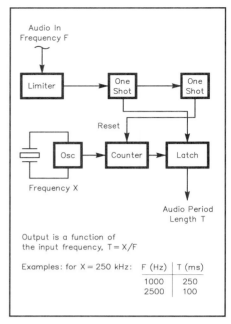

Fig 12.53—Block diagram of a digital SSTV demodulator.


```
Set line number, L, to 1
Repeat:
    Wait for sync
    Wait for end of sync
    If it was vertical sync, set L = 1

    Gather 128 pixels
    Display pixels on line L
    Increment L
    If L > 120, set L = 1
```

Fig 12.54—An outline of typical software written to display an SSTV frame from received digital picture information.

Fig 12.55—An example SSTV image.

The output of one filter is rectified to become a positive voltage; the output of the other is rectified to become a negative voltage; then the voltages are summed. A low-pass filter, with a 1-kHz cutoff, removes the audio carrier ripple while passing the slower video signal. With careful design, the result is a voltage that is fairly proportional to the input frequency. Finally, an analog to digital (A/D) converter processes the signal for the computer.

Another frequency-measuring approach uses digital circuitry to measure the period of each audio cycle (see **Fig 12.53**). When the signal amplitude crosses zero, a counter is reset. It then proceeds to count pulses from a crystal controlled oscillator. At the end of the audio cycle, the counter content is snatched, the counter is reset and the process starts all over again.

The digital approach offers a few advantages over the analog approach. A single chip can contain the counter and handle several other functions as well. The analog approach requires a handful of op amps, resistors, capacitors, diodes and an analog to digital (A/D) converter. The digital approach has crystal controlled accuracy and no adjustments are required. The frequency-to-voltage transfer function of the analog version isn't exactly linear and can change with temperature, power-supply variations and component aging.

Digital Signal Processing (DSP) is an exciting possibility for SSTV demodulators. With DSP, a high speed A/D converter is used to sample the audio input. After that, it's all software. DSP can be used to construct filters that are more flexible, accurate, stable and reproducible than their analog counterparts.

Once you have the tone-frequency information, the real work begins. The next step is to separate the composite signal into the sync and video components. To reduce the effects of noise, the sync pulses are cleaned up with a low pass filter and Schmitt trigger. Then, sync is used to control the timing of pixel sampling. **Fig 12.54** contains a high level outline of a program used to receive an 8-s B&W picture. Receiving colors isn't much more difficult. For nonRobot modes, gather the R, G, and B scans for each line, combine them and display a line in color. Robot modes require considerably more calculation to undo their encoding.

AN INEXPENSIVE SSTV SYSTEM

Here is a color SSTV/FAX480/weatherfax (**Fig 12.55**) system for IBM PCs and compatibles that is essentially 99% software! (It most recently appeared in July 1998 *QST*, pp 23-26.) And this system *transmits*,

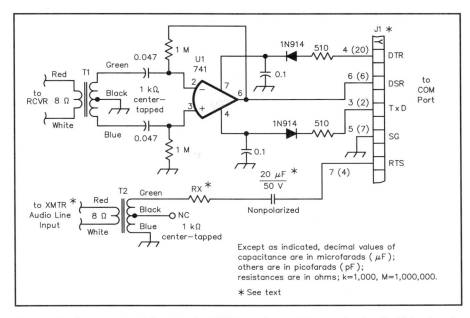

Fig 12.56—Schematic of the simple SSTV receive and transmit circuit. This circuit appears on page 34 of July 1998 *QST*. T1 and T2 are RadioShack 273-1380 audio-output transformers; the 20-μF, 50-V capacitor is a parallel combination of two RadioShack 272-999 10-μF, 50-V nonpolarized capacitors; equivalent parts can be substituted. Unless otherwise specified, resistors are 1/4-W, 5%-tolerance carbon composition or film units. An optional low-pass filter can be used between the output of the computer and the transmitter's audio-input line (see text). At J1, numbers in parentheses are for 25-pin connectors; other numbers are for 9-pin connectors.

too! The software is available from *ARRLWeb*. See page viii.

Ben Vester's, K3BC, work is aimed at the experimentally inclined, so if you're not familiar with BASIC programming, be prepared to learn a little about it if you want to maximize the utility of this system.

Hardware

Fig 12.56 shows a simple circuit used for receiving and transmitting. Connect the output of T2 to the phone patch input (often labeled LINE INPUT) of your transceiver. If you already have a phone patch, you can eliminate T2, and connect the line directly to the patch's phone-line terminals. Nearly all patches employ transformer isolation, but a simple ohmmeter check will verify that is true of *your* patch. (Avoid using the transceiver's mike input because of possible RF feedback problems.) R3 is set to the proper level for the audio going to the transmitter. SSTV has a 100% duty cycle signal, so you must set the audio signal to the transceiver at a level it can handle without overheating.

There is no low-pass filtering in the audio line between the computer output and transmitter audio input. On-the-air checks with many stations reveal that no additional external filtering is required when using SSB transmitters equipped with mechanical or crystal filters. If you intend to use this circuit with an AM or phasing-type SSB rig (or with

VHF/UHF FM transmitters), add audio filtering to provide the required spectral purity. An elliptical low-pass filter such as described by Campbell[1] should be adequate for most cases.

Circuit component values aren't critical nor is the circuit's physical construction. *Do* use a socket for the IC. A PC board is available from FAR Circuits,[2] but perfboard construction employing short leads works fine.

The Computer

The most important piece of hardware is the computer, which should have an 80286 (or better) microprocessor; a '386 machine running at 16 or 33 MHz definitely gives better results. You need a VGA color monitor that can provide a 640×480, 256-color noninterlaced display and a VGA (usually identified as SVGA) video adapter card that offers a 640×480×256-color mode.[3] The software directly addresses six of the most

1 R. Campbell, "High-Performance, Single-Signal Direct-Conversion Receivers," *QST*, Jan 1993, pp 32-40. See also Feedback, *QST*, Apr 1993, p 75.
2 FAR Circuits (see Address List in **References** chapter). The PC board is $4.50, plus $1.50 shipping.
3 Picture quality is degraded with an interlaced display. Few, if any, newer displays are interlaced at 640×480.

Commercial SSTV Products

All software and computer interfaces are for IBM PC with VGA display unless otherwise noted. Contact information for each of these sources appears in the Address List in the **References** chapter.

Scan Converters

DFM 1200 USA—PC Boards and instructions to build a Robot 1200C clone. The builder must collect EPROM and other parts. Muneki also supplies some of the hard-to-find parts.

> Donald P. Lucarell, K8SQL
> Felipe Rojas
> Muneki Yamafuzi, JF3GOH

SUPERSCAN 2001—Similar to 1200C but with many new features such as: four image memories, built-in mouse interface, on-screen help messages, and battery back up of CMOS memory to save system parameters when power is turned off. EPROMs developed by Martin Emmerson. Available assembled and in various semi-kit options.

> Jad Bashour

Tasco (TSC-70U)—A stand-alone color slow-scan TV converter. Receives and transmits color slow-scan without a PC. For picture storage, a PC interface is an optional module.

Replacement EPROMs—A brand new 1200C was capable of only the "Robot" modes. Martin supplies replacement EPROMs which add the Martin, Scottie, AVT, Wraase, and fax modes and other interesting features such as an "oscilloscope" tuning indicator. Product Reviews: Jul 1991 *73 Amateur Radio Today*, p 46 (version 4.0); *IVCA Newsletter* Fall 1991 (version 4.1); *IVCA Newsletter* Spring 1993 (version 4.2)

> Martin H. Emmerson MSc, G3OQD

Computer-based SSTV Systems

BMK-MULTY—Software for transmitting and receiving AMTOR, RTTY, CW, PACTOR, Audio Spectrum Analyzer, HF WEFAX, and SSTV.

> Schnedler Systems, AC4IW

MFJ-1278B—MCP for packet radio, RTTY, AMTOR, CW and so on. It is also capable of sending and receiving most popular SSTV modes with the MultiCom software.

> MFJ Enterprises Inc

Pasokon TV—Interface to send and receive SSTV fits inside expansion slot of computer. Software supports all popular modes, automatic receive mode selection from VIS code, up to 32k simultaneous colors on screen, graphical user interface with mouse support. Article: Jan 1993 *QST*, p 20. A free demo version, called EZSSTV, is available in many of the ham radio software depositories.

> Absolute Value Systems

PC SSTV 5—Compact separate send and receive interfaces plug into a serial port. Software supports the most popular modes, reads/writes popular image file formats, built-in text generating capability.

> Software Systems Consulting

Slow Scan II—Software to send and receive SSTV using the popular Sound Blaster (or compatible) sound card instead of interface dedicated to SSTV. Details: May 1993 *QEX*. A free demo version of the software is available on CompuServe: Go HAMNET, Library 6, search for "SSTV", "SLOWSC.ZIP".

> Harlan Technologies

SSTV Explorer—Low cost, receive-only system for most popular modes. Compact interface plugs into serial port. Has graphical user interface with mouse support, automatic receive mode selection, super VGA support with up to 32768 colors. Product Review: April 1994 *QST*, p 80.

> Radioware

Viewport VGA—External interface to send and receive plugs into printer port. Software (shareware by KA2PYJ) supports most popular modes. Construction article: *73* Aug 1992.

> A&A Engineering

Accessories and Related Software

ART (Amiga Robot Terminal)—Hardware interface and software to control Robot 1200C from Amiga computer. Contains paint program, multifont text, and many image processing functions. Supports Martin and Scottie EPROMs.

> Thomas M. Hibben, KB9MC

Audio Analyzer—Software for use with the Sound Blaster. Produces frequency vs time plots of audio signals. Useful for studying SSTV signals.

> Harlan Technologies

DFM SSTV Bandpass Filter—A bandpass filter especially designed for SSTV.

> Donald Lucarell

GEST—"all-in-one" SSTV utility package for the Robot 1200C. Includes paint program, text generation, special effects and image processing tools. Graphical user interface supports CGA, EGA, VGA and mouse. Controls the 1200C through parallel port.

> Torontel
> Royal Electronics (Canada)

HiRes—Paint program for use with the Robot 1200C. Has many impressive special effects and character fonts.

> Tom Jenkins, N9AMR

HiRes 32—New version of HiRes designed specifically for use with PC-based SSTV systems. Requires VGA display adapter capable of 32768 simultaneous colors.

> Tom Jenkins, N9AMR

Robot Helper—Robot 1200C control program for Microsoft Windows and OS/2 environments. Some features include: thumbnail previews of images on disk, dual image preview windows, fast image load and save to Robot, support for Robot or Martin EPROMs.

> William Montgomery, VE3EC

SCAN—Software for use with Robot 1200C.

> Bert Beyt, W5ZR

---

common SVGA chip types and also includes a VESA standard choice. If your video adapter card doesn't match one of the six, you'll need a VESA driver for your specific card. If you have trouble finding a driver, try checking on the Internet.

Software

GWBASIC is the programming tool. Although the guts of the program are contained in assembly language code (.ASM files), this code is available to the program (and you)

through BASIC. All of the modifications to the core programs (.ASM files) that adapt them to the multitude of SSTV/FAX modes are accomplished using BASIC POKEs. This allows experimenters with even a limited knowledge of BASIC programming to make modifications that add other modes, and so on. In deference to a few friends who complained about learning any BASIC, the programs include a system configuration list. The program uses this list to determine which POKEs to make. This system is

strictly keyboard controlled. The software uses a unique technique to get wider color definition than is normally available with a 256-color video card.

Some Program Details

One of the common SSTV practices is to retransmit a picture you just received so other SSTVers not copying the originating station can see the image. This capability is included.

RT.BAS is the receive and retransmit

program. On receive, you simply choose the mode from a menu, and wait for the picture transmission to complete. As of this writing, Robot 36 and 72 modes are available in either a synchronous or a line-synced mode. Other modes (all synchronous) are Scottie 1 and 2, Martin 1 and 2, AVT90, AVT94, Wraase 96, FAX480 and weatherfax.

When receiving, if you fail to get the mode selection made in time to catch the frame sync, you can go directly to copying by pressing the keyboard's spacebar. On all but the AVT modes, the next line sync is picked up and starts the picture. The AVT modes copy out of sync. Because the program allows you to scroll horizontally across the RGB color frames, you can resync after the picture has been received. A few images have nonstandard color registration, so the program can adjust color registration after the picture is received. You also can save the picture—usually after you have scrolled the picture so the CRT screen frames just the part you want to keep.

TX.BAS is used for transmitting any picture file. When queried, you provide the mode and the file name, and after a brief pause while the picture loads, press **G**(o) to transmit. To avoid additional switching complexity, VOX transmitter switching is used.

VU.BAS allows you to view a picture. It has the same adjustments available as RT.BAS. One feature (applicable only to the Robot modes) is the ability to "retune" the picture (in 10-Hz increments) as you view its color balance.

SLIDESHO.BAS gives you the vehicle to display a bunch of pictures as a slide show. Place SLIDESHO.BAS in a directory contained in your PATH statement so it can be called up from anywhere.

TIFCONV.BAS converts 640×480, 24-bit color, TIFF pictures into a format that can be transmitted by any of the supported SSTV modes except Robot. TIFF is a common format used to transfer higher-resolution pictures between programs. This program works with the Computer Eyes/RT[4] and Software Systems Consulting[5] frame grabbers. The picture output from this program can be

---

[4] ComputerEyes R/T by Digital Vision (see Address List in **References** chapter).
[5] Software Systems Consulting (see Address List in **References** chapter).

---

viewed with VU.BAS and, of course, is bound by 320×240 with 18-bit color.

LABEL.BAS allows you to add call signs and other text to the SSTV pictures. It takes any black-and-white TIFF (that is, 1-bit) file and creates a mask cutout where the black is. You can superimpose the cutout over an SSTV picture either in any color you want, or transfer a cutout of any background file you find interesting. The letters will then look like they were cut out of the background picture. Obviously, you can use squares or circles in addition to fonts to transfer a piece of one file onto another one. Use a cheap hand scanner to capture interesting fonts you find. You can get a three-dimensional effect by painting a color through the mask, then moving the mask a few pixels and rerunning the data through LABEL with a background file or another color. Or, run several different masks through LABEL in sequence to obtain different colors or patterns on different letters.

Work with Ben Vester's system continues. Look at articles by Vester in the SSTV Bibliography and watch *QST* for more discussion.

SUMMARY

For decades there was a convenient excuse for not trying SSTV: it cost kilobucks to buy a specialized piece of equipment. But you can't use that excuse anymore. There are several free programs that only require trivial interfaces to receive pictures. Once you get hooked, there are plenty of other home-brew projects and commercial products available at affordable prices. You need not be a computer wizard to install and use these systems.

SSTV is a rapidly changing area of Amateur Radio. Although it is still supported, the once-popular Robot 1200C has been discontinued. Many new products have been introduced.

For More Information

Contact information for each of these sources appears in the Address List in the **References** chapter.

Weekly nets:

Saturdays, 1500 UTC 14.230 MHz.
Saturdays, 1800 UTC 14.230 MHz.

SSTV Newsletter:

VISION from International Visual Communications Association (IVCA)

Magazines specializing in ATV:

Amateur Television Quarterly

ATV Today!
CQ-TV from British Amateur Television Club
The SPEC-COM Journal

Old A5 magazine reprints:

ESF copy services

Handbook:

Slow Scan Television Explained, by Mike Wooding, G6IQM (available from British Amateur Television Club and *Amateur Television Quarterly*).

SSTV Bibliography

Abrams, C., and R. Taggart, "Color Computer SSTV," *73*, Nov and Dec 1984.

Battles, B. and S. Ford, "Smile—You're on Ham Radio!" *QST*, Oct 1992, p 42.

Bodson, D., W4PWF, and Karty, S., N5SK, "FAX 480 and SSTV Interface and Software," *QST*, Jul 1998, pp 32-36.

Cameroni, G., I2CAB, and G. Morellato, I2AED (translated by Jim Grubbs, K9EI), "Get on SSTV—with the C-64," *Ham Radio*, Oct 1986, p 43.

Churchfield, T., K3HKR, "Amiga AVT System," *73 Amateur Radio*, Jul 1989, p 29.

Goodman, D., WA3USG, "SSTV with the Robot 1200C Scan Converter and the Martin Emmerson EPROM Version 4.0," *73 Amateur Radio Today*, Jul 1991, p 46.

Langner, J., WB2OSZ, "Color SSTV for the Atari ST," *73 Amateur Radio*, Dec 1989, p 38; Jan 1990, p 41.

Langner, J., WB2OSZ, "SSTV—The AVT System Secrets Revealed," *CQ-TV* 149 (Feb 1990), p 79.

Langner, J. WB2OSZ, "Slow Scan Television—It isn't expensive anymore," *QST*, Jan 1993, p 20.

Montalbano, J., KA2PYJ, "The ViewPort VGA Color SSTV System," *73*, Aug 1992, p 8.

Schick, M., KA4IWG, "Color SSTV and the Atari Computer," *QST*, Aug 1985, p 13.

Taggart, R., WB8DQT, "The Romscanner," *QST*, Mar 1986, p 21.

Taggart, R., WB8DQT, "A New Standard for Amateur Radio Facsimile," *QST*, Feb 1993, p 31.

Vester, B., K3BC "Vester SSTV/FAX80/ Fax System Upgrades," Technical Correspondence, *QST*, Jun 1994, pp 77-78.

Vester, B., K3BC, "SSTV: An Inexpensive System Continues to Grow," Dec 1994 *QST*, pp 22-24.

Vester, B., K3BC, "K3BC's SSTV Becomes TRUSCAN," Technical Correspondence, *QST*, Jul 1996, pp 66-67.

FAST-SCAN TELEVISION

Fast-scan amateur television (FSTV or just ATV) is a wide-band mode that uses standard broadcast, or NTSC, television scan rates. It is called "fast scan" only to differentiate it from slow-scan TV. In fact, no scan conversions or encoder/decoders are necessary with FSTV. Any standard TV set can display the amateur video and audio. Standard (1 V P-P into 75 Ω) composite video from home camcorders, cameras, VCRs or computers is fed directly into an AM ATV transmitter. The audio has a separate connector and goes through a 4.5 MHz FM subcarrier generator which is mixed with the video. This section was written by Tom O'Hara, W6ORG.

Amateurs regularly show themselves in the shack, zoom in on projects, show home video tapes, computer programs and just about anything that can be shown live or by tape (see **Figs 12.57** and **12.58**). Whatever the camera "sees" and "hears" is faithfully transmitted, including color and sound information. Picture quality is about equivalent to that of a VCR, depending on video signal level and any interfering carriers. All of the sync and signal-composition information is present in the composite-video output of modern cameras and camcorders. Most camcorders have an accessory cable or jacks that provide separate video and audio outputs. Audio output may vary from one camera to the next, but usually it has been amplified from the built-in microphone to between 0.1 to 1 V P-P (into a 10-kΩ load).

ATV transmitters have been carried by helium balloons to above 100,000 ft, to the edge of space. The result is fantastic video transmissions, showing the curvature of the Earth, that have been received as far as 500 miles from the balloon. Small cameras have been put into the cockpits of R/C model airplanes to transmit a pilot's-eye view. Many ATV repeaters retransmit Space Shuttle video and audio from NASA during missions. This is especially exciting for schools involved with SAREX. ATV is used for public service events such as parades, races, Civil Air Patrol searches and remote damage assessment.

Emergency service coordinators have found that live video from a site gives a better understanding of a situation than is possible from voice descriptions alone. Weather-radar video, WEFAX, or other computer generated video has also been carried by ATV transmitters for RACES groups during significant storms. This use enables better allocation of resources by presenting real-time information about the storm track. Computer graphics and video special effects are often transmitted to dazzle the viewers.

How Far Does ATV Go?

The theoretical snow-free line-of-sight distance for 10 W, given 15.8-dBd antennas and 2-dB feed-line loss at both ends, is 91 miles. (See **Table 12.16**.) However, except for temperature-inversion skip conditions, reflections, or through high hilltop repeaters, direct line-of-sight ATV contacts seldom exceed 25 miles. The RF horizon over flat terrain with a 50-ft tower is 10 miles. For best DX use low loss feed line and a broadband high-gain antenna, up as high as possible. The antenna system is the most important part of an ATV system because it affects both receive and transmit signal strength.

A snow-free, or "P5," picture rating (see **Fig 12.59**) requires at least 200 µV (−61 dBm) of signal at the input of the ATV receiver, depending on the system noise figure and bandwidth. The noise floor increases with bandwidth. Once the receiver system gain and noise figure reaches this floor, no additional gain will increase sensitivity. At 3-MHz bandwidth the noise floor is 0.8 µV (−109 dBm) at standard temperature. If you compare this to an FM voice receiver with 15 kHz bandwidth; there is a 23 dB difference in the noise floor. However the eye, much like the ear of experienced CW opera-

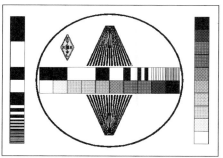

P5 —Excellent

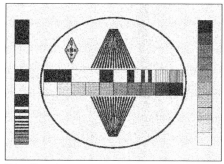

P4—Good

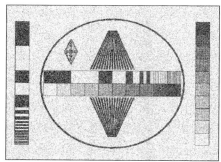

P3—Fair

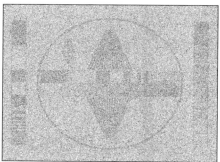

P2—Poor

P1—Barely perceptible
Fig 12.59—An ATV quality reporting system.

Fig 12.57—Students enjoy using ATV to communicate between science and computer classes.

Fig 12.58—The ATV view shows the aft end of the Space Shuttle cargo bay during mission STS-9.

Table 12.16

Line-of-Sight Snow-Free 70-cm ATV Communication Distances

This table relates transmit and receive station antenna gains to communication distances in miles for 1/10/100 W PEP at 440 MHz. To find the possible snow-free distance under line-of-sight conditions, select the column that corresponds to transmit antenna gain and the row for the receive antenna gain. Read the distance where the row and column intersect. Multiply the result by 0.5 for 902 MHz and 0.33 for 1240 MHz. The table assumes 2 dB of feed-line loss, a 3 dB system noise figure at both ends and snow-free is greater than 40 dB picture:noise ratio (most home cameras give 40 to 45 dB picture:noise; this is used as the limiting factor to define snow-free ATV pictures). The P unit picture rating system goes down about 6 dB per unit. For instance, P4 pictures would be possible at double the distances in the table.

	TX Antenna			
	0 dBd	4 dBd	9 dBd	15.8 dBd
RX Antenna				
0 dBd	0.8/2.5/8	1/3.5/11	2/7/22	5/15/47
4 dBd	1/3.5/11	2/6/19	3.5/11/34	7.5/23/75
9 dBd	2/7/22	3.5/11/34	6/19/60	13/42/130
15.8 dBd	5/15/47	7.5/23/75	13/42/130	29/91/290

tors, can pick out sync bars in the noise below the noise floor. Sync lock and large well contrasted objects or lettering can be seen between 1 and 2 μV. Color and subcarrier sound come out of the noise between 2 and 8 μV depending on their injection level at the transmitter and TV-set differences.

Two-meter FM is used to coordinate ATV contacts. Operators must take turns transmitting on the few available channels and the 2-m link allows full-duplex audio from many receiving stations to the ATV transmitting station, who is speaking on the sound subcarrier. This is great for interactive show and tell. Also it is much easier to monitor a squelched 2-m channel using an omni antenna rather than searching out each station with a beam. Depending on the third-harmonic relationship to the video on 70 cm, 144.34 MHz and 146.43 MHz (simplex) are the most popular frequencies; they are often mixed with the subcarrier sound on ATV repeater outputs.

Getting the Picture

Since the 70-cm band corresponds to cable channels 57 through 61, seeing your first ATV picture may be as simple as connecting a good outside 70-cm antenna (aligned for the customary local polarization) to a cable-ready TV set's antenna input jack. Cable channel 57 is 421.25 MHz, and each channel is progressively 6 MHz higher. (Note that cable and broadcast UHF channel frequencies are different.) Check the *ARRL Repeater Directory* for a local ATV repeater output that falls on one of these cable channels. Cable-ready TVs may not be as sensitive as a low-noise downconverter designed just for ATV, but this technique is well worth a try.

Most stations use a variable tuned downconverter specifically designed to convert the whole amateur band down to a VHF TV channel. Generally the 400 and 900 MHz

bands are converted to TV channel 3 or 4, whichever is not used in the area. For 1200 MHz converters, channels 7 through 10 are used to get more image rejection. The downconverter consists of a low-noise preamp, mixer and tunable or crystal controlled local oscillator. Any RF at the input comes out at the lower frequencies. All signal processing is done in the TV set. A complete receiver with video and audio output would require all the TV sets circuitry, less the sweep and CRT components. There is no picture-quality gain by going direct from a receiver to a video monitor (as compared with a TV set) because IF and detector bandwidth are still the limiting factors.

A good low-noise amateur downconverter with 15 dB gain ahead of a TV set will give sensitivity close to the noise floor. A preamp located in the shack will not significantly increase sensitivity, but rather will reduce dynamic range and increase the probability of intermod interference. Sensitivity can be increased by increasing antenna-system gain: reducing feed-line loss, increasing antenna gain or adding an antenna mounted preamp (which will eliminate the coax loss plus any loss through transmit linear amplifier TR relays). Remember that each 6 dB increase in combination of transmitted power, reduced coax loss, antenna gain or receiver sensitivity can double the line-of-sight distance.

Foliage greatly attenuates the signal at UHF, so place antennas above the tree tops for the best results. Beams made for 432-MHz weak-signal work or 440-MHz FM may not have enough SWR bandwidth to cover all the ATV frequencies for transmitting, but they are okay for reception. A number of manufacturers now make ATV beam antennas to cover the whole band from 420 to 450 MHz. Use low-loss coax (such as Belden 9913: 2.5 dB/100 ft at 400 MHz) or

Hardline for runs over 100 ft. All outside connectors must be weatherproofed with tape or coax sealer; any water that gets inside the coax will greatly increase the attenuation. Almost all ATV antennas use N connectors, which are more resistant to moisture contamination than other types.

Antenna polarization varies from area to area. Technically, the polarization should be chosen to give additional isolation (up to 20 dB) from other users near the channel. It is more common to find that the polarity was determined by the first local ATV operators (which antennas they had in place for other modes). Generally, those on 432 MHz SSB and weak-signal DX have horizontally polarized antennas, and those into FM, public service or repeaters will have vertical antennas. Check with local ATV operators before permanently locking down the antenna-mast clamps. Circularly polarized antennas let you work all modes, including satellites, with only 3 dB sacrificed when working a fixed polarity.

ATV Frequencies

Standard broadcast TV channels are 6 MHz wide to accommodate the composite video, 3.58 MHz color and 4.5 MHz sound subcarriers. (See **Fig 12.60**.) Given the NTSC 525 horizontal line and 30 frames per second scan rates, the resulting horizontal resolution bandwidth is 80 lines per MHz. Therefore, with the typical TV set's 3-dB rolloff at 3 MHz (primarily in the IF filter), up to 240 vertical black lines can be seen. Color bandwidth in a TV set is less than this, resulting in up to 100 color lines. Lines of resolution are often confused with the number of horizontal scan lines per frame. The video quality should be every bit as good as on a home video recorder.

The lowest frequency amateur band wide enough to support a TV channel is 70 cm (420 - 450 MHz), and it is the most popular. With transmit power, antenna gains

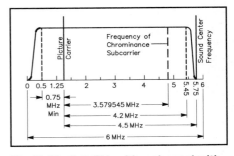

Fig 12.60—A 6-MHz video channel with the video carrier 1.25 MHz up from the lower edge. The color subcarrier is at 3.58 MHz and the sound subcarrier at 4.5 MHz above the video carrier.

and coax losses equal, decreasing frequency increases communication range. The 33-cm band goes half the distance that 70 cm does, but this can be made up to some extent with high-gain antennas, which are physically smaller at the higher frequency. A Technician class or higher license is required to transmit ATV on this band, and Novices can transmit ATV only in the 1270 to 1295 MHz segment of the 23-cm band.

Depending on local bandplan options, there is room for no more than two simultaneous ATV channels in the 33- and 70-cm bands without interference. Unlike cable channels, broadcast TV signals must skip a channel to keep a strong adjacent channel signal from interfering with a weaker on-channel signal. Cable companies greatly filter and equalize the signal amplitudes in order to use every channel.

Generally, because only two channels are available in the 70-cm band, an ATV repeater input on 439.25 or 434.0 MHz is shared with simplex. 421.25 MHz is the most popular in-band repeater output frequency. At least 12 MHz of separation is necessary for in-band repeaters because of filter-slope attenuation characteristics and TV-set adjacent-channel rejection. Some repeaters have their output on the 33-cm or 23-cm (the 923.25 and 1253.25 MHz output frequencies are most popular) bands which frees up a channel on 70 cm for simplex. Such crossband repeaters also make it easier for the transmitting operator to monitor the repeated video with only proper antenna separation to prevent receiver desensitization. 426.25 MHz is used for simplex, public service and R/C models in areas with crossband repeaters, or as alternative to the main ATV activities on 434.0 or 439.25 MHz. Before transmitting, check with local ATV operators, repeater owners and frequency coordinators listed in the *ARRL Repeater Directory* for the coordinated frequencies used in your area.

Since a TV set receives a 6-MHz bandwidth, ATV is more susceptible to interference from many other sources than are narrower modes. Interference 40 dB below the desired signal can be seen in video. Many of our UHF (and above) amateur bands are shared with radar and other government radio positioning services. These show up as horizontal bars in the picture. Interference from amateurs who are unaware of the presence of the ATV signal (or in the absence of a technically sound and publicized local band plan) can wipe out the sound or color or put diagonal lines in the picture.

DSB and VSB Transmission

While most ATV is double sideband (DSB) with the widest component being the sound subcarrier out plus and minus 4.5 MHz, over 90% of the spectrum power is in the first 1 MHz on both sides of the carrier for DSB or VSB (vestigial sideband). As can be seen in **Fig 12.61**, the video power density is down more than 30 dB at frequencies greater than 1 MHz from the carrier. DSB and VSB are both compatible with standard TV receivers, but the lower sound and color subcarriers are rejected in the TV IF filter as unnecessary. In the case of VSB, less than 5% of the lower sideband energy is attenuated. The other significant energy frequencies are the sound (set in the ATV transmitter at 15 dB below the peak sync) and the color at 3.58 MHz (greater than 22 dB down).

Narrow-band modes operating greater than 1 MHz above or below the video carrier are rarely interfered with or know that the ATV transmitter is on unless the narrow-band signal is on one of the subcarrier frequencies or the stations are too near one another. If the band is full and the lower sideband color and sound subcarrier frequencies need to be used by a dedicated link or repeater, a VSB filter in the antenna line can attenuate them another 20 to 30 dB, or the opposite antenna polarization can be used for more efficient packing of the spectrum. Since all amateur linear amplifiers re-insert the lower sideband to within 10 dB of DSB, a VSB filter in the antenna line is the only cost-effective way to reduce the unnecessary lower sideband subcarrier energy if more than 1 W is used. In the more populated areas, 2-m calling or coordination frequencies are often used to work out operating time shifts, and so on, between all users sharing or overlapping the same segment of the band.

ATV Identification

ATV identification can be on video or the sound subcarrier. A large high-contrast call-letter sign on the wall behind the operating table in view of the camera is the easiest way to fulfill the requirement. Transmitting stations fishing for DX during band openings often make up call-ID signs using fat black letters on a white background to show up best in the snow. Their city and 2-m monitoring frequency are included at the bottom of the sign to make beam alignment and contact confirmation easier.

Quite often the transmission time exceeds 10 minutes, especially when transmitting demonstrations, public-service events, space-shuttle video, balloon flights or a video tape. A company by the name of Intuitive Circuits makes a variety of boards that will overlay text on any video looped through them. Call letters and other information can be programmed into the board's non-volatile memory by on-board push buttons or an RS-232 line from a computer (depending on the version and model of the OSD board). There is even a model that will accept NMEA-0183 GPRMC data from a GPS receiver and overlay latitude, longitude, altitude, direction and speed as well as call letters on the applied camera video. This is ideal for ATV rockets, balloons and R/C vehicles. The overlaid ID can be selected to be on, off or flashed on for a few seconds every 10 minutes to automatically satisfy the ID requirement of 97.119 (see **Fig 12.62**). The PC Electronics VOR-2 board has an automatic nine minute timer, but it also has an end-of-transmission hang timer that switches to another video source for ID.

A 20-W ATV Transceiver

Many newcomers to ATV start out by buying an inexpensive downconverter board just to check out the local simplex or repeater activity. Once they see a picture it isn't long before they want to transmit. The downconverter board can be kept

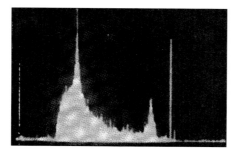

Fig 12.61—A spectrum-analyzer photo of a color ATV signal. Each vertical division represents 10 dB; horizontal divisions are 1 MHz. Spectrum power density varies with picture content, but typically 90% of the sideband power is within the first 1 MHz.

Fig 12.62—A photo of an ATV image of the Space Shuttle interior with K6KMN's repeater ID overlaid. Automatic video overlay in the picture easily solves the 10-minute ID requirement for Space Shuttle retransmissions and other long transmissions.

separate or put in a larger chassis with transmitter boards to make one convenient package, as shown in **Fig 12.63**. All the modules shown here are available wired and tested from PC Electronics and are also functionally representative of what is available from other suppliers. **Fig 12.64** shows a block diagram of this transceiver.

The complete 20-W ATV transceiver consists of the

- *TVC-2G* downconverter (420 - 450 MHz in, TV channel 2, 3 or 4 out)
- *TXA5-70* 80 mW exciter/modulator
- *FMA5-F* 4.5-MHz sound subcarrier generator
- *PA5-70* 20-W brick linear amplifier
- *DMTR* video detector, video monitor driver and TR relay modules.

The modules must be mounted in an aluminum enclosure for RF shielding and heat sinking. A 2.5×7×7-inch or larger aluminum chassis and bottom cover will make a nice transceiver. The Hammond 1590F diecast aluminum box makes a more rugged and RF tight enclosure. Lay all the modules in the selected chassis to position for best fit before drilling the mounting holes. Board wiring and mounting layouts come with each module.

Mount the PA5-70 amplifier and DMTR TR relay on the back panel, with the Mitsubishi M57716 RF power module

as low as possible for best air flow. Unscrew the power module and its board from the heatsink and poke through the four mounting holes and a piece of paper with a pencil. Use this as a template to center punch the drill locations on the chassis from the outside. Make sure the heatsink will mount at least $^1/_8$" above the bottom edge of the chassis. Drill the $^3/_{16}$-diam holes and carefully debur each side. The M57716 must be on a perfectly flat surface or the ceramic substrate could crack when its mounting bolts are tightened. Use a thin layer of heatsink compound under both the M57716 and the heatsink. Mount the M57716 and its board inside the chassis, and the heatsink outside by running the four screws from the M57716 side through the chassis into the heatsink.

The DMTR TR relay board mounts directly on a flange N UG58 chassis connector. Use RG-174 (small 50-Ω coax) for the RF leads to the amplifier and downconverter modules. To minimize RF coupling inside the chassis, carefully dress the coax braid back over its outer insulation (no more than $^1/_4$ inch) and solder the shield directly to the board ground planes. When soldering, make sure there are no bends or stress on the coax. Do not twist the braid into a "pig tail" at UHF.

A four-pin mic jack is used for the +13.8 V dc power connector. It is wired through a 4 A fast-acting fuse to the SPST POWER switch. The two unused pins can be used to control or power external devices such as a camera. A 1N4745A 16-V, 1-W Zener diode is connected from the transceiver side of the fuse to ground to help protect the circuits in case of accidental or reverse voltage. The downconverter, exciter and subcarrier generator can be mounted inside the chassis with #4-40×$^1/_2$ screws with double nuts for spacers (see module board mounting detail). Again, keep the exposed length of the interconnecting RG-174 center conductor less than $^1/_4$ inch. Solder the coax carefully and check with an ohmmeter for shorts. Use #18 wire for the amplifier power leads and #22 solid wire for all of the other wiring. Dress all dc leads away from the RF coaxes and the power module. The video and audio leads, and the panel-pot connections, can be #22 twisted pair (up to 6 inches long). Use RG-174 for longer runs.

You may want to remove and change some of the board mounted trimpots to panel mounted potentiometers to make adjustments easier. (For example, the video gain on the exciter, the mic and line gain on the sound subcarrier board, and the downconverter frequency tuning may be changed.) Remove the trimpots and run three wires from the mounting holes to their respective carbon (no wire wounds as they

(A)   **(B)**

Fig 12.63—A is the front view of a complete ATV transceiver and B is the inside view. This complete 20-W 70-cm ATV transceiver is assembled from readily available built and tested modules and mounted in a Hammond 1590F die-cast aluminum enclosure. On the box floor, left to right: TVC-2G downconverter, FMA5-F 4.5-MHz sound subcarrier generator and TXA5-70 80-mW exciter/video modulator. On the back (top left) is the downconverter-to-TV F connector and a 4-pin mic jack (which serves as the +13.8 V dc input). To the right is the DMTR TR relay board mounted to a flanged N connector. On the inside in front of the heatsink is the PA5 20-W power-amplifier module using a Mitsubishi M57716.

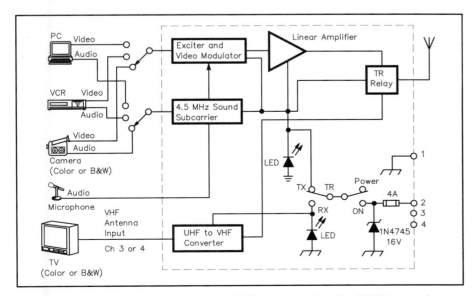

Fig 12.64—Block diagram of a complete ATV station using the 20-W transceiver.

are inductive at video frequencies) panel potentiometers. 100-Ω carbon panel controls for the video gain are difficult to find, but they are available from PC Electronics.

For RF purposes, bypass each video input connector (100-pF ceramic disc capacitor) and each audio connector (220-pF disc) directly at the connector with short leads.

Most camcorders use phono jacks for the composite-video and line-audio connections. A low-impedance mic with push-to-talk can be used in parallel with the camera or VCR audio, which is mixed in the sound subcarrier board and the transmit receive toggle switch. An F connector on the back panel supplies downconverter output to the TV set antenna input. Use 75-Ω coax for the line to the TV. (300-Ω twin lead picks up too much interference from strong adjacent-channel broadcast TV stations.) Do not put any other boards inside the chassis that might be RF susceptible.

Transceiver Checkout

Use an ohmmeter to verify that there are no short circuits in the coax or +13.8 V dc leads. (The antenna input will show a short because of the stripline tuned circuit.) Connect a good resonant 70-cm antenna, do not run a piece of wire or other band antenna just to try it out. With the transceiver off, connect the downconverter output coax to the TV set antenna jack. Switch the TV set on and select a channel that is not used in your area, usually 3 or 4. Adjust the fine tuning for minimum adjacent-channel interference.

Then turn on the transceiver and adjust the downconverter tuning for a known nearby ATV station that you have contacted on 2 m. Peak the input trimmer cap on the TR relay board for minimum snow.

Next, with no video connected, switch the transmitter on for no more than 10 seconds at a time while verifying that you have less than 1 W of reflected power (as shown by an RF power meter in the antenna line). Continued transmission into an SWR of more than 2:1 can damage the SAU4 power module. If the SWR is low, peak the trimmer cap on the DMTR board for maximum output, then proceed to set the blanking pedestal pot on the TXA5-70 exciter.

ATV is a complex waveform that requires that the video to sync ratio remains constant throughout all of the linear amplifiers and with camera contrast changes (see **Fig 12.65**). The modulator contains a blanking clamp circuit that also acts as a sync stretcher to compensate for amplifier gain compression. To set this level, the pedestal control is set to maximum power output and then backed off to 60% of that value. The sync tip, which is the peak power, is constant at the maximum power read and the blanking level is the 60% point. This procedure must be repeated anytime a different power amplifier is added or applied voltage is changed by more than 0.5 V. Any other RF power measurements with an averaging power meter under video modulation are meaningless.

The camera video can now be connected, and the video gain set for best picture as described by the receiving station (or by

observing a video monitor connected to the output jack on the DMTR board). Be careful not to overmodulate. Overmodulation is indicated by white smearing in the picture and sync buzz in the audio.

Connect a low-impedance (150 Ω to 600 Ω) dynamic mic (Radio Shack has some tape recorder replacements with a push-to-talk switch) into the mic jack and adjust the audio gain to a comfortable level as described by the receiving station. Electret mics are not good for this application because they are more susceptible to RF pickup (symptom: sync buzz in the audio). RF pickup may also be a problem with inadequately shielded mic cords. For example, it may be necessary to replace a cord having a spiral wrapped shield with one that has a braided shield, in order to improve shielding at UHF. The FMA5-F board has a soft limiter that comes in at the standard 25-kHz deviation.

The line-audio input has an independent volume control for the camcorder amplified mic or VCR audio, which is mixed with the low impedance mic input. This feature is great for voice-over commenting during video tapes.

Driving Amplifiers with ATV

Wide-band AM video requires some special design considerations for linear amplifiers (as compared to those for FM and SSB amplifiers). Many high-power amateur amplifiers would oscillate (and possibly self destruct) from high gain at low frequencies if they were not protected by feedback networks and power RF chokes. These same stability techniques can affect some of the 5-MHz video bandwidth. Sync, color and sound can be very distorted unless the amplifier has been carefully designed for both stability and AM video modulation.

Mirage, Teletec and Down East Microwave either make special ATV amplifiers or offer standard models that were designed for all modes, including ATV. Basically the collector and base bias supplies have a range of capacitors to keep the voltage constant under modulation while at the same time using the minimum-value low-resistance series inductors or chokes to prevent self oscillation.

Almost all amateur linear power amplifiers have gain compression from half to their full rated peak envelope power. To compensate for this, the ATV exciter/modulator has a sync stretcher to maintain the proper transmitted video to sync ratio (see **Fig 12.66**). With both video and sound subcarrier disconnected, the pedestal control is set for maximum power output. Peak sync should first be set to 90% of the rated peak envelope power. (This is necessary to give some head room for the 4.5 MHz sound that is mixed

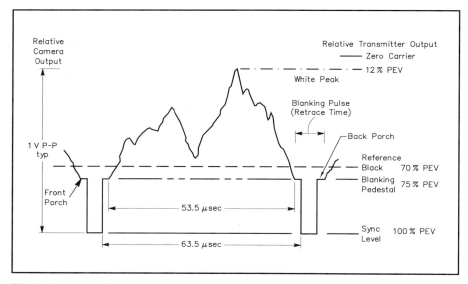

Fig 12.65—An ATV waveform. Camera and corresponding transmitter RF output power levels during one horizontal line scan for black-and-white TV. (A color camera would generate a "burst" of 8 cycles at 3.58 MHz on the back porch of the blanking pedestal.) Note that "black" corresponds to a higher transmitter output power than does "white." For the purposes of blanking pedestal setup with a RF power meter rather than an oscilloscope, the 75% PEV corresponds to slightly less than 60% power.

and adds with the video waveform.) The TXA5-70 exciter/modulator has a RF power control to set this. Once this is done, the blanking pedestal control can be set to 60% of the peak sync value. For example, a 100-W amplifier would first be set for 90 W with the RF power control and then 54 W with the pedestal control. Then the sound subcarrier can be turned back on and the video plugged in and adjusted for best picture. If you could read it on a peak-reading power meter made for video, the power is between 90 and 100 W PEP. On a dc oscilloscope connected to a RF diode detector in the antenna line, it can be seen that the sync and blanking pedestal power levels remain constant at their set levels regardless of video gain setting or average picture contrast. On an averaging meter like a Bird 43, however, it is normal to read something less than the pedestal set up power.

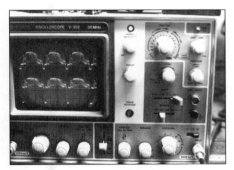

Fig 12.66—An oscilloscope used to observe a video waveform. The lower trace is the video signal as it comes out of the sync stretcher. The upper trace is the signal from the Mirage D1010-N amplifier.

ATV Repeaters

Basically there are two kinds of ATV repeaters: in band and cross band. 70-cm in-band repeaters are more difficult to build and use, yet they are more popular because equipment is more available and less expensive. Indeed, cable-ready TV sets tune the 70-cm band with no modifications.

Why are 70-cm repeaters more difficult to build? The wide bandwidth of ATV makes for special filter requirements. Response across the 6-MHz passband must be as flat as possible with minimum insertion loss, but also must sharply roll off to reject other users as little as 12 MHz away. Special multipole interdigital or combline VSB filters are used to meet the requirement. An ATV duplexer can be used to feed one broadband omnidirectional antenna, but an additional VSB filter is needed in the transmitter line for sufficient attenuation of noise and IMD products.

A cross-band repeater, because of the great frequency separation between the input and output, requires less sophisticated filtering to isolate the transmitter and receiver. In addition, a cross-band repeater makes it easier for users to see their own video (no duplexer is needed, only sufficient antenna spacing). Repeater linking is easier too, if the repeater outputs alternate between the 23- and 33-cm bands.

Fig 12.67 shows a block diagram for a simple 70-cm in-band repeater. No duplexer is shown because the antennas and VSB filters provide adequate isolation. The repeater transmitter power supply should be separate from the receiver and exciter supply. ATV is amplitude modulated, therefore the current varies greatly from maximum at the sync tip

to minimum during white portions of the picture. Power supplies are not generally made to hold tight regulation with such great current changes at rates up to several megahertz. Even the power supply leads become significant inductors at video frequencies; they will develop a voltage across them that can be transferred to other modules on the same power-supply line.

To prevent unwanted key up from other signal sources, ATV repeaters use a video operated relay (VOR). The VOR senses the horizontal sync at 15,734 Hz in much the same manner that FM repeaters use CTCSS tones. Just as in voice repeaters, an ID timer monitors VOR activity and starts the repeater video ID generator every nine minutes or a few seconds after a user stops transmitting.

Frequency Modulated ATV (FMATV)

While AM is the most popular mode because of greater equipment availability, lower cost, less occupied bandwidth and use of a standard TV set, FMATV is gaining interest among experimenters and also repeater owners for links. FM on the 1200-MHz band is the standard in Europe because there is little room for video in their allocated portion of the 70-cm band. FMATV occupies 17 to 21 MHz depending on deviation and sound subcarrier frequency. The US 70-cm band is wide enough but has great interference potential in all but the less populated areas. Most available FMATV equipment is made for the 1.2, 2.4 and 10.25-GHz bands. **Fig 12.68** is a block diagram of an FMATV receiver.

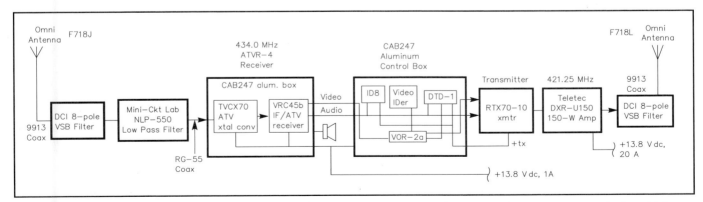

Fig 12.67—A block diagram of a 70-cm in-band ATV repeater. The antennas are Diamond omnidirectional verticals, which require 20 ft (minimum) of vertical separation to prevent receiver desensitization. The VSB filters are made by DCI; they have the proper band-pass characteristics and only 1 dB insertion loss. A low pass filter on the receiver is also necessary because cavity type filters repeat a passband at odd harmonics and the third-harmonic energy from the transmitter may not be attenuated enough. The receiver, 10-W transmitter and VOR are made by PC Electronics. The Communications Specialists DTD-1 DTMF decoder and ID8 Morse identifier (optional if a video ID is used) are used to remotely turn the repeater transmitter on or off and to create a CW ID, respectively. Alternatively, an Intuitive Circuits ATVR-4 ATV repeater controller board can do all the control box functions as well as remotely select from up to four video sources.

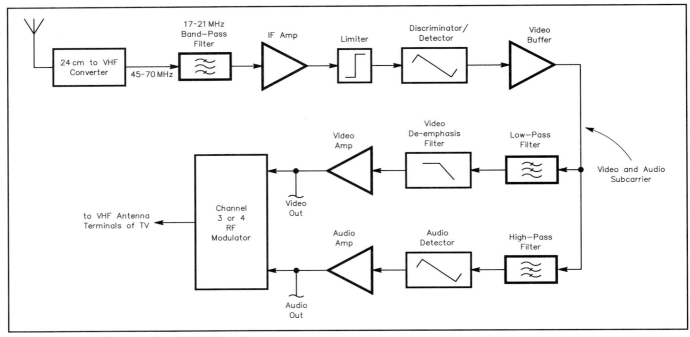

Fig 12.68—Block diagram of an FMATV receiver.

The US standard for FMATV is 4 MHz deviation with the 5.8-MHz sound subcarrier set to 10 dB below the video level. 1252 or 1255 MHz are suggested frequencies in order to stay away from FM voice repeaters and other users higher in the band while keeping sidebands above the 1240-MHz band edge. Using the US standard with Carsons rule for FM occupied bandwidth, it comes out to just under 20 MHz. So 1250 MHz would be the lowest possible frequency. Almost all modern FMATV equipment is synthesized, but if yours is not, use a frequency counter to monitor the frequency for warm up drift. Check with local frequency coordinators before transmitting because the band plan permits other modes in that segment.

Experimentally, using the US standard, FMATV gives increasingly better picture-to-noise ratios than AMATV at receiver input signals greater than 5 µV. Because of the wider noise bandwidth and FM threshold effect, AM video can be seen in the noise well before FM. For DX work, it has been shown that AM signals are recognizable signals in the snow at four times (12 dB) greater distance than FM signals with all other factors equal. Above the FM threshold, however, FM rapidly overtakes AM; snow-free pictures occur above 50 µV, or 4 times farther away than with AM signals. The crossover point is near the signal level where sound and color begin to appear for both systems. **Fig 12.69** compares AM and FMATV across a wide range of signal strengths.

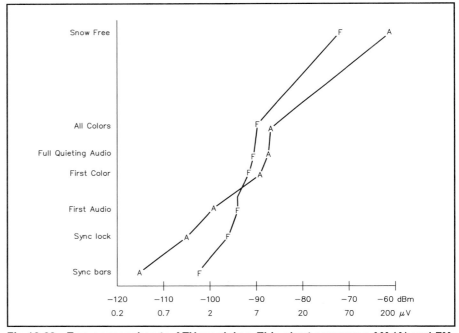

Fig 12.69—Two approaches to ATV receiving. This chart compares AM (A) and FM (F) ATV as seen on a TV receiver and monitor. Signal levels are into the same downconverter with sufficient gain to be at the noise floor. The FM receiver bandwidth is 17 MHz, using the US standard.

There are a variety of methods to receive FMATV. Older satellite receivers have a 70 or 45-MHz input and require a down converter with 40 to 50 dB gain ahead of them. Also satellite receivers are made for wider deviation and need some video gain to give the standard 1 volt peak to peak video output when receiving a signal with the standard 4-MHz deviation. Current satellite receivers directly tune anywhere from 900 to 2150 MHz and they only need a preamp added at the antenna for use on the 33 and 23-cm ham bands. The additional video gain can often be had by adjusting an internal pot or changing the gain with a resistor.

Some of the inexpensive Part 15 license-free wireless video receivers in the 33-cm band are 4 MHz deviation FM video, and most of the 2.4 GHz ones are FM, which can be used directly. However, they may or may not have the standard de-emphasis video network which may have to be added. On 2.4 GHz, some of the Part 15 frequencies are outside the band and care should be taken to use only those inside the 2390 to 2450 MHz ham band if modified. Wavecom Jr has been the most popular 2.4 GHz license-free video transmitter and receiver (available from ATV Research) and have been modified for higher power and other features as well as having all 4 of the channels in the ham band using interface boards from PC Electronics.

Gunnplexers on 10.4 GHz make inexpensive point to point ATV links for public-service applications or between repeaters. A 10-mW Gunnplexer with 17-dB horn can cover over 2 miles line-of-sight when received on a G8OZP low noise 3-cm LNB and satellite receiver. An application note for construction of the 3-cm transmitter comes with the GVM-1 Gunnplexer video modulator board from PC Electronics.

For short distance ATV from R/C vehicles, low-power FM ATV modules with 50 to 100-mW output in the 33, 23 or 13-cm bands are often used. These offer less desense possibility to the R/C receiver. An example can be seen on the model Humvee in the photo.

For greater distance such as with R/C aircraft, use up to a 1-W ATV transmitter board operating in the 70-cm band. Since R/C receivers at 50 or 72 MHz were not designed to be placed right next to a transmitter, it is necessary to shield the R/C receiver and put a simple 3-pole 100-MHz low-pass filter at the antenna input. An application note is available from PC Electronics.

Further ATV Reading

Amateur Television Quarterly Magazine.
CQ-TV, British ATV Club, a quarterly publication available through *Amateur Television Quarterly Magazine.*
Ruh, *ATV Secrets for the Aspiring ATVers*, Vol 1, 1991 and Vol 2, 1992. Available through *Amateur Television Quarterly Magazine.*
Taggart, "An Introduction to Amateur Television," April, May and June 1993 *QST* .

ATV Equipment Sources

Contact information for these sources appears in the Address List in the **References** chapter.
Advanced Receiver Research
ATV Research
Digital Communications, Inc (DCI)
Down East Microwave
Elktronics
Intuitive Circuits
Mini-Circuits Labs
PC Electronics
Phillips-Tech Electronics
Spectrum International
TX/RX Systems
Teletec
Wyman Research Inc

N8QPJ mounted an ATV setup aboard this model Humvee.

Radio Control

Amateur Radio gave birth to the radio control (R/C) hobby as we know it today. Part 97 of the FCC regulations (§97.215) specifically permits "remote control of model craft" as a licensed amateur station activity. Station identification is not required for R/C, and the transmitter power is limited to 1 W. Before 1950, development of telecommand radio systems small enough to be used for remote radio control of model aircraft, cars and boats, was primarily an Amateur Radio activity. In the early 1950s, the FCC licensed R/C transmitter operation on nonham frequencies, without an operator license examination. The invention of the transistor and the subsequent increase in R/C development activity lead to the sophisticated electronic control systems in use

today. This section was contributed by H. Warren Plohr, W8IAH.

The simplest electronic control systems are currently used in low-cost toy R/C models. These toys often use simple on/off switching control that can be transmitted by on/off RF carrier or tone modulation. More expensive toys and R/C hobby models use more sophisticated control techniques. Several simultaneous proportional and switching controls are available, using either analog or digital coding on a single RF carrier.

R/C hobby sales records show that control of model cars is the most popular segment of the hobby. Battery powered cars like that shown in **Fig 12.70** are the most popular. Other popular types include mod-

els powered by small internal combustion "gas" engines.

R/C model aircraft are next in the line of popularity and include a wide range of styles and sizes. Fixed-wing models like those shown in **Fig 12.71** are the most popular. They can be unpowered (gliders) or powered by either electric or "gas" motors. The basic challenge for a new model pilot is to operate the model in flight without crashing. Once this is achieved, the challenge extends to operating detailed scaled models in realistic flight, performing precision aerobatics, racing other models or engaging in model-to-model combat. The challenge for the R/C glider pilot is to keep the model aloft in rising air currents. The most popular rotary-wing aircraft models are helicopters. The

Fig 12.70—Photo of three R/C model electric cars.

Fig 12.71—Photo of two R/C aircraft models.

(A)

(B)

Fig 12.72—A, photo of Futaba's Conquest R/C aircraft transmitter. B shows the matching airborne system.

sophistication of model helicopters and their control systems can only be appreciated when one sees a skilled pilot perform a schedule of precision flight maneuvers. The most exotic maneuver is sustained inverted flight, a maneuver not attainable by a full-scale helicopter.

R/C boats are another facet of the hobby. R/C water craft models can imitate full-scale ships and boats. From electric motor powered scale warships that engage in scale battles, to "gas" powered racing hydroplanes, model racing yachts and even submarines.

Most R/C operation is no longer on Amateur Radio frequencies. The FCC currently authorizes 91 R/C frequencies between 27 MHz and 76 MHz. Some frequencies are for all models, some are for aircraft only and others for surface (cars, boats) models only. Some frequencies are used primarily for toys and others for hobbyist models. Amateur Radio R/C operators use the 6-m band almost exclusively. Spot frequencies in the upper part of the band are used in geographical areas where R/C operation is compatible with 6-m repeater operation and TV Channel-2 signals that can interfere with control. Eight spot frequencies, 53.1 to 53.8 MHz, spaced 100 kHz apart, are used. There is also a newer 200 kHz R/C band from 50.8 to 51.0 MHz providing ten channels spaced 20 kHz apart. The close channel spacing in this band requires more selective receivers than do the 53-MHz channels. The AMA *Membership Manual* provides a detailed

list of all R/C frequencies in current use.[7] The *ARRL Repeater Directory* lists current Amateur Radio R/C frequencies.

Fig 12.72 shows a typical commercial R/C system, consisting of a hand-held aircraft transmitter (A), a multiple-control receiver, four control servos and a battery (B). This particular equipment is available for any of the ten R/C frequencies in the 50.80 to 51.00 MHz band. Other commercially available control devices include relays (solid-state and mechanical) and electric motor speed controllers.

Some transmitters are tailored to specific kinds of models. A helicopter, for example, requires simultaneous control of both collective pitch and engine throttle. A model helicopter pilot commands this response with a linear motion of a single transmitter control stick. The linear control stick signal is conditioned within the transmitter to provide the encoder with a desired combination of nonlinear signals. These signals then command the two servos that control the vertical motion of the helicopter.

Transmitter control-signal conditioning is provided by either analog or digital circuitry. The signal conditioning circuitry is often designed to suit a specific type of model, and it is user adjustable to meet an individual model's control need. (Low-cost transmitters use analog circuitry.) They are available for helicopters, sailplanes and pattern (aerobatic) aircraft.

More expensive transmitters use digital microprocessor circuitry for signal conditioning. **Fig 12.73** shows a transmitter that uses a programmable microprocessor. It is available on any 6-m Amateur Radio R/C frequency with switch-selectable PPM or PCM coding. It can be programmed to suit the needs of a helicopter, sailplane or pattern aircraft. Nonvolatile memory retains up to four user-programmed model configurations.

Many R/C operators use the Amateur Radio channels to avoid crowding on the nonham channels. Others do so because they can operate home-built or modified R/C transmitters without obtaining FCC

type acceptance. Still others use commercial R/C hardware for remote control purposes around the shack. Low-cost R/C servos are particularly useful for remote actuation of tuners, switches and other devices. Control can be implemented via RF or hard wire, with or without control multiplexing.

R/C RF MODULATION

The coded PPM or PCM information for R/C can modulate an RF carrier via either amplitude- or frequency-modulation techniques. Commercial R/C systems use both AM and FM modulation for PPM, but use FM exclusively for PCM.

The AM technique used by R/C is 100% "down modulation." This technique switches the RF carrier off for the duration of the PPM pulse, usually 250 to 350 μs. A typical transmitter design consists of a third-overtone transistor oscillator, a buffer amplifier and a power amplifier of about $^1/_2$ W output. AM is achieved by keying the 9.6-V supply to the buffer and final amplifier.

Fig 12.73—Photo of Airtronics Infinity 660 R/C aircraft transmitter.

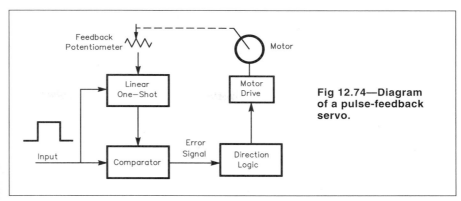

Fig 12.74—Diagram of a pulse-feedback servo.

The FM technique used by R/C is frequency shift keying (FSK). The modulation is applied to the crystal-oscillator stage, shifting the frequency about 2.5 or 3.0 kHz. The direction of frequency shift, up or down with a PPM pulse or PCM code, can be in either direction, as long as the receiver detector is matched to the transmitter. R/C manufacturers do not standardize, so FM receivers from different manufacturers may not be compatible.

SIGNALING TECHNIQUES

Background

Radio control (R/C) of models has used many different control techniques in the past. Experimental techniques have included both frequency- and time-division multiplexing, using both electronic and mechanical devices. Most current systems use time-division multiplexing of pulse-width information. This signaling technique, used by hobbyist R/C systems, sends pulse-width information to a remotely located pulse-feedback servo-mechanism. Servos were initially developed for R/C in the 1950s and are still used today in all but low-cost R/C toys.

Fig 12.74 is a block diagram of a pulse-feedback servo. The leading edge of the input pulse triggers a linear one-shot multivibrator. The width of the one-shot output pulse is compared to the input pulse. Any pulse width difference is an error signal that is amplified to drive the motor. The motor drives a feedback potentiometer that controls the one-shot timing. When this feedback loop reduces the error signal to a few microseconds, the

drive motor stops. The servo position is a linear function of the input pulse width. The motor-drive electronics are usually timed for pulse repetition rates of 50 Hz or greater and a pulse width range of 1 to 2 ms. A significantly slower repetition rate reduces the servomechanism slew rate but not the position accuracy.

In addition to motor driven servos, the concept of pulse-width comparison can be used to operate solid-state or mechanical relay switches. The same concept is used in solid-state proportional electric motor speed controllers. These speed controllers are used to operate the motors powering model cars, boats and aircraft. Currently available model speed controllers can handle tens of amperes of direct current at voltages up to 40 V dc using MOSFET semiconductor switches.

Requirements

The signaling technique required by R/C is the transmission of 1- to 2-ms-wide pulses with an accuracy of ±1 μs at repetition rates of about 50 Hz. A single positive-going dc pulse of 3 to 5 V amplitude can be hard wire transmitted successfully to operate a single control servomechanism. If such a pulse is sent as modulation of an RF carrier, however, distortion of the pulse width in the modulation/demodulation process is often unacceptable. Consequently, the pulse-width information is usually coded for RF transmission. In addition, most R/C systems require pulse-width information for more than one control. Time-division multiplexing of each control provides this multichannel capability. Two coding techniques are used to transfer the pulse-width information for multiple control channels, pulse-position modulation (PPM) and pulse-code modulation (PCM).

Pulse-Position Modulation

PPM is analog in nature. The timing

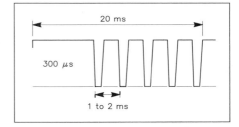

Fig 12.75—Diagram of a four-channel PPM RF envelope.

between transmitted pulses is an analog of the encoded pulse width. A train of pulses encodes multiple channels of pulse-width information as the relative position or timing between pulses. Therefore the name, pulse-position modulation. The transmitted pulse is about 300 μs in width and uses slow rise and fall times to minimize the transmitter RF bandwidth. The shape of the received waveform is unimportant because the desired information is in the timing between pulses. **Fig 12.75** diagrams a frame of five pulses that transmits four control channels of pulse-width information. The frame of modulation pulses is clocked at 50 Hz for a frame duration of 20 ms. Four multiplexed pulse widths are encoded as the times between five 300-μs pulses. The long period between the first and the last pulse is used by the decoder for control-channel synchronization.

PPM is often incorrectly called digital control because it can use digital logic circuits to encode and decode the control pulses. A block diagram of a typical encoder is shown in **Fig 12.76**. The 50-Hz clock frame generator produces the first 300-μs modulation pulse and simultaneously triggers the first one-shot in a chain of multivibrators. The trailing edge of each one-shot generates a 300-μs modulation pulse while simultaneously triggering the succeeding multivibrator one-shot.

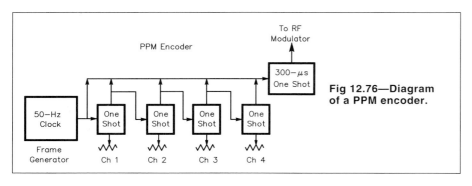

Fig 12.76—Diagram of a PPM encoder.

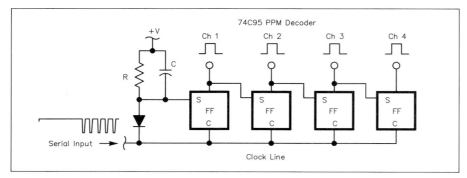

Fig 12.77—Diagram of a 74C95 PPM decoder.

In a four-channel system the fifth modulation pulse, which indicates control of the fourth channel, is followed by a modulation pause that is dependent on the frame rate. The train of 300-µs pulses are used to modulate the RF.

Received pulse decoding can also use digital logic semiconductors. **Fig 12.77** shows a simple four-control-channel decoder circuit using a 74C95 CMOS logic IC. The IC is a 4-bit shift register operated in the right-shift mode. Five data pulses spaced 1 to 2 ms apart, followed by a synchronization pause, contain the encoded pulse-width information in one frame. During the sync pause, the RC circuit dis-

charges and sends a logic-one signal to the 74C95 serial input terminal. Subsequent negative going data pulses remove the logic-one signal from the serial input and sequentially clock the logic one through the four D-flip-flops. The output of each flip-flop is a positive going pulse, with a width corresponding to the time between the clocking pulses. The output of each flip-flop is a demultiplexed signal that is used to control the corresponding servo.

Pulse Code Modulation

PCM uses true digital code to transfer R/C signals. The pulse width data of each control channel is converted to a binary word. The digital word information of each control channel is coded and multiplexed to permit transmission of multiple channels of control on a single RF carrier. On the receiving end, the process is reversed to yield the servo control signals.

There is no standard for how the digital word is coded for transmission. Therefore PCM R/C transmitters and receivers from different makers are not interchangeable. Some older PCM systems provide only 256 discrete positions for 90° of servo motion, thereby limiting servo resolution. Newer systems use more digital bits for each word and provide smooth servo motion with 512 and 1024 discrete positions. All PCM and PPM systems use the same servo input-signal and supply voltages. Therefore the servos of different manufacture are interchangeable once compatible wiring connectors have been installed.

Spread Spectrum

This introduction to spread spectrum communications was written by André Kesteloot, N4ICK. *The ARRL Spread Spectrum Sourcebook* contains a more complete treatment of the subject.

A Little History

Spread spectrum has existed at least since the mid 1930s. Despite the fact that John Costas, W2CRR, published a paper on nonmilitary applications of spread spectrum communications in 1959,[8] spread spectrum was used almost solely for military purposes until the late 1970s. In 1981, the FCC granted the Amateur Radio Research and Development Corporation (AMRAD) a Special Temporary Authorization to conduct Amateur Radio

[8]"Poisson, Shannon and the Radio Amateur," *Proceedings of the IRE*, Dec 1959.

spread spectrum experiments. In June 1986, the FCC authorized all US amateurs to use spread spectrum above 420 MHz.

Why Spread Spectrum

Faced with increasing noise and interference levels on most RF bands, traditional wisdom still holds that the narrower the RF bandwidth, the better the chances that "the signal will get through." This is not so.

In 1948, Claude Shannon published his famous paper, "A Mathematical Theory of Communication" in the *Bell System Technical Journal*, followed by "Communications in the Presence of Noise" in the *Proceedings of the IRE* for January 1949. A theorem that follows Shannon's, known as the Shannon-Hartley theorem, states that the channel capacity C of a band-limited gaussian channel is

$$C = W \log_2 (1 + S/N) \text{ bits/s} \qquad (6)$$

where

  W is the bandwidth,
  S is the signal power and
  N is the noise within the channel bandwidth.

This theorem states that should the channel be perfectly noiseless, the capacity of the channel is infinite. It should be noted, however, that making the bandwidth W of the channel infinitely large does *not* make the capacity infinite, because the channel noise increases proportionately to the channel bandwidth.

Within reason, however, one can trade power for bandwidth. In addition, the power density at any point of the occupied bandwidth can be very small, to the point that it may be well *below* the noise floor of the receiver. The US Navy Global Positioning System (GPS) is an excellent example of the use of what is called direct-sequence spread spectrum. The average signal at the GPS receiver's antenna terminals is approximately −160 dBW (for the C/A code). Since most sources of interference are relatively narrow-band, spread-spectrum users will also benefit, as narrow-band interfering signals are rejected automatically during the de-spreading process, as will be explained later in this section.

These benefits are obtained at the cost of fairly intricate circuitry: The transmitter must spread its signal over a wide bandwidth in accordance with a certain prearranged code, while the receiver must somehow synchronize on this code and recombine the received energy to produce a usable signal. To generate the code, use is made of pseudo-noise (PN) generators. The PN generators are selected for their correlation properties. This means that when two similar PN sequences are compared out of phase their correlation is nil (that is, the output is 0), but when they are exactly in phase their correlation produces a huge peak that can be used for synchronization purposes.

This synchronization process has been (and still is) the major complicating factor in any spread spectrum link, for how can one synchronize on a signal that can be well below the receiver's noise floor? Because of the cost associated with the complicated synchronization processes, spread spectrum applications were essentially military-related until the late 1970s. The development of ICs then allowed for the replacement of racks and racks of tube equipment by a few plug-in PC boards, although the complexity level itself did not improve. Amateur Radio operators could not afford such levels of complexity and had to find simpler solutions, at the cost of robustness in the presence of interference.

Spread-Spectrum Transmissions

A transmission can be called "spread spectrum" if the RF bandwidth used is (1) much larger than that needed for traditional modulation schemes and (2) independent of the modulation content. Although numerous spread spectrum modulation schemes are in existence, only two, frequency-hopping (FH) and direct-sequence spread spectrum (DSSS) are specifically authorized by the FCC for use by the Amateur Radio community.

To understand FH, let us assume a transmitter is able to transmit on any one of 100 discrete frequencies F1 through F100. We now force this equipment to transmit for 1 second on each of the frequencies, but in an apparently random

pattern (for example, F1, F62, F33, F47...; see **Fig 12.78**). Should some source interfere with the receiver site on three of those discrete frequencies, the system will still have achieved reliable transmission 97% of the time. Because of the built-in redundancy in human speech, as well as the availability of error-correcting codes in data transmissions, this approach is particularly attractive for systems that must operate in heavy interference.

In a DSSS transmitter, an RF carrier and a pseudo-random pulse train are mixed in a doubly balanced mixer (DBM). In the process, the RF carrier disappears and is replaced by a noise-like wide-band transmission, as shown in **Fig 12.79**. At the receiver, a similar pseudo-random signal is reintroduced and the spread spectrum signal is correlated, or despread, while narrow-band interference is spread simultaneously by the same process.

The technical complexity mentioned above is offset by several important advantages for military and space applications:

• *Interference rejection*. If the interference is not synchronized with the original spread spectrum signal, it will not appear after despreading at the receiver.

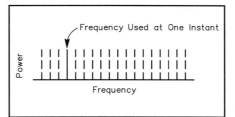

Fig 12.78—Power vs frequency for frequency-hopping spread spectrum signals. Emissions jump around to discrete frequencies in pseudo-random fashion.

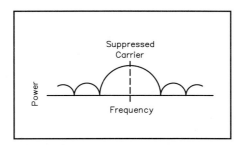

Fig 12.79—Power vs frequency for a direct-sequence-modulated spread spectrum signal. The envelope assumes the shape of a (sin x/x)² curve. With proper modulating techniques, the carrier is suppressed.

• *Security*. The length and sophistication of the pseudo-random codes used can be such as to make unauthorized recovery difficult if not impossible.
• *Power density*. Low power density makes for easy hiding of the RF signal and a resulting lower probability of detection.

As far as the Amateur Radio community is concerned, particular benefit will be derived from the interference rejection just mentioned, as it offers both robustness and reliability of transmissions, as well as low probability of interference to other users. Additionally, spread spectrum has the potential to allow better utilization of the RF spectrum allocated to amateurs. There is a limit as to how many conventional signals can be placed in a given band before serious transmission degradation takes place. Additional spread spectrum signals will not cause severe interference, but may instead only raise the background noise level. This becomes particularly important in bands shared with other users and in our VHF and UHF bands increasingly targeted by would-be commercial users. The utilization of a channel by many transmitters is essentially the concept behind CDMA (Code Division Multiple Access), a system in which several DSSS transmissions can share the same RF bandwidth, provided they utilize orthogonal pseudo-random sequences.

Amateur Spread Spectrum

When radio amateurs (limited in both financial resources and time available for experimentation) decided to try their hand at spread spectrum transmissions, they had to attack the problem by simplifying several assumptions. Security and privacy, the primary goals of the military, were sacrificed in favor of simplicity of design and implementation.

Experimentation sponsored by AMRAD began in 1981 and continues to this day. These experiments have lead to the design and construction of a practical DSSS UHF link. This project was described in May 1989 *QST* and was reprinted in *The ARRL Spread Spectrum Sourcebook*. In it, N4ICK offered a simple solution to the problem of synchronization. (Because of its simplicity, this solution does not offer all the anti-jamming properties of more sophisticated systems, but this should not be of concern to Amateur Radio operators.) The block diagram is shown in **Fig 12.80**. **Fig 12.81** shows the RF signals at the transmitter output, at the receiver antenna terminals and the recovered signal after correlation. James Vincent, G1PVZ, replaced the original FM scheme with a continuously variable delta modulation system, or CVSD. A description of his work can be found in the September and October 1993 issues of the British magazine *Electronics World & Wireless World*.

In addition to *The ARRL Spread Spectrum Sourcebook*, interested readers may want to pay particular attention to Robert Dixon's text, *Spread Spectrum Systems*. Additional information can be found in the publications and magazines listed below.

SS References

Dixon, *Spread Spectrum Systems*, second edition, 1984, Wiley Interscience, New York.

Dixon, *Spread Spectrum Techniques*, IEEE.

Golomb, *Shift Register Sequences*, 1982, Aegean Park Press, Laguna Hills, California.

Hershey, *Proposed Direct Sequence Spread Spectrum Voice Techniques for Amateur Radio Service*, 1982, US Department of Commerce, NTIA Report 82-111.

Holmes, *Coherent Spread Spectrum Systems*, 1982, Wiley Interscience, New York.

Kesteloot, Ed., *The ARRL Spread Spectrum Sourcebook* (Newington, CT: ARRL, 1990). Includes Hershey, *QST* and *QEX* material listed separately here.

The *AMRAD Newsletter* carries a monthly column on spread spectrum and reviews ongoing AMRAD experiments. Contact information appears in the Address List in the **References** chapter.

The following articles have appeared in Amateur Radio publications. All of the articles from *QST* and *QEX* are reproduced in *The ARRL Spread Spectrum Sourcebook*.

Feinstein, "Spread Spectrum—A report from AMRAD," *73*, November 1981.

Feinstein, "Amateur Spread Spectrum Experiments," *CQ*, July 1982.

Kesteloot, "Practical Spread Spectrum: A Simple Clock Synchronization Scheme," *QEX*, October 1986.

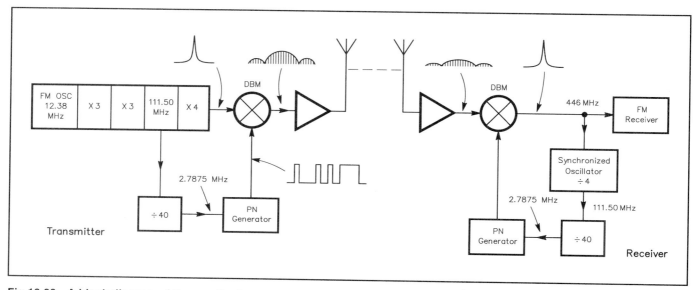

Fig 12.80—A block diagram of the practical spread spectrum link. The success of this arrangement lies in the use of a synchronized oscillator (right) to recover the transmitter clock signal at the receiving site.

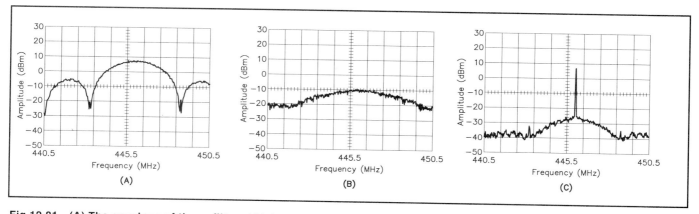

Fig 12.81—(A) The envelope of the unfiltered biphase-modulated spread spectrum signals as viewed on a spectrum analyzer. In this practical system, band-pass filtering is used to confine the spread spectrum signal to the amateur band. (B) At the receiver end of the line, the filtered spread spectrum signal is apparent only as a 10-dB hump in the noise floor. (C) The despread signal at the output of the receiver DBM. The original carrier—and any modulation components that accompany it—has been recovered. The peak carrier is about 45 dB above the noise floor—more than 30 dB above the hump shown at B. (These spectrograms were made at a sweep rate of 0.1 s/division and an analyzer bandwidth of 30 kHz; the horizontal scale is 1 MHz/division.)

Kesteloot, "Experimenting with Direct Sequence Spread Spectrum," *QEX*, December 1986.

Kesteloot, "Extracting Stable Clock Signals from AM Broadcast Carriers for Amateur Spread Spectrum Applications," *QEX*, October 1987.

Kesteloot, "Practical Spread Spectrum Achieving Synchronization with the Slip-Pulse Generator," *QEX*, May 1988.

Kesteloot, "A Practical Direct Sequence Spread-Spectrum UHF Link," *QST*, May 1989.

Kesteloot, "Practical Spread Spectrum Clock Recovery With the Synchronous Oscillator," *QEX*, June 1989.

Rohde, "Digital HF Radio A Sampling of Techniques," *Ham Radio*, April 1985.

Rinaldo, "Spread Spectrum and the Radio Amateur," *QST*, November 1980.

Sabin, "Spread Spectrum Applications in Amateur Radio," *QST*, July 1983.

Williams, "A Digital Frequency Synthesizer," *QST*, April 1984.

Williams, "A Microprocessor Controller for the Digital Frequency Synthesizer," *QST*, February 1985.

Contents

RF Power Amplifiers and Projects
13

This chapter describes the design and construction of power RF amplifiers for use in an Amateur Radio station. Dick Ehrhorn, W4ETO, contributed materially to this section.

An amplifier may be required to develop as much as 1500 W of RF output power, the legal maximum in the United States.

The voltages and currents needed to perform this feat are much higher than those found in other amateur equipment—the voltage and current levels are potentially lethal, in fact.

Every component in an RF power amplifier must be carefully selected to endure these high electrical stress levels without

failing. Large amounts of heat are produced in the amplifier and must be dissipated safely. Generation of spurious signals must be minimized, not only for legal reasons, but also to preserve good neighborhood relationships. Every one of these challenges must be overcome to produce a loud, clean signal from a safe and reliable amplifier.

Types of Power Amplifiers

Power amplifiers are categorized by their power level, intended frequencies of operation, device type, class of operation and circuit configuration. Within each of these categories there almost always are two or more options available. Choosing the most appropriate set of options from all those available is the fundamental concept of design.

SOLID STATE VERSUS VACUUM TUBES

With the exception of high-power amplifiers, nearly all items of amateur equipment manufactured commercially today use solid-state (semiconductor) devices exclusively. Semiconductor diodes, transistors and integrated circuits (ICs) offer several advantages in designing and fabricating equipment:

- Compact design—Even with their heat sinks, solid-state devices are smaller than functionally equivalent tubes, allowing smaller packages.
- "No-tune-up" operation—By their nature, transistors and ICs lend themselves to low impedance, broadband operation. Fixed-tuned filters made with readily available components can be used to suppress harmonics and other spurious signals. Bandswitching of such filters is

easily accomplished when necessary; it often is done using solid-state switches. Tube amplifiers, on the other hand, usually must be retuned on each band, and even for significant frequency movement within a band.

- Long life—Transistors and other semiconductor devices have extremely long lives if properly used and cooled. When employed in properly designed equipment, they should last for the entire useful life of the equipment—commonly 100,000 hours or more. Vacuum tubes wear out as their filaments (and sometimes other parts) deteriorate with time in normal operation; the useful life of a typical vacuum tube may be on the order of 10,000 to 20,000 hours.
- Manufacturing ease—Most solid-state devices are ideally suited for printed-circuit-board fabrication. The low voltages and low impedances that typify transistor and IC circuitry work very well on printed circuits (some circuits use the circuit board traces themselves as circuit elements); the high impedances found with vacuum tubes do not. The IC or transistor's physical size and shape also lends itself well to printed circuits and the devices usually can be soldered right to the board.

These advantages in fabrication mean reduced manufacturing costs. Based on all these facts, it might seem that there would be no place for vacuum tubes in a solid state world. Transistors and ICs do have significant limitations, however, especially in a practical sense. Individual RF power transistors available today cannot develop more than approximately 150 W output; this figure has not changed much in the past two decades.

Individual present-day transistors cannot generally handle the combination of current and voltage needed, nor can they safely dispose of the amount of heat dissipated, for RF amplification to higher power levels. So pairs of transistors, or even pairs of pairs, are usually employed in practical power amplifier designs, even at the 100-W level. Beyond the 300-W output level, somewhat exotic (at least for most radio amateurs) techniques of power combination from multiple amplifiers ordinarily must be used. Although this is commonly done in commercial equipment, it is an expensive proposition.

It also is far easier to ensure safe cooling of vacuum tubes, which operate satisfactorily at surface temperatures as high as 150-200°C and may be cooled by simply blowing sufficient ambient air past or through their relative large cooling surfaces. The very small cooling surfaces of power transistors should be held to 75-100°C to avoid drastically shortening their life expectancy. Thus, assuming worst-case 50°C ambient air temperature, the large cooling surface of a vacuum tube can be allowed to rise 100-150°C above ambient, while the small surface of a transistor must not be allowed to rise more than about 50°C. Moreover, power tubes are considerably more likely than transistors to survive, without significant damage, the rare instance of severe overheating.

Furthermore, RF power transistors are much less tolerant of electrical abuse than are most vacuum tubes. An overvoltage spike lasting only microseconds can—and is likely to—destroy transistors costing $75 to $150 each. A comparable spike is unlikely to have any effect on a tube. So the important message is this: designing with expensive RF power transistors demands using extreme caution to ensure that adequate thermal and electrical protection is provided. It is an area best left to knowledgeable designers.

Even if one ignores the challenge of the RF portions of a high-power transistor amplifier, there is the dc power supply to consider. A solid-state amplifier capable of delivering 1 kW of RF output might require regulated (and transient-free) 50 V at more than 40 A. Developing that much current is a challenging and expensive task. These limitations considered, solid-state amplifiers have significant practical advantages up to a couple of hundred watts output. Beyond that point, and certainly at the kilowatt level, the vacuum tube still reigns for amateur constructors because of its cost-effectiveness and ease of equipment design.

CLASSES OF OPERATION

The class of operation of an amplifier stage is defined by its conduction angle, the angular portion of each RF drive cycle, in degrees, during which plate current (or collector or drain current in the case of transistors) flows. This, in turn, determines the amplifier's gain, efficiency, linearity and input and output impedances.

- Class A: The conduction angle is 360°. DC bias and RF drive level are set so that the device is not driven to output current cutoff at any point in the driving-voltage cycle, so some device output current flows throughout the complete 360° of the cycle (see **Fig 13.1A**). Output voltage is generated by the variation of output current flowing through the load resistance. Maximum linearity and gain are achieved in a Class A amplifier, but the efficiency of

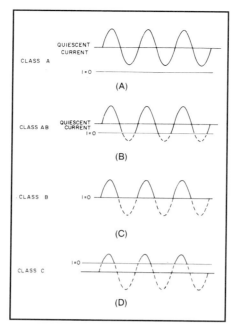

Fig 13.1—Amplifying device output current for various classes of operation. All assume a sinusoidal drive signal.

the stage is low. Maximum theoretical efficiency is 50%, but 25 to 30% is more common in practice.

- Class AB: The conduction angle is greater than 180° but less than 360° (see Fig 13.1B). In other words, dc bias and drive level are adjusted so device output current flows during appreciably more than half the drive cycle, but less than the whole drive cycle. Efficiency is much better than Class A, typically reaching 50-60% at peak output power. Class AB linearity and gain are not as good as that achieved in Class A, but are very acceptable for even the most rigorous high-power SSB applications in Amateur Radio.

Class AB vacuum tube amplifiers are further defined as class AB1 or AB2. In class AB1, the grid is not driven positive so no grid current flows. Virtually no drive power is required, and gain is quite high, typically 15-20 dB. The load on the driving stage is relatively constant throughout the RF cycle. Efficiency typically exceeds 50% at maximum output.

In Class AB2, the grid is driven positive on peaks and some grid current flows. Efficiency commonly reaches 60%, at the expense of greater demands placed on the driving stage and slightly reduced linearity. Gain commonly reaches 15 dB.

- Class B: Conduction angle = 180°. Bias

and RF drive are set so that the device is just cut off with no signal applied (see Fig 13.1C), and device output current flows during one half of the drive cycle. Efficiency commonly reaches as high as 65%, with fully acceptable linearity.

- Class C: The conduction angle is much less than 180°—typically 90°. DC bias is adjusted so that the device is cut off when no drive signal is applied. Output current flows only during positive crests in the drive cycle (see Fig 13.1D), so it consists of pulses at the drive frequency. Efficiency is relatively high—up to 80%—but linearity is extremely poor. Thus Class C amplifiers are not suitable for amplification of amplitude-modulated signals such as SSB or AM, but are quite satisfactory for use in on-off keyed stages or with frequency or phase modulation. Gain is lower than for the previous classes of operation, typically 10-13 dB.

- Classes D through H use various switched mode techniques and are not commonly found in amateur service. Their prime virtue is high efficiency, and they are used in a wide range of specialized audio and RF applications to reduce power-supply requirements and dissipated heat. These classes of RF amplifiers require fairly sophisticated design and adjustment techniques, particularly at high-power levels. The additional complexity and cost could rarely if ever be justified for amateur service.

Class of operation is independent of device type and circuit configuration (see **Analog Signal** chapter). The active amplifying device and the circuit itself must be uniquely applied for each operating class, but amplifier linearity and efficiency are determined by the class of operation. Clever amplifier design cannot improve on these fundamental limits. Poor design and implementation, though, can certainly prevent an amplifier from approaching its potential in efficiency and linearity.

MODELING THE ACTIVE DEVICE

It is very useful to have a model for the active devices used in a real-world RF power amplifier. Although the actual active device used in an amplifier might be a vacuum tube, a transistor or an FET, each model has certain common characteristics.

See **Fig 13.2A**, where a vacuum tube is modeled as a current generator in parallel with a *dynamic plate resistance* R_p and a *load resistance* R_L. In this simplified model, any residual reactances (such as the inductance of connecting leads and the output capacity of the tube) are not specifically shown. The control-grid voltage in a vacuum tube controls the stream of electrons moving between the cathode and the plate. An important measure for a tube is its *transconductance*, which is the change in plate current caused by a change in grid-cathode voltage. The plate current is:

$$i_p = g_m \times e_g \qquad (1)$$

where

i_p = plate current
g_m = transconductance (also called mutual conductance) of tube $\Delta i_p / \Delta e_g$
e_g = grid RF voltage.

The concept of dynamic plate resistance is sometimes misunderstood. It is a measure of how the plate current changes with a change in plate voltage, given a constant grid voltage. The control-grid voltage is by far the major determinant of the plate current in a triode. In a tetrode or pentode vacuum tube, the screen grid "screens" the plate current even further from the effect of changes in the plate voltage. For small-signal operation (where the plate voltage does not swing below the screen voltage) the plate current in a pentode or tetrode changes remarkably little when the plate voltage is changed. Thus the dynamic plate resistance is very high in a tetrode or pentode that is operating linearly, and only somewhat less for a triode. The plate current delivered into the load resistance R_L creates RF power.

An FET operates much like the vacuum-tube model. Obviously, there is no vacuum inside the case of an FET, and the FET electrodes are called gate, drain and source instead of grid, plate and cathode, but the current-generator model is just as viable for an FET as for a vacuum tube.

In a transistor, the base current controls the flows of electrons (or holes) in the collector circuit. See Fig 13.2B. A transistor operating in a linear fashion resembles the operation of a tetrode or pentode vacuum tube since the equivalent collector dynamic output resistance is also high. This is so because the collector current is not affected greatly by the collector voltage—it is mainly determined by the base current. The collector current in the current-generator model for a transistor is:

$$i_c = \beta \times i_b \qquad (2)$$

where

i_c = collector current
β = current gain of transistor
i_b = base current.

IMPEDANCE TRANSFORMATION —"MATCHING NETWORKS"

Over the years, some confusion in the

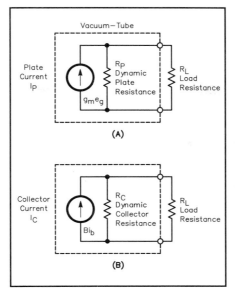

Fig 13.2—At A, the current-generator model for a vacuum-tube amplifier operating linearly. Typical values for R_p and R_L for small-signal vacuum tubes are 100 kΩ and 5 kΩ respectively. The plate current I_p is equal to the product of the tube tranconductance g_m times the grid voltage. At B, the current-generator model for a transistor. Typical values for R_C and R_L are on the same order as those for a small-signal vacuum tube.

amateur ranks has resulted from imprecise use of the terms *matching* and *matching network*. The term "matching" was first used in the technical literature in connection with transmission lines. When a matching network such as a *Transmatch* is tuned properly, it "matches" (that is, makes equal) a particular load impedance to the fixed characteristic impedance of the transmission line used at the Transmatch input.

In this chapter, we are concerned with using active devices to generate useful RF power. For a given active device, RF power is generated most efficiently, and with the least distortion for a linear amplifier, when it delivers RF current into an *optimum value of load resistance*. For an amplifier, the output network transforms the load impedance (such as an antenna) into an optimum value of load resistance for the active device. In part to differentiate active power amplifiers from passive transmission lines, we prefer to call such a transforming network an *output network*, rather than a matching network.

Output Networks and Class AB, B, and C Amplifiers

In Class AB, B and C amplifiers, we

select a load resistance that will keep the tube or transistor from dissipating too much power or, in the case of Class AB or B amplifiers, to achieve the desired linearity.

In these classes of amplifiers, the device output current is zero for large parts of the RF cycle. Because of this, the effective source resistance is no longer the simple dynamic plate resistance of a Class A amplifier. In fact, the value of R_p varies with the drive level. This means that, since the load resistance (of an antenna, for example) is constant, the efficiency of the amplifier also varies with the drive level.

It may at first appear contradictory that Class AB and B amplifiers use nonlinear devices but achieve "linear" operation nevertheless. The explanation is that the peak amplitude of device output current faithfully follows that of the drive voltage, even though its waveform does not. In tuned amplifiers, the flywheel effect of the resonant output network restores the missing part of each RF input cycle, as well as its sinusoidal waveform. In broadband transistor amplifiers, balanced push-pull circuitry commonly is used to restore the missing RF cycles, and low-pass filters on the output remove harmonics and thereby restore the sinusoidal RF waveform. The result in both cases is linear amplification of the input signal—by the clever application of nonlinear devices.

The usual practice in RF power amplifier design is to select an optimum load resistance that will provide the highest power output consistent with required linearity, while staying within the amplifying device's ratings. The optimum load resistance is determined by the amplifying device's current transfer characteristics and the amplifier's class of operation. For a transistor amplifier, the optimum load resistance is approximately:

$$R_L = \frac{V_{CC}^2}{2P_O} \qquad (6)$$

where

R_L = the load resistance
V_{CC} = the collector dc voltage
P_O = the amplifier power output in watts.

Vacuum tubes have complex current transfer characteristics, and each class of operation produces different RMS values of RF current through the load impedance. The optimum load resistance for vacuum-tube amplifiers can be approximated by the ratio of the dc plate voltage to the dc plate current at maximum signal, divided by a constant appropriate to each class of operation. The load resistance, in turn, determines the maximum power output and efficiency the amplifier can provide.

The optimum tube load resistance is

$$R_L = \frac{V_P}{K \times I_P} \qquad (7)$$

where

R_L = the appropriate load resistance, in ohms

V_P = the dc plate potential, in V

I_P = the dc plate current, in A

K = a constant that approximates the RMS current to dc current ratio appropriate for each class. For the different classes of operation:

Class A, $K \approx 1.3$
Class AB, $K \approx 1.5 - 1.7$
Class B, $K \approx 1.57 - 1.8$
Class C, $K \approx 2$.

Graphical or computer-based analytical methods may be used to calculate more precisely the optimum plate load resistance for specific tubes and operating conditions, but the above "rules of thumb" generally provide satisfactory results for design.

The ultimate load for an RF power amplifier usually is a transmission line connected to an antenna or the input of another amplifier. It usually isn't practical, or even possible, to modify either of these load impedances to the optimum value needed for high-efficiency operation. An output network is thus used to transform the real load impedance to the optimum load resistance for the amplifying device. Two basic types of output networks are found in RF power amplifiers: tank circuits and transformers.

TANK CIRCUITS

Parallel-resonant circuits and their equivalents have the ability to store energy. Capacitors store electrical energy in the electric field between their plates; inductors store energy in the magnetic field induced by the coil winding. These circuits are referred to as *tank circuits*, since they act as storage "tanks" for RF energy.

The energy stored in the individual tank circuit components varies with time. Consider for example the tank circuit shown in **Fig 13.4**. Assuming that R is zero, the tank circuit dissipates no power. Therefore, no power need be supplied by the source; hence no line current I_{LINE} flows. Only circulating current I_{CIRC} flows, and it is exactly the same through both L and C at any instant. Similarly, the voltage across L and C is always exactly the same. At some point the capacitor is fully charged, and the current through both the capacitor and inductor is zero. So the inductor has no magnetic field and therefore no energy stored in its field. All the energy in the tank is stored in the capacitor's electric field.

At this instant, the capacitor starts to discharge through the inductor. The current flowing in the inductor creates a magnetic field, and energy transferred from the capacitor is stored in the inductor's magnetic field. Still assuming there is no loss in the tank circuit, the increase in energy stored in the inductor's magnetic field is exactly equal to the decrease in energy stored in the capacitor's electric field. The total energy stored in the tank circuit stays constant; some is stored in the inductor, some in the capacitor. Current flow into the inductor is a function of both time and of the voltage applied by the capacitor, which decreases with time as it discharges into the inductor. Eventually, the capacitor's charge is totally depleted and all the tank circuit's energy is stored in the magnetic field of the inductor. At this instant, current flow through L and C is maximum and the voltage across the terminals of both L and C is zero.

Since energy no longer is being transferred to the inductor, its magnetic field begins to collapse and becomes a source of current, still flowing in the same direction as when the inductor was being driven by the capacitor. When the inductor becomes a current source, the voltage across its terminals reverses and it begins to recharge the capacitor, with opposite polarity from its previous condition. Eventually, all energy stored in the inductor's magnetic field is depleted as current decreases to zero. The capacitor is fully charged, and all the energy is then stored in the capacitor's electric field. The exchange of energy from capacitor to inductor and back to capacitor is then repeated, but with opposite voltage polarities and direction of current flow from the previous exchange. It can be shown mathematically that the "alternating" current and voltage produced by this process are sinusoidal in waveform, with a frequency of

$$f = \frac{1}{2\pi \sqrt{LC}} \qquad (8)$$

which of course is the resonant frequency of the tank circuit. In the absence of a load or any losses to dissipate tank energy, the tank circuit current would oscillate forever.

In a typical tank circuit such as shown in Fig 13.4, the values for L and C are chosen so that the reactance (X_L) of L is equal to the reactance (X_C) of C at the frequency of the signal generated by the ac voltage source. If R is zero (since X_L is equal to X_C), the line current I_{LINE} measured by M1 is close to zero. However, the circulating current in the loop made up of L, R and C is definitely not zero. Examine what would happen if the circuit were sud-

denly broken at points A and B. The circuit is now made up of L, C and R, all in series. X_L is equal to X_C, so the circuit is resonant. If some voltage is applied between points A and B, the magnitude of circulating current is limited only by resistance R. If R were equal to zero, the circulating current would be infinite!

The Flywheel Effect

A tank circuit can be likened to a flywheel—a mechanical device for storing energy. The energy in a flywheel is stored in the angular momentum of the wheel.

As soon as a load of some sort is attached, the wheel starts to slow or even stop. Some of the energy stored in the spinning flywheel is now transferred to the load. In order to keep the flywheel turning at a constant speed, the energy drained by the load must be replenished. Energy has to be added to the flywheel from some external source. If sufficient energy is added to the flywheel, it maintains its constant rotational speed.

In the real world, of course, flywheels and tank circuits suffer from the same fate; system losses dissipate some of the stored energy without performing any useful work. Air resistance and bearing friction slow the flywheel. In a tank circuit, resistive losses drain energy.

Tank Circuit Q

In order to quantify the ability of a tank circuit to store energy, a quality factor, Q, is defined. Q is the ratio of energy stored in a system during one complete RF cycle to energy lost.

$$Q = 2\pi \frac{W_S}{W_L} \qquad (9)$$

where

W_S = is the energy stored

W_L = the energy lost to heat and the load.

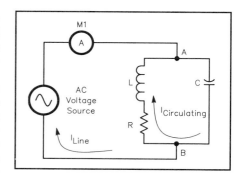

Fig 13.4—There are two currents in a tank circuit: the line current (I_{LINE}) and the circulating current (I_{CIRC}). The circulating current is dependent on tank Q.

By algebraic substitution and appropriate integration, the Q for a tank circuit can be expressed as

$$Q = \frac{X}{R} \qquad (10)$$

where

X = the reactance of either the inductor or the capacitor

R = the series resistance.

Since both circulating current and Q are proportional to 1/R, circulating current is therefore proportional to Q. The tank circulating current is equal to the line current multiplied by Q. If the line current is 100 mA and the tank Q is 10, then the circulating current through the tank is 1 A. (This implies, according to Ohm's Law, that the voltage potentials across the components in a tank circuit also are proportional to Q.)

When there is no load connected to the tank, the only resistances contributing to R are the losses in the tank circuit. The *unloaded Q* (Q_U) in that case is:

$$Q_U = \frac{X}{R_{LOSS}} \qquad (11)$$

where

X = the reactance of either the inductor or capacitor

R_{Loss} = the effective series loss resistance in the circuit.

A load connected to a tank circuit has exactly the same effect on tank operation as circuit losses. Both consume energy. It just happens that energy consumed by circuit losses becomes heat rather than useful output. When energy is coupled out of the tank circuit into a load, the *loaded Q* (Q_L) is:

$$Q_L = \frac{X}{R_{Loss} + R_{Load}} \qquad (12)$$

where R_{Load} is the load resistance. Energy dissipated in R_{Loss} is wasted as heat. Ideally, all the tank circuit energy should be delivered to R_{Load}. This implies that R_{Loss} should be as small as practical, to yield the highest reasonable value of unloaded Q.

Tank Circuit Efficiency

The efficiency of a tank circuit is the ratio of power delivered to the load resistance (R_{Load}) to the total power dissipated by losses (R_{Load} and R_{Loss}) in the tank circuit. Within the tank circuit, R_{Load} and R_{Loss} are effectively in series, and the circulating current flows through both. The power dissipated by each is therefore proportional to its resistance. The loaded tank efficiency can therefore be defined as

$$\text{Tank Efficiency} = \frac{R_{Load}}{R_{Load} + R_{Loss}} \times 100 \qquad (13)$$

where efficiency is stated as a percentage. By algebraic substitution, the loaded tank efficiency can also be expressed as

$$\text{Tank Efficiency} = \left(1 - \frac{Q_L}{Q_U}\right) \times 100 \qquad (14)$$

where

Q_L = the tank circuit loaded Q

Q_U = the unloaded Q of the tank circuit.

It follows then that tank efficiency can be maximized by keeping Q_L low, which keeps the circulating current low and the I^2R losses down. Q_U should be maximized for best efficiency; this means keeping the circuit losses low.

The selectivity provided by a tank circuit helps suppress harmonic currents generated by the amplifier. The amount of harmonic suppression is dependent upon circuit loaded Q_L, so a dilemma exists for the amplifier designer. A low Q_L is desirable for best tank efficiency, but yields poorer harmonic suppression. High Q_L keeps amplifier harmonic levels lower at the expense of some tank efficiency. At HF, a compromise value of Q_L can usually be chosen such that tank efficiency remains high and harmonic suppression is also reasonable. At higher frequencies, tank Q_L is not always readily controllable, due to unavoidable stray reactances in the circuit. However, unloaded Q_U can always be maximized, regardless of frequency, by keeping circuit losses low.

Tank Output Circuits

Tank circuit output networks need not take the form of a capacitor connected in parallel with an inductor. A number of equivalent circuits can be used to match the impedances normally encountered in a power amplifier. Most are operationally more flexible than a parallel-resonant tank. Each has its advantages and disadvantages for specific applications, but the final choice usually is based on practical construction considerations and the component values needed to implement a particular network. Some networks may require unreasonably high or low inductance or capacitance values. In that case, use another network, or a different value of Q_L. Several different networks may be investigated before an acceptable final design is reached.

The impedances of RF components and amplifying devices frequently are given in terms of a parallel combination of a resistance and a reactance, although

it is often easier to use a series R-X combination to design networks. Fortunately, there is a series impedance equivalent to every parallel impedance and vice versa. The equivalent circuits, and equations for conversion from one to the other, are given in **Fig 13.7**. In order to use most readily available design equations for computing matching networks, the parallel impedance must first be converted to its equivalent series form.

The Q_L of a parallel impedance can be derived from the series form as well. Substitution of the usual formula for calculating Q_L into the equations from Fig 13.7 gives

$$Q_L = \frac{R_P}{X_P} \qquad (15)$$

where

R_P = the parallel equivalent resistance

X_P = the parallel equivalent reactance.

Several impedance-matching networks are shown in the **Receivers** chapter. A low-pass T network and two low-pass L networks are possible matching networks. Both types of matching networks provide good harmonic suppression. The pi network is also commonly used for amplifier matching. Harmonic suppression of a pi network is a function of the impedance transformation ratio and the Q_L of the circuit. Second-harmonic attenuation is approximately 35 dB for a load impedance of 2000 Ω in a pi network with a Q_L of 10. The third harmonic is typically 10 dB lower and the fourth approximately 7 dB below that. A typical pi network as used in the output circuit of a tube amplifier is shown in **Fig 13.8**.

You can calculate Pi-network matching-circuit values using the following equations. These equations are from Elmer (W5FD) Wingfield's August 1983 *QST*

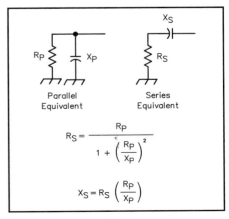

Fig 13.7—Parallel and series equivalent circuits and the formulas used for conversion.

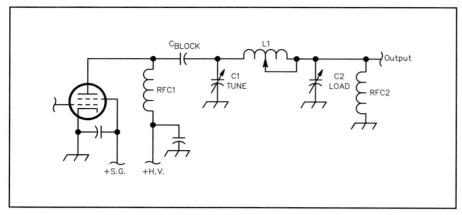

Fig 13.8—A pi matching network used at the output of a tetrode power amplifier. RFC2 is used for protective purposes in the event C$_{BLOCK}$ fails.

Pi-Network Equations at C$_{Min}$

Wingfield's equations are a great improvement because they solve for the desired component values in terms of Q_0, the desired output Q. When the overall circuit capacitance C$_{Min}$ at the plate is too great (common at higher frequencies), the normal equations do not work. Use the following procedure for this case.

$$Q_1 = \frac{R_1}{X_{CMin}} \qquad (A)$$

where X$_{CMin}$ is the reactance at minimum capacitance at the plate, including strays, such as the plate output capacitance and the minimum capacitance of variable C$_1$.

$$X_{C2} = \sqrt{\frac{R_1 \times R_2}{Q_1^2 + 1 - \frac{R_1}{R_2}}} \qquad (B)$$

$$Q_2 = \frac{R_2}{X_{C2}} \qquad (C)$$

$$X_L = \frac{Q_1^2 + 1}{Q_1^2} X_{CMin} + \frac{Q_2^2}{Q_2^2 + 1} X_{C2} \qquad (D)$$

$$Q_0 = Q_1 + Q_2 \qquad (E)$$

article, "New and Improved Formulas for the Design of Pi and Pi-L Networks," and Feedback in January 1984 *QST*. (See the Bibliography at the end of this chapter.)

Table 13.1 shows some data from a computer program Wingfield wrote to calculate these values. This program (PI-CMIN.EXE) and a similar program to calculate Pi-L network values (PI-LCMIN.EXE) are available from *ARRL Web* (see page viii), along with several other useful Wingfield programs. The programs are for IBM PC and compatible computers. A more complete set of tables is also available from ARRL as a template package. See the **References** chapter for ordering information.

The computer programs take into account the minimum practical capacitance (C$_{min}$) you can expect to achieve with your circuit, based on your knowledge of the tube output capacitance, stray circuit capacitance, the minimum capacitance of the variable tuning capacitors and a reasonable amount of capacitance for tuning. (Start with a minimum capacitance of about 35 pF for vacuum variable capacitors and about 45 to 50 pF for air variable capacitors.) If the following equations lead to a capacitor value less than the minimum capacitance you expect to achieve, use the minimum value to recalculate the other quantities as shown in Q$_1$ Based Pi-Network Equations. This will result in a final circuit operating Q value that is larger than the selected value. (Wingfield uses Q_0 to represent this output Q, which is the same as Q$_L$ referred to earlier in this chapter. We will use Q_0 in the equations.) The program output includes this new calculated Q_0 value.

Use the following equations to calculate specific component values for a Pi-network matching circuit. Select the desired circuit operating Q, Q_0, to satisfy these relationships, depending on whether the load resistance is higher or lower than the transformed resistance presented to

the plate:

$$Q_0^2 > \frac{R_1}{R_2} - 1 \quad \text{and} \quad Q_0^2 > \frac{R_2}{R_1} - 1 \qquad (16)$$

where:
R$_1$ is the input resistance to be matched, in ohms
R$_2$ is the load (output) resistance to be matched, in ohms.

Calculate the value of the input Q, Q$_1$:

$$Q_1 = \frac{R_1 Q_0 - \sqrt{R_1 R_2 Q_0^2 - (R_1 - R_2)^2}}{R_1 - R_2} \qquad (17)$$

We will work through an example as the equations are presented. Let's select $Q_0 = 12$, $R_1 = 1500 \ \Omega$ and $R_2 = 50 \ \Omega$.

$$Q_1 = \frac{1500 \times 12 - \sqrt{1500 \times 50 \times 12^2 - (1500 - 50)^2}}{1500 - 50}$$

$$Q_1 = \frac{1.80 \times 10^4 - \sqrt{8.6975 \times 10^6}}{1450} = 10.38$$

Next calculate the value of the output Q, Q$_2$:

$$Q_2 = Q_0 - Q_1 \qquad (18)$$

$$Q_2 = 12 - 10.38 = 1.62$$

Now calculate the reactance of the input capacitor, output capacitor and inductor.

$$X_{C1} = \frac{R_1}{Q_1} \qquad (19)$$

$$X_{C1} = \frac{1500}{10.38} = 144.5 \ \Omega$$

$$X_{C2} = \frac{R_2}{Q_2} \qquad (20)$$

$$X_{C2} = \frac{50}{1.62} = 30.86 \ \Omega$$

$$X_L = \frac{R_1 Q_0}{Q_1^2 + 1} \qquad (21)$$

$$X_L = \frac{1500 \times 12}{10.38^2 + 1} = \frac{1.80 \times 10^4}{108.74} = 165.5 \ \Omega$$

Finally, calculate the component values:

$$C_1 = \frac{1}{2\pi \ f \ X_{C1}} \qquad (22)$$

where f is in Hz and X$_{C1}$ is in ohms.

For our example, let's find the component values at 3.75 MHz.

Table 13.1
Pi-Network Values for Various Plate Impedances
(Sample Output from PI-CMIN.EXE by W5FD)

C in pF and L in µH
Pi-Net Values
$R2 = 50\ \Omega$, $Q_0 = 12$, $C(min) = 35$ pF

R1=1500 ohms

Band	C1	C2	L	
160	580	2718	13.9	
80	294	1378	7.0	
40	154	721	3.7	
30	109	511	2.7	
20	78	364	1.86	$Q_0=12.0$
17	61	285	1.45	$Q_0=12.0$
15	52	243	1.24	$Q_0=12.0$
12	44	207	1.06	$Q_0=12.0$
10	38	179	0.91	$Q_0=12.0$

R1=1600 ohms

Band	C1	C2	L	
160	547	2619	14.6	
80	278	1328	7.4	
40	145	695	3.9	
30	103	492	2.8	
20	73	351	1.96	$Q_0=12.0$
17	57	274	1.53	$Q_0=12.0$
15	49	234	1.31	$Q_0=12.0$
12	42	199	1.11	$Q_0=12.0$
10	36	172	0.96	$Q_0=12.0$

R1=1700 ohms

Band	C1	C2	L	
160	518	2527	15.4	
80	263	1281	7.8	
40	137	671	4.1	
30	97	475	2.9	
20	69	338	2.06	$Q_0=12.0$
17	54	265	1.61	$Q_0=12.0$
15	46	226	1.38	$Q_0=12.0$
12	39	192	1.17	$Q_0=12.0$
10	35	173	0.99	$Q_0=12.3$

R1=1800 ohms

Band	C1	C2	L	
160	491	2441	16.1	
80	249	1238	8.2	
40	130	648	4.3	
30	92	459	3.0	
20	66	327	2.16	$Q_0=12.0$
17	51	256	1.69	$Q_0=12.0$
15	44	218	1.44	$Q_0=12.0$
12	37	186	1.23	$Q_0=12.0$
10	35	180	0.99	$Q_0=13.0$

R1=1900 ohms

Band	C1	C2	L	
160	468	2360	16.9	
80	237	1197	8.6	
40	124	626	4.5	
30	88	443	3.2	
20	63	316	2.26	$Q_0=12.0$
17	49	247	1.77	$Q_0=12.0$
15	42	211	1.51	$Q_0=12.0$
12	36	180	1.29	$Q_0=12.0$
10	35	186	0.99	$Q_0=13.7$

R1=2000 ohms

Band	C1	C2	L	
160	446	2284	17.6	
80	226	1158	8.9	
40	118	606	4.7	
30	84	429	3.3	
20	60	306	2.36	$Q_0=12.0$
17	47	239	1.85	$Q_0=12.0$
15	40	204	1.58	$Q_0=12.0$
12	35	184	1.29	$Q_0=12.5$
10	35	193	0.98	$Q_0=14.4$

R1=2100 ohms

Band	C1	C2	L	
160	427	2213	18.4	
80	216	1122	9.3	
40	113	587	4.9	
30	80	416	3.5	
20	57	296	2.46	$Q_0=12.0$
17	45	232	1.92	$Q_0=12.0$
15	38	198	1.64	$Q_0=12.0$
12	35	189	1.30	$Q_0=13.0$
10	35	199	0.98	$Q_0=15.1$

R1=2200 ohms

Band	C1	C2	L	
160	409	2145	19.1	
80	207	1088	9.7	
40	109	569	5.1	
30	77	403	3.6	
20	55	287	2.56	$Q_0=12.0$
17	45	232	2.00	$Q_0=12.0$
15	37	192	1.71	$Q_0=12.0$
12	35	197	1.29	$Q_0=13.7$
10	35	205	0.98	$Q_0=15.8$

R1=2300 ohms

Band	C1	C2	L	
160	392	2081	19.8	
80	199	1055	10.1	
40	104	552	5.3	
30	74	391	3.7	
20	53	279	2.65	$Q_0=12.0$
17	41	218	2.08	$Q_0=12.0$
15	35	186	1.77	$Q_0=12.0$
12	35	210	1.30	$Q_0=12.0$
10	35	211	0.98	$Q_0=16.5$

R1=2400 ohms

Band	C1	C2	L	
160	377	2020	20.5	
80	191	1024	10.4	
40	100	536	5.5	
30	71	379	3.9	
20	51	270	2.75	$Q_0=12.0$
17	40	212	2.15	$Q_0=12.0$
15	35	192	1.78	$Q_0=12.5$
12	35	207	1.30	$Q_0=14.8$
10	35	216	0.98	$Q_0=17.2$

R1=2500 ohms

Band	C1	C2	L	
160	363	1961	21.3	
80	184	994	10.8	
40	96	520	5.6	
30	68	368	4.0	
20	49	262	2.85	$Q_0=12.0$
17	38	205	2.23	$Q_0=12.0$
15	35	198	1.78	$Q_0=13.0$
12	35	215	1.29	$Q_0=15.5$
10	35	222	0.98	$Q_0=17.9$

$$C_1 = \frac{1}{2\pi\ 3.75\times10^6 \times 144.5} = 294\ \text{pF}$$

$$C_2 = \frac{1}{2\pi\ f\ X_{C2}} \tag{23}$$

$$C_2 = \frac{1}{2\pi\ 3.75\times10^6 \times 30.86} = 1375\ \text{pF}$$

$$L = \frac{X_L}{2\pi\ f} \tag{24}$$

$$L = \frac{165.5}{2\pi\ 3.75\times10^6} = 7.02\ \mu\text{H}$$

As an alternate method, after selecting the values for Q_0, R_1 and R_2, you can use the following equations:

$$X_L = \tag{25}$$

$$\frac{Q_0(R_1+R_2)+2\sqrt{R_1 R_2 (Q_0^2+4)-(R_1+R_2)^2}}{Q_0^2+4}$$

$$X_L =$$

$$\frac{12(1500+50)+2\sqrt{1500\times50\ (12^2+4)-(1500+50)^2}}{12^2+4}$$

$$X_L =$$

$$\frac{1.86\times10^4+2\sqrt{1.11\times10^7-2.4025\times10^6}}{148} = 165.5\ \Omega$$

$$Q_1 = \sqrt{\frac{Q_0 R_1}{X_L}-1} \tag{26}$$

$$Q_1 = \sqrt{\frac{12\times1500}{165.5}-1} = 10.38$$

$$Q_2 = Q_0 - Q_1$$

or

$$Q_2 = \sqrt{\frac{Q_0 R_2}{X_L}-1} \tag{27}$$

$$Q_2 = \sqrt{\frac{12\times50}{165.5}-1} = 1.62$$

Use equations 19 and 20 to calculate the reactances of capacitors C_1 and C_2. Equations 22, 23 and 24 give the capacitance and inductance values for the pi network.

The pi-L network is a combination of a pi network followed by an L network. The pi network transforms the load resistance to an intermediate impedance level called the image impedance. Typically, the image impedance is chosen to be between 300 and 700 Ω. The L section then transforms from the image impedance down to 50 Ω. The output capacitor of the pi network is combined with the input capacitor for the L network, as shown in **Fig 13.9**. The pi-L configuration attenuates harmonics better than a pi network. Second harmonic level for a pi-L network with a Q_L of 10 is approximately 52 dB below the fundamental. The third harmonic is attenuated 65 dB and the fourth harmonic approximately 75 dB.

The following equations help you calculate pi-L matching-network values. Select an image resistance value (R_m) that the L network will supply as a load for the pi network. This value must be between the desired pi-L network input resistance (R_1) and the output load resistance (R_2). For example, you can use the value given:

$$R_m = \sqrt{R_1 R_2} \qquad (28)$$

The computer program, PI-LCMIN. EXE, uses 300 Ω for R_m in its calculations. Changing the image resistance results in a different network solution. Use this equation to compute the L network Q value, Q_L:

$$Q_L = \sqrt{\frac{R_m}{R_2} - 1} \qquad (29)$$

We will work through an example, using $R_1 = 1500\ \Omega$, $R_2 = 50\ \Omega$ and the desired pi-L network output Q, $Q_0 = 12$.

$$Q_L = \sqrt{\frac{300}{50} - 1} = 2.24$$

Use equations 30 and 31 to calculate the L-network reactances.

$$X_{L2} = Q_L R_2 \qquad (30)$$

$$X_{L2} = 2.24 \times 50 = 112\ \Omega \qquad (31)$$

$$X_{P2} = \frac{R_m}{Q_L} \qquad (32)$$

$$X_{P2} = \frac{300}{2.24} = 134\ \Omega$$

Next calculate the desired Q of the pi-network section ($Q_{0\pi}$).

$$Q_{0\pi} = Q_0 - Q_L \qquad (33)$$

$$Q_{0\pi} = 12 - 2.24 = 9.76$$

Use equations 17 through 21 or 25 through 29 to calculate the pi-network reactances, X_{C1}, X_{L1} and X_{P1} as shown in Fig 13.9. Be sure to use the value specified for R_m as R_2 in these calculations. Also use the value just calculated for $Q_{0\pi}$ as Q_O. Notice that X_{P1} is X_{C2} in equation 23.

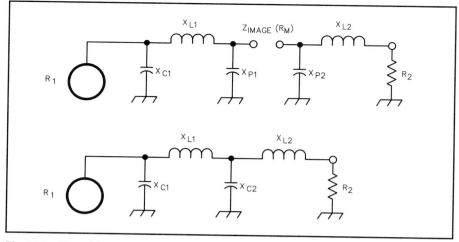

Fig 13.9—The pi-L network uses a pi network to transform the input impedances (R1) to the image impedance (Z_{IMAGE}). An L network transforms Z_{IMAGE} to R2.

$$Q_1 = \frac{R_1 Q_{0\pi} - \sqrt{R_1 R_m Q_{0\pi}^2 - (R_1 - R_m)^2}}{R_1 - R_m}$$

$$Q_1 =$$

$$\frac{1500 \times 9.76 - \sqrt{1500 \times 300 \times 9.76^2 - (1500 - 300)^2}}{1500 - 300}$$

$$Q_1 =$$

$$\frac{1.464 \times 10^4 - \sqrt{4.287 \times 10^7 - 1.44 \times 10^6}}{1200} = 6.84$$

$$Q_2 = Q_{0\pi} - Q_1 = 9.76 - 6.84 = 2.92$$

$$X_{C1} = \frac{R_1}{Q_1} = \frac{1500}{6.84} = 219.3\ \Omega$$

$$X_{P1} = \frac{R_m}{Q_2} = \frac{300}{2.92} = 102.7\ \Omega$$

$$X_{L1} = \frac{R_1 Q_{0\pi}}{Q_1^2 + 1} = \frac{1500 \times 9.76}{6.84^2 + 1} = 306.3\ \Omega$$

Combine the two parallel capacitors, X_{P1} and X_{P2} to find the Pi-L network X_{C2} value.

$$X_{C2} = \frac{X_{P1} X_{P2}}{X_{P1} + X_{P2}}$$

$$X_{C2} = \frac{102.7 \times 134}{102.7 + 134} = 58.3\ \Omega$$

Finally, calculate the capacitance and inductance values using equations 22 through 24. **Table 13.2** shows some data from Wingfield's program, PI-LCMIN. For the sample calculation shown here, we choose a frequency of 3.75 MHz.

$$C_1 = \frac{1}{2\pi\, f\, X_{C1}} = \frac{1}{2\pi \times 3.75 \times 10^6 \times 219.3}$$

$$C_1 = 193.5\ \text{pF}$$

$$C_2 = \frac{1}{2\pi\, f\, X_{C2}} = \frac{1}{2\pi\, 3.75 \times 10^6 \times 58.3}$$

$$C_2 = 730\ \text{pF}$$

$$L_1 = \frac{X_{L1}}{2\pi\, f} = \frac{306.3}{2\pi\, 3.75 \times 10^6} = 13.0\ \mu\text{H}$$

$$L_2 = \frac{X_{L2}}{2\pi\, f} = \frac{112}{2\pi\, 3.75 \times 10^6} = 4.75\ \mu\text{H}$$

The values for L and C in Tables 13.1 and 13.2 are based on purely resistive load impedances and assume ideal capacitors and inductors. Any other circuit reactances will modify these values.

Stray circuit reactances, including tube capacitances and capacitor stray inductances, should be included as part of the matching network. It is not uncommon for such reactances to render the use of certain matching circuits impractical, because they require either unacceptable loaded Q values or unrealistic component values. If all matching network alternatives are investigated and found unworkable, some compromise solution must be found.

Above 30 MHz, transistor and tube reactances tend to dominate circuit impedances. At the lower impedances found

Table 13.2
Pi-L Network Values for Various Plate Impedances
(Sample Output from PI-LCMIN.EXE by W5FD)

C in pF and L in μH
Pi-L Network Values
Rm = 300 Ω, Q_0 = 12, R2 = 50 Ω
$C_{(Min)}$ = 35 pF

R1=1500 ohms

Band	C1	C2	L1	L2	
160	382	1443	25.7	9.38	
80	194	732	13.0	4.76	
40	102	383	6.83	2.49	
30	72	270	4.82	1.76	
20	51	193	3.44	1.26	Q_0=12.0
17	40	151	2.69	0.98	Q_0=12.0
15	35	131	2.25	0.84	Q_0=12.2
12	35	123	1.64	0.71	Q_0=14.0
10	35	118	1.24	0.62	Q_0=15.9

R1=1600 ohms

Band	C1	C2	L1	L2	
160	362	1423	26.9	9.38	
80	184	722	13.6	4.76	
40	96	378	7.13	2.49	
30	68	267	5.04	1.76	
20	48	190	3.60	1.26	Q_0=12.0
17	38	149	2.81	0.98	Q_0=12.0
15	35	134	2.23	0.84	Q_0=12.8
12	35	126	1.63	0.71	Q_0=14.7
10	35	120	1.23	0.62	Q_0=16.7

R1=1700 ohms

Band	C1	C2	L1	L2	
160	344	1404	28.0	9.38	
80	175	712	14.2	4.76	
40	92	373	7.44	2.49	
30	65	263	5.25	1.76	
20	46	188	3.75	1.26	Q_0=12.0
17	36	147	2.94	0.98	Q_0=12.0
15	35	136	2.22	0.84	Q_0=13.4
12	35	129	1.62	0.71	Q_0=15.4
10	35	123	1.22	0.62	Q_0=17.5

R1=1800 ohms

Band	C1	C2	L1	L2	
160	328	1387	29.2	9.38	
80	166	703	14.8	4.76	
40	87	368	7.74	2.49	
30	61	260	5.47	1.76	
20	44	186	3.90	1.26	Q_0=12.0
17	35	147	3.01	0.98	Q_0=12.2
15	35	139	2.21	0.84	Q_0=13.9
12	35	131	1.61	0.71	Q_0=16.0
10	35	125	1.21	0.62	Q_0=18.2

R1=1900 ohms

Band	C1	C2	L1	L2	
160	313	1371	30.3	9.38	
80	159	695	15.4	4.76	
40	83	364	8.04	2.49	
30	59	257	5.68	1.76	
20	42	184	4.06	1.26	Q_0=12.0
17	35	149	2.99	0.98	Q_0=12.7
15	35	141	2.20	0.84	Q_0=14.5
12	35	133	1.60	0.71	Q_0=16.7
10	35	128	1.20	0.62	Q_0=19.0

R1=2000 ohms

Band	C1	C2	L1	L2	
160	300	1356	31.4	9.38	
80	152	687	15.9	4.76	
40	80	360	8.34	2.49	
30	56	254	5.89	1.76	
20	40	181	4.20	1.26	Q_0=12.0
17	35	152	2.97	0.98	Q_0=13.2
15	35	143	2.18	0.84	Q_0=15.1
12	35	136	1.59	0.71	Q_0=17.4
10	35	130	1.19	0.62	Q_0=19.7

R1=2100 ohms

Band	C1	C2	L1	L2	
160	288	1341	32.5	9.38	
80	146	680	16.5	4.76	
40	76	356	8.63	2.49	
30	54	251	6.09	1.76	
20	39	180	4.35	1.26	Q_0=12.0
17	35	154	2.97	0.98	Q_0=13.6
15	35	146	2.17	0.84	Q_0=15.6
12	35	138	1.58	0.71	Q_0=18.0
10	35	132	1.19	0.62	Q_0=20.5

R1=2200 ohms

Band	C1	C2	L1	L2	
160	277	1327	33.6	9.38	
80	140	673	17.0	4.76	
40	73	352	8.92	2.49	
30	52	249	6.30	1.76	
20	37	178	4.50	1.26	Q_0=12.0
17	35	156	2.95	0.98	Q_0=14.1
15	35	148	2.16	0.84	Q_0=16.2
12	35	140	1.57	0.71	Q_0=18.7
10	35	134	1.18	0.62	Q_0=21.3

R1=2300 ohms

Band	C1	C2	L1	L2	
160	266	1315	34.7	9.38	
80	135	667	17.6	4.76	
40	71	349	9.21	2.49	
30	50	246	6.50	1.76	
20	36	176	4.65	1.26	Q_0=12.0
17	35	158	2.93	0.98	Q_0=14.6
15	35	150	2.15	0.84	Q_0=16.7
12	35	142	1.57	0.71	Q_0=19.3
10	35	137	1.17	0.62	Q_0=22.0

R1=2400 ohms

Band	C1	C2	L1	L2	
160	257	1302	/35.8	9.38	
80	130	660	18.2	4.76	
40	68	346	9.50	2.49	
30	48	244	6.71	1.76	
20	35	176	4.71	1.26	Q_0=12.2
17	35	161	2.92	0.98	Q_0=15.0
15	35	152	2.13	0.84	Q_0=17.3
12	35	145	1.56	0.71	Q_0=20.0
10	35	139	1.17	0.62	Q_0=22.8

R1=2500 ohms

Band	C1	C2	L1	L2	
160	248	1291	36.9	9.38	
80	126	654	18.7	4.76	
40	66	343	9.79	2.49	
30	46	242	6.91	1.76	
20	35	178	4.70	1.26	Q_0=12.6
17	35	163	2.91	0.98	Q_0=15.5
15	35	154	2.13	0.84	Q_0=17.8
12	35	147	1.55	0.71	Q_0=20.6
10	35	141	1.16	0.62	Q_0=23.5

in transistor circuits, the standard networks can be applied so long as suitable components are used. Above 50 MHz, capacitors often exhibit values far different from their marked values because of stray internal reactances and lead inductance, and this requires compensation. Tuned circuits are frequently fabricated in the form of strip lines or other transmission lines in order to circumvent the problem of building "pure" inductances and capacitances. The choice of components is often more significant than the type of network used.

The high impedances encountered in VHF tube-amplifier plate circuits are not easily matched with typical networks. Tube output capacitance is usually so large that most matching networks are unsuitable. The usual practice is to resonate the tube output capacitance with a low-loss inductance connected in series or parallel. The result can be a very high Q tank circuit. Component losses must be kept to an absolute minimum in order to achieve reasonable tank efficiency. Output impedance transformation is usually performed by a link inductively coupled to the tank circuit or by a parallel transformation of the output resistance using a series capacitor.

(A)

(B)

Fig 13.10—The two methods of constructing the transformers outlined in the text. At A, the one-turn loop is made from brass tubing; at B, a piece of coaxial cable braid is used for the loop.

Transformers

Broadband transformers are often used in matching to the input impedance or optimum load impedance in a power amplifier. Multioctave power amplifier performance can be achieved by appropriate application of these transformers. The input and output transformers are two of the most critical components in a broadband amplifier. Amplifier efficiency, gain flatness, input SWR, and even linearity all are affected by transformer design and application. There are two basic RF transformer types, as described elsewhere in this *Handbook*: the conventional transformer and the transmission-line transformer.

The conventional transformer is wound much the same way as a power transformer. Primary and secondary windings are wound around a high-permeability core, usually made from a ferrite or powdered-iron material. Coupling between the secondary and primary is made as tight as possible to minimize leakage inductance. At low frequencies, the coupling between windings is predominantly magnetic. As the frequency rises, core permeability decreases and leakage inductance increases; transformer losses increase as well.

Typical examples of conventional transformers are shown in **Fig 13.10**. In Fig 13.10A, the primary windings consist of brass or copper tubes inserted into ferrite sleeves. The tubes are shorted together at one end by a piece of copper-clad circuit board material. The secondary winding is threaded through the tubes. Since the low-impedance winding is only a single turn, the transformation ratio is limited to the squares of integers; for example, 1, 4, 9, 16, and so on. The lowest effective transformer frequency is determined by the inductance of the one-turn winding. It should have a reactance, at the lowest frequency of intended operation, at least four times greater than the impedance it is connected to.

The coupling coefficient between the two windings is a function of the primary tube diameter and its length, and the diameters and insulation thickness of the wire used in the high-impedance winding. High impedance ratios, greater than 36:1, should use large-diameter secondary windings. Miniature coaxial cable (using only the braid as the conductor) works well. Another use for coaxial cable braid is illustrated in Fig 13.10B. Instead of using tubing for the primary winding, the secondary winding is threaded through copper braid. Performance of the two units is almost identical.

The cores used must be large enough so the core material will not saturate at the power level applied to the transformer.

Core saturation can cause permanent changes to the core permeability, as well as overheating. Transformer nonlinearity also develops at core saturation. Harmonics and other distortion products are produced, clearly an undesirable situation. Multiple cores can be used to increase the power capabilities of the transformer.

Transmission-line transformers are similar to conventional transformers, but can–be used over wider frequency ranges. In a conventional transformer, high-frequency performance deterioration is caused primarily by leakage inductance, which rises with frequency. In a transmission-line transformer, the windings are arranged so there is tight capacitive coupling between the two. A high coupling coefficient is maintained up to considerably higher frequencies than with conventional transformers.

Output Filtering

Amplifier output filtering is sometimes necessary to meet spurious signal requirements. Broadband amplifiers, by definition, provide little if any inherent suppression of harmonic energy. Even amplifiers using output tank circuits often require further attenuation of undesired harmonics. High-level signals from one transmitter, particularly at multiple transmitter sites, can be intercepted by an antenna connected to another transmitter, conducted down the feed line and mixed in a power amplifier, causing spurious outputs. For example, an HF transceiver signal radiated from a triband beam may be picked up by a VHF FM antenna on the same mast. The signal saturates the low-power FM transceiver output stage, even with power off, and is reradiated by the VHF antenna. Proper use of filters can reduce such spurious energy considerably.

The filter used will depend on the application and the level of attenuation needed. Band-pass filters attenuate spurious signals above and below the passband for which they are designed. Low-pass filters attenuate only signals above the cutoff frequency, while high-pass filters reduce energy below the design cutoff frequency.

The **Filters** chapter includes detailed information about designing suitable filters. Tables of component values in the **References** chapter allow you to select a particular design and scale the values for different frequencies and impedance ranges as needed.

TRANSMITTING DEVICE RATINGS

Plate Dissipation

The ultimate factor limiting the power-handling capability of a tube often (but not always) is its maximum plate dissipation rating. This is the measure of how many watts of heat the tube can safely dissipate, if it is cooled properly, without exceeding critical temperatures. Excessive temperature can damage or destroy internal tube components or vacuum seals, resulting in tube failure. The same tube may have different voltage, current and power ratings depending on the conditions under which it is operated, but its safe temperature ratings must not be exceeded in any case! Important cooling considerations are discussed in more detail in the Amplifier Cooling section of this chapter.

The efficiency of a power amplifier may range from approximately 25% to 75%, depending on its operating class, adjustment, and circuit losses. The efficiency indicates how much of the dc power supplied to the stage is converted to useful RF output power; the rest is dissipated as heat, mostly by the plate. By knowing the plate-dissipation limit of the tube and the efficiency expected from the class of operation selected, the maximum power input and output levels can be determined. The maximum safe power output is

$$P_{OUT} = \frac{P_D N_P}{100 - N_P} \qquad (34)$$

where

P_{OUT} = the power output in W
P_D = the plate dissipation in W
N_P = the efficiency (10% = 10).

The dc input power would simply be

$$P_{IN} = \frac{100 \, P_D}{100 - N_P} \qquad (35)$$

Almost all vacuum-tube power amplifiers in amateur service today operate as linear amplifiers (Class AB or B) with efficiencies of approximately 50% to 65%. That means that a useful power output of approximately 1 to 2.0 times the plate dissipation generally can be achieved. This requires, or course, that the tube is cooled enough to realize its maximum plate dissipation rating and that no other tube rating, such as maximum plate current or grid dissipation, is exceeded.

Type of modulation and duty cycle also influence how much output power can be achieved for a given tube dissipation. Some types of operation are less efficient than others, meaning that the tube must dissipate more heat. Some forms of modulation, such as CW or SSB, are intermittent in nature, causing less average heating than modulation formats such as RTTY in which there is continuous transmission. Power-tube manufacturers use two different rating systems to allow for the variations in service. CCS (Continuous Commercial Service) is the more conservative rating and is used for specifying tubes that are in constant use at full power. The second rating system is based on intermittent, low-duty-cycle operation, and is known as ICAS (Intermittent Commercial and Amateur Service). ICAS ratings are normally used by commercial manufacturers and individual amateurs who wish to obtain maximum power output consistent with reasonable tube life in CW and SSB service. CCS ratings should be used for FM, RTTY and SSTV applications. (Plate power transformers for amateur service are also rated in CCS and ICAS terms.)

Maximum Ratings

Tube manufacturers publish sets of maximum values for the tubes they produce. No maximum rated value should ever be exceeded. As an example, a tube might have a maximum plate-voltage rating of 2500 V, a maximum plate-current rating of 500 mA, and a maximum plate dissipation rating of 350 W. Although the plate voltage and current ratings might seem to imply a safe power input of 2500 V × 500 mA = 1250 W, this is true only if the dissipation rating will not be exceeded. If the tube is used in class AB2 with an expected efficiency of 60%, the maximum safe dc power input is

$$P_{IN} = \frac{100\,P_D}{100 - N_D} = \frac{100 \times 350}{100 - 60} = 875\,W$$

In this case, any combination of plate voltage and current whose product does not exceed 875 W (and which allows the tube to achieve the expected 60% efficiency) is acceptable. A good compromise might be 2000 V and 437 mA: 2000 × 0.437 = 874 W input. If the maximum plate voltage of 2500 is used, then the plate current should be limited to 350 mA (not 500 mA) to stay within the maximum plate dissipation rating of 350 W.

TRANSISTOR POWER DISSIPATION

RF power-amplifier transistors are limited in power-handling capability by the amount of heat the device can safely dissipate. Power dissipation for a transistor is abbreviated P_D. The maximum rating is based on maintaining a case temperature of 25°C (77°F), which is seldom possible if a conventional air-cooled heat sink is used in an ambient air temperature of 70° F or higher. For higher temperatures, the device must be derated (in terms of milliwatts or watts per degree C) as specified by the manu-

facturer for that particular device. The efficiency considerations described earlier in reference to plate dissipation apply here also. A rule of thumb for selecting a transistor suitable for a given RF power output level is to choose one that has a maximum dissipation (with the heat sink actually to be used) of twice the desired output power.

MAXIMUM TRANSISTOR RATINGS

Transistor data sheets specify the maximum operating voltage for several conditions. Of particular interest is the V_{CEO} specification (collector to emitter voltage, with the base open). In RF amplifier service the collector to emitter voltage can rise to twice the dc supply potential. Thus, if a 12-V supply is used, the transistor should have a V_{CEO} of 24 V or greater to preclude damage.

The maximum collector current is also specified by the manufacturer. This specification is actually limited by the current-carrying capabilities of the internal bonding wires. Of course, the collector current must stay below the level that generates heat higher than the allowable device power dissipation. Many transistors are also rated for the load mismatch they can safely withstand. A typical specification might be for a transistor to tolerate a 30:1 SWR at all phase angles.

Transistor manufacturers publish data sheets that describe all the appropriate device ratings. Typical operating results are also given in these data sheets. In addition, many manufacturers publish application notes illustrating the use of their devices in practical circuits. Construction details are usually given. Perhaps owing to the popularity of Amateur Radio among electrical engineers, many of the notes describe applications especially suited to the Amateur Service. Specifications for some of the more popular RF power transistors are found in the **Component Data** chapter.

PASSIVE COMPONENT RATINGS

Output Tank Capacitor Ratings

The tank capacitor in a high-power amplifier should be chosen with sufficient spacing between plates to preclude high-voltage breakdown. The peak RF voltage present

across a properly loaded tank circuit, without modulation, may be taken conservatively as being equal to the dc plate or collector voltage. If the dc supply voltage also appears across the tank capacitor, this must be added to the peak RF voltage, making the total peak voltage twice the dc supply voltage. At the higher voltages, it is usually desirable to design the tank circuit so that the dc supply voltages do not appear across the tank capacitor, thereby allowing the use of a smaller capacitor with less plate spacing. Capacitor manufacturers usually rate their products in terms of the peak voltage between plates. Typical plate spacings are given in **Table 13.3**.

Output tank capacitors should be mounted as close to the tube as temperature considerations will permit, to make possible the shortest path with the lowest possible inductive reactance from plate to cathode. Especially at the higher frequencies, where minimum circuit capacitance becomes important, the capacitor should be mounted with its stator plates well spaced from the chassis or other shielding. In circuits in which the rotor must be insulated from ground, the capacitor should be mounted on ceramic insulators of a size commensurate with the plate voltage involved and—most important of all, from the viewpoint of safety to the operator—a well-insulated coupling should be used

Table 13.3
Typical Tank-Capacitor Plate Spacings

Spacing Inches	Peak Voltage	Spacing Inches	Peak Voltage	Spacing Inches	Peak Voltage
0.015	1000	0.07	3000	0.175	7000
0.02	1200	0.08	3500	0.25	9000
0.03	1500	0.125	4500	0.35	11000
0.05	2000	0.15	6000	0.5	13000

Table 13.4
Copper Conductor Sizes for Transmitting Coils for Tube Transmitters

Power Output (Watts)	Band (MHz)	Minimum Conductor Size
1500	1.8-3.5	10
	7-14	8 or 1/8"
	18-28	6 or 3/16"
500	1.8-3.5	12
	7-14	10
	18-28	8 or 1/8"
150	1.8-3.5	16
	7-14	12
	18-28	10

*Whole numbers are AWG; fractions of inches are tubing ODs.

between the capacitor shaft and the knob. The section of the shaft attached to the control knob should be well grounded. This can be done conveniently by means of a metal shaft bushing at the panel.

Tank Coils

Tank coils should be mounted at least half their diameter away from shielding or other large metal surfaces, such as blower housings, to prevent a marked loss in Q. Except perhaps at 24 and 28 MHz, it is not essential that the coil be mounted extremely close to the tank capacitor. Leads up to 6 or 8 inches are permissible. It is more important to keep the tank capacitor, as well as other components, out of the immediate field of the coil.

The principal practical considerations in designing a tank coil usually are to select a conductor size and coil shape that will fit into available space and handle the required power without excessive heating. Excessive power loss as such is not necessarily the worst hazard in using too-small a conductor: it is not uncommon for the heat generated to actually unsolder joints in the tank circuit and lead to physical damage or failure. For this reason it's extremely important, especially at power levels above a few hundred watts, to ensure that all electrical joints in the tank circuit are secured mechanically as well as soldered. **Table 13.4** shows recommended conductor sizes for amplifier tank coils, assuming loaded tank circuit Qs of 15 or less on the 24 and 30 MHz bands and 8 to 12 on the lower frequency bands.

In the case of input circuits for screen-grid tubes where driving power is quite small, loss is relatively unimportant and almost any physically convenient wire size and coil shape is adequate.

The conductor sizes in Table 13.4 are based on experience in continuous-duty amateur CW, SSB, and RTTY service and assume that the coils are located in a reasonably well ventilated enclosure. If the tank area is not well ventilated and/or if significant tube heat is transferred to the coils, it is good practice to increase AWG wire sizes by two (for example, change from AWG 12 to AWG 10) and tubing sizes by $1/16$ inch.

Larger conductors than required for current handling are often used to maximize unloaded Q, particularly at higher frequencies. Where skin depth effects increase losses, the greater surface area of large diameter conductors can be beneficial. Small-diameter copper tubing, up to $3/8$ inch outer diameter, can be used successfully for tank coils up through the lower VHF range. Copper tubing in sizes suitable for constructing high-power coils

is generally available in 50 ft rolls from plumbing and refrigeration equipment suppliers. Silver plating the tubing further reduces losses. This is especially true as the tubing ages and oxidizes. Silver oxide is a much better conductor than copper oxides, so silver-plated tank coils maintain their low-loss characteristics even after years of use.

At VHF and above, tank circuit inductances do not necessarily resemble the familiar coil. The inductances required to resonate tank circuits of reasonable Q at these higher frequencies are small enough that only strip lines or sections of transmission line are practical. Since these are constructed from sheet metal or large-diameter tubing, current-handling capabilities normally are not a relevant factor.

RF Chokes

The characteristics of any RF choke vary with frequency. At low frequencies the choke presents a nearly pure inductance. At some higher frequency it takes on high impedance characteristics resembling those of a parallel-resonant circuit. At a still higher frequency it goes through a series-resonant condition, where the impedance is lowest—generally much too low to perform satisfactorily as a shunt-feed plate choke. As frequency increases further, the pattern of alternating parallel and series resonances repeats. Between resonances, the choke will show widely varying amounts of inductive or capacitive reactance.

In series-feed circuits, these characteristics are of relatively small importance because the RF voltage across the choke is negligible. In a shunt-feed circuit such as is used in most high-power amplifiers, however, the choke is directly in parallel with the tank circuit, and is subject to the full tank RF voltage. If the choke does not present a sufficiently high impedance, enough power will be absorbed by the choke to burn it out. To avoid this, the choke must have a sufficiently high reactance to be effective at the lowest frequency (*at least* equal to the plate load resistance), and yet have no series resonances near any of the higher frequency bands. A resonant-choke failure in a high-power amplifier can be very dramatic and damaging!

Thus any choke intended for shunt-feed use should be carefully investigated with a dip meter. The choke must be shorted end-to-end with a direct, heavy braid or strap. Because nearby metallic objects affect the resonances, it should be mounted in its intended position, but disconnected from the rest of the circuit. A dip meter coupled an inch or two away from one end of the choke nearly always will show a

deep, sharp dip at the lowest series-resonant frequency and shallower dips at higher series resonances.

Any choke to be used in an amplifier for the 1.8 to 28 MHz bands requires careful (or at least lucky!) design to perform well on *all* amateur bands within that range. Most simply put, the challenge is to achieve sufficient inductance that the choke doesn't "cancel" a large part of tuning capacitance on 1.8 MHz. At the same time, try to position all its series resonances where they can do no harm. In general, close wind enough #20 to #24 magnet wire to provide about 135 µH inductance on a $3/4$ to 1-inch diameter cylindrical form of ceramic, Teflon, or fiberglass. This gives a reactance of 1500 Ω at 1.8 MHz and yet yields a first series resonance in the vicinity of 25 MHz. Before the advent of the 24.9 MHz band this worked fine. But trying to "squeeze" the resonance into the narrow gaps between the 21, 24, and/or 28 MHz bands is quite risky unless sophisticated instrumentation is available. If the number of turns on the choke is selected to place its first series resonance at 23.2 MHz, midway between 21.45 and 24.89 MHz, the choke impedance will typically be high enough for satisfactory operation on the 21, 24 and 28 MHz bands. The choke's first series resonance should be measured very carefully as described above using a dip meter and calibrated receiver or RF impedance bridge, with the choke mounted in place on the chassis.

Investigations with a vector impedance meter have shown that "trick" designs, such as using several shorter windings spaced along the form, show little if any improvement in choke resonance characteristics. Some commercial amplifiers circumvent the problem by bandswitching the RF choke. Using a larger diameter (1 to 1.5 inch) form does move the first series resonance somewhat higher for a given value of basic inductance. Beyond that, it is probably easiest for an all-band amplifier to add or subtract enough turns to move the first resonance to about 35 MHz and settle for a little less than optimum reactance on 1.8 MHz.

Blocking Capacitors

A series capacitor is usually used at the input of the amplifier output circuit. Its purpose is to block dc from appearing on matching circuit components or the antenna. As mentioned in the section on tank capacitors, output-circuit voltage requirements are considerably reduced when only RF voltage is present.

To provide a margin of safety, the voltage rating for a blocking capacitor should be at least 25 to 50% greater than the dc voltage applied. A large safety margin is desirable, since blocking capacitor failure can bring catastrophic results.

To avoid affecting the amplifier's tuning and matching characteristics, the blocking capacitor should have a low impedance at all operating frequencies. Its reactance at the lowest operating frequency should be not more than about 5% of the plate load resistance.

The capacitor also must be capable of handling, without overheating or significantly changing value, the substantial RF current that flows through it. This current usually is greatest at the highest frequency of operation where tube output capacitance constitutes a significant part of the total tank capacitance. A significant portion of circulating tank current therefore flows through the blocking capacitor. As a conservative and very rough rule of thumb, the maximum RF current in the blocking capacitor (at 28 MHz) is

$$I_{CBlock} \approx I_p + 0.15 \times C_{OUT} \times V_{dc} \qquad (36)$$

where

I_{CBlock} = maximum RMS current through blocking capacitor, in A

C_{OUT} = output capacitance of the output tubes, in pF

V_{dc} = dc plate voltage, in kV

I_p = dc plate current at full output, in A.

Transmitting capacitors are rated by their manufacturers in terms of their RF current-carrying capacity at various frequencies. Below a couple hundred watts at the high frequencies, ordinary disc ceramic capacitors of suitable voltage rating work well in high-impedance tube amplifier output circuits. Some larger disk capacitors rated at 5 to 8 kV also work well for higher power levels at HF; for example, two inexpensive Centralab type DD-602 discs (0.002 µF, 6 kV) in parallel have proved to be a reliable blocking capacitor for 1.5-kW amplifiers operating at plate voltages to about 2.5 kV. At very high power and voltage levels and at VHF, ceramic "doorknob" transmitting capacitors are needed for their low losses and high current handling capabilities. So-called "TV doorknobs" may break down at high RF current levels and should be avoided.

The very high values of Q_L found in many VHF and UHF tube-type amplifier tank circuits often require custom fabrication of the blocking capacitor. This can usually be accommodated through the use of a Teflon "sandwich" capacitor. Here, the blocking capacitor is formed from two parallel plates separated by a thin layer of Teflon. This capacitor often is part of the tank circuit itself, forming a very low-loss blocking capacitor. Teflon is rated for a minimum break-down voltage of 2000 V per mil of thickness, so voltage breakdown should not be a factor in any practically realized circuit. The capacitance formed from such a Teflon sandwich can be calculated from the information presented elsewhere in this *Handbook* (use a dielectric constant of 2.1 for Teflon). In order to prevent any potential irregularities caused by dielectric thickness variations (including air gaps), Dow-Corning DC-4 silicone grease should be evenly applied to both sides of the Teflon dielectric. This grease has properties similar to Teflon, and will fill in any surface irregularities that might cause problems.

The very low impedances found in transistorized amplifiers present special problems. In order to achieve the desired low blocking-capacitor impedance, large-value capacitors are required. Special ceramic chips and mica capacitors are available that meet the requirements for high capacitance, large current carrying capability and low associated inductance. These capacitors are more costly than standard disk-ceramic or silver-mica units, but their level of performance easily justifies their price. Most of these special-purpose capacitors are either leadless or come with wide straps instead of normal wire leads. Disc-ceramic and other wire-lead capacitors are generally not suitable for transistor power-amplifier service.

SOURCES OF OPERATING VOLTAGES

Tube Filament or Heater Voltage

The heater voltage for the indirectly heated cathode- tubes found in low-power classifications may vary 10% above or below rating without seriously reducing the life of the tube. A power vacuum tube can use either a directly heated filament or an indirectly heated cathode. The filament voltage for either type should be held within 5% of rated voltage. Because of internal tube heating at UHF and higher, the manufacturers' filament voltage rating often is reduced at these higher frequencies. The derated filament voltages should be followed carefully to maximize tube life. Series dropping resistors may be required in the filament circuit to attain the correct voltage. The voltage should be measured at the filament pins of the tube socket while the amplifier is running. The filament choke and interconnecting wiring all have voltage drops associated with them. The high current drawn by a power-tube heater circuit causes substantial voltage drops to occur across even small resistances. Also, make sure that the plate power drawn from the power line does not cause the filament voltage to drop below the proper value when plate power is applied.

Thoriated filaments lose emission when the tube is overloaded appreciably. If the overload has not been too prolonged, emission sometimes may be restored by operating the filament at rated voltage, with all other voltages removed, for a period of 30 to 60 minutes. Alternatively, you might try operating the tube at 20% above rated filament voltage for five to ten minutes.

Vacuum-Tube Plate Voltage

DC plate voltage for the operation of RF amplifiers is most often obtained from a transformer-rectifier-filter system (see the **Power Supplies** chapter) designed to deliver the required plate voltage at the required current. It is not unusual for a power tube to arc over internally (generally from the plate to the screen or control grid) once or twice, especially soon after it is first placed into service. The flashover by itself is not normally dangerous to the tube, provided that instantaneous maximum plate current to the tube is held to a safe value and the high-voltage plate supply is shut off very quickly.

A good protective measure against this is the inclusion of a high-wattage power resistor in series with the plate high-voltage circuit. The value of the resistor, in ohms, should be approximately 10 to 15 times the no-load plate voltage in kV. This will limit peak fault current to 67 to 100 A. The series resistor should be rated for 25 or 50 W power dissipation; vitreous enamel coated wire-wound resistors of the common Ohmite or Clarostat types have been found to be capable of handling repeated momentary fault-current surges without damage. Aluminum-cased resistors such as some made by Dale are not recommended for this application. Each resistor also must be large enough to safely handle the maximum value of normal plate current; the wattage rating required may be calculated from $P = I^2R$. If the total filter capacitance exceeds 25 µF, it is a good idea to use 50-W resistors in any case. Even at high plate-current levels, the addition of the resistors does little to affect the dynamic regulation of the plate supply.

Since tube (or other high-voltage circuit) arcs are not necessarily self-extinguishing, a fast-acting plate overcurrent relay or primary circuit breaker also is recommended to quickly shut off ac power to the HV supply when an arc begins. Using this protective system, a mild HV flashover may go undetected, while a more severe one will remove ac power from the HV supply. (The cooling blower should remain energized however, since the tube may be hot when the HV is removed due to an arc.) If effective protection is not provided, however, a "normal" flashover, even in a new tube, is

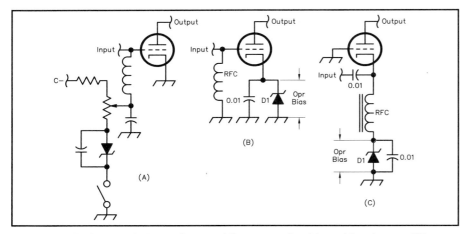

Fig 13.11—Various techniques for providing operating bias with tube amplifiers.

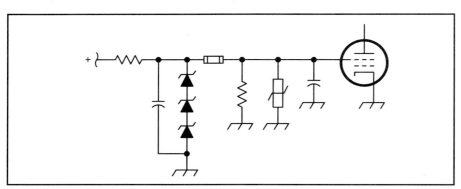

Fig 13.12—A Zener-regulated screen supply for use with a tetrode. Protection is provided by a fuse and a varistor.

likely to damage or destroy the tube, and also frequently destroys the rectifiers in the power supply as well as the plate RF choke. A power tube that flashes over more than about 3 to 5 times in a period of several months likely is defective and will have to be replaced before long.

Grid Bias

The grid bias for a linear amplifier should be highly filtered and well regulated. Any ripple or other voltage change in the bias circuit modulates the amplifier. This causes hum and/or distortion to appear on the signal. Since most linear amplifiers draw only small amounts of grid current, these bias-supply requirements are not difficult to achieve.

Fixed bias for class AB1 tetrode and pentode amplifiers is usually obtained from a variable-voltage regulated supply. Voltage adjustment allows setting bias level to give the desired resting plate current. **Fig 13.11A** shows a simple Zener-diode-regulated bias supply. The dropping resistor is chosen to allow approximately 10 mA of Zener current. Bias is then reasonably well regulated for all drive conditions up to 2 or 3 mA of grid current. The

potentiometer allows bias to be adjusted between Zener and approximately 10 V higher. This range is usually adequate to allow for variations in the characteristics of different tubes. Under standby conditions, when it is desirable to cut off the tube entirely, the Zener ground return is interrupted so the full bias supply voltage is applied to the grid.

In Fig 13.11B and C, bias is obtained from the voltage drop across a Zener diode in the cathode (or filament center-tap) lead. Operating bias is obtained by the voltage drop across D1 as a result of plate (and screen) current flow. The diode voltage drop effectively raises the cathode potential relative to the grid. The grid is therefore negative with respect to the cathode by the Zener voltage of the diode. The Zener-diode wattage rating should be twice the product of the maximum cathode current times the rated zener voltage. Therefore, a tube requiring 15 V of bias with a maximum cathode current of 100 mA would dissipate 1.5 W in the Zener diode. To allow a suitable safety factor, the diode rating should be 3 W or more. The circuit of Fig 13.11C illustrates how D1 would be used with a cathode driven

(grounded grid) amplifier as opposed to the grid driven example at B.

In all cases, the Zener diode should be bypassed by a 0.01-μF capacitor of suitable voltage. Current flow through any type of diode generates shot noise. If not bypassed, this noise would modulate the amplified signal, causing distortion in the amplifier output.

Screen Voltage For Tubes

Power tetrode screen current varies widely with both excitation and loading. The current may be either positive or negative, depending on tube characteristics and amplifier operating conditions. In a linear amplifier, the screen voltage should be well regulated for all values of screen current. The power output from a tetrode is very sensitive to screen voltage, and any dynamic change in the screen potential can cause distorted output. Zener diodes are commonly used for screen regulation.

Fig 13.12 shows a typical example of a regulated screen supply for a power tetrode amplifier. The voltage from a fixed dc supply is dropped to the Zener stack voltage by the current-limiting resistor. A screen bleeder resistor is connected in parallel with the zener stack to allow for the negative screen current developed under certain tube operating conditions. Bleeder current is chosen to be roughly 10 to 20 mA greater than the expected maximum negative screen current, so that screen voltage is regulated for all values of current between maximum negative screen current and maximum positive screen current. For external-anode tubes in the 4CX250 family, a typical screen bleeder current value would be 20 mA. For the 4CX1000 family, a screen-bleeder current of 70 mA is required.

Screen voltage should never be applied to a tetrode unless plate voltage and load also are applied; otherwise the screen tends to act like an anode and will draw excessive current. Supplying the screen through a series dropping resistor from the plate supply affords a measure of protection, since the screen voltage only appears when there is plate voltage. Alternatively, a fuse can be placed between the regulator and the bleeder resistor. The fuse should not be installed between the bleeder resistor and the tube, because the tube should never be operated without a load on the screen. Without a load, the screen potential tends to rise to the anode voltage. Any screen bypass capacitors or other associated circuits are likely be damaged by this high voltage.

In Fig 13.12, a varistor is connected from screen to ground. If, because of some circuit failure, the screen voltage should rise substantially above its nominal level, the varis-

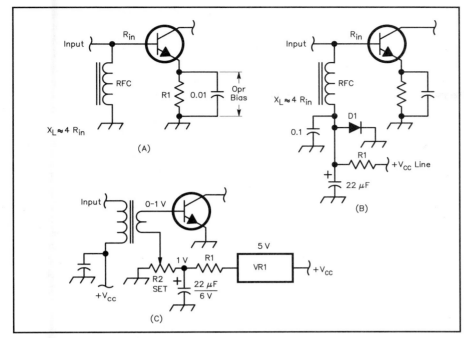

Fig 13.13—Biasing methods for use with transistor amplifiers.

tor will conduct and clamp the screen voltage to a low level. If necessary to protect the varistor or screen dropping resistors, a fuse or overcurrent relay may be used to shut off the screen supply so that power is interrupted before any damage occurs. The varistor voltage should be approximately 30 to 50% higher than normal screen voltage.

Transistor Biasing

Solid-state power amplifiers generally operate in Class C or AB. When some bias is desired during Class C operation (**Fig 13.13A**), a resistance of the appropriate value can be placed in the emitter return as shown. Most transistors will operate in Class C without adding bias externally, but in some instances the amplifier efficiency can be improved by means of emitter bias. Reverse bias supplied to the base of the Class C transistor should be avoided because it will lead to internal breakdown of the device during peak drive periods. The damage is frequently a cumulative phenomenon, leading to gradual destruction of the transistor junction.

A simple method for Class AB biasing is shown in Fig 13.13B. D1 is a silicon diode that acts as a bias clamp at approximately 0.7 V. This forward bias establishes linear-amplification conditions. That value of bias is not always optimum for a specified transistor in terms of IMD. Variable bias of the type illustrated in Fig 13.13C permits the designer sufficient flexibility to position the operating point for best linearity. The diode clamp or the reference sensor for another type of regulator is usually thermally bonded to the power transistor or its heat sink. The bias level then tracks the thermal characteristics of the output transistor. Since a transistor's current transfer characteristics are a function of temperature, thermal tracking of the bias is necessary to maintain device linearity and, in the case of bipolar devices, to prevent thermal runaway and the subsequent destruction of the transistor.

AMPLIFIER COOLING

Tube Cooling

Vacuum tubes must be operated within the temperature range specified by the manufacturer if long tube life is to be achieved. Tubes having glass envelopes and rated at up to 25-W plate dissipation may be used without forced-air cooling if the design allows a reasonable amount of convection cooling. If a perforated metal enclosure is used, and a ring of $1/4$ to $3/8$-inch-diameter holes is placed around the tube socket, normal convective air flow can be relied on to remove excess heat at room temperatures.

For tubes with greater plate dissipation ratings, and even for very small tubes operated close to maximum rated dissipation, forced-air cooling with a fan or blower is needed. Most manufacturers rate tube cooling requirements for continuous-duty operation. Their literature will indicate the required volume of air flow, in cubic feet per minute (CFM), at some particular back pressure. Often this data is given for several different values of plate dissipation, ambient air temperature and even altitude above sea level.

One extremely important consideration is often overlooked by power-amplifier designers and users alike: a tube's plate dissipation rating is only its maximum *potential* capability. The power that it can *actually* dissipate safely depends directly on the cooling provided. The actual power capability of virtually all tubes used in high-power amplifiers for amateur service

Table 13.5

Specifications of Some Popular Tubes, Sockets and Chimneys

Tube	CFM	Back Pressure (inches)	Socket	Chimney
3-500Z	13	0.13	SK-400, SK-410	SK-416
3CX800A7	19	0.50	SK-1900	SK-1906
3CX1200A7	31	0.45	SK-410	SK-436
3CX1200Z7	42	0.30	SK-410	—
3CX1500/8877	35	0.41	SK-2200, SK-2210	SK-2216
4-400A/8438	14	0.25	SK-400, SK-410	SK-406
4-1000A/8166	20	0.60	SK-500, SK-510	SK-506
4CX250R/7850	6.4	0.59	SK-600, SK-600A, SK602A, SK-610, SK-610A SK-611, SK-612, SK-620, SK-620A, SK-621, SK-630	SK-626
4CX400/8874	8.6	0.37	SK1900	SK606
4CX400A	8	0.20	SK2A	—
4CX800A	20	0.50	SK1A	—
4CX1000A/8168	25	0.20	SK-800B, SK-810B, SK-890B	SK-806
4CX1500B/8660	34	0.60	SK-800B, SK-1900	SK-806
4CX1600B	36	0.40	SK3A	CH-1600B

These values are for sea-level elevation. For locations well above sea level (5000 ft/1500 m, for example), add an additional 20% to the figure listed.

Table 13.6

Blower Performance Specifications

Wheel Dia	Wheel Width	RPM	Free Air CFM	CFM for Back Pressure (inches)					Cutoff	Stock No.
				0.1	0.2	0.3	0.4	0.5		
2"	1"	3160	15	13	4	—	—	—	0.22	2C782
3"	1-15/32"	3340	54	48	43	36	25	17	0.67	4C012
3"	1-7/8"	3030	60	57	54	49	39	23	0.60	4C440
3"	1-7/8"	2880	76	70	63	56	45	8	0.55	4C004
3-13/16"	1-7/8"	2870	100	98	95	90	85	80	0.80	4C443
3-13/16"	2-1/2"	3160	148	141	135	129	121	114	1.04	4C005

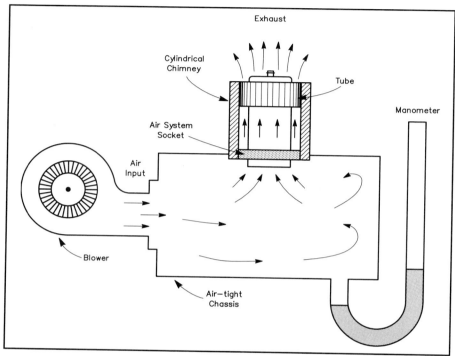

Fig 13.14—Air is forced into the chassis by the blower and exits through the tube socket. The manometer is used to measure system back pressure, which is an important factor in determining the proper size blower.

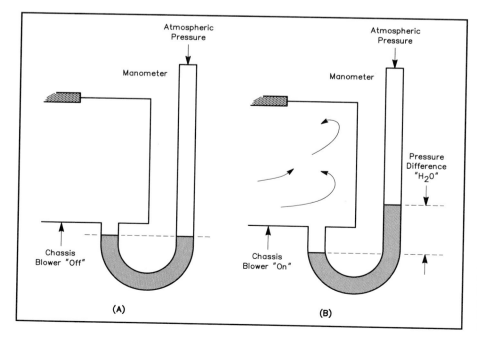

(A)

(B)

depends on the volume of air forced through the tube's cooling structure.

This requirement usually is given in terms of cubic feet of air per minute, (CFM), delivered into a "back pressure" representing the resistance of the tube cooler to air flow, stated in inches of water. Both the CFM of air flow required and the pressure needed to force it through the cooling system are determined by ambient air temperature and altitude (air density), as well as by the amount of heat to be dissipated. The cooling fan or blower must be capable of delivering the specified air flow into the corresponding back pressure. As a result of basic air flow and heat transfer principles, the volume of air flow required through the tube cooler increases considerably faster than the plate dissipation, and back pressure increases even faster than air flow. In addition, blower air output decreases with increasing back pressure until, at the blower's so-called "cut-off pressure," actual air delivery is zero. Larger and/or faster-rotating blowers are required to deliver larger volumes of air at higher back pressure.

Values of CFM and back pressure required to realize maximum rated plate dissipation for some of the more popular tubes, sockets and chimneys (with 25°C ambient air and at sea level) are given in **Table 13.5**. Back pressure is specified in inches of water and can be measured easily in an operational air system as indicated in **Figs 13.14** and **13.15**. The pressure differential between the air passage and atmospheric pressure is measured with a device called a manometer. A manometer is nothing more than a piece of clear tubing, open at both ends and fashioned in the shape of a "U." The manometer is temporarily connected to the chassis and is removed after the measurements are completed. As shown in the diagrams, a small amount of water is placed in the tube. At **Fig 13.15A** the blower is "off" and the water seeks its own level, because the air pressure (ordinary atmospheric pressure) is the same at both ends of the manometer tube. At B, the blower is "on" (socket, tube and chimney in place) and the pressure difference, in terms of inches of water, is measured. For most applications a standard ruler used for measurement will yield sufficiently accurate results.

Table 13.6 gives the performance specifications for a few of the many Dayton blowers which are available through Grainger catalog outlets in all 50 states. Other blowers having wheel diameters,

Fig 13.15—At A the blower is "off" and the water will seek its own level in the manometer. At B the blower is "on" and the amount of back pressure in terms of inches of water can be measured as indicated.

widths and rotational speeds similar to any in Table 13.6 likely will have similar flow and back-pressure characteristics. If in doubt about specifications, consult the manufacturer. Tube temperature under actual operating conditions is the ultimate criterion for cooling adequacy and may be determined using special crayons or lacquers which melt and change appearance at specific temperatures. The setup of Fig 13.15, however, nearly always gives sufficiently accurate information.

As an example, consider the cooling design of a linear amplifier to use one 3CX800A7 tube, to operate near sea level with the air temperature not above 25°C. The tube, running 1150-W dc input, easily delivers 750-W continuous output, resulting in 400 W plate dissipation ($P_{Dis} = P_{IN} - P_{OUT}$). According to the manufacturer's data, adequate tube cooling at 400 W P_D requires at least 6 CFM of air at 0.09 inches of water back pressure. Referring to Table 13.6, a Dayton no. 2C782 will do the job with a good margin of safety.

If the same single tube were to be operated at 2.3 kW dc input to deliver 1.5 kW output (substantially exceeding its maximum electrical ratings!), P_{IN} would be about 2300 W and $P_D \approx 800$ W. The minimum cooling air required would be about 19 CFM at 0.5 inches of water pressure—doubling P_{DIS}. more than tripling the CFM of air flow required and increasing back pressure requirements on the blower by a factor of 5.5!

However, two 3CX800A7 tubes are needed to deliver 1.5 kW of continuous maximum legal output power in any case. Each tube will operate under the same conditions as in the single-tube example above, dissipating 400-W. The total cooling air requirement for the two tubes is therefore 12 CFM at about 0.09 inches of water, only two-thirds as much air volume and one-fifth the back pressure required by a single tube. While this may seem surprising, the reason lies in the previously mentioned fact that both the airflow required by a tube and the resultant back pressure increase much more rapidly than P_D of the tube. Blower air delivery capability, conversely, decreases as back pressure is increased. Thus a Dayton 2C782 blower can cool two 3CX800A7 tubes dissipating 800 W total, but a much larger (and probably noisier) no. 4C440 would be required to handle the same power with a single tube.

In summary, three very important considerations to remember are these:

- A tube's actual safe plate dissipation capability is *totally dependent* on the amount of cooling air forced through its cooling system. Any air-cooled power tube's maximum plate dissipation rat-

ing is meaningless unless the specified amount of cooling air is supplied.
- Two tubes will always safely dissipate a given power with a significantly smaller (and quieter) blower than is required to dissipate the same power with a single tube of the same type. A corollary is that a given blower can virtually always dissipate more power when cooling two tubes than when cooling a single tube of the same type.
- Blowers vary greatly in their ability to deliver air against back pressure so blower selection should not be taken lightly.

A common method for directing the flow of air around a tube involves the use of a pressurized chassis. This system is shown in Fig 13.14. A blower attached to the chassis forces air around the tube base, often through holes in its socket. A chimney is used to guide air leaving the base area around the tube envelope or anode cooler, preventing it from dispersing and concentrating the flow for maximum cooling.

A less conventional approach that offers a significant advantage in certain situations is shown in **Fig 13.16**. Here the anode compartment is pressurized by the blower. A special chimney is installed between the anode heat exchanger and an exhaust hole in the compartment cover. When the blower pressurizes the anode compartment, there are two parallel paths for air flow: through the anode and its chimney, and through the air system socket. Dissipation, and hence cooling air required, generally is much greater for the anode than for the tube base. Because high-volume anode airflow need not be forced through restrictive air channels in the base area, back pressure may be very significantly reduced with certain tubes and sockets. Only airflow actually needed is bled through the base area. Blower back pressure requirements may sometimes be reduced by nearly half through this approach.

Table 13.5 also contains the part numbers for air-system sockets and chimneys available for use with the tubes that are listed. The builder should investigate which of the sockets listed for the 4CX250R, 4CX300A, 4CX1000A and 4CX1500A best fit the circuit needs. Some of the sockets have certain tube elements grounded internally through the socket. Others have elements bypassed to ground through capacitors that are integral parts of the sockets.

Depending on one's design philosophy and tube sources, some compromises in the cooling system may be appropriate. For example, if glass tubes are available inexpensively as broadcast pulls, a shorter life span may be acceptable. In such a case, an

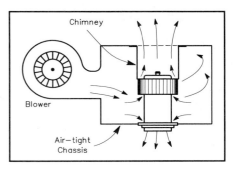

Fig 13.16—Anode compartment pressurization may be more efficient than grid compartment pressurization. Hot air exits upwards through the tube anode and through the chimney. Cool air also goes down through the tube socket to cool tube's pins and the socket itself.

increase of convenience and a reduction in cost, noise, and complexity can be had by using a pair of "muffin" fans. One fan may be used for the filament seals and one for the anode seal, dispensing with a blower and air-system socket and chimney. The air flow with this scheme is not as uniform as with the use of a chimney. The tube envelope mounted in a cross flow has flow stagnation points and low heat transfer in certain regions of the envelope. These points become hotter than the rest of the envelope. The use of multiple fans to disturb the cross air flow can significantly reduce this problem. Many amateurs have used this cooling method successfully in low-duty-cycle CW and SSB operation but it is not recommended for AM, SSTV or RTTY service. The true test of the effectiveness of a forced air cooling system is the amount of heat carried away from the tube by the air stream. The power dissipated can be calculated from the air flow temperatures. The dissipated power is

$$P_D = 169 \, Q_A \left[\frac{T_2}{T_1} - 1 \right] \qquad (37)$$

where
P_D = the dissipated power, in W
Q_A = the air flow, in CFM (cubic feet per minute)
T_1 = the inlet air temperature, kelvins (normally quite close to room temperature)
T_2 = the amplifier exhaust temperature, kelvins.

The exhaust temperature can be measured with a cooking thermometer at the air outlet. The thermometer should not be placed inside the anode compartment because of the high voltage present.

Transistor Cooling

Transistors used in power amplifiers dissipate significant amounts of power, and the heat so generated must be effectively removed to maintain acceptable device temperatures. Some bipolar power transistors have the collector connected directly to the case of the device, as the collector creates most of the heat generated when the transistor is in operation. Others have the emitter connected to the case. However, if operated close to maximum rated dissipation, even the larger case designs cannot normally conduct heat away fast enough to keep the operating temperature of the device within the safe area—the maximum temperature that a device can stand without damage. Safe area is usually specified in a device data sheet, often in graphical form. Germanium power transistors theoretically may be operated at internal temperatures up to 100°C, while silicon devices may be run at up to 200°C. However, to assure long device lifetimes much lower case temperatures—not greater than 50° to 75°C for germanium and 75° to 100°C for silicon—are highly desirable. Leakage currents in germanium devices can be very high at elevated temperatures; thus, silicon transistors are preferred for most power applications.

A properly chosen heat sink often is essential to help keep the transistor junction temperature in the safe area. For low-power applications a simple clip-on heat sink will suffice, while for 100 W or higher input power a massive cast-aluminum finned radiator usually is necessary. The appropriate size heat sink can be calculated based on the thermal resistance between the transistor case and ambient air temperature. The first step is to calculate the total power dissipated by the transistor:

$$P_D = P_{DC} + P_{RFin} - P_{RFout} \qquad (38)$$

where

   P_D = the total power dissipated by the transistor in W

   P_{DC} = the dc power into the transistor, in W

   P_{RFin} = the RF (drive) power into the transistor in W

   P_{RFout} = the RF output power from the transistor in W.

The value of P_D is then used to obtain the θ_{CA} value from

$$\theta_{CA} = (T_C - T_A)/P_D \qquad (39)$$

where

   θ_{CA} = the thermal resistance of the device case to ambient

   T_C = the device case temperature in °C

   T_A = the ambient temperature (room temperature) in °C.

A suitable heat sink, capable of transferring all of the heat generated by the transistor to the ambient air, can then be chosen from the manufacturer's specifications for θ_{CA}. A well-designed heat-sink system minimizes thermal path lengths and maximizes their cross-sectional areas. The contact area between the transistor and heat sink should have very low thermal resistance. The heat sink's mounting surface must be flat and the transistor firmly attached to the heat sink so intimate contact—without gaps or air voids—is made between the two. The use of silicone-based heat sink compounds can provide considerable improvement in thermal transfer. The thermal resistance of such grease is considerably lower than that of air, but not nearly as good as that of copper or aluminum. The quantity of grease should be kept to an absolute minimum. Only enough should be used to fill in any small air gaps between the transistor and heat sink mating surfaces. The maximum temperature rise in the transistor junction may easily be calculated by using the equation

$$T_J = (\theta_{JC} + \theta_{CA}) P_D + T_A \qquad (40)$$

where

   T_J = the transistor junction temperature in °C

   θ_{JC} = the manufacturer's published thermal resistance of the transistor

   θ_{CA} = the thermal resistance of the device case to ambient

   P_D = the power dissipated by the transistor

   T_A = the ambient temperature in °C.

The value of T_J should be kept well below the manufacturer's recommended maximum to prevent premature transistor failure. Measured values of the ambient temperature and the device case temperature can be used in the preceding formulas to calculate junction temperature. **The Real World Components** chapter contains a more detailed discussion of transistor cooling.

Design Guidelines and Examples

Most of the problems facing an amplifier designer are not theoretical, but have to do with real-world component limitations. **The Real World Components** chapter discusses the differences between ideal and real components

A simplified equivalent schematic of an amplifying device is shown in **Fig 13.17A**. The input is represented by a series (parasitic) inductance feeding a resistance in parallel with a capacitance. The output consists of a current generator in parallel with a resistance and capacitance, followed by a series inductance. This is a reasonably accurate description of both transistors and vacuum tubes, regardless of circuit configuration (as demonstrated in Figs 13.17B and C). Both input and output impedances have a resistive component in parallel with a reactive component. Each also has a series inductive reactance, which represents connecting leads within the device. These inductances, un-

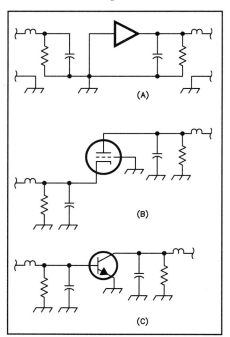

like the other components of input and output impedance, often are not characterized in manufacturers' device specifications.

The amplifier input and output matching networks must transform the complex impedances of the amplifying device to the source and load impedances (often 50-Ω transmission lines). Impedances associated with other parts of the amplifier circuit, such as a dc-supply choke, must also be considered in designing the matching networks. The matching networks and other circuit

Fig 13.17—The electrical equivalents for power amplifiers. At A, the input is represented by a series stray inductance, then a resistor in parallel with a capacitor. The output is a current source in parallel with a resistor and capacitor, followed by a series stray inductance. These effects are applied to tubes and transistors in B and C.

components are influenced by each other's presence, and these mutual effects must be given due consideration.

Perhaps the best way to clarify the considerations that enter into designing various types of RF power amplifiers is through example. The following examples illustrate common problems associated with power-amplifier design. They are not intended as detailed construction plans, but only demonstrate typical approaches useful in designing similar projects.

DESIGN EXAMPLE 1: A HIGH-POWER VACUUM TUBE HF AMPLIFIER

Most popular HF transceivers produce approximately 100-W output. The EIMAC 8877 can deliver 1500-W output for approximately 60 W of drive when used in a grounded grid circuit. Grounded-grid operation is usually the easiest tube amplifier circuit to implement. Its input impedance is relatively low, often close to 50 Ω. Input/output shielding provided by the grid and negative feedback inherent in the grounded-grid circuit configuration reduce the likelihood of amplifier instability and provide excellent linearity without critical adjustments. Fewer supply voltages are needed in this configuration compared to others: often just high-voltage dc for the plate and low-voltage ac for the filament.

The first step in the amplifier design process is to verify that the tube is actually capable of producing the desired results while remaining within manufacturer's ratings. The plate dissipation expected during normal operation of the amplifier is computed first. Since the amplifier will be used for SSB, a class of operation producing linear amplification must be used. Class AB2 provides a very good compromise between linearity and good efficiency, with effective efficiency typically exceeding 60%. Given that efficiency, an input power of 2500 W is needed to produce the desired 1500-W output. Operated under these conditions, the tube will dissipates about 1000 W—well within the manufacturer's specifications, provided adequate cooling airflow is supplied.

The grid in modern high-mu triodes is a relatively delicate structure, closely spaced to the cathode and carefully aligned to achieve high gain and excellent linearity. To avoid shortening tube life or even destruction of the tube, the specified maximum grid dissipation must not be exceeded for more than a few milliseconds under any conditions. For a given power output, the use of higher plate voltages tends to result in lower grid dissipation. It is important to use a plate voltage which is

high enough to result in safe grid current levels at maximum output. In addition to maximum ratings, manufacturers' data sheets often provide one or more sets of "typical operation" parameters. This makes it even easier for the builder to achieve optimum results.

The 8877, operating at 3500 V, can produce 2075 W of RF output with excellent linearity and 64 W of drive. Operating at 2700 V it can deliver 1085 W with 40 W of drive. To some extent, the ease and cost of constructing a high-power amplifier, as well as its ultimate reliability, are enhanced by using the lowest plate voltage which will yield completely satisfactory performance. Interpolating between the two sets of typical operating conditions suggests that the 8877 can comfortably deliver 1.5 kW output with a 3100-V plate supply and 50 to 55 W of drive. Achieving 2500-W input power at this plate voltage requires 800 mA of plate current—well within the 8877's maximum rating of 1.0 A.

The next step in the design process is to calculate the optimum plate load resistance at this plate voltage and current for Class AB2 operation and design an appropriate output matching network. From the earlier equations, R_L is calculated to be 2200 Ω.

Several different output networks might be used to transform the nominal 50-Ω resistance of the actual load to the 2200-Ω load resistance required by the 8877, but experience shows that pi and pi-L networks are most practical. Each can provide reasonable harmonic attenuation, is relatively easy to

build mechanically and uses readily available components. The pi-L gives significantly greater harmonic attenuation than the pi and usually is the better choice—at least in areas where there is any potential for TVI or crossband interference. In a multiband amplifier, the extra cost of using a pi-L network is the "L" inductor and its associated bandswitch section.

To simplify and avoid confusion with terminology previously used in the pi and pi-L network design tables, in the remainder of this chapter Q_{IN} is the loaded Q of the amplifier's input matching tank, Q_{OUT} is the loaded Q of the output pi-L tank, Q_{PI} is the loaded Q of the output pi section only, and Q_L is the loaded Q of the output L section only.

The input impedance of a grounded-grid 8877 is typically on the order of 50 to 55 Ω, shunted by input capacitance of about 38 pF. While this average impedance is close enough to 50 Ω to provide negligible input SWR, the instantaneous value varies greatly over the drive cycle—that is, it is nonlinear. This nonlinear impedance is reflected back as a nonlinear load impedance at the exciter output, resulting in increased intermodulation distortion, reduced output power, and often meaningless exciter SWR meter indications. In addition, the tube's parallel input capacitance, as well as parasitic circuit reactances, often are significant enough at 28 MHz to create significant SWR. A tank circuit at the amplifier input can solve both of these problems by tuning out the stray reactances and stabilizing (linearizing) the tube input impedance through its flywheel effect.

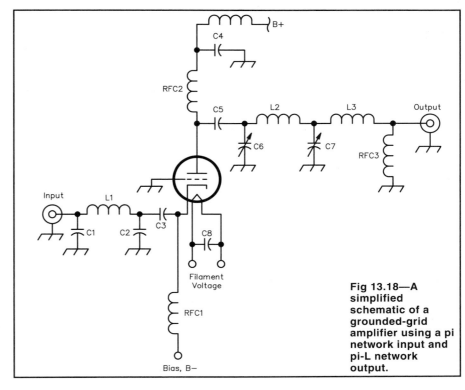

Fig 13.18—A simplified schematic of a grounded-grid amplifier using a pi network input and pi-L network output.

The input tank should have a loaded Q (called Q_{Lin} in this discussion) of at least two for good results. Increasing Q_{Lin} to as much as five results in a further small improvement in linearity and distortion, but at the cost of a narrower operating bandwidth. Even a Q_{Lin} of 1.0 to 1.5 yields significant improvement over an untuned input. A pi network commonly is used for input matching at HF.

Fig 13.18 illustrates these input and output networks applied in the amplifier circuit. The schematic shows the major components in the amplifier RF section, but with band-switching and cathode dc-return circuits omitted for clarity. C1 and C2 and L1 form the input pi network. C3 is a blocking capacitor to isolate the exciter from the cathode dc potential. Note that when the tube's average input resistance is close to 50 Ω, as in the case of the 8877, a simple parallel-resonant tank often can successfully perform the tuning and flywheel functions, since no impedance transformation is necessary. In this case it is important to minimize stray lead inductance between the tank and tube to avoid undesired impedance transformation.

The filament or "heater" in indirectly heated tubes such as the 8877 must be very close to the cathode to heat the cathode efficiently. A capacitance of several picofarads exists between the two. Particularly at very high frequencies, where these few picofarads represent a relatively low reactance, RF drive intended for the cathode can be capacitively coupled to the lossy filament and dissipated as heat. To avoid this, above about 50 MHz, the filament must be kept at a high RF impedance above ground. The high impedance (represented by choke RFC1 in Fig 13.18) minimizes RF current flow in the filament circuit so that RF dissipated in the filament becomes negligible. The choke's low-frequency resistance should be kept to a minimum to lessen voltage drops in the high-current filament circuit.

The choke most commonly used in this application is a pair of heavy-gauge insulated wires, bifilar-wound over a ferrite rod. The ferrite core raises the inductive reactance throughout the HF region so that a minimum of wire is needed, keeping filament-circuit voltage drops low. The bifilar winding technique assures that both filament terminals are at the same RF potential.

Below 30 MHz, the use of such a choke seldom is necessary or beneficial, but actually can introduce another potential problem. Common values of cathode-to-heater capacitance and heater-choke inductance often are series resonant in the 1.8 to 29.7 MHz HF range. A capacitance of 5 pF and an inductance of 50 μH, for example, resonate

at 10.0 MHz; the actual components are just as likely to resonate near 7 or 14 MHz. At resonance, the circuit constitutes a relatively low impedance shunt from cathode to ground, which affects input impedance and sucks out drive signal. An unintended resonance like this near any operating frequency usually increases input SWR and decreases gain on that one particular band. While aggravating, the problem rarely completely disables or damages the amplifier, and so is seldom pursued or identified.

Fortunately, the entire problem is easily avoided—below 30 MHz the heater choke can be deleted. At VHF-UHF, or wherever a heater isolation choke is used for any reason, the resonance can be moved below the lowest operating frequency by connecting a sufficiently large capacitance (about 1000 pF) between the tube cathode and one side of the heater. It is good practice also to connect a similar capacitor between the heater terminals. It also would be good practice in designing other VHF/UHF amplifiers, such as those using 3CX800A7 tubes, unless the builder can insure that the actual series resonance is well outside of the operating frequency range.

Plate voltage is supplied to the tube through RFC2. C5 is the plate blocking capacitor. The output pi-L network consists of tuning capacitor C6, loading capacitor C7, pi coil L2, and L coil L3. RFC3 is a high-inductance RF choke placed at the output for safety purposes. Its value, usually 100 μH to 2 mH, is high enough so that it appears as an open circuit across the output connector for RF. However, should the plate blocking capacitor fail and allow high voltage onto the output matching network, RFC3 would short the dc to ground and blow the power-supply fuse or breaker. This prevents dangerous high voltage from appearing on the feed line or antenna. It also prevents electrostatic charge—from the antenna or from blocking capacitor leakage—from building up on the tank capaci-

tors and causing periodic dc discharge arcs to ground. If such a dc discharge occurs while the amplifier is transmitting, it can trigger a potentially damaging RF arc.

Our next step is designing the input matching network. As stated earlier, tube input impedance varies moderately with plate voltage and load resistance as well as bias, but is approximately 50 to 55-Ω paralleled by C_{IN} of 38 pF, including stray capacitance. A simple parallel-resonant tank of $Q_{IN} = 2$ to 3 can provide an input SWR not exceeding 1.5:1, provided all wiring from RF input connector to tank to cathode is heavy and short. On each band a Q_{IN} between 2 to 3 requires an $X_{Ctot} = X_{Lin}$ between 25 and 17 Ω.

A more nearly perfect match, with greater tolerance for layout and wiring variations, may be achieved by using the pi input tank as shown in Fig 13.18. Design of this input matching circuit is straightforward. Component values are computed using a Q_{IN} between 2 or 3. Higher Q_{IN} values reduce the network's bandwidth, perhaps even requiring a front-panel tuning control for the wider amateur bands. The purpose of this input network is to present the desired input impedance to the exciter, not to add selectivity. As with a parallel tank, the value of the capacitor at the tube end of the pi network should be reduced by 38 pF; stray capacity plus tube C_{IN} is effectively in parallel with the input pi network's output.

The output pi-L network must transform the nominal 50-Ω amplifier load to a pure resistance of 2200 Ω. We previously calculated that the 8877 tube's plate must see 2200 Ω for optimum performance. In practice, real antenna loads are seldom purely resistive or exactly 50 Ω; they often exhibit SWRs of 2:1 or greater on some frequencies. It's desirable that the amplifier output network be able to transform any complex load impedance corresponding to an SWR up to about 2:1 into a resistance of 2200 Ω. The network also must compen-

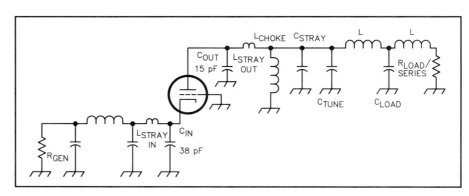

Fig 13.19—The effective reactances for the amplifier in Fig 13.18.

sate for tube C_{OUT} and other stray plate-circuit reactances, such as those of interconnecting leads and the plate RF choke. These reactances, shown in **Fig 13.19**, must be taken into account when designing the matching networks. Because the values of most stray reactances are not accurately known, the most satisfactory approach is to estimate them, and then allow sufficient flexibility in the matching network to accommodate modest errors.

Fig 13.19 shows the principal reactances in the amplifier circuit. C_{OUT} is the actual tube output capacitance of 10 pF plus the stray capacitance between its anode and the enclosure metalwork. This stray C varies with layout; we will approximate it as 5 pF, so C_{OUT} is roughly 15 pF. L_{OUT} is the stray inductance of leads from the tube plate to the tuning capacitor (internal to the tube as well as external circuit wiring.) External-anode tubes like the 8877 have essentially no internal plate leads, so L_{OUT} is almost entirely external. It seldom exceeds about 0.3 µH and is not very significant below 30 MHz. L_{CHOKE} is the reactance presented by the plate choke, which usually is significant only below 7 MHz. C_{STRAY} represents the combined stray capacitances to ground of the tuning capacitor stator and of interconnecting RF plate circuit leads. In a well constructed, carefully thought out power amplifier, C_{STRAY} can be estimated to be approximately 10 pF. Remaining components C_{TUNE}, C_{LOAD}, and the two tuning inductors, form the pi-L network proper.

The tables presented earlier in this chapter greatly simplify the task of selecting output circuit values. Both the pi and pi-L design tables are calculated for a Q_{OUT} value of 12. A pi network loaded Q much lower than 10 does not provide adequate harmonic suppression; a value much higher than 15 increases matching network losses caused by high circulating currents. For pi networks, a Q_{OUT} of 12 is a good compromise between harmonic suppression and circuit losses. In practice, it often is most realistic and practical with both pi and pi-L output networks to accept somewhat higher Q_{OUT} values on the highest HF frequencies—perhaps as large as 18 or even 20 at 28 MHz. When using a pi-L on the 1.8 and 3.5 MHz bands, it often is desirable to choose a moderately lower Q_{OUT}, perhaps 8 to 10, to permit using a more reasonably-sized plate tuning capacitor.

Nominal pi-L network component values for 2200-Ω plate impedance can be taken directly from Table 13.2. These values can then be adjusted to allow for circuit reactances outside the pi-L proper. First, low-frequency component values

should be examined. At 3.5 MHz, total tuning capacitance C1 value from Table 13.2 is 140 pF. From Fig 13.19 we know that three other stray reactances are directly in parallel with C_{TUNE} (assuming that L_{OUT} is negligible at the operating frequency, as it should be). The tube's internal and external plate capacitance to ground, C_{OUT}, is about 15 pF. Strays in the RF circuit, C_{STRAY}, are roughly 10 pF.

The impedance of the plate choke, X_{CHOKE}, is also in parallel with C_{TUNE}. Plate chokes with self-resonance characteristics suitable for use in amateur HF amplifiers typically have inductances of about 90 µH. At 3.5 MHz this is an inductive reactance of +1979 Ω. This appears in parallel with the tuning capacitance, effectively canceling an equal value of capacitive reactance. At 3.5 MHz, an X_C of 1979 Ω corresponds to 23 pF of capacitance—the amount by which tuning capacitor C_{TUNE} must be increased at 3.5 MHz to compensate for the effect of the plate choke.

The pi-L network requires an effective capacitance of 140 pF at its input at 3.5 MHz. Subtracting the 25 pF provided by C_{OUT} and C_{STRAY} and adding the 23 pF canceled by X_{CHOKE}, the actual value of C_{TUNE} must be $140 - 25 + 23 = 138$ pF. It is good practice to provide at least 10% extra capacitance range to allow matching loads having SWRs up to 2:1. So, if 3.5 MHz is the lower frequency limit of the amplifier, a variable tuning capacitor with a maximum value of at least 150 to 160 pF should be used.

Component values for the high end of the amplifier frequency range also must be examined, for this is where the most losses will occur. At 29.7 MHz, the values in Table 13.2 are chosen to accommodate a minimum pi-L input capacitance of 35 pF, yielding $Q_{OUT} = 21.3$. Since C_{OUT} and C_{STRAY} contribute 25 pF, C_{TUNE} must have a minimum value no greater than 10 pF. A problem exists, because this value is not readily achievable with a 150 to 160 pF air variable capacitor suitable for operation with a 3100 V plate supply. Such a capacitor typically has a minimum capacitance of 25 to 30 pF. Usually, little or nothing can be done to reduce the tube's C_{OUT} or the circuit C_{STRAY}, and in fact the estimates of these may even be a little low. If 1.8 MHz capability is desired, the maximum tuning capacitance will be at least 200 to 250 pF, making the minimum-capacitance problem at 29.7 MHz even more severe.

There are three potential solutions to this dilemma. We could accept the actual minimum value of pi-L input capacitance, around 50 to 55 pF, realizing that this will

raise the pi-L network's loaded Q to about 32. This results in very large values of circulating tank current. To avoid damage to tank components—particularly the bandswitch and pi inductor—by heat due to I²R losses, it will be necessary to either use oversize components or reduce power on the highest-frequency bands. Neither option is appealing.

A second potential solution is to reduce the minimum capacitance provided by C_{TUNE}. We could use a vacuum variable capacitor with a 300-pF maximum and a 5-pF minimum capacity. These are rated at 5 to 15 kV, and are readily available. This reduces the minimum effective circuit capacitance to 30 pF, allowing use of the pi-L table values for $Q_{OUT} = 12$ on all bands from 1.8 through 29.7 MHz. While brand new vacuum variables are quite expensive, suitable models are widely available in the surplus and used markets for prices not much higher than the cost of a new air variable. A most important caveat in purchasing a vacuum capacitor is to ensure that its vacuum seal is intact and that it is not damaged in any way. The best way to accomplish this is to "hi-pot" test the capacitor throughout its range, using a dc or peak ac test voltage of 1.5 to 2 times the amplifier plate supply voltage. For all-band amplifiers using plate voltages in excess of about 2500 V, the initial expense and effort of securing and using a vacuum variable input tuning capacitor often is well repaid in efficient and reliable operation of the amplifier.

A third possibility is the use of an additional inductance connected in series between the tube and the tuning capacitor. In conjunction with C_{OUT} of the tube, the added inductor acts as an L network to transform the impedance at the input of the pi-L network up to the 2200-Ω load resistance needed by the tube. This is shown in **Fig 13.20A**. Since the impedance at the input of the main pi-L matching network is reduced, the loaded Q_{OUT} for the total capacitance actually in the circuit is lower. With lower Q_{OUT}, the circulating RF currents are lower, and thus tank losses are lower.

C_{OUT} in Fig 13.20 is the output capacitance of the tube, including stray C from the anode to metal enclosure. X_L is the additional series inductance to be added. As determined previously, the impedance seen by the tube anode must be a 2200 Ω resistance for best linearity and efficiency, and we have estimated C_{OUT} of the tube as 15 pF. If the network consisting of C_{OUT} and X_L is terminated at A by 2200-Ω, we can calculate the equivalent impedance at point B, the input to the pi-L network, for various values of series X_L. The pi-L

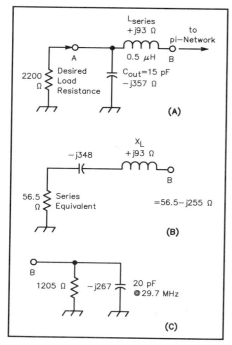

Fig 13.20—The effect of adding a small series inductance in vacuum tube output circuit. At A, a 0.5 μH coil L_{SERIES} is connected between anode and the output pi network, and this represents a reactance of + $j93\,\Omega$ at 29.7 MHz. The 15 pF output capacity (C_{OUT}) of the tube has a reactance of $-j357\,\Omega$ at 29.7 MHz. At B, the equivalent series network for the parallel 2200-Ω desired load resistance and the $-j357\,\Omega$ C_{OUT} is 56.5 Ω in series with $-j348\,\Omega$. At C, this series-equivalent is combined with the series +$j93\,\Omega$ X_{SERIES} and converted back to the parallel equivalent, netting an equivalent parallel network of 1205 Ω shunted by a 20 pF capacitor. The pi tuning network must transform the load impedance (usually 50 Ω) into the equivalent parallel combination and absorb the 20-pF parallel component. The series inductor has less effect as the operating frequency is lowered from 29.7 MHz.

network must then transform the nominal 50-Ω load at the transmitter output to this equivalent impedance.

We work backwards from the plate of the tube towards the C_{TUNE} capacitor. First, calculate the series-equivalent impedance of the parallel combination of the desired 2200-Ω plate load and the tube X_{OUT} (15 pF at 29.7 MHz = $-j357\,\Omega$). The series-equivalent impedance of this parallel combination is 56.5 − $j348\,\Omega$, as shown in Fig 13.20B. Now suppose we use a 0.5 μH inductor, having an impedance of 93 Ω + $j93\,\Omega$ at 29.7 MHz, as the series inductance X_L. The resulting series-equivalent impedance is 56.5 − $j348$ + $j93$, or 56.5 − $j255\,\Omega$. Converting back to the

parallel equivalent gives the network of Fig 13.20C: 1205 Ω resistance in parallel with − $j267\,\Omega$, or 20 pF at 29.7 MHz. The pi-L tuning network must now transform the 50-Ω load to a resistive load of 1205 Ω at B, and absorb the shunt capacity of 20 pF.

Using the pi-L network formulas in this chapter for R1 = 1205 and Q_{OUT} = 15 at 29.7 MHz yields a required total capacitive reactance of 1205/15 = 80.3 Ω, which is 66.7 pF at 29.7 MHz. Note that for the same loaded Q_{OUT} for a 2200-Ω load line, the capacitive reactance is 2200/15 = 146.7 Ω, which is 36.5 pF. When the 20 pF of transformed input capacity is subtracted from the 66.7 pF total needed, the amount of capacity is 46.7 pF. If the minimum capacity in C_{TUNE} is 25 pF and the stray capacity is 10 pF, then there is a margin of 46.7 − 35 = 10.7 pF beyond the minimum capacity for handling SWRs greater than 1:1 at the load.

The series inductor should be a high-Q coil wound from copper tubing to keep losses low. This inductor has a decreasing, yet significant effect, on progressively lower frequencies. A similar calculation to the above should be made on each band to determine the transformed equivalent plate impedance, before referring to Table 13.2. The impedance-transformation effect of the additional inductor decreases rapidly with decreasing frequency. Below 21 MHz, it usually may be ignored and pi-L network values taken directly from the pi-L tables for R1 = 2200 Ω.

The nominal 90-μH plate choke remains in parallel with C_{TUNE}. It is rarely possible to calculate the impedance of a real HF plate choke at frequencies higher than about 5 MHz because of self-resonances. However, as mentioned previously, the choke's reactance should be sufficiently high that the tables are useful if the choke's first series-resonance is at 23.2 MHz.

This amplifier is made operational on multiple bands by changing the values of inductance at L2 and L3 for different bands. The usual practice is to use inductors for the lowest operating frequency, and short out part of each inductor with a switch, as necessary, to provide the inductance needed for each individual band. Wiring to the switch and the switch itself add stray inductance and capacitance to the circuit. To minimize these effects at the higher frequencies, the unswitched 10-m L2 should be placed closest to the high-impedance end of the network at C6. Stray capacitance associated with the switch then is effectively in parallel with C7, where the impedance level is around 300 Ω. The effects of stray capacitance are relatively insignificant at this low impedance level. This configuration also mini-

mizes the peak RF voltage that the switch insulation must withstand.

Pi and L coil tap positions that yield desired values of inductance may be determined with fairly good accuracy by using a dip meter and a small mica capacitor of 5% tolerance. The pi and L coils and bandswitch should be mounted in the amplifier and their common point connected only to the bandswitch rotors. Starting at the highest-frequency switch position, lightly tack solder a short length of copper braid or strap to the pi or L switch stator terminal for that band. Using the shortest leads possible, tack a 50 to 100 pF, 5% dipped mica capacitor between the braid and a trial tap position on the appropriate coil. Lightly couple the dip meter and find the resonant frequency. The inductance then may be calculated from the equation

$$L = \frac{1,000,000}{C\,(2\pi f)^2} \qquad (41)$$

where

L = inductance in μH
C = capacitor value in pF
f = resonant frequency in MHz.

As each tap is located, it should be securely wired with strap or braid and the process repeated for successively lower bands.

The impedance match in both the input and output networks can be checked without applying dc voltage, once the amplifier is built. In operation, the tube input and output resistances are the result of current flow through the tube. Without filament power applied, these resistances are effectively infinite but C_{IN} and C_{OUT} are still present because they are passive physical properties of the tube. The tube input resistance can be simulated by an ordinary 5% 1/4-W to 2-W composition or film resistor (don't use wirewound, though; they are more inductive than resistive at RF). A resistor value within 10% of the tube input resistance, connected in parallel with the tube input, presents approximately the same termination resistance to the matching network as the tube does in operation.

With the input termination resistor temporarily soldered in place using very short leads, input matching network performance can be determined by means of a noise bridge or an SWR meter that does not put out more RF power than the temporary termination resistor is capable of dissipating. Any good self- or dipper-powered bridge or analyzer should be satisfactory. Connect the bridge to the amplifier input and adjust the matching network, as necessary, for lowest SWR. Be sure to remove the terminating resistor before powering up the amplifier!

The output matching network can be evaluated in exactly the same fashion, even though the plate load resistance is not an actual resistance in the tube like the input resistance. According to the reciprocity principle, if the impedance presented at the output of the plate matching network is 50-Ω resistive when the network input is terminated with R_L, then the tube plate will "see" a resistive load equal to R_L at the input when the output is terminated in a 50-Ω resistance (and vice versa). In this case, a suitable 2200-Ω resistor should be connected as directly as possible from the tube plate to chassis. If the distance is more than a couple of inches, braid should be used to minimize stray inductance. The bridge is connected to the amplifier output. If coil taps have been already been established as described previously, it is a simple matter to evaluate the output network by adjusting the tune and load capacitors, band by band, to show a perfect 50-Ω match on the SWR bridge.

When these tests are complete, the amplifier is ready to be tested for parasitic oscillations in preparation for full-power operation. Refer to Amplifier Stabilization, later in this chapter.

DESIGN EXAMPLE 2: A MEDIUM-POWER 144-MHZ AMPLIFIER

For several decades the 4CX250 family of power tetrodes has been used successfully up through 500 MHz. They are relatively inexpensive, produce high gain and lend themselves to relatively simple amplifier designs. In amateur service at VHF, the 4CX250 is an attractive choice for an amplifier. Most VHF exciters used now by amateurs are solid state and often develop 10 W or less output. The drive requirement for the 4CX250 in grounded cathode, Class AB operation ranges between 2 and 8 W for full power output, depending on frequency. At 144 MHz, manufacturer's specifications suggest an available output power of over 300 W. This is clearly a substantial improvement over 10 W, so a 4CX250B will be used in this amplifier.

The first design step is the same as in the previous example: verify that the proposed tube will perform as desired while staying within the manufacturer's ratings. Again assuming a basic amplifier efficiency of 60% for Class AB operation, 300 W of output requires a plate input power of 500 W. Tube dissipation is rated at 250 W, so plate dissipation is not a problem, as the tube will only be dissipating 200 W in this amplifier. If the recommended maximum plate potential of 2000 V is used, the plate current for 500-W input will be 250 mA, which is within the manufacturer's ratings. The plate load resistance

can now be calculated. Using the same formula as before, the value is determined to be 5333 Ω.

The next step is to investigate the output circuit. The manufacturer's specification for C_{OUT} is 4.7 pF. The inevitable circuit strays along with the tuning capacitor add to the circuit capacitance. A carefully built amplifier might only have 7 pF of stray capacitance, and a specially made tuning capacitor can be fabricated to have a midrange value of 3 pF. The total circuit capacitance adds up to about 15 pF. At 144 MHz this represents a capacitive reactance of only 74 Ω. The Q_L of a tank circuit with this reactance with a plate load resistance of 5333 Ω is 5333/74 = 72. A pi output matching network would be totally impractical, because the L required would be extremely small and circuit losses would be prohibitive. The simplest solution is to connect an inductor in parallel with the circuit capacitance to form a parallel-resonant tank circuit.

To keep tank circuit losses low with such a high Q_L, an inductor with very high unloaded Q must be used. The lowest-loss inductors are formed from transmission line sections. These can take the form of either coaxial lines or strip lines. Both have their advantages and disadvantages, but the strip line is so much easier to fabricate that it is almost exclusively used in VHF tank circuits today.

The reactance of a terminated transmission line section is a function of both its characteristic impedance and its length (see the **Transmission Lines** chapter). The reactance of a line terminated in a short circuit is

$$X_{IN} = Z_0 \tan \ell \qquad (42)$$

where

X_{IN} = is the circuit reactance
Z_0 = the line's characteristic impedance
ℓ = the transmission line length in degrees.

For lines shorter than a quarter wavelength (90°) the circuit reactance is inductive. In order to resonate with the tank circuit capacitive reactance, the transmission line reactance must be the same value, but inductive. Examination of the formula for transmission-line circuit reactance suggests that a wide range of lengths can yield the same inductive reactance, as long as the line Z_0 is appropriately scaled. Based on circuit Q considerations, the best bandwidth for a tank circuit results when the ratio of Z_0 to X_{IN} is between one and two. This implies that transmission line lengths between 26.5° and 45° give the best bandwidth. Between these two limits, and with some adjustment of Z_0, practical trans-

mission lines can be designed. A transmission line length of 35° is 8 inches long at 144 MHz, a workable dimension mechanically. Substitution of this value into the transmission-line equation gives a Z_0 of 105 Ω.

The width of the strip line and its placement relative to the ground planes determine the line impedance. Other stray capacitances such as mounting standoffs also affect the impedance. Accurate calculation of the line impedance for most physical configurations requires extensive application of Maxwell's equations and is beyond the scope of this book. The specialized case in which the strip line is parallel to and located halfway between two ground planes has been documented in *Reference Data for Radio Engineers* (see Bibliography). According to charts presented in that book, a 105-Ω strip line impedance is obtained by placing a line with a width of approximately 0.4 times the ground plane separation halfway between the ground planes. Assuming the use of a standard 3-inch-deep chassis for the plate compartment, this yields a strip-line width of 1.2 inches. A strip line 1.2 inches wide located 1.5 inches above the chassis floor and grounded at one end has an inductive reactance of 74 Ω at 144 MHz.

The resulting amplifier schematic diagram is shown in **Fig 13.21**. L2 is the strip-line inductance just described. C3 is the tuning capacitor, made from two parallel brass plates whose spacing is adjustable. One plate is connected directly to the strip line while the other is connected to ground through a wide, low-inductance strap. C2 is the plate blocking capacitor. This can be either a ceramic doorknob capacitor such as the Centralab 850 series or a homemade "Teflon sandwich." Both are equally effective at 144 MHz.

Impedance matching from the plate resistance down to 50 Ω can be either through an inductive link or through capacitive reactance matching. Mechanically, the capacitive approach is simpler to implement. **Fig 13.22** shows the development of reactance matching through a series capacitor (C4 in Fig 13.21). By using the parallel equivalent of the capacitor in series with the 50-Ω load, the load resistance can be transformed to the 5333-Ω plate resistance. Substitution of the known values into the parallel- to-series equivalence formulas reveals that a 2.15 pF capacitor at C4 matches the 50-Ω load to the plate resistance. The resulting parallel equivalent for the load is 5333 Ω in parallel with 2.13 pF. The 2.13-pF capacitor is effectively in parallel with the tank circuit.

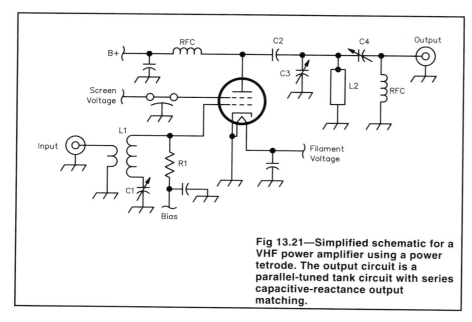

Fig 13.21—Simplified schematic for a VHF power amplifier using a power tetrode. The output circuit is a parallel-tuned tank circuit with series capacitive-reactance output matching.

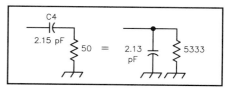

Fig 13.22—Series reactance matching as applied to the amplifier in Fig 13.21.

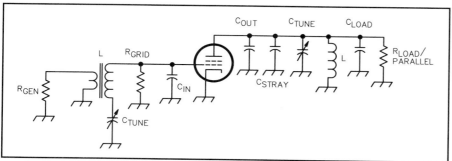

Fig 13.23—The reactances and resistances for the amplifier in Fig 13.21.

A new plate line length must now be calculated to allow for the additional capacitance. The equivalent circuit diagram containing all the various reactances is shown in **Fig 13.23**. The total circuit capacitance is now just over 17 pF, which is a reactance of 64 Ω. Keeping the strip-line width and thus its impedance constant at 105 Ω dictates a new resonant line length of 31°. This calculates to be 7.14 inches for 144 MHz.

The alternative coupling scheme is through the use of an inductive link. The link can be either tuned or untuned. The length of the link can be estimated based on the amplifier output impedance, in this case, 50 Ω. For an untuned link, the inductive reactance of the link itself should be approximately equal to the output impedance, 50 Ω. For a tuned link, the length depends on the link loaded Q, Q_L. The link Q_L should generally be greater than two, but usually less than five. For a Q_L of three this implies a capacitive reactance of 150 Ω, which at 144 MHz is just over 7 pF. The self-inductance of the link should of course be such that its impedance at 144 MHz is 150 Ω (0.166 μH).

Adjustment of the link placement determines the transformation ratio of the circuit line. Some fine adjustment of this parameter can be made through adjust-

ment of the link series tuning capacitor. Placement of the link relative to the plate inductor is an empirical process.

The input circuit is shown in Figs 13.21 and 13.23. C_{IN} is specified to be 18.5 pF for the 4CX250. This is only $-j60$ Ω at 2 m, so the pi network again is unsuitable. Since a surplus of drive is available with a 10-W exciter, circuit losses at the amplifier input are not as important as at the output. An old-fashioned "split stator" tuned input can be used. L1 in Fig 13.21 is series tuned by C_{IN} and C1. The two capacitors are effectively in series (through the ground return). A 20-pF variable at C1 set to 18.5 pF gives an effective circuit capacitance of 9.25 pF. This will resonate at 144 MHz with an inductance of 0.13 μH at L1. L1 can be wound on a toroid core for mechanical convenience. The 50-Ω input impedance is then matched by link coupling to the toroid. The grid impedance is primarily determined by the value for R1, the grid bias feed resistor.

DESIGN EXAMPLE 3: A BROADBAND HF SOLID-STATE AMPLIFIER

Linear power amplifier design using transistors at HF is a fundamentally simple process, although a good understanding of

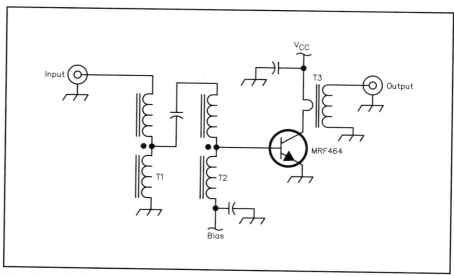

Fig 13.24—A simplified schematic of a broadband HF transistorized power amplifier. T1 and T2 are 4:1 broadband transformers to match the low input impedance of the transistor.

application techniques is important to insure that the devices are effectively protected against damage or destruction due to parasitic self-oscillations, power transients, load mismatch and/or overdrive.

An appropriate transistor meeting the desired performance specifications is selected on the basis of dissipation and power output. Transistor manufacturers greatly simplify the design by specifying each type of power transistor according to its frequency range and power output. The amplifier designer need only provide suitable impedance matching to the device input and output, along with appropriate dc bias currents to the transistor.

The Motorola MRF464 is an RF power transistor capable of 80 W PEP output with low distortion. Its usable frequency range extends through 30 MHz. At a collector potential of 28 V, a collector efficiency of 40% is possible. **Fig 13.24** shows the schematic diagram of a 2 to 30-MHz broadband linear amplifier using the MRF464. The input impedance of the transistor is specified by the manufacturer to be $1.4 - j0.30 \ \Omega$ at 30 MHz and decreases to $9.0 - j5.40 \ \Omega$ at 2 MHz. Transformers T1 and T2 match the 50-Ω amplifier input impedance to the median value of the transistor input impedance. They are both 4:1 stepdown ratio transmission-line transformers. A single 16:1 transformer could be used in place of T1 and T2, but 16:1 transformers are more difficult to fabricate for broadband service.

The specified transistor output resistance is approximately 6 Ω (in parallel with a corresponding output capacitance) across the frequency range. T3 is a ferrite-loaded conventional transformer with a step-up ratio of approximately 8:1. This matches the transistor output to 50 Ω.

The amplifier has a falling gain characteristic with rising frequency. To flatten out gain across the frequency range, negative feedback could be applied. However, most power transistors have highly reactive input impedances and large phase errors would occur in the feedback loop. Instability could potentially occur.

A better solution is to use an input correction network. This network is used as a frequency-selective attenuator for amplifier drive. At 30 MHz, where transistor gain is least, the input power loss is designed to be minimal (less than 2 dB). The loss increases at lower frequencies to compensate for the increased transistor gain. The MRF464 has approximately 12 dB more gain at 1.8 MHz than at 30 MHz; the compensation network is designed to have 12 dB loss at 1.8 MHz. A properly designed compensation network will result in an overall gain flatness of approximately 1 dB.

AMPLIFIER STABILIZATION

Stable Operating Conditions

Purity of emissions and the useful life (or even survival) of the active devices in a tube or transistor circuit depend heavily on stability during operation. Oscillations can occur at the operating frequency or far from it, because of undesired positive feedback in the amplifier. Unchecked, these oscillations pollute the RF spectrum and can lead to tube or transistor over-dissipation and subsequent failure. Each type of oscillation has its own cause and its own cure.

In a linear amplifier, the input and output circuits operate on the same frequency. Unless the coupling between these two circuits is kept to a small enough value, sufficient energy from the output may be coupled in phase back to the input to cause the amplifier to oscillate. Care should be used in arranging components and wiring of the two circuits so that there will be negligible opportunity for coupling external to the tube or transistor itself. A high degree of shielding between input and output circuits usually is required. All RF leads should be kept as short as possible and particular attention should be paid to the RF return paths from input and output tank circuits to emitter or cathode.

In general, the best arrangement using a tube is one in which the input and output circuits are on opposite sides of the chassis. Individual shielded compartments for the input and output circuitry add to the isolation. Transistor circuits are somewhat more forgiving, since all the impedances are relatively low. However, the high currents found on most amplifier circuit boards can easily couple into unintended circuits. Proper layout, the use of double-sided circuit board (with one side used as a ground plane and low-inductance ground return), and heavy doses of bypassing on the dc supply lines often are sufficient to prevent many solid-state amplifiers from oscillating.

VHF and UHF Parasitic Oscillations

RF power amplifier circuits contain parasitic reactances that have the potential to cause so-called parasitic oscillations at frequencies far above the normal operating frequency. Nearly all vacuum-tube amplifiers designed for operation in the 1.8 to 29.7 MHz frequency range exhibit tendencies to oscillate somewhere in the VHF-UHF range—generally between about 75 and 250 MHz depending on the type and size of tube. A typical parasitic resonant circuit is highlighted by bold lines in **Fig 13.25**. Stray inductance between the tube plate and the output tuning capacitor forms

a high-Q resonant circuit with the tube's C_{OUT}. C_{OUT} normally is much smaller (higher X_C) than any of the other circuit capacitances shown. The tube's C_{IN} and the tuning capacitor C_{TUNE} essentially act as bypass capacitors, while the various chokes and tank inductances shown have high reactances at VHF. Thus the values of these components have little influence on the parasitic resonant frequency.

Oscillation is possible because the VHF resonant circuit is an inherently high-Q parallel-resonant tank that is not coupled to the external load. The load resistance at the plate is very high, and thus the voltage gain at the parasitic frequency can be quite high, leading to oscillation.

The parasitic frequency, f_r, is approximately:

$$f_r = \frac{1000}{2\pi\sqrt{L_P \, C_{OUT}}} \qquad (43)$$

where
f_r = parasitic resonant frequency in MHz
L_P = total stray inductance between tube plate and ground via the plate tuning capacitor (including tube internal plate lead) in μH
C_{OUT} = tube output capacitance in pF.

In a well-designed HF amplifier, L_p might be in the area of 0.2 μH and C_{OUT} for an 8877 is about 10 pF. Using these figures, the equation above yields a potential parasitic resonant frequency of

$$f_r = \frac{1000}{2\pi\sqrt{0.2 \times 10}} = 112.5 \text{ MHz}$$

For a smaller tube, such as the 3CX800A7 with C_{OUT} of 6 pF, f_r = 145 MHz. Circuit details affect f_r somewhat, but these results do in fact correspond closely to actual parasitic oscillations experienced with these tube types. VHF-UHF parasitic oscillations can be prevented (*not* just minimized!) by reducing the loaded Q of the parasitic resonant circuit so that gain at its resonant frequency is insufficient to support oscillation. This is possible with any common tube, and it is especially easy with modern external-anode tubes like the 8877, 3CX800A7, and 4CX800A.

Z1 of Fig 13.25B is a parasitic suppressor. Its purpose is to add loss to the parasitic circuit and reduce its Q enough to prevent oscillation. This must be accomplished without significantly affecting normal operation. L_z should be just large enough to constitute a significant part of the total parasitic tank inductance (originally represented by L_P), and located right at the tube plate terminal(s). If L_z is made quite lossy, it will reduce the Q of the parasitic circuit as desired.

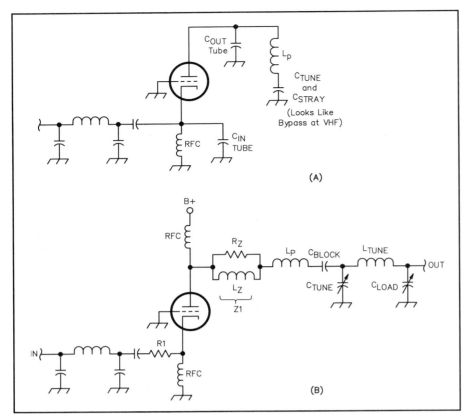

Fig 13.25—At A, typical VHF/UHF parasitic resonance in plate circuit. The HF tuning inductor in the pi network looks like an RF choke at VHF/UHF. The tube's output capacity and series stray inductance combine with the pi-network tuning capacity and stray circuit capacity to create a VHF/UHF pi network, presenting a very high impedance to the plate, increasing its gain at VHF/UHF. At B, Z1 lowers the Q and therefore gain at parasitic frequency.

The inductance and construction of L_z depend substantially on the type of tube used. Popular glass tubes like the 3-500Z and 4-1000A have internal plate leads made of wire. This significantly increases L_p when compared to external-anode tubes. Consequently, L_z for these large glass tubes usually must be larger in order to constitute an adequate portion of the total value of L_p. Typically a coil of 3 to 5 turns of #10 wire, 0.25 to 0.5 inches in diameter and about 0.5 to 1 inches long is sufficient. For the 8877 and similar tubes it usually is convenient to form a "horseshoe" in the strap used to make the plate connection. A "U" about 1 inch wide and 0.75 to 1 inch deep usually is sufficient. In either case, L_z carries the full operating-frequency plate current; at the higher frequencies this often includes a substantial amount of circulating tank current, and L_z must be husky enough to handle it without overheating even at 29 MHz.

Regardless of the form of L_z, loss may be introduced as required by shunting L_z with one or more suitable noninductive resistors. In high-power amplifiers, two composition or metal film resistors, each 100 Ω, 2 W, connected in parallel across L_z usually are adequate. For amplifiers up to perhaps 500 W a single 47-Ω, 2-W resistor may suffice. The resistance and power capability required to prevent VHF/UHF parasitic oscillations, while not overheating as a result of normal plate circuit current flow, depend on circuit parameters. Operating-frequency voltage drop across L_z is greatest at higher frequencies, so it is important to use the minimum necessary value of L_z in order to minimize power dissipation in R_z.

The parasitic suppressors described above very often will work without modification, but in some cases it will be necessary to experiment with both L_z and R_z to find a suitable combination. Some designers use nichrome or other resistance wire for L_z, but there is no credible evidence of any fundamental difference in performance as a result. Amplifier manufacturer W4ETO has never seen an HF amplifier using modern tubes that could not be made completely free of VHF parasitics by using one of the simple parasitic suppressor constructions described above.

In exceptionally difficult cases, particularly when using glass tetrodes or pentodes, additional parasitic suppression may be attained by connecting a low value resistor (about 10 to 15 Ω) in series with the tube input, near the tube socket. This is illustrated by R1 of Fig 13.25B. If the tube has a relatively low input impedance, as is typical of grounded-grid amplifiers and some grounded-cathode tubes with large C_{IN}, R1 may dissipate a significant portion of the total drive power.

Testing Tube Amplifiers for VHF-UHF Parasitic Oscillations

Every high-power amplifier should be tested before being placed in service, to insure that it is free of parasitic oscillations. For this test, nothing is connected to either the RF input or output terminals, and the bandswitch is first set to the lowest-frequency range. If the input is tuned and can be bandswitched separately, it should be set to the highest-frequency band. The amplifier control system should provide monitoring for both grid current and plate current, as well as a relay, circuit breaker, or fast-acting fuse to quickly shut off high voltage in the event of excessive plate current. To further protect the tube grid, it is a good idea to temporarily insert in series with the grid current return line a resistor of approximately 1000 Ω to prevent grid current from soaring in the event a vigorous parasitic oscillation breaks out during initial testing.

Apply filament and bias voltages to the amplifier, leaving plate voltage off and/or cutoff bias applied until any specified tube warmup time has elapsed. Then apply the lowest available plate voltage and switch the amplifier to transmit. Some idling plate current should flow. If it does not, it may be necessary to increase plate voltage to normal or to reduce bias so that at least 100 mA or so does flow. Grid current should be zero. Vary the plate tuning capacitor slowly from maximum capacitance to minimum, watching closely for any grid current or change in plate current, either of which would indicate a parasitic oscillation. If a tunable input network is used, its capacitor (the one closest to the tube if a pi circuit) should be varied from one extreme to the other in small increments, tuning the output plate capacitor at each step to search for signs of oscillation. If at any time either the grid or plate current increases to a large value, shut off plate voltage immediately to avoid damage! If moderate grid current or changes in plate current are observed, the frequency of oscillation can be determined by loosely coupling an RF absorption meter or a spectrum analyzer to the plate area. It will then be necessary to experiment with parasitic suppression measures until no signs of oscillation can be detected under any conditions. This process should be repeated using each bandswitch position.

When no sign of oscillation can be found, increase the plate voltage to its normal operating value and calculate plate dissipation (idling plate current times plate voltage). If dissipation is at least half of, but not more than, its maximum safe value, repeat the previous tests. If plate dissipation is much less than half of maximum safe value, it is desirable (but not absolutely essential) to reduce bias until it is. If no sign of oscillation is detected, the temporary grid resistor should be removed and the amplifier is ready for normal operation.

Parasitic Oscillations in Solid-State Amplifiers

In low-power solid-state amplifiers, parasitic oscillations can be prevented by using a small amount of resistance in series with the base or collector lead, as shown in **Fig 13.26A**. The value of R1 or R2 typically should be between 10 and 22 Ω. The use of both resistors is seldom necessary, but an empirical determination must be made. R1 or R2 should be located as close to the transistor as practical.

At power levels in excess of approximately 0.5 W, the technique of parasitic suppression shown in Fig 13.26B is effective. The voltage drop across a resistor would be prohibitive at the higher power levels, so one or more ferrite beads placed over connecting leads can be substituted (Z1 and Z2). A bead permeability of 125 presents a high impedance at VHF and above without affecting HF performance. The beads need not be used at both circuit locations. Generally, the terminal carrying the least current is the best place for these suppression devices. This suggests that the resistor or ferrite beads should be connected in the base lead of the transistor.

C3 of **Fig 13.27** can be added to some power amplifiers to dampen VHF/UHF parasitic oscillations. The capacitor should be low in reactance at VHF and UHF, but must present a high reactance at the operating frequency. The exact value selected will depend upon the collector impedance. A reasonable estimate is to use an X_C of 10 times the collector impedance at the operating frequency. Silver-mica or ceramic chip capacitors are suggested for this application. An additional advantage is the resultant bypassing action for VHF and UHF harmonic energy in the collector circuit. C3 should be placed as close to the collector terminal as possible, using short leads.

The effects of C3 in a broadband amplifier are relatively insignificant at the operating frequency. However, when a narrowband collector network is used, the added

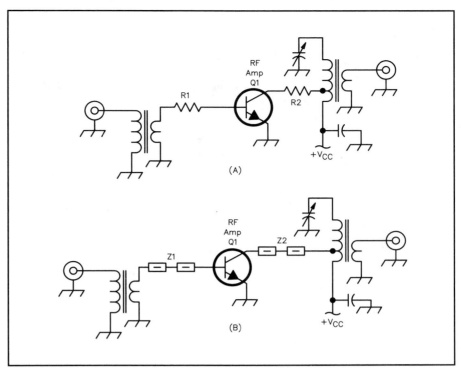

Fig 13.26—Suppression methods for VHF and UHF parasitics in solid-state amplifiers.

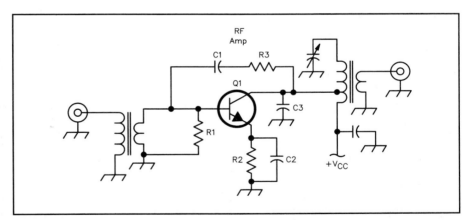

Fig 13.27—Illustration of shunt feedback in a transistor amplifier. C1 and R3 make up the feedback network.

capacitance of C3 must be absorbed into the network design in the same manner as the C_{OUT} of the transistor.

Low-Frequency Parasitic Oscillations

Bipolar transistors exhibit a rising gain characteristic as the operating frequency is lowered. To preclude low-frequency instabilities because of the high gain, shunt and degenerative feedback are often used. In the regions where low-frequency self-oscillations are most likely to occur, the feedback increases by nature of the feedback network, reducing the amplifier gain. In the circuit of Fig 13.27, C1 and R3 provide

negative feedback, which increases progressively as the frequency is lowered. The network has a small effect at the desired operating frequency but has a pronounced effect at the lower frequencies. The values for C1 and R3 are usually chosen experimentally. C1 will usually be between 220 pF and 0.0015 μF for HF-band amplifiers while R3 may be a value from 51 to 5600 Ω.

R2 of Fig 13.27 develops emitter degeneration at low frequencies. The bypass capacitor, C2, is chosen for adequate RF bypassing at the intended operating frequency. The impedance of C2 rises progressively as the frequency is lowered, thereby increasing the degenerative feedback caused by R2. This lowers the amplifier gain. R2 in a power stage is seldom greater than 10 Ω, and may be as low as 1 Ω. It is important to consider that under some operating and layout conditions R2 can cause instability. This form of feedback should be used only in those circuits in which unconditional stability can be achieved.

R1 of Fig 13.27 is useful in swamping the input of an amplifier. This reduces the chance for low-frequency self oscillations, but has an effect on amplifier performance in the desired operating range. Values from 3 to 27 Ω are typical. When connected in shunt with the normally low base impedance of a power amplifier, the resistors lower the effective device input impedance slightly. R1 should be located as close to the transistor base terminal as possible, and the connecting leads must be kept short to minimize stray reactances. The use of two resistors in parallel reduces the amount of inductive reactance introduced compared to a single resistor.

Although the same concepts can be applied to tube-type amplifiers, the possibility of self-oscillations at frequencies lower than VHF is significantly lower than in solid-state amplifiers. Tube amplifiers will usually operate stably as long as the input-to-output isolation is greater than the stage gain. Proper shielding and dc-power-lead bypassing essentially eliminate feedback paths, except for those through the tube itself.

On rare occasions tube-type amplifiers will oscillate at frequencies in the range of about 50 to 500 kHz. This is most likely with high-gain tetrodes using shunt feed of dc voltages to both grid and plate through RF chokes. If the resonant frequency of the grid RF choke and its associated coupling capacitor occurs close to that of the plate choke and its blocking capacitor, conditions may support a tuned-plate tuned-grid oscillation. For example, using typical values of 1 mH and 1000 pF, the expected parasitic frequency would be around 160 kHz.

Make sure that there is no low-impedance, low-frequency return path to ground through inductors in the input matching networks in series with the low impedances reflected by a transceiver output transformer. Usually, oscillation can be prevented by changing choke or capacitor values to insure that the input resonant frequency is much lower than that of the output.

Amplifier Neutralization

Depending on stage gain and interelectrode capacitances, sufficient positive feedback may occur to cause oscillation at the operating frequency. This should not occur in well-designed grounded-grid amplifiers, nor with tetrode or pentodes

operating at gains up to about 15 dB as is current practice at HF where 50 to 100 W of drive is almost always available. If triodes are grid-driven, however, and under certain other circumstances, neutralization may be necessary because of output energy capacitively coupled back to the input as shown in **Fig 13.28**. Neutralization involves coupling a small amount of output energy back to the amplifier input out of phase, to cancel the unwanted in-phase (positive) feedback. A typical circuit is given in **Fig 13.29**. L2 provides a 180° phase reversal because it is center tapped. C1 is connected between the plate and the lower half of the grid tank. C1 is then adjusted so that the energy coupled from the tube output through the neutralization circuit is equal in amplitude and exactly 180° out of phase with the energy coupled from the output back through the tube. The two signals then cancel and oscillation is impossible.

The easiest way to adjust a neutralization circuit is to connect a low-level RF source to the amplifier output tuned to the amplifier operating frequency. A sensitive RF detector like a receiver is then connected to the amplifier input. The amplifier must be turned off for this test. The amplifier tuning

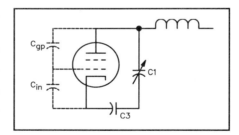

Fig 13.28—The equivalent feedback path due to the internal capacitance of the tube grid-plate structure in a power amplifier. Also see Fig 13.30.

and loading controls, as well as any input network adjustments are then peaked for maximum indication on the RF detector connected at the input. C1 is then adjusted for minimum response on the detector. This null indicates that the neutralization circuit is canceling energy coupled from the amplifier output to its input through tube, transistor, or circuit capacitances.

Screen-Grid Tube Stabilization

The plate-to-grid capacitance in a screen-grid tube is reduced to a fraction of a picofarad by the interposed grounded screen. Nevertheless, the power gain of these tubes may be so great in some circuits that only a very small amount of feedback is necessary to start oscillation. To assure a stable tetrode amplifier, it is usually necessary to load the grid circuit, or to use a neutralizing circuit.

Grid Loading

The need for a neutralizing circuit may often be avoided by loading the grid circuit to reduce stage gain, provided that the driving stage has some power capacity to spare. Loading by tapping the grid down on the grid tank coil, or by placing a "swamping" resistor from grid to cathode, is effective to stabilize an amplifier. Either measure reduces the gain of the amplifier, lessening the possibility of oscillation. If a swamping resistor is connected between grid and cathode with very short leads, it may help reduce any tendency toward VHF-UHF parasitic oscillations as well. In a class AB1 amplifier, which draws no grid current, a swamping resistor can be used to replace the bias supply choke if parallel feed is used.

Often, reducing stage gain to the value required by available drive power is sufficient to assure stability. If this is not practical or effective, the bridge neutralizing system for screen-grid tubes shown in **Fig 13.30** may be used. C1 is the neutralizing

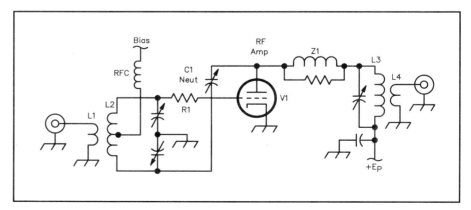

Fig 13.29—Example of neutralization of a single-ended RF amplifier.

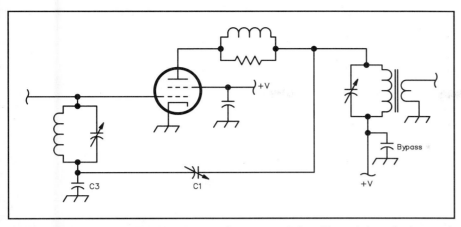

Fig 13.30—A neutralization circuit uses C1 to cancel the effect of the tube internal capacitance.

capacitor. The value of C1 should be chosen so that at some adjustment of C1,

$$\frac{C1}{C3} = \frac{C_{gp}}{C_{IN}} \qquad (44)$$

where

C_{gp} = tube grid-plate capacitance
C_{IN} = tube input capacitance.

The grid-to-cathode capacitance must include all strays directly across the tube capacitance, including the capacitance of the tuning capacitor.

THE SUNNYVALE/SAINT PETERSBURG KILOWATT-PLUS

This article describes a modern 1500-W output linear amplifier for the amateur HF bands. It uses a relatively recent arrival on the transmitting tube scene in the US, a 4CX1600B power tetrode made by Svetlana in Saint Petersburg, Russia. The amplifier was designed and constructed by George T. Daughters, K6GT, who lives in Sunnyvale, California—hence the name "Sunnyvale/Saint Petersburg" for this project. **Fig 13.31** shows the completed amplifier and the power supply cabinet.

Power tetrodes such as the 4CX1600B feature higher power gain than do the power triodes (such as the 3-500Z or 8877) often used in linear amplifiers. The increased power gain gives the designer additional flexibility, at the expense of a somewhat more complex dc supply design. This amplifier operates in the grounded-cathode configuration, with a 50-Ω resistor from control grid to ground. This provides a good load for the trans-

ceiver driving the amplifier, promotes amplifier stability and also eliminates the need for switched-input tuned circuits. The advantages of such a *passive-grid*, grounded-cathode design outweigh the cost and complication of the screen-grid supply needed by the tetrode tube.

The Svetlana 4CX1600B is designed with a "striped-cathode," where emission takes place mainly in the spaces between parallel control-grid wires. This reduces the number of electrons intercepted by the control grid under normal drive conditions. (The Eimac 4CX1500B is also designed this way.) However, the linearity of such a high-gain tetrode falls off rapidly if the control grid is allowed to draw any current at all. Even a small positive voltage at the control grid can cause a large current to flow in the grid.

Note that the control grid in this type of

high-gain tetrode is only rated at 2 W dissipation. (The first versions of the data sheet for the 4CX1600B specified the grid dissipation as 100 *milliwatts*!) By comparison, the control grid dissipation of the venerable, but much lower-gain, 4-1000A tetrode is 25 W. Any circumstance where measurable control grid current flows in the 4CX1600B will result in nonlinear operation, resulting not only in splatter, but also in possible damage to the control grid. It is thus important to provide some sort of *grid current prevention* scheme or, at the very least, a *grid current warning* alarm, for an amplifier using the 4CX1600B.

The grid of the 4CX1600B in this amplifier is tapped down on the input resistor. With 100 W of drive, the grid voltage cannot swing positive enough to result in significant grid current. Deliberate cathode degeneration (negative feedback) is also

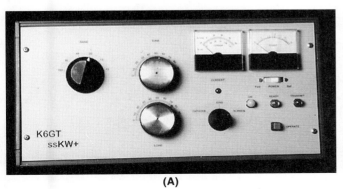

(A)

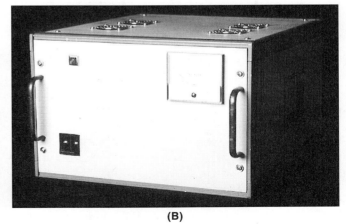

(B)

Fig 13.31—At A, photo of Sunnyvale/Saint Petersburg Kilowatt-Plus amplifier RF Deck. At B, the Power Supply cabinet.

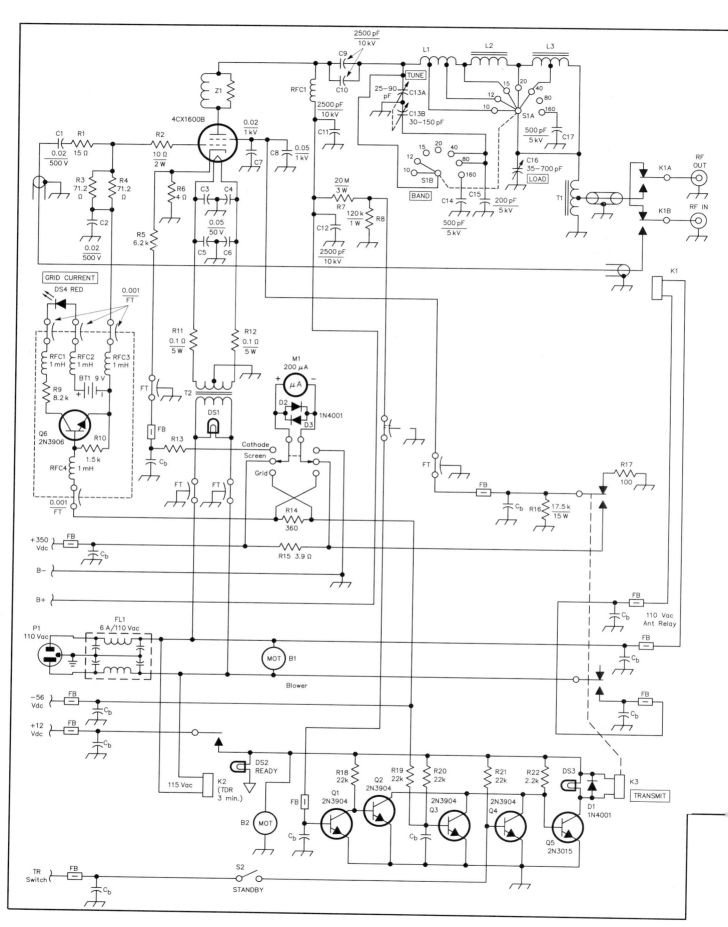

B1—Squirrel-cage blower capable of 36 cfm at 0.4 inches of water back pressure (Dayton 4C753 or similar)

B2—"Biscuit" blower, 12 V dc, 130 mA (Rotron BD12A3 or similar) mounted inside the pressurized RF deck to aid cooling the input grid resistor R1.

BT1—9 V transistor radio battery.

C1, C2—0.02 µF, 500 V disc ceramic.

C3, C4, C5, C6—0.05 µF, 50 V disc ceramic.

C7—Screen bypass capacitor (0.02 µF, 1 kV disc ceramic at the screen terminal on the socket in parallel with the internal bypass capacitor, which is part of the Svetlana SK-3A socket).

C8—0.05 µF, 1 kV disc ceramic.

C9, C10—parallel 2500 pF, 10 kV ceramic doorknob, Newark #46F5253.

C11, C12—2500 pF, 10 kV ceramic doorknob, Newark #46F5253.

C13—Plate tuning capacitor; front section is 30-150 pF; rear section is 25-90 pF (Command Technologies P/N 73-2-100-41).

C14, C17—500 pF, 5 kV ceramic doorknob.

C15—200 pF, 5 kV ceramic doorknob.

C16—Plate loading capacitor, 35-700 pF (Command Technologies P/N 73-1-45-65).

D1—1N4001.

DS1, DS2, DS3—Indicator lamps (green: 120 V ac; amber: 12 V; and red: 12 V).

DS4—Jumbo red LED.

FL1—IEC 110 V ac connector with 6 A line filter.

FT—0.001 µF, 1000 V feedthrough capacitors.

FB, C_B—RF decoupling components used in multiple places; ferrite beads FB-43-1801 and 0.01 µF, 1 kV disc-ceramic capacitors.

K1—110 V ac DPDT antenna changeover relay.

K2—115 V ac 3-minute time delay (Macromatic SS-6262-KK).

K3—12 V dc relay, DPST.

L1—Plate tank inductor; ¼-inch diameter, silver-plated copper tubing, 6 turns with inside diameter of 1¼ inches, followed by 4½ turns with inside diameter of 1¾ inches. Tap for 10 (and 12) meters is 4 turns from small-diameter end; tap for 15 (and 17) is 2 turns further down. All of L1 is used for 20 meters.

L2—Toroid coil; 5 turns #10 PTFE wire (40 inches long, overall) on two T-225-8 cores.

L3—Toroid coil; 6 groups of 3 each #10 PTFE wires (150 inches long, overall) on three T-225-28 cores.

M1—200 µA meter movement, internal resistance 2000 Ω.

P1—IEC power cable to J1 on Fig 13.34.

Q1 to Q6—2N3904 or similar (Silicon, general purpose, NPN).

Q5—2N3015 or similar (Silicon, low V_{ce}(Sat), NPN).

R1—15 Ω, Caddock MP-850, mounted on heat sink with R3 and R4.

R2—10 Ω, 2 W composition.

R3, R4—71.2 Ω Caddock MP-850, mounted on heat sink with R1.

R5, R13—6.2 kΩ, 1 W. (R5 is part of the cathode current meter multiplier, as is R13. Their values were chosen to provide 1.3 A full-scale reading on the meter used.)

R6—4 Ω, 12 W (4 each 16 Ω, 3 W, noninductive metal-oxide-film, in parallel on 4CX1600B tube socket).

R7—20 MΩ, 3 W (Caddock MX430).

R8—120 kΩ, 1 W composition.

R11, R12—Filament dropping resistors; 0.1 Ω, 5 W.

R16—Screen bleeder; 17.5 kΩ, 15 W (two 25 kΩ, 5 W in parallel, in series with 5 kΩ, 5 W).

RFC—1 mH RF choke.

RFC1—Plate choke, 91 turns #26 enamel on 1-inch diameter × 3.75 inch delrin form (Command Technologies P/N RFC-1).

T1—Broadband 2:1 transformer; 13 bifilar turns #12 PTFE (120 inches, overall) on three FT-240-61 cores. Note that plate tank inductors, bandswitch, plate RF choke, and toroidal RF transformer are part of Command Technologies HF-2500 plate tank circuit.

T2—Filament transformer, 12.6 V ac (center-tapped), 6A (Triad F-182).

V1—Svetlana 4CX1600B power tetrode in modified Svetlana SK-3A socket. The anode connector is a Svetlana AC-2, and the chimney and the chimney extension are each a Svetlana CH-1600B.

Z1—Parasitic suppressor; two turns of tinned copper strap (0.032-inch thick × 0.313-inch wide) over three 91 Ω, 2 W composition resistors in parallel.

and keeps the tube cut off to avoid the generation of any shot noise. In transmit, a 17.5-kΩ, 15-W resistor to ground is switched into the screen grid circuit to keep a constant load of 20 mA on the series regulator. This allows the regulator to function properly with up to −20 mA of screen current. (Negative screen current is a condition common to these types of power tetrodes under some load conditions.) The 20-mA constant load is indicated on the screen-current meter as "zero," so that the meter reads actual screen current from −20 to + 80 mA.

BUILDING IT

The heart of the amplifier consists of the RF deck, the control and metering circuitry and the cooling system. These are all mounted in a surplus 19-inch rack-mount cabinet of the sort picked up at surplus stores and hamfests. The power supply is built into another cabinet.

Fig 13.32 shows the schematic diagram of the RF deck. The 4CX1600B is mounted in the Svetlana SK-3A socket, modified as described below (to allow the cathode to operate above ground potential for negative feedback). Svetlana's CH-1600B chimney routes the cooling airflow through the anode cooling fins. An additional CH-1600B acts as a chimney extension, discharging the air through the top of the RF deck's cabinet. The cooling fan is a squirrel-cage blower. According to the 4CX1600B data sheet at 1600 W of plate dissipation, the blower should deliver at least 36 cfm (cubic feet per minute) of cooling air at an ambient temperature of 25°C, at a back pressure of 0.4 inches of water.

The low-cost filament transformer specified in **Fig 13.32** produces 13.5 V ac (with nominal mains voltage), so two 0.1-Ω, 5-W resistors were added to drop the voltage at the filament terminals of the 4CX1600B to the 12.6 V ac recommended by the tube manufacturer.

The input grid resistor is 51.6 Ω, with a dissipation capability exceeding 100 W. It consists of three Caddock MP850 resistors—two 71.2-Ω resistors in parallel, in series with 15 Ω, all mounted on a surplus heat sink (5.0 × 5.5 × 0.75 inch or 12.7 × 14.0 × 2.0 cm). This passive grid resistor is mounted below the chassis, near the SK-3A socket, and has its own small cooling "biscuit" fan. While the air below the chassis is pressurized by the main blower to provide cooling of the tube, the auxiliary fan cools the input resistors and keeps the air stirred up to prevent any stagnant hot air below the chassis.

The grid of the 4CX1600B is tapped at the 35.6-Ω point of the input resistive divider. As a further aid to stability, a

used to help prevent grid-current flow. This is accomplished by placing a noninductive resistor between the cathode and ground. In addition, a sensitive grid-current meter is provided, reading 1.3 mA at full-scale deflection. Finally, a simple, yet sensitive, grid-current-activated warning is also included in this design, using a red LED on the front panel as a warning lamp.

In receive, a 100 Ω resistor is switched into the screen grid circuit to chassis ground. This removes the screen voltage

10-Ω, 2-W composition resistor is placed in series with the control-grid lead. This arrangement results in an input SWR of 1.0:1 at 1.9 MHz, increasing to just over 1.6:1 at 29.6 MHz, mainly due to the reactance of the 86 pF input capacitance of the 4CX1600B. No frequency compensation was deemed necessary. The cathode resistor is made up of four 16-Ω, 3-W noninductive metal-oxide film resistors from the cathode terminal ring on the socket to each of the four socket mounting screws.

The plate tank circuit components include a heavy-duty bandswitch, a silver-plated inductor for the high bands, powdered iron toroidal inductors for the low bands and a plate choke wound on a Delrin form. These components are those used in a Command Technologies HF-2500 amplifier but other suitable components could be utilized. (As it is currently configured, the plate tank cannot be tuned to 30 meters. Operation at full power on this band would require another position on the bandswitch and another tap on the tank coil or compromises on other bands. These are options which the author considered to be unnecessary and undesirable, since US hams have a power limit of 200 W on 30 meters.)

The anode connector is a Svetlana AC-2, and the plate parasitic choke is two turns of tinned copper strap (0.032-inch thick × 0.188-inches wide, or 0.8 mm × 4.8 mm) over three 91-Ω, 2-W composition resistors in parallel. (Any value from 47 to 100 Ω will be satisfactory.) The antenna change-over relay has a 115 V ac coil (12 V dc would be fine also). The author's relay had wide, gold-plated contacts.

CONTROL CIRCUITRY

The control circuitry is shown in Fig 13.32. The amplifier is turned on with the main switch/breaker on the power-supply cabinet. When the switch is thrown, all voltages are ready (after the step-start delay in the plate supply). The 4CX1600B filament begins to heat; the cooling fans go on; the time delay starts and anode voltage is applied to the 4CX1600B. After the mandatory three minutes for filament warmup, the +12 V dc control voltage is enabled by the time-delay relay. At this time, the control circuitry (consisting of transistors Q1 to Q5) determines whether screen voltage can be applied to the 4CX1600B and whether to activate the antenna changeover relay. Q5 is the main switch activating T/R relay K2 whenever 12 V is available (that is, after the 3-minute warmup period). Screen voltage will thus be supplied to the tube only when all of the following conditions are met:

1. The anode voltage for the 4CX1600B is available. This is sensed in the RF deck by the resistive divider R7/R8 shown in Fig 13.32. If the HV sense line is low, then Q1 and Q2 hold the base of Q5 at a low level.
2. The negative control-grid bias is present. If this voltage is near zero, transistor Q3 is saturated, and again Q5 is turned off.
3. The T/R switch from the exciter has pulled the base of Q4 low, allowing its collector to rise.

THE POWER SUPPLY

Remember that almost every voltage inside a power supply for a high-power linear amplifier is lethal! Turn it off, unplug it, and short it out before you touch anything! Always apply the "one hand in the pocket" principle when working on anything above 24 V!

The high-voltage power supply uses a Peter W. Dahl ARRL-002 transformer, weighing 46 pounds. As shown in **Fig 13.34**, a simple *step-start circuit* using K1 and K2 limits the current surge charging the filter capacitors when power is first applied. The transformer's output is rectified by a bridge of K2AW's Silicon Alley 10-kV diode arrays, and the filter capacitor is made up of a string of ten 470 μF, 400-V electrolytic capacitors. These were removed from a laser power supply board, which was available at a local surplus store (Alltronics, Santa Clara, CA) for $14.95. The voltage is divided equally across the capacitor string by 25-kΩ, 25-W resistors that also serve as the power supply bleeder. (This divider results in a considerably higher bleeder current than the typical 100 kΩ resistors often seen. The result is a stiffer power supply, but more heat is generated.)

The author's junk box produced a transformer with output windings of 275 V ac at 60 mA, 6.3 V ac at 2 A, and 35 V ac at 150 mA. These windings were dedicated to a regulated 350-V screen supply, a regulated 12-V dc supply for relay and indicator lamps (using a full-wave doubler and a three-terminal IC regulator), and the control-grid bias supply. The circuitry for these supplies is very straightforward. These supplies were built in the same cabinet as the plate high-voltage supply.

All power supplies are cooled by a muffin fan on the rear panel of the cabinet. Although the fan probably isn't necessary, cool components are sure to last longer. The major source of heat in this cabinet is the bleeder-resistor chain, which dissipates about 36 W when the plate voltage is 3000 V. High voltage is monitored with a 200 μA surplus meter movement through a Caddock MX430 20 MΩ multiplier resistor.

All power to the RF deck is supplied from the power supply cabinet. There is a standard IEC 120-V ac cable for the 4CX1600B filament transformer and the antenna changeover relay, an auxiliary power cable and a high-voltage line for the anode voltage. The shielded auxiliary power cable carries the screen and control-grid bias voltages and the 12-V dc and the ground. The high-voltage line is a 40-kV #18 wire obtained from a local surplus store, with Millen 37001 connectors at each end.

In this design it is possible to plug in and turn on the HV supply without any connection to the RF deck. If you should forget to connect the ground wire and only connect the HV cable by itself, then a potentially *unsafe* condition exists, with high voltage on the RF deck chassis with respect to the power supply chassis. You can avoid this in several ways: Use a special high-voltage cable/connector that incorporates a chassis ground connection together with the HV lead. Or you could use an interlock system, with an additional high-current relay in the 240 V ac line that is activated only when an interlock cable is connected. (The interlock cable would contain a direct inter-chassis ground connection.) Finally, a simple but effective approach is to bundle the HV cable with the other inter-cabinet cables, with a distinctive bright warning label to remind the operator to make sure all connections are made between the power supply and the RF deck.

Because no control-grid current flows, the control-grid bias voltage (nominally −56 V) is provided by a simple half-wave voltage doubler, with low-power zener diodes and a potentiometer to allow grid bias adjustment for the desired no-signal cathode current. The common practice of using a zener diode in the cathode circuit to provide operating bias was rejected because of the need for actual resistance between the cathode and ground for negative feedback.

The screen supply provides a dc voltage of 350 V by means of a series electronic regulator. The regulator has a current-limiting feature, where the output voltage falls if the screen draws more than 60 mA. This prevents the screen grid dissipation from exceeding its maximum rating of 20 W.

MODIFYING THE SK-3A SOCKET

Because the stock Svetlana socket has the cathode tied directly to chassis ground (through the socket's mounting plate) and because an internal bypass capacitor for the screen grid is placed between the screen grid and the cathode, you must modify the socket for this application. You

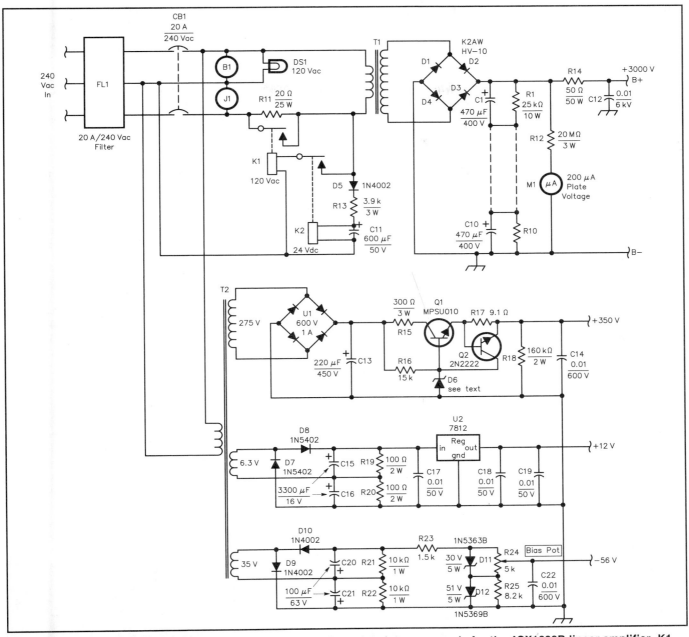

Fig 13.34—Schematic diagram of the high-voltage plate and regulated screen supply for the 4CX1600B linear amplifier. K1, K2 and associated circuitry provide a "step-start" characteristic to limit the power-on surge of charging current for the filter capacitors. Resistors are 1/2 W unless noted. Capacitors are disc ceramic unless noted and those marked with a + are electrolytic. Addresses for parts suppliers are given in the References chapter.

B1—Muffin fan (Rotron SU2A1 or similar).
C1 to C10—Filter capacitors; 470 μF, 400 V electrolytic.
C11—600 μF, 50 V electrolytic.
C12—0.01 μF, 6 kV disc ceramic.
C13—220 μF, 450 V electrolytic.
C14, C22—0.01 μF, 600 V disc ceramic.
C15, C16—3300 μF, 16 V electrolytic.
C17, C18, C19—0.01 μF, 50 V disc ceramic.
C20, C21—100 μF, 63 V electrolytic.
CB1—2 × 20 A, 240 V ac circuit breaker.
D1 to D4—K2AW's HV-10 rectifier diodes.
D5—1N4002.
D6—Zener diodes, three 1N4764A and one 1N5369B to total

approximately 350 V dc.
D7, D8—1N5402.
D9, D10—1N4002.
D11—Zener diode, 1N5363B (30 V, 5 W).
D12—Zener diode, 1N5369B (51 V, 5 W).
DS1—120 V ac indicator lamp (red).
FL1—240 V ac/20 A EMI filter.
J1—110 V ac, 15 A receptacle for plug P1 on Fig 13.32.
K1—120 V ac DPDT relay; both poles of 240 V ac/15 A contacts in parallel.
K2—24 V dc relay; 120 V ac/5 A contacts.
M1—200 μA meter movement.
Q1—MPSU010.
Q2—2N2222.
R1 to R10—Bleeder resistors; 25 kΩ, 10 W.
R11—20 Ω, 25 W.

R12—20 MΩ, 3 W (Caddock MX430).
R13—3.9 kΩ, 3 W.
R14—50 Ω, 50 W mounted on standoff insulators.
R15—300 Ω, 3 W.
R18—160 kΩ, 2 W composition.
R19, R20—100 Ω, 2 W composition.
R21, R22—10 kΩ, 1 W composition.
R24—5 kΩ potentiometer; sets control grid bias for desired no-signal cathode current.
T1—Plate transformer (Peter W. Dahl No. ARRL-002).
T2—Power transformer, 120 V / 275 V at 0.06 A, 6.3 V at 2 A, 35 V at 0.15 A.
U1—600 V, 1 A rectifier bridge.
U2—7812, +12 V IC voltage regulator.

will need four insulating shoulder washers (Teflon or other insulating material), made for 4-40 screws.

1. Drill out the four rivets holding the screen ring to the screen contactors at the very top of the socket.
2. At the bottom of the socket, remove the four nuts from the machine screws holding the socket assembly together.
3. Disassemble the socket:
   a) First remove the cathode contact ring. Be sure to mark its position relative to the underlying bakelite layer.
   b) Remove the bakelite socket layer, which has the factory markings and serial number, also marking its position relative to the socket mounting plate. (This is the 0.060-inch [1.5-mm] silver-plated brass plate.)
   c) Carefully remove the screen contactor assembly, freeing the contactor "ears" by springing them outward. Don't drop the screen capacitor! It is the ceramic annulus with silver plating on each side and it is very brittle.
   d) Finish removing the spring plate, the capacitor and the other spring plate, if they didn't already come out with the screen contactor assembly in step (c) above.
   e) Remove the mounting plate assembly, marking its position relative to the remaining socket assembly.
4. Drill out the four holes in the mounting plate assembly using a #14 drill (0.180 inches). These are the second set of holes in from the outer edge, through which the socket assembly screws pass. (The screws should still be in the top layer of the socket, with heater, grid, and cathode contactors.)
5. Put the new Teflon shoulder washers on the screws. When the socket is reassembled, the cathode will be isolated from the main mounting plate and the screen bypass capacitor.
6. Replace the capacitor assembly in the following order: spring, capacitor and spring. Now replace the screen contactor assembly and the bakelite bottom section, taking care to align this section with your previous mark. Carefully guide the socket solder tabs through the bakelite bottom without bending them.

Fig 13.35—Inside view of RF deck. Tank components are from a Command Technologies HF-2500 amplifier.

7. Cut the outer tabs off the cathode ring contact. After all of this work, you don't want this ring (the cathode terminal) to be grounded when you mount the socket in the chassis! Place the modified cathode contact ring over the screws.
8. Replace the washers and nuts on the socket assembly machine screws and tighten each a little at a time, until the assembly is snug.

This completes the socket conversion. The screen ring on the 4CX1600B is contacted exactly as before. The internal screen bypass capacitor still appears between the screen grid and ground (through the socket mounting plate). The heater, control grid, and screen contacts function exactly as in the original.

The cathode annulus on the 4CX1600B is contacted exactly as before, but the electrical connection for the cathode is now isolated from the chassis. The cathode contact on the socket is now made through the thin cathode ring on the bottom of the socket. (The ring is silver-plated and easily soldered, convenient for an application like the present one, which requires multiple contacts.)

METERING

The author obtained some attractive meters with 200 µA movements from a local surplus store. The internal resistance was 2000 Ω. One meter became a voltmeter on the anode power supply (0 to 4 kV); one became a triple-purpose multimeter to measure anode current (0 to 1.3A), screen-grid current (–20 to +80 mA), and control-grid current (0 to 1.3 mA). The third meter, not shown in the schematic, indicates forward (0 to 1500 W) and reflected power (0 to 150 W) at the output connector. After dc calibration against a digital multimeter, he carefully removed the cover and face of each movement and attached a homemade laser-printed scale.

GRID CURRENT WARNING

The circuitry for the grid-current warning indicator light is very simple and is shown in Fig 13.32 also. When control-grid current flows, it develops a voltage across R10. This causes the collector cur-

Fig 13.36—Another view of RF deck during construction.

rent of Q6 to light a red LED indicator brightly when grid current is about 1.0 mA. (Although the battery is always connected to the circuit of transistor Q6, the current drain due to collector-emitter leakage current is negligible, so battery life should be very long. If you don't like the floating 9-V battery, a small dc power supply could be included or a small "wall-wart" type of dc supply could be built right into the cabinet. It must however, be capable of floating at the grid potential, about 60 V away from chassis ground potential.)

When the grid-current warning LED flickers on voice peaks, it's time to back off the transceiver's RF output control to reduce the drive. In CW mode, many transceivers will put out a high-power spike on initial key closure, even when the RF output control is set to quite low values. If this happens with your transceiver, the warning blink from the LED will alert you to the problem. The circuitry for the grid-current warning indicator is built into a small aluminum minibox that uses feedthrough capacitors and RF

Table 13.7

4CX1600B, Class AB1, Passive Grid-Driven Service

	Zero Signal	Maximum Signal
Plate Voltage	3200 V	3040 V
Control Grid Bias Voltage	−56 V	−56 V
Screen Grid Voltage	350 V	350 V
DC Plate Current	280 mA	800 mA
Approx. Plate Load	—	2400 Ω
Drive Power	0 W	66 W
Power Output	0 W	1500 W
Intermodulation Distortion Products		
3rd order	—	−35 dB
5th order	—	−43 dB
7th order	—	−47 dB

chokes to eliminate stray RF.

RESULTS

The zero-signal plate current is about 280 mA, resulting in a zero-signal plate dissipation of about 900 W. At full 1.5 kW output on 40 meters, the plate current is about 0.8 A and the anode dissipation is less than 1000 W. (Until the TR switch is activated, the screen voltage is zero and the tube is effectively cut off, so there is no plate dissipation except during transmit periods.) After a heavy period of operating the amp-lifier, let the fan run for a few minutes in standby mode to cool the tube before turning the amplifier off.

Performance figures for the amplifier are presented in **Table 13.7**.

A 6-METER KILOWATT AMPLIFIER USING THE SVETLANA 4CX1600B

The Svetlana 4CX1600B tube has attracted a lot of attention lately because of its potent capabilities and relatively low cost. Because of its high gain and its large anode dissipation capabilities, the tube has relatively large input and output capacitances—85 pF at the input and 12 pF at the output. Stray capacitance of about 10 pF must be added in as well. On bands lower than 50 MHz, these capacitances can be dealt with satisfactorily with a broadband 50-Ω input resistor and conventional output tuning circuitry.

Fig 13.37 — Photo of the front panel of W1QWJ's 6-meter 4CX1600B amplifier.

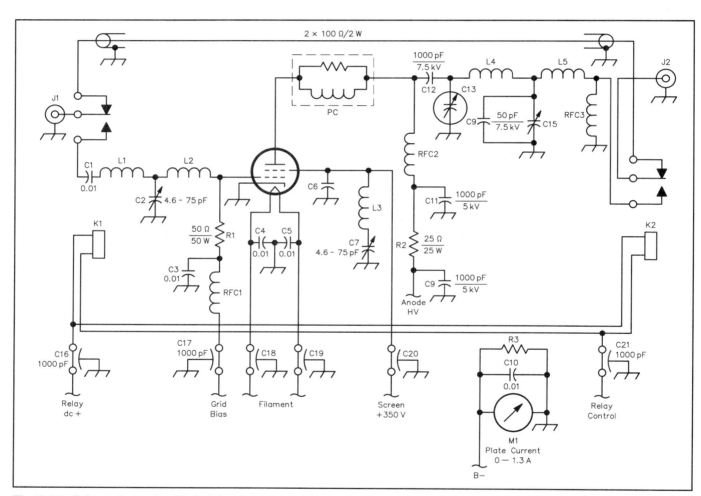

Fig 13.38—Schematic for the RF deck for the 6-meter 4CX1600B amplifier. Capacitors are disc ceramic unless noted. Addresses for parts suppliers are given in the References chapter.

C2, C7—4.6-75 pF, 500-V air-variable trimmer capacitor, APC style.
C6—Screen bypass capacitor, built into SK-3A socket.
C13—1-45 pF, 5 kV, Jennings CHV1-45-5S vacuum-variable capacitor.
C14—50 pF, 7.5 kV, NPO ceramic doorknob capacitor.
C15—4-102 pF, 1100V, HFA-100A type air-variable capacitor.
C16, 17, 18, 19, 20, 21—1000 pF, 1 kV feedthrough capacitors.

L1—11 turns, #16, 3/8-inch diameter, 1-inch long.
L2—9 turns #16, 3/8-inch diameter, close-wound.
L3—8 turns #16, 3/8-inch diameter, 7/8-inch long.
L4—1/4-inch copper tubing, 4 1/2 turns, 1 1/4 inches diameter, 4 1/2 inches long.
L5—5 turns #14, 1/2 inch diameter, 1 3/8 inches long.
M1—0-1.3 A meter, with homemade

shunt resistor, R3, across 0-10 mA movement meter.
PC—Parasitic suppressor, 2 turns #14, 1/2 inch diameter, shunted by two 100-Ω, 2-W carbon composition resistors in parallel.
RFC1—10 µH, grid-bias choke.
RFC2—Plate choke, 40 turns #20, 1/2 inch diameter, close-wound.
RFC3—Safety choke, 20 turns #20, 3/8 inch diameter.

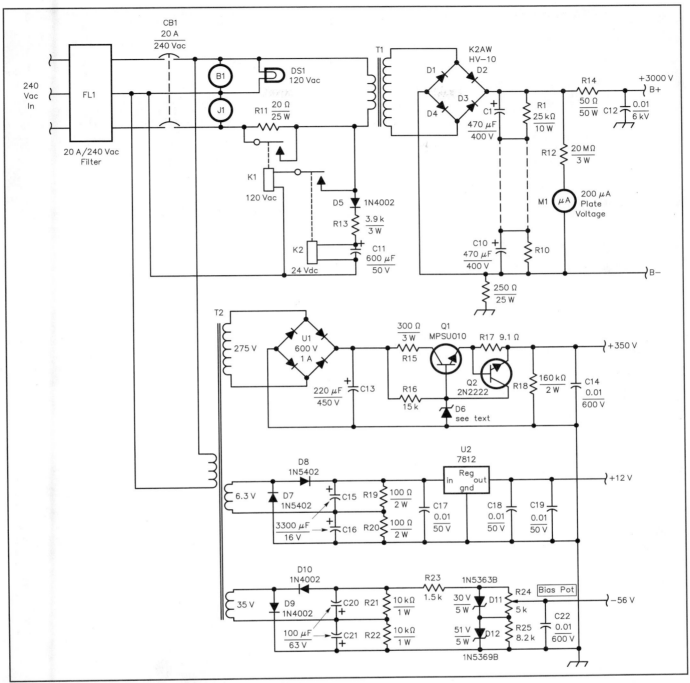

Fig 13.39—Partial schematic of K6GT HV power supply (see Fig 13.34), showing modification with 250-Ω, 25-W power resistor to ground on B– line, allowing for metering of the plate current in the amplifier.

See the article by George Daughters, K6GT, "The Sunnyvale/Saint Petersburg Kilowatt-Plus" earlier in this chapter for details on suitable control and power-supply circuitry. This 6-meter amplifier uses the same basic design as K6GT's, except for modified input and output circuits in the RF deck. See **Fig 13.37**, a photograph of the front panel of the 6-meter amplifier.

On the 50-MHz band the tube's high input capacitance must be tuned out. Author Dick Stevens, W1QWJ, used a T network so that the input impedance looks

like a nonreactive 50 Ω to the transceiver. To keep the output tuning network's loaded Q low enough for efficient power generation, he used a 1.5 to 46 pF Jennings CHV1-45-5S vacuum-variable capacitor, in a Pi-L configuration to keep harmonics low. You should use a quarter-wave shorted coaxial stub in parallel with the output RF connector to make absolutely sure that the second harmonic is reduced well below the FCC specification limits.

To guarantee stability, the author had to make sure the screen grid was kept as close

as possible to RF ground. This allows the screen to do its job "screening"—this minimizes the capacitance between the control grid and the anode. He used the Svetlana SK-3A socket, which includes a built-in screen bypass capacitor, and augmented that with a 50-MHz series-tuned circuit to ground. In addition, to prevent VHF parasitics, he used a parasitic suppressor in the anode circuit.

Unlike the K6GT HF amplifier, this 6-meter amplifier uses no cathode degeneration. W1QWJ wanted maximum stable

Fig 13.40—Close-up photo of the anode tank circuit for 6-meter kW amplifier. The air-cooling chimney has been removed in this photo.

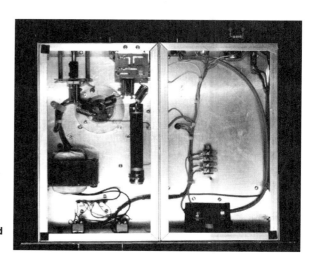

Fig 13.41—Underneath the 6-meter kW amplifier RF deck, showing on the left the tube socket and input circuitry.

power gain, with less drive power needed on 6 meters. He left the SK-3A socket in stock form, with the cathode directly grounded. This amplifier requires about 25 W of drive power to produce full output.

Fig 13.38 is a schematic of the RF deck built by W1QWJ. The control and power supply circuitry are basically the same as that used in Fig 13.32 and Fig 13.34 in the K6GT HF amplifier, except that plate current is monitored with a meter in series with the B– lead, since the cathode in this amplifier is grounded directly. The K6GT power supply is modified by inserting a 250-Ω, 25-W power resistor to ground in place of the direct ground connection. See **Fig 13.39**. In Fig 13.38, C1 blocks grid-bias dc voltage from appearing at the transceiver, while L1, L2 and C2 make up the T-network that tunes out the input capacitance of V1. R1 is a non-reactive 50-Ω 50-W resistor.

C6 is the built-in screen bypass capacitor in the SK-3A socket, while L3 and C7 make up the series-tuned screen bypass circuit. RFC3 is a safety choke, in case blocking capacitor C12 should break down and short, which would otherwise place high voltage at the output connector.

CONSTRUCTION

Like the K6GT amplifier, this W1QWJ amplifier is constructed in two parts: an RF deck and a power supply. Two aluminum chassis boxes bolted together and mounted to a front panel are used to make the RF deck. **Fig 13.40** shows the 4CX1600B tube and the 6-meter output tank circuit. **Fig 13.41** shows the underside of the RF deck, with the input circuitry shown in

more detail in **Fig 13.42**. The 50-Ω, 50-W noninductive power resistor is shown at the bottom of Fig 13.42. Note that the tuning adjustment for the input circuit is accessed from the rear of the RF deck.

AMPLIFIER ADJUSTMENT

The tune-up adjustments can be done without power applied to the amplifier and with the top and bottom covers removed. You can use readily available test instruments: an MFJ-259 SWR Analyzer and a VTVM with RF probe.

1. Activate the antenna changeover relay, either mechanically or by applying control voltage to it. Connect a 2700-Ω, $^1/_2$-W carbon composition resistor from anode to ground using short leads. Connect the SWR analyzer, tuned to 50 MHz, to the output connector. Adjust plate tuning and loading controls for a 1:1 SWR. You are using the Pi-L network in reverse this way.

2. Now, connect the MFJ-259 to the input connector and adjust the input T-network for a 1:1 SWR. Some spreading of the turns of the inductor may be required.

3. Disconnect the Pi-L output network from the tube's anode, leaving the 2700-Ω carbon composition resistor from the anode still connected. Connect the RF probe of the VTVM to the anode and run your exciter at low power into the amplifier's input connector. Tune the screen series-tuned bypass circuit for a distinct dip on the VTVM. The dip will be sharp and the VTVM reading should go to zero.

4. Now, disconnect the 2700-Ω carbon resistor from the anode and replace the covers. Connect the power supply and

control circuitry. When you apply power to the amplifier, you should find that only a slight tweaking of the output controls will be needed for final adjustment.

Fig 13.42—Close-up photo of the input circuitry for the 6-meter kW amplifier. Input tuning capacitor C2 is adjusted from the rear panel during operation, if necessary. The series-tuning capacitor C7 used to thoroughly ground the screen for RF is shown at the lower right. It is adjusted through a normally plugged hole in the rear panel during initial adjustment only.

A 144-MHZ AMPLIFIER USING THE 3CX1200Z7

This 2-m, 1-kW amplifier uses the EIMAC 3CX1200Z7 triode. The original article by Russ Miller, N7ART, appeared in December 1994 *QST*. The tube requires a warmup of about 10 s after applying filament voltage—no more waiting for three agonizingly long minutes until an amplifier can go on-line!

The 3CX1200Z7 is different from the earlier 3CX1200A7 by virtue of its external grid ring, redesigned anode assembly and a 6.3-V ac filament. One advantage to the 3CX1200Z7 is the wide range of plate voltages that can be used, from 2000 to 5500 V. This amplifier looks much like the easily duplicated W6PO design. The RF deck is a compact unit, designed for table-top use (See **Fig 13.44** and schematic in **Fig 13.45.**).

Table 13.9 gives some data on the 3CX1200Z7 and **Table 13.10** lists CW operating performance for this amplifier.

Input Circuit

Tuning is easy and docile. Grid bias is provided by an 8.2-V, 50-W Zener diode. Cutoff bias is provided by a 10-kΩ, 25-W resistor. A relay on the control board shorts out the cutoff-bias resistor, to place the amplifier in the TRANSMIT mode.

The author didn't use a tube socket. Instead, he bolted the tube directly to the top plate of the subchassis, using the four holes (drilled to clear a #6 screw) in the grid flange. Connections to the heater pins are via drilled and slotted brass rods. The

Fig 13.44—This table-top 2-m power amplifier uses a quick-warm-up tube, a real plus when the band suddenly opens for DX and you want to join in.

input circuit is contained within a 3¹/₂ × 6 × 7¹/₄-inch (HWD) subchassis (**Fig 13.46**).

Control Circuit

The control circuit (**Fig 13.47**) is a necessity. It provides grid overcurrent protection, keying control and filament surge control. To protect the tube filament from stressful surge current, a timer circuit places a resistor in series with the primary of the filament transformer. After four seconds, the timer shorts the resistor, allowing full filament voltage to be applied. C2 and R4 establish the time delay.

Another timer inhibits keying for a total of 10 s, to give the internal tube temperatures a chance to stabilize. C1 and R3 determine the time constant of this timer. After 10 s, the amplifier can be keyed by grounding the keying line. When the amplifier is not keyed, it draws no plate current. When keyed, idle current is approximately 150 mA, and the amplifier only requires RF drive to produce output. A safety factor is built in: the keying circuit requires +12 V from the high-voltage supply. This feature ensures that high voltage is present before the amplifier is driven.

The grid overcurrent circuit should be set to trip if grid current reaches 200 mA. When it trips, the relay latches and the NORMAL LED extinguishes. Restoration requires the operator to press the RESET switch.

Plate Circuit

Fig 13.48 shows an interior view of the plate compartment. A 4×2¹/₄-inch tuning capacitor plate and a 2×2-inch output coupling plate are centered on the anode collet. See **Fig 13.49**. Sufficient clearance in the collet hole for the 3CX1200Z7 anode must be left for the fingerstock. The hole diameter will be approximately 3⁵/₈ inches. **Fig 13.50** is a drawing of the plate line, **Fig 13.51** is a drawing of the plate tuning capacitor assembly, and **Fig 13.52** shows the output coupling assembly.

Cooling

The amplifier requires an air exhaust through the top cover, as the plate compartment is pressurized. Fashion a chimney from a 3¹/₂-inch waste-water coupling (black PVC) and a piece of ¹/₃₂-inch-thick Teflon sheet. The PVC should extend down from the underside of the amplifier cover plate by 1¹/₈ inches, with the Teflon sheet extending down ³/₄ inch from the bottom of the PVC.

The base of the 3CX1200Z7 is cooled using bleed air from the plate compart-

ment. This is directed at the tube base, through a ⁷/₈-inch tube set into the subchassis wall at a 45° angle. The recommended blower will supply more than enough air for any temperature zone. A smaller blower is not recommended, as it is doubtful that the base area will be cooled adequately. The 3CX1200Z7 filament draws 25 A at 6.3 V! It alone generates a great deal of heat around the tube base seals and pins, so good air flow is critical.

Construction

The amplifier is built into a 12×12×10-inch enclosure. A 12×10-inch partition is installed 7¹/₄ inches from the rear panel. The area between the partition and the front panel contains the filament transformer, control board, meters, switches, Zener diode and miscellaneous small parts. Wiring between the front-panel area and the rear panel is through a ¹/₂-inch brass tube, located near the shorted end of the right-hand plate line.

High voltage is routed from an MHV jack on the rear panel, through a piece of solid-dielectric RG-59 (*not* foam dielectric!), just under the shorted end of the left-hand plate line. The cable then passes through the partition to a high-voltage standoff insulator made from nylon. This insulator is fastened to the partition near the high-voltage feedthrough capacitor. A 10-Ω, 25-W resistor is connected between the insulator and the feedthrough capacitor.

Table 13.9
3CX1200Z7 Specifications
Maximum Ratings

Plate voltage: 5500 V
Plate current: 800 mA
Plate dissipation: 1200 W
Grid dissipation: 50 W

Table 13.10
CW Operating Data

Plate voltage: 3200 V
Plate current (operating): 750 mA
Plate current (idling): 150 mA
Grid current: 165 mA
DC Power input: 2400 W
RF Power output: 1200 W
Plate dissipation: 1200 W
Efficiency: 50%
Drive power: 85 W
Input reflected power: 1 W

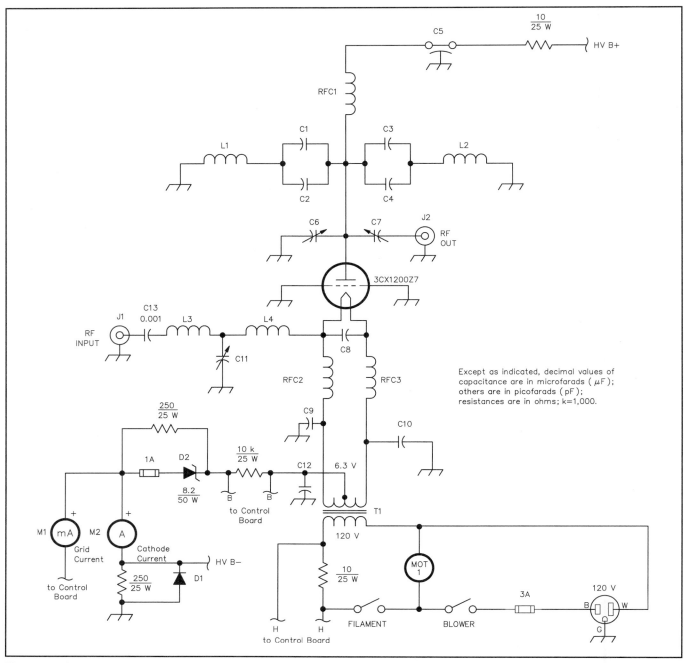

Fig 13.45—Schematic diagram of the 2-m amplifier RF deck. For supplier addresses, see the Address List in the **References** chapter.

C1-C4—100 pF, 5 kV, Centralab 850.
C5—1000 pF, 5 kV.
C6—Anode-tuning capacitor; see text and Fig 13.51 for details.
C7—Output-loading capacitor; see text and Fig 13.52 for details.
C8-C10, C13—1000-pF silver mica, 500 V.
C11—30-pF air variable.
C12—0.01 μF, 1 kV.
D1—1000 PIV, 3-A diode, 1N5408 or equiv.
D2—8.2-V, 50-W Zener diode, ECG 5249A.
J1—Chassis-mount BNC connector.

J2—Type-N connector fitted to output coupling assembly (see Fig 13.52).
L1, L2—Plate lines; see text and Fig 13.50 for details.
L3—5 t #14 enameled wire, ½-inch diameter, close wound.
L4—3 t #14, ⅝-inch diameter, ¼-inch spacing.
RFC1—7 t #14, ⅝-inch diameter, 1⅜ inch long.
RFC2, RFC3—10 t #12, ⅝-inch diameter, 2 inches long.

T1—Filament transformer. Primary: 120 V; secondary: 6.3 V, 25 A, center tapped. Available from Avatar Magnetics; part number AV-539.
M1—Grid milliammeter, 200 mA dc full scale.
M2—Cathode ammeter, 2 A dc full scale.
MOT1—140 free-air cfm, 120-V ac blower, Dayton 4C442 or equivalent.

Sources for some of the "hard to get parts" include Fair Radio Sales and Surplus Sales of Nebraska.

Fig 13.46—This view of the cathode-circuit compartment shows the input tuned circuit and filament chokes.

Fig 13.47—(below) Schematic diagram of the amplifier-control circuits.

C3—0.47-μF, 25-V tantalum capacitor.
D1-D5—1N4001 or equiv.
D6—1N4007 or equiv.
DS1—Yellow LED.
DS2—Green LED.
DS3—Red LED.
K1—Keying-inhibit relay, DPDT, 12-V dc coil, 1-A contact rating (Radio Shack 275-249 or equiv).
K2—Amplifier keying relay, SPDT, 12-V dc coil, 2-A contact rating (Radio Shack 275-248 or equiv).
K3—Filament delay relay, SPST, 12-V dc coil, 2-A contact rating (Radio Shack 275-248 or equiv).
K4—Grid-overcurrent relay, DPDT, 12-V dc coil, 1-A contact rating (Radio Shack 275-249 or equiv).
Q1, Q2, Q5—2N2222A or equiv.
Q3—MPF102 or equiv.
Q4—2N3819 or equiv.
S1—Normally closed, momentary pushbutton switch (Radio Shack 275-1549 or equiv).
T1—Power transformer, 120-V primary, 18-V, 1-A secondary.
U1—+12 V regulator, 7812 or equiv.

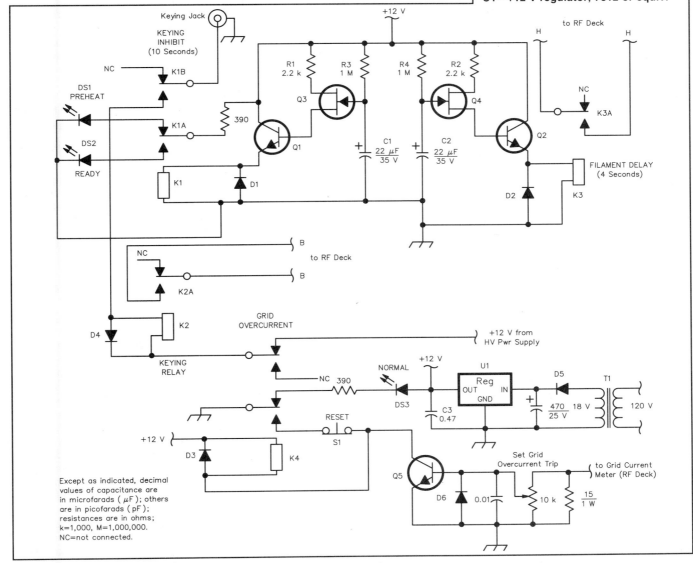

Fig 13.48—This top view of the plate compartment shows the plate-line arrangement, C1-C4 and the output coupling assembly.

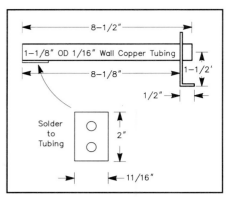

Fig 13.50—Plate line details.

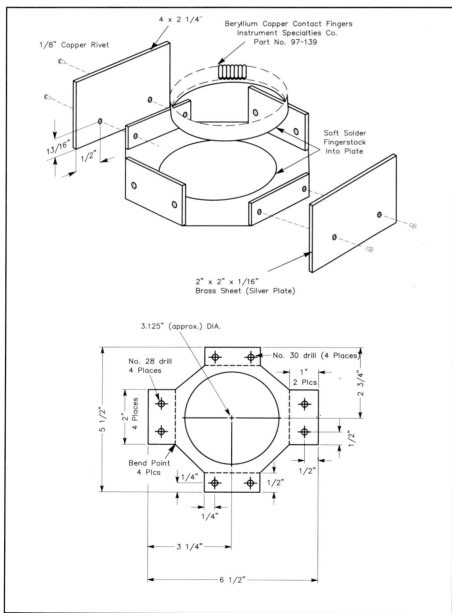

Fig 13.49—Anode collet details.

The plate lines are connected to the dc-blocking capacitors on the plate collet with $1^3/_4$×2-inch phosphor-bronze strips. The bottom of the plate lines are attached to the sides of the subchassis, with the edge of the L-shaped mounting bracket flush with the bottom of the subchassis.

When preparing the subchassis top plate for the 3CX1200Z7, cut a $2^{11}/_{16}$-inch hole in the center of the plate. This hole size allows clearance between the tube envelope and the top plate, without putting stress on the envelope in the vicinity of the grid flange seal.

Exercise care in placing the movable tuning plate and the movable output coupling disc, to ensure they cannot touch their fixed counterparts on the plate collet.

Operation

When the amplifier is first turned on, it cannot be keyed until:
• 10 s has elapsed
• High voltage is available, as confirmed by presence of +12 V to the keying circuit

Connect the amplifier to a dummy load through an accurate power meter capable of indicating 1500 W full scale. Key the amplifier and check the idling plate current. With 3200-V plate voltage, it should be in the vicinity of 150 mA. Now, apply a small amount of drive and adjust the input tuning for maximum grid current. Adjust the output tuning until you see an indication of RF output. Increase drive and adjust the output coupling and tuning for the desired output. Do not overcouple the output; once desired output is reached, do not increase loading. Insert the hold-down screw to secure the output coupling capacitor from moving. One setting is adequate for tuning across the 2-m band if

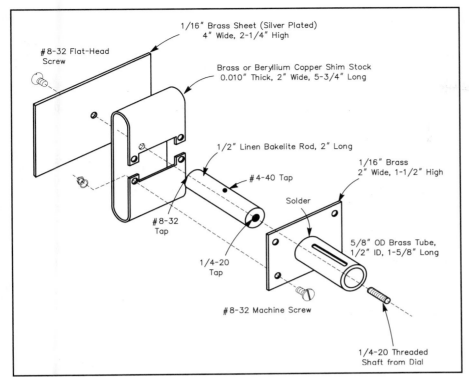

Fig 13.51—Plate tuning capacitor details.

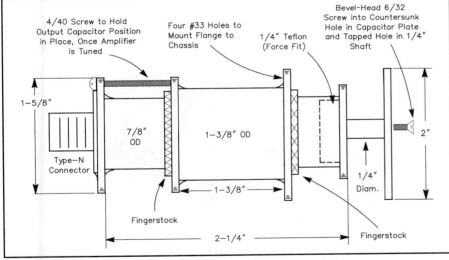

Fig 13.52—Details of the output coupling assembly.

Table 13.11
Power Supply Specifications
High voltage: 3200 V
Continuous current: 1.2 A
Intermittent current: 2 A
Step/Start delay: 2 s

the SWR on the transmission line is reasonably low.

When you shut down the amplifier, leave the blower running for at least three minutes after you turn off the filament voltage. The 3CX1200Z7 is an excellent tube. The author tried it with excessive drive, plate-current saturation, excessive plate dissipation—all the abuse it's likely to encounter in amateur applications. There were no problems, but that doesn't mean you should repeat these torture tests!

A Companion Power Supply

A good, solid-state high-voltage power supply is a necessity to ensure linearity in SSB operation. Specifications of the power supply are given in **Table 13.11**. The schematic and parts list for the author's power supply are in the **Power Supplies and Projects** chapter.

Conclusion

This amplifier is a reliable and cost-effective way to generate a big 2-m signal—almost as quickly as a solid-state amplifier. To ensure that the output of the amplifier meets current spectral purity requirements, a high-power output filter, as shown in **Fig 13.53**, should be used. The author reports that he can run full output while his wife watches TV in a nearby room.

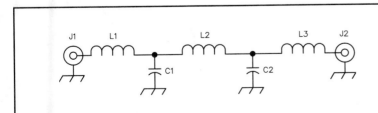

Fig 13.53—Schematic diagram of output harmonic filter.
C1, C2—27-pF Centralab 850 series ceramic transmitting capacitor.
J1, J2—Female chassis-mount N connector (UG-58 or equiv).
L1, L3—2 t #14 wire, 0.3125 inch ID, 0.375 inch long.
L2—3 t #14 wire, 0.3125 inch ID, 0.4375 inch long.

A 2-M BRICK AMP FOR HANDHELDS

Perhaps you've been looking for a fun weekend project and need a bit more output from your HT while operating mobile. This Brick Amp project may be exactly what you're looking for—construction is easy and all the parts are readily available. The following was contributed by ARRL Laboratory Engineer, Mike Gruber, W1MG.

The Brick Amp is easily driven at the low-output power setting of most handhelds. The same design is used for either a 25 or a 50 W version—only the amplifier module is changed. See **Fig 13.54** for a view of what's inside a typical module, alongside the finished amplifier.

The low-power 25 W Brick Amp complies with the bioeffects guidelines set forth in the **Safety** chapter. The full 50 W version can be built when more output is required. (Note: The bioeffects guidelines recommend that field-strength measurements be made in mobile installations of greater than 25 W output. Be sure to consult the **Safety** chapter before building the 50 W version.)

Circuit Details

The heart of this project is a Toshiba amplifier module. The S-AV7 is used for the 25 W output, while the mechanically identical S-AV17 is used in the 50 W version. Both modules are biased as class-C amplifiers, keeping their efficiency up and making them ideally suited for FM or CW use. Since they are not linear, however, they are not useful for other modes, such as SSB or AM.

From a builder's standpoint, these modules keep the parts count down to a minimum and construction simple. All you need to add for circuitry is input drive attenuation, if required, transmit/receive switching, an output filter and the usual dc filter and decoupling components. Beginners and seasoned veterans alike will no doubt appreciate this Brick Amp's simplicity!

DC Filtering and Decoupling

See the schematic diagram in **Fig 13.55**. C1 through C6 and chokes L1-L2 provide dc filtering and decoupling. D5 provides reverse-polarity protection, by blowing F1 should the +13.8 V line be wrongly connected.

The Input and Output Circuitry

The Brick Amp's input circuitry consists of a resistive T-pad formed by R1, R2 and R3. The pad attenuates the HT's output to the input level required by the module. Select the proper pad values based upon the HT output and the amplifier module selected. Refer to **Table 13.12** for pad resistor values for typical HT output levels.

Reduced drive power results in less than full rated output power, while excessive drive can result in exceeding the limits of the module. Proper T-pad selection is essential for full rated output power without exceeding design limits. Be sure to use only noninductive resistors, such as carbon or metal oxide, with the speci-

Table 13.12
T-Pad Values

25 W Module, S-AV7

HT Power (W)	Attenuation (dB)	R1,R3 (Ω)	R2 (Ω)
0.5	1	2.9	430
0.8	2	5.6	220
1.0	4	12	100
1.5	6	16	68
2.0	7	18	56
2.5	8	22	47
3.0	9	24	39
4.0	10	27	36
5.0	11	56/56*	62/62*

50 W Module, S-AV17

HT Power (W)	Attenuation (dB)	R1,R3 (Ω)	R2 (Ω)
0.5	4	12	100
0.8	6	16	68
1.0	7	20	56
1.5	9	24	39
2.0	10	27	36
2.5	11	27	30
3.0	12	30	27
4.0	13	33	24
5.0	14	62/62*	43/43*

Note: For power inputs up to and including 4.0 W, use 2 W resistors for R1 and R3, $\frac{1}{2}$ W resistors for R2. All resistors are carbon composition or metal oxide.
* Use parallel connected resistors: 2 W for R1 and R3; 1 W for R2.

(A)

(B)

Fig 13.54—At A, photo of 2-m Brick Amp assembly and at B, a 25 W power module with cover removed.

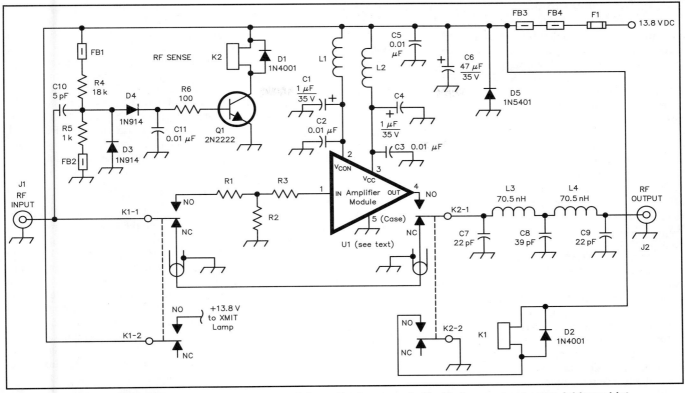

Fig 13.55—Schematic and parts list. Supplier contact information appears in the References chapter Address List.

Case—Hammond cat. no. 1590D.
Heat sink—5×7 inches, RF Parts.
C1, C4—1.0 µF, 35 V dipped tantalum (RS 272-1434).
C2, C3, C5, C11—0.01 µF, 500 V ceramic disc (RS 272-131).
C6—47 µF, 25 V electrolytic (RS 272-1027).
C7, C9—22 pF, DM-15 dipped mica.
C8—39 pF, DM-15 dipped mica.
C10—5 pF, DM-15 dipped mica.
D1, D2—1N4001 diode.

D3, D4—1N914 diode.
D5—1N5401 diode.
F1—10 A for 25 W, 15 A for 50 W.
K1, K2—221D012 relay, RF Parts.
FB1, FB2, FB3, FB4—56590-65/3B Ferroxcube, Communications Concepts Inc (CCI).
L1, L2—VK200-20/4B Ferroxcube ferrite choke, CCI.
L3, L4—70.5 nH, 7 t #20 AWG, 0.125 inch ID, 0.33 inch long, 0.10 inch leads.

Q1—2N2222.
R1, R2, R3—see Table 13.12.
S1—20 A at 12 V.
U1—Power module, Toshiba S-AV7 (25 W) or S-AV17 (50 W).
DS1, DS2—Pilot and transmit indicator lamps, 13.8 V.
J1, J2—connectors, BNC or UHF as desired.
Coax—RG-8X, 2 ft.

fied power ratings for the T pad.

CAUTION: Some HTs can generate a momentary high power output pulse when keyed in the low power mode, especially when first keyed. Such a spike could exceed the amplifier module specified input limits if a low-power T-pad is selected. Observe the HT output on an oscilloscope or check with its manufacturer to make sure it doesn't exhibit this characteristic.

The output circuitry is a low-pass filter consisting of C7, C8, C9, L3 and L4. **Fig 13.56** shows the filtered output to be better than –60 dBc, the FCC requirement for spectral purity for a transmitter at this frequency and power class.

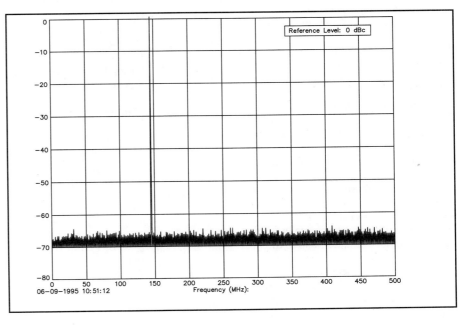

Fig 13.56—Plot of output spectrum for 25 W Brick Amp showing that it meets current FCC specifications for output purity, with second harmonic reduced by more than 60 dB.

The TR Switching Circuit

While in the receive mode, signals from the antenna are applied to the receiver through the normally closed relay contacts of K2-1 and K1-1. The low-pass filter used in transmit remains in the circuit for receive. This is a useful feature, since many HTs are prone to overload from strong out-of-band signals, such as from nearby UHF-TV transmitters. Further, harmonics generated by the TR switching circuit are suppressed to better than 60 dBc in the bypass mode. When the HT is keyed, RF is applied to voltage doubler D3 and D4. The junction of D3 and D4 is biased to approximately 0.5 V by R4 and R5 to facilitate diode turn-on at low RF levels. Once transistor Q1 is turned on, K2 is energized. The K2-2 contacts then energize K1 by applying +13.8 V to its coil. The K1-2 relay contacts provide power for an optional transmit light. This two-relay system provides two features not possible with a single relay:

1. Improved isolation between the input and output relay contacts. Coupling between these contacts is sufficiently small to prevent the amplifier from oscillating. (Initial experiments in the Lab with a single two-pole relay resulted in an excellent 2-m oscillator.)

2. "Hot switching" the output relay is eliminated. Since K2 is activated first, RF cannot appear across the K1 output relay contacts until K2 is already closed. Contact bounce with output power applied is thus eliminated, resulting in improved switching reliability and enhanced contact life.

This sequence is reversed when switching from transmit to receive. Once the RF is removed, Q1 is cut off and relay K2 is deenergized. Diodes D1 and D2 protect against voltage spikes created by the relay coils as their magnetic fields collapse upon deactivation. K1 is returned to the receive mode by the opening of the K2-2 contacts.

Construction Details

A surface-mount circuit board was selected for the Brick Amp, with traces on the top of the board and ground plane on the underside. Components connected to RF ground (such as the output filter capacitors C7, C8 and C9), have their ground leads soldered on both top and bottom sides to provide good RF ground continuity. Otherwise, component leads are soldered directly to top-side PC traces. The PC board is mounted above the heat sink surface, using metal #6 flat washers as spacers. The amplifier module pins and their associated PC board pads are in close vertical alignment, eliminating excessive

bending of the pins at solder time.

The heat sink and case were selected on the basis of availability, ruggedness and heat dissipation ability. The heat sink may be overkill, especially for the 25 W version. It is, however, readily available and adequate for the task, even in a hot car on a summer day after a long-winded transmission (such as the author has been known to make on occasion). If you intend to mount the Brick Amp in a car trunk, the sharp edges of a heat sink could be hazardous to its other contents. Be sure to give your mounting options careful consideration before making your final decision.

The case is die-cast aluminum, strong enough to withstand the most severe abuse in a car trunk, or any other mounting spot you may select. If you are budget minded and have a big junk box, here is where cost saving substitutions may be made. The only critical aspects of the heat sink is that it present a flat surface upon which to mount the module, large enough to mount the PC board, and that it meet the heat dissipating requirements for the conditions in which you intend to use the Brick Amp.

CAUTION: Before considering a heat sink, make sure that the module lies perfectly flat against the sink's mounting surface. Attempts to mount a module on a surface that is not flat can cause permanent damage to the module!

Begin construction by mounting the module at the center of the heat sink. Using the module as a template, drill two holes with a #36 drill bit. Remove any burrs around the holes with an oversized drill, and thread them both with a #6 tap. Clean the holes and heat sink mounting surface with a rag. Lightly rub 400 to 600 grit emery cloth (or fine steel wool) across the module and heat sink surfaces. It is not necessary to remove the black finish from the aluminum of the heat sink shown in Fig 13.54. Clean both surfaces and screw holes with a suitable solvent, such as denatured alcohol or flux remover, to remove any dirt, grease or oil. Wipe the exposed surfaces with a clean cloth and let dry.

Apply a very thin coat of thermal conducting grease to module and heat-sink mating surfaces. Place two #6 mounting screws through two flat washers and two #6 solder lugs (with internal lock teeth) pointing toward the PC board, through the mounting flanges on each side of the module. Alternately increase the torque on each screw until full torque is achieved and wipe off excess thermal compound with a rag. Be careful not to bend the module leads during this process.

Next, slide the board in place, line up the pins with the correct pads, center the board and mark the four mounting holes onto the heat sink. Drill holes with a #36 bit and tap them for #6 screws. Deburr and clean with solvent as before. Next, solder jumper wires through the 6 holes so indicated. Bend as shown in the inset to the layout in **Fig 13.57**, and solder the wire on the top and bottom of the PC board. These connections tie the underside ground plane foil on the board to ground. Install the components with ground connection through their holes now.

Mount the board to the heat sink using #6 screws and two washers as spacers. Make sure that the soldered wires do not touch the heat sink as torque is applied to the screws. More washers may be added as necessary to prevent the solder connections from touching the heat sink, but you must use an equal amount for each screw. Solder the module pins to the appropriate PC board pads, but be sure to leave sufficient free pin length to account for flexing from temperature changes and vibration.

Solder the ground lug at the input side of the module to ground on top of the PCB. Solder a short piece of braid from the underside of the PCB to the ground lug at the output side of the module.

Carefully bend the pins on each of the relays outward. Pliers may be used to accomplish this, or you may wish to try gently pressing all four lead tips against a hard surface. It is not necessary that they be at right angles to the relay, but they must be sufficiently bent to permit surface mount soldering. Avoid any unnecessary reworking or bending of these pins. Before soldering the relays onto the board, make certain they are oriented correctly. Carefully line up each relay pin with its pad and solder it in place.

Continue soldering the components on the board as shown in Fig 13.57. L3 and L4 can both be wound on the shank of a $^1/_8$ inch drill bit. Other construction data for these coils is given in the parts list. Finally, install the coax and +12 V dc jumpers.

Fuse the Brick Amp with either a ready-made cable having an in-line fuse pair, or a fuse holder in the case. Temporarily connect the dc power cable (with fuses installed) to the proper PC board pads and test the Brick Amp for operation and function.

Assuming the Brick Amp works correctly, install its case. Cut a square hole in the case large enough to accommodate the PC board and module. Position the hole so that the heat sink will be centered on the case. Mark and drill and tap (#6) holes for four mounting screws.

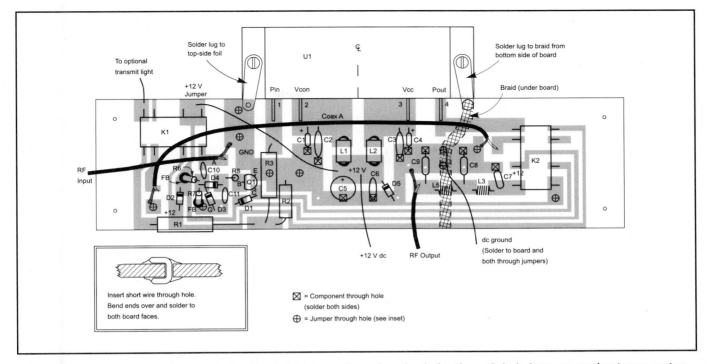

Fig 13.57—Part-placement diagram for the 2-m brick. Inset shows how leads for through-hole jumpers are bent over on top and bottom of PC board to provide good continuity for RF ground currents.

Install one or both of the optional lights, if desired. Make holes for both input and output connectors, dc power cable and on/off switch. (NOTE: A BNC for the input, and a UHF connector for the output facilitates coax connection to the HT and prevents accidental input/output cable reversal.) Tailor the input/output connectors and cabinet layout to suit your requirements. Install the heat sink and other components in the cabinet. Be sure to install the ferrite beads before soldering the wires and cables!

A PC-board template is in Chapter 30, **References**.

BIBLIOGRAPHY

Belcher, "RF Matching Techniques, Design and Example," *QST*, October 1972, pp 24-30.

Feynman, *Lectures on Physics, Vol. 1*, Addison-Wesley Publishing Co, 1977.

Goodman, "My Feed Line Tunes My Antenna," *QST*, April 1977, pp 40-42.

Granberg, "Build This Solid-State Titan," *QST*, June 1977, pp 27-31 (Part 1); *QST*, July 1977, pp 27-29 (Part 2). Granberg, "One KW-Solid-State Style," *QST*, April 1976, pp 11-14 (Part 1); *QST*, May 1976, pp 28-30 (Part 2).

Hejhall, "Broadband Solid-State Power Amplifiers for SSB Service," *QST*, March 1972, pp 36-43.

Johnson and Artigo, "Fundamentals of Solid-State Power-Amplifier Design," *QST*, September 1972, pp 29-36 (Part 1); *QST*, November 1972, pp 16-20 (Part 2); *QST*, April 1973, pp 28-34 (Part 3).

Johnson, "Heat Losses in Power Transformers," *QST*, May 1973, pp 31-34.

Knadle, "A Strip-line Kilowatt Amplifier 432 MHz," *QST*, April 1972, pp 49-55.

Meade, "A High-Performance 50-MHz Amplifier," *QST*, September 1975, pp 34-38.

Meade, "A 2-KW PEP Amplifier for 144 MHz," *QST*, December 1973, pp 34-38.

Measures, "Improved Parasitic Suppression for Modern Amplifier Tubes," *QST*, October 1988, pp 36-38, 66, 89.

Measures, "Parasitics Revisited," *QST*, September 1990, pp 15-18, October 1990, pp 32-35.

Olsen, "Designing Solid-State RF Power Circuits," *QST*, August 1977, pp 28-32 (Part 1); *QST*, September 1977, pp 15-18 (Part 2); *QST*, October 1977, pp 22-24 (Part 3).

Orr, *Radio Handbook*, 22nd Ed, Howard W. Sams & Co, Inc, 1981.

Potter and Fich, *Theory of Networks and Lines*, Prentice-Hall, Inc, 1963.

Reference Data for Radio Engineers, ITT, Howard & Sams Co, Inc.

RF Data Manual, Motorola, Inc, 1982.

Simpson, *Introductory Electronics for Scientists and Engineers*, Allyn and Bacon, Inc, 1975.

Solid State Power Circuits, RCA Designer's Handbook, 1972.

Terman, *Electronic and Radio Engineering*, McGraw-Hill Book Company, Inc. This book provides a lucid explanation of vacuum-tube amplifier design, explaining the current and voltage-source models very well.

White, "Thermal Design of Transistor Circuits," *QST*, April 1972, pp 30-34.

Wingfield, "New and Improved Formulas for the Design of Pi and Pi-L Networks," *QST*, August 1983, pp 23-29. (Feedback, *QST*, January 1984, p 49.)

Wingfield, "A Note on Pi-L Networks," *QEX*, December 1983, pp 5-9.

Contents

AC/RF Sources
(Oscillators and Synthesizers)

14

Just say in public that oscillators are one of the most important, fundamental building blocks in radio technology and you will immediately be interrupted by someone pointing out that *tuned-RF* (*TRF*) receivers can be built without any form of oscillator at all. This is certainly true, but it shows how some things can be taken for granted. What use is any receiver without signals to receive? All intentionally transmitted signals trace back to some sort of signal generator — an oscillator or frequency synthesizer. In contrast with the TRF receivers just mentioned, a modern, all-mode, feature-laden, MF/HF transceiver may contain in excess of a dozen RF oscillators and synthesizers, while a simple QRP CW transmitter may consist of nothing more than a single oscillator. (This chapter was written by David Stockton, GM4ZNX.)

In the 1980s, the main area of progress in the performance of radio equipment was the recognition of receiver intermodulation as a major limit to our ability to communicate, with the consequent development of receiver front ends with improved ability to handle large signals. So successful was this campaign that other areas of transceiver performance now require similar attention. One indication of this is any equipment review receiver dynamic range measurement qualified by a phrase like "limited by oscillator phase noise." A plot of a receiver's effective selectivity can provide another indication of work to be done: An IF filter's high-attenuation region may appear to be wider than the filter's published specifications would suggest — almost as if the filter characteristic has grown sidebands! In fact, in a

way, it has: This is the result of local-oscillator (LO) or synthesizer *phase noise* spoiling the receiver's overall performance. Oscillator noise is the prime candidate for the next major assault on radio performance.

The sheer number of different oscillator circuits can be intimidating, but their great diversity is an illusion that evaporates once their underlying pattern is seen. Almost all RF oscillators share one fundamental principle of operation: an amplifier and a filter operate in a loop (**Fig 14.1**). There are plenty of filter types to choose from:

- LC
- Quartz crystal and other piezoelectric materials
- Transmission line (stripline, microstrip, troughline, open-wire, coax and so on)
- Microwave cavities, YIG spheres, dielectric resonators
- Surface-acoustic-wave (SAW) devices

Should any new forms of filter be invented, it's a safe guess that they will also be applicable to oscillators. There is an equally large range of amplifiers to choose from:

- Vacuum tubes of all types
- Bipolar junction transistors
- Field effect transistors (JFET, MOSFET, GaAsFET, in all their varieties)
- Gunn diodes, tunnel diodes and other negative-resistance generators

It seems superfluous to state that anything that can amplify can be used in an oscillator, because of the well-known pro-

pensity of all prototype amplifiers to oscillate! The choice of amplifier is widened further by the option of using single- or multiple-stage amplifiers and discrete devices versus integrated circuits. Multiply all of these options with those of filter choice and the resulting set of combinations is very large, but a long way from complete. Then there are choices of how to couple the amplifier into the filter and the filter into the amplifier. And then there are choices to make in the filter section: Should it be tuned by variable capacitor, variable inductor or some form of sliding cavity or line?

Despite the number of combinations that are possible, a manageably small number of types will cover all but very special requirements. Look at an oscillator circuit and "read" it: What form of filter — *resonator* — does it use? What form of amplifier? How have the amplifier's input and output been coupled into the filter? How is the filter tuned? These are simple, easily answered questions that put oscillator types into appropriate catego-

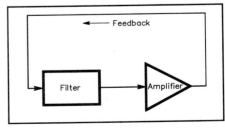

Fig 14.1 — Reduced to essentials, an oscillator consists of a filter and an amplifier operating in a feedback loop.

ries and make them understandable. The questions themselves may make more sense if we understand the mechanics of oscillation, in which *resonance* plays a major role.

HOW OSCILLATORS WORK

Maintained Resonance

The pendulum, a good example of a resonator, has been known for millennia and understood for centuries. It is closely analogous to an electronic resonator, as shown in **Fig 14.2**. The weighted end of the pendulum can store energy in two different forms: the kinetic energy of its motion and the potential energy of it being raised above its rest position. As it reaches its highest point at the extreme of a swing, its velocity is zero for an instant as it reverses direction. This means that it has, at that instant, no kinetic energy, but because it is also raised above its rest position, it has some extra potential energy.

At the center of its swing, the pendulum is at its lowest point with respect to gravity and so has lost the extra potential energy. At the same time, however, it is moving at its highest speed and so has its greatest kinetic energy. Something interesting is happening: The pendulum's stored energy is continuously moving between potential and kinetic forms. Looking at the pendulum at intermediate positions shows that this movement of energy is smooth. Newton provided the keys to understanding this. It took his theory of gravity and laws of motion to explain the behavior of a simple weight swinging on the end of a length of string and calculus to perform a quantitative mathematical analysis. Experiments had shown the period of a pendulum to be very stable and predictable. Apart from side effects of air drag and friction, the length of the period should not be affected by the mass of the weight, nor the amplitude of the swing.

A pendulum can be used for timing events, but its usefulness is spoiled by the action of drag or friction, which eventually stops it. This problem was overcome by the invention of the *escapement*, a part of a clock mechanism that senses the position of the pendulum and applies a small push in the right direction and at the right time to maintain the amplitude of its swing or oscillation. The result is a mechanical oscillator: The pendulum acts as the filter, the escapement acts as the amplifier and a weight system or wound-up spring powers the escapement.

Electrical oscillators are closely analogous to the pendulum, both in operation and in development. The voltage and current in the tuned circuit — often called *tank circuit* because of its energy-storage ability — both vary sinusoidally with time and are 90° out of phase. There are instants when the current is zero, so the energy stored in the inductor must be zero, but at the same time the voltage across the capacitor is at its peak, with all of the circuit's energy stored in the electric field between the capacitor's plates. There are also instants when the voltage is zero and the current is at a peak, with no energy in the capacitor. Then, all of the circuit's energy is stored in the inductor's magnetic field.

Just like the pendulum, the energy stored in the electrical system is swinging smoothly between two forms; electric field and magnetic field. Also like the pendulum, the tank circuit has losses. Its conductors have resistance, and the capacitor dielectric and inductor core are imperfect. Leakage of electric and magnetic fields also occurs, inducing currents in neighboring objects and just plain radiating energy off into space as radio waves. The amplitudes of the oscillating voltage and current decrease steadily as a result. Early intentional radio transmissions, such as those of Heinrich Hertz's experiments, involved abruptly dumping energy into a tuned circuit and letting it oscillate, or *ring*, as shown in **Fig 14.3**. This was done by applying a spark to the resonator. Hertz's resonator was a gapped ring, a good choice for radiating as much of the energy as possible. Although this looks very different from the LC tank of Fig 14.2, it has inductance *distributed* around its length and capacitance distributed across it and across its gap, as opposed to the *lumped* L and C values in

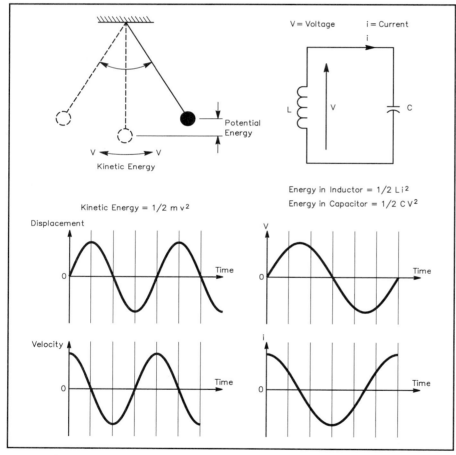

Fig 14.2 — A resonator lies at the heart of every oscillatory mechanical and electrical system. A mechanical resonator (here, a pendulum) and an electrical resonator (here, a tuned circuit consisting of L and C in parallel) share the same mechanism: the regular movement of energy between two forms — potential and kinetic in the pendulum, electric and magnetic in the tuned circuit. Both of these resonators share another trait: Any oscillations induced in them eventually die out because of losses — in the pendulum, due to drag and friction; in the tuned circuit, due to the resistance, radiation and inductance. Note that the curves corresponding to the pendulum's displacement vs velocity and the tuned circuit's voltage vs current, differ by one quarter of a cycle, or 90°.

Fig 14.2. The gapped ring therefore works just the same as the LC tank in terms of oscillating voltages and currents. Like the pendulum and the LC tank, its period, and therefore the frequency at which it oscillates, is independent of the magnitude of its excitation.

Making a longer-lasting signal with the Fig 14.3 arrangement merely involves repeating the sparks. The problem is that a truly continuous signal cannot be made this way. The sparks cannot be applied often enough or always timed precisely enough to guarantee that another spark reexcites the circuit at precisely the right instant. This arrangement amounts to a crude spark transmitter, variations of which served as the primary means of transmission for the first generation of radio amateurs. The use of damped waves is now entirely forbidden by international treaty because of their great impurity. Damped waves look a lot like car-ignition waveforms and sound like car-ignition interference when received.

What we need is a *continuous wave* (CW) oscillation — a smooth, sinusoidal signal of constant amplitude, without phase jumps, a "pure tone." To get it, we must add to our resonator an equivalent of the clock's escapement — a means of synchronizing the application of energy and a fast enough system to apply just enough energy every cycle to keep each cycle at the same amplitude.

Amplification

A sample of the tank's oscillation can be extracted, amplified and reinserted. The gain can be set to exactly compensate the tank losses and perfectly maintain the oscillation. The amplifier usually need only give low gain, so active devices can be used in oscillators not far below their unity (unity = 1) gain frequency. The amplifier's output must be lightly coupled into the tank — the aim is just to replace lost energy, not forcibly drive the tank. Similarly, the amplifier's input should not heavily load the tank. It is a good idea to think of *coupling* networks rather than *matching* networks in this application, because a matched impedance extracts the maximum available energy from a source, and this would certainly spoil an oscillator.

Fig 14.4A shows the block diagram of an oscillator. Certain conditions must be met for oscillation. The criteria that separate oscillator loops from stable loops are often attributed to Barkhausen by those aiming to produce an oscillator and to Nyquist by those aiming for amplifier stability, although they boil down to the same boundary. Fig 14.4B shows the loop

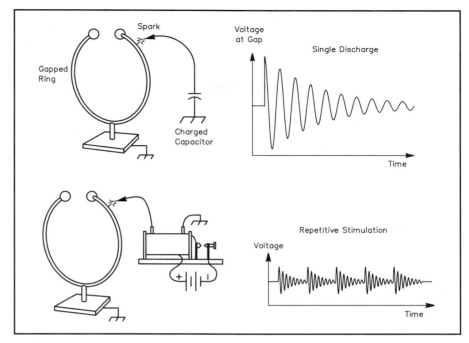

Fig 14.3 — Stimulating a resonance, 1880s style. Shock-exciting a gapped ring with high voltage from a charged capacitor causes the ring to oscillate at its resonant frequency. The result is a damped wave, each successive alternation of which is weaker than its predecessor because of resonator losses. Repetitively stimulating the ring produces trains of damped waves, but oscillation is not continuous.

broken and a test signal inserted. (The loop can be broken anywhere; the amplifier input just happens to be the easiest place to do it.) The criterion for oscillation says that at a frequency at which the phase shift around the loop is exactly zero, the net gain around the loop must equal or exceed unity (that is, one). (Later, when we design phase-locked loops, we must revisit this concept in order to check that *those* loops *cannot* oscillate.) Fig 14.4C shows what happens to the amplitude of an oscillator if the loop gain is made a little higher or lower than one.

The loop gain has to be precisely one if we want a stable amplitude. Any inaccuracy will cause the amplitude to grow to clipping or shrink to zero, making the oscillator useless. Better accuracy will only slow, not stop this process. Perfect precision is clearly impossible, yet there are enough working oscillators in existence to prove that we are missing something important. In an amplifier, nonlinearity is a nuisance, leading to signal distortion and intermodulation, yet nonlinearity is what makes stable oscillation possible. All of the vacuum tubes and transistors used in oscillators tend to reduce their gain at higher signal levels. With such components, only a tiny change in gain can shift the loop's operation between amplitude

growth and shrinkage. Oscillation stabilizes at that level at which the gain of the active device sets the loop gain at exactly one.

Another gain-stabilization technique involves biasing the device so that once some level is reached, the device starts to turn off over part of each cycle. At higher levels, it cuts off over more of each cycle. This effect reduces the effective gain quite strongly, stabilizing the amplitude. This badly distorts the signal (true in most common oscillator circuits) in the amplifying device, but provided the amplifier is lightly coupled to a high-Q resonant tank, the signal in the tank should not be badly distorted.

Many radio amateurs now have some form of circuit-analysis software, usually running on a PC. Attempts to analyze oscillators by this means often fail by predicting growing or shrinking amplitudes, and often no signal at all, in circuits that are known to work. Computer analysis of oscillator circuits can be done, but it requires a sophisticated program with accurate, nonlinear, RF-valid models of the devices used, to be able to predict operating amplitude. Often even these programs need some special tricks to get their modeled oscillators to start. Such software is likely to

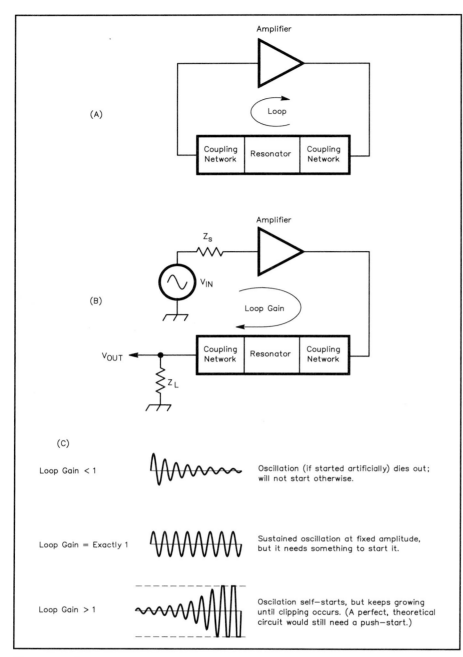

Fig 14.4 — The bare-bones oscillator block diagram of Fig 14.1 did not include two practical essentials: networks to couple power in and out of the resonator (A). Breaking the loop, inserting a test signal and measuring the loop's overall gain (B) allows us to determine whether the system can oscillate, sustain oscillation or clip (C).

be priced higher than most private users can justify, and it still doesn't replace the need for the user to understand the circuit. With that understanding, some time, some parts and a little patience will do the job, unassisted.

Textbooks give plenty of coverage to the frequency-determining mechanisms of oscillators, but the amplitude-determining mechanism is rarely covered. It is often not even mentioned. There is a

good treatment in Clarke and Hess, *Communications Circuits: Analysis and Design*.

Start-Up

Perfect components don't exist, but if we could build an oscillator from them, we would naturally expect perfect performance. We would nonetheless be disappointed. We could assemble from our perfect components an oscillator that exactly

met the criterion for oscillation, having slightly excessive gain that falls to the correct amount at the target operating level and so is capable of sustained, stable oscillation. But being capable of something is not the same as doing it, for there is another stable condition. If the amplifier in the loop shown in **Fig 14.4** has no input signal, and is perfect, it will give no output! No signal returns to the amplifier's input via the resonator, and the result is a sustained and stable *lack* of oscillation. Something is needed to start the oscillator.

This fits the pendulum-clock analogy: A wound-up clock is stable with its pendulum at rest, yet after a push the system will sustain oscillation. The mechanism that drives the pendulum is similar to a Class C amplifier: It does not act unless it is driven by a signal that exceeds its threshold. An electrical oscillator based on a Class C amplifier can sometimes be kicked into action by the turn-on transient of its power supply. The risk is that this may not always happen, and also that should some external influence stop the oscillator, it will not restart until someone notices the problem and cycles the power. This can be very inconvenient!

A real-life oscillator whose amplifier does not lose gain at low signal levels can self-start due to noise. **Fig 14.5** shows an oscillator block diagram with the amplifier's noise shown, for our convenience, as a second input that adds with the true input. The amplifier amplifies the noise. The resonator filters the output noise, and this signal returns to the amplifier input.. The importance of having slightly excessive gain until the oscillator reaches operating amplitude is now obvious. If the loop gain is slightly above one, the recirculated noise must, within the resonator's bandwidth, be larger than its original level at the input. More noise is continually summed in as a noise-like signal continuously passes around the loop, undergoing amplification and filtering as it does. The level increases, causing the gain to reduce. Eventually, it stabilizes at whatever level is necessary to make the net loop gain equal to one.

So far, so good. The oscillator is running at its proper level, but something seems very wrong. It is not making a proper sine wave; it is recirculating and filtering a noise signal. It can also be thought of as a Q multiplier with a controlled (high) gain, filtering a noise input and amplifying it to a set level. Narrow-band filtered noise approaches a true sine wave as the filter is narrowed to zero width. What this means is that we cannot make a true sine-wave signal — all we can do is make narrow-band filtered noise as an approximation to one. A high-quality, low-noise oscillator is merely one that does tighter filtering. Even a kick-started Class-C-amplifier oscillator has noise continuously entering its circulating signal, and so behaves similarly.

A small-signal gain greater than one is absolutely critical for reliable starting, but having too much gain can make the final operating level unstable. Some oscillators are designed around limiting amplifiers to make their operation predictable. AGC systems have also been used, with an RF detector and dc amplifier used to servo-control the amplifier gain. It is notoriously difficult to design reliable crystal oscillators that can be published or mass-produced without having occasional individuals refuse to start without some form of shock.

Mathematicians have been intrigued by "chaotic systems" where tiny changes in initial conditions can yield large changes in outcome. The most obvious example is meteorology, but much of the necessary math was developed in the study of oscillator start-up, because it is a case of chaotic activity in a simple system. The equations that describe oscillator start-up are similar to those used to generate many of the popular, chaotic fractal images.

PHASE NOISE

Viewing an oscillator as a filtered-noise generator is relatively modern. The older approach was to think of an oscillator making a true sine wave with an added, unwanted noise signal. These are just different ways of visualizing the same thing: They are equally valid views, which are used interchangeably, depending which best makes some point clear. Thinking in terms of the signal-plus-noise version, the noise surrounds the carrier, looking like sidebands and so can also be considered to be equivalent to random-noise FM and AM on the ideal sine-wave signal. This gives us a third viewpoint. Strangely, these noise sidebands are called *phase noise*. If we consider the addition of a noise voltage to a sinusoidal voltage, we must take into account the phase relationship. A *phasor diagram* is the clearest way of illustrating this. **Fig 14.6A** represents a clean sine wave as a rotating vector whose length is equal to the peak amplitude and whose frequency is equal to the number of revolutions per second of its rotation. Moving things are difficult to depict on paper, so phasor diagrams are usually drawn to show the dominant signal as stationary, with other components drawn relative to this.

Noise contains components at many

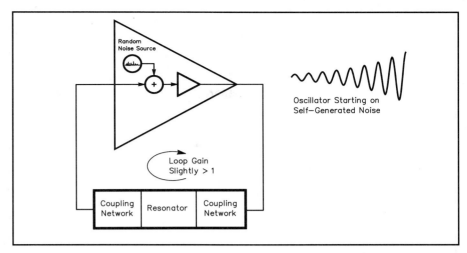

Fig 14.5 — An oscillator with noise. Real-world amplifiers, no matter how quiet, generate some internal noise; this allows real-world oscillators to self-start.

frequencies, so its phase with respect to the dominant, theoretically pure signal — the "carrier" — is random. Its amplitude is also random. Noise can only be described in statistical terms because its voltage is constantly and randomly changing, yet it does have an average amplitude that can be expressed in RMS volts. Fig 14.6B shows noise added to the carrier phasor, with the noise represented as a fuzzy, uncertain region in which the sum phasor wanders randomly. The phase of the noise is uniformly random — no direction is more likely than any other — but the instantaneous magnitude of the noise obeys a probability distribution like that shown, higher values being progressively rarer. Fig 14.6C shows how the extremities of the noise region can be considered as extremes of phase and amplitude variation from the normal values of the carrier.

Phase modulation and frequency modulation are closely related. Phase is the integral of frequency, so phase modulation resembles frequency modulation in which deviation decreases with increasing modulating frequency. Thus, there is no need to talk of "frequency noise" because *phase noise* already covers it.

Fig 14.6C clearly shows AM noise as the random variation of the length of the sum phasor, yet "amplitude noise" is rarely discussed. The oscillator's amplitude control mechanism acts to reduce the AM noise by a small amount, but the main reason is that the output is often fed into some form of limiter that strips off AM components just as the limiting IF amplifier in an FM receiver removes any AM on incoming signals. The limiter can be obvious, like a circuit to convert the signal to logic levels, or it can

be implicit in some other function. A diode ring mixer may be driven by a sine-wave LO of moderate power, yet this signal drives the diodes hard on and hard off, approximating square-wave switching. This is a form of limiter, and it removes the effect of any AM on the LO. Fig 14.6D shows the result of passing a signal with noise through a limiting amplifier. For these reasons, AM noise sidebands are rarely a problem in oscillators, and so are normally ignored. There is one subtle problem to beware of, however. If a sine wave drives some form of switching circuit or limiter, and the threshold is offset from the signal's mean voltage, any level changes will affect the exact switching times and cause *phase jitter*. In this way, AM sidebands are translated into PM sidebands and pass through the limiter. This is usually called *AM-to-PM conversion* and is a classic problem of limiters.

Effects of Phase Noise

You would be excused for thinking that phase noise is a recent discovery, but all oscillators have always produced it. Other changes have elevated an unnoticed characteristic up to the status of a serious impairment. Increased crowding and power levels on the ham bands, allied with greater expectations of receiver performance as a result of other improvements, have made phase noise more noticeable, but the biggest factor has been the replacement of VFOs in radios by frequency synthesizers. It is a major task to develop a synthesizer that tunes in steps fine enough for SSB and CW while competing with the phase-noise performance of a reasonable-quality LC VFO. Many synthesizers have fallen far short of that target. Phase noise is worse in higher-frequency oscillators

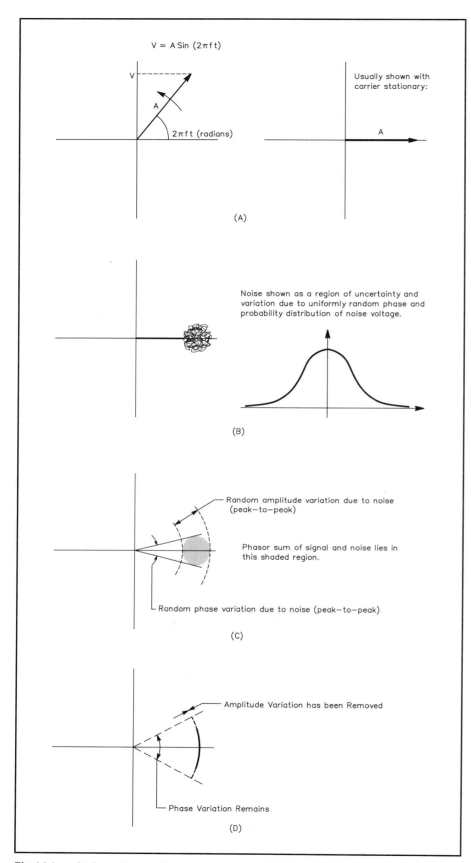

Fig 14.6 — At A, a phasor diagram of a clean (ideal) oscillator. Noise creates a region of uncertainty in the vector's length and position (B). AM noise varies the vector's length; PM noise varies the vector's relative angular position (C) Limiting a signal that includes AM and PM noise strips off the AM and leaves the PM (D).

and the trend towards general-coverage, upconverting structures has required that local oscillators operate at higher and higher frequencies. **Fig 14.7A** shows sketches of the relative phase-noise performance of some oscillators. The very high Q of the quartz crystal in a crystal oscillator gives it the potential for much lower phase noise than LC oscillators. A medium-quality synthesizer has close-in phase noise performance approaching that of a crystal oscillator, while further from the carrier frequency, it degrades to the performance of a modest-quality LC oscillator. There may be a small bump in the noise spectrum at the boundary of these two zones. A bad synthesizer can have extensive noise sidebands extending over many tens (hundreds in extreme cases) of kilohertz at high levels.

Phase noise on a transmitter's local oscillators passes through its stages, is amplified and fed to the antenna along with the intentional signal. The intentional signal is thereby surrounded by a band of noise. This radiated noise exists in the same proportion to the transmitter power as the phase noise was to the oscillator power if it passes through no narrow-band filtering capable of limiting its bandwidth. This radiated noise makes for a noisier band. In bad cases, nearby stations can be unable to receive over many tens of kilohertz. Fig 14.7B illustrates the difference between clean and dirty transmitters.

The effects of receiver-LO phase noise are more complicated, but at least it doesn't affect other stations' reception. The process is called *reciprocal mixing*.

Reciprocal Mixing

This is an effect that occurs in all mixers, yet despite its name, reciprocal mixing is an LO, not a mixer, problem. Imagine that the outputs of two supposedly unmodulated signal generators are mixed together and the mixer output is fed into an FM receiver. The receiver produces sounds, indicating that the resultant signal is modulated nonetheless. Which signal generator is responsible? This is a trick question, of course. A moment spent thinking about how a change in the frequency of either input signal affects the output signal will show that FM or PM on either input reaches the output. The best answer to the trick question is therefore "either or both."

The modulation on the mixer output is the combined modulations of the inputs. This means that modulating a receiver's local oscillator is indistinguishable from using a clean LO and having the same modulation present on *all* incoming signals. (This is also true for AM provided that a fully linear multiplier is used as the mixer, but mixers are commonly driven into switching, which strips any AM off the LO signal. This is the chief reason why the phase component of oscillator noise is more important than any AM component.)

The word *indistinguishable* is important in the preceding paragraph. It does not mean that the incoming signals are themselves modulated, but that the signals in the receiver IF and the noise in the IF, sound exactly as if they were. What really happens is that the noise components of the LO are extra LO signals that are offset from the carrier frequency. Each of them mixes other signals that are appropriately offset from the LO carrier into the receiver's IF. Noise is the sum of an infinite number of infinitesimal components spread over a range of frequencies, so the signals it mixes into the IF are spread into an infinite number of small replicas, all at different frequencies. This amounts to scrambling these other signals into noise. It is tedious to look at the effects of receiver LO phase noise this way. The concept of reciprocal mixing gives us an easier, alternative view that is much more digestible and produces identical results.

A poor oscillator can have significant noise sidebands extending out many tens of kilohertz on either side of its carrier. This is the same, as far as the signals in the receiver IF are concerned, as if the LO was clean and every signal entering the mixer had these noise sidebands. Not only will the wanted signal (and its noise sidebands) be received, but the noise sidebands added by the LO to signals near, but outside, the receiver's IF passband will overlap it. If the band is busy, each of many signals present will add its set of noise sidebands — and the effect is cumulative. This produces the appearance of a high background-noise level on the band. Many hams tend to accept this, blaming "conditions."

Hams now widely understand reception problems due to intermodulation, and almost everyone knows to apply RF attenuation until the signal gets no clearer. Intermodulation is a nonlinear effect, and the levels of the intermod products fall by greater amounts than the reduction in the intermodulating signals. The net result is less signal, but with the intermodulation products dropped still further. This improvement reaches a limit when more attenuation pushes the desired signal too close to the receiver's noise floor.

Reciprocal mixing is a linear process, and the mixer applies the same amount of noise "deviation" to incoming signals as that present on the LO. Therefore the ratio of noise-sideband power to signal power

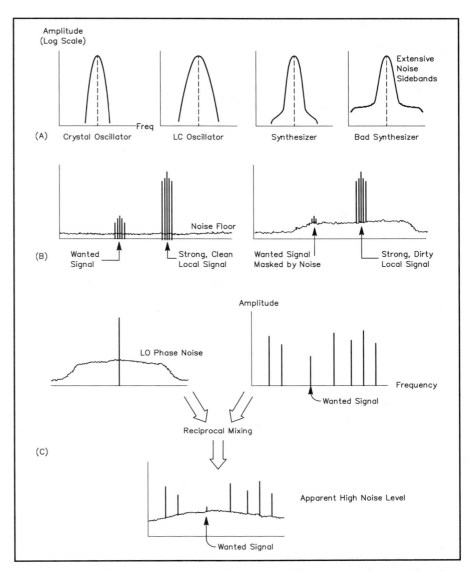

Fig 14.7 — The effects of phase noise. At A, the relative phase-noise spectra of several different signal-generation approaches. At B, how transmitted phase noise degrades the weak-signal performance even of nearby receivers with phase-quiet oscillators, raising the effective noise floor. What is perhaps most insidious about phase noise in Amateur Radio communication is that its presence in a receiver LO can allow strong, clean transmitted signals to degrade the receiver's noise floor just as if the transmitted signals were dirty. This effect, reciprocal mixing, is shown at C.

is the same for each signal, and the same as that on the LO. Switching in RF attenuation reduces the power of signals entering the mixer, but the reciprocal mixing process still adds the noise sidebands at the same *relative* power to each. Therefore, no reception improvement results. Other than building a quieter oscillator, the only way of improving things is to use narrow preselection to band-limit the receiver's input response and reduce the number of incoming signals subject to reciprocal mixing. This reduces the number of times the phase noise sidebands get added into the IF signal. Commercial *tracking*

preselectors — selective front-end circuits that tune in step with a radio's band changes and tuning, are expensive, but one that is manually tuned would make a modest-sized home-brew project and could also help reduce intermodulation effects. When using a good receiver with a linear front end and a clean LO, amateurs accustomed to receivers with poor phase-noise performance report the impression is of a seemingly emptier band with gaps between signals — and then they begin to find readable signals in some of the gaps.

Fig 14.7C shows how a noisy oscillator

affects transmission and reception. The effects on reception are worst in Europe, on 40 m, at night. Visitors from North America, and especially Asia, are usually shocked by the levels of background noise. In ITU Region 1, the Amateur Radio 40-m allocation is 7.0 to 7.1 MHz; above this, ultra-high-power broadcasters operate. The front-end filters in commercial ham gear are usually fixed band-pass designs that cover the wider 40-m allocations in the other regions. This allows huge signals to reach the mixer and mix large levels of LO phase noise into the IF. Operating colocated radios, on the same band, in a multioperator contest, requires linear front ends, preselection and state-of-the-art phase-noise performance. Outside of amateur circles, only warship operation is more demanding, with kilowatt transmitters and receivers sharing antennas on the same mast.

A Phase Noise Demonstration

Healthy curiosity demands some form of demonstration so the scale of a problem can be judged "by ear" before measurements are attempted. We need to be able to measure the noise of an oscillator alone (to aid in the development of quieter ones) and we also need to be able to measure the phase noise of the oscillators in a receiver (a transmitter can be treated as an oscillator). Conveniently, a receiver contains most of the functions needed to demonstrate its own phase noise.

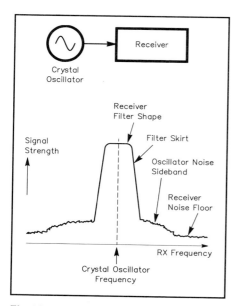

Fig 14.8 — Tuning in a strong, clean crystal-oscillator signal can allow you to hear your receiver's relative phase-noise level. Listening with a narrow CW filter switched in allows you to get better results closer to the carrier.

No mixer has perfect port-to-port isolation, and some of its local-oscillator signal leaks through into the IF. If we tune a general-coverage receiver, with its antenna disconnected, to exactly 0 Hz, the local oscillator is exactly at the IF center frequency, and the receiver acts as if it is tuned to a very strong unmodulated carrier. A typical mixer might give only 40 dB of LO isolation and have an LO drive power of at least 10 mW. If we tune away from 0 Hz, the LO carrier tunes away from the IF center and out of the passband. The apparent signal level falls. Although this moves the LO carrier out of the IF passband, some of its noise sidebands will not be, and the receiver will respond to this energy as an incoming noise signal. To the receiver operator, this sounds like a rising noise floor as the receiver is tuned toward 0 Hz. To get good noise floor at very low frequencies, some professional/military receivers, like the Racal RA1772, use very carefully balanced mixers to get as much port-to-port isolation as possible, and they also may switch a crystal notch filter into the first mixer's LO feed.

This demonstration cannot be done if the receiver tunes amateur bands only. As it is, most general-coverage radios inhibit tuning in the LF or VLF region. It could be suggested by a cynic that how low manufacturers allow you to tune is an indication of how far they think their phase-noise sidebands could extend!

The majority of amateur transceivers with general-coverage receivers are programmed not to tune below 30 to 100 kHz, so means other than the "0 Hz" approach are needed to detect LO noise in these radios. Because reciprocal mixing adds the LO's sidebands to clean incoming signals, in the same proportion to the incoming carrier as they exist with respect to the LO carrier, all we need do is to apply a strong, clean signal wherever we want within the receiver's tuning range. This signal's generator must have lower phase noise than the radio being evaluated. A general-purpose signal generator is unlikely to be good enough; a crystal oscillator is needed.

It's appropriate to set the level into the receiver to about that of a strong broadcast carrier, say S9 + 40 dB. Set the receiver's mode to SSB or CW and tune around the test signal, looking for an increasing noise floor (higher hiss level) as you tune closer towards the signal, as shown in **Fig 14.8**. Switching in a narrow CW filter allows you to hear noise closer to the carrier than is possible with an SSB filter. This is also the technique used to measure a receiver's effective selectivity, and some equipment reviewers kindly publish their plots in this

format. *QST* reviews, done by the ARRL Lab, often include the results of specific phase-noise measurements.

Measuring Receiver Phase Noise

There are several different ways of measuring phase noise, offering different trade-offs between convenience, cost and effort. Some methods suit oscillators in isolation, others suit them in situ (in their radios).

If you're unfamiliar with noise measurements, the units involved may seem strange. One reason for this is that a noise signal's power is spread over a frequency range, like an infinite number of infinitesimal sinusoidal components. This can be thought of as similar to painting a house. The area that a gallon of paint can cover depends on how thinly it's spread. If someone asks how much paint was used on some part of a wall, the answer would have to be in terms of paint volume per square foot. The wall can be considered to be an infinite number of points, each with an infinitesimal amount of paint applied to it. The question of what volume of paint has been applied at some specific point is unanswerable. With noise, we must work in terms of *power density*, of watts per hertz. We therefore express phase-noise level as a ratio of the carrier power to the noise's power density. Because of the large ratios involved, expression in decibels is convenient. It has been a convention to use *dBc* to mean "decibels with respect to the carrier."

For phase noise, we need to work in terms of a standard bandwidth, and 1 Hz is the obvious candidate. Even if the noise is measured in a different bandwidth, its equivalent power in 1 Hz can be easily calculated. A phase-noise level of 120 dBc in a 1-Hz bandwidth (often written as *120 dBc/Hz*) translates into each hertz of the noise having a power of 10^{-12} of the carrier power. In a bandwidth of 3 kHz, this would be 3000 times larger.

The most convenient way to measure phase noise is to buy and use a commercial phase noise test system. Such a system usually contains a state-of-the-art, low-noise frequency synthesizer and a low-frequency spectrum analyzer, as well as some special hardware. Often, a second, DSP-based spectrum analyzer is included to speed up and extend measurements very close to the carrier by using the Fast Fourier Transform (FFT). The whole system is then controlled by a computer with proprietary software. With a good system like this costing about $100,000, this is not a practical method for amateurs, although a few fortunate individuals have access to them at work. These systems are also overkill for our needs, because we are not particularly interested in determining phase-noise levels very close to and very far from the carrier.

It's possible to make respectable receiver-oscillator phase-noise measurements with less than $100 of parts and a multimeter. Although it's time-consuming, the technique is much more in keeping with the amateur spirit than using a $100k system! An ordinary multimeter will produce acceptable results, a meter capable of indicating "true RMS" ac voltages is preferable because it can give correct readings on sine waves *and* noise. **Fig 14.9** shows the setup. Measurements can only be made around the frequency of the crystal oscillator, so if more than one band is to be tested, crystals must be changed, or else a set of appropriate oscillators is needed. The oscillator should produce about +10 dBm (10 mW) and be *very* well shielded. (To this end, it's advisable to build the oscillator into a die-cast box and power it from internal batteries. A noticeable shielding improvement results even from avoiding the use of an external power switch; a reed-relay element inside the box can be positioned to connect the battery when a small permanent magnet is placed against a marked place outside the box.)

Likewise, great care must be taken with attenuator shielding. A total attenuation of around 140 dB is needed, and with so much attenuation in line, signal leakage can easily exceed the test signal that reaches the receiver. It's not necessary to be able to switch-select all 140 dB of the attenuation, nor is this desirable, as switches can leak. (The 1995 and older editions of this *Handbook* contain a step attenuator that's satisfactory.) All of the attenuators' enclosure seams must be soldered. A pair of boxes with 30 dB of fixed attenuation each are needed to complete the set. With 140 dB of attenuation, coax cable leakage is also a problem. The only countermeasure against this is to minimize all cable lengths and to interconnect test-system modules with BNC plug-to-plug adapters (UG-491As) where possible.

Ideally, the receiver could simply be tuned across the signal from the oscillator and the response measured using its signal-strength (S) meter. Unfortunately, receiver S meters are notoriously imprecise, so an equivalent method is needed that does not rely on the receiver's AGC system.

The trick is not to measure the response to a fixed level signal, but to measure the changes in applied signal power needed to give a fixed response. Here is a step-by-

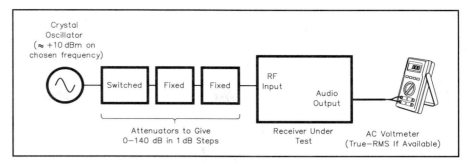

Fig 14.9 — Setup for measuring receiver-oscillator phase noise.

step procedure based on that described by John Grebenkemper, KI6WX, in March and April 1988 *QST*:

1. Connect the equipment as shown in Fig 14.9, but with the crystal oscillator off. Set the step attenuator to maximum attenuation. Set the receiver for SSB or CW reception with its narrowest available IF filter selected. Switch out any internal preamplifiers or RF attenuators. Select AGC off, maximum AF and RF gain. It may be necessary to reduce the AF gain to ensure the audio amplifier is at least 10 dB below its clipping point. The ac voltmeter or an oscilloscope on the AF output can be used to monitor this.

2. To measure noise, it is important to know the bandwidth being measured. A true-RMS ac voltmeter measures the power in the noise reaching it. To calculate the noise density, we need to divide by the receiver's *noise bandwidth*. The receiver's –6-dB IF bandwidth can be used as an approximation, but purists will want to plot the top 20 dB of the receiver's bandwidth on linear scales and integrate the area under it to find the width of a rectangle of equal area and equal height. This accounts properly for the noise in the skirt regions of the overall selectivity. (The very rectangular shape of common receiver filters tends to minimize the error of just taking the approximation.)

Switch on the test oscillator and set the attenuators to give an AF output above the noise floor and below the clipping level with the receiver peaked on the signal. Tune the receiver off to each side to find the frequencies at which the AF voltage is half that at the peak. The difference between these is the receiver's –6-dB bandwidth. High accuracy is not needed: 25% error in the receiver bandwidth will only cause a 1-dB error in the final result. The receiver's published selectivity specifications will be close enough. The benefit of integration is greater if the receiver has a very rounded, low-ringing or low-order filter.

3. Retune the receiver to the peak.

Switch the oscillator off and note the noise-floor voltage. Turn the oscillator back on and adjust the attenuator to give an AF output voltage 1.41 times (3 dB) larger than the noise floor voltage. This means that the noise power and the test signal power at the AF output are equal — a value that's often called the *MDS* (*minimum discernible signal*) of a receiver. Choosing a test-oscillator level at which to do this test involves compromise. Higher levels give more accurate results where the phase noise is high, but limit the lowest level of phase noise that can be measured because better receiver oscillators require a greater input signal to produce enough noise to get the chosen AF-output level. At some point, either we've taken all the attenuation out and our measurement range is limited by the test oscillator's available power, or we overload the receiver's front end, spoiling the results.

Record the receiver frequency at the peak, (f_0), the attenuator setting (A_0) and the audio output voltage (V_0). These are the carrier measurements against which all the noise measurements will be compared.

4. Now you must choose the offset frequencies — the spacings from the carrier — at which you wish to make measurements. The receiver's skirt selectivity will limit how close to the carrier noise measurements can be made. (Any measurements made too close in are valid measurements of the receiver selectivity, but because the signal measured under these conditions is sinusoidal and not noise like, the corrections for noise density and noise bandwidth are not appropriate.) It is difficult to decide where the filter skirt ends and the noise begins, and what corrections to apply in the region of doubt and uncertainty. A good practical approach is to listen to the audio and tune away from the carrier until you can't distinguish a tone in the noise. The ear is superb at spotting sine tones buried in noise, so this criterion, although subjective, errs on the conservative side.

Table 14.1

SSB Phase Noise of ICOM IC-745 Receiver Section

Oscillator output power = –3 dBm (0.5 mW)
Receiver bandwidth (Δf) = 1.8 kHz
Audio noise voltage –0.070 V
Audio reference voltage (V_0) = 0.105 V
Reference attenuation (A_0) = 121 dB

Offset Frequency (kHz)	Attenuation (A_1) (dB)	Audio V_1 (volts)	Audio V_2 (volts)	Ratio V_2/V_1	SSB Phase Noise (dBc/Hz)
4	35	0.102	0.122	1.20	–119
5	32	0.104	0.120	1.15	–122*
6	30	0.104	0.118	1.13	–124*
8	27	0.100	0.116	1.16	–127*
10	25	0.106	0.122	1.15	–129*
15	21	0.100	0.116	1.16	–133*
20	17	0.102	0.120	1.18	–137
25	14	0.102	0.122	1.20	–140
30	13	0.102	0.122	1.20	–141
40	10	0.104	0.124	1.19	–144
50	8	0.102	0.122	1.20	–146
60	6	0.104	0.124	1.19	–148
80	4	0.102	0.126	1.24	–150
100	3	0.102	0.126	1.24	–151
150	3	0.102	0.124	1.22	–151
200	0	0.104			–154
250	0	0.100			–154
300	0	0.98			–154
400	0	0.96			–154
500	0	0.96			–154
600	0	0.97			–154
800	0	0.96			–154
1000	0	0.96			–154

*Asterisks indicate measurements possibly affected by receiver overload (see text).

Tune the receiver to a frequency offset from f_0 by your first chosen offset and adjust the attenuators to get an audio output voltage as close as possible to V_0. Record the total attenuation, A_1 and the audio output voltage, V_1. The SSB phase noise (qualified as *SSB* because we're measuring the phase noise on only one side of the carrier, whereas some other methods cannot segregate between upper and lower noise sidebands and measure their sum, giving *DSB* phase noise) is now easy to calculate:

$$L(f) = A_1 - A_0 - 10 \log(BW_{noise}) \qquad (1)$$

where

$L(f) = $ SSB phase noise in dBc/Hz
$BW_{noise} = $ receiver noise bandwidth, Hz.

5. It's important to check for overload. Decrease the attenuation by 3 dB, and record the new audio output voltage, V_2. If all is well, the output voltage should increase by 22% (1.8 dB); if the receiver is operating nonlinearly, the increase will be less. (An 18% increase is still acceptable for the overall accuracy we want.) Record V_2/V_1 as a check: a ratio of 1.22:1 is ideal, and anything less than 1.18:1 indicates a bad measurement.

If too many measurements are bad, you may be overdriving the receiver's AF

Transmitter Phase-Noise Measurement in the ARRL Lab

Here is a brief description of the technique used in the ARRL Lab to measure transmitter phase noise. The system essentially consists of a direct-conversion receiver with very good phase-noise characteristics. As shown in Fig B, we use an attenuator after the transmitter, a Mini-Circuits ZAY-1 mixer, a Hewlett-Packard 8640B signal generator, a band-pass filter, an audio-frequency low-noise amplifier and a spectrum analyzer (HP 8563E) to make the measurements.

The transmitter signal is mixed with the output of the signal generator, and signals produced in the mixing process that are not required for the measurement process are filtered out. The spectrum analyzer then displays the transmitted phase-noise spectrum. The 100 mW output of the HP 8640B is barely enough to drive the mixer—the setup would work better with 200 mW of drive. To test the phase noise of an HP 8640B, we use a second '8640B as a reference source. It is quite important to be sure that the phase noise of the reference source is lower than that of the signal under test, because we are really measuring the combined phase-noise output of the signal generator and the transmitter. It would be quite embarrassing to publish phase-noise plots of the reference generator instead of the transmitter under test! The HP 8640B has much cleaner spectral output than most transmitters.

A sample phase-noise plot for an amateur transceiver is shown in Fig A. It was produced with the test setup shown in Fig B. Measurements from multiple passes are taken and averaged. A 3-Hz video bandwidth also helps average and smooth the plots. These plots do not necessarily reflect the phase-noise characteristics of all units of a particular model.

The log reference level (the top horizontal line on the

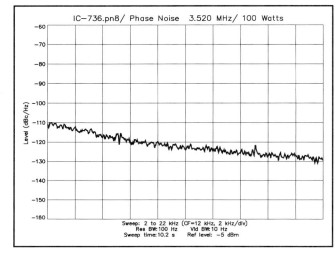

Fig A —Sample phase noise plot for an amateur transceiver.

amplifier, so try reducing the AF gain and starting again back at Step 3. If this doesn't help, reducing the RF gain and starting again at Step 3 should help if the compression is occurring late in the IF stages.

6. Repeat Steps 4 and 5 at all the other offsets you wish to measure. If measurements are made at increments of about half the receiver's bandwidth, any discrete (non-noise) spurs will be found. A noticeable tone in the audio can indicate the presence of one of these. If it is well clear of the noise, the measurement is valid, but the noise bandwidth correction should be ignored, giving a result in dBc.

Table 14.1 shows the results for an ICOM IC-745 as measured by KI6WX,

and **Fig 14.10** shows this data in graphic form. His oscillator power was only −3 dBm, which limited measurements to offsets less than 200 kHz. More power might have allowed noise measurements to lower levels, although receiver overload places a limit on this. This is not important, because the real area of interest has been thoroughly covered. When attempting phase-noise measurements at large offsets, remember that any front-end selectivity, before the first mixer, will limit the maximum offset at which LO phase-noise measurement is possible.

Measuring Oscillator and Transmitter Phase Noise

Measuring the composite phase noise of

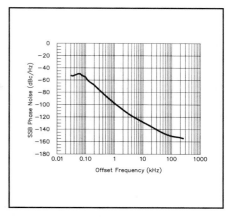

Fig 14.10 — The SSB phase noise of an ICOM IC-745 transceiver (serial number 01528) as measured by KI6WX.

---

scale in the plot) represents −60 dBc/Hz. It is common in industry to use a 0-dBc log reference, but such a reference level would not allow measurement of phase-noise levels below −80 dBc/Hz. The actual measurement bandwidth used on the spectrum analyzer is 100 Hz, but the reference is scaled for a 1-Hz bandwidth. This allows phase-noise levels to be read directly from the display in dBc/Hz. Because each vertical division represents 10 dB, the plot shows the noise level between −60 dBc/Hz (the top horizontal line) and −140 dBc/Hz (the bottom horizontal line). The horizontal scale is 2 kHz per division. The offsets shown in the plots are 2 through 20 kHz.

What Do the Phase-Noise Plots Mean?

Although they are useful for comparing different radios, plots can also be used to calculate the amount of interference you may receive from a nearby transmitter with known phase-noise characteristics. An approximation is given by

$$A_{QRM} = NL + 10 \times \log(BW)$$

where
A_{QRM} = Interfering signal level, dBc
NL = noise level on the receive frequency, dBc
BW = receiver IF bandwidth, in Hz

For instance, if the noise level is −90 dBc/Hz and you are using a 2.5-kHz SSB filter, the approximate interfering signal will be −56 dBc. In other words, if the transmitted signal is 20 dB over S9, and each S unit is 6 dB, the interfering signal will be as strong as an S3 signal.

The measurements made in the ARRL Lab apply only to transmitted signals. It is reasonable to assume that the phase-noise characteristics of most transceivers are similar on transmit and receiver because the same oscillators are generally used in local-oscillator (LO) chain.

In some cases, the receiver may have better phase-noise characteristics than the transmitter. Why the possible difference? The most obvious reason is that circuits often perform less than optimally in strong RF fields, as anyone who has experienced RFI problems can tell you. A less obvious reason results from the way that many high-dynamic-range receivers work. To get good dynamic range, a sharp crystal filter is often

placed immediately after the first mixer in the receive line. This filter removes all but a small slice of spectrum for further signal processing. If the desired filtered signal is a product of mixing an incoming signal with a noisy oscillator, signals far away from the desired one can end up in this slice. Once this slice of spectrum is obtained, however, unwanted signals cannot be reintroduced, no matter how noisy the oscillators used in further signal processing. As a result, some oscillators in receivers don't affect phase noise.

The difference between this situation and that in transmitters is that crystal filters are seldom used for reduction of phase noise in transmitting because of the high cost involved. Equipment designers have enough trouble getting smooth, click-free break-in operation in transceivers without having to worry about switching crystal filters in and out of circuits at 40-wpm keying speeds!—*Zack Lau, KH6CP/1, ARRL Laboratory Engineer*

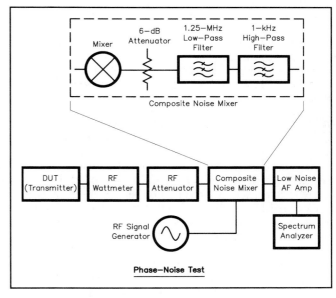

Fig B — ARRL transmitter phase-noise measurement setup.

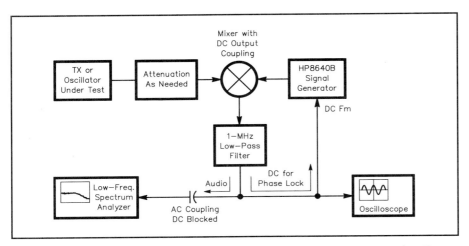

Fig 14.11 — Arrangement for measuring phase noise by directly converting the signal under test to audio. The spectrum analyzer views the signal's noise sidebands as audio; the signal's carrier, converted to dc, provides a feedback signal to phase-lock the Hewlett-Packard HP8640B signal generator to the signal under test.

a receiver's LO requires a clean test oscillator. Measuring the phase noise of an incoming signal, whether from a single oscillator or an entire transmitter, requires the use of a clean receiver, with lower phase noise than the source under test. The sidebar, "Transmitter Phase-Noise Measurement in the ARRL Lab," details the method used to measure composite noise (phase noise and amplitude noise, the practical effects of which are indistinguishable on the air) for *QST* Product Reviews. Although targeted at measuring high power signals from entire transmitters, this approach can be used to measure lower-level signals simply by changing the amount of input attenuation used.

At first, this method — using a low-frequency spectrum analyzer and a low-phase-noise signal generator — looks unnecessarily elaborate. A growing number of radio amateurs have acquired good-quality spectrum analyzers for their shacks since older model Tektronix and Hewlett-Packard instruments have started to appear on the surplus market at affordable prices. The obvious question is, "Why not just use one of these to view the signal and read phase-noise levels directly off the screen?" Reciprocal mixing is the problem. Very few spectrum analyzers have clean enough local oscillators not to completely swamp the noise being measured. Phase-noise measurements involve the measurement of low-level components very close to a large carrier, and that carrier will mix the noise sidebands of *the analyzer's LO* into its IF. Some way of notching out the carrier is needed, so that the analyzer need only handle the noise sidebands. A crystal filter could be designed to do the job, but this would be expensive, and

one would be needed for every different oscillator frequency to be tested. The alternative is to build a direct-conversion receiver using a clean LO like the Hewlett-Packard HP8640B signal generator and spectrum-analyze its "audio" output with an audio analyzer. This scheme mixes the carrier to dc; the LF analyzer is then ac-coupled, and this removes the carrier. The analyzer can be made very sensitive without overload or reciprocal mixing being a problem. The remaining problem is then keeping the LO — the HP8640B in this example — at exactly the carrier frequency. 8640s are based on a shortened-UHF-cavity oscillator and can drift a little. The oscillator under test will also drift. The task is therefore to make the 8640B track the oscillator under test. For once we get something for free: The HP8640B's FM input is dc coupled, and we can use this as an electronic fine-tuning input. As a further bonus, the 8640B's FM deviation control acts as a sensitivity control for this input. We also get a phase detector for free, as the mixer output's dc component depends on the phase relationship between the 8640B and the signal under test (remember to use the dc coupled port of a diode ring mixer as the output). Taken together, the system includes everything needed to create a crude phase-locked loop that will automatically track the input signal over a small frequency range. **Fig 14.11** shows the arrangement.

The oscilloscope is not essential for operation, but it is needed to adjust the system. With the loop unlocked (8640B FM input disconnected), tune the 8640 off the signal frequency to give a beat at the mixer output. Adjust the mixer drive levels to get an undistorted sine wave on the scope. This

ensures that the mixer is not being overdriven. While the loop is off-tuned, adjust the beat to a frequency within the range of the LF spectrum analyzer and use it to measure its level, "A_c," in dBm. This represents the carrier level and is used as the reference for the noise measurements. Connect the FM input of the signal generator, and switch on the generator's dc FM facility. Try a deviation range of 10 kHz to start with. When you tune the signal generator toward the input frequency, the scope will show the falling beat frequency until the loop jumps into lock. Then it will display a noisy dc level. Fine tune to get a mean level of 0 V. (This is a very-low-bandwidth, very-low-gain loop. Stability is not a problem, careful loop design is not needed. We actually want as slow a loop as possible; otherwise, the loop would track and cancel the slow components of the incoming signal's phase noise, preventing their measurement.)

When you first take phase-noise plots, it's a good idea to duplicate them at the generator's next lower FM-deviation range and check for any differences in the noise level in the areas of interest. Reduce the FM deviation range until you find no further improvement. Insufficient FM deviation range makes the loop's lock range narrow, reducing the amount of drift it can compensate. (It's sometimes necessary to keep gently trimming the generator's fine tune control.)

Set up the LF analyzer to show the noise. A sensitive range and 100-Hz resolution bandwidth are appropriate. Measure the noise level, "A_n," in dBm. We must now calculate the noise density that this represents. Spectrum-analyzer filters are normally *Gaussian*-shaped and bandwidth-specified at their –3-dB points. To avoid using integration to find their true-noise power bandwidth, we can reasonably assume a value of $1.2 \times BW$. A spectrum analyzer logarithmically compresses its IF signal ahead of the detectors and averaging filter. This affects the statistical distribution of noise voltage and causes the analyzer to read low by 2.5 dB. To produce the same scale as the ARRL Lab photographs, the analyzer reference level must be set to –60 dBc/Hz, which can be calculated as:

$$A_{ref} = A_c - 10 \log (1.2 \times BW) + 62.5 \text{ dBm}$$
(2)

where

A_{ref} = analyzer reference level, dBm
A_c = carrier amplitude, dBm.

This produces a scale of –60 dBc/Hz at the top of the screen, falling to –140 dBc/Hz at the bottom. The frequency scale is 0 to 20 kHz with a resolution bandwidth (BW in the above equation) of 100 Hz. This method combines the power of *both* sidebands and so measures DSB phase noise. To calculate the equivalent SSB phase noise, subtract 3 dB for noncoherent noise (the general "hash" of phase noise) and subtract 6 dB for coherent, discrete components (that is, single-frequency spurs). This can be done by setting the reference level 3 to 6 dB higher.

Low-Cost Phase Noise Testing

All that expensive equipment may seem far beyond the means of the average Amateur Radio experimenter. With careful shopping and a little more effort, alternative equipment can be put together for pocket money. (All of the things needed — parts for a VXO, a surplus spectrum analyzer and so on — have been seen on sale cheap enough to total less than $100.) The HP8640B is good and versatile, but for use at one oscillator frequency, you can build a VXO for a few dollars. It will only cover one oscillator frequency, but a VXO can provide even better phase-noise performance than the 8640B.

You can also get away without an audio spectrum analyzer. A spectrum-analyzer forerunner, the *wave analyzer*, may suffice. A wave analyzer neither scans nor plots signals. It has simple, hand-cranked tuning calibrated with a mechanical scale or frequency counter and displays received-signal level on a meter. The telephone companies used some variants of these called *selective level measuring sets* (*SLMS*s) to measure channel powers and pilot tones on their systems, and some of these instruments covered the frequency range of interest. Surplus dealers see little demand for them; they get treated as "boat anchors" and sold cheaply for stripping. One of these instruments will do just fine, but you'll have to plot your noise graph manually, point-by-point.

The mixer need not be an expensive, packaged unit. A Mini-Circuits Labs SRA-2H (see Address List in the **References** chapter) or any other medium-level diode-ring unit will do just as well.

OSCILLATOR CIRCUITS AND CONSTRUCTION

LC Oscillator Circuits

The LC oscillators used in radio equipment are usually arranged as *variable frequency oscillators* (*VFO*s). Tuning is achieved by either varying part of the capacitance of the resonator or, less commonly, by using a movable magnetic core to vary the inductance. Since the early days of radio, there has been a huge quest for the ideal, low-drift VFO. Amateurs and professionals have spent immense effort on this

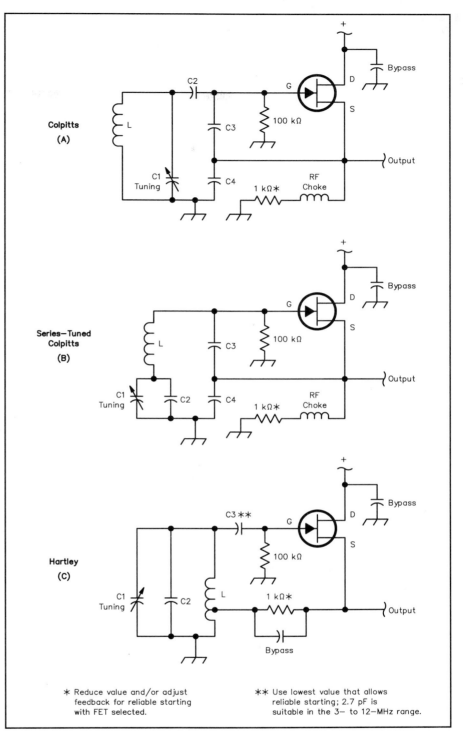

Fig 14.12 — The Colpitts (A), series-tuned Colpitts (B) and Hartley oscillator circuits. Rules of thumb: C3 and C4 at A and B should be equal and valued such that their X_C = 45 Ω at the operating frequency; for C2 at A, X_C = 100 Ω. For best stability, use C0G or NP0 units for all capacitors associated with the FETs' gates and sources. Depending on the FET chosen, the 1-kΩ source-bias-resistor value shown may require adjustment for reliable starting.

pursuit. A brief search of the literature reveals a large number of designs, many accompanied by claims of high stability. The quest for stability has been solved by the development of low-cost frequency synthesizers, which can give crystal-controlled stability. Synthesizers have other problems though, and the VFO still has much to offer in terms of signal cleanliness, cost and power consumption, making it attractive for home construction. No one VFO circuit has any overwhelming advantage over any other — component quality, mechanical design and the care taken in assembly are much more important.

Fig 14.12 shows three popular oscillator circuits stripped of any unessential frills so they can be more easily compared. The original Colpitts circuit (Fig 14.12A) is now often referred to as the *parallel-tuned Colpitts* because its series-tuned derivative (Fig 14.12B) has become com-

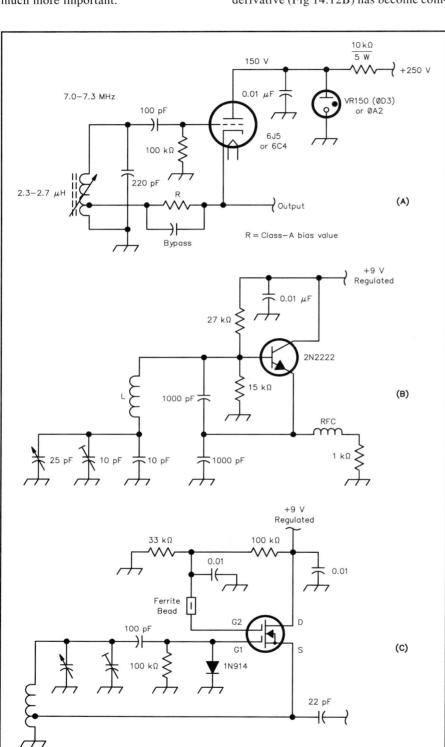

Fig 14.13 — Three more oscillator examples: at A, a triode-tube Hartley; at B, a bipolar junction transistor in a series-tuned Colpitts; at C, a dual-gate MOSFET Hartley.

mon. All three of these circuits use an amplifier with a voltage gain less than unity, but large current gain. The N-channel JFET source follower shown appears to be the most popular choice nowadays. In the parallel-tuned Colpitts, C3 and C4 are large values, perhaps 10 times larger than typical values for C1 and C2. This means that only a small fraction of the total tank voltage is applied to the FET, and the FET can be considered to be only lightly coupled into the tank. The FET is driven by the sum of the voltages across C3 and C4, while it drives the voltage across C4 alone. This means that the tank operates as a resonant, voltage-step-up transformer compensating for the less-than-unity-voltage-gain amplifier. The resonant circuit consists of L, C1, C2, C3 and C4. The resonant frequency can be calculated by using the standard formulas for capacitors in series and parallel to find the resultant capacitance effectively connected across the inductor, L, and then use the standard formula for LC resonance:

$$f = \frac{1}{2\pi\sqrt{LC}} \qquad (3)$$

where

f = frequency in hertz
L = inductance in henries
C = capacitance in farads.

For a wide tuning range, C2 must be kept small to reduce the effect of C3 and C4 swamping the variable capacitor C1.

The series-tuned Colpitts circuit works in much the same way. The difference is that the variable capacitor, C1, is positioned so that it is well-protected from being swamped by the large values of C3 and C4. In fact, *small* values of C3, C4 would act to limit the tuning range. Fixed capacitance, C2, is often added across C1 to allow the tuning range to be reduced to that required, without interfering with C3 and C4, which set the amplifier coupling. The series-tuned Colpitts has a reputation for better stability than the parallel-tuned original. Note how C3 and C4 swamp the capacitances of the amplifier in both versions.

The Hartley is similar to the parallel-tuned Colpitts, but the amplifier source is tapped up on the tank inductance instead of the tank capacitance. A typical tap placement is 10 to 20% of the total turns up from the "cold" end of the inductor. (It's usual to refer to the lowest-signal-voltage end of an inductor as *cold* and the other, with the highest signal voltage as *hot*.) C2 limits the tuning range as required; C3 is reduced to the minimum value that allows reliable starting. This is necessary because the Hartley's lack of the

Colpitts's capacitive divider would otherwise couple the FET's capacitances to the tank more strongly than in the Colpitts, potentially affecting the circuit's frequency stability.

In all three circuits, the 1-kΩ source resistor aids frequency stability by keeping the source higher above ground than the gate. Some commercial circuits accomplish this by connecting the gate to a high-value voltage divider (for example, 1 MΩ from the positive rail to gate, 1 MΩ from gate to ground). The gate-to-ground clamping diode (1N914 or similar) long used by radio amateurs as a means of avoiding gate-source conduction has been shown by Ulrich Rohde, KA2WEU, to degrade oscillator phase-noise performance, and its use is virtually unknown in professional circles.

Fig 14.13 shows some more VFOs to illustrate the use of different devices. The triode Hartley shown includes *permeability tuning*, which has no sliding contact like that to a capacitor's rotor and can be made reasonably linear by artfully spacing the coil turns. The slow-motion drive can be done with a screw thread. The disadvantage is that special care is needed to avoid backlash and eccentric rotation of the core. If a nonrotating core is used, the slides have to be carefully designed to prevent motion in unwanted directions. The Collins Radio Company made extensive use of tube-based permeability tuners, and a semiconductor version can still be found in a number of Ten-Tec radios.

Vacuum tubes cannot run as cool as competitive semiconductor circuits, so care is needed to keep the tank circuit away from the tube heat. In many amateur and commercial vacuum-tube oscillators, oscillation drives the tube into grid current at the positive peaks, causing rectification and producing a negative grid bias. The oscillator thus runs in Class C, in which the conduction angle reduces as the signal amplitude increases until the amplitude stabilizes. As in the FET circuits of Fig 14.12, better stability and phase-noise performance can be achieved in a vacuum-tube oscillator by moving it out of true Class C — that is, by including a bypassed cathode-bias resistor (the resistance appropriate for Class A operation is a good starting value). A small number of people still build vacuum-tube radios partly to be different, partly for fun, but the semiconductor long ago achieved dominance in VFOs.

The voltage regulator (*VR*) tube shown in **Fig 14.13** has a potential drawback: It is a gas-discharge device with a high striking voltage at which it starts to conduct

and a lower extinguishing voltage, at which it stops conducting. Between these extremes lies a region in which decreasing voltage translates to increasing current, which implies negative resistance. When the regulator strikes, it discharges any capacitance connected across it to the extinguishing voltage. The capacitance then charges through the source resistor until the tube strikes again, and the process repeats. This *relaxation oscillation* demonstrates how negative resistance can cause oscillation. The oscillator translates the resultant sawtooth modulation of its power supply into frequency and amplitude variation. Because of VR tubes' ability to support relaxation oscillation, a traditional rule of thumb is to keep the capacitance directly connected across a VR tube to 0.1 µF or less. A value much lower than this can provide sufficient bypassing in Fig 14.13A because the dropping resistor acts as a decoupler, increasing the bypass's effectiveness.

There is a related effect called *squegging*, which can be loosely defined as oscillation on more than one frequency at a time, but which may also manifest itself as RF oscillation interrupted at an AF rate, as in a superregenerative detector. One form of squegging occurs when an oscillator is fed from a power supply with a high source impedance. The power supply charges up the decoupling capacitor until oscillation starts. The oscillator draws current and pulls down the capacitor voltage, starving itself of power until oscillation stops. The oscillator stops drawing current, and the decoupling capacitor then recharges until oscillation restarts. The process, the low-frequency cycling of which is a form of relaxation oscillation, repeats indefinitely. The oscillator output can clearly be seen to be pulse modulated if an oscilloscope is used to view it at a suitable time-base setting. This fault is a well-known consequence of poor design in battery-powered radios. As dry cells become exhausted, their internal resistance rises quickly and circuits they power can begin to misbehave. In audio stages, such misbehavior may manifest itself in the *putt-putt* sound of the slow relaxation oscillation called *motorboating*.

Compared to the frequently used JFET, bipolar transistors, Fig 14.13B, are relatively uncommon in oscillators because their low input and output impedances are more difficult to couple into a high-Q tank without excessively loading it. Bipolar devices do tend to give better sample-to-sample amplitude uniformity for a given oscillator circuit, however, as JFETs of a given type tend to vary more in their characteristics.

The dual-gate MOSFET, Fig 14.13C, is very rarely seen in VFO circuits. It imposes the cost of the components needed to bias and suppress VHF parasitic oscillation at the second gate, and its inability to generate its own AGC through gate-source conduction forces the addition of a blocking capacitor, resistor and diode at the first gate for amplitude control.

VFO Components and Construction

Tuning Capacitors and Reduction Drives

As most commercially made radios now use frequency synthesizers, it has become increasingly difficult to find certain key components needed to construct a good VFO. The slow-motion drives, dials, gearboxes, associated with names like Millen, National, Eddystone and Jackson are no longer available. Similarly, the most suitable silver-plated variable capacitors with ball races at both ends, although still made, are not generally marketed and are expensive. Three approaches remain: scavenge suitable parts from old equipment; use tuning diodes instead of variable capacitors — an approach that, if uncorrected through phase locking, generally degrades stability and phase-noise performance; or use two tuning capacitors, one with a capacitance range $^1/_5$ to $^1/_{10}$ that of the other, in a bandset/bandspread approach.

Assembling a variable capacitor to a chassis and its reduction drive to a front panel can result in *backlash* — an annoying tuning effect in which rotating the capacitor shaft deforms the chassis and/or panel rather than tuning the capacitor. One way of minimizing this is to use the reduction drive to support the capacitor, and use the capacitor to support oscillator circuit board.

Fixed Capacitors

Traditionally, silver-mica fixed capacitors have been used extensively in oscillators, but their temperature coefficient is not as low as can be achieved by other types, and some silver micas have been known to behave erratically. Polystyrene film has become a proven alternative. One warning is worth noting: Polystyrene capacitors exhibit a permanent change in value should they ever be exposed to temperatures much over 70°C; they do not return to their old value on cooling. Particularly suitable for oscillator construction are the low-temperature-coefficient ceramic capacitors, often described as *NP0* or *C0G* types. These names are actually temperature-coefficient codes. Some ceramic capacitors are available with deliberate, controlled temperature coefficients so that they can be used to compensate for other causes of frequency drift with temperature. For example, the code N750 denotes a part with a temperature coefficient of –750 parts per million per degree Celsius. These parts are now somewhat difficult to obtain, so other methods are needed.

In a Colpitts circuit, the two large-value capacitors that form the voltage divider for the active device still need careful selection. It would be tempting to use any available capacitor of the right value, because the effect of these components on the tank frequency is reduced by the proportions of the capacitance values in the circuit. This reduction is not as great as the difference between the temperature stability of an NP0 ceramic part and some of the low-cost, decoupling-quality X7R-dielectric ceramic capacitors. It's worth using low-temperature coefficient parts even in the seemingly less-critical parts of a VFO circuit — even as decouplers. Chasing the cause of temperature drift is more challenging than fun. Buy critical components like high-stability capacitors from trustworthy sources.

Inductors

Ceramic coil forms can give excellent results, as can self-supporting air-wound coils (Miniductor). If you use a magnetic core, make it powdered iron, never ferrite, and support it stably. Stable VFOs have been made using toroidal cores, but again, ferrite must be avoided. Micrometals mix number 6 has a low temperature coefficient and works well in conjunction with NP0 ceramic capacitors. Coil forms in other materials have to be assessed on an individual basis.

A material's temperature stability will not be apparent until you try it in an oscillator, but you can apply a quick test to identify those nonmetallic materials that are lossy enough to spoil a coil's Q. Put a sample of the coil-form material into a microwave oven along with a glass of water and cook it about 10 s on low power. *Do not include any metal fittings or ferromagnetic cores.* Good materials will be completely unaffected; poor ones will heat and may even melt, smoke, or burst into flame. (This operation is a fire hazard if you try more than a tiny sample of an unknown material. Observe your experiment continuously and do not leave it unattended.)

Wes Hayward, W7ZOI, suggests annealing toroidal VFO coils after winding. Roy Lewallen, W7EL reports achieving success with this method by boiling his coils in water and letting them cool in air.

Voltage Regulators

VFO circuits are often run from locally regulated power supplies, usually from resistor/Zener diode combinations. Zener diodes have some idiosyncrasies which could spoil the oscillator. They are noisy, so decoupling is needed down to audio frequencies to filter this out. Zener diodes are often run at much less than their specified optimum bias current. Although this saves power, it results in a lower output voltage than intended, and the diode's impedance is much greater, increasing its sensitivity to variations in input voltage, output current and temperature. Some common Zener types may be designed to run at as much as 20 mA; check the data sheet for your diode family to find the optimum current.

True Zener diodes are low-voltage devices; above a couple of volts, so-called Zener diodes are actually avalanche types. The temperature coefficient of these diodes depends on their breakdown voltage and crosses through zero for diodes rated at about 5 V. If you intend to use nothing fancier than a common-variety Zener, designing the oscillator to run from 5 V and using a 5.1-V Zener, will give you a free advantage in voltage-versus-temperature stability. There are some diodes available with especially low temperature coefficients, usually referred to as *reference* or *temperature-compensated diodes*. These usually consist of a series pair of diodes designed to cancel each other's temperature drift. Running at 7.5 mA, the 1N829A gives 6.2 V ±5% and a temperature coefficient of just ±5 parts per million (ppm) maximum per degree Celsius. A change in bias current of 10% will shift the voltage less than 7.5 mV, but this increases rapidly for greater current variation. The 1N821A is a lower-grade version, at ±100 ppm/°C. The LM399 is a complex IC that behaves like a superb Zener at 6.95 V, ±0.3 ppm/°C. There are also precision, low-power, three-terminal regulators designed to be used as voltage references, some of which can provide enough current to run a VFO. There are comprehensive tables of all these devices between pages 334 and 337 of Horowitz and Hill, *The Art of Electronics*, 2nd ed.

Oscillator Devices

The 2N3819 FET, a classic from the 1960s, has proven to work well in VFOs, but, like the MPF102 also long-popular with ham builders, it's manufactured to wide tolerances. Considering an oscillator's importance in receiver stability, you should not hesitate to spend a bit more on a better device. The 2N5484, 2N5485 and 2N5486 are worth considering; together, their transconductance ranges span

that of the MPF102, but each is a better-controlled subset of that range. The 2N5245 is a more recent device with better-than-average noise performance that runs at low currents like the 2N3819. The 2N4416/A, also available as the plastic-cased PN4416, is a low-noise device, designed for VHF/UHF amplifier use, that has been featured in a number of good oscillators up to the VHF region. Its low internal capacitance contributes to low frequency drift. The J310 (plastic; the metal-cased U310 is similar) is another popular JFET in oscillators.

The 2N5179 (plastic, PN5179 or MPS5179) is a bipolar transistor capable of good performance in oscillators up to the top of the VHF region. Care is needed because its absolute-maximum collector-emitter voltage is only 12 V, and its collector current must not exceed 50 mA. Although these characteristics may seem to convey fragility, the 2N5179 is sufficient for circuits powered by stabilized 6-V power supplies.

VHF-UHF devices are not really necessary in LC VFOs because such circuits are rarely used above 10 MHz. Absolute frequency stability is progressively harder to achieve with increasing frequency, so free-running oscillators are rarely used to generate VHF-UHF signals for radio communication. Instead, VHF-UHF radios usually use voltage-tuned, phase-locked oscillators in some form of synthesizer. Bipolar devices like the BFR90 and MRF901, with f_Ts in the 5-GHz region and mounted in stripline packages, are needed at UHF.

Integrated circuits have not been mentioned until now because few specific RF-oscillator ICs exist. Some consumer ICs — the NE602, for example — include the active device(s) necessary for RF oscillators, but often this is no more than a single transistor fabricated on the chip. This works just as a single discrete device would, although using such on-chip devices may result in poor isolation from the rest of the circuits on the chip. There is one specialized LC-oscillator IC: the Motorola MC1648. This device has been made since the early 1970s and is a surviving member of a long-obsolete family of emitter-coupled-logic (ECL) devices. Despite the MC1648's antiquity, it has no real competition and is still widely used in current military and commercial equipment. Market demand should force its continued production for some more years to come. Its circuitry is complex for an oscillator, with a multitransistor oscillator cell controlled by a detector and amplifier in an on-chip ALC system. The MC1648's first problem is that the ECL families use only about a 1-V swing between logic levels. Because the oscillator is made using the same ECL-optimized semiconductor manufacturing processes and circuit design techniques, this same limitation applies to the signal in the oscillator tuned circuit. It is possible to improve this situation by using a tapped or transformer-coupled tank circuit to give improved Q, but this risks the occurrence of the device's second problem.

Periodically, semiconductor manufacturers modernize their plants and scrap old assembly lines used to make old products. Any surviving devices then must undergo some redesign to allow their production by the new processes. One common result of this is that devices are shrunk, when possible, to fit more onto a wafer. All this increases the f_T of the transistors in the device, and such evolution has rendered today's MC1648s capable of operation at much higher frequencies than the specified 200-MHz limit. This allows higher-frequency use, but great care is needed in the layout of circuits using it to prevent spurious oscillation. A number of old designs using this part have needed reengineering because the newer parts generate spurious oscillations that the old ones didn't, using PC-board traces as parasitic tuned circuits.

The moral is that a UHF-capable device requires UHF-cognizant design and layout even if the device will be used at far lower frequencies. **Fig 14.14** shows the MC1648 in a simple circuit and with a tapped resonator. These more complex circuits have a greater risk of presenting a stray resonance within the device's operating range, risking oscillation at an unwanted frequency. This device is *not* a prime choice for an HF VFO because the physical size of the variable capacitor and the inevitable lead lengths, combined with the need to tap couple to get sufficient Q for good noise performance, makes spurious oscillation difficult to avoid. The MC1648 is really intended for tuning-diode control in phase-locked loops operating at VHF. This difficulty is inherent in wideband devices, especially oscillator circuits connected to their tank by a single "hot" terminal, where there is simply no isolation between the amplifier's input and output paths. Any resonance in the associated circuitry can control the frequency of oscillation.

The popular NE602 mixer IC has a built-in oscillator and can be found in many published circuits. This device has separate input and output pins to the tank and has proved to be quite tame. It's still a relatively new part and may not have been "improved" yet (so far, it has pro-

gressed from the NE602 to the NE602A, the A version affording somewhat higher dynamic range than the original NE602). It might be a good idea for anyone laying out a board using one to take a little extra care to keep PCB traces short in the oscillator section to build in some safety margin so that the board can be used reliably in the future. Experienced (read: "bitten") professional designers know that their designs are going to be built for possibly more than 10 years and have learned to make allowances for the progressive improvement of semiconductor manufacture.

Three High-Performance HF VFOs

The G3PDM Vackar VFO

The Vackar VFO shown in **Fig 14.15** was developed over 20 years ago by Peter Martin, G3PDM, for the Mark II version of his high-dynamic range receiver. This can be found in the Radio Society of Great Britain's *Radio Communication Handbook*, with some further comments on the oscillator in RSGB's *Amateur Radio Techniques*. This is a prime example of an oscillator that has been successfully optimized for maximum frequency stability. Not only does it work extremely well, but it still represents the highest stability that can be achieved. Its developer commented

on a number of points, which also apply to the construction of different VFO circuits.

- Use a genuine Vackar circuit, with C1/(C4+C6) and C3/C2 = 6.
- Use a strong box, die-cast or milled from solid metal.
- Use a high-quality variable capacitor (double ball bearings, silver plated).
- Adjust feedback control C2, an air-dielectric trimmer, so that circuit just oscillates.
- Thoroughly clean all variable capacitors in an ultrasonic bath.
- Use an Oxley "Tempatrimmer," a fixed capacitor whose temperature coefficient is variable over a wide range, for adjustable temperature compensation. (The "Thermatrimmer" is a lower cost, smaller range alternative.)
- C1, C3 and C6 are silver-mica types glued to a solid support to reduce sensitivity to mechanical shock.
- The gate resistor is a 4.7-kΩ, 2-W carbon-composition type to minimize self-heating.
- The buffer amplifier is essential.
- Circuits using an added gate-ground diode seem to suffer increased drift.
- Its power supply should be well-regulated. Make liberal use of decoupling capacitors to prevent unintentional feedback via supply rails.

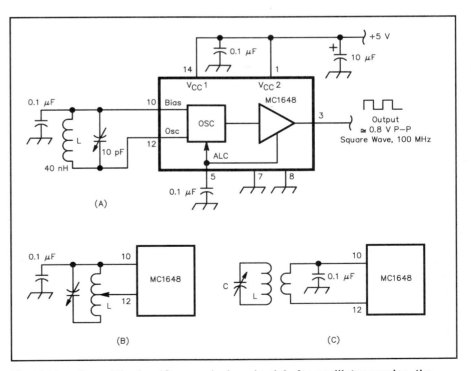

Fig 14.14 — One of the few ICs ever designed solely for oscillator service, the ECL Motorola MC1648 (A) requires careful design to avoid VHF parasitics when operating at HF. Keeping its tank Q high is another challenge; B and C show means of coupling the IC's low-impedance oscillator terminals to the tank by tapping up on the tank coil (B) or with a link

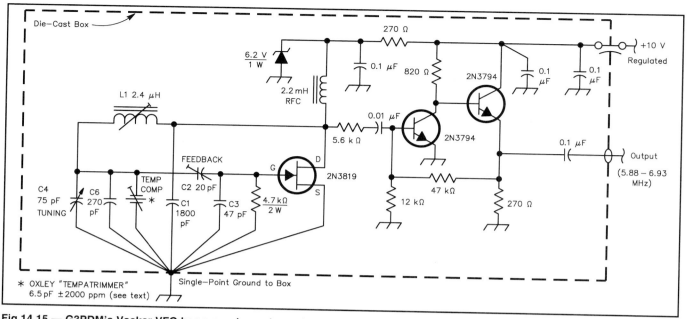

Fig 14.15 — G3PDM's Vackar VFO has proved popular and successful for two decades. The MPF102 can be used as a substitute for the 2N3819. Generally, VFOs can be adapted to work at other frequencies (within the limits of the active device). To do so, compute an adjustment factor: f_{old} / f_{new}. Multiply the value of each frequency-determining or feedback L or C by the factor. As frequency increases, it may help to increase feedback even more than indicated by the factor.

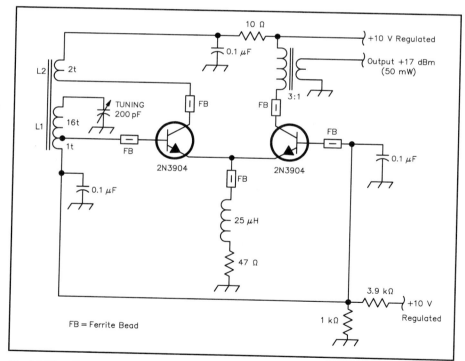

Fig 14.16 — This low-noise oscillator design by K7HFD operates at an unusually high power level to achieve a high C/N (carrier-to-noise) ratio. Need other frequencies? See Fig 14.15 (caption) for a frequency-scaling technique.

- Single-point grounding for the tank and FET is important. This usually means using one mounting screw of the tuning capacitor.
- The inductor used a ceramic form with a powdered-iron core mounted on a spring-steel screw.

- Short leads and thick solid wire (#16 to #18 gauge) are essential for mechanical stability.

Operating in the 6-MHz region, this oscillator drifted 500 Hz during the first minute as it warmed up and then drifted at

2 Hz in 30 minutes. It must be stressed that such performance does not indicate a wonderful circuit so much as care in construction, skillful choice of components, artful mechanical layout and diligence in adjustment. The 2-W gate resistor may seem strange, but 20 years ago many components were not as good as they are today. A modern, low-inductance component of low temperature coefficient and more modest power rating should be fine.

The buffer amplifier is an important part of a good oscillator system as it serves to prevent the oscillator seeing any impedance changes in the circuit being driven, which could otherwise affect the frequency.

The Oxley Tempatrimmer used by G3PDM was a rare and expensive component, used commercially only in very high-quality equipment, and it may no longer be made. An alternative temperature compensator can be constructed from currently available components, as will be described shortly. G3PDM also referred to his "standard mallet test," where a thump with a wooden hammer produced an average shift of 6 Hz, as an illustration of the benefits of solid mechanical design.

The K7HFD Low-Noise Oscillator

The other high performance oscillator example, shown in **Fig 14.16**, was designed for low-noise performance by Linley Gumm, K7HFD, and appears on page 126 of ARRL's *Solid State Design for the Radio Amateur*. Despite its publication in the home-brewer's bible, this circuit seems to have been overlooked by many builders. It uses no unusual components and looks simple, yet it is a subtle and sophisticated circuit. It represents the antithesis of G3PDM's VFO: In the pursuit of low noise sidebands, a number of design choices have been made that will degrade the stability of frequency over temperature.

The effects of oscillator noise have already been covered, and Fig 14.6 shows the effect of limiting on the signal from a noisy oscillator. Because AM noise sidebands can get translated into PM noise sidebands by imperfect limiting, there is an advantage to stripping off the AM as early as possible, in the oscillator itself. An ALC system in the oscillator will counteract and cancel only the AM components within its bandwidth, but an oscillator based on a limiter will do this over a broad bandwidth. K7HFD's oscillator uses a differential pair of bipolar transistors as a limiting amplifier. The dc bias voltage at the bases and the resistor in the common emitter path to ground establishes a controlled dc bias current. The ac voltage between the bases switches this current between the two collectors. This applies a rectangular pulse of current into link winding L2, which drives the tank. The output impedance of the collector is high in both the current on and current off states. Allied with the small number of turns of the link winding, this presents a very high impedance to the tank circuit, which minimizes the damping of the tank Q. The input impedance of the limiter is also quite high and is applied across only a one-turn tap of L1, which similarly minimizes the effect on the tank Q. The input transistor base is driven into conduction only on one peak of the tank waveform. The output transformer has the inverse of the current pulse applied to it, so the output is not a low distortion sine wave, although the output harmonics will not be as extensive as simple theory would suggest because the circuit's high output impedance allows stray capacitances to attenuate high-frequency components. The low-frequency transistors used also act to reduce the harmonic power.

With an output of +17 dBm, this is a power oscillator, running with a large dc input power, so appreciable heating results that can cause temperature-induced drift. The circuit's high-power operation is a deliberate ploy to create a high signal-to-noise ratio by having as high a signal power as possible. This also reduces the problem of the oscillator's broadband noise output. The limitation on the signal level in the tank is the transistors' base-emitter-junction breakdown voltage. The circuit runs with a few volts peak-to-peak across the one-turn tap, so the full tank is running at over 50 V P-P. The single easiest way to damage a bipolar transistor is to reverse bias the base-emitter junction until it avalanches. Most devices are only rated to withstand 5 V applied this way, the current needed to do damage is small, very little power is needed. If the avalanche current is limited to less than that needed to perform immediate destruction of the transistor, it is likely that there will be some degradation of the device, a reduction in its bandwidth and gain along with an increase in its noise. These changes are irreversible and cumulative. Small, fast signal diodes have breakdown voltages of over 30 V and less capacitance than the transistor bases, so one possible experiment would be to try the effect of adding a diode in series with the base of each transistor and running the circuit at even higher levels.

The amplitude must be controlled by the drive current limit. The voltage on L2 must never allow the collector of the transistor driving it to go into saturation, if this happens, the transistor presents a very low impedance to L2 and badly loads the tank, wrecking the Q and the noise performance. The circuit can be checked to verify the margin from saturation by probing the hot end of L2 and the emitter with an oscilloscope. Another, less obvious, test is to vary the power-supply voltage and monitor the output power. While the circuit is under current control, there is very little change in output power, but if the supply is low enough to allow saturation, the output power will change significantly.

The use of the 2N3904 is interesting, as it is not normally associated with RF oscillators. It is a cheap, plain, general-purpose type more often used at dc or audio frequencies. There is evidence that suggests some transistors, which have good noise performance *at RF*, have worse noise performance at low frequencies, and that the low-frequency noise they create can modulate an oscillator, creating noise sidebands. Experiments with low-noise audio transistors may be worthwhile, but many such devices have very low f_T and high junction capacitances. In the description of this circuit in *Solid State Design for the Radio Amateur*, the results of a phase-noise test made using a spectrum analyzer with a crystal filter as a preselector are given. Ten kilohertz out from the carrier, in a 3-kHz measurement bandwidth, the noise was over 120 dB below the carrier level. This translates into better than −120 − 10 log (3000), which equals −154.8 dBc/Hz, SSB. At this offset, −140 dBc is usually considered to be excellent. This is state-of-the-art performance by today's standards — in a 1977 publication.

A JFET Hartley VFO

Fig 14.17 shows an 11.1-MHz version of a VFO and buffer closely patterned after that used in 7-MHz transceiver designs published by Roger Hayward, KA7EXM, and Wes Hayward, W7ZOI (The Ugly Weekender) and Roy Lewallen, W7EL (the Optimized QRP Transceiver). In it, a 2N5486 JFET Hartley oscillator drives the two-2N3904 buffer attributed to Lewallen. This version diverges from the originals in that its JFET uses source bias (the bypassed 910-Ω resistor) instead of a gate-clamping diode and is powered from a low-current 7-V regulator IC instead of a Zener diode and dropping resistor. The 5-dB pad sets the buffer's output to a level appropriate for "Level 7" (+7-dBm-LO) diode ring mixers.

The circuit shown was originally built with a gate clamping diode, no source bias and a 3-dB output pad. Rebiasing the oscillator as shown increased its output by 2 dB without degrading its frequency stability (200 to 300-Hz drift at power up, stability within ±20 Hz thereafter at a constant room temperature).

Temperature Compensation

The general principle for creating a high-stability VFO is to use components with minimal temperature coefficients in circuits that are as insensitive as possible to changes in components' secondary characteristics. Even after careful minimization of the causes of temperature sensitivity, further improvement can still be desirable. The traditional method was to split one of the capacitors in the tank so that it could be apportioned between low-temperature-coefficient parts and parts with deliberate temperature dependency. Only a limited number of different, controlled temperature coefficients are available, so the proportioning between low coefficient and controlled coefficient parts was varied to "dilute" the temperature sensitivity of a part more sensitive than needed. This was a tedious process, involving much trial and error, an undertaking made more complicated by the difficulty of arranging means of heating and cooling the unit being compensated. (Hayward described such a means in December 1993 *QST*.) As commercial and military equipment have been based on frequency synthesizers for some time, supplies of capacitors with controlled temperature sensitivity are drying up. An alternative approach is needed.

A temperature-compensated crystal oscillator (TCXO) is an improved-stability version of a crystal oscillator, that is used widely in industry. Instead of using controlled-temperature coefficient-capacitors, most TCXOs use a network of thermistors and normal resistors to control the bias of a tuning

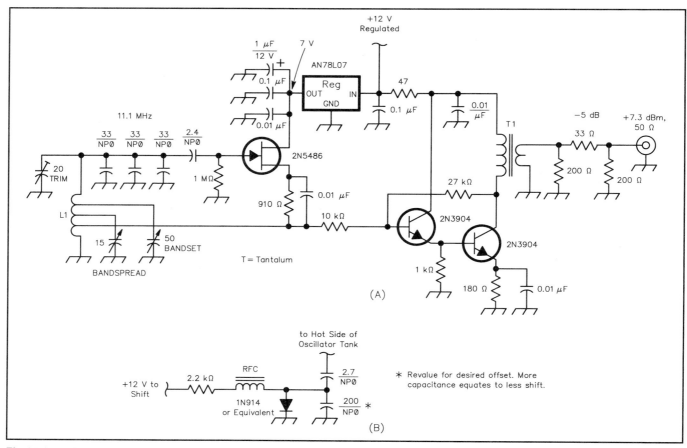

Fig 14.17 — Incorporating ideas from KA2WEU, KA7EXM, W7ZOI and W7EL, the oscillator at A achieves excellent stability and output at 11.1 MHz without the use of a gate-clamping diode, as well as end-running the shrinking availability of reduction drives through the use of bandset and bandspread capacitors. L1 consists of 10 turns of B & W #3041 Miniductor (#22 tinned wire, ⁵⁄₈ inch in diameter, 24 turns per inch). The source tap is 2¹⁄₂ turns above ground; the tuning-capacitor taps are positioned as necessary for bandset and bandspread ranges required. T1's primary consists of 15 turns of #28 enameled wire on an FT-37-72 ferrite core; its secondary, 3 turns over the primary. B shows a system for adding fixed TR offset that can be applied to any LC oscillator. The RF choke consists of 20 turns of #26 enameled wire on an FT-37-43 core. Need other frequencies? See Fig 14.15 (caption) for a frequency-scaling technique.

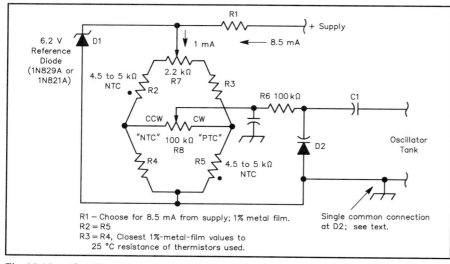

R1 – Choose for 8.5 mA from supply; 1% metal film.
R2 = R5
R3 = R4, Closest 1%-metal-film values to 25 °C resistance of thermistors used.

Fig 14.18 — Oscillator temperature compensation has become more difficult because of the scarcity of negative-temperature-coefficient diodes. This circuit, by GM4ZNX, uses a bridge containing two identical thermistors to steer a tuning diode for drift correction. The 6.2-V Zener diode used (a 1N821A or 1N829A) is a temperature-compensated part; just any 6.2-V Zener will not do.

diode. Manufacturers measure the temperature vs frequency characteristic of sample oscillators, and use a computer program to calculate the optimum normal resistor values for production. This can reliably achieve at least a tenfold improvement in stability. We are not interested in mass manufacture, but the idea of a thermistor tuning a varactor is worth stealing. The parts involved are likely to be available for a long time.

Browsing through component suppliers' catalogs shows ready availability of 4.5- to 5-kΩ-bead thermistors intended for temperature-compensation purposes, at less than a dollar each. **Fig 14.18** shows a circuit based on these. Commonly available thermistors have negative temperature coefficients, so as temperature rises, the voltage at the counterclockwise (ccw) end of R8 increases, while that at the clockwise (cw) end drops. Somewhere near the center, there is no change. Increasing the voltage on the tuning diode decreases its capacitance, so settings toward R8's ccw end simulate a negative-temperature-coefficient capacitor; toward its clockwise end, a positive-temperature-coefficient part. Choose R1 to pass 8.5 mA from whatever supply voltage is available to the 6.2-V reference diode, D1. The 1N821A/1N829A-family diode used has a very low temperature coefficient and needs 7.5 mA bias for best performance; the bridge takes 1 mA. R7 and R8 should be good-quality multiturn trimmers. D2 and C1 are best determined via trial and error; practical components aren't known well enough to rely on analytical models. Choose the least capacitance that provides enough compensation range. This reduces the noise added to the oscillator. (It is possible, though tedious, to solve for the differential varactor voltage with respect to R2 and R5, via differential calculus and circuit theory. The equations in Hayward's 1993 article can then be modified to accommodate the additional capacitors formed by D2 and C1.) Use a single ground point near D2 to reduce the influence of ground currents from other circuits. Use good-quality metal-film components for the circuit's fixed resistors.

The circuit requires two adjustments, one at each of two different temperatures, and achieving them requires a stable frequency counter that can be kept far enough from the radio so that the radio, not the counter, is subjected to the temperature extremes. (Using a receiver to listen to the oscillator under test can speed the adjustments.) After connecting the counter to the oscillator to be corrected, run the radio containing the oscillator and compensator in a room-temperature, draft-free environment until the oscillator's frequency reaches its stable operating temperature rise over ambient. Lock its tuning, if possible. Adjust R7 to balance the bridge. This causes a drop of 0 V across R8, a condition you can reach by winding R8 back and forth across its range while slowly adjusting R7. When the bridge is balanced and 0 V appears across R8, adjusting R8 causes no frequency shift. When you've found this R7 setting, leave it there, set R8 is to the exact center of its range and record the oscillator frequency.

Run the radio in a hot environment and allow its frequency to stabilize. Adjust R8 to restore the frequency to the recorded value. The sensitivity of the oscillator to temperature should now be significantly reduced between the temperatures at which you performed the adjustments.

You will also have somewhat improved the oscillator's stability outside this range.

For best results with any temperature-compensation scheme, it's important to group all the oscillator and compensator components in the same box, avoiding differences in airflow over components. A good oscillator should not dissipate much power, so it's feasible, even advisable, to mount all of the oscillator components in an unventilated box. In the real world, temperatures change, and if the components being compensated and the components doing the compensating have different thermal time constants, a change in temperature can cause a temporary change in frequency until the slower components have caught up. One cure for this is to build the oscillator in a thick-walled metal box that's slow to heat or cool, and so dominates and reduces the possible rate of change of temperature of the circuits inside. This is sometimes called a *cold oven*.

Shielding and Isolation

Oscillators contain inductors running at moderate power levels and so can radiate strong enough signals to cause interference with other parts of a radio, or with other radios. Oscillators are also sensitive to radiated signals. Effective shielding is therefore important. A VFO used to drive a power amplifier and antenna (to form a simple CW transmitter) can prove surprisingly difficult to shield well enough. Any leakage of the power amplifier's high-level signal back into the oscillator can affect its frequency, resulting in a poor transmitted note. If the radio gear is in the station antenna's near field, sufficient shielding may be even more difficult. The following rules of thumb continue to serve ham builders well:

- Use a complete metal box, with as few holes as possible drilled in it, with good contact around surface(s) where its lid(s) fit(s) on.
- Use feedthrough capacitors on power and control lines that pass in and out of the VFO enclosure, and on the transmitter or transceiver enclosure as well.
- Use *buffer amplifier* circuitry that amplifies the signal by the desired amount *and* provide sufficient attenuation of signal energy flowing in the reverse direction. This is known as *reverse isolation* and is a frequently overlooked loophole in shielding. Figs 14.15 and 14.17 include buffer circuitry of proven performance. As another (and higher-cost) option, consider using a high-speed buffer-amplifier IC (such as the LM6321N by National Semiconductor, a part that combines the high input impedance of an op amp with the ability to drive 50-Ω loads directly up into the VHF range).

- Use a mixing-based frequency-generation scheme instead of one that operates straight through or by means of multiplication. Such a system's oscillator stages can operate on frequencies with no direct frequency relationship to its output frequency.
- Use the time-tested technique of running your VFO at a *subharmonic* of the output signal desired — say, 3.5 MHz in a 7-MHz transmitter — and *multiply* its output frequency in a suitably nonlinear stage for further amplification at the desired frequency.

Quartz Crystals in Oscillators

Because crystals afford Q values and frequency stabilities that are orders of magnitude better than those achievable with LC circuits, fixed-frequency oscillators are usually on quartz-crystal resonators. Master references for frequency counters and synthesizers are always based on crystal oscillators.

So glowing is the executive summary of the crystal's reputation for stability that newcomers to radio experimentation naturally believe that the presence of a crystal in an oscillator will force oscillation at the frequency stamped on the can. This impression is usually revised after the first few experiences to the contrary! There is no sure-fire crystal oscillator circuit (although some are better than others); reading and experience soon provide a learner with plenty of anecdotes to the effect that:

- Some circuits have a reputation of being temperamental, even to the point of not always starting.
- Crystals sometimes mysteriously oscillate on unexpected frequencies.

Even crystal manufacturers have these problems, so don't be discouraged from building crystal oscillators. The occasional uncooperative oscillator is a nuisance, not a disaster, and it just needs a little individual attention. Knowing how a crystal behaves is the key to a cure.

Quartz and the Piezoelectric Effect

Quartz is a crystalline material with a regular atomic structure that can be distorted by the simple application of force. Remove the force, and the distorted structure springs back to its original form with very little energy loss. This property allows acoustic waves — sound — to propagate rapidly through quartz with very little attenuation, because the velocity of an acoustic wave depends on the elasticity and density (mass/volume) of the medium through which the wave travels.

If you heat a material, it expands. Heating may cause other characteristics of a material to change — such as elasticity, which affects the speed of sound in the material. In quartz, however, expansion and change in the speed of sound are very small and tend to cancel, which means that the transit time for sound to pass through a piece of quartz is very stable.

The third property of this wonder material is that it is *piezoelectric*. Apply an electric field to a piece of quartz and the crystal lattice distorts just as if a force had been applied. The electric field applies a force to electrical charges locked in the lattice structure. These charges are captive and cannot move around in the lattice as they can in a semiconductor, for quartz is an insulator. A capacitor's dielectric stores energy by creating physical distortion on an atomic or molecular scale. In a piezoelectric crystal's lattice, the distortion affects the entire structure. In some piezoelectric materials, this effect is sufficiently pronounced that special shapes can be made that bend *visibly* when a field is applied.

Consider a rod made of quartz. Any sound wave propagating along it eventually hits an end, where there is a large and abrupt change in acoustic impedance. Just as when an RF wave hits the end of an unterminated transmission line, a strong reflection occurs. The rod's other end similarly reflects the wave. At some frequency, the phase shift of a round trip will be such that waves from successive round trips exactly coincide in phase and reinforce each other, dramatically increasing

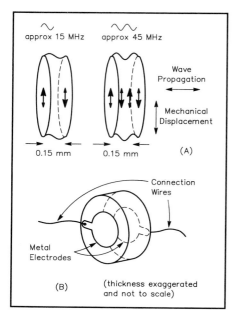

Fig 14.19 — Thickness-shear vibration at a crystal's fundamental and third overtone (A); B shows how the modern crystals commonly used by radio amateurs consist of etched quartz discs with electrodes deposited directly on the crystal surface.

the wave's amplitude. This is resonance.

The passage of waves in opposite directions forms a standing wave with antinodes at the rod ends. Here we encounter a seeming ambiguity: not just one, but a *family* of different frequencies, causes standing waves — a family fitting the pattern of $^1/_2$, $^3/_2$, $^5/_2$, $^7/_2$ and so on, wavelengths into the length of the rod. And this *is* the case: A quartz rod *can* resonate at any and all of these frequencies.

The lowest of these frequencies, where the crystal is $^1/_2$ wavelength long, is called the *fundamental* mode. The others are named the third, fifth, seventh and so on, *overtones*. There is a small phase-shift error during reflection at the ends, which causes the frequencies of the overtone modes to differ slightly from odd integer multiplies of the fundamental. Thus, a crystal's third overtone is very close to, but not exactly, three times, its fundamental frequency. Many people are confused by overtones and harmonics. Harmonics are additional signals at *exact* integer multiples of the fundamental frequency. Overtones are not signals at all; they are additional resonances that can be exploited if a circuit is configured to excite them.

The crystals we use most often resonate in the 1- to 30-MHz region and are of the *AT cut, thickness shear* type, although these last two characteristics are rarely mentioned. A 15-MHz-fundamental crystal of this type is about 0.15 mm thick. Because of the widespread use of reprocessed war-surplus, pressure-mounted *FT-243* crystals, you may think of crystals as small rectangles on the order of a half inch in size. The crystals we commonly use today are discs, etched and/or doped to their final dimensions, with metal electrodes deposited directly on the quartz. A crystal's diameter does not directly affect its frequency; diameters of 8 to 15 mm are typical.

AT cut is one of a number of possible standard designations for the orientation at which a crystal disc is sawn from the original quartz crystal. The crystal lattice atomic structure is asymmetric, and the orientation of this with respect to the faces of the disc influences the crystal's performance. *Thickness shear* is one of a number of possible orientations of the crystal's mechanical vibration with respect to the disc. In this case, the crystal vibrates perpendicularly to its thickness. This is not easy to visualize, and diagrams don't help much, but **Fig 14.19A** is an attempt at illustrating this. Place a moist bathroom sponge between the palms of your hands, move one hand up and down, and you'll see thickness shear in action.

There is a limit to how thin a disc can be made, given requirements of accuracy and price. Traditionally, fundamental-mode

crystals have been made up to 20 MHz, although 30 MHz is now common at a moderately raised price. Using techniques pioneered in the semiconductor industry, crystals have been made with a central region etched down to a thin membrane, surrounded by a thick ring for robustness. This approach can push fundamental resonances to over 100 MHz, but these are more lab curiosities than parts for everyday use. The easy solution for higher frequencies is to use a nice, manufacturably thick crystal on an overtone mode. All crystals have all modes, so if you order a 28.060-MHz, third-overtone unit for a little QRP transmitter, you'll get a crystal with a fundamental resonance somewhere near 9.353333 MHz, but its manufacturer will have adjusted the thickness to plant the third overtone exactly on the ordered frequency. An accomplished manufacturer can do tricks with the flatness of the disc faces to make the wanted overtone mode a little more active and the other modes a little less active. (As some builders discover, however, this does not *guarantee* that the wanted mode is the most active!)

Quartz's piezoelectric property provides a simple way of driving the crystal electrically. Early crystals were placed between a pair of electrodes in a case. This gave amateurs the opportunity to buy surplus crystals, open them and grind them a little to reduce their thickness, thus moving them to higher frequencies. The frequency could be reduced very slightly by loading the face with extra mass, such as by blackening it with a soft pencil. Modern crystals have metal electrodes deposited directly onto their surfaces (Fig 14.19B), and such tricks no longer work.

The piezoelectric effect works both ways. Deformation of the crystal produces voltage across its electrodes, so the mechanical energy in the resonating crystal can also be extracted electrically by the same electrodes. Seen electrically, at the electrodes, the mechanical resonances look like electrical resonances. Their Q is very high. A Q of 10,000 would characterize a *poor* crystal nowadays; 100,000 is often reached by high-quality parts. For comparison, a Q of over 200 for an LC tank is considered good.

Accuracy

A crystal's frequency accuracy is as outstanding as its Q. Several factors determine a crystal's frequency accuracy.

First, the manufacturer makes parts with certain tolerances: ±200 ppm for a low-quality crystal for use as in a microprocessor clock oscillator, ±10 ppm for a good-quality part for professional radio use. Anything much better than this starts to get expensive! A crystal's resonant frequency is influenced by the impedance presented to its terminals, and manufacturers assume that once a crystal is brought within several parts per million of the nominal frequency, its user will perform fine adjustments electrically.

Second, a crystal ages after manufacture. Aging could give increasing or decreasing frequency; whichever, a given crystal usually keeps aging in the same direction. Aging is rapid at first and then slows down. Aging is influenced by the care in polishing the surface of the crystal (time and money) and by its holder style. The cheapest holder is a soldered-together, two-part metal can with glass bead insulation for the connection pins. Soldering debris lands on the crystal and affects its frequency. Alternatively, a two-part metal can be made with flanges that are pressed together until they fuse, a process called *cold-welding*. This is much cleaner and improves aging rates roughly fivefold compared to soldered cans. An all-glass case can be made in two parts and fused together by heating in a vacuum. The vacuum raises the Q, and the cleanliness results in aging that's roughly ten times slower than that achievable with a soldered can. The best crystal holders borrow from vacuum-tube assembly processes and have a *getter*, a highly reactive chemical substance that traps remaining gas molecules, but such crystals are used only for special purposes.

Third, temperature influences a crystal. A reasonable, professional quality part might be specified to shift not more than ±10 ppm over 0 to 70°C. An AT-cut crystal has an *S*-shaped frequency-versus-temperature characteristic, which can be varied by slightly changing the crystal cut's orientation. **Fig 14.20** shows the general shape and the effect of changing the cut angle by only a few seconds of arc. Notice how all the curves converge at 25°C. This is because this temperature is normally chosen as the reference for specifying a crystal. The temperature stability specification sets how accurate the manufacturer must make the cut. Better stability may be needed for a crystal used as a receiver frequency standard, frequency counter clock and so on. A crystal's temperature characteristic shows a little hysteresis. In other words, there's a bit of offset to the curve depending whether temperature is increasing or decreasing. This is usually of no consequence except in the highest-precision circuits.

It is the temperature of the quartz which is important, and as the usual holders for crystals all give effective thermal insulation, only a couple of milliwatts dissipation by the crystal itself can be tolerated before self-heating becomes troublesome. Because such heating occurs in the quartz itself and does not come from the surrounding environment, it defeats the effects of temperature

compensators and ovens.

The techniques shown earlier for VFO for temperature compensation can also be applied to crystal oscillators. An after-compensation drift of 1 ppm is routine and 0.5 ppm is good. The result is a *temperature-compensated crystal oscillator* (*TCXO*). Recently, oscillators have appeared with built-in digital thermometers, microprocessors and ROM look-up tables customized on a unit-by-unit basis to control a tuning diode via a digital-to-analog converter (DAC) for temperature compensation. These *digitally temperature-compensated oscillators* (*DTCXOs*) can reach 0.1 ppm over the temperature range. With automated production and adjustment, they promise to become the cheapest way to achieve this level of stability.

Oscillators have long been placed in temperature-controlled *ovens*, which are typically held at 80°C. Stability of several parts per billion can be achieved over temperature, but this is a limited benefit as aging can easily dominate the accuracy. These are usually called *oven-controlled crystal oscillators* (*OCXOs*).

Fourth, the crystal is influenced by the impedance presented to it by the circuit in which it is used. This means that care is needed to make the rest of an oscillator circuit stable, in terms of impedance and phase shift.

Gravity can slightly affect crystal resonance. Turning an oscillator upside down usually produces a small frequency shift, usually much less than 1 ppm; turning the oscillator back over reverses this. This effect is quantified for the highest-quality reference oscillators.

The Equivalent Circuit of a Crystal

Because a crystal is a passive, two-terminal device, its electrical appearance is that of an impedance that varies with frequency. **Fig 14.21A** shows a very simplified sketch of the magnitude (phase is ignored) of the impedance of a quartz

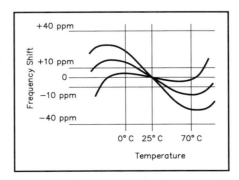

Fig 14.20 — Slight changes in a crystal cut's orientation shift its frequency-versus-temperature curve.

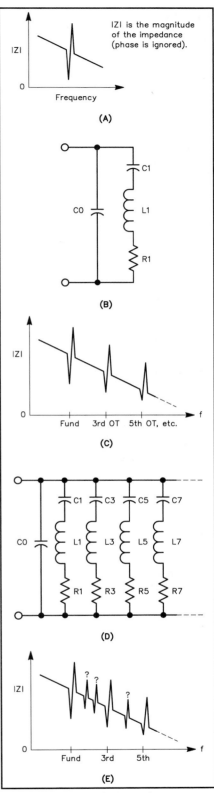

Fig 14.21 — **Exploring a crystal's impedance (A) and equivalent circuit (B) through simplified diagrams. C and D extend the investigation to include overtones; E, to spurious responses not easily predictable by theory or controllable through manufacture. A crystal may oscillate on any of its resonances under the right conditions.**

Table 14.2

Typical Equivalent Circuit Values for a Variety of Crystals

Crystal Type	Series L	Series C (pF)	Series R (Ω)	Shunt C (pF)
1-MHz fundamental	3.5 H	0.007	340	3
10-MHz fundamental	9.8 mH	0.026	7	6.3
30-MHz third overtone	14.9 mH	0.0018	27	6.2
100-MHz fifth overtone	4.28 mH	0.0006	45	7

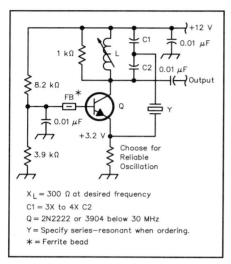

X$_L$ = 300 Ω at desired frequency
C1 = 3X to 4X C2
Q = 2N2222 or 3904 below 30 MHz
Y = Specify series—resonant when ordering.
✱ = Ferrite bead

Fig 14.22 — A basic series-mode crystal oscillator. A 2N5179 can be used in this circuit if a lower supply voltage is used; see text.

crystal. The general trend of dropping impedance with increasing frequency implies capacitance across the crystal. The sharp fall to a low value resembles a series-tuned tank, and the sharp peak resembles a parallel-tuned tank. These are referred to as series and parallel resonances. Fig 14.21B shows a simple circuit that will produce this impedance characteristic. The impedance looks purely resistive at the exact centers of both resonances, and the region between them has impedance increasing with frequency, which looks inductive.

C1 (sometimes called *motional capacitance*, C_m, to distinguish it from the lumped capacitance it approximates) and L1 (*motional inductance*, L_m) create the series resonance, and as C0 and R1 are both fairly small, the impedance at the bottom of the dip is very close to R1. At parallel resonance, L1 is resonating with C1 and C0 in series, hence the higher frequency. The impedance of the parallel tank is immense, the terminals are connected to a capacitive tap, that causes them to see only a small fraction of this, which is still a very large impedance. The

overtones should not be neglected, so Figs 14.21C and 14.21D show them included. Each overtone has series and parallel resonances and so appears as a series tank in the equivalent circuit. C0 again provides the shifted parallel resonance.

This is still simplified, because real-life crystals have a number of spurious, unwanted modes which add yet more resonances, as shown in **Fig 14.21E**. These are not well controlled and may vary a lot even between crystals made to the same specification. Crystal manufacturers work hard to suppress these spurs and have evolved a number of recipes for shaping crystals to minimize them. Just where they switch from one design to another varies from manufacturer to manufacturer.

Always remember that the equivalent circuit is just a representation of crystal behavior and does not represent circuit components actually present. Its only use is as an aid in designing and analyzing circuits using crystals. **Table 14.2** lists typical equivalent-circuit values for a variety of crystals. It is impossible to build a circuit with 0.026 to 0.0006-pF capacitors; such values would simply be swamped by strays. Similarly, the inductor must have a Q orders of magnitude better than is practically achievable, and impossibly low stray C in its winding.

The values given in Table 14.2 are nothing more than rough guides. A crystal's frequency is tightly specified, but this still allows inductance to be traded for capacitance. A good manufacturer could hold these characteristics within a ±25% band or could vary them over a 5:1 range by special design. Similarly marked parts from different sources vary widely in motional inductance and capacitance.

Quartz is not the only material that behaves in this way, but it is the best. Resonators can be made out of lithium tantalate and a group of similar materials that have lower Q, allowing them to be *pulled* over a larger frequency range in VCXOs. Much more common, however, are ceramic resonators based on the technology of the well-known ceramic IF filters. These have much lower Q than quartz

and much poorer frequency precision. They serve mainly as clock resonators for cheap microprocessor systems in which every last cent must be saved. A ceramic resonator could be used as the basis of a wide range, cheap VXO, but its frequency stability would not be as good as a good LC VFO.

Crystal Oscillator Circuits

Crystal oscillator circuits are usually categorized as series- or parallel-mode types, depending on whether the crystal's low- or high-impedance resonance comes into play at the operating frequency. The series mode is now the most common; parallel-mode operation was more often used with vacuum tubes. **Fig 14.22** shows a basic series-mode oscillator. Some people would say that it is an overtone circuit, used to run a crystal on one of its overtones, but this is not necessarily true. The tank (L-C1-C2) tunes the collector of the common-base amplifier. C1 is larger than C2, so the tank is tapped in a way that transforms to a lower impedance, decreasing signal voltage, but increasing current. The current is fed back into the emitter via the crystal. The common-base stage provides a current gain of less than unity, so the transformer in the form of the tapped tank is essential to give loop gain. There are *two* tuned circuits, the obvious collector tank and the series-mode one "in" the crystal. The tank kills the amplifier's gain away from its tuned frequency, and the crystal will only pass current at the series resonant frequencies of its many modes. The tank resonance is much broader than any of the crystal's modes, so it can be thought of as the crystal setting the frequency, but the tank selecting which of the crystal's modes is active. The tank could be tuned to the crystal's fundamental, or one of its overtones.

Fundamental oscillators can be built without a tank quite successfully, but there is always the occasional one that starts up on an overtone or spurious mode. Some simple oscillators have been known to change modes while running (an effect triggered by changes in temperature or loading) or to not always start in the same mode! A series-mode oscillator should present a low impedance to the crystal at the operating frequency. In Fig 14.22, the tapped collector tank presents a transformed fraction of the 1-kΩ collector load resistor to one end of the crystal, and the emitter presents a low impedance to the other. To build a practical oscillator from this circuit, choose an inductor with a reactance of about 300 Ω at the wanted frequency and calculate C1 in series with C2 to resonate with it. Choose C1 to be 3 to 4 times larger than C2. The amplifier's quiescent ("idling") current sets the gain and hence the operating level. This is not easily calculable, but can be found by experiment. Too little quiescent current and the oscillator will not start reliably; too much and the transistor can drive itself into saturation. If an oscilloscope is available, it can be used to check the collector waveform; otherwise, some form of RF voltmeter can be used to allow the collector voltage to be set to 2 to 3 V RMS. 3.3 kΩ would be a suitable starting point for the emitter bias resistor. The transistor type is not critical; 2N2222A or 2N3904 would be fine up to 30 MHz; a 2N5179 would allow operation as an overtone oscillator to over 100 MHz (because of the low collector voltage rating of the 2N5179, a supply voltage lower than 12 V is required). The ferrite bead on the base gives some protection against parasitic oscillation at UHF.

If the crystal is shorted, this circuit should still oscillate. This gives an easy way of adjusting the tank; it is even better to temporarily replace the crystal with a small-value (tens of ohms) resistor to simulate its *equivalent series resistance* (ESR), and adjust L until the circuit oscillates close to the wanted frequency. Then restore the crystal and set the quiescent current. If a lot of these oscillators were built, it would sometimes be necessary to adjust the current individually due to the different equivalent series resistance of individual crystals. One variant of this circuit has the emitter connected directly to the C1/C2 junction, while the crystal is a decoupler for the transistor base (the existing capacitor and ferrite bead not being used). This works, but with a greater risk of parasitic oscillation.

We commonly want to trim a crystal oscillator's frequency. While off-tuning the tank a little will pull the frequency slightly, too much detuning spoils the mode control and can stop oscillation (or worse, make the circuit unreliable). The answer to this is to add a trimmer capacitor, which will act as part of the equivalent series tuned circuit, in series with the crystal. This will shift the frequency in one way only, so the crystal frequency must be respecified to allow the frequency to be varying around the required value. It is common to specify a crystal's frequency with a standard load (30 pF is commonly specified), so that the manufacturer grinds the crystal such that the series resonance

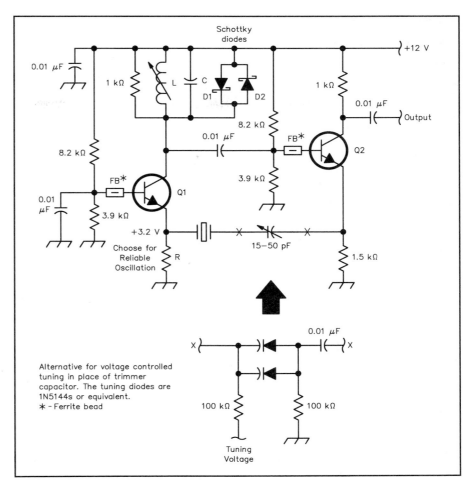

Fig 14.23 — A Butler crystal oscillator.

of the specified mode is accurate when measured with a capacitor of this value in series. A 15- to 50-pF trimmer can be used in series with the crystal to give fine frequency adjustment. Too little capacitance can stop oscillation or prevent reliable starting. The Q of crystals is so high that marginal oscillators can take several seconds to start!

This circuit can be improved by reducing the crystal's driving impedance with an emitter follower as in **Fig 14.23**. This is the *Butler* oscillator. Again the tank controls the mode to either force the wanted overtone or protect the fundamental mode. The tank need not be tapped because Q2 provides current gain, although the circuit is sometimes seen with C split, driving Q2 from a tap. The position between the emitters offers a good, low-impedance environment to keep the crystal's in-circuit Q high. R, in the emitter of Q1, is again selected to give reliable oscillation. The circuit has been shown with a capacitive load for the crystal, to suit a unit specified for a 30-pF load. An alternative circuit to give electrical fine tuning is also shown. The

diodes across the tank act as limiters to stabilize the operating amplitude and limit the power dissipated in the crystal by clipping the drive voltage to Q2. The tank should be adjusted to peak at the operating frequency, not used to trim the frequency. The capacitance in series with the crystal is the proper frequency trimmer.

The Butler circuit works well, and has been used in critical applications to 140 MHz (seventh-overtone crystal, 2N5179 transistor). Although the component count is high, the extra parts are cheap ones. Increasing the capacitance in series with the crystal reduces the oscillation frequency but has a progressively diminishing effect. Decreasing the capacitance pulls the frequency higher, to a point at which oscillation stops; before this point is reached, start-up will become unreliable. The possible amount of adjustment, called *pulling range*, depends on the crystal; it can range from less than ten to several hundred parts per million. Overtone crystals have much less pulling range than fundamental crystals on the same frequency; the reduction in pulling is roughly proportional to the square of the overtone number.

Low-Noise Crystal Oscillators

Fig 14.24A shows a crystal operating in its series mode in a series-tuned Colpitts circuit. Because it does not include an LC tank to prevent operation on unwanted modes, this circuit is intended for fundamental mode operation only and relies on that mode being the most active. If the crystal is ordered for 30-pF loading, the frequency trimming capacitor can be adjusted to compensate for the loading of the capacitive divider of the Colpitts circuit. An unloaded crystal without a trimmer would operate slightly off the exact series resonant frequency in order to create an inductive impedance to resonate with the divider capacitors. Ulrich Rohde, KA2WEU, in Fig 4-47 of his book *Digital PLL Frequency Synthesizers — Theory and Design*, published an elegant alternative method of extracting an output signal from this type of circuit, shown in Fig 14.24B. This taps off a signal from the current in the crystal itself. This can be thought of as using the crystal as a band-pass filter for the oscillator output. The RF choke in the emitter keeps the emitter bias resistor from loading the tank and degrading the Q. In this case (3-MHz operation), it has been chosen to resonate close to 3 MHz with the parallel capacitor (510 pF) as a means of forcing operation on the

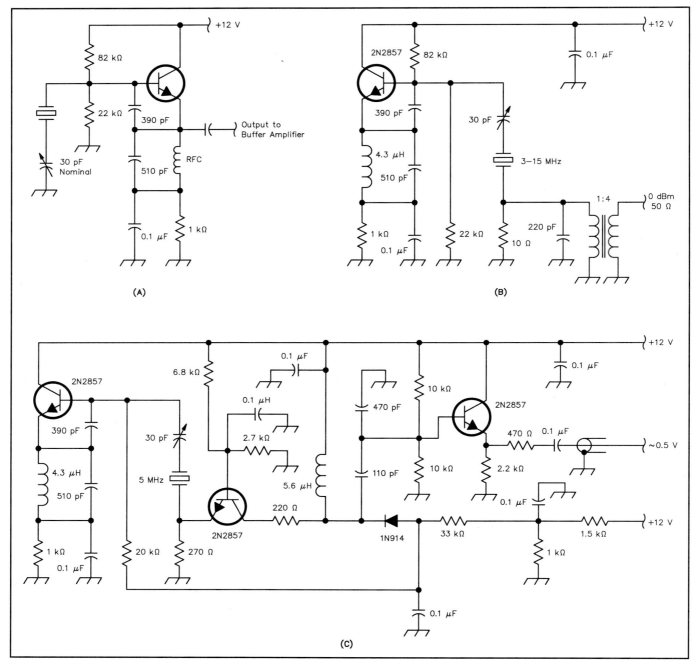

Fig 14.24 — The crystal in the series-tuned Colpitts oscillator at A operates in its series-resonant mode. B shows KA2WEU's low-noise version, which uses the crystal as a filter and features high harmonic suppression. The circuit at C builds on the B version by adding a common-base output amplifier and ALC loop.

wanted mode. The 10-Ω resistor and the transformed load impedance will reduce the in-circuit Q of the crystal, so a further development substituted a common base amplifier for the resistor and transformer. This is shown in Fig 14.24C. The common-base amplifier is run at a large quiescent current to give a very low input impedance, its collector is tuned to give an output with low harmonic content, and an emitter follower is used to buffer this from the load. This oscillator sports a simple ALC system, in which the amplified and rectified signal is used to reduce the bias voltage on the oscillator transistor's base. This circuit is described as achieving a phase noise level of 168 dBc/Hz a few kilohertz out from the carrier. This may seem far beyond what may ever be needed, but frequency multiplication to high frequencies, whether by classic multipliers or by frequency synthesizers, multiplies the deviation of any FM/PM sidebands as well as the carrier frequency. This means that phase noise worsens by 20 dB for each tenfold multiplication of frequency. A clean crystal oscillator and a multiplier chain is still the best way of generating clean microwave signals for use with narrow-band modulation schemes.

It has already been mentioned that overtone crystals are much harder to pull than fundamental ones. This is another way of saying that overtone crystals are less influenced by their surrounding circuit, which is helpful in a frequency standard oscillator like this one. Even though 5 MHz is in the main range of fundamental-mode crystals and this circuit will work well with them, an overtone crystal has been used. To further help stability, the power dissipated in the crystal is kept to about 50 μW. The common-base stage is effectively driven from a higher impedance than its own input impedance, under which conditions it gives a very low noise figure.

VXOs

Some crystal oscillators have frequency trimmers. If the trimmer is replaced by a variable capacitor as a front-panel control, we have a *variable crystal oscillator* (VXO): a crystal-based VFO with a narrow tuning range, but good stability and noise performance. VXOs are often used in small, simple QRP transmitters to tune a few kilohertz around common calling frequencies. Artful constructors, using optimized circuits and components, have achieved 1000-ppm tuning ranges. Poor-quality "soft" crystals are more pullable than high-Q ones. Overtone crystals are

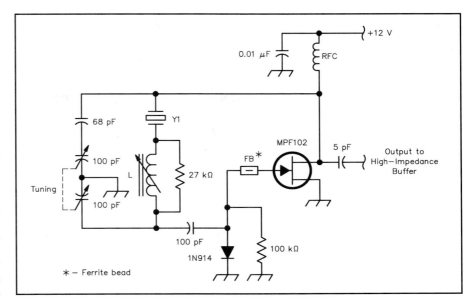

Fig 14.25 — A wide-range variable-crystal oscillator (VXO).

not suited to VXOs. For frequencies beyond the usual limit for fundamental mode crystals, use a fundamental unit and frequency multipliers.

ICOM and Mizuho made some 2 m SSB transceivers based on multiplied VXO local oscillators. This system is simple and can yield better performance than many expensive synthesized radios. SSB filters are available at 9 or 10.7 MHz, to yield sufficient image rejection with a single conversion. Choice of VXO frequency depends on whether the LO is to be above or below signal frequency and how much multiplication can be tolerated. Below 8 MHz multiplier filtering is difficult. Above 15 MHz, the tuning range per crystal narrows. A 50-200 kHz range per crystal should work with a modern front-end design feeding a good 9 MHz IF, for a contest quality 2 m SSB Receiver.

The circuit in **Fig 14.25** is a JFET VXO from Wes Hayward, W7ZOI, and Doug DeMaw, W1FB, optimized for wide-range pulling. Published in *Solid State Design for the Radio Amateur*, many have been built and its ability to pull crystals as far as possible has been proven. Ulrich Rohde, KA2WEU, has shown that the diode arrangement as used here to make signal-dependent negative bias for the gate confers a phase-noise disadvantage, but oscillators like this that pull crystals as far as possible need any available means to stabilize their amplitude and aid start-up. In this case, the noise penalty is worth paying. This circuit can achieve a 2000-ppm tuning range with amenable crystals. If you have some overtone crystals in your junk box whose fundamental frequency is close to the wanted value, they are worth trying.

This sort of circuit doesn't necessarily stop pulling at the extremes of the possible tuning range, sometimes the range is set by the onset of undesirable behavior such as jumping mode or simply stopping oscillating. L was a 16-μH slug-tuned inductor for 10-MHz operation. It is important to minimize the stray and interwinding capacitance of L as this dilutes the range of impedance that is presented to the crystal.

One trick that can be used to aid the pulling range of oscillators is to tune out the C0 of the equivalent circuit with an added inductor. **Fig 14.26** shows how. L is chosen to resonate with C0 for the individual crystal, turning it into a high impedance parallel-tuned circuit. The Q of this circuit is orders of magnitude less than the Q of the true series resonance of the crystal, so its tuning is much broader. The value of C0 is usually just a few picofarads, so L has to be a fairly large value considering the frequency it is resonated at. This means that L has to have low stray capacitance or else it will self-resonate at a lower frequency. The tolerance on C0 and the variations of the stray C of the inductor means that individual adjust-ment is needed. This technique can also work wonders in crystal ladder filters.

Logic-Gate Crystal Oscillators

A 180° phase-shift network and an inverting amplifier can be used to make an oscillator. A single stage RC low-pass network cannot introduce more than 90° of phase shift and only approaches that as the signal becomes infinitely attenuated. Two stages can only approach 180° and then have immense attenuation. It takes three stages to give 180° and not destroy the loop gain. **Fig 14.27A** shows the basic form of the *phase shift* oscillator; Fig 14.27B is an example the phase-shift oscillator as commonly used in commercial Amateur Radio transceivers as an audio sidetone oscillator. The frequency-determining network of an RC oscillator has a Q of less than one, which is a massive disadvantage compared to an LC or crystal resonator. The Pierce crystal oscillator is a converted phase shift oscillator with the crystal taking the place of one series resistor, Fig 14.27C. At the exact series resonance, the crystal looks resistive, so by suitable choice of the capacitor values, the circuit can be made to oscillate. The crystal has a far steeper phase/frequency relationship than the rest of the

network, so the crystal is the dominant controller of the frequency. The Pierce circuit is rarely seen in this full form, instead, a cut-down version has become the most common circuit for crystal clock oscillators for digital systems. Fig 14.27D shows this minimalist Pierce, using a logic inverter as the amplifier. R_{bias} provides dc negative feedback to bias the gate into its linear region. At first sight it appears that this arrangement should not oscillate, but the crystal is a resonator (not a simple resistor) and oscillation occurs offset from the series resonance, where the crystal appears inductive, which makes up the missing phase shift. This is one circuit which *cannot* oscillate exactly on the crystal's series resonance. This circuit is included

in many microprocessors and other digital ICs that need a clock, it is also the usual circuit inside the miniature clock oscillator cans. It is not a very reliable circuit, as operation is dependent on the crystal's equivalent series resistance and the output impedance of the logic gate, occasionally the logic device or the crystal needs to be changed to start oscillation, sometimes playing with the capacitor values is necessary. It is doubtful whether these circuits are ever designed, values (the two capacitors) seem to be arrived at by experiment. Once going, the circuit is reliable, but its drift and noise are moderate, not good — acceptable for a clock oscillator. The commercial packaged oscillators are the same, but the manufacturers have handled the

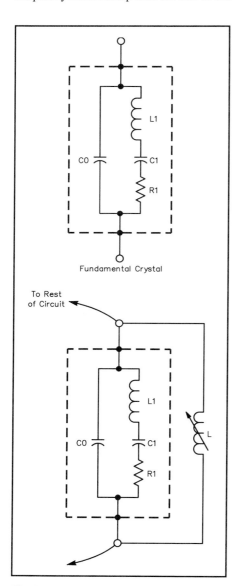

Fig 14.26 — Using an inductor to "tune out" C0 can increase a crystal oscillator's pulling range.

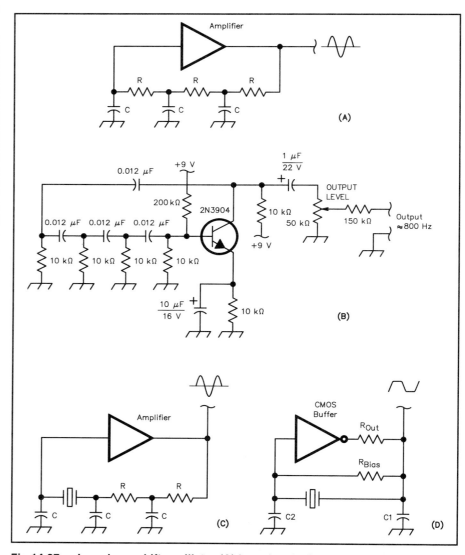

Fig 14.27 — In a phase-shift oscillator (A) based on logic gates, a chain of RC networks — three or more — provide the feedback and phase shift necessary for oscillation, at the cost of low Q and considerable loop loss. Many commercial Amateur Radio transceivers have used a phase-lead oscillator similar to that shown at B as a sidetone generator. Replacing one of the resistors in A with a crystal produces a Pierce oscillator (C), a cut-down version of which (D) has become the most common clock oscillator configuration in digital systems.

production foibles on a batch-by-batch basis.

RC Oscillators

Plenty of RC oscillators are capable of operating to several megahertz. Some of these are really constant-current-into-capacitor circuits which are easier to make in silicon. Like the phase-shift oscillator above, the timing circuit Q is less than one, giving very poor noise performance that's unsuitable even to the least demanding radio application. One example is the oscillator section of the CD4046 phase-locked-loop IC. This oscillator has poor stability over temperature, large batch-to-batch variation and a wide variation in its voltage-to-frequency relationship. It is not recommended that this sort of oscillator is used at RF in radio systems. (The '4046 phase detector section is very useful, however, as we'll see later.) These oscillators are best suited to audio applications.

VHF AND UHF OSCILLATORS

A traditional way to make signals at higher frequencies is to make a signal at a lower frequency (where oscillators are easier) and multiply it up to the wanted range. Multipliers are still one of the easiest ways of making a clean UHF/microwave signal. The design of a multiplier depends on whether the multiplication factor is an odd or even number. For odd multiplication, a Class-C biased amplifier can be used to create a series of harmonics; a filter selects the one wanted. For even multiplication factors, a full-wave-rectifier arrangement of distorting devices can be used to create a series of harmonics with strong even-order components, with a filter selecting the wanted component. At higher frequencies, diode-based passive circuits are commonly used. Oscillators using some of the LC circuits already described can be used in the VHF range, at UHF different approaches become necessary.

Fig 14.28 shows a pair of oscillators based on a resonant length of line. The first one is a return to basics: a resonator, an amplifier and a pair of coupling loops. The amplifier can be a single bipolar or FET device, or one of the monolithic microwave integrated circuit (MMIC) amplifiers. The second circuit is really a Hartley, and one was made as a test oscillator for the 70-cm band from a 10-cm length of wire suspended 10 mm over an unetched PC board as a ground plane, bent down and soldered at one end, a trimmer at the other end. The FET was a BF981 dual-gate device used as a source follower.

No free-running oscillator will be stable enough on these bands except for use with wideband FM or video modulation and AFC at the receiver. Oscillators in this range are almost invariably tuned with tuning diodes controlled by phase-locked-loop synthesizers, which are themselves controlled by a crystal oscillator.

There is one UHF oscillator which is extremely common, yet rarely applied intentionally. The answer to this riddle is a configuration that is sometimes deliberately built as a useful wide-tuning oscillator covering say, 500 MHz to 1 GHz — and is also the modus operandi of a very common form of *spurious* VHF/UHF oscillation in circuitry intended to process lower-frequency signals! This oscillator has no generally accepted name. It relies on the creation of a small negative resistance in series with a series resonant LC tank. **Fig 14.29A** shows the circuit in its simplest form. This circuit is well-suited to construction with printed-circuit inductors. Common FR4 glass-epoxy board is lossy at these frequencies; better performance can be achieved by using the much more expensive glass-Teflon board. If you can get surplus offcuts of this type of material, it has many uses at UHF and microwave, but it is difficult to use, as the adhesion between the copper and the substrate can be poor. A high-UHF transistor with a 5-GHz f_T like the BFR90 is suitable; the base inductor can be 30 mm of 1-mm trace folded into a hairpin shape (inductance, less than 10 nH). Analyzing this circuit using a comprehensive model of the UHF transistor reveals that the emitter presents an impedance that is small, resistive and negative to the outside world. If this is large enough to more than cancel the effective series resistance of the emitter tank, oscillation will occur. Fig 14.29B shows a very basic emitter-follower circuit with some capacitance to ground on both the input and output. If the capacitor shunting the input is a distance away from the transistor, the trace can look like an inductor — and small capacitors at audio and low RF can look like very good decouplers at hundreds of MHz. The length to the capacitor shunting the output will behave as a series resonator at a frequency where it is a ¼ wavelength long. This circuit is the same as in Fig 14.29A; it, too, can oscillate. The semiconductor manufacturers have steadily improved their small-signal transistors to give better gain and bandwidth so that any transistor circuit where all three electrodes find themselves decoupled to ground at UHF may oscillate at several hundred megahertz. The upshot of this is that there is no longer any branch of electronics where RF design and layout techniques can be safely ignored. A circuit must not just be de-

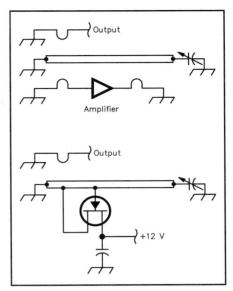

Fig 14.28 — Oscillators that use transmission-line segments as resonators. Such oscillators are more common than many of us may think, as Fig 14.29 reveals.

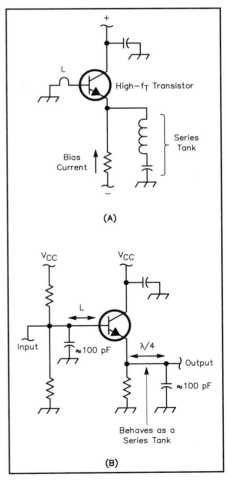

Fig 14.29 — High device gain at UHF and resonances in circuit-board traces can result in spurious oscillations even in non RF equipment.

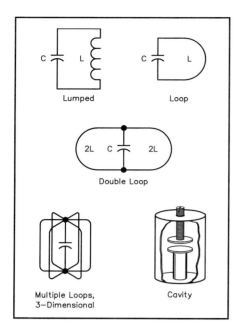

Fig 14.30 — Evolution of the cavity resonator.

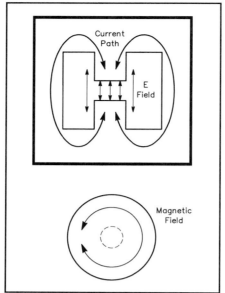

Fig 14.31 — Currents and fields in a cavity.

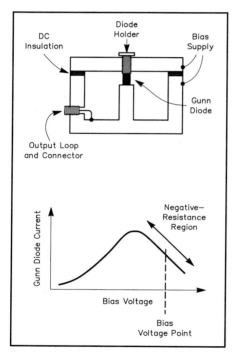

Fig 14.32 — A Gunn diode oscillator uses negative resistance and a cavity resonator to produce radio energy.

signed to do what it *should* do, it must also be designed so that it cannot do what it *should not* do.

There are three ways of taming such a circuit; adding a small resistor, perhaps 50 to 100 Ω in the collector lead, close to the transistor, or adding a similar resistor in the base lead, or by fitting a ferrite bead over the base lead under the transistor. The resistors can disturb dc conditions, depending on the circuit and its operating currents. Ferrite beads have the advantage that they can be easily added to existing equipment and have no effect at dc and low frequencies. Beware of some electrically conductive ferrite materials that can short transistor leads. If an HF oscillator uses beads to prevent any risk of spurs (Fig 14.16), the beads should be anchored with a spot of adhesive to prevent movement that can cause small frequency shifts. Ferrite beads of Fair-Rite no. 43 material are especially suitable for this purpose; they are specified in terms of impedance, not inductance. Ferrites at frequencies above their normal usable range become very lossy and can make a lead look not inductive, but like a few tens of ohms, resistive.

Microwave Oscillators

Low-noise microwave signals are still best made by multiplying a very-low-noise HF crystal oscillator, but there are a number of oscillators that work directly at microwave frequencies. Such oscillators can be based on resonant lengths of stripline or microstrip, and are simply scaled-down versions of UHF oscillators, using microwave transistors and printed striplines on a low-loss substrate, like alumina. Techniques for printing metal traces on substrates to form filters, couplers, matching networks and so on, have been intensively developed over the past two decades. Much of the professional microwave community has moved away from

waveguide and now uses low-loss coaxial cable with a solid-copper shield — *semi-rigid cable* — to connect circuits made flat on ceramic or Teflon based substrates. Semiconductors are often bonded on as unpackaged chips, with their bond-wire connections made directly to the traces on the substrate. At lower microwave frequencies, they may be used in standard surface-mount packages. From an Amateur Radio viewpoint, many of the processes involved are not feasible without access to specialized furnaces and materials. Using ordinary PC-board techniques with surface-mount components allows the construction of circuits up to 4 GHz or so. Above this, structures get smaller and accuracy becomes critical; also PC-board materials quickly become very lossy.

Older than stripline techniques and far more amenable to home construction, cavity-based oscillators can give the highest possible performance. The dielectric constant of the substrate causes stripline structures to be much smaller than they would be in free air, and the lowest-loss substrates tend to have very high dielectric constants. Air is a very low-loss dielectric with a dielectric constant of 1, so it gives high Q and does not force excessive miniaturization. **Fig 14.30** shows a series of structures used by G. R. Jessup, G6JP, to illustrate the evolution of a cavity from a tank made of lumped components. All cavities have a number of different modes of resonance, the orientation of the currents and fields are shown in **Fig 14.31**. The cavity can take different shapes, but that shown has proven to well-suppress unwanted modes. The gap need not be central and is often right at the top. A screw can be fitted through the top, protruding into the gap, to adjust the frequency.

To make an oscillator out of a cavity, an amplifier is needed. Gunn and tunnel diodes have regions in their characteristics where their current *falls* with increasing bias voltage. This is negative resistance. If such a device is mounted in a loop in a cavity and bias applied, the negative resistance can more than cancel the effective loss resistance of the cavity, causing oscillation. These diodes are capable of operating at extremely high frequencies and were discovered long before transistors were developed that had any gain at microwave frequencies.

A Gunn-diode cavity oscillator is the basis of many of the Doppler-radar modules used to detect traffic or intruders. **Fig 14.32** shows a common configuration. The coupling loop and coax output connector could be replaced with a simple aperture to couple into waveguide or a mixer cavity. **Fig 14.33** shows a transistor cavity oscillator version using a modern microwave transistor. FET or bipolar devices can be used. The two coupling loops are completed by the capacitance of the feedthroughs.

The *dielectric-resonator oscillator* (*DRO*) may soon become the most common microwave oscillator of all, as it is used in the downconverter of satellite TV receivers. The dielectric resonator itself is a ceramic cylinder, like a miniature hockey puck, several millimeters in diameter. The ceramic has a very high dielectric constant, so the surface (where ceramic meets air in an abrupt mismatch) reflects electromagnetic waves and makes the ceramic body act as a resonant cavity. It is mounted on a substrate and coupled to the active device of the oscillator by a stripline which runs past it. At 10 GHz, a FET made of gallium arsenide (GaAsFET), rather than silicon, is normally used. The dielectric resonator elements are made at frequencies appropriate to mass applications like satellite TV. The set-up charge to manufacture small quantities at special frequencies is likely to be prohibitive for the foreseeable future. The challenge with these devices is to devise new ways of using oscillators on industry standard frequencies. Their chief attraction is their low cost in large quantities and compatibility with microwave stripline (microstrip) techniques. Frequency stability and Q are competitive with good cavities, but are inferior to that achievable with a crystal oscillator and chain of frequency multipliers. Satellite TV downconverters need free running oscillators with less than 1 MHz of drift at 10 GHz.

There are a number of thermionic (vacuum-tube) microwave sources, klystrons, magnetrons and *backwards wave oscillators* (*BWO*s), but available devices are either very old or designed for very high power.

The *yttrium-iron garnet* (*YIG*) oscillator was developed for a wide tuning range as a solid-state replacement for the BWO, and many of them can be tuned over more than an octave. They are complete, packaged units that appear to be a heavy block of metal with low-frequency connections for power supplies and tuning, and an SMA connector for the RF output. The manufacturer's label usually states the tuning range and often the power supply voltages. This is very helpful because, with new units priced in the kilodollar range, it is important to be able to identify surplus units. The majority of YIGs are in the 2- to 18-GHz region, although units

down to 500 MHz and up to 40 GHz are occasionally found. In this part of the spectrum there is no octave-tunable device that can equal their cleanliness and stability. A 3-GHz unit drifting less than 1 kHz per second gives an idea of typical stability. This seems very poor — until we realize that this is 0.33 ppm per second. Nevertheless, any YIG application involving narrow-band modulation will usually require some form of frequency stabilizer.

Good quality, but elderly, RF spectrum analyzers have found their way into the workshops of a number of dedicated constructors. A 0- to 1500-MHz analyzer usually uses a 2- to 3.5-GHz YIG as its first local oscillator, and its tuning circuits are designed around it. A reasonable understanding of YIG oscillators will help in troubleshooting and repair.

YIG spheres are resonant at a frequency controlled not only by their physical dimensions, but also by any applied magnetic field. A YIG sphere is carefully oriented within a

coupling loop connected to a negative-resistance device and the whole assembly is placed between the poles of an electromagnet. Negative-resistance diodes have been used, but transistor circuits are now common. The support for the YIG sphere often contains a thermostatically controlled heater to reduce temperature sensitivity.

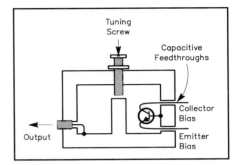

Fig 14.33 — A transistor can also directly excite a cavity resonator.

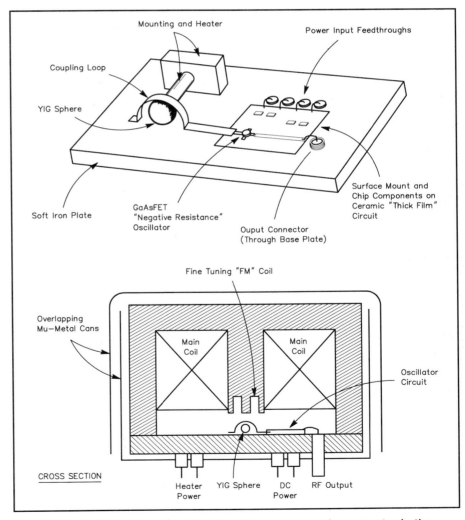

Fig 14.34 — A yttrium-iron-garnet (YIG) sphere serves as the resonator in the sweep oscillators used in many spectrum analyzers.

The first problem with a magnetically tuned oscillator is that magnetic fields, especially at low frequencies, are extremely difficult to shield; the tuning will be influenced by any local fields. Varying fields will cause frequency modulation. The magnetic core must be carefully designed to be all-enclosing in an attempt at self-shielding and then one or more nested mu-metal cans are fitted around everything. It is still important to site the unit away from obvious sources of magnetic noise, like power transformers.

Cooling fans are less obvious sources of fluctuating magnetic fields, yet some are 20 dB worse than a well designed 200-W 50/60-Hz transformer.

The second problem is that the oscillator's internal tuning coils need significant current from the power supply to create strong fields. This can be eased by adding a permanent magnet as a fixed "bias" field, but the bias will shift as the magnet ages. The only solution is to have a coil with many turns, but high inductance, which will require a high supply voltage to permit rapid tuning, thus increasing the power consumption. The usual compromise is to have dual coils: One with many turns allows slow tuning over a wide range; a second with much fewer turns allows fast tuning or FM over a limited range. The main coil can have a sensitivity in the 20 MHz/mA range; and the "FM" coil, perhaps 500 kHz/mA. The frequency/current relationship can have excellent inearity. **Fig 14.34** shows the construction of a YIG oscillator.

Frequency Synthesizers

Narrow-band modes on HF and all modes on VHF upwards, place difficult to impossible demands on VFO stability and setting accuracy. (Wideband modes, such as WBFM, ATV and higher-speed data, may not.) Some radios are only required to tune to a few channels, and an obvious solution is to use switched crystals. This means a pair of crystals for each channel, because the transmitter and receiver usually need separate LOs to create the IF offset needed even for simplex operation. This approach soon becomes unacceptably expensive and large as the number of channels is increased; also, there is labor involved in periodically checking and adjusting each channel individually. Commercial designers were desperate for any way of reducing the size of crystal bank needed. One way, once commonly used in aircraft radios, was to have a multisuperhet, with a switched group of crystals for each LO. The last LO would have crystal frequencies chosen to give steps equal to the required channel spacing, the next earlier LO would have its crystal frequencies chosen to give steps equal to the range of the last LO + 1 channel and so on.

If a step-tuned MF/HF SSB/CW radio must sound as if it tunes as continuously as an LC VFO, a 10-Hz step size is needed. For general coverage, close to a 30-MHz LO tuning range is needed, giving a total of 3 million channels. We handle this with a first LO with 3 crystals at 10-MHz increments, a second LO with 10 crystals at 1 MHz increments, a third LO with 10 crystals at 100-kHz increments and so on to a seventh LO with 10 crystals at 10-Hz increments. Sixty-three crystals are a big improvement over three million, but are still too many, and seven frequency conversions are also excessive. Using a VXO for fine tuning could eliminate the 10- and 100-Hz stages, but would still leave a total of 43 crystals.

A way was invented of having only one crystal bank and reusing it for successive stages. Additionally, the frequency-generation system was divorced from the receiver structure. **Fig 14.35** shows what has become known as a *direct synthesizer*. This has been greatly oversimplified to show just the general principle. In many designs, two stages of mixing were used in place of each mixer shown.

The first divide-by-10 reduces the increment size from the first switch tenfold. After the second divider, the influence of the first switch has been reduced one hundredfold and that of the second switch, tenfold. Notice how the structure has repeated cells, although the last cell has no divider. Careful choice of frequencies makes the cells, even the filters, absolutely identical. To get finer output step size, usually called finer *resolution*, just add more cells in the chain. This was revolutionary because it allows the generation of as many steps or channels as could be wanted from 11 crystal oscillators. (It is not necessary to use a decimal structure with 10 components; other numerical bases can be used, although reducing the number of components will require more cells for the same number of possible output channels.)

The disadvantage is that this method requires excellent shielding, for each cell operates in roughly the same frequency range, and any signal leaking between cells can reach the output as a spurious signal. Not only must shielding be used to prevent coupling by radiation, but power buses must be well-filtered as they enter each cell, and (less obviously) isolation amplifiers are necessary to prevent signals leaking between cells along the component frequency buses. These early frequency synthesizers could change frequency extremely rapidly if diode RF switches were used to select components and could be extended to microwave frequencies.

Phase-Locked Loops

A totally different approach, based on phase-locked loops, long ago came to dominate frequency synthesis. It is still dominant today, although it is now often combined with other techniques to form hybrids.

The principle of the *phase-locked loop* (*PLL*) synthesizer is very simple indeed. An oscillator can be built to cover the required frequency range, so what is needed is a system to keep its tuning correct. This

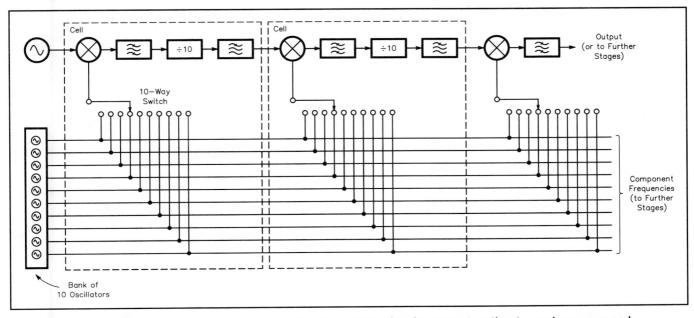

Fig 14.35 — Direct frequency synthesis involves mixing two or more signal sources together to produce sums and differences at desired output frequencies. Great elaboration (and, with it, great likelihood of spurious outputs) is necessary to achieve fine tuning steps. This direct synthesizer is simplified to emphasize the principles involved.

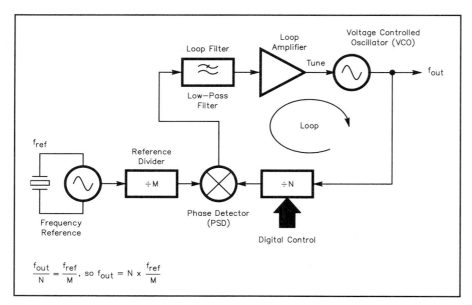

Fig 14.36 — A basic phase-locked-loop (PLL) synthesizer acts to keep the divided-down signal from its voltage-controlled oscillator (VCO) phase-locked to the divided-down signal from its reference oscillator. Fine tuning steps are therefore possible without the complication of direct synthesis.

is done by continuously comparing the phase of the oscillator to a stable reference, such as a crystal oscillator and using the result to steer the tuning. If the oscillator frequency is too high, the phase of its output will start to lead that of the reference by an increasing amount, which is detected and used to decrease the frequency of the oscillator. Too low a frequency causes an increasing phase lag, which is detected and used to increase the frequency of the oscillator. In this way, the oscillator is locked into a fixed phase relationship with the reference, which means that their frequencies are held exactly equal.

The oscillator is now under control, but is locked to a fixed frequency. **Fig 14.36** shows the next step. The phase detector does not simply compare the oscillator frequency with the reference oscillator; both signals have now been passed through frequency dividers. Advances in digital integrated circuits have made frequency dividers, up to low microwave frequencies, commonplace. The divider on the reference path has a fixed division factor, but that in the VCO path is programmable, the factor being entered digitally as a (usually) binary word. The phase detector is now operating at a lower frequency, which is a submultiple of both the reference and output frequencies.

The phase detector, via the loop amplifier, steers the oscillator to keep both its inputs equal in frequency. The reference frequency, divided by M, is equal to the output frequency divided by N. The out-

put frequency equals the reference frequency ×N/M. N is programmable (and an integer), so this synthesizer is capable of tuning in steps of F_{ref}/M.

As an example, to make a 2-m radio covering 144 to 148 MHz, having a 10.7-MHz IF, we need a local oscillator covering 154.7-158.7 MHz. If the channel spacing is 20 kHz, then F_{ref}/M is 20 kHz, so N is 7735 to 7935. There is a free choice of F_{ref} or M, but division by a round binary number is easiest. Crystals are readily available for 10.24 MHz, and one of these (divided by 512) will give 20 kHz at low cost. (ICs are readily available containing most of the circuitry necessary for the reference oscillator — less the crystal — the programmable divider and the phase detector.)

The phase detector in this example compares the relative timing of 5-V pulses (CMOS logic level) at 20 kHz. Inevitably, the phase detector output will contain strong components at 20 kHz and its harmonics in addition to the wanted "steering" signal. The loop filter must block these unwanted signals; otherwise the loop amplifier will amplify them and apply them to the VCO, generating unwanted FM. No filtering can be perfect and the VCO is very sensitive, so most synthesizers have measurable sidebands spaced at the phase detector operating frequency (and harmonics) away from the carrier. These are called *reference-frequency sidebands*. (Exactly which is the reference frequency is ambiguous: Do we mean the frequency of the reference oscillator, or

the frequency applied to the reference input of the phase detector? You must look carefully at context whenever *reference frequency* is mentioned.)

The loop filter is not usually built as a single block of circuitry, it is often made up of three areas: some components directly after the phase detector, a shaped frequency response in the loop amplifier and some more components between the loop amplifier and the VCO.

The PLL is like a feedback amplifier, although the "signal" around its loop is represented by frequency in some places, by phase in others and by voltage in others. Like any feedback amplifier, there is the risk of instability and oscillations traveling around the loop, which can be seen as massive FM on the output and a strong ac signal on the tuning line to the VCO. The loop's filtering and gain has to be designed to prevent this. Good design makes allowance for component variation and the variation in a VCO's sensitivity across its tuning range, as well as from unit to unit — and then leaves plenty of safety margin!

Like any feedback amplifier, the PLL has a frequency response and gain. The following example illustrates: Imagine that we shift the reference oscillator 1 Hz. The reference applied to the phase detector would shift 1/M Hz, and the loop would respond by shifting the output frequency to produce a matching shift at the other phase detector port, so the output is shifted by N/M Hz. Imagine now that we apply a very small amount of FM to the reference oscillator. The amount of deviation will be amplified by N/M — but this is only true for low modulating frequencies. If the modulating frequency is increased, eventually the loop filter starts to reduce the gain around the loop, the loop ceases to track the modulation and the deviation at the output falls. This is referred to as the *closed-loop frequency response* or *closed-loop bandwidth*. Poorly designed loops can have poor closed-loop responses, for example, the gain can have large peaks above N/M at some reference modulation frequencies, indicating marginal stability and excessively amplifying any noise at those offsets from the carrier. The design of the loop filtering and the gain around the loop sets the closed-loop performance.

In a single-loop synthesizer, there is a trade-off between how fine the step size can be versus the performance of the loop. A loop with a very fine step size runs the phase detector at a very low frequency, so the loop bandwidth has to be kept very, very low to keep the reference-frequency sidebands low. Also, the reference oscillator usually has much better phase-noise performance than the VCO, and (within the loop bandwidth) the loop acts to oppose the low-frequency components of the VCO phase-noise sidebands. This very useful cleanup activity is lost when loops have to be narrowed in bandwidth to allow narrow steps. Low-bandwidth loops are slow to respond — they exhibit overly long *settling times* — to the changes in N necessary to change channels. Absolutely everything seems to become impossible in any attempt to get fine resolution from a phase-locked loop.

Single-loop synthesizers are okay for the 2-m FM example for a number of reasons: Channel spacings in this band are not too small, FM is not as critical of phase noise as other modes, 2-m FM is rarely used for weak-signal DXing and the channelization involved does away with the desirability of simulating fast, smooth, continuous tuning. None of these excuses apply to MF/HF radios, however, and ways to circumvent the problems are needed.

A clue was given earlier, by carefully referring to *single*-loop synthesizers. It's possible to use *several* PLLs as components in a larger structure, dividing the frequency of one loop and adding it to the frequency of another that has not been divided. This represents a form of hybrid between the old direct synthesizer and the PLL. A form of PLL containing a mixer in place of the programmable divider can be used to perform the frequency addition.

Voltage-Controlled Oscillators

Voltage-controlled oscillators are commonly merely referred to as *VCOs*. (Voltage-tuned oscillator, VTO, more accurately describes the circuitry most commonly used in VCOs, but tradition is tradition! An exception to this is the YIG oscillator, sometimes used in UHF and microwave PLLs, which is *current*-tuned.) In all the oscillators described so far, except for permeability tuning, the frequency is controlled or trimmed by a variable capacitor. These are modified for voltage tuning by using a tuning diode.

When a normal silicon diode is reverse-biased, the mobile charge carriers are stripped out of a zone either side of the junction. This is called the *depletion region* (**Fig 14.37**); lacking any movable charges, it is a good insulator. The two outer regions have plenty of charge carriers, so are good conductors, and are connected to the diode's leads. This insulator surrounded by conductors is the classic recipe for a capacitor. If the reverse bias voltage is increased, the depletion region widens, reducing the capacitance. The capacitance depends on the cross-sec-

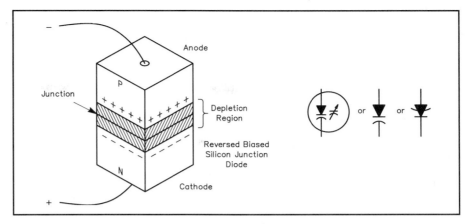

Fig 14.37 — Modern VCOs use the voltage-variable characteristic of diode PN junctions for tuning.

tional area of the diode, the dielectric constant of depleted silicon and the width of the depletion region (in just the same way as for a normal capacitor). The width of the depletion region depends on the doping profile of the semiconductor and the bias voltage. All semiconductor diodes do this. Common small-signal and power diodes can be used, but their capacitance range is small, as is true of any junction of a bipolar transistor or JFET. Deliberately made tuning diodes are carefully designed to give much higher capacitance ranges. These diodes go by a wide range of names: Varactors, Varicaps, Epicaps, tuning diodes, voltage-variable capacitors.

This effect is fast enough so that when a signal voltage exists across the diode, superimposed on the bias voltage, the capacitance varies following the instantaneous total voltage. This is the basis of the old tuning-diode frequency multipliers (once used to convert the output of a 2-m transmitter to a 70-cm band) and for *parametric amplifiers*. These applications are rare, however. Modulation of a diode's capacitance by the signal is usually a nuisance, distorting the signal and creating harmonics. This means that oscillators tuned by diodes can have significantly more harmonic content than those using normal capacitors, especially when run at high levels, as is often done to get best noise performance. (The collector-base junction of a transistor is a reverse-biased diode and also exhibits signal-dependent capacitance. The effect of this capacitance, including signal-dependent variation, is multiplied by the Miller effect and can cause rising distortion at increasing frequency in an amplifier.)

Tuning diodes cannot rival the Q of normal fixed and variable capacitors, and their Q degrades rapidly at the low voltage/high capacitance end of their range. In

ordinary LC oscillators it is normal for the capacitors to have a Q an order of magnitude greater than the inductor, making coil losses the dominant limitation. Tuning-diode Q, however, is comparable with that of the inductor, and care is therefore needed to choose the most suitable diodes and operate them in bias regions that afford favorable Q across the capacitance range needed. The capacitance-voltage curve becomes much steeper in the low bias/high capacitance region, and the tuning characteristic of oscillators will become very nonlinear at the low-frequency end of the range. It is a good idea to avoid running tuning diodes below about 2 V of reverse bias.

When choosing a diode, read its data sheet carefully. Often, capacitance ratios over 10:1 are proclaimed, but achieving the low-capacitance end of a given such range usually requires a higher-voltage bias supply than are often available. For a high-performance synthesizer, it is worth adding a higher-voltage power supply to run varactors in their best regions, possibly using parallel connected diodes to make up for the reduction in diode capacitance at higher reverse voltages.

Fig 14.38 shows a VCO used in the PLL frequency summer later in this chapter. This is a robust circuit, rather than an attempt at the ultimate in low-noise design, but it illustrates a few useful tricks. Instead of a single tuning diode, a cathode-to-cathode pair are used. The signal voltage increases the bias on one diode while it decreases that on the other. This gives a partial cancellation of the nonlinear effects on the signal, giving lower harmonic content from the oscillator. This is a very common arrangement, and dual diodes like that used here are becoming popular. The inductor helpfully provides a dc ground path for the anode of the top

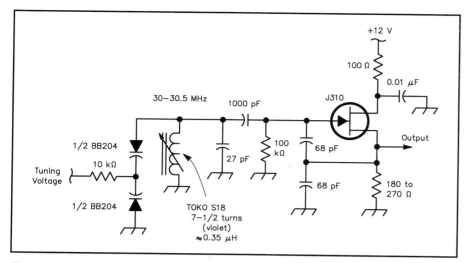

Fig 14.38 — A practical VCO. The tuning diodes are halves of a BB204 dual, common-cathode tuning diode (capacitance per section at 3 V, 39 pF) or equivalent. The ECG617, NTE617 and MV104 are suitable dual-diode substitutes, or use pairs of 1N5451s (39 pF at 4 V) or MV2109s (33 pF at 4 V).

diode, so only one resistor is needed to apply the tuning bias. (In high-performance, low noise synthesizers, it is normal to use an RF choke as the bias feed, to avoid modulating the oscillator with the noise generated in the high-value resistor.) The tuning range is small, 30.0 to 30.5 MHz, so the tuning diodes constitute only a part of the capacitance across the coil; the rest is in the form of high-Q fixed components. This limits the reduction of tank Q due to tuning-diode losses. When this oscillator is used in a PLL, the PLL adjusts the tuning voltage to completely cancel any drift in the oscillator (provided it does not drift further than it can be tuned) regardless of its cause, so no special care is needed to compensate a PLL VCO for temperature effects. Adjusting the inductor core will not change the frequency; the PLL will adjust the tuning voltage to hold the oscillator on frequency. This adjustment is important, though. It's set so that the tuning voltage neither gets too close to the maximum available, nor too low for acceptable Q, as the PLL is tuned across its full range.

In a high-performance, low-noise synthesizer, the VCO may be replaced by a bank of switched VCOs, each covering a section of the total range needed, each VCO having better Q and lower noise because the lossy diodes constitute only a smaller part of the total tank capacity than they would in a single, full-range oscillator. Another method uses tuning diodes for only part of the tuning range and switches in other capacitors (or inductor taps) to extend the range. The performance of wide-range VCOs can be improved by using a large

number of varactors in parallel in a high-C, low-L tank.

Programmable Dividers

Designing your own programmable divider is now a thing of the past, as complete systems have been made as single ICs for over 10 years. The only remaining reason to do so is when an ultra-low-noise synthesizer is being designed.

It may seem strange to think of a digital counter as a source of random noise, but digital circuits have propagation delays caused by analog effects, such as the charging of stray capacitances. Digital circuits are composed of the same sorts of components as analog circuits; there are no such things as digital and analog electrons! Signals are made of voltages and currents. The prime difference between analog and digital electronics lies in the different ways meaning is assigned to the magnitude of a signal. The differences in circuit design are consequences of this, not causes.

Thermal noise in the components of a digital circuit, added to the signal voltage will slightly change the times at which thresholds are crossed. As a result, the output of a digital circuit will have picked up some timing jitter. Jitter in one or both of the signals applied to a phase detector can be viewed as low-level random phase modulation. Within the loop bandwidth, the phase detector steers the VCO so as to oppose and cancel it, so phase jitter or noise is applied to the VCO in order to cancel out the jitter added by the divider. This makes the VCO noisier. The noise performance of the high-speed CMOS logic normally used in programmable di-

viders is good. For ultra-low-noise dividers, ECL devices are sometimes used.

A programmable divider (**Fig 14.39A**) consists of a loadable counter and some control circuitry. The counter is designed to count downward. The programmed division factor is loaded into the counter, and the incoming frequency clocks the downward counting. When the count reaches zero, the counter is quickly reloaded with the division factor before the next clock edge. The maximum frequency of operation is limited by the minimum time needed to reliably perform the loading operation. One improvement is to have a circuit to recognize a state a few clock cycles before the zero count is reached, so the reload can be performed at a more leisurely pace during those cycles. This imposes a minimum division ratio, but that is rarely a problem. An output pulse is given during the loading cycle. In this way a frequency is divided by the number loaded.

The maximum input frequency that can be handled is set by the speed of available logic. The synchronous reset cycle forces the entire divider to be made of equally fast logic. Just adding a fast *prescaler* — a fixed-ratio divider ahead of the programmable one — will increase the maximum frequency that can be used, but it also scales up the loop step size. Equal division has to be added to the reference input of the phase detector to restore the step size, reducing the phase detector's operating frequency. The loop bandwidth has to be reduced to restore the filtering needed to suppress reference frequency sidebands. This makes the loop much slower to change frequency and degrades the noise performance.

Dual-Modulus Prescalers

A plain programmable divider for VHF use would need to be built from ECL devices. It would be expensive, hot and power-hungry. The idea of a fast front end ahead of a CMOS programmable divider would be perfect if these problems could be circumvented. Consider a fast divide-by-ten prescaler ahead of a programmable divider where division by 947 is required. If the main divider is set to divide by 94, the overall division ratio is 940. The prescaler goes through its cycle 94 times and the main divider goes through its cycle once, for every output pulse. If the prescaler is changed to divide by 11 for 7 of its cycles for every cycle of the entire divider system, the overall division ratio is now $[(7 \times 11) + (87 \times 10)] = 947$. At the cost of a more elaborate prescaler and the addition of a slower programmable counter to control it, this prescaler does not multiply the step size and avoids all the problems of fixed prescaling. Fig 14.39B shows the general block diagram of a dual-modulus prescaled divider. The down-counter controls the modulus of the prescaler. The numerical example, just given, used decimal arithmetic, although binary is now usual. Each cycle of the system begins with the last output pulse having loaded the frequency control word, into both the main divider and the prescaler controller. If the part of the word loaded into the prescaler controller is not zero, the prescaler is set to divide by 1 greater than its normal ratio. Each cycle of the prescaler clocks the down-counter. Eventually, it reaches zero, and two things happen: The counter is designed to freeze at zero (and it will remain frozen until it is next reloaded) and the prescaler is switched back to its normal ratio, at which it will remain until the next reload. One way of visualizing this is to think of the prescaler as just being a divider of its normal ratio, but with the ability to "steal" a number of input pulses controlled by the data loaded into its companion down counter. Note that a dual-modulus prescaler system has a *minimum* division ratio, needed to ensure there are enough cycles of the prescaler to allow enough input pulses to be stolen.

Dual-modulus prescaler ICs are widely used and widely available. Devices for use to a few hundred megahertz are cheap, and devices in the 2-GHz region are becoming available, although fixed prescaling has been more common at such frequencies. Common prescaler IC division ratio pairs are: 8-9, 10-11, 16-17, 32-33, 64-65 and so on. Many ICs containing programmable dividers are available in versions with and without built-in prescaler controllers.

Phase Detectors

A phase detector (PSD) produces an output voltage that depends on the phase relationship between its two input signals. If two signals, *in phase on exactly the same frequency*, are mixed together in a conventional diode-ring mixer with a dc-coupled output port, one of the products is direct current (0 Hertz). If the phase relationship between the signals changes, the mixer's dc output voltage changes. With both signals in phase, the output is at its most positive; with the signals 180° out of phase, the output is at its most negative.

Fig 14.39 — Mechanism of a programmable frequency divider (A). B shows the function of a dual-modulus prescaler. The counter is reloaded with N when the count reaches 0 or 1, depending on the sequencer action. →

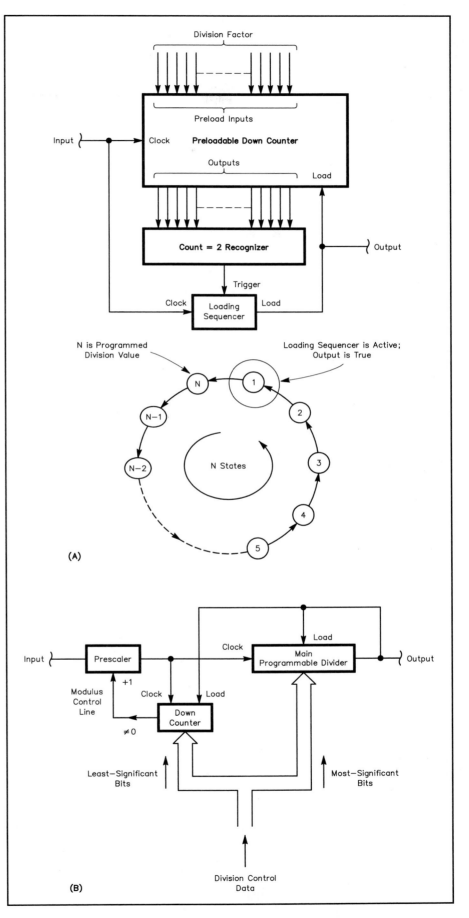

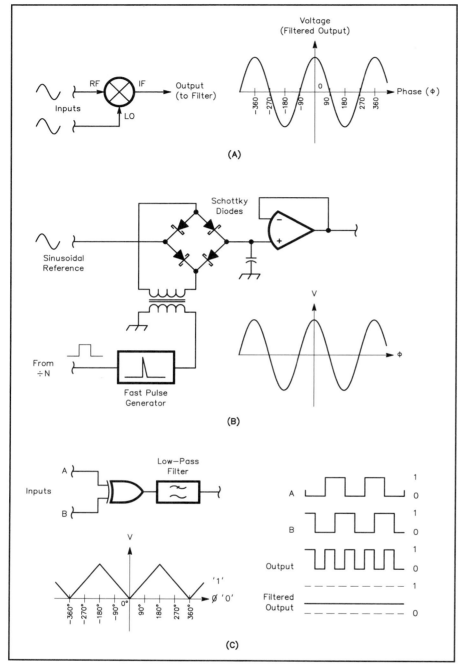

Fig 14.40 — Simple phase detectors: a mixer (A), a sampler (B) and an exclusive-or gate (C).

can be used as phase detectors, as can active mixers like the MC1496. Mixer manufacturers make some parts (Mini-Circuits RPD-1 and so on) that are specially optimized for phase-detector service. All these devices can make excellent, low-noise phase detectors but are not commonly used in ham equipment. A high-speed sample-and-hold circuit, based on a Schottky-diode bridge, can form a very low-noise phase detector and is sometimes used in specialized instruments. This is just a variant on the basic mixer; it produces a similar result, as shown in Fig 14.40B.

The most commonly used simple phase detector is just a single exclusive-OR (XOR) logic gate. This circuit gives a logic 1 output only when *one* of its inputs is at logic 1; if both inputs are the same, the output is a logic 0. If inputs A and B in Fig 14.40C are almost in phase, the output will be low most of the time, and its average filtered value will be close to the logic 0 level. If A and B are almost in opposite phase, the output will be high most of the time, and its average voltage will be close to logic 1. This circuit is very similar to the mixer. In fact, the internal circuit of ECL XOR gates is the same transistor tree as found in the MC1496 and similar mixers, with some added level shifting. Like the other simple phase detectors, it produces a cyclic output, but because of the square-wave input signals, produces a triangular output signal. To achieve this circuit's full output-voltage range, it's important that the reference and VCO signals applied operate at a 50% duty cycle.

Phase-Frequency Detectors

All the simple phase detectors described so far are really specialized mixers. If the loop is out of lock and the VCO is far off frequency, such phase detectors give a high-frequency output midway between zero and maximum. This provides no information to steer the VCO towards lock, so the loop remains unlocked. Various solutions to this problem, such as crude relaxation oscillators that start up to sweep the VCO tuning until lock is acquired, or laboriously adjusted "pretune" systems in which a DAC, driven from the divider control data, is used to coarse tune the VCO to within locking range, have been used in the past. Many of these solutions have been superseded, although pretune systems are still used in synthesizers that must change frequency very rapidly.

The *phase-frequency detector* is the usual solution to lock-acquisition problems. It behaves like a simple phase detector over an extended phase range, but its characteristic is not repetitive. Because its output voltage stays high or low, de-

When the phase difference is 90° (the signals are said to be *in quadrature*), the output is 0 V.

Applying sinusoidal signals to a phase detector causes the detector's output voltage to vary sinusoidally with phase angle, as in **Fig 14.40A**. This nonlinearity is not a problem, as the loop is usually arranged to run with phase differences close to 90°. What might seem to be a more serious complication is that the detector's phase-voltage characteristic repeats every 180°,

not 360°. Two possible input phase differences can therefore produce a given output voltage. This turns out not to be a problem, because the two identical voltage points lie on opposing slopes on the detector's output-voltage curve. One direction of slope gives positive feedback (making the loop unstable and driving the VCO away from what otherwise would be the lock angle) over to the other slope, to the true and stable lock condition.

MD108, SBL-1 and other diode mixers

pending on which input is higher in frequency, this PSD can steer a loop towards lock from anywhere in its tuning range. **Fig 14.41** shows the internal logic of the phase-frequency detector in the CD4046 PLL chip. (The CD4046 also contains an XOR PSD). When the phase of one input leads that of the other, one of the output MOSFETs is pulsed on with a duty cycle proportional to the phase difference. Which MOSFET receives drive depends on which input is leading. If both inputs are in phase, the output will include small pulses due to noise, but their effects on the average output voltage will cancel. If one signal is at a higher frequency than the other, its phase will lead by an increasing amount, and the detector's output will be held close to either V_{DD} or ground, depending on which input signal is higher in frequency. To get a usable voltage output, the MOSFET outputs can be terminated in a high-value resistor to $V_{DD}/2$, but the CD4046's output stage was really designed to drive current pulses into a capacitive load, with pulses of one polarity charging the capacitor and pulses of the other discharging it; the capacitor integrates the pulses.

Simple phase detectors are normally used to lock their inputs in quadrature, but phase-frequency detectors are used to lock their input signals in phase.

Traditional textbooks and logic-design courses give extensive coverage to avoiding *race hazards* caused by near-simultaneous signals racing through parallel paths in a structure to control a single output. Avoiding such situations is important in many circuits because the outcome is strongly dependent on slight differences in gate speed. The phase-frequency detector is the one circuit whose entire function depends on its built-in ability to *make* both its inputs race. Consequently, it's not easy to home-brew a phase-frequency detector from ordinary logic parts, but fortunately they are available in IC form, usually combined with other functions. The MC4044 is an old stand-alone TTL phase-frequency detector; the MC12040 is an ECL derivative. CMOS versions can be found in the CD4046 PLL chip and in almost all current divider-PLL synthesizer chips.

The race-hazard tendency does cause one problem in phase-frequency detectors: A device's delays and noise, rather than its input signals, control its phase-voltage characteristic in the zero-phase-difference region. This degrades the loop's noise performance and makes its phase-to-voltage coefficient uncertain and variable in a small region — unfortunately, the detector's normal operating range! It's therefore normal to bias operation slightly

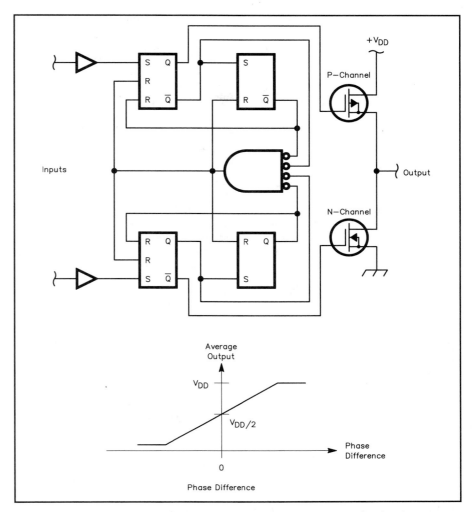

Fig 14.41 — Input signals very far off frequency can confuse a simple phase-detector; a phase-frequency detector solves this problem.

away from *exact* phase equality to avoid this problem.

PLL Synthesizer ICs

Simple PLL ICs have been available since the early 1970s, but most of these contain a crude low-Q VCO and a phase detector. They were intended for general use as tone detectors and demodulators, not as major elements in communication-quality frequency synthesizers. One device well worth noting in this group, however, is the CD4046, which contains a VCO and a pair of phase detectors. The CD4046's VCO is useless for our purposes, but its phase-frequency detector is quite good. Better yet, the CD4046 is a low-cost part and one of the cheapest ways of getting a good phase-frequency detector. Its VCO-disable pin is a definite design plus. Later CD4046 derivatives, the 74HC4046 (CMOS input levels) and 74HCT4046 (TTL compatible input levels) are usable to much higher frequencies

and seem to be more robust, but note that these versions are for +5 V supplies only, whereas the original CD4046 can be used up to 15 V.

More complex ICs specifically intended for frequency synthesizers have since appeared. They normally contain a programmable divider, a phase-frequency detector and a reference divider that usually allows a small choice of division ratios. Usually, the buffer amplifier on these parts' reference input is arranged so that it can be used as part of a simple crystal oscillator. This is adequate for modest frequency accuracy, but an independent TCXO or OCXO is better.

Outside of Japanese equipment, Motorola's range of CMOS synthesizer ICs seem by far the most widely used. The MC145151 brings all of its division-control bits out to individual pins and needs no sequencing for control. It is the best choice if only a few output frequencies are needed, because they can be programmed via a diode matrix. The MC145151's divide-by-N range is 3 to 16383, controlled by a 14-bit in binary format; the reference can be divided by 8, 128, 256, 512, 1024, 2048, 2410 and 8192. It's possible to operate the MC145151 to 30 MHz, and its phase-frequency detector is similar to that in the CD4046, lacking the MOSFET "Tri-State" output, but with the added benefit of a lock-detector output. The MC145152 is a variant that includes control circuitry for an external dual-modulus prescaler. The choice of external prescaler sets the maximum operating frequency, as well as maximum and minimum division ratios. For some unknown reason, the reference division choices differ from those of the MC145151: 8, 64, 128, 256, 512, 1024, 1160 and 2048. There is also a Tri-State phase-frequency detector output.

The MC145146 is simply an MC145152 with its control data formatted as a 4-bit bus writing to eight addresses. This reduces the number of pins required and allows more data to be input, so the reference divider is fully programmable from 3 to 4095.

Closing the Loop

The design of a PLL's loop amplifier and loop filtering are inextricably linked. The first task is to recognize the system as a feedback loop and take action to ensure that the loop is stable. The second task is to produce a closed-loop characteristic that best suits the application.

Phase-locked loops have acquired an evil reputation with radio amateurs because some PLL designers are more careful than others. A number of commercially made radios have included poorly designed synthesizers that produced excessive noise sidebands, or wouldn't lock reliably. Some irreproducible designs have been published for home construction that could not be made to work. Many experimenters have sweated over an unstable loop desperately trying anything to get it to lock stably. So the PLL has earned its shady reputation. Because of all of this aversion therapy, very few amateurs are now prepared to attempt to build a PLL.

Note carefully that we are designing a *loop*. All oscillators are loops, and any loop will oscillate when certain conditions are met. Look at the RC phase shift oscillator from which the "Pierce" crystal oscillator is derived (Fig 14.27A). Our PLL is a loop, containing an amplifier and a number of RC sections. It has all the parts needed to make an oscillator. If a loop is unstable and oscillates, there will be an oscillatory voltage superimposed on the VCO tuning voltage. This will produce massive unwanted FM on the output of the

PLL. This is absolutely undesirable, but PLLs have become notorious for the difficulty of fixing their instability problems.

Good loops *can* be designed. Trial and error, intuitive component choice, or "reusing" a loop amplifier design from a different synthesizer, will lead to a very low probability of success. We must resort to mathematics and take a custom approach to each new loop design.

Before you turn to a less demanding chapter, consider this: It's one of Nature's jokes that easy procedures often hide behind a terror-inducing facade. The math you need to design a good PLL may look weird when you see it for the first time and seem to involve some obviously impossible concepts, but it ends in a very simple procedure that allows you to calculate the response and stability of a loop. Professional designers must handle these things daily, and because they like differential equations no more than anyone else does, they use a graphical method to generate PLL designs. (We'll leave proofs in the textbooks, where they belong! Incidentally, if you can get the math required for PLL design under control, as a free gift you will also be ready to do modern filter synthesis and design beam antennas, since the math they require is essentially the same.)

The easy solutions to PLL problems can only be performed during the design phase. We must deliberately design loops with sufficient stability safety margins so that all forseeable variations in component tolerances, from part-to-part, over temperature and across the tuning range, cannot take the loop anywhere near the threshold of oscillation. More than this, it is important to have a large stability margin because loops with lesser margins will exhibit amplified phase-noise sidebands, and that is another notorious PLL problem that can be designed out.

Back at the beginning of the chapter, near Fig 14.4, the criterion for stability/oscillation of a loop was mentioned: Barkhausen's criterion for oscillation or Nyquist's stability criterion. This time Harry Nyquist is our hero.

A PLL is a negative feedback system, with the feedback opposing the input, this opposition or inversion around the loop amounts to a 180° phase shift. This is the frequency-*independent* phase shift round the loop, but there are also frequency-*dependent* shifts that will add in. These will inevitably give an increasing, lagging phase with increasing frequency. Eventually an extra 180° phase shift will have occurred, giving a total of 360° around the loop at some frequency. We are in trouble if the gain around the loop has not dropped

below unity (0 dB) by this frequency.

Note that we are not concerned only with the loop operating frequency. Consider a 30-MHz low-pass filter passing a 21-MHz signal. Just because there is no signal at 30 MHz, our filter is no less a 30-MHz circuit. Here we use the concept of frequency to describe "what would happen *if* a signal was applied at that frequency."

The next sections show how to calculate the gain and phase response around a loop. They also give a reliable recipe for a loop with sufficient stability margin to be dependable. Note that it does not matter if the phase goes very close to the limit at lower frequencies (it is the crossover point that counts) and that the recipe loop design brings the phase back to 30 to 45° of safety margin before the loop gain passes unity. Let's define some terms first:

Poles and Zeros

We all think of frequency as stretching from zero to infinity, but let's imagine that additional numbers could be used. This is pure imagination — we can't make signals at *complex* values (with *real* and *imaginary* components) of hertz — but if we try using complex numbers for frequency, some "impossible" things happen. For example, take a simple RC low-pass network shown in **Fig 14.42**. The frequency response of such a network is well-known, but its phase response is *not* so well-known. The equation shown for the *transfer function*, the "gain" of this simple circuit, is pretty standard, although the j is included to make it complete and accurately describe the output in phase as well as magnitude (j is an impossible number, the square root of −1, usually called an *imaginary* number and used to represent a 90° phase shift).

Although the equation shown in Fig 14.42 was constructed to show the behavior of the circuit at real-world frequencies, there is nothing to stop us from using it to explore how the circuit would behave at impossible frequencies, just out of curiosity. At one such frequency, the circuit goes crazy. At $f = -j2\pi RC$, the denominator of the equation is zero, and the gain is infinite! Infinite gain is a pretty amazing thing to achieve with a passive circuit — but because this can only happen at an impossible frequency, it cannot happen in the real world. This frequency is equal to the network's −3 dB frequency multiplied by j; it is called a *pole*. Other circuits, which produce real amplitude responses that start to *rise* with frequency, act similarly, but their crazy effect is called a *zero* — an imaginary frequency at which gain goes to zero. The frequency of a circuit's zero is again j times its 3-dB frequency (+3 dB in this case).

These things are clearly impossible, but we can map all the poles and zeros of a complex circuit and look at the patterns to determine if the circuit possesses unwanted gain and/or phase bumps across a frequency range of interest. Poles are associated with 3-dB roll-off frequencies and 45° phase lags, zeros are associated with 3-dB "roll-up" frequencies and 45° phase leads. We can plot the poles and zeros on a two-dimensional chart of real and imaginary frequency. (The names of this chart and its axes are results of its origins in *Laplace transforms*, which we don't need to touch in this discussion.) The traditional names for the axes are used, but we can add labels with clearer meanings.

What is a Pole? — A pole is associated with a bend in a frequency response plot where attenuation with increasing frequency increases by 6 dB per octave (20 dB per decade; an octave is a 2:1 frequency ratio, a decade is a 10:1 frequency ratio).

There are four ways to identify the existence and frequency of a pole:

1. A downward bend in a gain vs frequency plot, the pole is at the −3 dB point for a single pole. If the bend is more than 6 dB/octave, there must be multiple poles at this frequency.

2. A 90° change in a phase vs frequency plot, where lag increases with frequency. The pole is at the point of 45° added lag. Multiple poles will add their lags, as above.

3. On a circuit diagram, a single pole looks like a simple RC low-pass filter. The pole is at the −3 dB frequency ($1/(2\pi RC)$ Hz) Any other circuit that gives the same response will produce a pole at the same frequency.

4. In an equation for the transfer function of a circuit, a pole is a theoretical value of frequency, which would make the equation predict infinite gain. This is clearly impossible, but as the value of frequency will either be absolute zero, or will have an imaginary component, it is impossible to make a signal at a pole frequency.

What is a Zero? — A zero is the complement of a pole. Each zero is associated with an upward bend of 6 dB per octave in a gain vs frequency plot. The zero is at the +3 dB point. Each zero is associated with a transition on a phase-vs-frequency plot that *reduces* the lag by 90°. The zero is at the 45° point of this S-shaped transition.

In math, it is a frequency at which the transfer-function equation of a circuit predicts zero gain. This is not impossible in real life (unlike the pole), so zeros can be found at real-number frequencies as well

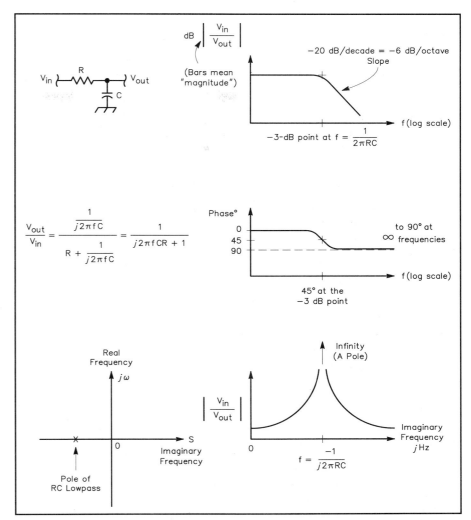

Fig 14.42 — A simple RC filter is a "pole."

as complex-number frequencies.

On a circuit diagram, a pure zero would need gain that increases with frequency forevermore above the zero frequency. This implies active circuitry that would inevitably run out of gain at some frequency, which implies one or more poles up there. In real-world circuits, zeros are usually not found making gain go up, but rather in conjunction with a pole, giving a gain slope between two frequencies and flat gain beyond them. Real-world zeros are only found chaperoned by a greater or equal number of poles.

Consider a classic RC high-pass filter. The gain increases at 6 dB per octave from 0 Hz (so there must be a zero at 0 Hz) and then levels off at $1/(2\pi RC)$ Hz. This leveling off is really a pole, it adds a 6 dB per octave roll off to cancel the roll-up of the zero.

Poles and Zeros in the Loop Amplifier — The loop-amplifier circuit used in the example loop has a blocking capacitor

in its feedback path. This means there is no dc feedback. At higher frequencies the reactance of this capacitor falls, increasing the feedback and so reducing the gain. This is an integrator. The gain is immense at 0 Hz (dc) and falls at 6 dB per octave. This points to a pole at 0 Hz. This is true whatever the value of the series resistance feeding the signal into the amplifier inverting input and whatever the value of the feedback capacitor. The values of these components scale the gain rather than shape it. The shape of the integrator gain is fixed at −6 dB per octave, but these components allow us to move it to achieve some wanted gain at some wanted frequency.

At high frequencies, the feedback capacitor in an ordinary integrator will have very low reactance, giving the circuit very low gain. In our loop amplifier, there is a resistor in series with the capacitor. This limits how low the gain can go, in other words the integrator's downward (with

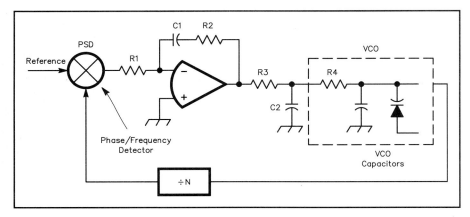

Fig 14.43 — A common loop-amplifier/filter arrangement.

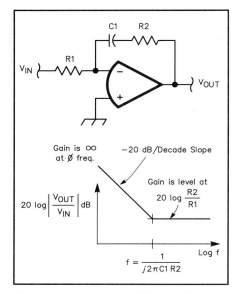

Fig 14.44 — Loop-amplifier detail.

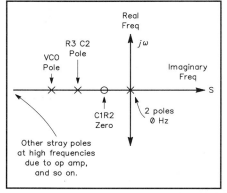

Fig 14.45 — The loop's resulting pole-zero "constellation."

increasing frequency) slope is leveled off. This resistor and capacitor make a zero, and its +6 dB per octave slope cancels the –6 dB per octave slope created by the pole at 0 Hz.

Recipe for A PLL Pole-Zero Diagram

We can choose a well-tried loop-amplifier circuit and calculate values for its components to make it suit our loop. **Fig 14.43** shows a common synthesizer loop arrangement. The op amp operates as an integrator, with R2 added to level off its falling gain. An integrator converts a dc voltage at its input to a ramping voltage at its output. Greater input voltages yield faster ramps. Reversing the input polarity reverses the direction of the ramp.

The system's phase-frequency detector (connected to work in the right sense) steers the integrator to ramp in the direc-

tion that tunes the VCO toward lock. As the VCO approaches lock, the phase detector reduces the drive to the integrator, and the ramping output slows and settles on the right voltage to give the exact, locked output frequency. Once lock is achieved, the phase detector outputs short pulses that "nudge" the integrator to keep the divided VCO frequency exactly locked in phase with the reference. Now we'll take a look around the entire loop and find the poles and zeros of the circuits.

The integrator produces a –20 dB per decade roll off from 0 Hz, so it has a pole at 0 Hz. It also includes R2 to cancel this slope, which is the same as adding a rising slope that exactly offsets the falling one. This implies a zero at $f = 1/(2\pi C1R2)$ Hz, as **Fig 14.44** shows. R3 and C2 make another pole at $f = 1/(2\pi C2R3)$ Hz. A VCO usually includes a series resistor that conveys the control voltage to the tuning diode, which also is loaded by various capacitors. This creates another pole.

The VCO generates frequency, while the phase detector responds to phase. Phase is the integral of frequency, so to-

gether they act as another integrator and add another pole at 0 Hz.

What About the Frequency Divider? — The frequency divider is like a simple attenuator in its effect on loop response.

Imagine a signal generator that is switched between 10 and 11 MHz on alternate Tuesdays. Imagine that we divide its output by 10. The output of the divider will change between 1 and 1.1 MHz, still on alternate Tuesdays. The divider divides the deviation of the frequency (or phase) modulated signal, but it cannot affect the modulating frequency.

A possible test signal passing around the example loop is in the form of frequency modulation as it passes through the divider, so it is simply attenuated. A divide by N circuit will give $20 \log_{10}(N)$ dB of attenuation (reduction of loop gain). This completes the loop. We can plot all this information on one *s-plane*, **Fig 14.45**, a plot sometimes referred to as a PLL system's *pole-zero constellation*.

Open-Loop Gain and Phase

Just putting together the characteristics of the blocks around the loop, without allowing for the loop itself, gives us the system's *open-loop* characteristic, which is all we really need.

What Is Stability? — Here "open-loop response" means the gain around the loop that would be experienced by a signal at some frequency *if* such a signal were inserted. We do not insert such signals in the actual circuit, but the concept of frequency dependent gain is still valid. We need to ensure that there is insufficient gain, at *all* frequencies, to ensure that the loop cannot create a signal and begin oscillating. More than this, we want a good safety margin to allow for component variations and because loops that are *close* to instability perform poorly.

The loop gain and phase are doing interesting things on our plots at frequencies in the AF part of the spectrum. This does not mean that there is visible operation at those frequencies. Imagine a bad, unstable loop. It will have a little too much loop gain at some frequency, and unavoidable noise will build up until a strong signal is created. The amplitude will increase until nonlinearities limit it. A 'scope view of the tuning voltage input to the VCO will show a big signal, often in the hertz to tens of kilohertz region, often large enough to drive the loop amplifier to the limits of its output swing, close to its supply voltages. As this big signal is applied to the tuning voltage input to the VCO, it will modulate the VCO frequency across a wide range. The output will look like that of a sweep generator on a spectrum analyzer. What is wrong? How can we fix this loop? Well, the problem may be excessive loop gain, or improper loop time constants. Rather than work directly with the loop time constants, it is far easier to work with the pole and zero frequencies of the loop response. It's just a different view of the same things.

Now imagine a good, stable loop, with an adequate stability safety margin. There will be some activity around the loop: the PSD (phase detector) will demodulate the phase noise of the VCO and feed the demodulated noise through the loop amplifier in such a way as to cancel the phase noise. This is how a good PLL should give *less* phase noise than the VCO alone would suggest. In a good loop, this noise will be such a small voltage that a 'scope will not show it. In fact, connecting test equipment in an attempt to measure it can usually *add* more noise than is there.

Finally, imagine a poor loop that is only just stable. With an inadequate safety margin, the action will be like that of a Q multiplier or a regenerative receiver close to the point of oscillation: There will be an amplified noise peak at some frequency. This spoils the effect of the phase detector trying to combat the VCO phase noise and gives the opposite effect. The output spectrum will show prominent bumps of exaggerated phase noise, as the excess noise frequency modulates the VCO.

Now, let's use the pole-zero diagram as a graphic tool to find the system open-loop gain and phase. As we do so, we need to keep in mind that the frequency we have been discussing in designing a loop response is the frequency of a theoretical test signal passing around the loop. In a real PLL, the loop signal exists *in two forms*: as sinusoidal voltage between the output of the phase detector and the input of the VCO, and as a sinusoidal modulation of the VCO frequency in the remainder of the loop.

With our loop's pole-zero diagram in hand, we can pick a frequency at which we want to know the system's open-loop gain and phase. We plot this value on the graph's vertical (Real Frequency) axis and draw lines between it and each of the poles and zeros, as shown in **Fig 14.46**. Next, we measure the lengths of the lines and the "angles of elevation" of the lines. The loop gain is proportional to the product of the lengths of all the lines to zeros *divided by* the product of all the lengths of the lines to the poles. The phase shift around the loop (lagging phase equates to a positive phase shift) is equal to the sum of the pole angles minus the sum of the zero angles. We can repeat this calculation for a number of different frequencies and draw graphs of the loop gain and phase versus frequency.

All the lines to poles and zeros are hy-potenuses of right triangles, so we can use Pythagoras's rule and the tangents of angles to eliminate the need for scale drawings. Much tedious calculation is involved because we need to repeat the whole business for each point on our open-loop response plots. This much tedious calculation is an ideal application for a computer.

The procedure we've followed so far gives only *proportional* changes in loop gain, so we need to calculate the loop gain's *absolute* value at some (chosen to be easy) frequency and then relate everything to this. Let's choose 1 Hz as our reference. (Note that it's usual to express angles in radians, not degrees, in these calculations and that this normally renders frequency in peculiar units of *radians per second*. We can keep frequencies in hertz if we remember to include factors of 2π in the right places. A frequency of 1 Hz = 2π radians/second, because 2π radians = $360°$.)

We must then calculate the loop's proportional gain at 1 Hz from the pole-zero diagram, so that the constant of proportion can be found. For starters, we need a reasonable estimate of the VCO's *voltage-to-frequency gain* — how much it changes frequency per unit change of tuning voltage. As we are primarily interested in stability, we can just take this number as the slope, in Hertz per volt, at the steepest part of voltage-vs-frequency tuning characteristic, which is usually at the low-frequency end of the VCO tuning range. (You can characterize a VCO's voltage-versus-frequency gain by varying the bias on its tuning diode with an adjustable power supply and measuring its tuning characteristic

with a voltmeter and frequency counter.)

The loop divide-by-N stage divides our theoretical modulation — the tuning corrections provided through the phase detector, loop amplifier and the VCO tuning diode — by its programmed ratio. The worst case for stability occurs at the divider's lowest N value, where the divider's "attenuation" is least. The divider's voltage-vs-frequency gain is therefore 1/N, which, in decibels, equates to $-20 \log(N)$.

The change from frequency to phase has a voltage gain of one at the frequency of 1 radian/second — $1/(2\pi)$, which equates to -16 dB, at 1 Hz. The phase detector will have a specified "gain" in volts per radian. To finish off, we then calculate the gain of the loop amplifier, including its feedback network, at 1 Hz.

There is no need even to draw the pole-zero diagram. **Fig 14.47** gives the equations needed to compute the gain and phase of a system with up to four poles and one zero. They can be extended to more singularities and put into a simple computer program. Alternatively, you can type them into your favorite spreadsheet and get printed plots. The computational power needed is trivial, and a listing to run these equations on a Hewlett-Packard HP11C (or similar pocket calculator, using RPN) is available from ARRL HQ for an SASE. (Contact the Technical Department Secretary and ask for the 1995 *Handbook* PLL design program.)

Fig 14.48 shows the sort of gain- and phase-versus-frequency plots obtained from a "recipe" loop design. We want to know where the loop's phase shift equals, or becomes more negative than, $-180°$.

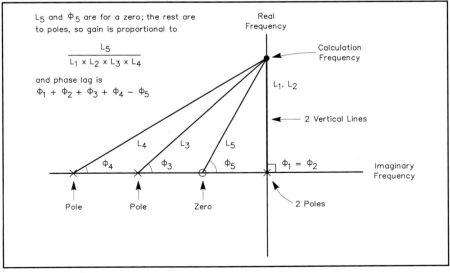

Fig 14.46 — **Calculating loop gain and phase characteristics from a pole-zero diagram.**

$$\text{Open loop gain (dB)} = 20 \log \left[\frac{\sqrt{P_1^2+1} \ \sqrt{P_2^2+1} \ \sqrt{P_3^2+1} \ \sqrt{P_4^2+1} \ \sqrt{Z^2+f^2}}{\sqrt{P_1^2+f^2} \ \sqrt{P_2^2+f^2} \ \sqrt{P_3^2+f^2} \ \sqrt{P_4^2+f^2} \ \sqrt{Z^2+1}} \times 10^{\frac{\text{unity freq gain}}{20}} \right]$$

$$\text{Phase (lead)} = \tan^{-1}\left(\frac{f}{Z}\right) - \tan^{-1}\left(\frac{f}{P_1}\right) - \tan^{-1}\left(\frac{f}{P_2}\right) - \tan^{-1}\left(\frac{f}{P_3}\right) - \tan^{-1}\left(\frac{f}{P_4}\right)$$

f is the frequency of the point to be characterized. P_1, P_2, P_3, P_4, Z are the frequencies of the poles and zero (all in same units!)

Fig 14.47 — Pole-zero frequency-response equations capable of handling up to four poles and one zero.

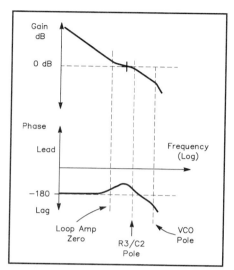

Fig 14.48 — An open-loop gain/phase plot.

Added to the –180° shift inherent in the loop's feedback (the polarity of which must be negative to ensure that the phase detector drives the VCO toward lock instead of away from it), this will be the point at which the loop itself oscillates — *if* any loop gain remains at this point. The game is to position the poles and zero, and set the unit frequency gain, so that the loop gain falls to below 0 dB before the –180° line is crossed.

At the low-frequency end of the Fig 14.48 plots, the two poles have had their effect, so the gain falls at 40 dB/decade, and the phase remains just infinitesimally on the safe side of –180°. The next influence is the zero, which throttles the gain's fall back to 20 dB/decade and peels the phase line away from the –180° line. The zero would eventually bring the phase back to 90° of lag, but the next pole bends it back before that and returns the gain slope back to a fall of 20 dB/decade. The last pole, that attributable to the VCO, pushes the phase over the –180° line.

It's essential that there not be a pole

between the 0-Hz pair and the zero, or else the phase will cross the line early. In this example, the R3-C2 pole and the zero do all the work in setting this critical portion of the loop's response. Their frequency spacing should create a phase bump of 30 to 45°, and their particular frequency positions are constrained as a compromise between sufficient loop bandwidth and sufficient suppression of reference-frequency sidebands. We know the frequency of the loop amplifier zero (that contributed by R1 and C1), so all we need to do now is design the loop amplifier to exhibit a frequency response such that the open-loop gain passes through 0 dB at a frequency close to that of the phase-plot bump.

This loop-design description may have been hard to follow, but a loop amplifier can be designed in under 30 minutes with a pocket calculator and a little practice. The high output impedance of commonly used CMOS phase detectors favors loop amplifiers based on FET-input op amps and allows the use of high-impedance RC networks (large capacitors can therefore be avoided in their design). LF356 and TL071 op amps have proven successful in loop-amplifier service and have noise characteristics suited to this environment.

To sum up, the recipe gives a proven pole zero pattern (listed in order of increasing frequency):

1. Two poles at 0 Hz (one is the integrator, the other is implicit in all PLs)

2. A zero controlled by the RC series in the loop amplifier feedback path.

3. A simple RC low-pass pole

4. A second RC low-pass pole formed by the resistor driving the capacitive load of the tuning voltage inputs of the VCO

To design a loop: Design the VCO, divider, PSD and find their coefficients to get the loop gain at unit frequency (1 Hz), then, keeping the poles and zeros in the above order, move them about until you get the same 45° phase bump at a frequency close to your desired bandwidth. Work out how much loop-amplifier unit-

frequency gain is needed to shift the gain-vs-frequency plot so that 0-dB loop gain occurs at the center of the phase bump. Then calculate R and C values to position the poles and zero and the feedback-RC values to get the required loop amplifier gain at 1 Hz.

To Cheat — You can scale the example loop to other frequencies: Just take the reactance values at the zero and pole frequencies and use them to scale the components, you get the nice phase plot bump at a scaled frequency. Even so, you still must do the loop-gain design.

Don't forget to do this for your lowest division factor, as this is usually the least stable condition because there is less attenuation. Also, VCOs are usually most sensitive at the low end of their tuning range.

Noise in Phase-Locked Loops

Differences in Q usually make the phase-noise sidebands of a loop's reference oscillator much smaller than those of the VCO. Within its loop bandwidth, a PLL acts to correct the phase-noise components of its VCO and impose those of the reference. Dividing the reference oscillator to produce the reference signal applied to the phase detector also divides the deviation of the reference oscillator's phase-noise sidebands, translating to a 20-dB reduction in phase noise per decade of division, a factor of 20 log(M) dB, where M is the reference divisor. Offsetting this, within its loop bandwidth the PLL acts as a frequency *multiplier*, and this multiplies the deviation, again by 20 dB per decade, a factor of 20 log(N) dB, where N is the loop divider's divisor. Overall, the reference sidebands are increased by 20 log (N/M) dB. Noise in the dividers is, in effect, present at the phase detector input, and so is increased by 20 log (N) dB. Similarly, op-amp noise can be calculated into an equivalent phase value at the input to the phase detector, and this can be increased by 20 log(N) to arrive at the effect it has on the output.

Phase noise can be introduced into a PLL by other means. Any amplifier stages between the VCO and the circuits that follow it (such as the loop divider) will contribute some noise as will microphonic effects in loop- and reference-filter components (such as those due to the piezo-electric properties of ceramic capacitors and the crystal filters sometimes used for reference-oscillator filtering). Noise on the power supply to the system's active components can modulate the loop. The fundamental and harmonics of the system's ac line supply can be coupled into the VCO directly or by means of ground loops.

Fig 14.49 shows the general shape of the PLL's phase noise output. The dashed curve shows the VCO's noise performance when unlocked; the solid curve shows how much locking the VCO to a cleaner reference improves its noise performance. The two noise bumps are a classic characteristic of a phase-locked loop. If the loop is poorly designed and has a low stability margin, the bumps may be exaggerated — a sign of noise amplification due to an overly peaky loop response.

Exaggerated noise bumps can also occur if the loop bandwidth is less than optimum. Increasing the loop bandwidth in such a case would widen the band over which the PLL acts, allowing it to do a better job of purifying the VCO — but this might cause other problems in a loop that's deliberately bandwidth-limited for better suppression of reference-frequency sidebands. Immense loop bandwidth is not desirable, either: Farther away from the carrier, the VCO may be so quiet on its own that widening the loop would make it *worse*.

Multiloop Synthesizers

The trade-off between small step size (resolution) and all other PLL performance parameters has already been mentioned. The *multiloop synthesizer*, a direct synthesizer constructed from two or more PLL synthesizers, is one way to break away from this trade-off. **Fig 14.50**, the block diagram of a five-loop synthesizer, reflects the complexity found in some professional receivers. All of its synthesis loops run with a 100-kHz step size, which allows about a 10-kHz loop bandwidth. Such a system's noise performance, settling time after a frequency change and reference-sideband suppression can all be very good.

Cost concerns generally render such elaboration beyond reach for consumer-grade equipment, so various cut-down versions, all of which involve trade-offs in performance, have been used. For ex-

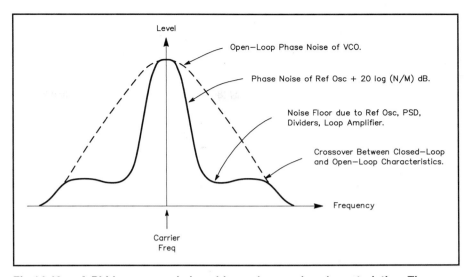

Fig 14.49 — A PLL's open- and closed-loop phase-noise characteristics. The noise bumps at the crossover between open- and closed-loop characteristics are typical of PLLs; the severity of the bumps reflects the quality of the system's design.

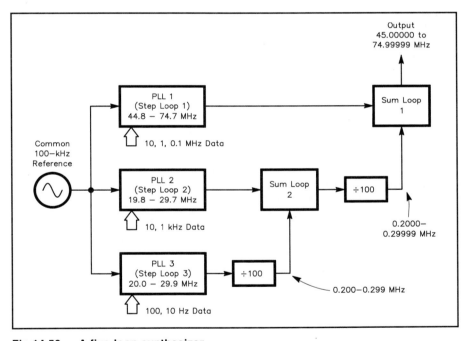

Fig 14.50 — A five-loop synthesizer.

ample, a three-loop machine can be made by replacing the lower three loops with a single loop that operates with a 1-kHz step size and accepting the slower frequency stepping determined by the narrow loop bandwidth this involves.

A Summing-Loop Synthesizer

Radio amateurs have tended to refer only to relatively modern radio gear as "synthesized," but any means of generating a frequency by adding and/or subtracting the frequencies of two or more signal sources also qualifies as frequency synthesis. It's therefore true to say that radio amateurs have used frequency-synthesized equipment for decades, with the synthesizer — in radios such as the R. L. Drake 4-line (in which the technique was termed *premixing*) — usually taking the form of a mixer, a crystal oscillator (one crystal per band segment) and a bank of switched filters.

Hams who build their own equipment have also successfully used this form of frequency synthesis. Among them, Ian

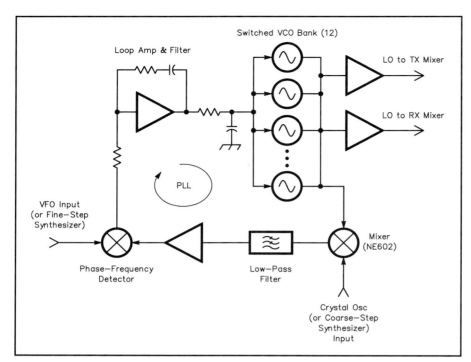

Fig 14.51 — In a summing-loop synthesizer, a mixer takes the place of the divider. The drawing diagrams a design by G3ROO and GM4ZNX that uses a 5- to 5.5-MHz VFO and crystal oscillators to provide LO signals appropriate for ham-band operation from 1.8 through 29.7 MHz with a 9-MHz IF.

Fig 14.52 — The G3ROO/GM4ZNX summing-loop PLL phase-locks its VCO to the frequency difference between a crystal oscillator and 5- to 5.5-MHz VFO (these circuits are not shown; see text). Table 14.3 lists the conversion-crystal frequency, VCO tuned-circuit padding capacitance (C_X) and VCO inductor data required for each band. Any 5- to 5.5-MHz VFO capable at least 2.5 mW (4 dBm) output with a 50-Ω load — 1 V P-P — can drive the circuit's VFO input.

C_X, C2, C3 — NP0 or C0G ceramic, 10% or tighter tolerance.

C65, C66, C67 — NP0, C0G or general-purpose ceramic, 10% tolerance or tighter.

D0A, D0B — BB204 dual, common-cathode tuning diode (capacitance per section at 3 V, 39 pF) or equivalent. The ECG617, NTE617 and MV104 are suitable dual-diode substitutes, or use pairs of 1N5451s (39 pF at 4 V) or MV2109s (33 pF at 4V).

D1 — BA244 switching diode. The 1N4152 is a suitable substitute.

L1 — Variable inductor; see Table 14.3 for value.

L13, L14 — 22-μH choke, 20% tolerance or better (Miller 70F225AI, 78F270J, 8230-52, 9250-223; Mouser 43LQ225; Toko 144LY-220J [Digi-Key TK4232] suitable).

T1 — 6 bifilar turns of #28 enameled wire on an FT-37-72 ferrite toroid ($\approx$30 μH per winding).

U2 — 74HC4046 or 74HCT4046 PLL IC (CD4046 unsuitable; see text).

Keyser, G3ROO, has designed and built several entire, multimode transceivers for the HF bands, of which many copies have been built. All have been based on the proven formula of a 9-MHz IF and a 5- to 5.5-MHz VFO upconverted to the proper LO frequency (desired signal + 9 MHz) by signals from a bank of switched crystal oscillators.

Misadjusting the nontrivial post-mixer filtering in such systems can cause spurious signals and out-of-band operation, and few home equipment builders have the test equipment necessary to adjust such filters easily and surely. G3ROO therefore decided to try *summing loop* synthesis, the means taken in a number of popular commercial ham transceivers from the mid- 1970s to mid-1980s, in his latest transceiver as a means of avoiding these difficulties.

A summing-loop synthesizer uses a mixer instead of a programmable divider, producing its output frequency by adding (or subtracting) the frequency of a high-stability HF VFO, which may be a VXO, VFO or PLL-synthesized, to (from) the frequency of a crystal oscillator. A circuit designed by G3ROO and GM4ZNX does this job in the latest G3ROO transceiver. **Fig 14.51** shows its block diagram

This loop can be easily aligned with just a DMM. In an attempt to counter the bad reputation of PLLs, it has been designed to have very dependable stability. Although it does not attempt to set a record for low phase noise, it is a circuit designed to be robust, easy to set up and to build the confidence of constructors venturing into a new area. The circuit (**Fig 14.52**) was first published in the G QRP Club's journal, *SPRAT* (No. 70, Spring 1992). At least 10 successful copies, all trouble-free, were built before the Radio Society of Great Britain's *Radio Communication* carried a construction article about it in December 1993. This loop was originally used to combine a 5- to 5.5-MHz VFO with the output from one of a bank of crystal oscillators. Suitable crystal oscillators and VFOs can be found earlier in this chapter. With a coil change, the G3PDM VFO (Fig 14.15) can be used to cover 5 to 5.5 MHz. Ian has chosen to operate the crystal oscillators above the desired output frequency, which makes the VFO tune "backwards" — increasing VFO frequency equates to decreasing loop output frequency — because crystals are commonly used on these frequencies in many of the older transceivers (the prototype used the crystal oscillator board from a Yaesu FT-707). **Table 14.3** lists the necessary conversion-crystal frequency, VCO tuned-circuit padding capacitance and VCO-inductor winding data on a band-by-band basis.

The summing loop's phase-frequency detector (U2, a 74HC4046 or 74HCT4046) steers the integrator and tunes whichever VCO is selected to keep its inputs (pin 3 and 14) locked equal in frequency. The VFO, applied to pin 14, serves as a variable reference. Via Q13, pin 3 of the phase detector receives the output of U1, an NE602AN, which mixes signals from the VCO and crystal oscillator appropriate to the band in use. U1's output contains sum and difference frequencies, but the phase detector responds only to the *difference* ($f_{XO} - f_{VCO}$, where f_{XO} is the crystal oscillator frequency and f_{VCO} is the VCO frequency) because of its close proximity to its reference signal (the VFO). Consequently, the difference between the VCO and crystal oscillator frequencies must equal the VFO frequency, and the loop operates as a *subtractor*.

You can use this loop to *add* frequencies (adding the signal from a 13-MHz crystal to the 5-MHz VFO for output at 18 MHz, for instance) by changing the crystal frequency and transposing the phase detector input signals (VFO to pin

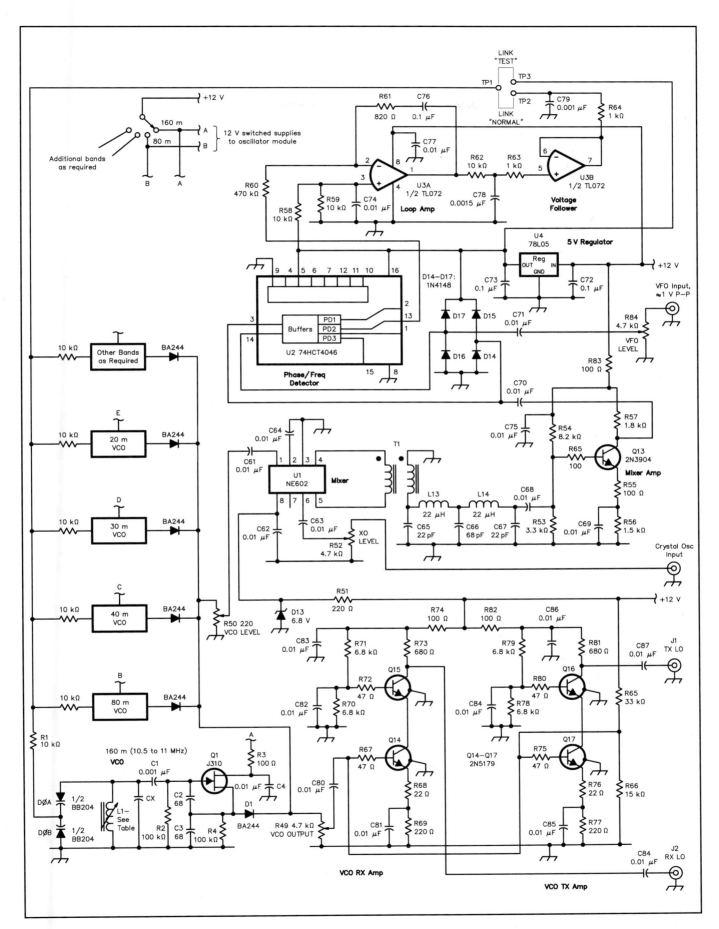

Table 14.3

Band-Specific Component Data for the Summing Loop

Band (MHz)	Output Range (MHz)	Crystal Frequency (MHz)	C_x (pF)	VCO Coil
1.5-2.0	10.5-11.0	16.0	0	20 turns on Toko 10K-series inductor form (≈ 4.29 μH)
3.5-4.0	12.5-13.0	18.0	0	16 turns on Toko 10K-series inductor form (≈ 3.03 μH)
7.0-7.5	16.0-16.5	21.5	0	12 turns on Toko 10K-series inductor form (≈ 1.85 μH)
10.0-10.5	19.0-19.5	24.5	27	8 turns on Toko 10K-series inductor form (≈ 0.87 μH)
14.0-14.5	23.0-23.5	28.5	56	8½ turns white Toko S-18-series (≈ 0.435 μH)
18.0-18.5	27.0-27.5	32.5	39	7½ turns violet Toko S-18-series (≈ 0.375 μH)
21.0-21.5	30.0-30.5	35.5	27	7½ turns violet Toko S-18-series (≈ 0.350 μH)
24.5-25.0	33.5-34.0	39.0	22	6½ turns blue Toko S-18-series (≈ 0.300 μH)
28.0-28.5	37.0-37.5	42.5	22	5½ turns green Toko S-18-series (≈ 0.245 μH)
28.5-29.0	37.5-38.0	43.0	22	5½ turns green Toko S-18-series (≈ 0.239 μH)
29.0-29.5	38.0-38.5	43.5	22	5½ turns green Toko S-18-series (≈ 0.232 μH)
29.5-30.0	38.5-39.0	44.0	22	5½ turns green Toko S-18 series (≈ 0.227 μH)

The Toko 10K-series forms have four-section bobbins. The VCO-coil windings for 160 through 30 m are therefore split into four equal sections (for example, 5 + 5 + 5 + 5 turns for the 160-m coil).

3, U2 output [via Q13], to pin 14). The phase detector inputs must be transposed for adding to keep the loop feedback negative; otherwise, the phase detector would drive the VCO in the wrong direction — *away* from lock instead of toward it. For this reason, you cannot have some crystals above and some crystals below the output frequency driving the same loop — unless you also build in means to transpose the phase detector inputs as necessary.

The loop's phase/frequency detector is a high-speed CMOS IC, the 74HC4046 or 74HCT4046. This part, which is faster, more robust and better behaved than the original CD4046, has a power-supply range of 2 to 6 V. (The CD4046 is not recommended in this circuit because its limited high-frequency response may make it unreliable.) The internal VCO in this chip is not worth using for our purposes, so we disable it by tying pin 5 high. Both inputs can be driven by ac-coupled signals, but many CMOS parts are prone to damage if their inputs exceed their supply voltage or go below ground potential, so the Fig 14.52 circuit includes clipping diodes to protect U2 against transients at turn on. Assuming a VFO input level of 1 V P-P, R84, VFO level, can be set to maximum. (You may need to turn it down for higher VFO input levels, but its setting is uncritical.)

The loop amplifier uses a JFET-input op amp (a TL072) with both inputs biased to +2.5 V. R61 and C76 give a zero in the open-loop response at 2 kHz. C76 and R60 set the integrator gain to 3.4 at 1 Hz. (This gain is much lower than the usual value for a loop with a programmable divider, because this loop does not have to make up for reduction in loop gain that occurs in division.) R62 and C78 give the next pole,

at 10 kHz. R63 is included as a result of the use of earlier-generation PC-board versions of the circuit. Its value is noncritical and can be anything from 0 to the value shown in Fig 14.52 because it merely connects the R62/C78 network to U3B, which isolates these parts from R64 and C79. They form the fourth pole, at 160 kHz. (The poles in the VCOs themselves are all much higher and can be ignored.) The overall loop design conforms exactly to the "recipe" described earlier and was designed using an HP11C pocket calculator program implementing the equations in Fig 14.47.

In normal operation, TP1 and TP2 are joined, completing the loop. For troubleshooting, TP1 can be connected to TP3 instead, breaking the loop and placing a fixed 5-V tuning voltage on all the tuning diodes. This should put the VCO frequencies somewhere within their normal operating range, so that the oscillators, amplifiers and mixer can be checked. The phase detector should drive the loop amplifier to one end of its output-voltage swing, which end depending on whether the VCO frequency is too high or too low. Tuning the VFO across its range should cause this to change, and the output voltage of the loop amplifier/filter system should change over to the other extreme.

All of the VCOs remain connected to the tuning-voltage line at all times. The VCOs are selected merely by switching +12 V to the desired one, applying power to its JFET and turning on its output switching diode, which connects its output to R49, VCO OUTPUT.

R49 allows adjustment of the TX LO and RX LO output level to suit whatever the system is driving. The TX LO and RX LO amplifiers are capable of 500 mV RMS

output, which is sufficient for driving a Plessey SL6440 mixer directly. Power amplification will be needed to drive a diode-ring mixer, however. The output amplifiers are designed to keep signals from *entering* the system and to prevent unwanted signals at either output from reaching the other output. This may sound odd, but if you use, say, the TX LO output to drive something noisy, like a frequency counter, its amplifier's reverse isolation provides valuable protection against unwanted signals reaching whatever is connected to the RX LO output — say, a receiver mixer. The output amplifiers' output impedance is roughly equal to their collector resistances (680 Ω), but coax terminated in 50 Ω can be driven successfully.

R50, VCO LEVEL, sets the VCO signal applied to the NE602. It can be set at maximum if the VCO circuit in Fig 14.52 is used, but has been included in case anyone wants to experiment with different devices and values in the VCO circuits, which could change their output level. VCO drive to the NE602 should be roughly 200 mV P-P.

The NE602's full differential output is used to avoid the need for a gain stage between it and the phase detector. The low-pass filter (C65, C66, C67, L13 and L14) passes the VFO tuning range (5 to 5.5 MHz) while rolling off the higher-frequency VCO signals. Ordinary 20%-tolerance chokes with Qs of 30 or more will work adequately at L13 and L14.

To adjust the unit, connect TP1 to TP3. Select each band in turn and adjust the frequency of the appropriate VCO to about the middle of the output frequency range for that band. Reconnect TP1 to TP2 to close the loop. With the loop locked, for each band in turn, tune the VFO to the center of its range (5.25 MHz) and check that the voltage on TP1 is roughly 5 V. This confirms that all ranges can lock. Now we need only check that the diode tuning voltage doesn't go too low or too high at the band edges. For each range in turn, tune the VFO from 5.5 to 5 MHz while monitoring the tuning voltage on TP1. It must not go below 3 V or above 9 V or there will not be insufficient margin for aging and temperature effects. (Ideally, the voltage should be 4 V at the low-frequency end of each band.) Adjust the cores of the VCO inductors as necessary to optimize the tuning-voltage swing.

Summing loops have a serious drawback: If the VCO is ever, even momentarily, tuned to a frequency on the opposite side of the crystal oscillator, the sense of the feedback via the phase/frequency detector changes, and positive feedback forces the VCO even further from the correct output frequency. The loop can latch-up far from lock. This design avoids that risk by using a bank of narrow-range VCOs — an approach with the added advantage that narrow-range VCOs can give better noise performance than wide-range ones. Designing a summing loop around a lower-frequency VFO than that used here would move the VCO frequencies closer to the crystal oscillator frequencies, requiring careful design to avoid latch-up.

A PLL's tuning voltage is a superb indicator of its health. A quick check with a voltmeter as the loop is tuned across its range will show if aging has brought the loop close to unlocking and when the VCOs need readjusting.

Lock Detectors

It's a good idea, especially on a transmitter, to have some warning if a synthesizer loses lock. Many phase-detector chips have auxiliary outputs that, with a little filtering and amplification, can be used to light warning lights and inhibit transmission.

Another method is to monitor the tuning voltage, where the loop amplifier is an integrator. The rest of the loop can be thought of as a dc feedback path around the loop amplifier — and the loop amplifier has no local feedback at dc. If the loop fails, the amplifier will drift off into saturation close to either its + or – supply voltage. Window comparator circuitry can be used to continuously monitor the tuning voltage to confirm that it stays within thresholds set between the normal tuning-voltage range and the limits of the op amp's output swing.

DIRECT DIGITAL SYNTHESIS

Direct digital synthesis (DDS) is covered in this *Handbook*'s **Digital Signal Processing** chapter, but as good consumer-affordable DDSs don't yet cover the full tuning range required for even an MF/HF transceiver, we need to consider how to exploit them in conjunction with phase-locked loops. The VFO in the summing loop just described could be replaced by a DDS covering 5 to 5.5 MHz. A single loop with a programmable divider could be used in place of the crystal-oscillator bank. The result would be very close to the approach used in the latest generation of commercial Amateur Radio gear. That manufacturers advertise them as using direct digital synthesis has given some people the impression that their synthesizers consist entirely of DDS. In fact, the DDS in modern ham transceivers replaces the lower-significance "interpolation" loops in what otherwise would be regular multiloop synthesizers.

This is not to say that the DDS in our current ham gear is not a great improvement over the size, complexity and cost of what it replaces. Its random noise sidebands are usually excellent, and it can execute fast, clean frequency changes. The latest devices do 32-bit phase arithmetic and so offer over a billion frequency steps, which translates into a frequency resolution of a few *millihertz* with the usual clock rates for our applications — so the old battle for better resolution and low cost has already been won. Direct digital synthesis is not without a few problems, but fortunately, hybrid structures using PLL and DDS can allow one technique to compensate for the weaknesses of the other.

The prime weakness of the direct digital synthesizer is its quantization noise. A DDS cannot construct a *perfect* sine wave because each sample it outputs must be the nearest available voltage level from the set its DAC can make. So a DDS's output waveform is really a series of steps that only approximate a true sine wave.

We can view a DDS's output as an ideal sine wave plus an irregular "error" waveform. The spectrum of the error waveform is the set of unwanted frequency components found on the output of the DDS. Quantization noise is not the only source of unwanted DAC outputs. DACs can also give large output spikes, called *glitches*, as they transit from one level to another. In some DACs, glitches cause larger unwanted components than quantization. These components are scattered over the full frequency range passed by the DDS's low-pass filter, and their frequencies shift as the DDS tunes to different frequencies. At some frequencies, a number of components may coincide and form a single, larger component.

A summing loop PLL acts as a tracking filter, with a bandwidth measured in kilohertz. Acting on a DDS's output, a summing loop passes only quantization components close to the carrier. A system designer seeking to minimize the DDS's noise contribution must choose between using a loop bandwidth narrow enough to filter the DDS (thereby reducing the loop's ability to purify its VCO) or a more expensive low "glitch energy" DAC with more bits of resolution (allowing greater loop bandwidth and better VCO-noise control).

Complex ICs containing most of a DDS have been on the market for over 10 years, steadily getting cheaper, adding functions and occasionally taking onboard the functions of external parts (first the sine ROM, now the DAC as well). The Analog Devices AD7008 is an entire DDS (just add a low-pass filter . . .) on one CMOS chip. It includes a 10-bit DAC (fast enough for the whole system to clock at 50 MHz) and some digital-modulation hardware.

For the home-brewing amateur, DDS's first drawback is that these devices must receive their frequency data in binary form, loaded via a serial port, and this really forces the use of a microprocessor system in the radio. DDS's second drawback for experimenters is that surface-mount packages are becoming the norm for DDS ICs. Offsetting this, the resulting simplification of an entire synthesizer to a few ICs should cause more people to experiment with them.

EXPLORING THE SYNTHESIZER IN A COMMERCIAL MF/HF TRANSCEIVER

Few people would contemplate building an entire synthesized transceiver, but far more will need to understand enough of a commercially built one to be able to fix it or modify it. Choosing a radio to use as an example was not too difficult. The ICOM IC-765 received high marks for its clean synthesizer in its *QST* Product Review, so it certainly has a synthesizer worth examining. We are grateful to ICOM America for their permission to reprint the IC-765's schematic in the dis-

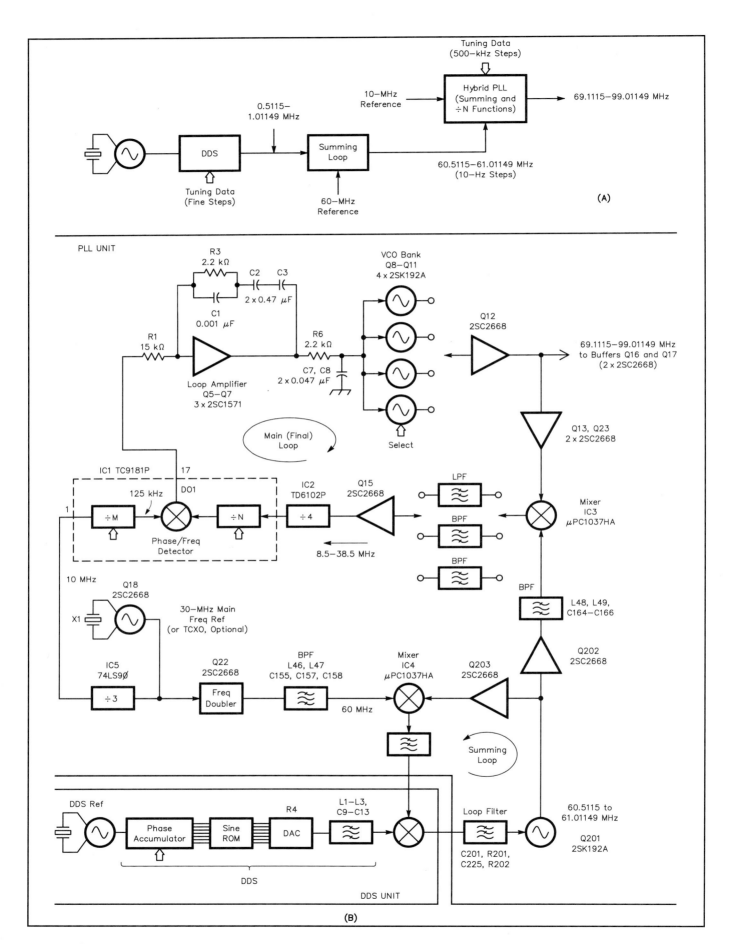

Fig 14.53 — Simplified (A) and detailed (B) block diagrams of the ICOM IC-765 frequency synthesizer.

cussion to follow.

Fig 14.53A shows a simplified block diagram of the IC-765's synthesizer. It contains one DDS and two PLLs. Notice that the DDS has its own frequency-reference oscillator. DDSs usually use binary arithmetic in their phase accumulators, so their step size is equal to their clock (reference) frequency divided by a large, round, binary number. The latest 32-bit machines give a step size of $1/(2^{32})$ of their clock frequency, that is, a ratio of 1/4,294,967,296. This means that if we want, say, a 10-Hz step size, we must have a peculiar reference frequency so that the increment of the DDS is a submultiple of 10 Hz. This is what ICOM designers chose. One alternative would be to use a convenient reference frequency (say 10 MHz) and accept a strange synthesizer step size. A 32-bit machine clocked at 10 MHz will give a step size of 0.002328... Hz. It would be simple to have the radio's microprocessor select the nearest of these very fine steps to the frequency set by the user, giving the user the appearance of a 10-Hz step size. An error of ±1.2 millihertz is trivial compared to the accuracy of all usual reference oscillators.

Installing the IC-765's optional high-stability 30-MHz TCXO does nothing to improve the accuracy of the radio's DDS section as it uses its own reference. The effect of the stability and accuracy of the DDS reference on the overall tuned frequency, however, is smaller than that of the main reference.

The DDS runs at a comfortably low frequency (0.5115 to 1.01149 MHz) to ease the demands placed on DAC settling time. The final loop needs a signal near 60 MHz as a summing input. Just mixing the DDS with 60 MHz would require a complex filter to reject the image. The IC-765 avoids this by using a summing loop. The summing loop VCO must only tune from 60.5115 to 61.01149 MHz, plus some margin for aging and temperature, but care is needed in this circuit because, as in the home-brew summing loop described earlier, the loop may latch up if this VCO goes below 60 MHz.

The final loop is a normal PLL with a programmable divider, but instead the VCO is not fed directly into the divider — it is mixed down by the 60.5115- to 61.01149-MHz signal first. The final loop uses a bank of four switched VCOs, each

one covering a one fourth of the system's full output range of 69.1115 to 99.01149 MHz. This mix-down feeds an 8.5- to 38.5-MHz signal into the programmable divider. To remove unwanted mixer outputs over this range, three different, switched filters are needed before the signal is amplified to drive the divider.

Fig 14.53B shows the synthesizer in greater detail. The programmable divider/PSD chip used (IC1, a TC9181P) will not operate up to 38 MHz, so a prescaler (IC2, a TD6102P) has been added before its input. This is a fixed divide-by-four prescaler, which forces the phase detector to be run at one-fourth of the step size, at 125 kHz. (A dual-modulus prescaler would have been very desirable here because it would allow the PSD to run at 500 kHz, easing the trade-off between reference-frequency suppression and loop bandwidth. Unfortunately, the lowest division factor necessary is too low for the simple application of any of the common ICs with prescaler controllers.)

Fig 14.54 shows the full circuit diagram of the IC-765's synthesizer. The DDS is on a small subboard, DDS unit (far right). IC1, an SC-1051, appears to be a custom IC, incorporating the summing loop's phase accumulator and phase detector. The DDS's sine ROM is split into two parts: IC2, an SC-1052; and IC3, an SC-1053. IC4 and IC5, both 74HCT374s, are high-speed-CMOS flip-flops used as latches to minimize glitches by closely synchronizing all 12 bits of data coming from the ROMs. The DAC is simply a binary-weighted resistor ladder network, R4, which relies on the CMOS latch outputs all switching exactly between the +5 V supply and ground.

Look at the output of the main loop phase detector (pin 17 of IC1 on the PLL unit), find Q5, Q6 and Q7 and look at the RC network around them. It's the recipe loop design again! R3 creates the zero, but notice C1 across it. This is a neat and economical way of creating one of the other poles. R6 and C7 create the final pole.

Fig 14.55 shows the phase noise of an IC-765 measured on a professional phase-noise-measurement system. The areas shown in Fig 14.49 can be seen clearly. The slope above 500 Hz is the phase noise of the final VCO and exactly tracks the *Good* curve in KI6WX's *QST* article on the effects of phase noise. Below the noise peak, the phase noise falls to levels 20 dB better than KI6WX's *Excellent* curve. This low-noise area complements the narrow CW filter skirts and the narrow notch filter.

Many technically inclined amateurs may be wondering what could be done to

improve the IC-765's synthesizer, or to design an improved one. The peak in the IC-765's phase noise at 500 Hz is quite prominent. It does not appear to be due to a marginally unstable loop design, but rather to a mismatch between the choice of loop bandwidth and the noise performance of the oscillators involved. The DDS's resistor-based DAC is unlikely to perform as well as a purpose-designed 12-bit DAC, and it also lets power-supply noise modulate the output. The system's designers appear to have deliberately reduced the loop bandwidth to better filter DDS spurs. If we were free to increase the cost and complexity of the unit, we could buy a high-performance DAC designed for low DDS spurs and trade this improvement off against noise by increasing the loop bandwidth — redesigning the loop poles, zero and gain. To help with the loop bandwidth, and get a 12-dB improvement in any noise from the main loop's phase detector, divider and loop amplifier, we could remove the fixed prescaler (IC2, the TD6102P) and operate the PSD at 500 kHz. This would require the design of a faster programmable divider that can handle 38 MHz.

Finally, the IC-765's VCOs could be improved. Their tuned circuits could be changed to low-L, high-C, multiple-tuning-diode types. If we added a higher-voltage power supply to allow higher diode tuning voltages than the 5.6-V level at which the design now operates (only little current is needed), higher-Q tuning diodes can be used — and we can avoid using them at lower tuning voltages, where tuning-diode Q degrades. This could reduce the phase noise above the noise peak by several decibels. Changing the loop bandwidth would reduce the noise bump's height and move it farther from the carrier.

It's important to keep all of these what-ifs in perspective: The IC-765's synthesizer was one of the best available in amateur MF/HF transceivers when this chapter was written, and its complexity is about half that of old multiloop, non-DDS examples. Like any product of mass production, its design involved trade-offs between cost, performance and component availability. As better components become less expensive, we can expect excellent synthesizer designs like the IC-765's to come down in price and even better synthesizer designs to become affordable in Amateur Radio gear.

SYNTHESIZERS: THE FUTURE

It's notoriously dangerous to predict the future in print, but there are some new developments being used in professional

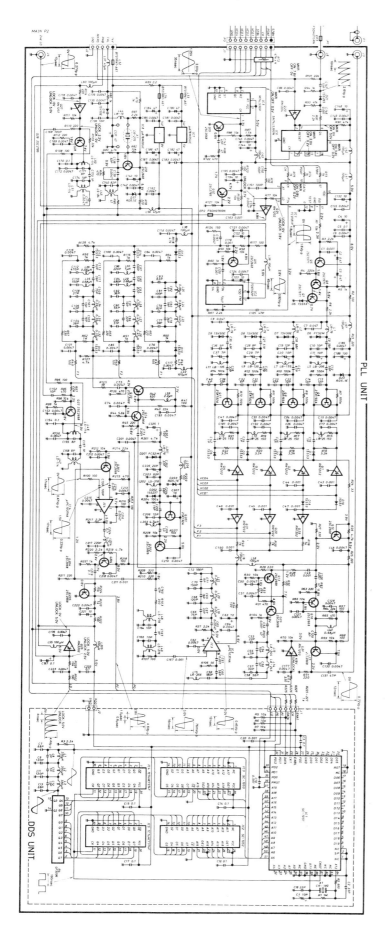

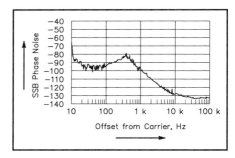

Fig 14.54 — The IC-765 frequency synthesizer down to the component level. The key ICs in the radio's DDS (IC1, IC2 and IC3, right) appear to be custom components made especially for this application.

Fig 14.55 — The phase noise of an ICOM IC-765 (serial no. 03077) as measured by a Hewlett-Packard phase-noise measurement system.

test equipment, on sale as this chapter is being written, that can reasonably be expected to percolate through into radio equipment.

Fractional-N Synthesis

A single loop would be capable of any step size if its programmable divider wasn't tied to integer numbers. Some designers long ago tried to use such *fractional-N* values by switching the division ratio between two integers, with a duty cycle that set the fractional part. This whole process was synchronized with the divider's operation. Averaged over many cycles, the frequency really did come out as wanted, allowing interpolation between the steps mandated by integer-N division. This approach was largely abandoned because it added huge sidebands (at the fractional frequency and its harmonics) to the loop's output. One such synthesizer design — which has been applied in a large amount of equipment and remains in use — uses a hybrid digital/analog system to compute a sawtooth voltage waveform of just the right amplitude, frequency and phase that can be added to the VCO tuning voltage to cancel the fractional-frequency FM sidebands. This system is complex, however, and like all cancellation processes, it can never provide complete cancellation. It is appreciably sensitive to changes due to tolerances, aging and alignment. Even when applied in a highly developed form using many tight-tolerance components, such a system cannot reduce its fractional frequency sidebands to a level much below −70 to −80 dBc.

A new approach, which has been described in a few articles in professional and trade journals, does not try to cancel its sidebands at all. Its basic principle is delightfully simple. A digital system switches the programmable divider of a normal single loop around a set of division values. This set of values has two properties: First, its average value is controllable in very small steps, allowing fine interpolation of the integer steps of the loop; second, the FM it applies to the loop is huge, but the resultant sideband energy is strongly concentrated in very-high-order sidebands. The loop cannot track such fast FM and so filters off this modulation!

The key to this elegant approach is that it deliberately shapes the spectrum of its loop's unwanted components such that they're small at frequencies at which the loop will pass them and large at frequencies filterable by the loop. The result is a reasonably clean output spectrum.

BIBLIOGRAPHY AND REFERENCES

Clarke and Hess, *Communications Circuits; Analysis and Design* (Addison-Wesley, 1971; ISBN 0-201-01040-2). Wide coverage of transistor circuit design, including techniques suited to the design of integrated circuits. Its age shows, but it is especially valuable for its good mathematical treatment of oscillator circuits, covering both frequency- and amplitude-determining mechanisms. Look for a copy at a university library or initiate an interlibrary loan.

J. Grebenkemper, "Phase Noise and Its Effects on Amateur Communications," *Part 1, QST*, Mar 1988, pp 14-20; *Part 2*, Apr 1988, pp 22-25. Also see Feedback, *QST*, May 1988, p 44. This material also appears in volumes 1 (1991) and 2 (1993) of *The ARRL Radio Buyer's Sourcebook*. Covers the effects of phase noise and details the measurement techniques used in the ARRL Lab. Those techniques are also described in a sidebar earlier in this chapter.

W. Hayward and D. DeMaw, *Solid State Design for the Radio Amateur* (Newington, CT: ARRL, 1986). A good sourcebook of RF design ideas, with good explanation of the reasoning behind design decisions.

"Spectrum Analysis...Random Noise Measurements," Hewlett-Packard Application Note 150-4. Source of correction factors for noise measurements using spectrum analyzers. HP notes should be available through local HP dealers.

U. Rohde, *Digital PLL Frequency Synthesizers* (Englewood Cliffs, NJ: Prentice-Hall, Inc). Now available solely from Compact Software, 201 McLean Blvd, Paterson, NJ 07504, this book contains the textbook-standard mathematical analyses of frequency synthesizers combined with unusually good insight into what makes a better synthesizer and a *lot* of practical circuits to entertain serious constructors. A good place to look for low-noise circuits and techniques.

U. Rohde, "Key Components of Modern Receiver Design," *Part 1, QST*, May 1994, pp 29-32; *Part 2, QST*, Jun 1994, pp 27-31; *Part 3, QST,* Jul 1994, pp 42-45. Includes discussion on phase-noise reduction techniques in synthesizers and oscillators.

Pappenfus and others, *Single Sideband Circuits and Systems* (McGraw-Hill). A book on HF SSB transmitters, receivers and accoutrements by Rockwell-Collins staff. Contains chapters on synthesizers and frequency standards. The frequency synthesizer chapter predates the rise of the DDS and other recent techniques, but good information about the effects of synthesizer performance on communications is spread throughout the book.

Motorola Applications Note AN-551, "Tuning Diode Design Techniques," included in Motorola RF devices databooks. A concise explanation of varactor diodes, their characteristics and use.

I. Keyser, "An Easy to Set Up Amateur Band Synthesiser," *RADCOM* (RSGB), Dec 1993, pp 33-36.

Contents

Mixers, Modulators and Demodulators

15

At base, radio communication involves translating information into radio form, letting it travel for a spell, and translating it back again. Translating information into radio form entails the process we call *modulation*, and *demodulation* is its reverse. One way or another, every transmitter used for radio communication, from the simplest to the most complex, includes a means of modulation; one way or another, every receiver used for radio communication, from the simplest to the most complex, includes a means of demodulation.

Modulation involves varying one or both of a radio signal's basic characteristics—amplitude and frequency (or phase)—to convey information. A circuit, stage or piece of hardware that modulates is called a *modulator*.

Demodulation involves reconstructing the transmitted information from the changing characteristic(s) of a modulated radio wave. A circuit, stage or piece of hardware that demodulates is called a *demodulator*.

Many radio transmitters, receivers and transceivers also contain *mixers*—circuits, stages or pieces of hardware that combine two or more signals to produce additional signals at sums of and differences between the original frequencies. Amateur Radio textbooks have traditionally handled mixers separately from modulators and demodulators, and modulators separately from demodulators.

This chapter, by David Newkirk, WJ1Z, and Rick Karlquist, N6RK, examines mixers, modulators and demodulators together because the job they do is essentially the same. Modulators and demodulators translate information into radio form and back again; mixers translate one frequency to others and back again. All of these translation processes can be thought of as forms of frequency translation or frequency shifting—the function traditionally ascribed to mixers. We'll therefore begin our investigation by examining what a mixer is (and isn't), and what a mixer does.

THE MECHANISM OF MIXERS AND MIXING

What is a Mixer?

Mixer is a traditional radio term for a circuit that shifts one signal's frequency up or down by combining it with another signal. The word *mixer* is also used to refer to a device used to blend multiple audio inputs together for recording, broadcast or sound reinforcement. These two mixer types differ in one very important way: A radio mixer makes new frequencies out of the frequencies put into it, and an audio mixer does not.

Mixing Versus Adding

Radio mixers might be more accurately called "frequency mixers" to distinguish them from devices such as "microphone mixers," which are really just signal *combiners*, *summers* or *adders*. In their most basic, ideal forms, both devices have two inputs and one output. The combiner simply *adds* the instantaneous voltages of the two signals together to produce the output at each point in time (**Fig 15.1**). The mixer, on the other hand, *multiplies* the instantaneous voltages of the two signals together to produce its output signal from instant to instant (**Fig 15.2**). Comparing the output spectra of the combiner and mixer, we see that the combiner's output contains only the frequencies of the two inputs, and nothing else, while the mixer's output contains *new* frequencies. Because it combines one energy with another, this process is sometimes called *heterodyning*, from the Greek words for *other* and *power*.

Mixing as Multiplication

Since a mixer works by means of multiplication, a bit of math can show us how they work. To begin with, we need to represent the two signals we'll mix, A and B, mathematically. Signal A's instantaneous amplitude equals

$$A_a \sin 2\pi f_a t \qquad (1)$$

in which A is peak amplitude, f is frequency, and t is time. Likewise, B's instantaneous amplitude equals

$$A_b \sin 2\pi f_b t \qquad (2)$$

Since our goal is to show that multiplying two signals generates sum and difference frequencies, we can simplify these

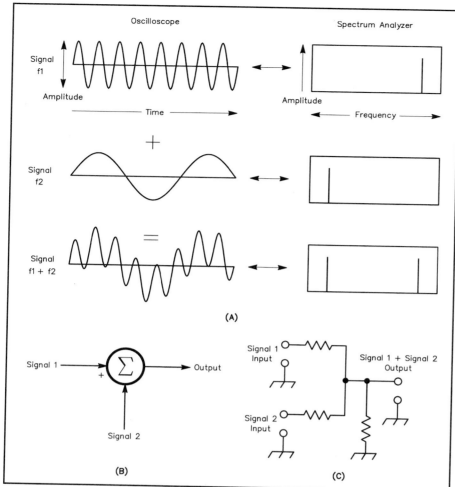

Fig 15.1—Adding or summing two sine waves of different frequencies (f1 and f2) combines their amplitudes without affecting their frequencies. Viewed with an **oscilloscope** (a real-time graph of amplitude versus time), adding two signals appears as a simple superimposition of one signal on the other. Viewed with a **spectrum analyzer** (a real-time graph of signal amplitude versus frequency), adding two signals just sums their spectra. The signals merely coexist on a single cable or wire. All frequencies that go into the adder come out of the adder, and no new signals are generated.

Drawing B, a block diagram of a summing circuit, emphasizes the stage's mathematical operation rather than showing circuit components. Drawing C shows a simple summing circuit, such as might be used to combine signals from two microphones. In audio work, a circuit like this is often called a mixer—but it does not perform the same function as an RF mixer.

signal definitions by assuming that the peak amplitude of each is 1. The equation for Signal A then becomes

$$a(t) = A \sin (2\pi f_a t) \qquad (3)$$

and the equation for Signal B becomes

$$b(t) = B \sin (2\pi f_b t) \qquad (4)$$

Each of these equations represents a sine wave and includes a subscript letter to help us keep track of where the signals go.

Merely combining Signal A and Signal B by letting them travel on the same wire develops nothing new:

$$a(t) + b(t) = \\ A \sin (2\pi f_a t) + B \sin (2\pi f_b t) \qquad (5)$$

As needlessly reflexive as equation 5 may seem, we include it to highlight the fact that multiplying two signals is a quite different story. From trigonometry, we know that multiplying the sines of two variables can be expanded according to the relationship

$$\sin x \sin y = \frac{1}{2}\left[\cos (x - y) - \cos (x + y)\right] \qquad (6)$$

Conveniently, Signals A and B are both sinusoidal, so we can use equation 6 to determine what happens when we multiply Signal A by Signal B. In our case, $x = 2\pi f_a t$ and $y = 2\pi f_b t$, so plugging them into equation 6 gives us

$$a(t) \bullet b(t) = \\ \frac{AB}{2}\cos\left(2\pi\left[f_a - f_b\right]t\right) - \frac{AB}{2}\cos\left(2\pi\left[f_a + f_b\right]t\right)$$

$$(7)$$

Now we see two momentous results: a sine wave at the frequency *difference* between Signal A and Signal B $2\pi(f_a - f_b)t$, and a sine wave at the frequency *sum* of Signal A and Signal B $2\pi(f_a + f_b)t$. (The products are cosine waves, but since equivalent sine and cosine waves differ only by a phase shift of 90°, both are called *sine waves* by convention.)

This is the basic process by which we translate information into radio form and translate it back again. If we want to transmit a 1-kHz audio tone by radio, we can feed it into one of our mixer's inputs and feed an RF signal—say, 5995 kHz—into the mixer's other input. The result is two radio signals: one at 5994 kHz (5995 − 1) and another at 5996 kHz (5995 + 1). We have achieved modulation.

Converting these two radio signals back to audio is just as straightforward. All we do is feed them into one input of another mixer, and feed a 5995-kHz signal into the mixer's other input. Result: a 1-kHz tone. We have achieved demodulation; we have communicated by radio.

The key principle of a radio mixer is that, in mixing multiple signal voltages together, it adds and subtracts their frequencies to produce new frequencies. (In the field of signal processing, this process, *multiplication in the time domain*, is recognized as equivalent to the process of *convolution in the frequency domain*. Those interested in this alternative approach to describing the generation of new frequencies through mixing can find more information about it in the many textbooks available on this fascinating subject.) The difference between the mixer we've been describing and any mixer, modulator or demodulator that you'll ever use is that it's ideal. We put in two signals and got just two signals out. *Real* mixers, modulators and demodulators, on the other hand, also produce *distortion* products that make their output spectra "dirtier" or "less clean," as well as putting out some energy at input-signal frequencies and their harmonics. Much of the art and science of making good use of multiplication in mixing, modulation and demodulation goes into minimizing these unwanted multiplication products (or their effects) and making multipliers do their frequency translations as efficiently as possible.

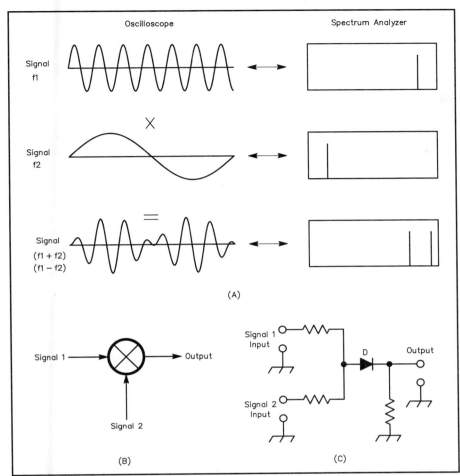

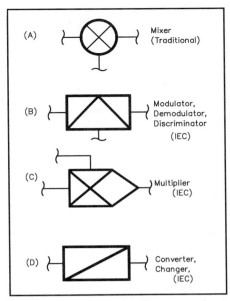

Fig 15.3—**We commonly symbolize mixers with a circled *X* (A) out of tradition, but other standards sometimes prevail (B, C and D). Although the converter/changer symbol (D) can conceivably be used to indicate frequency changing through mixing, the three-terminal symbols are arguably better for this job because they convey the idea of two signal sources resulting in a new frequency. (*IEC* stands for *International Electrotechnical Commission*.)**

Fig 15.2—*Multiplying* two sine waves of different frequencies produces a new output spectrum. Viewed with an oscilloscope, the result of multiplying two signals is a composite wave that seems to have little in common with its components. A spectrum-analyzer view of the same wave reveals why: The original signals disappear entirely and are replaced by two new signals—at the *sum* and *difference* of the original signals' frequencies. Drawing B diagrams a multiplier, known in radio work as a mixer. The *X* emphasizes the stage's mathematical operation. (The circled *X* is only one of several symbols you may see used to represent mixers in block diagrams, as Fig 15.3 explains.)

Drawing C shows a very simple multiplier circuit. The diode, D, does the mixing. Because this circuit does other mathematical functions and adds them to the sum and difference products, its output is more complex than f1 + f2 and f1 − f2, but these can be extracted from the output by filtering.

Putting Multiplication to Work

Piecing together a coherent picture of how multiplication works in radio communication isn't made any easier by the fact that traditional terms applied to a given multiplication approach and its products may vary with their application. If, for instance, you're familiar with standard textbook approaches to mixers, modulators and demodulators, you may be wondering why we didn't begin by working out the math involved by examining *amplitude modulation*, also known as *AM*. "Why not tell them about the *carrier* and how to get rid of it in a *balanced modulator*?" A transmitter enthusiast may ask

"Why didn't you mention *sidebands* and how we conserve spectrum space and power by getting rid of one and putting all of our power into the other?" A student of radio receivers, on the other hand, expects any discussion of the same underlying multiplication issues to touch on the topics of *LO feedthrough*, *mixer balance* (*single* or *double*?), *image rejection* and so on.

You likely expect this book to spend some time talking to you about these things, so it will. But *this* radio-amateur-oriented discussion of mixers, modulators and demodulators will take a look at their common underlying mechanism *before* turning you loose on practical mixer, modulator and

demodulator circuits. Then you'll be able to tell the forest from the trees. **Fig 15.3** shows the block symbol for a traditional mixer along with several IEC symbols for other functions mixers may perform.

It turns out that the mechanism underlying multiplication, mixing, modulation and demodulation is a pretty straightforward thing: Any circuit structure that *nonlinearly distorts* ac waveforms acts as a multiplier to some degree.

Nonlinear Distortion?

The phrase *nonlinear distortion* sounds redundant, but isn't. Distortion, an externally imposed change in a waveform, can be linear; that is, it can occur independently of signal amplitude. Consider a radio receiver front-end filter that passes only signals between 6 and 8 MHz. It does this by *linearly distorting* the single complex waveform corresponding to the wide RF spectrum present at the radio's antenna terminals, reducing the amplitudes of frequency components below 6 MHz and above 8 MHz relative to those between 6 and 8 MHz. (Considering multiple signals on a wire as one complex waveform is just as valid, and sometimes handier, than considering them as separate signals. In this case, it's a bit easier to think of distortion as something that happens to a waveform rather than something that happens to separate signals relative to each other. It would be just as valid—and certainly more in keeping with the consensus view—to say merely that the filter attenuates signals at frequencies below 6 MHz and above 8 MHz.) The filter's output waveform certainly differs from its input waveform; the waveform has been distorted. But because this distortion occurs independently of signal level or polarity, the distortion is linear. No new frequency components are created; only the amplitude relationships among the wave's existing frequency components are altered. This is *amplitude* or *frequency* distortion, and all filters do it or they wouldn't be filters.

Phase or *delay distortion*, also linear, causes a complex signal's various component frequencies to be delayed by different amounts of time, depending on their frequency but independently of their amplitude. No new frequency components occur, and amplitude relationships among existing frequency components are not altered. Phase distortion occurs to some degree in all real filters.

The waveform of a nonsinusoidal signal can be changed by passing it through a circuit that has only linear distortion, but only *nonlinear distortion* can change the waveform of a simple sine wave. It can also produce an output signal whose output waveform changes as a function of the input amplitude, something not possible with linear distortion. Nonlinear circuits often distort excessively with overly strong signals, but the distortion can be a complex function of the input level.

Nonlinear distortion may take the form of *harmonic distortion*, in which integer multiples of input frequencies occur, or *intermodulation distortion (IMD)*, in which different components multiply to make new ones.

Any departure from absolute linearity results in some form of nonlinear distortion, and this distortion can work for us or against us. Any so-called linear amplifier distorts nonlinearly to some degree; any device or circuit that distorts nonlinearly can work as a mixer, modulator, demodulator or frequency multiplier. An amplifier optimized for linear operation will nonetheless mix, but inefficiently; an amplifier biased for nonlinear amplification may be practically linear over a given tiny portion of its input-signal range. The trick is to use careful design and component selection to maximize nonlinear distortion when we want it, and minimize it when we don't. Once we've decided to maximize nonlinear

distortion, the trick is to minimize the distortion products we don't want, and maximize the products we desire.

Keeping Unwanted Distortion Products Down

Ideally, a mixer multiplies the signal at one of its inputs by the signal at its other input, but does not multiply a signal at the same input by itself, or multiple signals at the same input by themselves or by each other. (Multiplying a signal by itself—squaring it—generates harmonic distortion [specifically, *second-harmonic* distortion] by adding the signal's frequency to itself per equation 7. Simultaneously squaring two or more signals generates simultaneous harmonic and intermodulation distortion, as we'll see later when we explore how a diode demodulates AM.)

Consider what happens when a mixer must handle signals at two different frequencies (we'll call them f_1 and f_2) applied to its first input, and a signal at a third frequency (f_3) applied to its other input. Ideally, a mixer multiplies f_1 by f_3 and f_2 by f_3, but does not multiply f_1 and f_2 by each other. This produces output at the sum and difference of f_1 and f_3, and the sum and difference of f_2 and f_3, but *not* the sum and difference of f_1 and f_2. **Fig 15.4** shows that feeding two signals into one input of a mixer results in the same output as if f_1 and f_2 are each first mixed with f_3 in two separate mixers, and the outputs of these mixers are combined. This shows that a mixer, even though constructed with nonlinearly distorting components, actually behaves as a *linear frequency shifter*.

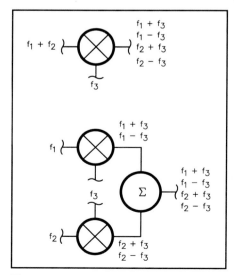

Fig 15.4—Feeding two signals into one input of a mixer results in the same output as if f₁ and f₂ are each first mixed with f₃ in two separate mixers, and the outputs of these mixers are combined.

Traditionally, we refer to this process as mixing and to its outputs as *mixing products*, but we may also call it frequency *conversion*, referring to a device or circuit that does it as a *converter*, and to its outputs as *conversion products*.

Real mixers, however, at best act only as *reasonably* linear frequency shifters, generating some unwanted IMD products—spurious signals, or *spurs*—as they go. Receivers are especially sensitive to unwanted mixer IMD because the signal-level spread over which they must operate without generating unwanted IMD is often 90 dB or more, and includes infinitesimally weak signals in its span. In a receiver, IMD products so tiny that you'd never notice them in a transmitted signal can easily obliterate weak signals. This is why receiver designers apply so much effort to achieving "high dynamic range."

The degree to which a given mixer, modulator or demodulator circuit produces unwanted IMD is often *the* reason why we use it, or don't use it, instead of another circuit that does its wanted-IMD job as well or even better.

Other Mixer Outputs

In addition to desired sum-and-difference products and unwanted IMD products, real mixers also put out some energy at their input frequencies. Some mixer implementations may *suppress* these outputs—that is, reduce one or both of their input signals by a factor of 100 to 1,000,000, or 20 to 60 dB. This is good because it helps keep input signals at the desired mixer-output sum or difference frequency from showing up at the IF terminal—an effect reflected in a receiver's *IF rejection* specification. Some mixer types, especially those used in the vacuum-tube era, suppress their input-signal outputs very little or not at all.

Input-signal suppression is part of an overall picture called *port-to-port isolation*. Mixer input and output connections are traditionally called *ports*. By tradition, the port to which we apply the shifting signal is the *local-oscillator (LO)* port. The convention for naming the other two ports (one of which must be an output, and the other of which must be an input) is usually that the higher-frequency port is called the *RF (radio frequency)* port and the lower-frequency port is called the *IF (intermediate frequency)* port. If a mixer's output frequency is lower than its input frequency, then the RF port is an input and the IF port is an output. If the output frequency is higher than the input frequency, the IF port may be the input and the RF port may be the output. (We hedge with *may be* because usage varies. When in doubt, check a diagram carefully to determine which port is the "gozinta" and which port is the "gozouta.")

It's generally a good idea to keep a mixer's input signals from appearing at its output port because they represent energy that we'd rather not pass on to subsequent circuitry. It therefore follows that it's usually a good idea to keep a mixer's LO-port energy from appearing at its RF port, or its RF-port energy from making it through to the IF port. But there are some notable exceptions.

Mixers and Amplitude Modulation

Now that we've just discussed what a fine thing it is to have a mixer that doesn't let its input signals through to its output port, we can explore a mixing approach that outputs one of its input signals so strongly that the fed-through signal's amplitude at least equals the combined amplitudes of the system's sum and difference products! This system, *amplitude modulation*, is the oldest means of translating information into radio form and back again. It's a frequency-shifting system in which the original unmodulated signal, traditionally called the *carrier*, emerges from the mixer along with the sum and difference products, traditionally called *sidebands*.

We can easily make the carrier pop out of our mixer along with the sidebands merely by building enough *dc level shift* into the information we want to mix so that its waveform never goes negative. Back at equation 1 and 2, we decided to keep our mixer math relatively simple by setting the peak voltage of our mixer's input signals directly equal to their sine values. Each input signal's peak voltage therefore varies between +1 and −1, so all we need to do to keep our modulating-signal term (provided with a subscript m to reflect its role as the modulating or information waveform) from going negative is add 1 to it. Identifying the carrier term with a subscript c, we can write

$$\text{AM signal} = \left(1 + m \sin 2\pi f_m t\right) \sin 2\pi f_c t \tag{8}$$

Notice that the modulation $(2\pi f_m t)$ term has company in the form of a coefficient, m. This variable expresses the modulating signal's varying amplitude—variations that ultimately result in amplitude modulation. Expanding equation 8 according to equation 6 gives us

$$\text{AM signal} = \sin 2\pi f_c t$$
$$+ \frac{1}{2} m \cos\left(2\pi f_c - 2\pi f_m\right)t$$
$$- \frac{1}{2} m \cos\left(2\pi f_c + 2\pi f_m\right)t \tag{9}$$

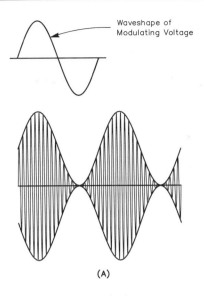

(A)

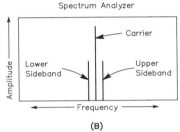

(B)

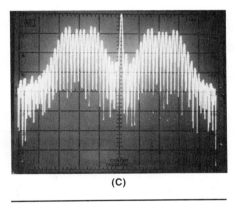

(C)

Fig 15.5—Graphed in terms of amplitude versus time (A), the *envelope* of a properly modulated AM signal exactly mirrors the shape of its modulating waveform, which is a sine wave in this example. This AM signal is modulated as fully as it can be—100%—because its envelope *just* hits zero on the modulating wave's negative peaks. Graphing the same AM signal in terms of amplitude versus frequency (B) reveals its three spectral components: Carrier, upper sideband and lower sideband. B shows sidebands as single-frequency components because the modulating waveform is a sine wave. With a complex modulating waveform, the modulator's sum and difference products really do show up as *bands* on either side of the carrier (C).

The modulator's output now includes the carrier ($\sin 2\pi f_c t$) in addition to sum and difference products that vary in strength according to m. According to the conventions of talking about modulation, we call the sum product, which comes out at a frequency higher than that of the carrier, the *upper sideband (USB)*, and the difference product, which comes out a frequency lower than that of the carrier, the *lower sideband (LSB)*. We have achieved amplitude modulation.

Why We Call It Amplitude Modulation

We call the modulation process described in equation 8 *amplitude modulation* because the complex waveform consisting of the sum of the sidebands and carrier varies with the information signal's magnitude (m). Concepts long used to illustrate AM's mechanism may mislead us into thinking that the *carrier* varies in strength with modulation, but careful study of equation 9 shows that this doesn't happen. The carrier, $\sin 2\pi f_c t$, goes into the modulator—we're in the modulation business now, so it's fitting to use the term *modulator* instead of *mixer*—as a sinusoid with an unvarying maximum value of |1|. The modulator multiplies the carrier by the dc level (+1) that we added to the information signal (m $\sin 2\pi f_m t$). Multiplying $\sin 2\pi f_c t$ by 1 merely returns $\sin 2\pi f_c t$. We have proven that the carrier's amplitude does not vary as a result of amplitude modulation. The carrier is, however, used by many circuits as a reference signal.

Overmodulation

Since an AM signal's information content resides entirely in its sidebands, it follows that the more energetic we make the sidebands, the more information energy will be available for an AM receiver to "recover" when it demodulates the signal. Even in an ideal modulator, there's a practical limit to how strong we can make an AM signal's sidebands relative to its carrier, however. Beyond that limit, we severely distort the waveform we want to translate into radio form.

We reach AM's distortion-free modulation limit when the sum of the sidebands and carrier at the modulator output *just reaches zero* at the modulating waveform's most negative peak (**Fig 15.5**). We call this condition *100% modulation*, and it occurs when m equals 1. (We enumerate *modulation percentage* in values from 0 to 100%. The lower the number, the less information energy in the sidebands. You may also see modulation enumerated in terms of a *modulation factor* from 0 to 1, which directly equals m; a modulation factor of 1 is the same as 100% modulation.)

Equation 9 shows that each sideband's voltage is half that of the carrier. Power varies as the square of voltage, so the power in each sideband of a 100%-modulated signal is therefore $(1/2)^2$ times, or 1/4, that of the carrier. A transmitter capable of 100% modulation when operating at a carrier power of 100 W therefore puts out a 150-W signal at 100% modulation, 50 W of which is attributable to the sidebands. (The *peak* envelope power [PEP] output of a double-sideband, full-carrier AM transmitter at 100% modulation is four times its carrier PEP. This is why our solid-state, "100-W" MF/HF transceivers are usually rated for no more than about 25 W carrier output at 100% amplitude modulation.)

One-hundred-percent modulation is a brick-wall limit because an amplitude modulator can't reduce its output to less than zero. Trying to increase modulation beyond the 100% point results in *overmodulation* (**Fig 15.6**), in which the modulation envelope no longer mirrors the shape of the modulating wave (Fig 15.6A). An overmodulated wave contains more energy than it did at 100% modulation, but some of the added energy now exists as *harmonics of the modulating waveform* (Fig 15.6B). This distortion makes the modulated signal take up more spectrum space than it needs. In voice operation, overmodulation commonly happens only on syllabic peaks, making the distortion products sound like crashy, transient noise we refer to as *splatter*.

Modulation Linearity

If we increase an amplitude modulator's

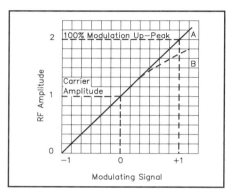

Fig 15.7—An ideal AM transmitter exhibits a straight-line relationship (A) between its instantaneous envelope amplitude and the instantaneous amplitude of its modulating signal. Distortion, and thus an unnecessarily wide signal, results if the transmitter cannot respond linearly across the modulating signal's full amplitude range.

modulating-signal input by a given percentage, we expect a proportional modulation increase in the modulated signal. We expect good *modulation linearity*. Suboptimal amplitude modulator design may not allow this, however. Above some modulation percentage, a modulator may fail to increase modulation in proportion to an increase its input signal (**Fig 15.7**). Distortion, and thus an unnecessarily wide signal, results.

Using AM to Send Morse Code

Fig 15.6A closely resembles what we see when a properly adjusted CW transmitter sends a string of dots (**Fig 15.8**). Keying a carrier on and off produces a wave that varies in amplitude and has double (upper and lower) sidebands that vary in spectral composition according to the duration and envelope shape of the on-off transitions. The emission mode we call *CW* is therefore a form of AM. The concepts of modulation percentage and overmodulation are usually not applied to generating an on-off-keyed Morse signal, however. This is related to how we copy CW by ear, and the fact that, in CW radio communication, we usually don't translate the received signal all the way back into its original pre-modulator (*baseband*) form, as a closer look at the process reveals.

In CW transmission, we usually open and close a keying line to make dc transitions that turn the transmitted carrier on and off. CW reception usually does not entirely reverse this process, however. Instead of demodulating a CW signal all

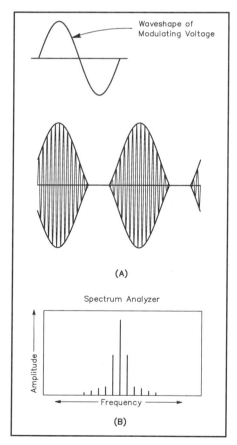

Fig 15.6—Overmodulating an AM transmitter results in a modulation envelope (A) that doesn't faithfully mirror the modulating waveform. This distortion creates additional sideband components that broaden the transmitted signal (B).

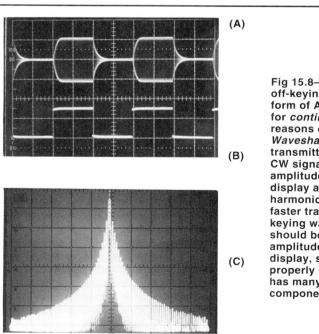

Fig 15.8—Telegraphy by on-off-keying a carrier is also a form of AM, called *CW* (short for *continuous wave*) for reasons of tradition. *Waveshaping* in a CW transmitter often causes a CW signal's RF envelope (the amplitude-versus-time display at A) to contain less harmonic energy than the faster transitions of its keying waveform (B) suggest should be the case. C, an amplitude-versus-frequency display, shows that even a properly shaped CW signal has many sideband components.

the way back to its baseband self—a shifting dc level—we want the presences and absences of its carrier to create long and short audio tones. Because the carrier is RF and not AF, we must mix it with a locally generated RF signal—from a *beat-frequency oscillator (BFO)*—that's close enough in frequency to produce a difference signal at AF. What goes into our transmitter as shifting dc comes out of our receiver as thump-delimited tone bursts of dot and dash duration. We have achieved CW communication.

The dots and dashes of a CW signal must start and stop abruptly enough so we can clearly distinguish the carrier's presences and absences from noise, especially when fading prevails. The keying sidebands, which sound like little more than thumps when listened to on their own, help our brains be sure when the carrier tone starts and stops.

It so happens that we always need to hear one or more harmonics of the fundamental keying waveform for the code to sound sufficiently crisp. If the transmitted signal will be subject to propagational fading—a safe assumption for any long-distance radio communication—we *harden* our keying by making the transmitter's output rise and fall more quickly. This puts more energy into more keying sidebands and makes the signal more copiable in the presence of fading—in particular, *selective fading*, which linearly distorts a modulated signal's complex waveform and randomly changes the sidebands' strength and phase relative to the carrier and each other. The appropriate keying hardness also depends on the keying speed. The faster the keying in WPM, the faster the on-off times—the harder the keying—must be for the signal to remain ear-readable through noise and fading.

Instead of thinking of this process in terms of modulation percentage, we just ensure that a CW transmitter produces sufficient keying-sideband energy for solid reception. Practical CW transmitters rarely do their keying with a modulator stage as such. Instead, one or more stages are turned on and off to modulate the carrier with Morse, with rise and fall times set by R and C values associated with the stages' keying and/or power supply lines. A transmitter's CW *waveshaping* is therefore usually hardwired to values appropriate for reasonably high-speed sending (35 to 55 WPM or so) in the presence of fading. As a result, we generally cannot vary keying hardness at will as we might vary a voice transmitter's modulation with a front-panel control. Rise and fall times of 1 to 5 ms (5 ms rise and fall times equate to a keying speed of 36 WPM in the presence of fading and 60 WPM if fading is absent) are common.

The faster a CW transmitter's output changes between zero and maximum, the more bandwidth its carrier and sidebands occupy. Making a CW signal's keying too hard is therefore spectrum-wasteful and unneighborly because it makes the signal wider than it needs to be. Keying sidebands that are stronger and wider than necessary are traditionally called *clicks* because of what they sound like on the air. There is a more detailed discussion of keying waveforms in the **Transceivers** chapter.

The Many Faces of Amplitude Modulation

We've so far examined mixers, multipliers and modulators that produce complex output signals of two types. One, the action of which equation 7 expresses, produces only the frequency sum of and frequency difference between its input signals. The other, the amplitude modulator characterized by equations 8 and 9, produces carrier output in addition to the frequency sum of and frequency difference between its input signals. Exploring the AM process led us to a discussion on-off-keyed CW, which is also a form of AM.

Amplitude modulation is nothing more and nothing less than varying an output signal's amplitude according to a varying voltage or current. All of the output signal types mentioned above are forms of amplitude modulation, and there are others. Their names and applications depend on whether the resulting signal contains a carrier or not, and both sidebands or not. Here's a brief overview of AM-signal types, what they're called, and some of the jobs you may find them doing:

- *Double-sideband (DSB), full-carrier AM* is often called just *AM*, and often what's meant when radio folk talk about just *AM*. (When the subject is broadcasting, *AM* can also refer to broadcasters operating in the 525- to 1705-kHz region, generically called *the AM band* or *the broadcast band* or *medium wave*. These broadcasters used only double-sideband, full-carrier AM for many years, but many now use combinations of amplitude modulation and *angle modulation*, which we'll explore shortly, to transmit stereophonic sound.) Equations 8 and 9 express what goes on in generating this signal type. What we call *CW*—Morse code done by turning a carrier on and off—is a form of DSB, full-carrier AM.

- *Double-sideband, suppressed-carrier AM* is what comes out of a circuit that does what equation 7 expresses—a sum (upper sideband), a difference (lower

sideband) and no carrier. We didn't call its sum and difference outputs upper and lower sidebands earlier in equation 7's neighborhood, but we'd do so in a transmitting application. In a transmitter, we call a circuit that suppresses the carrier while generating upper and lower sidebands a *balanced modulator*, and we quantify its *carrier suppression*, which is always less than infinite. In a receiver, we call such a circuit a *balanced mixer*, which may be *single-balanced* (if it lets either its RF signal or its LO [carrier] signal through to its output) or *double-balanced* (if it suppresses both its input signal and LO/carrier in its output), and we quantify its *LO suppression* and *port-to-port isolation*, which are always less than infinite. (Mixers [and amplifiers] that afford no balance whatsoever are sometimes said to be *single-ended*.) Sometimes, DSB suppressed-carrier AM is called just *DSB*.

- *Vestigial sideband (VSB), full-carrier AM* is like the DSB variety with one sideband partially filtered away for bandwidth reduction. Commercial television systems that transmit AM video use VSB AM.

- *Single-sideband, suppressed-carrier AM* is what you get when you generate a DSB, suppressed carrier AM signal and throw away one sideband with filtering or phasing. We usually call this signal type just *single sideband (SSB)* or, as appropriate, *upper sideband (USB)* or *lower sideband (LSB)*. In a modulator or demodulator system, the *unwanted sideband*—that is, the sum or difference signal we don't want—may be called just that, or it may be called the *opposite sideband*, and we refer to a system's *sideband rejection* as a measure of how well the opposite sideband is suppressed. In receiver mixers not used for demodulation and transmitter mixers not used for modulation, the unwanted sum or difference signal, or the input signal that produces the unwanted sum or difference, is the *image*, and we refer to a system's *image rejection*. A pair of mixers specially configured to suppress either the sum or the difference output is an *image-reject mixer (IRM)*. In receiver demodulators, the unwanted sum or difference signal may just be called the opposite sideband, or it may be called the *audio image*. A receiver capable of rejecting the opposite sideband or audio image is said to be capable of *single-signal* reception.

- *Single-sideband, full-carrier AM* is akin to full-carrier DSB with one sideband missing. Commercial and military communicators may call it *AM equivalent (AME)* or *compatible AM (CAM)*—com-

patible because it can be usefully demodulated in AM and SSB receivers and because it occupies about the same amount of spectrum space as SSB.)

- *Independent sideband (ISB) AM* consists an upper sideband and a lower sideband containing different information (a carrier of some level may also be present). Radio amateurs sometimes use ISB to transmit simultaneous slow-scan-television and voice information; international broadcasters sometimes use it for point-to-point audio feeds as a backup to satellite links.

Mixers and AM Demodulation

Translating information from radio form back into its original form—demodulation—is also traditionally called *detection*. If the information signal we want to detect consists merely of a baseband signal frequency-shifted into the radio realm, almost any low-distortion frequency-shifter that works according to equation 7 can do the job acceptably well.

Sometimes we recover a radio signal's information by shifting the signal right back to its original form with no intermediate frequency shifts. This process is called *direct conversion*. More commonly, we first convert a received signal to an *intermediate frequency* so we can amplify, filter and level-control it prior to detection. This is *superheterodyne* reception, and most modern radio receivers work in this way. Whatever the receiver type, however, the received signal ultimately makes its way to one last mixer or demodulator that completes the final translation of information back into audio, or into a signal form suitable for device control or computer processing. In this last translation, the incoming signal is converted back to recovered-information form by mixing it with one last RF signal. In heterodyne or *product* detection, that final frequency-shifting signal comes from a BFO. The incoming-signal energy goes into one mixer input port, BFO energy goes into the other, and audio (or whatever form the desired information takes) results.

If the incoming signal is full-carrier AM and we don't need to hear the carrier as a tone, we can modify this process somewhat, if we want. We can use the carrier itself to provide the heterodyning energy in a process called *envelope detection*.

Envelope Detection and Full-Carrier AM

Fig 15.5 graphically represents how a full-carrier AM signal's *modulation envelope* corresponds to the shape of the modulating wave. If we can derive from the modulated signal a voltage that varies ac-

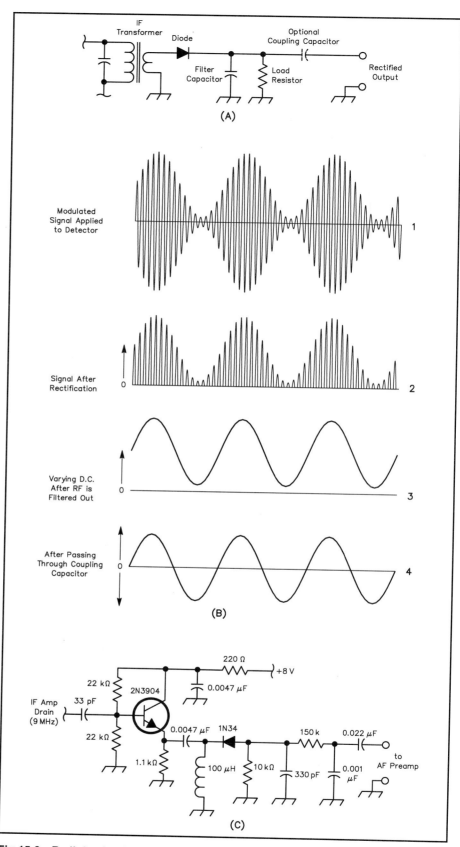

Fig 15.9—Radio's simplest demodulator, the diode rectifier (A), demodulates an AM signal by multiplying its carrier and sidebands to produce frequency sums and differences, two of which sum into a replica of the original modulation (B). Modern receivers often use an emitter follower to provide low-impedance drive for their diode detectors (C).

cording to the modulation envelope, we will have successfully recovered the information present in the sidebands. This process is called envelope detection, and we can achieve it by doing nothing more complicated than half-wave-rectifying the modulated signal with a diode (**Fig 15.9**).

That a diode demodulates an AM signal by allowing its carrier to multiply with its sidebands may jar those long accustomed to seeing diode detection ascribed merely to "rectification." But a diode is certainly nonlinear. It passes current in only one direction, and its output voltage is (within limits) proportional to the square of its input voltage. These nonlinearities allow it to multiply.

Exploring this mathematically is tedious with full-carrier AM because the process squares three summed components (carrier, lower sideband and upper sideband). Rather than fill the better part of a page with algebra, we'll instead characterize the outcome verbally: In "just rectifying" a DSB, full-carrier AM signal, a diode detector produces

- Direct current (the result of rectifying the carrier);
- A second harmonic of the carrier;
- A second harmonic of the lower sideband;
- A second harmonic of the upper sideband;
- Two difference-frequency outputs (upper sideband minus carrier, carrier minus lower sideband), each of which is equivalent to the modulating waveform's frequency, and both of which sum to produce the recovered information signal; and
- A second harmonic of the modulating waveform (the frequency difference between the two sidebands).

Three of these products are RF. Low-pass filtering, sometimes little more than a simple RC network, can remove the RF products from the detector output. A capacitor in series with the detector output line can block the carrier-derived dc component. That done, only two signals remain: the recovered modulation and, at a lower level, its second harmonic—in other words, second-harmonic distortion of the desired information signal.

Mixers and Angle Modulation

Amplitude modulation served as our first means of translating information into radio form because it could be implemented as simply as turning an electric noise generator on and off. (A spark transmitter consisted of little more than this.) By the 1930s, we had begun experimenting with translating information into radio

form and back again by modulating a radio wave's angular velocity (frequency or phase) instead of its overall amplitude. The result of this process is *frequency modulation (FM)* or *phase modulation (PM)*, both of which are often grouped under the name *angle modulation* because of their underlying principle.

A change in a carrier's frequency or phase for the purpose of modulation is called *deviation*. An FM signal deviates according to the amplitude of its modulating waveform, independently of the modulating waveform's frequency; the higher the modulating wave's amplitude, the greater the deviation. A PM signal deviates according to the amplitude *and frequency* of its modulating waveform; the higher the modulating wave's amplitude *and/or frequency*, the greater the deviation.

An angle-modulated signal can be mathematically represented as

$$f_c(t) = \cos\left(2\pi f_c t + m \sin\left(2\pi f_m t\right)\right)$$
$$= \cos\left(2\pi f_c t\right) \cos\left(m \sin\left(2\pi f_m t\right)\right) \quad (10)$$
$$- \sin\left(2\pi f_c t\right) \sin\left(m \sin\left(2\pi f_m t\right)\right)$$

In it, we see the carrier frequency ($2\pi f_c t$) and modulating signal ($\sin 2\pi f_m t$) as in the equation for AM (equation 8). We again see the modulating signal associated with a coefficient, m, that relates to degree of modulation. (In the AM equation, *m* is the modulation factor; in the angle-modulation equation, m is the *modulation index* and, for FM, equals the deviation divided by the modulating frequency.) We see that angle-modulation occurs as the cosine of the sum of the carrier frequency ($2\pi f_c t$) and the modulating signal ($\sin 2\pi f_m t$) times the modulation index (m). In its expanded form, we see the appearance of sidebands above and below the carrier frequency.

Angle modulation is a multiplicative process, so, like AM, it creates sidebands on both sides of the carrier. Unlike AM, however, angle modulation creates an *infinite* number of sidebands on either side of the carrier! This occurs as a direct result of modulating the carrier's angular velocity, to which its frequency and phase directly relate. If we continuously vary a wave's angular velocity according to another periodic wave's cyclical amplitude variations, the rate at which the modulated wave repeats *its* cycle—its frequency—passes through an infinite number of values. (How many individual amplitude points are there in one cycle of the modulating wave? An infinite number. How many corresponding discrete frequency or phase values does the cor-

responding angle-modulated wave pass through as the modulating signal completes a cycle? An infinite number!) In AM, the carrier frequency stays at one value, so AM produces two sidebands—the sum of its carrier's unchanging frequency value and the modulating frequency, and the difference between the carrier's unchanging frequency value and the modulating frequency. In angle modulation, the modulating wave shifts the frequency or phase of the carrier through an infinite number of different frequency or phase values, resulting in an infinite number of sum and difference products.

Wouldn't the appearance on the air of just a few such signals result in a bedlam of mutual interference? No, because most of angle modulation's uncountable sum and difference products are vanishingly weak in practical systems, and because they don't show up just anywhere in the spectrum. Rather, they emerge from the modulator spaced from the average ("resting," unmodulated) carrier frequency by integer multiples of the modulating frequency (**Fig 15.10**). The strength of the sidebands relative to the carrier, and the strength and phase of the carrier itself,

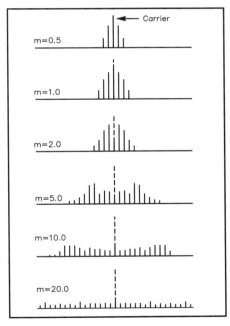

Fig 15.10—Angle-modulation produces a carrier and an infinite number of upper and lower sidebands spaced from the average ("resting," unmodulated) carrier frequency by integer multiples of the modulating frequency. (This drawing is a simplification because it only shows relatively strong, close-in sideband pairs; space constraints prevent us from extending it to infinity.) The relative amplitudes of the sideband pairs and carrier vary with modulation index, *m*.

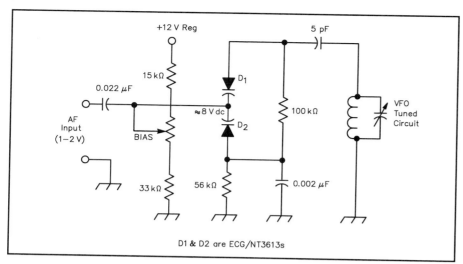

Fig 15.11—One or more tuning diodes can serve as the variable reactance in a reactance modulator. This HF reactance modulator circuit uses two diodes in series to ensure that the tuned circuit's RF-voltage swing cannot bias the diodes into conduction. D1 and D2 are "30-volt" tuning diodes that exhibit a capacitance of 22 pF at a bias voltage of 4. The BIAS control sets the point on the diode's voltage-versus-capacitance characteristic around which the modulating waveform swings.

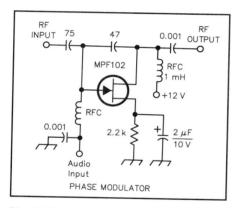

Fig 15.12—A series reactance modulator acts as a variable shunt around a reactance—in this case, a 47-pF capacitor—through which the carrier passes.

vary with the degree of modulation—the modulation index. (The *overall* amplitude of an angle-modulated signal does not change with modulation, however; when energy goes out of the carrier, it shows up in the sidebands, and vice versa.) In practice, we operate angle-modulated transmitters at modulation indexes that make all but a few of their infinite sidebands small in amplitude. (A mathematical tool called *Bessel functions*, part of the set of which is shown graphically in the **Modulation Sources** chapter, lets us determine the relative strength of the carrier and sidebands according to modulation index.) Selectivity in transmitter and receiver circuitry further modify this relationship, especially for

sidebands far away from the carrier.

Angle Modulators

Vary a reactance in or associated with an oscillator's frequency-determining element(s), and you vary the oscillator's frequency. Vary the tuning of a tuned circuit through which a signal passes, and you vary the signal's phase. A circuit that does this is called a *reactance modulator*, and can be little more than a tuning diode or two connected to a tuned circuit in an oscillator or amplifier (**Fig 15.11**). Varying a reactance through which the signal passes (**Fig 15.12**) is another way of doing the same thing.

The difference between FM and PM depends solely on how, and not how much, deviation occurs. A modulator that causes deviation in proportion to the modulating wave's amplitude and frequency is a phase modulator. A modulator that causes deviation only in proportion to the modulating signal's amplitude is a frequency modulator.

Increasing Deviation by Frequency Multiplication

Maintaining modulation linearity is just as important in angle modulation as it is in AM, because unwanted distortion is always our enemy. A given angle-modulator circuit can frequency- or phase-shift a carrier only so much before the shift stops occurring in strict proportion to the amplitude (or, in PM, the amplitude and frequency) of the modulating signal.

If we want more deviation than an angle modulator can linearly achieve, we can

operate the modulator at a suitable subharmonic—submultiple—of the desired frequency, and process the modulated signal through a series of *frequency multipliers* to bring it up to the desired frequency. The deviation also increases by the overall multiplication factor, relieving the modulator of having to do it all directly. A given FM or PM radio design may achieve its final output frequency through a combination of mixing (frequency shift, no deviation change) and frequency multiplication (frequency shift *and* deviation change).

The Truth About "True FM"

Something we covered a bit earlier bears closer study:

An FM signal deviates according to the amplitude of its modulating waveform, independently of the modulating waveform's frequency; the higher the modulating wave's amplitude, the greater the deviation. A PM signal deviates according to the amplitude *and frequency* of its modulating waveform; the higher the modulating wave's amplitude *and/or frequency*, the greater the deviation.

The practical upshot of this excerpt is that we can use a phase modulator to generate FM. All we need to do is run a PM transmitter's modulating signal through a low-pass filter that (ideally) halves the signal's amplitude for each doubling of frequency (a reduction of "6 dB per octave," as we sometimes see such responses characterized) to compensate for its phase modulator's "more deviation with higher frequencies" characteristic. The result is an FM, not PM, signal. FM achieved with a phase modulator is sometimes called *indirect FM* as opposed to the *direct FM* we get from a frequency modulator.

We sometimes see radio gear manufacturers claim that one piece of gear is better than another solely because it generates "true FM" as opposed to indirect FM. We can immunize ourselves against such claims by keeping in mind that direct and indirect FM *sound exactly alike in a receiver* when done correctly.

Depending on the nature of the modulation source, there *is* a practical difference between a frequency modulator and a phase modulator. Answering two questions can tell us whether this difference matters: Does our modulating signal contain a dc level or not? If so, do we need to accurately preserve that dc level through our radio communication link for successful communication? If both answers are *yes*, we must choose our hardware and/or information-encoding

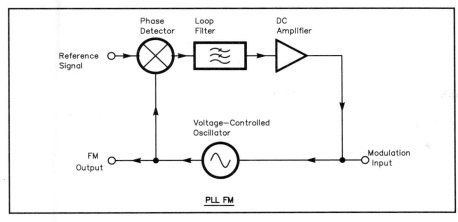

Fig 15.13—Frequency modulation, PLL-style.

approach carefully, because a frequency modulator can convey shifts in its modulating wave's dc level, while a phase modulator, which responds only to instantaneous changes in frequency and phase, cannot.

Consider what happens when we want to frequency-modulate a phase-locked-loop-synthesized transmitted signal. **Fig 15.13** block-diagrams a PLL frequency modulator. Normally, we modulate a PLL's VCO because it's the easy thing to do. As long as our modulating frequency results in frequency excursions too fast for the PLL to follow and correct—that is, as long as our modulating frequency is outside the PLL's *loop bandwidth*—we achieve the FM we seek. Trying to modulate a dc level by pushing the VCO to a particular frequency and holding it there fails, however, because a PLL's loop response includes dc. The loop sees the modulation's dc component as a correctable error and dutifully "fixes" it. FMing a PLL's VCO therefore can't buy us the dc response "true FM" is supposed to allow.

We *can* dc-modulate a PLL modulator, but we must do so by modulating the loop's *reference*. The PLL then adjusts the VCO to adapt to the changed reference, and our dc level gets through. In this case, the modulating frequency must be *within* the loop bandwidth—which dc certainly is—or the VCO won't be corrected to track the shift.

Mixers and Angle Demodulation

With the awesome prospect of generating an infinite number of sidebands still fresh in our minds, we may be a bit disappointed to learn that we commonly demodulate angle modulation by doing little more than turning it into AM and then envelope- or product-detecting it! But this is what happens in many of our FM receivers and transceivers, and we can get a handle on this process by realizing that a form of angle-modulation-to-AM conversion begins quite early in an angle-modulated signal's life because of distortion of the modulation by amplitude-linear circuitry—something that happens to angle-modulated signals, it turns out, in any linear circuit that doesn't have an amplitude-versus-frequency response that's utterly flat out to infinity.

Think of what happens, for example, when we sweep a constant-amplitude signal up in frequency—say, from 1 kHz to 8 kHz—and pass it through a 6-dB-per-octave filter (**Fig 15.14A**). The filter's rolloff causes the output signal's amplitude to decrease as frequency increases. Now imagine that we linearly sweep our constant-amplitude signal *back and forth* between 1 kHz and 8 kHz at a constant rate of 3 kHz per second (Fig 15.14B). The filter's output *amplitude* now varies cyclically over time as the input signal's *frequency* varies cyclically over time. Right before our eyes, a frequency change turns into an amplitude change. The process of converting angle modulation to amplitude modulation has begun.

This is what happens whenever an angle-modulated signal passes through circuitry with an amplitude-versus-frequency response that isn't flat out to infinity. As the signal deviates across the frequency-response curves of whatever circuitry passes it, its angle modulation is, to some degree, converted to AM—a form of crosstalk between the two modulation types, if we wish to look at it that way. (Variations in system phase linearity also cause distortion and FM-to-AM conversion, because the sidebands do not have the proper phase relationship with respect to each other and with respect to the carrier.)

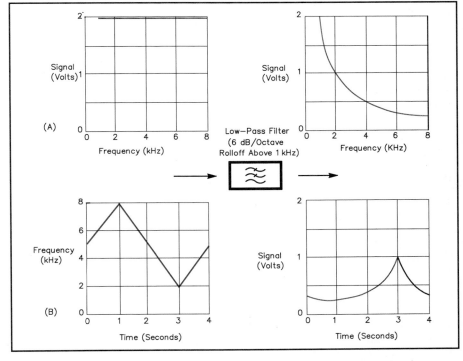

Fig 15.14—Frequency-sweeping a constant-amplitude signal and passing it through a low-pass filter results in an output signal that varies in amplitude with frequency (A). Sweeping the input signal back and forth between two frequency limits causes the output signal's amplitude to vary between two limits (B). This is the principle behind the angle-demodulation process called *frequency discrimination*.

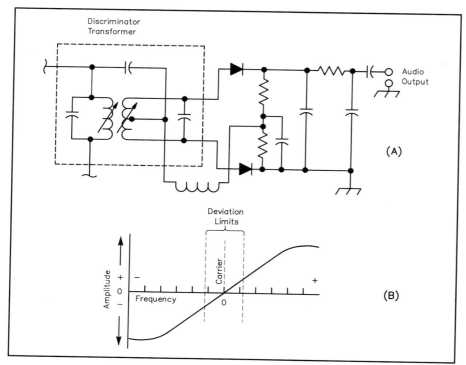

Fig 15.15—A *discriminator* (A) converts an angle-modulated signal's deviation into an amplitude variation (B) and envelope-detects the resulting AM signal. For undistorted demodulation, the discriminator's amplitude-versus-frequency characteristic must be linear across the input signal's deviation. A *crystal discriminator* uses a crystal as part of its frequency-selective circuitry.

All we need to do to put this effect to practical use is develop a circuit that does this frequency-to-amplitude conversion linearly (and, since more output is better, steeply) across the frequency span of the modulated signal's deviation. Then we envelope-demodulate the resulting AM, and we're in.

Fig 15.15 shows such a circuit—a *discriminator*—and the sort of amplitude-versus-frequency response we expect from it. (It's possible to use an AM receiver to recover understandable audio from a narrow angle-modulated signal by "off-tuning" the signal so its deviation rides up and down on one side of the receiver's IF selectivity curve. This *slope detection* process served as an early, suboptimal form of frequency discrimination.)

Quadrature Detection

It's also possible to demodulate an angle-modulated signal merely by multiplying it with a time-delayed copy of itself in a double-balanced mixer (**Fig 15.16**). For simplicity's sake, we'll represent the mixer's RF input signal as just a sine wave with an amplitude, A

$$A \sin (2\pi ft) \qquad (11)$$

and its time-delayed twin, fed to the mixer's LO input, as a sine wave with an amplitude, A, and a time delay of d:

$$A \sin [2\pi f(t + d)] \qquad (12)$$

Setting this special mixing arrangement into motion, we see

$$A \sin (2\pi ft) \cdot A \sin (2\pi ft + d)$$

$$= \frac{A^2}{2} \cos (2\pi fd)$$

$$- \frac{A^2}{2} \cos (2\pi fd) \cos (2 \cdot 2\pi ft) \qquad (13)$$

$$+ \frac{A^2}{2} \sin (2\pi fd) \sin (2 \cdot 2\pi ft)$$

Two of the three outputs—the second and third terms—emerge at twice the input frequency; in practice, we're not interested in these, and filter them out. The remaining term—the one we're after—varies in amplitude and sign according to how far and in what direction the carrier shifts away from its resting or center frequency (at which the time delay, d, causes the mixer's RF and LO inputs to be exactly 90° out of phase—in *quadrature*—with each other). We can examine this effect by replacing f in equations 11 and 12 with the sum term $f_c + f_s$, where f_c is the center frequency and f_s is the frequency shift. A 90° time delay is the same as a quarter cycle of f_c, so we can restate d as

$$d = \frac{1}{4f_c} \qquad (14)$$

The first term of the detector's output then becomes

$$\frac{A^2}{2} \cos \left(2\pi (f_c + f_s) d \right)$$

$$= \frac{A^2}{2} \cos \left(2\pi (f_c + f_s) \frac{1}{4f_c} \right) \qquad (15)$$

$$= \frac{A^2}{2} \cos \left(\frac{\pi}{2} + \frac{\pi f_s}{2f_c} \right)$$

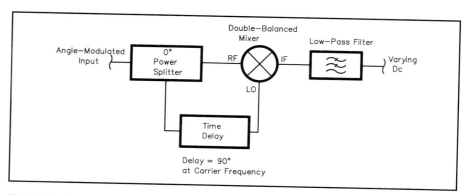

Fig 15.16—In *quadrature detection*, an angle-modulated signal multiplies with a time-delayed copy of itself to produce a dc voltage that varies with the amplitude and polarity of its phase or frequency excursions away from the carrier frequency. A practical quadrature detector can be as simple as a 0° power splitter (that is, a power splitter with in-phase outputs), a diode double-balanced mixer, a length of coaxial cable 1/4-λ (electrical) long at the carrier frequency, and a bit of low-pass filtering to remove the detector output's RF components. IC quadrature detectors achieve their time delay with one or more resistor-loaded tuned circuits (Fig 15.17).

When f_s is zero (that is, when the carrier is at its center frequency), this reduces to

$$\frac{A^2}{2} \cos\left(\frac{\pi}{2}\right) = 0 \qquad (16)$$

As the input signal shifts higher in frequency than f_c, the detector puts out a positive dc voltage that increases with the shift. When the input signal shifts lower in frequency than f_c, the detector puts out a negative dc voltage that increases with the shift. The detector therefore recovers the input signal's frequency or phase modulation as an amplitude-varying dc voltage that shifts in sign as f_s varies around f_c—in other words, as ac. We have demodulated FM by means of quadrature detection.

An ideal quadrature detector puts out 0 V dc when no modulation is present (with the carrier at f_c). The output of a real quadrature detector may include a small *dc offset* that requires compensation. If we need the detector's response all the way down to dc, we've got it; if not, we can put a suitable blocking capacitor in the output line for ac-only coupling.

Quadrature detection is more common than frequency discrimination nowadays because it doesn't require a special discriminator transformer, and because the necessary balanced-detector circuitry can easily be implemented in IC structures along with limiters and other receiver circuitry. The synchronous AM detector project later in this chapter uses such a chip, the Philips Components-Signetics NE604A, to do limiting and phase detection as part of a phase-locked loop (PLL); **Fig 15.17** shows another example.

PLL Angle Demodulation

Back at Fig 15.14, we saw how a PLL can be used as an angle modulator. A PLL also makes a fine angle *de*modulator. Applying an angle-modulated signal to a PLL keeps its phase detector and VCO hustling to maintain loop lock through the input signal's angle variations. The loop's error voltage therefore tracks the input signal's modulation, and its variations mirror the modulation signal. Turning the loop's varying dc error voltage into audio is just a blocking capacitor away.

Although we can't convey a dc level by directly modulating the VCO in a PLL angle modulator, a PLL demodulator can respond down to dc quite nicely. A constant frequency offset from f_c (a dc component) simply causes a PLL demodulator to swing its VCO over to the new input frequency, resulting in a proportional dc offset on the VCO control-voltage line. Another way of looking at the difference between a PLL angle modulator and a PLL angle demodulator is that a PLL demodulator works with a varying reference signal (the input signal), while a PLL angle modulator generally doesn't.

Amplitude Limiting Required

By now, it's almost household knowledge that FM radio communication systems are superior to AM in their ability to suppress and ignore static, manmade electrical noise and (through an angle-modulation-receiver characteristic called *capture effect*) cochannel signals sufficiently weaker than the desired signal. AM-noise immunity is not intrinsic to angle modulation, however; it must be designed into the angle-modulation receiver.

If we note the progress of A from the left to the right side of the equal sign in equation 13, we realize that the amplitude of a quadrature detector's input signal affects the amplitude of a quadrature detector's three output signals. A quadrature detector therefore responds to AM, and so does a frequency discriminator. To achieve FM's storied noise immunity, then, these angle demodulators must be preceded by *limiting* circuitry that removes all amplitude variations from the incoming signal.

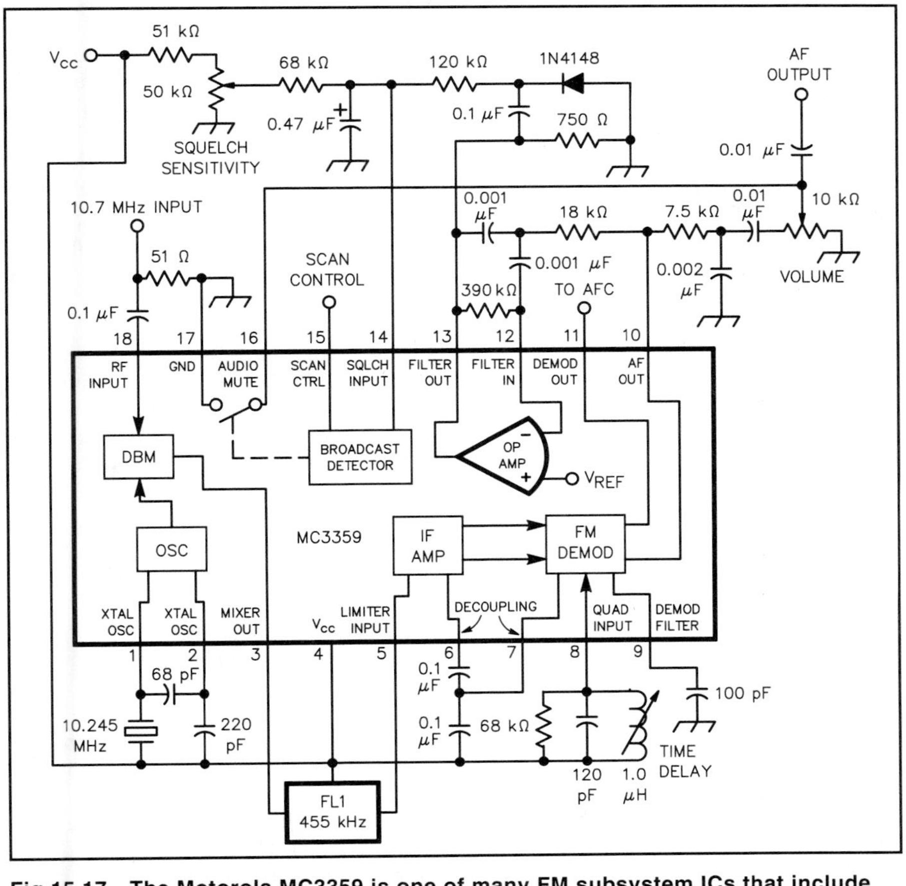

PRACTICAL BUILDING BLOCKS FOR MIXING, MODULATION AND DEMODULATION

So far, we've tended to look at mixing as a process that frequency-shifts one sinusoidal wave by mixing it with another. We need to expand our thinking to other cases, however, since it turns out that many practical mixers work best with *square-wave* signals applied to their LO inputs.

Sine-Wave Mixing, Square-Wave Mixing

Thinking of mixers as multiplying sine waves implies that mixers act like tiny analog computers performing millions of multiplications per second. It's certainly possible to build mixers this way by using an IC mixer circuit (**Fig 15.18**) conceived in 1967 by Barrie Gilbert and widely known as the Gilbert cell. (Gilbert himself was not responsible for this eponym; indeed, he has noted that a prior art search at the time found that essentially the same idea—used as a "synchronous detector" and not as true mixer—had already been

Fig 15.17—The Motorola MC3359 is one of many FM subsystem ICs that include limiter and quadrature-detection circuitry. The TIME DELAY coil is adjusted for minimum recovered-audio distortion.

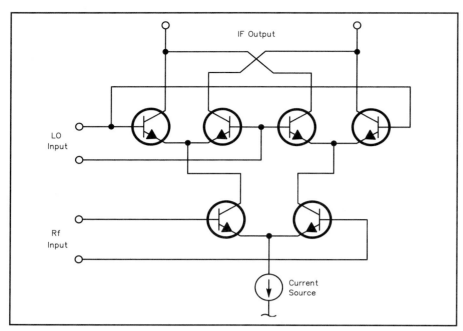

Fig 15.18—The Gilbert cell mixer. The Motorola MC1496 and Philips Components-Signetics NE602A are based on this circuit.

patented by H. E. Jones.) A Gilbert cell consists of two differential transistor pairs whose bias current is controlled by one of the input signals. The other signal drives the differential pairs' bases, but only after being "predistorted" in a diode circuit. (This circuit distorts the signal equally and oppositely to the inherent distortion of the differential pair.) The resulting output signal is an accurate multiplication of the input voltages.

Early Gilbert-cell ICs, such as the Motorola MC1495 multiplier, had a number of disadvantages, including critical external adjustments, narrow bandwidth, and limited dynamic range. Modern Gilbert cells, such as the Burr-Brown MPY600 and Analog Devices AD834, overcome most of these disadvantages, and have led to an increase in usage of analog multipliers as mixers. Most practical radio mixers do not work exactly as analog multipliers, however. In practice, they act more like fast analog *switches*.

In using a mixer as a fast switching device, we feed its LO input with a single-frequency square wave rather than a sine wave, and feed sine waves, audio, or other complex signals to the mixer's RF input. The RF port serves as the mixer's "linear" input, and therefore must preferably exhibit low intermodulation and harmonic distortion. Feeding a ±1-V square wave into the LO input alternately multiplies the linear input by +1 or −1. Multiplying the RF-port signal by +1 just transfers it to the output with no change. Multiplying the

RF-port signal by −1 does the same thing, except that the signal inverts (flips 180° in phase). The LO port need not exhibit low intermodulation and harmonic distortion; all it has to do is preserve the switching signal's fast rise and fall times.

Using square-wave LO drive allows us to simplify the Gilbert multiplier by dispensing with its predistortion circuitry. The Motorola MC1496, still in wide use despite its age, is an example of this. The Philips Components-Signetics NE602A and its relatives, popular with Amateur Radio experimenters, are modern MC1496 descendants. The vast majority of Gilbert-cell mixers in current use are square-wave LO-drive types. In practice, though, we don't have to square the LO signals we apply to them to make them work well. All we need to do is drive their LO inputs with a sine wave of sufficient amplitude to overdrive the associated transistors' bases. This clips the LO waveform, effectively resulting in square-wave drive.

Reversing-Switch Mixers

We can multiply a signal by a square wave without using an analog multiplier at all. All we need is a pair of balun transformers and four diodes (**Fig 15.19A**).

With no LO energy applied to the circuit, none of its diodes conduct (Fig 15.19B). RF-port energy (1) can't make it to the LO port because there's no direct connection between the secondaries of T1 and T2, and (2) doesn't produce

IF output because T2's secondary balance results in energy cancellation at its center tap, and because no complete IF-energy circuit exists through T2's secondary with both of its ends disconnected from ground.

Applying a square wave to the LO port biases the diodes so that, 50% of the time (Fig 15.19C), D1 and D2 are on and D3 and D4 are reverse-biased off. This unbalances T2's secondary by leaving its upper wire floating and connecting its lower wire to ground through T1's secondary and center tap. With T2's secondary unbalanced, RF-port energy emerges from the IF port.

The other 50% of the time (Fig 15.19D), D3 and D4 are on and D1 and D2 are reverse-biased off. This unbalances T2's secondary by leaving its lower wire floating, and connects its upper wire to ground through T1's secondary and center tap. With T2's secondary unbalanced, RF-port energy again emerges from the IF port—shifted 180° relative to the first case because T2's active secondary wires are now, in effect, transposed relative to its primary.

A reversing switch mixer's output spectrum is the same as the output spectrum of a multiplier fed with a square wave. This can be analyzed by thinking of the square wave in terms of its Fourier series equivalent, which consists of the sum of sine waves at the square wave frequency and all of its odd harmonics. The amplitude of the equivalent series' fundamental sine wave is $4/\pi$ times (2.1 dB greater than) the amplitude of the square wave. The amplitude of each harmonic is inversely proportional to its harmonic number, so the third harmonic is only 1/3 as strong as the fundamental (9.5 dB below the fundamental), the 5th harmonic is only 1/5 as strong (14 dB below the fundamental) and so on. The input signal mixes with each harmonic separately from the others, as if each harmonic were driving its own separate mixer, just as we illustrated with two sine waves in Fig 15.4. Normally, the harmonic outputs are so widely removed from the desired output frequency that they are easily filtered out, so a reversing-switch mixer is just as good as a sine-wave-driven analog multiplier for most practical purposes, and usually better—for radio purposes—in terms of dynamic range and noise.

An additional difference between multiplier and switching mixers is that a switching mixer's signal flow is reversible. It really only has one dedicated input (the LO input). The other terminals can be thought of as I/O (input/output) ports, since either one can be the input as long as the other is the output.

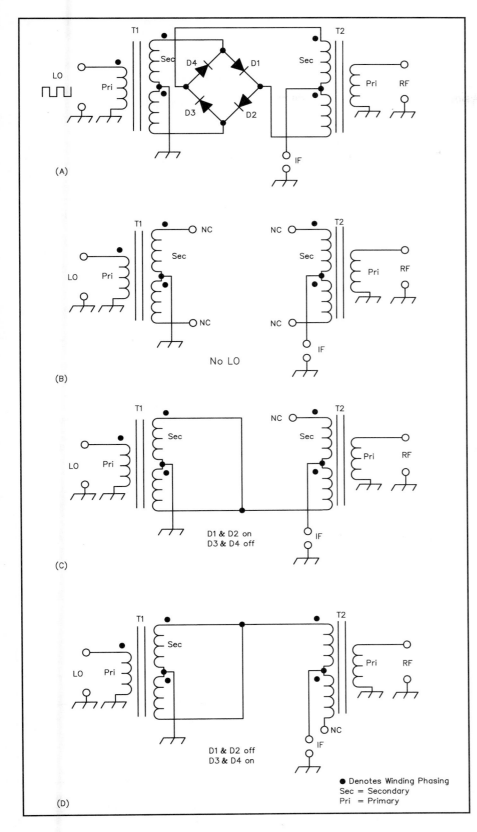

Fig 15.19—A reversing-switch mixer uses fast switching devices, balanced transformers and square-wave LO drive to reverse the RF-port signal's phase at a rate determined by the LO frequency. (Diodes are shown, but BJTs and FETs can also be used, as we'll soon see.) With no LO drive (B), no RF-port energy shows up at the IF port because T2's secondary has no ground return. When the LO waveform goes positive, D1 and D2 turn on and D3 and D4 turn off, unbalancing T2's secondary and giving the IF port a ground return through T1's secondary and its center tap. When the LO waveform goes negative, D1 and D2 turn off and D3 and D4 turn on, again unbalancing T2's secondary and again giving the IF port a ground return. RF-port energy gets through to the IF port in both states; alternately operating T2's secondary halves results in 180° phase reversal from state to state.

3.9 dB. This is 2.1 dB better than an ideal multiplier, in which the peak output voltages of the sum and difference products are each half that of (6 dB lower than) the input signal, as reflected in the $^{1}/_{2}$ coefficients in the right-hand terms of equation 7.

The switches and transformers in an ideal reversing-switch mixer dissipate no power, so an ideal reversing-switch mixer is lossless. If the LO driving signal contained no harmonics, half of the mixer's output power would show up in the sum product and half in the difference product, resulting in a 3-dB *conversion loss* relative to the RF-port signal. Some of the mixer's output goes into the sum and difference products of harmonic mixing, however; hence, the additional 0.9 dB of conversion loss, or 3.9 dB overall for an ideal reversing switch mixer. In practical mixers, switch and transformer losses increase conversion loss to a typical value of 6 dB.

The Diode Double-Balanced Mixer: A Basic Building Block

The most common implementation of a reversing switch mixer is the diode *double-balanced mixer* (DBM). DBMs can serve as mixers (including image-reject types), modulators (including single- and double-sideband, phase, biphase, and quadrature-phase types) and demodulators, limiters, attenuators, switches, phase detectors, and frequency doublers. In some of these applications, they work in conjunction with power dividers, combiners and hybrids.

The Basic DBM Circuit

We have already seen the basic diode DBM circuit (Fig 15.19). In its simplest form, a DBM contains two or more unbal-

Conversion Loss

At the instant of peak output voltage, a switching mixer's waveform is a full-wave-rectified sine wave. Thus its average output voltage equals that of a full-wave-rectified sine wave, which we know from power supply theory to be $2/\pi$, or 0.636, times the peak of the sine wave. The conversion loss of an ideal reversing-switch mixer is therefore 20 log $2/\pi$, or

anced-to-balanced transformers and a Schottky diode ring consisting of 4 × n diodes, where n is the number of diodes in each leg of the ring. Each leg commonly consists of up to four diodes.

As we've seen, the degree to which a mixer is *balanced* depends on whether either, neither or both of its input signals (RF and LO) emerge from the IF port along with mixing products. An unbalanced mixer suppresses neither its RF nor its LO; both are present at its IF port. A single-balanced mixer suppresses its RF or LO, but not both. A double-balanced mixer suppresses its RF *and* LO inputs. Diode and transformer uniformity in the Fig 15.19 circuit results in equal LO potentials at the center taps of T1 and T2. The LO potential at T1's secondary center tap is zero (ground); therefore, the LO potential at the IF port is zero.

Balance in T2's secondary likewise results in an RF null at the IF port. The RF potential between the IF port and ground is therefore zero—except when the DBM's switching diodes operate, of course!

The Fig 15.19 circuit normally also affords high RF-IF isolation because its balanced diode switching precludes direct connections between T1 and T2. A diode DBM can be used as a current-controlled switch or attenuator by applying dc to its IF port, albeit with some distortion. This causes opposing diodes (D2 and D4, for

instance) to conduct to a degree that depends on the current magnitude, connecting T1 to T2.

One extension of the single-diode-ring DBM is a *double* double-balanced mixer (DDBM) with high dynamic range and larger signal handling capability than a single-ring design. **Fig 15.20** diagrams such a DDBM, which uses transmission-line transformers and two diode rings. This type of mixer has higher 1-dB compression point (usually 3 to 4 dB lower than the LO drive) than a DBM. Low distortion is a typical characteristic of DDBMs. Depending on the ferrite core material used (ferrites with a magnetic permeability—μ— of 100 to 15,000), frequencies as low as a few hundred hertz and as high as a few gigahertz can be covered.

Diode DBM Components

Commercially manufactured diode DBMs generally consist of: (A) a supporting base; (B) a diode ring; (C) two or more ferrite-core transformers commonly wound with two or three twisted-pair wires; (D) encapsulating material; and (E) an enclosure.

Diodes

Hot-carrier (Schottky) diodes are the devices of choice for diode-DBM rings because of their low ON resistance, although ham-built DBMs for noncritical

MF/HF use commonly use switching diodes like the 1N914 or 1N4148. The forward voltage drop, V_f, across each diode in the ring determines the mixer's optimum local-oscillator drive level. Depending on the forward voltage drop of each of its diodes and the number of diodes in each ring leg, a diode DBM may be categorized as a Level 0, 3, 7, 10, 13, 17, 23 or 27 device. The numbers indicate the mixer's optimal LO drive level in dBm. As a rule of thumb, the LO signal must be must 20 dB larger than the RF and IF signals for proper operation. This ensures that the LO signal, rather than the RF or IF signals, switches the mixer's diodes on and off—a critical factor in minimizing IMD and maximizing dynamic range.

Schottky diodes are characterized by loss and contact resistance (R_s), junction capacitance (C_j), and forward voltage drop (V_f) at a known current, typically 1 mA or 10 mA. The lower the diode-to-diode V_f difference in millivolts, the better the diode match at dc. (Some early diode DBM designs used diodes in series with a parallel resistor/capacitor combination for automatic biasing.) Better diode matching (in V_f and C_j) results in higher isolation among the ports. Diodes capable of operating at higher frequencies have lower junction capacitance and lower parasitic inductance. **Fig 15.21** shows the equivalent circuit for Schottky diodes of three package types.

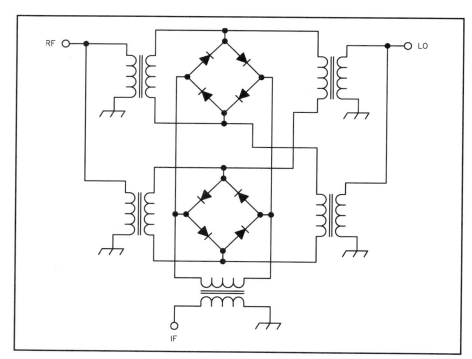

Fig 15.20—Five transmission-line transformers and two Schottky-quad rings form this double double-balanced mixer (DDBM). Such designs can provide lower distortion, better signal-handling capability and higher interport isolation than single-ring designs.

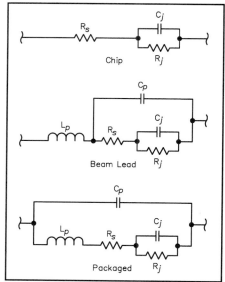

Fig 15.21—Its semiconductive properties aside, a Schottky diode can be represented as a network consisting of resistance, capacitance and/or inductance. Of these, the junction capacitance (C_j) plays an especially critical role in a double-balanced mixer's high-end response. (R_j=junction resistance, R_s=contact resistance, L_p=parasitic inductance and C_p=parasitic capacitance.)

Manufacturers of diodes suitable for DBMs characterize their diodes as low-barrier, medium-barrier, high-barrier and very-high-barrier (usually two or more diodes in each leg), with typical V_f values of 220 mV, 350 mV, 600 mV and 1 V or more, respectively. **Fig 15.22** shows a typical current-voltage (I-V) characteristic for a low-barrier Schottky quad capable of operating up to 4 GHz. Note that as current through the diodes increases, the V_f difference among the ring's diodes also increases, affecting the balance.

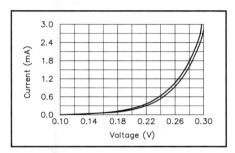

Fig 15.22—Current-voltage (I-V) characteristic for Schottky diode quad, showing worst-case voltage imbalance (the spread between the two curves) among the four diodes.

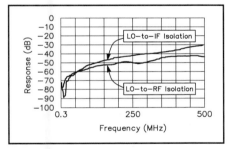

Fig 15.23—A diode DBM's port-to-port isolation depends on how well its diodes match and how well its transformers are balanced. This graph shows LO-IF and LO-RF isolation versus frequency for a Synergy Microwave CLP-4A3 mixer.

At higher frequencies, diode packaging becomes critical and expensive. As the frequency of operation increases, the effect of junction capacitance and package capacitance cannot be ignored. Part or all of the capacitance can be compensated at the mixer's highest operating frequency by properly designing the unbalanced-to-balanced transformers. The transformer inductance and diode junction capacitance form a low-pass network with its cutoff frequency higher than the frequency of operation. Compensated in this way, diodes with a junction capacitance of 0.2 pF can be used up to 8 to 10 GHz.

Transformers

From the DBM schematic shown in Fig 15.19, it's clear that the LO and RF transformers are unbalanced on the input side and balanced on the diode side. The diode ends of the balanced ports are 180° out of phase throughout the frequency range of interest. This property causes signal cancellations that result in higher port-to-port isolation. **Fig 15.23** plots LO-RF and LO-IF isolation versus frequency for Synergy Microwave's CLP-4A3 DBM. Isolations on the order of 70 dB occur at lower end of the band as a direct result of the balance among the four diode-ring legs and the RF phasing of the balanced ports.

As we learned in our discussion of generic switching mixers, transformer efficiency plays an important role in determining a mixer's conversion loss and drive-level requirement. Core loss, copper loss and impedance mismatch all contribute to transformer losses.

Ferrite in toroidal, bead, balun (multihole), or rod form can serve as DBM transformer cores. Radio amateurs commonly use Fair-rite Mix 43 ferrite (μ=950), but if the mixer will be used over a wide temperature range, the core material must be evaluated in terms of temperature coefficient and curie temperature (the temperature at which a ferromagnetic ma-

terial loses its magnetic properties). In some materials, μ may change drastically across the desired temperature range, causing a frequency-response shift with temperature. Once a suitable core material and form have been selected, frequency requirements determine the necessary core size. For a given core shape and size, the number of turns, wire size, and the number of twists determine transformer performance. Wire placement also plays an important role.

RF transformers combine lumped and distributed capacitance and inductance. The interwinding capacitance and characteristic impedance of a transformer's twisted wires sets the transformer's high-frequency response. The core's μ and size, and the number of winding turns, determine the transformer's lower frequency limit. Covering a specific frequency range requires a compromise in the number of turns used with a given core. Increasing a transformer's core size and number of turns improves its low-frequency response. Cores may be stacked to meet low-frequency performance specs.

Inexpensive mixers operating up to 2 GHz most commonly use twisted trifilar (three-wire) windings made of a wire size between #36 and #32. The number of twists per unit length of wire determines a winding's characteristic impedance. Twisted wires are analogous to transmission lines, and can be analyzed in terms of distributed interwinding capacitance. Decreasing the number of twists lowers the interwinding capacitance and increases the frequency of operation. On the other hand, using fewer twists per inch than four makes winding difficult because the wires tend to separate instead of behaving as a single cable.

The transmission-line effect predominates at the higher end of a transformer's frequency range. If two impedances, Z_1 and Z_2, need to be matched through a transmission line of characteristic impedance, Z_0, then

$$Z_0 = \sqrt{Z_1 \times Z_2} \qquad (17)$$

Fig 15.24 shows two types of transformers using twisted wires: (A) a three-wire type in which the primary winding is isolated from the secondary winding with a center tap, and (B) a two-wire (transmission-line) type in which two sets of transmission lines are interconnected to form a center tap at the secondary with a direct connection between primary and secondary. The primary-to-secondary turns ratio determines the impedance match as shown in equation 17. The properties of these two transformer types can be summarized as follows:

1. By virtue of its construction, the

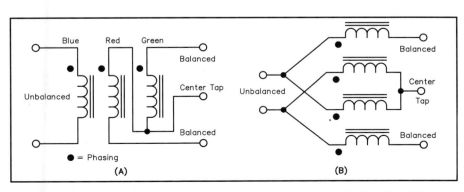

Fig 15.24—Transformers for DBMs: three-wire (A), and transmission-line (B).

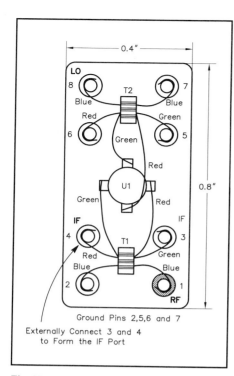

Ground Pins 2,5,6 and 7

Externally Connect 3 and 4
to Form the IF Port

Fig 15.25—How a typical commercial DBM is wired. The use of different wire colors for the transformers' various windings speeds assembly and minimizes error. U1, a Schottky-diode quad, contains D1-D4 of Fig 15.19.

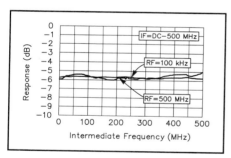

Fig 15.26—Conversion loss versus frequency for a typical diode DBM. The LO drive level is +7 dBm.

three-wire transformer is less symmetrical at higher frequencies than the transmission-line type.

2. The transformers' lower cutoff frequency (f_L) is determined by the equation

$$\omega L > 4R \qquad (18)$$

where

L = inductance of the winding
R = system impedance; for example, 50 Ω, 75 Ω and so on; and
$\omega = 2\pi f_L$.

3. The transmission-line transformer's upper cutoff frequency (f_H) is determined

by the highest frequency at which its wires' twists (that is, the coupling between them) allow it to function as a transmission line of the proper characteristic impedance.

4. Transformers convert one impedance, Z_1 (primary) to another, Z_2 (secondary) according to the relationship

$$Z_2 = Z_1 (N)^2 \qquad (19)$$

where N = secondary to primary turns ratio. Within certain limits, if Z_1 is varied, Z_2 also varies to a new value multiplied by N^2. Thus, a mixer designed for a 50-Ω system may work in a 75-Ω system with minor modifications.

Diode DBMs in Practice

Fig 15.25 shows the wiring of a typical commercial DBM made with toroidal cores. The wires are wrapped around the package pins and diode leads, and then soldered. In this unit, the LO transformer's primary winding connects across pins 7 and 8; the RF-transformer primary, across pins 1 and 2. The pin pairs 3-4 and 5-6 are connected externally to form the transformers' secondary center taps, one of which (5-6, that of the LO transformer) connects to a common ground point while the other (3-4, that of the RF transformer) serves as the IF port.

The DBM shown in Fig 15.25 has a dc-coupled IF port. If necessary, this DBM can be operated at a particular polarity (positive or negative) by appropriately connecting the LO, RF, IF and common ground points.

DBM Specifications

Most of the parameters important in building or selecting a diode DBM also apply to other mixer types. They include: conversion loss and amplitude flatness across the required IF bandwidth; variation of conversion loss with input frequency; variation of conversion loss with LO drive, 1-dB compression point; LO-RF, LO-IF and RF-IF isolation; intermodulation products; noise figure (usually within 1 dB of conversion loss); port SWR; and dc offset, which is directly related to isolation among the RF, LO and IF ports.

Conversion Loss

Fig 15.26 shows a plot of conversion loss versus intermediate frequency in a typical DBM. The curves show conversion loss for two fixed RF-port signals, one at 100 kHz and the another at 500 MHz, while varying the LO frequency from 100 kHz to 500 MHz.

Fig 15.27 plots a diode DBM's simulated output spectrum. Note that the RF

input is −20 dBm and the IF output (the frequency difference between the RF and LO signals) is −25 dBm, implying a conversion loss of 5 dB. This figure also applies to the sum of both signals (RF + LO).

We minimize a diode DBM's conversion loss, noise figure and intermodulation by keeping its LO drive high enough to switch its diodes on fully and rapidly. **Fig 15.28** plots noise figure for the DDBM shown in Fig 15.20. Its 4-dB noise figure assumes ideal transformers and somewhat idealized diodes; typical mixers have a noise figure of 5 to 6 dB. **Fig 15.29** plots conversion gain (loss) for the same mixer circuit.

Insufficient LO drive results in increased noise figure and conversion loss. IMD also increases because RF-port signals have a greater chance to control the mixer diodes when the LO level is too low.

Dynamic Range: Compression, Intermodulation and More

The output of a linear stage—including a mixer, which we want to act as a linear frequency shifter—tracks its input signal decibel by decibel, every 1-dB change in its input signal(s) corresponding to an identical 1-dB output change. This is the stage's *first-order* response.

Because no device is perfectly linear, however, two or more signals applied to it generate sum and difference frequencies. These IMD products occur at frequencies and amplitudes that depend on the order of the IMD response as follows:

- *Second-order* IMD products change 2 dB for every decibel of input-signal change (this figure assumes that the IMD comes from equal-level input signals), and appear at frequencies that result from the simple addition and subtraction of input-signal frequencies. For example, assuming that its input bandwidth is sufficient to pass them, an amplifier subjected to signals at 6 and 8 MHz will produce second-order IMD products at 2 MHz (8−6) and 14 MHz (8+6).

- *Third-order* IMD products change 3 dB for every decibel of input-signal change (this also assumes equal-level input signals), and appear at frequencies corresponding to the sums and differences of twice one signal's frequency plus or minus the frequency of another. Assuming that its input bandwidth is sufficient to pass them, an amplifier subjected to signals at 14.02 MHz (f_1) and 14.04 MHz (f_2) produces third-order IMD products at 14.00 ($2f_1 − f_2$), 14.06 ($2f_2 − f_1$), 42.08 ($2f_1 + f_2$) and 42.10 ($2f_2 + f_1$) MHz. The subtrac-

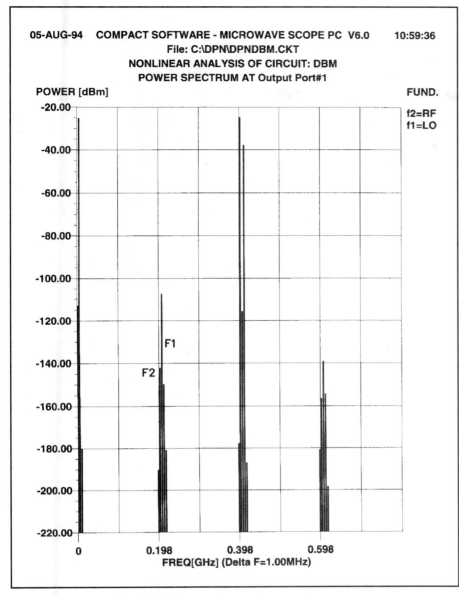

File: C:\DPN\DPNDBM.CKT
NONLINEAR ANALYSIS OF CIRCUIT: DBM
POWER SPECTRUM AT Output Port#1

POWER [dBm]

FUND.

f2=RF
f1=LO

F1

F2

FREQ[GHz] (Delta F=1.00MHz)

Fig 15.27—Simulated diode-DBM output spectrum. Note that the desired output products (the highest two products, RF − LO and RF + LO) emerge at a level 5 dB below the mixer's RF input (−20 dBm). This indicates a mixer conversion loss of 5 dB. (*Microwave SCOPE* simulation)

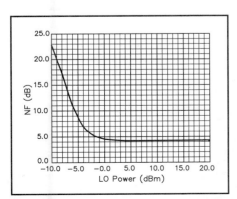

Fig 15.28—Noise figure versus LO drive for a DDBM built along the lines of Fig 15.20.

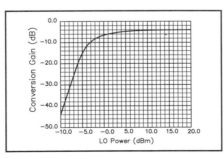

Fig 15.29—Conversion gain (loss) for a DDBM. Increasing a mixer's LO level beyond that sufficient to turn its switching devices all the way on merely makes them dissipate more LO power and does not improve performance.

tive products (the 14.02- and 14.04-MHz products in this example) are close to the desired signal and can cause significant interference. Thus, third-order-IMD performance is of great importance in receiver mixers and RF amplifiers.

It can be seen that the IMD order determines how rapidly IMD products change level per unit change of input level. Nth-order IMD products therefore change by n dB for every decibel of input-level change.

IMD products at orders higher than three can and do occur in communication systems, but the second- and third-order products are most important in receiver front ends. In transmitters, third- and higher-odd-order products are important because they widen the transmitted signal.

Intercept Point

The second type of dynamic range concerns the receiver's *intercept point*, sometimes simply referred to as *intercept*. Intercept point is typically measured by applying two or three signals to the antenna input, tuning the receiver to count the number of resulting spurious responses, and measuring their level relative to the input signal.

Because a device's IMD products increase more rapidly than its desired output as the input level rises, it might seem that steadily increasing the level of multiple signals applied to an amplifier would eventually result in equal desired-signal and IMD levels at the amplifier output. Real devices are incapable of doing this, however. At some point, every device *overloads*, and changes in its output level no longer equally track changes at its input. The device is then said to be operating in *compression*; the point at which its first-order response deviates from linearity by 1 dB is its *1-dB compression point*. Pushing the process to its limit ultimately leads to *saturation*, at which point input-signal increases no longer increase the output level.

The power level at which a device's second-order IMD products equal its first-order output (a point that must be extrapolated because the device is in compression by this point) is its *second-order intercept point*. Likewise, its *third-order intercept point* is the power level at which third-order responses equal the desired signal. **Fig 15.30** graphs these relationships.

Input filtering can improve second-order intercept point; device nonlinearities determine the third, fifth and higher-odd-number intercept points. In preamplifiers, third-order intercept point is directly related to dc input power; in mixers, to the local-oscillator power applied.

Testing and Calculating Intermodulation Distortion in Receivers

Second and third-order IMD can be measured using the setup of **Fig A**. The outputs of two signal generators are combined in a 3-dB hybrid coupler. Such couplers are available from various companies, and can be homemade. The 3-dB coupler should have low loss and should itself produce negligible IMD. The signal generators are adjusted to provide a known signal level at the output of the 3-dB coupler, say, −20 dBm for each of the two signals. This combined signal is then fed through a calibrated variable attenuator to the device under test. The shielding of the cables used in this system is important: At least 90 dB of isolation should exist between the high-level signal at the input of the attenuator and the low-level signal delivered to the receiver.

The measurement procedure is simple: adjust the variable attenuator to produce a signal of known level at the frequency of the expected IMD product ($f_1 \pm f_2$ for second-order, $2f_1 - f_2$ or $2f_2 - f_1$ for third-order IMD).

To do this, of course, you have to figure out what equivalent input signal level at the receiver's operating frequency corresponds to the level of the IMD product you are seeing. There are several ways of doing this. One way—the way used by the ARRL Lab in their receiver tests—uses the minimum discernible signal. This is defined as the signal level that produces a 3-dB increase in the receiver audio output power. That is, you measure the receiver output level with no input signal, then insert a signal at the operating frequency and adjust the level of this input signal until the output power is 3 dB greater than the no-signal power. Then, when doing the IMD measurement, you adjust the attenuator of Fig A to cause a 3-dB increase in receiver output. The level of the IMD product is then the same as the MDS level you measured.

There are several things I dislike about doing the measurement this way. The problem is that you have to measure noise power. This can be difficult. First, you need an RMS voltmeter or audio power meter to do it at all. Second, the measurement varies with time (it's noise!), making it difficult to nail down a number. And third, there is the question of the audio response of the receiver; its noise output may not be flat across the output spectrum. So I prefer to measure, instead of MDS, a higher reference level. I use the receiver's S meter as a reference. I first determine the input signal level it takes to get an S1 reading. Then, in the IMD measurement, I adjust the attenuator to again give an S1 reading. The level of the IMD product signal is now equal to the level I measured at S1. Note that this technique gives a different IMD level value than the MDS technique. That's OK, though. What we are trying to determine is the *difference* between the level of the signals applied to the receiver input and the level of the IMD product. Our calculations will give the same result whether we measure the IMD product at the MDS level, the S1 level or some other level.

An easy way to make the reference measurement is with the setup of Fig A. You'll have to switch in a lot of attenuation (make sure you have an attenuator with enough range), but doing it this way keeps all of the possible variations in the measurement fairly constant. And this way, the difference between the reference level and the input level needed to produce the desired IMD product signal level is simply the difference in attenuator settings between the reference and IMD measurements.

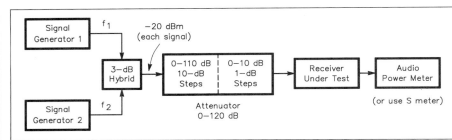

Fig A—Test setup for measurement of IMD performance. Both signal generators should be types such as HP 608, HP 8640, or Rhode & Schwarz SMDU, with phase-noise performance of −140 dBc/Hz or better at 20 kHz from the signal frequency.

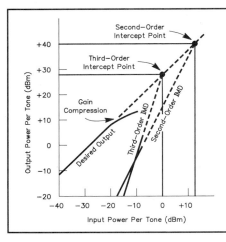

Fig 15.30—A linear stage's output tracks its input decibel by decibel on a 1:1 slope—its *first-order* response. *Second-order intermodulation distortion (IMD) products* produced by two equal-level input signals ("tones") rise on a 2:1 slope—2 dB for every 1 dB of input increase. *Third-order IMD products* likewise increase 3 dB for every 1 dB of increase in two equal tones. For each IMD order *n*, there is a corresponding *intercept point* IP_n at which the stage's first-order and *n*th order products are equal in amplitude. The first-order output of real amplifiers and mixers falls off (the device overloads and goes into *compression*) before IMD products can intercept it, but intercept point is nonetheless a useful, valid concept for comparing radio system performance. The higher an amplifier or mixer's intercept point, the stronger the input signals it can handle without overloading. The input and output powers shown are for purposes of example; every device exhibits its own particular IMD profile. (After W. Hayward, *Introduction to Radio Frequency Design*, Fig 6.17)

Calculating Intercept Points

Once we know the levels of the signals applied to the receiver input and the level of the IMD product, we can easily calculate the intercept point using the following equation:

$$IP_n = \frac{n \cdot P_A - P_{IM_n}}{n - 1} \tag{A}$$

Here, n is the order, P_A is the receiver input power (of one of the input signals), P_{IM_n} is the power of the IMD product signal, and IP_n is the nth-order intercept point. All powers should be in dBm. For second and third-order IMD, equation A results in the equations:

$$IP_2 = \frac{2 \cdot P_A - P_{IM_2}}{2 - 1} \tag{B}$$

$$IP_3 = \frac{3 \cdot P_A - P_{IM_3}}{3 - 1} \tag{C}$$

You can measure higher-order intercept points, too.

Example Measurements

To get a feel for this process, it's useful to consider some actual measured values.

The first example is a Rohde & Schwarz model EK085 receiver with digital preselection. For measuring second-order IMD, signals at 6.00 and 8.01 MHz, at −20 dBm each, were applied at the input of the attenuator. The difference in attenuator settings between the reference measurement and the level needed to produce the desired IMD product signal level was found to be 125 dB. The calculation of the second-order IP is then:

$$IP_2 = \frac{2(-20\,dBm) - (-20\,dBm - 125\,dB)}{2 - 1}$$
$$= -40\,dBm + 20\,dBm + 125\,dB$$
$$= +105\,dBm$$

For IP_3, we set the signal generators for 0 dBm at the attenuator input, using frequencies of 14.00 and 14.01 MHz. The difference in attenuator settings between the reference and IMD measurements was 80 dB, so:

$$IP_3 = \frac{3(0\,dBm) - (0\,dBm - 80\,dB)}{3 - 1}$$
$$= \frac{0\,dBm + 80\,dB}{2} = +40\,dBm$$

We also measured the IP_3 of a Yaesu FT-1000D at the same frequencies, using attenuator-input levels of −10 dBm. A difference in attenuator readings of 80 dB resulted in the calculation:

$$IP_3 = \frac{3(-10\,dBm) - (-10\,dBm - 80\,dB)}{3 - 1}$$
$$= \frac{-30\,dBm + 10\,dBm + 80\,dB}{2} = \frac{-20\,dBm + 80\,dB}{2}$$
$$= +30\,dBm$$

Synthesizer Requirements

To be able to make use of high third-order intercept points at these close-in spacings requires a low-noise LO synthesizer. You can estimate the required noise performance of the synthesizer for a given IP_3 value. First, calculate the value of receiver input power that would cause the IMD product to just come out of the noise floor, by solving equation A for P_A, then take the difference between the calculated value of P_A and the noise floor to find the dynamic range. Doing so gives the equation:

$$ID_3 = \frac{2}{3}\left(IP_3 + P_{min}\right) \tag{D}$$

Where ID_3 is the third-order IMD dynamic range in dB and P_{min} is the noise floor in dBm. Knowing the receiver bandwidth, BW (2400 Hz in this case) and noise figure, NF (8 dB) allows us to calculate the noise floor, P_{min}:

$$P_{min} = -174\,dBm + 10\log(BW) + NF$$
$$= -174\,dBm + 10\log(2400) + 8$$
$$= -132\,dBm$$

The synthesizer noise should not exceed the noise floor when an input signal is present that just causes an IMD product signal at the noise floor level. This will be accomplished if the synthesizer noise is less than:

$$ID + 10\log(BW) = 114.7\,dB + 10\log(2400)$$
$$= 148.5\,dBc/Hz$$

in the passband of the receiver. Such synthesizers hardly exist.—*Dr. Ulrich L. Rohde, KA2WEU*

Applying Diode DBMs

At first glance, applying a diode DBM is easy: We feed the signal(s) we want to frequency-shift (at or below the maximum level called for in the mixer's specifications, such as −10 dBm for the Mini-Circuits SBL-1 and Synergy Microwave S-1, popular Level 7 parts) to the DBM's RF port, feed the frequency-shifting signal (at the proper level) to the LO port, and extract the sum and difference products from the mixer's IF port.

There's more to it than that, however, because diode DBMs (along with most other modern mixer types) are *termination-sensitive*. That is, their ports—particularly their IF (output) ports—must be resistively terminated with the proper impedance (commonly 50 Ω, resistive). A wideband, resistive output termination is particularly critical if a mixer is to achieve its maximum dynamic range in receiving applications. Such a load can be achieved by

- terminating the mixer in a 50-Ω resistor or attenuator pad (a technique usually avoided in receiving applications because it directly degrades system noise figure);
- terminating the mixer with a low-noise, high-dynamic-range *post-mixer amplifier* designed to exhibit a wideband resistive input impedance; or
- terminating the mixer in a *diplexer*, a frequency-sensitive signal splitter that appears as a two-terminal resistive load at its input while resistively dissipating unwanted outputs and passing desired outputs through to subsequent circuitry.

Termination-insensitive mixers are available, but this label can be misleading. Some termination-insensitive mixers are nothing more than a termination-sensitive mixer packaged with an integral post-mixer amplifier. True

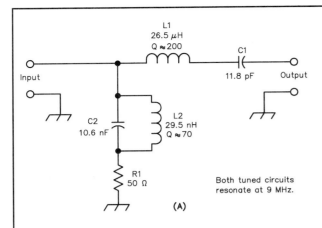

15-AUG-94   COMPACT SOFTWARE - ARRL Radio Designer 1.0   16:43:29
File: c:\data\word\books\hbk95\mixers\fig15-31.ckt
Simulation, Input Match (MS11) and Gain (MS21) of 9-MHz Diplexer

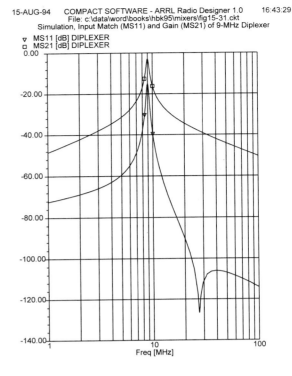

(C)

15-AUG-94   COMPACT SOFTWARE - ARRL Radio Designer 1.0   16:43:32
File: c:\data\word\books\hbk95\mixers\fig15-31.ckt
Simulation, Diplexer Gain (MS21) and Input Match in 9-MHz Region

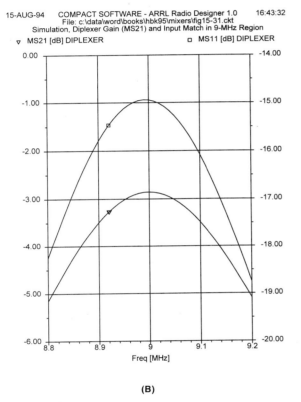

(B)

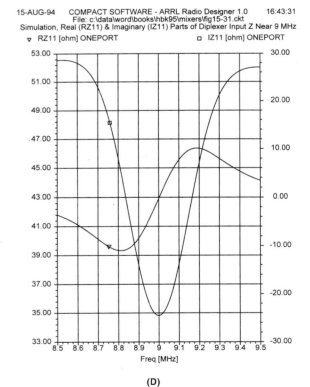

(D)

Fig 15.31—A diplexer resistively terminates energy at unwanted frequencies while passing energy at desired frequencies. This band-pass diplexer (A) uses a series-tuned circuit as a selective pass element, while a high-C parallel-tuned circuit keeps the network's terminating resistor R1 from dissipating desired-frequency energy. Computer simulation of the diplexer's response with *ARRL Radio Designer 1.0* characterizes the diplexer's insertion loss and good input match from 8.8 to 9.2 MHz (B) and from 1 to 100 MHz (C); and the real and imaginary components of the diplexer's input impedance from 8.8 to 9.2 MHz with a 50-Ω load at the diplexer's output terminal (D). The high-C, low-L nature of the L2-C2 circuit requires that C2 be minimally inductive; a 10,000-pF chip capacitor is recommended. This diplexer was described by Ulrich L. Rohde and T. T. N. Bucher in *Communications Receivers*: *Principles and Design*.

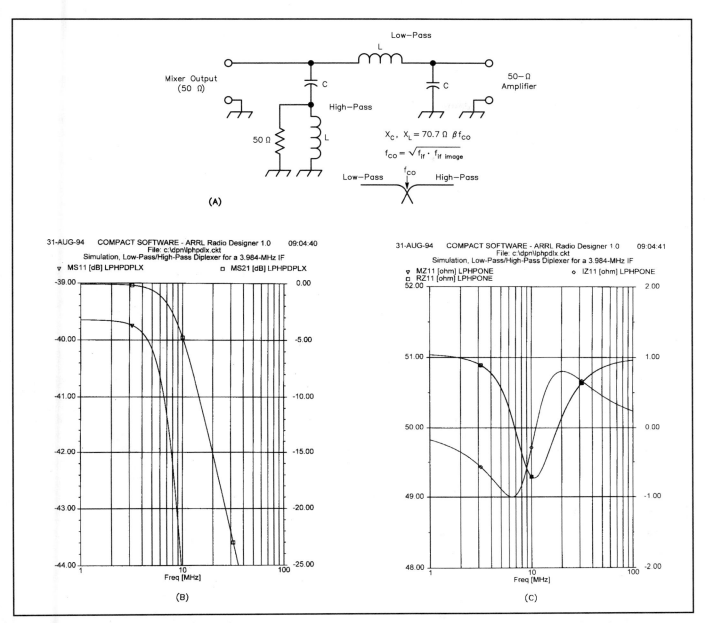

Fig 15.32—All of the inductors and capacitors in this high-pass/low-pass diplexer (A) exhibit a reactance of 70.7 Ω at its tuned circuits' 3-dB cutoff frequency (the geometric mean of the IF and IF image). B and C show *ARRL Radio Designer* simulations of this circuit configured for use in a receiver that converts 7 MHz to 3.984 MHz using a 10.984-MHz LO. The IF image is at 17.984 MHz, giving a 3-dB cutoff frequency of 8.465 MHz. The inductor values used in the simulation were therefore 1.33 μH (Q = 200 at 25.2 MHz); the capacitors, 265 pF (Q = 1000). This drawing shows idler load and "50-Ω Amplifier" connections suitable for a receiver in which the IF image falls at a frequency *above* the desired IF. For applications in which the IF image falls below the desired IF, interchange the 50-Ω idler load resistor and the diplexer's "50-Ω Amplifier" connection so the idler load terminates the diplexer low-pass filter and the 50-Ω amplifier terminates the high-pass filter.

termination-insensitive mixers are less common and considerably more elaborate. Amateur builders will more likely use one of the many excellent termination-sensitive mixers available in connection with a diplexer, post-mixer amplifier or both.

Fig 15.31 shows one diplexer implementation. In this approach, L1 and C1 form a series-tuned circuit, resonant at the desired IF, that presents low impedance between the diplexer's input and output terminals at the IF. The high-impedance parallel-tuned cir-

cuit formed by L2 and C2 also resonates the desired IF, keeping desired energy out of the diplexer's 50-Ω load resistor, R1.

The preceding example is called a *bandpass diplexer*. **Fig 15.32** shows another type: a *high-pass/low-pass diplexer* in which each inductor and capacitor has a reactance of 70.7 Ω at the 3-dB cutoff frequency. It can be used after a "difference" mixer (a mixer in which the IF is the difference between the signal frequency and LO) if the desired IF and its image frequency are far enough apart

so that the image power is "dumped" into the network's 51-Ω resistor. (For a "summing" mixer—a mixer in which the IF is the sum of the desired signal and LO—interchange the 50-Ω idler load resistor and the diplexer's "50-Ω Amplifier" connection.) Richard Weinreich, KØUVU, and R. W. Carroll described this circuit in November 1968 *QST* as one of several absorptive TVI filters.

Fig 15.33 shows a BJT post-mixer amplifier design made popular by Wes Hayward, W7ZOI, and John Lawson, K5IRK.

K5IRK. RF feedback (via the 1-kΩ resistor) and emitter degeneration (the ac-coupled 5.6-Ω emitter resistor) work together to keep the stage's input impedance low and uniformly resistive across a wide bandwidth.

Amplitude Modulation with a DBM

We can generate DSB, suppressed-carrier AM with a DBM by feeding the carrier to its RF port and the modulating signal to the IF port. This is a classical *balanced modulator*, and the result—sidebands at radio frequencies corresponding to the carrier signal plus audio and the RF signal minus audio—emerges from the DBM's LO port. If we also want to transmit some carrier along with the sidebands, we can dc-bias the IF port (with a current of 10 to 20 mA) to upset the mixer's balance and keep its diodes from turning all the way off. (This technique is sometimes used for generating CW with a balanced modulator otherwise intended to generate DSB as part of an SSB-generation process.) **Fig 15.34** shows a more elegant approach to generating full-carrier AM with a DBM.

As we saw earlier when considering the many faces of AM, two DBMs, used in conjunction with carrier and audio phasing, can be used to generate SSB, suppressed-carrier AM. Likewise, two DBMs can be used with RF and LO phasing as an image-reject mixer.

Phase Detection with a DBM

As we saw in our exploration of quadrature detection, applying two signals of equal frequency to a DBM's LO and RF ports produces an IF-port dc output proportional to the cosine of the signals' phase difference (**Fig 15.35**). (This assumes that the DBM has a dc-coupled IF port, of course. If it doesn't—and some DBMs don't—phase-detector operation is out.) Any dc output offset introduces error into this process, so critical phase-detection applications use low-offset DBMs optimized for this service.

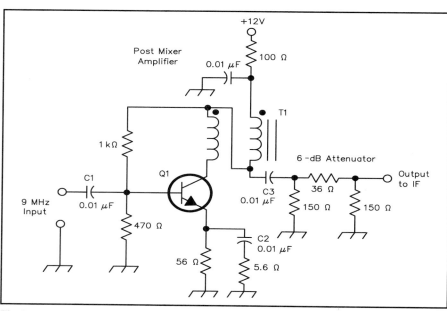

Fig 15.33—The post-mixer amplifier from Hayward and Lawson's Progressive Communications Receiver (November 1981 *QST*). This amplifier's gain, including the 6-dB loss of the attenuator pad, is about 16 dB; its noise figure, 4 to 5 dB; it output intercept, 30 dBm. The 6-dB attenuator is essential if a crystal filter follows the amplifier; the pad isolates the amplifier from the filter's highly reactive input impedance. This circuit's input match to 50 Ω below 4 MHz can be improved by replacing 0.01-μF capacitors C1, C2 and C3 with low-inductance 0.1-μF units (chip capacitors are preferable).

Q1—TO-39 CATV-type bipolar transistor, f_T=1 GHz or greater. 2N3866, 2N5109, 2SC1252, 2SC1365 or MRF586 suitable. Use a small heat sink on this transistor.

T1—Broadband ferrite transformer, ≈42 μH per winding: 10 bifilar turns of #28 enameled wire on an FT 37-43 core.

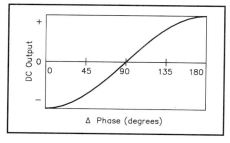

Fig 15.35—A phase detector's dc output is the cosine of the phase difference between its input and reference signals.

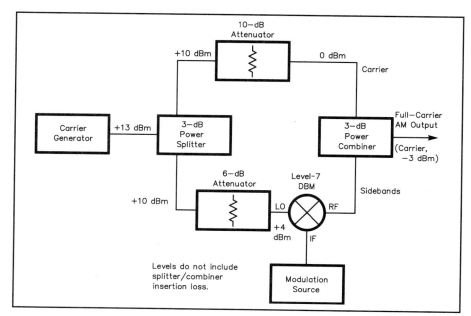

Fig 15.34—Generating full-carrier AM with a diode DBM.

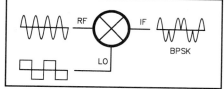

Fig 15.36—Mixing a carrier with a square wave generates biphase-shift keying (BPSK), in which the carrier phase is shifted 180° for data transmission. In practice, as in this drawing, the carrier and data signals are phase-coherent so the mixer switches only at carrier zero crossings.

Biphase-Shift Keying Modulation with a DBM

Back in our discussion of square-wave mixing, we saw how multiplying a switching mixer's linear input with a square wave causes a 180° phase shift during the negative part of the square wave's cycle. As **Fig 15.36** shows, we can use this effect to produce *biphase-shift keying (BPSK)*, a digital system that conveys data by means of carrier phase reversals. A related system, *quadrature phase-shift keying (QPSK)* uses two DBMs and phasing to convey data by phase-shifting a carrier in 90° increments.

Transistors as Mixer Switching Elements

We've covered diode DBMs in depth because their home-buildability, high performance and suitability for direct connection into 50-Ω systems makes them attractive to Amateur Radio builders. The abundant availability of high-quality manufactured diode mixers at resonable prices make them excellent candidates for home-construction projects. Although diode DBMs are common in telecommunications as a whole, their conversion loss and relatively high LO power requirement have usually driven the manufacturers of high-performance MF/HF Amateur Radio receivers and transceivers to other solutions. Those solutions have generally involved single- or double-balanced FET mixers—MOSFETs in the late 1970s and early 1980s, JFETs from the early 1980s to date. Many of the JFET designs are variations of a single-balanced mixer circuit introduced to *QST* readers in 1970!

Fig 15.37 shows the circuit as it was presented by William Sabin in "The Solid-State Receiver," *QST*, July 1970. Two 2N4416 JFETs operate in a common-source configuration, with push-pull RF input and parallel LO drive. **Fig 15.38** shows a similar circuit as implemented in the ICOM IC-765 transceiver. In this version, the JFETs (2SK125s) operate in common-gate, with the LO applied across a 220-Ω resistor between the gates and ground.

Bipolar junction transistors can also work well in switching mixers. In June 1994 *QST*, Ulrich Rohde, KA2WEU, published a medium-frequency mixer well-suited to shortwave applications and home-built projects. Shown in **Fig 15.39**, it consists of two transistors in a push-pull, single-balanced configuration. Because of the degenerative feedback introduced by the 20-Ω emitter resistors, the two transistors need not be matched. This mixer's advantage lies in its achievement of a 33-dBm output intercept with only 17 dBm of local oscillator drive. (Typically, a diode DBM with the same IMD performance requires 25 to 27 dBm of LO drive.) Tests indicate that the upper frequency limit of this mixer lies in the 500-MHz region. The circuit's lower frequency limit depends on the transformer inductances and the ferrites used for the transformer cores.

Examining the state of the art in the 1990s, however, we find that the best receive-mixer dynamic ranges are achieved with quads of RF MOSFETs operating as passive switches, with no drain voltage applied. The best of these techniques involves following a receiver's first mixer with a diplexer and low-loss roofing crystal filter, rather than terminating the mixer in a strong wideband amplifier.

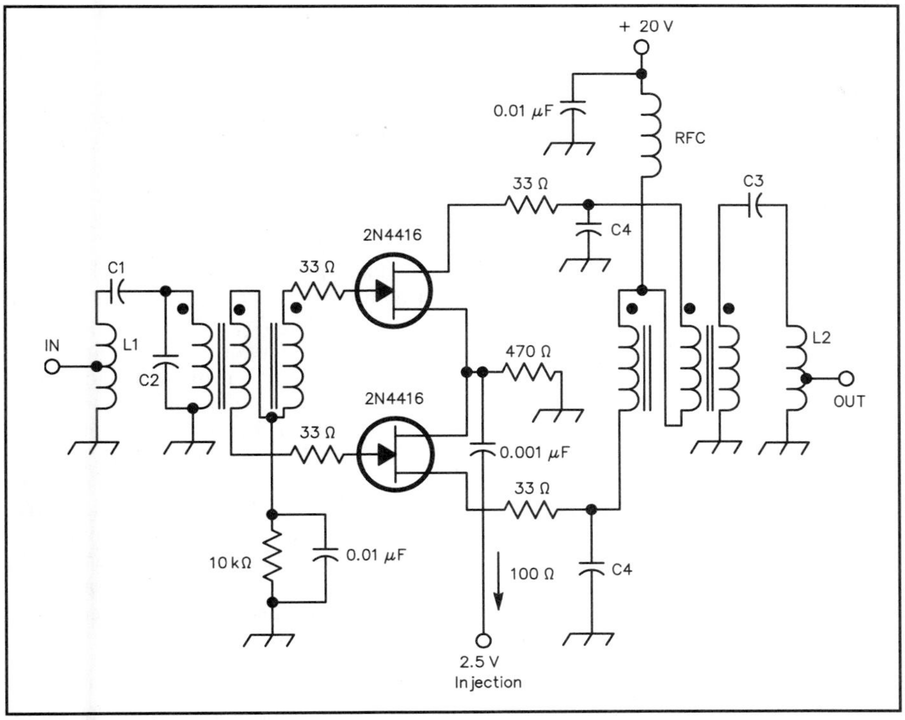

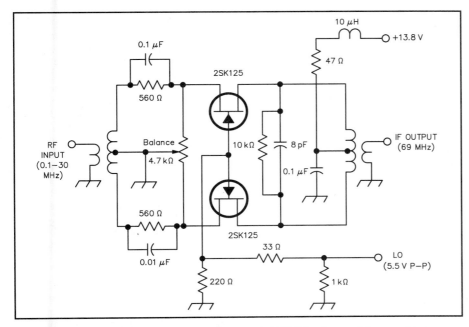

Fig 15.37—Two 2N4416 JFETs provide high dynamic range in this mixer circuit from Sabin 1970. L1, C1 and C2 form the input tuned circuit; L2, C3 and C4 tune the mixer output to the IF. The trifilar input and output transformers are broadband transmission-line types.

Fig 15.38—The ICOM IC-765's single-balanced 2SK125 mixer achieves a high dynamic range (per *QST* Product Review, an IP$_3$ of 10.5 dBm at 14 MHz with preamp off) with arrangement very much like Sabin's. The first receive mixer in many commercial Amateur Radio transceiver designs of the 1980s and 1990s used a 2SK125 pair in much this way.

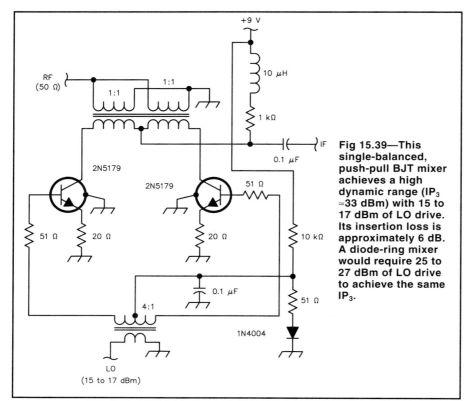

Fig 15.39—This single-balanced, push-pull BJT mixer achieves a high dynamic range (IP₃ ≈33 dBm) with 15 to 17 dBm of LO drive. Its insertion loss is approximately 6 dB. A diode-ring mixer would require 25 to 27 dBm of LO drive to achieve the same IP₃.

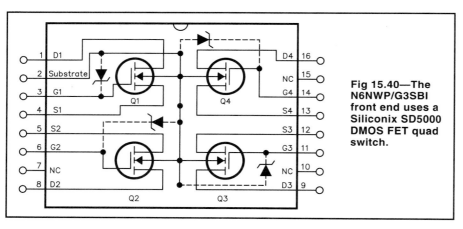

Fig 15.40—The N6NWP/G3SBI front end uses a Siliconix SD5000 DMOS FET quad switch.

A HIGH-DYNAMIC-RANGE MF/HF FRONT END

Jacob Mahkinson, N6NWP, achieved outstanding results with a Calogic SO8901 mixer and CMOS flip-flop. Colin Horrabin, G3SBI, found that a Siliconix SD5000 DMOS FET quad switch driven by a local oscillator signal from an advanced CMOS flip-flop (74AC74N) gave even better performance. His adaptation of Mahkinson's circuit is shown in Figs 15.40, 15.41, 15.42, 15.43 and 15.44. The Siliconix SD5000 (Fig 15.40) serves as the core of this system, which was experimentally built for 14-MHz reception with a 9-MHz IF, and later tested from 1.8 to 30 MHz. Mahkinson's system was described in February 1993 *QST*. Horrabin's experiments have been the subject of an ongoing discussion in

the Technical Topics column of The Radio Society of Great Britain's *Radio Communication* since September 1993.

The Mixer

Fig 15.41 shows the mixer/diplexer schematic. T1 matches the 50-Ω source (the preamplifier) to U2's RF port, and T2 matches U2's IF port to a 50-Ω load—the post-mixer amplifier.

In a commutation mixer, there is a compromise between conversion loss and intermodulation distortion. The impedance ratios of transmission-line transformers T1 and T2 are chosen to achieve low intermodulation distortion while keeping conversion loss at an acceptable level.

The LO waveform applied to U2 is of great importance. To achieve a high inter-

cept point, the LO drive must approach an ideal square wave with a 50% duty cycle. Flip-flop U1 divides the LO signal frequency by two and provides 50% duty cycle square waves at its complimentary outputs. The offset bias adjustment potentiometers, R6 and R7, allow compensation for mismatch among the SD5000's MOSFETs and ensure that the commutation mixer switches operate in a 50% duty cycle mode.

The mixer's intercept point also depends on the LO drive level. Zener diode D1 establishes the power supply voltage for U1 and, therefore, the voltage swing at the mixer LO port. In the interest of obtaining a high intercept point, a 9.1-V diode is used at D1.

A band-pass diplexer follows the mixer. It comprises L1, C8, C9, L2, C10, C11, R10 and R11. The 9-MHz signal passes through this network with minimum attenuation, while out-of-passband signals over a wide frequency range are dissipated. Resistor R10 presents a 50-Ω impedance to the mixer, while R11 presents a 50-Ω impedance to the input of the postmixer amplifier. C7 cancels the inductive reactance of the mixer's input source.

The measured performance of the mixer/diplexer circuit gives input intercepts of at least 42 dBm (HF bands) and 46 dBm (1.8 MHz).

The Preamplifier

The push-pull preamplifier (**Fig 15.42**) uses MRF586s in a "noiseless feedback" configuration. BIAS controls R12 and R14 allow the collector current of each transistor to be adjusted to the desired value of 25 mA. T3 matches the amplifier input to the 50-Ω band-pass input filter; T4 matches the preamp output to the mixer signal port.

Transformers T5 and T6 couple part of the collector signal back into the emitter (negative feedback), set the gain to 8 dB, and set the input and output impedances to 50 Ω. C12 cancels the inductive reactance of the mixer's source impedance at 14 MHz. (In a multiband receiver, this capacitor may be part of switchable preselector filters.) The measured performance of the circuit is: third-order output intercept, 48 dBm; 1-dB output compression point, 25 dBm; gain, 8.0 dB; noise figure, 2.0 dB; –1-dB frequency response 1 to 40 MHz. This performance fully meets the preamplifier design goals.

Post-Mixer Amplifier

The post-mixer amplifier (**Fig 15.43**) uses the same basic configuration as the preamplifier. This is Colin Horrabin's adaptation of Jacob Mahkinson's original post-mixer amplifier. The MRF580A transistors have a collector current bias of 60 mA per device. This helps to overcome the negative effects of load on the postmixer amplifier's intercept point. T11, a

Parts List for the N6NWP High-Dynamic Range Front End

Unless indicated otherwise, all of the circuits' resistors are ¼-W 1% tolerance, and all of the capacitors are 20%-tolerance ceramic. C7-C12 and C20 are dipped mica, polyester film or epoxy-coated ceramic.

U1—74HAC74N CMOS flip-flop.
U2—Siliconix SD5000 DMOS FET Quad switch.
U3—LM317MP adjustable voltage regulator IC.
U4—LM337MP adjustable voltage regulator IC.
Q1-Q2—Motorola MRF586 RF transistor.
Q3-Q4—Motorola MRF580A RF transistor. Use heat sink

(Thermalloy 2228B or Aavid 3257).
T1, T2—Mini-Circuits T4-1 4:1 RF transformer.
T3, T4—Mini-Circuits T1-1 1:1 RF transformer.
T11—Mini-Circuits T9-1 9:1 RF transformer.
L3-L5—100-µH RF choke.
L1, L2—Toko 332PN-T1012Z (Digi-Key TK5130) 1.0-µH variable inductor.
R6, R7—2-kΩ multiturn potentiometer.

R12, R14—200-Ω multiturn potentiometer.
D1—9.1-V, 0.5-W Zener diode, 1N5239 suitable. See text.
C7—39 pF ±10%.
C8, C10—330 pF ±5%.
C9, C11—22 pF ±5%.
C12—27 pF ±10%.
C20—100 pF ±10%.
T5, T6—homemade trifilar transformer wound with #32 enameled wire on Amidon BN-43-2402 core (Fair-Rite 2843002402) as per Fig 15.44.

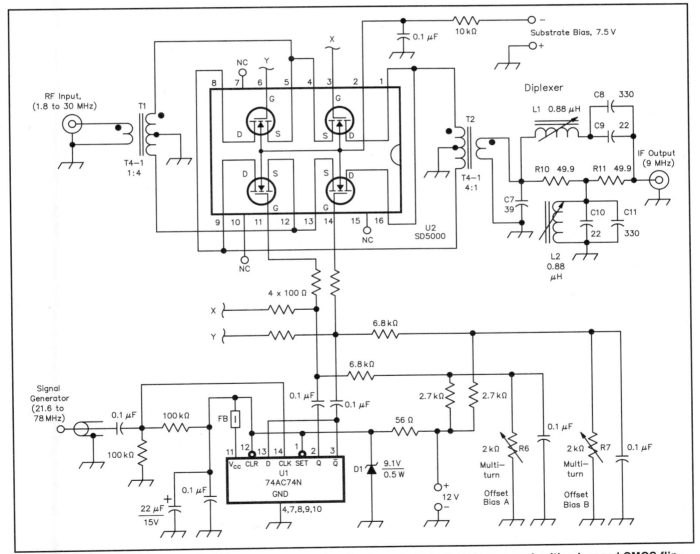

Fig 15.41—Colin Horrabin's DMOS FET quad switching mixer implements Mahkinson's approach with advanced CMOS flip-flops (74AC74Ns operated within their supply ratings) and the Siliconix SD5000 quad DMOS FET. Substrate bias (−7.5 V) is necessary to keep the SD5000's built-in diodes turned off. The ferrite bead on the 74AC74N's V_CC pin, and the 100-Ω resistors in series with the SD5000's gate-drive lines, maximize the mixer's dynamic range by damping ringing, thus keeping the mixer's LO waveform square. The OFFSET BIAS trimmers are set for best mixer balance at a gate-bias level of about 4.5 V.

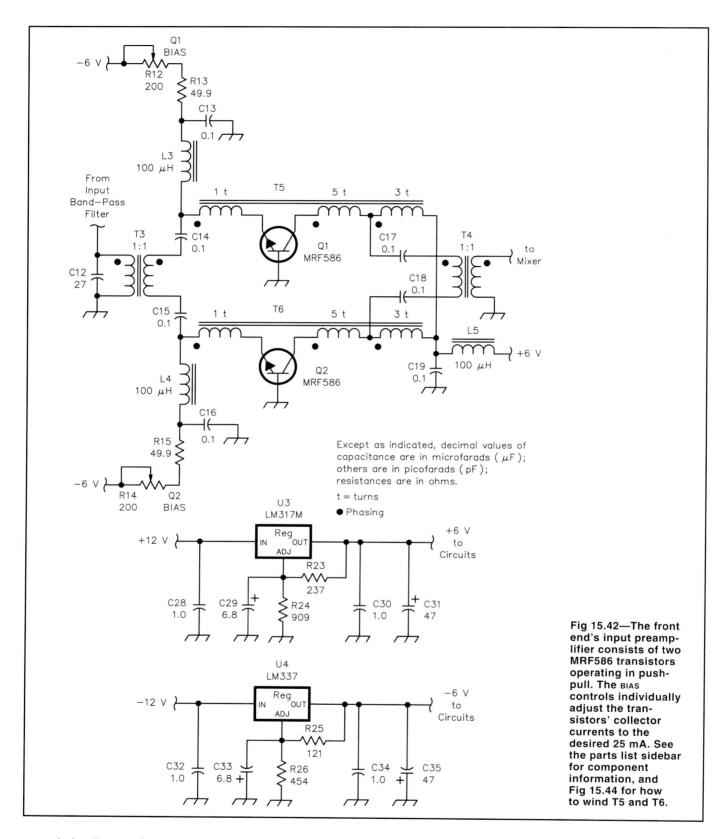

Fig 15.42—The front end's input preamplifier consists of two MRF586 transistors operating in push-pull. The BIAS controls individually adjust the transistors' collector currents to the desired 25 mA. See the parts list sidebar for component information, and Fig 15.44 for how to wind T5 and T6.

transmission-line transformer, matches the crystal filter's impedance (500 Ω in this case) to the amplifier's 50-Ω output impedance. C20 cancels the inductive reactance of the amplifier's input source. The measured performance of the post-mixer amplifier is: output intercept, +56 dBm; gain, 8.8 dB; and noise figure, 0.5 dB.

More High-Performance Mixer Experiments

Colin Horrabin, G3SBI, continued experimenting with variations of Jacob Mahkinson's original high-performance mixer circuit. This led to the development of a new mixer configuration, called an H mode mixer. This name comes from the signal path through the circuit. (See

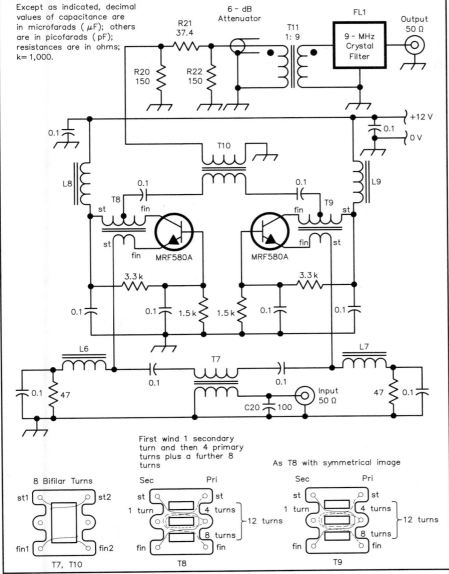

Fig 15.43 — G3SBI adapted the N6NWP design to build this post mixer amplifier using MRF580A RF transistors. All resistors are ¼ W metal-film units. All 0.1 μF capacitors are monolithic ceramic components. L8 and L9 have 4 turns of #28 AWG enameled wire on a ferrite bead. L6, L7, T7, T8, T9 and T10 are wound on Fair-Rite 28-43002402 (material 43) balun cores. L6 and L7 are 6 turns of #28 AWG enameled wire. T7 to T10 use #36 AWG enameled wire. Take two glass fiber Cambion 14-pin DIP sockets, cut each in half and bend the leads out at 90°. Mount the balun cores to a socket half with double sided tape and then wind the transformers as shown. The amplifier was constructed using ground-plane construction.

Fig 15.45A.) Horrabin is a professional scientist/engineer at the Science and Engineering Research Council's Daresbury Laboratory, which has supported his investigative work on the H-mode switched-FET mixer, and consequently holds intellectual title to the new mixer. This does not prevent readers from taking the development further or using the information presented here.

Inputs A and B are complementary square-wave inputs derived from the sine-wave local oscillator at twice the required square-wave frequency. If A is ON, then FETs F1 and F3 are ON and F2 and F4 are OFF. The direction of the RF signal across T1 is given by the E arrows. When B is ON, FETs F2 and F4 are ON and F1 and F3 are OFF. The direction of the RF signal across T1 reverses, as shown by the F arrows.

This is still the action of a switching mixer, but now the source terminal of each FET switch is grounded, so that the RF signal switched by the FET cannot modulate the gate voltage. In this configuration the transformers are important:

T1 is a Mini-Circuits type T4-1 and T2 is a pair of these same transformers with their primaries connected in parallel.

The Ubiquitous NE602: A Popular Gilbert Cell Mixer

Introduced as the NE602 in the mid-1980s, the Philips Components-Signetics NE602A mixer-oscillator IC has become greatly popular with amateur experimenters for transmit mixers, receive mixers and balanced modulators. **Fig 15.46** shows its equivalent circuit. The NE602A's typical current drain is 2.4 mA; its supply voltage range is 4.5 to 8.0.

As we learned in exploring sine-wave versus square-wave mixing, the NE602's mixer is a Gilbert cell multiplier. Its inputs (RF) and outputs (IF) can be single- or double-ended (balanced) according to design requirements (**Fig 15.47**). Each input's equivalent ac impedance is approximately 1.5 kΩ in parallel with 3 pF; each output's resistance is 1.5 kΩ. The mixer can typically handle signals up to 500 MHz. At 45 MHz, its noise figure is typically 5.0 dB; its typical conversion gain, 18 dB. Considering the NE602A's low current drain, its input IP_3 (measured at 45 MHz with 60-kHz spacing) is usefully good at −15 dBm. Factoring in the mixer's conversion gain results in an equivalent output IP_3 of about 5 dBm.

The NE602A's on-board oscillator can operate up to 200 MHz in LC and crystal-controlled configurations (**Fig 15.48** shows three possibilities). Alternatively, energy from an external LO can be applied to the chip's pin 6 via a dc blocking capacitor. At least 200 mV P-P of external LO drive is required for proper mixer operation.

NE602A Usage Notes

The '602 was intended to be used as the second mixer in double-conversion FM cellular radios, in which the first IF is typically 45 MHz, and the second IF is typically 455 kHz. Such a receiver's second mixer can be relatively weak in terms of dynamic range because of the adjacent-signal protection afforded by the high selectivity of the first-IF filter preceding it. When used as a first mixer, the '602 can provide a two-tone third-order dynamic range between 80 and 90 dB, but this figure is greatly diminished if a preamplifier is used ahead of the '602 to improve the system's noise figure.

When the '602 is used as a second mixer, the sum of the gains preceding it should not exceed about 10 dB. NE602 product detection therefore should not follow a high-gain IF section unless appropriate attenuation is inserted between the '602 and the IF strip.

The '602 is generally *not* a good choice for VHF and higher-frequency mixers because of its input noise and diminishing

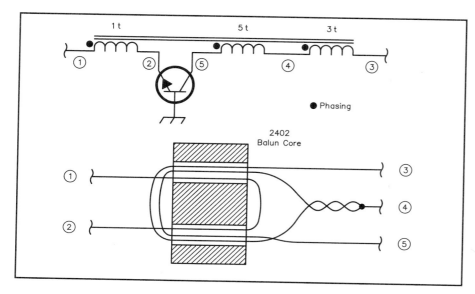

Fig 15.44—Transformers T5 and T6 are wound with #32 wire on two-hole ferrite cores (see the parts list sidebar for core information). Wind the 1-turn winding (1-2) first, the 3-turn winding (4-3) second and the 5-turn winding (5-4) third.

R2, BFO TUNING, drives D1 to provide manual detector tuning without upsetting the D2's control-voltage optimization. S2, BFO OFFSET, presets R2's tuning range into the optimum regions for LSB (−2 kHz), DSB AM (±0 kHz) and USB (+2 kHz). (These offsets are for a radio in which received sidebands are inverted relative to their on-air sense.) The BFO TUNING control therefore provides fine adjustment for detector lock around coarse receiver tuning steps (1 kHz in the Sangean AT-808 with which this circuit was originally used).

U3's phase detector requires a 90° phase shift between the incoming and reference phases to give the correct zero-phase output (again, approximately 2.16 V). The all-pass stage, Q1, operates as an isolation stage and adjustable phase shifter to generate the 90° phase shift.

U3 is an NE604N FM IF subsystem IC that contains limiting-amplifier stages (total gain, 101 dB) and a quadrature detector. Band limiting can be inserted between the limiter stages, but experiments with various RC and LC filters brought no improvement, and instead led to increased delay that upset the carrier/sideband phase relationships necessary for good quasi-synchronous detection. The Fig 15.49 circuit uses U3's quadrature detector as a phase detector that outputs control voltage for D2.

The most difficult aspect of the phase-locking chain is the selection of a time constant for the locking loop. Signal fading, in conjunction the relative absence or presence of phase-modulation components in the transmitted signal, play important roles in detector lock. Were fading not a problem, a short time constant—one allowing fast locking—would suffice for DSB AM. For SSB AM with carrier (which includes a PM component at all modulation frequencies), however, and DSB AM with fading (during which a fast PLL may unlock on the sudden phase shifts that can accompany fast, deep carrier fades), a long time constant is necessary. Particularly for SSB with carrier, the loop bandwidth must be reduced to below the lowest expected modulation frequency. C1 and R4 set the PLL time constant in Fig 15.49.

The received signal strength indicator (RSSI) output at pin 5 of U3 follows the input level logarithmically, giving an output of 1.1 V on noise only (RF INPUT shorted in Fig 15.49) and 3.3 V at an RF INPUT level of 3 mV. The RSSI output is adaptable as an AGC-detector output, making the NE604AN attractive for

IMD performance at high frequencies. There are applications, however, where 6-dB noise figures and 60- to 70-dB dynamic range performance is adequate. If your target specifications exceed these numbers, you should consider other mixers at VHF and up.

NE602A Relatives

The NE602A (SA602A for operation over a wider temperature range) began life as the NE602/SA602, a part with a slightly lower IP$_3$ than the A version. The pinout-identical NE612/SA612A costs less as a result of wider tolerances. All of these parts should nonetheless work satisfactorily in most "NE602" experimenter projects. The same mixer/oscillator topology, modified for slightly higher dynamic range at the expense of somewhat less mixer gain, is also available in the Philips Components-Signetics mixer/oscillator/FM IF chips NE/SA605 (input IP$_3$, typically −10 dBm) and NE/SA615 (input IP$_3$, typically −13 dBm).

A SYNCHRONOUS AM DETECTOR FOR 455 KHZ

Much like switching a receiver or transceiver to SSB and receiving AM as SSB, its carrier at zero beat, synchronous detection overcomes fading-related distortion by supplying an unfading carrier at the receiver. The difference between synchronous AM detection and normal SSB product detection is that synchronous detection *phase locks* its carrier to that of the incoming signal. No tuning error occurs if the received signal happens to fall between a synthesized receiver's tuning steps, and the detector's PLL compensates for modest tuning drift. The result is a dramatic fidelity improvement over diode-detected AM. This circuit, designed by Jukka Vermasvuori, OH2GF, and originally published in July 1993 QST, requires only a digital voltmeter for alignment.

The Synchronous Detector Circuit

See **Fig 15.49**. The unit uses popular NE602AN (mixer/oscillator) and NE604AN (FM subsystem) ICs to provide both synchronous and quasi-synchronous detection. (This discussion refers to U1 and U2 as NE602ANs, but NE602Ns, SA602Ns and SA602ANs will work equally well in this application. Likewise, an NE604N, NE604AN, SA604N or SA604AN will work well at U3 in this application.) Operating at a supply voltage of 6, the circuit draws 10 mA. U1, an NE602AN, acts as the BFO and product detector necessary for synchronous detection. Feeding U1's balanced inputs in push-pull helps keep BFO energy from backing out of the input pins and into U3's limiting circuitry. To take advantage of the chip's internal biasing, the input transformer (T1) is isolated with dc blocking capacitors. (U1 also supplies balanced audio output, but usefully reducing this to a single-ended output would have required an operational amplifier. Doing so would reduce even-harmonic distortion in recovered audio, but would not justify the increased circuit complexity and power consumption.)

U1's oscillator amplitude is optimized to 660 mV P-P (as measured across L1) by the 220-Ω resistor at the oscillator output at pin 7. The BFO frequency is adjusted by two variable-capacitance diodes (D1 and D2) in addition to the tank coil, L1. D2 receives its control voltage via switch S3 (BFO MODE), which selects control voltage from either U3's phase detector (SYNC) or a constant voltage from a resistive divider (CW/SSB/TUNE).

The fixed CW/SSB/TUNE voltage (2.16, set by the ratio of R5 and R6) corresponds to the phase detector's optimum output voltage at lock.

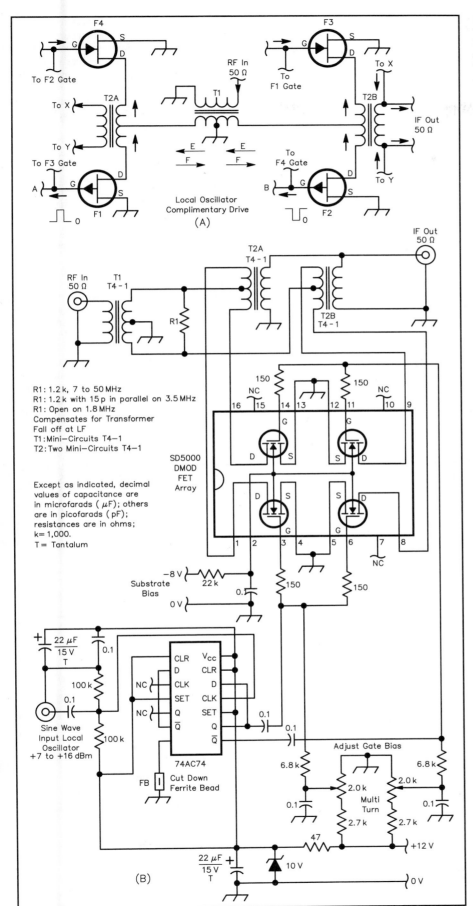

Fig 15.45 — A shows the operation of the H mode switched mixer developed by Colin Horrabin, G3SBI. B shows the actual mixer circuit implemented with the Siliconix SD5000 DMOS FET quad switch IC and a 74AC74 flip-flop.

simple IF-AGC designs. Because of the NE604AN's high gain, circuit layout can be critical and requires short leads and physically small bypass capacitors. Coupling must be minimized between pin 9 of U3 and the U1 oscillator components.

U2, an NE602AN, operates as a quasi-synchronous detector. The BFO energy it requires is readily available as a square wave at pin 9 of U3. Except for the fact that its BFO input is derived from limited input signal instead of a VCO, U2 functions the same way as U1.

Construction

Two evaluation models were constructed using ground-plane construction, mounting the ICs upside down and soldering their ground pins directly to ground with minimal lead length. The later version is constructed onto a long and narrow piece of circuit board intended to be the bottom plate of an add-on box to be fixed underside of a Sangean ATS-808 receiver. (**Fig 15.50** shows this version's general layout.)

With the circuitry arranged per

Quasi-Synchronous Detection

Synchronous detection can be mimicked by amplifying and limiting the AM signal sufficiently (at IF) so that only carrier remains, and substituting this signal for the BFO at the product detector. This *quasi-synchronous detection* acts much like envelope ("diode") detection and works best when the received signal does not fall to zero, as can often occur with SSB and, with AM, during fading. As the signal fades and the carrier-to-noise (C/N) ratio decreases, noise renders the detector's switching action inconsistent, and detection quality deteriorates rapidly. Thus, under conditions of low C/N ratio, quasi-synchronous detection exhibits a distinct detection threshold, as does a diode detector. The chief advantage of quasi-synchronous detection over simple diode rectification is its much lower threshold compared to a diode. The detector circuit presented in Fig 15.49 includes a quasi-synchronous detector for flexibility and A/B comparison with the synchronous circuit.—*Jukka Vermasvuori, OH2GF*

Fig 15.50 and receiver-detector IF-AF connections made with small-diameter co-axial cable to avoid crosstalk, BFO-signal leakage is unmeasurable at U3; that is, the voltage at RSSI does not change when the BFO is temporarily disabled under no-signal conditions.

Obtaining IF Drive

A simple BJT emitter follower (**Fig 15.51**)

can connect the synchronous detector to a solid-state transceiver. The detector can also be driven from the unbypassed cathode resistor of a vacuum-tube receiver's final IF stage (Fig 15.51B). If test equipment is not available to allow accurate measurement of the detector's input level, just keep the detector drive comfortably below that at which distortion begins.

IF signal can be obtained from the ATS-808 receiver via a 56-pF capacitor connected to the hot end of the '808's transformer T9 (at pin 16 of the '808's U1, a TA7758P IC). Connecting the detector cable detunes T9, which, though difficult to reach, must be retuned by turning its slug outwards a few turns to obtain maximum audio output. The detector's audio output (AF OUT) returns to the AT-808 by means of the '808's TONE switch, which can be rewired to select audio from the outboard detector.

The key to success with this circuit is getting *interference-free* IF drive. U3's RSSI output can be of critical importance in scoping out possible BFO leakage and/or unwanted signal input. With the detector's RF input shorted, a voltmeter connected to RSSI should indicate about 1.1. If it doesn't, U1's BFO signal may be getting into U3. The RSSI indication shouldn't be much higher than this with the detector connected to the receiver, with the receiver's RF gain control turned all the way down for minimal noise input to the detector. (The receiver used must be in "AM" mode—*BFO off*.) If the RSSI volt-

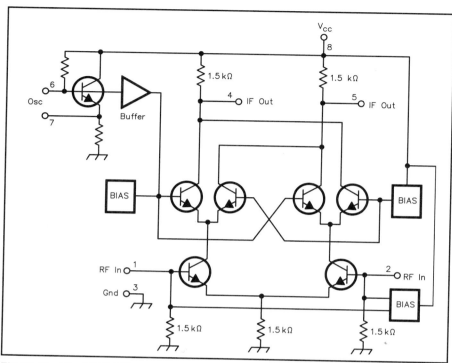

Fig 15.46—The NE602A's equivalent circuit reveals its Gilbert-cell heritage.

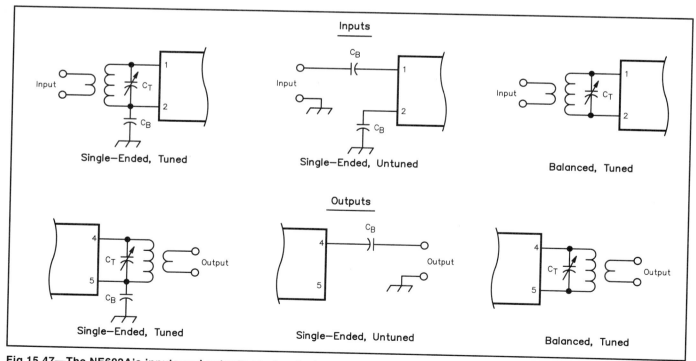

Fig 15.47—The NE602A's inputs and outputs can be single- or double-ended (balanced). The balanced configurations minimize second-order IMD and harmonic distortion, and unwanted envelope detection in direct-conversion service. C_T tunes its inductor to resonance; C_B is a bypass or dc-blocking capacitor. The arrangements pictured don't show all the possible input/output configurations; for instance, you can also use a center-tapped broadband transformer to achieve a balanced, untuned input or output.

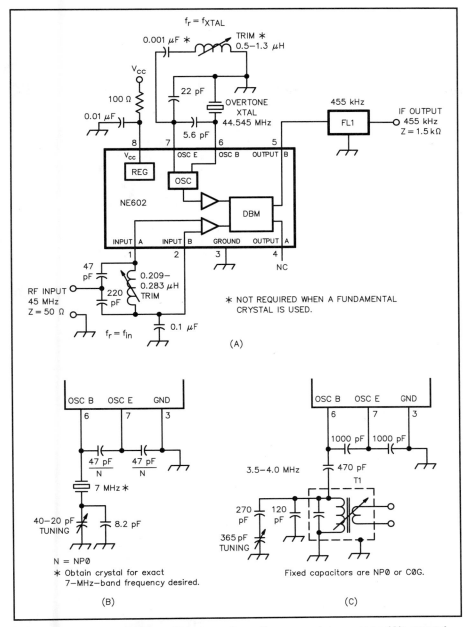

Fig 15.48—Three NE602A oscillator configurations: crystal overtone (A); crystal fundamental (B); and LC-controlled (C). T1 in C is a Mouser 10.7-MHz IF transformer, green core, 7:1 turns ratio, part no. 42IF123.

Accurately tune the receiver to a strong, pure carrier, such as a beacon. Adjust R2, BFO TUNING, for a wiper voltage of 2.00 with S2, BFO OFFSET, set to ±0 kHz. Mark as CENTER this point in its knob's travel. With the BFO MODE, switch in the SSB/CW/TUNING position, use a nonmetallic tool to adjust L1, the VCO coil, for zero beat with the incoming carrier. Returning S3 to the SYNC position should allow carrier lock if R1, SHIFT ADJ, is reasonably near adjustment. Adjust SHIFT ADJ for carrier lock if necessary. This completes coarse adjustment of SHIFT ADJ. Return the BFO MODE switch to the SSB/CW/TUNING position; the BFO should still be at or very close to zero beat with the incoming signal.

Return the BFO MODE switch to the SYNC position. After the detector locks, fine-tune SHIFT ADJ to minimize detected low-frequency hiss. (If a sufficiently strong unmodulated local signal is not available off-air, transmit into a dummy antenna with a PLL-synthesized transceiver and make this adjustment by listening to its signal. It should be possible to find a SHIFT ADJ setting at which detected hiss distinctly nulls. As a less desirable alternative, tune in an unfading AM signal modulated with a 1-kHz tone and adjust SHIFT ADJ for maximum tone recovery.) Once this is done, the detector's carrier phase is within exactly 0° or 180° of the BFO signal applied to the amplitude detector (U1), and the detector's response to phase noise has been minimized. Also significantly, this alignment procedure sets the detector to lock in a range centered on the control voltage that corresponds to optimum locking sensitivity *and* minimum phase noise demodulation. This completes alignment.

To zero-beat and lock a given station: Set the BFO TUNING control to its center (2.00 V) position, BFO MODE switch to CW/SSB/TUNE and BFO OFFSET to match the sideband(s)—LSB, USB or both—desired. Tune the receiver as close to zero beat as its tuning steps allow. Adjust BFO TUNING for zero beat. Switch the BFO MODE switch to SYNC to lock the detector.

Toggling S1, DETECTOR, between ENVELOPE and SYNC allows easy comparison of the effects of detection mode under adverse propagation conditions. The synchronous mode will be considerably superior much of the time. The quasi-synchronous (ENVELOPE) mode may give crisper audio under average or poor signal conditions; this effect may be due to increasing distortion as the signal approaches the noise floor, however.

Properly adjusted, the synchronous detector operates at less than 1% total harmonic distortion. (The quasi-synchronous detector provides comparable performance—but only on a nonfading test signal.) Measurements confirm the importance of setting R1, SHIFT ADJ, properly: Improper adjustment can increase low-order harmonic-distortion products by 3 to 12 dB in

age is above the 1-V range at this point, a receiver oscillator or some other signal, however inaudible, may be driving U3's limiter. *The detector will be unable to achieve and hold lock if anything other than the receiver IF signal captures its limiter.* The NE604's limiter stages, specified to work up to at least 21 MHz, are capable of 101 dB of overall gain and specified to be several decibels into limiting with *as little as 3 μV (−92 dBm) applied across a* 50-Ω load at the '604's input!

Operation

After checking the circuit, connect it to a 6-V power supply. The total current

consumption should be approximately 10 mA. Switch the DETECTOR switch to ENVELOPE; band noise should now be audible. Tune in a strong AM signal, switch S3, BFO MODE, to SSB/CW/TUNE, and set the DETECTOR switch to SYNC. The detector may sound very quiet at this point. Adjust L1's core until the signal swoops into audibility. This proves that the BFO is oscillating. If possible, measure the BFO level across L1 with an oscilloscope and 10:1 probe; it should be about 660 mV. (If test-equipment unavailability disallows this measurement, go to the next paragraph.) If it's not, experiment with R3's value to make it so.

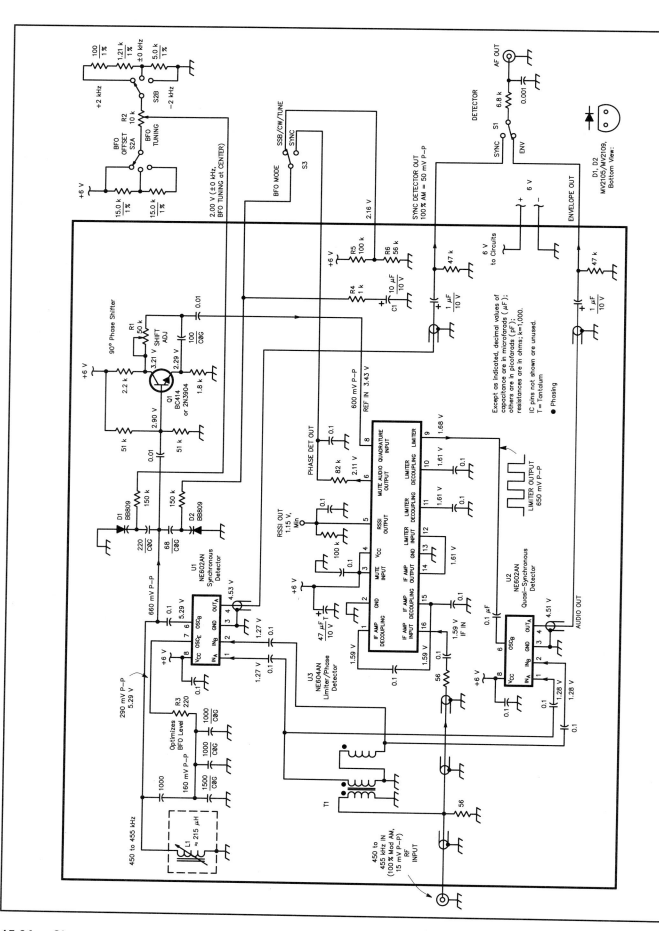

Fig 15.49—The synchronous detector operates in the 450- to 455-kHz region. Except as otherwise specified, its fixed-value resistors are ¼-W, 5%-tolerance units, and its capacitors' working voltages can be 10 or higher. See the References chapter for a list of part suppliers addresses that includes addresses of the firms mentioned below.

D1, D2—BB809 or BB409 tuning diode. Each of these, a "28-V" diode, exhibits approximately 33 pF at 2 V and an unusually high voltage-versus-capacitance slope of 10. Two paralleled 30-V Motorola tuning diodes (one MV2109 [≈45 pF at 2 V] and one MV2105 [≈18 pF at 2 V], both with a slope of 3) may serve as a substitute for each BB809 or BB409.

L1—Approximately 215 µH. Toko RWRS-T1019Z (nominally 220 µH, Q of 100 at 796 kHz, available as Digi-Key Corp no. TK1223), suitable.

R1—50-kΩ trimmer.

R2—10-kΩ linear control.

T1—13 trifilar turns of #28 enameled wire, twisted, on an FT-37-77 toroidal ferrite core.

U1, U2—Signetics NE602N, NE602AN, SA602N, SA602AN mixer/oscillator IC.

U3—Signetics NE604N, NE604AN, SA604N or SA604AN FM receiver subsystem IC.

addition to increasing the detector's sensitivity to phase noise.

REFERENCES

J. Dillon, "The Neophyte Receiver," *QST*, Feb 1988, pp 14-18. Describes a VFO-tuned, NE602-based direct-conversion receiver for 80 and 40 m.

B. Gilbert, "Demystifying the Mixer," self-published monograph, 1994.

P. Hawker, ed, "Super-Linear HF Receiver Front Ends," Technical Topics, *Radio Communication*, Sep 1993, pp 54-56.

W. Hayward, *Introduction to Radio Frequency Design* (Newington, CT: ARRL, 1994).

P. Horowitz and W. Hill, *The Art of Electronics*, 2nd ed (New York: Cambridge University Press, 1989).

S. Joshi, "Taking the Mystery Out of Double-Balanced Mixers," *QST*, Dec 1993, pp 32-36.

D. Kazdan "What's a Mixer?" *QST*, Aug 1992, pp 39-42.

J. Makhinson, "High Dynamic Range MF/HF Receiver Front End," *QST*, Feb 1993, pp 23-28. Also see Feedback, *QST*, Jun 1993, p 73.

RF/IF Designer's Handbook (Brooklyn, NY: Scientific Components, 1992).

RF Communications Handbook (Philips Components-Signetics, 1992).

L. Richey, "W1AW at the Flick of a Switch," New Ham Horizons, *QST*, Feb 1993, pp 56-57. Describes a crystal-controlled NE602 receiver buildable for 80, 40 and 20 m.

U. Rohde and T. Bucher, *Communications Receivers: Principles and Design* (New York: McGraw-Hill Book Co, 1988).

U. Rohde, "Key Components of Modern Receiver Design," Part 1, *QST*, May 1994, pp 29-32; Part 2, *QST*, Jun 1994, pp 27-31; Part 3, *QST*, Jul 1994, 42-45.

U. Rohde, "Testing and Calculating Intermodulation Distortion in Receivers," *QEX*, Jul 1994, pp 3-4.

W. Sabin, "The Solid-State Receiver," *QST*, Jul 1970, pp 35-43.

Synergy Microwave Corporation product handbook (Paterson, NJ: Synergy Microwave Corp, 1992).

R. Weinreich and R. W. Carroll, "Absorptive Filter for TV Harmonics," *QST*, Nov 1968, pp 20-25.

R. Zavrel, "Using the NE602," Technical Correspondence, *QST*, May 1990, pp 38-39.

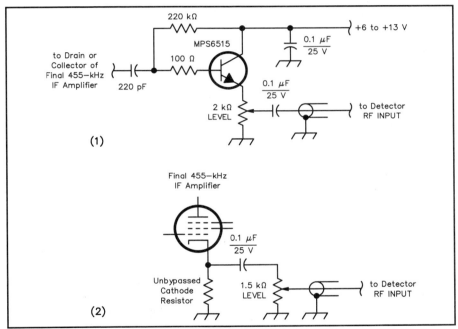

Fig 15.51—Simple circuitry can connect the synchronous detector to a solid-state transceiver (1) or a vacuum-tube receiver (2) if a 455-kHz output tap is not already available.

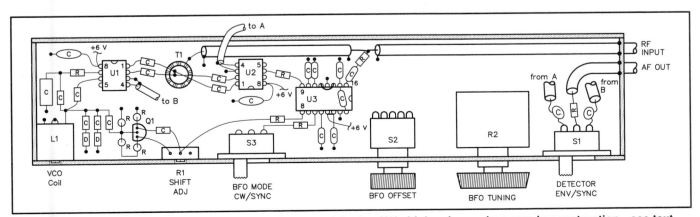

Fig 15.50—One recommended layout for the synchronous detector. U3's high gain requires care in construction—see text.

Contents

Filters and Projects

16

This chapter contains basic design information and examples of the most common filters used by radio amateurs. It was prepared by Reed Fisher, W2CQH, and includes a number of design approaches, tables and filters by Ed Wetherhold, W3NQN, and others. The chapter is divided into two major sections. The first section contains a discussion of filter theory with some design examples. It includes the tools needed to predict the performance of a candidate filter before a design is started or a commercial unit purchased. Extensive references are given for further reading and design information. The second section contains a number of selected practical filter designs for immediate construction.

Basic Concepts

A filter is a network that passes signals of certain frequencies and rejects or attenuates those of other frequencies. The radio art owes its success to effective filtering. Filters allow the radio receiver to provide the listener with only the desired signal and reject all others. Conversely, filters allow the radio transmitter to generate only one signal and attenuate others that might interfere with other spectrum users.

The simplified SSB receiver shown in **Fig 16.1** illustrates the use of several common filters. Three of them are located between the antenna and the speaker. They provide the essential receiver filter functions. A preselector filter is placed between the antenna and the first mixer. It passes all frequencies between 3.8 and 4.0 MHz with low loss. Other frequencies, such as out-of-band signals, are rejected to prevent them from overloading the first mixer (a common problem with shortwave broadcast stations). The preselector filter is almost always built with LC filter technology.

An intermediate frequency (IF) filter is placed between the first and second mixers. It is a band-pass filter that passes the desired SSB signal but rejects all others. The age of the receiver probably determines which of several filter technologies is used. As an example, 50 kHz or 455 kHz

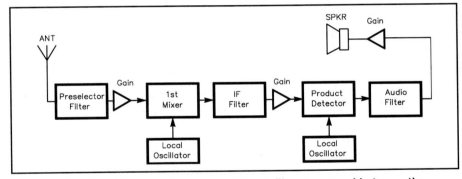

Fig 16.1—One-band SSB receiver. At least three filters are used between the antenna and speaker.

LC filters and 455 kHz mechanical filters were used through the 1960s. Later model receivers usually use quartz crystal filters with center frequencies between 3 and 9 MHz. In all cases, the filter bandwidth must be less than 3 kHz to effectively reject adjacent SSB stations.

Finally, a 300 Hz to 3 kHz audio band-pass filter is placed somewhere between the detector and the speaker. It rejects unwanted products of detection, power supply hum and noise. Today this audio filter is usually implemented with active filter technology.

The complementary SSB transmitter block diagram is shown in **Fig 16.2**. The same array of filters appear in reverse order.

First is a 300-Hz to 3-kHz audio filter, which rejects out-of-band audio signals such as 60-Hz power supply hum. It is placed between the microphone and the balanced mixer.

The IF filter is next. Since the balanced mixer generates both lower and upper sidebands, it is placed at the mixer output to pass only the desired lower (or upper) sideband. In commercial SSB transceivers this filter is usually the same as the IF filter used in the receive mode.

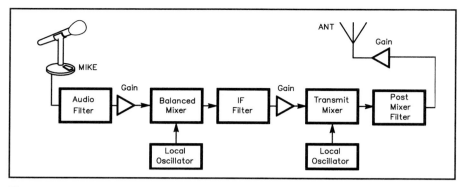

Fig 16.2—One-band transmitter. At least three filters are needed to ensure a clean transmitted signal.

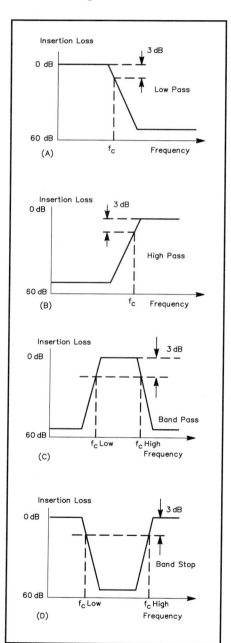

Fig 16.3—Idealized filter responses. Note the definition of f_c is 3 dB down from the break points of the curves.

Finally, a 3.8 to 4.0-MHz band-pass filter is placed between transmit mixer and antenna to reject unwanted frequencies generated by the mixer and prevent them from being amplified and transmitted.

This chapter will discuss the four most common types of filters: low-pass, high-pass, band-pass and band-stop. The idealized characteristics of these filters are shown in their most basic form in **Fig 16.3**.

A low-pass filter permits all frequencies below a specified cutoff frequency to be transmitted with small loss, but will attenuate all frequencies above the cutoff frequency. The "cutoff frequency" is usually specified to be that frequency where the filter loss is 3 dB.

A high-pass filter has a cutoff frequency above which there is small transmission loss, but below which there is considerable attenuation. Its behavior is opposite to that of the low-pass filter.

A band-pass filter passes a selected band of frequencies with low loss, but attenuates frequencies higher and lower than the desired passband. The passband of a filter is the frequency spectrum that is conveyed with small loss. The transfer characteristic is not necessarily perfectly uniform in the passband, but the variations usually are small.

A band-stop filter rejects a selected band of frequencies, but transmits with low loss frequencies higher and lower than the desired stop band. Its behavior is opposite to that of the band-pass filter. The stop band is the frequency spectrum in which attenuation is desired. The attenuation varies in the stop band rising to high values at frequencies far removed from the cutoff frequency.

FILTER FREQUENCY RESPONSE

The purpose of a filter is to pass a desired frequency (or frequency band) and reject all other undesired frequencies. A simple single-stage low-pass filter is shown in **Fig 16.4**. The filter consists of an inductor, L. It is placed between the volt-

age source e_g and load resistance R_L. Most generators have an associated "internal" resistance, which is labeled R_g.

When the generator is switched on, power will flow from the generator to the load resistance R_L. The purpose of this low-pass filter is to allow maximum power flow at low frequencies (below the cutoff frequency) and minimum power flow at high frequencies. Intuitively, frequency filtering is accomplished because the inductor has reactance that vanishes at dc but becomes large at high frequency. Thus, the current, I, flowing through the load resistance, R_L, will be maximum at dc and less at higher frequencies.

The mathematical analysis of Fig 16.4 is as follows: For simplicity, let $R_g = R_L = R$.

$$i = \frac{e_g}{2R + jX_L} \tag{1}$$

where
$$X_L = 2\pi f L$$
$$f = \text{generator frequency.}$$

Power in the load, P_L, is:

$$P_L = \frac{e_g^2 R_L}{4R^2 + X_L^2} \tag{2}$$

Available (maximum) power will be delivered from the generator when:

$$X_L = 0 \text{ and } R_g = R_L$$

$$P_O = \frac{E_g^2}{4R_g} \tag{3}$$

The filter response is:

$$\frac{P_L}{P_O} = \frac{\text{power in the load}}{\text{available generator power}} \tag{4}$$

The filter cutoff frequency, called f_c, is the generator frequency where

$$2R = X_L \text{ or } f_c = \frac{R}{\pi L} \tag{5}$$

As an example, suppose $R_g = R_L = 50 \, \Omega$

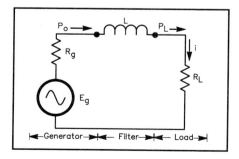

Fig 16.4—A single-stage low-pass filter consists of a series inductor. DC is passed to the load resistor unattenuated. Attenuation increases (and current in the load decreases) as the frequency increases.

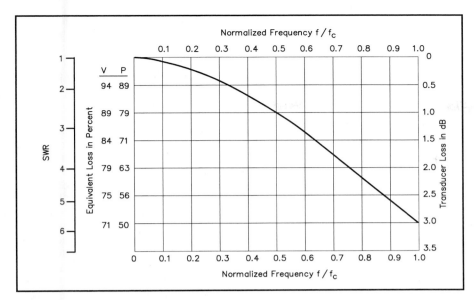

Fig 16.5—Transmission loss of a simple filter plotted against normalized frequency. Note the relationship between loss and SWR.

and the desired cutoff frequency is 4 MHz. Equation 4 states that the cutoff frequency is where the inductive reactance $X_L = 100\ \Omega$. At 4 MHz, using the relationship $X_L = 2\pi f L$, $L = 4\ \mu H$. If this filter is constructed, its response should follow the curve in **Fig 16.5**. Note that the gentle rolloff in response indicates a poor filter. To obtain steeper rolloff a more sophisticated filter, containing more reactances, is necessary. Filters are designed for a specific value of purely resistive load impedance called the *terminating resistance*. When such a resistance is connected to the output terminals of a filter, the impedance looking into the input terminals will equal the load resistance throughout most of the passband. The degree of mismatch across the passband is shown by the SWR scale at the left-hand side of Fig 16.5. If maximum power is to be extracted from the generator driving the filter, the generator resistance must equal the load resistance. This condition is called a "doubly terminated" filter. Most passive filters, including the LC filters described in this chapter, are designed for double termination. If a filter is not properly terminated, its passband response changes.

Certain classes of filters, called "transformer filters" or "matching networks" are specifically designed to work between unequal generator and load resistances. Band-pass filters, described later, are easily designed to work between unequal terminations.

All passive filters exhibit an undesired nonzero loss in the passband due to unavoidable resistances associated with the

reactances in the ladder network. All filters exhibit undesired transmission in the stop band due to leakage around the filter network. This phenomenon is called the "ultimate rejection" of the filter. A typical high-quality filter may exhibit an ultimate rejection of 60 dB.

Band-pass filters perform most of the important filtering in a radio receiver and transmitter. There are several measures of their effectiveness or *selectivity*. Selectivity is a qualitative term that arose in the 1930s. It expresses the ability of a filter (or the entire receiver) to reject unwanted adjacent signals. There is no mathematical measure of selectivity.

The term *Q* is quantitative. A band-pass filter's *quality factor* or Q is expressed as Q = (filter center frequency)/(3-dB bandwidth). *Shape factor* is another way some filter vendors specify band-pass filters. The shape factor is a ratio of two filter bandwidths. Generally, it is the ratio (60-dB bandwidth) / (6-dB bandwidth), but some manufacturers use other bandwidths. An ideal or *brick-wall* filter would have a shape factor of 1, but this would require an infinite number of filter elements. The IF filter in a high-quality receiver may have a shape factor of 2.

POLES AND ZEROS

In equation 1 there is a frequency called the "pole" frequency that is given by $f_p = 0$.

In equation 1 there also exists a frequency where the current i becomes zero. This frequency is called the *zero frequency* and is given by: $f_0 =$ infinity. Poles and zeros are intrinsic properties of all net-

works. The poles and zeros of a network are related to the values of inductances and capacitances in the network.

Poles and zero locations are of interest to the filter theorist because they allow him to predict the frequency response of a proposed filter. For low-pass and high-pass filters the number of poles equals the number of reactances in the filter network. For band-pass and band-stop filters the number of poles specified by the filter vendors is usually taken to be half the number of reactances.

LC FILTERS

Perhaps the most common filter found in the Amateur Radio station is the inductor-capacitor (LC) filter. Historically, the LC filter was the first to be used and the first to be analyzed. Many filter synthesis techniques use the LC filter as the mathematical model.

LC filters are usable from dc to approximately 1 GHz. Parasitic capacitance associated with the inductors and parasitic inductance associated with the capacitors make applications at higher frequencies impractical because the filter performance will change with the physical construction and therefore is not totally predictable from the design equations. Below 50 or 60 Hz, inductance and capacitance values of LC filters become impractically large.

Mathematically, an LC filter is a linear, lumped-element, passive, reciprocal network. Linear means that the ratio of output to input is the same for a 1-V input as for a 10-V input. Thus, the filter can accept an input of many simultaneous sine waves without intermodulation (mixing) between them.

Lumped-element means that the inductors and capacitors are physically much smaller than an operating wavelength. In this case, conductor lengths do not contribute significant inductance or capacitance, and the time that it takes for signals to pass through the filter is insignificant. (Although the different times that it takes for different frequencies to pass through the filter—known as group delay— is still significant for some applications.)

The term *passive* means that the filter does not need any internal power sources. There may be amplifiers before and/or after the filter, but no power is necessary for the filter's equations to hold. The filter alone always exhibits a finite (nonzero) insertion loss due to the unavoidable resistances associated with inductors and (to a lesser extent) capacitors. Active filters, as the name implies, contain internal power sources.

Reciprocal means that the filter can pass power in either direction. Either end of the filter can be used for input or output.

TIME DOMAIN VS FREQUENCY DOMAIN

Humans think in the time domain. Life experiences are measured and recorded in the stream of time. In contrast, Amateur Radio systems and their associated filters are often better understood when viewed in the frequency domain, where frequency is the relevant system parameter. *Frequency* may refer to a sine-wave voltage, current or electromagnetic field. The sine-wave voltage, shown in **Fig 16.6**, is a waveform plotted against time with equation $V = A \sin(2\pi f t)$. The sine wave has a peak amplitude A (measured in volts) and frequency, f (measured in cycles/second or Hertz). A graph showing frequency on the horizontal axis is called a spectrum. A filter response curve is plotted on a spectrum graph.

Historically, radio systems were best analyzed in the frequency domain. The radio transmitters of Hertz (1865) *and* Marconi (1895) consisted of LC resonant circuits excited by high-voltage spark gaps. The transmitters emitted packets of damped sine waves. The low-frequency (200-kHz) antennas used by Marconi were found to possess very narrow bandwidths, and it seemed natural to analyze antenna performance using sine-wave excitation. In addition, the growing use of 50 and 60-Hz alternating current (ac) electric

power systems in the 1890s demanded the use of sine-wave mathematics to analyze these systems. Thus engineers trained in ac power theory were available to design and build the early radio systems.

In the frequency domain, the radio world is imagined to be composed of many sine waves of different frequencies flowing endlessly in time. It can be shown by the Fourier transform (Ref 7) that all periodic waveforms can be represented by summing sine waves of different frequencies. For example, the square-wave voltage shown in **Fig 16.7** can be represented by a "fundamental" sine wave of frequency $f = 1/t$ and all its odd harmonics: 3f, 5f, 7f and so on. Thus, in the frequency domain a sine wave is a *narrowband* signal (zero bandwidth) and a square wave is a "wideband" signal.

If the square-wave voltage of Fig 16.7 is passed through a low-pass filter, which removes some of its high-frequency components, the waveform of **Fig 16.8** results. The filtered square wave now has a rise time, which is the time required to rise from 10% to 90% of its peak value (A). The rise time is approximately:

$$\tau_R = \frac{0.35}{f_c} \qquad (6)$$

where f_c is the cutoff frequency of the low-pass filter.

Thus a filter distorts a time-domain signal by removing some of its high-frequency components. Note that a filter cannot distort a sine wave. A filter can only change the amplitude and phase of sine waves. A linear filter will pass multiple sine waves without producing any intermodulation or "beats" between frequencies—this is the definition of *linear*.

The purpose of a radio system is to convey a time-domain signal originating at a source to some distant point with minimum distortion. Filters within the radio system transmitter and receiver may intentionally or unintentionally distort the source signal. A knowledge of the source signal's frequency-domain bandwidth is required so that an appropriate radio system may be designed.

Table 16.1 shows the minimum necessary bandwidth of several common source signals. Note that high-fidelity speech and music requires a bandwidth of 20 Hz to 15 kHz, which is that transmitted by high-quality FM broadcast stations. However, telephone-quality speech requires a bandwidth of only 200 Hz to 3 kHz. Thus, to

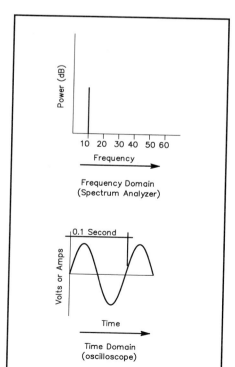

Fig 16.6—Ideal sine-wave voltage. Only one frequency is present.

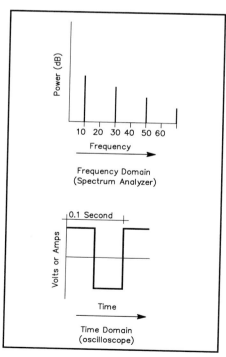

Fig 16.7—Square-wave voltage. Many frequencies are present, including f = 1/t and odd harmonics 3f, 5f, 7f with decreasing amplitudes.

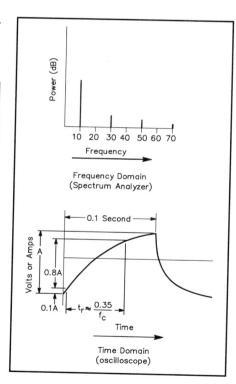

Fig 16.8—Square-wave voltage filtered by a low-pass filter. By passing the square wave through a filter, the higher frequencies are attenuated. The rectangular shape (fast rise and fall items) are *rounded* because the amplitude of the higher harmonics are decreased.

Table 16.1

Table 16.1

Typical Filter Bandwidths for Typical Signals.

Source	Required Bandwidth
High-fidelity speech and music	20 Hz to 15 kHz
Telephone-quality speech	200 Hz to 3 kHz
Radiotelegraphy (Morse code, CW)	200 Hz
HF RTTY	1000 Hz (varies with frequency shift)
NTSC television	60 Hz to 4.5 MHz
SSTV	200 Hz to 3 kHz
1200 bit/s packet	200 Hz to 3 kHz

minimize transmit spectrum, as required by the FCC, filters within amateur transmitters are required to reduce the speech source bandwidth to 200 Hz to 3 kHz at the expense of some speech distortion. After modulation the transmitted RF bandwidth will exceed the filtered source bandwidth if inefficient (AM or FM) modulation methods are employed. Thus the post-modulation *emission bandwidth* may be several times the original filtered source bandwidth. At the receiving end of the radio link, band-pass filters are required to accept only the desired signal and sharply reject noise and adjacent channel interference.

As human beings we are accustomed to operation in the time domain. Just about all of our analog radio connected design occurs in the frequency domain. This is particularly true when it comes to filters. Although the two domains are convertible, one to the other, most filter design is performed in the frequency domain.

Filter Synthesis

The image-parameter method of filter design was initiated by O. Zobel (Ref 1) of Bell Labs in 1923. Image-parameter filters are easy to design and design techniques are found in earlier editions of the ARRL *Handbook*. Unfortunately, image parameter theory demands that the filter terminating impedances vary with frequency in an unusual manner. The later addition of "m-derived matching half sections" at each end of the filter made it possible to use these filters in many applications. In the intervening decades, however, many new methods of filter design have brought both better performance and practical component values for construction.

MODERN FILTER THEORY

The start of modern filter theory is usually credited to S. Butterworth and S. Darlington (Refs 3 and 4). It is based on this approach: Given a desired frequency response, find a circuit that will yield this response.

Filter theorists were aware that certain known mathematical polynomials had "filter like" properties when plotted on a frequency graph. The challenge was to match the filter components (L, C and R) to the known polynomial poles and zeros. This pole/zero matching was a difficult task before the availability of the digital computer. Weinberg (Ref 5) was the first to publish computer generated tables of normalized low-pass filter component values. ("Normalized" means 1-Ω resistor terminations and cutoff frequency $\omega_c = 2\pi f_c = 1$ radian/s.)

An ideal low-pass filter response shows no loss from zero frequency to the cutoff frequency, but infinite loss above the cutoff frequency. Practical filters may approximate this ideal response in several different ways.

Fig 16.9 shows the Butterworth or "maximally-flat" type of approximation. The Butterworth response formula is:

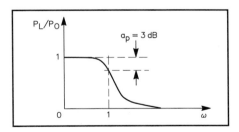

Fig 16.9—Butterworth approximation of an ideal low-pass filter response. The 3-dB attenuation frequency (f_C) is normalized to 1 radian/s.

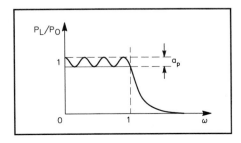

Fig 16.10—Chebyshev approximation of an ideal low-pass filter. Notice the ripple in the passband.

$$\frac{P_L}{P_O} = \frac{1}{1 + \left(\dfrac{\omega}{\omega_c}\right)^{2n}} \quad (7)$$

where

ω = frequency of interest,
ω_c = cutoff frequency
n = number of poles (reactances)
P_L = power in the load resistor
P_O = available generator power.

The passband is exceedingly flat near zero frequency *and* very high attenuation is experienced at high frequencies, but the approximation for both pass and stop bands is relatively poor in the vicinity of cutoff.

Fig 16.10 shows the Chebyshev approximation. Details of the Chebyshev response formula can be found in Ref 24. Use of this reference as well as similar references for Chebyshev filters requires detailed familiarity with Chebyshev polynomials.

IMPEDANCE AND FREQUENCY SCALING

Fig 16.11A shows normalized component values for Butterworth filters up to ten poles. Fig 16.11B shows the schematic diagrams of the Butterworth low-pass filter. Note that the first reactance in Fig 16.11B is a shunt capacitor C1, whereas in Fig 16.11C the first reactance is a series inductor L1. Either configuration can be used, but a design using fewer inductors is usually chosen.

In filter design, the use of *normalized* values is common. Normalized generally

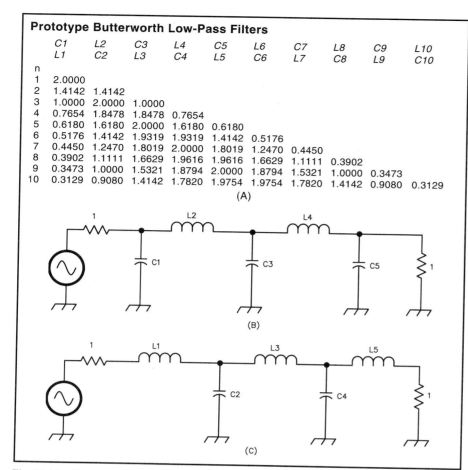

Prototype Butterworth Low-Pass Filters

n	C1 / L1	L2 / C2	C3 / L3	L4 / C4	C5 / L5	L6 / C6	C7 / L7	L8 / C8	C9 / L9	L10 / C10
1	2.0000									
2	1.4142	1.4142								
3	1.0000	2.0000	1.0000							
4	0.7654	1.8478	1.8478	0.7654						
5	0.6180	1.6180	2.0000	1.6180	0.6180					
6	0.5176	1.4142	1.9319	1.9319	1.4142	0.5176				
7	0.4450	1.2470	1.8019	2.0000	1.8019	1.2470	0.4450			
8	0.3902	1.1111	1.6629	1.9616	1.9616	1.6629	1.1111	0.3902		
9	0.3473	1.0000	1.5321	1.8794	2.0000	1.8794	1.5321	1.0000	0.3473	
10	0.3129	0.9080	1.4142	1.7820	1.9754	1.9754	1.7820	1.4142	0.9080	0.3129

(A)

(B)

(C)

Fig 16.11—Component values for Butterworth low-pass filters. Greater values of n require more stages.

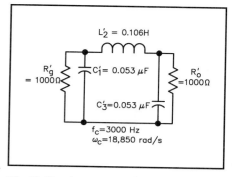

Fig 16.13—A 3-pole Butterworth filter scaled to 3000 Hz.

The normalized prototype, taken from Fig 16.11B is shown in **Fig 16.12**. The new (desired) inductor value is:

$$L' = \left(\frac{1000\ \Omega}{1\ \Omega}\right)\left(\frac{1\,\text{radian}/\text{second}}{2\pi(3000)\,\text{Hz}}\right)2\ \text{H}$$

or L' = 0.106 H.

The new (desired) capacitor value is:

$$C' = \left(\frac{1\ \Omega}{1000\,\Omega}\right)\left(\frac{1\,\text{radian}/\text{second}}{2\pi(3000)\,\text{Hz}}\right)1\ \text{F}$$

or C' = 0.053 µF.

The final denormalized filter is shown in **Fig 16.13**. The filter response, in the passband, should obey curve n = 3 in **Fig 16.14**. To use the normalized frequency response curves in Fig 16.14 calculate the frequency ratio f/f_c where f is the desired frequency and f_c is the cutoff frequency. For the filter just designed, the loss at 2000 Hz can be found as follows: When f is 2000 Hz, the frequency ratio is:

means a design based on 1-Ω terminations and a cutoff frequency (passband edge) of 1 radian/second. A filter is *denormalized* by applying the following two equations:

$$L' = \left(\frac{R'}{R}\right)\left(\frac{\omega}{\omega'}\right)L \tag{8}$$

$$C' = \left(\frac{R}{R'}\right)\left(\frac{\omega}{\omega'}\right)C \tag{9}$$

where

L', C', ω' and R' are the new (desired) values

L and C are the values found in the filter tables

R = 1 Ω

ω = 1 radian/s.

For example, consider the design of a 3-pole Butterworth low-pass filter for a transmitter speech amplifier. Let the desired cutoff frequency be 3000 Hz and the desired termination resistances be 1000 Ω.

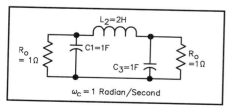

Fig 16.12—A 3-pole Butterworth filter designed for a normalized frequency of 1 radian/s.

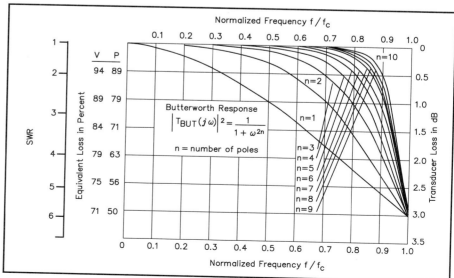

Fig 16.14—Passband loss of Butterworth low-pass filters. The horizontal axis is normalized frequency (see text).

$f/f_c = 2000/3000 = 0.67$. Therefore the predicted loss (from the n = 3 curve) is about 0.37 dB.

When f is 4000 Hz, the filter is operating in the stop-band (Fig 16.17). The resulting frequency ratio is: $f/f_c = 4000/3000 = 1.3$. Therefore the expected loss is about 8 dB. Note that as the number of reactances (poles) increases the filter response approaches the low-pass response of Fig 16.3A.

BAND-PASS FILTERS— SIMPLIFIED DESIGN

The design of band-pass filters may be directly obtained from the low-pass prototype by a frequency translation. The low-pass filter has a "center frequency" (in the parlance of band-pass filters) of 0 Hz. *The frequency translation from 0 Hz to the band-pass filter center frequency, f, is obtained by replacing in the low-pass prototype all shunt capacitors with parallel tuned circuits and all series inductors with series tuned circuits.*

As an example, suppose a band-pass filter is required at the front end of a home-brew 40-m QRP receiver to suppress powerful adjacent broadcast stations. The proposed filter has these characteristics:

- Center frequency, f_c = 7.15 MHz
- 3-dB bandwidth = 360 kHz
- terminating resistors = 50 Ω
- 3-pole Butterworth characteristic.

Start the design for the normalized 3-pole Butterworth low-pass filter (shown in Fig 16.11). First determine the center frequency from the band-pass limits. This frequency, f_O, is found by determining the geometric mean of the band limits. In this case the band limits are 7.15 + 0.360/2 = 7.33 MHz and 7.15 −0.360/2 = 6.97 MHz; then

$$f_O = \sqrt{f_{lo} \times f_{hi}} = \sqrt{6.97 \times 7.33} = 7.14 \text{ MHz} \quad (10)$$

where

f_{lo} = low frequency end of the band-pass (or band-stop)

f_{hi} = high frequency end of the band-pass (or band-stop)

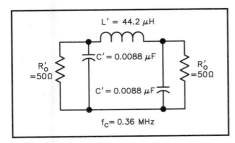

Fig 16.15—Interim 3-pole Butterworth low-pass filter designed for cutoff at 0.36 MHz.

[Note that in this case there is little difference between 7.15 (bandwidth center) and 7.147 (band-edge geometric mean) because the bandwidth is small. For wide-band filters, however, there can be a significant difference.]

Next, denormalize to a new interim low-pass filter having R' = 50 Ω and f' = 0.36 MHz.

$$L' = \left(\frac{50}{1}\right)\left(\frac{1}{2 \times \pi \times 0.36 \times 10^6}\right)2\,H = 44.2\,\mu H$$

$$C' = \left(\frac{1}{50}\right)\left(\frac{1}{2 \times \pi \times 0.36 \times 10^6}\right)1\,F = 0.0088\,\mu F$$

This interim low-pass filter, shown in **Fig 16.15**, has a cutoff frequency $f_c = 0.36$ MHz and is terminated with 50-Ω resis-

tors. The desired 7.147-MHz band-pass filter is achieved by parallel resonating the shunt capacitors with inductors and series resonating the series inductor with a series capacitor. All resonators must be tuned to the center frequency. Therefore, variable capacitors or inductors are required for the resonant circuits. Based on the L' and C' just calculated the parallel-resonating inductor values are:

$$L1 = L3 = \frac{1}{C'(2 \times \pi\, f_o)^2} = 0.056\,\mu H$$

The series-resonating capacitor value is:

$$C2 = \frac{1}{L'(2 \times \pi \times f_O)^2} = 11\,pF$$

The final band-pass filter is shown in **Fig 16.16**. The filter should have a 3-dB bandwidth of 0.36 MHz. That is, the

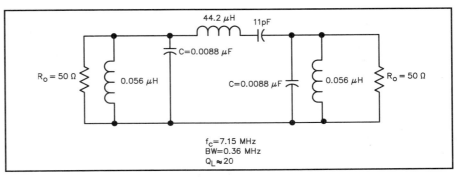

Fig 16.16—Final filter design consists of the low-pass filter scaled to a center frequency of 7.15 MHz.

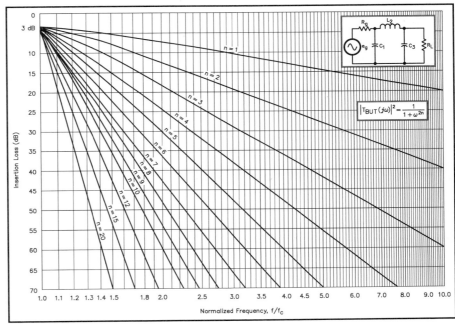

Fig 16.17—Stop-band loss of Butterworth low-pass filters. The almost vertical angle of the lines representing filters with high values of n (10, 12, 15, 20) show the slope of the filter will be very high (sharp cutoff).

3-dB loss frequencies are 6.97 MHz and 7.33 MHz. The filter's loaded Q is: Q = 7.147/0.36 or approximately 20.

The filter response, in the passband, falls on the "n = 3" curve in **Fig 16.17**. To use the normalized frequency response curves, calculate the frequency ratio f/f_c. For this band-pass case, f is the difference between the desired attenuation frequency and the center frequency, while f_c is the upper 3-dB frequency minus the center frequency. As an example the filter loss at 7.5 MHz is found by using the normalized frequency ratio given by:

$$\frac{f}{f_c} = \frac{7.5 - 7.147}{7.33 - 7.147} = 1.928$$

Therefore, from Fig 16.17 the expected loss is about 17 dB.

At 6 MHz the loss may be found by:

$$\frac{f}{f_c} = \frac{7.147 - 6}{7.33 - 7.147} = 6.26$$

The expected loss is approximately 47 dB. Unfortunately, awkward component values occur in this type of band-pass filter. The series resonant circuit has a very large LC ratio and the parallel resonant circuits have very small LC ratios. The situation worsens as the filter loaded Q_L ($Q_L = f_0/BW$) increases. Thus, this type of band-pass filter is generally used with a loaded Q less than 10.

Good examples of low-Q band-pass filters of this type are demonstrated by W3NQN's High Performance CW Filter and Passive Audio Filter for SSB in the 1995 and earlier editions of this handbook.

[Note: This analysis used the geometric f_c with the assumption that the filter response is symmetrical about f_c, which it is not. A more rigorous analysis yields 16.9 dB at 7.5 MHz and 50.7 dB at 6 MHz.—Ed.]

Q Restrictions—Band-pass Filters

Most filter component value tables assume lossless reactances. In practice, there are always resistance losses associated with capacitors and inductors (especially inductors). Lossy reactances in low-pass filters modify the response curve. There is finite loss at zero frequency and the cutoff "knee" at f_c will not be as sharp as predicted by theoretical response curves.

The situation worsens with band-pass filters. As loaded Q is increased the midband insertion loss may become intolerable. Therefore, before a band-pass filter design is started, estimate the expected loss.

An approximate estimate of band-pass filter midband response is given by:

$$\frac{P_L}{P_O} = \left(1 - \frac{Q_L}{Q_U}\right)^{2N} \quad (11A)$$

where:

P_L = power delivered to load resistor R_L

P_O = power available from generator:

$$P_O = \frac{e_g^2}{4R_L} \quad (11B)$$

Q_U = unloaded Q of inductor:

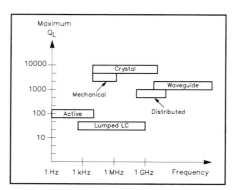

Fig 16.18—Frequency range and maximum loaded Q of band-pass filters. Crystal filters are shown with the highest Q_L and LC filters the lowest.

$$Q_U = \frac{2\pi \times f_0 \times L}{R} \quad (11C)$$

R = inductor series resistance
L = inductance
Q_L = filter loaded Q

$$Q_L = \frac{f_0}{BW_3} \quad (11D)$$

BW_3 = 3 dB bandwidth
N = number of filter stages.

This equation assumes that all losses are in the inductors. For example, the expected loss of the 7.15-MHz filter shown in Fig 16.16 is found by assuming $Q_U = 150$. Q_L is found by equation 11D to be = 7.147/0.36 = 19.8 or approximately 20. Since N = 3 then:

$$\frac{P_L}{P_O} = \left(1 - \frac{20}{150}\right)^6$$

from equation (11A), which equals 0.423. Expressed as dB this is equal to 10 log (0.423) = – 3.73 dB.

Therefore this filter may not be suitable for some applications. If the insertion loss is to be kept small there are severe restrictions on Q_L/Q_U. With typical lumped inductors Q_U seldom exceeds 200. Therefore, LC band-pass filters are usually designed with Q_L not exceeding 20 as shown in **Fig 16.18**.

This loss vs bandwidth trade-off is usually why the final intermediate frequency (IF) in older radio receivers was very low. These units used the equivalent of LC filters in their IF coupling. Generally, for SSB reception the desired receiver bandwidth is about 2.5 kHz. Then 50 kHz was often chosen as the final IF since this implies a loaded Q_L of 20. AM broadcast receivers require a 10-kHz bandwidth and use a 455-kHz IF, which results in $Q_L = 45$. FM broadcast receivers require a 200-kHz bandwidth and use a 10.7-MHz IF and $Q_L = 22$.

Filter Design Using Standard Capacitor Values

Practical filters must be designed using commercially available components. Therefore a set of tables, based upon *standard value capacitors* (SVC), has been generated to facilitate this real design process. The procedure presented here uses eight computer-calculated tables of performance parameters and component values for 5- and 7-branch Chebyshev and 5-branch elliptic 50-Ω filters. The tables permit the quick and easy selection of an equally terminated passive LC filter for applications where the attenuation response is of primary interest. All of the capacitors in the Chebyshev designs and the three nonresonating capacitors in the elliptic designs have standard, off-the-shelf values to simplify construction. Although the tables cover only the 1 to 10-MHz frequency range, a simple scaling procedure gives standard-value capacitor (SVC) designs for any impedance level and virtually any cutoff frequency.

The full tables are printed in the **References** chapter of this *Handbook*. Extracts from the tables are reprinted in this section to illustrate the design procedure.

The following text by Ed Wetherhold, W3NQN, is adapted from his paper entitled *Simplified Passive LC Filter Design for the EMC Engineer*. It was presented at an IEEE International Symposium on Electromagnetic Compatibility in 1985.

The approach is based upon the fact that for most nonstringent filtering applications, it is not necessary that the actual cutoff frequency exactly match the desired cutoff frequency. A deviation of 5% or so between the actual and desired cutoff frequencies is acceptable. This permits the use of design tables based on standard capacitor values instead of passband ripple attenuation or reflection coefficient.

STANDARD VALUES IN FILTER DESIGN CALCULATIONS

Capacitors are commercially available in special series of preferred values having designations of E12 (10% tolerance) and E24 (5% tolerance; Ref 22) The reciprocal of the E-number is the power to which 10 is raised to give the step multiplier for that particular series.

First the normalized Chebyshev and elliptic component values are calculated based on many ratios of standard capacitor values. Next, using a 50-Ω impedance level, the parameters of the designs are calculated and tabulated to span the 1-10 MHz decade. Because of the large number of standard-value capacitor (SVC) designs in this decade, the increment in cutoff frequency from one design to the next is sufficiently small so that virtually any cutoff frequency requirement can be satisfied. Using such a table, the selection of an appropriate design consists of merely scanning the cutoff frequency column to find a design having a cutoff frequency that most closely matches the desired cutoff frequency.

CHEBYSHEV AND ELLIPTIC FILTERS

Low-pass and high-pass 5- and 7-element Chebyshev and 5-branch elliptic designs were selected for tabulation because they are easy to construct and will satisfy the majority of nonstringent filtering requirements where the amplitude response is of primary interest. The pre-calculated 50-Ω designs are presented in eight tables of five low-pass and three high-pass designs with cutoff frequencies covering the 1-10 MHz decade. The applicable filter configuration and attenuation response curve accompany each table. In addition to the component values, attenuation vs frequency data and SWR are also included in the table. The passband attenuation ripples are so low in amplitude that they are swamped by the filter losses and are not measurable. For this reason, they are not shown in the response curves.

LOW-PASS TABLES

Fig 16.19 is an extract from the tables for the low-pass 5- and 7-element Chebyshev capacitor input/output configuration in the **References** chapter of this *Handbook*. This filter configuration is generally preferred to the alternate inductor input/output configuration because it requires fewer inductors. Generally, decreasing input impedance with increasing frequency in the stop band presents no problems. **Fig 16.20** shows the corresponding information for low-pass applications, but with an inductor input/output configuration. This configuration is useful when the filter input impedance in the stop band must rise with increasing frequency. For example, some RF transistor amplifiers may become unstable when terminated in a low-pass filter having a stop-band response with a decreasing input impedance. In this case, the inductor-input configuration may eliminate the instability. (Ref 23) Because only one capacitor value is required in the designs of Fig 16.20, it was feasible to have the inductor value of L1 and L5 also be a standard value. **Fig 16.21** is extracted from the table

---

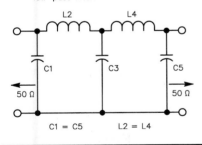

The schematic for a 5-element capacitor input/output Chebyshev low-pass filter.

Fig 16.19—A portion of a 5-element Chebyshev low-pass filter design table for 50-Ω impedance, C-in/out and standard E24 capacitor values. The full table is printed in the **References** chapter.

5-Element Chebyshev 50-Ω Lowpass SVC-Filter Designs C-In/Out, E24 Capacitor Values

No.	F_{CO}	—FREQUENCY (MHz)— 3 dB	20 dB	40 dB	MAX SWR	C1,5 (pF)	L2,4 (μH)	C3 (pF)
1	1.01	1.15	1.53	2.25	1.355	3600	10.8	6200
2	1.02	1.21	1.65	2.45	1.212	3000	10.7	5600
3	1.15	1.29	1.71	2.51	1.391	3300	9.49	5600
4	1.10	1.32	1.81	2.69	1.196	2700	9.88	5100
5	1.25	1.41	1.88	2.75	1.386	3000	8.67	5100
6	1.04	1.37	1.94	2.94	1.085	2200	9.82	4700
7	1.15	1.41	1.95	2.92	1.155	2400	9.37	4700

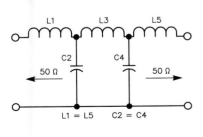

Schematic for a 5-element inductor-input/output Chebyshev low-pass filter. See the corresponding table in References chapter for the attenuation response curve.

Fig 16.20—A portion of a 5-element Chebyshev low-pass filter design table for 50-Ω impedance, L-in/out and standard-value L and C. The full table is printed in the **References** chapter.

5-Element Chebyshev 50-Ω Low-pass Filters with L-In/Out and Standard-value L and C

No.	F_{CO}	—FREQUENCY (MHz)— 3 dB	20 dB	40 dB	MAX SWR	L1,5 (μH)	C2,4 (pf)	L3 (μH)
1	0.744	1.15	1.69	2.60	1.027	5.60	4700	13.7
2	0.901	1.26	1.81	2.76	1.055	5.60	4300	12.7
3	1.06	1.38	1.94	2.93	1.096	5.60	3900	11.8
4	1.19	1.47	2.05	3.07	1.138	5.60	3600	11.2
5	1.32	1.58	2.17	3.23	1.192	5.60	3300	10.6
6	0.911	1.39	2.03	3.12	1.030	4.70	3900	11.4
7	1.08	1.50	2.16	3.29	1.056	4.70	3600	10.6
8	1.25	1.63	2.30	3.48	1.092	4.70	3300	9.92
9	1.42	1.77	2.46	3.68	1.142	4.70	3000	9.32
10	1.61	1.92	2.63	3.90	1.209	4.70	2700	8.79
11	1.05	1.64	2.41	3.72	1.025	3.90	3300	9.63

Fig 16.21—A portion of a 5-element elliptic low-pass filter design table for 50-Ω impedance, Standard E12 capacitor values for C1, C3 and C5. The full table is printed in the References chapter.

Schematic for a 5-branch elliptic low-pass filter.

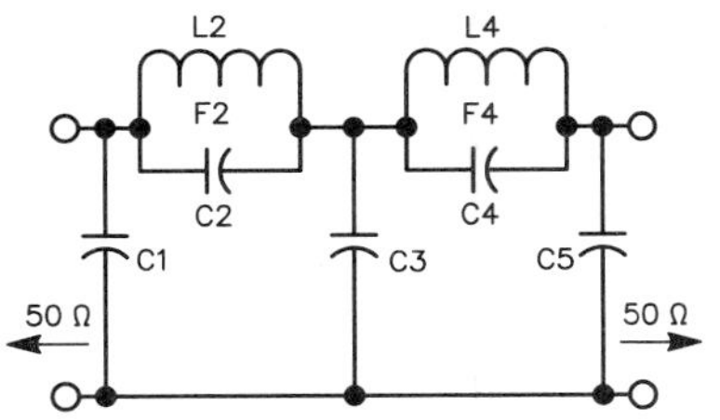

No.	(MHz) F_{co}	F-3dB	F-A$_s$	A$_s$ (dB)	Max SWR	(pF) C1	C3	C5	C2	C4	(µH) L2	L4	(MHz) F2	F4
1	0.795	0.989	1.57	47.4	1.092	2700	5600	2200	324	937	12.1	10.1	2.54	1.64
2	1.06	1.20	1.77	46.2	1.234	2700	4700	2200	341	982	9.36	7.56	2.82	1.85
3	1.47	1.57	2.15	45.4	1.586	2700	3900	2200	364	1045	6.32	4.88	3.32	2.23
4	0.929	1.18	1.91	48.0	1.077	2200	4700	1800	257	743	10.2	8.59	3.11	1.99
5	1.27	1.45	2.17	46.7	1.215	2200	3900	1800	271	779	7.85	6.39	3.45	2.26
6	1.69	1.82	2.54	45.9	1.489	2200	3300	1800	287	821	5.64	4.42	3.96	2.64
7	1.12	1.44	2.41	49.8	1.071	1800	3900	1500	192	549	8.45	7.25	3.95	2.52
8	1.49	1.73	2.70	48.8	1.183	1800	3300	1500	200	570	6.75	5.62	4.33	2.81
9	2.11	2.27	3.27	47.8	1.506	1800	2700	1500	213	604	4.55	3.64	5.12	3.40
10	1.28	1.66	2.63	46.3	1.064	1500	3300	1200	192	561	7.20	6.00	4.28	2.74
11	1.79	2.06	2.99	44.8	1.195	1500	2700	1200	204	592	5.52	4.42	4.75	3.11
12	2.52	2.70	3.63	43.8	1.525	1500	2200	1200	220	636	3.71	2.82	5.58	3.76

for the low-pass 5-branch elliptic filter with the capacitor input/output configuration, in which the nonresonating capacitors (C1, C3 and C5) are standard values. The alternate inductor input/output elliptic configuration is seldom used and therefore it is not included.

HIGH-PASS TABLES

A high-pass 5-element Chebyshev capacitor input/output configuration is shown in the table extract of **Fig 16.22**. Because the inductor input/output configuration is seldom used, it was not included. High-pass tables for elliptical filters appear in the **References** chapter.

SCALING TO OTHER FREQUENCIES AND IMPEDANCES

The tables shown are for the 1-10 MHz decade and for a 50-Ω equally terminated

impedance. The designs are easily scaled to other frequency decades and to other equally terminated impedance levels, however, making the tables a universal design aid for these specific filter types.

Frequency Scaling

To scale the frequency and the component values to the 10-100 or 100-1000 MHz decades, multiply all tabulated frequencies by 10 or 100, respectively. Then divide all C and L values by the same number. The A$_s$ and SWR data remain unchanged. To scale the filter tables to the 0.1-1 kHz, 1-10 kHz or the 10-100 kHz decades, divide the tabulated frequencies by 1000, 100 or 10, respectively. Next multiply the component values by the same number. By changing the "MHz" frequency headings to "kHz" and the "pF" and "µH" headings to "nF" and

"mH," the tables are easily changed from the 1-10 MHz decade to the 1-10 kHz decade and the table values read directly. Because the impedance level is still at 50 Ω, the component values may be awkward, but this can be corrected by increasing the impedance level by ten times using the impedance scaling procedure described below.

Impedance Scaling

All the tabulated designs are easily scaled to impedance levels other than 50 Ω, while keeping the convenience of standard-value capacitors and the "scan mode" of design selection. If the desired new impedance level differs from 50 Ω by a factor of 0.1, 10 or 100, the 50-Ω designs are scaled by shifting the decimal points of the component values. The other data remain unchanged. For example, if the

Fig 16.22—A portion of a 5-element Chebyshev high-pass filter design table for 50-Ω impedance, C-in/out and standard E24 capacitor values. The full table is printed in the References chapter.

No.	FREQUENCY (MHz) F_{co}	3 dB	20 dB	40 dB	Max SWR	C1,5 (pF)	L2,4 (H)	C3 (pF)
1	1.04	0.726	0.501	0.328	1.044	5100	6.45	2200
2	1.04	0.788	0.554	0.366	1.081	4300	5.97	2000
3	1.17	0.800	0.550	0.359	1.039	4700	5.85	2000
4	1.07	0.857	0.615	0.410	1.135	3600	5.56	1800
5	1.17	0.877	0.616	0.406	1.076	3900	5.36	1800
6	1.33	0.890	0.609	0.397	1.034	4300	5.26	1800
7	1.12	0.938	0.686	0.461	1.206	3000	5.20	1600
8	1.25	0.974	0.693	0.461	1.109	3300	4.86	1600
9	1.38	0.994	0.691	0.454	1.057	3600	4.71	1600
10	1.54	1.00	0.683	0.444	1.028	3900	4.67	1600

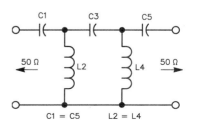

The schematic for a 5-element capacitor input/output Chebyshev high-pass filter.

impedance level is increased by ten or one hundred times (to 500 or 5000 Ω), the decimal point of the capacitor is shifted to the left one or two places and the decimal point of the inductor is shifted to the right one or two places. With increasing impedance the capacitor values become smaller and the inductor values become larger. The opposite is true if the impedance decreases.

When the desired impedance level differs from the standard 50-Ω value by a factor such as 1.2, 1.5 or 1.86, the following scaling procedure is used:

1. Calculate the impedance scaling ratio:

$$R = \frac{Z_x}{50} \qquad (12)$$

where Z_x is the desired new impedance level, in ohms.

2. Calculate the cutoff frequency (f_{50co}) of a "trial" 50-Ω filter,

$$f_{50co} = R \times f_{xco} \qquad (13)$$

where R is the impedance scaling ratio and f_{xco} is the desired cutoff frequency of the filter at the new impedance level.

3. From the appropriate SVC table select a design having its cutoff frequency closest to the calculated f_{50co} value. The tabulated capacitor values of this design are taken directly, but the frequency and inductor values must be scaled to the new impedance level.

4. Calculate the exact f_{xco} values, where

$$f_{xco} = \frac{f'_{50co}}{R} \qquad (14)$$

and f'_{50co} is the tabulated cutoff frequency of the selected design. Calculate the other frequencies of the design in the same way.

5. Calculate the inductor values for the new filter by multiplying the tabulated inductor values of the selected design by the square of the scaling ratio, R.

For example, assume a 600-Ω elliptic low-pass filter is desired with a cutoff frequency of 1.0 kHz. The elliptic low-pass table is frequency scaled to the 1-10 kHz decade by changing the table headings to kHz, nF and mH. A suitable design is then selected for scaling to 60 Ω. The 60-Ω design is then scaled to 600 Ω by shifting the decimal point to complete the scaling procedure. The calculations for this example follow, using the five steps outlined above and using the

table extract in Fig 16.21:

1. $R = \dfrac{Z_x}{50} = \dfrac{60}{50} = 1.2$

2. $f_{50co} = 1.2 \times 1.0 \text{ kHz} = 1.2 \text{ kHz}$

3. From the elliptic low-pass table (Fig 16.21), designs 5 and 10 have cutoff frequencies closest to the F_{50co} of 1.2 kHz. Either design is suitable and design 5 is chosen because of its better selectivity. The tabulated capacitor values of 2200 nF, 3900 nF, 1800 nF, 271 nF and 779 nF are copied directly.

4. All frequencies of the final design are calculated by dividing the tabulated frequencies (in kHz) of design 5 by the impedance scaling ratio, 1.2:

$$f_{co} = \frac{1.27}{1.2} = 1.06$$

$$f_{3dB} = \frac{1.45}{1.2} = 1.21$$

$$f_{AS} = \frac{2.17}{1.2} = 1.88$$

Note that a cutoff frequency of 1.0 kHz was desired, but a 1.06-kHz cutoff frequency will be accepted in exchange for the convenience of using an SVC design.

5. The L2 and L4 inductor values of design 5 are scaled to 60 Ω by multiplying them by the square of the impedance ratio, where $R = 1.2$ and $R^2 = 1.44$:

$$L2 = 1.44 \times 7.85 \text{ mH} = 11.3 \text{ mH}$$

$$L4 = 1.44 \times 6.39 \text{ mH} = 9.20 \text{ mH}$$

The 60-Ω design is now impedance scaled to 600 Ω by shifting the decimal points of the capacitor values to the left and the decimal points of the inductor values to the right. The final scaled component values for the 600-Ω filter are:

C1 = 0.22 μF
C3 = 0.39 μF
C5 = 0.18 μF
C2 = 27.1 nF
C4 = 77.9 nF
L2 = 113 mH
L4 = 92.0 mH

How to Use the Filter Tables

1. 50-Ω impedance level: Before selecting a filter design, the important parameters of the filter must be known, such as type (low-pass or high-pass), cutoff frequency, impedance level, preferred input

element (for low-pass only) and an approximation of the required stop band attenuation. It is obvious which tables to use for low-pass or high-pass applications, but it is not so obvious which one design of the many possible choices is optimum for the intended application.

Generally, the Chebyshev will be preferred over the elliptic because the Chebyshev does not require tuning of the inductors. If the relatively gradual attenuation rise of the Chebyshev is not satisfactory, however, then the elliptic should be considered. For audio filtering, the elliptic designs with high values of SWR are preferred because these designs have a much more abrupt attenuation rise than the Chebyshev. For RF applications, SWR values less than 1.2 are recommended to minimize undesired reflections. Low SWR is also important when cascading high-pass and low-pass filters to achieve a band-pass response more than two octaves wide. Each filter will operate as expected if it is correctly terminated, but this will occur only if both designs have the relatively constant terminal impedance that is associated with low SWR.

Once you know the filter type and response needed, select the table of designs most appropriate for the application on a trial basis. From the chosen table, scale the 1-10 MHz data to the desired frequency decade and search the cutoff frequency column for a value nearest the desired cutoff frequency. After finding a possible design, check the stop-band attenuation levels to see if they are satisfactory. Then check the SWR to see if it is appropriate for the application. Finally, check the component values to see if they are convenient. For example, in the audio-frequency range, the capacitor values probably will be in the microfarad range and capacitors in this size are available only in the E12 series of standard values. Then connect the components in accordance with the diagram shown in the table from which the design was selected.

2. Impedance levels other than 50 Ω: First calculate a "trial" filter design using the impedance scaling procedure previously explained. Then search the appropriate table for the best match to the trial filter and scale the selected design to the desired impedance level. In this way, the convenient scan mode of filter selection is used regardless of the desired impedance level.

Chebyshev Filter Design (Normalized Tables)

The figures and tables in this section provide the tools needed to design Chebyshev filters including those filters for which the previously published *standard value capacitor* (SVC) designs might not be suitable. **Table 16.2** lists normalized low-pass designs that, in addition to low-pass filters, can also be used to calculate high-pass, band-pass and band-stop filters in either the inductor or capacitor input/output configurations for equal impedance terminations. **Table 16.3** provides the attenuation for the resultant filter.

This material was prepared by Ed Wetherhold, W3NQN, who has been the author of a number of articles and papers on the design of LC filters. It is a complete revision of his previously published filter design material and provides both insight to the design and actual designs in just a few minutes.

For a given number of elements (N), increasing the filter reflection coefficient (RC or ρ) causes the attenuation slope to increase with a corresponding increase in both the passband ripple amplitude (a_p) and SWR and with a decrease in the filter return loss. All of these parameters are mathematically related to each other. If one is known, the others may be calculated. Filter designs having a low RC are preferred because they are less sensitive to component and termination impedance variations than are designs having a higher RC. The RC percentage is used as the independent variable in Table 16.2 because it is used as the defining parameter in the more frequently used tables, such as those by Zverev and Saal (see Refs 17 and 18).

The return loss is tabulated instead of passband ripple amplitude (a_p) because it is easy to measure using a return loss bridge. In comparison, ripple amplitudes less than 0.1 dB are difficult to measure accurately. The resulting values of attenuation are contained in Table 16.3 and corresponding values of a_p and SWR may be found by referring to the Equivalent Values of Reflection Coefficient, Attenuation, SWR and Return Loss table in the **References** chapter. The filter used (low pass, high-pass, band-pass and so on) will depend on the application and the stop-band attenuation needed.

Table 16.2

Element values of Chebyshev low-pass filters normalized for a ripple cutoff frequency (Fa_p) of one radian/sec ($1/2\pi$ Hz) and 1-Ω terminations.

Use the top column headings for the low-pass C-in/out configuration and the bottom column headings for the low-pass L-in/out configuration. Fig 16.23 shows the filter schematics.

N	RC (%)	Ret Loss (dB)	F3/F$_{ap}$ Ratio	C1 (F)	L2 (H)	C3 (F)	L4 (H)	C5 (F)	L6 (H)	C7 (F)	L8 (H)	C9 (F)
3	1.000	40.00	3.0094	0.3524	0.6447	0.3524						
3	1.517	36.38	2.6429	0.4088	0.7265	0.4088						
3	4.796	26.38	1.8772	0.6292	0.9703	0.6292						
3	10.000	20.00	1.5385	0.8535	1.104	0.8535						
3	15.087	16.43	1.3890	1.032	1.147	1.032						
5	0.044	67.11	2.7859	0.2377	0.5920	0.7131	0.5920	0.2377				
5	0.498	46.06	1.8093	0.4099	0.9315	1.093	0.9315	0.4099				
5	1.000	40.00	1.6160	0.4869	1.050	1.226	1.050	0.4869				
5	1.517	36.38	1.5156	0.5427	1.122	1.310	1.122	0.5427				
5	2.768	31.16	1.3892	0.6408	1.223	1.442	1.223	0.6408				
5	4.796	26.38	1.2912	0.7563	1.305	1.577	1.305	0.7563				
5	6.302	24.01	1.2483	0.8266	1.337	1.653	1.337	0.8266				
5	10.000	20.00	1.1840	0.9732	1.372	1.803	1.372	0.9732				
5	15.087	16.43	1.1347	1.147	1.371	1.975	1.371	1.147				
7	1.000	40.00	1.3004	0.5355	1.179	1.464	1.500	1.464	1.179	0.5355		
7	1.427	36.91	1.2598	0.5808	1.232	1.522	1.540	1.522	1.232	0.5808		
7	1.517	36.38	1.2532	0.5893	1.241	1.532	1.547	1.532	1.241	0.5893		
7	3.122	30.11	1.1818	0.7066	1.343	1.660	1.611	1.660	1.343	0.7066		
7	4.712	26.54	1.1467	0.7928	1.391	1.744	1.633	1.744	1.391	0.7928		
7	4.796	26.38	1.1453	0.7970	1.392	1.748	1.633	1.748	1.392	0.7970		
7	8.101	21.83	1.1064	0.9390	1.431	1.878	1.633	1.878	1.431	0.9390		
7	10.000	20.00	1.0925	1.010	1.437	1.941	1.622	1.941	1.437	1.010		
7	10.650	19.45	1.0885	1.033	1.437	1.962	1.617	1.962	1.437	1.033		
7	15.087	16.43	1.0680	1.181	1.423	2.097	1.573	2.097	1.423	1.181		
9	1.000	40.00	1.1783	0.5573	1.233	1.550	1.632	1.696	1.632	1.550	1.233	0.5573
9	1.517	36.38	1.1507	0.6100	1.291	1.610	1.665	1.745	1.665	1.610	1.291	0.6100
9	2.241	32.99	1.1271	0.6679	1.342	1.670	1.690	1.793	1.690	1.670	1.342	0.6679
9	2.512	32.00	1.1206	0.6867	1.357	1.688	1.696	1.808	1.696	1.688	1.357	0.6867
9	4.378	27.17	1.0915	0.7939	1.419	1.786	1.712	1.890	1.712	1.786	1.419	0.7939
9	4.796	26.38	1.0871	0.8145	1.427	1.804	1.713	1.906	1.713	1.804	1.427	0.8145
9	4.994	26.03	1.0852	0.8239	1.431	1.813	1.712	1.913	1.712	1.813	1.431	0.8239
9	8.445	21.47	1.0623	0.9682	1.460	1.936	1.692	2.022	1.692	1.936	1.460	0.9682
9	10.000	20.00	1.0556	1.025	1.462	1.985	1.677	2.066	1.677	1.985	1.462	1.025
9	15.087	16.43	1.0410	1.196	1.443	2.135	1.617	2.205	1.617	2.135	1.443	1.196
N	RC (%)	Ret Loss (dB)	F3/F$_{ap}$ Ratio	L1 (H)	C2 (F)	L3 (H)	C4 (F)	L5 (H)	C6 (F)	L7 (H)	C8 (F)	L9 (H)

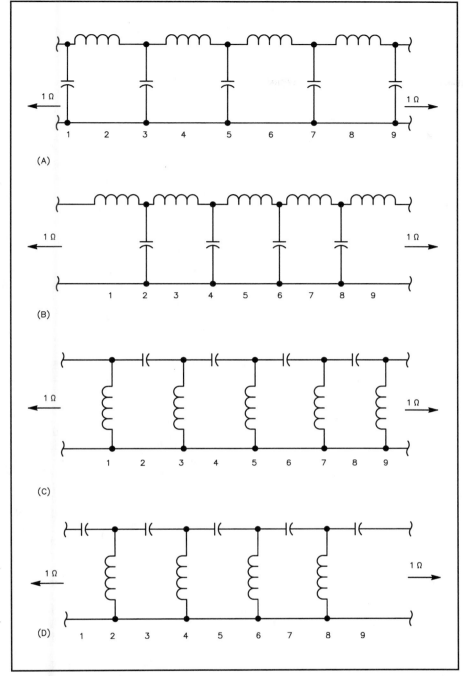

Fig 16.23—The schematic diagrams shown are low-pass and high-pass Chebyshev filters with the C-in/out and L-in/out configurations. For all normalized values see Table 16.2.

A: C-in/out low-pass configuration. Use the C and L values associated with the top column headings of the Table.

B: L-in/out low-pass configuration. For normalized values, use the L and C values associated with the bottom column headings of the Table.

C: L-in/out high-pass configuration is derived by transforming the C-in/out low-pass filter in A into an L-in/out high-pass by replacing all Cs with Ls and all Ls with Cs. The reciprocals of the lowpass component values become the highpass component values. For example, when n = 3, RC = 1.00% and C1 = 0.3524 F, L1 and L3 in C become 2.838 H.

D: The C-in/out high-pass configuration is derived by transforming the L-in/out low-pass in B into a C-in/out high-pass by replacing all Ls with Cs and all Cs with Ls. The reciprocals of the low-pass component values become the high-pass component values. For example, when n = 3, RC = 1.00% and L1 = 0.3524 H, C1 and C3 in D become 2.838 F.

The filter schematic diagrams shown in **Fig 16.23** are for low-pass and high-pass versions of the Chebyshev designs listed in Table 16.2. Both low-pass and high-pass equally terminated configurations and component values of the C-in/out or L-in/out filters can be derived from this single table. By using a simple procedure, the low-pass and high-pass designs can be transformed into corresponding band-pass and band-stop filters. The normalized element values of the low-pass C-in/out and L-in/out designs, Fig 16.23A and B, are read directly from the table using the values associated with either the top or bottom column headings, respectively.

The first four columns of Table 16.2 list N (the number of filter elements), RC (reflection coefficient percentage), *return loss* and the ratio of the 3-dB-to-F_{ap} frequencies. The passband maximum ripple amplitude (a_p) is not listed because it is difficult to measure. If necessary it can be calculated from the reflection coefficient. The $F3/F_{ap}$ ratio varies with N and RC; if both of these parameters are known, the $F3/F_{ap}$ ratio may be calculated. The remaining columns list the normalized Chebyshev element values for equally terminated filters for Ns from 3 to 9 in increments of 2.

The Chebyshev passband ends when the passband attenuation first exceeds the maximum ripple amplitude, a_p. This frequency is called the "ripple cutoff frequency, F_{ap}" and it has a normalized value of unity. All Chebyshev designs in Table 16.2 are based on the ripple cutoff frequency instead of the more familiar 3-dB frequency of the Butterworth response. However, the 3-dB frequency of a Chebyshev design may be obtained by multiplying the ripple cutoff frequency by the $F3/F_{ap}$ ratio listed in the fourth column.

The element values are normalized to a ripple cutoff frequency of 0.15915 Hz (one radian/sec) and 1-Ω terminations, so that the low-pass values can be transformed directly into high-pass values. This is done by replacing all Cs and Ls in the low-pass configuration with Ls and Cs and by replacing all the low-pass element values with their reciprocals. The normalized values are then multiplied by the appropriate C and L scaling factors to obtain the final values based on the desired ripple cutoff frequency and impedance level. The listed C and L element values are in farads and henries and become more reasonable after the values are scaled to the desired cutoff frequency and impedance level.

The normalized designs presented are a mixture: Some have integral values of *reflection coefficient (RC)* (1% and 10%) while others have "integral" values of

Table 16.3

Normalized Frequencies at Listed Attenuation Levels for Chebyshev Low-Pass Filters with N = 3, 5, 7 and 9.

N	RC(%)	Attenuation Levels (dB)										
		1.0	3.01	6.0	10	20	30	40	50	60	70	80
3	1.000	2.44	3.01	3.58	4.28	6.33	9.27	13.59	19.93	29.25	42.92	63.00
3	1.517	2.15	2.64	3.13	3.74	5.52	8.08	11.83	17.35	25.46	37.36	54.83
3	4.796	1.56	1.88	2.20	2.60	3.79	5.53	8.08	11.83	17.35	25.45	37.35
3	10.000	1.31	1.54	1.78	2.08	3.00	4.34	6.33	9.26	13.57	19.90	29.20
3	15.087	1.20	1.39	1.59	1.85	2.63	3.79	5.52	8.06	11.81	17.32	25.41
5	1.000	1.46	1.62	1.76	1.94	2.39	2.97	3.69	4.62	5.79	7.27	9.13
5	1.517	1.38	1.52	1.65	1.80	2.22	2.74	3.41	4.26	5.33	6.69	8.40
5	4.796	1.19	1.29	1.39	1.50	1.82	2.22	2.74	3.41	4.26	5.33	6.69
5	6.302	1.16	1.25	1.34	1.44	1.74	2.12	2.61	3.24	4.04	5.05	6.34
5	10.000	1.11	1.18	1.26	1.35	1.61	1.95	2.39	2.96	3.69	4.61	5.78
5	15.087	1.07	1.13	1.20	1.28	1.51	1.82	2.22	2.74	3.41	4.25	5.33
7	1.000	1.23	1.30	1.37	1.45	1.65	1.89	2.18	2.53	2.95	3.44	4.04
7	1.517	1.19	1.25	1.32	1.39	1.57	1.80	2.07	2.39	2.79	3.25	3.81
7	4.796	1.10	1.15	1.19	1.25	1.39	1.57	1.80	2.07	2.39	2.79	3.25
7	8.101	1.07	1.11	1.15	1.19	1.32	1.49	1.69	1.94	2.24	2.60	3.03
7	10.000	1.05	1.09	1.13	1.18	1.30	1.45	1.65	1.89	2.18	2.53	2.94
7	15.087	1.04	1.07	1.10	1.14	1.25	1.39	1.57	1.80	2.07	2.39	2.78
9	1.000	1.13	1.18	1.22	1.26	1.38	1.51	1.67	1.85	2.07	2.32	2.61
9	1.517	1.11	1.15	1.19	1.23	1.34	1.46	1.61	1.78	1.99	2.22	2.50
9	4.796	1.06	1.09	1.11	1.15	1.23	1.34	1.46	1.61	1.78	1.99	2.22
9	8.445	1.04	1.06	1.09	1.11	1.19	1.28	1.40	1.53	1.69	1.88	2.10
9	10.000	1.03	1.06	1.08	1.10	1.18	1.27	1.38	1.51	1.67	1.85	2.07
9	15.087	1.02	1.04	1.06	1.08	1.15	1.23	1.34	1.46	1.61	1.78	1.99

passband ripple amplitude (0.001, 0.01 and 0.1 dB). These ripple amplitudes correspond to reflection coefficients of 1.517, 4.796 and 15.087%, respectively. By having tabulated designs based on integral values of both reflection coefficient and passband ripple amplitude, the correctness of the normalized component values may be checked against those same values published in filter handbooks whichever parameter, RC or a_p, is used.

In addition to the customary normalized design listings based on integral values of reflection coefficient or ripple amplitude, Table 16.2 also includes unique designs having special element ratios that make them more useful than previously published tables. For example, for N = 5 and RC = 6.302, the ratio of C3/C1 is 2.000. This ratio allows 5-element low-pass filters to be realized with only one capacitor value because C3 may be obtained by using parallel-connected capacitors each having the same value as C1 and C5.

In a similar way, for N = 7 and RC = 8.101, C3/C1 and C5/C1 are also 2.000. Another useful N = 7 design is that for RC = 1.427%. Here the L4/L2 ratio is 1.25, which is identical to 110/88. This means a seventh-order C-in/out low-pass audio filter can be realized with four surplus 88-mH inductors. Both L2 and L6 can be 88 mH while L4 is made up of a series connection of 22 mH and 88 mH. The

22-mH value is obtained by connecting the two windings of one of the four surplus inductors in parallel. Other useful ratios also appear in the N = 9 listing for both C3/C1 and L4/L2.

Except for the first two N = 5 designs, all designs were calculated for a reflection coefficient range from 1% to about 15%. The first two N = 5 designs were included because of their useful L3/L1 ratios. Designs with an RC of less than 1% are not normally used because of their poor selectivity. Designs with RC greater than 15% yield increasingly high SWR values with correspondingly increased objectionable reflective losses and sensitivity to termination impedance and component value variations.

Low-pass and high-pass filters may be realized in either a C-input/output or an L-input/output configuration. The C-input/output configuration is usually preferred because fewer inductors are required, compared to the L-input/output configuration. Inductors are usually more lossy, bulky and expensive than capacitors. The selection of the filter order or number of filter elements, N, is determined by the desired stop-band attenuation rate of increase and the tolerable reflection coefficient or SWR. A steeper attenuation slope requires either a design having a higher reflection coefficient or more circuit elements. Consequently, to

select an optimum design, the builder must determine the amount of attenuation required in the stop band and the permissible maximum amount of reflection coefficient or SWR.

Table 16.3 shows the theoretical normalized frequencies (relative to the ripple cutoff frequency) for the listed attenuation levels and reflection coefficient percentages for Chebyshev low-pass filters of 3, 5, 7 and 9 elements. For example, for N = 5 and RC = 15.087%, an attenuation of 40 dB is reached at 2.22 times the ripple cutoff frequency (slightly more than one octave). The tabulated data are also applicable to high-pass filters by simply taking the reciprocal of the listed frequency. For example, for the same previous N and RC values, a high-pass filter attenuation will reach 40 dB at 1/2.22 = 0.450 times the ripple cutoff frequency.

The attenuation levels are theoretical and assume perfect components, no coupling between filter sections and no signal leakage around the filter. A working model should follow these values to the 60 or 70-dB level. Beyond this point, the actual response will likely degrade somewhat from the theoretical.

Fig 16.24 shows four plotted attenuation vs normalized frequency curves for N = 5 corresponding to the normalized frequencies in Table 16.3 At two octaves above the ripple cutoff fre-

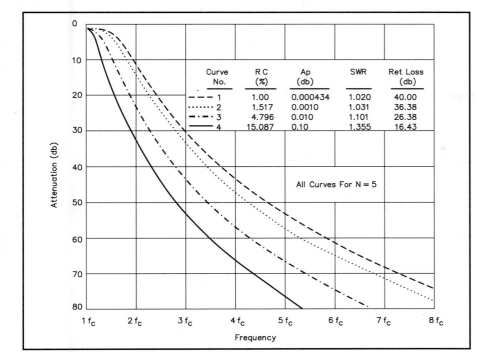

Fig 16.24—The graph shows attenuation vs frequency for four 5-element low-pass filters designed with the information obtained from Table 16.2. This graph demonstrates how reflection coefficient percentage (RC), maximum passband ripple amplitude (a_p), SWR, return loss and attenuation rolloff are all related. The exact frequency at a specified attenuation level can be obtained from Table 16.3.

Curve No.	RC (%)	Ap (db)	SWR	Ret Loss (db)
– – – 1	1.00	0.000434	1.020	40.00
· · · · 2	1.517	0.0010	1.031	36.38
– · – · 3	4.796	0.010	1.101	26.38
——— 4	15.087	0.10	1.355	16.43

All Curves For N = 5

L2, L6 = 1.392×1.989 µH = 2.77 µH
L4 = 1.633×1.989 µH = 3.25 µH

High-Pass Filter

The procedure for calculating a high-pass filter is similar to that for a low-pass filter, except a low-pass-to-high-pass transformation must first be performed. Assume a 50-Ω high-pass filter is needed to give more than 40 dB of attenuation one octave below ($f_c/2$) a ripple cutoff frequency of 4.0 MHz. Referring to Table 16.3, we see from the 40-dB column that a 7-element low-pass filter with RC of 4.796% will give 40 dB of attenuation at $1.8f_c$. If this filter is transformed into a high-pass filter, the 40-dB level is reached at $f_c/1.80$ or at $0.556f_c = 2.22$ MHz. Since the 40-dB level is reached before one octave from the 4-MHz cutoff frequency, this design will be satisfactory.

From Fig 16.23, we choose the low-pass L-in/out configuration in B and transform it into a high-pass filter by replacing all inductors with capacitors and all capacitors with inductors. Fig 16.23D is the filter configuration after the transformation. The reciprocals of the low-pass values become the high-pass values to complete the transformation. The high-pass values of the filter shown in Fig 16.23D are:

$$C1, C7 = \frac{1}{0.7970} = 1.255 \text{ F}$$

$$L2, L6 = \frac{1}{1.392} = 0.7184 \text{ H}$$

$$C3, C5 = \frac{1}{1.748} = 0.5721 \text{ F}$$

and

$$L4 = \frac{1}{1.633} = 0.6124 \text{ H}$$

Using the previously calculated C and L scaling factors, the high-pass component values are calculated the same way as before:

C1, C7 = 1.255×795.8 pF = 999 pF
C3, C5 = 0.5721×795.8 pF = 455 pF
L2, L6 = 0.7184×1.989 µH = 1.43 µH
L4 = 0.6124×1.989 µH = 1.22 µH

BAND-PASS FILTERS

Band-pass filters may be classified as either narrowband or broadband. If the ratio of the upper ripple cutoff frequency to the lower cutoff frequency is greater than two, we have a wideband filter. For wideband filters, the band-pass filter (BPF) requirement may be realized by simply cascading separate high-pass and low-pass filters having the same design

quency, f_c, the attenuation slope gradually becomes 6 dB per octave per filter element.

LOW-PASS AND HIGH-PASS FILTERS

Low-Pass Filter

Let's look at the procedure used to calculate the capacitor and inductor values of low-pass and high-pass filters by using two examples. Assume a 50-Ω low-pass filter is needed to give more than 40 dB of attenuation at $2f_c$ or one octave above the ripple-cutoff frequency of 4.0 MHz. Referring to Table 16.3, we see from the 40-dB column that a filter with 7 elements (N = 7) and a RC of 4.796% will reach 40 dB at 1.80 times the cutoff frequency or $1.8 \times 4 = 7.2$ MHz. Since this design has a reasonably low reflection coefficient and will satisfy the attenuation requirement, it is a good choice. Note that no 5-element filters are suitable for this application because 40 dB of attenuation is not achieved one octave above the cutoff frequency.

From Table 16.2, the normalized component values corresponding to N = 7 and RC = 4.796% for the C-in/out configuration are: C1, C7 = 0.7970 F, L2, L6 = 1.392 H, C3, C5 = 1.748 F and

L4 = 1.633 H. See Fig 16.23A for the corresponding configuration. The C and L normalized values will be scaled from a ripple cutoff frequency of one radian/sec and an impedance level of 1 Ω to a cutoff frequency of 4.0 MHz and an impedance level of 50 Ω. The C_s and L_s scaling factors are calculated:

$$C_S = \frac{1}{2\pi R f} \qquad (15)$$

$$L_S = \frac{R}{2\pi f} \qquad (16)$$

where:
R = impedance level
f = cutoff frequency.

In this example:

$$C_S = \frac{1}{2\pi R f} = \frac{1}{2\pi \times 50 \times 4 \times 10^6} = 795.8 \times 10^{-12}$$

$$L_S = \frac{R}{2\pi f} = \frac{50}{2\pi \times 4 \times 10^6} = 1.989 \times 10^{-6}$$

Using these scaling factors, the capacitor and inductor normalized values are scaled to the desired cutoff frequency and impedance level:

C1, C7 = 0.797×795.8 pF = 634 pF
C3, C5 = 1.748×795.8 pF = 1391 pF

impedance. (The assumption is that the filters maintain their individual responses even though they are cascaded.) *For this to be true, it is important that both filters have a relatively low reflection coefficient percentage (less than 5%) so the SWR variations in the passband will be small.*

For narrowband BPFs, where the separation between the upper and lower cutoff frequencies is less than two, it is necessary to transform an appropriate low-pass filter into a BPF. That is, we use the low-pass normalized tables to design narrowband BPFs.

We do this by first calculating a low-pass filter (LPF) with a cutoff frequency equal to the desired bandwidth of the BPF. The LPF is then transformed into the desired BPF by resonating the low-pass components at the geometric center frequency of the BPF.

For example, assume we want a 50-Ω BPF to pass the 75/80-m band and attenuate all signals outside the band. Based on the passband ripple cutoff frequencies of 3.5 and 4.0 MHz, the geometric center frequency = $(3.5 \times 4.0)^{0.5} = (14)^{0.5} = 3.741657$ or 3.7417 MHz. Let's slightly extend the lower and upper ripple cutoff frequencies to 3.45 and 4.058 MHz to account for possible component tolerance variations and to maintain the same center frequency. We'll evaluate a low-pass 3-element prototype with a cutoff frequency equal to the BPF passband of $(4.058-3.45)$ MHz = 0.608 MHz as a possible choice for transformation.

Further, assume it is desired to attenuate the second harmonic of 3.5 MHz by at least 40 dB. The following calculations show how to design an N = 3 filter to provide the desired 40-dB attenuation at 7 MHz and above.

The bandwidth (BW) between 7 MHz on the upper attenuation slope (call it "f+") of the BPF and the corresponding frequency at the same attenuation level on the lower slope (call it "f−") can be calculated based on $(f+)(f-) = (f_c)^2$ or

$$f- = \frac{14}{7} = 2 \text{ MHz}$$

Therefore, the bandwidth at this unknown attenuation level for 2 and 7 MHz is 5 MHz. This 5-MHz BW is normalized to the ripple cutoff BW by dividing 5.0 MHz by 0.608 MHz:

$$\frac{5.0}{0.608} = 8.22$$

We now can go to Table 16.3 and search for the corresponding normalized frequency that is closest to the desired normalized BW of 8.22. The low-pass design of N = 3 and RC = 4.796% gives 40 dB for

a normalized BW of 8.08 and 50 dB for 11.83. Therefore, a design of N = 3 and RC = 4.796% with a normalized BW of 8.22 is at an attenuation level somewhere between 40 and 50 dB. Consequently, a low-pass design based on 3 elements and a 4.796% RC will give slightly more than the desired 40 dB attenuation above 7 MHz. The next step is to calculate the C and L values of the low-pass filter using the normalized component values in Table 16.2.

From this table and for N = 3 and RC = 4.796%, C1, C3 = 0.6292 F and L2 = 0.9703 H. We calculate the scaling factors as before and use 0.608 MHz as the ripple cutoff frequency:

$$C_s = \frac{1}{2\pi Rf} = \frac{1}{2\pi \times 50 \times 0.608 \times 10^6} = 5235 \times 10^{-12}$$

$$L_s = \frac{R}{2\pi f} = \frac{50}{2\pi \times 0.608 \times 10^6} = 13.09 \times 10^{-6}$$

C1, C3 = 0.6292×5235 pF = 3294 pF and L2 = 0.9703×13.09 μH = 12.70 μH.

The LPF (in a pi configuration) is transformed into a BPF with 3.7417-MHz center frequency by resonating the low-pass elements at the center frequency. The resonating components will take the same identification numbers as the components they are resonating.

$$L1,L3 = \frac{25330}{\left(F_c^2 \times C1\right)}$$

$$= \frac{25330}{\left(14 \times 3294\right)} = 0.5493 \text{ μH}$$

$$C2 = \frac{25330}{\left(F_c^2 \times L2\right)}$$

$$= \frac{25330}{\left(14 \times 12.7\right)} = 142.5 \text{ pF}$$

where L, C and f are in μH, pF and MHz, respectively.
The BPF circuit after transformation (for N = 3, RC = 4.796%) is shown in **Fig 16.25**.

The component-value spread is

$$\frac{12.7}{0.549} = 23$$

and the reactance of L1 is about 13 Ω at the center frequency. For better BPF performance, the component spread should be reduced and the reactance of L1 and L3 should be raised to make it easier to achieve the maximum possible Q for these two inductors. This can be easily done by designing the BPF for an impedance level of 200 Ω and then using the center taps

on L1 and L3 to obtain the desired 50-Ω terminations. The result of this approach is shown in **Fig 16.26**.

The component spread is now a more reasonable

$$\frac{12.7}{2.20} = 5.77$$

and the L1, L3 reactance is 51.6 Ω. This higher reactance gives a better chance to achieve a satisfactory Q for L1 and L3 with a corresponding improvement in the BPF performance.

As a general rule, keep reactance values between 5 Ω and 500 Ω in a 50-Ω circuit. When the value falls below 5 Ω, either the equivalent series resistance of the inductor or the series inductance of the capacitor degrades the circuit Q. When the inductive reactance is greater than 500 Ω, the inductor is approaching self-resonance and circuit Q is again degraded. In practice, both L1 and L3 should be bifilar wound on a powdered-iron toroidal core to assure that optimum coupling is obtained between turns over the entire

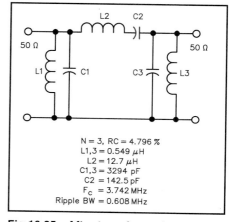

N = 3, RC = 4.796 %
L1,3 = 0.549 μH
L2 = 12.7 μH
C1,3 = 3294 pF
C2 = 142.5 pF
F_c = 3.742 MHz
Ripple BW = 0.608 MHz

Fig 16.25—After transformation of the band-pass filter, all parallel elements become parallel LCs and all series elements become series LCs.

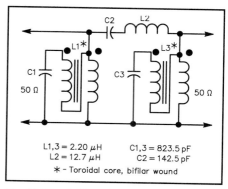

L1,3 = 2.20 μH   C1,3 = 823.5 pF
L2 = 12.7 μH   C2 = 142.5 pF
∗ - Toroidal core, bifilar wound

Fig 16.26—A filter designed for 200-Ω source and load provides better values. By tapping the inductors, we can use a 200-Ω filter design in a 50-Ω system.

Side-Slope Attenuation Calculations

The following equations allow the calculation of the frequencies on the upper and lower sides of a BPF response curve at any given attenuation level if the bandwidth at that attenuation level and the geometric center frequency of the BPF are known:

$$f_{lo} = -X + \sqrt{f_c{}^2 + X^2} \qquad (17)$$

$$f_{hi} = f_{lo} + BW \qquad (18)$$

where

BW = bandwidth at the given attenuation level,

f_c = geometric center frequency

$$X = \frac{BW}{2}$$

For example, if f = 3.74166 MHz and BW = 5 MHz, then

$$\frac{BW}{2} = X = 2.5$$

and:

$$f_{lo} = -2.5 + \sqrt{3.74166^2 + 2.5^2} = 2.00 \text{ MHz}$$

$$f_{hi} = f_{lo} + BW = 2 + 5 = 7 \text{ MHz}$$

BAND-STOP FILTERS

Band-stop filters may be classified as either narrowband or broadband. If the ratio of the upper ripple cutoff frequency to the lower cutoff frequency is greater than two, the filter is considered wideband. A wideband band-stop filter (BSF) requirement may be realized by simply paralleling the inputs and outputs of separate low-pass and high-pass filters having the same design impedance and with the low-pass filter having its cutoff frequency one octave or more below the high-pass cutoff frequency.

In order to parallel the low-pass and high-pass filter inputs and outputs without one affecting the other, it is essential that each filter have a high impedance in that portion of its stop band which lies in the passband of the other. *This means that each of the two filters must begin and end in series branches.* In the low-pass filter, the input/output series branches must consist of inductors and in the high-pass filter, the input/output series branches must consist of capacitors.

When the ratio of the upper to lower cutoff frequencies is less than two, the BSF is considered to be narrowband, and a calculation procedure similar to that of the narrowband BPF design procedure is used. However, in the case of the BSF, the design process starts with the design of a high-pass filter having the desired impedance level of the BSF and a ripple cutoff frequency the same as that of the desired ripple bandwidth of the BSF. After the HPF design is completed, every high-pass element is resonated to the center frequency of the BSF in the same manner as if it were a BPF, except that all shunt branches of the BSF will consist of series-tuned circuits, and all series branches will consist of parallel-tuned circuits—just the opposite of the resonant circuits in the BPF. The reason for this becomes obvious when the impedance characteristics of the series and parallel circuits at resonance are considered relative to the intended purpose of the filter, that is, whether it is for a band-pass or a band-stop application.

Quartz Crystal Filters

Practical inductor Q values effectively set the minimum achievable bandwidth limits for LC band-pass filters. Higher-Q circuit elements must be employed to extend these limits. These high-Q resonators include PZT ceramic, mechanical and co-axial devices. However, the quartz crystal provides the highest Q and best stability with temperature and time of all available resonators. Quartz crystals suitable for filter use are fabricated over a frequency range from audio to VHF.

The quartz resonator has the equivalent circuit shown in **Fig 16.31**. L_s, C_s and R_s represent the *motional* reactances and loss resistance. C_p is the parallel plate capacitance formed by the two metal electrodes separated by the quartz dielectric. Quartz has a dielectric constant of 3.78. **Table 16.4** shows parameter values for typical moderate-cost quartz resonators. Q_U is the resonator unloaded Q.

$$Q_U = 2\pi f_s r_s \qquad (19)$$

Q_U is very high, usually exceeding 25,000. Thus the quartz resonator is an ideal component for the synthesis of a high-Q band-pass filter.

A quartz resonator connected between

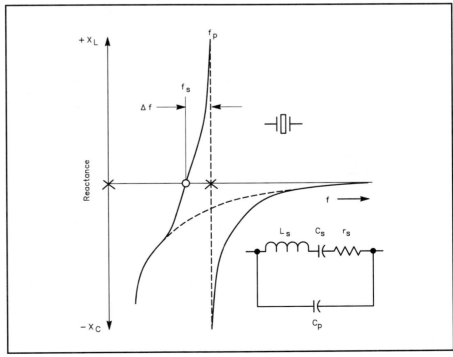

Fig 16.31—Equivalent circuit of a quartz crystal. The curve plots the crystal reactance against frequency. At f_p, the resonance frequency, the reactance curve goes to infinity.

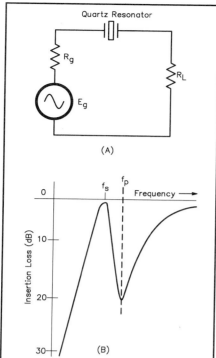

(A)

(B)

Fig 16.32—A: Series test circuit for a crystal. In the test circuit the output of a variable frequency generator, e_g, is used as the test signal. The frequency response in B shows the highest attenuation at resonance (f_p). See text.

Table 16.4
Typical Parameters for AT-Cut Quartz Resonators

Freq (MHz)	Mode n	rs (Ω)	Cp (pF)	Cs (pF)	L (mH)	Q_U
1.0	1	260	3.4	0.0085	2900	72,000
5.0	1	40	3.8	0.011	100	72,000
10.0	1	8	3.5	0.018	14	109,000
20	1	15	4.5	0.020	3.1	26,000
30	3	30	4.0	0.002	14	87,000
75	3	25	4.0	0.002	2.3	43,000
110	5	60	2.7	0.0004	5.0	57,000
150	5	65	3.5	0.0006	1.9	27,000
200	7	100	3.5	0.0004	2.1	26,000

Courtesy of Piezo Crystal Co, Carlisle, Pennsylvania

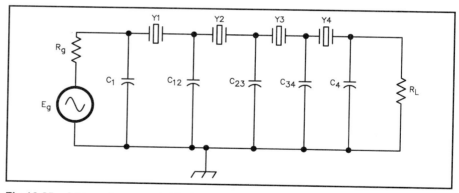

Fig 16.34—A half-lattice crystal filter. No phasing capacitor is needed in this circuit.

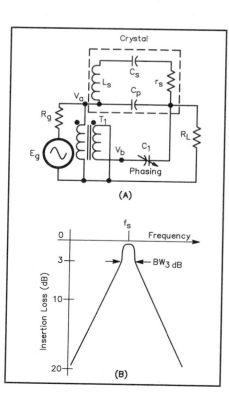

(A)

(B)

Fig 16.33—The practical one-stage crystal filter in A has the response shown in B. The phasing capacitor is adjusted for best response (see text).

Fig 16.35—A four-stage crystal ladder filter. The crystals must be chosen properly for best response.

generator and load, as shown in **Fig 16.32A**, produces the frequency response of Fig 16.32B. There is a relatively low loss at the series resonant frequency f_s and high loss at the parallel resonant frequency f_p. The test circuit of Fig 16.32A is useful for determining the parameters of a quartz resonator, but yields a poor filter.

A crystal filter developed in the 1930s is shown in **Fig 16.33A**. The disturbing effect of C_p (which produces f_p) is canceled by the *phasing capacitor,* C1. The voltage reversing transformer T1 usually consists of a bifilar winding on a ferrite core. Voltages V_a and V_b have equal magnitude but 180° phase difference. When C1 = C_p, the effect of C_p will disappear and a well-behaved single resonance will occur as shown in Fig 16.33B. The band-pass filter will exhibit a loaded Q given by:

$$Q_L = \frac{2\pi f_S L_S}{R_L} \tag{20}$$

This single-stage "crystal filter," operating at 455 kHz, was present in almost all high-quality amateur communications receivers up through the 1960s. When the filter was switched into the receiver IF amplifier the bandwidth was reduced to a few hundred Hz for Morse code reception.

The half-lattice filter shown in **Fig 16.34** is an improvement in crystal filter design. The quartz resonator parallel-plate capacitors, C_p, cancel each other. Remaining series resonant circuits, if prop-erly offset in frequency, will produce an approximate 2-pole Butterworth or Chebyshev response. Crystals A and B are usually chosen so that the parallel resonant frequency (f_p) of one is the same as the series resonant frequency (f_s) of the other.

Half-lattice filter sections can be cascaded to produce a composite filter with many poles. Until recently, most vendor-supplied commercial filters were lattice types. Ref 11 discusses the computer de-sign of half-lattice filters.

Many quartz crystal filters produced to-day use the ladder network design shown in **Fig 16.35**. In this configuration, all resonators have the same series resonant frequency f_s. Interresonator coupling is provided by shunt capacitors such as C12 and C23. Refs 12 and 13 provide good ladder filter design information. A test set for evaluating crystal filters is presented in the **Projects** section of this chapter.

Monolithic Crystal Filters

A monolithic (Greek: one-stone) crystal filter has two sets of electrodes deposited on the same quartz plate, as shown in **Fig 16.36**. This forms two resonators with acoustic (mechanical) coupling between them. If the acoustic coupling is correct, a 2-pole Butterworth or Chebyshev response will be achieved. More than two resonators can be fabricated on the same plate yielding a multipole response. Monolithic crystal filter technology is popular because it produces a low parts count, single-unit filter at lower cost than a lumped-element equivalent. Monolithic crystal filters are typically manufactured in the range from 5 to 30 MHz for the fundamental mode and up to 90 MHz for the third-overtone mode. Q_L ranges from 200 to 10,000.

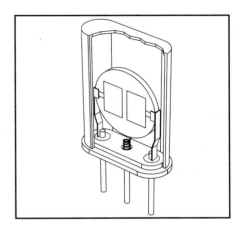

Fig 16.36—Typical two-pole monolithic crystal filter. This single small ($^1/_2$ to $^3/_4$-inch) unit can replace 6 to 12, or more, discrete components.

SAW Filters

The resonators in a monolithic crystal filter are coupled together by bulk acoustic waves. These acoustic waves are generated and propagated in the interior of a quartz plate. It is also possible to launch, by an appropriate transducer, acoustic waves which propagate only along the surface of the quartz plate. These are called "surface-acoustic-waves" because they do not appreciably penetrate the interior of the plate.

A surface-acoustic-wave (SAW) filter consists of thin aluminum electrodes, or fingers, deposited on the surface of a piezoelectric substrate as shown in **Fig 16.37**. Lithium Niobate ($LiNbO_3$) is usually favored over quartz because it yields less insertion loss. The electrodes make up the filter's transducers. RF voltage is applied to the input transducer and generates electric fields between the fingers. The piezoelectric material vibrates launching an acoustic wave along the surface. When the wave reaches the output transducer it produces an electric field between the fingers. This field generates a voltage across the load resistor.

Since both input and output transducers are not entirely unidirectional, some acoustic power is lost in the acoustic absorbers located behind each transducer. This lost acoustic power produces a midband electrical insertion loss typically greater than 10 dB. The SAW filter frequency response is determined by the choice of substrate material and finger pattern. The finger spacing, (usually one-quarter wavelength) determines the filter center frequency. Center frequencies are available from 20 to 1000 MHz. The number and length of fingers determines the filter loaded Q and shape factor.

Loaded Qs are available from 2 to 100, with a shape factor of 1.5 (equivalent to a dozen poles). Thus the SAW filter can be made broadband much like the LC filters that it replaces. The advantage is substantially reduced volume and possibly lower cost. SAW filter research was driven by military needs for exotic amplitude-response and time-delay requirements. Low-cost SAW filters are presently found in television IF amplifiers where high midband loss can be tolerated.

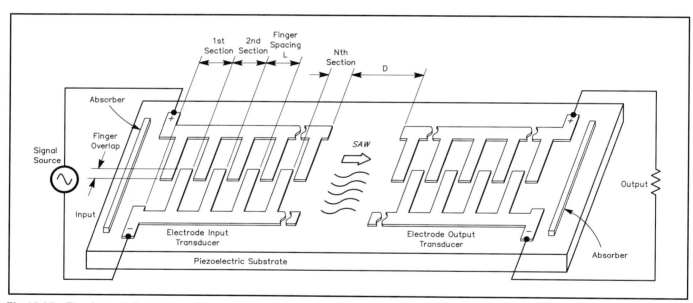

Fig 16.37—The *interdigitated* transducer, on the left, launches SAW energy to a similar transducer on the right (see text).

Transmission-Line Filters

LC filter calculations are based on the assumption that the reactances are *lumped*—the physical dimensions of the components are considerably less than the operating wavelength. Therefore the unavoidable interturn capacitance associated with inductors and the unavoidable series inductance associated with capacitors are neglected as secondary effects. If careful attention is paid to circuit layout and miniature components are used, lumped LC filter technology can be used up to perhaps 1 GHz.

Transmission-line filters predominate from 500 MHz to 10 GHz. In addition they are often used down to 50 MHz when narrowband ($Q_L > 10$) band-pass filtering is required. In this application they exhibit considerably lower loss than their LC counterparts.

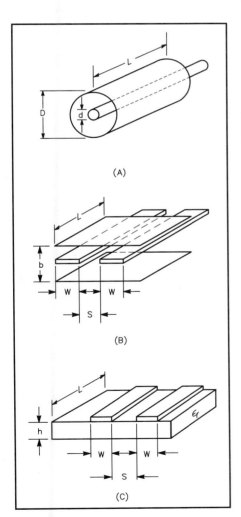

Fig 16.38—Transmission lines.
A: Coaxial line. B: Coupled stripline, which has two ground planes.
C: Microstripline, which has only one ground plane.

Replacing lumped reactances with selected short sections of TEM transmission lines results in transmission-line filters. In TEM, or *Transverse Electromagnetic Mode*, the electric and magnetic fields associated with a transmission line are at right angles (transverse) to the direction of wave propagation. Coaxial cable, stripline and microstrip are examples of TEM components. Waveguides and waveguide resonators are not TEM components.

Transmission Lines for Filters

Fig 16.38 shows three popular transmission lines used in transmission-line filters. The circular coaxial transmission line (coax) shown in Fig 16.38A consists of two concentric metal cylinders separated by dielectric (insulating) material. Coaxial transmission line possesses a characteristic impedance given by:

$$Z_0 = \frac{138}{\sqrt{\varepsilon}} \log\left(\frac{D}{d}\right) \tag{21}$$

A plot of Z_0 vs D/d is shown in **Fig 16.39**. At RF, Z_0 is an almost pure resistance. If the distant end of a section of coax is terminated in Z_0, then the impedance seen looking into the input end is also Z_0 at all frequencies. A terminated section of coax is shown in **Fig 16.40A**. If the distant end is not terminated in Z_0, the input impedance will be some other value. In Fig 16.40B the distant end is short circuited and the length is less than $1/4\ \lambda$. The input impedance is an inductive reactance as seen by the notation $+j$ in the equation in part B of the figure.

The input impedance for the case of the open-circuit distant end, is shown in Fig 16.40C. This case results in a capacitive reactance ($-j$). Thus, short sections of coaxial line (stubs) can replace the inductors and capacitors in an LC filter. Coax line inductive stubs usually have lower loss than their lumped counterparts.

X_L vs ℓ for shorted and open stubs is shown in **Fig 16.41**. There is an optimum value of Z_0 that yields lowest loss, given by

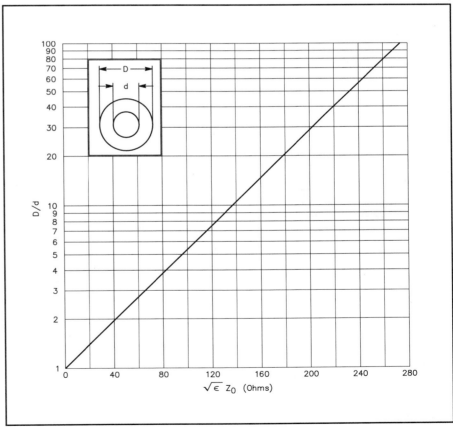

Fig 16.39—Coaxial-line impedance varies with the ratio of the inner- and outer-conductor diameters. The dielectric constant, ε, is 1.0 for air and 2.32 for polyethylene.

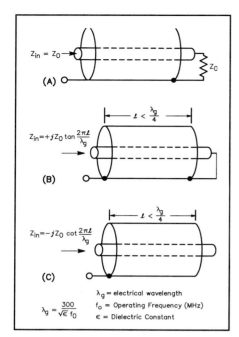

Fig 16.40—Transmission line stubs. A: A line terminated in its characteristic impedance. B: A shorted line less than ¹/₄-λ long is an *inductive* stub. C: An open line less than ¹/₄-λ long is a *capacitive* stub.

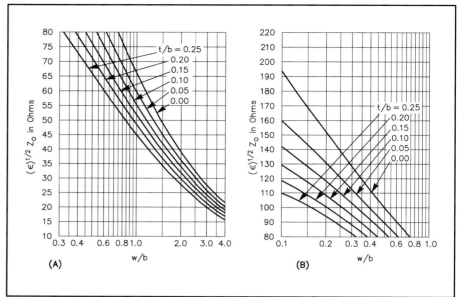

Fig 16.42—The Z_0 of stripline varies with *w*, *b* and *t* (conductor thickness). See Fig 16.38B. The conductor thickness is *t* and the plots are normalized in terms of *t/b*.

	Shorted Stub	Open Stub
$\dfrac{\ell}{\lambda_g}$	X_L Ω	X_C Ω
0	0	∞
0.05	16.2	154
0.10	36.3	68.8
0.125	50	50
0.15	68.8	36.3
0.20	154	16.2
0.25	∞	0

Fig 16.41—Stub reactance for various lengths of transmission line. Values are for $Z_0 = 50\ \Omega$. For $Z_0 = 100\ \Omega$, **double the tabulated values.**

$$Z_0 = \frac{75}{\sqrt{\varepsilon}} \qquad (22)$$

If the dielectric is air, $Z_0 = 75\ \Omega$. If the dielectric is polyethylene ($\varepsilon = 2.32$) $Z_0 = 50\ \Omega$. This is the reason why polyethylene dielectric flexible coaxial cable is usually manufactured with a 50-Ω characteristic impedance.

The first transmission-line filters were built from sections of coaxial line. Their mechanical fabrication is expensive and it is difficult to provide electrical coupling between line sections. Fabrication difficulties are reduced by the use of shielded strip transmission line (stripline) shown in Fig 16.38B. The outer conductor of stripline consists of two flat parallel metal plates (ground planes) and the inner conductor is a thin metal strip. Sometimes the inner conductor is a round metal rod. The dielectric between ground planes and strip can be air or a low-loss plastic such as polyethylene. The outer conductors (ground planes or shields) are separated from each other by distance *b*.

Striplines can be easily coupled together by locating the strips near each other as shown in Fig 16.38B. Stripline Z_0 vs width (*w*) is plotted in **Fig 16.42**. Air-dielectric stripline technology is best for low bandwidth ($Q_L > 20$) band-pass filters.

The most popular transmission line is microstrip (unshielded stripline), shown in Fig 16.38C. It can be fabricated with standard printed-circuit processes and is the least expensive configuration. Unfortunately, microstrip is the most lossy of the three lines, therefore it is not suitable for narrow band-pass filters. In microstrip the outer conductor is a single flat metal ground-plane. The inner conductor is a thin metal strip separated from the ground-plane by a solid dielectric substrate. Typical substrates are 0.062-inch G-10 fiber-glass ($\varepsilon = 4.5$) for the 50 MHz to 1-GHz frequency range and 0.031-inchTeflon ($\varepsilon = 2.3$) for frequencies above 1 GHz.

Conductor separation must be minimized or free-space radiation and unwanted coupling to adjacent circuits may become problems. Microstrip characteristic impedance and the effective dielectric constant (ε) are shown in **Fig 16.43**. Unlike coax and stripline, the effective dielectric constant is less than that of the substrate since a portion of the electromagnetic wave propagating along the microstrip "sees" the air above the substrate.

$Z_0\ \Omega$	$\varepsilon = 1$ (AIR) W/h	$\varepsilon = 2.3$ (RT/Duroid) W/h	$\sqrt{\varepsilon_e}$	$\varepsilon = 4.5$ (G–10) W/h	$\sqrt{\varepsilon_e}$
25	12.5	7.6	1.4	4.9	2.0
50	5.0	3.1	1.36	1.8	1.85
75	2.7	1.6	1.35	0.78	1.8
100	1.7	0.84	1.35	0.39	1.75
	$\sqrt{\varepsilon} = 1$				

Fig 16.43—Microstrip parameters (after H. Wheeler, *IEEE Transactions on MTT*, March 1965, p 132). ε_e is the effective ε.

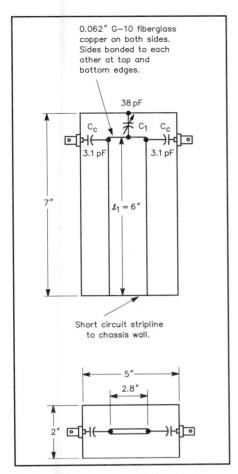

Fig 16.44—This 146-MHz stripline band-pass filter has been measured to have a Q$_L$ of 63 and a loss of approximately 1 dB.

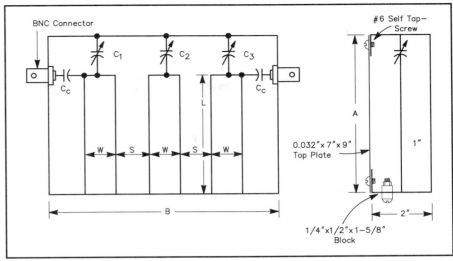

Fig 16.45—This Butterworth filter is constructed in combline. It was originally discussed by R. Fisher in December 1968 QST.

Dimension	52 MHz	146 MHz	222 MHz
A	9"	7"	7"
B	7"	9"	9"
L	7³/₈"	6"	6"
S	1"	1¹/₁₆"	1³/₈"
W	1"	1⁵/₈"	1⁵/₈"

Capacitance (pF)			
C1	110	22	12
C2	135	30	15
C3	110	22	12
C$_c$	35	6.5	2.8
Q$_L$	10	29	36

Performance			
BW3 (MHz)	5.0	5.0	6.0
Loss (dB)	0.6	0.7	—

The least-loss characteristic impedance for stripline and microstrip-lines is not 75 Ω as it is for coax. Loss decreases as line width increases, which leads to clumsy, large structures. Therefore, to conserve space, filter sections are often constructed from 50-Ω stripline or microstrip stubs.

Transmission-Line Band-Pass Filters

Band-pass filters can also be constructed from transmission-line stubs. At VHF the stubs can be considerably shorter than a quarter wavelength yielding a compact filter structure with less midband loss than its LC counterpart. The single-stage 146-MHz stripline band-pass filter shown in **Fig 16.44** is an example. This filter consists of a single inductive 50-Ω stripline stub mounted into a 2×5×7-inch aluminum box. The stub is resonated at 146 MHz with the "APC" variable capacitor,

C1. Coupling to the 50-Ω generator and load is provided by the coupling capacitors C$_c$. The measured performance of this filter is: f$_o$ = 146 MHz, BW = 2.3 MHz (Q$_L$ = 63) and midband loss = 1 dB.

Single-stage stripline filters can be coupled together to yield multistage filters. One method uses the capacitor coupled band-pass filter synthesis technique to design a 3-pole filter. Another method allows closely spaced inductive stubs to magnetically couple to each other. When the coupled stubs are grounded on the same side of the filter housing, the structure is called a "combline filter." Three examples of combline band-pass filters are shown in **Fig 16.45**. These filters are constructed in 2×7×9-inch chassis boxes.

Quarter-Wave Transmission-Line Filters

Fig 16.41 shows that when ℓ = 0.25 λ$_g$, the shorted-stub reactance becomes infi-

nite. Thus, a ¹/₄-λ shorted stub behaves like a parallel-resonant LC circuit. Proper input and output coupling to a ¹/₄-λ resonator yields a practical band-pass filter. Closely spaced ¹/₄-λ resonators will couple together to form a multistage band-pass filter. When the resonators are grounded on opposite walls of the filter housing, the structure is called an "interdigital filter" because the resonators look like interlaced fingers. Two examples of 3-pole UHF interdigital filters are shown in **Fig 16.46**. Design graphs for round-rod interdigital filters are given in Ref 16. The ¹/₄-λ resonators may be tuned by physically changing their lengths or by tuning the screw opposite each rod.

If the short-circuited ends of two ¹/₄-λ resonators are connected to each other, the resulting ¹/₂-λ stub will remain in resonance, even when the connection to ground-plane is removed. Such a floating ¹/₂-λ microstrip line, when bent into a

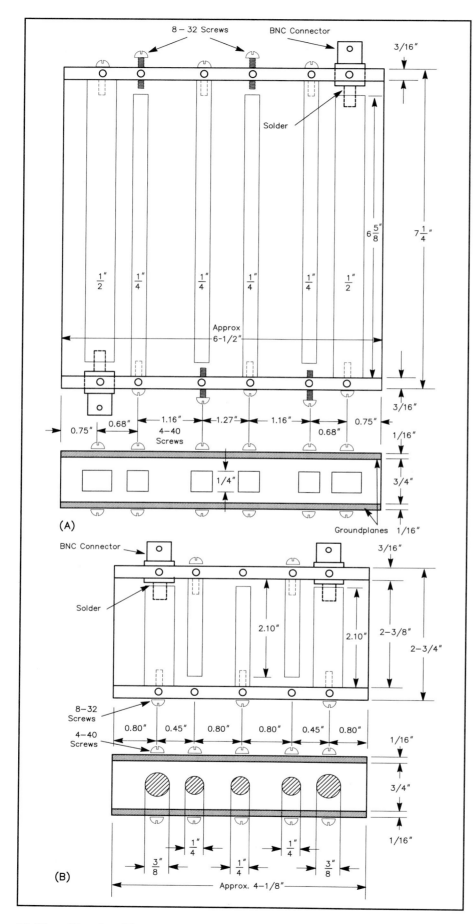

U-shape, is called a "hairpin" resonator. Closely coupled hairpin resonators can be arranged to form multistage band-pass filters. Microstrip hairpin band-pass filters are popular above 1 GHz because they can be easily fabricated using photo-etching techniques. No connection to the ground-plane is required.

Transmission-Line Filters Emulating LC Filters

Low-pass and high-pass transmission-line filters are usually built from short sections of transmission lines (stubs) that emulate lumped LC reactances. Sometimes low-loss lumped capacitors are mixed with transmission-line inductors to form a hybrid component filter. For example, consider the 720-MHz 3-pole microstrip low-pass filter shown in **Fig 16.47A** that emulates the LC filter shown in Fig 16.47B. C1 and C3 are replaced with 50-Ω open-circuit shunt stubs ℓ_C long. L2 is replaced with a short section of 100-Ω line ℓ_L long. The LC filter, Fig 16.47B, was designed for f_c = 720 MHz. Such a filter could be connected between a 432-MHz transmitter and antenna to reduce harmonic and spurious emissions. A reactance chart shows that X_C is 50 Ω, and the inductor reactance is 100 Ω at f_c. The microstrip version is constructed on G-10 fiberglass 0.062-inch thick, with ε = 4.5. Then, from Fig 16.43, w is 0.11 inch and l_C = 0.125 λ_g for the 50-Ω capacitive stubs. Also, from Fig 16.43, w is 0.024 inch and ℓ_L is 0.125 ℓ_g for the 100-Ω inductive line. The inductive line length is approximate because the far end is not a short circuit. ℓ_g is 300/(720)(1.75) = 0.238 m, or 9.37 inches. Thus ℓ_C is 1.1 inch and ℓ_L is 1.1 inch.

This microstrip filter exhibits about 20 dB of attenuation at 1296 MHz. Its response rises again, however, around 3 GHz. This is because the fixed-length transmission-line stubs change in terms of wavelength as the frequency rises. This particular filter was designed to eliminate third-harmonic energy near 1296 MHz from a 432-MHz transmitter and does a better job in this application than the Butterworth filter in Fig 16.46, which has spurious responses in the 1296-MHz band.

Fig 16.46—These 3-pole Butterworth filters (upper: 432 MHz, 8.6 MHz bandwidth, 1.4 dB pass-band loss; lower: 1296 MHz, 110 MHz bandwidth, 0.4 dB pass-band loss) are constructed as interdigitated filters. The material is from R. E. Fisher, March 1968 *QST*.

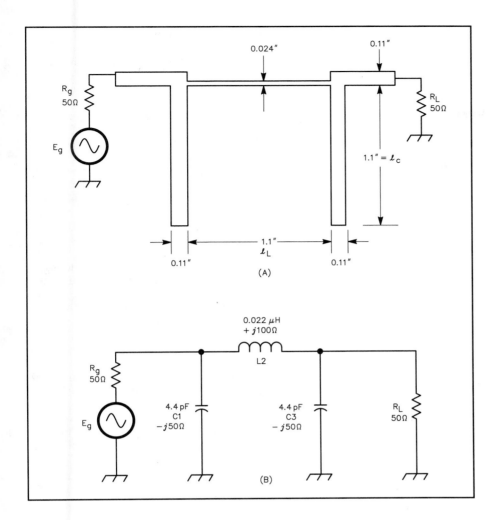

Fig 16.47—A microstrip 3-pole emulated-Butterworth low-pass filter with a cutoff frequency of 720 MHz. A: Microstrip version built with G-10 fiberglass board ($\varepsilon = 4.5$, h = 0.062 inches). B: Lumped LC version of the same filter. To construct this filter with lumped elements very small values of L and C must be used and stray capacitance and inductance must be reduced to a tiny fraction of the component values.

Helical Resonators

Ever-increasing occupancy of the radio spectrum brings with it a parade of receiver overload and spurious responses. Overload problems can be minimized by using high-dynamic-range receiving techniques, but spurious responses (such as the image frequency) must be filtered out before mixing occurs. Conventional tuned circuits cannot provide the selectivity necessary to eliminate the plethora of signals found in most urban and many suburban neighborhoods. Other filtering techniques must be used.

Helical resonators are usually a better choice than $1/4$-λ cavities on 50, 144 and 222 MHz to eliminate these unwanted inputs. They are smaller and easier to build. In the frequency range from 30 to 100 MHz it is difficult to build high-Q inductors and coaxial cavities are very large. In this frequency range the helical resonator is an excellent choice. At 50 MHz for example, a capacitively tuned, $1/4$-λ coaxial

cavity with an unloaded Q of 3000 would be about 4 inches in diameter and nearly 5 ft long. On the other hand, a helical resonator with the same unloaded Q is about 8.5 inches in diameter and 11.3 inches long. Even at 432 MHz, where coaxial cavities are common, the use of helical resonators results in substantial size reductions.

The helical resonator was described by W1HR in a *QST* article as a coil surrounded by a shield, but it is actually a shielded, resonant section of helically wound transmission line with relatively high characteristic impedance and low axial propagation velocity. The electrical length is about 94% of an axial $1/4$-λ or 84.6°. One lead of the helical winding is connected directly to the shield and the other end is open circuited as shown in **Fig 16.48**. Although the shield may be any shape, only round and square shields will be considered here.

Design

The unloaded Q of a helical resonator is determined primarily by the size of the shield. For a round resonator with a copper coil on a low-loss form, mounted in a copper shield, the unloaded Q is given by

$$Q_U = 50D\sqrt{f_o} \qquad (23)$$

where

D = inside diameter of the shield, in inches

f_O = frequency, in MHz.

D is assumed to be 1.2 times the width of one side for square shield cans. This formula includes the effects of losses and imperfections in practical materials. It yields values of unloaded Q that are easily attained in practice. Silver plating the shield and coil increases the unloaded Q by about 3% over that predicted by the equation. At VHF and UHF, however, it is

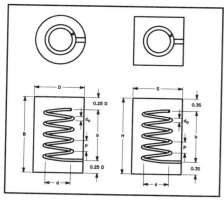

Fig 16.48—Dimensions of round and square helical resonators. The diameter, D (or side, S) is determined by the desired unloaded Q. Other dimensions are expressed in terms of D or S (see text).

more practical to increase the shield size slightly (that is, increase the selected Q_U by about 3% before making the calculation). The fringing capacitance at the open-circuit end of the helix is about 0.15 D pF (that is, approximately 0.3 pF for a shield 2 inches in diameter). Once the required shield size has been determined, the total number of turns, N, winding pitch, P and characteristic impedance, Z_0, for round and square helical resonators

with air dielectric between the helix and shield, are given by:

$$N = \frac{1908}{f_0 D} \qquad (24A)$$

$$P = \frac{f_0 D^2}{2312} \qquad (24B)$$

$$Z_0 = \frac{99,000}{f_0 D} \qquad (24C)$$

$$N = \frac{1590}{f_0 S} \qquad (24D)$$

$$P = \frac{f_0 S^2}{1606} \qquad (24E)$$

$$Z_0 = \frac{82,500}{f_0 S} \qquad (24F)$$

In these equations, dimensions D and S are in inches and f_0 is in megahertz. The design nomograph for round helical resonators in **Fig 16.49** is based on these formulas.

Although there are many variables to consider when designing helical resonators, certain ratios of shield size to length and coil diameter to length, provide optimum results. For helix diameter, d = 0.55 D or d = 0.66 S. For helix length, b = 0.825D or b = 0.99S. For shield length, B = 1.325 D and H = 1.60 S.

Fig 16.50 simplifies calculation of these dimensions. Note that these ratios result in a helix with a length 1.5 times its diameter, the condition for maximum Q. The shield is about 60% longer than the helix—although it can be made longer—to completely contain the electric field at the top of the helix and the magnetic field at the bottom.

The winding pitch, P, is used primarily to determine the required conductor size. Adjust the length of the coil to that given by the equations during construction. Conductor size ranges from 0.4 P to 0.6 P for both round and square resonators and are plotted graphically in **Fig 16.51**.

Obviously, an area exists (in terms of frequency and unloaded Q) where the designer must make a choice between a conventional cavity (or lumped LC circuit) and a helical resonator. The choice is affected by physical shape at higher frequencies. Cavities are long and relatively small in diameter, while the length of a helical resonator is not much greater than its diameter. A second consideration is that point where the winding pitch, P, is less than the radius of the helix (otherwise the structure tends to be nonhelical). This condition occurs when the helix has fewer than three turns (the "upper limit" on the

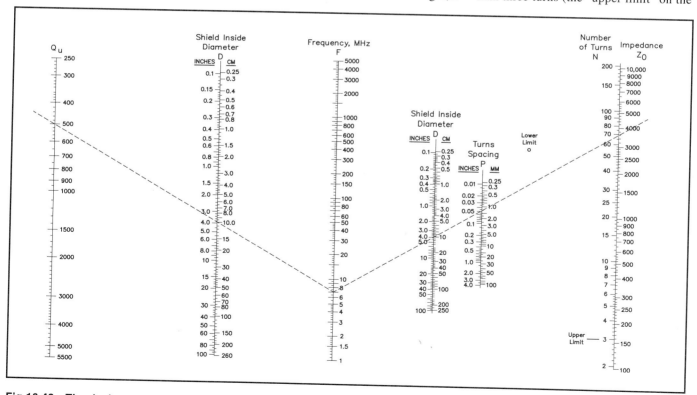

Fig 16.49—The design nomograph for round helical resonators starts by selecting Q_U and the required shield diameter. A line is drawn connecting these two values and extended to the frequency scale (example here is for a shield of about 3.8 inches and Q_U of 500 at 7 MHz). Finally the number of turns, N, winding pitch, P, and characteristic impedance, Z_0, are determined by drawing a line from the frequency scale through selected shield diameter (but this time to the scale on the right-hand side. For the example shown, the dashed line shows P ≈ 0.047 inch, N = 70 turns, and Z_n = 3600 Ω).

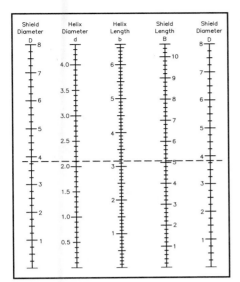

Fig 16.50—The helical resonator is scaled from this design nomograph. Starting with the shield diameter, the helix diameter, d, helix length, b, and shield length, B, can be determined with this graph. The example shown has a shield diameter of 3.8 inches. This requires a helix mean diameter of 2.1 inches, helix length of 3.1 inches, and shield length of 5 inches.

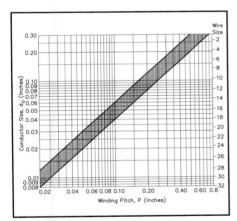

Fig 16.51—This chart provides the design information of helix conductor size vs winding pitch, P. For example, a winding pitch of 0.047 inch results in a conductor diameter between 0.019 and 0.028 inch (#22 or #24 AWG).

design nomograph of Fig 16.49).

Construction

The shield should not have any seams parallel to the helix axis to obtain as high an unloaded Q as possible. This is usually not a problem with round resonators because large-diameter copper tubing is used for the shield, but square resonators require at least one seam and usually more. The effect on unloaded Q is minimum if the seam is silver soldered carefully from one end to the other.

Results are best when little or no dielectric is used inside the shield. This is usually no problem at VHF and UHF because the conductors are large enough that a supporting coil form is not required. The lower end of the helix should be soldered to the nearest point on the inside of the shield.

Although the external field is minimized by the use of top and bottom shield covers, the top and bottom of the shield may be left open with negligible effect on frequency or unloaded Q. Covers, if provided, should make electrical contact with the shield. In those resonators where the helix is connected to the bottom cover, that cover must be soldered solidly to the shield to minimize losses.

Tuning

A carefully built helical resonator designed from the nomograph of Fig 16.49 will resonate very close to the design frequency. Slightly compress or expand the helix to adjust resonance over a small range. If the helix is made slightly longer than that called for in Fig 16.50, the resonator can be tuned by pruning the open end of the coil. However, neither of these methods is recommended for wide frequency excursions because any major deviation in helix length will degrade the unloaded Q of the resonator.

Most helical resonators are tuned by means of a brass tuning screw or high-quality air-variable capacitor across the open end of the helix. Piston capacitors also work well, but the Q of the tuning capacitor should ideally be several times the unloaded Q of the resonator. Varactor diodes have sometimes been used where remote tuning is required, but varactors can generate unwanted harmonics and other spurious signals if they are excited by strong, nearby signals.

When a helical resonator is to be tuned by a variable capacitor, the shield size is based on the chosen unloaded Q at the operating frequency. Then the number of turns, N *and* the winding pitch, P, are based on resonance at $1.5\ f_0$. Tune the resonator to the desired operating frequency, f_0.

Insertion Loss

The insertion loss (dissipation loss), I_L, in decibels, of all single-resonator circuits is given by

$$I_L = 20\log_{10}\left(\frac{1}{1 - \dfrac{Q_L}{Q_U}}\right) \qquad (25)$$

where
$\qquad Q_L = $ loaded Q
$\qquad Q_U = $ unloaded Q

This is plotted in **Fig 16.52**. For the most practical cases ($Q_L > 5$), this can be closely approximated by $I_L \approx 9.0\ (Q_L/Q_U)$ dB. The selection of Q_L for a tuned circuit is dictated primarily by the required selectivity of the circuit. However, to keep dissipation loss to 0.5 dB or less (as is the case for low-noise VHF receivers), the unloaded Q must be at least 18 times the Q_L.

Coupling

Signals are coupled into and out of helical resonators with inductive loops at the bottom of the helix, direct taps on the coil or a combination of both. Although the correct tap point can be calculated easily, coupling by loops and probes must be determined experimentally.

The input and output coupling is often provided by probes when only one resonator is used. The probes are positioned on opposite sides of the resonator for maximum isolation. When coupling loops are used, the plane of the loop should be perpendicular to the axis of the helix and separated a small distance from the bottom of the coil. For resonators with only a few turns, the plane of the loop can be tilted slightly so it is parallel with the slope of the adjacent conductor.

Helical resonators with inductive coupling (loops) exhibit more attenuation to signals above the resonant frequency (as compared to attenuation below resonance), whereas resonators with capacitive coupling (probes) exhibit more attenuation below the passband, as shown for a typical 432-MHz resonator in **Fig 16.53**. Consider this characteristic when choosing a coupling method. The passband can be made more symmetrical by using a combination of coupling methods (inductive input and capacitive output, for example).

If more than one helical resonator is required to obtain a desired band-pass characteristic, adjacent resonators may be coupled through apertures in the shield

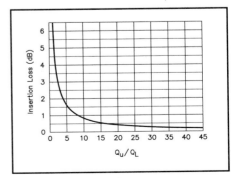

Fig 16.52—The ratio of loaded (Q_L) to unloaded (Q_U) Q determines the insertion loss of a tuned resonant circuit.

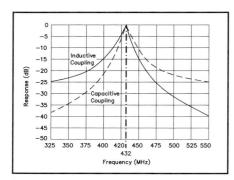

Fig 16.53—This response curve for a single-resonator 432-MHz filter shows the effects of capacitive and inductive input/output coupling. The response curve can be made symmetrical on each side of resonance by combining the two methods (inductive input and capacitive output, or vice versa).

wall between the two resonators. Unfortunately, the size and location of the aperture must be found empirically, so this method of coupling is not very practical unless you're building a large number of identical units.

Since the loaded Q of a resonator is determined by the external loading, this must be considered when selecting a tap (or position of a loop or probe). The ratio of this external loading, R_b, to the characteristic impedance, Z_0, for a $1/4$-λ resonator is calculated from:

$$K = \frac{R_b}{Z_0} = 0.785\left(\frac{1}{Q_L} - \frac{1}{Q_U}\right) \qquad (26)$$

Even when filters are designed and built properly, they may be rendered totally ineffective if not installed properly. Leakage around a filter can be quite high at VHF and UHF, where wave-lengths are short. Proper attention to shielding and good grounding is mandatory for minimum leakage. Poor coaxial cable shield connection into and out of the filter is one of the greatest offenders with regard to filter leakage. Proper dc-lead bypassing throughout the receiving system is good practice, especially at VHF and above. Ferrite beads placed over the dc leads may help to reduce leakage. Proper filter termination is required to minimize loss.

Most VHF RF amplifiers optimized for noise figure do not have a 50-Ω input impedance. As a result, any filter attached to the input of an RF amplifier optimized for noise figure will not be properly terminated and filter loss may rise substantially. As this loss is directly added to the RF amplifier noise figure, carefully choose and place filters in the receiver.

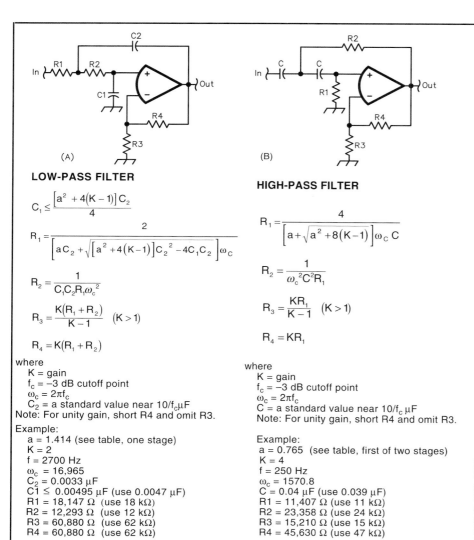

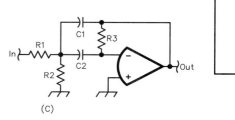

(A)

LOW-PASS FILTER

$$C_1 \leq \frac{\left[a^2 + 4(K-1)\right]C_2}{4}$$

$$R_1 = \frac{2}{\left[aC_2 + \sqrt{\left[a^2 + 4(K-1)\right]C_2{}^2 - 4C_1C_2}\right]\omega_c}$$

$$R_2 = \frac{1}{C_1 C_2 R_1 \omega_c{}^2}$$

$$R_3 = \frac{K(R_1 + R_2)}{K-1} \quad (K > 1)$$

$$R_4 = K(R_1 + R_2)$$

where
  K = gain
  f_c = −3 dB cutoff point
  $\omega_c = 2\pi f_c$
  C_2 = a standard value near $10/f_c$ μF
Note: For unity gain, short R4 and omit R3.

Example:
  a = 1.414 (see table, one stage)
  K = 2
  f = 2700 Hz
  ω_c = 16,965
  C_2 = 0.0033 μF
  C1 ≤ 0.00495 μF (use 0.0047 μF)
  R1 = 18,147 Ω (use 18 kΩ)
  R2 = 12,293 Ω (use 12 kΩ)
  R3 = 60,880 Ω (use 62 kΩ)
  R4 = 60,880 Ω (use 62 kΩ)

(B)

HIGH-PASS FILTER

$$R_1 = \frac{4}{\left[a + \sqrt{a^2 + 8(K-1)}\right]\omega_c C}$$

$$R_2 = \frac{1}{\omega_c{}^2 C^2 R_1}$$

$$R_3 = \frac{KR_1}{K-1} \quad (K > 1)$$

$$R_4 = KR_1$$

where
  K = gain
  f_c = −3 dB cutoff point
  $\omega_c = 2\pi f_c$
  C = a standard value near $10/f_c$ μF
Note: For unity gain, short R4 and omit R3.

Example:
  a = 0.765 (see table, first of two stages)
  K = 4
  f = 250 Hz
  ω_c = 1570.8
  C = 0.04 μF (use 0.039 μF)
  R1 = 11,407 Ω (use 11 kΩ)
  R2 = 23,358 Ω (use 24 kΩ)
  R3 = 15,210 Ω (use 15 kΩ)
  R4 = 45,630 Ω (use 47 kΩ)

(C)

BAND-PASS FILTER

Pick K, Q, $\omega_o = 2\pi f_c$
where f_c = center freq.
Choose C

Then $R1 = \dfrac{Q}{K_o \omega_o C}$

$$R2 = \frac{Q}{\left(2Q^2 - K_o\right)\omega_o C}$$

$$R3 = \frac{2Q}{\omega_o C}$$

Example:
  K = 2, f_o = 800 Hz, Q = 5 and
  C = 0.022 μF
  R1 = 22.6 kΩ (use 22 kΩ)
  R2 = 942 Ω (use 1000 Ω)
  R3 = 90.4 kΩ (use 91 kΩ or
  100 kΩ)

Fig 16.54—Equations for designing a low-pass RC active audio filter are given at A. B, C and D show design information for high-pass, band-pass and band-reject filters, respectively. All of these filters will exhibit a Butterworth response. Values of K and Q should be less than 10.

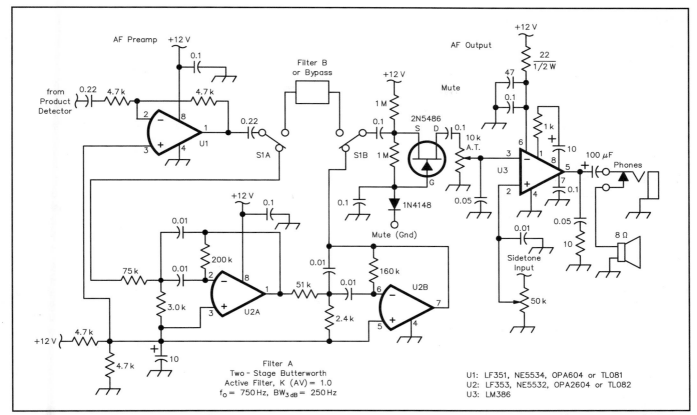

Fig 16.55—Typical application of a two-stage active filter in the audio chain of a QRP CW tranceiver. The filter can be bypassed, or another filter can be switched in by S1.

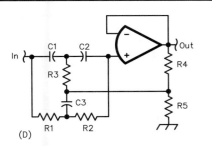

(D)

BAND-REJECT FILTER

$$F_0 = \frac{1}{2\pi R1C1}$$

$$K = 1 - \frac{1}{4Q}$$

$$R \gg (1-K) R1$$

where

$$C1 = C2 = \frac{C3}{2} = \frac{10\,\mu F}{f_0}$$

R1 = R2 = 2R3
R4 = (1 – K)R
R5 = K × R
Example:
  f_0 = 500 Hz, Q=10
  K = 0.975
  C1 = C2 = 0.02 μF (or use 0.022 μF)
  C3 = 0.04 μF (or use 0.044 μF)
  R1 = R2 = 15.92 kΩ (use 15 kΩ)
  R3 = 7.96 kΩ (use 7.5 kΩ)
  R ≫ 375 Ω (1 kΩ)
  R4 = 25 Ω (use 27Ω)
  R5 = 975 Ω (use 1 kΩ)

ACTIVE FILTERS

Passive HF filters are made from combinations of inductors and capacitors. These may be used at low frequencies, but the inductors often become a limiting factor because of their size, weight, cost and losses. The active filter is a compact, low-cost alternative made with op amps, resistors and capacitors. They often occupy a fraction of the space required by an LC filter. While active filters have been traditionally used at low and audio frequencies, modern op amps with small-signal bandwidths that exceed 1 GHz have extended their range into MF and HF.

Active filters can perform any common filter function: low pass, high pass, bandpass, band reject and all pass (used for phase or time delay). Responses such as Butterworth, Chebyshev, Bessel and ellip-

tic can be realized. Active filters can be designed for gain, and they offer excellent stage-to-stage isolation.

Despite the advantages, there are also some limitations. They require power, and performance may be limited by the op amp's finite input and output levels, gain and bandwidth. While LC filters can be designed for high-power applications, active filters usually are not.

The design equations for various filters are shown in **Fig 16.54**. A program (ACTFIL.EXE), useful for designing 1-4 stage filters, is available from *ARRLWeb* (see page viii). **Fig 16.55** shows a typical application of a two-stage, bandpass filter. A two-stage filter is considered the minimum acceptable for CW, while three or four stages will prove more effective under some conditions of noise and interference.

Factor "a" for Low- and High-Pass Filters

No. of Stages	Stage 1	Stage 2	Stage 3	Stage 4
1	1.414	-	-	-
2	0.765	1.848	-	-
3	0.518	1.414	1.932	-
4	0.390	1.111	1.663	1.962

These values are truncated from those of Appendix C of Ref 21, for even-order Butterworth filters.

CRYSTAL-FILTER EVALUATION

Crystal filters, such as those described earlier in this chapter, are often constructed of surplus crystals or crystals whose characteristics are not exactly known. Randy Henderson, WI5W, developed a swept frequency generator for testing these filters. It was first described in March 1994 *QEX*. This test instrument adds to the ease and success in quickly building filters from inexpensive microprocessor crystals.

A template, containing additional information, is available from the *ARRLWeb*, **http://www.arrl.org/notes**.

An Overview

The basic setup is shown in **Fig 16.56A**. The VCO is primarily a conventional LC-tuned Hartley oscillator with its frequency tuned over a small range by a varactor diode (MV2104 in part B of the figure). Other varactors may be used as long as the capacitance specifications aren't too different. Change the 5-pF coupling capacitor to expand the sweep width if desired.

The VCO signal goes through a buffer amplifier to the filter under test. The filter is followed by a wide-bandwidth amplifier and then a detector. The output of the detector is a rectified and filtered signal. This varying dc voltage drives the vertical input of an oscilloscope. At any particular time, the deflection and sweep circuitry commands the VCO to "run at this frequency." The same deflection voltage causes the oscilloscope beam to deflect left or right to a position corresponding to the frequency.

Any or all of these circuits may be eliminated by the use of appropriate commercial test equipment. For example, a commercial sweep generator would eliminate the need for everything but the wide-band amplifier and detector. Motorola, Mini-Circuits Labs and many others sell devices suitable for the wide-band amplifiers and detector.

The generator/detector system covers approximately 6 to 74 MHz in three ranges. Each tuning range uses a separate RF oscillator module selected by switch S1. The VCO output and power-supply input are multiplexed on the "A" lead to each oscillator. The tuning capacitance for each VCO is switched into the appropriate circuit by a second set of contacts on S1. C_T is the coarse tuning adjustment for each oscillator module.

Two oscillator coils are wound on PVC plastic pipe. The third, for the highest frequency range, is self supporting #14 copper wire. Although PVC forms with Super Glue dope may not be "state of the art" technology, frequency stability is completely adequate for this instrument.

The oscillator and buffer stage operate at low power levels to minimize frequency drift caused by component heating. Crystal filters cause large load changes as the frequency is swept in and out of the passband. These large changes in impedance tend to "pull" the oscillator frequency and cause inaccuracies in the passband shape depicted by the oscilloscope. Therefore a buffer amplifier is a necessity. The wide-band amplifier in **Fig 16.57** is derived from one in ARRL's *Solid State Design for the Radio Amateur*.

S2 selects a 50-Ω 10-dB attenuator in the input line. When the attenuator is in the line, it provides a better output match for the filter under test. The detector uses some forward bias for D2. A simple unbiased diode detector would offer about 50 dB of dynamic range. Some dc bias increases the dynamic range to almost 70 dB. D3, across the detector output (the scope input), increases the vertical-amplifier sensitivity while compressing or limiting the response to high-level signals. With this arrangement, high levels of attenuation (low-level signals) are easier to observe and low attenuation levels are still visible on the CRT. The diode only kicks in to provide limiting at higher signal levels.

The horizontal-deflection sweep circuit uses a dual op-amp IC (see **Fig 16.58**). One section is an oscillator; the other is an integrator. The integrator output changes linearly with time, giving a uniform brightness level as the trace is moved from side to side. Increasing C1 decreases the

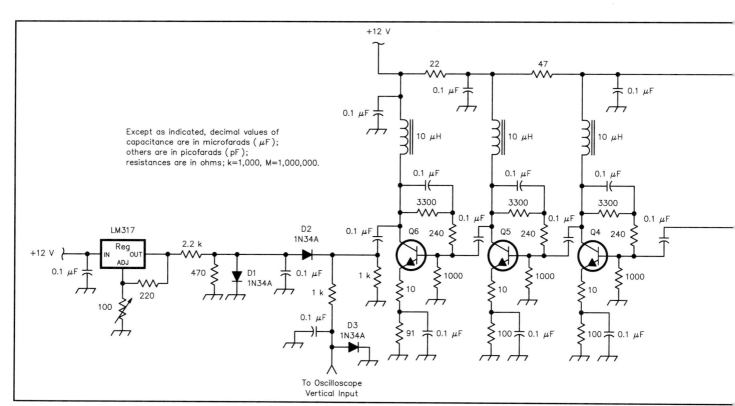

Except as indicated, decimal values of capacitance are in microfarads (μF); others are in picofarads (pF); resistances are in ohms; k=1,000, M=1,000,000.

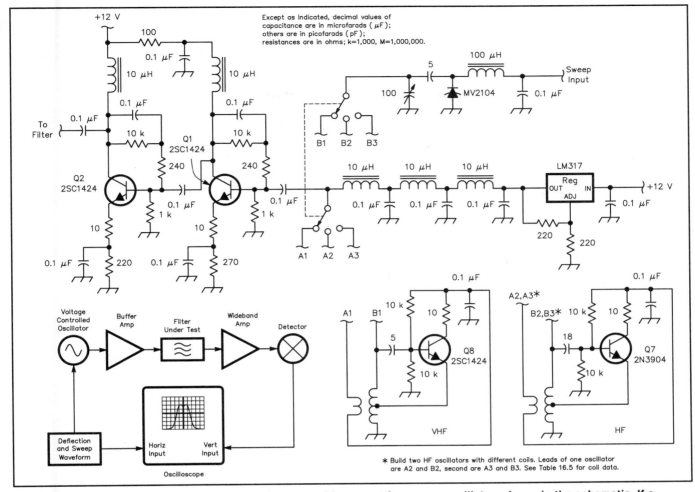

Fig 16.56—The test set block diagram, lower left, starts with a swept frequency oscillator, shown in the schematic. If a commercial swept-frequency oscillator is available, it can be substituted for the circuit shown.

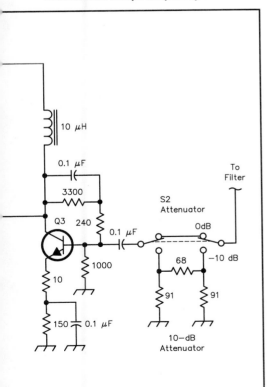

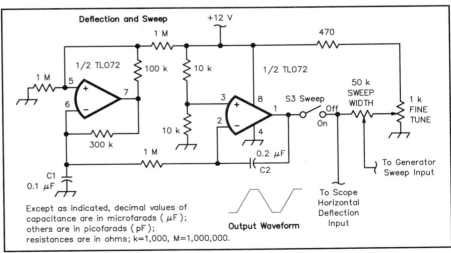

Fig 16.58—The sweep generator provides both an up and down sweep voltage (see text) for the swept frequency generator and the scope horizontal channel.

Fig 16.57—The filter under test is connected to Q3 on the right side of the schematic. The detector output, on the left side, connects to the oscilloscope vertical input. A separate voltage regulator, an LM 317, is used to power this circuit. Q3, Q4, Q5 and Q6 are 25C1424 or 2N2857.

sweep rate. Increasing C2 decreases the slope of the output waveform ramp.

Operation

The CRT is swept in both directions, left to right and right to left. The displayed curve is a result of changes in frequency, not time. Therefore it is unnecessary to incorporate the usual right-to-left, snap-back and retrace blanking used in oscilloscopes.

S3 in Fig 16.58 disables the automatic sweep function when opened. This permits manual operation. Use a frequency counter to measure the VCO output, from which bandwidth can be calculated. Turn the fine-tune control to position the CRT beam at selected points of the passband curve. The difference in frequency readings is the bandwidth at that particular point or level of attenuation.

Substitution of a calibrated attenuator for the filter under test can provide reference readings. These reference readings

may be used to calibrate an otherwise uncalibrated scope vertical display in dB.

The buffer amplifier shown here is set up to drive a 50-Ω load and the wide-bandwidth amplifier input impedance is about 50 Ω. If the filter is not a 50-Ω unit, however, various methods can be used to accommodate the difference.

Table 16.5
VCO Coils

Coil	Inside Diameter (inches)	Length (inches)	Turns, Wire	Inductance (μH)
large	0.85	1.1	18 t, #28	5.32
medium	0.85	0.55	7 t, #22	1.35
small	0.5	0.75	5 t, #14	0.27

The two larger coils are wound on ¾-inch PVC pipe and the smaller one on a ½-inch drill bit. Tuning coverage for each oscillator is obtained by squeezing or spreading the turns before gluing them in place. The output windings connected to A1, A2 and A3 are each single turns of #14 wire spaced off the end of the tapped coils.

References

A. Ward, "Monolithic Microwave Integrated Circuits," Feb 1987 *QST*, pp 23-29.

Z. Lau, "A Logarithmic RF Detector for Filter Tuning," Oct 1988 *QEX*, pp 10-11.

BAND-PASS FILTERS FOR 144 OR 222 MHz

Spectral purity is necessary during transmitting. Tight filtering in a receiving system ensures the rejection of out-of-band signals. Unwanted signals that lead to receiver overload and increased intermodulation-distortion (IMD) products result in annoying in-band "birdies." One solution is the double-tuned band-pass filters shown in **Fig 16.64**. They were designed by Paul Drexler, WB3JYO. Each includes a resonant trap coupled between

the resonators to provide increased rejection of undesired frequencies.

Many popular VHF conversion schemes use a 28-MHz intermediate frequency (IF), yet proper filtering of the image frequency is often overlooked in amateur designs. The low-side injection frequency used in 144-MHz mixing schemes is 116 MHz and the image frequency, 88 MHz, falls in TV channel 6. Inadequate rejection of a broadcast carrier at this frequency results in a strong, wideband signal at the low end of the 2-m band. A similar problem on the transmit side can cause TVI. These band-pass filters have effectively suppressed undesired mixing products. See **Fig 16.65** and **16.66**.

The circuit is constructed on a double-sided copper-clad circuit board. Minimize

component lead lengths to eliminate resistive losses and unwanted stray coupling. Mount the piston trimmers through the board with the coils soldered to the opposite end, parallel to the board. The shield between L1 and L3 decreases mutual coupling and improves the frequency response. Peak C1 and C3 for optimum response.

L1, C1, L3 and C3 form the tank circuits that resonate at the desired frequency. C2 and L2 reject the undesired energy while allowing the desired signal to pass. The tap points on L1 and L3 provide 50-Ω matching; they may be adjusted for optimum energy transfer. Several filters have been constructed using a miniature variable capacitor in place of C2 so that the notch frequency could be varied.

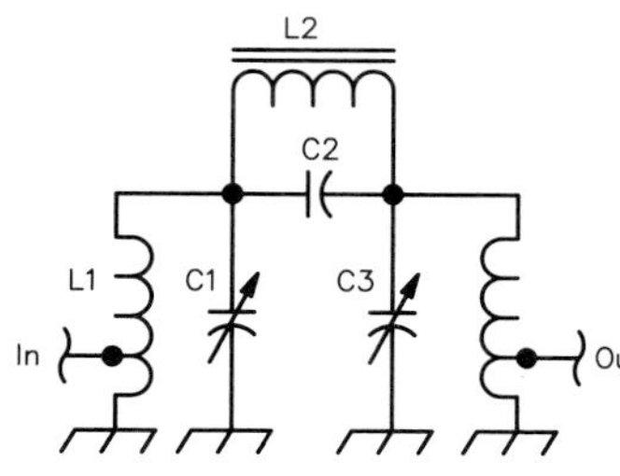

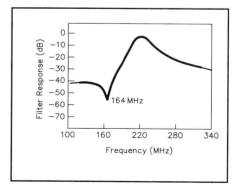

Fig 16.64—Schematic of the band-pass filter. Components must be chosen to work with the power level of the transmitter.

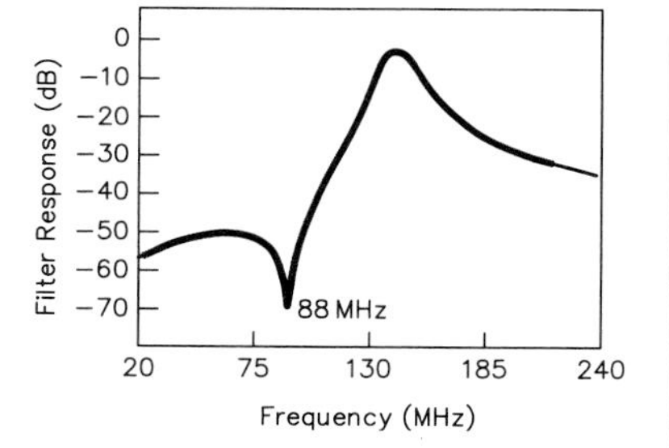

Component Values

	144 MHz	220 MHz
C2	1 pF	1 pF
C1, C3	1-7 pf piston	1-7 pF piston
L2	27t no. 26 enam on T37-10	15t no. 24 enam on T44-10
L1, L3	7t no. 18, ¼-in ID, tap 1½t	4t no. 18, ¼-in ID, tap 1½t

Fig 16.65—Filter response plot of the 144-MHz band-pass filter, with an image-reject notch for a 28 MHz IF.

Fig 16.66—Filter response plot of the 222-MHz band-pass filter, with an image-reject notch for a 28 MHz IF.

Switched Capacitor Filters

The *switched capacitor filter,* or SCF, uses an IC to synthesize a high-pass, low-pass, band-pass or notch filter. The performance of multiple-pole filters is available, with Q and bandwidth set by external resistors. An external clock frequency sets the filter center frequency, so this frequency may be easily changed or digitally controlled. Dynamic range of 80 dB, Q of 50, 5-pole equivalent design and maximum usable frequency of 250 kHz are available for such uses as audio CW and RTTY filters. In addition, all kinds of digital tone signaling such as DTMF and modem encoding and decoding are being designed with these circuits.

A CONTINUOUSLY VARIABLE BANDWIDTH AUDIO FILTER

Active audio filters are a very popular way to enhance the selectivity, especially on CW, of many rigs. Passive LC filters, usually based on 88-mH toroids, were used for many years, followed by active filters built with one or more opamps. Switched capacitor filters, such as this one designed by Denton Bramwell, K7OWJ, offer additional performance, with fewer chips. This project was first described in the July 1995 issue of *QST*.

General Description

Maxim[1] markets a series of useful filter (and other) chips that exhibit excellent performance. The filter described here rolls off at an impressive 96 dB per octave. Best of all, the upper cutoff frequency *can be continuously varied* to accommodate any reasonable desired bandwidth! The values given here provide a 3-dB bandwidth ranging from 450 Hz for CW to 2700 Hz for voice. The filter's bandwidth is determined by merely adjusting a potentiometer.

Maxim's switched-capacitor-chip family provides Bessel, Butterworth and elliptic low-pass designs. This Butterworth version has a flat passband—which is excellent if you're cascading stages—and is based on the MAX 295 chip.

These ICs are exceptionally easy to apply (see Table 1). If you want but one stage (48 dB per octave) and you don't need to vary the filter's bandwidth, all that's required are ±5-V supplies and a single capacitor. Working with an oscillator internal to the IC, the capacitor sets the clock rate. Attach input and output lines, add a few inexpensive components, a single-voltage power supply and you will have a working one-stage filter.

The '295 will not operate on voltages much higher than ±5 (split supply) or +10 V (a single supply). A potential of 12 V between the supply leads is the rated absolute maximum, so if you try to use this voltage, you may well pop the chip.

Circuit Description

Fig 16.67 shows the incoming audio from the receiver passing through a simple RC high-pass filter, and on to U1. U1's output is coupled through another simple RC high-pass filter to the input of U2. U2's output is filtered by two more stages of active RC high-passing using 741 op amps (U3 and U4). This fixes a −3 dB point about 300 Hz on the low-frequency side of the filter and a completely adjustable −3 dB point on the high side.

The bypassing shown uses 220-μF capacitors for decoupling the stages. Good decoupling is absolutely necessary to achieve good ultimate rejection. (Dual and quad op amps can't be decoupled from each other because they all use the same supply!) The 220-μF capacitors have a reactance of 2.9 Ω at 250 Hz. You might get by with smaller capacitance values.

Up to 5 mV of clock feedthrough can appear with the audio at the output of the '295. Maxim provides an op amp within the '295 for RC active filtering to suppress clock leakage. This is accomplished by C7, C8, R7, R8 and R9, which are chosen for a corner frequency of 3.4 kHz.

For continuously variable bandwidth service, it's necessary to use an external clock to feed the MAX295s (U1 and U2). A 555 timer (U5) is configured as an astable multivibrator. The output frequency is controlled by a potentiometer (R20). The corner frequency of the '295 is $1/50$ of the clock frequency. With the components shown, U5 delivers 37 kHz to 180 kHz, for a corner frequency ranging from 740 to 3,600 Hz. The 555 output frequency range will change with supply voltage. Be prepared to adjust the values if you use a supply other than ± 5 V. The resistance of most potentiometers is usually not tightly controlled, so the actual clock range you obtain may be a bit different from the values above

The supply shown for this filter provides ±5-V supplies—the negative supply must deliver about 25 mA. If you want to operate this filter from a single

Table 1
Selected Maxim Filter ICs

Part Number	Filter Type	Unit Price ($US)	Rolloff Characteristics	Notes
MAX291 MAX295	8th-order Butterworth	6	48 dB/octave; about 110 dB ultimate	Maximum flatness in the passband; excellent ultimate rejection
MAX293 MAX297	8th-order elliptic	6	−80 dB at 1.5 × the corner frequency; about 80 dB ultimate	Probably an excellent choice for a single-IC, fixed-frequency filter. A MAX297 can directly replace a MAX295, if steeper rolloff is needed and more in-band ripple is acceptable.

Note: For single-chip, fixed-frequency filters using the MAX295 and MAX297, the corner frequency can be set by a single capacitor connected between pin 1 and ground. The internal clock frequency, F_1, is determined by the value of the clock capacitor, C, according to the formula: $fl_{(kHz)} = 10^5 \div 3C_{(pF)}$. The corner frequency equals $1/50$ of the clock frequency.

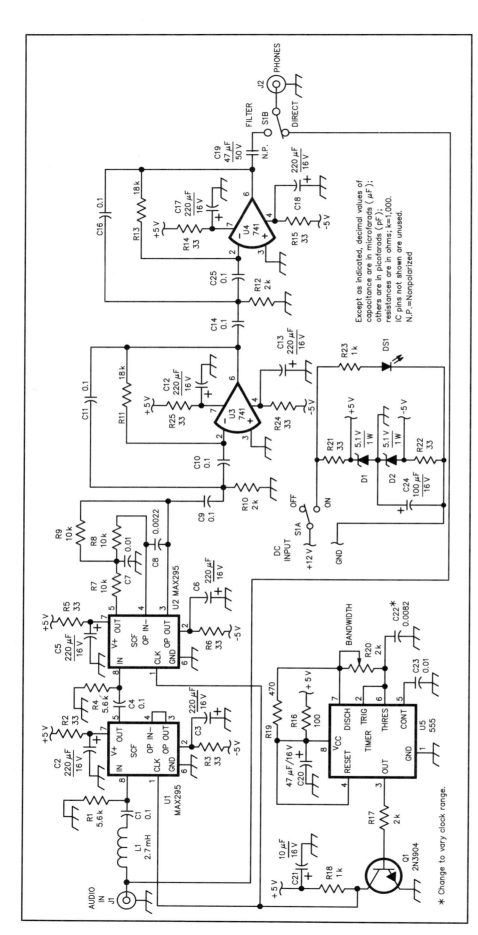

Fig 16.67—Schematic diagram of the continuously variable audio filter. Unless otherwise specified, resistors are ¼-W, 5%-tolerance carbon-composition or film units. Equivalent parts can be substituted.

C2, C3, C5, C6, C12, C13, C17, C18—
   220 µF, 16 V electrolytic
C19—47 µF, 50 V nonpolarized
C20—47 µF, 16 V electrolytic or
   tantalum
C21—10 µF, 16 V electrolytic or
   tantalum
C24—100 µF, 16 V electrolytic or
   tantalum
D1, D2—5.1-V, 1-W Zener diode
   (1N4733)
DS1—LED
J1, J2—Phono jack
L1—2.7 mH (optional)
Q1—2N3904
R20—2-kΩ linear-taper potentiometer
S1—DPDT toggle
U1, U2—MAX295
U3, U4—741 op amp
U5—555 timer

12-V supply, you can create a virtual ground by stacking two Zener diodes (see the inset of Fig 16.67). With this approach (rather than using separate positive and negative supplies, *be sure that the common (ground) line of the filter's PC board never meets the ground for the rest of your station!* The filter's ground must float. If you enclose your filter in a plastic box as I did (and the jacks are thus mounted in plastic), it's okay for the incoming audio and the 12-V supply to share the same return (C1 provides dc decoupling). Since the output is a jack to a set of headphones, the actual ground is not important—unless you add an outboard audio amplifier. Then you must add an audio transformer for isolation.

Construction

Layout is generally non-critical. The author built his filter on a general-purpose prototyping board using point-to-point wiring, but PC boards are available.[2]

In Use

With S1 off, the filter receives no power and the headphones are connected directly to the receiver output. With S1 on, power is applied to the filter and the headphones are connected to the filter output. Because the filter has unity gain, the input and output audio levels are equal, so there's no need to adjust your receiver's audio gain control when switching the filter in and out of the line.

Summary

To test the filter on-the-air, tune to a busy region of a band and listen as the interfering signals drop into oblivion as you rotate R20, the BANDWIDTH control. With an attenuation of 96 dB per octave, only a small adjustment of the potentiometer moves a signal from very bothersome to below the noise. By itself, the audio filter provides skirts as deep and as steep as better-quality IF filters.

The unit's distributed high-pass filtering makes a quite noticeable reduction in low-frequency grunts and rumbles. It also colors the atmospheric noise a bit. With the pot set for maximum width, you'll hear the apparent pitch of the background noise shift slightly upward when you turn on the filter. This shift represents noise you no longer have to cope with when copying a station.

Notes

[1]Maxim ICs are available from several sources including Digi-Key Corp. See the **References** chapter.

[2]PC boards are available from FAR Circuits. The price: $6 plus $1.50 shipping. FAR Circuits is listed in the **References** chapter Address List. A PC-board template package is available on *ARRLWeb*.

Filter Response with the Clock Set at 37.6 kHz (A) and 149.4 kHz (B)

With the clock frequency set for 37.6 kHz, curve (A) shows a 10 dB down response of roughly 260 Hz to 800 Hz, and 60 dB down response of about 100 Hz to 1180 Hz. You can calibrate the front panel control to whatever bandwidth you like—3 dB down, 6 dB down, or 60 dB down.

Curve (B), taken with the clock set to 149.4 kHz, shows about 250 Hz to 3050 Hz at 10 dB down and about 100 to 4400 Hz at 60 dB down.

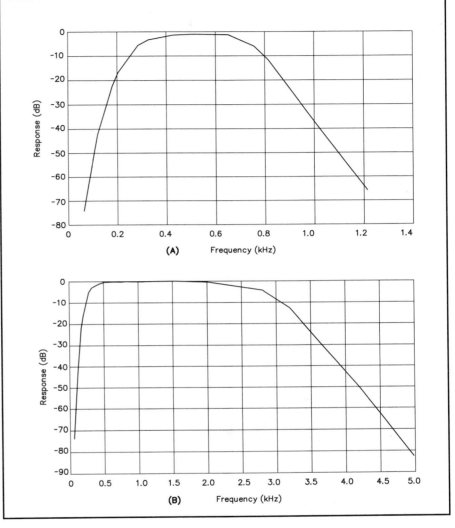

A BC-BAND ENERGY-REJECTION FILTER

Inadequate front-end selectivity or poorly performing RF amplifier and mixer stages often result in unwanted cross-talk and overloading from adjacent commercial or amateur stations. The filter shown is inserted between the antenna and receiver. It attenuates the out-of-band signals from broadcast stations but passes signals of interest (1.8 to 30 MHz) with little or no attenuation.

The high signal strength of local broadcast stations requires that the stop-band attenuation of the high-pass filter also be high. This filter provides about 60 dB of stop-band attenuation with less than 1 dB of attenuation above 1.8 MHz. The filter input and output ports match 50 Ω with a maximum SWR of 1.353:1 (reflection coefficient = 0.15). A 10-element filter yields adequate stop-band attenuation and a reasonable rate of attenuation rise.

The design uses only standard-value capacitors.

Building the Filter

The filter parts layout, schematic diagram, response curve and component values are shown in **Fig 16.68**. The standard capacitor values listed are within 2.8% of the design values. If the attenuation peaks (f2, f4 and f6) do not fall at 0.677, 1.293 and 1.111 MHz, tune the series-resonant circuits by slightly squeezing or separating the inductor windings.

Construction of the filter is shown in **Fig 16.69**. Use Panasonic NP0 ceramic disk capacitors (ECC series, class 1) or equivalent for values between 10 and 270 pF. For values between 330 pF and 0.033 μF, use Panasonic P-series polypropylene (type ECQ-P) capacitors. These capacitors are available through Digi-Key

(see the Address List in the **References** chapter) and other suppliers. The powdered-iron T-50-2 toroidal cores are available through Amidon, Palomar Engineers and others.

For a 3.4-MHz cutoff frequency, divide the L and C values by 2. (This effectively doubles the frequency-label values in Fig 16.68.) For the 80-m version, L2 through L6 should be 20 to 25 turns each, wound on T-50-6 cores. The actual turns required may vary one or two from the calculated values. Parallel-connect capacitors as needed to achieve the nonstandard capacitor values required for this filter.

Filter Performance

The measured filter performance is shown in Fig 16.68. The stop-band attenuation is more than 58 dB. The measured cutoff frequency (less than 1 dB attenuation) is under 1.8 MHz. The measured passband loss is less than 0.8 dB from 1.8 to 10 MHz. Between 10 and 100 MHz, the insertion loss of the filter gradually increases to 2 dB. Input impedance was measured between 1.7 and 4.2 MHz. Over the range tested, the input impedance of the filter remained within the 37 to 67.7 Ω input-impedance window (equivalent to a maximum SWR of 1.353:1).

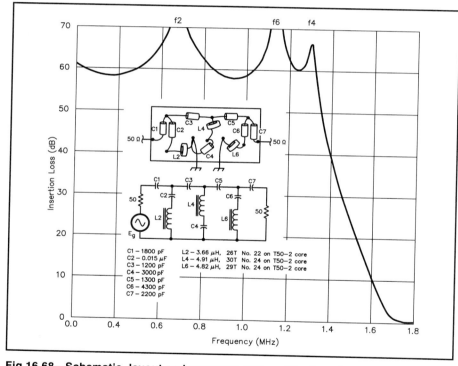

C1 – 1800 pF
C2 – 0.015 μF
C3 – 1200 pF
C4 – 3000 pF
C5 – 1300 pF
C6 – 4300 pF
C7 – 2200 pF

L2 – 3.66 μH, 26T No. 22 on T50-2 core
L4 – 4.91 μH, 30T No. 24 on T50-2 core
L6 – 4.82 μH, 29T No. 24 on T50-2 core

Fig 16.68—Schematic, layout and response curve of the broadcast band rejection filter.

Fig 16.69—The filter fits easily in a 2×2×5-inch enclosure. The version in the photo was built on a piece of perfboard.

SECOND-HARMONIC-OPTIMIZED (CWAZ) LOW-PASS FILTERS

The FCC requires transmitter spurious outputs below 30 MHz to be attenuated by 40 dB or more for power levels between 5 and 500 W. For power levels greater than 5 W, the typical second-harmonic attenuation (40-dB) of a seven-element Chebyshev low-pass filter (LPF) is marginal. An additional 10 dB of attenuation is needed to assure compliance with the FCC requirement.

Jim Tonne, WB6BLD, solved the problem of significantly increasing the second-harmonic attenuation of the seven-element Chebyshev LPF while maintaining an acceptable return loss (> 20 dB) over the amateur passband. Jim's idea was presented in February 1999 *QST* by Ed Wetherhold, W3NQN. These filters are most useful with single-band, single device transmitters. Common medium-power multiband transceivers use push-pull power amplifiers because such amplifiers inherently suppress the second harmonic.

Tonne modified a seven-element Chebyshev standard-value capacitor (SVC) LPF to obtain an additional 10 dB of stop-band loss at the second-harmonic frequency. He did this by adding a capacitor across the center inductor to form a resonant circuit. Unfortunately, return loss (RL) decreased to an unacceptable level, less than 12.5 dB. He needed a way to add the resonant circuit, while maintaining an acceptable RL level over the passband.

The typical LPF, and the Chebyshev SVC designs listed in this chapter all have acceptable RL levels that extend from the filter ripple-cutoff frequency down to dc. For many Amateur Radio application, we need an acceptable RL only over the amateur band for which the LPF is designed. We can trade RL levels below the amateur band for improved RL in the passband, and simultaneously increase the stop-band loss at the second-harmonic frequency.

THE CWAZ LOW-PASS FILTER

This new eight-element LPF has a topology similar to that of the seven-element Chebyshev LPF, with two exceptions: The center inductor is resonated at the second harmonic in the filter stop band, and the component values are adjusted to maintain a more than acceptable RL across the amateur passband. To distinguish this new LPF from the SVC Chebyshev LPF, Wetherhold named it the "Chebyshev with Added Zero" or "CWAZ" LPF design.

You should understand that CWAZ LPFs are *output filters for single-band*

transmitters. They provide optimum second and higher harmonic attenuation while maintaining a suitable level of return loss over the amateur band for which they're designed.

Fig 16.70 shows a schematic diagram of a CWAZ LPF design. **Table 16.6** lists suggested capacitor and inductor values for all amateur bands from 160 through 10 meters. If you want to calculate CWAZ values for different bands, simply divide the first-row C and L values (for 1 MHz) by the start frequency of the desired band.

For example, C1, 7 for the 160-meter design is equal to 2986/1.80 = 1659 pF. The other component values for the 160-meter LPF are calculated in a similar manner.

CWAZ VERSUS SEVENTH-ORDER SVC

The easiest way to demonstrate the superiority of a CWAZ LPF over the Chebyshev LPF is to compare the RL and insertion-loss responses of these two designs. **Fig 16.71** shows a 20-meter SVC Chebyshev LPF design based on the tables

Table 16.6

CWAZ 50-Ω Low-Pass Filters

Designed for second-harmonic attenuation in amateur bands below 30 MHz.

Band (m)	Start Frequency (MHz)	C1,7 (pF)	C3,5 (pF)	C4 (pF)	L2,6 (µH)	L4 (µH)	F4 (MHz)
—	1.00	2986	4556	680.1	9.377	8.516	2.091
160	1.80	1659 / 1450 + 220 / 1500 + 150	2531 / 2100 + 470 / 2200 + 330	378 / 330 + 47	5.21	4.73	3.76 / 3.78
80	3.50	853 / 470 + 390	1302 / 1150 + 150 / 1200 + 100	194 / 150 + 47	2.68	2.43	7.32 / 7.27
40	7.00	427 / 330 + 100	651 / 330 + 330	97.2 / 100	1.34	1.22	14.6 / 14.4
30	10.1	296 / 150 + 150	451 / 470	67.3 / 68	0.928	0.843	21.1 / 21.0
20	14.0	213 / 220	325 / 330	48.6 / 47	0.670	0.608	29.3 / 29.8
17	18.068	165 / 82 + 82	252 / 100 + 150	37.6 / 39	0.519	0.471	37.8 / 37.1
15	21.0	142 / 150	217 / 220	32.4 / 33	0.447	0.406	43.9 / 43.5
12	24.89	120 / 120	183 / 180	27.3 / 27	0.377	0.342	52.0 / 52.4
10	28.0	107 / 100	163 / 82 + 82	24.3 / 27	0.335	0.304	58.5 / 55.6

NOTE: The CWAZ low-pass filters are designed for a single amateur band to provide more than 50 dB attenuation to the second harmonic of the fundamental frequency and to the higher harmonics. All component values for any particular band are calculated by dividing the 1-MHz values in the first row (included for reference only) by the start frequency of the selected band. The upper capacitor values in each row show the calculated design values obtained by dividing the 1-MHz capacitor values by the amateur-band start frequency in megahertz. The lower standard-capacitor values are suggested as a convenient way to realize the design values. The middle capacitor values in the 160 and 80-meter-band designs are suggested values when the high-value capacitors (greater than 1000 pF) are on the low side of their tolerance range. The design F4 frequency (see upper value in the F4 column) is calculated by multiplying the 1-MHz F4 value by the start frequency of the band. The lower number in the F4 column is the F4 frequency based on the suggested lower capacitor value and the listed L4 value.

in this chapter. **Fig 16.72** shows the computer-calculated return- and insertion-loss responses of the LPF shown in Fig 16.71. The plotted responses were made using Jim Tonne's *ELSIE* filter design and analysis software. This DOS-based program is available from Trinity Software.[1]

Fig 16.73 shows the computer-calculated return- and insertion-loss responses of a CWAZ LPF intended to replace the seven-element 20-meter Chebyshev SVC LPF. The stop-band attenuation of the CWAZ LPF in the second-harmonic band is more than 60 dB and is substantially greater than that of the Chebyshev LPF. Also, the pass-band RL of the CWAZ LPF is quite satisfactory, at more than 25 dB. The disadvantages of the CWAZ design are that an extra capacitor is needed across L4, and several of the designs listed in

Table 16.6 require paralleled capacitors to realize the design values. Nevertheless, these disadvantages are minor in comparison to the increased second-harmonic stop-band attenuation that is possible with a CWAZ design.

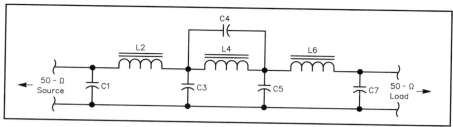

Fig 16.70—Schematic diagram of a CWAZ low-pass filter designed for maximum second-harmonic attenuation. See Table 16.6 for component values of CWAZ 50-Ω designs. L4 and C4 are tuned to resonate at the F4 frequency given in Table 16.6. For an output power of 10 W into a 50-Ω load, the RMS output voltage is $\sqrt{10 \times 50} = 22.4$ V. Consequently, a 100 V dc capacitor derated to 60 V (for RF filtering) is adequate for use in these LPFs if the load SWR is less than 2.5:1.

[1]Those seriously interested in passive LC filter design may experience the capabilities of *ELSIE* through a demo disk. The demo is restricted to LC filters of the third-order only. Contact Trinity Software (see the **References** chapter Address List for contact information).

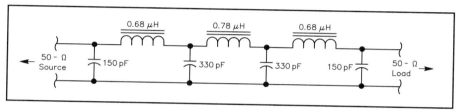

Fig 16.71—Schematic diagram of a 20-meter SVC Chebyshev LPF.

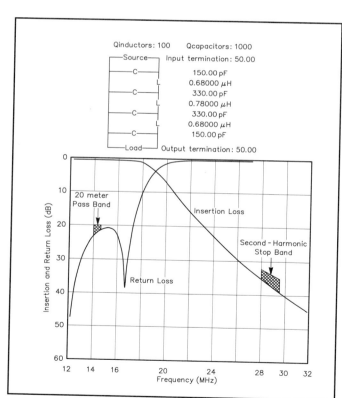

Fig 16.72—The plots show the *ELSIE* computer-calculated return- and insertion-loss responses of the seventh-order Chebyshev SVC low-pass filter shown in Fig 16.71. The 20-meter passband RL is about 21 dB, and the insertion loss over the second-harmonic frequency band ranges from 35 to 39 dB. A listing of the component values is included.

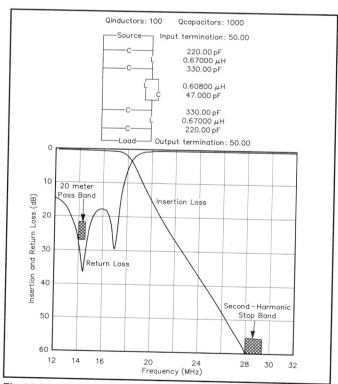

Fig 16.73—The plots show the *ELSIE* computer-calculated return- and insertion-loss responses of the eight-element low-pass filter using the CWAZ capacitor and inductor values listed in Table 16.6 for the 20-meter low-pass filter. Notice that the calculated attenuation to second-harmonic signals is greater than 60 dB, while RL over the 20-meter passband is greater than 25 dB.

THE DIPLEXER FILTER

This section, covering diplexer filters, was written by William E. Sabin, WØIYH. The diplexer is helpful in certain applications, and Chapter 15 shows them used as frequency mixer terminations.

Diplexers have a constant filter-input resistance that extends to the stop band as well as the passband. Ordinary filters that become highly reactive or have an open or short-circuit input impedance outside the passband may degrade performance.

Fig 16.74 shows a *normalized* prototype 5-element, 0.1-dB Chebyshev low-pass/high-pass (LP/HP) filter. This idealized filter is driven by a voltage generator with zero internal resistance, has load resistors of 1.0 Ω and a cutoff frequency of 1.0 radian per second (0.1592 Hz). The LP prototype values are taken from standard filter tables.[1] The first element is a series inductor. The HP prototype is found by:

a) replacing the series L (LP) with a series C (HP) whose value is 1/L, and

b) replacing the shunt C (LP) with a shunt L (HP) whose value is 1/C.

For the Chebyshev filter, the return loss is improved several dB by multiplying the prototype LP values by an experimentally derived number, K, and dividing the HP values by the same K. You can calculate the LP values in henrys and farads for a 50-Ω RF application with the following formulas:

$$L_{LP} = \frac{K L_{P(LP)} R}{2\pi f_{CO}}; \quad C_{LP} = \frac{K C_{P(LP)}}{2\pi f_{CO} R}$$

where

$L_{P(LP)}$ and $C_{P(LP)}$ are LP prototype values

K = 1.005 (in this specific example)

R = 50 Ω

f_{CO} = the cutoff (−3-dB response) frequency in Hz.

For the HP segment:

$$L_{HP} = \frac{L_{P(HP)} R}{2\pi f_{CO} K}; \quad C_{HP} = \frac{C_{P(HP)}}{2\pi f_{CO} K R}$$

where $L_{P(HP)}$ and $C_{P(HP)}$ are HP prototype values.

Fig 16.75 shows the LP and HP responses of a diplexer filter for the 80 meter band. The following items are to be noted:

- The 3 dB responses of the LP and HP meet at 5.45 MHz.
- The input impedance is close to 50 Ω at all frequencies, as indicated by the high value of return loss (SWR <1.07:1).
- At and near 5.45 MHz, the LP input reactance and the HP input reactance are conjugates, therefore they cancel and produce an almost perfect 50-Ω input

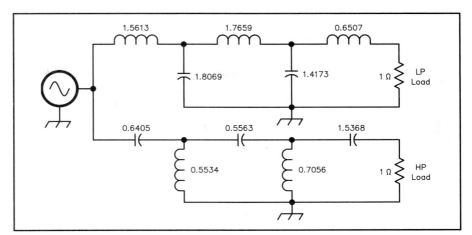

Fig 16.74—Low-pass and high-pass prototype diplexer filter design. The low-pass portion is at the top, and the high-pass at the bottom of the drawing. See text.

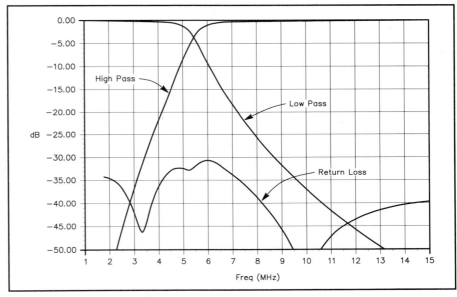

Fig 16.75—Response for the low-pass and high-pass portions of the 80-meter diplexer filter. Also shown is the return loss of the filter.

resistance in that region.

- Because of the way that the diplexer filter is derived from synthesis procedures, the transfer characteristic of the filter is pretty much independent of the actual value of the amplifier's dynamic output impedance.[2] This is a useful feature, since the RF power amplifier's output impedance is usually not known or specified.
- The 80-meter band is well within the LP response.
- The HP response is down more than 20 dB at 4 MHz.
- The second harmonic of 3.5 MHz is down only 18 dB at 7.0 MHz. Because the second harmonic attenuation of the

LP is not great, it is necessary that the amplifier itself be a well-balanced push-pull design that greatly rejects the second harmonic. In practice this is not a difficult task.

- The third harmonic of 3.5 MHz is down almost 40 dB at 10.5 MHz.

Fig 16.76A shows the unfiltered of a solid-state push-pull power amplifier for the 80-meter band. In the figure you can see that:

- The second harmonic has been suppressed by a proper push-pull design.
- The third harmonic is typically only 15 dB or less below the fundamental.

The amplifier output goes through our diplexer filter. The desired output comes

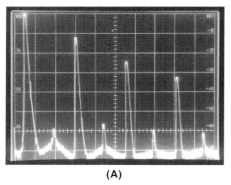 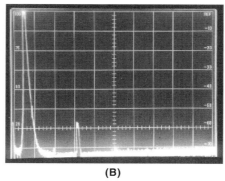 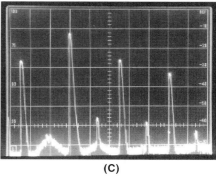

(A) (B) (C)

Fig 16.76—At A, the output spectrum of a push-pull 80-meter amplifier. At B, the spectrum after passing through the low-pass filter. At C, the spectrum after passing through the high-pass filter.

from the LP side, and is shown in Fig 16.76B. In it we see that:

- The fundamental is attenuated only about 0.2 dB.
- The LP has some harmonic content, however, the attenuation exceeds FCC requirements for a 100-W amplifier.

Fig 16.76C shows the HP output of the diplexer which terminates in the HP load or *dump* resistor. A small amount of the fundamental frequency (about 1%) is also lost in this resistor. Within the 3.5 to 4.0 MHz band the filter input resistance almost perfectly matches the 50-Ω amplifier output impedance. This is because power that would otherwise be *reflected* back to the amplifier is absorbed in the dump resistor.

Solid state power amplifiers tend to have stability problems that can be difficult to debug.[3] These problems may be evidenced by level changes in: load impedance, drive, gate or base bias, B+, etc. Problems may arise from:

- The reactance of the low-pass filter outside the desired passband. This is especially true for transistors that are designed for high-frequency operation.
- Self resonance of a series inductor at some high frequency.
- A stopband impedance that causes voltage, current and impedance reflections back to the amplifier, creating instabilities within the output transistors.

Intermodulation performance can also be degraded by these reflections. The strong third harmonic is especially bothersome for these problems.

The diplexer filter is an approach that can greatly simplify the design process, especially for the amateur with limited PA-design experience and with limited home-lab facilities. For these reasons, the amateur homebrew enthusiast may want to consider this solution, despite its slightly greater parts count and expense.

The diplexer is a good technique for narrowband applications such as the HF

amateur bands.[4] From Fig 16.75, we see that if the signal frequency is moved beyond 4.0 MHz the amount of desired signal lost in the dump resistor becomes large. For signal frequencies below 3.5 MHz the harmonic reduction may be inadequate. A single filter will not suffice for all the HF amateur bands.

This treatment provides you with the information to calculate your own filters. A *QEX* article has detailed instructions for building and testing a set of six filters for a 120-W amplifier. These filters cover all nine of the MF/HF amateur bands.[5] Check *ARRLWeb* at: **http://www.arrl.org/qex/**.

You can use this technique for other filters such as Bessel, Butterworth, linear phase, Chebyshev 0.5, 1.0, etc.[6] However, the diplexer idea does *not* apply to the elliptic function types.

The diplexer approach is a resource that can be used in any application where a constant value of filter input resistance over a wide range of passband and stopband frequencies is desirable for some reason. The *ARRL Radio Designer* program is an ideal way to finalize the design before the actual construction.[7] The coil dimensions and the dump resistor wattage need to be determined from a consideration of the power levels involved, as illustrated in Fig 16.76.

Another significant application of the diplexer is for elimination of EMI, RFI and TVI energy. Instead of being reflected and very possibly escaping by some other route, the unwanted energy is dissipated in the dump resistor.[7]

Notes
[1]Williams, A. and Taylor, F., *Electronic Filter Design Handbook*, any edition, McGraw-Hill.
[2]Storer, J.E., *Passive Network Synthesis*, McGraw-Hill 1957, pp 168-170. This book shows that the input resistance is ideally constant in the passband and the stopband and that the filter transfer characteristic is

ideally independent of the generator impedance.
[3]Sabin, W. and Schoenike, E., *HF Radio Systems and Circuits*, Chapter 12, Noble Publishing, 1998. This publication is available from ARRL as Order no. 7253. It can be ordered at: **http://www.arrl.org/catalog/**. Also the previous edition of this book, *Single-Sideband Systems and Circuits*, McGraw-Hill, 1987 or 1995.
[4]Dye, N. and Granberg, H., *Radio Frequency Transistors, Principles and Applications*, Butterworth-Heinemann, 1993, p 151.
[5]Sabin, W.E. WØIYH, "Diplexer Filters for the HF MOSFET Power Amplifier," *QEX*, Jul/Aug, 1999. Also check *ARRLWeb* at: **http://www.arrl.org/qex/**.
[6]See note 1. *Electronic Filter Design Handbook* has LP prototype values for various filter types, and for compexities from 2 to 10 components.
[7]Weinrich, R. and Carroll, R.W., "Absorptive Filters for TV Harmonics," *QST*, Nov 1968, pp 10-25.

OTHER FILTER PROJECTS

Filters for specific applications may be found in other chapters of this *Handbook*. Receiver input filters, transmitter filters, interstage filters and others can be separated from the various projects and built for other applications. Since filters are a first line of defense against *electromagnetic interference* (EMI) problems, the following filter projects appear in the EMI chapter:

- Differential-mode high-pass filter for 75-Ω coax (for TV reception)
- *Brute-force* ac-line filter
- Loudspeaker common-mode choke
- LC filter for speaker leads
- Audio equipment input filter

REFERENCES

1. O. Zobel, "Theory and Design of Electric Wave Filters," *Bell System Technical Journal*, Jan 1923.
2. ARRL *Handbook*, 1968, p 50.
3. S. Butterworth, "On the Theory of Filter

Amplifiers," *Experimental Wireless and Wireless Engineer*, Oct 1930, pp 536-541.

4. S. Darlington, "Synthesis of Reactance 4-Poles Which Produce Prescribed Insertion Loss Characteristics," *Journal of Mathematics and Physics*, Sep 1939, pp 257-353.

5. L. Weinberg, "Network Design by use of Modern Synthesis Techniques and Tables," *Proceedings of the National Electronics Conference*, vol 12, 1956.

6. Laplace Transforms: P. Chirlian, *Basic Network Theory,* McGraw Hill, 1969.

7. Fourier Transforms: *Reference Data for Engineers*, Chapter 7, 7th edition, Howard Sams, 1985.

8. Cauer Elliptic Filters: *The Design of Filters Using the Catalog of Normalized Low-Pass Filters,* Telefunken, 1966. Also Ref 7, pp 9-5 to 9-11.

9. M. Dishal, "Top Coupled Band-pass Filters," *IT&T Handbook*, 4th edition, American Book, Inc, 1956, p 216. Also, P. Geffe, *Simplified Modern Filter Design*, J. F. Rider, 1963, pp 42-48.

10. W. E. Sabin, WØIYH, "Designing Narrow Band-Pass Filters with a BASIC Program," May 1983 *QST*, pp 23-29.

11. U. R. Rohde, DJ2LR, "Crystal Filter Design with Small Computers" May 1981 *QST*, p 18.

12. J. A. Hardcastle, G3JIR, "Ladder Crystal Filter Design," Nov 1980 *QST*, p 20.

13. W. Hayward, W7ZOI, "A Unified Approach to the Design of Ladder Crystal Filters," May 1982 *QST*, p 21.

14. R. Fisher, W2CQH, "Combline VHF Band-pass Filters," Dec 1968 *QST*, p 44.

15. R. Fisher, W2CQH, "Interdigital Band-pass Filters for Amateur VHF/UHF Applications," Mar 1968 *QST*, p 32.

16. W. S. Metcalf, "Graphs Speed Interdigitated Filter Design," *Microwaves*, Feb 1967.

17. A. Zverev, *Handbook of Filter Synthesis*, John Wiley and Sons.

18. R. Saal, *The Design of Filters Using the Catalog of Normalized Low-Pass Filters*, Telefunken.

19. P. Geffe, *Simplified Modern Filter Design* (New York: John F. Rider, a division of Hayden Publishing Co, 1963).

20. *A Handbook on Electrical Filters* (Rockville, Maryland: White Electromagnetics, 1963).

21. A. B. Williams, *Electronic Filter Design Handbook* (New York: McGraw-Hill, 1981).

22. *Reference Data for Radio Engineers*, 6th edition, Table 2, p 5-3 (Indianapolis, IN: Howard W. Sams & Co, 1981).

23. R. Frost, "Large-Scale S Parameters Help Analyze Stability," *Electronic Design*, May 24, 1980.

24. E. Wetherhold, W3NQN, "Modern Design of a CW Filter Using 88 and 44-mH Surplus Inductors," Dec 1980 *QST*, pp 14-19. See also Feedback in Jan 1981 *QST*, p 43.

Contents

Receivers, Transmitters, Transceivers and Projects

17

In this chapter, William E. Sabin, WØIYH, discusses the "system design" of Amateur Radio receivers, transmitters and transceivers. "A Single-Stage Building Block" reviews briefly a few of the basic properties of the various individual building block circuits, described in detail in other chapters, and the methods that are used to combine and interconnect them in order to meet the requirements of the completed equipment. "The Amateur Radio Communication Channel" describes the relationships between the equipment system design and the electromagnetic medium that conveys radio signals from transmitter to receiver. This understanding helps to put the radio equipment mission and design requirements into perspective. Then we discuss receiver, transmitter, transceiver and transverter design techniques in general terms. At the end of the theory discussion is a list of references for further study on the various topics. The projects section contains several hardware descriptions that are suitable for amateur construction and use on the ham bands. They have been selected to illustrate system-design methods. The emphasis in this chapter is on analog design. Those functions that can be implemented using digital signal processing (DSP) can be explored in other chapters, but an initial basic appreciation of analog methods and general system design is very valuable.

A SINGLE-STAGE BUILDING BLOCK

We start at the very beginning with **Fig 17.1**, a generic single-stage module that would typically be part of a system of many stages. A signal source having an "open-circuit" voltage V_{gen} causes a current I_{gen} to flow through Z_{gen}, the impedance of the generator, and Z_{in}, the input impedance of the stage. This input current is responsible for an open-circuit output voltage V_d (measured with a high-impedance voltmeter) that is proportional to I_{gen}. V_d produces a current I_{out} and a voltage drop across Z_{out}, the output impedance of the stage and Z_{load}, the load impedance of the stage. Observe that the various Zs may contain reactance and resistance in various combinations. Let's first look at the different types of gain and power relationships that can be used to describe this stage.

Actual Power Gain

Current I_{gen} produces a power dissipation P_{in} in the resistive component of Z_{in} that is equal to $I_{gen}^2 R_{in}$. The current I_{out} produces a power dissipation P_{load} in the resistive component of Z_{load} that is equal to $I_{out}^2 R_{load}$. The actual power gain in dB is $10 \log (P_{load} / P_{in})$. This is the conventional usage of dB, to describe a power ratio.

Voltage Gain

The current I_{gen} produces a voltage drop across Z_{in}. V_d produces a current I_{out} and a voltage drop V_{out} across Z_{load}. The voltage gain is the ratio

$$V_{out} / V_{in} \qquad (1)$$

In decibels (dB) it is

$$20 \log (V_{out} / V_{in}) \qquad (2)$$

This alternate usage of dB, to describe a voltage ratio, is common practice. It is *different* from the power gain mentioned in the previous section because it does *not* take into account the power ratio or the resistance values involved. It is a voltage ratio only. It is used in troubleshooting and other instances where a rough indication of operation is needed, but precise measurement is unimportant. Voltage gain is often used in high-impedance circuits such as pentode vacuum tubes and is also sometimes convenient in solid-state circuits. Its improper usage often creates errors in radio circuit design because many calculations, for correct answers, require power ratios rather than voltage ratios. We will see several examples of this throughout this chapter.

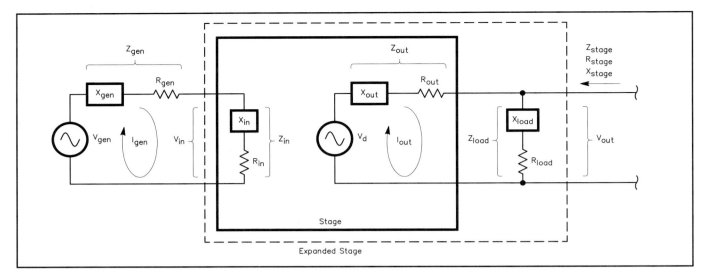

Fig 17.1—A single-stage building-block signal processor. The properties of this stage are discussed in the text.

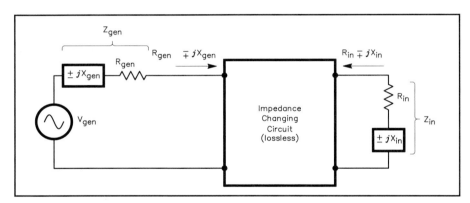

Fig 17.2—The conjugate impedance match of a generator to a stage input. The network input impedance is $R_{gen} \mp jX_{gen}$ and its output impedance is $R_{in} \mp jX_{in}$ (where either R term may represent a dynamic impedance). Therefore the generator and the stage input are both impedance matched for maximum power transfer.

Available Power

The maximum power, in watts, that can be obtained from the generator is $V_{gen}^2 / (4\ R_{gen})$. To see this, suppose temporarily that X_{gen} and X_{in} are both zero. Then let R_{in} increase from zero to some large value. The maximum power in R_{in} occurs when $R_{in} = R_{gen}$ and the power in R_{in} then has the value mentioned above (plot a graph of power in R_{in} vs R_{in} to verify this, letting $V_{gen} = 1$ V and $R_{gen} = 50$ W). This is called the "available power" (we're assuming sine-wave signals). If X_{gen} is an inductive (or capacitive) reactance and if X_{in} is an equal value of capacitive (or inductive) reactance, the net series reactance is nullified and the above discussion holds true. If the net reactance is not zero the current I_{gen} is reduced and the power in R_{in} is less than maximum. The process of tuning out the reactance and then transforming the resis-

tance of R_{in} is called "conjugate matching." A common method for doing this conjugate matching is to put an impedance transforming circuit of some kind, such as a transformer or a tuned circuit, between the generator and the stage input that "transforms" R_{in} to the value R_{gen} (as seen by the generator) and at the same time nullifies the reactance.

Fig 17.2 illustrates this idea and later discussion gives more details about these interstage networks. A small amount of power is lost within any lossy elements of the matching network. This same technique can be used between the output of the stage in Fig 17.1 and the load impedance Z_{load}. In this case, the stage delivers the maximum amount of power to the load resistance. If both input and output are processed in this way, the stage utilizes the generator signal to the maximum extent

possible. It is very important to note, however, that in many situations we do not want this maximum utilization. We deliberately "mismatch" in order to achieve certain goals that will be discussed later (Ref 1).

The dBm Unit of Power

In low-level radio circuitry, the watt (W) is inconveniently large. Instead, the milliwatt (mW) is commonly used as a reference level of power. The dB with respect to 1 mW is defined as

$$dBm = 10 \log (P_W / 0.001) \quad (3)$$

where

   dBm = Power level in dB with respect to 1 mW
   P_W = Power level, watts.

For example, 1 W is equivalent to 30 dBm. Also

$$P_W = 0.001 \times 10\ dB^{m/10} \quad (4)$$

Maximum Available Power Gain

The ratio of the power that is available from the stage, $V_d^2 / (4\ R_{out})$, to the power that is available from the generator, $V_{gen}^2 / (4\ R_{gen})$, is called the maximum available power gain. In some cases the circuit is adjusted to achieve this value, using the conjugate-match method described above. In many cases, as mentioned before, less than maximum gain is acceptable, perhaps more desirable.

Available Power Gain

Consider that in Fig 17.1 the stage and its output load Z_{load} constitute an "expanded" stage as defined by the dotted box. The power available from this new stage is determined by V_{out} and by R_{stage}, the resistive part of Z_{stage}. The available power gain is then $V_{out}^2 / (4\ R_{stage})$ divided by $V_{gen}^2 / (4\ R_{gen})$. This value of gain is used in a number of design procedures. Note that Z_{load} can be a physical network of some kind, or it may be partly or entirely the input impedance of the stage following the one shown in Fig 17.1.

In the latter case it is sometimes convenient to "detach" this input impedance from the next stage and make it part of the expanded first stage, as shown in Fig 17.1, but we note that Z_{out} is still the generator (source) impedance that the input of the next stage "sees."

Transducer Power Gain

The transducer gain is defined as the ratio of the power actually delivered to R_{load} in Fig 17.1 to the power that is available from the generator V_{gen} and R_{gen}. In other words, how much more power does the stage deliver to the load than the generator could deliver if the generator were impedance matched to the load? We will discuss how to use this kind of gain later.

Feedback (Undesired)

One of the most important properties of the single-stage building block in Fig 17.1 is that changes in the load impedance Z_{load} cause changes in the input impedance Z_{in}. Changes in Z_{gen} also affect Z_{out}. These effects are due to reverse coupling, within the stage, from output to input. For many kinds of circuits (such as networks, filters, attenuators, transformers and so on) these effects cause no unexpected problems.

But, as the chapter on **RF Power Amplifiers** explains in detail, in active circuits such as amplifiers this reverse coupling within one stage can have a major impact not only on that stage but also on other stages that follow and precede. It is the effect on system performance that we discuss here. In particular, if a stage is expected to have certain gain, noise factor and distortion specifications, all of these can be changed either by reverse coupling (undesired feedback) within the stage or adjacent stages. For example, internal feedback can cause the input impedance of a certain stage "A" (Fig 17.1) to become very large. If this impedance is the load impedance for the preceding stage, the gain of the preceding stage can become excessive, creating problems in both stages. This same feedback can cause the gain of stage "A" to become greater, thereby causing the next stage to be driven into heavy distortion. A very common event is that stage "A" goes into oscillation. All of these occurrences are common in poorly designed radio equipment. Changes in temperature and variations in

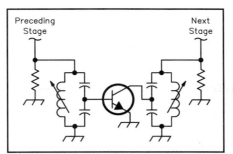

Fig 17.3—A double tuned transistor amplifier circuit that may oscillate due to excessive amplification and reverse coupling.

component tolerances are major contributors to these problems.

One particular example is shown in **Fig 17.3**, a transistor amplifier, shown in skeleton form, with sharply tuned resonators at input and output.

Because of reverse coupling, the two tuned circuits interact, making adjustments difficult or even impossible. The likelihood of oscillation is very high. There are two solutions: drastically reduce the gain of the amplifier, or use an amplifier circuit that has very little reverse coupling. Usually, both methods are used simultaneously (in the right amount) in order to get predictable performance. The object lesson for the system designer is that a combination of reduced gain and low reverse coupling is the safe way to go when designing a radio system. More stages may be required, but the price is well worthwhile. The cascode amplifier, grounded-gate amplifier, dual-gate FET and many types of IC amplifiers are examples of circuits that have little reverse coupling and good stability. "Neutralization" methods are used to cancel reverse coupling that

causes instability. All such circuits are said to be "unilateral," which means "in one direction" and both input and output can be independently tuned as in Fig 17.3 if the gain is not too high.

Feedback (Desired)

The **RF Power Amplifiers** chapter explains how negative feedback (good feedback) can be used to stabilize a circuit and make it much more predictable over a range of temperature and component tolerances. Here we wish to point out some system implications of negative feedback. One is that the gain, noise-figure and distortion performances within a stage are made much more constant and predictable. Therefore a system designer can put building blocks together with more confidence and less guesswork.

There are some problems, though. In some circuits the amount of feedback depends on both the output impedance of the driving circuit and the input impedance of the next stage. A classic example is the cascadable amplifier shown in **Fig 17.4**.

In this circuit, if the output load impedance becomes very low the amplifier input impedance becomes high, and vice versa (a "teeter-totter" effect). Other amplifier properties also can change. With amplifiers of this type it is important to maintain the correct impedances at the input and output interfaces. Any building block should be examined for effects of this kind. Data sheets frequently specify the reverse transfer values as well as those for forward transfer. Often, lab measurements are needed. Apply a signal to the output and measure the reverse coupling to the input. Where varying load and source impedances are involved, look for a circuit that is less vulnerable (that is, has less reverse coupling).

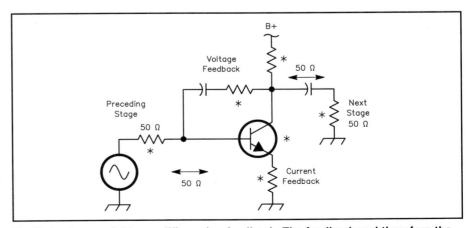

Fig 17.4—A cascadable amplifier using feedback. The feedback and therefore the amplifier performance depends on the load and driving-stage impedances.

Another problem is that feedback networks often add thermal noise sources to a circuit and so degrade its noise figure. In systems where this is a consideration, use so-called "lossless feedback" circuits. These circuits use very efficient transformers instead of resistors or lossy networks that introduce thermal noise into a system.

Noise Factor and Noise Figure

The output resistance of the signal generator that drives a typical signal processing block such as shown in Fig 17.4 is a source of thermal noise power, which is a natural phenomenon occurring in the resistive component of any impedance. It is caused by random motion of electrons within a conducting (or semiconducting) material. Note that the reactive part of an impedance is not a source of thermal noise power because the voltage across a pure reactance and the current through the reactance are in phase quadrature (90°) at any one frequency. The average value of the product of these two (the power) is zero. If this is true at any frequency, then it is true at all frequencies. Also, a purely "dynamic resistance" such as R_e, the dynamic resistance ($\Delta V/\Delta I$) of a perfect forward conducting PN junction, is also not a source of thermal noise. However, the junction is a source of "shot noise" power that, by the way, is only 50% as great as the thermal noise that R_e would have if it were an actual resistor (Ref 2).

Each "*" in Fig 17.4 indicates a noise source. Passive elements generate thermal noise. Active components such as transistors generate thermal noise and other types, such as shot noise and flicker (1/f) noise, internally. These "excess" noises are all superimposed on the signal from the generator. Therefore the noise factor of a single stage is a measure of how much the signal to noise ratio is degraded as a signal passes through that stage.

Refer now to the diagram and equations in **Fig 17.5**. F is noise factor and S_i/N_i is the input signal to noise ratio from the signal generator. S_o/N_o is signal to noise ratio at the output and kTB is the thermal noise power that is available from any value of resistance (kT = –174 dBm in a 1-Hz bandwidth at room temperature). G is S_o/S_i, the available power gain of the stage and B is the noise bandwidth at the *output* of the stage, assumed to be not wider than the noise bandwidth at the input. The case where the output noise bandwidth is wider will be considered in a later section.

Noise bandwidth is defined in Fig 17.5. An ideal rectangular frequency response has a maximum value that is defined at the reference frequency. The area under the rectangle is the same as the area under the actual filter response, therefore the noise within the rectangle and within the actual filter response are equal. The width of the rectangle is called the noise bandwidth. Various kinds of filters have certain ratios of signal bandwidth to noise bandwidth that can be measured or calculated.

Part of the output noise is amplified thermal noise from the signal generator. To find the noise that is generated within the stage, we must subtract the amplified signal generator noise from the total output noise. Fig 17.5 shows the equation that performs this operation and the quantity (F–1)kTBG is the excess noise that the stage contributes.

In general, the excess noise of a stage is the output noise minus the amplified noise

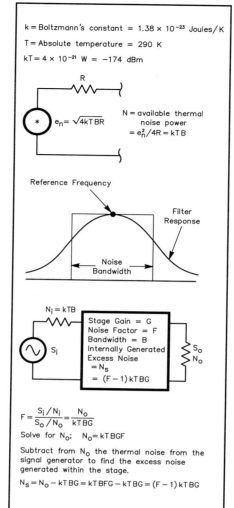

k = Boltzmann's constant = 1.38×10^{-23} Joules/K

T = Absolute temperature = 290 K

$kT = 4 \times 10^{-21}$ W = –174 dBm

$e_n = \sqrt{4kTBR}$

N = available thermal noise power
$= e_n^2/4R = kTB$

Reference Frequency

Filter Response

Noise Bandwidth

$N_i = kTB$

Stage Gain = G
Noise Factor = F
Bandwidth = B
Internally Generated
Excess Noise
$= N_s$
$= (F-1)kTBG$

S_i

S_o
N_o

$F = \dfrac{S_i/N_i}{S_o/N_o} = \dfrac{N_o}{kTBG}$

Solve for N_o; $N_o = kTBGF$

Subtract from N_o the thermal noise from the signal generator to find the excess noise generated within the stage.

$N_s = N_o - kTBG = kTBFG - kTBG = (F-1)kTBG$

Fig 17.5—Diagram and equations that explain F, the noise factor of a single stage. The excess noise generated within the stage is also indicated. The definition of noise bandwidth is included.

from the previous stage. A thorough understanding of this concept is very important for any one who designs radio systems that employ low-level signals. Finally, noise figure, NF, is 10 times the logarithm of F, the noise factor (Ref 3).

Noise Factor of a Passive Device

Often the stage in Fig 17.5 is a filter, an attenuator or some other passive device (no amplification) that contains only thermal noise sources. In a device of this kind, the output noise is thermal noise of the same value as the thermal noise of the generator alone. That is all the thermal noise sources inside the device and also the generator resistance can be combined into a single resistor whose available noise power is kTB, the same as that of the generator alone. Therefore no additional noise is added by the device. But the available signal power is reduced by the attenuation (loss of signal) of the device. Therefore, using the equation for noise factor in Fig 17.5, the noise factor F of the device is numerically equal to its attenuation.

For example, a 3-dB attenuator has a 3-dB noise figure, or a noise factor of 2. This important fact is very useful in radio design. It applies only when there is no amplification and no shot noise or 1/f noise sources within the device. This discussion assumes that all components and the generator are at the same temperature. If not, a slightly more complicated procedure involving Equivalent Noise Temperatures (T_E), to be discussed later in this chapter, can be used.

Sensitivity

Closely related to the concept of noise figure (or noise factor) is the idea of sensitivity. Suppose a circuit, a component or a complete system has a noise figure NF (dB) and therefore a noise output N_o (dBm). Then the value in dBm of a signal generator input that increases the total output (signal + noise) by 10 dB is defined as the "sensitivity." That is, $10 \times \log[(\text{signal} + \text{noise})/(\text{noise})] = 0$ dB. The ratio (signal) / (noise) is then equal to 9.54 dB (= $10 \times \log(9)$). N_o is equal to kTBFG as shown in Fig 17.5. B is noise bandwidth. Using this information, the sensitivity is

$$S(\text{dBm}) = 174\,(\text{dBm}) + 9.54\,(\text{dB}) + NF\,(\text{dB}) + 10\log(B)\,(\text{dB}) \qquad (5)$$

In terms of the "open circuit voltage" from a 50-Ω signal generator (twice the reading of the generator's voltmeter) the sensitivity is

$$E\,(\text{volts open circuit}) = 0.4467 \times 10^{S/20} \qquad (6)$$

N_o is 9.54 dB below the sensitivity

value. This is sometimes referred to in specifications as the "noise floor." The signal level that is equal to the noise floor is sometimes referred to as the minimum detectable signal (MDS). Also associated with N_o is the concept of "noise temperature" which we discuss later under Microwave Receivers.

Distortion in a Single Stage

Suppose the input to a stage is called X. If the stage is perfectly linear the output is Y, and Y = AX, where A is a constant of proportionality. That is, Y is a perfect replica of X, possibly changed in size. But if the stage contains something nonlinear such as a diode, transistor, magnetic material or other such device, then Y = AX + BX 2 + CX 3 +....... The additional terms are "distortion" terms that deliver to the output artifacts that were not present in the generator. Without getting too mathematical at this point, if the input is a pure sine wave at frequency f, the output will contain "harmonic distortion" at frequencies 2f, 3f and so on. If the input contains two signals at f1 and f2, the output contains *intermodulation distortion* (IMD) products at f1+f2, f1–f2, 2f1+f2, 2f2–f1, just to name a few. All semiconductors, vacuum tubes and magnetic materials create distortion and the radio designer's job is to limit the distortion products to acceptable levels. We wish to look at distortion from a system-design standpoint.

There are several ways to reduce distortion. One is to use a high-power device operated well below its maximum ratings. This leads to devices that dissipate more power in the form of heat. Unfortunately, these devices also tend to be noisier; so high power levels and low noise tend to be incompatible goals in most cases. (Some modern devices, such as certain GaAsFETs, achieve improved values of dynamic range.) Also, a large reduction in distortion is not always assured with this method, especially in transmitters.

Second, reduce the signal level into the device. This allows a lower power device to be used that will tend to be less noisy. To get the same output level, though, we must increase the gain of the stage. Then we run into another problem: if the signal at the output of this lower power stage becomes too large, distortion is generated at the output. Also, as mentioned before, high-gain stages tend to be unstable at RF.

Third, reduce the stage gain. But then we must add another stage in order to get the required output level. This additional stage turns out to be a high-power stage. The addition of another stage adds more noise and distortion contamination to the signal.

Fourth, use negative feedback. This is a powerful technique that is discussed in detail in the **RF Power Amplifiers** chapter. In general, if we increase the stage gain and perhaps make it more powerful, we can use feedback to reduce distortion and stabilize performance with respect to component variations. The feedback stage may be noisier, although the use of loss-less feedback can improve this situation. Negative feedback is the preferred method for reducing distortion in radio design, but the gain reduction due to feedback means that more stages are needed. This tends to reintroduce some noise and distortion.

A fifth way reduces distortion by increasing selectivity. For example, harmonics of an RF amplifier can be eliminated by a tuned circuit. Products such as f1 + f2 and f1 – f2 can often also be eliminated. Third-order products such as 2f1 – f2 and so on (and higher odd-order products) frequently are sufficiently close to f1 and f2 that selectivity does not help much, but if they are somewhat removed in frequency these so-called "adjacent channel" products can be greatly reduced.

A sixth way is to use push-pull circuits (see the **RF Power Amplifiers** chapter) that tend to greatly reduce "even-order" products such as 2f, 4f, f1 + f2, f1 – f2, and so on. But "odd-orders" such as 3f, 5f, 2f1 + f2, 2f1 – f2 are not reduced by this method except as noted later in the Modules in Combination section.

A seventh way uses diplexers to absorb undesired harmonics or other spurious products. So there are compromises to be made. The designer must look for the compromise that gets the job done in an acceptable manner and is optimal in some sense. For example, devices are available that are optimized for linearity

IMD Ratio

If a pair of equal-amplitude signals creates IMD products, the IMD ratios (IMR) are the differences, in dB, between each of the two tones and each of the IMD products (see **Fig 17.6**).

Intercept Point

The intercept point is a figure of merit that is commonly used to describe the IMD performance of an individual stage or a complete system. For example, third-order products increase at the rate of 3:1. That is, a 1-dB increase in the level of each of the two-tone input signals produces (ideally, but not always exactly true) a 3-dB increase in third-order IMD products. As the input levels increase, the distortion products seen at the output on a spectrum analyzer could catch up to, and equal, the level of the two desired

signals, if the circuit did not go into a limiting process (see next topic). The input level at which this occurs is the input intercept point. Fig 17.6 shows the concept graphically, and also derives from the geometry an equation that relates signal level, distortion and intercept point. A similar process is used to get a second-order intercept point for second-order IMD. These formulas are very useful in designing radio systems and circuits. If the input intercept point (dBm) and the gain of the stage (dB) are added the result is an output intercept point (dBm). Receivers are specified by input intercept point, referring distortion back to the receive antenna input. Transmitter specifications use output intercept, referring distortion to the transmit antenna output.

Gain Compression

The gain of a circuit that is linear and has little distortion products deteriorates rapidly when the instantaneous input or output level reaches a critical point where the peak or trough of the waveform begins to "clip" or "saturate." The 1-dB compression point occurs when the output is 1 dB

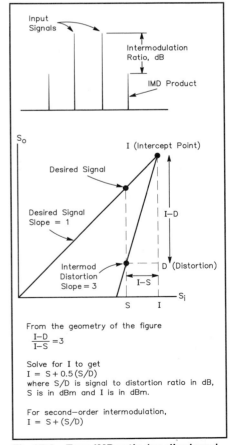

From the geometry of the figure

$$\frac{I-D}{I-S} = 3$$

Solve for I to get
I = S + 0.5 (S/D)
where S/D is signal to distortion ratio in dB, S is in dBm and I is in dBm.

For second-order intermodulation,
I = S + (S/D)

Fig 17.6—Top: IMD ratio (as displayed on a spectrum analyzer). Bottom: intercept point.

less than it would be if the stage were still linear. Some circuits do not need to be linear (and should not be linear), and we will look at several examples. In many applications linearity is necessary, especially in SSB receivers and transmitters. The situation for a linear circuit is optimum when the input and output become nonlinear simultaneously. This means that the gain, bias and load impedance are all properly coordinated. We will study this more closely in later sections.

Dynamic Range

There is a relationship between noise factor, IMD, gain compression and bandwidth in a building block stage. In general, an active circuit that has a low noise factor tends to have a poor intercept point and vice versa. A well-designed transistor or circuit tries to achieve the best of both worlds. Dynamic range is a measure of this capability. Suppose that a circuit has a third-order input intercept of +10 dBm, a noise factor of 6 dB and a noise bandwidth of 1000 Hz. We want to determine its dynamic range. At a certain level per tone of a two-tone input signal the third-order IMD products are equal to the noise level in the 1000-Hz band. The ratio, in dB, of each of the two tones to the noise level is called the "spurious free dynamic range" (SFDR). **Fig 17.7** illustrates the problem and derives the proper formula. Note that the bandwidth is an important player. For the example above, the dynamic range is DR = 0.67 (10 − (174 + 10 log(1000) + 6)) = 99 dB. Often the dynamic range is calculated using a 1.0-Hz bandwidth. This is called "normalization." Another kind of dynamic range compares the 1-dB compression level with the noise level. This is the CFDR (compression-free dynamic range). Fig 17.7 illustrates this also.

Modules in Combination

Quite often the performance of a single stage can be greatly improved by combining two identical modules. Because the input power is split evenly between the two modules the drive source power can be twice as great and the output power will also be twice as great. In transmitters, especially, this often works better than a single transistor with twice the power rating. Or, for the same drive and output power, each module need supply only one-half as much power, which usually means better distortion performance. Often, the total number of stages can be reduced in this manner, with resulting cost savings. If the combining is performed properly, using hybrid transformers, the modules interact with each other much less, which can avoid certain problems. These are the

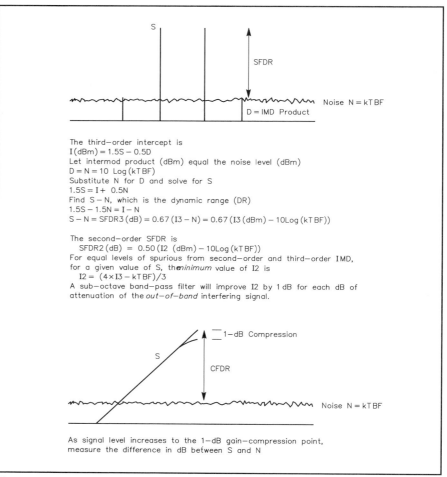

The third−order intercept is
I (dBm) = 1.5S − 0.5D
Let intermod product (dBm) equal the noise level (dBm)
D = N = 10 Log (kTBF)
Substitute N for D and solve for S
1.5S = I + 0.5N
Find S − N, which is the dynamic range (DR)
1.5S − 1.5N = I − N
S − N = SFDR3 (dB) = 0.67 (I3 − N) = 0.67 (I3 (dBm) − 10Log (kTBF))

The second−order SFDR is
SFDR2 (dB) = 0.50 (I2 (dBm) − 10Log (kTBF))
For equal levels of spurious from second−order and third−order IMD, for a given value of S, the *minimum* value of I2 is
I2 = (4×I3 − kTBF)/3
A sub−octave band−pass filter will improve I2 by 1 dB for each dB of attenuation of the *out−of−band* interfering signal.

As signal level increases to the 1−dB gain−compression point, measure the difference in dB between S and N

Fig 17.7—The definitions of spurious-free dynamic range (SFDR) and compression-free dynamic range (CFDR). The derivation yields a very useful equation for SFDR.

system-design implications of module combining.

Three methods are commonly used to combine modules: parallel (0°), push-pull (180°) and quadrature (90°). In RF circuit design, the combining is often done with special types of "hybrid" transformers called splitters and combiners. These are both the same type of transformer that can perform either function. The splitter is at the input, the combiner at the output. We will only touch very briefly on these topics in this chapter and suggest that the reader consult the **RF Power Amplifiers** chapter and the very considerable literature for a deeper understanding and for techniques used at different frequency ranges.

Fig 17.8 illustrates one example of each of the three basic types. In a 0° hybrid splitter at the input the tight coupling between the two windings forces the voltages at A and B to be equal in amplitude and also equal in phase if the two modules are identical. The 2R resistor between points A and B greatly reduces the transfer of power between A and B via the trans-

former, but only if the generator resistance is closely equal to R. The output combiner separates the two outputs C and D from each other in the same manner, if the output load is equal to R, as shown. No power is lost in the 2R resistor if the module output levels are identical.

The 180° hybrid produces push-pull operation. The advantages of push pull were previously discussed. The horizontal transformers, 1:1 balun transformers, allow one side of the input and output to be grounded. The R/2 resistors improve isolation between the two modules if the 2R resistors are accurate, and dissipate power if the two modules are not identical.

In a 90° hybrid splitter, if the two modules are identical but their identical input impedance values may not be equal to R, the hybrid input impedance is nevertheless R Ω, a fact that is sometimes very useful in system design. The power that is "reflected" from the mismatched module input impedance is absorbed in RX, the "dump" resistor, thus creating a virtual input impedance equal to R. The two module inputs are 90° apart. At the

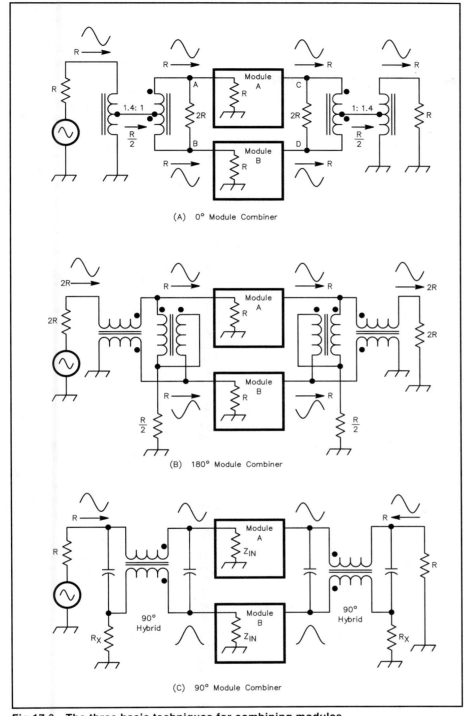

Fig 17.8—The three basic techniques for combining modules.

of situations. Further study of this chapter will reveal how these methods can be adapted to various situations. We will consider typical receiving circuits and typical transmitting circuits.

Properties of Cascaded Stages

Fig 17.9 shows a simple receiver "front end" circuit consisting of a preselector filter, an RF amplifier, a second filter and a double balanced diode mixer. We want to know the gain, bandwidth, noise factor, second and third-order intercept points, SFDR and CFDR for this combination, when the circuitry following these stages has the values shown. Let's consider one item at a time.

Gain of Cascaded Stages

The antenna tuned circuit L1, C1, C2 has some resistive loss; therefore the power that is available from it is less than the power that is available from the generator. Let's say this loss is 2.0 dB. Next, find the available power gain of the RF amplifier. First, note that the generator voltage V_s is transformed up to a larger voltage V_g by the input tuned circuit, according to the behavior of this kind of circuit. This step-up increases the gain of the RF amplifier because the FET now has a larger gate voltage to work with. (A bit of explanation: The FET has a high input impedance therefore, since the generator resistance R_s is only 50 Ω, a voltage step-up will utilize the FET's capabilities much better. But an excessive step-up opens up the possibility that the FET and other "downstream" circuits can be overdriven by a moderately large signal. So this step-up process should not be carried too far). The gain also depends on the drain load resistance, which is the mixer input impedance, stepped up by the circuit L2, C3, C4. Again, there is some loss within this tuned circuit, say 2.0 dB. If the drain load is too large the FET drain voltage swing can become excessive, creating distortion. The RF amplifier can become unstable due to excessive gain. Note also that the unbypassed source resistor provides negative feedback, to help make the RF amplifier more predictable. Dual-gate FETs have relatively little reverse coupling.

We come now to the mixer, whose available gain is about –6 dB. This is the difference between its available IF output power and its available RF input power. This is a fairly low-level mixer, so it can be easily overdriven if the RF gain is too high. Harmonic IMD and two-tone IMD can become excessive (see later discussion in this chapter). On the other hand, as we will discuss later, too little RF gain will yield

output, the two identical signals, 90° apart, are combined as shown and the output resistance is also R. This basic hybrid is a narrowband device, but methods for greatly extending the frequency range are in the literature (Ref 3). One advantage of the 90° hybrid is that catastrophic failure in one module causes a loss of only one half of the power output.

MULTISTAGE SYSTEMS

As the next step in studying system design we will build on what we've learned about single stages, and look at the methods for organizing several building block circuits and their interconnecting networks so that they combine and interact in a desirable and predictable manner. These methods are applicable to a wide variety

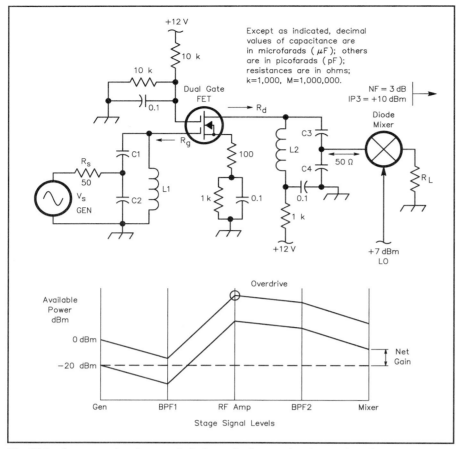

Fig 17.9—An example of cascaded stage design, a simple receiver front end.

Selectivity of Cascaded Stages

The simplified receiver example of Fig 17.9 shows two resonant circuits (filters) tuned to the signal frequency. They attenuate strong signals on adjacent frequencies so that these signals will not disturb the reception of a desired weak signal at center frequency. **Fig 17.10** shows the response of the first filter and also the composite response of both filters at the mixer input.

Consider first the situation at the output of the first filter. If a strong signal is present, somewhat removed from the center frequency, the selectivity of the first filter may just barely prevent excessive signal level in the RF amplifier. When this signal is amplified and filtered again by the second filter, its level at the input of the mixer may be excessive. Our system design problem is to coordinate the amplifier gain and second filter selectivity so that the mixer level is not too great. (A computer simulation tool, such as *ARRL Radio Designer,* can be instructive and helpful.) Then we can say that for that level of undesired signal at that frequency offset the cascade is properly designed.

The decisions regarding the "expected" maximum level and minimum frequency offset of the undesired signal are based on the operating environment for the equipment, with the realistic understanding that occasionally both of these values may be violated. If improvement is needed, it may then be necessary to (a) improve the selectivity, (b) use a more robust amplifier and mixer or (c) reduce amplifier gain. Very often, increases in cost, complexity and system noise factor are the byproducts of these measures.

Cascaded signal filters are often used to obtain a selectivity shape that has a flat top response and rapid or deep attenuation beyond the band edges. This method is often preferred over a single, more complex "brick wall" filter that has a very steep rate of attenuation outside the passband.

Noise Factor of a Cascade

In the example of Fig 17.9, the overall noise factor of the two-stage circuit is defined in the same way as for a single stage. It is the degradation in signal-to-noise ratio (S/N) from the signal generator to the output. This total noise factor can be found by direct measurement or by a stage-by-stage analysis. If we wish to optimize the total noise factor or look for trade-offs between it and other things such as gain and distortion, a stage-by-stage analysis is needed.

The definition of noise factor for a single stage applies as well to each stage

a poor receiver noise figure.

The concepts of available gain and transducer gain were introduced earlier. If we multiply the available gains of the input filter, RF amplifier, interstage filter and mixer, we have the available gain of the entire combination. The transducer gain is the ratio of the power actually delivered to R_L to the power that is available from the generator. To get the transducer gain of the combination, multiply the available gain of the first three circuits by the transducer gain of the last circuit (the mixer). This concept may require some thought on your part, but it is one that is frequently used and it adds understanding to how circuits are cascaded. One example, the transducer gain of a receiver, compares the signal power available from the antenna with the power into the loudspeaker (a perfectly linear receiver is assumed).

Fig 17.9 also shows an example of a commonly used graphical method for the available gain of a cascade. The loss or increase of available power at each step is shown. As the input increases the other values follow. But at some point, measurements of linearity or IMD will show that

some circuit is being driven excessively, as the example indicates. To improve performance at that point, we may want to make gain changes or take some other action. If the overload is premature, a more powerful amplifier or a higher-level mixer may be needed. It may be possible to reduce the gain of the RF amplifier by reducing the step-up in the input LC circuit or the drain load circuit, but this may degrade noise figure too much. This is where the "optimization" process begins.

A method that is often used in the lab is to plot the voltage levels at various points in the system. These voltages are easily measured with an RF voltmeter or spectrum analyzer, using a high-impedance probe. This is a convenient way to make comparative measurements, with the understanding that voltage values are not the same thing as power-gain values, although they may be mathematically related. Many times, these voltage measurements quickly locate excessive or deficient drive conditions during the design or troubleshooting process. Comparisons of measured values with previous measurements of the same kind on properly functioning

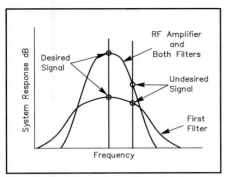

Fig 17.10—The gain and cumulative selectivity, between the generator and the mixer, of the example circuit in Fig 17.9.

in the chain. For each stage there is a signal and thermal noise generator, internal sources of excess noise and a noise bandwidth. In a cascade, the signal and thermal noise sources for a particular stage are found in the previous stage, as shown in **Fig 17.11A**. But this thermal noise source has already been accounted for as part of the excess noise for the previous stage. Therefore, this thermal noise must not be counted twice in the calculation. On this basis, Fig 17.11A derives the formula for the noise factor of a two-stage system. This formula can then be used to find the noise factor of the multisegment system in

Fig 17.11B by applying it repetitively, first to stage N + 1 and N, then to N and N − 1, then to N − 1 and N − 2 and so on, where N is the total number of stages and N + 1 is the rest of the system after the last stage. Fig 17.11B shows the cumulative noise figure (dB) at each point in the example of Fig 17.9. The graphical method aids the analysis visually.

Note that the diode mixer's noise figure approximately equals its gain loss. In applying the formula, if G1 is a lossy device, not an amplifier, then F1 equals its attenuation factor and G1 = 1/F1. Also, observe the critical role that values of RF amplifier gain and noise figure play in establishing the overall noise factor (or noise figure), despite the high noise figure that follows it. In Fig 17.11A and B, we assumed that the noise bandwidth does not increase toward the output. If the noise bandwidth does increase toward the output a complication occurs. Fig 17.11C provides a modified formula that is more accurate under these conditions. This situation is often encountered in practice, as we will see, especially in the discussion of receiver design (Ref 4).

Distortion in Cascaded Circuits

The IMD created in one stage combines with the distortion generated in following stages to produce a cumulative effect at the output of the cascade. The phase relationships between the distortion products of one stage and those of another stage can vary from 0° (full addition) to 180° (full subtraction). It is customary to assume that they add in-phase as a worst case. Under these conditions, **Fig 17.12** shows how to determine distortion at the input of a stage. Formulas are given for finding the third-order and second-order input intercept points in dBm. These formulas can be applied repetitively, in a manner similar to the noise-factor formula, to get the cumulative intercept point at each stage of the cascade. The output intercept point, in dBm, of a stage is equal to its input intercept point, in dBm, plus the gain, in dB, of the stage. When a purely passive, linear stage is part of the analysis, use a large value of intercept such as 100 dBm (10^7 W, Ref 5).

THE AMATEUR RADIO COMMUNICATION CHANNEL

In order to design radio equipment it is first necessary to know what specifications the equipment must have in order to establish and maintain communication. This is a very large and complex subject that we cannot fully explore here, however, it is possible to point out certain properties of the communication channel,

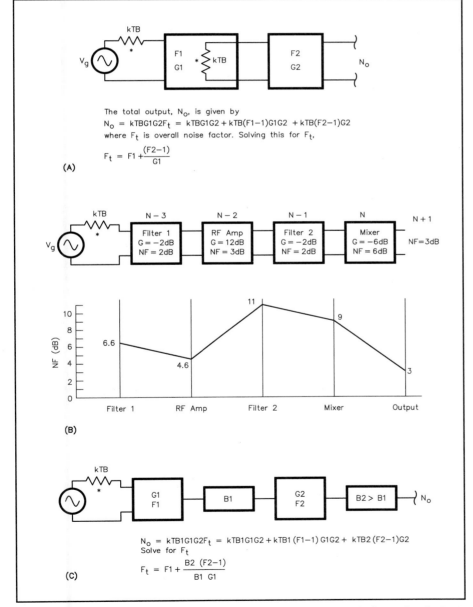

The total output, N_o, is given by
$$N_o = kTBG1G2F_t = kTBG1G2 + kTB(F1-1)G1G2 + kTB(F2-1)G2$$
where F_t is overall noise factor. Solving this for F_t,

$$F_t = F1 + \frac{(F2-1)}{G1}$$

(A)

(B)

$$N_o = kTB1G1G2F_t = kTB1G1G2 + kTB1\,(F1-1)\,G1G2 + kTB2\,(F2-1)G2$$
Solve for F_t

$$F_t = F1 + \frac{B2\,(F2-1)}{B1\,G1}$$

(C)

Fig 17.11—A: the noise factor of a two-stage network. B: cumulative noise factor for the example in Fig 17.10. C: noise factor when the bandwidth increases toward the output.

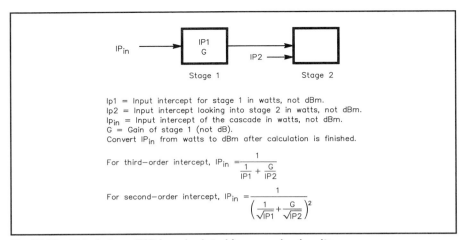

Ip1 = Input intercept for stage 1 in watts, not dBm.
Ip2 = Input intercept looking into stage 2 in watts, not dBm.
Ip$_{in}$ = Input intercept of the cascade in watts, not dBm.
G = Gain of stage 1 (not dB).
Convert IP$_{in}$ from watts to dBm after calculation is finished.

For third−order intercept, $IP_{in} = \dfrac{1}{\dfrac{1}{IP1} + \dfrac{G}{IP2}}$

For second−order intercept, $IP_{in} = \dfrac{1}{\left(\dfrac{1}{\sqrt{IP1}} + \dfrac{G}{\sqrt{IP2}}\right)^2}$

Fig 17.12—This is how IMD is calculated in cascade circuits.

especially as it pertains to Amateur Radio, and to discuss equipment requirements for successful communication. The "channel" is:

- the frequency band that is being transmitted and to which the distant receiver is tuned, and
- the electromagnetic medium that conveys the signal.

The Amateur Radio bands are, in fact, a very difficult arena for communications and a severe test of radio-equipment design. The very wide range of received signal levels, the high density of signals whose channels often overlap or are closely adjacent, the relatively low power levels and the randomness (the lack of formal operating protocols) are the main challenges for Amateur Radio equipment designers. An additional challenge is to design the equipment for moderate cost, which often implies technical specifications that are somewhat below commercial and military standards. These relaxed standards sometimes add to the amateur's problems.

Received Noise Levels

There are three major sources of noise arriving at the receive antenna:

- atmospheric noise generated by disturbances in the Earth's environment,
- galactic noise from outer space and
- noise from transmitters other than the desired signal.

Let's briefly discuss each of these kinds of noise.

Atmospheric noise (including manmade noise) is maximum at frequencies below 10 MHz, where it has *average* values about 40 dB above the thermal noise at 290 K (K = kelvins, absolute temperature). Above 10 MHz, its strength de-

creases at 20 dB per *octave*. At VHF and above, it is of little importance (Ref 6).

However, various studies have found that at certain times and locations and in certain directions this noise approaches the level of thermal noise at 290 K, even at the lower frequencies. Therefore the conventional wisdom that a low receiver noise figure is not important at low HF is not completely true. Amateurs, in particular, exploit these occurrences, and most amateur HF receivers have noise figures in the 8 to 12-dB range for this reason, among others. A very efficient antenna at a low frequency can modify this conclusion, though, because of its greater signal and noise gathering power (for example, a half-wave dipole gathers about 12 dB more power at 1.8 MHz than a half-wave dipole at 30 MHz (Ref 7)). When the noise level is high, an attenuator in the antenna lead can reduce receiver vulnerability to strong interfering signals without reducing the S/N ratio of weaker signals. In other words, the system (that is, receiver plus noisy antenna) dynamic range is improved (the receiver intercept point increases and the system noise is reduced). The antenna noise, after attenuation, should be several dB above the receiver internal noise. This is a typical example of a communication-link design consideration that may not be necessary if the receiver is of high quality.

Receive Antenna Directivity

If the receive antenna has gain, and can be aimed in a certain direction, it often happens that atmospheric noise is less in that direction. A lower receiver noise figure may then help. Or, if the noise arrives uniformly from all directions but the desired signal is increased by the antenna gain, then the S/N ratio is increased. That is, the noise is constant but the signal is

greater. (Explanation: if the noise is the same from all directions the high-gain antenna receives more noise from the desired direction but rejects noise from other directions; therefore the total received noise tends to remain constant.) This is one of the advantages of the HF rotary beam antenna. The same gain can also cause strong undesired signals to challenge the receiver's dynamic range (or null out an undesired signal).

Galactic Noise

The *average* noise level from outer space is about 20 dB above that of thermal noise at 20 MHz and decreases at about 20 dB per *decade* of frequency (Ref 6). But at microwave frequencies, high-gain antennas with very-low-noise-figure receivers are able to locate sources of relatively intense (and very low) galactic noise. Amateurs working at microwave frequencies up to 10 GHz go to great lengths to get their antenna gains and receiver sensitivities good enough to take advantage of the high and low noise levels.

Transmitter Noise

Fig 17.13A shows the spectral output of a typical amateur transmitter. The desired modulation lies within a certain well defined bandwidth, which is determined by the type of modulation. Because of unavoidable imperfections in transmitter design, there are some out-of-band modulation artifacts such as high-order IMD products. The signal filter (SSB, CW and so on) response also has some slope outside the passband. There is also a region of phase noise generated in the various mixers and local oscillators (LOs). These phase noise sidebands are "coherent." That is, the upper frequency sidebands have a definite phase relationship to the lower frequency sidebands. At higher values of frequency offset, a noncoherent "additive" noise shelf may become greater than phase noise and it can extend over a considerable frequency band. Other outputs such as harmonics and other transmitter-generated spurious emissions are problems.

The general design goals for the transmitter are:

1. Make the unavoidable out-of-band distortion products as small as technology and equipment cost and complexity will reasonably allow,
2. Design the synthesizers and other local oscillators and mixers so that phase noise, as measured in a bandwidth equal to that of the desired modulation, is less than the out-of-band distortion products in goal 1 and,
3. Make the wideband noise sufficiently

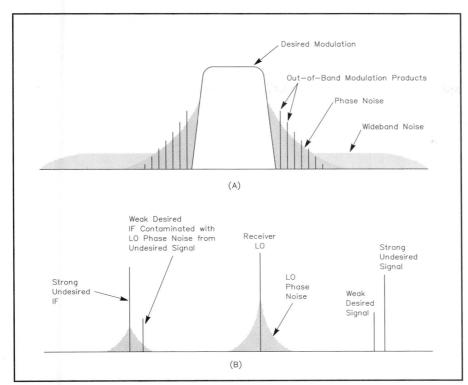

Fig 17.13—A: transmitter spectrum with discrete out-of-band products, phase noise and white noise. B: reciprocal mixing of LO phase noise onto an incoming signal.

small that the noise will be less than any unavoidable receiver noise at nearby receivers with the same bandwidth.

If the additive noise is very small, LO phase noise may come back into the picture. In narrowband systems such as Morse code (CW) it can be very difficult or impractical to make transmitted phase noise less than the normal Morse code sidebands (see later discussion of this topic). The general method to reduce wideband noise from the transmitter is to place the narrowband modulation band-pass filter at as high a signal level as possible and to follow that with a high-level mixer and then a low-noise first-stage RF amplifier.

Phase-noise amplitude varies with modulation. That is, the LO phase noise is modulated onto the outgoing signal by the "reciprocal mixing" process (the signal becomes the "LO" and the LO phase noise becomes the "signal"). If the actual LO to phase noise ratio is X dB, the ratio of the transmit signal to its phase noise is also X dB. In SSB the magnitude of the phase-noise sidebands is maximum only on modulation peaks. In CW it exists only when the transmitter is "key down." The additive noise, on the other hand, may be much more constant. If the power amplifi-

ers are Class A or Class AB, additive noise does not require any actual signal and tends to remain more nearly constant with modulation.

In a communication link design, the receiver's culpability must also be considered. The receiver's LOs also generate phase noise that is modulated onto an adjacent-channel signal (reciprocal mixing) to produce an in-band noise interference, as shown in Fig 17.13B. In view of this, the transmitter and receiver share equal responsibility regarding phase noise, and there is little point in making either one a great deal better unless the other is improved also. Nevertheless, high-quality receivers with low phase noise exist, and they are vulnerable to transmitter phase noise. The converse situation also exists; receiver phase noise can contaminate a clean incoming signal (Ref 8).

Receiver Gain and Transmitter Power Requirements

The minimum level of a received signal is a function of the antenna noise level and the bandwidth. As just one example, for an HF SSB system with a 2.0-kHz bandwidth and a noise level 10 dB above thermal (–131 dBm in a 2.0-kHz band) the minimum readable signal, say 3 dB above the noise

level, would be –128 dBm. Assume the receiver-generated noise is negligible. If the audio output to a loudspeaker is, say +20 dBm (0.1 W), then the total required receiver gain is 148 dB, which is an enormous amount of amplification. For a CW receiver with 200-Hz bandwidth, the minimum signal would be –138 dBm and the gain would be 158 dB. Receivers at other frequencies with lower noise levels can require even higher gains to get the desired audio output level. If the transmission path attenuation can be predicted or calculated, the required transmitter power can be estimated. These kinds of calculations are often done in UHF and microwave amateur work, but less often at HF (see the microwave receiver section for an example).

We do, however, get some "feel" for the receiver gain requirements, how the receiver interacts with the "channel," and that the minimum signal power is an almost incredibly small 1.6×10^{-16} W. On the other hand, amateur receiver S-meters are calibrated up to an input signal level of –13 dBm (60 dB above 100 µV from a 50-Ω source). Therefore the receiver must deal with a desired signal range of at least 115 dB (128 – 13) for the SSB example or at least 125 dB for the CW example (assuming that AGC limits the signal levels within the receiver).

Fading

Radio signals very often experience changes in strength due either to reflections from nearby objects (multipath) or, in the case of HF, to multiple reflections in the ionosphere. At a particular frequency and at a certain time, a signal arriving by multipath may decrease severely. The effect is noticed over a narrow band of frequencies called the "fading bandwidth." At HF, the center frequency of this fade band drifts slowly across the spectrum. Communication links are degraded by these effects, so equipment design and various communication modes are used to minimize them. For example, SSB is less vulnerable than conventional AM. In AM, loss of carrier or phase shift of the carrier, relative to the sidebands, causes distortion and reduces audio level.

The UHF/Microwave Channel

At frequencies above about 300 MHz, we need to account for the interaction of the Earth environment with the transmitted and received signals. Here are some of the things to consider:

1. Line-of-sight communications distance, as a function of receiver and transmitter antenna height.
2. Losses from atmospheric gasses and

water vapor (above several gigahertz).

3. Temperature effects on paths: reflections, refractions, diffractions and transmission "ducts."
4. Atmospheric density inversions due to atmospheric pressure variations and weather fronts.

5. Tropospheric reflections and scattering.
6. Meteor scattering (mostly at VHF but occasionally at UHF).
7. Receive-antenna sky temperature.

Competitive amateur operators who are active at these frequencies become proficient at recognizing and dealing with these communication channel effects and learn how they affect equipment design. They become proficient at estimating channel performance, including path loss, receive system noise figure (or noise temperature), antenna radiation patterns and gain.

Receiver Design Techniques

We will now look at the various kinds of receivers that are used by amateurs and at specific circuit designs that are commonly used in these receivers. The emphasis is entirely on analog approaches. Methods that use digital signal processing (DSP) for various signal processing functions are covered in the **DSP** chapter.

EARLY RECEIVER DESIGN METHODS

Fig 17.14 shows some early types of receivers. We will look briefly at each. Each discussion contains information that has wider applicability in modern circuit design, and is therefore not merely of historical interest. A lot of good old ideas are still around, with new faces.

The Crystal Set

In Fig 17.14A the antenna circuit (capacitive at low frequencies) is series resonated by the primary coil to maximize the current through both, which also maximizes the voltage across the secondary. The semiconductor crystal rectifier then demodulates the AM signal. This demodulation process utilizes the carrier of the AM signal as a "LO" and frequency translates (mixes) the RF signal down to baseband (audio). The rectifier and its output load impedance (the headphones) constitute a loading effect on the tuned circuit. For maximum audio output, a certain amount of coupling to the primary coil provides an optimum impedance match between the rectifier circuit and the tuning circuit. The selectivity is then somewhat less than the maximum obtainable. To improve the selectivity, reduce the secondary L/C ratio and/or decrease the coupling to the primary. Some decrease in audio output will usually result. This basic mechanism for demodulat-

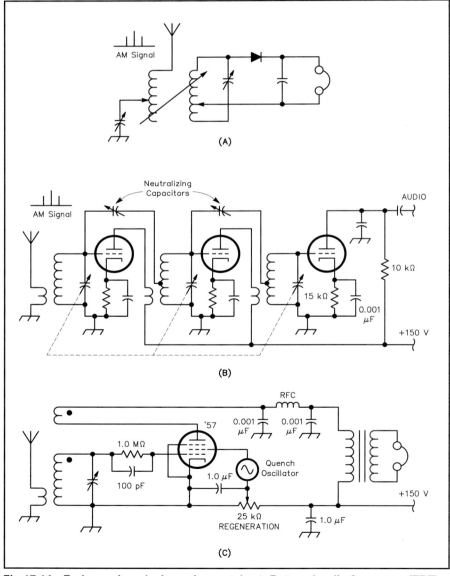

Fig 17.14—Early receiver designs. A: crystal set. B: tuned radio-frequency (TRF) receiver. C, a regenerative receiver.

ing an AM signal by using a rectifier is identical to that used in nearly all modern AM receivers. One important feature of this rectifier is a signal level "threshold effect" below which rectification quickly ceases. Therefore the crystal set, without RF amplification, is not very good for very weak signals. Early crystal receivers used large antennas to partially solve this problem, but they were vulnerable to strong signals (their dynamic range was not very good). However, a large antenna does make greater tuner selectivity possible (if needed) because looser coupling can be used in the tuner. That is, the loading of the secondary resonator by the antenna and rectifier can both be reduced somewhat.

There is one other interesting property of this detector. The two AM sidebands add in phase (coherently) at the audio output, but noise above and below the carrier frequency add in random phase (noncoherently). Therefore the detector provides a 3-dB improvement in signal-to-noise ratio, for a given average sideband power (Ref 9).

The Tuned Radio Frequency (TRF) Receiver

Fig 17.14B shows a TRF receiver. One, or possibly many, tuned RF stages are followed by a vacuum-tube implementation of the crystal rectifier (infinite-impedance square-law detector) described in the previous section. The RF amplification overcomes the threshold effect of the detector and the multiple tuned circuits (called "synchronous tuning"), usually isolated from each other by amplifier devices, allow much better selectivity. This selectivity is greatly reduced at the high end of the tuning range because the tuning capacitance becomes small. At the low end of the tuning dial, modulation sidebands may be rolled off or "clipped" due to excessive selectivity. This variation in selectivity, and also in gain, are the TRF's main drawbacks, which helped popularize the superheterodyne approach. The figure shows triode-tube amplifiers (they could also be three lead transistors: single-gate FETs or bipolar) that are neutralized to prevent self oscillation. This receiver is called a "Neutrodyne." Multigrid tubes or dual gate FETs do not need neutralization.

The Regenerative Receiver

Fig 17.14C (with the quench oscillator inactive) is a simple example of a regenerative receiver. The basic principle is that positive RF feedback, via the plate winding, is used to increase the RF gain up to and slightly beyond the point of self oscillation. With no signal, the internal shot and thermal noise peaks and a small bias voltage on the grid capacitor (caused by a

very slight grid current) combine to produce a stable and self-limiting oscillation that is similar to the behavior of an ordinary oscillator, except that right at the peak of the oscillation cycle the amplification is extremely large and the detector is therefore very sensitive. The Q of the tuned circuit is greatly increased by the introduction of negative resistance (see the **AC/RF Sources** chapter), but the small grid-current loading limits the increase. With greater or lesser feedback, or if a strong signal appears, the gain and Q drop rapidly.

The self oscillation heterodynes with an incoming CW or SSB signal to produce an audio beat note. The main advantages of the "regen" are the beat note, the absence of a weak-signal threshold effect and amplification. Slightly below the oscillation point, makeshift AM reception is possible. From about 1920 to 1935, the regen was the favorite HF receiver among amateurs. It was the subject of much design and development by them. It also required considerable operating skill (Ref 10).

The Superregenerative Receiver

Fig 17.14C (with the quench oscillator active) is a simple superregenerative receiver. The idea here is that the output of the 20-kHz quench oscillator modulates the detector and drives it through the point of oscillation and through the point of maximum gain and sensitivity at a super-audible rate. This results in an audio signal that is free of the audible heterodyne that the ordinary regen produces. Therefore, the receiver is more useful for AM or FM voice reception. Amplitude limitations, distortion and poor selectivity are inherent in superregenerative receivers—this is the price paid for relative simplicity.

Super regens were used for many years for 30-100 MHz voice reception, where the high ratio of signal frequency to quench frequency made the circuit more manageable. The classic amateur article on this subject by Ross Hull (at ARRL) appears in the July 1931 *QST*. See also any edition of the famous *Radiotron Designer's Handbook* by Langford-Smith, circa 1953.

MODERN RECEIVER DESIGN METHODS

The superheterodyne and the direct-conversion receiver are the most popular modern receivers and the chief topic of this discussion. Both were conceived in the 1915-1922 time frame. Direct conversion was used by Bell Labs in 1915 SSB experiments; it was then called the "homodyne" detector. E. H. Armstrong devised the superhet in about 1922, but for about 12 years it was considered too expensive for the (at that time) financially

strapped amateur operators. The advent of "single signal reception," pioneered by J. Lamb at ARRL, and the gradual end of the Great Depression era brought about the demise of the regenerative receiver. We begin with a discussion of the direct conversion receiver, which has been rediscovered by amateur equipment builders and experimenters in recent years.

Direct Conversion (D-C) Receivers

The direct conversion (D-C) receiver, in its simplest form, has some similarities to the regenerative receiver:

- The signal frequency is converted to audio in a single step,
- An oscillator very near the signal frequency produces an audible beat note,
- Signal bandwidth filtering is performed at baseband (audio),
- Signals and noise (both receiver noise and antenna noise) on both sides of the oscillation frequency appear equally in the audio output. The image (on the other side of zero beat) noise is an "excess" noise that degrades the noise factor and dynamic range.

There are three major differences favoring the D-C receiver:

- There is no delicate state of regeneration involved. A low-gain or passive mixer of high stability is used instead.
- The oscillator is a separate and very stable circuit that is buffered and coupled to the mixer.
- The D-C (that uses modern circuit design) has much better dynamic range.

The regen has enormous RF gain (and Q multiplication) and therefore little audio gain is needed. The D-C delivers a very low-level audio that must be greatly amplified and filtered. RF amplification, band-pass filtering and automatic gain control (AGC) can be easily placed ahead of the mixer with beneficial results.

An enhancement of the D-C concept can perform a fairly large reduction of the signal and noise image responses mentioned above. It is a major technical problem, however, to get a degree of reduction over a wide frequency range, say 1.6-30 MHz, that compares with that easily obtainable using superheterodyne methods.

D-C Receiver Design Example

Fig 17.15 is a schematic diagram of a simple D-C receiver that utilizes all of the principles mentioned above except image rejection. The emphasis is on simplicity for both SSB and CW reception on the 14-MHz band. The LO is standard *Handbook* circuitry and is not shown. **Fig 17.16**

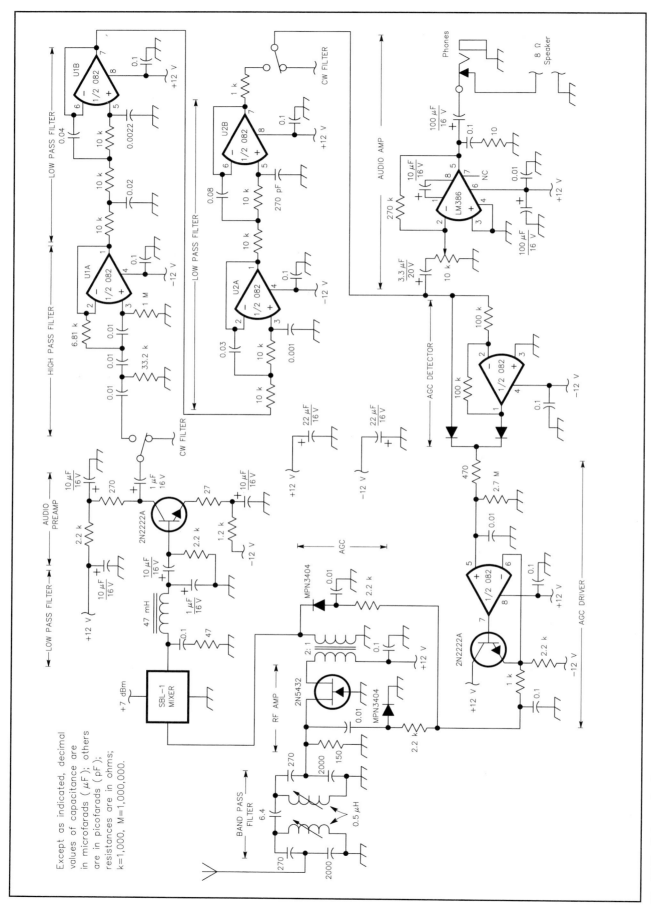

Fig 17.15—Schematic diagram of a simple D-C receiver for 20 m SSB and CW. Image cancellation is not used, but RF amplification and audio derived AGC are included.

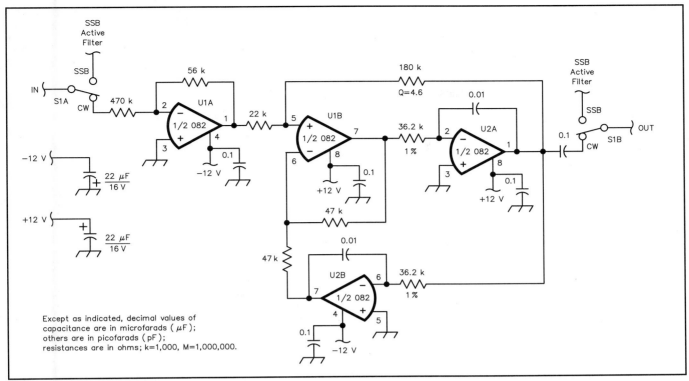

Fig 17.16—Schematic diagram of an active CW band-pass filter for the D-C receiver in Fig 17.15.

shows a simple example of an active CW band-pass filter centered at 450 Hz.

The input RF filter shields the receiver from large out-of-ham-band signals and has a noise figure of 2 dB. The grounded-gate RF amplifier has a gain of 8 dB, a noise figure of 3 dB and an input third-order-intercept point of 18 dBm. Its purpose is to improve a 14-dB noise figure (at the antenna input) without the amplifier to about 8.5 dB. It also eliminates any significant LO conduction to the antenna and provides opportunities for RF AGC. The total gain ahead of the mixer is about 6 dB, which degrades the IMD performance of the mixer and subsequent circuitry somewhat. However, the receiver still has a third-order intercept (IP3) of about 6 dBm for two tones within the range of the audio filter. The IP3 is 11 dBm (quite respectable) when one of the two tones is outside the range of the low-pass audio filter that precedes the first audio amplifier. The low-pass filter protects the audio amplifier from wideband signals and noise. The intercept point could be improved by eliminating the RF amplifier, but the antenna input noise figure would then be much worse. This is a common trade-off decision that receiver designers must make.

The above analysis would be correct for a conventional receiver, but in this case there is a small complication that we will mention only briefly. The noise sources ahead of the mixer, both thermal and excess, that are on the image side (the side of the carrier opposite a weak desired signal) are translated to the baseband and appear as an increased noise level at the input of the first audio amplifier. If the SFDR (previously defined) in a 1000-Hz bandwidth were ordinarily 95 dB, using the above numbers, the actual SFDR would be perhaps 2.5 dB less. An image-reject mixer would correct this problem. Observe that the low noise figure of the RF amplifier minimizes the gain needed to get the desired overall noise figure. Also, its good intercept point minimizes strong-signal degradation contributed by the amplifier. The input BPF eliminates problems from second-order IMD. Flicker-effect (1/f) noise in the mixer audio output may also be a problem, which the RF amplification reduces and the mixer design should minimize.

The preceding analysis illustrates the kind of thinking that goes into receiver design. If we can quantify performance in this manner we have a good idea of how well we have designed the receiver.

For the circuit in Fig 17.15 and the numbers given above, the gain ahead of the first audio amplifier is 0 dB. As stated before, this amplifier is protected from wideband interference by the 2-element low-pass audio filter ahead of it, which attenuates at a rate of 12 dB per octave. This filter could have more elements if

desired. By minimizing front-end gain, the tendency for the audio stages to overload before the earlier stages do so is minimized—if the audio circuitry is sufficiently "robust." This should be checked out using two-tone and gain-compression tests on the audio circuits. Audio-derived AGC helps prevent signal-path overload by strong desired (in-band) signals. Additional AGC can be applied in the audio section by using a variable-gain audio op amp (MC3340P).

The audio SSB and CW band-pass filters are simplified active op-amp filters that could be improved, if desired, by using methods mentioned in some of the references. In an advanced design, digital signal processing (DSP) could be used. Good shape factor is the main requirement for good adjacent-channel rejection. Good transient response (maximally flat group delay) would be a "runner-up" consideration. Digital FIR filters and analog elliptic filters are good choices.

Image Rejection in the D-C Receiver

Rejection of noise and signals on one side of the LO is a major enhancement and also a major complication of the D-C receiver (Refs 11, 12). **Fig 17.17** shows two correct ways to build an image canceling mixer and one incorrect way. The third way does not perform the required phase cancellations for image reduction. In practice, two ±45° phase shifters are used, rather than one 90° stage. As mentioned before, it is very difficult to get close phase tracking over a wide band of signal and LO frequencies. In amateur equipment, front panel "tweaker" controls would be practical.

The block diagram in **Fig 17.18** is a typical approach to an image-canceling D-C receiver. The two channels, including RF, mixers and audio must be very closely matched in amplitude and phase. The audio phase-shift networks must have equal gain and very close to a 90° phase difference. **Fig 17.19** relates phase error in degrees and amplitude error in dB to the rejection in dB of the opposite (image) sideband. For 30 or 40 dB of rejection, the need for close matching is apparent.

AUDIO PHASE SHIFTERS

Fig 17.20A shows an example of an audio phase-shift network. The stage in Fig 17.20B is one section, an active "all-pass" network that has these properties:

- The gain is exactly 1.0 at all frequencies and
- The phase shift changes from 180° at very low frequency to 0° at very high frequency.

The shift of this single stage is +90° at $f = 1/(2\pi RC)$. By cascading several of these with carefully selected values of RC the set of stages has a smooth phase shift across the audio band. A second set of stages is chosen such that the phase difference between the two sets is very close to 90°. The choices of R and C values have been worked out using computer methods; you can also find them in other handbooks

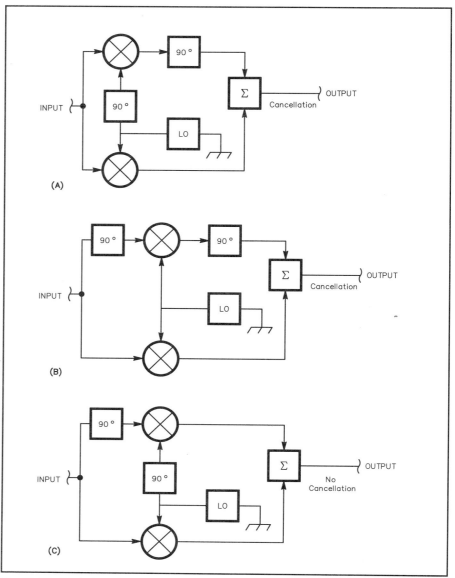

Fig 17.17—Both A and B are workable image cancelling mixer stages. The scheme at C will not cancel image signals.

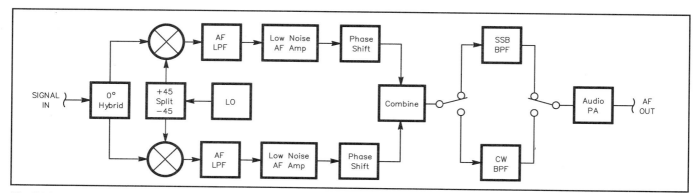

Fig 17.18—Typical block diagram of an image cancelling D-C receiver.

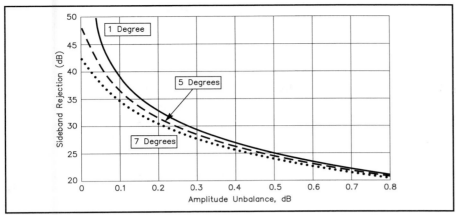

Fig 17.19—A plot of sideband rejection versus phase error and amplitude unbalance.

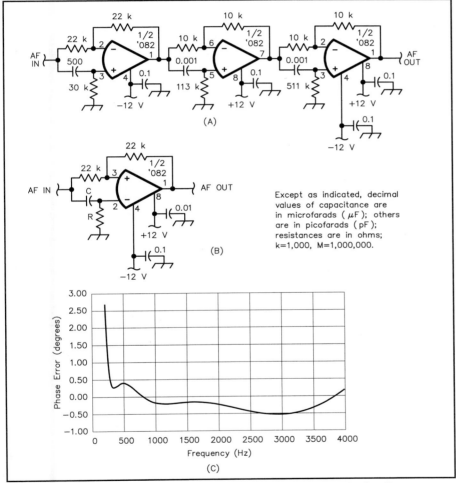

Fig 17.20—A: an example of an audio phase-shift circuit. B: a single all-pass stage. C: phase error vs audio frequency for a pair of circuits like that in A with appropriate values of R and C.

(Ref 13). Fig 17.20C shows the phase error for two circuits like the one shown in Fig 17.20A. Note the rapid increase in error at very low audio frequencies (an improvement would be desirable for CW work). These frequencies should be greatly attenuated by the audio band-pass filters that follow.

D-C Receiver Problem Areas

Because of the high audio gain, microphonic reactions due to vibration of low-level audio stages are common. Good, solid construction is necessary. Another problem involves leakage of the LO into the RF signal path by conduction and/or radiation. The random fluctuations in phase of the leakage signal interact with the LO to produce some unpleasant modulation and microphonic effects. Hum in the audio can be caused by interactions between the LO and the power supply; good bypassing and lead filtering of the power supply are needed. A small amount of RF amplification is beneficial for all of these problems.

The Superheterodyne Receiver

GENERAL DISCUSSION

The superheterodyne ("superhet") method is by far the most widely used approach to receiver and transmitter design. **Fig 17.21A** shows the basic elements as applied to an SSB/CW receiver, which we will consider first. We will consider a superhet transmitter later in this chapter.

RF from the antenna is filtered (preselected) by a band-pass filter of some kind to reduce certain kinds of spurious responses and then (possibly) RF amplified. A mixer, or frequency converter circuit, *multiplies* (in the time domain) its two inputs, the signal and the LO. The result of this multiplication process is a pair of output intermediate frequencies (IFs) that are the sum and difference of the signal and LO frequencies.

If the mixer is a perfect multiplier, as the equation in Fig 17.21 suggests, it is a linear mixer, these are the only output frequencies present and it has all the properties of any other linear circuit except for the change of frequency. If the mixer is a commutating (switching) mode mixer it is still a perfect mixer but additional frequencies of lesser amplitude are present. See the **Mixers, Modulators and Demodulators** chapter for a detailed discussion.

One of these outputs is selected to be the "desired" IF by the designer. It is then band-pass filtered and amplified. The bandwidths and shape factors of these filters are optimized for the kind of signal being received (AM, SSB, CW, FM, digital data). Two of the main attributes of the superhet are that this signal filtering band

shape and also the IF amplification are constant for any value of the receive signal frequency. An excessively narrow preselector filter could, however, have some effect on the desired signal, as we saw in the case of the TRF receiver.

A second mixer, or "detector" as it is usually called, translates the IF signal to baseband (audio) where it is further amplified, possibly filtered, and applied to an output transducer (headphones, loudspeaker, some other signal processor or display).

A superhet receiver may also contain multiple frequency conversions (IFs). Later discussion will focus on strategies used to select these IFs. Let's begin with a detailed discussion of the classic down-conversion superhet. Almost all of the topics apply as well to the various other kinds of receiver designs in subsequent sections.

Superhet Characteristics: A Down-Conversion Example

The desirability of the superhet approach is offset somewhat by certain penalties and problem areas. As a vehicle for mentioning these difficulties, seeing how to deal with them and discussing analysis and design methods; we use the tutorial example in **Fig 17.23**. That is a "down converting" single-conversion 14-MHz superhet with a 1.5-MHz IF. This receiver is simple and capable of fairly good performance in the 1.8 to 30-MHz frequency range. Fig 17.23 is intentionally incomplete and meant for instructional purposes only; do not attempt

to duplicate it as a project.

Block Diagram

The block labels of Fig 17.23 show that a preselector and RF amplifier are followed by downward frequency conversion to 1.5 MHz. This is followed by IF amplification and crystal filtering, a product detector, audio band-pass filters and an audio power output stage. Equal emphasis is given to SSB and CW. AGC circuitry is included. The audio and AGC circuits are the same as those in Fig 17.15 and Fig 17.16. As a first step, let's look at spurious responses of the mixer.

Mixer Spurious Responses

Mixers and their spurious responses are covered in detail in the Mixers chapter, but we will present a brief overview of the subject for our present purposes. We then will see how this information is used in the design process.

The mixer is vulnerable to RF signals other than the desired signal. Various harmonics of any undesired RF signal and harmonics of the LO combine to produce spurious IF outputs (called harmonic IMD). If these spurious outputs are within the IF passband they appear at the receiver output. The strength of these outputs depends on: harmonic number and strength of the RF signal as it appears at the mixer input, the harmonic number of the LO, the LO power rating (7 dBm, 17 dBm, and so on) and the design of the mixer.

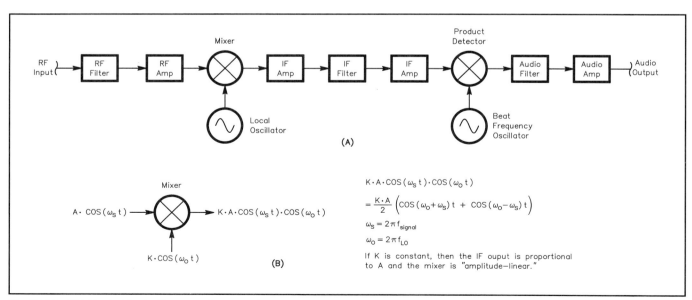

Fig 17.21—(A) Basic block diagram of a superhet receiver. (B) Showing how the input signal and a constant LO input produce a linear mixing action.

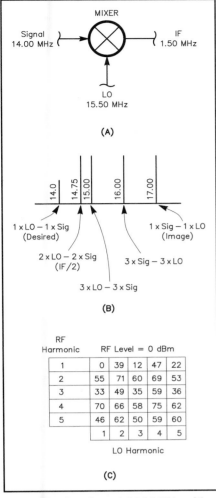

MIXER

Signal
14.00 MHz → IF 1.50 MHz

LO
15.50 MHz

(A)

14.0 14.75 15.00 16.00 17.00

1 × LO − 1 × Sig (Desired)   1 × Sig − 1 × LO (Image)

2 × LO − 2 × Sig (IF/2)   3 × Sig − 3 × LO

3 × LO − 3 × Sig

(B)

RF Harmonic	RF Level = 0 dBm				
1	0	39	12	47	22
2	55	71	60	69	53
3	33	49	35	59	36
4	70	66	58	75	62
5	46	62	50	59	60
	1	2	3	4	5

LO Harmonic

(C)

Fig 17.22—A: mixer at 14.00 MHz with LO at 15.50 and IF at 1.50. B: locations of strong signals that interfere with desired signal at 14.00 MHz due to harmonic IMD. C: typical chart of harmonic IMD products for a double balanced diode mixer with 0 dBm signal and 7 dBm LO.

Commercially available double-balanced diode mixers are so convenient, easy-to-use and of such low cost and high quality that they are used in many Amateur Radio receiver and transmitter projects. These mixers also do a good job of rejecting certain kinds of spurious responses. Our numerical examples will be based on typical published data for one of these mixers (Ref 14).

Fig 17.22A shows an example for the mixer tuned to a desired signal at 14.00 MHz with the LO at 15.50 MHz. The locations of undesired signals that cause a spurious response are shown in Fig 17.22B; they are at 14.75, 15.00, 16.00 and 17.00 MHz (there are many others of lesser importance). Each of these undesired signals produces a 1.50-MHz output from the mixer. The figure indicates the

harmonics of the undesired signal and the harmonics of the LO that are involved in each instance. The "order" of the spurious product is the sum of these harmonic numbers, for example the one at 16.00 MHz is a sixth-order product. The spurious at 17.00 MHz is called the "image" because it is also 1.50 MHz away from the LO, just as the 14.00 desired signal is 1.5 MHz away from the LO. It is a second-order response, as is the response at 14.00 MHz.

Fig 17.22C is a chart that shows the relative responses for various orders of harmonic IMD products for a signal level (desired or undesired) of 0 dBm and an LO level of +7 dBm. The values are typical for a great many +7-dBm mixers having various brand names and they improve greatly for higher level mixers (at the same RF levels). The second-order (desired and image) both have a reference value of 0 dB and the others are in dB below those two.

We can now consider the receiver design that suppresses these spurious responses so that they do not interfere with a weak desired signal at 14.00 MHz. If an interfering signal is reduced in amplitude at the mixer RF input by 1.0 dB, the suppression of that spur is improved by 1.0 × Signal Harmonic Number dB. This is true in principle, but in reality the reduction may be somewhat less. For example, the spur produced by 15.00 MHz is reduced 3 dB for each dB that we reduce its level. We accomplish this task by choosing the right mixer, limiting the amount of RF amplification and designing adequate selectivity into the preselector circuitry. With respect to selectivity, though, note that in many other mixing schemes the interfering signal is so close to the desired frequency that selectivity does little good. Then we must use a mixer with a higher LO level and/or reduce RF gain.

The design method is illustrated by the following numerical example. Suppose that a signal at 14.75 MHz (the IF/2 spur) is at −20 dBm (very strong) at the antenna and −10 dBm at the mixer RF port. From the chart, this spur will be reduced by 71 + 2 (0 − (−10)) = 91 dB to a level of −10 − 91 = −101 dBm. If this is not enough then a preselector will help. If the preselector attenuates 14.75 MHz by 5 dB, the total spur reduction will be 71 + 2(0 − (−15)) = 101 dB to a level of −15 − 101 = −116 dBm, a 15-dB improvement. Notice that spurs involving high harmonics of the signal frequency attenuate more quickly as the input RF level is reduced.

On the other hand, if we consider the image signal at 17 MHz, all of the reduction of this spur must come from the preselector. In other words, selectivity is

the only way to reduce the image response unless an image reducing mixer circuit is used. In this example, additional spur reduction is obtained by using a preselector circuit topology that has improved attenuation *above* the passband.

In designing *any* receiver we must be reasonable about spur and image reduction. Receiver cost and complexity can increase dramatically if we are not willing to accept an occasional spurious response due to some very strong and seldom occurring signal. In the case of a certain persistent interference some specific cure for that source can usually be devised. A sharply tuned "trap" circuit, a special preselector or a temporary antenna attenuator are a few examples. In practice, for down-conversion superhets, 90 dB of image reduction is excellent and 80 dB is usually plenty good enough for amateur work.

In classical down-conversion superhets, the preselection circuits are tuned and bandswitched in unison with the LO. They must all "track" each other across the dial. The cost and complexity of this arrangement have made this approach prohibitive in modern commercial multiband designs (Ref 15). For amateur work the approach in **Fig 17.23** is more practical, using switched or even plug-in band-pass preselectors and oscillator coils. A frequency counter, offset by the 1.5 MHz IF and connected to the LO, eliminates the need for a calibrated dial.

Two-Tone Intermodulation Distortion

Another important mixer spurious response is two-tone IMD. This distortion has been covered previously in this chapter, and the **Mixers** chapter gives more detail. From a system design standpoint, the trade-offs between receiver noise figure and IMD have been covered in this chapter, and the choices of mixer, RF gain (if any) and selectivity are decided in a study exercise of performance, cost and complexity.

A receiver that has a 10 to 20-dBm third-order intercept point for two signals 20 kHz and 40 kHz removed is an excellent receiver in many applications. Some advanced experimenters have built receivers with 25 to 40-dBm values of IP3. Values of 40 dBm are near the state of the art (Ref 16).

A matter of considerable interest concerns the way that IMD varies as the separation between the two tones increases. In Fig 17.23, for example, if one tone is 1.0 kHz (or 100 kHz) above 14.00 the other is 2.0 kHz (or 200 kHz) above. We see that for very close tone separations the IF filter may not prevent the tones from reaching the circuits following the IF filter. As the separation increases, first one,

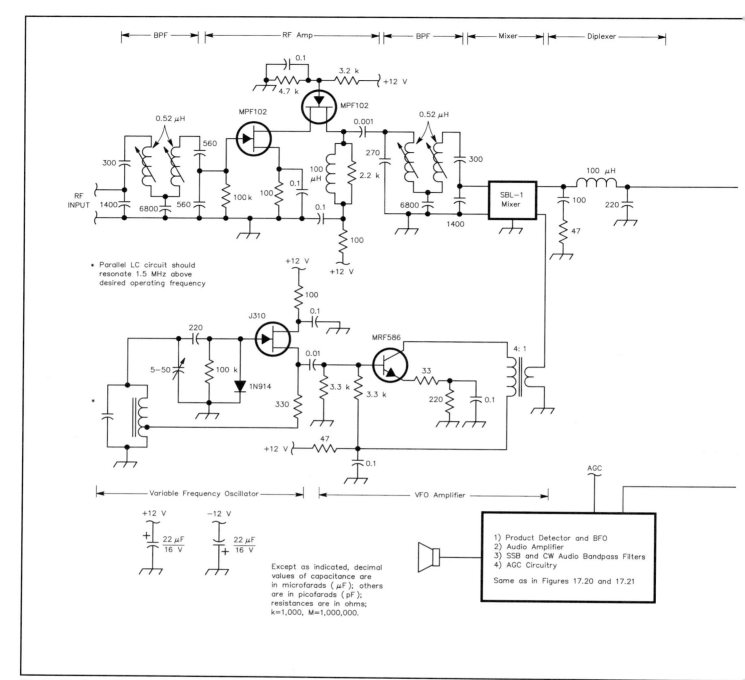

Fig 17.23—Specific example of a down-conversion superhet that is used to explain and analyze superhet behavior and design.

then both, tones fall outside the IF filter passband and the IMD becomes much less. However the mixer and the amplifier after the mixer are still vulnerable. At greater separations the preselector starts to protect these two stages, but the RF amplifier is not well protected by the first RF filter until the tone separation becomes greater, perhaps 200 kHz. It is a common procedure to plot a graph of receiver third-order input intercept point vs tone separation and then look for ways to improve the overall performance.

The stages after the IF filter are protected by AGC so that, hopefully, tones in the IF passband do not overdrive the circuits after the IF filter. But in the example of Fig 17.23 there is also a narrowband audio filter and the AGC is derived from the output of this filter. This means that circuits *after* the IF filter but *ahead of* the audio filter may not always be as well protected as we would like. Strong tones that get through the IF filter may be stopped by the audio filter and not affect the AGC. This particular example illustrates a very

common problem in all kinds of receivers that have *distributed* selectivity. It is also found universally in multiple conversion receivers, as we will discuss later.

GAIN AND NOISE FIGURE DISTRIBUTION

Based on the information given so far, the approach to designing a superhet receiver, whether a downconverter or any other kind, can now be summarized by the following guidelines:

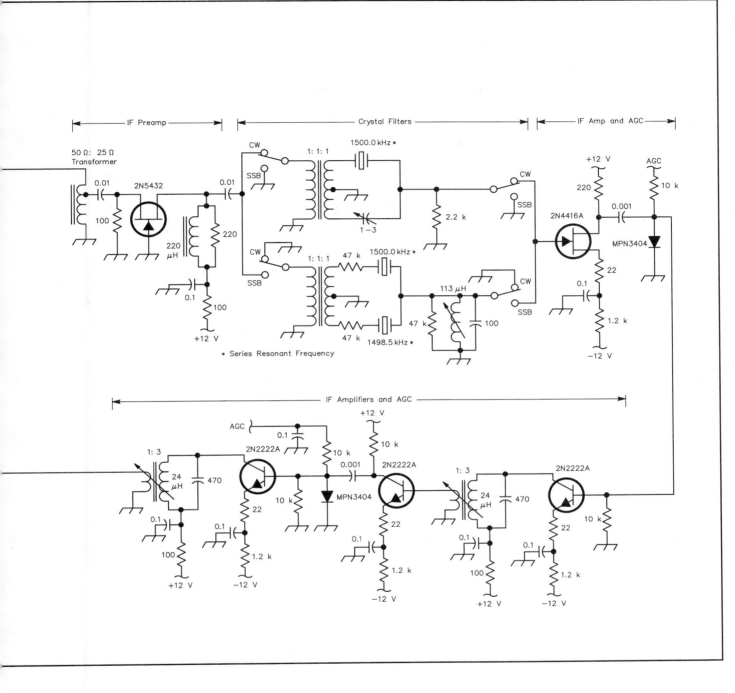

1. Try to keep the gain ahead of the mixer and the narrow band-pass filters (SSB, CW and so on) as low as possible. For a fixed components cost (such as mixers and amplifiers), this minimizes the IMD, both two tone and harmonic.

2. Reducing the gain implies that the noise figure may be a little higher. It is always best to avoid making the noise figure any lower than necessary. Noise figure is usually more important at microwave frequencies than at HF, and strong signal interference is usually less important. Where interference is a problem an increase in noise figure is almost always mandatory, except possibly when a higher-level mixer is used. A narrowband preselector, for example, will increase the noise figure (and also the intercept point) because of its passband attenuation.

3. Amplifier circuits and modules always involve a trade-off of some kind between intercept point and noise figure. Designers look for devices and circuits that optimize the SFDR for the particular kind of receiver under design.

4. If the receiver has distributed selectiv-ity, make the first IF filter good enough that the AGC/IF-overload problem mentioned above is mini-mized.

5. To *minimize* the gain ahead of the mixer, *follow* the mixer with a low-noise, high-dynamic-range amplifier with no more gain than necessary, say 10 dB or so (see Fig 17.23).

6. Terminate the mixer in such a way that its IMD is minimized. Fig 17.23 shows a simple IF diplexer that ab-sorbs the output image at 29.5 MHz (14.0 + 15.5).

7. The RF terminal of the mixer should be short circuited at the image frequency so that noise at the image frequency (from the preceding circuitry) is minimized.

8. Because a large amount of overall gain is needed, reducing front-end gain implies that the gain after the first IF filter must be very large. The problem of IF and audio noise then arises. It is very desirable to use a low-noise amplifier right after the first IF filter (see Fig 17.23) and to restrict the bandwidth of the IF/AF amplifiers. A second IF/AF filter downstream, and also possibly an image-reducing product detector, are excellent ways to accomplish this. This step also minimizes the degradation of receiver noise figure that can be caused by this wideband noise.

9. The LO must have very low phase noise to reduce reciprocal mixing. Also, the mixer must have good balance (meaning isolation or rejection) from LO port to RF and IF ports so that broadband additive noise from the LO amplifiers does not degrade the mixer noise figure. This is especially important when the RF amplifier gain has been minimized. If the mixer is not balanced in this sense at the LO port, a band-pass filter between LO and mixer is very desirable.

AUTOMATIC GAIN CONTROL (AGC)

The amplitude of the desired signal at each point in the receiver is controlled by AGC. Each stage has a distortion vs signal-level characteristic that must be known, and the stage input level must not become excessive. The signal being received has a certain signal-to-distortion ratio that must not be degraded too much by the receiver. For example, if an SSB signal has −30 dB distortion products the receiver should have −40 dB quality. A correct AGC design ensures that each stage gets the right input level. It is often necessary to redesign some stages in order to accomplish this (Ref 17).

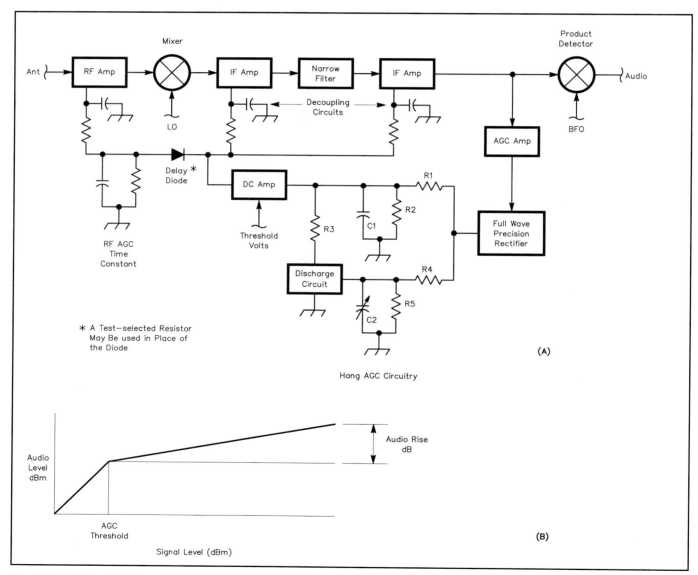

Fig 17.24—AGC principles. A: typical superhet receiver with AGC applied to multiple stages of RF and IF. B: audio output as a function of antenna signal level.

The AGC Loop

Fig 17.24A shows a typical AGC loop that is often used in amateur receivers. The AGC is applied to the stages through RF decoupling circuits that prevent the stages from interacting with each other. The AGC amplifier helps to provide enough AGC loop gain so that the gain-control characteristic of Fig 17.24B is achieved. The AGC action does not begin until a certain level, called the AGC threshold, is reached. The Threshold Volts input in Fig 17.24A serves this purpose. After that level is exceeded, the audio level slowly increases. The audio rise beyond the threshold value is usually in the 5 to 10-dB range. Too much or too little audio rise are both undesirable for most operators.

As an option, the AGC to the RF amplifier is held off, or "delayed," by the 0.6-V forward drop of the diode so that the RF gain does not start to decrease until larger signals appear. This prevents a premature increase of the receiver noise figure. Also, a time constant of one or two seconds after this diode helps keep the RF gain steady for the short term.

Fig 17.25 is a typical plot of the signal levels at the various stages of a certain ham band receiver. Each stage has the proper level and a 115-dB change in input level produces a 10-dB change in audio level. A manual gain control would produce the same effect.

AGC Time Constants

In Fig 17.24, following the precision rectifier, R1 and C1 set an "attack" time, to prevent excessively fast application of AGC. One or two milliseconds is a good value for the R1 × C1 product. If the antenna signal suddenly disappears, the AGC loop is opened because the precision rectifier stops conducting. C1 then discharges through R2 and the C1 × R2 product can be in the range of 100 to 200 ms. At some point the rectifier again becomes active, and the loop is closed again.

An optional modification of this behavior is the "hang AGC" circuit (Ref 18). If we make R2 × C1 much longer, say 3 seconds or more, the AGC voltage remains almost constant until the R5, C2 circuit decays with a switch selectable time constant of 100 to 1000 ms. At that time R3 quickly discharges C1 and full receiver gain is quickly restored. This type of control is appreciated by many operators because of the lack of AGC "pumping" due to modulation, rapid fading and other sudden signal level changes.

AGC Loop Problems

If the various stages have the property that each 1-V change in AGC voltage

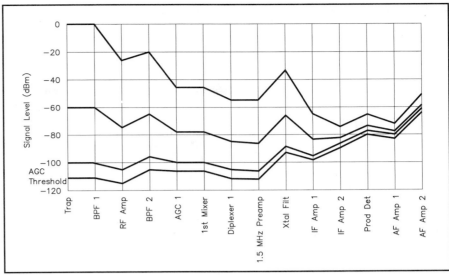

Fig 17.25—Gain control of a ham-band receiver using AGC. A manual gain control could produce the same result.

changes the gain by a constant amount (in dB), the AGC loop is said to be "log linear" and regular feedback principles can be used to analyze and design the loop. But there are some difficulties that complicate this textbook model. One has already been mentioned, that when the signal is rapidly decreasing the loop becomes "open loop" and the various capacitors discharge in an open-loop manner. When the signal is increasing beyond the threshold, or if it is decreasing slowly enough, the feedback theory applies more accurately. In SSB and CW receivers rapid changes are the rule and not the exception.

Another problem involves the narrow band-pass IF filter. The group delay of this filter constitutes a time lag in the loop that can make loop stabilization difficult. Moreover, these filters nearly always have much greater group delay at the edges of the passband, so that loop problems are aggravated at these frequencies. Over-shoots and undershoots, called "gulping," are very common. Compensation networks that advance the phase of the feedback help to offset these group delays. The design problem arises because some of the AGC is applied before the filter and some after the filter. It is a good idea to put as much fast AGC as possible after the filter and use a slower decaying AGC ahead of the filter. The delay diode and RC in Fig 17.24A are helpful in that respect. Complex AGC designs using two or more compensated loops are also in the literature. If a second cascaded narrow filter is used in the IF it is usually a lot easier to leave the second or "downstream" filter out of the AGC loop.

Another problem is that the control characteristic is often not log-linear. For example, dual-gate MOSFETs tend to have much larger dB/V at large values of gain reduction. Many IC amplifiers have the same problem. The result is that large signals cause instability because of excessive loop gain. There are variable gain op amps and other ICs available that are intended for gain control loops.

Audio frequency components on the AGC bus can cause problems because the amplifier gains are modulated by the audio and distort the desired signal. A hang AGC circuit can reduce or eliminate this problem.

Finally, if we try to reduce the audio rise to a very low value, the required loop gain becomes very large, and stability problems become very difficult. It is much better to accept a 5 to 10 dB variation of audio output.

Because many parameters are involved and many of them are not strictly log-linear, it is best to achieve good AGC performance through an initial design effort and finalize the design experimentally. Use a signal generator, attenuator and a signal pulser (2-ms rise and fall times, adjustable pulse rate and duration) at the antenna and a synchronized oscilloscope to look at the IF envelope. Tweak the time constants and AGC distribution by means of resistor and capacitor decade boxes. Be sure to test throughout the passband of each filter. The final result should be a smooth and pleasant sounding SSB/CW response, even with maximum RF gain and strong signals. Patience and experience are helpful.

Audio-Derived AGC

The example in Fig 17.15 shows audio-derived AGC. There is a problem with this approach also. At low audio frequencies the AGC can be slow to develop. That is, low-frequency audio sine waves take a long time to reach their peaks. During this time the RF/IF/AF stages can be overdriven. If the RF and IF gains are kept at a low level this problem can be reduced. Also, attenuating low audio frequencies prior to the first audio amplifier should help. With audio AGC, it is important to avoid so-called "charge pump" rectifiers or other slow-responding circuits that require multiple cycles to pump up the AGC voltage. Instead, use a peak-detecting circuit that responds accurately on the first positive or negative transition.

AGC Circuits

Fig 17.26 shows some gain controllable circuits. Fig 17.26A shows a two-stage 455-kHz IF amplifier with PIN-diode gain control. This circuit is a simplified adaptation from a production receiver, the Collins 651S. The IF amplifier section shown is preceded and followed by selectivity circuits and additional gain stages with AGC. The 1.0-μF capacitors aid in loop compensation. The favorable thing about this approach is that the transistors remain biased at their optimum operating point. Right at the point where the diodes start to conduct, a small increase in IMD may be noticed, but that goes away as diode current increases slightly. Two or more diodes can be used in series, if this is a problem (it very seldom is).

Fig 17.26B is an audio-derived AGC circuit using a full-wave rectifier that responds to positive or negative excursions of the audio signal. The RC circuit follows the audio closely.

Fig 17.26C shows a typical circuit for the MC1350P RF/IF amplifier. The graph of gain control vs AGC volts shows the change in dB/V. If the control is limited to the first 20 dB of gain reduction this chip should be favorable for good AGC transient response and good IMD performance. Use multiple low-gain stages rather than a single high-gain stage for these reasons. The gain control within the MC1350P is accomplished by diverting signal current from the first amplifier stage into a "current sink." This is also known as the "Gilbert multiplier" architecture. Another chip of this type is the NE/SA5209. This type of approach is simpler to implement than discrete-circuit approaches, such as dual-gate MOSFETs that are now being replaced by IC designs.

Fig 17.26D shows the high-end perfor-

mance Comlinear CLC520AJP (14-pin DIP plastic package) voltage controlled amplifier. It is specially designed for accurate log-linear AGC from 0 to 40 dB with respect to a preset maximum voltage gain from 6 to 40 dB. Its frequency range is dc to 150 MHz. It costs about $11.50 in small quantities and is an excellent IF amplifier for high-performance receiver or transmitter projects.

IF FILTERS

There are some aspects of IF-filter design that influence the system design of receivers and transmitters. The influence of group delay, especially at the band-pass edges, on AGC-loop performance has been mentioned. Shape factor is also significant (the ratio of two bandwidths, usually 60-dB:6-dB widths). To get good adjacent-channel rejection, the transition-band response should fall very quickly. Unfortunately, this goal aggravates group-delay problems at the passband edges. It also causes poor transient response, especially in CW filters. Another filter phenomenon can cause problems: at sharp passband edges signals and noise produce a raspy sound that is annoying and interferes with weak signals.

A desirable filter response would be slightly rounded at the edges of the passband, say to –6 dB, with a steep rolloff after that. This is known as a "transitional filter" (Ref 19). Cascaded selectivity with two filters, each having fewer "poles" (than a single filter would) is also a good approach. Both methods have a smoother group delay across the passband and reduce the problems mentioned above.

Ultimate Attenuation

In a high-gain receiver with as much as 110 dB of AGC the ultimate attenuation of the filter is important. Low-level leakage through or around the filter produces high-pitch interference that is especially noticeable on CW. Give special attention to parts layout, wiring and shielding. (Filter selector switches are often leakage culprits.) Cascaded IF filters also help very considerably.

Audio Filter Supplement

An audio band-pass filter can be used to supplement IF filtering. This can help to improve signal-to-noise ratio and reduce adjacent-channel interference. Supplementary audio filtering also helps reduce the high-frequency leakage problem mentioned above. Another significant problem: If AGC is made in the IF section, strong signals inside the IF passband but outside the audio passband can "pump" or modulate the AGC, rendering weak de-

sired signals hard to copy. This is especially noticeable during periods of high band activity, such as in a contest. These filters can use analog (see Fig 17.15 and Fig 17.16) or digital (DSP) technology.

Some Simple Crystal Filters

Fig 17.27 and **Fig 17.28** present two crystal filters to consider for a simple down-conversion receiver with a 1.5-MHz IF (see Fig 17.23). The crystals are a set of three available from JAN Crystals. (See the Address List in the **References** chapter for their current address. Mention the ARRL *Handbook*.) The filters are both driven from a low-impedance source (200 Ω, for example).

CW Filter

Fig 17.27A is a "semi-lattice" filter using a single crystal for CW work (Ref 19). Capacitor C_c balances the bridge circuit at the crystal's parallel-resonant frequency because it is equal to the holder capacitance C_0 of the crystal. The response is then symmetrical around the series resonant frequency of the crystal. The selectivity is determined by the value of R_{out}. As the value decreases the selectivity sharpens as shown in Fig 17.27B. If this filter is combined with an audio band-pass filter as in Fig 17.16, pretty good CW selectivity is possible. In Fig 17.27C, the capacitor is increased to 8.3 pF and a notch appears at –1.7 kHz. This is the "single signal" adjustment. Also, note that the response on the high side is degraded quite a bit. The notch can be located above or below center frequency by adjusting the capacitor value; the degradation is on the opposite side of center.

SSB Filter

Fig 17.28 is a "half-lattice" filter (Ref 19). The schematic diagram shows the LCR values and the series resonant frequencies of the two crystals. One of these (1.4998 MHz) is the same type as the one used in the CW filter. The trimmer capacitor equalizes the two values of C_0, the crystal shunt capacitance (very important) in case they are not already closely matched. Place the trimmer across the crystal that has the lowest value of C_0. The response curve shows good symmetry and modest adjacent-channel rejection. The output tuned circuit absorbs load capacitance to get a pure R_{load} (also important). The follow-up audio speech band-pass filter in Fig 17.15 will improve the overall response considerably.

Mechanical Filters

Mechanical filters use transducers and the magnetostriction principle of certain materials to obtain a multiresonator nar-

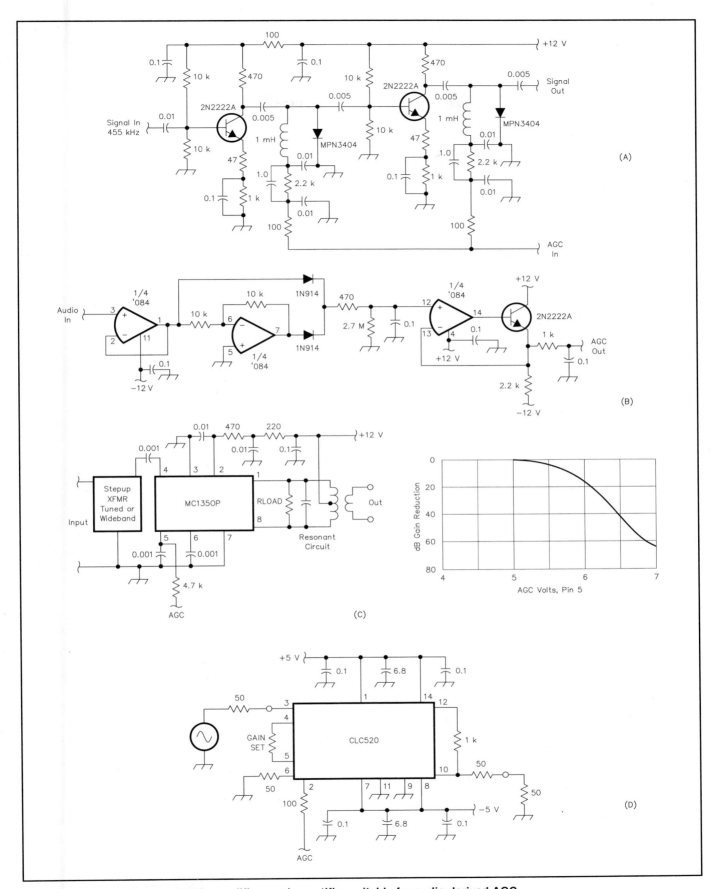

Fig 17.26—Some gain controllable amplifiers and a rectifier suitable for audio derived AGC.

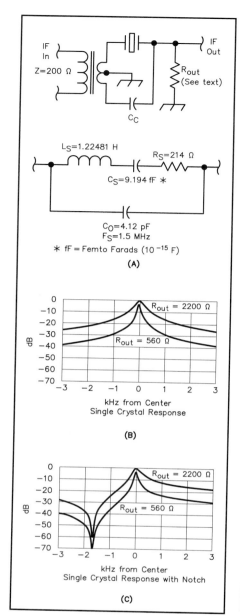

Fig 17.27—A single-crystal filter circuit for a simple CW receiver design. See also Fig 17.23.

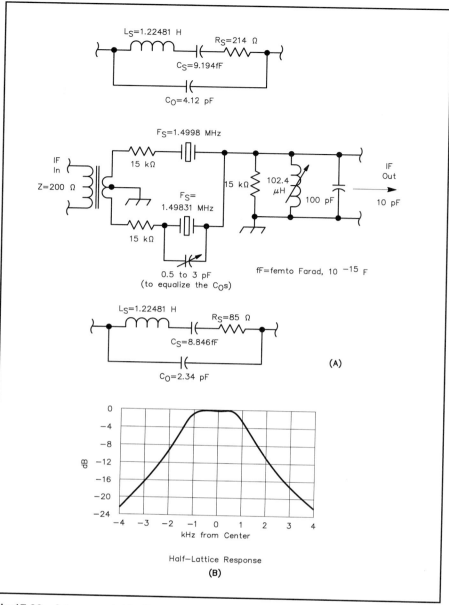

Fig 17.28—A two-crystal half-lattice filter for a simple SSB receiver. See also Fig 17.23.

row band-pass filter in the 100 to 500-kHz range. They are very frequency stable, accurate and reliable. An interesting example, for radio amateurs, is the Rockwell-Collins (Costa Mesa, California) "Low Cost Series" of miniaturized torsional-mode filters for 455.0 kHz. They come in four styles with 3 dB/60 dB bandwidths of 0.3/0.5/1.5/2.0, 2.5/5.2 and 5.5/11 kHz. In small quantities (1-4), they sell for $92 each (1997 price—mention the ARRL *Handbook* article), including tax and shipping. Used filters are sometimes available from various sources (Ref 15).

Multielement Crystal Filters

A discussion of more complex crystal filters appears in the **Filters and Projects** chapter of this *Handbook*. In this chapter we have considered only two very simple examples that might appeal especially to student designers and builders of a receiver that downconverts to an IF less than 2 MHz or so.

From the system design standpoint, note that for voice reception amateurs often use optional IF filters with less than the conventional bandwidth for SSB (for example 1.8 kHz), even though they reduce higher frequency speech components. This helps to improve adjacent-channel interference, which is a severe problem on some amateur bands.

It is common practice to use multipole crystal filters in the range from 5 to 10 MHz, because they can be economically designed for that frequency range. It is also common to cascade these filters with other types, such as mechanical or LC filters, at lower IFs (more about this later).

Filter Switching

Filter switching for different modes

(AM, SSB, CW, RTTY and so on) requires some careful design to prevent impedance mismatching, leakage (discussed before) and spurious coupling to other circuitry. There are three general methods for switching: mechanical, relays, solid-state (diodes or transistors). **Fig 17.29** shows examples of relay and diode switching that work quite well. The relays can be inexpensive miniature RadioShack 275-241 SPDT units, one at each input and one at each output. The diodes can be inexpensive Motorola MPN3404 PIN diodes. These circuits assume that all filters are terminated with the same impedance values (Ref 15).

In PIN diode applications, IMD *can* be a problem with inadequate bias or excessive signal levels. The application (PIN diode and circuit) should be tested at the highest expected signal level.

One major problem involves high-level IF-output-signal leakage or BFO leakage into the input of the filter, which can produce high passband ripple and other unpleasant problems such as AGC malfunctions.

THE VLF IF RECEIVER

An approach to IF selectivity that has been used frequently over the years in both home-built and factory-made amateur receivers uses a second down conversion from an IF at, say 4 or 5 MHz or even 455 kHz, to a very low frequency, usually 50 to 85 kHz. At these frequencies several double tuned LC filters, separated by amplifier stages, make possible excellent improvements in SSB/CW frequency response and ultimate attenuation along with a relatively flat group delay. These amplifier stages can also have AGC. Low-cost (four-pole) crystal filters (SSB and CW) at the higher IF followed by two lower-IF channels (SSB and CW) make a very desirable combination. This is also an effective way to assure a narrow noise bandwidth for the overall receiver. One requirement is that the circuitry ahead of the VLF downconverter must provide good rejection of an image frequency that is only 100 to 170 kHz away (Ref 20).

AM DEMODULATION

There is some interest among amateurs in double-sideband AM reception on the HF broadcast bands. Coherent AM detection is a way to reduce audio distortion that is caused by a temporary reduction of the carrier. This "selective fading" is due to phase cancellations caused by multipath propagation. By inserting a large, locally created carrier onto the signal this effect is reduced. The term "exalted carrier reception" is sometimes used. In reception of a double-sideband AM signal, the phase of the inserted carrier must be identical to that of the incoming carrier. If not, reduced audio and also audio distortion result. Therefore the common method is to use a phase-locked loop (PLL) to coordinate the phases of the incoming carrier and the locally generated carrier. This requires a Type II PLL, which drives, or integrates, the phase difference to zero degrees (Ref 21).

MULTIPLE CONVERSION SUPERHETS

There are a couple of drawbacks to the downconverting receiver just described.

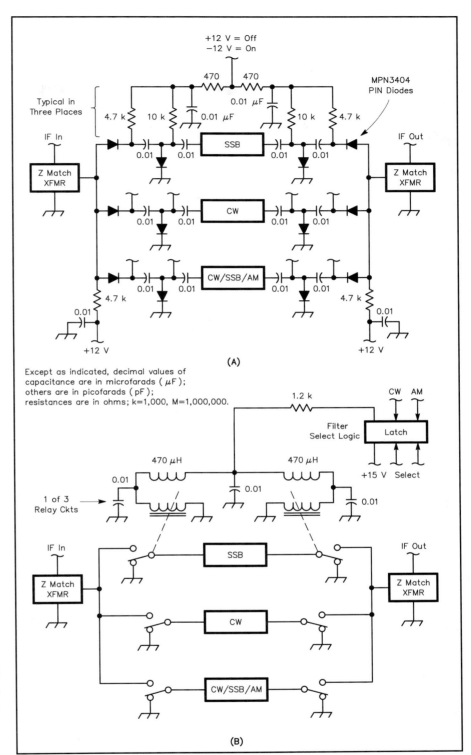

Except as indicated, decimal values of capacitance are in microfarads (μF); others are in picofarads (pF); resistances are in ohms; k=1,000, M=1,000,000.

Fig 17.29—IF filter switching using relays or PIN diodes.

First, the LO must be bandswitched. Also, its tuning must track with the preselector tuning even though the preselector is offset from the signal frequency by the amount of the IF. A tuning dial scale is required for each band, and the receiver must be fitted to it at the factory. This adds a lot of cost and complexity.

A solution to these problems is shown in **Fig 17.30**. A crystal controlled first mixer is preceded by a gang-tuned preselector and is followed by a wideband first IF that is 200-kHz wide. The second mixer has a VFO that tunes a 200-kHz range. To change bands, the crystal is switched and the preselector is band-switched. An additional tuned circuit removes the wideband additive noise from the crystal LO, so that it does not degrade the noise figure of the *unbalanced* mixer circuit.

One of the main design problems is to select the first IF, its bandwidth and the second mixer design so that harmonic IMD products (involving the signal, crystal frequency, first IF, second IF and VFO frequency) do not cause appreciable interference. In the example of Fig 17.30, a first IF at 2.9275 MHz (the signal frequency would be 14.2275 MHz) and a VFO at 2.7 MHz produce a fourth-order spurious response at 455 kHz, therefore the first IF filter must attenuate 2.9275 MHz sufficiently and the second mixer must reject the fourth-order response sufficiently. We have discussed the fourth-order (IF/2) response previously.

One of the main bonuses of this approach is that the tunable second LO can be very stable and accurately calibrated. This calibration is the same for any signal band. Another advantage is that the first crystal LO is very stable and has little

phase noise. A third bonus is that the high value of the first IF simplifies the preselector design for good image rejection in the first mixer (Ref 22).

The second mixer is vulnerable to two-tone IMD caused by strong interfering signals that lie within, or near, the 200-kHz-wide first-IF bandwidth, and that have been amplified by the circuitry preceding it. They do not make AGC because they are outside the narrow signal filters.

This cascaded-selectivity problem, which we have discussed previously, makes it necessary to very carefully control the gain and noise-figure distribution ahead of the second mixer. Also, put the narrow signal filter right after the second mixer and follow that with a low-noise IF amplifier, so that "front end" gain can be minimized. In more expensive receivers of this kind, the first IF is sharply gang-tuned along with the second LO in order to reduce this problem (Ref 23).

This general approach has been extended in order to make a general-coverage receiver that has acceptable spurious responses. The first IF can be switched between two different frequency ranges and various combinations of up conversion and down conversion are used. This subject is interesting, but more complex than we can cover here. This approach is also not frequently used at this time.

THE UP CONVERSION SUPERHET

The most common approach to superhet design today is the "up converter." This designation is reserved for receivers in which the first IF is greater than the highest receive frequency. First IF values can be as low as 35 MHz for low-cost HF receivers or as high as 3 GHz for wideband

receivers (and spectrum analyzers) that cover the 1 MHz to 2.5 GHz range. Let's begin by discussing the general properties of all up conversion receivers.

An Up Converter Example

The block diagram in **Fig 17.31** is one example for HF amateur SSB/CW use. The input circuit responds uniformly to a wide frequency band, 1.8 to 30 MHz. A 1.8 to 30-MHz band-pass filter is at the input. The absence of any narrow preselection is typical, but in difficult environments an electronically tuned or electromechanically tuned preselector is often used. Another option is a set of "half octave" (2 to 3 MHz, 3 to 4.5 MHz and so on) filters switched by PIN diodes or relays. This type of filter eliminates second-order IMD. For example, if we are listening to a weak signal at 2.00 MHz, two strong signals at 2.01 and 4.01 MHz would not create a spur at 2.00 MHz because the one at 4.01 MHz would be greatly attenuated.

Wideband Interference

The wideband circuitry in the front end is vulnerable to strong signals over the entire frequency range if no preselection is used. Therefore the strong-signal performance is a major consideration. Total receiver noise figure is usually allowed to increase somewhat in order to achieve this goal. Double balanced passive (or often active) mixers with high intercept points (second and third-order) and high LO levels are common. A typical high-quality up conversion HF receiver has a third-order intercept (IP3) of 20 to 30 dBm and a noise figure of 10 to 14 dB. High-end performers will have an IP3 of 32 to 40 dBm and a noise figure of 8 to 12 dB.

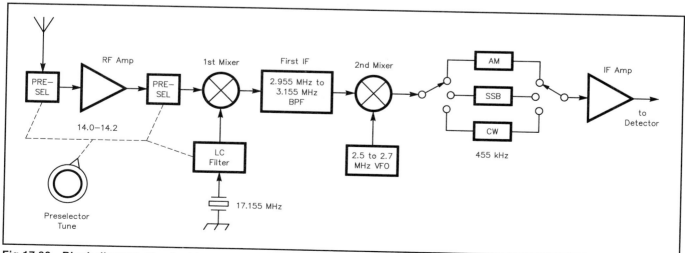

Fig 17.30—Block diagram of a double-conversion superhet that eliminates some of the tracking problems of the conventional superhet.

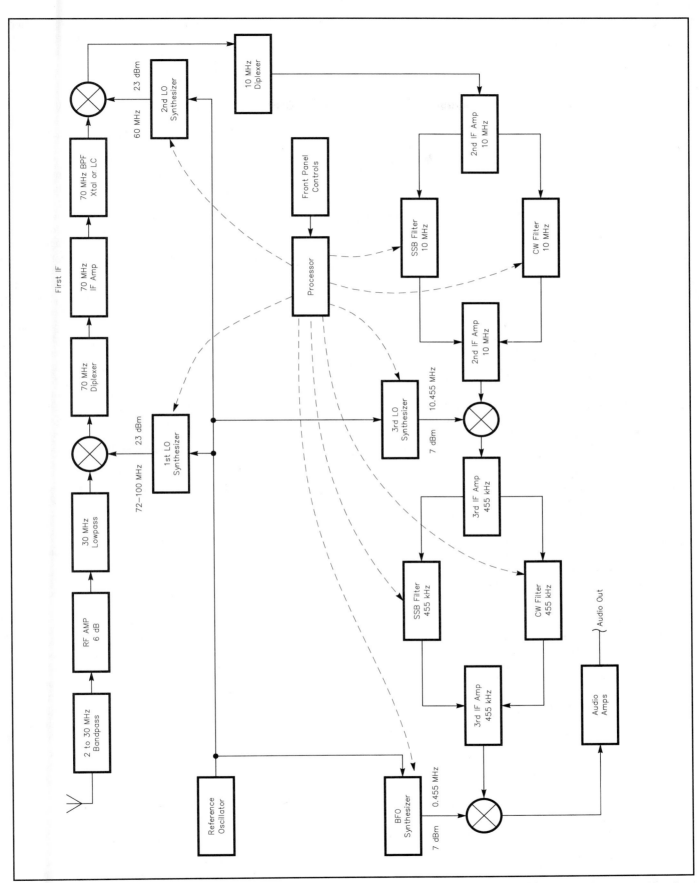

Fig 17.31—Block diagram of an HF up-conversion receiver for SSB and CW. Microprocessor control of receiver functions is included. LOs are from synthesizers.

As a practical matter, in all but the most severe situations with collocated transmitters, there is very little need in Amateur Radio for the most advanced receiver specifications. One reason for this involves statistics. To get two-tone IMD interference on a *high-quality* receiver at some particular frequency there must be two strong signals, or perhaps one very strong and a second weaker signal, on just the right pair of frequencies and at the same time. In nearly all cases, the "chances" of this are small. In Amateur Radio contest situations, these kinds of interactions are more probable. For persistent cases, other remedies are usually available.

After the Mixer

We would really like to go from the mixer directly into an SSB or CW filter, but at the high frequency of the first IF this is not realistic. Therefore we run into a major compromise: It is necessary to have at least one additional wide-band frequency conversion before getting to the narrow filters. The first IF filter can be as narrow as cost and technology will permit. In the 35 to 110-MHz range crystal filters with bandwidths of 10 to 20 kHz are available, but they are somewhat expensive in small quantities. Fig 17.31 shows an option with far less cost. The LC filter in the first IF is about 1.0-MHz wide but it has enough attenuation at 50 MHz to yield excellent image rejection in the second mixer. If we use a high-input-intercept, low-gain, low-noise amplifier followed by a strong second mixer (minimize the gain ahead of the second mixer and let the receiver noise figure go up a couple of dB) the overall receiver performance will be excellent, especially with the kinds of efficient antennas that amateurs use.

Terminating the Mixers

In the upconversion receiver, getting a pure wideband resistive termination for the mixer IF port is a problem. The output of the first mixer in Fig 17.31 contains undesired frequencies. For example, a 10-MHz signal produces 70-MHz (desired IF), 90-MHz (image) and 80-MHz (LO leakage). For a 2-MHz signal there would be 70, 72 and 74-MHz outputs. A filter that passes 70-MHz, rejects the others and at the same time terminates the mixer resistively over a wide band is a complicated band-pass diplexer.

Usually the termination is an amplifier input impedance plus a much simpler band-pass diplexer. The amplifier input should be a pure resistance, and it may then be required to deal with the vector sum of all three products. Diode upconverter mixers have typically 30 to 40 dB LO-to-IF isolation. If the

LO level is 23 dBm, the amplifier may be looking at –7 to –17 dBm of LO feedthrough, which is fairly strong. The output of the amplifier and also the next stage must deal with these amplified values. The second mixer is much easier to terminate with a diplexer, as Fig 17.31 shows.

At lower signal and LO levels (7 dBm or less), MMIC amplifiers like the MAV-11 may provide a good termination across a wide bandwidth. However, susceptibility to IMD must be checked carefully.

At the RF terminal of the mixer, any noise at the image frequency from previous stages (such as RF amplifier, antenna or even thermal noise) must not be allowed to enter the mixer because it degrades the mixer noise figure. The RF terminal should be short circuited at the image frequency if possible.

Choosing the First IF

The choice of the first IF is a compromise between cost and performance. First, consider harmonic IMD. Published data for several high-level diode-mixer models show that if the IF is greater than three times the highest signal frequency (greater than 90 MHz for a 0 to 30-MHz receiver) the rejection of harmonics of the signal frequency increases considerably. For example 3 times an interferer at 33 MHz produces a 99 MHz IF. The input 30-MHz low-pass filter would attenuate the 33-MHz signal and so would help considerably. On the other hand, 24.75 times 4 is also 99 MHz, but the mixer does a better job of rejecting this fourth harmonic. Other spurious responses tend to improve also.

However, other factors are involved, most important of which are the LO designs for the first and second mixers. In up-converters, the LOs are invariably synthesizers whose output frequencies are phase locked to a low-frequency reference crystal oscillator. As the LO frequencies increase, two other things increase: cost and quantity of high-frequency synthesizer components, and synthesizer phase noise. Also, the exact choice is interwoven with the details of the synthesizer design. Special IFs, such as 109.350 MHz, are chosen after complex trade-off studies. The cost of the first-IF signal-path components, especially filters, tends to increase also.

For all of these reasons, the IF is quite often chosen at a lower frequency. In Fig 17.31, a 70-MHz IF is shown. Crystal filters at this frequency are widely available at reasonable cost. LC filters with a 1.0 MHz (or less) bandwidth are easy to construct and get working. A 45-MHz IF is also popular. Helical resonator filters are excellent candidates at higher IF frequencies, although they can be a bit large. The problems associated with lower IFs

can be greatly improved by using a higher performance mixer. The costs are a better mixer and a more powerful LO amplifier.

The "Gray Area"

In the upconverter front end we encounter the "cascaded selectivity" problem that was mentioned previously in this chapter. Strong signals that are *within* the first IF passband but *outside* the SSB/CW passband (the gray area) do not make AGC and therefore are not controlled by AGC. These signals intermodulate in the mixers and in the amplifiers that precede the SSB/CW filters. It is important to realize that the receiver gray-area IMD performance is the composite of all these stages, and not just the first mixer alone. It is not until the first IF filter takes effect that things get a lot better. This degradation for strong adjacent-channel signals is an artifact of the upconverting superhet receiver. In practice, it often happens that reciprocal mixing with the adjacent channel signals (caused by receiver synthesizer phase noise) also gets into the act. Phase noise on the interfering signals is also not uncommon. The design problem is to make IMD in the gray area as small as reasonably possible. The example of Fig 17.31 is an acceptable compromise in this respect for a fairly high quality receiver.

Triple Conversion

The block diagram of Fig 17.31 shows a 10-MHz second IF. This IF is selected for two reasons:

- There are narrowband crystal filters available (including homemade ladder filters) in that vicinity.
- The image frequency for the second mixer is at 50 MHz, which can be highly attenuated by the first IF filter.

It is entirely possible to let the rest of the IF remain at this frequency, and many receivers do just that. Nonetheless, there are some advantages to having a third conversion to a lower IF. For example, it is desirable to get the large amount of IF gain needed at two different frequencies, so that stability problems and leakage from the output to input are reduced. Also, a wide variety of excellent filters are readily available at 455 kHz and other low frequencies.

Note that there is a cascade of the SSB and CW filters at the two frequencies. The desirability of this was discussed earlier and it is a powerful concept for narrow-band receiver design. Placing cascaded filters at two different frequencies has another advantage that we will look into presently. For the third mixer the image frequency is at 10.910 MHz, therefore the

10-MHz filter must have good rejection at that frequency. If the first IF filter has good rejection at 70.910 MHz it will help reduce this image also. We are looking for 90 to 100 dB rejection of this potentially serious image problem. If the first IF uses a 1.0 MHz wide LC filter, as we have suggested, some additional 10 MHz LC filtering will probably be needed.

The 50 kHz IF

The third IF could be the cascaded LC filter circuits at VLF, one for SSB and one for CW, which we discussed earlier. This is an excellent approach that has been used often. The image frequency for the third mixer is only 100 kHz away, so that problem needs careful attention. Image canceling mixers have been used to help get the required 90 to 100 dB (Ref 24).

Local Oscillator (LO) Leakage

It is easy to see that with four LOs running, some at levels of 0.2 W or more, interactions between the mixers can occur. In a multiple-conversion receiver mechanical packaging, shielding, circuit placement and lead filtering are very critical areas. As one example of a problem, if the 60-MHz second LO leaks into the first mixer, a vulnerability to strong 10-MHz input signals results. It is called "IF feedthrough" because the second IF is 10 MHz. Other audible "tweets" are very common; they occur at various frequencies that involve harmonics of LOs beating together in various mixers. It is a major exercise to devise a "frequency scheme" that minimizes tweets, or at least puts them where they do not cause too much trouble (for amateurs, outside the ham bands). After that the "dog work" of reducing the remaining tweets below the noise level begins. It is a very educational experience. Synthesizers produce numerous artifacts that can also be very troublesome. It is a very common dilemma to build a receiver using "cheap" construction and poorly conceived packaging, and then try to bully the thing into good behavior.

Frequency Tuning

The synthesized first and second LOs present several different ways to tune the receiver frequency. This chapter cannot get into the details of synthesizer design, so these are only a few brief remarks. Let's discuss two options from a system-design standpoint:

- Do all of the tuning in the first LO. If steps of 10 Hz (or 1 Hz) are needed, a single-loop synthesizer that tunes in, say 500-kHz or 1.0-MHz steps, can be used. Then, a direct digital synthesizer (DDS) that tunes in 10-Hz (or 1-Hz) steps is included in the main loop in what is termed a "translation loop." The DDS frequency is added into the loop in such a manner that its imperfections are not increased by frequency multiplication. Because the reference frequency for the loop is high (500 kHz or 1.0 MHz) the phase noise of the main loop is quite small, if the loop and the circuitry are correctly designed and if the LO frequency is not extremely high. The digital frequency readout is obtained from the bits that program the synthesizer. A simpler approach might use a free-running VFO plus a low-cost frequency counter instead of the DDS. The counter can be designed to display the receiver signal frequency.
- If the first IF filter is sufficiently wideband it is possible to tune the first LO in steps of 500 kHz or 1.0 MHz and tune the second LO in 10-Hz steps. This may be a simpler method because the second LO need only be tuned over a small range.

With this second approach the first LO could be a crystal oscillator with switched overtone crystals, one for each ham band. The second LO could be a combination of crystal and VFO. One disadvantage (not extremely serious for amateur work) is that the LOs are not locked to a very accurate reference. Another is that a separate crystal (easy to get) is needed for each band. A frequency counter on the second LO could be used to get a close approximation to the signal frequency. This approach might be of interest to the home builder who is not yet ready to get involved with synthesizers. A 500-kHz crystal calibrator in the receiver would mark the band edges accurately.

Passband Tuning

While listening to a desired signal in the presence of another partially overlapping and interfering signal, whether in SSB or CW mode, it is often possible to "move" the interference at least somewhat out of the receiver passband without affecting the tune frequency (pitch) of the desired signal. In Fig 17.31, if the processor has independent fine tuning control of the second and third LOs and also the BFO, it is a matter of software design to accomplish this. It is done by controlling the overlap or intersection of the passbands of the 10-MHz filters and the 455-kHz filters. There are three things that can be done: the bandwidth can be decreased, the center frequency of the passband can be moved and both can be done simultaneously.

This scheme works best when both SSB or both CW filters are of high quality and have the same bandwidth (for example, 2.5 kHz and 500 Hz), fairly flat response and the same shape factor. As the passband is made narrower by decreasing the overlap, however, the composite shape factor is degraded somewhat (it gets larger). For CW especially, this is not detrimental. A very steep-sided response at narrow bandwidths is not desirable from a transient-response standpoint. The effect is not serious for SSB either.

Later discussion in the Transceivers section of this chapter will present another method of passband tuning using a variable frequency mixer scheme rather than software control. This method is commonly found in manufactured equipment.

Noise Blanking

The desire to eliminate impulse noise from the receiver audio output has led to the development of special IF circuits that detect the presence of a noise impulse and open the signal path just long enough to prevent the impulse from getting through. Most often, a diode switch is used to open the signal path. An important design requirement is that the desired IF signal must be delayed slightly, ahead of the switch, so that the switch is opened precisely when the noise arrives at the switch. The circuitry that detects the impulse and operates the switch has a certain time delay, so the signal in the mainline IF path must be delayed also. The Transceivers section of this chapter describes how a noise blanker is typically implemented. (See also Ref 17.)

VHF and UHF Receivers

The basic ideas presented in previous sections all operate equally well in receivers that are intended for the VHF and UHF regions. One difference, however, is that narrow-band frequency modulation (NBFM) is commonly used. In recent times, however, SSB has increased in popularity because of its potentially narrower channel spacing. The commercial use of SSB at VHF has been encouraged by the use of a −10 dB pilot carrier combined with amplitude compandoring. This is known as amplitude-compandored SSB or ACSSB. Improvements in synthesizer design and frequency stability have also contributed to the growth of VHF SSB. We will focus the differences between VHF/UHF and HF receivers.

Narrowband FM (NBFM) Receivers

Fig 17.32A is a block diagram of an NBFM receiver for the VHF/UHF amateur bands.

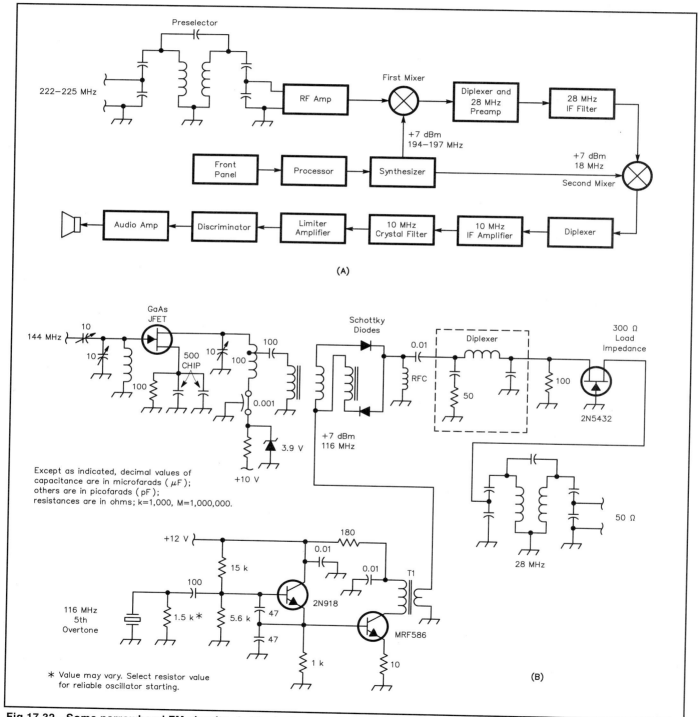

Fig 17.32—Some narrowband FM circuits. A: block diagram of a typical NBFM receiver. B: a front-end circuit with preselection and down conversion.

Front End

A low-noise front end is desirable because of the decreasing atmospheric noise level at these frequencies and also because portable gear uses short rod antennas at ground level. Nonetheless, the possibilities for gain compression and harmonic IMD, multitone IMD and cross modulation are also substantial. So dynamic range is an important design consideration, especially if large, high-gain antennas are used. FM limiting should not occur until after the NBFM signal filter. Because of the high occupancy of the VHF/UHF spectrum by powerful broadcast transmitters and nearby two-way radio services, front-end preselection is desirable, so that low noise figure can be achieved economically within the amateur band. Fig 17.32B is an example of a simple front end for the 144- to 148-MHz band.

Downconversion

Downconversion to the final IF can occur in one or two stages. Favorite IFs are in the 5 to 10-MHz region, but at the higher frequencies rejection of the image 10 to 20-MHz away can be difficult, requiring considerable preselection. So at the higher frequencies an intermediate IF in the 30 to 50-MHz region is a better choice. Fig 17.32A shows dual down-conversion.

IF Filters

The customary frequency deviation in amateur NBFM is about 5 kHz RMS (7-kHz peak) and the audio speech band extends to 5 kHz. This defines a minimum modulation index (defined as the deviation ratio) of 7/5 = 1.4. An inspection of the Bessel function plots shows that this condition confines most of the 300 to 5000-Hz speech information sidebands within a 12 to 15-kHz-wide bandwidth filter. With this bandwidth, channel separations of 20 or 25 kHz are achievable.

Many amateur NBFM transceivers are channelized in steps that can vary from 5 to 25 kHz. For low distortion of the audio output (after FM detection), this filter should have good phase linearity across the bandwidth. This would seem to preclude filters with very steep descent outside the passband, that tend to have very nonlinear phase near the band edges. But since the amount of energy in the higher speech frequencies is naturally less, the actual distortion due to this effect may be acceptable for speech purposes. A possible qualifier to this may be when pre-emphasis of the higher speech frequencies occurs at the transmitter and de-emphasis compensates at the receiver (a commonly found feature).

Limiting

After the filter, hard limiting of the IF is needed to remove any amplitude components. In a high-quality receiver, special attention is given to any nonlinear phase shift that might result from the limiter circuit design. This is especially important in phase-coherent data receivers. In amateur receivers for speech it may be less important. Also, the "ratio detector" (see the **Mixers, Modulators and Demodulators** chapter) largely eliminates the need for a limiter stage, although the limiter approach is probably still preferred.

FM Detection

The discussion of this subject is deferred to the **Mixers, Modulators and Demodulators** chapter. Quadrature detection is used on some popular NBFM multistage ICs. An example receiver chip will be presented later. Also see the Transceivers section of this chapter.

NBFM Receiver Weak-Signal Performance

The noise bandwidth of the IF filter is not much greater than twice the audio bandwidth of the speech modulation, as it would be in wideband FM. Therefore such things as capture effect, the threshold effect and the noise quieting effect so familiar to wideband FM are still operational, but somewhat less so, in NBFM. For NBFM receivers, sensitivity is specified in terms of a SINAD (see the **Test Procedures** chapter) ratio of 12 dB. Typical values are -110 to -125 dBm, depending on the low-noise RF preamplification that often can be selected or deselected (in strong signal environments).

LO Phase Noise

In an FM receiver, LO phase noise superimposes phase modulation, and therefore frequency modulation, onto the desired signal. This reduces the ultimate signal-to-noise ratio within the passband. This effect is called "incidental FM (IFM)." The power density of IFM (W/Hz) is equal to the phase noise power density (W/Hz) multiplied by the square of the frequency offset from the carrier (the familiar parabolic effect in FM). If the receiver uses high-frequency deemphasis at the audio output (-6 dB per octave from 300 to 3000 Hz, a common practice), the IFM level at higher audio frequencies can be reduced. Levels of total (integrated) IFM from 10 to 50 Hz are high quality for amateur voice work. Ordinarily, as the signal increases the noise would be "quieted" (that is, "captured") in an FM receiver, but in this case the signal and the phase noise riding "piggy back" on the signal increase

in the same proportion. As the signal becomes large the signal-to-noise ratio therefore approaches some final value (Ref 8). A similar ultimate SNR effect occurs in SSB receivers. On the other hand, a perfect AM receiver tends to suppress LO phase noise (Ref 9).

NBFM ICs

A wide variety of special ICs for NBFM receivers are available. Many of these were designed for "cordless" or cellular telephone applications and are widely used. **Fig 17.33** shows some popular versions for a 50-MHz NBFM receiver. One is an RF amplifier chip (NE/SA5204A) for 50-Ω input to 50-Ω output with 20 dB of gain. This gain should be reduced to perhaps 8 dB. The second chip (NE/SA602A) is a front-end device (Ref 25) with an RF amplifier, mixer and LO. The third is an IF amplifier, limiter and quadrature NBFM detector (NE/SA604A) that also has a very useful RSSI (logarithmic received signal strength indicator) output and also a "mute" function. The fourth is the LM386, a widely used audio-amplifier chip. Another NBFM receiver chip, complete in one package, is the MC3371P.

The NE/SA5204A plus the two tuned circuits help to improve image rejection. An alternative would be a single double tuned filter with some loss of noise figure. The Mini-Circuits MAR/ERA series of MMIC amplifiers are excellent devices also. The crystal filters restrict the noise bandwidth as well as the signal bandwidth. A cascade of two low-cost filters is suggested by the vendors. Half-lattice filters at 10 MHz are shown, but a wide variety of alternatives, such as ladder networks, are possible.

Another recent IC is the MC13135, which features double conversion and two IF amplifier frequencies. This allows more gain on a single chip with less of the cross-coupling that can degrade stability. This desirable feature of multiple downconversion was mentioned previously in this chapter.

The diagram in Fig 17.33 is (intentionally) only a general outline that shows how chips can be combined to build complete equipment. The design details and specific parts values can be learned from a careful study of the data sheets and application notes provided by the IC vendors. Amateur designers should learn how to use these data sheets and other information. The best places to learn about data sheets are data books and application notes.

UHF Techniques

The Ultra High Frequency spectrum is

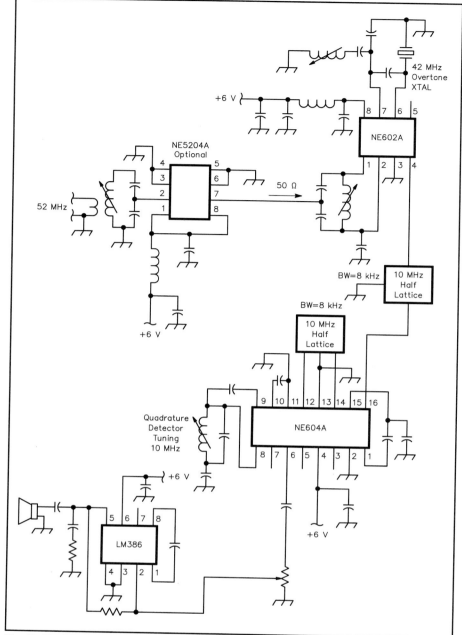

Fig 17.33—The NE/SA5204A, NE/SA602A, NE/SA604A NBFM ICs and the LM386 audio amplifier in a typical amateur application for 50 MHz.

range of discrete components. For example, the literature shows methods of building LC filters at as high as 2 GHz, using chip capacitors and tiny helical inductors. Commercially available amplifier and mixer circuits operate at 2 GHz, using these types of components on ceramic substrates.

ICs for UHF

In recent years a wide variety of highly miniaturized monolithic microwave ICs (MMIC) have become available at reasonable cost. Among these are the Avantek MODAMP and the Mini Circuits MAR and MAV/ERA lines. Designer kits containing a wide assortment of these are available at reasonable cost. They come in a wide variety of gains, intercepts and noise figures for frequency ranges from dc to 2 GHz. A more expensive option, for more sophisticated receiver applications, are the hybrid "cascadable" amplifiers, built on ceramic substrates and mounted in TO5 or TO8 metal cans. Most of these circuits are intended for a 50-Ω to 50-Ω interface.

A wide variety of hybrid amplifiers, designed for the Cable TV industry, are available for the frequency range from 1 to 1200 MHz (for example, the Motorola CA series, in type 714x packages). These have gains from 15 to 35 dB, output 1-dB compression points from 22 to 30 dBm and noise figures from 4.5 to 8.5 dB. Such units are excellent alternatives to discrete home-brew circuits for many applications where very low noise figures are not needed. In small quantities they may be a bit expensive sometimes, but the total cost of a home-built circuit, including labor (even the amateur experimenter's time is not really "free") is often at least as great. Home-built circuits do, however, have very important educational value.

UHF Design Aids

Circuit design and evaluation at the higher frequencies usually require some kind of minimal lab facilities, such as a signal generator, a calibrated noise generator and, hopefully, some kind of simple (or surplus) spectrum analyzer. This is true because circuit behavior and stability depend on a number of factors that are difficult to "guess at," and intuition is often unreliable. The ideal instrument is a vector network analyzer with all of the attachments (such as an S-parameter measuring setup), but very few amateurs can afford this. Another very desirable thing would be a circuit design and analysis program for the personal computer. Software packages created especially for UHF and microwave circuit design are available. They tend to be somewhat expensive, but

considered to be the range from 300 MHz to 3 GHz. All of the basic principles of radio system design and circuit design that have been discussed so far apply as well in this range, but the higher frequencies require some special thinking about the methods of circuit design and the devices that are used.

GaAs FET Preamp for 430 MHz

Fig 17.34 shows the schematic diagram and the physical construction of a typical RF circuit at 430 MHz. It is a GaAsFET preamplifier intended for low noise Earth-Moon-Earth or satellite reception. The

construction uses chip capacitors, small helical inductors and a stripline surface-mount GaAsFET, all mounted on a G10 (two layers of copper) glass-epoxy PC board. The very short length of interconnection leads is typical. The bottom of the PC board is a ground plane. At this frequency, lumped components are still feasible, while microstrip circuitry tends to be rather large.

At higher frequencies, microstrip methods become more desirable in most cases because of their smaller dimensions. However, the advent of tiny chip capacitors and chip resistors have extended the frequency

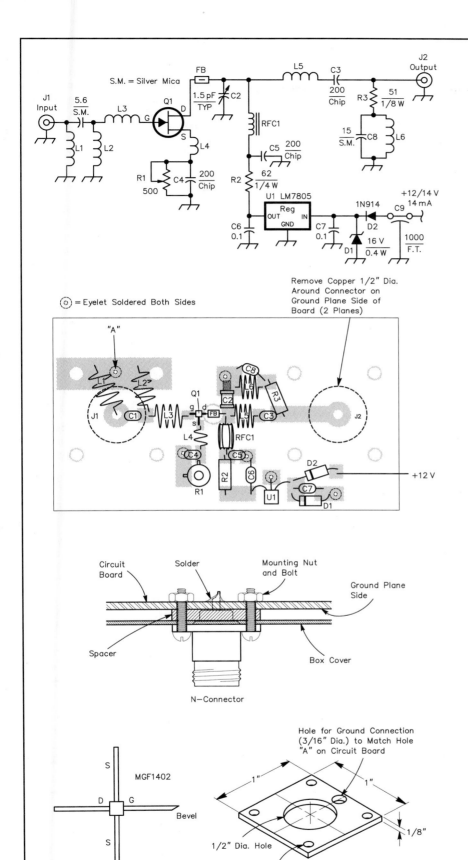

Fig 17.34—GaAsFET preamplifier schematic and construction details for 430 MHz. Illustrates circuit, parts layout and construction techniques suitable for 430-MHz frequency range.

C1—5.6-pF silver-mica capacitor or same as C2.

C2—0.6- to 6-pF ceramic piston trimmer capacitor (Johanson 5700 series or equiv).

C3, C4, C5—200-pF ceramic chip capacitor.

C6, C7—0.1-mF disc ceramic capacitor, 50 V or greater.

C8—15-pF silver-mica capacitor.

C9—500- to 1000-pF feedthrough capacitor.

D1—16- to 30-V, 500-mW Zener diode (1N966B or equiv).

D2—1N914, 1N4148 or any diode with ratings of at least 25 PIV at 50 mA or greater.

J1, J2—Female chassis-mount Type-N connectors, PTFE dielectric (UG-58 or equiv).

L1, L2—3t, #24 tinned wire, 0.110-inch ID spaced 1 wire diam.

L3—5t, #24 tinned wire, 3/16-inch ID, spaced 1 wire diam. or closer. Slightly larger diameter (0.010 inch) may be required with some FETs.

L4, L6—1t #24 tinned wire, 1/8-inch ID.

L5—4t #24 tinned wire, 1/8-inch ID, spaced 1 wire diam.

Q1—Mitsubishi MGF1402.

R1—200- or 500-Ω cermet potentiometer (initially set to midrange).

R2—62-Ω, 1/4-W resistor.

R3—51-Ω, 1/8-W carbon composition resistor, 5% tolerance.

RFC1—5t #26 enameled wire on a ferrite bead.

U1—5-V, 100-mA 3-terminal regulator (LM78L05 or equiv. TO-92 package).

worthwhile for a serious designer. Inexpensive SPICE programs are a good compromise. *ARRL Radio Designer* is an excellent, low cost choice.

A 902 to 928 MHz (33 cm) Receiver

Fig 17.35A is a block diagram of a 902-MHz down-converting receiver. A cavity resonator at the antenna input provides high selectivity with low loss. The first RF amplifier is a GaAsFET. Two additional 902-MHz band-pass microstrip filters and a BFR96 transistor provide more gain and image rejection (at RF–56 MHz) for the Mini Circuits SRA12 mixer. The output is at 28.0 MHz.

Cumulative Noise Figure

Fig 17.35B shows the cumulative noise figure (NF) of the signal path, including the 28-MHz receiver. The 1.5-dB cumulative NF of the input cavity and first-RF-amplifier combination, considered by itself, is degraded to 1.9 dB by the rest of the system following the first RF amplifier. The NF values of the various components for this example are reasonable, but may vary somewhat for actual hardware. Also, losses prior to the input, such as transmission-line losses (very important), are not included. They would be part of the complete receive-system analysis, however. It is common practice to place a low-noise preamp outdoors, right at the antenna, to overcome coax loss (and to permit use of inexpensive coax).

Local Oscillator (LO)Design

The +7 dBm LO at 874 to 900 MHz is derived from a set of crystal oscillators and frequency multipliers, separated by band-pass filters. These filters prevent a wide assortment of spurious frequencies from appearing at the mixer LO port. They also enhance the ability of the doubler stage to generate the second harmonic. That is, they have a very low impedance at the input frequency, thereby causing a large current to flow at the fundamental frequency. This increases the nonlinearity of the circuit, which increases the second-harmonic component. The higher filter impedance at the second harmonic produces a large harmonic output.

For very narrow-bandwidth use, such as EME, the crystal oscillators are often oven controlled or otherwise temperature compensated. The entire LO chain must be of low-noise design and the mixer should have good isolation from LO port to RF port (to minimize noise transfer from LO to RF).

A phase-locked loop using GHz-range prescalers (as shown in Fig 17.35C) is an alternative to the multiplier chain. The divide-by-N block is a simplification; in practice, an auxiliary dual-modulus divider in a "swallow count" loop would be involved in this segment. The cascaded 902-MHz band-pass filters in the signal path should attenuate any image frequency noise (at RF–56 MHz) that might degrade the mixer noise figure.

Summary

This example is fairly typical of receiver design methods for the 500 to 3000 MHz range, where down-conversion to an existing HF receiver is the most-convenient and cost-effective approach for amateurs. At higher frequencies a double down conversion with a first IF of 200 MHz or so, to improve image rejection, might be necessary. Usually, though, the presence of strong signals at image frequencies is less likely. Image-reducing mixers plus down conversion to 28 MHz is also coming into use, when strong interfering signals are not likely at the image frequency.

"No-Tune" Techniques

In recent years, a series of articles have appeared that emphasize simplicity of construction and adjustment. The use of

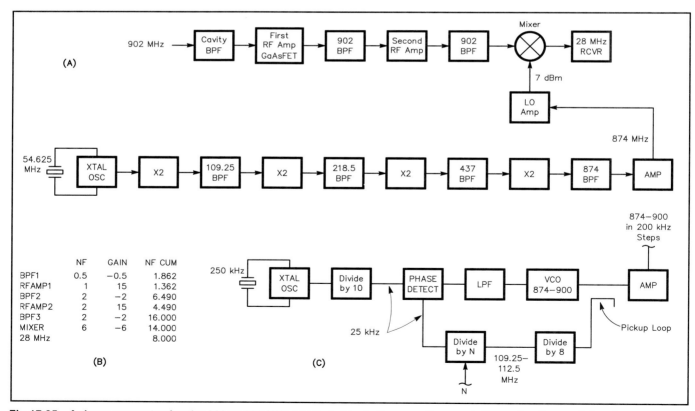

Fig 17.35—A down converter for the 902 to 928 MHz band. A: block diagram; B: cumulative noise figure of the signal path; C: alternative LO multiplier using a phase locked loop.

printed-circuit microstrip filters that require little or no adjustment, along with IC or MMIC devices, or discrete transistors, in precise PC-board layouts that have been carefully worked out, make it much easier to "get going" on the higher frequencies. Several of the **References** at the end of this discussion show how these UHF and microwave units are designed and built (Refs 26, 27). See also the projects at the end of this chapter.

Microwave Receivers

The world above 3 GHz is a vast territory with a special and complex technology and an "art form" that are well beyond the scope of this chapter. We will scratch the surface by describing a specific receiver for the 10-GHz frequency range and point out some of the important special features that are unique to this frequency range.

A 10-GHz Preamplifier

Fig 17.36A is a schematic and parts list, B is a PC-board parts layout and C is a photograph for a 10-GHz preamp, designed by Zack Lau at ARRL HQ. With very careful design and packaging techniques a noise figure approaching the 1 to 1.5-dB range was achieved. This depends on an accurate 50-Ω generator impedance and noise matching the input using a microwave circuit-design program such as *Touchstone* or *Harmonica*. Note that microstrip capacitors, inductors and transmission-line segments are used almost exclusively. The circuit is built on a 15-mil Duroid PC board. In general, this kind of performance requires some elegant measurement equipment that few amateurs have. A detailed discussion appears in Ref 28; it is recommended reading. On

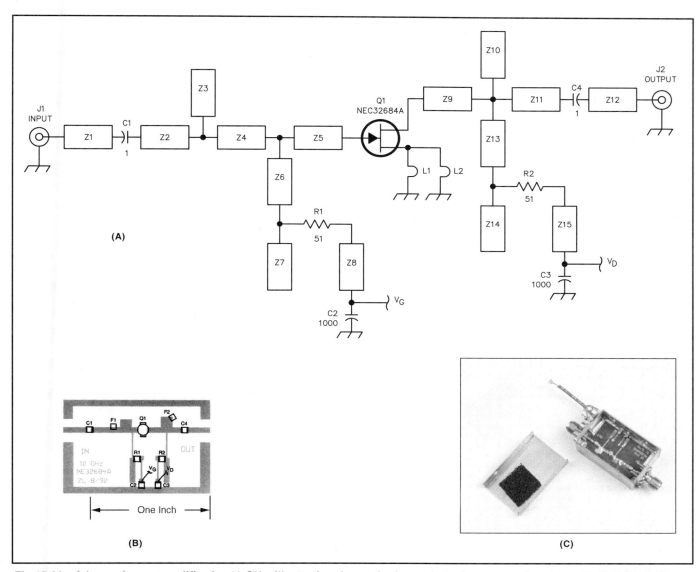

Fig 17.36—A low-noise preamplifier for 10 GHz, illustrating the methods used at microwaves. A: schematic. B: PC board layout. Use 15-mil 5880 Duroid, dielectric constant of 2.2 and a dissipation factor of 0.0011. For a negative of the board write, phone or e-mail the Technical Department Secretary at ARRL HQ and request the template from the December 1992 *QEX* "RF" column. C: A photograph of the completed preamp.

C1, C4—1-pF ATC 100-A chip capacitors. C1 must be very low loss.
C2, C3—1000-pF chip capacitors. (Not critical.) The ones from Mini Circuits work fine.

F1, F2—Pieces of copper foil used to tune the preamp.
J1, J2—SMA jacks. Ideally these should be microstrip launchers. The pin should be flush against the board.

L1, L2—The 15-mil lead length going through the board to the ground plane.
R1, R2—51-Ω chip resistors.
Z1-Z15—Microstriplines etched on the PC board.

the other hand, preamp noise figures in the 2 to 4-dB range are much easier to get (with simple test equipment) and are often satisfactory for terrestrial communication.

Articles written by those with expertise and the necessary lab facilities almost always include PC board patterns, parts lists and detailed instructions that are easily duplicated by readers. So it is possible to "get going" on microwaves using the material supplied in the articles. Complete kits are available from suppliers listed in the ARRL Address List in the **References** chapter. Microwave ham clubs and their publications are a good way to get started in microwave amateur technology.

Because of the frequencies involved, dimensions of microstrip circuitry must be very accurate. Dimensional stability and dielectric constant reliability of the boards must be very good.

System Performance

At microwaves, an estimation of system performance can often be performed using known data about the signal path terrain, atmosphere, transmitter and receivers systems.

Fig 17.37 shows a simplified example of how this works. This example is adapted from Dec 1980 *QST* (also see ARRL *UHF/ Microwave Experimenter's Manual*, p 7-55). In the present context of receiver design we wish to establish an approximate goal for the receiver system, including the antenna and transmission line.

A more detailed analysis includes terrain variations, refraction effects, the Earth's curvature, diffraction effects and interactions with the atmosphere's chemical constituents and temperature gradients. *The ARRL UHF/Microwave Experimenter 's Manual* is a good text for these matters.

In microwave work, where very low noise levels and low noise figures are encountered, experimenters like to use the "effective noise temperature" concept, rather than noise factor. The relationship between the two is given by

$$T_E = 290 (F - 1) \tag{7}$$

T_E is a measure, in terms of temperature, of the "excess noise" of a component (such as an amplifier). A resistor at 290 +

T_E would have the same available noise power as the device (referred to the device's input) specified by T_E. For a lossy device (such as a lossy transmission line) T_E is given by $T_E = 290 (L - 1)$, where L is the loss factor (same as its noise factor). The cascade of noise temperatures is similar to the Friis formula for cascaded noise factors and its use is illustrated in Fig 17.42.

$$T_S = T_G + T_{E1} + T_{E2}/G_1 + T_{E3}/(G_1G_2) \\ + T_{E4}/(G_1G_2G_3) + \ldots \tag{8}$$

where T_S is the system noise temperature (including the generator, which may be an antenna) and T_G is the temperature of the antenna.

The 290 number in the formulas for T_E is the standard ambient temperature (kelvins) at which the noise factor of a two-port transducer is defined and measured, according to an IEEE recommendation. So those formulas relate a noise factor F, measured at 290 K, to the temperature T_E. In general, though, it is perfectly correct to say that the ratio $(S_I/N_I)/(S_O/N_O)$ can be thought of as the ratio of total system output noise to that system output noise attributed to the "generator" alone, regardless of the temperature of the equipment or the nature of the generator, which may be an antenna at some arbitrary temperature, for example. This ratio is, in fact, a special "system noise factor (or figure), F_S" that need not be tied to any particular temperature such as 290 K. The use of the F_S notation avoids any confusion. As the example of Fig 17.37 shows, the value of this system noise factor F_S is just the ratio of the total system temperature to the antenna temperature.

Having calculated a system noise temperature, the receive system noise floor (that is, the antenna input level of a signal that would exactly equal system noise, both observed at the receiver output) associated with that temperature is:

$$N = k\, T_S\, B_N \tag{9}$$

where
$$k = 1.38 \times 10^{-23} \text{ and}$$
B_N = noise bandwidth.

The system noise figure F_S is indicated in the example also. It is higher than the sum of the receiver and coax noise figures.

The example includes a loss of 3 dB in the receiver transmission line. The formula for T_S in the example shows that this loss has a double effect on the system noise temperature, once in the second term (288.6) and again in the third term (2.0). If the receiver (or high-gain preamp with a 6 dB NF) were mounted at the antenna, the receive-system noise temperature would be reduced to 1064.5 K and a system noise

---

Analysis of a 10.368 GHz communication link with SSB modulation:
Free space path loss (FSPL) over a 50 mile line-of-sight path (S) at F = 10.368 GHz:
FSPL = 36.6 (dB) + 20 log F (MHz) + 20 log S (Mi) = 36.6 + 80.3 + 34 = 150.9 dB.

Effective isotropic radiated power (EIRP) from transmitter:
EIRP (dBm) = P_{XMIT} (dBm) + Antenna Gain (dBi)

The antenna is a 2 ft diameter (D) dish whose gain GA (dBi) is:
GA = 7.0 + 20 log D (ft) + 20 log F (GHz) = 7.0 + 6.0 + 20.32 = 33.3 dBi

Assume a transmission-line loss L_T, of 3 dB
The transmitter power P_T = 0.5 (mW PEP) = −3 (dBm PEP)
P_{XMIT} = P_T (dBm PEP) − L_T (dB) = −3 − 3 = −6 (dBm PEP)
EIRP = P_{XMIT} + G_A = −6 + 33.3 = 27.3 (dBm PEP)
Using these numbers the received signal level is:
P_{RCVD} = EIRP (dBm) − Path loss (dB)
= 27.3 (dBm PEP) − 150.9 (dB) = −123.6 (dBm PEP)
Add to this a receive antenna gain of 17 dB. The received signal is then
P_{RCVD} = −123.6 + 17 = −106.6 dBm

Now find the receiver's ability to receive the signal:
The antenna noise temperature T_A is 200 K. The receiver noise figure NF_R is 6 dB (FR=3.98, noise temperature T_R = 864.5 K) and its noise bandwidth (B) is 2400 Hz. The feedline loss L_L is 3 dB (F = 2.00, noise temperature T_L = 288.6 K). The system noise temperature is:
T_S = T_A + T_L + (L_L) (T_R)
T_S = 200 + 288.6 + (2.0) (864.5) = 2217.6 K
N_S = kT_SB = 1.38 × 10^{-23} × 2217.6 × 2400 = 7.34 × 10^{-17} W = −131.3 dBm
This indicates that the PEP signal is −106.6 −(−131.3) = 24.7 dB above the noise level. However, because the average power of speech, using a speech processor, is about 8 dB less than PEP, the average signal power is about 16.7 dB above the noise level.

To find the system noise factor F_S we note that the system noise is proportional to the system temperature T_S and the "generator" (antenna) noise is proportional to the antenna temperature T_A. Using the idea of a "system noise factor": F_S = T_S / T_A = 2217.6 / 200 = 11.09 = 10.45 dB.
If the antenna temperature were 290 K the system noise figure would be 9.0 dB, which is precisely the sum of receiver and receiver coax noise figures (6.0 + 3.0).

Fig 17.37—Example of a 10-GHz system performance calculation. Noise temperature and noise factor of the receiver are considered in detail.

figure, FS, of 7.26 dB, a very substantial improvement. Thus, it is the common practice to mount a preamp at the antenna.

Microwave Receiver for 10 GHz

Ref 29 provides a good example of modern amateur experimenter techniques for the 10-GHz band. The intended use for the radio is narrowband CW and SSB work, which requires extremely good frequency stability in the LO. Here, we will discuss the receiver circuit.

Block Diagram

Fig 17.38 is a block diagram of the receiver. Here are some important facets of the design.

1. The antenna should have sufficient gain. At 10 GHz, gains of 30 dBi are not difficult to get, as the example of Fig 17.37 demonstrates. A 4-ft dish might be difficult to aim, however.

2. For best results a very low-noise preamp at the antenna reduces loss of system sensitivity when antenna temperature is low. For example, if the antenna temperature at a quiet direction of the sky is 50 K and the receiver noise figure is 4 dB (due in part to transmission-line loss), the system temperature is 488 K for a system noise figure of 9.9 dB. If the receiver noise figure is reduced to 1.5 dB by adding a preamp at the antenna the system temperature is reduced to 170 K for a system noise figure of 3.4 dB, which is a very big improvement.

3. After two stages of RF amplification using GaAsFETs, a probe-coupled cavity resonator attenuates noise at the mixer's image frequency, which is $10.368 - 0.288 = 10.080$ GHz. An image reduction of 15 to 20 dB is enough to prevent image frequency noise generated by the RF amplifiers from affecting the mixer's noise figure.

4. The singly balanced diode mixer uses a "rat-race" 180° hybrid. Each terminal of the ring is $1/4$ wavelength (90°) from its closest neighbors. So the anodes of the two diodes are 180°

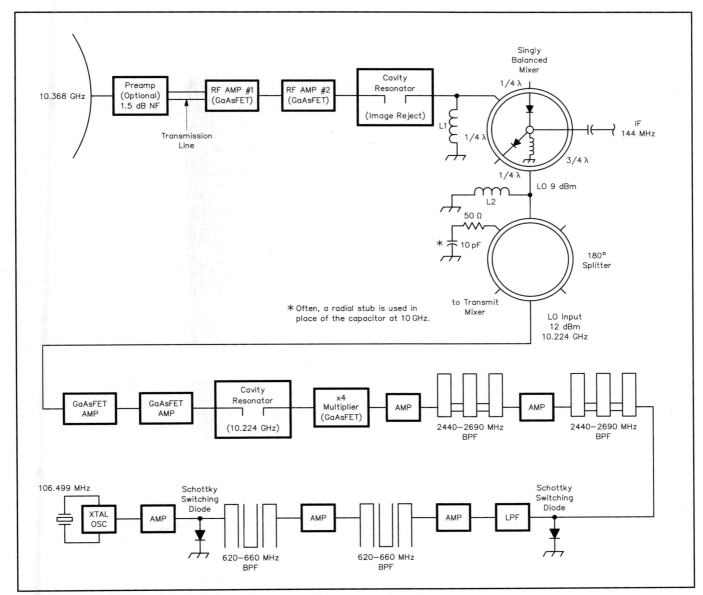

Fig 17.38—A block diagram of the microwave receiver discussed in the text.

(½ wavelength) apart with respect to the LO port, but in-phase with respect to the RF port. The inductors (L1, L2) connected to ground present a low impedance at the IF frequency. The mixer microstrip circuit is carefully "tweaked" to improve system performance. Use the better mixer in the transmitter.

5. The crystal oscillator is a fifth-overtone Butler circuit that is capable of high stability. The crystal frequency error and drift are multiplied 96 times (10.224/0.1065), so for narrowband SSB or CW work it may be difficult to get on (and stay on) the "calling frequency" at 10.368 GHz. One acceptable (not perfect) solution might be to count the 106.5 MHz with a frequency counter whose internal clock is constantly compared with WWV. Adjust to 106.5 MHz as required. At times there may be a small Doppler shift on the WWV signal. It may be necessary to switch to a different WWV frequency, or WWV's signals may not be strong enough. Surplus frequency standards of high quality are sometimes available. Many operators just "tune" over the expected range of uncertainty.

6. The frequency multiplier chain has numerous band-pass filters to "purify" the harmonics by reducing various frequency components that might affect the signal path and cause spurious responses. The final filter is a tuned cavity resonator that reduces spurs from previous stages. Oscillator phase noise amplitude is multiplied by 96.0 also, so the oscillator must have very good short term stability to prevent contamination of the desired signal.

7. A second hybrid splitter provides an LO output for the transmitter section of the radio. The 50-Ω resistor improves isolation between the two output ports. The two-part *QST* article (Ref 29) is recommended reading for this very interesting project, which provides a fairly straightforward (but not extremely simple) way to get started on 10 GHz.

Transmitter Design

TRANSMITTER DESIGN VS RECEIVER DESIGN

Many of the building blocks used in transmitter design are either identical to or very similar to those used in receiver design. Such things as mixers, oscillators, low-level RF/IF/AF amplifiers are the same. There is one major difference in the usage of these items, though. In a transmitter, the ratio of maximum to minimum signal levels for each of these is much less than in a receiver, where a very large ratio exists routinely. In a transmitter the signal, as it is developed to its final frequency and power level, is carefully controlled at each stage so that the stage is driven close to some optimum upper limit. The noise figures and dynamic ranges of the various stages are somewhat important, but not as important as in a receiver.

The transmitter design is concerned with the development of the desired high level of output power as cleanly, efficiently and economically as possible. Spurious outputs that create interference are a major concern. Protection circuitry that prevents self-destruction in the event of parts failures or mishandling by the operator help the reliability in ways that are unimportant in receivers.

THE SUPERHET SSB/CW TRANSMITTER

The same mixing schemes, IF frequencies and IF filters that are used for superhet receivers can be, and very often are, used for a transmitter. **Fig 17.39** is a block diagram of one approach. Let's discuss the various elements in detail, starting at the microphone.

Microphones (Mics)

A microphone is a transducer that converts sound waves into electrical signals. For speech, its frequency response should be as flat as possible from below 200 to above 3500 Hz. Response peaks in the microphone can increase the peak to average ratio of speech, which then degrades (increases) the peak to average ratio of the transmitted signal. If a transmitter uses speech processing, most microphones pick up a lot of background ambient noises because the speech amplification, whether it be at audio or IF/RF, may be as much as 20 dB greater than without speech processing. A "noise canceling" microphone is recommended to reduce this background pickup. Microphone output levels vary, depending on the microphone type. Typical amateur mics produce about 10 to 100 mV.

Ceramic

Ceramic mics have high output impedances but low level outputs. They require a high-resistance load (usually about 50 kΩ) for flat frequency response and lose low-frequency response as this resistance is reduced (electrically, the mic "looks like" a small capacitor). These mics vary widely in quality, so a "cheap" mic is not a good bargain because of its effect on the transmitted power level and speech quality.

Dynamic

A dynamic mic resembles a small loudspeaker, with an impedance of about 680 Ω and an output of about 12 mV on voice peaks. In many cases a transformer (possibly built-in) transforms the impedance to 100 kΩ or more and delivers about 100 mV on voice peaks. Dynamic mics are widely used by amateurs.

Electret

"Electret" mics use a piece of special insulator material that contains a "trapped" polarization charge (Q) at its surfaces and a capacitance (C). Sound waves modulate the capacitance (dC) of the material and cause a voltage change (dV) according to the law $dV/V = -dC/C$. For small changes (dC) the change (dV) is almost linear. A polarizing voltage of about 4 V is required to maintain the charge. The mic output level is fairly low, and a preamp is sometimes required. These mics have been greatly improved in recent years.

Microphone Amplifiers

The balanced modulator and (or) the audio speech processor need a certain optimum level, which can be in the range of 0.3 to 0.6 V ac into perhaps 1 kΩ to 10 kΩ. Excess noise generated within the microphone amplifier should be minimized, especially if speech processing is used. The circuit in **Fig 17.40** uses a low-noise BiFET op amp. The 620-Ω resistor is selected for a low impedance microphone, and switched out of the circuit for high-impedance mics. The amplifier gain is set by the 100-kΩ potentiometer.

It is also a good idea to experiment with the low-and high-frequency responses of the mic amplifier to compensate for the frequency response of the mic and the voice of the operator.

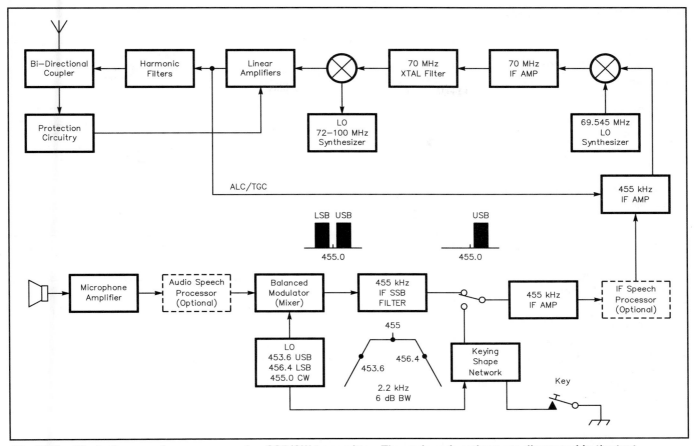

Fig 17.39—Block diagram of an up conversion SSB/CW transmitter. The various functions are discussed in the text.

Audio Speech Clipping

If the audio signal from the microphone amplifier is further amplified, say by as much as 12 dB, and then if the peaks are clipped (sometimes called "slicing") by 12 dB by a speech clipper, the output peak value is the same as before the clipper, but the average value is increased considerably. The resulting signal contains harmonics and IMD but the speech intelligibility, especially in a white-noise background, is improved by 5 or 6 dB.

The clipped waveform frequently tends to have a square-wave appearance, especially on voice peaks. It is then band-pass filtered to remove frequencies below 300 and above 3000 Hz. The filtering of this signal can create a "repeaking" effect. That is, the peak value tends to increase noticeably above its clipped value.

An SSB generator responds poorly to a square-wave audio signal. The Hilbert Transform effect, well known in mathematics, creates significant peaks in the RF envelope. These peaks cause out-of-band splatter in the linear amplifiers unless Automatic Level Control (ALC, to be discussed later) cuts back on the RF gain. The peaks increase the peak-to-average

ratio and the ALC reduces the average SSB power output, thereby reducing some of the benefit of the speech processing. The square-wave effect is also reduced by band-pass filtering (300 to 3000 Hz) the input to the clipper as well as the output.

Fig 17.41 is a circuit for a simple audio

speech clipper. A CLIP LEVEL potentiometer before the clipper controls the amount of clipping and an OUTPUT LEVEL potentiometer controls the drive level to the balanced modulator. The correct adjustment of these potentiometers is done with a two-tone audio input or by

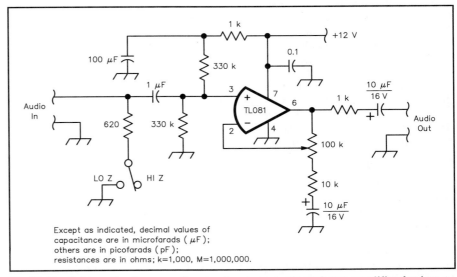

Except as indicated, decimal values of capacitance are in microfarads (μF); others are in picofarads (pF); resistances are in ohms; k=1,000, M=1,000,000.

Fig 17.40—Schematic diagram of a simple op-amp microphone amplifier for low- and high-impedance microphones.

talking into the microphone, rather than a single tone, because single tones don't exhibit the repeaking effect.

Audio Speech Compression

Although it is desirable to keep the voice level as high as possible, it is difficult to maintain constant voice intensity when speaking into the microphone. To overcome this variable output level, it is possible to use an automatic gain control that follows the average variations in speech amplitude. This can be done by rectifying and filtering some of the audio output and applying the resultant dc to a control terminal in an early stage of the amplifier.

The circuit of **Fig 17.42A** works on this AGC principle. One section of a Signetics 571N IC is used. The other section can be connected as an expander to restore the dynamic range of received signals that have been compressed in transmission. Operational transconductance amplifiers such as the CA3080 are also well suited for speech compression.

When an audio AGC circuit derives control voltage from the output signal the system is a closed loop. If short attack time is necessary, the rectifier-filter bandwidth must be opened up to allow syllabic modulation of the control voltage. This allows some of the voice frequency signal to enter the control terminal, causing distortion and instability. Because the syllabic frequency and speech-tone frequencies have relatively small separation, the simpler feedback AGC systems compromise fidelity for fast response.

Problems with loop dynamics in audio AGC can be side-stepped by eliminating the loop and using a forward-acting system. The control voltage is derived from the input of the amplifier, rather than from the output. Eliminating the feedback loop allows unconditional stability, but the trade-off between response time and fidelity remains. Care must be taken to avoid excessive gain between the signal input and the control voltage output. Otherwise the transfer characteristic can reverse; that is, an increase in input level can cause a decrease in output. A simple forward-acting compressor is shown in Fig 17.42B.

Balanced Modulators

A balanced modulator is a mixer. A more complete discussion of balanced modulator design is provided in the **Mixers** chapter. Briefly, the IF frequency LO (455 kHz in the example of Fig 17.39) translates the audio frequencies up to a pair of IF frequencies, the LO plus the audio frequency and the LO minus the audio frequency. The balance from the LO port to the IF output causes

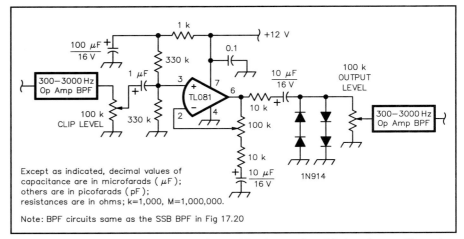

Fig 17.41—A simple audio speech clipper. The input signal is band pass filtered, amplified by 20 dB, clipped and band pass filtered again.

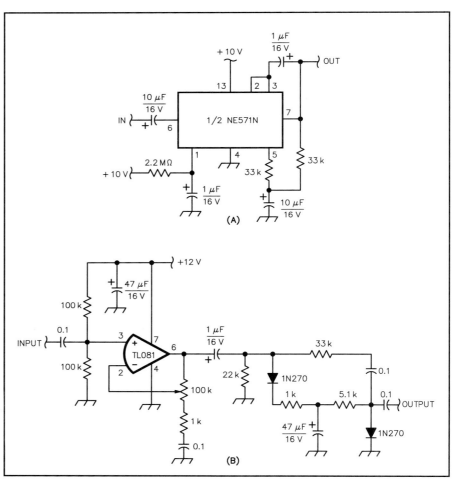

Fig 17.42—Typical solid-state compressor circuits. The circuit at A works on the AGC principle, while that at B is a forward-acting compressor.

the LO frequency to be suppressed by 30 to 40 dB. Adjustments are provided to improve the LO null.

The filter method of SSB generation uses an IF band-pass filter to pass one of the sidebands and block the other. In Fig 17.39 the filter is centered at 455.0 kHz. The LO is offset to 453.6 kHz or 456.4 kHz so that the upper sideband or the lower sideband (respectively) can pass through the filter. This creates a problem for the other LOs in the radio, because they must now be properly offset so that the final transmit output's carrier (suppressed) frequency coincides with the frequency readout on the front panel of the radio. Various schemes have been used to do this. One method uses two crystals for the 69.545-MHz LO that can be selected. In synthesized radios the programming of the microprocessor control moves the various LOs. Some synthesized radios use two IF filters at two different frequencies, one for USB and one for LSB, and a 455.0-kHz LO. These radios are often designed to transmit two independent sidebands (ISB, Ref 17).

In times past, balanced modulators using diodes, balancing potentiometers and numerous components spread out on a PC board were universally used. These days it doesn't make sense to use this approach. ICs and packaged diode mixers do a much better job and are less expensive. The most famous modulator IC, the MC1496, has been around for more than 20 years and is still one of the best and least expensive. **Fig 17.43** is a typical balanced modulator circuit using the MC1496.

The data sheets for balanced modulators and mixers specify the maximum level of audio for a given LO level. Higher audio levels create excessive IMD. The IF filter after the modulator removes higher-order IMD products that are outside its passband but the in-band IMD products should be at least 40 dB below each of two equal test tones. Speech clipping (AF or IF) can degrade this to 10 dB or so, but in the absence of speech processing the signal should be clean, in-band.

IF Filters

The desired IF filter response is shown in **Fig 17.44A**. The reduction of the carrier frequency is augmented by the filter response. It is common to specify that the filter response be down 20 dB at the carrier frequency. Rejection of the opposite sideband should (hopefully) be 60 dB, starting at 300 Hz below the carrier frequency, which is the 300-Hz point on the opposite sideband. The ultimate attenuation should be at least 70 dB. This would represent a very good specification for a high quality transmitter. The filter passband should be as flat as possible (ripple less than 1 dB or so).

Special filters, designated as USB or LSB, are designed with a steeper rolloff on the carrier frequency side, in order to improve rejection of the carrier and opposite

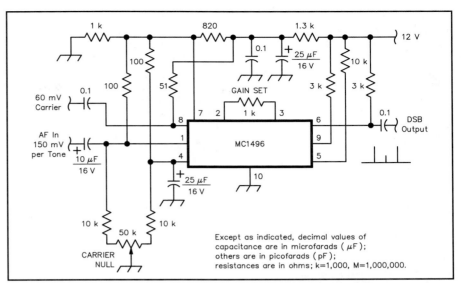

Fig 17.43—An IC balanced modulator circuit using the MC1496. The resistor from pin 2 to pin 3 sets the conversion gain.

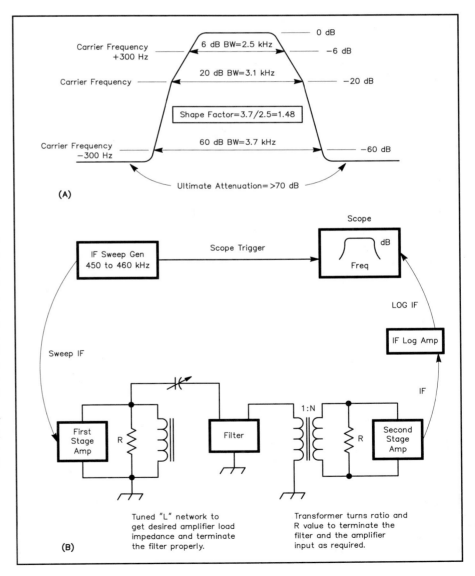

Fig 17.44—A: desired response of a SSB IF filter. B: one method of terminating a mechanical filter that allows easy and accurate tuning adjustment and also a possible test setup for performing the adjustments.

sideband. Mechanical filters are available that do this. Crystal-ladder filters (see the **Filters** chapter) are called "single-sideband" filters because they also have this property. The steep skirt can be on the low side or the high side, depending on whether the crystals are across the signal path or in series with the signal path, respectively.

Filters require special attention to their terminations. The networks that interface the filter with surrounding circuits should be accurate and stable over temperature. They should be easy to adjust. One very good way to adjust them is to build a narrowband sweep generator and look at the output IF envelope with a logarithmic amplifier, as indicated in Fig 17.44B. There are three goals: The driver stage must see the desired load impedance: the stage after the filter must see the desired source (generator) impedance and the filter must be properly terminated at both ends. Fig 17.44B shows two typical approaches. This kind of setup is a very good way to make sure the filters and other cir-cuitry are working properly.

Finally, overdriven filters (such as crystal or mechanical filters) can become nonlinear and generate distortion. So it is necessary to heed the manufacturer's instructions. Magnetic core materials used in the tuning networks must be sufficiently linear at the signal levels encountered. They should be tested for IMD separately.

IF Speech Clipper

Audio-clipper speech processors generate a considerable amount of in-band harmonics and IMD (involving different simultaneously occurring speech frequencies). The total distortion detracts somewhat from speech intelligibility. Other problems were mentioned in the section on audio processing. IF clippers overcome most of these problems, especially the Hilbert Transform problem (Ref 17).

Fig 17.45A is a diagram of a 455-kHz IF clipper using high-frequency op amps. 20 dB of gain precedes the diode clippers. A second amplifier establishes the desired output level. The clipping produces a wide band of IMD products close to the IF freuqency. Harmonics of the IF frequency are generated that are easily rejected by subsequent selectivity. "Close-in" IMD distortion products are band limited by the 2.5-kHz-wide IF filter so that out-of-band splatter is eliminated. The in-band IMD products are at least 10 dB below the speech tones.

Fig 17.46 shows oscilloscope pictures of an IF clipped two-tone signal at various levels of clipping. The level of clipping in a radio can be estimated by comparing with these photos. Listening tests verify that the IMD does not sound nearly as bad as harmonic distortion. In fact, processed speech sounds relatively clean and crisp. Tests also verify that speech intelligibility in a noise background is improved by 8 dB.

The repeaking effect from band-pass filtering the clipped IF signal occurs, and must be accounted for when adjusting the output level. A two-tone audio test signal or a speech signal should be used. The ALC circuitry (discussed later) will cut back the IF gain to prevent splattering in

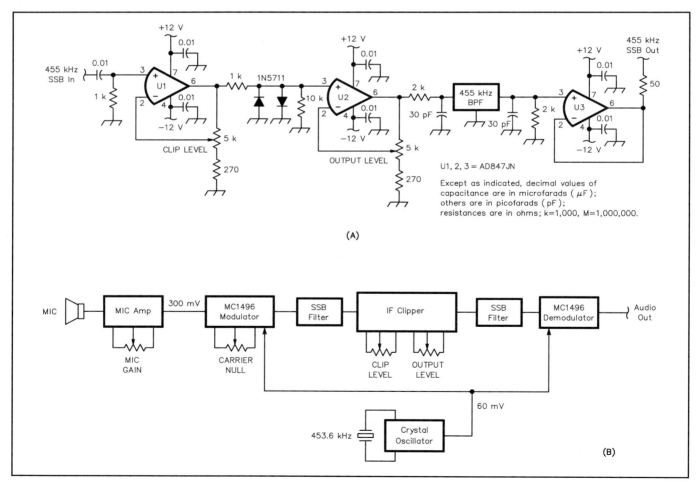

(A)

(B)

Fig 17.45—IF speech clipping. A: an IF clipper circuit approach. B: the audio signal is translated to 455 kHz, processed, and translated back to audio.

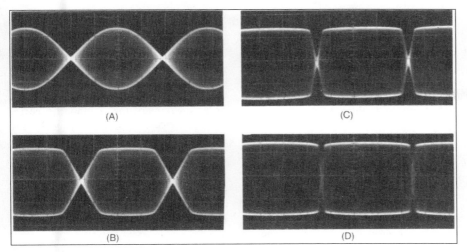

Fig 17.46—Two-tone envelope patterns with various degrees of RF clipping. All envelope patterns are formed using tones of 600 and 1000 Hz. At A, clipping threshold; B, 5 dB of clipping; C, 10 dB of clipping; D, 15 dB of clipping (from "RF Clippers for SSB," by W. Sabin, July 1967 QST, pp 13-18).

the power amplifiers. If the IF filter is of high quality and if subsequent amplifiers are "clean," the transmitted signal is of very high quality, very effective in noisy situations and often also in "pile-ups."

The extra IF gain implies that the IF signal entering the clipper must be free of noise, hum and spurious products. The cleanup filter also helps reduce the carrier frequency, which is outside the passband.

An electrically identical approach to the IF clipper can be achieved at audio frequencies. If the audio signal is translated to, say, 455 kHz, processed as described, and translated back to audio, all the desirable effects of IF clipping are retained. This output then plugs into the transmitter's microphone jack. Fig 17.45B shows the basic method. The mic amplifier and the MC1496 circuits have been previously shown and the clipper circuit is the same as in Fig 17.45A.

Another method, performed at audio, synthesizes mathematically the function of the IF clipper. This method is mentioned in Ref 17, and was an accessory for the Collins KWM380 transceiver.

The interesting operating principle in all of these examples is that the characteristics of the IF-clipped (or equivalent) speech signal do not change during frequency translation, even when translated down to audio and then back up to IF in a balanced modulator.

IF Linearity and Noise

Fig 17.39 indicates that after the last SSB filter, whether it is just after the SSB modulator or after the IF clipper, subsequent BPFs are considerably wider. For example, the 70-MHz crystal filter may be

15 to 30 kHz wide. This means that there is a "gray region" in the transmitter just like the one that we saw in the up conversion receiver, where out-of-band IMD that is generated in the IF amplifiers and mixers can cause adjacent-channel interference.

A possible exception, not shown in Fig 17.39, is that there may be an intermediate IF in the 10-MHz region that also contains a narrow filter, as we saw in the triple-conversion receiver in Fig 17.31.

The implication is that special attention must be paid to the linearity of these circuits. It's the designer's job to make sure that distortion in this gray area is much less than distortion generated by the PA and also less than the phase noise generated by the final mixer. Recall also that the total IMD generated in the exciter stages is the resultant of several amplifier and mixer stages in cascade; therefore, each element in the chain must have at least 40 to 50 dB IMD quality. The various drive levels should be chosen to guarantee this. This requirement for multistage linearity is one of the main technical and cost burdens of the SSB mode.

Of interest also in the gray region are white, additive thermal and excess noises originating in the first IF amplifier after the SSB filter and highly magnified on their way to the output. This noise can be comparable to the phase noise level if the phase noise is low, as it would be in a high-quality radio. Recall also that phase noise is at its worst on modulation peaks, but additive noise may be (and often is) present even when there is no modulation. This is a frequent problem in colocated transmitting and receiving environments. Many transmitter designs do not have the

benefit of the narrow filter at 70 MHz, so the amplified noise can extend over a much wider frequency range.

CW Mode

Fig 17.39 shows that in the CW mode a carrier is generated at the center of the SSB filter passband. There are two ways to make this carrier available. One way is to unbalance the balanced modulator so that the LO can pass through. Each kind of balanced modulator circuit has its own method of doing this. The approach chosen in Fig 17.39 is to go around the modulator and the SSB filter.

A shaping network controls the envelope of the IF signal to accomplish two things: control the shape of the Morse code character in a way that limits wideband spectrum emissions that can cause interference, and makes the Morse code signal easy and pleasant to copy.

RF Envelope Shaping

On-off keying is a special kind of low-level amplitude modulation (a low-signal-level stage is turned on and off). It generates numerous sidebands around the carrier frequency whose amplitudes are influenced by the shapes of the RF envelope rise and fall intervals. Consider an unprocessed keying waveform, a string of equal-length rectangular RF pulses separated by equal-length "dead times." Its spectrum consists of the carrier frequency, a pair of the usual modulation sidebands at the carrier frequency plus and minus the repetition rate of the pulses, and numerous other sidebands at multiples of the repetition rate. The higher frequency sidebands can create "key clicks" that extend many kHz either side of the carrier frequency.

Adjusting an exponentially shaped rise and fall time to about 5 ms (a value recommended by ARRL) controls key clicks, yet allows a wide range of practical keying speeds.

Fig 17.47A is a computer simulation of a single dot-space that is part of a continuous sequence, 20 ms on and 20 ms off (a 60 wpm rate), that has been shaped in this manner. This fast dot sequence is probably a "worst case" as far as Morse code interference is concerned. Fig 17.47B shows the line spectrum (Discrete Fourier Transform, DFT) of this periodic sequence. The carrier component is reduced because the on-off duty cycle is 50%, and some power is transferred from the carrier to the sidebands. (Note: This is a little different from ordinary AM, in which an additional source of power is used to create sidebands and the carrier remains constant.) The first pair of spectrum lines are ±25 Hz from the carrier. These are the normal modulation

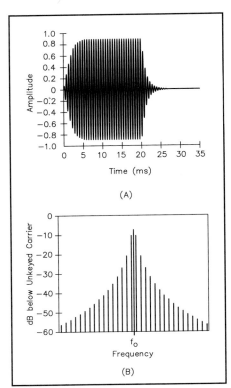

Fig 17.47—A: envelope, and B: spectrum of a 20-ms dot that has been shaped to 5 ms rise and fall times.

sidebands, determined only by the keying rate (25 Hz = 1/(40 ms)). The subsequent pairs are odd-order sidbands starting at ±75 kHz from the carrier (the symmetry of the on-off reduces even-order sidebands).

If the duration of the dot is held constant but the time between dots is greatly increased (the keying rate is greatly reduced) many more lines, much closer together and now including the even-order (50, 100, 150 Hz, and so on) sidebands, would appear in Fig 17.47B. But the general shape and also the bandwidth of the spectrum remain the same as shown. For example, a single, unrepeated dot would have essentially the same spectrum shape and width as in Fig 17.47B but would be a continuous spectrum (the Fourier Transform, FT) instead of the line spectrum (DFT) shown. In all cases the bandwidth of the interference, measured at the –20 dB points, is approximately 150 Hz. Finally, a single "key-down-and-hold' or "key-up-and-hold" operation would produce a similar spectrum of brief duration (a single key click) due to the shaping of the leading or trailing edge.

The rolloff rate of the higher sidebands in the shaped pulse is much faster than for a rectangular pulse. The example shown could be improved further by making the transitions "sinusoidal." The higher sidebands are still present but they are further

reduced, as verified by computer simulations. This waveform (or other kinds of "windowing") could be generated with DSP or with an op-amp state-variable analog computer in the Key Shape block of Fig 17.39. The linearity of the SSB amplifier chain will then preserve this envelope shape right up to the antenna.

Keying Circuits

The circuit in **Fig 17.48** is one example of an approach that will provide accurate control of the IF envelope. The PIN diodes are connected either to the signal source or to ground. They are turned on or off through U2 according to the exponential charging or discharging of the capacitor C1 through R1. R1 connects to +12 V or –12 V through U1. Some experimentation with R1 and R2 will be necessary, depending on the IF drive level, to get the desired rise and fall characteristics. Look at the IF envelope on an oscilloscope while sending a string of dots. Try an input IF level of 100 mV. The transformers are optimized for the IF in use.

Another approach is to use gain controllable IC amplifiers of the type discussed in connection with receiver AGC. See, for example, Fig 17.26. The key-up "turnoff" must be complete to avoid "backwave," residual output.

Wideband Noise

In the block diagram of Fig 17.39 the last mixer and the amplifiers after it are wideband circuits that are limited only by the harmonic filters and by any selectivity that may be in the antenna system. Wideband phase noise transferred onto the transmitted modulation by the last LO can extend over a wide frequency range, therefore LO (almost always a synthesizer of some kind) cleanliness is always a matter of great concern (Ref 8).

The amplifiers after this mixer are also sources of wide-band "white" or additive noise. This noise can be transmitted even during times when there is no modulation, and it can be a source of local interference. To reduce this noise: use a high-level mixer with as much signal output as possible, and make the noise figure of the first amplifier stage after the mixer as low as possible.

Transmitters that are used in close proximity to receivers, such as on shipboard, are always designed to control wideband emissions of both additive noise and pahse noise, referred to as "composite" noise.

Transmit Mixer Spurious Signals

The last IF and the last mixer LO in Fig 17.39 are selected so that, as much as possible, harmonic IMD products are far enough away from the operating fre-

quency that they fall outside the passband of the low-pass filters and are highly attenuated. This is difficult to accomplish over the transmitter's entire frequency range. It helps to use a high-level mixer and a low enough signal level to minimize those products that are unavoidable. Low-order crossovers that cannot be sufficiently reduce are unacceptable, however; the designer must "go back to the drawing board."

The **Regulations** chapter gives information regarding FCC requirements for harmonic and spurious levels. The **Filters** chapter describes the design of harmonic filters.

Automatic Level Control (ALC)

The purpose of ALC is to prevent the various stages in the transmitter from being overdriven. Over-drive can generate too much out-of-band distortion or cause excessive power dissipation, either in the amplifiers or in the power supply. ALC does this by sampling the peak amplitude of the modulation (the envelope variations) of the output signal and then developing a dc gain-control voltage that is applied to an early amplifier stage, as suggested in Fig 17.39.

ALC is usually derived from the last stage in a transmitter. This ensures that this last stage will be protected from overload. However, other stages prior to the last stage may not be as well protected; they may generate excessive distortion. It is possible to derive a composite ALC from more than one stage in a way that would prevent this problem. But designers usually prefer to design earlier stages conservatively enough so that, given a temperature range and component tolerances, the last stage can be the one source of ALC. The gain control is applied to an early stage so that all stages are aided by the gain reduction.

Speech Processing with ALC

A fast response to the leading edge of the modulation is needed to prevent a transient overload. After a strong peak, the control voltage is "remembered" for some time as the voltage in a capacitor. This voltage then decays partially through a resistor between peaks. An effective practice provides two capacitors and two time constants. One capacitor decays quickly with a time constant of, say 100 ms, the other with a time constant of several seconds. With this arrangement a small amount of speech processing, about 1 or 2 dB, can be obtained. (Explanation: The dB of improvement mentioned has to do with the improvement in speech intelligibility in a random noise back-ground.

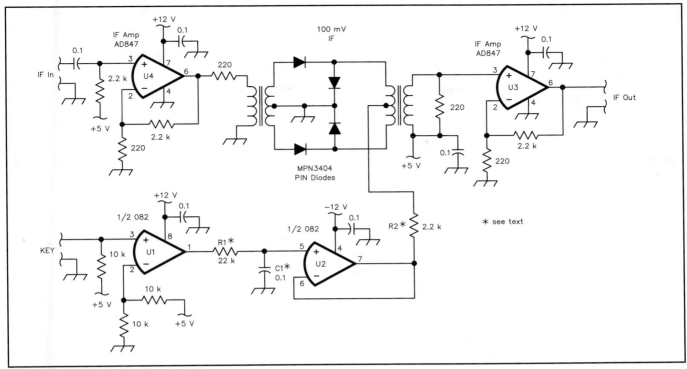

Fig 17.48—A keying circuit for the transmitter diagram in Fig 17.39.

This improvement is equivalent to what could be achieved if the transmit power were increased that same number of dB). The gain rises a little between peaks so that weaker speech components are enhanced. But immediately after a peak it takes awhile for the enhancement to take place, so weak components right after a strong peak are not enhanced very much. **Fig 17.49A** shows a complete ALC circuit that performs speech processing.

ALC in Solid-State Power Amplifiers

Fig 17.49B shows how a dual directional coupler can be used to provide ALC for a solid-state power amplifier (PA). The basic idea is to protect the PA transistors from excessive SWR and dissipation by monitoring both the forward power and the reflected power. The projects section of this chapter includes a 50-W amplifier project; the protection circuitry is discussed in detail there. Thermal protection is also included.

Transmit Gain Control (TGC)

This is a widely used feature in commercial and military equipment. A calibrated "tune-up" test carrier of a certain known level is applied to the transmitter. The output carrier level is sampled, using a diode detector. The resulting dc voltage is used to set the gain of a low-level stage. This control voltage is digitized and stored in memory so that it is semipermanent. A new voltage may be generated and stored

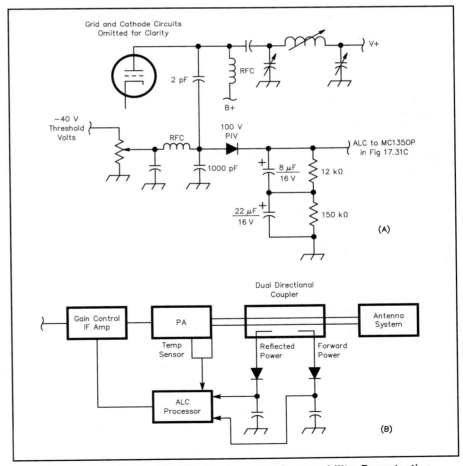

Fig 17.49—A: an ALC circuit with speech processing capability. B: protection method for a solid-state transmitter.

after each frequency change, or the stored value may be "fetched." A test signal is also used to do automatic antenna tuning. A dummy load is used to set the level and a low-level signal (a few mW) is used for the antenna tune-up.

Transmitter Output Load Impedance

The following logical processes are used to tune and load the final PA of a transmitter:

1. The RF input power requirement to the input terminal of the PA has been determined.
2. The desired load impedance of the plate/collector/drain of the PA has been determined, either graphically or by calculation, from the power to be delivered to the load, the dc power supply voltage and the ac voltage on the plate/collector/drain (see **RF Power Amplifiers** chapter).
3. The input impedance, looking toward the antenna, of the transmission line that is connected to the transmitter is adjusted by a network of some kind to its Z_0 value (if it is not already equal to that value).
4. A network of some kind is designed, which transforms the transmission-line Z_0 to the impedance required in step 2. This may be a sharply tuned resonator with impedance transforming capability, or it may be a wideband transformer of some kind.

Under these conditions the PA is performing as intended. Note that a knowledge of the output impedance of the PA is not needed to get these results. That is, we are interested mostly in the actual power gain of the PA, which does not require a knowledge of the amplifier's output impedance.

The output impedance, looking backward from the plate/collector/drain termi-nal of the network, in step 4, will have some influence on the selectivity of a resonant tuned circuit or the frequency response of a low-pass filter. This must (or should) be considered during the design process, but it is not needed during the "tune and load" process.

A Classic Amateur Vacuum-Tube Transmitter

Cascaded stage transmitter designs became very popular after about 1930. Home-brew solid-state CW transmitters often use the same general idea today. Fig **17.50** is a typical example. The first stage is an oscillator, which may be either crystal controlled or a variable frequency oscillator (VFO) of high stability. The oscillator is followed by a combination of buffer amplifiers and frequency multipliers, usually doublers, to arrive at the final frequency and power level. Mixer circuits and crystal-VFO "mix-master" frequency generator arrangements are not used, so spurious mixer products are not a problem.

Each stage provides the power input that the next stage needs and each stage increases in size and power dissipation. The vacuum tube '47 crystal osc, '46 doubler, 210 ("five watter") buffer, 203A ("fifty watter") final (200 W input) was a famous CW "rig" during the mid 1930s. HF CW DXers would usually have one favorite crystal that was just several Hz inside the band edge.

The process of frequency multiplication also multiplies the frequency drift, frequency/phase modulation and microphonic effects of the VFO. So for many years amateurs relied on crystal control, until they learned how to design very stable, low-noise VFOs. The frequency multiplier is an item worthy of discussion.

Frequency Multipliers

A passive multiplier using diodes is shown in **Fig 17.51A**. The full-wave rectifier circuit can be recognized, except that the dc component is shorted to ground. If the fundamental frequency ac input is 1.0 V RMS the second harmonic is 0.42 V RMS or 8 dB below the input, including some small diode losses. This value is found by calculating the Fourier Series coefficients for the full-wave-rectified sine wave, as shown in many textbooks.

Transistor and vacuum-tube frequency multipliers operate on the following principle: if a sine wave input causes the plate/collector/drain current to be distorted (not a sine wave) then harmonics of the input are generated. If an output resonant circuit is tuned to a harmonic the output at the harmonic is emphasized and other frequencies are attenuated. For a particular harmonic the current pulse should be distorted in a way that maximizes that harmonic. For example, for a doubler the current pulse should look like a half sine wave (180° of conduction). A transistor with Class B bias would be a good choice. For a tripler use 120° of conduction (Class C).

An FET, biased at a certain point, is very nearly a "square law" device. That is, the drain-current change is proportional to the square of the gate-voltage change. It is then an efficient frequency doubler that also deemphasizes the fundamental.

A push-push doubler is shown in Fig 17.51B. The FETs are biased in the square-law region and the BALANCE potentiometer minimizes the fundamental frequency. Note that the gates are in push-pull and the drains are in parallel. This causes second harmonics to add in-phase at the output and fundamental components to cancel.

Fig 17.51C shows an example of a bipolar-transistor doubler. The efficiency of a doubler of this type is typically 50%, a tripler 33% and a quadrupler 25%. Harmonics other than the one to which the

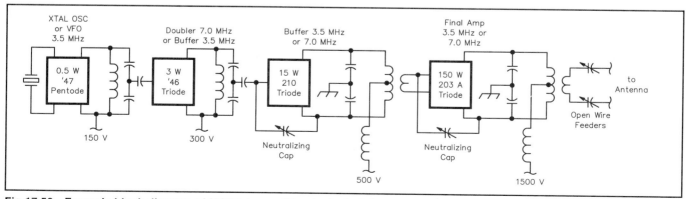

Fig 17.50—Example block diagram of MOPA transmitter, typical of the 1930s. Crystal control is shown, but VFOs replaced them during the '40s.

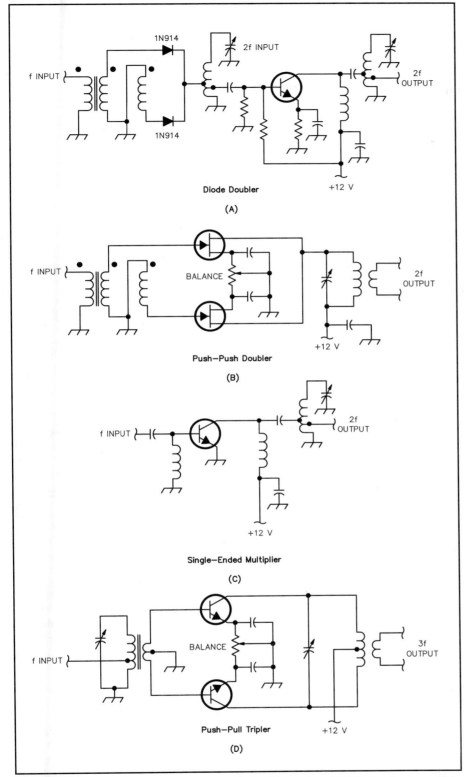

Fig 17.51—A: diode doubler. B: push-push doubler using JFETS. C: single-ended multiplier using a BJT. D: push-pull tripler using BJTs.

output tank is tuned will appear in the output unless effective band-pass filtering is applied. The collector tap on L1 is placed at the point that offers the best compromise between power output and spectral purity.

A push-pull tripler is shown in Fig 17.51D. The input and output are both push-pull. The balance potentiometer minimizes even harmonics. Note that the transistors have no bias voltage in the base

circuit; this places the transistors in Class C for efficient third-harmonic production. Choose an input drive level that maximizes harmonic output.

The step recovery diode (SRD) is an excellent device for harmonic generation, especially at microwave frequencies. The basic idea of the SRD is as follows: When the diode is forward conducting, a charge is stored in the diode's diffusion capacitance; and if the diode is quickly reverse-biased, the stored charge is very suddenly released into an LC harmonic-tuned circuit. The circuit is also called a "comb generator" because of the large number of harmonics that are generated. (The spectral display looks like a comb.) Phase-locked loops (PLLs) can be made to lock onto these harmonics. A typical low-cost SRD is the HP 5082-0180, found in the HP Microwave & RF Designer's Catalog. **Fig 17.52A** is a typical schematic. For more information regarding design details there are two References: Hewlett-Packard application note AN-920 and Ref 30.

The varactor diode can also be used as a multiplier. Fig 17.52B shows an example. This circuit depends on the fact that the capacitance of a varactor changes with the instantaneous value of the RF excitation voltage. This is a nonlinear process that generates harmonic currents through the diode. Power levels up to 25 W can be generated in this manner.

IMPEDANCE TRANSFORMATION BETWEEN CASCADED CIRCUITS

One of the most common tasks the electronics designer encounters is to correctly interface between the output of one circuit and the input of an adjacent circuit, so that both are operating in the desired manner. We introduced this in Figs 17.1 and 17.2. In this segment we will present a unified overview of the general topic that should be helpful to the designer of RF circuits. This is a very large topic, so we must stick to basic ideas and give References. The networks we will consider are in two categories, broadband and narrowband. The modern trend in Amateur Radio is to employ the personal computer, using software such as *ARRL Radio Designer*, *SPICE*, *Mathcad*, the Smith Chart and associated design programs.

Impedance "Matching" and "Transformation"

Fig 17.53A shows a network connected between a generator with internal resistance R_{IN} and a load R_{OUT}. In this case the load and source are "matched" because each sees itself, looking into the network. As Figs 17.1 and 17.2 explained, this idea

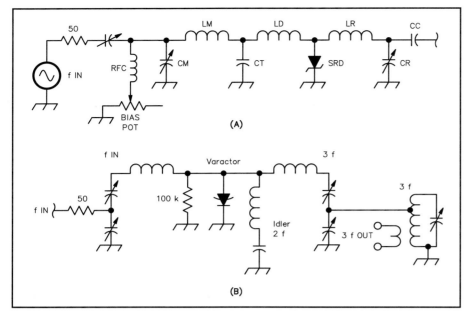

Fig 17.52—Diode frequency multipliers. A: step-recovery diode multiplier. B: varactor diode multiplier.

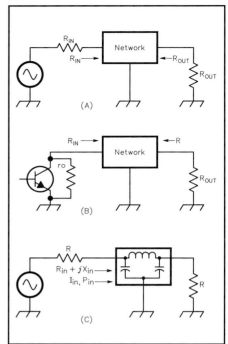

Fig 17.53—At A, matching network driven by generator. At B, matching network driven by transistor. At C. lowpass filter network.

extends to the idea of "conjugate match." There are many circuits that require this kind of impedance matching for textbook operation. In Fig 17.53B, showing a transistor amplifier, a different situation may exist. The transistor sees the R_{IN} that it needs for correct operation of the transistor.

That is, for a certain value of collector DC voltage and current and a certain allowed maximum value of collector RF voltage and current, a certain value of R_{IN} is required. However, the load R_{OUT}, looking back, may not see itself, but something much different, R, because the output resistance r_o of the transistor is not guaranteed to be the same as R_{IN} in Fig 17.53A. In this situation we could say that the load R_{OUT} is "transformed" to R_{IN}. This difference is important to understand in many applications. In some small-signal transistor circuits the dynamic output resistance r_o of the transistor and R_{OUT} are actually "matched" by a network as in Fig 17.53A. [Note: "dynamic" means that r_o is not a physical resistor but an internal, lossless *negative-feedback* property of the transistor]. In many other situations correct circuit performance requires that R_{OUT} see some specific value other than itself, looking into the output of the network.

A frequent practice for networks, including the *resonant* coupling networks (see Fig 17.59) is to combine a physical resistor with r_o so that the network is properly terminated at the input end. If the resistor is used, the transistor then sees the network and resistor combination; this

combination should be the load that we want. For example, suppose r_o is 10 kΩ and an R_{IN} of 1 kΩ (including r_o) is required. A 2 kΩ network and a 2.5 kΩ resistor in parallel could be used and the network is then correctly double-terminated as in Fig 17.53a.

Of course, some signal power will be lost in the 2.5 kΩ resistor, but in low-level circuits we often accept that. An elegant alternative is to use additional negative feedback in the transistor circuit to reduce r_o to 1 kΩ and use a 1 kΩ network. In some circuits, especially RF power amplifiers, the dissipation in an extra resistor cannot be tolerated, and in this case the network internal impedance R, looking into the output terminals, is usually ignored and the main goal is to get the right R_{IN}.

The **RF Power Amplifiers** chapter shows how Pi and Pi-L networks for power amplifiers are calculated. Use simulation to verify that the circuit is behaving as you want and to get network component losses. Simulations also show that the loading by a resistor or tube/transistor dynamic r_o at the R_{IN} end changes the selectivity, and this can be monitored by actual measurement. The Q and LC values can be modified as needed.

Mismatch in Lossless Networks

Fig 17.53C shows a generator with resistance R and a lowpass filter (LPF) connected to an R load resistor. At very low frequencies the filter is "transparent" and the maximum power is delivered to the load. The inductances and capacitances

are assumed to be lossless and the generator sees the R load. But at high frequencies the reactances cause the output power to be attenuated. The input impedance is $R_{IN} \pm j\, X_{IN}$. The power (real power) delivered to the filter input is $P_{IN} = I_{IN}^2 \times R_{IN}$.

Because L and C are lossless, this power P_{IN} must be identical to the power that is actually delivered to the R load. I_{IN}, X_{IN} and R_{IN} in the low-pass filter (LPF) are modified by the reactances in such a way that P_{IN}, therefore P_{LOAD}, is reduced. These changes are equivalent to an impedance mismatch between the generator and the filter. Or, we can say the generator no longer sees the correct load resistance R. This is the basic mechanism by which lossless networks control the frequency response.

BROADBAND TRANSFORMERS

Conventional Transformers

Fig 17.54A shows a push-pull amplifier that we will use to point out the main properties and the problems of conventional transformers. The medium of signal transfer from primary to secondary is magnetic flux in the core. If the core material is ferromagnetic then this is basically a nonlinear process that becomes increasingly nonlinear if the flux becomes too large or if there is a dc current through the winding that biases the core into a nonlinear region. Nonlinearity causes harmonics and IMD.

Push-pull operation eliminates the dc biasing effect if the stage is symmetrical. The magnetic circuit can be made more linear by adding more turns to the windings. This reduces the ac volts per turn, increases the reactance of the windings and therefore reduces the flux. For a given physical size, however, the wire resistance, distributed capacitance and leakage reactance all tend to increase as turns are added. This reduces efficiency and bandwidth. Higher permeability core materials and special winding techniques can improve things up to a point, but eventually linearity becomes more difficult to maintain.

Fig 17.54B is an approximate equivalent circuit of a typical transformer. It shows the leakage reactance and winding capacitance that affect the high-frequency response and the coil inductance that affects the low-frequency response. Fig 17.54C shows how these elements determine the frequency response, including a resonant peak at some high frequency.

The transformers in a system are correctly designed and properly coordinated when the total distortion caused by them is at least 10 dB less than the total distortion due to all other nonlinearities in the system. Do not over-design them in relation to the rest of the equipment. During the design process, distortion measurements are made on the transformers to verify this.

The main advantage of the conventional transformer, aside from its ability to transform between widely different impedances over a fairly wide frequency band, is the very high resistance between the windings. This isolation is important in many applications and it also eliminates coupling capacitors, which can sometimes be large and expensive.

In radio-circuit design, conventional transformers with magnetic cores are often used in high-impedance RF/IF amplifiers, in high-power solid-state amplifiers and in tuning networks such as antenna couplers. They are seldom used any more in audio circuits. Hybrid transformers, such as those in Fig 17.8, are often "conventional."

Fig 17.54D considers a typical application of a conventional transformer in a linear Class-A RF power amplifier. The load is 50 Ω and the maximum allowable transistor collector voltage and current excursions for linear operation are shown. The value of DC current and the sinewave limits are determined by studying the collector voltage-current curves (or constant-current curves) in the data manual to find the most linear region. To deliver this power to the 50 Ω load, the turns ratio is calculated from the equations. In an RF amplifier a ferrite or pow-

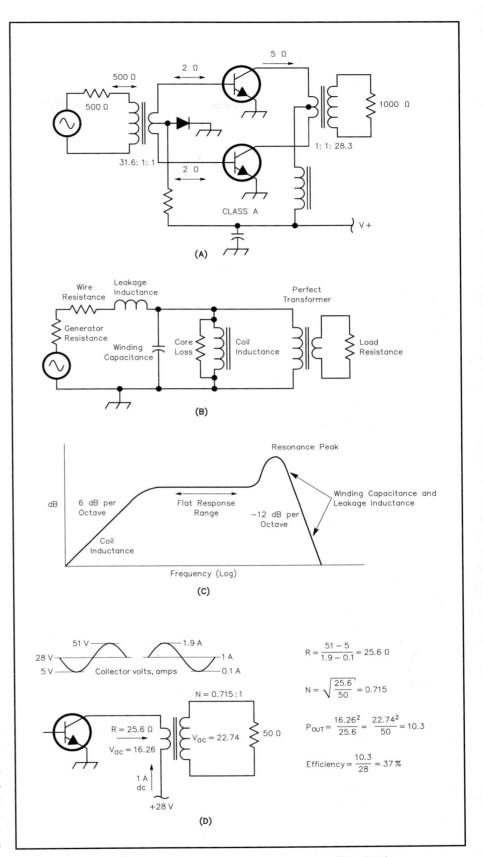

Fig 17.54—Conventional transformers in an RF power amplifier. Leakage reactances, stray capacitances and core magnetizations limit the bandwidth and linearity, and also create resonant peaks. D shows design example of transformer-coupled RF power amplifier.

dered-iron core would be used. The efficiency in this example is 38%.

Transmission Line Transformers

The basic transmission line transformer, from which other transformers are derived, is the 1:1 choke (or current) balun, shown in **Fig 17.55A**. We consider the following basic properties:

- A pair of close-spaced wires or a length of coax (ie, a transmission line) wraps around a ferrite rod or toroid or through a number of beads. For the 3.5 to 29.7 MHz band, type 43 ferrite (μ = 850), or equivalent, is usually preferred. Other types such as 77 (at 1.8 MHz, μ = 2000) or 61 (at VHF bands, μ = 120) are used. The Z_0 of the line should equal R.
- Because of the ferrite, a large impedance exists between points A and C and a virtually identical impedance between B and D. This is true for parallel wires and it is also true for coax. The ferrite affects the B to D impedance of the coax inner conductor equally.
- The conductors (two wires or coax braid and center-wire) are tightly coupled by electromagnetic fields and therefore constitute a good conventional transformer with a turns ratio of 1:1. The voltage from A to C is equal to and in-phase with that from B to D. These are called the *common-mode voltages* (CM).
- A common-mode (CM) current is one that has the same value and direction in both wires (or braid and center wire). Because of the ferrite, the CM current encounters a high impedance that acts to reduce (choke) the current. The normal differential-mode (DM) signal does not encounter this CM impedance because the electromagnetic fields due to equal and opposite currents in the two conductors cancel each other at the ferrite, so the magnetic flux in the ferrite is virtually zero.
- The main idea of the transmission line transformer is that although the CM impedance may be very large, the DM signal is virtually unopposed, especially if the line length is a small fraction of a wavelength.
- A common experience is a CM current that flows on the outside of a coax braid due to some external field, such as a nearby antenna or noise source. The balun reduces (chokes) the CM current due to these sources. But it is very important to keep in mind that the common-mode voltage across the ferrite winding that is due to this current is efficiently coupled to the center wire by conventional transformer action, as mentioned before and easily verified. This equality of CM voltages, and also

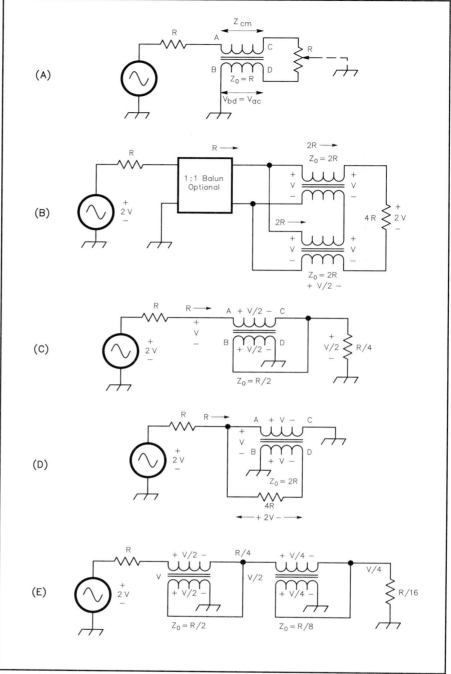

Fig 17.55—At A, basic balun. At B, 1:4 Guanella transformer. At C, Ruthroff transformer, 4:1 unbalanced. At D, Ruthroff 1:4 balanced transformer. At E, Ruthroff 16:1 unbalanced transformer.

CM impedances, reduces the *conversion* of a CM signal to an *undesired* DM signal that can interfere with the *desired* DM signal in both transmitters and receivers.

- The CM current, multiplied by the CM impedance due to the ferrite, produces a CM voltage. The CM impedance has L and C reactance and also R. So L, C and R cause a broad parallel self-resonance at some frequency. The R component

also produces some dissipation (heat) in the ferrite. This dissipation is an excellent way to dispose of a small amount of unwanted CM power.

- The main feature of the ferrite is that the choke is effective over a bandwidth of one, possibly two decades of frequency. In addition to the ferrite choke balun, straight or coiled lengths of coax (no core and almost no CM dissipation) are used within narrow frequency bands. A one-quarter-wave length of transmission line is a good choke balun at a single frequency or within a narrow band.

- The two output wires of the balun in Fig 17.55A have a high impedance with respect to, and are therefore "isolated" from, the generator. This feature is very useful because now any point of R at the output can be grounded. In a well-designed balun circuit *almost* all of the current in one conductor returns to the generator through the other conductor, despite this ground connection. Note also that the ground connection introduces some CM voltage across the balun cores and this has to be taken into account. This CM voltage is maximum if point C is grounded. If point D is grounded and if all "ground" connections are at the same potential, which they often are not, the CM voltage is zero and the balun may no longer be needed. In a coax balun the return current flows on the inside surface of the braid.

We now look briefly at a transmission line transformer that is based on the choke balun. Fig 17.55B shows two identical choke baluns whose inputs are in parallel and whose outputs are in series. The output voltage amplitude of each balun is identical to the common input, so the two outputs add in-phase (equal time delay) to produce twice the input voltage. It is the high CM impedance that makes this voltage addition possible. If the power remains constant the load current must be one-half the generator current, and the load resistor is $2V/0.5I = 4V/I = 4R$.

The CM voltage in each balun is $V/2$, so there is some flux in the cores. The right side floats. This is named the *Guanella* transformer. If Z_0 of the lines equals $2R$ and if the load is pure resistance $4R$ then the input resistance R is independent of line length. If the lines are exactly one-quarter wavelength, then $Z_{IN} = (2R)^2/Z_L$, an impedance inverter, where Z_{IN} and Z_L are complex. The quality of balance can often be improved by inserting a 1:1 balun (Fig 17.55A) at the left end so that both ends of the 1:4 transformer are floating and a ground is at the far left side as shown. The Guanella can also be operated from a grounded right end to a floating left end. The 1:1 balun at the left then allows a grounded far left end.

Fig 17.55C is a different kind, the *Ruthroff* transformer. The input voltage V is divided into two equal in-phase voltages AC and BD (they are tightly coupled), so the output is $V/2$. And because power is constant, $I_{OUT} = 2I_{IN}$ and the load is $R/4$. There is a CM voltage $V/2$ between A and C and between B and D, so in normal operation the core is not free of magnetic flux. The input and output both return to ground so it can also be operated from right to left for a 1:4 impedance stepup. The Ruthroff is often used as an amplifier interstage transformer, for example between 200 Ω and 50 Ω. To maintain low attenuation the line length should be much less than one-fourth wavelength at the highest frequency of operation, and its Z_0 should be $R/2$. A balanced version is shown in Fig 17.55D, where the CM voltage is V, not $V/2$, and transmission is from left-to-right only. Because of the greater flux in the cores, no different than a conventional transformer, this is not a preferred approach, although it could be used with air-wound coils (for example in antenna tuner circuits) to couple 75 Ω unbalanced to 300 Ω balanced. The tuner circuit could then transform 75 Ω to 50 Ω.

Fig 17.56 illustrates, in skeleton form, how transmission-line transformers can be used in a push-pull solid state power amplifier. The idea is to maintain highly balanced stages so that each transistor shares equally in the amplification in each stage. The balance also minimizes even-order harmonics so that low-pass filtering of the output is made much easier. In the diagram, T1 and T5 are current (choke) baluns that convert a grounded connection at one end to a balanced (floating) connection at the other end, with a high impedance to ground at both wires. T2 transforms the 50 Ω generator to

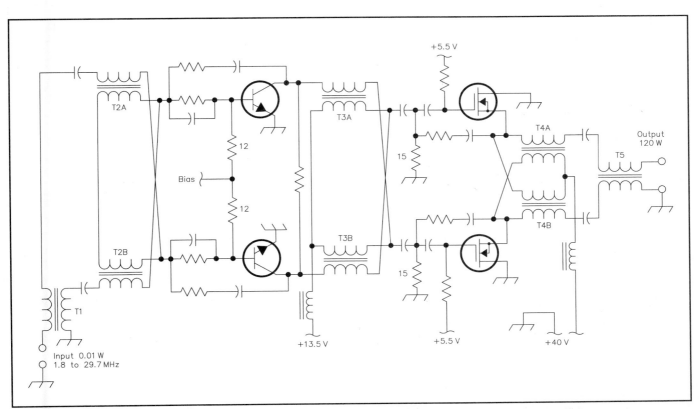

Fig 17.56—This illustrates how transmission-line transformers can be used in a push-pull power amplifier.

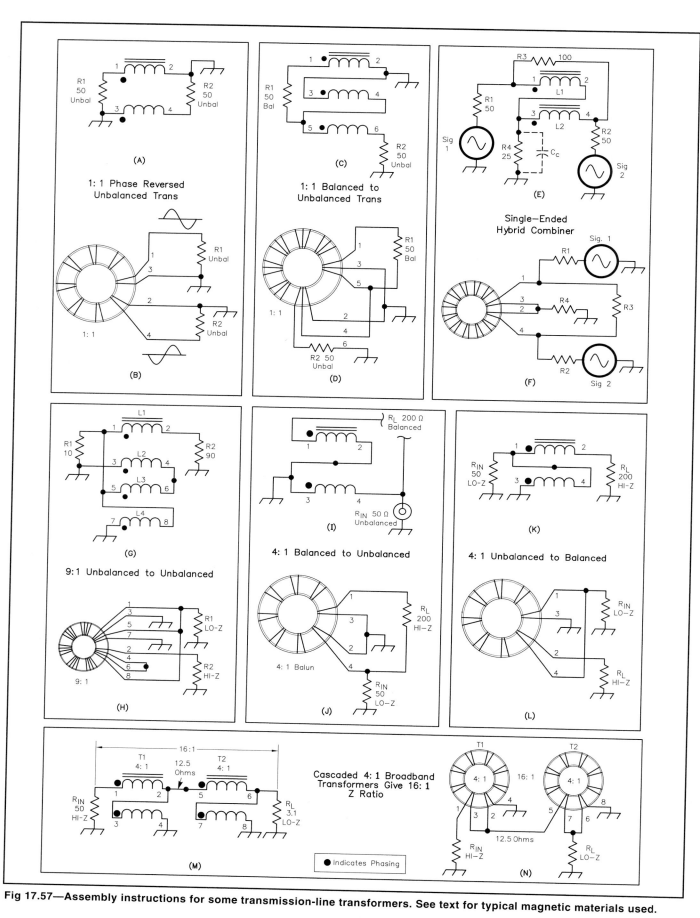

Fig 17.57—Assembly instructions for some transmission-line transformers. See text for typical magnetic materials used.

the 12.5 Ω (4:1 impedance) input impedance of the first stage. T3 performs a similar step-down transformation from the collectors of the first stage to the gates of the second stage. The MOSFETs require a low impedance from gate to ground. The drains of the output stage require an impedance step up from 12.5 Ω to 50 Ω, performed by T4. Note how the choke baluns and the transformers collaborate to maintain a high degree of balance throughout the amplifier. Note also the various feedback and loading networks that help keep the amplifier frequency response flat.

Tips on Toroids and Coils

Some notes about toroid coils: Toroids do have a small amount of leakage flux, despite rumors to the contrary. Toroid coils are wound in the form of a helix (screw thread) around the circular length of the core. This means that there is a small component of the flux from each turn that is perpendicular to the circle of the toroid (parallel to the axis through the hole) and is therefore not adequately linked to all the other turns. This effect is responsible for a small leakage flux and the effect is called the "one-turn" effect, since the result is equivalent to one turn that is wound around the outer edge of the core and not through the hole. Also, the inductance of a toroid can be adjusted, also despite rumors to the contrary. If the turns can be pressed closer together or separated a little, inductance variations of a few percent are possible.

A grounded aluminum shield between adjacent toroidal coils can eliminate any significant capacitive or inductive (at high frequencies) coupling. These effects are most easily noticed if a network analyzer is available during the checkout procedure, but how many of us are that lucky? Spot checks with an attenuator ahead of a receiver that is tunable to the harmonics are also very helpful.

There are many transformer schemes that use the basic ideas of Fig 17.55. Several of them, with their toroid winding instructions, are shown in **Fig 17.57**. Because of space limitations, for a comprehensive treatment we suggest Jerry Sevick's books *Transmission Line Transformers* and *Building and Using Baluns and Ununs*, both available from ARRL. For applications in solid-state RF power amplifiers, see Sabin and Schoenike, *HF Radio Systems and Circuits*, Chapter 12, also available from ARRL.

Tuned (Resonant) Networks

There is a large class of LC networks that utilize resonance at a single frequency to transform impedances over a narrow band. In many applications the circuitry that the network connects to has internal

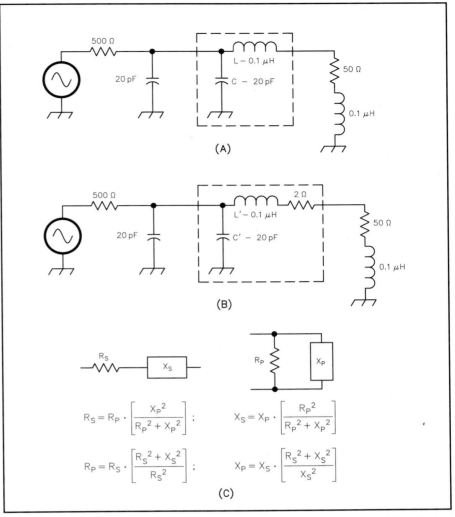

Fig 17.58—At A, impedance transformation, first iteration. At B, second iteration compensates L and C values for coil resistance. At C, series-parallel conversions.

reactances, inductive or capacitive, combined with resistance. We want to absorb these reactances, if possible, to become an integral part of the network design. By looking at the various available network possibilities we can identify those that will do this at one or both ends of the network. Some networks must operate between two different values of resistance, others can also operate between equal resistances. As mentioned before, nearly all networks also allow a choice of selectivity, or Q, where Q is (approximately) the resonant frequency divided by the 3-dB bandwidth.

As a simple example that illustrates the method, consider the generator and load of **Fig 17.58A**. We want to absorb the 20 pF and the 0.1 μH into the network. We use the formulas to calculate L and C for a 500 Ω to 50 Ω L-network, then subtract 20 pF from C and 0.1 μH from L. As a second iteration we can improve the design by considering the resistance of the L that we just

found. Suppose it is 2 Ω. We can recalculate new values L′ and C′ for a network from 500 Ω to 52 Ω, as shown in Fig 17.58B.

Further iterations are possible but usually trivial. More complicated networks and more difficult problems can use a computer to expedite absorbing process. Always try to absorb an inductance into a network L and a capacitance into a network C in order to minimize spurious LC resonances and undesired frequency responses. Inductors and capacitors can be combined in series or in parallel as shown in the example. Fig 17.58C shows useful formulas to convert series to parallel and vice versa to help with the designs.

A set of 14 simple resonant networks, and their equations, is presented in **Fig 17.59**. Note that in these diagrams RS is the low impedance side and RL is the high impedance side and that the X values are calculated in the top-down order given. The improved version of the program *NBMATCH.BAS* (lo-

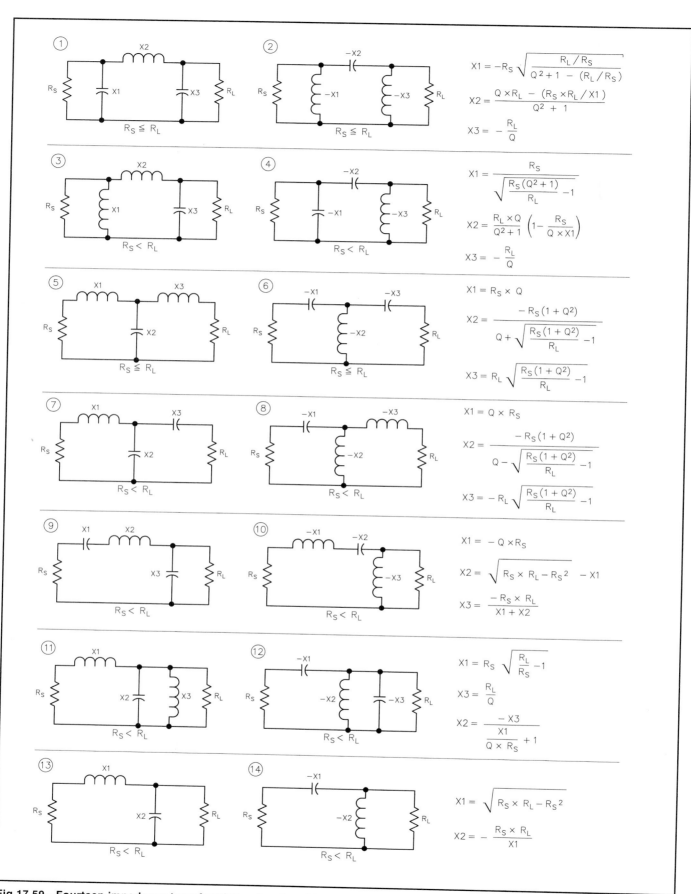

Fig 17.59—Fourteen impedance transforming networks with their design equations (for lossless components).

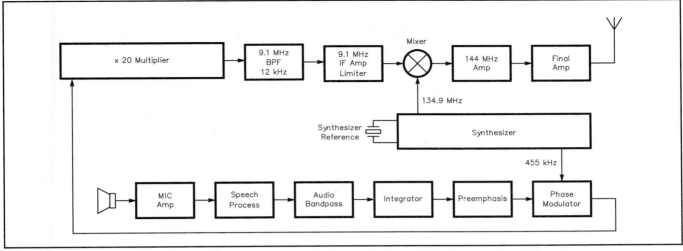

Fig 17.60—Block diagram of a VHF/UHF NBFM transmitter using the indirect FM (phase modulation) method.

cated on the CD-ROM in the back of this book) can perform the calculations.

ARRL Radio Designer can also help a lot with special circuit-design problems and some approaches to resonant network design. It can graph the frequency response, compute insertion loss and also tune the capacitances and inductances across a frequency band. You may select the selectivity (Q) in such programs based on frequency-response requirements. The program can also be trimmed to help realize realistic or standard component values. A math program such as *Mathcad* can also make this a quick and easy process. You can find additional information for Pi and Pi-L networks in the **RF Power Amplifiers** chapter.

NBFM Transmitter Block Diagram

Fig 17.60 shows the phase-modulation method, also known as indirect FM. It is the most widely used approach to NBFM. Phase modulation is performed at low IF, say 455 kHz. Prior to the phase modulator, speech filtering and processing are performed to achieve four goals:

1. Convert phase modulation to frequency modulation (see below),
2. Preemphasize higher speech frequencies for improved signal-to-noise ratio at the receive end,
3. Perform speech processing to emphasize the weaker speech components and
4. Adjust for the microphone's frequency response and possibly also the operator's voice characteristics.

Multiplier stages move the signal to some desired higher IF and also multiply the frequency deviation to the desired final value. If the FM deviation generated in the 455-kHz modulator is 250 Hz, the de-

viation at 9.1 MHz is 20 × 250, or 5 kHz. A second conversion to the final output frequency is then performed. Prior to this final translation, IF band-pass filtering is performed in order to minimize adjacent-channel interference that might be caused by excessive frequency deviation. This filter neds good phase linearity to assure that the FM sidebands maintain the correct phase relationships. If this is not done, an AM component is introduced to the signal, which can cause nonlinear distortion problems in the PA stages. The final frequency translation retains a constant value of FM deviation for any value of the output signal frequency.

The IF/RF amplifiers are Class C amplifiers because the signal in each amplifier contains, at any one instant, only a single value of instantaneous frequency and not multiple simultaneous frequencies as in SSB. These amplifiers are not sources of IMD, so they need not be "linear." The sidebands that appear in the output are a result only of the FM process (the Bessel functions).

In phase modulation, the frequency deviation is directly proportional to the frequency of the audio signal. To make the deviation independent of the audio frequency, an audio-frequency response that rolls off at 6 dB per octave is needed. An op-amp integrator circuit in the audio amplifier accomplishes this. This process converts phase modulation to frequency modulation. In addition, audio speech processing helps to maintain a constant value of speech amplitude, therefore constant IF deviation, with respect to audio speech levels. Also, preemphasis of the speech frequencies (6 dB per octave from 300 to 3000 Hz) is commonly used to improve the signal-to-noise ratio at the receive end.

Analysis shows that this is especially effective in FM systems when the corresponding deemphasis is used at the receiver (Ref 31).

An IF limiter stage may be used to ensure that any amplitude changes created during the modulation process are removed. The indirect-FM method allows complete frequency synthesis to be used in all the transmitter LOs, so that the channelization of the output frequency is very accurate. The IF and RF amplifier stages are operated in a highly efficient Class-C mode, which is helpful in portable equipment operating on small internal batteries.

NBFM is more tolerant of frequency misalignments, between the transmitter and receiver, than is SSB. In commercial SSB communication systems, this problem is solved by transmitting a pilot carrier that is 10 or 12 dB below PEP. The receiver phase locks to this pilot carrier. The pilot carrier is also used for squelch and AGC purposes. A short-duration "memory" feature in the receiver bridges across brief pilot-carrier dropouts, caused by multipath nulls.

"Direct FM" frequency modulates a high-frequency (say, 9 MHz or so) crystal oscillator by varying the voltage on a varactor. The audio is preemphasized and processed ahead of the frequency modulator. The Transceivers section of this chapter describes such a system.

TRANSVERTERS

At VHF, UHF and microwave frequencies, transverters that interact with factory-made transceivers in the HF or VHF range are common and are often home-built. These units convert the transceiver transmit signal up to a higher frequency and convert the receive frequency down to the transceiver receive frequency. The resulting performance and signal quality at the higher frequencies are enhanced by the frequency stability and the signal processing capabilities of the transceiver. For example, SSB and narrowband CW from 1.2 to 10 GHz are feasible, and becoming more popular. Some HF and VHF transceivers have special provisions such as connectors, signal-path switching and T/R switching that facilitate use with a transverter.

VHF Transverters

The methods of individual circuit design for a transverter are not much different than methods that have already been described. The most informative approach would be to study carefully an actual project description.

The interface between the transceiver and transverter requires some careful planning. For example, the transceiver power output must be compatible with the transverter's input requirements. This may require an attenuator or some modifications to a particular transverter or transceiver.

When receiving, the gain of the transverter must not be so large that the transceiver front-end is overdriven (system IMD is seriously degraded). On the other hand, the transverter gain must be high enough and its noise figure low enough so that the overall system noise figure is within a dB or so of the transverter's own noise figure. The formulas in this chapter for cascaded noise figure and cascaded third-order intercept points should be used during the design process to assure good system performance. The transceiver's performance should be either known or measured to assist in this effort.

Microwave Transverters

The microwave receiver section of this cahpter discussed a 10-GHz transverter project and gave references to the *QST* articles that give a detailed description. The reader is encouraged to refer to these articles and to review the previous material in this chapter.

Other Information

The *ARRL UHF/Microwave Experimenter's Manual* and ARRL's *Microwave Projects* contain additional interesting and valuable descriptions of transverter and transponder requirements.

TRANSCEIVERS

In recent years the transceiver has become the most popular type of purchased-equipment among amateurs. The reasons for this popularity are:

Fig 17.61—Photograph of Ten-Tec Omni VI Plus HF transceiver.

1. It is economical to use LOs (especially synthesizers), IF amplifiers and filters, power supplies, DSP modules and microprocessor controls for both transmit and receive.
2. It is simpler to perform transmit-receive (T/R) switching functions smoothly and with the correct timing within the same piece of equipment.
3. It is convenient to set a receive frequency and the identical (or properly offset) transmit frequency simultaneously.

In addition, transceivers have acquired very impressive arrays of operator aids that help the operator to communicate more easily and effectively. The complex design, numerous features and the very compact packaging have made the transceiver essentially a "store bought" item that is extremely difficult for the individual amateur to duplicate at home. The complexity of the work done by teams of design specialists at the factories is incompatible with the technical backgrounds of nearly all individual amateur operators.

The result of this modern trend is that amateur home-built equipment tends to be simpler, with less power output and more specialized (one-band, QRP, CW only, direct conversion, no-tune, receive only, transmit only and so on). Or, the amateur designs and builds add-on devices such as antenna couplers, active adaptive filters, computer interfaces and such.

Transceiver Example

As a way of providing a detailed, in-depth description of modern high-quality transceiver design, let's discuss one recent example, the Ten-Tec Omni VI Plus, an HF ham-band-only solid-state 100 W (output) transceiver, shown in **Fig 17.61**. Ref 32 is a Product Review of this radio. Let's consider first the signal-path block diagram in **Fig 17.62**, one section at a time.

Receiver Front End

The receive antenna can be either the same as the transmitting antenna or an auxiliary

receive antenna. A 20-dB attenuator can be switched in as needed. A 1.6-MHz high-pass filter attenuates the broadcast band. A 9.0-MHz trap attenuates any very strong signals at 9.0 MHz that might create interference in the form of blocking or harmonic IMD, especially when tuned to the 10.1-MHz (30-m) or 7-MHz (40-m) bands.

A set of band-pass filters, one for each HF amateur band, eliminates image responses and other spurs in the first mixer. These filters are also used in the low-level transmit stages. A low-noise, high-dynamic-range, grounded-gate JFET RF amplifier with about 9 dB of gain precedes the double balanced diode mixer, which uses 17 dBm of LO in a high-side mix

First IF

The first IF is 9.0 MHz. Because the LO is on the high side, there is a sideband inversion (USB becomes LSB and so on) after the first mixer. A grounded-gate, low-noise JFET amplifier terminates the first mixer in a resistive load and provides 6 dB of gain. This preamp helps to establish the receiver sensitivity (0.15 µV) with minimum gain preceding the mixer. The preamp is followed by a 15-kHz-wide two-pole filter, which is used for NBFM reception. It is also a roofing filter for the IF amplifier and the noise-blanker circuit that follow it.

The noise blanker gathers impulse energy from the 15-kHz filter, amplifies and rectifies it, and opens a balanced diode noise gate. The IF signal ahead of the gate is delayed slightly by a two-pole filter so that the IF noise pulse and the blanking pulse arrive at the gate at the same time (Ref 17).

The standard IF filter for SSB/CW has 8 poles, is centered at 9.0015 MHz and is 2.4 kHz wide at the –6 dB points. Following this filter, two optional 9.0-MHz filters with the following bandwidths can be installed:

Fig 17.62—Signal path block diagram, receive and transmit, for the Omni VI Plus transceiver.

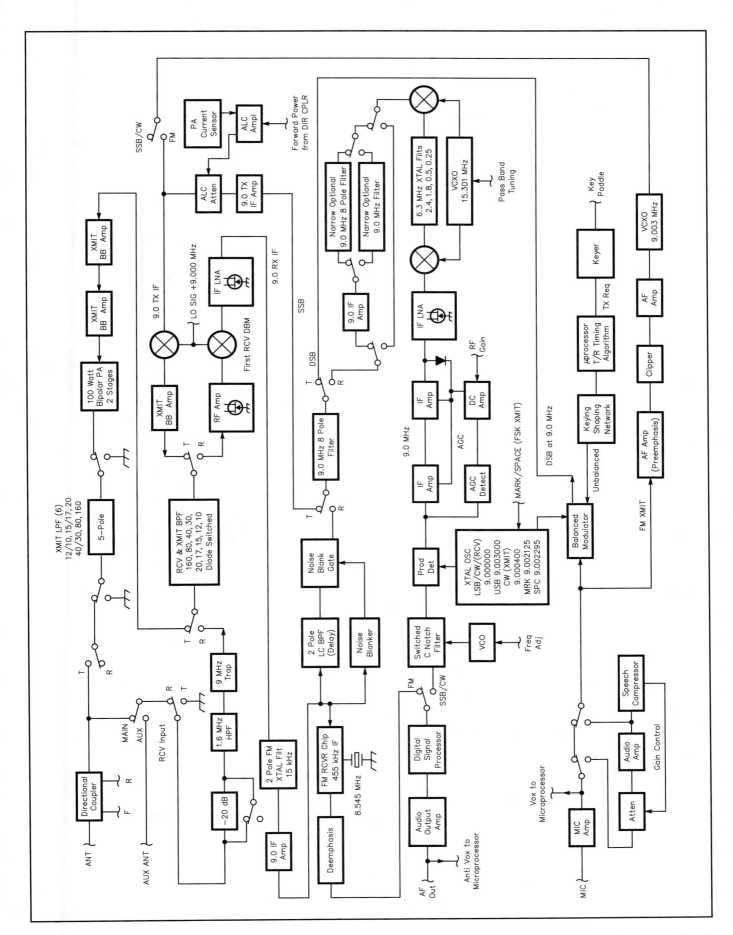

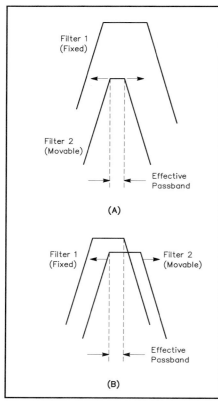

Fig 17.63—Explanation of passband tuning. A: wide first filter and narrow second filter. B: two narrow filters.

1.8 kHz, 500 Hz, 250 Hz or a 500 Hz RTTY filter. The optional filters are in cascade with the standard filter, for improved ultimate attenuation.

Passband Tuning Section

A mixer converts 9.0 MHz to 6.3 MHz and drives a standard 2.4 kHz wide filter. One of three optional filters, 1.8 kHz, 0.5 kHz (CW) or 250 Hz can be selected instead. A second mixer translates back to the 9.0-MHz frequency. A voltage tuned crystal oscillator at 15.300 MHz (tunable ± 1.5 kHz) is the LO for both mixers. This choice of LO and the 6.3-MHz IF results in very low levels of harmonic IMD products that might cross over the signal frequency and cause spurious outputs. The passband can be tuned ± 1.5 kHz.

The composite passband is the intersection of the fixed 9.0-MHz passband and the tunable 6.3-MHz passband. If the first filter is wide and the second much narrower the passband width remains constant over most of the adjustment range. If both have the same bandwidth the resultant bandwidth narrows considerably as the second filter is adjusted. This can be especially helpful in CW mode. **Fig 17.63** shows how passband tuning works.

IF Amplifiers after Passband Tuning

A low-noise grounded-gate JFET amplifier, with PIN-diode AGC, establishes a low noise figure and a low level of IF noise after the last IF filter. Two IC IF amplifiers (MC1350P) provide most of the receive IF gain. These three stages provide all of the AGC for the receiver. The AGC loop does not include the narrow-band IF filters. Two AGC recovery times (Fast and Slow) are available. AGC can be switched off for manual RF gain control as well. The AGC drives the S-meter, which is calibrated at 50 μV for S9 and 0.8 μV for S3.

Product Detector

The IC product detector (CA3053E) uses LO frequencies of 9.000 MHz for LSB and CW (in receive only), 9.003 MHz for USB. When switching between USB and LSB, for a constant value of signal carrier frequency (such as 14.20000 MHz), the LO of the first mixer is moved 3.00 kHz in order to keep the signal within the passband of the IF SSB filters. More about this later.

Audio Notch Filter

In the CW and digital modes, a switched-capacitor notch filter (MF5CN) places a narrow notch in the audio band. The location of the notch is determined by the clock rate applied to the chip. This is determined by a VCO (CD4046BE) whose frequency is controlled by the front panel NOTCH control.

NBFM Reception

After the 15 kHz wide IF filter at 9.0 MHz and before the noise blanker, the IF goes to the NBFM receiver chip (MC3371P). A mixer (8.545-MHz LO) converts it to 455 kHz. The signal goes through an off-chip ceramic band-pass filter, then goes back on-chip to the limiter stages and a quadrature detector. A received signal strength indicator (RSSI) output provides a dc voltage that is proportional to the dB level of the signal. This voltage goes to the front panel meter when in the NBFM mode. A squelch function (NBFM only) is controlled from a potentiometer on the front panel.

Audio Digital Signal Processing (DSP)

The DSP is based on the Analog Devices ADSP 2105 processor. The DSP program is stored in an EPROM and loaded into the 2105's RAM on power-up. DSP can be used in both SSB and CW. In USB or LSB (not CW or data) the DSP automatically locates and notches out one or several interfering carriers. In SSB or CW the manual audio notch filter described previously is also available, either as a notch filter or to reduce high frequency response (hiss filter). In the CW mode the DSP can be instructed to low-pass filter the audio with several

corner-frequency values. A DSP noise reduction function tracks desired signals and attenuates broadband noise by as much as 15 dB, depending on conditions.

Audio Output

The 1.5-W audio output uses a TDA2611 chip. Either FM audio or SSB/CW audio or, in transmit, a CW sidetone, can be fed to the speaker or headphones. The sidetone level (a software adjustment) is separate from the volume control. The audio output, after A/D conversion, is also fed to the Anti-VOX algorithm in the microprocessor.

Transmit Block Diagram

Now, let's look at the path from microphone or key to the antenna, one stage at a time.

Microphone Amplifier

The suggested microphone is 200 Ω to 50 kΩ at 5 mV (–62 dB). A polarizing voltage for electret mics is provided. The Mic Amp drives the balanced modulator, either directly or through the speech compressor. It also supplies VOX information to the microprocessor, via an A/D converter. The microprocessor software sets Vox hang time and sensitivity, as well as the Anti-VOX, via the keypad. Timing and delays for T/R switching are also in the software.

Speech Processor

The audio speech processor is a compressor, as discussed in a previous section of this chapter. A dc voltage that is proportional to the amount of compression is sent to the front panel meter so that compression can be set to the proper level. Clipper diodes limit any fast transients that might overdrive the signal path momentarily.

Balanced Modulator

The balanced modulator generates a double-sideband, suppressed carrier IF at 9.0 MHz. The LO is that used for the receive product detector. There is a carrier nulling adjustment. In CW and FSK modes, the modulator is unbalanced to let the LO pass through. A built-in iambic keyer (Curtis style A or B) is adjustable from 10 to 50 wpm. An external key or keyer can also be plugged in.

IF Filter

The standard 9.0015 MHz, 2.4-kHz-wide 8-pole filter (also used in receive) removes either the lower or upper sideband.The output is amplified at 9.0 MHz.

ALC

The forward-power measurement from the PA output directional coupler is used for ALC, which is applied at the output of the first 9.0-MHz IF amplifier by a PIN

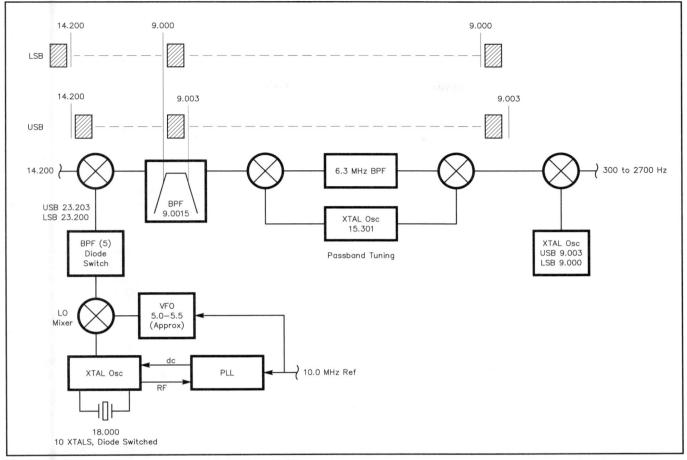

Fig 17.64—LO frequency management in USB and LSB.

diode attenuator. A front panel LED lights when ALC is operating. An additional circuit monitors dc current in the PA and cuts back on RF drive if the PA current exceeds 22 A.

Output Mixer

This mixer translates to the output signal frequency. The same LO frequency is used for each transmit mode (USB/LSB) and the same 3.00000-kHz frequency shift is used to assure that the frequency readout is always correct.

Band-pass Filters

The mixer output contains the image, at 9.0 plus LO and low values of harmonics and harmonic IMD products. A band-pass filter for each ham band, the same ones used in receive, eliminates all out of band products from the transmitter output.

PAs

Four stages of amplification, culminating in push-pull bipolar MRF454s in Class AB, supply 100 W output, CW or PEP from 160 to 10 m. Temperature sensing of the transistors in the last two stages helps to prevent

thermal damage. Full output can be maintained, key down, for 20 minutes. Forced air heat-sink cooling allows unlimited on-time.

LO Frequency Management

The LO goals are to achieve low levels of phase noise, high levels of frequency stability and at the same time keep equipment cost within the reach of as many amateurs as possible. As part one of the LO analysis we look at the method used to adjust the LOs in order to keep the speech spectrum of a USB or LSB signal within the 9.0-MHz IF filter passband. This method is somewhat typical for many equipment designs. Refer to **Fig 17.64**.

First Mixer and Product Detector

An SSB signal whose carrier frequency is at 14.20000 MHz, which may be LSB or USB, is translated so that the modulation (300 to 2700 Hz) in either case falls within the passband of the 9.0015-MHz 8-pole crystal filter. For a USB signal this is accomplished by increasing the first LO 3 kHz, as indicated. Since the LO is on the high side of the signal frequency there is a sideband "inversion" at the first mixer.

The passband tuning module does not change this relationship. At the product detector, the LO is increased 3 kHz in USB so that it is the same as the carrier (suppressed) frequency of the IF signal. Note that the designators "USB" or "LSB" at the product detector LO refer to the antenna signal, not the IF signal. The jog of the first mixer LO is accomplished partly within the crystal oscillator and partly within the VFO. The microprocessor sends the frequency instructions to both of these oscillators. Despite the frequency offsets, the digital readout displays the correct carrier frequency, in this example 14.20000 MHz. And, of course, the same procedures apply to the transmit mode.

Another interesting idea involves the first LO mixer. The 18-MHz crystal and the VFO are added to get 23.0 to 23.5 MHz, for the 20-m band. But the output of the LO mixer also contains the difference, 12.5 to 13.0 MHz, which is just right for the 80-m band. For the 20-m band one BPF selects 23.0 to 23.5 MHz. For the 80-m band the 12.5 to 13.0-MHz BPF is selected. A problem occurs, though, because now the direction of frequency

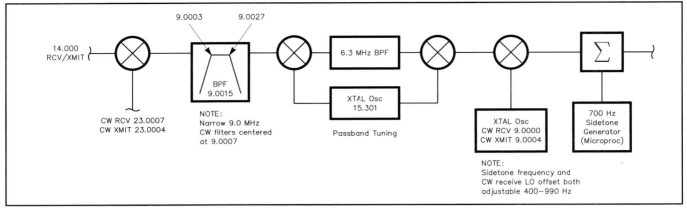

Fig 17.65—LO management in the CW receive and transmit modes.

tuning is reversed, from high to low. The microprocessor corrects this by reversing the direction of the tuning knob (an optical encoder). Other "book keeping" is performed so that the operation is transparent to the operator. A similar trick is used on the 17-m band and the 28.5 to 29.0-MHz segment of the 10-m band.

CW-Mode LO Frequencies

In CW mode the "transceiver problem" shows up. See **Fig 17.65** for a discussion of this problem. If the received carrier is on exactly 14.00000 MHz and if we want to transmit our carrier on that same exact frequency then the transmitter and the receiver are both "zero beat" at 14.00000 MHz. In receive we would have to retune the receiver, say up 700 Hz, to get an audible 700-Hz beat note. But then when we transmit we are no longer on 14.0000 MHz but are at 14.00070 MHz. We would then have to reset to 14.00000 when we transmit.

The transceiver's microprocessor performs all of these operations automatically. Fig 17.65 shows that in receive the first-LO frequency is increased 700 Hz. This puts the first IF at 9.0007 MHz, which is inside the passband of the 9.0 MHz IF filter. The BFO is at 9.0000 MHz and an audio beat note at 700 Hz is produced. This 700-Hz pitch is compared to a 700-Hz audio oscillator (from the microprocessor). When the two pitches coincide the signal frequency display of our transceiver coincides almost exactly with the frequency of the received signal. The digital frequency display reads "14.00000" at all times. The value of the 700-Hz reference beat can be adjusted between 400 and 950 Hz by the user. The receive LO shift matches that value.

When the optional 500/250 Hz CW 9.0 MHz IF filters are used, these are centered at 9.0007 MHz. These filters are used in receive but not in transmit.

When we transmit, the transmit fre-

quency is that which the frequency display indicates, 14.00000 MHz. However, there is a slight problem. The 9.0-MHz transmit IF must be increased slightly to get the speech signal within the passband of the 9.0 MHz IF filter. The transmit BFO is therefore at 9.0004 MHz. The mixer LO is also moved up 400 Hz so that the transmit output frequency will be exactly 14.0000 MHz.

In addition to the above actions, the RIT (receive incremental tuning) and the XIT (transmit incremental tuning) knobs permit up to ±9.9 kHz of independent control of the receive and transmit frequencies, relative to the main frequency readout.

Local Oscillators

Fig 17.64 indicates that the crystal oscillator for the first mixer is phase locked. Each of the 10 crystals is locked to a 100-kHz reference inside the PLL chip. This reference is derived from a 10-MHz system reference.

Let's go into some detail regarding the very interesting 5.0 to 5.5 MHz VFO circuitry. **Fig 17.66** is a block diagram. The VCO output, 200 to 220 MHz, is divided by 40 to get 5.0 to 5.5 MHz. The reference frequency for the PLL loop is 10 kHz, so each increment of the final output is 10 kHz / 40, which is 250 Hz. Phase noise in the PLL is also divided by 40, which is equivalent to 32 dB ($20 \times \log(40)$).

To get 10-Hz steps at the output, the voltage-tuned crystal oscillator at 39.94 MHz is tuned in 200-Hz steps (the division ratio from oscillator to final output is 20 instead of 40 because of the frequency doubler). To get 200-Hz steps in this oscillator, serial data from the microprocessor is fed into a latch. This data is sent from a RAM lookup table that has the correct values to get the 200-Hz increments very accurately. The outputs of the latch are fed into an R/2R ladder (D/A conversion) and the dc voltage tunes

the VCXO. Adjustment potentiometers calibrate the tuning range of the oscillator over 5000 Hz in 200-Hz increments. At the final output, this tuning range fills in the 10-Hz steps between the 250-Hz increments of the PLL. Although this circuit is not phase locked to a reference (it's an open-loop), the resulting frequency steps are very accurate, especially after the division by 20. This economical approach reduces the complexity and cost of the VFO considerably, but performs extremely well (very low levels of phase noise and frequency drift).

CW Break-In (Fast QSK) Keying

The radio is capable of break-in keying at rates up to 25 wpm when it is in the FAST QSK mode. This mode is also used for AMTOR FSK. **Fig 17.67** explains the action and the timing involved in this T/R switching. The sequence of events is as follows:

1. The key is pressed.
2. A Transmit Request is sent to the microprocessor.
3. The microprocessor changes the LO and BFO to their transmit frequencies.
4. After a 0.25-ms delay a Transmit Out logic level is sent to a jack on the rear panel.
5. The Transmit Out signal is jumpered to the TRANSMIT ENABLE jack. If an external Fast QSK PA is being used, an additional short delay is introduced while it is being switched to the transmit mode.
6. The Transmit Enable signal starts the keying-waveform circuit, which ramps up in 3 ms.
7. Near the bottom of this ramp, the "T" (transmit B+) voltage goes high and the "R" (receive B+) voltage goes low. The T/R reed relay (very fast) at the Omni VI PA output switches to transmit.
8. The shaped keying waveform goes to the balanced modulator and the RF envelope builds up. There is a very

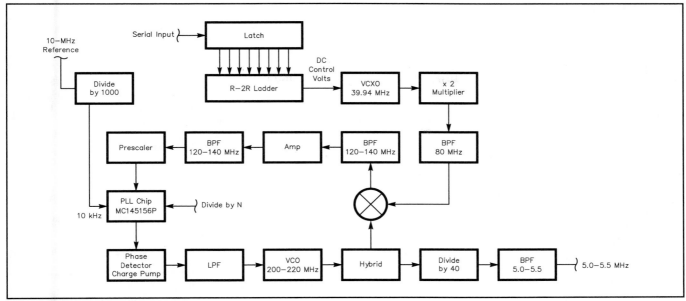

Fig 17.66—Block diagram of the VFO.

brief delay from balanced modulator to Omni VI output. The T/R relay is switched before the RF arrives.

9. The key is opened.
10. The Transmit Out and Transmit Enable lines go high.
11. The keying waveform ramps down. RF ramps down to zero.
12. After 5 ms (a fixed delay set by the microprocessor) "T" goes low and "R" goes high and the microprocessor returns the LO and BFO to their receive frequencies.

Slow QSK

In the Slow QSK mode the action is as described above. The radio reverts quickly to the receive mode. However, the receive audio is muted until the end of an extended (adjustable) delay time.

There is also a relay in the Omni VI that can be used to T/R switch a conventional (not Fast QSK) external PA. This option is selected from the operator's menu and is available only in the Slow QSK mode. The relay is held closed for the duration of the delay time in the slow QSK mode. When using this option the operator must ensure that the external PA switches fast enough on the first "key-down" so that it is not "hot switched" or that the first dot is missed. Many older PAs do not respond well at high keying rates. If the PA is slow, we can still use the Fast QSK mode with the Omni VI ("barefoot") and the PA will be bypassed because the optional relay is not energized in the Fast QSK mode.

VOX

In SSB mode, VOX and PTT perform

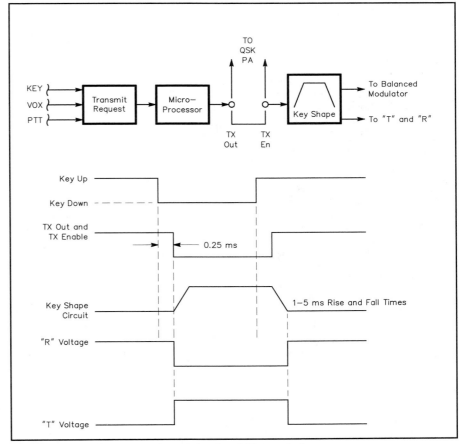

Fig 17.67—Fast QSK operation of the Omni VI transceiver.

the same functions and in the same manner as the CW Slow QSK described above. The VOX hang-time adjustment is separate from the CW hang-time adjustment. A MUTE jack allows manual switching (foot switch, and so on) to enable the transmit mode without applying RF. The key or the VOX then subsequently applies RF to the system. This arrangement helps if the external PA has slow T/R switching.

Operating Features

Modern transceivers have, over the years, acquired a large ensemble of operator's aids that have become very popular. Here are some descriptions of them:

Key Pad

The front panel key pad is the means for configuring a wide assortment of operating preferences and for selecting bands and modes.

Frequency Change

1. Use the tuning knob. The tuning rate can be programmed to 5.12, 2.56, 1.71, 1.28, 1.02 or 0.85 kHz per revolution. The knob has adjustable drag.
2. UP and DOWN arrows give 100 kHz per step.
3. Band selection buttons.
4. Keyboard entry of an exact frequency.

Mode Selection

1. *Tune*: puts the rig in CW mode, key down, for various "tune-up" operations.
2. *CW*: An optional DSP low-pass filter can be selected. Cutoff frequencies of 600, 800, 1000, 1200 or 1400 Hz can be designated. A SPOT function generates a 700-Hz audio sidetone that can be used for precise frequency setting (the received signal pitch matches the 700-Hz tone). The pitch of the sidetone can be adjusted from 400 to 950 Hz. Audio level is adjustable also. FAST QSK and SLOW QSK are available as previously described. Cascaded CW filters are available, with passband tuning of one of the filters. CW filter options: 500/250 Hz at 6.3 MHz IF and 500 Hz at 9.0-MHz IF.
3. *USB or LSB* : Standard SSB IF bandwidth is 2.4 kHz (two 8-pole filters in cascade). The second filter can be passband-tuned ±1.5 kHz. Additional IF filters with 1.8-kHz BW are available.
4. *FSK and AFSK*: Special FSK filter for receive. AMTOR operation with FAST QSK capability. AFSK generator can be plugged into microphone jack.
5. *FM*:15-kHz IF filters at 9.0 MHz and 455 kHz. Quadrature detection, RSSI output and squelch. Adjustable transmit deviation. FM transmit uses the direct method, as shown in Fig 17.69.
6. *VOX*:VOX sensitivity and hang time adjustable via the key pad. Anti-VOX level adjustable.

Time of Day Clock

There is a digital readout on front panel.

Built-In Iambic Keyer

Curtis type A or B, front panel speed knob. Adjustable dot-dash ratio. Also external key or keyer.

Dimmer

Adjusts brightness of front panel display.

Dual VFOs

Select A or B. Independent frequency, mode, RIT and filter choices stored for each VFO. Used for split-frequency operation.

Receiver Incremental Tuning (RIT)

Each VFO has its own stored RIT value.

Frequency Offset Display

RIT value adjusted with knob. RIT can be toggled on and off, or cleared to zero.

Transmitter Incremental Tuning (XIT)

Same comments as RIT. Simultaneous RIT and XIT.

Cross-Band and Cross-Mode Operation

For cross-band operation, use PTT for SSB and manual switch for CW.

Scratch-Pad Memory

Stores a displayed frequency. Restores that frequency to the VFO on command.

Band Register

Allows toggling between two frequencies on each band.

Memory Store

Store 100 values of frequency, band, mode, filter, RIT, XIT. Memory channels can be recalled by channel number (key pad), "scrolling" the memory channels or "memory tune" using the main tuning knob.

Lock

Locks the main tuning dial.

User Option Menu

Enables configuration of the radio via the keypad.

Meter

Select between receive signal strength (on SSB/CW or NBFM), speech processor level, forward power, SWR and PA dc current.

AGC

Fast, slow, off and manual RF gain control.

FM Squelch Adjust

Passband Tuning Knob

Notch

Automatically notch out several heterodynes on SSB/FM or manually notch on CW/digital modes. Adjustable low-pass filter in CW mode.

Antenna Switch

Auxiliary antenna may be selected in receive mode.

Interface Port

25-pin D connector for interface to personal computer.

OTHER TRANSCEIVERS

Other transceivers vary in cost, complexity and features. The one just described is certainly one of the best, at a reasonable price. For reviews of other transceivers (to see the differences in cost, features and performance specs) refer to the Product Reviews in *QST* and ARRL's *Radio Buyer's Sourcebook* series.

Let's look quickly at a more expensive transceiver, to get some feel for other options and design approaches at higher prices.

The Kenwood TS-950SDX

This radio has a "subreceiver," containing an SSB filter and a 500-Hz CW filter. Simultaneous reception on both receivers is possible. Both receivers can be tuned independently.

A stored-message memory allows automatic transmission of short CW messages.

It performs harmonic sampling and DSP at a 100-kHz IF. Digital detection is followed by digital filtering. The 100-kHz IF also has a voltage-tuned notch filter. SSB is generated by a DSP phasing method.

The synthesizer uses a DDS that can tune in 1-Hz steps. The receiver covers from 100 kHz to 30 MHz. The radio uses up conversion to a first IF of 73 MHz, followed by IFs at 8.8 MHz, 455 kHz and 100 kHz. The PA uses FET technology. A cooling fan is provided. The radio also contains a built-in automatic antenna tuner with memory retention of tuner settings.

References

[1]G. Gonzalez, *Microwave Transistor Amplifiers*, Englewood Cliffs, NJ, 1984, Prentice-Hall.

[2]Motchenbacher and Fitchen, *Low-Noise Electronic Design*, pp 22-23, New York NY, 1973, John Wiley & Sons.

[3]W. Hennigan, W3CZ, "Broadband Hybrid Splitters and Summers," Oct 1979 *QST*.

[4]W. Sabin, "Measuring SSB/CW Receiver Sensitivity," Oct 1992, *QST*. See also Technical Correspondence, Apr 1993 *QST*.

[5]W. Sabin, WØIYH, "A BASIC Approach to Calculating Cascaded Intercept Points and Noise Figure," Oct 1981 *QST*.

[6]Howard Sams & Co, Inc, *Reference Data for Radio Engineers,* Indianapolis, IN, p 29-2.

[7]J. Kraus and K. Carver, *Electromagnetics,* second edition, 1973, section 14-5, McGraw-Hill, NY.

[8]J. Grebenkemper, KI6WX, "Phase Noise and its Effects on Amateur Communications," Mar and Apr 1988 *QST*.

[9]W. Sabin, "Envelope Detection and AM Noise-Figure Measurement," Nov 1988 *RF Design* , p 29.

[10]H. Hyder, "A 1935 Ham Receiver," Sep 1986 *QST*, p 27.

[11]G. Breed, "A New Breed of Receiver," Jan 1988 *QST*, p 16.

[12]R. Campbell, "High-Performance, Single-Signal Direct-Conversion Receivers," Jan 1993 *QST* , p 32.

[13]A. Williams and F. Taylor, *Electronic Filter Design Handbook,* second edition, Chapter 7, McGraw-Hill, NY 1988.

[14]*RF Designer's Handbook,* Mini-Circuits Co, Brooklyn, NY.

[15]W. Sabin, "The Mechanical Filter in HF Receiver Design," Mar 1996 *QEX*.

[16]J. Makhinson, "A High-Dynamic-Range MF/HF Receiver Front End," Feb 1993 *QST*.

[17]W. Sabin and E. Schoenike, Eds., *Single-Sideband Systems and Circuits,* McGraw-Hill, 1987.

[18]B. Goodman, "Better AGC For SSB and Code Reception," Jan 1957 *QST*.

[19]A. Zverev, "Handbook of Filter Synthesis," Wiley & Sons, 1967.

[20]R. Pittman and G. Summers, "The Ultimate CW Receiver," Sep 1952 *QST*.

[21]J. Vermasvuori, "A Synchronous Detector for AM Reception," Jul 1993 *QST*.

[22]"The Collins 75S-3 Receiver," Product Review, Feb 1962 *QST*.

[23]"The 75A-4 Receiver," Product Review, Apr 1955 *QST*.

[24]S. Prather, "The Drake R-8 Receiver," Fall 1992 *Communications Quarterly*. Also see J. Kearman, "The Drake R-8 Shortwave Receiver," Mar 1992 *QST*, p 72.

[25]R. Zavrel, "Using the NE602," Technical Correspondence, May 1990 *QST*.

[26]R. Campbell, "A Single-Board, No-Tune 902 MHz Transverter," Jul 1991 *QST*.

[27]Z. Lau, "A No-Tune 222 MHz Transverter," Jul 1993 *QEX*.

[28]Z. Lau, "The Quest for 1 dB NF on 10 GHz," Dec 1992 *QEX*.

[29]Z. Lau, "Home-Brewing a 10-GHz SSB/CW Transverter," May and Jun 1993 *QST*.

[30]ARRL, *UHF/Microwave Experimenter's Manual,* 1990, p 6-50.

[31]M. Schwartz, *Information Transmission, Modulation and Noise,* third edition, McGraw-Hill, 1980.

[32]R. Healy, "The Omni VI Transceiver," Product Review, Jan 1993 *QST*, p 65.

A Rock-Bending Receiver for 7 MHz

This simple receiver by Randy Henderson, WI5W, originally published in Aug 1995 *QST*, is a direct-conversion type that converts RF directly to audio. Building a stable oscillator is often the most challenging part of a simple receiver. This one uses a tunable crystal-controlled oscillator that is both stable and easy to reproduce. All of its parts are readily available from multiple sources and the fixed-value capacitors and resistors are common components available from many electronics parts suppliers.

THE CIRCUIT

This receiver works by mixing two radio-frequency signals together. One of them is the signal you want to hear, and the other is generated by an oscillator circuit (Q1 and associated components) in the receiver. In **Fig 17.68**, mixer U1 puts out sums and differences of these signals and their harmonics. We don't use the sum of the original frequencies, which comes out of the mixer in the vicinity of 14 MHz. Instead, we use the frequency *difference* between the incoming signal and the receiver's oscillator—a signal in the audio range if the incoming signal and oscillator frequencies are close enough to each other. This signal is filtered in U2, and amplified in U2 and U3. An audio transducer (a speaker or headphones) converts U3's electrical output to audio.

How the Rock Bender Bends Rocks

The oscillator is a tunable crystal oscillator—a variable crystal oscillator, or V*XO*. Moving the oscillation frequency of a crystal like this is often called *pulling*. Because crystals consist of precisely sized pieces of quartz, crystals have long been called *rocks* in ham slang—and receivers, transmitters and transceivers that can't be tuned around due to crystal frequency control have been said to be *rockbound*. Widening this rockbound receiver's tuning range with crystal pulling made *rock bending* seem just as appropriate!

L2's value determines the degree of pulling available. Using FT-243-style crystals and larger L2 values, the oscillator reliably tunes from the frequency marked on the holder to about 50 kHz below that point with larger L2 values. (In the author's receiver a 25-kHz tuning range was achieved.) The oscillator's frequency stability is very good.

Inductor L2 and the crystal, Y1, have more effect on the oscillator than any other components. Breaking up L2 into two or three series-connected components often works better than using one RF choke. (The author used three molded RF chokes in series—two 10-µH chokes and one 2.7-µH unit.) Making L2's value too large makes the oscillator stop.

The author tested several crystals at Y1. Those in FT-243 and HC-6-style holders seemed more than happy to react to adjustment of C7 (TUNING). Crystals in the smaller HC-18 metal holders need more inductance at L2 to obtain the same tuning range. One tiny HC-45 unit from International Crystals needed 59 µH to eke out a mere 15 kHz of tuning range.

Input Filter and Mixer

C1, L1, and C2 form the receiver's input filter. They act as a peaked *low-pass* network to keep the mixer, U1, from responding to signals higher in frequency than the 40-meter band. (This is a good idea because it keeps us from hearing video buzz from local television transmitters, and signals that might mix with harmonics of the receiver's VXO.) U1, a Mini-Circuits SBL-1, is a passive diode-ring mixer. Diode-ring mixers usually perform better if the output is terminated properly. R11 and C8 provide a resistive termination at RF without disturbing U2A's gain or noise figure.

Audio Amplifier and Filter

U2A amplifies the audio signal from U1. U2B serves as an active low-pass filter. The values of C12, C13 and C14 are appropriate for listening to CW signals. If you want SSB stations to sound better, make the changes shown in the caption for Fig 17.68.

U3, an LM386 audio power amplifier IC, serves as the receiver's audio output stage. The audio signal at U3's output is more than a billion times more powerful than a weak signal at the receiver's input, so don't run the speaker/earphone leads near the circuit board. Doing so may cause a squealy audio oscillation at high volume settings.

CONSTRUCTION

If you're already an accomplished builder, you know that this project can be built using a number of construction techniques, so have at it! If you're new to building, you should consider building the Rock-Bending Receiver on a printed circuit (PC) board. (The parts list tells where you can buy one ready-made.) See **Fig 17.69** for details on the physical layout of several important components used in the receiver. **Fig 17.70** shows photos of two different receivers using two different approaches to construction — one using a PC board and the other using "ugly" techniques.

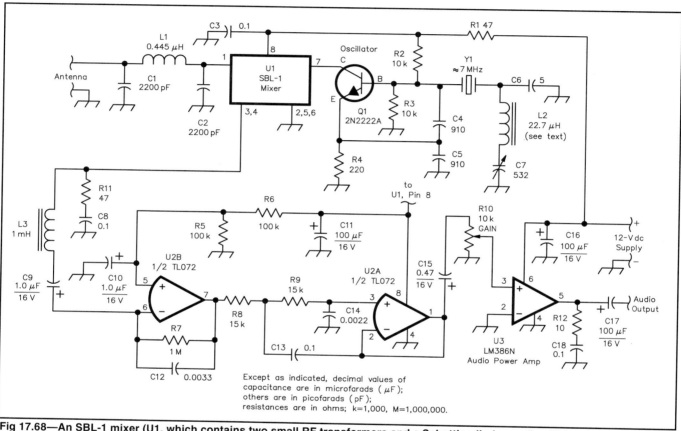

Fig 17.68—An SBL-1 mixer (U1, which contains two small RF transformers and a Schottky-diode quad), a TL072 dual op-amp IC (U2) and an LM386 low-voltage audio power amplifier IC (U3) do much of the Rock-Bending Receiver's magic. Q1, a variable crystal oscillator (VXO), generates a low-power radio signal that shifts incoming signals down to the audio range for amplification in U2 and U3. All of the circuit's resistors are ¼-W, 5%-tol types; the circuit's polarized capacitors are 16-V electrolytics, except C10, which can be rated as low as 10 V. The 0.1-μF capacitors are monolithic or disc ceramics rated at 16 V or higher.

C1, C2—Ceramic or mica, 10% tolerance.
C4, C5, and C6—Polystyrene, dipped silver mica, or C0G (formerly NP0) ceramic, 10% tolerance.
C7—Dual-gang polyethylene-film variable (266 pF per section) available as #24TR218 from Mouser Electronics (800-346-6873, 817-483-4422). Screws for mounting C7 are Mouser #48SS003. a rubber equipment foot serves as a knob. (Any variable capacitor with a maximum capacitance of 350 to 600 pF can be substituted; the wider the capacitance range, the better.)
C12, C13, C14—10% tolerance. For SSB, change C12, C13 and C14 to 0.001 μF.
U2—TL072CN or TL082CN dual JFET op amp.
L1—4 turns of AWG #18 wire on ¾-inch

PVC pipe form. Actual pipe OD is 0.85 inch. The coil's length is about 0.65 inch; adjust turns spacing for maximum signal strength. Tack the turns in place with cyanoacrylic adhesive, coil dope or Duco cement. (As a substitute, wind 8 turns of #18 wire around 75% of the circumference of a T-50-2 powdered-iron core. Once you've soldered the coil in place and have the receiver working, expand and compress the coil's turns to peak incoming signals, and then cement the winding in place.)
L2—Approximately 22.7 μH; consists of one or more encapsulated RF chokes in series (two 10-μH chokes [Mouser #43HH105 suitable] and one 2.7-μH choke [Mouser #43HH276 suitable]

used by author). See text
L3—1-mH RF choke. As a substitute, wind 34 turns of #30 enameled wire around an FT-37-72 ferrite core.
Q1—2N2222, PN2222 or similar small-signal, silicon NPN transistor.
R10—5 or 10-kΩ audio-taper control (RadioShack No. 271-215 or 271-1721 suitable).
U1—Mini-Circuits SBL-1 mixer.
Y1—7-MHz fundamental-mode quartz crystal. Ocean State Electronics carries 7030, 7035, 7040, 7045, 7110 and 7125-kHz units.

PC boards for this project are available from FAR Circuits (see the Address List in the References chapter). Price $5, plus $1.50 shipping (for up to four boards).

If you use a homemade double-sided circuit board based on the PC pattern in the **References** chapter, you'll notice that it has more holes than it needs to. The extra holes (indicated in the part-placement diagram with square pads) allow you to connect its ground plane to the ground traces on its foil side. (Doing so reduces the inductance of some of the board's ground paths.) Pass a short length of bare wire (a clipped-off component lead is fine) into each of these holes and solder on both sides. Some of the circuit's components (C1, C2 and others) have grounded leads accessible on both sides of the board. Solder these leads on both sides of the board.

Another important thing to do if you use a homemade double-sided PC board is to countersink the ground plane to clear all ungrounded holes. (Countersinking clears copper away from the holes so components won't short-circuit to the ground plane.) A 1/4-inch-diameter drill bit works well for this. Attach a control knob to the bit's shank and you can safely use the bit as a manual countersinking tool. If you countersink your board in a drill press, set it to about 300 rpm or less, and use very light pressure on the feed handle.

Mounting the receiver in a metal box or cabinet is a good idea. Plastic enclosures can't shield the TUNING capacitor from the presence of your hand, which may slightly affect the receiver tuning. You don't have to completely enclose the receiver—a flat aluminum panel screwed to a wooden base is an acceptable alternative. The panel supports the tuning capacitor, GAIN control and your choice of audio connector. The base can support the circuit board and antenna connector.

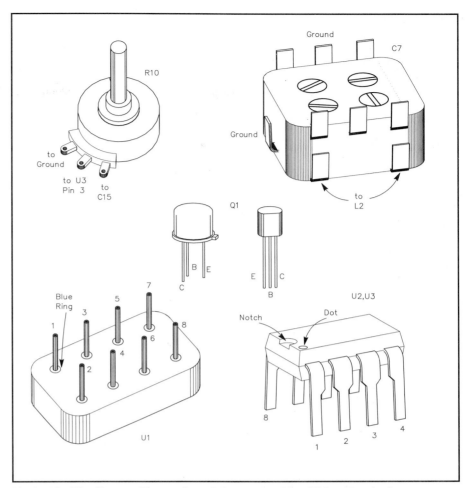

Fig 17.69—The Mouser Electronics part suggested for C7 has terminal connections as shown here. (You can use any variable capacitor with a maximum capacitance of 350 to 600 pF for C7, but its terminal configuration will differ from that shown here.) Two Q1-case styles are shown because plastic or metal transistors will work equally well for Q1. If you build your Rock-Bending Receiver using a prefab PC board, you should mount the ICs in 8-pin mini-DIP sockets rather than just soldering the ICs to the board.

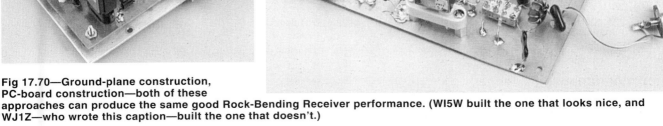

Fig 17.70—Ground-plane construction, PC-board construction—both of these approaches can produce the same good Rock-Bending Receiver performance. (WI5W built the one that looks nice, and WJ1Z—who wrote this caption—built the one that doesn't.)

CHECKOUT

Before connecting the receiver to a power source, thoroughly inspect your work to spot obvious problems like solder bridges, incorrectly inserted components or incorrectly wired connections. Using the schematic (and PC-board layout if you built your receiver on a PC board), recheck every component and connection one at a time. If you have a digital voltmeter (DVM), use it to measure the resistance between ground and everything that should be grounded. This includes things like pin 4 of U2 and U3, pins 2, 5, 6 of U1, and the rotor of C7.

If the grounded connections seem all right, check some supply-side connections with the meter. The connection between pin 6 of U3 and the positive power-supply lead should show less than 1 Ω of resistance. The resistance between the supply lead and pin 8 of U1 should be about 47 Ω because of R1.

If everything seems okay, you can apply power to the receiver. The receiver will work with supply voltages as low as 6 V and as high as 13.5 V, but it's best to stay within the 9 to 12-V range. When first testing your receiver, use a current-limited power supply (set its limiting between 150 and 200 mA) or put a 150-mA fuse in the connection be- tween the receiver and its power source. Once you're sure that everything is working as it should, you can remove the fuse or turn off the current limiting.

If you don't hear any signals with the antenna connected, you may have to do some troubleshooting. Don't worry; you can do it with very little equipment.

TROUBLE?

The first clue to look for is noise. With the GAIN control set to maximum, you should hear a faint rushing sound in the speaker or headphones. If not, you can use a small metallic tool and your body as a sort of test-signal generator. (If you have any doubt about the safety of your power supply, power the Rock-Bending Receiver from a battery during this test.) Turn the GAIN control to maximum. Grasp the metallic part of a screwdriver, needle or whatever in your fingers, and use the tool to touch pin 3 of U3. If you hear a loud scratchy popping sound, that stage is working. If not, then something directly related to U3 is the problem.

You can use this technique at U2 (pin 3, then pin 5) and all the way to the antenna. If you hear loud pops when touching either end of L3 but not the antenna connec- tor, the oscillator is probably not working. You can check for oscillator activity by putting the receiver near a friend's transceiver (both must be in the same room) and listening for the VXO. Be sure to adjust the TUNING control through its range when checking the oscillator.

The dc voltage at Q1's base (measured without the RF probe) should be about half the supply voltage. If Q1's collector voltage is about equal to the supply voltage, and Q1's base voltage is about half that value, Q1 is probably okay. Reducing the value of L2 may be necessary to make some crystals oscillate.

OPERATION

Although the Rock-Bending Receiver uses only a handful of parts and its features are limited, it performs surprisingly well. Based on tests done with a Hewlett-Packard HP 606A signal generator, the receiver's minimum discernible signal (by ear) appears to be 0.3 μV. The author could easily copy 1-μV signals with his version of the Rock-Bending Receiver.

Although most HF-active hams use transceivers, there are advantages in using separate receivers and transmitters. This is especially true if you are trying to assemble a simple home-built station.

A Wideband MMIC Preamp

This project illustrates construction techniques used in the microwave region (at and beyond 1 GHz). It also results in a neat "dc to daylight" preamplifier with many uses around your shack, not the least of which is monitoring the downlinks from Amateur Radio satellites. The original article was written by William Parmley, KR8L, in Nov 1997 *QST*.

The preamplifier uses the MAR-6 monolithic microwave integrated circuit (MMIC) manufactured by Mini-Circuits Labs. The MAR-6 is a four terminal, sur- face mount device (SMD) with an operat- ing frequency range from dc to 2 GHz, a noise figure of 3 dB, a gain of up to 20 dB, and input and output impedances of 50 Ω. The basic concept for the preamplifier and the construction techniques used to build it came from *The ARRL UHF/Microwave Experimenter's Manual*. The parts and cir- cuit board material in this project are readily available from sources such as Ocean State Electronics.

CIRCUIT DESCRIPTION

Fig 17.71 is the schematic for the preamplifier. C1 and C2 are dc blocking

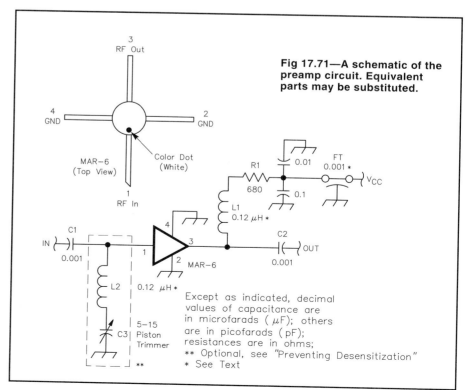

Fig 17.71—A schematic of the preamp circuit. Equivalent parts may be substituted.

capacitors. The device receives V_{CC} at the output lead, through RF choke L1 and limiting resistor R1. The only other components used are the bypass capacitors on the V_{CC} lead. C1 and C2 should present a low impedance at the lowest signal frequency of interest. The author designed his preamplifier for 435 MHz, a downlink frequency for many amateur satellites. Two 220 pF disc ceramic capacitors were used for C1 and C2. To use the preamplifier at 29 MHz for downlink signals from Russian RS-series satellites, C1 and C2 become 0.001 µF disc ceramic capacitors.

The power-supply voltage determines R1's value. The MAR-6 draws about 16 mA, and needs a Vcc of about 3.5 V. Use Ohm's Law to calculate the necessary voltage drop from your power supply voltage down to 3.5 V. The author's power supply provided about 14.6 V, so a 680 Ω, $^1/_2$-W resistor was used for R1. RF choke L1 helps isolate the power supply from the MMIC output. L1 is a homemade 0.12 µH choke, consisting of 8 turns of #30 enameled wire around the shank of a 3/16-inch drill bit, spaced for a total length of 0.3 inches. (Remove the drill bit; it's only a winding mandrel!)

This value of L1 was left in place when the preamplifier was used at lower frequencies. The remaining three essential parts are bypass capacitors. Because capacitors have self-resonant frequencies (resulting from unavoidable inductances in the devices and their leads), it is a common practice to use capacitors of several different values in parallel. This design uses a 0.001 µF feedthrough capacitor

passing through the circuit board ground plane. The parallel 0.01 µF and 0.1 µF capacitors are disc ceramics. L2 and C3 are optional components for 432 MHz, used to provide some selectivity against desensitization when transmitting on 144 MHz for satellite Mode J, at the expense of wideband coverage, of course.

CIRCUIT CONSTRUCTION

Fig 17.72 shows the circuit-board layout. The material used is double-sided, glass-epoxy board with a thickness of 0.0625 inches, known as FR-4 or G-10. This is the least expensive board material suited for microwave use. (The board I used is a product of GC/Thorsen in Rockford, Illinois.) Notice that most of the top of the board, and all the bottom side of the board, serves as circuit ground.

The signal-conducting part of the circuit is a "microstrip." (That is a strip-type transmission line: a conductor above or between extended conducting surfaces.—*Ed.*) The line width, board thickness and board dielectric constant determine the microstrip's characteristic impedance. A 0.1-inch-wide line and the ground plane on 0.0625-inch-thick G-10 form a 50-Ω transmission line, which matches the MMIC's input and output impedance.

The author fabricated his board by laying out the traces with a machinist's rule. Then he cut through the copper foil with a knife and lifted off the unwanted copper areas while heating them with a 100-W soldering gun. You could etch the board if you prefer, or use any other method you like. The single mounting pad was

"etched" by grinding away the copper with a hand-held grinder.

The MMIC is tiny. Connect it to the traces with the shortest possible distances between the traces and the body. (The author managed to achieve about 0.03 inch.) Also, the device leads are very delicate—if possible, do not bend them at all. To fit the MMIC leads flat on the PC-board traces without bending, a small depression was ground in the board dielectric for the MMIC body. Remember that, viewed from the top, the colored dot (white on the MAR-6) on the body marks pin 1, the input lead. The other leads are numbered counterclockwise; pin 3 is the output lead.

Mount the blocking capacitors as close to the board as possible. To do this, the capacitor leads were cut to about $^1/_{16}$-inch long. Both the capacitor leads and circuit traces were tinned and then the capacitors were soldered in place. This method of mounting minimizes lead inductance.

The author installed N connectors for his unit. To achieve a "zero lead length," he notched the ends of the board to fit the profile of the connectors and installed the connectors directly to the board. The center pin was laid on top of the microstrip and soldered. Then the connector body was soldered to the ground foil in four places: two on the top of the board and two on the bottom. Another very good technique is to drill a hole in the microstrip and insert the center pin from the bottom of the board. The center pin is then soldered to the microstrip, and the body is soldered to the ground foil or mounted with machine screws. (If you do this, be sure to remove a bit of foil from around the hole on the bottom side so the center pin doesn't short to ground.) The latter approach is much better if you want to mount the preamplifier into a box. You can mount the board on the inside of the lid with the connectors projecting through.

It is important that all portions of the ground foil be at equal potential, particularly near the MMIC and the board edges. To achieve this, wrap the long edges of the board with pieces of 0.003-inch-thick brass shim stock and solder them on both top and bottom. Thinner or thicker material is suitable (up to about 0.005 inch), as is copper flashing. Two small holes were drilled on either side of each MMIC ground lead, and a small Z-shaped wire was soldered to each side of the board. (A Z wire is a short, small-gauge, solid copper wire bent 90°, inserted through the hole, bent 90° again and soldered on both sides of the board.)

The inductor is also mounted using minimal-length leads. One lead connects to the microstrip and the other to the square

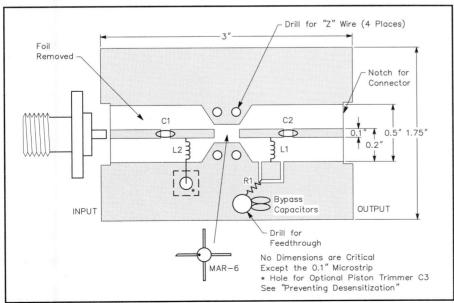

Fig 17.72—A part-placement diagram for KR8L's MMIC preamp. Dark areas are copper on the component (near) side. The reverse is a copper ground plane.

pad. The resistor connects from the pad to the feedthrough capacitor, and the other two bypass capacitors connect from the feed-through to the ground foil.

HOOKUP AND OPERATION

For the basic preamplifier design there is nothing to align or adjust. Simply connect the preamplifier between your antenna and receiver and apply power. If you connect the preamp to a transceiver, take precautions to prevent transmitting through the preamp! This preamplifier is very handy for many uses: adding gain to an older 10-meter receiver or scanner, boosting signal-generator output or for casual monitoring of the Amateur Radio satellites on 29, 145 and 435 MHz. A commercial metal box, home-made PC board or thin sheet-metal boxes make suitable cabinets for this project.

A Binaural I-Q Receiver

This little receiver was designed and built by Rick Campbell, KK7B. It was first described in the March 1999 issue of *QST*. It replaces the narrow filters and interference-fighting hardware and software of a conventional radio with a wide-open *binaural I-Q detector*. If you liken a conventional receiver to a high-powered telescope, this receiver is a pair of bright, wide-field binoculars. The receiver's classic junk-box-available-parts construction approach achieves better RF integrity than that of much commercial ham gear. A PC board and parts kit is available for those who prefer to duplicate a proven design.[1] The total construction time was only 17 hours. There are a number of toroids to wind, and performance was not compromised to simplify construction or reduce parts count. **Fig 17.73** is a photo of the front panel built by KK7B.

BINAURAL I-Q RECEPTION

Modern receivers use a combination of band-pass filters and digital signal processing (DSP) to select a single signal that is then amplified and sent to the speaker or headphones. When DSP is used, the detector often takes the form shown in **Fig 17.74**. The incoming signal is split into two paths, then mixed with a pair of local oscillators (LOs) with a relative 90° phase shift. This results in two baseband signals: an in-phase, or *I* signal, and a quadrature, or *Q* signal. Each of the two baseband signals contains all of the information in the upper and lower sidebands. The baseband pair also contains all of the information needed to determine whether a signal is on the upper or lower sideband before multiplication. An analog signal processor consisting of a pair of audio phase-shift networks and a summer could be used to reject one sideband. In a DSP receiver, the *I* and *Q* baseband signals are digitized and the resulting sets of numbers are phase-shifted and added.

The human brain is a good processor for information presented in pairs. We have two eyes and two ears. Generally speaking, we prefer to observe with both eyes open, and listen with both ears. This gives us depth of field and three-dimensional hearing that allows us to sort out the environment around us. The ear/brain combination can be used to process the output of the I-Q detectors as shown in **Fig 17.75**.

The sound of CW signals on a binaural I-Q receiver is like listening to a stereo recording made with two identical microphones spaced about six inches apart. The same information is present on each channel, but the *relative phase* provides a stereo effect that is perceived as three-dimensional space. Signals on different sidebands—and at different frequencies—appear to originate at different points in space. Because SSB signals are composed of many audio frequencies, they sound a little spread in the perceived three-dimensional sound space. This spreading also

Fig 17.73—A receiver with presence . . . to fully appreciate this receiver, you've got to hear it! "Once my ears got used to the effect, they had to drag me away from this radio. This is one I gotta have!"—*Ed Hare, W1RFI, ARRL Lab Supervisor*

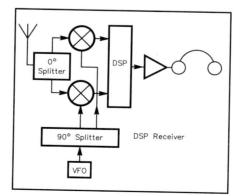

Fig 17.74—The simplified block diagram of a receiver using a DSP detector; see text.

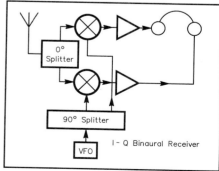

Fig 17.75—The block diagram of a binaural I-Q receiver that allows the ear/brain combination to process the detector output, resulting in stereo-like reception.

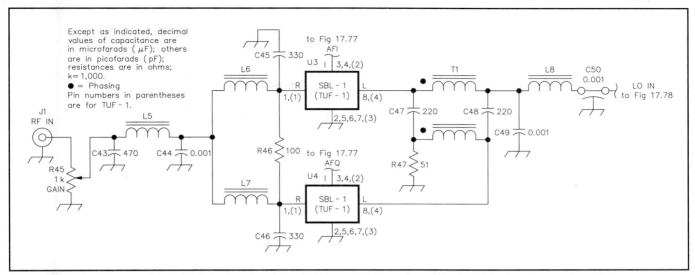

Fig 17.76—This diagram shows the front end and *I* and *Q* demodulators of the Binaural Weekender receiver. Unless otherwise specified, resistors are ¼ W, 5% tolerance carbon-composition or film units. Equivalent parts can be substituted. Pin connections for the SBL-1 and TUF-1 mixers at U3 and U4 are shown; the TUF-1 pin numbers are in parentheses. A kit is available (see Note 1). Parts are available from several distributors including Digi-Key Corp, Mouser Electronics, and Newark Electronics. See the References chapter for contact information.

C43—470 pF disc ceramic
C44, C49—0.001 µF metal polyester
C45, C46—330 pF disc ceramic
C47, C48—220 pF disc ceramic
C50—0.001 µF feed-through capacitor

J1—Chassis-mount female BNC connector
L5—1.6 µH, 24 turns #28 enameled wire on T-30-6 powdered-iron core
L6, L7—1.3 µH, 21 turns #28 enameled wire on T-30-6 powdered-iron core

L8—350 nH, 11 turns #28 enameled wire on T-30-6 powdered-iron core
R45—1 kΩ panel-mount pot
T1—17 bifilar turns #28 enameled wire on T-30-6 powdered-iron core
U3, U4—Mini-Circuits SBL-1 or TUF-1 mixer

occurs with most sounds encountered in nature, and is pleasant to hear.

To keep the receiver as simple as possible, a single-band direct-conversion (D-C) approach is used. A crystal-controlled converter can be added for operation on other bands, changing the receiver to a single-conversion superhet. Alternatively, the binaural I-Q detector can be used in a conventional superhet, with a tunable first converter and fixed-frequency BFO. If proper receiver design rules are followed, there is no advantage to either design over the other.

THE RECEIVER

Figs 17.76, **17.77** and **17.78** show the complete receiver schematic. In Fig 17.76, signals from the antenna are connected directly to a 1 kΩ GAIN pot on the front panel. J1 is a BNC antenna connector, popular with QRP builders. Adjusting the gain before splitting the signal path avoids the need for a two-gang volume control, and eliminates having to use separate RF and AF-gain adjustments. This volume-control arrangement leaves the "stereo background noise" constant and varies the signal-to-noise ratio. The overall gain is selected so that the volume is all the way up when the band is quiet. Resistor values R9 and R31 may be changed to modify the overall gain if required. After the volume

control, the signal is split with a Wilkinson divider and connected to two SBL-1 diode-ring mixers. (The TUF-1 is a better mixer choice, but I had more SBL-1s in my junk box.) The VFO signal is fed to the two mixers through a quadrature hybrid, described by Reed Fisher.[2] All of the circuitry under the chassis is broadband, and there are *no* tuning adjustments.

The audio-amplifier design of Fig 17.77 is derived from that used in the R1 High-Performance Direct-Conversion Receiver,[3] with appropriate simplifications. The R1 high-power audio output is not needed to drive headphones, the low-pass filter is eliminated, and the diplexer has fewer components. Distortion performance is not compromised—well over 60 dB of in-band two-tone dynamic range is available. The original article, and the additional notes in Technical Correspondence for February 1996,[4] describe the audio-amplifier chain in detail.

THE VFO

Fig 17.78 is the schematic of the receiver VFO, a JFET Hartley oscillator with a JFET buffer amplifier. Components for the VFO tuned circuit are chosen for linear tuning from 7.0 to 7.3 MHz with the available junk-box variable capacitor. Setting up the VFO is best done with a frequency counter, receiver and oscillo-

scope. The frequency counter makes it easy to select the parallel NP0 capacitors and squeezing and spreading the wire turns on L1 achieves the desired tuning range. After the tuning range is set, listen to the VFO signal with a receiver to make sure the VFO tunes smoothly and has a good note. Interrupt the power to hear its start-up chirp. The signal may sound ratty with the frequency counter on, so turn it off. The VFO is one area where craftsmanship pays off. Solid construction, a self-aligning variable-capacitor mounting, complete RF and air shielding and good capacitor bearings all contribute to a receiver that is a joy to tune.

Both connections to the VFO compartment are made with feed-through capacitors. The power supply connection is self-explanatory, but passing RF through a feed-through capacitor (at LO Out) may seem a bit unusual. Electrically, the capacitor is one element of a low-pass pi network. Using feed-through capacitors keeps local VHF signals (high-powered FM broadcast and TV signals near my location) out of the VFO compartment. A second pi network feeds the VFO signal to the detector circuit below the chassis. The use of VHF construction techniques in a 40 meter receiver may seem like overkill, but the present KK7B location is line-of-sight to broadcast towers serving the Port-

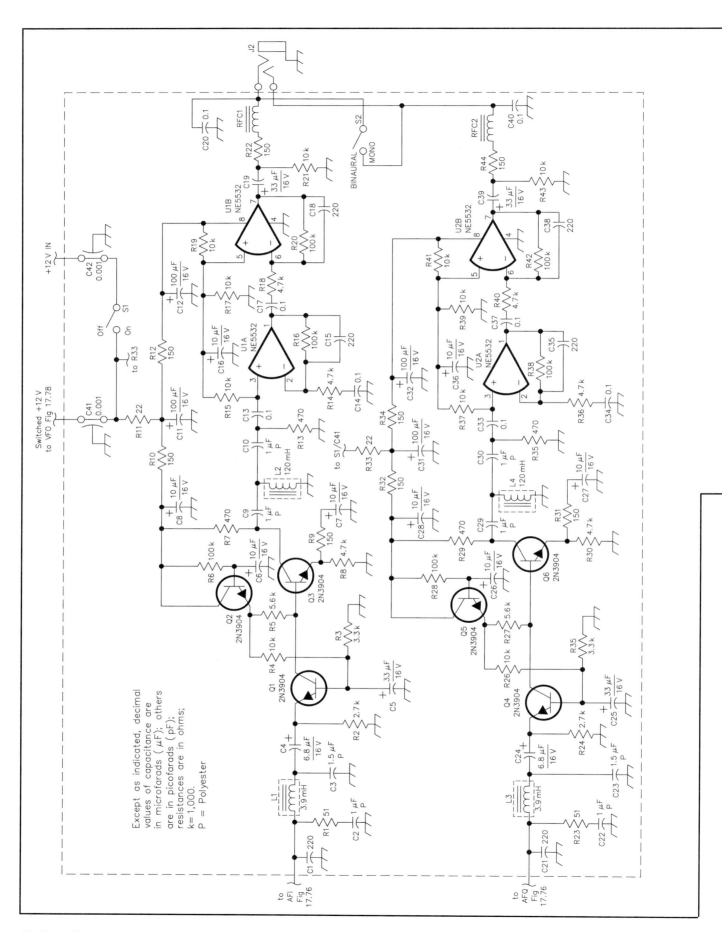

Fig 17.77—This diagram shows the receiver audio-amplifier design.

C1, C15, C18, C21, C35, C38—
220 pF disc ceramic

C2, C9, C10, C22, C29, C30—1 µF
metal polyester (Panasonic ECQ-
E(F) series)

C3, C23—1.5 µF metal polyester
(Panasonic ECQ-E(F) series)

C4, C24—6.8 µF, 16 V electrolytic
(Panasonic KA series)

C5, C19, C25, C39—33 µF, 16 V
electrolytic (Panasonic KA series)

C6, C7, C8, C16, C26, C27, C28,
C36—10 µF, 16 V electrolytic
(Panasonic KA series)

C11, C12, C31, C32—100 µF, 16 V
electrolytic (Panasonic KA series)

C13, C14, C17, C20, C33, C34, C37,
C40—0.1 µF metal polyester
(Panasonic V series)

C41, C42, C50—0.001 µF feedthrough capacitor

J2—1/8-inch stereo phone jack

L1, L3—3.9 mH Toko 10RB shielded
inductor

L2, L4—120 mH Toko 10RB shielded
inductor

Q1 through Q6—2N3904

RFC1, RFC2—10 turns #28
enameled wire on Amidon ferrite
bead FB 43-2401 (six-hole bead)

S1, S2—SPST toggle switch

U1, U2—NE5532 dual low-noise
high-output op amp

What Do You Hear?

Even the earliest solid-state direct-conversion (D-C) receivers had a *presence* or *clarity* that is rarely duplicated in more elaborate receivers. Many of us remember the first time we heard this crispness in a "homebrewed" D-C receiver. As we try to "enhance" our rigs through the addition of IF filters and other "features," we still hope that the result will be as clean as that first D-C receiver.

This binaural D-C receiver is such an experience—but even better. The binaural processing supplies the ears with additional information without compromising what was already there, enhancing the presence.

As you tune through a CW signal on a quiet band (best done with your eyes closed while sitting in a solid chair), a centered signal enters, but moves to the left background, undergoes circular motions at the back of your head as you tune through zero beat, repeats the previous gyrations on the right side, fades to the right background, and finally drops away in the center. Multiple signals within the receiver passband are distributed throughout this perceived space. With training, concentration on one signal allows it to be copied among the many. An SSB signal seems to occupy parts of the space, left and right, with clarity when properly tuned, leaving others vacant. Static crashes and white noise appear distributed throughout the entire space without well defined position. Receiver noise, although present, has no perceived position.

It's vital that this receiver include a front-panel switch to shift between binaural and monaural output. Although useful during the learning process, it becomes indispensable for the demonstrations that you will want to do. I used the switch to set up my son, Roger, KA7EXM, for the experience. We entered the shack and I handed him the headphones. He put one phone to just one ear, but I told him that he had to use both, that it would not work with just one. He put the phones on his head, casually tuned the receiver through the 40-meter CW band, removed the phones and commented, "Well, it sounds just like a direct-conversion receiver: A good one, but still just a direct-conversion receiver." I smiled and asked him to put the headphones on again. As I flipped the switch to the binaural position his hand reached out, seeking the support of the workbench. His facial expression became more serious. He eased into the chair and began tuning the receiver, very slowly at first. After a minute he took the headphones off, but remained speechless for a while—an unusual condition for Roger. Finally, he commented, "Wow! The appliance guys have never heard that!"

A builder of the Binaural Weekender should prepare for some truly unusual experiences.—*Wes Hayward, W7ZOI*

---

land, Oregon area. Using commercial HF gear with conventional bypassing under these circumstances provided disappointing results.

Fig 17.73 shows the prototype receiver front panel. Receiver controls are simple and intuitive. The ear/brain adjusts so naturally to binaural listening that I added a BINAURAL/MONO switch to provide a quick reminder of how signals sound on a conventional receiver. The switch acts much like the STEREO-MONO switch on an FM broadcast receiver—given the choice, it always ends up in the STEREO position!

The uses a pair of Koss SG-65 headphones with his receiver. They are not necessary, but have some useful features. First, at about $32, they are relatively inexpensive. Second, they have relatively high-impedance drivers, (90 Ω) so they can be driven at reasonable volume directly from an op amp. Finally, they make an attempt at low distortion. Other headphones in the same price bracket are acceptable, but some have much lower impedance and won't provide a very loud audio signal using the component values given in the schematic. Those $2.95 bubble-packed, throw-away headphones

are not a good choice! Audiophile headphones are fine, but don't really belong on an experimenter's bench. A stray clip-lead brushing across the wrong wire in the circuit can instantly burn out a driver and seriously ruin your day.

BUILDING A BINAURAL WEEKENDER

A few construction details are generally important, while others were determined by the components that happened to be in my junk box. The big reduction drive is delightful to use, but doesn't contribute to electrical performance. I purchased it at a radio flea market. The steel chassis provides a significant reduction in magnetic hum pickup, something that can be a problem if the receiver is operated near a power transformer. (Steel chassis are available from parts houses that cater to audio experimenters.) The VFO mounting and mushroom-can shield shown in **Fig 17.79**

are a simple way to eliminate mechanical backlash, keep radiated VFO energy off the antenna, prevent hand capacitance from shifting the tuning, and reduce VFO drift caused by air currents.

Experienced builders can duplicate this receiver simply using the schematic and construction techniques described here. Unlike a phasing receiver, there is no need to precisely duplicate the exact amplitudes and phases between the two channels. The ear/brain combination is the ultimate adaptive processor, and it quickly learns to focus on a desired signal and ignore interference. Small errors in phase and amplitude balance are heard as slight shifts in a signal's position. Standard-tolerance components may be used throughout.

One note about the kit version: A very good VFO can be built on an open PC board if the variable oscillator is not running on the desired output frequency. The Kanga kit VFO runs at one-half the

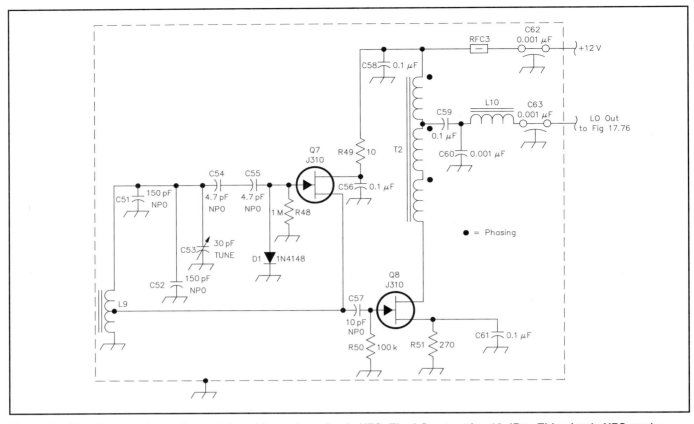

Fig 17.78—The diagram shows the prototype binaural receiver's VFO. The LO output is +10 dBm. This simple VFO works exceptionally well, but must be completely shielded for good D-C receiver performance. A receiver with an open PC-board VFO will work better if the variable oscillator is not running on the received frequency. As noted elsewhere, the kit version of the receiver uses a different VFO.

C51, C52—150 pF, NP0 disc ceramic
C53—30 pF air-dielectric variable
C54, C55—4.7 pF NP0 disc ceramic
C56, C57, C59, C61—0.1 μF metal polyester (Panasonic V series)
C57—10 pF NP0 disc ceramic
C60—0.001 pF metal polyester
C62, C63—0.001 μF feedthrough

capacitor
D1—1N4148
L9—1.5 μH, 22 turns #22 enameled wire on T-37-6 powdered-iron core; tap 5 turns from ground end.
L10—350 nH, 11 turns #28 on T-30-6 powdered-iron core

Q7, Q8—J310 (U310 used in prototype)
RFC3—10 turns #28 enameled wire on Amidon ferrite bead FB 43-2401 (six-hole bead used in prototype)
T2—10 trifilar turns #28 enameled wire on Amidon ferrite bead FB 43-2401 (six-hole bead used in prototype)

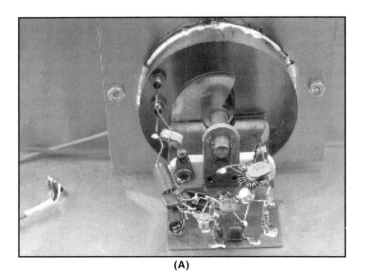
(A)

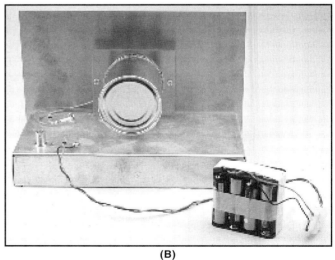
(B)

Fig 17.79 — A shows a close-up of the VFO. The simple VFO used in the prototype works exceptionally well, but must be completely shielded for good D-C receiver performance. B shows how an empty mushroom can live again as a VFO shield in the prototype receiver.

desired frequency, and is followed by a balanced frequency doubler and driver amplifier.

OTHER EXPERIMENTS

My earliest experiments with binaural detectors feeding stereo audio amplifiers were done in 1979, using two antennas. The technique works very well, but requires two antennas either physically spaced some distance apart, or of different polarization. Listening to OSCAR 13 on a binaural receiver with cross-polarized Yagis was an unsettling experience. The need for two antennas is a liability—these days most of us struggle to put up one. A number of experiments have also been done with binaural independent sideband (ISB) reception. These are profoundly interesting for AM broadcast reception, and could be used for amateur AM or DSB reception using a Costas Loop for carrier recovery. Binaural ISB detection of shortwave AM broadcasting can be analyzed as a form of spread spectrum with the ear/brain combination serving the despreading function, or as a form of frequency diversity, with the ear/brain as an optimal combiner.

The binaural techniques described here are analogous to binocular vision: They present the same information to each ear, but from a slightly different angle. This provides a very natural sound environment that the brain interprets as three-dimensional space. There are other "binaural" techniques that involve the use of different filter responses for the right and left ears. My experiments with different filter responses for the left and right ears have not been particularly interesting, and I have not pursued them.

SUMMARY

This little receiver is a joy to tune around the band. It is a serious *listening* receiver, and allows digging for weak signals in a whole new way. Digging for weak signals in a three-dimensional sound field is sometimes referred to as the "cocktail party effect." It is difficult to quantify the performance of a binaural receiver, because the final signal processing occurs in the brain of the listener—you. The experimental literature of psycho-acoustics suggests that the ear/brain combination provides a signal-to-noise advantage of approximately 3 dB when listening to speech or a single tone in the presence of uncorrelated binaural noise. The amount of additional noise in the opposite sideband is also 3 dB, so it appears that the binaural I-Q detector breaks even. In some applications, such as UHF weak-signal work, the binaural I-Q detector may have an advantage, as it permits listening to a larger slice of the band without a noise penalty. In other situations, such as CW sweepstakes, the "cocktail party" may get entirely out of hand. Binoculars and telescopes both have their place.

Notes

[1]The complete kit version, available from Kanga US, uses a different VFO circuit than the one shown here. The kit VFO runs at one-half the desired output frequency, and is followed by a balanced frequency doubler and driver amplifier.

Steel chassis such as the Hammond 1441-12 ($2 \times 7 \times 5$ inches [HWD]) with 1431-12 bottom plate and the Hammond 1441-14 ($2 \times 9 \times 5$ inches [HWD]) with 143-14 bottom plate are suitable enclosures. These chassis and bottom plates are not available in single quantities directly from Hammond, but are available from Allied Electronics and Newark Electronics. See the **References** chapter for contact information.

[2]Reed Fisher, W2CQH, "Twisted-Wire Quadrature Hybrid Directional Couplers," *QST*, Jan 1978, pp 21-23. See also IEEE Transactions MTT, Vol MTT-21, No. 5, May 1973, pp 355-357.

[3]Rick Campbell, KK7B, "High-Performance Direct-Conversion Receivers," *QST*, Aug 1992, pp 19-28.

[4]Rick Campbell, KK7B, "High-Performance, Single-Signal Direct-Conversion Receivers," *QST*, Jan 1993, pp 32-40. See also Feedback, *QST*, Apr 1993, p 75.

References

Campbell, Rick, KK7B, "Direct Conversion Receiver Noise Figure," Technical Correspondence, *QST*, Feb 1996, pp 82-85.

Campbell, Richard L., "Adaptive Array with Binaural Processor," *Proceedings of the IEEE Antennas and Propagation Society International Symposium*, Philadelphia, PA, June 1986, pp 953-956.

Campbell, Rick, KK7B, "Binaural Presentation of SSB and CW Signals Received on a Pair of Antennas," *Proceedings of the 18th Annual Conference of the Central States VHF Society*, Cedar Rapids, IA, July 1984, pp 27-33.

A Superregenerative VHF Receiver

Introduction

The complexity of many published receiver circuits has "scared off" some would-be home builders. Yet, making a receiver from scratch is an extremely rewarding experience. Charles Kitchin, N1TEV, put together this VHF receiver that can easily be built by the average person, and does not require any special components or test equipment. It covers roughly 118-170 MHz, and receives AM, WBFM and NBFM.

Don't expect to squeeze out the ultimate in selectivity or stability with this receiver, although it does provide a sensitivity of around 1 μV. It uses the principle of superregeneration for this high sensitivity with a low parts count. This design differs from other superregenerative circuits with an addition of a *quench waveform* control, which greatly helps to improve selectivity

for NBFM reception. Parts cost should be less than $20, and the receiver is powered by a standard 9-V battery.

This simple circuit is not going to compete with your modern digitally synthesized, multi-programmable transceiver but it is very useful (and fun) for monitoring your local 2-meter repeater, the aircraft band (118-137 MHz), marine radio, weather, police, snowplows, fire stations, telephone paging systems and many other types of local communications. As with any simple receiver, a certain amount of practice and patience is required in learning how to tune and adjust the set.

Regeneration and Superregeneration

Regenerative detectors are basically oscillators to which an input signal has been coupled. In a straight regenerative circuit, the input signal is coupled to the detector and then *regenerated* to very high levels by feeding back a portion of the output signal, in phase, back to its input, until just before or just at a critical point where a self sustaining oscillation begins. After that point, the circuit's gain stops increasing and starts going down, as most of the transistor's energy is devoted to generating its own oscillations.

The regeneration control is included to allow the operator to maintain the feedback at a point close to oscillation. With modern components, practical circuit gains of 20,000 are easily achieved in a single stage.

The superregenerative circuit uses an oscillating regenerative detector, which is periodically shut off or *quenched*.

Superregeneration allows the input signal to be regenerated over and over again, providing single-stage circuit gains of close to 1 million, even at UHF. Note the oscillations must be completely quenched each time, before being allowed to start-up again. Superregenerative detectors can use either a separate lower frequency oscillator to interrupt the detector (separately quenched) or as in this design, a single JFET is used to produce both oscillations (self quenched circuit).

Circuit Details

The receiver circuit shown in **Fig 17.80** uses a JFET RF stage with a separate dc return path for RF stage and detector. This uses a few more components than simply soldering a tap onto the coil, but it reduces the current drawn through the regeneration control. This allows the use of a Zener voltage regulator for the detector, which reduces the frequency drift due to changes in battery voltage.

The dual gang variable capacitor and toggle switch controls the band selected. A single section is used for the high band and the second section is simply switched in for low-band reception. The switch was soldered right onto the hot (not grounded)

terminal of the tuning capacitor. A series (mica) capacitor, with short leads was used to reduce the second section's total capacitance and set the desired frequency range of the low band.

Construction Details

Since this is a superregenerative receiver used at very high frequencies, stray circuit capacitances and multiple ground paths can prevent the detector from oscillating. It is important that the super-regenerative detector's tuning coil be physically located away from other objects—particularly chassis ground, the bottom and sides of the equipment box, if it is metal, and any shielding that may be present.

Avoid printed circuit boards: with all their components (especially the main tuning coil) mounted very close to the ground plane, the detector will be loaded down and may fail to oscillate properly, if it all. Instead, just use a piece of copper clad board and some standoffs. Suspend the components above the board on the standoffs or use the parts that have leads grounded to hold the other components above the board (this is known as the *dead bug* or *ugly* construction method). The parts are readily

available from most small parts suppliers listed in the **References** chapter.

Put the copper board inside a small box or use a block of wood and a piece of metal for the front panel. If you plan on placing the entire receiver inside a closed metal box, first build the circuit outside the box and be sure it oscillates before placing it inside.

Mount the tuning capacitor onto the copper board and pass its shaft through an oversized hole in the front panel. Mount all the other controls (except S1) to the front panel and connect them to the board using the shortest leads possible.

Receiver circuits are usually built backwards. Start with the audio stage. Wire it up as far back as the volume control and then test the stage by turning the control up halfway and placing your finger on the wiper (listen for a buzz). If it's not working, recheck the wiring or use a voltmeter to troubleshoot the problem. Be sure the supply voltage is present and that the voltage on pin 5 of the LM386 is at half supply.

After the audio stage is working, wire the detector and RF stage, but don't connect the two stages together (leave-out capacitor C2 in the schematic). Be sure to locate Q1 very close to L1 since C2 needs to be connected with very short

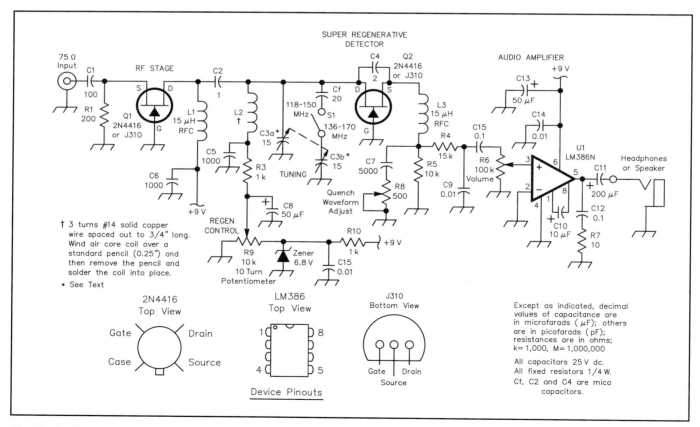

Fig 17.80—Schematic diagram of the receiver. C3 is a dual section, air-variable capacitor. C2 can be a fixed commercial capacitor, or a pair of twisted, insulated wires (see text). Cf can be increased to lower the low-frequency tuning range or decreased to increase this range. Compressing L2 will lower both tuning ranges.

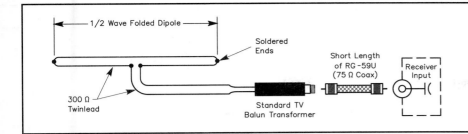

Fig 17.81—This simple dipole, made from 300-Ω TV line, will work fine. The TV line is attached to a coax feed line by a standard TV balun transformer. Make the dipole 39 inches long for the 2-meter band and a bit longer, 44.5 inches, for the aircraft band.

leads. Now, with no load on the detector, set R8 to mid position and turn up the regeneration control, R9, until oscillation occurs. You should hear a loud rushing noise. This indicates the detector is superregenerating.

L3 and C4 are the only components with critical values. Since your layout and construction will be different from the author's, some experimentation may be needed to get the detector to oscillate. Start with the values shown in the figure. Try changing the value of L3 first, followed by C4. Be sure the detector oscillates over the entire tuning range. R8 and R9 also affect oscillations. If you have any dead spots, you may want to move L2 and L3 farther away from ground. Once the values are set, no further changes should be needed. The C2 value shown, 1 pF, will work fine, but if you want to optimize the performance you will want to vary this. You also can use a 5-pF capacitor and tap down on the coil for best performance. Alternately, you can try using a pair of twisted, insulated wires for C2. Start with 4 turns, and increase the number until it sounds as though you are affecting the detector's oscillations.

Tuning the Receivers

For optimum sensitivity from these receivers, use "fresh" 9-V batteries. The frequency range can be adjusted by adding 10 or 15 pF in parallel with C3a (to move both the high range and low range down), in parallel with C3b (to move only the high range down) or reducing Cf to move the high range up. The turns on the main tuning coil, L1b, can be compressed or expanded to change both bands.

In these self-quenched circuits, the regeneration control also varies the quench frequency rate. For AM and wide-band FM reception this is a handy feature, since the operator only needs to adjust a single control. The setting for optimum sensitivity can usually be found by simply advancing the control past the detector's oscillation threshold and then to a point just below where the background (mush) noise suddenly begins to increase rapidly.

But for narrow-band FM reception, the second control is needed. Set R8 at midscale, then adjust the regeneration control for strong oscillation (high sensitivity). Next tune in the carrier of the desired station. After tuning to the center of the carrier, lower the regeneration level until the audio level increases sharply. If you turn the control too far down, the detector will squeal. The use of R8 creates a narrow-band "widow" on the regeneration control between the point where the

detector first begins to oscillate and the point where (narrow-band) audio begins to drop off rapidly. Increasing R8's resistance widens this region but lowers detector sensitivity. Because of their interaction, the regeneration control and the quench wave-form control both need to be adjusted for narrow-band FM reception.

Fig 17.81 shows a simple antenna for the receiver. Mount the dipole vertically, since most transmissions in this frequency range are vertically polarized.

Tuning Around

Yes, the construction technique is ugly. Yes, you will have to adjust R8 and R9 constantly for best reception. Yes, you will have to practice finding stations and calibrating the dial. But, you can build it in one or two evenings, hear hundreds of stations, AM, WBFM or NBFM, that transmit in this frequency range, and get surprisingly good performance from two FETs and one integrated circuit.

You can modify this receiver to cover the 6-m band by changing a few components. Connect both sections of C3 together, and increase L1 and L3 to 30 mH. Use 8 turns of #14 wire for L2. C2 becomes 2 pF and C4 5 pF. Compress or expand the turns of L2 to cover the band.

A 30/40 W SSB/CW 20-M Transceiver

This project is a unique *Handbook* offering. Unlike many projects, this one was not created to be duplicated by the home constructor. This project is for the confident and experienced builder who is looking for a challenge. The transceiver was created in the ARRL Lab by Zack Lau, W1VT (ex-KH6CP/1), to meet an unusual set of design goals.

Some old-timers may remember Project Goodwill. The project purchased and distributed QRP 20-m CW transceiver kits to prospective hams in many foreign countries with small ham populations. Those transceivers ran out a few years ago, and ARRL began development of a new transceiver.

The goal was to supply an inexpensive,

reliable, ready-built transceiver that provided moderate (not QRP) RF output in both the CW and SSB modes. **Fig 17.82** shows the transceiver block diagram. This transceiver was intended for inexpensive commercial manufacture. One of the important design goals was to eliminate as much tuning as possible, much like the microwave no-tune transverters. Of course, this conflicted with the next goal, which was to make the design as cheap as possible without sacrificing too much performance. Another goal was to accommodate a wide-tolerance PC-board process (single-sided with thick traces—50-mils minimum). Finally, the board was to use controls and connectors that are board-mounted.

In an attempt to simultaneously reduce the amount of tuning and keep costs down, a decision was made to use bilateral filters, mixers, and amplifiers where possible. It could be argued that the SBL-1s are too costly for an inexpensive transceiver project, but they offer several advantages. Being inherently bilateral, one needs only two mixers instead of four, reducing the number of parts needed. See the schematic in **Fig 17.83**. Also, they have pretty good carrier suppression, which eliminates a trimmer that would require adjustment, a very important consideration for this project. Finally, they are low-impedance devices, which makes broadband matching much easier. As a

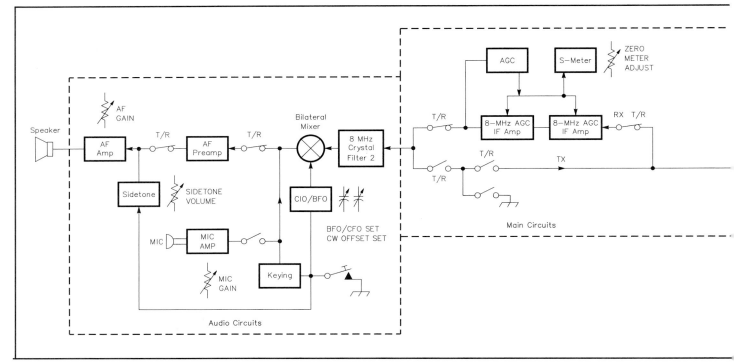

Fig 17.82—Block diagram of the transceiver.

bonus, they offer good dynamic range performance, sometimes outperforming NE602-based designs by 15 to 20 dB.

The no-tune requirement caused real problems in the power amplifier. The push-pull amplifier uses a pair of MRF477s that could probably be replaced by a single narrow-band MRF477 amplifier. A narrow-band amplifier would require adjustment, but one would get a significant savings. Additionally, a narrow-band amplifier would give more gain, allowing the driver stage to be run with a bit more feedback, increasing stability. The final amplifier shown is only conditionally stable, it will oscillate with severe mismatches. Fortunately, these transistors are pretty rugged, though you may wish to add some sort of SWR foldback circuitry for added insurance. To keep the costs down, such circuitry wasn't developed. A more complex bias circuit that independently biased the transistors might be better, but the simple circuit shown seems to work adequately with matched pairs of RF transistors.

The biasing circuit shows an interesting change in technology—it's probably more cost effective to use an integrated circuit and a variable resistor than to use the old alternative—a hand-selected 2-W resistor. Of course, if you just happen to have a junk box full of 2-W resistors, you're in good shape.

An 8-MHz IF was chosen, which allows the use of cheap microprocessor crystals and a 6-MHz VFO. Matching transformers are used because the mixer and amplifiers present close to 50-Ω terminations. Two separate four-pole filters are used to get a total of eight poles of filtering.

The cost cutting is most apparent in the simple AGC circuit. Initially, a hang-AGC circuit was used with good results, but it was removed to trim the cost as much as possible. An inferior audio-derived AGC circuit was used instead. One advantage of the audio-derived circuit is that it's less susceptible to BFO leakage, something that can easily be a problem when putting almost all the circuitry for an SSB transceiver on a single, single-sided PC board.

The audio amplifier is an old standby LM380 with an appropriate RC stabilization network on the output. While this IC does not have as much gain, or output, as newer chips, like the Signetics TDA 1015, it does seem easier to use. That's an important consideration when filling a big board with circuits that are bound to interact if you give them a chance. 2N5486 JFETs are used for audio switching. While it's certainly possible to reduce the parts count by using quad switching chips instead of transistors, ICs are more difficult to lay out properly in a complicated circuit. The sidetone is a triangle wave.

The VFO uses a JFET and a MOSFET buffer amplifier. Unfortunately MOSFETs are apparently being phased out, but you can still get them from a variety of surplus sources.

Zack built an "ugly" prototype on several unetched PC boards, and that circuit went through many changes as he worked with the manufacturer. What you see here is the circuit as it was when development stopped. It is

presented in the *Handbook* because there has been a consistent demand for an SSB transceiver with more that a few watts of output. Consider this a work in progress. If you build it, drop the *Handbook* Editor a note telling your experience and ideas. With a little more development, this could be a useful and widely popular project.

Construction

If the circuits are all working properly, alignment should be pretty easy. While there are two interacting adjustments for the band-pass filter, the BFO/carrier insertion oscillator and the VFO alignment, they aren't terribly difficult. Other adjustments are the S-meter zero, the sidetone volume, transmit gain, MRF477 bias setting, and mic-gain adjustment.

The bias setting for the finals is a trade-off, as one does get more gain and better IMD performance with a higher bias current, but the amount of heat generated is higher. A setting that results in a total current of 1 A during transmit seems to work well.

The mic gain should be adjusted for 30 W on voice peaks for good linearity, while 40 W is available on CW. One of the prototypes has a noise floor of −129 dBm.

1996 Updates

Lance, WS2B, tells us that Motorola recently discontinued the MRF 476 (Q23). NTE Cross Reference software gives the NTE236 as an alternative device. Dan's Small Parts may have the 40673 and the VN10KM.

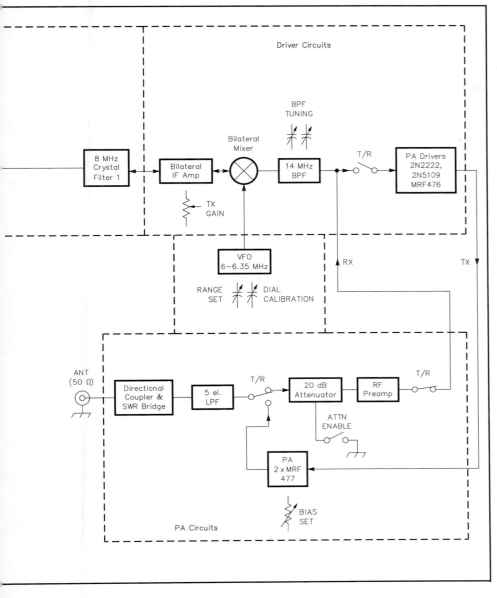

Fig 17.83—(see next two pages) Schematic diagram of the 20 m SSB/CW transceiver.

C1-5,C41-45—100-pF silver mica.
C7, C8, C65, C66—70-pF maximum ceramic or film trimmer capacitors.
C83—24-pF air trimmer (Johnson 189-509 used).
C84—50-pF tuning capacitor.
D6, D7, D8, D9, D10, D20, D21—1N4007 rectifier diodes used as PIN diodes.
D26, 27—1N4001 rectifier diodes thermally coupled to the nearest MRF477.
D28, D29—1N34 germanium diodes.

K1—Omron G2U-112-US SPDT relay or equiv.
L1, L2—17 t #22 enameled wire on a T-50-6 core.
L3—35 t #28 enameled wire on a T-50-6 core. Tap 8 turns up from ground end.
L4—13 t #28 enameled wire on a T-37-6 core.
L5, L7—12 t #22 enameled wire on a T-44-2 core.
L6—15 t #22 enameled wire on T-68-6 toroid core.

Q1, Q2, Q3, Q14, Q16, Q19, Q20—2N3904 NPN BJT.
Q4, Q5, Q12, Q15, Q18—2N3906 PNP BJT.
Q6, Q7, Q8, Q25—2N5486 N-channel JFET.
Q13—VN10KM VMOS FET, or equiv.
Q17—TIP 30 PNP BJT.
Q21—2N2222A metal cased NPN BJT.
Q22, Q24—2N5109 NPN BJT with heat sinks.
Q23—MRF476 NPN BJT with heat sinks.
Q26—40673 dual-gate MOSFET, or equiv.
Q27, Q28—MRF477 NPN BJT with appropriate heat sinks.
RFC1—1 mH toroidal RF choke. FT-37-72 toroid filled with one layer of #28 enameled wire.
RFC2—8 t #18 enameled wire on an FT-50-43 core.
T1, T2, T6, T7—12 t #28 enameled wire primary feeds crystal filter. 5 t #28 enameled wire secondary on an FT-37-43 core.
T3—20 t #28 enameled wire on an FT-37-43 core. Tap 7 t from collector end. Secondary is 4 t of #28 enameled wire over the primary.
T4, T5—18 t #28 enameled wire center-tapped primary on an FT-37-43 toroid. Secondary, 2 t #28 enameled wire. (One way of reducing the IF gain is to use only one half of the primary.)
T8, T9, T10, T13—5 t bifilar wound #28 enameled on an FT-37-43 toroid.
T11—5 t #24 enameled wire primary on an FT-37-43. Secondary, 3 t #22 enameled wire (to MRF 476).
T12—3 t #23 enameled wire primary on an FT-37-43 toroid. 5 t #22 enameled wire secondary (output).
T14—18 t #28 enameled wire primary. 5 t #28 secondary on an FT-37-43 core.
T15—Input is 3 t #24 enameled wire on a BLN-202-73 balun core. Output to transistors is 1 t center-tapped #24 enameled wire.
T16—8 t #20 bifilar wound on an FT-50-43 ferrite toroid.
T17—Output transformer. $^3/_{16}$-inch brass tubing and unetched circuit-board stock (double-sided) primary through 4 FB-63-43 ferrite beads. 4 t #20 enameled wire secondary through tubing.
T18, T19—31 t #24 enameled wire on a T-50-3 toroid. 1 t #24 enameled wire secondary.
U1, U9—SBL-1 double-balanced mixer.
U2—NE5514 quad op amp.
U3—LM380 audio amplifier.
U4, U5—MC1350 IF amplifier.
U6—LM358 dual op amp.
U10—78L05 regulator.
U8—LM393 dual comparator.
U11—LM317T adjustable regulator.
Y1, Y2, Y3, Y4, Y6, Y7, Y8, Y9—Matched 8-MHz microprocessor crystals (HC-18/U case). 250-Hz matching should be adequate.
Y5—8-MHz microprocessor crystal.

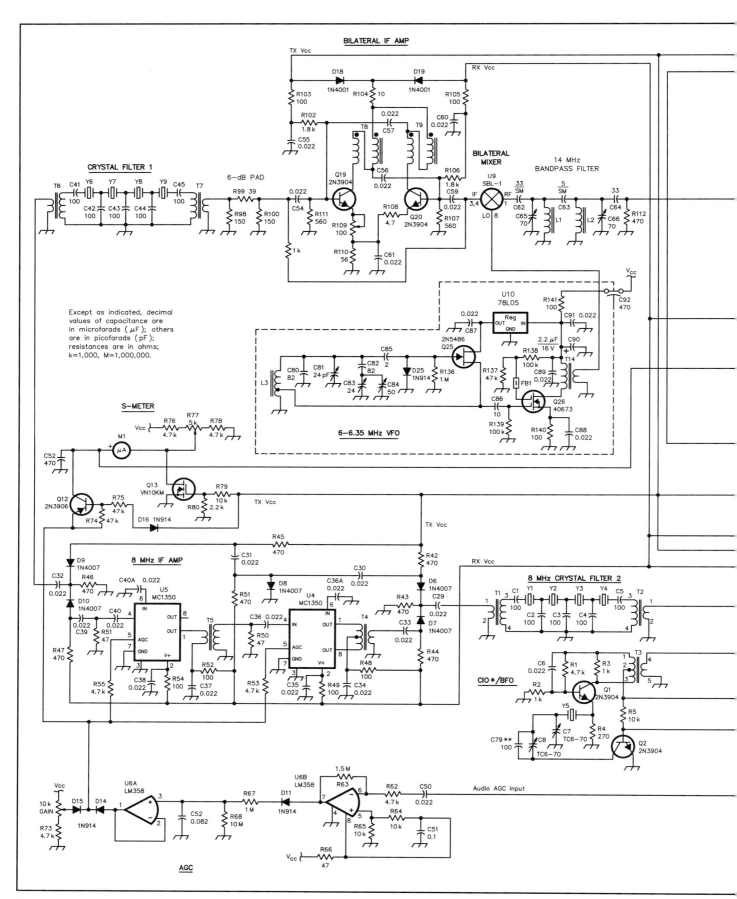

Fig 17.83—See previous page for details.

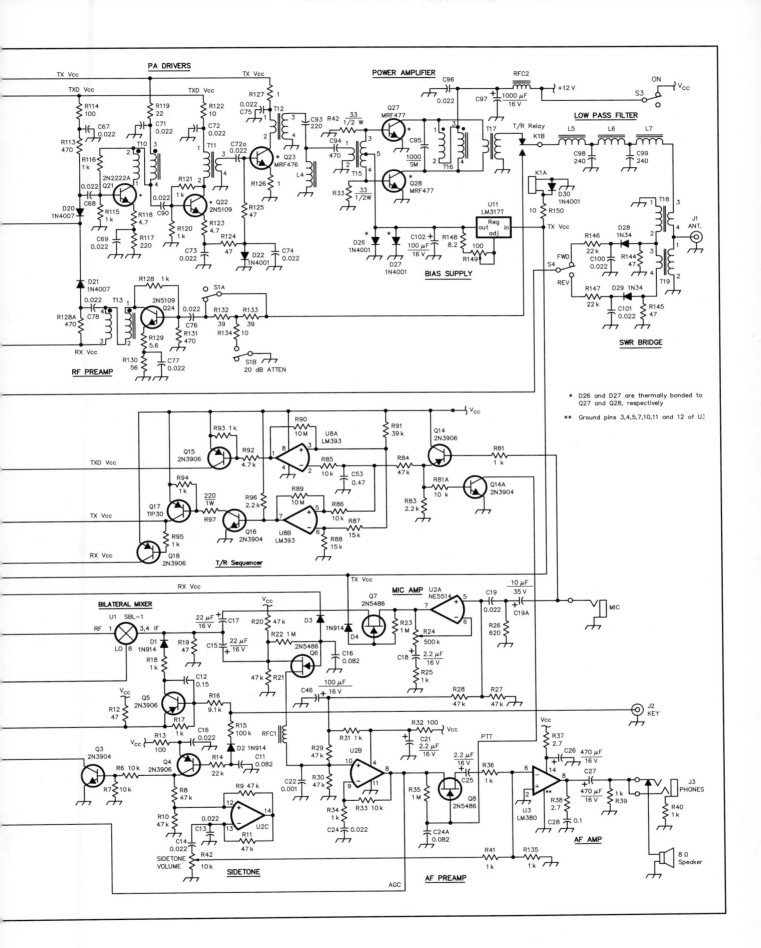

The NORCAL Sierra: An 80-15 m CW Transceiver

Most home-built QRP transceivers cover a single band, for good reason: complexity of the circuit and physical layout can increase dramatically when two or more bands are covered. This holds for most approaches to multiband design, including the use of multipole switches, transverters and various forms of electronic switching.[1]

If the designer is willing to give up instant band switching, then plug-in band modules can be used. Band modules are especially appropriate for a transceiver that will be used for extended portable operation, for example: back-packing. The reduced circuit complexity improves reliability, and the extra time it takes to change bands usually isn't a problem. Also, the operator need take only the modules needed for a particular outing.

The Sierra transceiver shown in **Fig 17.84** uses this technique, providing coverage of all bands from 80 through 15 m with good performance and relative simplicity.[2] The name Sierra was inspired by the mountain range of the same name—a common hiking destination for West Coast QRPers. The transceiver was designed and built by Wayne Burdick, N6KR, and field tested by members of NorCal, the Northern California QRP Club.[3]

Features

One of the most important features of the Sierra for the portable QRP operator is its low current drain. Because it has no relays, switching diodes or other active band-switching circuitry, the Sierra draws only 30 mA on receive.[4] Another asset for field operation is the Sierra's low-frequency VFO and premixing scheme, which provides 150 kHz of coverage and good frequency stability on all bands.

The receiver is a single-conversion superhet with audio-derived AGC and RIT. It has excellent sensitivity and selectivity, and will comfortably drive a speaker. Transmit features include full break-in keying, shaped keying and power output averaging 2 W, with direct monitoring of the transmitted signal in lieu of sidetone. Optional circuitry allows monitoring of relative power output and received signal strength.

Physically, the Sierra is quite compact—the enclosure is $2.7 \times 6.2 \times 5.3$ inches (HWD)—yet there is a large amount of unused space both inside and on the front and rear panels. This results from the use of PC board-mounted controls and connectors. The top cover is secured by quick-release plastic latches, which provide easy access to the inside of the enclosure. Band changes take only a few seconds.

Circuit Description

Fig 17.85 is a block diagram of the Sierra. The diagram shows specific signal frequencies for operation on 40 m. **Table 17.1** provides a summary of crystal oscillator and premix frequencies for all bands for all bands. The schematic is shown in **Fig 17.86**. See **Table 17.2** for band-module component values.

On all bands, the VFO range is 2.935 MHz to 3.085 MHz. The VFO tunes "backwards": At the low end of each band, the VFO frequency is 3.085 MHz. U7 is the premixer and crystal oscillator, while Q8 buffers the premix signal prior to injection into the receive mixer (U2) and transmit mixer (U8).

A low-pass filter, three band-pass filters and a premix crystal make up each band module. To make the schematic easier to follow, this circuitry is integrated into Fig 17.86, rather than drawn separately. J5 is the band module connector (see the note on the schematic).

The receive mixer is an NE602, which draws only 2.5 mA and requires only about 0.6 V (P-P) of oscillator injection at pin 6. An L network is used to match the receive mixer to the first crystal filter (X1-X4). This filter has a bandwidth of less than 400 Hz. The single-crystal second filter (X5) removes some of the noise generated by the IF amplifier (U7), a

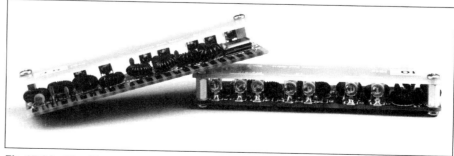

Fig 17.84—The Sierra transceiver. One band module is plugged into the center of the main PC board; the remaining boards are shown below the rig. Quick-release latches on the top cover of the enclosure make it easy to change bands.

Table 17.1

Crystal Oscillator and Premix (PMO) Frequencies in MHz

The premixer (U7) subtracts the VFO (2.935 to 3.085 MHz) from the crystal oscillator to obtain the PMO range shown. The receive mixer (U2) subtracts the RF input from the PMO signal, yielding 4.915 MHz. The transmit mixer (U8) subtracts 4.915 MHz from the PMO signal to produce an output in the RF range.

RF Range	Crystal Oscillator	PMO Range
3.500-3.650	11.500	8.415-8.565
7.000-7.150	15.000	11.915-12.065
10.000-10.150	18.000	14.915-15.065
14.000-14.150	22.000	18.915-19.065
18.000-18.150	26.000	22.915-23.065
21.000-21.150	29.000	25.915-26.065

technique W7ZOI described.[5] This second filter also introduces enough loss to prevent the IF amplifier from overdriving the product detector (U4).

The output of the AF amplifier (U3) is dc-coupled to the AGC detector. U3's output floats at $V_{cc}/2$, about 4 V, which happens to be the appropriate no-signal AGC voltage for the IF amplifier when it is operated at 8 V. C26, R5, R6, C76 and R7 provide AGC loop filtering. Like all audio-derived AGC schemes, this circuit suffers from pops or clicks at times.

Transmit signal monitoring is achieved by means of a separate 4.915 MHz oscillator for the transmitter; the difference between this oscillator and the BFO determines the AF pitch. Keying is exponentially shaped, with the rise time set by the turn-on delay of transmit mixer U8 and the fall time determined by C51, in the emitter of driver Q6.

Construction

The Sierra's physical layout and packaging make it relatively easy to build and align, although this isn't a project for the first-time builder. The boards and custom enclosure described here are included as part of an available kit.[6] Alternative construction methods are discussed below.

With the exception of the components on the band module, all of the circuitry for the Sierra is mounted on a single 5 × 6 inch PC board. This board contains not only the components, but all of the controls and connectors as well. The board is double-sided with plated-through holes, which permits flexible arrangement of the circuitry while eliminating nearly all hand-wiring. The only two jumpers on the board, W1 and W2, are short coaxial cables between the RF GAIN control and the receiver input filters.

A dual-row edge connector (J5) provides the interface between the main board and the band module. The 50 pins of J5 are used in pairs, so there are actually only 25 circuits (over half of which are ground connections).

The band module boards are 1.25 × 4 inches (HW). They, too, are double-sided, maximizing the amount of ground plane. Because the band modules might be inserted and removed hundreds of times over the life of the rig, the etched fingers that mate with J5 are gold-plated. Each etched finger on the front is connected to the corresponding finger on the back by a plated through hole, which greatly

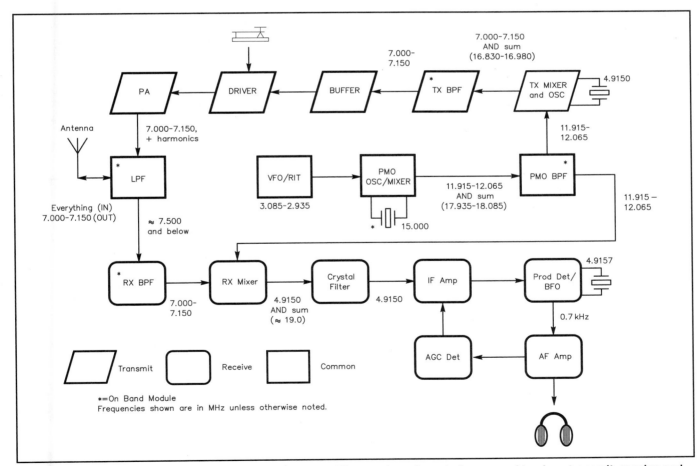

Fig 17.85—Block diagram of the Sierra transceiver. Three different-shaped symbols are used to show transmit, receive and common blocks. Those blocks with an asterisk (*) are part of the band module. Signal frequencies shown are for 40 m; see Table 17.1 for a list of crystal oscillator and premix frequencies for all bands.

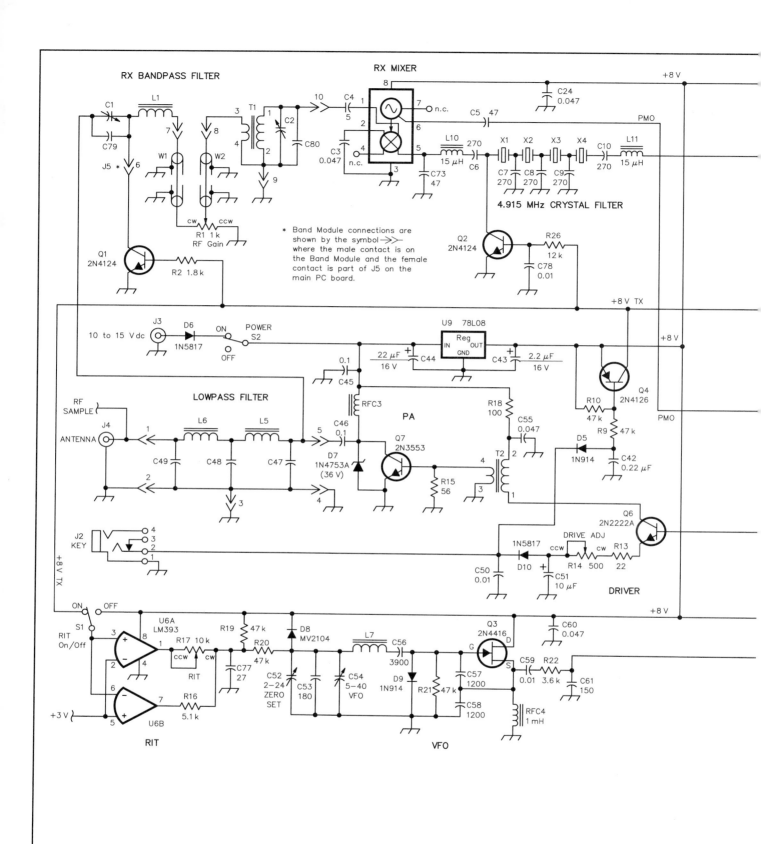

Fig 17.86—(See page 17.86 for details.)

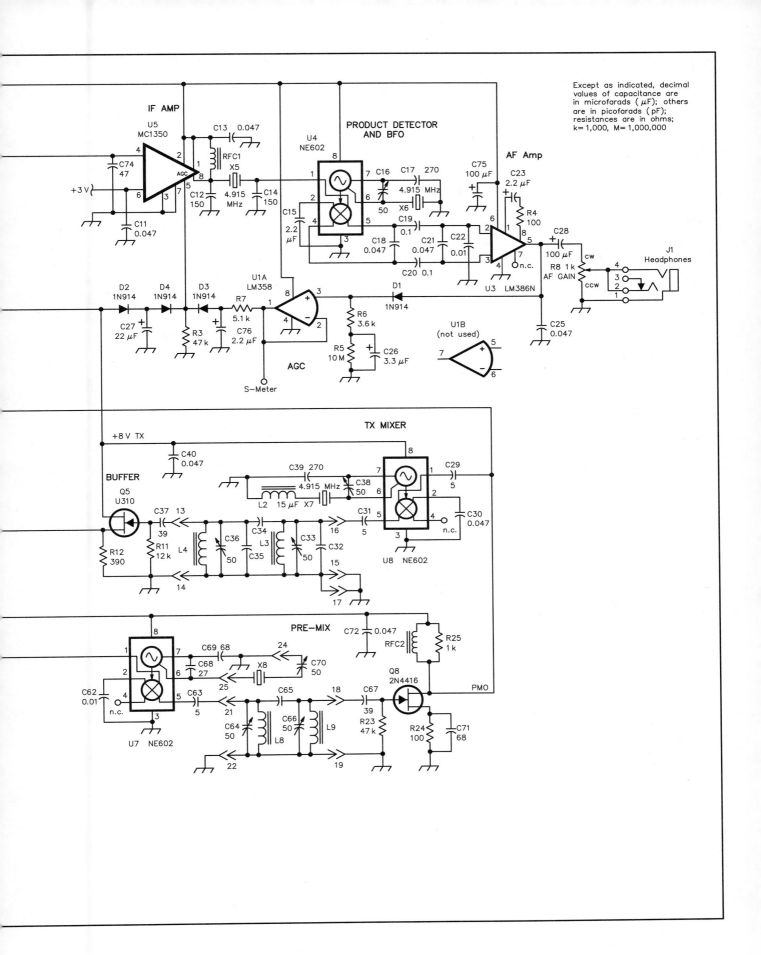

Fig 17.86—(see previous two pages) Schematic of the Sierra transceiver. Parts that change for each band are shown in Table 17.5.

C1, C2, C33, C36, C64, C66, C70— 9-50 pF right-angle-mount ceramic trimmer (same for all band modules, Mouser 24AA084)

C16, C38—Ceramic trimmer, 8-50 pF (Mouser 24AA024).

C52—Air variable, 2-24 pF (Mouser 530-189-0509-5).

C53—Disc, 180 pF, 5%, NP0.

C54—5-40 pF air variable with 8:1 vernier drive.

C56—Polystyrene, 3900 pF, 5%.

C57, C58—Polystyrene, 1200 pF, 5%.

D6, D10—1N5817, 1N5819 or similar .

D7—36 V, 1 W Zener diode (Mouser 333-1N4753A).

D8—MV2104 varactor diode, or equivalent.

J1, J2—PC-mount 3.5-mm stereo jack with switch (Mouser 161-3500).

J3—2.1-mm dc power jack (Mouser 16PJ031).

J4—PC-mount BNC jack (Mouser 177-3138).

J5—50 PIN, dual-row edgeboard connector with 0.156-inch spacing (Digi-Key S5253-ND).

L10, L11—18 µH; 18 t #28 enameled wire on an FT-37-61 toroid.

L2—Miniature RFC, 15 µH (Mouser, 43LS185).

L7—19 µH; 58 t #28 enameled wire on a T-68-7 toroid.

Q5—U310, J310, 2N4416 or other high-transconductance device.

R1, R8—PC-mount 1-kΩ pot (Mouser 31CW301).

R14, R101—500 Ω trimmer (Mouser 323-4295P-500).

R17—PC-mount 10-kΩ pot (Mouser 31CW401).

RFC1—3.5 µH; 8 t #26 enameled wire on an FT-37-61 toroid.

RFC2—7 µH; 4 t #26 enameled wire on an FT-37-43 toroid.

RFC3—34 µH; 9 t #26 enameled wire on an FT-37-43 toroid.

RFC4—Miniature RFC, 1 mH (Mouser 43LS103).

S1, S2—SPDT, PC mount, right angle toggle switch with threaded bushing. C&K 7101SDAV2QE is used in the kit; Digi-Key CKN1059-ND will work but does not have a threaded bushing.

T2—Primary: 12 t #26 enameled wire; secondary: 3 t on an FT-37-43 toroid.

U1—LM358N dual op-amp IC.

U2, U4, U7, U8—NE602AN mixer-oscillator IC.

U3—LM386N-1 audio amplifier IC.

U5—MC1350P IF amplifier IC.

U6—LM393N dual comparator IC.

U9—8 V regulator, TO-92 package (Digi-Key AN78L08-ND).

W1, W2—RG-174 coaxial jumper, about 3 inches long (see text).

X1-X7—4.915 MHz, HC-49 (Digi-Key CTX050). X1 through X5 should be matched (their series-resonant frequencies within 50 Hz).

improves reliability over that of a single finger contact.

Each band module requires eight toroids: two for the low-pass filter, and two each for the receive, transmit and premix band-pass filters. The builder can secure the toroids to the band module with silicone adhesive or Q-dope. Right-angle-mount trimmer capacitors allow alignment from above the module. Each band module has a top cover made of PC board material. The cover protects the components during insertion, removal and storage.

The VFO capacitor is a 5-40 pF unit with a built-in 8:1 vernier drive. The operating frequency is read from a custom dial fabricated from 0.060-inch Lexan. The dial mounts on a hub that comes with the capacitor.

The Sierra's custom 0.060-inch aluminum enclosure offers several benefits in both construction and operation. Its top and bottom covers are identical U-shaped pieces. The bottom is secured to the main board by two 0.375-inch standoffs, while the top is secured to the bottom by two long-life, quick-release plastic latches. As a result, the builder can easily remove both covers to make "live" adjustments or signal measurements without removing any controls, connectors or wires. The front and rear panels attach directly to the controls and connectors on the main board. This keeps the panels rigid and properly oriented.

Table 17.2
Band Module Components

All crystals are fundamental, 15-pF load capacitance, 0.005% frequency tolerance, in HC-49 holders. Fixed capacitors over 5 pF are 5% tolerance. All coils are wound with enameled wire.

			Band			
Part	80 m	40 m	30 m	20 m	17 m	15 m
C32, C35	33 pF, 5%	47 pF, 5%	not used	not used	not used	not used
C34	5 pF, 5%	5 pF, 5%	2 pF, 5%	2 pF, 5%	2 pF, 5%	2 pF, 5%
C47, C49	820 pF, 5%	330 pF, 5%	330 pF, 5%	220 pF, 5%	150 pF, 5%	150 pF, 5%
C48	1800 pF, 5%	820 pF, 5%	560 pF, 5%	470 pF, 5%	330 pF, 5%	330 pF, 5%
C65	5 pF, 5%	5 pF, 5%	2 pF, 5%	1 pF, 5%	1 pF, 5%	1 pF, 5%
L1	50 µH, 30 t #28 on FT-37-61	14 µH, 16 t #26 on FT-37-61	5.2 µH, 36 t #28 on T-37-2	2.9 µH, 27 t #28 on T-37-2	1.7 µH, 24 t #28 on T-37-6	1.9 µH, 25 t #28 on T-37-6
L3, L4	32 µH, 24 t #26 on FT-37-61	5.2 µH, 36 t #28 on T-37-2	4.4 µH, 33 t #28 on T-37-2	2.9 µH, 27 t #28 on T-37-2	1.7 µH, 24 t #28 on T-37-6	1.9 µH, 25 t #28 on T-37-6
L5, L6	2.1 µH, 23 t #26 on T-37-2	1.3 µH, 18 t #26 on T-37-2	1.0 µH, 16 t #26 on T-37-2	0.58 µH, 12 t #26 on T-37-2	0.43 µH, 12 t #26 on T-37-6	0.36 µH, 11 t #26 on T-37-6
L8, L9	8.0 µH, 12 t #26 on FT-37-61	2.5 µH, 25 t #28 on T-37-2	1.6 µH, 20 t #28 on T-37-2	1.3 µH, 18 t #26 on T-37-2	0.97 µH, 18 t #26 on T-37-6	0.87 µH, 17 t #28 on T-37-6
T1 (Sec same as L1)	Pri: 2 t #26 on FT-37-61	Pri: 1 t #26 on FT-37-61	Pri: 3 t #26 on T-37-2	Pri: 2 t #26 on T-37-2	Pri: 2 t #26 on T-37-6	Pri: 2 t #26 on T-37-6
X8	11.500 MHz (ICM 434162)	15.000 MHz (ICM 434162)	18.000 MHz (ICM 434162)	22.000 MHz (ICM 435162)	26.000 MHz (ICM 436162)	29.000 MHz (ICM 436162)

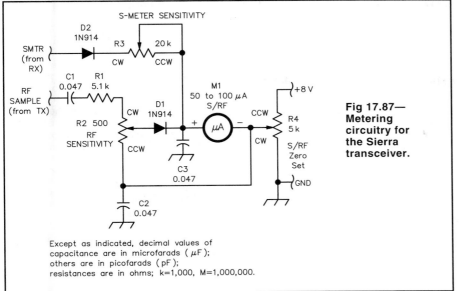

S-METER SENSITIVITY

Fig 17.87— Metering circuitry for the Sierra transceiver.

Except as indicated, decimal values of capacitance are in microfarads (μF); others are in picofarads (pF); resistances are in ohms; k=1,000, M=1,000,000.

As can be seen in the photograph, the interior of the rig is uncluttered. NorCal QRP Club members have taken advantage of this, building in keyers, frequency counters and other accessories—and even storing up to four band modules in the top cover. One popular addition is an S/RF meter, the circuit shown in **Fig 17.87**.

The construction techniques described above represent only one way to build the Sierra; other physical layouts may better suit your needs. For example: If no built-ins are needed, the rig could be built in a smaller enclosure. You could replace the VFO capacitor with a small 10-turn pot and a varactor diode. If necessary, eliminate RIT and metering.

If a different physical layout is required, determine the orientation and mounts for the band module connector first, then arrange the various circuit blocks around it. Use short leads and good ground-plane techniques to avoid instability, especially on the band modules. Point-to-point or "dead-bug" construction are possible, but in some cases shields and additional decoupling may be required. Use a reliable connector if band modules will be repeatedly inserted and removed.

Alignment

The minimum recommended equipment for aligning the rig is a DMM with home-made RF probe and a ham-band transceiver. Better still is a general-coverage receiver or frequency counter.[7] Start with a 40- or 20-m module; these are usually the easiest to align.

First, set the VFO to the desired band edge by adjusting C52. If exactly 150 kHz of range is desired, squeeze or spread the windings of L7 and readjust C52 iteratively until this range is obtained. RIT operation can also be checked at this time. Reduce the value of R19 if more RIT range is desired.

Prepare each band module for alignment by setting all of its trim caps to midrange. (The final settings will be close to midpoint in most cases.)

Receiver alignment is straightforward. Set BFO trimmer C16 to midrange, RF GAIN (R1) to maximum and AF GAIN (R8) so that noise can be heard on the phones or speaker. On the band module, peak the premix trimmers (C64 and C66) for maximum signal level measured at Q8's drain. Set the fine frequency adjustment (C70) by lightly coupling a frequency counter to U7, pin 7. Next, connect an antenna to J4 and adjust the receiver filter trimmers (C1 and C2) for maximum signal. The AGC circuitry normally requires no adjustment, but the no-signal gain of the IF amplifier can be increased by decreasing the value of R3.

Before beginning transmitter alignment, set the drive-level control, R14, to minimum. Key the rig while monitoring the transmitted signal on a separate receiver and peak the transmit band-pass filter using C33 and C36. Then, with a dummy load or well-matched antenna connected to J4, set R14 to about 90% of maximum and check the output power level. It may be necessary to stagger-tune C33 and C36 on the lower bands in order to obtain constant output power across the desired tuning range. On 80 m the –3 dB transmit band-

width will probably be less than 150 kHz.

Typically, output on 80, 40 and 20 m is 2.0-2.5 W, and on the higher bands 1.0-2.0 W. Some builders have obtained higher outputs on all bands by modifying the band-pass filters. However, filter modification may compromise spectral purity of the output, so the results should be checked with a spectrum analyzer. Also, note that the Sierra was designed to be a 2-W rig: additional RF shielding and decoupling may be required if the rig is operated at higher power levels.

PERFORMANCE

The Sierra design uses a carefully selected set of compromises to keep complexity low and battery life long. An example is the use of NE602 mixers, which affects both receive and transmit performance. On receive, the RF gain will occasionally need reduction when strong signals overload the receive mixer. On transmit, ARRL Lab tests show that the rig complies with FCC regulations for its power and frequency ranges.

Aside from the weak receive mixer, receiver performance is very good. There are no spurious signals (birdies) audible on any band. ARRL Lab tests show that the Sierra's receiver has a typical MDS of about –139 dBm, blocking dynamic range of up to 112 dB and two-tone dynamic range of up to 90 dB. AGC range is about 70 dB.

The Sierra's transmitter offers smooth break-in keying, along with direct transmit signal monitoring. There are two benefits to direct monitoring:

- the clean sinusoidal tone is easier on the ears than most sidetone oscillators and
- the pitch of the monitor tone is the correct receive-signal pitch to listen for when calling other stations.

The TR mute delay capacitor, C27, can be reduced to as low as 4.7 μF to provide faster break-in keying if needed.

The prototype Sierra survived its christening at Field Day, 1994, where members of the Zuni Loop Expeditionary Force used it on 80, 40, 20 and 15 m. There, Sierra compared favorably to the Heath HW-9 and several older Ten-Tec rigs, having as good or better sensitivity and selectivity—and in most cases better-sounding sidetone and break-in keying. While the other rigs had higher output power, they couldn't touch the Sierra's small size, light weight and low power consumption. The Sierra has consistently received high marks from stations worked too, with reports of excellent keying and stability.

CONCLUSION

At the time this article was written, over 100 Sierras had been built. Many have been used extensively in the field, where the rig's unique features are an asset. For some builders, the Sierra has become the primary home station rig.

The success of the Sierra is due, in large part, to the energy and enthusiasm of the members of NorCal, who helped test and refine early prototypes, procured parts for the field-test units and suggested future modifications.[8] This project should serve as a model for other clubs who see a need for an entirely new kind of equipment, perhaps something that is not available commercially.

Notes

[1]One of N6KR's previous designs, the Safari-4, is a good example of how complex a band-switched rig can get. See "The Safari-4...." Oct through Dec 1990 *QEX*.

[2]Band modules for 160, 12 and 10 m have also been built. Construction details for these bands are provided in the Sierra information packet available from the ARRL. Go to the ARRL Web page: **http://www.arrl.org/notes/** to download a template package or write to the ARRL Technical Department Secretary and request the '96 Handbook Sierra template package.

[3]For information about NorCal, write to Jim Cates, WA6GER, who is in the **References** Address List. Please include an SASE.

[4]Most multiband rigs draw from 150 to 500 mA on receive, necessitating the use of a larger battery. A discussion of battery life considerations can be found in "A So-lar-Powered Field Day," May 1995 *QST*.

[5]Solid-State Design, p 87.

[6]Full and partial kits are available. The full kit comes with all components, controls, connectors, and a detailed assembly manual. Complete band modules kits are available for 80, 40, 30, 20, 17 and 15 m. For information, write to Wilderness Radio (see Address List in the **References** chapter).

[7]The alignment procedure given here is necessarily brief. More complete instructions are provided with the ARRL template package and the kit. See note 2 to obtain a template.

[8]The author would like to acknowledge the contributions of several NorCal members: Doug Hendricks, KI6DS; Jim Cates, WA6GER; Bob Dyer, KD6VIO; Dave Meacham, W6EMD; Eric Swartz, WA6HHQ, Bob Warmke, W6CYX; Stan Cooper, K4DRD; Vic Black, AB6SO; and Bob Korte, KD6KYT.

A Broadband HF Amplifier Using Low-Cost Power MOSFETs

Many articles have been written encouraging experimenters to use power MOSFETs to build HF RF amplifiers.[1-8] That's because power MOSFETs—popular in the design of switching power supplies—cost as little as $1 each, whereas RF MOSFET prices start at about $35 each!

Mike Kossor, WA2EBY, designed and built this amplifier after hundreds of hours experimenting with power MOSFETs. The construction projects described in Notes 1 to 8, provide useful information about MOSFETs and general guidelines for working with them, including biasing, parasitic-oscillation suppression, broadband impedance-matching techniques and typical amplifier performance data.

With the design described here, 1 W of input power produces over 40 W of output (after harmonic filtering) from 160 through 10 meters. In addition to the basic amplifier, there is an RF-sensed TR relay and a set of low-pass filters designed to suppress harmonic output and comply with FCC requirements. The amplifier is built on double-sided PC board and requires *no tuning*. Another PC board contains the low-pass filters. Power-supply requirements are 28 V dc at 5 A, although the amplifier performs well at 13.8 V dc.

There are no indications of instability, no CW key clicks and no distortion on SSB has been reported by stations contacted while using the amplifier. To make it easy for you to duplicate this project, PC boards and parts kits are available, all at a cost of about $100![9] Etching patterns and parts-placement diagrams are included in the **References** chapter.

AN OVERVIEW OF MOSFETS

MOSFETs operate very differently from bipolar transistors. MOSFETs are voltage-controlled devices and exhibit a very high input impedance at dc, whereas bipolar transistors are current-controlled devices and have a relatively low input impedance.

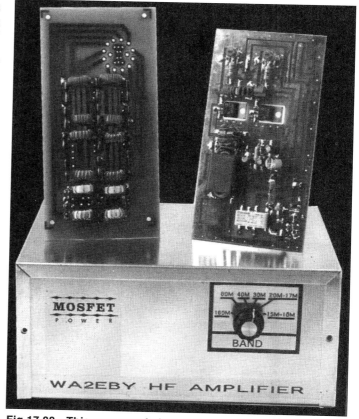

Fig 17.88—This rear-panel view of the amplifier shows the heat sink. The filter board mounts on the front panel.

Biasing a MOSFET for linear operation only requires applying a fixed voltage to its gate via a resistor. With MOSFETs, no special bias or feedback circuitry is required to maintain the bias point over temperature as is required with bipolar transistors to prevent thermal runaway.[10] With MOSFETs, the gate-threshold voltage increases with increased drain current. This works to turn off the device, especially at elevated temperatures as transconductance decreases and R_{DSon} (static drain-to-source *on* resistance) increases.

These built-in self-regulating actions prevent MOSFETs from being affected by thermal runaway. MOSFETs do not require negative feedback to suppress low-frequency gain as is often required with bipolar RF transistors. Bipolar transistor gain increases as frequency decreases. Very high gain at dc and low frequencies can cause unwanted, low-frequency oscillation to occur in bipolar transistor RF amplifiers unless negative feedback is employed to prevent it. Low-frequency oscillation can damage bipolar transistors by causing excess power dissipation, leading to thermal runaway. This rear-panel view of the amplifier shows the heat sink. The filter board mounts on the front panel.

MOSFET LIMITATIONS

Of course, MOSFETs do have their limitations. The high gate impedance and the device structure make them susceptible to electrostatic discharge (ESD) damage. Some easily applied precautions prevent this: Use a soldering iron with grounded tip; use a wrist strap connected to ground through a 1 MΩ resistor to bleed off excess body charge while handling MOSFETs and do all work on an anti-static mat connected to ground via a 1 MΩ resistor.

The sensitivity of a MOSFET's gate to static and high-voltage spikes also makes it vulnerable to damage resulting from parasitic oscillation. This undesired self-oscillation could result in excessive gate-to-source voltage that permanently damages the MOSFET's gate insulation. Another MOSFET limitation is gate capacitance. This parameter limits the frequency at which a MOSFET can operate effectively as an RF amplifier. The author recommends reviewing the referents of Notes 1 to 3 if you are interested in more detailed information about MOSFETs.

POWER MOSFET RF AMPLIFIERS

The author built several power MOSFET amplifiers to check their performance. His experiments underscore the need to observe *exact* construction techniques and physical layout if similar performance is to be expected. Although he used PC board construction, his results differed significantly in several of the experiments because the circuit layout was not the same as the original layout. A photo of the WA2EBY amplifier is shown in **Fig 17.88**.

Considerable experimentation (and he does mean considerable!) with several designs resulted in the circuit shown in **Fig 17.89**. This amplifier consists of two power MOSFETs operating in push-pull, and employs an RF-sensed TR relay.

During receive, TR relay K1 is de-energized. Signals from the antenna are connected to J2 and routed through K1 to a transceiver connected to J1. (This path loss is less than 0.3 dB from 1.8 MHz through 30 MHz.) In transmit, RF voltage from the transceiver is sampled by C17 and divided by R6 and R7. D2 and D3 rectify the RF voltage and charge C16. Q3 begins to conduct when the detected RF voltage across C16 reaches approximately 0.7 V. This energizes K1, which then routes the transmitted RF signal from J1 to the amplifier input and switches the amplifier output to the low-pass filter block and then to the antenna at J2. RF-sensed relay response is very fast. No noticeable clipping of the first CW character has been reported.

An RF attenuator (consisting of R8, R9 and R10) allows you to adjust the amplifier input power to 1 W. (The parts list contains resistor values to reduce the output of 2 or 5 W drivers to 1 W.) The 1 W signal is then applied to the primary of T1 via an input impedance-matching network consisting of L3.

T1 is a 1:1 balun that splits the RF signal into two outputs 180° out of phase. One of these signals is applied by C1 to the gate of Q1. The other signal is routed via C2 to the gate of Q2. The drains of Q1 and Q2 are connected to the primary of output transformer T3, where the two signals are recombined in phase to produce a single output. T3 also provides impedance transformation from the low output impedance of the MOSFETs to the 50-Ω antenna port. DC power is provided to the drains of Q1 and Q2 by a phase-reversal choke, T2. This is a very effective method to provide power to Q1 and Q2 while presenting a high impedance to the RF signal over a broad range of frequencies. The drain chokes for Q1 and Q2 are wound on the same core, and the phase of one of the chokes (see the phasing-dot markings on T2) is reversed. C9 increases the bandwidth of the impedance transformation provided by T3, especially at 21 MHz.

The 5 V bias supply voltage is derived from 28 V by Zener diode D1 and current-limiting resistor R11. Bypass capacitors C3, C4, C5, C6 and C13 remove RF voltages from the bias supply voltage. Gate bias for Q1 and Q2 is controlled independently. R1 adjusts the gate-bias voltage to Q1 via R3 and L1. R2 works similarly for Q2 via R4 and L2.

At low frequencies, the amplifier input impedance is essentially equal to the series value of R3 and R4. L1 and L2 improve the input-impedance match at higher frequencies. The low value of series resistance provided by R3 and R4 also reduces the Q of impedance-matching inductors L1 and L2, which improves stability. DC blocking capacitors C1 and C2 prevent loading the gate bias-supply voltage.

C14 keeps transistor Q3 conducting and K1 energized between SSB voice syllables or CW elements. Without C14, K1 would chatter in response to the SSB modulation envelope and fast keying. Increasing the value of C14 increases the time K1 remains energized during transmit. The reverse voltage generated by K1 when the relay is deenergized is clamped to a safe level by D4. D5 drops the 28 V supply to 13 V to power 12 V relay K1. D5 can be replaced with a jumper if K1 has a 28 V dc coil or if you intend to operate the amplifier with a 13.8 V dc supply.

HARMONIC FILTERING

Although biased for class AB linear operation, this amplifier (like others of its type) exhibits some degree of non-linearity, resulting in the generation of harmonics. This push-pull amplifier design cancels even-order harmonics (2f, 4f, 6f, etc) in the output transformer, T3. Odd-order harmonics are not canceled. Second-order harmonics generated by the amplifier are typically less than –30 dBc (30 dB below the carrier) whereas third-order harmonics are typically only –10 dBc. FCC regulations require all HF RF-amplifier harmonic output power to be at least –40 dBc at power levels between 50 to 500 W. To meet this requirement, it is common practice for HF amplifiers to use low-pass filters. Separate low-pass filters are needed for the 160, 80, 40 and 30 meter bands. The 20 and 17 meter bands can share the same low-pass filter. So, too, the 15, 12 and 10 meter bands can share a common low-pass filter; see Fig 17.89.

Switching between the six filters can be a messy wiring problem, especially on the higher-frequency bands where lead lengths should be kept short for optimum performance. This problem is solved by mounting all six low-pass filters on a PC board. A two-pole, six-position rotary switch (S1) mounted directly on the same PC board manages all filter interconnections. One pole of S1 connects the amplifier output to

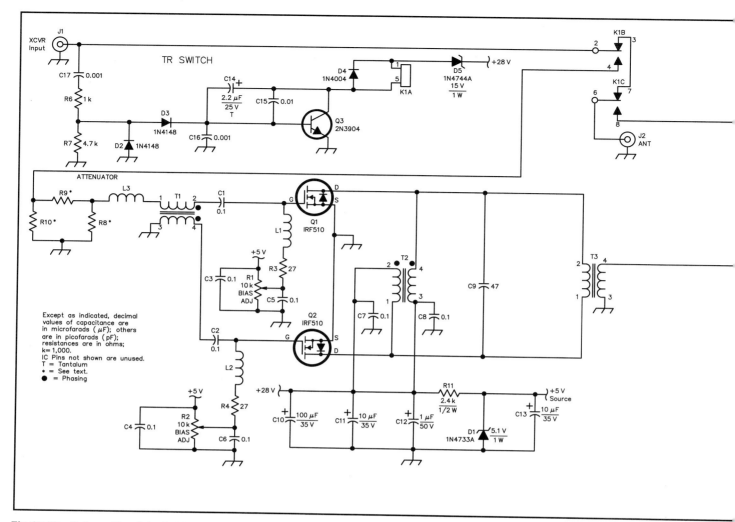

Fig 17.89—Schematic of the MOSFET all-band HF amplifier. Unless otherwise specified, resistors are ¼ W, 5% tolerance carbon-composition or film units. The low-pass filter section shows some filter component values that differ from the calculated values of a standard 50 Ω-input filter. Such differences improve the impedance matching between the amplifier and the load. Capacitors in the filter section are all dipped mica units. Equivalent parts can be substituted. Part numbers in parentheses are Mouser; see Note 9 and the References chapter for contact information.

C1-C8—0.1 µF chip (140-CC502Z104M)
C9—47 pF chip (140-CC502N470J)
C10—100 µF, 35 V (140-HTRL35V100)
C11, C13—15 µF, 35 V (140MLR35V10)
C12—1 µF, 50 V (140-MLRL50V1.0)
C14—2.2 µF, 35 V tantalum (581-2.2M35V)
C15—0.01 µF chip (140-CC502B103K)
C16, C17—0.001 µF chip (140-CC502B102K)
C18, C20, C22—1500 pF (5982-19-500V1500)
C19—2700 pF (5982-19-500V2700)
C21, C23, C25—820 pF (5982-19-500V820)
C24, C26—430 pF (5982-15-500V430)
C27, C29, C31—330 pF (5982-19-500V330)
C28—560 pF (5982-19-500V560)
C30, C34—180 pF (5982-15-500V180)
C32—200 pF (5982-15-500V200)
C33, C35—100 pF (5982-10-500V100)
D1—1N4733A, 5.1 V, 1 W Zener diode (583-1N4733A)
D4—1N4004A (583-1N4004A)
D2, D3—1N4148 (583-1N4148)

D5—1N4744A, 15 V, 1 W Zener diode (583-1N4744A)
J1, J2—SO-239 UHF connector (523-81-120) or BNC connector (523-31-10)
K1—12 V DPDT, 960 Ω coil, 12.5 mA (431-OVR-SH-212L)
L1, L2—9½ turns #24 enameled wire, closely wound 0.25-in. ID
L3—3½ turns #24 enameled wire, closely wound 0.190-in. ID
Q1, Q2—IRF510 power MOSFET (570-IRF510)
Q3—2N3904 (610-2N3904)
R1, R2—10 kΩ trim pot (323-5000-10K)
R3, R4—27 Ω, ½ W (293-27)
R6—1 kΩ chip (263-1K)
R7—4.7 kΩ chip (263-4.7K)
R8—130 Ω, 1 W (281-130); for 7 dB pad (5 W in, 1 W out)
R9—43 Ω, 2 W (282-43); for 7 dB pad (5 W in, 1 W out)
R10—130 Ω, 3 W (283-130); for 7 dB pad (5 W in, 1 W out)

R8, R10—300 Ω, ½ W (273-300); for 3 dB pad (2 W in, 1 W out)
R9—18 Ω, 1 W (281-18); for 3 dB pad (2 W in, 1 W out)
R11—2.4 kΩ, ½ W (293-2.4K)
S1—2 pole, 6 position rotary (10YX026)
T1—10 bifilar turns #24 enameled wire on an FT-50-43 core.
T2—10 bifilar turns #22 enameled wire on two stacked FT-50-43 cores.
T3—Pri 2 turns, sec 3 turns #20 Teflon-covered wire on BN-43-3312 balun core.
Misc: Aluminum enclosure 3.5×8×6 inches (HWD) (537-TF-783), two TO-220 mounting kits (534-4724), heat-sink compound (577-1977), amplifier and low-pass filter PC boards (see Note 9), heat sink (AAVID [Mouser 532-244609B02]; see text), about two feet of RG-58 coax, #24 enameled wire and #20 Teflon-insulated wire.

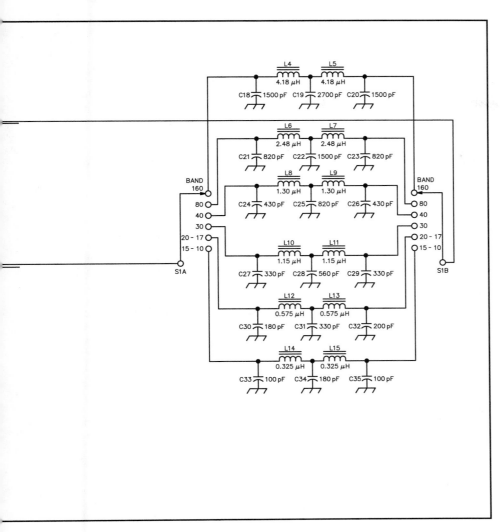

and add the secondary winding over the primary. Be sure to use Teflon-insulated wire for these windings; the high operating temperatures encountered will likely melt standard hook-up wire insulation.

Heat Sinking

Together, Q1 and Q2 dissipate up to 59 W. A suitable heat sink is required to prevent the transistors from overheating and damage. I used an AAVID 244609B02 heat sink originally designed for dc-to-dc power converters. The amplifier PC board and heat sink are attached to an aluminum enclosure by two #4-40 screws drilled through the PC board, enclosure and heat sink at diagonally opposite corners. See **Fig 17.90.** A rectangular cutout in the enclosure allows Q1 and Q2 direct access to the heat sink. This is essential because of the large thermal impedance associated with the TO-220 package (more on this topic later). Mark the locations of the transistor-tab mounting-hole location in the center of the heat sink, between the cooling fins. Disassemble the heat sink to drill 0.115 inch holes for #4-40 mounting screws, or tap #4-40 mounting holes in the center of the heat-sink fins.

Use mica insulators and grommets when mounting Q1 and Q2 to prevent the #4-40 mounting screws from shorting the TO-220 package drain connections (tabs) to ground. Coat both sides of the mica insulator with a *thin* layer of thermal compound to improve the thermal conduction between the transistor tab and the heat sink. Be sure to install the mica insulator on the heat sink *before* assembling the amplifier PC board to the enclosure and heat sink. The mica insulators are larger than the cut outs in the PC board, making it impossible to install them after the PC board is mounted.

LOW-PASS FILTER CONSTRUCTION

Inductor winding information for the low-pass filters is provided in **Table 17.3.**

Single Band

A PC-board trace is available on the amplifier PC board next to amplifier output (J3) to allow the installation of a single-band low-pass filter between the terminals of J3 and the J4 input to K1. This is handy if you intend to use the amplifier on one band only. The input inductor of the low-pass filter connects from J3 to the single PC trace adjacent to J3. The output inductor connects in series between the single PC trace to J4. The three filter capacitors connect from J3, J4 and the PC-board trace near J3 to ground. *This single trace is not used when multiple filters are*

one of the six filter inputs, while the other pole of S1 simultaneously connects the corresponding filter output to the TR relay, K1. Only two coaxial-cable connections are required between the RF amplifier and the low-pass filter board.

AMPLIFIER CONSTRUCTION

The amplifier is constructed on a double-sided PC board with plated through holes to provide top-side ground connections. Chip resistors and capacitors were used to simplify construction, but leaded capacitors may work if lead lengths are kept short. First, assemble all chip capacitors and resistors on the PC board. Tweezers help to handle chip components (or see the vacuum handler for chip components described in the **Station Setup and Accessory Projects** chapter. Work with only one component value at a time. (Chip caps and resistors can be very difficult to identify!) Chip capacitor and resistor mounting is simplified by tinning one side of the PC board trace with solder before positioning the capacitor or resistor. Touch the soldering iron tip to the capacitor

or resistor to tack it in place. Finish mounting by soldering the opposite side of the component. *Don't apply too much heat to chip capacitors.* The metalized contacts on the capacitor can be damaged or completely removed if too much heat is applied. Use a 15 to 20 W soldering iron and limit soldering time to five seconds.

Mount axial-leaded resistors, diodes and remaining capacitors next. To avoid damaging them, mount inductors and transformers last. L1 and L2 are wound on a 1/4-inch drill-bit shaft. By wrapping the wire around the shaft 10 times, you'll get 9 1/2 turns. The last turn arcs only a half-turn before entering the PC board. L3 is wound on a 0.190-inch diameter drill bit with 3 1/2 turns wound the same way as L1 and L2. Mounting K1 is simplified by first bending all its leads 90° outward so it lies flat on the PC board. Be sure to follow the anti-static procedures mentioned at the beginning of this project while handling MOSFETs. The gate input can be damaged by electrostatic discharge!

When winding T3, wind the primary first

required. Remember to remove the single trace adjacent to J3 on the amplifier PC board before attaching the amplifier board between the RF connectors on the rear panel of the enclosure.

Multiple-Band Filters

Using the amplifier on more than one band requires a different approach. A set of six low-pass filters is built on a double-sided PC board with plated through holes to provide top-side ground connections. A PC-board mount, two-pole, six-position rotary switch does all low-pass filter selection. Silver-mica, leaded capacitors are used in all the filters. On 160 through 30 meters, T-50-2 toroids are used in the inductors. T-50-6 toroids are used for inductors on 20 through 10 meters. The number of turns wound on a toroid core are counted on the toroid's OD as the wire passes through the core center. (The **Circuit Construction** chapter provides complete details for winding toroids.) Assemble one filter section at a time starting with the 160, 80, 40 and 30-meter filters. With the switch mounting position at your upper left, the 160-m filter input (C18) is near the top edge of the board and the filter output (C20) is near the bottom edge. *The last two filters are out of sequence;* the 15-10 meter filter comes *before* the 20-17 meter filter) and the inputs/outputs are reversed to simplify the PC-board layout. The input capacitors, C30 and C33, are mounted on the board *bottom edge,* and output capacitors, C32 and C35, are on the *top edge*.

Use care when assembling the rotary switch. All 14 terminals must fit through the PC board without damaging or bending the pins. Make sure there are no bent pins before you attempt assembly. Insert the rotary switch into the PC board. Do *not* press the rotary switch all the way into the PC-board holes flush with the ground plane! If you do, the top flange of the signal pins may short to the ground plane.

BIAS ADJUSTMENT

The biasing procedure is straightforward and requires only a multimeter to complete. First, set R1 and R2 fully counterclockwise, (0 V on the gates of Q1 and Q2). Terminate the RF input and output with 50-Ω loads. Next, connect the 28 V supply to the amplifier in series with a multimeter set to the 0-200 mA current range. Measure and record the idling current drawn by the 5 V bias supply. The value should be approximately 9.5 mA (28 − 5.1 V) / 2.4 kΩ = 9.5 mA). Set the Q1 drain current to 10 mA by adjusting R1 until the 28 V supply current increases by 10 mA above the idling current

Table 17.3
Low-Pass Filter Inductor Winding Information

(Refer to Fig 17.89)

Inductor Number	No. of Turns	Core
L4, L5	30	T-50-2
L6, L7	22	T-50-2
L8, L9	16	T-50-2
L10, L11	14	T-50-2
L12, L13	11	T-50-6
L14, L15	8	T-50-6

Note: All inductors are wound with #22 enameled wire except for L4 to L7, which are wound with #24 enameled wire.

(9.5 + 10 = 19.5 mA). Next, adjust R2 for a Q2 drain current of 10 mA. This is accomplished by adjusting R2 until the 28 V supply current increases by an additional 10 mA (to 29.5 mA).

AMPLIFIER PERFORMANCE

With a 28 V power supply and 1 W of drive, the RF output power of this amplifier exceeds 40 W from 1.8 MHz through 28 MHz. Peak performance occurs at 10 MHz, providing about 75 W after filtering! A performance graph for this amplifier is shown in **Fig 17.91A**.

As shown in Fig 17.91B, this amplifier achieves an efficiency of better than 50% over its frequency range, except at 7 MHz where the efficiency drops to 48%.

Fig 17.91C shows the input SWR of the amplifier. It exceeds 2:1 above 14 MHz.

The input SWR can be improved to better than 2:1 on all bands by adding a 3 dB pad (R8-R10 of Fig 17.89) at the input and supplying 2 W to the pad input. This keeps the amplifier drive at 1 W.

Fig 17.91D graphs the amplifier RF output power as a function of drain supply voltage. During this test, the amplifier RF drive level was kept constant at 1 W. As you can see, even when using a 13.8 V dc supply, the amplifier provides over 10 W output (a gain of more than 10 dB) from 1.8 to 30 MHz.

OPERATION

The amplifier requires no tuning while operating on any HF amateur band. You must, however, *be sure to select the proper low-pass filter prior to transmitting*. If the wrong low-pass filter is selected, damage to the MOSFETs may result. Damage will likely result if you attempt to operate the amplifier on a band with the low-pass filter selected for a lower frequency. For example, driving the amplifier with a 21 MHz signal while the 1.8 MHz low-pass filter is selected will likely destroy Q1 and/or Q2.

The amplifier can also be damaged by overheating. This limitation is imposed by the TO-220 packages in which Q1 and Q2 are housed. The thermal resistance from junction to case is a whopping 3.5°C/W. This huge value makes it virtually impossible to keep the junction temperature from exceeding the +150°C target for good reliability. Consider the following

Fig 17.90

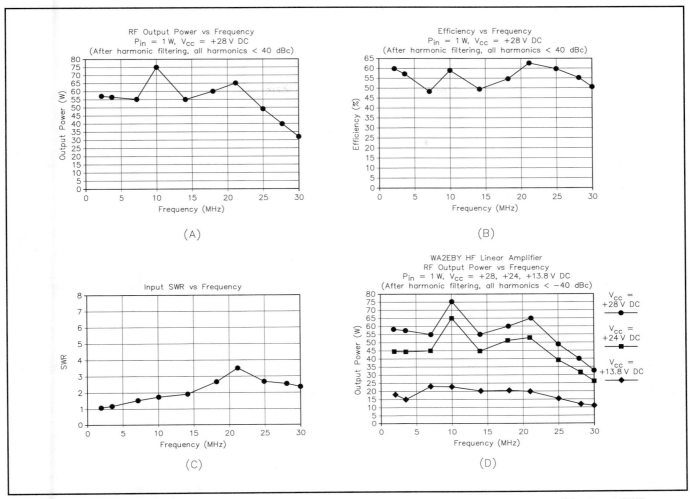

Fig 17.91—A shows the amplifier RF output power. B shows the amplifier efficiency. C shows the amplifier input SWR. D shows the amplifier RF output power versus supply voltage.

conditions: key down, 1 W input, 53 W output on 7 MHz (worst-case band for efficiency). The amplifier consumes 28 V × 4 A = 112 W, of which 53 W are sent to the antenna, so 59 W (112 W – 53 W = 59 W) are dissipated in Q1 and Q2. Assuming equal current sharing between Q1 and Q2, each transistor dissipates 29.5 W. To keep the transistor junction temperature below +150°C requires preventing the transistor case temperature from exceeding 46.8°C (150 –[3.5 × 29.5]) while dissipating 29.5 W. Also, there is a temperature rise across the mica insulator between the transistor case and heat sink of 0.5°C/W. That makes the maximum allowable heat-sink temperature limited to 46.8 (0.5 × 29.5) = 32°C. In other words, the heat sink must dissipate 59 W (29.5 from each transistor) with only a 7°C rise above room temperature (25°C). Even if the junction temperatures were allowed to

reach the absolute maximum of 175°C, the heat sink temperature must not exceed 57°C. Accomplishing this requires a heat sink with a thermal resistance of (57 – 25) /59 = 0.54°C/W. This is far less than the 1.9°C/W rating of the AAVID 244609B02 heat sink I used. The situation may seem bleak, but all is not lost. These calculations make it clear that the amplifier should not be used for AM, FM or any other continuous-carrier operation. The amplifier should be used only for CW and SSB operation where the duty cycle is significantly reduced.

Thermal performance of the amplifier is illustrated in **Fig 17.92A**. Data was taken under dc operating conditions with power-dissipation levels set equal to conditions under RF operation. A RadioShack brushless 12 V dc fan (RS 273-243A) blows air across the heat sink. Key down, the maximum rated junction temperature is reached in as little as five seconds. Pro-

longed key-down transmissions should be avoided for this reason.

Under intermittent CW conditions, the situation is very different. Transistor-case temperatures reached 66°C after operating four minutes under simulated CW conditions at 20 WPM (60 ms on, 60 ms off). The corresponding junction temperature is +141°C (based on an equivalent RMS power dissipation of 21.7 W per transistor). This keeps the junction temperature under the 150°C target (see Fig 17.92B). One simple way to reduce power dissipation is to reduce the power-supply voltage to 24 V. RF output power will decrease about 10 W from the maximum levels achieved with a 28 V supply.

From a thermal standpoint, the IRF510 power MOSFET is a poor choice for this RF amplifier application. Although I must say I am impressed with the robustness of these devices considering the times I spent testing them key down, five minutes at a

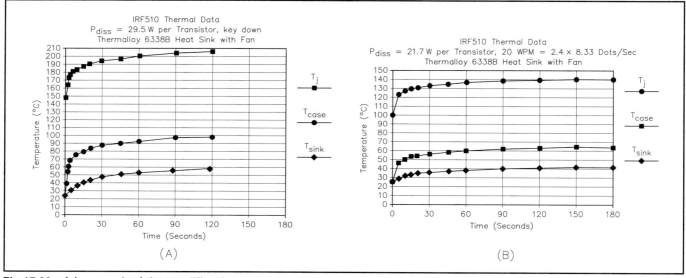

Fig 17.92—A is a graph of the amplifier thermal performance during key-down conditions. B is a graph of the amplifier thermal performance during simulated CW conditions.

time, without failure. Q1 and/or Q2 may need to be replaced after a year or so of operation because of the compromise in reliability. Considering their low cost, that is not a bad trade-off.

STABILITY

High gain, broad bandwidth and close input/output signal routing (within the TR relay) all work against stability. With a good load (< 2:1 SWR) the amplifier is stable from 1.8 MHz through 39 MHz. Oscillation was observed when the transmitter frequency was increased to 40 MHz. The output load match also affects stability. I spent a great deal of time trying to make this design unconditionally stable even with loads exceeding 3:1 SWR without sacrificing output power (gain) at 28 MHz without success. I did identify some reasonable compromises.

One of the easiest ways to improve stability and the input SWR seen by the RF source is to add an RF attenuator (pad) at the amplifier input. An attenuator is absolutely required if the transmitter (driver) provides more than 1 W to the amplifier. R8, R9 and R10 form an RF attenuator that attenuates the transmitter drive level, but does not attenuate received signals because it is only in the circuit when K1 is energized. To drive this amplifier with a 2-W-output transmitter requires use of a 3-dB pad. The pad improves the amplifier input SWR and the isolation between the amplifier's input and output. The drawback is that 1 W is wasted in the pad. Likewise, a 5-W driver requires use of a 7-dB pad, but 4 W are wasted in the pad. (Val-

ues for R8, R9 and R10 to make a 3-dB pad and a 7-dB pad are given in the parts list of the caption for Fig 17.94.) Installing a pad requires cutting the PC-board trace *under R9*, otherwise R9 would be shorted out by the trace. Make a small cut (0.1 inch wide) in the trace under R9 before soldering R9 in position. R8 and R10 have the same values, but may have different power ratings. Connect R10 between the RF input side of R9 and ground. Install R8 between the amplifier side of R9 and ground.

An impedance mismatch between the output of a 1-W driver and the amplifier input can be a source of instability. (Obviously, if the driving transmitter output power is only 1 W, you can't use a pad as described earlier.) If you encounter stability problems, try these remedies: Place a resistor in parallel with L1 and L2 to decrease the Q of the amplifier matching network (try values between 50 and 220 Ω). Try reducing the value of L3 or eliminating L3 entirely. Both of these modifications improve stability, but reduce the amplifier output power above 21 MHz.

SUMMARY

This project demonstrates how inexpensive power MOSFETs can be used to build an all-band linear HF power amplifier. Frequency of operation is extended beyond the limits of previous designs using the IRF510 and improved input-impedance matching. Long-term reliability is recognized as a compromise because of the poor thermal performance of the low-cost TO-220 package.

If you have been thinking about adding an amplifier to your QRP station, this project is a good way to experiment with amplifier design and is an excellent way to become familiar with surface-mount "chip" components. Mouser Electronics and Amidon, Inc provide parts kits for this project (see Note 9). These parts kits make it very easy to get started and more economical to "homebrew" this project.

ACKNOWLEDGMENTS

The author thanks the following individuals associated with this project: Harry Randel, WD2AID, for his untiring support in capturing the schematic diagram and parts layout of this project; Al Roehm, W2OBJ, for his continued support and encouragement in developing, testing, editing and publishing this project; Larry Guttadore, WB2SPF, for building, testing and photographing the project; Dick Jansson, WD4FAB, for thermal-design suggestions; Adam O'Donnell, N3RCS, for his assistance building prototypes; and his wife, Laura, N2TDL, for her encouragement and support throughout the project.

Notes

[1] Doug DeMaw, W1FB, "Power-FET Switches as RF Amplifiers," *QST*, Apr 1989, pp 30-33. See also Feedback, *QST*, May 1989, p 51.

[2] Wes Hayward, W7ZOI, and Jeff Damm, WA7MLH, "Stable HEXFET RF Power Amplifiers," Technical Correspondence, *QST*, Nov 1989, pp 38-40; also see Feedback, *QST*, Mar 1990, p 41.

[3] Jim Wyckoff, AA3X, "1 Watt In, 30 Watts Out with Power MOSFETs at 80 Meters," Hints

and Kinks, *QST,* Jan 1993, pp 50-51.

[4]Doug DeMaw, W1FB, "Go Class B or C with Power MOSFETs," *QST,* March 1983, pp 25-29.

[5]Doug DeMaw, W1FB, "An Experimental VMOS Transmitter", *QST,* May 1979, pp 18-22.

[6]Wes Hayward, W7ZOI, "A VMOS FET Transmitter for 10-Meter CW," *QST,* May 1979, pp 27-30.

[7]Ed Oxner, KB6QJ (ex-W9PRZ), "Build a Broadband Ultralinear VMOS Amplifier," *QST,* May 1979, pp 23-26.

[8]Gary Breed, K9AY, "An Easy-to-Build 25-Watt MF/HF Amplifier," *QST,* Feb 1994, pp 31-34.

[9]Parts for this project are available in five modular kits. The following three kits are available from Mouser Electronics: Amplifier components (Mouser P/N 371-HFAMP1) consisting of the amplifier PC board and all PC-board-mounted components (except for the ferrite cores). Price: $35, plus shipping. Amplifier hardware kit (Mouser P/N 371-HFAMP2) consisting of the aluminum enclosure, two UHF connectors, two TO-220 mounting kits, AAVID heat sink and one container of heat sink compound. Price: $30 plus shipping. Low-pass filter kit (Mouser P/N 371-HFAMP3) consisting of the low-pass filter PC board, rotary switch and all PC-board-mounted capacitors (inductor cores are *not* included). Price: $35, plus shipping. Part-placement diagrams accompany the PC boards. See the References chapter for contact information. PC boards only are available from Mouser Electronics: HF amplifier board (#371-AMPPWB-2); filter PC board (#371-LPPWB-2). Price $15 each, plus shipping.

The following two kits are available from Amidon Inc: Amplifier ferrite kit (Amidon P/N HFAFC) containing the ferrite cores, balun core and magnet and Teflon wire to wind the transformers for the HF amplifier. Price: $3.50 plus shipping. Low-pass filter cores kit (Amidon P/N HFFLT) containing all iron cores and wire for the low-pass filters. Price: $4.50 plus shipping. See the References chapter for contact information.

[10]Motorola Application Reports Q1/95, HB215, *Application Report AR346.* Thermal runaway is a condition that occurs with bipolar transistors because bipolar transistors conduct more as temperature increases, the increased conduction causes an increase in temperature, which further increases conduction, etc. The cycle repeats until the bipolar transistor overheats and is permanently damaged.

An Experimental ¹/₂-W CW Transmitter

Using the transmitter shown in **Fig 17.93**—a 74HC240 octal inverting buffer IC operating as a crystal oscillator and power amplifier—Lew Smith, N7KSB, has worked all continents and over 30 countries using just a roof-mounted ground-plane antenna. It is a simple, inexpensive means of having fun on 10, 15 or 20 m.

It's important to operate the circuit at a supply voltage from 7.8 to 8 V as a compromise between maximum power output and device safety. The logic chips have built-in input and output buffers. The extra gain provided by the extra stage makes it harder to get rid of key clicks. The ¹/₂-W rig's key-click filter therefore uses an unusually large (33-ms) time constant.

The output stages in 74HC devices are designed to have equal value pull-up and pull-down transistors. This minimizes even-order harmonics, simplifying the ¹/₂-W rig's output filtering.

The 74HC240 can directly drive a power MOSFET amplifier, as exemplified by the 74HC240-IRF510 transmitter described in "An Easy-to-Build 15-W Transmitter" (*Hambrew* magazine, Spring 1994).

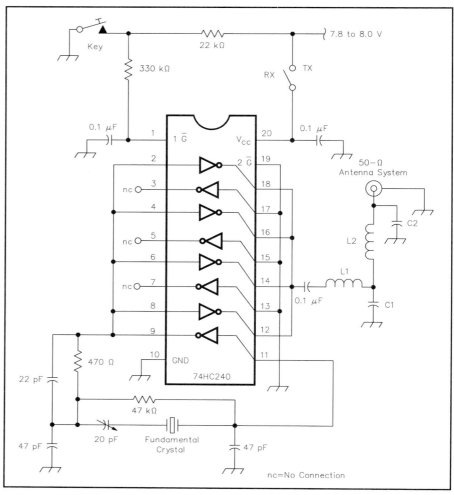

Fig 17.93—N7KSB's experimental ¹/₂-W CW transmitter uses a 74HC240 high-speed CMOS octal buffer: One section serves as a crystal oscillator; four sections amplify the signal; three sections are unused. (See Table 17.7 for output filter values.) Because the IC dissipates ¹/₂ W on 20 m and 0.9 W on 10 m, it needs a heat sink. (Epoxy the IC to the ground plane, dead-bug style.) L1 is sensitive to lead length, so plug-in filters (Lew uses phono plugs and jacks) may require removal of one or two turns to compensate for the extra lead length. (ARRL Lab tests indicate that this transmitter meets current FCC spectral-purity speci-fications for equipment in its power-output class and frequency range. The test version produced 0.51 W at 14 and 21 MHz, and 0.47 W at 28 MHz, with a 7.83-V supply.)

Table 17.7
Output Filter Component Values for the 74HC240 Transmitter

Band (m)	10	15	20
C1 (pF)	330	470	680
C2 (pF)	100	150	220
L1 (turns, length)	3 t, ⁵/₈"	4 t, ⁵/₈"	5.5 t, ⁵/₈"
L2	7 t, ⁵/₈"	10 t, 1"	12 t, 1"

C1 and C2 are mica or ceramic; L1 and L2 are #14 wire with air cores (³/₈" diam form).

A Drift-Free VFO

By following several design guidelines, Jacob Makhinson, N6NWP, built a low-cost, easy-to-construct LC VFO with a very low level of phase noise. The article originally appeared in December 1996 QST.

The method shown makes the oscillator essentially drift-free, with very little phase noise, VFOs built with these techniques are viable in applications where low overall noise level and wide dynamic range is of great importance. The technique can also spare VFO designers the drudgery of more conventional drift-compensating techniques.

Many VFO designs have appeared in the Amateur Radio literature, and the quest for a low-drift VFO hasn't ceased. If the frequency-stability requirements are stringent, the thermal-drift compensation can be very tedious. Wes Hayward's QST article[1] devoted to VFO drift compensation is an excellent example of this difficult pursuit.

DESIGN CRITERIA

To avoid degradation of the receiver's front end, several requirements should be imposed on the phase noise level of the VFO. An excessively high level of close-in phase noise (within the bandwidth of the SSB signal) may reduce the receiver's ability to separate closely spaced signals. As an example, a 14-pole crystal filter described in Note 2 provides adjacent-signal rejection of 103 dB at a 2-kHz offset. This requires the use of a VFO with –139 dBc/Hz phase noise at a 2-kHz offset.

$$Pn = P - 10\log(BW)$$
$$= -103 - 10\log(4000) = -139 \text{ dBc/Hz}$$

where

Pn = VFO phase-noise spectral density, in decibels relative to the carrier output power, in a 1-Hz bandwidth (dBc/Hz)

P = VFO power level (dBc) in a given bandwidth (BW)

BW = test bandwidth, in Hertz

In addition, excessive close-in phase noise may lead to reciprocal mixing, where the noise sidebands of a VFO mix with strong off-channel signals to produce unwanted IF signals. Excessive far-out phase noise may degrade the receiver dynamic range. In a properly designed receiver, the phase-noise-governed dynamic range (PNDR) should be equal to or better than the spurious-free dynamic range (SFDR). We can calculate the PNDR:[3]

$$PNDR = -Pn - 10\log(BW)$$

Assuming the PNDR equals the SFDR at 112 dB in a 2.5-kHz IF noise bandwidth, the required far-out phase noise level is –146 dBc/Hz:

$$Pn = -SFDR - 10\log(BW) =$$
$$-112 - 34 = -146 \text{ dBc/Hz}$$

Another form of VFO instability—frequency drift—has always been a nuisance and a great concern to the amateur community. The objective of this project was to keep the long-term frequency drift (seconds, minutes, hours) under 20 Hz. This includes thermal drift from both internal heating and environmental changes.

BLOCK DIAGRAM

The block diagram of **Fig 17.94** shows the LC VFO and the frequency stabilizer. The stabilizer monitors the VFO frequency and forms an error signal that is applied to the VFO to compensate for frequency drift. This technique, which is capable of stabilizing a VFO to within a few hertz, was devised by Klaas Spaargaren, PA0KSB, and first described in *RadCom* magazine in 1973.[4] This project builds upon Spaargaren's idea and presents a few refinements.

The stabilizer converts a free-running VFO into an oscillator that can be tuned in the usual fashion, but then locks to the nearest of a series of small frequency steps. Unlike traditional PLL frequency synthesizers, the stabilizer has no effect on the phase-noise performance of the VFO; it only compensates for thermal drift.

The timing signal (2.6 Hz) is derived from a crystal oscillator via a frequency divider. The timing signal drives a NAND gate to provide a crystal-controlled time window, during which the binary counter counts the VFO output. When the gate closes, the final digit of the count remains in the counter. For counts 0 to 3, the Q3 output of the counter is a logic 0; for counts of 4 to 7, a logic 1.

The result is stored in a D flip-flop memory cell: When the 2.6-Hz timing signal goes low, the first of three one-shots triggers. The second follows and clocks the binary counter Q3 output into the memory cell. The negative-going pulse from the third resets the counter for the next counting sequence.

The output of the memory cell is applied to an RC integrating circuit with a time constant of several minutes. This slowly changing dc voltage controls the VFO frequency via a couple of varicaps connected to a tap on the VFO coil.

If the counter output is 0, the memory-cell output is 1, which charges C and increases the VFO frequency. A counter output of 1 discharges C and decreases the VFO frequency. The stabilizer constantly searches for equilibrium, so the VFO frequency slowly swings a few hertz around the lock frequency. The circuit limits the frequency swing to a maximum of ±2 Hz, typically ±1 Hz.

A difficulty arises when the operator changes frequency because the control voltage is disturbed. If the memory-cell output connects directly to the RC integrator, the frequency correction that occurs immediately after tuning results in a frequency hop. To overcome this problem, an analog switch disconnects the integrator from the memory during tuning. The tuning detector—an infrared interrupter switch and a one-shot—controls the analog switch.

CIRCUIT DESCRIPTION VFO

The VFO is a tapped-coil Hartley oscillator that is optimized for low phase noise (see Fig 17.94B). It follows the design rules compiled by Ulrich Rohde, intended to minimize the phase noise in oscillators.[5]

The tank coil, L1, has an iron-powder toroidal core; coil Q exceeds 300. C1, C4, C5 and C7 are NP0 (C0G) ceramic capacitors (5% or 10% tolerance). C2 is the main tuning capacitor, and C3 is a small ceramic trimmer capacitor.

The VFO frequency range is set from 6.0 MHz to 6.4 MHz (to accommodate a 20-meter receiver with an 8-MHz IF). The loaded Q of the resonator is kept high by using a tapped coil and loose coupling to the gate of the FET through C7 (more than 8 kΩ at 6 MHz). The RF voltage swing across the resonator exceeds 50 V, P-P. Varicaps D1 and D2, which compensate for thermal drift, are connected across the coil's lower tap (less than 14% of the total turns) and have a negligible effect on overall phase-noise. J310 is the TO-92 version of U310—a very low-noise FET in HF applications.

An ALC loop limits the voltage swing. The signal is sampled at the primary of T1, rectified by the D5-C21 network and fed to the inverting input of an integrator, U1A, where it is compared against the reference voltage at the junction of R18 and R19. The dc voltage at the integrator output sets Q1's drain current so that the signal swing at T1's primary is always 2.5 V, P-P. The ALC loop also makes VFO performance independent of Q1's pinch-off voltage. The signal at Q1's source is a 6.5-V, P-P, sinusoid with almost no distortion.

Q2 is a high-impedance buffer that is loosely coupled to Q1. Q2's drain current is set to 3.4 mA (by the constant-current source, Q3-Q4) regardless of Q2's pinch-off voltage. Buffer 1 is a push-pull stage biased into slight conduction by resistors R11 and R12. It has excellent linearity and a very low output impedance, which is

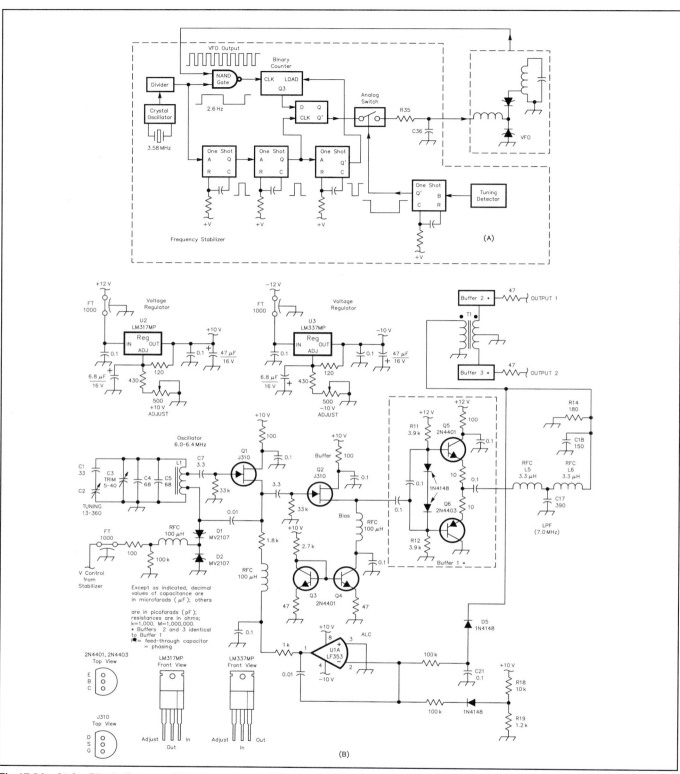

Fig 17.94—At A—Block diagram of the frequency stabilizer and VFO connections. At B—VFO schematic. Buffers 2 and 3 are identical to Buffer 1. Most of the parts are available from Mouser Electronics, Digi-Key Corporation or Allied Electronics. The cores for L1 and T1 are from Amidon Associates.[6] Use 1/4-W, 5%-tolerance carbon-composition or film resistors and ceramic, 20%-tolerance capacitors unless otherwise indicated. RF chokes or encapsulated inductors may be used for those labeled "RFC."

Q1, Q2—J310, N-channel JFET (Allied)
D1, D2—MV2107 or ECG/NTE613 tuning diode (Varicap, Allied)
L1—29 turns of #18 AWG enameled copper wire on a T-80-6 iron-powder toroidal core tapped at 4 turns and 20 turns from the cold end (Amidon)

T1—#32 AWG enameled copper wire on a BN-43-2402 two-hole ferrite balun core (Amidon) primary: 5 turns; secondary: 16 turns, center tapped
Vector part #8007 circuit board (Digi-Key)
Vector part #T44 terminals (Digi-Key)

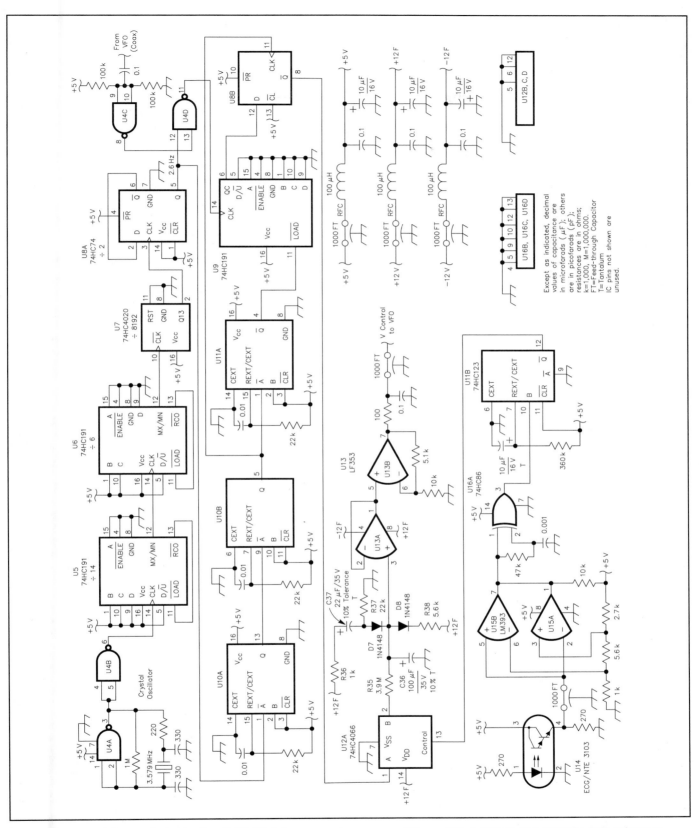

Figure 17.95—Stabilizer schematic. Use ¼-W, 5%-tolerance carbon-composition or film resistors and ceramic, 20%-tolerance capacitors unless otherwise indicated.

U4—74HC00 Quad NAND gate
U5, U6, U9—74HC191 presettable 4-bit binary counter
U7—74HC4020 14-bit binary ripple counter
U8—74HC74 dual D flip-flop

U10, U11—Dual 74HC123 one-shot
U12—4066 quad analog switch
U14—ECG/NTE3103 optical interrupter (Darlington output, Allied, see Note 6)

required to drive an LC filter. The filter (L5, L6, C17, C18 and R14) is a four-pole, 0.1-dB Chebyshev low-pass filter with a ripple frequency of 7 MHz. All harmonics at the VFO output are at least 45 dB below the fundamental.

T1 provides the two complementary outputs required for a commutation mixer and raises the voltage swing at the VFO output. Buffers 2 and 3 are electrically identical to buffer 1. They further decouple the VFO from its load and serve as low-distortion 50-Ω. drivers. The signal level at each output is 4 V, P-P, when driving a high-imped-ance load (eg, a CMOS gate), +10 dBm when driving a 50-Ω load.

FREQUENCY STABILIZER

NAND gates U4A and B (see **Fig 17.95**) comprise a Pierce crystal oscillator. The timing signal (2.6 Hz) appears at the output of the frequency divider (U5, U6, U7 and U8A). The exact frequency of the crystal and the timing signal is unimportant, but the stabilizer has been optimized for 2.3 to 2.7 Hz.

There are two requirements for the crystal oscillator: No harmonics should fall in the IF passband, and the crystal should have a low temperature coefficient. Crystal-oscillator thermal drift should not exceed 10 Hz within the temperature operating range. Crystals in HC-33 cases with frequencies between 2.0 and 3.58 MHz worked best for me. The frequency divider is sufficiently flexible to provide the desired timing-signal frequency.

U4C, biased into a linear range, converts the sinusoidal signal from one of the two VFO outputs into a square wave. U4D gates the VFO signal bursts into the clock input of the binary counter, U9. At the end of every burst, the final digit is held by the counter.

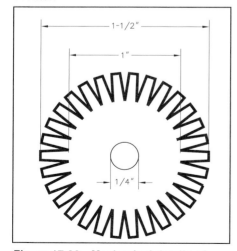

Figure 17.96—Mechanical details of interrupter wheel. Use a good reduction drive and make one tooth for every 20 to 40 Hz of frequency change. Use any rigid, opaque material.

The falling edge of the timing signal triggers U10A, the first of three cascaded one-shots. The pulse at the output of U10B clocks the data from the counter into U8B. The pulse at the output of U11A resets U9.

If the number of pulses in each successive burst is equal (no VFO drift), U9 constantly counts the same number, and the output of U8B never changes. In practice, however, U8B constantly toggles between two states. The integrating circuit, R35-C36 (time constant = 6.5 minutes), converts the toggling into a slowly changing voltage. Varicaps D1 and D2 transform a few millivolts of change into ±1 or 2 Hz change of VFO frequency.

U13A, a high-input-impedance buffer, prevents the discharge of C36. U13B, a noninverting amplifier with a gain of 1.5, ensures compliance between the control-voltage range and the capacitance-per-volt ratio of the Varicaps (1 to 6 V for best performance). Network R36, R37, C37 and D7 establishes the initial dc voltage applied to the varicaps; the value is set by the C37-C36 voltage divider.

An infrared interrupter switch, U14, serves as sensor in the tuning-detector circuit. The slotted interrupter detects the movements of a serrated disc (see **Fig 17.96**) on the VFO reduction-drive shaft. U15A and B, a two-level limit comparator, converts the signal at its input into pulses. U16A produces trigger pulses for the one-shot, U11B, by detecting both leading and falling edges of the signal at its input. U11B is retriggerable—its $\overline{Q}$ output stays low during manual tuning and for 3.6 seconds after tuning stops. Analog switch U12A disconnects C36 from the flip-flop during tuning, thus preserving the capacitor charge. This system does not provide for an RIT control.

CONSTRUCTION

The VFO and the stabilizer are in separate boxes. Mount components within the enclosures on the perf board's foil side. Make ground connections to the foil plane. Use Vector pins as terminal posts for the input and output signals.

The VFO box is a die-cast aluminum enclosure ($4^{11}/_{16} \times 3^{11}/_{16} \times 2^1/_{16}$ inches) to ensure mechanical rigidity. The two RF outputs exit the box via BNC connectors and coax. DC enters via feedthrough capacitors. Rigidly attach C2 to the enclosure wall. Cover L1 with a low-loss polystyrene Q dope and place it as far as possible from the ground plane and enclosure walls. The layout is not critical, but observe standard RF building methods: use short leads, dress them for minimum coupling and solder bypass capacitors directly to the ground plane close to the terminal they bypass.

The stabilizer is in a $5^1/_2 \times 3 \times 1^1/_4$-inch

LMB aluminum enclosure. Component placement and layout is not critical, but keep component leads short around the crystal oscillator. Use a BNC connector for the signal from the VFO module. Solder the ground pins of all ICs directly to the ground plane, and decouple each power-supply pin of ICs U4 through U9 to ground via a 0.1 μF capacitor. Route all dc voltages to the module via feedthrough capacitors to avoid RF leakage. Mount U14 so that the serrated disc is in the middle of the slot.

Mount C37 in a socket in case you need to adjust its value: For unknown or varying VFO drift direction, use a 22 μF capacitor to place the initial varicap control voltage at midrange ($V_c \approx 2.9$ V). Use 10-μF if VFO drift is predominantly negative ($V_c \approx 1.5$ V), and 33 μF if it's predominantly positive ($V_c \approx 4.0$ V).

MEASUREMENTS

The VFO thermal drift without the stabilizer was under 800 Hz at room temperature (after 90 minutes) and under 1500 Hz when the ambient temperature was raised 20°C. There was no attempt to compensate for thermal drift.

With the frequency stabilizer connected, the thermal drift did not exceed 10 Hz at room temperature, 20 Hz when raised 20°C. In one of the experiments, power was on for several days, and drift was under 10 Hz at room temperature. Frequency lock is attained in less than 10 seconds after the power is switched on.

With the components shown in the schematic, the stabilizer can compensate for a maximum 1800-Hz drift with a 25°C temperature rise. To compensate for a greater frequency drift, select varicaps with higher diode capacitances; the frequency swing will increase from ±1 or 2 Hz to a higher value.

Notes

[1]Wes Hayward, W7ZOI, "Measuring and Compensating Oscillator Frequency Drift," *QST*, Dec 1993, pp 37-41.

[2]Jacob Makhinson, N6NWP, "Designing and Building High-Performance Crystal Ladder Filters," *QEX*, Jan 1995, pp 3-17.

[3]Peter Chadwick, G3RZP, "Phase Noise Intermodulation and Dynamic Range," *Frequency Dividers and Synthesizers IC Handbook*, Plessey Semiconductors, 1988, p 151.

[4]Klaas Spaargaren, PA0KSB, "Technical Topics: Crystal-Stabilized VFO" *RadCom*, Jul 1973, pp 472-473. Comments followed in later "Technical Topics" columns. Also see, "Frequency Stabilization of L-C Oscillators," *QEX*, February 1996, pp 19-23.

[5]Ulrich Rohde, KA2WEU/DJ2LR, *Digital PLL Frequency Synthesizers* (Englewood Cliffs: Prentice-Hall, 1983) p 78.

[6]For contact information for Mouser Electronics, Digi-Key Corp, Allied Electronics and Amidon Inc, see the Address List in the **References** chapter.

Contents

Digital Signal Processing
18

Digital signal processing (DSP) is one of the great technological innovations of the last hundred years. It has found a permanent place not only in radio, but also in the exploration for oil and other fossil fuels, high-definition television (HDTV), compact-disc (CD) recording and many other facets of our lives. Its popularity stems from certain advantages: DSP filters do not need tuning and may be exactly duplicated from unit to unit; temperature variations are virtually non-existent; and DSP represents the ultimate in flexibility, since general-purpose DSP hardware can be programmed to perform many different functions, often eliminating other hardware. This chapter was written by Doug Smith, KF6DX.

DSP FUNDAMENTALS

In this chapter, you will see that DSP is about rapidly measuring analog signals, recording the measurements as a series of numbers, processing those numbers, then converting the new sequence back to analog signals. How we process the numbers depends on which of many possible functions we are performing. We will take a look at some of those functions and explore how real DSP systems are implemented in software and hardware.

Sampling

The process of generating a sequence of numbers that represent periodic measurements of a continuous analog waveform is called *sampling*. Each number in the sequence is a single measurement of the instantaneous amplitude of the waveform at a sampling time. When we make the measurements continually at regular intervals, the result is a sequence of numbers representing the amplitude of the signal at evenly spaced times.

This process is illustrated in **Fig 18.1**. Note that the frequency of the sine wave being sampled is much less than the *sampling frequency*, f_s. In other words, we are taking many samples during each cycle of the sine wave. The sampled waveform does not contain information about what

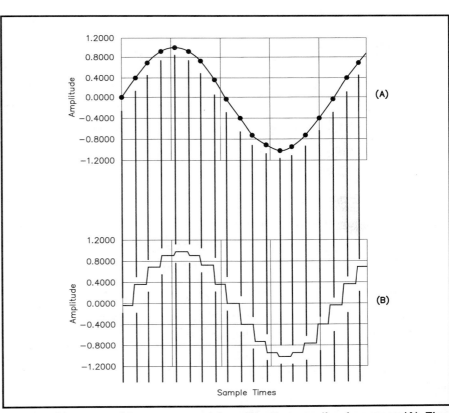

Fig 18.1—Sine wave of frequency much less than the sampling frequency (A). The sampled sine wave (B).

the analog signal did between samples, but it still roughly resembles the sine wave. Were we to feed the analog sine wave into a spectrum analyzer, we would see a single spike at the sine wave's frequency. Pretty obviously, the spectrum of the sampled waveform is not the same, since it is a stepwise representation.

The sampled signal's spectrum can be predicted and interpreted in the following way. The analog sine wave's spectrum is shown in **Fig 18.2A**, above the spectrum of the string of evenly spaced sampling impulses in Fig 18.2B. The sampled signal is just the *product* of the two signals; its spectrum is the *convolution* of the two input spectra, as shown in Fig 18.2C. The sampling process is equivalent to a mixing process: They each perform a multiplication of the two input signals.

Note that the sampled spectrum repeats at intervals of f_s. These repetitions are called *aliases* and are as real as the fundamental in the sampled signal. Each contains all the information necessary to fully describe the original signal. In general, we are only interested in the fundamental, but let's see what happens when the sampling frequency is *less than* that of the analog input.

Sine Wave, alias Sine Wave: Harmonic Sampling

Take the case wherein the sampling frequency is less than that of the analog sine wave. See **Fig 18.3**. The sampled output no longer matches the input waveform. Notice that the sampled signal retains the shape of a sine wave at a frequency lower than that of the input. Ordinarily, this would not be a happy situation.

A downward frequency translation is useful, though, in the design of IF-DSP receivers. In addition, lower sampling frequencies are good because they allow more time between samples for signal processing algorithms to do their work; that is, lower sampling rates ease the processing burden. Caution is required, though: An input signal near twice the sampling frequency would produce the same output as that of Fig 18.3. To use this technique, then, we must first limit the BW of the input: A BPF is called for. This is known as *harmonic sampling*. The BPF is referred to as an *anti-aliasing* filter.

Input signals must fall between the fundamental (or some harmonic) of the sampling frequency and the point half way to the next higher harmonic. A frequency translation will take place, but no information about the shape of the input signal will be lost. A spectral representation of harmonic sampling is shown in **Fig 18.4**. It reveals the basis for the often-misquoted

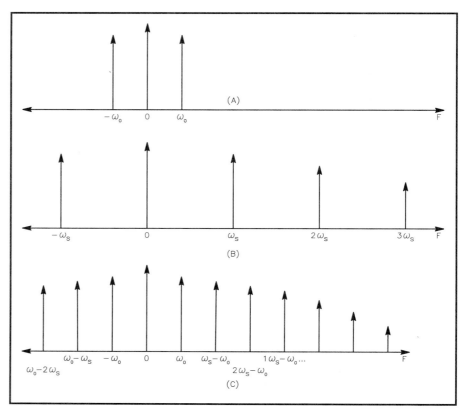

Fig 18.2—Spectrum of an analog sine wave (A). The spectrum of a string of evenly spaced sampling impulses (B). The spectrum of the sampled sine wave (C).

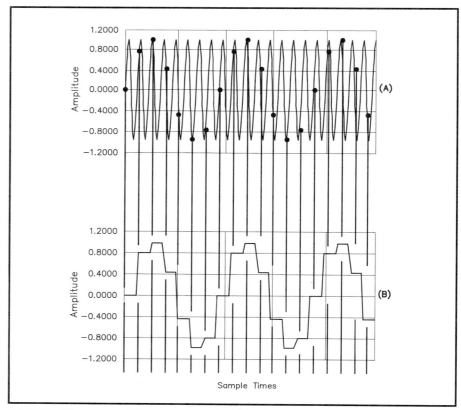

Fig 18.3—Sine wave of frequency greater than the sampling frequency (A). Harmonically sampled sine wave (B).

Nyquist sampling theorem: The sampling frequency must be at least twice the input BW to avoid aliasing. Such aliasing would destroy information; once incurred, nothing can remedy it.

Data Converters and Quantization Noise

The device used to perform sampling is called an *analog-to-digital converter* (ADC). An ADC produces a binary number that is directly proportional to the input amplitude, so only a limited number of amplitude values can be represented. An 8-bit ADC, for example, can only give one of 256 values. This means the amplitude reported is not the exact amplitude of the input, but only the closest value of those available. The difference is called the *quantization error*.

The amplitude reported by the ADC can, therefore, be thought of as the sum of two signals: the desired input and the quantization error. In a perfect ADC, the error cannot exceed ±1/2 of the value of the least-significant bit of the converter—this is the error signal's peak-to-peak ampli-tude. Assuming the desired input is changing and covers a large range of quantization levels, the error is just as likely to be negative as positive, and just as likely to be small as large. Hence, the error signal is pseudo-random and appears as *quantization noise*.

This noise is spread uniformly over the entire input BW of $f_s/2$. Taking this and the maximum signal the ADC can handle into account, the maximum signal-to-quantization-noise ratio produced by the ADC is:

$$\mathrm{SNR}_{max} \approx 6.02\,b + 1.76\ \mathrm{dB} \qquad (1)$$

where b is the number of bits used by the converter.

For a simple 16-bit ADC, the SNR cannot exceed about 98 dB. The reason we wrote that the quantization noise was pseudo-random and not truly random is the following: If there were a harmonic relationship between the input signal and the sampling frequency, the noise might tend to concentrate itself at discrete frequencies, yielding higher power levels than that indicated by Eq 1.

Aperture Jitter

In addition to quantization noise, noise is introduced in ADCs by slight variations in the exact times of sampling. Phase noise in the ADC's clock source, as well as other inaccuracies in the sampling mechanisms, produce undesired phase modulation of the sampled signal. Again, assuming it is uncorrelated with the input signal, this *aperture jitter noise* will be distributed across the entire input BW. Its amplitude is proportional to the squares of both the desired signal's frequency and the RMS time jitter in the sampling rate, and inversely proportional to the sampling rate itself. With contemporary crystal-derived clock sources, aperture jitter is usually not a significant factor until the sampling frequencies reach VHF; even at those frequencies, the effect may be small compared with quantization noise.

Over-Sampling and Sigma-Delta ADCs

The nature of the above-mentioned noise sources is such that if we could increase the sampling frequency by some factor N, then digitally filter the output back down to a lower rate, we could improve the SNR by almost the factor N. This is because the noise would be spread over a larger BW; much of the high-frequency noise would be eliminated by the digital filter. This technique is called *over-sampling*.

So-called *sigma-delta* converters use this method to achieve the best possible dynamic range. They employ one-bit quantizers at very high speed and digital decimation filters (described later) to reduce the sampling frequency, thus improving SNR. They represent the state of the art in ADC technology. Other factors, such as the noise figure of analog stages inside an ADC, tend to limit the SNR of real converters to within a few dB of that calculated by Eq 1.

Non-Linearity in ADCs

The quantization steps of a real converter are not perfectly spaced; conversion results are contaminated by the inaccuracy. In general, two types of non-linearity are characterized by manufacturers: *differential non-linearity* (DNL) and *integral non-linearity* (INL).

DNL is the measure of the output non-uniformity from one input step to the next. It is expressed as the maximum error in the output between adjacent input steps as measured over the entire input range of the device. The worst errors usually occur near the middle of the scale. Since we are talking about the accuracy of the smallest steps the converter can resolve, noisy low-order distortion products caused by this

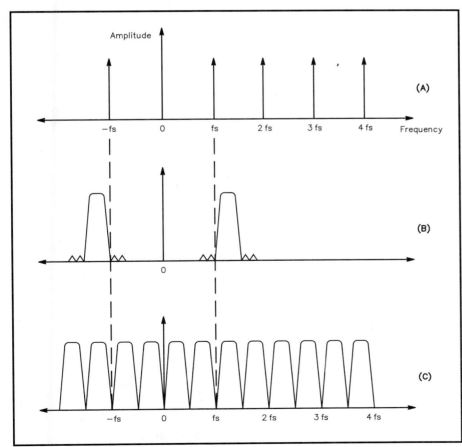

Fig 18.4—Spectrum of sampling impulses (A). Spectrum of a band of real signals (B). Spectrum of a harmonically sampled band of real signals (C).

effect limit dynamic range. Current technology uses correction systems to compensate for temperature variations that would otherwise further degrade performance.

An ADC is considered *monotonic* if a steady increase in the input signal always results in an increase in the output. Device manufacturers hold DNL to ± 0.5 bits or better so that monotonicity is maintained.

INL is a measure of an ADC's large-signal handling capability. To measure it, we first inject a signal of amplitude A and measure the output; when we inject a signal of amplitude 100A, we expect the output to grow in exact proportion. INL represents the maximum error in the output between *any two* input levels. Another way to think about this is to plot the input against the output and see how straight the line is. INL produces harmonic distortion and IMD; values for typical converters are ± 1 or 2 bits over the entire range.

Spurious-Free Dynamic Range and Dithering

Spurious-free dynamic range (SFDR) is defined as the ratio of the largest signal the converter can accurately handle to the largest source of noise and distortion caused by effects mentioned above. Quite often, undesired components may appear in unexpected parts of the input spectrum; spurious responses may be found without apparent explanation. It turns out there are explanations, of course, but we will defer that discussion. Suffice it to write here that manufacturers test for SFDR and usually specify it on their data sheets, especially for high-speed devices.

Sometimes noise and distortion effects conspire to add at discrete frequencies. It is found that the addition of random noise at the clock input helps dissipate these spurious responses. This technique is known as *dithering*. It may seem strange, but artificial noise—usually several bits in amplitude and high enough in frequency to be eliminated by the decimation filter—actually reduces quantization noise and improves performance rather than degrading it. This is discussed in more detail later.

Digital-to-Analog Converters: Additional Distortion Sources

Digital-to-analog converters (DACs) perform the conversion of binary numbers back into analog voltages—the inverse operation of ADCs. They suffer from all the inadequacies described earlier, as well as a few of their own. The first unique distortion of DACs is one of frequency response: *zero-order sample-and-hold distortion*.

Typical converters are sample-and-hold devices: They continue to output the last sampled value throughout the sample period. This effect acts as a low-pass filter having a frequency response:

$$H_f = \frac{\sin\left(\dfrac{\pi f}{f_s}\right)}{\left(\dfrac{\pi f}{f_s}\right)} \qquad (2)$$

The high-frequency roll-off is quite undesirable in many circumstances. For example, if the output frequency is one quarter the sampling frequency, an attenuation of about 1 dB will occur. Correction can be made for this, but an increase in sampling frequency reduces the attenuation. Interpolation of the sampled output signal (described later) is called for in many cases.

Settling Time and Glitch Energy

When the output of a DAC changes from one voltage to another, it obviously cannot do so instantaneously; a finite time is required for the voltage to reach its new value. This is known as the *settling time*. It is usually defined as the time required to settle to within some number of voltage-equivalent bits of the final value.

Glitch energy or *glitch area* is defined as the product of the voltage error during the settling time and the settling time itself. While volt-seconds are not units of energy, it is assumed the DAC is driving some kind of load; thus, these units can be translated into units of energy (watt-seconds), performing work on that load. The settling mechanism is an important factor in the production of spurious outputs in DACs. Manufacturers usually specify the glitch energy for their high-speed devices. It is an especially important number for direct-digital-synthesis (DDS) applications.

Note also that DACs produce aliases, again repeating at intervals of f_s. These must usually be removed using an analog LPF. Occasionally, a BPF may be used, and one of the aliases taken as the desired output. This can be a clever way of getting an upward frequency translation under certain conditions.

Reducing the Sampling Frequency: Decimation

As we have seen, sampling at high rates is beneficial because it eases the design of the analog filters we must use to avoid aliasing. It also reduces quantization noise and aperture jitter. We have also noted that lower sampling rates help reduce the computational burden in DSP systems. In addition, we will discover that when it is time to digitally filter some signals, making the filter's BW a large fraction of the sampling frequency makes it easier to build sharp-skirted filters—exactly what DSP is famous for.

Reduction of the sampling frequency is usually called *decimation*. Decimation is normally done by integer factors (although it does not have to be) and is equivalent to resampling the already-sampled signal at a lower rate. The resampled signal has a family of aliases, repeating at intervals of the lower sampling frequency; we have to reduce the BW to less than half this lower sampling frequency to avoid the aliasing that would destroy information.

The process of decimation is simple: Just throw away the unwanted samples. To decimate by two, for example, only every other sample is retained. A *decimation filter*, operating at the higher sampling rate, f_s, reduces signal BW to less than $f_s/4$ prior to discarding the samples to avoid aliasing. But why spend time computing filter outputs that we are only going to discard? We may compute only those we intend to keep. This is exactly the same as running the decimation filter at the lower rate. This method is typical of those used by DSP designers to save time and effort. See the chapter Appendix for a software project (Project A) that demonstrates decimation using Alkin's *PC-DSP* program. This program is included with the book listed in the Bibliography.

Increasing the Sampling Frequency: Interpolation

We learned that when it is time to convert back to analog, an artificial increase in sampling rate may be advantageous. It will push aliases higher in frequency where they are easier to remove by analog filtering, and it will relieve some of the sample-and-hold distortion. So, even having decimated the data at some earlier stage in our designs, we may later employ the process of *interpolation*.

Decimation was performed by deleting samples. Interpolation is performed by inserting them. The inserted samples have a value of zero and are placed between the existing samples. While this increases the sampling frequency, the information in the original samples is not destroyed; however, new information is added in the form of aliases, and an *interpolation filter* is usually required. This filter, most often a low-pass, operates at the higher sampling frequency, f_s, and eliminates components in the interpolated data above $f_s/4$.

The way numbers are represented in DSPs is a major consideration. Let's take a look at this before moving on to filtering algorithms.

Representation of Numbers: Floating-Point vs Fixed-Point

One of the things that makes general-purpose computers so useful is their ability to perform *floating-point* calculations. In this form of numeric representation, numbers are stored in two pieces: a fractional part, or *mantissa*, and an exponent. The mantissa is assumed to be a binary number representing an absolute value less than unity, and the exponent, a binary integer. This approach allows the computer to handle a large range of numbers, from very small to very large. Some DSP chips support floating-point calculations, but it is not as great an advantage in signal processing as it is in general-purpose computing because the range of values we are dealing with in DSP is limited anyway. For this reason, *fixed-point* processors are common in DSP.

A fixed-point processor treats numbers as just the mantissa and does away with the exponent. The radix point—the separation between the integer and fractional parts of a number—is usually assumed to reside to the left of the most-significant bit. This is convenient, since the product of two fractions less than unity is always another fraction less than unity. The *sum* of two fractions, though, may be greater than unity: *overflow* would be the result. Overflow is a constant concern for fixed-point DSP programmers and leads to considerations for *scaling* of data, as discussed further below, which may limit system dynamic range to less than the data converters' capabilities.

DSP ALGORITHMS FOR RADIO

Digital Filters

The ability to construct high-performance filters is probably the most important rationale for using DSP in radio transceivers. An expensive crystal or mechanical filter having a single BW can be replaced by a set of superior digital filters, offering as many BWs as the associated on-board memory can support.

As shape-factor requirements get more stringent, filters get more complex. As a filter gets more complex—with additional inductors and capacitors in the analog case, or additional delay elements in the digital case—the sensitivity of the filter's response to errors in the element values becomes more severe. Thus, for analog filters, precise values of resistance, inductance and capacitance must be maintained if the filter is to operate as designed. Establishing those values is difficult; holding them within tolerances over temperature variations and aging is more so.

DSP filters, on the other hand, are unchanging. The "component" values are numbers stored in a computer that are not susceptible to temperature changes or aging. Filters that would be impractical or impossible in the analog realm are easily implemented by DSP algorithms.

We can build digital filters having linear phase responses, which is very difficult in the analog world. This is an advantage mainly for digital communication modes such as FSK and PSK. Also, filters may be combined numerically to yield composite responses without the need for adding hardware. This is useful for passband tuning or graphic-equalizer applications.

DSP filters are usually characterized by their *impulse responses*. The impulse response of a digital filter is the output of the filter when the input is a one-sample, unity-amplitude impulse. Impulse response is directly related to frequency response by a *Fourier transform*, about which we will learn more later. Suffice it to write for now that digital filters may be broadly divided into two classes: finite impulse response (FIR) and infinite impulse response (IIR). The presence or absence of feedback separates the two.

FIR FILTERS

Take a look at the block diagram of the FIR filter shown in **Fig 18.5**. The string of boxes labeled z^{-1} is simply a delay line, with each box representing a one-sample delay. Programmers will note that with one input sample in each position, this is just a buffer of length five. Each buffer location may be referred to as a *tap* in the delay line. The datum at each tap, $x(n)$, is multiplied by one of the filter *coefficients*, $h(n)$. All the products are summed at each sample time to produce the filter output. At the next sample time, samples are shifted down the delay line by one position and the *multiply-and-accumulate*

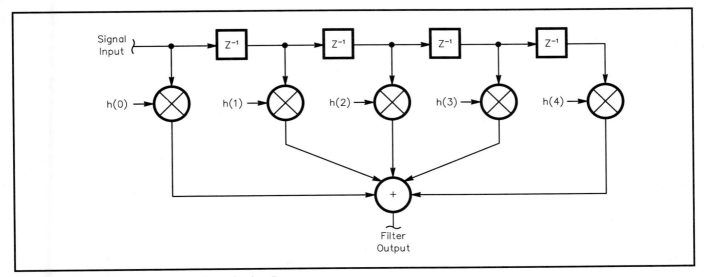

Fig 18.5—Block diagram of an FIR filter for L = 5.

(MAC) operation is performed again. Coefficients remain in place and do not shift. The mathematical expression describing this repetitive MAC operation is also called a *convolution sum*:

$$y(k) = \sum_{n=0}^{L-1} h(n) x(k-n) \qquad (3)$$

where x(k-n) represents the input data in the buffer.

Since the output depends only on past input values, the filter is said to be a *causal process*. Since no feedback is employed, it is unconditionally stable.

In an FIR filter, the set of coefficients, h(n), is identical to the impulse response of the filter. The trick, then, is to find the impulse response that gives us the frequency response we want. Almost any frequency response can be generated if we use enough taps. In general, low shape factors (steeper roll-offs) require more taps. Most filter-design methods begin with an estimate of the number of taps needed. Rabiner and Gold indicate the estimate may be taken as:

$$L = 1 - \frac{10 \log(\delta_1 \delta_2) - 15}{14 \left(\dfrac{f_T}{f_s} \right)} \qquad (4)$$

where δ_1 is the passband ripple, δ_2 is the stopband attenuation, f_T is the transition BW, f_s is the sampling frequency, and L, the number of taps, is called the length of the filter. This equation assumes that enough bits of resolution are used to achieve the required accuracy. In practice, filters of over 100 taps are used to realize shape factors of less than 1.15.

Normally, an FIR filter's impulse response has a symmetry about center; that is, h(0)=h(L−1), h(1)=h(L−2), and so forth. It turns out this is sufficient to ensure a linear phase response and flat group-delay characteristics. The total delay through an FIR filter of length L is:

$$t = \frac{L}{2f_s} \qquad (5)$$

As noted, this delay is *independent* of frequency. Remember that longer filters demand more processing than shorter filters.

When personal computers are used to design FIR filters, coefficients are usually represented in floating-point format to the full accuracy of the computer—often with 12 or more significant figures in the mantissa. Embedded, fixed-point DSP implementations ordinarily achieve only 16-bit accuracy. The *truncation* of coefficients and data to this accuracy affects the frequency response and ultimate attenua-

tion of filters, and may be the factor determining dynamic range. Also notice that when we multiply a 16-bit coefficient by a 16-bit datum, the product is a 32-bit number; we are then adding several 32-bit numbers in the final accumulator of an FIR filter. The result may grow by several more bits to 35 or so by the time we are done. At some stage, the result may overflow the accumulator, especially in FIR filters with small transition BWs (sharp skirts). The worst-case output can grow as large as the sum of the absolute value of all the coefficients:

$$y_{max} = \pm \sum_{n=0}^{L-1} |h(n)| \qquad (6)$$

We might have to scale the data, the coefficients, or both by the reciprocal of this number to avoid overflow.

The filter output at each sample time is usually *rounded* back down to the bit-resolution of the DAC; say, to 16 bits. The rounding operation introduces a small error in the result. This rounding error is directly analogous to quantization noise; it is computed in almost exactly the same way. A trade-off exists between the possibility of overflow, which is catastrophic, and loss of accuracy because of rounding. It is interesting to note that truncation of filter coefficients affects the frequency response of the filter, but not the amount of noise in the output. On the other hand, truncation and rounding of data do not affect the frequency response but add quantization noise to the output.

One FIR filter-design approach takes advantage of the fact that a filter's frequency response is the Fourier transform of its impulse response. Thus, we may start with a sampled version of the frequency response and apply an *inverse* Fourier transform to obtain the impulse response. All filter-design software is capable of using this method. Better designs may be obtained in many cases by using an algorithm developed by Parks and McClellan. This approach produces an *equi-ripple* design in which all of the passband ripples are the same amplitude, as are all the stopband ripples. Another popular algorithm is the *least-squares* method. Its claim to fame is that it minimizes the error in the desired frequency response.

Since finding coefficient sets for a given filter design is so computationally intensive, it is a good job for a computer program. DSP filter-design programs are readily available at low cost. Refer to the System Software section below for further discussion of filter design and the Bibliography at the end of this chapter for a list of software design tools. The article by Kossor

has a practical circuit example of a commutating BPF that employs principles of DSP. Also see Project B in the chapter Appendix for examples of FIR filter designs.

IIR Filters

While FIR filters have a lot going for them, they tend to require a large number of taps and a proportional amount of processing power. As opposed to that, an IIR filter can provide sharp transition BWs with relatively few calculations. What it will not provide, in general, is a linear phase response. In circumstances where the computational burden is of more concern than the phase response, IIR filters may be desirable.

Unlike FIR filters, IIR filters employ feedback: That is what makes their impulse responses infinite. For this reason, IIR filters are usually designed by converting traditional analog filter designs, such as Chebyshev and elliptical types. See the Filters and Projects chapter of this book for a description of those designs. The transfer function of an analog Chebyshev low-pass filter can be written as the ratio of a constant to an nth-order polynomial:

$$H_s = \frac{K}{a_0 s^n + a_1 s^{n-1} + a_2 s^{n-2} + \cdots + a_n s^0} \qquad (7)$$

Tables in the literature, such as in Zverev, list the values of the coefficients, a_n, related to the cutoff frequency; these are used to derive actual component values for the filter. The low-pass design can be transformed to band-pass or band-stop responses. Two popular methods exist for deriving the digital transfer function from the analog: These are known as the *impulse-invariant* and *bilinear transform* methods.

The impulse-invariant method assures that the digital filter will have an impulse response equivalent to its analog counterpart, and thus the same phase response. Problems arise, though, if the bands of interest are near half the sampling frequency; the digital filter's response can develop serious errors in this case. Because of this problem, the impulse-invariant method is not as good as the bilinear transform method. The bilinear transform method makes a convenient substitution for s in Eq 7 above. The filter output comes out as:

$$y(k) = \sum_{n=0}^{L-1} \alpha(n) x(k-n) - \sum_{n=1}^{L-1} \beta(n) y(k-n) \qquad (8)$$

This filter has L zeros and L−1 poles. The block diagram of such a filter for L = 5 is shown in **Fig 18.6**. Feedback is evident in the diagram: The paths involving coefficients ß loop back and are added to the signal path.

The *direct form* of Eq 8 may be factored into 2-pole sections and implemented in cascaded form. The output of each section serves as the input to the next. See **Fig 18.7** This configuration requires a few more multiplications than the direct form, but suffers less from instability problems that may plague IIR filters. Since feedback is being used, IIR filters are not necessarily unconditionally stable. They also tend to be prone to *limit cycles*, low-level oscillations that arise near the lower end of the dynamic range. For these and other reasons, data and coefficient storage should be cleared or set to zero before processing begins.

A Simple Digital Notch Filter

Along with common LPFs, HPFs and BPFs, radio designers are interested in one other type of filter: the notch. While most filter-design software can generate notch filters using FIR methods discussed above, Widrow and Stearns have described an unusual type in which the number of taps is minimized. In fact, they were able to prove that only two taps are needed for each frequency to be notched. This is great, since it reduces computation to almost nil. We will take a look at it here and touch briefly on some of the theory of *adaptive signal processing*, treated in depth later.

The situation is this: We want to copy a broadband signal, such as an SSB phone signal, and suddenly a dreadful carrier appears in the passband. Our notch filter will remove it and we will have complete control over the notch width, as well as a notch depth limited only by the bit resolution of our system. Dr. Widrow found that one can build a filtering system that minimizes repetitive signal energy by altering the filter coefficients "on the fly" using a certain algorithm. Known as the *least-mean-squares* (LMS) method, it describes a way to adjust filter coefficients over time to remove undesired, steady tones in the

input. A complex reference signal is used at the exact frequency of the offending tone. The algorithm then forms a BPF centered at the tone frequency whose output is subtracted from the input to create the notch. The block diagram of a two-tap system is shown in **Fig 18.8**.

The broadband input is called d(t). The reference input consists of two signals, $A\cos(\omega_0 t)$ and $A\sin(\omega_0 t)$. These signals feed multipliers having coefficients h(1) and h(2), which in turn feed an accumulator just as in a normal FIR filter. This is the BPF output; it is subtracted from the input to form the notch output, e(t). Note that the BPF output is also available at no additional overhead. While the initial values of the coefficients are unimportant to the steady state, the procedure for updating them with the LMS algorithm is:

$$h_{t+1}(1) = h_t(1) + 2\mu e(t)x_t(1)$$
$$h_{t+1}(2) = h_t(2) + 2\mu e(t)x_t(2) \qquad (9)$$

where $0 < \mu < 1$. Analysis shows that as the

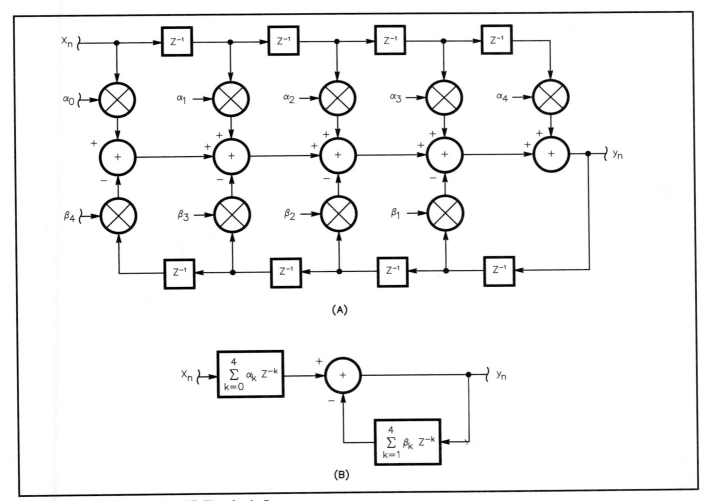

(A)

(B)

Fig 18.6—Block diagram of an IIR filter for L=5.

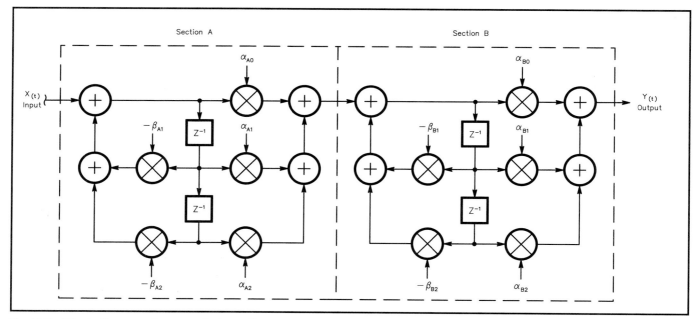

Fig 18.7—Block diagram of a cascade-form IIR filter.

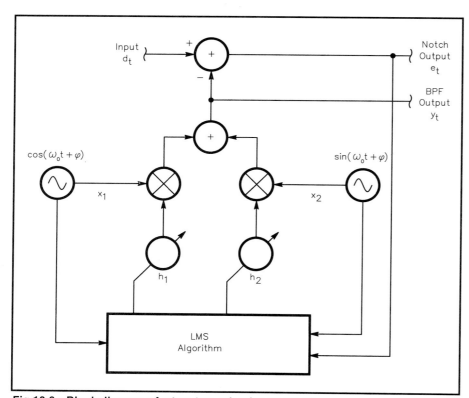

Fig 18.8—Block diagram of a two-tap, adaptive notch filter.

by varying the factor μ and the amplitude of the reference signal. The depth of the null is, in general, superior to that of a fixed filter because the algorithm tracks the correct phase relationship for ideal cancellation, even if the reference frequency is changing slowly with the offending tone. Each additional tone to be notched demands two additional taps in the filter. Noise in the input may cause us to have to add more taps to maintain sufficient accuracy. Additional detail of adaptive signal processing will be found below and in material shown in the Bibliography.

Lattice and Other Structures

While many filter-design software packages do not have the capability to work with them, *lattice structures* and other types of digital filters have seen use, especially in adaptive signal processing. Crystal and mechanical lattice filters are common elements of many transceivers. A digital lattice or ladder filter is a lot like its analog brother. The design of digital lattice filters is similar, as well. Digital lattice filters may be either FIR or IIR. Also note that from the IIR cascade form above, we can derive a *parallel form* that may be computationally beneficial in some cases. The design of this kind of filter is a very complicated session in partial fraction expansion. Widrow and Stearns provide more information on these and other exotic concepts.

reference inputs are sinusoidal, the system is linear and time-invariant for output e(t), although the coefficient values do not necessarily approach any fixed value. The 3-dB BW of the notch is:

$$BW = \frac{2\mu A^2}{t_s} \text{rad/s} \quad (10)$$

The Q of the filter may be readily computed. Thus, we have control over the BW

ANALYTIC SIGNALS AND MODULATION

DSP implementations of transceiver functions, such as modulation and demodulation, compel designers to examine the mathematics behind them. Computers are good at crunching numbers, but they do exactly what they are told! If we expect a DSP system to generate an SSB signal, for example, we had better know which calculations to perform and which to avoid.

Mathematics of Complex Signals

Because DSP makes it easy to build frequency-independent phase shifters—a fantasy in the analog world—the phasing or "I/Q" method has dominated other modulation techniques. Complex signals are not generally well understood and quite often form a stumbling block to those wishing to grasp DSP concepts. The idea of negative frequency is especially troublesome. The key to understanding these concepts lies in the theory of complex numbers. A real signal, such as a cosine wave, is normally thought of as a positive frequency. It can be transmitted and detected normally; however, we shall see that such a signal actually consists of positive *and* negative frequencies when examined in the complex domain.

A real cosine wave embodies the relation:

$$x(t) = \cos(\omega t) \qquad (11)$$

where $\omega = 2\pi f$, and t is time. In the complex domain, the cosine wave is really the sum of two complex signals:

$$x(t) \frac{1}{2}\left\{ \begin{array}{l} \left[\cos(\omega t) + j\sin(\omega t)\right] + \\ \left[\cos(\omega t) - j\sin(\omega t)\right] \end{array} \right\} \qquad (12)$$

This signal has both positive and negative frequency components. The real parts add and the imaginary parts cancel to make the equation true. In the complex plane, where the real part is one axis and the imaginary part the other, this signal can be represented as two vectors rotating in opposite directions. See **Fig 18.9**.

While this depiction is beautiful and elegant to the mathematician, what does it really mean to you and me? Well, it means that signals represented in complex form can have a one-sided spectrum—that is, only a positive or a negative frequency component. This is useful as we mix our signals upward to their final frequency positions in a modulator.

As our first example, let's select the task of taking a real input signal, such as the audio from a microphone, and converting it to an SSB signal that can be transmitted. We obviously have to translate the audio signal upward in frequency and preserve its spectral content within the band we want the transmitted signal to occupy. If we wish to produce an upper-sideband (USB) signal, we want the carrier and lower sideband to be suppressed as much as possible. Were we able to translate the spectrum of our cosine wave—with its symmetrical positive- and negative-frequency components—upward in frequency far enough, we would have two positive frequencies separated by twice the original signal's frequency. For a real signal, this is exactly what happens when it is applied to an analog mixer: Both sum and difference frequencies are generated. See the **Receivers, Transmitters, Transceivers and Projects** chapter for more detail of the operation of mixers as multipliers.

To move our sampled audio signal upward in frequency, we must multiply it by (mix it with) a local oscillator. The local-oscillator function can be implemented in DSP software using direct digital synthesis (DDS) techniques. In this case, though, the local oscillator must be complex; that is, it must have two outputs with a 90° phase relationship between them. This is the same as saying there must be both a sine and a cosine output from it. This will enable us to mix signals having a one-sided spectrum.

When we implement a complex mixer in DSP, we are multiplying complex numbers by complex numbers. Note that the calculations for the real and imaginary parts are carried out separately; each part is treated as if it were a single, real multiplication. Two complex numbers a + jb and c + jd, when multiplied, produce:

$$(a + jb)(c + jd) = (ac - bd) + j(ad + bc) \qquad (13)$$

Four real multiplications and two real additions are required.

Hilbert Transformers and an SSB Modulator

If we want to create a signal having a one-sided spectrum from a real input signal, such as from the microphone, we need to shift all the frequency components in the sampled signal by 90°. Fortunately, in DSP, we have a way to do that: the *Hilbert transformer*. Recall that an FIR filter with a symmetrical impulse response exhibits a constant, frequency-independent delay. It turns out a filter with an *anti-symmetrical* impulse response—that is, with h(0) = –h(L –1), h(1) = –h(L –2), and so forth—produces a linear phase response, too, but with a phase response exactly 90° different from the symmetrical-impulse-response filter. This is exactly the type of filter we need to generate the components of an *analytic signal*.

Fig 18.10 shows a system using a Hilbert transformer to create an analytic signal from the microphone audio. Since the Hilbert transformer includes not only a 90° phase shift, but also a fixed delay of L/2 sample periods, we need an L/2 delay in the leg that does not contain a phase shift. The delay through the two paths is then equal and the only difference between the two signals produced is the 90° phase shift. The non-phase-shifted signal is called I, the phase-shifted signal is called Q. Together, these signals form our analytic signal I + jQ. Now let's see what it looks like when we multiply this signal by a complex local oscillator.

In this case, we are performing the multiplication:

$$\begin{aligned} \left[\cos(\omega t) + j\sin(\omega t)\right]&\left[I(t) + jQ(t)\right] = \\ &\left[I(t)\cos(\omega t) - Q(t)\sin(\omega t)\right] \\ &+ j\left[I(t)\sin(\omega t) + Q(t)\cos(\omega t)\right] \end{aligned} \qquad (14)$$

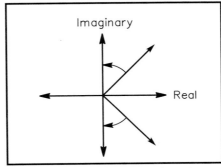

Fig 18.9—Vector representation of a real cosine wave.

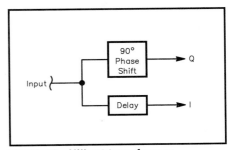

Fig 18.10—Hilbert transformer producing an analytic signal.

This is the equation for a USB signal. We are only interested in the real part of the result, since we only have one real channel on which to transmit. For this reason, the system is really a *half-complex mixer*.

A block diagram of such a mixer is shown in **Fig 18.11**. This is, in fact, the phasing method. Output signals are translated upward by the frequency of the local oscillator, ω radians per second, or ω/2π hertz. Most transmitter designs will translate signals to an IF significantly higher in frequency than audio, so it is wise to include an increase in the sampling rate prior to mixing. An interpolation filter is natu-

rally needed. It is particularly convenient to choose an interpolation factor of 4, because the cosine LO produces values of 1, 0, −1 and 0 during a full cycle; the sine LO produces values of 0, 1, 0 and −1. No actual multiplications need take place, saving time and accuracy. The Hilbert transformer can operate at the lower, original sampling rate, but we would like to include band-pass filtering to limit the spectrum to about 3 kHz BW. In fact, we can build a pair of DSP filters that provide the BPF response and the 90° phase relationship, as described below. Our SSB modulator then matches that shown in **Fig 18.12**.

Before discussing how to generate analytic filter pairs, it is worth noting a few properties of SSB signals created in this way. First, were we to add the I and Q signals instead of subtract them in the summation block of Fig 18.11, we would have an LSB signal instead of USB. It is not too hard to see that we could easily both add and subtract to produce a DSB, suppressed-carrier signal. We can even pre-add and subtract *two* audio signals to produce an independent-sideband (ISB) signal, as shown in **Fig 18.13**. More than two channels can be combined in this way. Second, since the amplitude of the carrier, $\cos(\omega_0 t) \pm j\sin(\omega_0 t)$, is constant, the amplitude of an SSB signal can be specified as some function of the modulating signal. If we think of the analytic audio signal as a vector in the complex plane, its length is equal to the signal's instantaneous amplitude:

$$A(t) = \left[I^2(t) + Q^2(t)\right]^{\frac{1}{2}} \quad (15)$$

Finally, the phase of the signal is the instantaneous angle of this rotating vector:

$$\phi(t) = \arctan\left[\frac{Q(t)}{I(t)}\right] \quad (16)$$

Now we can rewrite the real part of Eq 14 as:

$$A(t)\cos[\omega t + \phi(t)] \quad (17)$$

This expression clearly shows SSB to be a

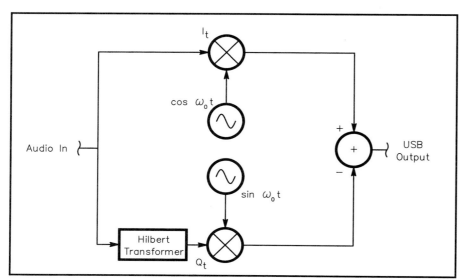

Fig 18.11—Block diagram of a half-complex mixer.

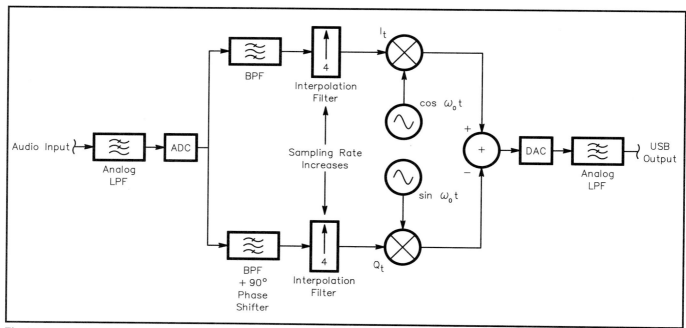

Fig 18.12—Block diagram of a digital SSB modulator.

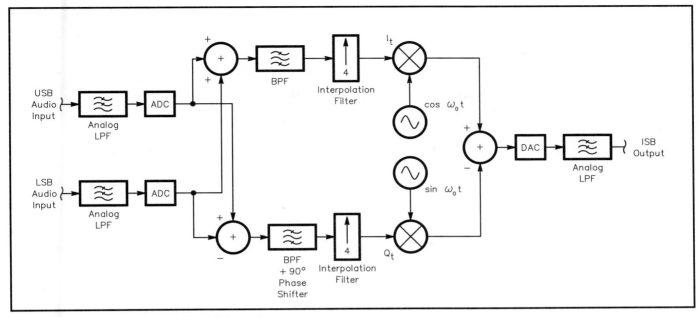

Fig 18.13—Block diagram of a digital ISB modulator.

hybrid of amplitude and phase modulation. Also, we can write the equation:

$$[I(t) + jQ(t)]$$
$$= A(t)\{\cos[\phi(t)] + j\sin[\phi(t)]\} \quad (18)$$

directly relating the envelope and phase to the analytic *baseband* signal. The amplitude and phase of the analytic signal are identical to those of the SSB wave so produced.

Analytic Filter-Pair Synthesis

We have seen how complex mixing translates signals in frequency with a one-sided spectrum. We will use this fact to our advantage in creating an analytic filter pair. Each filter will have the same frequency response as the other; they will differ only in their phase responses.

We begin by designing a low-pass filter having the desired transition-band characteristic, $H(\omega)$; we obtain its impulse response, $h(t)$. Multiplying the impulse response by a complex sinusoid of angular frequency ω_0 results in two sets of coefficients—one for the real part, and one for the imaginary part:

$$h_I(t) = h(t)\cos(\omega_0 t)$$
$$h_Q(t) = h(t)\sin(\omega_0 t) \quad (19)$$

The frequency response of either one of these filters is given by:

$$H_\omega = \frac{H_{(\omega - \omega_0)} + H_{(\omega + \omega_0)}}{2} \quad (20)$$

which is a BPF centered at ω_0. The I filter has a phase response differing 90° at every frequency from the Q filter. The frequency translation theorem works just as well on the responses of filters as it does on real signals. To perform this transformation of the L coefficients of the prototype LPF, we calculate new coefficients according to:

For $0 \le k \le L-1$,

$$h_I(k) = h(k)\cos\left[\omega_0\left(k - \frac{L}{2} + \frac{1}{2}\right)t_s\right] \text{ OR}$$

$$h_Q(k) = h(k)\sin\left[\omega_0\left(k - \frac{L}{2} + \frac{1}{2}\right)t_s\right]$$

$$(21)$$

where t_s is the sampling period. When the low-frequency transition band is placed near zero frequency, as we would like for SSB, the BW of each BPF is approximately twice that of the prototype LPF. A very interesting thing ordinarily happens when the number of taps is odd: The odd-numbered coefficients are zero. This allows reduction in computation by a factor of two. Refer to Project C in the Appendix for a practical example of how analytic filter pairs are generated.

We can alter the exciter's frequency response by convolving the impulse response of our analytic filter pair with that of a filter having the desired characteristic. New coefficients are calculated using the same convolution sum as in Eq 3. Graphic or parametric equalizers may be implemented in this way.

Demodulation: SSB

As in digital exciters, phasing methods prevail in receivers; the process is almost exactly the reverse of the modulator's. **Fig 18.14** presents the block diagram of a digital SSB receiver. After the IF signal is digitized, we wish to reduce the sampling rate and the filtered BW as soon as possible. This is because we need as much time as possible between input samples for the intense filtering and other computations we must perform. As noted above, reduced sampling rates also ease the design of the digital filters that provide the final selectivity. We therefore include a decimation filter and decimate by a factor of 4. Again, the LO signals take on only values of 1, 0, −1 and 0, eliminating multiplications. Digitized signals are translated to baseband using the complex mixing algorithms outlined above. Since the input signal, x(t), is real, only two multiplications are necessary:

$$I(t) = x(t)\cos(\omega t)$$
$$Q(t) = x(t)\sin(\omega t) \quad (22)$$

Now we have an analytic signal as before; the frequency of the BFO, ω rad/s, is chosen to beat the carrier frequency to zero hertz. An analytic filter pair precedes the summation in which we select the sideband we want. The equations work precisely in reverse: That is why they are Hilbert *transforms*.

AM Demodulation

One's first inclination is to demodulate an AM signal by rectifying it. A better way

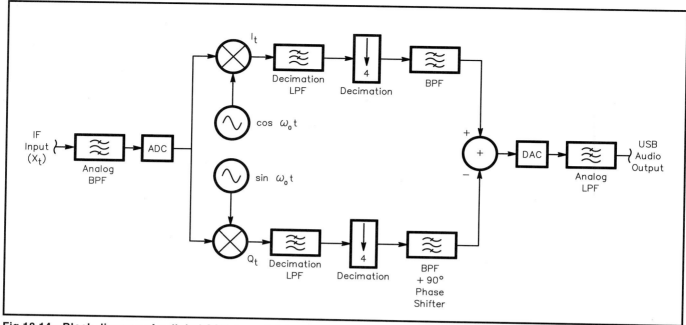

Fig 18.14—Block diagram of a digital SSB demodulator.

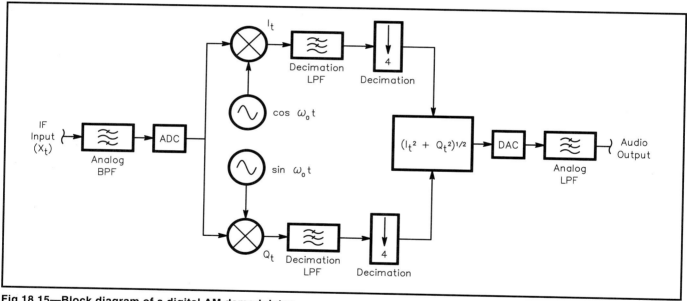

Fig 18.15—Block diagram of a digital AM demodulator.

is to use the I and Q signals we have already developed: Eq 15. Now we are stuck with computing square roots. Lucky for us, a fellow named Isaac Newton figured out a slick way almost 400 years ago. In the 17th century, these calculations were quite a burden—everything had to be done by hand! Because this is such a common problem in computing, a lot of additional effort has gone into finding faster algorithms since that time. A new, very fast look-up-table method is also presented here that may be more attractive where

enough memory is available.

An I/Q AM demodulator dodges problems associated with rectification methods. It also can use the decimation filters for final selectivity, obviating much of the computations found in the SSB demodulator. **Fig 18.15** shows the circuit. Newton's method for square roots goes like this: Take a crude guess at the square root of the number in question. Divide the number by the crude guess. Add the crude guess to this ratio and divide it all by 2. Use this result as the new crude guess and

repeat the process until the desired accuracy is obtained:

$$\text{let GUESS}_{new} = \left(\frac{\dfrac{\text{Number}}{\text{GUESS}_{old}} + \text{GUESS}_{old}}{2} \right)$$

$$\text{let GUESS}_{old} = \text{GUESS}_{new} \qquad (23)$$

REPEAT

In practice, the accuracy of the result reaches the limit of 16-bit representations in five or six iterations when the first guess is good. It is about half an order of magnitude slower than the following look-up table method, but is still among the best where memory is at a premium. Project D in the Appendix describes a *QuickBasic 4.5* example of Newton's method.

A new, very fast look-up-table method for computing integer square roots has been discovered. It employs a short (512-entry) table and first-order interpolation between table entries. First-order interpolation is described in detail in the DDS section below. To preserve accuracy, the algorithm also uses the process of argument normalization. The algorithm serves as our fifth software project in DSP.

The argument of this function—the number of which we must find the square root—is a 32-bit integer. The result is a 16-bit integer. Refer to **Fig 18.16**, a flow chart of the process. In the first step, the

argument is normalized to within the range 2^{22}-2^{24}. Arguments greater than 2^{24} are divided by an integral power of two, 2^k, where:

$$k = \Im\left[\log_2\left(\text{arg}\right) - 23\right] \qquad (24)$$

The script I indicates the integer part, and k—which takes on values of 0, 2, 4 or 6—is saved for later processing. Now the normalized argument is split into integer and fractional parts, with the radix point residing to the left of bit 15:

$$a = \Im\left(\frac{\text{arg}}{2^{k+8}}\right)$$

$$b = \mathcal{F}\left(\frac{\text{arg}}{2^{k+8}}\right) \qquad (25)$$

where a is the integer part and b is the fractional part. In other words, a comprises bits 16-23 of the normalized argument, and b is bits 0-15, as shown in the flow chart. Next, we use a as the address into the look-up table, fetching a 16-bit value, x_a. This value is the nearest table entry lower than the actual root. Fractional part b is used to interpolate between this value and the next higher table entry, x_{a+1}:

$$\text{root} = b\left(x_{a+1} - x_a\right) + x_a \qquad (26)$$

This is the square root of the normalized argument:

Finally, this result must be multiplied by the square root of 2^k, which is of course $2^{k/2}$. The result is then "de-normalized" and ready for use. Restricting k to an even integer (as we did) makes this a simple bit-shifting operation, as in the normalization process above. The 16-bit result produced by this algorithm is accurate to within several least-significant bits over the entire range of 32-bit arguments. It is quite a bit faster than the 5 or 6 iterations of Newton's method required for the same accuracy; this is because it avoids the divisions that Newton's method employs. Most DSPs take 3 or 4 times the processing time for a fractional division as they take for multiplication or look-up table indexing. Project E in the Appendix describes an assembly-language implementation of this square-root algorithm.

Additional threshold extension and distortion-avoidance procedures may be employed in an AM demodulator. Of particular interest is the *synchronous, exalted-carrier* demodulator. Synchronous, in this case, means that the demodulator's frequency standard is phase-locked to the received carrier. This forces the phases of modulation components into their correct relationships and therefore minimizes

phase distortion. A small advantage in SNR performance of up to 3 dB is also gained. DSP makes it relatively easy to build a narrow BPF, centered on the carrier, that strips the modulation prior to application to the PLL used to achieve lock. The exalted-carrier technique is a way of avoiding distortion caused by selective fading of the carrier. Ordinarily, when received carrier amplitude drops, the signal becomes over-modulated, even though it was not transmitted that way. Distortion can be severe. Exalted carrier strips the carrier from the signal using the narrow BPF; it is used only to drive the PLL. A copy of the limited carrier is then added back to the carrier-stripped signal prior to demodulation, at an amplitude that avoids over-modulation. See **Fig 18.17** for a block diagram of this type of demodulator. Refer to the chapter on Modulation Sources for more discussion of AM waveforms.

FM and PM Demodulation

Traditional FM and PM demodulators, such as discriminators (filters) and PLLs may be implemented in DSP; but again, the I/Q method carries distinct advantages as it exploits mathematical relationships. We already defined the phase of an analytic signal in Eq 16, and so we can build a PM demodulator directly by finding arctangents. Possibilities include look-up tables and Taylor series. For an FM demodulator, we would then differentiate the string of phase samples using the technique of *first differencing*. We simply take the difference between adjacent samples by subtracting them:

$$f(t) = \phi(t) - \phi(t - 1) \qquad (27)$$

and this is the FM demodulator's output.

One common analog technique that stands out among DSP implementations is the quadrature detector. It is certainly simple and convenient to generate delays and multipliers, such as are required. The input signal is multiplied by a time-delayed copy of itself to produce a voltage proportional to its phase excursions away from the center frequency. This voltage is also proportional to the amount of delay inserted. See **Fig 18.18**. When the delay is an odd integral multiple of one quarter the input period, the output is zero. Longer delays produce greater output-voltage sensitivities; that is, dV/dφ increases.

Digital BFO Generation: Direct Digital Synthesis

Synthesizers have come a long way since first becoming popular in HF transceivers of the 1970s. Availability of

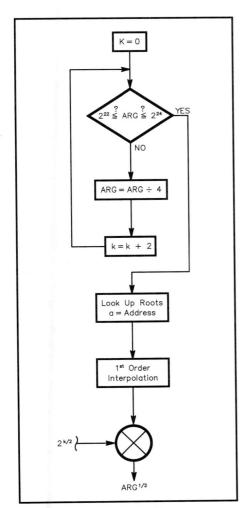

Fig 18.16—Flow chart of a fast square-root algorithm.

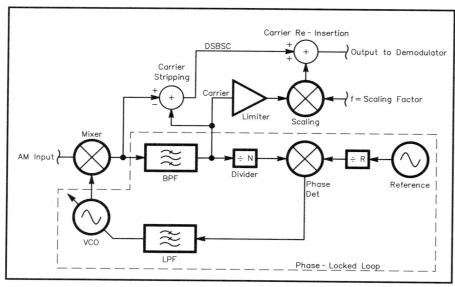

Fig 18.17—Block diagram of a synchronous, exalted-carrier demodulator.

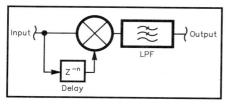

Fig 18.18—Block diagram of a digital quadrature detector.

components then lagged well behind the development of theory. Now, hardware capabilities have nearly caught up—which is the case for DSP in general—and are driving the very rapid advancement of equipment we are now experiencing. Paralleling breakthroughs in the microprocessor and data acquisition fields, progress in *direct digital synthesis* (DDS) has enabled performance levels only dreamed of a decade ago. Virtually all new designs may profit from this technology. Below, we will cover quite a few issues impacting transceiver performance: phase noise, spectral purity, frequency stability, lock times and tuning resolution. A DDS circuit using dedicated hardware is described; discussion of BFO and LO generation in software follows.

Synthesizer performance affects receiver dynamic range. Phase-noise and spectral-purity issues are in play. Phase noise is the unwanted phase modulation of transceiver frequency-control elements by circuit noise. It appears at and near the transmitter's output frequency and may cause interference to stations on adjacent frequencies. In addition, it may cause in-

terference in one's own receiver—even if the signals received are phase-noise free—through the process of *reciprocal mixing*. See the **AC/RF Sources (Oscillators and Synthesizers)** chapter for a discussion of this effect. The spectral purity of a synthesizer may also affect receiver dynamic range by introducing spurious responses where spurs exist on the synthesizer's output. This may be true especially for the first LO in a receiver across the entire range of frequencies present. It is extremely important, then, that this LO be clean.

Radio amateurs are free to operate anywhere within large frequency bands, so it might seem that frequency accuracy is not very critical. Prevalent narrow-band communication modes require it, though, and operators have come to expect excellent stability from their rigs. It is reasonable to expect ±20-Hz stability over a range of −10 to +50° C. Digital compensation techniques currently achieve this. We wish to attain a tuning speed that does not impose limitations on typical use. "Cross-band," or split-frequency operation ought to be considered. For a frequency step of ±600 kHz, an upper limit of 25 ms on the lock time of a synthesizer is a reasonable goal. Lock time is defined as the time required to settle within the stability limits we already set. The smallest frequency steps should be such that they do not impede performance. 10 Hz used to be good enough, but now certain digital modes benefit from finer tuning. In addition, the digital notch filter described before is so sharp that it occasionally needs to be within 1 Hz!

A DDS system generates digital

samples of a sine wave and converts them to an analog signal using a DAC. See **Fig 18.19**. In a DDS chip, a phase accumulator is incremented at each clock time; the phase information is used to look up a sine-wave amplitude from a table. This value is passed to the DAC, which outputs a step-wise sine wave. As we saw before, the spectrum of this sine wave is seasoned with aliases and contains other minor pollutants. Since the phase is represented by a binary number with a fixed number of bits, p, errors develop because the number is truncated to that number of bits. Truncation generates PM spurs in the DDS output. This occurs prior to the DAC. Further errors are related to the output resolution of the look-up table. Table values representing the amplitudes are truncated to some number of bits, a. This mechanism produces AM spurs in the output. According to Cercas *et al*, the largest PM spurs have amplitude:

$$P_{spur} = -(6.02p - 5.17)\text{dBc} \qquad (28)$$

and maximum AM spurs can rise to:

$$P_{spur} = -(6.02a + 1.75)\text{dBc} \qquad (29)$$

Phase noise at the output is that of the DDS clock source times the ratio of the output frequency to the clock frequency, as limited by divider noise. Spurious levels also tend to grow as the DDS output frequency approaches the Nyquist limit. Strange spurs at the output are usually related to IMD and harmonics of the desired signal and their aliases. Remember that frequencies exceeding half the sampling frequency "fold back" into the signal spectrum at a position determined by their frequency, modulo $f_s/2$. High-order harmonics are liable to find their way into one's band of interest. Traps at the DAC output have been known to suppress these responses. See Project F in the Appendix for the schematic and parts list of a DDS project.

In the analog signal we generate, the DAC introduces more AM spurs, harmonics and IMD because of its inherent nonlinearity, as discussed above. Spurs are also likely at the clock frequency, its harmonics and sub-harmonics. A higher-order LPF will take care of these, but we must see what we can do about the others. It turns out we may eliminate *all* the AM spurs by squaring the DDS output. We can do nothing about the remaining PM spurs. Cranking through Eq 28 will show that they can be made very low: −113 dBc for a 20-bit-address sine look-up table and 32-bit phase accumulator. This parameter is critical in case we want to use the DDS as the

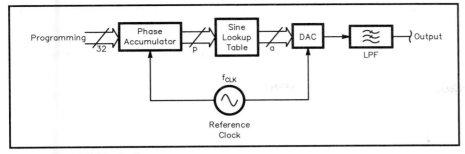

Fig 18.19—DDS block diagram.

reference to a high-frequency PLL circuit: The PLL will multiply the phase noise and PM spurs by the ratio of the PLL output frequency to the PLL reference frequency within the PLL loop BW. Outside the loop BW, the VCO itself is responsible for establishing spectral purity. So while dividing the DDS to the PLL reference frequency lowers phase noise and PM spurs, the PLL multiplies them back upward. A trade-off exists between spur levels and reference frequency, hence lock time.

A PLL reference frequency of near 100 kHz has been found to be sufficient for the desired lock times, with an output-to-reference ratio of 1000. Such a loop should achieve very fast lock times, as it can be expected to lock within 500 cycles of the reference input. The DDS tuning time is at least three orders of magnitude faster than this. In the example, the VCO output is near 100 MHz. DDS energy is injected at the reference input to the PLL chip, squaring it and dividing it by 10; the DDS runs near 1000 kHz. The block diagram of a PLL using a DDS as its reference is shown in **Fig 18.20**. Spurs and phase noise inside a loop BW of, say, 1 kHz are amplified by the PLL by the factor:

$$N = 20 \log \left(\frac{f_{VCO}}{f_{REF}} \right) = 40 \ \text{dB} \qquad (30)$$

Of course, we tune the hybrid synthesizer by programming the DDS; the PLL programming is fixed. Let's say we want 1-Hz tuning resolution at the VCO output. As the DDS frequency is 1/100 of the output, we must tune the DDS in 10 *millihertz* steps! Tuning resolution in a DDS circuit is determined by the phase accumulator's bit resolution, p, and the DDS clock's frequency, f_{clk}:

$$df_{DDS} = \frac{f_{CLK}}{2^P} \qquad (31)$$

A clock frequency around 10 MHz and p = 32 easily satisfy our conditions, producing a step size of 2.3 millihertz. As noted above, making the DDS output frequency a small fraction of the clock frequency makes it easier to get a clean output. A range of about half an octave eases the design of the LPF or BPF used at the DDS output to limit spurs, aliases and clock feed-through.

The phase-accumulator/look-up-table approach is equally useful in generating numeric BFOs in software. One of the first things to emerge when considering this

scheme is the potentially large size of the look-up table. To maintain the full dynamic range of a DSP system requires BFO phase and amplitude performance, as limited by Eqs 28 and 29, at least as good as the rest of the system. In 16-bit systems, we are shooting for about 90-100 dB of dynamic range. A table with $2^{16} = 65,536$ entries is not much of a problem for DDS chip manufacturers to include on-board, but it may tax available memory space in embedded systems.

Fortunately, a couple of ways around the problem have been uncovered. The first involves the process of interpolation, very much like the artificial increase of sampling frequencies we examined above. In this method, we restrict the number of table entries to some arbitrary number, $M \ll 2^{16}$, while keeping the bit-resolution of the entries themselves, a, high enough to satisfy the limits of Eq 29 for the spur levels we can tolerate. Take the case where $M = 2^8 = 256$ and a = 16. The phase accumulator, incremented at each sample time by an amount df that is directly proportional to the output frequency, forms the address into the look-up table. Let this address have bit-resolution p = 16. According to Eq 28, PM spurs will not exceed –91 dBc. Since there are only 256 table entries, we may use the most-significant byte (MSB) of the address to find the table entries that straddle the correct output value. We then use the least-significant byte (LSB) as an unsigned fraction to find out how far between the two table entries we must go to reach the correct output value. If, in order of increasing address in the table, our two adjacent table entries are d_1 and d_2, we may perform a first-order interpolation between the entries using:

$$d_{int} = d_1 \left(\frac{256 - LSB}{256} \right) + d_2 \left(\frac{LSB}{256} \right) \qquad (32)$$

This results in a linear, piece-wise representation of the data, as shown in **Fig 18.21**. The worst-case amplitude errors caused by this straight-line approximation place total harmonic distortion (THD) at the output at around 0.03% or –70 dBc. Much of this harmonic distortion is concentrated near half the sampling frequency, though, and may not be of much concern in actual systems. Doubling M would reduce THD to around 0.01%. Second and higher-order interpolation algorithms are available that out-perform the first-order approximations by a long way.

In systems where an even smaller look-up table must be used, computation of sines and cosines using Taylor series might be attractive. THD is less than

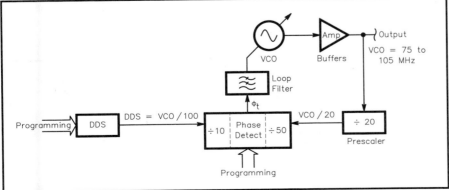

Fig 18.20—Block diagram of a DDS/PLL hybrid synthesizer.

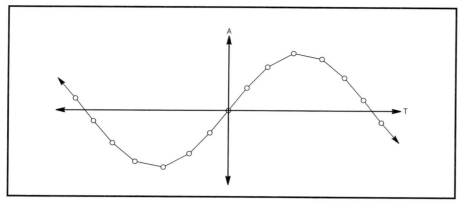

Fig 18.21—Linear piece-wise representation of data resulting from first-order interpolation.

0.008% when using four or five terms from the polynomials:

$$\sin(x) = x - \frac{1}{3!}x^3 + \frac{1}{5!}x^5 - \frac{1}{7!}x^7 \cdots \quad (33)$$

and

$$\cos(x) = 1 - \frac{1}{2!}x^2 + \frac{1}{4!}x^4 - \frac{1}{6!}x^6 + \frac{1}{8!}x^8 \cdots \quad (34)$$

DIGITAL SPEECH PROCESSING

Virtually all modern transmitters employ fast-attack, slow-decay RF compression: It is called automatic level control (ALC). Because transmitters are usually peak-power limited, some form of gain control is necessary to prevent overdrive of the final RF power amplifier.

RF Compression

A typical ALC system detects the transmitter's envelope with a rectifier and filter, applying this control signal to some gain-controlled stage or stages in the exciter. An increasing level from the envelope-detector results in decreasing gain such that the peak envelope power (PEP) is regulated. ALC is a servo loop employing negative feedback, usually developed only on voice peaks. As the decay time of the detector is decreased, some amplification of parts of speech falling between peaks is achieved. Enhancement cannot exceed the total gain reduction occurring at the voice peaks and usually falls in the range of 3-6 dB. The increase in the transmitter's average output power (talk power) may be quite a bit less than this depending on the characteristics of the voice, especially the *peak-to-average ratio*. In a digital exciter, we may eliminate the need for an analog gain-controlled stage by employing a numeric gain control factor in software and simply regulating the modulator's output level.

Human voices have peak-to-average ratios as high as 15 dB. This does not utilize a peak-limited transmitter very well in SSB mode: At the 100-W PEP level, the average output power might be as little as 3 W! RF compression raises the average output power and tends to further improve intelligibility by bringing out subtle parts

of speech. In a digital I/Q modulator, we have a distinct advantage in designing an RF compressor: The RF envelope can be calculated before the modulation is performed. Once the microphone audio has been sampled and converted to an analytic signal, Eq 15 may be used to compute the envelope. To avoid the time-consuming square-root calculation, we may use an approximation:

$$\left\{ \begin{array}{l} \text{For:} |I| > |Q|, \left(I^2 + Q^2\right)^{\frac{1}{2}} \approx |I| + 0.4|Q| \\ |Q| \ge |I|, \left(I^2 + Q^2\right)^{\frac{1}{2}} \approx |Q| + 0.4|I| \end{array} \right\} \quad (35)$$

The envelope signal is used to compress the range of baseband levels prior to modulation so that the peak-to-average ratio is reduced. A block diagram of this system is shown in **Fig 18.22**. The net effect of the system can be shown to be identical to that of a direct, RF compressor. This naturally involves distortion, since the transmitter is no longer linear; however, the distortion produced enhances the syllabic and formant energy in speech without introducing the "mushy" sound caused by heavy audio compression or clipping. As

the attack and decay times of an RF compressor are made faster, it approaches the performance of an RF clipper, known to be the most effective form of processing. Because the baseband audio is processed prior to filtering and modulation, occupied BW does not increase much; low-order IMD products will be created, though, that fall within the desired transmit BW. These products ultimately limit the effectiveness of the compressor. This technique may also be applied to receivers.

Audio Compression: Building an AM Transmitter

It has long been a problem to hold the carrier and modulation levels constant in AM transmitters covering several octaves of frequency, such as at HF. Because a baseband signal may not have symmetrical positive and negative amplitudes about its average value, a suitable analog ALC system would be incredibly complex.

In DSP, we may prevent *carrier shift* using adaptive techniques; we prevent over-modulation using an audio compressor. (Refer to **Fig 18.23**.) First, the ratio of drive level to output level, $d(t) / y(t)$, is easily computed by a DSP when the transmitter is on. From this, we can calculate

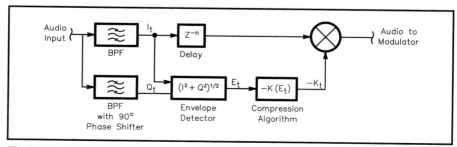

Fig 18.22—Digital RF compressor block diagram.

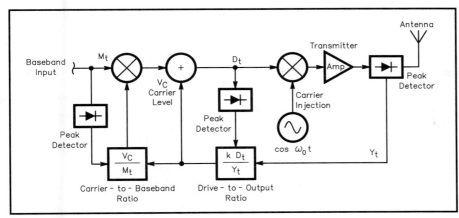

Fig 18.23—AM ALC block diagram.

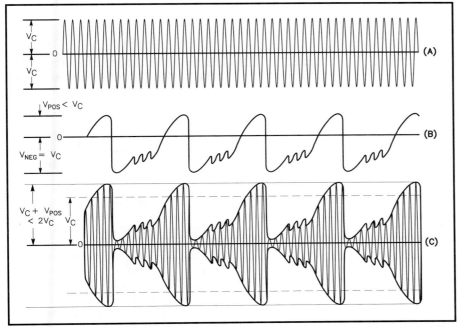

Fig 18.24—AM carrier (A). Baseband input with asymmetrical amplitudes (B). AM modulator output (C).

what drive level is required to reach exactly 25% of the peak-power setting. We want the carrier to have this amplitude, regardless of modulation (or lack of it). Second, the baseband signal applied to the modulator must have a maximum peak level equal to the carrier's drive level established above. When the carrier and compressed baseband levels are added, the result is a 100%-modulated AM signal.

Fig 18.24 shows this situation, using a baseband signal whose negative excursions are greater than its positive excursions about the average value. Now two servomechanisms are operating in our AM ALC: One continually computes the drive-to-output ratio and sets the carrier level; the other compresses the peak baseband signal to that same peak level. Since the baseband peak detector has to find either the highest negative or highest positive peak, asymmetrical audio inputs may produce an unexpected result: Either the upward or downward modulation may reach 100% before the other can do so. If the downward modulation limits baseband amplitude first, the upward modulation would not cause the transmitter to reach its set PEP level without introducing a carrier shift.

INTERFERENCE-REDUCTION TECHNIQUES

We touched on the idea of a manually tuned adaptive notch filter using the LMS algorithm. These principles are explored in more detail here, especially as they apply to interference- and noise-reduction systems. The nature of information-bearing signals is that they are in some way coherent; that is, they have some features that distinguish them from noise. For example, voice signals have attributes related to the pitch, syllabic content and impulse response of a person's voice.

Adaptive Filtering

We will find it possible to build an adaptive filter that accentuates those repetitive components and suppresses the non-repetitive (noise). Much research has been done about detection of a sinusoidal signal buried in noise. Adaptive filtering methods are based on the exploitation of the statistical properties of the sampled input signal, specifically, *autocorrelation*. Simply put, autocorrelation refers to how recent samples of a waveform resemble past input samples. We will build an *adaptive predictor*, which actually makes a reasonable guess at what the next sample will be based on past samples. This leads directly to an adaptive noise-reduction system.

An Adaptive Interference Canceler

Imagine we have some sampled input signal, $x(t)$, that we want to adaptively filter to enhance its repetitive content. In the case of a CW signal, all that is required is a BPF centered on the desired frequency. We know that this signal takes the form of a sine wave and that its amplitude will change markedly. Its frequency may not be absolutely constant, either, but we will assume it is fixed for now. We set up an FIR filter structure and an error-measurement system to compare a reference sine wave, $d(t)$, with the output of the filter, $y(t)$. See **Fig 18.25**. Sine wave $d(t)$ is the same frequency we expect the CW tone to be. The difference output, e(t), is known as the *error signal.*

Now imagine some person is watching the error signal and has their hands on the controls that change the filter coefficients. (See **Fig 18.26**.) Minimizing the error signal by tweaking the coefficients forces the filter to converge to a BPF centered at the frequency of d(t). The speed and accuracy of that convergence is going to depend on how well the person analyzes and reacts to the error data. If it is difficult to tell that a sine wave is present, then adjusting the filter will also be difficult. Further, if the sampling rate is high enough, a person will not be able to keep up; they can check the error only so often or can generate long-term averages of the error.

Using the typical processes of the human mind, the person will soon discover that if they turn the controls the wrong way, the error increases. This information is used to reverse the direction of adjustment; the person then turns the controls the other way. It soon emerges that the person is on a *performance surface*, with an "uphill" and a "downhill," and they know the goal is to go only downhill. So they trash about with the controls, sometimes making mistakes, but ultimately making headway overall down the hill. At some point, the error gets rather small: They know they are near the "bottom of the bowl." Once at the bottom, it is uphill no matter which way they go. The goal of minimizing the error e(t) has been achieved. They continue gently flailing about with the controls, but always staying near the bottom. This situation is analogous to aligning an analog BPF with an adjustment tool.

After doing this whole thing several times, the person finds that certain rules help speed up the process. First, there is a relationship between the magnitude of the error and the amount they must tweak the controls. If the total error is large, a lot of tweaking must be done; if small, then it is better to make small adjustments to stay near the bottom of the performance surface. Second, there is a correlation between the error, e(t); the input samples, x(t); and the coefficient set, h(t) they need to adjust. Derivation of algorithms providing for steepest descent down the hill is a long and tedious exercise in linear algebra. Let's just say the person goes to school, becomes an expert in matrix mathematics and discovers that one of the fastest and most accurate ways down the hill is to adjust coefficients at sample time t according to:

$$h_{t+1}(k) = h_t(k) + 2\mu e(t)x(t) \qquad (36)$$

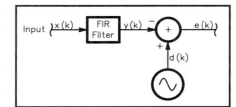

Fig 18.25—An adaptive modeling system.

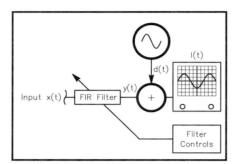

Fig 18.26—An adaptive modeling system, which requires a person at the filter controls.

This is the LMS algorithm. It was developed by Widrow and Hoff in the late 1950s.

Replacing the person with the LMS algorithm, as shown in **Fig 18.27**, we have our manually tuned adaptive interference canceler. Note that both the desired output, y(t), and the undesired, e(t), are available. This is nice in case we want to take only the broadband component and reject the tone. Performance issues of interest include the adjustment error near the bottom of the performance surface and the speed of adaptation. One of the first things we notice about the LMS algorithm is that each of these factors is directly proportional to μ. We select its value, which ranges from 0 to 1, to set the desired properties. A trade-off exists between speed and misadjustment. Large values of μ result in fast convergence, but large misadjustment in the steady state. Total

misadjustment is also proportional to the number of filter taps, L, and this may place a limitation on the complexity of the filter that may be used. The total delay through the filter also grows with its length; it may become unacceptably large under certain conditions. As in Eq 10, the 3-dB BW of the adaptive BPF is:

$$BW = \frac{2\mu A^2}{t_s} \, rad/s \qquad (37)$$

Small values of μ result in narrower filters that take longer to adapt. Attempts may be made to adjust μ on the fly by using a value that changes is proportion to the error, e(t). A large value is selected initially for rapid convergence, then it is decreased to minimize the long-term misadjustment. This works fine as long as the characteristics of the input signal are not rapidly changing.

An Adaptive Interference Canceler Without An External Reference: An Adaptive Predictor

In the above example, we knew pretty much what to expect at the output: a sine wave of known frequency. What happens when we do not know much about the nature of the input signal, except that it contains coherent components? Quite a few circumstances like this arise in practice. It might seem at first that adaptive processing could not be applied; however, if a delay, z^{-n} is inserted in the *primary input*, x(t), to create the *reference input*, d(t), periodic signals may be detected and thereby enhanced (or eliminated). See **Fig 18.28**. This delay forms an *auto-correlation offset*, representing the time difference used to compare past input samples with present samples. The amount of delay must be chosen so that the desired components in the input signal correlate with themselves, and the undesired components do not. This is an *adaptive predictor*: Predictable components are enhanced, while the unpredictable parts are removed. Experiments show that for any given value of μ, the filter converges quickest when

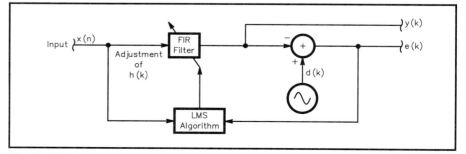

Fig 18.27— An adaptive interference canceler.

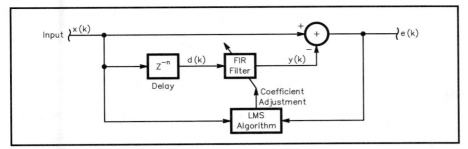

Fig 18.28—An adaptive predictor.

the delay, z^{-n}, is set between one half and one times the filter's total delay.

We may predict this circuit's noise-reduction performance using the ratio of the pre-filtered BW to that of the converged filter:

$$\Delta SNR = 10\log\left(\frac{BW_{input}}{BW_{filter}}\right)$$

$$= 10\log\left(\frac{BW_{input}}{2\mu A^2 f_s}\right) \quad (38)$$

As an example, for $\mu = 0.005$, $A = 1$, $BW_{input} = 3$ kHz and $f_s = 15$ kHz, the SNR improvement is about 13 dB. When adaptive filters with many taps are used, multiple tones may be either enhanced or notched. Under most conditions, the undesired components are large compared to the desired; enhancement of signals is needed most when the input SNR is low. This situation may not give us enough thrashing about to find our way down the performance surface to convergence.

Adding artificial noise to satisfy this condition is tempting, but it turns out we can alter the algorithm slightly to improve our lot without actually adding such noise. These additional terms in the algorithm are known as *leakage terms*.

The unique feature of *leaky LMS algorithms* is a continual "nudging" of the filter coefficients toward zero. The effect of a leakage term can be striking, especially when applied to noise-reduction of voice signals. The SNR increases because the filter coefficients tend toward a lower throughput gain in the absence of coherent input signals. More significantly, leakage helps the filter adapt under low-SNR conditions—exactly when we need noise-reduction the most. One way to implement leakage is to add a small constant of the appropriate sign to each coefficient at every sample time:

$$h_{t+1}(k) = h_t(k)+2\mu e(t)x(t)$$
$$-\lambda\left\{sign\left[h_t(k)\right]\right\} \quad (39)$$

The value of λ may be altered to vary

the amount of leakage. Large values prevent the filter from converging on *any* input components, and things get very quiet indeed. Small values are useful in extending the noise floor of the system. In the absence of coherent input signals, the coefficients move geometrically toward zero; during convergent conditions, the total misadjustment is increased to at least λ, but this is not usually serious enough to affect signal quality.

An alternate way to implement leakage is to scale the coefficients at each sample time by some factor, γ, thus also nudging them toward zero:

$$h_{t+1}(k) = \gamma h_t(k)+2\mu e(t)x(t) \quad (40)$$

For values of γ just less than unity, leakage is small; values near zero represent large leakage and again prevent the filter from converging. It can be shown that the leaky LMS algorithm is equivalent to adding normalized noise power to the input $x(t)$ equal to:

$$\sigma^2 = \frac{1-\gamma}{2\mu} \quad (41)$$

The leaky LMS algorithm must adapt to survive, much as a hummingbird must flap its wings. Were the factor μ suddenly set to zero, the coefficients would all die away, never to recover. Therefore, it is perhaps unwise to use these algorithms with adaptive values of μ. Although values for γ and μ of greater than unity have been tried, the inventors refer to these procedures as "the dangerous LMS algorithm." Enough said.

FOURIER TRANSFORMS

While Fourier transforms are not used exclusively for interference reduction, we present them under that heading here because they are generally superior to adaptive-filtering algorithms in that application. The penalty for this greater effectiveness is an increased computational burden. The relationship Joseph Fourier (pronounced **foor**-ee-ay, 1768-1830) formulated between the application of heat to a solid body and its propagation has direct analogy to the behavior of electrical signals as they pass though filters and other networks. The laws he wrote define the connection between time- and frequency-domain descriptions of signals. They form the basis for DSP spectral analysis, which

makes them extremely valuable tools for many functions, including digging signals out of the noise, as we will see.

A Fourier transform is a mathematical technique for determining the frequency content of a signal. Applied to a signal over some finite period of time, it produces an output that describes frequency content by assuming that the section of the signal being analyzed repeats itself indefinitely. Of course, when we analyze a real signal, such as a couple of seconds of speech, we know that those few seconds do not, in fact, repeat endlessly. So at best, the Fourier transform can give us only an approximation of the frequency content. If we looked at a large enough chunk of the sig-

nal, that approximation would be pretty good. Certain mathematical conditioning of the input data will help us control the error, even for relatively short analysis intervals.

Originally, the Fourier transform was developed for continuous signals. In DSP, we use a variant of it called the *discrete Fourier transform* (DFT). It is the discrete version because it operates on sampled signals. It is a *block transform* because it converts a block of N input samples into a block of N output *bins*. The input block may be any N contiguous samples. A DFT makes use of complex sinusoids and produces a complex result. When the input data are real, meaning they lack an imaginary part, half the output block consists of the *complex conjugates* of the other half, and so is redundant. When a complex input is used, none of the output bins is redundant.

We learned before that a complex sinusoid is just a pair of waves: a cosine wave and sine wave of the same frequency. Since we will be dragging around a lot of these in the equations below, we introduce a little mathematical shorthand for them called the *Euler identity*:

$$e^{j\omega t} = \cos(\omega t) + j\sin(\omega t) \qquad (42)$$

where e is base of natural logarithms. We will shorten this even more later. For each output bin k, where $0 \le k \le N-1$, the DFT is computed as:

$$X(k) = \sum_{n=0}^{N-1} x(n)e^{\frac{-j2\pi nk}{N}} \qquad (43)$$

Expanding Eq 43 using the Euler identity yields:

$$X(k) = \sum_{n=0}^{N-1} x(n)\cos\left(\frac{2\pi nk}{N}\right) - j\sum_{n=0}^{N-1} x(n)\sin\left(\frac{2\pi nk}{N}\right) \qquad (44)$$

So each bin has a real part and an imaginary part. Note that each part is calculated using the same convolution sum we saw in Eq 3. Eq 44 is in normal complex-number form: a + jb. These coefficients a and b yield the amplitude and phase of the signal x(t) at frequency $\omega = (k f_s)/N$:

$$A_k = \left(a_k^2 + b_k^2\right)^{\frac{1}{2}} \qquad (45)$$

$$\phi_k = \arctan\left(\frac{b}{a}\right) \qquad (46)$$

k is directly proportional to the frequency of its bin according to:

$$f_k = \frac{k f_s}{N}, \text{ for } k < \frac{N}{2} \qquad (47)$$

The bins are evenly spaced in frequency by the amount $f_1 = f_s/N$, but there are actually only N/2 real frequencies represented. As mentioned above, half the DFT bins produced from a real input are redundant. Complex inputs may analyze positive and negative frequencies separately.

Working in reverse, we may reconstruct time-domain signal x(t) by summing X(k) for all values of k:

$$x(t) = \frac{1}{N}\sum_{k=0}^{N-1} X(k)e^{\frac{j2\pi kn}{N}} \qquad (48)$$

This is the *inverse discrete Fourier transform* (IDFT or DFT^{-1}). It is important to note the duality of the DFT/IDFT relationship. The transforms are not really altering the signal in any way, they are only different ways of representing it mathematically. The strength of the DFT in noise-reduction systems is that it evaluates the amplitude and phase of each frequency component to the exclusion of others.

As far as we can reduce the *resolution BW*, f_s/N, we can eliminate additional noise by artificially zeroing frequency bins not meeting a pre-defined amplitude threshold. Finer resolution BW is obtained by increasing the number of bins, N, decreasing the sampling frequency, or both. Increasing the number of bins, N, involves taking a larger block of N input samples; the larger block represents a longer time span. Obviously we have to wait for N samples to be taken before we can Fourier transform a complete block: A delay of N samples is the result.

Since the DFT assumes the input block repeats indefinitely, we have discontinuities at the beginning and end of the block where the data have been chopped out of the continuous string of input samples. These abrupt discontinuities cause unexpected spectral components to appear, just as fast on-off keying of a CW transmitter does. This phenomenon is known as *spectral leakage*. Discrete signal components in the input "leak" some of their energy into adjacent frequency bins, smearing the spectrum slightly. Increasing the number of bins, N, helps alleviate this problem. Increasing N moves the bins closer together; a signal that falls between two bins will still cause leakage into adjacent bins, but since the bins are closer together, the spread in frequency will be less. Even so, input components are still spreading their energy over several bins and this overlap makes it difficult to determine their exact amplitudes and phases.

To minimize that problem, we use a technique known as *windowing* on the input data prior to transformation. The data block is multiplied by a *window function*, then used as input to the DFT normally. Window functions are chosen to shape the block of data by removing the sharp transitions in its envelope. Examples of window functions and their DFTs are shown in **Fig 18.29**.

The rectangular window is equivalent to not using a window at all, as all the samples are multiplied by unity. The other window functions achieve various amounts of side-lobe reduction. These window functions are also used to design filters using the Fourier transform method. In fact, these sequences can be used as the impulse responses of prototype LPFs, as should be evident from their frequency responses. Notice that they each involve a trade-off between transition BW and ultimate attenuation. Also note that in the figure, values of ultimate attenuation are plotted without regard to dynamic-range limitations which may be imposed by the bit-resolution of actual systems.

FAST FOURIER TRANSFORMS

In the years before computers, reduction of computational burden was extremely desirable. Many excellent mathematicians, including Runge, applied their wits to the problem of calculating DFTs more rapidly than the direct form of Eq 43. They recognized that the direct form requires N complex multiplications and additions per bin and that N bins are to be calculated, for a total computational burden proportional to N^2. The first breakthrough was achieved when they realized that the complex sinusoid $e^{-j2\pi kn/N}$ is periodic with period N, so a reduction in computations is possible through the symmetry property:

$$e^{\frac{-j2\pi k(N-n)}{N}} = e^{\frac{j2\pi kn}{N}} \qquad (49)$$

This led to the construction of algorithms that effectively break any N DFT computations of length N, into N computations of length $\log_2 N$. Thus, the computational burden is reduced to be proportional to $N\log_2 N$. Because even this much calculation was not practical by hand, the usefulness of the faster algorithms was overlooked until Cooley and Tukey revived it in the 1960s.

To exploit the symmetry referred to, we have to break the DFT computations of length N into successively smaller calculations. This is done by *decomposing* either the input or output sequence. Algorithms wherein the input sequence, x(t), is decomposed into smaller sub-sequences are called *decimation-in-time* FFT algorithms; output decompositions result in *decimation-in-frequency* FFTs. The decomposition is based on the fact that for some convenient number of samples, N, many of the sine and cosine values are the same and products can be combined prior to computing the convolution sums. In addition, other products have factors that are other sine and cosine values. It turns out that electing to decompose by successive factors of 2 produces a very compact and efficient algorithm: a *radix-2 FFT algorithm*.

Now for that additional bit of complex-sinusoidal shorthand mentioned earlier. Lots of complex sinusoids will appear in the diagrams to follow, so it sure would be nice to reduce the clutter a bit more. Let's follow the popular DSP text of Oppenheim

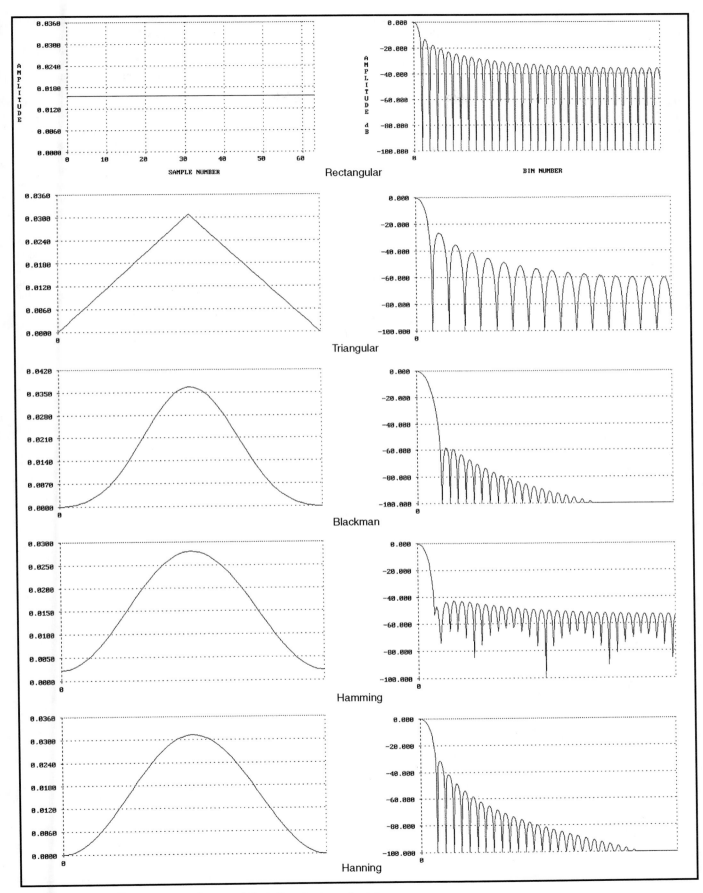

Fig 18.29—Various window functions and their Fourier transforms.

and Schafer and select the notation:

$$e^{\frac{-j2\pi kn}{N}} = W_N^{kn} \qquad (50)$$

This is used in **Fig 18.30** in a flow chart for a complete FFT calculation, for N=8. Multiplication symbols represent complex multiplications, addition symbols represent complex additions. Note that each complex multiplication requires 4 real multiplications and 2 real additions. Complex additions need 2 real additions.

We have 8 input points and 8 output points. Observe that the diagram could not be drawn without crossing many signal paths—there is a lot of calculation going on! Computations progress from left to right in $\log_2 N = 3$ stages; each stage requires N complex multiplications and additions, so the total burden is proportional to $N\log_2 N$. Further, each stage transforms N complex numbers into another set of N

complex numbers. This suggests we should use a complex array of size N to store the inputs and outputs of each stage as we go along.

An examination of the branching of terms in the diagram reveals that pairs of intermediate results are linked by pairs of calculations like that shown in **Fig 18.31**. Because of the appearance of this diagram, it is known as a *butterfly computation*.

Making use of another symmetry of complex sinusoids, we can reduce the total multiplications of the butterfly by another factor of 2. A modified butterfly flow diagram is shown in **Fig 18.32**. This calculation can be performed *in place* because of the one-to-one correspondence between the inputs and outputs of each butterfly. The nodes are connected horizontally on the diagram. The data from locations a and b are required to compute the new data to be stored in those same locations, hence only one array is needed during calculation. A complete

8-*point* FFT with the modified butterflies is shown in **Fig 18.33**.

An interesting result of our decomposition of the input sequence is that in the diagram, the input samples are no longer in ascending order; in fact, they are in *bit-reversed* order. It turns out this is a necessity for doing the calculation in place. To see why this is so, let's review briefly what happens in the decomposition process. We first separate the input samples into even- and odd-numbered samples. Naturally, all the even-numbered samples appear in the top half of the diagram, the odds in the bottom. Next, we separated each of these sets into their even- and odd-numbered parts. This process was repeated until we had N sub-sequences of length one. It resulted in the sorting of the input data in a bit-reversed way. This is not very convenient for us in setting up the calculation, but at least the output arrives in the correct order.

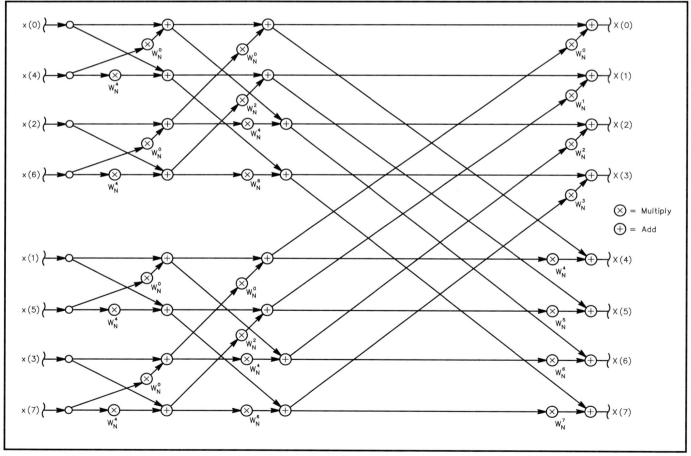

Fig 18.30—Flow chart of an 8-sample FFT.

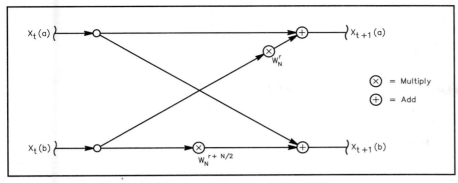

Fig 18.31—Butterfly calculation in a decimation-in-time FFT.

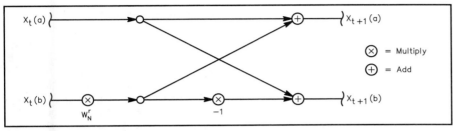

Fig 18.32—Modified butterfly calculation.

While we are on the subject, this business of bit-reversed indexing is the first thing that ties one's brain in knots during coding of these algorithms, so let's have at it. Several approaches are feasible to translate a normally ordered index to a bit-reversed one: a look-up table, the bit-polling method, reverse bit-shifting and the reverse counter approach.

The look-up table is perhaps the most straightforward approach. The table may be calculated ahead of time and the index used as an address into the table. Most systems do not require very large values of N, so the space taken by the table is not objectionable.

For more space-sensitive applications, the bit-polling method may be attractive. Since the bit-reversed indices were generated through successive divisions by two and determination of odd or even, a tree structure can be devised that leads us to the correct translation, based on bit-polling.

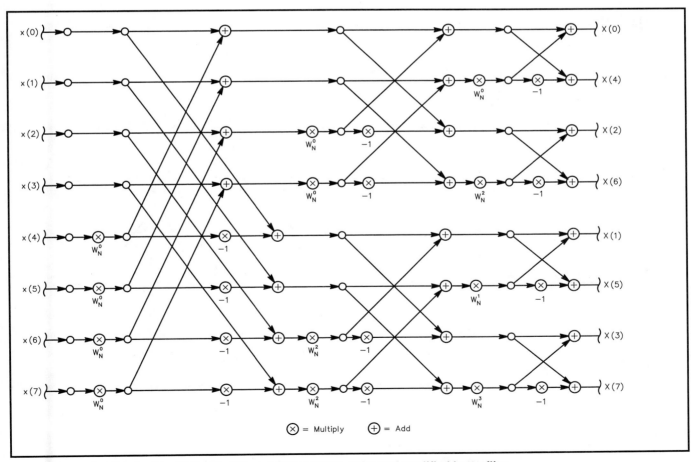

Fig 18.33—Decimation-in-time FFT with different input/output order and modified butterflies.

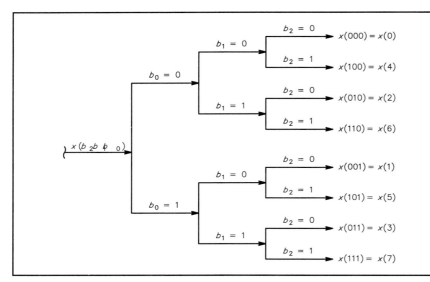

Fig 18.34—Polling tree for bit reversal.

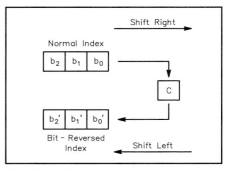

Fig 18.35—Register arrangement for bit-reversal shifting.

See **Fig 18.34**. The algorithm examines the least-significant bit, then branches either upward or downward based on the state of the bit. Then the second least-significant bit is examined and another branch taken, and so forth, until all bits have been examined.

The bit-shifting method requires about the same computation time as bit-polling. Two registers are used: one for the input index shifting right through the carry bit, the other shifting left through carry. After all the bits have been shifted, the left-shifting register contains the result. See **Fig 18.35**.

Finally, Gold and Rader have described a flow diagram for a bit-reversal counter than can be "decremented" each time the index is to change. If data are actually to be moved during sorting, the exchange is made between data at input index n and bit-reversed index m, but only once. That is, only N/2 exchanges need be performed.

During the actual calculations, indexing of data and coefficients requires attention to many details. In particular, several symme-

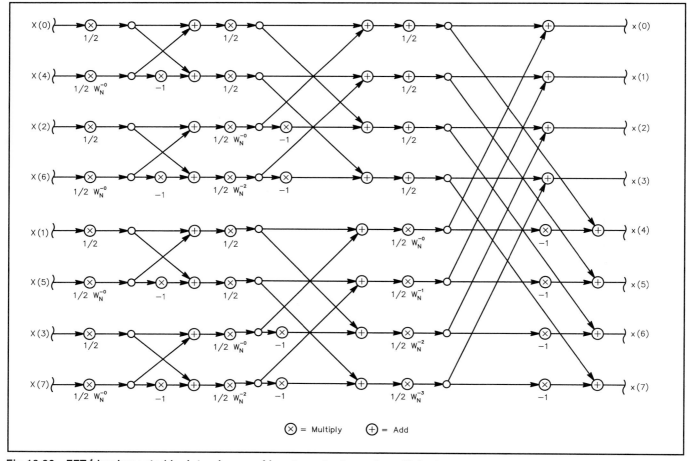

Fig 18.36—FFT⁻¹ implemented by interchange of inputs, outputs, and coefficients.

tries about offsets of the index may be exploited. At the first stage of Fig 18.32, all the multipliers are equal to $W_N^0 = 1$, so no actual multiplications need take place; all the butterfly inputs are adjacent elements of the input array x(t). At the second stage, all the multipliers are either W_N^0 or integral powers of $W_N^{N/4}$ and the butterfly inputs are two samples apart, and so forth.

The coefficients are indexed in ascending order. These are normally calculated ahead of time and stored in a table. Another way is to use a *recursion formula* to generate them on the fly, but this is discouraged because of numerical-accuracy effects that destroy the efficiency of the technique.

All those multiplications and additions take their toll on the numerical accuracy of our final result. Quantization noise is multiplied and added as well, and at the output of a DFT, the noise power grows by N times.

In an FFT calculation, the situation is roughly the same; however, the requirement to avoid overflow at intermediate stages may force us to scale the data, the coefficients, or both. This further reduces the dynamic range of any FFT. Results have been offered indicating noise increases in the vicinity of 12N. In addition, the quantization-noise contribution of the coefficients increases in inverse proportion to p, the number of bits used to represent them. This, in turn, means that the noise increase with respect to N is slow.

In FFT-based noise-reduction systems, we perform some modification of the frequency-domain data, such as zeroing bins not meeting a pre-defined amplitude threshold. Then we transform the modified data back to the time domain. The duality of the Fourier transform and its inverse can be shown in the flow diagram of a FFT^{-1} as in **Fig 18.36**. This diagram was produced from Fig 18.33 by simply substituting $\frac{1}{2}W_N^{-kn}$ for W_N^{kn} at each stage and, of course, using X(k) as the input to obtain x(t) as the output.

Alternatively, we may compute the FFT^{-1} by using the FFT flow diagram and swapping the inputs and outputs and reversing the direction of signal flow. It is important to note that this is a consequence of that fact that we can rearrange the nodes of the flow diagrams however we want, so long as we do not alter the result. The transforms work just as well in reverse as they do in the forward direction.

DAMN-FAST FOURIER TRANSFORMS

When it is necessary to compute Fourier transforms on a sample-by-sample basis, or where frequency resolution must be non-uniform across the sampling BW, even traditional FFTs may be too computationally intensive for the processing horsepower available. A class of algorithms that computes the next transform output very rapidly—based solely on current transform output and the next input sample—has been discovered. A method is included here for controlling its inherent divergence problem by brute force.

The derivation begins by looking at how the Fourier transform results change for each bin at each sample time. Say we start with some discrete Fourier transform output bins X (k) at sample time r. Then we compute the DFT for the next sample time r + 1 and examine the sequences to see what has changed. For r = 0, each DFT sequence expands to:

$$X_0(k) = W_N^{0k} x(0) + W_N^{1k} x(1) + W_N^{2k}$$
$$x(2) + ... + W_N^{(N-1)k} x(N-1)$$
$$X_1(k) = W_N^{0k} x(1) + W_N^{1k} x(2) + W_N^{2k} \quad (51)$$
$$x(3) + ... + W_N^{(N-1)k} x(N)$$

What is evident is that each input sample x(n) that was multiplied by W_N^{nk} in the summation for $X_0(k)$ is now multiplied by $W_N^{(n-1)k}$ in the summation for $X_1(k)$. The *ratio* of the two sequences is nearly:

$$\frac{X_1(k)}{X_0(k)} \approx \frac{W_N^{(n-1)k}}{W_N^{nk}}$$
$$= W_N^{-k} \quad (52)$$

We still have two terms "hanging out" of the relationship, namely the first and the last:

$$W_N^{0k} x(0) = x(0) \text{ and } W_N^{(N-1)k} x(N) \quad (53)$$

that have not been accounted for in the ratio. If we first subtract x(0) from $X_0(k)$ before taking the ratio, then add the new term $W_N^{(N-1)k}x(N)$ after, we have the correct result:

$$X_1(k) = W_N^{-k}\left[X_0(k) - x(0)\right]$$
$$+ W_N^{(N-1)} x(N) \quad (54)$$

Now this may be simplified a little, since:

$$W_N^{(N-1)k} = e^{\frac{-2\pi j(N-1)k}{N}}$$
$$= e^{\frac{-2\pi jN}{N}} \cdot e^{\frac{2\pi jk}{N}} \quad (55)$$
$$= W_N^{-k}$$

and substituting:

$$X_1(k) = W_N^{-k}\left[X_0(k) - x(0) + x(N)\right] \quad (56)$$

This is the damn-fast Fourier transform (DFFT). It means: For N values of k, we can compute the new DFT from the old with N complex multiplications and 2N complex additions, or a computational burden proportional to N. If we begin with $X_0(k)=0$ and take the first N value of x(n)=0, we can start the thing rolling. It saves computation over the FFT by a factor of:

$$\frac{N\log_2 N}{2N} = \frac{\log_2 N}{2} \quad (57)$$

which for large values of N is very significant indeed. For example, if N = 1024, the improvement is by a factor of five. Over the direct-form DFT, it is a factor of N^2/N faster. But there is a catch: An error term will grow in the output because the truncation and rounding noise discussed previously is cumulative. The error will continue to grow unless we do something about it.

The simplest way to handle the situation is to compute two DFFTs for all the output bins k, resetting every other block of N input samples to zero. In other words, one DFFT begins at some time with an input buffer that has been zeroed, the other continues to operate on the continuous stream of real input samples. As sample-taking continues, DFFT output is taken from the second calculation. As the buffer of the first DFFT gradually fills with real samples, the block of zeroes it originally held disappears. At this point, each DFFT produces the same result except for the greater error in the second DFFT because of truncation and rounding effects. Output is then taken from the first DFFT and the buffer of the second is zeroed; the calculations continue for another N iterations, at which time the exchange and reset are again done, and so forth, continually. This places an upper bound on the cumulative error to that associated with 2N iterations and increases the computational burden by a factor of two. Now the savings over the FFT is only:

$$\frac{\log_2 N}{4} \quad (58)$$

which for N > 16 still represents an improvement. DFFT output quantization noise is at least twice that of the DFT.

Frequency resolution of DFFTs is controlled by the block length, N, used in the calculations, just as in DFTs or FFTs. Resolution may be set differently, though, for each bin; further, not all bins need be computed to compute any particular bin, unlike the Cooley-Tukey FFT. Is there an inverse DFFT? Well, because inverse Fourier transforms map into the time domain, it is simple enough to just compute the next output sample rather than the next N output samples. The easiest output term to compute is x(0), since all coefficients are $W_N^0 = 1$. The output is then just:

$$x(0) = \frac{1}{N} \sum_{k=0}^{N-1} X(k) \qquad (59)$$

and only one multiplication is involved.

RADIO ARCHITECTURES FOR DSP

SUPERHETERODYNE WITH BASEBAND DSP

It is common these days to apply DSP techniques at audio or baseband, especially by using outboard processing units with older, analog receivers. A drawback to this approach is that while a receiver's selectivity may be improved this way, special gain-control settings must be employed. To see why this is so, let's look at a typical arrangement, shown in **Fig 18.37**. The receiver's BW is 3 kHz and we wish to use the outboard DSP unit to implement an RTTY filter with BW = 500 Hz. It follows that some of the signals we digitize are undesired and this raises a problem. When the desired signal is strong relative to the undesired, everything is fine; however, when a strong undesired signal appears within the receiver's BW but outside our DSP filter's BW, the receiver's AGC acts on the combination, reducing the level of our desired signal, as well as that of the undesired signal. We may elect to solve this in several ways; each involves digitally adjusting the gain applied to the desired signal to keep its level constant—the goal of any AGC system.

A block diagram of one such *digital AGC system* is shown in **Fig 18.38**. It consists of a gain-control block (multiplier) and a power-ratio detector. The detector computes the ratio of the sum of the undesired signal's peak amplitude, m, and the filtered signal's amplitude, n, to n:

$$k = \frac{m+n}{n} \qquad (60)$$

k is the factor by which the filtered output must be digitally boosted to keep its peak level constant. This detector includes a fast-attack, slow-decay filter. The decay rate is chosen to match that of the receiver's analog AGC. Analog AGCs usually have a decay that—when plotted as dB *vs* seconds—looks fairly close to a straight line. This exponential decay is achieved in DSP by multiplying the stored detector value by a constant near unity at each sample time. When the ratio k suddenly jumps upward, along with m + n, the stored value is updated immediately to get a fast attack.

The gain-controlled stage is simply a multiplier; its inputs are k and the filtered signal. Note that k ≥ 1 always, hence the multiplication is not the simple fractional type described above. We may now have a need to extend our fixed-point math to values greater than one. This is tedious but not too difficult. We just handle the integer and fractional parts separately. Additions and subtractions are straightforward, but we have to multiply k by a fractional decay factor, δ, then multiply k by another fraction—the filtered signal—at each sample time. Separating the integer and fractional parts by a radix point, we adopt the notation k = (a.b) where:

$$a \in \mathfrak{I}.b \in \mathcal{F} \qquad (61)$$

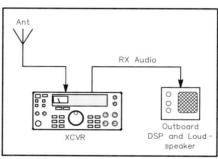

Fig 18.37—A typical use of an outboard, baseband DSP processor.

multiplier; its inputs are k and the filtered signal. Note that k ≥ 1 always, hence the meaning that we treat a as an integer, and b as a fraction. Numbers a and b are ordinarily represented in binary. A number whose absolute value is less than unity has a zero integer part: δ = (0.d). The result of the multiplication (a.b)(0.d) = e.f is:

$$(a.b)(0.d) = \left(\left[\mathfrak{I}ad + \mathfrak{I}bd \right] \times \left[\mathcal{F}ad + \mathcal{F}bd \right] \right) \qquad (62)$$

requiring four real multiplications, just as in complex math.

A delay is inserted in the path of input signal m to compensate the delay through the DSP filter. Scaling might be necessary to prevent overflow in the gain-controlled stage. Special attention must be paid to what happens during the attack time. Some receivers exhibit *AGC overshoot*, which may cause spikes on incoming signals, resulting in rapid gain excursions. A good approach seems to be to allow gain adjustment in proportion to the attack time of input signal m, but only if m persists at that level for several milliseconds, to avoid triggering on noise pulses.

In practice, baseband DSP filtering may be limited by *in-band IMD* and synthesizer *phase-noise* effects that plague the analog transceiver itself. These cause unwanted signals to appear in the passband, masking the desired signal. With a perfect receiver, performance is limited by the available

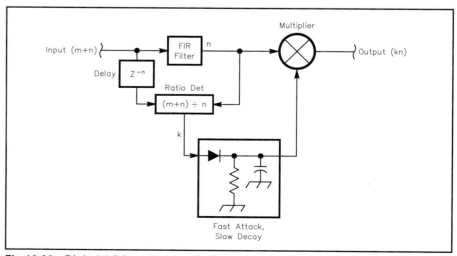

Fig 18.38—Digital AGC system block diagram.

SNR and SFDR of the ADC in use and by the phase noise of its clock. Noise-reduction algorithms may be very effective, though, even in the face of these margins. As we move the digitization point closer to the antenna, converter noise and phase-noise issues become more critical; other factors actually aid in the resolution of some of the problems outlined above as we go to IF-DSP.

Of course, a receiver's AGC may be disabled to avoid this kind of arrangement, but only with some degradation of dynamic range.

IF-DSP AT A LOW IF

The primary reason for wanting to digitize signals closer to the antenna is to eliminate expensive filters and other hardware whose functions can be performed in DSP. By going to a low IF, we can get rid of balanced modulators and multiple crystal or mechanical filters; demodulation, squelch, and digital AGC duties are also done in software; many things judged quite difficult or impossible in the analog world may be included, as well.

To do it in a receiver, we apply harmonic sampling and a fast, 16-bit sigma-delta ADC at an IF just above the audio range. This IF is selected to be comparable with the ultimate BW of the *roofing filters* used in the receiver's front end. Recall that in harmonic sampling, the sampling frequency may be as low as the IF minus half its BW. An IF BW of 15 kHz, for example, requires a sampling frequency of at least 30 kHz. The center IF itself may be almost anything greater than 7.5 kHz at this mini-

mum sampling frequency. We ought to consider what *image rejection* we are going to get based on such a low IF, though. Roofing and other analog filters will determine it by their attenuation at an offset from center equal to twice the IF. If we intend to use the same IF in transmit mode, the 2nd LO will appear at an offset equal to the IF. Quite a few poles of filtering in the analog sections are still required around this arrangement. See **Fig 18.39**.

From the antenna, signals are low-pass filtered to remove first-mixing image responses and to eliminate LO leakage. Then, they are mixed to a VHF first IF to dodge as many spurious responses as possible. A VHF first IF may be selected above twice or even three times the highest RF to get away from second and third-order mixing products. Six to eight poles of crystal BPF may be used in the strip, with several gain-controlled stages interspersed. A traditional IF analog AGC is employed. In any receiver design, it is best to distribute gain and loss evenly to avoid degradation of the SNR under reduced-gain conditions. We would like SNR to continue increasing as the input signal increases, as far as possible. Gain reduction, therefore, is usually made to occur at the stages closest to the antenna first, followed by subsequent stages.

First-IF signals are converted directly to the low IF, then amplified and possibly filtered further. Enough amplification is needed to raise the second-IF signal to within about 10 dB of the maximum ADC input level. This maximizes the SNR available from the ADC and the dynamic

range available for digital AGC operation, as described in the Baseband section above. A 10-dB margin leaves the *headroom* needed to accommodate analog AGC overshoot and noise spikes. Overload of the ADC is catastrophic and must never be allowed to occur. IF-DSP architectures that include analog AGC usually must also have digital AGC loops. Note that it is possible to build an analog front end that has the required dynamic range ($\geq$115 dB) and be rid of analog AGC, but this would entail some more-expensive hardware and tighter tolerances. Analog AGC makes it easier to keep the front end linear over the range of input signals expected. These days, receivers may be called on to handle input signals as large as one watt!

One way to handle the situation is to employ a gain-compensation scheme such as that described in the Baseband-DSP section above. We may, in addition, arrange for the DSP to monitor the analog AGC voltage to find out what it is doing. Knowledge of the receiver's AGC voltage-*vs.*-gain curve leads to a new algorithm: When the analog gain is changing rapidly, rapid changes in gain-control variable k are allowed, as in the baseband case. When time constants are set correctly, we do not even have to know the exact amount of analog gain reduction. If we want the digital AGC's threshold to reside above the receiver's noise floor, though, we *do* have to know the gain reduction accurately.

Traditionally, HF receivers have been designed with analog AGCs that have

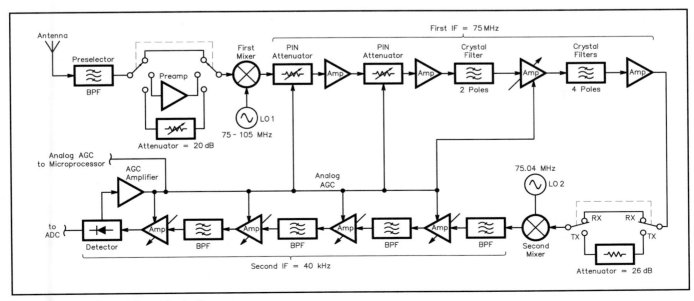

Fig 18.39—IF-DSP receiver block diagram.

thresholds or "knees" around 3 µV. This means that signals below that threshold are not gain-controlled; when input signals are low, the receiver gets quiet. Note that this parameter is unrelated to the SNR as determined by the noise floor of the receiver. Digital AGCs may be designed with any threshold—it may even be made variable via a front-panel control. The effect is much the same as that produced by an IF-gain control. A *knee-less* digital AGC system may result in operator fatigue, because receiver and atmospheric noise are boosted in the absence of strong input signals.

Another approach to analog AGC in a digital receiver involves generating the gain-control voltage using a detector in the digital portion. AGC voltage comes from a DAC controlled by the DSP. See **Fig 18.40**. The delay between detection of IF signals and the application of gain control causes the same problem as in traditional analog AGCs: The loop filter must be optimized with regard to amplitude and phase response so it can minimize overshoot. Delays encountered in DSP filters may require the use of a secondary ADC and detector, solely for AGC purposes. Some designers have even gone as far as to build separate IF strips for detection and application of analog AGC, all the way up to and including the receiver's first IF. Such architectures tend to rapidly become complicated: One of the other schemes detailed above should suffice for Amateur Radio applications.

Conversion plans used in IF-DSP receivers may also be used in the transmitter by simply swapping the LOs, inputs and outputs. A switching arrangement for this is shown in **Fig 18.41**. Isolation between the ports of the LO's DPDT switch must be set to the desired level of spectral purity. An example of such a switch is shown in **Fig 18.42** using PIN diodes at VHF. Switch control voltages swing between +5 V and –5 V; when the series diodes are on, the shunt diodes are off, and vice versa. This particular circuit was designed for a

75-105 MHz LO and achieves better than 80 dB of isolation between the ports. Switching of the first mixer's input is best achieved by a relay for HF circuits, PIN diodes for VHF and above. The second mixer's output may be switched using various commercial ICs, such as the Signetics NE630. Isolation in these switches is important because it reduces spurious responses in both receive and transmit modes.

Gain-controlled stages or step attenuators may have to be employed to provide

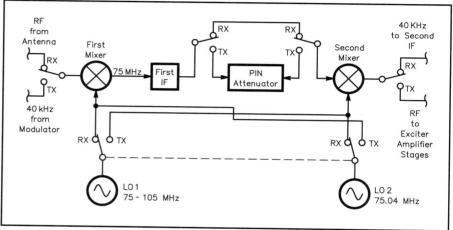

Fig 18.41—Block diagram of IF-DSP conversion scheme with T/R switching added.

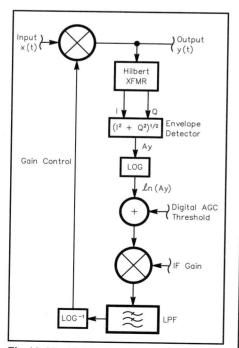

Fig 18.40—IF-DSP receiver with digitally derived analog AGC.

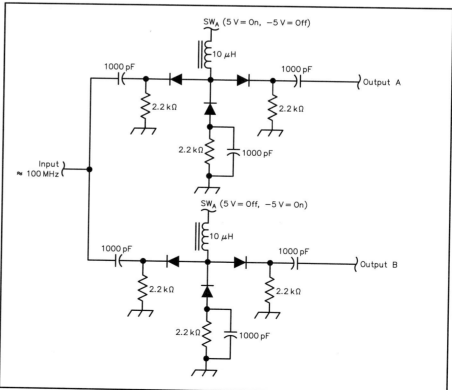

Fig 18.42—SPDT LO switch using PIN diodes. Diodes are Philips BA682 or equiv.

for a difference in IF-strip gain between receive and transmit. To see why this might be necessary, examine the difference in gain between a receiver and a transmitter in typical service. A receiver takes as little as −132 dBm from the antenna and amplifies it to around 1 W (+30 dBm) at the loudspeaker. The power gain is:

$$GAIN_{RX} = 30-(-132)dBm = 162 dB \quad (63)$$

In a transmitter, a typical dynamic microphone might produce 5 mV RMS into 600 Ω, or −44 dBm. To get to 100 W or +50 dBm, the gain is:

$$GAIN_{TX} = 50-(-44)dBm = 94 dB \quad (64)$$

The receiver has the far more difficult task, but the transmitter is still doing yeoman's duty. Considering a maximum path loss of:

$$LOSS_{PATH} = 50-(-132)dBm = 182 dB \quad (65)$$

it is a wondrously large amount of enhancement we get from our electronics, since the total power gain from the microphone on one end to the loudspeaker on the other must be:

$$GAIN_{TOTAL} = 162+94 = 256 dB \quad (66)$$

or a factor of 4×10^{25}!

DIRECT DIGITAL CONVERSION

In the ultimate digital receiver, signals are sampled directly from the antenna without any intervening stages. In practice, though, some gain is required ahead of the ADC because of the present limitations of data-conversion technology. As far as gain stages can be made with high dynamic range and good large-signal-handling capability, direct digital conversion (DDC) comes within reach. In this technique, signals are converted directly to baseband without any intermediate analog mixing stages. Refer to **Fig 18.43**. The LO is, in

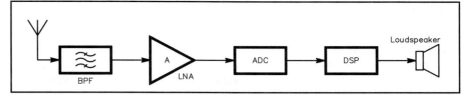

Fig 18.43—DDC receiver block diagram.

effect, placed very close to the desired signal and through harmonic sampling and decimation, translates it to baseband. The closeness of the LO to the signal accentuates phase-noise effects such as reciprocal mixing (discussed above) and makes short-term clock stability a major issue. Fortunately, low-noise, crystal-derived clock designs are becoming available. RMS clock jitter is usually specified in units of time (picoseconds rms), but a clock's phase-noise-*versus*-frequency characteristic tells the entire tale.

The Nyquist criterion compactly determines the sampling rate required for any given signal or group of signals. If the digitized BW is 50 kHz, the minimum sampling rate is 100 kHz, even if the signal's frequency is in the VHF range or beyond. Ancillary sample-and-hold devices may be employed in a DDC receiver to ease the BW requirements for the ADC. The digitized BW must remain within half the final sampling frequency to avoid aliasing; for this reason, interest in narrow pre-selector filters has been renewed.

In the example of a 50-kHz received BW, any increase in sampling rate above 100 kHz is called *over-sampling*. Over-sampling is important because it allows for an SNR gain by spreading quantization noise over a larger BW, then filtering it, as discussed above. We are using harmonic sampling, though, so we are also *under-sampling* our signal. We can

be both over-sampling and under-sampling at the same time since one is defined with respect to BW and the other by the frequencies of interest.

Frequency planning is of special concern in DDC architectures. Quite often, spurious responses appear in high-speed ADC and DAC outputs that we must plan to avoid. Those are largely responsible for establishing the SFDR of high-speed converters. Problems are also created in supposedly linear stages that generate significant harmonic content, since those harmonics show up as aliases in the digitized spectrum and may mix with other products. Careful selection of sampling frequency and IF can place these spurs where they are harmless: outside the band of interest. Over-sampling only moves us toward the goal by providing more spectrum into which spurs may harmlessly fall.

The technique known as dithering further improves SFDR by spreading the energy contained in spurs—those caused by the DNL of data converters—over greater BWs. Dithering artificially adds noise to the clock input of the ADC or DAC to achieve this spectral spreading. Typical values used are in the range of 10-30 bits peak-to-peak. Spurious reduction of over 20 dB has been attained with high-speed (40-Msample/s) converters.

ADVANCED DSP TOPICS IN COMMUNICATIONS

A few advanced subjects are worth briefly mentioning here. They represent exciting trends in DSP communications science that hold significant promise for the future.

ADAPTIVE BEAMFORMING

Thus far, we have considered the application of DSP techniques to processing of signals in the time and frequency domains. *Adaptive beamforming*—the creation of antenna systems with automatically varying patterns—extends the concept of adaptive DSP to the spatial domain. The main goal of "smart-antenna" methods is to condition the radiation pattern of an array so as to maximize received SNR of the signal of interest, and to minimize interference and noise on an adaptive basis. In this sense, adaptive means that the antenna array's pattern changes in response to changes in the strength and direction of desired and undesired signals.

An adaptive antenna array consists of two or more antennas separated in space, connected to a multi-channel, adaptive signal processor. At least one of the antennas in the array feeds the DSP through an adaptive filter. See **Fig 18.44** for an illustration of a two-antenna system. Such a system was developed by Howells and Applebaum around 1960. Two omnidirectional antennas are used: One is called the primary antenna, the other the reference antenna, in direct analogy to the adaptive-interference-canceling arrangement described before. Let's say we have one desired signal and one undesired signal present, originating from different directions of the compass. The output from each antenna contains both signals, but with different amplitudes and times of arrival. The LMS algorithm may use direction of arrival, strength, energy integrated over time, or one of many other criteria as the basis for interference cancellation. Discussion below is limited to the first case of directional adaptation.

Consider what happens when the interference is much stronger than the desired signal: The adaptive filter coefficients are determined almost exclusively by the interference. In the steady state, the adaptive filter's output contains a replica of the interference appearing at the primary antenna, canceling it in the summation. In real systems, each antenna feeds a separate RF receiver where signals are amplified and detected. Receiver and atmospheric noise have a significant effect on performance. The system just described works fine as long as the interference is strong compared with the desired signal. When it is not, leakage terms may bring

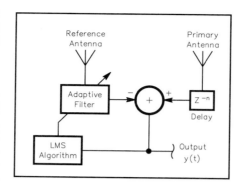

Fig 18.44—Block diagram of a two-antenna, spatial-diversity adaptive system.

remarkable improvements to performance, as in the case of adaptive noise reduction above.

Another type of adaptive beamformer developed by Widrow, Griffiths, Mantey and Goode uses a *pilot* signal to steer the array. The pilot signal forces the antenna array to *look* in a specified direction while still retaining the ability to form nulls where interference is present. This has the effect of removing the restriction that the interference be stronger than the desired signal. Others have developed simpler algorithms that accomplish the same thing, but the pilot-signal algorithm remains viable for certain applications. We can dream of many possible configurations for phased arrays. Each must be evaluated for the relationship between angle of arrival (or other criteria) and amplitude/phase of signals appearing at various elements in the array.

PERCEPTUAL SPEECH CODING FOR BW REDUCTION

Driven by the desire to digitally code audio at relatively low bit rates, a lot of effort has been expended on coding algorithms over the years. Much research has been done on the modeling of human speech and hearing toward that desire. A myriad of schemes have evolved for both modeling and coding—they are too numerous to list here. Let it suffice that these schemes may be divided into two camps: time-domain and frequency-domain. The former type of coding involves the audio waveform directly; the second implies analysis and manipulation of the spectral content of the signal over time.

Time-domain coders use the sample-by-sample characteristics of an analog signal to produce a coded digital signal, which is usually then stored (as on a CD) or transmitted (as through the telephone network). The public switched telephone network

(PSTN)—the universal communications medium to which the most people have access—uses a form of audio coding called *μ-law* coding. In it, signals are digitized according to a logarithmic (rather than linear) transfer function so as to extend dynamic range. 8-bit quantization is used at a sampling rate of 8 kHz, producing a 64-kbit/s bit stream. Other systems such as *delta modulation* also produce *toll-quality* audio at similar bit rates.

Frequency-domain coders manipulate audio signals by analyzing their spectral content in discrete time intervals. To achieve coding efficiency, these coders exploit the fact that the spectral content of speech signals does not change significantly over short intervals, on the order of 20-30 ms. Spectral signatures may be stored or transmitted using relatively few bits per second. A trade-off may be made between recovered speech quality and transmitted bit rate. To get an idea of what is possible, consider the following example that is a little outside the two classifications above.

It is evident that during normal speech at about 150 words per minute, a textual representation transmitted digitally requires a data rate of only (150 words/min) (5 letters/word) (7 bits/letter) = 5250 bits/minute = 87.5 bits/s. This may be transmitted in a very narrow BW. This system cannot convey any of the characteristics of the speaker's voice, though, such as loudness, timbre, or emotion: It is impossible to tell who the speaker is—unless they tell you! So we must conclude that additional BW is necessary to include such information and to make the recovered speech sound natural.

FFT-based spectral analysis and synthesis may be very effective where it is desired to transmit coded information in the analog domain. The same BW savings may be achieved when transmitting analog signals as when transmitting them digitally, as long as the transmission medium does not introduce deleterious distortions on the coded signals. Media such as HF radio introduce serious multi-path and phase distortion that have proven more damaging to digital modes than analog.

A coded signal may not be the final coder output. It might serve as the input to another algorithm that determines whether speech is present or not, that decides who is talking, or that determines what is being said. Speech-recognition algorithms use many of the same principles as speech-compression algorithms. This technology is being put to use in communications and elsewhere for voice control of equipment, computer dictation, and voice-print verification.

HARDWARE FOR EMBEDDED DSP SYSTEMS

What is it about a microprocessor that makes it a DSP? Well, DSPs are special because they include facilities uniquely designed for the type of calculations common in signal-processing algorithms. They are almost all 16-bit machines, or better, and so are very powerful even without their special facilities. DSPs may be classified primarily by their representation of numbers (fixed-point *vs.* floating-point), also by their data-path width (16-bit, 32-bit), by their programmability (general-purpose *vs.* dedicated co-processor), and their speed.

FIXED-POINT DSPS

Fixed-point DSPs are generally simpler than floating-point units, so they are typically less expensive. Fixed-point processors are common in embedded systems, especially for radio. Special software instructions and separate high-speed computational units are included to accelerate the processing of those common DSP calculations already mentioned. Perhaps the most-used operation is the convolution sum, performed as a series of MAC instructions (see the section on FIR Filters). Designers are interested in how many MACs per second a DSP can execute, because for anything beyond simple audio processing, only a small amount of time is available between samples for filtering and other functions.

A typical 16-bit, fixed-point DSP is shown in the block diagram of **Fig 18.45**. It employs what is called the *Harvard architecture*: It has separate program and data memory paths and also includes a *pipeline* for holding instructions waiting to be executed. This arrangement speeds things along because the CPU can fetch future instructions even when it is executing the current instruction or fetching data from another path.

Consider how this affects an FIR filter algorithm, for example. For each tap in the filter, the processor must multiply a constant (a filter coefficient) by a data value (a stored sample). When the processor can fetch both values simultaneously, an entire cycle time is saved. The subsequent addition of the product to the accumulator and the incrementing of indices for the next MAC instruction may also be executed in a single cycle. When large filters are being implemented, time savings quickly mount. Contrast this with the many cycles needed to perform the same operations in a general-purpose computer and you will see why specialized processors are so much more capable of handling sampled signals.

This business of execution speed is a large factor in the selection of a DSP for any particular use. System planning must begin by reckoning how many instructions can be executed between sample times. In a system with a 30-kHz sampling rate, only 33 μs are available, so a fixed-point DSP that can execute two million MACs per second (2 M-MACs/s) can only get 66 of these in the space between samples. For all but the simplest of systems, this is generally insufficient power for good filtering and other requirements and a separate filter *co-processor* must be employed. This is discussed further below. DSPs are now available having over 200 M-MACs/s performance.

Many fixed-point DSPs are available that also have undedicated parallel and serial input/output (I/O) on board. These may be very useful for embedded applications by obviating the need for other hardware. Processors embedded in radios have traditionally been shut off during times when no user input is present, stopping their clocks. This is done to eliminate the digital-circuit noise that otherwise would

be difficult to remove. With a DSP in critical signal paths, this luxury is not possible. Careful attention to shielding, grounding and bypassing must therefore be paid. A DSP and associated support components humming along at 25 MHz—or more—tend to generate lots of noise and discrete spectral elements. They also tend to draw significant current, although dissipations in the one-watt range are typical; for base-station equipment, this is not usually a big concern.

Fixed-point math brings with it a limitation on the range of numbers that can be represented, notwithstanding the extended integer/fractional representation demonstrated above. This limitation may form an obstacle to achieving the highest possible dynamic range. For this reason, floating-point DSPs are also widely available for use where greater boundaries must be set on the range of numbers handled. **Table 18.1** shows a listing of popular fixed-point DSPs, along with their floating-point cousins. Manufacturers supply evaluation boards, some of which include data converters and other support circuitry. Control software running on a desk-top computer is available for downloading *object code*—the DSP instructions that make up the program—as well as for debugging by use of tools such as break-points and register dumps.

FLOATING-POINT DSPS

Representation of numbers is a critical decision to be made early in the system design process. A decision to use a floating-point DSP, at generally higher cost than fixed-point, is usually made either to remove dynamic-range barriers or to grant greater flexibility to algorithms that require scaling of data and coefficients, such as the FFT algorithms discussed above. We saw that each floating-point number requires two storage locations: one for the mantissa and one for the exponent. One would expect the processing of these numbers to be slowed by having to handle twice the data, but floating-point architectures are devised in such a way as to minimize or even eliminate this apparent handicap.

Multiplying two floating-point numbers involves multiplying the mantissas, then adding the exponents and any carry (or borrow) from the multiplication. Since multiplications generally require more time than additions, summing the exponents does not really slow the machine very much. Adding two floating-point numbers, though, requires the addition of the mantissas and a possible adjustment to

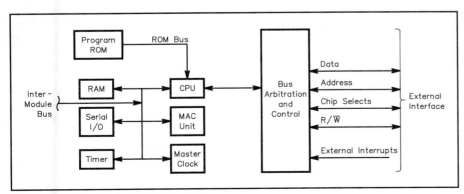

Fig 18.45—Fixed-point DSP block diagram.

Table 18.1

Fixed and Floating-Point DSPs

Part Number	Manufacturer	# of bits	Fixed/Floating
TMS320Cxx	Texas Instr.	16	Fixed
DSP320Cxx	Microchip	16	"
DSP16	ATT	16	"
ADSP21xx	Analog Dev.	16	"
MC68HC16	Motorola	16	"
MC5600x	Motorola	24	"
MB862xx	Fujitsu	24	Floating
MC9600x	Motorola	32	"
DSP32x	ATT	32	"
TMS320Cxx	Texas Instr.	32	"

the exponent, and this is always a bit slower than can be done on fixed-point numbers. With an optimized MAC unit, even this restriction can be removed for the bulk of calculations in typical DSP applications. Other than for these points, the block diagram of a floating-point DSP does not look very different from that of the fixed-point unit in Fig 18.45.

SELECTING DATA CONVERTERS

Complete DSP systems almost always include data converters in the form of one or more ADCs and DACs. Selection of these devices for any particular application is made with regard to cost, bit-resolution, speed, SFDR, and digital interface. Manufacturers characterize devices on these bases and obviously, we must choose them so they will handle the highest sampling rate at our analog interface. In general, bit-resolution and speed determine SFDR. Dual 16-bit ADCs and DACs are now very common because they are used in compact-disc (CD) recorders and playback units at a sampling rate of 44.1 kHz. Note that 44.1 kilosamples/s of two channels in a stereo system is equal to $(2) \times (44,100) \times (16) \approx 1.41$ megabits per second. Devices with 20 and even 24-bit capability are catching on. This is a lot of data

and the bit-resolution of data converters is most often chosen to match that of the DSP, although there may be advantages in having slightly more bit-resolution in the DSP to mitigate round-off errors, as noted in the FIR Filters section above.

We noted before that over-sampling of input signals brings significant advantages for the DSP designer. For this reason, sigma-delta ADCs are the "top of the crop" for use in IF-DSP and DDC receivers. As sampling frequencies increase, over-sampling becomes more difficult to achieve. Engineers working in cellular radio and similar technologies deal with much wider BWs than most of those found in Amateur Radio, and so must grapple with reduced dynamic range; fortunately, they also require less. ADCs that handle 12 to 16 bits at speeds exceeding 65 MHz are available. Viable DDC designs are finding their ways into many commercial services worldwide.

Converters must interface with DSPs through a high-speed digital connection of some kind. Parallel transfer—all 16 bits at once, for example—is more common among DACs than ADCs. High-speed, three-line serial interfaces are popular among converter manufacturers and several standards have evolved. Some of these

are compatible with one another. Bearing in mind the amount of data being transferred, realize that these serial links may run at clock speeds in excess of 100 MHz. ADC/DAC evaluation boards may be connected to DSP evaluation boards to form a prototype DSP system. Some data converters are listed in **Table 18.2**.

EXTRA PROCESSING POWER: DSP CO-PROCESSORS

Quite often, a single, general-purpose DSP by itself is not sufficient to handle the computational load in a project. This may be determined early in the system design by evaluating the number of MACs required by filters and other algorithms. Several solutions present themselves: adding one or more general-purpose DSPs, adding specialized co-processor chips, or designing a custom co-processor using programmable-logic chips.

More than one general-purpose DSP may be used to augment net data capacity. The trend these days, though, is to use dedicated co-processor chips that are optimized for the function they are to perform. This is especially true of FFT and other operations that do not lend themselves well to the MAC procedures for which general-purpose chips are optimized. Whatever the algorithm, it seems that multiplication of numbers takes the most time, so a co-processor that incorporates a fast-multiplication algorithm is desirable. A lot of effort has gone into fast multipliers since the 1980s and for the IF-DSP or DDC designer, a knowledge of how it is done may bring plentiful results.

The multiplication of two binary numbers may be decomposed into an addition of several binary numbers. We know that fast binary addition is readily achieved by relatively simple logic. Let's take a look at this, since it forms the basis for most fast multipliers. Shown in **Fig 18.46** is the long multiplication of two 4-bit binary numbers. It is performed in base two the same way as it is in base ten: First, take the least-significant digit of the lower multiplicand and multiply it by the other multiplier. Since in binary, this digit is either one or zero, the digits we write under the line is either a copy of the top multiplicand, or all zeros. Then, the next-significant digit of the lower multiplicand is used, with the result written below the first and shifted one digit to the left. This process continues until all bits of the lower multiplicand have been used. Finally, all the interim results are added. This last result is the product of the two numbers. Note that the result may contain a number of bits as high as the sum of the number of bits in both the multiplicands. Project G in the Appendix

Table 18.2

Data Converters

Part Number	Manufacturer	# bits	Speed	ADC/DAC
HI1276	Harris/Intersil	8	500 Ms/s	ADC
AD7722	Analog Dev.	16	200 ks/s	"
ADC76	Burr Brown	16	50 ks/s	"
PCM1750	"	18	44 ks/s	dual ADC
CS5322	Crystal	24	2 ks/s	ADC
BT254	Brooktree	24	30 Ms/s	"
Note: Also see Maxim, National, Sipex, Analogic				
CA3338A	Harris/Int.	8	50 Ms/s	DAC
HI1171	"	8	40 Ms/s	"
HI5780	"	10	40 Ms/s	"
HI20201	"	10	160 Ms/s	"
PCM56	Burr Brown	16	93 ks/s	DAC
PCM66	"	16	44 ks/s	dual DAC
Note: See also National, Analog Dev., Maxim, etc.				

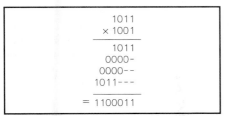

```
        1011
      × 1001
      ------
        1011
       0000-
      0000--
     1011---
      ------
   = 1100011
```

Fig 18.46—Long multiplication of two 4-bit binary numbers.

Table 18.3
Co-processors and DDC Chips

Part Number	Manufacturer	# bits	Speed	Function
HSP50016	Harris/Int.	16	52 Ms/s	DSP down-conv.
HSP50110	"	10	60 Ms/s	Quadr. tuner
HSP50210	"	10	52 Ms/s	DSP Costas loop
HSP50306	"	6	2 Mbit/s	QPSK demod
HSP43xxx	"	10-24	var.	DSP filters
510	Harris et al	16	10 Ms/s	Mult/Acc
LMA2010	Logic Dev, IDT	16	40 Ms/s	Mult/Acc
HSP4510x	Harris/Int.	20/32	33 Ms/s	DDS
Various	Xylinx, Altera, Atmel, etc.	8-32	>100 Ms/s	FPGAs

shows how simple logic is used to implement a fast multiplier. Pipe-lining and latency issues are discussed there.

Refinements of this technique that use look-up tables and combinatorial methods yield speed increases. Field-programmable gate array (FPGA) manufacturers have worked out the details of these algorithms and routinely provide them to users. FPGAs are available now in very-high-speed versions ($f_{clk} \geq 200$ MHz) that may be used for DSP co-processing.

FPGA designs may also employ the Harvard architecture using external, *dual-port memory* to provide a register-based interface to host DSPs. Normally, one sample is passed to the co-processor and one retrieved at each sample time. Filters exceeding 100 taps may be implemented this way, saving processing time in the host DSP for other housekeeping tasks.

Entire down-conversion and I/Q modulation sub-systems have been incorporated on a single chip. These chip sets may be advantageous where FPGA-based designs either do not meet requirements or are too expensive. A sampling of ready-to-use co-processors and DDC chips is given in **Table 18.3**. Also read some of the reference material listed at the end of this chapter for more information on dedicated DSP co-processors.

DSP SYSTEM SOFTWARE

ASSEMBLY LANGUAGE AND TIMING REQUIREMENTS

Embedded-DSP application software is most often written in *assembly language*, the native language of the DSP in use. Instructions to be executed are arranged in order, according to the *von Neumann model*, and entered as lines in a text file, using the mnemonics provided by the DSP manufacturer. When this *source code* is ready, an *assembler* program is invoked that translates the source code into object code—the numbers that the DSP understands as instructions. The object code is then transferred to the program memory of the target system for execution.

The reason assembly language is so prevalent in embedded applications is the critical timing involved. Programs compiled in high-level languages do not always handle interrupt-driven events well (the input or output samples) and may bog down. To minimize the required hardware speed, processing of some second-line tasks such as squelch and ALC must have reduced sampling rates to fit into the whole picture. Only a part of their processing burden may be performed at each sample time. This is a form of *time-distributed processing* and is just one in the DSP designer's bag of tricks.

Someone will always think of something

more for a transceiver to do and it is better to err on the side of higher speed and more memory at the start than to run out later. Even so, DSP designers must carefully evaluate all the functions included at the outset. Other shortcuts—like the assumption of only integer values by a BFO at one-fourth the sampling frequency—may present themselves, but one cannot always count on it; one must plan diligently to avoid roadblocks. In addition, *unexpected things can occur* if due thought is not given to quantization and scaling effects, especially where adaptive processing is applied, no matter the representation of numbers used. DSP-chip manufacturers provide assemblers and instruction details free of charge. Their applications engineers are ordinarily ready to assist. A plethora of information is available on the Web.

FILTER-DESIGN SOFTWARE

Several software packages for DSP filter design are listed at the end of this chapter. Many more are available. You can expect to find reasonably priced software that will design FIR and IIR filters, as well as let you perform convolution, multiplication, addition, logarithms and other calculations on numeric sequences.

FIR filters usually may be designed with a choice of method (Fourier, Parks-

McClellan, least-squares), length, frequency response, and ripple magnitude; they may use various window functions to achieve different shape factors and passband/stopband attenuations. Some are able to take coefficient and data quantization into account and some are not. Large filters may deviate significantly from their theoretical responses because of these effects, so if you are contemplating reasonably long filters, check into this capability.

IIR filter design usually includes a choice of various analog-filter prototypes. Software packages may vary in their ability to display, print, or plot responses and write coefficient files to disk. Filter coefficients are generally part of system firmware and must be transferred from the host DSP to a filter co-processor on demand. It must be possible to translate the filter-design software's output to a format the compiler software understands. A translation program may have to be written to accomplish this.

Longer and more-complex FIR filters may be implemented by convolving the impulse responses of several different filters. This allows the alteration of the frequency response of standard filters to include graphic or parametric equalization and IF shift. Such filtering systems are already being employed in Amateur Radio and commercial transceivers.

OTHER DSP DESIGN TOOLS

FPGA design software is generally available from chip manufacturers. In addition, many schematic-capture and PCB-layout software vendors provide interfaces to popular FPGAs and other programmable devices. Hardware Design Language (HDL) and Verilog Hardware Design Language (VHDL) have become popular for translating user requirements into programming code for FPGAs. Most FPGA programmers understand HDL or VHDL.

A rich variety of flow-chart software exists in both the public and private domains. It may be especially useful for time-sensitive applications in DSP.

BIBLIOGRAPHY

(Key: **D** = disk included, **A** = disk available, **F** = filter design software)

DSP Software Tools

Alkin, O., *PC-DSP*, Prentice Hall, Englewood Cliffs, NJ, 1990 (**DF**).

Kamas, A. and Lee, E., *Digital Signal Processing Experiments*, Prentice Hall, Englewood Cliffs, NJ, 1989 (**DF**).

Momentum Data Systems, Inc., *QEDesign*, Costa Mesa, CA, 1990 (**DF**).

Stearns, S. D. and David, R. A., *Signal Processing Algorithms in FORTRAN and C*, Prentice Hall, Englewood Cliffs, NJ, 1993 (**DF**).

DSP Textbooks

Frerking, M. E., *Digital Signal Processing in Communication Systems*, Van Nostrand Reinhold, New York, NY, 1994.

Ifeachor, E. and Jervis, B., *Digital Signal Processing: A Practical Approach*, Addison-Wesley, 1993 (**AF**).

Madisetti, V. K. and Williams, D. B., Editors, *The Digital Signal Processing Handbook*, CRC Press, Boca Raton, FL, 1998 (**D**).

Oppenheim, A. V. and Schafer, R. W., *Digital Signal Processing*, Prentice Hall, Englewood Cliffs, NJ, 1975.

Proakis, J. G. and Manolakis, D., *Digital Signal Processing*, Macmillan, New York, NY, 1988.

Proakis, J. G., Rader, C. M., et. al., *Advanced Digital Signal Processing*, Macmillan, New York, NY, 1992.

Rabiner, L. R. and Schafer, R. W., *Digital Processing of Speech Signals*, Prentice Hall, Englewood Cliffs, NJ, 1978.

Widrow, B. and Stearns, S. D., *Adaptive Signal Processing*, Prentice Hall, Englewood Cliffs, NJ, 1985.

Articles

Albert, J. and Torgrim, W., "Developing Software for DSP," *QEX*, March, 1994, pp 3-6.

Anderson, P. T., "A Simple SSB Receiver Using a Digital Down-Converter," *QEX*, March, 1994, pp 17-23.

Anderson, P. T., "A Faster and Better ADC for the DDC-Based Receiver," *QEX*, Sep/Oct 1998, pp 30-32.

Applebaum, S. P., "Adaptive arrays," *IEEE Transactions Antennas and Propagation*, Vol. PGAP-24, PP. 585-598, September, 1976

Ash, J. et al., "DSP Voice Frequency Compandor for Use in RF Communications," *QEX*, July, 1994, pp 5-10.

Beals, K., "A 10-GHz Remote-Control System for HF Transceivers," *QEX*, Mar/Apr, 1999, pp 9-15.

Bloom, J., "Measuring SINAD Using DSP," *QEX*, June, 1993, pp 9-18.

Bloom, J., "Negative Frequencies and Complex Signals," *QEX*, September, 1994.

Brannon, B., "Basics of Digital Receiver Design," *QEX*, Sep/Oct, 1999, pp 36-44.

Cahn, H., "Direct Digital Synthesis—An Intuitive Introduction," *QST*, August, 1994, pp 30-32.

Cercas, F. A. B., Tomlinson, M. and Albuquerque, A. A., "Designing With Digital Frequency Synthesizers," *Proceedings of RF Expo East*, 1990.

de Carle, B., "A Receiver Spectral Display Using DSP," *QST*, January, 1992, pp 23-29.

Dick, R., "Tune SSB Automatically," *QEX*, Jan/Feb, 1999, pp 9-18.

Emerson, D., "Digital Processing of Weak Signals Buried in Noise," *QEX*, January, 1994, pp 17-25.

Forrer, J., "Programming a DSP Sound Card for Amateur Radio," *QEX*, August, 1994.

Green, R., "The Bedford Receiver: A New Approach," *QEX*, Sep/Oct, 1999, pp 9-23.

Hale, B., "An Introduction to Digital Signal Processing," *QST*, September, 1992, pp 43-51.

Kossor, M., "A Digital Commutating Filter," *QEX*, May/Jun, 1999, pp 3-8.

Morrison, F., "The Magic of Digital Filters," *QEX*, February, 1993, pp 3-8.

Olsen, R., "Digital Signal Processing for the Experimenter," *QST*, November, 1994, pp 22-27.

Reyer, S. and Herschberger, D., "Using the LMS Algorithm for QRM and QRN Reduction," *QEX*, September, 1992, pp 3-8.

Rohde, D., "A Low-Distortion Receiver Front End for Direct-Conversion and DSP Receivers," *QEX*, Mar/Apr, 1999, pp 30-33.

Runge, C., *Z. Math. Physik*, Vol 48, 1903; also Vol 53, 1905.

Smith, D., "Signals, Samples and Stuff: A DSP Tutorial, Parts 1-4," *QEX*, Mar/Apr-Sep/Oct, 1998.

Smith, D., "PTC: Perceptual Transform Coding for Bandwidth Reduction of Speech in the Analog Domain, Part 1," *QEX*, May/Jun, 2000.

Ulbing, Sam, "Surface-Mount Technology—You Can Work With It! Parts 1-4," *QST*, April-July, 1999

Ward, R., "Basic Digital Filters," *QEX*, August, 1993, pp 7-8.

APPENDIX: DSP PROJECTS

Project A: Decimation
Project B: FIR Filter Design Variations
Project C: Analytic Filter-Pair Generation
Project D: Newton's Method for Square Roots in QuickBasic 4.5
Project E: A Fast Square-Root Algorithm Using a Small Look-Up Table in Assembly Language
Project F: A High-Performance DDS
Project G: A Fast Binary Multiplier in High-Speed CMOS Logic

PROJECT A: DECIMATION

This project illustrates the concept of decimation using Alkin's *PC-DSP* program, included with the book of that name listed in the Bibliography. First, generate 40 samples of the sinusoid y(n) = sin(n/4), where $0 \leq n \leq 39$. This sequence may be generated using the "Sine" function of the "Generate" sub-menu under the "Data" menu, with parameters Var1 = SIN, A = 1, B = 0.25, C = 0 and #Samples = 40. Press F2 to display the data, which should match **Fig 18.A1**.

Next, decimate the sequence by a factor of 2 using the "Decimate" function found in the "Process" sub-menu under the "Data" menu. Use parameters Var1 = SIN2, Var2 = SIN, Factor = 2. Display the new sequence by pressing F2. It should match **Fig 18.A2**.

PROJECT B: FIR FILTER DESIGN VARIATIONS

An FIR filter's ultimate attenuation and its transition BW are largely determined by the filter's length: the number of taps used in its design. Fourier and other design methods do not always readily optimize the trade-off among transition BW, ultimate attenuation and ripple. One way to achieve better ultimate attenuation at the expense of passband ripple is to convolve the impulse responses of two short filters to obtain a longer filter. The two impulse-response sequences are processed by precisely the same convolution sum that is used to compute FIR filter outputs (Eq 3 in the main text).

A filter obtained by convolving two filters of length L has length 2L −1. In one example, two LPFs of length 31 may be convolved to produce a filter of length 61. The resulting frequency response, plotted against that of a LPF designed with Fourier methods for an identical length of 61 taps, would show that the ultimate attenuation of the convolved filter is 20 dB or 10 times greater than that of the plain, Fourier-designed filter. Also, the convolved filter would have a greater passband ripple and a narrower transition region. Quite often, filters that were designed using different window functions may be convolved to get some of the benefits of each in the final filter.

A look back at Fig 18.29 reveals that different window functions achieve different transition BWs and values of ultimate attenuation. The rectangular window attains a narrow transition BW, but a poor ultimate attenuation; the Blackman window, on the other hand, has nearly optimal ultimate attenuation and a moderate transition BW. Let's see what happens when we convolve the impulse responses of filters designed using each method. We will constrain ourselves to filters with odd numbers of taps so that the convolved impulse response will also have an odd number of taps.

Using your favorite filter-design software, first design a LPF by the Fourier method with a length of 31, using a rectangular window, and a cut-off frequency (−6 dB point) of $0.25f_s$. Its frequency

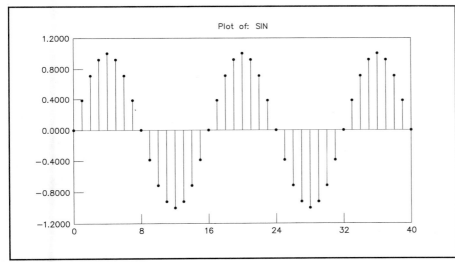

Fig 18.A1—A 40-sample sine wave.

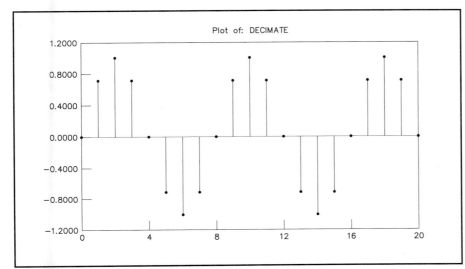

Fig 18.A2—Decimated, 20-sample sine wave.

response is shown in **Fig 18.B1A**. We produce a second filter having the same cutoff frequency of $0.25f_s$ using a Blackman window, whose response is shown in **Fig 18.B1B**. The response of the filter formed by the convolution of the two filters is shown in **Fig 18.B1C**, along with that of a standard Fourier-designed LPF. The final filter has length 61 taps. Notice that the filter obtains the benefits of the rectangular window's sharp transition region and those of the Blackman window's good ultimate attenuation.

A second advantage may be garnered by convolving two different filters in that their responses may be governed separately, while producing desired changes in frequency (or phase) response. An good example of this arises when it is desired to alter the audio response of an SSB transmitter (or receiver), but keep the ultimate attenuation characteristics the same. A long BPF with excellent transition properties may be convolved with a much shorter filter that is manipulated to provide the desired passband response.

FIR filters used in Amateur Radio transceivers must usually have at least 60 dB ultimate attenuation. This generally requires at least 63 taps. As our second FIR filter variation, let's consider a case wherein we want to customize an IF-DSP transmitter's frequency response without impacting opposite-sideband rejection. We will use a 99-tap BPF in each leg of a Hilbert transformer (as part of an SSB modulator) whose response is convolved with that of a 31-tap filter describing the variation in frequency response we want. The 99-tap fixed filter has the frequency response shown in **Fig 18.B2A**. The 31-tap filter has been designed using Fourier methods to have a 6 dB/octave rise in its frequency response, as shown in **Fig 18.B2B**.

The frequency response of the convolution of the two filters' impulse responses is shown in **Fig 18.B2C**. It is important to note that the net response is that of the *product* of the two filters' frequency responses; that is, if $H_1(\omega)$ and $H_2(\omega)$ are the two frequency response functions, the final response is simply:

$$H_{composite}(\omega) = H_1(\omega)H_2(\omega) \qquad \text{(B1)}$$

PROJECT C: ANALYTIC FILTER PAIR GENERATION

Frequency-translation properties of complex multiplication work just as well on the responses of filters as they do on real signals. In this project, we will explore just how these properties are applied to the generation of analytic filter pairs.

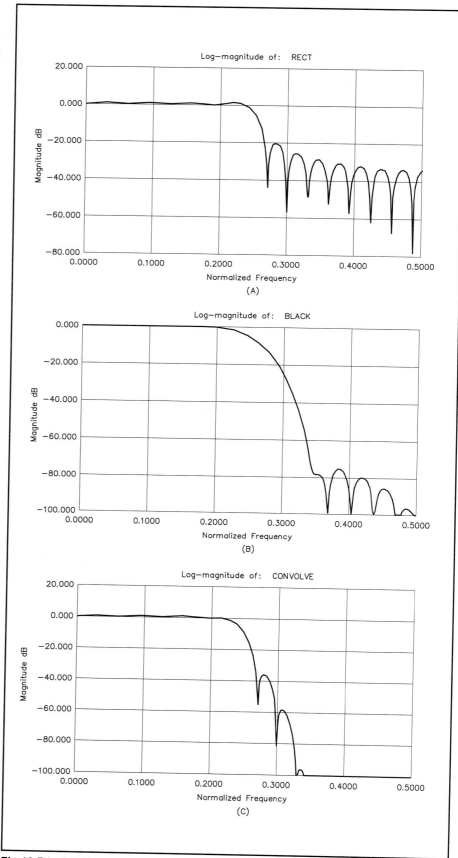

Fig 18.B1—LPF frequency response, rectangular window (A). LPF frequency response, Blackman window (B). LPF frequency response, convolution of filters shown in A and B (C).

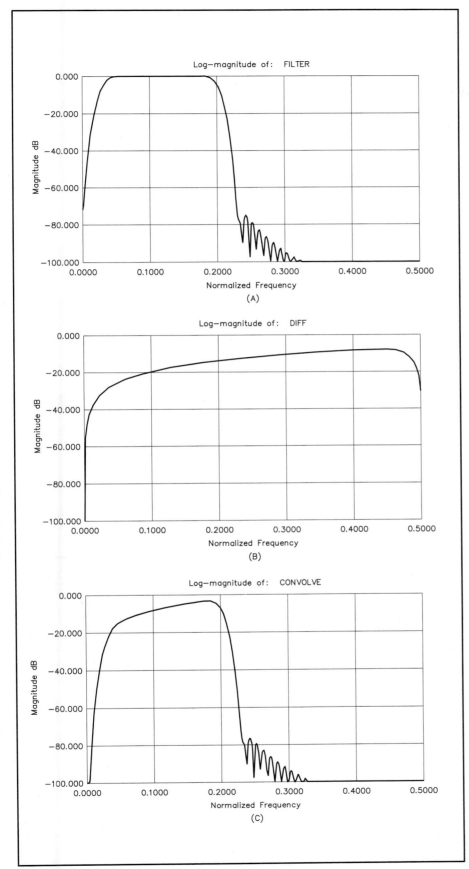

Fig 18.B2—BPF for SSB use, L=99 (A). LPF having rising frequency response, L=31 (B). Frequency response of convolution of filters shown in A and B (C).

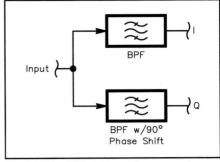

Fig 18.C1—Hilbert transformer using an analytic filter pair.

Analytic filter pairs are used to produce complex signals from real signals for the purposes of modulation, demodulation, and other processing algorithms.

An analytic filter pair consists of two filters (usually BPFs) whose frequency responses are identical, but whose phase responses differ at every frequency by 90°. These filters are used in legs of a Hilbert transformer, as shown in **Fig 18.C1**. The creation of these filters begins with the design of a LPF prototype having the desired passband, transition-band, and stopband characteristics. Such a prototype filter, as might suffice for an SSB receiver, would have a frequency response such as that shown in **Fig 18.C2A** (page 39).

The filter's impulse response (L = 63) is then multiplied by a sine-wave sequence (also L = 63) whose frequency represents the amount of upward translation applied to the LPF's frequency response. If the sine wave is high enough in frequency, the resulting impulse response is a BPF filter centered on ω_0, the sine wave's frequency. See **Fig 18.C2B**. Likewise, the prototype LPF's impulse response is multiplied by a cosine-wave sequence to produce a filter having the same frequency response as that of the sine-wave filter, but with a phase response differing by 90°. Sample-by-sample multiplication occurs according to Eq 21 in the main text.

When an analytic filter pair is used in a demodulator, IF shift may be included by varying the frequency of ω_0. In combination with various filter BWs, IF shift is useful in avoiding interference by modifying a receiver's frequency response. Further modification may be obtained by convolving each filter in the analytic pair with a filter having the desired characteristic. The phase relation between the filters in the pair will not be altered by the convolution.

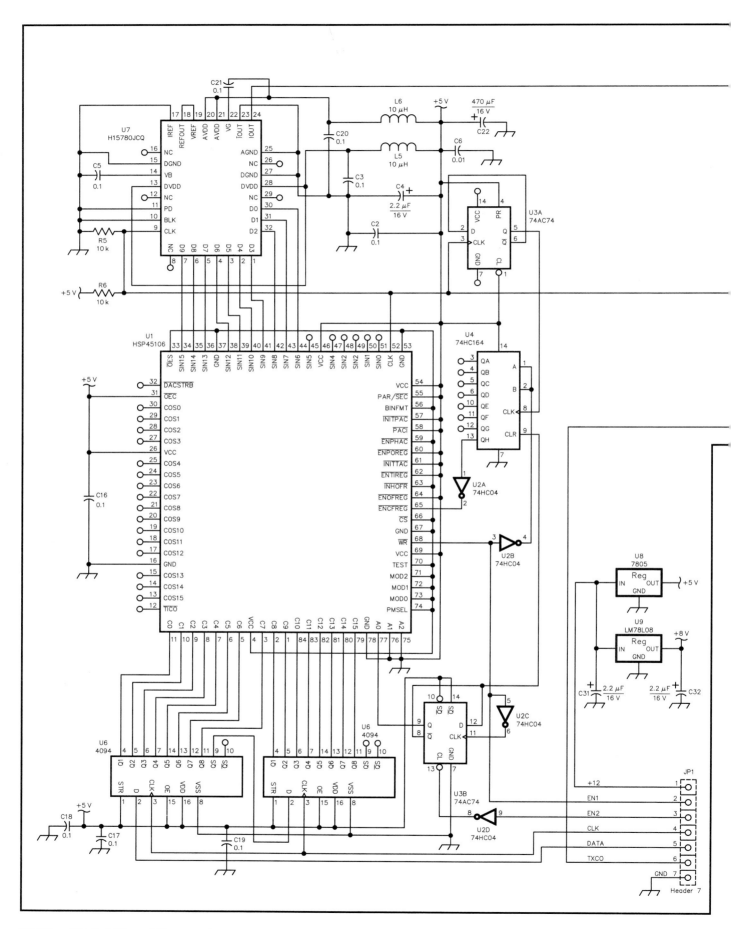

18.38   Digital Signal Processing

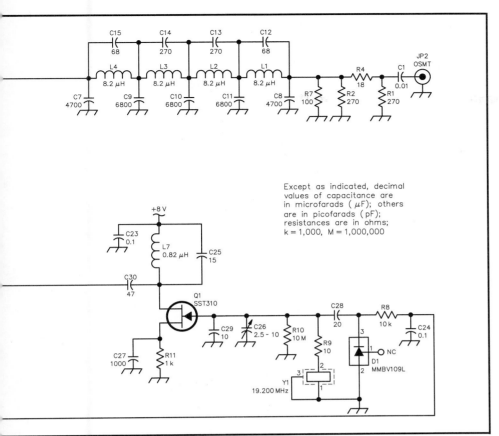

Fig 18.F1—High-performance DDS schematic diagram.

DERROR, initially defined to be one least-significant bit or $1/(2^{15}) \approx 30 \times 10^{-6}$. Note that if DERROR is small or zero, convergence may never be reached because of quantization noise. A loop counter, K, is established to count iterations. The program displays on the computer screen the argument, its root and the iteration count. Users may readily modify the program to use random numbers as arguments to time the number of roots per second it calculates.

The program is included in the *2001 ARRL Handbook* companion software. This software is available for free download from *ARRLWeb* at: **http://www.arrl.org/notes**.

PROJECT E: A FAST SQUARE-ROOT ALGORITHM USING A SMALL LOOK-UP TABLE

This project is a machine-language example of a fast square-root algorithm. The target processor in this case is the Motorola MC68HC16Z1, a 16-bit, fixed-point DSP. The method is depicted in **Fig 18.16** in the main text. Like the previous project, this is included in the *2001 ARRL Handbook* companion software, which is available for free download from *ARRLWeb* at **http://www.arrl.org/notes**.

PROJECT F: A HIGH-PERFORMANCE DDS

A DDS is described below that is used as a reference for a PLL. See **Fig 18.F1**. This DDS is designed to cover a small range of frequencies near 1 MHz. A crystal-oscillator clock at 19.2 MHz is applied to both the DDS, a Harris/Intersil HSP45106, and the DAC, a Harris/Intersil HI5780. Making the DDS output frequency a small fraction of the clock frequency makes it relatively easy to obtain excellent spurious performance. PM spurs are limited to –90 dBc and AM spurs to

PROJECT D: NEWTON'S METHOD FOR SQUARE ROOTS IN QUICKBASIC 4.5

In this example of Newton's method, a generic *BASIC* program is given that computes the root of a 32-bit integer to within an error margin, DERROR. The root of a 32-bit integer is naturally a 16-bit integer. Emphasis is placed in what follows on speed of execution and accuracy as influenced by truncation and rounding. 32-bit integer variables are defined DEFLONG, 16-bit integers are DEFINT. Integer math in *QuickBasic* is much faster than floating-point math.

As described in the AM Demodulation section in the main text, Newton's method iteratively converges on a result. Experience has shown that three to six iterations are necessary to obtain best accuracy for a 16-bit result, but here we execute as many iterations as necessary to obtain accuracy

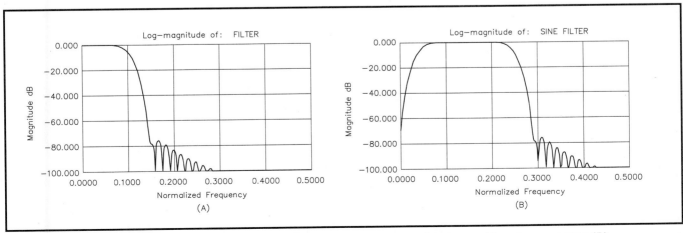

Fig 18.C2—LPF prototype frequency response (A). BPF frequency response of processed impulse response (B).

about –60 dBc. If the output is not squared at the input to a PLL chip, an external Schmitt-trigger squaring stage may be added, eliminating virtually all the AM spurs prior to the LPF.

The LPF at the output of the circuit is a 4-section elliptical type. Design impedance is 100 ohms. This filter cuts out many high-frequency spurs and stops clock feed-through. The DAC's 10 input lines are fed from the 10 most-significant bits of one of the DDS's outputs. The HSP45106 has two 16-bit outputs (sine and cosine) to accommodate the needs of complex-mixer designs, but only one is being used here.

The DDS chip itself is programmed using a 16-bit parallel interface. This is transformed into a serial interface by shift registers U5 and U6, divider U3 and counter U4. Each time the frequency is changed, an internal 32-bit phase-increment accumulator must be updated. The phase increment is just f_{out}/f_{clk}, expressed as a 32-bit, unsigned fraction. This value is written into the chip in two 16-bit segments, most-significant bit of the most-significant word first.

During serial programming, a data bit is placed on the DATA line by the host microprocessor; the clock line is toggled high, then low to shift the bit into the shift registers. After the first 16 bits have been shifted, they are written into the DDS by toggling the ENABLE line. Counter U4 supplies the necessary write pulse with appropriate timing. The remaining 16 bits are then shifted and written to the chip, completing the operation.

An example of the output spectrum of this circuit is shown in **Fig 18.F2**. Components are surface-mount types and care must be exercised during construction.

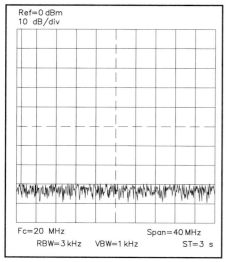

Fig 18.F2—Typical output spectrum of DDS.

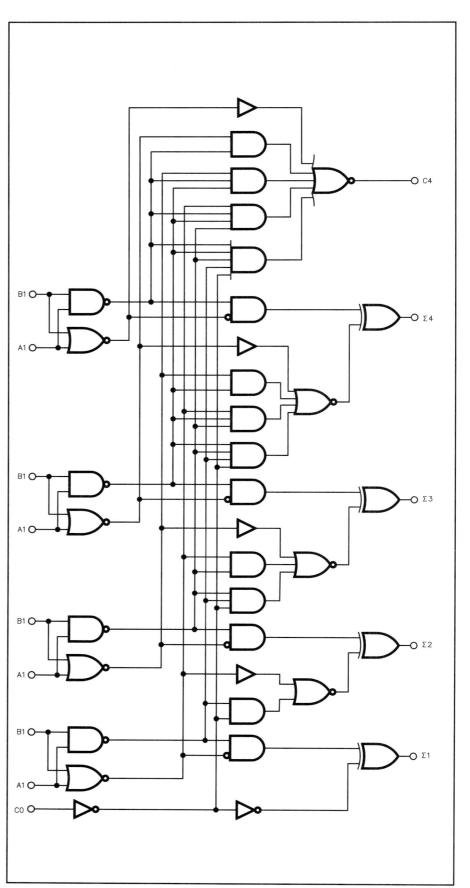

Fig 18.G1—A 4-bit adder schematic diagram.

18.40   Digital Signal Processing

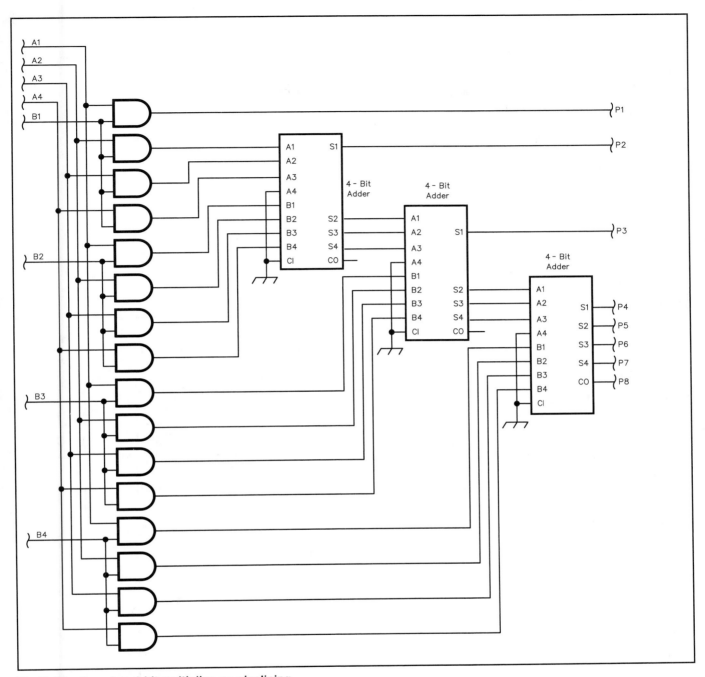

Fig 18.G2—Complete 4-bit multiplier, no pipelining.

See Ulbing's article in the Bibliography for information on surface-mount soldering techniques.

PROJECT G: A FAST BINARY MULTIPLIER IN HIGH-SPEED CMOS LOGIC

In this project, a fast 4-bit binary multiplier is described that may be constructed from 'HC-series logic gates or programmed into an FPGA. Two variations are explored: one without pipelining, and one with pipelining. Pipelining is employed where the propagation delays of gates limit throughput.

As seen in Fig 18.45 in the main text, a 4-bit multiplication may be broken into several 4-bit additions. In our circuit, 4-bit adders are used to add rows of bits in the summation, each one producing a single output bit. The diagram of a fast, 4-bit adder with look-ahead carry is shown in **Fig 18.G1**.

In this multiplier, 4-bit adders are used to add adjacent rows of bits in the traditional way. A multiplier connected this way is shown in **Fig 18.G2**. Not all bits in each addend have mates in the other, so

4-bit adders suffice. In the case where execution speed exceeds the reciprocal of the total propagation delay, pipelining must be employed to avoid error.

To use pipelining, we place storage registers between the stages of addition and one interim result is held by each stage at each clock time. See **Fig 18.G3**. The result is the same, but appears only after a latency of three clock times. When maximum gate delays are well known, this approach also yields more predictable performance because the latency is independent of the input data.

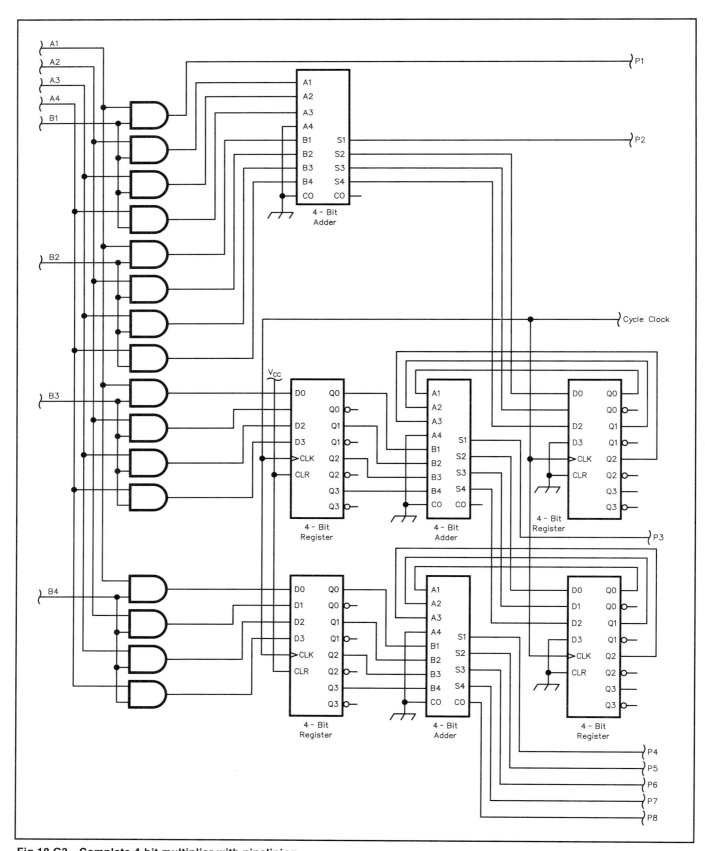

Fig 18.G3—Complete 4-bit multiplier with pipelining.

Contents

Transmission Lines

19

RF power is rarely generated right where it will be used. A transmitter and the antenna it feeds are a good example. To radiate effectively, the antenna should be high above the ground and should be kept clear of trees, buildings and other objects that might absorb energy. The transmitter, however, is most conveniently installed indoors, where it is out of the weather and is readily accessible. A *transmission line* is used to convey RF energy from the transmitter to the antenna. A transmission line should transport the RF from the source to its destination with as little loss as possible. This chapter was written by Dean Straw, N6BV.

There are three main types of transmission lines used by radio amateurs: coaxial lines, open-wire lines and waveguides. The most common type is the *coaxial* line, usually called *coax*. See **Fig 19.1A**. Coax is made up of a center conductor, which may be either stranded or solid wire, surrounded by a concentric outer conductor. The outer conductor may be braided shield wire or a metallic sheath. A flexible aluminum foil is employed in some coaxes to improve shielding over that obtainable from a woven shield braid. If the outer conductor is made of solid aluminum or copper, the coax is referred to as *Hardline*.

The second type of transmission line utilizes parallel conductors side by side, rather than the concentric ones used in coax. Typical examples of such *open-wire* lines are 300-Ω TV ribbon line and 450-Ω ladder line. See Fig 19.1B. Although open-wire lines are enjoying a sort of renais-sance in recent years due to their inher-ently lower losses in simple multiband antenna systems, coaxial cables are far more prevalent, because they are much more convenient to use.

The third major type of transmission line is the *waveguide*. While open-wire and coaxial lines are used from power-line frequencies to well into the microwave region, waveguides are used at microwave frequencies only. Waveguides will be covered at the end of this chapter.

TRANSMISSION LINE BASICS

In either coaxial or open-wire line, currents flowing in each of the two conductors travel in opposite directions. If the physical spacing between the two parallel

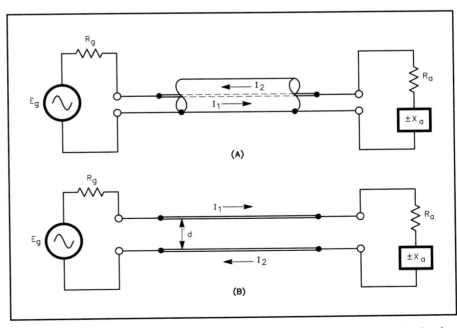

Fig 19.1—In A, coaxial cable transmission line connecting signal generator having source resistance R_g to reactive load $R_a \pm jX_a$, where X_a is either a capacitive (–) or inductive (+) reactance. Velocity factor (VF) and characteristic impedance (Z_0) are properties of the line, as discussed in the text. B shows open-wire balanced transmission line.

Table 19.1

Characteristics of Commonly Used Transmission Lines

RG or Type	Part Number	Z_0 Ω	VF %	Cap. pF/ft	Cent. Cond. AWG	Diel.	Shield	Jacket	OD in.	Max V (RMS)	Matched Loss (dB/100)			
											1 MHz	10	100	1000
RG-6	Belden 8215	75	66	20.5	#21 Solid	PE	FC	PE	0.275	2700	0.4	0.8	2.7	9.8
RG-8	TMS LMR400	50	85	23.9	#10 Solid	FPE	FC	PE	0.405	600	0.1	0.4	1.3	4.1
RG-8	Belden 9913	50	84	24.6	#10 Solid	ASPE	FC	P1	0.405	600	0.1	0.4	1.3	4.5
RG-8	WM CQ102	50	84	24.0	#9.5 Solid	ASPE	S	P2	0.405	600	0.1	0.4	1.3	4.5
RG-8	DRF-BF	50	84	24.5	#9.5 Solid	FPE	FC	PEBF	0.405	600	0.1	0.5	1.6	5.2
RG-8	WM CQ106	50	82	24.5	#9.5 Solid	FPE	FC	P2	0.405	600	0.2	0.6	1.8	5.3
RG-8	Belden 9914	50	82	24.8	#10 Solid	TFE	FC	P1	0.405	3700	0.1	0.5	1.6	6.0
RG-8	Belden 8237	52	66	29.5	#13 Flex	PE	S	P1	0.405	3700	0.2	0.6	1.9	7.4
RG-8X	TMS LMR240	50	84	24.2	#15 Solid	FPE	FC	PE	0.242	300	0.2	0.8	2.5	8.0
RG-8X	WM CQ118	50	82	25.0	#16 Flex	FPE	S	P2	0.242	300	0.3	0.9	2.8	8.4
RG-8X	Belden 9258	50	80	25.3	#16 Flex	TFE	S	P1	0.242	300	0.3	1.0	3.3	14.3
RG-9	Belden 8242	51	66	30.0	#13 Flex	PE	D	P2N	0.420	3700	0.2	0.6	2.1	8.2
RG-11	Belden 8213	75	78	17.3	#14 Solid	FPE	S	PE	0.405	600	0.2	0.4	1.5	5.4
RG-11	Belden 8238	75	66	20.5	#18 Flex	PE	S	P1	0.405	600	0.2	0.7	2.0	7.1
RG-58C	TMS LMR200	50	83	24.5	#17 Solid	FPE	FC	PE	0.195	300	0.3	1.0	3.2	10.5
RG-58	WM CQ124	53.5	66	28.5	#20 Solid	PE	S	P2N	0.195	1400	0.4	1.3	4.3	14.3
RG-58	Belden 8240	53.5	66	28.5	#20 Solid	PE	S	P1	0.193	1400	0.3	1.1	3.8	14.5
RG-58A	Belden 8219	50	78	26.5	#20 Flex	FPE	S	P1	0.198	300	0.4	1.3	4.5	18.1
RG-58C	Belden 8262	50	66	30.8	#20 Flex	PE	S	P2N	0.195	1400	0.4	1.4	4.9	21.5
RG-58A	Belden 8259	50	66	30.8	#20 Flex	PE	S	P1	0.193	1400	0.4	1.5	5.4	22.8
RG-59	Belden 8212	75	78	17.3	#20 Solid	TFE	S	PE	0.242	300	0.6	1.0	3.0	10.9
RG-59B	Belden 8263	75	66	20.5	#23 Solid	PE	S	P2N	0.242	1700	0.6	1.1	3.4	12.0
RG-62A	Belden 9269	93	84	13.5	#22 Solid	ASPE	S	P1	0.260	750	0.3	0.9	2.7	8.7
RG-62B	Belden 8255	93	84	13.5	#24 Solid	ASPE	S	P2N	0.260	750	0.3	0.9	2.9	11.0
RG-63B	Belden 9857	125	84	9.7	#22 Solid	ASPE	S	P2N	0.405	750	0.2	0.5	1.5	5.8
RG-142B	Belden 83242	50	69.5	29.2	#18 Solid	TFE	D	TFE	0.195	1400	0.3	1.1	3.9	13.5
RG-174	Belden 8216	50	66	30.8	#26 Solid	PE	S	P1	0.101	1100	1.9	3.3	8.4	34.0
RG-213	Belden 8267	50	66	30.8	#13 Flex	PE	S	P2N	0.405	3700	0.2	0.6	2.1	8.2
RG-214	Belden 8268	50	66	30.8	#13 Flex	PE	D	P2N	0.425	3700	0.2	0.6	1.9	8.0
RG-216	Belden 9850	75	66	20.5	#18 Flex	PE	D	P2N	0.425	3700	0.2	0.7	2.0	7.1
RG-217	M17/79-RG217	50	66	30.8	#9.5 Solid	PE	D	P2N	0.545	7000	0.1	0.4	1.4	5.2
RG-218	M17/78-RG218	50	66	29.5	#4.5 Solid	PE	S	P2N	0.870	11000	0.1	0.2	0.8	3.4
RG-223	Belden 9273	50	66	30.8	#19 Solid	PE	D	P2N	0.212	1700	0.4	1.2	4.1	14.5
RG-303	Belden 84303	50	69.5	29.2	#18 Solid	TFE	S	TFE	0.170	1400	0.3	1.1	3.9	13.5
RG-316	Belden 84316	50	69.5	29.0	#26 Solid	TFE	S	TFE	0.098	900	1.2	2.7	8.3	29.0
RG-393	M17/127-RG393	50	69.5	29.4	#12 Solid	TFE	D	TFE	0.390	5000	0.2	0.5	1.7	6.1
RG-400	M17/128-RG400	50	69.5	29.4	#20 Solid	TFE	D	TFE	0.195	1900	0.4	1.1	3.9	13.2
LMR500	TMS LMR500	50	85	23.9	#7 Solid	FPE	FC	PE	0.500	2500	0.1	0.3	0.9	3.3
LMR600	TMS LMR600	50	86	23.4	#5.5 Solid	FPE	FC	PE	0.590	4000	0.1	0.2	0.8	2.7
LMR1200	TMS LMR1200	50	88	23.1	#0 Tube	FPE	FC	PE	1.200	4500	0.04	0.1	0.4	1.3

Hardline

1/2"	CATV Hardline	50	81	25.0	#5.5	FPE	SM	none	0.500	2500	0.05	0.2	0.8	3.2
1/2"	CATV Hardline	75	81	16.7	#11.5	FPE	SM	none	0.500	2500	0.05	0.2	0.8	3.2
7/8"	CATV Hardline	50	81	25.0	#1	FPE	SM	none	0.875	4000	0.03	0.1	0.6	2.9
7/8"	CATV Hardline	75	81	16.7	#5.5	FPE	SM	none	0.875	4000	0.03	0.1	0.6	2.9
LDF4-50A	Heliax –1/2"	50	88	25.9	#5 Solid	FPE	CC	PE	0.630	1400	0.05	0.2	0.6	2.4
LDF5-50A	Heliax — 7/8"	50	88	25.9	0.355"	FPE	CC	PE	1.090	2100	0.03	0.10	0.4	1.3
LDF6-50A	Heliax – 1 1/4"	50	88	25.9	0.516"	FPE	CC	PE	1.550	3200	0.02	0.08	0.3	1.1

Parallel Lines

TV Twinlead		300	80	5.8	#20	PE	none	P1	0.500					
Transmitting Tubular		300	80	5.8	#20	PE	none	P1	0.500	8000	0.09	0.3	1.1	3.9
Window Line		450	91	4.0	#18	PE	none	P1	1.000	10000	0.02	0.08	0.3	1.1
Open Wire Line		600	92	1.1	#12	none	none	none	varies	12000	0.02	0.06	0.2	0.7

Approximate Power Handling Capability (1:1 SWR, 40°C Ambient):

	1.8 MHz	7	14	30	50	150	220	450	1 GHz
RG-58 Style	1350	700	500	350	250	150	120	100	50
RG-59 Style	2300	1100	800	550	400	250	200	130	90
RG-8X Style	1830	840	560	360	270	145	115	80	50
RG-8/213 Style	5900	3000	2000	1500	1000	600	500	350	250
RG-217 Style	20000	9200	6100	3900	2900	1500	1200	800	500
LDF4-50A	38000	18000	13000	8200	6200	3400	2800	1900	1200
LDF5-50A	67000	32000	22000	14000	11000	5900	4800	3200	2100
LMR500	12000	6000	4200	2800	2200	1200	1000	700	450
LMR1200	39000	19000	13000	8800	6700	3800	3100	2100	1400

Legend:

ASPE	Air Spaced Polyethylene	P1	PVC, Class
BF	Flooded direct bury	P2	PVC, Class 2
CC	Corrugated Copper	PE	Polyethylene
D	Double Copper Shields	S	Single Shield
		SM	Smooth Aluminum
DRF	Davis RF	TFE	Teflon
FC	Foil/Copper Shields	TMS	Times Microwave Systems
		WM	Wireman
FPE	Foamed Polyethylene	**	Not Available or varies
Heliax	Andrew Corp Heliax		
N	Non-Contaminating		

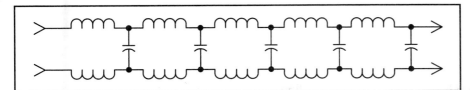

Fig 19.2—Equivalent of an infinitely long lossless transmission line using lumped circuit constants.

conductors in an open-wire line is small in terms of wavelength, the phase difference between the currents will be very close to 180°. If the two currents also have equal amplitudes, the field generated by each conductor will cancel that generated by the other, and the line will not radiate energy, even if it is many wavelengths long.

The equality of amplitude and 180° phase difference of the currents in each conductor in an open-wire line determine the degree of radiation cancellation. If the currents are for some reason unequal, or if the phase difference is not 180°, the line will radiate energy. How such imbalances occur and to what degree they can cause problems will be covered in more detail later.

In contrast to an open-wire line, the outer conductor in a coaxial line acts as a shield, confining RF energy within the line. Because of *skin effect* (see the **Real World** chapter in this *Handbook*), current flowing in the outer conductor of a coax does so mainly on the inner surface of the outer conductor. The fields generated by the currents flowing on the outer surface of the inner conductor and on the inner surface of the outer conductor cancel each other out, just as they do in open-wire line.

In a real (non-ideal) transmission line, the energy actually travels somewhat slower than the speed of light (typically from 65 to 97% of light speed), depending primarily on the dielectric properties of the insulating materials used in the construction of the line. The fraction of the speed of propagation in a transmission line compared to the speed of light in free space is called the *velocity factor* (VF) of the line. The velocity factor causes the line's *electrical* wavelength to be shorter than the wavelength in free space. Eq 1 describes the physical length of an electrical wavelength of transmission line.

$$\lambda = \frac{983.6}{f} \times VF \qquad (1)$$

where
λ = wavelength, in ft
f = frequency in MHz
VF = velocity factor.
Each transmission line has a character-

istic velocity factor, related to the specific properties of its insulating materials. The velocity factor must be taken into account when cutting a transmission line to a specific electrical length. **Table 19.1** shows various velocity factors for the transmission lines commonly used by amateurs. For example, if RG-8A, which has a velocity factor of 0.66, were used to make a quarter-wavelength line at 3.5 MHz, the length would be $(0.66 \times 983.6/3.5)/4 = 46.4$ ft long, instead of the free-space length of 70.3 ft. Open-wire line has a velocity factor of 0.97, close to unity, because it lacks a substantial amount of solid insulating material. Conversely, molded 300-Ω TV line has a velocity factor of 0.80 to 0.82 because it does use solid insulation between the conductors.

A perfectly lossless transmission line may be represented by a whole series of small inductors and capacitors connected in an infinitely long line, as shown in **Fig 19.2**. (We first consider this special case because we need not consider how the line is terminated at its end, since there is no end.)

Each inductor in Fig 19.2 represents the inductance of a very short section of one wire and each capacitor represents the capacitance between two such short sections. The inductance and capacitance values per unit of line depend on the size of the conductors and the spacing between them. The smaller the spacing between the two conductors and the greater their diameter, the higher the capacitance and the lower the inductance. Each series inductor acts to limit the rate at which current can charge the following shunt capacitor, and in so doing establishes a very important property of a transmission line: its *surge impedance*, more commonly known as its *characteristic impedance*. This is usually abbreviated as Z_0, and is approximately equal to $\sqrt{L/C}$, where L and C are the inductance and capacitance per unit length of line.

The characteristic impedance of an air-insulated parallel-conductor line, neglecting the effect of the insulating spacers, is given by

$$Z_0 = 276 \log_{10} \frac{2S}{d} \qquad (2)$$

where
Z_0 = characteristic impedance
S = center-to-center distance between conductors
d = diameter of conductor (in same units as S).

The characteristic impedance of an air-insulated coaxial line is given by

$$Z_0 = 138 \log_{10}\left(\frac{b}{a}\right) \qquad (3)$$

where
Z_0 = characteristic impedance
b = inside diameter of outer conductors
a = outside diameter of inner conductor (in same units as b).

It does not matter what units are used for S, d, a or b, so long as they are the same units. A line with closely spaced, large conductors will have a low characteristic impedance, while one with widely spaced, small conductors will have a relatively high characteristic impedance. Practical open-wire lines exhibit characteristic impedances ranging from about 200 to 800 Ω, while coax cables have Z_0 values between 25 to 100 Ω.

All practical transmission lines exhibit some power loss. These losses occur in the resistance that is inherent in the conductors that make up the line, and from leakage currents flowing in the dielectric material between the conductors. We'll next consider what happens when a real transmission line, which is not infinitely long, is terminated in real load impedances.

Matched-lines

Real transmission lines do not extend to infinity, but have a definite length. In use they are connected to, or *terminate* in, a load, as illustrated in **Fig 19.3A**. If the load is a pure resistance whose value equals the characteristic impedance of the line, the line is said to be *matched*. To current traveling along the line, such a load at the end of the line acts as though it were still more transmission line of the same characteristic impedance. In a matched transmission line, energy travels outward along the line from the source until it reaches the load, where it is completely absorbed.

Mismatched Lines

Assume now that the line in Fig 19.3B is terminated in an impedance Z_a which is not equal to Z_0 of the transmission line. The line is now a *mismatched* line. RF energy reaching the end of a mismatched line will not be fully absorbed by the load impedance. Instead, part of the energy will be reflected back toward the source. The amount of reflected versus absorbed energy depends on the degree of mismatch between the characteristic impedance of the line and the load impedance connected to its end.

The reason why energy is reflected at a discontinuity of impedance on a transmission line can best be understood by examining some limiting cases. First, consider the rather extreme case where the line is shorted at its end. Energy flowing to the load will encounter the short at the end, and the voltage at that point will go to zero, while the current will rise to a maximum. Since the current can't develop any power in a dead short, it will all be reflected back toward the source generator.

If the short at the end of the line is replaced with an open circuit, the opposite will happen. Here the voltage will rise to maximum, and the current will by definition go to zero. The phase will reverse, and all energy will be reflected back towards the source. By the way, if this sounds to you like what happens at the end of a half-wave dipole antenna, you are quite correct. However, in the case of an antenna, energy traveling along the antenna is lost by radiation on purpose, whereas a good transmission line will lose little energy to radiation because of field cancellation between the two conductors.

For load impedances falling between the extremes of short- and open-circuit, the phase and amplitude of the reflected wave will vary. The amount of energy reflected and the amount of energy absorbed in the load will depend on the difference between the characteristic impedance of the line and the impedance of the load at its end.

Now, what actually happens to the energy reflected back down the line? This energy will encounter another impedance discontinuity, this time at the generator. Reflected energy flows back and forth between the mismatches at the source and load. After a few such journeys, the reflected wave diminishes to nothing, partly as a result of finite losses in the line, but mainly because of absorption at the load. In fact, if the load is an antenna, such absorption at the load is desirable, since the energy is actually radiated by the antenna.

If a continuous RF voltage is applied to the terminals of a transmission line, the voltage at any point along the line will consist of a vector sum of voltages, the composite of waves traveling toward the load and waves traveling back toward the source generator. The sum of the waves traveling toward the load is called the *forward* or *incident* wave, while the sum of the waves traveling toward the generator is called the *reflected wave*.

Reflection Coefficient and SWR

In a mismatched transmission line, the ratio of the voltage in the reflected wave at any one point on the line to the voltage in the forward wave at that same point is defined as the *voltage reflection coefficient*. This has the same value as the current reflection coefficient. The reflection coefficient is a complex quantity (that is, having both amplitude and phase) and is generally designated by the Greek letter ρ (rho), or sometimes in the professional literature as Γ (Gamma). The relationship between R_a (the load resistance), X_a (the load reactance), Z_0 (the line characteristic impedance, whose real part is R_0 and whose reactive part is X_0), Z_0' is the complex conjugate of Z_0 and the complex reflection coefficient ρ is

$$\rho = \frac{Z_a - Z_0'}{Z_a + Z_0} = \frac{\left(R_a \pm jX_a\right) - \left(R_0 \mp jX_0\right)}{\left(R_a \pm jX_a\right) + \left(R_0 \pm jX_0\right)} \quad (4)$$

For most transmission lines the characteristic impedance Z_0 is almost completely resistive, meaning that $Z_0 = R_0$ and $X_0 \cong 0$. The magnitude of the complex reflection coefficient in Eq 4 then simplifies to:

$$|\rho| = \sqrt{\frac{\left(R_a - R_0\right)^2 + X_a^2}{\left(R_a + R_0\right)^2 + X_a^2}} \quad (5)$$

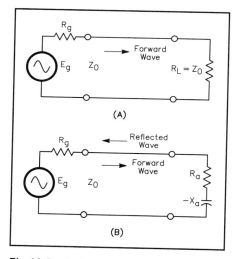

Fig 19.3—At A the coaxial transmission line is terminated with resistance equal to its Z_0. All power is absorbed in the load. At B, coaxial line is shown terminated in an impedance consisting of a resistance and a capacitive reactance. This is a mismatched line, and a reflected wave will be returned back down the line toward the generator. The reflected wave reacts with the forward wave to produce a standing wave on the line. The amount of reflection depends on the difference between the load impedance and the characteristic impedance of the transmission line.

For example, if the characteristic impedance of a coaxial line is 50 Ω and the load impedance is 120 Ω in series with a capacitive reactance of –90 Ω, the magnitude of the reflection coefficient is

$$|\rho| = \sqrt{\frac{(120-50)^2 + (-90)^2}{(120+50)^2 + (-90)^2}} = 0.593$$

Note that if R_a in Eq 4 is equal to R_0 and X_a is 0, the reflection coefficient, ρ, is 0. This represents a matched condition, where all the energy in the incident wave is transferred to the load. On the other hand, if R_a is 0, meaning that the load has no real resistive part, the reflection coefficient is 1.0, regardless of the value of R_0. This means that all the forward power is reflected since the load is completely reactive. The concept of reflection is often shown in terms of the *return loss,* which is the reciprocal of the reflection coefficient, in dB. In the example above, the return loss is 4.5 dB.

If there are no reflections from the load, the voltage distribution along the line is constant or *flat*. A line operating under these conditions is called either a *matched* or a *flat* line. If reflections do exist, a voltage *standing-wave* pattern will result from the interaction of the forward and reflected waves along the line. For a lossless transmission line, the ratio of the maximum peak voltage anywhere on the line to the minimum value anywhere on the line (which must be at least 1/4 λ) is defined as the *voltage standing-wave ratio,* or VSWR. Reflections from the load also produce a standing-wave pattern of currents flowing in the line. The ratio of maximum to minimum current, or ISWR, is identical to the VSWR in a given line.

In amateur literature, the abbreviation *SWR* is commonly used for standing-wave ratio, as the results are identical when taken from proper measurements of either current or voltage. Since SWR is a ratio of maximum to minimum, it can never be less than one-to-one. In other words, a perfectly flat line has an SWR of 1:1. The SWR is related to the magnitude of the complex reflection coefficient by

$$SWR = \frac{1 + |\rho|}{1 - |\rho|} \quad (6)$$

and conversely the reflection coefficient magnitude may be defined from a measurement of SWR as

$$|\rho| = \frac{SWR - 1}{SWR + 1} \quad (7)$$

The definitions in Eq 6 and 7 are valid for any line length and for lines which are lossy, not just lossless lines longer than ¼ λ at the frequency in use. Very often the load impedance is not exactly known, since an antenna usually terminates a transmission line, and the antenna impedance may be influenced by a host of factors, including its height above ground, end effects from insulators, and the effects of nearby conductors. We may also express the reflection coefficient in terms of forward and reflected power, quantities which can be easily measured using a directional RF wattmeter. The reflection coefficient may be computed as

$$\rho = \sqrt{\frac{P_r}{P_f}} \qquad (8)$$

where

P_r = power in the reflected wave
P_f = power in the forward wave.

If a line is not matched (SWR > 1:1) the difference between the forward and reflected powers measured at any point on the line is the net power going toward the load from that point. The forward power measured with a directional wattmeter (often referred to as a reflected power meter or *reflectometer*) on a mismatched line will thus always appear greater than the forward power measured on a flat line with a 1:1 SWR.

Losses in Transmission Lines

A real transmission line exhibits a certain amount of loss, caused by the resistance of the conductors used in the line and by dielectric losses in the line's insulators. The *matched-line loss* for a particular type and length of transmission line, operated at a particular frequency, is the loss when the line is terminated in a resistance equal to its characteristic impedance. The loss in a line is lowest when it is operated as a matched-line.

Line losses increase when SWR is greater than 1:1. Each time energy flows from the generator toward the load, or is reflected at the load and travels back toward the generator, a certain amount will be lost along the line. The net effect of standing waves on a transmission line is to increase the average value of current and voltage, compared to the matched-line case. An increase in current raises I²R (ohmic) losses in the conductors, and an increase in RF voltage increases E²/R losses in the dielectric. Line loss rises with frequency, since the conductor resistance is related to skin effect, and also because dielectric losses rise with frequency.

Matched-line loss is stated in decibels per hundred feet at a particular frequency. **Fig 19.4** shows the matched-line loss per hundred feet versus frequency for a number of common types of lines, both coaxial and open-wire balanced types. For example, RG-213 coax cable has a matched-line loss of 2.5 dB/100 ft at 100 MHz. Thus, 45 ft of this cable feeding a 50-Ω load at 100 MHz would have a loss of

$$\text{Matched line loss} = \frac{2.5 \text{ dB}}{100 \text{ ft}} \times 45 \text{ ft} = 1.13 \text{ dB}$$

If a line is not matched, standing waves will cause additional loss beyond the inherent matched-line loss for that line. On lines which are inherently lossy, the total line loss (the sum of matched-line

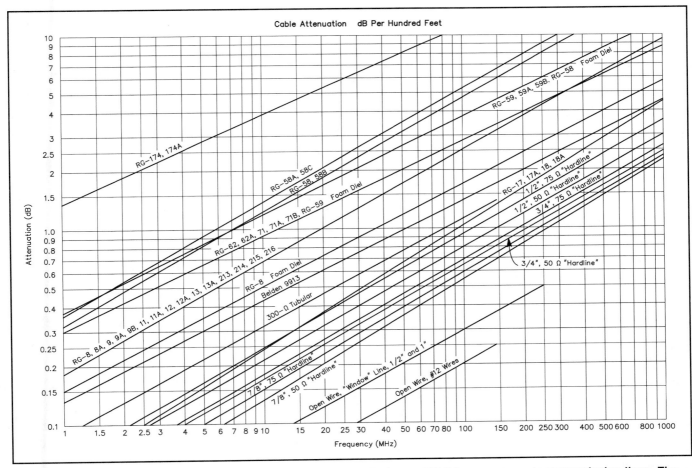

Fig 19.4—This graph displays the matched line attenuation in decibels per 100 ft for many popular transmission lines. The vertical axis represents attenuation and the horizontal axis frequency. Note that these loss figures are only accurate for properly matched transmission lines.

loss and additional loss due to SWR) can be surprisingly high for high values of SWR.

Total Mismatched-Line Loss (dB)

$$= 10 \log \left(\frac{a^2 - |\rho^2|}{a(1 - |\rho|)^2} \right) \qquad (9)$$

where

$a = 10^{ML/10}$ = matched-line ratio

$|\rho| = \dfrac{SWR - 1}{SWR + 1}$

ML = matched-line loss in dB
SWR = SWR measured at load

Because of losses in a transmission line, the measured SWR at the input of the line is less than the SWR measured at the load end of the line.

$$SWR \text{ at input} = \frac{a + |\rho|}{a - |\rho|} \qquad (10)$$

For example, RG-8A solid-dielectric coax cable exhibits a matched-line loss per 100 ft at 28 MHz of 1.18 dB. A 250-ft length of this cable has a matched-line loss of 2.95 dB. Assume that we measure the SWR at the load as 6:1.

$a = 10^{2.95/10} = 1.972$

$|\rho| = \dfrac{6.0 - 1}{6.0 + 1} = 0.714$

$\text{Total Loss} = 10 \log \dfrac{1.972^2 - 0.714^2}{1.972 \left(1 - 0.714^2\right)} = 5.4 \text{ dB}$

$SWR_{in} = \dfrac{1.972 + 0.714}{1.972 - 0.714} = 2.1$

The additional loss due to the 6:1 SWR at 28 MHz is 5.4 – 3.0 = 2.4 dB. The SWR at the input of the 250-ft line is only 2.1:1, because line loss has masked the true extent of the SWR (6:1) at the load end of the line.

The losses become larger if coax with a larger matched-line loss is used under the same conditions. For example, RG-58A coaxial cable is about one-half the diameter of RG-8A, and it has a matched-line loss of 2.5 dB/100 ft at 28 MHz. A 250-ft length of RG-58A has a total matched-line loss of 6.3 dB. With a 6:1 SWR at the load, the additional loss due to SWR is 3.0 dB, for a total loss of 9.3 dB. The additional cable loss due to the mismatch reduces the SWR at the input of the line to 1.4:1. An unsuspecting operator measuring the SWR at his transmitter might well believe that everything is just fine, when in truth only about 12% of the transmitter power is getting to the antenna! Be suspicious of very low SWR readings for an antenna fed

Table 19.2
Matched-Line Loss for 250 ft of Three Common Coaxial Cables

Comparisons of line losses versus frequency for 250-ft lengths of three different coax cable types: small-diameter RG-58A, medium-diameter RG-8A, and ¾-inch OD 50-Ω Hardline. At VHF, the losses for the small-diameter cable are very large, while they are moderate at 3.5 MHz.

Xmsn Line	3.5 MHz Matched-Line Loss, dB	3.5 MHz Loss, 6:1 SWR, dB	28 MHz Matched-Line Loss, dB	28 MHz Loss, 6:1 SWR, dB	146 MHz Matched-Line Loss, dB	146 MHz Loss 6:1 SWR, dB
RG-58A	1.9	4.0	6.3	9.3	16.5	19.6
RG-8A	0.9	2.2	3.0	5.4	7.8	10.8
¾" 50-Ω Hardline	0.2	0.5	0.7	1.8	2.1	4.2

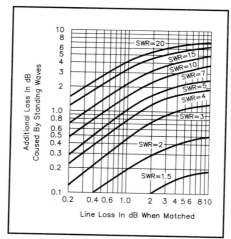

Fig 19.5—Increase in line loss because of standing waves (SWR measured at the load). To determine the total loss in decibels in a line having an SWR greater than 1, first determine the matched line loss for the particular type of line, length and frequency, on the assumption that the line is perfectly matched (from Fig 19.4). For example, Belden 9913 has a matched line loss of 0.49 dB/100 ft at 14 MHz. Locate 0.49 dB on the horizontal axis. For an SWR of 5:1, move up to the curve corresponding to this SWR. The increase in loss due to SWR is 0.65 dB beyond the matched line loss.

with a long length of coaxial cable, especially if the SWR remains low across a wide frequency range. Most antennas have narrow SWR bandwidths, and the SWR *should* change across a band.

On the other hand, if expensive ¾-inch diameter 50-Ω Hardline cable is used at 28 MHz, the matched-line loss is only 0.28 dB/100 ft. For 250 ft of Hardline the matched-line loss is 0.7 dB, and the additional loss due to a 6:1 SWR is 1.1 dB. The total loss is 1.8 dB. See **Table 19.2** for a summary of the losses for 250 ft of the three types of coax as a function of fre-

quency for matched-line and 6:1 SWR conditions.

At the upper end of the HF spectrum, when the transmitter and antenna are separated by a long transmission line, the use of bargain coax may prove to be a very poor cost-saving strategy. A 7.5 dB linear amplifier, to offset the loss in RG-58A compared to Hardline, would cost a great deal more than higher-quality coax. Furthermore, no *transmitter* amplifier can boost *receiver* sensitivity—loss in the line has the same effect as putting an attenuator in front of the receiver.

At the low end of the HF spectrum, say 3.5 MHz, the amount of loss in common coax lines is less of a problem for the range of SWR values typical on this band. For example, consider an 80-m dipole cut for the middle of the band at 3.75 MHz. It exhibits an SWR of about 6:1 at the 3.5 and 4.0 MHz ends of the band. At 3.5 MHz, 250 ft of RG-58A small-diameter coax has an additional loss of 2.1 dB for this SWR, giving a total line loss of 4.0 dB. If larger-diameter RG-8A coax is used instead, the additional loss due to SWR is 1.3 dB, for a total loss of 2.2 dB. This is an acceptable level of loss for most 80-m operators.

However, the loss situation gets dramatically worse as the frequency increases into the VHF and UHF regions. At 146 MHz, the total loss in 250 ft of RG-58A with a 6:1 SWR at the load is 16.5 dB, 10.8 dB for RG-8A, and 4.2 dB for ¾-inch 50-Ω Hardline. At VHF and UHF, a low SWR is essential to keep line losses low, even for the best coaxial cable. The length of transmission line must be kept as short as practical at these frequencies.

The effect of SWR on line loss is shown graphically in **Fig 19.5**. The horizontal axis is the attenuation, in decibels, of the line when perfectly matched. The vertical axis gives the additional attenuation due to SWR. If long coaxial-cable transmission lines are necessary, the matched loss of the coax used should be kept as low as

Table 19.3
Modeled Data for a 100-ft Flat-Top Antenna

100-ft long, 50-ft high, center-fed dipole over average ground, using coaxial or open-wire transmission lines. Antenna impedance computed using *NEC2* computer program, with ground relative permittivity of 13, ground conductivity of 5 mS/m and Sommerfeld/Norton ground model. Note the extremely reactive impedance levels at many frequencies, but especially at 1.8 MHz. If this antenna is fed directly with RG-8A coax, the losses are unacceptably large on 160 m, and undesirably high on most other bands also. The RF voltage at 3.8 MHz for high-power operation with open-wire line is extremely high also, and would probably result in arcing either on the line itself, or more likely in the Transmatch. Each transmission line is 100 ft long.

Frequency (MHz)	Antenna Impedance (Ohms)	SWR RG-8A Coax	Loss 100 ft RG-8A Coax	Loss 100 ft 450-Ω Line	Max Volt. RG-8A 1500 W	Max Volt. 450-Ω Line 1500 W
1.8 MHz	4.5 − j 1673	1818:1	25.9 dB	12.1 dB	1640	7640
3.8 MHz	38.9 − j 362	63:1	5.7 dB	0.9 dB	1181	3188
7.1 MHz	481 + j 964	49:1	5.8 dB	0.3 dB	981	1964
10.1 MHz	2584 − j 3292	134:1	10.4 dB	0.9 dB	530	2869
14.1 MHz	85.3 − j 123.3	6.0:1	1.9 dB	0.5 dB	530	1863
18.1 MHz	2097 + j 1552	65:1	9.0 dB	0.6 dB	780	2073
21.1 MHz	345 − j 1073	73:1	9.8 dB	0.8 dB	757	2306
24.9 MHz	202 + j 367	18:1	5.2 dB	0.4 dB	630	1563
28.4 MHz	2493 − j 1375	65:1	10.1 dB	0.7 dB	690	2051

possible, meaning that the highest-quality, largest-diameter cable should be used.

Choosing a Transmission Line

It is no accident that coaxial cable became as popular as it has since it was first widely used during World War II. Coax is mechanically much easier to use than open-wire line. Because of the excellent shielding afforded by its outer shield, coax can be run up a metal tower leg, taped together with numerous other cables, with virtually no interaction or crosstalk between the cables. At the top of a tower, coax can be used with a rotatable Yagi or quad antenna without worrying about shorting or twisting the conductors, which might happen with an open-wire line. Class 2 PVC non-contaminating outer jackets are designed for long-life outdoor installations. Class 1 PVC outer jacks are not recommended for outdoor installations. Coax can be buried underground, especially if it is run in plastic piping (with suitable drain holes) so that ground water and soil chemicals cannot easily deteriorate the cable. A cable with an outer jacket of polyethylene (PE) rather than polyvinyl chloride (PVC) is recommended for direct-bury installations.

Open-wire line must be carefully spaced away from nearby conductors, by at least several times the spacing between its conductors, to minimize possible electrical imbalances between the two parallel conductors. Such imbalances lead to line radiation and extra losses. One popular type of open-wire line is called *ladder line* because the insulators used to separate the two parallel, uninsulated conductors of the line resemble the steps of a ladder. Long lengths of ladder line can twist together in the wind

and short out if not properly supported.

Despite the mechanical difficulties associated with open-wire line, there are some compelling reasons for its use, especially in simple multiband antenna systems. Every antenna system, no matter what its physical form, exhibits a definite value of impedance at the point where the transmission line is connected. Although the input impedance of an antenna system is seldom known exactly, it is often possible to make a close estimate of its value, especially since sophisticated computer-modeling programs have become available to the radio amateur. As an example, **Table 19.3** lists the computed characteristics versus frequency for a multiband, 100-ft long center-fed dipole, placed 50 ft above average ground having a dielectric constant of 13 and a conductivity of 10 mS/m.

These values were computed using a complex program called *NEC2* (Numerical Electromagnetic Code), which incorporates a sophisticated Sommerfeld/Norton ground-modeling algorithm for antennas close to real earth. A nonresonant 100-ft length was chosen as an illustration of a practical size that many radio amateurs could fit into their backyards, although nothing in particular recommends this antenna over other forms. It is merely used as an example.

Examine Table 19.3 carefully in the following discussion. Columns three and four show the SWR on a 50-Ω RG-8A coaxial transmission line directly connected to the antenna, followed by the total loss in 100 ft of this cable. The impedance for this nonresonant, 100-ft long antenna varies over a very wide range for the nine operating frequencies. The SWR on a 50-Ω coax connected directly to this

antenna would be *extremely* high on some frequencies, particularly at 1.8 MHz, where the antenna is highly capacitive because it is much short of resonance. The loss for an SWR of 1818:1 in 100 ft of RG-8A at 1.8 MHz is a staggering 25.9 dB.

Contrast this to the loss in 100 ft of 450-Ω open-wire line. Here, the loss at 1.8 MHz is 12.1 dB. While 12.1 dB of loss is not particularly desirable, it is almost 14 dB better than the coax! Note that the RG-8A coax exhibits a good deal of loss on almost all the bands due to mismatch. Only on 14 MHz does the loss drop down to 1.9 dB, where the antenna is just past 3/2-λ resonance. From 3.8 to 28.4 MHz the open-wire line has a maximum loss of only 0.9 dB.

Columns six and seven in Table 19.3 list the maximum RMS voltage for 1500 W of RF power on the 50-Ω coax and on the 450-Ω open-wire line. The maximum RMS voltage for 1500 W on the open-wire line is extremely high, at 7640 V at 1.8 MHz. The voltage for a 100-W transmitter would be reduced by a ratio of $\sqrt{1500/100} = 3.87:1$. This is 1974 V, still high enough to cause arcing in many Transmatches.

In general, such a nonresonant antenna is a proven, practical multiband radiator when fed with 450-Ω open-wire ladder line connected to a Transmatch, although a longer antenna would be preferable for more efficient 160-m operation, even with open-wire line. The Transmatch and the line itself must be capable of handling the high RF voltages and currents involved for high-power operation. On the other hand, if such a multiband antenna is fed directly with coaxial cable, the losses on most frequencies are prohibitive. Coax is most suitable for antennas whose resonant feed-point impedances are close to the characteristic impedance of the feed line.

The Transmission Line as Impedance Transformer

If the complex mechanics of reflections, SWR and line losses are put aside momentarily, a transmission line can very simply be considered as an impedance trans-former. A certain value of load imped-ance, consisting of a resistance and reac-tance, at the end of the line is transformed into another value of impedance at the input of the line. The amount of transfor-mation is determined by the electrical length of the line, its characteristic imped-ance, and by the losses inherent in the line. The input impedance of a real, lossy trans-mission line is computed using the follow-ing equation

$$Z_{in} = Z_0 \times \frac{Z_L \cosh(\eta \ell) + Z_0 \sinh(\eta \ell)}{Z_L \sinh(\eta \ell) + Z_0 \cosh(\eta \ell)} \quad (11)$$

where
Z_{in} = complex impedance at input of line
$= R_{in} \pm jX_{in}$

Reflections on the Smith Chart

Although most radio amateurs have seen the Smith Chart, it is often regarded with trepidation. It is supposed to be complicated and subtle. However, the chart is ex-tremely useful in circuit analysis, especially when transmission lines are involved. The Smith Chart is not limited to transmission-line and antenna problems.

The basis for the chart is Eq 4 in the main text relating reflection coefficient to a terminating imped-ance. Eq 4 is repeated here:

$$\rho = \frac{Z - Z_0}{Z + Z_0} \quad (1)$$

where Z_0 is the characteristic impedance of the chart, and $Z = R + jX$ is a complex terminating imped-ance. Z might be the feed-point impedance of an antenna connected to a Z_0 transmission line.

It is useful to define a normalized impedance $z = Z/Z_0$. The normalized resistance and reactance become $r = R/Z_0$ and $x = X/Z_0$. Inserting these into Eq 1 yields:

$$\rho = \frac{z-1}{z+1} \quad (2)$$

where ρ and z are both complex, each having a magnitude and a phase when expressed in polar coordinates, or a real and an imaginary part in XY coordinates.

Eq 1 and 2 have some interesting and useful properties, characteristics that make them physically significant:

- Even though the components of z (and Z) may take on values that are very large, the reflection coefficient ρ, is restricted to always having a magnitude between zero and one if z has a real part, r, that is positive.
- If all possible values for ρ are examined and plotted in polar coordinates, they will lie within a circle with a radius of one. This is termed *the unit circle*. A plot is shown in **Fig A**.
- An impedance that is perfectly matched to Z_0, the characteristic value for the chart, will produce a ρ at the center of the unit circle.
- Real Z values, ones that have no reactance, "map" onto a horizontal line that divides the top from the bottom of the unit circle. By conven-tion, a polar variable with an angle of zero is on the x axis, to the right of the origin.
- Impedances with a reactive part produce ρ values away from the dividing line. Inductive impedances with the imaginary part greater than zero appear in the upper half of the chart, while capacitive impedances appear in the lower half.
- Perhaps the most interesting and exciting property of the reflection coefficient is the way it describes

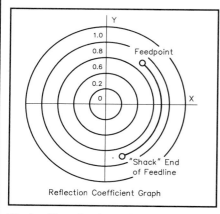

Fig A—Plot of polar reflection coefficient. Circles represent contours of constant ρ. The starting "feedpoint" value, 0.5 at +45°, represents an antenna impedance of 69.1 +j 65.1 Ω with Z_0 = 50 Ω. The arc represents a 15-ft section of 50-Ω, VF 0.66 transmission line at 7 MHz, yielding a shack ρ of 0.5 at −71.3°. The shack z is calculated as 40.3 − j 50.9 Ω.

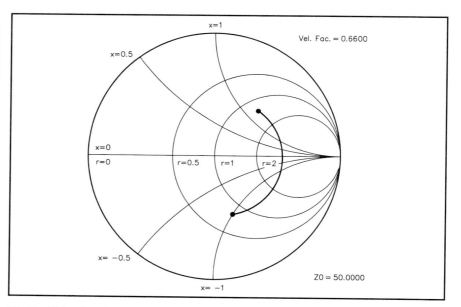

Fig B—This plot shows a Smith Chart. The circles now represent contours of constant normalized resistance or reactance. Note the arc with the markers: This illustrates the same antenna and line used in the previous figure. The plot is the same on the two charts; only the scale details have changed.

Z_L = complex load impedance at end of line = $R_a \pm jX_a$

Z_0 = characteristic impedance of line = $R_0 \pm jX_0$

η = complex loss coefficient = $\alpha + j\beta$

α = matched-line loss attenuation constant, in nepers/unit length (1 neper = 8.688 dB; most cables are rated in dB/100 ft)

β = phase constant of line in radians/unit length (related to physical length of line by the fact that 2π radians = 1 wavelength, and by Eq 1)

ℓ = electrical length of line in same units of length measurement as α or above.

Solving this equation manually is tedious, since it incorporates hyperbolic cosines and sines of the complex loss coefficient, but it may be solved using a traditional paper Smith Chart or a computer program. *The ARRL Antenna Book* has a chapter detailing the use of the Smith Chart. *MicroSmith* is a sophisticated graphical Smith Chart program written for the IBM PC, and is available through the ARRL. *TL* (Transmission Line) is another ARRL program that performs this transformation, but without Smith Chart graphics. *TL.EXE* is available from *ARRLWeb* (see page viii).

---

the impedance-transforming properties of a transmission line, presented in closed mathematical form in the main text as Eq 11. Neglecting loss effects, a transmission line of electrical length θ will transform a normalized impedance represented by ρ to another with the same magnitude and a new angle that differs from the original by -2θ. This rotation is clockwise.

Clearly, the reflection coefficient is more than an intermediate step in a mathematical development. It is a useful, alternative description of complex impedance. However, our interest is still focused on impedance; we want to know, for example, what the final z is after transformation with a transmission line. This is the problem that Phillip Smith solved in creating the Smith Chart. Smith observed that the unit circle, a graph of reflection coefficient, could be labeled with lines representing *normalized impedance*. A Smith Chart is shown in **Fig B**. All of the lines on the chart are complete or partial circles representing a line of constant normalized resistance and reactance.

How might we use the Smith Chart? A classic application relates antenna feed-point impedance to the impedance seen at the end of the "shack" end of the line. Assume that the antenna impedance is known, $Z_a = R_a + jX_a$. This complex value is converted to normalized impedance by dividing R_a and X_a by Z_0 to yield $r_a + jx_a$, and is plotted on the chart. A compass is then used to draw an arc of a circle centered at the origin of the chart. The arc starts at the normalized antenna impedance and proceeds in a clockwise direction for $2\theta°$, where θ is the electrical degrees, derived from the physical length and velocity factor of the transmission line. The end of the arc represents the normalized impedance at the end of the line in the

shack; it is denormalized by multiplying the real and imaginary parts by Z_0.

Antenna feedpoint Z can also be inferred from an impedance measurement at the shack end of the line. A similar procedure is followed. The only difference is that rotation is now in a counterclockwise direction. The Smith Chart is much more powerful than depicted in this brief summary. A detailed treatment is given by Phillip H. Smith in his classic book: *Electronic Applications of the Smith Chart* (McGraw-Hill, 1969). I also recommend his article "Transmission Line Calculator" in Jan 1939 *Electronics*. Joseph White presented a wonderful summary of the chart in a short but outstanding paper: "The Smith Chart: An Endangered Species?" Nov 1979 *Microwave Journal*. *MicroSmith* is available from ARRL for $39. The impedance matching tutorial is included.
—Wes Hayward, W7ZOI

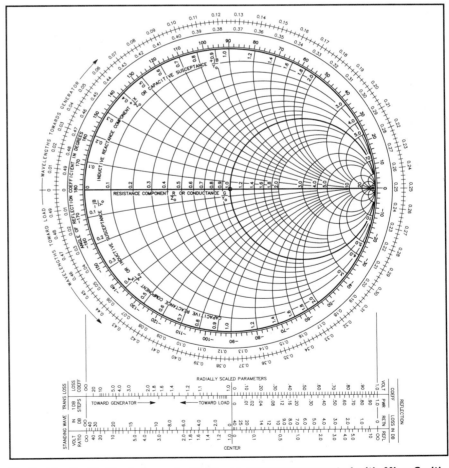

Fig C—The Smith Chart shown in Fig B was computer generated with *MicroSmith*. A much more detailed plot is presented here; this is the chart form used by Smith, suitable for graphic applications. Numbers are calculated by the computer in *MicroSmith*. This chart is used with the permission of Analog Instruments.

Lines as Stubs

The impedance-transformation properties of a transmission line are useful in a number of applications. If the terminating resistance is zero (that is, a short) at the end of a low-loss transmission line which is less than $1/4\lambda$, the input impedance consists of a reactance, which is given by a simplification of Eq 11.

$$X_{in} \cong Z_0 \tan \ell \qquad (12)$$

If the line termination is an open circuit, the input reactance is given by

$$X_{in} \cong Z_0 \cot \ell \qquad (13)$$

The input of a short (less than $1/4\lambda$) length of line with a short circuit as a terminating load appears as an inductance, while an open-circuited line appears as a capacitance. This is a useful property of a transmission line, since it can be used as a low-loss inductor or capacitor in matching networks. Such lines are often referred to as *stubs*.

A line that is an electrical quarter wavelength is a special kind of a stub. When a quarter-wave line is short circuited at its load end, it presents an open circuit at its input. Conversely, a quarter-wave line with an open circuit at its load end presents a short circuit at its input. Such a line inverts the sense of a short or an open circuit at the frequency for which the line is a quarter-wave long. This is also true for frequencies that are odd multiples of the quarter-wave frequency. However, for frequencies where the length of the line is a half wavelength, or integer multiples thereof, the line will duplicate the termination at its end.

For example, if a shorted line is cut to be a quarter wavelength at 7.1 MHz, the impedance looking into the input of the cable will be an open circuit. The line will have no effect if placed in parallel with a transmitter's output terminal. However, at twice the frequency, 14.2 MHz, that same line is now a half wavelength, and the line looks like a short circuit. The line, often dubbed a *quarter-wave stub* in this application, will act as a trap for not only the second harmonic, but also for higher even-order harmonics, such as the fourth or sixth harmonics.

Quarter-wave stubs made of good-quality coax, such as RG-213, offer a convenient way to lower transmitter harmonic levels. Despite the fact that the exact amount of harmonic attenuation depends on the impedance (often unknown) into which they are working at the harmonic frequency, a quarter-wave stub will typically yield 20 to 25 dB of attenuation of the second harmonic when placed directly at the output of a transmitter feeding common amateur antennas. Because different manufacturing runs of coax will have slightly different velocity factors, a quarter-wave stub is usually cut a little longer than calculated, and then carefully pruned by snipping off short pieces, while monitoring the response at the fundamental frequency, using a grid-dip meter or a receiver noise bridge. Because the end of the coax is an open circuit while pieces are being snipped away, the input of a quarter-wave line will show a short circuit exactly at the fundamental frequency. Once the coax has been pruned to frequency, a short jumper is soldered across the end, and the response at the second harmonic frequency is measured.

We will examine further applications of quarter-wave transmission lines later in the next section.

Matching the Antenna to the Line

When transmission lines are used with a transmitter, the most common load is an antenna. When a transmission line is connected between an antenna and a receiver, the receiver input circuit is the load, not the antenna, because the power taken from a passing wave is delivered to the receiver.

Whatever the application, the conditions existing at the load, and *only* the load, determine the reflection coefficient, and hence the standing-wave ratio, on the line. If the load is purely resistive and equal to the characteristic impedance of the line, there will be no standing waves. If the load is not purely resistive, or is not equal to the line Z_0, there will be standing waves. No adjustments can be made at the input end of the line to change the SWR at the load. Neither is the SWR affected by changing the line length, except as previously described when the SWR at the input of a lossy line is masked by the attenuation of the line.

Only in a few special cases is the antenna impedance the exact value needed to match a practical transmission line. In all other cases, it is necessary either to operate with a mismatch and accept the SWR that results, or else to bring about a match between the line and the antenna.

Technical literature sometimes uses the term *conjugate match* to describe the condition where the reactance seen looking toward the load from any point on the line is the complex conjugate of the impedance seen looking toward the source. A conjugate match is necessary to achieve the maximum power gain possible from a small-signal amplifier. For example, if a small-signal amplifier at 14.2 MHz has an output impedance of 25.8 − j11.0 Ω, then the maximum power possible will be generated from that amplifier when the output load is 25.8 + j11.0 Ω. The amplifier and load system is resonant because the ±11.0-Ω reactances cancel.

Now, assume that 100 ft of 50-Ω RG-213 coax at 14.2 MHz just happens to be terminated in an impedance of 115 − j25 Ω. Eq 11 calculates that the impedance looking into the input of the line is 25.8 − j11.0 Ω. If this transmission line is connected directly to the small-signal amplifier above, then a conjugate match is created, and the amplifier generates the maximum possible amount of power it can generate.

However, if the impedance at the output of the amplifier is not 25.8 − j11.0 Ω, then a matching network is needed between the amplifier and its load for maximum power gain. For example, if 50 ft of RG-213 is terminated in a 72 − j34 Ω antenna impedance, the impedance at the line input becomes 35.9 − j21.6 Ω. A matching network is designed to transform 35.9 − j21.6 Ω to 25.8 + j11.0 Ω, so that once again a conjugate match is created for the small-signal amplifier.

Now, let us consider what happens with amplifiers where the power level is higher than the milliwatt level of small-signal amplifiers. Most modern transmitters are designed to work into a 50-Ω load. Most will reduce power automatically if the load is not 50 Ω—this protects them against damage and ensures linear operation without distortion.

Many amateurs use an *antenna tuner* between their transmitter and the transmission line feeding the antenna. The antenna tuner's function is to transform the impedance, whatever it is, at the shack-end of the transmission line into the 50 Ω required by their transmitter. Note that the SWR on the transmission line between the antenna and the output of the antenna tuner is rarely exactly 1:1, even though the SWR on the short length of line between the tuner and the transmitter is 1:1.

Therefore, some loss is unavoidable: additional loss due to the SWR on the line, and loss in the antenna tuner itself. However, most amateur antenna installations use antennas that are reasonably close to resonance, making these types of losses small enough to be acceptable.

Despite the inconvenience, if the antenna tuner could be placed at the antenna rather than at the transmitter output, it can transform the 72 − j34 Ω antenna impedance to a nonreactive 50 Ω. Then the line SWR is 1:1.

Impedance matching networks can take

a variety of physical forms, depending on the circumstances.

Matching the Antenna to the Line, at the Antenna

This section describes methods by which a network can be installed at the antenna itself to provide matching to a transmission line. Having the matching system up at the antenna rather than down in the shack at the end of a long transmission line does seem intuitively desirable, but it is not always very practical, especially in multiband antennas.

If a highly reactive antenna can be tuned to resonance, even without special efforts to make the resistive portion equal to the line's characteristic impedance, the resulting SWR is often low enough to minimize additional line loss due to SWR. For example, the multiband dipole in Table 19.3 has an antenna impedance of $4.5 - j1673 \Omega$ at 1.8 MHz. Assume that the antenna reactance is tuned out with a network consisting of two symmetrical inductors whose reactance is $+836.5 \Omega$ each, with a Q of 200. The inductors are made up of 73.95 μH coils in series with inherent loss resistors of $836.5/200 = 4.2 \Omega$. The total series resistance is thus $4.5 + 2 \times (4.2) = 12.9 \Omega$, and the antenna reactance and inductor reactance cancel out. See **Fig 19.6**.

If this tuned system is fed with 50-Ω coaxial cable, the SWR is $50/12.9 = 3.88:1$, and the loss in 100 ft of RG-8A cable would be 0.47 dB. The radiation efficiency is $4.5/12.9 = 34.9\%$. Expressed another way, there is 4.57 dB of loss. Adding the 0.47 dB of loss in the line yields an overall system loss of 5.04 dB. Compare this to the loss of 17.1 dB if the RG-8A coax is used to feed the antenna directly, without any matching at the antenna. The use of a moderately high-Q resonator has yielded almost 12 dB of "gain" (that is, less loss) compared to the nonresonator case. The drawback of course is that the antenna is now resonated on only one frequency, but it certainly is a lot more efficient on that one frequency.

The Quarter-Wave Transformer or "Q" Section

The range of impedances presented to the transmission line is usually relatively small on a typical amateur antenna, such as a dipole or a Yagi when it is operated close to resonance. In such antenna systems, the impedance transforming properties of a quarter-wave section of transmission line are often utilized to match the transmission line at the antenna.

One example of this technique is an ar-ray of stacked Yagis on a single tower. Each antenna is resonant and is fed in parallel with the other Yagis, using equal lengths of coax to each antenna. A stacked array is used to produce not only gain, but also a wide vertical elevation pattern, suitable for coverage of a broad geographic area. (See *The ARRL Antenna Book* for details about Yagi stacking.) The feed-point impedance of two 50-Ω Yagis fed with equal lengths of feed line connected in parallel is 25 Ω (50 Ω/2); three in parallel yield 16.7 Ω; four in parallel yield 12.5 Ω. The nominal SWR for a stack of four Yagis is 4:1 (50 Ω/12.5 Ω). This level of SWR does not cause excessive line loss, provided that low-loss coax feed line is used. However, many station designers want to be able to select, using relays, any individual antenna in the array, without having the load seen by the transmitter change. Perhaps they might wish to turn one antenna in the stack in a different direction and use it by itself. If the load changes, the amplifier must be retuned, an inconvenience at best.

See **Fig 19.7**. If the antenna impedance and the characteristic impedance of a feed

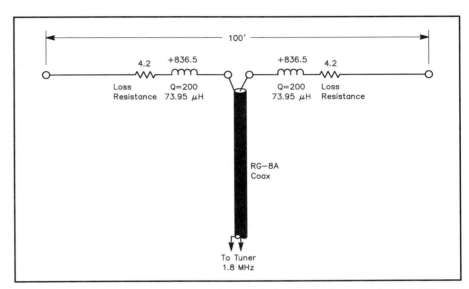

Fig 19.6—The efficiency of the dipole in Table 19.3 can be improved at 1.8 MHz with a pair of inductors inserted symmetrically at the feedpoint. Each inductor is assumed to have a Q of 200. By resonating the dipole in this fashion the system efficiency, when fed with RG-8A coax, is almost 20 dB better than using this same antenna without the resonator. The disadvantage is that the formerly multiband antenna can only be used on a single band.

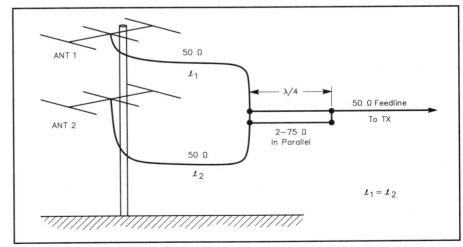

Fig 19.7—Array of two stacked Yagis, illustrating use of quarter-wave matching sections. At the junction of the two equal lengths of 50-Ω feed line the impedance is 25 Ω. This is transformed back to 50 Ω by the two paralleled 75-Ω, quarter-wave lines, which together make a net characteristic impedance of 37.5 Ω. This is close to the 35.4 Ω value computed by the formula $\sqrt{25 \times 50}$.

line to be matched are known, the characteristic impedance needed for a quarter-wave matching section of low-loss cable is expressed by another simplification of Eq 11.

$$Z = \sqrt{Z_1 Z_0} \qquad (14)$$

where

Z = characteristic impedance needed for matching section
Z_1 = antenna impedance
Z_0 = characteristic impedance of the line to which it is to be matched.

Example: To match a 50-Ω line to a Yagi stack consisting of two antennas fed in parallel to produce a 25-Ω load, the quarter-wave matching section would require a characteristic impedance of

$$Z = \sqrt{50 \times 25} = 35.4 \, \Omega$$

A transmission line with a characteristic impedance of 35 Ω could be closely approximated by connecting two equal lengths of 75-Ω cable (such as RG-11A) in parallel to yield the equivalent of a 37.5- Ω cable. Three Yagis fed in parallel would require a quarter-wave transformer made using a cable having a characteristic impedance of

$$\sqrt{16.7 \times 50} = 28.9 \, \Omega$$

This is approximated by using a quarter-wave section of 50-Ω cable in parallel with a quarter-wave section of 75-Ω cable, yielding a net impedance of 30 Ω, quite close enough to the desired 28.9 Ω. Four Yagis fed in parallel would require a quarter-wave transformer made up using cable with a characteristic impedance of 25 Ω, easily created by using two 50-Ω cables in parallel.

T- and Gamma-Match Sections

Many types of antennas exhibit a feed-point impedance lower than the 50-Ω characteristic impedance of commonly available coax cable. Both the so-called *T-Match* and the *Gamma-Match* are used extensively on Yagi and quad beam antennas to increase the antenna feed impedance to 50 Ω.

The method of matching shown in **Fig 19.8** is based on the fact that the impedance between any two points equidistant from the center along a resonant antenna is resistive, and has a value that depends on the spacing between the two points. It is therefore possible to choose a pair of points between which the impedance will have the right value to match a transmission line. In practice, the line cannot be connected directly at these points because the distance between them is much greater than the conductor spacing of a

practical transmission line. The T arrangement in Fig 19.8A overcomes this difficulty by using a second conductor paralleling the antenna to form a matching section to which the line may be connected.

The T is particularly well suited to use with parallel-conductor feed line. The operation of this system is somewhat complex. Each T conductor (Y in the drawing) forms a short section of transmission line with the antenna conductor opposite it. Each of these transmission-line sections can be considered to be terminated in the impedance that exists at the point of connection to the antenna. Thus, the part of the antenna between the two points carries a transmission-line current in addition to the normal antenna current. The two transmission-line matching sections are in series, as seen by the main transmission line.

If the antenna by itself is resonant at the operating frequency, its impedance will be purely resistive. In this case the matching-section lines are terminated in a resistive load. As transmission-line sections however, these matching sections are terminated in a short, and are shorter than a quarter wavelength. Thus their input impedance, the impedance seen by the main transmission line looking into the matching-section terminals, will be inductive as well as resistive. The reactive component of the input impedance must be tuned out before a proper match can be obtained.

One way to do this is to detune the antenna just enough, by shortening its length,

to cause capacitive reactance to appear at the input terminals of the matching section, thus canceling the reactance introduced. Another method, which is considerably easier to adjust, is to insert a variable capacitor in series with each matching section where it connects to the transmission line, as shown in the chapter on **Antennas**. The capacitors must be protected from the weather.

When the series-capacitor method of reactance compensation is used, the antenna should be the proper length for resonance at the operating frequency. Trial positions of the matching-section taps are then taken, each time adjusting the capacitor for minimum SWR, until the lowest possible SWR has been achieved. The unbalanced (gamma) arrangement in Fig 19.8B is similar in principle to the T, but is adapted for use with single coax line. The method of adjustment is the same.

The Hairpin Match

In beam antennas such as Yagis or quads, which utilize parasitic directors and reflectors to achieve directive gain, the mutual impedance between the parasitic and the driven elements lowers the resistive component of the driven-element impedance, typically to a value between 10 and 30 Ω. If the driven element is purposely cut slightly shorter than its half-wave resonant length, it will exhibit a capacitive reactance at its feedpoint. A shunt inductor as shown in **Fig 19.9** placed across the feed-point center insulator can be used to transform the antenna resistance to match the character-

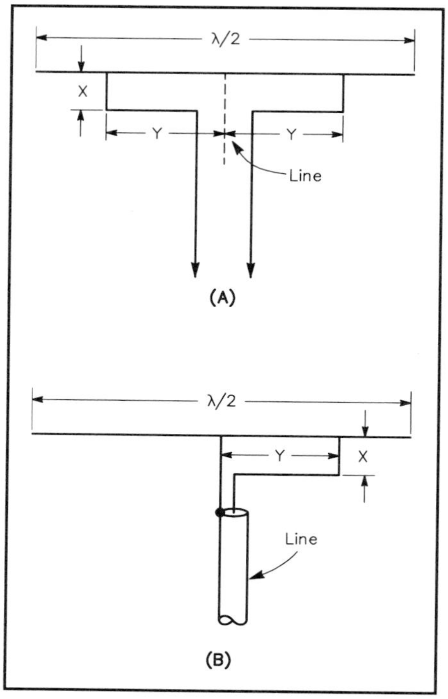

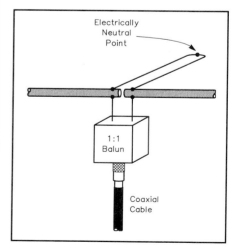

Fig 19.9—Hairpin match, sometimes called the Beta match. The "hairpin" is a shunt inductor, which together with the series capacitive reactance of an electrically short driven element, forms an L network. This L network transforms the antenna resistive component to 50 Ω.

Fig 19.8—The T match (A) and gamma match (B).

istic impedance of the transmission line, while canceling out the capacitive reactance simultaneously. The antenna's capacitive reactance and the hairpin shunt inductor form an L network.

For mechanical convenience, the shunt inductor is often constructed using heavy-gauge aluminum wire bent in the shape of a hairpin. The center of the hairpin, the end farthest from the driven element, is grounded to the boom, since this point in a balanced feed system is equidistant from the antenna feed terminals. This gives some protection against static buildup and a certain measure of lightning protection. The disadvantage of the Hairpin match is that it does require that the driven element be split and insulated at its center. Since only the length of the driven element and the value of shunt inductance can be varied in the Hairpin, the SWR often cannot be brought down to exactly 1:1 at a desired frequency in the band, as it can be with the T or gamma matches previously described.

Matching the Line to the Transmitter

So far we have been concerned mainly with the measures needed to achieve acceptable amounts of loss and a low SWR when real coax lines are connected to real antennas. Not only is feed-line loss minimized when the SWR is kept within reasonable bounds, but also the transmitter is able to deliver its rated output power, at its rated level of distortion, when it sees the load resistance it was designed to feed.

Most modern amateur transmitters use broadband, untuned solid-state final amplifiers designed to work into a 50-Ω load. Such a transmitter very often utilizes built-in protection circuitry, which automatically reduces output power if the SWR rises to more than about 2:1. Protective circuits are needed because many solid-state devices will willingly and almost instantly destroy themselves attempting to deliver power into low-impedance loads. Solid-state devices are a lot less forgiving than vacuum tube amplifiers, which can survive momentary overloads without being destroyed instantly. Pi networks used in vacuum-tube amplifiers typically have the ability to match a surprisingly wide range of impedances on a transmission line. See the **Amplifiers** chapter in this *Handbook*.

Besides the rather limited option of using only inherently low-SWR antennas to ensure that the transmitter sees the load for which it was designed, we radio amateurs have another alternative. We can use an antenna tuner. The function of an antenna tuner is to transform the impedance at the input end of the transmission line, whatever it may be, to the 50 Ω needed to keep the transmitter loaded properly. Do not forget: A tuner does not alter the SWR on the transmission line going to the antenna; it only keeps the transmitter looking into the load for which it was designed. Indeed, some solid-state transmitters incorporate (usually at extra cost) automatically tuned *antenna couplers* (another name for antenna tuner), so that they too can cope with practical antennas and transmission lines that are not perfectly flat. The range of impedances which can be matched is typically rather limited, however, especially at lower frequencies.

Over the years, radio amateurs have derived a number of circuits for use as tuners. At one time, when open-wire transmission line was more widely used, link coupled tuned circuits were in vogue. See **Fig 19.10**. With the increasing popularity of coaxial cable used as feed lines, other circuits have become more prevalent. The most common form of antenna tuner in recent years is some variation of a T configuration, as shown in **Fig 19.11A**.

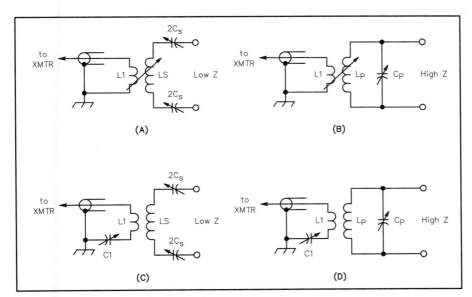

Fig 19.10—Simple antenna tuners for coupling a transmitter to a balanced line presenting a load different from the transmitter's design load impedance, usually 50 Ω. A and B, respectively, are series and parallel tuned circuits using variable inductive coupling between coils. C and D are similar but use fixed inductive coupling and a variable series capacitor, C1. A series tuned circuit works well with a low-impedance load; the parallel circuit is better with high impedance loads (several hundred ohms or more).

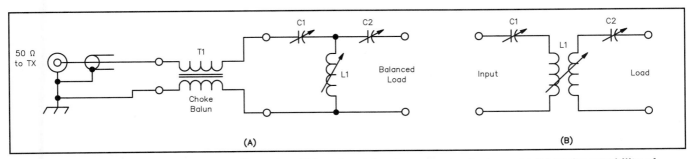

Fig 19.11—Antenna tuner network in T configuration. This network has become popular because it has the capability of matching a wide range of impedances. At A, the balun transformer at the input of the antenna tuner preserves balance when feeding a balanced transmission line. At B, the T configuration is shown as two L networks back to back.

The T network can be visualized as being two L networks back to front, where the common element has been conceptually broken down into two inductors in parallel. See Fig 19.11B. The L network connected to the load transforms the output impedance $R_a \pm jX_a$ into its parallel equivalent by means of the series output capacitor C2. The first L network then transforms the parallel equivalent back into the series equivalent and resonates the reactance with the input series capacitor C1.

Note that the equivalent parallel resistance R_p across the shunt inductor can be a very large value for highly reactive loads, meaning that the voltage developed at this point can be very high. For example, assume that the load impedance at 3.8 MHz presented to the antenna tuner is $Z_a = 20 - j1000$. If C2 is 300 pF, then the equivalent parallel resistance across L1 is 66326 Ω. If 1500 W appears across this parallel resistance, a peak voltage of 14106 V is produced, a very substantial level indeed. Highly reactive loads can produce very high voltages across components in a tuner.

The ARRL computer program *TL* calculates and shows graphically the Transmatch values for operator selected antenna impedances transformed through lengths of various types of practical transmission lines. The **Station Accessories** chapter includes an antenna tuner project, and *The ARRL Antenna Book* contains detailed information on tuner design and construction.

Myths About SWR

This is a good point to stop and mention that there are some enduring and quite misleading myths in Amateur Radio concerning SWR.

Despite some claims to the contrary, a high SWR *does not by itself* cause RFI, or TVI or telephone interference. While it is true that an antenna located close to such devices can cause overload and interference, the SWR on the feed line to that antenna has nothing to do with it, providing of course that the tuner, feed line or connectors are not arcing. The antenna is merely doing its job, which is to radiate. The transmission line is doing its job, which is to convey power from the transmitter to the radiator.

A second myth, often stated in the same breath as the first one above, is that a high SWR will cause excessive radiation from a transmission line. SWR has nothing to do with excessive radiation from a line. *Imbalances* in open-wire lines cause radiation, but such imbalances are not related to SWR. This subject will be covered more in the section on baluns.

A third and perhaps even more prevalent myth is that you can't "get out" if the SWR on your transmission line is higher than 1.5:1, or 2:1 or some other such arbitrary figure. On the HF bands, if you use reasonable lengths of good coaxial cable (or even better yet, open-wire line), the truth is that you need not be overly concerned if the SWR at the load is kept below about 6:1. This sounds pretty radical to some amateurs who have heard horror story after horror story about SWR. The fact is that if you can load up your transmitter without any arcing inside, or if you use a tuner to make sure your transmitter is operating into its rated load resistance, you can enjoy a very effective station, using antennas with feed lines having high values of SWR on them. For example, a 450-Ω open-wire line connected to the multiband dipole shown in Table 19.3 would have a 19:1 SWR on it at 3.8 MHz. Yet time and again this antenna has proven to be a great performer at many installations.

Fortunately or unfortunately, SWR is one of the few antenna and transmission-line parameters easily measured by the average radio amateur. Ease of measurement does not mean that a low SWR should become an end in itself! The hours spent pruning an antenna so that the SWR is reduced from 1.5:1 down to 1.3:1 could be used in far more rewarding ways—making QSOs, for example, or studying transmission-line theory.

Loads and Balancing Devices

Center-fed dipoles and loops are *balanced*, meaning that they are electrically symmetrical with respect to the feed point. A balanced antenna should be fed by a balanced feeder system to preserve this electrical symmetry with respect to ground, thereby avoiding difficulties with unbalanced currents on the line and undesirable radiation from the transmission line itself. Line radiation can be prevented by a number of devices which detune or *decouple* the line for currents radiated by the antenna back onto the line that feeds it, greatly reducing the amplitude of such *antenna currents*.

Many amateurs use center-fed dipoles or Yagis, fed with unbalanced coaxial line. Some method should be used for connecting the line to the antenna without upsetting the symmetry of the antenna itself. This requires a circuit that will isolate the balanced load from the unbalanced line, while still providing efficient power transfer. Devices for doing this are called *baluns* (a contraction for "balanced to unbalanced"). A balanced antenna fed with balanced line, such as two-wire ladder

line, will maintain its inherent balance, so long as external causes of unbalance are avoided. However, even they will require some sort of balun at the transmitter, since modern transmitters have unbalanced (coax) outputs.

If a balanced antenna is fed at the center through a coaxial line without a balun, as indicated in **Fig 19.12A**, the inherent symmetry and balance is upset because one side of the radiator is connected to the shield while the other is connected to the inner conductor. On the side connected to the shield, current can be diverted from flowing into the antenna, and instead can flow down over the outside of the coaxial shield. The field thus set up cannot be canceled by the field from the inner conductor because the fields inside the cable cannot escape through the shielding of the outer conductor. Hence currents flowing on the outside of the line will be responsible for some radiation from the line.

This is a good point to say that striving for perfect balance in a line and antenna system is not always absolutely mandatory. For example, if a nonresonant center fed dipole is fed with open-wire line and a tuner for multiband operation, the most desirable radiation pattern for general-purpose communication is actually an omnidirectional pattern. A certain amount of feed-line radiation might actually help fill in otherwise undesirable nulls in the azimuthal pattern of the antenna itself. Furthermore, the radiation pattern of a co-axial-fed dipole that is only a few tenths of a wavelength off the ground (50 ft high on the 80-m band, for example) is not very directional anyway, because of its severe interaction with the ground.

Purists may cry out in dismay, but there are many thousands of coaxial-fed dipoles in daily use worldwide that perform very effectively without the benefit of a balun. See **Fig 19.13A** for a worst-case comparison between a dipole with and without a balun at its feed point. This is with a 1-λ long feed line slanted downward 45° under one side of the antenna. Common-mode currents are radiated and conducted onto the braid of the feed line, which in turn radiates. The amount of pattern distortion is not particularly severe for a dipole, however. It is debatable whether the bother and expense of installing a balun for such an antenna is worthwhile.

However, some form of balun should be used to preserve the pattern of an antenna that is purposely designed to be highly directional, such as a Yagi or a quad. Fig 19.13B shows the distortion that can result from common-mode currents conducted and radiated back onto the feed

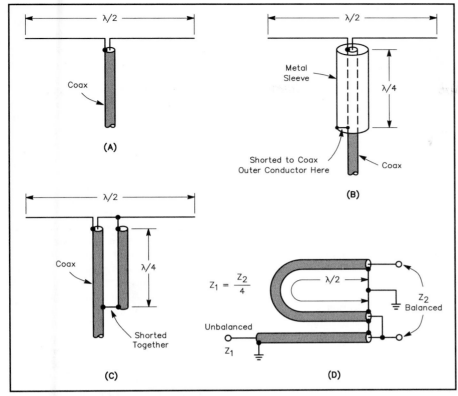

line for a 5-element Yagi. This antenna has purposely been designed for an excellent pattern but the common-mode currents seriously distort the rearward pattern and reduce the forward gain as well. A balun is highly desirable in this case.

Quarter-Wave Baluns

Fig 19.12B shows a balun arrangement known as a *bazooka*, which uses a sleeve over the transmission line. The sleeve, together with the outside portion of the outer coax conductor, forms a shorted quarter-wave line section. The impedance looking into the open end of such a section is very high, so the end of the outer conductor of the coaxial line is effectively isolated from the part of the line below the sleeve. The length is an electrical quarter wave, and because of the velocity factor may be physically shorter if the insulation between the sleeve and the line is not air. The bazooka has no effect on antenna impedance at the frequency where the quarter-wave sleeve is resonant. However, the sleeve adds inductive shunt reactance at frequencies lower, and capacitive shunt reactance at frequencies higher than the quarter-wave-resonant frequency. The bazooka is mostly used at VHF, where its physical size does not present a major

Fig 19.12—Quarter-wave baluns. Radiator with coaxial feed (A) and methods of preventing unbalanced currents from flowing on the outside of the transmission line (B and C). The ½ λ-phasing section shown at D is used for coupling to an unbalanced circuit when a 4:1 impedance ratio is desired or can be accepted.

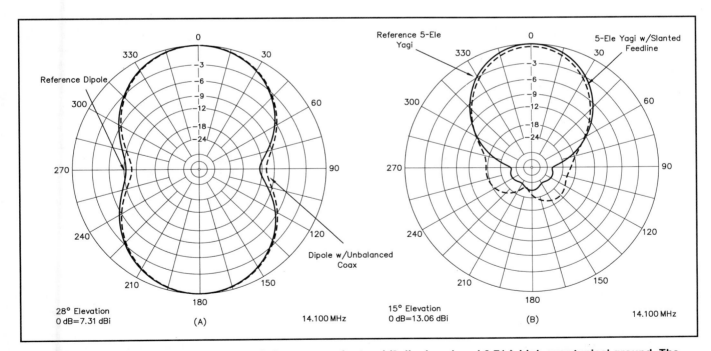

Fig 19.13—At A, computer-generated azimuthal responses for two λ/2 dipoles placed 0.71 λ high over typical ground. The solid line is for a dipole with no feed line. The dashed line is for an antenna with its feed line slanted 45° down to ground. Current induced on the outer braid of the 1-λ-long coax by its asymmetry with respect to the antenna causes the pattern distortion. At B, azimuthal response for two 5-element 20-meter Yagis placed 0.71 λ over average ground. Again, the solid line is for a Yagi without a feed line and the dashed line is for an antenna with a 45° slanted, 1-λ long feed line. The distortion in the radiated pattern is now clearly more serious than for a simple dipole. A balun is needed at the feed point, and most likely λ/4 way down the feed line from the feed point to suppress the common-mode currents and restore the pattern.

problem. On HF a quarter-wavelength rigid sleeve becomes considerably more challenging to construct, especially for a rotary antenna such as a Yagi.

Another method that gives an equivalent effect is shown at Fig 19.12C. Since the voltages at the antenna terminals are equal and opposite (with reference to ground), equal and opposite currents flow on the surfaces of the line and second conductor. Beyond the shorting point, in the direction of the transmitter, these currents combine to cancel out. The balancing section acts like an open circuit to the antenna, since it is a quarter-wave parallel-conductor line shorted at the far end, and thus has no effect on normal antenna operation. This is not essential to the line balancing function of the device, however, and baluns of this type are sometimes made shorter than a quarter wavelength to provide a shunt inductive reactance required in certain matching systems (such as the Hairpin).

Fig 19.12D shows a third balun, in which equal and opposite voltages, balanced to ground, are taken from the inner conductors of the main transmission line and half-wave phasing section. Since the voltages at the balanced end are in series while the voltages at the unbalanced end are in parallel, there is a 4:1 step down in impedance from the balanced to the unbalanced side. This arrangement is useful for coupling between a 300-Ω balanced line and a 75-Ω unbalanced coaxial line.

Broadband Baluns

At HF and even at VHF, broadband baluns are generally used nowadays. Examples of broadband baluns are shown in **Fig 19.14**.

Choke or current baluns force equal and opposite currents to flow. The result is that currents radiated back onto the transmission line by the antenna are effectively reduced, or "choked off," even if the antenna is not perfectly balanced. If winding inductive reactance becomes marginal at lower frequencies, the balun's ability to eliminate antenna currents is reduced, but (for the 1:1 balun) no winding impedance appears across the line.

If induced current on the line is a problem, perhaps because the feed line must be run in parallel with the antenna for some portion of its length, additional baluns can be placed at approximately $1/4$-λ intervals along the line. Current baluns are particularly useful for feeding asymmetrical antennas with balanced line.

Broadband Balun Construction

Either type of broadband balun can be constructed using a variety of techniques. Construction of choke (current) baluns is described here. The objective is to obtain a high impedance for currents that tend to flow on the line. Values from a few hundred to over a thousand ohms of inductive reactance are readily achieved. These baluns work best with antennas having resonant feed-point impedances less than 100 Ω or so (400 Ω for 4:1 baluns). This is because the winding inductive reactance must be high relative to the antenna impedance for effective operation. A rule of thumb is that the inductive reactance should be four times higher than the antenna impedance. High impedances are difficult to achieve over a wide frequency range. Any sort of transformer which is operated at impedances for which it was not designed can fail, sometimes spectacularly.

The simplest construction method for a 1:1 balun for coaxial line is simply to wind a portion of the line into a coil. See **Fig 19.15**. This type of choke balun is simple, cheap and effective. Currents on the outside of the line encounter the coil's impedance, while currents on the inside are unaffected. A flat coil (like a coil of rope) shows a broad resonance that easily covers three bands, making it reasonably effective over the entire HF range. If particular problems are encountered on a single band, a coil that is resonant at that band may be added. The coils shown in **Table 19.4** were constructed to have a high impedance at the indicated frequencies, as measured with an impedance meter. Many other geometries can also be effective. This construction technique is not effective with open-wire or twin-lead line because of coupling between adjacent turns. A 4:1 choke balun is shown in **Fig 19.14B**.

Ferrite-core baluns can provide a high impedance over the entire HF range. They may be wound either with two conductors in bifilar fashion, or with a single coaxial cable. Rod or toroidal cores may be used

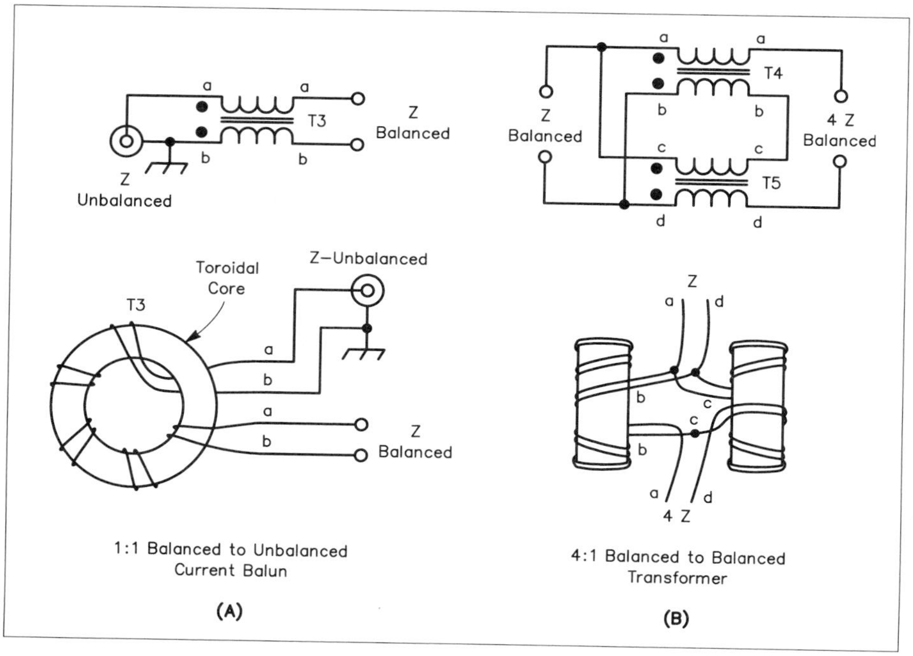

Fig 19.14—Broadband baluns. (A) 1:1 current balun and (B) 4:1 current transformer wound on two cores, which are separated. Use 12 bifilar turns of #14 enameled wire wound on FT240-43 cores for A and B. Distribute bifilar turns evenly around core.

Table 19.4
Effective Choke (Current Baluns)

Wind the indicated length of coaxial feed line into a coil (like a coil of rope) and secure with electrical tape. The balun is most effective when the coil is near the antenna. Lengths are not critical.

Single Band (Very Effective)

Freq MHz	RG-213, RG-8	RG-58
3.5	22 ft, 8 turns	20 ft, 6-8 turns
7	22 ft, 10 turns	15 ft, 6 turns
10	12 ft, 10 turns	10 ft, 7 turns
14	10 ft, 4 turns	8 ft, 8 turns
21	8 ft, 6-8 turns	6 ft, 8 turns
28	6 ft, 6-8 turns	4 ft, 6-8 turns

Multiple Band

Freq, MHz	RG-8, 58, 59, 8X, 213
3.5-30	10 ft, 7 turns
3.5-10	18 ft, 9-10 turns
14-30	8 ft, 6-7 turns

(**Fig 19.14A**). Current through a choke balun winding is the "antenna current" on the line; if the balun is effective, this current is small. Baluns used for high-power operation should be tested by checking for temperature rise before use. If the core overheats, add turns or use a larger or lower-loss core. It also would be wise to investigate the imbalance causing such high line antenna currents.

Type 72, 73 or 77 ferrite gives the greatest impedance over the HF range. Type 43 ferrite has lower loss, but somewhat less impedance. Core saturation is not a problem with these ferrites at HF, since they overheat due to loss at flux levels well below saturation. The loss occurs because there is insufficient inductive reactance at lower frequencies. Ten to twelve turns on a toroidal core or 10 to 15 turns on a rod are typical for the HF range. Winding impedance increases approximately as the square of the number of turns.

Another type of choke balun that is very effective was originated by M. Walter Maxwell, W2DU. A number of ferrite toroids are strung, like beads on a string, directly onto the coax where it is connected to the antenna. The "bead" balun in **Fig 19.16** consists of 50 FB73-2401 ferrite beads slipped over a 1-ft length of RG-303 coax. The beads fit nicely over the insulating jacket of the coax and occupy a total length of 9½ inches. Twelve FB-77-1024 or equivalent beads will come close to doing the same job using RG-8A or RG-213 coax. Type-73 material is recommended for 1.8 to 30 MHz use, but type-77 material may be substituted; use type-43 material for 30 to 250 MHz.

The cores present a high impedance to any RF current that would otherwise flow on the outside of the shield. The total impedance is in approximate proportion to the stacked length of the cores. The impedance stays fairly constant over a wide range of frequencies. Again, 70-series ferrites are a good choice for the HF range; use type-43 if heating is a problem. Type-43 or -61 is the best choice for the VHF range. Cores of various materials can be used in combination, permitting construction of baluns effective over a very wide frequency range, such as from 2 to 250 MHz.

WAVEGUIDES

A waveguide is a hollow, conducting tube, through which microwave energy is transmitted in the form of electromagnetic waves. The tube does not carry a current in the same sense that the wires of a two-conductor line do. Instead, it is a boundary that confines the waves to the enclosed space. Skin effect on the inside walls of

Fig 19.15—RF choke formed by coiling the feed line at the point of connection to the antenna. The inductance of the choke isolates the antenna from the remainder of the feed line.

Fig 19.16—W2DU bead balun consisting of 50 FB-73-2401 ferrite beads over a length of RG-303 coax. See text for details.

the waveguide confines electromagnetic energy inside the guide, in much the same manner that the shield of a coaxial cable confines energy within the coax. Microwave energy is injected at one end (either through capacitive or inductive coupling or by radiation) and is received at the other end. The waveguide merely confines the energy of the fields, which are propagated through it to the receiving end by means of reflections off its inner walls.

Evolution of a Waveguide

Suppose an open-wire line is used to carry UHF energy from a generator to a load. If the line has any appreciable length, it must be well insulated from the supports to avoid high losses. Since high-quality insulators are difficult to make for microwave frequencies, it is logical to support the transmission line with quarter-wavelength stubs, shorted at the far end. The open end of such a stub presents an infinite impedance to the transmission line, provided that the shorted stub is nonreactive. However, the shorting link has finite length and, therefore, some inductance. This inductance can be nullified by making the RF current flow on the surface of a plate rather than through a thin wire. If the plate is large enough, it will prevent the magnetic lines of force from encircling the RF current.

An infinite number of these quarter-wave stubs may be connected in parallel without affecting the standing waves of voltage and current. The transmission line may be supported from the top as well as the bottom, and when infinitely many supports are added, they form the walls of a waveguide at its cutoff frequency. **Fig 19.17** illustrates how a rectangular waveguide evolves from a two-wire parallel transmission line. This simplified analysis also shows why the cutoff dimension is a half wavelength.

While the operation of waveguides is usually described in terms of fields, current does flow on the inside walls, just as fields exist between the conductors of a two-wire transmission line. At the waveguide cutoff frequency, the current is concentrated in the center of the walls, and disperses toward the floor and ceiling as the frequency increases.

Analysis of waveguide operation is

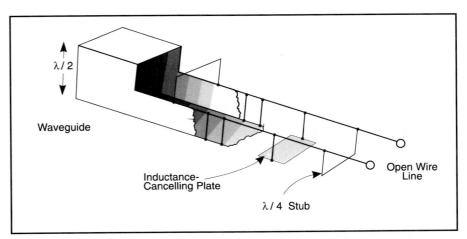

Fig 19.17—At its cutoff frequency a rectangular waveguide can be analyzed as a parallel two-conductor transmission line supported from top and bottom by an infinite number of quarter-wavelength stubs.

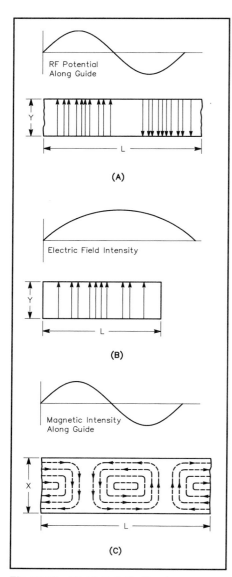

Fig 19.18—Field distribution in a rectangular waveguide. The TE$_{1,0}$ mode of propagation is depicted.

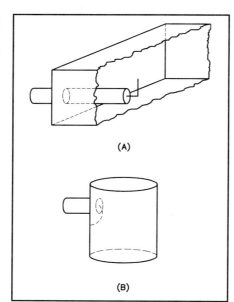

Fig 19.19—Coupling to waveguide and resonators. The probe at A is an extension of the inner conductor of coax line. At B an extension of the coax inner conductor is grounded to the waveguide to form a coupling loop.

Table 19.5
Wavelength Formulas for Waveguide

	Rectangular	Circular
Cut-off wavelength	2X	3.41R
Longest wavelength transmitted with little attenuation	1.6X	3.2R
Shortest wavelength before next mode becomes possible	1.1X	2.8R

based on the assumption that the guide material is a perfect conductor of electricity. Typical distributions of electric and magnetic fields in a rectangular guide are shown in **Fig 19.18**. The intensity of the electric field is greatest (as indicated by closer spacing of the lines of force in Fig 19.18B) at the center along the X dimension and diminishes to zero at the end walls. Zero field intensity is a necessary condition at the end walls, since the existence of any electric field parallel to any wall at the surface would cause an infinite current to flow in a perfect conductor, which is an impossible situation.

Modes of Propagation

Fig 19.18 represents a relatively simple distribution of the electric and magnetic fields. An infinite number of ways exist in which the fields can arrange themselves in a guide, as long as there is no upper limit to the frequency to be transmitted. Each field configuration is called a mode. All modes may be separated into two general groups. One group, designated TM (transverse magnetic), has the magnetic field entirely crosswise to the direction of propagation, but has a component of electric field in the propagation direction. The other type, designated TE (transverse electric) has the electric field entirely crosswise to the direction of propagation, but has a component of magnetic field in the direction of propagation. TM waves are sometimes called E waves, and TE waves are sometimes called H waves. The TM and TE designations are preferred, however.

The particular mode of transmission is identified by the group letters followed by subscript numbers; for example TE$_{1,1}$, TM$_{1,1}$ and so on. The number of possible modes increases with frequency for a given size of guide. There is only one possible mode (called the *dominant mode*) for the lowest frequency that can be transmitted. The dominant mode is the one normally used in practical work.

Waveguide Dimensions

In rectangular guides the critical dimension (shown as X in Fig 19.18C) must be more than one-half wavelength at the lowest frequency to be transmitted. In practice, the Y dimension is usually made about equal to $^1/_2$ X to avoid the possibility of operation at other than the dominant mode. Cross-sectional shapes other than rectangles can be used; the most important is the circular pipe.

Table 19.5 gives dominant-mode wavelength formulas for rectangular and circular guides. X is the width of a rectangular guide, and R is the radius of a circular guide.

Coupling to Waveguides

Energy may be introduced into or extracted from a waveguide or resonator by means of either the electric or magnetic field. The energy transfer frequently takes place through a coaxial line. Two methods for coupling are shown in **Fig 19.19**. The probe at A is simply a short extension of the inner conductor of the feed coaxial line, oriented so that it is parallel to the electric lines of force. The loop shown at B is arranged to enclose some of the magnetic lines of force. The point at which maximum coupling will be obtained depends on the particular mode of propagation in the guide or cavity; the coupling will be maximum when the coupling device is in the most intense field.

Coupling can be varied by rotating the probe or loop through 90°. When the probe is perpendicular to the electric lines the coupling will be minimum; similarly, when the plane of the loop is parallel to the magnetic lines, the coupling will be minimum. See *The ARRL Antenna Book* for more information on waveguides.

BIBLIOGRAPHY

J. Devoldere, *ON4UN's Low Band DXing*, 3rd Edition (Newington: ARRL, 1999).

W. Everitt, *Communication Engineering*, 2nd Edition (New York: McGraw-Hill, 1937).

H. Friis, and S. Schelkunoff, *Antennas: Theory and Practice* (New York: John Wiley and Sons, 1952).

M. Maxwell, *Reflections: Transmission*

Lines and Antennas (Newington, CT: ARRL, 1990) [out of print].

F. Regier, "Series-Section Transmission Line Impedance Matching," *QST,* Jul 1978, pp 14-16.

A. Roehm, W2OBJ, "Some Additional Aspects of the Balun Problem," *The ARRL Antenna Compendium Vol 2,* p 172.

W. Sabin, WØIYH, "Computer Modeling of Coax Cable Circuits," *QEX*, Aug 1996 pp 3-10.

P. Smith, "Transmission Line Calculator," *Electronics*, Jan 1939.

P. Smith, *Electronic Applications of the Smith Chart* (McGraw-Hill, 1969).

W.D. Stewart, Jr, N7WS, "Balanced Transmission Lines in Current Amateur Practice," *The ARRL Antenna Compendium Vol 6* (Newington: ARRL, 1999).

R. D. Straw, N6BV, Ed.,*The ARRL Antenna Book,* 18th Edition (Newington: ARRL, 1997).

F. Witt, "Baluns in the Real (and Complex) World," *The ARRL Antenna Compendium Vol 5* (Newington: ARRL, 1996).

Contents

Antennas & Projects

20

ANTENNA BASICS

Every ham needs at least one antenna, and most hams have built one. This chapter, by Chuck Hutchinson, K8CH, covers theory and construction of antennas for most radio amateurs. Here you'll find simple verticals and dipoles, as well as quad and Yagi projects and other antennas that you can build and use.

The amount of available space should be high on the list of factors to consider when selecting an antenna. Those who live in urban areas often must accept a compromise antenna for the HF bands because a city lot won't accommodate full-size wire dipoles, end-fed systems or high supporting structures. Other limitations are imposed by the amount of money available for an antenna system (including supporting hardware), the number of amateur bands to be worked and local zoning ordinances.

Operation objectives also come into play. Do you want to dedicate yourself to serious contesting and DXing? Are you looking for general-purpose operation that will yield short- and long-haul QSOs during periods of good propagation? Your answers should result in selecting an antenna that will meet your needs. You might want to erect the biggest and best collection of antennas that space and finances will allow. If a modest system is the order of the day, then use whatever is practical and accept the performance that follows. Practically any radiator works well under some propagation conditions, assuming the radiator is able to accept power and radiate it at some useful angle. Any antenna is a good one if it meets your needs!

In general, the height of the antenna above ground is the most critical factor at the higher end of the HF spectrum, that is from roughly 14 through 30 MHz. This is because the antenna should be clear of conductive objects such as power lines, phone wires, gutters and the like, plus high enough to have a low radiation angle. Lower frequency antennas, operating between 2 and 10 MHz, should also be kept well away from conductive objects and as high above ground as possible if you want good performance.

Antenna Polarization

Most HF-band antennas are either vertically or horizontally polarized, although circular polarization is possible, just as it is at VHF and UHF. *Polarization* is determined by the position of the radiating element or wire with respect to the earth. Thus a radiator that is parallel to the earth radiates horizontally, while an antenna at a right angle to the earth (vertical) radiates a vertical wave. If a wire antenna is slanted above earth, it radiates waves that have both a vertical and a horizontal component.

For best results in line-of-sight communications, antennas at both ends of the circuit should have the same polarization; cross polarization results in many decibels of signal reduction. It is not essential for both stations to use the same antenna polarity for ionospheric propagation (sky wave). This is because the radiated wave is bent and it tumbles considerably during its travel through the ionosphere. At the far end of the communications path the wave may be horizontal, vertical or somewhere in between at any given instant. On multihop transmissions, in which the signal is refracted more than once from the ionosophere, and subsequently reflected from the Earth's surface during its travel, considerable polarization shift will occur. For that reason, the main consideration for a good DX antenna is a low angle of radiation rather than the polarization.

Antenna Bandwidth

The *bandwidth* of an antenna refers generally to the range of frequencies over which the antenna can be used to obtain good performance. The bandwidth is often referenced to some SWR value, such as, "The 2:1 *SWR bandwidth* is 3.5 to 3.8 MHz." Popular amateur usage of the term "bandwidth" most often refers to the 2:1 SWR bandwidth. Other specific bandwidth terms are also used, such as the *gain bandwidth* and the *front-to-back ratio bandwidth*.

For the most part, the lower the operating frequency of a given antenna design, the narrower is the bandwidth. This follows the rule that the bandwidth of a resonant circuit doubles as the frequency of operation is doubled, assuming the Q is the same for each case. Therefore, it is often difficult to cover all of the 160 or 80-m band for a particular level of SWR with a dipole antenna. It is important to recognize that SWR bandwidth does not always relate directly to gain bandwidth. Depending on the amount of feed-line loss, an 80-m dipole with a relatively narrow 2:1 SWR bandwidth can still radiate a good signal at each end of the band, provided that an antenna tuner is used to allow the transmitter to load properly. Broadbanding techniques, such as fanning the far ends of a dipole to simulate a conical type of dipole, can help broaden the SWR response curve.

Current and Voltage Distribution

When power is fed to an antenna, the current and voltage vary along its length. The current is nearly zero (a current *node*)

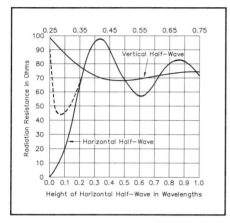

Fig 20.1—Curves showing the radiation resistance of vertical and horizontal half-wavelength dipoles at various heights above ground. The broken-line portion of the curve for a horizontal dipole shows the resistance over "average" real earth, the solid line for perfectly conducting ground.

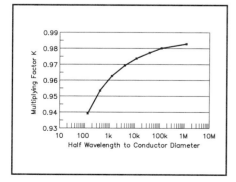

Fig 20.2—Effect of antenna diameter on length for half-wavelength resonance, shown as a multiplying factor, K, to be applied to the free-space, half-wavelength equation.

at the ends. The current does not actually reach zero at the current nodes, because of capacitance at the antenna ends. Insulators, loops at the antenna ends, and support wires all contribute to this capacitance, which is also called the "end effect." In the case of a half-wave antenna there is a current maximum (a current *loop*) at the center.

The opposite is true of the RF voltage. That is, there is a voltage loop at the ends, and in the case of a half-wave antenna there is a voltage minimum (node) at the center. The voltage is not zero at its node because of the resistance of the antenna, which consists of both the RF resistance of the wire (ohmic loss resistance) and the *radiation resistance*. The radiation resistance is the equivalent resistance that would dissipate the power the antenna radiates, with a current flowing in it equal to the antenna current at a current loop (maximum). The loss resistance of a half-wave antenna is ordinarily small, compared with the radiation resistance, and can usually be neglected for practical purposes.

Impedance

The *impedance* at a given point in the antenna is determined by the ratio of the voltage to the current at that point. For example, if there were 100 V and 1.4 A of RF current at a specified point in an antenna and if they were in phase, the impedance would be approximately 71 Ω.

Antenna impedance may be either resistive or complex (that is, containing resistance and reactance). This will depend on whether or not the antenna is *resonant* at the operating frequency. You need to know the impedance in order to match the feeder to the feedpoint. Some operators mistakenly believe that a mismatch, however small, is a serious matter. This is not true. The importance of a matched line is described in detail in the **Transmission Lines** chapter of this book. The significance of a perfect match becomes more pronounced only at VHF and higher, where feed-line losses are a major factor.

Some antennas possess a theoretical input impedance at the feedpoint close to that of certain transmission lines. For example, a 0.5-λ (or half-wave) center-fed dipole, placed at a correct height above ground, will have a feedpoint impedance of approximately 75 Ω. In such a case it is practical to use a 75-Ω coaxial or balanced line to feed the antenna. But few amateur half-wave dipoles actually exhibit a 75-Ω impedance. This is because at the lower end of the high-frequency spectrum the typical height above ground is rarely more than ¼ λ. The 75-Ω feed-point impedance is most likely to be realized in a practical installation when the

horizontal dipole is approximately ½, ¾ or 1 wavelength above ground. Coax cable having a 50-Ω characteristic impedance is the most common transmission line used in amateur work.

Fig 20.1 shows the difference between the effects of perfect ground and typical earth at low antenna heights. The effect of height on the radiation resistance of a horizontal half-wave antenna is not drastic so long as the height of the antenna is greater than 0.2 λ. Below this height, while decreasing rapidly to zero over perfectly conducting ground, the resistance decreases less rapidly with height over actual ground. At lower heights the resistance stops decreasing at around 0.15 λ, and thereafter increases as height decreases further. The reason for the increasing resistance is that more and more of the induction field of the antenna is absorbed by the earth as the height drops below ¼ λ.

Conductor Size

The impedance of the antenna also depends on the diameter of the conductor in relation to the wavelength, as indicated in **Fig 20.2**. If the diameter of the conductor is increased, the capacitance per unit length increases and the inductance per unit length decreases. Since the radiation resistance is affected relatively little, the decreased L/C ratio causes the Q of the antenna to decrease so that the resonance curve becomes less sharp with change in frequency. This effect is greater as the diameter is increased, and is a property of some importance at the very high frequencies where the wavelength is small.

Directivity and Gain

All antennas, even the simplest types, exhibit directive effects in that the intensity of radiation is not the same in all directions from the antenna. This property of radiating more strongly in some directions than in others is called the *directivity* of the antenna.

The *gain* of an antenna is closely related to its directivity. Because directivity is based solely on the shape of the directive pattern, it does not take into account any power losses that may occur in an actual antenna system. Gain takes into account those losses.

Gain is usually expressed in decibels, and is based on a comparison with a "standard" antenna—usually a dipole or an *isotropic radiator*. An isotropic radiator is a theoretical antenna that would, if placed in the center of an imaginary sphere, evenly illuminate that sphere with radiation. The isotropic radiator is an unambiguous standard, and so is frequently used as the comparison for gain measurements.

When the standard is the isotropic radiator in free space, gain is expressed in dBi. When the standard is a dipole, *also located in free space*, gain is expressed in dBd.

The more the directive pattern is compressed—or focused—the greater the power gain of the antenna. This is a result of power being concentrated in some directions at the expense of others. The directive pattern, and therefore the gain, of an antenna at a given frequency is determined by the size and shape of the antenna, and on its position and orientation relative to the Earth.

Elevation Angle

For HF communication, the vertical (elevation) angle of maximum radiation is of considerable importance. You will want to erect your antenna so that it radiates at desirable angles. **Tables 20.1, 20.2 and 20.3** show optimum elevation angles from locations in the continental US. These figures are based on statistical averages over all portions of the solar sunspot cycle.

Since low angles usually are most effective, this generally means that horizontal antennas should be high—higher is usually better. Experience shows that satisfactory results can be attained on the bands above 14 MHz with antenna heights between 40 and 70 ft. **Fig 20.3** shows this effect at work in horizontal dipole antennas.

Imperfect Ground

Earth conducts, but is far from being a perfect conductor. This influences the radiation pattern of the antennas that we use. The effect is most pronounced at high vertical angles (the ones that we're least interested in for long-distance communications) for horizontal antennas. The consequences for vertical antennas are greatest at low angles, and are quite dramatic as can be clearly seen in **Fig 20.4**, where the elevation pattern for a 40-m vertical half-wave dipole located over average ground is compared to one located over saltwater. At 10° elevation, the saltwater antenna has about 7 dB more gain than its landlocked counterpart.

A vertical antenna may work well at HF for a ham living in the area between Dallas, Texas and Lincoln, Nebraska. This area is pastoral, has low hills, and rich soil. Ground of this type has very good conductivity. By contrast, a ham living in New Hampshire, where the soil is rocky and a poor conductor, may not be satisfied with the performance of a vertical HF antenna.

Table 20.1
Optimum Elevation Angles to Europe

Band	Northeast	Southeast	Upper Midwest	Lower Midwest	West Coast
10 m	5°	3°	3°	7°	3°
12 m	5°	6°	4°	6°	5°
15 m	5°	7°	8°	5°	6°
17 m	4°	8°	7°	5°	5°
20 m	11°	9°	8°	5°	6°
30 m	11°	11°	11°	9°	8°
40 m	15°	15°	14°	14°	12°
75 m	20°	15°	15°	11°	11°

Table 20.2
Optimum Elevation Angles to Far East

Band	Northeast	Southeast	Upper Midwest	Lower Midwest	West Coast
10 m	4°	5°	5°	5°	6°
12 m	4°	8°	5°	12°	6°
15 m	7°	10°	10°	10°	8°
17 m	7°	10°	9°	10°	5°
20 m	4°	10°	9°	10°	9°
30 m	7°	13°	11°	12°	9°
40 m	11°	12°	12°	12°	13°
75 m	12°	14°	14°	12°	15°

Table 20.3
Optimum Elevation Angles to South America

Band	Northeast	Southeast	Upper Midwest	Lower Midwest	West Coast
10 m	5°	4°	4°	4°	7°
12 m	5°	5°	6°	3°	8°
15 m	5°	5°	7°	4°	8°
17 m	4°	5°	5°	3°	7°
20 m	8°	8°	8°	6°	8°
30 m	8°	11°	9°	9°	9°
40 m	10°	11°	9°	9°	10°
75 m	15°	15°	13°	14°	14°

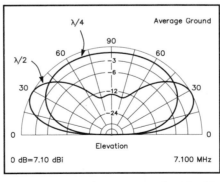

Fig 20.3—Elevation patterns for two 40-m dipoles over average ground (conductivity of 5 mS/m and dielectric constant of 13) at $1/4 \lambda$ (33 ft) and $1/2 \lambda$ (66 ft) heights. The higher dipole has a peak gain of 7.1 dBi at an elevation angle of about 26°, while the lower dipole has more response at high elevation angles.

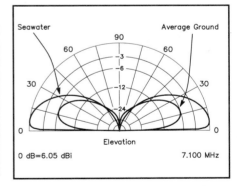

Fig 20.4—Elevation patterns for a vertical dipole over sea water compared to average ground. In each case the center of the dipole is just over $1/4 \lambda$ high. The low-angle response is greatly degraded over average ground compared to sea water, which is virtually a perfect ground.

Dipoles and the Half-Wave Antenna

A fundamental form of antenna is a wire whose length is half the transmitting wavelength. It is the unit from which many more complex forms of antennas are constructed and is known as a *dipole antenna*. The length of a half-wave in free space is

$$\text{Length (ft)} = \frac{492}{f(\text{MHz})} \quad (1)$$

The actual length of a resonant $1/2$-λ antenna will not be exactly equal to the half wavelength in space, but depends on the thickness of the conductor in relation to the wavelength. The relationship is shown in Fig 20.2, where K is a factor that must be multiplied by the half wavelength in free space to obtain the resonant antenna length. An additional shortening effect occurs with wire antennas supported by insulators at the ends because of the capacitance added to the system by the insulators (end effect). The following formula is sufficiently accurate for wire antennas for frequencies up to 30 MHz.

Length of half-wave antenna (ft)

$$= \frac{492 \times 0.95}{f(\text{MHz})} = \frac{468}{f(\text{MHz})} \quad (2)$$

Example: A half-wave antenna for 7150 kHz (7.15 MHz) is 468/7.15 = 65.45 ft, or 65 ft 5 inches.

Above 30 MHz use the following formulas, particularly for antennas constructed from rod or tubing. K is taken from Fig 20.2.

Length of half-wave antenna (ft)

$$= \frac{492 \times K}{f(\text{MHz})} \quad (3)$$

$$\text{length (in.)} = \frac{5904 \times K}{f(\text{MHz})} \quad (4)$$

Example: Find the length of a half-wave antenna at 50.1 MHz, if the antenna is made of $1/2$-inch-diameter tubing. At 50.1 MHz, a half wavelength in space is

$$\frac{492}{50.1} = 9.82 \text{ ft}$$

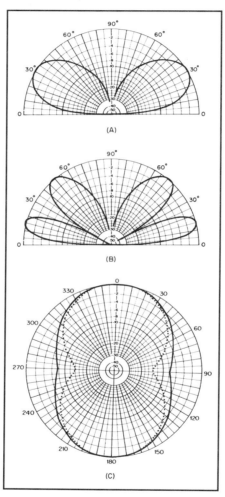

(A)

(B)

(C)

Fig 20.6—At A, elevation response pattern of a dipole antenna placed $1/2$ λ above a perfectly conducting ground. At B, the pattern for the same antenna when raised to one wavelength. For both A and B, the conductor is coming out of the paper at right angle. C shows the azimuth patterns of the dipole for the two heights at the most-favored elevation angle, the solid-line plot for the $1/2$-λ height at an elevation angle of 30°, and the broken-line plot for the 1-λ height at an elevation angle of 15°. The conductor in C lies along 90° to 270° axis.

From equation 1 the ratio of half wavelength to conductor diameter (changing wavelength to inches) is

$$\frac{(9.82 \times 12)}{0.5 \text{ inch}} = 235.7$$

From Fig 20.2, K = 0.965 for this ratio. The length of the antenna, from equation 3 is

$$\frac{492 \times 0.965}{50.1} = 9.48 \text{ ft}$$

or 9 ft 5$3/4$ inches. The answer is obtained directly in inches by substitution in equation 4

$$\frac{5904 \times 0.965}{50.1} = 113.7 \text{ inches}$$

The length of a half-wave antenna is also affected by the proximity of the dipole ends to nearby conductive and semiconductive objects. In practice, it is often necessary to do some experimental "pruning" of the wire after cutting the antenna to the computed length, lengthening or shortening it in increments to obtain a low SWR. When the lowest SWR is obtained for the desired part of an amateur band, the antenna is resonant at that frequency. The value of the SWR indicates the quality of the match between the antenna and the feed line. If the lowest SWR obtainable is too high for use with solid-state rigs, a Transmatch or line-input matching network may be used, as described in the **Transmission Lines** and **Station Setup** chapters.

Radiation Characteristics

The radiation pattern of a dipole antenna in free space is strongest at right angles to the wire (**Fig 20.5**). This figure-8 pattern appears in the real world if the dipole is $1/2$ λ or greater above earth and is not degraded by nearby conductive objects. This assumption is based also on a symmetrical feed system. In practice, a coaxial feed line may distort this pattern slightly, as shown in Fig 20.5. Minimum horizontal radiation occurs off the ends of the dipole if the antenna is parallel to the earth.

As an antenna is brought closer to ground, the elevation pattern peaks at a higher elevation angle as shown in Fig 20.3. **Fig 20.6** illustrates what happens to the directional pattern as antenna height changes. Fig 20.6C shows that there is significant radiation off the ends of a low horizontal dipole. For the $1/2$-λ height (solid line), the radiation off the ends is only 7.6 dB lower than that in the broadside direction.

Feed Methods

Most amateurs use either *coax* or *open-*

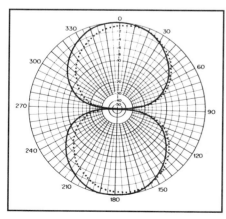

Fig 20.5—Response of a dipole antenna in free space, where the conductor is along 90° to 270° axis, solid line. If the currents in the halves of the dipole are not in phase, slight distortion of the pattern will occur, broken line. This illustrates case where balun is not used on a balanced antenna fed with unbalanced line.

wire transmission line. Coax is the common choice because it is readily available, its characteristic impedance is close to that of the antenna and it may be easily routed through or along walls and among other cables. The disadvantages of coax are increased RF loss and low working voltage (compared to that of open-wire line). Both disadvantages make coax a poor choice for high-SWR systems.

Take care when choosing coax. Use $1/4$-inch foam-dielectric cables only for low power (25 W or less) HF transmissions. Solid-dielectric $1/4$-inch cables are okay for 300 W if the SWR is low. For high-power installations, use $1/2$-inch or larger cables.

The most common two-wire transmission lines are *ladder line* and *twin lead*. Since the conductors are not shielded, two-wire lines are affected by their environment. Use standoffs and insulators to keep the line several inches from structures or other conductors. Ladder line has very low loss (twin lead has a little more), and it can stand very high voltages (SWR) as long as the insulators are clean.

Two-wire lines are usually used in balanced systems, so they should have a balun at the transition to an unbalanced transmitter or coax. A Transmatch will be needed to match the line input impedance to the transmitter.

Baluns

A balun is a device for feeding a balanced load with an unbalanced line, or vice versa (see the **Transmission Lines** chapter of this book). Because dipoles are balanced (electrically symmetrical about their feed-points), a balun should be used at the feed-point when a dipole is fed with coax. When coax feeds a dipole directly (as in **Fig 20.7**), current flows on the outside of the cable shield. The shield can conduct RF onto the transmitter chassis and induce RF onto metal objects near the system. Shield currents can impair the function of instruments connected to the line (such as SWR meters and SWR-protection circuits in the transmitter). The shield current also produces some feed-line radiation, which changes the antenna radiation pattern, and allows objects near the cable to affect the antenna-system performance.

The consequences may be negligible: A slight skewing of the antenna pattern usually goes unnoticed. Or, they may be significant: False SWR readings may cause the transmitter to shut down or destroy the output transistors; radiating coax near a TV feed line may cause strong local interference. Therefore, it is better to eliminate feed-line radiation whenever possible, and

a balun should be used at any transition between balanced and unbalanced systems. (The **Transmission Lines** chapter thoroughly describes baluns and their construction.) Even so, balanced or unbalanced systems without a balun often operate with no apparent problems. For temporary or emergency stations, do not let the lack of a balun deter you from operating.

Practical Dipole Antennas

A classic dipole antenna is $1/2$-λ long and fed at the center. The feed-point impedance is low at the resonant frequency, f_0, and odd harmonics thereof. The impedance is high near even harmonics. When fed with coax, a classic dipole provides a reasonably low SWR at f_0 and its odd harmonics.

When fed with ladder line (see **Fig 20.8A**) and a Transmatch, the classic dipole should be usable near f_0 and all harmonic frequencies. (With a wide-range Transmatch, it may work on all frequencies.) If there are problems (such as extremely high SWR or evidence of RF on objects at the operating position), change the feed-line length by adding or subtracting $1/8$ λ at the problem frequency. A few such adjustments should yield a workable solution. Such a system is sometimes called a "center-fed Zepp." A true "Zepp" antenna is an end-fed dipole that is matched by $1/4$ λ of open-wire feed line (see Fig 20.8B). The antenna was originally used on zeppelins, with the dipole trailing from the feeder, which hung from the airship cabin. It is intended for use on a single band, but should be usable near odd harmonics of f_0.

Most dipoles require a little pruning to reach the desired resonant frequency. Here's a technique to speed the adjustment.

How much to prune: When assembling the antenna, cut the wire 2 to 3% longer than the calculated length and record the length. When the antenna is complete, raise it to the working height and check the SWR at several frequencies. Multiply the frequency of the SWR minimum by the antenna length and divide the result by the desired f_0. The result is the finished length; trim both ends equally to reach that length and you're done.

Loose ends: Here's another trick, if you use nonconductive end support lines. When assembling the antenna, mount the end insulators in about 5% from the ends. Raise the antenna and let the ends hang free. Figure how much to prune and cut it from the hanging ends. If the pruned ends are very long, wrap them around the insulated line for support.

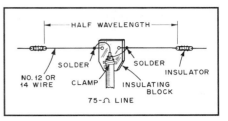

Fig 20.7—Method of affixing feed line to the center of a dipole antenna. A plastic block is used as a center insulator. The coax is held in place by a clamp. A balun is often used to feed dipoles or other balanced antennas to ensure that the radiation pattern is not distorted. See text for explanation.

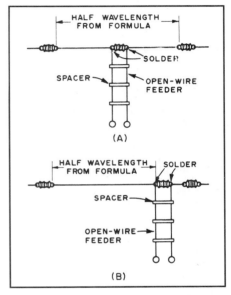

Fig 20.8—Center-fed multiband "Zepp" antenna (A) and an end-fed Zepp at (B).

Dipole Orientation

Dipole antennas need not be installed in a horizontal straight line. They are generally tolerant of bending, sloping or drooping as required by the antenna site. Remember, however, that dipole antennas are RF conductors. For safety's sake, mount all antennas away from conductors (especially power lines), combustibles and well beyond the reach of passersby.

A *sloping dipole* is shown in **Fig 20.9**. This antenna is often used to favor one direction (the "forward direction" in the figure). With a nonconducting support and poor earth, signals off the back are weaker than those off the front. With a nonconducting mast and good earth, the response is omnidirectional. There is no gain in any direction with a nonconducting mast.

A conductive support such as a tower acts as a parasitic element. (So does the coax shield, unless it is routed at 90° from the antenna.) The parasitic effects vary with earth quality, support height and other conductors on the support (such as a beam at the top). With such variables, performance is very difficult to predict.

Losses increase as the antenna ends approach the support or the ground. To prevent feed-line radiation, route the coax away from the feed-point at 90° from the antenna, and continue on that line as far as possible.

An *Inverted V* antenna appears in **Fig 20.10**. While "V" accurately describes the shape of this antenna, this antenna should not be confused with long-wire V antennas, which are highly directive. The radiation pattern and dipole impedance depend on the apex angle, and it is very important that the ends do not come too close to lossy ground.

Bent dipoles may be used where antenna space is at a premium. **Fig 20.11** shows several possibilities; there are many more. Bending distorts the radiation pattern somewhat and may affect the impedance as well, but compromises are acceptable when the situation demands them. When an antenna bends back on itself (as in Fig 20.11B) some of the signal is canceled; avoid this if possible.

Remember that current produces the radiated signal, and current is maximum at the dipole center. Therefore, performance is best when the central area of the antenna is straight, high and clear of nearby objects. Be safe! Keep any bends, sags or hanging ends well clear of conductors (especially power lines) and combustibles, and beyond the reach of persons.

Multiband Dipoles

There are several ways to construct coax-fed multiband dipole systems. These techniques apply to dipoles of all orientations. Each method requires a little more work than a single dipole, but the materials don't cost much.

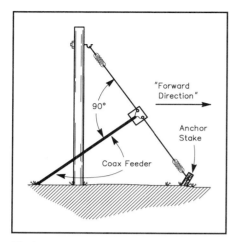

Fig 20.9—Example of a sloping ½-λ dipole, or "full sloper." On the lower HF bands, maximum radiation over poor to average earth is off the sides and in the "forward direction" as indicated, if a nonconductive support is used. A metal support will alter this pattern by acting as a parasitic element. How it alters the pattern is a complex issue depending on the electrical height of the mast, what other antennas are located on the mast, and on the configuration of guy wires.

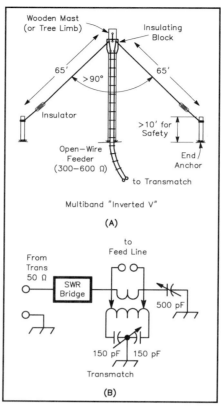

Fig 20.10—At A, details for an inverted V fed with open-wire line for multiband HF operation. A Transmatch is shown at B, suitable for matching the antenna to the transmitter over a wide frequency range. The included angle between the two legs should be greater than 90° for best performance.

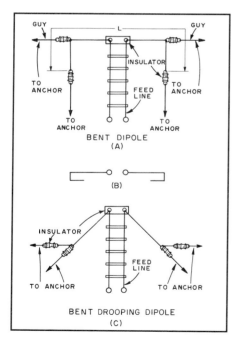

Fig 20.11—When limited space is available for a dipole antenna, the ends can be bent downward as shown at A, or back on the radiator as shown at B. The inverted V at C can be erected with the ends bent parallel with the ground when the available supporting structure is not high enough.

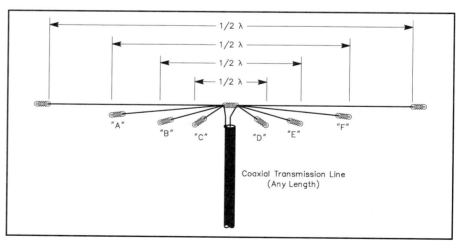

Fig 20.12—Multiband antenna using paralleled dipoles, all connected to a common 50 or 75-Ω coax line. The half-wave dimensions may be either for the centers of the various bands or selected for favorite frequencies in each band. The length of a half wave in feet is 468/frequency in MHz, but because of interaction among the various elements, some pruning for resonance may be needed on each band. See text.

Parallel dipoles are a simple and convenient answer. See **Fig 20.12**. Center-fed dipoles present low-impedances near f_0, or its odd harmonics, and high impedances elsewhere. This lets us construct simple multiband systems that automatically select the appropriate antenna. Consider a 50-Ω resistor connected in parallel with a 5-kΩ resistor. A generator connected across the two resistors will see 49.5 Ω, and 99% of the current will flow through the 50-Ω resistor. When resonant and nonresonant antennas are parallel connected, the nonresonant antenna takes little power and has little effect on the total feed-point impedance. Thus, we can connect several antennas together at the feed-point, and power naturally flows to the resonant antenna.

There are some limits, however. Wires in close proximity tend to couple and produce mutual inductance. In parallel dipoles, this means that the resonant length of the shorter dipoles lengthens a few percent. Shorter antennas don't affect longer ones much, so adjust for resonance in order from longest to shortest. Mutual inductance also reduces the bandwidth of shorter dipoles, so a Transmatch may be needed to achieve an acceptable SWR across all bands covered. These effects can be reduced by spreading the ends of the dipoles.

Also, the power-distribution mechanism requires that only one of the parallel dipoles is near resonance on any amateur band. Separate dipoles for 80 and 30 m should not be parallel connected because the higher band is near an odd harmonic of the lower band ($80/3 \approx 30$) and center-fed dipoles have low impedance near odd harmonics. (The 40 and 15-m bands have a similar relationship.) This means that you must either accept the lower performance of the low-band antenna operating on a harmonic or erect a separate antenna for those odd-harmonic bands. For example, four parallel-connected dipoles cut for 80, 40, 20 and 10 m (fed by a single Transmatch and coaxial cable) work reasonably on all HF bands from 80 through 10 m.

Trap dipoles provide multiband operation from a coax-fed single-wire dipole. **Fig 20.13** shows a two-band trap antenna. A trap is a parallel-resonant circuit that effectively disconnects wire beyond the trap at the resonant frequency. Traps may be constructed from coiled sections of coax or from discrete LC components.

Choose capacitors (C1 in the figure) that are rated for high current and voltage. Mica transmitting capacitors are good. Ceramic transmitting capacitors may work, but their values may change with temperature. Use large wire for the induc-

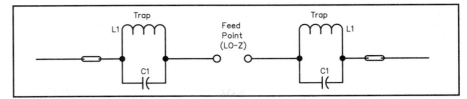

Fig 20.13—Example of a trap dipole antenna. L1 and C1 can be tuned to the desired frequency by means of a dip meter before they are installed in the antenna.

The Trusty Slingshot

Trees make excellent supports for wire antennas, but how do you get a rope over tall branches? Some hams use bows and arrows with good success, but many prefer a slingshot. Use a short section of leader (2 to 4 ft is adequate) between the weight and the line. The leader should be strong enough to withstand the shock of launching the weight, but must be rated at less tensile strength than the main line. Then, if the weight wraps around a limb or otherwise becomes stuck, pulling on the line will break the leader and free the main line. (Often, the weight breaks free and drops to the ground.) This arrangement works better than using one continuous piece of line and getting the whole mess hopelessly stuck in the tree.

Use a "Wrist Rocket" or equivalent, with 20 to 25-lb monofilament line (the main line) laid out neatly on the ground. Use an 8 to 12-lb monofilament line as the leader, with a 2 to 3-oz teardrop fishing weight. This combination is a low-cost method to shoot lines accurately over 100 to 125-ft fir trees. With a little patience, a 2 to 3-oz weight is sufficient to drop the line to ground level for attachment to support ropes, and so on.—*K7FL*

tors to reduce loss. Any reactance (X_L and X_C) above 100 Ω (at f_0) will work, but bandwidth increases with reactance (up to several thousand ohms).

Check trap resonance before installation. This can be done with a dip meter and a receiver. To construct a trap antenna, cut a dipole for the highest frequency and connect the pretuned traps to its ends. It is fairly complicated to calculate the additional wire needed for each band, so just add enough wire to make the antenna $1/2 \lambda$ and prune it as necessary. Because the inductance in each trap reduces the physical length needed for resonance, the finished antenna will be shorter than a simple $1/2$-λ dipole.

Shortened Dipoles

Inductive loading increases the electrical length of a conductor without increasing its physical length. Therefore, we can build physically short dipole antennas by placing inductors in the antenna. These are called "loaded antennas," and *The ARRL Antenna Book* shows how to design them. There are some trade-offs involved: Inductively loaded antennas are less efficient and have narrower bandwidths than full-size antennas. Generally they should not be shortened more than 50%.

Building Dipole Antennas

The purpose of this section is to offer information on the actual physical con-

struction of wire antennas. Because the dipole, in one of its configurations, is probably the most common amateur wire antenna, it is used in the following examples. The techniques described here, however, enhance the reliability and safety of all wire antennas.

Wire

Choosing the right type of wire for the project at hand is the key to a successful antenna—the kind that works well and stays up through a winter ice storm or a gusty spring wind storm. What gauge of wire to use is the first question to settle, and the answer depends on strength, ease of handling, cost, availability and visibility. Generally, antennas that are expected to support their own weight, plus the weight of the feed line should be made from #12 wire. Horizontal dipoles, Zepps, some long wires and the like fall into this category. Antennas supported in the center, such as inverted-V dipoles and delta loops, may be made from lighter material, such as #14 wire—the minimum size called for in the National Electrical Code.

The type of wire to be used is the next important decision. The wire specifications table in the **Component Data** chapter shows popular wire styles and sizes. The strongest wire suitable for antenna service is *copperclad steel*, also known as *copperweld*. The copper coating is necessary for RF service because steel is a rela-

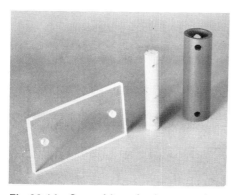

Fig 20.14—Some ideas for homemade antenna insulators.

tively poor conductor. Practically all of the RF current is confined to the copper coating because of skin effect. Copper-clad steel is outstanding for permanent installations, but it can be difficult to work with. Kinking, which severely weakens the wire, is a constant threat when handling any solid conductor. Solid-copper wire, either hard drawn or soft drawn, is another popular material. Easier to handle than copper-clad steel, solid copper is available in a wide range of sizes. It is generally more expensive however, because it is all copper. Soft drawn tends to stretch under tension, so

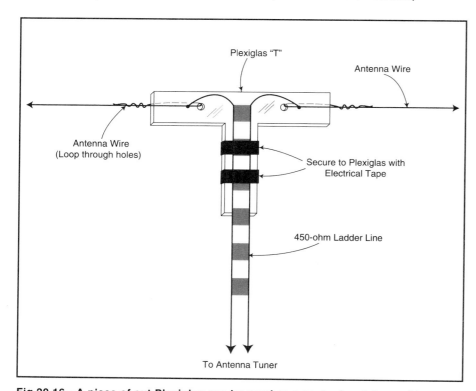

Fig 20.15—Some homemade dipole center insulators. The one in the center includes a built-in SO-239 connector. Others are designed for direct connection to the feed line. (See the Transmission Lines chapter for details on baluns.)

Plexiglas "T"

Antenna Wire

Antenna Wire
(Loop through holes)

Secure to Plexiglas with
Electrical Tape

450-ohm Ladder Line

To Antenna Tuner

Fig 20.16—A piece of cut Plexiglas can be used as a center insulator and to support a ladder-line feeder. The Plexiglas acts to reduce the flexing of the wires where they connect to the antenna.

periodic pruning of the antenna may be necessary in some cases. Enamel-coated *magnet-wire* is a good choice for experimental antennas because it is easy to manage, and the coating protects the wire from the weather. Although it stretches under tension, the wire may be prestretched before final installation and adjustment. A local electric motor rebuilder might be a good source for magnet wire.

Hook-up wire, speaker wire or even ac lamp cord are suitable for temporary installations. Almost any copper wire may be used, as long as it is strong enough for the demands of the installation. Steel wire is a poor conductor at RF; avoid it.

It matters not (in the HF region at least) whether the wire chosen is insulated or bare. If insulated wire is used, a 3 to 5% shortening beyond the standard 468/f length will be required to obtain resonance at the desired frequency, because of the increased distributed capacitance resulting from the dielectric constant of the plastic insulating material. The actual length for resonance must be determined experimentally by pruning and measuring because the dielectric constant of the insulating material varies from wire to wire. Wires that might come into contact with humans or animals should be insulated to reduce the chance of shock or burns.

Insulators

Wire antennas must be insulated at the ends. Commercially available insulators are made from ceramic, glass or plastic. Insulators are available from many Amateur Radio dealers. Radio Shack and local hardware stores are other possible sources.

Acceptable homemade insulators may be fashioned from a variety of material including (but not limited to) acrylic sheet or rod, PVC tubing, wood, fiberglass rod or even stiff plastic from a discarded container. **Fig 20.14** shows some homemade insulators. Ceramic or glass insulators will usually outlast the wire, so they are highly recommended for a safe, reliable, permanent installation. Other materials may tear under stress or break down in the presence of sunlight. Many types of plastic do not weather well.

Many wire antennas require an insulator at the feedpoint. Although there are many ways to connect the feed line, there are a few things to keep in mind. If you feed your antenna with coaxial cable, you have two choices. You can install an SO-239 connector on the center insulator and use a PL-259 on the end of your coax, or you can separate the center conductor from the braid and connect the feed line directly to the antenna wire. Although it costs less to connect direct, the use of con-

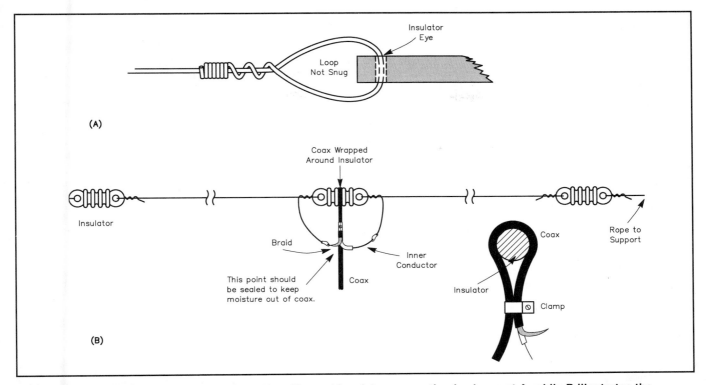

Fig 20.17—Details of dipole antenna construction. The end insulator connection is shown at A, while B illustrates the completed antenna. This is a balanced antenna and is often fed with a balun.

costs less to connect direct, the use of connectors offers several advantages.

Coaxial cable braid soaks up water like a sponge. If you do not adequately seal the antenna end of the feed line, water will find its way into the braid. Water in the feed line will lead to contamination, rendering the coax useless long before its normal lifetime is up. It is not uncommon for water to drip from the end of the coax inside the shack after a year or so of service if the antenna connection is not properly waterproofed. Use of a PL-259/SO-239 combination (or connector of your choice) makes the task of waterproofing connections much easier. Another advantage to using the PL-259/SO-239 combination is that feed line replacement is much easier, should that become necessary.

Whether you use coaxial cable, ladder line, or twin lead to feed your antenna, an often-overlooked consideration is the mechanical strength of the connection. Wire antennas and feed lines tend to move a lot in the breeze, and unless the feed line is attached securely, the connection will weaken with time. The resulting failure can range from a frustrating intermittent electrical connection to a complete separation of feed line and antenna. **Fig 20.15** illustrates several different ways of attaching the feed line to the antenna. An idea for supporting ladder line is shown in **Fig 20.16**.

Putting It Together

Fig 20.17 shows details of antenna construction. Although a dipole is used for the examples illustrated here, the techniques illustrated here apply to any type of wire antenna. **Table 20.5** shows dipole lengths for the amateur HF bands.

How well you put the pieces together is second only to the ultimate strength of the materials used in determining how well your antenna will work over the long term. Even the smallest details, such as how you connect the wire to the insulators (Fig 20.17A), contribute significantly to antenna longevity. By using plenty of wire at the insulator and wrapping it tightly, you will decrease the possibility of the wire pulling loose in the wind. There is no need to solder the wire once it is wrapped. There is no electrical connection here, only mechanical. The high heat needed for soldering can anneal the wire, significantly weakening it at the solder point.

Similarly, the feed-line connection at the center insulator should be made to the antenna wires after they have been secured to the insulator (Fig 20.17B). This way, you will be assured of a good electrical connection between the antenna and feed line without compromising the mechanical strength. Do a good job of soldering the antenna and feed-line connections. Use a heavy iron or a torch, and be sure to

Table 20.5

Dipole Dimensions for Amateur Bands

Freq MHz	Overall Length	Leg Length
28.4	16' 6"	8' 3"
21.1	22' 2"	11' 1"
18.1	25' 10"	12' 11"
14.1	33' 2"	16' 7"
10.1	46' 4"	23' 2"
7.1	65' 10"	32' 11"
3.6	130' 0"	65' 0"

clean the materials thoroughly before starting the job. Proper planning should allow you to solder indoors at a workbench, where the best possible joints may be made. Poorly soldered or unsoldered connections will become headaches as the wire oxidizes and the electrical integrity degrades with time. Besides degrading your antenna performance, poorly made joints can even be a cause of TVI because of rectification. Spray paint the connections with acrylic for waterproofing.

If made from the right materials, the dipole should give a builder years of maintenance-free service—unless of course a tree falls on it. As you build your antenna, keep in mind that if you get it right the first time, you won't have to do it again for a long time.

A 135-FT MULTIBAND CENTER-FED DIPOLE

An 80-m dipole fed with ladder line is a versatile antenna. If you add a wide-range matching network, you have a low-cost antenna system that works well across the entire HF spectrum. Countless hams have used one of these in single-antenna stations and for Field Day operations.

For best results place the antenna as high as you can, and keep the antenna and ladder line clear of metal and other conductive objects. Despite significant SWR on some bands, system losses are low. (See the **Transmission Lines** chapter.) You can

make the dipole horizontal, or you can install in as an inverted V. ARRL staff analyzed a 135-ft dipole at 50 ft above typical ground and compared that to an inverted V with the center at 50 ft, and the ends at 10 ft. The results show that on the 80-m band, it won't make much difference which configuration you choose. (See **Fig 20.18**.) The inverted V exhibits additional losses because of its proximity to ground.

Fig 20.19 shows a comparison between

a 20-m flat-top dipole and the 135-ft flat-top dipole when both are placed at 50 ft above ground. At a 10° elevation angle, the 135-ft dipole has a gain advantage. This advantage comes at the cost of two deep, but narrow, nulls that are broadside to the wire.

Fig 20.20 compares the 135-ft dipole to the inverted-V configuration of the same antenna on 14.1 MHz. Notice that the inverted-V pattern is essentially omnidirec-

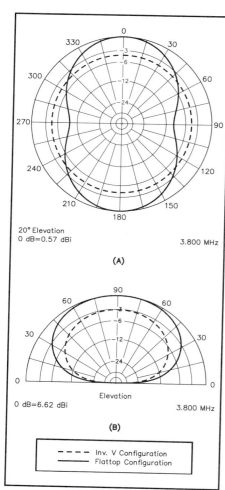

Fig 20.18—Patterns on 80 m for 135-ft, center-fed dipole erected as a horizontal dipole at 50 ft, and as an inverted V with the center at 50 ft and the ends at 10 ft. The azimuth pattern is shown at A, where conductor lies in the 90° to 270° plane. The elevation pattern is shown at B, where conductor comes out of paper at right angle. At the fundamental frequency the patterns are not markedly different.

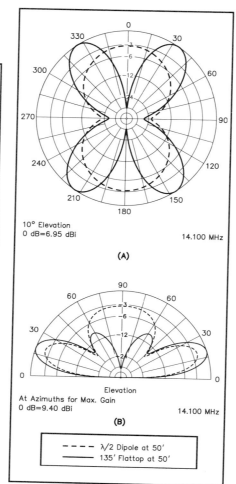

Fig 20.19—Patterns on 20 m comparing a standard 1/2-λ dipole and a multiband 135-ft dipole. Both are mounted horizontally at 50 ft. The azimuth pattern is shown at A, where conductors lie in the 90° to 270° plane. The elevation pattern is shown at B. The longer antenna has four azimuthal lobes, centered at 35°, 145°, 215°, and 325°. Each is about 2 dB stronger than the main lobes of the 1/2-λ dipole. The elevation pattern of the 135-ft dipole is for one of the four maximum-gain azimuth lobes, while the elevation pattern for the 1/2-λ dipole is for the 0° azimuthal point.

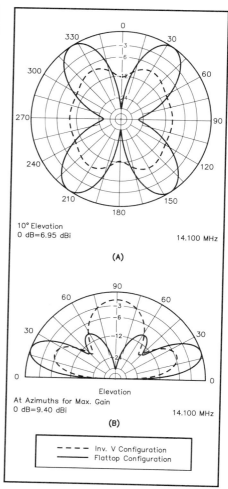

Fig 20.20—Patterns on 20 m for two 135-ft dipoles. One is mounted horizontally as a flat-top and the other as an inverted V with 120° included angle between the two legs. The azimuth pattern is shown at A, and the elevation pattern is shown at B. The inverted V has about 6 dB less gain at the peak azimuths, but has a more uniform, almost omnidirectional, azimuthal pattern. In the elevation plane, the inverted V has a fat lobe overhead, making it a somewhat better antenna for local communication, but not quite so good for DX contacts at low elevation angles.

tional. That comes at the cost of gain, which is less than that for a horizontal flat-top dipole.

As expected, patterns become more complicated at 28.4 MHz. As you can see in **Fig 20.21**, the inverted V has the advantage of a pattern with slight nulls, but with reduced gain compared to the flat-top configuration.

Installed horizontally, or as an inverted V, the 135-ft center-fed dipole is a simple antenna that works well from 3.5 to 30 MHz. Bandswitching is handled by a Transmatch that is located near your operating position.

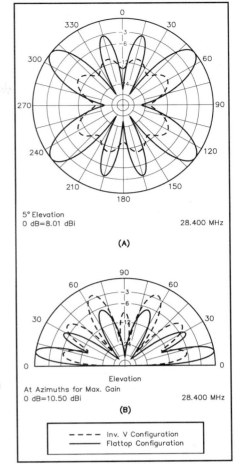

(A)

(B)

Fig 20.21—Patterns on 10 m for 135-ft dipole mounted horizontally and as an inverted V, as in Fig 20.20. The azimuth pattern is shown at A, and the elevation pattern is shown at B. Once again, the inverted-V configuration yields a more omnidirectional pattern, but at the expense of almost 8 dB less gain than the flat-top configuration at its strongest lobes.

Antenna Modeling by Computer

Modern computer programs have made it a *lot* easier for a ham to evaluate antenna performance. The elevation plots for the 135-ft long center-fed dipole were generated using a sophisticated computer program known as *NEC*, short for "Numerical Electromagnetics Code." *NEC* is a general-purpose antenna modeling program, capable of modeling almost any antenna type, from the simplest dipole to extremely complex antenna designs. Various mainframe versions of *NEC* have been under continuous development by US government researchers for several decades.

But because it is a general-purpose program, *NEC* can be very slow when modeling some antennas—such as long-boom, multi-element Yagis. There are other, specialized programs that work on Yagis much faster than *NEC*. Indeed, *NEC* has developed a reputation for being accurate (if properly applied!), but decidedly difficult to learn and use. A number of commercial software developers have risen to the challenge and created more "user-friendly" versions. Check the ads in *QST*.

NEC uses a "Method of Moments" algorithm. The mathematics behind this algorithm are pretty formidable to most hams, but the basic principle is simple. An antenna is broken down into a set of straight-line wire "segments." The fields resulting from the current in each segment and from the mutual interaction between segments are vector-summed in the far field to create azimuth and elevation-plane patterns.

The most difficult part of using a *NEC*-type of modeling program is setting up the antenna's geometry—you must condition yourself to think in three-dimensional coordinates. Each end point of a wire is represented by three numbers: an x, y and z coordinate. An example should help sort things out. See **Fig A**, showing a "model" for a 135-foot center-fed dipole, made of #14 wire placed 50 ft above flat ground. This antenna is modeled as a single, straight wire.

For convenience, ground is located at the *origin* of the coordinate system, at (0, 0, 0) feet, directly under the center of the dipole. The dipole runs parallel to, and above, the y-axis. Above the origin, at a height of 50 feet, is the dipole's feedpoint. The "wingspread" of the dipole goes toward the left (that is, in the "negative y" direction) one-half the overall length, or −67.5 ft. Toward the right, it goes +67.5 ft. The "x" dimension of our dipole is zero. The dipole's ends are thus represented by two points, whose coordinates are: (0, −67.5, 50) and (0, 67.5, 50) ft. The thickness of the antenna is the diameter of the wire, #14 gauge.

To run the program you must specify the number of segments into which the dipole is divided for the method-of-moments analysis. The guideline for setting the number of segments is to use at least 10 segments per half-wavelength. In Fig A, our dipole has been divided into 11 segments for 80-m operation. The use of 11 segments, an odd rather than an even number such as 10, places the dipole's feedpoint (the "source" in *NEC*-parlance) right at the antenna's center and at the center of segment number six.

Since we intend to use our 135-foot long dipole on all HF amateur bands, the number of segments used actually

(Continued on next page)

should vary with frequency. The penalty for using more segments in a program like *NEC* is that the program slows down roughly as the square of the segments—double the number and the speed drops to a fourth. However, using too few segments will introduce inaccuracies, particularly in computing the feed-point impedance. The commercial versions of *NEC* handle such nitty-gritty details automatically.

Let's get a little more complicated and specify the 135-ft dipole, configured as an inverted-V. Here, as shown in **Fig B**, you must specify *two* wires. The two wires join at the top, (0, 0, 50) ft. Now the specification of the source becomes more complicated. The easiest way is to specify two sources, one on each end segment at the junction of the two wires. If you are using the "native" version of *NEC*, you may have to go back to your high-school trigonometry book to figure out how to specify the end points of our "droopy" dipole, with its 120° included angle. Fig B shows the details, along with the trig equations needed.

So, you see that antenna modeling isn't entirely a cut-and-dried procedure. The commercial programs do their best to hide some of the more unwieldy parts of *NEC*, but there's still some art mixed in with the science. And as always, there are trade-offs to be made—segments versus speed, for example.

However, once you do figure out exactly how to use them, computer models are wonderful tools. They can help you while away a dreary winter's day, designing antennas on-screen—without having to risk life and limb climbing an ice-covered tower. And in a relatively short time a computer model can run hundreds, or even thousands, of simulations as you seek to optimize an antenna for a particular parameter. Doesn't that sound better than trying to optimally tweak an antenna by means of a thousand cut-and-try measurements, all the while hanging precariously from your tower by a climbing belt?!—*R. Dean Straw, N6BV, Senior Assistant Technical Editor*

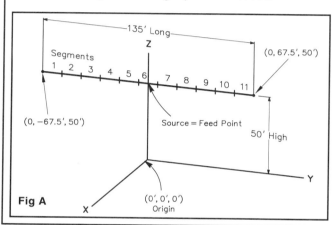

Fig A

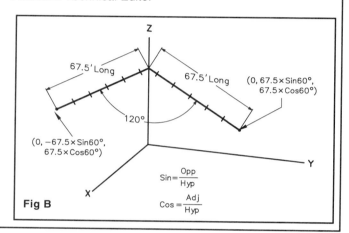

Fig B

Computer-aided Design of Loaded Short-doublet Antennas

By Richard Ellers, K8JLK, 426 Central Pkwy, SE, Warren, OH 44483-6213

◊ I've devised a BASIC program[1] you can use to design shortened doublet (dipole) antennas, using loading coils of known inductance. The program uses common BASIC commands and should run as-is on any IBM PC-compatible or Apple II personal computer. This program is an offshoot and adaptation of previously published charts, a formula and a BASIC program. The charts and formula, published in *QST*,[2] and the original BASIC program, published in *CQ*,[3] were devised to calculate the inductances required to resonate a short doublet, given the overall antenna length, the coil spacing and the element diameter.

The development of my program began when I decided to build a 2-element phased array of short doublets, using four surplus mobile coils I'd bought at a hamfest. Checking through my ham magazines, I found Dick (K5QY) Sander's ingenious BASIC version of Jerry Hall's (then K1PLP, now K1TD) comprehensive work on the design of off-center-loaded antennas. However, I was in the opposite position: *I had the inductance values in hand* and needed instead to figure *where to place the coils* in an antenna of a certain size.

Not being very adept at higher math, I didn't even attempt to rewrite Jerry's formula or K5QY's program. Instead, I revised K5QY's program by adding a computer routine known as a *binary search*. Although more often used to seek data in large files, my program uses a binary search as a form of computerized empirical determination (more commonly known as cut-and-try). The search is accurate—and rapid. In just seconds, the routine finds the coil spacing that matches the desired

antenna to the existing inductance.

I wrote the program to find the coil spacing for a single antenna (and for a series of 10 antennas) between two specified lengths. Incidentally, because the program actually calculates the center-to-coil distance (as did K5QY's version), it can be used to design short vertical antennas, but you must enter *twice the desired overall length*. As with K5QY's program, mine handles antenna elements of wire or tubing, the diameters being entered in decimal fractions of an inch where necessary. To keep the program short and simple, I omitted K5QY's optional wire table. (You can find such a wire table in Chapter 24, **Component Data**, or use the following as a guide: 22-gauge wire has a diameter of 0.025 inch; 10-gauge wire's diameter is 0.101 inch.)

Program error traps catch (and explain) parameter combinations that don't match the given inductor. One trap catches dimensions that would normally put the program in an endless loop. I've tested my program for many configurations and frequencies using inductances first determined by K5QY's program: The results matched *every* time.

Notes

[1] The software contains ASCII QBASIC files for both IBM-compatible and Mac computers in file ELLERS.EXE. It is available from the ARRL BBS and internet sites. See page viii.
[2] J. Hall, "Off-Center-Loaded Antennas," *QST*, Sep 1974, pp 28-34 and 58.
[3] D. Sander, "A Computer Designed Loaded Dipole Antenna," *CQ*, Dec 1981, p 44.

A TRAP DIPOLE FOR 40, 80 AND 160 M

This antenna was designed for amateurs with limited space who also wanted to operate the low bands. It was first described in July 1992 *QST* by A. C. Buxton, W8NX, and features innovative coaxial-cable traps.

Fig 20.25 shows the antenna layout; it is resonant at 1.865, 3.825, and 7.225 MHz. The antenna is made of #14 stranded wire and two pairs of coaxial traps. Construction is conventional in most respects, except for the high inductance-to-capacitance (L/C) ratio that results from the unique trap construction.

The traps use two series-connected coil layers, wound in the same direction using RG-58 coaxial cable's center conductor, together with the insulation over the center conductor. The black outer jacket from the cable is stripped and discarded. The shield braid is also removed from the cable (pushing is easier than pulling the shield off). No doubt you will want to save the braid for use in other projects. RG-58 with a stranded center conductor is best for this project. **Fig 20.26** shows the traps. The 3.8-MHz trap is shown with the weatherproofing cover of electrical tape removed to show construction details.

Precautions and Trap Specifications

With this trap-winding configuration, there are two thicknesses of coax dielectric material between adjacent turns, which doubles the breakdown voltage of the traps. The transformer action of the two windings gives a second doubling of the trap-voltage rating, bringing it to 5.6 kV.

The 7-MHz traps have 33 µH of inductance and 15 pF of capacitance, and the 3.8-MHz traps have 74 µH of inductance and 24 pF of capacitance. The trap Qs are over 170 at their design frequencies.

These traps are suitable for high-power operation. Do not use RG-8X or any other foam-dielectric cable for making the traps. Breakdown voltage is less for foam dielectric, and the center conductor tends to migrate through the foam when there is a short turn radius.

Loading caused by the traps causes a reduced bandwidth for any trap dipole compared to a half-wave dipole. This antenna covers 65 kHz of 160 m, 75 kHz of 80 m, and the entire 40-m band with less 2:1 SWR.

Construction

Although these traps are similar in many ways to other coaxial-cable traps, the shield winding of the common coax-cable trap has been replaced by a top winding that fits snugly into the grooves formed by the bottom layer. Capacitance is reduced to 7.1 pF per ft, compared to 28.5 pF per ft with conventional coax traps made from RG-58. Trap reactance can be up to four times greater than that provided by con-

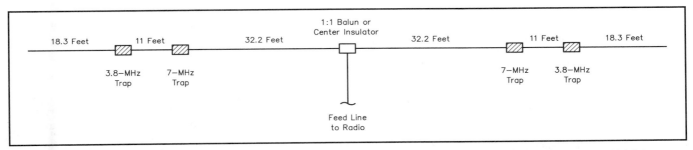

Fig 20.25—The W8NX trap dipole resonates in the SSB portions of the 40, 80 and 160-m bands. The antenna is 124 ft long.

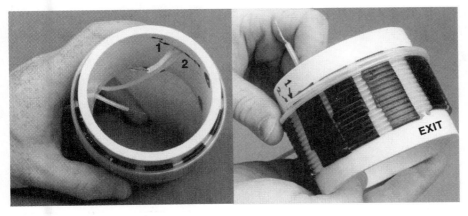

Fig 20.26—The W8NX coaxial-cable traps use two layered windings in series to provide an unusually high inductance-to-capacitance ratio, higher Q, and twice the breakdown voltage of single-layer traps. At A is shown an inside view of a W8NX two-layer trap, showing how the windings enter and exit the form. Two holes at each end of the PVC form pass the windings in and out of the form. At B, an outside view of a partially assembled W8NX trap. The bottom winding starts at hole "1" and reenters the form just below the "EXIT" hole. The wire then comes back through the inside of the form to hole "2," where it comes back out to make the second winding in the same direction, on top of the first. It reenters the form at the "EXIT" hole. The black electrical tape holds the bottom winding in place from spreading as you wind the top layer over it. Other holes drilled in the ends of the form provide convenient points for the antenna wires to connect mechanically to the traps.

ventional coax-cable traps.

The coil forms are cut from PVC pipe. The 7-MHz trap form is made from 2-inch-ID pipe with an outer diameter of 2.375 inches. The 3.8-MHz trap form is made from 3-inch pipe with an outer diameter of 3.5 inches. The 7-MHz trap uses a 12.3-turn bottom winding and an 11.4-turn top winding. The 3.8-MHz trap uses a 14.3-turn bottom winding and a 13.4-turn top winding. All turns are close wound. The 40-m trap frequency is 7.17 MHz and the 80-m trap frequency is 3.85 MHz.

Use a #30 (0.128-inch) diameter drill for the feed-through holes in the PVC coil forms. The start and end holes of the 7-MHz traps are spaced 1.44 inches center to center, measured parallel to the trap center line. The holes in the 3.8-MHz traps are 1.66 inches apart. Wind the traps with a single length of coax center conductor. The lengths are 17.55 ft for the 7-MHz traps and 28.45 ft for the 3.8-MHz traps. These lengths include the trap pigtails and a few inches for fine tuning.

Use electrical tape to keep the turns of the inner-layer winding closely spaced during the winding process. This counteracts the tendency of the tension in the outer-layer winding to spread the bottom-layer turns. Stick the tape strips directly to the coil form before winding and then tightly loop them over and around the bottom layer before winding the outer layer. Use six or more tape strips for each trap. Other smaller holes may be drilled in the ends of the traps to provide a place for the antenna wires to make secure mechanical connections to the traps.

80-M BROADBAND DIPOLE WITH COAXIAL RESONATOR MATCH

This material has been condensed from an article by Frank Witt, AI1H, that appeared in April 1989 *QST*. A full technical description appears in *The ARRL Antenna Compendium, Volume 2*.

Fig 20.27 shows the detailed dimensions of the 3.5-MHz coaxial resonator match broadband dipole. Notice that the coax is an electrical quarter wavelength, has a short at one end, an open at the other end, a strategically placed crossover, and is fed at a tee junction. (The crossover is made by connecting the shield of one coax segment to the center conductor of the adjacent segment and by connecting the remaining center conductor and shield in a similar way.) At AI1H, the antenna is constructed as an inverted-V dipole with a 110° included angle and an apex at 60-ft. The measured SWR vs frequency is shown in **Fig 20.28**. Also in Fig 20.28 is the SWR characteristic for an uncompensated inverted-V dipole made from the same materials and positioned exactly as was the broadband version.

The antenna is made from RG-8 coaxial cable and #14 AWG wire, and is fed with 50-Ω coax. The coax should be cut so that the stub lengths of Fig 20.27 are within 1/2 inch of the specified values. PVC plastic pipe couplings and SO-239 UHF chassis connectors can be used to make the T and crossover connections, as shown in **Fig 20.29** at A and B. Alternatively, a standard UHF T connector and coupler can be used for the T, and the crossover may be a soldered connection (Fig 20.29C). Witt used RG-8 because of its ready availability, physical strength, power handling capability and moderate loss.

Cut the wire ends of the dipole about three ft longer than the lengths given in Fig 20.27. If there is a tilt in the

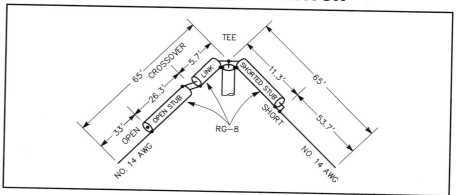

Fig 20.27—Coaxial-resonator-match broadband dipole for 3.5 MHz. The coax segment lengths total 1/4 λ. The overall length is the same as that of a conventional inverted-V dipole.

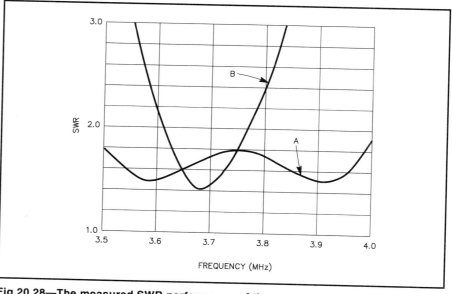

Fig 20.28—The measured SWR performance of the antenna of Fig 20.27, curve A. Also shown for comparison is the SWR of the same dipole without compensation, curve B.

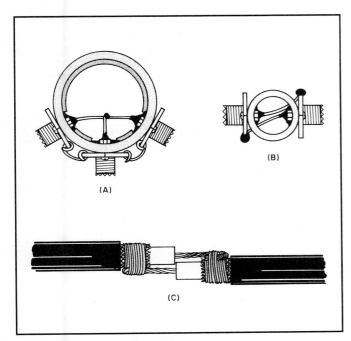

Fig 20.29—T and crossover construction. At A, a 2-inch PVC pipe coupling can be used for the T, and at B, a 1-inch coupling for the crossover. These sizes are the nominal inside diameters of the PVC pipe which is normally used with the couplings. The T could be made from standard UHF hardware (an M-358 T and a PL-258 coupler). An alternative construction for the crossover is shown at C, where a direct solder connection is made.

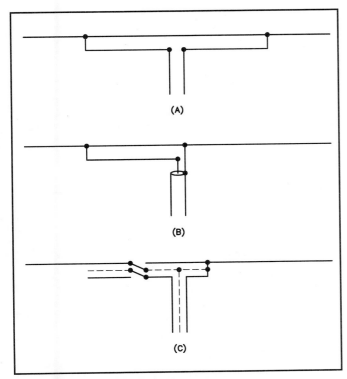

Fig 20.30—Dipole matching methods. At A, the T match; at B, the gamma match; at C, the coaxial resonator match.

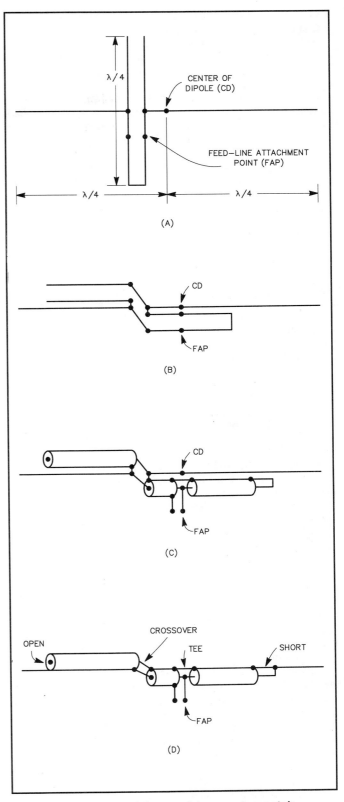

Fig 20.31—Evolution of the coaxial-resonator-match broadband dipole. At A, the resonant transformer is used to match the feed line to the off-center-fed dipole. The match and dipole are made collinear at B. At C, the balanced transmission-line resonator/transformer of A and B is replaced by a coaxial version. Because the shield of the coax can serve as a part of the dipole radiator, the wire adjacent to the coax match may be eliminated, D.

SWR-frequency curve when the antenna is first built, it may be "flattened" to look like the shape given in Fig 20.28 by increasing or decreasing the wire length. Each end should be lengthened or shortened by the same amount.

A word of caution: If the coaxial cable chosen is not RG-8 or equivalent, the dimensions will need to be modified. The following cable types have about the same characteristic impedance, loss and velocity factor as RG-8 and could be substituted: RG-8A, RG-10, RG-10A, RG-213 and RG-215. If the Q of the dipole is particularly high or the radiation resistance is unusually low because of different ground characteristics, antenna height, surrounding objects and so on, then different segment lengths will be required. In fact, if the dipole Q is too high, broadbanding is possible, but an SWR under 2:1 over the whole band cannot be achieved.

What is the performance of this broadband antenna relative to that of a conventional inverted-V dipole? Apart from the slight loss (about 1 dB at band edges, less elsewhere) because of the nonideal matching network, the broadband version will behave essentially the same as a dipole cut for the frequency of interest. That is, the radiation patterns for the two cases will be virtually the same. In reality, the dipole itself is not "broadband," but the coaxial resonator match provides a broadband match between the transmission line and the dipole antenna. This match is a remarkably simple way to broaden the SWR response of a dipole.

The Coaxial Resonator Match

The coaxial resonator match performs the same function as the T match and the gamma match; that is, matching a transmission line to a resonant dipole. These familiar matching devices as well as the coaxial resonator match are shown in **Fig 20.30**. The coaxial resonator match has some similarity to the gamma match in that it allows connection of the shield of the coaxial feed line to the center of the dipole, and it feeds the dipole off center. The coaxial resonator match has a further advantage: It can be used to broadband the antenna system while it is providing an impedance match.

The coaxial resonator match is a resonant transformer made from a quarter-wave long piece of coaxial cable. It is based on a technique used at VHF and UHF to realize a low-loss impedance transformation.

The Coaxial Resonator Match Broadband Dipole

Fig 20.31 shows the evolution of the

broadband dipole. Now it becomes clear why coaxial cable is used for the quarter-wave resonator/transformer; interaction between the dipole and the matching network is minimized. The effective dipole feedpoint is located at the crossover. In effect, the match is physically located "inside" the dipole. Currents flowing on the inside of the shield of the coax are associated with the resonator; currents flowing on the outside of the shield of the coax are the usual dipole currents. Skin effect provides a degree of isolation and allows the coax to perform its dual function. The wire extensions at each end make up the remainder of the dipole, making the overall length equal to one half-wave.

A useful feature of an antenna using the

coaxial resonator match is that the entire antenna is at the same dc potential as the feed line, thereby avoiding charge buildup on the antenna. Hence, noise and the potential of lightning damage are reduced.

A Model for DXers

The design of Fig 20.27 may be modified to yield a "3.5-MHz DX Special." In this case the band extends from 3.5 MHz to 3.85 MHz. Over that band the SWR is better than 1.6:1 and the matching network loss is less than 0.75 dB. See **Fig 20.32** for measured performance of a 3.5-MHz DX Special built and used by Ed Parsons, K1TR. Design dimensions for the DX Special are given in **Fig 20.33**.

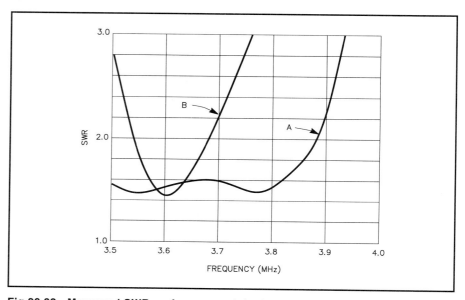

Fig 20.32—Measured SWR performance of the 3.5-MHz DX Special, curve A. Note the substantial broadbanding relative to a conventional uncompensated dipole, curve B.

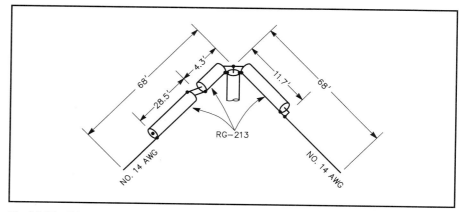

Fig 20.33—Dimensions for the 3.5-MHz DX Special, an antenna optimized for the phone and CW DX portions of the 3.5-MHz band.

A 40-M AND 15-M DUAL-BAND DIPOLE

Two popular ham bands, especially for Novice and Technician class operators, are those at 7 and 21 MHz. As mentioned earlier, dipoles have harmonic resonances at odd multiples of their fundamental resonances. Because 21 MHz is the third harmonic of 7 MHz, 7-MHz dipoles are harmonically resonant in the popular ham band at 21 MHz. This is attractive because it allows you to install a 40-m dipole, feed it with coax, and use it without an antenna tuner on both 40 and 15 m.

But there's a catch: The third harmonic of the Novice 40-m allocation (7100-7150 kHz) begins at 21,300 kHz; yet the Novice segment of 15 m is 21,100-21,200 kHz. As a result of this and other effects, a 40-m dipole does not provide a low SWR in the 40 *and* 15-m Novice segments without a tuner.

An easy fix for this, as shown in **Fig 20.34**, is to capacitively load the antenna about a quarter wavelength (at 21.1 MHz) away from the feedpoint in both wires. Known as *capacitance hats*, the simple loading wires shown lower the antenna's resonant frequency on 15 m without substantially affecting resonance on 40 m.

To put this scheme to use, first measure, cut and adjust the dipole to resonance at the desired 40-m frequency. Then, cut two 2-ft-long pieces of stiff wire (such as #12 or #14 house wire) and solder the ends of each one together to form two loops. Twist the loops in the middle to form figure-8s, and strip and solder the wires where they cross. Install these capacitance hats on the dipole by stripping the antenna wire (if necessary) and soldering the hats to the dipole about a third of the way out from the feedpoint (placement isn't critical) on each wire. To resonate the antenna on 15 m, adjust the loop shapes (*not while you're transmitting!*) until the SWR is acceptable in the desired segment of the 15-m band.

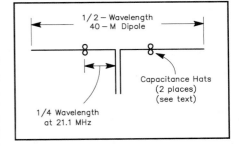

Fig 20.34—Figure-8-shaped capacitance hats made and placed as described in the text, can make your 40-m dipole resonate anywhere you like in the 15-m band.

A RESONANT FEED-LINE DIPOLE

This antenna, first described by James Taylor, W2OZH in August 1991 *QST*, uses a section of the feed line as part of the antenna. Taylor's design takes advantage of the fact that separate currents can flow on the inside and outside of the shield of a coaxial cable. Current flows from the feedpoint back along the outside of the coax's shield. At $1/4 \lambda$ from the feedpoint, an RF choke effectively stops current flow, and thus the dipole is formed.

The antenna, also known as an RFD, is shown in **Fig 20.35**. Dipole dimensions can be taken from Table 20.5. Length of coax and number of turns for the choke are given in **Table 20.7**.

The RFD has the advantage of being end fed. That means no feed line supported by the antenna. This makes it easy to erect, and makes the RFD particularly handy for portable operation. Like any dipole, the RFD should be installed as high, and in the clear, as possible.

Table 20.7

Choke Dimensions for RFD Antenna

Freq	RG-213, RG-8	RG-58
3.5	22 ft, 8 turns	20 ft, 6-8 turns
7	22 ft, 10 turns	15 ft, 6 turns
10	12 ft, 10 turns	10 ft, 7 turns
14	10 ft, 4 turns	8 ft, 8 turns
21	8 ft, 6-8 turns	6 ft, 8 turns
28	6 ft, 6-8 turns	4 ft, 6-8 turns

Wind the indicated length of coaxial feed line into a coil (like a coil of rope) and secure with electrical tape. Lengths are not highly critical.

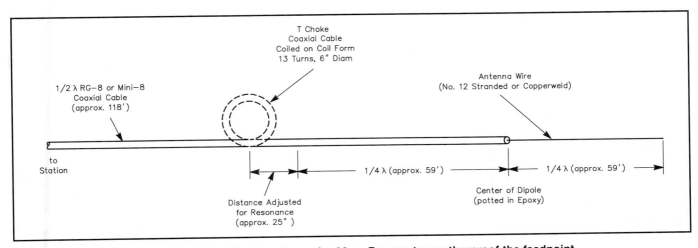

Fig 20.35—The RFD (resonant feed-line dipole) antenna for 80 m. Be sure to weatherproof the feedpoint.

A Simple Quad for 40 Meters

Many amateurs yearn for a 40-meter antenna with more gain than a simple dipole. While two-element rotary 40-meter beams are available commercially, they are costly and require fairly hefty rotators to turn them. This low-cost, single-direction quad is simple enough for a quick Field Day installation, but will also make a home station very competitive on the 40-meter band.

This quad uses a 2-inch outside diameter, 18-foot boom, which should be mounted no less than 60 feet high, preferably higher. (Performance tradeoffs with height above ground will be discussed later.) The basic design is derived from the N6BV 75/80-meter quad described in *The ARRL Antenna Compendium, Vol 5*. However, since this simplified 40-meter version is unidirectional and since it covers only one portion of the band (CW or Phone, but not both), all the relay-switched components used in the larger design have been eliminated.

The layout of the simple 40-meter quad at a boom height of 70 feet is shown in **Fig 20.36**. The wires for each element are pulled out sideways from the boom with black ⅛-inch Dacron rope designed specifically to withstand both abrasion and UV radiation. The use of the proper type of rope is very important—using a cheap substitute is not a good idea. You will not enjoy trying to retrieve wires that have become, like Charlie Brown's kite, hopelessly entangled in nearby trees, all because a cheap rope broke during a windstorm! At a boom height of 70 feet, the quad requires a "wingspread" of 140 feet for the side ropes. This is the same wingspread needed by an inverted-V dipole at the same apex height with a 90° included angle between the two legs.

The shape of each loop is rather unusual, since the bottom ends of each element are brought back close to the supporting tower. (These element ends are insulated from the tower and from each other). Having the elements near the tower makes fine-tuning adjustments much easier—after all, the ends of the loop wires are not 9 feet out, on the ends of the boom! The feed-point resistance with this loop configuration is close to 50 Ω, meaning that no matching network is necessary. By contrast, a more conventional diamond or square quad-loop configuration exhibits about a 100-Ω resistance.

Another bonus to this loop configuration is that the average height above ground is higher, leading to a slightly lower angle of radiation for the array and less loss because the bottom of each element is raised higher above lossy ground. The drawback to this unusual layout is that four more "tag-line" stay ropes are necessary to pull the elements out sideways at the bottom, pulling against the 10-foot separator ropes shown in Fig 20.36.

CONSTRUCTION

You must decide before construction whether you want coverage on CW (centered on 7050 kHz) or on Phone (centered on 7225 kHz), with roughly 120 kHz of coverage between the 2:1 SWR points. If the quad is cut for the CW portion of the band, it will have less than about a 3.5:1 SWR at 7300 kHz, as shown in **Fig 20.37**. The pattern will deteriorate to about a 7 dB F/B at 7300 kHz, with a reduction in gain of almost 3 dB from its peak in the CW band. It is possible to use a quad tuned for CW in the phone band if you use an antenna tuner to reduce the SWR and if you can take the reduction in performance. To put things in perspective, a quad tuned for CW but operated in the phone band will still work about as well as a dipole.

Next, you must decide where you want

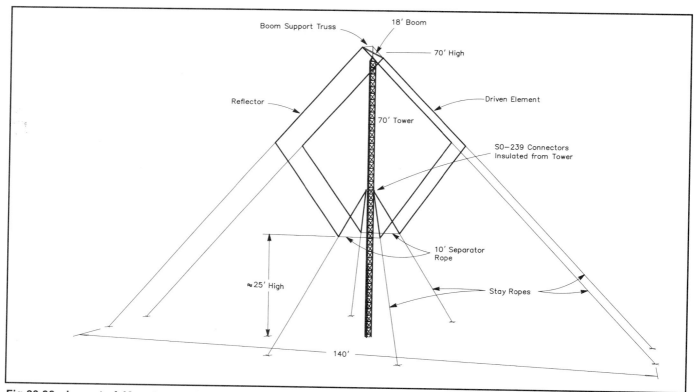

Fig 20.36—Layout of 40-meter quad with a boom height of 70 feet. The four stay ropes on each loop pull out each loop into the desired shape. Note the 10-foot separator rope at the bottom of each loop, which helps it hold its shape. The feed line is attached to the driven element through a choke balun, consisting of 10 turns of coax in a 1-foot diameter loop. You could also use large ferrite beads over the feed-line coax, as explained in Chapter 19. Both the driven element and reflector loops are terminated in SO-239 connectors tied back to (but insulated from) the tower. The reflector SO-239 has a shorted PL-259 normally installed in it. This is removed during fine-tuning of the quad, as explained in the text.

to point the quad. A DXer or contester in the USA might want to point this single-direction design to cover Europe and North Africa. For Field Day, a group operating on the East Coast would simply point it west, while their counterparts on the West Coast would point theirs east.

The mechanical requirements for the boom are not severe, especially since a top truss support is used to relieve stress on the boom due to the wires pulling on it from below. The boom is 18 feet long, made of 2-inch diameter aluminum tubing. You can probably find a suitable boom from a scrapped triband or monoband Yagi. You will need a suitable set of U-bolts and a mounting plate to secure the boom to the face of a tower. Or perhaps you might use lag screws to mount the boom temporarily to a suitable tree on Field Day! On a 70-foot high tower, the loop wires are brought back to the tower at the 37.5-foot level and tied there using insulators and rope. The lowest points of

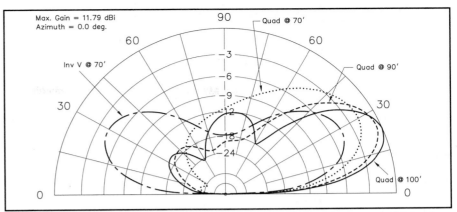

Fig 20.39—Comparisons of the elevation patterns for quads at boom heights of 70, 90 and 100 feet, referenced to an inverted-V dipole at 70 feet.

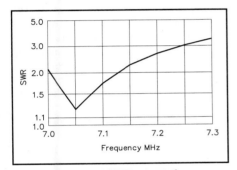

Fig 20.37—Plot of SWR versus frequency for a quad tuned for CW operation.

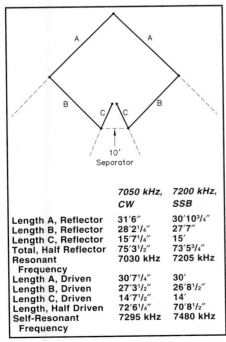

Fig 20.38—Dimensions of each loop, for CW or Phone operation.

	7050 kHz, CW	7200 kHz, SSB
Length A, Reflector	31′6″	30′10³⁄₄″
Length B, Reflector	28′2¹⁄₄″	27′7″
Length C, Reflector	15′7¹⁄₄″	15′
Total, Half Reflector	75′3¹⁄₂″	73′5³⁄₄″
Resonant Frequency	7030 kHz	7205 kHz
Length A, Driven	30′7¹⁄₄″	30′
Length B, Driven	27′3¹⁄₂″	26′8¹⁄₂″
Length C, Driven	14′7¹⁄₂″	14′
Length, Half Driven	72′6¹⁄₄″	70′8¹⁄₂″
Self-Resonant Frequency	7295 kHz	7480 kHz

the loops are located about 25 feet above ground for a 70-foot tower. **Fig 20.38** gives dimensions for the driven element and reflector for both the CW and the Phone portions of the 40-meter band.

Guy Wires

Anyone who has worked with quads knows they are definitely three-dimensional objects! You should plan your installation carefully, particularly if the supporting tower has guy wires, as most do. Depending on where the guys are located on the tower and the layout of the quad with reference to those guys, you will probably have to string the quad loops over certain guys (probably at the top of the tower) and under other guys lower down.

It is very useful to view the placement of guy wires using the "View Antenna" function in the *EZNEC* modeling program. This allows you to visualize the 3-D layout of an antenna. You can "rotate" yourself around the tower to view various aspects of the layout. *EZNEC* will complain about grounding wires directly but will still allow you to use the View Antenna function. Note also that it is best to insulate guy wires to prevent interaction between them and the antennas on a tower, but this may not be necessary for all installations.

FINE TUNING, IF NEEDED

We specify stranded #14 hard-drawn copper wire for the elements. During the course of installation, however, the loop wires could possibly be stretched a small amount as you pull and yank on them, trying to clear various obstacles. This may shift the frequency response and the performance slightly, so it is useful to have a tuning procedure for the quad when it is finally up in the air.

The easiest way to fine-tune the quad while on the tower is to use a portable, battery-operated SWR indicator (such as the Autek RF-1 or the MFJ-259) to adjust the reflector and the driven element

lengths for specific resonant frequencies. You can eliminate the influence of mutual coupling to the other element by open-circuiting the other element.

For convenience, each quad loop should be connected to an SO-239 UHF female connector that is insulated from but tied close to the tower. You measure the driven element's resonant frequency by first removing the shorted PL-259 normally inserted into the reflector connector. Similarly, the reflector's resonant frequency can be determined by removing the feed line normally connected to the driven element's feed point.

Obviously, it's easiest if you start out with extra wire for each loop, perhaps 6 inches extra on each side of the SO-239. You can then cut off wire in ¹⁄₂-inch segments equally on each side of the connector. This procedure is easier than trying to splice extra wire while up on the tower. Alligator clips are useful during this procedure, but just don't lose your hold on the wires! You should tie safety strings from each wire back to the tower. Prune the wire lengths to yield the resonant frequencies (±5 kHz) shown in Fig 20.38 and then solder things securely. Don't forget to reinsert the shorted PL-259 into the reflector SO-239 connector to turn it back into a reflector.

HIGHER IS BETTER

This quad was designed to operate with the boom at least 60 feet high. However, it will work considerably better for DX work if you can put the boom up even higher. **Fig 20.39** shows the elevation patterns for four antennas: a reference inverted-V dipole at 70 feet (with a 90° included angle between the two legs), and three quads, with boom height of 70, 90 and 100 feet respectively. At an elevation angle of 20°, typical for DX work on 40 meters, the quad at 100 feet has about a 5 dB advantage over an inverted-V dipole at 70 feet, and about a 3 dB advantage over a quad with a boom height of 70 feet.

Vertical Antennas

One of the more popular amateur antennas is the *vertical*. Although amateurs often use the term "vertical" somewhat loosely, it usually refers to a single radiating element placed vertically over the ground. A vertical is usually ¹/₄-λ long electrically and constructed out of wire or tubing. Many amateurs live on city lots with zoning restrictions preventing them from putting up tall towers with large Yagi arrays. Amateurs are also creative and ingenious—a vertical disguised to look like a flagpole can be both patriotic and a good radiator of RF!

Vertical antennas are omnidirectional radiators. This can be beneficial or detrimental, depending on the exact situation. On transmission there are no nulls in any direction, unlike most horizontal antennas. However, QRM on receive can't be nulled out from the directions that are not of interest, unless multiple verticals are used in an array.

When compared to horizontal antennas, verticals also suffer more acutely from two main types of losses—*ground return losses* for currents in the near field, and *far-field ground losses*. Ground losses in the near field can be minimized by using many ground radials. This is covered in detail in the following section.

Far-field losses are highly dependent on the conductivity and dielectric constant of the earth around the antenna, extending out as far as 100 λ from the base of the antenna. There is very little that someone can do to change the character of the ground that far away—other than moving to a small island surrounded by saltwater! Far-field losses greatly affect low-angle

radiation, causing the radiation patterns of practical vertical antennas to fall far short of theoretical patterns over "perfect ground," often seen in classical texts. **Fig 20.41** shows the elevation pattern response for two different 40-m quarter-wave verticals. One is placed over a theoretical infinitely large, infinitely conducting ground. The second is placed over an extensive radial system over average soil, having a conductivity of 5 mS/m and a dielectric constant of 13. This sort of soil is typical of heavy clay found in pastoral regions of the US mid-Atlantic states. At a 10° elevation angle, the real antenna losses are almost 6 dB compared to the theoretical one; at 20° the difference is about 3 dB. See *The ARRL Antenna Book* chapter on the effects of the earth for further details.

While real verticals over real ground are not a magic method to achieve low-angle radiation, cost versus performance and ease of installation are incentives that inspire many antenna builders. For use on the lower frequency amateur bands—notably 160 and 80 m—it is not always practical to erect a full-size vertical. At 1.8 MHz, a full-sized quarter-wave vertical is 130 ft high. In such instances it is often necessary to accept a shorter radiating element and use some form of *loading*.

Fig 20.42 provides curves for the physical height of verticals in wavelength versus radiation resistance and reactance. Although the plots are based on perfectly conducting ground, they show general trends for installations where many radials have been laid out to make a ground screen. As the radiator is made shorter, the

radiation resistance decreases—with 6 Ω being typical for a 0.1-λ high antenna. The lower the radiation resistance, the more the antenna efficiency depends on ground conductivity and the effectiveness of the ground screen. Also, the bandwidth decreases markedly as the length is reduced toward the left of the scale in Fig 20.42. It can be difficult to develop suitable matching networks when radiation resistance is very low.

Ground Systems

The importance of an effective ground system for vertical antennas cannot be emphasized too strongly. However, it is not always possible to install a radial network that approaches the ideal. The AM broadcast industry uses 120 buried radials as a standard radial ground system for their antennas. This is rarely practical for most amateurs, and a compromise ground is certainly better than no ground at all. The amateur should experiment with whatever is physically possible when working with vertical antennas. Even modest radial systems can produce contacts, even if they don't consistently crack big pileups!

Although the matter of less-than-optimum ground systems could be debated almost endlessly, some practical rules of thumb are in order for those wishing to erect vertical antennas. Generally a large number of shorter radials offers a better ground system than a few longer ones. For example, 8 radials of ¹/₈ λ are preferred over 4 radials of ¹/₄ λ. If the physical height of the vertical is an ¹/₈ λ, the radial wires should be of the same length and dispersed uniformly from the base.

The conductor size of the radials is not especially significant. Wire gauges from #4 to #20 have been used successfully by amateurs. Copper wire is preferred, but where soil is low in acid (alkali), aluminum wire can be used. The wires may be bare or insulated, and they can be laid on the earth's surface or buried a few inches below ground. Insulated wires will have greater longevity by virtue of reduced corrosion and dissolution from soil chemicals. The amateur should bury as much ground wire as time and budget permit. Some operators have literally miles of wire buried radially beneath their vertical antennas.

When property dimensions do not allow a classic installation of equally spaced radial wires, they can be placed on the ground as space permits. They may run away from the antenna in only one or two compass directions. They may be bent to fit on your property.

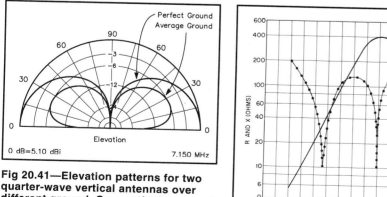

Fig 20.41—Elevation patterns for two quarter-wave vertical antennas over different ground. One vertical is placed over "perfect" ground, and the other is placed over average ground. The far-field response at low elevation angles is greatly affected by the quality of the ground—as far as 100 λ away from the vertical antenna.

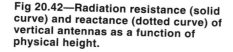

Fig 20.42—Radiation resistance (solid curve) and reactance (dotted curve) of vertical antennas as a function of physical height.

A single ground rod, or group of them bonded together, is seldom as effective as a collection of random-length radial wires. In some instances a group of short radial wires can be used in combination with ground rods driven into the soil near the base of the antenna. Bear in mind, though, that RF currents at MF and HF seldom penetrate the earth more than several inches. If a metal fence skirts the property it can be used as part of the ground system. Rolls of galvanized "chicken wire" fencing fanned out from the base of the vertical make good ground systems, especially when used in conjunction with longer wire radials. Six 30-ft long by 3-ft wide rolls of chicken wire make a good ground screen for a 160-m vertical, backed up with 6 or more 130-ft long radials. A good rule is to use anything that will serve as a ground when developing a radial ground system.

All radial wires should be connected together at the base of the vertical antenna. The electrical bond needs to be of low resistance. Best results will be obtained when the wires (and chicken-wire screen, if used) are soldered together at the junction point. When a grounded vertical is used, the ground wires should be affixed securely to the base of the driven element. A lawn edging tool is excellent for cutting slits in grass sod or in soil when laying radial wires.

Ground return losses are lower when vertical antennas and their radials are elevated above ground, a point that is well-known by those using *ground plane* antennas on their roofs. Even on 160 or 80 m, effective vertical antenna systems can be made with as few as four quarter-wave long radials elevated 10 to 20 ft off the ground.

Full-Size Vertical Antennas

When it is practical to erect a full-size $1/4 \lambda$ vertical antenna, the forms shown in **Fig 20.43** are worthy of consideration. The example at A is the well-known *vertical ground plane*. The ground system consists of four above-ground radial wires. The length of the radials and the driven element is derived from the standard equation

$$L \text{ (ft)} = \frac{234}{f\text{(MHz)}} \qquad (6)$$

With four equidistant radial wires drooped at approximately 45° (Fig 20.43A), the feed-point impedance is roughly 50 Ω. When the radials are at right angles to the radiator (Fig 20.43B) the impedance approaches 36 Ω. Besides minimizing ground return losses, another major advantage in this type of vertical antenna over a ground-mounted type is

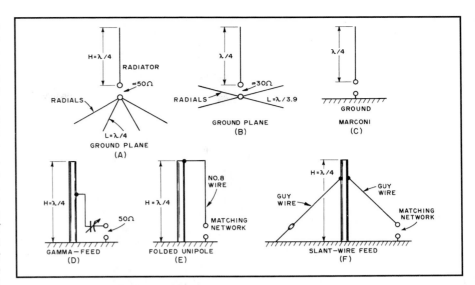

Fig 20.43—Various types of vertical antennas.

that the system can be elevated well above nearby conductive objects (power lines, trees, buildings and so on). When drooping radials are used, they can also serve as guy wires for the mast that supports the antenna. The coax shield braid is connected to the radials, and the center conductor to the driven element.

The *Marconi* vertical antenna shown in Fig 20.43C is the classic form taken by a ground-mounted vertical. It can be grounded at the base and shunt fed, or it can be isolated from ground, as shown, and series fed. As always, this vertical antenna depends on an effective ground system for efficient performance. If a perfect ground were located below the antenna, the feed impedance would be near 36 Ω. In a practical case, owing to imperfect ground, the impedance is more apt to be in the vicinity of 50 to 75 Ω.

A gamma feed system for a grounded $1/4$-λ vertical is presented in Fig 20.43D. Some rules of thumb for arriving at workable gamma-arm and capacitor dimensions are to make the rod length 0.04 to 0.05 λ, its diameter $1/3$ to $1/2$ that of the driven element and the center-to-center spacing between the gamma arm and the driven element roughly 0.007 λ. The capacitance of C1 at a 50-Ω matched condition will be about 7 pF per meter of wavelength. The absolute value of C1 will depend on whether the vertical is resonant and on the precise value of the radiation resistance. For best results, make the radiator approximately 3% shorter than the resonant length.

Amateur antenna towers lend themselves to use as shunt-fed verticals, even though an HF-band beam antenna is usu-

ally mounted on the tower. The overall system should be close to resonance at the desired operating frequency if a gamma feed is used. The HF-band beam will contribute somewhat to *top loading* of the tower. The natural resonance of such a system can be checked by dropping a #12 or #14 wire from the top of the tower (making it common to the tower top) to form a folded unipole (Fig 20.43E). A four- or five-turn link can be inserted between the lower end of the drop wire and the ground system. A dip meter is then inserted in the link to determine the resonant frequency. If the tower is equipped with guy wires, they should be broken up with strain insulators to prevent unwanted loading of the vertical. In such cases where the tower and beam antennas are not able to provide $1/4$-λ resonance, portions of the top guy wires can be used as top-loading capacitance. Experiment with the guy-wire lengths (using the dip-meter technique) while determining the proper dimensions.

A folded-unipole is depicted at E of Fig 20.43. This system has the advantage of increased feed-point impedance. Furthermore, a Transmatch can be connected between the bottom of the drop wire and the ground system to permit operation on more than one band. For example, if the tower is resonant on 80 m, it can be used as shown on 160 and 40 m with reasonable results, even though it is not electrically long enough on 160. The drop wire need not be a specific distance from the tower, but you might try spacings between 12 and 30 inches.

The method of feed shown at Fig 20.43F is commonly referred to as "slant-wire feed." The guy wires and the tower com-

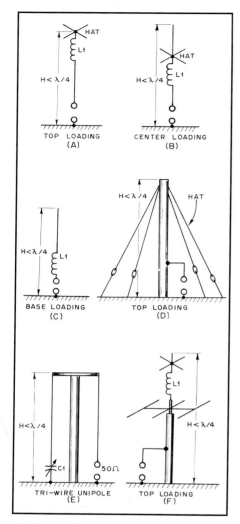

Fig 20.44—Vertical antennas that are less than one-quarter wavelength in height.

bine to provide quarter-wave resonance. A matching network is placed between the lower end of one guy wire and ground and adjusted for an SWR of 1:1. It does not matter at which level on the tower the guy wires are connected, assuming that the Transmatch is capable of effecting a match to 50 Ω.

Physically Short Verticals

A group of short vertical radiators is presented in **Fig 20.44**. Illustrations A and B are for top and center loading. A capacitance hat is shown in each example. The hat should be as large as practical to increase the radiation resistance of the antenna and improve the bandwidth. The wire in the loading coil is chosen for the largest gauge consistent with ease of winding and coil-form size. The larger wire diameters will reduce the resistive (I^2R) losses in the system. The coil-form material should have a medium or high dielectric constant. Phenolic or fiberglass tubing is entirely adequate.

A base-loaded vertical is shown at C of Fig 20.44. The primary limitation is that the high current portion of the vertical exists in the coil rather than the driven element. With center loading, the portion of the antenna below the coil carries high current, and in the top-loaded version the entire vertical element carries high current. Since the high-current part of the antenna is responsible for most of the radiating, base loading is the least effective of the three methods. The radiation resistance of the coil-loaded antennas shown is usually less than 16 Ω.

A method for using guy wires to top load a short vertical is illustrated in Fig 20.44D. This system works well with gamma feed. The loading wires are trimmed to provide an electrical quarter wavelength for the overall system. This method of loading will result in a higher radiation resistance and greater bandwidth than the systems shown at A through C. If an HF-band or VHF array is at the top the tower, it will simply contribute to the top loading.

A three-wire unipole is shown at E. Two #8 drop wires are connected to the top of the tower and brought to ground level. The wires can be spaced any convenient distance from the tower—normally 12 to 30 inches from one side. C1 is adjusted for best SWR. This type of vertical has a fairly narrow bandwidth, but because C1 can be motor driven and controlled from the operating position, frequency changes can be accomplished easily. This technique will not be suitable for matching to 50-Ω line unless the tower is less than an electrical quarter wavelength high.

A different method for top loading is shown at F. Barry Boothe, W9UCW, described this method in December 1974 *QST*. An extension is used at the top of the tower to effect an electrical quarter-wavelength vertical. L1 is a loading coil with sufficient inductance to provide antenna resonance. This type of antenna lends itself nicely to operation on 160 m.

A method for constructing the top-loading shown in Fig 20.44F is illustrated in **Fig 20.45**. Pipe section D is mated with the mast above the HF-band beam antenna. A loading coil is wound on solid Plexiglas rod or phenolic rod (item C), then clamped inside the collet (B). An aluminum slug (part A) is clamped inside item B. The top part of A is bored and tapped for a $3/8 \times 24$ stud. This permits a standard 8-ft stainless-steel mobile whip to be threaded into item A above the loading coil. The capacitance hat (Fig 20.45B) can be made from a ¼-inch-thick brass or aluminum plate. It may be round or square. Lengths of ⅛-inch brazing rod can be threaded and screwed into the edge of the aluminum plate. The plate contains a row of holes along its perimeter, each having been tapped for a 6-32 thread. The capacitance hat is affixed to item A by means of the 8-ft whip antenna. The whip will increase the effective height of the vertical antenna.

Cables and Control Wires on Towers

Most vertical antennas of the type shown in Fig 20.43 consist of towers, usually with HF or VHF beam antennas at the top. The rotator control wires and the coaxial feeders to the top of the tower will not affect antenna performance adversely. In fact, they become a part of the compos-

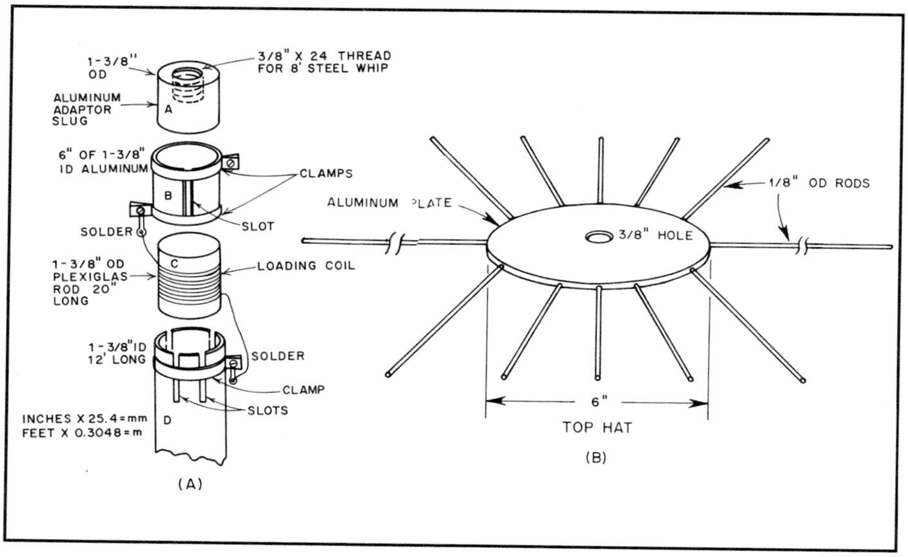

Fig 20.45—At A are the details for the tubing section of the loading assembly. Illustration B shows the top hat and its spokes. The longer the spokes, the better.

ite antenna. To prevent unwanted RF currents from following the wires into the shack, simply dress them close to the tower legs and bring them to ground level. This decouples the wires at RF. The wires should then be routed along the earth surface (or buried underground) to the operating position. It is not necessary to use bypass capacitors or RF chokes in the rotator control leads if this is done, even when maximum legal power is employed.

Trap Verticals

The 2-band trap vertical antenna of **Fig 20.46** operates in much the same manner as a trap dipole or trap Yagi. The notable difference is that the vertical is one half of a dipole. The radial system (inground or above-ground) functions as a ground plane for the antenna, and represents the missing half of the dipole. Once again, the more effective the ground system, the better will be the antenna performance.

Trap verticals usually are adjusted as $1/4$-λ radiators. The portion of the antenna below the trap is adjusted as a $1/4$-λ radiator at the higher proposed operating frequency. That is, a 20/15-m trap vertical would be a resonant quarter wavelength at 15 m from the feedpoint to the bottom of the trap. The trap and that portion of the antenna above the trap (plus the 15-m section below the trap) constitute the complete antenna during 20-m operation. But because the trap is in the circuit, the overall physical length of the vertical antenna will be slightly less than that of a single-band, full-size 20-m vertical.

Traps

The trap functions as the name implies: It traps the 15-m energy and confines it to the part of the antenna below the trap.

During 20-m operation it allows the RF energy to reach all of the antenna. The trap in this example is tuned as a parallel resonant circuit to 21 MHz. At this frequency it divorces the top section of the vertical from the lower section because it presents a high impedance (barrier) at 21 MHz. Generally, the trap inductor and capacitor have a reactance of 100 to 300 Ω. Within that range it is not critical.

The trap is built and adjusted separately from the antenna. It should be resonated at the center of the portion of the band to be operated. Thus, if one's favorite part of the 15-m band is between 21.0 and 21.1 MHz, the trap should be tuned to 21.05 MHz.

Resonance is checked by using a dip meter and detecting the dipper signal in a calibrated receiver. Once the trap is adjusted it can be installed in the antenna, and no further adjustment will be required. It is easy, however, to be misled after the system is assembled: Attempts to check the trap with a dip meter will suggest that the trap has moved much lower in frequency (approximately 5 MHz lower in a 20/15-m vertical). This is because the trap is part of the overall antenna, and the resultant resonance is that of the total antenna. Measure the trap separate from the rest of the antenna.

Multiband operation is quite practical by using the appropriate number of traps and tubing sections. The construction and adjustment procedure is the same, regardless of the number of bands covered. The highest frequency trap is always closest to the feed end of the antenna, and the next to lowest frequency trap is always the farthest from the feedpoint. As the operating frequency is progressively lowered, more traps and more tubing sections become a functional part of the antenna.

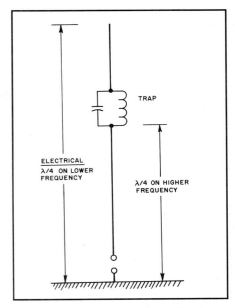

Fig 20.46—A two-band trap vertical antenna. The trap should be resonated by itself as a parallel resonant circuit at the center of the operating range for the higher frequency band. The reactance of either the inductor or the capacitor range from 100 to 300 Ω. At the lower frequency the trap will act as a loading inductor, adding electrical length to the total antenna.

Traps should be weatherproofed to prevent moisture from detuning them. Several coatings of high dielectric compound, such as Polystyrene Q Dope, are effective. Alternatively, a protective sleeve of heat-shrink tubing can be applied to the coil after completion. The coil form for the trap should be of high dielectric quality and be rugged enough to sustain stress during periods of wind.

TIPS ON INSTALLING AND CONNECTING TO GROUND RODS

◊ Driving a ground rod 8 feet into the ground with a sledgehammer batters the rod end into an ugly flare. Some types of ground clamps can't open far enough to slip over the enlarged rod. Of course, you can file, grind or saw off the flared end, but doing all of these things at ground level can be difficult.

Alternatively, you could slip the clamp over the rod before driving it into the ground; or use a clamp that opens far enough to pass over the flare. In the case of $1/2$-inch ground rods, however, clamps wide enough to pass the flare may not tighten adequately.

After considering these problems, I attached my shack ground wire to a $1/2$-inch ground rod as follows. I drilled a tap-size hole, about $3/4$ inch deep, into the rod top. I tapped this hole for a $1/4$-inch, standard thread. Driving in

a hex-head bolt permitted firm attachment of the wire to the rod end.

Although I did this by drilling only one hole, drilling a pilot hole—say, about $1/8$ inch in diameter—*before* driving in the rod would assist. Doing so would allow you to put the rod in a vise for stability and accurate drilling. Your sledgehammer may obliterate this hole, but you should be able to relocate it by probing with a center punch.

In any such drilling and thread cutting, use a sharp drill and lubricate it and your tap often while cutting. One more tip: If you have a welder friend, consider having him or her weld a $1/4$-inch bolt to the ground-rod top—after you've driven it in—to provide a stud for connections.—*A. W. Edwards, K5CN, Corpus Christi, Texas*

DUAL-BAND VERTICALS FOR 17/40 OR 12/30 M

Thanks to the harmonic relationships between the HF ham bands, many antennas can be made to do "double duty." The simple verticals described here cover two bands at once. Here's how to turn a 30-m $1/4$-λ vertical into a 0.625-λ vertical for the 12-m band, and a 40-m $1/4$-λ vertical into a 0.625-λ vertical for the 17-m band. These verticals were designed and constructed

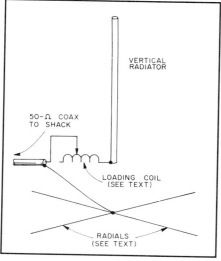

Fig 20.47—Dual-band vertical. Use a switch or relay to remove the loading coil from the circuit for lower frequency operation. Adjust the coil tap for best SWR on the higher-frequency band. The radial system should be as extensive as possible. See *The ARRL Antenna Book* for more information on ground systems for vertical antennas.

Table 20.8

Specifications for Dual-Band Verticals

Bands	Height	Required Matching Inductance (μH)
12 m & 30 m	23' 5"	0.99
17 m & 40 m	32' 3"	1.36

by John J. Reh, K7KGP. The write-up first appeared in April 1989 *QST*.

Construction Details

For the 30 and 12-m vertical, an old aluminum multiband vertical was cut to a length of 25 ft, 3 inches. This corresponds to a design frequency of 24.95 MHz. The length-to-diameter ratio is approximately 460. The input impedance of a vertical that is substantially longer than a $1/4$-λ (in this case 0.625 λ) is particularly sensitive to the λ/D ratio of the radiating element. If this antenna is duplicated with materials having a significantly different λ/D ratio, the results may be different.

After installing a good ground system, the input impedance was measured and found to have a resistance of about 50 Ω, and a capacitance of about -155 Ω (at 24.95 MHz). At 10.125 MHz, the input impedance was just under 50 Ω, and purely resistive. To tune out the reactance at 24.95 MHz, a series inductor is installed (see **Fig 20.47**) and tapped to resonance at the design frequency. The easiest way to find resonance is by measuring the antenna SWR. Use a good-quality coil for the series inductor. The recommended coil has a diameter of $2^{1/2}$ inches, and has 6 turns per inch (B&W stock no. 3029). Resonance on 12 m was established with $3^{1/4}$ turns. The SWR on 12 m is 1.1:1, and on 30 m, 1:1. To change bands from 12 to 30 m, move the coil tap to the end of the coil closest to the vertical element. Alternatively, a single-pole switch or remotely operated relay can be installed at the base of the vertical for bandswitching.

The Ground System

Maximum RF current density—and therefore maximum ground losses—for $1/4$-λ verticals occurs in the immediate area of the base of the antenna. Maximum return current ground loss for a 0.625-λ vertical occurs about $1/2$ λ away from the base of the antenna. It's important to have the

lowest possible losses in the immediate area for both types of verticals. In addition to a ground radial system, 6×6-ft aluminum ground screen is used at the base of the antenna. The screen makes a good tie point for the radials and conducts ground currents efficiently. Seventeen wire radials, each about 33 ft long, are spaced evenly around the antenna. More radials would probably work better. Each radial is bolted to the screen using corrosion-resistant #10-24 hardware. (Do not attempt to connect copper directly to aluminum. The electrical connection between the two metals will quickly deteriorate.) The radials can be made of bare or insulated wire. Make sure the ground screen is bolted to the ground side of the antenna with heavy-gauge wire. Current flow is fairly heavy at this point.

Table 20.8 gives specifications for the dual-band vertical. If your existing 40-m vertical is a few inches longer than 32 ft, 3 inches, try using it anyway—a few inches isn't too critical to performance on 17 m.

Automatic Bandswitching

In October 1989 *QST*, James Johnson, W8EUI, presented this scheme for automatic bandswitching of the 40/17-m vertical. Johnson shortened his 40-m vertical approximately 12 inches and found an inductance that gave him 40 and 17-m band operation with an SWR of less than 1.4:1 across each band. He used an inductor made from B&W air-wound coil stock (no. 3033). This coil is 3 inches in diameter, and has $3^{1/8}$ turns of #12 wire wound at 6 turns per inch, providing an inductance of about 2.8 μH. Johnson experimentally determined the correct tap position.

For the 30/12-m version, start with the vertical radiator 9 inches shorter than the value given in the table. In both cases, radiator height and inductance should be adjusted for optimum match on the two bands covered.

FERRITE SHIELD-CURRENT CHOKES CURE STRAY RF ON VERTICAL-ANTENNA TRANSMISSION LINES

◊ A vertical antenna can induce current on the shield of its coax feeder. The antenna induces current on the shield in the same manner that it induces current in the ground-system radials.

Since the shield is connected to station ground, RF current flowing on the feeder shield also flows on the outer surfaces of the equipment when the transmitter operates. This current can induce undesirable voltages in interconnecting cables, causing erratic operation of computer keyboards and power supplies. RF feedback also can occur

through the transmitter microphone circuit.

These problems occurred when I operated my kilowatt power amplifier and Kenwood TS-940S transceiver. (RF feedback in the TS-940 microphone preamplifier has been recognized by others and can be solved with bypass capacitors and RF chokes.) A better solution—one that will address all station problems caused by coax-shield RF currents—is to prevent the RF from getting into the station from the coax shield. Gary Peterson, K0CX, suggested that I do this by putting a ferrite-bead shield-current choke in my

A TREE-MOUNTED HF GROUNDPLANE ANTENNA

A tree-mounted, vertically polarized antenna may sound silly. But is it, really? Perhaps engineering references do not recommend it, but such an antenna does not cost much, is inconspicuous, and it works. This idea was described by Chuck Hutchinson, K8CH, in *QST* for September 1984.

The antenna itself is simple, as shown in **Fig 20.48**. A piece of RG-58 cable runs to the feedpoint of the antenna, and is attached to a porcelain insulator. Two radial wires are soldered to the coax-line braid at this point. Another piece of wire forms the radiator. The top of the radiator section is suspended from a tree limb or other convenient support, and in turn supports the rest of the antenna.

The dimensions for the antenna are given in **Fig 20.49**. (You can use the values given for leg lengths in Table 20.5.) All three wires of the antenna are $1/4$-λ long. This generally limits the usefulness of the antenna for portable operation to 7 MHz and higher bands, as temporary supports higher than 35 or 40 ft are difficult to come by. Satisfactory operation can be accomplished on 3.5 MHz with an inverted-L configuration of the radiator, if you can overcome the accompanying difficulty of erecting the antenna at the operating site. The tree-mounted vertical idea can also be used for fixed station installations to make an "invisible" antenna. Shallow trenches can be slit for burying the coax feeder and the radial wires. The radiator itself is difficult to see unless you are standing right next to the tree.

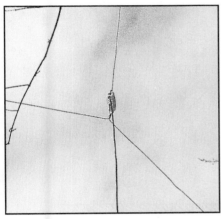

Fig 20.48—The feedpoint of the tree-mounted groundplane antenna. The opposite ends of the two radial wires may be connected to stakes or other convenient anchor points. Make sure that the radials are high enough so that people cannot come in contact with them.

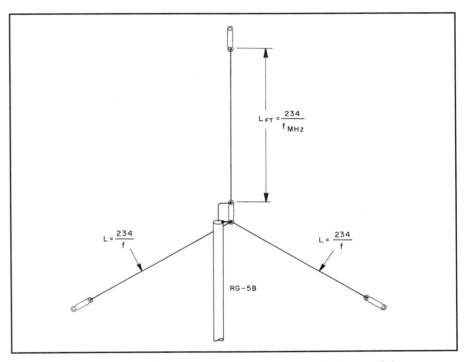

$$L_{FT} = \frac{234}{f_{MHz}}$$

$$L = \frac{234}{f}$$

$$L = \frac{234}{f}$$

RG-58

Fig 20.49—Dimensions and construction of the tree-mounted groundplane antenna.

---

antenna feed line where it enters my station.

A good source of the required components is The Wireman, Inc (see Address List in **References**). The required components are a package of 50 type FB73-2401 ferrite beads (Wireman part #912B), 15 inches of RG-303 coax, and two coax connectors (UHF connectors [PL-259s; the Wireman's #702] with inserts [UG-175s; the Wireman's #704] for installation on RG-58 coax, or N connectors intended for installation on RG-58 [UG-536B/Us, available as the Wireman's #737). Although the Wireman sells these parts bundled as part of a W2DU Balun kit [#833], the coax in the kit is too short to accommodate the connectors, so I recommend ordering the parts individually instead of buying them as The Wireman's W2DU Balun kit.

To assemble the choke, install one connector on the cable. Because the UG-536B/U connector is made for RG-58 (which has a larger diameter than RG-303), you must heat-shrink a piece of 1-inch-long, $3/16$-inch-diameter plastic tubing on the end of the coax before you install the connector. This will make the connector fit the cable tightly.

Next, put a $1/2$-inch-long piece of heat-shrink tubing on the cable about 1 inch from the connector to hold the beads in place. Then place the 50 beads on the cable. Put on another $1/2$-inch piece of tubing to anchor the other end of the beads, and then shrink a piece of $3/8$-inch-diameter tubing over the beads. Install the other connector on the cable. Install the choke on the transmission line where it leaves your station.

This should largely do away with problems related to stray RF traveling into your station on your feed line's coax shield. An idea for clubs: Build one or more chokes that members can borrow to test this fix's effectiveness before building their own chokes.

If your station is not in the immediate vicinity of your antenna and you have problems with stray RF with a dipole or inverted **V**, the problem may be due to lack of a balun or use of an ineffective balun. If this is the case, install the choke at the antenna feedpoint as discussed in W2DU's book *Reflections* rather than using the choke at the station end of your feed line.—*Bruce R. Palmer, KØWM, Edmond, Oklahoma*

Inverted L and Sloper antennas

This section covers variations on the vertical antenna. **Fig 20.50A** shows a flat-top T vertical. Dimension H should be as tall as possible for best results. The horizontal section, L, is adjusted to a length that provides resonance. Maximum radiation is polarized vertically despite the horizontal top-loading wire. A variation of the T antenna is depicted at B of Fig 20.50. This antenna is commonly referred to as an *inverted L*. Vertical member H should be as long as possible. L is added to provide an electrical quarter wavelength overall.

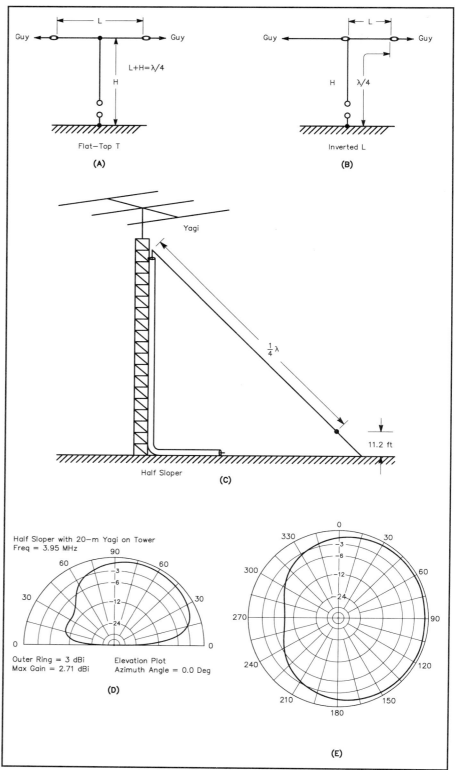

The Half-Sloper Antenna

Many hams have had excellent results with *half-sloper* antennas, while others have not had such luck. Work by ARRL Technical Advisor John S. Belrose, VE2CV, has brought some insight to the situation through computer modeling with *ELNEC* and antenna-range tests. The following is taken from VE2CV's Technical Correspondence in Feb 1991 *QST*, pp 39 and 40. Essentially, the half sloper is a top-fed vertical antenna worked against a ground plane (such as a grounded Yagi antenna) at the top of the tower. The tower acts as a reflector.

For half slopers, the input impedance, the resonant length of the sloping wire and the antenna pattern all depend on the tower height, the angle (between the sloper and tower) the type of Yagi and the Yagi orientation. Here are several configurations extracted from VE2CV's work:

At 160 m—use a 40-m beam on top of a 95-ft tower with a 55° sloper apex angle. The radiation pattern varies little with Yagi type. The pattern is slightly cardioid with about 8 dB front-to-back ratio at a 25° takeoff angle (see Fig 20.50D and E). Input impedance is about 50 Ω.

At 80 m—use a 20-m beam on top of a 50-ft tower with a 55° sloper apex angle. The radiation pattern and input impedance are similar to those of the 160-m half sloper.

At 40 m—use a 20-m beam on top of a 50-ft tower with a 55° sloper apex angle. The radiation pattern and impedance depend strongly on the azimuth orientation of the Yagi. Impedance varies from 76 to 127 Ω depending on Yagi direction.

Fig 20.50—Some variations in vertical antennas. D is the vertical radiation pattern in the plane of a half sloper, with the sloper to the right. E is the azimuthal pattern of the half sloper (90° azimuth is the direction of the sloping wire). Both patterns apply to 160- and 80-m antennas described in the text.

1.8-MHz INVERTED L

The antenna shown in **Fig 20.51** is simple and easy to construct. It is a good antenna for the beginner or the experienced 1.8 MHz DXer. Because the overall electrical length is greater than $1/4 \lambda$, the feed-point resistance is on the order of 50 Ω, with an inductive reactance. That reactance is canceled by a series capacitor, which for power levels up to the legal limit can be a air-variable capacitor with a voltage rating of 1500 V. Adjust antenna length and variable capacitor for lowest SWR.

A yardarm or a length of line attached to a tower can be used to support the vertical section of the antenna. (Keep the inverted L as far from the tower as is practical. Certain combinations of tower height and Yagi top loading can interact severely with the Inverted-L antenna—a 70-ft tower and a 5-element Yagi, for example.) For best results the vertical section should be as long as possible. A good ground system is necessary for good results.

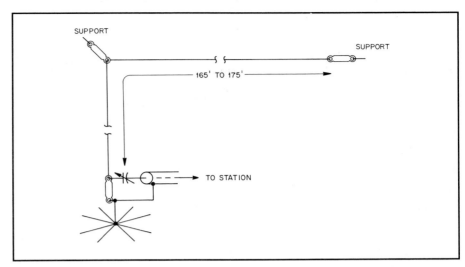

Fig 20.51—The 1.8-MHz inverted L. Overall wire length is 165 to 175 ft. The variable capacitor has a maximum capacitance of 500 to 800 pF.

THE AE6C DUAL-BAND INVERTED-L ANTENNA

In July 1991 *QST*, Dennis Monticelli, AE6C, described a dual-band inverted L. A drawing of his installation is shown in **Fig 20.52**. Dimensions and values are given in **Table 20.9**. For ease of explanation, only the 80/40-m version is described here. On 80 m the antenna is 0.375 λ long, which raises the antenna's radiation resistance and feed-point impedance, and thus decreases the effect of ground losses. (The feed-point exhibits inductive reactance—which is canceled with series capacitance.) For 40 m the antenna is 0.75 λ long, and is therefore resonant on that band. The pattern on 40 m, depending on how you install the antenna, resembles a combination of a vertical and a low dipole.

If you use a modest radial system with this antenna, the feed-point impedance is roughly 100 Ω on both bands. **Fig 20.53** and **Fig 20.54** show details of the matching system and the 1:2 (50 to 100 Ω) broadband transformer. The matching capacitor, C1, is not used on the higher frequency band, and so it is shorted out to operate there.

To tune the antenna, start with the dimensions given in Table 20.9. With the

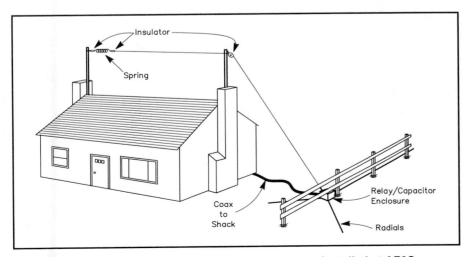

Fig 20.52—Drawing of the Dual-Band Inverted-L antenna installed at AE6C.

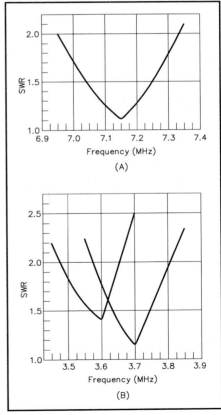

Fig 20.55—At A, 40-m SWR curve. At B, measured SWR curves for 80 m. Curves representing two different series capacitor values are shown.

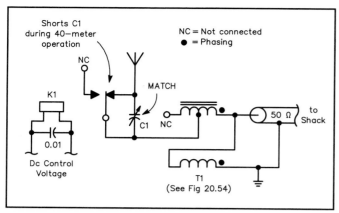

Fig 20.53—The resonating, impedance-matching, and band-switching circuitry required at the base of the inverted L. Fig 20.54 shows details of T1.

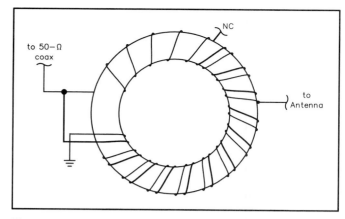

Fig 20.54—Winding details for constructing broadband bifilar transformer T1, which is wound on T-200-2 core. The primary is 16 turns of #14 enameled wire, and the secondary is 10 turns of #14 enameled wire tapped at about the eighth turn from the feed-line end.

Table 20.9

Recommended Wire Lengths and Capacitor Values*

Bands	Vertical Length (ft)	Horizontal Length (ft)	Total Length (ft)	Series Capacitor
80/40	32	64	96	≈ 100 pF
160/80	64	128	192	≈ 200 pF

* The total length is important, but the portions allocated to the vertical and horizontal members aren't critical.

series capacitor shorted, trim the wire length for best SWR at your 40-m frequency of interest. Unshort the capacitor and adjust it for best SWR at your frequency of interest on 80 m. You can further adjust SWR by changing transformer taps. (This will affect both bands.) As shown in **Fig 20.55**, a single capacitance value yields a 2:1 SWR bandwidth of approximately 225 kHz on 80 m. Different capacitance values can be used to cover other parts of the 80-m band.

You can make you own fixed-value, high-voltage, high-current capacitor from a piece of coax. For example, RG-8 (solid dielectric) exhibits 29.5 pF per ft and can withstand 4 kV at several amperes. Connect it by attaching one end of the inner conductor to the antenna and the braid (at the same end) to the feed system; leave the other end open. Start with about 4 ft of RG-8 and trim the coax for minimum SWR. (Make your cuts when you're not transmitting!) Other solid-dielectric cables are also suitable for this application. The **Transmission Lines** chapter gives capacitance per unit length in its table of coaxial-cable characteristics. The ends of the coaxial capacitors must be sealed to prevent water ingress.

Cork Your SO-239 Connectors!

◊ Those little red plastic caps (CAPLUGS) that are used to protect unused coax (SO-239) connectors from moisture, dirt and damage are expensive if you are lucky enough to find them. An alternative is to use plastic caps from inexpensive champagne bottles (Fig A). These caps fit perfectly.—*Roy Berkowitz, K3NFU, Monroeton, Pennsylvania*

Fig A—Champagne corks may be recycled into protective caps for SO-239 connectors.

SIMPLE, EFFECTIVE, ELEVATED GROUND-PLANE ANTENNAS

This article describes a simple and effective means of using a grounded tower, with or without top-mounted antennas, as an elevated ground-plane antenna for 80 and 160 m. It first appeared in a June 1994 *QST* article by Thomas Russell, N4KG.

FROM SLOPER TO VERTICAL

Recall the quarter-wavelength sloper, also known as the *half-sloper*. It consists of an isolated quarter wavelength of wire, sloping from an elevated feedpoint on a grounded tower. Best results were usually obtained when the feedpoint was somewhere below a top-mounted Yagi antenna. You feed a sloper by attaching the center conductor of a coaxial cable to the sloping wire and the braid of the cable to the tower leg. Now, imagine four (or more) slopers, but instead of feeding each individually, connect them together to the center conductor of a single feed line. *Voilà!* Instant elevated ground plane.

Now, all you need to do is determine how to tune the antenna to resonance. With no antennas on the top of the tower, the tower can be thought of as a fat conductor and should be approximately 4% shorter than a quarter wavelength in free space. Calculate this length and attach four insulated quarter-wavelength radials at this distance from the top of the tower. For 80 m, a feedpoint 65 ft below the top of an unloaded tower is called for. The tower guys must be broken up with insulators for all such installations. For 160 m, 130 ft of tower above the feedpoint is needed.

What can be done with a typical grounded-tower-and-Yagi installation? A top-mounted Yagi acts as a large capacitance hat, top loading the tower. Fortunately, top loading is the most efficient means of loading a vertical antenna.

The examples in **Fig 20.56** should give us an idea of how much top loading might be expected from typical amateur antennas. The values along the horizontal axis

tell us the approximate vertical height replaced by the antennas listed in a top-loaded vertical antenna. To arrive at the remaining amount of tower needed for resonance, subtract these numbers from the nonloaded tower height needed for resonance. Note that for all but the 10-m antennas, the equivalent loading equals or exceeds a quarter wavelength on 40 m. For typical HF Yagis, this method is best used only on 80 and 160 m.

CONSTRUCTION EXAMPLES

Consider this example: A TH7 triband Yagi mounted on a 40-ft tower. The TH7 has approximately the same overall dimensions as a full-sized 3-element 20-m beam, but has more interlaced elements. Its equivalent loading is estimated to be 40 ft. At 3.6 MHz, 65 ft of tower is needed without loading. Subtracting 40 ft of equivalent loading, the feedpoint should be 25 ft below the TH7 antenna.

Ten quarter-wavelength (65-ft) radials

were run from a nylon rope tied between tower legs at the 15-ft level, to various supports 10 ft high. Nylon cord was tied to the insulated, stranded, #18 wire, without using insulators. The radials are all connected together and to the center of an exact half wavelength (at 3.6 MHz) of RG-213 coax, which will repeat the antenna feed impedance at the other end. **Fig 20.57** is a drawing of the installation. The author used a Hewlett-Packard low-frequency impedance analyzer to measure the input impedance across the 80-m band.

An exact resonance (zero reactance) was seen at 3.6 MHz, just as predicted. The radiation resistance was found to be 17 Ω. The next question is, how to feed and match the antenna.

One good approach to 80-m antennas is to tune them to the low end of the band, use a low-loss transmission line, and switch an antenna tuner in line for operation in the higher portions of the band. With a 50-Ω line, the 17-Ω radiation resistance represents a 3:1 SWR, meaning that an antenna tuner should be in-line for all

Antenna	Boom Length (feet)	Equivalent Loading (feet)
3L 20	24	39
5L 15	26	35
4L 15	20	31
3L 15	16	28
5L 10	24	28
4L 10	18	24
3L 10	12	20
TH7	24	40 (estimated)
TH3	14	27 (estimated)

Fig 20.56— Effective loading of common Yagi antennas.

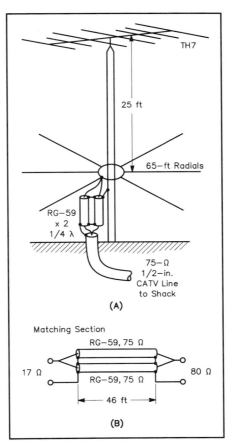

Fig 20.57—At A, an 80-m top-loaded, reverse-fed elevated ground plane, using a 40-ft tower carrying a TH7 triband Yagi antenna. At B, dimensions of the 3.6-MHz matching network, made from RG-59.

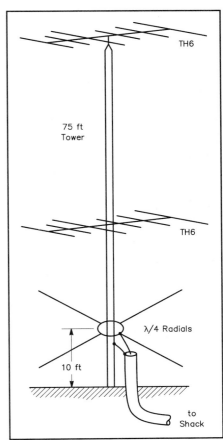

Fig 20.58—A 160-m antenna using a 75-ft tower carrying stacked triband Yagis.

frequencies. For short runs, it would be permissible to use RG-8 or RG-213 directly to the tuner. If you have a plentiful supply of low-loss 75-Ω CATV rigid coax, you can take another approach.

Make a quarter-wave (70 feet × 0.66 velocity factor = 46 ft) 37-Ω matching line by paralleling two pieces of RG-59 and connecting them between the feedpoint and a run of the rigid coax to the transmitter. The magic of quarter-wave matching transformers is that the input impedance (R_i) and output impedance (R_o) are related by:

$$Z_o^2 = R_i \times R_o \qquad \text{(Eq 1)}$$

For $R_i = 17\ \Omega$ and $Z_o = 37\ \Omega$, $R_o = 80\ \Omega$, an almost perfect match for the 75-Ω CATV coax. The resulting 1.6:1 SWR at the transmitter is good enough for CW operation without a tuner.

160-M OPERATION

On the 160-m band, a resonant quarter-wavelength requires 130 ft of tower above the radials. That's a pretty tall order. Subtracting 40 ft of top loading for a 3-element 20-m or TH7 antenna brings us to a more reasonable 90 ft above the radials. Additional top loading in the form of more antennas will reduce that even more.

Another installation, using stacked TH6s on a 75-ft tower, is shown in **Fig 20.58**. The radials are 10 ft off the ground.

North Shadow

By Irvin L. McNally, K6WX, 26119 Fairlane Dr, Sun City, CA 92586

◊ Determining which way is north is essential for setting up a satellite dish or beam antenna. Articles I've read have shown different ways of making this determination. One method covers the use of a magnetic compass and a knowledge of the local magnetic variation (the difference between true north and magnetic north). However, an accurate value of the local magnetic variation is not always available. And with small hand-held compasses, it's difficult to read the exact bearing because the scale isn't graduated in degrees.

Another method is to observe the north star, Polaris, and sight from the observer to a vertical plumb-bob line to the star. The latter step isn't easy to do in the dark. Furthermore, unless a *Nautical Almanac* is available to determine the exact time of Polaris' meridian transit, there will be a maximum error of about 1°.

Homeowners can try another source of information: the surveyor's plot plan may have noted the magnetic or true bearing relative to the property line.

Mariners use the noon sight to determine their position at sea. With a sextant and chronometer, the altitude and time of the sun's meridian crossing is determined. Combining this data with the declination of the sun, equation of time,[2] sextant correction, refraction and dip, the position can readily be calculated.

I figured that this method could be used in a reverse calculation to find north. Knowing your exact longitude and the equation of time, the local time of the sun's meridian crossing can be determined. This is the time when the shadow of a vertical pole or tower points true north.

Because the equation of time varies from day to day, its value must be taken from the *Nautical Almanac* or an appropriate table. This difference is shown in the *Nautical Almanac* for each day of the year. World globes used to have an analema printed on a barren area of the Pacific Ocean. The analema is a figure-eight pattern that represents the year-long cycle of the sun's declination and the equation of time. Most globe users fail to grasp its significance.

I've written an IBM BASIC computer program[1] that determines the local time of the north shadow for any day of the year. The required information is: local longitude, month number, day number and local time zone number. A *Nautical Almanac* isn't needed because the equation of time, taken from Mixter's *Primer of Navigation,*[2] is in the program. The maximum error using this table is about 20 seconds, with a possible error of 2 seconds or less, four times a year.

Notes

[1]The program *SHADOW.EXE* is available from *ARRLWeb*. See page viii.

[2]G. Mixter, *Primer of Navigation* (New York: Van Nostrand Reinhold Co., 5th ed.), 1967, Chapter 21, Table 10, p 539. ISBN: 0-442-05443-2.

Yagi and Quad Directive Antennas

Most antennas described earlier in this chapter have unity gain compared to a dipole, or just slightly more. For the purpose of obtaining gain and directivity it is convenient to use a Yagi-Uda or quad beam antenna. The former is commonly called a *Yagi*, and the latter is referred to as a *quad*.

Most operators prefer to erect these antennas for horizontal polarization, but they can be used as vertically polarized arrays merely by rotating the elements by 90°. In effect, the beam antenna is turned on its side for vertical polarity. The number of elements used will depend on the gain desired and the limits of the supporting structure. Many amateurs obtain satisfactory results with only two elements in a beam antenna, while others have four or five elements operating on a single amateur band.

Regardless of the number of elements used, the height-above-ground considerations discussed earlier for dipole antennas remain valid with respect to the angle of radiation. This is demonstrated in **Fig 20.60** at A and B where a comparison of radiation characteristics is given for a 3-element Yagi at one-half and one wavelength above average ground. It can be seen that the higher antenna (Fig 20.60B) has a main lobe that is more favorable for DX work (roughly 15°) than the lobe of the lower antenna in Fig 20.60A (approximately 30°). The pattern at B shows that some useful high-angle radiation exists also, and the higher lobe is suitable for short-skip contacts when propagation conditions dictate the need.

The azimuth pattern for the same antenna is provided in **Fig 20.61**. Most of the power is concentrated in the *main lobe* at 0° azimuth. The lobe directly behind the main lobe at 180° is often called the *backlobe*. Note that there are small *sidelobes* at approximately 110° and 260° in azimuth. The peak power difference, in decibels, between the "nose" of the main lobe at 0° and the strongest rearward lobe is called the *front-to-rear ratio (F/R)*. In this case the worst-case rearward lobe is at 180°, and the F/R is 12 dB. It is infrequent that two 3-element Yagis with different element spacings and tuning will yield the same lobe patterns. The pattern of Fig 20.61 is shown only for illustrative purposes.

Parasitic Excitation

In most of these arrangements the additional elements receive power by induction or radiation from the driven element and reradiate it in the proper phase relationship to give the desired effect. These elements are called *parasitic elements*, as contrasted to *driven elements*, which receive power directly from the transmitter through the transmission line.

The parasitic element is called a *director* when it reinforces radiation on a line pointing to it from the driven element, and a *reflector* when the reverse is the case. Whether the parasitic element is a director or reflector depends on the parasitic element tuning, which is usually adjusted by changing its length.

Gain, Front-to-Rear Ratio and SWR

The gain of an antenna with parasitic elements varies with the spacing and tuning of the elements. Element tuning is a function of length, diameter and taper schedule if the element is constructed with telescoping tubing. For any given spacing, there is a tuning condition that will give maximum gain at this spacing. However, the maximum front-to-rear ratio seldom, if ever, occurs at the same condition that gives maximum forward gain. The impedance of the driven element in a parasitic array, and thus the SWR, also varies with the tuning and spacing.

It is important to remember that all these parameters change as the operating frequency is varied. For example, if you operate both the CW and phone portions of the 20-m band with a Yagi or quad antenna, you probably will want an antenna that "spreads out" the performance over most of the band. Such designs typically must sacrifice a little gain in order to achieve good F/R and SWR performance across the band. The longer the boom of a Yagi or a quad, and the more elements that are placed on that boom, the better will be the overall performance over a given amateur band. For the lower HF bands, the size of the antenna quickly becomes impractical for truly "optimal" designs, and compromise is necessary.

Two-Element Beams

A 2-element beam is useful where space or other considerations prevent the use of a three element, or larger, beam. The general practice is to tune the parasitic element as a reflector and space it about 0.15 λ from the driven element, although some successful antennas have been built with 0.1-λ spacing and director tuning. Gain vs element spacing for a 2-element antenna is given in **Fig 20.62** for the special case where the parasitic element is resonant. It is indicative of the performance to be expected under maximum-gain tuning conditions. Changing the tuning of the driven element in a Yagi or quad will not materially affect the gain or F/R. Thus, only the spacing and the tuning of the single parasitic element have any effect on the performance of a 2-element Yagi or quad. Most 2-element Yagi designs achieve a compromise F/R of about 10 dB, together with acceptable SWR and gain across a frequency band with a percentage bandwidth less than about 4%. A 2-element quad can achieve better F/R, gain and SWR across a band, at the expense of greater mechanical complexity compared to a Yagi.

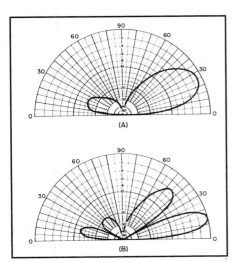

Fig 20.60—Elevation-plane response of a 3-element Yagi placed ½ λ above perfect ground at (A) and the same antenna spaced 1 λ above ground at (B).

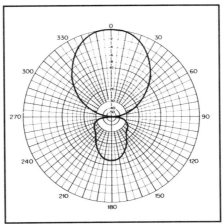

Fig 20.61—Azimuth-plane pattern of a typical three-element Yagi in free space. The Yagi's boom is along the 0° to 180° axis.

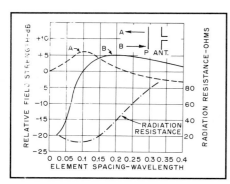

Fig 20.62—Gain vs element spacing for a 2-element Yagi, having one driven and one parasitic element. The reference point, 0 dB, is the field strength from a half-wave antenna alone. The greatest gain is in the direction A at spacings of less than 0.14 λ, and in direction B at greater spacings. The front-to-rear ratio is the difference in decibels between curves A and B. Variation in radiation resistance of the driven element is also shown. These curves are for the special case of a self-resonant parasitic element, but are representative of how a 2-element Yagi works. At most spacings the gain as a reflector can be increased by slight lengthening of the parasitic element; the gain as a director can be increased by shortening. This also improves the front-to-rear ratio.

Three-Element Beams

A theoretical investigation of the 3-element case (director, driven element and reflector) has indicated a maximum gain of about 9.7 dBi. A number of experimental investigations have shown that the spacing between the driven element and reflector for maximum gain is in the region of 0.15 to .25 λ. With 0.2-λ reflector spacing, **Fig 20.63** shows that the gain variation with director spacing is not especially critical. Also, the overall length of the array (boom length in the case of a rotatable antenna) can be anywhere between 0.35 and 0.45 λ with no appreciable difference in the maximum gain obtainable.

If maximum gain is desired, wide spacing of both elements is beneficial because adjustment of tuning or element length is less critical and the input resistance of the driven element is generally higher than with close spacing. A higher input resistance improves the efficiency of the antenna and makes a greater bandwidth possible. However, a total antenna length, director to reflector, of more than 0.3 λ at frequencies of the order of 14 MHz introduces difficulty from a construction standpoint. Lengths of 0.25 to 0.3 λ are therefore used frequently for this band, even

though they are less than optimum from the viewpoint of maximum gain.

In general, Yagi antenna gain drops off less rapidly when the reflector length is increased beyond the optimum value than it does for a corresponding decrease below the optimum value. The opposite is true of a director. It is therefore advisable to err, if necessary, on the long side for a reflector and on the short side for a director. This also tends to make the antenna performance less dependent on the exact frequency at which it is operated: An increase above the design frequency has the same effect as increasing the length of both parasitic elements, while a decrease in frequency has the same effect as shortening both elements. By making the director slightly short and the reflector slightly long, there will be a greater spread between the upper and lower frequencies at which the gain starts to show a rapid decrease.

We recommend "plumbers delight"

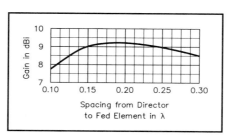

Fig 20.63—General relationship of gain of 3-element Yagi vs director spacing, the reflector being fixed at 0.2 λ. This antenna is tuned for maximum forward gain.

construction, where all elements are mounted directly on, and grounded to, the boom. This puts the entire array at dc ground potential, affording better lightning protection. A gamma- or T-match section can be used for matching the feed line to the array.

Computer-Optimized Yagis

Yagi designers are now able to take advantage of powerful personal computers and software to optimize their designs for the parameters of gain, F/R and SWR across frequency bands. ARRL Senior Assistant Technical Editor Dean Straw, N6BV, has designed a family of Yagis for HF bands. These can be found in **Tables 20.10** through **Table 20.14**, for the 10, 12, 15, 17 and 20-m amateur bands.

For 12 through 20 m, each design has been optimized for better than 20 dB F/R, and an SWR of less than 2:1 across the entire amateur frequency band. For the 10-m band, the designs were optimized for the lower 800 kHz of the band, from 28.0 to 28.8 MHz. Each Yagi element is made of telescoping 6061-T6 aluminum tubing, with 0.058 inch thick walls. This type of element can be telescoped easily, using techniques shown in **Fig 20.64**. Measuring each element to an accuracy of $^1/_8$ inch results in performance remarkably consistent with the computations, without any need for "tweaking" or fine-tuning when the Yagi is on the tower.

Each element is mounted above the boom with a heavy rectangular aluminum plate, by means of galvanized U-bolts with

Table 20.10

10-m Optimized Yagi Designs

	Spacing Between Elements (in.)	Seg 1 Length (in.)	Seg 2 Length (in.)	Seg 3 Length (in.)	Midband Gain F/R
310-08					
Refl	0	24	18	66.750	7.2 dBi
DE	36	24	18	57.625	22.9 dB
Dir 1	54	24	18	53.125	
410-14					
Refl	0	24	18	64.875	8.4 dBi
DE	36	24	18	58.625	30.9 dB
Dir 1	36	24	18	57.000	
Dir 2	90	24	18	47.750	
510-24					
Refl	0	24	18	65.625	10.3 dBi
DE	36	24	18	58.000	25.9 dB
Dir 1	36	24	18	57.125	
Dir 2	99	24	18	55.000	
Dir 3	111	24	18	50.750	

Note: For all antennas, the tube diameters are: Seg 1=0.750 inch, Seg 2=0.625 inch, Seg 3=0.500 inch.

Table 20.11

12-m Optimized Yagi Designs

	Spacing Between Elements (in.)	Seg 1 Length (in.)	Seg 2 Length (in.)	Seg 3 Length (in.)	Midband Gain F/R
312-10					
Refl	0	36	18	69.000	7.5 dBi
DE	40	36	18	59.125	24.8 dB
Dir 1	74	36	18	54.000	
412-15					
Refl	0	36	18	66.875	8.5 dBi
DE	46	36	18	60.625	27.8 dB
Dir 1	46	36	18	58.625	
Dir 2	82	36	18	50.875	
512-20					
Refl	0	36	18	69.750	9.5 dBi
DE	46	36	18	61.750	24.9 dB
Dir 1	46	36	18	60.500	
Dir 2	48	36	18	55.500	
Dir 3	94	36	18	54.625	

Note: For all antennas, the tube diameters are: Seg 1 = 0.750 inch, Seg 2 = 0.625 inch, Seg 3 = 0.500 inch.

Table 20.12

15-m Optimized Yagi Designs

	Spacing Between Elements (in.)	Seg 1 Length (in.)	Seg 2 Length (in.)	Seg 3 Length (in.)	Seg 4 Length (in.)	Midband Gain F/R
315-12						
Refl	0	30	36	18	61.375	7.6 dBi
DE	48	30	36	18	49.625	25.5 dB
Dir 1	92	30	36	18	43.500	
415-18						
Refl	0	30	36	18	59.750	8.3 dBi
DE	56	30	36	18	50.875	31.2 dB
Dir 1	56	30	36	18	48.000	
Dir 2	98	30	36	18	36.625	
515-24						
Refl	0	30	36	18	62.000	9.4 dBi
DE	48	30	36	18	52.375	25.8 dB
Dir 1	48	30	36	18	47.875	
Dir 2	52	30	36	18	47.000	
Dir 3	134	30	36	18	41.000	

Note: For all antennas, the tube diameters (in inches) are: Seg 1 = 0.875, Seg 2 = 0.750, Seg 3 = 0.625, Seg 4 = 0.500.

Table 20.13

17-m Optimized Yagi Designs

	Spacing Between Elements (in.)	Seg 1 Length (in.)	Seg 2 Length (in.)	Seg 3 Length (in.)	Seg 4 Length (in.)	Seg 5 Length (in.)	Midband Gain F/R
317-14							
Refl	0	24	24	36	24	60.125	8.1 dBi
DE	65	24	24	36	24	52.625	24.3 dB
Dir 1	97	24	24	36	24	48.500	
417-20							
Refl	0	24	24	36	24	61.500	8.5 dBi
DE	48	24	24	36	24	54.250	27.7 dB
Dir 1	48	24	24	36	24	52.625	
Dir 2	138	24	24	36	24	40.500	

Note: For all antennas, tube diameters (inches) are: Seg 1=1.000, Seg 2=0.875, Seg 3=0.750, Seg 4=0.625, Seg 5=0.500.

Table 20.14

20-m Optimized Yagi Designs

	Spacing Between Elements (in.)	Seg 1 Length (in.)	Seg 2 Length (in.)	Seg 3 Length (in.)	Seg 4 Length (in.)	Seg 5 Length (in.)	Seg 6 Length (in.)	Midband Gain F/R
320-16								
Refl	0	48	24	20	42	20	69.625	7.3 dBi
DE	80	48	24	20	42	20	51.250	23.4 dB
Dir 1	106	48	24	20	42	20	42.625	
420-26								
Refl	0	48	24	20	42	20	65.625	8.6 dBi
DE	72	48	24	20	42	20	53.375	23.4 dB
Dir 1	60	48	24	20	42	20	51.750	
Dir 2	174	48	24	20	42	20	38.625	

Note: For all antennas, tube diameters (inches) are: Seg 1=1.000, Seg 2=0.875, Seg 3=0.750, Seg 4=0.625, Seg 5=0.500.

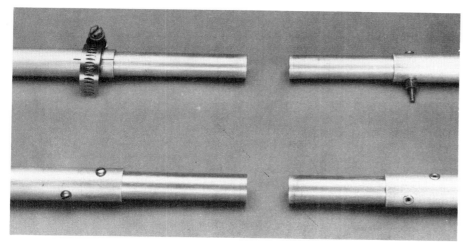

Fig 20.64—Some methods of connecting telescoping tubing sections to build beam elements. See text for a discussion of each method.

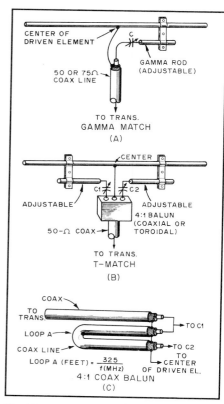

Fig 20.66—Illustrations of gamma and T matching systems. At A, the gamma rod is adjusted along with C until the lowest SWR is obtained. A T match is shown at B. It is the same as two gamma-match rods. The rods and C1 and C2 are adjusted alternately for a best SWR. A coaxial 4:1 balun transformer is shown at C. A toroidal balun can be used in place of the coax model shown. The toroidal version has a broader frequency range than the coaxial one. The T match is adjusted for 200 Ω and this balun steps this balanced value down to 50 Ω, unbalanced. Or the T match can be set for 300 Ω, and the balun used to step this down to 75 Ω unbalanced. Dimensions for the gamma and T match rods will depend on the tubing size used, and the spacing of the parasitic elements of the beam. Capacitors C, C1 and C2 can be 140 pF for 14-MHz beams. Somewhat less capacitance will be needed at 21 and 28 MHz.

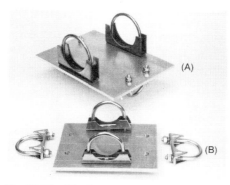

Fig 20.65—The boom-to-element plate at A uses muffler-clamp-type U-bolts and saddles to secure the round tubing to the flat plate. The boom-to-mast plate at B is similar to the boom-to-element plate. The main difference is the size of materials used.

saddles, as shown in **Fig 20.65**. This method of element mounting is rugged and stable, and because the element is mounted away from the boom, the amount of element detuning due to the presence of the boom is minimal. The element dimensions given in each table already take into account any element detuning due to the boom-to-element mounting plate. The element-to-boom mounting plate for all the 10-m Yagis is a 0.250-inch thick flat aluminum plate, 4 inches wide by 4 inches long. For the 12 and 15-m Yagis, a 0.375-inch thick flat aluminum plate, 5 inches wide by 6 inches long is used, and for the 17 and 20-m Yagis, a 0.375-inch thick flat aluminum plate, 6 inches wide by 8 inches long is used. Where the plate is rectangular, the long dimension is in line with the element.

Each design table shows the dimensions

for *one-half* of each element, mounted on one side of the boom. The other half of each element is the same, mounted on the other side of the boom. Use a tubing sleeve inside the center portion of the element so that the element is not crushed by the mounting U-bolts. Each telescoping section is inserted 3 inches into the next size of tubing. For example, in the 310-08.YAG design (3 elements on an 8-ft boom), the reflector tip, made out of $^1/_2$-inch OD tubing, sticks out 66.75 inches from the $^5/_8$-inch OD tubing. For each 10-m element, the overall length of each $^5/_8$-inch OD piece of tubing is 21 inches, before insertion into the $^3/_4$-inch piece. Since the $^3/_4$-inch OD tubing is 24 inches long on each side of the boom, the center portion of each element is actually 48 inches of uncut $^3/_4$-inch OD tubing.

The boom for all these antennas should be constructed with at least 2-inch-OD tubing, with 0.065-inch wall thickness. Because each boom has 3 inches extra space at each end, the reflector is actually placed 3 inches from one end of the boom. For the 310-08.YAG, the driven element is placed 36 inches ahead of the reflector, and the director is placed 54 inches ahead of the driven element.

Each antenna is designed with a driven element length appropriate for a gamma or T matching network, as shown in **Fig 20.66**. The variable gamma or T capacitors can be housed in small plastic cups for weatherproofing; receiving-type variable capacitors with close plate spacing can be used at powers up to a few hundred watts. Maximum capacitance required is usually 140 pF at 14 MHz and proportionally less at the higher frequencies.

The driven-element's length may require slight readjustment for best match,

particularly if a different matching network is used. *Do not change either the lengths or the telescoping tubing schedule of the parasitic elements*—they have been optimized for best performance and will not be affected by tuning of the driven element.

Tuning Adjustments

Preliminary matching adjustments can be done on the ground. The beam should be set up so the reflector element rests on the earth, with the beam pointing upward. The matching system is then adjusted for best SWR. When the antenna is raised to its operating height, only slight touch-up of the matching network may be required.

Construction of Yagis

Most beams and verticals are made from sections of aluminum tubing. Compromise beams have been fashioned from less-expensive materials such as electrical conduit (steel) or bamboo poles wrapped with conductive tape or aluminum foil. The steel conduit is heavy, is a poor conductor and is subject to rust. Similarly, bamboo with conducting material attached to it will deteriorate rapidly in the weather. The dimensions shown for the Yagis in the preceding section are designed for specific telescoping aluminum elements, but the elements may be scaled to different sizes by using the information about tapering and scaling in Chapter 2 of *The ARRL Antenna Book*, although with a likelihood of deterioration in performance over the whole frequency band.

For reference, **Table 20.15** details the standard sizes of aluminum tubing, available in many metropolitan areas. Dealers may be found in the Yellow Pages under "Aluminum." Tubing usually comes in 12-ft lengths, although 20-ft lengths are available in some sizes. Your aluminum dealer will probably also sell aluminum plate in various thicknesses needed for boom-to-mast and boom-to-element connections.

Aluminum is rated according to its hardness. The most common material used in antenna construction is grade 6061-T6. This material is relatively strong and has good workability. In addition, it will bend without taking a "set," an advantage in antenna applications where the pieces are constantly flexing in the wind. The softer grades (5051, 3003 and so on) will bend much more easily, while harder grades (7075 and so on) are more brittle.

Wall thickness is of primary concern when selecting tubing. It is of utmost importance that the tubing fits snugly where the element sections join. Sloppy joints will make a mechanically unstable antenna. The "magic" wall thickness is 0.058 inch. For example (from Table 20.15), 1-inch outside diameter (OD) tubing with a 0.058-inch wall has an inside diameter (ID) of 0.884 inch. The next smaller size of tubing, 7/8 inch, has an OD of 0.875 inch. The 0.009-inch difference provides just the right amount of clearance for a snug fit.

Fig 20.64 shows several methods of fastening antenna element sections together. The slot and hose clamp method shown in Fig 20.64A is probably the best for joints where adjustments are needed. Generally, one adjustable joint per element half is sufficient to tune the antenna—usually the tips at each end of an element are made adjustable. Stainless steel hose clamps (beware—some "stainless steel" models do not have a stainless screw and will rust) are recommended for longest antenna life.

Fig 20.64B, C and D show possible fastening methods for joints that are not adjustable. At B, machine screws and nuts hold the elements in place. At C, sheet metal screws are used. At D, rivets secure the tubing. If the antenna is to be assembled permanently, rivets are the best choice. Once in place, they are permanent.They will never work free, regardless of vibration or wind. If aluminum rivets with aluminum mandrels are employed, they will never rust. Also, being aluminum, there is no danger of corrosion from interaction between dissimilar metals. If the antenna is to be disassembled and moved periodically, either B or C will work. If machine screws are used, however, take precautions to keep the nuts from vibrating free. Use of lock washers, lock nuts and flexible adhesive such as silicone bathtub sealant will keep the hardware in place.

Use of a conductive grease at the element joints is essential for long life. Left untreated, the aluminum surfaces will oxidize in the weather, resulting in a poor connection. Some trade names for this conductive grease are Penetrox, Noalox and Dow Corning Molykote 41. Many electrical supply houses carry these products.

Boom Material

The boom size for a rotatable Yagi or quad should be selected to provide stability to the entire system. The best diameter for the boom depends on several factors, but mostly the element weight, number of elements and overall length. Two-inch-diameter booms should not be made any longer than 24 ft unless additional support is given to reduce both vertical and horizontal bending forces. Suitable reinforcement for a long 2-inch boom can consist of a truss or a truss and lateral support, as shown in **Fig 20.67**.

A boom length of 24 ft is about the point where a 3-inch diameter begins to be very worthwhile. This dimension provides a considerable amount of improvement in overall mechanical stability as well as increased clamping surface area for element hardware. The latter is extremely important to prevent rotation of elements around the boom if heavy icing is commonplace. Pinning an element to the boom with a

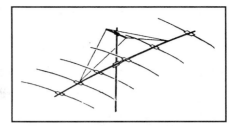

Fig 20.67—A long boom needs both vertical and horizontal support. The crossbar mounted above the boom can support a double truss, which will help keep the antenna in position.

large bolt helps in this regard. On smaller diameter booms, however, the elements sometimes work loose and tend to elongate the pinning holes in both the element and the boom. After some time the elements shift their positions slightly (sometimes from day to day) and give a ragged appearance to the system, even though this may not harm the electrical performance.

A 3-inch-diameter boom with a wall thickness of 0.065 inch is very satisfactory for antennas up to about a 5-element, 20-m array that is spaced on a 40-ft boom. A truss is recommended for any boom longer than 24 ft. One possible source for large boom material is irrigation tubing sold at farm supply houses.

Putting It Together

Once you assemble the boom and elements, the next step is to fasten the elements to the boom securely and then fasten the boom to the mast or supporting structure. Be sure to leave plenty of material on either side of the U-bolt holes on the element-to-boom mounting plates. The U-bolts selected should be a snug fit for the tubing. If possible, buy muffler-clamp U-bolts that come with saddles.

The boom-to-mast plate shown in Fig 20.65B is similar to the boom-to-element plate. The size of the plate and number of U-bolts used will depend on the size of the antenna. Generally, antennas for the bands up through 20 m require only two U-bolts each for the mast and boom. Longer antennas for 15 and 20 m (35-ft booms and up) and most 40-m beams should have four U-bolts each for the boom and mast because of the torque that the long booms and elements exert as the antennas move in the wind. When tightening the U-bolts, be careful not to crush the tubing. Once the wall begins to collapse, the connection begins to weaken. Many aluminum suppliers sell 1/4-inch or 3/8-inch plates just right for this application. Often they will shear pieces to the correct size on

Table 20.15
Standard Sizes of Aluminum Tubing
6061-T6 (61S-T6) Round Aluminum Tube in 12-ft Lengths

OD (in.)	Wall Thickness (in.)	Wall Thickness stubs ga	ID (in.)	Approx Weight (lb) per ft	Approx Weight (lb) per length	OD (in.)	Wall Thickness (in.)	Wall Thickness stubs ga	ID (in.)	Approx Weight (lb) per ft	Approx Weight (lb) per length
3/16	0.035	no. 20	0.117	0.019	0.228	1 1/4	0.035	no. 20	1.180	0.155	1.860
	0.049	no. 18	0.089	0.025	0.330		0.049	no. 18	1.152	0.210	2.520
1/4	0.035	no. 20	0.180	0.027	0.324		0.058	no. 17	1.134	0.256	3.072
	0.049	no. 18	0.152	0.036	0.432		0.065	no. 16	1.120	0.284	3.408
	0.058	no. 17	0.134	0.041	0.492		0.083	no. 14	1.084	0.357	4.284
5/16	0.035	no. 20	0.242	0.036	0.432	1 3/8	0.035	no. 20	1.305	0.173	2.076
	0.049	no. 18	0.214	0.047	0.564		0.058	no. 17	1.259	0.282	3.384
	0.058	no. 17	0.196	0.055	0.660	1 1/2	0.035	no. 20	1.430	0.180	2.160
3/8	0.035	no. 20	0.305	0.043	0.516		0.049	no. 18	1.402	0.260	3.120
	0.049	no. 18	0.277	0.060	0.720		0.058	no. 17	1.384	0.309	3.708
	0.058	no. 17	0.259	0.068	0.816		0.065	no. 16	1.370	0.344	4.128
	0.065	no. 16	0.245	0.074	0.888		0.083	no. 14	1.334	0.434	5.208
7/16	0.035	no. 20	0.367	0.051	0.612		*0.125	1/8"	1.250	0.630	7.416
	0.049	no. 18	0.339	0.070	0.840		*0.250	1/4"	1.000	1.150	14.823
	0.065	no. 16	0.307	0.089	1.068	1 5/8	0.035	no. 20	1.555	0.206	2.472
1/2	0.028	no. 22	0.444	0.049	0.588		0.058	no. 17	1.509	0.336	4.032
	0.035	no. 20	0.430	0.059	0.708	1 3/4	0.058	no. 17	1.634	0.363	4.356
	0.049	no. 18	0.402	0.082	0.948		0.083	no. 14	1.584	0.510	6.120
	0.058	no. 17	0.384	0.095	1.040	1 7/8	0.058	no. 17	1.759	0.389	4.668
	0.065	no. 16	0.370	0.107	1.284	2	0.049	no. 18	1.902	0.350	4.200
5/8	0.028	no. 22	0.569	0.061	0.732		0.065	no. 16	1.870	0.450	5.400
	0.035	no. 20	0.555	0.075	0.900		0.083	no. 14	1.834	0.590	7.080
	0.049	no. 18	0.527	0.106	1.272		*0.125	1/8"	1.750	0.870	9.960
	0.058	no. 17	0.509	0.121	1.452		*0.250	1/4"	1.500	1.620	19.920
	0.065	no. 16	0.495	0.137	1.644	2 1/4	0.049	no. 18	2.152	0.398	4.776
3/4	0.035	no. 20	0.680	0.091	1.092		0.065	no. 16	2.120	0.520	6.240
	0.049	no. 18	0.652	0.125	1.500		0.083	no. 14	2.084	0.660	7.920
	0.058	no. 17	0.634	0.148	1.776	2 1/2	0.065	no. 16	2.370	0.587	7.044
	0.065	no. 16	0.620	0.160	1.920		0.083	no. 14	2.334	0.740	8.880
	0.083	no. 14	0.584	0.204	2.448		*0.125	1/8"	2.250	1.100	12.720
7/8	0.035	no. 20	0.805	0.108	1.308		*0.250	1/4"	2.000	2.080	25.440
	0.049	no. 18	0.777	0.151	1.810	3	0.065	no. 16	2.870	0.710	8.520
	0.058	no. 17	0.759	0.175	2.100		*0.125	1/8"	2.700	1.330	15.600
	0.065	no. 16	0.745	0.199	2.399		*0.250	1/4"	2.500	2.540	31.200
1	0.035	no. 20	0.930	0.123	1.467						
	0.049	no. 18	0.902	0.170	2.040						
	0.058	no. 17	0.884	0.202	2.424						
	0.065	no. 16	0.870	0.220	2.640						
	0.083	no. 14	0.834	0.281	3.372						
1 1/8	0.035	no. 20	1.055	0.139	1.668						
	0.058	no. 17	1.009	0.228	2.736						

*These sizes are extruded; all other sizes are drawn tubes.

Shown here are standard sizes of aluminum tubing that are stocked by most aluminum suppliers or distributors in the United States and Canada.

request. As with tubing, the relatively hard 6061-T6 grade is a good choice for mounting plates.

The antenna should be put together with good-quality hardware. Stainless steel is best for long life. Rust will attack plated steel hardware after a short while, making nuts difficult, if not impossible, to remove. If stainless muffler clamps are not available, the next best thing is to have them plated. If you can't get them plated, then at least paint them with a good zinc-chromate primer and a finish coat or two. Good-quality hardware is more expensive initially, but if you do it right the first time, you won't have to take the antenna down after a few years and replace the hardware. Also, when repairing or modifying an installation, nothing is more frustrating than fighting rusty hardware at the top of a tower.

Quad Antennas

One of the more effective DX arrays is called a *quad* antenna. It consists of two or more loops of wire, each supported by a bamboo or fiberglass cross-arm assembly. The loops are a quarter wavelength per side (full wavelength overall). One loop is driven and the other serves as a parasitic element—usually a reflector. A variation of the quad is called the *delta loop*. The electrical properties of both antennas are the same. Both antennas are shown in **Fig 20.68**. They differ mainly in their physical properties, one being of plumber's delight construction, while the other uses insulating support members. One or more directors can be added to either antenna if additional gain and directivity are desired, though most operators use the 2-element arrangement.

It is possible to interlace quads or "deltas" for two or more bands, but if this is done the formulas given in Fig 20.68

may have to be changed slightly to compensate for the proximity effect of the second antenna. For quads the length of the full-wave loop can be computed from

$$\text{Full} - \text{wave loop} = \frac{1005}{f(\text{MHz})} \qquad (7)$$

If multiple arrays are used, each antenna should be tuned separately for maximum forward gain, or best front-to-rear ratio, as noted on a field-strength meter. The reflector stub on the quad should be adjusted for this condition. The gamma match should be adjusted for best

SWR. The resonance of the antenna can be found by checking the frequency at which the lowest SWR occurs. The element length (driven element) can be adjusted for resonance in the most-used portion of the band by lengthening or shortening it.

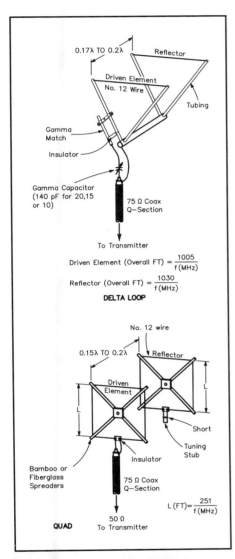

Fig 20.68—Information on building a quad or a delta-loop antenna. The antennas are electrically similar, but the delta-loop uses plumber's delight construction. The λ/4 length of 75-Ω coax acts as a Q-section transformer from approximate 100-Ω feedpoint impedance of quad to 50-Ω feed line coax.

FEEDING A QUAD WITH OPEN-WIRE LINE

◊ After purchasing a GEM quad antenna, I determined that the best place for the antenna on my property was 375 feet from my operating position. Even low-loss coax could easily result in a 3-dB loss at 28 MHz.

At the encouragement of Jay Kolinsky, NE2Q, I looked at using open-wire transmission line. Using readily available 450-Ω line, the losses would be low and the line cost went down to only 10 cents per foot. As we learned later, there were even bigger benefits.

Feeding the Antenna

I've talked to a lot of people using quads, and I couldn't find anyone who had fed their quad with open-wire line. A call to the antenna manufacturer produced some suggestions, but not firm guidance on how to do the job.

The antenna was set up for four bands: 20, 17, 15 and 10 m. Jay and I decided to take the easy way out and tie all of its driven elements together, spacing their terminals with a 1 1/2-inch porcelain insulator, and feed all four driven elements with the open-wire line as shown in Figure 2. We ignored the driven elements' sum feed-point impedance, since using open-wire line meant using an antenna tuner between the transmitter and the line, anyway. We did not attempt to tune the antenna's driven elements, deciding instead to carefully adjust the quad's front-to-back ratio on each band and let the tuner take care of the rest.

Unexpected Results

Careful front-to-back ratio adjustments yielded results better than those claimed by the antenna manufacturer. NT5E went on the air

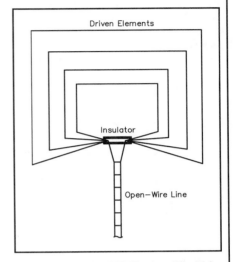

Figure 2—Curtis, NT5E, simplified his quad feed by connecting its elements' feed points together and feeding all three simultaneously with open-wire line.

with a new antenna, and the performance proved well worth the work. Using the tuner, I was able to operate anywhere in the bands with a 1:1 SWR. The only drawback was that I still needed antennas for 7, 10 and 24 MHz—or did I?

A quick check proved that the system would load up at 10 and 24 MHz. More careful adjustments were required for tuning up at 7 MHz. We roughly measured its radiation pattern as similar to that of a half-wave loop. My antenna had grown—at least in terms of frequency coverage! Further rough measurements at 7 MHz showed that there was a slight front-to-back ratio, possibly because of the reflectors present for 14, 18, 21 and 28 MHz.—*Curtis Robb, NT5E, Boerne, Texas*

A FIVE-BAND, TWO-ELEMENT HF QUAD

Two quad designs are described in this article, both nearly identical. One was constructed by KC6T from scratch, and the other was built by Al Doig, W6NBH, using modified commercial triband quad hardware. The principles of construction and adjustment are the same for both models, and the performance results are also essentially identical. One of the main advantages of this design is the ease of (relatively) independent performance adjustments for each of the five bands. These quads were described by William A. Stein, KC6T, in *QST* for April 1992. Both models use 8-ft-long, 2-inch diameter booms, and conventional X-shaped spreaders (with two sides of each quad loop parallel to the ground).

The Five-Band Quad as a System

Unless you are extraordinarily lucky, you should remember one general rule: Any quad must be adjusted for maximum performance after assembly. Simple quad designs can be tuned by pruning and re-stringing the elements to control front-to-rear ratio and SWR at the desired operat-ing frequency. Since each element of this quad contains five concentric loops, this adjustment method could lead to a nervous breakdown!

Fig 20.69 shows that the reflectors and driven elements are each independently adjustable. After assembly, adjustment is simple, and although gamma-match components on the driven element and capaci-tors on the reflectors add to the antenna's parts count, physical construction is not difficult. The reflector elements are pur-posely cut slightly long (except for the 10-m reflector), and electrically shortened by means of a tuning capacitor. The driven-element gamma matches set the lowest SWR at the desired operating frequency.

As with most multiband directive an-tennas, the designer can optimize any two of the following three attributes at the expense of the third: forward gain, front-to-rear ratio and bandwidth (where the SWR is less than 2:1). These three charac-teristics are related, and changing one changes the other two. The basic idea be-hind this quad design is to permit (without resorting to trimming loop lengths, spac-ing or other gross mechanical adjust-ments):

- The forward gain, bandwidth and front-to-rear ratio may be set by a simple adjustment after assembly. The adjust-ments can be made on a band-by-band basis, with little or no effect on previ-ously made adjustments on the other bands.
- Setting the minimum SWR in any por-tion of each band, with no interaction with previously made front-to-back or SWR adjustments.

The first of the two antennas described, the KC6T model, uses aluminum spread-ers with PVC insulators at the element attachment points. (The author elected not to use fiberglass spreaders because of their high cost.) The second antenna, the W6NBH model, provides dimensions and adjustment values for the same antenna, but using standard triband-quad fiberglass spreaders and hardware. If you have a triband quad, you can easily adapt it to this design. When W6NBH built his antenna, he had to shorten the 20-m reflector be-cause the KC6T model uses a larger 20-m reflector than W6NBH's fiberglass spreaders would allow. Performance is essentially identical for both models.

Mechanical Considerations

Even the best electrical design has no value if its mechanical construction is lacking. Here are some of the things that contribute to mechanical strength: The gamma-match capacitor KC6T used was a small, air-variable, chassis-mount capaci-tor mounted in a plastic box (see **Fig 20.70**). A male UHF connector was mounted to the box, along with a screw terminal for connection to the gamma rod.

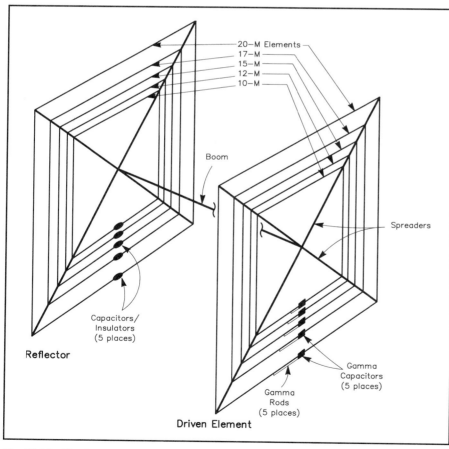

Fig 20.69—Mechanical layout of the five-band quad. The boom is 8 ft long; see Table 20.16 for all other dimensions.

Fig 20.70—Photo of one of the feed-point gamma-match capacitors.

The terminal lug and wire are for later connection to the driven element. The box came from a local hobby shop, and the box lid was replaced with a piece of $1/32$-inch ABS plastic, glued in place after the capacitor, connector and wiring had been installed. The capacitor can be adjusted with a screwdriver through an access hole. Small vent (drain) holes were drilled near corresponding corners of each end.

Enclose the gamma-match capacitor in such a manner that you can tape unwanted openings closed so that moisture can't be directly blown in during wind and rainstorms. Also, smaller boxes and sturdy mounts to the driven element ensure that you won't pick up gamma capacitor assemblies along with the leaves after a wind storm.

Plastic gamma-rod insulators/standoffs were made from $1/32$-inch ABS, cut $1/2$-inch wide with a hole at each end. Use a knife to cut from the hole to the side of each insulator so that one end can be slipped over the driven element and the other over the gamma rod. Use about four such insulators for each gamma rod, and mount the first insulator as close to the capacitor box as possible. Apply five-minute epoxy to the element and gamma rod at the insulator hole to keep the insulators from sliding. If you intend to experiment with gamma-rod length, perform this gluing operation after you have made the final gamma-rod adjustments.

Element Insulators

As shown in Fig 20.69, the quad uses insulators in the reflectors for each band to break the loop electrically, and to allow reflector adjustments. Similar insulators were used to break up each driven element so that element impedance measurements could be made with a noise bridge. After the impedance measurements, the driven-element loops are closed again. The insulators are made from $1/4 \times 2 \times 3/4$-inch phenolic stock. The holes are $1/2$-inch apart. Two terminal lugs (shorted together at the center hole) are used in each driven element. They offer a convenient way to open the loops by removing one screw. **Fig 20.71** shows these insulators and the gamma-match construction schematically. **Table 20.16** lists the component values, element lengths and gamma-match dimensions.

Element-to-Spreader Attachment

Probably the most common problem with quad antennas is wire breakage at the element-to-spreader attachment points. There are a number of functional attachment methods; **Fig 20.72** shows one of them. The attachment method with both

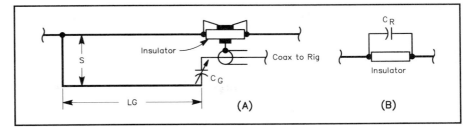

Fig 20.71—Gamma-match construction details (A) and reflector-tuning capacitor (C_R) attachment schematic (B). The gamma matches consist of matching wires (one per band) with series capacitors (C_g). See Table 20.16 for lengths and component specifications.

Table 20.16

Element Lengths and Gamma-Match Specifications of the KC6T and W6NBH Five-Band Quads

KC6T Model

Band (MHz)	Driven Element	Length (in.)	Spacing	C_g (pF)	Reflector Length (in.)	C_R (pF)
			Gamma Match			
14	851.2	33	2	125	902.4	68
18	665.6	24	2	110	705.6	47
21	568	24	1.5	90	604.8	43
24.9	483.2	29.75	1	56	514.4	33
28	421.6	26.5	1	52	448.8	(jumper)

W6NBH Model

Band (MHz)	Driven Element	Length (in.)	Spacing	C_g (pF)	Reflector Length (in.)	C_R (pF)
			Gamma Match			
14	851.2	31	2	117	890.4	120
18	665.6	21	2	114	705.6	56
21	568	26	1.5	69	604.8	58
24.9	483.2	15	1	75.5	514.4	54
28	421.6	18	1	41	448.8	(jumper)

KC6T and W6NBH spreaders is the same, even though the spreader constructions differ. The KC6T model uses #14 AWG, 7-strand copper wire; W6NBH used #18, 7-strand wire. At the point of element attachment (see **Fig 20.73**), drill a hole through both walls of the spreader using a #44 (0.086-inch) drill. Feed a 24-inch-long piece of antenna wire through the hole and center it for use as an attachment wire.

After fabricating the spider/spreader assembly, lay the completed assembly on a flat surface and cut the element to be installed to the correct length, starting with the 10-m element. Attach the element ends to the insulators to form a closed loop before attaching the elements to the spreaders. Center the insulator between the spreaders on what will become the bottom side of the quad loop, then carefully measure and mark the element-mounting-points with fingernail polish (or a similar substance). Do *not* depend on the at-rest position of the spreaders to guarantee that the mounting

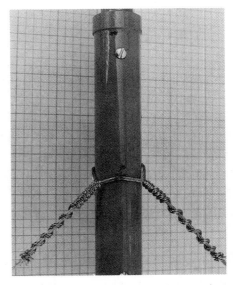

Fig 20.72—Attaching quad wires to the spreaders must minimize stress on the wires for best reliability. This method (described in the text) cuts the chances of wind-induced wire breakage by distributing stress.

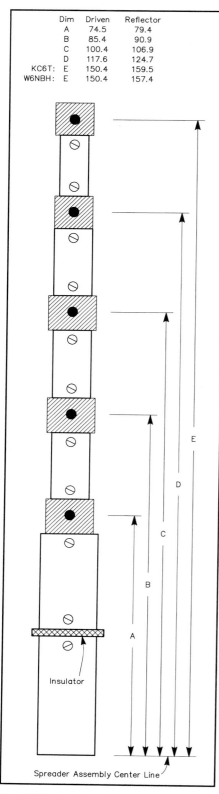

Dim	Driven	Reflector
A	74.5	79.4
B	85.4	90.9
C	100.4	106.9
D	117.6	124.7
KC6T: E	150.4	159.5
W6NBH: E	150.4	157.4

Insulator

Spreader Assembly Center Line

Fig 20.73—Spreader-drilling diagram and dimensions (in.) for the five-band quad. These dimensions apply to both spreader designs described in the text, except that most commercial spreaders are only a bit over 13 ft (156 inches) long. This requires compensation for the W6NBH model's shorter 20-m reflector as described in the text.

points will all be correct.

Holding the mark at the centerline of the spreader, tightly loop the attachment wire around the element and then gradually space out the attachment-wire turns as shown. The attachment wire need not be soldered to the element. The graduated turn spacing minimizes the likelihood that the element wire will flex in the same place with each gust of wind, thus reducing fatigue-induced wire breakage.

Feeding the Driven Elements

Each driven element is fed separately, but feeding five separate feed lines down the tower and into the shack would be costly and mechanically difficult. The ends of each of these coax lines also require support other than the tension (or lack of thereof) provided by the driven element at the feed-point. It is best to use a remote coax switch on the boom approximately 1 ft from the driven-element spider-assembly attachment point.

At installation, the cables connecting the gamma-match capacitors and the coax switch help support the driven elements and gamma capacitors. The support can be improved by taping the cables together in several places. A single coaxial feed line (and a control cable from the remote coax switch, if yours requires one) is the only required cabling from the antenna to the shack.

The KC6T Model's Composite Spreaders

If you live in an area with little or no wind, spreaders made from wood or PVC are practical but, if you live where winds can reach 60 to 80 mi/h, strong, lightweight spreaders are a must. Spreaders constructed with electrical conductors (in this case, aluminum tubing) can cause a myriad of problems with unwanted resonances, and the problem gets worse as the number of bands increases.

To avoid these problems, this version uses composite spreaders made from machined PVC insulators at the element-attachment points. Aluminum tubing is inserted into (or over) the insulators 2 inches on each end. This spreader is designed to withstand 80 mi/h winds. The overall insulator length is designed to provide a 3-inch center insulator clear of the aluminum tubing. The aluminum tubing used for the 10-m section (inside dimension "A" in Fig 20.73) is 1$\frac{1}{8}$-inch diameter × 0.058-inch wall. The next three sections are $\frac{3}{4}$-inch diameter × 0.035-inch wall, and the outer length is made from $\frac{1}{2}$ inch diameter × 0.035-inch wall. The dimensions shown in Fig 20.73 are *attachment point* dimensions only.

Attach the insulators to the aluminum using #6 sheet metal screws. Mechanical strength is provided by Devcon no. S 220 Plastic Welder Glue (or equivalent) applied liberally as the aluminum and plastic parts are joined. Paint the PVC insulators before mounting the elements to them. Paint protects the PVC from the harmful effects of solar radiation. As you can see from Fig 20.73, an additional spreader insulator located about halfway up the 10-m section (inside dimension "A") removes one of the structure's electrical resonances not eliminated by the attachment-point insulators. Because it mounts at a relatively high-stress point in the spreader, this insulator is fabricated from a length of heavy-wall fiberglass tubing.

Composite spreaders work as well as fiberglass spreaders, but require access to a well-equipped shop, including a lathe. The main objective of presenting the composite spreader is to show that fiberglass spreaders aren't a basic requirement—there are many other ways to construct usable spreaders. If you can lay your hands on a used multiband quad, even one that's damaged, you can probably obtain enough spreaders to reduce construction costs considerably.

Gamma Rod

The gamma rod is made from a length of #12 solid copper wire (W6NBH used #18, 7-strand wire). Dimensions and spacings are shown in Table 20.16. If you intend to experiment with gamma-rod lengths and capacitor settings, cut the gamma-rod lengths about 12 inches longer than the length listed in the table. Fabricate a sliding short by soldering two small alligator clips back-to-back such that they can be clipped to the rod and the antenna element and easily moved along the driven element. Note that gamma-rod spacing varies from one band to another. When you find a suitable shorting-clip position, mark the gamma rod, remove the clip, bend the gamma rod at the mark and solder the end to the element.

The W6NBH Model

As previously mentioned, this model uses standard 13-ft fiberglass spreaders, which aren't quite long enough to support the larger 20-m reflector specified for the KC6T model. The 20-m W6NBH reflector loop is cut to the dimensions shown in Table 20.16, 12 inches shorter than that for the KC6T model. To tune the shorter reflector, a 6-inch-long stub of antenna wire (spaced 2 inches) hangs from the reflector insulator, and the reflector tuning capacitor mounts on another insulator at the end of this stub.

Gamma-Match and Reflector-Tuning Capacitor

Use an air-variable capacitor of your choice for each gamma match. Approximately 300 V can appear across this capacitor (at 1500 W), so choose plate spacing appropriately. If you want to adjust the capacitor for best match and then replace it with a fixed capacitance, remember that several amperes of RF will flow through the capacitance. If you choose disc-ceramic capacitors, use a parallel combination of at least four 1-kV units of equal value. Any temperature coefficient is acceptable. NP0 units are not required. Use similar components to tune the reflector elements.

Adjustments

Well, here you are with about 605 ft of wire. Your antenna will weigh about 45 pounds (the W6NBH version is slightly lighter) and have about 9 square ft of wind area. If you chose to, you can use the dimensions and capacitance values given, and performance should be excellent. If you adjust the antenna for minimum SWR at the band centers, it should cover all of the lower four bands and 28 to 29 MHz with SWRs under 2:1; front-to-rear ratios are given in **Table 20.17**.

Instead of building the quad to the dimensions listed and hoping for the best, you can adjust your antenna to account for most of the electrical environment variables of your installation. The adjustments are conceptually simple: First adjust the reflector's electrical length for maximum front-to-rear ratio (if you desire good gain, but are willing to settle for a narrower than maximum SWR bandwidth), or accept some compromise in front-to-rear ratio that results in the widest SWR bandwidth. You can make this adjustment by placing an air-variable capacitor (about 100-pF maximum) across the open reflector loop ends, one band at a time, and adjusting the capacitor for the desired front-to-rear ratio. The means of doing this will be discussed later.

During these reflector adjustments, the driven-element gamma-match capacitors may be set to any value and the gamma rods may be any convenient length (but the sliding-short alligator clips should be installed somewhere near the lengths specified in Table 20.16). After completing the front-to-rear adjustments, the gamma capacitors and rods are adjusted for minimum SWR at the desired frequency.

Adjustment Specifics

Adjust each band by feeding it separately. You can make a calibrated variable capacitor (with a hand-drawn scale and wire pointer). Calibrate the capacitor using your receiver, a known-value inductor and a dip meter (plus a little calculation).

To adjust front-to-rear ratio, simply clip the (calibrated) air-variable capacitor across the open ends of the desired reflector loop. Connect the antenna to a portable receiver with an S meter. Point the back of the quad at a signal source, and slowly adjust the capacitor for a dip in the S-meter reading.

Table 20.17
Measured Front-to-Rear Ratios

Band	KC6T Model	W6NBH Model
14	25 dB	16 dB
18	15 dB	10 dB
21	25 dB	>20 dB
24.9	20 dB	>20 dB
28	20 dB	>20 dB

After completing the front-to-rear adjustments, replace the variable capacitor with an appropriate fixed capacitor sealed against the weather. Then move to the driven-element adjustments. Connect the coax through the SWR bridge to the 10-m gamma-match capacitor box. Use an SWR bridge that requires only a watt or two (not more than 10 W) for full-scale deflection in the calibrate position on 10 m. Using the minimum necessary power, measure the SWR. Go back to receive and adjust the capacitor until (after a number of transmit/receive cycles) you find the minimum SWR. If it is too high, lengthen or shorten the gamma rod by means of the sliding alligator-clip short and make the measurements again.

Stand away from the antenna when making transmitter-on measurements. The adjustments have minimal effect on the previously made front-to-rear settings, and may be made in any band order. After making all the adjustments and sealing the gamma capacitors, reconnect the coax harness to the remote coax switch.

HF Mobile Antennas

This section is by Jack Kuecken, KE2QJ. Jack is an antenna engineer who has written a number of articles for ARRL publications.

An ideal HF mobile antenna is:

1. Sturdy. Stays upright at highway speeds.
2. Mechanically stable. Sudden stops or sharp turns do not cause it to whip about, endangering other vehicles.
3. Flexibly mounted. Permits springing around branches and obstacles at slow speeds.
4. Weatherproof. Handles the impact of wind, snow and ice at high speed.
5. Tunable to all of the HF bands without stopping the vehicle.
6. Mountable without altering the vehicle in ways which lower the resale value.
7. Efficient as possible.
8. Easily removed for sending the car through a car wash, etc.

For HF mobile operation, the ham must use an electrically small antenna. The possibility that the antenna might strike a fixed object places a limitation on its height. On Interstate highways, an antenna tip at 11.5 feet above the pavement is usually no problem. However, on other roads you may encounter clearances of 9.5 or 10 feet. You should be able to easily *tie down* the antenna for a maximum height of about 7 feet to permit passage through low-clearance areas. The antenna should be usable while in the tied-down position.

If the base of an antenna is 1 foot above the pavement and the tip is at 11.5 feet, the length is 10.5 feet which is 0.1 λ at 9.37 MHz, and 0.25 λ at 23.4 MHz. That means that the antenna will require a matching network for all of the HF bands except 10 and 12 meters

The power radiated by the antenna is equal to the radiation resistance times the square of the antenna current. The radiation resistance of an electrically small antenna is given by:

$$Rr = 395 \times (h/\lambda)^2$$

where

h = radiator height in meters
λ = wavelength in meters = 300/Freq in MHz

The capacitance in pF of an electrically small antenna is given approximately by:

$$C = (55.78 \times h)/((den1) \times (den2))$$

where

(den1) = (ln(h/r)-1)
(den2) = (1 − (F × h/75)²)
ln = natural logarithm
r = conductor radius in meters
F = frequency in MHz

Characteristics of a 10.5-foot (3.2 meter) whip with a 0.003 m radius and, assuming a base loading coil with a Q of 200 and coil stray capacitance of 2 pF, are given in **Table 20.19**.

Radiation resistance rises in a nonlinear fashion and the capacitance drops just as dramatically with increase in the ratio h/λ. **Fig 20.82** shows the relationship of capacitance to height. This can be used for estimating antenna capacitance for other heights.

Fig 20.83 shows that capacitance is not very sensitive to frequency for h/λ less than 0.075, 8 MHz in this case. However, the sensitivity increases rapidly thereafter.

Table 20.19 shows that at 3.5 MHz an inductance of 62.5 μH will cancel the capacitive reactance. This results in an impedance of 7.43 Ω which means that additional matching is required. In this case the radiation efficiency of the system is only 0.074 or 7.4%. In other words, nearly 93% of energy at the terminals is wasted in heating the matching coil.

System Q is controlled by the Q of the coil. The bandwidth between 2:1 SWR points of the system = 0.36 × F/Q. In this case, bandwidth = 0.36 × 3.5/200 = 6.3 kHz

If we could double the Q of the coil, the efficiency would double and the bandwidth would be halved. The converse is also true. In the interest of efficiency, the highest possible Q should be used!

Another significant factor arises from the high Q. Let's assume that we deliver

Table 20.19
Characteristics of a 10.5-foot whip antenna

F (MHz)	C (pF)	Rr	Impedance	Efficiency	L (μH)
1.8	30.1	0.146	13.72 −j2716	0.01064	240
3.5	30.6	0.55	7.43 −j1375	0.074	62.5
7	32.8	2.2	7.04 −j644	0.312	14.6
10	36.5	4.5	6.5 −j408	0.692	6.49
14	46.5	8.8	10 −j232	0.88	2.64

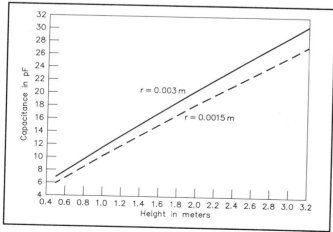

Fig 20.82—Relationship at 3.5 MHz between vertical radiator length and capacitance. The two curves show that the capacitance is not very sensitive to radiator diameter.

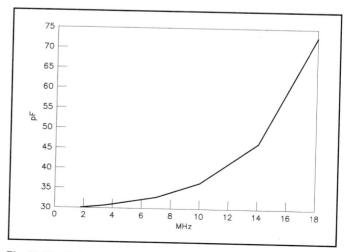

Fig 20.83—Relationship between frequency and capacitance for a 3.2-meter vertical whip.

100 watts to the 7.43 Ω at the antenna terminals. The current is 3.67 A and flows through the 1375-Ω reactance of the coil giving rise to 1375 × 3.67= 5046 VRMS (7137 Vpeak) across the coil.

With only 30.6 pF of antenna capacitance, the presence of significant stray capacitance at the antenna base shunts currents away from the antenna. RG-58 has about 21 pF/foot. A 1.5-foot length would halve the radiation efficiency of our example antenna. For cases like the whip at 3.5 MHz, the matching network has to be right at the antenna!

BASE, CENTER OR DISTRIBUTED LOADING

There is no clear-cut advantage in terms of radiation performance for either base or center-loaded antennas for HF mobile. Antennas with distributed (or continuous) loading have appeared in recent use. How do they compare?

BASE LOADING

In the design procedure, one estimates the capacitance, capacitive reactance and radiation resistance as shown previously. One then calculates the expected loss resistance of the loading coil required to resonate the antenna. There is generally additional resistance amounting to about half of the coil loss which must be added in. As a practical matter, it is usually not possible to achieve a coil Q in excess of 200 for such applications.

Using the radiation resistance plus 1.5 times the coil loss and the power rating desired for the antenna, one may select the wire size. For high efficiency coils, a current density of 1000 A/inch^2 is a good compromise. For the 3.67 A of the example we need a wire 0.068-inch diameter, which roughly corresponds to #14 AWG. Higher current densities can lead to a melted coil.

Design the coil with a pitch equal to twice the wire diameter and the coil diameter approximately equal to the coil length. These proportions lead to the highest Q in air core coils.

The circuit of **Fig 20.84** will match essentially all practical HF antennas on a car or truck. The circuit actually matches the antenna to 12.5 Ω and the transformer boosts it up to 50 Ω. Actual losses alter the required values of both the shunt inductor and the series capacitor. At a frequency of 3.5 MHz with an antenna impedance of 0.55 −j1375 Ω and a base capacitance of 2 pF results in the values shown in **Table 20.20**. Inductor and capacitor values are highly sensitive to coil Q. Furthermore, the inductor values are considerably below the 62.5 µH required

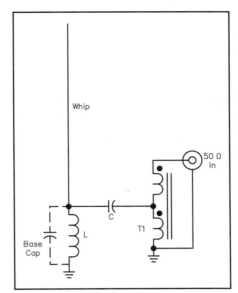

Fig 20.84—The base-matched mobile whip antenna.

to resonate the antenna.

This circuit has the advantage that the tuning elements are all at the base of the antenna. The whip radiator itself has minimal mass and wind resistance. In addition, the rig is protected by the fact that there is a dc ground on the radiator so any accidental discharge or electrical contact is kept out of the cable and rig. Variable tuning elements allow the

antenna to be tuned to other frequencies.

Connect the antenna, L and C. Start with less inductor than required to resonate the antenna. Tune the capacitor to minimum SWR. Increase the inductance and tune for minimum SWR. When the values of L and C are right, the SWR will be 1:1.

For remote or automatic tuning the drive motors for the coil and capacitor and the limit switches can be operated at RF ground potential. Mechanical connections to the RF components should be through insulated couplings.

CENTER LOADING

Center loading increases the current in the lower half of the whip as shown in **Fig 20.85**. One can start by calculating the capacitance for the section above the coil just as done for the base loaded antenna. This permits the calculation of the loading inductance. The center loaded antenna is often operated without any base matching in which case the resistive component can be assumed to be 50 Ω for purposes of calculating the current rating and selecting wire size for the inductor.

The reduced size top section results in reduced capacitance which requires a much larger loading inductor. Center loading requires twice as much inductive reactance as base loading. For equal coil Qs, loss resistance is twice as great for center loading. If the coil is above the center, the inductance must be even larger, and the loss resistance increases accordingly. These

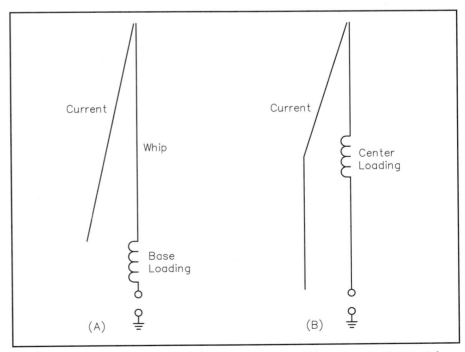

Fig 20.85—Relative current distribution on a base-loaded antenna is shown at A and for a center-loaded antenna at B.

factors tend to negate the advantage of the improved current distribution.

Because of the high value of inductance required, optimum Q coils are very large . One manufacturer of this type coil does not recommend their use in rain or inclement weather. The large wind resistance necessitates a very sturdy mount for operation at highway speed. Owing to the Q of these large coils the use of a base matching element in the form of either a tapped inductor or a shunt capacitor is usually needed to match to 50 Ω.

Another manufacturer places the coil above the center and uses a small extendable *wand* for tuning. To minimize wind resistance, the coil lengths are several times their diameters. These antenna coils are usually close wound with enameled wire. The coils are covered with a heat-shrink sleeve. If used in heavy rain or snow for extended periods water may get under the sleeving and seriously detune and lower the Q of the coils. These antennas usually do not require a base matching element. The resistance seems to come out close enough to 50 Ω.

It is possible to make a center-loaded antenna that is remotely tunable across the HF bands; however, this requires a certain amount of mechanical sophistication. The drive motor, limit switches and position sensor can be located in a box at the antenna base and drive the coil tuning mechanism through an electrically isolated shaft. Alternatively, the equipment could be placed adjacent to the loading coil requiring all of the electrical leads to be choked off to permit RF feeding of the base. The latter choice is probably the most difficult to realize.

CONTINUOUSLY LOADED ANTENNAS

Antennas consisting of a fiberglass sleeve with the radiator wound in a continuous spiral to shorten a CB antenna from 8.65 feet to 5 or 6 feet have been on the market for many years. This modest shortening has little impact on the efficiency but does narrow the bandwidth.

One line of mobile antennas uses periodic loading on a relatively small diameter tube. A series of taps along the length are used to select among the HF bands. An adjustable tip allows one to move about a single band. Because the length to diameter ratio is so large the loading coil Q is relatively low. The antenna is most effective above 20 meters.

The Screwdriver Antenna

The screwdriver antenna consists of a top whip attached to a long slender coil about 1.5 inches in diameter. The coil screws itself out of a base tube which has

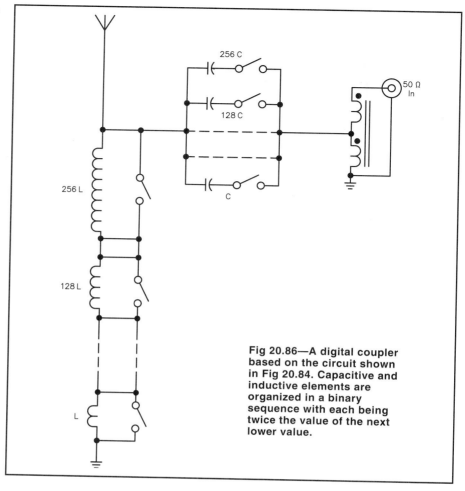

Fig 20.86—A digital coupler based on the circuit shown in Fig 20.84. Capacitive and inductive elements are organized in a binary sequence with each being twice the value of the next lower value.

a set of contact fingers at the top. For lower frequencies more of the coil is screwed out of the base tube and at maximum frequency the coil is entirely *swallowed* by the base tube.

The antenna is tunable over a wide range of frequencies by remote control. It has the advantage that the drive mechanism is operated at ground potential with RF isolation in the mechanical drive shaft. On the other hand, the antenna is not easily extended to 10.5 foot length for maximum efficiency on 80 and 40 meters. Because of its shape, coil Q will not be very high.

Table 20.20

Values of L and C for the Circuit of Fig 20.84 on 3.5 MHz

Coil Q	L (µH)	C (pF)	System Efficiency
300	44	11.9	0.083
200	29.14	35	0.0372
100	22.2	58.1	0.014

DIGITAL VERSUS ANALOG COUPLERS

Digital HF antenna couplers were first used by the military about 1960 for radios with Automatic Link Establishment. In this mode, the military radio has a list of frequencies ranging 2 to 30 MHz. It will try these in some sequence and will *lock* on the frequency giving the best reception. During the search, frequencies change much too fast to permit the use of conventional roller coils and motor driven vacuum capacitors. By comparison the digital coupler can jump from one memory setting to another in milliseconds.

For matching a mobile whip, the circuit shown in **Fig 20.86** will suffice. The inductor and capacitor can each be made up of about 8 binary sequenced steps. For example, at 3.5 MHz, the 10.5-foot antenna has an impedance of about $0.55 - j1684$ Ω. From Table 20.20 we see that we could use a series inductance sequence of 20, 10, 5, 2.5, 1.25, 0.625, 0.32 and 0.16 µH. We can use a relay to short unwanted elements. In this way we could theoretically produce any value of inductance between 0 and 39.84 µH in steps of 0.16 µH. In reality you will never reach a zero inductance. With all of the relays shorted, the wiring inductance and contact inductance of 8 relays appear in series. Also, each of the coils will have the open circuit capacitance of a relay contact across it in addition to the normal stray capacitance.

With most relays it does not make sense to switch less than 2 pF. For that reason, the capacitance chain would consist of 2, 4, 8, 16, 32, 64, 128 and possibly 256 pF. This would give a maximum of 510 pF and a step size of 2 pF. Each relay has an open circuit capacitance of about 1 pF, and that gives a minimum capacitance of 8 to 9 pF. As a practical matter, there is also the stray capacitance between the relay contacts and the coil windings.

In a high Q matching circuit that handles 100 W, the individual relays must handle 4 or 5 kV with the contacts open and several amperes of RF with the contacts closed. If we can unkey the transmitter so that the coupler will not have to switch under power, we'll still need some sizeable relays. If the inductors have lower Q, the voltages and currents will be correspondingly lower. Some military couplers use Jennings vacuum latching relays. This is expensive, as each of the 16 or 17 relays costs more than $100.

If coil and antenna Qs are kept or forced low, the voltages and currents become more reasonable. However, if the antenna

size is restricted this reduction comes only at the cost of decreased efficiency. A commercially available ham/marine digital coupler employs RF reed relays rated for 5 kV and 1 A, and restricts the power at low frequencies if the antenna is small. Another ham/marine unit uses small relays in series where voltage requirements are great and in parallel where current requirements are great—not good engineering practice. A third offering is not too specific about the power rating with very high Q loads.

There are no successful examples of 100 W plus couplers that use PIN diode switching. Their use is highly problematic given the high-Q loads they would handle.

A REMOTELY TUNED ANALOG ANTENNA COUPLER

KE2QJ built an antenna coupler designed for 100-W continuous-duty operation that will tune an antenna 10 feet or longer to any frequency from 3.5 to 30 MHz. With longer antennas, the power rating is higher and the lowest frequency is lower. The design requires only hand tools to build; however, access to a drill press and a lathe could save labor.

The roller coils and air variable capacitors to be used are not widely manufactured these days. Tube-type linear amplifiers still use air variable capacitors but these are generally built on order for the manufacturer and are not readily available to consumers in small quantity.

Until the 1970s, E. F. Johnson manu-

factured roller coils and air variable capacitors that were suitable for kilowatt amplifier finals and high power antenna couplers. On occasion one or more of these may be found in the original box, but they tend to be expensive. Ten Tec and MFJ both manufacture antenna couplers and offer some components in small quantities.

The following data refer to generic motors, capacitors and inductors. The descriptions are intended to aid the builder in selecting items from surplus, hamfest flea market offerings or salvage of old equipment.

THE MOTORS

Two motors are required, one to drive the inductor and one to drive the capacitor. The design employs permanent magnet dc gearhead motors with a nominal 12-V rating. A permanent magnet (PM) motor can be reversed by simply reversing the polarity of the drive voltage, and its speed can be controlled over a wide range by pulsing the power on and off with a variable duty cycle. The motor should have an output shaft speed on the order of 60 to 180 r/min (1 to 3 r/s) although this is not critical. New, such motors, can cost as much as $65 to $150 in small quantities. However, they can be found surplus and in repair shops for a few dollars.

The motor you are looking for is 1 to 1.5 inches in diameter and perhaps 2.5 inches long. It might be rated 12 or 24 V and have a $1/4$-inch diameter output shaft. At 12 V it

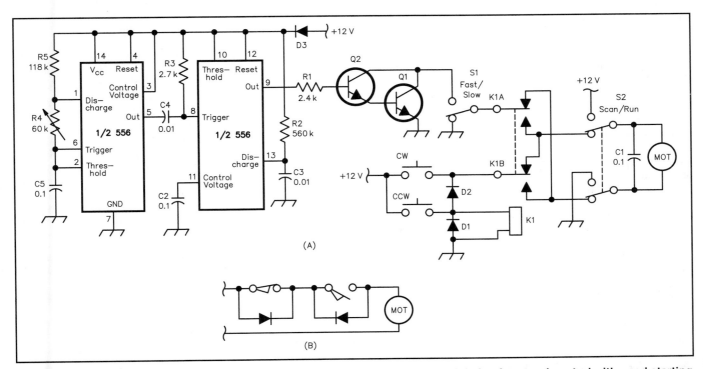

Fig 20.87—At A, motor control circuit used by KE2QJ. This circuit uses pulse modulation for speed control with good starting torque. Direction of rotation is controlled by the relay. At B, how to add limit switches to the circuit. See text.

should have enough torque to make it hard to stop the shaft with your fingers. Tape recorders, fax machines, film projectors, windshield wipers and copiers often use this type of motor.

LIMIT SWITCHES

On a remotely operated unit it is usually necessary to have limit switches to prevent the device from *crashing* into the ends. On an external roller coil these can be microswitches with paddles mounted on each end of the coil. As the coil is wound to one end, the roller operates the paddle and opens the limit switch which stops the motor.

Fig 20.87A illustrates a simple motor control circuit. Relay K1 is arranged as a DPDT polarity reversing switch. If switch CCW is pressed, the motor rotates CCW and the steering diode prevents the relay from operating. If CW is pressed, relay K1 operates reversing the polarity at the motor. The motor is energized through the steering diode.

Fig 20.87B shows how to add limit switches. The diodes across the switches are called anti-jam diodes. When a switch opens, the diode permits current to flow in the reverse direction and the motor to move the roller away from the open switch.

The photograph of **Fig 20.88** shows the mounting of the switches on the coil. The diode should be a power rectifier type rated for several times the motor current and at least 60 V.

POSITION READER

While not necessary, it is worthwhile to have a way to determine inductor position. An easy way to do this is to couple a 10-turn potentiometer to the coil shaft or drive gears. Make sure that the potentiometer turns less than 10 turns between limits. Don't try to make it come out exact.

Because the potentiometer is a light mechanical load, a belt drive reduction works well and won't slip if properly tensioned. Fig 20.88 shows the potentiometer and the gear drive.

You may be able to find suitable gears. However, a belt drive requires less precise shaft positioning than fine tooth gears. With a lathe, pulleys can be made in almost any ratio. Vacuum cleaner belts and O rings make handy belts.

COUPLINGS

In this coupler circuit, both ends of the capacitor are *hot* with RF although the end adjacent to the transformer is at relatively low voltage. Nevertheless, the capacitor shaft must be insulated from the motor shaft. The coil can be driven from the

Fig 20.88—Photo of the inductor drive assembly from KE2QJ's antenna coupler.

grounded end. Insulation is not necessary, but use a coupler between the motor shaft and the coil to compensate for any misalignment. Universal joints and insulating couplings are available from most electronics supply houses. You can make a coupling from a length of flexible plastic tubing which fits snugly over the shafts. Clamp the tubing to the shafts to avoid slippage.

THE CAPACITOR

The easiest capacitor to use is an air variable. It should have a range of approximately 10 to 250 pF. The plate spacing should be 2 mm ($^{1}/_{16}$ inch) or more, and the plate edges should be smooth and rounded. The capacitor should be capable of continuous 360° rotation, and it would be nice if it had ball bearings. The straight-line capacitance design is best for this application. Several capacitors of this type are available in military surplus ARC-5 series transmitters. These are approximately $2 \times 2 \times 3$ inches.

The capacitor must be mounted on stand-off insulators although high voltage will not be present on the frame. A cam that briefly operates a microswitch when the capacitor goes through minimum can be used to flash an LED on the remote control panel. This provides an indication that the capacitor is turning.

THE INDUCTOR

As calculated earlier, and assuming an inductor Q a bit under 300 is attainable, the roller inductor for this coupler should have a maximum inductance on the order of 40 µH. The wire should be at least #14 AWG wound about 8 t/inch.

You can use **Table 20.21** as a guide to buy a roller coil at a hamfest. The seller may not know the inductance of the coil. The antenna loading coil from an ARC-5 transmitter will work, but the wire is a bit small.

You could make the loading coil by threading 2, 2.5 or 3-inch diameter white, thick-wall PVC pipe with 8 t/inch. If the pipe is threaded in a lathe, the wire can be

wound into the threads under considerable tension. This helps to prevent the wires from coming loose with wear or temperature.

THE TRANSFORMER

The transformer consists of a bifilar winding on an Amidon FT-114-61 core. Start with two 2-foot lengths of #18 insulated wire; Teflon insulation is preferable. Twist the wire with a hand drill until there are about 5 t/inch (not critical). Wind 12 turns onto the core. This should about fill it up. Attach the starting end of one wire to the finish end of the other. This is the 12.5-Ω tap. One of the free ends is grounded and the other is the 50-Ω tap. Mount the coil on a plastic or wooden post through the center of the coil. A metal screw can be used as long as it does not make a complete turn around the core.

CONSTRUCTION

For ease of service, mount the inductor, its drive motor, position sensing potentiometer and limit switch assembly on a single aluminum plate. A plug and socket assembly permits rapid disconnection and removal. Make a similar assembly for the capacitor, its drive motor, transformer and the interrupter. Both assemblies should be made on $^{1}/_{16}$ to $^{1}/_{8}$-inch thick aluminum. These individual assemblies make it easier to fix problems.

The chassis shown in **Fig 20.89A** is made of a single piece $^{1}/_{16}$ to $^{3}/_{32}$-inch aluminum bent in an L shape. Two chassis-stiffening braces are riveted in place. Alternatively, the chassis can be made of flat sheets with aluminum angles riveted around the edge.

Mount the coil and capacitor assemblies parallel to the long leg of the L. Punch a 1-inch hole in the center of the short end of the L. Cover the hole with an insulator made of PVC, Teflon or other suitable material.

The rest of the case is a 4-sided wooden assembly as shown in Fig 20.89B. The back wall of the box is drilled to accept the two pivot pins. The box is slid over the chassis and the pivot pins engaged. The tie-down screw secures the box. For service, remove the tie-down screw and

Table 20.21
Data for 40 µH Coils

Diameter	Length	Turns
2.3 inch	5.625 inch	45
(58 mm)	(143 mm)	
2.8 inch	4.25 inch	34
(71 mm)	(108 mm)	
3.3 inch	3.375 inch	27
(84 mm)	(86 mm)	

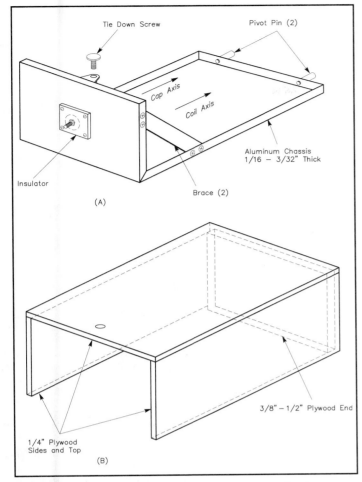

Fig 20.89—At A, the chassis for the coupler mounting box. At B, the box cover.

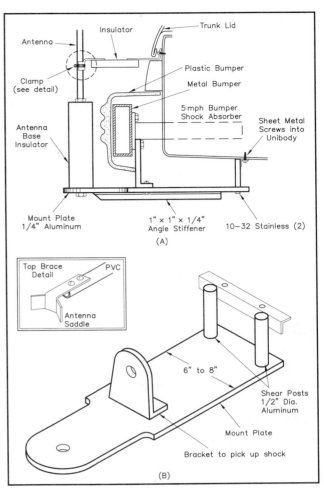

Fig 20.90—Antenna mounting detail. At A, the overall plan. At B, detail of the mount plate.

slide off the cover. The works of the coupler are very easy to get at!

The box is made of $1/4$-inch exterior grade plywood except for the back plate, which is $3/8$ or $1/2$-inch plywood. The sides, top and back should overlap the flanges on the chassis by $1/2$ inch. The inside corner seams of the box should be reinforced with $1/2$ or $3/4$-inch square strips. Assembly can be with any water resistant glue.

Finish the box, inside and out, with several coats of clear urethane varnish, sanding lightly between coats. This leaves a smooth plastic finish. This can be sprayed with an exterior paint that matches your car's color.

If the sides of the box fit closely over the flanges, no fastening beside the tie-down screw is required. A nearly perfect seal will leak out hot air when the sun shines on it and will draw in cold damp air in the evening, trapping moisture inside. A moderate fit will keep rain and snow out and permit the box to *breathe* freely, thereby keeping the inside dry.

MOUNTING THE WHIP AND THE BOX

Plastics in bumpers and bodies makes the mounting of a mobile whip antenna problematic. Modern bumpers are covered with plastic and the bumper is attached to the car unibody through a *5-MPH* shock absorber. The latter item is an unreliable ground.

The arrangement of **Fig 20.90** solves many of these problems. It uses a $1/4$-inch aluminum plate 6 to 8 inches wide and long enough to fit between a reasonably strong place on the unibody and the place behind the bumper where the antenna wants to be. This plate is fitted with an angle bracket for the lower bolt on the shock/bumper mounting. This plate is stiffened with a length of $1 \times 1 \times 1/4$-inch aluminum angle bolted in several places.

Near the forward edge of the plate, two $1/2$-inch diameter aluminum shear posts are fitted. The bottom of each is tapped 10-32 and bolted through the mount plate with a stainless 10-32 screw. At the top of these shear posts another piece of $1 \times 1 \times 1/2$-inch

angle is attached which is screwed to the unibody with three or four #10 stainless sheet metal screws. A bracket attaches the mount plate to the bumper's shock absorber. The angle bracket may either be welded to the plate or bolted with angle stock. In the event that the car is hit from behind or backs into an obstacle, the two 10-32 screws will shear off, thereby preventing the mount from defeating the 5 MPH crushable shock absorber. The part protruding behind the bumper may be cut down in width to 3 inches and rounded for appearance and safety.

Any type of base insulator may be used, but try to bring the base of the antenna to the height of the coupler output terminal. You can make a good base insulator from thick-wall white PVC $1 1/2$ inch pipe. Reinforce each end. Start with a $1 1/2$ inch length of pipe. Remove a $5/8$-inch wide strip so the remaining portion can be rolled and pressed into the open end of the insulator. Apply PVC pipe glue just before pressing in the piece; this gives the

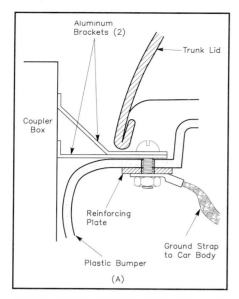

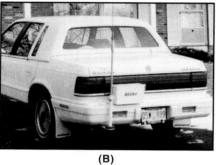

(B)

Fig 20.91—Box mounting detail. At A, mounting-bracket design. At B, photo of KE2QJ's installation.

insulator a double wall thickness at each end. Aluminum plugs can be turned for a snug fit and tapped for $^3/_8$-24 hardware. The plugs can be held in place with 8-32 stainless screws.

The upper antenna brace has an aluminum plate at one end that goes under the trunk lid (see Fig 20.90). A length of $^1/_2$-inch diameter heavy-wall white PVC pipe, which serves as an insulator, is screwed to this. At the other end of the insulator, another aluminum piece is bent to form a saddle for the antenna which is clamped to the saddle. This clamp should be as high as convenient above the mount plate, preferably not less than a foot. The mount plate should be sturdy enough for you to stand on and with the brace will easily hold a whip upright at 70 MPH or more.

The coupler box mounting is shown in **Fig 20.91**. Brackets can be made of $^1/_8 \times$ 2-inch aluminum with a brace going perhaps 2 inches from the corner. The brackets bolt or rivet to the chassis. The bracket reaches through the gap between the trunk lid and the plastic top of the bumper. For reinforcement, a pair of reinforcement plates $1.5 \times 2 \times ^1/_4$-inch

thick are bolted to the plastic on the under side of the bumper. Ground the reinforcement plates to the unibody with some $^3/_4$ or 1 inch ground braid.

Two 10-32 stainless screws hold each reinforcement plate to the plastic bumper and a central $^1/_4$-20 tapped hole holds down the box bracket. One need only remove two screws to get the box off the car for car wash, etc. You have to open the trunk to remove the antenna coupler, and this provides a measure of security.

THE SPRING AND WHIP

A section of 1-inch diameter aluminum tubing extends from the top of the insulator to the base of the spring. It's usually best to have the spring about 4 feet above the pavement. The type used for CB whips works well. A 7-foot whip brings the top to about 11.5 feet above the pavement. The 7-foot whip can be a cut down CB unit. Don't use the type with helical winding. When the antenna is tied down, the bow of the whip should be about 7 feet above the pavement.

TUNING

It is best to initially tune the antenna using low power. A power attenuator just after the transceiver will limit SWR, but your SWR indicator must be on the antenna side of the attenuator.

For a first tune-up set the capacitor control to SCAN and slowly advance the inductor from minimum inductance toward maximum. As the inductor approaches the correct value the SWR will start to kick down. At this point take the capacitor off of SCAN and JOG it to a best tune. Next, JOG the inductor and repeat; the SWR should go down. Continue until a 1:1 SWR is obtained. Record the potentiometer setting. The next time you want to use this frequency run the coil directly to the logged setting.

In the SCAN position the capacitor motor runs at full voltage. When you JOG the capacitor for low SWR you will find the speed far too fast for sharp tuning. The slow speed tuning is provided by using duty-factor modulation of the motor current. The circuit of Fig 20.87A supplies fixed width pulses with variable timing. At the slowest speed, the unit will supply about one pulse per second and the motor shaft will rotate one degree or so per pulse. The full voltage pulse provides good starting torque.

If the SWR cannot be brought to 1:1, examine the coil and capacitor to see whether either is at maximum or minimum. At high frequencies above 24 MHz it may be necessary to place a capacitor between the coupler and the antenna base.

RADIATION PATTERNS

At the lower frequencies the pattern

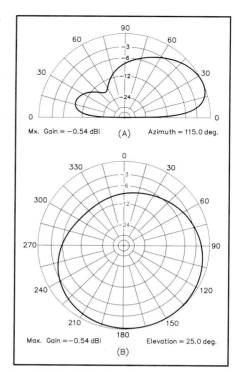

Fig 20.92—At A, elevation pattern of the KE2QJ mobile antenna. The pattern is in the plane that runs diagonal through the car. At B, azimuth pattern at 25° elevation for the same antenna. The operating frequency is 18.130 MHz.

tends to be essentially round in azimuth. At 20 meters the pattern tends to become more and more directive. The patterns in **Fig 20.92** were calculated using *EZNEC*. The frequency is 18.13 MHz and the antenna is mounted at the left rear corner of a mid-size sedan. It may be seen that the pattern has more than 10 dB maximum-to-minimum ratio with the broad maximum along the diagonal of the vehicle occupied by the antenna. If the antenna were mounted in the center of the vehicle, the omnidirectional characteristics would be improved. However, the antenna would have to be much shorter to stay under 11.5 feet. The shorter antenna would likely be weaker in its best direction than the taller antenna is in its worst.

References

J. S. Belrose, VE3BLW, "Short Antennas for Mobile Operation," *QST,* Sept 1953, pp 30-35, 109.

J. Kuecken, KE2QJ, *Antennas and Transmission Lines*, MFJ Publishing, Ch 25.

J. Kuecken, KE2QJ, "A High Efficiency Mobile Antenna Coupler," *The ARRL Antenna Compendium, Vol 5*, pp 182-188.

J. Kuecken, KE2QJ, "Easy Homebrew Remote Controls," *The ARRL Antenna Compendium, Vol 5*, pp 189-193.

J. Kuecken, KE2QJ, "A Remote Tunable Center Loaded Antenna," *The ARRL Antenna Compendium, Vol 6*.

VHF/UHF Antennas

INTRODUCTION

Improving an antenna system is one of the most productive moves open to the VHF enthusiast. It can increase transmitting range, improve reception, reduce interference problems and bring other practical benefits. The work itself is by no means the least attractive part of the job. Even with high-gain antennas, experimentation is greatly simplified at VHF and UHF because an array is a workable size, and much can be learned about the nature and adjustment of antennas. No large investment in test equipment is necessary.

Whether we buy or build our antennas, we soon find that there is no one "best" design for all purposes. Selecting the antenna best suited to our needs involves much more than scanning gain figures and prices in a manufacturer's catalog. The first step should be to establish priorities.

Gain

As has been discussed previously, shaping the pattern of an antenna to concentrate radiated energy, or received signal pickup, in some directions at the expense of others is the only possible way to develop gain. Radiation patterns can be controlled in various ways. One is to use two or more driven elements, fed in phase. Such arrays provide gain without markedly sharpening the frequency response, compared to that of a single element. More gain per element, but with some sacrifice in frequency coverage, is obtained by placing parasitic elements into a Yagi array.

Radiation Pattern

Antenna radiation can be made omnidirectional, bidirectional, practically unidirectional, or anything between these conditions. A VHF net operator may find an omnidirectional system almost a necessity but it may be a poor choice otherwise. Noise pickup and other interference problems tend to be greater with omnidirectional antennas. Maximum gain and low radiation angle are usually prime interests of the weak-signal DX aspirant. A clean pattern, with lowest possible pickup and radiation off the sides and back, may be important in high-activity areas, where the noise level is high, or when challenging modes like EME (Earth-Moon-Earth) are employed.

Height Gain

In general, the higher a VHF antenna is installed, the better will be the results. If raising the antenna clears its view over nearby obstructions, it may make dramatic improvements in coverage. Within reason, greater height is almost always worth its cost, but height gain must be balanced against increased transmission-line loss. Line losses can be considerable at VHF, and they increase with frequency. The best available line may be none too good, if the run is long in terms of wavelength. Consider line losses in any antenna planning.

Physical Size

A given antenna design for 432 MHz, say a 5-element Yagi on a 1-λ boom, will have the same gain as one for 144 MHz, but being only one-third the size it will intercept only one-ninth as much energy in receiving. Thus, to be equal in communication effectiveness, the 432-MHz array should be at least equal in physical size to the 144-MHz one, requiring roughly three times the number of elements. With all the extra difficulties involved in going higher in frequency, it is well to be on the big side in building an antenna for the UHF bands.

DESIGN FACTORS

Having sorted out objectives in a general way, we face decisions on specifics, such as polarization, type of transmission line, matching methods and mechanical design.

Polarization

Whether to position the antenna elements vertically or horizontally has been a question since early VHF pioneering. Tests show little evidence on which to set up a uniform polarization policy. On long paths there is no consistent advantage, either way. Shorter paths tend to yield higher signal levels with horizontal in some kinds of terrain. Man-made noise, especially ignition interference, tends to be lower with horizontal. Verticals, however, are markedly simpler to use in omnidirectional systems and in mobile work.

Early VHF communication was largely vertical, but horizontal gained favor when directional arrays became widely used. The major trend to FM and repeaters, particularly in the 144-MHz band, has tipped the balance in favor of verticals in mobile work and for repeaters. Horizontal predominates in other communication on 50 MHz and higher frequencies. It is well to check in advance in any new area in which you expect to operate, however, as some localities may use vertical polarization. A circuit loss of 20 dB or more can be expected with cross-polarization.

Transmission Lines

There are two main categories of transmission lines used at HF through UHF: balanced and unbalanced. Balanced lines include *open-wire lines* separated by insulating spreaders, and *twin-lead*, in which the wires are embedded in solid or foamed insulation. Unbalanced lines are represented by the family of coaxial cables, commonly called *coax*. Line losses in either types of line result from ohmic resistance, radiation from the line and deficiencies in the insulation.

Large conductors, closely spaced in terms of wavelength, and using a minimum of insulation, make the best balanced lines. Characteristic impedances are between 300 to 500 Ω. Balanced lines work best in straight runs, but if bends are unavoidable, the angles should be as gentle as possible. Care should also be taken to prevent one wire from coming closer to metal objects than the other.

Properly built open-wire line can operate with very low loss in VHF and even UHF installations. A total line loss under 2 dB per hundred ft at 432 MHz is readily obtained. A line made of #12 wire, spaced $^3/_4$ inch or less with Teflon spreaders, and running essentially straight from antenna to station, can be better than anything but the most expensive "Hardline" coax, at a fraction of the cost. This assumes the use of high-quality baluns to match into and out of the balanced line, with a short length of low-loss coax for the rotating section from the top of the tower to the antenna. A similar 144-MHz setup could have a line loss under 1 dB.

Small coax such as RG-58 or RG-59 should never be used in VHF work if the run is more than a few feet. Half-inch lines (RG-8 or RG-11) work fairly well at 50 MHz, and are acceptable for 144-MHz runs of 50 ft or less. If these lines have foam rather than solid insulation they are about 30% better. Aluminum-jacket "Hardline" coaxial cables with large inner conductors and foam insulation are well worth their cost. Hardline can sometimes even be obtained for free from local Cable TV operators as "end runs"—pieces at the end of a roll. The most common CATV variety is ½-inch OD 75-Ω Hardline. Waterproof commercial connectors for Hardline are fairly expensive, but enterprising amateurs have "home-brewed" low-cost connectors. If they are properly waterproofed, connectors and Hardline can last almost indefinitely. Of course, a disadvantage implied by their name is that Hardline must not be bent too sharply, because it will kink. See *The ARRL Antenna Book* for details on Hardline connectors.

Effects of weather on transmission lines should not be ignored. A well-constructed open-wire line works well in nearly any weather, and it stands up well. TV type twin-lead is almost useless in heavy rain, wet snow or icing. The best grades of coax are impervious to weather. They can be run underground, fastened to metal towers without insulation, or bent into almost any convenient position, with no adverse effects on performance. However, beware of "bargain" coax. Lost transmitter power can be made up to some extent by increasing power, but once lost, a weak signal can never be recovered in the receiver.

Impedance Matching

Theory and practice in impedance matching are given in detail in the **Transmission Lines** chapter, and in theory, at least, is the same for frequencies above 50 MHz. Practice may be similar, but physical size can be a major modifying factor in choice of methods.

Delta Match

Probably the first impedance match was made when the ends of an open line were fanned out and tapped onto a half-wave antenna at the point of most efficient power transfer, as in **Fig 20.93A**. Both the side length and the points of connection either side of the center of the element must be adjusted for minimum reflected power in the line, but the impedances need not be known. The delta makes no provision for tuning out reactance, so the length of the dipole is pruned for best SWR.

Once thought to be inferior for VHF applications because of its tendency to radiate if adjusted improperly, the delta has come back to favor now that we have good methods for measuring the effects of matching. It is very handy for phasing multiple-bay arrays with low-loss open lines, and its dimensions in this use are not particularly critical.

Gamma and T Matches

The gamma match is shown in Fig 20.93C, and the T match is shown in Fig 20.93D. These matches are covered in more detail in the chapter on **Transmission Lines**. There being no RF voltage at the center of a half-wave dipole, the outer conductor of the coax is connected to the element at this point, which may also be the junction with a metallic or wooden boom. The inner conductor, carrying the RF current, is tapped out on the element at the matching point. Inductance of the arm is canceled by means of C1. Both the point of contact with the element and the setting of the capacitor are adjusted for zero reflected power, with a bridge connected in the coaxial line.

The capacitor can be made variable temporarily, then replaced with a suitable fixed unit when the required capacitance value is found, or C1 can be mounted in a waterproof box. Maximum capacitance should be about 100 pF for 50 MHz and 35 to 50 pF for 144 MHz. The capacitor and arm can be combined with the arm connecting to the driven element by means of a sliding clamp, and the inner end of the arm sliding inside a sleeve connected to the inner conductor of the coax. It can be constructed from concentric pieces of tubing, insulated by plastic sleeving or shrink tubing. RF voltage across the capacitor is low, once the match is adjusted properly, so with a good dielectric, insulation presents no great problem, if the initial adjustment is made with low power. A clean, permanent, high-conductivity bond between arm and element is important, as the RF current is high at this point.

Because it is inherently somewhat unbalanced, the gamma match can sometimes introduce pattern distortion, particularly on long-boom, highly directive Yagi arrays. The T-match, essentially two gamma matches in series creating a balanced feed system, has become popular for this reason. A coaxial balun like that shown in Fig 20.93B is used from the bal-

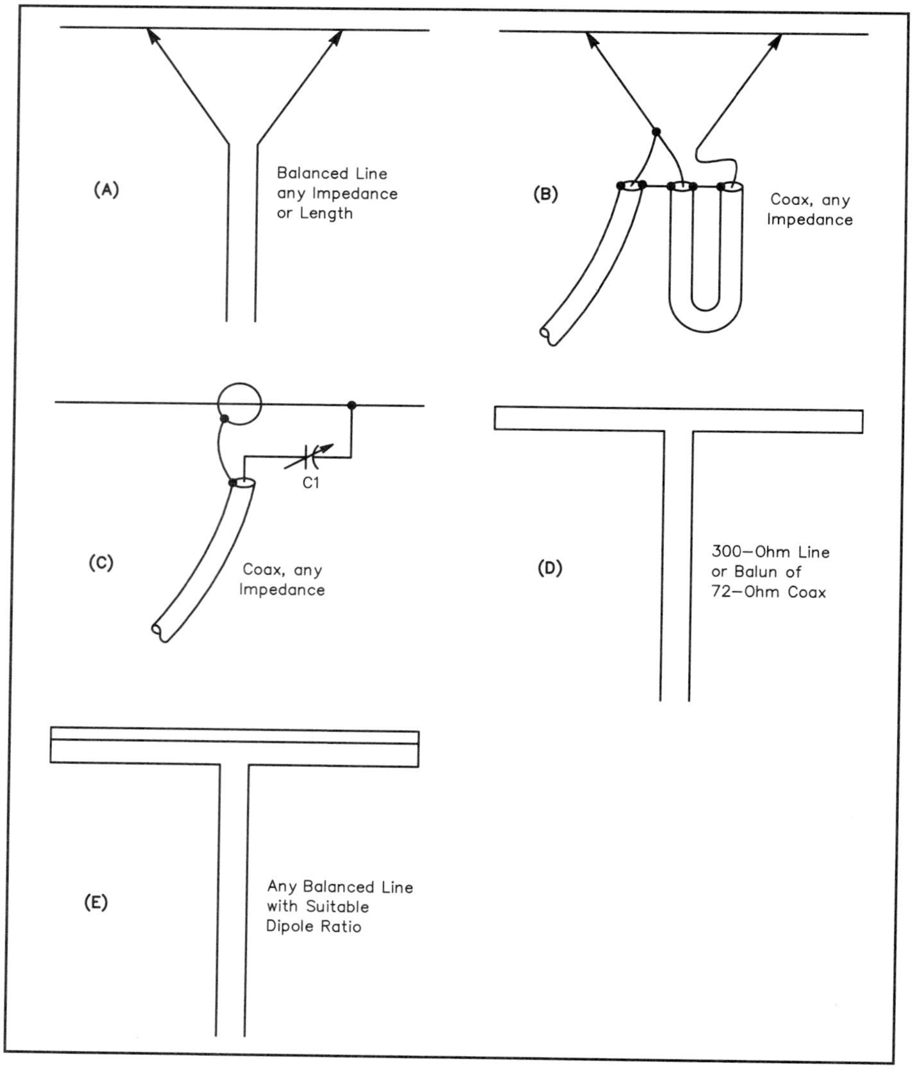

Fig 20.93—Matching methods commonly used in VHF antennas. In the delta match, A and B, the line is fanned out to tap on the dipole at the point of best impedance match. The gamma match, C, is for direct connection of coax. C1 tunes out inductance in the arm. Folded dipole of uniform conductor size, D, steps up antenna impedance by a factor of four. Using a larger conductor in the unbroken portion of the folded dipole, E, gives higher orders of impedance transformation.

anced T-match to the unbalanced coaxial line going to the transmitter. See K1FO's Yagi designs later in this chapter for details.

Folded Dipole

The impedance of a half-wave antenna broken at its center is 72 Ω. If a single conductor of uniform size is folded to make a half-wave dipole, as shown in Fig 20.93D, the impedance is stepped up four times. Such a folded dipole can thus be fed directly with 300-Ω line with no appreciable mismatch. Coaxial line of 70 to 75 Ω impedance may also be used if a 4:1 balun is added. Higher impedance step-up can be obtained if the unbroken portion is made larger in cross-section than the fed portion, as in Fig 20.93E.

Baluns

Conversion from balanced loads to unbalanced lines, or vice versa, can be performed with electrical circuits, or their equivalents made of coaxial line. A balun made from flexible coax is shown in **Fig 20.94A**. The looped portion is an electrical half-wave. The physical length depends on the propagation factor of the line used, so it is well to check its resonant frequency, as shown at B. The two ends are shorted, and the loop at one end is coupled to a dip-meter coil. This type of balun gives an impedance step-up of 4:1, 50 to 200 Ω, or 75 to 300 Ω typically.

Coaxial baluns giving a 1:1 impedance transfer are shown in **Fig 20.95**. The coaxial sleeve, open at the top and connected to the outer conductor of the line at the lower end (A) is the preferred type. A conductor of approximately the same size as the line is used with the outer conductor to form a quarter-wave stub, in B. Another piece of coax, using only the outer conductor, will serve this purpose. Both baluns are intended to present an infinite impedance to any RF current that might otherwise tend to flow on the outer conductor of the coax.

Stacking Yagis

Where suitable provision can be made for supporting them, two Yagis mounted one above the other and fed in phase may be preferable to one long Yagi having the same theoretical or measured gain. The pair will require a much smaller turning space for the same gain, and their lower radiation angle can provide interesting results. On long ionospheric paths a stacked pair occasionally may show an apparent gain much greater than the 2 to 3 dB that can be measured locally as the gain from stacking.

Optimum spacing for Yagis with booms

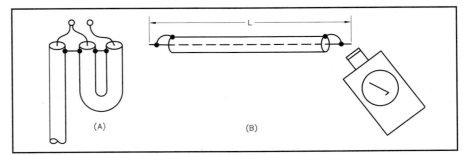

Fig 20.94—Conversion from unbalanced coax to a balanced load can be done with a half-wave coaxial balun, A. Electrical length of the looped section should be checked with a dip meter, with ends shorted, B. The half-wave balun gives a 4:1 impedance step up.

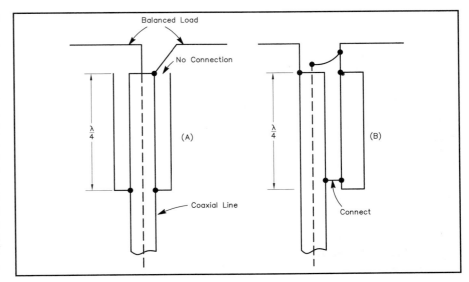

Fig 20.95—The balun conversion function, with no impedance change is accomplished with quarter-wave lines, open at the top and connected to the coax outer conductor at the bottom. The coaxial sleeve, A, is preferred.

longer than 1 λ is one wavelength, but this may be too much for many builders of 50-MHz antennas to handle. Worthwhile results are possible with as little as $^1/_2$ λ (10 ft), but $^5/_8$ λ (12 ft) is markedly better. The difference between 12 and 20 ft may not be worth the added structural problems involved in the wider spacing, at 50 MHz at least.

The closer spacings give lowered measured gain, but the antenna patterns are cleaner (less power in the high-angle elevation lobes) than with one-wavelength spacing. Extra gain with wider spacings is usually the objective on 144 MHz and higher bands, where the structural problems are not quite as severe as on 50 MHz.

One method for feeding two 50-Ω antennas, as might be used in a stacked Yagi array, is shown in **Fig 20.96**. The transmission lines from each antenna, with a balun feeding each antenna (not shown in the drawing for simplicity), to the common feedpoint must be equal in length and an odd multiple of a quarter wavelength. This line acts as an quarter-wave (Q-sec-

tion) impedance transformer and raises the feed impedance of each antenna to 100 Ω. When the coaxes are connected in parallel at the coaxial T fitting, the resulting impedance is close to 50 Ω.

Circular Polarization

Polarization is described as "horizontal" or "vertical," but these terms have no meaning once the reference of the Earth's surface is lost. Many propagation factors can cause polarization change—reflection or refraction and passage through magnetic fields (Faraday rotation), for example. Polarization of VHF waves is often random, so an antenna capable of accepting any polarization is useful. Circular polarization, generated with helical antennas or with crossed elements fed 90° out of phase, will respond to any linear polarization.

The circularly polarized wave in effect threads its way through space, and it can be left- or right-hand polarized. These polarization senses are mutually exclusive, but either will respond to any plane (horizontal or vertical) polarization. A wave generated

Fig 20.96—A method for feeding a stacked Yagi array. Note that baluns at each antenna are not specifically shown. Modern-day practice is to use current ("choke") baluns made up of ferrite beads slipped over the outside of the coax and taped to prevent movement. See Transmission Lines chapter for details.

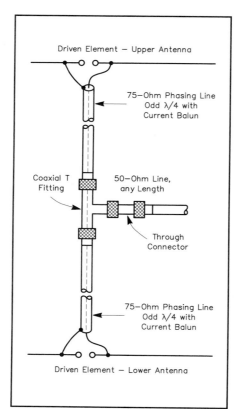

with right-hand polarization, when reflected from the moon, comes back with left-hand plarization, a fact to be borne in mind in setting up EME circuits. Stations communicating on direct paths should have the same polarization sense.

Both senses can be generated with crossed dipoles, with the aid of a switchable phasing harness. With helical arrays, both senses are provided with two antennas wound in opposite directions.

Hardwood Dowels Strengthen Antenna Elements

There is a very simple and inexpensive way to strengthen your beam antenna prior to hoisting it to the top of the tower. Most well-stocked hardware stores stock assorted sizes of hardwood dowel. Select dowels that can be snugly inserted inside the boom and elements and placed where they are attached to each other. Carefully redrill any holes blocked by the dowels and assemble the beam as instructed. You will then have a more solid attachment point and can really tighten the U-bolts and fittings without fear of deforming the tubing. Lightweight beams can be strengthened by inserting the dowel in the entire length of the boom. This method can be used to straighten elements that have been damaged. These dowels add very little weight to the antenna. I have done this for over 25 years and am completely satisfied with the results.—H. A. "Tony" Miller, W5BWA, Alexandria, Louisiana

A QUICK ANTENNA FOR 223 MHz

Here is a functional antenna for 223 MHz that you can make in less than an hour (Fig A). To build it, you'll need 9 feet of #10 copper wire, 6 inches of small-diameter copper tubing, and a 10-foot length of PVC pipe or some other physical support.

Bend the antenna from one piece of wire. Slide the copper tubing over the top end of the antenna, and adjust how far it extends beyond the wire to get the lowest SWR. (Don't handle the antenna while transmitting—make adjustments only while receiving.) For more precision, you can move the coaxial feed line's taps on the antenna's matching stub (the 12-inch section at the bottom) about an eighth of an inch at a time. My antenna shows an SWR of 1.2 at 223 MHz.—William Bruce Cameron, WA4UZM, Temple Terrace, Florida

Fig A—WA4UZM's quick 223-MHz antenna gets you going on 1¼ meters in a hurry. (The text explains how to adjust it for minimum SWR.) To support the antenna, lash it to a piece of PVC pipe with nylon cable ties.

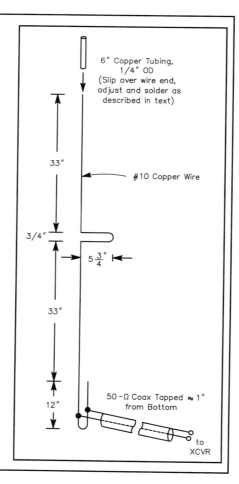

Car-Engine Heater Keeps Rotator Lubricant Flowing

My antenna rotator manual says that the rotator's lubricant should flow at temperatures below −20 °F, but during a −28° spell last winter, nothing moved. I attached a magnetically mounted car engine heater to the bottom of the rotator mounting plate, warmed up the rotator, and got it moving again.— *Richard Mollentine, WA0KKC, Overland Park, Kansas*

¹/₄-λ ANTENNAS FOR HOME, CAR AND PORTABLE USE

Quarter-wave vertical antennas are useful for local communications when size, cost and ease of construction are important. For theoretical information about quarter-wave vertical antennas see earlier parts of this chapter or *The ARRL Antenna Book*.

Construction Materials

The antennas shown in the following sections are built on a coaxial connector. Use UHF or N connectors for the fixed station antennas. BNC connectors are good for mobile and portable antennas. BNC and N connectors are better than PL-259 connectors for VHF/UHF outdoor use because: (1) they provide a constant impedance over the frequencies of interest, and (2) they are weatherproof when the appropriate connector or cap is attached. The ground-plane antennas require a panel-mount connector (it has mounting holes to hold the radials).

If the antenna is sheltered from weather, copper wire is sufficiently rigid for the element and radials. Antennas exposed to the weather should be made from ¹/₁₆- to ¹/₈-inch brass or stainless-steel rod.

Radials may be made from ³/₁₆-inch aluminum rod or tubing and mounted on an aluminum sheet. Do not use aluminum for the antenna element because it cannot be easily soldered to the coaxial-connector center pin.

Where the figures call for #4-40 hardware, stainless steel or brass is best. Use cadmium-plated hardware if stainless steel or brass is not available.

Fixed-Station Antennas

The ground-plane antenna in **Fig 20.97** uses female chassis-mount connectors to support the element and four radials. If you have chosen large-diameter wire or tubing for the radials refer to **Fig 20.98**. Cut a metal sheet as shown (size is not critical, and the mounting tab is optional). Drill the sheet to accept the coaxial connector on hand (usually ¹¹/₁₆ inch) and the 4-40 hardware for the radials and connector. Bend the plate or radials as shown with the aid of a bench vise. Mount the coaxial connector and radials to the plate.

Small diameter (¹/₁₆-inch) radials may be attached directly to the mounting lugs of the coaxial connector with 4-40 hardware. To install ³/₃₂- or ¹/₈-inch radials, bend a hook at one end of each radial for insertion through the connector lug. (You may need to enlarge the lug holes slightly for ¹/₈-inch rod.) Solder the radials (and hardware, if used) to the connector using a large soldering iron or propane torch.

Solder the element to the center pin of the connector. If the element does not fit inside the solder cup, use a short section of brass tubing as a coupler (a slotted ¹/₈-inch-ID tube will fit over an SO-239 or N-receptacle center pin).

One mounting method for fixed-station antennas appears in Fig 20.97. The method shown is probably the easiest and strongest. Alternatively, a tab (Fig 20.98D) or "L" bracket could be fastened to the side of a mast with a hose clamp. Once the antenna is mounted and tested, thoroughly seal the open side of the coaxial connector with RTV sealant, and weatherproof the connections with rust-preventative paint.

Mobile Applications

In order to achieve a perfect omnidirectional radiation pattern, mobile vertical antennas should be located in the center of the vehicle roof. Practically speaking however, vertical antennas work well anywhere on a metal auto body (although the radiation pattern may not be omnidirectional). In the interest of RF safety, antennas that are not mounted on the roof should be placed as far from the vehicle occupants as is practical.

The mobile antenna shown here is based on a mating pair of BNC connectors. (You could use similar techniques with N connectors.) If you plan to remove the antenna element, obtain a matching connector cap to protect the open jack from weather. The jack may be mounted directly through the vehicle body if you wish. If you do so, select a body panel with sufficient strength to support the antenna, and be sure that there is adequate access to both sides of the panel for connector and cable installation.

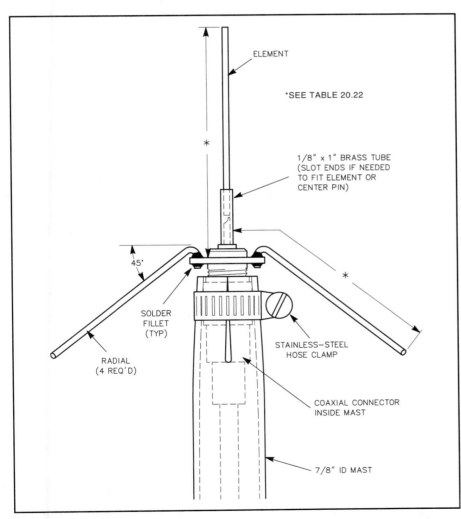

ELEMENT

*SEE TABLE 20.22

1/8" x 1" BRASS TUBE
(SLOT ENDS IF NEEDED
TO FIT ELEMENT OR
CENTER PIN)

45°

SOLDER
FILLET
(TYP)

RADIAL
(4 REQ'D)

STAINLESS–STEEL
HOSE CLAMP

COAXIAL CONNECTOR
INSIDE MAST

7/8" ID MAST

Fig 20.97—A simple groundplane antenna for the 144, 222 and 440-MHz bands. The feed line and connector are inside the mast, and a hose clamp squeezes the slotted mast end to tightly grip the plug body. See Table 20.22 for element and radial measurements

Most homemade VHF and UHF mobile antennas are mounted on an $\frac{1}{16}$-inch-thick aluminum L bracket, which is fastened to the side of the hood or trunk opening (see **Fig 20.99A** and **Fig 20.99B**). If the opening lip slopes at the antenna location, the lower edge of the L bracket should match that slope. Hold the sheet against the inside of the opening lip, orient the sheet so that its top edge is parallel with the ground, and mark where the opening lip crosses the edges. Connect the marks with a straight line, and cut the sheet on that line.

Drill the sheet as required for the coaxial connector, and bend as indicated.

Better Adhesion for Suction-Cup Mounted Antennas

I have used a suction-cup-mounted Squalo antenna on my car for over 60,000 miles without the suction cups coming loose—after I put silicone grease on the edge of the suction cups. (Apparently, the grease fills voids in the rubber.) The silicone grease has not damaged my car's finish.—*Roger Gibson, K4KLK, Raleigh, North Carolina*

Mount the coaxial connector, and install the cable. Hold the bracket against the side of the opening, drill and secure it with two #6 or #8 sheet-metal screws and lock washers.

The antenna shown here was made for the 2-m band. The element is about $19\frac{1}{4}$ inches of $\frac{1}{8}$-inch brass rod (from a welding-supply store). Refer to Fig 20.99C and D and follow this procedure to install the element in the male BNC connector:

1) Prepare a special insulator from a 1-inch piece of RG-58 cable by removing all of the jacket and shield. Also remove the center-conductor insulation as shown.

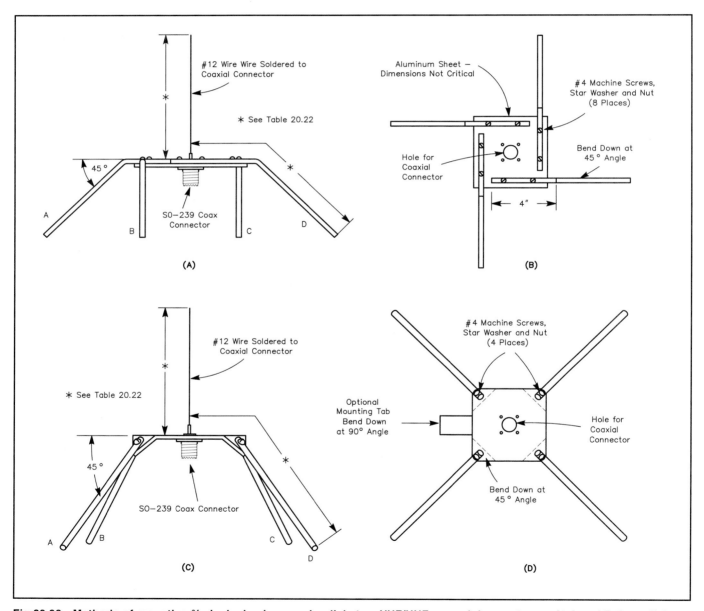

Fig 20.98—Methods of mounting $\frac{3}{16}$-inch aluminum-rod radials to a VHF/UHF groundplane antenna. At A and B the radials are made approximately $1\frac{1}{2}$ inches longer than ℓ, then bent (45°) and attached to a flat aluminum sheet. At C and D, the radials are somewhat shorter than ℓ, and the corners of the aluminum sheet are bent to provide the 45° angle. In both cases, ℓ is measured from the radial tip to the element. The size of the aluminum sheet is not critical. The mounting tab shown at D is optional; it could be added to the sheet in A and B if desired.

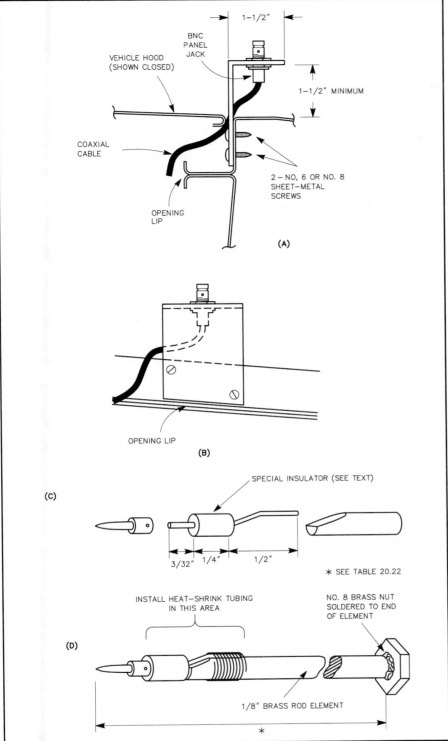

Table 20.22

¼-Wavelength Vertical Antenna Element and Radial Sizes

	Band		
	144 MHz	222 MHz	440 MHz
Lengths			
ℓ†	19¼"	12½"	6⁵/₁₆"
Diameters			
Brass rod	1/8"	3/32"	1/16"
Stainless-steel rod	3/32"	1/16"	1/16"

$$^†\ell = \frac{234}{f}$$

where
  ℓ = length, in ft
  f = frequency, in MHz

2) File or grind an angle on the end of the brass rod.

3) Place the special insulator against the end of the brass rod and lash them together with fine bare wire.

4) Solder the wire, lashing and BNC center pin in place. (Use a light touch; too much heat causes the insulation to swell and makes it difficult to assemble the connector.)

5) Apply heat-shrink tubing to cover the special insulator and lashing. Place the tubing so that at least ¹/₈ inch of the special insulator remains exposed at the center pin.

6) Place the element in the connector body and fill the back of the connector with epoxy to support the element. Install the BNC nut before the epoxy sets. (The BNC washer, clamp and gasket are discarded.)

7) Solder a #8 brass nut (or other rounded brass object) to the end of the element to serve as eye protection.

This general procedure can be adapted to all of the element materials recommended and to N connectors as well.

Fig 20.99—Details of the mobile vertical mount and BNC center-pin connection. A shows a cross section of a typical vehicle hood opening with the hood closed. B shows the lower edge of the L bracket trimmed to match the slope of the hood (this ensures that the antenna is plumb). C shows the BNC center pin, the special insulator made from RG-58 and the end of the element shaped for connection. At D, the special insulator has been lashed to the element with fine bare wire and the center pin positioned. The center pin and lashing should be soldered, and the area indicated should be covered with heat-shrink tubing to prevent contact with the plug body when it is installed. The brass nut serves as eye protection.

AN ALL-COPPER 2-M J-POLE

Rigid copper tubing, fittings and assorted hardware can be used to make a really rugged J-pole antenna for 2 m. When copper tubing is used, the entire assembly can be soldered together, ensuring electrical integrity, and making the whole antenna weatherproof. This material came from an article by Michael Hood, KD8JB, in *The ARRL Antenna Compendium, Vol. 4.*

No special hardware or machined parts are used in this antenna, nor are insulating materials needed, since the antenna is always at dc ground. Best of all, even if the parts aren't on sale, the antenna can be built for less than $15. If you only build one antenna, you'll have enough tubing left over to make most of a second antenna.

Construction

Copper and brass is used exclusively in this antenna. These metals get along together, so dissimilar metal corrosion is eliminated. Both metals solder well, too. See **Fig 20.100**. Cut the copper tubing to the lengths indicated. Item 9 is a 1 1/4-inch nipple cut from the 20-inch length of 1/2-inch tubing. This leaves 18 3/4 inches for the $\lambda/4$-matching stub. Item 10 is a 3 1/4-inch long nipple cut from the 60-inch length of 3/4-inch tubing. The 3/4-wave element should measure 56 3/4 inches long. Remove burrs from the ends of the tubing after cutting, and clean the mating surfaces with sandpaper, steel

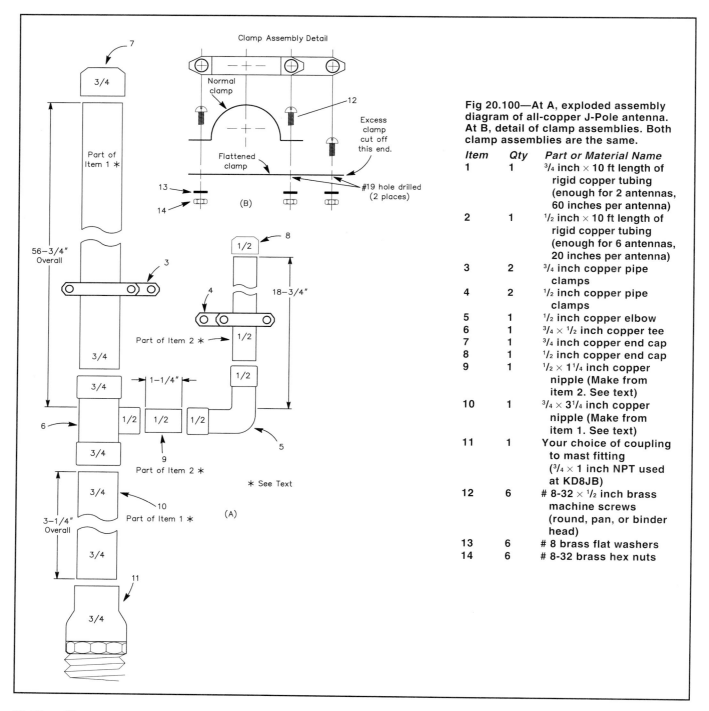

Fig 20.100—At A, exploded assembly diagram of all-copper J-Pole antenna. At B, detail of clamp assemblies. Both clamp assemblies are the same.

Item	Qty	Part or Material Name
1	1	3/4 inch × 10 ft length of rigid copper tubing (enough for 2 antennas, 60 inches per antenna)
2	1	1/2 inch × 10 ft length of rigid copper tubing (enough for 6 antennas, 20 inches per antenna)
3	2	3/4 inch copper pipe clamps
4	2	1/2 inch copper pipe clamps
5	1	1/2 inch copper elbow
6	1	3/4 × 1/2 inch copper tee
7	1	3/4 inch copper end cap
8	1	1/2 inch copper end cap
9	1	1/2 × 1 1/4 inch copper nipple (Make from item 2. See text)
10	1	3/4 × 3 1/4 inch copper nipple (Make from item 1. See text)
11	1	Your choice of coupling to mast fitting (3/4 × 1 inch NPT used at KD8JB)
12	6	# 8-32 × 1/2 inch brass machine screws (round, pan, or binder head)
13	6	# 8 brass flat washers
14	6	# 8-32 brass hex nuts

wool, or emery cloth.

After cleaning, apply a very thin coat of flux to the mating elements and assemble the tubing, elbow, tee, endcaps and stubs. Solder the assembled parts with a propane torch and rosin-core solder. Wipe off excess solder with a damp cloth, being careful not to burn yourself. The copper tubing will hold heat for a long time after you've finished soldering. After soldering, set the assembly aside to cool.

Flatten one each of the $1/2$-inch and $3/4$-inch pipe clamps. Drill a hole in the flattened clamp as shown in Fig 20.100B. Assemble the clamps and cut off the excess metal from the flattened clamp using the unmodified clamp as a template. Disassemble the clamps.

Assemble the $1/2$-inch clamp around the $1/4$-wave element and secure with two of the screws, washers, and nuts as shown in Fig 20.100B. Do the same with the $3/4$-inch clamp around the $3/4$-wave element. Set the clamps initially to a spot about 4 inches above the bottom of the "J" on their respective elements. Tighten the clamps only finger tight, since you'll need to move them when tuning.

Tuning

The J-Pole can be fed directly from 50 Ω

coax through a choke balun (3 turns of the feed coax rolled into a coil about 8 inches in diameter and held together with electrical tape). Before tuning, mount the antenna vertically, about 5 to 10 ft from the ground. A short TV mast on a tripod works well for this purpose. When tuning VHF antennas, keep in mind that they are sensitive to nearby objects—such as your body. Attach the feed line to the clamps on the antenna, and make sure all the nuts and screws are at least finger tight. It really doesn't matter to which element ($3/4$-wave element or stub) you attach the coaxial center lead. The author has done it both ways with no variation in performance. Tune the antenna by moving the two feed-point clamps equal distances a small amount each time until the SWR is minimum at the desired frequency. The SWR will be close to 1:1.

Final Assembly

The final assembly of the antenna will determine its long-term survivability. Perform the following steps with care. After adjusting the clamps for minimum SWR, mark the clamp positions with a pencil and then remove the feed line and clamps. Apply a very thin coating of flux to the inside of the clamp and the corresponding

surface of the antenna element where the clamp attaches. Install the clamps and tighten the clamp screws.

Solder the feed line clamps where they are attached to the antenna elements. Now, apply a small amount of solder around the screw heads and nuts where they contact the clamps. Don't get solder on the screw threads! Clean away excess flux with a non-corrosive solvent.

After final assembly and erecting/mounting the antenna in the desired location, attach the feed line and secure with the remaining washer and nut. Weatherseal this joint with RTV. Otherwise, you may find yourself repairing the feed line after a couple years.

On-Air Performance

Years ago, prior to building the first J-Pole antenna for this station, the author used a standard $1/4$-wave ground plane vertical antenna. While he had no problem working various repeaters around town with a $1/4$-wave antenna, simplex operation left a lot to be desired. The J-Pole performs just as well as a Ringo Ranger, and significantly better than the $1/4$-wave ground-plane vertical.

VHF/UHF Yagis

Without doubt, the Yagi is king of homestation antennas these days. Today's best designs are computer optimized. For years amateurs as well as professionals designed Yagi arrays experimentally. Now we have powerful (and inexpensive) personal computers and sophisticated software for antenna modeling. These have brought us antennas with improved performance, with little or no element pruning required.

A more complete discussion of Yagi design can be found earlier in this chapter. For more coverage on this topic and on stacking Yagis, see the most recent edition of *The ARRL Antenna Book*.

3 AND 5-ELEMENT YAGIS FOR 6 M

Boom length often proves to be the de-

ciding factor when one selects a Yagi design. ARRL Senior Assistant Technical Editor Dean Straw, N6BV, created the designs shown in **Table 20.23**. Straw generated the designs in the table for convenient boom lengths (6 and 12 ft). The 3-element design has about 8 dBi gain, and the 5-element version has about 10 dBi gain. Both antennas exhibit better than 22 dB front-to-rear ratio, and both cover 50 to 51 MHz with better than 1.6:1 SWR.

Element lengths and spacings are given in the table. Elements can be mounted to the boom as shown in **Fig 20.101**. Two muffler clamps hold each aluminum plate to the boom, and two U bolts fasten each element to the plate, which is 0.25 inches thick and 4×4 inches square. Stainless steel is the best choice for hardware, however, galvanized hardware can be substituted. Automotive muffler clamps do not

work well in this application, because they are not galvanized and quickly rust once exposed to the weather.

The driven element is mounted to the boom on a Bakelite plate of similar dimension to the other mounting plates. A 12-inch piece of Plexiglas rod is inserted into the driven element halves. The Plexiglas allows the use of a single clamp on each side of the element and also seals the center of the elements against moisture. Self-tapping screws are used for electrical connection to the driven element.

Refer to **Fig 20.102** for driven element and Hairpin match details. A bracket made from a piece of aluminum is used to mount the three SO-239 connectors to the driven element plate. A 4:1 transmission-line balun connects the two element halves, transforming the 200-Ω resistance at the Hairpin match to 50 Ω at the center connector. Note

Table 20.23

Optimized 6-m Yagi Designs (See page 20.34)

	Spacing From Reflector (in.)	Seg 1 Length (in.)	Seg 2 Length (in.)	Midband Gain F/R
306-06				
Refl	0	36	22.500	8.1 dBi
DE	24	36	16.000	28.3 dB
Dir 1	66	36	15.500	
506-12				
OD		0.750	0.625	
Refl	0	36	23.625	10.0 dBi
DE	24	36	17.125	26.8 dB
Dir 1	36	36	19.375	
Dir 2	80	36	18.250	
Dir 3	138	36	15.375	

Note: For all antennas, telescoping tube diameters (in inches) are: Seg1=0.750, Seg2=0.625. See page 20.34 for element details.

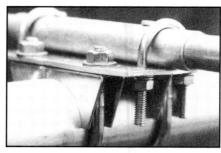

Fig 20.101—The element-to-boom clamp. Galvanized U bolts are used to hold the element to the plate, and 2-inch galvanized muffler clamps hold the plates to the boom.

that the electrical length of the balun is λ/2, but the physical length will be shorter due to the velocity factor of the particular coaxial cable used. The Hairpin is connected directly across the element halves. The exact center of the hairpin is electrically neutral and should be fastened to the boom. This has the advantage of placing the driven element at dc ground potential.

The Hairpin match requires no adjustment as such. However, you may have to change the length of the driven element slightly to obtain the best match in your preferred portion of the band. Changing the driven-element length will not adversely affect antenna performance. *Do not adjust the lengths or spacings of the other elements—they are optimized already.* If you decide to use a gamma match, add 3 inches to each side of the driven element lengths given in the table for both antennas.

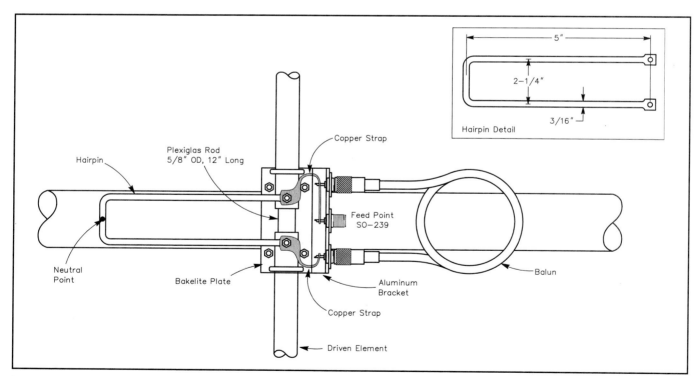

Fig 20.102—Detailed drawing of the feed system used with the 50-MHz Yagi. Balun lengths: For cable with 0.80 velocity factor—7 ft, 10³/₈ in. For cable with 0.66 velocity factor—6 ft, 5³/₄ in.

A PORTABLE 3-ELEMENT 2-M BEAM

In April 1993 *QST*, Nathan Loucks, WBØCMT, described the 2-m beam shown in **Fig 20.103**. The boom and mast are made from ³/₄-inch PVC plumber's pipe. The three pieces of PVC pipe are held together with a PVC T joint and secured by screws. Elements can be made from brass brazing or hobby rods. (If you can't find a 40-inch rod for the reflector, you can solder wire extensions to obtain the full length.)

Drill holes that provide a snug fit to the elements approximately ¹/₄ inch or so from the boom ends. Epoxy the director and reflector in place after centering them in these holes. A pair of holes spaced ¹/₄ inch and centered 16 inches from the reflector hold the two-piece driven element. The short ends of the element halves should extend about ¹/₄ inch through the boom. Solder the 50-Ω feed line to the driven element as shown in **Fig 20.104**.

Loucks used a pair of 4-inch pieces held in place by #12 or #14 jam screws (electrical connectors) to extend and adjust the driven element to allow for operation in various parts of the 2-m band. You can trim the driven element to length for operation in the desired portion of the band if you prefer.

The figures show the beam assembled for vertical polarization. You may want to turn the boom pieces 90° for horizontal polarization for SSB or CW operation.

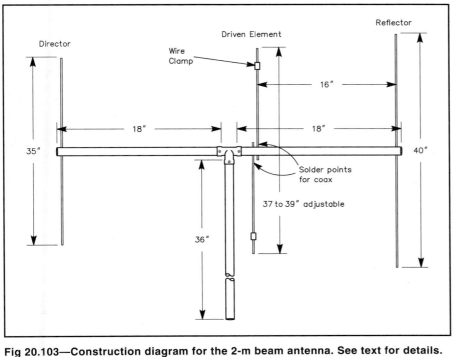

Fig 20.103—Construction diagram for the 2-m beam antenna. See text for details.

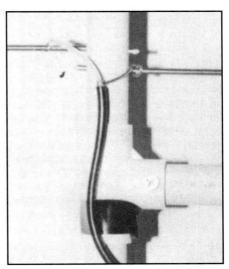

Fig 20.104—Solder the coaxial cable to the driven element pieces as shown.

High-Performance VHF/UHF Yagis

This construction information is presented as an introduction to the three high-performance VHF/UHF Yagis that follow. All were designed and built by Steve Powlishen, K1FO.

For years the design of long Yagi antennas seemed to be a mystical black art. The problem of simultaneously optimizing 20 or more element spacings and element lengths presented an almost unsolvable set of simultaneous equations. With the unprecedented increase in computer power and widespread availability of antenna analysis software, we are now able to quickly examine many Yagi designs and determine which approaches work and which designs to avoid.

At 144 MHz and above, most operators desire Yagi antennas two or more wavelengths in length. This length (2 λ) is where most classical designs start to fall apart in terms of gain per boom length, bandwidth and pattern quality. Extensive computer and antenna range analysis has proven that the best possible design is a Yagi that has both varying element spacings and varying element lengths.

This logarithmic-design approach (pioneered by Gunter Hoch, DL6WU, and others) starts with closely spaced directors. The director spacings gradually increase until a constant spacing of about 0.4 λ is reached. Conversely, the director lengths start out longest with the first director and decrease in length in a decreasing rate of change until they are virtually constant in length. This method of construction results in a wide gain bandwidth. A bandwidth of 7% of the center frequency at the −1 dB forward-gain points is typical for these Yagis even when they are longer than 10 λ. The log-taper design also reduces the rate of change in driven-element impedance vs frequency. This allows the use of simple dipole driven elements while still obtaining acceptable driven-element SWR over a wide frequency range. Another benefit is that the resonant frequency of the Yagi changes very little as the boom length is increased. The driven-element impedance also changes moderately with boom length. The tapered approach creates a Yagi with a very clean radiation pattern. Typically, first side lobe levels of ≈17 dB in the E plane, ≈15 dB in the H plane, and all other lobes at ≈20 dB or more are possible on designs from 2 λ to more than 14 λ.

The actual rate of change in element lengths is determined by the diameter of the elements (in wavelengths). The spacings can be optimized for an individual boom length or chosen as a best compromise for most boom lengths.

The gain of long Yagis has been the subject of much debate. Recent measurements and computer analysis by both amateurs and professionals indicates that given an optimum design, doubling a Yagi's boom length will result in a maximum theoretical gain increase of about 2.6 dB. In practice, the real gain increase may be less because of escalating resistive losses and the greater possibility of construction error. **Fig 20.105** shows the maximum possible gain per boom length expressed in decibels, referenced to an isotropic radiator. The actual number of directors does not play an important part in determining the gain vs boom length as long as a reasonable number of directors are used. The use of more directors per boom length will normally give a wider gain bandwidth; however, a point exists where too many directors will adversely affect all performance aspects.

While short antennas (< 1.5 λ) may show increased gain with the use of quad or loop elements, long Yagis (> 2 λ) will not exhibit measurably greater forward gain or pattern integrity with loop type elements. Similarly, loops used as driven elements and reflectors will not significantly change the properties of a long log-taper Yagi. Multiple-dipole driven-element assemblies will also not result in any significant gain increase per given boom length when compared to single-dipole feeds.

Once a long-Yagi director string is properly tuned, the reflector becomes relatively noncritical. Reflector spacings between 0.15 λ and 0.2 λ are preferred. The spacing can be chosen for best pattern and driven-element impedance. Multiple-reflector arrangements will not significantly increase the forward gain of a Yagi that has its directors properly optimized for forward gain. Many multiple-reflector schemes such as trireflectors and corner reflectors have the disadvantage of lowering the driven-element impedance compared to a single optimum-length reflector. The plane or grid reflector, shown in **Fig 20.106**, may however reduce the intensity of unwanted rear lobes. This can be used to reduce noise pickup on EME or satellite arrays. This type of reflector will usually increase the driven-element impedance compared to a single reflector. This sometimes makes driven-element

matching easier. Keep in mind that even for EME, a plane reflector will add considerable wind load and weight for only a few tenths of a decibel of receive signal-to-noise improvement.

Yagi Construction

Normally, aluminum tubing or rod is used for Yagi elements. Hard-drawn enamel-covered copper wire can also be used on Yagis above 420 MHz. Resistive losses are inversely proportional to the square of the element diameter and the square root of its conductivity.

Element diameters of less than ³⁄₁₆ inch or 4 mm should not be used on any band. The size should be chosen for reasonable strength. Half-inch diameter is suitable for 50 MHz, ³⁄₁₆ to ³⁄₈ inch for 144 MHz and 3/16 inch is recommended for the higher bands. Steel, including stainless steel and unprotected brass or copper wire, should not be used for elements.

Boom material may be aluminum tubing, either square or round. High-strength aluminum alloys such as 6061-T6 or 6063-T651 offer the best strength-to-weight advantages. Fiberglass poles have been used (where available as surplus). Wood is a popular low-cost boom material. The wood should be well seasoned and free from knots. Clear pine, spruce and Douglas fir are often used. The wood should be well treated to avoid water absorption and warping.

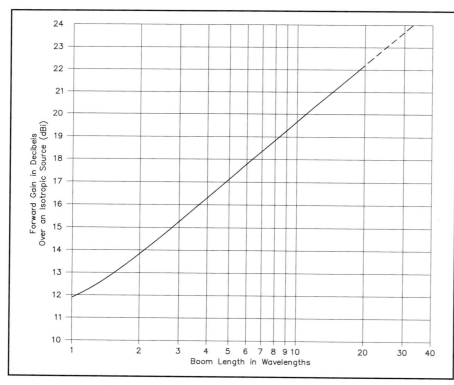

Fig 20.105—This chart shows maximum gain per boom length for optimally designed long Yagi antennas.

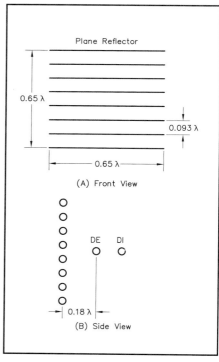

Fig 20.106—Front and side views of a plane-reflector antenna.

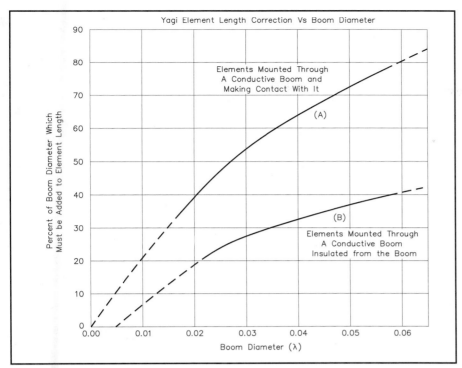

Fig 20.107—Yagi element correction vs boom diameter. Curve A is for elements mounted through a round or square conductive boom, with the elements in mechanical contact with the boom. Curve B is for insulated elements mounted through a conductive boom, and for elements mounted on top of a conductive boom (elements make electrical contact with the boom). The patterns were corrected to computer simulations to determine Yagi tuning. The amount of element correction is not affected by element diameter.

Elements may be mounted insulated or uninsulated, above or through the boom. Mounting uninsulated elements through a metal boom is the least desirable method unless the elements are welded in place. The Yagi elements will oscillate, even in moderate winds. Over several years this element oscillation will work open the boom holes. This will allow the elements to move in the boom and create noise (in your receiver) when the wind blows. Eventually the element-to-boom junction will corrode (aluminum oxide is a good insulator). This loss of electrical contact between the boom and element will reduce the boom's effect and change the resonant frequency of the Yagi.

Noninsulated elements mounted above the boom will perform fine as long as a good mechanical connection is made. Insulating blocks mounted above the boom will also work, but they require additional fabrication. One of the most popular construction methods is to mount the elements through the boom using insulating shoulder washers. This method is lightweight and durable. Its main disadvantage is difficult disassembly, making this method of limited use for portable arrays.

If a conductive boom is used, element lengths must be corrected for the mounting method used. The amount of correction is dependent upon the boom diameter in wavelengths. See **Fig 20.107**. Elements mounted through the boom and not insulated require the greatest correction. Mounting on top of the boom or through the boom on insulated shoulder washers requires about half of the through-the-boom correction. Insulated elements mounted at least one element diameter above the boom require no correction over the free-space length.

The three following antennas have been optimized for typical boom lengths on each band.

A HIGH-PERFORMANCE 432-MHz YAGI

This 22-element, 6.1-λ, 432-MHz Yagi was originally designed for use in a 12-Yagi EME array built by K1FO. A lengthy evaluation and development process preceded its construction. Many designs were considered and then analyzed on the computer. Next, test models were constructed and evaluated on a home-made antenna range. The resulting design is based on W1EJ's computer-optimized spacings.

The attention paid to the design process has been worth the effort. The 22-element Yagi not only has exceptional forward gain (17.9 dBi), but has an unusually "clean" radiation pattern. The measured E-plane pattern is shown in **Fig 20.108**. Note that a 1-dB-per-division axis is used to show pattern detail. A complete description of the design process and construction methods appears in December 1987 and January 1988 *QST*.

Like other log-taper Yagi designs, this one can easily be adapted to other boom lengths. Versions of this Yagi have been built by many amateurs. Boom lengths ranged between 5.3 λ (20 elements) and 12.2 λ (37 elements).

The size of the original Yagi (169 inches long, 6.1 λ) was chosen so the antenna could be built from small-diameter boom material (⁷/₈ inch and 1 inch round 6061-T6 aluminum) and still survive high winds and ice loading. The 22-element Yagi weighs about 3.5 pounds and has a wind load of approximately 0.8 square ft. This allows a high-gain EME array to be built with manageable wind load and weight. This same low wind load and weight lets the tropo operator add a high-performance 432-MHz array to an existing tower without sacrificing antennas on other bands.

Table 20.24 lists the gain and stacking specifications for the various length Yagis. The basic Yagi dimensions are shown in **Table 20.25**. These are free-space element lengths for ³/₁₆-inch-diameter elements. Boom corrections for the element mounting method must be added in. The element-length correction column gives the length that must be added to keep the Yagi's center frequency optimized for use at 432 MHz. This correction is required to use the same spacing pattern over a wide range of boom lengths. Although any length Yagi will work well, this design is at its best when made with 18 elements or more (4.6 λ). Element material of less than ³/₁₆-inch diameter is not recommended because resistive losses will reduce the gain by about 0.1 dB, and wet-weather performance will be worse.

Quarter-inch-diameter elements could be used if all elements are shortened by 3 mm. The element lengths are intended for use with a slight chamfer (0.5 mm) cut into the element ends. The gain peak of the array is centered at 437 MHz. This allows acceptable wet-weather performance,

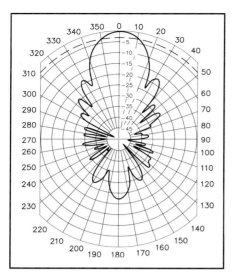

Fig 20.108—Measured E-plane pattern for the 22-element Yagi. Note: This antenna pattern is drawn on a linear dB grid, rather than on the standard ARRL log-periodic grid, to emphasize low sidelobes.

Table 20.24
Specifications for 432-MHz Yagi Family

No. of El	Boom Length (λ)	Gain (dBi)*	FB Ratio (dB)	DE Impd (Ω)	Beamwidth E / H (°)	Stacking E / H (in.)
15	3.4	15.67	21	23	30 / 32	53 / 49
16	3.8	16.05	19	23	29 / 31	55 / 51
17	4.2	16.45	20	27	28 / 30	56 / 53
18	4.6	16.8	25	32	27 / 29	58 / 55
19	4.9	17.1	25	30	26 / 28	61 / 57
20	5.3	17.4	21	24	25.5 / 27	62 / 59
21	5.7	17.65	20	22	25 / 26.5	63 / 60
22	6.1	17.9	22	25	24 / 26	65 / 62
23	6.5	18.15	27	30	23.5 / 25	67 / 64
24	6.9	18.35	29	29	23 / 24	69 / 66
25	7.3	18.55	23	25	22.5 / 23.5	71 / 68
26	7.7	18.8	22	22	22 / 23	73 / 70
27	8.1	19.0	22	21	21.5 / 22.5	75 / 72
28	8.5	19.20	25	25	21 / 22	77 / 75
29	8.9	19.4	25	25	20.5 / 21.5	79 / 77
30	9.3	19.55	26	27	20 / 21	80 / 78
31	9.7	19.7	24	25	19.6 / 20.5	81 / 79
32	10.2	19.8	23	22	19.3 / 20	82 / 80
33	10.6	19.9	23	23	19 / 19.5	83 / 81
34	11.0	20.05	25	22	18.8 / 19.2	84 / 82
35	11.4	20.2	27	25	18.5 / 19.0	85 / 83
36	11.8	20.3	27	26	18.3 / 18.8	86 / 84
37	12.2	20.4	26	26	18.1 / 18.6	87 / 85
38	12.7	20.5	25	25	18.9 / 18.4	88 / 86
39	13.1	20.6	25	23	18.7 / 18.2	89 / 87
40	13.5	20.8	26	21	17.5 / 18	90 / 88

*Gain is approximate real gain based upon gain measurements made on six different-length Yagis.

while reducing the gain at 432 MHz by only 0.05 dB.

The gain bandwidth of the 22-element Yagi is 31 MHz (at the –1 dB points). The SWR of the Yagi is less than l.4:1 between 420 and 440 MHz. **Fig 20.109** is a network analyzer plot of the driven-element SWR vs frequency. These numbers indicate just how wide the frequency response of a log-taper Yagi can be, even with a simple dipole driven element. In fact, at one antenna gain contest, some ATV operators conducted gain vs frequency measurements from 420 to 440 MHz. The 22-element Yagi beat all entrants including those with so-called broadband feeds.

To peak the Yagi for use on 435 MHz (for satellite use), you may want to shorten all the elements by 2 mm. To peak it for use on 438 MHz (for ATV applications), shorten all elements by 4 mm. If you want to use the Yagi on FM between 440 MHz and 450 MHz, shorten all the elements by 10 mm. This will provide 17.6 dBi gain at 440 MHz, and 18.0 dBi gain at 450 MHz. The driven element may have to be adjusted if the element lengths are shortened.

Although this Yagi design is relatively broadband, pay close attention to copying the design exactly. Metric dimensions are used because they are convenient for a Yagi sized for 432 MHz. Element holes should be drilled within ±2 mm. Element lengths should be kept within ±0.5 mm. Elements can be accurately constructed if they are first

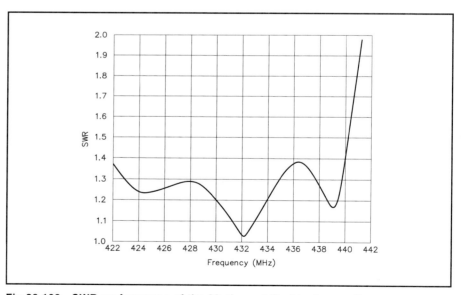

Fig 20.109—SWR performance of the 22-element Yagi in dry weather.

rough cut with a hack saw and then held in a vise and filed to the exact length.

Larger arrays require more attention to making all Yagis identical. Elements are mounted on shoulder insulators and run through the boom (see **Fig 20.110**). The element retainers are stainless-steel push nuts.

These are made by several companies, including Industrial Retaining Ring Co in Irvington, New Jersey, and AuVeco in Ft Mitchell, Kentucky. Local industrial hardware distributors can usually order them for you. The element insulators are not critical. Teflon or black polyethylene are probably

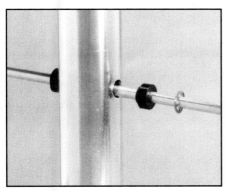

Fig 20.110—Element-mounting detail. Elements are mounted through the boom using plastic insulators. Stainless steel push-nut retaining rings hold the element in place.

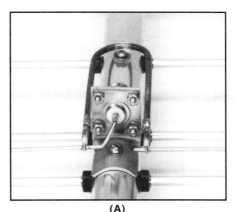

(A)

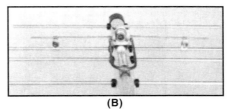

(B)

Fig 20.111—Several views of the driven element and T match.

the best materials. The Yagi in the photographs is made with black Delryn insulators, available from Rutland Arrays in New Cumberland, Pennsylvania.

The driven element uses a UG-58 connector mounted on a small bracket. The UG-58 should be the type with the press-in center pin. UG-58s with center pins held in by "C" clips will usually leak water. Some connectors use steel retaining clips, which will rust and leave a conductive

stripe across the insulator. The T-match wires are supported by the UT-141 balun. RG-303 or RG-142 Teflon-insulated cable could be used if UT-141 cannot be ob-

tained. **Fig 20.111** shows details of the driven-element construction. Driven element dimensions are given in **Fig 20.112**.

Dimensions for the 22-element Yagi are listed in **Table 20.26**. **Fig 20.113** details the Yagi's boom layout. Element material can be either $^3/_{16}$-inch 6061-T6 aluminum rod or hard aluminum welding rod.

A 24-ft-long, 10.6-λ, 33-element Yagi was also built. The construction methods used were the same as the 22-element Yagi. Telescoping round boom sections of 1, 1$^1/_8$, and 1$^1/_4$ inches in diameter were used. A boom support is required to keep boom sag acceptable. At 432 MHz, if boom sag is much more than two or three inches, H-plane pattern distortion will occur. Greater amounts of boom sag will reduce the gain of a Yagi. **Table 20.27** lists the proper dimensions for the antenna when built with the previously given boom diameters. The boom layout is shown in **Fig 20.114**, and the driven element is described in **Fig 20.115**. The 33-element Yagi exhibits the same clean pattern traits as the 22-element Yagi (see **Fig 20.116**). Measured gain of the 33-element Yagi is 19.9 dBi at 432 MHz. A measured gain sweep of the 33-element Yagi gave a ≈1 dB gain bandwidth of 14 MHz with the ≈1 dB points at 424.5 MHz and 438.5 MHz.

A HIGH-PERFORMANCE 144-MHz YAGI

This 144-MHz Yagi design uses the latest log-tapered element spacings and lengths. It offers near-theoretical gain per boom length, an extremely clean pattern and wide bandwidth. The design is based upon the spacings used in a 4.5-λ 432-MHz computer-developed design by W1EJ. It is quite similar to the 432-MHz Yagi described elsewhere in this chapter. Refer to that project for additional construction diagrams and photographs.

Mathematical models do not always directly translate into real working examples. Although the computer design

provided a good starting point, the author, Steve Powlishen, K1FO, built several test models before the final working Yagi was obtained. This hands-on tuning included changing the element-taper rate in order to obtain the flexibility that allows the Yagi to be built with different boom lengths.

The design is suitable for use from 1.8 λ (10 elements) to 5.1 λ (19 elements). When elements are added to a Yagi, the center frequency, feed impedance and front-to-back ratio will range up and down. A modern tapered design will mini-

mize this effect and allow the builder to select any desired boom length. This Yagi's design capabilities per boom length are listed in **Table 20.28**.

The gain of any Yagi built around this design will be within 0.1 to 0.2 dB of the maximum theoretical gain at the design frequency of 144.2 MHz. The design is intentionally peaked high in frequency (calculated gain peak is about 144.7 MHz). It has been found that by doing this, the SWR bandwidth and pattern at 144.0 to 144.3 MHz will be better, the Yagi will be less affected by weather and its performance in arrays will be more predictable. This design starts to drop off in performance if built with fewer than 10 elements. At less than 2 λ, more traditional designs perform well.

Table 20.29 gives free-space element lengths for $^1/_4$-inch-diameter elements. The use of metric notation allows for much easier dimensional changes during the design stage. Once you become familiar with the metric system, you'll probably find that construction is easier without the burden of cumbersome English fractional units. For $^3/_{16}$-inch-diameter elements, lengthen all parasitic elements by 3 mm. If $^3/_8$-inch-diameter elements are used,

Table 20.28

Specifications for the 144-MHz Yagi Family

No. of El	Boom Length (λ)	Gain (dBd)	DE Impd (Ω)	FB Ratio (dB)	Beamwidth E / H (°)	Stacking E / H (ft)
10	1.8	11.4	27	17	39 / 42	10.2 / 9.5
11	2.2	12.0	38	19	36 / 40	11.0 / 10.0
12	2.5	12.5	28	23	34 / 37	11.7 / 10.8
13	2.9	13.0	23	20	32 / 35	12.5 / 11.4
14	3.2	13.4	27	18	31 / 33	12.8 / 12.0
15	3.6	13.8	35	20	30 / 32	13.2 / 12.4
16	4.0	14.2	32	24	29 / 30	13.7 / 13.2
17	4.4	14.5	25	23	28 / 29	14.1 / 13.6
18	4.8	14.8	25	21	27 / 28.5	14.6 / 13.9
19	5.2	15.0	30	22	26 / 27.5	15.2 / 14.4

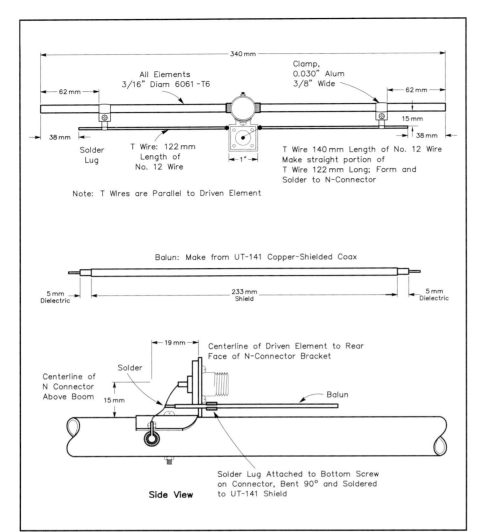

Table 20.26

Dimensions for the 22-Element 432-MHz Yagi

Element Number	Element Position (mm from rear of boom)	Element Length (mm)	Boom Diam (in.)
REF	30	346	
DE	134	340	
D1	176	321	
D2	254	311	
D3	362	305	7/8
D4	496	301	
D5	652	297	
D6	828	295	
D7	1020	293	
D8	1226	291	
D9	1444	289	
D10	1672	288	
D11	1909	286	
D12	2152	285	1
D13	2403	284	
D14	2659	283	
D15	2920	281	
D16	3184	280	
D17	3452	279	7/8
D18	3723	278	
D19	3997	277	
D20	4272	276	

Fig 20.112—Details of the driven element and T match for the 22-element Yagi. Lengths are given in millimeters to allow precise duplication of the antenna. See text. Use care soldering to connector center pin to prevent damage.

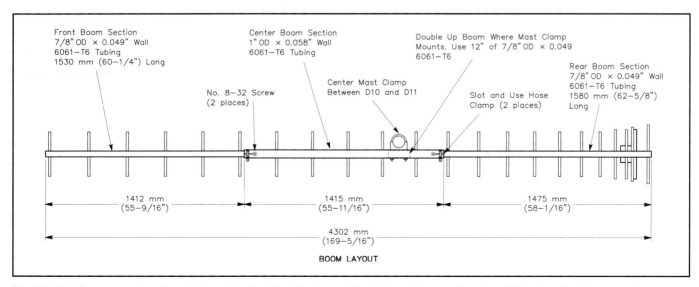

Fig 20.113—Boom-construction information for the 22-element Yagi. Lengths are given in millimeters to allow precise duplication of the antenna. See text.

Table 20.25

Table 20.25
Free-Space Dimensions for 432-MHz Yagi Family

Element lengths are for 3/16-inch-diameter material.

El No.	Element Position (mm from rear of boom)	Element Length (mm)	Element Correction*	El No.	Element Position (mm from rear of boom)	Element Length (mm)	Element Correction*
REF	0	340		D21	4520	269	0
DE	104	334		D22	4798	269	0
D1	146	315		D23	5079	268	0
D2	224	306		D24	5360	268	+1
D3	332	299		D25	5642	267	+1
D4	466	295		D26	5925	267	+1
D5	622	291		D27	6209	266	+1
D6	798	289		D28	6494	266	+1
D7	990	287		D29	6779	265	+2
D8	1196	285		D30	7064	265	+2
D9	1414	283		D31	7350	264	+2
D10	1642	281	−2	D32	7636	264	+2
D11	1879	279	−2	D33	7922	263	+2
D12	2122	278	−2	D34	8209	263	+2
D13	2373	277	−2	D35	8496	262	+2
D14	2629	276	−2	D36	8783	262	+2
D15	2890	275	−1	D37	9070	261	+3
D16	3154	274	−1	D38	9359	261	+3
D17	3422	273	−1				
D18	3693	272	0				
D19	3967	271	0				
D20	4242	270	0				

*Element correction is the amount to shorten or lengthen all elements when building a Yagi of that length.

Table 20.27
Dimensions for the 33-Element 432-MHz Yagi

Element Number	Element Position (mm from rear of boom)	Element Length (mm)	Boom Diam (in.)
REF	30	348	
DE	134	342	
D1	176	323	
D2	254	313	
D3	362	307	
D4	496	303	1
D5	652	299	
D6	828	297	
D7	1020	295	
D8	1226	293	
D9	1444	291	
D10	1672	290	
D11	1909	288	
D12	2152	287	1 1/8
D13	2403	286	
D14	2659	285	
D15	2920	284	
D16	3184	284	
D17	3452	283	
D18	3723	282	1 1/4
D19	3997	281	
D20	4272	280	
D21	4550	278	
D22	4828	278	
D23	5109	277	1 1/8
D24	5390	277	
D25	5672	276	
D26	5956	275	
D27	6239	274	
D28	6524	274	
D29	6809	273	1
D30	7094	273	
D31	7380	272	

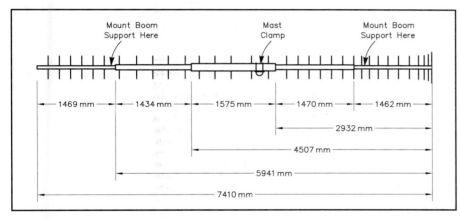

Fig 20.114—Boom-construction information for the 33-element Yagi. Lengths are given in millimeters to allow precise duplication of the antenna.

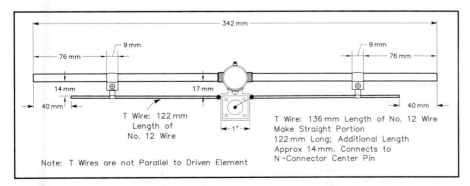

Fig 20.115—Details of the driven element and T match for the 33-element Yagi. Lengths are given in millimeters to allow precise duplication of the antenna.

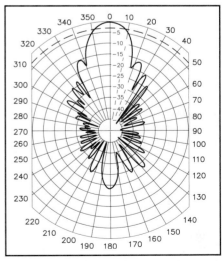

Fig 20.116—E-plane pattern for the 33-element Yagi. This pattern is drawn on a linear dB grid scale, rather than the standard ARRL log-periodic grid, to emphasize low sidelobes.

Table 20.29
Free-Space Dimensions for the 144-MHz Yagi Family

Element Diameter is ¹/₄-inch.

El No.	Element Position (mm from rear of boom)	Element Length (mm)
REF	0	1038
DE	312	955
D1	447	956
D2	699	932
D3	1050	916
D4	1482	906
D5	1986	897
D6	2553	891
D7	3168	887
D8	3831	883
D9	4527	879
D10	5259	875
D11	6015	870
D12	6786	865
D13	7566	861
D14	8352	857
D15	9144	853
D16	9942	849
D17	10744	845

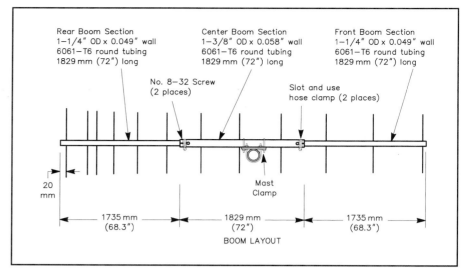

Fig 20.117—Boom layout for the 12-element 144-MHz Yagi. Lengths are given in millimeters to allow precise duplication.

Table 20.30
Dimensions for the 12-Element 2.5-λ Yagi

Element Number	Element Position (mm from rear of boom)	Element Length (mm)	Boom Diam (in.)
REF	0	1044	
DE	312	955	
D1	447	962	1¹/₄
D2	699	938	
D3	1050	922	
D4	1482	912	
D5	1986	904	
D6	2553	898	1³/₈
D7	3168	894	
D8	3831	889	
D9	4527	885	1¹/₄
D10	5259	882	

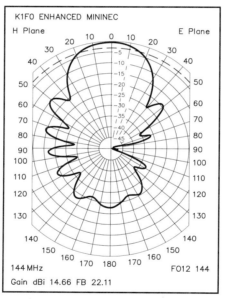

Fig 20.119—H- and E-plane pattern for the 12-element 144-MHz Yagi.

shorten all of the directors and the reflector by 6 mm. The driven element will have to be adjusted for the individual Yagi if the 12-element design is not adhered to.

For the 12-element Yagi, ¹/₄-inch-diameter elements were selected because smaller-diameter elements become rather flimsy at 2 m. Other diameter elements can be used as described previously. The 2.5-λ boom was chosen because it has an excellent size and wind load vs gain and pattern trade-off. The size is also convenient; three 6-ft-long pieces of aluminum tubing can be used without any waste. The relatively large-diameter boom sizes (1¹/₄ and 1³/₈ inches) were chosen, as they provide

an extremely rugged Yagi that does not require a boom support. The 12-element 17-ft-long design has a calculated wind survival of close to 120 mi/h! The absence of a boom support also makes vertical polarization possible.

Longer versions could be made by telescoping smaller-size boom sections into the last section. Some sort of boom support will be required on versions longer than 22 ft. The elements are mounted on shoulder insulators and mounted through the boom. However, elements may be mounted, insulated or uninsulated, above

or through the boom, as long as appropriate element length corrections are made. Proper tuning can be verified by checking the depth of the nulls between the main lobe and first side lobes. The nulls should be 5 to 10 dB below the first side-lobe level at the primary operating frequency. The boom layout for the 12-element model is shown in **Fig 20.117**. The actual corrected element dimensions for the 12-element 2.5-λ Yagi are shown in **Table 20.30**.

The design may also be cut for use at 147 MHz. There is no need to change element spacings. The element lengths should be shortened by 17 mm for best operation between 146 and 148 MHz. Again, the driven element will have to be adjusted as required.

The driven-element size (¹/₂-inch diameter) was chosen to allow easy impedance matching. Any reasonably sized driven element could be used, as long as appropriate length and T-match adjustments are made. Different driven-element dimensions are required if you change the boom length. The calculated natural driven-element impedance is given as a guideline. A balanced T-match was chosen because it's easy to adjust for best SWR and provides a balanced radiation pattern. A 4:1 half-wave coaxial balun is used, although impedance-transforming quarter-wave sleeve baluns could also be used. The calculated natural impedance will be useful in determining what impedance transformation will be required at the 200-Ω balanced feedpoint. *The ARRL Antenna Book* contains information on calculating folded-dipole and T-match driven-element parameters. A balanced feed is important for best operation on this antenna.

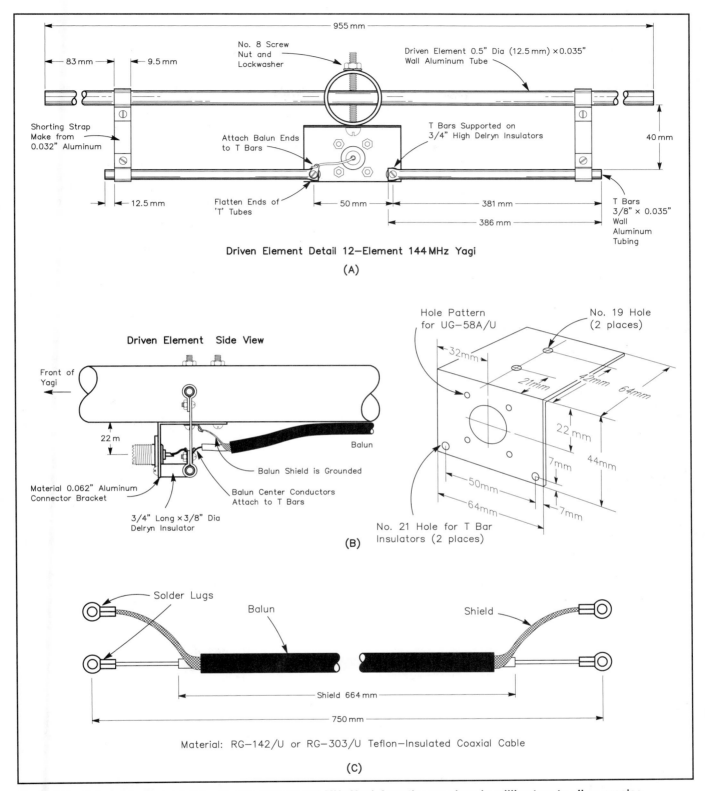

Fig 20.118—Driven-element detail for the 12-element 144-MHz Yagi. Lengths are given in millimeters to allow precise duplication. Use care soldering to connector center to prevent pin damage.

Gamma matches can severely distort the pattern balance. Other useful driven-element arrangements are the Delta match and the folded dipole, if you're willing to sacrifice some flexibility. **Fig 20.118** de-

tails the driven-element dimensions.

A noninsulated driven element was chosen for mounting convenience. An insulated driven element may also be used. A grounded driven element may be less af-

fected by static build-up. On the other hand, an insulated driven element allows the operator to easily check his feed lines for water or other contamination by the use of an ohmmeter from the shack.

Fig 20.119 shows computer-predicted E- and H-plane radiation patterns for the 12-element Yagi. The patterns are plotted on a 5-dB-per-division linear scale instead of the usual ARRL polar-plot graph. This expanded scale plot is used to show greater pattern detail. The pattern for the 12-element Yagi is so clean that a plot done in the standard ARRL format would be almost featureless, except for the main lobe and first sidelobes.

The excellent performance of the 12-element Yagi is demonstrated by the reception of Moon echoes from several of the larger 144-MHz EME stations with only one 12-element Yagi. Four of the 12-element Yagis will make an excellent starter EME array, capable of working many EME QSOs while being relatively small in size. The advanced antenna builder can use the information in Table 20.28 to design a "dream" array of virtually any size.

A SMALL LOOP FOR 160 M

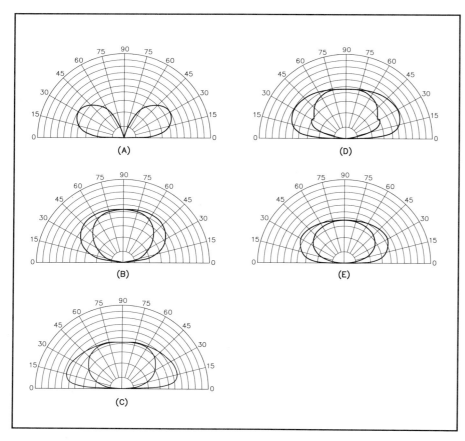

Fig 20.120—Relative comparisons of small loop elevation-plane radiation patterns to those of dipoles and verticals. At A, a vertical; B, a dipole; C, loop over radials; D, loop over short radials; E, loop over ground (no radials). In B, the inner pattern is that in the plane parallel to the antenna axis. In C through E, the inner patterns are those in the plane perpendicular to the antenna axis.

For amateurs with limited space, a small loop antenna is ideal for 160 m. Because small loops have high Q, they exhibit very narrow bandwidth, suppress harmonics and give a significant receiving noise reduction compared to dipoles. This project by Charles J. Mozzochi, W1LYQ, appeared in the June 1993 issue of QST. More information on theory and design considerations can be found in The ARRL Antenna Book.

Mozzochi found that for distances of 700 to 1000 miles, his loop performed the same as a full-size inverted-V antenna (62 ft high at the midpoint and fed with open-wire transmission line). Contacts with Europe report signals from the loop essentially the same as that from the inverted V. However, some report signals from the loop slightly stronger than from the inverted V. Both antennas are oriented to favor Europe. Also, the inverted V is more than 90% efficient, whereas the loop's efficiency is close to 50% ($\approx$3 dB)

Radiation Pattern

Fig 20.120 compares the vertical radiation patterns of the loop with those of a vertical and a dipole. The inner-pattern axis is perpendicular to that of the outer patterns. Note that the loop performs better at low angles, making it a good DX antenna. Fig 20.121 compares horizontal patterns at four elevation angles.

Physical Description

Fig 20.122 shows the antenna schematically and Fig 20.123 shows the base section of the antenna. The outer loop is made in the

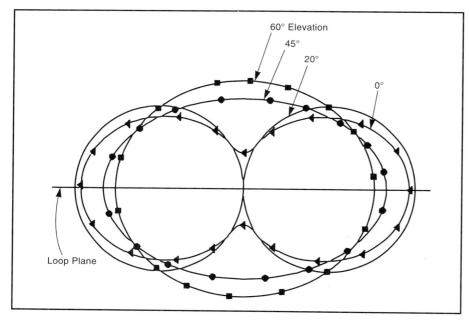

Fig 20.121—Electrically small loop antenna azimuth-plane radiation patterns at several wave angles.

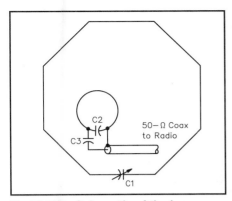

Fig 20.122—Schematic of the loop antenna.

Fig 20.123—Feed-point detail of the W1LYQ loop. C1, a vacuum variable, is secured to the top of a wooden box. C2 is made of ten "doorknob" capacitors in parallel, and C3 is one large doorknob. The loops are tied off to nearby objects for support.

form of a hexagon with eight 10-ft sections of 1-inch copper pipe joined with 45° couplers. The inner loop is a circle, 15 ft, 6 inches in circumference, made from ⁵/₈-inch copper pipe. Choose values and ratings for C1, C2 and C3 as discussed later.

The antenna is mounted vertically, approximately a foot off the ground, although it would be wise to mount it higher or install a protective wooden fence around the antenna to keep away people or pets. Very high voltages occur on small transmitting loops! Mozzochi installed 12 radials, ¹/₄-λ long, under the loop. These radials are joined under the loop, but are not electrically connected to it. The radials act as a reflective screen to help reduce ground loss. The use of radials is not absolutely necessary, but they can help reduce losses.

Construction

Because the outer loop's radiation resistance is on the order of 0.1 Ω, every effort must be made to minimize losses. Antenna efficiency increases with the diameter of the pipe used in both loops. One-inch copper pipe is used as a compromise in the outer loop; larger pipe would provide better efficiency, but would weigh considerably more. Mozzochi supports the outer loop with nylon rope strung between two trees. Smaller nylon ropes guy the lower section between trees and a fence. The inner loop is hung from the top of the outer loop with nylon rope, and nylon rope is used to guy and position the inner loop.

Solder all of the loop joints with rosin-core tin/lead or tin/lead/silver solder. Do not use acid-core solder.

Capacitors

C1 is a 200-pF vacuum-variable unit. A split-stator capacitor would work in place of a vacuum variable; however, under no circumstances should a conventional variable capacitor be used at C1 because the loss in the wiper contacts is significant compared to the outer loop's radiation re-

sistance. C2 is composed of ten 100-pF "doorknob" capacitors connected in parallel via copper straps, and C3 is a single 740-pF doorknob.

The outer loop radiates and the inner loop functions as a low loss matching and coupling device. This antenna puts very high voltages across C1 and C2 and passes unusually high currents through them. The formulas in **Table 20.31** give approximate currents and voltages for C1, C2 and C3, where P is the applied power, in watts.

The formulas in Table 20.31 are based on a steady-state analysis, under the assumption that the radiation resistance and the loss resistance in the outer loop are each 0.1 Ω. These resistances can differ significantly from this value depending on how the loop is constructed and located. Furthermore, substantial transient currents and voltages can occur under certain situations. It is wise to incorporate a safety factor of two to three after calculating the capacitor voltages and currents from these equations.

Tuning and Matching

To adjust the antenna to resonance at 1.85 MHz, follow this procedure:

1) Place the inner loop in either lower corner area of the outer loop, but *not touching it*, in a plane parallel to the outer loop and approximately 20 inches from the outer loop.

2) Set the transmitter to 1.85 MHz.

3) Adjust C1 for minimum SWR.

By making minor adjustments in the loop position and C1's setting, you should easily get the SWR down to 1:1. Once a 1:1 SWR is achieved at 1.85 MHz, you need only adjust C1 for other frequencies between 1.8 and 2 MHz. The achievable SWR will be 1:1 from 1.8 to 1.85 MHz,

Table 20.31
Calculating Capacitor Ratings

$V_{C1} = 1800 \sqrt{P}$
$V_{C2} = 18 \sqrt{P}$
$V_{C3} = 17 \sqrt{P}$
$I_{C1} = 2.2 \sqrt{P}$
$I_{C2} = 0.21 \sqrt{P}$
$I_{C3} = 0.15 \sqrt{P}$

Example: if P = 1000 watts, then V_{C1} = 18,000; V_{C2} = 180; V_{C3} = 170; I_{C1} = 22 A; I_{C2} = 2.1 A; and I_{C3} = 1.5 A.

and beginning at 1.85 MHz the SWR will climb slowly but steadily to 2:1 at 2 MHz. Changing frequency by more than 2 to 3 kHz requires readjusting C1 for minimum SWR. Keep in mind that a change of 1 pF in C1 makes a change of 3 to 5 kHz in the antenna's resonant frequency. Be sure that you use a motor drive with sufficient gear reduction.

For optimum performance, C1 should be mounted at the top of the loop. Also, it is possible to eliminate C2 and C3 by a suitable choice of the size and position of the inner loop. Consider this if you're planning to use high power with the loop.

Warning

Operating a loop antenna indoors or close to dwellings can raise the risk of interference to consumer devices and ham gear. Also, it's prudent to minimize RF exposure to people who may be near the antenna by using the minimum necessary transmitter power to carry on the desired communications. Whenever possible, mount your antennas as far as possible from people and dwellings.

A Simple Gain Antenna for 28 MHz

With the large number of operators and wide availability of inexpensive, single-band radios, the 10-m band could well become the hangout for local ragchewers that it was before the advent of 2-m FM, even at a low point in the solar cycle.

This simple antenna provides gain over a dipole or inverted V. It is a resonant loop with a particular shape. It provides 2.1 dB gain over a dipole at low radiation angles when mounted well above ground. The antenna is simple to feed—no matching network is necessary. When fed with 50-Ω coax, the SWR is close to 1:1 at the design frequency, and is less than 2:1 from 28.0-28.8 MHz for an antenna resonant at 28.4 MHz.

The antenna is made from #12 AWG wire (see **Fig 20.124**) and is fed at the center of the bottom wire. Coil the coax into a few turns near the feedpoint to provide a simple balun. A coil diameter of about a foot will work fine. You can support the antenna on a mast with spreaders made of bamboo, fiberglass, wood, PVC or other nonconducting material. You can also use aluminum tubing both for support and conductors, but you'll have to readjust the antenna dimensions for resonance.

This rectangular loop has two advantages over a resonant square loop. First, a square loop has just 1.1 dB gain over a dipole. This is a power increase of only 29%. Second, the input impedance of a square loop is about 125 Ω. You must use a matching network to feed a square loop with 50-Ω coax. The rectangular loop achieves gain by compressing its radiation pattern in the elevation plane. The azimuth plane pattern is slightly wider than that of a dipole (it's about the same as that of an inverted V). A broad pattern is an advantage for a general-purpose, fixed antenna. The rectangular loop provides a bidirectional gain over a broad azimuth region.

Mount the loop as high as possible. To provide 1.7 dB gain at low angles over an inverted V, the top wire must be at least 30 ft high. The loop will work at lower heights, but its gain advantage disappears. For example, at 20 ft the loop provides the same gain at low angles as an inverted V.

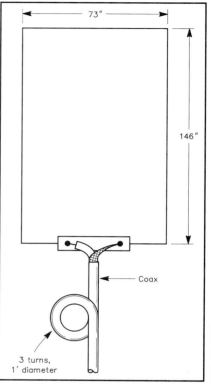

Fig 20.124—Construction details of the 10-m rectangular loop antenna.

BIBLIOGRAPHY

A. Christman, "Elevated Vertical Antenna Systems," *QST*, Aug 1988, pp 35-42.

A. C. Doty, Jr, J. A. Frey and H. J. Mills, "Efficient Ground Systems for Vertical Antennas," *QST*, Feb 1983, pp 20-25.

A. C. Doty, Jr, J. A. Frey and H. J. Mills, "Vertical Antennas: New Design and Construction Data," in *The ARRL Antenna Compendium, Volume 2* (Newington: ARRL, 1989), pp 2-9.

R. Fosberg, "Some Notes on Ground Systems for 160 Meters," *QST*, Apr 1965, pp 65-67.

J. Sevick, "The Ground-Image Vertical Antenna," *QST*, Jul 1971, pp 16-22.

J. Sevick, "The W2FMI Ground-Mounted Short Vertical," *QST*, Mar 1973, pp 13-18, 41.

J. Sevick, "A High Performance 20-, 40- and 80-Meter Vertical System," *QST*, Dec 1973, pp 14-18.

J. Sevick, "Short Ground-Radial Systems for Short Verticals," *QST*, Apr 1978, pp 30-33.

J. Stanley, "Optimum Ground Systems for Vertical Antennas," *QST*, Dec 1976, pp 13-15.

Textbooks on Antennas

C. A. Balanis, *Antenna Theory, Analysis and Design* (New York: Harper & Row, 1982).

D. S. Bond, *Radio Direction Finders*, 1st ed. (New York: McGraw-Hill Book Co).

W. N. Caron, *Antenna Impedance Matching* (Newington: ARRL, 1989).

R. S. Elliott, *Antenna Theory and Design* (Englewood Cliffs, NJ: Prentice Hall, 1981).

K. Henney, *Principles of Radio* (New York: John Wiley and Sons, 1938), p 462.

H. Jasik, *Antenna Engineering Handbook*, 1st ed. (New York: McGraw-Hill, 1961).

W. C. Johnson, *Transmission Lines and Networks*, 1st ed. (New York: McGraw-Hill Book Co, 1950).

R. C. Johnson and H. Jasik, *Antenna Engineering Handbook*, 2nd ed. (New York: McGraw-Hill, 1984).

E. C. Jordan and K. G. Balmain, *Electromagnetic Waves and Radiating Systems*, 2nd ed. (Englewood Cliffs, NJ: Prentice-Hall, Inc, 1968).

R. W. P. King, *Theory of Linear Antennas* (Cambridge, MA: Harvard Univ. Press, 1956).

R. W. P. King, H. R. Mimno and A. H. Wing, *Transmission Lines, Antennas and Waveguides* (New York: Dover Publications, Inc, 1965).

King, Mack and Sandler, *Arrays of Cylindrical Dipoles* (London: Cambridge Univ Press, 1968).

J. D. Kraus, *Electromagnetics* (New York: McGraw-Hill Book Co).

J. D. Kraus, *Antennas*, 2nd ed. (New York: McGraw-Hill Book Co, 1988).

E. A. Laport, *Radio Antenna Engineering* (New York: McGraw-Hill Book Co, 1952).

J. L. Lawson, *Yagi-Antenna Design*, 1st ed. (Newington: ARRL, 1986).

M. W. Maxwell, *Reflections—Transmission Lines and Antennas* (Newington: ARRL, 1990).

G. M. Miller, *Modern Electronic Communication* (Englewood Cliffs, NJ: Prentice Hall, 1983).

V. A. Misek, *The Beverage Antenna Handbook* (Hudson, NH: V. A. Misek, 1977).

More Wire Antenna Classics (Newington, CT: ARRL, 1999).

T. Moreno, *Microwave Transmission Design Data* (New York: McGraw-Hill, 1948).

L. A. Moxon, *HF Antennas for All Locations* (Potters Bar, Herts: Radio Society of Great Britain, 1982), pp 109-111.

S. A. Schelkunoff, *Advanced Antenna Theory* (New York: John Wiley & Sons, Inc, 1952).

S. A. Schelkunoff and H. T. Friis, *Antennas Theory and Practice* (New York: John Wiley & Sons, Inc, 1952).

J. Sevick, *Transmission Line Transformers*, 3rd ed. (Atlanta: Noble Publishing, 1996).

F. E. Terman, *Radio Engineers' Handbook*, 1st ed. (New York, London: McGraw-Hill Book Co, 1943).

F. E. Terman, *Radio Engineering*, 3rd ed. (New York: McGraw-Hill, 1947).

The ARRL Antenna Book, 19th Ed. (Newington, CT: ARRL, 2000)

Radio Communication Handbook, 5th ed. (London: RSGB, 1976).

Vertical Antenna Classics (Newington, CT: ARRL, 1996)

Wire Antenna Classics (Newington, CT: ARRL, 1999-2000).

Contents

Propagation

21

R adio waves, like light waves and all other forms of electromagnetic radiation, normally travel in straight lines. Obviously this does not happen all the time, because long-distance communication depends on radio waves traveling beyond the horizon. How radio waves propagate in other than straight-line paths is a complicated subject, but one that need not be a mystery. This chapter, by Emil Pocock, W3EP, provides basic understanding of the principles of electromagnetic radiation, the structure of the Earth's atmosphere and solar-terrestrial interactions necessary for a working knowledge of radio propagation. More detailed discussions and the underlying mathematics of radio propagation physics can be found in the references listed at the end of this chapter.

FUNDAMENTALS OF RADIO WAVES

Radio belongs to a family of electromagnetic radiation that includes infrared (radiation heat), visible light, ultraviolet, X-rays and the even shorter-wavelength gamma and cosmic rays. Radio has the longest wavelength and thus the lowest frequency of this group. See **Table 21.1**. Electromagnetic waves result from the interaction of an electric and a magnetic field. An oscillating electric charge in a piece of wire, for example, creates an electric field and a corresponding magnetic field. The magnetic field in turn creates an

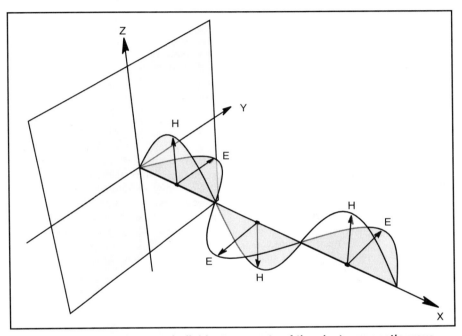

Fig 21.1—Electric and magnetic field components of the electromagnetic wave. The polarization of a radio wave is the same direction as the plane of its electric field.

electric field, which creates another magnetic field, and so on.

These two fields sustain themselves as a composite *electromagnetic wave*, which propagates itself into space. The electric and magnetic components are oriented at right angles to each other and 90° to the direction of travel. The polarization of a

radio wave is usually designated the same as its electric field. This relationship can be visualized in **Fig 21.1**. Unlike sound waves or ocean waves, electromagnetic waves need no propagating medium, such as air or water. This property enables electromagnetic waves to travel through the vacuum of space.

Velocity

Radio waves, like all other electromagnetic radiation, travel nearly 300,000 km (186,400 mi) per second in a vacuum. Radio waves travel more slowly through any other medium. The decrease in speed through the atmosphere is so slight that it is usually ignored, but sometimes even this small difference is significant. The speed

Table 21.1
The Electromagnetic Spectrum

Radiation	Frequency	Wavelength
X-ray	3×10^5 THz and higher	10 Å and shorter
Ultraviolet	800 THz - 3×10^5 THz	4000 - 10 Å
Visible light	400 THz - 800 THz	8000 - 4000 Å
Infrared	300 GHz - 400 THz	1 mm - .0008 mm
Radio	10 kHz - 300 GHz	30,000 km - 1 mm

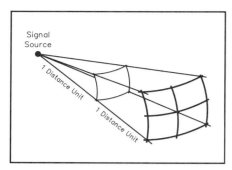

Fig 21.2—Radio energy disperses as the square of the distance from its source. For the change of one distance unit shown the signal is only one quarter as strong. Each spherical section has the same surface area.

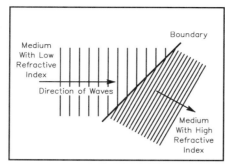

Fig 21.3—Radio waves are refracted as they pass at an angle between dissimilar media. The lines represent the crests of a moving wave front and the distance between them is the wavelength. The direction of the wave changes because one end of the wave slows down before the other as it crosses the boundary between the two media. The wavelength is simultaneously shortened, but the wave frequency (number of crests that pass a certain point in a given unit of time) remains constant.

of a radio wave in a piece of wire, by contrast, is about 95% that of free space, and the speed can be even slower in other media.

The speed of a radio wave is always the product of wavelength and frequency, whatever the medium. That relationship can be stated simply as:

$$c = f\lambda$$

where

  c = speed in m/s
  f = frequency in hertz
  λ = wavelength in m

The *wavelength (λ)* of any radio frequency can be determined from this simple formula. In free space, where the speed is 3×10^8 m/s, the wavelength of a 30-MHz radio signal is thus 10 m. Wavelength decreases in other media because the propagating speed is slower. In a piece of wire, the wavelength of a 30-MHz signal shortens to about 9.5 m. This factor must be taken into consideration in antenna designs and other applications.

Wave Attenuation and Absorption

Radio waves weaken as they travel, whether in the near vacuum of cosmic space or within the Earth's atmosphere. *Free-space attenuation* results from the dispersal of radio energy from its source. See **Fig 21.2**. Attenuation grows rapidly with distance because signals weaken with the square of the distance traveled. If the distance between transmitter and receiver is increased from 1 km to 10 km (0.6 to 6 mi), the signal will be only one-hundredth as strong. Free-space attenuation is a major factor governing signal strength, but radio signals undergo a variety of other losses as well.

Energy is lost to *absorption* when radio waves travel through media other than a vacuum. Radio waves propagate through the atmosphere or solid material (like a wire) by exciting electrons, which then reradiate energy at the same frequency. This process is not perfectly efficient, so some radio energy is transformed into heat and retained by the medium. The amount of radio energy lost in this way depends on the characteristics of the medium and on the frequency. Attenuation in the atmosphere is minor from 10 MHz to 3 GHz, but at higher frequencies, absorption due to water vapor and oxygen can be high.

Radio energy is also lost during refraction, diffraction and reflection—the very phenomena that allow long-distance propagation. Indeed, any form of useful propagation is accompanied by attenuation. This may vary from the slight losses encountered by refraction from sporadic-E clouds near the maximum usable frequency, to the more considerable losses involved with tropospheric forward scatter or D-Layer absorption in the lower HF bands. These topics will be covered later. In many circumstances, total losses can become so great that radio signals become too weak for communication.

Refraction

Electromagnetic waves travel in straight lines until they are deflected by something. Radio waves are *refracted*, or bent, slightly when traveling from one medium to another. Radio waves behave no differently from other familiar forms of electromagnetic radiation in this regard. The apparent bending of a pencil partially immersed in a glass of water demonstrates this principle quite dramatically.

Refraction is caused by a change in the velocity of a wave when it crosses the boundary between one propagating medium and another. If this transition is made at an angle, one portion of the wavefront slows down (or speeds up) before the other, thus bending the wave slightly. This is shown schematically in **Fig 21.3**.

The amount of bending increases with the ratio of the *refractive indices* of the two media. Refractive index is simply the velocity of a radio wave in free space divided by its velocity in the medium. Radio waves are commonly refracted when they travel through different layers of the atmosphere, whether the highly charged ionospheric layers 100 km (60 mi) and higher, or the weather-sensitive area near the Earth's surface. When the ratio of the refractive indices of two media is great enough, radio waves can be *reflected*, just like light waves striking a mirror. The Earth is a rather lossy reflector, but a metal surface works well if it is several wavelengths in diameter.

Scattering

The direction of radio waves can also be altered through *scattering*. The effect seen by a beam of light attempting to penetrate fog is a good example of light-wave scattering. Even on a clear night, a highly directional search light is visible due to a small amount of atmospheric scattering perpendicular to the beam. Radio waves are similarly scattered when they encounter randomly arranged objects of wavelength size or smaller, such as masses of electrons or water droplets. When the density of scattering objects becomes great enough, they behave more like a propagating medium with a characteristic refractive index.

If the scattering objects are arranged in some alignment or order, scattering takes place only at certain angles. A rainbow provides a good analogy for *field-aligned scattering* of light waves. The arc of a rainbow can be seen only at a precise angle away from the sun, while the colors result from the variance in scattering across the light-wave frequency range. Ionospheric electrons can be field-aligned by magnetic forces in auroras and under other unusual circumstances. Scattering in such cases is best perpendicular to the Earth's magnetic field lines.

Reflection

At amateur frequencies above 30 MHz, reflections from a variety of large objects, such as water towers, buildings, airplanes, mountains and the like can provide a useful means of extending over-the-horizon paths several hundred km. Two stations need only beam toward a common reflector, whether stationary or moving. Contrary to common sense notions, the best position for a reflector is not midway between two stations. Signal strength increases as the reflector approaches one end of the path, so the most effective reflectors are those closest to one station or the other.

Maximum range is limited by the radio line-of-sight distance of both stations to the reflector and by reflector size and shape. The reflectors must be many wavelengths in size and ideally have flat surfaces. Large airplanes make fair reflectors and may provide the best opportunity for long-distance contacts. The calculated limit for airplane reflections is 900 km (560 mi), assuming the largest jets fly no higher than 12,000 m (40,000 ft), but actual airplane reflection contacts are likely to be considerably shorter.

Knife-Edge Diffraction

Radio waves can also pass behind solid objects with sharp upper edges, such as a mountain range, by *knife-edge diffraction*. This is a common natural phenomenon that affects light, sound, radio and other coherent waves, but it is difficult to comprehend. **Fig 21.4** depicts radio signals approaching an idealized knife-edge. The portion of the radio waves that strike the base of the knife-edge is entirely blocked, while that portion passing several wavelengths above the edge travel on relatively unaffected. It might seem at first glance that a knife-edge as large as a mountain, for example, would completely prevent radio signals from appearing on the other side but that is not quite true. Something quite unexpected happens to radio signals that pass just over a knife-edge.

Normally, radio signals along a wave front interfere with each other continuously as they propagate through unobstructed space, but the overall result is a uniformly expanding wave. When a portion of the wave front is blocked by a knife-edge, the resulting interference pattern is no longer uniform. This can be understood by visualizing the radio signals right at the knife-edge as if they constituted a new and separate transmitting point, but in phase with the original source. The signals adjacent to the knife-edge still interact with signals passing above the edge, but they cannot interact with signals that have been obstructed below the edge. The resulting *interference pattern* no longer creates a uniformly expanding wave front, but rather appears as a pattern of alternating strong and weak bands of waves that spread in a nearly 180° arc behind the knife-edge.

The crest of a range of hills or mountains 50 to 100 wavelengths long can serve as a reasonable knife-edge diffractor at radio frequencies. Hill crests that are clearly defined and free of trees, buildings and other clutter make the best edges, but even rounded hills may serve as a diffracting edge. Alternating bands of strong and weak signals, corresponding to the interference pattern, will appear on the surface of the Earth behind the mountain, known as the *shadow zone*. The phenomenon is generally reciprocal, so that two-way communication can be established under optimal conditions. Knife-edge diffraction might make it possible to complete paths of 100 km or more that might otherwise be entirely obstructed by mountains or seemingly impossible terrain.

Ground Waves

A *ground wave* is the result of a special form of diffraction that primarily affects longer-wavelength vertically polarized radio waves. It is most apparent in the 80- and 160-m amateur bands, where practical ground-wave distances may extend beyond 200 km (120 mi). The term ground wave is often mistakenly applied to any short-distance communication, but the actual mechanism is unique to the longer-wave bands.

Radio waves are bent slightly as they pass over a sharp edge, but the effect extends to edges that are considerably rounded. At medium and long wavelengths, the curvature of the Earth looks like a rounded edge. Bending results when the lower part of the wave front loses energy due to currents induced in the ground. This slows down the lower part of the wave, causing the entire wave to tilt forward slightly. This tilting follows the curvature of the Earth, thus allowing low- and medium-wave radio signals to propagate over distances well beyond line of sight.

Ground wave is most useful during the day at 1.8 and 3.5 MHz, when D-layer absorption makes skywave propagation more difficult. Vertically polarized antennas with excellent ground systems provide the best results. Ground-wave losses are reduced considerably over saltwater and are worst over dry and rocky land.

SKY-WAVE PROPAGATION AND THE SUN

The Earth's atmosphere is composed primarily of nitrogen (78%), oxygen (21%) and argon (1%), with smaller amounts of a dozen other gases. Water vapor can account for as much as 5% of the atmosphere under certain conditions. This ratio of gases is maintained until an altitude of about 80 km (50 mi), when the mix begins to change. At the highest levels, helium and hydrogen predominate.

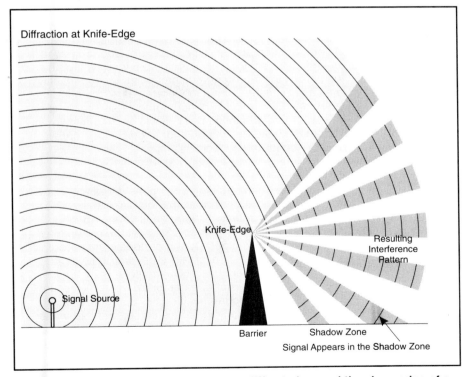

Fig 21.4—Radio, light and other waves are diffracted around the sharp edge of a solid object that is large in terms of wavelengths. Diffraction results from interference between waves right at the knife-edge and those that are passing above it. Some signals appear behind the knife-edge as a consequence of the interference pattern. Hills or mountains can serve as natural knife-edges at radio frequencies.

Propagation Summary, by Band

Medium Frequencies (300 kHz-3 MHz)

The only amateur medium-frequency band is situated just above the domestic AM broadcast band. Ground wave provides reliable communication out to 150 km (90 mi) or during the day, when no other form of propagation is available. Long-distance paths are made at night via the F_2 layer.

1.8-2.0 MHz (160 m)

The top band, as it is sometimes called, suffers from extreme daytime D-layer absorption. Even at high radiation angles, virtually no signal can pass through to the F layer, so daytime communication is limited to ground-wave coverage. At night, the D layer quickly disappears and worldwide 160-m communication becomes possible via F_2-layer skip. Atmospheric and man-made noise limit propagation. Tropical and midlatitude thunderstorms cause high levels of static in summer, making winter evenings the best time to work DX at 1.8 MHz. A proper choice of receiving antenna can often significantly reduce the amount of received noise while enhancing desired signals.

High Frequencies (3-30 MHz)

A wide variety of propagation modes are useful on the HF bands. The lowest two bands in this range share many daytime characteristics with 160 m. The transition between bands primarily useful at night or during the day appears around 10 MHz. Most long-distance contacts are made via F_2-layer skip. Above 21 MHz, more exotic propagation, including TE, sporadic E, aurora and meteor scatter, begin to be practical.

3.5-4.0 MHz (80 m)

The lowest HF band is similar to 160 m in many respects. Daytime absorption is significant, but not quite as extreme as at 1.8 MHz. High-angle signals may penetrate to the E and F layers. Daytime communication range is typically limited to 400 km (250 mi) by ground-wave and skywave propagation. At night, signals are often propagated halfway around the world. As at 1.8 MHz, atmospheric noise is a nuisance, making winter the most attractive season for the 80-m DXer.

7.0-7.3 MHz (40 m)

The popular 40-m band has a clearly defined skip zone during the day. D-layer absorption is not as severe as on the lower bands, so short-distance skip via the E and F layers is possible. During the day, a typical station can cover a radius of approximately 800 km (500 mi). Ground-wave propagation is not important. At night, reliable worldwide communication via F_2 is common on the 40-m band.

Atmospheric noise is less troublesome than on 160 and 80 m, and 40-m DX signals are often of sufficient strength to override even high-level summer static. For these reasons, 40 m is the lowest-frequency amateur band considered reliable for DX communication in all seasons. Even during the lowest point in the solar cycle, 40 m may be open for worldwide DX throughout the night.

10.1-10.15 MHz (30 m)

The 30-m band is unique because it shares characteristics of both daytime and nighttime bands. D-layer absorption is not a significant factor. Communication up to 3000 km (1900 mi) is typical during the daytime, and this extends halfway around the world via all-darkness paths. The band is generally open via F_2 on a 24-hour basis, but during a solar minimum, the MUF on some DX paths may drop below 10 MHz at night. Under these conditions, 30 m adopts the characteristics of the daytime bands at 14 MHz and higher. The 30-m band shows the least variation in conditions over the 11-year solar cycle, thus making it generally useful for long-distance communication anytime.

14.0-14.35 MHz (20 m)

The 20-m band is traditionally regarded as the amateurs' primary long-haul DX favorite. Regardless of the 11-year solar cycle, 20 m can be depended on for at least a few hours of worldwide F_2 propagation during the day. During solar maximum periods, 20 m will often stay open to distant locations throughout the night. Skip distance is usually appreciable and is always present to some degree. Daytime E-layer propagation may be detected along very short paths. Atmospheric noise is not a serious consideration, even in the summer. Because of its popularity, 20 m tends to be very congested during the daylight hours.

18.068-18.168 MHz (17 m)

The 17-m band is similar to the 20-m band in many respects, but the effects of fluctuating solar activity on F_2 propagation are more pronounced. During the years of high solar activity, 17 m is reliable for daytime and early-evening long-range communication, often lasting well after sunset. During moderate years, the band may open only during sunlight hours and close shortly after sunset. At solar minimum, 17 m will open to middle and equatorial latitudes, but only for short periods during midday on north-south paths.

21.0-21.45 MHz (15 m)

The 15-m band has long been considered a prime DX band during solar cycle maxima, but it is sensitive to changing solar activity. During peak years, 15 m is reliable for daytime F_2-layer DXing and will often stay open well into the night. During periods of moderate solar activity, 15 m is basically a daytime-only band, closing shortly after sunset. During solar minimum periods, 15 m may not open at all except for infrequent north-south transequatorial circuits. Sporadic E is observed occasionally in early summer and mid-winter, although this is not common and the effects are not as pronounced as on the higher frequencies.

24.89-24.99 MHz (12 m)

This band offers propagation that combines the best of the 10- and 15-m bands. Although 12 m is primarily a daytime band during low and moderate sunspot years, it may stay open well after sunset during the solar maximum. During years of moderate solar activity, 12 m opens to the low and middle latitudes during the daytime hours, but it seldom remains open after sunset. Periods of low solar activity seldom cause this band to go completely dead, except at higher latitudes. Occasional daytime openings, especially in the lower latitudes, are likely over north-south paths. The main sporadic-E season on 24 MHz lasts from late spring through summer and short openings may be observed in mid-winter.

28.0-29.7 MHz (10 m)

The 10-m band is well known for extreme variations in characteristics and variety of propagation modes. During solar maxima, long-distance F_2 propagation is so efficient that very low power can produce loud signals halfway around the globe. DX is abundant with modest equipment. Under these conditions, the band is usually open from sunrise to a few hours past sunset. During periods of moderate solar activity, 10 m usually opens only to low and transequatorial latitudes around noon. During the solar minimum, there may be no F_2 propagation at any time during the day or night.

Sporadic E is fairly common on 10 m, especially May through August, although it may appear at any time. Short skip, as it is sometimes called on the HF bands, has little relation to the solar cycle and occurs regardless of F-layer conditions. It provides single-hop communication from 300 to 2300 km (190 to 1400 mi) and multiple-hop opportunities of 4500 km (2800 mi) and farther.

Ten meters is a transitional band in that it also shares some of the propagation modes more characteristic of VHF. Meteor scatter, aurora, auroral E and trans-equatorial spread-F provide the means of making contacts out to 2300 km (1400 mi) and farther, but these modes often go unnoticed at 28 MHz. Techniques similar to those used at VHF can be very effective on 10 m, as signals are usually stronger and more persistent. These exotic modes can be more fully exploited, especially during the solar minimum when F_2 DXing has waned.

Very High Frequencies (30-300 MHz)

A wide variety of propagation modes are useful in the VHF range. F-layer skip appears on 50 MHz during solar cycle peaks. Sporadic E and several other E-layer phenomena are most effective in the VHF range. Still other forms of VHF ionospheric propagation, such as field-aligned irregularities (FAI) and transequatorial spread F (TE), are rarely observed at HF. Tropospheric propagation, which is not a factor at HF, becomes increasingly important above 50 MHz.

50-54 MHz (6 m)

The lowest amateur VHF band shares many of the characteristics of both lower and higher frequencies. In the absence of any favorable ionospheric propagation conditions, well-equipped 50-MHz stations work regularly over a radius of 300 km (190 mi) via tropospheric scatter, depending on terrain, power, receiver capabili-

ties and antenna. Weak-signal troposcatter allows the best stations to make 500-km (310-mi) contacts nearly any time. Weather effects may extend the normal range by a few hundred km, especially during the summer months, but true tropospheric ducting is rare.

During the peak of the 11-year sunspot cycle, worldwide 50-MHz DX is possible via the F_2 layer during daylight hours. F_2 backscatter provides an additional propagation mode for contacts as far as 4000 km (2500 mi) when the MUF is just below 50 MHz. TE paths as long as 8000 km (5000 mi) across the magnetic equator are common around the spring and fall equinoxes of peak solar cycle years.

Sporadic E is probably the most common and certainly the most popular form of propagation on the 6-m band. Single-hop E-skip openings may last many hours for contacts from 600 to 2300-km (370 to 1400 mi), primarily during the spring and early summer. Multiple-hop E_s provides transcontinental contacts several times a year, and contacts between the US and South America, Europe and Japan via multiple-hop E-skip occur nearly every summer.

Other types of E-layer ionospheric propagation make 6 m an exciting band. Maximum distances of about 2300 km (1400 mi) are typical for all types of E-layer modes. Propagation via FAI often provides additional hours of contacts immediately following sporadic E events. Auroral propagation often makes its appearance in late afternoon when the geomagnetic field is disturbed. Closely related auroral-E propagation may extend the 6-m range to 4000 km (2500 mi) and sometimes farther across the northern states and Canada, usually after midnight. Meteor scatter provides brief contacts during the early morning hours, especially during one of the dozen or so prominent annual meteor showers.

144-148 MHz (2 m)

Ionospheric effects are significantly reduced at 144 MHz, but they are far from absent. F-layer propagation is unknown except for TE, which is responsible for the current 144-MHz terrestrial DX record of nearly 8000 km (5000 mi). Sporadic E occurs as high as 144 MHz less than a tenth as often as at 50 MHz, but the usual maximum single-hop distance is the same, about 2300 km (1400 mi). Multiple-hop sporadic-E contacts greater than 3000 km (1900 mi) have occurred from time to time across the continental US, as well as across Southern Europe.

Auroral propagation is quite similar to that found at 50 MHz, except that signals are weaker and more Doppler-distorted. Auroral-E contacts are rare. Meteor-scatter contacts are limited primarily to the periods of the great annual meteor showers and require much patience and operating skill. Contacts have been made via FAI on 144 MHz, but its potential has not been fully explored.

Tropospheric effects improve with increasing frequency, and 144 MHz is the lowest VHF band at which

(Continued on page 21.6)

weather plays an important propagation role. Weather-induced enhancements may extend the normal 300- to 600-km (190- to 370-mi) range of well-equipped stations to 800 km (500 mi) and more, especially during the summer and early fall. Tropospheric ducting extends this range to 2000 km (1200 mi) and farther over the continent and at least to 4000 km (2500 mi) over some well-known all-water paths, such as that between California and Hawaii.

222-225 MHz (135 cm)

The 135-cm band shares many characteristics with the 2-m band. The normal working range of 222-MHz stations is nearly as good as comparably equipped 144-MHz stations. The 135-cm band is slightly more sensitive to tropospheric effects, but ionospheric modes are more difficult to use. Aurora and meteor-scatter signals are somewhat weaker than at 144 MHz, and sporadic-E contacts on 222 MHz are extremely rare. FAI and TE may also be well within the possibilities of 222 MHz, but reports of these modes on the 135-cm band are uncommon. Increased activity on 222-MHz will eventually reveal the extent of the propagation modes on the highest of amateur VHF bands.

Ultra-High Frequencies (300-3000 MHz) and Higher

Tropospheric propagation dominates the bands at UHF and higher, although some forms of E-layer propagation are still useful at 432 MHz. Above 10 GHz, atmospheric attenuation increasingly becomes the limiting factor over long-distance paths. Reflections from airplanes, mountains and other stationary objects may be useful adjuncts to propagation at 432 MHz and higher.

420-450 MHz (70 cm)

The lowest amateur UHF band marks the highest frequency on which ionospheric propagation is commonly observed. Aurora signals are weaker and more Doppler distorted; the range is usually less than at 144 or 222 MHz. Meteor scatter is much more difficult than on the lower bands, because bursts are significantly weaker and of much shorter duration. Although sporadic E and FAI are unknown as high as 432 MHz and prob-ably impossible, TE may be possible.

Well-equipped 432-MHz stations can expect to work over a radius of at least 300 km (190 mi) in the absence of any propagation enhancement. Tropospheric refraction is more pronounced at 432 MHz and provides the most frequent and useful means of extended range contacts. Tropospheric ducting supports contacts of 1500 km (930 mi) and farther over land. The current 432-MHz terrestrial DX record of more than 4000 km (2500 mi) was accomplished via ducting over water.

902-928 MHz (33-cm) and Higher

Ionospheric modes of propagation are nearly unknown in the bands above 902 MHz. Aurora scatter may be just within amateur capabilities at 902 MHz, but signal levels will be well below those at 432 MHz. Doppler shift and distortion will be considerable, and the signal bandwidth may be quite wide. No other ionospheric propagation modes are likely, although high-powered research radars have received echoes from auroras and meteors as high as 3 GHz.

Almost all extended-distance work in the UHF and microwave bands is accomplished with the aid of tropospheric enhancement. The frequencies above 902 MHz are very sensitive to changes in the weather. Tropospheric ducting occurs more frequently than in the VHF bands and the potential range is similar. At 1296 MHz, 2000-km (1200-mi) continental paths and 4000-km (2500-mi) paths between California and Hawaii have been spanned many times. Contacts of 1000 km (620 mi) have been made on all bands through 10 GHz in the US and over 1600 km (1000 mi) across the Mediterranean Sea. Well-equipped 903- and 1296-MHz stations can work reliably up to 300 km (190 mi), but normal working ranges generally shorten with increasing frequency.

Other tropospheric effects become evident in the GHz bands. Evaporation inversions, which form over very warm bodies of water, are usable at 3.3 GHz and higher. It is also possible to complete paths via scattering from rain, snow and hail in the lower GHz bands. Above 10 GHz, attenuation caused by atmospheric water vapor and oxygen become the most significant limiting factors in long-distance communication.

Solar radiation acts directly or indirectly on all levels of the atmosphere. Adjacent to the surface of the Earth, solar warming controls all aspects of the weather, powering wind, rain and other familiar phenomena. *Solar ultraviolet (UV) radiation* creates small concentrations of ozone (O_3) molecules between 10 and 50 km (6 and 30 mi). Most UV radiation is absorbed by this process and never reaches the Earth.

At even higher altitudes, UV and *X-ray radiation* partially ionize atmospheric gases. Electrons freed from gas atoms eventually recombine with positive ions to recreate neutral gas atoms, but this takes some time. In the low-pressure environ-ment at the highest altitudes, atoms are spaced far apart and the gases may remain ionized for many hours. At lower altitudes, recombination happens rather quickly, and only constant radiation can keep any appreciable portion of the gas ionized.

Structure of the Earth's Atmosphere

The atmosphere, which reaches to more than 600 km (370 mi) altitude, is divided into a number of regions, shown in **Fig 21.6**. The weather-producing *tropo-sphere* lies between the surface and an average altitude of 10 km (6 mi). Between 10 and 50 km (6 and 30 mi) are the strato-sphere and the imbedded ozonosphere, where ultraviolet absorbing ozone reaches its highest concentrations. About 99% of atmospheric gases are contained within these two lowest regions.

Above 50 km to about 600 km (370 mi) is the ionosphere, notable for its effects on radio propagation. At these altitudes, atomic oxygen and nitrogen predominate under very low pressure. High-energy solar UV and X-ray radiation ionize these gases, creating a broad region where ions are created in relative abundance. The ionosphere is subdivided into distinctive D, E and F regions.

The magnetosphere begins around 600 km (370 mi) and extends as far as 160,000 km (100,000 mi) into space. The

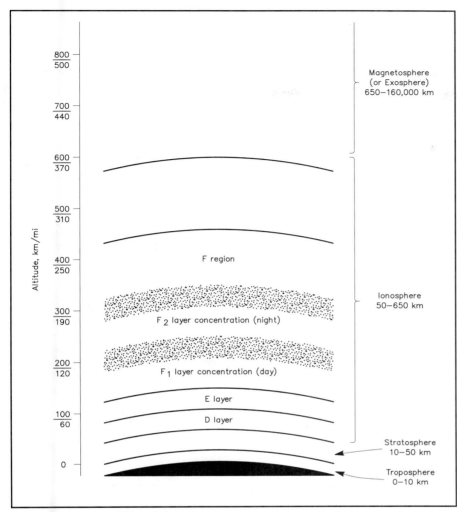

Fig 21.6—Regions of the atmosphere.

predominant component of atmospheric gases gradually shifts from atomic oxygen, to helium and finally to hydrogen at the highest levels. The lighter gases may reach escape velocity or be swept off the atmosphere by the solar wind. At about 3,200 and 16,000 km (2000 and 9900 mi), the Earth's magnetic field traps energetic electrons and protons in two bands, known as the Van Allen belts. These have only a minor effect on terrestrial radio propagation.

The Ionosphere

The ionosphere plays a basic role in long-distance communication in all the amateur bands from 1.8 MHz to 30 MHz. Ionospheric effects are less apparent in the very high frequencies (30-300 MHz), but they persist at least through 432 MHz. As early as 1902, Oliver Heaviside and Arthur E. Kennelly independently suggested the existence of a layer in the upper atmosphere

that could account for the long-distance radio transmissions made the previous year by Guglielmo Marconi and others. Edward Appleton confirmed the existence of the Kennelly-Heaviside layer during the early 1920s and used the letter E on his diagrams to designate the electric waves that were apparently reflected from it.

In 1924, Appleton discovered two additional layers in the *ionosphere*, as he and Robert Watson-Watt named this atmospheric region, and noted them with the letters D and F. Appleton was reluctant to alter this arbitrary nomenclature for fear of discovering yet other layers, so it has stuck to the present day. The basic physics of ionospheric propagation was largely worked out by the 1920s, yet both amateur and professional experimenters made further discoveries through the 1930s and 1940s. Sporadic E, aurora, meteor scatter and several types of field-aligned scattering were among additional ionospheric

phenomena that required explanation.

Ionospheric Refraction

The refractive index of an ionospheric layer increases with the density of free-moving electrons. In the most dense regions of the F layer, that density can reach a trillion electrons per cubic meter (10^{12} e/m^3). Even at this high level, radio waves are refracted gradually over a considerable vertical distance, usually amounting to tens of km. Radio waves become useful for terrestrial propagation only when they are refracted enough to bring them back to Earth. See **Fig 21.7**.

Although refraction is the primary mechanism of ionospheric propagation, it is usually more convenient to think of the process as a reflection. The *virtual height* of an ionospheric layer is the equivalent altitude of a reflection that would produce the same effect as the actual refraction. The virtual height of any ionospheric layer can be determined using an ionospheric sounder, or *ionosonde*, a sort of vertically oriented radar. The ionosonde sends pulses that sweep over a wide frequency range, generally from 2 MHz to 6 MHz or higher, straight up into the ionosphere. The frequencies of any echoes are recorded against time and then plotted as distance on an ionogram. **Fig 21.8** depicts a simple ionogram.

The highest frequency that returns echoes at vertical incidence is known as the *vertical incidence* or *critical frequency*. The critical frequency is almost totally a function of ion density. The higher the ionization at a particular altitude, the higher becomes the critical frequency. Physicists are more apt to call this the *plasma* frequency, because technically gases in the ionosphere are in a plasma, or partially ionized state. F-layer critical frequencies commonly range from about 1 MHz to as high as 15 MHz.

Maximum and Lowest Usable Frequencies

When the frequency of a vertically incident signal is raised above the critical frequency of an ionospheric layer, that portion of the ionosphere is unable to refract the signal back to Earth. However, a signal above the critical frequency may be returned to Earth if it enters the layer at an oblique angle, rather than at vertical incidence. This is fortunate because it permits two widely separated stations to communicate on significantly higher frequencies than the critical frequency. See **Fig 21.9**.

The highest frequency supported by the ionosphere between two stations is the *maximum usable frequency* (MUF) for that path. If the separation between the stations

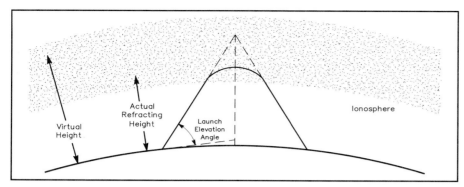

Fig 21.7—Gradual refraction in the ionosphere allows radio signals to be propagated long distances. It is often convenient to imagine the process as a reflection with an imaginary reflection point at some virtual height above the actual refracting region. The other figures in this chapter show ray paths as equivalent reflections, but you should keep in mind that the actual process is a gradual refraction.

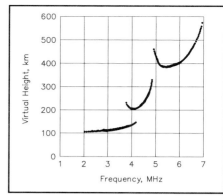

Fig 21.8—Simplified vertical incidence ionogram showing echoes returned from the E, F_1 and F_2 layers. The critical frequencies of each layer (4.1, 4.8 and 6.8 MHz) can be read directly from the ionogram scale.

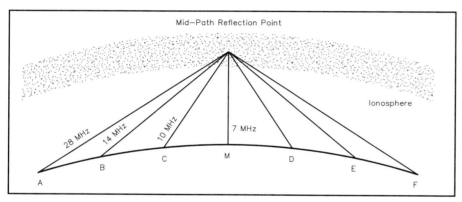

Fig 21.9—The relationships between critical frequency, maximum usable frequency (MUF) and skip zone can be visualized in this simplified, hypothetical case. The critical frequency is 7 MHz, allowing frequencies below this to be used for short-distance ionospheric communication by stations in the vicinity of point M. These stations cannot communicate by the ionosphere at 14 MHz. Stations at points B and E (and beyond) can communicate because signals at this frequency are refracted back to Earth because they encounter the ionosphere at an oblique angle of incidence. At greater distances, higher frequencies can be used because the MUF is higher at the larger angles of incidence (low launch angles). In this figure, the MUF for the path between points A and F, with a small launch angle, is shown to be 28 MHz. Each pair of stations can communicate at or below the MUF of the path between them, but not below the LUF—see text.

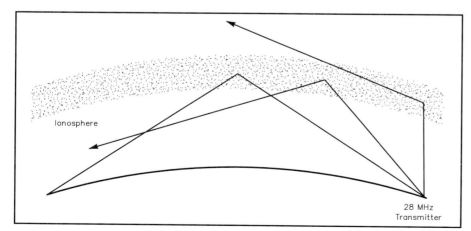

Fig 21.10—Signals at the MUF propagated at a low angle to the horizon provide the longest possible one-hop distances. In this example, 28-MHz signals entering the ionosphere at higher angles are not refracted enough to bring them back to Earth.

is increased, a still higher frequency can be supported at lower launch angles. The MUF for this longer path is higher than the MUF for the shorter path. When the distance is increased to the maximum one-hop distance, the launch angle of the signals between the two stations is zero (that is, the ray path is tangential to the Earth at the two stations) and the MUF for this path is the highest that can be supported by that layer of the ionosphere at that location. This maximum distance is about 4000 km (2500 mi) for the F_2 layer and about 2300 km (1400 mi) for the E layer.

The MUF is a function of path, time of day, season, location, solar UV and X-ray radiation levels and ionospheric disturbances. For vertically incident waves, the MUF is the same as the critical frequency. For path lengths at the limit of one-hop propagation, the MUF can be several times the critical frequency. See **Table 21.2**. The ratio between the MUF and the critical frequency is known as the *maximum usable frequency factor* (MUFF).

The term *skip zone* is closely related to MUF. When two stations are unable to communicate with each other on a particular frequency because the ionosphere is unable to refract the signal from one to the other through the required angle — that is, the frequency is below the MUF — the stations are said to be in the skip zone for that frequency. Stations within the skip zone may be able to work each other at a lower frequency, or by ground wave if they are close enough. There is no skip zone at frequencies below the critical frequency.

The MUF at any time on a particular path is just that — the *maximum* usable frequency. Frequencies below the MUF will also propagate along the path, but ionospheric absorption and noise at the receiving location (perhaps due to thun-

Table 21.2
Maximum Usable Frequency Factors (MUFF)

Layer	Maximum Critical Frequency (MHz)	MUFF	Useful Operating Frequencies (MHz)
F_2	15.0	3.3-4.0	1-60
F_1*	5.5	4.0	10-20
E*	4.0	4.8	5-20
E_s	30.0	5.3	20-160
D*	Not observed	—	None

* Daylight only

derstorms, local or distant) may make the received signal-to-noise ratio too low to be usable. In this case, the frequency is said to be below the *lowest usable frequency* (LUF). This occurs most frequently below 10 MHz, where atmospheric and man-made noises are most troublesome. The LUF can be lowered somewhat by the use of high power and directive antennas, or through the use of communications modes that permit reduced receiver bandwidth or are less demanding of SNR — CW instead of SSB, for example. This is not true of the MUF, which is limited by the physics of ionospheric refraction, no matter how high your transmitter power or how narrow your receiver bandwidth. The LUF can be higher than the MUF, in which case there is no frequency that supports communication on the particular path at that time.

Ionospheric Fading

HF signal strengths typically rise and fall over periods of a few seconds to several minutes, and rarely hold at a constant level for very long. Fading is generally caused by the interaction of several radio waves from the same source arriving along different propagation paths. Waves that arrive in phase combine to produce a stronger signal, while those out of phase cause destructive interference and a lower net signal strength. Short-term variations in ionospheric conditions may change individual path lengths or signal strengths enough to cause fading. Even signals that arrive primarily over a single path may vary as the propagating medium changes. Fading may be most notable at sunrise and sunset, especially near the MUF, when the ionosphere undergoes dramatic transformations. Other ionospheric traumas, such as auroras and geomagnetic storms, also produce severe forms of HF fading.

The 11-Year Solar Cycle

The density of ionospheric layers depends on the amount of solar radiation reaching the Earth, but solar radiation is not constant. Variations result from daily and seasonal motions of the Earth, the sun's own 27-day rotation and the 11-year cycle of solar activity. One visual indicator of both the sun's rotation and the solar cycle is the periodic appearance of dark spots on the sun, which have been observed continuously since the mid18th century. On average, the number of *sunspots* reaches a maximum every 10.7 years, but the period has varied between 7 and 17 years. Cycle 19 peaked in 1958, with an average sunspot number of over 200, the highest recorded to date. **Fig 21.11** shows average monthly sunspot numbers for the past four cycles.

Sunspots are cooler areas on the sun's surface associated with high magnetic activity. Active regions adjacent to sun-spot groups, called *plages*, are capable of producing great flares and sustained bursts of radiation in the radio through X-ray spectrum. During the peak of the 11-year solar cycle, average solar radiation increases along with the number of flares and sunspots. The ionosphere becomes more intensely ionized as a consequence, resulting in higher critical frequencies, particularly in the F_2 layer. The possibilities for long-distance communications are considerably improved during solar maxima, especially in the higher-frequency bands.

One key to forecasting F-layer critical frequencies, and thus long-distance propagation, is the intensity of ionizing UV and X-ray radiation. Until the advent of satellites, UV and X-ray radiation could not be measured directly, because they were almost entirely absorbed in the upper atmosphere. The sunspot number provided the most convenient approximation of general solar activity. The sunspot number is not a simple count of the number of visual spots, but rather the result of a complicated formula that takes into consideration size, number and grouping. The sunspot number varies from near zero during the solar-cycle minimum to over 200.

Another method of gauging solar activity is the *solar flux*, which is a measure of the intensity of 2800-MHz (10.7 cm) radio noise coming from the sun. The 2800-MHz radio flux correlates well with the intensity of ionizing UV and X-ray radiation and provides a convenient alternative to sunspot numbers. It commonly varies on a scale of 60-300 and can be related to sunspot numbers, as shown in **Fig 21.12**. The Dominion Radio Astrophysical Observatory, Penticton, British Columbia, measures the 2800-MHz solar flux daily at local noon. (Prior to June 1991, the Algonquin Radio Observatory, Ontario, made the measurements.) Radio station WWV broadcasts the latest solar-flux index at 18 minutes after each hour; WWVH does the same at 45 minutes after the hour. The Penticton solar flux is employed in a wide variety of other applications. Daily, weekly, monthly and even 13-month smoothed average solar flux readings are commonly used in propagation predictions.

High flux values generally result in higher MUFs, but the actual procedures for predicting the MUF at any given hour and path are quite complicated. Solar flux is not the sole determinant, as the angle of the sun to the Earth, season, time of day, exact location of the radio path and other factors must all be taken into account. MUF forecasting a few days or months ahead involves additional variables and even more uncertainties.

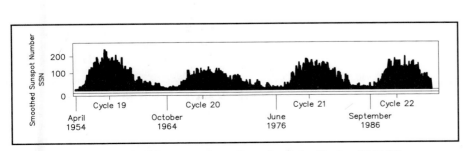

Fig 21.11—Average monthly sunspot numbers for Solar Cycles 19 to 22.

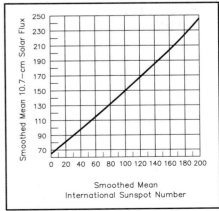

Fig 21.12—Approximate conversion between solar flux and sunspot number.

The Sun's 27-Day Rotation

Sunspot observations also reveal that the sun rotates on its own axis. The sun is composed of extremely hot gases and does not turn uniformly. At the equator, the period is just over 25 days, but it approaches 35 days at the poles. Sunspots that affect the Earth's ionosphere, which appear almost entirely within 35° of the sun's equator, take about 26 days for one rotation. After taking into account the Earth's movement around the sun, the apparent period of solar rotation is about 27 days.

Active regions must face the Earth in the proper orientation to have an impact on the ionosphere. They may face the Earth only once before rotating out of view, but they often persist for several solar rotations. The net effect is that solar activity often appears in 27-day cycles corresponding to the sun's rotation, even though the active regions themselves may last for several solar rotations.

Solar-Ionospheric Disturbances

The sun's surface sometimes erupts into cataclysmic events known as *solar flares*. Solar flares release huge amounts of energy in the form of electromagnetic radiation in a wide frequency range from HF radio to X-rays and cosmic rays, as well as ejecting particles, including electrons and protons. The first indications of a solar flare reach the Earth in eight minutes as a visible brightness near a sunspot group, increases in UV and X-ray radiation and high levels of noise in the VHF radio band. The flare and associated radiation may last only a short time, but its effects can be dramatic.

The extra X-ray radiation causes an immediate increase in D- and E-layer ionization known as a *sudden ionospheric disturbance* (SID). Extreme D-layer absorption may cause a short-term blackout of all HF

communications on the sun-facing side of the Earth. Signals in the 2 to 30-MHz range may completely disappear. In extreme cases, nearly all background noise will be gone as well. SIDs may last up to an hour, when ionospheric conditions temporarily return to normal. Cosmic rays penetrate deep into the ionosphere at the poles. This radiation produces intense ionization and consequent absorption of HF signals known as a *polar cap absorption* (PCA) event. A PCA event may last for days.

When a solar flare occurs, most of the time the electrons and protons ejected from the sun do not reach the Earth, because their trajectory takes them in another direction. If they do reach Earth, they do so 20 to 40 hours after the flare. As these relatively low-energy charged particles sweep past, they distort the Earth's geomagnetic field, causing a *geomagnetic storm*. This results in acceleration of the particles to energy levels that permit them to penetrate into the ionosphere at the poles. This tremendous influx causes auroral displays at mid-latitudes and can disrupt HF communications for several hours or longer. Extraordinary radio noise and interference can accompany geomagnetic storms and associated

Table 21.3
Geomagnetic Indices

Description	Typical K	A Range
Quiet	0-1	0-7
Unsettled	2	8-15
Active	3	16-29
Minor storm	4	30-49
Major storm	5	50-99
Severe storm	6-9	100 and greater

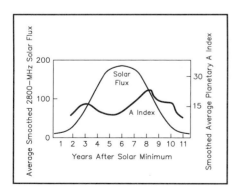

Fig 21.13—Geomagnetic activity (measured as the A-index) also follows an 11-year cycle. Average values over the past few cycles show that geomagnetic activity peaks before and after the peak of solar flux.

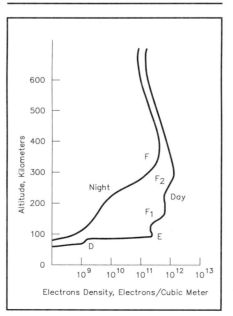

Fig 21.14—Typical electron densities for the various ionospheric regions.

auroras, especially at HF. Radio emissions from solar flares may be heard as sudden increases in noise on the VHF bands.

Many geomagnetic storms result from particle clouds unrelated to solar flares. These clouds come from coronal holes, disappearing filaments and coronal mass ejections from the sun.

Effects of ionospheric storms at HF vary considerably. Communications may be temporarily blacked out during a sudden ionospheric disturbance, but ionospheric paths may be generally noisy, weakened or disrupted for several days. Transpolar signals at 14 MHz and higher may be considerably attenuated and take on a hollow multipath sound. The number of geomagnetic storms varies considerably from year to year, with peak geomagnetic activity following the peak of solar activity. See **Fig 21.13**.

Geomagnetic activity is monitored by devices known as magnetometers, which may be as simple as a magnetic compass rigged to record its movements. Small variations in the geomagnetic field are scaled to two measures known as the K and A indices. The *K index* provides an indication of magnetic activity during the previous three hours on a finite scale of 0-9. Very quiet conditions are reported as 0 or 1, while geomagnetic storm levels begin at 4. Daily geomagnetic conditions are summarized in an *A index* that corresponds to daily K index values. See **Table 21.3**. A worldwide network of magnetometers monitors the Earth's magnetic field, because magnetic activity varies with location.

At 18 minutes past the hour, radio stations WWV and WWVH broadcast the latest solar flux number, the average planetary A-Index and the latest Boulder K-Index. In addition, they broadcast a descriptive account of the condition of the geomagnetic field and a forecast for the next three hours. You should keep in mind that the A-Index is a description of what happened *yesterday*. Strictly speaking, the K-Index is valid only for Boulder, Colorado. However, the trend of the K-Index is very important for propagation analysis and forecasting. A rising K foretells worsening HF propagation conditions, particularly for transpolar paths. At the same time, a rising K alerts VHF operators to the possibility of enhanced auroral activity, particularly when the K-Index rises above 3.

D-Layer Propagation

The *D layer* is the lowest region of the ionosphere, situated between 55 and 90 km (30 and 60 mi). See **Fig 21.14**. It is ionized primarily by the strong ultraviolet emission of solar hydrogen and short X-rays, both of which penetrate through the upper atmosphere. The D layer exists only during daylight, because constant radiation is needed to replenish ions that quickly recombine into neutral molecules. The D layer abruptly disappears at night so far as amateur MF and HF signals are concerned. D-layer ionization varies a small amount over the solar cycle. It is unsuitable as a refracting medium for any radio signals.

Daytime D-Layer Absorption

Nevertheless, the D layer plays an important role in HF communications. During daylight hours, radio energy as high as 5 MHz is effectively absorbed by the D layer, severely limiting the range of daytime 1.8- and 3.5-MHz signals. Signals at 7 MHz and 10 MHz pass through the D layer and on to the E and F layers only at relatively high angles. Low-angle waves, which must travel a much longer distance through the D layer, are subject to greater absorption. As the frequency increases above 10 MHz, radio waves pass through the D layer with increasing ease.

Nighttime D Layer

D-layer ionization falls 100-fold as soon as the sun sets and the source of ionizing radiation is removed. Low-band HF signals are then free to pass through to the E layer (also greatly diminished at night) and on to the F layer, where the MUF is almost always high enough to propagate 1.8- and 3.5-MHz signals half way around the world. Long-distance propagation at 7 and 10 MHz generally improves at night as well, because absorption is less and low-angle waves are able to reach the F layer.

D-Layer Ionospheric Forward Scatter

Radio signals in the 25-100 MHz range can be scattered by ionospheric irregularities, turbulence and stratification in the D and lower reaches of the E layers. Signals propagated by ionospheric forward scatter undergo very high losses, so signals are apt to be very weak. Typical scatter distances at 50 MHz are 800-1500 km (500-930 mi). This is not a common mode of propagation, but under certain conditions, ionospheric forward scatter can be very useful.

Ionospheric forward scatter is best during daylight hours from 10 AM to 2 PM local time, when the sun is highest in the sky and D-layer ionization peaks. It is worst at night. Scattering may be marginally more effective during the summer and during the solar cycle maximum due to somewhat higher D-layer ionization. The maximum path length of about 2000 km (1200 mi) is limited by the height of the scattering region, which is centered about 70 km (40 mi). Ionospheric scatter signals are typically weak, fluttery and near the noise level. Ionization from meteors sometimes temporarily raises signals well out of the noise for up to a few seconds at a time.

This mode may find its greatest use when all other forms of propagation are absent, primarily because ionospheric scatter signals are so weak. For best results at 28 and 50 MHz, a 3-element Yagi or larger, several hundred watts of power and a sensitive receiver are required. The paths are direct. CW is preferred, although, under optimal conditions, ionospheric scatter signals may be consistent enough to support SSB communications. Scattering is not efficient below 25 MHz. The very best-equipped pairs of 144-MHz stations may also be able to complete ionospheric scatter contacts.

E-Layer Propagation

The *E layer* lies between 90 and 150 km (60 and 90 mi) altitude, but a narrower region centered at 95 to 120 km (60 to 70 mi) is more important for radio propagation. E-layer nitrogen and oxygen atoms are ionized by short UV and long X-ray radiation. The normal E layer exists primarily during daylight hours, because like the D layer, it requires a constant source of ionizing radiation. Recombination is not as fast as in the denser D layer and absorption is much less. The E layer has a daytime critical frequency that varies between 3 and 4 MHz with the solar cycle. At night, the normal E layer all but disappears.

Daytime E Layer

The E layer plays a small role in propagating HF signals but can be a major factor limiting propagation during daytime hours. Its usual critical frequency of 3 to 4 MHz, with a maximum MUF factor of about 4.8, suggests that single-hop E-layer skip might be useful between 5 and 20 MHz at distances up to 2300 km (1400 mi). In practice this is not the case, because the potential for E-layer skip is severely limited by D-layer absorption. Signals radiated at low angles at 7 and 10 MHz, which might be useful for the longest-distance contacts, are largely absorbed by the D layer. Only high-angle signals pass through the D layer at these frequencies, but high-angle E-layer skip is typically limited to 1200 km (750 mi) or so. Signals at 14 MHz penetrate the D layer at lower angles at the cost of some absorption, but the casual operator may not be able to distinguish between signals propagated by the E layer or higher-angle F-layer propagation.

An astonishing variety of other propagation modes finds their home in the E layer, and this perhaps more than makes up for its ordinary limitations. Each of these other modes—sporadic E, field-aligned irregularities, aurora, auroral E and meteor scatter—are aberrant forms of propagation with unique characteristics. They are primarily useful only on the highest HF and lower VHF bands.

Sporadic E

Short skip, long familiar on the 10-m band during the summer months, affects the VHF bands as high as 222 MHz. *Sporadic E* (E_s), as this phenomenon is properly called, commonly propagates 28, 50 and 144-MHz radio signals between 500 and 2300 km (300 and 1400 mi). Signals are apt to be exceedingly strong, allowing even modest stations to make E_s contacts. At 21 MHz, the skip distance may only be a few hundred km. During the most intense E_s events, skip may shorten to less than 200 km (120 mi) on the 10-m band and disappear entirely on 15 m. Unusual multiple-hop E_s has supported contacts up to 10,000 km (6200 mi) on 28 and 50 MHz and more than 3,000 km (1900 mi) on 144 MHz. The first confirmed 220-MHz E_s contact was made in June 1987, but such contacts are likely to remain very rare.

Sporadic E at midlatitudes (roughly 15 to 45°) may occur at any time, but it is most common in the Northern Hemisphere during May, June and July, with a less-intense season at the end of December and early January. Its appearance is independent of the solar cycle. Sporadic E is most likely to occur from 9 AM to noon local time and again early in the evening between 5 PM and 8 PM. Midlatitude E_s events may last only a few minutes to many hours. In contrast, sporadic E is an almost constant feature of the polar regions at night and the equatorial belt during the day.

Efforts to predict midlatitude E_s have not been successful, probably because its causes are complex and not well understood. Studies have demonstrated that thin and unusually dense patches of ionization in the E layer, between 100 and 110 km (60 and 70 mi) altitude and 10 to 100 km (6 to 60 mi) in extent, are responsible for most E_s reflections. Sporadic-E clouds may form suddenly, move quickly from their birthplace, and dissipate within a few hours. Professional studies have recently focused on the role of heavy metal ions, probably of meteoric origin, and wind shears as two key factors in creating the dense patchy regions of E-layer ionization.

Sporadic-E clouds exhibit an MUF that can rise from 28 MHz through the 50-MHz band and higher in just a few minutes. When the skip distance on 28 MHz is as short as 400 or 500 km (250 or 310 mi), it is an indication that the MUF has reached 50 MHz for longer paths at low launch angles. Contacts at the maximum one-hop sporadic-E distance, about 2300 km (1400 mi), should then be possible at 50 MHz. E-skip contacts as short as 700 km (435 mi) on 50 MHz, in turn, may indicate that 144-MHz contacts in the 2300-km (1400 mi) range can be completed. See **Fig 21.15**. Sporadic-E openings occur about a tenth as often at 144 MHz in comparison to 50 MHz and for much shorter periods.

Sporadic E can also have a detrimental effect on HF propagation by masking the F_2 layer from below. HF signals may be prevented from reaching the higher levels of the ionosphere and the possibilities of long F_2 skip. Reflections from the tops of sporadic-E clouds can also have a masking effect, but they may also lengthen the F_2 propagation path with a top-side intermediate hop that never reaches the Earth.

E-Layer Field-Aligned Irregularities

Amateurs have experimented with a little-known scattering mode known as *field-aligned irregularities* (FAI) at 50 and 144 MHz since 1978. FAI commonly appear directly after sporadic-E events and may persist for several hours. Oblique-angle scattering becomes possible when electrons are compressed together due to the action of high-velocity ionospheric acoustic (sound) waves. The resulting irregularities in the distribution of free electrons are aligned parallel to the Earth's magnetic field, in something like moving vertical rods. A similar process of electron field-alignment takes place during radio aurora, making the two phenomena quite similar.

Most reports suggest that 8 PM to midnight may be the most productive time for FAI. Stations attempting FAI contacts point their antennas toward a common scattering region that corresponds to an active or recent E_s reflection point. The best direction must be probed experimentally, for the result is rarely along the great-circle path. Stations in south Florida, for example, have completed 144-MHz FAI contacts with north Texas when participating stations were beamed toward a common scattering region over northern Alabama.

FAI-propagated signals are weak and fluttery, reminiscent of aurora signals.

Doppler shifts of as much as 3 kHz have been observed in some tests. Stations running as little as 100 W and a single Yagi should be able to complete FAI contacts during the most favorable times, but higher power and larger antennas may yield better results. Contacts have been made on 50 and 144 MHz and 222-MHz FAI seems probable as well. Expected maximum distances should be similar to other forms of E-layer propagation, or about 2300 km (1400 mi).

Aurora

Radar signals as high as 3000 MHz have been scattered by the *aurora borealis* or northern lights (*aurora australis* in the Southern Hemisphere), but amateur aurora contacts are common only from 28 through 432 MHz. By pointing directional antennas generally north toward the center of aurora activity, oblique paths between stations up to 2300 km (1400 mi) apart can be completed. See **Fig 21.16**. High power and large antennas are not necessary. Stations with small Yagis and as little as 10 W output have used auroras on frequencies as high as 432 MHz, but contacts at 902 MHz and higher are exceedingly rare. Aurora propagation works just as well in the Southern Hemisphere, in which case antennas must be pointed south.

The appearance of auroras is closely linked to solar activity. During massive geomagnetic storms, high-energy particles flow into the ionosphere near the polar regions, where they ionize the gases of the E layer and higher. This unusual ionization produces spectacular visual auroral displays, which often spread southward into the midlatitudes. Auroral ionization in the E layer scatters radio signals in the VHF and UHF ranges.

In addition to scattering radio signals, auroras have other effects on worldwide radio propagation. Communication below 20 MHz is disrupted in high latitudes, primarily by absorption, and is especially noticeable over polar and near polar paths. Signals on the AM broadcast band through the 40-m band late in the afternoon may become weak and watery. The 20-m band may close down altogether. Satellite operators have also noticed that 144-MHz downlink signals are often weak and distorted when satellites pass near the polar regions. At the same time, the MUF in equatorial regions may temporarily rise dramatically, providing transequatorial paths at frequencies as high as 50 MHz.

Auroras occur most often around the spring and fall equinoxes (March-April and September-October), but auroras may appear in any month. Aurora activity generally peaks about two years before and

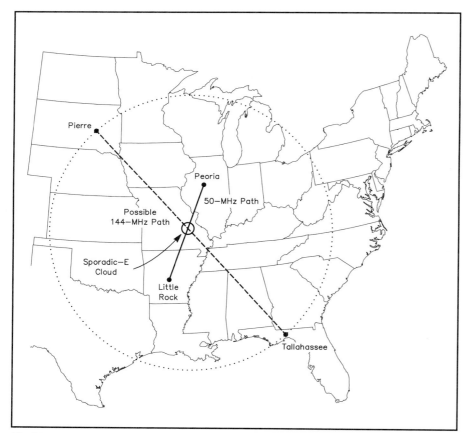

Fig 21.15—50 MHz sporadic-E contacts of 700 km (435 mi) or shorter (such as between Peoria and Little Rock) indicate that the MUF on longer paths is above 144 MHz. Using the same sporadic-E region reflecting point, 144-MHz contacts of 2200 km (1400 mi), such as between Pierre and Tallahassee, should be possible.

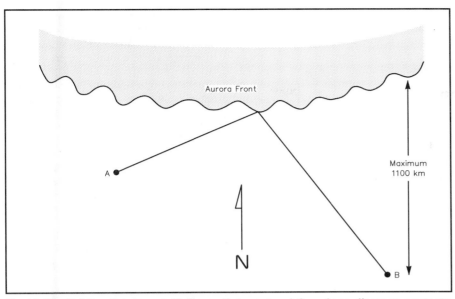

Fig 21.16—Point antennas generally north to make oblique long-distance contacts on 28 through 432 MHz via aurora scattering. Optimal antenna headings may shift considerably to the east or west depending on the location of the aurora.

after solar cycle maximum. Radio aurora activity is usually heard first in late afternoon and may reappear later in the evening. Auroras may be anticipated by following the A- and K-index reports on WWV. A K index of five or greater and an A index of at least 30 are indications that a geomagnetic storm is in progress and an aurora likely. The probability, intensity and southerly extent of auroras increase as the two index numbers rise. Stations north of 42° latitude in North America experience many aurora openings each year, while those in the Gulf Coast states may hear aurora signals no more than once a year, if that often.

Aurora-scattered signals are easy to identify. On 28- and 50-MHz SSB, signals sound very distorted and somewhat wider than normal; at 144 MHz and above, the distortion may be so severe that only CW is useful. Aurora CW signals have a distinctive note variously described as a buzz, hiss or mushy sound. This characteristic aurora signal is due to Doppler broadening, caused by the movement of electrons within the aurora. An additional Doppler shift of 1 kHz or more may be evident at 144 MHz and several kilohertz at 432 MHz. This second Doppler shift is the result of massive electrical currents that sweep electrons toward the sun side of the Earth during magnetic storms. Doppler shift and distortion increase with higher frequencies, while signal strength dramatically decreases.

It is not necessary to see an aurora to make aurora contacts. Useful auroras may be 500-1000 km (310-620 mi) away and

below the visual horizon. Antennas should be pointed generally north and then probed east and west to peak signals, because auroral ionization is field aligned. This means that for any pair of stations, there is an optimal direction for aurora scatter. Offsets from north are usually greatest when the aurora is closest and often provide the longest contacts. There may be some advantage to antennas that can be elevated, especially when auroras are high in the sky.

Auroral E

Radio auroras may evolve into a propagation mode known as *auroral E* at 28, 50 and rarely 144 MHz. Doppler distortion disappears and signals take on the characteristics of sporadic E. The most effective antenna headings shift dramatically away from oblique aurora paths to direct great-circle bearings. The usual maximum distance is 2300 km (1400 mi), typical for E-layer modes, but 28- and 50-MHz auroral-E contacts of 5000 km (3100 mi) are sometimes made across Canada and the northern US, apparently using two hops. Contacts at 50 MHz between Alaska and the east coasts of Canada and the northern US have been completed this way. Transatlantic 50-MHz auroral-E paths are also likely, although only one such contact has been reported.

Typically, 28- and 50-MHz auroral E appears across the northern third of the US and southern Canada when aurora activity is diminishing. This usually happens after midnight on the eastern end of the path. Auroral-E signals sometimes have a

slightly hollow sound to them and build slowly in strength over an hour or two, but otherwise they are indistinguishable from sporadic E. Auroral-E paths are almost always east-west oriented, perhaps because there are few stations at very northern latitudes to take advantage of this propagation.

Auroral E may also appear while especially intense auroras are still in progress, as happened during the great aurora of March 1989. On that occasion, 50-MHz propagation shifted from Doppler-distorted aurora paths to clear-sounding auroral E over a period of a few minutes. Many 6-m operators as far south as Florida and Southern California made single- and double-hop auroral-E contacts across the country. At about the same time, the MUF reached 144 MHz for stations west of the Great Lakes to the Northeast, the first time auroral E had been reported so high in frequency. At least two other rare instances of 2-m auroral E have been reported.

Meteor Scatter

Contacts between 800 and 2300 km (500 and 1400 mi) can be made at 28 through 432 MHz via reflections from the ionized trails left by meteors as they travel through the ionosphere. The kinetic energy of meteors no larger than grains of rice are sufficient to ionize a column of air 20 km (12 mi) long in the E layer. The particle itself evaporates and never reaches the ground, but the ionized column may persist for a few seconds to a minute or more before it dissipates. This is enough time to make very brief contacts by reflections from the ionized trails. Millions of meteors enter the Earth's atmosphere every day, but few have the required size, speed and orientation to the Earth to make them useful for meteor-scatter propagation.

Radio signals in the 30- to 100-MHz range are reflected best by meteor trails, making the 50-MHz band prime for meteor-scatter work. The early morning hours around dawn are usually the most productive, because the Earth's rotation contributes to the high speed of meteors heading into the Earth's path. Meteor contacts ranging from a second or two to more than a minute can be made nearly any morning at 28 or 50 MHz. Meteor-scatter contacts at 144 MHz and higher are more difficult because reflected signal strength and duration drop sharply with increasing frequency. A meteor trail that provides 30 seconds of communication at 50 MHz will last only a few seconds at 144 MHz, and less than a second at 432 MHz.

Meteor scatter opportunities are somewhat better during July and August be-

Table 21.4
Major Annual Meteor Showers

Name	Peak Dates	Approximate Rate (meteors/hour)
Quadrantids	Jan 3	50
Arietids	Jun 7-8	60
Perseids	Aug 11-13	80
Orionids	Oct 20-22	20
Geminids	Dec 12-13	60

cause the average number of meteors entering the Earth's atmosphere peaks during those months. The best times are during one of the great annual meteor showers, when the number of useful meteors may increase ten-fold over the normal rate of five to ten per hour. See **Table 21.4**. A meteor shower occurs when the Earth passes through a relatively dense stream of particles, thought to be the remnants of a comet, that are also in orbit around the sun. The most-productive showers are relatively consistent from year to year, although several can produce great storms periodically.

Because meteors provide only fleeting moments of communication even during one of the great meteor showers, special operating techniques are often used to increase the chances of completing a contact. Prearranged schedules between two stations establish times, frequencies and precise operating standards. Usually, each station transmits on alternate 15-second periods until enough information is pieced together a bit at a time to confirm contact. Nonscheduled random meteor contacts are common on 50 MHz and 144 MHz, but short transmissions and alert operating habits are required.

It is helpful to run several hundred watts to a single Yagi, but meteor-scatter can be used by modest stations under optimal conditions. During the best showers, a few watts and a small directional antenna are sufficient at 28 or 50 MHz. At 144 MHz, at least 100 W output and a long Yagi are needed for consistent results. Proportionately higher power is required for 222 and 432 MHz even under the best conditions.

F-Layer Propagation

The region of the F layers, from 150 km (90 mi) to over 400 km (250 mi) altitude, is by far the most important for long-distance HF communications. F-region oxygen atoms are ionized primarily by ultraviolet radiation. During the day, ionization reaches maxima in two distinct layers. The F_1 layer forms between 150 and 250 km (90 and 160 mi) and disappears at night. The F_2 layer extends above 250 km (160 mi), with a peak of ionization around

300 km (190 mi). At night, F-region ionization collapses into one broad layer at 300-400 km (190-250 mi) altitude. Ions recombine very slowly at these altitudes, because molecular density is relatively low. Maximum ionization levels change significantly with time of day, season and year of the solar cycle.

F_1 Layer

The daytime F_1 layer is not important to HF communication. It exists only during daylight hours and is largely absent in winter. Radio signals below 10 MHz are not likely to reach the F_1 layer, because they are either absorbed by the D layer or refracted by the E layer. Signals higher than 20 MHz that pass through both of the lower ionospheric regions are likely to pass through the F_1 layer as well, because the F_1 MUF rarely rises above 20 MHz. Absorption diminishes the strength of any signals that continue through to the F_2 layer during the day. Some useful F_1-layer refraction may take place between 10 and 20 MHz during summer days, yielding paths as long as 3000 km (1900 mi), but these would be practically indistinguishable from F_2 skip.

F_2 and Nighttime F Layers

The F_2 layer forms between 250 and 400 km (160 and 250 mi) during the daytime and persists throughout the night as a single consolidated F region 50 km (30 mi) higher in altitude. Typical ion densities are the highest of any ionospheric layer, with the possible exception of some unusual E-layer phenomenon. In contrast to the other ionospheric layers, F_2 ionization varies considerably with time of day, season and position in the solar cycle, but it is never altogether absent. These two characteristics make the F_2 layer the most important for long-distance HF communications.

The F_2-layer MUF is nearly a direct function of UV solar radiation, which in turn follows closely the solar cycle. During the lowest years of the cycle, the daytime MUF may climb above 14 MHz for only a few hours a day. In contrast, the MUF may rise beyond 50 MHz during peak years and stay above 14 MHz throughout the night. The virtual height of F_2 averages 330 km (210 mi), but varies between 200 and 400 km (120 and 250 mi). Maximum one-hop distance is about 4000 km (2500 mi), but it may be effectively longer when the MUF is above 40 MHz. Near-vertical incidence skywave propagation just below the critical frequency provides reliable coverage out to 200-300 km (120-190 mi) with no skip zone. It is most often observed on 7 MHz during the day.

In general, both F_2-layer ionization and MUF build rapidly at sunrise, usually reach a maximum in the afternoon, and then decrease to a minimum prior to sunrise. Depending on the season, the MUF is generally highest within 20° of the equator and lower toward the poles. For this reason, transequatorial paths may be open at a particular frequency when all other paths are closed.

In contrast to all the other ionospheric layers, daytime ionization in the winter F_2 layer averages four times the level of the summer at the same period in the solar cycle, doubling the MUF. This so-called *winter anomaly* is caused by the Earth moving closer to the Sun and tilting. Wintertime F_2 conditions are much superior to those in summer, because the MUF is much higher.

Multihop F-Layer Propagation

Most HF communication beyond 4000 km (2500 mi) takes place via multiple ionospheric hops. Radio signals are reflected from the Earth back toward space for additional ionospheric refractions. A series of ionospheric refractions and terrestrial reflections commonly create paths half-way around the Earth. Each hop involves additional attenuation and absorption, so the longest-distance signals tend to be the weakest. Even so, it is possible for signals to be propagated completely around the world and arrive back at their originating point. Multiple reflections within the F layer may bypass ground reflections altogether, creating what are known as *chordal hops*, with lower total attenuation. It takes a radio signal about 0.15 second to make a round-the-world trip.

Multihop paths can take on many different configurations, as shown in the examples of **Fig 21.17**. E-layer (especially sporadic E) and F-layer hops may be mixed. In practice, multihop signals arrive via many different paths, which often increases the problems of fading. Analyzing multihop paths is complicated by the effects of D- and E-layer absorption, possible reflections from the tops of sporadic-E layers, disruptions in the auroral zone and other phenomena.

F-Layer Long Path

Most HF communication takes place along the shortest great-circle path between two stations. Short-path propagation is always less than 20,000 km (12,000 mi)—halfway around the Earth. Nevertheless, it may be possible at times to make the same contact in exactly the opposite direction via the *long path*. The long-path distance will be 40,000 km (25,000 mi) minus the short-path length. Signal strength via the long path is usually considerably less than the more direct short-path. When both paths are open simultaneously, there may be a distinctive sort of echo on received signals. The time interval of the echo represents the difference between the short-path and long-path distances.

Sometimes there is a great advantage to using the long path when it is open, because signals can be stronger and fading less troublesome. There are times when the short path may be closed or disrupted by E-layer blanketing, D-layer absorption or F-layer gaps, especially when operating just below the MUF. Long paths that predominantly cross the night side of the Earth, for example, are sometimes useful because they generally avoid blanketing and absorption problems. Daylight-side long paths may take advantage of higher F-layer MUFs that occur over the sunlit portions of the Earth.

F-Layer Gray-Line

Gray-line paths can be considered a special form of long-path propagation that take into account the unusual ionospheric configuration along the twilight region between night and day. The gray line, as the twilight region is sometimes called, extends completely around the world. It is not precisely a line, for the distinction between daylight and darkness is a gradual transition due to atmospheric scattering. On one side, the gray line heralds sunrise and the beginning of a new day; on the opposite side, it marks the end of the day and sunset.

The ionosphere undergoes a significant transformation between night and day. As day begins, the highly absorbent D and E layers are recreated, while the F-layer MUF rises from its pre-dawn minimum. At the end of the day, the D and E layers quickly disappear, while the F-layer MUF continues its slow decline from late afternoon. For a brief period just along the gray-line transition, the D and E layers are not well formed, yet the F_2 MUF usually remains higher than 5 MHz. This provides a special opportunity for stations at 1.8 and 3.5 MHz.

Normally, long-distance communication on the lowest two amateur bands can take place only via all-darkness paths because of daytime D-layer absorption. The gray-line propagation path, in contrast, extends completely around the world. See **Fig 21.18**. This unusual situation lasts less than an hour at sunrise and sunset when the D-layer is largely absent, and may support contacts that are difficult or impossible at other times.

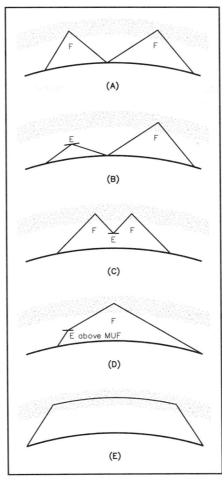

Fig 21.17—Multihop paths can take many different configurations, including a mixture of E- and F-layer hops. (A) Two F-layer hops. Five or more consecutive F-layer hops are possible. (B) An E-layer hookup to the F layer. (C) A top-side E-layer reflection can shorten the distance of two F-layer hops. (D) Refraction in the E layer above the MUF is insufficient to return the signal to Earth, but it can go on to be refracted in the F layer. (E) The Pedersen ray, which originates from a signal launched at a relatively high angle above the horizon into the E or F region, may result in a single-hop path, 5000 km (3100 mi) or more. This is considerably further than the normal 4000-km (2500 mi) maximum F-region single-hop distance, where the signal is launched at a very low takeoff angle. The Pedersen ray can easily be disrupted by any sort of ionospheric gradient.

The gray line generally runs north-south, but it varies by 23° either side of true north as measured at the equator over the course of the year. This variation is caused by the tilt in the Earth's axis. The gray line is exactly north-south through the poles at the equinoxes (March 21 and September 21) and is at its 23° extremes

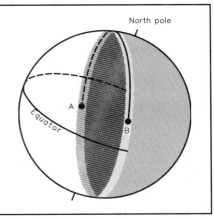

Fig 21.18—The gray line encircles the Earth, but the tilt at the equator to the poles varies over 46° with the seasons. Long-distance contacts can often be made halfway around the Earth along the gray line, even as low as 1.8 and 3.5 MHz. The strength of the signals, characteristic of gray-line propagation, indicates that multiple Earth-ionosphere hops are not the only mode of propagation, since losses in many such hops would be very great. Chordal hops, where the signals are confined to the ionosphere for at least part of the journey, are involved.

on June 21 and December 21. Over a one-year period, the gray line crosses a 46° sector of the Earth north and south of the equator, providing optimum paths to slightly different parts of the world each day. Many commonly available computer programs plot the gray line on a flat map or globe. The *ARRL Operating Manual* provides sunrise and sunset times over the entire year for several hundred worldwide locations. The position of the gray line on any date can also be plotted manually on a globe from these data.

F-Layer Backscatter and Sidescatter

Special forms of F-layer scattering can create unusual paths within the skip zone. *Backscatter* and *sidescatter* signals are usually observed just above the MUF for the direct path and allow communications not normally possible by other means. They also provide a useful clue to propagation conditions, because reception of backscatter or sidescatter signals suggests that the MUF is just below the operating frequency. Backscattered signals are generally weak and have a characteristic hollow sound. Useful communication distances range from 100 km (60 mi) to the normal one-hop distance of 4000 km (2500 mi).

Backscatter and sidescatter are closely related and the terminology does not precisely distinguish between the two. Back-

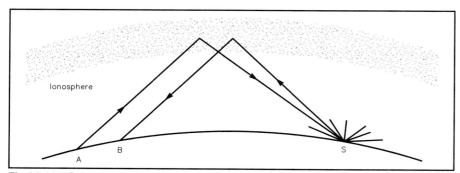

Fig 21.19—Schematic of a simple backscatter path. Stations A and B are too close to make contact via normal F-layer ionospheric refraction. Signals scattered back from a distant point on the Earth's surface (S), often the ocean, may be accessible to both and create a backscatter circuit.

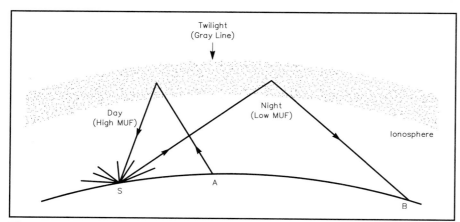

Fig 21.20—Backscatter path across the gray line. Stations A and B are too close to make contact via normal ionospheric refraction, but may hear each other's signals scattered from point S. Station A makes use of a high-angle refraction on the day side of the gray line, where the MUF is high. Station B makes use of a night-time refraction, with a lower MUF and lower angle of propagation. Note that station A points away from B to complete the circuit.

scatter usually refers to single-hop signals that have been scattered by the Earth or the ocean at some distant point back toward the transmitting station. Two stations spaced a few hundred km apart can often communicate via a backscatter path near the MUF. See **Fig 21.19**.

Sidescatter usually refers to a circuit that is oblique to the normal great-circle path. Two stations can make use of a common side-scattering region well off the direct path, often toward the south. European and North American stations sometimes complete 28-MHz contacts via a scattering region over Africa. US and Finnish 50-MHz operators observed a similar effect early one morning in November 1989 when they made contact by beaming off the coast of West Africa.

When backscattered signals cross an area where there is a sharp gradient in ionospheric density, such as between night and day, the path may take on a different geometry, as shown in **Fig 21.20**. In this case, stations can communicate because backscattered signals return via the day side ionosphere on a shorter hop than the night side. This is possible because the dayside MUF is higher and thus the skip distance shorter. The net effect is to create a backscatter path between two stations within the normal skip zone.

Transequatorial Spread-F

Discovered in 1947, *transequatorial spread-F* (TE) supports propagation between 5000 and 8000 km (3100 and

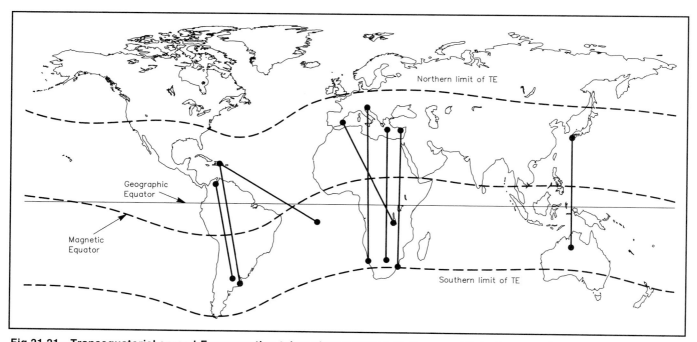

Fig 21.21—Transequatorial spread-F propagation takes place between stations equidistant across the geomagnetic equator. Distances up to 8000 km (5000 mi) are possible on 28 through 432 MHz. Note the geomagnetic equator is considerably south of the geographic equator in the Western Hemisphere.

5000 mi) across the equator from 28 MHz to as high as 432 MHz. Stations attempting TE contacts must be nearly equidistant from the geomagnetic equator. Many contacts have been made at 50 and 144 MHz between Europe and South Africa, Japan and Australia and the Caribbean region and South America. Fewer contacts have been made on the 222-MHz band. TE signals have been heard at 432 MHz, but so far, no two-way contacts have resulted.

Unfortunately for most continental US stations, the geomagnetic equator dips south of the geographic equator in the Western Hemisphere, as shown in **Fig 21.21**, making only the most southerly portions of Florida and Texas within TE range. TE contacts from the southeastern part of the country may be possible with Argentina, Chile and even South Africa.

Transequatorial spread-F peaks between 5 PM and 10 PM during the spring and fall equinoxes, especially during the peak years of the solar cycle. The lowest probability is during the summer. Quiet geomagnetic conditions are required for TE to form. Signals have a rough aurora-like note, sometimes termed *flutter fading*. High power and large antennas are not required to work TE, as VHF stations with 100 W and single long Yagis have been successful.

The best explanation of TE propagation suggests that the F_2 layer near the equator bulges and intensifies slightly, particularly during solar maxima. Irregular field-aligned ionization forms shortly after sunset in an area 100-200 km (60-120 mi) north and south of the geomagnetic equator and 500-3000 km (310-1900 mi) wide. For this reason, the mode is sometimes called *transequatorial field-aligned irregularities*. It moves west with the setting sun. The MUF may increase to twice its normal level 15° either side of the geomagnetic equator.

Field alignment of ionospheric irregularities favors refraction along magnetic field lines, that is north-south. VHF and UHF signals are refracted twice over the geomagnetic equator at angles that normally would be insufficient to bring the signals back toward Earth. See **Fig 21.22**. The geometry is such that two shallow reflections in the F_2 layer can create north-south terrestrial paths up to 8000 km (5000 mi).

Spread-F propagation also occurs over the polar regions, but because of low population densities, amateurs have rarely reported making use of it. Near the northern magnetic pole (located in extreme northeastern Canada), spread-F is a nearly permanent feature of winter. During summer, it appears most summer nights and at least

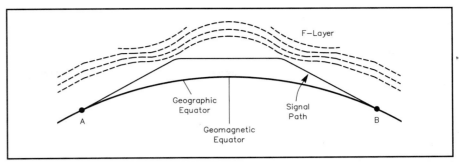

Fig 21.22—Cross-section of a transequatorial spread-F signal path, showing the effects of ionospheric bulging and a double refraction above the normal MUF.

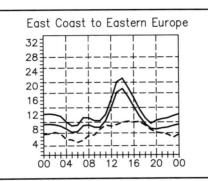

Fig 21.23—Propagation prediction chart for East Coast to Europe that appeared in *QST* for December 1994. An average 2800-MHz (10.7-cm) solar flux of 82 was assumed for the mid-December to mid-January period. On 10% of these days, the highest frequency propagated was predicted at least as high as the uppermost curve (the Highest Possible Frequency, or HPF, approximately 21 MHz), and for 50% of the days as high as the middle curve, the MUF. The broken lines show the Lowest Usable Frequency (LUF) for a 1500-W CW transmitter.

Table 21.5
Shortwave Broadcasting Bands

Frequency (MHz)	Band (m)
2.300-2.495	120
3.200-3.400	90
3.900-4.000	75
4.750-5.060	60
5.959-6.200	49
7.100-7.300	41
9.500-9.900	31
11.650-12.050	25
13.600-13.800	22
15.100-15.600	19
17.550-17.900	16
21.450-21.850	13
25.600-26.100	11

half the time during the day. There is a greater probability of polar spread-F appearing during the equinox periods and during the solar cycle maximum. Field-alignment in the polar regions suggests that some form of backscatter signals, similar to aurora, would be most likely.

MUF PREDICTION

F-layer MUF prediction is key to forecasting HF communications paths at particular frequencies, dates and times, but forecasting is complicated by several variables. Solar radiation varies over the course of the day, season, year and solar cycle. These regular intervals provide the main basis for prediction, yet recurrence is far from reliable. In addition, forecasts are predicated on a quiet geomagnetic field, but the condition of the Earth's magnetic field is most difficult to predict weeks

or months ahead. For professional users of HF communications, uncertainty is a nuisance for maintaining reliable communications paths, while for many amateurs it provides an aura of mystery and chance that adds to the fun of DXing. Nevertheless, many amateurs want to know what to expect on the HF bands to make best use of available on-the-air time, plan contest strategy, ensure successful net operations or engage in other activities.

MUF Forecasts

Long-range forecasts several months ahead, such as those published in *QST* and other journals, provide only the most general form of prediction. A series of 30 charts, similar to **Fig 21.23**, forecast average propagation for a one-month period over specific paths each month in *QST*. The charts assume a single average solar flux value for the entire period. Early editions of *The ARRL Operating Manual* included a similar series of such charts for three representative sunspot values and a variety of paths.

The uppermost curve shows the highest frequency that will be propagated on at least 10% of the days. The given values might be exceeded considerably on a few rare days. On at least half the days, propagation will be possible as high as the

MUF Prediction on the Home Computer

Like predicting the weather, predicting propagation—even with the best computer software available—is *not* an exact science. The processes occurring as a signal is propagated from one point on the Earth to another are enormously complicated and subject to an incredible number of variables. Experience and a knowledge of propagation conditions (as related to solar activity) are needed when you actually get on the air to check out the bands. Keep in mind, too, that ordinary computer programs are written mainly to calculate propagation for great-circle paths via the F layer. Scatter, skew-path, auroral and other such propagation modes may provide contacts when computer predictions indicate no contacts are possible.

Brief information about prediction programs for the IBM PC and compatible computers follows. The programs can be divided roughly into two categories. The first includes programs best suited for quick, on-the-fly predictions. Programs such as MINIMUF, IONSOUND or MINIPROP are excellent tools for assessing what bands are likely to be open in the near future, usually using recent propagation indices broadcast on WWV as data input—to see, for example, whether 21 MHz is likely to be open tomorrow morning on the path from Kansas City to Berlin, given a solar flux averaging 95 over the last several days.

The second category includes programs designed for long-term analysis and station planning. These programs require more investment in computer hardware, and will often take a considerable period of time to do their complex calculations. Most long-term planning programs benefit greatly from the use of a math coprocessor, although most of them will run, even if slowly, without a math coprocessor in the system. ASAPS, IONCAP and CAPMAN typify this category of heavy-duty propagation programs.

Table 21.7 summarizes the features and attributes of these programs. Each program is copyrighted unless otherwise indicated.

ASAPS V2.2

ASAPS, for Advanced Stand-Alone Prediction System, was developed in Australia. It rivals IONCAP (see below) in its analysis capability but performs calculations in significantly less time. It is also interactive with the user; transmit power levels, antennas and other parameters may be changed and the new results viewed almost instantly without further menu entries. Available from: IPS Radio and Space Services.

IONCAP, Version PC.27; CAPMAN, Version 3.0

IONCAP, short for Ionospheric Communications Analysis and Prediction, was written by an agency of the US government. This program is considered by many amateurs and professionals alike as the most comprehensive and best HF prediction program available. The program has been under development for almost 30 years, and was ported to PCs from a mainframe environment. IONCAP offers no menu; rather, an ASCII input file containing instructions and data must be prepared for program execution. CAPMAN is a "user-friendly" version of IONCAP that has been tailored for ham use. IONCAP is available from: National Technical Information Service. CAPMAN is available from Kangaroo Tabor Software.

IONSOUND HDX, IONSOUND, IONSOUND PRO

There are now three versions of IONSOUND, at price levels from free to $75. The low-end IONSOUND HDX program is tailored specifically for the locations formerly shown in the "How's DX?" column in *QST*. It provides calculated data that one would not expect from a program of its price class. Its bigger brothers, of course, provide more features—IONSOUND PRO is the top of the line model. Graphs present mode chirp plots (frequency versus delay time versus intensity), much as an ionsonde oblique-incidence sounder might produce. For each path calculation, the user must first answer several screens of questions, such as noise environment at the terminal points, receiver bandwidth, required S/N ratio, order of layer modes and so on. Available from Skywave Technologies.

MINIMUF, Version 3.5

Written in BASIC, this was the first prediction program to become available for use on home computers. The program was published in Dec 1982 *QST* (R. B. Rose, "MINIMUF, A Simplified MUF-Prediction Program for Microcomputers," pp 36-38). This is public-domain software that has been customized by many subsequent software writers. The core algorithm's accuracy suffers outside the range from 250 to 6000 miles. MINIMUF calculations do not consider the E region, further limiting its accuracy.

---

middle curve. Propagation will exceed the lowest curve on at least 90% of the days. The MUF on any particular day cannot be determined from the charts, but the calculated time of the MUF is reliable.

Short-range forecasts a few days ahead still depend on predictions of solar radiation and geomagnetic indices, but the previous 27-day history and current conditions improve these estimates considerably. Daily forecasts are even more reliable, as they are based on measurements of solar and geomagnetic activity no more than a few hours old. Forecasts can be made at home using one of several popular programs for the personal computer, including CAPMAN, IONCAP, IONSOUND, MINIMUF and MINIPROP PLUS. These are described in more detail in the accompanying sidebar.

Direct Observation

Propagation conditions can be determined directly by listening to the HF band. The simplest method is to tune higher in frequency until no more long-distance stations are heard. This point is roughly just above the MUF to anywhere in the world at that moment. The highest usable amateur band would be the next lowest one. If HF stations seem to disappear around 23 MHz, for example, the 15-m band at 21 MHz might make a good choice for DXing. By carefully noting station locations as well, the MUF in various directions can also be determined quickly.

The shortwave broadcast bands (see **Table 21.5**) are most convenient for MUF browsing, because there are many high-powered stations on regular schedules. Take care to ensure that programming is actually transmitted from the originating country. A Radio Moscow or BBC program, for example, may be relayed to a transmitter outside Russia or England for retransmission. An excellent guide to shortwave broadcast stations is the *World Radio TV Handbook*, available through the ARRL.

MINIPROP PLUS, Version 2.0

MINIPROP has undergone several revisions since it first appeared. It was written primarily for the amateur community, and has an excellent user interface, with great graphics. In addition to the customary propagation data, it provides sunrise-sunset and gray-line information, along with a world map showing either long- or short-path propagation graphically. MINIPROP PLUS also produces a unique "DX Compass" showing the MUF in 12 azimuth directions for a given time of day. Available from Sheldon C. Shallon, W6EL.

Application Tips

Because of the lag in F-layer response to a rapid increase in solar activity, it is best to use either a 5, 15 or 90-day running average of the 2800-MHz (10.7-cm) solar flux for prediction calculations. The type of application determines which is best. The 5-day mean is a short-term dynamic input; the 90-day mean is appropriate for long-term planning. The ultimate test, of course, of a prediction program is to get on the air and listen to the signals arriving from the part of the world you just modeled!

Table 21.7

Features and Attributes of Propagation Prediction Programs

	ASAPS V. 4	VOACAP Windows	HFx 1.06	MINIPROP PLUS 2.5	CAPMan	WinCAP Wizard 2
User Friendliness	Good	Good	Excellent	Good	Good	Good
Review data	Yes	Yes	Yes	Yes	Yes	Yes
User library of QTHs	Yes	Yes	Yes	Yes	Yes	Yes
Bearings, distances	Yes	Yes	Yes	Yes	Yes	Yes
MUF calculation	Yes	Yes	Yes	Yes	Yes	Yes
LUF calculation	Yes	Yes	Yes	No	Yes	Yes
Wave angle calculation	Yes	Yes	Yes	Yes	Yes	Yes
Vary minimum wave angle	Yes	Yes	Yes	Yes	Yes	Yes
Path regions and hops	Yes	Yes	Yes	Yes	Yes	Yes
Multipath effects	Yes	Yes	Yes	No	Yes	Yes
Path probability	Yes	Yes	Yes	Yes	Yes	Yes
Signal strengths	Yes	Yes	Yes	Yes	Yes	Yes
S/N ratios	Yes	Yes	Yes	No	Yes	Yes
Long path calculation	Yes	Yes	Yes	Yes	Yes	Yes
Antenna selection	Yes	Yes	Yes	Indirectly	Yes	Isotropic
Vary antenna height	Yes	Yes	Yes	Indirectly	Yes	No
Vary ground characteristics	Yes	Yes	No	No	Yes	No
Vary transmit power	Yes	Yes	Yes	Indirectly	Yes	Yes
Graphic displays	Yes	Yes	Yes	Yes	Yes	Yes
UT-day graphs	Yes	Yes	Yes	Yes	Yes	Yes
Color monitor support	Yes	Yes	Yes	Yes	Yes	Yes
Hard disk required	Yes	Yes	Yes	No	Yes	Yes
Save data to disk	Yes	Yes	Yes	No	Yes	Yes
Area Mapping	No	Yes	Yes	Yes	Yes	No
Documentation	Yes	On-line	On-line	Yes	Yes	Yes
Price class	$275+	free†	$129	$60	$89+	$29.95+

"Review data" indicates ability to review previous program display screens.
Price classes are for early 1999 and subject to change.
†Available on the World Wide Web at: **http://elbert.its.bldrdoc.gov/hf.html**
+Shipping and handling extra.

WWV and WWVH

The standard time stations WWV (Ft Collins, Colorado) and WWVH (Kauai, Hawaii), which transmit on 2.5, 5, 10, 15 and 20 MHz, are also popular for propagation monitoring. They transmit 24 hours a day. Daily monitoring of these stations for signal strength and quality can quickly provide a good basic indication of propagation conditions. In addition, each hour they broadcast the geomagnetic A and K indices, the 2800-MHz (10.7-cm) solar flux, and a short forecast of conditions for the next day. These are heard on WWV at 18 minutes past each hour and on WWVH at 45 minutes after the hour. The same information is also available by telephoning the recorded message at 303-497-3235. The K index is updated every three hours, while the A index and solar flux are updated after 2100 UTC. These data are useful for making predictions on home computers, especially when averaged over several days of solar flux observations.

Beacons

Automated beacons in the higher amateur bands can also be useful adjuncts to propagation watching. Beacons are ideal for this purpose because most are designed to transmit 24 hours a day. Among the best organized beacon system is one designed by the Northern California DX Foundation for the 20-m band. Nine beacons on five continents transmit in successive one-minute intervals, with the tenth minute silent. More on this system, along with a longer list of HF, VHF and UHF beacons, can be found in *The ARRL Operating Manual*. Other interested groups publish updated lists of beacons with call sign, frequency, location, transmitter mode, power, and antenna. Beacons often include location as part of their automated message, and many can be located from their call sign. Thus, even casual scanning of beacon subbands can be useful.

Table 21.6 provides the frequencies where beacons useful to HF propagation are most commonly placed.

PROPAGATION IN THE TROPOSPHERE

All radio communication involves propagation through the troposphere for at least part of the signal path. Radio waves traveling through the lowest part of the atmosphere are subject to refraction, scattering and other phenomena, much like ionospheric effects. Tropospheric conditions are rarely significant below 30 MHz, but they are very important at 50 MHz and higher. Much of the long-distance work on the VHF, UHF and microwave bands depends on some form of tropospheric propagation. Instead of watching solar activity and geomagnetic indices, those who use tropospheric propagation are much more concerned about the weather.

Line of Sight

At one time it was thought that communications in the VHF range and higher would be restricted to line-of-sight paths. Although this has not proven to be the case even in the microwave region, the concept of line of sight is still useful in understanding tropospheric propagation. In the vacuum of space or in a completely homogeneous medium, radio waves do travel essentially in straight lines, but these conditions are almost never met in terrestrial propagation.

Radio waves traveling through the troposphere are ordinarily refracted slightly earthward. The normal drop in temperature, pressure and water-vapor content with increasing altitude change the index of refraction of the atmosphere enough to cause refraction. Under average conditions, radio waves are refracted toward Earth enough to make the horizon appear 1.15 times farther away than the visual horizon. Under unusual conditions, tropospheric refraction may extend this range significantly.

A simple formula can be used to estimate the distance to the radio horizon under average conditions:

$$d = \sqrt{2h}$$

where

d = distance to the radio horizon, miles

h = height above average terrain, ft,

$$d = \sqrt{17h}$$

where

d = distance to the radio horizon, km

h = height above average terrain, m.

The distance to the radio horizon for an antenna 30 m (98 ft) above average terrain is thus 22.6 km (14 mi), a station on top of a 1000-m (3280-ft) mountain has a radio horizon of 130 km (80 mi).

Atmospheric Absorption

Atmospheric gases, most notably oxygen and water vapor, absorb radio signals, but neither is a significant factor below 10 GHz. Attenuation from rain becomes important at 3.3 GHz, where signals passing through 20 km (12 mi) of heavy showers incur an additional 0.2 dB loss. That same rain would impose 12 dB additional loss at 10 GHz and losses continue to increase with frequency. Heavy fog is similarly a problem only at 5.6 GHz and above. More detailed information about atmospheric absorption in the microwave bands can be found in the *ARRL UHF/Microwave Experimenter's Manual.*

Tropospheric Scatter

Contacts beyond the radio horizon out to a working distance of 100 to 500 km (60 to 310 mi), depending on frequency, equipment and local geography, are made every day without the aid of obvious propagation enhancement. At 1.8 and 3.5 MHz, local communication is due mostly to ground wave. At higher frequencies, especially in the VHF range and above, the primary mechanism is scattering in the troposphere, or *troposcatter.*

Most amateurs are unaware that they use troposcatter even though it plays an essential role in most local communication. Radio signals through the VHF range are scattered primarily by small gradients in the index of refraction of the lower atmosphere due to turbulence, along with changes in temperature. Radio signals in the microwave region can also be scattered by rain, snow, fog, clouds and dust. That tiny part that is scattered forward and toward the Earth creates the over-the-horizon paths. Troposcatter path losses are considerable and increase with frequency.

The maximum distance that can be linked via troposcatter is limited by the height of a scattering volume common to two stations, shown schematically in **Fig 21.24.** The highest altitude for which scattering is efficient at amateur power levels is about 10 km (6 mi). An application of the distance-to-the-horizon formula yields 800 km (500 mi) as the limit for troposcatter paths, but typical maxima are more like half that. Tropospheric scatter varies little with season or time of day, but it is difficult to assess the effect of weather on troposcatter alone. Variations in tropospheric refraction, which is very sensitive to the weather, probably account

Table 21.6
Popular Beacon Frequencies

Frequencies (MHz) *Comments*

Frequencies (MHz)	Comments
14.100	Northern California DX Foundation beacons
28.2-28.3	Several dozen beacons worldwide
50.0-50.1	Most US beacons are within 50.06-50.08 MHz
70.03-70.13	Beacons in England, Ireland, Gibraltar and Cyprus

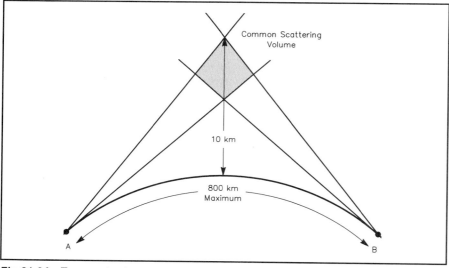

Fig 21.24—Tropospheric-scatter path geometry. The lower boundary of the common scattering volume is limited by the take-off angle of both stations. The upper boundary of 10 km (6 mi) altitude is the limit of efficient scattering in the troposphere. Signal strength increases with the scattering volume.

for most of the observed day-to-day differences in troposcatter signal strength.

Troposcatter does not require special operating techniques or equipment, as it is used unwittingly all the time. In the absence of all other forms of propagation, especially at VHF and above, the usual working range is essentially the maximum troposcatter distance. Ordinary working range increases most dramatically with antenna height, because that lowers the take-off angle to the horizon. Working range increases less quickly with antenna gain and transmitter power. For this reason, a mountaintop is the choice location for extending ordinary troposcatter working distances.

Rain Scatter in the Troposphere

Scatter from raindrops is a special case of troposcatter practical in the 1296-MHz to 10-GHz range. Stations simply point their antennas toward a common area of rain. A certain portion of radio energy is scattered by the raindrops, making possible over-the-horizon or obstructed-path contacts, even with low power. The theoretical range for rain scatter is as great as 600 km (370 mi), but the experience of amateurs in the microwave bands suggests that expected distances are less than 200 km (120 mi). Snow and hail make less efficient scattering media unless the ice particles are partially melted. Smoke and dust particles are too small for extraordinary scattering, even in the microwave bands.

Refraction and Ducting in the Troposphere

Radio waves are refracted by natural gradients in the index of refraction of air with altitude, due to changes in temperature, humidity and pressure. Refraction under standard atmospheric conditions extends the radio horizon somewhat beyond the visual line of sight. Favorable weather conditions further enhance normal tropospheric refraction, lengthening the useful VHF and UHF range by several hundred kilometers and increasing signal strength. Higher frequencies are more sensitive to refraction, so its effects may be observed in the microwave bands before they are apparent at lower frequencies.

Ducting takes place when refraction is so great that radio waves are bent back to the surface of the Earth. When tropospheric ducting conditions exist over a wide geographic area, signals may remain very strong over distances of 1500 km (930 mi) or more. Ducting results from the gradient created by a sharp increase in temperature with altitude, quite the opposite of normal atmospheric conditions. A simultaneous drop in humidity contributes

to increased refractivity. Useful temperature inversions form between 250 and 2000 m (800-6500 ft) above ground. The elevated inversion and the Earth's surface act something like the boundaries of a natural open-ended waveguide. Radio waves of the right frequency range caught inside the duct will be propagated for long distances with relatively low losses. Several common weather conditions can create temperature inversions.

Radiation Inversions in the Troposphere

Radiation inversions are probably the most common and widespread of the various weather conditions that affect propagation. Radiation inversions form only over land after sunset as a result of progressive cooling of the air near the Earth's surface. As the Earth cools by radiating heat into space, the air just above the ground is cooled in turn. At higher altitudes, the air remains relatively warmer, thus creating the inversion. A typical radiation-inversion temperature profile is shown in **Fig 21.25**.

The cooling process may continue through the evening and predawn hours, creating inversions that extend as high as 500 m (1500 feet). Deep radiation inversions are most common during clear, calm, summer evenings. They are more distinct in dry climates, in valleys and over open ground. Their formation is inhibited by wind, wet ground and cloud cover. Although radiation inversions are common and widespread, they are rarely strong enough to cause true ducting. The enhanced conditions so often observed after sunset during the summer are usually a result of this mild kind of inversion.

High-Pressure Weather Systems

Large, sluggish, high-pressure systems (or anticyclones) create the most dramatic and widespread tropospheric ducts due to subsidence. Subsidence inversions in high-pressure systems are created by air that is sinking. As air descends, it is compressed and heated. Layers of warmer air—temperature inversions—often form between 500 and 3000 m (1500-10,000 ft) altitude, as shown in **Fig 21.26**. Ducts usually intensify during the evening and early morning hours, when surface temperatures drop and suppress the tendency for daytime ground-warmed air to rise. In the Northern Hemisphere, the longest and strongest radio paths usually lie to the south of high-pressure centers. See **Fig 21.27**.

Sluggish high-pressure systems likely to contain strong temperature inversions are common in late summer over the east-

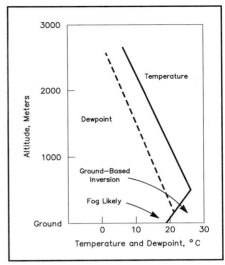

Fig 21.25—Temperature and dewpoint profile of an early-morning radiation inversion. Fog may form near the ground. The midday surface temperature would be at least 30°C.

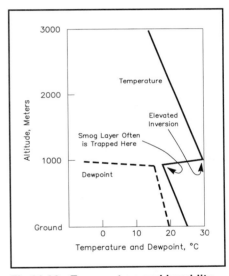

Fig 21.26—Temperature and humidity profile across an elevated duct at 1000 m altitude. Such inversions typically form in summertime high-pressure systems. Note the air is very dry in the inversion.

ern half of the US. They generally move southeastward out of Canada and linger for days over the Midwest, providing many hours of extended propagation. The southeastern part of the country and the lower Midwest experience the most high-pressure openings; the upper Midwest and East Coast somewhat less frequently; the western mountain regions rarely.

Semipermanent high-pressure systems, which are nearly constant climatic features in certain parts of the world, sustain the longest and most exciting ducting paths. The

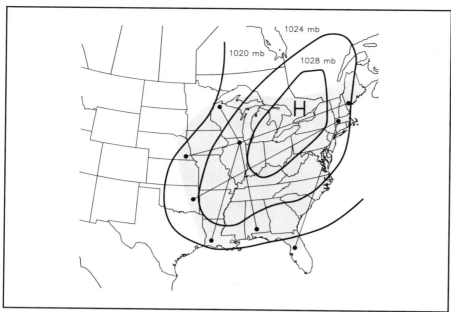

Fig 21.27—Surface weather map for September 13, 1993, shows that the eastern US was dominated by a sprawling high-pressure system. The shaded portion shows the area in which ducting conditions existed on 144 through 1286 MHz and higher.

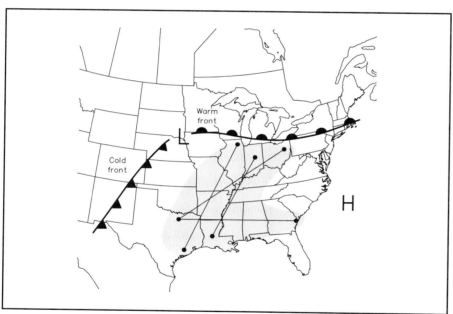

Fig 21.28—Surface weather map for June 2, 1980, with a typical spring wave cyclone over the southeastern quarter of the US. The shaded portion shows where ducting conditions existed.

Eastern Pacific High, which migrates northward off the coast of California during the summer, has been responsible for the longest ducting paths reported to date. Countless contacts in the 4000-km (2500 mi) range have been made from 144 MHz through 5.6 GHz between California and Hawaii. The Bermuda High is a nearly permanent feature of the Caribbean area, but during the summer it moves north and often covers the southeastern US. It has supported contacts in excess of 2800 km (1700 mi) from Florida and the Carolinas to the West Indies, but its full potential has not been exploited. Other semipermanent highs lie in the Indian Ocean, the western Pacific and off the coast of western Africa.

Wave Cyclone

The wave cyclone is a more dynamic weather system that usually appears during the spring over the middle part of the American continent. The wave begins as a disturbance along a boundary between cooler northern and warmer southern air masses. Southwest of the disturbance, a cold front forms and moves rapidly eastward, while a warm front moves slowly northward on the eastward side. When the wave is in its open position, as shown in **Fig 21.28**, north-south radio paths 1500 km (930 mi) and longer may be possible in the area to the east of the cold front and south of the warm front, known as the warm sector. East-west paths nearly as long may also open in the southerly parts of the warm sector.

Wave cyclones are rarely productive for more than a day in any given place, because the eastward-moving cold front eventually closes off the warm sector. Wave cyclone temperature inversions are created by a southwesterly flow of warm, dry air above 1000 m (3200 ft) that covers relatively cooler and moister gulf air flowing northward near the Earth's surface. Successive waves spaced two or three days apart may form along the same frontal boundary.

Warm Fronts and Cold Fronts

Warm fronts and cold fronts sometimes bring enhanced tropospheric conditions, but rarely true ducting. A warm front marks the surface boundary between a mass of warm air flowing over an area of relatively cooler and more stationary air. Inversion conditions may be stable enough several hundred kilometers ahead of the warm front to create extraordinary paths.

A cold front marks the surface boundary between a mass of cool air that is wedging itself under more stationary warm air. The warmer air is pushed aloft in a narrow band behind the cold front, creating a strong but highly unstable temperature inversion. The best chance for enhancement occurs parallel to and behind the passing cold front.

Other Conditions Associated With Ducts

Certain kinds of wind may also create useful inversions. The *Chinook* wind that blows off the eastern slopes of the Rockies can flood the Great Plains with warm and very dry air, primarily in the springtime. If the ground is cool or snow-covered, a strong inversion can extend as far as Canada to Texas and east to the Mississippi River. Similar kinds of *foehn* winds, as these mountain breezes are called, can be found in the Alps, Caucasus Mountains and other places.

The land breeze is a light, steady, cool wind that commonly blows up to 50 km (30 mi) inland from the oceans, although the distance may be greater in some circumstances. Land breezes develop after sunset on clear summer evenings. The land cools more quickly than the adjacent ocean. Air cooled over the land flows near the surface of the Earth toward the ocean to displace relatively warmer air that is rising. See **Fig 21.29**. The warmer ocean air, in turn, travels at 200-300 m (600-1000 ft) altitude to replace the cool surface air. The land-sea circulation of cool air near the ground and warm air aloft creates a mild inversion that may remain for hours. Land-breeze inversions often bring enhanced conditions and occasionally allow contacts in excess of 800 km (500 mi) along coastal areas.

In southern Europe, a hot, dry wind known as the *sirocco* sometimes blows northward from the Sahara Desert over relatively cooler and moister Mediterranean air. Sirocco inversions can be very strong and extend from Israel and Lebanon westward past the Straits of Gibraltar. Sirocco-type inversions are probably responsible for record-breaking microwave contacts in excess of 1500 km (930 mi) across the Mediterranean.

Marine Boundary Layer Effects

Over warm water, such as the Caribbean and other tropical seas, evaporation inversions may create ducts that are useful in the microwave region between 3.3 and 24 GHz. This inversion depends on a sharp drop in water-vapor content rather than on an increase in temperature to create ducting conditions. Air just above the surface of water at least 30°C is saturated because of evaporation. Humidity drops significantly within 3 to 10 m (10 to 30 ft) altitude, creating a very shallow but stable duct. Losses due to water vapor absorption may be intolerable at the highest ducting frequencies, but breezes may raise the effective height of the inversion and open the duct to longer wavelengths. Stations must be set up right on the beaches to ensure being inside an evaporation inversion.

Tropospheric Fading

Tropospheric turbulence and small changes in the weather are responsible for most fading at VHF and higher. Local weather conditions, such as precipitation, warm air rising over cities and the effects of lakes and rivers, can all contribute to tropospheric instabilities that affect radio propagation. Fast-flutter fading at 28 MHz and above is often the result of an airplane that temporarily creates a second propagation path. Flutter results as the phase relationship between the ordinary tropospheric signal and that reflected by the airplane change with the airplane's movement.

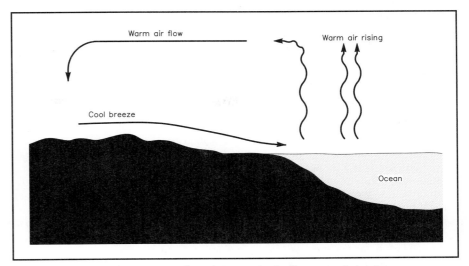

Fig 21.29—Land-breeze convection along a coast after sunset creates a temperature inversion over the land.

EXTRATERRESTRIAL PROPAGATION

Communication of all sorts into space has become increasingly important. Amateurs confront extraterrestrial propagation when accessing satellite repeaters or using the moon as a reflector. Special propagation problems arise from signals that travel from the Earth through the ionosphere (or a substantial portion of it) and back again. Tropospheric and ionospheric phenomena, so useful for terrestrial paths, are unwanted and serve only as a nuisance for space communication. A phenomenon known as Faraday rotation may change the polarization of radio waves traveling through the ionosphere, presenting special problems to receiving weak signals. Cosmic noise also becomes an important factor when antennas are intentionally pointed into space.

Faraday Rotation

Magnetic and electrical forces rotate the polarization of radio waves passing through the ionosphere. For example, signals that leave the Earth as horizontally polarized, and return after a reflection from the moon may not arrive with the same polarization. An additional 20 dB of path loss is incurred when polarization is shifted by 90°, an intolerable amount when signals are marginal.

Faraday rotation is difficult to predict and its effects change over time and with operating frequency. At 144 MHz, the polarization of space waves may shift back into alignment with the antenna within a few minutes, so often just waiting can solve the Faraday problem. At 432 MHz, it may take half an hour or longer for the polarization to become realigned. Use of circular polarization completely eliminates this problem, but creates a new one for EME paths. The sense of circularly polarized signals is reversed with reflection, so two complete antenna systems are normally required, one with left-hand and one with right-hand polarization.

Earth-Moon-Earth

Amateurs have used the moon as a reflector on the VHF and UHF bands since 1960. Maximum allowable power and large antennas, along with the best receivers, are normally required to overcome the extreme free-space and reflection losses involved in Earth-Moon-Earth (EME) paths. More modest stations make EME contacts by scheduling operating times when the Moon is at perigee on the horizon. The Moon, which presents a target only one-half degree wide, reflects only 7% of the radio signals that reach it. Techniques have to be designed to cope with Faraday rotation, cosmic noise, Doppler shift (due to the Moon's movements) and other difficulties. In spite of the problems involved, hundreds of stations have made contacts via the Moon on all bands from 50 MHz to 10 GHz. The techniques of EME communication are discussed in the chapter on **Repeaters, Satellites, EME and DFing.**

Satellites

Accessing amateur satellites generally does not involve huge investments in antennas and equipment, yet station design does have to take into account special challenges of space propagation. Free-space loss is a primary consideration, but it is manageable when satellites are only a few hundred kilometers distant. Free-space path losses to satellites in high Earth orbits are considerably greater, and appropriately larger antennas and higher powers are needed.

Satellite frequencies below 30 MHz can be troublesome. Ionospheric absorption and refraction may prevent signals from reaching space, especially to satellites at very low elevations. In addition, man-made and natural sources of noise are high. VHF and especially UHF are largely immune from these effects, but free-space path losses are greater. Problems related to polarization, including Faraday rotation, intentional or accidental satellite tumbling and the orientation of a satellite's antenna in relation to terrestrial antennas, are largely overcome by using circularly polarized antennas. More on using satellites can be found in the chapter on **Repeaters, Satellites, EME and DFing.**

NOISE AND PROPAGATION

Noise simply consists of unwanted radio signals that interfere with desired communications. In some instances, noise imposes the practical limit on the lowest usable frequencies. Noise may be classified by its sources: man-made, terrestrial and cosmic. Interference from other transmitting stations on adjacent frequencies is not usually considered noise and may be controlled, to a some degree anyway, by careful station design.

Man-Made Noise

Many unintentional radio emissions result from man-made sources. Broadband radio signals are produced whenever there is a spark, such as in contact switches, electric motors, gasoline engine spark plugs and faulty electrical connections. Household appliances, such as fluorescent lamps, microwave ovens, lamp dimmers and anything containing an electric motor may all produce undesirable broadband radio energy. Devices of all sorts, especially computers and anything controlled by microprocessors, television receivers and many other electronics also emit radio signals that may be perceived as noise well into the UHF range. In many cases, these sources are local and can be controlled with proper measures. See the **EMI** chapter.

High-voltage transmission lines and associated equipment, including transformers, switches and lightning arresters, can generate high-level radio signals over a wide area, especially if they are corroded or improperly maintained. Transmission lines may act as efficient antennas at some frequencies, adding to the noise problem. Certain kinds of street lighting, neon signs and industrial equipment also contribute their share of noise.

Lightning

Static is a common term given to the ear-splitting crashes of noise commonly heard on nearly all radio frequencies, although it is most severe on the lowest frequency bands. Atmospheric static is primarily caused by lightning and other natural electrical discharges. Static may result from close-by thunderstorms, but most static originates with tropical storms. Like any radio signals, lightning-produced static may be propagated over long distances by the ionosphere. Thus static is generally higher during the summer, when there are more nearby thunderstorms, and at night, when radio propagation generally improves. Static is often the limiting factor on 1.8 and 3.5 MHz, making winter a more favorable time for using these frequencies.

Precipitation Static and Corona Discharge

Precipitation static is an almost continuous hash-type noise that often accompanies various kinds of precipitation, including snowfall. Precipitation static is caused by rain drops, snowflakes or even wind-blown dust, transferring a small electrical charge on contact with an antenna. Electrical fields under thunderstorms are sufficient to place many objects such as trees, hair and antennas, into corona discharge. Corona noise may sound like a harsh crackling in the radio—building in intensity, abruptly ending, and then building again, in cycles of a few seconds to as long as a minute. A corona charge on an antenna may build to some critical level and then discharge in the atmosphere with an audible pop before recharging. Precipitation static and corona discharge can be a nuisance from LF to well into the VHF range.

Cosmic Sources

The sun, distant stars, galaxies and other cosmic features all contribute radio noise well into the gigahertz range. These cosmic sources are perceived primarily as a more-or-less constant background noise at HF. In the VHF range and higher, specific sources of cosmic noise can be identified and may be a limiting factor in terrestrial and space communications. The sun is by far the greatest source of radio noise, but its effects are largely absent at night. The center of our own galaxy is nearly as noisy as the sun. Galactic noise is especially noticeable when high-gain VHF and UHF antennas, such as may be used for satellite or EME communications, are pointed toward the center of the Milky Way. Other star clusters and galaxies are also radio hot-spots in the sky. Finally, there is a much lower cosmic background noise that seems to cover the entire sky.

FURTHER READING

J. E. Anderson, "MINIMUF for the Ham and the IBM Personal Computer," *QEX*, Nov 1983, pp 7-14.

B. R. Bean and E. J. Dutton, *Radio Meteorology* (New York: Dover, 1968).

K. Davies, *Ionospheric Radio* (London: Peter Peregrinus, 1989). Excellent, though highly technical text on propagation.

G. Grayer, "VHF/UHF Propagation," Ch 2 of *The VHF/UHF DX Book* (Buckingham, England: DIR, 1992).

G. Jacobs, T. Cohen, R. Rose, *The NEW Shortwave Propagation Handbook*, CQ Communications, Inc. (Hicksville, NY: 1995)

L. F. McNamara, *Radio Amateur's Guide to the Ionosphere* (Malabar, Florida: Krieger Publishing Company, 1994). Excellent, quite-readable text on HF propagation.

C. Newton, *Radio Auroras* (Potters Bar, England: Radio Society of Great Britain, 1991).

E. Pocock, "UHF and Microwave Propagation," Ch 3 of *The ARRL UHF/Microwave Experimenter's Manual* (Newington, Connecticut: ARRL, 1990).

E. Pocock, Ed., *Beyond Line of Sight: A History of VHF Propagation from the Pages of QST* (Newington, Connecticut: ARRL, 1992).

Contents

Station Setup and Accessory Projects

22

Although many hams never try to build a major project, such as a transmitter, receiver or amplifier, they do have to assemble the various components into a working station. There are many benefits to be derived from assembling a safe, comfortable, easy-to-operate collection of radio gear, whether the shack is at home, in the car or in a field. This chapter, written by Wally Blackburn, AA8DX, covers the many aspects of setting up an efficient station.

This chapter will detail some of the "how tos" of setting up a station for fixed, mobile and portable operation. Such topics as station location, finding adequate power sources, station layout and cable routing are covered, along with some of the practical aspects of antenna erection and maintenance.

Regardless of the type of installation you are attempting, good planning greatly increases your chances of success. Take the time to think the project all the way through, consider alternatives, and make rough measurements and sketches during your planning and along the way. You will save headaches and time by avoiding "shortcuts." What might seem to save time now may come back to haunt you with extra work when you could be enjoying your shack.

One of the first considerations should be to determine what type of operating you intend to do. While you do not want to strictly limit your options later, you need to consider what you want to do, how much you have to spend and what room you have to work with. There is a big difference between a casual operating position and a "big gun" contest station, for example.

Fixed Stations

SELECTING A LOCATION

Selecting the right location for your station is the first and perhaps the most important step in assembling a safe, comfortable, convenient station. The exact location will depend on the type of home you have and how much space can be devoted to your station. Fortunate amateurs will have a spare room to devote to housing the station; some may even have a separate building for their exclusive use. Most must make do with a spot in the cellar or attic, or a corner of the living room is pressed into service.

Examine the possibilities from several angles. A station should be comfortable; odds are good that you'll be spending a lot of time there over the years. Some unfinished basements are damp and drafty—not an ideal environment for several hours of leisurely hamming. Attics have their drawbacks, too; they can be stifling during warmer months. If possible, locate your station away from the heavy traffic areas of your home. Operation of your station should not interfere with family life. A night of chasing DX on 80 m may be exciting to you, but the other members

Fig 22.1—Danny, KD4HQV, appreciates the simplicity that his operating position affords. *(Photo courtesy Conard Murray, WS4S)*

Fig 22.2—VE6AFO's QSL card reveals an impressive array of gear. Although many hams would appreciate having this much space to devote to a station, most of us must make do with less.

Fig 22.3—Scott, KA9FOX, operated this well laid-out station, W9UP, during a recent contest. *(Photo courtesy NØBSH)*

of your household may not share your enthusiasm.

Keep in mind that you must connect your station to the outside world. The location you choose should be convenient to a good power source and an adequate ground. If you use a computer and modem, you may need access to a telephone jack. There should be a fairly direct route to the outside for running antenna feed lines, rotator control cables and the like.

Although most homes will not have an "ideal" space meeting all requirements, the right location for you will be obvious after you scout around. The amateurs whose stations are depicted in **Figs 22.1 through 22.3** all found the right spot for them. Weigh the trade-offs and decide which features you can do without and which are necessary for your style of operation. If possible pick an area large enough for future expansion.

THE STATION GROUND

Grounding is an important factor in overall station safety, as detailed in the **Safety** chapter. An effective ground system is necessary for every amateur station. The mission of the ground system is two-fold. First, it reduces the possibility of electrical shock if something in a piece of equipment should fail and the chassis or cabinet becomes "hot." If connected to a properly grounded outlet, a three-wire electrical system grounds the chassis. Much amateur equipment still uses the ungrounded two-wire system, however. A ground system to prevent shock hazards is generally referred to as *dc ground*.

The second job the ground system must perform is to provide a low-impedance path to ground for any stray RF current inside the station. Stray RF can cause equipment to malfunction and contributes to RFI problems. This low-impedance path is usually called *RF ground*. In most stations, dc ground and RF ground are provided by the same system.

Ground Noise

Noise in ground systems can affect our sensitive radio equipment. It is usually related to one of three problems:

1) Insufficient ground conductor size
2) Loose ground connections
3) Ground loops

These matters are treated in precise scientific research equipment and certain industrial instruments by attention to certain rules. The ground conductor should be at least as large as the largest conductor in the primary power circuit. Ground conductors should provide a solid connection to both ground and to the equipment being grounded. Liberal use of lock washers and star washers is highly recommended. A loose ground connection is a tremendous source of noise, particularly in a sensitive receiving system.

Ground loops should be avoided at all costs. A short discussion of what a ground loop is and how to avoid them may lead you down the proper path. A ground loop is formed when more than one ground current is flowing in a single conductor. This commonly occurs when grounds are "daisy-chained" (series linked). The correct way to ground equipment is to bring all ground conductors out radially from a common point to either a good driven earth ground or a cold-water system. If one or more earth grounds are used, they should be bonded back to the service entrance panel. Details appear in the **Safety** chapter.

Ground noise can affect transmitted and received signals. With the low audio levels required to drive amateur transmitters, and the ever-increasing sensitivity of our receivers, correct grounding is critical.

STATION POWER

Amateur Radio stations generally require a 120-V ac power source. The 120-V ac is then converted to the proper ac or dc levels required for the station equipment. Power supply theory is covered in the **Power Supplies** chapter, and safety issues are covered in the **Safety** chapter. If your station is located in a room with electrical outlets, you're in luck. If your station is located in the basement, an attic or another area without a convenient 120-V source, you will have to run a line to your operating position.

Surge Protection

Typically, the ac power lines provide an adequate, well-regulated source of electrical power for most uses. At the same time, these lines are fraught with frequent power surges that, while harmless to most household equipment, may cause damage to more sensitive devices such as computers or test equipment. A common method of protecting these devices is through the use of surge protectors. More information on these and lightning protection is in the **Safety** chapter.

STATION LAYOUT

Station layout is largely a matter of personal taste and needs. It will depend mostly on the amount of space available, the equipment involved and the types of operating to be done. With these factors in mind, some basic design considerations apply to all stations.

The Operating Table

The operating table may be an office or

Fig 22.4—The basement makes a good location if it is dry. A ready-to-assemble computer desk makes an ideal operating table at a reasonable price. This setup belongs to WK8H. *(Photo courtesy AA8DX)*

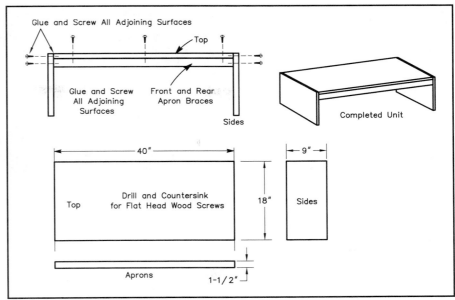

Fig 22.5—A simple but strong equipment shelf can be built from readily available materials. Use ³/₄-inch plywood along with glue and screws for the joints for adequate strength.

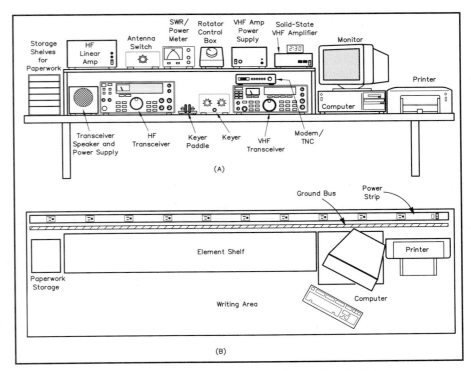

Fig 22.6—Example station layout as seen from the front (A) and the top (B). The equipment is spaced far enough apart that air circulates on all sides of each cabinet.

computer desk, a kitchen table or a custom-made bench. What you use will depend on space, materials at hand and cost. The two most important considerations are height and size of the top. Most commercial desks are about 29 inches above the floor. This is a comfortable height for most adults. Heights much lower or higher than this may cause an awkward operating position.

The dimensions of the top are an important consideration. A deep (36 inches or more) top will allow plenty of room for equipment interconnections along the back, equipment about midway and room for writing toward the front. The length of the top will depend on the amount of equipment being used. An office or computer desk makes a good operating table. These are often about 36 inches deep and 60 inches wide. Drawers can be used for storage of logbooks, headphones, writing materials, and so on. Desks specifically designed for computer use often have built-in shelves that can be used for equipment stacking. Desks of this type are available ready-to-assemble at most discount and home improvement stores. The low price and adaptable design of these desks make them an attractive option for an operating position. An example is shown in **Fig 22.4.**

Stacking Equipment

No matter how large your operating table is, some vertical stacking of equipment may be necessary to allow you to reach everything from your chair. Stacking pieces of equipment directly on top of one another is not a good idea because most amateur equipment needs air flow around it for cooling. A shelf like that shown in **Fig 22.5** can improve equipment layout in many situations. Dimensions of

the shelf can be adjusted to fit the size of your operating table.

Arranging the Equipment

When you have acquired the operating table and shelving for your station, the next task is arranging the equipment in a convenient, orderly manner. The first step is to provide power outlets and a good ground as described in a previous section. Be conservative in estimating the number of power outlets for your installation; radio equipment has a habit of multiplying with time, so plan for the future at the outset.

Fig 22.6 illustrates a sample station layout. The rear of the operating table is spaced about 1 1/2 ft from the wall to allow easy access to the rear of the equipment. This installation incorporates two separate operating positions, one for HF and one for VHF. When the operator is seated at the HF operating position, the keyer and transceiver controls are within easy reach. The keyer, keyer paddle and transceiver are the most-often adjusted pieces of equipment in the station. The speaker is positioned right in front of the operator for the best possible reception. Accessory equipment not often adjusted, including the amplifier, antenna switch and rotator control box, is located on the shelf above the transceiver. The SWR/power meter and clock, often consulted but rarely touched, are located where the operator can view them without head movement. All HF-related equipment can be reached without moving the chair.

This layout assumes that the operator is right-handed. The keyer paddle is operated with the right hand, and the keyer speed and transceiver controls are operated with the left hand. This setup allows the operator to write or send with the right hand without having to cross hands to adjust the controls. If the operator is left-handed, some repositioning of equipment is necessary, but the idea is the same. For best results during CW operation, the paddle should be weighted to keep it from "walking" across the table. It should be oriented such that the operator's entire arm from wrist to elbow rests on the table top to prevent fatigue.

Some operators prefer to place the station transceiver on the shelf to leave the table top clear for writing. This arrangement leads to fatigue from having an unsupported arm in the air most of the time. If you rest your elbows on the table top, they will quickly become sore. If you rarely operate for prolonged periods, however, you may not be inconvenienced by having the transceiver on the shelf. The real secret to having a clear table top for logging, and so on, is to make the operating table deep enough that your entire arm from elbow to wrist rests on the table with the front panels of the equipment at your fingertips. This leaves plenty of room for paperwork, even with a microphone and keyer paddle on the table.

The VHF operating position in this station is similar to the HF position. The amplifier and power supply are located on the shelf. The station triband beam and VHF beam are on the same tower, so the rotator control box is located where it can be seen and reached from both operating positions. This operator is active on packet

Fig 22.7—It was back to basics for Elias, K4IX, during a recent Field Day.

Fig 22.8—Richard, WB5DGR, uses a homebrew 1.5-kW amplifier to seek EME contacts from this nicely laid out station.

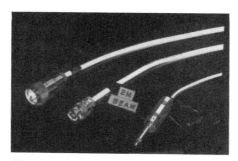

Fig 22.9—Labels on the cables make it much easier to rearrange things in the station. Labeling ideas include masking tape, cardboard labels attached with string and labels attached to fasteners found on plastic bags (such as bread bags).

radio on a local VHF repeater, so the computer, printer, terminal node controller and modem are all clustered within easy reach of the VHF transceiver.

This sample layout is intended to give you ideas for designing your own station. Study the photos of station layouts presented here, in other chapters of this *Handbook* and in *QST*. Visit the shacks of amateur friends to view their ideas. Station layout is always changing as you acquire

new gear, dispose of old gear, change operating habits and interests or become active on different bands. Configure the station to suit your interests, and keep thinking of ways to refine the layout. **Figs 22.7** and **22.8** show station arrangements tailored for specific purposes.

Equipment that is adjusted frequently sits on the table top, while equipment requiring infrequent adjustment is perched on a shelf. All equipment is positioned so the operator does not have to move the chair to reach anything at the operating position.

Aids for Hams with Disabilities

A station used by an amateur with physical disabilities or sensory impairments may require adapted equipment or particular layout considerations. The station may be highly customized to meet the operator's needs or just require a bit of "tweaking."

The myriad of individual needs makes describing all of the possible adaptive methods impractical. Each situation must be approached individually, with consideration to the operator's particular needs.

However, many types of situations have already been encountered and worked through by others, eliminating the need to start from scratch in every case.

An excellent resource is the Courage Handi-Ham System. The Courage Handi-Ham System, a part of the Courage Center, provides a number of services to hams (and aspiring hams) with disabilities. These include study materials, equipment loans, adapted equipment, a newsletter and much more. Information needed to reach the Courage Handi-Hams is in the **References** chapter.

INTERCONNECTING YOUR EQUIPMENT

Once you have your equipment and get it arranged, you will have to interconnect it all. No matter how simple the station, you will at least have antenna, power and microphone or key connections. Equipment such as amplifiers, computers, TNCs and so on add complexity. By keeping your equipment interconnections well organized and of high quality, you will avoid problems later on.

Often, ready-made cables will be available. But in many cases you will have to make your own cables. A big advantage of making your own cables is that you can customize the length. This allows more flexibility in arranging your equipment and avoids unsightly extra cable all over the place. Many manufacturers supply connectors with their equipment along with pinout information in the manual.

Fig 22.10—The back of this Ten-Tec Omni VI HF transceiver shows some of the many types of connectors encountered in the amateur station. Note that this variety is found on a single piece of equipment. *(Photo courtesy AA8DX)*

Fig 22.11—The wires on one side of this terminal block have connectors; the others do not. The connectors make it possible to secure different wire sizes to the strip and also make it much easier to change things around.

This allows you to make the necessary cables in the lengths you need for your particular installation.

Always use high quality wire, cables and connectors in your shack. Take your time and make good mechanical and electrical connections on your cable ends. Sloppy cables are often a source of trouble. Often the problems they cause are intermittent and difficult to track down. You can bet that they will crop up right in the middle of a contest or during a rare DX QSO! Even worse, a poor quality connection could cause RFI or even create a fire hazard. A cable with a poor mechanical connection could come loose and short a power supply to ground or apply a voltage where it should not be. Wire and cables should have good quality insulation that is rated high enough to prevent shock hazards.

Interconnections should be neatly bundled and labeled. Wire ties, masking tape or paper labels with string work well. See **Fig 22.9**. Whatever method you use, proper labeling makes disconnecting and reconnecting equipment much easier. **Fig 22.10** illustrates the number of potential interconnections in a modern, full-featured transceiver.

Wire and Cable

The type of wire or cable to use depends on the job at hand. The wire must be of sufficient size to carry the necessary current. Use the tables in the **Component Data** chapter to find this information. Never use underrated wire; it will be a fire hazard. Be sure to check the insulation too. For high-voltage applications, the insulation must be rated at least a bit higher than the intended voltage. A good rule of thumb is to use a rating at least twice what is needed.

Use good quality coaxial cable of sufficient size for connecting transmitters,

transceivers, antenna switches, antenna tuners and so on. RG-58 might be fine for a short patch between your transceiver and SWR bridge, but is too small to use between your legal-limit amplifier and Transmatch.

Hookup wire may be stranded or solid. Generally, stranded is a better choice since it is less prone to break under repeated flexing. Many applications require shielded wire to reduce the chances of RF getting into the equipment. RG-174 is a good choice for control, audio and some low-power applications. Shielded microphone or computer cable can be used where more conductors are necessary. For more information, see the **Transmission Lines** chapter.

Connectors

While the number of different types of connectors is mind-boggling, many manufacturers of amateur equipment use a few standard types. If you are involved in any group activities such as public service or emergency-preparedness work, check to see what kinds of connectors others in the group use and standardize connectors wherever possible. Assume connectors are not waterproof, unless you specifically buy one clearly marked for outdoor use (and assemble it correctly).

Audio, Power and Control Connectors

The simplest form of connector is found on terminal blocks. Although it is possible to strip the insulation from wire and wrap it around the screw, this method is not ideal. The wire tends to "squirm" out from under the screw when tightening, allowing strands to hang free, possibly shorting to other screws.

Terminal lugs, such as those in **Fig 22.11**, solve the problem. These lugs may be crimped (with the proper tool), soldered

(A)

(B)

(C)

Fig 22.12—The plugs shown at A are often used to connect equipment to remote power supplies. The multipin connectors at B are used for control, signal and power lines. The DIN plug at C offers shielding and is often used for connecting accessories to transceivers.

or both. Terminal lugs are available in different sizes. Use the appropriate size for your wire to get the best results.

Some common multipin connectors are shown in **Fig 22.12**. The connector in Fig 22.12A is often referred to as a "Cinch-Jones connector." It is frequently used for connections to power supplies from various types of equipment. Supplying from two to eight conductors, these connectors are keyed so that they go together only one way. They offer good mechanical and electrical connections, and the pins are large enough to handle high current. If your cable is too small for the strain relief fitting, build up the outer jacket with a few layers of electrical tape until the strain relief clamps securely. The strain relief will keep your wires from breaking away under flexing or from a sudden tug on the cable.

The plug in Fig 22.12B is usually called a "molex" connector. This plug consists of an insulated outer shell that houses the individual male or female "fingers." Each finger is individually soldered or crimped onto a conductor of the cable and inserted in the shell, locking into place. These connectors are used on many brands of amateur gear for power and accessory connections.

Fig 22.12C shows a DIN connector. Commonly having five to eight pins, these connectors are a European standard that have found favor with amateur equipment manufacturers around the world. They are generally used for accessory connections. A smaller version, the Miniature DIN, is becoming popular. It is most often used in portable gear but can be found on some full-size equipment as well.

Various types of phone plugs are shown in **Fig 22.14**. The ¹/₄-inch (largest) is usually used on amateur equipment for headphone and Morse key connections. They are available with plastic and metal bodies. The metal is usually a better choice because it provides shielding and is more durable.

Fig 22.14 also shows the ¹/₈-inch phone plug. These plugs, sometimes called miniature phone plugs, are used for earphone, external speaker, key and control lines. There is also a subminiature (³/₃₂-inch) phone plug that is not common on amateur gear.

The phono, or RCA, plug shown in **Fig 22.15** is popular among amateurs. It is used for everything from amplifier relay-control lines, to low-voltage power lines, to low-level RF lines, to antenna lines. Several styles are available, but the best choice is the shielded type with the screw-on metal body. As with the phone plugs, the metal bodies provide shielding and are very durable.

Nowhere is there more variation than among microphone connectors. Manufacturers seem to go out of their way to use

incompatible connectors! The most popular types of physical connectors are the four- and eight-pin microphone connectors shown in **Fig 22.16**. The simplest connectors provide three connections: audio, ground and push-to-talk (PTT). More complex connectors allow for such things as control lines from the microphone for frequency changes or power to the microphone for a preamplifier. When connecting a microphone to your rig, especially an after-market one, consult the manual. Follow the manufacturer's recommendations for best results.

If the same microphone will be used for multiple rigs with incompatible connectors, one or more adapters will be necessary. Adapters can be made with short pieces of cable and the necessary connectors at each end.

RF Connectors

There are many different types of RF connectors for coaxial cable, but the three most common for amateur use are the *UHF*, *Type N* and *BNC* families. The type of connector used for a specific job depends on the size of the cable, the frequency of operation and the power levels involved.

The so-called UHF connector is found on most HF and some VHF equipment. It is the only connector many hams will ever see on coaxial cable. PL-259 is another name for the UHF male, and the female is also known as the SO-239. These connectors are rated for full legal amateur power at HF. They are poor for UHF work because they do not present a constant impedance, so the UHF label is a misnomer. PL-259 connectors are designed to fit RG-8 and RG-11 size cable (0.405-inch

OD). Adapters are available for use with smaller RG-58, RG-59 and RG-8X size cable. UHF connectors are not weatherproof.

Fig 22.17 shows how to install the solder type of PL-259 on RG-8 cable. Proper preparation of the cable end is the key to success. Follow these simple steps. Measure back about ³/₄-inch from the cable end and slightly score the outer jacket around its circumference. With a sharp knife, cut through the outer jacket, through the braid, and through the dielectric, right down to the center conductor. Be careful not to score the center conductor. Cutting through all outer layers at once keeps the braid from separating. Pull the severed outer jacket, braid and dielectric off the end of the cable as one piece. Inspect the area around the cut, looking for any strands of braid hanging loose and snip them off. There won't be any if your knife was sharp enough. Next, score the outer jacket about ⁵/₁₆-inch back from the first cut. Cut through the jacket lightly; do not score the braid. This step takes practice. If

Fig 22.15—Phono plugs have countless uses around the shack. They are small and shielded; the type with the metal body is easy to grip. Be careful not to use too much heat when soldering the ground (outer) conductor—you may melt the insulation.

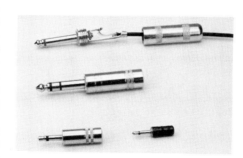

Fig 22.14—The phone-plug family. The ¹/₄-inch type is often used for headphone and key connections on amateur equipment. The three-circuit version is used with stereo headphones. The mini phone plug is commonly used for connecting external speakers to receivers and transceivers. A submini-phone plug is shown in the foreground for comparison. The shielded style with metal barrel is more durable than the plastic style.

Fig 22.16—The four-pin mike connector is common on modern transmitters and receivers. More elaborate rigs use the eight-pin type. The extra conductors may be used for switches to remotely control the frequency or to power a preamplifier built into the mike case.

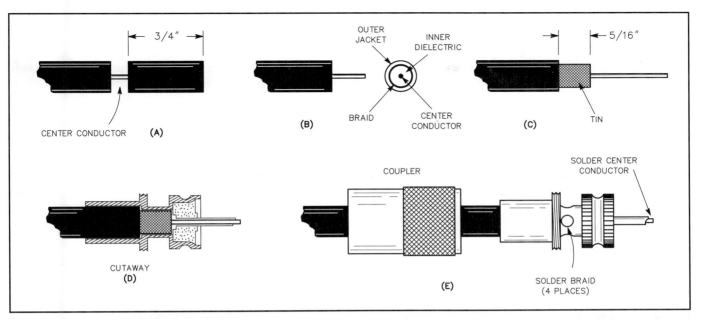

Fig 22.17—The PL-259, or UHF, connector is almost universal for amateur HF work and is popular for equipment operating in the VHF range. Steps A through E are described in detail in the text.

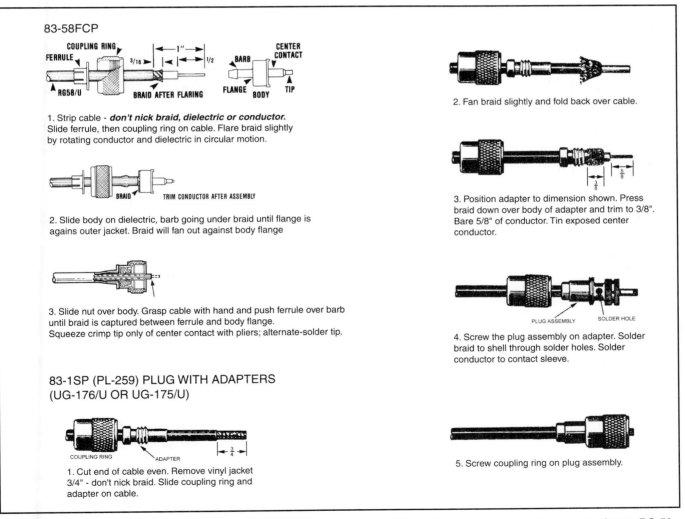

Fig 22.18—Crimp-on connectors and adapters for use with standard PL-259 connectors are popular for connecting to RG-58 and RG-59 type cable. *(Courtesy Amphenol Electronic Components, RF Division, Bunker Ramo Corp)*

you score the braid, start again. Remove the outer jacket.

Tin the exposed braid and center conductor, but apply the solder sparingly and avoid melting the dielectric. Slide the coupling ring onto the cable. Screw the connector body onto the cable. If you prepared the cable to the right dimensions, the center conductor will protrude through the center pin, the braid will show through the solder holes, and the body will actually thread onto the outer cable jacket.

Solder the braid through the solder holes. Solder through all four holes; poor connection to the braid is the most common form of PL-259 failure. A good connection between connector and braid is just as important as that between the center conductor and connector. Use a large soldering iron for this job. With practice, you'll learn how much heat to use. If you use too little heat, the solder will bead up, not really flowing onto the connector body. If you use too much heat, the dielectric will melt, letting the braid and center conductor touch. Most PL-259s are nickel plated, but silver-plated connectors are much easier to solder and only slightly more expensive.

Solder the center conductor to the center pin. The solder should flow on the inside, not the outside, of the center pin. If you wait until the connector body cools off from soldering the braid, you'll have less trouble with the dielectric melting. Trim the center conductor to be even with the end of the center pin. Use a small file to round the end, removing any solder that built up on the outer surface of the center pin. Use a sharp knife, very fine sandpaper or steel wool to remove any solder flux from the outer surface of the center pin. Screw the coupling ring onto the body, and you're finished.

Fig 22.18 shows two options available

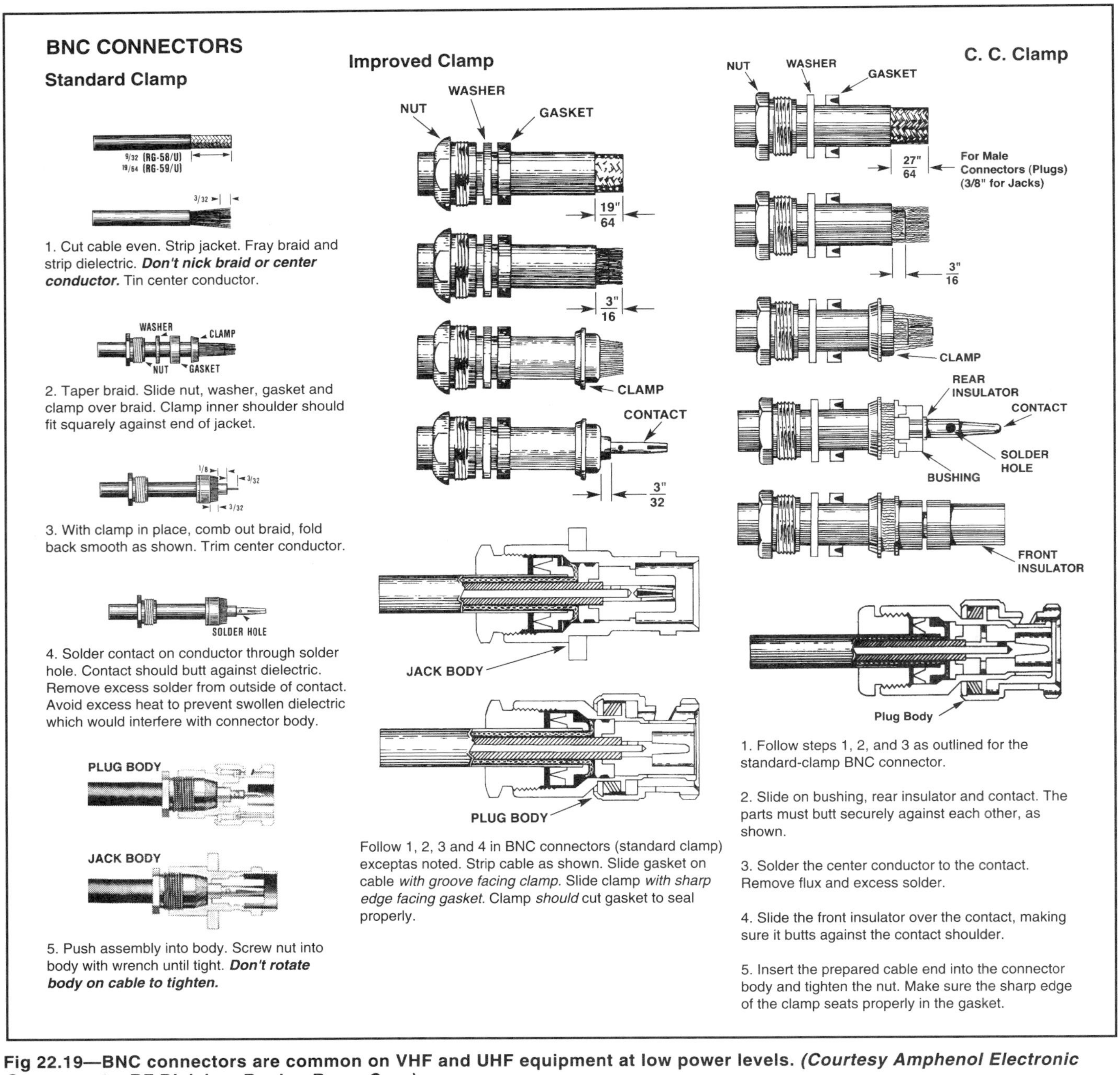

BNC CONNECTORS
Standard Clamp
9/32 (RG-58/U)
19/64 (RG-59/U)
3/32
1. Cut cable even. Strip jacket. Fray braid and strip dielectric. Don't nick braid or center conductor. Tin center conductor.
WASHER
CLAMP
NUT
GASKET
2. Taper braid. Slide nut, washer, gasket and clamp over braid. Clamp inner shoulder should fit squarely against end of jacket.
1/8
3/32
3/32
3. With clamp in place, comb out braid, fold back smooth as shown. Trim center conductor.
SOLDER HOLE
4. Solder contact on conductor through solder hole. Contact should butt against dielectric. Remove excess solder from outside of contact. Avoid excess heat to prevent swollen dielectric which would interfere with connector body.
PLUG BODY
JACK BODY
5. Push assembly into body. Screw nut into body with wrench until tight. Don't rotate body on cable to tighten.
Improved Clamp
WASHER
NUT
GASKET
19"/64
3"/16
CLAMP
CONTACT
3"/32
JACK BODY
PLUG BODY
Follow 1, 2, 3 and 4 in BNC connectors (standard clamp) exceptas noted. Strip cable as shown. Slide gasket on cable with groove facing clamp. Slide clamp with sharp edge facing gasket. Clamp should cut gasket to seal properly.
C. C. Clamp
NUT
WASHER
GASKET
27"/64
For Male Connectors (Plugs) (3/8" for Jacks)
3"/16
CLAMP
REAR INSULATOR
CONTACT
SOLDER HOLE
BUSHING
FRONT INSULATOR
Plug Body
1. Follow steps 1, 2, and 3 as outlined for the standard-clamp BNC connector.
2. Slide on bushing, rear insulator and contact. The parts must butt securely against each other, as shown.
3. Solder the center conductor to the contact. Remove flux and excess solder.
4. Slide the front insulator over the contact, making sure it butts against the contact shoulder.
5. Insert the prepared cable end into the connector body and tighten the nut. Make sure the sharp edge of the clamp seats properly in the gasket.

Fig 22.19—BNC connectors are common on VHF and UHF equipment at low power levels. ***(Courtesy Amphenol Electronic Components, RF Division, Bunker Ramo Corp)***

if you want to use RG-58 or RG-59 size cable with PL-259 connectors. The crimp-on connectors manufactured specially for the smaller cable work very well if installed correctly. The alternative method involves using adapters for the smaller cable with standard RG-8 size PL-259s. Prepare the cable as shown. Once the braid is prepared, screw the adapter into the PL-259 shell and finish the job as you would a PL-259 on RG-8 cable.

The BNC connectors illustrated in **Fig 22.19** are popular for low power levels at VHF and UHF. They accept RG-58 and RG-59 cable, and are available for cable

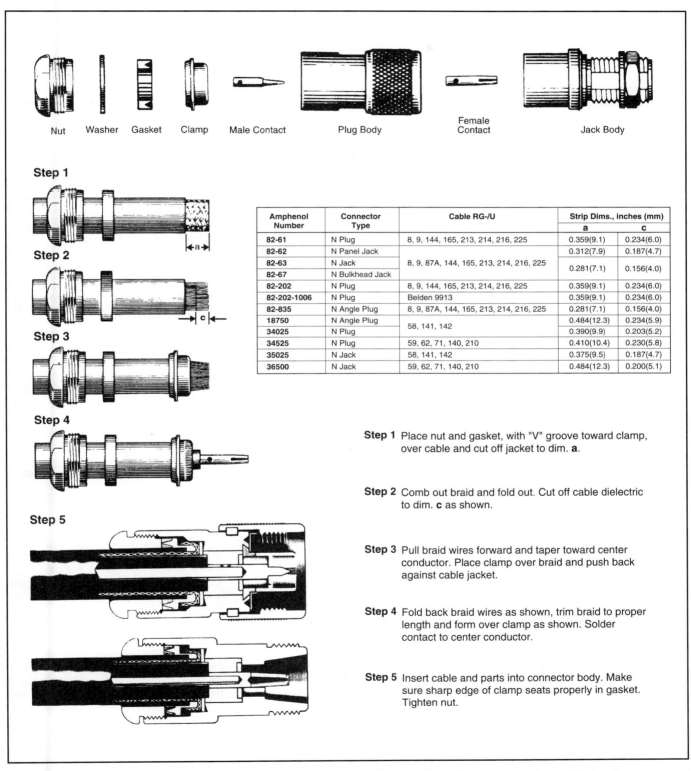

Amphenol Number	Connector Type	Cable RG-/U	Strip Dims., inches (mm)	
			a	c
82-61	N Plug	8, 9, 144, 165, 213, 214, 216, 225	0.359(9.1)	0.234(6.0)
82-62	N Panel Jack		0.312(7.9)	0.187(4.7)
82-63	N Jack	8, 9, 87A, 144, 165, 213, 214, 216, 225	0.281(7.1)	0.156(4.0)
82-67	N Bulkhead Jack			
82-202	N Plug	8, 9, 144, 165, 213, 214, 216, 225	0.359(9.1)	0.234(6.0)
82-202-1006	N Plug	Belden 9913	0.359(9.1)	0.234(6.0)
82-835	N Angle Plug	8, 9, 87A, 144, 165, 213, 214, 216, 225	0.281(7.1)	0.156(4.0)
18750	N Angle Plug	58, 141, 142	0.484(12.3)	0.234(5.9)
34025	N Plug		0.390(9.9)	0.203(5.2)
34525	N Plug	59, 62, 71, 140, 210	0.410(10.4)	0.230(5.8)
35025	N Jack	58, 141, 142	0.375(9.5)	0.187(4.7)
36500	N Jack	59, 62, 71, 140, 210	0.484(12.3)	0.200(5.1)

Step 1 Place nut and gasket, with "V" groove toward clamp, over cable and cut off jacket to dim. **a**.

Step 2 Comb out braid and fold out. Cut off cable dielectric to dim. **c** as shown.

Step 3 Pull braid wires forward and taper toward center conductor. Place clamp over braid and push back against cable jacket.

Step 4 Fold back braid wires as shown, trim braid to proper length and form over clamp as shown. Solder contact to center conductor.

Step 5 Insert cable and parts into connector body. Make sure sharp edge of clamp seats properly in gasket. Tighten nut.

Fig 22.20—Type N connectors are a must for high-power VHF and UHF operation. *(Courtesy Amphenol Electronic Components, RF Microwave Operations)*

mounting in both male and female versions. Several different styles are available, so be sure to use the dimensions for the type you have. Follow the installation instructions carefully. If you prepare the cable to the wrong dimensions, the center pin will not seat properly with connectors of the opposite gender. Sharp scissors are a big help for trimming the braid evenly.

The Type N connector, illustrated in **Fig 22.20**, is a must for high-power VHF and UHF operation. N connectors are available in male and female versions for cable mounting and are designed for RG-8 size cable. Unlike UHF connectors, they are designed to maintain a constant impedance at cable joints. Like BNC connectors, it is important to prepare the cable to the right dimensions. The center pin must be positioned correctly to mate with the center pin of connectors of the opposite gender. Use the right dimensions for the connector style you have.

Computer Connectors

As if the array of connectors related to amateur gear were not enough, the prevalence of the computer in the shack has brought with it another set of connectors to consider. Most connections between computers and their peripherals are made with some form of multiconductor cable. Examples include shielded, unshielded and ribbon cable. Common connectors used are the 9- and 25-pin D-Subminiature connector, the DIN and Miniature DIN and the 36-pin Amphenol connector. Various edge-card connectors are used internally (and sometimes externally) on many computers. **Fig 22.21** shows a variety of computer connectors.

EIA-232 Serial Connections

The serial port on a computer is arguably the most used, and often most troublesome, connector encountered by the amateur. The serial port is used to connect modems, TNCs, computer mice and some printers to the computer. As the name implies, the data is transmitted serially.

The EIA-232-D (commonly referred to as RS-232) standard defines a system used to send data over relatively long distances. It is commonly used to send data anywhere from a few feet to 50 feet or more. The standard specifies the physical connection and signal lines. The serial ports on most computers comply with the EIA-232-D standard only to the degree necessary to operate with common peripherals. **Fig 22.22** shows the two most common connectors used for computer serial ports. A 9-pin connector can be adapted to a 25-pin by connecting like signals. Earth ground is not provided in the 9-pin version.

Fig 22.21—Various computer connectors.

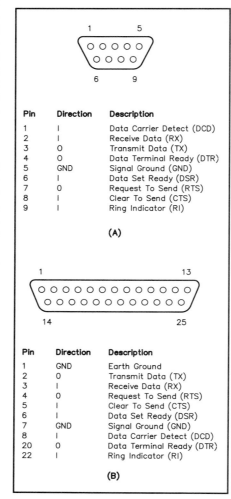

Fig 22.22—The two most common implementations of EIA-232-D serial connections on personal computers use 9- and 25-pin connectors.

Equipment connected via EIA-232-D is usually classified in one of two ways: DTE (data terminal equipment) or DCE (data communication equipment). Terminals and computers are examples of DTE, while modems and TNCs are DCE.

The binary data is represented by specific voltage levels on the signal line. The EIA-232-D standard specifies that a bi-

nary one is represented by a voltage ranging from –3 to –25 V. A binary zero ranges from 3 to 25 V. ±12 V is a common level in many types of equipment, but anything within the specified ranges is just as valid.

The RTS (request to send), CTS (clear to send), DTR (data terminal ready) and DSR (data set ready) lines are used for handshaking signals. These signals are used to coordinate the communication between the DTE and DCE. The RTS and DTR line are used by the DTE to indicate to the DCE that it is ready to receive data from the DCE. The DCE uses the CTS/DSR lines to signal the DTE as to whether or not it is ready to accept data. DCD (data carrier detect) is also sometimes used by the DCE to signal the DTE that an active carrier is present on the communication line. A +12-V signal represents an active handshaking signal. The equipment "drops" the line to –12 V when it is unable to receive data.

You may notice that the name "ready to send" is sort of a misnomer for the DTE since it actually uses it to signal that it is ready to receive. This is a leftover from when communication was mostly one-way—DTE to DCE. Note also that the signal names really only make sense from the DTE point of view. For example, pin 2 is called TD on both sides, even though the DCE is receiving data on that pin. This is another example of this one-way terminology.

It would be much too simple if all serial devices implemented all of the EIA-232-D specifications. Some equipment ignores some or all of the handshaking signals. Other equipment expects handshaking signals to be used as specified. Connecting these two types of equipment together will result in a frustrating situation. One side will blindly send data while the other side blindly ignores all data sent to it!

Fig 22.23 shows the different possible ways to connect equipment. Fig 22.23A shows how to connect a "normal" DTE/DCE combination. This assumes both sides correctly implement all of the handshaking signals. If one or both sides ignore handshaking signals, the connections shown in Fig 22.23B will be necessary. In this scheme, each side is sending the handshaking signals to itself. This little bit of deceit will almost always work, but handshaking signals that are present will be ineffective.

Null Modem Connections

Some equipment does not fall completely in the DTE or DCE category. Some serial printers, for example, act as DCE while others act as DTE. Whenever a DTE/

DTE or DCE/DCE connection is needed a special connection, known as a *null modem* connection, must be made. An example might be connecting two computers together so they can transmit data back and forth. A null modem connection simply crosses the signal and handshaking lines. **Fig 22.24** shows a normal null modem connection (A) and one for equipment that ignores handshaking (B).

Parallel Connections

Another common computer port is the parallel port. The most popular use for the parallel port by far is for printer connections. As the name implies, data is sent in a parallel fashion. There are eight data lines accompanied by a number of control and handshaking lines. A parallel printer connection typically uses a 25-pin D-Subminiature connector at the computer end and a 36-pin Amphenol connector (often called an Epson connector) on the printer.

Connecting Computers to Amateur Equipment

Most modern transceivers provide a serial connection that allows external control of the rig, typically with a computer. Commands sent over this serial control line can cause the rig to change frequency, mode and other parameters. Logging and contest software running on the computer often takes advantage of this capability.

The serial port of most radios operates with the TTL signal levels of 0 V for a binary 0 and 5 V for a binary 1. This is incompatible with the ±12 V of the serial port on the computer. For this reason, level shifting is required to connect the radio to the computer.

A couple of level shifter projects appear in the projects section at the end of this chapter. One of these two examples will work in most rig control situations, although some minor modifications may be necessary. Use the manual and technical documentation to find out what signals your radio requires and choose the circuit that fits the bill. The important factors to note are whether handshaking is implemented and what polarity the radio expects for the signals. In some cases, a 5-V level represents a logic 1 (active high) and in others a logic 0 (active low).

CSMA/CD Bus

Some equipment, notably ICOM rigs with the CI-V interface and recent Ten-Tec gear, use a CSMA/CD (carrier-sense multiple access/collision detect) bus that can interconnect a number of radios and computers simultaneously. This bus basically consists of a single wire, on which the devices transmit and receive

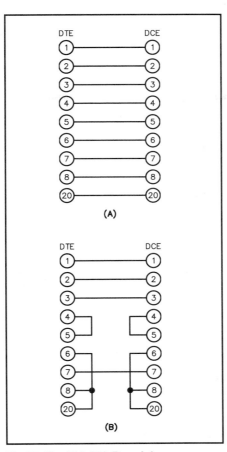

Fig 22.23—EIA-232-D serial connections for normal DTE/DCE (A) and those that ignore handshaking (B).

data, and a ground wire. **Fig 22.25** illustrates the CSMA/CD scheme.

Each device connected to the bus has its own unique digital address. A radio comes from the manufacturer with a default address that can be changed if desired, usually by setting dip switches inside the radio. Information is sent on the bus in the form of packets that include the control data and the address of the device (radio or computer) for which they are intended. A device receives every packet but only acts on the data when its address is embedded in the packet.

A device listens to the bus before transmitting to make sure it is idle. A problem occurs when both devices transmit on the bus at the same time: They both listen, hear nothing and start to send. When this happens, the packets garble each other. This is known as a collision. That is where the CSMA/CD bus collision-detection feature comes in: The devices detect the collision and each sender waits a random amount of time before resending. The sender waiting the shorter random time will get to send first.

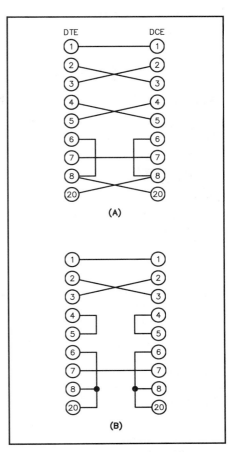

Fig 22.24—EIA-232-D null modem serial connection.

Computer/TNC Connections

TNCs (terminal node controllers) also connect to the computer (or terminal) via the serial port. A TNC typically implements handshaking signals. Therefore, a connector like the one in Fig 22.23A will be necessary. Connectors at the TNC end vary with manufacturer. The documentation included with the TNC will provide details for hooking the TNC to the computer.

DOCUMENTING YOUR STATION

An often neglected but very important part of putting together your station is properly documenting your work. Ideally, you should diagram your entire station from the ac power lines to the antenna on paper and keep the information in a special notebook with sections for the various facets of your installation. Having the station well documented is an invaluable aid when tracking down a problem or planning a modification. Rather than having to search your memory for information on what you did a long time ago, you'll have

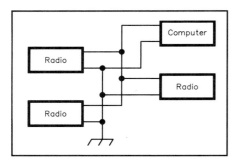

Fig 22.25—The basic two-wire bus system that ICOM and some Ten-Tec radios share among several radios and computers. In its simplest form, the network would include only one radio and one computer.

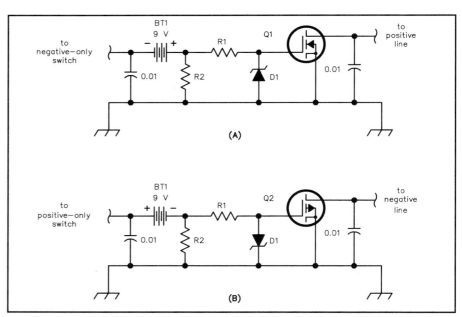

Fig 22.26—Level-shifter circuits for opposite input and output polarities. At A, from a negative-only switch to a positive line; B, from a positive-only switch to a negative line.

BT—9-V transistor-radio battery.
D1—15-V, 1-W Zener diode (1N4744 or equiv).
Q1—IRF620.

Q2—IRF220 (see text).
R1—100 Ω, 10%, ¼ W.
R2—10 kΩ, 10%, ¼ W.

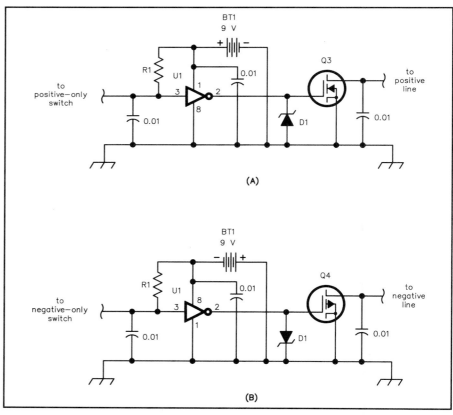

Fig 22.27—Circuits for same-polarity level shifters. At A, for positive-only switches and lines; B, for negative-only switches and lines.

BT1—9-V transistor-radio battery.
D1—15-V, 1-W Zener diode (1N4744 or equiv).
Q3—IRF620.
Q4—IRF220 (see text).

R1—10 kΩ, 10%, ¼ W.
U1—CD4049 CMOS inverting hex buffer, one section used (unused sections not shown; pins 5, 7, 9, 11 and 14 tied to ground).

the facts on hand.

Besides recording the interconnections and hardware around your station, you should also keep track of the performance of your equipment. Each time you install a new antenna, measure the SWR at different points in the band and make a table or plot a curve. Later, if you suspect a problem, you'll be able to look in your records and compare your SWR with the original performance.

In your shack, you can measure the power output from your transmitter(s) and amplifier(s) on each band. These measurements will be helpful if you later suspect you have a problem. If you have access to a signal generator, you can measure receiver performance for future reference.

INTERFACING HIGH-VOLTAGE EQUIPMENT TO SOLID-STATE ACCESSORIES

Many amateurs use a variety of equipment manufactured or home brewed over a considerable time period. For example, a ham might be keying a '60s-era tube rig with a recently built microcontroller-based electronic keyer. Many hams have modern solid-state radios connected to high-power vacuum-tube amplifiers.

Often, there is more involved in connecting HV (high-voltage) vacuum-tube gear to solid-state accessories than a cable and the appropriate connectors. The solid-state switching devices used in some equipment will be destroyed if used to switch the HV load of vacuum tube gear. The polarity involved is important too. Even if the voltage is low enough, a key-line might bias a solid-state device in such a way as to cause it to fail. What is needed is another form of level converter.

MOSFET Level Converters

While relays can often be rigged to interface the equipment, their noise, slow speed and external power requirement make them an unattractive solution in some cases. An alternative is to use power MOSFETs. Capable of handling substantial voltages and currents, power MOSFETs have become common design items. This has made them inexpensive and readily available.

Nearly all control signals use a common ground as one side of the control line. This leads to one of four basic level-conversion scenarios when equipment is interconnected:

1) A positive line must be actuated by a negative-only control switch.

2) A negative line must be actuated by a positive-only control switch.

3) A positive line must be actuated by a positive-only control switch.

4) A negative line must be actuated by a negative-only control switch.

In cases 3 and 4 the polarity is not the problem. These situations become important when the control-switching device is incapable of handling the required open-circuit voltage or closed-circuit current.

Case 1 can be handled by the circuit in **Fig 22.26A**. This circuit is ideal for interfacing keyers designed for grid-block keying to positive CW key lines. A circuit suitable for case 2 is shown in **Fig 22.26B**. This circuit is simply the mirror image of that in Fig 22.26A with respect to circuit

polarity. Here, a P-channel device is used to actuate the negative line from a positive-only control switch.

Cases 3 and 4 require the addition of an inverter, as shown in **Fig 22.27**. The inverter provides the logic reversal needed to drive the gate of the MOSFET high, activating the control line, when the control switch shorts the input to ground.

Almost any power MOSFET can be used in the level converters, provided the voltage and current ratings are sufficient to handle the signal levels to be switched. A wide variety of suitable devices is available from most large mail-order supply houses.

Mobile and Portable Installations

Time and again, radio amateurs have been pressed into service in times of need. New developments outside of Amateur Radio (cellular phones, for example) often bring with them predictions that amateurs will no longer be needed to provide emergency communications. Just as often, a disaster proves beyond doubt the falseness of that exclamation. When the call for emergency communication is voiced by government and disaster relief organizations, mobile and portable equipment is pressed into service where needed. In addition to the occasional emergency or disaster type of communications, mobile and portable operation under normal conditions can challenge and reward the amateur operator.

Most mobile operation today is carried out by means of narrow-band repeaters.

Major repeater frequencies reside in the 146 and 440-MHz bands. As these bands become increasingly congested, the 222 and 1240-MHz bands are being used for this reliable service mode as well. Many amateurs also enjoy mobile and portable HF operation because of the challenge and possibilities of worldwide communication.

MOBILE STATIONS

Installation and setup of mobile equipment can be considerably more challenging than for a fixed station. Tight quarters, limited placement options and harsher environments require innovation and attention to detail for a successful installation. The equipment should be placed so that operation will not interfere with driving. Driving safely is always the primary consideration; operating radio equipment is

secondary. See **Fig 22.28** for one neat solution. If your vehicle has an airbag, be sure it can deploy unimpeded.

Mobile operation is not confined to lower power levels than in fixed stations. Many modern VHF FM transceivers are capable of 25 to 50 W of output. Compact HF rigs usually have outputs in the 100-W range and run directly from the 13.6-V supply.

If a piece of equipment will draw more than a few amps, it is best to run a heavy cable directly to the battery. Few circuits in an automobile electrical system can safely carry the more than 20 A required for a 100-W HF transceiver. Check the table in the **Component Data** chapter to verify the current handling capabilities of various gauges of wire and cable. Adequate and well-placed fuses are necessary to prevent fire hazards. For maximum safety, fuse both the hot and ground lines near the battery. Automobile fires are costly and dangerous.

The limited space available makes antennas for mobile operation quite different than those for fixed stations. This is

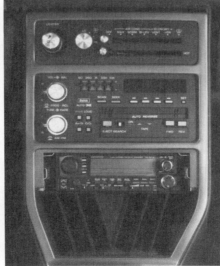

Fig 22.28—N8KDY removed the ashtray for his mobile installation. The old faceplate for the ashtray is used as a cover for the rig when not in use. This may help reduce the temptation for would-be thieves. *(Photo courtesy AA8DX)*

Fig 22.34—Photograph of a commercial dc-to-ac inverter that operates from 6 to 12 V dc and delivers 120 V ac (square wave) at 175 W.

especially true for HF antennas. The **Antennas** chapter contains information for building and using mobile antennas.

Interference

In the past, interference in mobile installations almost always concerned interference to the radio equipment. Examples include ignition noise and charging-system noise. Modern automobiles—packed with arrays of sensors and one or more onboard computers—have made interference a two-way street. The original type of interference (to the radio) has also increased with the proliferation of these devices. An entire chapter in the *ARRL RFI Book* is devoted to ways to prevent and cure this problem.

PORTABLE STATIONS

Many amateurs experience the joys of portable operation once each year in the annual emergency exercise known as Field Day. Setting up an effective portable station requires organization, planning and some experience. For example, some knowledge of propagation is essential to picking the right band or bands for the intended communications link(s). Portable operation is difficult enough without dragging along excess equipment and antennas that will never be used.

Some problems encountered in portable operation that are not normally experienced in fixed-station operation include finding an appropriate power source and erecting an effective antenna. The equipment used should be as compact and lightweight as possible. A good portable setup is simple. Although you may bring gobs of gear to Field Day and set it up the day before, during a real emergency speed is of the essence. The less equipment to set up, the faster it will be operational.

Portable AC Power Sources

There are two popular sources of ac power for use in the field. One is referred to as a dc-to-ac converter, or more commonly, an inverter. The ac output of an inverter is a square wave. Therefore, some types of equipment cannot be operated from the inverter. Certain types of motors are among those devices that require a sine-wave output. **Fig 22.34** shows a typical commercial inverter. This model delivers 120 V of ac at 175 W continuous power rating. It requires 6 or 12 V dc input.

Besides having a square-wave output, inverters have some other traits that make them less than desirable for field use. Commonly available models do not provide a great deal of power. The 175-W model shown in Fig 22.34 could barely power a few light bulbs, let alone a number of trans-

ceivers. Higher-power models are available but are quite expensive. Another problem is that the batteries supplying the inverter with primary power are discharged as power is drawn from the inverter.

Popularity and a number of competing manufacturers have caused gasoline generators to come down considerably in price. For a reliable, adequate source of ac (with sine-wave output), the gasoline-engine-driven generator is the best choice. (While still referred to as generators, practically all modern units actually use alternators to generate ac power.) Generators have become smaller and lighter as manufacturers have used aluminum and other lightweight materials in their construction. **Fig 22.35** shows the type of generator often used during Field Day.

Generators in the 3 to 5-kW range are easily handled by two people and can provide power for a relatively large multioperator field site. Most generators provide 12 V dc output in addition to 120/240 V ac.

Generator Maintenance

Proper maintenance is necessary to obtain rated output and a decent service life from a gasoline generator. A number of simple measures will prolong the life of the equipment and help maintain reliability.

It is a good idea to log the dates the unit is used and the operating time in hours. Many generators have hour-meters to make this simple. Include dates of maintenance and the type of service performed. The manufacturer's manual should be the primary source of maintenance information and the final word on operating procedures and safety. The manual should be thoroughly covered by all persons who will operate and maintain the unit.

Particular attention should be paid to fuel quality and lubricating oil. A typical gasoline generator is often used at or near its rated capacity. The engine driving the alternator is under a heavy load that varies with the operation of connected electrical equipment. For these reasons, the demands on the lubricating oil are usually greater than for most gasoline-engine powered equipment such as lawn mowers, tractors and even automobiles. Only the grades and types of oil specified in the manual should be used. The oil should be changed at the specified intervals usually given as a number of operating hours.

Fuel should be clean, fresh and of good quality. Many problems with gasoline generators are caused by fuel problems. Examples include dirt or water in the fuel and old, stale fuel. Gasoline stored for any length of time changes as the more volatile

Fig 22.35—Modern gasoline engine-powered generators offer considerable ac power output in a relatively compact and lightweight package.

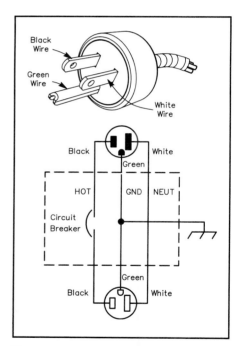

Fig 22.36—A simple accessory that provides overload protection for generators that do not have such provisions built in.

components evaporate. This leaves excess amounts of varnish-like substances that will clog carburetor passages. If the generator will be stored for a long period, it is a good idea to run it until all of the fuel is burned. Another option is the use of fuel stabilizers added to the gasoline before storage.

Spark plugs should be changed as specified. Faulty spark plugs are a common cause of ignition problems. A couple of spare spark plugs should always be

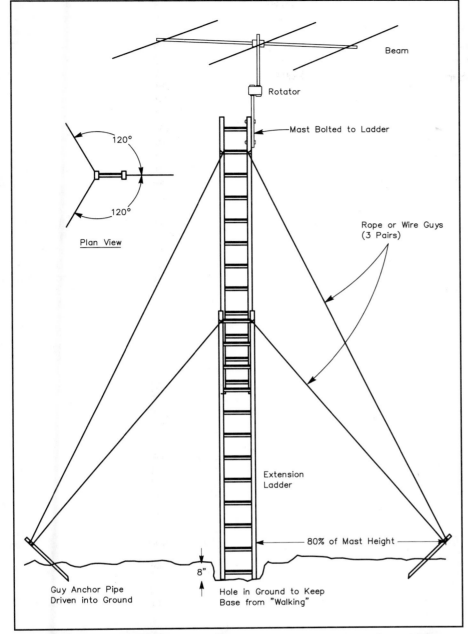

Beam

Rotator

Mast Bolted to Ladder

120°

120°

Plan View

Rope or Wire Guys
(3 Pairs)

Extension
Ladder

80% of Mast Height

8"

Guy Anchor Pipe
Driven into Ground

Hole in Ground to Keep
Base from "Walking"

Fig 22.38—An aluminum extension ladder makes a simple but sturdy portable antenna support. Attach the antenna and feed lines to the top ladder section while it is nested and laying on the ground. Push the ladder vertical, attach the bottom guys and extend the ladder. Attach the top guys. Do not attempt to climb this type of antenna support.

Portable Antennas

An effective antenna system is essential to all types of operation. Effective portable antennas, however, are more difficult to devise than their fixed-station counterparts. A portable antenna must be light, compact and easy to assemble. It is also important to remember that the portable antenna may be erected at a variety of sites, not all of which will offer ready-made supports. Strive for the best antenna system possible because operations in the field are often restricted to low power by power supply and equipment considerations. Some antennas suitable for portable operation are described in the **Antennas** chapter.

Antenna Supports

While some amateurs have access to a truck or trailer with a portable tower, most are limited to what nature supplies, along with simple push-up masts. Select a portable site that is as high and clear as possible. Elevation is especially important if your operation involves VHF. Trees, buildings, flagpoles, telephone poles and the like can be pressed into service to support wire antennas. Drooping dipoles are often chosen over horizontal dipoles because they require only one support.

An aluminum extension ladder makes an effective antenna support, as shown in **Fig 22.38**. In this installation, a mast, rotator and beam are attached to the top of the second ladder section with the ladder near the ground. The ladder is then pushed vertical and the lower set of guy wires attached to the guy anchors. When the first set of guy wires is secured, the ladder may be extended and the top guy wires attached to the anchors. Do not attempt to climb a guyed ladder.

Figs 22.39 and **22.40** illustrate two methods for mounting portable antennas described by Terry Wilkinson, WA7LYI. Although the antennas shown are used for VHF work, the same principles can be applied to small HF beams as well.

In Fig 22.39, a 3-ft section of Rohn 25 tower is welded to a pair of large hinges, which in turn are welded to a steel plate measuring approximately 18×30 inches. One of the rear wheels of a pickup truck is "parked" on the plate, ensuring that it will not move. In Fig 22.39, quad array antennas for 144 and 222 MHz are mounted on a Rohn 25 top section, complete with rotator and feed lines. The tower is then pushed up into place using the hinges, and guy ropes anchored to heavy-duty stakes driven into the ground complete the installation. This method of portable tower installation offers an exceptionally easy-to-erect, yet sturdy, antenna support. Towers installed in this

kept with the unit, along with tools needed to change them. Always use the type of spark plug recommended by the manufacturer.

Generator Ground

A proper ground for the generator is absolutely necessary for both safety reasons and to ensure proper operation of equipment powered from the unit. Most generators are supplied with a three-wire

outlet, and the ground should connect to the plug as shown in **Fig 22.36**. Some generators require that the frame be grounded also. An adequate pipe or rod should be driven into the ground near the generator and connected to the provided clamp or lug. If no connection is provided, a clamp can be used to connect the ground lead to the frame of the generator. As always, follow the manufacturer's recommendations.

manner may be 30 or 40 ft high; the limiting factor is the number of "pushers" and "rope pullers" needed to get into the air. A portable station located in the bed of the pickup truck completes the installation.

The second method of mounting portable beams described by WA7LYI is shown in Fig 22.40. This support is in-

(B)

Fig 22.39—The portable tower mounting system by WA7LYI. At A, a truck is "parked" on the homemade base plate to weigh it down. At B, the antennas, mast and rotator are mounted before the tower is pushed up. Do not attempt to climb a temporary tower installation.

tended for use with small or medium-sized VHF and UHF arrays. The is available from any dealer selling television antennas; tripods of this type are usually mounted on the roof of a house. Open the tripod to its full size and drive a pipe into the ground at each leg. Use a hose clamp or small U-bolt to anchor each leg to its pipe.

The rotator mount is made from a 6-inch-long section of 1 1/2-inch-diameter pipe welded to the center of an "X" made from two 2-ft-long pieces of concrete reinforcing rod (rebar). The rotator clamps onto the pipe, and the whole assembly is placed in the center of the tripod. Large rocks placed on the rebar hold the rotator in place, and the antennas are mounted on a 10 or 15-ft mast section. This system is easy to make and set up.

Tips for Portable Antennas

Any of the antennas described in the **Antennas** chapter or available from commercial manufacturers may be used for portable operation. Generally, though, big or heavy antennas should be passed over in favor of smaller arrays. The couple of decibels of gain a 5-element, 20-m beam may have over a 3-element version is insignificant compared to the mechanical considerations. Stick with arrays of reasonable size that are easily assembled.

Wire antennas should be cut to size and tuned prior to their use in the field. Be careful when coiling these antennas for transport, or you may end up with a tangled mess when you need an antenna in a hurry. The

coaxial cable should be attached to the center insulator with a connector for speed in assembly. Use RG-58 for the low bands and RG-8X for higher-band antennas. Although these cables exhibit higher loss than standard RG-8, they are far more compact and weigh much less for a given length.

Beam antennas should be assembled and tested before taking them afield. Break the beam into as few pieces as necessary for transportation and mark each joint for speed in reassembly. Hex nuts can be replaced with wing nuts to reduce the number of tools necessary.

Ground Rod Installation

A large sledgehammer, a small step ladder and a lot of elbow grease. That's the usual formula for driving in an 8-foot ground rod. Michael Goins, WB5YKX, reports success using fluid hydraulics to ease the task in very dense clay soil.

He suggests digging a small hole, about a foot deep, just enough to hold a few gallons of water. When the hole is complete, pour water into the hole, and then push the ground rod in as far as it will go. Next, pull it out completely. Some of the water will run into the smaller hole made by the ground rod.

Repeat the process, allowing water to run into the small hole each time you remove the rod. Continue pushing and removing until the rod is sunk as far as you want. About 6 inches of rod above the ground is usually enough to allow convenient connection of bonding clamps.

(A)

(B)

(C)

Fig 22.40—The portable mast and tripod by WA7LYI. At A, the tripod is clamped to stakes driven into the ground. The rotator is attached to a homemade pipe mount. At B, rocks piled on the rotator must keep the rotator from twisting and add weight to stabilize the mast. At C, a 10-ft mast is inserted into the tripod/rotator base assembly. Four 432-MHz Quagis are mounted at the top.

THE TiCK-2—A TINY CMOS KEYER 2

TiCK-2 stands for "Tiny CMOS Keyer 2." It is based on an 8-pin DIP microcontroller from Microchip Corporation, the PIC 12C509. This IC is a perfect candidate for all sorts of Amateur Radio applications because of its small size and high performance capabilities. This project was described fully in Oct 1997 *QST* by Gary M. Diana, Sr, N2JGU, and Bradley S. Mitchell, WB8YGG. The keyer has the following features:

- One memory message—a single memory message, capable of at least playing back "CQ CQ DE *callsign callsign* K"
- Mode A and B iambic keying
- Low current requirement—to support portable use.
- Low parts count—consistent with a goal for small physical size.
- Simple interfaces—this includes the rig and user interfaces. The user interface must be simple; ie, the operator shouldn't need a manual. The rig interface should be simple as well: paddles in, key line out.
- Sidetone—supply an audible sidetone for user-feedback functions and to support transceivers that do not have a built-in sidetone.
- Paddle select—allow the operator to swap the dot and dash paddles without having to rewire the keyer (or flip the paddles upside down!).
- Manual keying—permit interfacing a straight key (or external keyer) to the TiCK.

DESIGN

The PIC 12C509 has two pages of 512 bytes of program read-only memory (ROM) and 41 bytes of random-access memory (RAM). This means that all the keyer *functions* have to fit within the ROM. The keyer *settings* such as speed, paddle selection, iambic mode and sidetone enable are stored in RAM. The RAM in this microcontroller is *volatile*, that is, the values stored in memory are lost if the power to the chip is cycled off, then on—but there's not much of a need to do that because of the low power requirements for the chip.

A 12C509 has eight pins. Two pins are needed for the dc input and ground connections. The IC requires a clock signal. Several clock-source options are available; you can use: a crystal, RC (resistor and capacitor) circuit, resonator, or the IC's internal oscillator. The authors chose the internal 4-MHz oscillator to reduce the external parts count. Two I/O lines are used for the paddle input. One output feeds the key line, another output is required for the audio feedback (sidetone) and a third I/O line is assigned to a pushbutton.

USER INTERFACE

Using the two paddles and a pushbutton, you can access all of the TiCK's functions. Certain user-interface functions need to be more easily accessible than others; a prioritized list of functions (from most to least accessible) is presented in **Table 22.1**.

The TiCK employs a *single button interface* (SBI). This simplifies the TiCK PC board, minimizes the part count and makes for ease of use. Most other electronic keyers have multibutton user interfaces, which, if used infrequently, make it difficult to remember the commands. Here, a single button push takes you through the functions, one at a time, at a comfortable pace (based on the current speed of the keyer). Once the code for the desired function is heard, you simply let up on the button. The TiCK then executes the appropriate function, and/or waits for the appropriate input, either from the paddles or the pushbutton itself, depending on the function in question. Once the function is complete, the TiCK goes back into keyer mode, ready to send code through the key line.

The TiCK-2 IC generates a sidetone signal that can be connected to a piezoelectric element or fed to the audio chain of a transceiver. The latter option is rig-specific, but can be handled by more experienced builders.

THE TiCK LIKES TO SLEEP

To meet the low-current requirement, the authors took advantage of the 12C509's ability to *sleep*. In sleep mode, the processor shuts down and waits for input from either of the two paddles. While sleeping, the TiCK-2 consumes just a few microamperes. The TiCK-2 doesn't wait long to go to sleep either: As soon as there is no input from the paddles, it's snoozing! This feature should be especially attractive to amateurs who want to use the TiCK-2 in a portable station.

ASSEMBLING THE TiCK

The TiCK-2's PC board size (1×1.2 inches) supports its use as an embedded and stand-alone keyer. The PC board has two dc input ports, one at J2 for 7 to 25 V and another (J4, AUXILIARY) for 2.5 to 5.5 V. The input at J2 is routed to an on-board 5 V regulator (U2), while the AUXILIARY input feeds the TiCK-2 directly. When making the dc connections, observe proper polarity: There is no built-in reverse-voltage protection at either dc input port.

Table 22.1
TiCK-2 User Interface Description

Action	TiCK-1 and 2 Response	Function
Press pushbutton	S (dit-dit-dit)	Speed adjust: Press dit to decrease, dah to increase speed.
Hold pushbutton down	M (dah-dah)	Memory playback: Plays the message from memory, using the key line and sidetone (if enabled).
Hold pushbutton down	T (dah)	Tune: To unkey rig, press either paddle or pushbutton.
Hold pushbutton down	A (dit-dah)	ADMIN mode: Allows access to various TiCK-2 IC setup parameters.
Hold pushbutton down	I (dit-dit)	Input mode: Allows message entry. Press pushbutton when input is complete.
Hold pushbutton down	P (dit-dah-dah-dit)	Paddle select: Press paddle desired to designate as dit paddle.
Hold pushbutton down	A (dit-dah)	Audio select: Press dit to enable sidetone, dah to disable. Default: enabled.
Hold pushbutton down	SK (dit-dit-dit dah-dit-dah)	Straight key select: Pressing either paddle toggles the TiCK to/from straight key/keyer mode. Default: keyer mode.
Hold pushbutton down	M (dah-dah)	Mode select: Pressing the dit paddle puts the TiCK into iambic mode A; dah selects iambic mode B (the default).
Hold pushbutton down	K (dah-dit-dah)	Keyer mode: If pushbutton is released, the keyer returns to normal operation.
Hold pushbutton down	S (dit-dit-dit)	Cycle repeats with Speed adjust.

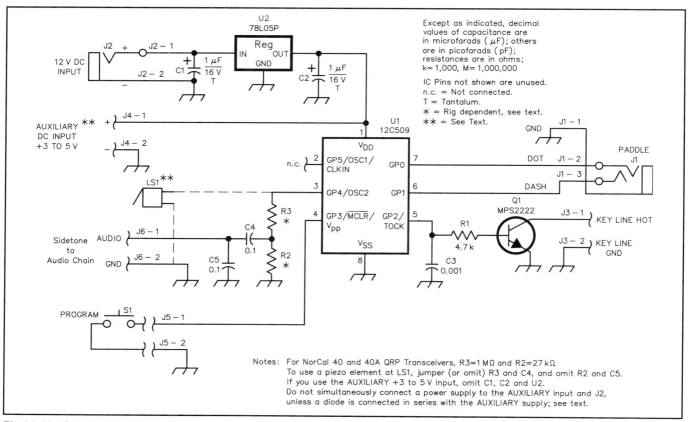

Fig 22.42—Schematic of the TiCK-2 keyer. Equivalent parts can be substituted. Unless otherwise specified, resistors are ¼ W, 5%-tolerance carbon-composition or film units. The PIC12C509 IC must be programmed before use; see Note 2. RS part numbers in parentheses are RadioShack; M = Mouser (Mouser Electronics, see the Address List in the References chapter).

C1, C2—1 µF, 16 V tantalum
 (RS 272-1434; M 581-1.0K35V).
J1—3-circuit jack (RS 274-249;
 M 161-3402).
J2—2-circuit jack (RS 274-251) or
 coaxial (RS 274-1563 or 274-1576).
J3—2-circuit jack (RS 274-251;
 M16PJ135).
LS1—Optional piezo element
 (RS 273-064).
Q1—MPS2222A, 2N2222, PN2222, NPN
 (RS 276-2709; M 333-PN2222).
S1—Normally open pushbutton
 (RS 275-1571; M 10PA011).

U1—Programmed PIC 12C509, available from Embedded Research. (See the Address List in the References chapter.) The TiCK-2 chip/data sheet, $10; TiCK-2 programmed IC, PC board and manual, $15; TiCK-2 programmed IC, PC-board, parts and manual, $21. All prices are postpaid within the continental US. Canadian residents please add $5; all others add $6 for shipping. New York state residents please add 8% sales tax. The DIP and SOIC (surface-mount)

chips and kits are the same price. Please specify which one you prefer when ordering. Note: Components *not* included with the SMD version kit are the voltage regulator and voltage divider components for the audio output. Source code is not available.
U2—5 V, 100 mA regulator
 (RS 276-1770; M 333-ML78L05A).
Misc: PC board, 8-pin DIP socket, hardware, wire (use stranded #22 to #28, Teflon insulated for heat/solder resistance).

The voltage regulator's bias current is quite high and will drain a 9 V battery quickly, even though the TiCK itself draws very little current. For this reason, the "most QRP way" to go may be to power the chip via the AUXILIARY power input and omit U2, C1 and C2.

When using the AUXILIARY dc input port, or if *both* dc inputs are used, connect a diode between the power source and the AUXILIARY power input pin, attaching the diode anode to the power source and the cathode to the AUXILIARY power input pin. This provides IC and battery protection and can also be used to deliver battery backup for your keyer settings.

SIDETONE

A piezo audio transducer can be wired directly to the TiCK-2's audio output: pads and board space are available for voltage-divider components. This eliminates the need to interface the TiCK-2 with a transceiver's audio chain. Use a piezo *element*, not a piezo *buzzer*. A piezo *buzzer* contains an internal oscillator and requires only a dc voltage to generate the sound, whereas a piezo *element* requires an external oscillator signal (available at pin 3 of U1).

If you choose to embed the keyer in a rig and want to hear the keyer's sidetone instead of the rig's sidetone, you may choose to add R2, R3, C4 and C5. Typi-

cally R3 should be 1 MΩ to limit current. R2's value is dictated by the amount of drive required. A value of 27 kΩ is a good start. C4 and C5 values of 0.1 µF work quite well. C4 and C5 soften the square wave and capacitively couple J6-1 to the square-wave output of pin 3. Decreasing the value of R2 decreases the amount of drive voltage, especially below 5 V. Use a 20 kΩ to 30 kΩ trimmer potentiometer at R2 when experimenting.

IN USE

To avoid RF pickup, keep all leads to and from the TiCK as short as possible. The authors tested the TiCK in a variety of

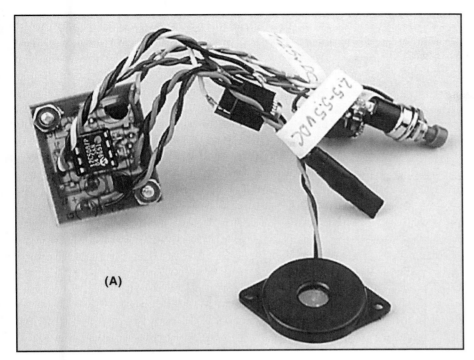

RF environments and found it to be relatively immune to RF. Make sure your radio gear is well grounded and avoid situations that cause an RF-hot shack.

The TiCK keys low-voltage positive lines, common in today's solid-state rigs. Don't try to directly key a tube rig because you will likely—at a minimum—ruin output transistor Q1.

To use the TiCK as a code-practice oscillator, connect a piezo element to the audio output at pin 3. If more volume is needed, use an audio amplifier, such as RadioShack's 277-1008.

The higher the power-supply voltage (within the specified limits), the greater the piezo element's volume. Use 5 V (as opposed to 3 V) if more volume is desired. Also, try experimenting with the location of the piezo element to determine the proper mounting for maximum volume.

JUST WHEN YOU THOUGHT IT WAS SMALL ENOUGH...

In addition to the DIP version of the TiCK-2, there is also a surface-mount version of the keyer. This uses a 12C509 IC, which resides in a medium-size SOIC package measuring roughly 5×5 mm! This is approximately *two-thirds* the size of the DIP version of the IC. (For simplicity, the surface-mount keyer does not include the regulator circuitry.)

The surface-mount version of the TiCK PC board has no provisions for standoffs, but it can easily be mounted in an enclosure (or on the back of a battery!) using double-stick foam tape. The authors used this method to put the surface-mount TiCK into some really tiny enclosures. To assist the builder, the pads on the TiCK board are larger than necessary.

Fig 22.43—At A, photo of DIP version, and at B, SMD version of the TiCK-2.

VINTAGE RADIO T/R ADAPTER

This T/R Adapter provides automatic transmit/receive switching for many vintage transmitters and receivers. It provides time-sequenced antenna transfer, receiver muting and transmitter keying in semi and full break-in CW modes or with push-to-talk systems in AM or SSB.

General Description

The vintage radio adapter consists of two assemblies, the control unit and a remotely located antenna relay. The control unit accepts key, keyer or push-to-talk input and produces time-sequenced outputs to transfer the antenna between receiver and transmitter, mute the receiver and key

either cathode keyed or grid block keyed transmitters. The antenna relay is remotely mounted to aid in running coaxial cable and to minimize relay noise. The antenna relay is rated to handle over 100 W. If high power is contemplated, an antenna relay with a higher power rating is needed. A complete circuit board kit is available, making the unit very easy to duplicate. If you choose to build the unit from scratch, a template package that includes the PC board layout and part placement diagram is available from ARRL HQ.[1]

How it Works

The control circuit of the adapter is based on the C1V keyer/controller chip from

Radio Adventures Corp. See **Fig 22.44A**. In this circuit, the controller features of the chip are used to provide the sequencing outputs to various circuits. By adding a few components, the keyer functions of the C1V can be utilized if desired.

When the KEY input line is pulled low by a key, bug, keyer or push-to-talk switch, the C1V controller chip immediately raises the ANT signal pin, pin 9, and SEMI signal, pin 3, switching the antenna from receiver to transmitter and muting the receiver. About 5.5 ms later the C1V raises the TX signal pin, pin 10, keying the transmitter. The delay provides time for the relays to transfer and for the receiver to quiet.

When the KEY line is released, the C1V

controller delays 5.5 ms and then lowers the TX signal, pin 10, unkeying the transmitter. This 5.5-ms delay compensates for the 5.5-ms transmit delay at the start of the keying sequence, hence preserving the keying waveform. 5.5 ms after lowering the TX pin, the C1V controller lowers the ANT pin, pin 9, switching the antenna back to the receiver and unmuting the receiver. This delay allows time for the transmitter power to decay before switching the antenna, preventing "hot" switching the antenna relay. This process provides full break-in keying up to about 40 WPM. See **Fig 22.45A**.

Although ANT signal, pin 9, follows the keying, SEMI, pin 3, delays on release of the key. The amount of delay is determined by the setting of the DELAY potentiometer, R1. The delay can be varied from about 10 ms to about 1.5 seconds. If KEY line is again lowered before the delay is completed, the delay timer is reset and the full delay time is available after KEY is released. The output of SEMI, pin 3, provides semi break-in timing for use in situations where QSK is not desired.

Level shifter transistor Q1 can be connected to either ANT or SEMI as desired. Q1 drives relay amplifier Q3, which in turn energizes relay K1 and the remote antenna relay (see **Fig 22.44B**).

Level shifter transistor Q2 drives Q4,

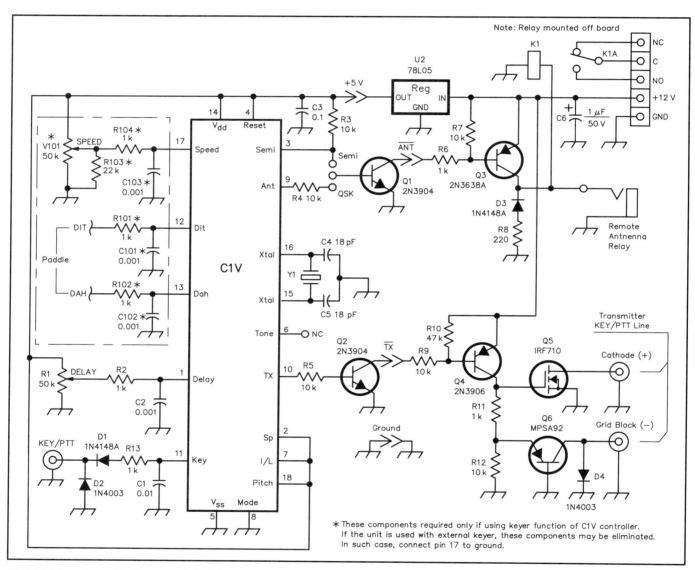

Fig 22.44A—Schematic diagram of the T/R switch. The circuit within the dashed lines is optional—it is used with the keyer function of the C1V controller. If an external keyer is used, this circuit can be eliminated; in this case, connect pin 17 to ground.

C1—0.01-μF, 50-V disc capacitor (Mouser CD50Z6-103M).
C2—0.001-μF, 50-V disc capacitor (Mouser CD50P6-102M).
C3—0.1-μF disc capacitor (Mouser CD100U5-104M).
C4, C5—18-pF, 100-V, 10% NP0 disc capacitor (Mouser 100N2-018J).
C6—1-μF, 50-V, vertical electrolytic (Mouser XRL50V1.0).
D1, D3—1N4148A switching diode.

D2, D4—1N4003 power diode.
K1—SPDT relay, 2 A, 12-V dc coil (Radio Adventures BAS111DC12).
Q1, Q2—2N3904 NPN, TO-92.
Q3—2N3638A PNP, TO-92.
Q4—2N3906 PNP, TO-92.
Q5—IRF710 HEXFET, TO-220.
Q6—PNP, TO-92 (MPSA92).
R1—50-kΩ potentiometer, linear taper (Mouser 31CN405).
R2, R6, R11, R13—1 kΩ, 1/4-W, 5% carbon film.

R3, R4, R5, R7, R9, R12—10-kΩ, 1/4-W, 5% carbon film.
R8—220-Ω, 1/4-W, 5% carbon film.
R10—47 kΩ, 1/4-W, 5% carbon film.
U1—C1V keyer/controller chip (Radio Adventures).
U2—78L05 5-V voltage regulator TO-92 (Mouser NJM78L05A).
Y1—2.0-MHz resonator (supplied with C1V).
Misc—Vintage Radio Adapter PC board (Radio Adventures 090-0112).

which provides drive for the HEXFET transistor used for cathode (positive) keying and the high voltage PNP transistor used for grid block (negative) keying.

The HEXFET power transistor selected for the cathode keying output will key currents beyond 500 mA and open circuit voltages of beyond 200 V. These ratings should be adequate for most cathode-keyed rigs.

The PNP transistor is rated at over 200 V and 10 mA. If your application requires higher voltage or current, a transistor with adequate ratings must be substituted and emitter resistor R11 must be reduced in value to provide sufficient current. The value of the resistor can be found with the following formula: R11 = 10/I. For example, to switch 20 mA, R11 = 10/0.02 = 500 Ω, a 470-Ω resistor would be used.

Construction

Construction is straightforward if the circuit board kit is used. If you choose to hand wire your unit on perf board or lay out your own circuit board, layout is not critical. The finished boards can be mounted in an enclosure of your choice. It is convenient to mount the antenna relay in its own enclosure to reduce the possibility of RF getting into the logic circuits and to make routing of coaxial cables more direct. Note that the relays are mounted on small carrier PC boards and are suspended by their leads only. This method of mounting greatly reduces relay noise that occurs when the relays are mounted on the main PC board.

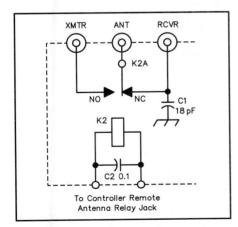

Fig 22.44B—The remote antenna relay schematic diagram.

C1—18-pF disc ceramic, #100N2-018J.
C2—0.1-μF disc ceramic, #CD100U5-104M.
K2—SPDT relay, 2 A, 12 V dc coil.
Misc:
  Bud Econobox, 1¹/₁₆×3⁵/₈×1¹/₂, #CU-123. SO-239 coaxial connectors.

Installation

Power requirements for the adapter are 12 V dc at approximately 120 mA. Antenna connections are straightforward and need no further discussion. Select either the cathode or grid block output for connection to your transmitter as appropriate. See **Fig 22.45B**.

There are a couple of options for muting the receiver. **Fig 22.45C** (left) shows a typical circuit for muting the receiver in the cathode or RF gain circuit. Many receivers bring

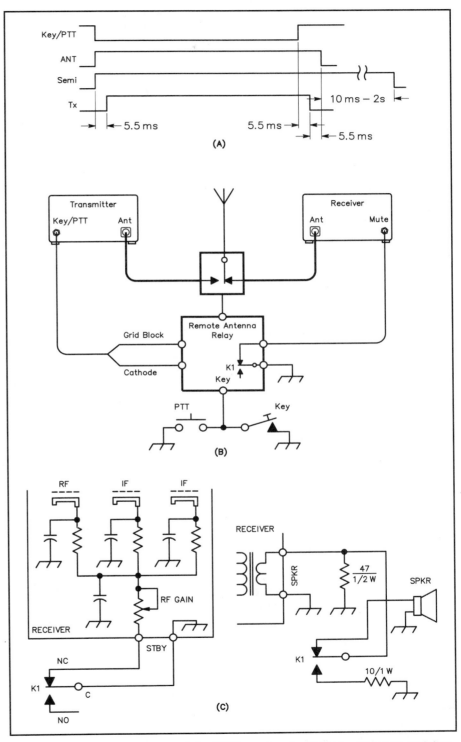

Fig 22.45—(A) Timing diagram of the C1V keyer/controller chip. The text explains how the control circuit works. (B) Block diagram showing how the antenna switch can connect to a vintage station. (C) Sample receiver mute circuits. Left: The receiver can be muted in the cathode or RF gain circuit. Right: If you prefer, the mute relay can also be connected to the speaker circuit.

the mute terminals to the rear apron. In some cases it may be necessary to go inside the receiver to bring out the cathode circuit.

If your receiver does not provide mute terminals and/or you do not want to mute in the cathode circuit, you can connect the mute relay K1 into the speaker circuit as shown in Fig 22.45C (right). The resistors ensure that the speaker terminals of the receiver are always terminated by a load close to 8 Ω, preventing possible damage to the receiver if the resistors were not used.

Note: Some receivers, especially those built in the '30s and '40s, put the receiver into a standby condition by opening the center tap of the power transformer high-voltage winding. *Do not* attempt to mute your receiver by this method. Use cathode or speaker muting instead.

For AM or SSB systems, connect the receiver as discussed, choose the QSK connection fo2r level shifter transistor Q1 and connect the microphone PTT switch to the key/PTT input.

Note
[1]See the **References** chapter for template.

QUICK AND EASY CW WITH YOUR PC

A couple of chips and a few hours work will yield this CW only terminal for a PC. Designed by Ralph Taggart, WB8DQT, the software transforms a computer into a Morse machine that's a full-function CW keyboard *and* a receive display terminal.

The circuit works with IBM-compatible PCs and uses the printer port to communicate with the computer. Parts cost is generally less than $50, and a printed-circuit board is available to make construction easier.

Circuit Description

Each stage of the circuit in **Fig 22.46A** is labeled with its function. Ferrite beads are used to keep RF from entering the unit. K1 provides isolation, so any transmitter may be keyed without worrying about polarity. Fig 22.46B shows the power and computer interconnections for the circuit board.

Power Supply Options

Three voltage sources are required (+12 V, +5 V and −9 V) at relatively low current. The simplest approach is to use a wall-mount power transformer/supply (200 mA minimum) to provide the +12 V. A 7805 voltage regulator chip (U6) produces +5 V from the +12 V bus for the 74LS TTL ICs. Since the −9 V current requirements are very low, a 9 V alkaline transistor battery was used in WB8DQT's unit. This battery is switched in and out using one set of contacts on the **POWER** switch and will last a long time—unless you forget to turn the unit off between operating sessions!

The Computer Connection

The circuit connects to the PC parallel printer port, which is usually a DB-25F (female) connector on the rear of the computer. A standard cable with a mating DB-25M (male) connector on one end and DB-25F on the other end could be used. This cable is available at any computer store, but it would require a DB-25M to be mounted on your project box.

Unfortunately, DB-25 connectors need an odd-shaped mounting hole, which is difficult to make with standard shop tools. Since only four conductors are needed (ground, printer data bits 0 and 1, and the strobe data bit), it's easier to make a cable. Use a 4-pin microphone connector at one end and a DB-25M at the other end. Wire the cable as follows:

Function	Microphone Plug	DB-25M Connector
Ground	1	25
Printer data 0	2	2
Printer data 1	3	3
Printer strobe	4	1

Drill a 5/8-inch round hole on the rear apron of the project enclosure for the mating chassis-mount 4-pin microphone socket.

Keying Options

For equipment with a positive, low-voltage keying line, point K on the board can be connected directly to the keying jack. In this case, omit K1 and its 1N4004 diode. For a wider range of transmitting equipment, use the keying relay. Mount it anywhere in the cabinet using a dab of silicone adhesive or a piece of double-sided foam mounting tape.

Construction

The simplest way to construct this circuit is on a single PC board. The PC-board pattern and parts overlay are available from the ARRL.[1] Make a PC board, or use the overlay to wire the circuit using perf-board. An etched and drilled PC board, with a silk-screened parts layout, is available from FAR Circuits.[2]

Any cabinet or enclosure that can ac-

[1]See the **References** chapter for a template.
[2]A circuit board is available for $5 (plus $1.50) from FAR Circuits (see the Address List in the **References** chapter).

Fig 22.46—Schematic of the CW interface. All fixed value resistors are 1/4 W, 5%-tolerance carbon film. Capacitance values are in microfarads (μF). RS indicates RadioShack part numbers. IC sections not shown are not used.

C1-C3, C5, C7-C13—0.1 μF monolithic or disc ceramic, 50 V.
C4—0.047 μF Polypropylene (dipped Mylar), 50 V.
C6—0.22 μF Polypropylene (dipped Mylar), 50 V.
C14—1 μF Tantalum or electrolytic, 50 V.
C15—0.47 μF Tantalum or electrolytic, 50 V.
C16—10 μF Tantalum or electrolytic, 50 V.
D1, D3—1N4004.
D2—1N270 germanium.
DS1—Green panel-mount LED.
DS2—Red panel-mount LED.
FB—Ferrite beads (11 total).
K1—12 V dc SPST reed relay (RS-275-233).
J1, J2—RCA phono jacks.
P1—4-pin microphone jack (RS-274-002).
Q1, Q3—2N4401.
Q2—MPF102.
R1—1 kΩ.
R2, R3—10 kΩ.
U1—NE567CN PLL tone decoder (8 pin).
U2—74LS14N hex Schmitt trigger (14 pin).
U3, U4—LM741CN op amp (8 pin).

U5—74LS00N quad NAND gate (14 pin).
4-pin microphone plug (RS-274-001).
4-pin microphone socket (chassis mount).
DB-25M Connector (RS-276-1547).
DB-25 Shell (RS-276-1549).
Coaxial power connector (RS-274-1563).
8-pin DIP IC sockets.
14-pin DIP IC sockets.
DPDT miniature toggle switch.

J3 is a panel-mounting coaxial power jack to match your wall-mount/transformer power supply. BT1 is a 9-V alkaline battery. See text. C17 and C18 are 0.1 μF, 50 V monolithic or disc ceramic bypass capacitors. The +5 V regulator chip should be mounted to the grounded wall of the cabinet. Off-board components are duplicated in section B of this drawing (J1, J2, P1, the CW LED indicator, and K1). The CW and POWER indicators are panel-mounting LED indicators (red for POWER and green for CW). FB indicates optional ferrite beads used to prevent RF interference with the interface circuits.

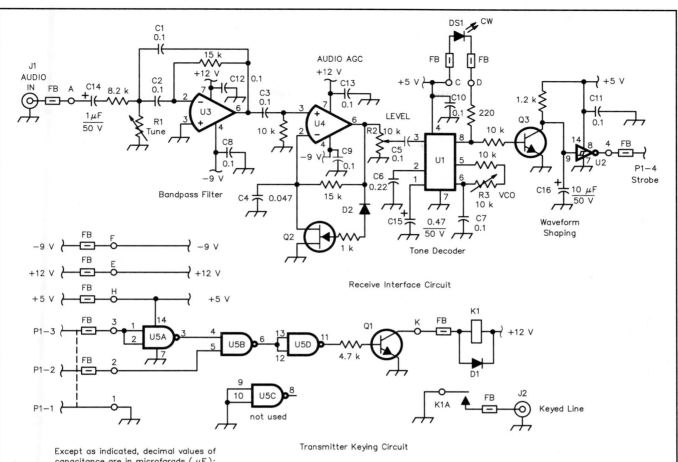

Except as indicated, decimal values of
capacitance are in microfarads (μF);
others are in picofarads (pF);
resistances are in ohms; k=1,000, M=1,000,000.
Labeled circles (A-K, 1-4) indicate PC board connections.

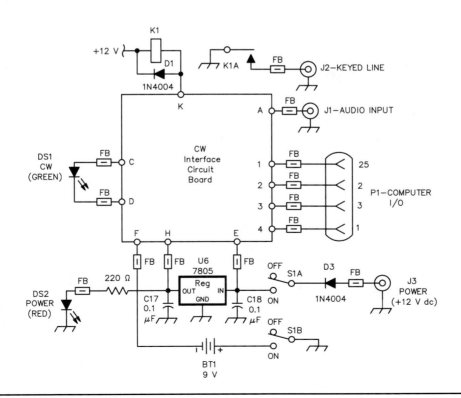

commodate the circuit board can be used. The POWER switch and POWER and CW LED indicators are the only front-panel items. J2 (KEYED LINE), J1 (AUDIO IN), J3 (+12 V DC POWER) and P1 (COMPUTER) are on the rear apron of the enclosure.

Alignment

There are three alignment adjustments, all of which are for the receive mode. Start by loading (and running) the software and turning the unit on. Switch the receiver to a dummy antenna to eliminate any interfering signals, and tune the receiver to a strong signal from a frequency calibrator or any other stable signal source. Carefully adjust the receiver for peak audio output. You may need a *Y* connector so the receiver can feed the interface and a speaker.

Connect a pair of headphones to the junction of the 0.1-µF capacitor and 10-kΩ resistor at pin 3 of U4. Adjust the Tune (R1) control on the PC board for the loudest signal. The filter is sharp, so make the adjustment carefully.

Set the PC board Level pot (R2) to midrange and adjust the VCO pot (R3) until the CW LED (DS1) comes on. Decrease the Level setting slightly (adjust the control in a counterclockwise direction) and readjust the VCO pot, if required, to cause the CW LED to light. Continue to reduce the Level setting in small steps, each time readjusting the VCO setting, until you reach the point where operation of the CW indicator becomes erratic.

Now turn the Level control back (clockwise) to just past the point where the LED comes on with no sign of erratic operation.

The Level threshold setting is critical for best operation of the receive demodulator. If the control is advanced too far, the LED will trigger on background noise and copy will be difficult. If you reduce the setting too far, the interface will trigger erratically, even with a clean beat note. If you have a reasonably good CW receiver (CW bandwidth crystal filters and/or good audio filtering), you can back down the Level control until the LED stops flickering on all but the strongest noise pulses, but where it will still key reliably on a properly tuned CW signal.

Software Installation

The software for this project is available from *ARRLWeb* (see page viii in the front of this book). The distribution files include MORSE.EXE, a sample set-up file (CW.DAT), a sample logging file (LOG.DAT), the HELP text file (CWHELP.DAT) and the program Quick-BASIC source code (MORSE.BAS). To run the program log into the directory holding these files and type *MORSE <CR>*. The symbol <CR> stands for *Return* or *Enter*, depending on your keyboard.

The program menu permits you to enter or change the following items:

SPEED—Select a transmitting speed from 5 to 60 WPM. The program autocalibrates to your computer clock speed, and transmitting speeds are accurate to within 1%. On receive, the system automatically tracks the speed of the station you are copying up to 50 or 60 WPM.

YOUR CALL—You can enter your call sign so you never have to type it in

routine exchanges. The call can be changed at any time if you want to use the program for contests, special events, or any other situation where you will be using another call.

OTHER CALL—If you enter the call of the station you are working (or would like to work), you can send all standard call exchanges at the beginning and end of a transmission with a single keystroke.

CQ OPTIONS—Select one of two CQ formats. The "standard" format is a 3x3 call using your call sign. The program also lets you store a custom CQ format, which is useful for contests.

MESSAGE BUFFERS—There are two message buffers. Either can be used for transmitting.

SIDETONE—Select on or off and a frequency of 400 to 1200 Hz.

WEIGHTING—Variable from 0.50 through 1.50.

DEFAULT SETUP—All the information discussed up to this point can be saved into a default disk file (CW.DAT). These choices will then be selected whenever you boot the program. Any setup can be saved at anytime.

LOGGING—The program supports a range of logging functions. It even includes the ability to check the log and let you know if you have worked that station before. If you have fully implemented the logging options, it will tell you the operator's name and QTH.

HELP FILES—If you forget how to use a function or are using the program for the first time, you can call up on-screen HELP files that explain every function.

AN EXPANDABLE HEADPHONE MIXER

From time to time, active amateurs find themselves wanting to listen to two or more rigs simultaneously with one set of headphones. For example, a DXer might want to comb the bands looking for new ones while keeping an ear on the local 2-m DX repeater. Or, a contester might want to work 20 m in the morning while keeping another receiver tuned to 15 m waiting for that band to open. There are a number of possible uses for a headphone mixer in the ham shack.

The mixer shown in **Figs 22.47** and **22.48** will allow simultaneous monitoring of up to three rigs. Level controls for each channel allow the audio in one channel to be prominent, while the others are kept in the background. Although this project was built for operation with three different rigs, the builder may vary the number of input sections to suit particular station requirements. This mixer was built in the ARRL Lab by Mark Wilson, K1RO.

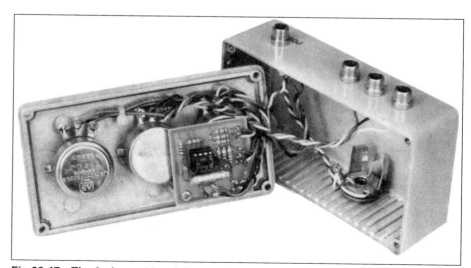

Fig 22.47—The 3-channel headphone mixer is built on a small PC board. Lead length was kept to a minimum to aid stability.

CIRCUIT DETAILS

The heart of the mixer is an LM386 low-power audio amplifier IC. This 8-pin device is capable of up to 400-mW output at 8 Ω—more than enough for headphone listening. The LM386 will operate from 4- to 12-V dc, so almost any station power supply, or even a battery, will power it.

As shown in Fig 22.48, the input circuitry for each channel consists of an 8.2-Ω resistor (R1-R3) to provide proper termination for the audio stage of each transceiver, a 5000-Ω level control (R4-R6) and a 5600-Ω resistor (R7-R9) for isolation between channels. C1 sets the gain of the LM386 to 46 dB. With pins 1 and 8 open, the gain would be 26 dB. Feedback resistor R10 was chosen experimentally for minimum amplifier total harmonic distortion (THD). C2 and R11 form a "snubber" to prevent high-frequency oscillation, adding to amplifier stability. None of the parts values are particularly critical, except R1-R3, which should be as close to 8 Ω as possible.

CONSTRUCTION

Most of the components are arranged on a small PC board.[1] Perfboard will work fine also, but some attention to detail is

[1]See the **References** chapter for a template.

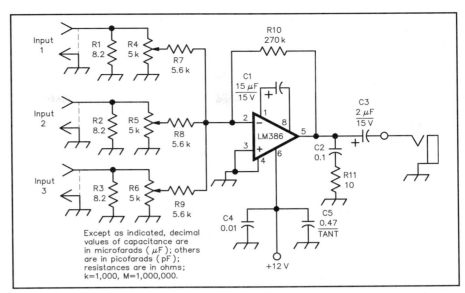

Fig 22.48—Schematic diagram of the LM386 headphone mixer. All resistors are ¼ W. Capacitors are disc ceramic unless noted.

necessary because of the high gain of the LM386. Liberal use of ground connections, short lead lengths and a bypass capacitor on the power-supply line all add to amplifier stability.

The mixer was built in a small diecast box. Tantalum capacitors and ¼-W resistors were used to keep size to a minimum. The '386 IC is available from RadioShack (cat. no. 276-1731). A 0.01-µF capacitor

and a ferrite bead on the power lead help keep RF out of the circuit. In addition, shielded cable is highly recommended for all connections to the mixer. The output jack is wired to accept stereo headphones.

Output power is about 250 mW at 5% THD into an 8-Ω load. The output waveform faithfully reproduces the input waveform, and no signs of oscillation or instability are apparent.

A VACUUM MANIPULATOR FOR CHIP COMPONENTS

Many builders could benefit from a simple machine to manipulate tiny chip components. This vacuum handler, designed and built by Dave Reynolds, KE7QF, accurately places those parts on your circuit board and holds them in place as you solder. You supply the vacuum — by sucking on a piece of plastic tubing such as that used with aquarium pumps. Other vacuum sources are also available. You might try a long piece of line connected to the vacuum system on your car. An old refrigerator compressor motor (with the refrigerant safely disposed) can also be used as a simple vacuum pump.

A common automotive universal joint serves as the swing bearing. It's strong, precise and inexpensive; a miracle of mechanical evolution. All the materials needed to build the manipulator are common items, but its performance is first class, and it definitely looks cool on the workbench. **Table 22.2** lists most of the necessary parts and materials.

The 12 inch square base was cut from pre-finished shelf material, and it is elevated by pieces of ³/₄ × 1¹/₂ inch wood scrap cut to length for each side. The base is not critical; it only needs to be solid and level.

Build your manipulator with the arm extending inward from your off-dominant

Table 22.2
Parts List

Item	Cost
One slotted shelf bracket	$2.00
Two auto universal joints	15.00
A pre-finished shelf board	5.00
Three conduit hangers	.90
One ball pen refill	1.00
1¹/₂ × 1¹/₂ inch aluminum angle	3.00
Bolts, wood scraps, sheet metal	Scrounged
Total — a very reasonable	$26.90

side. I'm right-handed so the open side of my manipulator is on the right. My soldering iron is on the right for the same reason. My unit includes a soldering iron holder made from a discarded speaker grill. That's it at the upper right of the lead photo. A ³/₄ inch trash hole in front of the iron has a small removable container below.

My probe is made from a ball pen refill with the ball carefully ground off, then emptied by prying out the plug and wick, and running water through backwards under pressure. Soaking the refill in a test tube of acetone helped loosen up the mess, but I warn you, mess is a key word. The stainless pen part looks really neat, but unless you're ready to challenge the messy aspects, you might want to use small diameter aluminum tubing.

The probe needs to be solid, non-solderable, and the right size. That means about 0.040 or 0.050 inch across the opening. Spin the probe in a drill and smooth the end where you ground off the ball. It has to make a good vacuum seal with the parts you want to pick up.

The arm and also the swing-bearing backboard were cut from ³/₄ inch wood, shaped, finished and painted. You could use pine, poplar or just about any wood you happen to have.

I never need more than about four inches of working radius for my boards, so the arm is about 5 inches long. See the lead photo for a view of the arm, with the pen refill installed through the head. Carefully drill vertically through the wooden probe head. I drilled a second hole in the end of the head, then threaded a thumb screw from a mobile speaker mount into the head to secure the probe. Any small bolt will do the job.

The new universal joints cost $6.95 each. I chose one for a 1970s Datsun; GMB part number 1052. Any joint will work, but I chose this one for the grease fitting hole in the center of one end cap. A 2¹/₂ inch long #10 bolt goes through the grease fitting hole, through the body of the cross piece and out through the second end cap. The bolt is tightened just enough to ensure slack-free swing of the arm. See **Fig 22.49**. Originally I had planned to drill the other bolt hole in the opposite end cap. After ruining several bits trying to drill this hole, I bought another $6.95 joint just to get a second bearing cap with the grease fitting hole in its center. Those caps are hardened and extremely difficult to drill. I recommend you buy two of the joints on your first trip to the auto parts store. Universal joints for some other vehicles also have the grease fitting centered in one end cap. Ask your friendly auto parts dealer for help identifying a suitable universal joint.

The right and left bearings of the U-joint are clamped to the backboard, and they

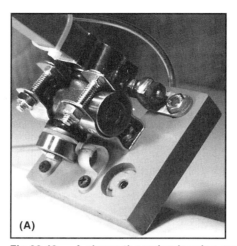 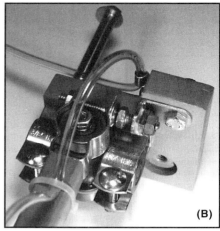

(A)   (B)

Fig 22.49 — A shows the swing bearing assembly with tensioner, backstop adjustment, and thumbscrew captive nut. B shows a view of the swing bearing, with good detail of the backstop adjustment, the double-nutted bolt through the universal joint body and the hose routing. The cutout section of the back board allows clearance for the mounting bolt as the assembly swings around to pick up components.

provide the up and down swing of the arm. The bearings at the top and bottom provide the side to side swing. The bearing holders shown in the photos are called #0 electrical conduit hangers; they're usually used to secure ¹/₂ inch electrical conduit to a wall. I formed the curves a little closer to the 0.984 inch diameter of the bearing cap by squeezing on them with a vise while the bearing cap was in place.

Stretched across the bottom bearing is my state-of-the-art "elastomeric excursion biasing means" (EEBM)—a rubber band. This pulls back on the lower bearing, resulting in a slight downward tension on the vacuum tip. Holding down on the tiny parts keeps them in place and prevents "tombstoning."

The backboard, bearing and probe arm are a subassembly, as shown in Fig 22.49. A small angle bracket and adjustment bolt is mounted at the top right of the backboard. The top bearing clamp contacts this bolt as a "rear limit." In use, you hold the probe at this rear-limit, and position the circuit board so the probe is exactly over where you want the part. Then, when you return with the part, it ends up precisely where you want it, ready to solder.

The backboard assembly is secured to the base with a thumb screw through a slotted hole in a piece of angle aluminum on the base. You need about ³/₈ inch side adjustment to re-center the probe for different tape widths. (I later found that by removing the swing bearing and arm, the base became an excellent ESD controlled working area for other projects.)

At the rear of the backboard, a long bolt with washers serves as a solder roll holder. It reduces clutter and helps you concentrate on assembly. See Fig 22.49B.

Much of the precision is determined

where the part is picked up. It needs to be in correct angular position and the probe needs to center on it. The arm should swing over to the pickup point in a wide arc of less than 90°. Mine goes about 70°.

If your parts are not confined in tape, you can devise a part holder that's moveable instead of adjusting the backboard. I did that for a while. SM parts like to bounce and slide off the work surface, always into a hiding place. Just try finding a SOT-23 transistor in a rug! That's one reason why I added the metal rail around the edge of the machine. The rail is grounded to help with ESD control.

BUILDING A TAPE FEEDER

I still lost parts, so I built the aluminum magazine shown in **Fig 22.50** to feed taped parts accurately. I purchased inexpensive *cut tape* rolls of a few hundred of the component values that I always seem to design around. When I encounter a part I don't have on tape, I slide a length of old tape into the magazine, and it positions the odd part.

The magazine only needs to stabilize about a 4 inch by ¹/₂ inch wide roll of parts tape. No center axle is required. Tapes for big or small parts differ, so you want to make the feed-out guides adjustable for different widths and different thicknesses. See the photos of Fig 22.50 for details. Tapes that you may plan to control will be on the order of 0.040 × 0.300 inches up to perhaps 0.080 × 0.500 for SOT223 parts.

The magazine slides into a slot cut in the floor of the fixture, and a bolt holds it securely in place. The same bolt secures the tape's side adjustment plate. Parts should be picked up at the same elevation as the circuit board, so the height of the board holder is the height you need for the pick up area. I used a spacer made from the

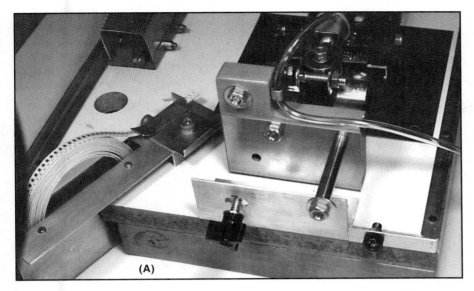

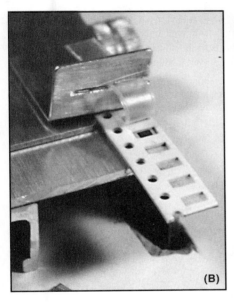

Fig 22.50 — A shows the backboard subassembly detached from the mounting bracket and adjustment slot, with the parts-tape holder magazine feeding parts tape into guides. B shows a close-up view of the tape holder. C shows the tape exiting the magazine guides. Note the adjustment nuts at the top and right side, and the "peeling guide" for the tape cover.

about $5/8$ inch wide by $1/2$ inch high. It is adjusted by a wing nut on a 10-24 through bolt, and a short piece of rubber hose provides spring tension to hold the board firmly by its edges. **Fig 22.51** shows the board holder. It does not fasten to the assembly floor. You move the board and holder to position the appropriate mounting location under the probe tip. You need a board holder like this to allow for parts and connections on the bottom side. It also bleeds off electrical charges, so any substitute material should be metal.

USING THE VACUUM HANDLER

After you have built all the pieces, using the vacuum handler is a simple process. First, swing the arm over the work area until the top clamp hits the rear limit screw. Now position the circuit board (and holder) so the exact part location is under the probe tip. Now swing the arm over to the tape holder and let it down to the component. Suck on the end of the plastic tubing like you were drinking through a straw (or turn on your vacuum source) to pick up the part. See **Fig 22.52A.**

Swing the arm back over the work are a until it hits the stop and lower the component into position. The rubber band and the weight of the arm will hold everything in position while you solder. See Fig 22.52B. Solder both sides, and you're done. Position the board for the next component and go pick it up.

MISCELLANEOUS RAMBLINGS

Surface mount implies no leads through drilled holes. While many SM parts are plentiful and inexpensive, others, like inductors and large capacitors, can be difficult to obtain. I use a combination of chip parts and conventional parts with the leads formed for soldering to the surface. Conventional ICs with leads bent inward make excellent SM parts.

I made a fixture to form the pins evenly by using a couple of elevated sheet metal edges to confine the "dead bug" legs. Large washers space the sheet metal to the appropriate height. The holes through the sheet metal are a bit large, to allow some adjustment when clamping the IC to the fixture. A form made from a 0.3 inch wide piece of 0.060 inch thick circuit board material is placed between the legs, and the appropriate legs are bent flat inward. Then the form is removed and the ends of the bent legs are bent the rest of the way down to the body. This results in "knees" that contact the pads for soldering. Pins are left straight when they need to go through a hole for contact on the other side of the circuit board. Ground this fixture for working with CMOS. **Fig 22.53** shows the details of this fixture and its use.

Chip resistors and capacitors come in

same shelf hanger that provided the circuit board holder, to be described later.

The side adjuster needs to be about 0.040 inch thick and adjustable to keep the tape confined to a straight line of movement. Both adjustments should be firm, but not tight. In use, the tape is advanced by hand, and excess tape and cover is snipped off as required.

The sheet metal guide that holds down on the tape includes a section bent upward at the front, with a narrow horizontal slot. The slot is for the tape's paper cover to be peeled back and put through, out of the way. That solves a nuisance problem. The cutoffs are pushed into the trash hole.

AN ELECTROSTATIC-DISCHARGE-PROTECTED WORK SURFACE

A square sheet-metal assembly floor is placed where the work is done. The straight edges give visual reference of whether or not the board, and consequently the part, is positioned squarely. All metal parts are electrically connected on the bottom of the base, and a banana jack on the rail is provided to connect to the workbench electrostatic ground system. If you open and install all static sensitive parts in this environment, you shouldn't have ESD problems.

Observe standard safety practices, including a resistor (usually 1 MΩ) in the ground circuit to protect you from serious electrical shock in the case of unintended contact with line voltage. It hurts; I've been there. It's a good idea to measure the ac voltages between "line neutral" points on your bench and earth ground. I once found 60 V at 80 mA!

The circuit board holder is made from two small lengths of metal shelf hanger material which has a U type section of

size numbers such as 2512, 1210, 1206, 0805, 0603, or 0402. A 1206 size may be considered as 0.12 inches long and 0.06 inches wide. Applying tenths of an inch as above, the tiny 0402 is about 0.040 × 0.020. I wouldn't recommend the tiny ones for beginners. I like the 1206 because it gives good size reduction and works out well on a $^1/_{10}$ inch grid.

An hour spent installing surface mount components will convert you. It's faster, cheaper, and parts are widely available — more available than many of the parts we loved in the "good old days." Probably ten thousand resistors fit in the space of an 813. You could shrink your cluttered workshop into a briefcase. There are no leads to cut or form. Lots less holes to drill. Industry likes it and you will too.

If this article encourages you to begin designing circuits and boards to enter surface mount, bear in mind that a computer isn't necessary. Use your brain. Draw your *proven circuit* on both sides of quadrille paper with a pencil. Erase and rearrange as required; it becomes an interesting puzzle and it's rewarding at the end.

I found a deal on some 0.015 inch diameter 63-37 solder at a hamfest, and it's much easier than the 0.032 inch diameter solder that I started with. My soldering iron is a good quality unit, without bells or whistles. It has a 0.060 inch chisel tip and it's rated at 25 W. I use a grounded frame isolation transformer and stay clear of the house "ground neutral."

Some suppliers insist on selling reels of ten thousand pieces, but they are *not* your only source. Others now sell single parts at about the same price as through-hole parts. A tape roll of a hundred resistors for several dollars is reasonable for a serious builder. Most designers tend to use just a few values, and good coverage of the common ones probably needn't exceed ten or so different values.

Then there's scrounging. SM parts are easily removed, and there's a huge supply of discarded phones, audio, computer boards and on and on. Leads don't get shortened, and many recovered parts are better than anything you can buy. You learn about design practice, and that can be applied to your projects.

So build a fixture something like this and you'll have more fun than those guys who are just sitting around saying "Nope, ya just can't build stuff no more." They speak for themselves.

Some sources of catalogs that include SM parts (this is not a complete listing) are DC Electronics, Digi-Key, Jameco, Mouser and Tech America. See the **References** chapter for contact information.

Fig 22.51 — This photo shows the construction details for the board holder. Notice that SM and conventional parts share board space on this project.

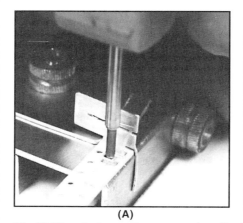 

(A)　　　　　　　　　　　　　　(B)

Fig 22.52 — A shows the vacuum handler lifting a component off the tape package. B shows a DO-35 diode, ready to solder onto the homemade board. Two quick dabs of solder and the job is done.

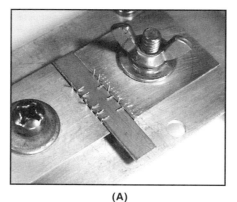 

(A)　　　　　　　　　　　　　　(B)

Fig 22.53 — A shows the pin former in action. Selected pins are bent flat onto the 0.060 × 0.3 inch center guide. The grounding jack is just visible in the upper right corner. B shows the ends of the pins bent down after the center guide was removed. Several pins were left straight for contrast here.

AUDIO BREAK-OUT BOX

Two integrated circuits and a small PC board are all you need to solve the problem of feeding one receiver into several add-ons, such as a TNC, a PC interface or a speaker. Ben Spencer, G4YNM, described this project in March 1995 *QST*. It takes the audio output from a receiver and applies it to the inputs of four identical, independent, low-level AF amplifiers and one high-level (1-W output) AF amplifier.

Each low-level output channel can provide up to 20 dB of gain that's independently adjustable. You can apply audio to each of your accessories at a selected level without changing the level to the other accessories. In addition you can set the level to the speaker independently. Turn the speaker volume up to tune in the signal, and then turn the speaker volume down once tuning is finished and the mode is operating.

Circuit Description

Four identical low-level channels, each feeding an amplifier (U1A, B, C and D), are shown in **Fig 22.56**. Using the top channel as an example, C1 connects the input jack J1 to the noninverting input of U1A. R3 and R4 set U1A's voltage gain.

Fig 22.56—This audio break-out box requires less than 500-mA from a 12-VDC supply. All resistors are 1/4-W, 5%-tolerance carbon-composition or film units unless otherwise specified. RS numbers in parentheses are RadioShack stock numbers.

C1, C3, C4, C6, C7, C9, C10, C12, C13, C15, C17—100 µF, 16-V radial electrolytic or tantalum (RS 272-1028).

C2, C5, C8, C11—1 µF, 16-V radial electrolytic or tantalum (RS 272-1434).

C14, C16—0.1 µF, 50 V disc ceramic (RS 272-135).

R1, R2, R5, R6, R9, R10, R13, R14—100 kΩ.

R3, R7, R11, R15—10 kΩ.

R4, R8, R12, R16, R17—100-kΩ log or audio taper, panel-mount potentiometer (RS 271-1722) or PC-board vertical-mount trimmer potentiometer; see text.

R18—2.7 Ω, 1/2 W.

U1—TL084, TL074, or LM324 quad op amp (RS 276-1711).

U2—LM380N 2-W audio power amplifier. The LM380 is available in several packages. Be sure to use the 14-pin DIP if you are going to build this project on the PC board from FAR or from the ARRL template.

Misc: Single-sided PC board (see Notes 1 and 2), enclosure, knobs, IC sockets, input and output connectors of choice, hook-up wire.

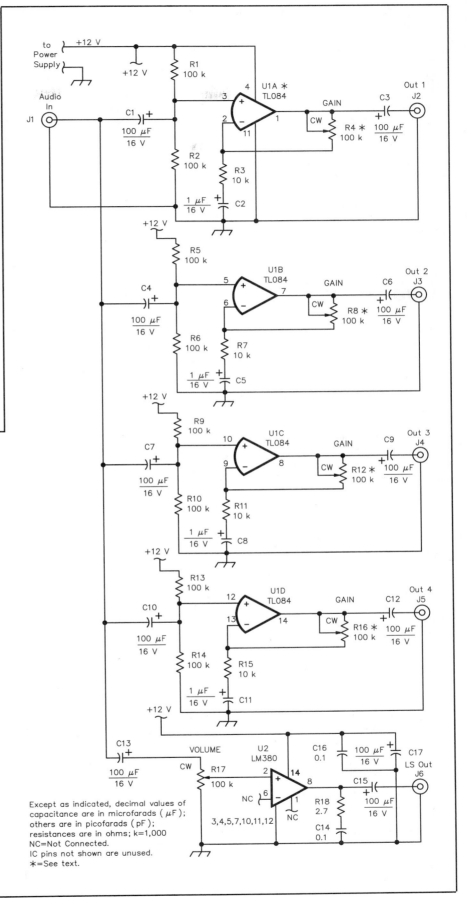

Except as indicated, decimal values of capacitance are in microfarads (µF); others are in picofarads (pF); resistances are in ohms; k=1,000
NC=Not Connected.
IC pins not shown are unused.
∗=See text.

R4 is the gain control, and when set fully clockwise (maximum resistance), the amplifier's gain is 10 (20 dB). At a counterclockwise (minimum resistance) setting, the amplifier's gain is 1 (0 dB).

The lower cut-off frequency (set by C2 and R3) is 16 Hz. The upper cut-off frequency of each channel is well beyond the audio frequency range. Each channel's output is dc isolated from its load; for example, U1A's output is dc isolated by C3.

R17 is the volume control for AF power amplifier U2 This stage will drive a

[1]PC boards are available for $6, plus $1.50 shipping, from FAR Circuits (see the Address List in the **References** chapter).
[2]See the **References** chapter for a template.

low-impedance load such as a loudspeaker (4 to 16 Ω) at a level up to 1 W.

Construction

A single-sided PC board is available,[1] but the unit will work equally well built on perf-board. A template available from the ARRL[2] includes a PC board layout and a parts layout. This parts layout also can be used as a guide for construction on perf-board. The PC board directly accepts vertical and horizontal-mount single-turn potentiometers, but you can run wires from the mounting holes to front-panel-mount potentiometers. Since the project uses high-gain audio circuits, enclose it in a metal box. Place the input and output jacks

on the rear panel to keep the interconnecting leads out of the way.

Checkout

After rechecking your wiring and soldering, connect the circuit to a 12-V power supply. The current drawn should be less than 50 mA when no audio is applied. Connect J1 to the AF output of your receiver and a speaker to J6. Adjust R17, VOLUME, for a comfortable listening level. Next check the operation of the low-power outputs by connecting J2, J3, J4 and J5 to a small earplug. Vary the four gain controls to check their operation. Each gain control can now be set to provide the audio level needed for each add-on.

AN SWR DETECTOR AUDIO ADAPTER

This SWR detector audio adapter is designed specifically for blind or vision-impaired amateurs, but anyone can use it. The basic circuit can be adapted to any application where you want to use an audio tone rather than a meter to give an indication of the value of a dc voltage.

Usually a meter (or meters) is used to display SWR by measuring the feed line forward and reflected voltages. This adapter generates two tones with frequencies that are proportional to these voltages. The tones are fed to a pair of

Fig 22.57—Schematic of the SWR detector audio-adapter circuit. Unless otherwise specified, resistors are ¼-W, 5%-tolerance carbon-composition or film units. All capacitors are disc ceramic unless otherwise stated. The circuits of A and B are identical, each driving one earphone of an 8- to 32-Ω stereo headset. At A, the forward-voltage circuit; at B, the reflected-voltage circuit. A voltage regulator that provides 5-V dc is shown at C.
R5A, R5B—10-kΩ horizontal-mount trimmer potentiometer; optionally, a 25-kΩ dual-gang, panel-mount potentiometer can be used.
R6—25-kΩ dual-gang, panel-mount potentiometer.
U1, U3—LM358N dual op amp (available from Jameco. See the Address List in the References chapter) or substitute an NTE928M (available from Hosfelt Electronics).
U2—74LS629 dual VCO (available from Jameco) or a NTE74LS629 (available from Hosfelt Electronics).
Misc: PC board, stereo headphone jack, 8 to 32-Ω stereo headphones, mounting hardware.

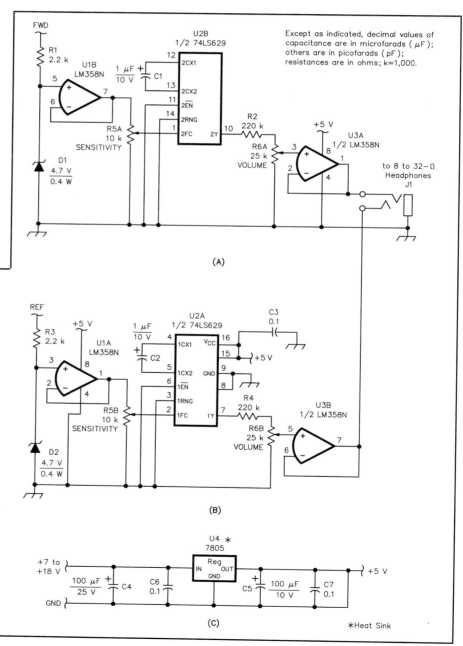

stereo headphones (the miniature types are ideal) so one ear hears the forward-voltage tone and the other ear hears the reflected-voltage tone. Ben Spencer, G4YNM, described this system in the July 1994 *QST*. He connected the forward voltage tone to his left earphone and reverse voltage tone to his right earphone. Thus, tuning up a transmitter is simply a matter of tuning for the highest pitched tone in the left ear, and the lowest pitched tone in the right ear.

The PC board can be installed in existing SWR detectors, and the forward and reflected voltages obtained by tapping into the lines that currently connect the voltage sensors to the existing meters or meter selector switch.

Circuit Description

The audio-adapter circuit is shown in **Fig 22.57**. Each half of the adapter circuit operates identically. Most SWR detectors consist of two RF voltage sensors, one for forward voltage and one for reverse. These voltages are diode-rectified. The resulting dc voltages are fed to meters that indicate relative forward and reflected power. With this circuit, the forward and reflected voltages are applied to the audio adapter board and drive voltage-controlled oscillators (VCOs).

The forward dc voltage from the SWR detector is routed to R1, buffered by U1B,

fed to Sensitivity control R5A and applied to VCO U2B. As the voltage on pin 1 of U2B increases, so does the frequency of the tone output at U2B pin 10.

Adjusting the Sensitivity control sets the range of audio tones produced by the audio adapter. This signal is fed via VOLUME control R6A to the audio amplifier (U3A) to drive the left headphone. Zener diode D1 limits the maximum input voltage, partly to protect U1B, but also to limit the upper VCO frequency to about 3 kHz.

Without any dc input, each VCO runs at a low frequency (approximately 380 Hz) to tell you the unit is operating. Increased voltage on the transmission line— even from a low-power transmitter— is sufficient to cause the tone frequency to increase noticeably. As the voltage decreases, so does the frequency of the tone.

Construction

A single-sided PC board and template package are available.[1,2] The PC board is small enough to fit inside most existing SWR detectors and the circuit can be battery operated if required. Mount a stereo headphone jack on the SWR detector's

[1]PC boards are available for $2.50, plus $1.50 shipping, from FAR Circuits (see the address list in the **References** chapter).
[2]See the **References** chapter for a template.

front panel to accept the headphone plug.

R6A and R6B are parts of a dual-section, panel-mount potentiometer. R5A and R5B are PC-board mounted trimmer potentiometers. For those who want a panel-mounted Sensitivity control, a dual-gang potentiometer can be substituted for R5A and R5B.

Testing and Calibration

Once the unit is installed in an SWR detector, connect a 5-V power supply to the audio adapter board. When power is applied. you should hear two identical low frequency tones in the headphones. Adjust the VOLUME controls to provide a comfortable listening level.

Next, connect your transmitter to a dummy load via the SWR detector. When you key your transmitter, the tone in the left earpiece should increase in frequency quite dramatically, representing increasing forward power. Theoretically, with a matched line and load, there should be no reflected voltage and therefore, the right-headphone tone shouldn't change. In all probability, however, the tone frequency will increase, but only slightly.

If you use an antenna tuner for matching your antenna system, you'll hear the two tones change frequency according to the degree of mismatch. The best match is indicated by the forward headphone tone reaching its maximum frequency while the reflected (right) headphone tone frequency decreases to its minimum.

PC VOLTMETER AND SWR BRIDGE

Personal computers are very good at doing arithmetic. To use this capability around the shack, the first thing to do is convert whatever you want to measure (voltage, power, SWR) to numbers. Next you have to find a way to put these numbers into your computer. Paul Danzer, N1II, took a single chip A/D (analog to digital converter) and built this unit to connect to a computer printer port. Construction and testing is just a few evenings' work. The software to run the chip is included with the *Handbook* companion software, available for downloading from the ARRL Internet site (see page viii).

Circuit Description

The circuit consists of a single-chip A/D converter, U2, and a DB-25 male plug (**Fig 22.58**). Pins 2 and 3 are identical voltage inputs, with a range from 0 to slightly less than the supply voltage V_{CC} (+5 V). R1, R2, C3 and C4 provide some input isolation and RF bypass. There are

four signal leads on U2—DO is the converted data from the A/D out to the computer, DI and CS are control signals from the computer and CLK is a computer generated clock signal sent to pin 7 of U2.

The +5 V supply is obtained from a +12 V source and regulator U1. One favorite accident, common in many ham shacks, is to connect power supply leads backwards. Diode D1 prevents any damage from this action. Current drain is usually less than 20 mA, so any 5-V regulator may be used for U1. The power supply ground, circuit ground and computer ground are all tied together.

In this form the circuit gives two identical dc voltmeters. To extend the range, a 2:1 divider, using 50-kΩ resistors, is shown. Resistor accuracy is not important, since the circuit is calibrated in the accompanying software.

The breadboard circuit, built on a universal PC board (RadioShack 276-150), is shown in **Fig 22.59**. The voltage regulator is

on the top left and the converter chip, U2, in an 8-pin socket. Power is brought in through a MOLEX plug which follows the standard suggested earlier in this chapter. Signal input and ground are on the wire stubs. Two strips of soft aluminum, bent into L-shapes, hold the male DB-25 connector (RadioShack 276-1547) to the PC board.

Use It As An SWR Bridge

Most analog SWR measuring devices use a meter, which has a nonlinear scale calibration. An SWR of 3:1 is usually close to center scale, and values above this are rarely printed. To use the PC voltmeter as an SWR bridge indicator, move jumpers WA and WB from the A1 and B1 positions to the A2 and B2 positions Disconnect the cathodes (banded end) of the diode detectors in your SWR bridge and connect them to J1 and J2.

The current that flows out of these diodes, and into J1 and J2, goes through the 25 kΩ resistors R7 and R8, to provide voltages of less than 5 V. These voltages are proportional to the forward and reverse voltages developed in the directional coupler. The software in the PC takes the sum and difference of these forward and reverse voltages, and calculates the SWR.

Software

The software, including a voltmeter function and an SWR function, is written in GW-BASIC and saved as an ASCII file. Therefore you can read it on any word processor, but if you modify it make sure you resave it as an ASCII file. It can be imported into QBasic and most other BASIC dialects.

It was written to be understandable rather than to be most efficient. Each line of basic code has a comment or explanation. It can be modified for most computers. The printer port used is LPT1, which is at a hex address of 378, 379 and 37A. If you wish to use LPT2 (printer port 2) try changing these addresses to 278, 279 and 27A.

Gary Sutcliffe, W9XT, wrote a small BASIC program to help you find the addresses of your printer ports. Run FINDLPT.BAS, which is included with the *Handbook* companion software, available for downloading from the ARRL Internet site (see page viii).

The CONV.BAS program was written to run on computers as slow as 4.7-MHz PC/XTs. If you get erratic results with a much faster computer, set line 1020 (CD=1) to a higher value to increase the width of the computer generated clock pulses.

The CONV.BAS program operates by first reading the value of voltage at point A into the computer, followed by the voltage at point B. It then prints on the screen these two values, and computes their sum and difference to derive the SWR. If you use the project as a voltmeter, simply ignore the SWR reading on the screen or suppress it by deleting lines 2150, 2160 and 2170. If the two voltages are very close to each other (within 1 mV) the program declares a bad reading for SWR.

Calibration

Lines 120 and 130 in the program independently set the calibration for the two voltage inputs. To calibrate a channel, apply a known voltage to the input point A. Read the value on the PC screen. Now multiply the constant in line 120 by the correct value and divide the result by the value you previously saw on the screen. Repeat the procedure for input point B and line 130.

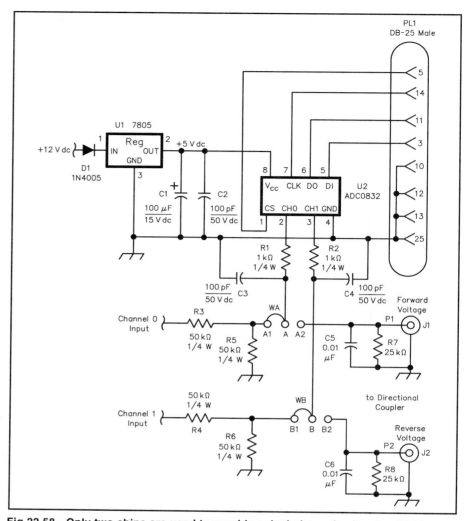

Fig 22.58—Only two chips are used to provide a dual-channel voltmeter. PL1 is connected through a standard 25-pin cable to a computer printer port. U2 requires an 8-pin IC socket. All resistors are ¼ W. You can use the A/D as an SWR display by connecting it to a sensor such as the one used in the Tandem Match described in this chapter (see text). A few more resistors are all that are needed to change the voltmeter scale. The 50 kΩ resistors form 2:1 voltage dividers, extending the voltmeter scale (on both channels) to almost 10-V dc.

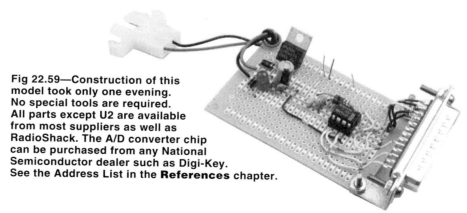

Fig 22.59—Construction of this model took only one evening. No special tools are required. All parts except U2 are available from most suppliers as well as RadioShack. The A/D converter chip can be purchased from any National Semiconductor dealer such as Digi-Key. See the Address List in the **References** chapter.

THE TANDEM MATCH—AN ACCURATE DIRECTIONAL WATTMETER

Most SWR meters are not very accurate at low power levels because the detector diodes do not respond to low voltage in a linear fashion. This design uses a compensating circuit to cancel diode nonlinearity. It also provides peak detection for SSB operation and direct SWR readout that does not vary with power level. **Fig 22.65** is a photo of the completed project. The following information is condensed from an article by John Grebenkemper, KI6WX, in January 1987 *QST*. Some modifications by KI6WX were detailed in the "Technical Correspondence" column of July 1993 *QST*. A PC Board is available from FAR Circuits.[1]

CIRCUIT DESCRIPTION

A directional coupler consists of an input port, an output port and a coupled port. Ideally, a portion of the power flowing from the input to the output appears at the coupled port, but *none* of the power flowing from the output to the input appears at the coupled port.

The coupler used in the Tandem Match consists of a pair of toroidal transformers connected in tandem. The configuration was patented by Carl G. Sontheimer and Raymond E. Fredrick (US Patent no. 3,426,298, issued February 4, 1969). It has been described by Perras, Spaudling (see bibliography) and others. With coupling factors of 20 dB greater, this coupler is suitable to sample both forward and reflected power.

The configuration used in the Tandem Match works well over the frequency range of 1.8 to 54 MHz, with a nominal coupling factor of 30 dB. Over this range, insertion loss is less than 0.1 dB. The coupling factor is flat to within ±0.1 dB from 1.8 to 30 MHz, and increases to only ±0.3 dB at 50 MHz. Directivity exceeds 35 dB from 1.8 to 30 MHz and exceeds 26 dB at 50 MHz.

The low-frequency limit of this directional coupler is determined by the inductance of the transformer secondary windings. The inductive reactance should be greater than 150 Ω (three times the line characteristic impedance) to reduce insertion loss. The high-frequency limit of this directional coupler is determined by the length of the transformer windings. When the winding length approaches a significant fraction of a wavelength, coupler performance deteriorates.

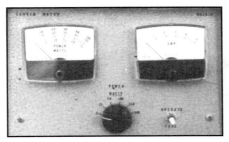

Fig 22.65—The Tandem Match uses a pair of meters to display net forward power and true SWR simultaneously.

The coupler described here may overheat at 1500 W on 160 m (because of the high circulating current in the secondary of T2). The problem could be corrected by using a larger core or one with greater permeability. A larger core would require longer windings; that option would decrease the high-frequency limit.

Most amateur directional wattmeters use a germanium-diode detector to minimize the forward voltage drop. Detector voltage drop is still significant, however, and an uncompensated diode detector does not respond to small signals in a linear fashion. Many directional wattmeters compensate for diode nonlinearity by adjusting the meter scale.

The effect of underestimating detected power worsens at low power levels. Under these conditions, the ratio of the forward power to the reflected power is overestimated because the reflected power is always less than the forward power. This results in an instrument that underestimates SWR, particularly as power is reduced. A directional wattmeter can be checked for this effect by measuring SWR at several power levels. The SWR should be independent of power level.

The Tandem Match uses a feedback circuit to compensate for diode nonlinearity. Transmission-line SWR is displayed on a linear scale. Since the displayed SWR is not affected by changes in transmitter power, a matching network can be simply adjusted to minimize SWR. Transmatch adjustment requires only a few watts.

CONSTRUCTION

The schematic diagram for the Tandem Match is shown in **Fig 22.70**. The circuit is designed to operate from batteries and draws very little power. Much of the circuitry is of high impedance, so take care to isolate it from RF fields. House the circuit in a metal case. Most problems in the prototype were caused by stray RF in the op-amp circuitry.

The schematic shows two construction options. Connect jumpers W1, W2 and W3 to use the circuit as it was originally designed (with two 9-V batteries and TLC27L4 or TLC27M4 op amps). By omitting these jumpers, any quad FET-input op amps can be used instead of the TLC27x4s. Possible substitutes include the TL064, TL074, TL084, LF347 and LF444. In that case you should also omit the 9-V batteries and the automatic turn-on circuitry of Q1, Q2 and Q3 (everything to the left of the jumpers on the top row of the diagram). Now you will have to connect an external + 15 V supply between the + V line and chassis ground and a –15 V supply to the – V line.

The FAR Circuits Tandem Match circuit board is double sided, but does not have plated-through holes. The component side is mainly the chassis and circuit ground planes, although there are a few signal traces. You will have to install "jumper posts" in a few locations, and solder them to both sides of the board to connect these traces. Carefully follow the schematic diagram and parts-placement diagram supplied with the board to identify these "posts." Check the board carefully to ensure that none of the ground traces pass too close to a circuit lead. You may have to scrape a bit of foil away from a few places around the component holes. This is easy with an X-ACTO knife.

The trimmer pots must be square multiturn units with top adjustment screws for use with the FAR Circuits board.

Table 22.4
Performance Specifications for the Tandem Match

Power range:	1.5 to 1500 W
Frequency range:	1.8 to 54 MHz
Power accuracy:	Better than ±10% (±0.4 dB)
SWR accuracy:	Better than ±5%
Minimum SWR:	Less than 1.05:1
Power display:	Linear, suitable for use with either analog or digital meters
SWR display:	Linear, suitable for use with either analog or digital meters
Calibration:	Requires only an accurate voltmeter

[1]See the Address List in the **References** chapter. PC board template packages are also available from FAR Circuits.

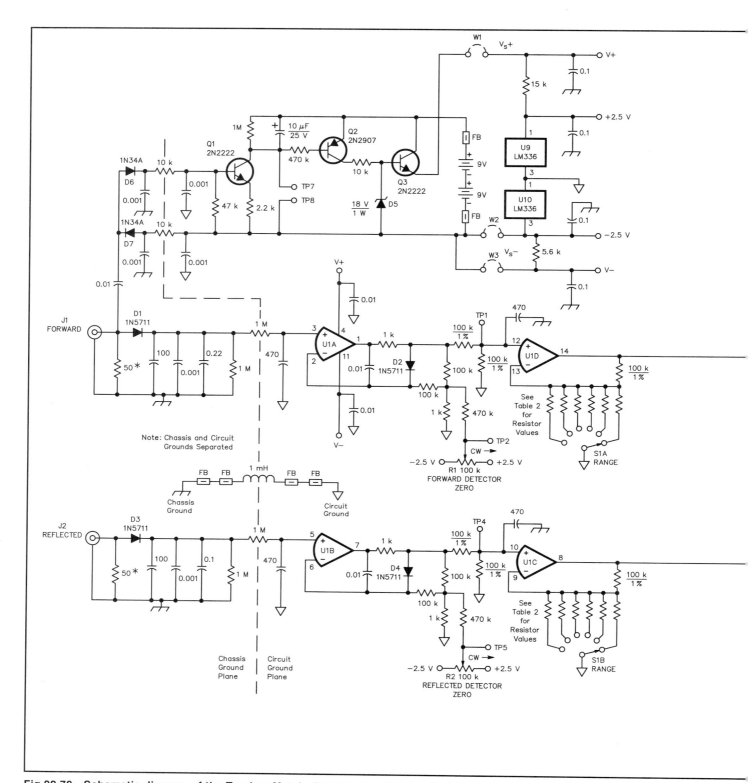

Fig 22.70—Schematic diagram of the Tandem Match directional wattmeter. Parts identified as RS are from RadioShack. Contact information for parts suppliers appears in the References chapter.

D1-D4—1N5711
D6, D7—1N34A or 1N271
D8-D14—1N914.
FB—Ferrite bead, Amidon FB-73-101 or equiv.
J1, J2—SO-239 connector.

J3, J4—Open-circuit jack.
M1, M2—1 mA panel meter.
Q1, Q3, Q4—2N2222 metal case only.
Q2—2N2907 metal case or equiv.
R1, R2, R5—100-kΩ, 10-turn cermet Trimpot.

R3, R4—10-kΩ, 10-turn, cermet Trimpot.
U1-U3—TLC27M4 op amp
U4—TLC27L2 or TLC27M2.
U5-U7—CA3146.
U9, U10—LM336.

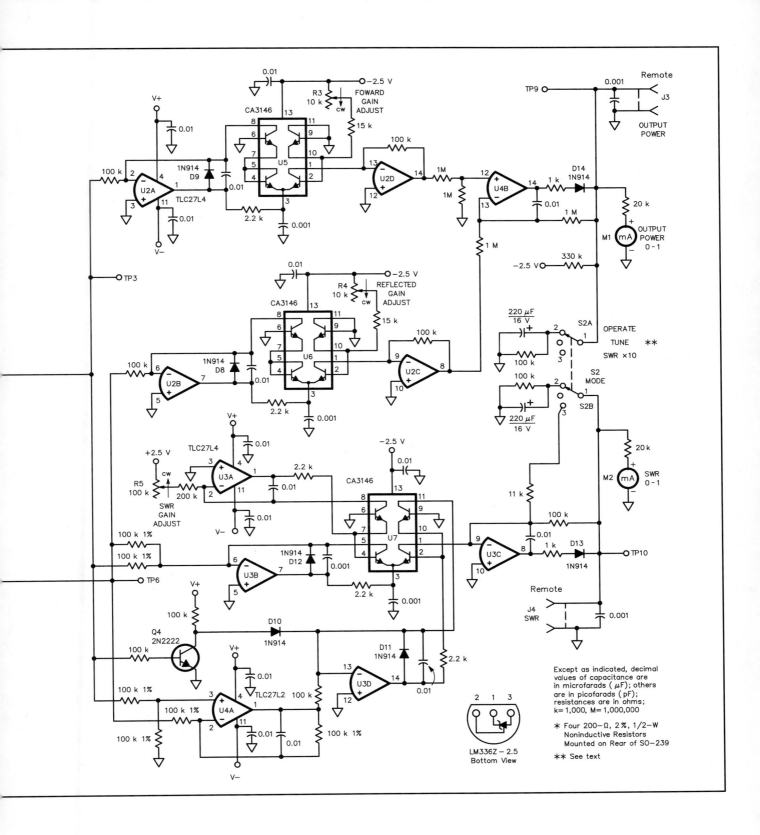

Mount the ferrite beads so they don't touch any board trace; the beads have sufficient leakage to cause problems in the high impedance parts of the circuit. Before mounting the SO-239 connectors to the circuit board, enlarge the center location holes to ⁵/₈ inch diameter to accept the connector body. The components connected to the SO-239 are soldered directly between the center pin and the board traces. See Fig 22.72.

DIRECTIONAL COUPLER

The directional coupler is constructed in its own small (2³/₄ × 2³/₄ × 2¹/₄-inch) aluminum box (see **Fig 22.71**). Two pairs of SO-239 connectors are mounted on opposite sides of the box. A piece of PC board is run diagonally across the box to improve coupler directivity. The pieces of RG-8X coaxial cable pass through holes in

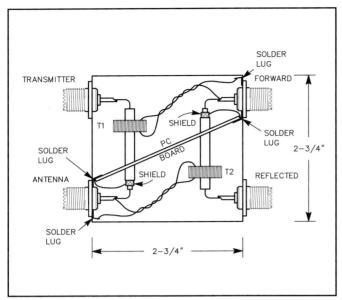

Fig 22.71 —Construction details for the directional coupler. A metal case is required.

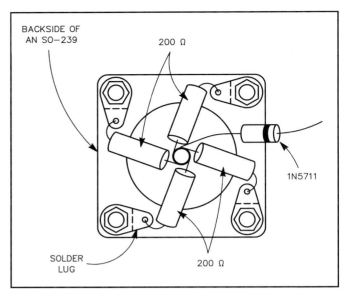

Fig 22.72—The parallel load resistors mounted on an SO-239 connector. Four 200-Ω resistors are mounted in parallel to provide a 50-Ω detector load.

the PC board. (Note: Some brands of "mini 8" cable have extremely low breakdown voltage ratings and are unsuitable to carry even 100 W when the SWR exceeds 1:1. See "High-Power Operation" for details of a coupler made with RG-8 cable.)

Begin by constructing T1 and T2, which are identical except for their end connections. (Refer to Fig 22.71.) The primary for each transformer is the center conductor of a length of RG-8X coaxial cable. Cut two cable lengths sufficient for mounting as shown in the figure. Strip the cable jacket, braid and dielectric as shown. The cable braid is used as a Faraday shield between the transformer windings, so it is only grounded at one end. *Important— connect the braid only at one end or the directional-coupler circuit will not work properly!* Wind two transformer secondaries, each 31 turns of #24 enameled wire on a T-50-3 iron-powder core. Slip each core over one of the prepared cable pieces (including both the shield and the outer insulation). Mount and connect the transformers as shown in Fig 22.71, with the wires running through separate holes in the copper-clad PC board.

The directional coupler can be mounted separately from the rest of the circuitry if desired. If so, use two coaxial cables to carry the forward- and reflected-power signals from the directional coupler to the detector inputs. Be aware, however, that any losses in the cables will affect power readings.

This directional coupler has not been used at power levels in excess of 100 W. For more information about using Tandem

Match at high power levels, see "High-Power Operation."

DETECTOR AND SIGNAL-PROCESSING CIRCUITS

The detector and signal-processing circuits were constructed on a perforated, copper-clad circuit board. These circuits use two separate grounds—*it is extremely important to isolate the grounds as shown in the circuit diagram.* Failure to do so may result in faulty circuit operation. Separate grounds prevent RF currents on the cable shield from affecting the op-amp circuitry.

The directional coupler requires good 50-Ω loads. They are constructed on the back of the female UHF chassis connectors where the cables from the directional coupler enter the wattmeter housing. Each load consists of four 200-Ω resistors connected from the center conductor of the UHF connector to the four holes on the mounting flange, as shown in **Fig 22.72**. The detector diode is then mounted from the center conductor of the connector to the 100-pF and 1000-pF bypass capacitors, which are located next to the connector. The response of this load and detector combination measures flat to beyond 500 MHz.

Schottky-barrier diodes (type 1N5711) were used in this design because they were readily available. Any RF-detector diode with a low forward voltage drop (less than 300 mW) and reverse breakdown voltage greater than 30 V could be used. (Germanium diodes could be used in this circuit, but performance will suffer. If germanium diodes are used, reduce the

values of the detector-diode and feedback-diode load resistors by a factor of 10.)

The rest of the circuit layout is not critical, but keep the lead lengths of 0.001- and 0.01-μF bypass capacitors short. The capacitors provide additional bypass paths for the op-amp circuitry.

D6 and D7 form a voltage doubler to detect the presence of a carrier. When the forward power exceeds 1.5 W, Q3 switches on and stays on until about 10 seconds after the carrier drops. (A connection from TP7 to TP9 forces the unit on, even with no carrier present.) The regulated references of +2.5 V and –2.5 V generated by the LM334 and LM336 are critical. Zener-diode substitutes would significantly degrade performance.

The four op amps in U1 compensate for nonlinearity of the detector diodes. D1-D2 and D3-D4 are the matched diode pairs discussed above. A RANGE switch selects the meter range. (A six-position switch was used here because it was handy.) The resistor values for the RANGE switch are shown in **Table 22.5** Full-scale input power gives an output at U1C or U1D of 7.07 V. The forward- and reflected-power detectors are zeroed with R1 and R2.

The forward- and reflected-detector voltages are squared by U2, U5 and U6 so that the output voltages are proportional to forward and reflected power. The gain constants are adjusted using R3 and R4 so that an input of 7.07 V to the squaring circuit gives an output of 5 V. The difference between these two voltages is used by U4B to yield an output that is proportional to the power delivered to the

Table 22.5

Range-Switch Resistor Values

Full-Scale Power Level (W)	Range Resistor (1% Precision) (kΩ)
1	2.32
2	3.24
3	4.02
5	5.23
10	7.68
15	9.53
20	11.0
25	12.7
30	15.0
50	18.7
100	28.7
150	37.4
200	46.4
250	54.9
300	63.4
500	100.0
1000	237.0
1500	649.0
2000	open

transmission line. This voltage is peak detected (by an RC circuit connected to the OPERATE position of the MODE switch) to indicate and hold the maximum power measurement during CW or SSB transmissions.

SWR is computed from the forward and reflected voltages by U3, U4 and U7. When no carrier is present, Q4 forces the SWR reading to be zero (that is, when the forward power is less than 2% of the full-scale setting of the RANGE switch). The SWR computation circuit gain is adjusted by R5. The output is peak detected in the OPERATE mode to steady the SWR reading during CW or SSB transmissions.

Transistor arrays (U5, U6 and U7) are used for the log and antilog circuits to guarantee that the transistors will be well matched. Discrete transistors may be used, but accuracy may suffer.

A three-position toggle switch selects the three operating modes. In the OPERATE mode, the power and SWR outputs are peak detected and held for a few seconds to allow meter reading during actual transmissions. In the TUNE mode, the meters display instantaneous output power and SWR.

A digital voltmeter is used to obtain more precise readings than are possible with analog meters. The output power range is 0 to 5 V (0 V = 0 W and 5 V = full scale). SWR output varies from 1 V (SWR = 1:1) to 5 V (SWR = 5:1). Voltages above 5 V are unreliable because of voltage limiting in some of the op amp circuits.

CALIBRATION

The directional wattmeter can be cali-brated with an accurate voltmeter. All calibration is done with dc voltages. The directional-coupler and detector circuits are inherently accurate if correctly built. To calibrate the wattmeter, use the following procedure:

1) Set the MODE switch to TUNE and the RANGE switch to 100 W or less.

2) Jumper TP7 to TP8. This turns the unit on.

3) Jumper TP1 to TP2. Adjust R1 for 0 V at TP3.

4) Jumper TP4 to TP5. Adjust R2 for 0 V at TP6.

5) Adjust R1 for 7.07 V at TP3.

6) Adjust R3 for 5.00 V at TP9, or a full-scale reading on M1.

7) Adjust R2 for 7.07 V at TP6.

8) Adjust R4 for 0 V at TP9, or a zero reading on M1.

9) Adjust R2 for 4.71 V at TP6.

10) Adjust R5 for 5.00 V at TP10, or a full-scale reading on M2.

11) Set the RANGE switch to its most sensitive scale.

12) Remove jumpers from TP1 to TP2 and TP4 to TP5.

13) Adjust R1 for 0 V at TP3.

14) Adjust R2 for 0 V at TP6.

15) Remove jumper from TP7 to TP8.

This completes the calibration procedure. This procedure has been found to equal calibration with expensive laboratory equipment. The directional wattmeter should now be ready for use.

ACCURACY

Performance of the Tandem Match has been compared to other well-known directional couplers and laboratory test equipment, and it equals any amateur directional wattmeter tested. Power measurement accuracy of the Tandem Match compares well to a Hewlett-Packard HP-436A power meter. The HP meter has a specified measurement error of less than ±0.05 dB. The Tandem Match tracked the 436A within ±0.5 dB from 10 mW to 100 W and within ±0.1 dB from 1 W to 100 W. The unit was not tested above 1200 W because a transmitter with a higher power rating was not available.

SWR performance was equally good when compared to the SWR calculated from measurements made with 436A and a calibrated directional coupler. The Tandem Match tracked the calculated SWR within ±5% for SWR values from 1:1 to 5:1. SWR measurements were made at 8 W and 100 W.

OPERATION

Connect the Tandem Match in the 50-Ω line between the transmitter and the antenna matching network (or antenna if no matching network is used). Set the RANGE switch to a range greater than the transmitter output rating and the MODE switch to TUNE. When the transmitter is keyed, the Tandem Match automatically switches on and indicates both power delivered to the antenna and SWR on the transmission line. When no carrier is present, the output power and SWR meters indicate zero.

The OPERATE mode includes RC circuitry to momentarily hold the peak-power and SWR readings during CW or SSB transmissions. The peak detectors are not ideal, so there could be about 10% variation from the actual power peaks and the SWR reading. The SWR×10 mode increases the maximum readable SWR to 50:1. This range should be sufficient to cover any SWR value that occurs in amateur use. (A 50-ft open stub of RG-8 yields a measured SWR of only 43:1, or less, at 2.4 MHz because of cable loss. Higher frequencies and longer cables exhibit a smaller maximum SWR.)

It is easy to use the Tandem Match to adjust an antenna-matching network. Adjust the transmitter for minimum output power (at least 1.5 W). With the carrier on and the MODE switch set to TUNE or SWR×10, adjust the matching network for minimum SWR. Once minimum SWR is obtained, set the transmitter to the proper operating mode and output power. Place the Tandem Match in the OPERATE mode.

PARTS

Few parts suppliers carry all the components needed for these couplers. Each may stock different parts. Good sources include Digi-Key, Surplus Sales of Nebraska, Newark Electronics and Anchor Electronics. See the Address List in the **References** Chapter.

HIGH-POWER OPERATION

This material was condensed from a letter by Frank Van Zant, KL7IBA, that appears in July 1989 *QST* (pp 42-43). In April 1988, Zack Lau, W1VT, described a directional-coupler circuit (based on the same principle as Grebenkemper's circuit) for a QRP transceiver (see the bibliography at the end of this chapter). The main advantage of Lau's circuit is very low parts count.

Grebenkemper uses complex log-antilog amplifiers to provide good measurement accuracy. This application gets away from complex circuitry, but retains reasonable measurement accuracy over the 1 to 1500-W range. It also forfeits the SWR-computation feature.

Lau's coupler uses ferrite toroids. It works great at low power levels, but the ferrite toroids heat excessively with high power, causing erratic meter readings and the potential for burned parts.

The Revised Design

Powdered-iron toroids are used for the transformers in this version of Lau's basic circuit. The number of turns on the secondaries was increased to compensate for the lower permeability of powdered iron.

Two meters display reflected and forward power (see **Fig 22.73**). The germanium detector diodes (D1 and D2—1N34) provide fairly accurate meter readings particularly if the meter is calibrated (using R3, R4 and R5) to place the normal transmitter output at midscale. If the winding sense of the transformers is reversed, the meters are transposed (the forward-power meter becomes the reflected-power meter, and vice versa).

Construction

Fig 22.74 shows the physical layout of

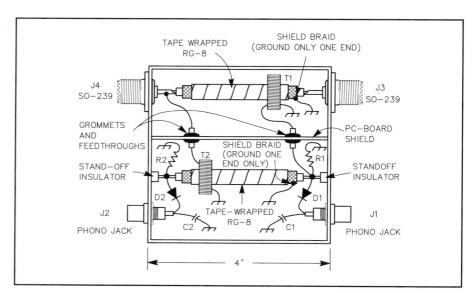

Fig 22.74—Directional-coupler construction details. Grommets or standoff insulators can be used to route the secondary windings of T1 and T2 through the PC-board shield. A 3¹/₂×3¹/₂×4-inch metal box serves as the enclosure.

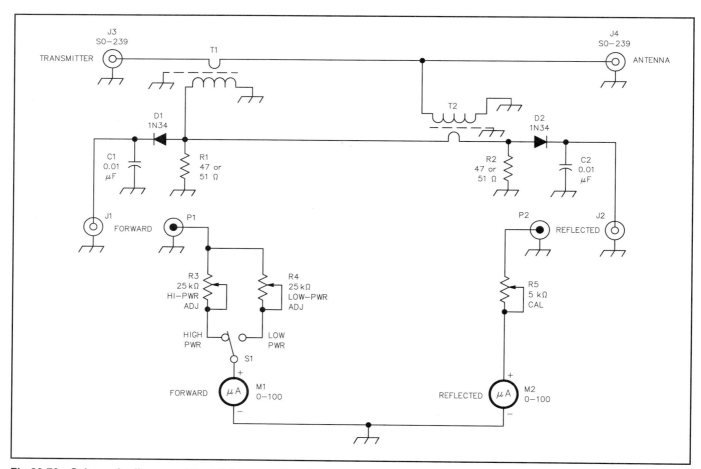

Fig 22.73—Schematic diagram of the high-power directional coupler. D1 and D2 are germanium diodes (1N34 or equiv). R1 and R2 are 47- or 51-Ω ¹/₂-W resistors. C1 and C2 have 500-V ratings. The secondary windings of T1 and T2 each consist of 40 turns of #26 to 30 enameled wire on T-68-2 powdered-iron toroid cores. If the coupler is built into an existing antenna tuner, the primary of T1 can be part of the tuner coaxial output line. The remotely located meters (M1 and M2) are connected to the coupler box at J1 and J2 via P1 and P2.

this coupler. The pickup unit is mounted in a 3¹/₂×3¹/₂×4-inch box. The meters, PC-mount potentiometers and HIGH/LOW power switch are mounted in a separate box or a compartment in an antenna tuner.

The primary windings of T1 and T2 are constructed much as Grebenkemper described, but use RG-8 with its jacket removed so that the core and secondary winding may fit over the cable. The braid is wrapped with fiberglass tape to insulate it from the secondary winding. An excellent alternative to fiberglass tape—with even higher RF voltage-breakdown characteristics—is ordinary plumber's Teflon pipe tape, available at most hardware stores.

The transformer secondaries are wound on T-68-2 powdered-iron toroid cores. They are 40 turns of #26 to 30 enameled wire spread evenly around each core. By using #26 to 30 wire on the cores, the cores slip over the tape-wrapped RG-8 lines. With #26 wire on the toroids, a single layer of tape (slightly more with Teflon tape) over the braid provides an extremely snug fit for the core. Use care when fitting the cores onto the RG-8 assemblies. After the toroids are mounted on the RG-8 sections,

coat the assembly with General Cement Corp Polystyrene Q Dope, or use a spot or two of RTV sealant to hold the windings in place and fix the transformers on the RG-8 primary windings.

Mount a PC-board shield in the center of the box. between T1 and T2, to minimize coupling between transformers. Suspend T1 between SO-239 connectors and T2 between two standoff insulators. The detector circuits (C1, C2, D1, D2, R1 and R2) are mounted inside the coupler box as shown.

Calibration, Tune up and Operation

The coupler has excellent directivity. Calibrate the meters for various power levels with an RF ammeter and a 50-Ω dummy load. Calculate I^2R for each power level, and mark the meter faces accordingly. Use R3, R4 and R5 to adjust the meter readings within the ranges. Diode nonlinearities are thus taken into account, and Grebenkemper's signal-processing circuits are not needed for relatively accurate power readings.

Start the tune-up process using about

10 W, adjust the antenna tuner for minimum reflected power, and increase power while adjusting the tuner to minimize reflected power.

This circuit has been built into several antenna tuners with good success. The bridge worked well at 1.5-kW output on 1.8 MHz. It also worked fine from 3.5 to 30 MHz with 1.2- and 1.5-kW output. The antenna is easily tuned for a 1:1 SWR using the null indication provided.

Amplifier settings for a matched antenna, as indicated with the wattmeter, closely agreed with those for a 50-Ω dummy load. Checks with a Palomar noise bridge and a Heath Antenna Scope also verified these findings. This circuit should handle more than 1.5 kW, *as long as the SWR on the feed line through the wattmeter is kept at or near 1:1.* (On one occasion high power was applied while the antenna tuner was not coupled to a load. Naturally the SWR was extremely high, and the output transformer secondary winding opened like a fuse. This resulted from the excessively high voltage across the secondary. The damage was easily and quickly repaired.)

AN EXTERNAL AUTOMATIC ANTENNA SWITCH FOR USE WITH YAESU OR ICOM RADIOS

This antenna-switching-control project involves a combination of ideas from several earlier published articles.[1, 2, 3] This system was designed to mount the antenna relay box outside the shack, such as on a tower. With this arrangement, only a single antenna feed line needs to be brought into the shack. The lead photo shows the control unit and relay box, designed and built by Joe Carcia, NJ1Q. As the W1AW chief operator, Joe has plans for the switch at W1AW. Either an ICOM or Yaesu HF radio will automatically select the proper antenna. In addition, a manual switch can override the ICOM automatic selection. That feature also provides a way to use the antenna with other radios. The antenna switch is not a two-radio switch, though. It will only work with one radio at a time.

Many builders may want to use only the ICOM or only the Yaesu portion of the interface circuitry, depending on the brand of radio they own. The project is a "hacker's dream." It can be built in a variety of forms, with the only limitations being the builder's imagination.

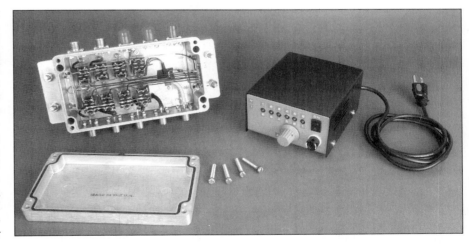

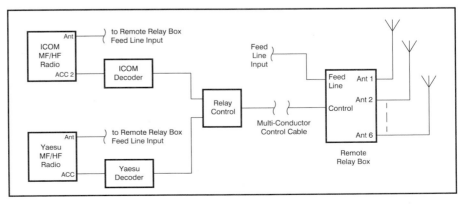

Fig 22.75 — Block diagram of the remotely controlled automatic antenna switch.

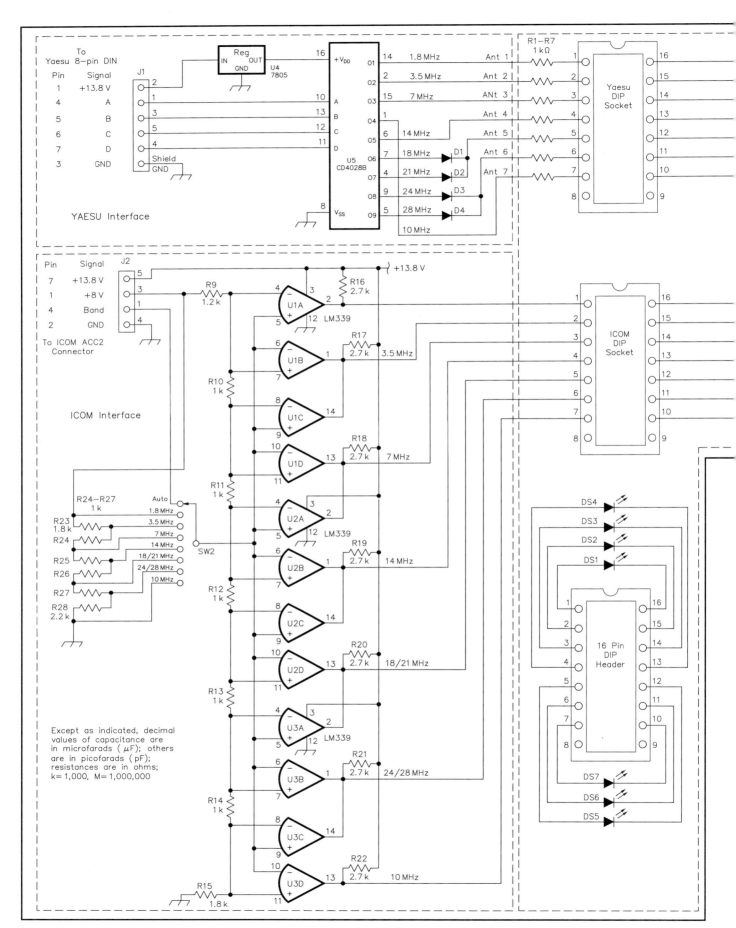

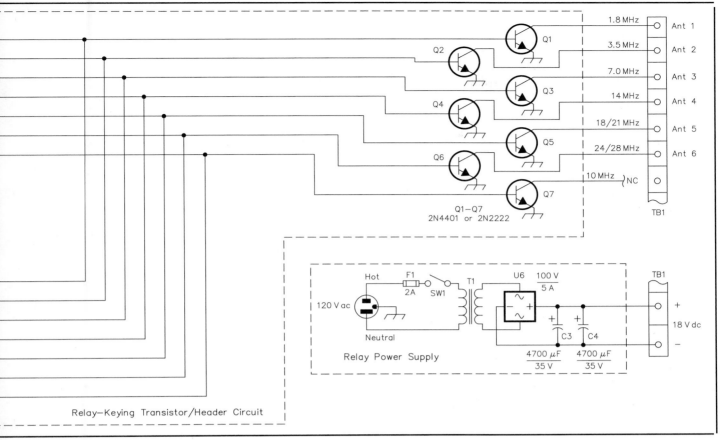

Fig. 22.76 — This schematic diagram shows the circuitry inside the main control box. The circuit is divided into four main sections: The *Icom interface circuit*, the *Yaesu interface circuit*, the *Relay-keying transistor/header board* and the *Relay power supply*. Most of the components are available from RadioShack. Resistors are ¼ W, 5% tolerance carbon-composition. The rotary switch allows manual antenna selection, which would be useful for rigs other than ICOM or Yaesu. A stable 8-V reference source must be made available to use the manual switch with non-ICOM rigs. The DIP sockets are used by the LED header to select the appropriate interface circuit for the radio being used. Note that only the Yaesu interface selector socket has resistors in line. Use flexible, stranded wire for the LED header and LEDs. This part of the circuit can be hard wired if only one type of rig will be used.

C1 — 10 µF, 16 V electrolytic
C2 — 0.1 µF, 50 V
C3, C4 — 4700 µF, 35 V
D1-D4 — 1N4001
DS1-DS7 — Red LED
F1 — 2A fuse
J1, J2 — Five-pin DIN socket (panel mount)
Q1-Q7 — 2N4401 or 2N2222

R1-R7 — 1 kΩ, ½ W
R9—1.2 kΩ
R10-R14 — 1 kΩ
R15, R23 — 1.8 kΩ
R16-R22, R28 — 2.7 kΩ
R24-R27 — 1 kΩ
SW1 — SPST, 250 V, 15 A
SW2 — 1 pole, 8 position rotary
T1 — 12.6 V ac C.T. 2A (or equivalent)

TB1 — 8 position terminal barrier strip
U1-U3 — LM339 comparator
U4 — LM7805 +5 V regulator
U5 — CD4028B (or equivalent) BCD decoder
U6 — 100 V, 5 A (or better) bridge rectifier
16 pin DIP sockets
16 pin DIP header

CIRCUIT DESCRIPTION

Fig 22.75 is a block diagram of the complete system. An ICOM or Yaesu HF radio connects to the appropriate decoder via the accessory connector on the back of the radio. Some other modern rigs have an accessory connector used for automatic bandswitching of amplifiers, tuners and other equipment. For example, Ten-Tec radios apply a 10 to 14-V dc signal to pins on the DB-25 interface connector for the various bands. Other radios use particular voltages on one of the accessory-connector pins to indicate the selected band. Check the owner's manual of your radio

for specific information, or contact the manufacturer's service department for more details. You may be able to adapt the ideas presented in this project for use with other radios.

A single length of coax and a multiconductor control cable run from the rig and decoder/control box to the remotely located switch unit. The remote relay box is equipped with SO-239 connectors for the input as well as the output to each antenna. You can use any type of connectors, though.

ICOM radios use an 8-V reference and a voltage divider system to provide a stepped band-data output voltage. **Table 22.6**

Table 22.6

ICOM Accessory Connector Output Voltages By Band

Band (MHz)	Output Voltage
1.8	7 – 8.0
3.5	6 – 6.5
7	5 – 5.5
14	4 – 4.5
18, 21	3 – 3.5
24, 28	2 – 2.5
10	0 – 1.2

Note: The voltage step between bands is not constant, but close to 1.0 V, and the 10-MHz band is not in sequence with the others.

shows the output voltage at the accessory socket when the radio is switched to the various bands. Notice that seven voltage steps can be used to select different antennas. The ICOM accessory connector pin assignments needed for this project are:

Pin 1   +8 V reference
Pin 2   Ground
Pin 4   Band signal voltage
Pin 7   +12 (13.8) V supply

Yaesu radios provide the band information as binary coded decimal (BCD) data on four lines. Nine different BCD values allow you to select a different antenna for each of the MF/HF bands. Table 22.6 also shows the BCD data from Yaesu radios for the various bands. The Yaesu 8-pin DIN accessory connector pin assignments needed for this project are:

Pin 1   +12 (13.8) V supply
Pin 3   Ground
Pin 4   Band Data A
Pin 5   Band Data B
Pin 6   Band Data C
Pin 7   Band Data D

Fig 22.76 is the schematic diagram for the control box. We will discuss each part of the control circuit later in this description. First, let's turn our attention to the external antenna box.

EXTERNAL ANTENNA BOX

Only the number of control lines going out to the relay box limits the number of antennas this relay box will switch. The unit shown in the lead photo has ten SO-239 connectors, to switch the common feed line to any of nine antennas. Many hams will use an eight-conductor rotator cable (such as *Belden 9405*) to the relay box. Using eight wires, we can control seven relays (six for antennas and one to ground the feed line for lightning protection) with the relay coil B+ supply, as well as a ground lead. The lead photo shows eight relays, and I plan to add two additional relays so I can select between all nine antennas when I install the unit at W1AW. The box contains a 12 V dc voltage regulator (LM7812 or equivalent), which I bolted to the aluminum box. I used an insulating spacer (TO-220 mounting hardware) between the back of the regulator and the box, and applied a layer of heatsink compound on both sides of the insulating wafer. There is also a connector for power and control lines. I used a DB-15 connector because I plan to add more relays and control lines later, so I can switch between 9 antennas. A DB-9 connector would be suitable for use with the eight-conductor control cable, or you may wish to use a weatherproof connector. **Fig 22.77** shows the relay box schematic diagram.

Since the box will be located outside,

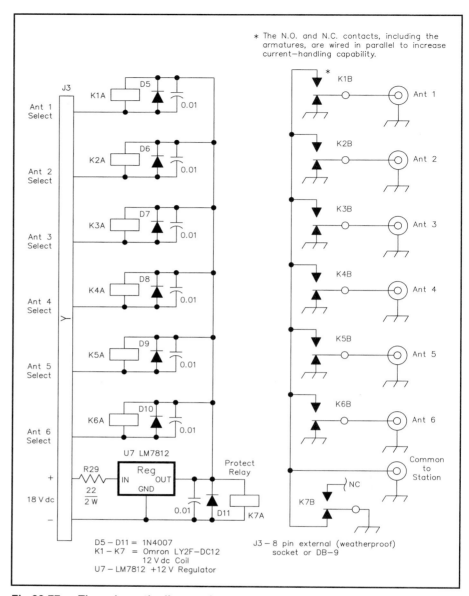

Fig 22.77 — The schematic diagram for the external antenna relay box. All relays are DPDT, 250 V ac, 15 A contacts. R29 is used to limit the regulator current. Mount the regulator using TO-220 mounting hardware, with heatsink compound. With the exception of the normally closed and normally open contacts, all wiring is #22 solid copper wire.

D5-D11 — 1N4007
J3 — 8 pin external weatherproof connector (or DB-9 with appropriate weather sealant).

K1-K7 — Omron LY2F-DC12 with 12 V coil (Allied Electronics Stock number 821-2019)
U7 — LM7812 +12 V regulator

I used a weatherproof metal box—a *Hammond Manufacturing*, type 1590Z150, watertight aluminum box. It's about $8^1/_2 \times 4^3/_4 \times 3^1/_8$ inches. This is a rather hefty box, meant to be exposed to years of various weather conditions. You can, however, use almost anything.

The coax connectors are mounted so each particular antenna connector is close to the relay, without too much crowding. I used flange-mount SO-239s (although the single hole type will also work). For added weather protection (and conductivity), I applied Penetrox® to the connector flange mount, including the threads of the mounting screws. On the power/control line connector, I used Coax Seal®.

I attached aluminum angle stock on either side of the box to mount it on a tower leg. The U-bolts should be of the proper size to fit the tower leg. They should also be galvanized or made of stainless steel.

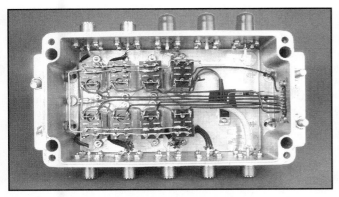

Photo A — This photo shows the external relay box. The LM7812 regulator is mounted to the bottom of the box. The relay normally open and normally closed contacts are wired in parallel using #12 solid copper. Joe Carcia plans to install two additional relays to use this project at W1AW, so he included the extra flange-mount SO-239 connectors when he was drilling holes in the box. Unused SO-239s can be capped off.

Photo B — This photo shows the inside of the control box. A RadioShack aluminum enclosure holds all of the components. The ICOM interface is built on a two-section RadioShack Universal Project Board (276-159B) and another single section of the same board. The bottom board in the enclosure as well as the right half of the middle section hold the ICOM circuit. The Yaesu interface is built on another section of the Universal Project Board, and is on the left side of the middle section. The top circuit board is a RadioShack Universal Project Board (276-150), which holds the DIP sockets and relay-selection transistors. The circuit boards use point-to-point wiring. All high voltage leads are insulated. The LEDs are mounted in holders on the front panel. The 7805 5-V regulator is mounted on the back panel using TO-220 mounting hardware, with heatsink compound.

Yaesu Band Data Voltage Output (BCD)

Band	A (1)	B (2)	C (4)	D (8)	(BCD Equiv.)
1.8	5V	0V	0V	0V	1
3.5	0V	5V	0V	0V	2
7.0	5V	5V	0V	0V	3
10.1	0V	0V	5V	0V	4
14	5V	0V	5V	0V	5
18.068	0V	5V	5V	0V	6
21	5V	5V	5V	0V	7
24.89	0V	0V	0V	5V	8
28	5V	0V	0V	5V	9

ANTENNA RELAYS

One of the more difficult parts of this project was the modification of the relays (DPDT Omron LY2F-DC12). To improve isolation, the moveable contacts (armature) are wired in parallel and the connecting wire is routed through a hole in the relay case.

Remove the relay from its plastic case. Unsolder and remove the small wires from the armatures. Carefully solder a jumper across the armature lugs. I used #20 solid copper. Then solder a piece of very flexible wire (such as braid from RG-58 cable) to either armature lug. Obviously, the location of the wire depends on which side you wish to connect the SO-239. You will also need to make a hole in the plastic case that is large enough to accommodate the armature wire without placing any strain on the free movement of the armature. I slipped a length of insulating tubing over this wire to prevent it from shorting to the aluminum box.

The normally open and normally closed contacts are also wired in parallel. This can be done on the lugs themselves. For this, I used no. 12 solid copper wire.

I mounted the relays in the aluminum box, oriented so they could be wired together without difficulty. (See **Photo A.**) With the exception of the wire used for the relay coils (no. 22 solid wire), I used no. 12 solid copper wire for the rest of the connections.

To eliminate the possibility of spikes or "back emf," a 1N4007 diode is soldered across the coil contacts of each relay. In addition, 0.01-μF capacitors across the diodes will reduce the possibility of stray RF causing problems with the relay operation.

Since the cable run from the shack to the tower can be quite long, consideration has to be given to the voltage drop that may occur. The relays require 12 V dc. As such, I installed a 12 V dc regulator in the box, and fed it with 18 V dc (at 2 amps) from the control box. If the cable run is not that long, however, you could just use a 12-V supply.

One of the relays is used for lightning protection. When not in use, the relay grounds the line coming in from the shack. When the control box is activated, it applies power to this relay, thus removing the ground on the station feed line. All the antenna lines are grounded through the normally closed relay contacts. They remain grounded until the relay receives power from the control box.

CONTROL BOX

This is the heart of the system. The 18 V dc power supply for the relays is located in this box, in addition to the Yaesu and ICOM decoder circuits and the relay-control circuitry. All connections to the relay box are made via an 8-position terminal barrier strip mounted on the back of the control box.

The front of the box has LEDs that indicate the selected antenna. A rotary switch can be used for manual antenna selection. The power switch and fuse are also located on the front panel.

The wiring schemes on the Yaesu and ICOM ACC sockets are so different, I opted to have a 5-pin DIN connector for each rig on the control box. Since there is only one set of LEDs, I used an 8-pin DIP header to select the appropriate control circuit for each radio. See **Photo B.**

ICOM CIRCUITRY

This circuit originally appeared in April 1993 *QST*.[2] I've modified the circuit slightly to fit my application. The original circuit allowed for switching between seven antennas (from 160 to 10 meters). The Band Data signal from the ICOM radios goes to a string of LM339 comparators. Resistors R9 through R15 divide the 8-V reference signal from the rig to provide midpoint references between the band signal levels. The LM339 comparators decide which band the radio is on. A single comparator selects the 1.8 or 10 MHz band because those bands are at opposite ends of the range. The other bands each use two comparators. One determines if the band signal is above the band level and the other determines if it is below the band level. If the signal is between those two levels, the appropriate LED and relay-selection transistor switch is turned on.

I used point-to-point wiring on RadioShack Universal Project Boards to build the various circuit sections. The ICOM interface uses both sections of a 276-159B project board shown at the bottom of the stack in Photo B for U1 and U2. Another section of project board holds U3, located on the right side of the middle section.

The ICOM circuit allows for manual antenna selection. The 8-V reference is normally taken directly from the ICOM ACC socket. If this circuit is to be used with other equipment, then a regulated 8-V source should be provided.

YAESU CIRCUITRY

The neat thing about Yaesu band data is that it's in a binary format. This means you can use a simple BCD decoder for band switching. The BCD output ranges from 1 to 9. In essence, you can switch between 9 antennas (or bands). Since the relay box switches just six antennas, I incorporated steering diodes (D1 through D4 in Fig 22.76) so I can use one antenna connection for multiple bands. In this regard, I opted to use one antenna connection for 17 and 15 meters, and another connection for 12 and 10 meters because the ICOM band data combines those bands. I did not include the control line or relay for a 30 meter antenna with this version of the project.

One section of the RadioShack 276-159B project board holds the Yaesu interface circuit. That board is shown on the left side of the middle layer of the stack shown in Photo B.

DIP SOCKETS AND HEADER

A RadioShack Universal Project Board, 276-150 holds the DIP sockets along with the relay keying transistors. This board is shown as the top layer in Photo B. The Yaesu socket has 1-kΩ resistors wired in series with each input pin. The other header connects directly to the ICOM circuitry.

The DIP header is used to switch the keying transistors between the ICOM and Yaesu circuitry. The LEDs are used to indicate antenna number. Use stranded wire (for its flexibility) when connecting to the LEDs.

RELAY KEYING TRANSISTORS

Both circuits use the same transistor-keying scheme, so I only needed one set of transistors. Each transistor collector connects to the terminal barrier strip. The emitters are grounded, and the bases are wired in parallel to the two 16-pin DIP sockets. The band data turns on one of the transistors, effectively grounding that relay-control lead. Current flows through the selected relay coil, switching that relay to the normally open position and connecting the station feed line to the proper antenna.

POWER SUPPLY

The power supply is used strictly for the relays. Other power requirements are taken from the rig used. There is room here for variations on the power supply theme. In this case, I used a 12.6 V, center-tapped, 2 A power transformer. I feed the output to a bridge rectifier, and two 4700 µF, 35-V electrolytics. (I happened to have these parts on hand.)

NOTES

[1]"An Antenna Switching System for Multi-Two and Single-Multi Contesting," by Tony Brock-Fisher, K1KP, January 1995 *NCJ*.

[2]"A Remotely Controlled Antenna Switch," by Nigel Thompson, April 1993 *QST*.

[3]"NA Logging Program© Section 11"

A TRIO OF TRANSCEIVER/COMPUTER INTERFACES

Virtually all modern Amateur Radio transceivers (and many general-coverage receivers) have provisions for external computer control. Most hams take advantage of this feature using software specifically developed for control, or primarily intended for some other purpose (such as contest logging), with rig control as a secondary function.

Unfortunately, the serial port on most radios cannot be directly connected to the serial port on most computers. The problem is that most radios use TTL signal levels while most computers use RS-232-D.

The interfaces described here simply convert the TTL levels used by the radio to the RS-232-D levels used by the computer, and vice versa. Interfaces of this type are often referred to as level shifters. Two basic designs, one having a couple of variations, cover the popular brands of radios.

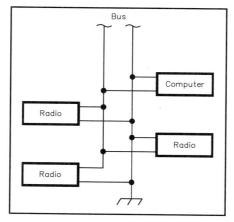

Fig 22.79—The basic two-wire bus system that ICOM and newer Ten-Tec radios share among several radios and computers. In its simplest form, the bus would include only one radio and one computer.

This article, by Wally Blackburn, AA8DX, first appeared in February 1993 *QST*.

TYPE ONE: ICOM CI-V

The simplest interface is the one used for the ICOM CI-V system. This interface works with newer ICOM and Ten-Tec rigs. **Fig 22.79** shows the two-wire bus system used in these radios.

This arrangement uses a CSMA/CD (carrier-sense multiple access/collision detect) bus. This refers to a bus that a number of stations share to transmit and receive data. In effect, the bus is a single wire and common ground that interconnect a number of radios and computers.

The single wire is used for transmitting and receiving data. Each device has its own unique digital address. Information is transferred on the bus in the form of packets that include the data and the address of

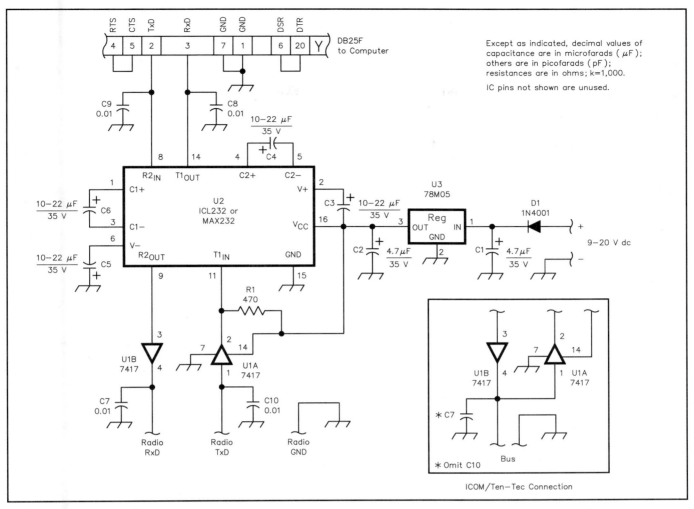

Fig 22.80—ICOM/Ten-Tec/Yaesu interface schematic. The insert shows the ICOM/Ten-Tec bus connection, which simply involves tying two pins together and eliminating a bypass capacitor.

C7-C10—0.01-μF ceramic disc.   U1—7417 hex buffer/driver.   U2—Harris ICL232 or Maxim MAX232.

the intended receiving device.

The schematic for the ICOM/Ten-Tec interface is shown in **Fig 22.80**. It is also the Yaesu interface. The only difference is that the transmit data (TxD) and receive data (RxD) are jumpered together for the ICOM/Ten-Tec version.

The signal lines are active-high TTL. This means that a logical one is represented by a binary one (+5 V). To shift this to RS-232-D it must converted to –12 V while a binary zero (0 V) must be converted to +12 V. In the other direction, the opposites are needed: –12 V to +5 V and +12 V to 0 V.

U1 is used as a buffer to meet the interface specifications of the radio's circuitry and provide some isolation. U2 is a 5-V-powered RS-232-D transceiver chip that translates between TTL and RS-232-D levels. This chip uses charge pumps to obtain ±10 V from a single +5-V supply. This device is used in all three interfaces.

A DB25 female (DB25F) is typically used at the computer end. Refer to the discussion of RS-232-D earlier in the chapter for 9-pin connector information. The interface connects to the radio via a ¹/₈-inch phone plug. The sleeve is ground and the tip is the bus connection.

It is worth noting that the ICOM and Ten-Tec radios use identical basic command sets (although the Ten-Tec includes additional commands). Thus, driver software is compatible. The manufacturers are to be commended for working toward standardizing these interfaces somewhat. This allows Ten-Tec radios to be used with all popular software that supports the ICOM CI-V interface. When configuring the software, simply indicate that an ICOM radio (such as the IC-735) is connected.

TYPE TWO: YAESU INTERFACE

The interface used for Yaesu rigs is iden-

tical to the one described for the ICOM/ Ten-Tec, except that RxD and TxD are not jumpered together. Refer to Fig 22.80. This arrangement uses only the RxD and TxD lines; no flow control is used.

The same computer connector is used, but the radio connector varies with model. Refer to the manual for your particular rig to determine the connector type and pin arrangement.

TYPE THREE: KENWOOD

The interface setup used with Kenwood radios is different in two ways from the previous two: Request-to-Send (RTS) and Clear-to-Send (CTS) handshaking is implemented and the polarity is reversed on the data lines. The signals used on the Kenwood system are active-low. This means that 0 V represents a logic one and +5 V represents a logic zero. This characteristic makes it easy to fully isolate the radio and the computer since a signal line only has to be grounded to assert it. Optoisolators can be used to simply switch the line to ground.

The schematic in **Fig 22.81** shows the Kenwood interface circuit. Note the different grounds for the computer and the radio. This, in conjunction with a separate power supply for the interface, provides excellent isolation.

The radio connector is a 6-pin DIN plug. The manual for the rig details this connector and the pin assignments.

Some of the earlier Kenwood radios require additional parts before their serial connection can be used. The TS-440S and R-5000 require installation of a chipset and some others, such as the TS-940S require an internal circuit board.

CONSTRUCTION AND TESTING

The interfaces can be built using a PC board, breadboarding, or point-to-point wiring. PC boards and MAX232 ICs are available from FAR Circuits. See the address list in the **References** chapter. The PC board template is available in the **References** chapter.

It is a good idea to enclose the interface in a metal case and ground it well. Use of a separate power supply is also a good idea. You may be tempted to take 13.8 V from your radio—and it works well in many cases: but you sacrifice some isolation and may have noise problems. Since these interfaces draw only 10 to 20 mA, a wall transformer is an easy option.

The interface can be tested using the data

Table 22.7
Kenwood Interface Testing

Apply	Result
GND to Radio-5	−8 to −12 V at PC-5
+5 V to Radio-5	+8 to +12 V at PC-5
+9 V to PC-4	+5 V at Radio-4
−9 V to PC-4	0 V at Radio-4
GND to Radio-2	−8 to −12 at PC-3
+5 V to Radio-2	+8 to +12 V at PC-3
+9 V to PC-2	+5 V to Radio-3
−9 V to PC-2	0 V to Radio-3

in **Tables 22.7, 8 and 9**. Remember, all you are doing is shifting voltage levels. You will need a 5-V supply, a 9-V battery and a voltmeter. Simply supply the voltages as described in the corresponding table for

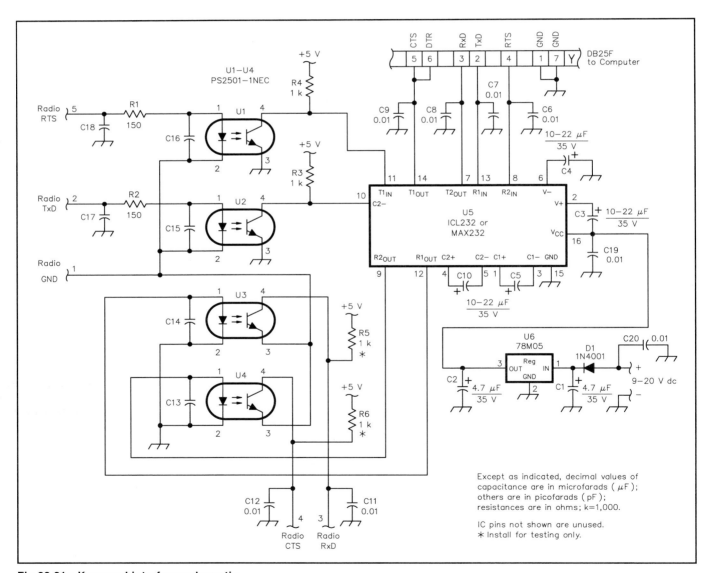

Fig 22.81—Kenwood interface schematic.
C6-C9, C11, C12, C17, C18—0.01-μF ceramic disc.

C13-C16, C19-C21—0.01 μF ceramic disc.
U1-U4—PS2501-1NEC (available from Digi-Key).

U5—Harris ICL232 or Maxim MAX232.

Table 22.8
ICOM/Ten-Tec Interface Testing

Apply	Result
GND to Bus	+8 to +12 V at PC-3
+5 V to Bus	−8 to −12 V at PC-3
−9 V to PC-2	+5 V on Bus
+9 V to PC-2	0 V on Bus

Table 22.9
Yaesu Interface Testing

Apply	Result
GND to Radio TxD	+8 to +12 V at PC-3
+5 V to Radio TxD	−8 to −12 V at PC-3
+9 V to PC-2	0 V at Radio RxD
−9 V to PC-2	+5 V at Radio RxD

your interface and check for the correct voltage on the other side. When an input of –9 V is called for, simply connect the positive terminal of the battery to ground.

During normal operation, the input signals to the radio float to 5 V because of pullup resistors inside the radio. These include RxD on the Yaesu interface, the bus on the ICOM/Ten-Tec version, and RxD and CTS on the Kenwood interface. To simulate this during testing, these lines must be tied to a 5-V supply through 1-kΩ resistors. Connecting these to the supply without current-limiting resistors will damage the interface circuitry. R5 and R6 in the Kenwood schematic illustrate this. They are not shown (but are still needed) in the ICOM/Ten-Tec/Yaesu schematic. Also, be sure to note the separate grounds on the Kenwood interface during testing.

Another subject worth discussing is the radio's communication configuration. The serial ports of both the radio and the computer must be set to the same baud rate, parity, and number of start and stop bits. Check your radio's documentation and configure your software or use the PC-DOS/MS-DOS MODE command as described in the computer manual.

A COMPUTER-CONTROLLED TWO-RADIO SWITCHBOX

This versatile computer-controlled two-radio switchbox was designed by Dean Straw, N6BV, who made it primarily for contest operations using one of the popular computer logging programs, such as *CT*, *NA* or *TR*. The switchbox was built into two boxes, a main unit and a hard-wired remote head. **Fig 22.82** shows the back of the main unit, and **Fig 22.83** shows the small wired-remote head. The remote head is compact enough to place almost anywhere on a crowded operating desk. Besides toggle switches, it uses red and green LED annunciators to tell the operator exactly what is happening.

RadioShack components were used throughout the project as much as possible so that parts availability should not be a hurdle for potential builders. The overall cost using all-new parts was about $160.

OVERVIEW OF FEATURES

The switchbox controls both transmitting and receiving functions for either phone or CW modes. (Data modes that connect through the transceiver's microphone input or that use direct FSK could also be controlled through the switchbox, using additional external switching.) This particular switchbox was built to work with two ICOM IC-765 HF transceivers, but you can easily wire the microphone, PTT, headphone and CW key-line connections to match your own radios.

Receiving Features

For this discussion, assume that Radio A is located to the left in front of the operator and Radio B is to the right. Assume also that the two radios are connected to separate antennas (and perhaps linear amplifiers), and that interaction and overload between the two radios has been mini-

Fig 22.82—A view of the rear panel of the main box of the Two-Radio Switchbox.

Fig 22.83—The remote head for the Two-Radio Switchbox.

mized by good engineering. In other words, Radio B can receive effectively on one frequency band, even while Radio A is transmitting full power on another band—and vice versa. Here we'll assume that you are using stereo headphones. You can select:

1. Radio A in both ears (monaural)—for both transmit and receive, in the RX A switch position.

2. Radio B in both ears (monaural)—for both transmit and receive, in the RX B switch position.

3. Radio A in the left ear; Radio B in the right ear—for both transmit and receive, in the STEREO switch position.

4. Radio A in both ears in receive; Radio B in both ears while Radio A is transmitting, toggling automatically while in the AUTO TX switch position.

5. Radio B in both ears in receive; Radio A in both ears while Radio B is transmitting, toggling automatically while in the AUTO TX switch position.

6. Green LEDs on the remote head give instant indication of the source(s) of audio

in the stereo headphones.

The AUTO TX facility in 4 and 5 above allows you to call CQ on one radio, while devoting full attention to listening to the second radio. You could, for example, look for new multipliers in a contest or to check whether another band is open or not. Late in a contest, you can easily become mesmerized listening to your own voice from a voice recorder calling CQ, or listening to the computer automatically calling CQ on CW. The AUTO TX facility forces you to pay attention to the second radio—but this function can be switched off, if you like, from the remote head. If you choose the STEREO receiving mode, the AUTO TX function is automatically disabled, since it would make it pretty confusing to have the right and left audio sources shift automatically.

Another useful feature in STEREO is a BLEND control on the main box. This allows you to shift the apparent position of the right-hand receive audio somewhere between full-right and near the middle of your spatial hearing range. Some opera-

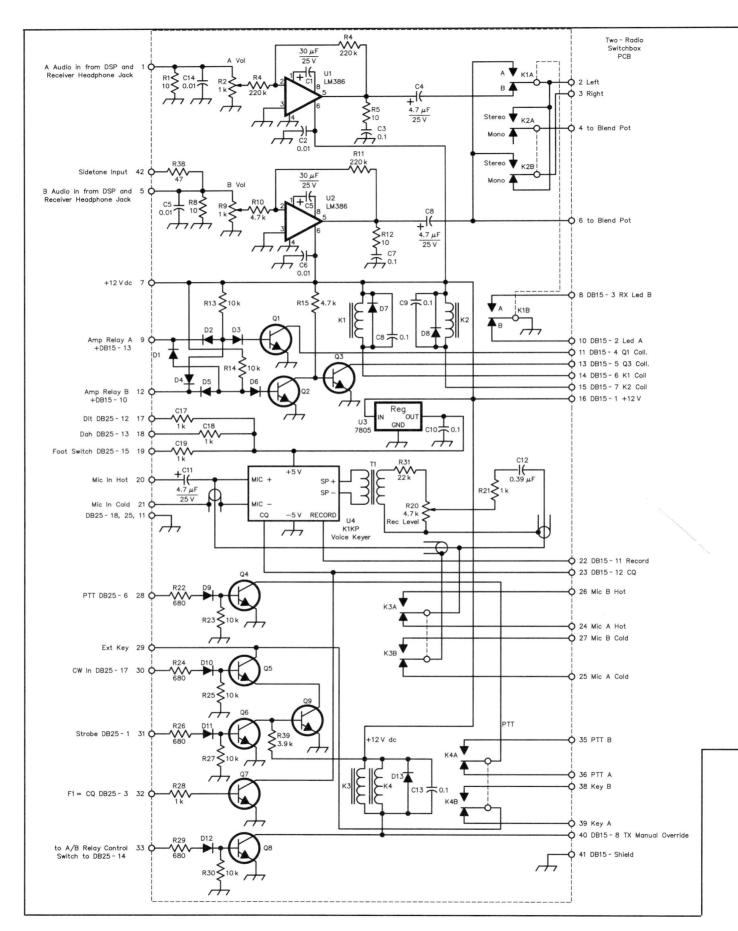

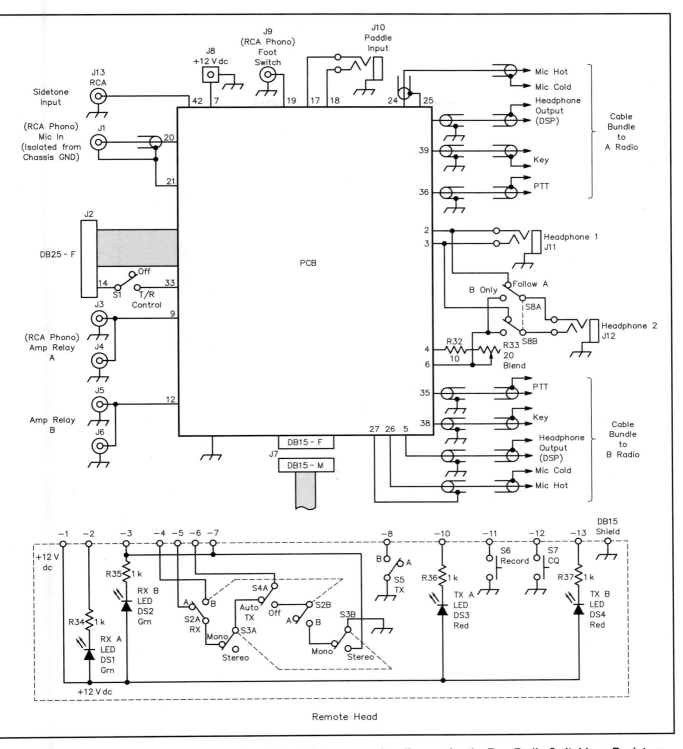

Fig 22.84—At A, schematic diagram for the PCB. At B, the interconnection diagram for the Two-Radio Switchbox. Resistors are ¼ W. Capacitors are disc ceramic. Capacitors marked with polarity are electrolytic.

D1-D13—General-purpose silicon switching diodes, such as 1N4148 or 1N914 (RS 276-1620).

DS1, DS2—Green LED (RS 276-069A).

DS3, DS4—Red LED (RS 276-068A).

J1—Insulated RCA phono jack for microphone input.

J3-6, J6, J9, J13—Chassis-mounting, grounded RCA phono jack (RS 274-346).

J10-12—¼-inch stereo phone jack (RS 274-312B).

J8—DC power jack (RS 274-1565A).

K1-K4—DPDT 12-V PC-mount relay (RS 275-249A).

Q1-Q9—General-purpose NPN switching transistor, 2N2222 or 2N3904 style (RS 276-1617).

R2, R9—1 kΩ PCB potentiometer (RS 271-280).

R21—4.7 kΩ PCB potentiometer (RS 271-281).

S1-S8—Flat lever switch, DPDT (RS 275-636B).

S6, S7—Momentary contact SPST (RS 275-1556A).

U1, U2—LM386 audio amplifier (RS 276-1731).

U3—7805 5-V regulator (RS 271-281).

U4—Voice record/playback module (RS 276-1326). Remove R5 and R6 from PCB.

tors claim that this helps cut down on fatigue during long operating sessions when using stereo reception.

There is a second stereo headphone jack on the main box, with a switch labelled FOLLOW A or B ONLY. A second operator can either monitor what the first operator is doing (perhaps for training or coordination during a contest), or else the second operator can pay full attention to the second receiver.

Each audio channel incorporates an LM-386 IC audio amplifier that provides more-than-adequate power to drive any types of headphones, regardless of their impedance. Separate level controls for each LM-386 equalize audio levels. These amplifiers are very useful for radios that have marginal or inadequate internal stereo headphone amplifiers. Each input is shunted by a 10-Ω resistor to provide a proper load for the radio driving it.

The switchbox also has a separate SIDETONE input jack for audio from an external keyer. This connects to the B channel so that you can still hear sidetone when you have selected AUTO TX (with Radio A as transmitter) and use the paddle instead of the computer to send CW, perhaps to send a fill for a missed report.

The wiring bundles going to each radio are set up to accommodate external DSP filters, something that can bring a high-quality older transceiver up to "modern" status, comparable to the newer radios with all their DSP bells and whistles.

Transmitting Features

1. The microphone "hot" line, the microphone "cold" line, the CW key line and the PTT line are all switched between Radio A and Radio B. Both microphone "hot" and "cold" leads are switched to reduce the possibility of ground-loop induced 60-Hz hum on your transmitted signal.

2. In the TX A switch position on the remote head, Radio A is selected manually, or the computer program can control transmitter selection through one of its parallel ports. Placing the switch in position TX B overrides computer control and selects Radio B manually.

3. A manual T/R CONTROL switch, S5, on the main box can disable automatic computer control of the transmitter selection. While the *TR* program allows this function to be set by software control, *CT* doesn't have this ability, so S5 was added.

4. An external paddle may be connected to J10 to send CW, using the *TR* computer program as a keyer.

5. Two sets of paralleled, diode-isolated RCA phono jacks (J3 through J6) are mounted on the main box. You can connect the AMPLIFIER RELAY control lines from each transceiver to both the switchbox and two external amplifiers

Fig 22.85—Inside the main box. With service loops for the cable bundles going to the two radios, the inside of the main box is cramped and a larger box would be a good idea!

without needing Y-connectors. By the way, only one radio can transmit at a time with this switchbox. This keeps you completely legal in any single-operator or multi-single contest category.

6. An RCA phono connector J9 is available for an external foot-switch input to the *TR* program.

7. A K1KP-style RadioShack voice-keyer module (see the project "A Simple Voice Keyer" in this chapter) is built into the main box. The voice keyer can be controlled either by an external computer or from the front panel of the remote head.

8. A separate EXT KEYER RCA phono connector J13 mounts on the main box. This parallels the LPT CW keyer output from the computer program.

9. Red LEDs on the remote head indicate which transmitter is active while it is transmitting.

THE SCHEMATIC

Fig 22.84A shows the schematic of the PCB used in the two-radio switchbox, and Fig 22.84B shows the interconnection diagram from the PCB to the other components in the main box and the remote head. Switching relays K1 and K2 are the heart of the receive-audio circuitry. K1A selects audio from either Receiver A or B (buffered by the LM-386 audio ampliers U1 and U2) and applies it to the left-ear terminal of headphone output jack J11. Relay K2B normally connects the right-ear terminal of J11 in parallel with the left-ear terminal, for monaural operation. In the STEREO mode, K2B puts the output from Receiver B into the right ear. Note that S3A on the remote head disables K1 when STEREO mode is selected, so that Receiver A

output remains in the left ear, with Receiver B output in the right ear.

Relay contacts K1B are used to turn on the Green receive LEDs in the control head to indicate whether audio is coming from Receiver A alone, B alone, or both A and B together in stereo. Many contest operators have trained themselves to listen to two radios simultaneously in STEREO mode. But many of us are only endowed with a single brain, and we get easily distracted in stereo! Thus the AUTO CQ feature was added to the switchbox.

In this mode, when S4 in the control head is set to AUTO TX and S3 is set to MONO, closure to ground on the AMPLIFIER RELAY line by either transceiver will toggle between the audio outputs from Receiver A and B automatically. When the radio stops transmitting, the receive audio will toggle back. Both AMPLIFIER RELAY control lines are isolated from each other by summing diodes.

Just so you won't be surprised, both the *CT* and *TR* software programs energize "Radio 2" as what we call "Radio A," and "Radio 1" as "Radio B." This is slightly non-intuitive but it seems to be a function of the default state of the computer parallel port. The operator, however, quickly becomes accustomed to this and toggles between the radios, while watching the computer screen to see which frequency band is active.

The author borrowed most of the transmitting selection circuitry from an N6TR design. (N6TR is the creator of the *TR* program.) Transistor Q7 was added to control the K1KP-style voice keyer from the computer program's F1 function key when in phone mode.

CONSTRUCTION

Fig 22.85 shows a view of the inside of the completed main unit box. The author chose to use a standard Bud LMB 6×4× 3¹/₂-inch aluminum box, but things got pretty crowded as you can see. You will probably want to use a bigger box—and you will probably choose to use smaller cables in the bundles going to each transceiver. N6BV taped together three standard "zip cords" for the key-line, PTT and receiver audio cables, plus a shielded cable for the microphone line. This bundle was definitely overkill in terms of current-handling capacity and made the service loops inside the main box quite bulky. Smaller-gauge "speaker wire" would have been far easier to handle.

Most of the switchbox circuitry was built on a single-sided copper PCB using "wired-traces construction" (also called the "Lazy PC Board" technique), described in the CIRCUIT CONSTRUCTION chapter. A large-diameter drill was used to ream out the copper around holes where the ground plane was supposed to be removed. The RadioShack 273-1326 voice-keyer board was stuck to the tops of relays K3 and K4 using double-sided sticky tape.

A 25-pin DB25F connector was mounted on the main box for J2 and an inexpensive 6-foot long 25-conductor DB25M-to-DM25M cable (bought at a local computer store) went to the computer's parallel port. Much of the point-to-point wiring to the DB25 and DB15 connectors used ribbon wires from a scavenged surplus computer.

The remote-control head required a total of 12 wires from the main box, and a 15-pin DB15M connector J7 was used so that an inexpensive commercial VGA computer DB15F cable could be used. The connector at one end was cut off the 6-foot long cable, which was then hard-wired into the remote head. Tie wraps were used to provide mechanical strain reliefs for cables entering into the main box and into the remote head.

The author created labels using a word processing program in 12-point Times Roman typeface. These were laser printed onto a thin mylar sheet used for creating overhead transparency films. Clear nail polish ("Hard As Nails") was brushed lightly over the labels on the mylar sheet to protect them from wear. After the polish had dried, the labels were cut out using a paper cutter and they were then stuck onto the boxes using more clear nail polish as glue.

OPERATION

The switchbox and remote head can be operated entirely manually, with no connection to a computer, if you like. However, it is a lot more fun when you can control each radio from the computer keyboard, especially if you've interfaced the radio to read and write the frequency to the transceivers! Then you can "point-and-shoot" from packet DX spots or you can type in a desired frequency and press the [Enter] key to tune your radio.

You must set up your computer program to control the two-radio switchbox through a parallel (LPT) port. If you control your radios' frequencies, you will do that through serial ports. Follow the directions for your software carefully. In general, you've got to get everything *exactly right* in order for all functions to work properly, particularly the frequency and mode controls for your radio. Connect the switchbox to a source of +12 V dc that can source about 1 A. The ICOM radios have a jack on the back that will provide this.

With the control head switches set to TX A, MONO and TX AUTO OFF, toggle between Radio 2 and Radio 1. In the *CT* program press ALT-, (press the ALT key and the comma key together) or in *TR* press ALT-R to toggle between transmitter A and B. You should hear relays changing in the switchbox and when you go into transmit the red Transmit LEDs on the remote head will light for the appropriate transmitter.

Now, switch between the RX A and the RX B positions. The green receiver RX A and RX B LEDs on the control head should alternately light. Try the STEREO switch to see both green LEDs light up. Switch back to MONO and then switch the TX AUTO switch on and key the transmitter. The RX A and RX B LEDs should toggle as you key and unkey the transmitter.

Adjust the receiver front-panel volume controls for normal audio levels, probably with the front-panel headphone plugs from the switchbox pulled out so that the speakers in the radios are enabled. Then plug both headphones into the radios and adjust the 1 kΩ level pots on the PCB so that the levels are equal for Radio A and Radio B when the switch is changed between RX A and RX B positions on the remote head. Now, check out STEREO headphone operation and adjust the BLEND pot for the spacial placement you prefer.

Now, put one of the radios into SSB mode, after connecting the RF output to a dummy load so you don't interfere with anyone on the air. Speak into the microphone while holding down the RECORD button on the remote head and record a message into the voice keyer module. Now, press the Voice CQ button and adjust the 5- kΩ level pot on the PCB to equalize the level between the microphone and the voice keyer.

You should now be all set to enjoy computer-controlled two-radio operation!

TR TIME-DELAY GENERATOR

If you've ever blown up your new GaAsFET preamp or hard-to-find coaxial relay, or are just plain worried about it, this transmit/receive (TR) time-delay generator is for you. This little circuit makes it simple to put some reliability into your present station or to get that new VHF or UHF transverter on the air fast, safe and simple. Its primary application is for VHF/UHF transverter, amplifier and antenna switching, but it can be used in any amplifier-antenna scheme. An enable signal to the TR generator will produce sequential output commands to receive relay, a TR relay, an amplifier and a transverter—automatically. All you do is

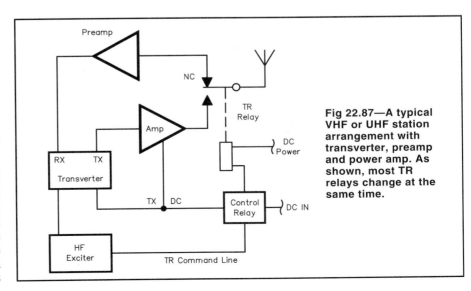

Fig 22.87—A typical VHF or UHF station arrangement with transverter, preamp and power amp. As shown, most TR relays change at the same time.

sit back and work DX! This project was designed and built by Chip Angle, N6CA.

WHY SEQUENCE?

Several problems may arise in stations using transverters, extra power amplifiers and external antenna-mounted TR relays. The block diagram of a typical station is shown in **Fig 22.87**. When the HF exciter is switched into transmit by the PTT or VOX line, it immediately puts out a ground (or in some cases a positive voltage) command for relay control, and an RF signal.

If voltage is applied to the transverter, amplifier and antenna relays simultaneously, RF can be applied as the relay contacts bounce. In most cases, RF will be applied before a relay can make full closure. This can easily arc contacts on dc and RF relays and cause permanent

damage. In addition, if the TR relay is not fully closed before RF from the power amplifier is applied, excessive RF may leak into the receive side of the relay. The likely result—preamplifier failure!

Fig 22.88 is a block diagram of a station with a remote-mounted preamp and antenna relays. The TR time-delay generator supplies commands, one after another, going into transmit and going back to receive from transmit, to turn on all station relays in the right order, eliminating the problems just described.

CIRCUIT DETAILS

Here's how it works. See the schematic diagram in **Fig 22.89**. Assume we're in receive and are going to transmit. A ground command to Q2 (or a positive voltage command to Q1) turns Q2 off.

This allows C1 to charge through R1 plus 1.5 kΩ. This rising voltage is applied to all positive (+) inputs of U1, a quad comparator. The ladder network on all negative (−) inputs of U1 sets the threshold point of each comparator at a successively higher level. As C1 charges up, each comparator, starting with U1A, will sequentially change output states.

The comparator outputs are fed into U2, a quad exclusive-OR gate. This was included in the design to allow "state programming" of the various relays throughout the system. Because of the wide variety of available relays, primarily coaxial, you may be stuck with a relay that's exactly what you need—except its contacts are open when it's energized. To use this relay, you merely invert the output state of the delay generator by using a

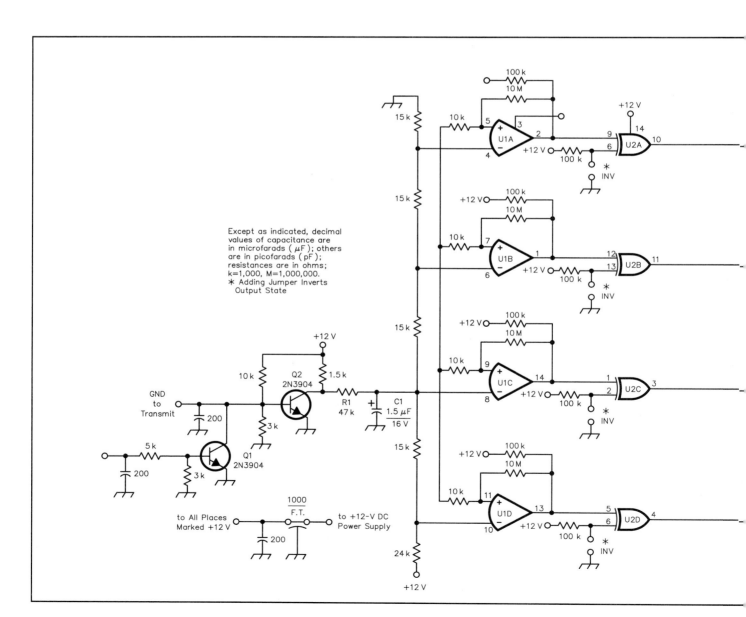

Except as indicated, decimal values of capacitance are in microfarads (μF); others are in picofarads (pF); resistances are in ohms; k=1,000, M=1,000,000.
∗ Adding Jumper Inverts Output State

jumper between the appropriate OR-gate input and ground. Now, the relay will be "on" during receive and "off" during transmit. This might seem kind of strange; however, high-quality coaxial relays are hard to come by and if "backwards" relays are all you have, you'd better use them.

The outputs of U2 drive transistors Q3-Q6, which are "on" in the receive mode. Drive from the OR gates turns these transistors "off." This causes the collectors of Q3-Q6 to go high, allowing the base-to-emitter junctions of Q7-Q10 to turn on the relays in sequential order. The LEDs serve as built-in indicators to check performance and sequencing of the generator. This is convenient if any state changes are made.

When the output transistors (Q7-Q10) are turned on, they pull the return side of the relay coils to ground. These output transistors were selected because of their high beta, a very low saturation voltage (V_{CE}) and low cost. They can switch (and have been tested at) 35 V at 600 mA for many days of continuous operation. If substitutions are planned, test one of the new transistors with the relays you plan to use to be sure that the transistor will be able to power the relay for long periods.

To go from transmit to receive, the sequencing order is reversed. This gives additional protection to the various system components. C1 discharges through R1 and Q2 to ground.

Fig 22.90 shows the relative states and duration of the four output commands when enabled. With the values specified for R1 and C1, there will be intervals of 30 to 50 milliseconds between the four output commands. Exact timing will vary because of component tolerances. Most likely everything will be okay with the values shown, but it's a good idea to check the timing with an oscilloscope just to be sure. Minor changes to the value of R1 may be necessary.

Most relays, especially coaxial, will require about 10 ms to change states and stop

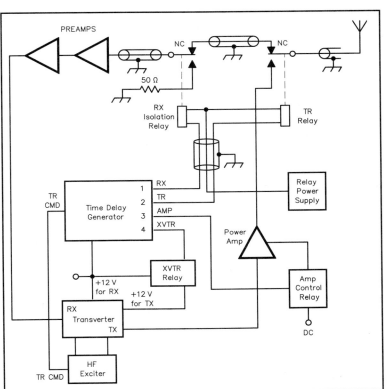

Fig 22.88—Block diagram of the VHF/UHF station with a remote-mounted preamp and antenna relays. The TR time-delay generator makes sure that everything switches in the right order.

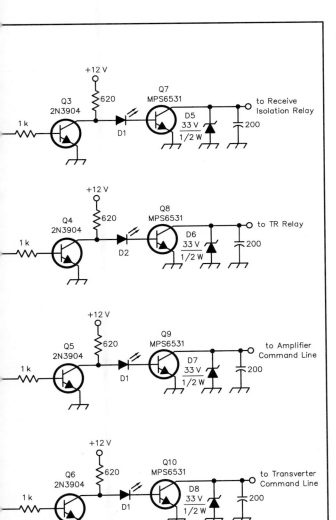

Fig 22.89—Schematic diagram of the TR time-delay generator. Resistors are ¼ W. Capacitors are disc ceramic. Capacitors marked with polarity are electrolytic.

D1-D4—Red LED (MV55, HP 5082-4482 or equiv.)
D5-D8—33-V, 500-mW Zener diode (1N973A or equiv.)
C1—1.5-µF, 16-V or greater, axial-lead electrolytic capacitor. See text.
Q1-Q6—General purpose NPN transistor (2N3904 or equiv.)
Q7-Q10—Low-power NPN amplifier transistor, MPS6531 or equiv.
 Must be able to switch up to 35 V at 600 mA continuously. See text.
R1—47-kΩ, ¼-W resistor. This resistor sets the TR delay time constant and may have to be varied slightly to achieve the desired delay. See text.
U1—Quad comparator, LM339 or equiv.
U2—Quad, 2-input exclusive or gate (74C86N, CD4030A or equiv.)

bouncing. The 30-ms delay will give adequate time for all closures to occur.

CONSTRUCTION AND HOOKUP

One of the more popular antenna changeover schemes uses two coaxial relays: one for actual TR switching and one for receiver/preamplifier protection. See Fig 22.88.

Many RF relays have very poor isolation especially at VHF and UHF frequencies. Some of the more popular surplus relays have only 40-dB isolation at 144 MHz or higher. If you are running high power, say 1000 W (+60 dBm) at the relay, the receive side of the relay will see +20 dBm (100 mW) when the station is transmitting. This power level is enough to inflict fatal damage on your favorite preamplifier.

Adding a second relay, called the RX isolation relay here, terminates the preamp in a 50-Ω load during transmission and increases the isolation significantly. Also, in the event of TR relay failure, this extra relay will protect the receive preamplifier.

As shown in Fig 22.88, both relays can be controlled with three wires. This scheme provides maximum protection for the receiver. If high-quality relays are used and verified to be in working order, relay losses can be kept well below 0.1 dB, even at 1296 MHz. The three-conductor cable to the remote relays should be shielded to eliminate transients or other interference.

By reversing the RX-TX state of the TR relays (that is, connecting the transmitter Hardline and 50-Ω preamp termination to the normally open relay ports instead of the normally closed side), receiver protection can be provided. When the station is not in use and the system is turned off, the receive preamplifier will be terminated in 50 Ω instead of being connected to the antenna. The relays must be energized to receive. This might seem a little backward; however, if you are having static-charge-induced preamplifier failures, this may solve your problem.

Most coaxial relays aren't designed to be energized continuously. Therefore, adequate heat sinking of coaxial relays must be considered. A pair of Transco Y relays can be energized for several hours when mounted to an aluminum plate 12 inches square and ¼ inch thick. Thermal paste will give better heat transfer to the plate. For long-winded operators, it is a good idea to heat sink the relays even when they are energized only in transmit.

Fig 22.91 shows typical HF power amplifier interconnections. In this application, amplifier in/out and sequencing are all provided. The amplifier will always have an antenna connected to its output before drive is applied.

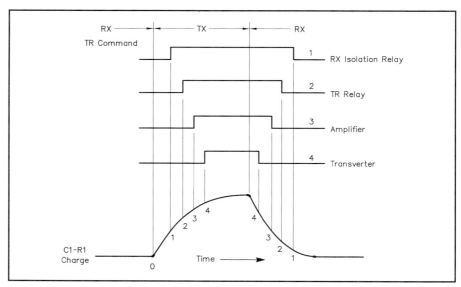

Fig 22.90—The relative states and durations of the four output commands when enabled. This diagram shows the sequence of events when going from receive to transmit and back to receive. The TR delay generator allows about 30 to 50 ms for each relay to close before activating the next one in line.

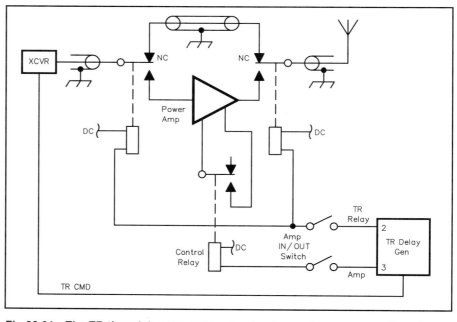

Fig 22.91—The TR time-delay generator can also be used to sequence the relays in an HF power amplifier.

Fig 22.92—The completed time-delay generator fits in a small aluminum box.

Many TR changeover schemes are possible depending on system requirements. Most are easily satisfied with this TR delay generator.

The TR delay generator is built on a 2½ × 3¼-inch PC board.[2] See **Fig 22.92**. Connections to the rest of the system are made through feedthrough capacitors. Do not use feedthrough capacitors larger than 2000 pF because peak current through the output switching transistors may be excessive.

A SIMPLE 10-MINUTE ID TIMER

This project was originally described in "Hints and Kinks" in the November 1993 issue of *QST* by John Conklin, WDØO. It is an update to an earlier WDØO design for which parts are no longer available.

This simple and effective timer can be built in an evening and uses inexpensive and easily obtained parts. Its timing cycle is independent of supply voltage and resets automatically upon power-up, as well as at the end of each cycle.

CONSTRUCTION AND ADJUSTMENT

Assembly is straightforward and parts layout is not critical. The circuit of Fig 22.94 can be built on a small piece of perfboard and housed in an inexpensive enclosure. To calibrate the time, set R1, TIME ADJ, at midpoint initially, and then adjust it by trial and error to achieve a 10-minute timing cycle. The buzzer will sound for about 1 second at the end of each cycle.

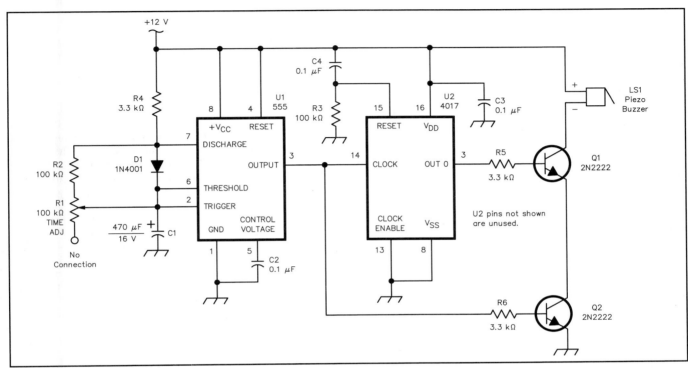

Fig 22.94—The new and improved 10-minute reminder uses easy-to-get components. Part numbers in parentheses are RadioShack; equivalent parts can be substituted. Resistors are $1/4$-W, 5% or 10% tolerance units.

C1—470-µF electrolytic, 16 V or more (272-957).
C2, C3, C4—0.1 µF, 16 V or more (272-109).
D1—1N4001, 50 PIV, 1 A (276-1101).

LS1—12-V piezo buzzer (273-074).
Q1, Q2—2N2222 or MPS2222A (276-2009).
R1—100-kΩ trimmer potentiometer (271-284).

R2, R3—100 kΩ (271-1347).
R4, R5, R6—3.3 kΩ (271-1328)
U1—555 timer IC (276-1723).
U2—4017 decade counter IC (276-2417).

HIGH-POWER ARRL ANTENNA TUNER FOR BALANCED OR UNBALANCED LINES

Only rarely does a transmission line connect at one end to a real-world antenna that has an impedance of exactly 50 Ω. An antenna tuner is often used to transform whatever impedance results at the input to the transmission line to the 50 Ω needed by a modern transceiver. Generally, only when a transceiver is working into the load for which it was designed can it deliver its rated power, at its rated level of distortion. Many transceivers have built-in antenna tuners capable of handling a modest range of impedance mismatches. Most are rated for SWRs up to 3:1 on an unbalanced coax line. Such a built-in tuner will probably work fine when you use the transceiver by itself. Thus, if your transceiver has a built-in antenna tuner and if you use coax-fed antennas, you probably don't need an external antenna tuner.

REASONS FOR USING AN ANTENNA TUNER

If you use a linear amplifier, however, you may find that it can't load some coax-fed antennas with even moderate SWRs, particularly on 160 or 80 meters. This is usually due to a loading capacitor that is marginal in capability. Some amplifiers even have protective circuits that prevent you from using the amplifier when the

SWR is higher than about 2:1. For this situation you may well need a high-power antenna tuner. Bear in mind that although an antenna tuner will bring the SWR down to 1:1 at the amplifier—that is, it presents a 50-Ω load to the amplifier—it will not change the actual SWR condition on the transmission line going to the antenna itself. Fortunately, most amateur HF coax-fed antennas are operated close to resonance and any additional loss on the line due to SWR is not a big problem. If you wish to operate a single-wire antenna on multiple frequency bands, an antenna tuner also will be needed. As an example, if you choose a 130-foot long dipole for this task, fed in the center with 450-Ω ladder line, the feed-point impedance of this antenna over the 1.8 to 29.7-MHz range will vary drastically! Further, the antenna and the feed line are both balanced, requiring a balanced type of antenna tuner. What you need is a balanced antenna tuner that can handle a very wide range of impedances, all without arcing or overheating internally.

DESIGN PHILOSOPHY BEHIND THE ARRL HIGH-POWER TUNER

Dean Straw, N6BV, designed this antenna tuner with three objectives in mind:

First, it would operate over a wide range of loads, at full legal power. Second, it would be a high efficiency design, with minimal losses, including losses in the balun. This leads to the third objective: Include a balun operating within its design impedances. Often, a balun is added to the output of a tuner. If it is designed as a 4:1 unit, it expects to see 200 Ω on its output. Connect it to ladder line and let it see a 1000-Ω load, and spectacular arcing can occur even at moderate (100 W) power levels.

For that reason this unit was designed with the balun on the input of the tuner. This antenna tuner is designed to handle full legal power from 160 to 10 meters, matching a wide range of either balanced or unbalanced impedances. The network configuration is a high-pass T-network, with two series variable capacitors and a variable shunt inductor. See **Fig 22.95** for the schematic of the tuner. Note that the schematic is drawn in a somewhat unusual fashion. This is done to emphasize that the common connection of the series input and output capacitors and the shunt inductor is actually the subchassis used to mount these components away from the tuner's cabinet. The subchassis is insulated from the main cabinet using four heavy-duty

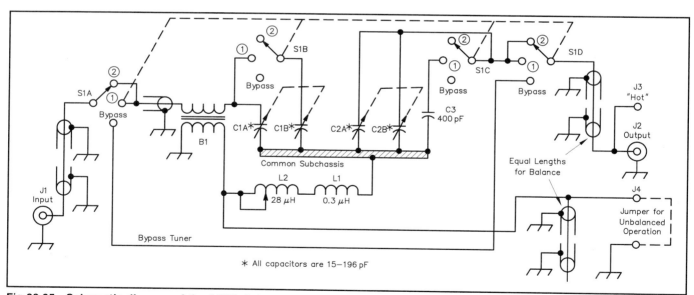

Fig 22.95—Schematic diagram of the ARRL Antenna Tuner.

B1—Balun, 12 turns bifilar wound #10 Formvar wire side-by-side on 2.4-inch OD Type 43 core.

C1, C2—15-196 pF transmitting variable with voltage rating of 3000 V peak, such as the E. F. Johnson 154-507-1.

C3—Home-made 400 pF capacitor; more than 10 kV voltage breakdown.

Made from plate glass from a "5×7-inch" picture frame, sandwiched in between a 4×6-inch, 0.030-inch thick aluminum plate and the electrically floating subchassis that also forms the common connection between C1, C2 and L1.

L1—Fixed inductor, approximately

0.3 μH, 4 turns of 1/4-inch copper tubing formed on 1-inch OD tubing.

L2—Rotary inductor, 28 μH inductance, Cardwell E. F. Johnson 229-203, with steatite coil form.

S1—HV switch, Radio Switch Corp R862E033000.

2-inch steatite cones.

While a T-network type of tuner can be very lossy if care isn't taken, it is very flexible in the range of impedances it can match. Special attention has been paid to minimize power loss in this tuner—particularly for low-impedance loads on the lower-frequency amateur bands. Preventing arcing or excessive power dissipation for low-impedance loads on 160 meters represents the most challenging conditions for an antenna tuner designer! To see the computed range of impedances it can handle, look over the tables in the ASCII file called TUNER.SUM available from the ARRL BBS and internet sites (see page viii). The tables were created using the program AAT, included with the 18th Edition of *The ARRL Antenna Book*. They show the percentage of power lost for 253 individual impedances, for each amateur band from 1.8 to 29.7 MHz. Where a match cannot be achieved a blank is shown; where the capability is limited due to voltage breakdown of one of the components a "V" is placed. Where the minimum capacitance of a tuning capacitor is too large for a match, a "C−" is placed. Where an "L+" appears, there is insufficient inductance available to match that impedance. A "P" indicates that the power loss at that impedance exceeds 20%. For example, assume that the load at 1.8 MHz is $12.5 + j\, 0\ \Omega$. For this example, the output capacitor C3 in this example is set by the program to 750 pF. This dictates the values for the other two components. At 1.8 MHz, for typical values of component unloaded Q (200 for the coil), 7.9% of the power delivered to the input of the network is lost as heat. For 1500 W at the input, the loss in the network is thus 119 W. Of this, 98 W ends up in the inductor, which must be able to handle this without melting or detuning. The T-network must be used judiciously, lest it burn itself up or arc over internally!

One of the techniques used to minimize power lost in this tuner is the use of a relatively large output capacitor. (The output variable capacitor has a maximum capacitance of approximately 400 pF, including an estimated 20 pF of stray capacitance.) An additional 400 pF of fixed capacitor can be switched across the output variable capacitor. At 750 pF output capacitance, enough heat is generated at 1500 W input to make the inductor uncomfortably warm to the touch after 30 seconds of full-power key-down operation, but not enough to destroy the coil for a 12.5-Ω load.

For a variable capacitor used in a T-network tuner, there is a trade-off between the range of minimum to maximum capacitance and the voltage rating. This tuner uses two identical Cardwell-Johnson dual-section 154-507-1 air-variable ca-

pacitors, rated at 3000 V. Each section of the capacitor ranges from 15 to 196 pF, with an estimated 10 pF of stray capacitance associated with each section. Both sections are wired in parallel for the output capacitor, while they are switched in or out using switch S1B for the input capacitor. This strategy allows the minimum capacitance of the input capacitor to be smaller to match high-impedance loads at the higher frequencies.

The roller inductor is a high-quality Cardwell 229-203-1 unit, with a steatite body to enable it to dissipate heat without damage. The roller inductor is augmented with a series 0.3 µH coil made of four turns of 1/4-inch copper tubing formed on a 1-inch OD form (which is then removed). This fixed coil can dissipate more heat when low values of inductance are needed for low-impedance loads at high frequencies. Both variable capacitors and the roller inductor use ceramic-insulated shaft couplers, since all components are "hot" electrically. Each shaft goes through a grounded bushing at the front panel to make sure none of the knobs is hot.

The balun allowing operation with balanced loads is placed at the input of this antenna coupler, rather than at the output where it is commonly placed in other designs. Putting the balun at the input stresses the balun less, since it is operating into its design resistance of 50 Ω, once the network is tuned. For unbalanced (coax) operation, the common point at the bottom of the roller inductor is grounded at the feedthrough insulator at the rear of the cabinet. In the prototype antenna tuner, the balun was wound using 12 turns of #10 formvar insulated wire, wound side-by-side in bifilar fashion on a 2.4-inch OD core of type 43 material. After 60 seconds of key-down operation at 1500 W on 29.7 MHz, the wire becomes warm to the touch, although the core itself remains cool. We estimated that 25 W was being dissipated in the balun.

Alternatively, if you don't intend to use the tuner for balanced lines, you can delete the balun altogether. A piece of RG-213 coax is used to connect the output coaxial socket (in parallel with the "hot" insulated feedthrough insulator) to S1D common. This adds approximately 15 pF fixed capacity to ground. An equal length of RG-213 is used at the "cold" feed-through insulator so that the circuit remains balanced to ground when used with balanced transmission lines. When the cold terminal is jumpered to ground for unbalanced loads (that is, using the coax connector), the extra length of RG-213 is shorted out and is thus out of the circuit.

CONSTRUCTION

The prototype antenna tuner was mounted in a Hammond model 14151 heavy-duty, painted steel cabinet. This is an exceptionally well-constructed cabinet that does not flex or jump around on the operating table when the roller inductor shaft is rotated vigorously. The electrical components inside were spaced well away from the steel cabinet to keep losses down, especially in the variable inductor. There is also lots of clearance between components and the chassis itself to prevent arcing and stray capacity to ground. See **Figs 22.96** and **22.97** showing the layout inside the cabinet of the prototype tuner. **Fig 22.98** shows a view of the front panel. The turns-counter dial for the roller inductor was bought from Surplus Sales of Nebraska.

The 400 pF fixed capacitor is constructed using low-cost plate glass from a 5×7-inch picture frame, together with an approximately 4×6-inch flat piece of sheet aluminum that is 0.030-inch thick. The tuner's 10 1/2×8-inch subchassis forms the other plate of this homebrew capacitor. For mechanical rigidity, the subchassis uses two 1/16-inch thick aluminum plates. The 1/16-inch thick glass is epoxied to the bottom of the subchassis. The 4×6-inch aluminum sheet forming the second plate of the 400-pF fixed capacitor is in turn epoxied to the glass to make a stable, high-voltage, high-current fixed capacitor. Two strips of wood are screwed down over the

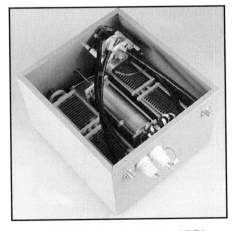

Fig 22.96—Interior view of the ARRL Antenna Tuner. The balun is mounted near the input coaxial connector. The two feedthrough insulators for balanced-line operation are located near the output coaxial unbalanced connector. The Radioswitch Corporation high-voltage switch is mounted to the front panel. Ceramic-insulated shaft couplers through ground 1/4-inch panel bushings couple the variable components to the knobs.

Fig 22.97—The subchassis, showing the four white insulators used to isolate the subchassis from the cabinet. The homemade 400-pF fixed capacitor C3 is epoxied to the bottom of the subchassis, sandwiching a piece of plate glass as the dielectric between the subchassis and a flat piece of aluminum.

Fig 22.99—Bottom view of subchassis, showing the two strips of wood ensuring mechanical stability of the C3 capacitor assembly.

assembly underneath the subchassis to make sure the capacitor stays in place. The estimated breakdown voltage is 12,000 V. See **Fig 22.99** for a bottom view of the subchassis.

Note: The dielectric constant of the glass in a cheap ($2 at Walmart) picture frame varies. The final dimensions of the aluminum sheet secured to the glass with one-hour epoxy was varied by sliding it in and out until 400 pF was reached, while the epoxy was still wet, using an Autek RF-1 as a capacitance meter. Don't let epoxy slop over the edges—this can arc and burn.

S1 is bolted directly to the front of the cabinet. S1 is a special high-voltage RF switch from Radio Switch Corporation, with four poles and three positions. It is not inexpensive, but we wanted to have no weak points in the prototype unit. A more frugal ham might want to substitute two more common surplus DPDT switches for S1. One would bypass the tuner when the operator desires to do that. The other would switch the additional 400 pF fixed capacitor across variable C3 and also parallel both sections of C1 together for the lower frequencies. Both switches would have to be capable of handling high RF voltages, of course.

OPERATION

The ARRL Antenna Tuner shown here is designed to handle the output from transmitters that operate up to 1.5 kW. An external SWR indicator is used between the transmitter and the antenna tuner to show when a matched condition is attained. Most often the SWR meter built

Fig 22.98—Front panel view of the ARRL Antenna Tuner. The high-quality turns counter dial is from Surplus Sales of Nebraska.

into the transceiver is used to tune the tuner and then the amplifier is switched on. The builder may want to integrate an SWR meter in the tuner circuit between J1 and the arm of S1A. Never "hot switch" an antenna tuner, as this can damage both transmitter and tuner. For initial setting below 10 MHz, set S1 to position 2 and C1 at midrange, C2 at full mesh. With a few watts of RF, adjust the roller inductor for a decrease in reflected power. Then adjust C1 and L2 alternately for the lowest possible SWR, also adjusting C2 if necessary. If a satisfactory SWR cannot be achieved, try S1 at position 3 and repeat the steps above. Finally, increase the transmitter power to maximum and touch up the tuner's controls if necessary. When tuning, keep your transmissions brief and identify your station.

For operation above 10 MHz, again initially use S1 set to position 2, and if SWR

cannot be lowered properly, try S1 set to position 3. This will probably be necessary for 24 or 28-MHz operation. In general, you want to set C2 for as much capacitance as possible, especially on the lower frequencies. This will result in the least amount of loss through the antenna tuner. The first position of S1 permits switched-through operation direct to the antenna when the antenna tuner is not needed.

FURTHER COMMENTS ABOUT THE ARRL ANTENNA TUNER

Surplus coils and capacitors are suitable for use in this circuit. L2 should have at least 25 μH of inductance and be constructed with a steatite body. There are roller inductors on the market made with Delrin plastic bodies but these are very prone to melting under stress and should be avoided. The tuning capacitors need to have 200 pF or more of capacitance per section at a breakdown voltage of at least 3000 V. You could save some money by using a single-section variable capacitor for the output capacitor, rather than the dual-section unit we used. It should have a maximum capacitance of 400 pF and a voltage rating of 3000 V.

Measured insertion loss for this antenna tuner is low. The worst-case load tested was four 50-Ω dummy loads in parallel to make a 12.5-Ω load at 1.8 MHz. Running 1500 W keydown for 30 seconds heated the variable inductor enough so that you wouldn't want to keep your hand on it for long. None of the other components became hot in this test.

At higher frequencies (and into a 50-Ω

load at 1.8 MHz), the coil was only warm to the touch at 1500 W keydown for 30 seconds. The #10 balun wire, as mentioned previously, was the warmest component in the antenna tuner for frequencies above 14 MHz, although it was far from catastrophic.

REFERENCES

1. Frank Witt, "How to Evaluate Your Antenna Tuner—*Parts 1 and 2*," *QST*, April and May 1995, pp 30-34 and pp 33-37 respectively.

2. Frank Witt, "Baluns in the Real (and Complex) World," *The ARRL Antenna Compendium, Vol 5* (Newington, ARRL: 1996), pp 171-181.

3. "*QST* Compares: Four High-Power Antenna Tuners," *QST*, Mar 1997, pp 73-77.

4. *The ARRL Antenna Book*, 18th Edition (Newington, ARRL: 1997). See Chapters 25 and 26.

USING PIC MICROCONTROLLERS IN AMATEUR RADIO PROJECTS

This article by John A. Hansen, W2FS first appeared in October 1998 *QST*. It is an introduction to using one of the simplest and cheapest microprocessors, Microchip Technology's Peripheral Interface Controllers—PICs—in Amateur Radio applications. It provides you with the necessary background to begin using PIC chips and give you some pointers on where to learn more about them. Most importantly, it will describe how to build a circuit that allows *you* to program PIC chips yourself—at a cost of less than $5! The next time you want to build a project that includes a PIC chip, if the source code is available, you'll be able to *program your own chip* instead of paying someone to do it for you!

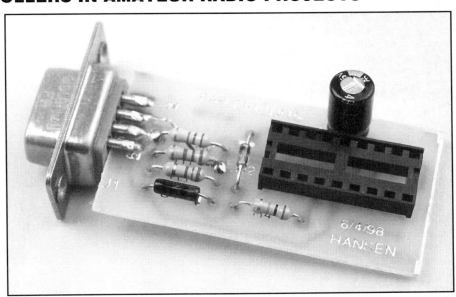

WHAT EMBEDDED MICROCONTROLLERS DO

An embedded microcontroller is a tiny computer that receives data, makes calculations or decisions based on that data and then acts in response. It interacts with the rest of the world through pins that can be configured as inputs *and* outputs. Configuring a pin as an *input* means that the microcontroller can *read* the pin to determine whether the voltage on it is high or low. When a pin is *high*, it means that 5 V is applied to it. When the pin is *low*, it means the pin is at ground potential. When a pin is configured as an *output*, it means that the microcontroller itself can make the pin high (+5 V) or low (0 V).

The microcontroller acts on a set of instructions (a *program*) that determine how the microcontroller converts these input signals into output signals. The range of functions that these microcontrollers can perform is incredible! That's especially true when you consider that the chip's behavior is limited solely to finding out whether input pins are high or low, then setting output pins either high or low in response!

Microcontroller inputs might be pushbuttons (or arrays of buttons in a keypad), sensors of various types such as temperature, pressure, or acceleration (fed through an analog-to-digital converter — ADC), or a serial or parallel data stream from any device capable of generating serial or parallel data (DTMF decoders, radio computer ports, PCs, etc). As long as the information can be presented to the microcontroller as a high or low signal on one or more input pins, the controller can recognize the information and perform predefined functions in response.

By using transistor switches or relays, the microcontroller outputs can switch external devices on and off, generate sounds, or send serial or parallel data to control other devices. Text or data from the microcontroller can be displayed on an LCD panel, or sent to a speech synthesis chip to be read aloud. Microcontroller projects in *QST* include a repeater controller[1], a CW Ider[2] and a remote base controller.[3] Each of these projects would have been *vastly* more complex, less capable and much more expensive if embedded microcontrollers had not been available. And think of it: We have only begun to scratch the surface of things that can be done with these ICs!

THE MICROCHIP PIC

Microchip Technology's series of PIC microcontrollers are among the most widely used by experimenters.[4] They are quite cheap and relatively easy to use. Microchip makes an assortment of these processors that hold differing program sizes and amounts of data. Some of these ICs include such features as on-board ADCs, serial ports, large numbers of input or output pins and multiple timers. To program these chips, you need a special programmer. To erase *most* of them, you need an ultraviolet eraser. In addition to the cost of the eraser (about $50), erasing one of these PICs takes about four minutes. So, if you are debugging a program by running it, checking how it functions and then making changes, getting to the finished project can be painfully slow.

Fortunately, Microchip has developed one series of chips that is *electrically* programmable and erasable. These chips can be programmed and erased over and over again—electrically. Not only does this make the process of programming and re-programming easier, it also saves the time and expense of a UV eraser.

One common version of these electrically programmable and erasable products was Microchip's 16C84. It has been largely superseded by their 16F84. In the 'F84, Microchip replaced the 16C84's EEPROM with flash memory. Operationally, both ICs are quite similar.[5] The 16F84 is nearly pin and code compatible with the 'C84, and has more room for data. In single quantities, the 16F84 costs about $6. If you buy 25 of them at a time, you can cut that cost by a third. Because of its ease of use, the 16F84 is an ideal chip to use when learning about PIC microcontrollers.

A ROAD MAP TO THE 16F84

Fig 22.100 shows a very simple circuit for experimenting with the 16F84 PIC: an LED and a piezo speaker. The PIC's pins labeled RB0-RB7 and RA0-RA4 are general input/output pins that the IC's program can specify as receiving or sending signals. In this simple circuit, we use only two outputs: RB0 and RB1. Through RB0, we can make the PIC light an LED. RB1 has been hooked to a small piezo speaker to allow experimentation with sound. The piezo speaker is an excellent choice for this application because it can be hooked directly to the output pin of the PIC to provide a fairly loud sound.

The rest of the parts in this circuit are needed in every PIC circuit to make it work. Positive 5 V is always applied to the V_{DD} pin and ground to the V_{SS} pin. Fig 22.100 shows how to use a 7805 voltage regulator to convert a 9 to 16 V supply to 5 V. An easy alternative is to buy a battery holder that holds four AA cells and use them to power the circuit. Although four AA cells deliver 6 V rather than 5 V, the PIC will handle that voltage level without difficulty.

The PIC needs a clock source connected to pins **OSC1** and **OSC2**. We have a number of clock options. The simplest is an RC timing circuit (see insert A of Fig 22.100), but its clock rate is slow and it does not provide a high-stability timing source. If you simply want to light an LED, the RC circuit is fine, but it is inappropriate for tasks that require more accurate timing (such as serial data communication). You can also use an external oscillator, but this is a more expensive solution than is really necessary. A crystal can be connected between the **OSC1** and **OSC2** (see insert B of Fig 22.100). Microchip recommends including small-value ca-

pacitors on both crystal leads, but most users have found this to be unnecessary. TV colorburst crystals (3.5795 MHz) are readily available, inexpensive and work well in this application.

The cheapest way to provide a stable clock, however, is to use a ceramic resonator (shown connected to U2 in Fig 22.100). These resonators are extremely small parts (smaller than crystals) and generally cost less than $1 each. (Make sure you get resonators with integrated capacitors.) To use a resonator, simply hook its outside pins to the PIC's **OSC1** and **OSC2** pins and the resonator's middle pin to ground.

Finally, you need to apply 5 V dc to the PIC's **MCLR** pin. You could connect 5 V dc directly to this pin, but it's better to apply it through a 10 kΩ current-limiting resistor; then you can restart the PIC's program by simply shorting U2 pins 4 and

5. If you try this without the resistor, you'll create a short circuit across your power supply. You will find that 10 kΩ resistors are extremely common in PIC circuits. They are used to limit current. When you apply 5 V to a 10 kΩ resistor, there is a current of 0.5 mA (E/R = 5/10000 = 0.0005). This is sufficient current for the chip to detect the signal, offers minimal power supply drain and will not damage the PIC. If you are going to do extensive work with PICs, buy 10 kΩ resistors in bulk. Because the PIC is a CMOS device, it is a good idea to tie each *unused* input pin to +5 V through a 10 kΩ resistor in your final design. For experimentation purposes, this isn't necessary.

I'll get to the programming of the PIC later on. (First, we need a programmer for the PIC.) When it comes time to build the circuit of Fig 22.100, I suggest you do so using a solderless breadboard

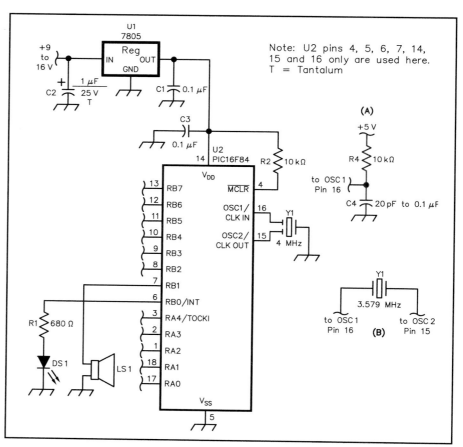

Fig 22.100 — Schematic of the PIC sound-experimentation circuit. Unless otherwise specified, resistors are ¼ W, 5% tolerance carbon-composition or film units; equivalent parts can be substituted. RS part numbers in parentheses are RadioShack; DK numbers are Digi-Key.

C2—1 µF, 25 V tantalum (DK P2059;
  (RS 272-1434)
LS1 — Piezo speaker element
  (RS 273-091)
U1 — 7805 12 V, 1 A positive regulator
  (DK NJM7805FA; RS 276-1770)

U2 — PIC16F84 microcontroller
  (see PIC Resources sidebar)
Y1 — 4 MHz resonator (DK PX400) or
  3.579 MHz crystal (DK CTX049, HC-49
  holder)

(RadioShack and other suppliers have them.) This will provide you with a test bed for experiments and trial designs for your circuits. You will be able to build a circuit with any combination of inputs and outputs you would like. Once you have finalized your design, you can transfer it to a protoboard or a PC board.

PROGRAMMING HARDWARE

Before you can use a PIC, you must program it for the task you want it to accomplish. In some instances, the program might be written for you. Authors of many *QST* projects that use PICs (and other micros) make the source code available. So, even if you have no interest in writing your own PIC programs, it may be useful to have a PIC programmer to burn chips using code written by others. There are a number of commercially made PIC programmers available, some of which cost more than $150.

For those who like to roll their own, **Fig 22.101** shows an incredibly simple PIC 16C84/16F84 programming circuit. It is based on a design by Ludwig Catta, with

a few changes made to ensure that all parts can be purchased at RadioShack. The device is powered directly from your computer's serial port, so no other power supply is required. If all the parts are purchased new, this programmer still costs less than $5 to build. You will need a socket to hold the PIC while it is being programmed. You can use a standard 18 pin DIP socket if you're simply burning a single chip based on someone else's program code. If you're planning to use the programmer to do your own development work (requiring frequent insertion and extraction of the PIC), however, buy a zero-insertion-force (ZIF) socket to minimize the wear and tear on the chips.

SOFTWARE NEEDED TO USE THE PROGRAMMER

The code (program) that is eventually loaded into the PIC consists of a series of hexadecimal numbers collectively called *machine language*. Generally speaking, it's extremely difficult to actually write programs in machine language, so virtually all designers use an intermediate step:

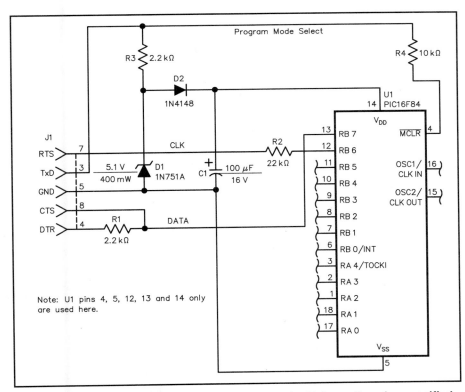

Fig 22.101 — The simple PIC programmer schematic. Unless otherwise specified, resistors are 1/4 W, 5% tolerance carbon-composition or film units. Equivalent parts can be substituted. RS part numbers in parentheses are RadioShack; DK numbers are Digi-Key.

C1 — 100 μF, 16 V electrolytic (DK P1119)
D1 — 5.1 V, 400 mW Zener diode (DK 1N5231BDICT; RS 276-565)
D2 — 1N4148 or 1N914 (DK 1N4148DICT; RS 276-1122)
J1 — D89
U1 — PIC16F84 microcontroller (see PIC Resources sidebar)

PIC Resources
Books

Benson, David, *Easy PIC'n: A Beginner's Guide to using PIC 16/17 Microcontrollers* (Kelseyville, California: Square 1 Electronics, 1996). It's a good introduction to PIC chips and assembly programming.

Benson, David, *PIC'n Up the Pace: PIC 16/17 Microcontroller Applications Guide* (Kelseyville, California: Square 1 Electronics, 1997). Essential reading if you are serious about learning assembler. If you plan to use a *C* or *BASIC* compiler, you need not have this book.

Predko, Myke, *Programming and Customizing the PIC Microcontroller* (New York, McGraw-Hill, 1998). Somewhat more advanced, this book contains a good discussion of the PIC architecture and lots of projects.

Peatman, John B., *Design With PIC Microcontrollers* (New York: Prentice Hall, 1997). This is a well written college-level text on PIC microcontrollers.

Compilers

PCM, a *C* compiler available from Custom Computer Services, Inc. (See the Address List in the **References** chapter for contact information.)

Hi-Tech's PIC *C* compiler. Although this compiler is too expensive for most of us ($850), a working demo is available for free that will compile small projects. (See the Address List in the **References** chapter for contact information.)

PicBasic Compiler, a *BASIC* compiler available from microEngineering Labs. (See the Address List in the **References** chapter for contact information.)

PIC Chips and Other Parts

Digi-Key Corporation. (See the Address List in the **References** chapter for contact information.) Digi-Key also stocks ceramic resonators and ZIF sockets.

ITU Technologies, see information above.

JDR Microdevices. (See the Address List in the **References** chapter for contact information.)

[Author's note: This is not an exhaustive list of resources. There is a wide range of resources available for PICs including complete development environments that sell for over $2000. I have focused here only on those resources that are within the budget of a typical amateur.—*John Hansen, W2FS*]

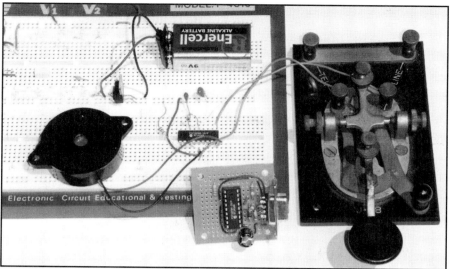

Fig 22.104 — CPO on a breadboard—keyed by one of the famous J-38s. A perfboard version of the Ludwig Catta (Ludi) PIC programmer is in the foreground. *Photo by John Martinson, WB2WXN*

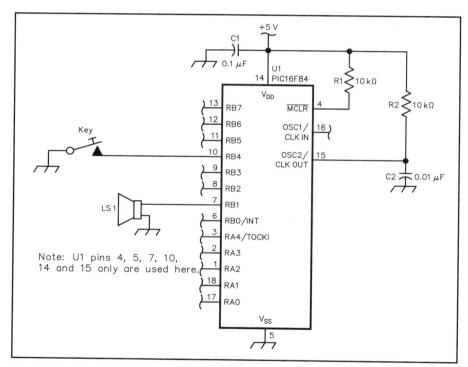

Fig 22.103 — Code-practice oscillator schematic using a PIC microcontroller. Unless otherwise specified, resistors are ¼ W, 5% tolerance carbon-composition or film units. Equivalent parts can be substituted.

LS1 — Piezo speaker element (RS 273-091)

U1 — PIC16F84 microcontroller (see PIC Resources sidebar)

They write the software in *assembler* or a *high-level language* (such as *BASIC* or *C*). They then use an *assembler* or *compiler* to convert the code to the machine language that the PIC can understand. You can usually determine which type of code is in a file by the file name's extension:

- HEX—machine language files that can be loaded directly into the PIC
- ASM—files written in assembler. They must be converted into HEX before they can be loaded into the PIC.

- C, H—files written in *C*. They must be compiled before they can be used. Depending on the compiler used, they might be converted into .ASM files, or directly to HEX.
- BAS—files written in *BASIC*. They must be compiled before they can be used.

For *BASIC* and *C* files, you cannot use just any compiler to convert them into HEX files. You must use a compiler *specifically designed* to generate instructions that can be read by the PIC.

The most popular tool used to convert assembly language files to HEX files is available from Microchip itself. Called *MPASM*, it is easy to use—best of all, it is *free*! You can obtain *MPASM* from Microchip's Web site.[6] You can write the assembly language instructions using any editor that handles plain ASCII text (such as *Windows Notepad* or the DOS *Edit* program), then use *MPASM* to compile the code into a HEX file.

You also need software to load the HEX file into the PIC. The best program I've found for doing this is called *PIX*.[7] It supposedly is a DOS program, but the author had trouble getting it to run when his computer was booted in MS-DOS! *PIX* runs fine when he runs it in a DOS box within *Windows*, however (that's a switch!). To get *PIX* to work with the programmer of Fig 22.101, first edit the *PIX.CFG* file that comes with the program. Find the two lines that say:

Port=LPT1

and

Programmer = Shaer

Place a semicolon at the start of each line. Then find the line that says:

;Programmer=Ludi

and remove the semicolon. Next, add a line that says:

Port=COM*x*

where *x* is the number of the serial port to which your programmer is connected.

It is important to use a *short* serial cable to connect your programmer to your computer. In building the programmer in Fig 22.101, I simply glued a female DB-9 connector onto the programmer itself and then plugged the programmer directly into the serial port on my computer. In any case, do not use a cable longer than about a foot. See **Fig 22.102**. FAR Circuits makes PC boards for the programmer and the other projects in this article. See the Address List in the **References** chapter.

When you run the program with the simple PIC programmer hooked up to your serial port, you will probably see a dialog box that says, **MODEM DETECTED** or

Table 22.10

An Assembly-Language Program to Create a Code Practice Oscillator (CPO.ASM)

```
        list      p=16F84
        __config  0x3FF3
portb   equ       0x06
        org       0x000
        movlw     0x00
        option
        movlw     0xFD
        tris      portb
start   btfsc     portb,4
        goto      start
        bsf       portb,1
        bcf       portb,1
        goto      start
        end
```

No/Bad Hardware. **Not True Continue**. Highlight the answer **YES** and press **ENTER**. The program will then start. This program allows you to read and write code to PICs and erase PICs. It will even disassemble the HEX code to show you the assembly language instructions. Pressing function key **F3** allows you to load your HEX file into the *PIX* program. When you press the **F9** key, the program loads into the PIC.

In summary, you use an ASCII text editor to write your assembly language instructions and save the file with an ASM extension. Then, use *MPASM* to assemble the ASM file into a HEX file. Finally, use *PIX* to load the HEX file into your PIC. To test your program, simply plug the programmed PIC into the completed target circuit and apply power. It should automatically begin executing its program.

BUILDING YOUR FIRST PROGRAM

Learning to write assembly-language programs for the PIC is no trivial matter. This section is not a tutorial in assembler, but will give you a sense of what it is like.[8]

The circuit of **Fig 22.103** can be used to construct a simple code-practice oscillator (CPO). Sure, there are simpler ways to build a CPO than using a computer-on-a-chip, but this application does provide a good introduction to programming PICs. Furthermore, you can build this project using only two resistors, two capacitors, the 16F84 PIC and a piezo speaker, making this just about as inexpensive a way to build a CPO as the more traditional methods. The CPO circuit uses an RC timing circuit for the clock. The key is connected between pin RB4 and ground. The speed at which the processor runs can be altered by changing the value of C2. Any capacitance value between 220 pF to 0.01 μF will work, with lower capacitance values re-

sulting in a higher-pitched oscillator.

Using your text editor, enter the code given in **Table 22.10** and name the saved file CPO.ASM. Enter the code in three columns as shown. You don't need to have the exact spacing shown between columns, but you do need to have your code in three columns. Note the two underscores preceding the word "config." In Table 22.10, they appear as one long underline.

The first line of code tells the assembler which PIC is being used. In the second line, the internal configuration of the clock and the timers is set up. Here we specify that we are using an RC circuit for the clock. If we wanted to use a colorburst crystal or a ceramic resonator, we would have specified 0x3FF1.

The third line makes the code easier to read. The internal location for the Port B I/O lines (RB0 through RB7) is 0x06. This line of code specifies that we will call this location "portb" instead. The "org" instruction in the next line says where the program should start; in this case at the first instruction entered in the chip (at location 0).

The next two lines enable internal pull-up resistors on all of the Port B pins that are used as inputs. This means that 5 V is applied to each of these pins through a current-limiting resistor. This happens inside the chip and requires no external components. As a result, each of the input lines sees a 5 V signal, unless you short the pin to ground. This is the mechanism used for keying the CPO. When you key the CPO, it momentarily connects pin RB4 to ground. The program detects that this pin is low, and generates a tone as long as the pin is grounded.

The next two lines determine which of the PIC's pins will be inputs, and which will be outputs.[9] The PIC assumes the pins are inputs, unless it is specifically told to make them outputs. In this case, the value FD (11111101 in binary) changes RB1 to an output. The next line is labeled "start" so that the program can loop back up to this line when it needs to. Once it gets to this point, all the program does is repeatedly execute the remaining lines of the program (except the end statement). This forms what is called an *infinite loop* because the program just continues doing this until you shut off the power to the microprocessor. In most programming environments infinite loops are avoided at all costs — they are often the things that cause computers to "hang" when they occur unintentionally. With PICs, however, infinite loops are very common. They are used whenever you want the PIC to continue running the same program over and

Fig 22.102 — Here's the homemade PIC programmer. The DB9 connector is glued to the perf board for direct connection to the PC Com port. *Photo by John Martinson, WB2WXN*.

over again until the power is shut off.

The line labeled "start" says that if pin four on Port B (RB4) is low (grounded), skip the next line of code. That next line just sends the program back to "start." Thus, if pin RB4 is high (+ 5 V), the program continues to alternate between these two lines of code until pin RB4 is grounded. When pin RB4 is grounded, the instruction to go back to "start" is skipped. The next two lines of code take pin RB1 high (bsf) and low (bcf) again. Then the program goes back up to "start." The effect of this is that if pin RB4 is grounded, the program causes pin RB1 to alternate between 5 V and ground at the same rate that the microprocessor clock is running. This produces a rectangular wave (not quite square) at about 700 Hz. Because we are using an RC circuit to clock the chip, it runs at a relatively low frequency. If we had used a 4 MHz crystal instead (which runs the microprocessor clock at 1 MHz), it would have been necessary to insert additional delay instructions to slow down the rate at which pin RB1 alternates between high and low. A CPO that runs at 1 MHz is not very useful! You won't hear it!

That's all there is to it. After you have saved the CPO.ASM file, use *MPASM* to compile it into a HEX file using the command: **MPASM CPO**

If any errors are reported during the compiling process, it means you have mistyped something. By viewing the *CPO.ERR* file, you can find out which lines contain the errors. After the program assembles without error, use *PIX* to load the HEX file into your 16F84 PIC.

When you get this project running, try changing the code to lower the tone. Or, try having the PIC light an LED (as in Fig

22.100) *and* generate a tone. Each time you change the program, you need to rerun *MPASM* to reassemble the code and reload it into the chip with *PIX*. By experimenting with variations of this basic circuit, you can better understand how assembly-language instructions work.

ISN'T THERE AN EASIER WAY?

Of course, there is an easier way to do this. You can use a high-level language such as *C* or *BASIC* to write your program. For that, however, you need a compiler that converts your *BASIC* or *C* code into assembler or machine language. The cheapest of these compilers costs about $100. There are indications that some share-ware compilers are beginning to

---

Table 22.11

An Assembly-Language Program for a PIC Morse Code Generator

```
        list      p=16f84                                          enddit  MOVLW   3C          ;add a small delay
        radix     hex                                                      MOVWF   10
        __config  0x3FF1                                                   CALL    time
dahlen  equ       d'99'     ; <==controls code speed..increase            RETLW   00
ditlen  equ       d'33'     ;<==(though not over 255) for
        org       0x000     ;slower code, decrease for       lspace                 0xB4      ;subroutine to make a
        MOVLW     00        ;faster code. Make sure the                            ;letter space
        MOVWF     0A.       ;top number is three times ;the bottom     MOVLW   10
        GOTO      start                                             CALL    time
        NOP                                                         RETLW   00
        NOP
time    MOVF      10,W                                      start   CLRF    04          ;MAIN PROGRAM
        BTFSC     03,2                                                                  ;STARTS HERE
        GOTO      endtime                                           MOVLW   0xFD
uptop   MOVLW     01                                                TRIS    6
        MOVWF     0D
upagn   CLRF      0C                                        top     CALL    dit         ;modify this code to
doagn   DECFSZ    0C,F                                                                  ;xmit the
        GOTO      doagn                                             CALL    dah         ;CW you want to send
        DECFSZ    0D,F                                              CALL    dah
        GOTO      upagn                                             CALL    lspace      ;lspace is a pause for
        MOVLW     4A                                                                    ;the space between
        MOVWF     0C                                                CALL    dit         ;letters. Include it after
upone   DECFSZ    0C,F                                                                  ;each letter.
        GOTO      upone                                             CALL    dit         ;As it appears here, the
        DECFSZ    10,F                                                                  ;IDer will
        GOTO      uptop                                             CALL    dah         ;transmit W2FS
endtime RETLW     00                                                CALL    dah
                                                                    CALL    dah
dah     MOVLW     01        ;subroutine to do a dah                 CALL    lspace
        MOVWF     0E                                                CALL    dit
agn3dah MOVLW     dahlen                                            CALL    dit
        SUBWF     0E,W                                              CALL    dah
        BTFSC     03,0                                              CALL    dit
        GOTO      enddah                                            CALL    lspace
        BSF       06,1      ;turn on pin B1                         CALL    dit
        MOVLW     01                                                CALL    dit
        MOVWF     10                                                CALL    dit
        CALL      time      ;wait 1 millisecond
        BCF       06,1      ;turn off pin B1                        MOVLW   01
        MOVLW     01                                                MOVWF   0E
        MOVWF     10                                                                    ;this code programs the
        CALL      time      ;wait 1 millisecond                                         ;delay between IDs
        MOVF      0E,W                                      loop    MOVLW   d'150'      ;<= 4 x this number =
        INCF      0E,F                                                                  ;number of seconds
        GOTO      agn3dah   ;loop up to do it again.                 SUBWF   0E,W        ;between IDs when
enddah  MOVLW     3C        ;add a small delay                                          ;using a 4 MHz resonator.
        MOVWF     10                                                BTFSC   03,0
        CALL      time                                              GOTO    bottom
        RETLW     00                                                MOVLW   10
                                                                    MOVWF   0F
dit     MOVLW     01        ;subroutine to do a dit         again   MOVLW   0xFA
        MOVWF     0E                                                MOVWF   10
agn3dit MOVLW     ditlen                                            CALL    time
        SUBWF     0E,W                                              DECFSZ  0F,F
        BTFSC     03,0                                              GOTO    again
        GOTO      enddit                                            MOVF    0E,W
        BSF       06,1      ;turn on pin B1                         INCF    0E,F
        MOVLW     01                                                GOTO    loop
        MOVWF     10                                        bottom  GOTO    top
        CALL      time      ;wait 1 millisecond                     END
        BCF       06,1      ;turn off pin B1
        MOVLW     01
        MOVWF     10
        CALL      time      ;wait 1 millisecond
        MOVF      0E,W
        INCF      0E,F
        GOTO      agn3dit   ;loop up and do it again.
```

come into the market, however. I've had very good luck with the Custom Computer Services, Inc compiler called *PCM*. It costs just under $100, and has built-in routines to make serial communication particularly easy. If you are going to do a lot of work with PICs, spending the money on a compiler may be worth considering

CREATING A PIC-BASED IDER

Table 22.11 contains the assembly-language code that makes a 16F84 PIC generate Morse code. The code is designed to work with the circuit of Fig 22.100, using a ceramic resonator or a 3.5795-MHz colorburst crystal. Audio output is obtained at pin RB1. The program is designed to be easy to modify to insert your choice of call sign, Morse code speed and transmit interval. To change the Morse code speed, alter the lines labeled "dahlen" and "ditlen." To change the ID interval, change the number between the apostrophes in the line labeled "loop." Change the lines that start with the label "top" to change the call sign (or other text) that is to be transmitted by the chip.

Follow the pattern shown in the program and make sure you put an "lspace" after the end of each letter. If you plan to transmit text other than just a call sign, you can do this by adding extra "lspace" instructions to create longer delays between the words. When altering this code, be sure to keep the label "top" on the same line as the first code element to be sent.

If you type in the program yourself, you may leave out the comments on each line by dropping the semicolon and the text that follows it. Blank lines can either be left in or deleted as you prefer. Alternatively, you can download this file from the ARRL's FTP site and edit it. The program may look long, but it only uses about 10% of the program capacity of a 16F84.

Notes

[1] Jeff Otterson, N1KDO, Peter Gailunas, KA1OKQ, Richard Cox, N1LTL, "Build a $60 Talking Repeater Controller," *QST*, Feb 1997, pp 37-40.
[2] Bob Anding, AA5OY, "A PIC of an IDer,"

QST, Jan 1998, pp 36-38.
[3] John Hansen, W2FS, "An Inexpensive, Remote-Base Station Controller Using the Basic Stamp," *QST*, May 1998, pp 33-37.
[4] Microchip Technology (see the Address List in the **References** chapter).
[5] Because these chips are so similar, any project you find that uses a 16C84 (such as the Talking Repeater Controller; see Note 1) can be built with a 16F84.
[6] Microchip Technology (see the Address List in the **References** chapter). At their Web site, you will also find a free program called *MPLAB*. It is a complete development environment including an editor, a simulator and an assembler.
[7] Available at **http://home5.swipnet.se/~w-53783/**.
[8] For a good beginner's tutorial in assembler for the PIC, see David Benson, *Easy PIC'n: A Beginner's Guide to using PIC 16/17 Microcontrollers*, (Kelseyville, California: Square 1 Electronics, 1996).
[9] Microchip considers the *option* and *tris* instructions outdated. However, they will work in most applications. Because the alternative methods for configuring the I/O pins and pull-up resistors are somewhat more complicated, in the interest of simplicity, the first approach is selected for this project.

Contents

Repeaters, Satellites, EME and Direction Finding

23

Repeaters

This section was written by Paul M. Danzer, N1II.

In the late 1960s two events occurred that changed the way radio amateurs communicated. The first was the explosive advance in solid state components — transistors and integrated circuits. A number of new "designed for communications" integrated circuits became available, as well as improved high-power transistors for RF power amplifiers. Vacuum tube-based equipment, expensive to maintain and subject to vibration damage, was becoming obsolete.

At about the same time, in one of its periodic reviews of spectrum usage, the Federal Communications Commission (FCC) mandated that commercial users of the VHF spectrum reduce the deviation of truck, taxi, police, fire and all other commercial services from 15 kHz to 5 kHz. This meant that thousands of new narrowband FM radios were put into service and an equal number of wideband radios were no longer needed.

As the new radios arrived at the front door of the commercial users, the old radios that weren't modified went out the back door, and hams lined up to take advantage of the newly available "commercial surplus." Not since the end of World War II had so many radios been made available to the ham community at very low or at least acceptable prices. With a little tweaking, the transmitters and re-

ceivers were modified for ham use, and the great repeater boom was on.

WHAT IS A REPEATER?

Trucking companies and police departments learned long ago that they could get much better use from their mobile radios by using an automated relay station called a repeater. Not all radio dispatchers are located near the highest point in town or have access to a 300-ft tower. But a re-

peater, whose basic idea is shown in **Fig 23.1**, can be more readily located where the antenna system is as high as possible and can therefore cover a much greater area.

Types of Repeaters

The most popular and well-known type of amateur repeater is an FM voice system on the 29, 144, 222 or 440-MHz bands. Tens of thousands of hams use small 12-V

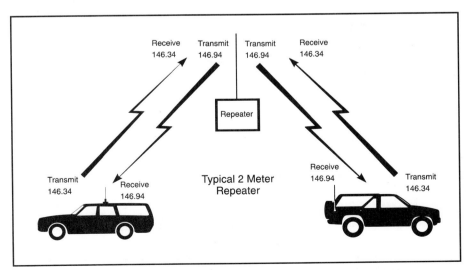

Fig 23.1 — Typical 2-m repeater, showing mobile-to-mobile communication through a repeater station. Usually located on a hill or tall building, the repeater amplifies and retransmits the received signal on a different frequency.

Table 23.1

Types of Repeaters

ATV — Amateur TV	Same coverage advantages as voice repeaters to hams using wideband TV in the VHF and UHF bands. Often consist of pairs of repeaters — one for the ATV and the other for the voice coordination.
AM and SSB	There is no reason to limit repeaters to FM. There are a number of other modulation-type repeaters, some experimental and some long-established.
Digipeaters	Digital repeaters used primarily for packet communications (see the **Modes** chapter). Can use a single channel (single port) or several channels (multi-port) on one or more VHF and UHF bands.
Multi-channel (wideband)	Amateur satellites are best-known examples. Wide bandwidth (perhaps 50 to 200 kHz) is selected to be received and transmitted so all signals in bandwidth are heard by the satellite (repeater) and retransmitted, usually on a different VHF or UHF band. Satellites are discussed elsewhere in this chapter.
	Although not permitted or practical for terrestrial use in the VHF or UHF spectrum, there is no reason wideband repeaters cannot be established in the microwave region where wide bandwidths are allowed. This would be known as frequency multiplexing.

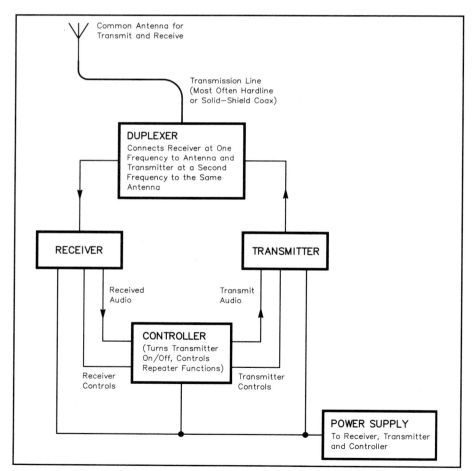

Fig 23.2 — The basic components of a repeater station. In the early days of repeaters, many were home-built. These days, most are commercial, and are far more complex than this diagram suggests.

powered radios in their vehicles for both casual ragchewing and staying in touch with what is going on during heavy traffic or commuting times. Others have low-power battery-operated hand-held units, known as "handi-talkies" or "HTs" for 144, 222 or 440 MHz. Some mobile and hand-held transceivers operate on two bands. But there are several other types of ham radio repeaters. **Table 23.1** describes them.

FM is the mode of choice, as it was in commercial service, since it provides a high degree of immunity to mobile noise pulses. Operations are *channelized* — all stations operate on the same transmit frequency and receive on the same receive frequency. In addition, since the repeater receives signals from mobile or fixed stations and retransmits these signals simultaneously, the transmit and receive frequencies are different, or *split*. Direct contact between two or more stations that listen and transmit on the same frequency is called operating *simplex*.

Individuals, clubs, amateur civil defense support groups and other organizations all sponsor repeaters. Anyone with a valid amateur license for the band can establish a repeater in conformance with the FCC rules. No one owns specific repeater frequencies, but nearly all repeaters are *coordinated* to minimize repeater-to-repeater interference. Frequency coordination and interference are discussed later in this chapter.

Block Diagrams

Repeaters normally contain at least the sections shown in **Fig 23.2**. After this, the sky is the limit on imagination. As an example, a remote receiver site can be used to try to eliminate interference (**Fig 23.3**).

The two sites can be linked either by telephone ("hard wire") or a VHF or UHF link. Once you have one remote receiver site it is natural to consider a second site to better hear those "weak mobiles" on the other side of town (**Fig 23.4**). Some of the stations using the repeater are on 2 m while others are on 440? Just link the two repeaters! (**Fig 23.5**).

Want to help the local Civil Air Patrol (CAP)? Add a receiver for aircraft emergency transmitters (ELT). Tornadoes? It is now legal to add a weather channel receiver (**Fig 23.6**).

The list goes on and on. Perhaps that is why so many hams have put up repeaters.

Repeater Terminology

Here are some definitions of terms used in the world of Amateur Radio FM and repeaters:

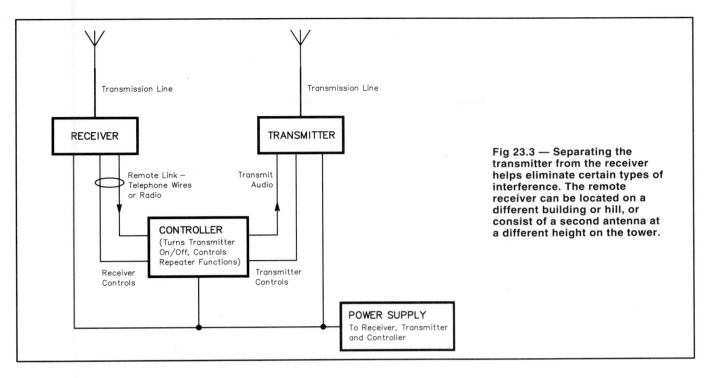

Fig 23.3 — Separating the transmitter from the receiver helps eliminate certain types of interference. The remote receiver can be located on a different building or hill, or consist of a second antenna at a different height on the tower.

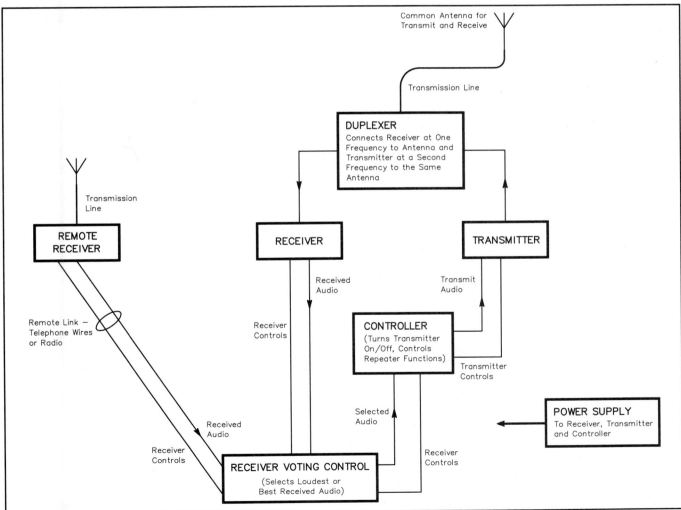

Fig 23.4 — A second remote receiver site can provide solid coverage on the other side of town.

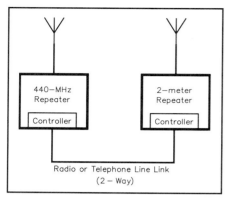

Fig 23.5 — Two repeaters using different bands can be linked for added convenience.

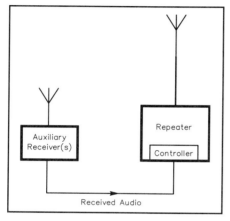

Fig 23.6 — For even greater flexibility, you can add an auxiliary receiver.

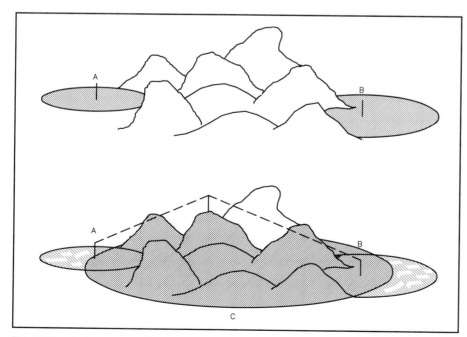

Fig 23.7 — In the upper diagram, stations A and B cannot communicate because their mutual coverage is limited by the mountains between them. In the lower diagram, stations A and B can communicate because the coverage of each station falls within the coverage of repeater C, which is on a mountaintop.

access code — one or more numbers and/or symbols that are keyed into the repeater with a DTMF tone pad to activate a repeater function, such as an autopatch.

autopatch — a device that interfaces a repeater to the telephone system to permit repeater users to make telephone calls. Often just called a "patch."

break — the word used to interrupt a conversation on a repeater *only* to indicate that there is an emergency.

carrier-operated relay (COR) — a device that causes the repeater to transmit in response to a received signal.

channel — the pair of frequencies (input and output) used by a repeater.

closed repeater — a repeater whose access is limited to a select group (see *open repeater*).

control operator — the Amateur Radio operator who is designated to "control" the operation of the repeater, as required by FCC regulations.

courtesy beep — an audible indication that a repeater user may go ahead and transmit.

coverage — the geographic area within which the repeater provides communications.

CTCSS — abbreviation for continuous tone-controlled squelch system, a series of subaudible tones that some repeaters use to restrict access. (see **closed repeater**)

digipeater — a packet radio (digital) repeater.

DTMF — abbreviation for dual-tone multifrequency, the series of tones generated from a keypad on a ham radio transceiver (or a regular telephone).

duplex or **full duplex** — a mode of communication in which a user transmits on one frequency and receives on another frequency simultaneously (see *half duplex*).

duplexer — a device that allows the repeater transmitter and receiver to use the same antenna simultaneously.

frequency coordinator — an individual or group responsible for assigning frequencies to new repeaters without causing interference to existing repeaters.

full quieting — a received signal that contains no noise.

half duplex — a mode of communication in which a user transmits at one time and receives at another time.

hand-held — a small, lightweight portable transceiver small enough to be carried easily; also called HT (for Handie-Talkie, a Motorola trademark).

hang time — the short period following a transmission that allows others who want to access the repeater a chance to do so; a *courtesy beep* sounds when the repeater is ready to accept another transmission.

input frequency — the frequency of the repeater's receiver (and your transceiver's transmitter).

intermodulation distortion (IMD) — the unwanted mixing of two strong RF signals that causes a signal to be transmitted on an unintended frequency.

key up — to turn on a repeater by transmitting on its input frequency.

machine — a repeater system.

magnetic mount or **mag-mount** — an antenna with a magnetic base that permits quick installation and removal from a motor vehicle or other metal surface.

NiCd — a nickel-cadmium battery that may be recharged many times; often used to power portable transceivers. Pronounced "NYE-cad."

open repeater — a repeater whose access is not limited.

output frequency — the frequency of the repeater's transmitter (and your transceiver's receiver).

over — a word used to indicate the end of a voice transmission.

Repeater Directory — an annual ARRL publication that lists repeaters in the US, Canada and other areas.

separation or **split** — the difference (in kHz) between a repeater's transmitter and receiver frequencies. Repeaters that use unusual separations, such as 1 MHz on 2 m, are sometimes said to have "oddball splits."

simplex — a mode of communication in which users transmit and receive on the same frequency.

time-out — to cause the repeater or a repeater function to turn off because you have transmitted for too long.

timer — a device that measures the length of each transmission and causes the repeater or a repeater function to turn off after a transmission has exceeded a certain length.

tone pad — an array of 12 or 16 numbered keys that generate the standard telephone dual-tone multifrequency (*DTMF*) dialing signals. Resembles a standard telephone keypad. (see *autopatch*)

Advantages of Using a Repeater

When we use the term *repeater* we are almost always talking about transmitters and receivers on VHF or higher bands, where radio-wave propagation is normally line of sight. Sometimes a hill or building in the path will allow refraction or other types of edge effects, reflections and bending. But for high quality, consistently solid communications, line of sight is the primary mode.

We know that the effective range of VHF and UHF signals is related to the height of each antenna. Since repeaters can usually be located at high points, one great advantage of repeaters is the extension of coverage area from low-powered mobile and portable transceivers.

Fig 23.7 illustrates the effect of using a repeater in areas with hills or mountains. The same effect is found in metropolitan areas, where buildings provide the primary blocking structures.

Siting repeaters at high points can also have disadvantages. When two nearby repeaters use the same frequencies, your transceiver might be able to receive both. But since it operates FM, the *capture effect* usually ensures that the stronger signal will capture your receiver and the weaker signal will not be heard — at least as long as the stronger repeater is in use.

It is also simpler to provide a very sensitive receiver, a good antenna system, and a slightly higher power transmitter at just one location — the repeater — than at each mobile, portable or home location. A superior repeater system compensates for the low power (5 W or less), and small, inefficient antennas that many hams use to operate through them. The repeater maintains the range or coverage we want, despite our equipment deficiencies. If both the hand-held transceiver and the

Fig 23.8 — In the Rocky Mountain west, hand-held transceivers can often cover great distances, thanks to repeaters located atop high mountains. *(photo courtesy WBØKRX and NØIET)*

repeater are at high elevations, for example, communication is possible over great distances, despite the low output power and inefficient antenna of the transceiver (see **Fig 23.8**).

Repeaters also provide a convenient meeting place for hams with a common interest. It might be geographic — your town — or it might be a particular interest such as DX or passing traffic. Operation is channelized, and usually in any area you can find out which channel — or repeater — to pick to ragchew, get highway information, or whatever your need or interest is. The fact that operation is channelized also provides an increased measure of driving safety — you don't have to tune around and call CQ to make a contact, as on the HF bands. Simply call on a repeater frequency — if someone is there and they want to talk, they will answer you.

Emergency Operations

When there is a weather-related emergency or a disaster (or one is threatening), most repeaters in the affected area immediately spring to life. Emergency operation and traffic always take priority over other ham activities, and many repeaters are equipped with emergency power sources just for these occasions.

Almost all Amateur Radio emergency organizations use repeaters to take advantage of their extended range, uniformly good coverage and visibility. Most repeaters are well known — everyone active in an area with suitable equipment knows the local repeater frequencies. For those who don't, many transceivers provide the ability to scan for a busy frequency. See **Fig 23.9**.

Repeaters and the FCC

The law in the United States changes over time to adapt to new technology and changing times. Since the early 1980s, the trend has been toward deregulation, or more accurately in the case of radio amateurs, self-regulation. Hams have established band plans, calling frequencies, digital protocols and rules that promote efficient communication and interchange of information.

Originally, repeaters were licensed separately with detailed applications and control rules. Repeater users were forbidden to use their equipment in any way that could be interpreted as commercial. In some cases, even calling a friend at an office where the receptionist answered with the company name was interpreted as a problem.

The rules have changed, and now most nonprofit groups and public service events can be supported and businesses can be called — as long as the participating radio

Fig 23.9 — During disasters like the Mississippi River floods of 1993, repeaters over a wide area are used solely for emergency-related communication until the danger to life and property is past. *(photo courtesy WA9TZL)*

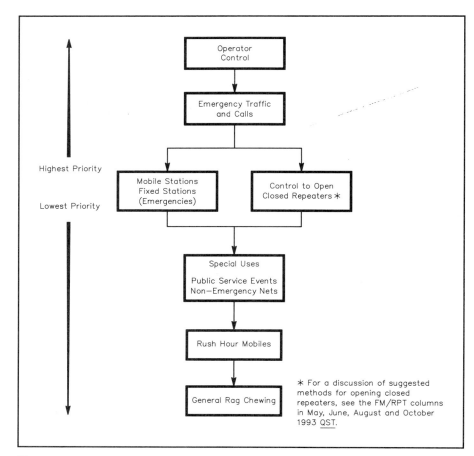

Fig 23.10 — The chart shows recommended repeater operating priorities. Note that, in general, priority goes to mobile stations.

amateurs are not earning a living from this specific activity.

We can expect this trend to continue. For the latest rules and how to interpret them, see *QST* and *The FCC Rule Book*, published by the ARRL.

FM REPEATER OPERATION AND EQUIPMENT

Operating Techniques

There are almost as many operating procedures in use on repeaters as there are repeaters. Only by listening can you determine the customary procedures on a particular machine. A number of common operating techniques are found on many repeaters, however.

One such common technique is the transmission of *courtesy tones.* Suppose several stations are talking in rotation — one following another. The repeater detects the end of a transmission of one user, waits a few seconds, and then transmits a short tone or beep. The next station in the rotation waits until the beep before transmitting, thus giving any other station wanting to join in a brief period to transmit their call sign. Thus the term *courtesy tone* — you are politely pausing to allow other stations to join in the conversation.

Another common repeater feature that encourages polite operation is the *repeater timer.* Since repeater operation is channelized — allowing many stations to use the same frequency — it is polite to keep your transmissions short. If you forget this little politeness many repeaters simply cut off your transmission after 2 or 3 minutes of continuous talking. After the repeater "times out," the timer is reset and the repeater is ready for the next transmission. The timer length is often set to 3 minutes or so during most times of the day and 1 or 1½ minutes during commuter rush hours when many mobile stations want to use the repeater.

A general rule, in fact law — both internationally and in areas regulated by the FCC — is that emergency transmissions always have priority. These are defined as relating to life, safety and property damage. Many repeaters are voluntarily set up to give mobile stations priority, at least in checking onto the repeater. If there is going to be a problem requiring help, the request will usually come from a mobile station. This is particularly true during rush hours; some repeater owners request that fixed stations refrain from using the repeater during these hours. Since fixed stations usually have the advantages of fixed antennas and higher power, they can operate simplex more easily. This frees the repeater for mobile stations that need it.

A chart of suggested operating priorities is given in **Fig 23.10**. Many but not all repeaters conform to this concept, so it can be used as a general guideline.

The figure includes a suggested priority control for *closed repeaters*. These are repeaters whose owners wish, for any number of reasons, not to have them listed as available for general use. Often they require transmission of a *subaudible* or *CTCSS* tone (discussed later). Not all repeaters requiring a CTCSS tone are closed. Other closed repeaters require the transmission of a coded telephone push-button *(DTMF)* tone sequence to turn on. It is desirable that all repeaters, including generally closed repeaters, be made available at least long enough for the presence of emergency information to be made known.

Repeaters have many uses. In some areas they are commonly used for formal traffic nets, replacing or supplementing the nets usually found on 75-m SSB. In other areas they are used with tone alerting for severe-weather nets. Even when a particular repeater is generally used for ragchewing it can be linked for a special purpose. As an example, an ARRL volunteer official may hold a periodic section meeting across her state, with linked re-

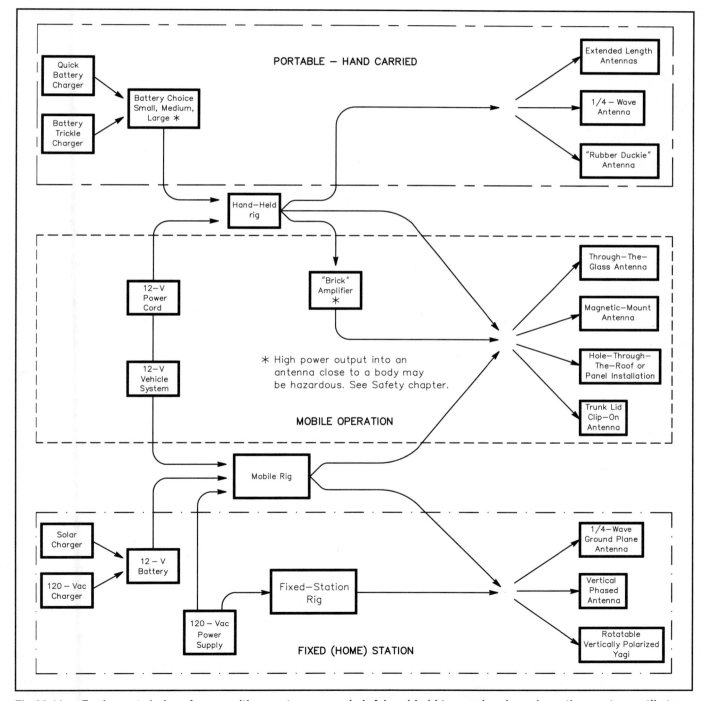

Fig 23.11 — Equipment choices for use with repeaters are varied. A hand-held transceiver is perhaps the most versatile type of radio, as it can be operated from home, from a vehicle and from a mountaintop.

peaters allowing both announcements and questions directed back to her.

One of the most common and important uses of a repeater is to aid visiting hams. Since repeaters are listed in the *ARRL Repeater Directory* and other directories, hams traveling across the country with mobile or hand-held radios often check into local repeaters asking for travel route, restaurant or lodging information. Others just come on the repeater to say hello to the local group. In most areas courtesy prevails — the visitor is given priority to say hello or get the needed help.

Detailed information on repeater operating techniques is included in a full chapter of the *ARRL Operating Manual*.

Home and Mobile Equipment

There are many options available in equipment used on repeaters—both home-built and commercial. It is common to use the same radio for both home station and mobile, or mobile and hand-held use. A number of these options are shown in **Fig 23.11**.

Hand-Held Transceivers

A basic hand-held radio with 100 mW to 5 W output can be mounted in an automobile with or without a booster amplifier or "brick."

Several types of antennas can be used in the hand-held mode. The smallest and most convenient is a rubber flex antenna, known as a "rubber duckie," a helically wound antenna encased in a flexible tube. Unfortunately, to obtain the small size the use of a wire helix or coil often produces a very low efficiency.

A quarter-wave whip, which is about 19 inches long for the 2-m band, is a good choice for enhanced performance. The rig and your hand act as a ground plane and a reasonably efficient result is obtained. A longer antenna, consisting of several electrical quarter-wave sections in series, is also commercially available. Although this antenna usually produces extended coverage, the mechanical strain of 30 or more inches of antenna mounted on the radio's antenna connector can cause problems. After several months, the strain may require replacement of the connector.

Selection of batteries will change the output power from the lowest generally available — 0.1 or 0.5 W — to the 5-W level. Charging is accomplished either with a "quick" charger in an hour or less or with a trickle charger overnight.

Power levels higher than 7 W may cause a safety problem on hand-held units, since the antenna is usually close to the operator's head and eyes. See the **Safety** chapter for more information.

For mobile operation, a 12-V power cord plugs into the auto cigarette lighter. In addition, commercially available brick amplifiers — available either assembled or as kits — can be used to raise the output power level of the hand-held radio to 10 to 70 W. These amplifiers often come with transmit-receive sensing and optional preamplifiers. One such unit is shown in block form in **Fig 23.12**.

Mobile Equipment

Mobile antennas range from quick and easy "clip-it-on" mounting to "drill through the car roof" assemblies. The four general classes of mobile antennas shown in the center section of Fig 23.11 are the most popular choices. Before experimenting with antennas for your vehicle, there are some precautions to be taken.

Through-the-glass antennas: Rather than trying to get the information from your dealer or car manufacturer, test any such antenna first using masking tape or some other temporary technique to hold the antenna in place. Some windshields are metallicized for defrosting, tinting and AM car radio reception. Having this metal in the way of your through-the-glass antenna will seriously decrease its efficiency.

Magnet-mount antennas are convenient, but only if your car has a metal roof. The metal roof serves as the ground plane.

Through-the-roof antenna mounting: Drilling a hole in your car roof may not be the best option unless you intend to keep the car for the foreseeable future. This mounting method provides the best efficiency, however, since the (metal) roof serves as a ground plane. Before you drill, carefully plan and measure how you intend to get the antenna cable down under the interior car headliner to the radio.

Trunk lid and clip-on antennas: These antennas are good compromises. They are usually easy to mount and they perform acceptably. Cable routing must be planned. If you are going to run more than a few watts, do not mount the antenna close to one of the car windows — a significant portion of the radiated power may enter the car interior.

Mobile rigs used at home can be powered either from rechargeable 12-V batteries or fixed power supplied from the 120-V ac line. Use of 12-V batteries has the advantage of providing back-up communications ability in the event of a power interruption. When a storm knocks down power lines and telephone service, it is common to hear hams using their mobile or 12-V powered rigs making autopatch calls to the power and telephone company to advise them of loss of service.

Home Station Equipment

The general choice of fixed-location antennas is also shown in Fig 23.11. A rotatable Yagi is normally not only unnecessary but undesirable for repeater use, since it has the potential of extending your transmit range into adjacent area repeaters on the same frequency pair. All antennas used to communicate through repeaters should be vertically polarized for best performance.

Both commercial and home-made $^1/_4$-λ and larger antennas are popular for home use. A number of these are shown in the **Antennas** chapter. Generally speaking,

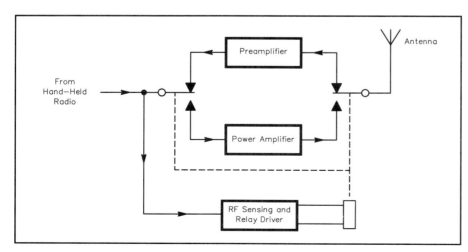

Fig 23.12 — This block diagram shows how a "brick" amplifier can be used with a receiver preamplifier. RF energy from the transceiver is detected, turns on the relay, and puts the RF power amplifier in line with the antenna. When no RF is sensed from the transceiver, the receiving preamp is in line.

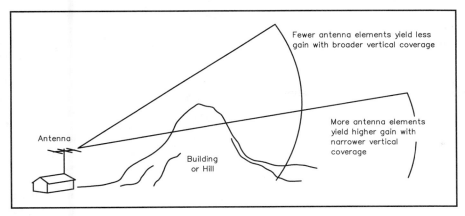

Fig 23.13 — As with all line-of-sight communications, terrain plays an important role in how your signal gets out.

Table 23.2

Standard Telephone (DTMF) Tones

Low Tone Group

| | High Tone Group | | | |
	1209 Hz	1336 Hz	1477 Hz	1633 Hz
697 Hz	1	2	3	A
770 Hz	4	5	6	B
852 Hz	7	8	9	C
941 Hz	*	0	#	D

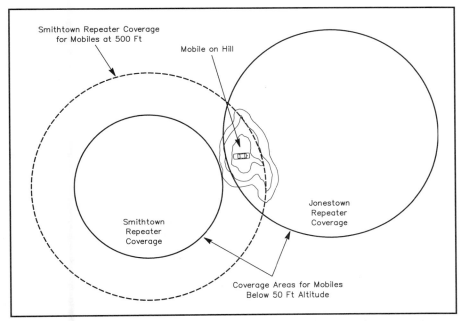

Fig 23.14 — When two repeaters operate on the same frequencies, a well-situated operator can key up both repeaters simultaneously. Frequency coordination prevents this occurrence.

¼-λ sections may be stacked up to provide more gain on any band. As you do so, however, more and more power is concentrated toward the horizon. This may be desirable if you live in a flat area. See **Fig 23.13**.

While most hams do not to try to build transceivers for use on repeaters, accessories provide a fertile area for construction and experimentation. What mobile operator has not wished that there was no need to hold the microphone continuously? The **Station Accessories** chapter includes a boom microphone unit that can be adapted to mobile use.

A single-pole, single-throw switch can be mounted in a small box and Velcro used to attach it temporarily to your seat. Flip the lever on to transmit and flip it off to listen — in the meantime your hands are free.

Autopatches and Tones

One of the most attractive features of repeaters is the availability of autopatch services. This allows the mobile or portable station to use a standard telephone key pad to connect the repeater to the local telephone line and make outgoing calls.

Table 23.2 shows the tones used for these services. Some keyboards provide the standard 12 sets of tones corresponding to the digits 0 through 9 and the special signs # and *. Others include the full set of 16 pairs, providing special keys A through D. The tones are arranged in two groups, usually called the low tones and high tones. Two tones, one from each group, are required to define a key or digit. For example, pressing 5 will generate a 770-Hz tone and a 1336-Hz tone simultaneously.

The standards used by the telephone company require the amplitudes of these two tones to have a certain relationship. Fortunately, most tone generators used for this purpose have the amplitude relationship as part of their construction. Initially, many hams used surplus telephone company keypads. These units were easily installed — usually just two or three wires were connected. Unfortunately they were constructed with wire contacts and their reliability was not great when used in a moving vehicle.

Many repeaters require pressing a code number sequence or the special figures * or # to turn the autopatch on and off. Out-of-area calls are usually locked out, as are services requiring the dialing of the prefix 0 or 1. "Speed dial" is often available, although occasionally this can conflict with the use of * or # for repeater control, since these special symbols are used by the telephone company for its own purposes.

Some repeaters require the use of *subaudible* or CTCSS tones to utilize the autopatch, while others require these tones just to access the repeater in normal use. Taken from the commercial services, subaudible tones are not generally used to keep others from using a repeater but rather are a method of minimizing interference from users of the same repeater frequency.

For example, in **Fig 23.14** a mobile station on hill A is nominally within the normal coverage area of the Jonestown repeater (146.16/76). The Smithtown repeater, also on the same frequency pair, usually cannot hear stations 150 miles away but since the mobile is on a hill he is in the coverage area of both Jonestown and Smithtown. Whenever the mobile transmits he is heard by both repeaters.

The common solution to this problem, assuming it happens often enough, is to equip the Smithtown repeater with a CTCSS decoder and require all users of the repeater to transmit a CTCSS tone to access the repeater. Thus, the mobile station on the hill does not come through the Smithtown repeater, since he is not transmitting the required CTCSS tone.

Table 23.3
CTCSS (PL) Tone Frequencies

The purpose of CTCSS (PL) is to reduce cochannel interference during band openings. CTCSS (PL) equipped repeaters would respond only to signals having the CTCSS tone required for that repeater. These repeaters would not respond to weak distant signals on their inputs and correspondingly not transmit and repeat to add to the congestion. The standard ANSI/EIA frequency codes, in hertz, with their Motorola alphanumeric designators, are as follows:

67.0—XZ	151.4—5Z
69.3—WZ	156.7—5A
71.9—XA	159.8
74.4—WA	162.2—5B
77.0—XB	165.5
79.7—WB	167.9—6Z
82.5—YZ	171.3
85.4—YA	173.8—6A
88.5—YB	177.3
91.5—ZZ	179.9—6B
94.8—ZA	183.5
97.4—ZB	186.2—7Z
100.0—1Z	189.9
103.5—1A	192.8—7A
107.2—1B	199.5
110.9—2Z	203.5—M1
114.8—2A	206.5—8Z
118.8—2B	210.7—M2
123.0—3Z	218.1—M3
127.3—3A	225.7—M4
131.8—3B	229.1—9Z
136.5—4Z	233.6—M5
141.3—4A	241.8—M6
146.2—4B	250.3—M7
	254.1—0Z

Table 23.3 shows the available CTCSS tones. They are usually transmitted by adding them to the transmitter audio but at an amplitude such that they are not readily heard by the receiving station. It is common to hear the tones described by their code designators — a carryover from their use by Motorola in their commercial communications equipment.

Listings in the *ARRL Repeater Directory* include the CTCSS tone required, if any.

Frequency Coordination and Band Plans

Since repeater operation is channelized, with many stations sharing the same frequency pairs, the amateur community has formed coordinating groups to help minimize conflicts between repeaters and among repeaters and other modes. Over the years, the VHF bands have been divided into repeater and nonrepeater subbands. These frequency-coordination groups maintain lists of available frequency pairs in their areas. A complete list of frequency coordinators, band plans and repeater pairs is included in the *ARRL Repeater Directory*.

Each VHF and UHF repeater band has been subdivided into repeater and nonrepeater channels. In addition, each band has a specific *offset* — the difference between the transmit frequency and the receive frequency for the repeater. While most repeaters use these standard offsets, others use "oddball splits." These non-

---

Table 23.4
Standard Frequency Offsets for Repeaters

Band	Offset
29 MHz	100 kHz
52 MHz	1 MHz
144 MHz	600 kHz
222 MHz	1.6 MHz
440 MHz	5 MHz
902 MHz	12 MHz
1240 MHz	12 MHz

---

standard repeaters are generally also coordinated through the local frequency coordinator. **Table 23.4** shows the standard frequency offsets for each repeater band.

The 10-m repeater band offers an additional challenge for repeater users. It is the only repeater band where ionospheric propagation is a regular factor. Coupled with the limited number of repeater frequency assignments available, the standard in this band is to use CTCSS tones on a regional basis. **Table 23.5** lists the coordinated tone assignments. As can be seen, 10-m repeaters in the 4th call area will use either the 146.2 or 100.0 (4B or 1Z) CTCSS tone.

AT THE REPEATER SITE

For details on the many elements that go into planning and installing a repeater at a particular site, request the 96 Handbook Repeater template from the ARRL Technical Secretary. (There is a nominal charge for postage and handling.)

Table 23.5
10-M CTCSS Frequencies

In 1980 the ARRL Board of Directors adopted the 10-m CTCSS (PL) tone-controlled squelch frequencies listed below for voluntary incorporation into 10-m repeater systems to provide a uniform national system.

Call Area	Tone 1		Tone 2	
W1	131.8 Hz-3B		91.5 Hz-ZZ	
W2	136.5	-4Z	94.8	-ZA
W3	141.3	-4A	97.4	-ZB
W4	146.2	-4B	100.0	-1Z
W5	151.4	-5Z	103.5	-1A
W6	156.7	-5A	107.2	-1B
W7	162.2	-5B	110.9	-2Z
W8	167.9	-6Z	114.8	-2A
W9	173.8	-6A	118.8	-2B
W0	179.9	-6B	123.0	-3Z
VE	127.3	-3A	88.5	-YB

Satellites

Unless propagation enhancements are used, radio communication distances are essentially limited by the curvature of the Earth. Propagation effects that are dependent upon the atmosphere or ionosphere can be conditionally (and sometimes unpredictably) used to transmit radio signals around the Earth's curvature, thus thwarting the straight-line radio range concept, even at VHF and UHF frequencies. Communicating beyond line-of-sight distances, however, may require the use of high power and gain antennas. These types of communications are defined as "terrestrial communications."

Because objects in space are visible from a number of locations on the Earth at the same time, it is possible to predict communications between stations within this "circle of visibility." This can be achieved by using the space object as a passive reflector for radio energy, or if the space object contains a transponding radio transmitter/receiver, it can act as a radio relay. The predictable signal path to the space object and back avoids the uncertain attenuation inherent in terrestrial propagation.

Amateur Radio space communications have two major facets: artificial satellites and our natural satellite, the Moon. Together, they make VHF and higher frequencies usable for amateur intercontinental communications and push today's technology to the limit. This section, written by Robert Diersing, N5AHD, covers communication from and through artificial spacecraft. EME or moonbounce communication is covered later in this chapter.

THE AMATEUR SATELLITES

The Amateur Radio satellite program began with the design, construction and launch of OSCAR I in 1961 under the auspices of the Project OSCAR Association in California. The acronym "OSCAR," which has been attached to almost all Amateur Radio satellite designations on a worldwide basis, stands for *Orbiting Satellite Carrying Amateur Radio*. Project OSCAR was instrumental in organizing

---

Current Amateur Satellites

OSCAR 10, the second Phase 3 satellite, was launched on June 16, 1983, aboard an ESA Ariane rocket, and was placed in an elliptical orbit. OSCAR 10 carries Mode B and Mode L transponders. Due to internal damage, it is currently uncontrollable, but still active.

OSCAR 11, a scientific/educational low-orbit satellite, was built at the University of Surrey in England and launched on March 1, 1984. This UoSAT spacecraft has also demonstrated the feasibility of store-and-forward packet digital communications and is fully operational.

OSCAR 16, also known as PACSAT, was launched in January 1990. A digital store-and-forward packet radio file server, it has an experimental S-band beacon at 2401.143 MHz.

OSCAR 19, also known as LUSAT, was sponsored by AMSAT Argentina. Launched in January 1990, it is nearly identical to OSCAR 16.

OSCAR 20, launched into low Earth orbit in February 1990, is the second amateur satellite designed and built in Japan. It carries Mode J and Mode JD (digital store-and-forward) transponders. Its digital functions are no longer operational.

OSCAR 22, another of the UoSAT series for both amateur and commercial services, was launched in July 1991. UO-22 now operates in amateur store-and-forward service as well as a 110°-wide CCD camera viewing the Earth.

OSCAR 25, known as KITSAT-B, was launched in September 1993.

OSCAR 27 is a companion module aboard the commercial EyeSat-A microsat. Launched in September 1993,

OSCAR 27 is an experimental platform designed by AMRAD. At the time of this writing, it is being used primarily as an FM voice repeater.

RS 15, launched in December 1994, is a Mode A spacecraft; its uplink is on the 2-m band, and its downlink is on 10 m.

OSCAR 29, launched from Japan in 1996, is similar to OSCAR 20 with the exception that its packet BBS has 9600-baud capability.

OSCAR 31, launched in July 1998, is the first Thai microsat. Known as TMSat, it was constructed at the University of Surrey by Thai engineers and the UoS staff. Similar in construction to KITSAT-OSCAR 25, it has a Mode JD 9600-baud FSK digital transponder.

OSCAR 34, launched from the shuttle *Discovery* in October 1998, is also known as PANSAT, for Petite Amateur Navy Satellite. It carries a spread-spectrum communication package fabricated by student officers and faculty at the Naval Postgraduate School in Monterey, California. PANSAT is used for store-and-forward digital packet communication using direct sequence spread spectrum modulation. Not yet available for Amateur Radio use.

OSCAR 35, launched in February 1999, is also called SUNSAT. Designed and built at the University of Stellenbosch in South Africa, it includes digital store-and-forward capability and a voice "parrot" repeater used primarily for educational purposes. SUNSAT also carried two NASA experiments and an experimental pushbroom imager capable of taking pictures of Earth.

the construction of the next three Amateur Radio satellites — OSCARs II, III and IV. *The Radio Amateur's Satellite Handbook*, published by ARRL has details of the early days of the amateur space program.

In 1969, the Radio Amateur Satellite Corporation (AMSAT) was formed in Washington, DC. AMSAT has participated in the vast majority of amateur satellite projects, both in the United States and internationally, beginning with the launch of OSCAR 5. Now, many countries have their own AMSAT organizations, such as AMSAT-UK in England, AMSAT-DL in Germany, BRAMSAT in Brazil and AMSAT-LU in Argentina. All of these organizations operate independently but may cooperate on large satellite projects and other items of interest to the worldwide Amateur Radio satellite community. Because of the many AMSAT organizations now in existence, the US AMSAT organization is frequently designated AMSAT-NA.

Beginning with OSCAR 6, amateurs started to enjoy the use of satellites with lifetimes measured in years as opposed to weeks or months. The operational lives of OSCARs 6, 7, 8 and 9, for example, ranged between four and eight years. All of these satellites were low Earth orbiting (LEO) with altitudes approximately 800-1200 km. LEO Amateur Radio satellites have also been launched by other groups not associated with any AMSAT organization such as the Radio Sputniks 1-8 and the ISKRA 2 and 3 satellites launched by the former Soviet Union.

The short-lifetime LEO satellites (OSCARs I through IV and 5) are sometimes designated the *Phase I* satellites, while the long-lifetime LEO satellites are sometimes called the *Phase II* satellites. There are other conventions in satellite naming that are useful to know. First, it is common practice to have one designation for a satellite before launch and another after it is successfully launched. Thus, OSCAR 10 (discussed later) was known as Phase 3B before launch. Next, the AMSAT designator may be added to the name, for example, AMSAT-OSCAR 10, or just AO-10 for short. Finally, some other designator may replace the AMSAT designator such as the case with Japanese-built Fuji-OSCAR 29 (FO-29).

In order to provide wider coverage areas for longer time periods, the high-altitude Phase 3 series was initiated. Phase 3 satellites often provide 8-12 hours of communications for a large part of the Northern Hemisphere. After losing the first satellite of the Phase 3 series to a launch vehicle failure in 1980, AMSAT-OSCAR 10 was successfully launched and became operational in 1983. AMSAT-OSCAR 13, the followup to the AO-10 mission, was launched in 1988 and re-entered the atmosphere in 1996. AO-10 provides some wide-area communications capability at certain times of the year despite the failure of its onboard computer memory. The successor to AO-13, Phase 3D, is awaiting launch at this writing.

With the availability of the long access time and wide coverage of satellites like AO-10 and the upcoming Phase 3D, it may seem that the lower altitude orbits and shorter access times of the Phase II series would be obsolete. This certainly might be true were it not for the incorporation of digital store-and-forward technology into many current satellites operating in low Earth orbit. Satellites providing store-and-forward communication services using packet radio techniques are generically called *PACSATs*. Files stored in a PACSAT message system can be anything from plain ASCII text to digitized pictures and voice.

The first satellite with a digital store-and-forward feature was UoSAT-OSCAR 11. UO-11's Digital Communications Experiment (DCE) was not open to the general Amateur Radio community although it was utilized by designated "gateway" stations. The first satellite with store-and-forward capability open to all amateurs was the Japanese Fuji-OSCAR 12 satellite, launched in 1986. FO-12 was succeeded by FO-20, launched in 1990, and FO-29, launched in 1996. In addition to providing digital store-and-forward service. FO-20 and FO-29 also have analog linear transponders for CW and SSB communication.

By far the most popular store-and-forward satellites are the *PACSATs* utilizing the PACSAT Broadcast Protocol. These PACSATs fall into two general categories — the *Microsats*, based on technology developed by AMSAT-NA, and the *UOSATs*, based on technology developed by the University of Surrey in the UK. While both types are physically small spacecraft, the Microsats represent a truly innovative design in terms of size and capability. A typical Microsat is a cube measuring 23 cm (9 in) on a side and weighing about 10 kg (22 lb). The satellite will contain an onboard computer, enough RAM for the message storage, two to three transmitters, a multichannel receiver, telemetry system, batteries and the battery charging/power conditioning system.

Amateur Radio satellites have evolved to provide two primary types of communication services — analog transponders for real-time CW and SSB communication and digital store-and-forward

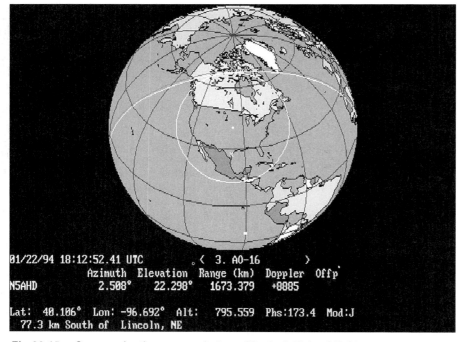

01/22/94 18:12:52.41 UTC   °< 3. AO-16   >
        Azimuth Elevation Range (km) Doppler Offp`
N5AHD   2.508°   22.298° 1673.379   +8885

Lat: 40.106° Lon: -96.692° Alt: 795.559 Phs:173.4 Mod:J
  77.3 km South of Lincoln, NE

Fig 23.15— Communication range circle or "footprint" for AO-16.

for non real-time communication. Which of the two types interest you the most will probably depend on your current Amateur Radio operating habits. If you enjoy real-time DX QSOs on the HF bands, you may be most interested in the high-altitude wide-coverage satellites such as the Phase 3D satellite. On the other hand, if you are a computer and terrestrial packet radio enthusiast you may be more interested in the digital store-and-forward satellites like AO-16, UO-22 and KO-25. Whatever your preference, the remainder of this section should provide the information to help you make a successful entry into the specialty of amateur satellite communications.

Basic Operations and Terminology

Since both low and high Earth orbit (LEO and HEO) satellites are available for use, it would be a good idea to acquire a mental picture of the communication range for each type of orbit. In **Fig 23.15**, the white circle, centered roughly on the United States, is a typical footprint for a low Earth orbit satellite like AO-16. Stations within the footprint can store and/or retrieve messages to/from the store-and-forward message system. For satellites that are used for real-time SSB and CW QSOs, only stations that are in the footprint simultaneously can communicate.

In **Fig 23.16**, for the Phase 3D satellite at apogee, the range circle is quite substantial. Keep in mind that for LEO satellites like AO-16 the footprint is moving quickly, and for HEO satellites it is moving slowly. LEO satellites will typically have access times of 12 to 20 minutes while HEO satellites can have access times as long as 10 to 12 hours. For a more complete discussion of orbital mechanics and other topics in this section, see *The Radio Amateur's Satellite Handbook* published by the ARRL.

When accessing an Amateur Radio satellite, the ground station receiver is tuned to the satellite's *downlink* frequency. If the particular satellite supports two-way communication, the ground station transmits on the satellite's *uplink* frequency. The uplink and downlink frequencies will be in different bands, and each combination of bands used will have a *mode* designator. For example, the combination of an uplink in the 2-m band and a downlink in the 10-m band is called Mode A. More discussion of operating modes can be found in the next two sections, but you may wish to look at **Tables 23.6** and **23.7** for some examples of the different modes available.

The exact manner in which satellite up-

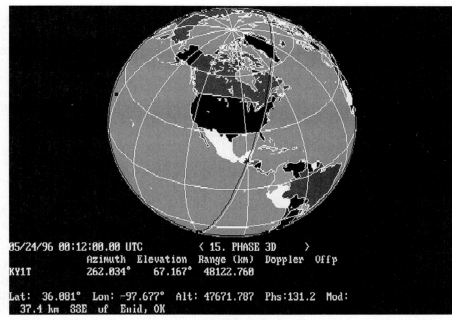

05/24/96 00:12:00.00 UTC   < 15. PHASE 3D >
      Azimuth   Elevation   Range (km)   Doppler   Offp
KY1T   262.034°   67.167°   48122.760
Lat: 36.081°   Lon: -97.677°   Alt: 47671.787   Phs:131.2   Mod:
  37.4 km SSE of Enid, OK

Fig 23.16—The Earth as seen from the Phase 3D satellite at apogee, depicted by *InstantTrack* satellite-tracking software. The broad footprint brings nearly all of the US into range at the same time.

Table 23.6
Analog Transponder Frequencies

RS Satellites

	RS-13	*RS-15*
Mode A		
Uplink	145.960-146.000	145.858-145.898
Downlink	29.460-29.500	29.354-29.394
Beacons	29.458/29.504	29.353/29.399
Mode A Robot		
Uplink	145.840	
Downlink	29.504	
Mode K		
Uplink	21.260-21.300	
Downlink	29.460-29.500	
Beacons	29.458/29.504	
Mode K Robot		
Uplink	21.138	
Downlink	29.504	
Mode T		
Uplink	21.260-21.300	
Downlink	145.960-146.000	
Beacons	145.862/145.908	
Mode T Robot		
Uplink	21.138	
Downlink	145.908	

Phase 3 Satellites

Satellite	Mode	Uplink (MHz)	Downlink (MHz)
AO-10	B	435.030-435.180	145.825-145.975
	Beacon		145.810

Other Satellites

Satellite	Mode	Uplink (MHz)	Downlink (MHz)
FO-20	J(A)	145.900-146.000	435.800-435.900
	Beacon		435.795
FO-29	J(A)	145.900-146.000	435.800-435.900
	Beacon		435.795

Glossary of Satellite Terminology

AMSAT — A registered trademark of the Radio Amateur Satellite Corporation, a nonprofit scientific/educational organization located in Washington, DC. It builds and operates Amateur Radio satellites and has sponsored the OSCAR program since the launch of OSCAR 5. (AMSAT, PO Box 27, Washington, DC 20044.)

Anomalistic period — The elapsed time between two successive perigees of a satellite.

AO-# — The designator used for AMSAT OSCAR spacecraft in flight, by sequence number.

AOS — Acquisition of signal. The time at which radio signals are first heard from a satellite, usually just after it rises above the horizon.

Apogee — The point in a satellite's orbit where it is farthest from Earth.

Area coordinators — An AMSAT corps of volunteers who organize and coordinate amateur satellite user activity in their particular state, municipality, region or country. This is the AMSAT grassroots organization set up to assist all current and prospective OSCAR users.

Argument of perigee — The polar angle that locates the perigee point of a satellite in the orbital plane; drawn between the ascending node, geocenter, and perigee; and measured from the ascending node in the direction of satellite motion.

Ascending node — The point on the ground track of the satellite orbit where the sub-satellite point (SSP) crosses the equator from the Southern Hemisphere into the Northern Hemisphere.

Az-el mount — An antenna mount that allows antenna positioning in both the azimuth and elevation planes.

Azimuth — Direction (side-to-side in the horizontal plane) from a given point on Earth, usually expressed in degrees. North = 0° or 360°; East = 90°; South = 180°; West = 270°.

Circular polarization (CP) — A special case radio energy emission where the electric and magnetic field vectors rotate about the central axis of radiation. As viewed along the radiation path, the rotation directions are considered to be right-hand (RHCP) if the rotation is clockwise, and left-hand (LHCP) if the rotation is counterclockwise.

Descending node — The point on the ground track of the satellite orbit where the sub-satellite point (SSP) crosses the equator from the Northern Hemisphere into the Southern Hemisphere.

Desense — A problem characteristic of many radio receivers in which a strong RF signal overloads the receiver, reducing sensitivity.

Doppler effect — An apparent shift in frequency caused by satellite movement toward or away from your location.

Downlink — The frequency on which radio signals originate from a satellite for reception by stations on Earth.

Earth station — A radio station, on or near the surface of the Earth, designed to transmit or receive to/from a spacecraft.

Eccentricity — The orbital parameter used to describe the geometric shape of an elliptical orbit; eccentricity values vary from e = 0 to e = 1, where e = 0 describes a circle and e = 1 describes a straight line.

EIRP — Effective isotropic radiated power. Same as ERP except the antenna reference is an isotropic radiator.

Elliptical orbit — Those orbits in which the satellite path describes an ellipse with the Earth at one focus.

Elevation — Angle above the local horizontal plane, usually specified in degrees. (0° = plane of the Earth's surface at your location; 90° = straight up, perpendicular to the plane of the Earth).

Epoch — The reference time at which a particular set of parameters describing satellite motion (*Keplerian elements*) are defined.

EQX — The reference equator crossing of the ascending node of a satellite orbit, usually specified in UTC and degrees of longitude of the crossing.

ERP — Effective radiated power. System power output after transmission-line losses and antenna gain (referenced to a dipole) are considered.

ESA — European Space Agency. A consortium of European governmental groups pooling resources for space exploration and development.

FO-# — The designator used for Japanese amateur satellites, by sequence number. Fuji-OSCAR 12 and Fuji-OSCAR 20 were the first two such spacecraft.

Geocenter — The center of the Earth.

Geostationary orbit — A satellite orbit at such an altitude (approximately 22,300 miles) over the equator that the satellite appears to be fixed above a given point.

Groundtrack — The imaginary line traced on the surface of the Earth by the subsatellite point (SSP).

Inclination — The angle between the orbital plane of a satellite and the equatorial plane of the Earth.

Increment — The change in longitude of ascending node between two successive passes of a specified satellite, measured in degrees West per orbit.

Iskra — Soviet low-orbit satellites launched manually by cosmonauts aboard Salyut missions. Iskra means "spark" in Russian.

JAMSAT — Japan AMSAT organization.

Keplerian Elements — The classical set of six orbital element numbers used to define and compute satellite orbital motions. The set is comprised of inclination, Right Ascension of Ascending Node (RAAN), eccentricity, argument of perigee, mean anomaly and mean motion, all specified at a particular epoch or reference year, day and time. Additionally, a decay rate or drag factor is usually included to refine the computation.

LHCP — Left-hand circular polarization.

LOS — Loss of signal — The time when a satellite passes out of range and signals from it can no longer be heard. This usually occurs just after the satellite goes below the horizon.

Mean anomaly (MA) — An angle that increases uniformly with time, starting at perigee, used to indicate where a satellite is located along its orbit. MA is usually specified at the reference epoch time where the Keplerian elements are defined. For AO-10 the orbital time is divided into 256 parts, rather than degrees of a circle, and MA (sometimes called phase) is specified from 0 to 255. Perigee is therefore at MA = 0 with apogee at MA = 128.

Mean motion — The Keplerian element to indicate the complete number of orbits a satellite makes in a day.

Microsat — Collective name given to a series of small amateur satellites having store-and-forward capability (OSCARs 14-19, for example).

NASA — National Aeronautics and Space Administration, the US space agency.

Nodal period — The amount of time between two successive ascending nodes of satellite orbit.

Orbital elements — See **Keplerian Elements**.

Orbital plane — An imaginary plane, extending throughout space, that contains the satellite orbit.

OSCAR — Orbiting Satellite Carrying Amateur Radio.

PACSAT — Packet radio satellite (see **Microsat** and **UoSAT-OSCAR**).

Pass — An orbit of a satellite.

Passband — The range of frequencies handled by a satellite translator or transponder.

Perigee — The point in a satellite's orbit where it is closest to Earth.

Period — The time required for a satellite to make one complete revolution about the Earth. See **Anomalistic period** and **Nodal period**.

Phase I — The term given to the earliest, short-lived OSCAR satellites that were not equipped with solar cells. When their batteries were depleted, they ceased operating.

Phase 2 — Low-altitude OSCAR satellites. Equipped with solar panels that powered the spacecraft systems and recharged their batteries, these satellites have been shown to be capable of lasting up to five years (OSCARs 6, 7 and 8, for example).

Phase 3 — Extended-range, high-orbit OSCAR satellites with very long-lived solar power systems (OSCARs 10 and Phase 3D, for example).

Phase 4 — Proposed OSCAR satellites in geostationary orbits.

Precession — An effect that is characteristic of AO-10 and Phase 3 orbits. The satellite apogee SSP will gradually change over time.

Project OSCAR — The California-based group, among the first to recognize the potential of space for Amateur Radio; responsible for OSCARs I through IV.

QRP days — Special orbits set aside for very low power uplink operating through the satellites.

RAAN — Right Ascension of Ascending Node. The Keplerian element specifying the angular distance, measured eastward along the celestial equator, between the vernal equinox and the hour circle of the ascending node of a spacecraft. This can be simplified to mean roughly the longitude of the ascending node.

Radio Sputnik — Russian Amateur Radio satellites (see **RS #**).

Reference orbit — The orbit of Phase II satellites beginning with the first ascending node during that UTC day.

RHCP — Right-hand circular polarization.

RS # — The designator used for most Russian Amateur Radio satellites (RS-1 through RS-15, for example).

Satellite pass — Segment of orbit during which the satellite "passes" nearby and in range of a particular ground station.

Sidereal day — The amount of time required for the Earth to rotate exactly 360° about its axis with respect to the "fixed" stars. The sidereal day contains 1436.07 minutes (see **Solar day**).

Solar day — The solar day, by definition, contains exactly 24 hours (1440 minutes). During the solar day the Earth rotates slightly more than 360° about its axis with respect to "fixed" stars (see **Sidereal day**).

Spin modulation — Periodic amplitude fade-and-peak resulting from the rotation of a satellite's antennas about its spin axis, rotating the antenna peaks and nulls.

SSC — Special service channels. Frequencies in the downlink passband of AO-10 that are set aside for authorized, scheduled use in such areas as education, data exchange, scientific experimentation, bulletins and official traffic.

SSP — Subsatellite point. Point on the surface of the Earth directly between the satellite and the geocenter.

Telemetry — Radio signals, originating at a satellite, that convey information on the performance or status of onboard subsystems. Also refers to the information itself.

Transponder — A device onboard a satellite that receives radio signals in one segment of the spectrum, amplifies them, translates (shifts) their frequency to another segment of the spectrum and retransmits them. Also called linear translator.

UoSAT-OSCAR (UO #) — Amateur Radio satellites built under the coordination of radio amateurs and educators at the University of Surrey, England.

Uplink — The frequency at which signals are transmitted from ground stations to a satellite.

Window — Overlap region between acquisition circles of two ground stations referenced to a specific satellite. Communication between two stations is possible when the subsatellite point is within the window.

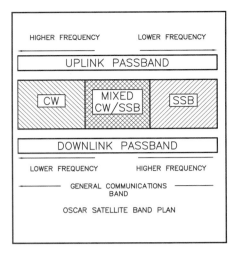

Fig 23.17 — The OSCAR satellite band plan allows for CW-only, mixed CW/SSB, and SSB-only operation. Courteous operators observe this voluntary band plan at all times.

Table 23.7
Uplink and Downlink Frequencies for the Phase-3D Satellite

Uplinks

Band	Digital (MHz)	Analog (MHz)	Center (MHz)
15 m	N/A	21.210-21.250	21.230
12 m	N/A	24.920-24.960	24.940
2 m	145.800-145.840	145.840-145.990	145.915
70 cm	435.300-435.550	435.550-435.800	435.675
23 cm(1)	1269.000-1269.250	1269.250-1269.500	1269.375
23 cm(2)	1268.075-1268.325	1268.325-1268.575	1268.450
13 cm(1)	2400.100-2400.350	2400.350-2400.600	2400.475
13 cm(2)	2446.200-2446.450	2446.450-2446.700	2446.575
6 cm	5668.300-5668.550	5668.550-5668.800	5668.675

Downlinks

Band	Digital (MHz)	Analog (MHz)	Center (MHz)
2 m	145.955-145.990	145.805-145.955	145.880
70 cm	435.900-436.200	435.475-435.725	435.600
13 cm	2400.650-2400.950	2400.225-2400.475	2400.350
3 cm	10451.450-10451.750	10451.025-10451.275	10451.150
1.5 cm	24048.450-24048.750	24048.025-24048.275	24048.150

All downlink passbands are inverted from the uplink passbands.

Beacons Band	General Beacon (MHz)	Middle Beacon (MHz)	Engineering Beacon (MHz)
2 m	N/A	145.880	N/A
70 cm	435.450	435.600	435.850
13 cm (1)	2400.200	2400.350	2400.600
13 cm (2)	2401.200	2401.350	2401.600
3 cm	10451.000	10451.150	10451.400
1.5 cm	24048.000	24048.150	24048.400

Note: The absence of a 2-m beacon is due strictly to characteristics of the IF Matrix and the limited bandwidth available on that band. The beacons on the other bands are for various purposes, including providing spacecraft engineering data to the command stations. All beacons can be modulated with 400 bits per second BPSK and possibly other formats.

links and downlinks are utilized depends on whether the primary purpose of the satellite is to provide analog or digital communication services. Satellites make use of *transponders*. Transponders regenerate all signals appearing in their input (uplink) frequency band on their output (downlink) frequency band. CW, SSB and FM signals appearing at the input will appear as CW, SSB, or FM signals on the output. Depending on the design of the transponder, USB on the input may appear either as USB or LSB on the output. The low-to-high frequency relationship of the uplink and downlink frequency bands may also differ. Note in Table 23.7, for example, that all downlink passbands are inverted from the uplink passbands. This means that a signal at the low end of the uplink will be retransmitted at the high end of the downlink. On the other hand, RS-15 Mode A uplink is 145.860 to 146.000 MHz, while the downlink is 29.360 to 29.400 MHz. Consequently, a signal at the low end of the uplink band will appear at the low end of the downlink band. The band plan used on most OSCAR satellites is given in **Fig 23.17**. FM is rarely used on amateur satellite transponders because of the energy required for the 100% duty cycle. There are exceptions, however, such as AO-27 and SO-35, each of which has a single FM voice channel using computer-reconstituted voice modulation for their downlink transmissions.

In contrast to a satellite such as RS-10, whose primary mission is providing linear transponders for CW and SSB communications, uplink and downlink frequencies on digital communications satellites are usually channelized. **Table 23.9** shows

that specific frequencies are used for both uplink and downlink. The reason for multiple uplinks and a single downlink on some satellites is the uncoordinated *Aloha* access used by ground stations. Generally, the satellite can handle requests from more than one ground station without overloading its own downlink.

Remember that the ground station will experience *Doppler* shift of the downlink frequency as the satellite moves with respect to the observer. For satellites with linear transponders, operating procedures have been established to minimize interference to other stations in the passband while staying tuned to the desired station. For digital satellites, the modem usually tunes the receiver frequency to compensate for Doppler shift.

Satellites with Analog Transponders

Table 23.6 is a list of frequencies for all Amateur Radio satellites providing linear transponder communication facilities, and Table 23.7 contains the Phase 3D frequencies. Both were accurate as of 1999.

A sensible approach for getting started in amateur satellite communication is to choose one of the low Earth orbit satellites (RS-15, for example) operating on frequencies for which you already have equipment. Even though the access times will be much shorter than with the higher orbit satellites, experience can be gained using existing equipment and simple antennas. Then, if the bug bites hard, assemble a station to work the wider coverage birds.

There is so much emphasis on the wide-area coverage of high altitude satellites, that the low Earth orbit (LEO) satellites often do not receive proper attention. There is a great amount of satisfaction to be gained from working other stations via LEO satellites, however. Moreover, such contacts provide practice at tracking and tuning that will prove valuable no matter which satellite is eventually used.

A first attempt at amateur satellite communication should be undertaken as inexpensively as possible. Successful operation on LEO satellites can be realized using omnidirectional antennas, an uplink

Table 23.9

Uplink and Downlink Frequencies for Satellites with Two-Way Digital Communications Links

Satellite	Mode	Uplink Freq (MHz)	Modulation	Rate (bps)	Downlink Freq (MHz)	Modulation	Rate (bps)
AO-16	JD	145.900 145.920 145.940 145.960	Manchester Encoded AFSK		437.051 437.026 2401.143	RCBPSK	1200
LO-19	JD	145.840 145.860 145.880 145.900	Manchester Encoded AFSK	1200	437.154 437.126	BPSK RCBPSK	1200 1200
UO-22	JD	145.900 145.975	FSK	9600	435.120	FSK	9600
KO-23	JD JD	145.850 145.900	FSK	9600	435.175	FSK	9600
KO-25	JD	145.980	FSK	9600	436.500	FSK	9600
TO-31	JD	TBA	—	—	436.923	—	—
TO-32	JD	TBA	—	—	435.325/ 435.225	—	—

Fig 23.19 — Simple ground plane and Yagi antennas can be used for low-Earth-orbit (LEO) satellite contacts.

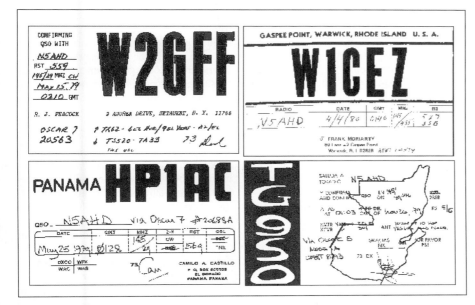

Fig 23.18 — Some stations worked from a QTH on the lower Texas Gulf Coast using simple antennas.

power in the area of 100 W EIRP and a good receiver. If Mode A is used, a 10-m receive preamp might prove useful. Similarly, if Mode J is used, a 70-cm preamp could be beneficial. One goal of an entry-level approach is to eliminate the complexity of high-gain steerable antennas. A power level of 50-100 W into an omnidirectional antenna is more than adequate for CW QSOs and at times will support SSB QSOs. The author has had many satellite contacts using the approach described here. **Fig 23.18** shows a few QSL cards from contacts made from his QTH on the lower Texas Gulf Coast. **Fig 23.19** shows the ground plane and small Yagi antennas used on the 2-m and 70-cm bands. On 10 m, either a dipole or wire loop antenna was used.

Assembling a Mode B Station

If antenna installation restrictions and budget constraints are not a problem, you may want to try the high-orbit satellites using Mode B (430 MHz up, 145 MHz down). This section gives the important considerations for assembling this type of station. Obviously, these requirements can be realized in many different system configurations. More information can be found in the section on equipping a station.

A typical station will use a 145-MHz circularly polarized antenna of at least 13 dBic having switchable polarization sense between RHCP and LHCP. If switchable polarization is not available, then RHCP is the preferred choice. Even though antennas with gain exceeding 18 dBi are available, they are not cost effective because the local noise floor becomes the limiting factor. Keep in mind that this antenna is used for both the Mode B downlink and the Mode J uplink.

Similarly, the 435-MHz antenna should be circularly polarized and have at least 13 dBic gain. Higher gains, in the range of 14 to 18 dBic, are preferred. Switchable polarization is even more desirable because the 435-MHz antenna serves as the uplink on Modes B and S as well as the downlink for Mode J. Increased antenna gain is more usable because the local noise floor is not usually a limiting factor as it is at 145 MHz.

The 1269-MHz Mode-L uplink antenna should have at least 20-dBic gain. Due to the short wavelength (23 cm), the required gain is achievable with relatively small antenna arrays such as four phased helices. Arrays of loop Yagis and standard horizontally polarized Yagis can be used but there will be a 3-dB penalty for polarization mismatch.

The 2400-MHz Mode-S downlink antenna should have at least 26 dBic gain. However, the required gain is easily achievable using a 4-ft parabolic dish or a quad helix array.

The antenna array must be steerable in both azimuth and elevation. The elevation rotator boom must be made of a nonmetallic material such as fiberglass. Feed-line loss should be held below 2 dB, and less than 1 dB is preferable.

The EIRP of the 435 and 145-MHz transmitting system should be no more than 1000 W and adjustable, allowing the lowest required power level to be used. For Modes B and J most communications can be conducted using 100 to 300 W EIRP. A higher EIRP may be needed on Mode S, but the requirement will still be below the 1000 W EIRP level. For Mode L, the required EIRP at 1269 MHz is between 3000 and 5000 W.

The 145-MHz receiving system should have a noise figure no greater than 2 dB but less than 1 dB is probably not usable even if it can be achieved. The 435 and 2400-MHz receiving system noise figures should be less than 1 dB.

Using high altitude satellites should be considered weak signal work. Always improve the receiving system first before increasing transmit EIRP. Once Phase 3D is operating, the downlink gain requirements will be much lower due to the higher transmitter powers used at the spacecraft.

ASSEMBLING A STATION FOR PHASE 3D

This section was written by ARRL Lab Engineer Zack Lau, W1VT.

If all goes well, Phase 3D will be similar to OSCAR 13, but much more "user friendly" for voice users. Thus, a station that did well with OSCAR 13 has little to worry about; the equipment will be more than enough for Phase 3D. Those building new stations can take advantage of technology improvements in the satellite, and get acceptable performance with more modest SSB/CW stations. Due to the laws of physics, those expecting loud signals like those of low Earth orbit satellites will still be disappointed. A station 100 times farther away is 40 dB weaker (400 km vs 40,000 km). Thus, digital users won't see any signal strength improvement compared to low Earth orbit satellites currently in use—the extra distance will eliminate improvements in power and antennas.

Perhaps the biggest change is the orbit—it will repeat every two days. Thus, manually rotated or even fixed antennas will become much more practical, possi-

bly eliminating the need for an expensive rotator system, for those who just want to maintain a schedule with another station. Stations without rotators may wish to use smaller antennas to maximize their operating time. This generally requires more power and better receivers to compensate. Mast mounting equipment near the antenna will reduce needed antenna size. Using circularly polarized antennas as opposed to higher gain linearly polarized antennas will help considerably toward optimizing your satellite time if the antennas don't move. However, rotatable linear arrays are probably preferred for local use. Horizontal polarization is the standard for terrestrial SSB/CW.

The next biggest change for most users will be on the 2401-MHz downlink—there will be a linear 60-W PEP transponder, as opposed to the 1-W experiment aboard OSCAR 13. Thus, it becomes possible to use a simple RHCP 16-turn helix instead of a 2-ft dish. However, it may be more cost effective to use a 2-ft dish with a 1.8 dB NF receiver than to obtain a 0.6-dB NF receiver for the helix. Since these are system noise figures, it isn't unusual to need a 0.4-dB NF preamp to get that 0.6-dB NF receiver. The dish with a no-tune preamp and receive converter makes a lot more sense for a builder with minimal test equipment. The antenna gain to receiver temperature ratio (G/T) to shoot for is 0.53/kelvin. There will also be a 13-cm uplink—plan on +27-dBWic. This is 5 W at the feed of a 20-dBic 2-ft dish. 10 W to a loop Yagi would also work; the extra 3 dB compensates for the polarization mismatch. However, the satellite won't be capable of in-band full duplex—a band can only be used on transmit or receive, not both simultaneously. Thus, since the 13-cm downlink is expected to be used heavily, the uplink is likely to get little use.

The 436-MHz uplink will need about 20 dBWic—10 W to a 5-turn 3-ft boom helix or a 5-ft boom circularly polarized Yagi. Slightly larger antennas can compensate for feed-line loss. Chances are, there will be little benefit to running a big amplifier—there will be an automatic notcher called LEILA to prevent stations from hogging the transponder. Hopefully, this will force stations to improve receive capability, when they find it difficult to hear themselves on the satellite.

On 436-MHz receive, you want a gain to system noise temperature ratio of 0.032/kelvin. A 12-dB antenna has a gain of 16. Thus, for a 12-dBic antenna you need a noise temperature under 500 kelvins, or 4 dB. Earth noise, feed-line noise and antenna noise all add to the re-

ceiver noise. A mast-mounted preamplifier and a small Yagi will work quite well. If the feed-line run is short, perhaps 50 ft, a larger antenna would allow having the preamplifier near the operating position.

The 1269-MHz uplink will need about 26-dBWic—8 W to a 12-ft boom loop Yagi, 10 W to a RHCP 16-turn helix, or 6 W to a 3-ft dish. Current rules prohibit having a 1269-MHz downlink, so this isn't planned for any of the satellites.

The 146-MHz uplink will need about 18 dBWic—10 W to a 5-ft boom circularly polarized Yagi or 3-turn helix. Again, this is power at the feed of the antenna.

The 146-MHz downlink depends heavily on your local noise level. Amateurs in rural areas can do just fine with a 2-dB system noise figure and an 8-dBic antenna. The predicted G/T needed is 0.008/kelvin. Those in heavily populated urban areas may be disappointed with the results—even with a big circularly polarized beam and a mast-mounted preamplifier. These amateurs should consider using a quieter band for the downlink.

Two meters does have a distinct advantage in one area—less attenuation through trees. As the frequency goes up, so does the attenuation through trees. Thus, while it is possible to hear the 2.4-GHz downlink indoors, tree blockages often degrade signals. It gets worse if you are using a small antenna with a low-noise preamplifier. Not only does a tree block more of a small antenna, but it also acts as a warm noise source. This noise adds to the system noise figure, degrading signals even more.

Amateurs attempting to contact the satellite on the horizon with microwaves may notice two degradations to the path. Atmospheric loss can add another 1.6 dB of path loss at 2.4 GHz, increasing to 3 dB at 10 GHz, though this is typically under 0.1 dB at 10 GHz for vertical paths. An antenna fixed on the horizon will also see noise from the warm Earth, reducing system sensitivity. A more serious problem may be finding excellent locations where one can worry about such details.

It is easy to overestimate the ease of obtaining a low noise figure, particularly at microwaves. A single bad connector, adapter or piece of coax can stop the system from meeting expectations. Avoid cheap connectors and coax. Getting all the pieces to work properly together can be a challenge. Fortunately, MMICs and computer aided design have resulted in designs that reduce potential problems. Still, it is possible to have pieces that work fine by themselves, but poorly as a system. People have even had problems with poorly designed power supplies generating spurs or modulating received signals. Fortunately

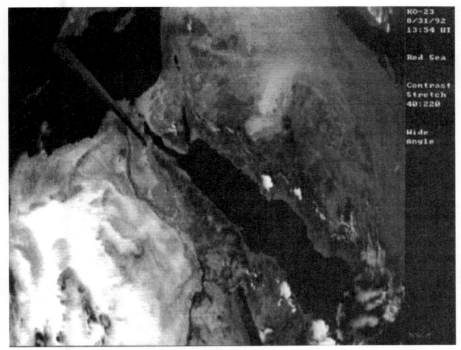

Fig 23.20 — This image of the Red Sea was downloaded from KITSAT-OSCAR 23. *(photo courtesy Harold Price, NK6K)*

microwave ovens have not interfered with 2.4-GHz amateur satellite work. Similarly, it is easy to underestimate the ease of obtaining low angle radiation at 2 m. The antenna height required may not be practical. It is often wise to have a bit of excess capability, often called link margin. If you do have excess uplink power, you should have a method of easily scaling it back.

While they are not expected to be as popular as the lower bands, a 5668-MHz uplink (34 dBWic), a 60-W PEP 10451-MHz downlink (G/T = 13/kelvin) and a 24048-MHz downlink will be included. Even with state-of-the-art equipment, it is likely that the latter will not be heard on long LOS paths due to atmospheric absorption.

Simple antennas like the turnstile work well with the satellite overhead, but not so well near the horizon. A simple vertical works better near the horizon. It makes a lot of sense to match the orbit track with the antenna pattern, keeping in mind that some computer simulations aren't accurate with wires close to real ground. Ionized atmospheric layers can significantly disturb satellite communications by blocking signals to and from the satellite. There may also be a 21-MHz uplink, but it is likely that it will not work well for many users, due to the high

galactic noise, and the modest antenna on the satellite.

The picture isn't quite so rosy for digital users—the 146-MHz uplink will need +22 dBWic, 10 W to a 12-ft boom circularly polarized Yagi. The 1270-MHz uplink will need +34 dBWic, or 10 W to a 6-ft dish.

The 436-MHz downlink will require a G/T of 0.12/kelvin, or a 13 dBic antenna with a 1 dB preamp (allowing 50 kelvins for sky and antenna noise, and 0.5 dB extra receiver noise).

DIGITAL COMMUNICATIONS SATELLITES

The amateur satellite enthusiast with an interest in digital communications will find a multitude of satellites with which to experiment. All of the digital communications satellites currently operating in the Amateur Satellite Service are in low Earth orbit. At first, it might seem that the short access times of LEO satellites would not support useful communications services. But, as will be seen shortly, this is certainly not the case.

There are three general categories of Amateur Radio satellites having digital communications links. First, there are those that transmit telemetry and other information of interest using digital codes

but do not provide store-and-forward message service. Satellites in this category include DOVE-OSCAR 17 (DO-17) and UoSAT-OSCAR 11 (UO-11). Also, KITSAT-OSCAR 23 transmits images of the Earth (see **Fig 23.20**). Satellites such as DO-17 and UO-11 provide an excellent opportunity to learn the mechanics of tracking LEO spacecraft and decoding their digital transmissions. At the same time, study of the captured telemetry data will provide an appreciation of many aspects of spacecraft engineering. A listing of the special-purpose digital satellites can be found in Table 23.8.

Another class of LEO satellites are those that provide store-and-forward message services via a user interface similar to those found on terrestrial packet radio bulletin board systems.

Finally, there are many satellites that provide store-and-forward services using the PACSAT Broadcast Protocol developed by Ward and Price. Satellites using the PACSAT Broadcast Protocol include: AMSAT-OSCAR 16 (AO-16), LUSAT-OSCAR 19 (LO-19), UoSAT-OSCAR 22 (UO-22), KITSAT-OSCAR 23 (KO-23) and KITSAT-OSCAR 25 (KO-25). In addition to these satellites, there are other projects in the design and construction stages that will also use the PACSAT Broadcast Protocol (PBP). Table 23.9 contains a list of the digital store-and-forward satellites operating at the time of publication.

Satellites Transmitting Digital Telemetry Data Only

Monitoring satellites transmitting digital telemetry data provides an excellent receive-only introduction to amateur satellite operations for the computer enthusiast. Of course, one could monitor telemetry from any of the digital satellites, but this section will deal with UO-11 because it does not provide two-way communication capabilities.

UoSAT-OSCAR 11

UO-11 transmits various kinds of data on its 2-m downlink (145.825 MHz) and most of it is plain-text ASCII, including bulletins and spacecraft telemetry. It is important to note that UO-11 transmits plain text and not packets such as those used in terrestrial 2-m packet radio networks. This means that a packet radio TNC is not required at the ground station.

Fig 23.21 shows a typical equipment configuration for receiving UO-11 transmissions. As can be seen, all that is necessary is to connect the receiver audio output to the demodulator input and the serial data output from the demodulator to the com-

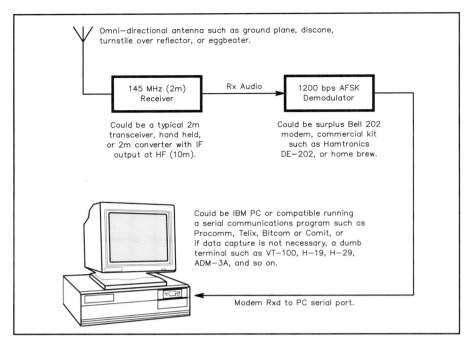

Fig 23.21 — Equipment needed to monitor digital transmissions from UoSAT-OSCAR 11.

```
UOSAT-2          0005026192842
005126014674022066033981040567050432060237070557080495090035F
104423113560120003130680C14228D15355716182C17585E185419195638
203890211893226600230001240006250007261005273503285294295196
305116310367322850E33590C340007352859363431374473384953395294
408095411235426440430007441690E45000146000247514348528349491 1
50548C51100552697F537032547215550000560003575227585181595180
6083E3615BD4621E0B633341644402651D0F66DFEC67000168000E69000F
```

Fig 23.22 — A UoSAT-OSCAR 11 (UoSAT-2) telemetry frame as monitored.

puter serial port. If a modem is purchased at a flea market or other used equipment outlet, be sure that it a Bell 202 standard as opposed to Bell 212. The type of modem is usually obvious from the model number but the 212 is much more common than the 202. Kits for Bell 202 demodulators are available commercially and construction plans have also been published in *QST*.

In a minimal equipment configuration it is also possible to eliminate the expense of a computer and substitute a serial terminal instead. However, capturing the received telemetry for later decoding or real-time telemetry decoding will require a computer. In this regard, remember that 80286 and earlier PCs are now sold at very reasonable prices and even discarded outright. These machines are entirely adequate to serve as substitutes for serial terminals and can perform the telemetry capture and decoding functions as well.

Fig 23.22 shows a typical UO-11 raw

telemetry frame while **Fig 23.23** shows a telemetry frame decoded to engineering data. Telemetry capture and decoding software for UO-11 is available from AMSAT-NA and AMSAT-UK.

Satellites Utilizing the PACSAT File Broadcast Protocol

Many Amateur Radio operators make use of terrestrial packet radio bulletin board systems (PBBS). These PBBSs are the Amateur Radio counterpart of microcomputer-based bulletin board systems, which use the public telephone network. However, much of the activity on terrestrial PBBSs is taken up with the repeated retrieval of information of general interest. For example, Amateur Radio operators interested in satellite operations may access a PBBS to obtain the reference elements used in their orbital prediction programs. Consequently, the exact same information may be transmitted many times.

On terrestrial networks, repeated transmission of the same data can be tolerated because the system capacity is available by virtue of the 24-hour per day access time and multiple stations providing PBBS service. Such capacity obviously does not exist in the case of a LEO satellite-based system with limited visibility time at any particular ground station location. What is needed is a way for multiple users to benefit from the same transmission of a particular file since many users are within the satellite footprint at the same time.

Satellites such as AO-16 and UO-22 are in low Earth orbits at an average altitude of 800 km. From that vantage point, over populated areas such as the continental United States, hundreds and perhaps even thousands of potential users are within the satellite's footprint. Although at any given ground station location (in the middle latitudes) there will be only 50 to 60 minutes of access time per day, there is still sufficient time for any individual station to receive a large amount of data. For example, with AO-16 operating at 1200 bps, it is possible to receive approximately 500 kbytes of data per day. For UO-22, operating at 9600 bps, about 4 Mbytes of data could be received. This assumes, of course, that the ground station is in operation for all times the satellite is visible.

Based on the nature of the system components and on the experience gained with the UoSAT-2 Digital Communications Experiment (DCE)[1], Ward and Price have developed the PACSAT Protocol Suite, which is fully documented in Notes 2-6. Look to these references for the complete details of the protocol implementation. The data-link layer protocol used is AX.25.[7] The PACSAT Protocol Suite implements a file broadcast mode and a file server mode using a common file format. Each of these two modes will be described briefly. The hardware and software required to access the satellite will be presented in the next section.

PACSAT File Header

Files being transmitted in broadcast mode and files being uploaded in file server mode make use of the PACSAT file header. **Fig 23.27** shows an example of the information contained in the file header. In broadcast mode, an individual data-link layer frame information field contains only the file number (ID), file type and the offset to the location in the file where the data belongs. The other information needed to identify the file and its attributes are contained in the file header. User software (PFHADD) has been provided to add

```
UoSAT-2    ORBIT NO. 10335   DATE: 86/02/07  -  Friday  -  86.038   TIME:  02.04.00

   CHANNEL          PARAMETER                  RAW VALUE    ACTUAL    UNITS
     00       Solar Array Current  -Y            512        7.600     mA
     01       Nav Magnetometer   X Axis          467        1.350     uT
     02       Nav Magnetometer   Y Axis          206      -37.926     uT
     03       Nav Magnetometer   Z Axis          398       -9.021     uT
     04       Sun Sensor   No. 1                 056        0.000
     05       Sun Sensor   No. 2                 043        0.000
     06       Sun Sensor   No. 3                 023        0.000
     07       Sun Sensor   No. 4                 055        0.000
     08       Sun Sensor   No. 5                 049        0.000
     09       Sun Sensor   No. 6                 035        0.000
     10       Solar Array Current  +Y            442      140.600     mA
     11       Nav Magentometer   Temp            356       -7.536     Degrees  C
     12       Horizon Sensor                     000        0.000
     13       Spare                              068        0.000
     14       DCE RAMUNIT Current                228       23.522     mA
     15       DCE CPU Current                    355       83.950     mA
     16       DCE GMEM Current                   182       28.905     mA
     17       Facet Temperature   +X             585      -21.000     Degrees  C
     18       Facet Temperature   +Y             541      -12.200     Degrees  C
     19       Facet Temperature   +Z             563      -16.600     Degrees  C
     20       Solar Array Current  -X            389      241.300     mA
     21       +10 Volt Line Current              189      183.330     mA
     22       PCM Voltage  +10V                  660        9.900     Volts
     23       P/W Logic Current  (+5V)           000        0.000     mA
     24       P/W Geiger Current  (+14V)         000        0.000     mA
     25       P/W Elec sp.curr  (+10V)           000        0.000     mA
     26       P/W Elec sp.curr  (-10V)           100        9.300     mA
     27       Facet Temperature   -X             350       26.000     Degrees  C
     28       Facet Temperature   -Y             529       -9.800     Degrees  C
     29       Facet Temperature   -Z             519       -7.800     Degrees  C
     30       Solar Array Current  +X            509       13.300     mA
     31       -10 Volt Line Current              036       17.280     mA
     32       PCM Voltage  -10V                  285       10.260     Volts
     33       1802 Computer Current  (+10V)      591      124.110     mA
     34       Digitalker  Current  (+5V)         000        0.000     mA
     35       145 MHz Beacon Power Output        283      432.500     mW
     36       145 MHz Beacon Current             342       75.240     mA
     37       145 MHz Beacon Temperature         447        6.600     Degrees  C
     38       Command Decoder Temperature  (+Y)  495       -3.000     Degrees  C
     39       Telemetry System Temperature  (+X) 529       -9.800     Degrees  C
     40       Solar Array Voltage  (+30V)        792       27.600     Volts
     41       +5 Volt Line Current               123      119.310     mA
     42       PCM Voltage  +5V                   644        5.410     Volts
     43       DSR Current  (+5V)                 000        0.000     mA
     44       Command Receiver Current           169      155.480     mA
     45       435 MHz Beacon Power Output        000        0.000     mW
     46       435 MHz Beacon Current             000        0.000     mA
     47       435 MHz Beacon Temperature         514       -6.800     Degrees  C
     48       P/W Temperature  (-X)              527       -9.400     Degrees  C
     49       BCR Temperature  (-Y)              489       -1.800     Degrees  C
     50       Battery Charge/Discharge  Current  540      237.600     mA
     51       +14 Volt Line Current              100      500.000     mA
     52       Battery Voltage  (+14V)            695      145.950     Volts
     53       Battery Cell Voltages  (MUX)       701        0.000
     54       Telemetry System Current           715       14.300     mA
     55       2401 MHz Beacon Power Output       000        0.000     mW
     56       2401 MHz Beacon Current            000        0.000     mA
     57       Battery Temperature                522       -8.400     Degrees  C
     58       2401 MHz Beacon Temperature        518       -7.600     Degrees  C
     59       CCD Imager Temperature             518       -7.600     Degrees  C
```

Fig 23.23 — A UoSAT-OSCAR 11 (UoSAT-2) telemetry frame decoded to engineering units.

a PACSAT file header to a file before it is uploaded and to remove or display a file header after the file has been downloaded (PHS).

File and Directory Broadcast Mode

The PACSAT Broadcast Protocol has the following attributes: (1) Any frame, when received independently, can be placed in the proper location within the file to which it belongs; (2) When all frames have been received, the receiving station can tell that the file is complete; and (3) For file types where it makes sense, partial files are usable. This implies that if a data compression scheme is used, it should be possible to incrementally decompress the file.

The broadcast mode transmits files in the message system memory and their directory entries by giving each file on the broadcast queue a certain amount of downlink time. The broadcasts continue in a round-robin fashion until the user's request has been filled.

File broadcasting is done as a series of AX.25[7] unnumbered information (UI) frames. UI frames are not acknowledged by the receiver and order of delivery is not guaranteed. The terminal node controller

(TNC) passes the frame on to the application program only if the frame is correctly received. Error checking of the frame is done via CRC-16 by the TNC. The format of the information field of a broadcast frame is shown below.

<flags> <file id> <file type> <offset> <data> <crc>

In the information field format above, the file ID field is a file number assigned by the file server system when the file is uploaded rather than an ASCII character string file name. The offset gives the position relative to the beginning of the file where the data belongs. The CRC shown is a check on the I-field contents only and is included to allow detection of errors on the serial link between the TNC and the computer.

Requests to place files in the broadcast queue are likewise done with UI frames. The spacecraft does respond to broadcast requests but not in terms of a data-link layer acknowledgment. It only sends a UI frame with "OK" in the information field to the station making a successful broadcast request. Error indications, such as broadcast queue full, are also transmitted as UI frames.

Even though a station may also access the satellite in a connected-mode transaction (described in the next section), the file and directory broadcast mode is the primary method of operation. Since multiple users in the satellite footprint may want to capture the same files and update their directories at the same time, downlink utilization is maximized when broadcasting is used. Individual users may request fills of specific "holes" in their captured files and directories, but the rebroadcast of entire files or directories for multiple users is eliminated.

File Server Mode

AO-16, UO-22 and similar satellites can also operate in file server mode, which is transaction oriented. Currently, the file server mode is used only for uploading files to the message system. An upload transaction can be resumed later if it was previously interrupted (by LOS, for example).

When using the file server mode, an AX.25 connection exists between the ground station and the spacecraft. Standard balanced-mode HDLC procedures control the exchange at the data-link layer. The transaction-oriented operation ensures that the availability of the uplinks is maximized.

PACSAT Ground Station Equipment

A typical equipment configuration for

utilizing AO-16 and UO-22 is shown in **Fig 23.28**. Even though the diagram shows a station set up to operate on both AO-16 and UO-22, a sensible approach would be to set up for AO-16 operation first and then progress to UO-22. This is particularly true if you had been operating on FO-20 because the radios and modems are already in place and attention can be focused on installing and using the PB and PG software. Even if you have not used FO-20 it is still easier to set up for AO-16 first: 1200 bps operation does not usually require any internal connections and/or modifications to the transmitter and receiver, whereas 9600 bps operation usually does require some internal connections.

```
file number        :   0x0
file name          :
file extension     :
file size          :   20270
create date        :   Tue Jan 08 04:56:58 1991
last modified      :   Tue Jan 08 04:57:26 1991
seu flag           :   0x00
file type          :   0x00
body checksum      :   0x662d
header checksum    :   0x1a24
body offset        :   186
source             :   n5ahd
ax25 uploader      :
upload time        :   uninitialized
download count     :   000
destination        :   wd5ivd
ax25 downloader    :
download time      :   uninitialized
expiry time        :   uninitialized
compression type   :   0x00
priority           :   000
user filename      :   ntc01.doc
title              :   article draft
keywords           :   NTC92
```

Fig 23.27 — PACSAT file header contents.

PACSAT Ground Station Software Capabilities and Operation

To use all of the communication facilities available, four computer programs, PB, PG, PHS and PFHADD, are available free of charge to the Amateur Radio community. PB allows files and directories being broadcast to be captured on the receiving station's computer. PB also allows a station to request the broadcast of hole fills in partially received files and directories. PG is used to upload files to the spacecraft for later broadcast. PFHADD adds the header required for uploading a user file. PHS will display or remove the file header after downloading a file. PB, PG, PFHADD and PHS run on IBM-PC/AT and compatible systems.

Recall that file and directory broadcasting is done in AX.25 unconnected mode and file uploads are done in AX.25 connected mode. Consequently, PB looks for UI frames from the spacecraft and places them in the proper location in the file being received if the user has requested that the file be captured. PG, on the other hand, establishes a connection with the file server and attempts to complete the transaction requested by the user. The following brief discussion of user software operation will provide some insight into the mechanics of utilizing the communications facilities of the satellites.

A user wishing to monitor files and directories being broadcast on the downlink would configure his/her station equipment as shown in Fig 23.28 and execute the PB program on the station computer. **Fig 23.29** shows a typical screen display from the PB program while monitoring UO-22 downlink traffic. The lower half of the screen shows certain informational messages exactly as they appear on the downlink. The upper left corner of the screen shows files for which capture is in progress (in this case none), and the upper right cor-

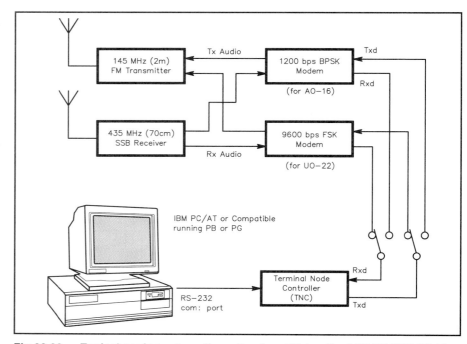

Fig 23.28 — Typical equipment configuration for utilizing the AMSAT-OSCAR 16 and UoSAT-OSCAR 22 satellites.

ner shows directory headers and message numbers being heard on the downlink.

The last line of the screen is a status line. "DIR: Part (05)" means that an updated directory has been partially received and there are five holes (missing pieces). "AUTO: Dir" shows that the ground station computer directory is being updated automatically from the monitored directory data. "Dir" could be replaced by a file number being downloaded. The values labeled "s:," "b:," "d:" and "e:" stand for

data rate in bytes per second for the last five seconds, number of bytes monitored from broadcast files, number of bytes monitored from broadcast directories and number of CRC errors between the TNC and the ground station computer.

The line beginning with "PB:" shows which stations have made requests for files or directories (or hole fills) to be broadcast. Station call signs with the suffix "\D" have made directory requests while the others have made file broadcast requests.

```
Download:  Priority Auto Grab Never  Fill  Dir      Info. View dir. Quit! Help.
Message    Holes       Size  Offset Rcvd  Dir   3a11 S:EISLOG       T:         F:
                                           Dir   5126 S:j2g         T:VK6AKI   F:VK6BMD
                                           Dir   512f S:Image view  T:VE1HD    F:SM5BVF
                                           Dir   50d6 S:AD920713    T:         F:
                                           Dir   50d9 S:BL920713    T:         F:
                                           Dir   5134 S:F I N N I   T:OH6LFG   F:OH7BY
                                           Dir   5133 S:OH1311T1.Z  T:JA6FTL   F:OH6SAT
                                           Message  4efc heard.
                                           Message  4e32 heard.
                                           Dir   50d8 S:TD920713    T:         F:
                                           Dir   50d7 S:AL920713    T:         F:
OK N0GIB
OK N0GIB
OK N0GIB
PB: WB7QKK KF5OJ\D KC2PH K8TL WB5EKW\D N0GIB\D K8YAH
HIT V2.16 PBP V2.05 DBP V1.00
Mon Jul 13 17:49:21 1992 Uptime= 92/22:9:12    EDAC=  2158   Fmem=4204
Lmem=2741 d:0 s:0.
Open  1 a : W5ERO
OK N5AHD
Open  1 a : W5ERO
OK VE8DX

DIR: Part  (05)    AUTO: Dir                  s:0427  b:007650  d:001505  e:
```

Fig 23.29 — PB display while receiving data from UoSAT-OSCAR 22 satellite downlink.

```
Keys: Prio Auto Grab Never Find aRchview  Quit Help Main   Select=All   Mail
Message   |S|         Subject              |  To  |  From | Posted at |  Size
    5140  |g| answers                      | W3TMZ | N6KK  | 07/13 17:56 |   420
    513f  |g| KCT/T                        |WB0NCR | N6KK  | 07/13 17:54 |   991
    513e  |g| 511                          | N4OUL | N6KK  | 07/13 17:54 |  1090
    5136  |■| CSDP Members                 | ALL CS|WB0NCR | 07/13 17:44 |   443
    5134  |g| F I N N I S H...             |OH6LFG | OH7BY | 07/13 17:29 |  1349
    50d9  |g| BL920713                     |       |       | 07/13 17:26 |  1340
    50d7  |g| AL920713                     |       |       | 07/13 17:23 | 14485
    512f  |g| Image viewer                 | VE1HD |SM5BVF | 07/13 15:55 |   695
    5126  |g| j2g                          |VK6AKI |VK6BMD | 07/13 15:21 | 46410
    5125  |g| VIDEO                        |VK3AHJ | VK8SO | 07/13 13:44 |   753
    5120  |g| eb3cdc.001                   |EB3CDC | EA4RJ | 07/13 12:49 |  1754
    511f  |■| images.hlp                   | ALL   | VE1HD | 07/13 12:48 |   725
    5118  |■| DSP-12 query                 | ALL   | DJ1KM | 07/13 12:41 |   737
    5108  |g| RE_EANET                     |EA5DOM |EB3CDC | 07/13 11:16 |  1796
    5101  |g| Supertrak  again             | ON6UG | ZL1WN | 07/13 10:16 |   810
    50ff  |g| Graphics Packet assist       |ZL1BIV |ZL2AMD | 07/13 10:16 | 10517
    50fd  |g| RE. NET-EA                   | EA4RJ |EB3CDC | 07/13 09:33 |  2826

All Mail AL BL
DIR: Up-To-Date    AUTO: Idle          s:0969  b:369695  d:135249  e:0002
```

Fig 23.30 — A display of a portion of a downloaded directory from UoSAT-OSCAR 22.

The message "Open: 1 a: W5ERO" shows that station W5ERO is a connected-mode user (probably doing a file upload) on uplink 2 and that uplink 1 is available for another user.

Fig 23.30 shows a portion of the ground station computer directory after it has been captured from the downlink traffic. The upper right corner of the screen shows the file broadcast selection criteria in the message "Select = All Mail." This means that message traffic addressed to "All" or traffic to this specific station's call sign will be downloaded automatically. These criteria can be changed to suit the station operator through selection equations employing relational and logical operators that test appropriate fields in the PACSAT

File Header. Consistent with the selection of "All," note that file numbers 5136, 511f and 5118 have a square block in the "S(tatus)" column. This indicates that these files, which have a "to" address of "ALL," have already been downloaded. At the lower left corner of the screen the message "All Mail AL BL" appears. These are the selection criteria for the directory display, as opposed to the criteria for automatic file downloading. Thus, the directory display will show AL (activity log) and BL (broadcast log) files in addition to files addressed to other satellite users.

A user wishing to upload a file would first attach the PACSAT file header to the file and then use the PG program to upload it to the satellite. PG is used only for file

uploading and operates in connected mode using the AX.25 data link layer protocol. When using PG, a connection is established, an upload transaction executed and the connection terminated as a result of a single operator command. The one-transaction-per-connection philosophy ensures maximum utilization of the uplinks in connected mode. There is no wasted time while a ground station operator executes a command and then pauses deciding what to do next.

[This section, including Figs 23.29 and 23.30, is reprinted with permission from "The Development of Low-Earth-Orbit Store-and-Forward Satellites in the Amateur Radio Service," *Proc IEEE International Phoenix Conference on Computers and Communications*, Tempe, AZ, March 23-26, 1993, pp 378-386, ©1993 IEEE.]

WiSP

The software package just described, consisting of the PB, PG, PHS and PFHADD programs, is the set of programs initially made available for accessing satellites utilizing the PACSAT Broadcast Protocol (PBP). More recently, considerable software development activity has resulted in several alternatives to the original program suite. The most significant of these new programs is the Windows application WiSP developed by Chris Jackson, ZL2TPO. An alternative to PB called XPB has been developed for the Linux X-Windows environment by John Melton, G0ORX/N6LYT, and Jonathan Naylor, G4KLX. Finally, a version of PB designed specifically for the IBM OS/2 environment is currently in development. The WiSP and XPB packages are available via FTP from several different sites including ftp.amsat.org. WiSP requires the payment of a registration fee to your national AMSAT organization while XPB falls under the GNU Public License. A brief explanation and a few examples of WiSP operation follow.

Fig 23.31 shows the display produced by the WiSP ground station control (GSC) program. Although the display shows only one satellite, the program may be configured to track multiple satellites with priorities assigned to each. **Fig 23.32** shows the graphical tracking feature of WiSP that may be invoked by the user if desired. When a satellite comes into view, a user-specified program can be run. This program could be something as simple as a terminal program to display raw downlink data. For the digital store-and-forward satellites like AO-16, LO-19, UO-22, KO-23 and KO-25, the MSPE program that is part of the WiSP package will usually be

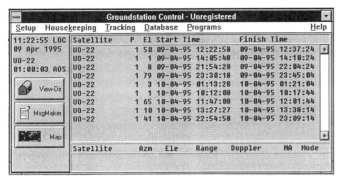

Fig 23.31—WiSP ground station control (GSC) screen showing the next visibility times for UO-22 along with the current clock time and the countdown to next AOS timer.

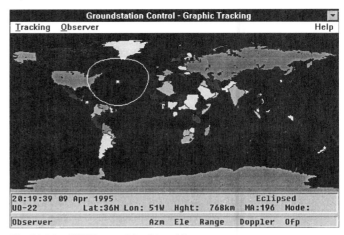

Fig 23.32—WiSP graphical tracking screen.

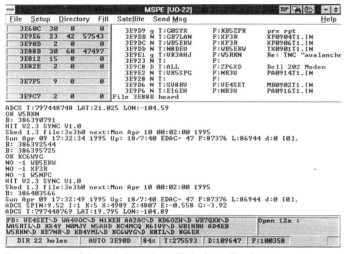

Fig 23.33—WiSP real-time downlink data display screen. This is the WiSP equivalent to Fig 23.29 for PB.

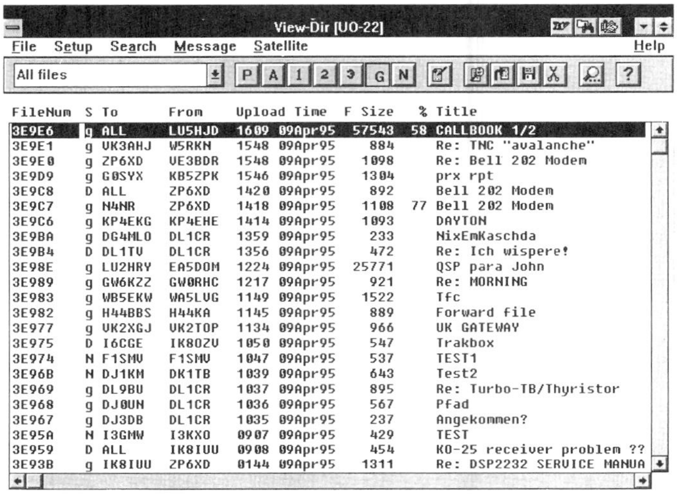

Fig 23.34—WiSP display produced by the View-DIR function. This is the WiSP equivalent to Fig 23.30 for PB.

run. **Fig 23.33** shows a typical screen produced by MSPE while monitoring UO-22. Notice that this is WiSP's equivalent to Fig 23.29 produced by the original PB program. Users may select which files should be automatically downloaded and processed for later reading. Finally, **Fig 23.34** shows a display produced by the View Dir(ectory) function. Once again there is a close parallel between the information shown my View-Dir and that shown in Fig 23.30 from PB. The WiSP package also has radio tuning and rotator control features. Sophisticated ground station software packages such as WiSP truly demonstrate the maturity of the ground segment that supports digital store-and-forward Amateur Radio satellites.

EQUIPPING A STATION

The previous sections have shown there are many satellites supporting Amateur Radio communications in many different modes. The satellite enthusiast must take into consideration his/her own desires, goals and financial resources when purchasing and assembling equipment for an amateur satellite ground station. Because there are so many different combinations of station equipment possible, it would be a good idea to define some broad categories that arise naturally from a combination of the available satellites, individual operating goals and required expenditures.

One possible set of categories for amateur satellite stations consists of: (1) receive-only stations; (2) stations to work LEO satellites with analog transponders; (3) stations to work LEO digital store-and-forward satellites; (4) stations to work HEO satellites with analog transponders; and (5) stations utilizing satellites with uplinks and/or downlinks in the microwave bands (above 450 MHz [70 cm]). As always, there are many trade-offs that can be made. Some of the common ones will be mentioned later.

Perhaps the biggest difference between terrestrial and satellite communications is that the latter is full-duplex operation. This means that you transmit and receive simultaneously. When communicating through an analog transponder, you can hear your own downlink signal while transmitting, as well as that of the station being worked. Full duplex provides the opportunity for a fully interactive conversation, as if the other station is in the very same room.

Successful satellite operation demands that you can locate and hear your own signal from the spacecraft. Choose equipment with this goal in mind. Equipping a station for full-duplex operation is not too difficult because the transmitter is on a different band than the receiver. Ground-station

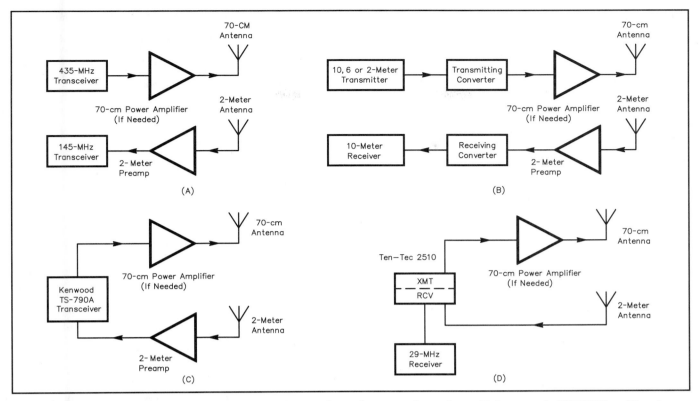

Fig 23.35 — Several different Mode-B satellite-station configurations are shown here. At A, separate VHF/UHF multimode transceivers are used for transmitting and receiving. The configuration shown at B uses transmitting and receiving converters or transverters with HF equipment. At C, a multimode, multiband transceiver can perform both transmitting and receiving function, full duplex, in one package. The Ten-Tec 2510 shown at D contains a 435-MHz transmitter and a 2 m to 10-m receiving converter.

configurations for high-altitude satellites vary according to the communications "mode" being used. **Figs 23.35, 36, 37** and **38** show several different configurations suitable for Modes B, J, L and S. An example of an entry-level receive-only station can be seen in **Fig 23.39**.

Computer System

The main reason for mentioning computer equipment in this section is that in many cases some other part of the station will be used in conjunction with a computer. For example, automatic antenna positioning may be done using a computer and appropriate rotator control interface.

One of the most common uses for a computer in the amateur satellite station is determining when and where a particular satellite will be visible. When considering this aspect of satellite operations, think carefully about what you really need and what "would be nice." If you are an entry-level operator and using omnidirectional antennas, a simple listing of AOS, LOS and position at 1-minute intervals is sufficient. Many different orbital prediction

programs are available; they range in complexity from those that produce simple time, heading and position printed output (see **Fig 23.40**) to those that produce graphical displays in real time (see Figs 23.15 and 16).

You may be better off with one computer dedicated to satellite-related functions rather than trying to do orbit prediction, antenna pointing, radio frequency control and your word processing and financial records all on the same machine. There is nothing more annoying than stopping in the middle of some other important work to get the right programs going for the next satellite contact. With every new generation of microprocessor, systems using the preceding technologies become more plentiful at reasonable prices. As your interest in amateur satellite operations solidifies, keep your eye open for a good computer buy at the next hamfest.

Other Microprocessor-Based Equipment

If you develop a serious interest in the

digital satellites, you will want to consider a DSP-based TNC. Since digital amateur satellites tend to use different modulation techniques than terrestrial packet radio, TNCs will usually require an additional external or internal modem. Although choosing a DSP-based TNC will result in a higher initial cost, the modem then becomes a matter of software rather than additional hardware. Consequently, as new modulation techniques require new modems, only software has to be changed, either by downloading from a PC or changing ROM chips in the TNC.

Antennas

Antennas for receive-only and entry-level stations will usually be simple fixed-position types such as ground plane, turnstile over reflector, dipole and wire loop. High-orbit satellites such as AO-10 require high-gain antennas that are movable in both azimuth and elevation. In between these two extremes are the antenna requirements for the typical LEO satellite station that has evolved beyond the entry level. In this case the gain requirements are not as

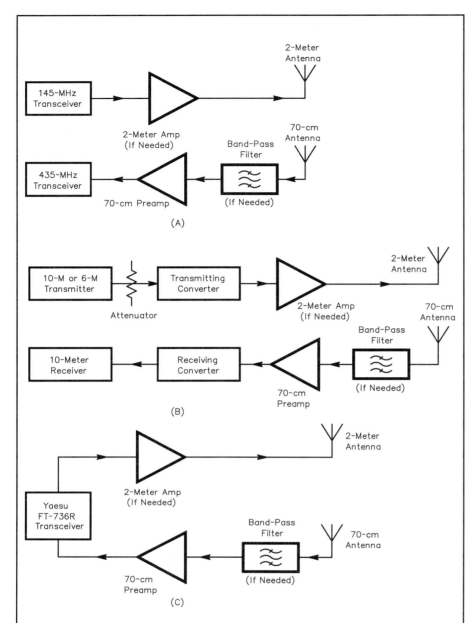

Fig 23.36 — Several different Mode-J satellite-station configurations are shown here. At A, separate VHF/UHF multimode transceivers are used for transmitting and receiving. The configuration shown at B uses transmitting and receiving converters or transverters with HF equipment. At C, a multimode, multiband transceiver can perform both transmitting and receiving functions, full duplex, in one package.

high as for AO-10 but for serious, dependable, day-to-day operation, automated azimuth and elevation positioning will become most desirable.

The best antennas for use in the amateur satellite service are circularly polarized (CP). The present trend in satellite arrays for 145 and 435 MHz is to use two complete Yagis mounted perpendicular to each other on the same boom. One set of elements is mounted $1/4$ wavelength ahead of the other. The antennas are fed in phase and are switchable from RHCP to LHCP. This is in contrast to using helical antennas. Circularly polarized Yagi antenna arrays are manufactured by Cushcraft, Telex/HyGain, M^2 and others. A typical set of crossed-Yagi antennas is shown in **Fig 23.41.**

Satellite antennas should be mounted as close to the station as possible. Height above ground makes no difference for satellite work, except that the antennas must be mounted high enough that trees and other obstructions do not block the view of the satellite at low elevations. A low mount allows use of shorter feed lines (lower losses) and often reduces noise pickup by the antennas. Many operators are able to set up

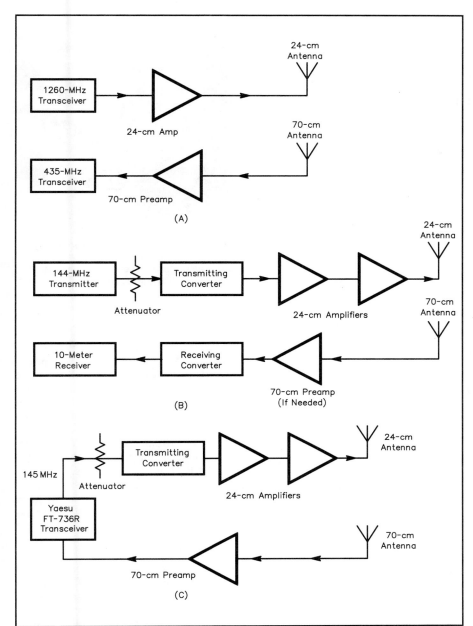

Fig 23.37 — Several different Mode-L satellite-station configurations are shown here. At A, separate VHF/UHF multimode transceivers are used for transmitting and receiving. The configuration shown at B uses transmitting and receiving converters or transverters with HF equipment. At C, a multimode, multiband transceiver can be used for full duplex receiving and transmitting (with the addition of a 2-m to 24-cm transmitting converter).

their antennas on a 10 to 15-ft mast right next to the shack and have only 20 ft of feed line. Plan to use good-quality, low-loss coaxial cable from the start, such as Belden 9913. Even better are runs of Hardline coaxial cable, available from a number of manufacturers.

Mode L transmitting antennas and Mode S receiving antennas have taken numerous forms, mostly based on the technology needed for EME communications. Loop Yagis are popular, as are the large parabolic reflector antennas seen in EME and TVRO services. While higher gain means lower transmitter power, the narrow beamwidths require the operator to reposition the antenna more often.

Although a practical CP Yagi for 24 cm has not yet been demonstrated, such an approach may be feasible. As Mode L is only operated near the satellite apogee, essentially on the satellite antenna pattern main lobe, RHCP operation is the only CP sense needed. This makes the use of a helical antenna attractive. A home-built Mode L helical antenna array is shown in Fig 23.41. Active Mode L operators have also found that a small parabolic dish (6 ft in diameter, or larger) with a circularly polarized feed can make a fine Mode L

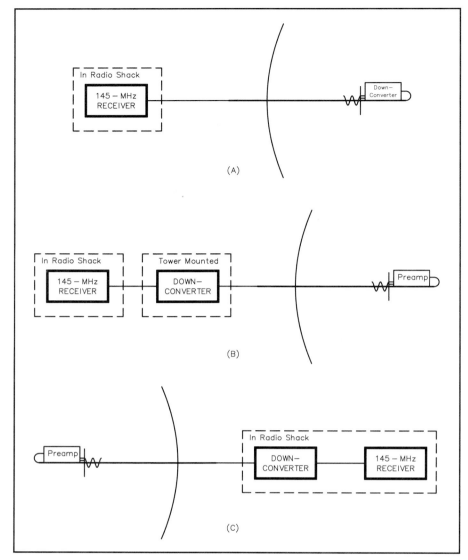

Fig 23.38 — Several different Mode-S satellite-station configurations are shown here.

Fig 23.39 — An entry-level receiving station using 2 m and 70-cm converters and a general coverage communications receiver.

antenna. A number of reasonably priced TVRO dishes are available; they require only the addition of a suitable feed for Mode L service. For TVRO dishes of 12 ft diameter, a dual-band (70 cm and 24 cm) feed for Mode L is a possibility, as the gain at 70 cm is sufficient for excellent reception, and the gain is very substantial for QRP 24-cm transmissions.

For Mode S reception, the most commonly used antennas are a small helix (16 turns) or a small parabolic dish (less than 3 ft).

Antenna Accessories

Long-boom Yagis, such as the KLM antennas shown in Fig 23.41, can suffer from boom sag that might cause pattern distortion and pointing errors. In addition, a boom support is desirable in areas where high winds or ice are a problem. To avoid possible interference with the antenna pattern, the boom brace must be made from nonconductive material such as Phillystran HPTG2100 guy cable. Details for the brace are shown in **Fig 23.42**.

The vertical boom-brace support member is nonconductive and made from a fiberglass fishing rod blank. A short piece of threaded stainless-steel rod inserted in the top of the tube is used to adjust tension on the boom brace. A 2-inch piece of $5/16$-inch copper tubing brazed across the threaded rod in a "T" fashion holds the Phillystran cable in place. Jam nuts secure the threaded rod once the boom is straight.

Experience with the exposed relays on the polarity switchers used on some commercial antennas has shown that they are prone to failure caused by an elusive

```
              ORBIT NO. 20988 EPOCH: 30.219444444384 DATE: 1/30/94

                       D/L     U/L                     SSP   SSP
      UTC     Az   El   Dplr    Dplr  Range  Height    Lat   Long   MA

    05:16:00  164   0   9761   -3259  3222    795       1     90     85
    05:17:00  164   4   9757   -3258  2821    796       4     91     87
    05:18:00  164   9   9694   -3237  2422    796       8     92     90
    05:19:00  164  14   9543   -3187  2029    797      12     92     93
    05:20:00  163  22   9235   -3083  1649    797      15     93     95
    05:21:00  162  33   8601   -2872  1295    797      19     94     98
    05:22:00  160  50   7232   -2415   997    798      22     95    100
    05:23:00  145  76   4274   -1427   821    798      26     96    103
    05:24:00    0  70    -634     211   847    798      29     97    105
    05:25:00  352  46   -5179   1728  1061    798      33     98    108
    05:26:00  350  31   -7660   2557  1376    799      36     99    110
    05:27:00  349  20   -8791   2934  1738    799      40    100    113
    05:28:00  348  13   -9319   3110  2121    799      43    101    115
    05:29:00  348   8   -9578   3197  2515    799      47    102    118
    05:30:00  348   3   -9700   3237  2915    799      50    104    121
    05:31:00  348   0   -9746   3253  3316    799      54    106    123
```

Fig 23.40 — Tabular output from an orbit prediction program showing time and position information for AO-16.

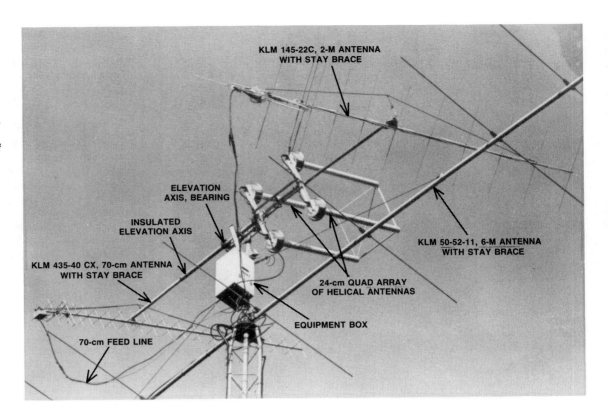

Fig 23.41 — A popular commercially manufactured antenna array for AO-10 Modes B and J, is a pair of KLM crossed Yagis. Shown also are Mode-L helical antennas. The large box on the mast contains a 2-m and 70-cm preamplifier, coaxial relays, a 24-cm transverter and power amplifier, and power-supply regulators.

KLM 145-22C, 2-M ANTENNA WITH STAY BRACE

ELEVATION AXIS, BEARING

INSULATED ELEVATION AXIS

KLM 435-40 CX, 70-cm ANTENNA WITH STAY BRACE

KLM 50-52-11, 6-M ANTENNA WITH STAY BRACE

24-cm QUAD ARRAY OF HELICAL ANTENNAS

EQUIPMENT BOX

70-cm FEED LINE

mechanism known as "diurnal pumping." The relay is covered with a plastic case, and the seam between the case and PC board is sealed with a silicone sealant. It is not hermetically sealed, however. As a result, the day/night temperature swings pump air and moisture in and out of the relay case. Under the right conditions of temperature and moisture content, moisture from the air will condense inside the relay case when the air cools. Water builds up inside the case, promoting extensive corrosion and unwanted electrical conduction, seriously degrading relay performance in a short time.

If you have antennas with sealed plastic relays, such as the KLM CX series, you can avoid problems by making the modifications shown in **Fig 23.43**. Relocate the 4:1 balun as shown and place a clear polystyrene plastic refrigerator container over the relay. Notch the container edges for the driven element and the boom so the container will sit down over the relay, sheltering it from the elements. Bond the container in place with a few dabs of RTV adhesive sealant.

Position the antenna in an "X" orientation, so neither set of elements is parallel to the ground. The switcher board should now be canted at an angle, and one side of

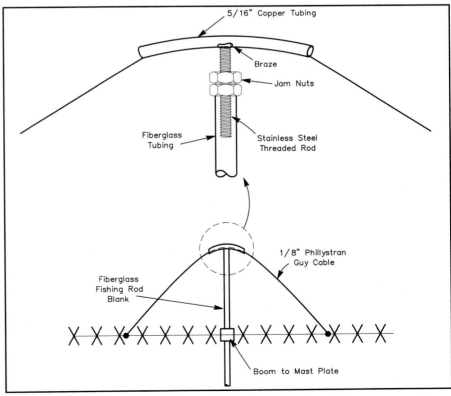

5/16" Copper Tubing

Braze

Jam Nuts

Fiberglass Tubing

Stainless Steel Threaded Rod

1/8" Phillystran Guy Cable

Fiberglass Fishing Rod Blank

Boom to Mast Plate

Fig 23.42 — A boom brace may be desirable for long Yagis. This arrangement is made from non-conductive material to prevent undesirable effects on the antenna pattern.

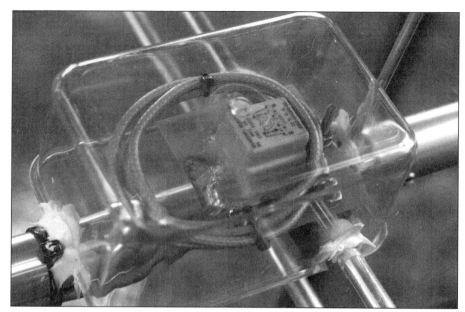

Fig 23.43 — KLM 2M-22C antenna CP switching relay with relocated balun and protective cover.

the relay case should be lower than the other. Carefully drill, by hand, a pair of $^3/_{32}$-inch holes through the low side to vent the relay case. The added cover keeps rain water off the relay, and the holes will prevent any build-up of condensation inside the relay case. Relays so treated have remained clean and operational over periods of years without any problems.

Antenna Rotators

Unlike stations located on the surface of the Earth, AO-10 and Phase 3D will be found somewhere in the sky above. Operators commonly aim antennas toward another station by changing the pointing angle, or *azimuth* (sometimes called az). Aiming antennas toward AO-10 and Phase 3D requires the control of antenna *elevation* (el). Satellite antennas must be able to rotate from side to side and up and down simultaneously. While the use of electrically controlled antenna rotators will be discussed here, it might be noted that Phase 3 satellite motions are slow enough that hand-operated, "armstrong" antenna control is feasible. At times, the antennas may not need repositioning for periods of up to four hours. On the other hand, the fast-moving LEO satellites such as AO-16 and UO-22 will almost certainly require an automatic positioning system.

Azimuth rotators are commonly used for positioning terrestrial HF and VHF antennas. Antennas for low-orbit satellites

can be on the smaller and lighter side, so light-duty TV-antenna rotators such as those sold by Alliance, Channel Master, Radio Shack and others could be used for the azimuth rotator. Today's high-gain satellite arrays are a bit large for these light-duty rotators. Look for something more robust, such as a rotator recommended for turning a small HF beam or VHF array. Various models manufactured by Alliance, Daiwa, Kenpro, Telex and others are advertised in *QST*.

Elevation rotator selection is more limited. Commercially manufactured models such as the Yaesu G-500A and G-5400B are available. The G-500A has been available in the past under different designations such as KLM and Kenpro KR-500 and is designed for elevating small-to medium-size VHF or UHF arrays. The G-5400B is a combined azimuth and elevation rotator. See **Fig 23.44**. Home-built elevation mounts can also be fabricated from TVRO antenna jack-screw motion controls or other similar muscular devices.

A lower cost alternative is the Alliance U110 TV-antenna rotator. Rotators of this type have been used successfully by satellite operators for many years. Despite its relatively light construction, the U110 will handle antenna loads weighing up to 40 pounds. The key to success is to achieve a static balance of the antenna mass so the rotator does not have to elevate a "dead" load. A highly attractive feature of the el-

evation rotators noted above is that the cross boom to be rotated passes completely through the rotator. This allows the mounting of one antenna on each side of center and the adjustment of their respective positions for a side-to-side balanced load.

Automatic Antenna Positioning

Several products are now available to automatically steer a satellite antenna array under computer control. Automatic antenna pointing is particularly useful when operating the OSCARs in low Earth orbit. Among the products available are the Kansas City Tracker/Tuner sold by L. L. Grace Co and the TrakBox sold by TAPR. There are other sources for similar products but these two represent the two general approaches to automated antenna control. The Kansas City Tracker is installed in an existing station computer while the TrakBox is a standalone controller with its own built-in microprocessor.

The Kansas City Tracker is available either as a tracker only or with the tuner option which sets and corrects the radio frequency throughout a pass. In either case, the tracking and/or tuning functions are carried out by the rotator control card in conjunction with Terminate and Stay Resident (TSR) programs running in the computer. Position and Doppler frequency correction information for the TSRs is supplied by the tracking program. Some rotators, such as the Yaesu G-5400B, already have a connector for the computer interface.

The TrakBox and similar units are specialized microprocessor-based control units. Usually, the position information is downloaded from a PC to the control box. After receiving the position information, the controller operates in a standalone mode. Devices such as the TrakBox have an LCD display that shows the operational status of the controller.

Whether to use a tracker and/or tuner that plug into the PC expansion bus or a standalone unit is a complicated question involving both technical matters and personal preference. If your station computer is using at least an 80386 class processor, the question is largely a matter of preference. Consider the situation of working one of the digital satellites like UO-22. With everything running in a single computer that machine must: update antenna position, set the radio frequency often enough to keep up with the Doppler shift, accept data from the TNC at a rate of 19,200 bps, update files and directories on disk as data is being captured, and update the screen display.

Finally, even though a tracker/tuner

Fig 23.44 — The Yaesu G-5400B azimuth-elevation rotator includes a DIN computer connection.

construction project has not been included in this section, it is an area where homebrew systems are certainly a possibility. Relay drivers can be built to control the rotators via signals from a parallel port. Analog-to-digital converters can be used to read the position indicating voltage from the meter circuit. Furthermore, most modern transceivers can be controlled by commands from a serial I/O port. Consequently, it is not beyond the capabilities of someone with programming and hardware experience to build an antenna and frequency control system.

Antenna Cross-Boom Construction

One requirement not commonly discussed is that of using a nonmetallic elevation axis boom for antennas that have their boom-to-mast mounting hardware in the center of the boom. A metal cross boom will seriously distort the beam pattern of a circularly polarized antenna, so it is important to make those portions of the cross boom nearest to the antennas from nonmetallic materials.

From a structural standpoint, the best nonmetallic material for this job is glass-epoxy composite tubing, because its stiffness is excellent. Lengths of this material may be found at an industrial supply house that specializes in plastics. Also, KLM sells lengths of 1 1/2 in. OD fiberglass

masting for this purpose, and Telex/HyGain includes fiberglass masting with their OSCAR antennas. If you have a rotator that will accept a 1 1/2 in. elevation boom, then your best bet is to use a single piece of this tubing. The 1 1/2 in. OD fiberglass tubing also slides into the ID of the 1 1/2 in. pipe used for the heavy-duty elevation axis.

A less expensive alternative is to make the cross boom from a combination of metallic and nonmetallic tubing. For strength and stiffness, use a short length of steel or aluminum tubing through the middle of the rotator. Let the metal tubing extend for about 6 in. on each side of the rotator. Then install nonmetallic masting, such as common PVC pipe, over the steel stubs.

The elevation boom pictured in Fig 23.41 was constructed with this method. The center piece that fits through the U110 is a 2-ft section of 1.33-in.-OD steel tubing that originally was part of the top support rail of a chain-link fence. Attached to the steel stub on each side of the rotator is a 4-ft length of 1 1/4 in., schedule-40 PVC pipe. This pipe slipped nicely over the center stub. The fit is perfect — no machining was needed.

Unlike glass-epoxy tubing, PVC pipe is not very stiff. The secret to making PVC pipe capable of supporting satellite Yagis is to insert a wooden dowel into the PVC

pipe, along its entire length. The finished dimension of 1 3/8 in. wooden clothes-rod dowel (the kind you might hang inside a closet) is just perfect for a slide fit into the pipe. This material is available from most lumber yards. Add a few 1/4 in. bolts to each side to secure the pieces, and you have a sturdy, inexpensive, nonmetallic elevation boom.

Receivers

The old adage "You can't work 'em if you can't hear 'em" especially applies to satellite operation. Receiving requirements for AO-10 are demanding, but pleasurable results can be achieved with the right kind of equipment. OSCAR operation is a weak-signal mode where contacts can be made with signals that are only 4 dB stronger than the noise. Conversational quality can be assured with signals that are 6 to 9 dB above the noise.

The first step to be taken before attempting to work such high altitude satellites as AO-10 is to assemble the best receiving setup possible. There is no point in getting transmitting capability until the satellite signals can be comfortably heard.

Amateurs active on 2 m with a multimode transceiver already have the basic building block for receiving Mode B and transmitting on Modes J and L (with an additional transmitting converter). If you currently have no VHF equipment, consider that a multimode, multiband transceiver will also allow you to explore the exciting world of terrestrial 2-m and 70-cm SSB operations. The basic requirements are that the rig includes SSB and CW modes and that it covers the entire 144-MHz band and (most of the) 420-MHz band. A multimode transceiver also makes an excellent replacement for an FM-only 144-MHz rig.

The equipment manufacturers listed in **Table 23.10** all make suitable transceivers, either single-band or multiband units. The current crop of base-station rigs includes the Kenwood single-band transceivers TS-711A (2 m) and TS-811 (70 cm), and multiband transceiver TS-471H and IC-475H (70 cm). Yaesu offers their multiband FT-726R and FT-736R transceivers. There are also several compact multimode radios intended for mobile use that will be quite usable. These include the Yaesu single-band FT-290R and FT-790R, Kenwood single-band TR-751A and TR-851A, and the ICOM single-band IC-290H. In addition, there are often good buys on the used market, if you're interested in an older radio. Gear such as the Kenwood TS-700 series, Yaesu FT-225RD and ICOM IC-

The Yaesu 736R is a multimode transceiver for 2 m (144-148 MHz) and 70 cm (430-450 MHz). It can be used for full-duplex receiving and transmitting on AO-10 Mode B and AO-13 Modes B and J. An optional 23-cm module covers 1230-1300 MHz (Mode L). Approximate power output: 20 W on 2 m and 70 cm, and 10 W on 23 cm.

251 are still popular. Many of these transceivers have been reviewed in *QST*.

Users of Mode L may want to consider the use of a full transceiver in the station for the 24-cm transmissions, as such a unit will also allow operations on the 23-cm band (1296 MHz). Kenwood and Yaesu offer 23-cm modules for their multiband transceivers for this service. ICOM also offers the single-band IC-271A. Alternatively, there are some 24-cm transmitting converters and transverters offered that employ a 2-m IF.

An excellent solution to receiving Modes B, J and L satellite signals can be found in the form of receiving converters used with a high-quality HF transceiver or receiver. The 2-m and 70-cm receiving converter consists of a mixer and a local oscillator and may contain a preamplifier. The local oscillator frequency is usually chosen so that signals will be converted to the 10-m band. In addition, a number of manufacturers offer transverters that include receiving and transmitting converters in the same package. Receiving converters are available from several suppliers listed in Table 23.10.

There are several advantages to using a receiving converter. Modern HF transceivers and receivers most likely have excellent frequency stability, a frequency readout in 1 kHz or smaller steps, good SSB and CW crystal filters, an effective noise blanker and high dynamic range. Chances are good that a multimode VHF transceiver will offer some, but not all, of these features. Cost is another factor. If you already own an HF rig, but are not interested in terrestrial VHF/UHF SSB operation (you don't need 2-m transmit capability for Mode B), the cost of building or buying a superior receiving converter will be significantly less than that of even an older multimode transceiver. One commercial example of a Mode-B-only unit is the Ten-Tec 2510B, providing an excellent 2-m receiving converter and a complete 70-cm SSB/CW transmitter, all with coupled VFOs for simultaneous tracking on both bands. While the '2510 is an economical way to get into satellite operation, it is limited to Mode B service only and cannot provide any help for Mode J and L services.

Experience has shown that daytime noise will often raise the practical 2-m receiver noise floor by 10 to 20 dB, thus making Mode B daytime communications difficult, at best. Weak downlink signals are often no match for the noise. In general, noise is not a problem on 70-cm (for Mode J and Mode L reception), but in some areas interference from airport radar can be troublesome. In addition, local FM repeaters may be heard in the satellite passband of the ground-based receiver because the VHF transceiver may offer poor rejection of strong nearby signals. Use of a high-dynamic-range receiving converter with a good HF transceiver has been shown to solve both of these problems. The lesson is that many VHF transceivers have noise blankers that are inadequate for VHF/UHF satellite operation and some VHF transceivers do not work well in areas with many nearby, strong signals. Consequently, in some cases, bet-

Table 23.10

Suppliers of Equipment of Interest to Satellite Operators

Contact information appears in the *Handbook* Address List in the **References** chapter. Send updates to the *Handbook* Editor at ARRL Headquarters.

Multimode VHF and UHF Transceivers and Specialty Equipment

ICOM America
Kenwood Communications
Yaesu USA

Converters, Transverters and Preamplifiers

Advanced Receiver Research
Angle Linear
Hamtronics
Henry Radio
The PX Shack
Radio Kit
RF Concepts
Spectrum International
SSB Electronics

Power Amplifiers

Alinco Electronics
Communications Concepts
Down East Microwave
Encomm
Falcon Communications
Mirage Communications
RF Concepts
TE Systems

Antennas

Cushcraft Corp
Down East Microwave
KLM Electronics
Telex Communications

Rotators

Alliance
Daiwa
Electronic Equipment Bank
Kenpro
M² Enterprises
Telex
Yaesu USA

Other Suppliers

AEA
ATV Research
Down East Microwave
Electronic Equipment Bank
Grove Enterprises
M² Enterprises
Microwave Components of Michigan
PacComm
SHF Microwave Parts
Tucson Amateur Packet Radio
 (TAPR)

Note: This is a partial list. The ARRL does not
 endorse specific products.

ter results may be achieved with a receiving converter than with a VHF multimode transceiver.

Receiving Accessories

For those stations using a receiving converter and an HF receiver for downlink reception, an in-line switchable attenuator, installed between the converter output and the antenna jack of the 10-m receiver, may prove useful. Such an attenuator can be used to lower the AGC level and improve the perceived signal-to-noise ratio. In addition, by adjusting the attenuator so that the S meter on the HF rig rests at zero at no signal, more accurate signal reports can be given. An attenuator circuit is shown in the **Test Equipment** chapter. A useful modified form may include only three steps — 5, 10 and 20 dB. These three settings allow attenuation in 5-dB steps from 0 to 35 dB.

Preamplifiers

No discussion of satellite receiving systems would be complete without mentioning preamplifiers. Good, low-noise preamplifiers are essential for receiving weak downlink signals. Multimode rigs and most transverters will hear much better with the addition of a GaAsFET preamplifier ahead of the receiver front end, albeit at the expense of a considerable reduction in the third-order intercept point of the receiver. While a preamplifier can be added right at the receiver in the station, it may not do much good there. Considerably better results can be obtained if the preamp is mounted near the antenna. Indeed, antenna mounting of a preamp is essential for UHF and higher operation. Losses in the feed line will seriously degrade the noise figure of even the best preamplifier mounted at the receiver, while an antenna-mounted preamplifier

can overcome nearly all of these noise figure problems.

Table 23.10 lists several sources of commercially built preamplifiers. These are available in several configurations. Some models are designed to be mounted in a receive-only line, for use with a receiving converter or transverter. Others, designed with multimode transceivers in mind, have built-in relays and circuitry that automatically switch the preamplifier out of the antenna line during transmit. Still others are housed, with relays, in weatherproof enclosures that mount right at the antenna. For the equipment builder, several suitable designs appear in *The Radio Amateur's Satellite Handbook*.

Tower-Mounted Preamplifiers

Mast-mounting of sensitive electronic equipment has been a fact of life for the serious VHF/UHFer for years, although it may seem to be a strange or difficult technology for many HF operators. To get the most out of your satellite station, you'll need to mount a low-noise preamplifier on the tower or mast, near the antenna, so that feed-line losses do not degrade low-noise performance. Feed-line losses ahead of the preamplifier add directly to receiver noise figure. A preamp with a 0.5-dB noise figure will not do you much good if there is 3 dB of feed-line loss between it and the antenna.

In Fig 23.41, note the large white box located below the elevation rotator. **Fig 23.45** shows the interior of this box. A close look shows a 2-m preamplifier for the Mode B downlink and relays to switch it in and out of the line to the antenna. This setup is designed to beat excessive feed-line losses for basic stations. Transmitting equipment for the 70-cm and 24-cm bands is also housed in the same enclosure. Normal installations may require only the receiving preamplifier (2 m for

Mode B and 70 cm for Mode J and L) to be mounted on the mast. The transmitting equipment is discussed further in the next section.

Transmitters

The AO-10 Mode B uplink requires a controllable 5 to 50 W of 435-MHz RF power at the antenna. This assumes a good antenna, which will be discussed later. Feed-line losses in a typical 435-MHz installation can easily run 3 dB, so you'll need anywhere between 10 and 100 W output from your transmitter.

Since there are many combinations of transmitter power and antenna gain that

Fig 23.45 — Interior of the tower-mounted equipment rack with the cover removed. The 70-cm equipment is on the left, while power-supply regulators, a 24-cm transmit converter and a 2-m preamp are mounted on the right.

Kenwood's TS-790A is a dual-band (144-148 MHz and 430-450 MHz) multimode transceiver. A 23-cm(1240-1300 MHz) module is optional. Power output is 35-45 W on 144 MHz, 30-40 W on 430 MHz and 10 W on 1240 MHz.

The ICOM IC-820 is a state-of-the-art transceiver designed especially for satellite use.

will result in a satisfactory signal through AO-10, satellite users generally talk about their uplink capability in terms of effective radiated power (ERP). ERP takes into account antenna gain, feed-line loss and RF output power. For example, a 10-W signal into a 3-dB-gain antenna will have an ERP of 20 W (3 dB greater than, or twice as strong as, 10 W). This assumes no loss in the feed line and all 10 W from the transmitter reaches the antenna. If the signal is 10 W into a 10-dB-gain antenna, the ERP is 100 W. The same 100-W ERP can be achieved with a 50-W transmitter and a 3-dB-gain antenna.

Stations with an uplink ERP as low as 10 W can be copied through AO-10, Modes B and J, but ERP levels of 100 to 400 W are the norm. No matter what your ERP, your signal on the downlink should never be stronger than the general beacon at 145.81 MHz (Mode B) and 435.65 MHz (Mode J). As a reminder, satellite service ground stations do not have to be as strong as the beacon to provide excellent communications; good operators will adjust their signals to just the level needed for the QSO. You must have a way of adjusting your uplink signal power so that your downlink is no stronger than the beacon. These points are discussed in detail in *The ARRL Operating Manual*.

If the Mode-B satellite ground station has a 10-W transmitter, a short run of low-loss feed line and good antenna gain, an additional amplifier would probably not be needed. If losses and gains do not add up to the required ERP, a 30 to 40-W amplifier may be needed. Some operators have 100-W amplifiers, but with the antennas available today, use of that much power is guaranteed to create an uplink signal that far exceeds the beacon level. This is considered by good operators to be an antisocial action. Considerate operators

with the 100-W amplifiers quickly reduce drive power to lower the ERP to acceptable levels.

Most satellite operators use UHF multimode transceivers to generate Mode B uplink signals. The manufacturers listed in Table 23.10 make 70-cm multimode transceivers that are similar to the 2-m units as described earlier. Although most of these transceivers provide 10-W output, some can deliver 25 W or more.

Earlier satellites in the AMSAT programs created and fostered the need for good multimode 2-m transceivers, as they formed the nucleus of the satellite station. Current and future satellite programs will find the 70-cm transceiver as the focal point of the satellite station, emphasizing the trend to even higher frequencies.

For Mode L transmitting, there are several transmitting converters, transverters, amplifiers and even a multimode transceiver available from the suppliers listed in Table 23.10.

Transmitting Accessories

The tower-mounted equipment rack shown in Fig 23.45 contains the 70-cm and 24-cm power amplifiers. Tower mounting of a transmitter is probably unnecessary for 70-cm, but becomes much more important for higher frequencies. Feed-line losses at 70 cm are generally twice those of 2 m, while those at 24 cm are about twice those of 70 cm, or four times those of 2 m. For an 80-ft feed line, the losses at 24 cm can easily reach 6 dB for even the best coaxial cable. Consequently, a 100-W amplifier in the shack will only yield 25 W at the antenna. A good alternative is to place the transmitting converter and a 20-W solid-state amplifier in the tower-mounted box near the antennas. The results are nearly the same, and you

avoid the time and money needed to generate high power that will just be lost in the feed line anyway.

A coaxial RF sampler is connected to the output of the 70-cm amplifier since it is good amateur practice to monitor the power at the antenna to be sure the transmitter is working properly. Coaxial relays are used for proper switching of the power amplifiers and 70-cm preamp for OSCAR Mode J and L operation as well as for terrestrial communications.

One very important aspect of using GaAsFET preamps with transmitting equipment is getting everything to switch at the proper time. If transmitters, amplifiers and antenna relays are keyed simultaneously, it's likely that RF will be applied to the feed line before the relays are fully connected to the antenna load. Such hot switching can easily arc the contacts on expensive coaxial relays. In addition, if the TR relay is not fully closed, RF may be applied to the preamplifier. Such bursts of RF energy are guaranteed to destroy the GaAsFET in the preamplifier. Many pieces of transmitting equipment (especially multimode transceivers) emit a short burst of RF power when switched on or off, so there is the risk of transmitting into your preamp even if you are careful to pause before keying.

Ideally, keying of a transmitter should follow a timing sequence that will ensure the safety of the equipment. When you switch into transmit from receive, the coaxial relays change state to remove the preamplifier from the line. Next, the power amplifier is keyed on. The last thing that happens is that the transmitter RF is enabled. When switching back to receive, the sequence is just the opposite. First, the transmitter RF is switched off, the power amplifier is dis-

The Yaesu FT-847's satellite mode provides the user with a full featured satellite transceiver. As a bonus, it also covers all the HF bands!

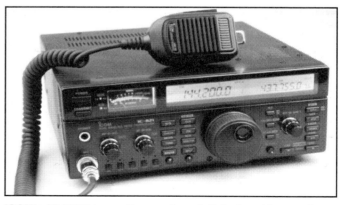

ICOM's IC-821H is another modern, full-featured transceiver designed for satellite use.

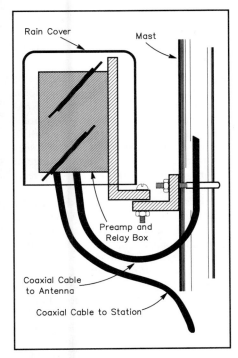

Fig 23.46 — Protection for tower-mounted equipment need not be elaborate. Be sure to dress the cables as shown so that water drips off the cable jacket before it reaches the enclosure.

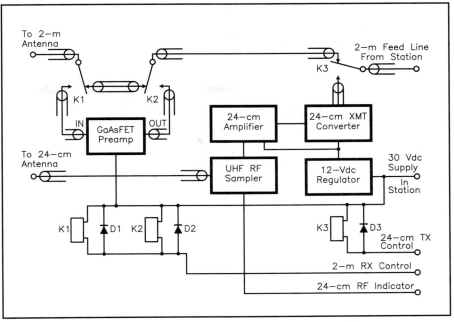

Fig 23.47 — Control circuitry for the mast-mounted 2-m preamplifier and 24-cm Mode-L transmitter. K1-K3 are surplus coaxial relays.

abled, and then the TR relays change state to place the preamp back in service. Solid-state sequencers to control station TR switching are shown in the **Station Accessories** chapter.

Specialized Transceivers

Separate transceivers or transmitting and receiving converters are no longer the only way to go. Modern equipment offerings by ICOM, Kenwood, Yaesu and Ten-Tec, tailored for the satellite user, do it all in one package.

The Yaesu FT-847, for example, covers all bands through 450 MHz—including HF. In satellite mode, it has duplex crossband capability. Its dual VFOs can be tied together, making it easy to track frequencies. This is particularly helpful when using satellites with inverting transponders, such as FO-29 and the upcoming Phase 3D. When the radio is in this configuration, tuning the uplink VFO causes the downlink VFO to tune automatically in the opposite direction.

ICOM's IC-821H is a dual-band (2-m/70 cm) multimode transceiver that also features a handy satellite mode. Like the FT-847, its dual-VFO design makes it easy to tune satellites with inverting transpon-

ders. In satellite mode, the Main (transmit) and Sub band frequencies follow each other, either in the same direction or in reverse.

The Kenwood TR-790A and Yaesu FT-726R and FT-736R units start out as 2-m multimode transceivers. They are, however, expandable to work on other bands with the addition of optional modules. The Mode-B satellite operator would most likely be interested in the TR-790A or FT-726R/FT-736R with the stock 144-MHz and 430-MHz modules. These same RF modules will also serve well for Mode J directly and Mode L using an outboard 23-cm transmitting converter. Both manufacturers also offer 23-cm transceive modules, for Mode L, for their multiband units. To tie it all together, these Kenwood and Yaesu transceivers also both offer satellite modules (stock or optional) that allow the amateur to transmit on one band for the uplink while receiving on another band for the downlink. This is full duplex operation; the effect is the same as having two separate radios in one box.

Tower-Mounted Equipment Shelters

A great many amateurs seem apprehensive about placing their valuable radio equipment outdoors. Such fears are unfounded if adequate care is taken to protect the equipment from the elements. The equipment shown in the photos has been outdoors

for years without any adverse effects.

Fig 23.46 shows the basic scheme for weatherproofing tower-mounted equipment. The fundamental concept is to provide a cover to shelter equipment from rain. A deep drawn aluminum pan can make an excellent shelter. A trip to the housewares section of the local department store will reveal a variety of plastic and aluminum trays and pans that can make suitable rain covers. Polyethylene plastic is not durable in the sunlight. Clear polystyrene refrigerator containers work better than those made of polyethylene, and aluminum is best of all. Choose a cover that is large enough for your equipment; remember to leave room for connecting cables.

The bottom of the rain cover is open to the elements. This is done on purpose and will not cause any problems. Do not try to hermetically seal the enclosure. By leaving the bottom open, adequate ventilation will prevent accumulation of water condensation. Just make sure that water cannot run into the enclosure by way of cables coming from above. Form the cables as shown in Fig 23.46 to provide drip loops. Adding a piece of window screen over the opening should be considered to avoid infestations of nesting insects, such as wasps, and the wire-hungry ravages of squirrels.

The mast-mounted enclosure shown earlier is a welded aluminum box purchased from a surplus dealer. It was used because it was available and the price was right. You

Fig 23.48 — The satellite station at WD4FAB, with a home-brew antenna controller mounted above a low-profile station controller. On the right is a Yaesu FT-726R transceiver, with a power meter, clock and 23-cm transverter mounted on top.

don't really need a big box like this if you just want to protect a preamp and relays.

Station Control

Fig 23.47 is a schematic diagram of the control circuitry for the 2-m side tower-mounted rack. Parts of this diagram will be helpful, even if only the preamp is mounted at the antenna. Note that this circuit is designed around the surplus coaxial relays that were available at the time. Your version will probably be different and will depend on the relays available to you.

This circuitry performs several functions. For starters, it places the preamp in the line only during receiving periods and takes it out of the line during transmitting periods as well as at those times when the station is not in use. This is needed if the satellite array is used for terrestrial transceive operation as well. The switching arrangement shown also protects the preamp from stray electromagnetic pulses (EMP), such as lightning strokes, when the station is not in use. EMP protection is desirable even if the antenna and preamp are used only for receiving satellite signals.

Fig 23.47 is only a little more complicated than the average mast-mounted preamp setup because it also allows 2-m RF to drive a 24-cm Mode L transmitting converter. An extra relay (K3) is used to switch between 2-m and 24-cm operation. K1, an SPDT transfer relay, switches the antenna between the input of the preamp and a through line to K2. K2, another SPDT relay, switches the feed line, from the shack, between the preamp output and K1. The relays are connected so that they must be energized to place the preamp in line, thus ensuring that the preamp is disconnected when not in use.

Depending on the complexity of your sat-

ellite station, you might want to combine most of the switching and control circuitry into a single box so that it provides ready access to all controls. **Fig 23.48** shows a Minibox cut to a low profile (beneath the multi-antenna rotator control unit) that contains all of the switches needed to control the station accessories. It houses a TR sequencer circuit board as well. This box controls the following functions: change antenna polarization from RHCP to LHCP on 2 m and 70 cm; switch the 2-m and 70-cm preamplifiers in and out of the circuit; and switch the power amplifier in or out of the line. The box also contains a high performance 2-m receive converter, for use with the station HF transceiver, when the Mode B conditions get to be really difficult. As shown, home-built control unit boxes can be dressed up with the addition of a plotted or printed label on white paper. The label is bonded to the panel with the use of thin double-sided adhesive tape. A covering of clear label tape protects the label from smudges and dirt.

Station Equipment Summary

Satellite communication, like any other facets of Amateur Radio, requires some specialized station equipment and accessories. Having the best equipment does not necessarily guarantee success. There are a number of "hints-and-kinks" type ideas that can make OSCAR operation far more satisfying. Some of the equipment items that have been discussed here provide capabilities beyond the bare minimum needed for successful satellite operation. They may also be of value in VHF and UHF terrestrial work. Design your station to suit your own needs. Some operators may continually tinker with their station equipment, making frequent improvements, as part of their participation in the Amateur Radio hobby.

SUPPORTING CONCEPTS AND THEORY

Previous sections have described the satellites and communication modes available as well as various types of equipment needed for successful satellite operations. As is often the case with other specialized communications systems, there are certain concepts and theory peculiar to satellite communication. The purpose of this section is to review some topics that range in importance from "essential" to "interesting to know."

Orbital mechanics and tracking will be examined first. The material presented here will be more descriptive than mathematical. For a complete treatment of this subject, the reader is directed to *The Radio Amateur's Satellite Handbook* published by the ARRL. The need to use circularly polarized antennas was mentioned in the discussion of equipping a station. This section includes a more complete discussion of circular polarization along with a related phenomenon whose effects can be mitigated somewhat by using circularly polarized antennas — spin modulation. The sections on path loss and link budget will help explain why certain combinations of receiving and transmitting equipment and antennas were recommended.

Tracking and Orbital Mechanics

In order to complete a QSO via an Amateur Radio satellite, the ground station operators must be able to predict when the satellite will be visible at their respective locations. Previous sections have recommended a stepwise approach to the complexities of amateur satellite operation. Expertise in the area of satellite tracking can be gained in a similar manner.

One method of tracking that is adequate for operations using omnidirectional antennas on LEO satellites uses the *Oscarlocator,* published by the ARRL as part of *The Satellite Handbook.* The *Oscarlocator* is a graphical tracking aid that consists of a satellite ground track indicator and a set of range circles overlaid on a polar projection map of the world. Since the ground track is calibrated in minutes, it is only necessary to position the track at the correct equator crossing location and note how many minutes later the satellite will cross inside the range circle.

There are, of course, many computer programs available for satellite tracking. These range in complexity from those that produce text output such as shown in Fig 23.40 to those that produce graphical output in real-time such as seen in Figs 23.15 and 16. It is also possible to obtain real-time tracking programs that will, in conjunction with the proper antenna controller, position your

Table 23.11

Polarization and Gain of AO-10 Antennas

High Gain Antennas			Omni Antennas	
Frequency (MHz)	Polarization	Gain (dBi)	Polarization	Gain (dBi)
146	RHCP	9.0	Linear	0.0
436	RHCP	9.5	Linear	2.1
1269	RHCP	12.0	RHCP	0.0

Fig 23.49 — This diagram shows the far-field radiation pattern (undeformed) for the 2-m high-gain antenna on AO-10 and AO-13. (Adapted from *The AMSAT-Phase III Satellite Operations Manual*, prepared by AMSAT and Project OSCAR.)

antennas automatically.

All of these programs rely on what is called a reference *Keplerian element set* for their operation. The Keplerian element sets for Amateur Radio satellites are distributed via on-the-air nets, the Internet and printed media.

The reference element sets used by the program need to be updated periodically. This is necessary because a future position is being computed based on a known previous position specified by the element set. As the length of time between the reference element set and the future prediction increases, the predictions are prone to inaccuracies due to perturbations of the orbit such as atmospheric drag.

How often the reference elements should be updated is dependent on the satellite. Reference elements for LEO satellites should probably be updated about once per month. Element sets for HEO satellites such as AO-10 and Phase 3D need not be updated nearly as often. This is particularly true if the element set in use has been produced by averaging previous element sets over a long period of time. Other special cases such as the US Space Shuttle SAREX missions can require very frequent updates of the elements sets because the orbit geometry can change significantly and frequently depending on the goals of the mission. Reference element sets are distributed by AMSAT every week by many media including HF and VHF packet radio networks, the packet radio satellites such as AO-16 and UO-22, as well as commercial information services such as CompuServe's HamNet.

Satellite tracking software is available for a wide variety of computers from many different sources. However, if you purchase your tracking program from AMSAT-NA, you can be sure that your contribution will help finance the next Amateur Radio satellite project. No matter where you purchase your program, you should be sure that it has the features you need and will run correctly on your computer system.

Circular Polarization

In the HF bands, polarization differences between antennas are not really noticeable because of the nonlinearities of ionospheric reflections. On the VHF and UHF bands, however, there is little ionospheric reflection. Cross-polarized stations (one using a vertical antenna, the other a horizontal antenna) often find considerable difficulty, with upwards of 20-dB loss. Such linearly polarized antennas are "horizontal" or "vertical" in terms of the antenna's position relative to the surface of the Earth, a reference that loses its meaning in space.

The need to use circularly polarized (CP) antennas for space communications is well established. If spacecraft antennas used linear polarization, ground stations would not be able to maintain polarization alignment with the spacecraft because of changing orientation. Ground stations using CP antennas are not as sensitive to the polarization motions of the spacecraft antenna, and therefore will maintain a better communications link.

All AO-10 gain antennas (for 2 m, 70 cm, 24 cm and 13 cm) are configured for RHCP operation along their maximum gain direction. See **Table 23.11** and **Fig 23.49**. Since this direction is also the main antenna lobe along the spacecraft +Z axis, the best communications with AO-10 will also be along that direction. Since the AO-10 radiations are RHCP, ground stations should also be RHCP for optimum communications.

There are times, however, when LHCP provides a better satellite link. AO-13 was designed so that the main antenna lobe was oriented toward the center of the Earth when the satellite was at apogee. AO-10 is not under active orientation control due to a failure of the flight computer memory. Thus, the AO-10 main-lobe orientation is a matter of chance.

Independently switchable RHCP/LHCP antenna circularity is necessary, especially for 70 cm. It is not required for 24 cm operation as Mode L operations are only conducted on the main lobe. Switchable RHCP/LHCP on 2 m is very convenient and useful for Mode B operations.

Spin Modulation

A characteristic of the AO-10 design is that the 2-m and 70-cm high-gain array patterns have three side lobes located along the spin axis that are only about 3 dB weaker than the primary axial lobe. The effective radiation pattern intended for this antenna is

Table 23.12

Summary of One-Way Transmission Losses for Communications Paths Between Earth and Phase 3 Spacecraft at Apogee in an 11.7-Hour Elliptical Orbit, as Viewed Near AOS-LOS Range

	Attenuation (dB)		
Loss Mechanism	2 m	70 cm	24 cm
Path Loss	168.07	177.57	186.86
Tropo/Ionospheric Refraction	0.002	0.0003	0.0002
Tropo Absorption	0.1	0.7	1.65
Ionospheric Absorption (by category)			
D Layer	0.12	0.013	0.002
F-Layer	0.12	0.013	0.002
Aurora	0.13	0.014	0.002
Polar Cap Absorption (a rare D-layer event)	0.47	0.053	0.006
Field-Aligned Irregularities	†	†	†

†Little data available; characterized by rapid amplitude and phase fluctuations.
(Adapted from *The AMSAT-Phase III Satellite Operations Manual* prepared by AMSAT and Project OSCAR)

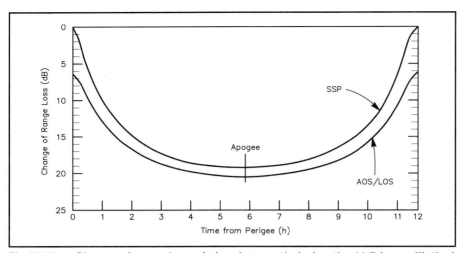

Fig 23.50 — Change of range loss of signal strength during the 11.7-hour elliptical orbit of AO-10. (Adapted from *The AMSAT-Phase III Satellite Operations Manual*, prepared by AMSAT and Project OSCAR.)

Table 23.13

Link Budget Computations for Mode-B and Mode-L Transponders on OSCAR 10 at Apogee and AOS-LOS Range

Ground Station Uplink

Symbol	Parameter	Mode B (70-cm uplink)	Mode L (24-cm uplink)
N_f	Sat Receiver Noise Floor	−136.7 dBm	−137.4 dBm
SNR	Avg Signal-to-Noise Ratio	15.0 dB	15.0 dB
P_s	Signal at Satellite	−121.7 dBm	−122.4 dBm
G_r	Satellite Antenna Gain	9.0 dBi	12.0 dBi
P_a	Signal at Sat Antenna	−130.7 dBm	−134.4 dBm
P_L	Path Loss	177.57 dB	186.86 dB
$L_{i,t}$	Iono/Tropo Loss	0.73 dB	1.65 dB
$L_{p,p}$	Pointing/Polarization Loss	1.5 dB	1.5 dB
P_t+G_t	Reqd Gnd Stn avg ERP	19.10 dBW	25.61 dBW
	or:	81.3 W	363.9 W
	Gnd Stn PEP ERP	25.10 dBW	31.51 dBW
	or:	323.6 W	1448.8 W
G_t	Gnd Station Antenna Gain	16.0 dBi	19.0 dBi
P_t	Gnd Stn PEP Power Output	9.1 dBW	13.6 dBW
	or:	8.1 W	18.2 W

Satellite Downlink

Symbol	Parameter	Mode B (2-m downlink)	Mode L (70-cm downlink)
P_t	Satellite Max. PEP Output	16.99 dBW	16.99 dBW
	Average Output	10.99 dBW	10.99 dBW
L_m	Multi-user Load Share	−15.0 dB	−20.0 dB
G_s	Satellite AGC Compression	0.0 dB	0.0 dB
G_t	Satellite Antenna Gain	9.0 dBi	9.5 dBi
P_u	Average User ERP Output	34.99 dBm	30.49 dBm
P_L	Path Loss	168.07 dB	177.57 dB
$L_{i,t}$	Iono/Tropo Loss	0.33 dB	0.73 dB
$L_{p,p}$	Pointing/Polarization Loss	1.5 dB	1.5 dB
G_r	Receiving Antenna Gain	12.0 dBi	16.0 dBi
P_s	Received Signal Level	−122.91 dBi	−133.31 dBm
P_n	Noise in Rcv Bandwidth	−137.76 dBm	−146.67 dBm
SNR	Avg Signal-to-Noise Ratio	14.8 dB	13.4 dB

Link computations are based on the following parameters:
 11.7 hour elliptical orbit
 Eccentricity = 0.6
 Slant range = 41.395 km at AOS/LOS
 2.4 kHz SSB Signal
 6 dB peak (PEP) to average signal ratio

Ground Station Receiving Noise Characteristics
 at 2 m: T = 505 K, NF = 4.4 dB
 at 70 cm: T = 65 K, NF = 0.9 dB

(Adapted from *The AMSAT Phase III Satellite Operations Manual*, prepared by AMSAT and Project OSCAR)

shown in Fig 23.49. Lobes in the off-axis antenna patterns produce amplitude modulation of signal strength as the spacecraft spins about its axis. Off-axis operations of AO-10 are an established fact of life, creating spin modulation frequencies of about 0.5 Hz from a 10 r/min spin rate.

Stations using CP antennas are much less prone to be affected by spin modulation than those using linearly polarized antennas. Good signals are maintainable even far off-axis from the main lobe, if the ground station is using switchable CP.

In free space, a station located at a distance from an RF source will receive an average power flow from that source that is inversely proportional to the square of the distance. Doubling the range will reduce the signal by 6 dB. Since Phase 3 satellites have substantially elliptical orbits, the changes in path length (slant range) and path loss is correspondingly sizable. **Fig 23.50** illustrates the changes in path length of AO-10/P3D signals for all conditions from apogee to perigee and for direct overhead passes (subsatellite point location) to the far-limb viewing at AOS or LOS.

Other physical processes introduce additional losses in signals traveling to and from a Phase 3 spacecraft through the troposphere and ionosphere. While many of these effects are frequency dependent, the additional path losses are small compared to the slant range path losses. **Table 23.12** summarizes the path losses for the 2-m, 70-cm and 24-cm bands.

Link Budget

Link budget computations can be deceiving to the unwary observer. The problems arise from the application of wide-bandwidth spacecraft transponders to handle a wide variety of complex, time-averaged, narrow-bandwidth QSOs. The analysis presented in **Table 23.13** employs a somewhat lower signal-to-noise ratio than the source information. Actual on-the-air experience has shown these lower values to be workable. The computation is also performed on the basis of averaged RF power levels. The conversion to PEP is done at the end, assuming a 6-dB relationship between PEP and average.

In the link computations, no allowances have been made for ground station transmission line attenuation. Information for attenuation of a variety of popular feed lines is shown in the **Transmission Lines** chapter. If you measure RF power near the antenna and have a tower-mounted, low-noise preamp, the link computations can be used as presented. The computation is based on two possible worst-case conditions: with the maximum slant range values at AOS/LOS, and with the spacecraft at apogee. An important assumption is that the spacecraft antenna pattern is on-axis to the communications path, a highly unusual spacecraft orientation condition. Of course, an infinite number of cases could be presented in such a tabular assessment.

Another highly variable parameter that must be taken into account is pointing and polarization loss. With the wide variety of AO-10 offset pointing situations that are seen in typical operation, the pointing and polarization losses can easily achieve values of 10 dB or more. Nevertheless, the presentation of Table 23.12 is in the ballpark for Mode B operations for those power levels and received signal conditions seen in practice.

THE 4 × 3 × 5 MHz FILTER FOR MODE J

If your 435-MHz receive system is sensitive enough to experience desense from your 2-m uplink, the filter shown in **Fig 23.51** should solve the problem. Most Mode J OSCAR users have experienced some difficulty getting satisfactory results on this mode. Being able to receive well is the secret. Adding this filter should narrow the passband enough to allow rejection of unwanted noise and birdies. Insert it before any preamp or converter in the antenna feed line. If the third-harmonic level is high, it may be necessary to use a similar filter built for the 144-MHz band after your uplink 144-MHz rig.

Most plumbing-supply outlets can supply you with the material for the 3/4- and 3-inch copper pipe. The only other item of cost is the type of coax receptacle you want to use. Make it adaptable to your system, without sacrificing loss. The filter should cost less than $10.

Be careful when you solder double-sided PC board. Direct the heat of your torch at the pipe and enough heat will transfer to the board to allow the solder to flow. It is not necessary to use PC board, as copper or brass that is thick enough to support the unit will do. If copper or brass is not available, a soup can you can solder to works well. The more stable the structure, the better.

The filter has a narrow passband, but with a good high-gain, low-noise system you should be able to peak up the noise with no signal. A low-power 145.050-MHz signal into a dummy load should give you a test signal to peak the filter.

The insertion loss measured in the ARRL Lab was around 0.4 to 0.5 dB. If you want to improve this figure, use silver braze and silver plating. This filter was fashioned after a design by Joe Reisert, W1JR, and was built by Jay Rusgrove, W1VD.

A similar filter is available as a kit from the Microwave Filter Co, Inc (see the **References** chapter). MFC developed their model 9397 filter specially as a kit for amateur Mode-J satellite operators. Dick Jansson, WD4FAB, describes the kit fully in September 1992 QEX.

Parts List

Piece No.

#		
1	Pipe, copper 3" diam, 5" long	Cut ends square. Drill or punch for connectors 3 3/4" from bottom.
2	Pipe, copper 3/4" diam, 4" long	AgSn (plumbing alloy) solder to center of 10.
3	Disk, copper 3/4" diam 1/16"-1/8" thick	Drill through center. Solder solid hook-up wire between disk and connector to space disk 3/16" from piece 2.
4	Disk, copper 3/4" diam 1/16"-1/8" thick	Drill through center. Solder solid hook-up wire between disk and connector to space disk 3/16" from piece 2.
5	Connector, coax	BNC, SMA or N type. Solder to prevent turning. For large connector, use chassis punch.
6	Same as 5.	
7	Nut, brass 1/4"-20 hex	
8	Same as 7.	
9	PC board, double sided. Top 4" × 4".	Drill hole in center to clear 1/4-20 bolt. Solder 7 and 8 each side of hole. (Use bolt 11 to hold nuts in place when soldering.)
10	PC board, double sided. Bottom 4" × 4".	Solder 2 in center.
11	Bolt, brass 1/4-20 × 3"	Insert through 12, then through 7 and 8.
12	Locking nut, brass, 1/4-20 hex	To hold piece 11 after resonance adjustment.

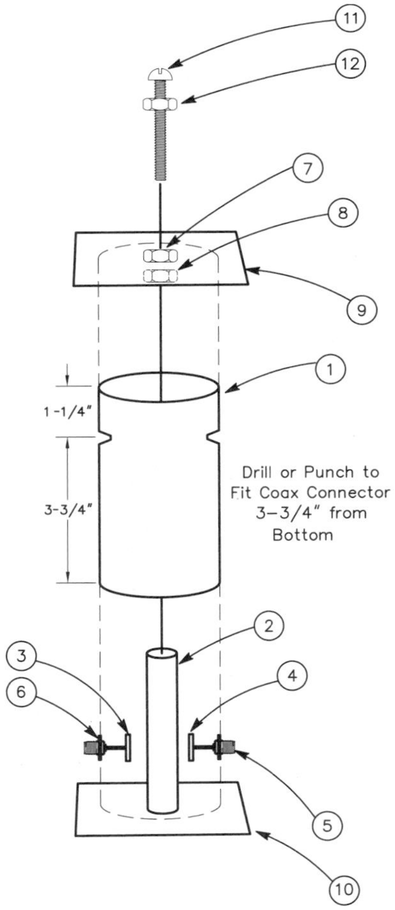

Fig 23.51—Parts list for the 4 × 3 × 5 MHz filter.

PARABOLIC REFLECTOR AND HELICAL ANTENNAS FOR MODE S

The Mode S transponder has become very popular for a variety of reasons. Among the reasons are: good performance can be realized with a physically small downlink antenna and good quality downconverters and preamps are available at reasonable prices. Increased operation on Mode S has long been advocated by a number of people including Bill McCaa, KØRZ, who led the team that designed and built the Mode S transponder[14] and James Miller, G3RUH, who operated one of the AO-13 command stations.[15] Ed Krome, K9EK, and James Miller have published many articles detailing the construction of preamps, downconverters, and antennas for Mode S.[16-22] The following is a condensation of several articles written by G3RUH describing two types of antennas that can be easily built and used for reception of the Phase 3 Mode S downlink.

Parabolic Reflector

There are three parts to the dish antenna — the parabolic reflector, the boom, and the feed. There are as many ways to accomplish the construction as there are constructors. It is not necessary to slavishly replicate every nuance of the design. The only critical dimensions occur in the feed system. When the construction is complete, you will have a 60-cm diameter S-band dish antenna with a gain of about 20 dBi with RHCP and a 3 dB beamwidth of 18°. Coupled with the proper downconverter, performance will be more than adequate for Mode S.

The parabolic reflector used for the original antenna was intended to be a lampshade. Several of these aluminum reflectors were located in department store surplus. The dish is 585 mm in diameter and 110 mm deep corresponding to an f/d ratio of 585/110×16 = 0.33 and a focal length of 0.33×585 = 194 mm. The f/d of 0.33 is a bit too concave for a simple feed to give optimal performance but the price was right, and the under-illumination keeps ground noise pickup to a minimum. The reflector already had a 40-mm hole in the center with three 4-mm holes around it in a 1-inch radius circle.

The boom passes through the center of the reflector and is made from 12.5 mm square aluminum tube. The boom must be long enough to provide for mounting to the rotator boom on the back side of the dish. The part of the boom extending through to the front of the dish must be long enough to mount the feed at the focus. If you choose to mount the downconverter or a preamp near the feed, some additional length will be necessary. Carefully check the requirements for your particular equipment.

A 3 mm thick piece of aluminum, 65 mm in diameter is used to support the boom at the center of the reflector. Once the center mounting plate is installed, the center boom is attached using four small angle brackets — two on each side of the reflector. See **Fig 23.52** for details of reflector and boom assembly.

A small helix is used for the S-band antenna feed. The reflector for the helix is made from a 125 mm square piece of 1.6 mm thick aluminum. The center of the reflector has a 13-mm hole to accommodate the square center boom described above. The type N connector is mounted to the reflector at 21.25 mm from the middle. This distance from the middle is, of course, the radius of a helical antenna for S band. Mount the N connector with spacers so that the back of the connector is flush with the reflector surface. The helix feed assembly is shown in **Fig 23.53.**

Copper wire about 3.3 mm in diameter is used to wind the helix. Wind four turns around a 40-mm diameter form. The turns are wound counterclockwise. This is because the polarization sense is reversed from RHCP when reflected from the dish surface. The wire helix will spring out slightly when winding is complete.

Once the helix is wound, carefully stretch it so that the turns are spaced 28 mm (±1 mm). Make sure the finished spacing of the turns is nice and even. Cut off the first half turn. Carefully bend the first quarter turn about 10° so it will be parallel to the reflector surface once the helix is attached

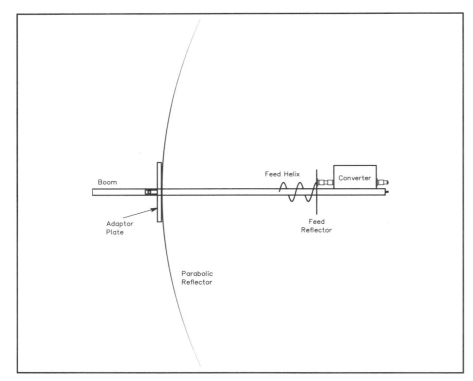

Fig 23.52 — Detail of 60-cm Mode-S dish and feed.

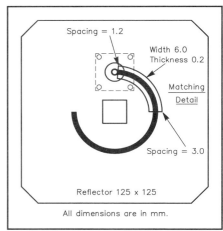

Fig 23.53 — Helix side of 60-cm Mode-S dish feed. The N-type connector is fixed with three screws, and is mounted on a 1.6-mm spacer to bring the PTFE molding flush with the reflector. Dimensions are in mm; 1 in. = 25.4 mm.

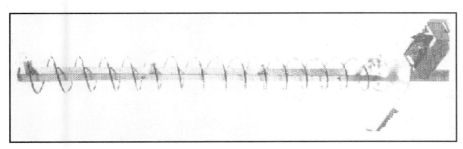

Fig 23.54 — 16-turn helix antenna for Mode S.

to the N connector. This quarter turn will form part of the matching section.

Cut a strip of brass 0.2 mm thick and 6 mm wide matching the curvature of the first quarter turn of the helix by using a paper pattern. Be careful to get this pattern and subsequent brass cutting done exactly right. Using a large soldering iron and working on a heat-proof surface, solder the brass strip to the first $^1/_4$ turn of the helix. Unless you are experienced at this type of soldering, getting the strip attached just right will require some practice. If it doesn't turn out right, just dismantle, wipe clean and try again.

After tack soldering the end of the helix to the type N connector, the first $^1/_4$ turn, with its brass strip in place, should be 1.2 mm above the reflector at its start (at the

N connector) and 3.0 mm at its end. Be sure to line up the helix so its axis is perpendicular to the reflector. Cut off any extra turns to make the finished helix have 2$^1/_4$ turns total. Once you are satisfied, apply a generous amount of solder at the point the helix attaches to the N connector. Remember this is all that supports the helix.

Once the feed assembly is completed, pass the boom through the middle hole and complete the mounting by any suitable method. The middle of the helix should be at the geometric focus of the dish. In the figures shown here, the feed is connected directly to the downconverter and then the downconverter is attached to the boom. You may require a slightly different configuration depending on whether you are attaching a downconverter, preamp, or just

a cable with connector. Angle brackets may be used to secure the feed to the boom in a manner similar to the boom-to-reflector mounting. Be sure to use some method of waterproofing if needed for your preamp and/or downconverter.

16-Turn Helix

The 16-turn helix described in this section was designed to be as physically small as possible and still allow reception of Mode S downlink signals.[21] The results of tests using the antenna while AO-13 was at apogee of 43,000 km can be found in Reference 22. When coupled with an adequate preamp/downconverter system, the 15.5 dBic gain of the 16-turn helix is adequate for CW operation, and under good conditions, may be adequate for SSB operation as well.

The 16-turn helix is shown in **Fig 23.54**. The helix and reflector plate are constructed as described for the parabolic dish above, except that the helix is wound right handed (clockwise). The matching section spacing from the reflector is 2 mm at the start and 8 mm at the end. The helix is supported at every fifth turn, starting with turn 3/4, using PTFE (Teflon) spacers screwed to the boom.

For additional information on constructing antennas for use at microwave frequencies, see *The ARRL UHF/Microwave Experimenter's Manual*.

MODE-S RECEIVE CONVERTER

This project, designed by Zack Lau, W1VT, in the ARRL Lab, was first published in July 1994 *QEX*, pp 25-30. Its goal is a simple yet high-performance 13-cm receive converter optimized for 2401-MHz OSCAR Mode-S reception. The design takes advantage of recent advances in PHEMT technology to simplify the circuitry while also improving performance. See **Fig 23.55**. A template package is available.[23] The finished unit checks out with a 0.33-dB NF and 31 dB of conversion gain, according to the ARRL Lab's HP 346A/ 8970 noise-figure meter. This low noise figure, combined with a 15-turn helix antenna, achieves a gain-to-noise-temperature ratio of around 1. The converter also is small enough to be mounted at the focus of a dish with minimal blockage.

The biggest design simplification comes from the use of Hewlett-Packard PHEMT GaAs MGA-86576 MMICs, which are almost ideal in this application.

The MGA-86576 has a relatively low noise figure of 1.5 dB, just the right output power of +7 dBm, a low current draw of 16 mA, and a gain of 23 dB that peaks in-band. The major disadvantage of the device is that it requires very good grounding for stability. Hewlett-Packard suggests four plated-through holes under each lead. Since this design is for amateurs to duplicate, rather than for commercial manufacturers, very thin circuit board, 15-mil 5880 Rogers Duroid, was used. Microwave Components of Michigan has been selling this material for many years. Excellent stability can be obtained with this thin board by bending the leads sharply and running them through the board. Unlike those of some surface-mount devices, the leads of these MMICs are long enough to go through and bend back against the board for a good mechanical attachment.

An advantage to PHEMT MMICs is

their significant power efficiency. The 564 to 2256-MHz multiplier draws only 23 mA, even when an inefficient LM317L linear regulator is used. See **Table 23.14**.

The mixer is a Mini-Circuits SYM-11, a surface mount device with a reasonable RF port impedance. Its 50-Ω input SWR is roughly 2:1.

Since the two MMICs and PHEMT draw only about 50 mA, the local oscillator

Table 23.14

Measured Performance of the 564 to 2256-MHz Multiplier

Input (dBm)	Output (dBm)
7.0	3.3
7.9	5.0
8.8	6.0
10.0	6.8
10.7	7.0
13.0	7.3

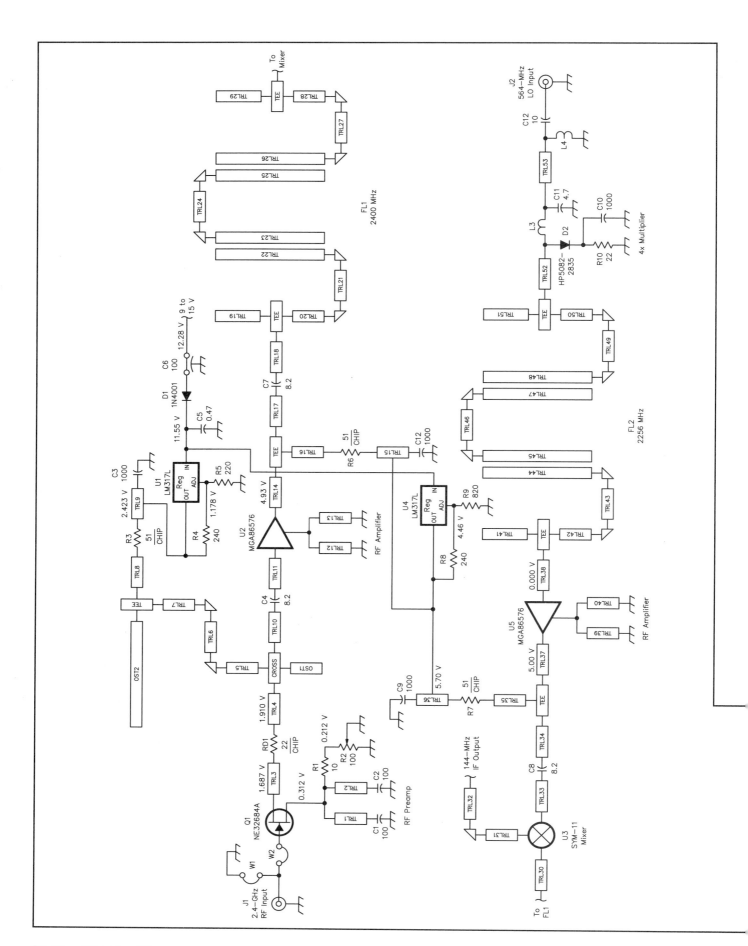

board was redesigned. See **Fig 23.56**. Size and current consumption were both cut in half. A major part of the size reduction was obtained by using smaller filters. This was done by capacitively loading the hairpin loops of the filter with 4.7-pF chip capacitors. This adds an interesting variable to no-tune designs. A disadvantage is that capacitor tolerances can skew the center frequency of the filter, perhaps unacceptably. On the other hand, it also can compensate for variations in board material. Thus, a bad batch of boards might well be salvaged merely by changing the value of the chip capacitors. This circuit might also be used for a 561.6 or 568-MHz LO. Multiply the former frequency by 10 for a 5616-MHz, 6-cm LO and the latter by 18 for a 10,224-MHz, 3-cm LO.

A major variable in dealing with G-10 or FR-4 glass-epoxy board is the board thickness. Variations can be as large as 14%—significant in determining the resonant frequency of a microstrip filter. Rogers advertises an available thickness tolerance of ±1.5%.

The local oscillator is the circuit used in the no-tune transverters.[24] This circuit is not recommended if you wish to set the oscillator to a precise frequency, however. Like many overtone circuits, this one may be difficult to get running properly, since there are at least three things that can go wrong. The most insidious is a parallel resonance in the tank bypass circuit at around 100 MHz. This may prevent the circuit from oscillating properly. This problem can be prevented by changing the value of the bypass capacitors or by changing the spacing between them so the stray inductance changes. The next possibility is that the tank circuit may not resonate at the desired frequency. The easiest solution is to install a 47-Ω resistor in place of

the crystal and resonating inductor to see what the tuning range of the tank circuit is. Be sure to verify that the oscillator is operating at 94 MHz. It is entirely possible that the output is near 564 MHz but the oscillator is operating at some other frequency, such as 80.6 MHz. Finally, the parallel resonating inductor must resonate near the desired overtone. Since the shunt capacitance across the crystal seems to vary, the inductance value also must vary. This shunt capacitance can be measured with a 1-kHz capacitance meter that reads a few picofarads accurately.

With a temperature compensating capacitor for C5, a home-built 561-MHz local oscillator drifted 368 Hz over a temperature variation of 41° (0 to 41°C). Unfortunately, temperature compensating capacitors of specific values are not readily available in small quantities, although they can be found in "bargain assortments" at Radio Shack.

CONSTRUCTION

After etching the Teflon board, cut slots in the board for ground foils and transistor source leads. These are marked with thin pads. Use a no. 10 X-acto knife with a sharp blade. The slot should just touch the outside of the pad, so the transistor will cover the copper pads. In addition to the four slots for the transistors, there are seven slots to ground pads with thin copper foil (roughly 1 mil thick). Next, cut the "U" for the RF preamplifier. This allows a piece of unetched circuit board to be soldered in place for the ground plane. Finally, three holes are needed for the IF and dc power connections. Coax can be run between the IF connection and a panel-mount connector. Copper can be cleared away from the holes with a hand-held drill bit. This takes a bit of practice with such

thin board material.

A disadvantage of 15-mil board is poor mechanical rigidity. Thus, the board should be mounted in a brass frame. 0.025×1-inch brass strip can be used. Attach the connectors to the brass strips with screws, then solder the strips to the circuit board. The strips without connectors are then soldered to the circuit board. An extra brass strip was added at the center of the enclosure near the mixer for extra stiffness.

The preamplifier construction is almost identical to the design published in Nov 1993 *QEX*.[25] Instead of an SMA connector, however, a nonstandard N connector was used—one with a panel-mount flange similar in size to a BNC connector.

R2 is set so the voltage across R1 is 0.10 V, making the bias current of Q1 10 mA. W1 and W2 may have to be tweaked for best noise figure. It is easy to damage Q1 via electrostatic discharge while doing this, so take proper precautions.

OPERATING HINTS

If this converter will be used with a typical transceiver, the mixer must be protected from the transmitter. When turned off, many modern transceivers transmit momentarily. Installation of a postamplifier and attenuator is recommended if you wish to connect this converter to the antenna port of a transceiver. This unit was really meant to be hooked up at the antenna, much like a mast-mounted preamplifier. There is little benefit to using an expensive GaAs FET preamplifier if it is fed with many feet of lossy coax. However, you can separate the preamp/MMIC amplifier and the rest of the converter, mounting the RF amplifiers at the antenna and having the converter in the station. Doing the calculations to determine the noise figure of the new system is recommended.

Fig 23.55—Schematic of the S-band converter.

C1, C2—100-pF ATC 100A chip capacitors (substitution not recommended).
C3, C9, C10—1000-pF chip capacitors.
C4, C7, C8—8.2-pF, 50-mil chip capacitors.
C5—0.1-μF or larger capacitor.
C6—100-pF feedthrough capacitor (any value from 10 pF to 0.1 μF should work fine).
C11—4.7-pF chip capacitor.
C12—10-pF chip capacitor.
D1—1N4001 rectifier diode.

D2—Hewlett-Packard 5082-2835 Schottky diode (or another Schottky switching diode).
L3, L4—4 turns #28 enameled wire, 1/16-inch ID.
RD1—22-Ω, 1/10-W chip resistor, 50×80 mils.
R3, R6, R7—51-Ω, 1/10-W chip resistor.
U1, U4—LM317L adjustable voltage regulator.
U2, U5—Hewlett-Packard MGA 86576 PHEMT MMIC.
U3—Mini-Circuits SYM-11 mixer.

W1—#32 silver-plated wire taken from 20-gauge Teflon stranded wire (8-mil diameter). This is a loop whose ends are 200 mils apart using 250 mils of wire (plus more for the connections). One end is 55 mils above ground; the other end is grounded.
W2—#32 silver-plated wire. This loop is between the center pin of the coax and the transistor gate. The gate lead is 20 mils long. Not counting connections, the length is 310 mils. The ends of the loop are 182 mils apart.

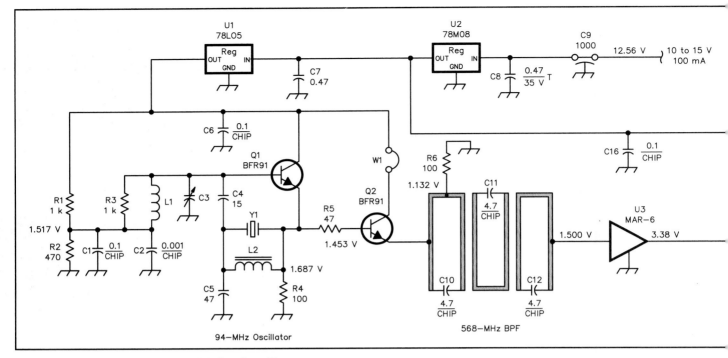

Fig 23.56—Schematic of the 564-MHz local oscillator.

C3—2 to 10-pF trimmer capacitor.
C4—15-pF NP0 capacitor.
C5—47-pF N1500 capacitor. A
commonly available NP0 could be
used, but the temperature
compensation improves stability.
(See text.)
C8—0.33-µF or larger capacitor to
prevent U2 from oscillating.
C9—1000-pF feedthrough capacitor.
Any value from 100 pF to 0.1 µF
should work well.
J1—SMA panel jack. At 564 MHz
connectors are optional.

L1—7 turns #28 enameled wire,
0.1-inch diameter, closewound.
8 turns may work if C3 tunes low
enough.
L2—14 turns #26 enameled wire on a
T-25-10 core. As many as 17 turns
may be needed on a T-25-6 core,
though 10 turns are usually
specified. It depends on the crystal's
shunt capacitance.
R7—270-Ω resistor. Replace with a
470-Ω resistor to run the MMIC from
+12 V (no 8-V regulator).
R8—82-Ω resistor. Replace with a
220-Ω, 1/2-W resistor to run the MMIC
from +12 V.

RFC1—8 turns #26 enameled wire,
0.10-inch ID, closewound.
U1—78L05 5-V regulator.
U2—78M08 8-V regulator.
U3—MAR-6, MSA-0685 MMIC.
U4—MAR-3, MSA-0385 MMIC.
W1—Wire jumper.
Y1—94-MHz crystal. International
Crystal Manufacturing part number
473390. For better temperature
stability, mount the crystal on the
ground-plane side of the board and
cover it with antistatic foam.

A SIMPLE JUNKBOX SATELLITE RECEIVER

This project, by John Reed, W6IOJ, appeared first in April 1994 *QEX*. Single-conversion receivers—including direct-conversion designs—have received a good deal of attention, primarily because of the article by Rick Campbell, KK7B, in August 1992 *QST*. The receiver described here has been configured for monitoring the 70-cm polar-orbiting PACSATs, and has demonstrated very good performance in spite of the limitations of single-conversion designs. In addition, it has retained the simplicity, compactness and versatility of single conversion. The receiver can operate in any part of the 70-cm band. The output is an IF signal having a 250-Hz to 2-MHz passband.

Fig 23.57 is a block diagram of the receiver. This diagram shows one particular PACSAT configuration—the receiver is being used with a 50-kHz IF filter/amplifier/FM discriminator, a 9600-baud modem and a TNC/computer.

GENERAL DESCRIPTION

The receiver's 3×5¼×5⅞-inch metal cabinet contains two 4¾×5-inch circuit boards. One is the UHF circuit board that has a 70-cm input filter, low-noise monolithic preamplifier, double-balanced mixer and IF preamplifier. See **Fig 23.58**. There is also an LO driver consisting of a tripler/filter arrangement operating from a 145-MHz, 6-mW source. The VFO circuit board has a 24-MHz varactor-tuned VFO and a 145-MHz frequency multiplier followed by a two-stage monolithic amplifier. This amplifier output is the LO

driver input. The VFO circuit board includes a 10-V regulator operating from a 12 to 20-V external source.

On the front panel there is a 12-position band selector switch covering 3.6 MHz of the 70-cm band in 300-kHz steps. Fine tuning within these steps is accomplished with a potentiometer covering a 500-kHz spread. The selector range can be placed in any part of the 70-cm band by trimmer adjustments located on the circuit boards. The back panel contains a BNC connector for the 70-cm input and jacks for the receiver output and the AFC input.

LIMITATIONS

Single-conversion receivers lack discrimination of the unused sideband. For example, in a PACSAT application with a

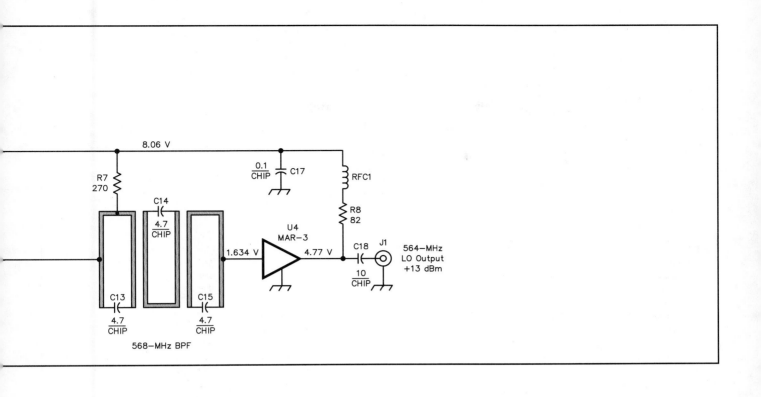

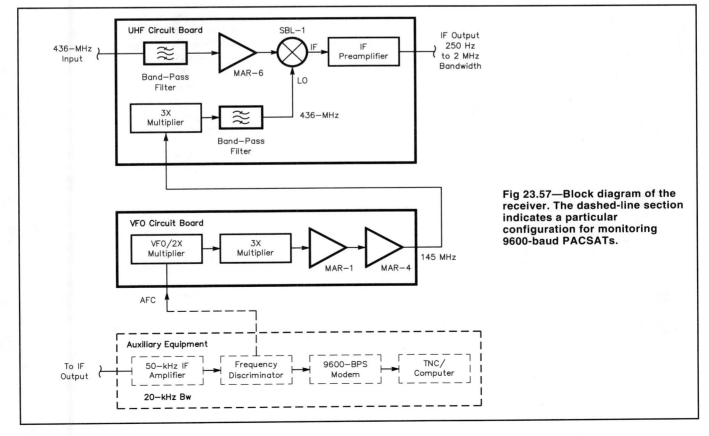

Fig 23.57—Block diagram of the receiver. The dashed-line section indicates a particular configuration for monitoring 9600-baud PACSATs.

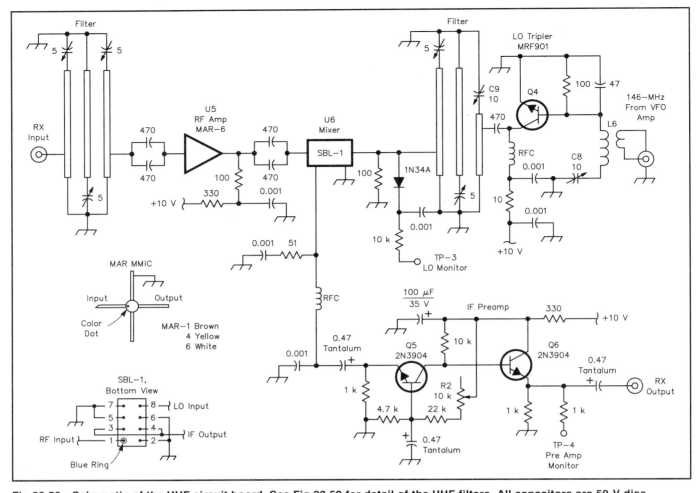

Fig 23.58—Schematic of the UHF circuit board. See Fig 23.59 for detail of the UHF filters. All capacitors are 50-V disc ceramics unless otherwise noted. All RFCs are 20 turns of #26 wire having an ID of 1/16 inch. C8 and C9 are 10-pF FILMTRIMS (Sprague-Goodman part #GYA10000). L6 is 7 turns of #18 wire, 1/4-inch ID, 3/8 inch long with a 1-turn link coupling coil connected to the VFO amplifier output coupling cable.

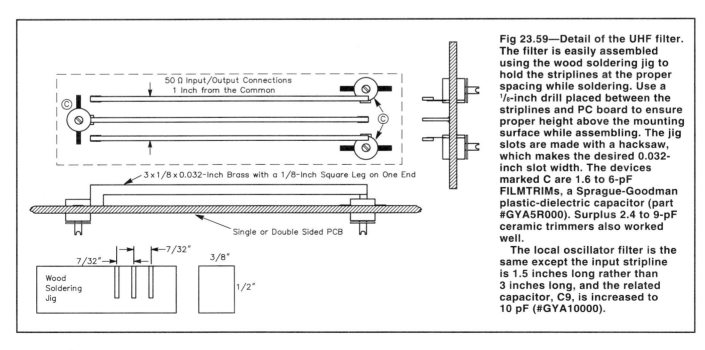

Fig 23.59—Detail of the UHF filter. The filter is easily assembled using the wood soldering jig to hold the striplines at the proper spacing while soldering. Use a 1/8-inch drill placed between the striplines and PC board to ensure proper height above the mounting surface while assembling. The jig slots are made with a hacksaw, which makes the desired 0.032-inch slot width. The devices marked C are 1.6 to 6-pF FILMTRIMs, a Sprague-Goodman plastic-dielectric capacitor (part #GYA5R000). Surplus 2.4 to 9-pF ceramic trimmers also worked well.

The local oscillator filter is the same except the input stripline is 1.5 inches long rather than 3 inches long, and the related capacitor, C9, is increased to 10 pF (#GYA10000).

50-kHz IF there will be an image frequency 100 kHz from the received signal. In actual operation, there has been no interference from a signal at this image frequency while monitoring PACSATs. In the rare case where there may be interference, you can tune to the opposite sideband, placing the image at a different frequency.

Probably of more importance is that unattenuated noise at the image frequency causes a 3-dB S/N degradation. Although this is clearly not optimal, typical signal variations of polar-orbiting satellites are so large that this loss does not represent a major compromise. A second possible limitation is 1/f noise originating from the diode mixer. But practically all 1/f noise is below 10 kHz. Therefore, PACSAT application, with its 50-kHz IF, is not affected. Even in applications requiring the use of lower frequencies, the receiver's RF amplifier will largely override the 1/f noise.

The IF preamplifier has been left wideband simply as a versatility feature. Although the dynamic ranges of both the mixer and IF preamplifier start to roll over at about the same input levels, a filter between the mixer and preamplifier will help avoid possible overloading effects of unwanted signals that pass through the input filter but are outside the useful passband. For example, a 50-kHz filter (20-kHz bandwidth) will improve the performance of the PACSAT configuration during some interference conditions.

Frequency stability is a major consideration of simple 70-cm local-oscillator design. One influencing factor in this case is that polar-orbiting satellites have total Doppler shifts of up to 20 kHz. This alone requires automatic frequency control— unless you are willing to keep one hand on the tuner! Of course, if AFC can compensate for Doppler, it can also compensate for some drift in the local oscillator.

CIRCUIT BOARDS

The circuit boards are assembled using a glue-down stripline technique that holds the components and acts as conducting RF links. Using this method, the printed-circuit board foil remains a solid groundplane, making it appropriate to use a single-sided board. A second feature is that the glued-down pads can be easily removed to accommodate layout changes. The component striplines are about 1/8-inch wide, and the lengths are determined by how many connections are desirable in a single line. The connecting pads are separated by foil notches made with a hacksaw, about 3/16-inch apart or longer, depending upon layout convenience.

The 50-Ω RF conducting lines are made 3/32-inch wide, assuming the use of standard glass-epoxy 0.059-inch material. The width is different from conventional etched striplines due to the raised gluedown stripline edge effect. In this particular application, the critical RF lines are so short that the type of PC board material is of little consequence. Elmer's Clear Household Cement can be used for fastening the striplines. The cement sets up enough to use the pad in a few minutes. Removal of a pad becomes difficult after setting-up for several weeks or more.

UHF CIRCUIT BOARD

The possibility of overloading of the preamplifier by off-frequency interference is minimized by first passing the input signals through a three-section stripline filter. It has an insertion loss of 0.7 dB. As shown in **Fig 23.59**, the filter is easy to build from readily available materials. Although the diagram specifies a Sprague capacitor, which is available from many sources (Digi-Key, for example), there are inexpensive surplus miniature ceramic trimmers that will work fine as long as the minimum capacitance is 2.5 pF or less.

The MAR-6 preamplifier MMIC has a typical gain of 18 dB with a noise figure of 3.0 dB. The critical operating characteristic is the 3.5-V bias voltage (measured at the MMIC output terminal). A 10-V V_{CC}, together with 430-Ω series resistors, sets the proper bias to allow the MMIC to operate near its nominal 16-mA current specification. Although chip coupling capacitors are recommended, standard disc ceramics offer little performance compromise. People with poor eyesight will find disc ceramics much easier to use. Two capacitors in parallel reduce possible compromising inductance. The two MMIC ground leads are raised above the board surface using strips of 1/16-inch thick brass to make them level with the input and output leads, which connect to the gluedown striplines.

The SBL 1 mixer is mounted in the conventional manner. Use of a drill to slightly ream each of the eight pin holes to avoid pin contact with the foil allows proper connection to the glue-down striplines. Grounded pins are soldered to the PC board with a small piece of soldering braid over the pin. This permits a relatively easy desoldering procedure if for some reason it becomes necessary to remove the mixer. The LO level into the mixer is monitored by a diode peak-reading detector. The nominal level is 5 mW, or about 0.8 V at TP3.

The LO driver filter is like the input fil-ter except the input stripline is made shorter and used with a larger value capacitor. This optimizes loading of Q4, the MRF 901 tripler. The LO driver has a maximum output of 20 mW. The output is reduced to the desired level by the drive control, R1, located on the VFO circuit board.

The IF preamplifier is similar to the one described in Campbell's *QST* article. The grounded-base stage, Q5, provides a 50-Ω load to the mixer and approximately 40 dB of gain. It is followed by an emitter follower to supply a low output impedance.

VFO CIRCUIT BOARD

See **Fig 23.60**. The VFO, Q1, is a 24-MHz JFET Colpitts oscillator. It is tuned by a 12-V Zener diode connected to operate like a varactor. It produces approximately a 10-pF capacitance change that provides the desired 436-MHz LO shift of 4 MHz (222-kHz shift at the VFO). The output is taken from the FET drain with a 48-MHz tuned circuit. This doubles the frequency while providing reasonable isolation from the VFO.

Fig 23.61 shows the results of tests made of various VFO components. Fig 23.61 also describes two acceptable component combinations. The dominant temperature-sensitive components are the inductor (L1) and the tuned circuit capacitors (C2, 3, 4). The band-set trimmers (both the capacitors and the variable ferrite), and the varactor are less sensitive. The two recommended combinations have an initial turn-on frequency shift of about 10 kHz during the first 15 minutes of operation. After that the shift is less than 10 kHz. An uncompensated configuration using all mica capacitors together with a Plexiglas coil form has a frequency shift of about 15 kHz per degree change of ambient temperature.

A resistive divider network, shown in **Fig 23.62**, provides the varactor tuning voltage and an input for the AFC or scanner function. A 12-position switch selects voltage divider resistors having values that compensate for the varactor nonlinearity, resulting in a 300-kHz frequency shift for each position. The voltage for the varactor will vary from 3.7 to 6.6 V or 3.4 to 6.0 V depending on the position of the fine-tuning potentiometer. The AFC/scanner input is nominally biased to 5 V. The frequency can be shifted up to 200 kHz either plus or minus by forcing the bias to 0 or 10 V. The bias resistance is 5 kΩ.

Two 10-turn potentiometers, one for bandset and the other for fine tuning, are an alternative method. Frequency calibra-

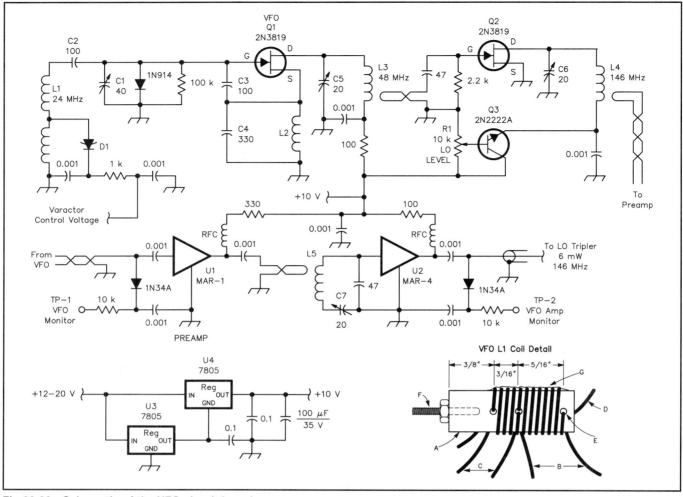

Fig 23.60—Schematic of the VFO circuit board.

C1—40-pF FILMTRIM (Sprague-Goodman #GYC40000).
C2—100-pF silver mica.
C3—100-pF polypropylene (Panasonic #ECQ-P1H101JZ).
C4—330-pF silver mica.
D1—1N4742, 12-V Zener diode (used as a varactor).
L2—10 turns #26 wire wound on a ¼-inch diam form (wood dowel).
L3—10 turns #26 wire wound on a ¼-inch diam form with a 2-turn link coupling coil made from #30 wire-wrapping wire. The twisted pair is about 2 inches long.

L4—5 turns #18 wire, ¼-inch ID, ⁵/₁₆-inch long with a 1-turn link coupling coil made from #22 hookup wire. The twisted pair is about 1 inch long.
L5—7 turns #18 wire, ¼ inch ID, ³/₈ inch long with a 1-turn link coupling coil made from #22 hookup wire. The twisted pair is about 1.5 inches long.
RFC—20 turns #26 wire, ¹/₁₆ inch ID. Output coupling to the tripler—about 1 ft of miniature microphone cable (RS 278-510).
A—³/₈ inch diam Plexiglas rod.

B—7 turns coil #26 gauge wire.
C—4 turns coil #26 gauge wire.
D—0.010-inch diam carpet thread wound with the coil for uniform turn spacing.
E—3¹/₁₆ inch diam holes to hold the coil.
F—#4-40 coil mounting stud. It is cemented or threaded into the Plexiglas form.
G—Place several lines of clear cement along the coil length (Elmer's Clear Household Cement).

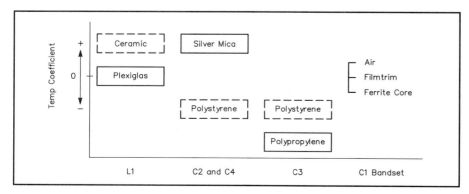

Fig 23.61—Relative temperature coefficients of the VFO components as observed during circuit operation. The solid lines indicate a configuration using a solid Plexiglas coil form and the dashed lines a thin-wall ceramic form. Both configurations will operate continuously for six hours together with a 10° F ambient temperature change with less than a 20-kHz shift in frequency. There is little performance difference between the three types of band-set methods as a function of temperature change.

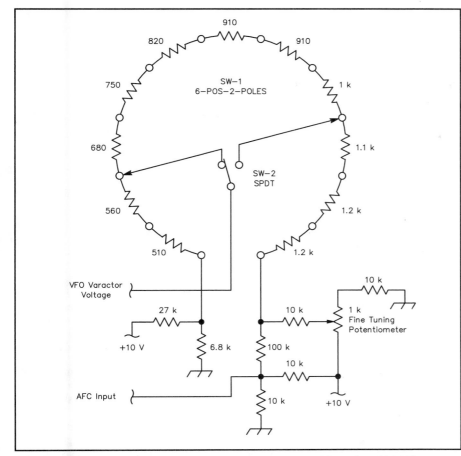

Fig 23.62—Schematic of the varactor voltage controller. There is a 300-kHz frequency shift between each SW-1 position. The fine tuning frequency shift is 500 kHz. The total tuning range is 4 MHz. The AFC input can shift the frequency up to ±200 kHz. It is biased to +5 V by a 5-kΩ resistive divider. Maximum shift occurs when it is forced to 0 or +10 V.

Table 23.15

Procedures for Final Alignment

1. Adjust all capacitors to minimum capacitance except C1, C5 and C8. Set these three at maximum capacitance. Set R1 for maximum VFO output.

2. Turn the tuner controls for midband response. SW1: pos 6, SW2: low pos. Fine tuning potentiometer: midway point.

3. Set the 2-m transceiver to 145.33 MHz and arrange conditions such that a rubber duck, or some other pick-up device, can be placed near L1.

4. Tune the VFO to 24 MHz by decreasing C1 until the VFO is heard on the receiver.

5. Tune L3/C5 to 48 MHz by decreasing C5 while peaking the 2-m receiver response.

6. Peak the 146-MHz multiplier response by decreasing C6 while monitoring TP1; it should read about 0.3 V.

7. Peak the VFO amp response by increasing C7 while monitoring TP2; it should read greater than 1 V.

8. Peak L6/C8 to 146 MHz by increasing C8 while monitoring TP2; it will null to about 0.8 V.

9. Peak the LO filter by decreasing C9 while monitoring TP3, then trim the remaining two 5-pF capacitors. TP3 should read about 1.2 V.

10. Correct the LO mixer input by adjusting R1 on the VFO circuit board for 0.8 V at TP3.

11. Normalize the IF preamp operation by adjusting R2 to make the dc voltage at TP4 2 V.

12. Peak the three input filter 5-pF capacitors for maximum response to the 2-m transmitter's third harmonic.

13. Optimize the S/N by tuning to a marginal input signal that is about +10 dB S/N (transmitter harmonic, noise, etc) and peaking the three filter 5-pF capacitors. S/N alignment requires the receiver input be terminated with a 50-Ω load.

Parts Suppliers

MMICs—Down East Microwave.

SBL1—Oak Hills Research.

Panasonic P-Series Polypropylene capacitors and Sprague-Goodman

FILMTRIMs—Digi-Key Corp.

Silver mica capacitors and ceramic trimmers—All Electronics.

Plexiglas—Check the Yellow Pages under Plastics.

Brass—Hobby shops usually stock small sheets of brass in various thicknesses.

tion is performed simply by monitoring the varactor input voltage with a meter. This method worked very well. It didn't compensate for the nonlinear varactor, but the tuning was fine enough to make that unimportant.

The VFO is followed by an FET tripler and a two-stage MMIC amplifier. The 145-MHz tripler has a maximum output of about 0.5 mW. This level is controlled by varying the FET drain voltage with the potentiometer R1, and monitored by a diode peak detector (TP1). The first MMIC MAR1 stage has a gain of about 18 dB and an output capability of +7 dBm in the VHF region. Its nominal operating current is 17 mA with a bias of 5 V. Although this output is probably enough to drive the LO tripler, the MAR4 was added as a safety factor to allow for circuit performance variations. It has a gain of about 8 dB with an output of +13 dBm. Its nominal operating current is 50 mA with a bias voltage of 6 V. The bias resistors, 300 and 100 Ω, together with the 10-V source, operate the MMICs close to their nominal values.

There is a relatively high-Q tuned circuit used in the interstage coupling between the two MMICs for filtering out unwanted VFO responses. There are three tuned circuits for this purpose, L4/C6, L5/C7 and L6/C8, to ensure a reasonably pure waveform for driving the LO tripler. Output of the VFO amplifier is monitored by a peak voltmeter at TP2. It is also used to initially align L6/C8. At resonance, TP2 will read a minimum value due to lowering of the load impedance. About one foot of miniature microphone cable is used to connect the amplifier RF output to the UHF circuit board. There is a bit of loss in using this cable, but there is power to spare and cable flexibility is an important consideration.

ALIGNMENT

The receiver can be aligned using nothing more than a multimeter and a 2-m transceiver. The 2-m receiver permits initial frequency set of the VFO, and the third harmonic of the 2-m transmitter provides a signal for initial test and alignment of the input filter and RF amplifier. The re-

quired alignment steps are detailed in **Table 23.15**.

Performance of the compact stripline filter is far from outstanding. However, it does allow me to monitor PACSATs while transmitting into the adjacent 2-m uplink antenna without interfering with the received signal (for the PACSAT full-duplex mode). The author has concluded that a more complex internal filter would not substantially improve the receiver's performance.

The receiver has been used to copy FM 9600-baud packet from UO-22 and KO-23, and 1200-baud PSK from AO-16. It is a simple assembly that is a pleasure to use.

AN INTEGRATED L-BAND SATELLITE ANTENNA AND AMPLIFIER

The new Phase 3D satellite offers a variety of bands and modes; literally something for everyone. It will have communications capability on every authorized amateur satellite frequency between 146 MHz and 24 GHz. Uplinks are on 146, 435, 1270, 2400 and 5670 MHz. Downlinks are on 146, 435, 2400 and 10450 MHz, and 24 GHz. All uplink receivers and downlink transmitters share a common intermediate frequency and are switched through a matrix. This allows any uplink to be paired with any downlink; even multiple transmitters and receivers may be paired. Frequency pairings will be selected based on operating considerations. Up and downlinks in the same band (such as on 146, 435 and 2400 MHz) will never be used simultaneously.

To demonstrate what can be done, using a typical commercially available amplifier module, Ed Krome, K9EK (ex-KA9LNV), built an amplifier for the L-band uplink. The amplifier is mounted on a boom, counterbalancing a helical antenna. The antenna design was taken from a program in the *ARRL UHF/Microwave Experimenter's Manual*. All of the work was done in the author's home workshop without any special tools or equipment. This project shows you can still home-brew equipment, even for the latest of the amateur satellites.

The amplifier/antenna combination takes a "system" approach to providing a convenient and practical method of generating a satisfactory Phase 3D uplink, while keeping cable and construction costs to a minimum.

It is likely that the same frequency combinations that have proven historically popular will be used frequently on Phase 3D. Much was learned about communications with AO-13. Mode L (to be termed Mode L/U), with its 1270 MHz uplink and 435 MHz downlink, was very popular and an excellent performer. Mode S (now Mode U/S; 2400 MHz downlink and 435 MHz uplink), was included as an experiment, and proved to be extremely successful. S-band downlinks with 30 inch diameter parabolic dish antennas could provide almost armchair SSB copy. It is expected that Phase 3D will see both of those modes frequently, as well as a combination that takes the best of both, Mode L/S.

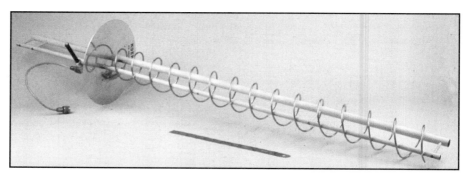

Photo A—This antenna was designed to be home-brewed without any special tools or shop equipment. It is mounted on two 4-foot long hardwood dowel rods. See the text for construction details.

Photo B—The matching brick amplifier can be mounted on the reverse side of the mast to counterbalance the antenna.

Antenna details
Construction parameters for a 1270 MHz, 15 turn helix antenna (from *helix.bas* by KA1GT) (all dimensions are in inches)

Length of wire in each turn = 10.445

Total length of wire required for entire antenna = 156.7

Coil diameter (center of wire) = 3.24

Spacing between turns (center to center) = 2.31

Circumference = 10.186

Total length of antenna coil = 34.7

For the 23 cm, or L band, uplink, both kits and commercial equipment are available. Commercial equipment, such as the ICOM IC-1271 series and the 23-cm plug-in module for the popular Yaesu FT-736R, provide about 10-W of output power. Ten watts, coupled to an antenna as small as a 12-turn helix, is predicted to be adequate for satisfactory SSB communications. However, microwave frequencies present different problems from the lower bands. In particular, 10-W in the shack does not necessarily mean you will see those 10 watts at the antenna. Common (and reasonably priced) coaxial feed lines are quite lossy at 1270 MHz. The old standby, RG-8U, loses almost 13 dB per 100 feet, which leaves $^1/_2$-W out from the 10-W input. Hardline is a good solution, but tends to be expensive.

AN INTEGRATED APPROACH

A much better solution may be to integrate the antenna and final amplifier, eliminating the feed-line losses between the two. Fortunately, an elegant solution exists for the "amplifier" part of this integration. This is the M57762 amplifier module, a readily available "hybrid" amplifier (referred to as a *brick*) designed for linear amplifier service on 23 cm. This 50-Ω impedance module requires only the addition of suitable dc power circuitry, input and output connectors and a heat sink to provide a 10 to 20-W, 13 dB gain amplifier.

Many antenna designs are available. One practical design for satellite communications is the helix antenna. Helix antennas are broadbanded, inherently circularly polarized and relatively easy to construct. A suitable helix antenna may be designed using KA1GT's *helix.bas* computer program from *The ARRL UHF/ Microwave Experimenter's Manual*. The antenna shown uses 15 turns of no. 8 copper wire and provides a calculated gain of 15 dB. This design was selected because it provides slightly more than the minimum required gain and physically fits on a 3 foot long boom.

The L band uplink amplifier and antenna shown carries the integration a step further. The amplifier is mounted on the back end of the helix antenna frame (on the opposite side of the clamp for the crossboom), thereby serving as a counterweight to partially offset the weight of the helix. This is desirable to reduce elevation rotor load.

The amplifier is fed from a shack-mounted 23-cm transmitter, but is rated for less input power than what the common transmitters (as previously noted) produce. To extend our development of an integrated arrangement even further, it may be possible to use the length of cable between the shack-mounted transmitter and the pole-mounted amplifier to attenuate the transmitter's signal down to the level required by the amplifier. The M57762 is rated at 1-W (+30 dBm) input. First, calculate the attenuation required between the transmitter and the amplifier to prevent damage to the amplifier. Then calculate the amount of your feed line required to provide that attenuation. If the run between the transmitter and amplifier is shorter than the cable required for attenuation, simply use the whole required length of cable and coil the excess up as a drip loop at the bottom of the tower.

ANTENNA CONSTRUCTION

Probably the easiest way to build the helix itself is to stretch out the required length of wire (adding a foot or two to hold on to, which will be cut off later), then mark the wire every "Length of Turn" distance. Calculate the cumulative length and mark from that to prevent errors. Then close wind the marked wire smoothly around a form slightly smaller than the noted diameter. It seems easier to "unwind" the helix while stretching it to size than to "wind" it. Stretch the helix along a rod until the overall length is achieved. Measure and equalize spacing between turns. Finally, cut off the ends to get the desired helix. The helix is effectively fed from its circumference through a matching transformer in the form of a *fin* soldered to the first ¼ turn of the helix. A ½ inch wide strip of 0.015 inch brass stock, cut to match the curve of the helix,

works fine. Position the fin back about ¼ inch from the connector, then solder it to the first quarter turn, parallel to the reflector plate. A useable method of connecting the large helix wire to the N connector center conductor is to flatten the last ¼ inch of the wire with a hammer and block, then drill a hole in the flattened end to fit over the center conductor. Finally, solder it in place. On the antenna shown, a return loss of 16 dB was attained when the end of the helix was spaced about 0.1 inch from the reflector.

The antenna frame was constructed from 4-foot long hardwood dowel rods. The frame shown uses a ¾ inch rod on top and ½ inch rod underneath, jointed together by ¼ inch dowel sections slipped into drilled holes and glued into place. Since the inside diameter of the helix is equal to the pitch diameter minus one times the wire diameter, this frame is 3.1 inches outside dimension. The spacers in the 36 inch helix section were installed first, then the free ends of the rods were passed through properly spaced holes drilled in the reflector plate, then the remaining spacers were installed (through holes drilled in the dowels) and glued in place. Varnish the frame before installing the helix. The reflector plate is secured to the frame with homemade angle brackets. The helix was slipped over the frame, stretched to the correct turn-to-turn spacing and held to the frame in several places with small tie-wraps. A homemade clamp plate is used to connect the top dowel rod to the antenna crossboom. Keep the crossboom attachment close to the back of the reflector to aid balance of the finished assembly. Since the amplifier is built with

all RF and dc connections on one end, it may be mounted connectors down on the end of the antenna frame and easily weatherproofed with an inverted plastic container. The cables are routed up into the amplifier in such a way as to provide drip loops and prevent water from getting into the amplifier.

A problem with helix antennas is that the helix itself is insulated from dc ground. Therefore, static electricity may build up on the helix until it damages the attached device. One method of preventing such buildup is to add a shorted ¼ wavelength stub to the antenna feed. Since a shorted ¼ wave stub presents an extremely high impedance at its non-shorted end, it is virtually "invisible" to the RF flowing between the antenna and amplifier at the design frequency but fully grounded for dc. The stub shown is actually ¾ wavelength since ¼ wavelength is too short to be physically practical. Three-quarter wavelength is measured from the center conductor of the main cable to the shorted end of the stub. With RG-213 cable, the length of the stub from the end of the male N connector to the short is 4⁹/₁₆ inches. Attach the stub to the antenna through a "T" fitting. Assembled return loss was measured at 20 dB.

AMPLIFIER CONSTRUCTION

The amplifier (**Fig 23.62A**) is constructed by mounting the module itself on a heat sink, and using an etched circuit board, slipped under the leads on the brick, to provide both RF and dc connections. The only things critical about the board are the width of the 50-Ω input and output lines (0.1 inch wide on 0.062 inch thick,

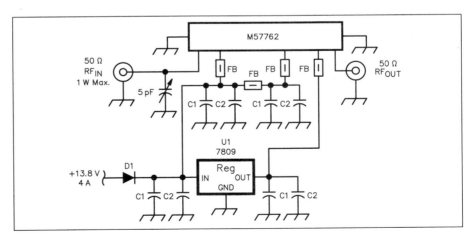

Fig 23.62A—Just a handful of parts are needed to connect the brick amplifier module. All capacitor pairs are 10 μF/35 V chip or tantalum units in parallel with 1000 pF chip capacitors. D1 is a 4-A (minimum), 50-V power rectifier such as Digi-Key GI820CT-ND. It prevents damage due to reverse connection of the power leads. U1 is a 7809 voltage regulator (9-V, 1-A). Check RF Parts and Down East Microwave (see the Chapter 30 Address List) for pricing and availability of the amplifier module.

G10 board). Keep all leads from the module to the board traces as short as possible. The connectors (type-N are recommended) should be mounted on the end of the heat sink in such a manner that the center conductors lay on the board traces. Keep everything short! No insulator is required between the module and the heat sink. Be sure to use thermal conductive grease between the brick and the heat sink. The circuit board also must be grounded to the heat sink. Ensure good grounding of the circuit board to the heat sink by drilling and tapping several holes through the circuit board as shown.

A template, with additional construction details and a PC board layout, is available from the ARRL. See Chapter 30, **References**, for ordering information.

SELECTED SATELLITE REFERENCES

Equipment Selection

D. DeMaw, "Trio-Kenwood TS-700S 2-Meter Transceiver," QST, Feb 1978, pp 31-32.

C. Hutchinson, "Ten-Tec 2510 Mode B Satellite Station," QST, Oct 1985, pp 41-43.

D. Ingram, "The Ten-Tec 2510 OSCAR Satellite Station/Converter," CQ Magazine, Feb 1985, pp 44-46.

J. Kleinman, "ICOM IC-290H All-Mode 2-Meter Transceiver," QST, May 1983, pp 36-37.

J. Lindholm, "ICOM IC-471A 70-cm Transceiver," QST, Aug 1985, pp 38-39.

R. Schetgen, Ed., The ARRL Radio Buyer's Sourcebook and The ARRL Radio Buyer's Sourcebook, Vol 2 (Newington: ARRL, 1991 and 1993). A compilation of product reviews from QST.

R. Roznoy, Ed. The ARRL VHF/UHF Radio Buyer's Sourcebook (Newington: ARRL, 1997). A compilation of VHF and UHF Product Reviews from QST.

M. Wilson, "ICOM IC-271A 2-Meter Multimode Transceiver," QST, May 1985, pp 40-41.

M. Wilson, "Yaesu Electronics Corp FT-726R VHF/UHF Transceiver," QST, May 1984, pp 40-42.

M. Wilson, "Yaesu FT-480R 2-Meter Multimode Transceiver," QST, Oct 1981, pp 46-47.

H. Winard and R. Soderman, "A Survey of OSCAR Station Equipment," Orbit, no. 16, Nov-Dec 1983, pp 13-16 and no. 18, Mar-Apr 1984, pp 12-16.

S. Ford, "PacComm PSK-1T Satellite Modem and TNC," QST, Jul 1993, p 46.

R. Healy, "Down East Microwave 432PA 432-MHz Amplifier Kit," QST, Mar 1993, p 66.

R. Healy, "Down East Microwave DEM432 No-Tune 432-MHz Transverter," QST, Mar 1993, p 64.

R. Jansson, "SSB Electronic SP-70 Mast-Mount Preamplifier," QST, Mar 1993, p 63.

S. Ford, "QST Compares: SSB Electronic UEK-2000S and Down East Microwave SHF-2400 2.4-GHz Satellite Down-converters," QST, Feb 1994, p 69.

S. Ford, "ICOM IC-821H VHF/UHF Multimode Transceiver," QST, Mar 1997, pp 70-73.

S. Ford, "Yaesu FT-847 HF/VHF/UHF Transceiver," QST, Aug 1998, pp 64-69.

Antennas

M. Davidoff, "Off-Axis Circular Polarization of Two Orthogonal Linearly Polarized Antennas," Orbit, no. 15, Sep-Oct 1983, pp 14-15.

J. L. DuBois, "A Simple Dish for Mode-L," Orbit, no. 13, Mar-Apr 1983, pp 4-6.

B. Glassmeyer, "Circular Polarization and OSCAR Communications," QST, May 1980, pp 11-15.

R. D. Straw, Ed., The ARRL Antenna Book (Newington: ARRL, 1997). Available from your local radio store or from ARRL.

R. Jansson, "Helical Antenna Construction for 146 MHz," Orbit, May-Jun 1981, pp 12-15.

R. Jansson, "KLM 2M-22C and KLM 435-40CX Yagi Antennas," QST, Oct 1985, pp 43-44.

R. Jansson, "70-Cm Satellite Antenna Techniques," Orbit, no. 1, Mar 1980, pp 24-26.

R. Messano, "An Indoor Loop for Satellite Work," QST, Jul 1999, p 55.

C. Richards, "The Chopstick Helical," Orbit, No. 5, Jan-Feb 1981, pp 8-9.

V. Riportella, "Amateur Satellite Communications," QST, May 1986, pp 70-71.

G. Schrick, "Antenna Polarization," The ARRL Antenna Compendium, Vol 1 (Newington: ARRL, 1985), pp 152-156.

A. Zoller, "Tilt Rather Than Twist," Orbit, No. 15, Sep-Oct 1983, pp 7-8.

R. Jansson, "M^2 Enterprises 2M-CP22 and 436-CP30 Satellite Yagi Antennas," QST, Nov 1992, p 69.

S. Ford, "M^2 Enterprises EB-144 Eggbeater Antenna," QST, Sep 1993, p 75.

Microsats

Mills, S.E., "Step Up to the 38,400 Bps Digital Satellites," QST, Apr 2000, pp 42-45.

Loughmiller, D and B. McGwier, "Microsat: The Next Generation of OSCAR Satellites." Part 1: QST, May 1989, pp 37-40. Part 2: QST, Jun 1989, pp 53-54, 70.

Loughmiller, D., "Successful OSCAR Launch Ushers in the '90s," QST, May 1990, pp 40-41.

Loughmiller, D., "WEBERSAT-OSCAR 18: Amateur Radio's Newest Eye in the Sky," QST, Jun 1990, pp 50-51.

Ford, S., "The Road Less Traveled," QST, Apr 1996, p 58.

Ford, S., "KITSAT-OSCAR 23 Reaches Orbit," QST, Oct 1992, p 93.

Ford, S., "Satellite on a String: SEDSAT-1," QST, Nov 1992, p 113.

Ford, S., "WEBERSAT—Step by Step," QST, Dec 1992, p 63.

Ford, S., "AMRAD-OSCAR 27," QST, Dec 1993, p 107.

Ford, S., "Two More PACSATs!" QST, Oct 1993, p 98.

Soifer, R., "AO-27: An FM Repeater in the Sky," QST, Jan 1998, pp 64-65.

Ford, S., "Meet the Multifaceted SUNSAT," QST, Mar 1999, p 90.

Ford, S., "Flash! Three New Satellites in Orbit!," QST, Mar 2000, p 92.

Soifer, R., "UO-14: A User-Friendly FM Repeater in the Sky," QST, Aug 2000, p 64.

Phase 3D

Coggins, B., "A Box for Phase 3D," QST, Jun 1997, p 94.

Ford, S. and Z. Lau, "Get Ready for Phase 3D!" Part 1: QST, Jan 1997, p 28; Part 2: QST, Feb 1997, p 50; Part 3: QST, Mar 1997, p 42; Part 4: QST, Apr 1997, p 45; Part 5: QST, May 1997, p 28.

Ford, S., "Phase 3D: The Ultimate EasySat," May 1995 QST, p 21.

Tynan, B. and R. Jansson, "Phase 3D—A Satellite for All." Part 1: QST, May 1993, p 49; Part 2: QST, Jun 1993, p 49.

Ford, S., "Phase 3D Update," Amateur Satellite Communication, QST, Dec 1994, p 105.

"Putting the Finishing Touches on Phase 3D," QST, Oct 1998, p 20.

Fuji-OSCARs

Ford, S., "K1CE's Secret OSCAR-20 Station," QST, Mar 1993, p 47.

Ford, S., "Fuji-OSCAR 20 Mode JA," QST, Sep 1993, p 104.

Software

S. Ford, "Short Takes—Nova for Windows 32," QST, Apr 2000, p 65.

Ford, S., "Will O' the WISP," QST, Feb 1995, p 90.

Ford, S., "Satellite-Tracking Software," QST, Dec 1993, p 89.

General Texts and Articles

M. Davidoff, The Radio Amateur's Satellite Handbook (Newington: ARRL, 2000).

S. Ford, "An Amateur Satellite Primer," QST, Apr 2000, p 36.

G. McElroy, "Keeping Track of OSCAR: A Short History," QST, Nov 1999, p 65.

The ARRL Satellite Anthology (Newington: ARRL, 1999).

The ARRL Operating Manual (Newington: ARRL, 2000).

The AMSAT-Phase III Satellite Operations Manual, prepared by Radio Amateur Satellite Corp and Project OSCAR, Inc, 1985. Available from AMSAT. (See Address List in **References** chapter.)

Satellite Notes

[1]R. J. Diersing and J. W. Ward, "Packet Radio in the Amateur Satellite Service," IEEE J. Selected Areas in Communications, Vol 7, no. 2, Feb 1989, pp 226-234.

[2]H. E. Price and J. W. Ward, "PACSAT Protocol Suite—An Overview," in Proc. 9th ARRL Computer Networking Conference, London, Ontario, Canada, 1990, pp 203-206.

[3]H. E. Price and J. W. Ward, "PACSAT Broadcast Protocol," in *Proc. 9th ARRL Computer Networking Conference*, London, Ontario, Canada, 1990, pp 232-244.

[4]H. E. Price and J. W. Ward, "PACSAT Data Specification Standards," in *Proc. 9th ARRL Computer Networking Conference*, London, Ontario, Canada, 1990, pp 207-208.

[5]J. W. Ward and H. E. Price, "PACSAT Protocol: File Transfer Level 0," in *Proc. 9th ARRL Computer Networking Conference*, London, Ontario, Canada, 1990, pp 209-231.

[6]J. W. Ward and H. E. Price, "PACSAT File Header Definition," in *Proc. 9th ARRL Computer Networking Conference*, London, Ontario, Canada, 1990, pp 245-252.

[7]T. L. Fox, *AX.25 Amateur Packet Radio Link-Layer Protocol*. Newington, CT: American Radio Relay League, 1984.

[14]McCaa, William D., "Hints on Using the AMSAT-OSCAR 13 Mode S Transponder," *The AMSAT Journal*, Vol 13, no. 1, Mar 1990, pp 21-22.

[15]Miller, James, "Mode S — Tomorrow's Downlink?" *The AMSAT Journal*, Vol 15, no. 4, Sep/Oct 1992, pp 14-15.

[16]Krome, Ed, "Development of a Portable Mode S Groundstation." *The AMSAT Journal*, Vol 16, no. 6, Nov/Dec 1993, pp 25-28.

[17]Krome, Ed, "S Band Reception: Building the DEM Converter and Preamp Kits," *The AMSAT Journal*, Vol 16, no. 2, Mar/Apr 1993, pp 4-6.

[18]Krome, Ed, "A Satellite Mode S Loop Yagi Antenna," *The AMSAT Journal*, Vol 16, no. 3, May/Jun 1993, pp 4-6.

[19]Krome, Ed, "Mode S: Plug and Play!" *The AMSAT Journal*, Vol 14, no. 1, Jan 1991, pp 21-23, 25.

[20]Miller, James, "A 60-cm S-Band Dish Antenna," *The AMSAT Journal*, Vol 16 no. 2, Mar/Apr 1993, pp 7-9.

[21]Miller, James, "Small is Best," *The AMSAT Journal*, Vol 16, no. 4, Jul/Aug 1993, p 12.

[22]Miller, James, "S-Band 16t Helix Update," *The AMSAT Journal*, Vol 16, no. 6, Nov/Dec 1993, p 29.

[23]A template package, including an etching pattern for the converter microwave circuitry, an etching pattern for the 564-MHz LO circuitry, a cutting and drilling guide for the Teflon microwave board, and parts-placement diagrams for the Teflon microwave board and the LO board is available from the ARRL. See Chapter 30, **References**, for details.

[24]Davey, Jim, WA8NLC, "A No-Tune Transverter for the 2304-MHz Band," *QST*, Dec 1992, pp 33-39.

[25]Lau, Zack, W1VT, "RF," *QEX*, Nov 1993, pp 20-23.

Earth-Moon-Earth (EME)

EME communication, also known as "moonbounce," has become a popular form of space communication. The concept is simple: The moon is used as a passive reflector for VHF and UHF signals. With a total path length of nearly 500,000 miles, EME is the ultimate DX. EME is a natural and passive propagation phenomenon, and EME QSOs count toward the WAS, DXCC and VUCC awards. EME opens up the VHF and UHF bands to a new universe of worldwide DX.

The first demonstration of EME capability was done by the US Army Signal Corps just after WW II. In the 1950s, using 400 MW of effective radiated power, the US Navy established a moon relay link between Washington, DC, and Hawaii that could handle four multiplexed Teletype (RTTY) channels. The first successful amateur reception of EME signals occurred in 1953 by W4AO and W3GKP.

It took until 1960 for two-way amateur communications to take place. Using surplus parabolic dish antennas and high-power klystron amplifiers, the Eimac Radio Club, W6HB, and the Rhododendron Swamp VHF Society, W1BU, accomplished this milestone in July 1960 on 1296 MHz. In the 1960s, the first wave of amateur EME enthusiasts established amateur-to-amateur contacts on 144 MHz and 432 MHz. In April 1964, W6DNG and OH1NL made the first 144-MHz EME QSO. 432-MHz EME experimentation was delayed by the 50-W power limit (removed January 2, 1963). Only one month after the first 144-MHz QSO was made, the 1000-ft-diameter dish at Arecibo, Puerto Rico, was used to demonstrate the viability of 432-MHz EME, when a contact was made between KB4BPZ and W1BU. The first amateur-to-amateur 432-MHz EME QSO occurred in July 1964 between W1BU and KH6UK.

The widespread availability of reliable low-noise semiconductor devices along with significant improvements in Yagi arrays ushered in the second wave of amateur activity in the 1970s. Contacts between stations entirely built by amateurs became the norm instead of the exception. In 1970, the first 220- and 2304-MHz EME QSOs were made, followed by the first 50-MHz EME QSO in 1972. 1970s activity was still concentrated on 144 and 432 MHz, although 1296-MHz activity grew.

As the 1980s approached, another quantum leap in receive performance occurred with the use of GaAsFET preamplifiers. This, and improvements in Yagi performance (led by DL6WU's log-taper design work), and the new US amateur power output limit of 1500 W have put EME in the grasp of most serious VHF and UHF operators. The 1980s saw 144- and 432-MHz WAS and WAC become a reality for a great number of operators. The 1980s also witnessed the first EME QSOs on 3456 MHz and 5760 MHz (1987), followed by EME QSOs on 902 MHz and 10 GHz (1988).

EME is still primarily a CW mode. As stations have improved, SSB is now more popular. Regardless of the transmission

Tommy Henderson, WD5AGO, pursues 144-MHz EME from his Tulsa, Oklahoma, QTH with this array. Local electronics students helped with construction.

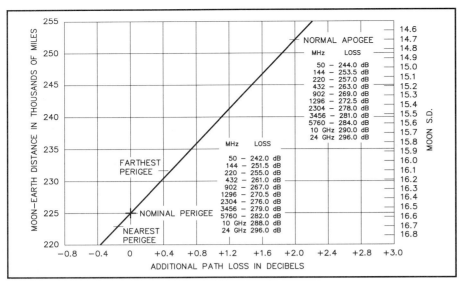

Fig 23.63 — Variations in EME path loss can be determined from this graph. SD refers to semi-diameter of the moon, which is indicated for each day of the year in *The Nautical Almanac*.

mode, successful EME operating requires:

1) As close to the legal power output as possible.

2) A fairly large array (compared to OSCAR antennas).

3) Accurate azimuth and elevation rotation.

4) Minimal transmission-line losses.

5) A low system noise figure, preferably with the preamplifier mounted at the array.

Choosing an EME Band

Making EME QSOs is a natural progression for many weak-signal terrestrial operators. Looking at EME path loss vs frequency (**Fig 23.63**), it may seem as if the lowest frequency is best, because of reduced path loss. This is not entirely true. The path-loss graph does not account for the effects of cosmic and man-made noise, nor does it relate the effects of ionospheric scattering and absorption. Both short- and long-term fading effects also must be overcome.

50-MHz EME is quite a challenge, as the required arrays are very large. In addition, sky noise limits receiver sensitivity at this frequency. Because of power and licensing restrictions, it is not likely that many foreign countries will be able to get on 50-MHz EME.

144 MHz is probably the easiest EME band to start on. It supports the largest number of EME operators. Commercial equipment is widely available; a 144-MHz EME station can almost be completely assembled from off-the-shelf equipment. 222 MHz is a good frequency for EME, but there are only a handful of active stations, and 222 MHz is available only

in ITU Region 2.

432 MHz is the most active EME band after 144 MHz. Libration fading (see Fig 23.118) is more of a problem than at 144 MHz, but sky noise is more than an order of magnitude less than on 144 MHz. The improved receive signal-to-noise ratio may more than make up for the more rapid fading. However, 432-MHz activity is most concentrated into the one or two weekends a month when conditions are expected to be best.

902 MHz and above should be considered if you primarily enjoy experimenting and building equipment. If you plan to operate at these frequencies, an unobstructed moon window is a must. The antenna used is almost certain to be a dish. 902 MHz has the same problem that 222 MHz has — it's not an international band. Equipment and activity are expected to be limited for many years.

1296 MHz currently has a good amount of activity from all over the world. Recent equipment improvements indicate 1296 MHz should experience a significant growth in activity over the next few years. 2300 MHz has received renewed interest. It suffers from nonaligned international band assignments and restrictions in different parts of the world.

Antenna Requirements

The tremendous path loss incurred over the EME circuit requires a high-power transmitter, a low-noise receiver and a high-performance antenna array. Although single-Yagi QSOs are possible, most new EME operators will rapidly become frustrated unless they are able to

work many different stations on a regular basis. Because of libration fading and the nature of weak signals, a 1- or 2-dB increase in array gain will often be perceived as being much greater. An important antenna parameter in EME communications is the antenna noise temperature. This refers to the amount of noise received by the array. The noise comes from cosmic noise (noise generated by stars other than the sun), Earth noise (thermal noise radiated by the Earth), and noise generate by manmade sources such as power-line leaks and other broadband RF sources.

Yagi antennas are almost universally used on 144 MHz. Although dish antennas as small as 24 ft in diameter have been successfully used, they offer poor gain-to-size trade-offs at 144 MHz. The minimum array gain for reliable operation is about 18 dBd (20.1 dBi). The minimum array gain should also allow a station to hear its own echoes on a regular basis. This is possible by using four 2.2-λ Yagis. The 12-element 2.5-λ Yagi described in the **Antennas and Projects** chapter is an excellent choice. When considering a Yagi design, you should avoid old-technology Yagis, that is, designs that use either constant-width spacings, constant-length directors or a combination of both. These old-design Yagis will have significantly poorer side lobes, a narrower gain bandwidth and a sharper SWR bandwidth than modern log-taper designs. Modern wideband designs will behave much more predictably when stacked in arrays, and, unlike many of the older designs, will deliver close to 3 dB of stacking gain.

222-MHz requirements are similar to those of 144 MHz. Although dish antennas are somewhat more practical, Yagis still predominate. The 16-element 3.8-λ Yagi described in the **Antennas and Projects** chapter is a good building block for 222-MHz EME. Four of these Yagis are adequate for a minimal 222-MHz EME station, but six or eight will provide a much more substantial signal.

At 432 MHz, parabolic-dish antennas become viable. The minimum gain for reliable 432-MHz EME operation is 24 dBi.

Yagis are also used on 432 MHz. The 22-element Yagi described in the **Antennas and Projects** chapter is an ideal 432-MHz design. Four of the 22-element Yagis meet the 24-dBi-gain criteria, and have been used successfully on EME. If you are going to use a fixed polarization Yagi array, you should plan on building an array with substantially more than 24-dBi gain if you desire reliable contacts with small stations. This extra gain is needed to overcome polarization misalignment.

At 902 MHz and above, the only antenna worthy of consideration is a parabolic dish. While it has been proven that Yagi antennas are capable of making EME QSOs at 1296 MHz, Yagi antennas, whether they use rod or loop elements, are simply not practical.

EME QSOs have been made at 1296 MHz with dishes as small as 6 ft in diameter. For reliable EME operation with similarly equipped stations, a 12-ft diameter dish (31 dBi gain at 1296 MHz) is a practical minimum. TVRO dishes, which are designed to operate at 3 GHz make excellent antennas, provided they have an accurate surface area. The one drawback of TVRO dishes is that they usually have an undesirable F/d ratio. More information on dish construction and feeds can be found in *The ARRL Antenna Book* and *The ARRL UHF/Microwave Experimenter's Manual*.

Polarization Effects

All of the close attention paid to operating at the best time, such as nighttime perigee, with high moon declination and low sky temperatures is of little use if signals are not aligned in polarization between the two stations attempting to make contact. There are two basic polarization effects. The first is called spatial polarization. Simply stated, two stations (using az-el mounts and fixed linear polarization) that are located far apart, will usually not have their arrays aligned in polarization as seen by the moon. Spatial polarization can easily be predicted, given the location of both stations and the position of the moon.

The second effect is Faraday rotation. This is an actual rotation of the radio waves in space, and is caused by the charge level of the Earth's ionosphere. At 1296 MHz and above, Faraday rotation is virtually nonexistent. At 432 MHz, it is believed that up to a 360° rotation is common. At 144 MHz, it is believed that the wavefront can actually rotate seven or more complete 360° revolutions. When Faraday rotation is combined with spatial polarization, there are four possible results:

1) Both stations hear each other and can QSO.
2) Station A hears station B, station B does not hear station A.
3) Station B hears station A, station A does not hear station B.
4) Neither station A not station B hear each other.

At 144 MHz, there are so many revolutions of the signal, and the amount of Faraday rotation changes so fast that, generally, hour-long schedules are arranged. At 432 MHz, Faraday rotation can take hours to change. Because of this, half-hour schedules are used. During the daytime, you can count on 90 to 180° of rotation. If both stations are operating during hours of darkness, there will be little Faraday rotation, and the amount of spatial polarization determines if a schedule should be attempted.

At 1296 MHz and above, circular polarization is standard. The predominant array is a parabolic reflector, which makes circular polarization easy to obtain. Although the use of circular polarization would make one expect signals to be constant, except for the effect of the moon's distance, long-term fading of 6 to 9 dB is frequently observed.

With improved long-Yagi designs, for years the solution to overcoming polarization misalignment has been to make the array larger. Making your station's system gain 5 or 6 dB greater than required for minimal EME QSOs will allow you to work more stations, simply by moving you farther down the polarization loss curve. After about 60° of misalignment, however, making your station large enough to overcome the added losses quickly becomes a lifetime project! See **Fig 23.64**.

At 432 MHz and lower, Yagis are widely used, making the linear polarization standard. Although circular polarization may seem like a simple solution to polarization problems, when signals are reflected off the moon, the polarization sense of circularly polarized radio wave is reversed, requiring two arrays of opposite polarization sense be used. Initially, crossed Yagis with switchable polarization may also look attractive. Unfortunately, 432-MHz Yagis are physically small enough that the extra feed lines and switching devices become complicated, and usually adversely affect array performance. Keep in mind that even at 144 MHz, Yagis cannot tolerate metal mounting masts and frames in line with the Yagi elements.

When starting out on EME, keep in mind that it is best to use a simple system. You will still be able to work many of the larger fixed-polarization stations and those who have polarization adjustment (only one station needs to have polarization control). Once you gain understanding and confidence in your simple array, a more complex array such as one with polarization rotation can be attempted.

Receiver Requirements

A low-noise receiving setup is essential for successful EME work. Many EME signals are barely, but not always, out of the noise. To determine actual receiver performance, any phasing line and feed line losses, along with the noise generated in the receiver, must be added to the array noise reception. When all losses are considered, a system noise figure of 0.5 dB (35 K) will deliver about all the performance that can be used at 144 MHz, even when low-loss phasing lines and a quiet array are used.

The sky noise at 432 MHz and above is low enough (cold sky is <15 K (kelvins) at 432 MHz, and 5 K at 1296 MHz) so the lowest possible noise figure is desired. Current high-performance arrays will have array temperatures near 30 K when unwanted noise pickup is added in. Phasing line losses must also be included, along with any relay losses. Even at 432 MHz, it is impossible to make receiver noise insignificant without the use of a liquid-cooled preamplifier. Current technology gives a minimum obtainable GaAsFET preampli-

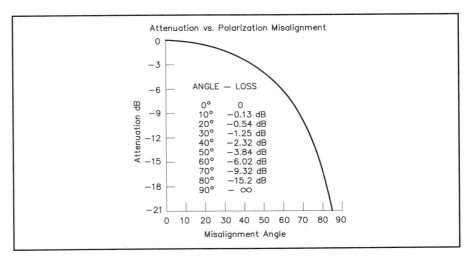

Fig 23.64 — The graph shows how quickly loss because of polarization misalignment increases after 45°. The curve repeats through 360°, showing no loss at 0° and maximum loss at 90° and 270°.

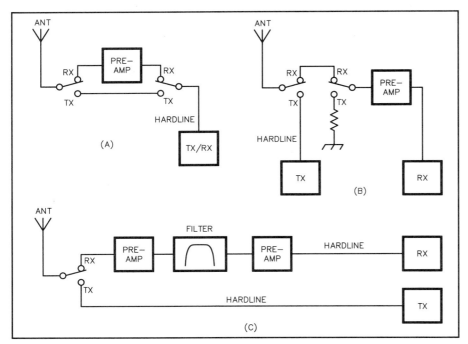

Fig 23.65 — Two systems for switching a preamplifier in and out of the receive line. At A, a single length of cable is used for both the transmit and receive line. At B is a slightly more sophisticated system that uses two separate transmission lines. At C, a high-isolation relay is used for TR switching. The energized position is normally used on receive.

Table 23.16
Transmitter Power Required for EME Success

Power at the array

50 MHz	1500 W
144 MHz	1000 W
222 MHz	750 W
432 MHz	500 W
902 MHz	200 W
1296 MHz	200 W
2300 MHz and above	100 W

fier noise figure, at room temperature, of about 0.35 dB (24 K). See **Fig 23.65**.

GaAsFET preamps have also been standard on 1296 MHz and above for several years. Noise figures range from about 0.4 dB at 1296 MHz (30 K) to about 2 dB (170 K at 10 GHz). HEMT devices are now available to amateurs, but are of little use below 902 MHz because of 1/f noise. At higher frequencies, HEMT devices have already shown impressively low noise figures. Current HEMT devices are capable of noise figures close to 1.2 dB at 10 GHz (93 K) without liquid cooling.

At 1296 MHz, a new noise-limiting factor appears. The physical temperature of the moon is 210 K. This means that just like the Earth, it is a black-body radiator. The additional noise source is the reflection of sun noise off the moon. Just as a full moon reflects sunlight to Earth, the rest of the electromagnetic spectrum is also reflected. On 144 and 432 MHz, the beamwidth of a typical array is wide enough (15° is typical for 144 MHz, 7° for 432 MHz) that the moon, which subtends a 0.5° area is small enough to be insignificant in the array's pattern. At 1296 MHz, beamwidths approach 2°, and moon-noise figures of up to 5 dB are typical at full moon. Stations operating at 2300 MHz and above have such narrow array patterns that many operators actually use moon noise to assure that their arrays are pointed at the moon!

A new weak-signal operator is encouraged to experiment with receivers and filters. A radio with passband tuning or IF-shift capability is desired. These features are used to center the passband and the pitch of the CW signal to the frequency at which the operator's ears perform best. Some operators also use audio filtering. Audio filtering is effective in eliminating high-frequency noise generated in the radio's audio or IF stages. This noise can be very fatiguing during extended weak signal operation. The switched-capacitor audio filter has become popular with many operators.

Transmitter Requirements

Although the maximum legal power (1500 W out) is desirable, the actual power required can be considerably less, depending on the frequency of operation and size of the array. Given the minimum array gain requirements previously discussed, the power levels recommended for reasonable success are shown in **Table 23.16**.

The amplifier and power supply should be constructed with adequate cooling and safety margins to allow extended slow-speed CW operation without failure. The transmitter must also be free from drift and chirp. The CW note must be pure and properly shaped. Signals that drift and chirp are harder to copy. They are especially annoying to operators who use narrow CW filters. A stable, clean signal will improve your EME success rate.

Calculating EME Capabilities

Once all station parameters are known, the expected strength of the moon echoes can be calculated given the path loss for the band in use (see Fig 23.61). The formula for the received signal-to-noise ratio is:

$$S/N = P_o - L_t + G_t - P_1 + G_r - P_n \qquad (1)$$

where

P_o = transmitter output power (dBW)
L_t = transmitter feed-line loss (dB)
G_t = transmitting antenna gain (dBi)
P_1 = total path loss (dB)
G_r = receiving antenna gain (dBi)
P_n = receiver noise power (dBW).

Receiver noise power, P_n, is determined by the following:

$$P_n = 10 \log_{10} KBT_s \qquad (2)$$

where

$K = 1.38 \times 10^{-23}$ (Boltzmann's constant)
B = bandwidth (Hz)
T_s = receiving system noise temperature (K).

Receiving system noise temperature, T_s, can be found from:

$$T_s = T_a + (L_r - 1) T_1 + L_r T_r \qquad (3)$$

where

T_a = antenna temperature (K)
L_r = receiving feed-line loss (ratio)
T_1 = physical temperature of feed line (normally 290 K)
T_r = receiver noise temperature (K).

An example calculation for a typical 432-MHz EME link is:

P_o = +30 dBW (1000 W)
L_t = 1.0 dB
G_t = 26.4 dBi (8 × 6.1-λ 22-el Yagis)
P_1 = 262 dB
G_r = 23.5 dBi (15 ft parabolic)
T_a = 60 K
L_r = 1.02 (0.1-dB preamp at antenna)
T_1 = 290 K
T_r = 35.4 K (NT = 0.5 dB)
T_s = 101.9 K
P_n = −188.5 dB
S/N = + 5.4 dB

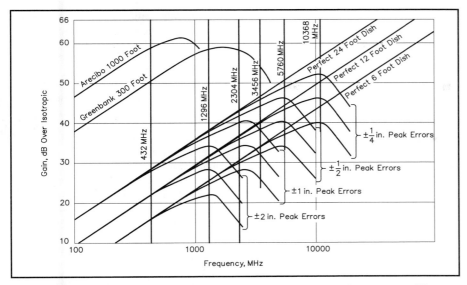

Fig 23.66— Parabolic-antenna gain vs size, frequency and surface errors. All curves assumed 60% aperture efficiency and 10-dB power taper. Reference: J. Ruze, British IEEE.

It is obvious that EME is no place for a compromise station. Even relatively sophisticated equipment provides less-than-optimum results.

Fig 23.66 gives parabolic dish gain for a perfect dish. The best Yagi antennas will not exceed the gain curve shown in the **Antennas and Projects** chapter. If you are using modern, log-taper Yagis, properly spaced, figure about 2.8 to 2.9 dB of stacking gain. For old-technology Yagis, 2.5 dB may be closer to reality. Any phasing line and power divider losses must also be subtracted from the array gain.

Locating the Moon

The moon orbits the Earth once in approximately 28 days, a lunar month. Because the plane of the moon's orbit is tilted from the Earth's equatorial plane by approximately 23.5°, the moon swings in a sinewave pattern both north and south of the equator. The angle of departure of the moon's position at a given time from the equatorial plane is termed declination (abbreviated decl). Declination angles of the moon, which are continually changing (a few degrees a day), indicate the latitude on the Earth's surface where the moon will be at zenith. For this presentation, positive declination angles are used when the moon is north of the equator, and negative angles when south.

The longitude on the Earth's surface where the moon will be at zenith is related to the moon's Greenwich Hour Angle, abbreviated G.H.A. or GHA. "Hour angle" is defined as the angle in degrees to the west of the meridian. If the GHA of the moon were 0°, it would be directly over the Greenwich meridian. If the moon's GHA were 15°, the moon would be directly over the meridian designated as 15° W longitude on a globe. As one can readily understand, the GHA of the moon is continually changing, too, because of both the orbital velocity of the moon and the Earth's rotation inside the moon's orbit. The moon's GHA changes at the rate of approximately 347° per day.

GHA and declination are terms that may be applied to any celestial body. *The Astronomical Almanac* (available from the Superintendent of Documents, US Government Printing Office) and other publications list the GHA and decl of the sun and moon (as well as for other celestial bodies that may be used for navigation) for every hour of the year. This information may be used to point an antenna when the moon is not visible. *Almanac* tables for the sun may be useful for calibrating remote-readout systems.

Using the Almanac

The Astronomical Almanac and other almanacs show the GHA and declination of the sun or moon at hourly intervals for every day of the period covered by the book. Instructions are included in such books for interpolating the positions of the sun or moon for any time on a given date. The orbital velocity of the moon is not constant, and therefore precise interpolations are not linear.

Fortunately, linear interpolations from one hour to the next, or even from one day to the next, will result in data that is entirely adequate for Amateur Radio purposes. If linear interpolations are made from 0000 UTC on one day to 0000 UTC on the next, worse-case conditions exist when apogee or perigee occurs near midday on the next date in question. Under such conditions, the total angular error in the position of the moon may be as much as a sixth of a degree. Because it takes a full year for the Earth to orbit the sun, the similar error for determining the position of the sun will be no more than a few hundredths of a degree.

If a polar mount (a system having one axis parallel to the Earth's axis) is used, information from the *Almanac* may be used directly to point the antenna array. The local hour angle (LHA) is simply the GHA plus or minus the observer's longitude (plus if east longitude, minus if west). The LHA is the angle west of the observer's meridian at which the celestial body is located. LHA and declination information may be translated to an EME window by taking local obstructions and any other constraints into account.

Azimuth and Elevation

An antenna system that is positioned in azimuth (compass direction) and elevation (angle above the horizon) is called an *az-el* system. For such a system, some additional work will be necessary to convert the almanac data into useful information. The GHA and decl information may be converted into azimuth and elevation angles with the mathematical equations that follow. A calculator or computer that treats trigonometric functions may be used. *CAUTION:* Most almanacs list data in degrees, minutes, and either decimal minutes or seconds. Computer programs generally require this information in degrees and decimal fractions, so a conversion may be necessary before the almanac data is entered.

Determining az-el data from equations follows a procedure similar to calculating great-circle bearings and distances for two points on the Earth's surface. There is one additional factor, however. Visualize two observers on opposite sides of the Earth who are pointing their antennas at the moon. Imaginary lines representing the boresights of the two antennas will converge at the moon at an angle of approximately 2°. Now assume both observers aim their antennas at some distant star. The boresight lines now may be considered to be parallel, each observer having raised his antenna in elevation by approximately 1°. The reason for the necessary change in elevation is that the Earth's diameter in comparison to its distance from the moon is significant. The same is not true for distant stars, or for the sun.

Equations for az-el calculations are:

$$\sin E = \sin L \sin D + \cos L \cos D \cos LHA$$
$$(4)$$

$$\tan F = \frac{\sin E - K}{\cos E} \tag{5}$$

$$\cos C = \frac{\sin D - \sin E \sin L}{\cos E \cos L} \tag{6}$$

where

E = elevation angle for the sun
L = your latitude (negative if south)
D = declination of the celestial body
LHA = local hour angle = GHA plus or minus your longitude (plus if east longitude, minus if west longitude)
F = elevation angle for the moon
K = 0.01657, a constant (see text that follows)
C = true azimuth from north if sin LHA is negative; if sin LHA is positive, then the azimuth = 360 − C.

Assume our location is 50° N latitude, 100° W longitude. Further assume that the GHA of the moon is 140° and its declination is 10°. To determine the az-el information we first find the LHA, which is 140 minus 100 or 40°. Then we solve equation 4:

$$\sin E = \sin 50 \sin 10 + \cos 50 \cos 10 \cos 40$$
$$\sin E = 0.61795 \text{ and } E = 38.2°$$

Solving equation 5 for F, we proceed. (The value for sin E has already been determined in equation 4.)

$$\tan F = \frac{0.61795 - 0.06175}{\cos 38.2}$$

$$= 0.76489$$

From this, F, the moon's elevation angle, is 37.4°.

We continue by solving equation 6 for C. (The value of sin E has already been determined.)

$$\cos C = \frac{\sin 10 - 0.61795 \sin 50}{\cos 38.2 \cos 50}$$

$$= 0.59308$$

C therefore equals 126.4°. To determine if C is the actual azimuth, we find the polarity for sin LHA, which is sin 40° and has a positive value. The actual azimuth then is 360 − C = 233.6°.

If az-el data is being determined for the sun, omit equation 5; equation 5 takes into account the nearness of the moon. The solar elevation angle may be determined from equation 4 alone. In the above example, this angle is 38.2°.

The mathematical procedure is the same for any location on the Earth's surface. Remember to use negative values for southerly latitudes. If solving equation 4 or 5 yields a negative value for E or F, this indicates the celestial body below the horizon.

These equations may also be used to determine az-el data for man-made satellites, but a different value for the constant, K, must be used. K is defined as the ratio of the Earth's radius to the distance from the Earth's center to the satellite.

The value for K as given above, 0.01657 is based on an average Earth-moon distance of 239,000 miles. The actual Earth-moon distance varies from approximately 225,000 to 253,000 mi. When this change in distance is taken into account, it yields a change in elevation angle of approximately 0.1° when the moon is near the horizon. For greater precision in determining the correct elevation angle for the moon, the moon's distance from the Earth may be taken as:

$$D = -15,074.5 \times SD + 474,332$$

where

D = moon's distance in miles
SD = moon's semi-diameter, from the almanac.

Computer Programs

As has been mentioned, a computer may be used in solving the equations for azimuth and elevation. For EME work, it is convenient to calculate az-el data at 30-minute intervals or so, and to keep the results of all calculations handy during the EME window. Necessary antenna-position corrections can then be made periodically.

A BASIC language program for the IBM PC is available from the ARRL Technical Secretary. Request the '95 *Handbook* EME template. This program provides azimuth and elevation information for half-hour intervals during a UTC day when the celestial body is above the horizon. The program makes a linear interpolation of GHA and declination values (discussed earlier) during the period of the UTC day.

Commercial, shareware and public-domain tracking programs are also available. See the **References** chapter for a list of some available programs. *RealTrak* prints out antenna azimuth and elevation headings for nearly any celestial object. It can be used with the Kansas City Tracker program described in the satellite section to track celestial objects automatically. *VHF PAK* provides real-time moon and celestial object position information. Two other real-time tracking programs are *EME Tracker* and the *VK3UM EME Planner*.

Libration Fading of EME Signals

One of the most troublesome aspects of receiving a moonbounce signal, besides the enormous path loss and Faraday rotation fading, is libration fading. This section will deal with libration (pronounced *lie-brayshun*) fading, its cause and effects, and possible measures to minimize it.

Libration fading of an EME signal is characterized in general as fluttery, rapid, irregular fading not unlike that observed in tropospheric scatter propagation. Fading can be very deep, 20 dB or more, and the maximum fading will depend on the operating frequency. At 1296 MHz the maximum fading rate is about 10 Hz, and, scales directly with frequency.

On a weak CW EME signal, libration fading gives the impression of a randomly keyed signal. In fact on very slow CW telegraphy the effect is as though the keying is being done at a much faster speed. On very weak signals only the peaks of libration fading are heard in the form of occasional short bursts or "pings."

Fig 23.67 shows samples of a typical EME echo signal at 1296 MHz. These recordings, made at W2NFA, show the wild fading characteristics with sufficient S/N ratio to record the deep fades. Circular polarization was used to eliminate Faraday fading; thus these recordings are of libration fading only. The recording bandwidth was limited to about 40 Hz to minimize the higher sideband-frequency components of libration fading that exist but are much smaller in amplitude. For those who would like a better statistical description, libration fading is Raleigh distributed. In the recordings shown in Fig 23.63, the average signal-return level computed from path loss and mean reflection coefficient of the moon is at about the +15 dB S/N level.

It is clear that enhancement of echoes far in excess of this average level is observed. This point should be kept clearly in mind when attempting to obtain echoes or receive EME signals with marginal equipment. The probability of hearing an occasional peak is quite good since random enhancement as much as 10 dB is possible. Under these conditions, however, the amount of useful information that can be copied will be near zero. Enthusiastic newcomers to EME communications will be stymied by this effect since they know they can hear the signal strong enough on peaks to copy but can't make any sense out of what they try to copy.

What causes libration fading? Very simply, multipath scattering of the radio waves from the very large (2000-mile diameter) and rough moon surface combined with the relative motion between Earth and moon called librations.

To understand these effects, assume first that the Earth and moon are stationary (no libration) and that a plane wave front arrives at the moon from your Earthbound station as shown in **Fig 23.68A**.

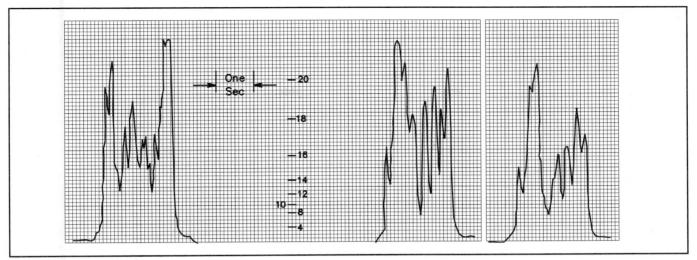

Fig 23.67 — Chart recording of moon echoes received at W2FNA on July 26, 1973, at 1630 UTC. Antenna gain 44 dBi, transmitting power 400 W and system temperature 400 K.

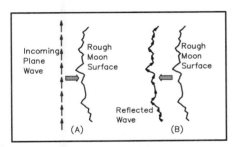

Fig 23.68 — How the rough surface of the moon reflects a plane wave as one having many field vectors.

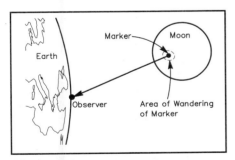

Fig 23.69 — The moon appears to "wander" in its orbit about the Earth. Thus a fixed marker on the moon's surface will appear to move about in a circular area.

The reflected wave shown in Fig 23.68B consists of many scattered contributions from the rough moon surface. It is perhaps easier to visualize the process as if the scattering were from many small individual flat mirrors on the moon that reflect small portions (amplitudes) of the incident wave energy in different directions (paths) and with different path lengths (phase). Those paths directed toward the moon arrive at your antenna as a collection of small wave fronts (field vectors) of various amplitudes and phases. The vector summation of all these coherent (same frequency) returned waves (and there is a near-infinite array of them) takes place at the feedpoint of your antenna (the collecting point in your antenna system). The level of the final summation as measured by a receiver can, of course, have any value from zero to some maximum. Remember that we assumed the Earth and moon were stationary, which means that the final summation of these multipath signal returns from the moon will be one fixed value. The condition of zero relative motion between Earth and moon is a rare event that will be discussed later in this section.

Consider now that the Earth and moon are moving relative to each other (as they are in nature), so the incident radio wave "sees" a slightly different surface of the moon from moment to moment. Since the lunar surface is very irregular, the reflected wave will be equally irregular, changing in amplitude and phase from moment to moment. The resultant continuous summation of the varying multipath signals at your antenna feed-point produces the effect called libration fading of the moon-reflected signal.

The term *libration* is used to describe small perturbations in the movement of celestial bodies. Each libration consists mainly of its diurnal rotation; moon libration consists mainly of its 28-day rotation which appears as a very slight rocking motion with respect to an observer on Earth. This rocking motion can be visualized as follows: Place a marker on the surface of the moon at the center of the moon disc, which is the point closest to the observer, as shown in **Fig 23.69**. Over time,

we will observe that this marker wanders around within a small area. This means the surface of the moon as seen from the Earth is not quite fixed but changes slightly as different areas of the periphery are exposed because of this rocking motion. Moon libration is very slow (on the order of 10^{-7} radians per second) and can be determined with some difficulty from published moon ephemeris tables.

Although the libration motions are very small and slow, the larger surface area of the moon has nearly an infinite number of scattering points (small area). This means that even slight geometric movements can alter the total summation of the returned multipath echo by a significant amount. Since the librations of the Earth and moon are calculable, it is only logical to ask if there ever occurs a time when the total libration is zero or near zero. The answer is yes, and it has been observed and verified experimentally on radar echoes that minimum fading rate (not depth of fade) is coincident with minimum total libration. Calculation of minimum total libration is at best tedious and can only be done successfully by means of a computer. It is a problem in extrapolation of rates of change in coordinate motion and in small differences of large numbers.

EME OPERATING TECHNIQUES

Many EME signals are near the threshold of readability, a condition caused by a combination of path loss, Faraday rotation and libration fading. This weakness and unpredictability of the signals has led to the development of techniques for the exchange of EME information that differ from those used for normal terrestrial work. The fading of EME signals chops

Table 23.17

Signal Reports Used on 144-MHz EME

T — Signal just detectable
M — Portions of call copied
O — Complete call set has been received
R — Both "O" report and call sets have
　 been received
SK — End of contact

Table 23.18

Signal Reports Used on 432-MHz EME

T — Portions of call copied
M — Complete calls copied
O — Good signal—solid copy (possibly
　 enough for SSB work)
R — Calls and reports copied
SK — End of contact

dashes into pieces and renders strings of dots incomplete. This led to the use of the "T M O R" reporting system. Different, but similar, systems are used on the low bands (50 and 144 MHz) and the high bands (432 MHz and above). **Tables 23.17** and **23.18** summarize the differences between the two systems.

As equipment and techniques have improved, the use of normal RST signal reports has become more common. It is now quite common for two stations working for the first time to go straight to RST reports if signals are strong enough. These normal reports let stations compare signals from one night to the next. EME QSOs are often made during the ARRL VHF contests. These contacts require the exchange of 4-digit grid locators. On 432 MHz and above, the sending of GGGG has come to mean "Please send me your grid square," or conversely, "I am now going to send my grid square."

The length of transmit and receive periods is also different between the bands. On 50 and 144 MHz, 2-minute sequences are used. That is, stations transmit for two full minutes, and then receive for two full minutes. One-hour schedules are used, with the eastern-most station (referenced to the international date line) transmitting first. **Table 23.19** gives the 2-minute se-

quence procedure. On 222 MHz, both the 144 and 432-MHz systems are used.

On 432 MHz and above, 2½-minute sequences are standard.

The longer period is used to let stations with variable polarization have adequate time to peak the signal. The last 30 seconds is reserved for signal reports only. **Table 23.20** provides more information on the 432-MHz EME QSO sequence. The western-most station usually transmits first. However, if one of the stations has variable polarization, it may elect to transmit second, to take the opportunity to use the first sequence to peak the signal. If both stations have variable polarization, the station that transmits first should leave its polarization fixed on transmit, to avoid "polarization chasing."

CW sending speed is usually in the 10 to 13-wpm range. It is often best to use greater-than-normal spacing between individual dits and dahs, as well as between complete letters. This helps to overcome libration fading effects. The libration fading rate will be different from one band to another. This makes the optimum CW speed for one band different from another. Keep in mind that characters sent too slowly will be chopped up by typical EME fading. Morse code sent too fast will simply be jumbled. Pay attention to the sending practices of the more successful stations, and try to emulate them.

Doppler shift must also be understood. As the moon rises or sets it is moving toward or away from objects on Earth. This leads to a frequency shift in the moon echoes. The amount of Doppler shift is directly proportional to frequency. At 144 MHz, about 500 Hz is the maximum shift. On 432 MHz, the maximum shift is 1.5 kHz. The shift is upward on moonrise and downward on moonset. When the moon is due south, your own echoes will have no Doppler shift, but stations located far away will still be affected. For scheduling, the accepted practice is to transmit zero beat on the schedule frequency, and tune to compensate for the Doppler shift. Be careful—most transmitters and transceivers have a built-in CW offset. Some radios read this offset when transmitting, and others don't. Find out how your transmitter operates and compensate as required.

Random operation has become popular in recent years. In the ARRL EME contest, many of the big guns will not even accept schedules during the contest periods, because they can slow down the pace of their contest contacts.

EME Operating Times

Obviously, the first requirement for EME operation is to have the moon visible by both EME stations. This requirement not only consists of times when the moon is above the horizon, but when it is actually clear of obstructions such as trees and buildings. It helps to know your exact EME operating window, specified in the form of beginning and ending GHAs (Greenwich Hour Angle) for different moon declinations. This information allows two different stations to quickly determine if they can simultaneously see the moon.

Once your moon window is determined, the next step is to decide on the best times during that window to schedule or operate. Operating at perigee is preferable because of the reduced path loss. Fig 23.61 shows that not all perigees are equal. There is about a 0.6-dB difference between the closest and farthest perigee points. The next concern is operating when the moon is in a quiet spot of the sky. Usually, northern declinations are preferred, as the sky is quietest at high declinations. If the moon is too close to the sun, your array will pick up sun noise and reduce the sensitivity off your receiver. Finally, choosing days with minimal libration fading is also desirable.

Perigee and apogee days can be determined from the *Astronomical Almanac* by inspecting the tables headed "S.D." (semi-diameter of the moon in minutes of arc). These semi-diameter numbers can be compared to Fig 23.63 to obtain the approximate moon distance. Many computer programs for locating the moon now give the moon's distance. The expected best weekends to operate on 432 MHz and the higher bands are normally printed well in advance in various EME newsletters.

When the moon passes through the galactic plane, sky temperature is at its maximum. Even on the higher bands this is one of the least desirable times to operate. The

Table 23.19

144-MHz Procedure — 2-Minute Sequence

Period	1½ minutes	30 seconds
1	Calls (W6XXX DE W1XXX)	
2	W1XXX DE WE6XXX	TTTT
3	W6XXX DE W1XXX	OOOO
4	RO RO RO RO	DE W1XXX K
5	R R R R R R	DE W6XXX K
6	QRZ? EME	DE W1XXX K

Table 23.20

432-MHz Procedure—2½-Minute Sequence

Period	2 minutes	30 seconds
1	VE7BBG DE K2UYH	
2	K2UYH DE VE7BBG	
3	VE7BBG DE K2UYH	TTT
4	K2UYH DE VE7BBG	MMM
5	RM RM RM RM	DE K2UYH K
6	R R R R R	DE VE7BBG SK

areas of the sky to avoid are the constellations of Orion and Gemini (during northern declinations), and Sagittarius and Scorpios (during southern declinations). The position of the moon relative to these constellations can be checked with information supplied in the *Astronomical Almanac* or *Sky and Telescope* magazine.

Frequencies and Scheduling

According to the ARRL-sponsored band plan, the lower edge of most bands is reserved for EME operation. On 144 MHz, EME frequencies are primarily between 144.000 and 144.080 MHz for CW, and 144.100 and 144.120 MHz for SSB. Random CW activity is usually between 144.000 and 144.020 MHz. In the US, 144.000 to 144.100 MHz is a CW subband, so SSB QSOs often take place by QSYing up 100 kHz after a CW contact has been established. Because of the large number of active 144-MHz stations, coordinating schedules in the small EME window is not simple. The more active stations usually have assigned frequencies for their schedules.

On 432 MHz, the international EME CW calling frequency is 432.010 MHz. Random SSB calling is done on 432.015 MHz. Random activity primarily takes place between 432.000 and 432.020 MHz. The greater Doppler shift on 432 MHz requires greater separation between schedule frequencies than on 144 MHz. Normally 432.000 MHz, 432.020 MHz and each 5-kHz increment up to 432.070 MHz are used for schedules.

Activity on 1296 MHz is centered between 1296.000 and 1296.040 MHz. The random calling frequency is 1296.010 MHz. Operation on the other bands requires more specific coordination. Activity on 33 cm is split between 902 and 903 MHz. Activity on 2300 MHz has to accommodate split-band procedures because of the different band assignments around the world.

EME Net Information

An EME net meets on 14.345 MHz on weekends for the purpose of arranging schedules and exchanging EME informa-

tion. The net meets at 1600 UTC. OSCAR satellites are becoming more popular for EME information exchange. When Mode B is available, a downlink frequency of 145.950 MHz is where the EME group gathers. On Mode L and Mode JL, the downlink frequency is 435.975 MHz.

Other Modes

Most EME contacts are still made on CW, although SSB has gained in popularity and it is now common to hear SSB QSOs on any activity weekend. The ability to work SSB can easily be calculated from Eq 1. The proper receiver bandwidth (2.3 kHz) is substituted. SSB usually requires a +3-dB signal-to-noise ratio, whereas slow-speed CW contacts can be made with a 0-dB signal-to-noise ratio. Slow-scan television and packet communication has been attempted between some of the larger stations. Success has been limited because of the greater signal-to-noise ratios required for these modes, and severe signal distortion from libration fading.

Radio Direction Finding

Far more than simply finding the direction of an incoming radio signal, radio direction finding (RDF) encompasses a variety of techniques for determining the exact location of a signal source. The process involves both art and science. RDF adds fun to ham radio, but has serious purposes, too.

This section was written by Joe Moell, KØOV.

RDF is almost as old as radio communication. It gained prominence when the British Navy used it to track the movement of enemy ships in World War I. Since then, governments and the military have developed sophisticated and complex RDF systems. Fortunately, simple equipment, purchased or built at home, is quite effective in Amateur Radio RDF.

In European and Asian countries, direction finding contests are foot races. The object is to be first to find four or five transmitters in a large wooded park. Young athletes have the best chance of capturing the prizes. This sport is known as *foxhunting* (after the British hill-and-dale horseback events) or *ARDF* (Amateur Radio direction finding).

In North America and England, most RDF contests involve mobiles—cars, trucks, vans, even motorcycles. It may be possible to drive all the way to the transmitter, or there may be a short hike at the

end, called a *sniff*. These competitions are also called foxhunting by some, while others use *bunny hunting, T-hunting* or the classic term *hidden transmitter hunting*.

In the 1950s, 3.5 and 28 MHz were the most popular bands for hidden transmitter hunts. Today, most competitive hunts worldwide are for 144-MHz FM signals, though other VHF bands are also used. Some international foxhunts include 3.5-MHz events.

Even without participating in RDF contests, you will find a knowledge of the techniques useful. They simplify the search for a neighborhood source of power-line interference or TV cable leakage. RDF must be used to track down emergency radio beacons, which signal the location of pilots and boaters in distress. Amateur Radio enthusiasts skilled in transmitter hunting are in demand by agencies such as the Civil Air Patrol and the US Coast Guard Auxiliary for search and rescue support.

The FCC's Field Operations Bureau has created an Amateur Auxiliary, administered by the ARRL Section Managers, to deal with interference matters. In many areas of the country, there are standing agreements between Local Interference Committees and district FCC offices, permitting volunteers to provide evidence leading to prosecution in serious cases of

malicious amateur-to-amateur interference. RDF is an important part of the evidence-gathering process.

The most basic RDF system consists of a directional antenna and a method of detecting and measuring the level of the radio signal, such as a receiver with signal strength indicator. RDF antennas range from a simple tuned loop of wire to an acre of antenna elements with an electronic beam-forming network. Other sophisticated techniques for RDF use the Doppler effect, or measure the time of arrival difference of the signal at multiple antennas.

All of these methods have been used from 2 to 500 MHz and above. However, RDF practices vary greatly between the HF and VHF/UHF portions of the spectrum. For practical reasons, high gain beams, Dopplers and switched dual antennas find favor on VHF/UHF, while loops and phased arrays are the most popular choices on 6 m and below. Signal propagation differences between HF and VHF also affect RDF practices. But many basic transmitter hunting techniques, discussed later in this chapter, apply to all bands and all types of portable RDF equipment.

RDF ANTENNAS FOR HF BANDS

Below 50 MHz, gain antennas such as Yagis and quads are of limited value for RDF. The typical installation of a tribander on a 70-ft tower yields only a general direction of the incoming signal, due to ground effects and the antenna's broad forward lobe. Long monoband beams at greater heights work better, but still cannot achieve the bearing accuracy and repeatability of simpler antennas designed specifically for RDF.

RDF Loops

An effective directional HF antenna can be as uncomplicated as a small loop of wire or tubing, tuned to resonance with a capacitor. When immersed in an electromagnetic field, the loop acts much the same as the secondary winding of a transformer. The voltage at the output is proportional to the amount of flux passing through it and the number of turns. If the loop is oriented such that the greatest amount of area is presented to the magnetic field, the induced voltage will be the highest. If it is rotated so that little or no area is cut by the field lines, the voltage induced in the loop is zero and a null occurs.

To achieve this transformer effect, the loop must be small compared with the signal wavelength. In a single-turn loop, the conductor should be less than 0.08 λ long. For example, a 28-MHz loop should be less than 34 inches in circumference, giving a diameter of approximately 10 inches. The loop may be smaller, but that will reduce its voltage output. Maximum output from a small loop antenna is in directions corresponding to the plane of the loop; these lobes are very broad. Sharp nulls, obtained at right angles to that plane, are more useful for RDF.

For a perfect bidirectional pattern, the loop must be balanced electrostatically with respect to ground. Otherwise, it will exhibit two modes of operation, the mode of a perfect loop and that of a nondirectional vertical antenna of small dimensions. This dual-mode condition results in mild to severe inaccuracy, depending on the degree of imbalance, because the outputs of the two modes are not in phase.

The theoretical true loop pattern is illustrated in **Fig 23.70A**. When properly balanced, there are two nulls exactly 180° apart. When the unwanted antenna effect is appreciable and the loop is tuned to resonance, the loop may exhibit little directivity, as shown in Fig 23.70B. By detuning the loop to shift the phasing, you may obtain a useful pattern similar to Fig 23.70C. While not symmetrical, and not necessarily at right angles to the plane of the loop, this pattern does exhibit a pair of nulls.

By careful detuning and amplitude balancing, you can approach the unidirectional pattern of Fig 23.70D. Even though there may not be a complete null in the pattern, it resolves the 180° ambiguity of Fig 23.70A. Korean War era military loop antennas, sometimes available on today's surplus market, use this controlled-antenna-effect principle.

An easy way to achieve good electrostatic balance is to shield the loop, as shown in **Fig 23.71**. The shield, represented by the dashed lines in the drawing, eliminates the antenna effect. The response of a well-constructed shielded loop is quite close to the ideal pattern of Fig 23.70A.

For 160 through 30 m, single-turn loops that are small enough for portability are usually unsatisfactory for RDF work. Multiturn loops are generally used instead. They are easier to resonate with practical capacitor values and give higher output voltages. This type of loop may also be shielded. If the total conductor length remains below 0.08 λ, the directional pattern is that of Fig 23.70A.

Ferrite Rod Antennas

Another way to get higher loop output is to increase the permeability of the medium in the vicinity of the loop. By winding a coil of wire around a form made of high-permeability material, such as ferrite rod, much greater flux is obtained in the coil without increasing the cross-sectional area.

Modern magnetic core materials make compact directional receiving antennas practical. Most portable AM broadcast receivers use this type of antenna, commonly called a *loopstick*. The loopstick is the most popular RDF antenna for portable/mobile work on 160 and 80 m.

As does the shielded loop discussed earlier, the loopstick responds to the magnetic field of the incoming radio wave, and not to the electrical field. For a given size of loop, the output voltage increases with increasing flux density, which is obtained by choosing a ferrite core of high permeability and low loss at the frequency of interest. For increased output, the turns may be wound over two rods taped together. A practical loopstick antenna is described later in this chapter.

A loop on a ferrite core has maximum signal response in the plane of the turns, just as an air core loop. This means that maximum response of a loopstick is broadside to the axis of the rod, as shown in **Fig 23.72**. The loopstick may be shielded to eliminate the antenna effect; a U-shaped or C-shaped channel of aluminum or other form of "trough" is best. The shield must not be closed, and its length should equal or slightly exceed the length of the rod.

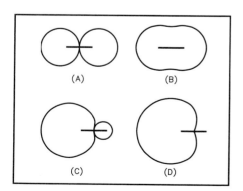

Fig 23.70 — Small loop field patterns with varying amounts of antenna effect — the undesired response of a loop acting merely as a mass of metal connected to the receiver antenna terminals. The horizontal lines show the plane of the loop turns.

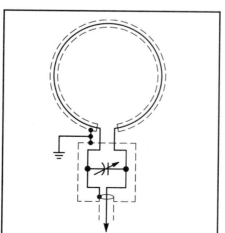

Fig 23.71 — Electrostatically shielded loop for RDF. To prevent shielding of the loop from magnetic fields, leave the shield unconnected at one end.

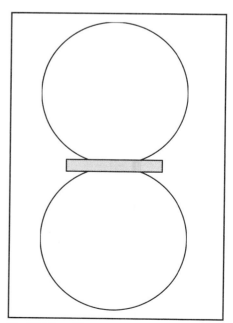

Fig 23.72 — Field pattern for a ferrite rod antenna. The dark bar represents the rod on which the loop turns are wound.

Sense Antennas

Because there are two nulls 180° apart in the directional pattern of a small loop or loopstick, there is ambiguity as to which null indicates the true direction of the target station. For example, if the line of bearing runs east and west from your position, you have no way of knowing from this single bearing whether the transmitter is east of you or west of you.

If bearings can be taken from two or more positions at suitable direction and distance from the transmitter, the ambiguity can be resolved and distance can be estimated by triangulation, as discussed later in this chapter. However, it is almost always desirable to be able to resolve the ambiguity immediately by having a unidirectional antenna pattern available.

You can modify a loop or loopstick antenna pattern to have a single null by adding a second antenna element. This element is called a sense antenna, because it senses the phase of the signal wavefront for comparison with the phase of the loop output signal. The sense element must be omnidirectional, such as a short vertical. When signals from the loop and the sense antenna are combined with 90° phase shift between the two, a heart-shaped (cardioid) pattern results, as shown in **Fig 23.73A**.

Fig 23.73B shows a circuit for adding a sense antenna to a loop or loopstick. For the best null in the composite pattern, signals from the loop and sense antennas must be of equal amplitude. R1 adjusts the level of the signal from the sense antenna.

In a practical system, the cardioid pattern null is not as sharp as the bidirectional null of the loop alone. The usual procedure when transmitter hunting is to use the loop alone to obtain a precise line of bearing, then switch in the sense antenna and take another reading to resolve the ambiguity.

Phased Arrays and Adcocks

Two-element phased arrays are popular for amateur HF RDF base station installations. Many directional patterns are possible, depending on the spacing and phasing of the elements. A useful example is two $^{1}/_{2}$-λ elements spaced $^{1}/_{4}$ λ apart and fed 90° out of phase. The resultant pattern is a cardioid, with a null off one end of the axis of the two antennas and a broad peak in the opposite direction. The directional frequency range of this antenna is limited to one band, because of the critical length of the phasing lines.

The best-known phased array for RDF is the Adcock, named after the man who invented it in 1919. It consists of two vertical elements fed 180° apart, mounted so the array may be rotated. Element spacing is not critical, and may be in the range from $^{1}/_{10}$ to $^{3}/_{4}$ λ. The two elements must be of identical lengths, but need not be self-resonant; shorter elements are commonly used. Because neither the element spacing nor length is critical in terms of wavelengths, an Adcock array may operate over more than one amateur band.

Fig 23.74 is a schematic of a typical Adcock configuration, called the H-Adcock because of its shape. Response to a vertically polarized wave is very similar to a conventional loop. The passing wave induces currents I1 and I2 into the vertical members. The output current in the transmission line is equal to their difference. Consequently, the directional pattern has two broad peaks and two sharp nulls, like the loop. The magnitude of the difference current is proportional to the spacing (d) and length (l) of the elements. You will get somewhat higher gain with larger dimen-

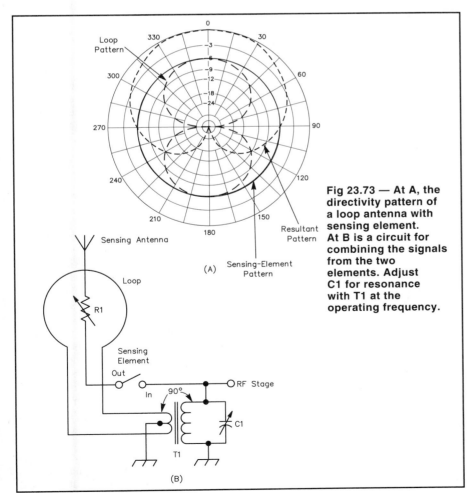

Fig 23.73 — At A, the directivity pattern of a loop antenna with sensing element. At B is a circuit for combining the signals from the two elements. Adjust C1 for resonance with T1 at the operating frequency.

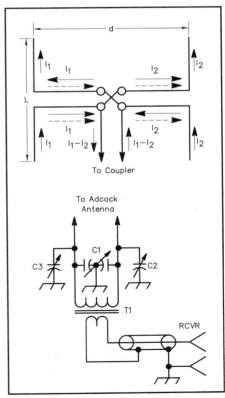

Fig 23.74 — A simple Adcock antenna and its coupler.

sions. The Adcock of **Fig 23.75**, designed for 40 m, has element lengths of 12 ft and spacing of 21 ft (approximately 0.15 λ).

Fig 23.76 shows the radiation pattern of the Adcock. The nulls are broadside to the axis of the array, becoming sharper with increased element spacings. When element spacing exceeds ³/₄ λ, however, the antenna begins to take on additional unwanted nulls off the ends of the array axis.

The Adcock is a vertically polarized antenna. The vertical elements do not respond to horizontally polarized waves, and the currents induced in the horizontal members by a horizontally polarized wave (dotted arrows in Fig 23.74) tend to balance out regardless of the orientation of the antenna.

Since the Adcock uses a balanced feed system, a coupler is required to match the

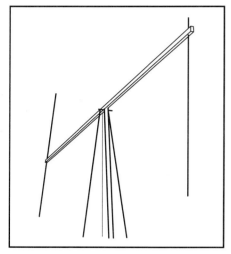

Fig 23.75 — An experimental Adcock antenna on a wooden frame.

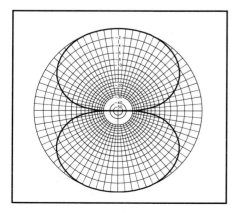

Fig 23.76 — The pattern of an Adcock array with element spacing of ¹/₂ wavelength. The elements are aligned with the vertical axis.

unbalanced input of the receiver. T1 is an air-wound coil with a two-turn link wrapped around the middle. The combination is resonated with C1 to the operating frequency. C2 and C3 are null-clearing capacitors. Adjust them by placing a low-power signal source some distance from the antenna and exactly broadside to it. Adjust C2 and C3 until the deepest null is obtained.

While you can use a metal support for the mast and boom, wood is preferable because of its nonconducting properties. Similarly, a mast of thick-wall PVC pipe gives less distortion of the antenna pattern than a metallic mast. Place the coupler on the ground below the wiring harness junction on the boom and connect it with a short length of 300-Ω twin-lead feed line.

Loops vs Phased Arrays

Loops are much smaller than phased arrays for the same frequency, and are thus the obvious choice for portable/mobile HF RDF. For base stations in a triangulation network, where the 180° ambiguity is not a problem, Adcocks are preferred. In general, they give sharper nulls than loops, but this is in part a function of the care used in constructing and feeding the individual antennas, as well as of the spacing of the elements. The primary construction considerations are the shielding and balancing of the feed line against unwanted signal pickup and the balancing of the antenna for a symmetrical pattern. Users report that Adcocks are somewhat less sensitive to proximity effects, probably because their larger aperture offers some space diversity.

Skywave Considerations

Until now we have considered the directional characteristics of the RDF loop only in the two-dimensional azimuthal plane. In three-dimensional space, the response of a vertically oriented small loop is doughnut-shaped. The bidirectional null (analogous to a line through the doughnut hole) is in the line of bearing in the azimuthal plane and toward the horizon in the vertical plane. Therefore, maximum null depth is achieved only on signals arriving at 0° elevation angle.

Skywave signals usually arrive at nonzero wave angles. As the elevation angle increases, the null in a vertically oriented loop pattern becomes more shallow. It is possible to tilt the loop to seek the null in elevation as well as azimuth. Some amateur RDF enthusiasts report success at estimating distance to the target by measurement of the elevation angle with a tilted loop and computations based on estimated height of the propagating ionospheric layer. This method seldom provides high

accuracy with simple loops, however.

Most users prefer Adcocks to loops for skywave work, because the Adcock null is present at all elevation angles. Note, however, that an Adcock has a null in all directions from signals arriving from overhead. Thus for very high angles, such as under-250-mile skip on 80 and 40 m, neither loops nor Adcocks will perform well.

Electronic Antenna Rotation

State-of-the-art fixed RDF stations for government and military work use antenna arrays of stationary elements, rather than mechanically rotatable arrays. The best known type is the Wullenweber antenna. It has a large number of elements arranged in a circle, usually outside of a circular reflecting screen. Depending on the installation, the circle may be anywhere from a few hundred feet to more than a quarter of a mile in diameter. Although the Wullenweber is not practical for most amateurs, some of the techniques it uses may be applied to amateur RDF.

The device which permits rotating the antenna beam without moving the elements has the classic name *radiogoniometer*, or simply *goniometer*. Early goniometers were RF transformers with fixed coils connected to the array elements and a moving pickup coil connected to the receiver input. Both amplitude and phase of the signal coupled into the pickup winding are altered with coil rotation in a way that corresponded to actually rotating the array itself. With sufficient elements and a goniometer, accurate RDF measurements can be taken in all compass directions.

Beam Forming Networks

By properly sampling and combining signals from individual elements in a large array, an antenna beam is electronically rotated or steered. With an appropriate number and arrangement of elements in the system, it is possible to form almost any desired antenna pattern by summing the sampled signals in appropriate amplitude and phase relationships. Delay networks and/or attenuation are added in line with selected elements before summation to create these relationships.

To understand electronic beam forming, first consider just two elements, shown as A and B in **Fig 23.77**. Also shown is the wavefront of a radio signal arriving from a distant transmitter. The wavefront strikes element A first, then travels somewhat farther before it strikes element B. Thus, there is an interval between the times that the wavefront reaches elements A and B.

We can measure the differences in arrival times by delaying the signal received at element A before summing it with that

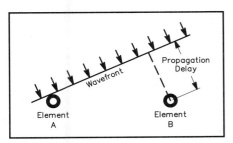

Fig 23.77 — One technique used in electronic beam forming. By delaying the signal from element A by an amount equal to the propagation delay, two signals are summed precisely in phase, even though the signal is not in the broadside direction.

from element B. If two signals are combined directly, the amplitude of the sum will be maximum when the delay for element A exactly equals the propagation delay, giving an in-phase condition at the summation point. On the other hand, if one of the signals is inverted and the two are added, the signals will combine in a 180° out-of-phase relationship when the element A delay equals the propagation delay, creating a null. Either way, once the time delay is determined by the amount of delay required for a peak or null, we can convert it to distance. Then trigonometry calculations provide the direction from which the wave is arriving.

Altering the delay in small increments steers the peak (or null) of the antenna. The system is not frequency sensitive, other than the frequency range limitations of the array elements. Lumped-constant networks are suitable for delay elements if the system is used only for receiving. Delay lines at installations used for transmitting and receiving employ rolls of coaxial cable of various lengths, chosen for the time delay they provide at all frequencies, rather than as simple phasing lines designed for a single frequency.

Combining signals from additional elements narrows the broad beamwidth of the pattern from the two elements and suppress unwanted sidelobes. Electronically switching the delays and attenuations to the various elements causes the formed beam to rotate around the compass. The package of electronics that does this, including delay lines and electronically switched attenuators, is the beam forming network.

METHODS FOR VHF/UHF RDF

Three distinct methods of mobile RDF are commonly in use by amateurs on VHF/UHF bands: directional antennas, switched dual antennas and Dopplers.

Each has advantages over the others in certain situations. Many RDF enthusiasts employ more than one method when transmitter hunting.

Directional Antennas

Ordinary mobile transceivers and hand-helds work well for foxhunting on the popular VHF bands. If you have a light-weight beam and your receiver has an easy-to-read S-meter, you are nearly ready to start. All you need is an RF attenuator and some way to mount the setup in your vehicle.

Amateurs seldom use fractional wave-length loops for RDF above 60 MHz because they have bidirectional characteristics and low sensitivity, compared to other practical VHF antennas. Sense circuits for loops are difficult to implement at VHF, and signal reflections tend to fill in the nulls. Typically VHF loops are used only for close-in sniffing where their compactness and sharp nulls are assets, and low gain is of no consequence.

Phased Arrays

The small size and simplicity of 2-element driven arrays make them a common choice of newcomers at VHF RDF. Antennas such as phased ground planes and ZL Specials have modest gain in one direction and a null in the opposite direction. The gain is helpful when the signal is weak, but the broad response peak makes it difficult to take a precise bearing.

As the signal gets stronger, it becomes possible to use the null for a sharper S-meter indication. However, combinations of direct and reflected signals (called *multipath*) will distort the null or perhaps obscure it completely. For best results with this type of antenna, always find clear locations from which to take bearings.

Parasitic Arrays

Parasitic arrays are the most common RDF antennas used by transmitter hunters in high competition areas such as Southern California. Antennas with significant gain are a necessity due to the weak signals often encountered on weekend-long T-hunts, where the transmitter may be over 200 miles distant. Typical 144-MHz installations feature Yagis or quads of three to six elements, sometimes more. Quads are typically home-built, using data from *The ARRL Antenna Book* and *Transmitter Hunting* (see Bibliography).

Two types of mechanical construction are popular for mobile VHF quads. The model of **Fig 23.78** uses thin gauge wire (solid or stranded), suspended on wood dowel or fiberglass rod spreaders. It is lightweight and easy to turn rapidly by

hand while the vehicle moves. Many hunters prefer to use larger gauge solid wire (such as AWG 10) on a PVC plastic pipe frame (**Fig 23.79**). This quad is more rugged and has somewhat wider frequency

Fig 23.78 — The mobile RDF installation of WB6ADC features a thin wire quad for 144 MHz and a mechanical linkage that permits either the driver or front passenger to rotate the mast by hand.

Fig 23.79 — KØOV uses this mobile setup for RDF on several bands, with separate antennas for each band that mate with a common lower mast section, pointer and 360° indicator. Antenna shown is a heavy gauge wire quad for 2 m.

range, at the expense of increased weight and wind resistance. It can get mashed going under a willow, but it is easily re-shaped and returned to service.

Yagis are a close second to quads in popularity. Commercial models work fine for VHF RDF, provided that the mast is attached at a good balance point. Light-weight and small-diameter elements are desirable for ease of turning at high speeds.

A well-designed mobile Yagi or quad installation includes a method of selecting wave polarization. Although vertical polarization is the norm for VHF-FM communications, horizontal polarization is allowed on many T-hunts. Results will be poor if a VHF RDF antenna is cross-polarized to the transmitting antenna, because multipath and scattered signals (which have indeterminate polarization) are enhanced, relative to the cross-polarized direct signal. The installation of Fig 23.78 features a slip joint at the boom-to-mast junction, with an actuating cord to rotate the boom, changing the polariza-tion. Mechanical stops limit the boom rotation to 90°.

Parasitic Array Performance for RDF

The directional gain of a mobile beam (typically 8 dB or more) makes it unex-celled for both weak signal competitive hunts and for locating interference such as TV cable leakage. With an appropriate receiver, you can get bearings on any sig-nal mode, including FM, SSB, CW, TV, pulses and noise. Because only the re-sponse peak is used, the null-fill problems and proximity effects of loops and phased arrays do not exist.

You can observe multiple directions of arrival while rotating the antenna, allow-ing you to make educated guesses as to which signal peaks are direct and which are from nondirect paths or scattering. Skilled operators can estimate distance to the transmitter from the rate of signal strength increase with distance traveled. The RDF beam is useful for transmitting, if necessary, but use care not to damage an attenuator in the coax line by transmitting through it.

The 3-dB beamwidth of typical mobile-mount VHF beams is on the order of 80°. This is a great improvement over 2-ele-ment driven arrays, but it is still not pos-sible to get pinpoint bearing accuracy. You can achieve errors of less than 10° by care-fully reading the S-meter. In practice, this is not a major hindrance to successful mobile RDF. Mobile users are not as con-cerned with precise bearings as fixed sta-tion operators, because mobile readings are used primarily to give the general di-

rection of travel to "home in" on the sig-nal. Mobile bearings are continuously up-dated from new, closer locations.

Amplitude-based RDF may be very dif-ficult when signal level varies rapidly. The transmitter hider may be changing power, or the target antenna may be moving or near a well-traveled road or airport. The resultant rapid S-meter movement makes it hard to take accurate bearings with a quad. The process is slow because the antenna must be carefully rotated by hand to "eyeball average" the meter readings.

Switched Antenna RDF Units

Three popular types of RDF systems are relatively insensitive to variations in sig-nal level. Two of them use a pair of verti-cal dipole antennas, spaced $^1/_2 \lambda$ or less apart, and alternately switched at a rapid rate to the input of the receiver. In use, the indications of the two systems are similar, but the principles are different.

Switched Pattern Systems

The switched pattern RDF set (**Fig 23.80**) alternately creates two cardioid antenna patterns with lobes to the left and the right. The patterns are generated in much the same way as in the phased arrays described above. PIN RF diodes select the alternating patterns. The combined an-tenna outputs go to a receiver with AM detection. Processing after the detector output determines the phase or amplitude difference between the patterns' responses to the signal.

Switched pattern RDF sets typically have a zero center meter as an indica-tor. The meter swings negative when the signal is coming from the user's left, and positive when the signal source is on the right. When the plane of the an-

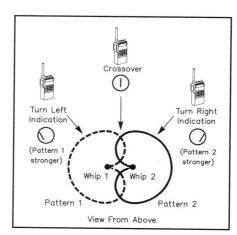

Fig 23.80 — In a switched pattern RDF set, the responses of two cardioid antenna patterns are summed to drive a zero center indicator.

tenna is exactly perpendicular to the direction of the signal source, the meter reads zero.

The sharpness of the zero crossing indi-cation makes possible more precise bear-ings than those obtainable with a quad or Yagi. Under ideal conditions with a well-built unit, null direction accuracy is within 1°. Meter deflection tells the user which way to turn to zero the meter. For example, a negative (left) reading requires turning the antenna left. This solves the 180° am-biguity caused by the two zero crossings in each complete rotation of the antenna system.

Because it requires AM detection of the switched pattern signal, this RDF system finds its greatest use in the 120-MHz air-craft band, where AM is the standard mode. Commercial manufacturers make portable RDF sets with switched pattern antennas and built-in receivers for field portable use. These sets can usually be adapted to the amateur 144-MHz band. Other designs are adaptable to any VHF receiver that covers the frequency of in-terest and has an AM detector built in or added.

Switched pattern units work well for RDF from small aircraft, for which the two vertical antennas are mounted in fixed po-sitions on the outside of the fuselage or simply taped inside the windshield. The left-right indication tells the pilot which way to turn the aircraft to home in. Since street vehicles generally travel only on roads, fixed mounting of the antennas on them is undesirable. Mounting vehicular switched-pattern arrays on a rotatable mast is best.

Time of Arrival Systems

Another kind of switched antenna RDF set uses the difference in arrival times of the signal wavefront at the two antennas. This narrow-aperture Time-Difference-of-Arrival (TDOA) technology is used for many sophisticated military RDF systems. The rudimentary TDOA implementation of **Fig 23.81** is quite effective for amateur use. The signal from transmitter 1 reaches antenna A before antenna B. Conversely, the signal from transmitter 3 reaches an-tenna B before antenna A. When the plane of the antenna is perpendicular to the sig-nal source (as transmitter 2 is in the fig-ure), the signal arrives at both antennas simultaneously.

If the outputs of the antennas are alter-nately switched at an audio rate to the re-ceiver input, the differences in the arrival times of a continuous signal produce phase changes that are detected by an FM dis-criminator. The resulting short pulses sound like a tone in the receiver output.

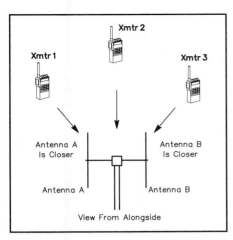

Fig 23.81 — A dual-antenna TDOA RDF system has a similar indicator to a switched pattern unit, but it obtains bearings by determining which of its antennas is closer to the transmitter.

The tone disappears when the antennas are equidistant from the signal source, giving an audible null.

The polarity of the pulses at the discriminator output is a function of which antenna is closer to the source. Therefore, the pulses can be processed and used to drive a left-right zero center meter in a manner similar to the switched pattern units described above. Left-right LED indicators may replace the meter for economy and visibility at night.

RDF operations with a TDOA dual antenna RDF are done in the same manner as with a switched antenna RDF set. The main difference is the requirement for an FM receiver in the TDOA system and an AM receiver in the switched pattern case. No RF attenuator is needed for close-in work in the TDOA case.

Popular designs for practical do-it-yourself TDOA RDF sets include the Simple Seeker (described elsewhere in this chapter) and the W9DUU design (see article by Bohrer in the Bibliography). Articles with plans for the Handy Tracker, a simple TDOA set with a delay line to resolve the dual-null ambiguity instead of LEDs or a meter, are listed in the Bibliography.

Performance Comparison

Both types of dual antenna RDFs make good on-foot "sniffing" devices and are excellent performers when there are rapid amplitude variations in the incoming signal. They are the units of choice for airborne work. Compared to Yagis and quads, they give good directional performance over a much wider frequency range. Their indications are more precise than those of beams with broad forward lobes.

Dual-antenna RDF sets frequently give inaccurate bearings in multipath situations, because they cannot resolve signals of nearly equal levels from more than one direction. Because multipath signals are a combined pattern of peaks and nulls, they appear to change in amplitude and bearing as you move the RDF antenna along the bearing path or perpendicular to it, whereas a non-multipath signal will have constant strength and bearing.

The best way to overcome this problem is to take large numbers of bearings while moving toward the transmitter. Taking bearings while in motion averages out the effects of multipath, making the direct signal more readily discernible. Some TDOA RDF sets have a slow-response mode that aids the averaging process.

Switched antenna systems generally do not perform well when the incoming signal is horizontally polarized. In such cases, the bearings may be inaccurate or unreadable. TDOA units require a carrier type signal such as FM or CW; they usually cannot yield bearings on noise or pulse signals.

Unless an additional method is employed to measure signal strength, it is easy to "overshoot" the hidden transmitter location with a TDOA set. It is not uncommon to see a TDOA foxhunter walk over the top of a concealed transmitter and walk away, following the opposite 180° null, because there is no display of signal amplitude.

Doppler RDF Sets

RDF sets using the Doppler principle are popular in many areas because of their ease of use. They have an indicator that instantaneously displays direction of the signal source relative to the vehicle heading, either on a circular ring of LEDs or a digital readout in degrees. A ring of four, eight or more antennas picks up the signal. Quarter-wavelength monopoles on a ground plane are popular for vehicle use, but half-wavelength vertical dipoles, where practical, perform better.

Radio signals received on a rapidly moving antenna experience a frequency shift due to the Doppler effect, a phenomenon well known to anyone who has observed a moving car with its horn sounding. The horn's pitch appears higher than normal as the car approaches, and lower as the car recedes. Similarly, the received radio frequency increases as the antenna moves toward the transmitter and vice versa. An FM receiver will detect this frequency change.

Fig 23.82 shows a $1/4$-λ vertical antenna being moved on a circular track around

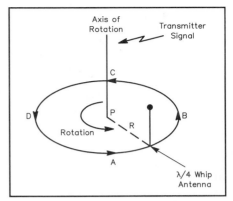

Fig 23.82 — A theoretical Doppler antenna circles around point P, continuously moving toward and away from the source at an audio rate.

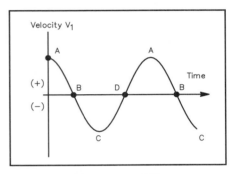

Fig 23.83 — Frequency shift versus time produced by the rotating antenna movement toward and away from the signal source.

point P, with constant angular velocity. As the antenna approaches the transmitter on its track, the received frequency is shifted higher. The highest instantaneous frequency occurs when the antenna is at point A, because tangential velocity toward the transmitter is maximum at that point. Conversely, the lowest frequency occurs when the antenna reaches point C, where velocity is maximum away from the transmitter.

Fig 23.83 shows a plot of the component of the tangential velocity that is in the direction of the transmitter as the antenna moves around the circle. Comparing Figs 23.82 and 23.83, notice that at B in Fig 23.83, the tangential velocity is crossing zero from the positive to the negative and the antenna is closest to the transmitter. The Doppler shift and resulting audio output from the receiver discriminator follow the same plot, so that a negative-slope zero-crossing detector, synchronized with the antenna rotation, senses the incoming direction of the signal.

The amount of frequency shift due to the Doppler effect is proportional to the RF frequency and the tangential antenna velocity. The velocity is a function of the radius of rotation and the angular velocity (rotation rate). The radius of rotation must be less than $1/4\ \lambda$ to avoid errors. To get a usable amount of FM deviation (comparable to typical voice modulation) with this radius, the antenna must rotate at approximately 30,000 RPM (500 Hz). This puts the Doppler tone in the audio range for easy processing.

Mechanically rotating a whip antenna at this rate is impractical, but a ring of whips, switched to the receiver in succession with RF PIN diodes, can simulate a rapidly rotating antenna. Doppler RDF sets must be used with receivers having FM detectors. The DoppleScAnt and Roanoke Doppler (see Bibliography) are mobile Doppler RDF sets designed for inexpensive home construction.

Doppler Advantages and Disadvantages

Ring-antenna Doppler sets are the ultimate in simplicity of operation for mobile RDF. There are no moving parts and no manual antenna pointing. Rapid direction indications are displayed on very short signal bursts.

Many units lock in the displayed direction after the signal leaves the air. Power variations in the source signal cause no difficulties, as long as the signal remains above the RDF detection threshold. A Doppler antenna goes on top of any car quickly, with no holes to drill. Many Local Interference Committee members choose Dopplers for tracking malicious interference, because they are inconspicuous (compared to beams) and effective at tracking the strong vertically polarized signals that repeater jammers usually emit.

A Doppler does not provide superior performance in all VHF RDF situations. If the signal is too weak for detection by the Doppler unit, the hunt advantage goes to teams with beams. Doppler installations are not suitable for on-foot sniffing. The limitations of other switched antenna RDFs also apply: (1) poor results with horizontally polarized signals, (2) no indication of distance, (3) carrier type signals only and (4) inadvisability of transmitting through the antenna.

Readout to the nearest degree is provided on some commercial Doppler units. This does not guarantee that level of accuracy, however. A well-designed four-monopole set is typically capable of ±5° accuracy on 2 m, if the target signal is vertically polarized and there are no multipath effects.

The rapid antenna switching can introduce cross modulation products when the user is near strong off-channel RF sources. This self-generated interference can tem-

porarily render the system unusable. While not a common problem with mobile Dopplers, it makes the Doppler a poor choice for use in remote RDF installations at fixed sites with high power VHF transmitters nearby.

Mobile RDF System Installation

Of these mobile VHF RDF systems, the Doppler type is clearly the simplest from

Fig 23.84 — A set of TDOA RDF antennas is light weight and mounts readily through a sedan window without excessive overhang.

Fig 23.85 — A window box allows the navigator to turn a mast mounted antenna with ease while remaining dry and warm. No holes in the vehicle are needed with a properly designed window box.

a mechanical installation standpoint. A four-whip Doppler RDF array is easy to implement with magnetic mount antennas. Alternately, you can mount all the whips on a frame that attaches to the vehicle roof with suction cups. In either case, setup is rapid and requires no holes in the vehicle.

You can turn small VHF beams and dual-antenna arrays readily by extending the mast through a window. Installation on each model vehicle is different, but usually the mast can be held in place with some sort of cup in the arm rest and a plastic tie at the top of the window, as in **Fig 23.84**. This technique works best on cars with frames around the windows, which allow the door to be opened with the antenna in place. Check local vehicle codes, which limit how far your antenna may protrude beyond the line of the fenders. Larger antennas may have to be put on the passenger side of the vehicle, where greater overhang is generally permissible.

The window box (**Fig 23.85**) is an improvement over through-the-window mounts. It provides a solid, easy-turning mount for the mast. The plastic panel keeps out bad weather. You will need to custom-design the box for your vehicle model. Vehicle codes may limit the use of a window box to the passenger side.

For the ultimate in convenience and versatility, cast your fears aside, drill a hole through the center of the roof and install a waterproof bushing. A roof-hole mount permits the use of large antennas without overhang violations. The driver, front passenger and even a rear passenger can turn the mast when required. The installation in Fig 23.79 uses a roof-hole bushing made from mating threaded PVC pipe adapters and reducers. When it is not in use for RDF, a PVC pipe cap provides a watertight cover. There is a pointer and 360° indicator at the bottom of the mast for precise bearings.

DIRECTION-FINDING TECHNIQUES AND PROJECTS

The ability to locate a transmitter quickly with RDF techniques is a skill you will acquire only with practice. It is very important to become familiar with your equipment and its limitations. You must also understand how radio signals behave in different types of terrain at the frequency of the hunt. Experience is the best teacher, but reading and hearing the stories of others who are active in RDF will help you get started.

Verify proper performance of your portable RDF system before you attempt to track signals in unknown locations. Of primary concern is the accuracy and symmetry of the antenna pattern. For instance, a lopsided figure-8 pattern with a loop, Adcock, or TDOA set leads to large bearing errors. Nulls should be exactly 180° apart and exactly at right angles to the loop plane or the array boom. Similarly, if feed-line pickup causes an off-axis main lobe in your VHF RDF beam, your route to the target will be a spiral instead of a straight line.

Perform initial checkout with a low-powered test transmitter at a distance of a few hundred feet. Compare the RDF bearing indication with the visual path to the transmitter. Try to "find" the transmitter with the RDF equipment as if its position were not known. Be sure to check all nulls on antennas that have more than one.

If imbalance or off-axis response is found in the antennas, there are two options available. One is to correct it, insofar as possible. A second option is to accept it and use some kind of indicator or correction procedure to show the true directions of signals. Sometimes the end result of the calibration procedure is a compromise between these two options, as a perfect pattern may be difficult or impossible to attain.

The same calibration suggestions apply for fixed RDF installations, such as a base station HF Adcock or VHF beam. Of course it does no good to move it to an open field. Instead, calibrate the array in its intended operating position, using a portable or mobile transmitter. Because of nearby obstructions or reflecting objects, your antenna may not indicate the precise direction of the transmitter. Check for imbalance and systemic error by taking readings with the test emitter at locations in several different directions.

The test signal should be at a distance of 2 or 3 miles for these measurements, and should be in as clear an area as possible during transmissions. Avoid locations where power lines and other overhead wiring can conduct signal from the transmitter to the RDF site. Once antenna adjustments are optimized, make a table of bearing errors noted in all compass directions. Apply these error values as corrections when actual measurements are made.

Preparing to Hunt

Successfully tracking down a hidden transmitter involves detective work — examining all the clues, weighing the evidence and using good judgment. Before setting out to locate the source of a signal, note its general characteristics. Is the frequency constant, or does it drift? Is the signal continuous, and if not, how long are transmissions? Do transmissions occur at regular intervals, or are they sporadic? Irregular, intermittent signals are the most difficult to locate, requiring patience and quick action to get bearings when the transmitter comes on.

Refraction, Reflections and the Night Effect

You will get best accuracy in tracking ground wave signals when the propagation path is over homogeneous terrain. If there is a land/water boundary in the path, the different conductivities of the two media can cause bending (refraction) of the wave front, as in **Fig 23.86A**. Even the most sophisticated RDF equipment will not indicate the correct bearing in this situation, as the equipment can only show the direction from which the signal is arriving. RDFers have observed this phenomenon on both HF and VHF bands.

Signal reflections also cause misleading bearings. This effect becomes more pronounced as frequency increases. T-hunt hiders regularly achieve strong signal bounces from distant mountain ranges on the 144-MHz band.

Tall buildings also reflect VHF/UHF signals, making midcity RDF difficult. Hunting on the 440-MHz and higher amateur bands is even more arduous because of the plethora of reflecting objects.

In areas of signal reflection and multipath, some RDF gear may indicate that the signal is coming from an intermediate point, as in Fig 23.86B. High gain VHF/UHF RDF beams will show direct and reflected signals as separate S-meter peaks, leaving it to the operator to determine which is which. Null-based RDF antennas, such as phased arrays and loops, have the most difficulty with multipath, because the multiple signals tend to make the nulls very shallow or fill them in entirely, resulting in no bearing indication at all.

If the direct path to the transmitter is masked by intervening terrain, a signal reflection from a higher mountain, building, water tower, or the like may be much stronger than the direct signal. In extreme cases, triangulation from several locations will appear to "confirm" that the transmitter is at the location of the reflecting object. The direct signal may not be detectable until you arrive at the reflecting point or another high location.

Objects near the observer such as concrete/steel buildings, power lines and chain-link fences will distort the incoming wavefront and give bearing errors. Even a dense grove of trees can sometimes have an adverse effect. It is always best to take readings in locations that are as open and clear as possible, and to take bearings from numerous positions for confirmation. Testing of RDF gear should also be done in clear locations.

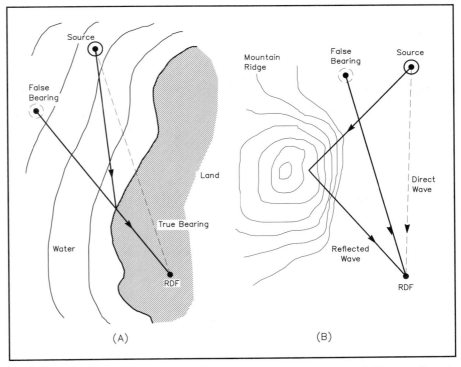

Fig 23.86 — RDF errors caused by refraction (A) and reflection (B). The reading at A is false because the signal actually arrives from a direction that is different from that to the source. At B, a direct signal from the source combines with a reflected signal from the mountain ridge. The RDF set may average the signals as shown, or indicate two lines of bearing.

Locating local signal sources on frequencies below 10 MHz is much easier during daylight hours, particularly with loop antennas. In the daytime, D-layer absorption minimizes skywave propagation on these frequencies. When the D layer disappears after sundown, you may hear the signal by a combination of ground wave and high-angle skywave, making it difficult or impossible to obtain a bearing. RDFers call this phenomenon the *night effect*.

While some mobile T-hunters prefer to go it alone, most have more success by teaming up and assigning tasks. The driver concentrates on handling the vehicle, while the assistant (called the "navigator" by some teams) turns the beam, reads the meters and calls out bearings. The assistant is also responsible for maps and plotting, unless there is a third team member for that task.

Maps and Bearing-Measurements

Possessing accurate maps and knowing how to use them is very important for successful RDF. Even in difficult situations where precise bearings cannot be obtained, a town or city map will help in plotting points where signal levels are high and low. For example, power line noise tends to propagate along the power line and radiates as it does so. Instead of a single source, the noise appears to come

from a multitude of sources. This renders many ordinary RDF techniques ineffective. Mapping locations where signal amplitudes are highest will help pinpoint the source.

Several types of area-wide maps are suitable for navigation and triangulation. Street and highway maps work well for mobile work. Large detailed maps are preferable to thick map books. Contour maps are ideal for open country. Aeronautical charts are also suitable. Good sources of maps include auto clubs, stores catering to camping/hunting enthusiasts and city/county engineering departments.

A *heading* is a reading in degrees relative to some external reference, such as your house or vehicle; a *bearing* is the target signal's direction relative to your position. Plotting a bearing on a hidden transmitter from your vehicle requires that you know the vehicle location, transmitter heading with respect to the vehicle and vehicle heading with respect to true north.

First, determine your location, using landmarks or a navigation device such as a loran or GPS receiver. Next, using your RDF equipment, determine the bearing to the hidden transmitter (0 to 359.9°) with respect to the vehicle. Zero degrees heading corresponds to signals coming from directly in front of the vehicle, signals from the right indicate

90°, and so on.

Finally, determine your vehicle's true heading, that is, its heading relative to true north. Compass needles point to magnetic north and yield magnetic headings. Translating a magnetic heading into a true heading requires adding a correction factor, called *magnetic declination*, which is a positive or negative factor that depends on your location.

Declination for your area is given on US Geological Survey (USGS) maps, though it undergoes long-term changes. Add the declination to your magnetic heading to get a true heading.

As an example, assume that the transmitted signal arrives at 30° with respect to the vehicle heading, that the compass indicates that the vehicle's heading is 15°, and the magnetic declination is +15°. Add these values to get a true transmitter bearing (that is, a bearing with respect to true north) of 60°.

Because of the large mass of surrounding metal, it is very difficult to calibrate an in-car compass for high accuracy at all vehicle headings. It is better to use a remotely mounted flux-gate compass sensor, properly corrected, to get vehicle headings, or to stop and use a hand compass to measure the vehicle heading from the outside. If you T-hunt with a mobile VHF beam or quad, you can use your manual compass to sight along the antenna boom for a magnetic bearing, then add the declination for true bearing to the fox.

Triangulation Techniques

If you can obtain accurate bearings from two locations separated by a suitable distance, the technique of *triangulation* will give the expected location of the transmitter. The intersection of the lines of bearing from each location provides a *fix*. Triangulation accuracy is greatest when stations are located such that their bearings intersect at right angles. Accuracy is poor when the angle between bearings approaches 0° or 180°.

There is always uncertainty in the fixes obtained by triangulation due to equipment limitations, propagation effects and measurement errors. Obtaining bearings from three or more locations reduces the uncertainty. A good way to show the probable area of the transmitter on the triangulation map is to draw bearings as a narrow sector instead of as a single line. Sector width represents the amount of bearing uncertainty. **Fig 23.87** shows a portion of a map marked in this manner. Note how the bearing from Site 3 has narrowed down the probable area of the transmitter position.

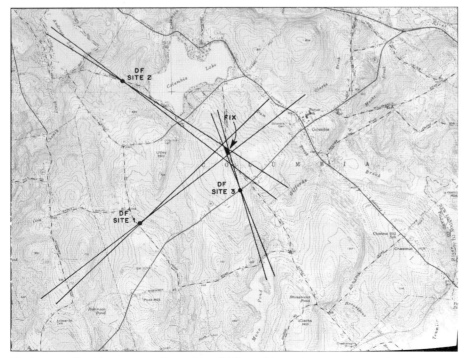

Fig 23.87 — Bearing sectors from three RDF positions drawn on a map for triangulation. In this case, bearings are from loop antennas, which have 180° ambiguity.

Computerized Transmitter Hunting

A portable computer is an excellent tool for streamlining the RDF process. Some T-hunters use one to optimize VHF beam bearings, generating a two-dimensional plot of signal strength versus azimuth. Others have automated the bearing-taking process by using a computer to capture signal headings from a Doppler RDF set, vehicle heading from a flux-gate compass, and vehicle location from a GPS receiver (**Fig 23.88**). The computer program can compute averaged headings from a Doppler set to reduce multipath effects.

Provided with perfect position and bearing information, computer triangulation could determine the transmitter location within the limits of its computational accuracy. Two bearings would exactly locate a fox. Of course, there are always uncertainties and inaccuracies in bearing and position data. If these uncertainties can be determined, the program can compute the uncertainty of the triangulated bearings. A "smart" computer program can evaluate bearings, triangulate the bearings of multiple hunters, discard those that appear erroneous, determine which locations have particularly great or small multipath problems and even "grade" the performance of RDF stations.

By adding packet radio connections to a group of computerized base and mobile RDF stations, the processed bearing data from each can be shared. Each station in the network can display the triangulated bearings of all. This requires a common map coordinate set among all stations. The USGS Universal Transverse Mercator (UTM) grid, consisting of 1×1-km grid squares, is a good choice.

The computer is an excellent RDF tool, but it is no substitute for a skilled "navigator." You will probably discover that using a computer on a high-speed T-hunt requires a full-time operator in the vehicle to make full use of its capabilities.

Skywave Bearings and Triangulation

Many factors make it difficult to obtain accuracy in skywave RDF work. Because of Faraday rotation during propagation, skywave signals are received with random polarization. Sometimes the vertical component is stronger, and at other times the horizontal. During periods when the vertical component is weak, the signal may appear to fade on an Adcock RDF system. At these times, determining an accurate signal null direction becomes very hard.

For a variety of reasons, HF bearing accuracy to within 1 or 2° is the exception rather than the rule. Errors of 3 to 5° are common. An error of 3° at a thousand miles represents

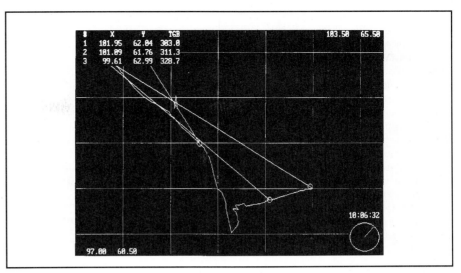

Fig 23.88 — Screen plot from a computerized RDF system showing three T-hunt bearings (straight lines radiating from small circles) and the vehicle path (jagged trace). The grid squares correspond to areas of standard topographic maps.

a distance of 52 miles. Even with every precaution taken in measurement, do not expect cross-country HF triangulation to pinpoint a signal beyond a county, a corner of a state or a large metropolitan area. The best you can expect is to be able to determine where a mobile RDF group should begin making a local search.

Triangulation mapping with skywave signals is more complex than with ground or direct waves because the expected paths are great-circle routes. Commonly available world maps are not suitable, because the triangulation lines on them must be curved, rather than straight. In general, for flat maps, the larger the area encompassed, the greater the error that straight line triangulation procedures will give.

A highway map is suitable for regional triangulation work if it uses some form of conical projection, such as the Lambert conformal conic system. This maintains the accuracy of angular representation, but the distance scale is not constant over the entire map.

One alternative for worldwide areas is the azimuthal-equidistant projection, better known as a great-circle map. True bearings for great-circle paths are shown as straight lines from the center to all points on the Earth. Maps centered on three or more different RDF sites may be compared to gain an idea of the general geographic area for an unknown source.

For worldwide triangulation, the best projection is the *gnomonic*, on which all great circle paths are represented by straight lines and angular measurements with respect to meridians are true. Gnomonic charts are custom maps prepared especially for government and military agencies.

Skywave signals do not always follow the great-circle path in traveling from a transmitter to a receiver. For example, if the signal is refracted in a tilted layer of the ionosphere, it could arrive from a direction that is several degrees away from the true great-circle bearing.

Another cause of signals arriving off the great-circle path is termed *sidescatter*. It is possible that, at a given time, the ionosphere does not support great-circle propagation of the signal from the transmitter to the receiver because the frequency is above the MUF for that path. However, at the same time, propagation may be supported from both ends of the path to some mutually accessible point off the great-circle path. The signal from the source may propagate to that point on the Earth's surface and hop in a sideways direction to continue to the receiver.

For example, signals from Central Europe have propagated to New England by hopping from an area in the Atlantic Ocean off the northwest coast of Africa, whereas the great-circle path puts the reflection point off the southern coast of Greenland. Readings in error by as much as 50° or more may result from sidescatter. The effect of propagation disturbances may be that the bearing seems to wander somewhat over a few minutes of time, or it may be weak and fluttery. At other times, however, there may be no telltale signs to indicate that the readings are erroneous.

Closing In

On a mobile foxhunt, the objective is usually to proceed to the hidden T with minimum time and mileage. Therefore, do not go far out of your way to get off-course bearings just to triangulate. It is usually better to take the shortest route along your initial line of bearing and "home in" on the signal. With a little experience, you will be able to gauge your distance from the fox by noting the amount of attenuation needed to keep the S-meter on scale.

As you approach the transmitter, the signal will become very strong. To keep the S-meter on scale, you will need to add an RF attenuator in the transmission line from the antenna to the receiver. Simple resistive attenuators are discussed in another chapter.

In the final phases of the hunt, you will probably have to leave your mobile and continue the hunt on foot. Even with an attenuator in the line, in the presence of a strong RF field, some energy will be coupled directly into the receiver circuitry. When this happens, the S-meter reading changes only slightly or perhaps not at all as the RDF antenna rotates, no matter how much attenuation you add. The cure is to shield the receiving equipment. Something as simple as wrapping the receiver in foil or placing it in a bread pan or cake pan, covered with a piece of copper or aluminum screening securely fastened at several points, may reduce direct pickup enough for you to get bearings.

Alternatively, you can replace the receiver with a field-strength meter as you close in, or use a heterodyne-type active attenuator. Plans for these devices are at the end of this chapter.

The Body Fade

A crude way to find the direction of a VHF signal with just a hand-held transceiver is the body fade technique, so named because the blockage of your body causes the signal to fade. Hold your HT close to your chest and turn all the way around slowly. Your body is providing a shield that gives the hand-held a cardioid sensitivity pattern, with a sharp decrease in sensitivity to the rear. This null indicates that the source is behind you (**Fig 23.89**).

If the signal is so strong that you can't find the null, try tuning 5 or 10 kHz off frequency to put the signal into the skirts of the IF passband. If your hand-held is dual-band (144/440 MHz) and you are hunting on 144 MHz, try tuning to the much weaker third harmonic of the signal in the 440-MHz band.

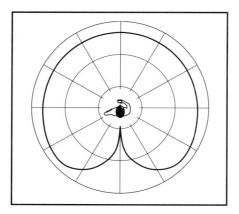

Fig 23.89 — When performing the body fade maneuver, a hand-held transceiver exhibits this directional pattern.

The body fade null, which is rather shallow to begin with, can be obscured by reflections, multipath, nearby objects, etc. Step well away from your vehicle before trying to get a bearing. Avoid large buildings, chain-link fences, metal signs and the like. If you do not get a good null, move to a clearer location and try again.

Air Attenuators

In microwave parlance, a signal that is too low in frequency to be propagated in a waveguide (that is, below the *cutoff frequency*) is attenuated at a predictable logarithmic rate. In other words, the farther inside the waveguide, the weaker the signal gets. Devices that use this principle to reduce signal strength are commonly known as *air attenuators*. Plans for a practical model for insertion in a coax line are in *Transmitter Hunting* (see Bibliography).

With this principle, you can reduce the level of strong signals into your hand-held transceiver, making it possible to use the body fade technique at very close range. Glen Rickerd, KC6TNF, documented this technique for *QST*. Start with a pasteboard mailing tube that has sufficient inside diameter to accommodate your hand-held. Cover the outside of the tube completely with aluminum foil. You can seal the bottom end with foil, too, but it probably will not matter if the tube is long enough. For durability and to prevent accidental shorts, wrap the foil in packing tape. You will also need a short, stout cord attached

Fig 23.90 — The air attenuator for a VHF hand-held in use. Suspend the radio by the wrist strap or a string inside the tube.

to the hand-held. The wrist strap may work for this, if long enough.

To use this air attenuation scheme for body fade bearings, hold the tube vertically against your chest and lower the hand-held into it until the signal begins to weaken (**Fig 23.90**). Holding the receiver in place, turn around slowly and listen for a sudden decrease in signal strength. If the null is poor, vary the depth of the receiver in the tube and try again. You do not need to watch the S-meter, which will likely be out of sight in the tube. Instead, use noise level to estimate signal strength.

For extremely strong signals, remove the "rubber duck" antenna or extend the wrist strap with a shoelace to get greater depth of suspension in the tube. The depth that works for one person may not work for another. Experiment with known signals to determine what works best for you.

THE SIMPLE SEEKER

The Simple Seeker for 144 MHz is the latest in a series of dual-antenna TDOA projects by Dave Geiser, WA2ANU. Fig 23.79 and accompanying text shows its principle of operation. It is simple to perform rapid antenna switching with diodes, driven by a free-running multivibrator. For best RDF performance, the switching pulses should be square waves, so antennas are alternately connected for equal times. The Simple Seeker uses a CMOS version of the popular 555 timer, which demands very little supply current. A 9-V alkaline battery will give long life. See **Fig 23.95** for the schematic diagram.

PIN diodes are best for this application because they have low capacitance and handle a moderate amount of transmit power. Philips ECG553, NTE-555, Motorola MPN3401 and similar types are suitable. Ordinary 1N4148 switching diodes are acceptable for receive-only use.

Off the null, the polarity of the switching pulses in the receiver output changes (with respect to the switching waveform), depending on which antenna is nearer the source. Thus, comparing the receiver output phase to that of the switching waveform determines which end of the null line points toward the transmitter. The common name for a circuit to make this comparison is a *phase detector*, achieved in this unit with a simple bridge circuit. A phase detector balance control is included, although it may not be needed. Serious imbalance indicates incorrect receiver tuning, an off-frequency target signal, or misalignment in the receiver IF stages.

Almost any audio transformer with approximately 10:1 voltage step-up to a center-tapped secondary meets the requirements of this phase detector. The output is a positive or negative indication, applied to meter M1 to indicate left or right.

Antenna Choices

Dipole antennas are best for long-distance RDF. They ensure maximum signal pickup and provide the best load for transmitting. **Fig 23.96** shows plans for a pair of dipoles mounted on an H frame of $1/2$-inch PVC tubing. Connect the 39-inch elements to the switcher with coaxial cables of *exactly* equal length. Spacing between dipoles is about 20 inches for

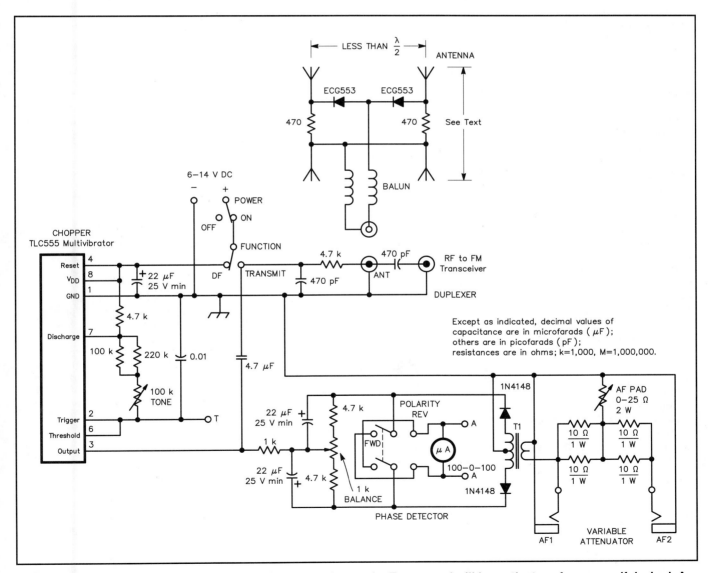

Fig 23.95 — Schematic of the Simple Seeker. A capacitor from point T to ground will lower the tone frequency, if desired. A single SPDT center-off toggle switch can replace separate power and function switches.

2 m, but is not critical. To prevent external currents flowing on the coax shield from disrupting RDF operation, wrap three turns (about 2 inch diameter) of the incoming coax to form a choke balun.

For receive-only work, dipoles are effective over much more than their useful transmit bandwidth. A pair of appropriately spaced 144-MHz dipoles works from 130 to 165 MHz. You will get greater tone amplitude with greater dipole spacing, making it easier to detect the null in the presence of modulation on the signal. But do not make the spacing greater than one-half free-space wavelength on any frequency to be used.

Best bearing accuracy demands that signals reach the receiver only from the switched antenna system. They should not arrive on the receiver wiring directly (through an unshielded case) or enter on wiring other than the antenna coax. The phase detecting system is less amplitude sensitive than systems such as quads and Yagis, but if you use small-aperture antennas such as "rubber duckies," a small signal leak may have a big effect. A wrap of aluminum foil around the receiver case helps block unwanted signal pickup, but tighter shielding may be needed.

Fig 23.97 shows a "sniffer" version of the unit with helix antennas. The added RDF circuits fit in a shielded box, with the switching pulses fed through a low-pass filter (the series 4.7-kΩ resistor and shunt 470-pF capacitor) to the receiver. The electronic switch is on a 20-pin DIP pad, with the phase detector on another pad (see **Fig 23.98**).

Because the phase detector may behave differently on weak and strong signals, the Simple Seeker incorporates an audio attenuator to allow either a full-strength audio or a lesser, adjustable received signal to feed the phase detector. You can plug headphones into jack AF2 and connect receiver audio to jack AF1 for no attenuation into the phase detector, or reverse the external connections, using the pad to control level to both the phones and the phase detector.

Convention is that the meter or other indicator deflects left when the signal is to the left. Others prefer that a left meter indication indicates that the antenna is rotated too far to the left. Whichever your choice, you can select it with the DPDT polarity switch. Polarity of audio output varies between receivers, so test the unit and receiver on a known signal source and mark the proper switch position on the unit

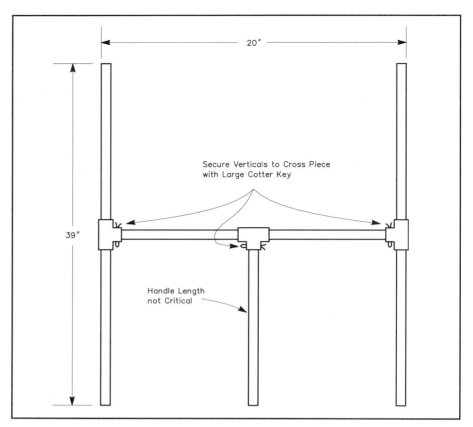

Fig 23.96 — "H" frame for the dual dipole Simple Seeker antenna set, made from ¹/₂-in. PVC tubing and tees. Glue the vertical dipole supports to the tees. Connect vertical tees and handle to the cross piece by drilling both parts and inserting large cotter pins. Tape the dipole elements to the tubes.

Fig 23.97 — Field version of the Simple Seeker with helix antennas.

Fig 23.98 — Interior view of the Simple Seeker. The multivibrator and phase detector circuits are mounted at the box ends. This version has a convenient built-in speaker.

before going into the field.

PIN diodes, when forward biased, exhibit low RF resistance and can pass up to approximately 1 W of VHF power without damage. The transmit position on the function switch applies steady dc bias to one of the PIN diodes, allowing communications from a hand-held RDF transceiver.

AN ACTIVE ATTENUATOR FOR VHF-FM

During a VHF transmitter hunt, the strength of the received signal can vary from roughly a microvolt at the starting point to nearly a volt when you are within inch of the transmitter, a 120-dB range. If you use a beam or other directional array, your receiver must provide accurate signal-strength readings throughout the hunt.

Zero to full scale range of S-meters on most hand-held transceivers is only 20 to 30 dB, which is fine for normal operating, but totally inadequate for transmitter hunting. Inserting a passive attenuator between the antenna and the receiver reduces the receiver input signal. However, the usefulness of an external attenuator is limited by how well the receiver can be shielded.

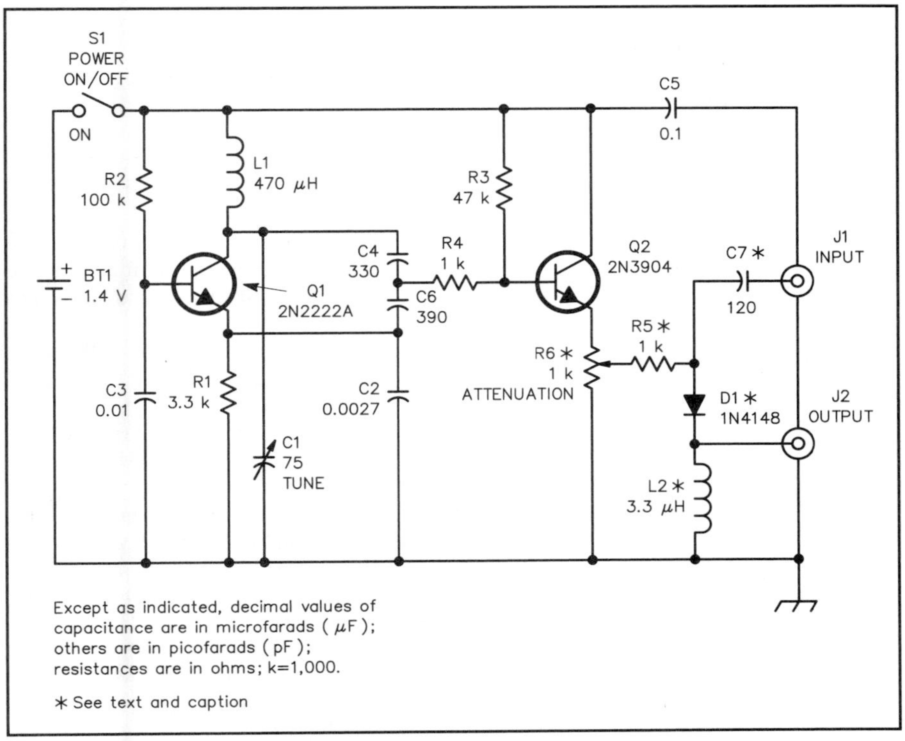

Anjo Eenhoorn, PA0ZR, has designed a simple add-on unit that achieves continuously variable attenuation by mixing the received signal with a signal from a 500-kHz oscillator. This process creates mixing products above and below the input frequency. The spacing of the closest products from the input frequency is equal to the local oscillator (LO) frequency. For example, if the input signal is at 146.52 MHz, the closest mixing products will appear at 147.02 and 146.02 MHz.

The strength of the mixing products varies with increasing or decreasing LO signal level. By DFing on the mixing product frequencies, you can obtain accurate headings even in the presence of a very strong received signal. As a result, any hand-held transceiver, regardless of how poor its shielding may be, is usable for transmitter hunting, up to the point where complete blocking of the receiver front end occurs. At the mixing product frequencies, the attenuator's range is greater than 100 dB.

Varying the level of the oscillator signal provides the extra advantage of controlling the strength of the input signal as it passes through the mixer. So as you close in on the target, you have the choice of monitoring and controlling the level of the input signal or the product signals, whichever provides the best results.

The LO circuit (**Fig 23.99**) uses the easy-to-find 2N2222A transistor. Trimmer capacitor C1 adjusts the oscillator's frequency. Frequency stability is only a minor concern; a few kilohertz of drift is tolerable. Q1's output feeds an emitter-follower buffer using a 2N3904 transistor, Q2. A linear-taper potentiometer (R6) controls the oscillator signal level present at the cathode of the mixing diode, D1. The diode and coupling capacitor C7 are in series with the signal path from antenna input to attenuator output.

This frequency converter design is unorthodox; it does not use the conventional configuration of a doubly balanced mixer, matching pads, filters and so on. Such sophistication is unnecessary here. This approach gives an easy to build circuit that consumes very little power. PA0ZR uses a tiny 1.4-V hearing-aid battery with a homemade battery clip. If your enclosure permits, you can substitute a standard AAA-size battery and holder.

Construction and Tuning

For a template for this project, including the PC board layout and parts overlay, see Chapter 30, **References**. A circuit board is available from FAR Circuits. The prototype (**Fig 23.100**) uses a

Except as indicated, decimal values of capacitance are in microfarads (μF); others are in picofarads (pF); resistances are in ohms; k=1,000.

＊ See text and caption

Fig 23.99 — Schematic of the active attenuator. Resistors are ¹/₄-W, 5%-tolerance carbon composition or film.

BT1 — Alkaline hearing-aid battery, Duracell SP675 or equivalent.
C1 — 75-pF miniature foil trimmer.
J1, J2 — BNC female connectors.
L1 — 470-μH RF choke.

L2 — 3.3-μH RF choke.
R6 — 1-kΩ, 1-W linear taper (slide or rotary).
S1 — SPST toggle.

Fig 23.100—Interior view of the active attenuator. Note that C7, D1 and L2 are mounted between the BNC connectors. R5 (not visible in this photograph) is connected to the wiper of slide pot R6.

plated enclosure with female BNC connectors for RF input and output. C7, D1, L2 and R5 are installed with point-to-point wiring between the BNC connectors and the potentiometer. S1 mounts on the rear wall of the enclosure.

Most hams will find the 500-kHz frequency offset convenient, but the oscillator can be tuned to other frequencies. If VHF/UHF activity is high in your area, choose an oscillator frequency that creates mixing products in clear portions of the band. The attenuator was designed for 144-MHz RDF, but will work elsewhere in the VHF/UHF range.

You can tune the oscillator with a frequency counter or with a strong signal of known frequency. It helps to enlist the aid of a friend with a hand-held transceiver a short distance away for initial tests. Connect a short piece of wire to J1, and cable your hand-held transceiver to J2. Select a simplex receive frequency and have your assistant key the test transmitter at its lowest power setting. (Better yet, attach the transmitter to a dummy antenna.)

With attenuator power on, adjust R6 for mid-scale S-meter reading. Now retune the hand-held to receive one of the mixing products. Carefully tune C1 and R6 until you hear the mixing product. Watch the S-meter and tune C1 for maximum reading.

If your receiver features memory channels, enter the hidden transmitter frequency along with both mixing product frequencies before the hunt starts. This allows you to jump from one to the other at the press of a button.

When the hunt begins, listen to the fox's frequency with the attenuator switched on. Adjust R6 until you get a peak reading. If the signal is too weak, connect your quad or other RDF antenna directly to your transceiver and hunt without the attenuator until the signal becomes stronger.

As you get closer to the fox, the attenuator will not be able to reduce the on-frequency signal enough to get good bearings. At this point, switch to one of the mixing product frequencies, set R6 for on-scale reading and continue. As you make your final approach, stop frequently to adjust R6 and take new bearings. At very close range, remove the RDF antenna altogether and replace it with a short piece of wire. It's a good idea to make up a short length of wire attached to a BNC fitting in advance, so you do not damage J1 by sticking random pieces of wire into the center contact.

While it is most convenient to use this system with receivers having S-meters, the meter is not indispensable. The active attenuator will reduce signal level to a point where receiver noise becomes audible. You can then obtain accurate fixes with null-seeking antennas or the "body fade" technique by simply listening for maximum noise at the null.

RDF BIBLIOGRAPHY

Bohrer, "Foxhunt Radio Direction Finder," *73 Amateur Radio*, Jul 1990, p 9.

Bonaguide, "HF DF—A Technique for Volunteer Monitoring," *QST*, Mar 1984, p 34.

DeMaw, "Maverick Trackdown," *QST*, Jul 1980, p 22.

Dorbuck, "Radio Direction Finding Techniques," *QST*, Aug 1975, p 30.

Eenhoorn, "An Active Attenuator for Transmitter Hunting," *QST*, Nov 1992, p 28.

Flanagan and Calabrese, "An Automated Mobile Radio Direction Finding System," *QST*, Dec 1993, p 51.

Geiser, "A Simple Seeker Direction Finder," *ARRL Antenna Compendium, Volume 3*, p 126.

Gillette, "A Fox-Hunting DF Twin-'Tenna," *QST*, Oct 1998, pp 41-44.

Johnson and Jasik, *Antenna Engineering Handbook*, Second Edition, New York: McGraw-Hill.

Kossor, "A Doppler Radio-Direction Finder," *QST*, Part 1: May 1999, pp 35-40; Part 2: June 1999, pp 37-40.

McCoy, "A Linear Field-Strength Meter," *QST*, Jan 1973, p 18.

Moell and Curlee, *Transmitter Hunting: Radio Direction Finding Simplified*, Blue Ridge Summit, PA: TAB/McGraw-Hill. (This book, available from ARRL, includes plans for the Roanoke Doppler RDF unit and in-line air attenuator, plus VHF quads and other RDF antennas.)

Moell, "Transmitter Hunting—Tracking Down the Fun," *QST*, Apr 1993, p 48 and May 1993, p 58.

Moell, "Build the Handy Tracker," *73 Magazine*, Sep 1989, p 58 and Nov 1989, p 52.

O'Dell, "Simple Antenna and S-Meter Modification for 2-Meter FM Direction Finding," *QST*, Mar 1981, p 43.

O'Dell, "Knock-It-Down and Lock-It-Out Boxes for DF," *QST*, Apr 1981, p 41.

Ostapchuk, "Fox Hunting is Practical and Fun!" *QST*, Oct 1998, pp 68-69.

Rickerd, "A Cheap Way to Hunt Transmitters," *QST*, Jan 1994, p 65.

The "Searcher" (SDF-1) Direction Finder, Rainbow Kits.

Contents

Component Data
24

None of us has the time or space to collect all the literature available on the many different commercially available manufactured components. Even if we did, the task of keeping track of new and obsolete devices would surely be formidable. Fortunately, amateurs tend to use a limited number of component types. This chapter, by Douglas Heacock, AAØMS, provides information on the components most often used by the Amateur Radio experimenter.

COMPONENT VALUES

Throughout this *Handbook*, composition resistors and small-value capacitors are specified in terms of a system of "preferred values." This system allows manufacturers to supply these components in a standard set of values, which, when considered along with component tolerances, satisfy the vast majority of circuit requirements.

The preferred values are based on a roughly logarithmic scale of numbers between 1 and 10. One decade of these values for three common tolerance ratings is shown in **Table 24.1**.

The Table represents the two significant digits in a resistor or capacitor value. Multiply these numbers by multiples of ten to get other standard values. For example, 22 pF, 2.2 µF, 220 µF, and 2200 µF are all standard capacitance values, available in all three tolerances. Standard resistor values include 3.9 Ω, 390 Ω, 39,000 Ω and 3.9 MΩ in ±5% and ±10% tolerances. All standard resistance values, from less than 1 Ω to about 5 MΩ are based on this table.

Each value is greater than the next smaller value by a multiplier factor that depends on the tolerance. For ±5% devices, each value is approximately 1.1 times the next lower one. For ±10% devices, the multiplier is 1.21, and for ±20% devices, the multiplier is 1.47. The resultant values are rounded to make up the series.

Tolerance refers to a range of acceptable values above and below the specified component value. For example, a 4700-Ω resistor rated for ±20% tolerance can have an actual value anywhere between 3760 Ω and 5640 Ω. You may always substitute a closer-tolerance device for one with a wider tolerance. For projects in this *Handbook*, assume a 10% tolerance if none is specified.

COMPONENT MARKINGS

The values, tolerances or types of most small components are typically marked with a color code or an alphanumeric code according to standards agreed upon by component manufacturers. The Electronic Industries Association (EIA) is a US agency that sets standards for electronic components, testing procedures, performance and device markings. The EIA co-operates with other standards agencies such as the International Electrotechnical Commission (IEC), a world-wide standards agency. You can often find published EIA standards in the engineering library of a college or university.

The standard EIA color code is used to identify a variety of electronic components. Most resistors are marked with color bands according to the code, shown in **Table 24.2**. Some types of capacitors and inductors are also marked using this color code.

Table 24.1

Standard Values for Resistors and Capacitors

±5%	±10%	±20%
1.0	1.0	1.0
1.1		
1.2	1.2	
1.3		
1.5	1.5	1.5
1.6		
1.8	1.8	
2.0		
2.2	2.2	2.2
2.4		
2.7	2.7	
3.0		
3.3	3.3	3.3
3.6		
3.9	3.9	
4.3		
4.7	4.7	4.7
5.1		
5.6	5.6	
6.2		
6.8	6.8	6.8
7.5		
8.2	8.2	
9.1		
10.0	10.0	10.0

Table 24.2

Resistor-Capacitor Color Codes

Color	Significant Figure	Decimal Multiplier	Tolerance (%)	Voltage Rating*
Black	0	1	-	-
Brown	1	10	1*	100
Red	2	100	2*	200
Orange	3	1,000	3*	300
Yellow	4	10,000	4*	400
Green	5	100,000	5*	500
Blue	6	1,000,000	6*	600
Violet	7	10,000,000	7*	700
Gray	8	100,000,000	8*	800
White	9	1,000,000,000	9*	900
Gold	-	0.1	5	1000
Silver	-	0.01	10	2000
No color	-	-	20	500

*Applies to capacitors only

Resistor Markings

Carbon-composition, carbon-film, and metal-film resistors are typically manufactured in roughly cylindrical cases with axial leads. They are marked with color bands as shown in **Fig 24.1A**. The first two bands represent the two significant digits of the component value, the third band represents the multiplier, and the fourth band (if there is one) represents the tolerance. Some units are marked with a fifth band that represents the percentage of resistance change per

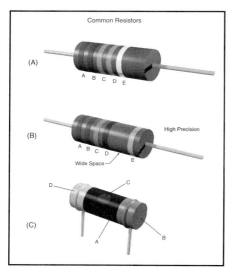

Fig 24.1—Color coding and body size for fixed resistors. The color code is given in Table 24.2. The colored areas have the following significance.

A—First significant figure of resistance in ohms.
B—Second significant figure.
C—Decimal multiplier.
D—Resistance tolerance in percent. If no color is shown the tolerance is ±20%.
E—Relative percent change in value per 1000 hours of operation; Brown, 1%; Red 0.1%; Orange 0.01%; Yellow 0.001%.

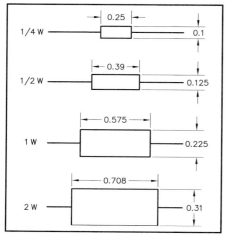

Fig 24.2—Typical carbon-composition resistor sizes.

Table 24.3
EIA Temperature Characteristic Codes for Ceramic Disc Capacitors

Minimum temperature		Maximum temperature		Maximum capacitance change over temperature range	
X	−55°C	2	+45°C	A	±1.0%
Y	−30°C	4	+65°C	B	±1.5%
Z	+10°C	5	+85°C	C	±2.2%
		6	+105°C	D	±3.3%
		7	+125°C	E	±4.7%
				F	±7.5%
				P	±10%
				R	±15%
				S	±22%
				T	−33%, +22%
				U	−56%, +22%
				V	−82%, +22%

Table 24.4
EIA Capacitor Temperature-Coefficient Codes

Industry	EIA	Industry	EIA
NP0	C0G	N330	S2H
N033	S1G	N470	U2J
N075	U1G	N1500	P3K
N150	P2G	N2200	R3L
N220	R2G		

Table 24.5
EIA Capacitor Tolerance Codes

Code	Tolerance
C	±1/4 pF
D	±1/2 pF
F	±1 pF or ±1%
G	±2 pF or ±2%
J	±5%
K	±10%
L	±15%
M	±20%
N	±30%
P or GMV*	−0%, +100%
W	−20%, +40%
Y	−20%, +50%
Z	−20%, +80%

*GMV = guaranteed minimum value.

1000 hours of operation: brown = 1%; red = 0.1%; orange = 0.01%; and yellow = 0.001%. Precision resistors (EIA Std RS-279, Fig 24.1B) and some mil-spec (MIL STD-1285A) resistors also use five color bands. On precision resistors, the first *three* bands are used for significant figures and the space between the fourth and fifth bands is wider than the others, to identify the tolerance band. On the military resistors, the fifth band indicates reliability information such as failure rate.

For example, if a resistor of the type shown in Fig 24.1A is marked with A = red; B = red; C = orange; D = no color, the significant figures are 2 and 2, the multiplier is 1000, and the tolerance is ±20%. The device is a 22,000-Ω, ±20% unit.

Some resistors are made with radial leads (Fig 24.1C) and are marked with a color code in a slightly different scheme. For example, a resistor as shown in Fig 24.1C is marked as follows: A (body) = blue; B (end) = gray; C (dot) = red; D (end) = gold. The significant figures are 6 and 8, the multiplier is 100, and the tolerance is ±5%; 6800 Ω with ±5% tolerance.

Resistor Power Ratings

Carbon-composition and metal-film resistors are available in standard power ratings of 1/10, 1/8, 1/4, 1/2, 1 and 2 W. The 1/10- and 1/8-W sizes are relatively expensive and difficult to purchase in small quantities. They are used only where miniaturization is essential. The 1/4, 1/2, 1, and 2-W composition resistor packages are drawn to scale in **Fig 24.2**. Metal-film resistors

are typically slightly smaller than carbon-composition units of the same power rating. Film resistors can usually be identified by a glossy enamel coating and an hourglass profile. Carbon-film and metal-film are the most commonly available resistors today, having largely replaced the less-stable carbon-composition resistors.

Capacitor Markings

A variety of systems for capacitor markings are in use. Some use color bands, some use combinations of numbers and letters. Capacitors may be marked with their value, tolerance, temperature characteristics, voltage ratings or some subset of these specifications. **Fig 24.3** shows several popular capacitor marking systems.

In addition to the value, ceramic disk capacitors may be marked with an alphanumeric code signifying temperature characteristics. **Table 24.3** explains the EIA code for ceramic-disk capacitor temperature characteristics. The code is made up of one character from each column in the table. For example, a capacitor marked Z5U is suitable for use between +10 and +85°C, with a maximum change in capacitance of −56% or +22%.

Capacitors with highly predictable temperature coefficients of capacitance are

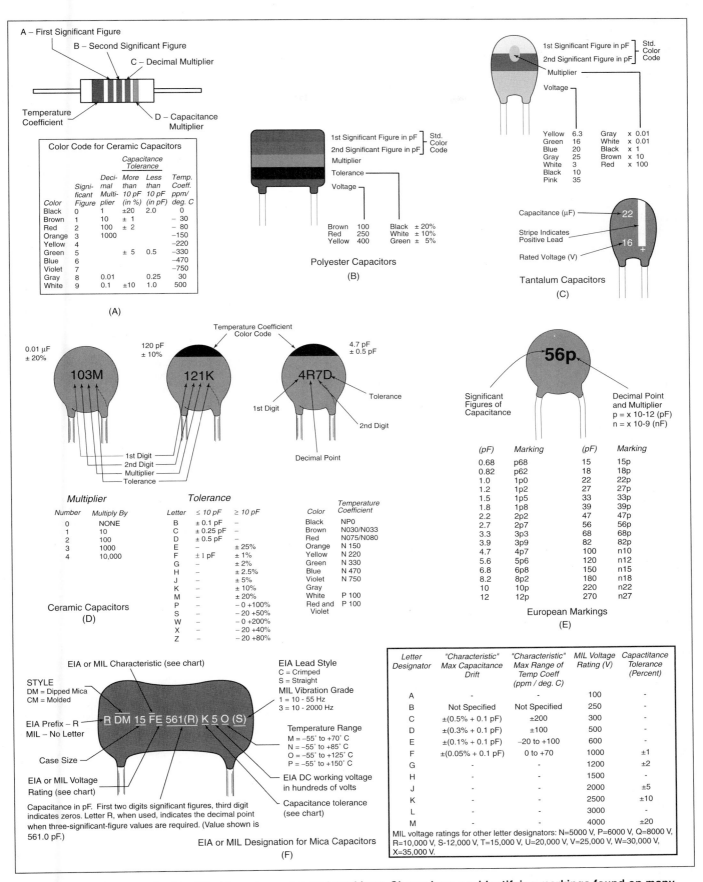

Color Code for Ceramic Capacitors (A)

| | | Capacitance Tolerance | | |
Color	Significant Figure	Decimal Multiplier	More than 10 pF (in %)	Less than 10 pF (in pF)	Temp. Coeff. ppm/ deg. C
Black	0	1	±20	2.0	0
Brown	1	10	± 1		− 30
Red	2	100	± 2		− 80
Orange	3	1000			−150
Yellow	4				−220
Green	5		± 5	0.5	−330
Blue	6				−470
Violet	7				−750
Gray	8	0.01		0.25	30
White	9	0.1	±10	1.0	500

Polyester Capacitors (B)

Brown	100	Black	± 20%
Red	250	White	± 10%
Yellow	400	Green	± 5%

Tantalum Capacitors (C)

Yellow	6.3	Gray	x 0.01
Green	16	White	x 0.01
Blue	20	Black	x 1
Gray	25	Brown	x 10
White	3	Red	x 100
Black	10		
Pink	35		

Ceramic Capacitors (D)

Multiplier

Number	Multiply By
0	NONE
1	10
2	100
3	1000
4	10,000

Tolerance

Letter	≤ 10 pF	≥ 10 pF
B	± 0.1 pF	−
C	± 0.25 pF	−
D	± 0.5 pF	−
E	−	± 25%
F	± 1 pF	± 1%
G	−	± 2%
H	−	± 2.5%
J	−	± 5%
K	−	± 10%
M	−	± 20%
P	−	− 0 +100%
S	−	− 20 +50%
W	−	− 0 +200%
X	−	− 20 +40%
Z	−	− 20 +80%

Color	Temperature Coefficient
Black	NP0
Brown	N030/N033
Red	N075/N080
Orange	N 150
Yellow	N 220
Green	N 330
Blue	N 470
Violet	N 750
Gray	
White	P 100
Red and Violet	P 100

European Markings (E)

Decimal Point and Multiplier
p = x 10-12 (pF)
n = x 10-9 (nF)

(pF)	Marking	(pF)	Marking
0.68	p68	15	15p
0.82	p62	18	18p
1.0	1p0	22	22p
1.2	1p2	27	27p
1.5	1p5	33	33p
1.8	1p8	39	39p
2.2	2p2	47	47p
2.7	2p7	56	56p
3.3	3p3	68	68p
3.9	3p9	82	82p
4.7	4p7	100	n10
5.6	5p6	120	n12
6.8	6p8	150	n15
8.2	8p2	180	n18
10	10p	220	n22
12	12p	270	n27

EIA or MIL Designation for Mica Capacitors (F)

R DM 15 FE 561(R) K 5 O (S)

STYLE
DM = Dipped Mica
CM = Molded

EIA Prefix – R
MIL – No Letter

EIA Lead Style
C = Crimped
S = Straight

MIL Vibration Grade
1 = 10 - 55 Hz
3 = 10 - 2000 Hz

Temperature Range
M = −55° to +70° C
N = −55° to +85° C
O = −55° to +125° C
P = −55° to +150° C

EIA DC working voltage in hundreds of volts

Capacitance tolerance (see chart)

Capacitance in pF. First two digits significant figures, third digit indicates zeros. Letter R, when used, indicates the decimal point when three-significant-figure values are required. (Value shown is 561.0 pF.)

Letter Designator	"Characteristic" Max Capacitance Drift	"Characteristic" Max Range of Temp Coeff (ppm / deg. C)	MIL Voltage Rating (V)	Capacitance Tolerance (Percent)
A	-	-	100	-
B	Not Specified	Not Specified	250	-
C	±(0.5% + 0.1 pF)	±200	300	-
D	±(0.3% + 0.1 pF)	±100	500	-
E	±(0.1% + 0.1 pF)	−20 to +100	600	-
F	±(0.05% + 0.1 pF)	0 to +70	1000	±1
G	-	-	1200	±2
H	-	-	1500	-
J	-	-	2000	±5
K	-	-	2500	±10
L	-	-	3000	-
M	-	-	4000	±20

MIL voltage ratings for other letter designators: N=5000 V, P=6000 V, Q=8000 V, R=10,000 V, S-12,000 V, T=15,000 V, U=20,000 V, V=25,000 V, W=30,000 V, X=35,000 V.

Fig 24.3—Capacitors can be identified by color codes and markings. Shown here are identifying markings found on many common capacitor types.

sometimes used in oscillators that must be frequency stable with temperature. If an application called for a temperature coefficient of –750 ppm/°C (N750), a capacitor marked U2J would be suitable. The older industry code for these ratings is being replaced with the EIA code shown in **Table 24.4**. NP0 (that is, N-P-zero) means "negative, positive, zero;" it is a characteristic often specified for RF circuits requiring temperature stability, such as VFOs. A capacitor of the proper value marked C0G is a suitable replacement for an NP0 unit.

Some capacitors, such as dipped silver-mica units, have a letter designating the capacitance tolerance. These letters are deciphered in **Table 24.5**.

Surface-Mount Resistor and Capacitor Markings

Many different types of electronic components, both active and passive, are now available in surface-mount packages. These are commonly-known as chip resistors and capacitors. The very small size of these components leaves little space for marking with conventional codes, so brief alphanumeric codes are used to convey the most information in the smallest possible space.

Surface-mount resistors are typically marked with a three- or four-digit value code and a character indicating tolerance. The nominal resistance, expressed in ohms, is identified by three digits for 2% (and greater) tolerance devices. The first two digits represent the significant figures; the last digit specifies the multiplier as the exponent of ten. (It may be easier to remember the multiplier as the number of zeros you must add to the significant figures.) For values less than $100\,\Omega$, the letter R is substituted for one of the significant digits and represents a decimal point. Here are some examples:

Resistor Code	Value
101	10 and 1 zero = 100 Ω
224	22 and 4 zeros = 220,000 Ω
1R0	1.0 and no zeros = 1 Ω
22R	22.0 and no zeros = 22 Ω
R10	0.1 and no zeros = 0.1 Ω

If the tolerance of the unit is narrower than ±2%, the code used is a four-digit code where the first three digits are the significant figures and the last is the multiplier. The letter R is used in the same way to represent a decimal point. For example, 1001 indicates a 1000-Ω unit, and 22R0 indicates a 22-Ω unit.

The tolerance rating for a surface-mount resistor is expressed with a single character at the end of the numeric value code, according to **Table 24.6**.

Surface-mount capacitors are marked with a two-character code consisting of a letter indicating the significant digits (see **Table 24.7**) and a number indicating the multiplier (see **Table 24.8**). The code represents the capacitance in picofarads. For

Table 24.6
SMT Resistor Tolerance Codes

Letter	Tolerance
D	±0.5%
F	±1.0%
G	±2.0%
J	±5.0%

Table 24.7
SMT Capacitor Significant Figures Code

Character	Significant Figures	Character	Significant Figures
A	1.0	T	5.1
B	1.1	U	5.6
C	1.2	V	6.2
D	1.3	W	6.8
E	1.5	X	7.5
F	1.6	Y	8.2
G	1.8	Z	9.1
H	2.0	a	2.5
J	2.2	b	3.5
K	2.4	d	4.0
L	2.7	e	4.5
M	3.0	f	5.0
N	3.3	m	6.0
P	3.6	n	7.0
Q	3.9	t	8.0
R	4.3	y	9.0
S	4.7		

Table 24.8
SMT Capacitor Multiplier Codes

Numeric Character	Decimal Multiplier
0	1
1	10
2	100
3	1,000
4	10,000
5	100,000
6	1,000,000
7	10,000,000
8	100,000,000
9	0.1

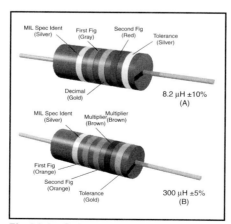

Fig 24.4—Color coding for tubular encapsulated RF chokes. At A, an example of the coding for an 8.2-μH choke is given. At B, the color bands for a 330-μH inductor are illustrated. The color code is given in Table 24.2.

example, a chip capacitor marked "A4" would have a capacitance of 10,000 pF, or 0.01 μF. A unit marked "N1" would be a 33-pF capacitor. If there is sufficient space on the device package, a tolerance code may be included (see Fig 24.3D for tolerance codes). Surface-mount capacitors can be very small; you may need a magnifying glass to read the markings.

INDUCTORS AND CORE MATERIALS

Inductors, both fixed and variable, are available in a wide variety of types and packages, and many offer few clues as to their values. Some coils and chokes are marked with the EIA color code shown in Table 24.2. See **Fig 24.4** for another marking system for tubular encapsulated RF chokes.

Most powdered-iron toroid cores that we amateurs use are manufactured by Micrometals, who uses paint to identify the material used in the core. The Micrometals color code is part of **Table 24.9**. **Table 24.10** gives the physical characteristics of powdered-iron toroids. Ferrite cores are not typically painted, so identification is more difficult. See **Table 24.11** for information about ferrite cores.

TRANSFORMERS

Many transformers, including power transformers, IF transformers and audio transformers, are made to be installed on PC boards, and have terminals designed for that purpose. Some transformers are manufactured with wire leads that are color-coded to identify each connection. When colored wire leads are present, the color codes in **Tables 24.12**, **24.13** and **24.14** usually apply.

In addition, many miniature IF transformers are tuned with slugs that are color-coded to signify their application. **Table 24.15** lists application vs slug color.

SEMICONDUCTORS

Most semiconductor devices are clearly marked with the part number and in some cases, a manufacturer's date code as well. Identification of semiconductors can be difficult, however, when the parts are "house-marked" (marked with codes used by an equipment manufacturer instead of the standard part numbers). In such cases, it is often possible to find the standard equivalent or a suitable replacement by using one of the semiconductor cross-reference directories available from various replacement-parts distributors. If you look up the house number and find the recommended replacement part, you can often find other standard parts that are replaced by that same part.

Diodes

Most diodes are marked with a part number and some means of identifying which lead is the cathode. Some diodes are marked with a color-band code (see **Fig 24.5**). Important diode parameters include maximum forward current, maximum peak inverse voltage (PIV) and the power-handling capacity.

Transistors

Some important parameters for transistor selection are voltage and current limits, power-handling capability, beta or gain characteristics and useful frequency range. The case style may also be an issue; some transistors are available in several different case styles.

Integrated Circuits

Integrated circuits (ICs) come in a variety of packages, including transistor-like metal cans, dual and single in-line packages (DIPs and SIPs), flat-packs and surface-mount packages. Most are marked with a part number and a four-digit manufacturer's date code indicating the year (first two digits) and week (last two digits) that the component

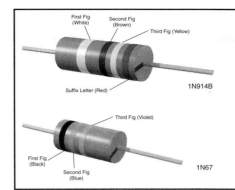

Fig 24.5—Color coding for semiconductor diodes. At A, the cathode is identified by the double-width first band. At B, the bands are grouped toward the cathode. Two-figure designations are signified by a black first band. The color code is given in Table 24.2. The suffix-letter code is A—Brown, B—red, C—orange, D—yellow, E—green, F—blue. The 1N prefix is understood.

Table 24.9
Powdered-Iron Toroid Cores: Magnetic Properties

Inductance and Turns Formula

The turns required for a given inductance or inductance for a given number of turns can be calculated from:

$$N = 100 \sqrt{\frac{L}{A_L}} \qquad\qquad L = A_L \left(\frac{N^2}{10,000} \right)$$

where N = number of turns; L = desired inductance (μH); A_L = inductance index (μH per 100 turns).*

A_L Values

					Mix						
Size	26**	3	15	1	2	7	6	10	12	17	0
T-12	na	60	50	48	20	18	17	12	7.5	7.5	3.0
T-16	145	61	55	44	22	na	19	13	8.0	8.0	3.0
T-20	180	76	65	52	27	24	22	16	10.0	10.0	3.5
T-25	235	100	85	70	34	29	27	19	12.0	12.0	4.5
T-30	325	140	93	85	43	37	36	25	16.0	16.0	6.0
T-37	275	120	90	80	40	32	30	25	15.0	15.0	4.9
T-44	360	180	160	105	52	46	42	33	18.5	18.5	6.5
T-50	320	175	135	100	49	43	40	31	18.0	18.0	6.4
T-68	420	195	180	115	57	52	47	32	21.0	21.0	7.5
T-80	450	180	170	115	55	50	45	32	22.0	22.0	8.5
T-94	590	248	200	160	84	na	70	58	32.0	na	10.6
T-106	900	450	345	325	135	133	116	na	na	na	19.0
T-130	785	350	250	200	110	103	96	na	na	na	15.0
T-157	870	420	360	320	140	na	115	na	na	na	na
T-184	1640	720	na	500	240	na	195	na	na	na	na
T-200	895	425	na	250	120	105	100	na	na	na	na

*The units of A_L (μH per 100 turns) are an industry standard; however, to get a correct result use A_L only in the formula above.
**Mix-26 is similar to the older Mix-41, but can provide an extended frequency range.

Magnetic Properties Iron Powder Cores

Mix	Color	Material	μ	Temp stability (ppm/°C)	f (MHz)	Notes
26	Yellow/white	Hydrogen reduced	75	825	dc - 1	Used for EMI filters and dc chokes
3	Gray	Carbonyl HP	35	370	0.05 - 0.50	Excellent stability, good Q for lower frequencies
15	Red/white	Carbonyl GS6	25	190	0.10 - 2	Excellent stability, good Q
1	Blue	Carbonyl C	20	280	0.50 - 5	Similar to Mix-3, but better stability
2	Red	Carbonyl E	10	95	2 - 30	High Q material
7	White	Carbonyl TH	9	30	3 - 35	Similar to Mix-2 and Mix-6, but better temperature stability
6	Yellow	Carbonyl SF	8	35	10 - 50	Very good Q and temp. stability for 20-50 MHz
10	Black	Powdered iron W	6	150	30 - 100	Good Q and stability for 40 - 100 MHz
12	Green/white	Synthetic oxide	4	170	50 - 200	Good Q, moderate temperature stability
17	Blue/yellow	Carbonyl	4	50	40 - 180	Similar to Mix-12, better temperature stability, Q drops about 10% above 50 MHz, 20% above 100 MHz
0	Tan	phenolic	1	0	100 - 300	Inductance may vary greatly with winding technique

Courtesy of Amidon Assoc and Micrometals
Note: Color codes hold only for cores manufactured by Micrometals, which makes the cores sold by most Amateur Radio distributors.

Table 24.10
Powdered-Iron Toroid Cores: Dimensions

Red E Cores—500 kHz to 30 MHz (μ = 10)			
No.	OD (in)	ID (in)	H (in)
T-200-2	2.00	1.25	0.55
T-94-2	0.94	0.56	0.31
T-80-2	0.80	0.50	0.25
T-68-2	0.68	0.37	0.19
T-50-2	0.50	0.30	0.19
T-37-2	0.37	0.21	0.12
T-25-2	0.25	0.12	0.09
T-12-2	0.125	0.06	0.05

Black W Cores—30 MHz to 200 MHz (μ=7)			
No.	OD (In)	ID (In)	H (In)
T-50-10	0.50	0.30	0.19
T-37-10	0.37	0.21	0.12
T-25-10	0.25	0.12	0.09
T-12-10	0.125	0.06	0.05

Yellow SF Cores—10 MHz to 90 MHz (μ=8)			
No.	OD (In)	ID (In)	H (In)
T-94-6	0.94	0.56	0.31
T-80-6	0.80	0.50	0.25
T-68-6	0.68	0.37	0.19
T-50-6	0.50	0.30	0.19
T-26-6	0.25	0.12	0.09
T-12-6	0.125	0.06	0.05

Number of Turns vs Wire Size and Core Size

Approximate maximum number of turns—single layer wound—enameled wire.

Wire Size	T-200	T-130	T-106	T-94	T-80	T-68	T-50	T-37	T-25	T-12
10	33	20	12	12	10	6	4	1		
12	43	25	16	16	14	9	6	3		
14	54	32	21	21	18	13	8	5	1	
16	69	41	28	28	24	17	13	7	2	
18	88	53	37	37	32	23	18	10	4	1
20	111	67	47	47	41	29	23	14	6	1
22	140	86	60	60	53	38	30	19	9	2
24	177	109	77	77	67	49	39	25	13	4
26	223	137	97	97	85	63	50	33	17	7
28	281	173	123	123	108	80	64	42	23	9
30	355	217	154	154	136	101	81	54	29	13
32	439	272	194	194	171	127	103	68	38	17
34	557	346	247	247	218	162	132	88	49	23
36	683	424	304	304	268	199	162	108	62	30
38	875	544	389	389	344	256	209	140	80	39
40	1103	687	492	492	434	324	264	178	102	51

Actual number of turns may differ from above figures according to winding techniques, especially when using the larger size wires. Chart prepared by Michel J. Gordon, Jr., WB9FHC
Courtesy of Amidon Assoc.

was made. ICs are frequently house-marked, and the cross-reference directories mentioned above can be helpful in identification and replacement.

Another very useful reference tool for working with ICs is *IC Master*, a master selection guide that organizes ICs by type, function and certain key parameters. A part number index is included, along with application notes and manufacturer's information for tens of thousands of IC devices. Some of the data from *IC Master* is also available on computer disks.

IC part numbers usually contain a few digits that identify the circuit die or function and several other letters and/or digits that identify the production process, manufacturer and package. For example, a '4066 IC contains four independent SPST switches. Harris (CD74HC4066, CD4066B and CD4066BE), National (MM74HC4066, CD4066BC and CD4066BM) and Panasonic (MN74HC4066 and MN4066B) all make similar devices (as do many other manufacturers) with slight differences. Among the numbers listed, "CD" (CMOS Digital), "MM" (MOS Monolithic), and "MN" indicate CMOS parts. "74" indicates a commercial quality product (for applications from 0°C to 70°C), which is pin compatible with the 74/54 TTL families. "HC" means high-speed CMOS family, which is as fast as the LS TTL family. The "B" suffix, as is CD4066B, indicates a buffered output. This is only a small example of the conventions used in IC part numbers. For more information look at data books from the various manufacturers. Base diagrams for many common ICs appear in *The ARRL Electronics Data Book*.

When choosing ICs that are not exact replacements, several operating needs and performance aspects should be considered. First, the replacement power requirements must be met: Some ICs require 5 V dc, others 12 V and some need both positive and negative supplies. Current requirements vary among the various IC families, so be sure that sufficient current is available from the power supply. If a replacement IC uses much more current than the device it replaces, a heat sink or blower may be needed to keep it cool.

Next consider how the replacement interacts with its neighboring components. Input capacitance and "fanout" are critical factors in digital circuits. Increased input capacitance may overload the driving circuits. Overload slows circuit operation, which may prevent lines from reaching the "high" condition. Fanout tells how many inputs a device can drive. The fanout of a replacement should be equal to, or greater than, that required in the circuit. Operating speed and propagation delay are also significant. Choose a replacement IC that operates at or above the circuit clock speed. (Although increased speed can increase EMI and cause other problems.) Some circuits may not function if the propagation delay varies much from the specified part. Look at the **Digital** chapter for details of how these operating characteristics relate to circuit performance.

Analog ICs have similar characteristics. Input and output capacities are often defined as how much current an analog IC can "sink" (accept at an input) or "source" (pass to a load). A replacement should be able to source or sink at least as much current as the device it replaces. Analog speed is sometimes listed as bandwidth (as in discrete-component circuits) or slew rate (common in op amps). Each of these quantities should meet or exceed that of the replaced component.

Some ICs are available in different operating temperature ranges. Op amps, for example, are commonly available in three standard ranges:

- Commercial 0°C to 70°C
- Industrial –25°C to 85°C
- Military –55°C to 125°C

In some cases, part numbers reflect the temperature ratings. For example, an LM301A op amp is rated for the commercial temperature range; an LM201A op amp for the industrial range and an LM101A for the military range.

When necessary, you can add interface circuits or buffer amplifiers that improve the input and output capabilities of replacement ICs, but auxiliary circuits cannot improve basic device ratings, such as speed or bandwidth.

An excellent source of information on many common ICs is *The ARRL Electronics Data Book*, which contains detailed data for digital ICs (CMOS and TTL), op amps and other analog ICs.

Table 24.11
Ferrite Toroids: A_L Chart (mH per 1000, turns) Enameled Wire

Core Size	63/67-Mix $\mu = 40$	61-Mix $\mu = 125$	43-Mix $\mu = 850$	77 (72) Mix $\mu = 2000$	J (75) Mix $\mu = 5000$
FT-23	7.9	24.8	188.0	396	980
FT-37	19.7	55.3	420.0	884	2196
FT-50	22.0	68.0	523.0	1100	2715
FT-82	22.4	73.3	557.0	1170	NA
FT-114	25.4	79.3	603.0	1270	3170

Number turns = $1000 \sqrt{\text{desired L (mH)} \div A_L \text{ value (above)}}$

Ferrite Magnetic Properties

Property	Unit	63/67-Mix	61-Mix	43-Mix	77 (72) Mix	J (75)-Mix
Initial perm (μ_i)		40	125	850	2000	5000
Maximum perm.		125	450	3000	6000	8000
Saturation flux density @ 10 oer	Gauss	1850	2350	2750	4600	3900
Residual flux density	Gauss	750	1200	1200	1150	1250
Curie temp.	°C	450	350	130	200	140
Vol. resistivity	ohm/cm	1×10^8	1×10^8	1×10^5	1×10^2	5×10^2
Resonant circuit frequency	MHz	15-25	0.2-10	0.01-1	0.001-1	0.001-1
Specific gravity		4.7	4.7	4.5	4.8	4.8
Loss factor	$\dfrac{1}{\mu_i Q}$	110×10^{-6} @25 MHz	32×10^{-6} @2.5 MHz	120×10^{-6} @1 MHz	4.5×10^{-6} @0.1 MHz	15×10^{-6} @0.1 MHz
Coercive force	Oer	2.40	1.60	0.30	0.22	0.16
Temp. Coef. of initial perm.	%/°C (20-70°C)	0.10	0.15	1.0	0.60	0.90

Ferrite Toroids—Physical Properties

Core Size	OD	ID	Height	A_e	l_e	V_e	A_S	A_W
FT-23	0.230	0.120	0.060	0.00330	0.529	0.00174	0.1264	0.01121
FT-37	0.375	0.187	0.125	0.01175	0.846	0.00994	0.3860	0.02750
FT-50	0.500	0.281	0.188	0.02060	1.190	0.02450	0.7300	0.06200
FT-82	0.825	0.520	0.250	0.03810	2.070	0.07890	1.7000	0.21200
FT-114	1.142	0.750	0.295	0.05810	2.920	0.16950	2.9200	0.43900

OD—Outer diameter (inches)
ID—Inner diameter (inches)
Hgt—Height (inches)
A_W—Total window area (in)^2

A_e—Effective magnetic cross-sectional area (in)^2
l_e—Effective magnetic path length (inches)
V_e—Effective magnetic volume (in)^3
A_S—Surface area exposed for cooling (in)^2

Courtesy of Amidon Assoc.

Table 24.12
Power-Transformer Wiring Color Codes

Non-tapped primary leads:	Black
Tapped primary leads:	Common: Black Tap: Black/yellow striped Finish: Black/red striped
High-voltage plate winding: Center tap:	Red Red/yellow striped
Rectifier filament winding: Center tap:	Yellow Yellow/blue striped
Filament winding 1: Center tap:	Green Green/yellow striped
Filament winding 2: Center tap:	Brown Brown/yellow striped
Filament winding 3: Center tap:	Slate Slate/yellow striped

Table 24.13
IF Transformer Wiring Color Codes

Plate lead:	Blue
B+ lead:	Red
Grid (or diode) lead:	Green
Grid (or diode) return:	Black

Note: If the secondary of the IF transformer is center-tapped, the second diode plate lead is green-and-black striped, and black is used for the center-tap lead.

Table 24.14
IF Transformer Slug Color Codes

Frequency	Application	Slug color
455 kHz	1st IF	Yellow
	2nd IF	White
	3rd IF	Black
	Osc tuning	Red
10.7 MHz	1st IF	Green
	2nd or 3rd IF	Orange, Brown or Black

Table 24.15
Audio Transformer Wiring Color Codes

Plate lead of primary	Blue
B+ lead (plain or center-tapped)	Red
Plate (start) lead on center-tapped primaries	Brown (or blue if polarity is not important)
Grid (finish) lead to secondary	Green
Grid return (plain or center tapped)	Black
Grid (start) lead on center tapped secondaries	Yellow (or green if polarity not important)

Note: These markings also apply to line-to-grid and tube-to-line transformers.

OTHER SOURCES OF COMPONENT DATA

There are many sources you can consult for detailed component data. Many manufacturers publish data books for the components they make. Many distributors will include data sheets for parts you order if you ask for them. Parts catalogs themselves are often good sources of component data. The following list is representative of some of the data resources available from manufacturers and distributors.

Motorola Small-Signal Transistor Data
Motorola RF Device Data
Motorola Linear and Interface ICs
Signetics: General Purpose/Linear ICs
NTE Technical Manual and Cross Reference
TCE SK Replacement Technical Manual and Cross Reference
National Semiconductor:
   Discrete Semiconductor Products Databook
   CMOS Logic Databook
   Linear Applications Handbook
   Linear Application-Specific ICs Databook
   Operational Amplifiers Databook

Copper Wire Specifications

Bare and Enamel-Coated Wire

Wire Size (AWG)	Diam (Mils)	Area (CM[1])	Enamel Wire Coating Turns / Linear inch[2]			Feet per Pound Bare	Ohms per 1000 ft 25∞ C	Current Carrying Capacity Continuous Duty[3]			Nearest British SWG No.
			Single	Heavy	Triple			at 700 CM per Amp[4]	Open air	Conduit or bundles	
1	289.3	83694.49				3.948	0.1239	119.564			1
2	257.6	66357.76				4.978	0.1563	94.797			2
3	229.4	52624.36				6.277	0.1971	75.178			4
4	204.3	41738.49				7.918	0.2485	59.626			5
5	181.9	33087.61				9.98	0.3134	47.268			6
6	162.0	26244.00				12.59	0.3952	37.491			7
7	144.3	20822.49				15.87	0.4981	29.746			8
8	128.5	16512.25				20.01	0.6281	23.589			9
9	114.4	13087.36				25.24	0.7925	18.696			11
10	101.9	10383.61				31.82	0.9987	14.834			12
11	90.7	8226.49				40.16	1.2610	11.752			13
12	80.8	6528.64				50.61	1.5880	9.327			13
13	72.0	5184.00				63.73	2.0010	7.406			15
14	64.1	4108.81	15.2	14.8	14.5	80.39	2.5240	5.870	32	17	15
15	57.1	3260.41	17.0	16.6	16.2	101.32	3.1810	4.658			16
16	50.8	2580.64	19.1	18.6	18.1	128	4.0180	3.687	22	13	17
17	45.3	2052.09	21.4	20.7	20.2	161	5.0540	2.932			18
18	40.3	1624.09	23.9	23.2	22.5	203.5	6.3860	2.320	16	10	19
19	35.9	1288.81	26.8	25.9	25.1	256.4	8.0460	1.841			20
20	32.0	1024.00	29.9	28.9	27.9	322.7	10.1280	1.463	11	7.5	21
21	28.5	812.25	33.6	32.4	31.3	406.7	12.7700	1.160			22
22	25.3	640.09	37.6	36.2	34.7	516.3	16.2000	0.914		5	22
23	22.6	510.76	42.0	40.3	38.6	646.8	20.3000	0.730			24
24	20.1	404.01	46.9	45.0	42.9	817.7	25.6700	0.577			24
25	17.9	320.41	52.6	50.3	47.8	1031	32.3700	0.458			26
26	15.9	252.81	58.8	56.2	53.2	1307	41.0200	0.361			27
27	14.2	201.64	65.8	62.5	59.2	1639	51.4400	0.288			28
28	12.6	158.76	73.5	69.4	65.8	2081	65.3100	0.227			29
29	11.3	127.69	82.0	76.9	72.5	2587	81.2100	0.182			31
30	10.0	100.00	91.7	86.2	80.6	3306	103.7100	0.143			33
31	8.9	79.21	103.1	95.2		4170	130.9000	0.113			34
32	8.0	64.00	113.6	105.3		5163	162.0000	0.091			35
33	7.1	50.41	128.2	117.6		6553	205.7000	0.072			36
34	6.3	39.69	142.9	133.3		8326	261.3000	0.057			37
35	5.6	31.36	161.3	149.3		10537	330.7000	0.045			38
36	5.0	25.00	178.6	166.7		13212	414.8000	0.036			39
37	4.5	20.25	200.0	181.8		16319	512.1000	0.029			40
38	4.0	16.00	222.2	204.1		20644	648.2000	0.023			
39	3.5	12.25	256.4	232.6		26969	846.6000	0.018			
40	3.1	9.61	285.7	263.2		34364	1079.2000	0.014			
41	2.8	7.84	322.6	294.1		42123	1323.0000	0.011			
42	2.5	6.25	357.1	333.3		52854	1659.0000	0.009			
43	2.2	4.84	400.0	370.4		68259	2143.0000	0.007			
44	2.0	4.00	454.5	400.0		82645	2593.0000	0.006			
45	1.8	3.10	526.3	465.1		106600	3348.0000	0.004			
46	1.6	2.46	588.2	512.8		134000	4207.0000	0.004			

Teflon Coated, Stranded Wire

(As supplied by Belden Wire and Cable)

Size	Strands[5]	Turns per Linear inch[2] UL Style No.		
		1180	1213	1371
16	19×29	11.2		
18	19×30	12.7		
20	7×28	14.7	17.2	
20	19×32	14.7	17.2	
22	19×34	16.7	20.0	23.8
22	7×30	16.7	20.0	23.8
24	19×36	18.5	22.7	27.8
24	7×32		22.7	27.8
26	7×34		25.6	32.3
28	7×36		28.6	37.0
30	7×38		31.3	41.7
32	7×40			47.6

Notes

[1]A circular mil (CM) is a unit of area equal to that of a one-mil-diameter circle ($\pi/4$ square mils). The CM area of a wire is the square of the mil diameter.

[2]Figures given are approximate only; insulation thickness varies with manufacturer.

[3]Maximum wire temperature of 212°F (100°C) with a maximum ambient temperature of 13°F (57°C) as specified by the manufacturer. The *National Electrical Code* or local building codes may differ.

[4]700 CM per ampere is a satisfactory design figure for small transformers, but values from 500 to 1000 CM are commonly used. The *National Electrical Code* or local building codes may differ.

[5]Stranded wire construction is given as "count"×"strand size" (AWG).

Color Code for Hookup Wire

Wire Color	Type of Circuit
Black	Grounds, grounded elements and returns
Brown	Heaters or filaments, off ground
Red	Power supply B plus
Orange	Screen grids and base 2 of transistors
Yellow	Cathodes and transistor emitters
Green	Control grids, diode plates, and base 1 of transistors
Blue	Plates and transistor collectors
Violet	Power supply, minus leads
Gray	Ac power line leads
White	Bias supply, B or C minus, AGC

Note: Wires with tracers are coded in the same manner as solid-color wires, allowing additional circuit identification over solid-color wiring. The body of the wire is white and the color band spirals around the wire lead. When more than one color band is used, the widest band represents the first color.

Crystal Holders

Note: Solder Seal, Cold Weld, and Resistance Weld sealing methods are commonly available. All dimensions are in inches

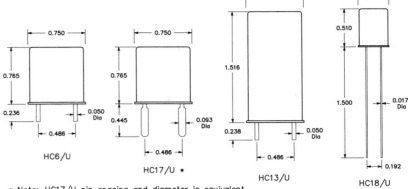

HC6/U   HC17/U *   HC13/U   HC18/U

* Note: HC17/U pin spacing and diameter is equivalent to the older FT−243 (32 pF) holder.

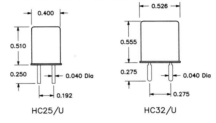

HC25/U   HC32/U

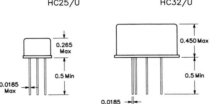

HC33/U

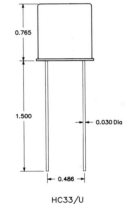

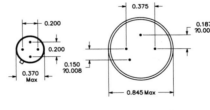

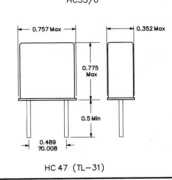

PIN	CONNECTION
1	No Connection
2	Crystal
3	Ground
4	Crystal

PIN	CONNECTION
1	No Connection
2	Crystal
3	Ground
4	Crystal

HC 35 (TO−5)   HC 40 (TL−90)   HC 47 (TL−31)

Aluminum Alloy Characteristics

Common Alloy Numbers

Type	Characteristic
2024	Good formability, high strength
5052	Excellent surface finish, excellent corrosion resistance, normally not heat treatable for high strength
6061	Good machinability, good weldability, can be brittle at high tempers
7075	Good formability, high strength

General Uses

Type	Uses
2024-T3	Chassis boxes, antennas, anything that will be bent or flexed repeatedly
7075-T3	
6061-T6	Mounting plates, welded assemblies or machined parts

Common Tempers

Type	Characteristics
T0	Special soft condition
T3	Hard
T6	Very hard, possibly brittle
TXXX	Three digit tempers—usually specialized high-strength heat treatments, similar to T6

Miniature Lamp Guide

BULB STYLES

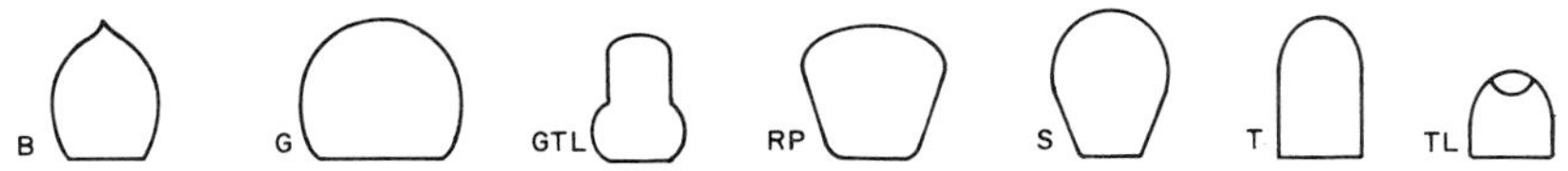

B   G   GTL   RP   S   T   TL

**Bulbs are described by a letter indicating shape and a number that is an approximation of diameter expressed in eighths of an inch. For example S-8 is "S" shape, 8 eighths or 1 inch in diam.

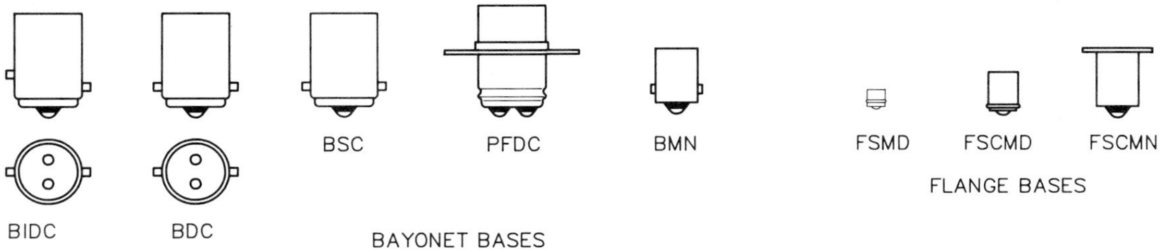

BSC   PFDC   BMN

FSMD   FSCMD   FSCMN

FLANGE BASES

BIDC   BDC

BAYONET BASES

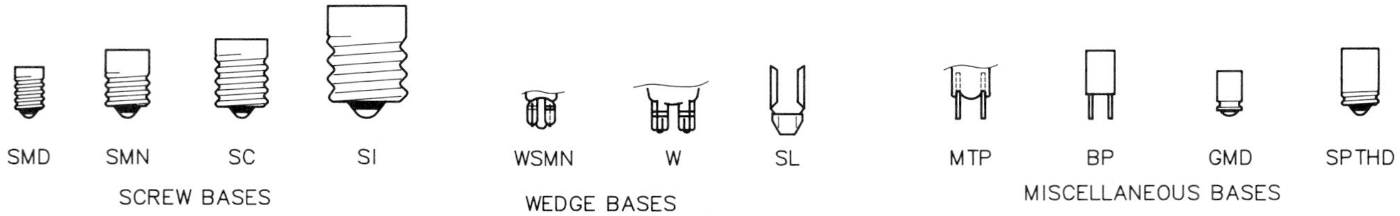

SMD   SMN   SC   SI

WSMN   W   SL

MTP   BP   GMD   SPTHD

SCREW BASES   WEDGE BASES   MISCELLANEOUS BASES

Lamp Base Legend

BDC	Bayonet, dual-contact	FSCMD	Flanged, midget single-contact	SI	Screw, intermediate
BIDC	Bayonet, indexed dual-contact	FSMD	Flanged, submidget	SMD	Screw, midget
BMN	Bayonet, miniature	GMD	Midget grooved	SMN	Screw, miniature
BP	Bipin	MTP	Miniature two-pin	SPTHD	Screw, special thread
BSC	Bayonet, single-contact	PFDC	Prefocused dual contact	W	Wedge
FSCMN	Flanged, single-contact,	SL	Slide (various sizes)	WSMN	Wedge, subminiature
	miniature	SC	Screw, candleabra	WT	Wire terminal

†k = 1000

Type	Bulb	Base	V	A	Life†	Type	Bulb	Base	V	A	Life†
PR2	B-3½	FSCMN	2.38	0.500	15	159	T-3¼	W	6.30	0.150	5K
PR3	B-3½	FSCMN	3.57	0.500	15	161	T-3¼	W	14.00	0.190	4K
PR4	B-3½	FSCMN	2.33	0.270	10	168	T-3¼	W	14.00	0.350	1.5K
PR6	B-3½	FSCMN	2.47	0.300	30	219	G-3½	BMN	6.30	0.250	5K
PR7	B-3½	FSCMN	3.70	0.300	30	222	TL-3	SMN	2.25	0.250	0.5
PR12	B-3½	FSCMN	5.95	0.500	15	239	T-3¼	BMN	6.30	0.360	5K
PR13	B-3½	FSCMN	4.75	0.500	15	240	T-3¼	BMN	6.30	0.360	5K
10	G-3½	MTP	2.50	0.500	3K	259	T-3¼	W	6.30	0.250	5K
12	G-3½	MTP	6.30	0.150	5K	268	T-1¾	FSCMD	2.50	0.350	10K
13	G-3½	SMN	3.70	0.300	15	305	S-8	BSC	28.00	0.510	300
14	G-3½	SMN	2.47	0.300	15	307	S-8	BSC	28.00	0.670	300
19	G-3½	MTP	14.40	0.100	1K	308	S-8	BDC	28.00	0.670	300
27	G-4½	SMN	4.90	0.300	30	313	T-3¼	BMN	28.00	0.170	500
37	T-1¾	WSMN	14.00	0.090	1.5K	323	T-1¼	SPTHD	3.00	0.190	350
40	T-3¼	SMN	6.30	0.150	3K	327	T-1¾	FSCMD	28.00	0.040	4K
43	T-3¼	BMN	2.50	0.500	3K	327AS15	T-1¾	FSCMD	28.00	0.040	4K
44	T-3¼	BMN	6.30	0.250	3K	328	T-1¾	FSCMD	6.00	0.200	1K
45	T-3¼	BMN	3.20	0.350	3K	330	T-1¾	FSCMD	14.00	0.080	1.5K
46	T-3¼	SMN	6.30	0.250	3K	331	T-1¾	FSCMD	1.35	0.060	3K
47	T-3¼	BMN	6.30	0.150	3K	334	T-1¾	GMD	28.00	0.040	4K
48	T-3¼	SMN	2.00	0.060	1K	335	T-1¾	SMD	28.00	0.040	4K
49	T-3¼	BMN	2.00	0.060	1K	336	T-1¾	GMD	14.00	0.080	1.5K
50	G-3½	SMN	7.50	0.220	1K	337	T-1¾	GMD	6.00	0.200	1K
51	G-3½	BMN	7.50	0.220	1K	338	T-1¾	FSCMD	2.70	0.060	6K
52	G-3½	SMN	14.40	0.100	1K	342	T-1¾	SMD	6.00	0.040	10K
53	G-3½	BMN	14.40	0.120	1K	344	T-1¾	FSCMD	10.00	0.014	50K
55	G-4½	BMN	7.00	0.410	500	345	T-1¾	FSCMD	6.00	0.040	10K
57	G-4½	BMN	14.00	0.240	500	346	T-1¾	GMD	18.00	0.040	10K
63	G-6	BSC	7.00	0.630	1K	349	T-1¾	FSCMD	6.30	0.200	5K
73	T-1¾	WSMN	14.00	0.080	15K	370	T-1¾	FSCMD	18.00	0.040	10K
74	T-1¾	WSMN	14.00	0.100	500	373	T-1¾	SMD	14.00	0.080	1.5K
82	G-6	BDC	6.50	1.020	500	375	T-1¾	FSCMD	3.00	0.015	10K
85	T-1¾	WSMN	28.00	0.040	7K	376	T-1¾	FSCMD	28.00	0.060	25K
86	T-1¾	WSMN	6.30	0.200	20K	380	T-1¾	FSCMD	6.30	0.040	20K
88	S-8	BDC	6.80	1.910	300	381	T-1¾	FSCMD	6.30	0.200	20K
93	S-8	BSC	12.80	1.040	700	382	T-1¾	FSCMD	14.00	0.080	15K
112	TL-3	SMN	1.20	0.220	5	385	T-1¾	FSCMD	28.00	0.040	10K
130	G-3½	BMN	6.30	0.150	5K	386	T-1¾	GMD	14.00	0.080	15K
131	G-3½	SMN	1.30	0.100	50	387	T-1¾	FSCMD	28.00	0.040	7K
158	T-3¼	W	14.00	0.240	500	388	T-1¾	GMD	28.00	0.040	7K

Type	Bulb	Base	V	A	Life†
397	T-1¾	GMD	10.00	0.040	5K
398	T-1¾	GMD	6.30	0.200	5K
399	T-1¾	SMD	28.00	0.040	7K
502	G-4½	SMN	5.10	0.150	100
555	T-3¼	W	6.30	0.250	3K
656	T-3¼	W	28.00	0.060	2.5K
680AS15	T-1	WT	5.00	0.060	60K
682AS15	T-1	FSMD	5.00	0.060	60K
683AS15	T-1	WT	5.00	0.060	25K
685AS15	T-1	FSMD	5.00	0.060	25K
715AS15	T-1	WT	5.00	0.115	40K
715AS25	T-1	WT	5.00	0.115	40K
718AS25	T-1	FSMD	5.00	0.115	40K
755	T-3¼	BMN	6.30	0.150	20K
756	T-3¼	BMN	14.00	0.080	15K
757	T-3¼	BMN	28.00	0.080	7.5K
1034	S-8	BIDC	14.00	0.590	5K
1073	S-8	BSC	12.80	1.800	200
1130	S-8	BDC	6.40	2.630	200
1133	RP-11	BSC	6.20	3.910	200
1141	S-8	BSC	12.80	1.440	1K
1143	RP-11	BSC	12.50	1.980	400
1184	RP-11	BDC	5.50	6.250	100
1251	G-6	BSC	28.00	0.230	2K
1445	G-3½	BMN	14.40	0.130	2K
1487	T-3¼	SMN	14.00	0.200	3K
1488	T-3¼	BMN	14.00	0.150	200
1490	T-3¼	BMN	3.20	0.160	3K
1493	S-8	BDC	6.50	2.750	100
1619	S-8	BSC	6.70	1.900	500
1630	S-8	PFDC	6.50	2.750	100
1691	S-8	BSC	28.00	0.610	1K
1705	T-1¾	WT	14.00	0.080	1.5K
1728	T-1¾	WT	1.35	0.060	3K
1730	T-1¾	WT	6.00	0.040	20K
1738	T-1¾	WT	2.70	0.060	6K
1762	T-1¾	WT	28.00	0.040	4K
1764	T-1¾	WT	28.00	0.040	4K
1767	T-1¾	SMD	2.50	0.200	500
1768	T-1¾	SMD	6.00	0.200	1K
1775	T-1¾	SMD	6.30	0.075	1K
1813	T-3¼	BMN	14.40	0.100	1K
1815	T-3¼	BMN	14.00	0.200	3K
1816	T-3¼	BMN	13.00	0.330	1K
1818	T-3¼	BMN	24.00	0.170	250
1819	T-3¼	BMN	28.00	0.040	2.5K
1820	T-3¼	BMN	28.00	0.100	1K
1821	T-3¼	SMN	28.00	0.170	500
1822	T-3¼	BMN	36.00	0.100	1K
1828	T-3¼	BMN	37.50	0.050	3K
1829	T-3¼	BMN	28.00	0.070	1K
1835	T-3¼	BMN	55.00	0.050	5K
1847	T-3¼	BMN	6.30	0.150	5K
1850	T-3¼	BMN	5.00	0.090	1.5K
1864	T-3¼	BMN	28.00	0.170	1.5K
1866	T-3¼	BMN	6.30	0.250	5K
1869	T-1¾	WT	10.00	0.014	50K
1891	T-3¼	BMN	14.00	0.240	500
1892	T-3¼	BMN	14.40	0.120	1K
1893	T-3¼	BMN	14.00	0.330	7.5K
1895	G-4½	BMN	14.00	0.270	2K
2102	T-1¾	WT	18.00	0.040	10K
2107	T-1¾	WT	10.00	0.040	5K
2158	T-1¾	WT	3.00	0.015	10K
2162	T-1¾	WT	14.00	0.100	10K
2169	T-1¾	WT	2.50	0.350	20K
2180	T-1¾	WT	6.30	0.040	20K
2181	T-1¾	WT	6.30	0.200	20K
2182	T-1¾	WT	14.00	0.080	40K
2187	T-1¾	WT	28.00	0.040	7K
2304	T-1¾	BP	3.00	0.300	1.5K
2307	T-1¾	BP	6.30	0.200	5K
2314	T-1¾	BP	28.00	0.050	1K
2316	T-1¾	BP	18.00	0.040	10K
2324	T-1¾	BP	28.00	0.040	4K
2335	T-1¾	BP	14.00	0.080	15K
2337	T-1¾	BP	6.30	0.200	20K
2342	T-1¾	BP	28.00	0.040	25K
3149	T-1¾	BP	5.00	0.060	5K
6803AS25	T-¾	WT	5.00	0.060	60K
6833AS15	T-¾	WT	5.00	0.060	25K
6838	T-1	WT	28.00	0.024	4K
6839	T-1	FSMD	28.00	0.024	4K
7001	T-1¾	BP	24.00	0.050	2K
7003	T-1¾	BP	24.00	0.050	2K
7153AS15	T-1¾	WT	5.00	0.115	40K
7265	T-1	BP	5.00	0.060	5K
7327	T-1¾	BP	28.00	0.040	4K
7328	T-1¾	BP	6.00	0.200	1K
7330	T-1¾	BP	14.00	0.080	1.5K
7344	T-1¾	BP	10.00	0.014	50K
7349	T-1¾	BP	6.30	0.200	5K
7361	T-1¾	BP	5.00	0.060	25K
7362	T-1¾	BP	5.00	0.115	40K
7367	T-1¾	BP	10.00	0.040	5K
7370	T-1¾	BP	18.00	0.040	10K
7371	T-1¾	BP	12.00	0.040	10K
7373	T-1¾	BP	14.00	0.100	10K
7374	T-1¾	BP	28.00	0.040	10K
7375	T-1¾	BP	3.00	0.015	10K
7376	T-1¾	BP	28.00	0.065	10K
7377	T-1¾	BP	6.30	0.075	1K
7380	T-1¾	BP	6.30	0.040	30K
7381	T-1¾	BP	6.30	0.200	20K
7382	T-1¾	BP	14.00	0.080	15K
7387	T-1¾	BP	28.00	0.040	7K
7410	T-1¾	BP	14.00	0.080	15K
7839	T-1	BP	28.00	0.025	4K
7876	T-1¾	BP	28.00	0.060	25K
7931	T-1¾	BP	1.35	0.060	3K
7945	T-1¾	BP	6.00	0.040	20K
7968	T-1¾	BP	2.50	0.200	500
8099	T-1	BP	18.00	0.020	16K
8362	T-1¾	SMD	14.00	0.080	15K
8369	T-1¾	SMD	28.00	0.065	10K

STANDARD LINE-VOLTAGE LAMPS

Type	V	W	Bulb	Base
10C7DC	115-125	10	C-7	BDC
3S6	120, 125	3	S-6	SC
6S6	30, 48, 115, 120, 125, 130, 135, 145, 155	6	S-6	SC
6S6/R	115-125	6	S-6 (red)	SC
6S6/W	115-125	6	S-6 (white)	SC
6T4½	120, 130	6	T-4½	SC
7C7	115-125	7	C-7	SC
7C7/W	115-125	7	C-7 (white)	SC
10C7	115-125	10	C-7	SC
10S6	120	10	S-6	SC
10S6/10	220, 230, 250	10	S-6	SC
6S6DC	30, 120, 125, 145	6	S-6	BDC
10S6/10DC	230, 250	10	S-6	BDC
40S11 N	115-125	40	S-11	SI
120MB	120	3	T-2½	BMN
120MB/6	120	6	T-2½	BMN
120PSB	120	3	T-2	SL
120PS	120	3	T-2	WT
120PS/6	120	6	T-2½	WT

INDICATOR LAMPS

Each has a T-2 bulb and a slide base.

Type	V	A	Life†
6PSB	6.00	0.140	20K
12PSB	12.00	0.170	12K
24PSB	24.00	0.073	10K
28PSB	28.00	0.040	5K
48PSB	48.00	0.050	10K
60PSB	60.00	0.050	7.5K
120PSB	120.00	0.025	7.5K

NEON GLOW LAMPS

Operating circuit voltage 105-125.

Type	Breakdown Voltage		Bulb	Base	W	External Resistance†
	AC	DC				
NE-2	65	90	T-2	WT	1/12	150K
NE-2A	65	90	T-2	WT	1/15	100K
NE-2D	65	90	T-2	FSCMD	1/12	100K
NE-2E	65	90	T-2	WT	1/12	100K
NE-2H	95	135	T-2	WT	1/4	30K
NE-2J	95	135	T-2	FSCMD	1/4	30K
NE-2V	65	90	T-2	WT	1/12	100K
NE-45	65	90	T-4 1/2	SC	1/4	NONE
NE-51	65	90	T-3 1/4	BMN	1/25	220K
NE-51H	95	135	T-3 1/4	BMN	1/7	47K
NE-84	95	135	T-2	SL	1/4	30K
NE-120PSB	95	95	T-2	SL	1/4	NONE

Metal-Oxide Varistor (MOV) Transient Suppressors†

Listed by voltage.

Type No.	ECG/NTE†† no.	V ac$_{RMS}$	Maximum Applied Voltage V ac$_{Peak}$	Maximum Energy (Joules)	Maximum Peak Current (A)	Maximum Power (W)	Maximum Varistor Voltage (V)
V180ZA1	1V115	115	163	1.5	500	0.2	285
V180ZA10	2V115	115	163	10.0	2000	0.45	290
V130PA10A		130	184	10.0	4000	8.0	350
V130PA20A		130	184	20.0	4000	15.0	350
V130LA1	1V130	130	184	1.0	400	0.24	360
V130LA2	1V130	130	184	2.0	400	0.24	360
V130LA10A	2V130	130	184	10.0	2000	0.5	340
V130LA20A	524V13	130	184	20.0	4000	0.85	340
V150PA10A		150	212	10.0	4000	8.0	410
V150PA20A		150	212	20.0	4000	15.0	410
V150LA1	1V150	150	212	1.0	400	0.24	420
V150LA2	1V150	150	212	2.0	400	0.24	420
V150LA10A	524V15	150	212	10.0	2000	0.5	390
V150LA20A	524V15	150	212	20.0	4000	0.85	390
V250PA10A		250	354	10.0	4000	0.85	670
V250PA20A		250	354	20.0	4000	7.0	670
V250PA40A		250	354	40.0	4000	13.0	670
V250LA2	1V250	250	354	2.0	400	0.28	690
V250LA4	1V250	250	354	4.0	400	0.28	690
V250LA15A	2V250	250	354	15.0	2000	0.6	640
V250LA20A	2V250	250	354	20.0	2000	0.6	640
V250LA40A	524V25	250	354	40.0	4000	0.9	640

†† ECG and NTE numbers for these parts are identical, except for the prefix. Add the "ECG" or "NTE" prefix to the numbers shown for the complete part number.

Voltage-Variable Capacitance Diodes†

Listed numerically by device

Device	CT Nominal Capacitance pF ±10% @ V_R = 4.0 V f = 1.0MHz	Capacitance Ratio 4-60 V Min.	Q @ 4.0 V 50 MHz Min.	Case Style
1N5441A	6.8	2.5	450	
1N5442A	8.2	2.5	450	
1N5443A	10	2.6	400	DO-7
1N5444A	12	2.6	400	
1N5445A	15	2.6	450	
1N5446A	18	2.6	350	
1N5447A	20	2.6	350	
1N5448A	22	2.6	350	DO-7
1N5449A	27	2.6	350	
1N5450A	33	2.6	350	
1N5451A	39	2.6	300	
1N5452A	47	2.6	250	
1N5453A	56	2.6	200	DO-7
1N5454A	68	2.7	175	
1N5455A	82	2.7	175	
1N5456A	100	2.7	175	
1N5461A	6.8	2.7	600	
1N5462A	8.2	2.8	600	
1N5463A	10	2.8	550	DO-7
1N5464A	12	2.8	550	
1N5465A	15	2.8	550	
1N5466A	18	2.8	500	
1N5467A	20	2.9	500	
1N5468A	22	2.9	500	DO-7
1N5469A	27	2.9	500	
1N5470A	33	2.9	500	

Device	CT Nominal Capacitance pF ±10% @ V_R = 4.0 V f = 1.0MHz	Capacitance Ratio 4-60 V Min.	Q @ 4.0 V 50 MHz Min.	Case Style
1N5471A	39	2.9	450	
1N5472A	47	2.9	400	
1N5473A	56	2.9	300	DO-7
1N5474A	68	2.9	250	
1N5475A	82	2.9	225	
1N5476A	100	2.9	200	
MV2101	6.8	2.5	450	
MV2102	8.2	2.5	450	
MV2103	10	2.0	400	TO-92
MV2104	12	2.5	400	
MV2105	15	2.5	400	
MV2106	18	2.5	350	
MV2107	22	2.5	350	
MV2108	27	2.5	300	TO-92
MV2109	33	2.5	200	
MV2110	39	2.5	150	
MV2111	47	2.5	150	
MV2112	56	2.6	150	
MV2113	68	2.6	150	TO-92
MV2114	82	2.6	100	
MV2115	100	2.6	100	

†For package shape, size and pin-connection information, see manufacturers' data sheets. Many retail suppliers offer data sheets to buyers free of charge on request. Data books are available from many manufacturers and retailers.

Zener Diodes

Volts	0.25	0.4	0.5	Power (Watts) 1.0	1.5	5.0	10.0	50.0
1.8	1N4614							
2.0	1N4615							
2.2	1N4616							
2.4	1N4617	1N4370,A	1N4370,A 1N5221,B 1N5985,B 1N5222B					
2.5								
2.6	1N702,A							
2.7	1N4618	1N4371,A	1N4371,A 1N5223,B 1N5839, 1N5986 1N5224B					
2.8								
3.0	1N4619	1N4372,A	1N4372 1N5225,B 1N5987					
3.3	1N4620	1N746,A 1N764,A 1N5518	1N746,A 1N5226,B 1N5988	1N3821 1N4728,A	1N5913	1N5333,B		
3.6	1N4621	1N747,A 1N5519	1N747A 1N5227,B 1N5989	1N3822 1N4729,A	1N5914	1N5334,B		
3.9	1N4622	1N748,A 1N5520	1N748A 1N5228,B 1N5844, 1N5990	1N3823 1N4730,A	1N5915	1N5335,B	1N3993A	1N4549,B 1N4557,B
4.1	1N704,A							
4.3	1N4623	1N749,A 1N5521	1N749,A 1N5229,B 1N5845 1N5991	1N3824 1N4731,A	1N5916	1N5336,B	1N3994,A	1N4550,B 1N4558,B
4.7	1N4624	1N750,A 1N5522	1N750A 1N5230,B 1N5846, 1N5992	1N3825 1N4732,A	1N5917	1N5337,B	1N3995,A	1N4551,B 1N4559,B
5.1	1N4625 1N4689	1N751,A 1N5523	1N751,A, 1N5231,B 1N5847 1N5993	1N3826 1N4733	1N5918	1N5338,B	1N3996,A	1N4552,B 1N4560,B
5.6	1N708A 1N4626	1N752,A 1N5524	1N752,A 1N5232,B 1N5848, 1N5994	1N3827 1N4734,A	1N5919	1N5339,B	1N3997,A	1N4553,B 1N4561,B
5.8	1N706A	1N762						
6.0			1N5233B 1N5849			1N5340,B		
6.2	1N709,1N4627 MZ605, MZ610 MZ620, MZ640	1N753,A 1N821,3,5,7,9;A	1N753,A 1N5234,B, 1N5850 1N5995	1N3828,A 1N4735,A	1N5920	1N5341,B	1N3998,A	1N4554,B 1N4562,B
6.4	1N4565-84,A							
6.8	1N4099	1N754,A 1N957,B 1N5526	1N754,A 1N757,B 1N5235,B 1N5851 1N5996	1N3016,B 1N3829 1N4736,A	1N3785 1N5921	1N5342,B	1N2970,B 1N3999,A	1N2804B 1N3305B 1N4555, 1N4563
7.5	1N4100	1N755,A 1N958,B 1N5527	1N755A, 1N958,B 1N5236,B 1N5852 1N5997	1N3017,A,B 1N3830 1N4737,A	1N3786 1N5922	1N5343,B	1N2971,B 1N4000,A	1N2805,B 1N3306,B 1N4556, 1N4564
8.0	1N707A							
8.2	1N712A 1N4101	1N756,A 1N959,B 1N5528	1N756,A 1N959,B 1N5237,B 1N5853 1N5998	1N3018,B 1N4738,A	1N3787 1N5923	1N5344,B	1N2972,B	1N2806,B 1N3307,B
8.4		1N3154-57,A	1N3154,A 1N3155-57					
8.5	1N4775-84,A		1N5238,B 1N5854					
8.7	1N4102					1N5345,B		
8.8		1N764						
9.0		1N764A	1N935-9;A,B					
9.1	1N4103	1N757,A 1N960,B 1N5529	1N757,A, 1N960,B 1N5239,B, 1N5855 1N5999	1N3019,B 1N4739,A	1N3788 1N5924	1N5346,B	1N2973,B	1N2807,B 1N3308,B
10.0	1N4104	1N758,A 1N961,B 1N5530,B	1N758,A, 1N961,B 1N5240,B, 1N5856 1N6000	1N3020,B 1N4740,A	1N3789 1N5925	1N5347,B	1N2974,B	1N2808,B 1N3309,A,B
11.0	1N715,A 1N4105	1N962,B 1N5531	1N962,B 1N5241,B 1N5857, 1N6001 1N941-4;A,B	1N3021,B 1N4741,A	1N3790 1N5926	1N5348,B	1N2975,B	1N2809,B 1N3310,B
11.7	1N716,A 1N4106							
12.0		1N759,A 1N963,B 1N5532	1N759,A, 1N963,B, 1N5242,B, 1N5858 1N6002	1N3022,B 1N4742,A	1N3791 1N5927	1N5349,B	1N2976,B	1N2810,B 1N3311,B
13.0	1N4107	1N964,B 1N5533	1N964,B 1N5243,B, 1N5859 1N6003	1N3023,B 1N4743,A	1N3792 1N5928	1N5350,B	1N2977,B	1N2811,B 1N3312,B

Zener Diodes—Continued

Volts	0.25	0.4	0.5	Power (Watts) 1.0	1.5	5.0	10.0	50.0
14.0	1N4108	1N5534	1N5244B 1N5860			1N5351,B	1N2978,B	1N2812,B 1N3313,B
15.0	1N4109	1N965,B 1N5535	1N965,B 1N5245,B, 1N5861, 1N6004	1N3024,B 1N4744A	1N3793 1N5929	1N5352,B	1N2979,A,B	1N2813,A,B 1N3314,B
16.0	1N4110	1N966,B 1N5536	1N966,B, 1N5246,B 1N5862, 1N6005	1N3025,B 1N4745,A	1N3794 1N5930	1N5353,B	1N2980,B	1N2814,B 1N3315,B
17.0	1N4111	1N5537	1N5247,B 1N5863			1N5354,B	1N2981B	1N2815,B 1N3316,B
18.0	1N4112	1N967,B 1N5538	1N967,B 1N5248,B 1N5864, 1N6006	1N3026,B 1N4746,A	1N3795 1N5931	1N5355,B	1N2982,B	1N2816,B 1N3317,B
19.0	1N4113	1N5539	1N5249,B 1N5865			1N5356,B	1N2983,B	1N2817,B 1N3318,B
20.0	1N4114	1N968,B 1N5540	1N968,B 1N5250,B 1N5866, 1N6007	1N3027,B 1N4747,A	1N3796 1N5932,A,B	1N5357,B	1N2984,B	1N2818,B 1N3319,B
22.0	1N4115	1N959,B 1N5541	1N969,B 1N5241,B 1N5867, 1N6008	1N3028,B 1N4748,A	1N3797 1N5933	1N5358,B	1N2985,B	1N2819,B 1N3320,A,B
24.0	1N4116	1N5542 1N9701B	1N970,B 1N5252,B, 1N586 1N6009	1N3029,B 1N4749,A	1N3798 1N5934	1N5359,B	1N2986,B	1N2820,B 1N3321,B
25.0	1N4117	1N5543	1N5253,B 1N5869			1N5360,B	1N2987B	1N2821,B 1N3322,B
27.0	1N4118	1N971,B	1N971 1N5254,B, 1N5870, 1N6010	1N3030,B 1N4750,A	1N3799 1N5935	1N5361,B	1N2988,B	1N2822B 1N3323,B
28.0	1N4119	1N5544	1N5255,B 1N5871			1N5362,B		
30.0	1N4120	1N972,B 1N5545	1N972,B 1N5256,B, 1N5872, 1N6011	1N3031,B 1N4751,A	1N3800 1N5936	1N5363,B	1N2989,B	1N2823,B 1N3324,B
33.0	1N4121	1N973,B 1N5546	1N973,B 1N5257,B 1N5873 1N6012	1N3032,B 1N4752,A	1N3801 1N5937	1N5364,B	1N2990,A,B	1N2824,B 1N3325,B
36.0	1N4122	1N974,B	1N974,B 1N5258,B 1N5874, 1N6013	1N3033,B 1N4753,A	1N3802 1N5938	1N5365,B	1N2991,B	1N2825,B 1N3326,B
39.0	1N4123	1N975,B	1N975,B, 1N5259,B 1N5875, 1N6014	1N3034,B 1N4754,A	1N3803 1N5939	1N5366,B	1N2992,B	1N2826,B 1N3327,B
43.0	1N4124	1N976,B	1N976,B 1N5260,B, 1N5876, 1N6015	1N3035,B 1N4755,A	1N3804 1N5940	1N5367,B	1N2993,A,B	1N2827,B 1N3328,B
45.0							1N2994B	1N2828B 1N3329B
47.0	1N4125	1N977,B	1N977,B, 1N5261,B 1N5877, 1N6016	1N3036,B 1N4756,A	1N3805 1N5941	1N5368,B	1N2996,B	1N2829,B 1N3330,B
50.0								1N2830B 1N3331B
51.0	1N4126	1N978,B	1N978,B, 1N5262,A,B 1N5878, 1N6017	1N3037,B 1N4757,A	1N3806 1N5942	1N5369,B	1N2997,B	1N2831,B 1N3332,B
52.0							1N2998B	1N3333
56.0	1N4127	1N979,B	1N979 1N5263,B 1N6018	1N3038,B 1N4758,A	1N3807 1N5943	1N5370,B	1N2999,B	1N2822,B 1N3334,B
60.0	1N4128		1N5264,A,B			1N5371,B		
62.0	1N4129	1N980,B	1N980 1N5265,A,B 1N6019	1N3039,B 1N4759,A	1N3808 1N5944	1N5372,B	1N3000,B	1N2833,B 1N3335,B
68.0	1N4130	1N981,B	1N981,B 1N5266,A,B 1N6020	1N3040,A,B 1N4760,A	1N3809 1N5945	1N5373,B	1N3001,B	1N2834,B 1N3336,B
75.0	1N4131	1N982,B	1N982 1N5267,A,B 1N6021	1N3041,B 1N4761,A	1N3810 1N5946	1N5374,B	1N3002,B	1N2835,B 1N3337,B
82.0	1N4132	1N983,B	1N983 1N5268,A,B 1N6022	1N3042,B 1N4762,A	1N3811 1N5947	1N5375,B	1N3003,B	1N2836,B 1N3338,B
87.0	1N4133		1N5269,B			1N5376,B		
91.0	1N4134	1N984,B	1N984 1N5270,B 1N6023	1N3043,B 1N4763,A	1N3812 1N5948	1N5377,B	1N3004,B	1N2837,B 1N3339,B
100.0	1N4135	1N985	1N985,B 1N5271,B 1N6024	1N3044,A,B 1N4764,A	1N3813 1N5949	1N5378,B	1N3005,B	1N2838,B 1N3340,B
105.0							1N3006B	1N2839,B 1N3341,B
110.0		1N986	1N986 1N5272,B 1N6025	1N3045,B 1M110ZS10	1N3814 1N5950	1N5379,B	1N3007A,B	1N2840,B 1N3342,B

Zener Diodes—Continued

Volts	0.25	0.4	0.5	Power (Watts) 1.0	1.5	5.0	10.0	50.0
120.0		1N987	1N987,B 1N5273,B 1N6026	1N3046,B 1M120ZS10	1N3815 1N5951	1N5380,B	1N3008A,B	1N2841,B 1N3343,B
130.0		1N988	1N988,B 1N5274,B 1N6027	1N3047,B 1M130ZS10	1N3816 1N5952	1N5381,B	1N3009,B	1N2842,B 1N3344,B
140.0		1N989	1N5275,B			1N5382B	1N3010B	1N3345B
150.0		1N990	1N989 1N5276,B 1N6028	1N3048,B 1M150ZS10	1N3817 1N5953	1N5383,B	1N3011,B	1N2843,B 1N3346,B
160.0		1N991	1N990 1N5277,B 1N6029	1N3049,B 1M160ZS10	1N3818 1N5954	1N5384,B	1N3012A,B	1N2844,B 1N3347,B
170.0		1N992	1N5278,B	1M170ZS10		1N5385,B		
175.0							1N3013B	1N3348B
180.0			1N991,B 1N5279,B 1N6030	1N3050,A,B 1M180ZS10	1N3819 1N5955	1N5386,B	1N3014,B	1N2845,B 1N3349,B
190.0			1N5280,B			1N5387,B		
200.0			1N992 1N5281,B 1N6031	1N3051,B 1M200ZS10	1N3820 1N5956	1N5388B	1N3015,B	1N2846,B 1N3350,B

Semiconductor Diode Specifications[†]

Listed numerically by device

Device	Type	Material	Peak Inverse Voltage, PIV (V)	Average Rectified Current Forward (Reverse) $I_O(A)(I_R(A))$	Peak Surge Current, I_{FSM} 1 s @ 25°C (A)	Average Forward Voltage, V_F (V)
1N34	Signal	Ge	60	8.5 m (15.0 μ)		1.0
1N34A	Signal	Ge	60	5.0 m (30.0 μ)		1.0
1N67A	Signal	Ge	100	4.0 m (5.0 μ)		1.0
1N191	Signal	Ge	90	5.0 m		1.0
1N270	Signal	Ge	80	0.2 (100 μ)		1.0
1N914	Fast Switch	Si	75	75.0 m (25.0 n)	0.5	1.0
1N1183	RFR	Si	50	40 (5 m)	800	1.1
1N1184	RFR	Si	100	40 (5 m)	800	1.1
1N2071	RFR	Si	600	0.75 (10.0 μ)		0.6
1N3666	Signal	Ge	80	0.2 (25.0 μ)		1.0
1N4001	RFR	Si	50	1.0 (0.03 m)		1.1
1N4002	RFR	Si	100	1.0 (0.03 m)		1.1
1N4003	RFR	Si	200	1.0 (0.03 m)		1.1
1N4004	RFR	Si	400	1.0 (0.03 m)		1.1
1N4005	RFR	Si	600	1.0 (0.03 m)		1.1
1N4006	RFR	Si	800	1.0 (0.03 m)		1.1
1N4007	RFR	Si	1000	1.0 (0.03 m)		1.1
1N4148	Signal	Si	75	10.0 m (25.0 n)		1.0
1N4149	Signal	Si	75	10.0 m (25.0 n)		1.0
1N4152	Fast Switch	Si	40	20.0 m (0.05 μ)		0.8
1N4445	Signal	Si	100	0.1 (50.0 n)		1.0
1N5400	RFR	Si	50	3.0 (500 μ)	200	
1N5401	RFR	Si	100	3.0 (500 μ)	200	
1N5402	RFR	Si	200	3.0 (500 μ)	200	
1N5403	RFR	Si	300	3.0 (500 μ)	200	
1N5404	RFR	Si	400	3.0 (500 μ)	200	
1N5405	RFR	Si	500	3.0 (500 μ)	200	
1N5406	RFR	Si	600	3.0 (500 μ)	200	
1N5408	RFR	Si	1000	3.0 (500 μ)	200	
1N5711	Schottky	Si	70	1 m (200 n)	15 m	0.41 @ 1 mA
1N5767	Signal	Si		0.1 (1.0 μ)		1.0
1N5817	Schottky	Si	20	1.0 (1 m)	25	0.75
1N5819	Schottky	Si	40	1.0 (1 m)	25	0.9
1N5821	Schottky	Si	30	3.0		
ECG5863	RFR	Si	600	6	150	0.9
1N6263	Schottky	Si	70	15 m	50 m	0.41 @ 1 mA
5082-2835	Schottky	Si	8	1 m (100 n)	10 m	0.34 @ 1 mA

Si = Silicon; Ge = Germanium; RFR = rectifier, fast recovery.

[†] For package shape, size and pin-connection information see manufacturers' data sheets. Many retail suppliers offer data sheets to buyers free of charge on request. Data books are available from many manufacturers and retailers.

European Semiconductor Numbering System (PRO Electron Code)

$$\boxed{B|F|R90}$$

First Letter (Material)	Second Letter (Type)	Third, Fourth, Fifth Character (Serial Code)
A Germanium	A Low-power diode, voltage-variable capacitor	Y## Industrial service (no letter "Z"). ## is a W## registration number from 10 to 99.
B Silicon	B Varicap	100 - Device for consumer or entertainment use. 999
C Compound materials such as cadmium sulfide or gallium arsenide used in semiconductor devices (Energy gap band of 1.3 or more electron-volts)	C Small-signal audio transistor	
D Materials with an energy gap band of less than 0.6 electron-volts such as indium antimonide	D Audio power transistor	
R Radiation detectors, photo-conductive cells, hall-effect generators and so on	E Tunnel diode	
	F Small-signal RF transistor	
	G Miscellaneous	
	H Field probe	
	K Hall generator	
	L RF-power transistor	
	M Hall modulators and multipliers	
	P Photodiode, phototransistor, photoconductive cell (LDR), radiation device	
	R Low-power controlled rectifier	
	S Low-power switching transistor	
	T Breakdown devices, high-power controlled rectifier, Schottky diode, Thyristor, pnpn diodes	
	U High-power switching transistor	
	X Multiplier diode	
	Y High-power rectifier (diode)	
	Z Zener diode	

Japanese Semiconductor Nomenclature

All transistors manufactured in Japan are registered with the Electronic Industries Association of Japan (EIAJ). In addition, the Japan industrial Standard JIS-C-7012 provides type numbers for transistors and thyristors.

Each transistor type number contains five elements.

i	ii	iii	iv	v
2	S	C	82D	A
Figure	Letter	Letter	Figure	Letter

i) Kind of device, indicating number of effective electrical connections minus one.

ii) For a semiconductor registered with the EIAJ this letter is always an S.

iii) This letter designates polarity and application, as follows:

Letter	Polarity and Application
A	PNP transistor, high frequency
B	PNP transistor, low frequency
C	NPN transistor, high frequency
D	NPN transistor, low frequency
E	P-gate thyristor
G	N-gate thyristor
H	N-base unijunction transistor
J	P-channel FET
K	N-channel FET
M	Bi-directional triode thyristor

iv) These figures designate the order of application for EIAJ registration, starting with 11.

v) This letter indicates the level of improvement. An improvement device may be used in place of a previous-generation device, but not necessarily the other way around.

Suggested Small-Signal FETs

Device No.	Type	Max Diss (mW)	Max V_{DS} (V)[3]	$VGS_{(off)}$ (V)[3]	Min gfs (µS)	Input C (pF)	Max ID (mA)[1]	f_{max} (MHz)	Noise Figure (typ)	Case	Base	Mfr[2]	Applications
2N4416	N-JFET	300	30	−6	4500	4	−15	450	400 MHz 4 dB	TO-72	1	S, M	VHF/UHF amp, mix, osc
2N5484	N-JFET	310	25	−3	2500	5	30	200	200 MHz 4 dB	TO-92	2	M	VHF/UHF amp, mix, osc
2N5485	N-JFET	310	25	−4	3500	5	30	400	400 MHz 4 dB	TO-92	2	S	VHF/UHF amp, mix, osc
2N5486	N-JFET	360	25	−2	5500	5	15	400	400 MHz 4 dB	TO-92	2	M	VHF/UHF amp, mix, osc
3N200 NTE222 SK3065	N-dual-gate MOSFET	330	20	−6	10,000	4-8.5	50	500	400 MHz 4.5 dB	TO-72	3	R	VHF/UHF amp, mix, osc
3N202 NTE454 SK3991	N-dual-gate MOSFET	360	25	−5	8000	6	50	200	200 MHz 4.5 dB	TO-72	3	S	VHF amp, mixer
MPF102 ECG451 SK9164	N-JFET	310	25	−8	2000	4.5	20	200	400 MHz 4 dB	TO-92	2	N, M	HF/VHF amp, mix, osc
MPF106 2N5484	N-JFET	310	25	−6	2500	5	30	400	200 MHz 4 dB	TO-92	2	N, M	HF/VHF/UHF amp, mix, osc
40673 NTE222 SK3050	N-dual-gate MOSFET	330	20	−4	12,000	6	50	400	200 MHz 6 dB	TO-72	3	R	HF/VHF/UHF amp, mix, osc
U304	P-JFET	350	−30	+10		27	−50	—	—	TO-18	4	S	analog switch chopper
U310	N-JFET	500 300	30 30	−6	10,000	2.5	60	450	450 MHz 3.2 dB	TO-52	5	S	common-gate VHF/UHF amp,
U350	N-JFET Quad	1W	25	−6	9000	5	60	100	100 MHz 7 dB	TO-99	6	S	matched JFET doubly bal mix
U431	N-JFET Dual	300	25	−6	10,000	5	30	100	$\dfrac{10nV}{\sqrt{Hz}}$	TO-99	7	S	matched JFET cascode amp and bal mix
2N5670	N-JFET	350	25	8	3000	7	20	400	100 MHz 2.5 dB	TO-92	2	M	VHF/UHF osc, mix, front-end amp
2N5668	N-JFET	350	25	4	1500	7	5	400	100 MHz 2.5 dB	TO-92	2	M	VHF/UHF osc, mix, front-end amp
2N5669	N-JFET	350	25	6	2000	7	10	400	100 MHz 2.5 dB	TO-92	2	M	VHF/UHF osc, mix, front-end amp
J308	N-JFET	350	25	6.5	8000	7.5	60	1000	100 MHz 1.5 dB	TO-92	2	M	VHF/UHF osc, mix, front-end amp
J309	N-JFET	350	25	4	10,000	7.5	30	1000	100 MHz 1.5 dB	TO-92	2	M	VHF/UHF osc, mix, front-end amp
J310	N-JFET	350	25	6.5	8000	7.5	60	1000	100 MHz 1.5 dB	TO-92	2	M	VHF/UHF osc, mix, front-end amp
NE32684A	HJ-FET	165	2.0	−0.8	45,000	—	30	20 GHz	12GHz 0.5 dB	84A		NE	Low-noise amp

Notes:

[1] 25°C.

[2] M = Motorola; N = National Semiconductor; NE=NEC; R = RCA; S = Siliconix.

[3] For package shape, size and pin-connection information, see manufacturers' data sheets. Many retail suppliers offer data sheets to buyers free of charge on request. Data books are available from many manufacturers and retailers.

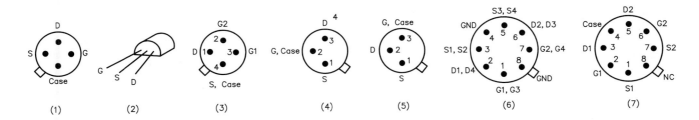

Low-Noise Transistors

Device	NF (dB)	F (MHz)	f_T (GHz)	I_C (mA)	Gain (dB)	F (MHz)	$V_{(BR)CEO}$ (V)	I_C (mA)	P_T (mW)	Case
MRF904	1.5	450	4	15	16	450	15	30	200	TO-206AF
MRF571	1.5	1000	8	50	12	1000	10	70	1000	Macro-X
MRF2369	1.5	1000	6	40	12	1000	15	70	750	Macro-X
MPS911	1.7	500	7	30	16.5	500	12	40	625	TO-226AA
MRF581A	1.8	500	5	75	15.5	500	15	200	2500	Macro-X
BFR91	1.9	500	5	30	16	500	12	35	180	Macro-T
BFR96	2	500	4.5	50	14.5	500	15	100	500	Macro-T
MPS571	2	500	6	50	14	500	10	80	625	TO-226AA
MRF581	2	500	5	75	15.5	500	18	200	2500	Macro-X
MRF901	2	1000	4.5	15	12	1000	15	30	375	Macro-X
MRF941	2.1	2000	8	15	12.5	2000	10	15	400	Macro-X
MRF951	2.1	2000	7.5	30	12.5	2000	10	100	1000	Macro-X
BFR90	2.4	500	5	14	18	500	15	30	180	Macro-T
MPS901	2.4	900	4.5	15	12	900	15	30	300	TO-226AA
MRF1001A	2.5	300	3	90	13.5	300	20	200	3000	TO-205AD
2N5031	2.5	450	1.6	5	14	450	10	20	200	TO-206AF
MRF4239A	2.5	500	5	90	14	500	12	400	3000	TO-205AD
BFW92A	2.7	500	4.5	10	16	500	15	35	180	Macro-T
MRF521*	2.8	1000	4.2	−50	11	1000	−10	−70	750	Macro-X
2N5109	3	200	1.5	50	11	216	20	400	2500	TO-205AD
2N4957*	3	450	1.6	−2	12	450	−30	−30	200	TO-206AF
MM4049*	3	500	5	−20	11.5	500	−10	−30	200	TO-206AF
2N5943	3.4	200	1.5	50	11.4	200	30	400	3500	TO-205AD
MRF586	4	500	1.5	90	9	500	17	200	2500	TO-205AD
2N5179	4.5	200	1.4	10	15	200	12	50	200	TO-206AF
2N2857	4.5	450	1.6	8	12.5	450	15	40	200	TO-206AF
2N6304	4.5	450	1.8	10	15	450	15	50	200	TO-206AF
MPS536*	4.5	500	5	−20	4.5	500	−10	−30	625	TO-226AA
MRF536*	4.5	1000	6	−20	10	1000	−10	−30	300	Macro-X

* denotes a PNP device

Complimentary devices

NPN	PNP
2N2857	2N4957
MRF904	MM4049
MRF571	MRF521

For package shape, size and pin-connection information, see manufacturers' data sheets. Many retail suppliers offer data sheets to buyers free of charge on request. Data books are available from many manufacturers and retailers.

VHF and UHF Class-A Transistors

The devices listed below are recommended for class-A linear applications, and include medium-power parts that are useful at frequencies from 100 MHz to 2 GHz.

Device	Frequency (MHz)	V_{CC} (V)	P_O @ 1 dB Compression (W)	Small Signal Gain/Frequency (MHz)	Bias Point (V_{dc}/A)	Package
MRA1000-3.5L	1000	19	3.5	10/1000	19/0.6	145A-09/1
MRA1000-7L	1000	19	7	9/1000	19/1.2	145A-09/1
MRA1000-14L	1000	19	14	8/1000	19/2.4	145A-09/1
MRF1029	1000	25	1.5	8/1000	25/0.2	244-04/1
MRF1030	1000	25	3	7.5/1000	25/0.4	244-04/1
MRF1031	1000	25	4.5	7/1000	25/0.6	244-04/1
MRF1032	1000	25	6	6.5/1000	25/0.85	244-04/1
MRF3094	2000	20	0.5	10.5/2000	20/0.12	328A-03/1
MRF3104	2000	20	0.5	10.5/2000	20/0.12	305A-01/1
MRF3095	2000	20	0.8	9/2000	20/0.12	328A-03/1
MRF3105	2000	20	0.8	9/2000	20/0.12	305A-01/1
MRF3096	2000	20	1.6	9/2000	20/0.24	328A-03/1
MRF3106	2000	20	1.6	9/2000	20/0.24	305A-01/1
MRF2000-5L	2000	20	5	7/2000	19/0.6	360A-01/1

For package shape, size and pin-connection information, see manufacturers' data sheets. Many retail suppliers offer data sheets to buyers free of charge on request. Data books are available from many manufacturers and retailers.

Monolithic Amplifiers (50 Ω)

Mini-Circuits Labs MMICs

Device	Freq Range (MHz)	Gain (dB) at 1000 MHz	Output Level 1 dB Comp (dBm)	NF (dB)	I_{max} (mA)	P_{max} (mW)
MAR-1	dc - 1000	15.5	+1.5	5.5	40	200
MAR-2	dc - 2000	12.0	+4.5	6.5	60	325
MAR-3	dc - 2000	12.0	+10.0	6.0	70	400
MAR-4	dc - 1000	8.0	+12.5	6.5	85	500
MAR-6	dc - 2000	16.0	+2.0	3.0	50	200
MAR-7	dc - 2000	12.5	+5.5	5.0	60	275
MAR-8	dc - 1000	22.5	+12.5	3.3	65	500
RAM-1	dc - 1000	15.5	+1.5	5.5	40	200
RAM-2	dc - 2000	11.8	+4.5	6.5	60	325
RAM-3	dc - 2000	12.0	+10.0	6.0	80	425
RAM-4	dc - 1000	8.0	+12.5	6.5	100	540
RAM-6	dc - 2000	16.0	+2.0	2.8	50	200
RAM-7	dc - 2000	12.5	+5.5	4.5	60	275
RAM-8	dc - 1000	23.0	+12.5	3.0	65	420
MAV-1	dc - 1000	15.0	+1.5	5.5	40	200
MAV-2	dc - 1500	11.0	+4.5	6.5	60	325
MAV-3	dc - 1500	11.0	+10.0	6.0	70	400
MAV-4	dc - 1000	7.5	+11.5	7.0	85	500
MAV-11	dc - 1000	10.5	+17.5	3.6	80	550

RAM-x, case VV105; MAR-x, case BBB123; MAV-x, case AF190[†]

Avantek MMICs

Device	Freq Range (MHz)	Typical Gain (dB)	Output Level 1 dB Comp (dBm)	NF (dB)	I_{max} (mA)	P_{max} (mW)
MSA-01xx	dc - 1300	18.5	1.5	5.5	40	200
MSA-02xx	dc - 2800	12.5	4.5	6.5	60	325
MSA-03xx	dc - 2800	12.5	10	6.0	80	425
MSA-04xx	dc - 4000	8.3	11.5	7.0	85	500
MSA-05xx	dc- 2800	7.0	19.0	6.5	135	1.5
MSA-06xx	dc - 800	19.5	2.0	3.0	50	200
MSA-07xx	dc - 2500	13.0	5.5	4.5	50	175
MSA-08xx	dc - 6000	32.5	12.5	3.0	65	500
MSA-09xx	dc - 6000	7.2	10.5	6.2	65	500
MSA-11xx	50-1300	12.0	17.5	3.6	80	550

Each listing represents a series of devices in different cases. Performance varies somewhat with the case (for example, the frequency range is often 30% less for a plastic package, as compared to that with a ceramic package).[†]

Hewlett-Packard MMIC[†]

Device	Freq Range (GHz)	Typical Gain (dB)	Output Level 1 dB Comp (dBm)	NF (dB)	I_{max} (mA)
MGA-86576	1.5-8	15.4	3.8	2.1	22

Motorola Hybrid Amplifiers (50 Ω)

Device type	Freq Range (MHz)	Gain (dB) min/typ	Supply voltage (V)	Output Level, 1 dB Comp (dBm)	NF at 250 MHz (dB)
MWA110	0.1 - 400	13/14	2.9	−2.5	4
MWA120	0.1 - 400	13/14	5	+8.2	5.5
MWA130	0.1 - 400	13/14	5.5	+18	7
MWA131	0 - 400	13/14	5.5	+20	5
MWA210	0.1 - 600	9/10	1.75	+1.5	6
MWA220	0.1 - 600	9/10	3.2	+10.5	6.5
MWA230	0.1 - 600	9/10	4.4	+18.5	7.5
MWA310	0.1 - 1000	7/8	1.6	+3.5	6.5
MWA320	0.1 - 1000	7/8	2.9	+11.5	6.7
MWA330	0.1 - 1000	na/6.2	4	+15.2	9

MWAxxx case 31A-03/2[†]

[†]For package shape, size and pin-connection information, see manufacturers' data sheets. Many retail suppliers offer data sheets to buyers free of charge on request. Data books are available from many manufacturers and retailers.

General Purpose Transistors†

Listed numerically by device

Device	Type	V_{CEO} Maximum Collector Emitter Voltage (V)	V_{CBO} - Maximum Emitter Base Voltage (V)	V_{EBO} Maximum Emitter Base Voltage (V)	I_c Maximum Collector Current (mA)	P_D Maximum Device Dissipation (W)	Minimum DC Current Gain h_{FE} I_C = 0.1 mA	Minimum DC Current Gain h_{FE} I_C = 150 mA	Current-Gain Bandwidth Product f_T* (MHz)	Noise Figure NF Maximum (dB)
2N918	NPN	15	30	3.0	50	0.200	20 (3 mA)	—	600	6.0
2N2102	NPN	65	120	7.0	1000	1.0	20	40	60	6.0
2N2218	NPN	30	60	5.0	800	0.8	20	40	250	
2N2218A	NPN	40	75	6.0	800	0.8	20	40	250	
2N2219	NPN	30	60	5.0	800	3.0	35	100	250	
2N2219A	NPN	40	75	6.0	800	3.0	35	100	300	4.0
2N2222	NPN	30	60	5.0	800	1.2	35	100	250	
2N2222A	NPN	40	75	6.0	800	1.2	35	100	200	4.0
2N2905	PNP	40	60	5.0	600	0.6	35	—	200	
2N2905A	PNP	60	60	5.0	600	0.6	75	100	200	
2N2907	PNP	40	60	5.0	600	0.400	35	—	200	
2N2907A	PNP	60	60	5.0	600	0.400	75	100	200	
2N3053	NPN	40	60	5.0	700	5.0	—	50	100	
2N3053A	NPN	60	80	5.0	700	5.0	—	50	100	
2N3563	NPN	15	30	2.0	50	0.600	20	—	800	
2N3904	NPN	40	60	6.0	200	0.625	40	—	300	5.0
2N3906	PNP	40	40	5.0	200	1.5	60	—	250	4.0
2N4037	PNP	40	60	7.0	1000	5.0	—	50		
2N4123	NPN	30	40	5.0	200	0.35	—	25(50 mA)	250	6.0
2N4124	NPN	25	30	5.0	200	0.350	120 (2 mA)	60(50 mA)	300	5.0
2N4125	PNP	30	30	4.0	200	0.625	50 (2 mA)	25(50 mA)	200	5.0
2N4126	PNP	25	25	4.0	200	0.625	120 (2 mA)	60(50 mA)	250	4.0
2N4401	NPN	40	60	6.0	600	0.625	20	100	250	
2N4403	PNP	40	40	5.0	600	0.625	30	100	200	
2N5320	NPN	75	100	7.0	2000	10.0	—	30(1 A)		
2N5415	PNP	200	200	4.0	1000	10.0	—	30(50 mA)	15	
MM4003	PNP	250	250	4.0	500	1.0	20 (10 mA)	—		
MPSA55	PNP	60	60	4.0	500	0.625	—	50 (0.1 A)	50	
MPS6531	NPN	40	60	5.0	600	0.625	60 (10 mA)	90 (0.1 A)		
MPS6547	NPN	25	35	3.0	50	0.625	20 (2 mA)	—	600	

* Test conditions: I_C = 20 mA dc; VCE = 20 V; f = 100 MHz

RF Power Amplifier Modules

Listed by frequency

Device	Supply (V)	Frequency Range (MHz)	Ouput Power (W)	Power Gain (dB)	Package†	Mfgr/ Notes
M57735	17	50-54	14	21	H3C	MI; SSB mobile
M57719N	17	142-163	14	18.4	H2	MI; FM mobile
S-AV17	16	144-148	60	21.7	5-53L	T, FM mobile
S-AV7	16	144-148	28	21.4	5-53H	T, FM mobile
MHW607-1	7.5	136-150	7	38.4	301K-02/3	MO; class C
BGY35	12.5	132-156	18	20.8	SOT132B	P
M67712	17	220-225	25	20	H3B	MI; SSB mobile
M57774	17	220-225	25	20	H2	MI; FM mobile
MHW720-1	12.5	400-440	20	21	700-04/1	MO; class C
MHW720-2	12.5	440-470	20	21	700-04/1	MO; class C
M57789	17	890-915	12	33.8	H3B	MI
MHW912	12.5	880-915	12	40.8	301R-01/1	MO; class AB
MHW820-3	12.5	870-950	18	17.1	301G-03/1	MO; class C

Manufacturer codes: MO = Motorola; MI = Mitsubishi; P = Philips; T = Toshiba.

†For package shape, size and pin-connection information, see manufacturers' data sheets. Many retail suppliers offer data sheets to buyers free of charge on request. Data books are available from many manufacturers and retailers.

General Purpose Silicon Power Transistors

TO-220 case*

NPN	PNP	I_c Max (A)	V_{CEO} Max (V)	h_{FE} Min	F_T (MHz)	Power Dissipation (W)
D44C8		4	60	100/220	50	30
	D45C8	−4	−60	40/120	50	30
TIP29		1	40	15/75	3	30
	TIP30A	1	40	15/75	3	30
TIP29A		1	60	15/75	3	30
	TIP30A	1	60	15/75	3	30
TIP29B		1	80	15/75	3	30
TIP29C		1	100	15/75	3	30
	TIP30C	1	100	15/75	3	30
TIP47		1	250	30/150	10	40
TIP48		1	300	30/150	10	40
TIP49		1	350	30/150	10	40
TIP50		1	400	30/150	10	40
TIP110		2	60	500	> 5	50
	TIP115	2	−60	500	> 5	50
TIP116		2	80	500	25	50
TIP31		3	40	25	3	40
	TIP32	3	40	25	3	40
TIP31A		3	60	25	3	40
	TIP32A	3	60	25	3	40
TIP31B		3	80	25	3	40
	TIP32B	3	80	25	3	40
TIP31C		3	100	25	3	40
	TIP32C	3	100	25	3	40
2N6124		4	45	25/100	2.5	40
2N6122		4	60	25/100	2.5	40
MJE13004		4	300	6/30	4	60
TIP120		5	60	1000	> 5	65
	TIP125	5	−60	1000	> 10	65
TIP42		6	15/75	3		65
TIP41A		6	60	15/75	3	65
TIP41B		6	80	15/75	3	65
2N6290		7	50	30/150	4	40
	2N6109	7	50	30/150	4	40
2N6292		7	70	30/150	4	40
	2N6107	7	70	30/150	4	40
MJE3055T		10	60	20/70	—	75
	MJE2955T	10	60	20/70	—	57
TIP140		10	60	500	> 5	125
	TIP145	10	−60	500	> 10	125

TO-204 case (TO-3)*

NPN	PNP	I_c Max (A)	V_{CEO} Max (V)	h_{FE} Min	F_T (MHz)	Power Dissipation (W)
2N6486		15	40	20/150	5	75
2N6488		15	80	20/150	5	75
2N6545		8	400	7/35	6	125
2N3789		10	60	15	4	150
2N3715		10	60	30	4	150
	2N3791	10	60	30	4	150
2N5875		10	60	20/100	4	150
2N3790		10	80	15	4	150
2N3716		10	80	30	4	150
	2N3792	10	80	30	4	150
2N3055		15	60	20/70	2.5	115
	MJ2955	15	60	20/70	2.5	115
2N3055A		15	60	20/70	0.8	115
2N5881		15	60	20/100	4	160
2N5880		15	80	20/100	4	160
2N6249		15	200	10/50	2.5	175
2N6250		15	275	8/50	2.5	175
2N6546		15	300	6/30	6-24	175
2N6251		15	350	6/50	2.5	175
2N5630		16	120	20/80	1	200
2N3773		16	140	15/60	4	200
2N5039		20	75	20/100	60	140
2N5303		20	80	15/60	2	200
2N6284		20	100	750/18K	—	160
	2N6287	20	100	750/18K	—	160
MJ15003		20	140	25/150	2	250
	MJ15004	20	140	25/150	2	250
2N5885		25	60	—	4	200
2N5886		25	80	20/100	4	200
	2N5884	25	80	20/100	4	200
MJ15024		25	250	15/60	5	250
2N3771		30	40	—	2	150
2N5301		30	40	15/60	2	200
2N5302		30	60	15/60	2	200
	2N4399	30	60	15/60	2	200
MJ802		30	100	25/100	2	200
	MJ4502	30	100	25/100	2	200

▒ = Complimentary pairs

*For package shape, size and pin-connection information, see manufacturers' data sheets. Many retail suppliers offer data sheets to buyers free of charge on request. Data books are available from many manufacturers and retailers.

RF Power Transistors

Device	Output Power (W)	Input Power (W)	Gain (dB)	Typ Supply Voltage (V)	Case†	Mfr
1.5 to 30 MHz, HF SSB/CW						
2SC2086	0.3		13	12	TO-92	MI
BLV10	1		18	12	SOT123	PH
BLV11	2		18	12	SOT123	PH
MRF476	3	0.1	15	12.5-13.6	221A-04/1	MO
BLW87	6		18	12	SOT123	PH
2SC2166	6		13.8	12	TO-220	MI
BLW83	10		20	26	SOT123	PH
MRF475	12	1.2	10	12.5-13.6	221A-04/1	MO
MRF433	12.5	0.125	20	12.5-13.6	211-07/1	MO
2SC3133	13		14	12	TO-220	MI
MRF485	15	1.5	10	28	221A-04/1	MO
2SC1969	16		12	12	TO-220	MI
BLW50F	16		19.5	45	SOT123	PH
MRF406	20	1.25	12	12.5-13.6	221-07/1	MO
SD1285	20	0.65	15	12.5	M113	SG
MRF426	25	0.16	22	28	211-07/1	MO
MRF427	25	0.4	18	50	211-11/1	MO
MRF477	40	1.25	15	12.5-13.6	211-11/1	MO
MRF466	40	1.25	15	28	211-07/1	MO
BLW96	50		19	40	SOT121	PH
2SC3241	75		12.3	12.5	T-45E	MI
SD1405	75	3.8	13	12.5	M174	SG
2SC2097	75		12.3	13.5	T-40E	MI
MRF464	80	2.53	10	28	211-11/1	MO
MRF421	100	10	10	12.5-13.6	211-11/1	MO
SD1487	100	7.9	11	12.5	M174	SG
2SC2904	100		11.5	12.5	T-40E	MI
SD1729	130	8.2	12	28	M174	SG
MRF422	150	15	10	28	211-11/1	MO
MRF428	150	7.5	13	50	211-11/1	MO
SD1726	150	6	14	50	M174	SG
PT9790	150	4.8	15	50	211-11/1	MO
MRF448	250	15.7	12	50	211-11/1	MO
MRF430	600	60	10	50	368-02/1	MO
50 MHz						
MRF475	4	0.4	10	125-13.6	221A-04/1	MO
MRF497	40	4	10	12.5-13.6	221A-04/2	MO
SD1446	70	7	10	12.5	M113	SG
MRF492	70	5.6	11	12.5-13.6	211-11/1	MO
SD1405	100	20	7	12.5	M174	SG
VHF to 175 MHz						
2N4427	0.7		8	7.5	TO-39	PH
2N3866	1		10	28	TO-39	PH
BFQ42	1.5		8.4	7.5	TO-39	PH
2SC2056	1.6		9	7.2	T-41	MI
2N3553	2.5	0.25	10	28	79-04/1	MO
BFQ43	3		9.4	7.5	TO-39	PH
SD1012	4	0.25	12	12.5	M135	SG
2SC2627	5		13	12.5	T-40	MI
2N5641	7	1	8.4	28	144B-05/1	MO
MRF340	8	0.4	13	28	221A-04/2	MO
BLW29	9		7.4	7.5	SOT120	PH
SD1143	10	1	10	12.5	M135	SG
2SC1729	14		10	13.5	T-31E	MI
SD1014-02	15	3.5	6.3	12.5	M135	SG
BLV11	15		8	13.5	SOT123	PH
2N5642	20	3	8.2	28	145A-09/1	MO
MRF342	24	1.9	11	28	221A-04/2	MO
BLW87	25		6	13.5	SOT123	PH
2SC1946	28		6.7	13.5	T-31E	MI
MRF314	30	3	10	28	211-07/1	MO
SD1018	40	14	4.5	12.5	M135	SG
2N5643	40	6.9	7.6	28	145A-09/1	MO
BLW40	40		10	12.5	SOT120	PH
MRF315	45	5.7	9	28	211-07/1	MO
PT9733	50	10	7	28	145A-09/1	MO
MRF344	60	15	6	28	221A-04/2	MO
2SC2694	70		6.7	12.5	T-40	MI
BLV75/12	75		6.5	12.5	SOT119	PH
MRF316	80	8	10	28	316-01/1	MO
SD1477	100	25	6	12.5	M111	SG
BLW78	100		6	28	SOT121	PH
MRF317	100	12.5	9	28	316-01/1	MO
TP9386	150	15	10	28	316-01/1	MO
220 MHz						
MRF207	1	0.15	8.2	12.5	79-04/1	MO
2N5109	2.5		11 .	12	TO-205AD	MO
MRF227	3	0.13	13.5	12.5	79-05/5	MO
MRF208	10	1	10	12.5	145A-09/1	MO
MRF226	13	1.6	9	12.5	145A-09/1	MO
2SC2133	30		8.2	28	T-40E	MI
2SC2134	60		7	28	T-40E	MI
2SC2609	100		6	28	T-40E	MI
UHF to 512 MHz						
2N4427	0.4		10	12.5	TO-39	PH
2SC3019	0.5		14	12.5	T-43	MI
MRF581	0.6	0.03	13	12.5	317-01/2	MO
2SC908	1		4	12.5	TO-39	MI
2N3866	1		10	28	TO-39	PH
2SC2131	1.4		6.7	13.5	TO-39	MI
BLX65E	2		9	12.5	TO-39	PH
BLW89	2		12	28	SOT122	PH
MRF586	2.5		16.5	15	79-04	MO
MRF630	3	0.33	9.5	12.5	79-05/5	MO
2SC3020	3	0.3	10	12.5	T-31E	MI
BLW80	4		8	12.5	SOT122	PH
BLW90	4		11	12.5	SOT122	PH
MRF652	5	0.5	10	12.5	244-04/1	MO
MRF587	5		16.5	15	244A-01/1	MO
2SC3021	7	1.2	7.6	12.5	T-31E	MI
BLW81	10		6	12.5	SOT122	PH
MRF653	10	2	7	12.5	244-04/1	MO
BLW91	10		9	28	SOT122	PH
MRF654	15	2.5	7.8	12.5	244-04/1	MO
2SC3022	18	6	4.7	12.5	T-31E	MI
BLU20/12	20		6.5	12.5	SOT119	PH
BLX94A	25		6	28	SOT48/2	PH
2SC2695	28		4.9	13.5	T-31E	MI
BLU30/12	30		6	12.5	SOT119	PH
BLU45/12	45		4.8	12.5	SOT119	PH
2SC2905	45		4.8	12.5	T-40E	MI
MRF650	50	15.8	5	12.5	316-01/1	MO
TP5051	50	6	9	24	333A-02/2	MO
BLU60/12	60		4.4	12.5	SOT119	PH
2SC3102	60	20	4.8	12.5	T-41E	MI
BLU60/28	60		7	28	SOT119	PH
MRF658	65	25	4.15	12.5	316-01/1	MO
MRF338	80	15	7.3	28	333-04/1	MO
SD1464	100	28.2	5.5	28	M168	SG
UHF to 960 MHz						
MRF581	0.6	0.06	10	12.5	317-01/2	MO
MRF8372	0.75	0.11	8	12.5	751-04/1	MO
MRF557	1.5	0.23	8	12.5	317D-02/2	MO
BLV99	2		9	24	SOT172	PH
SD1420	2.1	0.27	9	24	M122	SG
MRF839	3	0.46	8	12.5	305A-01/1	MO
MRF896	3	0.3	10	24	305-01/1	MO
MRF891	5	0.63	9	24	319-06/2	MO
2SC2932	6		7.8	12.5	T-31B	MI
SD1398	6	0.6	10	24	M142	SG
2SC2933	14	3	6.7	12.5	T-31B	MI
SD1400-03	14	1.6	9.5	24	M118	SG
MRF873	15	3	7	12.5	319-06/2	MO
SD1495-03	30	6	7	24	M142	SG
SD1424	30	5.3	7.5	24	M156	SG
MRF897	30	3	10	24	395B-01/1	MO
MRF847	45	16	4.5	12.5	319-06/1	MO
BLV101A	50		8.5	26	SOT273	PH
SD1496-03	55	10	7.4	24	M142	SG
MRF898	60	12	7	24	333A-02/1	MO
MRF880	90	12.7	8.5	26	375A-01/1	MO
MRF899	150	24	8	26	375A-01/1	MO

Manufacturer codes:
MI = Mitsubishi; MO = Motorola; PH = Philips; SG = SGE/Thomson

†For package shape, size and pin-connection information, see manufacturers' data sheets. Many retail suppliers offer data sheets to buyers free of charge on request. Data books are available from many manufacturers and retailers.

Power FETs

Device	Type	VDSS min (V)	RDS(on) max (W)	ID max (A)	PD max (W)	Case†	Mfr
BS250P	P-channel	45	14	0.23	0.7	E-line	Z
IRFZ30	N-channel	50	0.050	30	75	TO-220	IR
MTP50N05E	N-channel	50	0.028	25	150	TO-220AB	M
IRFZ42	N-channel	50	0.035	50	150	TO-220	IR
2N7000	N-channel	60	5	0.20	0.4	E-line	Z
VN10LP	N-channel	60	7.5	0.27	0.625	E-line	Z
VN10KM	N-channel	60	5	0.3	1	TO-237	S
ZVN2106B	N-channel	60	2	1.2	5	TO-39	Z
IRF511	N-channel	60	0.6	2.5	20	TO-220AB	M
MTP2955E	P-channel	60	0.3	6	25	TO-220AB	M
IRF531	N-channel	60	0.180	14	75	TO-220AB	M
MTP23P06	P-channel	60	0.12	11.5	125	TO-220AB	M
IRFZ44	N-channel	60	0.028	50	150	TO-220	IR
IRF531	N-channel	80	0.160	14	79	TO-220	IR
ZVP3310A	P-channel	100	20	0.14	0.625	E-line	Z
ZVN2110B	N-channel	100	4	0.85	5	TO-39	Z
ZVP3310B	P-channel	100	20	0.3	5	TO-39	Z
IRF510	N-channel	100	0.6	2	20	TO-220AB	M
IRF520	N-channel	100	0.27	5	40	TO-220AB	M
IRF150	N-channel	100	0.055	40	150	TO-204AE	M
IRFP150	N-channel	100	0.055	40	180	TO-247	IR
ZVP1320A	P-channel	200	80	0.02	0.625	E-line	Z
ZVN0120B	N-channel	200	16	0.42	5	TO-39	Z
ZVP1320B	P-channel	200	80	0.1	5	TO-39	Z
IRF620	N-channel	200	0.800	5	40	TO-220AB	M
MTP6P20E	P-channel	200	1	3	75	TO-220AB	M
IRF220	N-channel	200	0.400	8	75	TO-220AB	M
IRF640	N-channel	200	0.18	10	125	TO-220AB	M

Manufacturers: IR = International Rectifier; M = Motorola; S = Siliconix; Z = Zetex.

†For package shape, size and pin-connection information, see manufacturers' data sheets. Many retail suppliers offer data sheets to buyers free of charge on request. Data books are available from many manufacturers and retailers.

Logic IC Families

Type	Propagation Delay for C_L = 50 pF (ns) Typ	Max	Max Clock Frequency (MHz)	Power Dissipation ($CL = 0$) @ 1 MHz (mW/gate)	Output Current @ 0.5 V max (mA)	Input Current (Max mA)	Threshold Voltage (V)	Supply Voltage (V) Min	Typ	Max
CMOS										
74AC	3	5.1	125	0.5	24	0	V+/2	2	5 or 3.3	6
74ACT	3	5.1	125	0.5	24	0	1.4	4.5	5	5.5
74HC	9	18	30	0.5	8	0	V+/2	2	5	6
74HCT	9	18	30	0.5	8	0	1.4	4.5	5	5.5
4000B/74C (10 V)	30	60	5	1.2	1.3	0	V+/2	3	5 - 15	18
4000B/74C (5V)	50	90	2	3.3	0.5	0	V+/2	3	5 - 15	18
TTL										
74AS	2	4.5	105	8	20	0.5	1.5	4.5	5	5.5
74F	3.5	5	100	5.4	20	0.6	1.6	4.75	5	5.25
74ALS	4	11	34	1.3	8	0.1	1.4	4.5	5	5.5
74LS	10	15	25	2	8	0.4	1.1	4.75	5	5.25
ECL										
ECL III	1.0	1.5	500	60	—	—	−1.3	−5.19	−5.2	−5.21
ECL 100K	0.75	1.0	350	40	—	—	−1.32	−4.2	−4.5	−5.2
ECL100KH	1.0	1.5	250	25	—	—	−1.29	−4.9	−5.2	−5.5
ECL 10K	2.0	2.9	125	25	—	—	−1.3	−5.19	−5.2	−5.21
GaAs										
10G	0.3	0.32	2700	125	—	—	−1.3	−3.3	−3.4	−3.5
10G	0.3	0.32	2700	125	—	—	−1.3	−5.1	−5.2	−5.5

Source: Horowitz (W1HFA) and Hill, *The Art of Electronics—2nd edition,* page 570. © Cambridge University Press 1980, 1989. Reprinted with the permission of Cambridge University Press.

Three-Terminal Voltage Regulators

Listed numerically by device

Device	Description	Package	Voltage	Current (Amps)
317	Adj Pos	TO-205	+1.2 to +37	0.5
317	Adj Pos	TO-204,TO-220	+1.2 to +37	1.5
317L	Low Current Adj Pos	TO-205,TO-92	+1.2 to +37	0.1
317M	Med Current Adj Pos	TO-220	+1.2 to +37	0.5
338	Adj Pos	TO-3	+1.2 to +32	5.0
350	High Current Adj Pos	TO-204,TO-220	+1.2 to +33	3.0
337	Adj Neg	TO-205	-1.2 to -37	0.5
337	Adj Neg	TO-204,TO-220	-1.2 to -37	1.5
337M	Med Current Adj Neg	TO-220	-1.2 to -37	0.5
309		TO-205	+5	0.2
309		TO-204	+5	1.0
323		TO-204,TO-220	+5	3.0
140-XX	Fixed Pos	TO-204,TO-220	Note 1	1.0
340-XX		TO-204,TO-220		1.0
78XX		TO-204,TO-220		1.0
78LXX		TO-205,TO-92		0.1
78MXX		TO-220		0.5
78TXX		TO-204		3.0
79XX	Fixed Neg	TO-204,TO-220	Note 1	1.0
79LXX		TO-205,TO-92		0.1
79MXX		TO-220		0.5

Note 1—XX indicates the regulated voltage; this value may be anywhere from 1.2 V to 35 V. A 7815 is a positive 15-V regulator, and a 7924 is a negative 24-V regulator.

The regulator package may be denoted by an additional suffix, according to the following:

Package	Suffix
TO-204 (TO-3)	K
TO-220	T
TO-205 (TO-39)	H, G
TO-92	P, Z

For example, a 7812K is a positive 12-V regulator in a TO-204 package. An LM340T-5 is a positive 5-V regulator in a TO-220 package. In addition, different manufacturers use different prefixes. An LM7805 is equivalent to a µA7805 or MC7805.

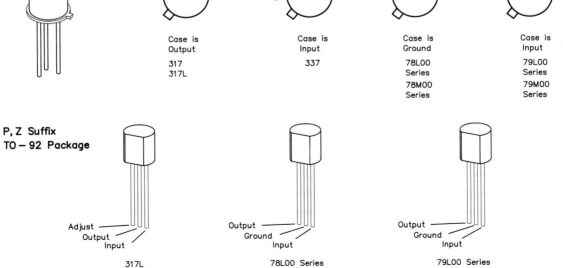

K Suffix
Metal TO — 204 Package

BOTTOM VIEW

Pins 1 and 2 Electrically
Isolated from Case.
Case is Third
Electrical Connection.

Adj V_{in}

V_{out}

Case is
Output

317
350

Adj V_{out}

V_{in}

Case is
Input

337

In   Out

Gnd

Case is
Ground

140 k-XX
340 k-XX
309
7800 Series
78T00 Series

Gnd   In

Out

Case is
Input

7900 Series

T Suffix
TO — 220 Package

Center Lead is Connected to the Heat Sink

Output   Input   Ground   Input

Adjust   Adjust   Input   Ground
  Output   Input   Ground   Input
    Input   Output   Output   Output

317   337   7800 Series   7900 Series
350   337M   78T00 Series   79M00 Series
                       78M00 Series
                       140T — XX
                       340T — XX

H, G Suffix
TO — 205 Package

BOTTOM VIEW

Adj   Out   Out   Out

In   Out   Adj   In   In   Gnd   Gnd   In

Case is   Case is   Case is   Case is
Output   Input   Ground   Input

317   337   78L00   79L00
317L   Series   Series
              78M00   79M00
              Series   Series

P, Z Suffix
TO — 92 Package

Adjust   Output   Output
  Output   Ground   Ground
    Input   Input   Input

317L   78L00 Series   79L00 Series

Op Amp ICs

Listed by device number

Device	Type	Freq Comp	Max Supply* (V)	Min Input Resistance (MΩ)	Max Offset Voltage (mV)	Min dc Open-Loop Gain (dB)	Min Output Current (mA)	Min Small-Signal Bandwidth (MHz)	Min Slew Rate (V/µs)	Notes
101A	Bipolar	ext	44	1.5	3.0	79	15	1.0	0.5	General purpose
108	Bipolar	ext	40	30	2.0	100	5	1.0		
124	Bipolar	int	32		5.0	100	5	1.0		Quad op amp, low power
148	Bipolar	int	44	0.8	5.0	90	10	1.0	0.5	Quad 741
158	Bipolar	int	32		5.0	100	5	1.0		Dual op amp, low power
301	Bipolar	ext	36	0.5	7.5	88	5	1.0	10	Bandwidth extendable with external components
324	Bipolar	int	32		7.0	100	10	1.0		Quad op amp, single supply
347	BiFET	ext	36	10^6	5.0	100	30	4	13	Quad, high speed
351	BiFET	ext	36	10^6	5.0	100	20	4	13	
353	BiFET	ext	36	10^6	5.0	100	15	4	13	
355	BiFET	ext	44	10^6	10.0	100	25	2.5	5	
355B	BiFET	ext	44	10^6	5.0	100	25	2.5	5	
356A	BiFET	ext	36	10^6	2.0	100	25	4.5	12	
356B	BiFET	ext	44	10^6	5.0	100	25	5.0	12	
357	BiFET	ext	36	10^6	10.0	100	25	20.0	50	
357B	BiFET	ext	36	10^6	5.0	100	25	20.0	30	
358	Bipolar	int	32		7.0	100	10	1.0		Dual op amp, single supply
411	BiFET	ext	36	10^6	2.0	100	20	4.0	15	Low offset, low drift
709	Bipolar	ext	36	0.05	7.5	84	5	0.3	0.15	
741	Bipolar	int	36	0.3	6.0	88	5	0.4	0.2	
741S	Bipolar	int	36	0.3	6.0	86	5	1.0	3	Improved 741 for AF
1436	Bipolar	int	68	10	5.0	100	17	1.0	2.0	High-voltage
1437	Bipolar	ext	36	0.050	7.5	90		1.0	0.25	Matched, dual 1709
1439	Bipolar	ext	36	0.100	7.5	100		1.0	34	
1456	Bipolar	int	44	3.0	10.0	100	9.0	1.0	2.5	Dual 1741
1458	Bipolar	int	36	0.3	6.0	100	20.0	0.5	3.0	
1458S	Bipolar	int	36	0.3	6.0	86	5.0	0.5	3.0	Improved 1458 for AF
1709	Bipolar	ext	36	0.040	6.0	80	10.0	1.0		
1741	Bipolar	int	36	0.3	5.0	100	20.0	1.0	0.5	
1747	Bipolar	int	44	0.3	5.0	100	25.0	1.0	0.5	Dual 1741

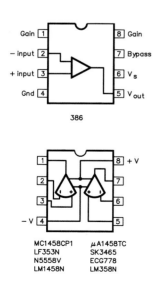

386

MC1458CP1 µA1458TC
LF353N   SK3465
N5558V   ECG778
LM1458N   LM358N

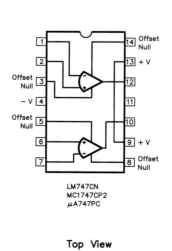

LM747CN
MC1747CP2
µA747PC

Top View

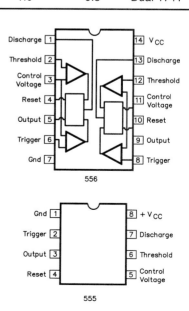

556

555

Device	Type	Freq Comp	Max Supply* (V)	Min Input Resistance (MΩ)	Max Offset Voltage (mV)	Min dc Open-Loop Gain (dB)	Min Output Current (mA)	Min Small-Signal Bandwidth (MHz)	Min Slew Rate (V/µs)	Notes
1748	Bipolar	ext	44	0.3	6.0	100	25.0	1.0	0.8	Noncompensated 1741
1776	Bipolar	int	36	50	5.0	110	5.0		0.35	Micro power, programmable
3140	BiFET	int	36	1.5×10^6	2.0	86	1	3.7	9	Strobable output
3403	Bipolar	int	36	0.3	10.0	80		1.0	0.6	Quad, low power
3405	Bipolar	ext	36		10.0	86	10	1.0	0.6	Dual op amp and dual comparator
3458	Bipolar	int	36	0.3	10.0	86	10	1.0	0.6	Dual, low power
3476	Bipolar	int	36	5.0	6.0	92	12		0.8	
3900	Bipbolar	int	32	1.0		65	0.5	4.0	0.5	Quad, Norton single supply
4558	Bipolar	int	44	0.3	5.0	88	10	2.5	1.0	Dual, wideband
4741	Bipolar	int	44	0.3	5.0	94	20	1.0	0.5	Quad 1741
5534	Bipolar	int	44	0.030	5.0	100	38	10.0	13	Low noise, can swing 20V P-P across 600
5556	Bipolar	int	36	1.0	12.0	88	5.0	0.5	1	Equivalent to 1456
5558	Bipolar	int	36	0.15	10.0	84	4.0	0.5	0.3	Dual, equivalent to 1458
34001	BiFET	int	44	10^6	2.0	94		4.0	13	JFET input
AD745	BiFET	int	±18	10^4	0.5	63	20	20	12.5	Ultra-low noise, high speed

LT1001   Precision op amp, low offset voltage (15 µV max), low drift (0.6 µV/°C max), low noise (0.3 µVp-p)
LT1007   Extremely low noise (0.06 µVp-p), very high gain (20 x 10^6 into 2 kΩ load)
LT1360   High speed, very high slew rate (800 V/µs), 50 MHz gain bandwidth, ±2.5 V to ±15 V supply range

Device	Type	Freq Comp	Max Supply* (V)	Min Input Resistance (MΩ)	Max Offset Voltage (mV)	Min dc Open-Loop Gain (dB)	Min Output Current (mA)	Min Small-Signal Bandwidth (MHz)	Min Slew Rate (V/µs)	Notes
NE5514	Bipolar	int	±16	100	1		10	3	0.6	
NE5532	Bipolar	int	±20	0.03	4	47	10	10	9	Low noise
OP-27A	Bipolar	ext	44	1.5	0.025	115		5.0	1.7	Ultra-low noise, high speed
OP-37A	Bipolar	ext	44	1.5	0.025	115		45.0	11.0	
TL-071	BiFET	int	36	10^6	6.0	91		4.0	13.0	Low noise
TL-081	BiFET	int	36	10^6	6.0	88		4.0	8.0	
TL-082	BiFET	int	36	10^6	15.0	99		4.0	8.0	Low noise
TL-084	BiFET	int	36	10^6	15.0	88		4.0	8.0	Quad, high-performance AF
TLC27M2	CMOS	int	18	10^6	10	44		0.6	0.6	Low noise
TLC27M4	CMOS	int	18	10^6	10	44		0.6	0.6	Low noise

*From –V to +V terminals

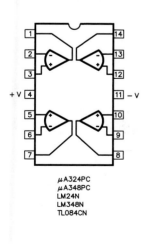

μA324PC
μA348PC
LM24N
LM348N
TL084CN

Top View

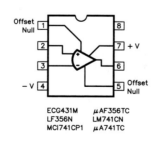

ECG431M   μAF356TC
LF356N   LM741CN
MCI741CP1   μA741TC

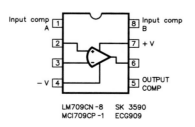

LM709CN-8   SK 3590
MCI709CP-1   ECG909

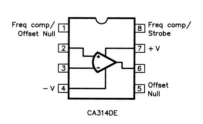

CA314DE

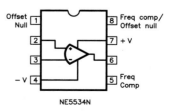

NE5534N

Triode Transmitting Tubes

Type	Maximum Ratings				Freq (MHz)	Ampl Factor	Cathode		Capacitances (pF)			Base Diagram	Service Class[1]	Typical Operation						
	Power Diss. (W)	Plate (V)	Plate (mA)	Grid dc (mA)			(V)	(I)	C_{IN}	C_P	C_{OUT}			Plate (V)	Grid (V)	Plate (mA)	Grid dc (mA)	Input (W)	P to P Load (kΩ)	Output (W)
5675	5	165	30	8	3000	20	6.3	0.135	2.3	1.3	0.09	Fig 21	GG0	120	-8	25	4	-	-	0.05
2C40	6.5	500	25	-	500	36	6.3	0.75	2.1	1.3	0.05	Fig 11	CTO	250	-5	20	0.3	-	-	0.075
5893	8.0	400	40	13	1000	27	6.0	0.33	2.5	1.75	0.07	Fig 21	CT	350	-33	35	13	2.4	-	6.5
													CP	300	-45	30	12	2.0	-	6.5
2C43	12	500	40	-	1250	48	6.3	0.9	2.9	1.7	0.05	Fig 11	CTO	470	-	38[7]	-	-	-	9[2]
811-A	65	1500	175	50	60	160	6.3	4.0	5.9	5.6	0.7	3G	CT	1500	-70	173	40	7.1	-	200
													CP	1250	-120	140	45	10.0	-	135
													B/GG	1250	0	27/175	28	12	-	165
													AB1	1250	0	27/175	13	3.0	-	155
812-A	65	1500	175	35	60	29	6.3	4.0	5.4	5.5	0.77	3G	CT	1500	-120	173	30	6.5	-	190
													CP	1250	-115	140	35	7.6	-	130
													B[2]	1500	-48	283/10	270[4]	5.0	13.2	340
3CX100A5[6]	100 / 70	1000 / 600	125[5] / 100[5]	50	2500	100	6.0	1.05	7.0	2.15	0.035	-	AGG	800	-20	80	30	6	-	27
													CP	600	-15	75	40	6	-	18
2C39	100	1000	60	40	500	100	6.3	1.1	6.5	1.95	0.03	-	G1C	600	-35	60	40	5.0	-	20
													CTO	900	-40	90	30	-	-	40
													CP	600	-150	100[5]	50	-	-	
AX9900 / 5866[6]	135	2500	200	40	150	25	6.3	5.4	5.8	5.5	0.1	Fig 3	CT	2500	-200	200	40	16	-	390
													CP	2000	-225	127	40	16	-	204
													B[2]	2500	-90	80/330	350[4]	14[3]	15.68	560
572B / T160L	160	2750	275	-	-	170	6.3	4.0	-	-	-	3G	CT	1650	-70	165	32	6	-	205
8873	200	2200	250	-	500	160	6.3	3.2	19.5	7.0	0.03	Fig 87	B/CG[2]	2400	-2.0	90/500	-	100	-	600
													AB2	2000	-	22/500	98[3]	27[3]	-	505
8875	300	2200	250	-	500	160	6.3	3.2	19.5	7.0	0.03	-	AB2	2000	-	22/500	98[3]	27[3]	-	505
833A	350 / 450[6]	3300 / 4000[6]	500	100	30 / 20[6]	35	10	10	12.3	6.3	8.5	Fig 41	CTO	2250	-125	445	85	23	-	780
													CP	3000	-160	335	70	20	-	800
													CP	2500	-300	335	75	30	-	635
													CP	3000	-240	335	70	26	-	800
8874	400	2200	350	-	500	160	6.3	3.2	19.5	7.0	0.03	-	B2	3000	-70	100/750	400[4]	20[4]	9.5	1650
													AB2	2000	-	22/500	98[3]	27[3]	-	505
3-400Z	400	3000	400	-	110	200	5	14.5	7.4	4.1	0.07	Fig 3	B/GG	3000	0	100/333	120	32	-	655
3-500Z	500	4000	400	-	110	160	5	14.5	7.4	4.1	0.07	Fig 3	B/GG	3000	0	370	115	30	5	750
3-600Z	600	4000	425	-	110	165	5	15.0	7.8	4.6	0.08	Fig 3	CT	3500	-75	300	115	22	-	850
3CX800A7	800	2250	600	60	350	200	13.5	1.5	26	-	6.1	Fig 87	B/GG	2200	-	400	118	33	-	810
													AB2.GG[7]	2200	-8.2	500	110	35	-	950
3-1000Z	1000	3000	800	-	110	200	7.5	21.3	17	6.9	0.12	Fig 3	B/GG	3000	0	180/670	36	65	-	1360
3CX1200A7	1200	5000	800	-	110	200	7.5	21.0	20	12	0.2	Fig 3	AB2.GG	3600	-10	700	300	85	-	1500
8877	1500	4000	1000	-	250	200	5.0	10	42	10	0.1	-	AB2	2500	-8.2	1000	230	57	-	1520

[1]KEY TO CLASS-OF-SERVICE ABBREVIATIONS

A₁ = Class-A₁ AF modulator.
AB₁ = Class-AB₁ push-pull AF modulator.
AB₂ = Class-AB₂ push-pull AF modulator.
B = Class-B push-pull AF modulator.
CM = Frequency multiplier.
CP = Class-C plate-modulated telephone.

CT = Class-C CW.
CTO = Class-C amplifier-oscillator.
AB₂-GG = Grounded-grid class AB₂ amplifier.
BGG = Grounded-grid class B amp. (single tone)
GGO = Grounded-grid oscillator.
GIC = Grid-isolation circuit.
GMA = Grid-modulated amplifier.

[2]Values are for two tubes in push-pull.
[3]Maximum signal value.
[4]Peak AF grid-to-grid volts.
[5]Maximum cathode current in mA.
[6]Forced-air cooling required.
[7]Key-down CW.

Tetrode and Pentode Transmitting Tubes

Type	Plate Diss (W)	Plate (V)	Screen Diss (W)	Screen (V)	Freq (MHz)	Cathode (V)	Cathode (A)	C_{IN}	C_{CP}	C_{OUT}	Base	Serv Class[1]	Plate (V)	Screen (V)	Supp (V)	Grid (V)	Plate (mA)	Screen (mA)	Grid (mA)	P_{in} (W)	P-to-P Load (kΩ)	P_{out} (W)
6146	25	750	3	250	60	6.3	1.25	13	0.24	8.5	7CK	CT	500	170	—	-66	135	9	2.5	0.2	—	48
6146A												CT	750	160	—	-62	120	11	3.1	0.2	—	70
8032	25	750	3	250	60	12.6	0.585	13	0.24	8.5	7CK	CT[6]	400	190	—	-54	150	10.4	2.2	3.0	—	35
6883												CP	400	150	—	-87	112	7.8	3.4	0.4	—	32
												CP	600	150	—	-87	112	7.8	3.4	0.4	—	52
6159B	25	750	3	250	60	26.5	0.3	13	0.24	8.5	7CK	AB2[8]	600	190	—	-48	28/270	1.2/20	2[2]	0.3	5	113
													750	165	—	-46	22/240	0.3/20	2.6[2]	0.4	7.4	131
												AB1[8]	750	195	—	-50	23/220	1/26	100[3]	0	8	120
807 807W 5933	30	750	3.5	300	60	6.3	0.9	12	0.2	7	5AW	CT	750	250	—	-45	100	6	3.5	0.22	—	50
												CP	600	275	—	-90	100	6.5	4	0.4	—	42.5
												AB1	750	300	—	-35	15/70	3/8	75[3]	0	—	72
1625	30	750	3.5	300	60	12.6	0.45	12	0.2	7	5AZ	B[5]	750	—	—	0	15/240	—	555[3]	5.3[2]	6.65	120
6146B 8298A	35	750	3	250	60	6.3	1.125	13	0.22	8.5	7CK	CT	750	200	—	-77	160	10	2.7	0.3	—	85
												CP	600	175	—	-92	140	9.5	3.4	0.5	—	62
												AB1	750	200	—	-48	25/125	6.3	—	—	3.5	61
813[7]	125	2500	20	800	30	10	5	16.3	0.25	14	5BA	CTO	1250	300	0	-75	180	35	12	1.7	—	170
												AB1	2250	400	0	-155	220	40	15	4	—	375
												AB1	2500	750	0	-95	25/145	27[2]	0	0	16	245
												AB2[2]	2000	750	0	-90	40/315	1.5/58	230[3]	0.1[2]	16	455
													2500	750	0	-95	35/360	1.2/55	235[3]	0.35[2]	17	650
4-400A	400[4]	4000	35	600	110	5	14.5	12.5	0.12	4.7	5BK	CT/CP	4000	300	—	-170	270	22.5	10	10	4.0	720
												GG	2500	0	—	0	80/270[9]	55[9]	100[9]	38[9]	—	325
												AB1	2500	750	—	-130	95/317	0/14	0	0	—	425
4CX400A	400	2500	8	400	500	6.3	3.2	24	.08	7	See Note 11	AB2 GD	2200	325	—	-30	100/270	22	2	9	—	405
												AB2 GD	2500	400	—	-35	100/400	18	1	13	—	610

Type	Plate Diss (W)	Plate (V)	Screen Diss (W)	Screen (V)	Freq (MHz)	Cathode (V)	Cathode (A)	C_{IN}	C_{CP}	C_{OUT}	Base	Serv Class[1]	Plate (V)	Screen (V)	Supp (V)	Grid (V)	Plate (mA)	Screen (mA)	Grid (mA)	P_{in} (W)	P-to-P Load (kΩ)	P_{out} (W)
4CX800A	800	2500	15	350	150	12.6	3.6	51	.9	11	See Note 12	AB2 GD	2200	350	—	-56	160/550	24	1	32	—	750
												AB2 GD	2200	300	—	-57	100/590	20	2	41	—	750
8166 4-1000A	1000	6000	75	1000	—	7.5	21	27.2	0.24	7.6	—	CT	3000	500	—	-150	700	146	38	11	—	1430
												CP	3000	500	—	-200	600	145	36	12	—	1390
												AB2	4000	500	—	-60	300/1200	0/95	11	11	7	3000
												GG	3000	0	—	0	100/700[9]	105[9]	170[9]	130[9]	2.5	1475
4CX1000A	1000	3000	12	400	400	6	12.5	35	0.15	12	—	AB1[8]	2000	325	—	-55	500/2000	-4/60	—	—	2.8	2160
													2500	325	—	-55	500/2000	-4/60	—	—	3.1	2920
4CX1600B	1600	3300	20	350	250	12.6	4.4	86	0.15	12	See Note 13	AB2 GD	3000	325	—	-55	500/1800	-4/60	—	—	3.85	3360
													2400	350	—	-53	500/1100	20	2	28	—	1600
													2400	350	—	-70	200/870	48	2	83[10]	—	1500
												AB2	3200	240	—	-57	200/740	21	1	33	—	1600

[1] SERVICE CLASS ABBREVIATIONS:
AB2GD = AB2 linear with 50-Ω passive grid driver circuit.
B = Class-B push-pull at modulator.
CM = Frequency multiplier.
CP = Class-C plate-modulated phone.
CT = Class-C telegraph.
CTO = Class-C amplifier-oscillator.
GG = Grounded-grid (grid and screen connected together).

[2] Maximum signal value.
[3] Peak grid-to-grid volts.
[4] Forced-air cooling required.
[5] Two tubes triode connected, G2 to G1 through 20kΩ Input to G2.
[6] Typical operation at 175 MHz.
[7] ±1.5 V.
[8] Values are for two tubes.
[9] Single tone.
[10] 24-Ω cathode resistance.
[11] Base same as 4CX250B. Socket is Russian SK2A.
[12] Socket is Russian SK1A.
[13] Socket is Russian SK3A.

TV Deflection Tubes

Type	Plate Dissipation Watts	Screen Dissipation Watts	Transconductance Micromhos	Heater (6.3V) Amperes	Cin pF	Cgp pF	Cout pF	Base	Class of Service	Plate Voltage	Screen Voltage	Grid Voltage	Plate Current mA	Screen Current mA	Grid Current mA	Approx. Driving Power Watts	Approx. Output Power Watts
						Capacitances			RF Operation (Up to 30 MHz)								
6DQ5	24	3.2	10.5k	2.5	23	0.5	11	8JC									
6DQ6B	18	3.6	7.3k	1.2	15	0.5	7	6AM	C	400	200	−40	100	12	1.5	0.1	25
6FH6	17	3.6	6k	1.2	33	0.4	8	6AM									
6GC6	17.5	4.5	6.6k	1.2	15	0.55	7	8JX									
6GJ5	17.5	3.5	7.1k	1.2	15	0.26	6.5	9NM	C	500	200	−75	180	15	5	0.43	63
									AB₁	500	200	−43	85	4			35
6HF5	28	5.5	11.3k	2.25	24	0.56	10	12FB	C	500	140	−85	232	12.5	8	0.76	77
									AB₁	500	140	−46	133	4.5			58
6JB6	17.5	3.5	7.1k	1.2	15	0.2	6	9QL	C	500	200	−75	180	13.3	5	0.43	63
									AB₁	500	200	−42	85	4.2			35
6JE6C	30	5	10.5k	2.5	24.3		14.5	9QL	C	500	125	−85	222	17	8	0.82	76
									AB₁	500	125	−44	110	3.9			47
6JG6A	17	3.5	10k	1.6	22	0.7	9	9QU	C	450	150	−80	202	20	8	0.75	63
									AB₁	450	150	−35	98	4.5			38
6JM6	17.5	3.5	7.3k	1.2	16	0.6	7	12FJ	C	500	200	−75	190	13.7	4	0.32	61
									AB₁	500	200	−42	85	4.4			37
6JN6	17.5	3.5	7.3k	1.2	16	0.34	7	12FK									
6JS6C	30	5.5		2.25	24	0.7	10	12FY									
6KD6	33	5	14k	2.85	40	0.8	16	12GW	GG	800	0	−11	150			12.5	82
6LB6	30	5	13.4k	2.25	33	0.4	18	12GJ									
6LG6	28	5	11.5k	2	25	0.8	13	12HL									
6LQ6	30	5	9.6k	2.5	22	0.46	11	9QL									
6MH6	38.5	7	14k	2.65	40	1.0	20	12GW									

Note: For AB₁ operation, input data is average
2-tone value. Output power is PEP.

EIA Vacuum-Tube Base Diagrams

Base diagrams correspond to the codes in "Base" columns of the tube-data tables. Bottom views are shown throughout.
Base connections are abbreviated as follows:

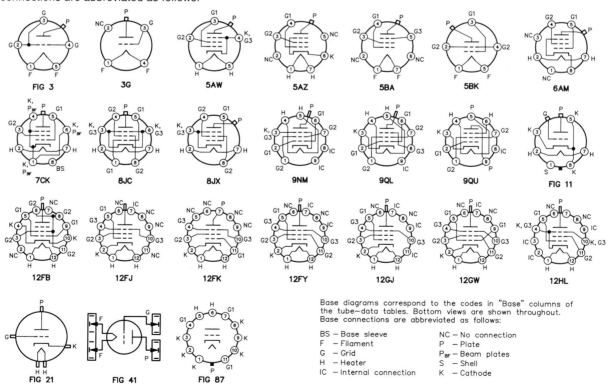

FIG 3 3G 5AW 5AZ 5BA 5BK 6AM

7CK 8JC 8JX 9NM 9QL 9QU FIG 11

12FB 12FJ 12FK 12FY 12GJ 12GW 12HL

FIG 21 FIG 41 FIG 87

Base diagrams correspond to the codes in "Base" columns of
the tube—data tables. Bottom views are shown throughout.
Base connections are abbreviated as follows:

BS — Base sleeve NC — No connection
F — Filament P — Plate
G — Grid P_BF — Beam plates
H — Heater S — Shell
IC — Internal connection K — Cathode

Alphabetical subscripts (D = diode, P = pentode, T = triode and HX = hexode) indicate structures in multistructure tubes. Subscript CT indicates filament or heater center tap.

Generally, when pin 1 of a metal-envelope tube (except all triodes) is shown connected to the envelope, pin 1 of a glass-envelope counterpart (suffix G or GT) is connected to an internal shield.

Properties of Common Thermoplastics

Polyvinyl Chloride (PVC)

Advantages:
—can be compounded with plasticizers, filters, stabilizers, lubricants and impact modifiers to produce a wide range of physical properties
—can be pigmented to almost any color
—Rigid PVC has good corrosion and stain resistance, thermal & electrical insulation, and weatherability

Disadvantages:
—base resin can be attacked by aromatic solvents, ketones, aldehydes, naphthalenes, and some chloride, acetate, and acrylate esters
—should not be used above 140°

Applications:
—conduit
—conduit boxes
—electrical fittings
—housings
—pipe
—wire and cable insulation

Polystyrene

Advantages:
—low cost
—moderate strength
—electrical properties only slightly affected by temperature and humidity
—sparkling clarity
—impact strength is increased by blending with rubbers, such as polybutadiene

Disadvantages:
—brittle
—low heat resistance

Applications:
—capacitors
—light shields
—knobs

Polyphenylene Sulfide (PPS)

Advantages:
—excellent dimensional stability
—strong
—high-temperature stability
—chemical resistant
—Inherently completely flame retardant
—completely transparent to microwave radiation.

Applications:
R3-R5 have various glass-fiber levels that are suitable for applications demanding high mechanical and impact strength as well as good dielectric properties.
R8 and R10 are suitable for high arc-resistance applications
R9-901 is suitable for encapsulation of electronic devices

Polypropylene

Advantages:
—low density
—good balance of thermal, chemical, and electrical properties
—moderate strength (increases significantly with glass-fiber reinforcement)

Disadvantages:
—Electrical properties affected to varying degrees by temperature (as temperature goes up, dielectric strength increases and volume resistivity decreases.)
—Inherently unstable in presence of oxidative and UV radiation

Applications:
—Automotive battery cases
—blower housings
—fan blades
—fuse housings
—insulators
—lamp housings
—supports for current-carrying electrical components.
—TV yokes

Polyethylene (PE)

Advantages: Low Density PE
—Good toughness
—excellent chemical resistance
—excellent electrical properties
—low coefficient of friction
—near zero moisture absorption
—easy to process
—relatively low heat resistance

Disadvantage
—susceptible to environmental and some chemical stress cracking
—wetting agents (such as detergents) accelerate stress cracking

Advantages: High Density PE
—Same as above, plus increased rigidity and tensile strength

Advantages: Ultra-High Molecular Weight PE
—outstanding abrasion resistance
—low coefficient of friction
—high impact strength
—excellent chemical resistance
—material does not break in impact strength tests using standard notched specimens

Applications:
—bearings
—components requiring maximum abrasion resistance, impact strength, and low coefficent of friction

Phenolic

Advantages:
—low cost
—superior heat resistance
—high heat-deflection temperatures
—good electrical properties
—good flame resistance
—excellent moldability
—excellent dimensional stability
—good water and chemical resistance

Applications:
—commutators and housings for small motors
—heavy duty electrical components
—rotary-switch wafers
—insulating spacers

Nylon

Advantages
—excellent fatigue resistance
—low coefficient of friction
—toughness a function of degree of crystallinity
—resists many fuels and chemicals
—good creep- and cold-flow resistance as compared to less rigid thermoplastics
—resists repeated impacts

Disadvantages:
—all nylons absorb moisture
—nylons that have not been compounded with a UV stabilizer are sensitive to UV light, and thus not suitable for extended outdoor use

Applications
—bearings
—housing and tubing
—rope
—wire coatings
—wire connectors
—wear plates

Properties of Common Thermoplastics

ASTM or UL test	Property	NYLONS (DRY, AS MOLDED) Type					PHENOLICS Type of compound						POLYETHYLENE			
		6/6	6	6/12	11	Castable	General purpose	impact	Non-bleeding	Electrical	Heat resistant	Special purpose*	Low density	Medium density	High density	Ultrahigh molecular weight
	PHYSICAL															
D792	Specific gravity	1.14	1.13	1.06	1.04	1.15-1.17	1.35-1.46	1.36-1.41	1.37-1.38	1.36-1.75	1.41-1.84	1.37-1.75	0.910-0.925	0.926-0.940	0.941-0.965	0.928-0.941
D792	Specific volume (in³/lb)	24.2	24.5	25.9	26.6	23.8							30.4-29.9	29.9-29.4	29.4-28.7	29.4
D570	Water absorption, 24 h. ⅛-in. thk (%)	1.2	1.6	0.25	0.4	0.9	0.6-0.7	0.6-0.9	0.8-0.9	0.05-0.20	0.30-0.35	0.20-0.40	<0.01	<0.01	<0.01	<0.01
	MECHANICAL															
D638	Tensile strength (psi)	12,000	11,800	8,800	8,500	11,000-14,000	6,500-7,000	6,000-7,000	6,000-7,000	5,000-7,000	5,000-6,000	7,000-9,000	600-2,300	1,200-3,500	3,100-5,500	4,000-6,000
D638	Elongation (%)	60	200	150	120	10-50	11-13	12	10	17-25	14	10	90-800	50-600	20-1,000	200-500
D638	Tensile modulus (10⁵ psi)	4.2	3.8	2.9	1.8	3.5-4.5							0.14-0.38	0.25-0.55	0.6-1.8	0.20-1.10
D785	Hardness, Rockwell ( )	121 (R)	119 (R)	114 (R)	—	112-120 (R)	70-95 (E)	82 (E)	82 (E)	75-88 (E)	94 (E)	76 (E)	10 (R)	15 (R)	65 (R)	55 (R)
D790	Flexural modulus (10⁵ psi)	4.1	3.9	2.9	1.5	—	11-14	12-25	10-12	12-25	11-23	10-19	0.08-0.60	0.60-1.15	1.0-2.0	1.0-1.7
D256	Impact strength, Izod (ft-lb/in of notch)	1.0	0.8	1.0	3.3	0.9	0.30-0.35	0.6-1.05	0.28	0.28-0.45	0.26	0.50	No break	0.5-16	0.5-20	No break
	THERMAL															
C177	Thermal conductivity (Btu-in/hr-ft²-°F)	1.7	1.7	1.5	—	1.7	7.1†	7.9†	—	16.0†	—	8.8†	8.0†	8.0-10.0†	11.0-12.4†	11.0†
D696	Coef of thermal expansion (10⁻⁵ in./in.-°F)	4.0	4.5	5.0	5.1	5.0	3.95	3.56	4.40	2.60	2.80	3.60	5.6-12.2	7.8-8.9	6.1-7.2	7.8
D648	Deflection temperature (°F) At 264 psi	194	152	194	118	300-425	275-360	270-500	370	310-400	330-380	360-430	90-105	105-120	110-130	118
	At 66 psi	455	365	356	154	400-425							100-121	120-165	140-190	170
UL 94	Flammability rating	V-2	V-2	V-2	—		V-1	HB	—	V-0	V-0	HB				
	ELECTRICAL															
D149	Dielectric strength (V/mil) Short time, ⅛ in. thk	600	400	400	425	500-600*	350	350-400	200	400	170	175	460-700	460-650	450-500	900*
D150	Dielectric constant At 1 kHz	3.9	3.7	4.0	3.3	3.7	5.2-5.3	5.2-5.4	—	4.9-6.5	11.7	7.8	2.25-2.35	2.30-2.35	2.30-2.35	—
D150	Dissipation factor At 1 kHz	0.02	0.02	0.02	0.03	0.02	0.04-0.05	0.04-0.06	—	0.025-0.10	0.15	0.12	0.0002	0.0002	0.0003	0.0002
D257	Volume resistivity (ohm-cm) At 73°F, 50% RH	10¹⁵	10¹⁵	10¹⁵	2×10¹³	—	10¹¹-10¹²	10¹¹-10¹²	10¹²	10¹¹-10¹³	10¹²	10¹¹	10¹⁵	10¹⁵	10¹⁵	10¹⁸
D495	Arc resistance (s)	116	—	121	—	—	100	50	—	184	181	—	135-160	200-235	—	—
	OPTICAL															
D542	Refractive Index												1.51	1.52	1.54	—
D1003	Transmittance (%)												4-50	4-50	10-50	—

*kV/cm.
*Chemical-resistant compound. †⅛-in. thick specimens.
*0.040 in. thick specimen
†(10⁻⁴ cal-cm/sec-cm²-°C)

ASTM or UL test	POLYPROPYLENE			POLYPHENYLENE SULFIDE[a]						POLYSTYRENE					POLYVINYL CHLORIDE	
				Glass reinforced			Glass and mineral filled			Polymers		Copolymers				
	Unmodified resin	Glass reinforced	Impact grade	R-3	R-4	R-8	R-9	R-10[b]	R-11	General purpose	Impact modified	Crystal clear	Impact modified	10-20% (wt.) Glass reinf*	Rigid	Flexible
D792	0.905	1.05-1.24	0.89-0.91	1.57	1.67	1.8	1.9	1.96-1.98	1.98	1.04-1.09	1.03-1.10	1.08-1.10	1.05-1.08	1.13-1.22	1.30-1.58	1.20-1.70
D792	30.8-30.4	24.5	30.8-30.5							26.0-25.6	28.1-25.2	–			20.5-19.1	–
D570	0.01-0.03	0.01-0.05	0.01-0.03	–	<0.05	0.03	–	–	–	0.03-0.10	0.05-0.6	0.1	0.1	0.08	0.04-0.4	0.15-0.75
D638	5,000	6,000-14,500	2,800-4,400	15,500	17,500	10,750	11,000	10,000-11,500	11,000	5,000-12,000	1,500-7,000	7,000-7,600	4,800-7,200	10,500-12,500	6,000-7,500	1,500-3,500
D638	10-20	2.0-3.6	350-500	1.1	1.25	0.47	0.5	0.5-0.6	0.6	0.5-2.0	2-60	1.4-1.7	2.0-20.0	1.3-2.0	40-80	200-450
D638	1.6	4.5-9.0	1.0-1.7	–	–	–	–	–	–	4.0-6.0	1.4-5.0	4.4-4.7	2.8-4.2	6.3-10.0	3.5-6.0	–
D785	80-110 (R)	110 (R)	50-85 (R)	–	123 (R)	121 (R)	–	120 (R)	–	65-80	10-90	108	80	101	65-85D (Shore)	50-100A (Shore)
D790	1.7-2.5	3.8-8.5	1.2-1.8	14	17	22	21	18	20	4.0-4.7	1.5-4.6	4.6-4.9	3.2-4.5	5.5-9.8	3.5	–
D256	0.5-2.2	1.0-5.0	1.0-15	1.0	1.1	0.59	0.7	0.6-1.0	0.8	0.2-0.45	0.5-4.0	0.3-0.5	0.5-4.4	1.8-2.6	0.4-20.0	–
C117	2.8†	–	3.0-4.0†	–	2.0	–	–	–	–	2.4-3.3	1.0-3.0	2.4-3.3	1.0-3.0	–	3.5-5.0†	3.0-4.0†
D696	3.2-5.7	1.6-2.9	3.3-4.7	–	2.2	1.6	1.1	–	–	3.3-4.4	1.9	3.5-3.7	3.5-3.7	2.0-2.2	2.8-5.6	3.9-13.9
D648	125-140 / 200-250	230-300 / 310	120-135 / 160-210	500 / 500	500 / 500	500 / 500	500 / 500	500 / 500	500	190-220 / 180-230	160-200 / 180-220	235-249	235-249	235-260	140-170 / 135-180	–
UL 94	HB[b]	HB[b]	HB[b]	V-0	V-0/5V	V-0/5V	V-0	V-0/5V	V-0	HB[b]	HB[b]	HB[b]	HB[b]	HB[b]	–	–
D149	500-660	475	500-650							500-700	300-600	500-700	300-600	–	350-500	300-400
D150	2.2-2.6	2.36	2.3	–	4.0*	4.3*	4.5*	4.8-6.1*	–	2.40-2.65	2.4-4.5	–	–	–	3.0-3.8	4.0-8.0
D150	0.0005-0.0018	0.0017	0.0003	–	0.0014*	0.016*	0.0072*	0.01-0.02*	–	0.0001-0.0003	0.0004-0.0020	–	–	–	0.009-0.017	0.07-0.16
D257	10^{17}	2×10^{18}	10^{18}	–	–	–	–	–	–	10^{17}–10^{19}	10^{16}	–	–	–	$>10^{16}$	10^{11}–10^{15}
D495	160	100	–	–	34	182	180	116-182	–	60-135	20-100	95	95	–	60-80	–
D542										1.60	–	1.59	–	–		
D1003										87-92	35.57	92	–	–		

[b]V-2, V-1, and V-0 grades are also available. *At 1.0 MHz

[a]Test specimen molding conditions, 275°F mold temperature.

[b]Representative of a series of various pigmented compounds.

Coaxial Cable End Connectors

UHF Connectors

Military No.	Style	Cable RG- or Description
PL-259	Str (m)	8, 9, 11, 13, 63, 87, 149, 213, 214, 216, 225
UG-111	Str (m)	59, 62, 71, 140, 210
SO-239	Pnl (f)	Std, mica/phenolic insulation
UG-266	Blkhd (f)	Rear mount, pressurized, copolymer of styrene ins.

Adapters		
PL-258	Str (f/f)	Polystyrene ins.
UG-224,363	Blkhd (f/f)	Polystyrene ins.
UG-646	Ang (f/m)	Polystyrene ins.
M-359A	Ang (m/f)	Polystyrene ins.
M-358	T (f/m/f)	Polystyrene ins.
Reducers		
UG-175		55, 58, 141, 142 (except 55A)
UG-176		59, 62, 71, 140, 210

Family Characteristics:
All are nonweatherproof and have a nonconstant impedance. Frequency range: 0-500 MHz. Maximum voltage rating: 500 V (peak).

N Connectors

Military No.	Style	Cable RG-	Notes
UG-21	Str (m)	8, 9, 213, 214	50 Ω
UG-94A	Str (m)	11, 13, 149, 216	70 Ω
UG-536	Str (m)	58, 141, 142	50 Ω
UG-603	Str (m)	59, 62, 71, 140, 210	50 Ω
UG-23, B-E	Str (f)	8, 9, 87, 213, 214, 225	50 Ω
UG-602	Str (f)	59, 62, 71, 140, 210	—
UG-228B, D, E	Pnl (f)	8, 9, 87, 213, 214, 225	—
UG-1052	Pnl (f)	58, 141, 142	50 Ω
UG-593	Pnl (f)	59, 62, 71, 140, 210	50 Ω
UG-160A, B, D	Blkhd (f)	8, 9, 87, 213, 214, 225	50 Ω
UG-556	Blkhd (f)	58, 141, 142	50 Ω
UG-58, A	Pnl (f)		50 Ω
UG-997A	Ang (f)		50 Ω 11/16"

Pnl mount (f) with clearance above panel

M39012/04-	Blkhd (f)		Front mount hermetically sealed
UG-680	Blkhd (f)		Front mount pressurized

N Adapters

Military No.	Style	Notes
UG-29,A,B	Str (f/f)	50 Ω, TFE ins.
UG-57A.B	Str (m/m)	50 Ω, TFE ins.
UG-27A,B	Ang (f/m)	Mitre body
UG-212A	Ang (f/m)	Mitre body
UG-107A	T (f/m/f)	—
UG-28A	T (f/f/f)	—
UG-107B	T (f/m/f)	—

Family Characteristics:
N connectors with gaskets are weatherproof. RF leakage: −90 dB min @ 3 GHz. Temperature limits: TFE: −67° to 390°F (−55° to 199°C). Insertion loss 0.15 dB max @ 10 GHz. Copolymer of styrene: −67° to 185°F (−55° to 85°C). Frequency range: 0-11 GHz. Maximum voltage rating: 1500 V P-P. Dielectric withstanding voltage 2500 V RMS. SWR (MIL-C-39012 cable connectors) 1.3 max 0-11 GHz.

BNC Connectors

Military No.	Style	Cable RG-	Notes
UG-88C	Str (m)	55, 58, 141, 142, 223, 400	
UG-959	Str (m)	8, 9	
UG-260,A	Str (m)	59, 62, 71, 140, 210	Rexolite ins.
UG-262	Pnl (f)	59, 62, 71, 140, 210	Rexolite ins.
UG-262A	Pnl (f)	59, 62, 71, 140, 210	nwx, Rexolite ins.
UG-291	Pnl (f)	55, 58, 141, 142, 223, 400	
UG-291A	Pnl (f)	55, 58, 141, 142, 223, 400	nwx
UG-624	Blkhd (f)	59, 62, 71, 140, 210	Front mount Rexolite ins.
UG-1094A	Blkhd		Standard
UG-625B	Receptacle		
UG-625			

BNC Adapters

Military No.	Style	Notes
UG-491,A	Str (m/m)	
UG-491B	Str (m/m)	Berylium, outer contact
UG-914	Str (f/f)	
UG-306	Ang (f/m)	
UG-306A,B	Ang (f/m)	Berylium outer contact
UG-414,A	Pnl (f/f)	# 3-56 tapped flange holes
UG-306	Ang (f/m)	
UG-306A,B	Ang (f/m)	Berylium outer contact
UG-274	T (f/m/f)	
UG-274A,B	T (f/m/f)	Berylium outer contact

Family Characteristics:
Z = 50 Ω. Frequency range: 0-4 GHz w/low reflection; usable to 11 GHz. Voltage rating: 500 V P-P. Dielectric withstanding voltage 500 V RMS. SWR: 1.3 max 0-4 GHz. RF leakage −55 dB min @ 3 GHz. Insertion loss: 0.2 dB max @ 3 GHz. Temperature limits: TFE: −67° to 390°F (−55° to 199°C); Rexolite insulators: −67° to 185°F (−55° to 85°C). "Nwx" = not weatherproof.

HN Connectors

Military No.	Style	Cable RG-	Notes
UG-59A	Str (m)	8, 9, 213, 214	
UG-1214	Str (f)	8, 9, 87, 213, 214, 225	Captivated contact
UG-60A	Str (f)	8, 9, 213, 214	Copolymer of styrene ins.
UG-1215	Pnl (f)	8, 9, 87, 213, 214, 225	Captivated contact
UG-560	Pnl (f)		
UG-496	Pnl (f)		
UG-212C	Ang (f/m)		Berylium outer contact

Family Characteristics:
Connector Styles: Str = straight; Pnl = panel; Ang = Angle; Blkhd = bulkhead. Z = 50 W. Frequency range = 0-4 GHz. Maximum voltage rating = 1500 V P-P. Dielectric withstanding voltage = 5000 V RMS SWR = 1.3. All HN series are weatherproof. Temperature limits: TFE: −67° to 390°F (−55° to 199°C); copolymer of styrene: −67° to 185°F (−55° to 85°C).

Cross-Family Adapters

Families	Description	Military No.
HN to BNC	HN-m/BNC-f	UG-309
N to BNC	N-m/BNC-f	UG-201,A
	N-f/BNC-m	UG-349,A
	N-m/BNC-m	UG-1034
N to UHF	N-m/UHF-f	UG-146
	N-f/UHF-m	UG-83,B
	N-m/UHF-m	UG-318
UHF to BNC	UHF-m/BNC-f	UG-273
	UHF-f/BNC-m	UG-255

Contents

Circuit Construction

25

Home construction of electronics projects can be a fun part of Amateur Radio. Some folks have said that hams don't build things nowadays; this just isn't so! An ARRL survey shows that 53% of active hams build some electronic projects. When you go to any ham flea market, you see row after row of dealers selling electronic components; people are leaving those tables with bags of parts. They must be doing something with them.

Even experienced constructors will find valuable tips in this chapter. It discusses tools and their uses, electronic construction techniques, tells how to turn a schematic into a working circuit and then summarizes common mechanical construction practices. This chapter was written by Ed Hare, W1RFI, Bruce Hale, KB1MW, Ian White, G3SEK, and Chuck Adams, K5FO.

SHOP SAFETY

All the fun of building a project will be gone if you get hurt. To make sure this doesn't happen, let's first review some safety rules.

- Read the manual! The manual tells all you need to know about the operation and safety features of the equipment you are using.
- Do not work when you are tired. You will be more likely to make a mistake or forget an important safety rule.
- Never disable any safety feature of any tool. If you do, sooner or later someone will make the mistake the safety feature was designed to prevent.
- Never fool around in the shop. Practical jokes and horseplay are in bad taste at social events; in a shop they are downright dangerous. A work area is a dangerous place at all times; even hand tools can hurt someone if they are misused.
- Keep your shop neat and organized. A messy shop is a dangerous shop. A knife left laying in a drawer can cut someone looking for another tool; a hammer left on top of a shelf can fall down at the worst possible moment; a sharp tool left on a chair can be a dangerous surprise for the weary constructor who sits down.
- Wear the proper safety equipment. Wear eye-protection goggles when working with chemicals or tools. Use earplugs or earphones when working near noise. If you are working with dangerous chemicals, wear the proper protective clothing.
- Make sure your shop is well ventilated. Paint, solvents, cleaners or other chemicals can create dangerous fumes. If you feel dizzy, get into fresh air immediately, and seek medical help if you do not recover quickly.
- Get medical help when necessary. Every workshop should contain a good first-aid kit. Keep an eye-wash kit near any dangerous chemicals or power tools that can create chips. If you become injured, apply first aid and then seek medical help if you are not sure that you are okay. Even a small burn or scratch on your eye can develop into a serious problem.
- Respect power tools. Power tools are not forgiving. A drill can go through your hand a lot easier than metal. A power saw can remove a finger with ease. Keep away from the business end of power tools. Tuck in your shirt, roll up your sleeves and remove your tie before using any power tool. If you have long hair, tie it back so it can't become entangled in power equipment.
- Don't work alone. Have someone nearby who can help if you get into trouble when working with dangerous equipment, chemicals or voltages.
- Think! Pay attention to what you are doing. No list of safety rules can cover all possibilities. Safety is always your responsibility. You must think about what you are doing, how it relates to the tools and the specific situation at hand.

TOOLS AND THEIR USES

All electronic construction makes use of tools, from mechanical tools for chassis fabrication to the soldering tools used for circuit assembly. A good understanding of tools and their uses will enable you to perform most construction tasks.

While sophisticated and expensive tools often work better or more quickly than simple hand tools, with proper use, simple hand tools can turn out a fine piece of equipment. **Table 25.1** lists tools indispensable for construction of electronic equipment. These tools can be used to perform nearly any construction task. Add tools to your collection from time to time, as finances permit.

Sources of Tools

Radio-supply houses, mail-order stores and most hardware stores carry the tools required to build or service Amateur Radio equipment. Bargains are available at ham flea markets or local neighborhood sales, but beware! Some flea-market bargains are really shoddy imports that won't work very well or last very long. Some used tools are offered for sale because the owner is not happy with their performance.

There is no substitute for quality! A high-quality tool, while a bit more expensive, will last a lifetime. Poor quality tools don't last long and often do a poor job even when brand new. You don't need to buy machinist-grade tools, but stay away from cheap tools; they are not the bargains they might appear to be.

Table 25.1
RECOMMENDED TOOLS AND MATERIALS

Simple Hand Tools

Screwdrivers
Slotted, 3-inch, 1/8-inch blade
Slotted, 8-inch, 1/8-inch blade
Slotted, 3-inch, 3/16-inch blade
Slotted, stubby, 1/4-inch blade
Slotted, 4-inch, 1/4-inch blade
Slotted, 6-inch, 5/16-inch blade
Phillips, 2-1/2-inch, #0 (pocket clip)
Phillips, 3-inch, #1
Phillips, stubby, #2
Phillips, 4-inch, #2
Phillips, 4-inch, #2
Long-shank screwdriver with holding clip on blade
Jeweler's set
Right-angle, slotted and Phillips

Pliers, Sockets and Wrenches
Long-nose pliers, 6- and 4-inch
Diagonal cutters, 6- and 4-inch
Channel-lock pliers, 6-inch
Slip-joint pliers
Locking pliers (Vise Grip or equivalent)
Socket nut-driver set, 3/16- to 1/2-inch
Set of socket wrenches for hex nuts
Allen (hex) wrench set
Wrench set
Adjustable wrenches, 6- and 10-inch
Tweezers, regular and reverse-action
Retrieval tool/parts holder, flexible claw
Retrieval tool, magnetic

Cutting and Grinding Tools
File set consisting of flat, round, half-round, and triangular. Large and miniature types recommended
Burnishing tool
Wire strippers
Wire crimper
Hemostat, straight
Scissors
Tin shears, 10-inch
Hacksaw and blades
Hand nibbling tool (for chassis-hole cutting)
Scratch awl or scriber (for marking metal)
Heavy-duty jackknife
Knife blade set (X-ACTO or equivalent)
Machine-screw taps, #4-40 through #10-32 thread
Socket punches, 1/2 in, 5/8 in, 3/4 in, 1-1/8 in, 1-1/4 in, and 1-1/2 in.
Tapered reamer, T-handle, 1/2-inch maximum width
Deburring tool

Miscellaneous Hand Tools
Combination square, 12-inch, for layout work
Hammer, ball-peen, 12-oz head
Hammer, tack
Bench vise, 4-inch jaws or larger
Center punch
Plastic alignment tools
Mirror, inspection
Flashlight, penlight and standard

Magnifying glass
Ruler or tape measure
Dental pick
Calipers
Brush, wire
Brush, soft
Small paintbrush
IC-puller tool

Hand-Powered Tools
Hand drill, 1/4-inch chuck or larger
High-speed drill bits, #60 through 3/8-inch diameter

Power Tools
Motor-driven emery wheel for grinding
Electric drill, hand-held
Drill press
Miniature electric motor tool (Dremel or equivalent) and accessory drill press

Soldering Tools and Supplies
Soldering pencil, 30-W, 1/8-inch tip
Soldering iron, 200-W, 5/8-inch tip
Solder, 60/40, resin core
Soldering gun, with assorted tips
Desoldering tool
Desoldering wick

Safety
Safety glasses
Hearing protector, earphones or earplugs
Fire extinguisher
First-aid kit

Useful Materials
Medium-weight machine oil
Contact cleaner, liquid or spray can
Duco modeling cement or equivalent
Electrical tape, vinyl plastic
Sandpaper, assorted
Emery cloth
Steel wool, assorted
Cleaning pad, Scotchbrite or equivalent
Cleaners and degreasers
Contact lubricant
Sheet aluminum, solid and perforated, 16- or 18-gauge, for brackets and shielding.
Aluminum angle stock, 1/2×1/2-inch
1/4-inch-diameter round brass or aluminum rod (for shaft extensions)
Machine screws: Round-head and flat head, with nuts to fit. Most useful sizes: 4-40, 6-32 and 8-32, in lengths from 1/4-inch to 1-1/2 inches. (Nickel-plated steel is satisfactory except in strong RF fields, where brass should be used.)
Bakelite, Lucite, polystyrene and copper-clad PC-board scraps.
Soldering lugs, panel bearings, rubber grommets, terminal-lug wiring strips, varnished-cambric insulating tubing, heat-shrinkable tubing.
Shielded and unshielded wire.
Tinned bare wire, #22, #14 and #12.
Enameled wire, #20 through #30

Care of Tools

The proper care of tools is more than a matter of pride. Tools that have not been cared for properly will not last long or work well. Dull or broken tools can be safety hazards. Tools that are in good condition do the work for you; tools that are misused or dull are difficult to use.

Store tools in a dry place. Tools do not fit in with most living-room decors, so they are often relegated to the basement or garage. Unfortunately, many basements or garages are not good places to store tools; dampness and dust are not good for tools. If your tools are stored in a damp place, use a dehumidifier. Sometimes you can minimize rust by keeping your tools lightly oiled, but this is a second-best solution. If you oil your tools, they may not rust, but you will end up covered in oil every time you use them. Wax or silicone spray is a better alternative.

Store tools neatly. A messy toolbox, with tools strewn about haphazardly, can be more than an inconvenience. You may waste a lot of time looking for the right tool and sharp edges can be dulled or nicked by tools banging into each other in the bottom of the box. As the old adage says, every tool should have a place, and every tool should be in its place. If you must search the workbench, garage, attic and car to find the right screwdriver, you'll spend more time looking for tools than building projects.

Sharpening

Many cutting tools can be sharpened. Send a tool that has been seriously dulled to a professional sharpening service. These services can resharpen saw blades, some files, drill bits and most cutting blades. Touch up the edge of cutting tools with a whetstone to extend the time between sharpenings.

Sharpen drill bits frequently to minimize the amount of material that must be removed each time. Frequent sharpening also makes it easier to maintain the critical surface angles required for best cutting with least wear. Most inexpensive drill-bit sharpeners available for shop use do a poor job, either from the poor quality of the sharpening tool or inexperience of the operator. Also, drills should be sharpened at different angles for different applications. Commercial sharpening services do a much better job.

Intended Purpose

Don't use tools for anything other than

their intended purpose! If you use a pair of wire cutters to cut sheet metal, a pliers as a vise or a screwdriver as a pry bar, you ruin a good tool. Although an experienced constructor can improvise with tools, most take pride in not abusing them.

Tool Descriptions and Uses

Specific applications for tools are discussed throughout this chapter. Hand tools are used for so many different applications that they are discussed first, followed by some tips for proper use of power tools.

Soldering Iron

Soldering is used in nearly every phase of electronic construction so you'll need soldering tools. A soldering tool must be hot enough to do the job and lightweight enough for agility and comfort. A 100-W soldering gun is overkill for printed-circuit work, for example. A temperature-controlled iron works well, although the cost is not justified for occasional projects. Get an iron with a small conical or chisel tip.

Soldering is not like gluing; solder does more than bind metal together and provide an electrically conductive path between them. Soldered metals and the solder combine to form an alloy.

You may need an assortment of soldering irons to do a wide variety of soldering tasks. They range in size from a small 25-W iron for delicate printed-circuit work to larger 100 to 300-W sizes used to solder large surfaces. Several manufacturers also sell soldering guns. Small "pencil" butane torches are also available, with optional soldering-iron tips. A small butane torch is available from the Solder-It Company. This company also sells a soldering kit that contains paste solders (in syringes) for electronics, pot metal and plumbing. See the Address List in the **References** chapter for the address.

Keep soldering tools in good condition by keeping the tips well tinned with solder. Do not run them at full temperature for long periods when not in use. After each period of use, remove the tip and clean off any scale that may have accumulated. Clean an oxidized tip by dipping the hot tip in sal ammoniac (ammonium chloride) and then wiping it clean with a rag. Sal ammoniac is somewhat corrosive, so if you don't wipe the tip thoroughly, it can contaminate electronic soldering.

If a copper tip becomes pitted, file it smooth and bright and then tin it immediately with solder. Modern soldering iron tips are nickel or iron clad and should not be filed.

The secret of good soldering is to use the right amount of heat. Many people who have not soldered before use too little heat, dabbing at the joint to be soldered and making little solder blobs that cause unintended short circuits.

Solders have different melting points, depending on the ratio of tin to lead. Tin melts at 450°F and lead at 621°F. Solder made from 63% tin and 37% lead melts at 361°F, the lowest melting point for a tin and lead mixture. Called 63-37 (or eutectic), this type of solder also provides the most rapid solid-to-liquid transition and the best stress resistance.

Solders made with different lead/tin ratios have a plastic state at some temperatures. If the solder is deformed while it is in the plastic state, the deformation remains when the solder freezes into the solid state. Any stress or motion applied to "plastic solder" causes a poor solder joint.

60-40 solder has the best wetting qualities. Wetting is the ability to spread rapidly and bond materials uniformly. 60-40 solder also has a low melting point. These factors make it the most commonly used solder in electronics.

Some connections that carry high current can't be made with ordinary tin-lead solder because the heat generated by the current would melt the solder. Automotive starter brushes and transmitter tank circuits are two examples. Silver-bearing solders have higher melting points, and so prevent this problem. High-temperature silver alloys become liquid in the 1100°F to 1200°F range, and a silver-manganese (85-15) alloy requires almost 1800°F.

Because silver dissolves easily in tin, tin bearing solders can leach silver plating from components. This problem can be greatly reduced by partially saturating the tin in the solder with silver or by eliminating the tin. Tin-silver or tin-lead-silver alloys become liquid at temperatures from 430°F for 96.5-3.5 (tin-silver), to 588°F for 1.0-97.5-1.5 (tin-lead-silver). A 15.0-80.0-5.0 alloy of lead-indium-silver melts at 314°F.

Never use acid-core solder for electrical work. It should be used only for plumbing or chassis work. For circuit construction, only use fluxes or solder-flux combinations that are labeled for electronic soldering.

The resin or the acid is a *flux*. Flux removes oxide by suspending it in solution and floating it to the top. Flux is not a cleaning agent! Always clean the work before soldering. Flux is not a part of a soldered connection — it merely aids the soldering process. After soldering, remove any remaining flux. Resin flux can be removed with isopropyl or denatured alcohol. A cotton swab is a good tool for applying the alcohol and scrubbing the excess flux away. Commercial flux-removal sprays are available at most electronic-part distributors.

The two key factors in quality soldering are time and temperature. Generally, rapid heating is desired, although most unsuccessful solder jobs fail because insufficient heat has been applied. Be careful; if heat is applied too long, the components or PC board can be damaged, the flux may be used up and surface oxidation can become a problem. The soldering-iron tip should be hot enough to readily melt the solder without burning, charring or discoloring components, PC boards or wires. Usually, a tip temperature about 100°F above the solder melting point is about right for mounting components on PC boards. Also, use solder that is sized appropriately for the job. As the cross section of the solder decreases, so does the amount of heat required to melt it. Diameters from 0.025 to 0.040 inches are good for nearly all circuit wiring.

Here's how to make a good solder joint. This description assumes that solder with a flux core is used to solder a typical PC board connection such as an IC pin.

• Prepare the joint. Clean all conductors thoroughly with fine steel wool or a plastic scrubbing pad. Do the circuit board at the beginning of assembly and individual parts such as resistors and capacitors immediately before soldering. Some parts (such as ICs and surface-mount components) cannot be easily cleaned; don't worry unless they're exceptionally dirty.

• Prepare the tool. It should be hot enough to melt solder applied to its tip quickly (half a second when dry, instantly when wet with solder). Apply a little solder directly to the tip so that the surface is shiny. This process is called "tinning" the tool. The solder coating helps conduct heat from the tip to the joint.

• Place the tip in contact with one side of the joint. If you can place the tip on the underside of the joint, do so. With the tool below the joint, convection helps transfer heat to the joint.

• Place the solder against the joint directly opposite the soldering tool. It should melt within a second for normal PC connections, within two seconds for most other connections. If it takes longer to melt, there is not enough heat for the job at hand.

• Keep the tool against the joint until the solder flows freely throughout the joint. When it flows freely, solder tends to form concave shapes between the conductors. With insufficient heat solder does not flow freely; it forms convex shapes—blobs. Once solder shape changes from convex to concave, remove the tool from the joint.

• Let the joint cool without movement at room temperature. It usually takes no more than a few seconds. If the joint is moved before it is cool, it may take on a dull, satin look that is characteristic of a "cold" solder joint. Reheat cold joints until the solder flows freely and hold them still until cool.

• When the iron is set aside, or if it loses its shiny appearance, wipe away any dirt with a wet cloth or sponge. If it remains dull after cleaning, tin it again.

Overheating a transistor or diode while soldering can cause permanent damage. Use a small heat sink when you solder transistors, diodes or components with plastic parts that can melt. Grip the component lead with a pair of pliers up close to the unit so that the heat is conducted away (be careful — it is easy to damage delicate component leads). A small alligator clip also makes a good heat sink.

Mechanical stress can damage components, too. Mount components so there is no appreciable mechanical strain on the leads.

Soldering to the pins of coil forms or male cable plugs can be difficult. Use a suitable small twist drill to clean the inside of the pin and then tin it with resin-core solder. While it is still liquid, clear the surplus solder from each pin with a whipping motion or by blowing through the pin from the inside of the form or plug. Watch out for flying hot solder! Next, file the nickel plate from the pin tip. Then insert the wire and solder it. After soldering, remove excess solder with a file, if necessary.

When soldering to the pins of plastic coil forms, hold the pin to be soldered with a pair of heavy pliers to form a heat sink. Do not allow the pin to overheat; it will loosen and become misaligned.

In order to remove components, you need to learn the art of desoldering — removing solder from components and PC boards so they can be separated easily. Use commercially made wicking material (braid) to soak up excess solder from a joint. Another useful tool is an air-suction solder remover. Another method is to heat the joint and "flick" the wet solder off. (Watch out for solder splashes!)

Soldering equipment gets *hot!* Be careful. Treat a soldering burn as you would any other. Handling lead or breathing soldering fumes is also hazardous. Observe these precautions to protect yourself and others:

• Properly ventilate the work area. If you can smell fumes, you are breathing them.
• Wash your hands after soldering, especially before handling food.
• Minimize direct contact with flux and flux solvents.

For more information about soldering hazards and the ways to make soldering safer, see "Making Soldering Safer," by Brian P. Bergeron, MD, NU1N (Mar 1991 *QST*, pp 28-30) and "More on Safer Soldering," by Gary E. Meyers, K9CZB (Aug 1991 *QST*, p 42).

Screwdrivers

For construction or repair, you need to have an assortment of screwdrivers. Each blade size is designed to fit a specific range of screw-head sizes. Using the wrong size blade usually damages the blade, the screw head or both. You may also need stubby sizes to fit into tight spaces. Right-angle screwdrivers are inexpensive and can get into tight spaces that can't otherwise be reached.

Electric screwdrivers are relatively inexpensive. If you have a lot of screws to fasten, they can save a lot of time and effort. They come with a wide assortment of screwdriver and nut-driver bits. An electric drill can also function as an electric screwdriver, although it may be heavy and over-powered for some applications.

Keep screwdriver blades in good condition. If a blade becomes broken or worn out, replace the screwdriver. A screwdriver only costs a few dollars; do not use one that is not in perfect condition. Save old screwdrivers to use as pry bars and levers, but use only good ones on screws. Filing a worn blade seldom gives good results.

Pliers and Vice Grips

Pliers and vice grips are used to hold or bend things. They are not wrenches! If pliers are used to remove a nut or bolt, the nut or the pliers is usually damaged. Pliers are not intended for heavy-duty applications. Use a metal brake to bend heavy metal; use a vice to hold a heavy component. To remove a nut, use a wrench or nut driver. There is one exception to this rule of thumb: To remove a nut that is stripped too badly for a wrench, use a pair of pliers, a vice grip or a diagonal cutter to bite into the nut and turn it a bit. If you do this, use an old tool or one dedicated to just this purpose; this technique is not good for the tool. If the pliers jaws or teeth become worn, replace the tool.

Wire Cutters

Wire cutters are primarily used to cut wires or component leads. The choice of diagonal blades (sometimes called

"dikes") or end-nip blades depends on the application. Diagonal blades are most often used to cut wires, while the end-nip blades are useful to cut off the ends of components that have been soldered into a printed-circuit board. Some delicate components can be damaged by cutting their leads with dikes. Scissors designed to cut wire can be used.

Wire strippers are handy, but you can usually strip wires using a diagonal cutter or a knife. This is not the only use for a knife, so keep an assortment handy.

Do not use wire cutters or strippers on anything other than wire! If you use a cutter to trim a protruding screw head, or cut a hardened-steel spring, you will usually damage the blades.

Files

Files are used for a wide range of tasks. In addition to enlarging holes and slots, they are used to remove burrs, shape metal, wood or plastic and clean some surfaces in preparation for soldering. Files are especially prone to damage from rust and moisture. Keep them in a dry place. The cutting edge of the blades can also become clogged with the material you are removing. Use file brushes (also called file cards) to keep files clean. Most files cannot be sharpened easily, so when the teeth become worn, the file must be replaced. A worn file is sometimes worse than no file at all. At best, a worn file requires more effort.

Drill Bits

Drill bits are made from carbon steel, high-speed steel or carbide. Carbon steel is more common and is usually supplied unless a specific request is made for high-speed bits. Carbon-steel drill bits cost less than high-speed or carbide types; they are sufficient for most equipment construction work. Carbide drill bits last much longer under heavy use. One disadvantage of carbide bits is that they are brittle and break easily, especially if you are using a hand-held power drill.

Twist drills are available in a number of sizes. Those listed in bold type in **Table 25.2** are the most commonly used in construction of amateur equipment. You may not use all of the drills in a standard set, but it is nice to have a complete set on hand. You should also buy several spares of the more common sizes. Although Table 25.2 lists drills down to #54, the series extends to number #80.

Specialized Tools

Most constructors know how to use common tools, such as screwdrivers, wrenches and hammers. Let's discuss

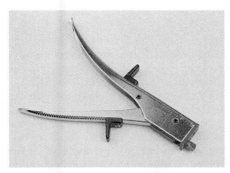

Fig 25.1 — A nibbling tool is used to remove small sections of sheet metal.

Fig 25.2 — A deburring tool is used to remove the burrs left after drilling a hole.

Fig 25.3 — A socket punch is used to easily punch a hole in sheet metal.

Table 25.2
Numbered Drill Sizes

No.	Diameter (Mils)	Will Clear Screw	Drilled for Tapping from Steel or Brass
1	228.0	12-24	—
2	221.0	—	—
3	213.0	—	14-24
4	209.0	12-20	—
5	205.0	—	—
6	204.0	—	—
7	201.0	—	—
8	199.0	—	—
9	196.0	—	—
10	193.5	—	—
11	191.0	10-24 10-32	—
12	189.0	—	—
13	185.0	—	—
14	182.0	—	—
15	180.0	—	—
16	177.0	—	12-24
17	173.0	—	—
18	169.5	—	—
19	166.0	8-32	12-20
20	161.0	—	—
21	159.0	—	10-32
22	157.0	—	—
23	154.0	—	—
24	152.0	—	—
25	149.5	—	10-24
26	147.0	—	—
27	144.0	—	—
28	140.0	6-32	—
29	136.0	—	8-32
30	128.5	—	—
31	120.0	—	—
32	116.0	—	—
33	113.0	4-40	—
34	111.0	—	—
35	110.0	—	—
36	106.5	—	6-32
37	104.0	—	—
38	101.5	—	—
39	099.5	3-48	—
40	098.0	—	—
41	096.0	—	—
42	093.5	—	—
43	089.0	—	4-40
44	086.0	2-56	—
45	082.0	—	—
46	081.0	—	—
47	078.5	—	3-48
48	076.0	—	—
49	073.0	—	—
50	070.0	—	2-56
51	067.0	—	—
52	063.5	—	—
53	059.5	—	—
54	055.0	—	—

other tools that are not so common.

A hand nibbling tool is shown in **Fig 25.1**. Use this tool to remove small "nibbles" of metal. It is easy to use; position the tool where you want to remove metal and squeeze the handle. The tool takes a small bite out of the metal. When you use a nibbler, be careful that you don't remove too much metal, clip the edge of a component mounted to the sheet metal or grab a wire that is routed near the edge of a chassis. Fixing a broken wire is easy, but something to avoid if possible. It is easy to remove metal but nearly impossible to put it back. Do it right the first time!

Deburring Tool

A deburring tool is just the thing to remove the sharp edges left on a hole after most drilling or punching operations. See **Fig 25.2**. Position the tool over the hole as shown and rotate it around the hole edge to remove burrs or rough edges. As an alternative, select a drill bit that is somewhat larger than the hole, position it over the hole, and spin it lightly to remove the burr.

Socket Punches

Greenlee is the most widely known of the socket-punch manufacturers. Most socket punches are round, but they do come in other shapes. To use one, drill a pilot hole large enough to clear the bolt that runs through the punch. Then, mount the punch as shown in **Fig 25.3**, with the cutter on one side of the sheet metal and the socket on the other. Tighten the nut with a wrench until the cutter cuts all the way through the sheet metal.

Useful Shop Materials

Small stocks of various materials are used when constructing electronics equipment. Most of these are available from hardware or radio-supply stores. A representative list is shown at the end of Table 25.1.

Small parts, such as machine screws, nuts, washers and soldering lugs can be economically purchased in large quantities (it doesn't pay to buy more than a life-time supply). For items you don't use often, many radio-supply stores or hardware stores sell small quantities and assortments.

A DELUXE SOLDERING STATION

The simple tool shown in **Figs 25.4** through **25.6** can enhance the usefulness and life of a soldering iron as well as make electronic assembly more convenient. It includes a protective heat sink and a tip-cleaning sponge rigidly attached to a sturdy base for efficient one-handed operation.

Fig 25.4 — A compact assembly of commonly available items, this soldering station makes soldering easier. Miniature toggle switches are used because they are easy to operate.

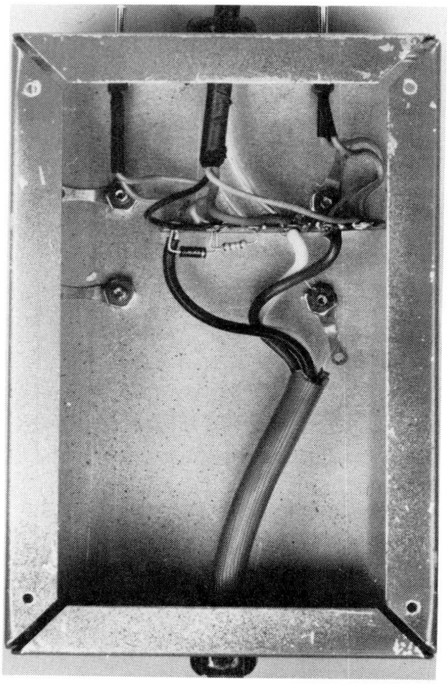

Soldering-iron tips and heating elements last longer if operated at a reduced temperature when not being used. Temperature reduction is accomplished by half-wave rectification of the applied ac. D1 conducts during only one-half of the ac cycle. With current flowing only in one direction, only one electrode of the neon bulb glows. Closing S1 short-circuits the diode and applies full power to the soldering iron, igniting both bulb electrodes brightly.

The base for the unit is a 2×6×4-inch (HWD) aluminum chassis (Bud AC-431 or equivalent). A 30- or 40-W soldering iron fits neatly on the chassis top. The holder has two mounting holes in each foot. A sponge tray nests between the feet and the case. In this model, a sardine tin is used for the sponge tray.

The tray and iron holder are secured to the chassis by 6-32 × ¹/₂-inch pan-head machine screws and nuts, with flat washers under the screw heads (sponge tray) and lock washers under the nuts (chassis underside). One of these nuts fastens a six-lug tie point strip to the chassis bottom. Use the soldering-iron holder base as a template for drilling the chassis and sponge tray. The floor of the sponge tray must be sealed around the screw heads to prevent moisture from leaking into the electrical components below the chassis. RTV compound was used for this purpose in the unit pictured.

Notice that the soldering iron and the soldering station use separate ac line cords. This ensures that the cord of the soldering iron will be long enough to do useful work. Bushings are used to anchor both cords. If these aren't available, grommets and cable clamps work well. Knotting the cords inside the chassis is a simple technique that normally provides adequate strain relief.

The underchassis assembly is shown in Fig 25.5. The neon bulb is installed in a ³/₁₆-inch-ID grommet. The leads are insulated with spaghetti insulation or heat-shrink tubing to prevent short circuits. If you mount the bulb in a fixture or socket, use a clear lens to ensure that the electrodes are distinctly visible. Install a cover on the bottom of the chassis to prevent accidental contact with the live ac wiring. Stick-on rubber feet prevent the bottom of the unit from scratching your work surface.

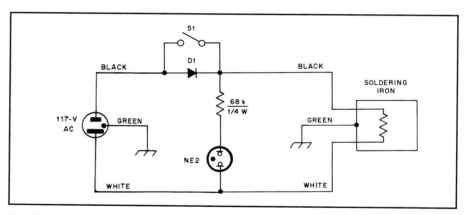

Fig 25.6 — Schematic diagram of the soldering station. D1 is a silicon diode, 1-A, 400-PIV. S1 is a miniature SPST toggle switch rated 3 A at 125 V. This circuit is satisfactory for use with irons having power ratings up to 100 W.

Fig 25.5 — View of the soldering-station chassis underside with the bottom plate removed. #24 hookup wire is adequate for all connections. Make sure no possibility of a short circuit exists.

SOLDERING-IRON TEMPERATURE CONTROL

A temperature control gives greater flexibility than the simple control just described. An incandescent-light dimmer can be used to control the working temperature of the tip. **Fig 25.7** shows a temperature control built into an electrical box. A dimmer and a duplex outlet are mounted in the box; the wiring diagram is shown in **Fig 25.8**. Only one of the two ac outlets is controlled by the dimmer. A jumper on the duplex outlet connects the hot terminals of both outlets together. This jumper must be removed. The hot terminal is narrower than the neutral one and its connecting screw is usually brass. Neutral terminals remain interconnected.

The dimmer shown in Fig 25.7 can be purchased at any hardware or electrical-supply store. The knob is capable of fine control of the soldering temperature.

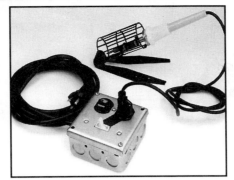

Fig 25.7 — An incandescent-light dimmer controls soldering-iron tip temperature. Only one of the duplex outlets is connected through the dimmer.

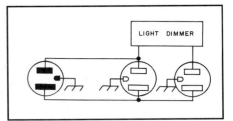

Fig 25.8 — Schematic diagram of the soldering-iron temperature control.

Electronic Circuits

Most of the construction projects undertaken by the average amateur involve electronic circuitry. The circuit is the "heart" of most amateur equipment. It might seem obvious, but in order for you to build it, the circuit must work! Don't always assume that a "cookbook" circuit that appears in an applications note or electronics magazine is flawless. These are sometimes design examples that have not always been thoroughly debugged. Many home-construction projects are "one-time" deals; the author has put one together and it worked. In some cases, component tolerances or minor layout changes might make it difficult to get a second unit working.

Protecting Components

You need to take steps to protect the electronic and mechanical components you use in circuit construction. Some components can be damaged by rough handling. Dropping a 1/4-W resistor causes no harm, but dropping a vacuum tube or other delicate subassemblies usually causes damage.

Some components are easily damaged by heat. Some of the chemicals used to clean electronic components (such as flux removers, degreasers or control-lubrication sprays) can damage plastic. Check them for safety before you use them.

Electrostatic Discharge

Some components, especially high-impedance components such as FETs and CMOS gates, can be damaged by electro-static discharge (ESD). Protect these parts from static charges. Most people are familiar with the static charge that builds up when one walks across a carpet then touches a metal object; the resultant spark can be quite lively. Walking across a carpet on a dry day can generate 35 kV! A worker sitting at a bench can generate voltages up to 6 kV, depending on conditions, such as when relative humidity is less than 20%.

You don't need this much voltage to damage a sensitive electronic component; damage can occur with as little as 30 V. The damage is not always catastrophic. A MOSFET can become noisy, or lose gain; an IC can suffer damage that causes early failure. To prevent this kind of damage, you need to take some precautions.

The energy from a spark can travel inside a piece of equipment to effect internal components. Protection of sensitive electronic components involves the prevention of static build-up together with the removal of any existing charges by dissipating any energy that does build up.

Several techniques can be used to minimize static build-up. First, remove any carpet in your work areas. You can replace it with special antistatic carpet, but this is expensive. It's less expensive to treat the carpet with antistatic spray, which is available from Chemtronics, GC Thorsen and other lines carried by electronics wholesalers.

Even the choice of clothing you wear can affect the amount of ESD. Polyester has a much greater ESD potential than cotton.

Many builders who have their workbench on a concrete floor use a rubber mat to minimize the risk of electric shocks from the ac line. Unfortunately, the rubber mat increases the risk of ESD. An antistatic rubber mat can serve both purposes.

Many components are shipped in antistatic packaging. Leave components in their conductive packaging. Other components, notably MOSFETs, are shipped with a small metal ring that temporarily shorts all of the leads together. Leave this ring in place until the device is fully installed in the circuit.

These precautions help reduce the build-up of electrostatic charges. Other techniques offer a slow discharge path for the charges or keep the components and the operator handling them at the same ground potential.

One of the best techniques is to connect the operator and the devices being handled to earth ground, or a common reference point. It is not a good idea to directly ground an operator working on electronic equipment, though; the risk of shock is too great. If the operator is grounded through a high-value resistor, ESD protection is still offered but there is no risk of shock.

The operator is usually grounded through a conductive wrist strap. 3M makes a grounding wrist band. This wrist band is equipped with a snap-on ground lead. A 1-MΩ resistor is built into the snap

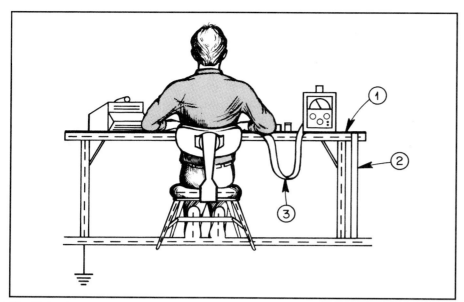

Fig 25.9—A work station that has been set up to minimize ESD features (1) a grounded dissipative work mat and (2) a wrist strap that (3) grounds the worker through high resistance.

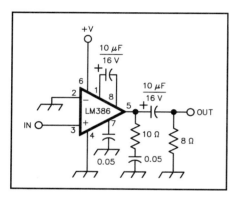

Fig 25.10 — Schematic diagram of the audio amplifier used as a design example of various construction techniques.

Fig 25.11 — The example audio amplifier of Fig 25.10 built using ground-plane construction.

of the strap to protect the user should a live circuit be contacted. Build a similar resistor into any homemade ground strap.

The devices and equipment being handled are also grounded, by working on a charge-dissipating mat that is connected to ground. The mat should be an insulator that has been impregnated with a resistance material. Suitable mats and wrist straps are made by 3M, GC Electronics and others; they are available from most electronics supply houses. Fig 25.9 shows a typical ESD-safe work station.

The work area should also be grounded, directly or through a conductive mat. Use a soldering iron with a grounded tip to solder sensitive components. Most irons that have three-wire power cords are properly grounded. When soldering static-sensitive devices, use two or three jumpers to ground you, the work and the iron. If the iron does not have a ground wire in the power cord, clip a jumper from the metal part of the iron near the handle to the metal box that houses the temperature control. Another jumper connects the box to the work. Finally, a jumper goes from the box to an elastic wrist band for static grounding.

Use antistatic bags to transport susceptible components or equipment. Keep your workbench free of objects such as paper, plastic and other static-generating items. Use conductive containers with a dissipative surface coating for equipment storage.

All of the antistatic products described above are available from Newark Electronics and other suppliers. See the Address List in the **References** chapter.

Electronics Construction Techniques

Several different point-to-point wiring techniques or printed-circuit boards (PC boards) can be used to construct electronic circuits. Most circuit projects use a combination of techniques. The selection of techniques depends on many different factors and builder preferences.

The simple audio amplifier shown in **Fig 25.10** will be built using various point-to-point or PC-board techniques. This shows how the different construction methods are applied to a typical circuit.

Point-to-Point Techniques

Point-to-point techniques include all circuit construction techniques that rely on tie points and wiring, or component leads, to build a circuit. This is the technique used in most home-brew construction projects. It is sometimes used in commercial construction, such as old vacuum-tube receivers and modern tube amplifiers.

Point-to-point is also used to connect the "off-board" components used in a printed-circuit project. It can be used to interconnect the various modules and printed-circuit boards used in more complex electronic systems. Most pieces of electronic equipment have at least some point-to-point wiring.

Ground-Plane Construction

A point-to-point construction technique that uses the leads of the components as tie points for electrical connections is known as "ground-plane construction," "dead-bug" or "ugly construction." (The term "ugly construction" was coined by Wes Hayward, W7ZOI.) "Dead-bug construction" gets its name from the appearance of an IC with its leads sticking up in the air. In most cases, this technique uses copper-clad circuit-board material as a foundation and ground plane on which to build a circuit using point-to-point wiring, so in this chapter it is called "ground-plane construction." An example is shown in **Fig 25.11**.

Ground-plane construction is quick and simple: You build the circuit on an unetched piece of copper-clad circuit board. Wherever a component connects to ground, you solder it to the copper board. Ungrounded connections between components are made point-to-point. Once you learn how to build with a ground-plane board, you can grab a piece of circuit board and start building any time you see an interesting circuit.

A PC board has strict size limits; the components must fit in the space allotted. Ground-plane construction is more flexible; it allows you to use the parts on hand. The circuit can be changed easily — a big help when you are experimenting. The greatest virtue of ground-plane construction is that it is fast.

Ground-plane construction is something like model building, connecting parts using solder almost —but not exactly — like glue. In ground-plane construction you build the circuit directly from the schematic, so it can help you get familiar with a circuit and how it works. You can build subsections of a large circuit on small ground-plane modules and string them together into a larger design.

Circuit connections are made directly, minimizing component lead length. Short lead lengths and a low-impedance ground conductor help prevent circuit instability. There is usually less intercomponent capacitive coupling than would be found between PC-board traces, so it is often better than PC-board construction for RF, high-gain or sensitive circuits.

Use circuit components to support other circuit components. Start by mounting one component onto the ground plane, building from there. There is really only one two-handed technique to mount a component to the ground plane. Bend one of the component leads at a 90° angle, then trim off the excess. Solder a blob of solder to the board surface, perhaps about 0.1 inch in diameter, leaving a small dome of solder. Using one hand, hold the component in place on top of the soldered spot and reheat the component and the solder. It should flow nicely, soldering the component securely. Remove the iron tip and hold the component perfectly still until the solder cools. You can then make connections to the first part.

Connections should be mechanically secure before soldering. Bend a small hook in the lead of a component, then "crimp" it to the next component(s). Do not rely only on the solder connections to provide mechanical strength; sooner or later one of these connections will fail, resulting in a dead circuit.

In most cases, each circuit has enough grounded components to support all of the components in the circuit. This is not always possible, however. In some circuits, high-value resistors can be used as stand-off insulators. One resistor lead is soldered to the copper ground plane, the other lead is used as a circuit connection point. You can use 1/4- or 1/2-W resistors in values from 1 to 10 MΩ. Such high-value resistors permit almost no current to flow, and in low-impedance circuits they act more like insulators than resistors. As a rule of thumb, resistors used as stand-off insulators should have a value that is at least 10 times the circuit impedance at that circuit point.

Fig 25.12A shows how to use the stand-off technique to wire the circuit shown at Fig 25.12C. Fig 25.12B shows how the resistor leads are bent before the stand-off component is soldered to the ground plane.

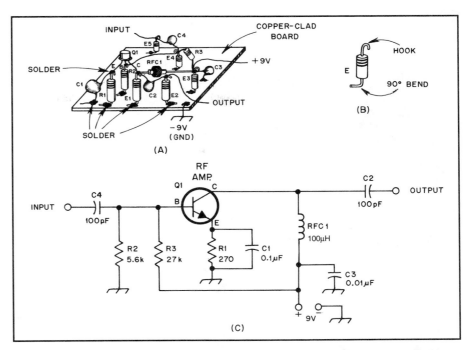

Fig 25.12 — Pictorial view of a circuit board that uses ground-plane construction is shown at A. A close-up view of one of the standoff resistors is shown at B. Note how the leads are bent. The schematic diagram at C shows the circuit displayed at A.

Components E1 through E5 are resistors that are used as stand-off insulators. They do not appear in the schematic diagram. The base circuitry at Q1 of Fig 25.12A has been stretched out to reduce clutter in the drawing. In a practical circuit, all of the signal leads should be kept as short as possible. E4 would, therefore, be placed much closer to Q1 than the drawing indicates.

No stand-off posts are required near R1 and R2 of Fig 25.12. These two resistors serve two purposes: They are not only the normal circuit resistances, but function as stand-off posts as well. Follow this practice wherever a capacitor or resistor can be employed in the dual role.

Wired Traces — the Lazy PC Board

If you already have a PC-board design, but don't want to copy the entire circuit — or you don't want to make a double-sided PC board — then the easiest construction technique is to use a bare board (or perfboard) and hard-wire the traces.

Drill the necessary holes in a piece of single-sided board, remove the copper ground plane from around the holes, and then wire up the back using component leads and bits of wire instead of etched traces (**Fig 25.13**).

To transfer an existing board layout, make a 1:1 photocopy and tape it to your piece of PC board. Prick through the holes with an automatic (one-handed) center

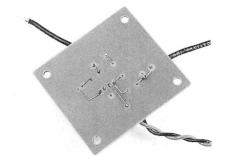

Fig 25.13 — The audio amplifier built using wired-traces construction.

punch or by firm pressure with a sharp scriber, remove the photocopy and drill all

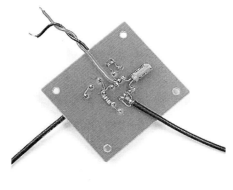

Fig 25.15 — The audio amplifier built using terminal-and-wire construction.

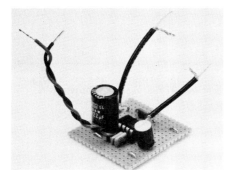

Fig 25.14 — The audio amplifier built on perforated board. Top view at A; bottom view at B.

Fig 25.16 — The audio amplifier built on a solderless prototyping board.

the holes. Holes for ground leads are optional — you generally get a better RF ground by bending the component lead flat to the board and soldering it down. Remove the copper around the rest of the holes by pressing a drill bit lightly against the hole and twisting it between your fingers. A drill press can also be used, but either way, don't remove too much board material. Then wire up the circuit beneath the board. The results look very neat and tidy — from the top, at least!

Circuits that contain components originally designed for PC-board mounting are good candidates for this technique. Wired

traces would also be suitable for circuits involving multipin RF ICs, double-balanced mixers and similar components. To bypass the pins of these components to ground, connect a miniature ceramic capacitor on the bottom of the board directly from the bypassed pin to the ground plane.

A wired-trace board is fairly sturdy, even though many of the components are only held in by their bent leads and blobs of solder. A drop of cyanoacrylate "super glue" can hold down any larger components, components with fragile leads or any long leads or wires that might move.

Perforated Construction Board

A simple approach to circuit building uses a perforated board (perfboard). Perfboard is available with many different hole patterns. Choose the one that suits your needs. Perfboard is usually unclad, although it is made with pads that facilitate soldering.

Circuit construction on perforated board is easy. Start by placing the compo-

nents loosely on the board and moving them around until a satisfactory layout is obtained. Most of the construction techniques described in this chapter can be applied to perfboard. The audio amplifier of Fig 25.10 is shown constructed with this technique in **Fig 25.14**.

Perfboard and accessories are widely available. Accessories include mounting hardware and a variety of connection terminals for solder and solderless construction.

Terminal and Wire

A perfboard is usually used for this technique (**Fig 25.15**). Push terminals are inserted into the hole in a perfboard. Components can then be easily soldered to the terminals. As an alternative, drill holes into a bare or copper-clad board wherever they are needed. The components are usually mounted on one side of the board and wires are soldered to the bottom of the board, acting as wired PC-board "traces." If a component has a reasonably rigid lead to which you can attach other components, use that instead of a push terminal, a modification of the ground-plane construction technique.

If you are using a bare board to provide a ground plane, drill holes for your terminals with a high-speed PC-board drill and drill press. Mark the position of the hole with a center punch to prevent the drill from skidding. The hole should provide a snug fit for the push terminal.

Mount RF components on top of the board, keeping the dc components and much of the interconnecting wiring underneath. Make dc feed-through connections with terminals having bypass capacitors on top of the board. Use small solder-in feedthrough capacitors for more critical applications.

Solderless Prototype Board

One construction alternative that works well for audio and digital circuits is the solderless prototype board (protoboard), shown in **Fig 25.16**. It is usually not suitable for RF circuits.

A protoboard has rows of holes with spring-loaded metal strips inside the board. Circuit components and hookup wire are inserted into the holes, making contact with the metal strips. Components that are inserted into the same row are connected together. Component and interconnection changes are easy to make.

Protoboards have some minor disadvantages. The boards are not good for building RF circuits; the metal strips add too much stray capacitance to the circuit. Large component leads can deform the metal strips.

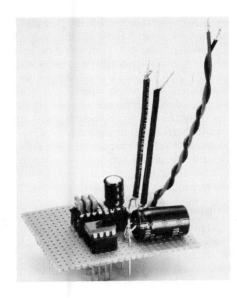

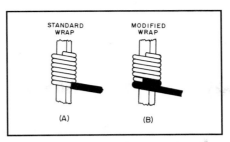

Fig 25.18 — Wire-wrap connections. Standard wrap is shown at A; modified wrap at B.

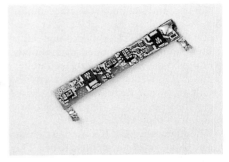

Fig 25.20 — A modern surface-mount IC, properly installed on a PC board.

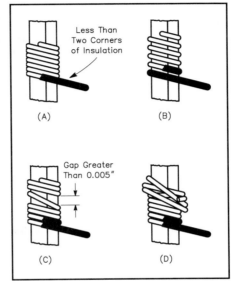

Fig 25.19 — Improper wire-wrap connections. Insufficient insulation for modified wrap is shown at A; a spiral wrap at B, where there is too much space between turns; an open wrap at C, where one or more turns are improperly spaced and an overwrap at D, where the turns overlap on one or more turns.

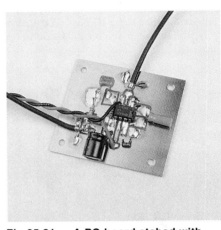

Fig 25.21 — A PC-board etched with copper "islands" for surface mounting of a standard DIP IC.

Fig 25.17 — The audio amplifier built using wire-wrap techniques.

Wire Wrap

Wire-wrap techniques can be used to quickly construct a circuit without solder. Low- and medium-speed digital circuits are often assembled on a wire-wrap board. The technique is not limited to digital circuits, however. **Fig 25.17** shows the audio amplifier built using wire wrap. Circuit changes are easy to make, yet the method is suitable for permanent assemblies.

Wire wrap is done by wrapping a wire around a small square post to make each connection. A wrapping tool resembles a thick pencil. Electric wire-wrap guns are convenient when many connections must be made. The wire is almost always #30 wire with thin insulation. Two wire-wrap methods are used: the standard and the modified wrap (**Fig 25.18**). The modified wrap is more secure. The wrap-post terminals are square (wire wrap works only on posts with sharp corners). They should be long enough for at least two connections. Fig 25.18 and **Fig 25.19** show proper and improper wire-wrap techniques. Mount

small components on an IC header plug. Insert the header into a wire-wrap IC socket as shown in Fig 25.17. The large capacitor in that figure has its leads soldered directly to wire-wrap posts.

Surface Mounting

Surface mounting is not new — it was an established ground-plane and professional technique for years before its appearance in consumer and amateur electronics. This technique is particularly suitable for PC-board construction, although it can be applied to many other construction techniques. Surface-mounted components take up very little space on a PC board.

Modern automated manufacturing techniques and surface-mount technology have evolved together; most modern ICs

are being made specifically for this technique. **Fig 25.20** shows a surface-mount IC soldered onto a board.

Chip resistors and capacitors are common in UHF and microwave designs. Chip devices have low stray inductance and capacitance, making them excellent components to use in this frequency range. Other components, such as transistors and diode arrays are also available in this space-saving format.

Surface-mount techniques are not limited to "surface-mount" ICs, however. This technique can be used to mount standard resistors, capacitors or ICs.

Two different ground-plane construction techniques work well for surface mounting components. One method is to cut out or etch small insulated islands in the PC board (see **Fig 25.21**). This works with either single- or double-sided board. Cut out the islands with a small hobby knife, making parallel cuts spaced as needed. Peel away the copper with the point of a hot soldering iron. An alternative is to use a hand-held hobbyist grinder with a cutting bit. You can also design a surface-mount PC-board pattern.

The second method of surface mounting is shown in **Fig 25.22**. Cut small patches of single-sided board and superglue them onto the copper ground plane. Although very effective, this technique is tedious for all but the simplest circuits. Don't glue your fingers to the PC board. Keep the glue solvent (usually acetone or nail-polish remover) nearby!

The surface-mount ICs used in industry are not easy for experimenters to use. They have tiny pins designed for precision PC boards. Most hams avoid these in homebrew designs. Sooner or later, you may need to replace one, though. If you do, don't try to get the old IC out in one piece! This will damage the IC beyond use anyway, and may damage the PC board in the process.

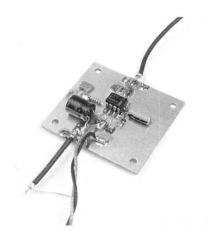

Fig 25.22 — A PC board with glued copper traces for surface mounting of a standard DIP IC.

Although it requires a delicate touch and small tools, it's possible to change a surface mount IC at home. To remove the old one, use small, sharp wire cutters to cut the IC pins flush with the IC. This usually leaves just enough old pin to grab with a tiny pair of needle-nose pliers or a hemostat. Heat the soldered connection with a small iron and use the pliers to gently pull the pin from the PC board.

To install a new part, apply a small blob of solder to one of the pads. Position the IC on the PC board and press it down while applying heat to the pin over the solder blob. With only that one pin soldered, inspect the position of each other pin relative to its pad. Reheat the first pin and reposition the part until all pins are in place. Then solder the remaining pins to their pads. Watch out for solder bridges. They are easy to make on traces with such small spacing. When you are done, inspect your work carefully.

Small "chip" surface-mount components are a bit trickier because they are too small to hold safely during soldering. Special care must be used when soldering chip components. **Fig 25.23** illustrates a good technique. All surfaces should be clean and lightly tinned before you solder the chip. Position the chip in place and hold it down with a toothpick. Do not use a screwdriver or tweezers, because the metal can easily damage the ceramic base of the component. Lay the chisel tip of a 15- to 20-W soldering pencil on the surface to which the chip is to be soldered, with the tip of the chisel just touching the chip component.

Touch and flow a minimum amount of solder between the chip and the soldering-pencil tip and pull the tip away at a low

angle to the surface. Repeat the procedure on the other side of the chip. If you don't use too much heat while soldering the second end, you may not have to hold it in place. Fig 25.23C shows how the connection should look when done. Inspect the solder and chip with a magnifier. The solder should have flowed properly and there should be no fractures or cracks. Don't overheat the chip or the metallization may separate from the ceramic and ruin the component. It requires a little practice, but the technique can be mastered.

Printed-Circuit Boards

Many builders prefer the neatness and miniaturization made possible by the use of etched printed-circuit boards (PC boards). Once designed, a PC board is easily duplicated, making PC boards ideal for group projects. To make a PC board, resist material is applied to a copper-clad bare PC board, which is then immersed into an acid etching bath to remove selected areas of copper. In a finished board, the conductive copper is formed into a pattern of conductors or "traces" that form the actual wiring of the circuit.

PC Board Stock

PC board stock consists of a sheet made from insulating material, usually glass epoxy or phenolic, coated with conductive copper. Copper-clad stock is manufactured with phenolic, FR-4 fiberglass and Teflon base materials in thicknesses up to $1/8$ inch. The copper thickness varies. It is usually plated from 1 to 2 oz per square foot of bare stock.

Resists

Resist is a material that is applied to a PC board to prevent the acid etchant from eating away the copper on those areas of the board that are to be used as conductors. There are several different types of resist materials, both commercial and home brew. When resist is applied to those areas of the board that are to remain as copper traces, it "resists" the acid action of the etchant.

The PC board stock must be clean before any resist is applied. This is discussed later in the chapter. After you have applied resist, by whatever means, protect the board by handling it only at its edges. Do not let it get scraped. Etch the board as soon as possible, to minimize the likelihood of oxidation, moisture or oils contaminating the resist or bare board.

Tape

To make a single PC board, Scotch, adhesive or masking tape, securely applied, makes a good resist. (Don't use drafting tape; its glue may be too weak to

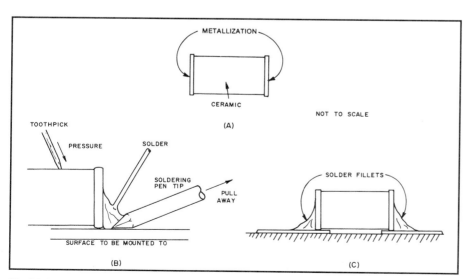

Fig 25.23 — A typical chip capacitor is shown at A. The proper technique for holding and soldering the chip is shown at B. At C, the final appearance of the component in the circuit. Check for good solder flow, no metallization separation and no cracks or fractures in the ceramic.

hold in the etching bath.) Apply the tape to the entire board, transfer the circuit pattern by means of carbon paper, then cut out and remove the sections of tape where the copper is to be etched away. An X-acto hobby knife is excellent for this purpose.

Resist Pens

Several electronics suppliers sell resist pens. Use a resist pen to draw PC-board artwork directly onto a bare board. Commercially available resist pens work well. Several types of permanent markers also function as resist, especially the "Sharpie" brand. They come in fine-point and regular sizes; keep two of each on hand.

Paint

Some paints are good resists. Exterior enamel works well. Nail polish is also good, although it tends to dry quickly so you must work fast. Paint the pattern onto the copper surface of the board to be etched. Use an artist's brush to duplicate the PC board pattern onto bare PC-board stock. Tape a piece of carbon paper to the PC-board stock. Tape the PC-board pattern to the carbon paper. Trace over the original layout with a ballpoint pen. The carbon paper transfers the outline of the pattern onto the bare board. Fill in the outline with the resist paint. After paint has been applied, allow it to dry thoroughly before etching.

Rub-On Transfer

Several companies, (Kepro Circuit Systems, DATAK Corp, GC Electronics) produce rub-on transfer material that can also be used as resist. Patterns are made with various width traces and for most components, including ICs. As the name implies, the pad or trace is positioned on the bare board and rubbed to adhere to the board.

Etchant

Etchant is an acid solution that is designed to remove the unwanted copper areas on PC-board stock, leaving other areas to function as conductors. Almost any strong acid bath can serve as an etchant, but some acids are too strong to be safe for general use. Two different etchants are commonly used to fabricate prototype PC boards: ammonium persulphate and ferric chloride. The latter is the more common of the two.

Ferric chloride etchant is usually sold ready-mixed. It is made from one part ferric chloride crystals and two parts water, by volume. No catalyst is required.

Etchant solutions become exhausted as they are used. Keep a supply on hand. Dispose of the used solution safely; follow the instructions of your local environmental protection authority.

Most etchants work better if they are hot. A board that takes 45 minutes to etch at room temperature will take only a few minutes if the etchant is hot. Use a heat lamp to warm the etchant to the desired temperature. A darkroom thermometer is handy for monitoring the temperature of the bath.

Be careful! Do not heat your etchant above the recommended temperature, typically 160°F. If it gets too hot, it will probably damage the resist. Hot or boiling etchant is also a safety hazard.

Insert the board to be etched into the solution and agitate it continuously to keep fresh chemicals near the board surface. This speeds up the etching process. Normally, the circuit board should be placed in the bath with the copper side facing up.

After the etching process is completed, remove the board from the tray and wash it thoroughly with water. Use medium-grade steel wool to rub off the resist.

WARNING: Use a glass or other nonreactive container to hold etching chemicals. Most etchants will react with a metal container. Etchant is caustic and can burn eyes or skin easily. Use rubber gloves and wear old clothing, or a lab smock, when working with any chemicals. If you get some on your skin, wash it with soap and cold water. Wear safety goggles (the kind that fit snugly on your face) when working with any dangerous chemicals. Read the safety labels and follow them carefully. If you get etchant in your eyes, wash immediately with large amounts of cool water and seek immediate medical help. Even a small chemical burn on your eye can develop into a serious problem.

Planning and Layout

A PC board can be a real convenience. If you want to build a project and a ready-made PC board is available, you can assemble the project quickly and expect it to work. This is true because someone else has done most of the real work involved — designing the PC board layout and fixing any "bugs" caused by intertrace capacitive coupling, ground loops and similar problems. In most cases, if a ready-made board is not available, ground-plane construction is a lot less work than designing, debugging and then making a PC board.

A later section of this chapter explains how to turn a schematic into a working circuit. It is not as simple as laying out the PC board just like the circuit is drawn on the schematic. Read that section before you design a PC board.

Rough Layout

Start by drawing a rough scale pictorial diagram of the layout. Draw the intercon-

necting leads to represent the traces that are needed on the board. Rearrange the layout as necessary to find an arrangement that completes all of the circuit traces with a minimum number of jumper-wire connections. In some cases, however, it is not possible to complete a design without at least a few jumpers.

Layout

After you have completed a rough layout, redraw the physical layout on a grid. Graph paper works well for this. Most IC pins are on 0.1-inch centers. Use graph paper that has 10 lines per inch to draw artwork at 1:1 and estimate the distance halfway between lines for 0.05-inch spacing. Drafting templates are helpful in the layout stage. Local drafting-supply stores should be able to supply them. The templates usually come in either full-scale or twice normal size.

To lay out a double-sided board, ensure that the lines on both sides of the paper line up (hold the paper up to the light). You can then use each side of the paper for each side of the board.

When using graph paper for a PC-board layout, include bolt holes, notches for wires and other mechanical considerations. Fit the circuit into and around these, maintaining clearance between parts.

Most modern components have leads on 0.1-inch centers. The rows of dual-inline-package (DIP) IC pins are spaced 0.3 or 0.4 inch. Measure the spacing for other components. Transfer the dimensions to the graph paper. It is useful to draw a schematic symbol of the component onto the layout.

Most IC specification sheets show a top view of the pin locations. If you are designing the "foil" side of a PC board, be sure to invert the pin out.

Draw the traces and pads the way they will look. Using dots and lines is confusing. It's okay to connect more than one lead per pad, or run a lead through a pad, although using more than two creates a complicated layout. In that case, there may be problems with solder bridges that form short circuits. Traces can run under some components; it is possible to put two or three traces between 0.4-inch centers for a $^1/_4$-W resistor, for example.

Leave power-supply and other dc paths for last. These can usually run just about anywhere, and jumper wires are fine for these noncritical paths.

Do not use traces less than 0.010 inch (10 mil) wide. If 1-oz stock is used, a 10-mil trace can safely carry up to 500 mA. To carry higher current, increase the width of the traces in proportion. (A trace should be 0.200 inch to carry 10 A, for example.) Allow 0.1 inch between traces for each kilovolt in the circuit.

When doing a double-sided board, use pads on both sides of the board to connect traces through the board. Home-brew PC boards do not use plated-through holes (a manufacturing technique that has copper and tin plating inside all of the holes to form electrical connections). Use a through hole and solder the associated component to both sides of the board. Make other through-hole connections with a small piece of bus wire providing the connection through the board; solder it on both sides. This serves the same purpose as the plated-through holes found in commercially manufactured boards.

After you have planned the physical design of the board, decide the best way to complete the design. For one or two simple boards, draw the design directly onto the board, using a resist pen, paint or rub-on resist materials. To transfer the design to the PC board, draw light, accurate pencil lines at 0.1- or 0.05-inch centers on the PC board. Draw both horizontal and vertical lines, forming a grid. You only need lines on one side. For single-sided boards, use this grid to transfer the layout directly onto the board surface. To make drilling easier, use a center punch to punch the centers of holes accurately. Do this before applying the resist so the grid is visible.

When drawing a pad with plenty of room around it, use a pad about 0.05 to 0.1 inch in diameter. For ICs, or other close quarters, make the pad as small as 0.03 inch or so. A "ring" that is too narrow invites soldering problems; the copper may delaminate from the heat. Pads need not be round. It's okay to shave one or more edges if necessary, to allow a trace to pass nearby.

Draw the traces next. A drafting triangle can help. It should be spaced about 0.1 inch above the table, to avoid smudging the artwork. Use a 9-inch or larger triangle, with a rubber grommet taped to each

corner (to hold it off the table). Select a sturdy triangle that doesn't bend easily.

Align the triangle with the grid lines by eye and make straight, even traces similar to the layout drawing. The triangle can help with angled lines, too. Practice on a few pieces of scrap board.

Make sure that the resist adheres well to the PC board. Most problems can be seen by eye; there can be weak areas or bare spots. If necessary, touch up problems with additional resist. If the board is not clean the resist will not adhere properly. If necessary, remove the resist, clean the board and start from the beginning.

Discard troublesome pens. Resist pens dry out quickly. Keep a few on hand, switch back and forth and put the cap back on each for a bit to give the pen a chance to recover.

Once all of the artwork on the board is drawn, check it against the original artwork. It is easy to leave out a trace. It is not easy to put copper back after a board is etched. In a pinch, replace the missing trace with a small wire.

Applied resist takes about an hour to dry at room temperature. Fifteen minutes in a 200°F oven is also adequate.

Special techniques are used to make double-sided PC boards. See the section on double-sided boards for a description.

Making a PC Board

Several techniques can be used to make PC boards. They usually start with a PC-board "pattern" or artwork. All of the techniques have one thing in common: this pattern needs to be transferred to the copper surface of the PC board. Unwanted copper is then removed by chemical or mechanical means.

Most variations in PC-board manufacturing technique involve differences in resist or etchant materials or techniques.

Cut the Board to Size

No matter what technique you use, you should determine the required size of the PC board, then cut the board to size. Trimming off excess PC-board material can be difficult after the components are installed.

Board Preparation

The bare (unetched) PC-board stock should be clean and dry before any resist is applied. (This is not necessary if you are using stock that has been treated with presensitized photoresist.) Wear rubber gloves when working with the stock to avoid getting fingerprints on the copper surface. Clean the board with soap and water, then scrub the board with #000 steel wool. Rinse the board thoroughly then dry it with a clean, lint-free cloth. Keep the board clean and free of fingerprints or for-

eign substances throughout the entire manufacturing process.

No-Etch PC Boards

The simplest way to make PC boards is to mechanically remove the unwanted copper. Use a grinding tool, such as the Moto-Tool manufactured by the Dremel Company (available at most hardware or hobby stores). Another technique is to score the copper with a strong, sharp knife, then remove unwanted copper by heating it with a soldering iron and lifting it off with a knife while it is still hot. This technique requires some practice and is not very accurate. It often fails with thin traces, so use it only for simple designs.

Photographic Process

Many magazine articles feature printed-circuit layouts. Some of these patterns are difficult to duplicate accurately by hand. A photographic process is the most efficient way to transfer a layout from a magazine page to a circuit board.

The resist ink, tape or dry-transfer processes can be time consuming and tedious for very complex circuit boards. As an alternative, consider the photo process. Not only does the accuracy improve, you need not trace the circuit pattern yourself!

A copper board coated with a light-sensitive chemical is at the heart of the photographic process. In a sense, this board becomes your photographic film.

Make a contact print of the desired pattern by transferring the printed-circuit artwork to special copy film. This film is attached to the copper side of the board and both are exposed to intense light. The areas of the board that are exposed to the light—those areas not shielded by the black portions of the artwork—undergo a chemical change. This creates a transparent image of the artwork on the copper surface.

Develop the PC board, using techniques and chemicals specified by the manufacturer. After the board is developed, etch it to remove the copper from all areas of the board that were exposed to the light. The result is a PC board that looks like it was made in a factory.

Kepro and GC Thorsen both sell materials and supplies for all types of PC-board manufacturing. Radio Shack also sells PC-board materials. See **Fig 25.24**. If you're looking for printed-circuit board kits, chemicals, tools and other materials, contact Ocean State Electronics. They carry products by Kepro, GC Thorsen, Datak and the Meadowlake Corporation.

Iron-On Resist

One company that makes an iron-on resist is the Meadowlake Corporation.

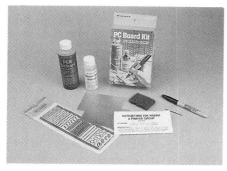

Fig 25.24 — PC-board materials are available from several sources. This kit is from Radio Shack.

Double-Sided PC Boards — by Hand!

Forget those nightmares about expensive photoresists that didn't work; forget that business of fifty bucks a board! You don't need computer-aided design to make a double-sided PC board; just improve on the basics, and keep it simple. Anyone can make low-cost double-sided boards with traces down to 0.020 inch, with perfect front-to-back hole registration.

To make a double-sided board, drill the holes before applying the resist artwork; that is the only way to assure good front-to-back registration. The artwork on both sides can then be properly positioned to the holes. PC-board drilling was discussed earlier in the text.

After you have drilled the board, clean its surface thoroughly. After that, wear clean rubber or cotton gloves to keep it clean. One fingerprint can really mess up the application of resist or the etchant.

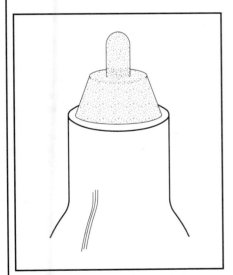

Fig A — Make a permanent marker into a specialized PC-board drawing tool. Simply press the marker point into a drilled hole to form a modified point as shown. More pressure produces a wider shoulder that makes larger pads on the PC board.

Tape the board to your work surface, making sure it can't move around. Transfer the artwork from your layout grid to the PC board, drawing by hand with a resist pen.

Allot enough time to finish at least one side of the artwork in one sitting. Start with the pads. To make a handy pad-drawing tool, press the tip of a regular-size Sharpie into one of the drilled PC-board holes. This "smooshes" the tip into the shape of the hole, leaving a flat shoulder to draw the pad. See **Fig A**. The diameter of the pad is determined by how hard the pen is pressed; pressing too hard forms a pad that is way too large for most applications. Practice on scrap board first. Use this modified pen to fill in all the holes and draw the pads at the same time. Use an unmodified resist pen to draw all of the traces and to touch up any voids or weak areas in the pads. For the rest of the drawing, the procedure described for single-sided boards applies to double-sided boards, too.

After the resist is applied to the first side, carefully draw the second side. Inspect the board thoroughly; you may have scratched or smudged the first side while you were drawing the second.

Etching a double-sided board is not much different than etching a single-sided board, except that you must ensure that the etchant is able to reach both sides of the board. If you dunk the board in and out of the etchant solution, both sides are exposed to the etchant. If you use a tray, put some spacers on the bottom and rest the board on the spacers. (The spacers must be put on the board edges, not where you want to actually etch.) This ensures that etchant gets to both sides. If you use this method, turn the board over once or twice during the process. — *Dave Reynolds, KE7QF*

Or Photo-Etched

You can also make double-sided boards at home without drawing the

layout by hand. This procedure can't produce results to match the finest professionally made double-sided boards, but it can make boards that are good enough for many moderately complex projects.

Start with the same sort of artwork used for single-sided boards, but leave a margin for taping at one edge. It is critical that the patterns for the two sides are accurately sized. The chief limiting factor in this technique is the requirement that matching pads on the two sides are positioned correctly. Not only must the two sides match each other, but they must also be the correct size for the parts in the project. Slight reproduction errors can accumulate to major problems in the length of a 40-pin DIP IC. One good tool to achieve this requirement is a photocopy machine that can make reductions and enlargements in 1% steps. Perform a few experiments to arrive at settings that yield accurately sized patterns.

Choose two holes at opposite corners of the etching patterns. Tape one of the two patterns to one side of the PC board. Choose some small wire and a drill bit that closely matches the wire diameter. For example, #20 enameled wire is a close match for a #62 or a #65 drill, depending on the thickness of the wire's enamel coating. Drill through the pattern and the board at the two chosen holes. Drill the chosen holes through the second pattern. Place two pieces of the wire through the PC board and slide the second pattern down these wire "pins" to locate the pattern on the board. Tape the second pattern in position and remove the pins. From this point on, expose and process each side of the board as if it were a single-sided board, but take care when exposing each side to keep the reverse side protected from light.
— *Bob Schetgen, KU7G*

Their products make an artwork positive using a standard photocopier. A clothes iron transfers the printed resist pattern to the bare PC board.

Some experimenters have reported satisfactory results using standard photocopier paper or the output from a laser printer. Apparently the toner makes a reasonable resist. Note that the artwork for this method must be reversed with respect to a normal etching pattern because the print must be placed with the toner against the copper.

To transfer the resist pattern onto the board, place the pattern on the board (image side toward the copper), then firmly press a hot iron onto the entire surface. Use plenty of heat and even pressure. This melts the resist, which then sticks to the bare PC board. This is not a perfect process; there will probably be bad areas on the resist. The amount of heat, the cleanliness of the bare board and the "skill" of the operator may affect the outcome.

The key to making high quality boards with the photocopy techniques is to be good at retouching the transferred resist. Fortunately, the problems are usually easy to retouch, if you have a bit of patience. A resist pen does a good job of reinforcing any spotty areas in large areas of copper.

Double-Sided PC Boards

All of the examples used to describe the above techniques were single-sided PC boards, with traces on one side of the board and either a bare board or a ground plane on the other side. PC boards can also have patterns etched onto both sides, or even have multiple layers. Most home-construction projects use single-sided boards, although some kit builders supply double-sided boards. Multilayer boards are rare in ham construction. One method for making double-sided boards is described in the sidebar, "Double-Sided PC Boards—by Hand."

Tin Plating

Most commercial PC boards are tin plated, to make them easier to solder.

Commercial tin-plating techniques require electroplating equipment not readily available to the home constructor. Immersion tin plating solutions can deposit a thin layer of tin onto a copper PC board. Using them is easy; put some of the solution into a plastic container and immerse the board in the solution for a few minutes. The chemical action of the tin-plating solution replaces some of the copper on the board with tin. The result looks nearly as good as a commercially made board. Agitate the board or solution from time to time. When the tinning is complete, take the board out of the solution and rinse it for five minutes under running water. If you don't remove all of the residue, solder may not adhere well to the surface. Kepro sells immersion tin plating solution.

Drilling a PC Board

After you make a PC board using one of the above techniques, you need to drill holes in the board for the components. Use a drill press, or at least improvise one. Boards can be drilled entirely "free hand"

with a hand-held drill but the potential for error is great. A drill press or a small Moto-Tool in an accessory drill press makes the job a lot easier. A single-sided board should be drilled after it is etched; the easiest way to do a double-sided board is to do it before the resist is applied.

To drill in straight lines, build a small movable guide for the drill press so you can slide one edge of the board against it and line up all of the holes on one grid line at a time. See **Fig 25.25**. This is similar to the "rip fence" set up by most woodworkers to cut accurately and repeatably with a table saw.

The drill-bit sizes available in hardware stores are too big for PC boards. You can use high-speed steel bits, but glass epoxy stock tends to dull these after a few hundred holes. (When your drill bit becomes worn, it makes a little "hill" around each drilled hole, as the worn bit pushes and pulls the copper rather than drilling it.) A PC-board drill bit, available from many electronic suppliers, will last for thousands of holes! If you are doing a lot of boards, it is clearly worth the investment.

Small drill bits are usually ordered by number. Here are some useful numbers and their sizes:

Number	Diameter
68	0.0310"
65	0.0350"
62	0.0380"
60	0.0400"

Use high RPM and light pressure to make good holes. Count the holes on both the board and your layout drawing to ensure that none are missed. Use a larger-size drill bit, lightly spun between your fingers, to remove any burrs. Don't use too much pressure; remove only the burr.

"Ready-Made" PC Boards

Utility PC Boards

"Utility" PC boards are an alternative to custom-designed etched PC boards. They offer the flexibility of perforated board construction and the mechanical and electrical advantages of etched circuit connection pads. Utility PC boards can be used to build anything from simple passive filter circuits to computers.

Circuits can be built on boards on which the copper cladding has been divided into connection pads. Power supply voltages can be distributed on bus strips. Boards like those shown in **Fig 25.26** are commercially available.

An audio amplifier constructed on a utility PC board is shown in **Fig 25.27**. Component leads are inserted into the board and soldered to the etched pads. Wire jumpers connect the pads together to

Fig 25.25 — This home-built drill fence makes it easy to drill PC-board holes in straight rows.

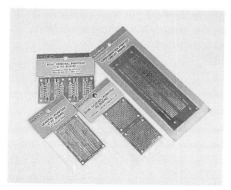

Fig 25.26 — Utility PC boards like these are available from many suppliers.

(A)

(B)

Fig 25.27 — The audio amplifier built on a multipurpose PC breadboard. Top view at A; bottom view at B.

complete the circuit.

Utility boards with one or more etched plugs for use in computer-bus, interface and general purpose applications are widely available. Connectors, mounting hardware and other accessories are also available. Check with your parts supplier for details.

PC-Board Assembly Techniques

Once you have etched and drilled a PC board you are ready to use it in a project. Several tools come in handy: needle-nose pliers, diagonal cutters, pocket knife, wire strippers, clip leads and soldering iron.

Cleanliness

Make sure your PC board and component leads are clean. Clean the entire PC board before assembly; clean each component before you install it. Corrosion looks dark instead of bright and shiny. Don't use sandpaper to clean your board. Use a piece of fine steel wool or a Scotchbrite cleaning pad to clean component leads or PC board before you solder them together.

Installing Components

In a construction project that uses a PC board, most of the components are installed on the board. Installing components is easy — stick the components in the right board holes, solder the leads, and cut off the extra lead length. Most construction projects have a parts-placement diagram that shows you where each component is installed.

Getting the components in the right holes is called "stuffing" the circuit board. Inserting and soldering one component at a time takes too long. Some people like to put the components in all at once, and then turn the board over and solder all the leads. If you bend the leads a bit (about 20°) from the bottom side after you push them through the board, the components are not likely to fall out when you turn the board over.

Start with the shortest components — horizontally mounted diodes and resistors. Larger components sometimes cover smaller components, so these smaller parts must be installed first. Use adhesive tape to temporarily hold difficult components in place while you solder.

PC-Board Soldering

To solder components to a PC board, bend the leads at a slight angle; apply the soldering iron to one side of the lead, and flow the solder in from the other side of the lead. See **Fig 25.28**. Too little heat causes a bad or "cold" solder joint; too much heat can damage the PC board. Prac-

tice a bit on some spare copper stock before you tackle your first PC board project. After the connection is soldered properly, clip the lead flush with the solder.

Special Concerns

Make sure you have the components in the right holes before you solder them. Components that have polarity, such as diodes, ICs and some capacitors must be oriented as shown on the parts-placement diagram.

FROM SCHEMATIC TO WORKING CIRCUIT

Some people don't know how to turn a schematic into a working circuit. One thing is usually true — you can't build it the way it looks on the schematic. Many design and layout considerations that apply in the real world of practical electronics don't appear on a schematic.

PC Boards — Always the Best Choice?

PC boards are everywhere — in all kinds of consumer electronics, in most of your Amateur Radio equipment. They are also used in most kits and construction projects. A newcomer to electronics might think that there is some unwritten law against building equipment in any other way!

The misconception that everything needs to be built on a printed-circuit board is often a stumbling block to easy project construction. In fact, a PC board is probably the worst choice for a one-time project. In actuality, a moderately complex project (like a QRP transmitter) can be built in much less time using other techniques. The additional design, layout and manufacturing is usually much more work than it would take to build the project by hand.

So why does everyone use PC boards? The most important reason is that they are reproducible. They allow many units to be mass-produced with exactly the same layout, reducing the time and work of conventional wiring and minimizing the possibilities of wiring errors. If you can buy a ready-made PC board or kit for your project, it can save a lot of construction time.

Using a PC board usually makes project construction easier by minimizing the risk of wiring errors or other construction blunders. Inexperienced constructors usually feel more confident when construction has been simplified to the assembly of components onto a PC board. One of the best ways to get started with home construction (to some the best part of Amateur Radio) is to start by assembling a few kits using PC boards. The ARRL Technical

Information Service (TIS) has prepared a list of kit manufacturers. If you would like one, send an SASE, with a request for the "Kits" TIS information package, to the ARRL Technical Department Secretary.

One-Time Projects

Kits are fun, but another facet of electronics construction is building and developing your own circuits, starting from circuit diagrams. For one-time construction, PC boards are really not necessary. It takes time to lay out, drill and etch a PC board. Alterations are difficult to make if you change your ideas or make a mistake. Most important, PC boards aren't always the best technique for building RF circuits.

Layout

A circuit diagram is a poor guide toward a proper layout. Circuit diagrams are drawn to look attractive on paper. They follow drafting conventions that have very little to do with the way the circuit works. On a schematic, ground and supply voltage symbols are scattered all over the place. The first rule of RF layout is — *do not wire RF circuits as they are drawn!* How a circuit works in practice depends on the layout. Poor layout can ruin the performance of even a well-designed circuit.

How to Design a Good Circuit Layout

The easiest way to explain good layout practices is to take you through an example. **Fig 25.29** is the circuit diagram of a two-stage receiver IF amplifier using dual-gate MOSFETs. It is only a design example, so the values are only typical. To analyze which things are important to the layout of this circuit, ask these questions:

- Which are the RF components, and which are only involved with LF or dc?

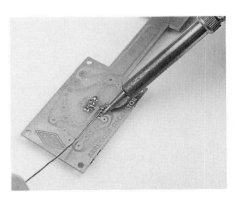

Fig 25.28 — This is how to solder a component to a PC board. Make sure that the component is flush with the board on the other side.

- Which components are in the main RF signal path?
- Which components are in the ground return paths?

Use the answers to these questions to plan the layout. The RF components that are in the main RF signal path are usually the most critical. The AF or dc components can usually be placed anywhere. The components in the ground return path should be positioned so they are easily connected to the circuit ground. Answer the questions, apply the answers to the layout, then follow these guidelines:

- Avoid laying out circuits so their inputs and outputs are close together. If a stage's output is too near a previous stage's input, the output signal can feedback into the input and cause problems.
- Keep component leads as short as practical. This doesn't necessarily mean as short as possible — just consider lead length as part of your design.
- Remember that metal transistor cases conduct, and that a transistor's metal case is usually connected to one of its leads. Prevent cases from touching ground or other components, unless called for in the design.

In our design example, the RF components are shown in heavy lines, though not

all of these components are in the main RF signal path. The RF signal path consists of T1/C1, Q1, T2/C4, C7, Q2, T3/C11. These need to be positioned in almost a straight line, to avoid feedback from output to input. They form the backbone of the layout, as shown in **Fig 25.30A**.

The question about ground paths requires some further thought — what is really meant by "ground" and "ground-return paths"? Some points in the circuit need to be kept at RF ground potential. The best RF ground potential on a PC board is a copper ground plane covering one entire side. Points in the circuit which cannot be connected directly to ground for dc reasons must be bypassed ("decoupled") to ground by capacitors that provide ground-return paths for RF.

In Fig 25.30, the components in the ground-return paths are the RF bypass capacitors C2, C3, C5, C8, C9 and C12. R4 is primarily a dc biasing component, but it is also a ground return for RF so its location is important.

The values of RF bypass capacitors are chosen to have a low reactance at the frequency in use; typical values would be 0.1 µF at LF, 0.01 µF at HF, and 0.001 µF or less at VHF. Not all capacitors are suitable for RF decoupling; the most common are disc ceramic capacitors. RF decoupling capacitors should always have

short leads.

Almost every RF circuit has an input, an output and a common ground connection. Many circuits also have additional ground connections, both at the input side and at the output side. Maintain a low-impedance path between input and output ground connections. The input ground connections for Q1 are the grounded ends of C1 and the two windings of T1. (The two ends of an IF transformer winding are generally not interchangeable; one is designated as the "hot" end, and the other must be connected or bypassed to RF ground.) The capacitor that resonates with the adjustable coil is often mounted inside the can of the IF transformer, leaving only two component leads to be grounded as shown in Fig 25.30B.

The RF ground for Q1 is its source connection via C3. Since Q1 is in a plastic package that can be mounted in any orientation, you can make the common ground either above or below the signal path in Fig 25.30B. Although the circuit diagram shows the source at the bottom. the practical circuit works much better with the source at the top, because of the connections to T2.

It's a good idea to locate the hot end of the main winding close to the drain lead of the transistor package, so the other end is toward the top of Fig 25.30B. If the source

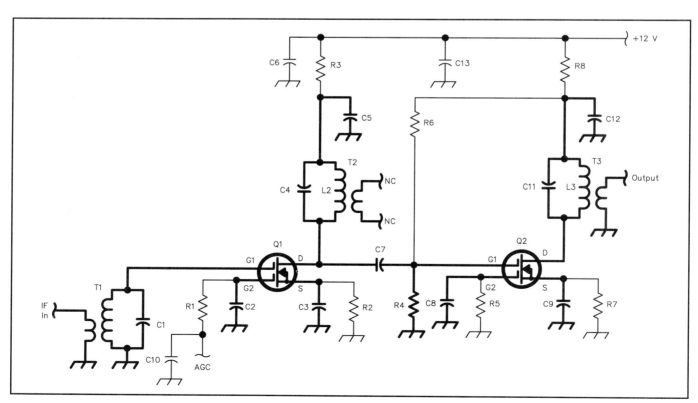

Fig 25.29 — The IF amplifier used in the design example. C1, C4 and C11 are not specified because they are internal to the IF transformers.

of Q1 is also toward the top of the layout, there is a common ground point for C3 (the source bypass capacitor) and the output bypass capacitor C5. Gate 2 of Q1 can safely be bypassed toward the bottom of the layout.

C7 couples the signal from the output of Q1 to the input of Q2. The source of Q2 should be bypassed toward the top of the layout, in exactly the same way as the source of Q1. R4 is not critical, but it should be connected on the same side as the other components. Note how the pinout of T3 has placed the output connection as far as possible from the input. With this layout for the signal path and the critical RF components, the circuit has an excellent chance of working properly.

DC Components

The rest of the components carry dc, so their layout is much less critical. Even so, try to keep everything well separated from the main RF signal path. One good choice is to put the 12-V connections along the top of the layout, and the AGC connection at the bottom. The source bias resistors R2 and R7 can be placed alongside C3 and C9. The gate-2 bias resistors for Q2, R5 and R6 are not RF components so their locations aren't too critical. R7 has to cross the signal path in order to reach C12, however, and the best way to avoid signal pickup would be to mount R7 on the opposite side of the copper ground plane from the signal wiring. Generally speaking, $1/8$-W or $1/4$-W metal-film or carbon-film resistors are best for low-level RF circuits.

Actually, it is not quite accurate to say that resistors such as R3 and R8 are not "RF" components. They provide a high impedance to RF in the positive supply lead. Because of R8, for example, the RF signal in T2 is conducted to ground through C5 rather than ending up on the 12-V line, possibly causing unwanted RF feedback. Just to be sure, C6 bypasses R3 and C13 serves the same function for R8. Note that the gate-2 bias resistor R6 is connected to C12 rather than directly to the 12-V supply, to take advantage of the extra decoupling provided by R8 and C13.

If you build something, you want it to work the first time, so don't cut corners! Some commercial PC boards take liberties with layout, bypassing and decoupling. Don't assume that you can do the same. Don't try to eliminate "extra" decoupling components such as R3, C6, R8 and C13, even though they might not all be absolutely necessary. If other people's designs have left them out, put them in again. In the long run it's far easier to take a little more time and use a few extra components, to build in some insur-

ance that your circuit will work. For a one-time project, the few extra parts won't hurt your pocket too badly; they may save untold hours in debugging time.

A real capacitor does not work well over a large frequency range. A 10-μF electrolytic capacitor cannot be used to bypass or decouple RF signals. A 0.1-μF capacitor will not bypass UHF or microwave signals. Choose component values to fit the range. The upper frequency limit is limited by the series inductance, L_S. In fact, at frequencies higher than the frequency at which the capacitor and its series inductance form a resonant circuit, the capacitor actually functions as an inductor. This is why it is a common practice to use two capacitors in parallel for bypassing, as shown in **Fig 25.31**. At first glance, this might appear to be unnecessary. However, the self-resonant frequency of C1 is usually 1 MHz or less; it cannot supply any bypassing above that frequency. However, C2 is able to bypass signals up into the lower VHF range.

Let's summarize how we got from Fig 25.29 to Fig 25.30B:

• Lay out the signal path in a straight line.
• By experimenting with the placement and orientation of the components in the

RF signal path, group the RF ground connections for each stage close together, without mixing up the input and output grounds.
• Place the non-RF components well clear of the signal path, freely using decoupling components for extra measure.

Practical Construction Hints

Now it's time to actually construct a project. The layout concepts discussed earlier can be applied to nearly any construction technique. Although you'll eventually learn from your own experience, the following guidelines give a good start:

• Divide the unit into modules built into separate shielded enclosures — RF, IF, VFO, for example. Modular construction improves RF stability, and makes the individual modules easier to build and test. It also means that you can make major changes without rebuilding the whole unit. RF signals between the modules can usually be connected using small coaxial cable.
• Use a full copper ground plane. This is your largest single assurance of RF stability and good performance.
• Keep inputs and outputs well separated

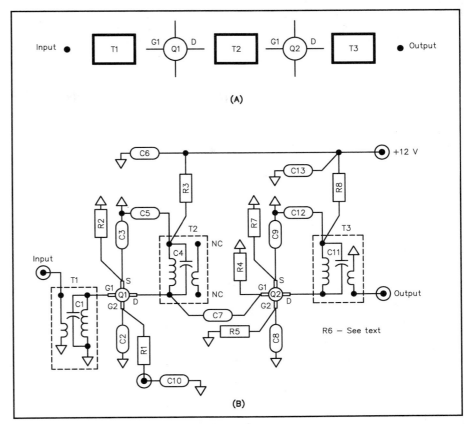

Fig 25.30 — Layout sketches. The preliminary line-up is shown in A; the final layout in B.

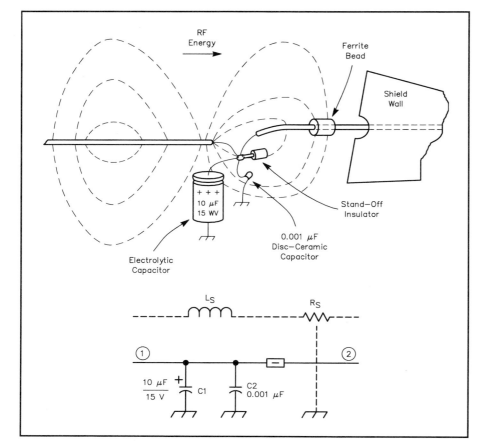

Fig 25.31 — Two capacitors in parallel afford better bypassing across a wide frequency range.

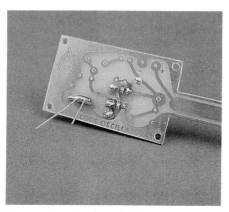

Fig 25.32 — A solder bridge has formed a short circuit between PC board traces.

Combination Techniques

You can use a mixture of construction techniques on the same board and in most cases you probably should. Even though you choose one style for most of the wiring, there will probably be places where other techniques would be better. If so, do whatever is best for that part of the circuit. The resulting hybrid may not be pretty (these techniques aren't called "ugly construction" for nothing), but it will work!

Mount dual-in-line package (DIP) ICs in an array of drilled holes, then connect them using wired traces as described earlier. It is okay to mount some of the components using a ground-plane method, push pins or even wire wrap. On any one board, you may use a combination of these techniques, drilling holes for some ICs, or gluing others upside down, then surface mounting some of the pins, and other techniques to connect the rest. These combination techniques are often found in a project that combines audio, RF and digital circuitry.

A Final Check

No matter what construction technique is chosen, do a final check before applying power to the circuit! Things do go wrong, and a careful inspection minimizes the risk of a project beginning and ending its life as a puff of smoke! Check wiring carefully. Make a photocopy of the schematic and mark each lead on the schematic with a red X when you've verified that it's connected to the right spot in the circuit.

Inspect solder connections. A bad solder joint is much easier to find before the PC board is mounted to a chassis. Look for any damage caused to the PC board by soldering. Look for solder "bridges" between adjacent circuit-board traces. Solder bridges (**Fig 25.32**) occur when solder accidentally connects two or more conductors that are supposed to be isolated. It is often difficult to distinguish a solder bridge from a conductive trace on a tin-plated board. If you find a bridge, remelt it and the adjacent trace or traces to allow the solder's surface tension to absorb it. Double check that each component is installed in the proper holes on the board and that the orientation is correct. Make sure that no component leads or transistor tabs are touching other components or PC board connections. Check the circuit voltages before installing ICs in their sockets. Ensure that the ICs are oriented properly and installed in the correct sockets.

for each stage, and for the whole unit. If possible, lay out all stages in a straight line. If an RF signal path doubles back or recrosses itself it usually results in instability.

- Keep the stages at different frequencies well-separated to minimize interstage coupling and spurious signals.
- Use interstage shields where necessary, but don't rely on them to cure a bad layout.
- Make all connections to the ground plane short and direct. Locate the common ground for each stage between the input and the output ground. Single-point grounding may work for a single stage, but it is rarely effective in a complex RF system.
- Locate frequency-determining components away from heat sources and mount them so as to maximize mechanical strength.
- Avoid unwanted coupling between tuned circuits. Use shielded inductors or toroids rather than open coils. Keep the RF high-voltage points close to the ground plane. Orient air-wound coils at right angles to minimize mutual coupling.
- Use lots of extra RF bypassing, especially on dc supply lines.
- Try to keep RF and dc wiring on opposite sides of the board, so the dc wiring is well away from RF fields.
- Compact designs are convenient, but don't overdo it! If the guidelines cited above mean that a unit needs to be bigger, make it bigger.

Microwave Construction Techniques

Microwave construction is becoming more popular, but at these frequencies the size of physical component leads and PC-board traces cannot be neglected. Microwave construction techniques either minimize these stray values or make them part of the circuit design.

Microwave construction does not always require tight tolerances and precision construction. A fair amount of error can often be tolerated if you are willing to tune your circuits, as you do at MF/HF. This usually requires the use of variable components that can be expensive and tricky to adjust.

Proper design and construction techniques, using high precision, can result in a "no-tune" microwave design. To build one of these no-tune projects, all you need do is buy the parts and install them on the board. The circuit tuning has been precisely controlled by the board and component dimensions so the project should work.

One tuning technique you can use with a microwave design, if you have the suitable test equipment, is to use bits of copper foil or EMI shielding tape as "stubs" to tune circuits. Solder these small bits of conductor into place at various points in the circuit to make reactances that can actually tune a circuit. After their position has been determined as part of the design, tuning is accomplished by removing or adding small amounts of conductor, or slightly changing the placement of the tuning stub. The size of the foil needed depends on your ability to determine changes in circuit performance, as well as the frequency of operation and the circuit board parameters. A precision setup that lets you see tiny changes allows you to use very small pieces of foil to get the best tuning possible.

From a mechanical accuracy point of view, the most tolerant type of construction is waveguide construction. Tuning is usually accomplished via one or more screws threaded into the waveguide. It becomes unwieldy to use waveguide on the amateur bands below 10 GHz because the dimensions get too large.

At 24 GHz and above, even waveguide becomes small and difficult to work with. At these frequencies, most readily available coax connectors work unreliably, so these higher bands are really a challenge. Special SMA connectors are available for use at 24 GHz.

Modular construction is a useful technique for microwave circuits. Often, circuits are tested by hooking their inputs and output to known 50-Ω sources and loads. Modules are typically kept small to prevent the chassis and PC board from acting as a waveguide, providing a feedback path between the input and output of a circuit, resulting in instability.

At microwave frequencies, the mechanical aspects and physical size of circuits become very much a part of the design. A few millimeters of conductor has significant reactance at these frequencies. This even affects VHF and HF designs! The traces and conductors used in an HF or VHF design resonate on microwave frequencies. If a high-performance FET has lots of gain in this region, a VHF preamplifier might also function as a 10-GHz oscillator if the circuit stray reactances were just right (or wrong!). You can prevent this by using shields between the input and output or by adding microwave absorptive material to the lid of the shielded module. (SHF Microwave sells absorptive materials. See the Address List in the **References** chapter.)

It is important to copy microwave circuits exactly, unless you really know what you are doing. "Improvements," such as better shielding or grounding can sometimes cause poor performance. It isn't usually attractive to substitute components, particularly with the active devices. It may look possible to substitute different grades of the same wafer, such as the ATF13135 and the AFT13335, but these are really the same transistor with different performance measurements. While two transistors may have exactly the same gain and noise figure at the desired operating frequency, often the impedances needed to maintain stability at other frequencies are be different. Thus, the "substitute" may oscillate, while the proper transistor would work just fine.

You can often substitute MMICs (monolithic microwave integrated circuits) for one another because they are designed to be stable and operate with the same input and output impedances (50 Ω).

The size of components used at microwaves can be critical—in some cases, a chip resistor 80 mils across is not a good substitute for one 60 mils across. Hopefully, the author of a construction project tells you which dimensions are critical, but you can't always count on this; the author may not know. It's not unusual for a person to spend years building just one prototype, so it's not surprising that the author might not have built a dozen different samples to try possible substitutions.

When using glass-epoxy PC board at microwave frequencies, the crucial board parameter is the thickness of the dielectric. It can vary quite a bit, in excess of 10%. This is not surprising; digital and lower-frequency analog circuits work just fine if the board is a little thinner or thicker than usual. Some of the board types used in microwave-circuit construction are a generic teflon PC board, Duroid 5870 and 5880. These boards are available from Microwave Components of Michigan. See the Address List in the **References** chapter.

Proper connectors are a necessary expense at microwaves. At 10 GHz, the use of the proper connectors is essential for repeatable performance. Do not hook up microwave circuits with coax and pigtails. It might work but it probably can't be duplicated. SMA connectors are common because they are small and work well. SMA jacks are sometimes soldered in place, although 2-56 hardware is more common. — *Zack Lau, KH6CP, ARRL Laboratory Engineer*

Other Construction Techniques

Wiring

Select the wire used in connecting amateur equipment by considering: the maximum current it must carry, the voltage its insulation must withstand and its use.

To minimize EMI, the power wiring of all transmitters should use shielded wire.

Receiver and audio circuits may also require the use of shielded wire at some points for stability or the elimination of hum. Coaxial cable is recommended for all 50-Ω circuits. Use it for short runs of high-impedance audio wiring.

When choosing wire, consider how much current it will carry. The Wire Table in the **References** chapter lists common wire sizes and current capacities. Stranded wire is usually preferred over solid wire because stranded wire better withstands the inevitable bending that is part of building and troubleshooting a circuit. Solid wire is more rigid than stranded wire; use it where mechanical rigidity is needed or desired.

Wire with typical plastic insulation is

How to Buy Parts for Electronics Projects

The number one question received by the ARRL Technical Information Service starts out "Where can I buy…" It seems that one of the most perplexing problems faced by the would-be constructor is where to get parts. Sometimes you are lucky—the circuit author has made a kit available. But not every project has a ready-made kit. If you would like to expand your construction horizons, you can learn to be your own "purchasing manager." That means searching out parts sources and dealing with them in person.

In reality, it is not all that difficult to find most parts. Unfortunately, though, the days of the local electronic parts supplier seem to be gone. This is not surprising. Years ago an electronics supplier had to stock a relatively small number of electronic components — resistors, capacitors, tube sockets, a few relays and variable resistors. Technology has increased the number of components by a few orders of magnitude. Nowadays, the number of integrated circuits alone is enough to fill a multivolume book. No single electronics supplier could possibly stock them all. It has become a mail-order world; the electronics world is no exception.

Although it is no longer always possible to purchase all of your electronic needs from a local electronics supplier, the good news is that you don't have to! For a few dollars, mail-order companies are willing to supply whatever you need. You only need do two things to obtain nearly any electronic component — make a phone call and write a check.

Become an electronic catalog collector. The **References** chapter has a list of electronic-component suppliers. Write to these companies and request their catalogs. Electronic suppliers also advertise in magazines that cater to ham-radio and electronics enthusiasts. If you are lucky, you may have a local source of electronics parts. Look in the Yellow Pages under "Electronic Equipment Suppliers" to find the local outlets. Radio Shack is one local source found nearly everywhere. They carry an assortment of the more common electronic parts. You'll probably need to order from more than one mail-order company. (It's almost a corollary to Murphy's Law: No matter how wide a selection you find in one mail-order catalog, there's always at least one part you must buy somewhere else!)

While you're waiting for your catalogs, look at the parts list for the project you want to build. Unfortunately, you can't just photocopy the list and send it off to a mail-order company with a note that says "please send me these parts." You need to convert the part list into a part-order list that shows the order number and quantity required of each component. This may require a similar list for each parts supplier where one supplier does not have all the parts.

Check the type, tolerance, power rating and other key characteristics of the parts. Group the parts by those parameters before grouping them by value. If all of the circuit's components are already grouped by value on the parts list, you can just count the number of each value. Each time you add parts to the order list, check them off the published parts list. Sometimes the parts list does not include common components like resistors and capacitors. If this is the case, make a copy of the schematic and check off the parts as you build your shopping list.

Although you'll probably be able to order exactly the right number of each part for a project, buy a few extras of some parts for your junk box. It's always good to have a few extra parts on hand; you may break a component lead during assembly, or damage a solid-state component with too much heat or by wiring it in backwards. If you don't have extras, you'll need to order another part. Even if you don't need the extras for this project, they may come in handy, and you'll be encouraged to build another project! Pick up an extra toroid or two as well.

Now's the time to decide whether you're going to build your project with ground-plane construction or PC board. If you need a PC-board, FAR Circuits and others have them for many ARRL book and *QST* projects. If you're going to use ground-plane construction, buy a good-sized piece of single-sided copper-clad board—glass-epoxy board if you can. Phenolic board is inferior because it is brittle and deteriorates rapidly with soldering heat.

Don't forget an enclosure for your project. This is often overlooked in parts lists for most projects, because different builders like different enclosures. Make sure there's room in the box for all of the components used in your project. Some people like to cram projects in the smallest possible box, but miniaturization can be extremely frustrating if you're not good at it.

There are almost always a few items you can't get from one company and most have minimum orders. You may need to distribute your order between two or more companies to meet minimum-order requirements. Some companies put out beautiful catalogs, but their minimum order is $25 or they charge $5 for shipping if you place a small order.

If you order enough parts, you'll soon find out which companies you like to deal with and which have slow service. It is frustrating to receive most of an order, then wait months for the parts that are on back-order. If you don't want the company to back-order your parts, write clearly on the order form, "Do not back-order parts." They will then ship the parts they have and leave you to order the rest from somewhere else.

If you are in a hurry, call the company to inquire about the availability of the parts in your order. Some companies take credit-card orders over the telephone. Some companies hold orders a few weeks to allow personal checks to clear.

If you are familiar with the catalogs and policies of electronic-component suppliers, you will find that getting parts is not difficult. Concentrate on the fun part — building the circuit and getting it working. —*Bruce Hale, KB1MW*

---

good for voltages up to about 500 V. Use Teflon-insulated or other high-voltage wire for higher voltages. Teflon insulation does not melt when a soldering iron is applied. This makes it particularly helpful in tight places or large wiring harnesses. Although Teflon-insulated wire is more expensive, it is often available from industrial surplus houses. Inexpensive wire strippers make the removal of insulation from hookup wire an easy job. Solid wire is often used to wire HF circuits. Bare soft-drawn tinned wire, #22 to #12 (depending on mechanical requirements) is suitable. Avoid kinks by stretching a piece 10 or 15 ft long and then cutting it into short, convenient lengths. Run RF wiring directly from point to point with a minimum of sharp bends and keep the wire well-spaced from the chassis or other grounded metal surfaces. Where the wiring must pass through the chassis or a partition, cut a clearance hole and line it with a rubber grommet. If insulation is necessary, slip spaghetti insulation or heat-shrink tubing over the wire. For power-supply leads, bring the wire through the chassis via a feedthrough capacitor.

In transmitters where the peak voltage does not exceed 500 V, shielded wire is satisfactory for power circuits. Shielded wire is not readily available for higher voltages — use point-to-point wiring instead. In the case of filament circuits carrying heavy current, it is necessary to use #10 or #12 bare or enameled wire. Slip the bare wire through spaghetti then cover it with copper braid pulled tightly over the spaghetti. Slide the shielding back over the insulation and flow solder into the end of the braid; the braid will stay in place, making it unnecessary to cut it back or secure it in place. Clean the braid first so solder will take with a minimum of heat.

For receivers, RF wiring follows the methods described above. At RF, most of the current flows on the surface of the wire (a phenomenon called "skin effect"). Hollow tubing is just as good a conductor at RF as solid wire.

High-Voltage Techniques

High-voltage wiring requires special care. You need to use wire rated for the voltage it is carrying. Most standard hookup wire is inadequate. High-voltage wire is usually insulated with Teflon or special multilayer plastic. Some coaxial cable is rated at up to 3700 V.

Air is a great insulator, but high voltage can break down its resistance and form an arc. You need to leave ample room between any circuit carrying voltage and any nearby conductors. At dc, leave a gap of at least 0.1 inch per kilovolt. The actual breakdown voltage of air varies with the frequency of the signal, humidity and the shape of the conductors.

High voltage is also prone to corona discharge, a bleeding off of charge, primarily from sharp edges. For this reason, all connections need to be soldered, leaving only rounded surfaces on the soldered connection. It takes a little practice to get a "ball" of solder on each joint, but for voltages above 5 kV it is important.

Be careful working near high-voltage circuits! Most high-voltage power supplies can deliver a lethal shock.

Cable Lacing and Routing

Where power or control leads run together for more than a few inches, they present a better appearance when bound

together in a single cable. Both plastic and waxed-linen lacing cords are available. You can also use a variety of plastic devices to bundle wires into cables and to clamp or secure them in place. **Fig 25.33** shows some of the products and techniques that you might use. Check with your local electronic parts supplier for items that are in stock.

To give a commercial look to the wiring of any unit, route any dc leads and shielded signal leads along the edge of the chassis. If this isn't possible, the cabled leads should then run parallel to an edge of the chassis. Further, the generous use of the tie points mounted parallel to an edge of the chassis, for the support of one or both ends of a resistor or fixed capacitor, adds to the appearance of the finished unit. In a similar manner, arrange the small components so that they are parallel to the panel or sides of the chassis.

Tie Points

When power leads have several branches in the chassis, it is convenient to use fiber-insulated multiple tie points as anchors for junction points. Strips of this kind are also useful as insulated supports for resistors, RF chokes and capacitors. Hold exposed points of high-voltage wiring to a minimum; otherwise, make them inaccessible to accidental contact.

Winding Coils

Winding coils seems so simple, yet many new constructors run into difficulty. Understanding the techniques prevents some of the frustration or construction errors associated with coil winding.

Close-wound coils are readily wound on the specified form by anchoring one end of the length of wire (in a vise or to a doorknob) and the other end to the coil form. Straighten any kinks in the wire and then pull to keep the wire under slight tension. Wind the coil to the required number of turns while walking toward the anchor, always maintaining a slight tension on the wire.

To space-wind the coil, wind the coil simultaneously with a suitable spacing medium (heavy thread, string or wire) in the manner described above. When the winding is complete, secure the end of the coil to the coil-form terminal and then carefully unwind the spacing material. If the coil is wound under suitable tension, the spacing material can be easily removed without disturbing the winding. Finish space-wound coils by judicious applications of Duco cement to hold the turns in place.

The "cold" end of a coil is the end at (or close to) chassis or ground potential. Wind coupling links on the cold end of a coil to minimize capacitive coupling.

Winding Toroidal Inductors

Toroidal inductors and transformers are specified for many projects in this *Handbook*. The advantages of these cores include compactness and a self-shielding property. **Figs 25.34** and **25.35** illustrate the proper way to wind and count turns on a toroidal core.

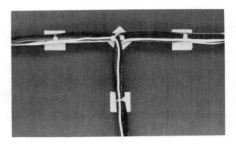

Fig 25.33 — **Adhesive-backed blocks and plastic wraps used to make wiring cables.**

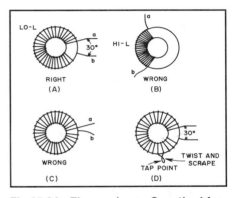

Fig 25.34—**The maximum-Q method for winding a single-layer toroid is shown at A. A 30° gap is best. Methods at B and C have greater distributed capacitance. D shows how to place a tap on a toroid coil winding.**

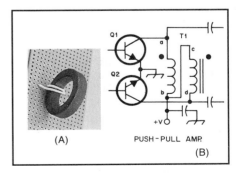

Fig 25.35—**A shows a toroid core with two turns of wire (see text). Large black dots, like those at T1 in B, indicate winding polarity (see text).**

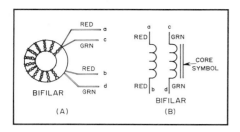

Fig 25.36 — Schematic and pictorial presentation of a bifilar-wound toroidal transformer.

When you wind a toroid inductor, count each pass of the wire through the toroid center as a turn. You can count the number of turns by counting the number of times the wire passes through the center of the core. See Fig 25.35A.

Multiwire Windings

A bifilar winding is one that has two identical lengths of wire, which when placed on the core result in the same number of turns for each wire. The two wires are wound on the core side by side at the same time, just as if a single winding were being applied. An easier and more popular method is to twist the two wires (8 to 15 turns per inch is adequate), then wind the twisted pair on the core. The wires can be twisted handily by placing one end of each in a bench vise. Tighten the remaining ends in the chuck of a small hand drill and turn the drill to twist the pair.

A trifilar winding has three wires, and a quadrifilar winding has four. The procedure for preparation and winding is otherwise the same as for a bifilar winding. **Fig 25.36** shows a bifilar toroid in schematic and pictorial form. The wires have been twisted together prior to placing them on the core. It is helpful, though by no means essential, to use wires of different color when multifilar-winding a core. It is more difficult to identify multiple windings on a core after it has been wound. Various colors of enamel insulation are available, but it is not easy for amateurs to find this wire locally or in small-quantity lots. This problem can be solved by taking lengths of wire (enameled magnet wire), cleaning the ends to remove dirt and grease, then spray painting them. Ordinary aerosol-can spray enamel works fine. Spray lacquer is not as satisfactory because it is brittle when dry and tends to flake off the wire.

The winding sense of a multifilar toroidal transformer is important in most circuits. Fig 25.35B illustrates this principle. The black dots (called phasing dots) at the top of the T1 windings indicate polarity. That is, points a and c are both start or finish ends of their respective windings. In this example, points a and d are of opposite phase (180° phase difference) to provide push-pull voltage output from Q1 and Q2.

After you wind a coil, scrape the insulation off the wire before you solder it into the circuit.

MECHANICAL FABRICATION

Electronic construction has many mechanical aspects. Cutting, bending and drilling are part of most electronic-construction projects.

Buy or Build a Chassis?

Most projects end up in some sort of an enclosure, and most hams choose to purchase a ready-made chassis for small projects, but some projects require a custom enclosure. Even a ready-made chassis may require a fabricated sheet-metal shield or bracket, so it's good to learn something about sheet-metal and metal-fabrication techniques.

Most often, you can buy a suitable enclosure. These are sold by Radio Shack and most electronics distributors. See the list in the **References** chapter.

Select an enclosure that has plenty of room. A removable cover or front panel can make any future troubleshooting or modifications easy. A project enclosure should be strong enough to hold all of the components without bending or sagging; it should also be strong enough to stand up to expected use and abuse.

Cutting and Bending Sheet Metal

Enclosures, mounting brackets and shields are usually made of sheet metal. Most sheet metal is sold in large sheets, 4×8 ft or larger. It must be cut to the size needed.

Most sheet metal is thin enough to cut with metal shears or a hacksaw. A jigsaw or bandsaw makes the task easier. If you use any kind of saw, select a blade that has teeth fine enough so that at least two teeth are in contact with the metal at all times.

If a metal sheet is too large to cut conveniently with a hacksaw, it can be scored and broken. Make scratches as deep as possible along the line of the cut on both sides of the sheet. Then, clamp it in a vise and work it back and forth until the sheet breaks at the line. Do not bend it too far before the break begins to weaken, or the edge of the sheet might bend. A pair of flat bars, slightly longer than the sheet being bent, make it easier to hold a sheet firmly in a vise. Use "C" clamps to keep the bars from spreading at the ends.

Smooth rough edges with a file or by sanding with a large piece of emery cloth or sandpaper wrapped around a flat block.

Finishing Aluminum

Give aluminum chassis, panels and parts a sheen finish by treating them in a caustic bath. Use a plastic container to hold the solution. Ordinary household lye can be dissolved in water to make a bath solution. Follow the directions on the container. A strong solution will do the job more rapidly.

Stir the solution with a stick of wood until the lye crystals are completely dissolved. If the lye solution gets on your skin, wash with plenty of water. If you get any in your eyes, immediately rinse with plenty of clean, room-temperature water and seek medical help. It can also damage your clothing, so wear something old. Prepare sufficient solution to cover the piece completely. When the aluminum is immersed, a very pronounced bubbling takes place. Provide ventilation to disperse the escaping gas. A half hour to two hours in the bath is sufficient, depending on the strength of the solution and the desired surface characteristics.

Chassis Working

With a few essential tools and proper procedure, building radio gear on a metal chassis is a relatively simple matter. Aluminum is better than steel, not only because it is a superior shielding material, but also because it is much easier to work and provides good chassis contact when used with secure fasteners.

Spend sufficient time planning a project to save trouble and energy later. The actual construction is much simpler when all details are worked out beforehand. Here we discuss a large chassis-and-cabinet project, such as a high-power amplifier. The techniques are applicable to small projects as well.

Cover the top of the chassis with a piece of wrapping paper or graph paper. Fold the edges down over the sides of the chassis and fasten them with adhesive tape. Place the front panel against the chassis front and draw a line there to indicate the chassis top edge.

Assemble the parts to be mounted on the chassis top and move them about to find a satisfactory arrangement. Consider that some will be mounted underneath the chassis and ensure that the two groups of components won't interfere with each other.

Place controls with shafts that extend through the cabinet first, and arrange them so that the knobs will form the desired pattern on the panel. Position the shafts perpendicular to the front chassis edge. Locate any partition shields and panel brackets next, then sockets and any other parts. Mark the mounting-hole centers of each part accurately on the paper. Watch out for capacitors with off-center shafts that do not line up with the mounting holes. Do not forget to mark the centers of socket holes and holes for wiring leads. Make the large center hole for a socket *before* the small mounting holes. Then use the socket itself as a template to mark the centers of the mounting holes. With all chassis holes marked, center-punch and drill each hole.

Next, mount on the chassis the capacitors and any other parts with shafts extending to the panel. Fasten the front panel to the chassis temporarily. Use a machinist's square to extend the line (vertical axis) of any control shaft to the chassis front and mark the location on the front panel at the chassis line. If the layout is complex, label each mark with an identifier. Also mark the back of the front panel with the locations of any holes in the chassis front that must go through the front panel. Remove the front panel.

PC-Board Materials

Much tedious sheet-metal work can be eliminated by fabricating chassis and enclosures from copper-clad printed-circuit board material. While it is manufactured in large sheets for industrial use, some hobby electronics stores and surplus outlets market usable scraps at reasonable prices. PC-board stock cuts easily with a small hacksaw. The nonmetallic base material isn't malleable, so it can't be bent. Corners are easily formed by holding two pieces at right angles and soldering the seam. This technique makes excellent RF-tight enclosures. If mechanical rigidity is required of a large copper-clad surface, solder stiffening ribs at right angles to the sheet.

Fig 25.37 shows the use of PC-board stock to make a project enclosure. This enclosure was made by cutting the pieces to size, then soldering them together. Start by laying the bottom piece on a workbench, then placing one of the sides in place at right angles. Tack-solder the second piece in two or three places, then start at one end and run a bead of solder down the entire seam. Use plenty of solder and plenty of heat. Continue with the rest of the pieces until all but the top cover is in place.

In most cases, it is better to drill all needed holes in advance. It can sometimes be difficult to drill holes after the enclosure is soldered together.

You can use this technique to build enclosures, subassemblies or shields. This technique is easy with practice; hone your skills on a few scrap pieces of PC-board stock.

Drilling Techniques

Before drilling holes in metal with a hand drill, indent the hole centers with a center punch. This prevents the drill bit from "walking" away from the center when starting the hole. Predrill holes greater than $1/2$-inch in diameter with a smaller bit that is large enough to contain the flat spot at the large bit's tip. When the metal being drilled is thinner than the depth of the drill-bit tip, back up the metal with a wood block to smooth the drilling process.

The chuck on the common hand drill is limited to $3/8$-inch bits. Some bits are much larger, with a $3/8$-inch shank. If necessary, enlarge holes with a reamer or round file. For very large or odd-shaped holes, drill a series of closely spaced small holes just inside of the desired opening. Cut the metal remaining between the holes with a cold chisel and file or grind the hole to its finished shape. A nibbling tool also works well for such holes.

Use socket-hole punches to make socket holes and other large holes in an aluminum chassis. Drill a guide hole for the punch center bolt, assemble the punch with the bolt through the guide hole and tighten the bolt to cut the desired hole. Oil the threads of the bolt occasionally.

Cut large circular holes in steel panels or chassis with an adjustable circle cutter ("flycutter"). Occasionally apply machine oil to the cutting groove to speed the job. Test the cutter's diameter setting by cutting a block of wood or scrap material first.

Remove burrs or rough edges that result from drilling or cutting with a burr-remover, round or half-round file, a sharp knife or chisel. Keep an old chisel sharpened and available for this purpose.

Rectangular Holes

Square or rectangular holes can be cut with a nibbling tool or a row of small holes as previously described. Large openings can be cut easily using socket-hole punches.

Construction Notes

If a control shaft must be extended or insulated, a flexible shaft coupling with adequate insulation should be used. Satisfactory support for the shaft extension, as well as electrical contact for safety, can be provided by means of a metal panel bushing made for the purpose. These can be obtained singly for use with existing shafts, or they can be bought with a captive extension shaft included. In either case the panel bushing gives a solid feel to the control. The use of fiber washers between ceramic insulation and metal brackets, screws or nuts will prevent the ceramic parts from breaking.

Fig 25.37 — A box made entirely from PC-board stock.

Painting

Painting is an art, but, like most arts, successful techniques are based on skills that can be learned. The surfaces to be painted must be clean to ensure that the paint will adhere properly. In most cases, you can wash the item to be painted with soap, water and a mild scrub brush, then rinse thoroughly. When it is dry, it is ready for painting. Avoid touching it with your bare hands after it has been cleaned. Your skin oils will interfere with paint adhesion. Wear rubber or clean cotton gloves.

Sheet metal can be prepared for painting by abrading the surface with medium-grade sandpaper, making certain the strokes are applied in the same direction (not circular or random). This process will create tiny grooves on the otherwise smooth surface. As a result, paint or lacquer will adhere well. On aluminum, one or two coats of zinc chromate primer applied before the finish paint will ensure good adhesion.

Keep work areas clean and the air free of dust. Any loose dirt or dust particles will probably find their way onto a freshly painted project. Even water-based paints produce some fumes, so properly ventilate work areas.

Select a paint suitable to the task. Some paints are best for metal, others for wood and so on. Some dry quickly, with no fumes; others dry slowly and need to be thoroughly ventilated. You may want to select a rust-preventative paint for metal surfaces that might be subjected to high moisture or salts.

Most metal surfaces are painted with some sort of spray, either from a spray gun or from spray cans of paint. Either way, follow the manufacturer's instructions for a high-quality job.

Summary

If you're like most amateurs, once you've got the building bug, you won't let your soldering iron stay cold for long. Starting is the hardest part. Now, the next time you think about adding another project to your station, you'll know where to start.

Contents

Test Procedures and Projects

26

This chapter, written by ARRL Technical Advisor Doug Millar, K6JEY, covers the test equipment and measurement techniques common to Amateur Radio. With the increasing complexity of amateur equipment and the availability of sophisticated test equipment, measurement and test procedures have also become more complex. There was a time when a simple bakelite cased volt-ohm meter (VOM) could solve most problems. With the advent of modern circuits that use advanced digital techniques, precise readouts and higher frequencies, test requirements and equipment have changed. In addition to the test procedures in this chapter, other test procedures appear in Chapters 14 and 15.

TEST AND MEASUREMENT BASICS

The process of testing requires a knowledge of what must be measured and what accuracy is required. If battery voltage is measured and the meter reads 1.52 V, what does this number mean? Does the meter always read accurately or do its readings change over time? What influences a meter reading? What accuracy do we need for a meaningful test of the battery voltage?

A Short History of Standards and Traceability

Since early times, people who measured things have worked to establish a system of consistency between measurements and measurers. Such consistency ensures that a measurement taken by one person could be duplicated by others — that measurements are reproducible. This allows discussion where everyone can be assured that their measurements of the same quantity would have the same result. In most cases, and until recently, consistent measurements involved an artifact: a physical object. If a merchant or scientist wanted to know what his pound weighed, he sent it to a laboratory where it was compared to the official pound. This system worked well for a long time, until the handling of the standard pound removed enough molecules so that its weight changed and measurements that compared in the past no longer did so.

Of course, many such measurements depended on an accurate value for the force of gravity. This grew more difficult with time because the outside environment — such things as a truck going by in the street — could throw the whole procedure off. As a result, scientists switched to physical constants for the determination of values. As an example, a meter was defined as a stated fraction of the circumference of the Earth over the poles.

Generally, each country has an office that is in charge of maintaining the integrity of the standards of measurement and is responsible for helping to get those standards into the field. In the United States that office is the National Institute of Standards and Testing (NIST), formerly the National Bureau of Standards. The NIST decides what the volt and other basic units should be and coordinates those units with other countries. For a modest fee, NIST will compare its volt against a submitted sample and report the accuracy of the sample. In fact, special batteries arrive there each day to be certified and returned so laboratories and industry can verify that their test equipment really does mean 1.527 V when it says so.

Basic Units: Frequency and Time

Frequency and time are the most basic units for many purposes and the ones known to the best accuracy. The formula for converting one to the other is to divide the known value into 1. Thus the time to complete a single cycle at 1 MHz = 0.000001 s.

The history of the accuracy of time keeping, of course, begins with the clock. Wooden clocks, water clocks and mechanical clocks were ancestors to our current standard: the electronic clock based on frequency. In the 1920s, quartz crystal controlled clocks were developed in the laboratory and used as a standard. With the advent of radio communication time intervals could be transmitted by radio, and a very fundamental standard of time and frequency could be used locally with little effort. Today transmitters in several countries broadcast time signals on standard calibrated frequencies. **Table 26.1** contains the locations and frequencies of some of these stations.

In the 1960s, Hewlett-Packard began selling self-contained time and frequency standards called cesium clocks. In a cesium clock a crystal frequency is generated and multiplied to microwave frequencies. That energy is passed through a chamber filled with cesium gas. The gas acts as a very narrow band-pass filter. The output signal is detected and the crystal oscillator frequency is adjusted automatically so that a maximum of energy is detected. The output of the crystal is thus linked to the stability of the cesium gas and is usually accurate to several parts in 10^{-12}. This is much superior to a crystal oscillator alone; but at close to $40,000

Table 26.1

Standard Frequency Stations

(Note: In recent years, frequent changes in these schedules have been common.)

Call Sign	Location	Frequency (MHz)
BSF	Taiwan	5, 15
CHU	Ottawa, Canada	3.330, 7.335, 14.670
FFH	France	2.500
IAM/IBF	Italy	5.000
JJY	Japan	2.5, 5, 8, 10, 15
LOL	Argentina	5, 10
RID	Irkutsk	5.004, 10.004, 15.004
RWM	Moscow	5, 4.996, 9.996, 14.996
WWV/WWVH	USA	2.5, 5, 10, 15, 20
VNG	Australia	2.5, 5
ZSC	South Africa	4.291, 8.461, 12.724 (part time)

each, cesium frequency standards are a bit extravagant for amateur use.

A rubidium frequency standard is an alternative to the cesium clock. They are not quite as accurate as the cesium, but they are much less expensive, relatively quick to warm up and can be quite small. Older models occasionally appear surplus. As with any precision instrument, it should be checked over and calibrated before use.

Most hams do not have access to cesium or rubidium standards—or need them. Instead we use crystal oscillators. Crystal oscillators provide three levels of stability. The least accurate is a single crystal mounted on a circuit board. The crystal frequency is affected by the temperature environment of the equipment, to the extent of a few parts per million (ppm) per degree Celsius. For example, the frequency of a 10-MHz crystal with temperature stability rated at 3 ppm might vary 60 Hz when temperature of the crystal changes by 2°C. If the crystal oscillator is followed by a frequency multiplier, any variation in the crystal frequency is also multiplied. Even so, the accuracy of a simple crystal oscillator is sufficient for most of our needs and most amateur equipment relies on this technique. For a discussion of crystal oscillators and temperature compensation, look in the **Oscillators** chapter of this book.

The second level of accuracy is achieved when the temperature around the crystal is stabilized, either by an "oven" or other nearby components. Crystals are usually designed to stabilize at temperatures far above any reached in normal operating environments. These oscillators are commonly good to 0.1 ppm per day and are widely used in the commercial two-way radio industry.

The third accuracy level uses a double oven with proportional heating. The two ovens compensate for each other automatically and provide excellent temperature stability. The ovens must be left on continuously, however, and warm-up requires several days to two weeks.

Crystal *aging* also affects frequency stability. Some crystals change frequency over time (age) so the circuit containing the crystal must contain components to compensate for this change. Other crystals become more stable over time and become excellent frequency standards. Many commercial laboratories go to the expense of buying and testing several examples of the same oscillator and select the best one for use. As a result, many surplus oscillators are surplus for a reason. Nevertheless, a good stable crystal oscillator can be accurate to 1×10^{-9} per day and very appropriate for amateur applications.

Time and Frequency Calibration

Many hams have digital frequency counters, which range from surplus lab equipment to new highly integrated instruments with nearly everything on one chip. Almost all of these are very precise and display nine or more digits. Many are even quite stable. Nonetheless, a 10-MHz oscillator accurate to 1 ppm per month can vary ±10 Hz in one month. This drift rate may be acceptable for many applications, but the question remains: How accurate is it?

This question can be answered by calibrating the oscillator. There are several ways to perform this calibration. The most accurate method compares the unit in question by leaving the oscillator operating, transporting it to an oscillator of known frequency and then making a comparison. A commonly used comparison method connects the output of the calibrated oscillator into the horizontal input of a high frequency oscilloscope, and the oscillator to be measured to the vertical input. It helps, but they need not be on the same frequency. By noting how long it takes the sine wave to travel one division at a given sweep speed, one can calculate the resulting drift in parts per million per minute (ppm/min).

Another technique of oscillator calibration uses a VLF phase comparator. This is a special direct-conversion receiver that picks up the signal from WWVB on 60 kHz. Phase comparison is used to compare WWVB with the divided frequency of the oscillator being tested. Many commercial units have a small strip chart printer attached and switches to determine the receiver frequency. Since these 60-kHz VLF Comparator receivers have been largely replaced by units that use Loran signals or rubidium standards, they can be found at very reasonable prices. A very effective 60-kHz antenna can be made by attaching an audio transformer with the low-impedance winding connected to the receiver antenna terminals by way of a series dc blocking capacitor. The high-impedance winding is then connected between ground and a random length of wire. A typical VLF Comparator can track an oscillator well into a few parts in 10^{-10}. This technique directly compares the oscillator with an NIST standard and can even characterize oscillator drift characteristics in ppm per day or week.

Another fairly direct method compares an oscillator with one of the WWV HF signals. The received signal is not immensely accurate, but if the oscillator of a modern HF transceiver is carefully compared, it will be accurate enough for all but the most demanding work.

The last and least accurate way to calibrate an oscillator is to compare it with another oscillator or counter owned by you or another local ham. Unless the calibration of the other oscillator or counter is known, this comparison could be very misleading. True accuracy is not determined by the label of a famous company or impressive looks. Metrologists (people who calibrate and measure equipment) spend more time calibrating oscillators than any other piece of equipment.

DC Instruments and Circuits

This section discusses the basics of analog and digital dc meters. It covers the design of range extenders for current, voltage and resistance; construction of a simple meter; functions of a digital voltmeter (DVM) and procedures for accurate measurements.

Basic Meters

In measuring instruments and test equipment suitable for amateur purposes, the ultimate readout is generally based on a measurement of direct current. There are two basic styles of meters: analog meters that use a moving needle display, and digital meters that display the measured values in digital form. The analog meter for measuring dc current and voltage uses a magnet and a coil to move a pointer over a calibrated scale in proportion to the current flowing through the meter.

The most common dc analog meter is the D'Arsonval type, consisting of a coil of wire to which the pointer is attached so that the coil moves (rotates) between the poles of a permanent magnet. When current flows through the coil, it sets up a magnetic field that interacts with the field of the magnet to cause the coil to turn. The design of the instrument normally makes the pointer move in direct proportion to the current.

Digital Multimeters

In recent years there has been a flood of inexpensive digital multimeters (DMMs) ranging from those built into probes to others housed in large enclosures. They are more commonly referred to as digital voltmeters (DVMs) even though they are multimeters; they usually measure voltage, current and resistance. After some years of refining circuits such as the "successive approximation" and "dual slope" methods, most meters now use the dual-slope method to convert analog voltages to a digital reading. DVMs have basically three main sections as shown in **Fig 26.1**.

The first section scales the voltage or current to be measured. It has four main circuits:

- a chain of multiplier resistors that reduce the input voltage to 0-1 V,
- a converter that changes 0-1 V ac to dc,
- an amplifier that raises signals in the 0-100 mV range to 0-1 V and
- a current driver that provides a constant

current to the multiplier chain for resistance measurements.

The second section is an integrator. It is usually based on an operational amplifier that is switched by a timing signal. The timing signal initially shorts the input of the integrator to provide a zero reference. Next a reference voltage is connected to charge the capacitor for a determined amount of time. Finally the last part of the timing cycle allows the capacitor to discharge. The time it takes the capacitor to discharge is proportional to either the input voltage (V_{in}, after it was scaled into the range of 0 to 1 V) or 1 minus V_{in}, depending on the meter design. This discharge time is measured by the next section of the DVM, which is actually a frequency counter. Finally, the output of the frequency counter is scaled to the selected range of voltage or current and sent to the final section of the DVM — the digital display.

Since the timing is quite fast and the capacitor is not used long enough to drift much in value, the components that most determine accuracy are the reference voltage source and the range multiplier resis-

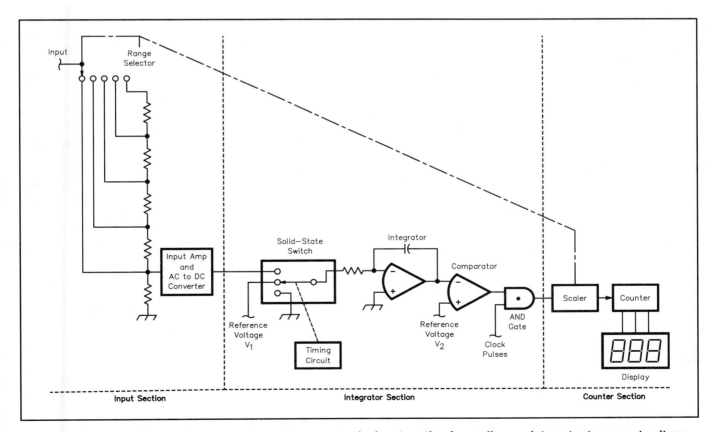

Fig 26.1 — A typical digital voltmeter consists of three parts: the input section for scaling, an integrator to convert voltage to pulse count, and a counter to display the pulse count representing the measured quantity.

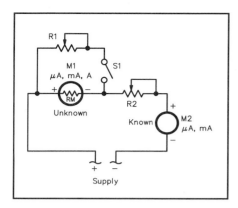

Fig 26.2 — This test setup allows safe measurement of a meter's internal resistance. See text for the procedure and part values.

tors. With the availability of integrated resistor networks that are deposited or diffused onto the same substrate, drift is automatically compensated because all branches of a divider drift in the same direction simultaneously. The voltage sources are generally Zener diodes on substrates with accompanying series resistors. Often the resistor and Zener have opposite temperature characteristics that cancel each other. In more complex DVMs, extensive digital circuitry can insert values to compensate for changes in the circuit and can even be automatically calibrated remotely in a few moments.

Liquid crystal displays (LCD) are commonly used for commercial DVMs. As a practical matter they draw little current and are best for portable and battery-operated use. The usual alternative, light emitting diode (LED) displays, draw much more current but are better in low-light environments. Some older surplus units use gas plasma displays (orange-colored digits). You may have seen plasma displays on gas-station pumps. They are not as bright as LEDs, but are easier to read. On the down side, plasma displays require high-voltage power supplies, draw considerable current and often fail after 10 years or so.

The advantages of DVMs are high input resistance (10 MΩ on most ranges), accurate and precise readings, portability, a wide variety of ranges and low price. There is one disadvantage, however: Digital displays update rather slowly, often only one to two times per second. This makes it very difficult to adjust a circuit for a peak (maximum) or null (minimum) response using only a digital display. The changing digits do not give any clue of the measurement

trend and it is easy to tune peak or null between display upaates. In answer, many new DVMs are built with an auxiliary bar-graph display that is updated constantly, thus providing instantaneous readings of relative value and direction of changes.

Current Ranges

The sensitivity of an analog meter is usually expressed in terms of the current required for full-scale deflection of the pointer. Although a very wide variety of ranges is available, the meters of interest in amateur work give maximum deflection with currents measured in microamperes or milliamperes. They are called microammeters and milliammeters, respectively.

Thanks to the relationships between current, voltage and resistance expressed by Ohm's Law, it is possible to use a single low-range instrument (for example, 1 mA or less for full-scale pointer deflection) for a variety of direct-current measurements. Through its ability to measure current, the instrument can also be used indirectly to measure voltage. In the same way, a measurement of both current and voltage will obviously yield a value of resistance. These measurement functions are often combined in a single instrument: the volt-ohm-milliammeter or VOM, a multirange meter that is one of the most useful pieces of test equipment an amateur can possess.

Accuracy

The accuracy of a D'Arsonval-movement dc meter is specified by the manufacturer. A common specification is ±2% of full scale, meaning that a 0-100 μA meter, for example, will be correct to within 2 μA at any part of the scale. There are very few cases in amateur work where accuracy greater than this is needed. When the instrument is part of a more complex measuring circuit, however, the design and components can each cause error that accumulates to reduce the overall accuracy.

Extending Current Range

Because of the way current divides between two resistances in parallel, it is possible to increase the range (more specifically to decrease the sensitivity) of a dc current meter. The meter itself has an inherent resistance (its internal resistance) which determines the full-scale current passing through it when its rated voltage is applied. (This rated voltage is on the order of a few millivolts.) When an external resistance is

n parallel with the meter, the current will divide between the two and the meter will respond only to that part of the current that flows through its movement. Thus, it reads only part of the total current; the effect makes more total current necessary for a full-scale meter reading. The added resistance is called a "shunt."

We must know the meter's internal resistance before we can calculate the value for a shunt resistor. Internal resistance may vary from a fraction of an ohm to a few thousand ohms, with greater resistance values associated with greater sensitivity. When this resistance is known, it can be used in the formula below to determine the required shunt for a given multiplication:

$$R = \frac{R_m}{n - 1} \qquad (1)$$

where

    R = shunt resistance, ohms
    R_m = meter internal resistance, ohms
    n = the factor by which the original meter scale is to be multiplied.

Often the internal resistance of a particular meter is unknown (when the meter is purchased at a flea market or is taken from a commercial piece of equipment, for example). Unfortunately, the internal resistance of a meter cannot be measured directly with an ohmmeter without risk of damage to the meter movement.

Fig 26.2 shows a method to safely measure the internal resistance of a linearly calibrated meter. It requires a calibrated meter that can measure the same current as the unknown meter. The system works as follows: S1 is switched off and R2 is set for maximum resistance. A supply of constant voltage is connected to the supply terminals (a battery will work fine) and R2 is adjusted so that the unknown meter reads exactly full scale. Note the current shown on M2. Close S1 and alternately adjust R1 and R2 so that the unknown meter (M1) reads exactly half scale and the known meter (M2) reads the same value as in the step above. At this point, half of the current in the circuit flows through M1 and half through R1. To determine the internal resistance of the meter, simply open S1 and read the resistance of R1 with an ohm-meter.

The values of R1 and R2 will depend on the meter sensitivity and the supply voltage. The maximum resistance value for R1 should be approximately twice the expected internal resistance of the meter. For highly sensitive meters (10 μA and less),

1 kΩ should be adequate. For less-sensitive meters, 100 Ω should suffice. Use no more supply voltage than necessary.

The value for minimum resistance at R2 can be calculated using Ohm's Law. For example, if the meter reads 0 to 1 mA and the supply is a 1.5-V battery, the minimum resistance required at R2 will be:

$$R2 = \frac{1.5}{0.001}$$

$$R2\,(min) = 1500\,\Omega$$

In practice a 2- or 2.5-kΩ potentiometer would be used.

Making Shunts

Homemade shunts can be constructed from several kinds of resistance wire or from ordinary copper wire if no resistance wire is available. The copper wire table in the **Component Data** chapter of this *Handbook* gives the resistance per 1000 ft for various sizes of copper wire. After computing the resistance required, determine the smallest wire size that will carry the full-scale current, again from the wire table. Measure off enough wire to give the required resistance. A high-resistance 1- or 2-W carbon-composition resistor makes an excellent form on which to wind the wire, as the high resistance does not affect the value of the shunt. If the shunt gets too hot, go to a larger diameter wire of a greater length.

VOLTMETERS

If a large resistance is connected in series with a meter that measures current, as shown in **Fig 26.3**, the current multiplied by the resistance will be the voltage drop across the resistance. This is known as a multiplier. An instrument used in this way is calibrated in terms of the voltage drop across the multiplier resistor and is called a voltmeter.

Sensitivity

Voltmeter sensitivity is usually expressed in ohms per volt (Ω/V), meaning that the meter full-scale reading multiplied by the sensitivity will give the total resistance of the voltmeter. For example, the resistance of a 1 kΩ/V voltmeter is 1000 times the full-scale calibration voltage. Then by Ohm's Law the current required for full-scale deflection is 1 milliampere. A sensitivity of 20 kΩ/V, a commonly used value, means that the instrument is a 50-μA meter.

As voltmeter sensitivity (resistance) increases, so does accuracy. Greater meter resistance means that less current is drawn from the circuit and thus the circuit under test is less affected by connection of the meter. Although a 1000-Ω/V meter can be used for some applications, most good meters are 20 kΩ/V or more. Vacuum-tube voltmeters (VTVMs) and their modern equivalent FET voltmeters (FETVOMs) are usually 10-100 MΩ/V and DVMs can go even higher.

Multipliers

The required multiplier resistance is found by dividing the desired full-scale voltage by the current, in amperes, required for full-scale deflection of the meter alone. To be mathematically correct, the internal resistance of the meter should be subtracted from the calculated value. This is seldom necessary (except perhaps for very low ranges) because the meter resistance is usually very low compared with the multiplier resistance. When the instrument is already a voltmeter with an internal multiplier, however, the meter resistance is significant. The resistance required to extend the range is then:

$$R = R_m\,(n - 1) \qquad (2)$$

where
R_m = total resistance of the instrument
n = factor by which the scale is to be multiplied

For example, if a 1-kΩ/V voltmeter having a calibrated range of 0 to 10 V is to be extended to 1000 V, R_m is 1000 × 10 = 10 kΩ, n is 1000/10 = 100 and R = 10,000 × (100 − 1) = 990 kΩ.

When extending the range of a voltmeter or converting a low-range meter into a voltmeter, the rated accuracy of the instrument is retained only when the multiplier resistance is precise. High-precision, hand-made and aged wire-wound resistors are used as multipliers of high-quality instruments. These are relatively expensive, but the home constructor can do well with 1% tolerance metal-film resistors. They should be derated when used for this purpose. That is, the actual power dissipated in the resistor should not be more than $^1/_{10}$ to $^1/_4$ the rated dissipation. Also, use care to avoid overheating the resistor body when soldering. These precautions will help prevent permanent change in the resistance of the unit.

Many DVMs use special resistor groups that have been etched on quartz or sapphire and laser trimmed to value. These resistors are very stable and often quite accurate. They can be bought new from various suppliers. It is also possible to "rescue" the divider/multiplier resistors

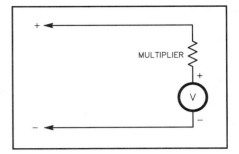

Fig 26.3 — A voltmeter is constructed by placing a current-indicating instrument in series with a high resistance, the "multiplier."

from an older DVM that no longer functions and use them as multipliers. Look for a series of four or five resistors that add up to 10 MΩ: .9,9,90,900,9,000,90,000 and 900,000 Ω. There is usually another 1-MΩ resistor in series to isolate the meter from the circuit under test. A few of these high-accuracy resistors in "odd" values can help calibrate less-expensive instruments.

DC Voltage Standards

For a long time NIST has statistically compared a bank of special Weston Cell or cadmium sulfate batteries to arrive at the standard volt. By using a special tapped resistor, a 1.08-V battery can be compared to other voltages and instruments compared. However, these are very high-impedance batteries that deliver almost no current and are relatively temperature sensitive. They are made up of a solution of cadmium and mercury in opposite legs of an "H" shaped glass container. You can read much more about them in *Calibration—Philosophy and Practice*, published by the John Fluke Co of Mount Lake Terrace, Washington.

Hams often use an ordinary flashlight battery as a convenient voltage reference. A fresh *D cell* usually provides 1.56 V under no load, as would be measured by a DVM. The Heath Company, which supplied thousands of kits to the ham community for many years, used such batteries as the calibration references for many of their kits.

Recently, NIST has been able to use a microwave to voltage converter called a "Josephson Junction" to determine the value of the volt. The converter transfers the accuracy of a frequency standard to the accuracy of the voltage that comes out of it. The converter generates only 5 mV, however, which then must be scaled to the standard 1-V level. One problem with high-accuracy measurements is stray noise (low-level voltages) that creates a floor below which measurements are

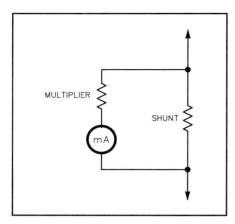

Fig 26.4 — A voltmeter can be used to measure current as shown. For reasonable accuracy, the shunt should be 5% of the circuit impedance or less, and the meter resistance should be 20 times the circuit impedance or more.

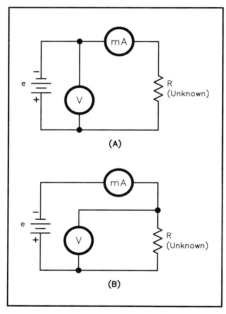

Fig 26.5 — Power or resistance can be calculated from voltage and current measurements. At A, error introduced by the ammeter is dominant. At B, error introduced by the voltmeter is dominant. The text gives an example.

meaningless. For that reason, meters with five or more digits must be very quiet and any comparisons must be made at a voltage high enough to be above the noise.

DC MEASUREMENT CIRCUITS

Current Measurement with a Voltmeter

A current-measuring instrument should have very low resistance compared with the resistance of the circuit being mea-

sured; otherwise, inserting the instrument will alter the current from its value when the instrument is removed. The resistance of many circuits in radio equipment is high and the circuit operation is affected little, if at all, by adding as much as a few hundred ohms in series. [Even better, use a resistor that is part of the working circuit if one exists. Unsolder one end of the resistor, measure its resistance, reinstall it and then make the measurement.—*Ed.*] In such cases the voltmeter method of measuring current in place of an ammeter, shown in **Fig 26.4**, is frequently convenient. A voltmeter (or low-range milliammeter provided with a multiplier and operating as a voltmeter) having a full-scale voltage range of a few volts is used to measure the voltage drop across a suitable value of resistance acting as a shunt.

The value of shunt resistance must be calculated from the known or estimated maximum current expected in the circuit (allowing a safe margin) and the voltage required for full-scale movement of the meter with its multiplier. For example, to measure a current estimated at 15 A on the 2-V range of a DVM, we need to solve Ohm's Law for the value of R:

$$R_{shunt} = \frac{2\,V}{15\,A} = 0.133\,\Omega$$

This resistor would dissipate $15^2 \times 0.133 = 29.92$ W. For a short-duration measurement, 30 1.0-Ω, 1-W resistors could be parallel connected in two groups of 15 (0.067 Ω per group) that are series connected to yield 0.133 Ω. For long-duration measurements, 2- to 5-W resistors would be better.

Power

Power in direct-current circuits is usually determined by measuring the current and voltage. When these are known, the power can be calculated by multiplying voltage, in volts, by the current, in amperes. If the current is measured with a milliammeter, the reading of the instrument must be divided by 1000 to convert it to amperes.

The setup for measuring power is shown in **Fig 26.5A**, where R is any dc load, not necessarily an actual resistor. In this measurement it is always best to use the lowest voltmeter or ammeter scale that allows reading the measured quantity. This results in the percentage error being less than if the meter was reading in the very lowest part of the selected scale.

Resistance

If both voltage and current are measured in a circuit such as that in Fig 26.5, the value of resistance R (in case it is un-

known) can be calculated from Ohm's Law. For reasonable results, two conditions should be met:

1. The internal resistance of the current meter should be less than 5% of the circuit resistance.

2. The input impedance of the voltmeter should be greater than 20 times the circuit resistance.

These conditions are important because both meters tend to load the circuit under test. The current meter resistance adds to the unknown resistance, while the voltmeter resistance decreases the unknown resistance as a result of their parallel connection.

Ohmmeters

Although Fig 26.5B suffices for occasional resistance measurements, it is inconvenient when we need to make frequent measurements over a wide range of resistance. The device generally used for this purpose is the ohmmeter. Its simplest form is a voltmeter (or milliammeter, depending on the circuit used) and a small battery. The meter is calibrated so that the value of an unknown resistance can be read directly from the scale. **Fig 26.6** shows some typical ohmmeter circuits. In the simplest circuit, Fig 26.6A, the meter and battery are connected in series with the unknown resistance. If a given movement of the meter's needle is obtained with terminals A-B shorted, inserting the resistance to be measured will cause the meter reading to decrease. When the resistance of the voltmeter is known, the following formula can be applied.

$$R = \frac{eR_m}{E} - R_m \qquad (3)$$

where

R = unknown resistance, ohms
e = voltage applied (A-B shorted)
E = voltmeter reading with R connected
R_m = resistance of the voltmeter.

The circuit of Fig 26.6A is not suited to measuring low values of resistance (less than 100 Ω or so) with a high-resistance voltmeter. For such measurements the circuit of Fig 26.6B is better. The unknown resistance is

$$R = \frac{I_2 R_m}{I_1 - I_2} \qquad (4)$$

where

R = unknown resistance, ohms
R_m = the internal resistance of the milliammeter, ohms
I_1 = current with R disconnected from terminals A-B, amps
I_2 = current with R connected, amps.

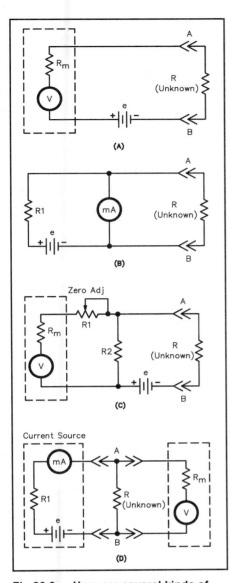

Fig 26.6 — Here are several kinds of ohmmeters. Each is explained in the text.

This formula is based on the assumption that the current in the complete circuit will be essentially constant whether or not the unknown terminals are short circuited. This requires that R1 be much greater than R_m. For example, 3000 Ω for a 1-mA meter with an internal resistance of perhaps 50 Ω. In this case, a 3-V battery would be necessary in order to obtain a full-scale deflection with the unknown terminals open. R1 can be an adjustable resistor, to permit setting the open-terminal current to exact full scale.

A third circuit for measuring resistance is shown in Fig 26.6C. In this case a high-resistance voltmeter is used to measure the voltage drop across a reference resistor, R2, when the unknown resistor is connected so that current flows through it, R2 and the battery in series. With suitable R2s

(low values for low-resistance, high values for high-resistance unknowns), this circuit gives equally good results for resistance values in the range from one ohm to several megohms. The voltmeter resistance, R_m, must be much greater (50 times or more) than that of R2. A 20-kΩ/V instrument (50-μA movement) is generally used. If the current through the voltmeter is negligible compared with the current through R2, the formula for the unknown is

$$R = \frac{eR2}{E} - R2 \qquad (5)$$

where

    R and R2 are in ohms

    e = voltmeter reading with R removed and A shorted to B.

    E = voltmeter reading with R connected.

R1 sets the voltmeter reading exactly to full scale when the meter is calibrated in ohms. A 10-kΩ pot is suitable with a 20-kΩ/V meter. The battery voltage is usually 3 V for ranges to 100 kΩ and 6 V for higher ranges.

Four-Wire Resistance Measurements

In situations where a very low resistance, like a 50-Ω dummy load, is to be measured, the resistance of the test leads can be significant. The average lead resistance is about 0.9 Ω through both leads, which would make a 50.5-Ω dummy load appear to be 51.4 Ω. To compensate for lead resistance, some meters allow for four-wire measurements. Briefly, two wires from the current source and two wires from the measuring circuit exit the meter case separately and connect directly to the unknown resistance (see Fig 26.6D). This eliminates the voltage drop in the current-source leads from the measurement. In practice, four-wire systems use special test clips that are similar to alligator clips, except that the jaws are insulated from each other and a meter lead is attached to each jaw. In some meters, an additional control allows the operator to short the test leads together and adjust the meter for a zero reading before making low-resistance measurements.

Bridge Circuits

Bridges are an important class of measurement circuits. They perform measurement by comparison with some known component or quantity, rather than by direct reading. VOMs, DVMs and other meters are convenient, but their accuracy is limited. The accuracy of manufactured analog meters is determined at the factory, while digital meters are accurate only to

some percentage ±1 in the least-significant digit. The accuracy of comparison measurements, however, is determined only by the comparison standard and bridge sensitivity.

Bridge circuits are useful across most of the frequency spectrum. Most amateur applications are at RF, as shown later in this chapter. The principles of bridge operation are easier to understand at dc, however, where bridge operation is simple.

The Wheatstone Bridge

A simple resistance bridge, known as the Wheatstone bridge, is shown in **Fig 26.7**. All other bridge circuits are based on this design. The four resistors, R1, R2, R3 and R4 in Fig 26.7A, are known as the bridge arms. For the voltmeter reading to be zero (null) the voltage dividers consisting of the pairs R1-R3 and R2-R4 must be equal. This means

$$\frac{R1}{R3} = \frac{R2}{R4} \qquad (6)$$

When this occurs the bridge is said to be *balanced*.

The circuit is usually drawn as shown at Fig 26.7B when used for resistance measurement. Equation 6 can be rewritten

$$RX = RS \left(\frac{R2}{R1} \right) \qquad (7)$$

RX is the unknown resistor. R1 and R2 are usually made equal; then the calibrated adjustable resistance (the standard), RS, will have the same value as RX when RS is set to show a null on the voltmeter.

Note that the resistance ratios, rather than the actual resistance values, determine the voltage balance. The values do have important practical effects on the sensitivity and power consumption, however. Bridge sensitivity is the ability of the meter to respond to slight unbalance near the null point; a sharper null means a more accurate setting of RS at balance.

The Wheatstone bridge is rarely used by amateurs for resistance measurement, since it is easier to measure resistances with VOMs and DVMs. Nonetheless, it is worthwhile to understand its operation as the basis of more complex bridges.

ELECTRONIC VOLTMETERS

We have seen that the resistance of a simple voltmeter (as in Fig 26.3) must be extremely high in order to avoid "loading" errors caused by the current that necessarily flows through the meter. The use of high-resistance meters tends to cause difficulty in measuring relatively low voltages because multiplier resistance progressively lessens as the voltage range is lowered.

Voltmeter resistance can be made inde-

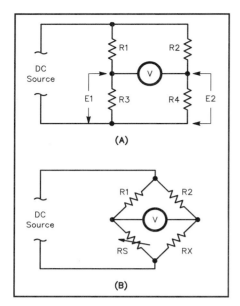

Fig 26.7 — A Wheatstone bridge circuit. A bridge circuit is actually a pair of voltage dividers (A). B shows how bridges are normally drawn.

pendent of voltage range by using vacuum tubes, FETs or op amps as dc amplifiers between the circuit under test (CUT) and the indicator, which may be a conventional meter movement or a digital display. Because the input resistance of the electronic devices mentioned is extremely high (hundreds of megohms) they have negligible loading effect on the CUT. They do, however, require a closed dc path in their input circuits (although this path can have very high resistance). They are also limited in the voltage level that their input circuits can handle. Because of this, the device

actually measures a small voltage across a portion of a high-resistance voltage divider connected to the CUT. Various voltage ranges are obtained from appropriate taps on the voltage divider.

In the design of electronic voltmeters it has become standard practice to use a voltage divider with a total resistance of 10 MΩ, tapped as required, in series with a 1-MΩ resistor incorporated in the meter. The total voltmeter resistance, including probe, is therefore 11 MΩ. The 1-MΩ resistor serves to isolate the voltmeter circuit from the CUT.

AC Instruments and Circuits

Most ac measurements differ from dc measurements in that the accuracy of the measurement depends on the purity of the sine wave. It is fairly easy to measure an ac voltage to between 1% and 5%, but getting down to 0.01% is difficult. Measurements to less than 0.01% must be left to precision laboratories. In general, amateurs measure ac voltages in household circuits, audio stages and RF power measurements, and 1% to 5% accuracy is usually close enough.

This section covers basic measurements, the nature of sine waves and meters. There are four common ways to measure ac voltage:

- Use a rectifier to change the ac to dc and then measure the dc.
- Heat a resistor in a Wheatstone bridge with the ac and measure the bridge unbalance.
- Heat a resistor surrounded by oil and measure the temperature rise.
- Use electronic circuits (such as multipliers and logarithmic amplifiers) with mathematical ac-to-dc conversion formulas. This method is not common, but it's interesting.

Calorimetric Meters

In a calorimetric meter, power is ap-

plied to a resistor that is immersed in the flow path of a special oil. This oil transfers the heat to another resistor that is part of a bridge. As the resistor heats, its resistance changes and the bridge becomes unbalanced. An attached meter registers the unbalance of the bridge as ac power. This type of meter is accurate for both dc and ac. They frequently operate from dc well into the GHz range. For calibration, an accurate dc voltage is applied and the reading is noted; a similar ac voltage is applied and the readings are compared. Some calorimetric meters are complicated, but others are simple.

Thermocouple Meters and RF Ammeters

In a thermocouple meter, alternating current flows through a low-resistance heating element. The power lost in the resistance generates heat that warms a thermocouple, which consists of a pair of junctions of two different metals. When one junction is heated a small dc voltage is generated in response to the difference in temperature of the two junctions. This voltage is applied to a dc milliammeter that is calibrated in suitable ac units. The heater-thermocouple/dc-meter combination is usually housed

in a regular meter case.

Thermocouple meters are available in ranges from about 100 mA to many amperes. Their useful upper frequency limit is in the neighborhood of 100 MHz. Amateurs use these meters mostly to measure current through a known load resistance and calculate the RF power delivered to the load.

RECTIFIER INSTRUMENTS

The response of a rectifier RF ammeter is proportional (depending on the design) to either the peak or average value of the rectified ac wave, but never directly to the RMS value. These meters cannot be calibrated in RMS without knowing the relationship that exists between the real reading and the RMS value. This relationship may not be known for the circuit under test.

Average-reading ac meters work best with pure sine waves and RMS meters work best with complicated wave forms. Since many practical measurements involve nonsinusoidal forms, it is necessary to know what your instrument is actually reading, in order to make measurements intelligently. Most VOMs and VTVMs use averaging techniques, while DVMs may use either one. In all cases, check the meter instruction manual to be sure what it reads.

Peak and Average with Sine-Wave Rectification

Peak, average and RMS values of ac waveforms are discussed in the **AC Theory** chapter. Because the positive and negative half cycles of the sine wave have the same shape, half-wave rectification of either the positive half or the negative half gives exactly the same result. With full-wave rectification, the peak reading is the same, but the average reading is doubled, because there are twice as many half cycles per unit of time.

Asymmetrical Wave Forms

A nonsinusoidal waveform is shown in **Fig 26.8A**. When the positive half cycles of this wave are rectified, the peak and average values are shown at B. If the polarity is reversed and the negative half cycles are rectified, the result is shown in Fig 26.8C. Full-wave rectification of such a lopsided wave changes the average value, but the peak reading is always the same as that of the half cycle that produces the highest peak in half-wave rectification.

Effective-Value Calibration

The actual scale calibration of commercially made rectifier voltmeters is very often (almost always, in fact) in terms of RMS values. For sine waves, this is satisfactory and useful because RMS is the standard measurement at power-line frequencies. It is also useful for many RF applications when the waveform is close to sinusoidal. In other cases, particularly in the AF range, the error may be considerable when the waveform is not pure.

Turn-Over

From Fig 26.8 it is apparent that the calibration of an average-reading meter will be the same whether the positive or negative sides are rectified. A half-wave peak-reading instrument, however, will indicate different values when its connections to the circuit are reversed (turn-over effect). Very often readings are taken both ways, in which case the sum of the two is the peak-to-peak (P-P) value, a useful figure in much audio and video work.

Average- vs Peak-Reading Circuits

For traditional analog displays, the basic difference between average- and peak-reading rectifier circuits is that the output is not filtered for averaged readings, while a filter capacitor is charged to the peak value of the output voltage in order to measure peaks. **Fig 26.9A** and **B** show typical average-reading circuits, one half-wave and the other full-wave. In the absence of dc filtering, the meter responds to wave forms such as those shown at B, C and D in Fig 26.8; and since the inertia of the pointer system makes it unable to follow the rapid variations in current, it averages them out mechanically.

In Fig 26.9A, D1 actuates the meter; D2 provides a low-resistance dc return in the meter circuit on the negative half cycles. R1 is the voltmeter multiplier resistance. R2 forms a voltage divider with R1 (through D1) that prevents more than a few ac volts from appearing across the rectifier-meter combination. A corresponding resistor can be used across the full-wave bridge circuit.

In these two circuits there is no provision to isolate the meter from any dc voltage in the circuit under test. The resulting errors can be avoided by connecting a large nonpolarized capacitor in series with the hot lead. The reactance must be low compared with the meter impedance (at the lowest frequency of interest, more on this later) in order for the full ac voltage to be applied to the meter circuit. Some meters may require as much as 1 μF at line (60 Hz) frequencies. Such capacitors are usually not included in VOMs.

Voltage doubler and shunt peak-reading circuits are shown in Fig 26.9C and D. In both circuits, C1 isolates the rectifier from dc voltage in the circuit under test. In the voltage-doubler circuit, the time constant of the C2-R1-R2 combination must be very large compared with the period of the lowest ac frequency to be measured; similarly with C1-R1-R2 in the shunt cir-

cuit. This is so because the capacitor is charged to the peak value (V_{P-P} in C, V_P in D) when the ac wave reaches its maximum and then must hold the charge (so it can register on a dc meter) until the next maximum of the same polarity. If the time constant is 20 times the ac period, the charge will have decreased by about 5% when the next charge occurs. The average voltage

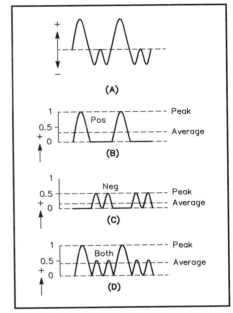

Fig 26.8— Peak vs average ac values for an asymmetrical wave form. Note that the peak values are different with positive or negative half-cycle rectification.

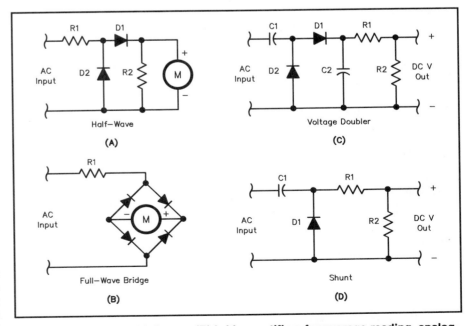

Fig 26.9 — Half (A) and full-wave (B) bridge rectifiers for average-reading, analog-display meters. Peak-reading circuits: a voltage doubler (C) and a shunt circuit (D). All circuits are discussed in the text.

drop will be smaller, so the error is appreciably less. The error will decrease rapidly with increasing frequency (if there is no change in the circuit values), but it will increase at lower frequencies.

In Fig 26.9C and D, R1 and R2 form a voltage divider that reduces the voltage to some desired value. For example, if R1 is 0Ω in the voltage doubler, the voltage across R2 is approximately V_{P-P}; if R1 = R2, the output is approximately V_P (as long as the waveform is symmetrical).

The most common application of the shunt circuit is an RF probe to read V_{RMS}. In that case, R2 is the input impedance of a VTVM or DVM: 11 MΩ. R1 is chosen so that 71% of the peak value appears across R2. This converts the peak reading to RMS for sine-wave ac. R1 is therefore approximately 4.7 MΩ, making the total resistance nearly 16 MΩ. A capacitance of 0.05 μF is sufficient for low audio frequencies under these conditions. Much smaller values of capacitance may be used at RF.

Voltmeter Impedance

The impedance of a voltmeter at the frequency being measured may have an effect on the accuracy similar to that caused by the resistance of a dc voltmeter, as discussed earlier. The ac meter is a resistance in parallel with a capacitance. Since the capacitive reactance decreases with increasing frequency, the impedance also decreases with frequency. The resistance does change with voltage level, particularly at very low voltages (10 V or less) depending on the sensitivity of the meter and the kind of rectifier used.

The ac load resistance represented by a diode rectifier is about one-half of its dc-load resistance. In Fig 26.9A the dc load is essentially the meter resistance, which is generally quite low compared with the multiplier resistance R1. Hence, the total resistance will be about the same as the multiplier resistance. The capacitance depends on the components and construction, test-lead length and location, and other such factors. In general, the capacitance has little or no effect at lower line and audio frequencies, but ordinary VOMs lose accuracy at high audio frequencies and are of little use at RF. Rectifiers with very low inherent capacitance are used at RF and they are usually located at the probe tip to reduce losses.

Similar limitations apply to peak-reading circuits. In the shunt circuit, the resistive part of the impedance is smaller than in the voltage-doubler circuit because the dc load resistance, R1/R2, is directly across the circuit under test and in parallel with the diode ac load resistance. In both peak-reading circuits the effective capacitance may range from 1 or 2 to a few hundred pF, with 100 pF typical in most instruments.

Scale Linearity

Fig 26.10 shows a typical current/voltage chart for a small semiconductor diode, which shows that the forward dynamic resistance of the diode is not constant, but rapidly decreases as the forward voltage increases from zero. The change from high to low resistance happens at much less than 1 V, but is in the range of voltage needed for a dc meter. With an average-reading circuit the current tends to be proportional to the square of the applied voltage. This makes the readings at the low end of the meter scale very crowded. For most measurement purposes, however, it is far more desirable for the output to be linear (that is, for the reading to be directly proportional to the applied voltage), which means that the markings on the meter are more evenly spaced.

To obtain that kind of linearity it is necessary to use a relatively large load resistance for the diode: Large enough that this resistance, rather than the diode resistance, will determine how much current flows. With this technique you can have a linear reading meter, but at the expense of sensitivity. The resistance needed depends on the type of diode; 5 kΩ to 50 kΩ is usually enough for a germanium rectifier,

---

Sources for RF Ammeters

When it comes to getting your own RF ammeter, there's good news and bad news. First, the bad news. New RF ammeters are expensive: about $70 to $200 (in 1994). AM radio stations are the main users of these today. The FCC defines the output power of AM stations based on the RF current in the antenna, so new RF ammeters are made mainly for that market. They are quite accurate, and their prices reflect that.

The good news is that used RF ammeters are often available. For example, Fair Radio Sales (see the Address List in the **References** chapter) has been a consistent RF-ammeter source. Ham flea markets are also worth trying. Some grubbing around in your nearest surplus store or some older ham's junk box may provide just the RF ammeter you need.

RF Ammeter Substitutes

Don't despair if you can't find a used RF ammeter. It's possible to construct your own. Both hot-wire and thermocouple units can be homemade.

Pilot lamps in series with antenna wires, or coupled to them in various ways, can indicate antenna current[*] or even forward and reflected power.[†]

Another approach is to use a small low-voltage lamp as the heat/light element and use a photo detector driving a meter as an indicator. (Your eyes and judgment can serve as the indicating part of the instrument.) A feed-line balance checker could be as simple as a couple of lamps with the right current rating and the lowest voltage rating available. You should be able to tell fairly well by eye which bulb is brighter or if they are about equal. You can calibrate a lamp-based RF ammeter with 60-Hz or dc power.

As another alternative, you can build an RF ammeter that uses a dc meter to indicate rectified RF from a current transformer that you clamp over a transmission line wire.[††]

Copper-Top Battery Testers as RF Ammeters

Finally, there are the *free* RF ammeters that come as the testers with Duracell batteries! For 1.5-V cells, these are actually 3 to 5-Ω resistors with built-in liquid-crystal displays. The resistor heats the liquid-crystal strip; the length of the "lighted" portion (heat turns the strip clear, exposing the fluorescent ink beneath) indicates the magnitude of the current.

Despite their "+" and "–" markings, these indicators are not polarized. Their resistance is low enough to have relatively little effect on a 50-Ω system. (For example, putting one in series with a 50-Ω dummy load would increase the system SWR from 1 to 1.1:1. These testers can measure about 200 to 400 mA. (You can achieve higher ranges by means of a shunt.) Best of all, if you burn out one of these "meters" during your tests, you can replace it at any drugstore, hardware store or supermarket for a few dollars, with some batteries thrown in free.

—*John Stanley, K4ERO*

* F. Sutter, "What, No Meters?" *QST*, Oct 1938, p 49.
† C. Wright, "The Twin-Lamp," *QST*, Oct 1947, pp 22-23, 110 and 112.
†† Z. Lau, "A Relative RF Ammeter for Open-Wire Lines," *QST*, Oct 1988, pp 15-17.

depending on the dc meter sensitivity, but several times as much may be needed for silicon diodes. Higher resistances require greater meter sensitivity; that is, the basic meter must be a microammeter rather than a low-range milliammeter.

Reverse Current

When semiconductor diodes are reverse biased, a small leakage current flows. This reverse current flows during the half cycle when the diode should appear open, and the current causes an error in the dc meter reading. The quantity of reverse current is indicated by a diode's *back resistance* specification. This back resistance is so high that reverse current is negligible with silicon diodes, but back resistance may be less than 100 kΩ for germanium diodes.

The practical effect of semiconductor back resistance is to limit the amount of resistance that can be used in the dc load. This in turn affects the linearity of the meter scale. For practical purposes, the back resistance of vacuum-tube diodes is infinite.

RF Voltage

Special precautions must be taken to minimize the capacitive component of the voltmeter impedance at RF. If possible, the rectifier circuit should be installed permanently at the point where the RF voltage is to be measured, using the shortest possible RF connections. The dc meter can be remotely located, however.

For general RF measurements an RF probe is used in conjunction with a 10 MΩ electronic voltmeter. The circuit of **Fig 26.11**, which is basically the shunt peak-reading circuit of Fig 26.9D, is gen-

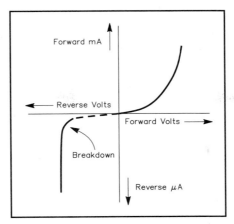

Fig 26.10 — Voltage vs current characteristics for a typical semiconductor diode. Actual values vary with different part numbers, but the forward current will always be increasing steeply with 1 V applied. Note that the forward current scale (mA) is 1000 times larger than that for reverse leakage current (μA). Breakdown voltage varies from 15 V to several hundred volts.

erally used. The series resistor, which is installed in the probe close to the rectifier, prevents RF from being fed through the probe cable to the electronic voltmeter. (In addition, the capacitance of the coaxial cable serves as a bypass for any RF on the lead.) This resistor, in conjunction with the 10-MΩ divider resistance of the electronic voltmeter, reduces the peak rectified voltage to a dc value equivalent to the RMS of the RF signal. Therefore, the RF readings are consistent with the regular dc calibration.

Of the diodes readily available to amateurs, the germanium point-contact or Schottky diode is preferred for RF-probe

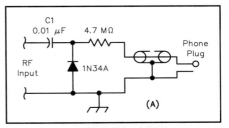

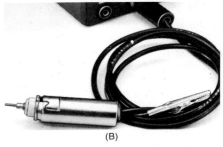

Fig 26.11 — At A is the schematic, at B a photo, of an RF probe for electronic voltmeters. The case of this probe is a seven-pin ceramic tube socket and a 2¹/₄-inch tube shield. A grommet protects the cable where it leaves the tube shield, and an alligator clip on the cable braid connects the probe to the ground of the circuit under test.

applications. It has low capacitance (on the order of 1 pF) and in high-back-resistance Schottky diodes, the reverse current is not serious. The principal limitation is that its safe reverse voltage is only about 50 to 75 V, which limits the applied voltage to 15 or 20 V RMS. Diodes can be series connected to raise the overall rating. At RF, however, it is more common to use capacitors or resistors as voltage dividers and apply the divider output to a single diode.

RF Power

RF power can be measured by means of an accurately calibrated RF voltmeter connected across a dummy load in which the power is dissipated. If the load is a known pure resistance, the power, by Ohm's Law, is equal to E^2/R, where E is the RMS voltage.

The Hewlett-Packard 410B/C VTVM

The Hewlett-Packard 410B and 410C VTVMs have been standards of bench measurement for industry, and they are now available as industrial surplus. These units are not only excellent VTVMs, but they are also good wide-range RF power meters. Both models use a vacuum-tube detector mounted in a low-loss probe for ac measurements. With an adapter that allows the probe to contact the center conductor of a transmission line, it will give very good RF voltage measurements from 50 mV to 300 V and from 20 Hz to beyond 500 MHz. Very few other measuring instruments provide this range in a single sensor/meter. In addition to the 410B or C, you will also need a probe adapter HP model #11042A.

Do not take the probe apart for inspection because that can change the calibration. You can quickly check the probe by feeling it after it has been warmed up for about 15 minutes. If the body of it feels warm it is probably working. Inside, the 410B is quite different from the C model, with the B being simpler. The meter scales are also different; the 410C offers better resolution and perhaps better accuracy.

To make RF power measurements, remove the ac probe tip and twist lock the probe into the 11042A probe adapter. Attach the output of the adapter to a dummy load and use the formula

$$P = \frac{E^2}{R} \qquad (8)$$

where

    P = power, in watts
    E = value given by the meter, in volts
    R = resistance of the dummy load, in ohms.

The resistance of the dummy load should be accurately known to at least ±0.1 Ω, preferably measured with a four-wire arrangement as described in the section on ohmmeters. For frequent measurements, make a chart of voltage vs power at your most often used wattages.

THE MICROWATTER

This simple, easy to build terminating microwattmeter by Denton Bramwell,

K7OWJ appeared in June 1997 *QST*.[1] It can measure power levels from below −50 dBm (10 nW) to −20 dBm (10 µW) with an accuracy of within 1 dB at frequencies from below the broadcast band up to 2 meters. Beyond that, it's still capable of making *relative* power measurements. The range of the meter can be extended by using an attenuator or by adding a broadband RF amplifier.

There are many uses for this meter. Combined with a signal generator, you can use it to characterize crystal and LC filters, for direct, on-air checks of field strength and measuring antenna patterns. If you are fortunate enough to have a 50-Ω oscilloscope probe (or a high-impedance probe), you can also use the Microwatter as an RF voltmeter for circuit testing.

Design

The circuit concept is quite simple (see **Fig 26.12**). Two nearly identical diodes (D1 and D2) are biased on by a small dc current. RF energy is coupled to one of the diodes, which then acts as a square-law detector. The difference between the voltages present on the two diodes is amplified by a differential amplifier (U1) and applied to an analog voltmeter equipped with hand-calibrated ranges. Getting adequate stability requires careful selection of parts.

D1 and D2 must have a low junction capacitance and a fast response in order to provide the desired bandwidth. Their temperature coefficients must also be extremely well matched. The solution is to use a very old part: the CA3039 diode array.[2] The diodes in this device are very fast silicon types, with exceptional thermal matching.[3] Any two diodes of the array—except the one tied to the substrate—can be used for D1 and D2. Unused IC pins can be cut off or bent out of the way. Similarly rated arrays of hot-carrier diodes may provide a better frequency response.

The Burr-Brown INA2128 provides two differential amplifiers with exceptional specifications: low thermal drift, superb common-mode rejection and low noise. In this circuit, one half of the INA2128 provides dc amplification of the detected voltage; the other half serves as a low-impedance offset-voltage source used for error nulling.

The RF-input terminating resistor is composed of two 100-Ω resistors (R3 and R4), providing a cleaner 50-Ω termination than a single resistor does. If you have them, use chip resistors. If not, clip the leads off two standard 100-Ω resistors and carefully scrape away any paint at the

ends. After installing the resistors, verify the input resistance with an ohmmeter.

A stable power source is required. Operating the circuit from an unregulated battery supply allows slight relative changes in the local ground as the terminal voltage drops, creating drift in the reading. The Zener diodes (D3 and D4) following regulator U2 provide a simple, dynamic means of splitting the 12-V supply into the three voltages required to power the circuit.

C1 and C11 should be as close as physically possible to U1 pins 1 and 2. In combination with R1 and R5, these capacitors provide good immunity to unwanted RF. R14, the **OFFSET NULL** potentiometer, allows nulling the dc amplifier offset and any offset from slight mismatches in the CA3039 diodes. Use a multiturn potentiometer for R14—10 turns or more. Whether you use a panel-mount pot, or a PC-board-mounted trimmer as I did, R14 must be accessible. On the most-sensitive range, the instrument must be zeroed before each use.

The forward voltage drop difference between any two diodes in the CA3039 is specified as "typically less than 0.5 mV." If your device is typical, the nominal values of 5.1 kΩ for R10 and 3.9 kΩ for R9 will do just fine. The manufacturer does not reject parts until the forward voltage drop difference is *5 mV*—so the 0.5 mV figure isn't totally dependable. You can always rotate the diode array so that pins 1, 2, 3 and 4 occupy the spots designated for pins 5, 6, 7 and 8. That gives you a second set of diodes from which to choose. The simplest general solution for a homebrew project is to use nonidentical values for R9 and R10.

If your diode array isn't typical, your Microwatter won't zero properly. The solution is simple. Remove R14 from the circuit. Temporarily connect one end of a 10-kΩ potentiometer to the +8.1 V supply and attach the other end to the −3.9 V supply. Connect the pot arm to the PC-board point of R14's arm. With this arrangement, you'll be able to zero your Microwatter, although the setting will be sensitive and might not produce an exact zero. Once you have the zero point, turn off the Microwatter and remove the 10-kΩ pot carefully without disturbing its setting. Measure the resistance between the pot arm and end that was connected to the +8.1 V point— this is the value of R10. R9's value is that measured between the pot arm and the end previously connected to the −3.9 V point. The combined value of these two resistors and that of R14 provides a total resistance of about 10.5 kΩ, and the resistance ratios

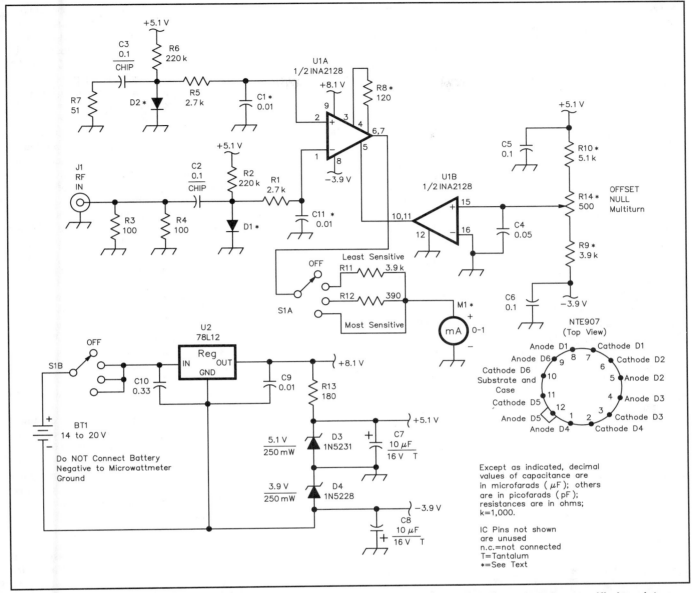

Fig 26.12—Schematic of the Microwatter circuit. Equivalent parts can be substituted. Unless otherwise specified, resistors are ¼-W, 5%-tolerance carbon-composition or film units.

D1, D2—Part of CA3039 or NTE907 diode array; see text and Notes 1 and 5. The NTE907 is available from Mouser Electronics (see References chapter Address List for contact information).

M1—1-mA meter movement, 50 Ω internal resistance; see text
R14—500 Ω, 10-turn (or more) potentiometer; see text
S1—2-pole, 4-position rotary switch
U1—Burr Brown amplifier INA2128P;

Digi-Key INA2128P; available from Digi-Key Corp (see References chapter Address List for contact information).
U2—78L12, positive 12-V, 100-mA voltage regulator; Digi-Key LM78L12ACZ.

will allow easy meter zeroing.

The Meter Movement

The specified meter movement has a 50-Ω internal resistance. Hence, on the most sensitive scale, a 50 mV output from U1 provides full-scale deflection. The middle and upper scales require 440 mV and 3.95 V, respectively, for full-scale deflection. Here's how you can use a meter that has a different internal resistance or current sensitivity.

Take, for example, a 25-µA meter movement with an internal resistance of 1910 Ω. A signal level of 47.8 mV (25 µA × 1910 Ω) provides full-scale deflection, so this meter can be substituted directly for the 1-mA, 50-Ω meter on the most sensitive scale. On the middle and top scales, the values of R12 and R11 should be 15.69 kΩ and

156.1 kΩ, or values close to those.

If the most sensitive meter you can find requires 80 to 100 mV to drive it to full scale deflection, don't worry. The Microwatter's most-sensitive scale will be compressed, but quite usable. The middle and top scales will still run full scale, probably not even requiring a change in the values of R11 and R12, if you use a 1-mA movement.

Construction and Calibration

For the prototype, components are mounted on double-sided PC board (see Note 1) with one side acting as a ground-plane. The cabinet is a 5×7×3-inch aluminum box, primed and painted gray (see **Fig 26.13**). Power is supplied by a set of NiCd batteries glued to the inside rear of the cabinet. The PC board is small enough to be mounted on the back of the input BNC connector without other support. Once you have completed PC-board assembly, remove the solder flux from the board.

To calibrate the Microwatter, you'll need a 50-Ω RF source with a known output level. A suitable signal source can be made from a crystal oscillator and an attenuator, using at least 6 dB of attenuation between the oscillator and the Microwatter at all times. The output of such a system can be calibrated with an oscilloscope, or with a simple RF meter. Any frequency in the mid-HF range is suitable for calibration—K7OWJ used 10 MHz.

Remove the meter-face cover and prepare the face for new markings. You can paint it white and use India ink to make the meter scale arcs and incremental marks. Before calibrating the Microwatter, let it stabilize at room temperature for an hour, and turn it on at least 15 minutes before use. (At these low power levels, even the heat generated by the small dc bias on the diodes needs time to stabilize.) Set the Microwatter to its most sensitive scale. Use R14 to set the meter needle to your chosen meter zero point and mark that point with a pencil. Then apply a –40 dBm

signal and mark the top end of the scale. Decrease the input signal level in 1-dB steps, marking each step. Recheck your meter zeroing, switch the Microwatter to the middle scale, apply a –30 dBm signal and mark this point. Again, decrease drive in 1-dB steps, marking each. Repeat with the least-sensitive scale, starting with –20 dBm. Once this is done, replace the meter-face cover and your Microwatter is ready to use.

DIRECTIONAL WATTMETERS

Directional wattmeters of varying quality are commonly used by the amateur community. The high quality standard is made by the Bird Electronic Corporation, who call their proprietary line THRULINE. The units are based on a sampling system built into a short piece of 50-Ω transmission line with plug-in elements for various power and frequency ranges.

AC BRIDGES

In its simplest form, the ac bridge is exactly the same as the Wheatstone bridge

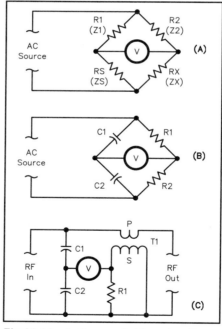

Fig 26.14—A shows a bridge circuit generalized for ac or dc use. B is a form of ac bridge for RF applications. C is an SWR bridge for use in transmission lines.

discussed earlier in the dc measurement section of this chapter. However, complex impedances can be substituted for resistances, as suggested by **Fig 26.14A**. The same bridge equation holds if Z (complex impedance) is substituted for R in each arm. For the equation to be true, however, both phase angles and magnitudes of the impedances must balance; otherwise, a true null voltage is impossible to obtain. This means that a bridge with all "pure" arms (pure resistance or reactance) cannot measure complex impedances; a combination of R and X must be present in at least one arm aside from the unknown.

The actual circuits of ac bridges take many forms, depending on the intended measurement and the frequency range to be covered. As the frequency increases, stray effects (unwanted capacitances and inductances) become more pronounced. At RF, it takes special attention to minimize them.

Most amateur built bridges are used for RF measurements, especially SWR measurements on transmission lines. The circuits at **Fig 26.14B** and C are favorites for this purpose.

Fig 26.14B is useful for measuring both transmission lines and lumped constant components. Combinations of resistance and capacitance are often used in one or more arms; this may be required for eliminating the effects of stray capacitance. The bridge shown in Fig 26.14C is used only on transmission lines and only on those lines having the characteristic impedance for which the bridge is designed.

SWR Measurement

The theory behind SWR measurement is covered in the **Transmission Lines** chapter and more fully in *The ARRL Antenna Book*. Projects to measure SWR appear in the **Station Setup and Accessory Projects** chapter of this *Handbook*.

Notes
[1] A PC board and some components for this project is available from FAR Circuits. A PC-board template package is not available.
[2] The NTE907 replacement is available from Mouser Electronics.
[3] The absolute value of the difference in forward drop between any two diodes is 1 mV/°C.

Fig 26.13—Photo of the Microwatter.

Frequency Measurement

The FCC Rules for Amateur Radio require that transmitted signals stay inside the frequency limits of bands consistent with the operator's license privileges. The exact frequency need not be known, as long as it is within the limits. On these limits there are no tolerances: Individual amateurs must be sure that their signal stays safely inside. The current limits for each license class can be found in the **References** chapter and in the current edition of *The FCC Rule Book*, published by the ARRL.

Staying within these limits is not difficult; many modern transceivers do so automatically, within limits. If your radio uses a PLL synthesized frequency source, just tune in WWV or another frequency standard occasionally.

Checks on older equipment require some simple equipment and careful adjusting. The equipment commonly used is the frequency marker generator and the method involves use of the station receiver, as shown in **Fig 26.15**.

FREQUENCY MARKER GENERATORS

A marker generator, in its simplest form, is a high-stability oscillator that generates a series of harmonic signals. When an appropriate fundamental is chosen, harmonics fall near the edges of the amateur frequency allocations.

Most US amateur band and subband limits are exact multiples of 25 kHz. A 25-kHz fundamental frequency will therefore produce the right marker signals if its harmonics are strong enough. But since harmonics appear at 25-kHz intervals throughout the spectrum, there is still a problem of identifying particular markers. This is easily solved if the receiver has reasonably good calibration. If not, most marker circuits provide a choice of fundamental outputs, say 100 and 50 kHz as well as 25 kHz. Then the receiver can be first set to a 100-kHz interval. From there, the desired 25-kHz (or 50-kHz) points can be counted. Greater frequency intervals are rarely required. Instead, tune in a signal from a station of known frequency and count off the 100-kHz points from there.

Transmitter Checking

To check transmitter frequency, tune in the transmitter signal on a calibrated receiver and note the dial setting at which it is heard. To start, reduce the transmitter to its lowest possible level to avoid receiver overload. Also, place a direct short across the receiver antenna terminals, reduce the

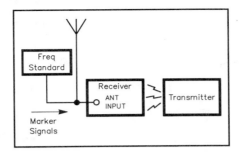

Fig 26.15 — A setup for checking transmitter frequency. Use care to ensure that the transmitter does not overload the receiver. False signals would result; see text.

RF gain to minimum and switch in any available receiver attenuators to help prevent receiver IMD, overload and possible false readings.

Place the transmitter on standby (not transmitting) and use the marker generator as a signal source. Tune in and identify the nearest marker frequencies above and below the transmitter signal. The transmitter frequency is between these two known frequencies. If the marker frequencies are accurate, this is all you need to know, except that the transmitter frequency must not be so close to a band (or subband) edge that sidebands extend past the edge.

If the transmitter signal is inside a marker at the edge of an assignment, to the extent that there is an audible beat note with the receiver BFO turned off, normal CW sidebands are safely inside the edge. (So long as there are no abnormal sidebands such as those caused by clicks and chirps.) For 'phone the safety allowance is usually taken to be about 3 kHz, the usual width of one sideband. A frequency difference of this much can be estimated by noting the receiver dial settings for the two 25-kHz markers that are either side of the signal and dividing 25 by the number of dial divisions between them. This will give the number of kilohertz per dial division. It is a prudent practice to allow an extra kHz margin when setting the transmitter close to a band or subband edge (5-kHz is a safe HF margin for most modes on modern transmitters).

Transceivers

The method described above is good when the receiver and transmitter are separate pieces of equipment. When a transceiver is used and the transmitting

frequency is automatically the same as that to which the receiver is tuned, setting the tuning dial to a spot between two known marker frequencies is all that is required. The receiver incremental tuning control (RIT) must be turned off.

The proper dial settings for the markers are those at which, with the BFO on, the signal is tuned to zero beat (the spot where the beat note disappears as tuning makes its pitch progressively lower). Exact zero beat can be determined by a very slow rise and fall of background noise, caused by a beat of a cycle or less per second. In receivers with high selectivity it may not be possible to detect an exact zero beat, because low audio frequencies from beat notes may be prevented from reaching the speaker or headphones.

Most commercial equipment has some way to match either the equipment's internal oscillator or marker generator with the signal received from WWV on one of its short-wave frequencies. It is a good idea to do this check on a new piece of gear. A recheck about a month later will show if anything has changed. Normal commercial equipment drifts less than 1 kHz after warm up.

Also check the dial linearity of equipment that has an analog dial or subdial. Often analog dials do not track frequency accurately across an entire band. Such radios usually provide for pointer adjustment so that dial error can be minimized at the most often used part of a band.

Frequency-Marker Circuits

The frequency in most amateur frequency markers is determined by a 100-kHz or 1-MHz crystal. Although the marker generator should produce harmonics every 25-kHz and 50-kHz, crystals (or other high-stability resonators) for frequencies lower than 100 kHz are expensive and rare. There is really no need for them, however, since it is easy to divide the basic frequency down to the desired frequency; 50- and 25-kHz steps require only two successive divisions by two (from 100 kHz). In the division process, the harmonics of the generator are strengthened so they are useful up to the VHF range. Even so, as frequency increases the harmonics weaken.

Current marker generators are based on readily available crystals. A 1 MHz basic oscillator would first be divided by 10 to produce 100 kHz and then followed by two successive divide-by-two stages to produce 50 kHz and 25 kHz.

A MARKER GENERATOR WITH SELECTABLE OUTPUT

Fig 26.16 shows a marker generator with selectable output for 100, 50 or 25-kHz intervals. It provides marker signals well up into the 2-m band. The project was first built by Bruce Hale, KB1MW, in the ARRL Lab. A more detailed presentation appeared in the **Station Accessories** chapter of *Handbooks* from 1987 through 1994. An etching pattern and parts-placement diagram are available for this project. See Chapter 30, **References**, for the Marker Generator template.

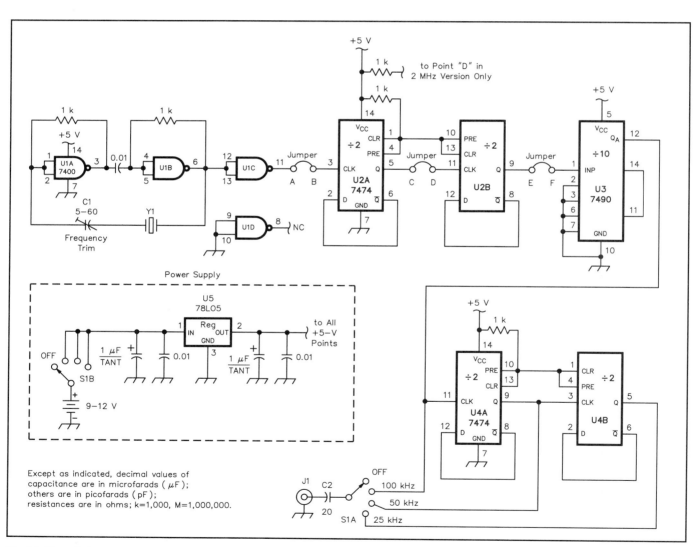

Fig 26.16 — Schematic diagram of the marker generator.

C1 — 5-60 pF miniature trimmer capacitor.
C2 — 20-pF disc ceramic.
U1 — 7400 or 74LS00 quad NAND gate.

U2, U4 — 7474 or 74LS74 dual D flip-flop.
U3 — 7490 or 74LS90 decade counter.

U5 — 78L05, 7805 or LM340-T5 5-V voltage regulator.
Y1 — 1, 2 or 4-MHz crystal in HC-25, HC-33 or HC-6 holder.

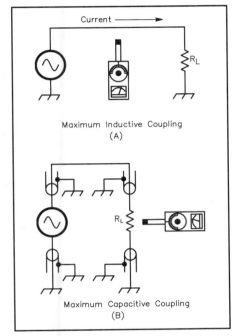

Fig 26.17 — Dip meter coupling. (A) uses inductive coupling and (B) uses capacitive coupling.

A 1, 2 or 4-MHz computer-surplus crystal is suitable for Y1. Several prototypes were built with such crystals and all could be tuned within 50 Hz at 100 kHz. The marker division ratio must be chosen once the crystal frequency is selected. This is accomplished by means of several jumpers that disable part or all of U2A, the dual flip-flop IC. When Y1 is a 1-MHz crystal, U2 may be omitted entirely. **Table 26.2** gives the jumper placement for each crystal frequency.

The prototypes were built with TTL logic ICs. Unused TTL gate inputs should always be connected either to ground or to Vcc through a pull-up resistor. If low-power Schottky (they have "LS" as part of their numerical designator) parts are available for U1 through U4, use them. They draw much less current than plain TTL, and will greatly extend battery life.

DIP METERS

This device is often called a transistor dip meter or a grid-dip oscillator (from vacuum-tube days). Most dip meters can also serve as absorption frequency meters (in this mode measurements are read at the current peak, rather than the dip). Further, some dip meters have a connection for headphones. The operator can usually hear signals that do not register on the meter. Because the dip meter is an oscillator, it can be used as a signal generator in certain cases where high accuracy or stability are not required.

A dip meter may be coupled to a circuit either inductively or capacitively. Inductive coupling results from the magnetic field generated by current flow. Therefore, inductive coupling should be used when a conductor with relatively high current is convenient. Maximum inductive coupling results when the axis of the pick-up coil is placed perpendicular to a nearby current path (see **Fig 26.17**).

Capacitive coupling is required when current paths are magnetically confined or shielded. (Toroidal inductors and coaxial cables are common examples of magnetic self shielding.) Capacitive coupling depends on the electric field produced by voltage. Use capacitive coupling when a point of relatively high voltage is convenient. (An example might be the output of a 12-V powered RF amplifier. *Do not* attempt dip-meter measurements on true high-voltage equipment such as vacuum-tube amplifiers or switching power supplies while they are energized.) Capacitive coupling is maximum when the end of the pick-up coil is near a point of high voltage (see Fig 26.17). In either case, the circuit under test is affected by the presence of the dip meter. Always use the minimum coupling that yields a noticeable indication.

Use the following procedure to make reliable measurements. First, bring the dip meter gradually closer to the circuit while slowly varying the dip-meter frequency. When a current dip occurs, hold the meter steady and tune for minimum current. Once the dip is found, move the meter away from the circuit and confirm that the dip comes from the circuit under test (the current reading should increase with distance from the circuit until the dip is gone). Finally, move the meter back toward the circuit until the dip is just noticed. Retune the meter for minimum current and read the dip-meter frequency with a calibrated receiver or frequency meter.

The current dip of a good measurement is smooth and symmetrical. A asymmetrical dip indicates that the dip-meter oscillator frequency is being significantly influenced by the test circuit. Such conditions do not yield usable readings.

A measurement of effective unloaded inductor Q can be made with a dip meter and an RF voltmeter (or a dc voltmeter with an RF probe). Make a parallel resonant circuit using the inductor and a capacitance equal to that of the application circuit. Connect the RF voltmeter across this parallel combination and measure the resonant frequency. Adjust the dip-meter/circuit coupling for a convenient reading on the voltmeter, then maintain this dip-meter/circuit relationship for the remainder of the test. Vary the dip meter frequency until the voltmeter reading drops to 0.707 times that at resonance. Note the frequency of the dip meter and repeat the process, this time varying the frequency on the opposite side of resonance. The difference between the two dip meter readings is the test-circuit bandwidth. This can be used to calculate the circuit Q:

$$Q = \frac{f_0}{BW} \qquad (9)$$

where
  f_0 = operating frequency,
  BW = measured bandwidth in the same units as the operating frequency.

When purchasing a dip meter, look for one that is mechanically and electrically stable. The coils should be in good condition. A headphone connection is helpful. Battery-operated models are convenient for antenna measurements.

A DIP METER WITH DIGITAL DISPLAY

An up-to-date dip meter was described by Larry Cicchinelli in the October 1993 issue of *QEX*. It consists of a dip meter with a three-digit frequency display. The analog portion of the circuit consists of an FET oscillator, voltage-doubler detector, dc-offset circuit and amplifier. The digital portion of the circuit consists of a high-impedance buffer, prescaler, counter, display driver, LED display and control circuit.

Circuit Description

The dip meter shown in **Fig 26.18** has four distinct functional blocks. The RF oscillator is a standard Colpitts using a common junction FET, Q1, as the active element. Its range

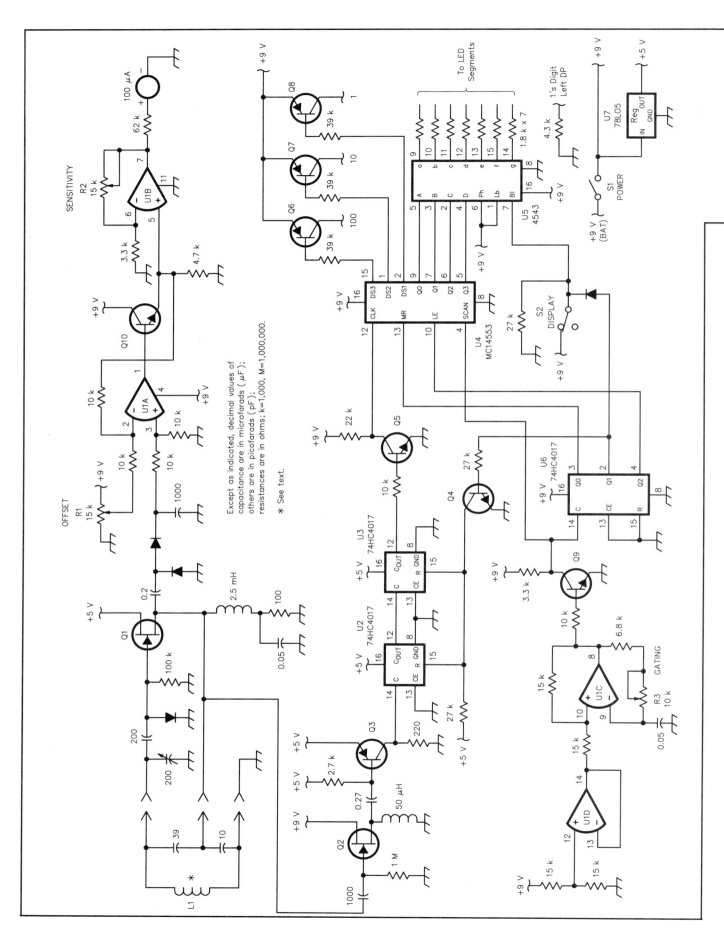

Table 26.3
Calculated Coil Data for the Dip Meter

Frequency (MHz)	L μH	Turns
1.7 to 3.1	48.6	52
2.8 to 5.9	16.3	23
5.6 to 11.9	4.0	9
9.7 to 20.7	1.3	5
19.0 to 45.0	0.3	2

is about 1.7 to 45 MHz. The 200-pF tuning capacitor gives a 2:1 tuning range. A 2:1 frequency range requires a capacitor with a 4:1 range. The sum of the minimum capacitance of the variable, the capacitors across the inductor and the strays must therefore be in the order of 70 pF. The values of L1 were determined experimentally by winding the coils and observing the lower and upper frequency values. [**Table 26.3** shows winding data calculated from the author's schematic. —Ed.] The coil sizes were experimentally determined, and the coils are constructed on 1¼-inch-diameter plug-in coil forms with #20 enameled wire. They are close wound. The number of turns shown are only a starting point; you may need to change them slightly in order to cover the desired frequency range.

The tapped capacitors are mounted inside the coil forms so that their values could be different for each band if required. The frequency spread of the lowest band is less than 2:1 because the tapped capacitor values are larger than those for the other bands. The frequency spread of the highest band is greater than 2:1 because its capacitors are smaller.

The analog display circuit begins with a voltage-doubler detector in order to get higher sensitivity. It drives a dc-offset circuit, U1A. R1 inserts a variable offset that is subtracted from the detector voltage. This allows the variable gain stage, U1B, to be more sensitive to variations in the detector output voltage. Q9 follows U1C to get an output gating voltage closer to ground. The resistor in series with the meter is chosen to limit the meter current to a safe value. For example, if a 1-mA meter is used, the resistor

should be 8.2 kΩ.

The prescaler begins with a high-impedance buffer and amplifier, Q2 and Q3. If you are going to use the meter for the entire frequency range described, take care in the layout of both the oscillator and buffer/amplifier circuits. The digital portion of the prescaler is a divide-by-100 circuit consisting of two divide-by-10 devices, U2 and U3. The devices used were selected because they were available. Any similar devices may be used as long as the reset circuit is compatible. Q5 is a level translator that shifts the 5-V signal to 9 V.

The first part of the digital display block is the oscillator circuit of U1C, which creates the gate time for the frequency counter. R3 adjusts the oscillator to a frequency of 500 Hz, yielding a 1-ms gate. The best way to set this frequency is to listen for the dip-meter output on a communications receiver and adjust R3 until the display agrees with the receiver. Once this calibration has been made for one of the bands, all the bands are calibrated. U1D gives a low-impedance voltage reference for U1C. Q9 was added to the output of the oscillator to remove a small glitch, which can cause the counter to trigger incorrectly. This type of oscillator has the advantage of simplicity. This circuit is fairly stable, easy to adjust and has a low parts count.

The digital system controller, U6, is a divide-by-10 counter. It has 10 decoded outputs, each of which goes high for one period of the input clock. The Q0 output is used to reset the frequency counter, U4. The Q1 output is used to enable the prescaler and disable the display and the Q2 output latches the count value into the frequency counter. Since the prescaler can only count while Q1 is high, it will be enabled for only 1 ms. Normally a 1-ms gate will yield 1-kHz resolution. Since the circuit uses a divide-by-100 prescaler, the resolution becomes 100 kHz.

The frequency counter, U4, is a three-digit counter with multiplexed BCD outputs. The clock input is driven from the prescaler, hence it is the RF oscillator frequency divided by 100. This signal is present for only 1 ms out of every 10 ms. The digit scanning is controlled by the 500-Hz oscillator of U1C.

U5 is a BCD-to-seven-segment decoder/ driver. Its outputs are connected to each of the three common-anode, seven-segment displays in parallel. Only the currently active digit will be turned on by the digit strobe outputs of U5, via Q6, Q7 and Q8. The diode connected to the blanking input of U5 disables the display while U4 is counting. U7 is a 5-V regulator that allows the use of a single 9-V battery for both the circuit and the LEDs. S2 turns on the displays once the unit has been adjusted for a dip.

The circuit draws about 20 mA with the LEDs off and up to 35 mA with the LEDs on.

Many of the resistor values are not critical, and those used were chosen based upon availability; the op-amp circuits depend primarily on resistance ratios. The resistor at the collector of Q3 is critical and should not be varied. Use 0.27-μF monolithic capacitors. They have the required good high-frequency characteristics over the range of the meter.

Most of the parts can be purchased from Digi-Key. They did not have the '4543 IC, which was purchased from Hosfelt Electronics. (See Address List in the **References** chapter.) The 74HC4017 may be substituted with a 74HCT4017. The circuit is built on a 4-inch-square perf board (with places for up to 12 ICs) and is housed in a 7×5×3-inch minibox.

Operation

To use the unit, set the gain, R2, fully clockwise for maximum sensitivity. With this setting, the output of the offset circuit (Q10 emitter) is at ground. As R1 is rotated, the voltage on the arm approaches and then becomes less than, the detector output. At this point the meter will start to deflect upward. Adjust R1 so that the meter reads about center scale. (Manual adjustment allows for variations in the output level of the RF oscillator.) As L1 is brought closer to the circuit under test, the meter will deflect downward as energy is absorbed by the circuit. For best results use the minimum possible coupling to the circuit being tested. If the dip meter is overcoupled to the test circuit, the oscillator frequency will be pulled.

FREQUENCY COUNTERS

One of the most accurate means of measuring frequency is a frequency counter. This instrument is capable of numerically displaying the frequency of the signal supplied to its input. For example, if an oscillator operating at 8.244 MHz is connected to

a counter input, 8.244 would be displayed. At present, counters are usable well up into the gigahertz range. Most counters that are used at high frequencies make use of a prescaler ahead of a basic low-frequency counter. A prescaler divides the high-fre-

quency signal by 10, 100, 1000 or some other fixed amount so that a low-frequency counter can display the operating frequency.

The accuracy of the counter depends on its internal crystal reference. A more accurate crystal reference yields more accurate readings. Crystals for frequency counters are manufactured to close tolerances. Most counters have a trimmer capacitor so that the crystal can be set exactly on frequency. Crystal frequencies of 1 MHz, 5 MHz or 10 MHz have become more or less standard. For calibration, harmonics of the crystal can be compared to a known reference station, such as those shown in Table 26.1, or other frequency standard and adjusted for zero beat.

Many frequency counters offer options to increase the accuracy of the counter timebase; this directly increases the counter accuracy. These options usually employ temperature-compensated crystal oscillators (TCXOs) or crystals mounted in constant temperature ovens that keep the crystal from being affected by changes in ambient (room) temperature. Counters with these options may be accurate to 0.1

ppm (part per million) or better. For example, a counter with a timebase accuracy of 5.0 ppm and a second counter with a TCXO accurate to 0.1 ppm are available to check a 436-MHz CW transmitter for satellite use. The counter with the 5-ppm timebase could have a frequency error of as much as 2.18 kHz, while the possible error of the counter with the 0.1 ppm timebase is only 0.0436 kHz.

Other Instruments and Measurements

This section covers a variety of test equipment that is useful in receiver and transmitter testing. It includes RF and audio generators, an inductance meter, oscilloscopes, spectrum analyzers, a calibrated noise source, a noise bridge, an advanced resonance indicator, combiners, attenuators and dummy loads. A number of applications of this equipment to basic transmitter and receiver testing is also included.

RF OSCILLATORS FOR CIRCUIT ALIGNMENT

Receiver testing and alignment uses equipment common to ordinary radio service work. Inexpensive RF signal generators are available, both complete and in kit form. However, any source of signal that is weak enough to avoid overloading the receiver usually will serve for alignment work. The frequency marker generator is a satisfactory signal source. In addition, its frequencies, although not continuously adjustable, are known far more precisely, since the usual signal-generator calibration is not highly accurate. An attenuator described later in this chapter can be added for relative dB measurements. When buying a used or inexpensive signal generator, look for these attributes: output level is calibrated, the output doesn't "ring" too badly when tapped, and doesn't drift too badly when warmed up. Many military surplus units are available that can work quite well. Commercial units such as the HP608 are

big and stable, and they may be inexpensive.

AUDIO-FREQUENCY OSCILLATORS

An audio signal generator should provide a reasonably pure sine wave. The best oscillator circuits for this use are RC

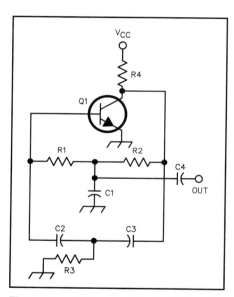

Fig 26.19 — Values for the twin-T audio oscillator circuit range from 18 kΩ for R1-R2 and 0.05 μF for C1 (750 Hz) to 15 kΩ and 0.02 μF for 1800 Hz. For the same frequency range, R3 and C2-C3 vary from 1800 Ω and 0.02 μF to 1500 Ω and 0.01 μF. R4 is 3300 Ω and C4, the output coupling capacitor, can be 0.05 μF for high-impedance loads.

coupled, operating as close to a class-A amplifier as possible. Variable frequencies covering the entire audio range are needed for determining frequency response of audio amplifiers.

An oscillator generating one or two frequencies with good waveform is sufficient for most phone-transmitter testing and simple troubleshooting in AF amplifiers. A two-tone (dual) oscillator is very useful for testing and adjusting sideband transmitters.

A circuit of a simple RC oscillator that is useful for general testing is given in **Fig 26.19**. This Twin-T arrangement gives a waveform that is satisfactory for most purposes. The oscillator can be operated at any frequency in the audio range by varying the component values. R1, R2 and C1 form a low-pass network, while C2, C3 and R3 form a high-pass network. As the phase shifts are opposite, there is only one frequency at which the total phase shift from collector to base is 180°: Oscillation will occur at this frequency. When C1 is about twice the capacitance of C2 or C3 the best operation results. R3 should have a resistance about 0.1 that of R1 or R2 (C2 = C3 and R1 = R2). Output is taken across C1, where the harmonic distortion is least. Use a relatively high impedance load — 100 kΩ or more.

Most small-signal AF transistors can be used for Q1. Either NPN or PNP types are satisfactory if the supply polarity is set correctly. R4, the collector load resistor may be changed a little to adjust the oscillator for best output waveform.

A WIDE-RANGE AUDIO OSCILLATOR

A wide-range audio oscillator that will provide a moderate output level can be built from a single 741 operational amplifier (**Fig 26.20**). Power is supplied by two 9-V batteries from which the circuit draws 4 mA. The frequency range is selectable from 8 Hz to 150 kHz. Distortion is approximately 1%. The output level under a light load (10 $k\Omega$) is 4 to 5 V. This can be increased by using higher battery voltages, up to a maximum of plus and minus 18 V, with a corresponding adjustment of R_F.

Pin connections shown are for the TO-5 case and the eight-pin DIP package. Variable resistor R_F is trimmed for an output level of about 5% below clipping as seen on an oscilloscope. This should be done for the temperature at which the oscillator will normally operate, as the lamp is sensitive to ambient temperature. This unit was originally described by Shultz in November 1974 *QST*; it was later modified by Neben as reported in June 1983 *QST*.

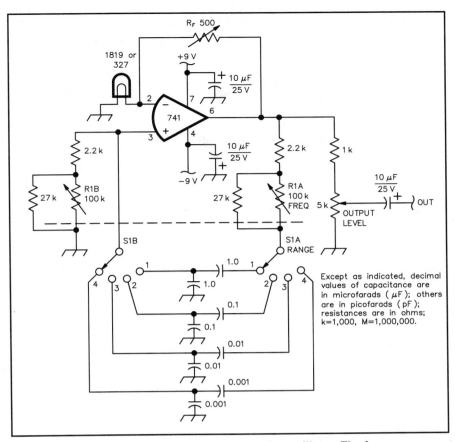

Fig 26.20 — A single IC (741 op amp) based audio oscillator. The frequency range is set by switch S1.

MEASURE INDUCTANCE AND CAPACITANCE WITH A DVM

Many of us have a DVM (Digital Voltmeter) or VOM (Volt Ohm Meter) in the shack, but few of us own an inductance or capacitance meter. If you have ever looked into your junk box and wanted to know the value of the unmarked parts, these simple circuits will give you the answer. They may be built in one evening (**Fig 26.23** and **Fig 26.24**), and will adapt your DVM or VOM to measure inductance or capacitance. The units are calibrated against a known part. Therefore, the overall accuracy depends only on the calibration values and not on the components used to build the circuits. If it is carefully calibrated, an overall accuracy of 10% may be expected if used with a DVM and slightly less with a VOM.

CONSTRUCTION

The circuits may be constructed on a small perf board (Radio Shack dual mini board, # 276-168), or if you prefer, on a PC board. A template and parts layout may be obtained from the ARRL.[1] Layout is non-critical—almost any construction technique will suffice. Wire-wrapping or point-to-point soldering may be used.

The circuits are available in kit form from Electronic Rainbow Inc. (see Address List in the **References** chapter). The IA inductance adapter kit is $14.95. A separate cabinet is available for $8.95. The PC board is 1.75×2.5 inches. The CA-1 capacitance adapter kit is $12.95 and comes with a 1.80×2.0-inch PC board.

INDUCTANCE ADAPTER FOR DVM/VOM
Description

The circuit shown in Fig 26.23 converts an unknown inductance into a voltage that

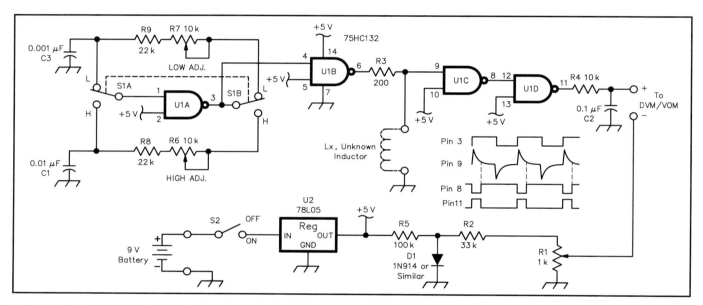

Fig 26.23—All components are 10% tolerance. 1N4148 or equivalent may be substituted for D1. A LM7805 may be substituted for the 78L05. All fixed resistors are ¼-W carbon composition. Capacitors are in μF.

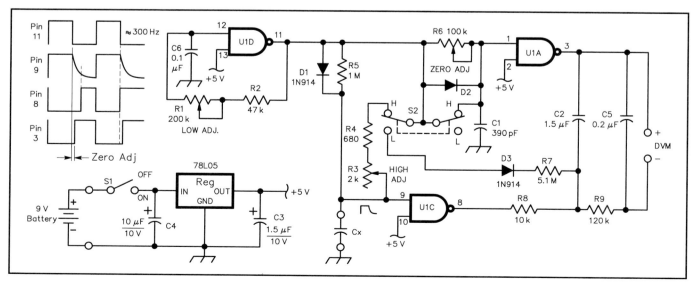

Fig 26.24—All components are 10% tolerance. A LM7805 may be substituted for the 78L05. All fixed resistors are ¼-W carbon composition. Capacitors are in μF unless otherwise indicated.

can be displayed on a DVM or VOM. Values between 3 and 500 µH are measured on the low range and from 100 µH to 7 mH on the high range. NAND gate U1A is a two frequency RC square-wave oscillator. The output frequency (pin 3) is approximately 60 kHz in the low range and 6 kHz in the high range. The square-wave output is buffered by U1B and applied to a differentiator formed by R3 and the unknown inductor, LX. The stream of spikes produced at pin 9 decay at a rate proportional to the time constant of R3-LX. Because R3 is a constant, the decay time is directly proportional to the value of LX. U1C squares up the positive going spikes, producing a stream of negative going pulses at pin 8 whose width is proportional to the value of LX.

They are inverted by U1D (pin 11) and integrated by R4-C2 to produce a steady dc voltage at the + output terminal. The resulting dc voltage is proportional to LX and the repetition rate of the oscillator. R6 and R7 are used to calibrate the unit by setting a repetition rate that produces a dc voltage corresponding to the unknown inductance. D1 provides a 0.7 volt constant voltage source that is scaled by R1 to produce a small offset reference voltage for zeroing the meter on the low inductance range.

When S1 is low, *mV* corresponds to *µH* and when high, *mV* corresponds to *mH*. A sensitive VOM may be substituted for the DVM with a sacrifice in resolution.

Test and Calibration

Short the LX terminals with a piece of wire and connect a DVM set to the 200-mV range to the output. Adjust R1 for a zero reading. Remove the short and substitute a known inductor of approximately 400 uH. Set S1 to the low (in) position and adjust R7 for a reading equal to the known inductance. Switch S1 to the high position and connect a known inductor of about 5 mH. Adjust R6 for the corresponding value. For instance, if the actual value of the calibration inductor is 4.76 mH, adjust R7 so the DVM reads 476 mV.

CAPACITANCE ADAPTER FOR DVM/VOM

Description

The circuit shown in Fig 26.24 measures capacitance from 2.2 to 1000 pF in the low range, from 1000 pF to 2.2 µF in the high range. U1D of the 74HC132 (pin 11) produces a 300 Hz square-wave clock. On the rising edge CX rapidly charges through D1. On the falling edge CX slowly discharges through R5 on the low range and through R3-R4 on the high range. This produces an asymmetrical waveform at pin 8 of U1C with a duty cycle proportional to the unknown capacitance, CX. This signal is integrated by R8-R9-C2 producing a dc voltage at the negative meter terminal proportional to the unknown capacitance. A constant reference voltage is produced at the positive meter terminal by

integrating the square-wave at U1A, pin 3. R6 alters the symmetry of this square-wave producing a small change in the reference voltage at the positive meter terminal. This feature provides a zero adjustment on the low range. The DVM measures the difference between the positive and negative meter terminals. This difference is proportional to the unknown capacitance.

Test and Calibration

Without a capacitor connected to the input terminals, set SW2 to the low range (out) and attach a DVM to the output terminals. Set the DVM to the 2-volt range and adjust R6 for a zero meter reading. Now connect a 1000 pF calibration capacitor to the input and adjust R1 for a reading of 1.00 volt. Switch to the high range and connect a 1.00 µF calibration capacitor to the input. Adjust R3 for a meter reading of 1.00 volts. The calibration capacitors do not have to be exactly 1000 pF or 1.00 µF, as long as you know their exact value. For instance, if the calibration capacitor is known to be 0.940 µF, adjust the output for a reading of 940 mV.

Notes

[1]See Chapter 30, **References**, for the template.

A SIX DIGIT PROGRAMMABLE FREQUENCY COUNTER AND DIGITAL DIAL

This six digit programmable frequency counter has multiple uses around the shack. It can be used as a general purpose frequency counter, a digital dial for home-built rigs, an add-on digital dial for commercial rigs, or it can be added to vintage tube-type receivers for precise frequency spotting. The counter displays frequencies greater than 50 MHz with 100-Hz resolution. It updates 40 times per second, providing near real time tuning response, and is programmable to account for the offset between a receiver local oscillator and the operating frequency. It also will display the correct operating frequency with reverse tuning local oscillators.

The counter is based on the PIC16C57, an eight bit CMOS Microcontroller by Microchip Technology Inc. This part is custom programmed by Radio Adventures Corp. (see Address List in the **References** chapter), and is supplied by them as the C5 fre-

quency counter chip. The easiest way to build this project is to order the complete kit (model A-2) including a silk-screened punched chassis from Radio Adventures for $79.95. The PC boards may be ordered separately, the counter board (BK-172) is $37.95 and the display board (BK-171) is $24.95. The programmed C5 chip alone is also available for $14.95. The kit comes with a metal case, instructions, and all parts including two high quality PC boards. If you choose to build the counter from scratch, a PC board layout and a parts location drawing are available from ARRL.[1]

Some of the advanced features of the counter include:

- an anti-jitter code that reduces last digit jitter
- five push-button selectable, non-volatile programmable offsets (total of 16)
- programmable reverse counting for re-

verse tuning VFOs
- selectable direct frequency readout
- programmable 100-Hz digit blanking
- programmable automatic display blanking for power conservation
- automatic display enable when the frequency changes
- leading zero blanking of megahertz digits.

The frequency counter requires 9 to 13-volts dc. Current requirements vary with operating frequency and the status of the display. At 5 MHz the operating current is approximately 100 mA with the display active, and approximately 40 mA with the display blanked. The main board is about 2.2×3.9-inches. The display board is a separate 1.4×3.9-inch assembly.

CIRCUIT DESCRIPTION

The block diagram (**Fig 26.25**) shows

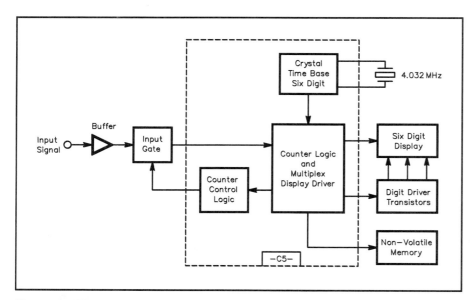

Fig 26.25—Block diagram showing the crystal time base, counter control logic, six digit counter control logic, multiplexed display driver and memory control for the Six Digit Programmable Frequency Counter.

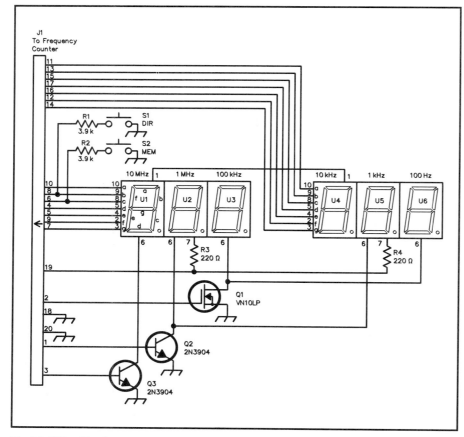

Fig 26.26B—Display circuitry schematic. J1 is connected to J5 through a 20-conductor ribbon cable. All fixed resistors are $^{1}/_{4}$-W, 5%-tolerance carbon film. Capacitance values are in microfarads (µF).

Q1—VN10LP VMOS power FET (Digi-Key VN10LP-ND)
Q2, Q3—2N3904 NP transistor
U1, U2, U6—9 mm green LED 7-Segment display (Kingbright SC36-11GWA)
U3, U4, U5—9 mm yellow LED 7-Segment display (Kingbright SC36-11YWA)

the crystal time base, counter control logic, six digit counter logic, multiplexed display driver and memory sections. The crystal is 4.032 MHz, which puts the fundamental and most lower order harmonics outside the ham bands.

The unit is built on two PC boards, the C5 frequency counter chip on one and the display on the other. A complete schematic is shown in **Fig 26.26**. Capacitor C5 is used to calibrate the counter. A two transistor buffer precedes the count gate and provides some amplification and buffering between the counter and the circuit being measured. Two switch selected input connections are provided—a 50-Ω termination for the digital dial functions and a high impedance termination for direct frequency measurement. The signal passes through U1A to the counter chip.

Using the pulse count stored in the C5 count registers, the chip makes necessary calculations and drives a common cathode display consisting of two groups of three digits each. The display and front panel switches are mounted on a separate PC board for ease of mounting and are connected to the main board by a ribbon cable. Current limiting resistors mounted on the main board are used to provide segment current limiting and provide protection to the C5 in case wiring to the display is accidentally shorted.

Programmed parameters such as offset, slope and auto-blanking are stored in U3, a non-volatile EEPROM memory chip. The programming and operating switches, which share pins with the display, are isolated by 3.9-kΩ resistors.

The C5 counter chip provides 16 channels of program storage. Each program stores frequency offset, normal/inverted counting, continuous/automatic blanking display and 100/1000 Hz resolution selection. The first five channels, 00 through 04, are selected by the front panel MEM switch. All sixteen channels may be selected through the binary jumper connected to J4 (MEMORY SELECT), on the main PC board. The terminals of J4 are pulled low to select a certain channel. For instance, to select memory channel five, terminals 1 and 4 would be connected to the ground pin.

Since the PGM and MODE switches are only used when the counter is being setup, they are located on the rear panel. The MEM and DIR frequency functions are used more often, therefore, they are accessible from the front panel.

The power supply for the counter should provide between 9 to 13 V of ripple-free dc. The counter main board contains a 5-V regulator (U4) that provides power to the counter, logic and memory chips. The 9 to 13-V power is passed through an RC filter to the two-transistor buffer. If volt-

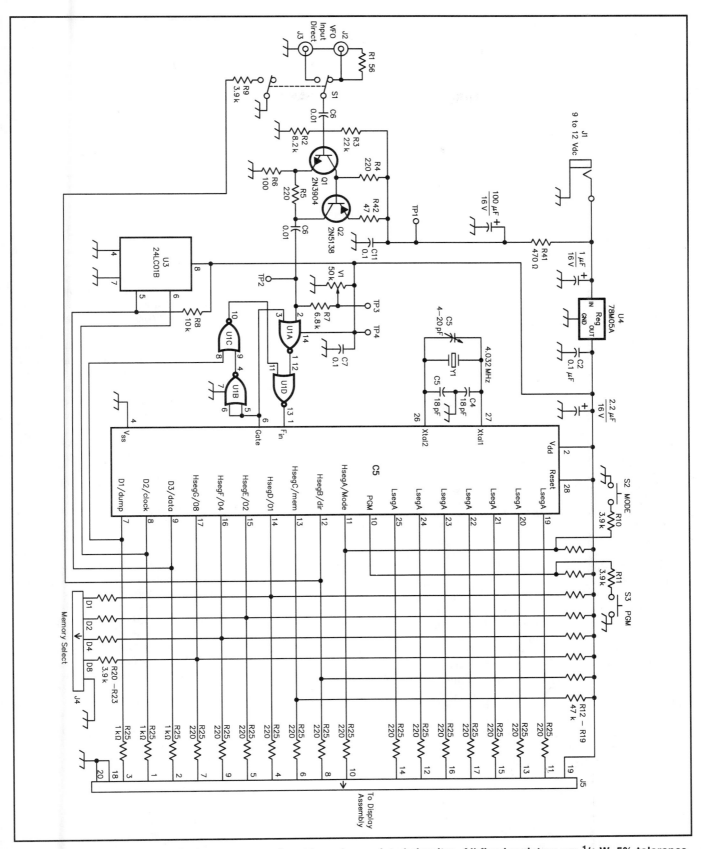

Fig 26.26A—Schematic of the C5 frequency counter chip and associated circuitry. All fixed resistors are ¹/4-W, 5%-tolerance carbon film. Capacitance values are in microfarads (μF).

U1—74HC02 quad 2 input NOR gate
(Digi-Key MM74HC02N-ND)
U2—C5 frequency counter IC (8 bit

microcontroller) see text
U3—24LC01B/P 128×8 EEPROM (Digi-
Key 24LC01B/P-ND)

Y1—4.032 MHz 20 pF HC-49/U crystal
(Mouser 520-HCA403-20)

ages below 9 V or above 13 V are used, the voltage dropping resistor, R41, can be changed such that the voltage at test point TP1 is about 6.5 V.

CONSTRUCTION

Construction is very straightforward if the circuit board kit is used. If you choose to hand wire your unit on perfboard or lay out your own PC board, use short connections in the front end of the counter (around J1, J2, Q1, Q2 and U1). The wiring to the display, memory and switches is less critical. The use of sockets for the counter, gate and memory ICs is strongly recommended. A good quality magnifying glass can be very helpful in identifying the various small components and checking the solder joints after assembly. A small 20 to 25 W soldering iron with a small tip will be necessary for making reliable solder joints without bridging. Use only electronic grade rosin core solder.

TEST AND CALIBRATION

After assembly, but before inserting the ICs into their sockets, apply dc power and check for +5 V ± 10% at the following pins:

U1, pin 14
U2, pins 2, 9, 10, 28
U3, pins 5, 8

The voltage at U1, pin 2, should vary between 0 and 5 V as potentiometer V1 is rotated. Set this voltage to 2.5 V. U2 should have 1.65 V ± 10% on pins 11, 12, 13, 14, 15, 16, 17.

All other pins on U1, U2 and U3 should measure 0 V. The voltage at TP1 should be 5-8 V and should be several volts less than the voltage powering the unit. If the unit fails one or more of these tests, there is a wiring error of some sort. Correct the error before proceeding.

When all is well with the above tests, remove the dc power and insert the ICs into their appropriate sockets. Re-apply power. The display should come alive. Position the input selector switch to the direct frequency position with no signal applied. The display should indicate 000.X where X may be a 0 or 1. If the display is indicating random numbers, adjust the trimmer pot V1 slightly to obtain the correct display. Adjusting V1 beyond the point where the display stabilizes will reduce the sensitivity of the counter. Attach a signal source with a known frequency to the direct frequency input. The signal should be in the range of 75 mV to 1 V rms. The higher the frequency, up to 50 MHz, the better. Adjust the trimmer ca-

pacitor C5 until the frequency displayed by the counter matches the applied frequency. The unit is now calibrated and ready for use.

OPERATION

The finished boards may be mounted in a stand alone enclosure or may be built into another piece of equipment. Since the counter is a digital device, it is possible that some noise may be heard in sensitive receivers on a quiet band. Very good results have been obtained when the counter is mounted in its own enclosure. If noise is present in built-in installations, some shielding may be necessary. Additional filtering of the supply voltage line may also help. Minimize the length of the ribbon cable connecting the main board and the display board.

In the digital dial mode, the counter requires about +10 dBm of signal over most of its operating range. Input to the counter is taken from the tunable oscillator or from the output of the pre-mix system. The general rule is to sample the signal at a low impedance point such as the emitter or source for transistors or the cathode in tube type equipment. Use small diameter coaxial cable, such as RG-174, to connect the digital dial to the equipment. A small capacitor, usually in the range of 10-100 pF, connected in series with the center conductor of the cable can be used to establish the proper signal level. Use the smallest value that gives reliable counter operation over the frequency range of interest. In broadband tube type equipment the oscillator signal amplitude may vary too much for reliable counting. In such cases, back-to-back parallel diodes may be connected to limit the signal level. If a broadband oscilloscope is available, observe the signal level at TP2 and adjust the value of the coupling capacitor to obtain between 0.6 and 2.5 V peak-to-peak.

Once a stable reading is obtained over the operating frequency range, the counter can be programmed with desired features and offsets. Programming is carried out in the following order. First, select the desired memory channel to be programmed, second, program the desired display mode, third, program the MHz offset and finally, program the kHz offset. There is no need to program unused memory channels. A detailed description of each programming step follows:

- Select the desired memory channel by closing the MEM switch and holding it until the desired channel number is displayed. Alternatively, switch the MEMORY SELECT lines to the binary

Table 26.5
Counter Display Modes

Mode	Invert	Blank	100 Hz
00			X
01	X		
20		X	X
30	X	X	X
40			
50	X		
60		X	
70	X	X	

value desired. Remember that the lines are active low. For instance, if you wanted to select memory channel 7, you would pull lines 01, 02 and 04 low.

- Program the mode by closing and holding the PGM switch until the display quits blinking *PROG*. Release the PGM switch. Close and hold the MODE switch until the display indicates the desired mode. Release the MODE switch. See **Table 26.5** to select desired mode. Store the selected mode by momentarily closing the PGM switch. The counter will now display frequency, indicating that the program mode has been exited.
- Program the MHz offset by closing and holding the PGM switch until the display quits blinking *PROG*. While the display is still blinking, momentarily close the DIR switch on the front panel. This action places the counter into the MHz program mode. When the display quits blinking release the PGM switch and use the DIR and MEM front panel switches to retard or advance the MHz display until it displays the frequency being monitored. Momentarily close the PGM switch to store the MHz offset.
- Program the kHz offset by closing and holding the PGM switch until the display quits blinking *PROG*. Release the PGM switch and use the DIR and MEM front panel switch to retard or advance the kHz display until it displays the frequency being monitored. Momentarily close the PGM switch to store the kHz offset.

This same procedure is used to program each memory channel.

Notes

[1]A template for the six digit programmable frequency counter is available at: **http://www.arrl.org/notes**.

OSCILLOSCOPES

Most engineers and technicians will tell you that the most useful single piece of test and design equipment is the triggered-sweep oscilloscope (commonly called just a "scope"). This section was written by Dom Mallozzi, N1DM.

Oscilloscopes can measure and display voltage relative to time, showing the waveforms seen in electronics textbooks. Scopes are broken down into two major classifications: analog and digital. This does not refer to the signals they measure, but rather to the methods used inside the scope to process signals for display.

ANALOG OSCILLOSCOPES

Fig 26.27 shows a simplified diagram of a triggered-sweep oscilloscope. At the heart of nearly all scopes is a cathode-ray tube (CRT) display. The CRT allows the visual display of an electronic signal by taking two electric signals and using them to move (deflect) a beam of electrons that strikes the screen. Unlike a television CRT, an oscilloscope uses electrostatic deflection rather than magnetic deflection. Wherever the beam strikes the phosphorescent screen of the CRT it causes a small spot to glow. The exact location of the spot is a result of the voltage applied to the vertical and horizontal inputs.

All of the other circuits in the scope are used to take the real-world signal and convert it to a form usable by the CRT. To trace how a signal travels through the oscilloscope circuitry start by assuming that the trigger select switch is in the INTERNAL position.

The input signal is connected to the input COUPLING switch. The switch allows selection of either the ac part of an ac/dc signal or the total signal. If you wanted to measure, for example, the RF swing at the collector of an output stage (referenced to the dc level), you would use the dc-coupling mode. In the ac position, dc is blocked from reaching the vertical amplifier chain so that you can measure a small ac signal superimposed on a much larger dc level. For example, you might want to measure a 25 mV 120-Hz ripple on a 13 Vdc power supply. Note that you should not use ac coupling at frequencies below 30 Hz, because the value of the blocking capacitor represents a considerable series impedance to very low-frequency signals.

After the coupling switch, the signal is connected to a calibrated attenuator. This is used to reduce the signal to a level that can be tolerated by the scope's vertical amplifier. The vertical amplifier boosts the signal to a level that can drive the CRT and also adds a bias component to locate the waveform on the screen.

A small sample of the signal from the vertical amplifier is sent to the trigger circuitry. The trigger circuit feeds a start pulse to the sweep generator when the input signal reaches a certain level. The sweep generator gives a precisely timed signal that looks like a triangle (see **Fig 26.28**). This triangular signal causes the scope trace to sweep from left to right, with the zero-voltage point representing the left side of the screen and the maximum voltage representing the right side of the screen.

The sweep circuit feeds the horizontal amplifier that, in turn, drives the CRT. It is also possible to trigger the sweep system from an external source (such as the system clock in a digital system). This is done by using an external input jack with the trigger select switch in the EXTERNAL position.

The trigger system controls the horizontal sweep. It looks at the trigger source (internal or external) to find out if it is positive- or negative-going and to see if the signal has passed a particular level. **Fig 26.29A** shows a typical signal and the dotted line on the figure represents the trigger level. It is important to note that once a trigger circuit is "fired" it cannot fire again until the sweep has moved all the way across the screen from left to right. In normal operation. the TRIGGER LEVEL control is manually adjusted until a stable display is seen. Some scopes have an AUTOMATIC position that chooses a level to lock the display in place without manual adjustment.

Fig 26.29B shows what happens when the level has not been properly selected. Because there are two points during a single cycle of the waveform that meet the triggering requirements, the trigger circuit will have a tendency to jump from one trigger point to another. This will make the waveform jitter from left to right. Adjustment of the TRIGGER control will fix this problem.

The horizontal travel of the trace is calibrated in units of time. If the time of one cycle is known, we can calculate the frequency of the waveform. In **Fig 26.30**, for example, if the SWEEP speed selector is set at 10 μs/division and we count the number of divisions (vertical bars) between peaks of the waveform (or any similar well defined points that occur once per cycle) we can find the period of one cycle. In this case it is 80 μs. This means that the frequency of the waveform is 12,500 Hz (1/80 μs). The accuracy of the measured frequency depends on the accuracy of the scope's sweep oscillator (usually approximately 5%) and the linearity of the ramp generator. This

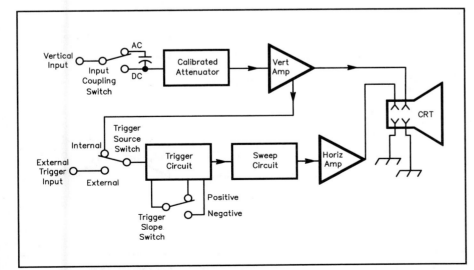

Fig 26.27 — Typical block diagram of a simple triggered-sweep oscilloscope.

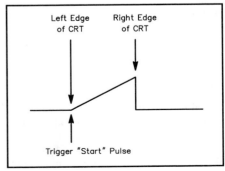

Fig 26.28 — The sweep trigger starts the ramp waveform that sweeps the CRT electron beam from side to side.

accuracy cannot compete with even the least-expensive frequency counter, but the scope can still be used to determine whether a circuit is functioning properly.

Dual-Trace Oscilloscopes

Dual-trace oscilloscopes can display two waveforms at once. This type of scope has two vertical input channels that can be displayed either alone, together or one after the other. **Fig 26.31** shows a simplified block diagram of a dual-trace oscilloscope. The only differences between this scope and the previous example are the additional vertical amplifier and the "channel switching circuit." This block determines whether we display channel A, channel B or both (simultaneously). The dual display is not a true dual display (there is only one electron gun in the CRT) but the dual traces are synthesized in the scope.

There are two methods of synthesizing a dual-trace display from a single-beam scope. These two methods are referred to as *chopped mode* and *alternate mode*. In the chopped mode a small portion of the channel A waveform is written to the CRT, then a corresponding portion of the channel B waveform is written to the CRT. This procedure is continued until both waveforms are completely written on the CRT. The chopped mode is especially useful where an actual measure of the phase difference between the two waveforms is required. The chopped mode is usually most useful on slow sweep speeds (times greater than a few microseconds per division).

In the alternate mode, the complete channel A waveform is written to the CRT followed immediately by the complete channel B waveform. This happens so quickly that it appears that the waveforms are displayed at the same time. This mode of operation is not useful at very slow sweep speeds, but is good at most other sweep speeds.

Most dual-trace oscilloscopes also have a feature called "X-Y" operation. This feature allows one channel to drive the horizontal amplifier of the scope (called the X channel) while the other channel (called Y in this mode of operation) drives the vertical amplifier. Some oscilloscopes also have an external Y input. X-Y operation allows the scope to display Lissajous patterns for frequency and phase comparison and to use specialized test adapters such as curve tracers or spectrum analyzer front ends. Because of frequency limitations of most scope horizontal amplifiers the X channel is usually limited to a 5 or 10-MHz bandwidth.

DIGITAL OSCILLOSCOPES

The classic analog oscilloscope just discussed has existed for over 50 years. In the last 15 years, the digital oscilloscope has advanced from a specialized laboratory device to a very useful general-purpose tool, with a price attractive to an active experimenter. It uses digital circuitry and microprocessors to enhance the processing and display of signals. These result in dramatically improved accuracy for both amplitude and time measurements. When configured as a digital storage oscilloscope (DSO) it can read a stored waveform for as long as you wish without time limitations incurred by an analog type of storage scope.

Examine the simplified block diagram shown in **Fig 26.32**. After the signal goes through the vertical input attenuators and amplifiers, it arrives at the analog-to-digital converter (ADC). The ADC assigns

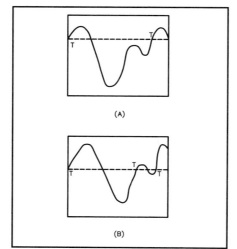

Fig 26.29 — In order to produce a stable display the selection of the trigger point is very important. Selecting the trigger point in A produces a stable display, but the trigger shown at B will produce a display that "jitters" from side to side.

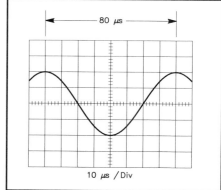

Fig 26.30 — Oscilloscopes with a calibrated sweep rate can be used to measure frequency. Here the waveform shown has a period of 80 microseconds (8 divisions × 10 μs per division) and therefore a period of 1/80 μs or 12.5 kHz.

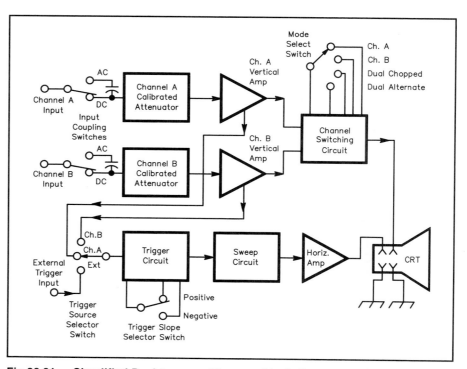

Fig 26.31 — Simplified Dual-trace oscilloscope block diagram. Note the two identical input channels and amplifiers.

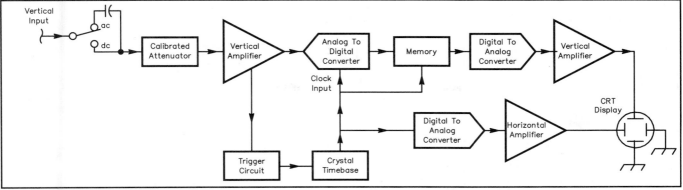

Fig 26.32 — Simplified block diagram of a digital oscilloscope. Note: The microprocessors are not shown for clarity after ABC's of Oscilloscopes copyright Fluke Corporation (Reproduced with Permission).

a digital value to the level of the analog input signal and puts this in a memory similar to computer RAM. This value is stored with an assigned time, determined by the trigger circuits and the crystal timebase. The digital oscilloscope takes discrete amplitude samples at regular time intervals. If you were to take this data directly from memory, it would put a series of dots on the screen. You would then have to connect the dots to reconstruct the original waveform. The digital scope's microprocessor does this for you by mathematically processing the signal while reading it back from the memory and driving a digital-to-analog converter

(DAC), which then drives the vertical deflection amplifier. A DAC also takes the digital stored time data and uses it to drive the horizontal deflection amplifier.

For the vertical signals you will see manufacturers refer to "8-bit digitizing," or perhaps "10-bit resolution." This is a measure of how many digital levels that are shown along the vertical (voltage) axis. More bits give you better resolution and accuracy of measurement. An 8-bit vertical resolution means each vertical screen has 2^8 (or 256) discrete values; similarly, 10 bits resolution yields 2^{10} (or 1024) discrete values.

It is important to understand some of the limitations resulting from sampling the signal rather than taking a continuous, analog measurement. When you try to reconstruct a signal from individual discrete samples, you must take samples at least twice as fast as the highest frequency signal being measured. If you digitize a 100-MHz sine wave, you should take samples at a rate of 200 million samples a second (referred to as 200 Megasamples/second). Actually, you really would like to take samples even more often, usually at a rate at least five times higher than the input signal.

If the sample rate is not high enough, very fast signal changes between sampling points will not appear on the display. For example, **Fig 26.33** shows one signal measured using both analog and digital scopes. The large spikes seen in the analog-scope display are not visible on the digital scope. The sampling frequency of the digital scope is not fast enough to store the higher frequency components of the waveform. If you take samples at a rate less than twice the input frequency, the reconstructed signal has a wrong apparent frequency; this is referred to as *aliasing*. In Fig 26.33 you can see that there is about one sample taken per cycle of the input waveform. This does not meet the 2:1 criteria established above. The result

is that the scope reconstructs a waveform with a different apparent frequency.

Many older digital scopes had potential problems with aliasing. Newer scopes use advanced techniques to check themselves. A simple manual check for aliasing is to use the highest practical sweep speed (shortest time per division) and then to change to other sweep speeds to verify that the apparent frequency doesn't change.

LIMITATIONS

Oscilloscopes have fundamental limits, primarily in frequency of operation and range of input voltages. For most purposes the voltage range of a scope can be expanded by the use of appropriate probes. The frequency response (also called the bandwidth) of a scope is usually the most important limiting factor. At the specified maximum response frequency, the response will be down 3 dB (0.707 voltage). For example, a 100-MHz 1-V sine wave fed into a 100-MHz bandwidth scope will read approximately 0.707 V on the scope display. The same scope at frequencies below 30 MHz (down to dc) should be accurate to about 5%.

A parameter called *rise time* is directly related to bandwidth. This term describes a scope's ability to accurately display voltages that rise very quickly. For example, a very sharp and square waveform may appear to take some time in order to reach a specified fraction of the input voltage level. The rise time is usually defined as the time required for the display to show a change from the 10% to 90% points of the input waveform, as shown in **Fig 26.34**. The mathematical definition of rise time is given by:

$$t_r = \frac{0.35}{BW} \tag{10}$$

where
  t_r = rise time, μs
  BW = bandwidth, MHz.

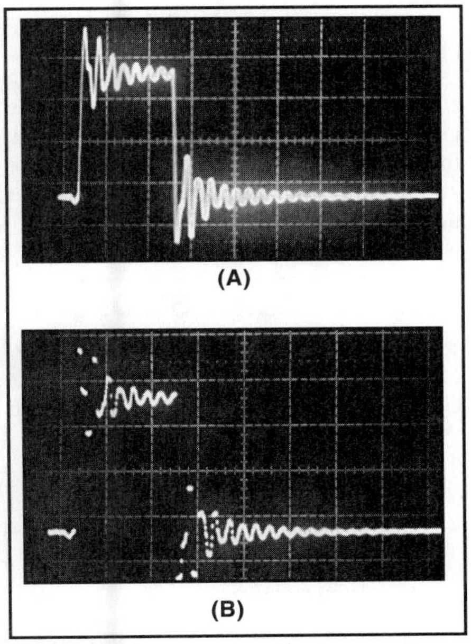

Fig 26.33 — Comparison of an analog scope waveform (A) and that produced by a digital oscilloscope (B). Notice that the digital samples in B are not continuous, which may leave the actual shape of the waveform in doubt for the fastest rise time displays the scope is capable of producing.

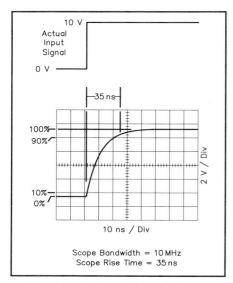

Fig 26.34 — The bandwidth of the oscilloscope vertical channel limits the rise time of the signals displayed on the scope.

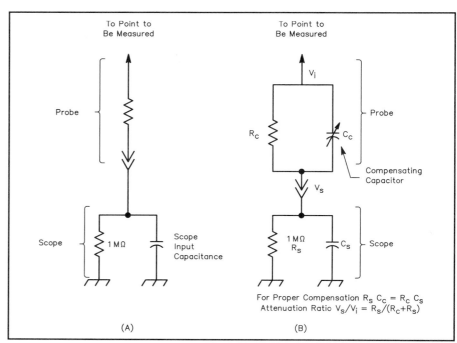

For Proper Compensation $R_S C_C = R_C C_S$
Attenuation Ratio $V_S/V_i = R_S/(R_C+R_S)$

(A)   (B)

Fig 26.35 — Uncompensated probes such as the one at A are sufficient for low-frequency and slow-rise-time measurements. However, for accurate display of fast rise times with high-frequency components the compensated probe at B must be used. The variable capacitor is adjusted for proper compensation (see text for details).

It is also important to note that all but the most modern (and expensive) scopes are not designed for precise measurement of either time or frequency. At best, they will not have better than 5% accuracy in these applications. This does not change the usefulness of even a moderately priced oscilloscope, however. The most important value of an oscilloscope is that it presents an image of what is going on in a circuit and quickly shows which component or stage is at fault. It can show modulation levels, relative gain between stages and oscillator output.

OSCILLOSCOPE PROBES

Oscilloscopes are usually connected to a circuit under test with a short length of shielded cable and a probe. At low frequencies, a piece of small-diameter coax cable and some sort of insulated test probe might do. Unfortunately, at higher frequencies the capacitance of the cable would produce a capacitive reactance much less than the one-megohm input impedance of the oscilloscope. In addition each scope has a certain built-in capacitance at its input terminals (usually between 5 and 35 pF). These two capacitances cause problems when probing an RF circuit with a relatively high impedance.

The simplest method of connecting a signal to a scope is to use a specially designed probe. The most common scope probe is a *×10 probe* (called a times ten probe). This probe forms a 10:1 voltage divider using the built-in resistance of the probe and the input resistance of the scope. When using a ×10 probe, all voltage

readings must be multiplied by 10. For example, if the scope is on the 1 V/division range and a ×10 probe was in use, the signals would be displayed on the scope face at 10 V/division.

Unfortunately a resistor alone in series with the scope input seriously degrades the scope's rise-time performance and therefore its bandwidth. Since the scope input looks like a parallel RC circuit, the series resistor feeding it causes a significant reduction in available charging current from the source. This may be corrected by using a compensating capacitor in parallel with the series resistor. Thus two dividers are formed: one resistive voltage divider and one capacitive voltage divider. With these two dividers connected in parallel and the RC relationships shown in **Fig 26.35**, the probe and scope should have a flat response curve through the whole bandwidth of the scope.

To account for manufacturing tolerances in the scope and probe the compensating capacitor is made variable. Most scopes provide a "calibrator" output that produces a known-frequency square wave for the purpose of adjusting the compensating capacitor in a probe. **Fig 26.36** shows possible responses when the probe is connected to the oscilloscope's calibrator jack.

If a probe cable is too short, do not attempt to extend the length of the cable

by adding a piece of common coaxial cable. The cable usually used for probes is much different than common 50 or 75-Ω coax. In addition the compensating capacitor in the probe is chosen to compensate for the provided length of cable. It usually will not have enough range to compensate for extra lengths.

The shortest ground lead possible should be used from the probe to the circuit ground. Long ground leads are inductors at high frequencies. In these circuits they cause ringing and other undesirable effects.

THE MODERN SCOPE

For many years a scope (even a so-called portable) was big and heavy. Computers and modern ICs have reduced

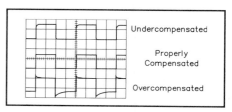

Fig 26.36 — Displays of a square-wave input illustrating undercompensated, properly compensated and overcompensated probes.

the size and weight. Modern scopes can take other forms than the traditional large cabinet with built-in CRT. Some modern scopes use an LCD display for true portability. Some scopes take the form of a card plugged into a PC, where they use the PC and monitor for display. Even if they don't use a PC for a display, many scopes can attach to a PC and download their data for storage and analysis using advanced mathematical techniques. Many high-end scopes now incorporate non-traditional functions, such as Fast Fourier Transforms (FFT). This allows limited spectrum analysis or other advanced mathematical techniques to be applied to the displayed waveform.

BUYING A USED SCOPE

Many hams will end up buying a used scope due to price. If you buy a scope and intend to service it yourself, be aware all scopes that use tubes or a CRT contain lethal voltages. Treat an oscilloscope with the same care you would use with a tube-type high-power amplifier. The CRT should be handled carefully because if dropped it will crack and implode, resulting in pieces of glass and other materials being sprayed everywhere in the immediate vicinity. You should wear a full-face safety shield and other appropriate safety equipment to protect yourself.

Another concern when servicing an older scope is the availability of parts. Many scopes made since about 1985 have used special ICs, LCDs and microprocessors. Some of these may not be available or may be prohibitive in cost. You should buy a used scope from a reputable vendor— even better yet, try it out before you buy it. Make sure you get the operators manual also.

AN HF ADAPTER FOR NARROW-BANDWIDTH OSCILLOSCOPES

Fig 26.37 shows the circuit of a simple piece of test equipment that will allow you to display signals that are beyond the normal bandwidth of an inexpensive oscilloscope. This circuit was built to monitor modulation of a 10-m signal on a scope that has a 5-MHz upper-frequency limit. This design features a Mini-Circuits Laboratory SRA-1 mixer. (See the Address List in the **References** chapter for contact information.) Any stable oscillator or VFO with an output of 10 dBm can be used for the local oscillator (LO), which mixes with the HF signal to produce an IF in the bandwidth of the oscilloscope.

The mixer can handle RF signal levels up to –3 dBm without clipping, so this was set as an upper limit for the RF input. A toroidal transformer coupler is constructed by winding a 31-turn secondary of #28-AWG wire on a 3E2A core, which has a 0.038-inch diameter. An FT-37-75 is suitable. The primary is a piece of coaxial cable passed through the core center. The coupler gives 30 dB of attenuation and has a flat response from 0.5 to 100 MHz. An additional 20-dB of attenuation was added for a total of 50 dB before the mixer. One-watt resistors will do fine for the attenuator. The completed adapter should be built into a shielded box.

This circuit, with a 25-MHz LO frequency, is useful on frequencies in the 20 to 30-MHz range with transmitters of up to 50-W power output. By changing the frequency of the LO, any frequency in the range of the coupler can be displayed on a 5-MHz-bandwidth oscilloscope. The frequency displayed will be the difference between the LO and the input signal. As an example, a 28.1-MHz input and a 25-MHz LO will be seen as a 3.1-MHz signal on the oscilloscope.

More attenuation will be required for higher-power transmitters. This circuit was described by Kenneth Stringham Jr., AE1X, in the Hints and Kinks column of February 1982 *QST*.

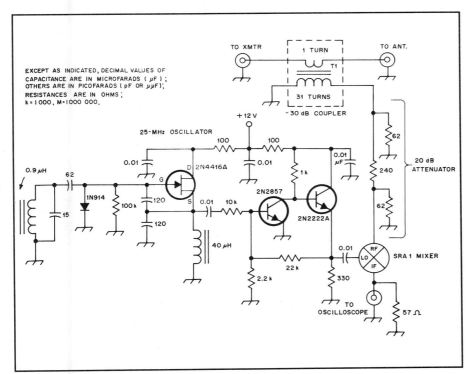

Fig 26.37 — This adapter displays HF signals on a narrow-bandwidth oscilloscope. It uses a 10-dBm 25-MHz LO, –30-dB coupler, 20-dB attenuator and diode-ring mixer. See text for further information.

A CALIBRATED NOISE SOURCE

NOISE FIGURE MEASUREMENT

One of the most important measurements in communications is the noise figure of a receiving set up. Relative measurements are often easy, while accurate ones are more difficult and expensive. One EME (moon bounce) station checks noise and system performance by measuring the noise of the sun reflected off the moon. While the measurement source (use of the sun and moon) is not expensive, the measuring equipment on 2 m consists of 48 antennas (each over 30 ft long). This measurement equipment is not for everyone!

The rest of us use more conventional noise sources and measuring techniques. Coverage of noise figure and its measurement appear in the **Transceivers** chapter of this *Handbook*.

Most calibrated and stable noise sources are expensive, but not this unit developed by Bill Sabin, WØIYH. It first appeared in May 1994 *QST*. When hams use a noise source, it is usually included in an RF bridge used to measure impedances and adjust antenna tuners. A somewhat different device (an *accurately calibrated and stable* noise source) is also useful. Combining a broadband RF noise source of known power output and a known output impedance with a true-RMS voltmeter, results in an excellent instrument for making interesting and revealing measurements on a variety of circuits hams commonly use. (Later on, some examples will be described.) The true-RMS voltmeter can be an RF voltmeter, a spectrum analyzer or an AF voltmeter at the output of a linear receiver.[1]

Calibrated noise generators and noise-figure meters are available at medium to astronomical prices. Here is a low-cost approach which can be used with reasonable confidence for many amateur applications where accuracy to tenths of a decibel is not needed, but where precision (repeatability) and comparative measurements are much more important. PC boards are available for this project.[2]

Semiconductor Noise Diodes

Any Zener diode can be used as a source of noise. If, however, the source is to be calibrated and used for reliable measurements, avalanche diodes specially designed for this purpose are preferable by far.[3] A good noise diode generates its noise through a carefully controlled *bulk avalanche* mechanism which exists *throughout* the PN junction, not merely at the junction surfaces where unstable and unreliable surface effects predominate due to local breakdown and impurity.[4] A true noise diode has a very low *flicker noise* (1/f) effect and tends to create a uniform level of truly *Gaussian noise* over a wide band of frequencies.[5] In order to maximize its bandwidth, the diode also has very low junction capacitance and lead inductance.

This project uses the NOISE/COM NC302L diode. It consists of a glass, axial-lead DO-35 package and is rated for use from 10 Hz to 3 GHz, if appropriate construction methods are followed. Prior to sale, the diodes are factory aged for 168 hours and are well stabilized. NOISE/COM has kindly agreed to make these diodes available to amateur experimenters for the special price of $10 each; the usual low-quantity price is about $25.[6]

Noise Source Design

The noise source presents two kinds of available output power. One is the thermal noise (−174 dBm/Hz at room temperature) when the diode is turned off. This is called N_{OFF}. The other is the sum of this same thermal noise and an "excess" noise, N_E, which is created by the diode when turned on, called N_{ON} (equivalent to $N_{OFF} + N_E$). For accurate measurements, the output impedance of the test apparatus must be the same (on or off) so that the device under test (DUT) always sees the same generator impedance. In Amateur Radio work, this impedance is usually 50 Ω, resistive.

The circuit design must guarantee this condition.

For maximum frequency coverage, a PC-board layout and coax connector suitable for use at microwaves are needed. For lower frequency usage, a less stringent approach can be employed. Two noise sources are presented here. One is for the 0.5 to 500-MHz region and uses conventional components that many amateurs already have. The other is for the 1-MHz to 2.5-GHz range; it uses chip components and an SMA connector.

Circuit Diagram and Construction

Figs 26.38 and **26.39** show the simple schematics of the two noise sources. In series with the diode is a 46.4-Ω resistor that combines with the dynamic resistance of the diode in the avalanche, noise generator mode (about 4 Ω) to total about 50 Ω. When the applied voltage polarity is reversed, the diode is forward conducting and its dynamic resistance is still about 4 Ω, but the avalanche noise is now turned off. As a result, the noise source output impedance is always about 50 Ω. The 5-dB pad reduces the effect of any small impedance differences, so that the output impedance is nearly constant from the *on* to the *off* condition, and the SWR is less than 2:1.

Consider the noise situation of the noise diode when it is forward conducting. The resistance of the forward biased PN junction is a *dynamic* resistance. This dynamic resistance is *not* a source of thermal noise, since it is not an actual physical resistance such as in a resistor or lossy network. However, the 0.6-V forward drop across the PN junction does produce a shot noise effect. The mathematics of this shot noise shows that the noise power associated with this effect is only about 50% of the thermal noise power that would be available from a physical resistor having the same value as the dynamic resistance. Therefore, the forward biased junction does *not* add excess noise to the system.[7] There is an 1/f noise effect associated with this shot noise in the diode, but its corner frequency is at about 100 kHz and of no importance at higher frequencies. Also, the small amount of bulk resistance contributes a little thermal noise.

In order to maximize the unit's flatness

---

[1]W. Sabin, "Measuring SSB/CW Receiver Sensitivity," *QST*, Oct 1992, pp 30-34. See also Technical Correspondence, *QST*, Apr 1993, pp 73-75.

[2]PC boards are available from FAR Circuits, See **References** chapter Address List; price, $3.75 plus $1.50 shipping. A PC board template for the Sabin noise source is available at: **http://www.arrl.org/notes**.

[3]The term *Zener diode* is commonly used to denote a diode that takes advantage of avalanche effect, even though the Zener effect and the avalanche effect are not exactly the same thing at the device-physics level.

[4]The term *bulk avalanche* refers to the avalanche multiplication effect in a PN junction. A carrier (electron or hole) with sufficient energy collides with atoms and causes more carriers to be knocked loose. This effect "avalanches" and it occurs throughout the volume of the PN junction. This mechanism is responsible for the high-quality noise generation in a true noise diode.

[5]*Gaussian noise* refers to the instantaneous values of a noise voltage. These values conform to the Gaussian probability density function of statistics.

[6]NOISE/COM Co, for contact information see the **References** chapter Address List.

[7]Motchenbacher and Fitchen, *Low Noise Electronic Design* (New York: Wiley & Sons, 1973), p 22.

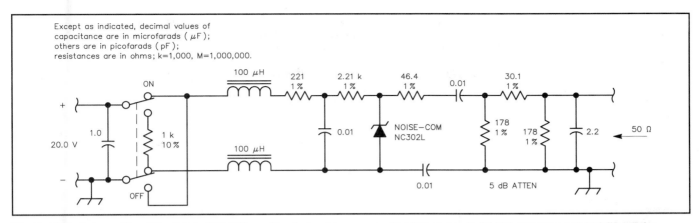

Fig 26.38 — Schematic of the 0.5 to 500-MHz calibrated noise source. Resistors are 1/8-W, 1%-tolerance metal-film units. 1% resistors are available from Digi-Key. See the References chapter for the address.

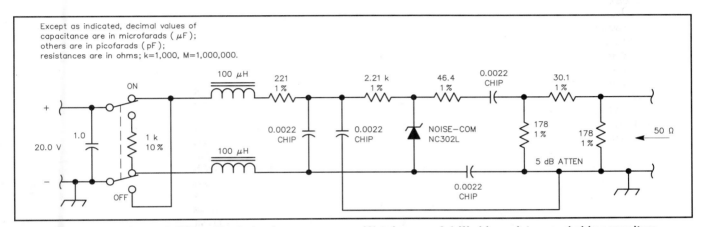

Fig 26.39 — The 1-MHz to 2.5-GHz calibrated noise source uses 1%-tolerance, 0.1-W chip resistors and chip capacitors. 1% resistors are available from Digi-Key. See the References chapter for the address.

Fig 26.40 — An inside view of the 0.5 to 500-MHz noise source.

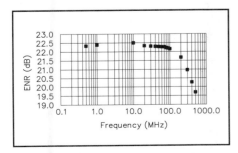

Fig 26.41 — Sample calibration chart for the 0.5 to 500-MHz noise source.

Fig 26.42 — A view inside of the 1-MHz to 2.5-GHz noise source.

and the coupling capacitors. The power-supply voltage must be clean, well by-passed and set accurately. **Fig 25.40** shows a 0.5 to 500-MHz unit. This construction method satisfies quite well the electrical requirements wanted for this model. At 500 MHz, the return loss with respect to 50 Ω at the output jack decreased to 10 dB. A calibration chart (**Fig 26.41**) is attached to the unit's top for easy reference. **Fig 26.42** shows the inside of the 1-MHz to 2.5-GHz noise source.

Calibrating the Noise Source

If the construction is solid, the calibration should last for a long time. There are two ways to calibrate the noise source. If the unit *has been carefully constructed and its correct operation verified*, NOISE/COM will calibrate home-built units over the desired frequency range for $25 plus return shipping charges. Note that one factory calibrated unit can be used as a reference for many home calibrated units. **Fig 26.43** shows the NOISE/COM calibra-

and frequency response bandwidth, noise-source construction methods should aim for RF circuit lead lengths as close to zero as possible as well as minimum inductance in the ground path

tion data for both models of prototype noise sources, including SWR data. The noise data is strictly valid only at room temperature, so it's necessary to avoid extreme temperature environments.

The second calibration method requires a signal generator with known output levels at the various desired calibration frequencies. One approach is to build a tunable weak-signal oscillator that can be compared to some accessible high-quality signal generator, using a sensitive receiver as a detector.[8] The level of the signal source in dBm is needed.

Access to a multistage attenuator is also desirable. Build the attenuator using the nearest 1% values of metal-film resistors, so that systematic errors are minimized. A total attenuation of 25 dB in 0.1-dB steps is desirable. Attenuator construction must be appropriate for use at the intended frequency range. In some cases, a high-frequency correction chart may be needed.

With the calibrated signal source and the attenuator feeding the receiver in an SSB or CW mode the techniques discussed in the reference of Note 1 should be used to determine the excess noise (N_E) of the noise source and the noise bandwidth (B_N) of the receiver.

Excess Noise Ratio

A few words about excess noise ratio (ENR) are needed. It is defined as the ratio of excess noise to thermal noise. That is,

$$ENR = \frac{N_{ON} - N_{OFF}}{N_{OFF}} = \frac{N_E}{N_{OFF}} \quad (12)$$

When the noise source is turned on, its output is $N_{OFF} + N_E$. The ratio of N_{ON} to N_{OFF} is then

$$\frac{N_{ON}}{N_{OFF}} = \frac{N_{OFF} + N_E}{N_{OFF}}$$

$$= 1 + \frac{N_E}{N_{OFF}} = 1 + ENR \quad (13)$$

Therefore, ENR is a measure of how much the noise increases and the noise generator can be calibrated in terms of its ENR.

Normalizing ENR to a 1-Hz bandwidth and converting to decibels, this is

$$ENR\,(dB) = 174\,(dBm/Hz) + \frac{N_E\,(dBm)}{B_N\,(Hz)} \quad (14)$$

Prepare a calibration chart and attach it to the top of the unit (see Fig 26.41). If the

[8]W. Hayward and D. DeMaw, *Solid State Design for the Radio Amateur* (Newington: ARRL, 1986).

Frequency (MHz)	0.5 to 500 MHz Unit		1.0 to 2500 MHz Unit	
	ENR (dB)	SWR	NR (dB)	SWR
0.5	22.33	1.03		
1	22.38	1.03	21.38	1.03
10	22.45	1.04	21.46	1.03
20	22.35	1.06		
30	22.32	1.06		
40	22.32	1.09		
50	22.30	1.11		
60	22.29	1.12		
70	22.25	1.15		
80	22.22	1.17		
90	22.20	1.20		
100	22.15	1.23	21.80	1.07
200	21.65	1.42		
300	20.96	1.62		
400	20.25	1.70		
500	19.60	1.90	20.71	1.44
1000			20.12	1.86
1500			20.00	2.06
2000			20.70	2.14
2500			21.51	1.88

Fig 26.43 — NOISE/COM calibration data for both prototype noise sources. The data is not universal; it varies from unit to unit.

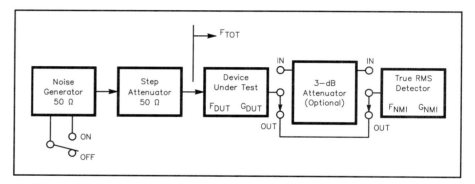

Fig 26.44 — Setup for measuring noise figure of a device under test (DUT).

unit is to be factory calibrated, first perform the calibration procedure to ensure everything is working properly. Remember, a factory calibrated unit can be used as a reference for other home calibrated units, once the calibration-transfer procedures have been worked out. This requires some careful thinking and proper techniques. Generally speaking, a NOISE/COM calibration is the best choice.

Noise-Figure Measurement

The thermal noise power available from the attenuator remains constant for any value of attenuator setting. But the excess noise and therefore the ENR (in dB) due to the noise diode is equal to the calibration point of the source minus the setting (in dB) of the attenuator.

The noise-figure measurement of a device under test (DUT) uses the Y method and the setup in **Fig 26.44**. If the DUT has a noise-generator input and a true-RMS noise-measuring instrument at the output,

then the total output noise (including the contribution of the measuring instrument) with the noise generator turned off is

$$N_{OFF(TOT)} = kTB_N F_{TOT} G_{DUT} G_{NMI} \quad (15)$$

where

kTB_N = thermal noise,
G_{DUT} = gain of the DUT,
G_{NMI} = gain of the noise-measuring instrument,
F_{TOT} = noise factor of the combination of the DUT and the noise-measuring instrument.

When the noise generator is turned on, the output noise is

$$N_{ON(TOT)} = kTB_N F_{TOT} G_{DUT} G_{NMI} +$$
$$(ENR)\,kTB_N\,G_{DUT}\,G_{NMI} \quad (16)$$

Where the last term is the contribution of excess noise by the noise generator. Note that none of these values is in dB or dBm.

If we divide equation 16 by equation 15 and say that the ratio

$$\frac{N_{ON(TOT)}}{N_{OFF(TOT)}} = Y \qquad (17)$$

then,

$$\frac{F_{TOT} + ENR}{F_{TOT}} = 1 + \frac{ENR}{F_{TOT}}$$

Note that $kTBN$, G_{DUT} and G_{NMI} disappear, so that these quantities need not be known to measure noise factor. If we solve equation 17 for F_{TOT}, we get the noise factor

$$F_{TOT} = \frac{ENR}{Y - 1} \qquad (18)$$

If the noise output doubles (increases by 3 dB) when we turn on the noise source, then $Y = 2$ and the noise factor is numerically equal to the excess noise ratio (ENR). If the attenuator steps are not fine enough or if the attenuator is not reliable over the entire frequency range, use equation 18 to get a better answer. (It's much simpler to use a good fine-step attenuator.) The value of F_{TOT} is that of the DUT in cascade with the noise-measuring instrument. To find F_{DUT}, we must know the noise factor F_{NMI} of the noise-measuring instrument and G_{DUT} and then use the Friis formula, unless G_{DUT} is very large (as it would be if the DUT were a high-gain receiver (see footnote 1).

$$F_{DUT} = F_{TOT} - \frac{F_{NMI} - 1}{G_{DUT}} \qquad (19)$$

The validity of equation 19 (if we need to use it) requires that the noise bandwidth of the noise-measuring instrument be less than the noise bandwidth of the DUT (see the referent of Note 1). Verify this before proceeding.

There's another advantage to using the power-doubling method. If the 3-dB attenuator of Fig 26.44 is used to maintain a constant noise level into the following stages and the RMS meter, this means that the noise factor, using the calibration scale and the input attenuator (without using equation 18), is

$$F_{DUT} = ENR + \frac{1}{G_{DUT}} \qquad (20)$$

If G_{DUT} is large, then the last term can be neglected. If G_{DUT} is small, we need to know its value. However, we do not need to know the noise factor F_{NMI} of the circuitry after the DUT, as we did in the previous discussion.

The 3-dB attenuator method also re-moves all restrictions regarding the type of noise measuring instrument, since the meter reading is now used only as a reference point. This last statement applies only when two noise (or two signal generator) inputs are being compared.

Frequency Response Measurements

The noise generator, in conjunction with a spectrum analyzer, is an excellent tool for measuring the frequency response of a DUT, if the noise source is much stronger than the internal noise of the DUT and that of the spectrum analyzer. Many spectrum analyzers are not equipped with tracking generators, which can be quite expensive for an amateur's budget.

The spectrum analyzer needs to be calibrated for a noise input, if accurate amplitude measurements are needed, because it responds differently to noise signals than to sine-wave signals. The envelope detection of noise, combined with the logarithmic amplification of the spectrum analyzer, creates an error of about 2.5 dB for a noise signal (the noise is that much greater than the instrument indicates). Also, the noise bandwidth of the IF filter is different from its resolution bandwidth. Some modern spectrum analyzers have internal DSP algorithms that make the corrections so that external noise sources and also carrier-to-noise ratios, normalized to some noise bandwidth like 1.0 Hz, can be measured with fair accuracy if the input noise is a few decibels above thermal. One example is the Tektronix Model 2712. If only relative response readings are needed, then these corrections are not needed.

Also, the noise source itself can be used to establish an accurate reference level (in dBm) on the screen. An accurate, absolute measurement with the DUT in place will then be this reference level (in dBm), plus the increment in decibels produced by the DUT.

The noise-generator output can be viewed as a collection of sine waves separated by, say, 1 Hz. Each separated frequency "bin" has its own Gaussian amplitude and random phase with respect to all the others. So the DUT is simultaneously looking at a collection or "ensemble," of input signals. As the spectrum analyzer frequency sweeps, it looks simultaneously at all of the DUT frequencies that fall within the spectrum analyzer's IF noise bandwidth. The spectrum display is thus the "convolution" of the IF filter frequency response and the DUT frequency response. If the DUT is a narrow filter, a very narrow resolution and a slow sweep are needed in the spectrum analyzer. In addition, the analyzer's video, or post-detection, filter has a narrow bandwidth and also requires some settling time to get an accurate reading. So, some experience and judgment are required to use a spectrum analyzer this way.

Using Your Station Receiver

Your station receiver can also be used as a spectrum analyzer. Place a variable attenuator between the DUT and the receiver. As you tune your receiver, in a narrow CW mode, adjust the attenuator for a constant reference level receiver output. The attenuator values are inversely related to the frequency response.

A calibrated noise source with an adjustable attenuator that can be easily switched into a receiver antenna jack is an excellent tool for measuring antenna noise level or incoming weak signal level (in dBm) or for establishing correct receiver operation.

The noise source can also be combined with a locally generated data-mode waveform of a known dBm value to get an approximate check on modem performance or to make adjustments that might assure correct operation of the system. The rigorous evaluation of system performance requires special equipment and techniques that may be unavailable at most amateur stations. Or, you could evaluate the intelligibility improvement of your SSB transmitter's speech processor in a noise background.

Summary

The calibrated, flat-spectrum noise generator described in this article is quite a useful instrument for amateur experimenters. Its simplicity and low cost make it especially attractive. Getting a good calibration is the main challenge, but once it is achieved, the calibration lasts a long time, if the right diode is used. The ENR of the units described here is in the range of 20 dB. Use of a high-quality, external, 10-dB attenuator barrel will get into the range of 10-dB ENR. If the unit is sent to NOISE/COM the attenuator should also be sent, with the request that it be included in the calibration. That attenuator then "belongs" to the noise source and should be so tagged. If the attenuator is of high quality, the output SWR will also be improved. NOISE/COM suggests periodic recalibration, at your discretion.

A NOISE BRIDGE FOR 1.8 THROUGH 30 MHz

The noise bridge, sometimes referred to as an antenna noise bridge or RX noise bridge, is an instrument that measures the impedances of antennas or other electrical circuits. The unit shown in **Fig 26.45** provides adequate accuracy for most measurements in the 1.8- through 30-MHz range. Battery operation and small physical size make this unit ideal for remote-location use. This classic bridge circuit was updated by Mark Shelhamer, WA3YNO. Additional information about using the noise bridge for transmission-line measurements appears in that article and in the **Transmission Line and Antenna Measurements** chapter of *The ARRL Antenna Book*. A detector, such as the station receiver, is required for operation. An etching pattern and parts placement diagram are in Chapter 30, **References**.

The noise bridge consists of two parts: the noise generator and the bridge circuitry. See **Fig 26.46**. A 6.8-V Zener diode serves as the noise source. The broadband noise signal is amplified by U1 and associated components to produce an approximate S9 signal in the receiver.

The bridge portion of the circuit consists of T1, C1 and R1. T1 is a ferrite core wound as shown in the schematic detail. This design eliminates phase shift and the ferrite core has sufficient permeability to eliminate low-frequency resistance shift. One winding of T1 couples noise energy into the bridge circuit. The remaining two windings are each in one arm of the bridge. C1 and R1 complete the known arm; the UNKNOWN circuit with C3 comprises the remainder of the bridge. The terminal labeled RCVR is for connection to the detector.

The reactance range of a noise bridge depends on several factors, including operating frequency, value of the series capacitor (C3 in the figure) and the range of the variable capacitor (C1 in the figure). The zero-reactance point occurs when C1 is either nearly fully meshed or fully unmeshed.

Construction

The noise bridge can be put in a home-made aluminum enclosure that measures $5 \times 2^3/_8 \times 3^3/_4$ inches. Many of the circuit components are mounted on a circuit board that is fastened to the rear wall of the cabinet. The circuit-board layout keeps the lead lengths to the board from the bridge and coaxial connectors to a minimum.

Potentiometer R1 must be mounted carefully. For accurate readings it must be very well insulated from ground. In the prototype the control is mounted on a piece of Plexiglas, which was fastened to the chassis with a piece of aluminum angle stock. Additionally, a ¼-inch control-shaft coupling and a length of phenolic rod were used to keep the control away from the front panel. Use a high-quality potentiometer to ensure good measurement results.

The variable capacitor is easier to mount because the rotor is grounded. It should be a high-quality unit. Two female RF fittings on the rear panel are connected to a detector (receiver) and to the UNKNOWN circuit. Plastic insulated phono connectors should *not* be used because they might influence bridge accuracy at higher frequencies. Miniature coaxial cable (RG-174) is used for the connection between the RCVR connector and circuit board. Attach one end of C3 to the circuit board and the other directly to the UNKNOWN circuit connector.

Bridge Compensation

Stray capacitance and inductance in the bridge circuit can affect impedance readings. If a very accurate bridge is required, use the next steps to make readings more accurate.

Good calibration loads are necessary to check the accuracy of the noise bridge. Four are needed here: a 0-Ω (short circuit) load, a 50-Ω load, a 180-Ω load and a variable-resistance load. The short-circuit and fixed-resistance loads are used to check the accuracy of the noise bridge; the variable-resistance load is used when measuring co-axial-cable loss.

Construction details of the loads are shown in **Fig 26.47**. Each load is constructed inside a connector. The leads should be kept as short as possible. The resistors must be noninductive (not wire wound). Carbon-composition (¼-W) resistors should work fine. The potentiometer in the variable-resistance load is a miniature PC-mount unit with a maximum resistance of 100 Ω, or less. The potentiometer wiper and one of the end leads are connected to the center pin of the connector; the other lead is connected to ground.

Stray Capacitance

Stray capacitance on the variable-resistor side of the bridge tends to be higher than that on the unknown side. This is because of parasitic capacitance in the variable resistor, R1.

The effect of parasitic capacitance is most easily detected using the 180-Ω load. Mea-

Fig 26.45 — Noise bridge construction details. Press-on lettering is used for the calibration marks. Note that the potentiometer must be isolated from ground.

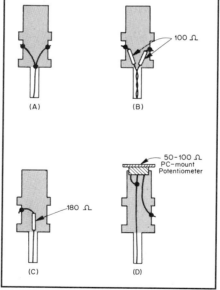

Fig 26.47 — Loads used to check and calibrate a noise bridge are built into a PL-259 shell. Leads should be kept as short as possible to minimize parasitic inductance. The connector shell is screwed in place after construction and is not shown in the figure. (A) is a short circuit; (B) depicts a 50-Ω load; (C) is a 180-Ω load; (D) shows a variable-resistance load used to determine the loss in a coaxial cable.

sure and record the actual resistance of the load, R_L. Connect the load to the UNKNOWN connector, tune the receiver to 1.8 MHz and null the bridge. (See "Finding the Null" below for tips.) Use an ohmmeter across R1 to measure its dc resistance. The magnitude of the stray capacitance can be calculated by:

$$C_P = C_S \left(\sqrt{\frac{R1}{R_L} - 1} \right) \qquad (21)$$

where

R_L = load resistance (as measured), ohms

R1 = resistance of the variable resistor, ohms

C_S = series capacitance (either C3 or C4, whichever is selected), pF.

We can compensate for C_P by placing a variable capacitor, C_C, in the side of the bridge with lesser stray capacitance. If R1 is greater than R_L, stray capacitance is greater on the variable resistor side of the bridge: Place C_c between point U (on the circuit board) and ground. If R1 is less than R_L, stray capacitance is greater on the unknown side: Place C_c between point B and ground.

If the needed compensating capacitance is only a few pF, you can use a gimmick capacitor (made by twisting two short pieces of insulated, solid wire together) for C_c. A gimmick capacitor is adjusted by trimming its length.

Compensate the bridge by setting the dc resistance of R1 equal to R_L. With the bridge at 1.8 MHz. alternately adjust C_c

and C1 to obtain a null.

Stray Inductance

Parasitic inductance, if present, should be only a few tens of nH. This represents a few ohms of inductive reactance at 30 MHz. The effect is best observed by reading the reactance of the 0-Ω test load at 1.8 MHz and 30 MHz; the indicated reactance should be the same at both frequencies.

If the reactance reading decreases as frequency is increased, parasitic inductance is greater in the known arm and compensating inductance is needed between point U and C3. If the reactance increases with frequency, the unknown-arm inductance is greater and compensating inductance should be placed between point B and R1.

Compensate for stray inductance by placing a single-turn coil, made from a 1 to 2-inch length of solid wire, in the appropriate arm of the bridge. Adjust the size of this coil until the reactance reading remains constant from 1.8 to 30 MHz.

Calibration

Good calibration accuracy is necessary for accurate noise-bridge measurements. Calibration of the resistance scale is straightforward. To do this, tune the receiver to a frequency near 10 MHz. Attach the 0-Ω load to the UNKNOWN connector and null the bridge. This is the zero-resistance point; mark it on the front-panel resistance scale. The rest of the resistance range is calibrated by adjusting R1, measuring R1 with an accurate ohmmeter, calculating the increase from the zero point and marking the increase on the front panel.

Most bridges have the reactance scale marked in capacitance because capacitance does not vary with frequency. Unfortunately, that requires calibration curves or complex calculations to find the load reactance. An alternative method is to mark the reactance scale in ohms at a reference frequency of 10 MHz. This method calibrates

Table 26.6

Noise Bridge Calibration with Coaxial Cable

This data is for Radio Shack RG-8M cable (R_0 = 52.5 Ω) cut to exactly λ/4 at 10 MHz; the reactances and capacitances shown correspond to this frequency.

Capacitance

C(pF)	f(MHz)	C(pF)	f(MHz)
10	9.798	−10	10.219
20	9.612	−20	10.459
30	9.440	−30	10.721
40	9.280	−40	11.010
50	9.130	−50	11.328
60	8.990	−60	11.679
70	8.859	−70	12.064
80	8.735	−80	12.484
90	8.618	−90	12.935
100	8.508	−100	13.407
110	8.403	−110	13.887
120	8.304	−120	14.357
130	8.209	−130	14.801
140	8.119	−140	15.211

Reactance

Xi	f(MHz)	Xi	f(MHz)
10	3.318	−10	19.376
20	4.484	−20	18.722
30	5.262	−30	18.048
40	5.838	−40	17.368
50	6.286	−50	16.701
60	6.647	−60	16.063
70	6.943	−70	15.472
80	7.191	−80	14.938
90	7.404	−90	14.459
100	7.586	−100	14.045
110	7.747	−110	13.683
120	7.884	−120	13.370
130	8.009	−130	13.097
140	8.119	−140	12.861
150	8.217	−150	12.654
160	8.306	−160	12.473
170	8.387	−170	12.313
180	8.460	−180	12.172
190	8.527	−190	12.045
200	8.588	−200	11.932
210	8.645	−210	11.831
220	8.697	−220	11.739
230	8.746	−230	11.655
240	8.791	−240	11.579
250	8.832	−250	11.510
260	8.872	−260	11.446
270	8.908	−270	11.387
280	8.942	−280	11.333
290	8.975	−290	11.283
300	9.005	−300	11.236
350	9.133	−350	11.045
400	9.232	−400	10.905
450	9.311	−450	10.798
500	9.375	−500	10.713

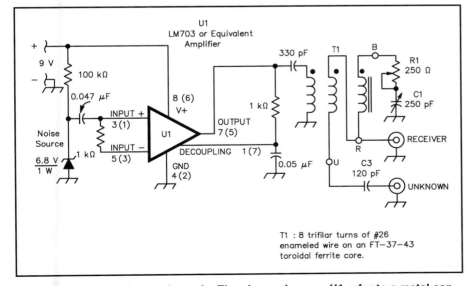

T1 : 8 trifilar turns of #26 enameled wire on an FT-37-43 toroidal ferrite core.

Fig 26.46 — Noise bridge schematic. The pin numbers on U1 refer to a metal can LM703 or the NTE/ECG replacement. Those in parentheses are for a National Semiconductor mini-DIP LM703N. See the text for placement of compensation C or L at points B, U or R.

the bridge near the center of its range and shows reactance directly, but it requires a simple calculation to scale the reactance reading for frequencies other than 10 MHz. The scaling equation is:

$$X_{u(f)} = X_{u(10)} \frac{10}{f} \qquad (22)$$

where

f = frequency, MHz
$X_{u(10)}$ = reactance of the unknown load at 10 MHz.
$X_{u(f)}$ = reactance of the unknown load at f.

A shorted piece of coaxial cable serves as a reactance source. (The reactance of a shorted, low-loss coaxial cable is dependent only on the cable length, the measurement frequency and the cable characteristic impedance.) Radio Shack RG-8M is used here because it is easy to get, has relatively low loss and has an almost purely resistive characteristic impedance. Prepare the calibration cable as follows:

1. Cut a length of coaxial cable that is slightly longer than λ/4 at 10 MHz (about 20 ft for RG-8M). Attach a suitable connector to one end of the cable; leave the other end open circuited.

2. Connect the 0-Ω load to the noise bridge UNKNOWN connector and set the receiver frequency to 10 MHz. Adjust the noise bridge for a null. Do not adjust the reactance control after the null is found.

3. Connect the calibration cable to the bridge UNKNOWN terminal. Null the bridge by adjusting only the variable resistor and the receiver frequency. The receiver frequency should be less than 10 MHz; if it is above 10 MHz, the cable is too short and you need to prepare a longer one.

4. Gradually cut short lengths from the end of the coaxial cable until you obtain a null at 10 MHz by adjusting only the resistance control. Then connect the cable center and shield conductors at the open end with a short length of braid. Verify that the bridge nulls with zero reactance at 20 MHz.

5. The reactance of the coaxial cable (normalized to 10 MHz) can be calculated from:

$$X_{i(10)} = R_0 \frac{f}{10} \tan\left(2\pi \frac{f}{40}\right) \qquad (23)$$

where

$X_{i(10)}$ = cable reactance at 10 MHz
R_0 = characteristic resistance of the coaxial cable (52.5 Ω for Radio Shack RG-8M)
f = frequency in MHz.

The results have less than 5% error for reactances less than 500 Ω, as long as the test-cable loss is less than 0.2 dB. This error becomes significantly less at lower reactances (2% error at 300 Ω for a 0.2-dB loss cable). The loss in 18 ft of RG-8M is 0.13 dB at 10 MHz. Reactance data for Radio Shack RG-8M is given in **Table 26.6**.

With the prepared cable and calibration values on hand, go on to calibrate the reactance scale. Tune the receiver to the appropriate frequency for the desired reactance (given in the table or found using the equation). Adjust the resistance and reactance controls to null the bridge. Mark the reactance reading on the front panel. Repeat this process until all desired reactance values have been marked. The resistance values needed to null the bridge during this calibration procedure may be

quite high (more than 100 Ω) at the higher reactances.

This calibration method is much more accurate than using fixed capacitors across the UNKNOWN connector. Also, you can calibrate a noise bridge in less than an hour using this method.

Finding the Null

In use, a receiver is attached to the RCVR connector and some load of unknown value is connected to the UNKNOWN terminal. The receiver allows us to hear the noise present across the bridge arms at the frequency the receiver is tuned to. The strength of the noise signal depends on the strength of the noise-bridge battery, the receiver bandwidth/sensitivity and the impedance difference between the known and unknown bridge arms. The noise is stronger and the null more obvious with wide receiver passbands. Set the receiver to the widest bandwidth AM mode available.

When the impedances of the known and unknown bridge arms are equal, the voltage across the receiver is minimized; this is a null. In use, the null may be difficult to find because it appears only when both bridge controls approach the values needed to balance the bridge.

To find the null, set C1 to midscale, sweep R1 slowly through its range and listen for a reduction in noise (it's also helpful to watch the S meter). If no reduction is heard, set R1 to midrange and sweep C1. If there is still no reduction, begin at one end of the C1 range and sweep R1. Change C1 by about 10% and sweep R1 with each change until some noise reduction appears. Once noise reduction begins, adjust C1 and R1 alternately for minimum signal.

A SIGNAL GENERATOR FOR RECEIVER TESTING

The oscillator shown in **Fig 26.48** and **Fig 26.49** was designed for testing high-performance receivers. Parts cost for the oscillator has been kept to a minimum by careful design. While the stability is slightly less than that of a well-designed crystal oscillator, the stability of the unit should be good enough to measure most amateur receivers. In addition, the ability to shift frequency is important when dealing with receivers that have spurious responses. More importantly, LC oscillators with high-Q components often have much better phase noise performance than crystal oscillators, because of power limitations in the crystal oscillators (crystals are easily damaged by excessive power).

The circuit is a Hartley oscillator followed by a class-A buffer amplifier. A 5-V regulator is used to keep the power supply output stable. The amplifier is cleaned up by a seven-element Chebyshev low-pass filter, which is terminated by a 6-dB attenuator. The attenuator keeps the filter working properly, even with a receiver that has an input impedance other than 50 Ω. A receiver designed to work with a 50-Ω system may not have a 50-Ω input impedance. The +4 dBm output is strong enough for most receiver measurements. It may even be too strong for some receivers. Note that sensitive components like crystal filters may require a step attenuator to lower the output level.

Construction

This unit is built in a box made of double-sided circuit board. Its inside dimensions are 1×2.2×5 inches (HWD). The copper foil of the circuit board makes an excellent shield, while the fiberglass helps temperature stability. Capacitor C2 should be soldered di-

rectly across L1 to ensure high Q. Since this is an RF circuit, leads should be kept short. While silver mica capacitors have slightly better Q, NP0 capacitors may offer better stability. Mounting the three inductors orthogonal (axis of the inductors 90° from each other) reduced the second-order harmonic by 2 dB when it was compared to the first unit that was made.

Alignment and Testing

The output of the regulator should be +5 V. The output of the oscillator should be +4 dBm (2.5 mW) into a 50-Ω load. Increasing the value of C3 will increase the power output to a maximum of about 10 mW. The frequency should be around 3.7 MHz. Additional capacitance across L1 (in parallel with C2) will lower the frequency if desired, while the trimmer capacitor (C1) specified will allow adjustment to a specific frequency. The drift

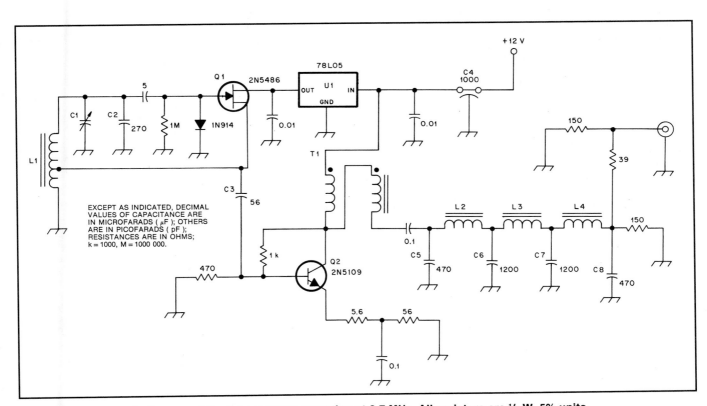

Fig 26.49 — Schematic diagram of the LC oscillator operating at 3.7 MHz. All resistors are ¼ W, 5% units.

C1 — 1.4 to 9.2-pF air trimmer (value and type not critical)
C2 — 270-pF silver-mica or NP0 capacitor. Value may be changed slightly to compensate for variations in L1.
C3 — 56-pF silver mica or NP0 capacitor. Value may be changed to adjust output power.
C4 — 1000 pF solder-in feedthrough

capacitor. Available from Microwave Components of Michigan (see References chapter).
C5-C8 — Silver-mica, NP0 disc or polystyrene capacitor.
D1 — 1N914, 1N4148.
L1 — 31t #18 enameled wire on T-94-6 core. Tap 8 turns from ground end (7.5 µH).
L2, L4 — 21t #22 enameled wire on a

T-50-2 core (2.5 µH, 2.43 µH ideal).
L3 — 23t #22 enameled wire on a T-50-2 core. (2.9 µH, 3.01 µH ideal).
T1 — 7t #22 enameled wire bifilar wound on an FT-37-43 core.
Q1 — 2N5486 JFET. MPF102 may give reduced output.
Q2 — 2N5109.
U1 — 78L05 low-current 5-V regulator.

of one of the first units made was 5 Hz over 25 minutes after a few minutes of warm up. If the warm-up drift is large, changing C2 may improve the situation somewhat. For most receivers, a drift of 100 Hz while you are doing measurements is not bad.

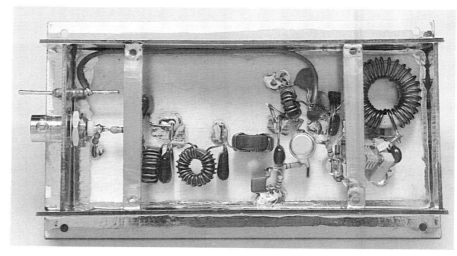

Fig 26.48 — A low-cost LC oscillator for receiver measurements. Toroidal cores are used for all of the inductances.

HYBRID COMBINERS FOR SIGNAL GENERATORS

Many receiver performance measurements require two signal generators to be attached to a receiver simultaneously. This, in turn, requires a combiner that isolates the two signal generators (to keep one generator from being frequency or phase modulated by the other). Commercially made hybrid combiners are available from Mini-Circuits Labs (see the Address List in the **References** chapter).

Alternatively, a hybrid combiner is not difficult to construct. The combiners described here (see **Fig 26.50**) provide 40 to 50 dB of isolation between ports (connections) while attenuating the desired signal paths (each input to output) by 6 dB. The 50-Ω impedance of the system is kept constant (very important if accurate measurements are to be made).

The combiners are constructed in small boxes made from double-sided circuit-board material. Each piece is soldered to the next one along the entire length of the seam. This makes a good RF-tight enclosure. BNC coaxial fittings are used on the units shown. However, any type of coaxial connector can be used. Leads must be kept as short as possible and precision resistors (or matched units from the junk box) should be used. The circuit diagram for the combiners is shown in **Fig 26.51**.

Fig 26.50 — The hybrid combiner on the left is designed to cover the 1 to 50-MHz range; the one on the right 50 to 500 MHz.

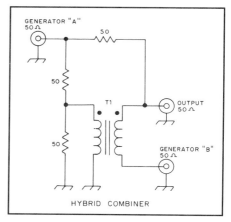

Fig 26.51 — A single bifilar wound transformer is used to make a hybrid combiner. For the 1 to 50-MHz model, T1 is 10 turns of #30 enameled wire bifilar wound on an FT-23-77 ferrite core. For the 50 to 500-MHz model, T1 consists of 10 turns of #30 enameled wire bifilar wound on an FT-23-63 ferrite core. Keep all leads as short as possible when constructing these units.

Return Loss Bridges

Return loss is a measure of how closely an impedance matches a reference impedance in phase angle and magnitude. If the reference impedance equals the measured impedance level with a 0° phase difference it has a return loss of infinity. **Fig A** shows basic return-loss measurement setups. Return-loss bridges are good for measuring filter response because return loss measurements are a more sensitive measure of passband response than insertion-loss measurements.

A 100 Hz to 100-kHz Return-Loss Bridge

Ed Wetherhold, W3NQN, has developed a low-frequency return-loss bridge (RLB) that can be adapted to different impedance levels. (See **Fig B**.) This bridge is used primarily for testing passive-LC filters that have been designed to work at a certain impedance level. Return-loss measurements require that the signal generator and RLB match the specific filter impedance level.

The characteristic impedance of this RLB is set by the values of four resistors (R1 = R2 = R3 = R4 = characteristic impedance). Ed mounted the four resistors on a plug-in module and placed their interconnections on the socket. Additional modules may be built for any impedance.

Table A provides computed values of return loss for several known loads and a 500-Ω RLB. If you build this RLB, use the values in the table to check its operation. For this frequency range, use an ac voltmeter in place of the power meter shown in Fig A. Choose one with good ac response well above 100 kHz.

Ed originally described this bridge circuit in an article published in 1993. That article provides complete construction details. Reprints are available from Ed Wetherhold, W3NQN (for $3). (For contact information, see the **References** chapter Address List.) The RLB shown is useful from 100 Hz to 100 kHz.

An RF Return-Loss Bridge

At HF and higher frequencies, return-loss bridges are used as shown in Fig A for making measurements in RF circuits. The schematic of a simple bridge is shown in **Fig C**. (Notice that the circuit is identical to that of a hybrid combiner.) It is built in a small box with short leads to the coax connectors. Either 49.9-Ω 1% metal-film or 51-Ω ¼-W carbon resistors may be used. The transformer is wound with 10 bifilar turns of #30 enameled wire on a high permeability ferrite core such as an FT-23-43 or similar.

Apply the output of the signal generator to the RF INPUT port of the RLB. It may be necessary to attenuate the generator output to avoid overloading the amplifier under test. Connect the bridge DETECTOR port to a power meter through a step attenuator and leave the UNKNOWN port of the bridge open circuited. Set the step attenuator for a relatively high level of attenuation and note the power meter indication.

Now connect the unknown impedance, Z_u, to the bridge. The power meter reading will decrease. Adjust the step attenuator to produce the same reading obtained when the UNKNOWN port was open circuited. The difference between the two settings of the attenuator is the return loss, measured in dB.

The unknown impedance measured by this technique is not limited to amplifier inputs. Coax cable attached to an antenna, a filter, or any other fixed impedance device can be characterized by return loss. Return

Fig A — A shows a setup to measure an unknown impedance with a return loss bridge. B is for measuring filter response.

Table A

Performance and Test Data for the 100-Hz to 100-kHz RLB

Bridge and signal-generator impedance = 500 Ω

Bridge Directivity

Frequency	Return Loss (dB, Unknown = 500-Ω)
100 Hz to 10 kHz	> 45
10 kHz to 80 kHz	40
80 kHz to 100 kHz	30

Return Loss of Known Resistive Loads

$$R_{LOAD} = LF \times Z_{BRIDGE}$$

where

R_{LOAD} = Load resistance, ohms
LF = Load factor
Z_{BRIDGE} = characteristic bridge impedance, ohms.

LF	Return Loss (dB)
5.848	3
3.009	6
1.925	10
1.222	20
1.065	30

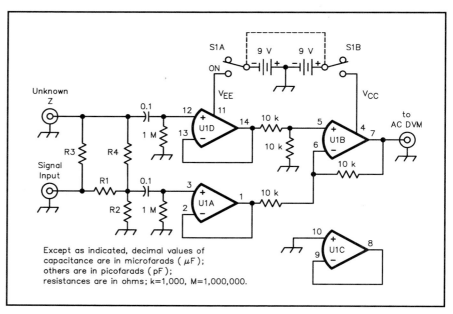

Fig B — An active low-frequency RLB. U1 is an LM324 quad op amp. R1, R2, R3 and R4 are ¼-W, 1% resistors with the same value as the output impedance of the signal generator and the input impedance of the test circuit.

loss is measured in dB, and it is related to a quantity known as the voltage reflection coefficient, ρ:

$$RL = -20 \log |\rho|$$

$$|\rho| = 10^{\frac{-RL}{20}}$$

where

RL = return loss, dB
ρ = voltage reflection coefficient.

The relationship of return loss to SWR is:

$$SWR = \frac{1 + |\rho|}{1 - |\rho|}$$

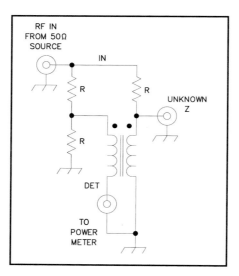

Fig C — An RLB for RF. Keep the lead lengths short. Wind the transformer on a high-permeability ferrite core. Use either 51-Ω carbon or 49.9-Ω 1% metal-film resistors.

Receiver Performance Tests

Comparing the performance of one receiver to another is difficult at best. The features of one receiver may outweigh a second, even though its performance under some conditions is not as good as it could be. Although the final decision on which receiver to purchase will more than likely be based on personal preference and cost, there are ways to compare receiver performance characteristics. Some of the more important parameters are sensitivity, blocking dynamic range and two-tone IMD dynamic range.

Instruments for measuring receiver performance should be of suitable quality and calibration. Always remember that accuracy can never be better than the tools used to make the measurements. Common instruments used for receiver testing include:

- Signal generators
- Hybrid combiner
- Audio ac voltmeter
- Distortion meter (FM measurements only)
- Noise figure meter (only required for noise figure measurements)
- Step attenuators (10 dB and 1 dB steps are useful)

Signal generators must be calibrated accurately in dBm or μV. The generators should have extremely low leakage. That is, when the output of the generator is switched off, no signal should be detected at the operating frequency with a sensitive receiver. Ideally, at least one of the signal generators should be capable of amplitude modulation. A suitable lab-quality piece would be the HP-8640B.

While most signal generators are calibrated in terms of microvolts, the real concern is not with the voltage from the generator but with the power available. The unit that is used for most low-level RF work is the milliwatt, and power is often specified in decibels with respect to 1 mW (dBm). Hence, 0 dBm would be 1 mW. The dBm level, in a 50-Ω load, can be calculated with the aid of the following equation:

$$dBm = 10 \log_{10} \left[20 \left(V_{RMS} \right)^2 \right] \qquad (24)$$

where
  dBm = power with respect to 1 mW
  V = RMS voltage available at the output of the signal generator.

The convenience of a logarithmic power unit such as the dBm becomes apparent when signals are amplified or attenuated. For example, a −107 dBm signal that is applied to an amplifier with a gain of 20 dB will result in an output increased by 20 dB. Therefore in this example (−107 dBm + 20 dB) = 87 dBm. Similarly, a −107 dBm signal applied to an attenuator with a loss of 10 dB will result in an output of (−107 dBm − 10 dB) or −117 dBm.

A hybrid combiner is a three-port device used to combine the signals from a pair of generators for all dynamic range measurements. It has the characteristic that signals applied at ports 1 or 2 appear at port 3 and are attenuated by 3 dB. However, a signal from port 1 is attenuated 30 or 40 dB when sampled at port 2. Similarly, signals applied at port 2 are isolated from port 1 some 30 to 40 dB. The isolating properties of the box prevent one signal generator from being frequency or phase modulated by the other. A second feature of a hybrid combiner is that a 50-Ω impedance level is maintained throughout the system.

Audio voltmeters should be calibrated in dB as well as volts. This facilitates easy measurements and eliminates the need for cumbersome calculations. Be sure that the step attenuators are in good working order and suitable for the frequencies involved. A distortion meter, such as the Hewlett-Packard 339A, is required for FM sensitivity measurements and a noise figure meter, such as the Hewlett-Packard 8970A, is excellent for certain kinds of sensitivity measurements.

Receiver Sensitivity

Several methods are used to determine receiver sensitivity. The mode under consideration often determines the best choice. One of the most common sensitivity measurements is minimum discernible signal (MDS) or noise floor. It is suitable for CW and SSB receivers.

This measurement indicates the minimum discernible signal that can be detected with the receiver. This level is defined as that which will produce the same audio-output power as the internally generated receiver noise. Hence, the term "noise floor."

To measure MDS, use a signal generator tuned to the same frequency as the receiver (see **Fig 26.56**). With the generator output at 0 or with maximum attenuation of its output note the voltmeter reading. Next increase the generator output level until the ac voltmeter at the receiver audio-output jack shows a 3-dB increase. The signal input at this point is the MDS. Be certain that the receiver is peaked on the generator signal. The filter bandwidth can affect the MDS. Always compare MDS readings taken with identical filter bandwidths. (A narrow bandwidth tends to improve MDS performance.) MDS can be expressed in μV or dBm.

In the hypothetical example of Fig 26.56, the output of the signal generator is −133 dBm and the step attenuator is set to 4 dB. Here is the calculation:

$$\begin{aligned} \text{Noise floor} &= -133 \text{ dBm} - 4 \text{ dB} \\ &= -137 \text{ dBm} \end{aligned} \qquad (25)$$

where the noise floor is the power available at the receiver antenna terminal and −4 dB is the loss through the attenuator.

Receiver sensitivity is also often expressed as 10 dB S+N/N (a 10-dB ratio of signal + noise to noise) or 10 dB S/N (signal to noise). The procedure and measurement are identical to MDS, except that the input signal is increased until the receiver output increases by 10 dB for 10 dB S+N/N and 9.5 dB for 10 dB S/N (often called "10 dB signal to noise ratio"). AM receiver sensitivity is usually expressed in this manner with a 30% modulated, 1-kHz test signal. (The modulation in this case is keyed on and off and the signal level is adjusted for the desired increase in audio output.)

SINAD is a common sensitivity measurement normally associated with FM receivers. It is an acronym for "*si*gnal plus *n*oise *a*nd *d*istortion." SINAD is a measure of signal quality:

$$SINAD = \frac{\text{signal} + \text{noise} + \text{distortion}}{\text{noise} + \text{distortion}} \qquad (26)$$

where SINAD is expressed in dB. In this

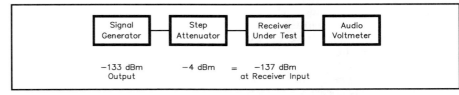

Fig 26.56 — A general test setup for measuring receiver MDS, or noise floor. Signal levels shown are for an example discussed in the text.

example, all quantities to the right of the equal sign are expressed in volts, and the ratio is converted to dB by multiplying the log of the fraction by 20.

SINAD(dB)

$$= 20 \log \left(\frac{\text{Signal(V)} + \text{Noise(V)} + \text{Distortion(V)}}{\text{Noise(V)} + \text{Distortion(V)}} \right)$$

Let's look at this more closely. We can consider distortion to be a part of the receiver noise because distortion, like noise, is an unwanted signal added to the desired signal by the receiving system. Then, if we assume that the desired signal is much stronger than the noise, SINAD closely approximates the signal to noise ratio. The common 12-dB SINAD specification therefore corresponds to a 4:1 S/N ratio (noise + distortion = 0.25 × signal).

The basic test setup for measuring SINAD is shown in **Fig 26.57**. The level of input signal is adjusted to provide 25% distortion (12 dB SINAD). Narrow-band FM signals, typical for amateur communications, usually have 3-kHz peak deviation when modulated at 1000 Hz.

Noise figure is another measure of receiver sensitivity. It provides a sensitivity evaluation that is independent of the system bandwidth. Noise figure is discussed further in the **Transceivers** chapter.

Dynamic Range

Dynamic range is the ability of the receiver to tolerate strong signals outside of its band-pass range. Two kinds will be considered:

Blocking dynamic range (blocking DR) is the difference, in dB, between the noise floor and a signal that causes 1 dB of gain compression in the receiver. It indicates the signal level, above the noise floor, that begins to cause desensitization.

IMD dynamic range (IMD DR) measures the impact of two-tone IMD on a receiver. IMD is the production of spurious responses that results when two or more signals mix. IMD occurs in any receiver when signals of sufficient magnitude are present. IMD DR is the difference, in dB, between the noise floor and the strength of two equal incoming signals that produce a third-order product 3 dB above the noise floor.

What do these measurements mean? When the IMD DR is exceeded, false signals begin to appear along with the desired signal. When the blocking DR is exceeded, the receiver begins losing its ability to amplify weak signals. Typically, the IMD DR is 20 dB or more below the blocking DR, so false signals appear well before sensitivity is significantly decreased. IMD DR is one of the most significant param-

eters that can be specified for a receiver. It is generally a conservative evaluation for other effects, such as blocking, which will occur only for signals well outside the IMD dynamic range of the receiver.

Both dynamic range tests require two signal generators and a hybrid combiner. When testing blocking DR (see **Fig 26.58**), one generator is set for a weak signal of roughly −110 dBm. The receiver is tuned to this frequency and peaked for maximum response. (ARRL Lab procedures require this level to be about 10 dB below the 1-dB compression point, if the AGC can be disabled. Otherwise, the level is set to 20 dB above the MDS.)

The second generator is set to a frequency 20 kHz away from the first and its level is increased until the receiver output drops by 1 dB, as measured with the ac voltmeter.

In the example shown, the output of the generator is −7 dBm, the loss through the

combiner is fixed at 3 dB and the step attenuator is set to 10 dB. The 1-dB compression level is calculated as follows:

$$\begin{aligned} \text{Blocking level} &= -7\,\text{dBm} - 3\,\text{dBm} - 10\,\text{dB} \\ &= -20\,\text{dBm} \end{aligned}$$
(27)

To express this as a dynamic range, the blocking level is referenced to the receiver noise floor (calculated earlier). Calculate it as follows:

$$\begin{aligned} \text{Blocking DR} &= \text{noise floor} - \text{blocking level} \\ &= -137\,\text{dBm} - (-20\,\text{dBm}) \\ &= -117\,\text{dB} \end{aligned}$$
(28)

This value is usually expressed as an absolute value: 117 dB.

Two-Tone IMD Test

The setup for measuring IMD DR is shown in **Fig 26.59**. Two signals of equal level, spaced 20-kHz apart are injected

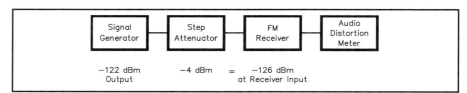

Fig 26.57 — FM SINAD test setup.

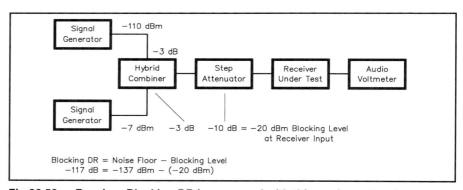

Fig 26.58 — Receiver Blocking DR is measured with this equipment and arrangement. Measurements shown are for the example discussed in the text.

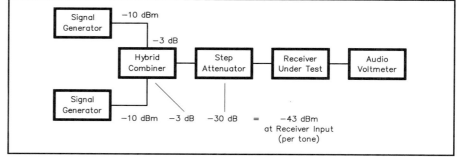

Fig 26.59 — Receiver IMD DR test setup. Signal levels shown are for the example discussed in the text.

into the receiver input. When we call these frequencies f1 and f2, the so-called third-order IMD products will appear at frequencies of (2f1 − f2) and (2f2 − f1). If the two input frequencies are 14.040 and 14.060 MHz, the third-order products will be at 14.020 and 14.080 MHz. Let's talk through a measurement with these frequencies.

First, set the generators for f1 and f2. Adjust each of them for an output of −10 dBm. Tune the receiver to either of the third-order IMD products. Adjust the step attenuator until the IMD product produces an output 3 dB above the noise level as read on the ac voltmeter.

For an example, say the output of the generator is −10 dBm, the loss through the combiner is 3 dB and the amount of attenuation used is 30 dB. The signal level at the receiver antenna terminal that just begins to cause IMD problems is calculated as:

$$\text{IMD level} = -10\,\text{dBm} - 3\,\text{dB} - 30\,\text{dB}$$
$$= -43\,\text{dBm} \qquad (29)$$

To express this as a dynamic range the IMD level is referenced to the noise floor as follows:

$$\text{IMD DR} = \text{noise floor} - \text{IMD level} \quad (30)$$
$$= -137\,\text{dBm} - (-43\,\text{dBm})$$
$$= -94\,\text{dB}$$

Therefore, the IMD dynamic range of this receiver would be 94 dB.

Third-Order Intercept

Another parameter used to quantify receiver performance is the third-order input intercept (IP^3). This is the point at which the desired response and the third-order IMD response intersect, if extended beyond their linear regions (see **Fig 26.60**). Greater IP^3 indicates better

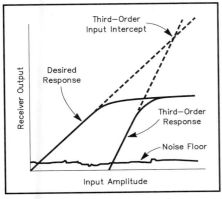

Fig 26.60 — A plot of the receiver characteristics that determine third-order input intercept, a measure of receiver performance.

receiver performance. Calculate IP^3 like this:

$$IP^3 = 1.5\,(\text{IMD dynamic range in dB})$$
$$+ (\text{MDS in dBm}) \qquad (31)$$

For our example receiver:

$$IP^3 = 1.5\,(94\,\text{dB}) + (-137\,\text{dBm}) = +4\,\text{dBm}$$

The example receiver we have discussed here is purely imaginary. Nonetheless, its performance is typical of contemporary communications receivers.

Evaluating the Data

Thus far, a fair amount of data has been gathered with no mention of what the numbers really mean. It is somewhat easier to understand exactly what is happening by arranging the data as shown in **Fig 26.61**. The base line represents power levels with a very small level at the left and a higher level (0 dBm) at the right.

The noise floor of our hypothetical receiver is at −137 dBm, the IMD level (the level at which signals will begin to create spurious responses) at −43 dBm and the blocking level (the level at which signals will begin to desensitize the receiver) at −20 dBm. The IMD dynamic range is some 23 dB smaller than the blocking dynamic range. This means IMD products will be heard long before the receiver begins to desensitize, some 23 dB sooner.

SPECTRUM ANALYZERS

A spectrum analyzer is similar to an oscilloscope. Both visually present an electrical signal through graphic representation. The oscilloscope is used to observe electrical signals in the time domain (amplitude as a function of time). The time domain, however, gives little information about the frequencies that make up complex signals. Amplifiers, mixers, oscillators, detectors, modulators and filters are best characterized in terms of their frequency response. This information is obtained by viewing electrical signals in the *frequency domain* (amplitude as a func-

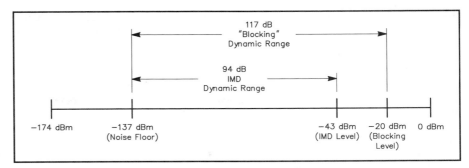

Fig 26.61 — Performance plot of the receiver discussed in the text. This is a good way to visualize the interaction of receiver-performance measurements.

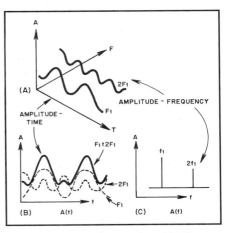

Fig 26.62 — A complex signal in the time and frequency domains. A is a three-dimensional display of amplitude, time and frequency. B is an oscilloscope display of time vs amplitude. C is spectrum analyzer display of the frequency domain and shows frequency vs amplitude.

tion of frequency). One instrument that can display the frequency domain is the spectrum analyzer.

Time and Frequency Domain

To better understand the concepts of time and frequency domain, see **Fig 26.62**. The three-dimensional coordinates show time (as the line sloping toward the bottom right), frequency (as the line rising toward the top right) and amplitude (as the vertical axis). The two discrete frequencies shown are harmonically related, so we'll refer to them as f1 and 2f1.

In the representation of time domain at B, all frequency components of a signal are summed together. In fact, if the two discrete frequencies shown were applied to the input of an oscilloscope, we would see the solid line (which corresponds to f1 + 2f1) on the display.

In the frequency domain, complex signals (signals composed of more than one frequency) are separated into their individual frequency components. A spectrum analyzer measures and displays the power level at each discrete frequency; this display is shown at C.

The frequency domain contains information not apparent in the time domain and therefore the spectrum analyzer offers advantages over the oscilloscope for certain measurements. As might be expected, some measurements are best made in the time domain. In these cases, the oscilloscope is a valuable instrument.

Spectrum Analyzer Basics

There are several different types of spectrum analyzers, but by far the most common is nothing more than an electronically tuned superheterodyne receiver. The receiver is tuned by means of a ramp voltage. This ramp voltage performs two functions: First, it sweeps the frequency of the analyzer local oscillator; second, it deflects a beam across the horizontal axis of a CRT display, as shown in **Fig 26.63**. The vertical axis deflection of the CRT beam is determined by the strength of the received signal. In this way, the CRT displays frequency on the horizontal axis and signal strength on the vertical axis.

Most spectrum analyzers use an up-converting technique so that a fixed tuned input filter can remove the image. Only the first local oscillator need be tuned to tune the receiver. In the up-conversion design, a wide-band input is converted to

an IF higher than the highest input frequency. As with most up-converting communications receivers, it is not easy to achieve the desired ultimate selectivity at the first IF, because of the high frequency. For this reason, multiple conversions are used to generate an IF low enough so that the desired selectivity is practical. In the example shown, dual conversion is used: The first IF is at 400 MHz; the second at 10.7 MHz.

In the example spectrum analyzer, the first local oscillator is swept from 400 MHz to 700 MHz; this converts the input (from nearly 0 MHz to 300 MHz) to the first IF of 400 MHz. The usual rule of thumb for varactor tuned oscillators is that the maximum practical tuning ratio (the ratio of the highest frequency to the lowest frequency) is an octave, a 2:1 ratio. In our example spectrum analyzer, the tuning ratio of the first local oscillator is 1.75:1, which meets this specification.

The image frequency spans 800 MHz to 1100 MHz and is easily eliminated using a low-pass filter with a cut-off frequency around 300 MHz. The 400-MHz first IF is converted to 10.7 MHz where the ultimate selectivity of the analyzer is obtained. The image of the second conversion, (421.4 MHz), is eliminated by the first IF filter. The attenuation of the image should be great, on the order of 60 to 80 dB. This requires a first IF filter with a high Q; this is achieved by using helical resonators, SAW resonators or

cavity filters. Another method of eliminating the image problem is to use triple conversion; converting first to an intermediate IF such as 50 MHz and then to 10.7 MHz. As with any receiver, an additional frequency conversion requires added circuitry and adds potential spurious responses.

Most of the signal amplification takes place at the lowest IF; in the case of the example analyzer this is 10.7 MHz. Here the communications receiver and the spectrum analyzer differ. A communications receiver demodulates the incoming signal so that the modulation can be heard or further demodulated for RTTY or packet or other mode of operation. In the spectrum analyzer, only the signal strength is needed.

In order for the spectrum analyzer to be most useful, it should display signals of widely different levels. As an example, signals differing by 60 dB, which is a thousand to one difference in voltage or a million to one in power, would be difficult to display. This would mean that if power were displayed, one signal would be one million times larger than the other (in the case of voltage one signal would be a thousand times larger). In either case it would be difficult to display both signals on a CRT. The solution to this problem is to use a logarithmic display that shows the relative signal levels in decibels. Using this technique, a 1000:1 ratio of voltage reduces to a 60-dB difference.

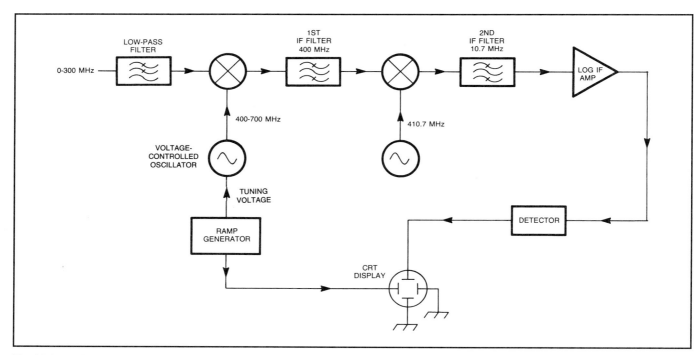

Fig 26.63 — A block diagram of a superheterodyne spectrum analyzer. Input frequencies of up to 300 MHz are up converted by the local oscillator and mixer to a fixed frequency of 400 MHz.

The conversion of the signal to a logarithm is usually performed in the IF amplifier or detector, resulting in an output voltage proportional to the logarithm of the input RF level. This output voltage is then used to drive the CRT display.

Spectrum Analyzer Performance Specifications

The performance parameters of a spectrum analyzer are specified in terms similar to those used for radio receivers, in spite of the fact that there are many differences between a receiver and a spectrum analyzer.

The sensitivity of a receiver is often specified as the minimum discernible signal, which means the smallest signal that can be heard. In the case of the spectrum analyzer, it is not the smallest signal that can be heard, but the smallest signal that can be seen. The dynamic range of the spectrum analyzer determines the largest and smallest signals that can be simultaneously viewed on the analyzer. As with a receiver, there are several factors that can affect dynamic range, such as IMD, second- and third-order distortion and blocking. IMD dynamic range is the maximum difference in signal level between the minimum detectable signal and the level of two signals of equal strength that generate an IMD product equal to the minimum detectable signal.

Although the communications receiver is an excellent example to introduce the spectrum analyzer, there are several differences such as the previously explained lack of a demodulator. Unlike the communications receiver, the spectrum analyzer is not a sensitive radio receiver. To preserve a wide dynamic range, the spectrum analyzer often uses passive mixers for the first and second mixers. Therefore, referring to Fig 26.63, the noise figure of the analyzer is no better than the losses of the input low-pass filter plus the first mixer, the first IF filter, the second mixer and the loss of the second IF filter. This often results in a combined noise figure of more than 20 dB. With that kind of noise figure the spectrum analyzer is obviously not a communications receiver for extracting very weak signals from the noise but a measuring instrument for the analysis of frequency spectrum.

The selectivity of the analyzer is called the resolution bandwidth. This term refers to the minimum frequency separation of two signals of equal level that can be resolved so there is a 3-dB dip between the two. The IF filters used in a spectrum analyzer differ from a communications receiver in that the filters in a spectrum analyzer have very gentle skirts and rounded passbands, rather than the flat passband and very steep skirts used on an IF filter in a high-quality communications receiver. This rounded passband is necessary because the signals pass into the filter passband as the spectrum analyzer scans the desired frequency range. If the signals suddenly pop into the passband (as they would if the filter had steep skirts), the filter tends to ring; a filter with gentle skirts is less likely to ring. This ringing, called scan loss, distorts the display and requires that the analyzer not sweep frequency too quickly. All this means that the scan rate must be checked periodically to be certain the signal amplitude is not affected by fast tuning.

Spectrum Analyzer Applications

Spectrum analyzers are used in situations where the signals to be analyzed are very complex and an oscilloscope display would be an indecipherable jumble. The spectrum analyzer is also used when the frequency of the signals to be analyzed is very high. Although high-performance oscilloscopes are capable of operation into the UHF region, moderately priced spectrum analyzers can be used well into the gigahertz region.

A spectrum analyzer can also be used to view very low-level signals. For an oscilloscope to display a VHF waveform, the bandwidth of the oscilloscope must extend from zero to the frequency of the waveform. If harmonic distortion and other higher-frequency distortions are to be seen the bandwidth of the oscilloscope must exceed the fundamental frequency of the waveform. This broad bandwidth can also admit a lot of noise power. The spectrum analyzer, on the other hand, analyzes the waveform using a narrow bandwidth; thus it is capable of reducing the noise power admitted.

Probably the most common application of the spectrum analyzer is the measurement of the harmonic content and other spurious signals in the output of a radio transmitter. **Fig 26.64** shows two ways to connect the transmitter and spectrum analyzer. The method shown at A should not be used for wide-band measurements since most line-sampling devices do not exhibit a constant-amplitude output over a broad frequency range. Using a line sampler is fine for narrow-band measurements, however. The method shown at B is used in the ARRL Lab. The attenuator must be capable of dissipating the transmitter power. It must also have sufficient attenuation to protect the spectrum analyzer input. Many spectrum analyzer mixers can be damaged by only a few milliwatts, so most analyzers have an adjustable input attenuator that will provide a reasonable amount of attenuation to protect the sensitive input mixer from damage. The power limitation of the attenuator itself is usually on the order of a watt or so, however. This means that 20 dB of additional attenuation is required for a 100-W transmitter, 30 dB for a 1000-W transmitter and so on, to limit the input to the spectrum analyzer to 1 W. There are specialized attenuators that are made for transmitter testing; these attenuators provide the necessary power dissipation and attenuation in the 20 to 30-dB range.

When using a spectrum analyzer it is very important that the maximum amount of attenuation be applied before a measurement is made. In addition, it is a good practice to start with maximum attenuation and view the entire spectrum of a signal before the attenuator is adjusted. The signal being viewed could appear to be at

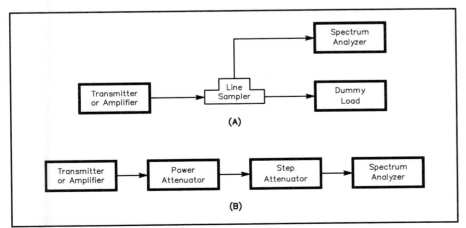

Fig 26.64 — Alternate bench setups for viewing the output of a high power transmitter or oscillator on a spectrum analyzer. A uses a line sampler to pick off a small amount of the transmitter or amplifier power. In B, most of the transmitter power is dissipated in the power attenuator.

a safe level, but another spectral component, which is not visible, could be above the damage limit. It is also very important to limit the input power to the analyzer when pulse power is being measured. The average power may be small enough so the input attenuator is not damaged, but the peak pulse power, which may not be readily visible on the analyzer display, can destroy a mixer, literally in microseconds.

When using a spectrum analyzer it is necessary to ensure that the analyzer does not generate additional spurious signals that are then attributed to the system under test. Some of the spurious signals that can be generated by a spectrum analyzer are harmonics and IMD. If it is desired to measure the harmonic levels of a transmitter at a level below the spurious level of the analyzer itself, a notch filter can be inserted between the attenuator and the spectrum analyzer as shown in **Fig 26.65**. This reduces the level of the fundamental signal and prevents that signal from generating harmonics within the analyzer, while still allowing the harmonics from the transmitter to pass through to the analyzer without attenuation. Use caution with this technique; detuning the notch filter or inadvertently changing the transmitter frequency will allow potentially high levels of power to enter the analyzer. In addition, use care when choosing filters; some filters (such as cavity filters) respond not only to the fundamental but notch out odd harmonics as well.

It is good practice to check for the generation of spurious signals within the spectrum analyzer. When a spurious signal is generated by a spectrum analyzer, adding attenuation at the analyzer input will cause the internally generated spurious signals to decrease by an amount greater than the added attenuation. If attenuation added ahead of the analyzer causes all of the visible signals to decrease by the same amount, this indicates a spurious-free display.

The input impedance for most RF spectrum analyzers is 50 Ω; not all circuits have convenient 50-Ω connections that can be accessed for testing purposes, however. Using a probe such as the one shown in **Fig 26.66** allows the analyzer to be used as a troubleshooting tool. The probe can be used to track down signals within a transmitter or receiver, much like an oscilloscope is used. The probe shown offers a 100:1 voltage reduction and loads the circuit with 5000 Ω. A different type of probe is shown in **Fig 26.67**. This inductive pickup coil (sometimes called a "sniffer") is very handy for troubleshooting. The coil is used to couple signals from the radiated magnetic field of a circuit into

the analyzer. A short length of miniature coax is wound into a pick-up loop and soldered to a larger piece of coax. The use of the coax shields the loop from coupling energy from the electric field component. The dimensions of the loop are not critical, but smaller loop dimensions make the loop more accurate in locating the source of radiated RF. The shield of the coax provides a complete electrostatic shield without introducing a shorted turn.

The sniffer allows the spectrum analyzer to sense RF energy without contacting the circuit being analyzed. If the loop is brought near an oscillator coil, the oscillator can be tuned without directly contacting (and thus disturbing) the circuit. The oscillator can then be checked for reliable starting and the generation of spurious sidebands. With the coil brought near the tuned circuits of amplifiers or frequency multipliers, those stages can be tuned using a similar technique.

Even though the sniffer does not contact the circuit being evaluated, it does extract some energy from the circuit. For this reason, the loop should be placed as far from the tuned circuit as is practical. If

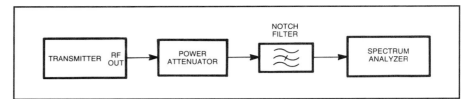

Fig 26.65 — A notch filter is another way to reduce the level of a transmitter's fundamental signal so that the fundamental does not generate harmonics within the analyzer. However in order to know the amplitude relationship between the fundamental and the transmitter's actual harmonics and spurs the attenuation of the fundamental in the notch filter must be known.

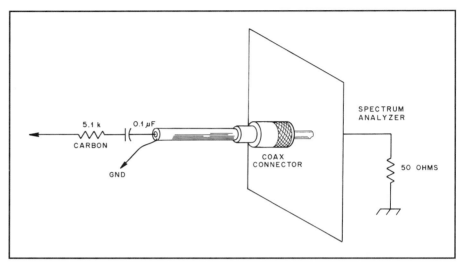

Fig 26.66 — A voltage probe designed for use with a spectrum analyzer. Keep the probe tip (resistor and capacitor) and ground leads as short as possible.

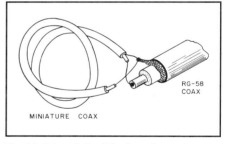

Fig 26.67 — A "sniffer" probe consisting of an inductive pick-up. It has an advantage of not loading the circuit under test. See text for details.

the loop is placed too far from the circuit, the signal will be too weak or the pick-up loop will pick up energy from other parts of the circuit and not give an accurate indication of the circuit under test.

The sniffer is very handy to locate sources of RF leakage. By probing the shields and cabinets of RF-generating equipment (such as transmitters) egress and ingress points of RF energy can be identified by increased indications on the analyzer display.

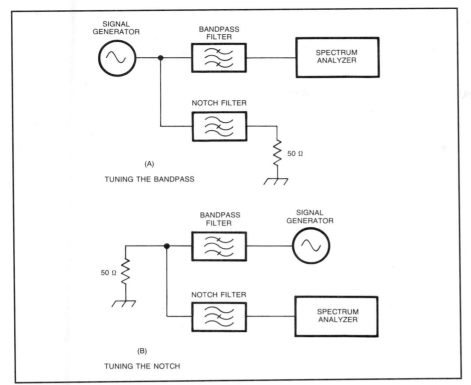

(A)

TUNING THE BANDPASS

(B)

TUNING THE NOTCH

Fig 26.68 — Block diagram of a spectrum analyzer and signal generator being used to tune the band-pass and notch filters of a duplexer. All ports of the duplexer must be properly terminated and good quality coax with intact shielding used to reduce leakage.

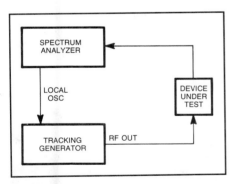

Fig 26.69 — A signal generator (shown in the figure as the "Tracking Generator") locked to the local oscillator of a spectrum analyzer can be used to determine filter response over a range of frequencies.

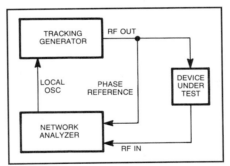

Fig 26.70 — A network analyzer is usually found in commercial communications development labs. It can measure both the phase and magnitude of the filter input and output signals. See text for details.

One very powerful characteristic of the spectrum analyzer is the instrument's capability to measure very low-level signals. This characteristic is very advantageous when very high levels of attenuation are measured. **Fig 26.68** shows the setup for tuning the notch and passband of a VHF duplexer. The spectrum analyzer, being capable of viewing signals well into the low microvolt region, is capable of measuring the insertion loss of the notch cavity more than 100 dB below the signal genera-tor output. Making a measurement of this sort requires care in the interconnection of the equipment and a well designed spectrum analyzer and signal generator. RF energy leaking from the signal generator cabinet, line cord or even the coax itself, can get into the spectrum analyzer through similar paths and corrupt the measurement. This leakage can make the measurement look either better or worse than the actual attenuation, depending on the phase relationship of the leaked signal.

Extensions of Spectral Analysis

What if a signal generator is connected to a spectrum analyzer so that the signal generator output frequency is exactly the same as the receiving frequency of the spectrum analyzer? It would certainly appear to be a real convenience not to have to continually reset the signal generator to the desired frequency. It is, however, more than a convenience. A signal generator connected in this way is called a tracking generator because the output frequency tracks the spectrum analyzer input frequency. The tracking generator makes it possible to make swept frequency measurements of the attenuation characteristics of circuits, even when the attenuation involved is large.

Fig 26.69 shows the connection of a tracking generator to a circuit under test. In order for the tracking generator to create an output frequency exactly equal to the input frequency of the spectrum analyzer, the internal local oscillator frequencies of the spectrum analyzer must be known. This is the reason for the interconnections between the tracking generator and the spectrum analyzer. The test setup shown will measure the gain or loss of the circuit under test. Only the magnitude of the gain or loss is available; in some cases, the phase angle between the input and output would also be an important and necessary parameter.

The spectrum analyzer is not sensitive to the phase angle of the tracking generator output. In the process of generating the tracking generator output, there are no guarantees that the phase of the tracking generator will be either known or constant. This is especially true of VHF spectrum analyzers/tracking generators where a few inches of coaxial cable represents a significant phase shift.

One effective way of measuring the phase angle between the input and output of a device under test is to sample the phase of the input and output of device under test and apply the samples to a phase detector. **Fig 26.70** shows a block diagram of this technique. An instrument that can measure both the magnitude and phase of a signal is called a vector network analyzer or simply a network analyzer. The magnitude and phase can be displayed either separately or together. When the magnitude and phase are displayed together the two can be presented as two separate traces, similar to the two traces on a dual-trace oscilloscope. A much more useful method of display is to present the magnitude and phase as a polar plot where the locus of the points of a vector having the length of the magnitude and the angle of the phase are displayed. Very sophisticated network analyzers can display all of the S parameters of a circuit in either a polar format or a Smith Chart format.

Transmitter Performance Tests

The test setup used in the ARRL Laboratory for measuring an HF transmitter or amplifier is shown in **Fig 26.71**. As can be seen, different power levels dictate different amounts of attenuation between the transmitter or amplifier and the spectrum analyzer.

Spurious Emissions

Fig 26.72 shows the broadband spectrum of a transmitter, showing the harmonics in the output. The horizontal (frequency) scale is 5 MHz per division; the main output of the transmitter at 7 MHz can be seen about 1.5 major divisions from the left of the trace. A very large apparent signal is seen at the extreme left of the trace. This occurs at what would be zero frequency and it is caused by the first local oscillator frequency being exactly the first

IF. All up-converting superheterodyne spectrum analyzers have this IF feedthrough; in addition, this signal is occasionally accompanied by a smaller spurious signal; generated within the analyzer. To determine what part of the displayed signal is a spurious response caused by IF feedthrough and what is an actual input signal, simply remove the input signal and observe the trace. It is not necessary or desirable that the transmitter be modulated for this broadband test.

Other transmitter tests that can be performed with a spectrum analyzer include measurement of two-tone IMD and SSB carrier and unwanted sideband suppression.

Two-Tone IMD

Investigating the sidebands from a modulated transmitter requires a narrow-band spectrum analysis and produces displays similar to that shown in **Fig 26.73**. In this example, a two-tone test signal is used to modulate the transmitter. The display shows the two test tones plus some of the IMD produced by the SSB transmitter. The test setup used to produce this display is shown in **Fig 26.74**.

In this example, a two-tone test signal with frequencies of 700 and 1900 Hz is used to modulate the transmitter. Set the transmitter output and audio input to the manufacturer's specifications. Each desired tone is adjusted to be equal in amplitude and centered on the display. The step attenuators and analyzer controls are then adjusted to set the two desired signals 6 dB below the 0-dB reference (top) line. The IMD products can then be read directly from the display in terms of "dB below Peak Envelope Power (PEP)." (In the example shown, the third-order products are 30 dB below PEP, the fifth-order products are 37 dB down, the seventh-order products are down 44 dB.)

Carrier and Unwanted Sideband Suppression

Single-tone audio input signals can be used with the same setup to measure unwanted sideband and carrier suppression of SSB signals. In this case, set the single tone to the 0-dB reference line. (Once the level is set, the audio can be disabled for carrier suppression measurements in order to eliminate IMD and other effects.)

Phase Noise

Phase/composite noise is also measured with spectrum analyzers in the ARRL Lab.

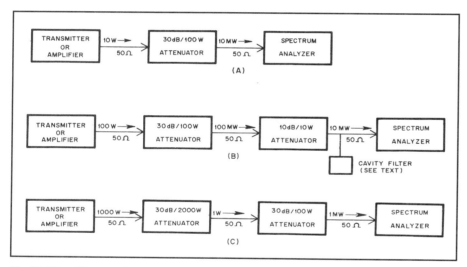

Fig 26.71 — These setups are used in the ARRL Laboratory for testing transmitters or amplifiers with several different power levels.

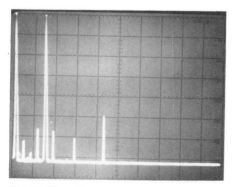
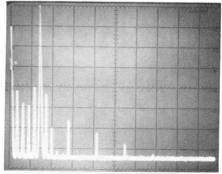

Fig 26.72 — Comparison of two different transmitters on the 40-m band as seen on a spectrum analyzer display. The photograph at A shows a relatively clean transmitted signal but the transmitter at B shows more spurious signal content. Horizontal scale is 5 MHz per division; vertical is 10 dB per division. According to current FCC spectral purity requirements both transmitters are acceptable.

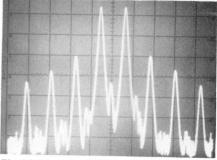

Fig 26.73 — An SSB transmitter two-tone test as seen on a spectrum analyzer. Each horizontal division represents 1 kHz and each vertical division is 10 dB. The third-order products are 30 dB below the PEP (top line), the fifth-order products are down 37 dB and seventh-order products are down 44 dB. This represents acceptable (but not ideal) performance.

This test requires specialized equipment and is included here for information purposes only.

The purpose of the Composite-Noise test is to observe and measure the phase and amplitude noise, as well as any close-in spurious signals generated by a transmitter. Since phase noise is the primary noise component in any well-designed transmitter, almost all of the noise observed during this test is phase noise.

This measurement is accomplished in the lab by converting the transmitter output down to a frequency band about 10 or 20 kHz above baseband. A mixer and a signal generator (used as a local oscillator)

are used to perform this conversion. Filters remove the 0-Hz component as well as any unwanted heterodyne components. A spectrum analyzer (see **Fig 26.75**) displays the remaining noise and spurious signals from 2 to 20 kHz from the carrier frequency (in the CW mode).

Tests in the Time Domain

Oscilloscopes are used for transmitter testing in the time domain. Dual-trace instruments are best in most cases, providing easy to read time-delay measurements between keying input and RF- or audio-output signals. Common transmitter measurements performed with 'scopes include

CW keying wave shape and time delay and SSB/FM transmit-to-audio turnaround tests (important for many digital modes).

A typical setup for measuring CW keying waveform and time delay is shown in **Fig 26.76**. A keying test generator is used to repeatedly key the transmitter at a controlled rate. The generator can be set to any reasonable speed, but ARRL tests are usually conducted at 20-ms on and 20-ms off (25 Hz, 50% duty cycle). **Fig 26.77** shows a typical display. The rise and fall times of the RF output pulse are measured between the 10% and 90% points on the leading and trailing edges, respectively. The delay times are measured between the 50% points of the keying and RF output waveforms. Look at the **Transceivers** chapter for further discussion of CW keying issues.

For voice modes (SSB/FM), a PTT-to-RF output test is similar to CW keying tests. It measures rise and fall times, as well as the on- and off-delay times just as in the CW test. See **Fig 26.78** for the test setup.

"Turn-around time" is the time it takes for a transceiver to switch from the 50% fall time of a keying pulse to 50% rise of audio output. The test setup is shown in **Fig 26.79**. Turn-around time measurements require extreme care with respect to transmitter output power, attenuation, signal-generator output and the maximum input signal that can be tolerated by the generator. The generator's specifications must not be exceeded and the input to the receiver must be at the required level, usually S9. Receiver AGC is usually off for this test, but experimentation with AGC and signal input level can reveal surprising variations. The keying rate must be

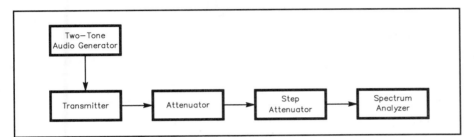

Fig 26.74 — The test setup used in the ARRL Laboratory to measure the IMD performance of transmitters and amplifiers.

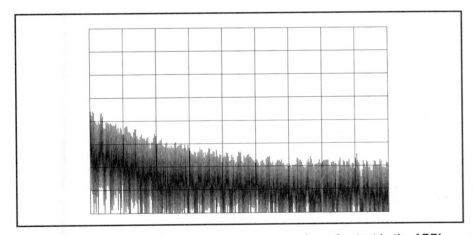

Fig 26.75 — The spectral-display results of a composite-noise test in the ARRL Lab. This display is for the ICOM IC-707 reviewed in April 94 *QST*. Power output is 100 W at 14 MHz. Vertical divisions are 10 dB; horizontal divisions are 2 kHz. The log reference level (the top horizontal line on the scale) represents –60 dBc/Hz and the baseline is –140 dBc/Hz. The carrier, off the left edge of the plot, is not shown. This plot shows composite transmitted noise 2 to 20 kHz from the carrier.

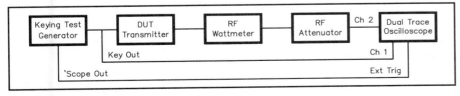

Fig 26.76 — CW keying waveform test setup.

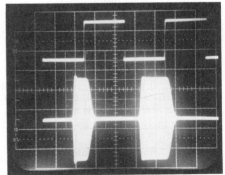

Fig 26.77 — Typical CW keying waveform test results. This display is for the ICOM IC-707 (semi-break-in mode) reviewed in April 94 *QST*. The upper trace is the actual key closure; the lower trace is the RF envelope. Horizontal divisions are 10 ms. The transceiver was being operated at 100 W output at 14 MHz.

considerably slower than the turn-around time; rates of 200-ms on/200-ms off or faster, have been used with success in Product Review tests at the ARRL Lab.

Turn-around time is an important consideration with some digital modes. AMTOR, for example, requires a turn-around time of 35 ms or less.

Oscilloscope Bibliography

R. vanErk, *Oscilloscopes, Functional Operation and Measuring Examples*, McGraw-Hill Book Co, New York, 1978.

V. Bunze, *Probing in Perspective—Application Note 152*, Hewlett-Packard Co, Colorado Springs, CO, 1972 (Pub No. 5952-1892).

The XYZs of Using a Scope, Tektronix, Inc, Portland, OR, 1981 (Pub No. 41AX-4758).

Basic Techniques of Waveform Measurement (Parts 1 and 2), Hewlett-Packard Co, Colorado Springs, CO, 1980 (Pub No. 5953-3873).

J. Millman, and H. Taub, *Pulse Digital and Switching Waveforms*, McGraw-Hill Book Co, New York, 1965, pp 50-54.

V. Martin, *ABCs of DMMs*, Fluke Corp, PO Box 9090, Everett, WA 98206.

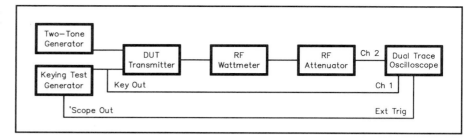

Fig 26.78 — PTT-to-RF-output test setup for voice-mode transmitters.

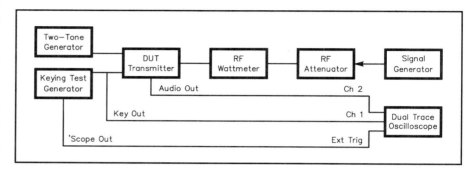

Fig 26.79 — Transmit-receive turn-around time test setup.

GLOSSARY

AC (Alternating current) — The polarity constantly reverses, as contrasted to dc (direct current) where polarity is fixed.

Analog — Signals which have a full set of values. If the signal varies between 0 and 10 V all values in this range can be found. Compare this to a *digital* system.

Attenuator — A device which reduces the amplitude of a signal

Average value — Obtained by recording or measuring *N* samples of a signal, adding up all of these values, and dividing this sum by *N*.

Bandwidth — A measure of how wide a signal is in frequency. If a signal covers 14,200 to 14,205 kHz its bandwidth is said to be 5 kHz.

BNC — A small connector used with coax cable.

Bridge circuit — Four passive elements, such as resistors, inductors, connected as a pair of voltage dividers with a meter or other measuring device across two opposite junctions. Used to indicate the relative values of the four passive elements. See the chapter discussion of Wheatstone bridges.

CMOS — A family of digital logic elements usually selected for their low power drain. See **Digital** chapter of this *Handbook*.

Coaxial cable (coax) — A cable formed of two conductors that share the same axis. The center conductor may be a single wire or a stranded cable. The outer conductor is called the shield. The shield may be flexible braid, foil, semi-rigid or rigid metal. For more information, look in the **Transmission Lines** chapter.

Combiners — See *Hybrid.*

D'Arsonval meter — A common mechanical meter consisting of a permanent magnet and a moving coil (with pointer attached).

DC (direct current) — The polarity is fixed for all time, as contrasted to ac (alternating current) where polarity constantly reverses.

Digital — A system that allows signals to assume a finite range of states. Binary logic is the most common example. Only two values are permitted in a binary system: one value is defined as a logical *1* and the other value as a logical *0*. See the *Handbook* chapter on **Digital Systems**.

Divider — A network of components that produce an output signal that is a fraction of the input signal. The ratio of the output to the input is the division factor. An analog divider divides voltage (a string of series connected resistors) or current (parallel connected resistors). Digital dividers divide pulse trains or frequency.

DMM (Digital multimeter) — A test instrument that usually measures at least: voltage, current and resistance, and displays the result on a numeric digit display, rather than an analog meter.

Dummy antenna or dummy load — A resistor or set of resistors used in place of an antenna to test a transmitter without radiating any electromagnetic energy into the air.

DVM (digital volt meter) — See *DMM.*

FET voltmeter — See also *VTVM*. An updated version of a VTVM using field effect transistors (FETs) in place of vacuum tubes.

Flip-flop — A digital circuit that has two stable states. See the chapter on **Digital Techniques**.

Frequency marker — Test signals generated at selected intervals (such as 25 kHz, 50 kHz, 100 kHz) for calibrating the dials of receivers and transmitters.

Fundamental — The first signal or frequency in a series of harmonically related signals. This term is often used to describe an oscillator or transmitter's desired signal.

Harmonic — A signal occurring at some integral multiple (such as two, three, four) of a *fundamental* frequency.

Hybrid (hybrid combiners) — A device used to connect two signal generators to one receiver for test purposes, without the two generators affecting each other.

IC (integrated circuit) — A complete circuit built into a single electronic component.

LCD (liquid crystal display) — A low-power display device utilizing the physics of liquid crystals. They usually need either ambient light or backlighting to be seen.

LED (light emitting diode) — A diode that emits light when an appropriate voltage (usually 1.5 V at about 20 mA) is connected. They are used either as tiny pilot lights or in bar shapes to display letters and numbers.

Loran — A navigation system using very-low-frequency transmitters.

Marker — See *Frequency marker.*

Multiplier — A circuit that purposely creates some desired *harmonic* of its input signal. For example, a frequency multiplier that takes energy from a 3.5-MHz exciter and puts out RF at 7 MHz is a two times multiplier, usually called a frequency doubler.

N — A type of coaxial cable connector common at UHF and higher frequencies.

NAND — A digital element that performs the *not-and* function. See the **Digital** chapter.

Noise (noise figure) — Noise is generated in all electrical circuits. It is particularly critical in those stages of a receiver that are closest to the antenna (RF amplifier and mixer), because noise generated in these stages can mask a weak signal. The noise figure is a measure of this noise generation. Lower noise figures mean that less noise is generated and weaker signals can be heard.

NOR — A digital element that performs the *not-or* function. See the **Digital** chapter.

Null (nulling) — The process of adjusting a circuit for a minimum reading on a test meter or instrument. At a perfect null there is null, or no, energy to be seen.

Ohmmeter — A meter that measures the value of resistors. Usually part of a multimeter. See *VOM* and *DMM.*

Peak value — The highest value of a signal during the measuring time. If a measured voltage varies in value from 1 to 10 V over a measuring period, the peak value would be the highest measured, 10 V.

PL-259 — A connector used for coaxial cable, usually at HF. It is also known as a male UHF connector. It is an inexpensive and common connector, but it is not weatherproof, nor is its impedance constant over frequency.

Prescaler — A circuit used ahead of a counter to extend the counter range to higher frequencies. A counter capable of operating up to 50 MHz can count up to 500 MHz when used with a ÷10 prescaler.

Q — The ratio of the reactance to the resistance of a component or circuit. It provides a measure of bandwidth. Lower resistive losses make for a higher Q, and a narrower bandwidth.

RMS (root mean square) — A measure of the value of a voltage or current obtained by taking values from successive small time slices over a complete cycle of the wave form, squaring those values, taking the mean of the squares, and then the square root of the mean. Very significant when working with good ac sine waves, where the RMS of the sine wave is 0.707 of the peak value.

Scope — Slang for oscilloscope. See the Oscilloscope section of this chapter.

Shunt — Elements connected in parallel.

Sinusoidal (sine wave) — The nominal waveform for unmodulated RF energy and many other ac voltages.

Spectrum — Used to describe a range of frequencies or wavelengths. The RF spectrum starts at perhaps 10 kHz and extends up to several hundred gigahertz. The light spectrum goes from infrared to ultraviolet.

Spurious emissions, or spurs — Unwanted energy generated by a transmitter or other circuit. These emissions include, but are not limited to, *harmonics.*

Thermocouple — A device made up of two different metals joined at two places. If one joint is hot and the other cold a voltage may be developed, which is a measure of the temperature difference.

Time domain — A measurement technique where the results are plotted or shown against a scale of time. In contrast to the frequency domain, where the results are plotted against a scale of frequency.

TTL (Transistor-transistor-logic) — A logic IC family commonly used with 5 V supplies. See the chapter on **Digital Techniques**.

Vernier dial or vernier drive — A mechanical system of tuning dials, frequently used in older equipment, where the knob might turn 10 times for each single rotation of the control shaft.

VOM (volt-ohm-meter) — A multimeter whose design predates multiple scale meters (see **DMM**).

VTVM (vacuum tube voltmeter) — A meter that was developed to provide a high input resistance and therefore low current drain (loading) from the circuit being tested. Now replaced by the FET meter.

Wheatstone bridge — See *Bridge circuit.*

Contents

Troubleshooting and Repair

27

Traditionally, the radio amateur has maintained a working knowledge of electronic equipment. This knowledge, and the ability to make repairs with whatever resources are available, keeps amateur stations operating when all other communications fail. This troubleshooting ability is not only a tradition; it is fundamental to the existence of the service.

This chapter, by Ed Hare, W1RFI, tells you what to do when you are faced with equipment failure or a circuit that doesn't work. It will help you ask and answer the right questions: "Should I fix it or send it back to the dealer for repair? What do I need to know to be able to fix it myself? Where do I start? What kind of test equipment do I need?" The best answers to these questions will depend on the type of test equipment you have available, the availability of a schematic or service manual and the depth of your own electronic and troubleshooting experience.

Not everyone is an electronics wizard; your set may end up at the repair shop in spite of your best efforts. The theory you learned for the FCC examinations and the information in this *Handbook* can help you decide if you can fix it yourself. If the problem is something simple (and most are), why not avoid the effort of shipping the radio to the manufacturer? It is gratifying to save time and money, but, even better, the experience and confidence you gain by fixing it yourself may prove even more valuable.

Although some say troubleshooting is as much art as it is science, the repair of electronic gear is not magic. It is more like detective work. A knowledge of complex math is not required. However, you must have, or develop, the ability to read a sche-matic diagram and to visualize signal flow through the circuit.

SAFETY FIRST

Always! Death is permanent. A review of safety must be the first thing discussed in a troubleshooting chapter. Some of the voltages found in amateur equipment can be fatal! Only 50 mA flowing through the body is painful; 100 to 500 mA is usually fatal. Under certain conditions, as little as 24 V can kill.

Make sure you are 100% familiar with all safety rules and the dangerous conditions that might exist in the equipment you are servicing. Remember, if the equipment is not working properly, dangerous conditions may exist where you don't expect them. Treat every component as potentially "live."

Some older equipment uses "ac/dc" circuitry. In this circuit, one side of the chassis is connected directly to the ac line. This is an electric shock waiting to happen.

A list of safety rules can be found in **Table 27.1**. You should also read the **Safety** chapter of this *Handbook* before you proceed.

GETTING HELP

Other hams may be able to help you with your troubleshooting and repair problems,

---

Table 27.1

Safety Rules

1. Keep one hand in your pocket when working on live circuits or checking to see that capacitors are discharged.
2. Include a conveniently located ground-fault current interrupter (GFCI) circuit breaker in the workbench wiring.
3. Use only grounded plugs and receptacles.
4. Use a GFCI protected circuit when working outdoors, on a concrete or dirt floor, in wet areas, or near fixtures or appliances connected to water lines, or within six feet of any exposed grounded building feature.
5. Use a fused, power limiting isolation transformer when working on ac/dc devices.
6. Switch off the power, *disconnect equipment from the power source, ground the output of the internal dc power supply,* and discharge capacitors when making circuit changes.
7. Do not subject electrolytic capacitors to excessive voltage, ac voltage or reverse voltage.
8. Test leads should be well insulated.
9. Do not work alone!
10. Wear safety glasses for protection against sparks and metal fragments.
11. Always use a safety harness when working above ground level.
12. Wear shoes with nonslip soles that will support your feet when climbing.
13. Wear rubber-sole shoes or use a rubber mat when standing on the ground or on a concrete floor.
14. Wear a hard hat when someone is working above you.
15. Be careful with tools that may cause short circuits.
16. Replace fuses only with those having proper ratings.

either with a manual or technical help. Check with your local club or repeater group. You may get lucky and find a troubleshooting "wizard." (On the other hand, you may get some advice that is downright dangerous, so be selective.) You can also place a classified ad in one of the ham magazines, looking for a rare manual.

Your fellow hams in the ARRL Field organization may also help. Technical Coordinators (TCs) and Technical Specialists (TSs) are volunteers who are willing to help hams with technical questions. For the name and address of a local TC or TS, contact your Section Manager (listed in the front of any recent issue of *QST*).

THEORY

To fix electronic equipment, you need to understand the system and circuits you are troubleshooting. A working knowledge of electronic theory, circuitry and components is an important part of the process. If necessary, review the electronic and circuit theory explained in the other chapters of this book. When you are troubleshooting, you are looking for the unexpected. Knowing how circuits are supposed to work will help you to look for things that are out of place.

TEST EQUIPMENT

Many of the steps involved in troubleshooting efficiently require the use of test equipment. We cannot see electrons flow. However, electrons do affect various devices in our equipment, with results we can measure.

Some people think they need expensive test instruments to repair their own equipment. This is not so! In fact, you probably already own the most important instruments. Some others may be purchased inexpensively, rented, borrowed or built at home. The test equipment available to you may limit the kind of repairs you can do, but you will be surprised at the kinds of repair work you can do with simple test equipment.

Senses

Although they are not "test equipment" in the classic sense, your own senses will tell you as much about the equipment you are trying to fix as the most-expensive spectrum analyzer. We each have some of these natural "test instruments."

Eyes — Use them constantly. Look for evidence of heat and arcing, burned components, broken connections or wires, poor solder joints or other obvious visual problems.

Ears — Severe audio distortion can be detected by ear. The "snaps" and "pops"

of arcing or the sizzling of a burning component may help you track down circuit faults. An experienced troubleshooter can diagnose some circuit problems by the sound they make. For example, a bad audio-output IC sounds slightly different than a defective speaker.

Nose — Your nose can tell you a lot. With experience, the smells of ozone, an overheating transformer and a burned carbon-composition resistor each become unique and distinctive.

Finger — Carefully use your fingers to measure low heat levels in components. Small-signal transistors can be fairly warm to the touch; anything hotter can indicate a circuit problem. (Be careful; some high-power devices or resistors can get downright hot during normal operation.)

Brain — More troubleshooting problems have been solved with a VOM and a brain than with the most expensive spectrum analyzer. You must use your brain to analyze data collected by other instruments.

"Internal" Equipment

Some "test equipment" is included in the equipment you repair. Nearly all receivers include a speaker. An S meter is usually connected ahead of the audio chain. If the S meter shows signals, it indicates that the RF and IF circuitry is probably functioning. Analyze what the unit is doing and see if it gives you a clue.

Some older receivers include a crystal frequency calibrator. The calibrator signal, which is rich in harmonics, is injected in the RF chain close to the antenna jack and may be used for signal tracing and alignment.

Bench Equipment

Here is a summary of test instruments and their applications. Some items serve several purposes and may substitute for others on the list. The list does not cover all equipment available, only the most common and useful instruments. The theory and operation of much of this test equipment is discussed in more detail in the **Test** chapter.

Multimeters — The multimeter is the most often used piece of test equipment. This group includes vacuum-tube voltmeters (VTVMs), volt-ohm-milliammeters (VOMs), field-effect transistor VOMs (FETVOMs) and digital multimeters (DMMs). Multimeters are used to read bias voltages, circuit resistance and signal level (with an appropriate probe). They can test resistors, capacitors (within certain limitations), diodes and transistors.

DMMs have become quite inexpensive.

Their high input impedance, accuracy and flexibility are well worth the cost. Many of them contain other test equipment as well, such as capacitance meters, frequency counters, transistor testers and even digital thermometers. Some DMMs are affected by RF, so most technicians keep an analog-display VOM on hand for use near RF equipment.

When buying an analog meter, look for one with an input impedance of 20 kΩ/V or better. Reasonably priced models are available with 30 kΩ/V ($35) and 50 kΩ/V ($40). The 10 MΩ or better input impedance of DMMs, FETVOMs, VTVMs and other electronic voltmeters makes them the preferred instruments for voltage measurements.

Test leads — Keep an assortment of wires with insulated, soldered alligator clips. Commercially made leads have a high failure rate because they use small wire that is not soldered to the clips; it is best to make your own.

Open wire leads (**Fig 27.1A**) are good for dc measurements, but they can pick up unwanted RF energy. This problem is reduced somewhat if the leads are twisted together (Fig 27.1B). A coaxial cable lead is much better, but its inherent capacitance can affect RF measurements.

The most common probe is the low-capacitance (×10) probe shown in Fig 27.1C. This probe isolates the oscilloscope from the circuit under test, preventing the 'scope's input and test-probe capacitance from affecting the circuit and changing the reading. A network in the probe serves as a 10:1 divider and compensates for frequency distortion in the cable and test instrument.

Demodulator probes (see the **Test** chapter and the schematic shown in Fig 27.1D) are used to demodulate or detect RF signals, converting modulated RF signals to audio that can be heard in a signal tracer or seen on a low-bandwidth 'scope.

You can make a probe for inductive coupling as shown in Fig 27.1E. Connect a two- or three-turn loop across the center conductor and shield before sealing the end. The inductive pick up is useful for coupling to high-current points.

RF power and SWR meters — Every shack should have one. It is used to measure forward and reflected RF power. A standing-wave ratio (SWR) meter can be the first indicator of antenna trouble. It can also be used between an exciter and power amplifier to spot an impedance mismatch.

Simple meters indicate relative power SWR and are fine for Transmatch adjustment and line monitoring. However, if you want to make accurate measurements a calibrated wattmeter with a directional coupler is required.

Dummy load — A dummy load is a necessity in any shack. Do not put a signal on the air while repairing equipment. Defective equipment can generate signals that interfere with other hams or other radio services. A dummy load also provides a known, matched load (usually 50 Ω) for use during adjustments.

When buying a dummy load, avoid used, oil-cooled dummy loads unless you can be sure that the oil does not contain PCBs. This biologically hazardous compound was common in transformer oil until a few years ago.

Dip meter — This device is often called a transistor dip meter or a grid-dip oscillator from vacuum-tube days.

Most dip meters can also serve as an absorption frequency meter. In this mode, measurements are read at the current peak, rather than the dip. Some meters have a connection for headphones. The operator can usually hear signals that do not register on the meter. Because the dip meter is an oscillator, it can be used as a signal generator in certain cases where high accuracy or stability are not required.

When purchasing a dip meter, look for

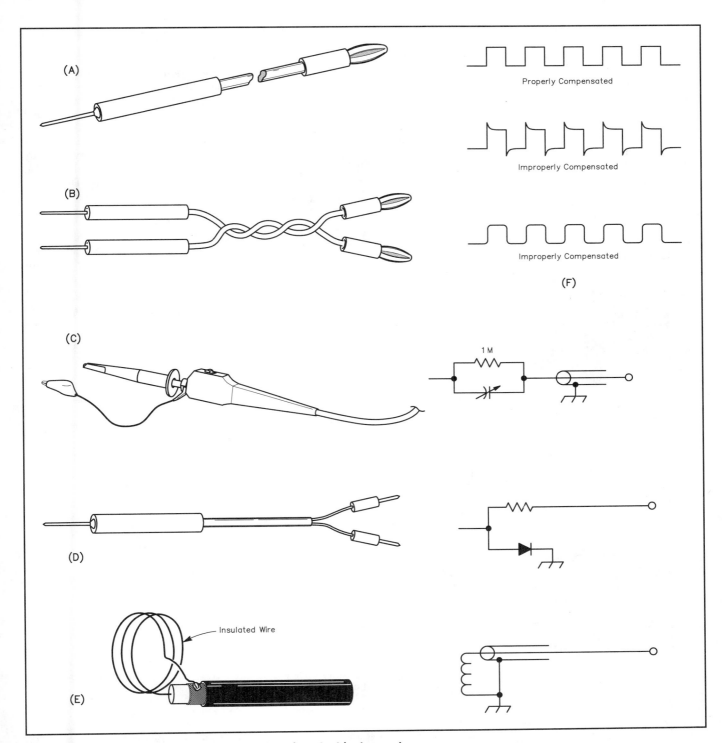

Fig 27.1—An array of test probes for use with various test instruments.

one that is mechanically and electrically stable. The coils should be in good condition. A headphone connection is helpful. Battery operated models are easier to use for antenna measurements. Dip meters are not nearly as common as they once were.[1]

Oscilloscope — The oscilloscope, or 'scope, is the second most often used piece of test equipment, although a lot of repairs can be accomplished without one. The trace of a 'scope can give us a lot of information about a signal at a glance.

The simplest way to display a waveform is to connect the vertical amplifier of the 'scope to a point in the circuit through a simple test lead. When viewing RF, use a low-capacitance probe that has been adjusted to match the 'scope. Select the vertical gain and time-base (horizontal scale, **Fig 27.2**) for the most useful displayed waveform.

A 'scope waveform shows voltage (if

[1]The ARRL has prepared a list of dip-meter sources. These are available on the World Wide Web (**http://www.arrl.org/notes/ 1832/templates/dips.pdf**).

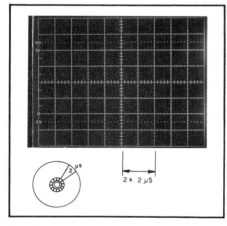

Fig 27.2—An oscilloscope display showing the relationship between time-base setting and graticule lines.

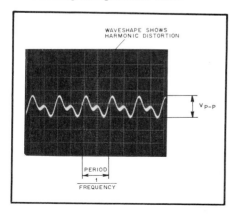

Fig 27.3—Information available from a typical oscilloscope display of a waveform.

calibrated), approximate period (frequency is the reciprocal of the period) and a rough idea of signal purity (see **Fig 27.3**). If the 'scope has dual-trace capability (meaning it can display two signals at once), a second waveform may be displayed and compared to the first. When the two signals are taken from the input and output of a stage, stage linearity and phase shift can be checked (see **Fig 27.4**).

An important specification of an oscilloscope is its amplifier bandwidth. This tells us the frequency at which amplifier response has dropped 3 dB. The instrument will display higher frequencies, but its accuracy at higher frequencies is not known. Even well below its rated bandwidth a 'scope is not capable of much more

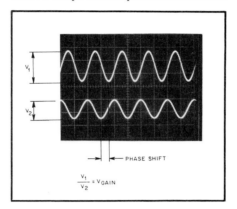

Fig 27.4—A dual-trace oscilloscope display of amplifier input and output waveforms.

[2]More information about the signal injector and signal sources appears in "Some Basics of Equipment Servicing," February 1982 *QST* (Feedback, May 1982).

than about 5% accuracy. This is adequate for most amateur applications.

An oscilloscope will show gross distortions of audio and RF waveforms, but it cannot be used to verify that a transmitter meets FCC regulations for harmonics and spurious emissions. Harmonics that are down only 20 dB from the fundamental would be illegal in most cases, but they would not change the oscilloscope waveform enough to be seen.

When buying a 'scope, get the greatest bandwidth you can afford. Old Hewlett-Packard or Tektronix 'scopes are usually quite good for amateur use.

Signal generator — Although signal generators have many uses, in troubleshooting they are most often used for signal injection (more about this later) and alignment.

An AF/RF signal-injector schematic is shown in **Fig 27.5**. If frequency accuracy is needed, the crystal-controlled signal source of **Fig 27.6** can be used. The AF/RF circuit provides usable harmonics up to 30 MHz, while the crystal controlled oscillator will function with crystals from 1 to 15 MHz. These two projects are not meant to compete with standard signal generators, but they are adequate for signal injection.[2] A better generator is

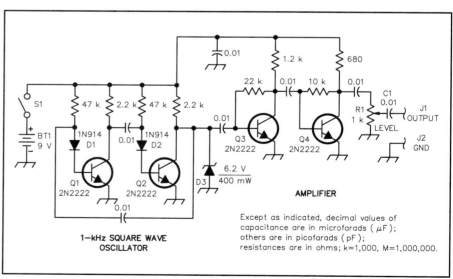

Fig 27.5—Schematic of the AF/RF signal injector. All resistors are 1/4-W, 5% carbon units, and all capacitors are disc ceramic.
BT1 — 9-V battery.
D1, D2 — Silicon switching diode, 1N914 or equiv.
D3 — 6.2-V, 400-mW Zener diode.
J1, J2 — Banana jack.
Q1-Q4 — General-purpose silicon NPN transistors, 2N2222 or similar.
R1 — 1-kΩ panel-mount control.
S1 — SPST toggle switch.

required for receiver alignment or for receiver quality testing.

When buying a generator, look for one that can generate a sine wave signal. A good signal generator is double or triple shielded against leakage. Fixed-frequency audio should be available for modulation of the RF signal and for injection into audio stages. The most versatile generators can generate amplitude and frequency modulated signals.

Good generators have stable frequency controls with no backlash. They also have multiposition switches to control signal level. A switch marked in dBm is a good indication that you have located a high-quality test instrument. The output jack should be a coaxial connector (usually a BNC or N), not the kind used for microphone connections.

Some older, high-quality units are common. Look for World War II surplus units of the URM series, Boonton, GenRad, Hewlett-Packard, Tektronix, Measurements Inc or other well-known brand names. Some home-built signal generators may be quite good, but make sure to check construction techniques, level control and shielding quality.

Signal tracer — Signals can be traced with a voltmeter and an RF probe, a dip meter with headphones or an oscilloscope, but there are some devices made especially for signal tracing. A signal tracer is primarily a high-gain audio amplifier. It may have a built-in RF detector, or rely on an external RF probe. Most convert the traced signal to audio through a speaker.

The tracer must function as a receiver and detector for each frequency range in the test circuit. A high-impedance tracer input is necessary to prevent circuit loading.

A general-coverage receiver can be used to trace RF or IF signals, if the receiver covers the necessary frequency range. Most receivers, however, have a low-impedance input that severely loads the test circuit. To minimize loading, use a capacitive probe or loop pickup. When the probe is held near the circuit, signals will be picked up and carried to the receiver. It may also pick up stray RF, so make sure you are listening to the correct signal by switching the circuit under test on and off while listening.

Tube tester — Vacuum-tube testers used to be found in nearly every drug or department store. They are scarce now because tubes are no longer used in modern consumer or (most) amateur equipment. Older tube gear is found in many ham shacks or flea markets, though. There are many aficionados of vintage gear who enjoy working with old vacuum-tube equipment.

Most simple tube testers measure the cathode emission of a vacuum tube. Each grid is shorted to the plate through a switch and the current is observed while the tube operates as a diode. By opening the switches from each grid to the plate (one at a time), we can check for opens and shorts. If the plate current does not drop slightly as a switch is opened, the element connected to that switch is either open or shorted to another element. (We cannot tell an open from a short with this test.) The emission tester does not necessarily indicate the ability of a tube to amplify.

Other tube testers measure tube gain (transconductance). Some transconductance testers read plate current with a fixed bias network. Others use an ac signal to drive the tube while measuring plate current.

Most tube testers also check interelement leakage. Contamination inside the tube envelope may result in current leakage between elements. The paths can have high resistance, and may be caused by gas or deposits inside the tube. Tube testers use a moderate voltage to check for leakage. Leakage can also be checked with an ohmmeter using the ×1M range, depending on the actual spacing of tube elements.

Transistor tester — Transistor testers are similar to transconductance tube testers. Device current is measured while the device is conducting or while an ac signal is applied at the control terminal. Commercial surplus units are often seen at ham flea markets. Some DMMs being sold today also include a built-in, simple transistor tester.

Most transistor failures appear as either an open or shorted junction. Opens and shorts can be found easily with an ohmmeter; a special tester is not required.

Transistor gain characteristics vary widely, however, even between units with the same device number. Testers can be used to measure the gain of a transistor. A tester that uses dc signals measures only transistor dc alpha and beta. Testers that apply an ac signal show the ac alpha or beta. Better testers also test for leakage.

In addition to telling you whether a transistor is good or bad, a transistor tester can help you decide if a particular transistor has sufficient gain for use as a replacement. It may also help when matched transistors are required. The final test is the repair circuit.

Frequency meter — Most frequency counters are digital units, often able to show frequency to a 1-Hz resolution. Some older "analog" counters are sometimes found surplus, but a low-cost digital counter will out-perform even the best of these old "classics."

Power supplies — A well-equipped test bench should include a means of varying the ac-line voltage, a variable-voltage regulated dc supply and an isolation transformer.

AC-line voltage varies slightly with load. An autotransformer with a movable tap lets you boost or reduce the line volt-

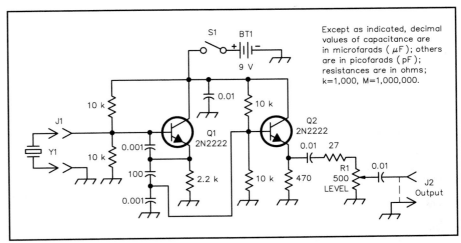

Fig 27.6 — Schematic of the crystal-controlled signal source. All resistors are ¼-W, 5% carbon units, and all capacitors are disc ceramic. A full-size etching pattern and parts-placement diagram are in Chapter 30, References.

BT1 — 9-V transistor radio battery.
J1 — Crystal socket to match the crystal type used.
J2 — RCA phono jack or equivalent.
Q1, Q2 — General-purpose silicon NPN transistors, 2N2222 or similar.

R1 — 500-Ω panel-mount control.
S1 — SPST toggle switch.
Y1 — 1 to 15-MHz crystal.

age slightly. This is helpful to test circuit functions with supply-voltage variations.

As mentioned earlier, ac/dc radios must be isolated from the ac line during testing and repair. Keep an isolation transformer handy if you want to work on table-model broadcast radios or television sets (check for other ac/dc equipment, too. Even some old phonographs or Amateur Radio transceivers used this dangerous circuit design).

A good multivoltage supply will help with nearly any analog or digital troubleshooting project. Several of the distributors listed in the **References** chapter stock bench power supplies. A variable-voltage dc supply may be used to power various small items under repair or provide a variable bias supply for testing active devices. Construction details for a laboratory power supply appear in the **Power Supplies** chapter.

If you want to work on vacuum-tube gear, the maximum voltage available from the dc supply should be high enough to serve as a plate or a bias supply for common tubes (about 300 to 400 V ought to do it).

Accessories — There are a few small items that may be used in troubleshooting. You may want to keep them handy.

Many circuit problems are sensitive to temperature. A piece of equipment may work well when first turned on (cold) but

fail as it warms up. In this case, a cold source will help you find the intermittent connection. When you cool the bad component, the circuit will suddenly start working again (or stop working). Cooling sprays are available from most parts suppliers.

A heat source helps locate components that fail only when hot. A small incandescent lamp can be mounted in a large piece of sleeve insulation to produce localized heat for test purposes.

A heat source is usually used in conjunction with a cold source. If you have a circuit that stops working when it warms up, heat the circuit until it fails, then cool the components one by one. When the circuit starts working again, the last component sprayed was the bad one.

A stethoscope (with the pickup removed — see **Fig 27.7**) or a long piece of sleeve insulation can be used to listen for arcing or sizzling in a circuit.

WHERE TO BEGIN

New Construction

In most repair work, the technician is aided by the knowledge that the circuit once worked. It is only necessary to find the faulty part(s) and replace it. This is not so with newly constructed equipment. Repair of equipment with no working history is a special, and difficult, case. You may be dealing with a defective component, construction error or even a faulty design. Carefully checking for these defects can save you hours.

All Equipment

Check the Obvious

Try the easy things first. If you are able to solve the problem by replacing a fuse or reconnecting a loose cable, you might be able to avoid a lot of effort. Many experienced technicians have spent hours troubleshooting a piece of equipment only to learn the hard way that the on/off switch was "off" or that they were not using the equipment properly.

Read the manual! Your equipment may be working as designed. Many electronic "problems" are caused by a switch that is set in the wrong position, or a unit that is being asked to do something it was not designed to do. Before you open up your equipment for major surgery, make sure you are using it correctly.

Next, make sure the equipment is plugged in, that the ac outlet does indeed have power, that the equipment is switched "on" and that all of the fuses are good. If the equipment uses batteries or an external power supply, make sure these are working.

Check that all wires, cables and accessories are working and plugged in to the right connectors or jacks. In a "system," it is often difficult to be sure which component or subsystem is bad. Your transmitter may not work on SSB because the transmitter is bad, but it could also be a bad microphone.

Connector faults are more common than component troubles. Consider poor connections as prime suspects in your troubleshooting detective work. Do a thorough inspection of the connections. Is the antenna connected? How about the speaker, fuses and TR switch? Are transistors and ICs firmly seated in their sockets? Are all interconnection cables sound and securely connected? Many of these problems are obvious to the eye, so look around carefully.

Simplify the Problem

If the broken equipment is part of a system, you need to find out exactly which part of the system is bad. For example, if your amateur station is not putting out any RF, you need to determine if it is a microphone problem, a transmitter problem, an amplifier problem or a problem somewhere in your station wiring. If you are trying to diagnose a bad channel on your home modular stereo system, it could be anything from a bad cable to a bad amplifier to a bad speaker.

Simplify the system as much as possible. To troubleshoot the "no-RF" problem, temporarily eliminate the amplifier from the station configuration. To diagnose the stereo system, start troubleshooting by checking just the amplifier with a set of known good headphones. Simplifying the problem will often isolate the bad component quickly.

Documentation

Once you have determined that a piece of equipment is indeed broken, you need to do some preparation before you diagnose and fix it. First, locate a schematic diagram and service manual. It is possible to troubleshoot without a service manual, but a schematic is almost indispensable.

The original equipment manufacturer is the best source of a manual or schematic. However, many old manufacturers have gone out of business. Several sources of equipment manuals are listed in the **References** chapter.

If all else fails, you can sometimes reverse engineer a simple circuit by tracing wiring paths and identifying components to draw your own schematic. If you have access to the databooks for the active devices used in the circuit, the pin-out diagrams and applications notes will

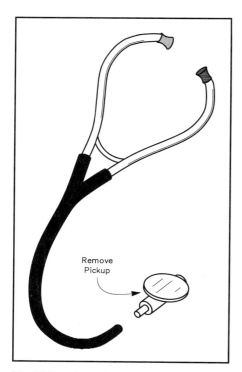

Fig 27.7 — A stethoscope, with the pickup removed, is used to listen for arcing in crowded circuits.

Remove Pickup

sometimes be enough to help you understand and troubleshoot the circuit.

Define Problems

To begin troubleshooting, define the problem accurately. Ask yourself these questions:

1. What functions of the equipment do not work as they should; what does not work at all?
2. What kind of performance can you realistically expect?
3. Has the trouble occurred in the past? (Keep a record of troubles and maintenance in the owner's manual or log book.)

Write the answers to the questions. The information will help with your work, and may help service personnel if their advice or professional service is required.

Take It Apart

All of the preparation work has been done. It is time to really dig in. You usually will have to start by taking the equipment apart. This is the part that can trap the unwary technician. Most experienced service technicians can tell you the tale of the equipment they took apart and were unable to easily put back together. Don't let it happen to you.

Take lots of notes about the way you take it apart. Take notes about each component you remove. Write down the order in which you do things, color codes, part placements, cable routings, hardware notes and anything else you think you might need to be able to reassemble the equipment weeks from now when the back-ordered part comes in.

Put all of the screws in one place. A plastic jar with a lid works well; if you drop it the plastic is not apt to break and the lid will keep all the parts from flying around the work area (you will never find them all). It may pay to have a separate labeled container for each subsystem.

Look Around

Many service problems are visible, if you look for them carefully. Many a technician has spent hours tracking down a failure, only to find a bad solder joint or burned component that would have been spotted in careful inspection of the printed-circuit board. Start troubleshooting by carefully inspecting the equipment.

It is time consuming, but you really need to look at every connector, every wire, every solder joint and every component. A connector may have loosened, resulting in an open circuit. You may spot broken wires or see a bad solder joint. Flexing the printed-circuit board or tugging on components a bit while looking at their solder joints will often locate a defective solder job. Look for scorched components.

Make sure all of the screws securing the printed-circuit board are tight and making good electrical contact. (Do not tighten the adjusting screws, however! You will ruin the alignment.) See if you can find evidence of previous repair jobs; these may not have been done properly. Make sure that each IC is firmly seated in its socket. Look for pins folded underneath the IC rather than making contact with the socket. If you are troubleshooting a newly constructed circuit, make sure each part is of the correct value or type number and is installed correctly.

If your careful inspection doesn't reveal

Look for the Obvious

The best example of how looking for the obvious can save a lot of repair time comes from my days as the manager of an electronics service shop. We had hired a young engineering graduate to work for us part time. He was the proud holder of a First-Class FCC Radiotelephone license (the predecessor to today's General Radiotelephone license). He was a likable sort, but, well . . . the chip on his young shoulder was a bit hard to take sometimes.

One day, I had asked him to repair a "tube-type" FM tuner. He had been poking around without success: hooking up a voltmeter, oscilloscope and signal generator, pretty much in that order. Finally, in total exasperation, he pronounced that the unit was beyond economical repair and suggested that I return it to the customer unfixed. The particular customer was a "regular," so I wanted to be sure of the diagnosis before I sent the tuner back. I told the tech I wanted to take a look at it before we wrote it off.

He started to expound loudly that there was no way that I, a lowly technician (even though I was also his boss) could find a problem that he, an engineering graduate and holder of a First Class . . . you get the idea. I did remind him gently that I was the boss, and he, realizing that I had him there, stepped aside, mumbling something about my suiting myself. He stepped back to gloat when I couldn't find it either.

I began by giving the tuner a thorough visual inspection. I looked it over carefully from stem to stern, while listening to our young apprentice proclaiming with certainty that one cannot fix electronic equipment by merely looking at it. I didn't see anything obviously wrong, so I decided to move wires and components around, looking for a bad solder joint or broken component. Of course, I had to listen to him telling me that one cannot possibly find bad components by touch. Unfortunately for our loud friend, he couldn't have been more wrong.

I grabbed hold of a ceramic bypass capacitor to give it a little wiggle, and much to my surprise it was hot enough to cause some real pain. I kept my composure; it was an opportunity for a good learning experience. Ceramic capacitors don't get very hot unless they are either shorted or very leaky. I kept silent and never let on that my finger "probe" had indeed located the bad part. I set the tuner down, sighed a bit, and then looked him right in the eye when I pointed to the capacitor and said "Change that part!"

They probably heard his bellowing in the next county! He went on and on about how there was just no way in the world that I could tell a good part from a bad part by just looking and touching things. He alternated between accusing me of pulling his leg and guessing, then back to just plain bellowing again. After letting this "source of great noise" run his course, I offered the ultimate shop challenge — I bet him a can of soda pop.

The traditional shop challenge did the trick. He smugly grabbed a replacement part from the bin and got out his soldering iron. In a matter of seconds (a new shop record, I believe) the capacitor was installed. He hooked the tuner up to a test amplifier and turned them on. After a couple of seconds, he smugly turned to me and started an "I told ya' so!" Just then, the last tube warmed up and the sounds of our local rock station blasted out of the speaker. He stopped in mid "toldya" and stared at the tuner in disbelief. The tempo and pitch of his voice jumped by an order of magnitude as he asked me how I managed to fix the tuner without using even an ohmmeter to test a fuse. It was weeks before I told him —the soda pop tasted especially good.

The moral of the story is clear; sophisticated test equipment and procedures are useful in troubleshooting, but they are no substitute for the experience of a veteran troubleshooter. — *Ed Hare, W1RFI, ARRL Laboratory Supervisor*

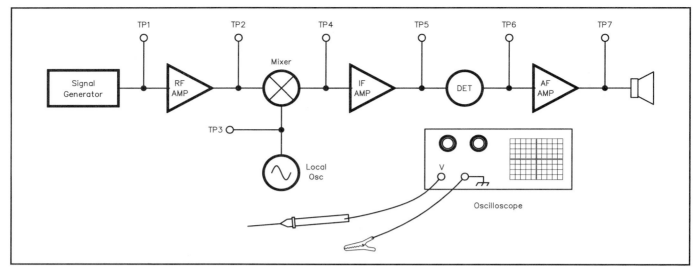

Fig 27.8 — Signal tracing in a simple receiver.

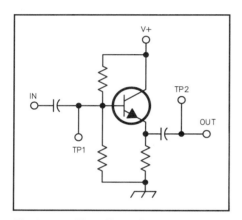

Fig 27.9 — The effect of circuit impedance on an oscilloscope display. Although the circuit functions as a current amplifier, the change in impedance from TP1 to TP2 results in the traces described. This is a common-collector amplifier.

anything, it is time to apply power to the unit under test and continue the process. Observe all safety precautions while troubleshooting equipment. There are voltages inside some equipment that can kill you. If you are not qualified to work safely with the voltages and conditions inside of the equipment, do not proceed. See Table 27.1 and the **Safety** chapter.

Other Senses

With power applied to the unit, listen for arcs and look and smell for smoke. If no problems are apparent, you will have to start testing the various parts of the circuit.

VARIOUS APPROACHES

There are two fundamental approaches to troubleshooting: the systematic approach and the instinctive approach. The systematic approach uses a defined process to analyze and isolate the problem. An instinctive approach relies on troubleshooting experience to guide you in selecting which circuits to test and which tests to perform. The systematic approach is usually chosen by beginning troubleshooters.

At the Block Level

The block diagram is a road map. It shows the signal paths for each circuit function. These paths may run together, cross occasionally or not at all. Those blocks that are not in the paths of faulty functions can be eliminated as suspects. Sometimes the symptoms point to a single block, and no further search is necessary.

In cases where more than one block is suspect, several approaches may be used. Each requires testing a block or stage. Signal injection, signal tracing, instinct or combination of all techniques may be used to diagnose and test electronic equipment.

Systematic Approaches

The instinctive approach works well for those with years of troubleshooting experience. Those of us who are new to this game need some guidance. A systematic approach is a disciplined procedure that allows us to tackle problems in unfamiliar equipment with a reasonable hope of success.

There are two common systematic approaches to troubleshooting at the block level. The first is signal tracing; the second is signal injection. The two techniques

are very similar. Differences in test equipment and the circuit under test determine which method is best in a given situation. They can often be combined.

Power Supplies

You may be able to save quite a bit of time if you test the power supply first. All of the other circuits may be dead if the power supply is not working. Power supply diagnosis is discussed in detail later in this chapter.

Signal Tracing

In signal tracing, start at the beginning of a circuit or system and follow the signal through to the end. When you find the signal at the input to a specific stage, but not at the output, you have located the defective stage. You can then measure voltages and perform other tests on that stage to locate the specific failure. This is much faster than testing every component in the unit to determine which is bad.

It is sometimes possible to use over-the-air signals in signal tracing, in a receiver for example. However, if a good signal generator is available, it is best to use it as the signal source. A modulated signal source is best.

Signal tracing is suitable for most types of troubleshooting of receivers and analog amplifiers. Signal tracing is the best way to check transmitters because all of the necessary signals are present in the transmitter by design. Most signal generators cannot supply the wide range of signal levels required to test a transmitter.

Equipment

A voltmeter, with an RF probe, is the

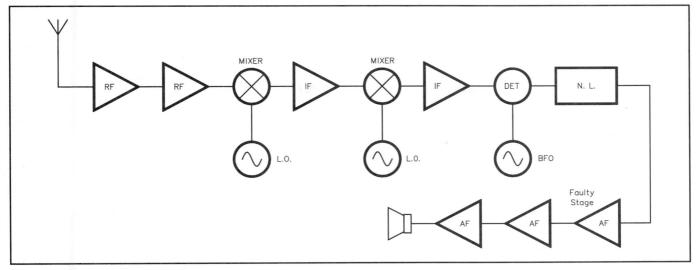

Fig 27.10 — The 14-stage receiver diagnosed by the "divide and conquer" technique.

most common instrument used for signal tracing. Low-level signals cannot be measured accurately with this instrument. Signals that do not exceed the junction drop of the diode in the probe will not register at all, but the presence, or absence, of larger signals can be observed.

A dedicated signal tracer can also be used. It is essentially an audio amplifier. An experienced technician can usually judge the level and distortion of the signal by ear. You cannot use a dedicated signal tracer to follow a signal that is not amplitude modulated (single sideband is a form of AM). Signal tracing is not suitable for tracing CW signals, FM signals or oscillators. To trace these, you will have to use a voltmeter and RF probe or an oscilloscope.

An oscilloscope is the most versatile signal tracer. It offers high input impedance, variable sensitivity, and a constant display of the traced waveform. If the oscilloscope has sufficient bandwidth, RF signals can be observed directly. Alternatively, a demodulator probe can be used to show demodulated RF signals on a low-bandwidth 'scope. Dual-trace scopes can simultaneously display the waveforms, including their phase relationship, present at the input and output of a circuit.

Procedure

First, make sure that the circuit under test and test instruments are isolated from the ac line by transformers. Set the signal source to an appropriate level and frequency for the unit you are testing. For a receiver, a signal of about 100 µV should be plenty. For other circuits, use the schematic, an analysis of circuit function and your own good judgment to set the signal level.

In signal tracing, start at the beginning

and work toward the end of the signal path. Switch on power to the test circuit and connect the signal-source output to the test-circuit input. Place the tracer probe at the circuit input and ensure that you can hear the test signal. Observe the characteristics of the signal if you are using a 'scope (see **Fig 27.8**). Compare the detected signal to the source signal during tracing.

Move the tracer probe to the output of the next stage and observe the signal. Signal level should increase in amplifier stages and may decrease slightly in other stages. The signal will not be present at the output of a "dead" stage.

Low-impedance test points may not provide sufficient signal to drive a high-impedance signal tracer, so tracer sensitivity is important. Also, in some circuits the output level appears low where there is an impedance change from input to output of a stage (see **Fig 27.9**). For example, the circuit in Fig 27.9 is a common-collector current amplifier with a high input impedance and low output impedance. The voltages at TP1 and TP2 are approximately equal and in phase.

There are two signals — the test signal and the local oscillator signal — present in a mixer stage. Loss of either one will result in no output from the mixer stage. Switch the signal source on and off repeatedly to make sure that the tracer reading varies (it need not disappear) with source switching.

Signal Injection

Like signal tracing, signal injection is particularly suited to some situations. Signal injection is a good choice for receiver troubleshooting because the receiver already has a detector as part of the design.

It is suitable for either high- or low-impedance circuits and can be used with vacuum tubes, transistors or ICs.

Equipment

If you are testing equipment that does not include a suitable detector as part of the circuit, some form of signal detector is required. Any of the instruments used for signal tracing are adequate.

Most of the time, your signal injector will be a signal generator. There are other injectors available, some of which are square-wave audio oscillators rich in RF harmonics (see Fig 27.5). These are usually built into a pen-sized case with a test probe at the end. These "pocket" injectors do have their limits because you can't vary their output level or determine their frequency. They are still useful, though, because most circuit failures are caused by a stage that is completely dead.

Consider the signal level at the test point when choosing an instrument. The signal source used for injection must be able to supply appropriate frequencies and levels for each stage to be tested. For example, a typical superheterodyne receiver requires AF, IF and RF signals that vary from 6 V at AF, to 0.2 µV at RF. Each conversion stage used in a receiver requires another IF from the signal source.

Procedure

If an external detector is required, set it to the proper level and connect it to the test circuit. Set the signal source for AF and inject a signal directly into the signal detector to test operation of the injector and detector. Move the signal source to the input of the preceding stage and observe the signal. Continue moving the signal source to the inputs of successive stages.

When you inject the signal source to the input of the defective stage, there will be no output. Prevent stage overload by reducing the level of the injected signal as testing progresses through the circuit. Use suitable frequencies for each tested stage.

Make a rough check of stage gain by injecting a signal at the input and output of an amplifier stage. You can then compare how much louder the signal is when injected at the input. This test may mislead you if there is a radical difference in impedance from stage input to output. Understand the circuit operation before testing.

Mixer stages present a special problem because they have two inputs, rather than one. A lack of output signal from a mixer can be caused by either a faulty mixer or a faulty local oscillator (LO). Check oscillator operation with a 'scope or absorption wavemeter, or by listening on another receiver. If none of these instruments are available, inject the frequency of the LO at the LO output. If a dead oscillator is the only problem, this should restore operation.

If the oscillator is operating, but off frequency, a multitude of spurious responses will appear. A simple signal injector that produces many frequencies simultaneously is not suitable for this test. Use a well-shielded signal generator set to an appropriate level at the LO frequency.

Divide and Conquer

Under certain conditions, the block search may be speeded by testing at the middle of successively smaller circuit sections. Each test limits the fault to one half of the remaining circuit (see **Fig 27.10**). Let's say the receiver has 14 stages and the fault is in stage 12. This approach requires only four tests to locate the faulty stage, a substantial saving of time.

This "divide and conquer" tactic cannot be used in equipment that splits the signal path between the input and the output. Test readings taken inside feedback loops are misleading unless you understand the circuit and the waveform to be expected at each point in the test circuit. It is best to consider all stages within a feedback loop as a single block during the block search.

Both signal tracing and signal injection procedures may be speeded by taking some diagnostic short cuts. Rather than check each stage sequentially, check a point halfway through the system. As an example:

An HF receiver is not working. There is absolutely no response from the speaker. First, substitute a suitable speaker — still no sound. Next, check the power supply — no problem there. No clues indicate any

particular stage. Signal tracing or injection must provide the answer.

Get out the signal generator and switch it on. Set the generator for a low-level RF signal, switch the signal off and connect the output to the receiver. Switch the signal on again and place a high-impedance signal-tracer probe at the antenna connection. Instantly, the tracer emits a strong audio note. Good; the test equipment is functioning.

Move the probe to the input of the receiver detector. As the tracer probe touches the circuit the familiar note sounds. Next, set the tracer for audio and place the probe halfway through the audio chain. It is silent! Move the probe halfway back to the detector, and the note appears once again. Yet, no signal is present at the output of the stage. You now know that the defect is somewhere between the two points tested. In this case, the third audio stage is faulty.

The Instinctive Approach

In an "instinctive" approach to troubleshooting, you rely on your judgment and experience to decide where to start testing, what and how to test. When you immediately check power supply voltages, or the ac fuse on a unit that is completely nonfunctional, that is an example of an instinctive approach. If you are faced with a receiver that has distorted audio and immediately start testing the speaker and audio output stage, or if you immediately start checking the filter and bypass capacitors in an audio stage that is oscillating or "motorboating" you are troubleshooting on instinct.

Most of our discussion on the instinctive approach is really a collection of tips and guidelines. Read them to build your troubleshooting skills.

The check for connector problems mentioned at the beginning of this section is a good idea. Experience has shown connector faults to be so common that they should be checked even before a systematic approach begins.

When instinct is based on experience, searching by instinct may be the fastest procedure. If your instinct is correct, repair time and effort may be reduced substantially. As experience and confidence grow, the merits of the instinctive approach grow with them. However, inexperienced technicians who choose this approach are at the mercy of chance.

TESTING WITHIN A STAGE

Once you have followed all of the troubleshooting procedures and have isolated your problem to a single defective stage or circuit, a few simple measure-

ments and tests will usually pinpoint one or more specific components that need adjustment or replacement.

First, check the parts in the circuit against the schematic diagram to be sure that they are reasonably close to the design values, especially in a newly built circuit. Even in a commercial piece of equipment, someone may have incorrectly changed them during attempted repairs. A wrong-value part is quite likely in new construction, such as a home-brew project.

Voltage Levels

Check the circuit voltages. If the voltage levels are printed on the schematic, this is easy. If not, analyze the circuit and make some calculations to see what the circuit voltages should be. Remember, however, that the printed or calculated voltages are nominal; measured voltages may vary from the calculations.

When making measurements, remember the following points:

- Make measurements at device leads, not at circuit-board traces or socket lugs.
- Use small test probes to prevent accidental shorts.
- Never connect or disconnect power to solid-state circuits with the switch on.
- Consider the effect of the meter on measured voltages. A 20-kΩ/V meter may load down a high-impedance circuit and change the voltage.

Voltages may give you a clue to what is wrong with the circuit. If not, check the active device. If you can check the active device in the circuit, do so. If not, remove it and test it, or substitute a known good device. After connections, most circuit failures are caused directly or indirectly by a bad active device. The experienced troubleshooter usually tests or substitutes these first.

Analyze the other components and determine the best way to test each. There is additional information about electronic components in the electronic-theory chapters and in the **Component Data** chapter.

There are two voltage levels in most circuits (V+ and ground, for example). Most component failures (opens and shorts) will shift dc voltages near one of these levels.

Typical failures that show up as incorrect dc voltages include: open coupling transformers; shorted capacitors; open, shorted or overheated resistors and open or shorted semiconductors.

Noise

A slight hiss is normal in all electronic circuits. This noise is produced whenever current flows through a conductor that is warmer than absolute zero. Noise is compounded and amplified by succeeding stages. Repair is necessary only when noise threatens to obscure normally clear signals.

Semiconductors can produce hiss in two ways. The first is normal — an even white noise that is much quieter than the desired signal. Faulty devices frequently produce excessive noise. The noise from a faulty device is usually erratic, with pops and crashes that are sometimes louder than the desired signal. In an analog circuit, the end result of noise is usually sound. In a control or digital circuit, noise causes erratic operation: unexpected switching and so on.

Noise problems usually increase with temperature, so localized heat may help you find the source. Noise from any component may be sensitive to mechanical vibration. Tapping various components with an insulated screwdriver may quickly isolate a bad part. Noise can also be traced with an oscilloscope or signal tracer.

Nearly any component or connection can be a source of noise. Defective components are the most common cause of crackling noises. Defective connections are a common cause of loud, popping noises.

Check connections at cables, sockets and switches. Look for dirty variable-capacitor wipers and potentiometers. Mica trimmer capacitors often sound like lightning when arcing occurs. Test them by installing a series 0.01-µF capacitor. If the noise disappears, replace the trimmer.

Potentiometers are particularly prone to noise problems when used in dc circuits. Clean them with spray cleaner and rotate the shaft several times.

Rotary switches may be tested by jumpering the contacts with a clip lead. Loose contacts may sometimes be repaired, either by cleaning, carefully rebending the switch contacts or gluing loose switch parts to the switch deck. Operate variable components through their range while observing the noise level at the circuit output.

Oscillations

Oscillations occur whenever there is sufficient positive feedback in a circuit that has gain. (This can even include digital devices.) Oscillation may occur at any frequency from a low-frequency audio buzz (often called "motorboating") well up into the RF region.

Unwanted oscillations are usually the result of changes in the active device (increased junction or interelectrode capacitance), failure of an oscillation suppressing component (open decoupling or

bypass capacitors or neutralizing components) or new feedback paths (improper lead dress or dirt on the chassis or components). It can also be caused by improper design, especially in home-brew circuits. A shift in bias or drive levels may aggravate oscillation problems.

Oscillations that occur in audio stages do not change as the radio is tuned because the operating frequency, and therefore the component impedances, do not change. However, RF and IF oscillations usually vary in amplitude as operating frequency is changed.

Oscillation stops when the positive feedback is removed. Locating and replacing the defective (or missing) bypass capacitor may effect an improvement. The defective oscillating stage can be found more reliably with a signal tracer or oscilloscope.

Amplitude Distortion

Amplitude distortion is the product of nonlinear operation. The resultant waveform contains not only the input signal, but new signals at other frequencies as well. All of the frequencies combine to produce the distorted waveform. Distortion in a transmitter gives rise to splatter, harmonics and interference.

Fig 27.11 shows some typical cases of distortion. Clipping (also called flattopping) is the consequence of excessive drive. The corners on the waveform show that harmonics are present. (A square wave contains the fundamental and all odd harmonics.) These odd harmonics would be heard well away from the operating frequency, possibly outside of amateur bands. Key clicks are similar to clipping.

Harmonic distortion produces radiation at frequencies far removed from the fundamental; it is a major cause of electromagnetic interference (EMI). Harmonics are generated in nearly every amplifier. When they occur in a transmitter, they are usually caused by insufficient transmitter filtering (either by design, or because of filter component failure).

Incorrect bias brings about unequal amplification of the positive and negative wave sections. The resultant waveform is rich in harmonics.

Frequency Distortion

If a "broadband" amplifier, such as an audio amplifier, doesn't amplify all frequencies equally, there is frequency distortion. In many cases, this "frequency distortion" is deliberate, as in a transmitter microphone amplifier that has been designed to pass only frequencies from 200 to 2000 Hz. In most cases, the amateur's ability to detect and measure

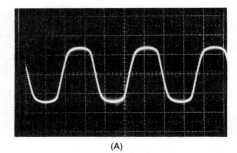

(A)

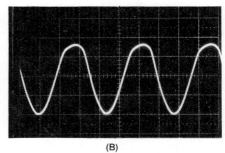

(B)

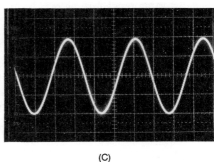

(C)

Fig 27.11 — Examples of distorted waveforms. The result of clipping is shown in A. Nonlinear amplification is shown in B. A pure sine wave is shown in C for comparison.

distortion is limited by available test equipment.

Distortion Measurement

A distortion meter is used to measure distortion of AF signals. A spectrum analyzer is the best piece of test gear to measure distortion of RF signals. If a distortion meter is not available, an estimation of AF distortion can sometimes be made with a function generator (sine and square waves) and an oscilloscope.

To estimate the amount of frequency distortion in an audio amplifier, set the generator for a square wave and look at it on the 'scope. (Use a low-capacitance probe.) The wave should show square corners and a flat top. Next, inject a square wave at the amplifier input and again look at the input wave on the 'scope. Any new distortion is a result of the test circuit loading the generator output. (If the wave shape is severely distorted, the test is not valid.) Now, move the test probe to the test circuit output and look at the waveform. Refer to **Fig 27.12** to evaluate square-wave distortion and its cause.

The above applies only to audio amplifiers without frequency tailoring. In RF gear, the transmitter may have a very narrow audio passband, so inserting a square wave into the microphone input may result in an output that is difficult to interpret. The frequency of the square wave will have a significant effect.

Anything that changes the proper bias of an amplifier can cause distortion. This includes failures in the bias components, leaky transistors or vacuum tubes with interelectrode shorts. These conditions may mimic AGC trouble. Improper bias often results from an overheated or open resistor. Heat can cause resistor values to permanently increase. Leaky, or shorted capacitors and RF feedback can also produce distortion by disturbing bias levels. Distortion is also caused by circuit imbalance in Class AB or B amplifiers.

Oscillations in an IF amplifier may produce distortion. They cause constant, full AGC action, or generate spurious signals that mix with the desired signal. IF oscillations are usually evident on the S meter, which will show a strong signal even with the antenna disconnected.

Alignment

Alignment is rarely the cause of an electronics problem. As an example, suppose an AM receiver suddenly begins producing weak and distorted audio. An inexperienced person frequently suspects poor alignment as a common problem. Even though the manufacturer's instructions and the proper equipment are not available, our "friend" (this would never be one of US!) begins "adjusting" the transformer cores. Before long, the set is hopelessly misaligned. Now our misguided ham must send the radio to a shop for an alignment that was not needed before repairs were attempted.

Alignment does not shift suddenly. A normal signal tracing procedure would have shown that the signal was good up to the audio-output IC, but badly distorted after that. The defective IC that caused the problem would have been easily found and quickly replaced.

Contamination

Contamination is another common service problem. Cold soda pop spilled into a hot piece of electronics is an extreme example (but one that does actually happen).

Conductive contaminants range from water to metal filings. Most can be removed by a thorough cleaning. Any of the residue-free cleaners can be used, but remember that the cleaner may also be conductive. Do not apply power to the circuit until the area is completely dry.

Keep cleaners away from variable-capacitor plates, transformers and parts that may be harmed by the chemical. The most common conductive contaminant is solder, either from a printed-circuit board "solder bridge" or a loose piece of solder deciding to surface at the most inconvenient time.

Solder "Bridges"

In a typical PC-board solder bridge, the solder that is used to solder one component has formed a short circuit to another PC-board trace or component. Unfortunately, they are common in both new construction and repair work. Look carefully for them after you have completed any soldering, especially on a PC-board. It is even possible that a solder bridge may exist in equipment you have owned for a long time, unnoticed until it suddenly decided to become a short circuit.

Related items are loose solder blobs, loose hardware or small pieces of component leads that can show up in the most awkward and troublesome places.

Arcing

Arcing is a serious sign of trouble. It may also be a real fire hazard. Arc sites are usually easy to find because an arc that generates visible light or noticeable sound also pits and discolors conductors.

Arcing is caused by component failure, dampness, dirt or lead dress. If the dampness is temporary, dry the area thoroughly and resume operation. Dirt may be cleaned from the chassis with a residue-free cleaner. Arrange leads so high-voltage conductors are isolated. Keep them away from sharp corners and screw points.

Arcing occurs in capacitors when the working voltage is exceeded. Air-dielectric variable capacitors can sustain occasional arcs without damage, but arcing indicates operation beyond circuit limits. Transmatches working beyond their ability may suffer from arcing. A failure or high SWR in an antenna circuit may also cause transmitter arcing.

Replacing Parts

If you have located a defective component within a stage, you need to replace it. When replacing socket mounted components, be sure to align the replacement part correctly. Make sure that the pins of the device are properly inserted into the socket.

Some special tools can make it easier to remove soldered parts. A chisel-shaped soldering tip helps pry leads from printed-circuit boards or terminals. A desoldering

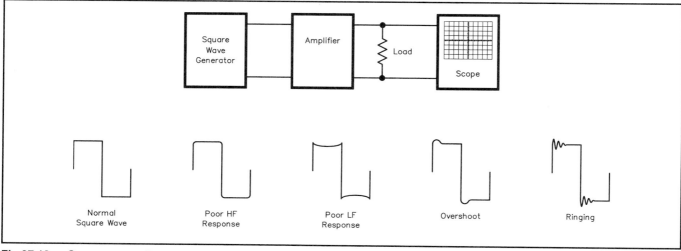

Fig 27.12 — Square-wave distortion and probable causes.

iron or bulb forms a suction to remove excess solder, making it easier to remove the component. Spring-loaded desoldering pumps are more convenient than bulbs. Desoldering wick draws solder away from a joint when pressed against the joint with a hot soldering iron.

In all cases, remember that soldering tools and melted solder can be hot and dangerous! Wear protective goggles and clothing when soldering. A full course in first aid is beyond the scope of this chapter, but if you burn your fingers, run the burn immediately under cold water and seek first aid or medical attention. Always seek medical attention if you burn your eyes; even a small burn can develop into serious trouble.

TYPICAL SYMPTOMS AND FAULTS

Power Supplies

Many equipment failures are caused by power-supply trouble. Fortunately, most power-supply problems are easy to find and repair (see **Fig 27.13**). First, use a voltmeter to measure output. Loss of output voltage is usually caused by an open circuit. (A short circuit draws excessive current that opens the fuse, thus becoming an open circuit.)

Most fuse failures are caused by a shorted diode in the power supply or a shorted power device (RF or AF) in the failed equipment. More rarely, one of the filter capacitors can short. If the fuse has opened, turn off the power, replace the fuse and measure the load-circuit dc resistance. The measured resistance should be consistent with the power-supply ratings. A short or open load circuit indicates a problem.

If the measured resistance is too low, check the load circuit with an ohmmeter to locate the trouble. (Nominal circuit resistances are included in some equipment manuals.) If the load circuit resistance is normal, suspect a defective regulator IC or problem in the rest of the unit. Electrolytic capacitors fail with long (two years) disuse; the electrolytic layer may be reformed as explained later in this chapter.

IC regulators can oscillate, sometimes causing failure. The small-value capacitors on the input, output or adjustment pins of the regulator prevent oscillations. Check or replace these capacitors whenever a regulator has failed.

AC ripple (hum) is usually caused by low-value filter capacitors in the power supply. Less likely, hum can also be caused by excessive load, a regulation problem or RF feedback in the power supply. Look for a defective filter capacitor (usually open or low-value), defective regulator or shorted filter choke. In older equipment, the defective filter capacitor will often have visible leaking electrolyte: Look for corrosion residue at the capacitor leads. In new construction projects make sure RF energy is not getting into the power supply.

Here's an easy filter-capacitor test: Temporarily connect a replacement capacitor (about the same value and working voltage) across the suspect capacitor. If the hum goes away, replace the bad component permanently.

Once the faulty component is found, inspect the surrounding circuit and consider what may have caused the problem. Sometimes one bad component can cause another to fail. For example, a shorted filter capacitor increases current flow and burns out a rectifier diode. While the defective diode is easy to find, the capacitor may show no visible damage.

Switching Power Supplies

Switching power supplies are quite different than conventional supplies. In a "switcher," a switching transistor is used to change dc voltage levels. They usually have AF oscillators and complex feedback paths. Any component failure in the rectifiers, switch, feedback path or load usually results in a completely dead supply. Every part is suspect. While active device failure is still the number one suspect, it pays to carefully test all components if a diagnosis cannot be made with traditional techniques.

Some equipment, notably TVs and monitors, derive some of the power-supply voltages from the proper operation of other parts of the circuit. In the case of a TV or monitor, voltages are often derived by adding secondary low-voltage windings to the flyback transformer and rectifying the resultant ac voltage (usually about 15 kHz). These voltages will be missing if there is any problem with the circuit they are derived from.

Amplifiers

Amplifiers are the most common circuits in electronics. The output of an ideal amplifier would match the input signal in every respect except magnitude: No distortion or noise would be added. Real amplifiers always add noise and distortion.

Gain

Gain is the measure of amplification. Gain is usually expressed in decibels (dB) over a specified frequency range, known as the bandwidth or passband of the ampli-

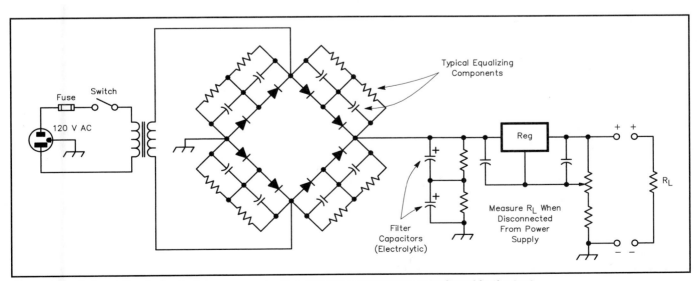

Fig 27.13 — Schematic of a typical power supply showing the components mentioned in the text.

fier. When an amplifier is used to provide a stable load for the preceding stage, or as an impedance transformer, there may be little or no voltage gain.

Amplifier failure usually results in a loss of gain or excessive distortion at the amplifier output. In either case, check external connections first. Is there power to the stage? Has the fuse opened? Check the speaker and leads in audio output stages, the microphone and push-to-talk (PTT) line in transmitter audio sections. Excess voltage, excess current or thermal runaway can cause sudden failure of semiconductors. The failure may appear as either a short, or open, circuit of one or more PN junctions.

Thermal runaway occurs most often in bipolar transistor circuits. If degenerative feedback (the emitter resistor reduces

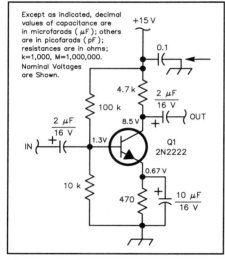

Fig 27.14 — The decoupling capacitor in this circuit is designated with an arrow.

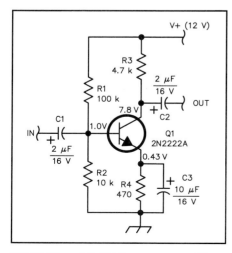

Fig 27.15 — A typical common-emitter audio amplifier.

base-emitter voltage as conduction increases) is insufficient, thermal runaway will allow excessive current flow and device failure. Check transistors by substitution, if possible.

Faulty coupling components can reduce amplifier output. Look for component failures that would increase series, or decrease shunt, impedance in the coupling network. Coupling faults can be located by signal tracing or parts substitution. Other passive component defects reduce amplifier output by shifting bias or causing active-device failure. These failures are evident when the dc operating voltages are measured.

In a receiver, a fault in the AGC loop may force a transistor into cutoff or saturation. Open the AGC line to the device and substitute a variable voltage for the AGC signal. If amplifier action varies with voltage, suspect the AGC-circuit components; otherwise, suspect the amplifier.

In an operating amplifier, check carefully for oscillations or noise. Oscillations are most likely to start with maximum gain and the amplifier input shorted. Any noise that is induced by 60-Hz sources can be heard, or seen with a 'scope synchronized to the ac line.

Unwanted amplifier RF oscillations should be cured with changes of lead dress or circuit components. Separate input leads from output leads; use coaxial cable to carry RF between stages; neutralize interelement or junction capacitance. Ferrite beads on the control element of the active device often stop unwanted oscillations.

Low-frequency oscillations ("motorboating") indicate poor stage isolation or inadequate power-supply filtering. Try a better lead-dress arrangement and/or check the capacitance of the decoupling network (see **Fig 27.14**). Use larger capacitors at the power-supply leads; increase the number of capacitors or use separate decoupling capacitors at each stage. Coupling capacitors that are too low in value can also cause poor low-frequency response. Poor response to high frequencies is usually caused by circuit design.

Amplifiers vs Switches

To help you hone your skills, let's analyze a few simple circuits. There is often a big difference in the performance of similar-looking circuits. Consider the differences between a common-emitter amplifier and a common-emitter switch circuit.

Common-Emitter Amplifier

Fig 27.15 is a schematic of a common-emitter transistor amplifier. The emitter,

base and collector leads are labeled e, b and c, respectively. Important dc voltages are measured at these points and designated V_e, V_b and V_c. Similarly, the important currents are I_e, I_b and I_c. V+ indicates the supply voltage.

First, analyze the voltages and signal levels in this circuit. The "junction drop," is the potential measured across a semiconductor junction that is conducting. It is typically 0.6 V for silicon and 0.2 V for germanium transistors.

This is a Class-A linear circuit. In Class-A circuits, the transistor is always conducting some current. R1 and R2 form a voltage divider that supplies dc bias (V_b) for the transistor. Normally, V_e is equal to V_b less the emitter-base junction drop. R4 provides degenerative dc bias, while C3 provides a low-impedance path for the signal. From this information, normal operating voltages can be estimated.

The bias and voltages will be set up so that the transistor collector voltage, V_c, is somewhere between V+ and ground potential. A good rule of thumb is that V_c should be about 0.5 V+, although this can vary quite a bit, depending on component tolerances. The emitter voltage is usually a small percentage of V_c, say about 10%.

Any circuit failure that changes I_c (ranging from a shorted transistor or a failure in the bias circuit) changes V_c and V_e as well. An increase of I_c lowers V_c and raises V_e. If the transistor shorts from collector to emitter, V_c drops to about 1.2 V, as determined by the voltage divider formed by R3 and R4.

You would see nearly the same effect if the transistor were biased into saturation by collector-to-base leakage, a reduction in R1's value or an increase in R2's value. All of these circuit failures have the same effect. In some cases, a short in C1 or C2 could cause the same symptoms.

To properly diagnose the specific cause of low V_c, consider and test all of these parts. It is even more complex; an increase in R3's value would also decrease V_c. There would be one valuable clue, however; if R3 increased in value, I_c would not increase; V_e would also be low.

Anything that decreases I_c increases V_c. If the transistor failed "open," R1 increased in value, R2 were shorted to ground or R4 opened, then V_c would be high.

Common-Emitter Switch

A common-emitter transistor switching circuit is shown in **Fig 27.16**. This circuit functions differently than the circuit shown in Fig 27.15. A linear amplifier is designed so that the output signal is a faithful reproduction of the input signal. Its input and output may have any value from V+ to ground.

The switching circuit of Fig 27.16, however, is similar to a "digital" circuit. The active device is either on or off, 1 or 0, just like digital logic. Its input signal level should either be 0 V or positive enough to switch the transistor on fully (saturate). Its output state should be either full off (with no current flowing through the relay), or full on (with the relay energized). A voltmeter placed on the collector will show either approximately +12 V or 0 V, depending on the input.

Understanding this difference in operation is crucial to troubleshooting the two circuits. If V_c were +12 V in the circuit in Fig 27.15, it would indicate a circuit failure. A V_c of +12 V in the switching circuit, is normal when V_b is 0 V. (If V_b measured 0.8 V or higher, V_c should be low and the relay energized.)

DC Coupled Amplifiers

In dc coupled amplifiers, the transistors are directly connected together without coupling capacitors. They comprise a unique troubleshooting case. Most often, when one device fails, it destroys one or more other semiconductors in the circuit. If you don't find all of the bad parts, the remaining defective parts can cause the installed replacements to fail immediately. To reliably troubleshoot a dc coupled circuit, you must test every semiconductor in the circuit and replace them all at once.

Oscillators

In many circuits, a failure of the oscillator will result in complete circuit failure. A transmitter will not transmit, and a superheterodyne receiver will not receive if you have an internal oscillator failure. (These symptoms do not always mean oscillator failure, however.)

Whenever there is weakening or complete loss of signal from a radio, check oscillator operation and frequency. There are several methods:

- Use a receiver with a coaxial probe to listen for the oscillator signal.
- A dip meter can be used to check oscillators. In the absorptive mode, tune the dip meter to within ±15 kHz of the oscillator, couple it to the circuit, and listen for a beat note in the dip-meter headphones.
- Look at the oscillator waveform on a 'scope. The operating frequency can't be determined with great accuracy, but you can see if the oscillator is working at all. Use a low capacitance (10×) probe for oscillator observations.
- Tube oscillators usually have negative

grid bias when oscillating. Use a high-impedance voltmeter to measure grid bias. The bias also changes slightly with frequency.

- Emitter current varies slightly with frequency in transistor oscillators. Use a sensitive, high-impedance voltmeter across the emitter resistor to observe the current level. (You can use Ohm's Law to calculate the current value.)

Many modern oscillators are phase-locked loops (PLLs). A PLL is a marriage of an analog oscillator and digital control circuitry. Read the Digital Circuitry section in this chapter and the **Oscillators** chapter of this book in order to learn PLL repair techniques.

To test for a failed oscillator tuned with inductors and capacitors, use a dip meter in the active mode. Set the dip meter to the oscillator frequency and couple it to the oscillator output circuit. If the oscillator is dead, the dip-meter signal will take its place and temporarily restore some semblance of normal operation. Tune the dip meter very slowly, or you may pass stations so quickly that they sound like "birdies."

Stability

We are spoiled; modern amateur equipment is very stable. Drift of several kilohertz per hour was once normal. You may want to modify old equipment for more stability, but drift that is consistent with the equipment design is not a defect. (This applies to new equipment as well as old.) It is normal for some digital displays to flash back and forth between two values for the least-significant digit.

Drift is caused by variations in the oscillator. Poor voltage regulation and heat are the most common culprits. Check regulation with a voltmeter (use one that is not affected by RF). Voltage regulators are usually part of the oscillator circuit. Check them by substitution.

Chirp is a form of rapid drift that is usually caused by excessive oscillator loading or poor power-supply regulation. The most common cause of chirp is poor design. If chirp appears suddenly in a working circuit, look for component or design defects in the oscillator or its buffer amplifiers. (For example, a shorted coupling capacitor increases loading drastically.) Also check lead dress, tubes and switches for new feedback paths (feedback defeats buffer action).

Frequency instability may also result from defects in feedback components. Too much feedback may produce spurious signals, while too little makes oscillator startup unreliable.

Sudden frequency changes are fre-

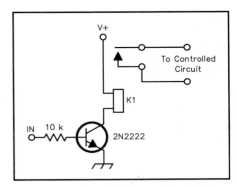

Fig 27.16 — A typical common-emitter switching amplifier.

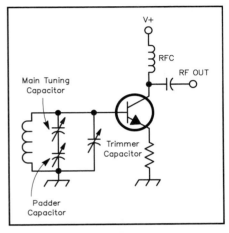

Fig 27.17 — A partial schematic of a simple oscillator showing the locations of the trimmer and padder capacitors.

quently the result of physical variations. Loose components or connections are probable causes. Check for arcing or dirt on printed-circuit boards, trimmers and variable capacitors, loose switch contacts, bad solder joints or loose connectors.

Frequency Accuracy

Dial tracking errors may be associated with oscillator operation. Misadjustments in the frequency-determining components make dial accuracy worse at the ends of the dial. Tracking errors that are constant everywhere in the passband can be caused by misalignment or by slippage in the dial drive mechanism or indicator. This is usually cured by calibration of a simple mechanical adjustment.

In LC oscillators, tracking at the high-frequency end of the dial is controlled by trimmer capacitors. A trimmer is a variable capacitor connected in parallel with the main tuning capacitor (see **Fig 27.17**). The trimmer represents a higher percentage of the total capacitance at the high end of the tuning range. It has relatively little effect on tuning characteristics at the low-frequency end of the dial.

Low-end tracking is adjusted by a padder capacitor. A padder is a variable capacitor that is connected in series with the main tuning capacitor. Padder capacitance has a greater effect at the low-frequency end of the dial. The padder capacitor is often eliminated to save money. In that case, the low-frequency tracking is adjusted by the main tuning coil.

Control Circuitry

Semiconductors have made it practical to use diodes for switching, running only a dc lead to the switching point. This eliminates problems caused by long analog leads in the circuit. Semiconductor switching usually reduces the cost and complexity of switching components. Switching speed is increased; contact corrosion and breakage are eliminated. In exchange, troubleshooting is complicated by additional components such as voltage regulators and decoupling capacitors (see **Fig 27.18**). The technician must consider many more components and symptoms when working with diode and transistor switched circuits.

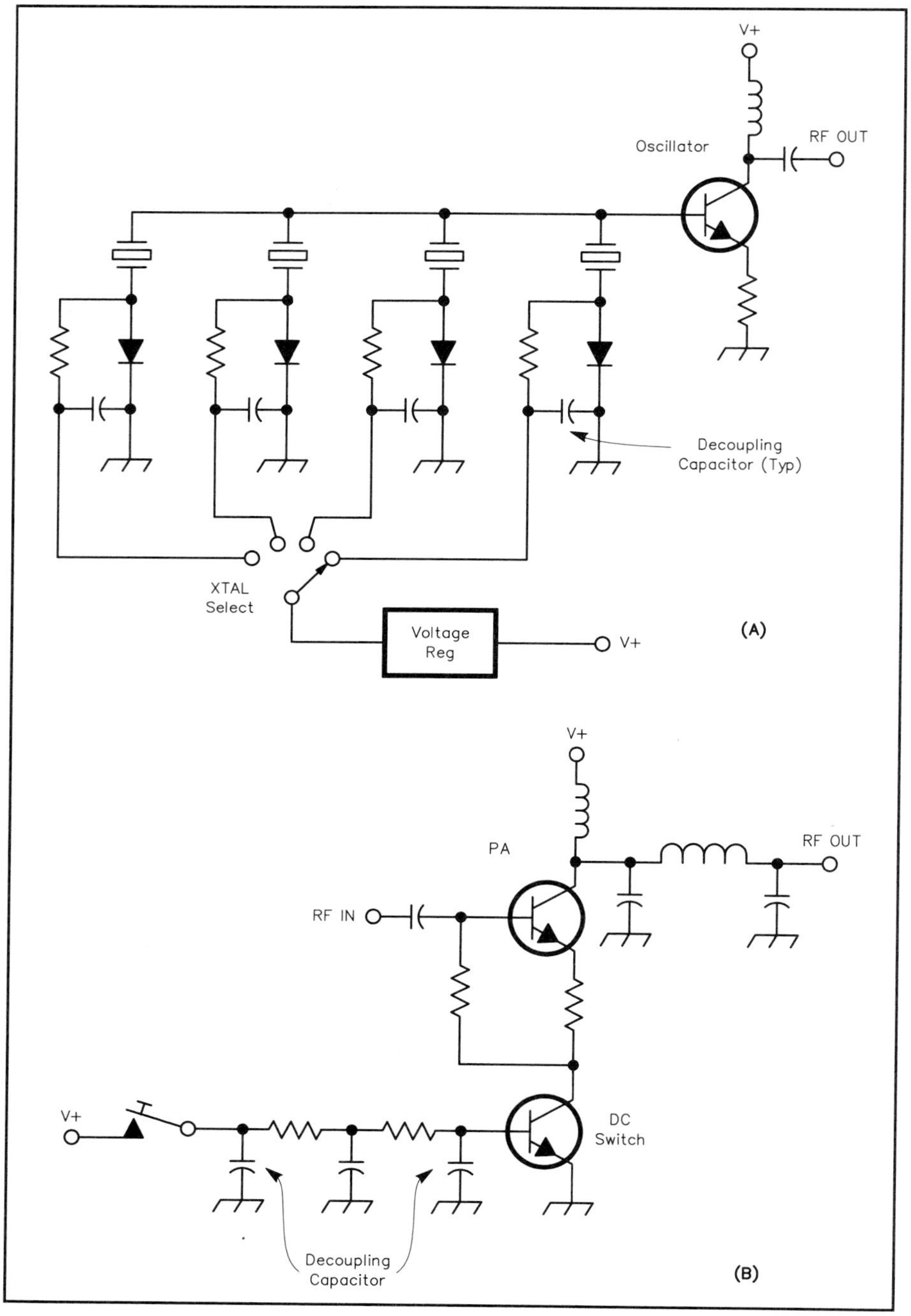

Mechanical switches are relatively rugged. They can withstand substantial voltage and current surges. The environment does not drastically affect them, and there is usually visible damage when they fail. Semiconductor switching offers inexpensive, high-speed operation. When subjected to excess voltage or current, however, most transistors and diodes silently expire. Occasionally, if the troubleshooter is lucky, one sends up a smoke signal to mark its passing.

Temperature changes semiconductor characteristics. A normally adequate control signal may not be effective when transistor beta is lowered by a cold environment. Heat may cause a control voltage regulator to produce an improper control signal.

A control signal is actually a bias for the semiconductor switch. Forward biased diodes and transistors act as closed switches; reverse biased components simulate open switches. If the control (bias) signal is not strong enough to completely saturate the semiconductor, conduction may not continue through a full ac cycle. Severe distortion can be the result.

When dc control leads provide unwanted feedback paths, switching transistors may become modulators or mixers. Additionally, any reverse biased semiconductor junction is a potential source of white noise.

Microprocessor Control

Nearly every new transceiver is controlled by a miniature computer. Entire books have been written about microprocessor (µP) control. Many of the techniques are discussed in the Digital Circuitry section. The **Digital** chapter of this book will also help you troubleshoot a µP problem. Many such problems end up back at the factory for service, however; the surface mounted components are just too difficult for most hams to replace. For successful repair of microprocessor controlled circuits, you should have the knowledge and test equipment necessary for computer repair. Familiarity with machine-language programming may also be desirable.

Fig 27.18 — Diode switching selects oscillator crystals at A. A transistor switch is used to key a power amplifier at B.

Digital Circuitry

The digital revolution has hit most ham shacks and amateur equipment. Microprocessors have brought automation to everything from desk clocks to ham transceivers and computer controlled EME antenna arrays. Although every aspect of their operation may be resolved to a simple 1 or 0, or tristate (an infinite impedance or open circuit), the symptoms of their failure are far more complicated. As with other equipment:

- Observe the operating characteristics.
- Study the block diagram and the schematic.
- Test.
- Replace defective parts.

Problems in digital circuits have two elementary causes. First, the circuit may give false counts because of electrical noise at the input. Second, the gates may lock in one state.

False counts from noise are especially likely in a ham shack. (A 15- to 20-μs voltage spike can trigger a TTL flip-flop.) Amateur Radio equipment often switches heavy loads; the attendant transients can follow the ac line or radiate directly to nearby digital equipment. Oscillation in the digital circuit can also produce false counts.

How these false counts affect a circuit is dependent on the design. A station clock may run fast, but a microprocessor controlled transceiver may "decide" that it is only a receiver. It might even be difficult to determine that there is a problem without a logic analyzer or a multitrace oscilloscope and a thorough understanding of circuit operation.

Begin by removing the suspect equipment from RF fields. If the symptoms stop when there is no RF energy around, you need to shield the equipment from RF.

In the mid '90s, microprocessors in general use clock speeds up to a few hundred megahertz. (They are increasing all the time.) It may be impossible to filter RF signals from the lines when the RF is near the clock frequency. In these cases, the best approach is to shield the digital circuit and all lines running to it.

If digital circuitry interferes with other nearby equipment, it may be radiating spurious signals. These signals can interfere with your Amateur Radio operation or other services. Digital circuitry can also be subject to interference from strong RF fields. Erratic operation or a complete "lock up" is often the result. *The ARRL RFI Book* has a chapter on computer and digital interference. That chapter discusses interference to and from digital devices and circuits.

Logic Levels

To troubleshoot a digital circuit, check for the correct voltages at the pins of each chip. The correct voltages may not always be known, but you should be able to identify the power pins (V_{cc} and ground).

The voltages on the other pins should be either a logic high, a logic low, or tristate (more on this later). In most working digital circuitry the logic levels are constantly changing, often at RF rates. A dc voltmeter may not give reliable readings. An oscilloscope or logic analyzer is usually needed to troubleshoot digital circuitry.

Most digital circuit failures are caused by a failed logic IC. In clocked circuits, listen for the clock signal with a coax probe and a suitable receiver. If the signal is found at the clock chip, trace it to each of the other ICs to be sure that the clock system is intact. Some digital circuits use VHF clock speeds; an oscilloscope must have a bandwidth of at least twice the clock speed to be useful. If you have a suitable scope, check the pulse timing and duration against circuit specifications.

As in most circuits, failures are catastrophic. It is unlikely that an AND gate will suddenly start functioning like an OR gate. It is more likely that the gate will have a signal at its input, and no signal at the output. In a failed device, the output pin will have a steady voltage. In some cases, the voltage is steady because one of the input signals is missing. Look carefully at what is going into a digital IC to determine what should be coming out. Keep manufacturers' data books handy. These data books describe the proper functioning of most digital devices.

Tristate Devices

Many digital devices are designed with a third logic state, commonly called tristate. In this state, the output of the device acts as if it weren't there at all. Many such devices can be connected to a common "bus," with the devices that are active at any given time selected by software or hardware control signals. A computer's data and address busses are good examples of this. If any one device on the bus fails by locking itself on in a 0 or 1 logic state, the entire bus becomes nonfunctional. These tristate devices can be locked "on" by inherent failure or a failure of the signal that controls them.

Simple Gate Tests

Logic gates, flip-flops and counters can be tested (see **Fig 27.19**) by triggering them manually, with a power supply (4 to 5 V is a safe level). Diodes may be checked with an ohmmeter. Testing of more complicated ICs requires the use of a logic analyzer, multitrace scope or a dedicated IC tester.

TROUBLESHOOTING HINTS

Receivers

A receiver can be diagnosed using any of the methods described earlier, but if there is not even a faint sound from the speaker, signal injection is not a good technique. If you lack troubleshooting experience, avoid following instinctive hunches. That leaves signal tracing as the best method.

The important characteristics of a receiver are selectivity, sensitivity, stability and fidelity. Receiver malfunctions ordinarily affect one or more of these areas.

Selectivity

Tuned transformers or the components used in filter circuits may develop a shorted turn, capacitors can fail and alignment is required occasionally. Such defects are accompanied by a loss of sensitivity. Except in cases of catastrophic failure (where either the filter passes all signals, or none), it is difficult to spot a loss of selectivity. Bandwidth and insertion-loss measurements are necessary to judge filter performance.

Sensitivity

A gradual loss of sensitivity results from gradual degradation of an active device or long-term changes in component values. Sudden partial sensitivity changes are usually the result of a component failure, usually in the RF or IF stages. Complete and sudden loss of sensitivity is caused by an open circuit anywhere in the signal path or by a "dead" oscillator.

Receiver Stability

The stability of a receiver depends on its oscillators. See the Oscillators section elsewhere in this chapter.

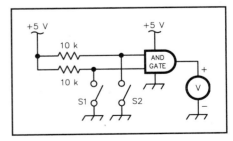

Fig 27.19 — This simple digital circuit can be tested with a few components. In this case, an AND gate is tested. Open and close S1 and S2 while comparing the voltmeter reading with a truth table for the device.

Distortion

Receiver distortion may be the effect of poor connections or faulty components in the signal path. AGC circuits produce many receiver defects that appear as distortion or insensitivity.

AGC

AGC failure usually causes distortion that affects only strong signals. All stages operate at maximum gain when the AGC influence is removed. An S meter can help diagnose AGC failure because it is operated by the AGC loop.

An open AGC bypass capacitor causes feedback through the loop. This often results in a receiver "squeal" (oscillation). Changes in the loop time constant affect tuning. If stations consistently blast, or are too weak for a brief time when first tuned in, the time constant is too fast. An excessively slow time constant makes tuning difficult, and stations fade after tuning. If the AGC is functioning, but the "timing" seems wrong, check the large-value capacitors found in the AGC circuit — they usually set the AGC time constants. If the AGC is not functioning, check the AGC-detector circuit. There is often an AGC voltage that is used to control several stages. A failure in any one stage could affect the entire loop.

Detector Problems

Detector trouble usually appears as complete loss or distortion of the received signal. AM, SSB and CW signals may be weak and unintelligible. FM signals will sound distorted. Look for an open circuit in the detector near the detector diodes. If tests of the detector parts indicate no trouble, look for a poor connection in the power-supply or ground lead. A BFO that is "dead" or off frequency prevents SSB and CW reception. In modern rigs, the BFO frequency is either crystal controlled, or derived from the PLL.

Receiver Alignment

Unfortunately, IF transformers are as enticing to the neophyte technician as a carburetor is to a shade-tree mechanic. In truth, radio alignment (and for that matter, carburetor repair) is seldom required. Circuit alignment may be justified under the following conditions:
- The set is very old and has not been adjusted in many years.
- The circuit has been subject to abusive treatment or environment.
- There is obvious misalignment from a previous repair.
- Tuned-circuit components or crystals have been replaced.
- An inexperienced technician attempted alignment without proper equipment. ("But all the screws in those little metal cans were loose!")
- There is a malfunction, but all other circuit conditions are normal. (Faulty transformers can be located because they will not tune.)

Even if one of the above conditions is met, do not attempt alignment unless you have the proper equipment. Receiver alignment should progress from the detector to the antenna terminals. When working on an FM receiver, align the detector first, then the IF and limiter stages and finally the RF amplifier and local oscillator stages. For an AM receiver, align the IF stages first, then the RF amplifier and oscillator stages.

Both AM and FM receivers can be aligned in much the same manner. Always follow the manufacturer's recommended alignment procedure. If one is not available, follow these guidelines:

1. Set the receiver RF gain to maximum, BFO control to zero or center (if applicable to your receiver) and tune to the high end of the receiver passband.

2. Disable the AGC.

3. Set the signal source to the center of the IF passband, with no modulation and minimum signal level.

4. Connect the signal source to the input of the IF section.

5. Connect a voltmeter to the IF output as shown in **Fig 27.20**.

6. Adjust the signal-source level for a slight indication on the voltmeter.

7. Peak each IF transformer in order, from the meter to the signal source. The adjustments interact; repeat steps 6 and 7

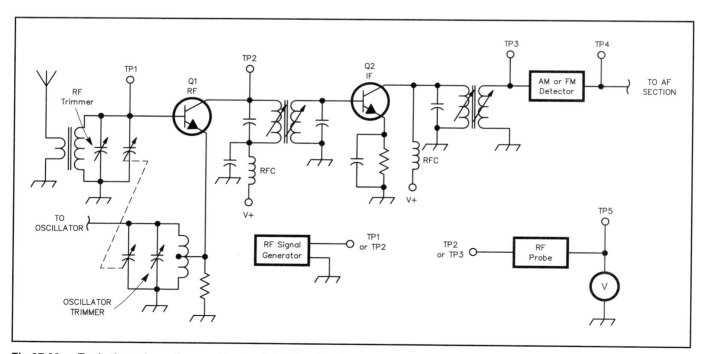

Fig 27.20 — Typical receiver alignment test points. To align the entire radio, connect a dc voltmeter at TP4. Inject an IF signal at TP2 and adjust the IF transformers. Move the signal generator to TP1 and inject an RF signal for alignment of the RF amplifier and oscillator stages. To align a single stage, place the generator at the input and an RF voltmeter (or demodulator probe and dc voltmeter) at the output: TP1/TP2 for RF, TP2/TP3 for IF.

until adjustment brings no noticeable improvement.

8. Remove the signal source from the IF-section input, reduce the level to minimum, set the frequency to that shown on the receiver dial and connect the source to the antenna terminals. If necessary, tune around for the signal — if the local oscillator is not tracking, it may be off.

9. Adjust the signal level to give a slight reading on the voltmeter.

10. Adjust the trimmer capacitor of the RF amplifier for a peak reading of the test signal. (Verify that you are reading the correct signal by switching the source on and off.)

11. Reset the signal source and the receiver tuning for the low end of the passband.

12. Adjust the local-oscillator padder for peak reading.

13. Steps 8 through 11 interact, so repeat them until the results are as good as you can get them.

Transmitters

Many potential transmitter faults are discussed in several different places in this chapter. There are, however, a few techniques used to ensure stable operation of RF amplifiers in transmitters that are not covered elsewhere.

High-power RF amplifiers often use parasitic chokes to prevent instability. Older parasitic chokes usually consist of a 51- to 100-Ω noninductive resistor with a coil wound around the body and connected to the leads. It is used to prevent VHF and UHF oscillations in a vacuum-tube amplifier. The suppressor is placed in the plate lead, close to the plate connection.

In recent years, problems with this style of suppressor have been discovered. Look at the **Amplifiers** chapter for information about suppressing parasitics.

Parasitic chokes often fail from excessive current flow. In these cases, the resistor is charred. Occasionally, physical shock or corrosion produces an open circuit in the coil. Test for continuity with an ohmmeter.

Transistor amplifiers are protected against parasitic oscillations by low-value resistors or ferrite beads in the base or collector leads. Resistors are used only at low power levels (about 0.5 W), and both methods work best when applied to the base lead. Negative feedback is used to prevent oscillations at lower frequencies. An open component in the feedback loop may cause low-frequency oscillation, especially in broadband amplifiers.

Keying

The simplest form of modulation is on/off keying. Although it may seem that there cannot be much trouble with such an elementary form of modulation, two very important transmitter faults are the result of keying problems.

Key clicks are produced by fast rise and decay times of the keying waveform. Most transmitters include components in the keying circuitry to prevent clicks. When clicks are experienced, check the keying filter components first, then the succeeding stages. An improperly biased power amplifier, or a Class C amplifier that is not keyed, may produce key clicks even though the keying waveform earlier in the circuit is correct. Clicks caused by a linear amplifier may be a sign of low-frequency parasitic oscillations. If they occur in an amplifier, suspect insufficient power-supply decoupling. Check the power-supply filter capacitors and all bypass capacitors.

The other modulation problem associated with on/off keying is called backwave. Backwave is a condition in which the signal is heard, at a reduced level, even when the key is up. This occurs when the oscillator signal feeds through a keyed amplifier. This usually indicates a design flaw, although in some cases a component failure or improper keyed-stage neutralization may be to blame.

Low Output Power

Some transmitters automatically reduce power in the TUNE mode. Check the owner's manual to see if the condition is normal. Check the control settings. Transmitters that use broadband amplifiers require so little effort from the operator that control settings are seldom noticed. The CARRIER (or DRIVE) control may have been bumped. Remember to adjust tuned amplifiers after a significant change in operating frequency (usually 50 to 100 kHz). Most modern transmitters are designed to reduce power if there is high (say 2:1) SWR. Check these obvious external problems before you tear apart your rig.

Power transistors may fail if the SWR protection circuit malfunctions. Such failures occur at the "weak link" in the amplifier chain: It is possible for the drivers to fail without damaging the finals. An open circuit in the "reflected" side of the sensing circuit leaves the transistors unprotected, a short "shuts them down."

Low power output in a transmitter may also spring from a misadjusted carrier oscillator or a defective SWR protection circuit. If the carrier oscillator is set to a frequency well outside the transmitter passband, there may be no measurable output. Output power will increase steadily as the frequency is moved into the passband.

Transceivers

Switching

Elaborate switching schemes are used in transceivers for signal control. Many transceiver malfunctions can be attributed to relay or switching problems. Suspect the switching controls when:

- The S meter is inoperative, but the unit otherwise functions. (This could also be a bad S meter.)
- There is arcing in the tank circuit. (This could also be caused by a bad antenna system.)
- Plate current is high during reception.
- There is excessive broadband PA noise in the receiver.

Since transceiver circuits are shared, stage defects frequently affect both the transmit and receive modes, although the symptoms may change with mode. Oscillator problems usually affect both transmit and receive modes, but different oscillators, or frequencies, may be used for different emissions. Check the block diagram.

For example, one particular transceiver uses a single carrier oscillator with three different crystals (see **Fig 27.21**). One crystal sets the carrier frequency for CW, AM and FSK transmit. Another sets USB transmit and USB/CW receive, and a third sets LSB transmit and LSB/FSK receive. This radio showed a strange symptom. After several hours of CW operation, the receiver produced only a light hiss on USB and CW. Reception was good in other modes, and the power meter showed full output during CW transmission. An examination of the block diagram and schematic showed that only one of the crystals (and seven support components) was capable of causing the problem.

VOX

Voice operated transmit (VOX) controls are another potential trouble area. If there is difficulty in switching to transmit in the VOX mode, check the VOX-SENSITIVITY and ANTI-VOX control settings. Next, see if the PTT and manual (MOX) transmitter controls work. If the PTT and MOX controls function, examine the VOX control diodes and amplifiers. Test the switches, control lines and control voltage if the transmitter does not respond to other TR controls.

VOX SENSITIVITY and ANTI-VOX settings should also be checked if the transmitter switches on in response to received audio. Suspect the ANTI-VOX circuitry next.

Unacceptable VOX timing results from a poor VOX-delay adjustment, or a bad resistor or capacitor in the timing circuit or VOX amplifiers.

Alignment

The mixing scheme of the modern SSB transceiver is complicated. The signal passes through many mixers, oscillators and filters. Satisfactory SSB communication requires accurate adjustment of each stage. Do not attempt any alignment without a copy of the manufacturer's instructions and the necessary test equipment.

Troubleshooting Charts

Tables 27.2 through 27.5 list some common problems and possible cures. These tables are not all-inclusive. They are a collection of hints and shortcuts that may save you some troubleshooting time. If you don't find your problem listed, continue with systematic troubleshooting.

COMPONENTS

Once you locate a defective part, it is time to select a replacement. This is not always an easy task. Each electronic component has a function. This section acquaints you with the functions, failure modes and test procedures of resistors, capacitors, inductors and other components. Test the components implicated by symptoms and stage-level testing. In most cases, a particular faulty component will be located by these tests. If a faulty component is not indicated, check the circuit adjustments. As a last resort, use a shotgun approach — replace all parts in the problem area with components that are known to be good.

Check the Circuit

Before you install a replacement component of any type, you should be sure that another circuit defect didn't cause the failure. Check the circuit voltages carefully before installing any new component. Check the potential on each trace to the bad component. The old part may have "died" as a result of a lethal voltage. Measure twice — repair once! (With apologies to the old carpenter.) Of course, circuit performance is the final test of any substitution.

Fuses

Most of the time, when a fuse fails, it is for a reason — usually a short circuit in the load. A fuse that has failed because of a short circuit usually shows the evidence of high current: a blackened interior with little blobs of fuse element everywhere. Fuses can also fail by fracturing the element at either end. This kind of failure is not visible by looking at the fuse. Check even "good" fuses with an ohmmeter. You may save hours of troubleshooting.

For safety reasons, always use *exact* replacement fuses. Check the current and voltage ratings. The fuse timing (fast, normal or slow blow) must be the same as the original. Never attempt to force a fuse that is not the right size into a fuse holder. The substitution of a bar, wire or penny for a fuse invites a "smoke party."

Wires

Wires seldom fail unless abused. Short circuits can be caused by physical damage to insulation or by conductive contamination. Damaged insulation is usually apparent during a close visual inspection of the conductor or connector. Look carefully

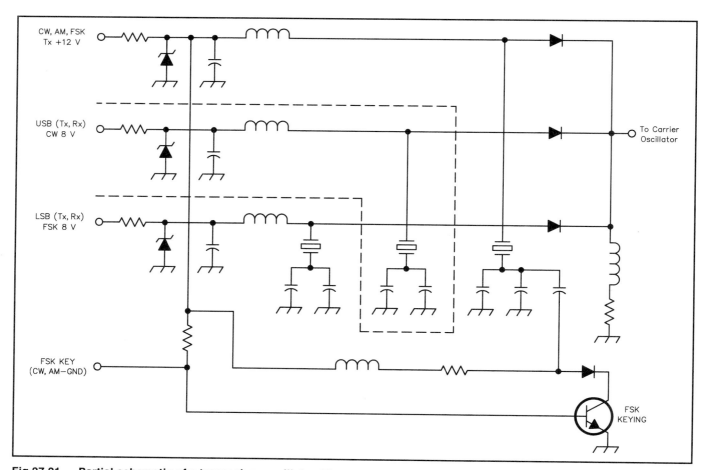

Fig 27.21 — Partial schematic of a transceiver oscillator. The symptoms described in the text are caused by one or more components inside the dashed lines or a faulty USB/CW control signal.

where conductors come close to corners or sharp objects. Repair worn insulation by replacing the wire or securing an insulating sleeve (spaghetti) or heat-shrink tubing over the worn area.

When wires fail, the failure is usually caused by stress and flexing. Nearly everyone has broken a wire by bending it back and forth, and broken wires are usually easy to detect. Look for sharp bends or bulges in the insulation.

When replacing conductors, use the same material and size, if possible. Substitute only wire of greater cross-sectional area (smaller gauge number) or material of greater conductivity. Insulated wire should be rated at the same, or higher, temperature and voltage as the wire it replaces.

Connectors

Connection faults are one of the most common failures in electronic equipment. This can range from something as simple as the ac-line cord coming out of the wall, to a connector having been put on the wrong socket, to a defective IC socket. Connectors that are plugged and unplugged frequently can wear out, becoming intermittent or noisy. Check connectors carefully when troubleshooting.

Connector failure can be hard to detect. Most connectors maintain contact as a result of spring tension that forces two conductors together. As the parts age, they become brittle and lose tension. Any connection may deteriorate because of nonconductive corrosion at the contacts. Solder helps prevent this problem but even soldered joints suffer from corrosion when exposed to weather.

The dissipated power in a defective connector usually increases. Signs of excess heat are sometimes seen near poor connections in circuits that carry moderate current. Check for short and open circuits with an ohmmeter or continuity tester. Clean those connections that fail as a result of contamination.

Occasionally, corroded connectors may be repaired by cleaning, but replacement of the conductor/connector is usually required. Solder all connections that may be subject to harsh environments and protect them with acrylic enamel, RTV compound or a similar coating.

Choose replacement connectors with consideration of voltage and current ratings. Use connectors with symmetrical pin arrangements only where correct insertion will not result in a safety hazard or circuit damage.

Resistors

Resistors usually fail by becoming an open circuit. More rarely they change value. This is usually caused by excess heat. Such heat may come from external sources or from power dissipated within the resistor. Sufficient heat burns the re-

Table 27.2

Symptoms and Their Causes for All Electronic Equipment

Symptom	Cause
Power Supplies	
No output voltage	Open circuit (usually a fuse or transformer winding)
Hum or ripple	Faulty regulator, capacitor or rectifier, low-frequency oscillation
Amplifiers	
Low gain	Transistor, coupling capacitors, emitter-bypass capacitor, AGC component, alignment
Noise	Transistors, coupling capacitors, resistors
Oscillations	Dirt on variable capacitor or chassis, shorted op-amp input
Untuned (oscillations do not change with frequency)	Audio stages
Tuned	RF, IF and mixer stages
Squeal	Open AGC-bypass capacitor
Static-like crashes	Arcing trimmer capacitors, poor connections
Static in FM receiver	Faulty limiter stage, open capacitor in ratio detector, weak RF stage, weak incoming signal
Intermittent noise	All components and connections, band-switch contacts, potentiometers (especially in dc circuits), trimmer capacitors, poor antenna connections
Distortion (constant)	Oscillation, overload, faulty AGC, leaky transistor, open lead in tab-mount transistor, dirty potentiometer, leaky coupling capacitor, open bypass capacitors, imbalance in tuned FM detector, IF oscillations, RF feedback (cables)
Distortion (strong signals only)	Open AGC line, open AGC diode
Frequency change	Physical or electrical variations, dirty or faulty variable capacitor, broken switch, loose compartment parts, poor voltage regulation, oscillator tuning (trouble when switching bands)
No Signals	
All bands	Dead VFO or heterodyne oscillator, PLL won't lock
One band only	Defective crystal, oscillator out of tune, band switch
No function control	Faulty switch, poor connection, defective switching diode or circuit
Improper Dial Tracking	
Constant error across dial	Dial drive
Error grows worse along dial	Circuit adjustment

Table 27.3

Transmitter Problems

Symptom	Cause
Key clicks	Keying filter, distortion in stages after keying
Modulation Problems	
Loss of modulation	Broken cable (microphone, PTT, power), open circuit in audio chain, defective modulator
Distortion on transmit	Defective microphone, RF feedback from lead dress, modulator imbalance, bypass capacitor, improper bias, excessive drive
Arcing	Dampness, dirt, improper lead dress
Low output	Incorrect control settings, improper carrier shift (CW signal outside of passband) audio oscillator failure, transistor or tube failure, SWR protection circuit
Antenna Problems	
Poor SWR	Damaged antenna element, matching network, feed line, balun failure (see below), resonant conductor near antenna, poor connection at antenna
Balun failure	Excessive SWR, weather or cold-flow damage in coil choke, broken wire
RFI	Arcing or poor connections anywhere in antenna system or nearby conductors

sistor until it becomes an open circuit.

Resistors can also fracture and become an open circuit as a result of physical shock. Contamination of a high-value resistor (100 kΩ or more) can cause a change in value through leakage. This contamination can occur on the resistor body, mounts or printed-circuit board. Resistors that have changed value should be replaced. Leakage is cured by cleaning the resistor body and surrounding area.

In addition to the problems of fixed-value resistors, potentiometers and rheostats can develop noise problems, especially in dc circuits. Dirt often causes intermittent contact between the wiper and resistive element. To cure the problem, spray electronic contact cleaner into the control, through holes in the case, and rotate the shaft a few times.

The resistive element in wire-wound potentiometers eventually wears and breaks from the sliding action of the wiper. In this case, the control needs to be replaced.

Replacement resistors should be of the same value, tolerance, type and power rating as the original. The value should stay within tolerance. Replacement resis-tors may be of a different type than the original, if the characteristics of the replacement are consistent with circuit requirements.

Substitute resistors can usually have a greater power rating than the original, except in high-power emitter circuits where the resistor also acts as a fuse or in cases where the larger size presents a problem.

Variable resistors should be replaced with the same kind (carbon or wire wound) and taper (linear, log, reverse log and so on) as the original. Keep the same, or better tolerance and pay attention to the power rating.

In all cases, mount high-temperature resistors away from heat-sensitive components. Keep carbon resistors away from heat sources. This will extend their life and ensure minimum resistance variations.

Capacitors

Capacitors usually fail by shorting, opening or becoming electrically (or physically) leaky. They rarely change value. Capacitor failure is usually caused by excess current, voltage, temperature or age. Leakage can be external to the ca-pacitor (contamination on the capacitor body or circuit) or internal to the capacitor.

Tests

The easiest way to test capacitors is out of circuit with an ohmmeter. In this test, the resistance of the meter forms a timing circuit with the capacitor to be checked. Capacitors from 0.01 μF to a few hundred μF can be tested with common ohmmeters. Set the meter to its highest range and connect the test leads across the discharged capacitor. When the leads are connected, current begins to flow. The capacitor passes current easily when discharged, but less easily as the charge builds. This shows on the meter as a low resistance that builds, over time, toward infinity.

The speed of the resistance build-up corresponds to capacitance. Small capacitance values approach infinite resistance almost instantly. A 0.01-μF capacitor checked with an 11-MΩ FETVOM would increase from zero to a two-thirds scale reading in 0.11 s, while a 1-μF unit would require 11 s to reach the same reading. If the tested capacitor does not reach infinity within five times the period taken to reach the two-thirds point, it has excess leakage. If the meter reads infinite resistance immediately, the capacitor is open. (Aluminum electrolytics normally exhibit high-leakage readings.)

Fig 27.22 shows a circuit that may be used to test capacitors. To use this circuit, make sure that the power supply is off, set S1 to CHARGE and S2 to TEST, then connect the capacitor to the circuit. Switch on the power supply and allow the capacitor to charge until the voltmeter reading stabilizes. Next, switch S1 to TEST and watch the meter for a few seconds. If the capacitor is good, the meter will show no potential. Any appreciable voltage indicates excess leakage. After testing, set S1 to CHARGE, switch off the power supply, and press the DISCHARGE button until the meter shows 0 V, then remove the capacitor from the test circuit.

Capacitance can also be measured with a capacitance meter, an RX bridge or a dip meter. Some DMMs (digital multimeters) measure capacitance. Capacitance measurements made with DMMs and dedicated capacitance meters are much more accurate than those made with RX bridges or dip meters. To determine capacitance with a dip meter, a parallel-resonant circuit should be constructed using the capacitor of unknown value and an inductor of known value. The formula for resonance is discussed in the **AC Theory** chapter of this book.

Table 27.4
Receiver Problems

Symptom	Cause
Low sensitivity	Semiconductor contamination, weak tube, alignment
Signals and calibrator heard weakly	
(low S-meter readings)	RF chain
(strong S-meter readings)	AF chain, detector
No signals or calibrator heard, only hissing	RF oscillators
Distortion	
On strong signals only	AGC fault
AGC fault	Active device cut off or saturated
Difficult tuning	AGC fault
Inability to receive	Detector fault
AM weak and distorted	Poor detector, power or ground connection
CW/SSB unintelligible	BFO off frequency or dead
FM distorted	Open detector diode

Table 27.5
Transceiver Problems

Symptom	Cause
Inoperative S meter	Faulty relay
PA noise in receiver	
Excessive current on receive	
Arcing in PA	
Reduced signal strength on transmit and receive	IF failure
Poor VOX operation	VOX amplifiers and diodes
Poor VOX timing	Adjustment, component failure in VOX timing circuits or amplifiers
VOX consistently tripped by receiver audio	AntiVOX circuits or adjustment

It is best to keep a collection of known components that have been measured on accurate L or C meters. Alternatively, a "standard" value can be obtained by ordering 1 or 2% components from an electronics supplier. A 10%-tolerance component can be used as a standard; however, the results will only be known to within 10%. The accuracy of tests made with any of these alternatives depends on the accuracy of the "standard" value component. Further information on this technique appears in Bartlett's article, "Calculating Component Values," in Nov 1978 *QST*.

Cleaning

The only variety of common capacitor that can be repaired is the air-dielectric variable capacitor. Electrical connection to the moving plates is made through a spring-wiper arrangement (see **Fig 27.23**). Dirt normally builds on the contact area, and they need occasional cleaning. Before cleaning the wiper/contact, use gentle air pressure and a soft brush to remove all dust and dirt from the capacitor plates. Apply some electronic contact cleaning fluid. Rotate the shaft quickly several times to work in the fluid and establish contact. Use the cleaning fluid sparingly, and keep it off the plates except at the contact point.

Replacements

Replacement capacitors should match the original in value, tolerance, dielectric, working voltage and temperature coefficient. Use only ac-rated capacitors for line service. If exact replacements are not available, substitutes may vary from the original part in the following respects: Bypass capacitors may vary from one to three times the capacitance of the original. Coupling capacitors may vary from one half to twice the value of the original. Capacitance values in tuned circuits (especially filters) must be exact. (Even then, any replacement will probably require circuit realignment.)

If the same kind of capacitor is not available, use one with better dielectric characteristics. Do not substitute polarized capacitors for nonpolarized parts. Capacitors with a higher working voltage may be used, although the capacitance of an electrolytic capacitor used significantly below its working voltage will usually increase with time.

The characteristics of each type of capacitor are discussed in the **Real World Components** chapter. Consider these characteristics if you're not using an exact replacement capacitor.

Inductors and Transformers

The most common inductor or trans-

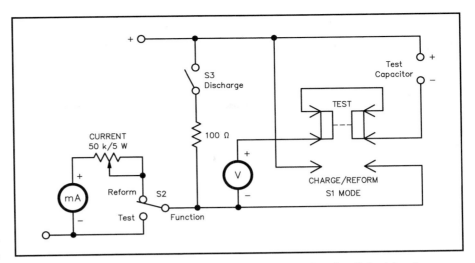

Fig 27.22 — A fixture for testing capacitors and reforming the dielectric of electrolytic capacitors. Use 12 V for testing the capacitor. Use the capacitor working voltage for dielectric reformation.

former failure is a broken conductor. More rarely, a short circuit can occur across one or more turns of a coil. In an inductor, this changes the value. In a transformer, the turns ratio and resultant output voltage changes. In high-power circuits, excessive inductor current can generate enough heat to melt plastics used as coil forms.

Inductors may be checked for open circuit failure with an ohmmeter. In a good inductor, dc resistance rarely exceeds a few ohms. Shorted turn and other changes in inductance show only during alignment or inductance measurement.

The procedure for measurement of inductance with a dip meter is the same as that given for capacitance measurement, except that a capacitor of known value is used in the resonant circuit.

Replacement inductors must have the same inductance as the original, but that is only the first requirement. They must also carry the same current, withstand the same voltage and present nearly the same Q as the original part. Given the original as a pattern, the amateur can duplicate these qualities for many inductors. Note that inductors with ferrite or iron-powder cores are frequency sensitive, so the replacement must have the same core material.

If the coil is of simple construction, with the form and core undamaged, carefully count and write down the number of turns and their placement on the form. Also note how the coil leads are arranged and connected to the circuit. Then determine the wire size and insulation used. Wire diameter, insulation and turn spacing are critical to the current and voltage ratings of an inductor. (There is little hope of matching coil characteristics unless the wire is duplicated exactly in the new part.) Next,

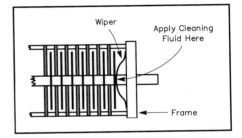

Fig 27.23 — Partial view of an air-dielectric variable capacitor. If the capacitor is noisy or erratic in operation, apply electronic cleaning fluid where the wiper contacts the rotor plates.

remove the old winding (be careful not to damage the form) and apply a new winding in its place. Be sure to dress all coil leads and connections in exactly the same manner as the original. Apply Q dope to hold the finished winding in place.

Follow the same procedure in cases where the form or core is damaged, except that a suitable replacement form or core (same dimensions and permeability) must be found.

Ready-made inductors may be used as replacements if the characteristics of the original and the replacement are known and compatible. Unfortunately, many inductors are poorly marked. If so, some comparisons, measurements and circuit analysis are usually necessary.

When selecting a replacement inductor, you can usually eliminate parts that bear no physical resemblance to the original part. This may seem odd, but the Q of an inductor depends on its physical dimensions and the permeability of the core material. Inductors of the same value, but of vastly different size or shape, will likely have a great difference in Q. The Q of the new inductor can be checked by installing it in the circuit, aligning the stage and performing the manufacturer's passband tests. Although this practice is all right in a pinch, it does not yield an accurate Q measurement. Methods to measure Q appear in the **Test Procedures** chapter.

Once the replacement inductor is found, install it in the circuit. Duplicate the placement, orientation and wiring of the original. Ground-lead length and arrangement should not be changed. Isolation and magnetic shielding can be improved by replacing solenoid inductors with toroids. If you do, however, it is likely that many circuit adjustments will be needed to compensate for reduced coupling and mutual inductance. Alignment is usually required whenever a tuned-circuit component is replaced.

A transformer consists of two inductors that are magnetically coupled. Transformers are used to change voltage and current levels (this changes impedance also). Failure usually occurs as an open circuit or short circuit of one or more windings.

Amateur testing of power transformers is limited to ohmmeter tests for open circuits and voltmeter checks of secondary voltage. Make sure that the power-line voltage is correct, then check the secondary voltage against that specified. There should be less than 10% difference between open-circuit and full-load secondary voltage.

Replacement transformers must match the original in voltage, volt-ampere (VA), duty cycle and operating-frequency ratings. They must also be compatible in size. (All transformer windings should be insulated for the full power-supply voltage.)

Relays

Although relays have been replaced by semiconductor switching in low-power circuits, they are still used extensively in high-power Amateur Radio equipment. Relay action may become sluggish. AC relays can buzz (with adjustment becoming impossible). A binding armature or weak springs can cause intermittent switching. Excessive use or hot switching ruins contacts and shortens relay life.

You can test relays with a voltmeter by jumpering across contacts with a test lead (power on, in circuit) or with an ohmmeter (out of circuit). Look for erratic readings across the contacts, open or short circuits at contacts or an open circuit at the coil.

Most failures of simple relays can be repaired by a thorough cleaning. Clean the contacts and mechanical parts with a residue-free cleaner. Keep it away from the coil and plastic parts that may be damaged. Dry the contacts with lint-free paper, such as a business card; then burnish them with a smooth steel blade. Do not use a file to clean contacts.

Replacement relays should match or exceed the original specifications for voltage, current, switching time and stray impedance (impedance is significant in RF circuits only). Many relays used in transceivers are specially made for the manufacturer. Substitutes may not be available from any other source.

Before replacing a multicontact relay, make a drawing of the relay, its position, the leads and their routings through the surrounding parts. This drawing allows you to complete the installation properly, even if you are distracted in the middle of the operation.

Semiconductors

Diodes

The primary function of a diode is to pass current in one direction only. They can be easily tested with an ohmmeter.

Signal or switching diodes — The most common diode in electronics equipment, they are used to convert ac to dc, to detect RF signals or to take the place of relays to switch ac or dc signals within a circuit. Signal diodes usually fail open, although shorted diodes are not rare. They can easily be tested with an ohmmeter.

Power-rectifier diodes — Most equipment contains a power supply, so power-rectifier diodes are the second-most common diodes in electronic circuitry. They usually fail shorted, blowing the power-supply fuse.

Other diodes — Zener diodes are made with a predictable reverse-breakdown voltage and used as voltage regulators. Varactor diodes are specially made for use as voltage controlled variable capacitors. (Any semiconductor diode may be used as a voltage-variable capacitance, but the value will not be as predictable as that of a varactor.) A Diac is a special-purpose diode that passes only pulses of current in each direction.

Diode tests — There are several basic tests for most diodes. First, is it a diode? Does it conduct in one direction and block current flow in the other? An ohmmeter is suitable for this test in most cases. An ohmmeter will read high resistance in one direction, low resistance in the other. Make sure the meter uses a voltage of more than 0.7 V and less than 1.5 V to measure resistance. Use a good diode to determine the meter polarity.

Diodes should be tested out of circuit. Disconnect one lead of the diode from the circuit, then measure the forward and reverse resistance. Diode quality is shown by the ratio of reverse to forward resistance. A ratio of 100:1 or greater is common for signal diodes. The ratio may go as low as 10:1 for old power diodes.

The first test is a forward-resistance test. Set the meter to read ×100 and connect the test probes across the diode. When the negative terminal of the ohmmeter battery is connected to the cathode, the meter will typically show about 200 to 300 Ω (forward resistance) for a good silicon diode, 200 to 400 Ω for a good germanium diode. The exact value varies quite a bit from one meter to the next.

Next, test the reverse resistance. Reverse the lead polarity and set the meter to ×1M (times one million, or the highest scale available on the meter) to measure diode reverse resistance. Good diodes should show 100 to 1000 MΩ for silicon and 100 kΩ to 1 MΩ for germanium. When you are done, mark the meter lead polarity for future reference.

This procedure measures the junction resistances at low voltage. It is not useful to test Zener diodes. A good Zener diode will not conduct in the reverse direction at voltages below its rating.

We can also test diodes by measuring the voltage drop across the diode junction while the diode is conducting. (A test circuit is shown in **Fig 27.24**.) To test, connect the diode, adjust the supply voltage until the current through the diode matches the manufacturer's specification and compare the junction drop to that specified. Silicon junctions usually show about 0.6 V, while germanium is typically 0.2 V. Junction voltage-drop increases with cur-

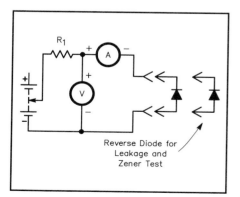

Fig 27.24 — A diode conduction, leakage and Zener-point test fixture. The ammeter should read mA for conduction and Zener point, μA for leakage tests.

rent flow. This test can be used to match diodes with respect to forward resistance at a given current level.

A final simple diode test measures leakage current. Place the diode in the circuit described above, but with reverse polarity. Set the specified reverse voltage and read the leakage current on a milliammeter. (The currents and voltages measured in the junction voltage-drop and leakage tests vary by several orders of magnitude.)

The most important specification of a Zener diode is the Zener (or avalanche) voltage. The Zener-voltage test also uses the circuit of Fig 27.24. Connect the diode in reverse. Set the voltage to minimum, then gradually increase it. You should read low current in the reverse mode, until the Zener point is reached. Once the device begins to conduct in the reverse direction, the current should increase dramatically. The voltage shown on the voltmeter is the Zener point of the diode. If a Zener diode has become leaky, it might show in the leakage-current measurement, but substitution is the only dependable test.

Replacement diodes — When a diode fails, check associated components as well. Replacement rectifier diodes should have the same current and peak inverse voltage (PIV) as the original. Series diode combinations are often used in high-voltage rectifiers, with resistor and capacitor networks to distribute the voltage equally among the diodes.

Switching diodes may be replaced with diodes that have equal or greater current ratings and a PIV greater than twice the peak-to-peak voltage encountered in the circuit. Switching time requirements are not critical except in RF, logic and some keying circuits. Logic circuits may require exact replacements to assure compatible switching speeds and load characteristics. RF switching diodes used near resonant circuits must have exact replacements as the diode resistance and capacitance will affect the tuned circuit.

Voltage, current and capacitance characteristics must be considered when replacing varactor diodes. Once again, exact replacements are best. Zener diodes should be replaced with parts having the same Zener voltage and equal or better current, power, impedance and tolerance specifications. Check the associated current-limiting resistor when replacing a Zener diode.

Bipolar Transistors

Transistors are primarily used to switch or amplify signals. Transistor failures occur as an open junction, a shorted junction, excess leakage or a change in amplification performance.

Most transistor failure is catastrophic. A transistor that has no leakage and amplifies at dc or audio frequencies will usually perform well over its design range. For this reason, transistor tests need not be performed at the planned operating frequency. Tests are made at dc or a low frequency (usually 1000 Hz). The circuit under repair is the best test of a potential replacement part. Swapping in a replacement transistor in a failed circuit will often result in a cure.

A simple and reliable bipolar-transistor test can be performed with the transistor in a circuit and the power on. It requires a test lead, a 10-kΩ resistor and a voltmeter. Connect the voltmeter across the emitter/collector leads and read the voltage. Then use the test lead to connect the base and emitter (**Fig 27.25A**). Under these conditions, conduction of a good transistor will be cut off and the meter should show nearly the entire supply voltage across the emitter/collector leads. Next, remove the clip lead and connect the 10-kΩ resistor from the base to the collector. This should bias the transistor into conduction and the emitter/collector voltage should drop (Fig 27.25B). (This test indicates transistor response to changes in bias voltage.)

Transistors can be tested (out of circuit) with an ohmmeter in the same manner as diodes. Look up the device characteristics before testing and consider the consequences of the ohmmeter-transistor circuit. Limit junction current to 1 to 5 mA for small-signal transistors. Transistor destruction or inaccurate measurements may result from careless testing.

Use the $\times 100$ Ω and $\times 1000$-Ω ranges for small-signal transistors. For high-power transistors use the $\times 1$ Ω and $\times 10$-Ω ranges. The reverse-to-forward resistance ratio for good transistors may vary from 30:1 to better than 1000:1.

Germanium transistors sometimes show high leakage when tested with an ohmmeter. Bipolar transistor leakage may be specified from the collector to the base, emitter to base or emitter to collector (with the junction reverse biased in all cases). The specification may be identified as I_{cbo}, I_{bo}, collector cutoff current or collector leakage for the base-collector junction, I_{ebo}, and so on for other junctions.[3] Leak-

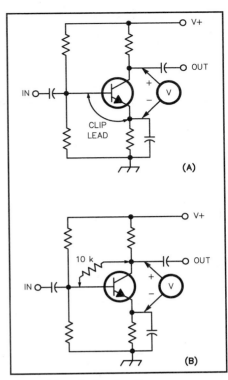

Fig 27.25 — An in-circuit semiconductor test with a clip lead, resistor and voltmeter. The meter should read V+ at (A). During test (B) the meter should show a decrease in voltage, ranging from a slight variation down to a few millivolts. It will typically cut the voltage to about half of its initial value.

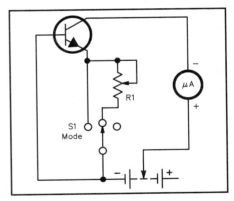

Fig 27.26 — A test circuit for measuring collector-base leakage with the emitter shorted to ground, open or connected to ground through a variable resistance, depending on the setting of S1. See the transistor manufacturer's instructions for test conditions and the setting of R1 (if used). Reverse battery polarity for PNP transistors.

[3]The term "I_{cbo}" means "Current from collector to base with emitter open." The subscript notation indicates the status of the three device terminals. The terminals measured are listed first, with the remaining terminal listed as "s" (shorted) or "o" (open).

age current increases with junction temperature.

A suitable test fixture for base-collector leakage measurements is shown in **Fig 27.26.** Make the required connections and set the voltage as stated in the transistor specifications and compare the measured leakage current with that specified. Small-signal germanium transistors exhibit I_{cbo}

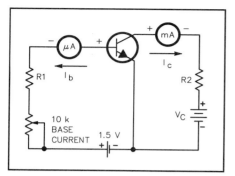

Fig 27.27 — A test circuit for measuring transistor beta. Values for R1 and R2 are dependent on the current range of the transistor tested. Reverse the battery polarity for PNP transistors.

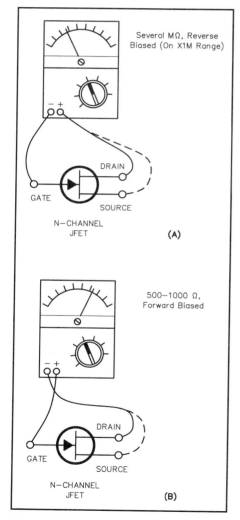

Fig 27.28 — Ohmmeter tests of a JFET. The junction is reverse biased at A and forward biased at B.

and I_{ebo} leakage currents of about 15 μA. Leakage increases to 90 μA or more in high-power components. Leakage currents for silicon transistors are seldom more than 1 μA. Leakage current tends to double for every 10°C increase above 25°C.

Breakdown-voltage tests actually measure leakage at a specified voltage, rather than true breakdown voltage. Breakdown voltage is known as BV_{cbo}, BV_{ces} or BV_{ceo}. Use the same test fixture shown for leakage tests, adjust the power supply until the specified leakage current flows, and compare the junction voltage against that specified.

A circuit to measure dc current gain is shown in **Fig 27.27**. Transistor gain can range from 10 to over 1000 because it is not usually well controlled during manufacture. Gain of the active device is not critical in a well-designed transistor circuit.

The test conditions for transistor testing are specified by the manufacturer. When testing, do not exceed the voltage, current (especially in the base circuit) or dissipated-power rating of the transistor. Make sure that the load resistor is capable of dissipating the power generated in the test.

While these simple test circuits will identify most transistor problems, RF devices should be tested at RF. Most component manufacturers include a test-circuit schematic on the data sheet. The test circuit is usually an RF amplifier that operates near the high end of the device frequency range.

Semiconductor failure is sometimes the result of environmental conditions. Open junctions, excess leakage (except with germanium transistors) and changes in amplification performance result from overload or excessive current. Electrostatic discharge can destroy a semiconductor in microseconds. Shorted junctions are caused by voltage spikes. Check surrounding parts for the cause of the transistor's demise, and correct the problem before installing a replacement.

JFETs

Junction FETs can be tested with an ohmmeter in much the same way as bipolar transistors (see text and **Fig 27.28**). Reverse leakage should be several megohms or more. Forward resistance should be 500 to 1000 Ω.

MOSFETs

MOS (metal-oxide semiconductor) layers are extremely fragile. Normal body static is enough to damage them. Even "gate protected" (a diode is placed across the MOS layer to clamp voltage)

MOSFETs may be destroyed by a few volts of static electricity.

Make sure the power is off, capacitors discharged and the leads of a MOSFET are shorted together before installing or removing it from a circuit. Use a voltmeter to be sure the chassis is near ground potential, then touch the chassis before and during MOSFET installation and removal. This assures that there is no difference of potential between your body, the chassis and the MOSFET leads. Ground the soldering-iron tip with a clip lead when soldering MOS devices. The FET source should be the first lead connected to and the last disconnected from a circuit. The insulating layers in MOSFETs prevent testing with an ohmmeter. Substitution is the only practical means for amateur testing of MOSFETs.

FET Considerations

Replacement FETs should be of the same kind as the original part: JFET or MOSFET, P-channel or N-channel, enhancement or depletion. Consider the breakdown voltage required by the circuit. The breakdown voltage should be at least two to four times the power-supply and signal voltages in amplifiers. Allow for transients of ten times the line voltage in power supplies. Breakdown voltages are usually specified as $V_{(BR)GSS}$ or $V_{(BR)GDO}$.

The gate-voltage specification gives the gate voltage required to cut off or initiate channel current (depending on the mode of operation). Gate voltages are usually listed as $V_{GS(OFF)}$, V_p(pinch off), V_{TH} (threshold) or $I_D(ON)$ or I_{TH}.

Dual-gate MOSFET characteristics are more complicated because of the interaction of the two gates. Cutoff voltage, breakdown voltage and gate leakage are the important traits of each gate.

Integrated Circuits

The basics of integrated circuits are covered in earlier chapters of this book. Amateurs seldom have the sophisticated equipment required to test ICs. Even a multitrace 'scope can view only their simplest functions. We must be content to check every other possible cause, and only then assume that the problem lies with an IC. Experienced troubleshooters will tell you that — most of the time anyway — if a defective circuit uses an IC, it is the IC that is bad.

Linear ICs — There are two major classes of ICs: linear and digital. Linear ICs are best replaced with identical units. Original equipment manufacturers are the best source of a replacement; they are the only source with a reason to stockpile obsolete or custom-made items. If substitution of an IC is unavoidable, first try the cross-reference guides published by several distributors. You can also look in manufacturers' databooks and compare pinouts and other specifications.

Digital ICs — It is usually not a good idea to substitute digital devices. While it may be okay to substitute an AB74LS00YZ from manufacturer "A" with a CD74LS00WX from a different manufacturer, you will usually not be able to replace an LS (low-power Schottky) device with an S (Schottky), C (CMOS) or any of a number of other families. The different families all have different speed, current-consumption, input and output characteristics. You would have to analyze the circuit to determine if you could substitute one type for another. The characteristics of various digital families are discussed in the **Digital** chapter.

Semiconductor Substitution

In all cases try to obtain exact replacement semiconductors. Specifications vary slightly from one manufacturer to the next. Cross-reference equivalents are useful, but not infallible. Before using an equivalent, check the specifications against those for the original part. When choosing a replacement, consider:

- Is it silicon or germanium?
- Is it a PNP or an NPN?
- What are the operating frequency and input/output capacitance?
- How much power does it dissipate (often less than $V_{max} \times I_{max}$)?
- Will it fit the original mount?
- Are there unusual circuit demands (low noise and so on)?
- What is the frequency of operation?

Remember that cross-reference equivalents are not guaranteed to work in every application. There may be cases where two dissimilar devices have the same part number, so it pays to compare the listed replacement specifications with the intended use. If "the book" says to use a diode in place of an RF transistor, it isn't going to work! Derate power specifications, as recommended by the manufacturer, for high-temperature operation.

Tubes

The most common tube failures in amateur service are caused by cathode depletion and gas contamination. Whenever a tube is operated, the coating on the cathode loses some of its ability to produce electrons. It is time to replace the tube when electron production (cathode current, I_c) falls to 50 - 60% of that exhibited by a new tube.

Gas contamination in a tube can often

be identified easily because there may be a greenish or whitish-purple glow between the elements during operation. (A faint deep-purple glow is normal in most tubes.) The gas reduces tube resistance and leads to runaway plate current evidenced by a red glow from the anode, interelectrode arcing or a blown power-supply fuse. Less common tube failures include an open filament, broken envelope and interelectrode shorts.

The best test of a tube is to substitute a new one. Another alternative is a tube tester; these are now rare. You can also do some limited tests with an ohmmeter. Tube tests should be made out of circuit so circuit resistance does not confuse the results:

Use an ohmmeter to check for an open filament (remove the tube from the circuit first). A broken envelope is visually obvious, although a cracked envelope may appear as a gassy tube. Interelectrode shorts are evident during voltage checks on the operating stage. Any two elements that show the same voltage are probably shorted. (Remember that some interelectrode shorts, such as the cathode-suppressor grid, are normal.)

Generally, a tube may be replaced with another that has the same type number. Compare the data sheets of similar tubes to assess their compatibility. Consider the base configuration and pinout, interelectrode capacitances (a small variation is okay except for tubes in oscillator service), dissipated power ratings of the plate and screen grid and current limitations (both peak and average). For example, the 6146A may be replaced with a 6146B (heavy duty), but not vice versa.

In some cases, minor type-number differences signify differences in filament voltages, or even base styles, so check all specifications before making a replacement. (Even tubes of the same model number, prefix and suffix vary slightly, in some respects, from one supplier to the next.)

AFTER THE REPAIRS

Once you have completed your troubleshooting and repairs, it is time to put the equipment back together. Take a little extra time to make sure you have done everything correctly.

All Units

Give the entire unit a complete visual inspection. Look for any loose ends left over from your troubleshooting procedures — you may have left a few components temporarily soldered in place or overlooked some other repair error. Look for cold solder joints and signs of damage

incurred during the repair. Double check the position, leads and polarity of components that were removed or replaced.

Make sure that all ICs are properly oriented in their sockets and all of the pins are properly inserted in the IC socket or printed-circuit board holes. Test fuse continuity with an ohmmeter and verify that the current rating matches the circuit specification.

Look at the position of all of the wires and components. Make sure that wires and cables will be clear of hot components, screw points and other sharp edges. Make certain that the wires and components will not be in the way when covers are installed and the unit is put back together.

Separate the leads that carry dc, RF, input and output as much as possible. Plug-in circuit boards should be firmly seated with screws tightened and lock washers installed if so specified. Shields and ground straps should be installed just as they were on the original.

For Transmitters Only

Since the signal produced by an HF transmitter can be heard the world over, a thorough check is necessary after any service has been performed. Do not exceed the transmitter duty cycle while testing. Limit transmissions to 10 to 20 seconds unless otherwise specified by the owner's manual.

1. Set all controls as specified in the operation manual, or at midscale.

2. Connect a dummy load and a power meter to the transmitter output.

3. Set the drive or carrier control for low output.

4. Switch the power on.

5. Transmit and quickly set the final-amplifier bias to specifications.

6. In narrowband equipment, slowly tune the output network through resonance. The current dip should be smooth and repeatable. It should occur simultaneously with the maximum power output. Any sudden jumps or wiggles of the current meter indicate that the amplifier is unstable. Adjust the neutralization circuit (according to the manufacturer's instructions) if one is present or check for oscillation. An amplifier usually requires neutralization whenever active devices, components or lead dress (that affect the output/input capacitance) are changed.

7. Check to see that the output power is consistent with the amplifier class used in the PA (efficiency should be about 25% for Class A, 50 to 60% for Class AB or B, and 70 to 75% for Class C).

8. Repeat steps 4 through 6 for each band of operation from lowest to highest frequency.

9. Check the carrier balance (in SSB transmitters only) and adjust for minimum power output with maximum RF drive and no microphone gain.

10. Adjust the VOX controls.

11. Measure the passband and distortion levels if equipment (wideband 'scope or spectrum analyzer) is available.

Other Repaired Circuits

After the preliminary checks, set the circuit controls per the manufacturer's specifications (or to midrange if specifications are not available) and switch the power on. Watch and smell for smoke, and listen for odd sounds such as arcing or hum. Operate the circuit for a few minutes, consistent with allowable duty cycle. Verify that all operating controls function properly.

Check for intermittent connections by subjecting the circuit to heat, cold and slight flexure. Also, tap or jiggle the chassis lightly with an alignment tool or other insulator.

If the equipment is meant for mobile or portable service, operate it through an appropriate temperature range. Many mobile radios do not work on cold mornings, or on hot afternoons, because a temperature-dependent intermittent was not found during repairs.

Button It Up

After you are convinced that you have repaired the circuit properly, put it all back together. If you followed the advice in this book, you have all the screws and assorted doodads in a secure container. Look at the notes you took while taking it apart; put it back together in the reverse order. Don't forget to reconnect all internal connections, such as ac-power, speaker or antenna leads.

Once the case is closed, and all appears well, don't neglect the final, important step — make sure it still works. Many an experienced technician has forgotten this important step, only to discover that some

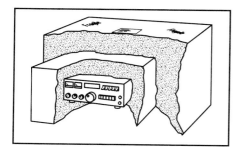

Fig 27.29 — Ship equipment packed securely in a box within a box.

minor error, such as a forgotten antenna connector, has left the equipment non-functional.

PROFESSIONAL REPAIRS

This chapter does not tell you how to perform all repairs. Repairs that deal with very complex and temperamental circuits, or that require sophisticated test equipment, should be passed on to a professional.

The factory authorized service personnel have a lot of experience. What seems like a servicing nightmare to you is old hat to them. There is no one better qualified to service your equipment than the factory.

If the manufacturer is no longer in business, check with your local dealer or look in the classified ads in electronics and Amateur Radio magazines. You can usually find one or more companies that service "all makes and models." Your local TV shop might be willing to tackle a repair, especially if you have located a schematic.

If you are going to ship your equipment somewhere for repair, notify the repair center first. Get authorization for shipping and an identification name or number for the package.

Packing It Up

You can always blame shipping damage on the shipper, but it is a lot easier for all concerned if you package your equipment properly for shipping in the first place. Firmly secure all heavy components, either by tying them down or blocking them off with shipping foam. Large vacuum tubes should be wrapped in packing material or shipped separately. Make sure that all circuit boards and parts are firmly attached.

Use a box within a box for shipping. (See **Fig 27.29**.) Place the equipment and some packing material inside a box and seal it with tape. Place that box inside another that is at least six inches larger in each dimension. Fill the gap with packing material, seal, address and mark the outer box. Choose a good freight carrier and insure the package.

Don't forget to enclose a statement of the trouble, a short history of operation and any test results that may help the service technician. Include a good description of the things you have tried. Be honest! At current repair rates you want to tell the technician everything to help ensure an efficient repair.

Even if you ended up sending it back to the factory, you can feel good about your experience. You learned a lot by trying,

and you have sent it back knowing that it really did require the services of a "pro." Each time you troubleshoot and repair a piece of electronic circuitry, you learn something new. The down side is that you may develop a reputation as a real electronics whiz. You may find yourself spending a lot of time at club meetings offering advice, or getting invited over to a lot of shacks for a late-evening pizza snack. There are worse fates.

References

J. Bartlett, "Calculating Component Values," *QST*, Nov 1978.

J.Carr, *How to Troubleshoot and Repair Amateur Radio Equipment*, Blue Ridge Summit, PA: TAB Books Inc, 1980.

D. DeMaw, "Understanding Coils and Measuring their Inductance," *QST*, Oct 1983.

H. Gibson, *Test Equipment for the Radio Amateur*, London, England: Radio Society of Great Britain, 1974.

C. Gilmore, *Understanding and Using Modern Electronic Servicing Test Equipment*, TAB Books, Inc, 1976.

F. Glass, *Owner Repair of Amateur Radio Equipment*, Los Gatos, CA: RQ Service Center, 1978.

R. Goodman, *Practical Troubleshooting with the Modern Oscilloscope*, TAB Books, Inc, 1979.

A. Haas, *Oscilloscope Techniques*, New York: Gernsback Library, Inc, 1958.

C. Hallmark, *Understanding and Using the Oscilloscope*, TAB Books, Inc, 1973.

A. Helfrick, *Amateur Radio Equipment Fundamentals*, Englewood Cliffs, NJ: Prentice-Hall Inc, 1982.

K. Henney, and C. Walsh, *Electronic Components Handbook*, New York: McGraw-Hill Book Company, 1957.

L. Klein, and K. Gilmore, *It's Easy to Use Electronic Test Equipment*, New York: John R. Rider Publisher, Inc (A division of Hayden Publishing Company), 1962.

J. Lenk, *Handbook of Electronic Test Procedures*, Prentice-Hall Inc, 1982.

G. Loveday, and A. Seidman, *Troubleshooting Solid-State Circuits*, New York: John Wiley and Sons, 1981.

A. Margolis, *Modern Radio Repair Techniques*, TAB Books Inc, 1971.

H. Neben, "An Ohmmeter with a Linear Scale," *QST*, Nov 1982.

H. Neben, "A Simple Capacitance Meter You Can Build," *QST*, Jan 1983.

F. Noble, "A Simple LC Meter," *QST*, Feb 1983.

J. Priedigkeit, "Measuring Inductance and Capacitance with a Reflection-Coefficient Bridge," *QST*, May 1982.

H. Sartori, "Solid Tubes — A New Life for Old Designs," *QST*, Apr 1977; "Questions on Solid Tubes Answered," Technical Correspondence, *QST*, Sep 1977.

B. Wedlock, and J. Roberge, *Electronic Components and Measurements*, Prentice-Hall, Inc, 1969.

"Some Basics of Equipment Servicing," series, *QST*, Dec 1981-Mar 1982; Feedback May 1982.

Contents

Electromagnetic Interference (EMI)
28

THE SCOPE OF THE PROBLEM

As our lives become filled with technology, the likelihood of electronic interference increases. Every lamp dimmer, garage-door opener or other new technical "toy" contributes to the electrical noise around us. Many of these devices also "listen" to that growing noise and may react unpredictably to their electronic neighbors.

Sooner or later, nearly every Amateur Radio operator will have a problem with interference. Most cases of interference can be cured! The proper use of "diplomacy" skills and standard cures will usually solve the problem.

This chapter, by Ed Hare, W1RFI, is only an overview. *The ARRL RFI Book* contains detailed information on the causes of and cures for nearly every type of interference problem.[1]

Important Terms

Bypass capacitor — a capacitor used to provide a low-impedance radio-frequency path around a circuit element.

Common-mode signals — signals that are in phase on both (or several) conductors in a system.

Conducted signals — signals that travel by electron flow in a wire or other conductor.

Decibel (dB) — a logarithmic unit of relative power measurement that expresses the ratio of two power levels.

Differential-mode signals — Signals that arrive on two or more conductors such that there is a 180° phase difference between the signals on some of the conductors.

Electromagnetic compatibility (EMC) — the ability of electronic equipment to be operated without performance degradation from interference.

Electromagnetic interference (EMI) — any electrical disturbance that interferes with the normal operation of electronic equipment.

Emission — electromagnetic energy propagated from a source by radiation.

Filter — a network of resistors, inductors and/or capacitors that offer little resistance to certain frequencies while blocking or attenuating other frequencies.

Fundamental overload — interference resulting from the fundamental signal of a radio transmitter.

Ground — a low-impedance electrical connection to the earth. Also, a common reference point in electronic circuits.

Harmonics — signals at exact multiples of the operating (or *fundamental*) frequency.

High-pass filter — a filter designed to pass all frequencies above a cutoff frequency, while rejecting frequencies below the cutoff frequency.

Induction — the transfer of electrical signals via magnetic coupling.

Interference — the unwanted interaction between electronic systems.

Intermodulation — the undesired mixing of two or more frequencies in a nonlinear device, which produces additional frequencies.

Low-pass filter — a filter designed to pass all frequencies below a cutoff frequency, while rejecting frequencies above the cutoff frequency.

Noise — any signal that interferes with the desired signal in electronic communications or systems.

Nonlinear — having an output that is not in linear proportion to the input.

Notch filter — a filter that rejects or suppresses a narrow band of frequencies within a wider band of frequencies.

Passband — the band of frequencies that a filter conducts with essentially no attenuation.

Radiated emission — radio-frequency energy that is coupled between two systems by electromagnetic fields.

Radio-frequency interference (RFI) — interference caused by a source of radio-frequency signals. This is a subclass of EMI.

Spurious emission — An emission, on frequencies outside the necessary bandwidth of a transmission, the level of which may be reduced without affecting the information being transmitted.

Susceptibility — the characteristic of electronic equipment that permits undesired responses when subjected to electromagnetic energy.

TVI — interference to television systems.

Pieces of the Problem

Every interference problem has two components — the equipment that is involved and the people who use it. A solution requires that we deal with both the equipment and the people effectively.

First, define the term "interference" without emotion. The ARRL recommends that the hams and their neighbors cooperate to find solutions. This view is shared by the FCC.

---

[1] ARRL Order no. 6834, available from ARRL Publication Sales or your local Amateur Radio equipment dealer.

Responsibility

When an interference problem occurs, we may ask "Who is to blame?" The ham and the neighbor often have different opinions. It is almost natural (but unproductive) to fix blame instead of the problem.

No amount of wishful thinking (or demands for the "other guy" to solve the problem) will result in a cure for interference. Each individual has a unique perspective on the situation, and a different degree of understanding of the personal and technical issues involved. On the other hand, each person has certain responsibilities to the other and should be prepared to address those responsibilities fairly.

FCC Regulations

A radio operator is responsible for the proper operation of the radio station. This responsibility is spelled out clearly in Part 97 of the FCC regulations. If interference is caused by a spurious emission from your station, you *must* correct the problem there.

Fortunately, most cases of interference are *not* the fault of the transmitting station. Most interference problems involve some kind of electrical noise or fundamental overload.

Personal Diplomacy

What happens when you first talk to your neighbor sets the tone for all that follows. Any technical solutions cannot help if you are not allowed in your neighbor's house to explain them! If the interference is not caused by spurious emissions from your station, however, you should be a locator of solutions, not a provider of solutions.

Your neighbor will probably *not* understand all of the technical issues — at least not at first. Understand that, regardless of fault, an interference problem is annoying to your neighbor. Let your neighbor know that you want to help find a solution and that you want to begin by talking things over.

Talk about some of the more important technical issues, in nontechnical terms. Interference can be caused by unwanted signals from your transmitter. Assure your neighbor that you will check your station thoroughly and correct any problems. You should also discuss the possible susceptibility of consumer equipment. If you have a copy of the consumer pamphlet "What to Do If You Have an Electronic Interference

Problem," give it to your neighbor.[2]

Here is a good analogy: If you tune your TV to channel 3, and see channel 8 instead, would you blame channel 8? No. You might check another set to see if it has the same problem, or call channel 8 to see if the station has a problem. If channel 8 was operating properly, you would likely decide that your TV set is broken. Now, if you tune your TV to channel 3, and see your local shortwave radio station (quite possibly Amateur Radio), don't blame the shortwave station without some investigation. In fact, many televisions respond to strong signals outside the television bands. They may be working as designed, but require added filters and/or shields to work properly near a strong, local RF signal.

Your neighbor will probably feel much better if you explain that you will help *find* a solution, even if the interference is *not* your fault. This offer can change your image from neighborhood villain to hero, especially if the interference is not caused by your station. (This is often the case.)

PREPARE YOURSELF

Learn About EMI

In order to troubleshoot and cure EMI, you need to learn more than just the basics. This is especially important when dealing with your neighbor. If you visit your neighbor's house and try a few dozen things that don't work (or make things worse), your neighbor may lose confidence in your ability to help cure the problem. If that happens, you may be asked to leave.

Local Help

If you are not an expert (and even experts can use moral support), you should find some local help. Fortunately, such help is often available from your Section Technical Coordinator (TC). The TC knows of any local RFI committees, and may have valuable contacts in the local utility companies. Even an expert can benefit from a TC's help.

---

[2]The ARRL "EMI/RFI Package" contains additional information about EMI, a list of EMI-filter sources, EMI-resistant telephones, telephone-company contacts and a pamphlet that explains interference in nontechnical terms. It is available from the ARRL Technical Department Secretary. The handling charge is $2 for ARRL members and $4 for non-members, prepaid.

The easiest way to find your TC is through your ARRL Section Manager (SM). There is a list of ARRL Officers and SMs on page 12 of any recent *QST*. Contact your SM through the address or telephone number listed. He or she can quickly put you in contact with the best source of local help.

Even if you can't secure the help of a local expert, a second ham can be a valuable asset. Often a second party can help defuse any hostility. It is also helpful to have someone to operate your station while you and your neighbor run through troubleshooting steps and try various cures.

Prepare Your Home

The first step toward curing an interference problem is to make sure your own signal is clean. You must eliminate all interference in your own house to be sure you are not causing the interference! This is also a valuable troubleshooting tool: If you know your station is clean, you have cut the size of the problem in half! If the FCC ever gets involved, you can demonstrate that you are not interfering with your own equipment.

Apply EMI cures to your own consumer electronics equipment. When your neighbor sees your equipment working well, it demonstrates that filters work and cause no harm.

To clean up your station, clean up the mess! A rat's nest of wires, unsoldered connections and so on in your station can contribute to EMI. To help build a better relationship, you may want to show your station to your neighbor. A clean station looks professional; it inspires confidence in your ability to solve the EMI problem.

Install a transmit filter (low-pass or band-pass) and a reasonable station ground. (If the FCC becomes involved, they will ask you about both items.) Show your neighbor that you have installed the necessary filter on your transmitter and explain that if there is still interference, it is necessary to try filters on the neighbor's equipment, too.

Operating practices and station-design considerations can affect EMI. Don't overdrive a transmitter or amplifier; that can increase its harmonic output. You can take steps to reduce the strength of your signal at the victim equipment. This might include reducing transmit power. Locate the antenna as far as possible from susceptible equipment or its wiring (ac line, telephone, cable TV). Antenna orientation may be important. For example, if your HF dipole at 30 ft is coupling into the neighbor's overhead cable-TV drop, that coupling could be reduced 20 dB by changing to a vertical antenna — even more by orienting the antenna so that the drop is off its end. Try different modes; CW or FM usually do not generate nearly as much telephone interference as AM or SSB, for example.

Call Your Neighbor

Now that you have learned more about EMI, located some local help (we'll assume it's the TC) and done all of your homework, make contact with your neighbor. First, arrange an appointment convenient for you, the TC and your neighbor. After you introduce the TC, allow him or her to explain the issues to your neighbor. Your TC will be able to answer most questions, but be prepared to assist with support and additional information as required.

Invite the neighbor to visit your station. Show your neighbor some of the things you do with your radio equipment. Point out any test equipment you use to keep your station in good working order. Of course, you want to show the filters you have installed on your transmitter.

Next, have the TC operate your station on several different bands. Show your neighbor that your home electronics equipment is working properly while your station is in operation. Point out the filters you have installed to correct any susceptibility problems.

At this point, tell your neighbor that the next step is to try some of these cures on his or her equipment. This is a good time to emphasize that the problem is probably not your fault, but that you and the TC will try to help find a solution anyway.

Table 28.1 is a list of the things needed to troubleshoot and solve most EMI problems. Decide ahead of time which of these items are needed and take them with you.

At Your Neighbor's Home

You and the TC should now visit the neighbor's home. Inspect the equipment installation and ask when the interference occurs, what equipment is involved and what frequencies or channels are affected. The answers are valuable clues. Next, either you or the TC should operate your station while the other observes the effects. Try all bands and modes that you use. Ask the neighbor to demonstrate the problem.

The tests may show that your station isn't involved at all. You may immediately recognize electrical noise or some kind of equipment malfunction. If so, explain your findings to the neighbor and suggest that he or she contact appropriate service personnel.

EMC Fundamentals

Knowledge is one of the most valuable tools for solving EMI problems. A successful EMI cure usually requires familiarity with the relevant technology and troubleshooting procedures.

SOURCE-PATH-VICTIM

All cases of EMI involve a *source* of electromagnetic energy, a device that responds to this electromagnetic energy (*victim*) and a transmission path that allows energy to flow from the source to the victim. Sources include radio transmitters, receiver local oscillators, computing devices, electrical noise, lightning and other natural sources.

There are three ways that EMI can travel from the source to the victim: radiation, conduction and induction. Radiated EMI propagates by electromagnetic radiation from the source, through space to the victim. A conducted signal travels over wires connected to the source and the victim. Induction occurs when two circuits

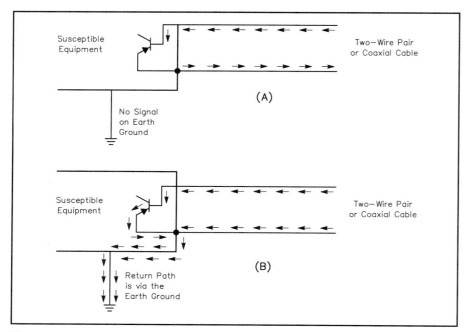

Fig 28.1 — A shows a differential-mode, while B shows a common-mode signal. The two kinds of signals are described in the text.

are magnetically coupled. Most EMI occurs via conduction, or some combination of radiation and conduction. For example, a signal is radiated by the source and picked up by a conductor attached to the victim (or directly by the victim's circuitry) and is then conducted into the victim. EMI from induction is rare.

DIFFERENTIAL VS COMMON-MODE

It is important to understand the differences between differential-mode and common-mode conducted signals (see **Fig 28.1**). Each of these conduction modes requires different EMI cures. Differential-mode cures, (the typical high-pass filter, for example) do not attenuate common-mode signals. On the other hand, a typical common-mode choke does not affect interference resulting from a differential-mode signal.

Differential-mode currents usually have two easily identified conductors. In a two-wire transmission line, for example, the signal leaves the generator on one line and returns on the other. When the two conductors are in close proximity, they form a transmission line and there is a 180° phase difference between their respective signals. It's relatively simple to build a filter that passes desired signals and shunts unwanted signals to the return line. Most *desired* signals, such as the TV signal inside a coaxial cable are differential-mode signals.

In a common-mode circuit, many wires of a multiwire system act as if they were a

single wire. The result can be a good antenna, either as a radiator or as a receptor of unwanted energy. The return path is usually earth ground. Since the source and return conductors are usually well separated, there is no reliable phase difference between the conductors and no convenient place to shunt unwanted signals. Toroid chokes are the answer to common-mode interference. (The following explanation applies to rod cores as well as toroids, but since rod cores may couple into nearby circuits, use them only as a last resort.)

Toroids work differently, but equally well, with coaxial cable and paired conductors. A common-mode signal on a coaxial cable is usually a signal that is present on the *outside* of the cable *shield*. When we wrap the cable around a ferrite-toroid core the choke appears as a reactance in series with the outside of the shield, but it has no effect on signals inside the cable because their field is (ideally) confined inside the shield. With paired conductors such as zip-cord, signals with opposite phase set up magnetic fluxes of opposite phase in the core. These "differential" fluxes cancel each other, and there is no net reactance for the differential signal. To common-mode signals, however, the choke appears as a reactance in series with the line.

Toroid chokes work less well with single-conductor leads. Because there is no return current to set up a canceling flux, the choke appears as a reactance in series with *both* the desired and undesired signals.

SOURCES OF EMI

The basic causes of EMI can be grouped into several categories:

- Fundamental overload effects
- External noise
- Spurious emissions from a transmitter
- Intermodulation distortion or other external spurious signals

As an EMI troubleshooter, you must determine which of these are involved in your interference problem. Once you do, it is easy to select the necessary cure.

Fundamental Overload

Most cases of interference are caused by fundamental overload. The world is filled with RF signals. Properly designed equipment should be able to select the desired signal, while rejecting all others. Unfortunately, because of design deficiencies such as inadequate shields or filters, some equipment is unable to reject strong out-of-band signals.

A strong fundamental signal can enter equipment in several different ways. Most commonly, it is conducted into the equipment by wires connected to it. Possible conductors include antennas and feed lines, interconnecting cables, power lines and ground wires. TV antennas and feed lines, telephone or speaker wiring and ac power leads are the most common points of entry.

The effect of an interfering signal is directly related to its strength. The strength of a radiated signal diminishes with the square of the distance from the source: When the distance from the source doubles, the strength of the electromagnetic field decreases to one-fourth of its strength at the original distance from the source. This characteristic can often be used to help solve EMI cases. You can often make a significant improvement by moving the victim equipment and the antenna farther away from each other.

External Noise

Most cases of interference reported to the FCC involve some sort of external noise source. The most common of these noise sources are electrical. External "noise" can also come from transmitters or from unlicensed RF sources such as computers, video games, electronic mice repellers and the like.

Electrical noise is fairly easy to identify by looking at the picture of a susceptible TV or listening on an HF receiver. A photo of electrical noise on a TV screen is shown in the TVI section of this chapter. On a receiver, it usually sounds like a buzz, sometimes changing in intensity as the arc or spark sputters a bit. If you determine the problem to be caused by external noise, it must be cured at the source. Refer to the Electrical Noise section of this chapter and the ARRL RFI book.

Spurious Emissions

All transmitters generate some (hopefully few) RF signals that are outside their allocated frequency bands. These out-of-band signals are called spurious emissions, or *spurs*. Spurious emissions can be discrete signals or wideband noise. Harmonics, the most common spurious emissions, are signals at exact multiples of the operating (or *fundamental*) frequency. Other discrete spurious signals are usually caused by the superheterodyne mixing process used in most modern transmitters. **Fig 28.2** shows the spectral output of a transmitter, including harmonics and mixing products.

Transmitters may also produce broadband noise and/or "parasitic" oscillations. (Parasitic oscillations are discussed in the **Amplifiers** chapter.) If these unwanted signals cause interference to another radio service, FCC regulations require the owner to correct the problem.

Troubleshooting EMI

Most EMI cases are complex. They involve a source, a path and a victim. Each of these main components has a number of variables: Is the problem caused by harmonics, fundamental overload, conducted emissions, radiated emissions or a combination of all of these factors? Should it be fixed with a low-pass filter, high-pass filter, common-mode chokes or ac-line filter? How about shielding, isolation transformers, a different ground or antenna configuration?

By the time you finish with these questions, the possibilities could number in the millions. You probably will not see your exact problem and cure listed in this book or any other. You must diagnose the problem!

Troubleshooting an EMI problem is a three-step process, and all three steps are equally important:

- Identify the problem

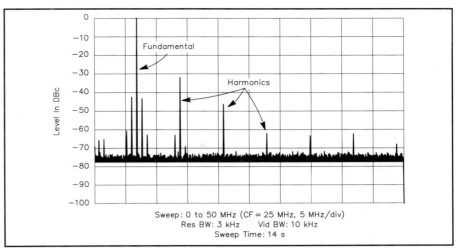

Fig 28.2 — The spectral output of a typical amateur transmitter. The fundamental is at 7 MHz. There are visible harmonics at 14, 21 and 28 MHz. Unlabeled lines are nonharmonic spurious emissions. This transmitter complies with the stringent FCC spectral-purity regulations regarding amateur transmitters with less than 5 W of RF output.

- Diagnose the problem
- Cure the problem.

IDENTIFY THE PROBLEM

Is It Really EMI? — Before trying to solve a suspected case of EMI, verify that the symptoms actually result from external causes. A variety of equipment malfunctions or external noise can look like interference. "Your" EMI problem might be caused by another ham or a radio transmitter of another radio service, such as a local CB or police transmitter.

Is It Your Station? — If it appears that your station is involved, operate your station on each band, mode and power level that you use. Note all conditions that produce interference. If no transmissions produce the problem, your station *may* not be the cause. (Although some contributing factor may have been missing in the test.) Have your neighbor keep notes of when and how the interference appears: what time of day, what station, what other appliances were in use, what was the weather? You should do the same whenever you operate. If you can readily reproduce the problem with your station, you

can start to troubleshoot the problem.

DIAGNOSE THE PROBLEM

Look Around — Aside from the brain, eyes are a troubleshooter's best tool. Look around. Installation defects contribute to many EMI problems. Look for loose connections, shield breaks in a cable-TV installation or corroded contacts in a telephone installation. Fix these first.

Problems that occur only on harmonics of the fundamental signal usually indicate the transmitter. Harmonics can also be generated in nearby semiconductors, such as an unpowered VHF receiver left connected to an antenna, or a corroded connection in a tower guy wire. Harmonics can also be generated in the front-end components of the TV or radio experiencing interference.

Is the wiring connected to the victim equipment resonant on one or more amateur bands? If so, a common-mode choke placed at the middle of the wiring may be an easy cure.

These are only a few of the questions you might need to ask. Any information you gain about the systems involved will help find the EMI cause and cure.

Cures

At Your Station — Make sure that your own station and consumer equipment are clean. This cuts the size of the problem in half! Once this is done, you won't need to diagnose or troubleshoot your station later. Also, any cures successful at your house may work at your neighbor's as well. If you do have problems in your own house,

refer first to the Transmitter section of this chapter, or continue through the troubleshooting steps and specific cures and take care of your own problem first.

Simplify the Problem — Don't tackle a complex system — such as a telephone system in which there are two lines running to 14 rooms — all at once. You could spend

the rest of your life running in circles and never find the true cause of the problem.

There's a better way. In our hypothetical telephone system, first locate the telephone jack closest to the telephone service entrance. Disconnect the lines to more remote jacks and connect one EMI-resistant telephone at the remaining jack.

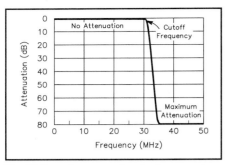

Fig 28.3 — An example of a low-pass filter response curve.

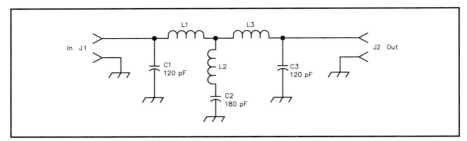

Fig 28.4 — A low-pass filter for amateur transmitting use. Complete construction information appears in the Transmitters chapter of the ARRL RFI book.

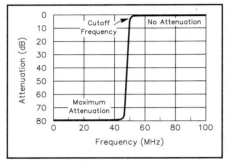

Fig 28.5 — An example of a high-pass filter response curve.

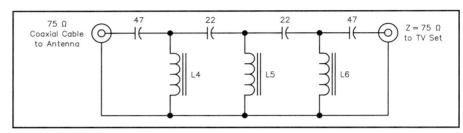

Fig 28.6 — A differential-mode high-pass filter for 75-Ω coax. It rejects HF signals picked up by a TV antenna or that leak into a cable-TV system. It is ineffective against common-mode signals. All capacitors are high-stability, low-loss, NP0 ceramic discs. Values are in pF. The inductors are all #24 enameled wire on T-44-0 toroid cores. L4 and L6 are each 12 turns (0.157 μH). L5 is 11 turns (0.135 μH).

If the interference remains, try cures until the problem is solved, then start adding lines and equipment back one at a time, fixing the problems as you go along. If you are lucky, you will solve all of the problems in one pass. If not, at least you can point to one piece of equipment as the source of the problem.

Multiple Causes — Many EMI problems have multiple causes. These are usually the ones that give new EMI troubleshooters the most trouble. If, for example, a TVI problem is caused by harmonics from the transmitter, an arc in the transmitting antenna, an overloaded TV preamp, differential-mode fundamental overload generating harmonics in the TV tuner, induced and conducted RF in the ac-power system and a common-mode signal picked up on the shield of the TV's coaxial feed line, you would never find a cure by trying only one at a time!

In this case, the solution requires that you apply all of the cures at the same time. When troubleshooting, if you try a cure, leave it in place. When you finally try a cure that really works, start removing the "temporary" attempts one at a time. If the interference returns, you know that there were multiple causes.

OVERVIEW OF TECHNIQUES

Shields

Shields are used to set boundaries for radiated energy. Thin conductive films, copper braid and sheet metal are the most common shield materials. Maximum shield effectiveness usually requires solid sheet metal that completely encloses the source or susceptible circuitry or equipment. Small discontinuities, such as holes or seams, decrease shield effectiveness.

Filters

A major means of separating signals relies on their frequency differences. Filters offer little opposition to certain frequencies while blocking others. Filters vary in attenuation characteristics, frequency characteristics and power-handling capabilities. The names given to various filters are based on their uses.

Low-pass filters pass frequencies below some cutoff frequency, while attenuating frequencies above that cutoff frequency. A typical low-pass filter curve is shown in **Fig 28.3**. A schematic is shown in **Fig 28.4**. These filters are difficult to construct properly so you should buy one. Many retail Amateur Radio stores that advertise in *QST* stock low-pass filters.

High-pass filters pass frequencies above some cutoff frequency while attenuating frequencies below that cutoff frequency. A typical high-pass filter curve is shown in **Fig 28.5**. **Fig 28.6** shows a schematic of a typical high-pass filter. Again, it is best to buy one of the commercially available filters.

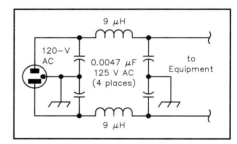

Fig 28.7 — A "brute-force" ac-line filter.

Bypass capacitors can be used to cure EMI problems. A bypass capacitor is usually placed between a signal or power lead and circuit ground. It provides a low-impedance path to ground for RF signals. Bypass capacitors for HF signals are usually 0.01 μF, while VHF bypass capacitors are usually 0.001 μF.

AC-line filters, sometimes called "brute-force" filters, are used to filter RF energy from power lines. A schematic is shown in **Fig 28.7**. Use ac-rated components as specified. We *strongly* recommend UL-listed, commercially made ac-line filters; the ac-power lines are no place for home-brew experimentation.

Common-Mode Chokes

Common-mode chokes may be the best-kept secret in Amateur Radio. The differential-mode filters described earlier are

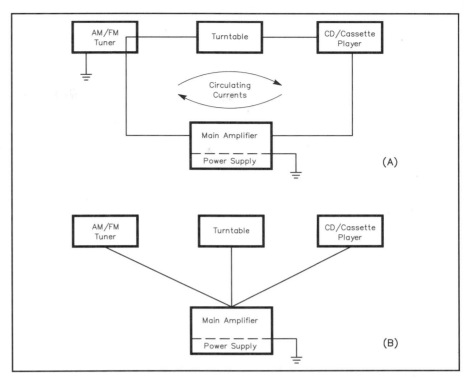

Fig 28.10 — A shows a stereo system grounded as an undesirable "ground loop." B is the proper way to ground a multiple-component system.

Fig 28.8 — Several styles of common-mode chokes.

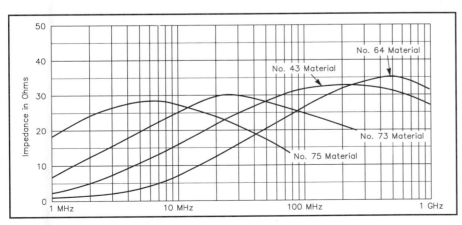

Fig 28.9 — Impedance vs frequency plots for "101" size ferrite beads.

not effective against common-mode signals. To eliminate common-mode signals properly, you need common-mode chokes. They may help nearly any interference problem, from cable TV to telephones to audio interference caused by RF picked up on speaker leads.

Common-mode chokes usually have ferrite core materials. These materials are well suited to attenuate common-mode currents. Several kinds of common-mode chokes are shown in **Fig 28.8**.

The optimum size and ferrite material are determined by the application and frequency. For example, an ac cord with a plug attached cannot be easily wrapped on a small ferrite core. The characteristics of ferrite materials vary with frequency, as shown by the graph in **Fig 28.9**.

Grounds

An electrical ground is not a huge sink that somehow swallows noise and unwanted signals. Ground is a *circuit* concept, whether the circuit is small, like a radio receiver, or large, like the propagation path between a transmitter and cable-TV installation. Ground forms a universal reference point between circuits.

This chapter deals with the EMC aspects of grounding. While grounding is not a cure-all for EMI problems, ground is an important safety component of any electronics installation. It is part of the lightning protection system in your station and a critical safety component of your house wiring. Any changes made to a grounding system must not compromise these important safety considerations. Refer to the **Safety** chapter for important information about grounding.

Many amateur stations have several grounds: a safety ground that is part of the ac-wiring system, another at the antenna for lightning protection and perhaps another at the station for EMI control. These grounds can interact with each other in ways that are difficult to predict.

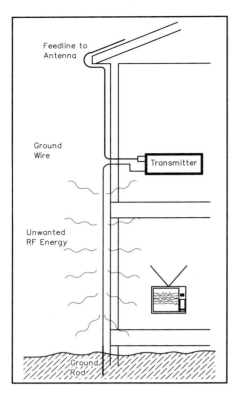

Fig 28.11 — When a transmitter is located on an upper floor, the ground lead may act as an antenna for VHF/UHF energy. Such stations may be better off without a normal ground.

Ground Loops

All of these station grounds can form a large ground loop. This loop can act as a large loop antenna, with increased susceptibility to lightning or EMI problems. **Fig 28.10** shows a ground loop and a proper single-point ground system.

When is Ground not a Ground?

In many stations, it is impossible to get a good RF connection to earth ground. Most practical installations require several feet of wire between the station ground connection and an outside ground rod. Many troublesome harmonics are in the VHF range. At VHF, a ground wire length can be several wavelengths long — a very effective long-wire antenna! Any VHF signals that are put on a long ground wire will be radiated. This is usually not the intended result of grounding.

Take a look at the station shown in **Fig 28.11**. In this case, the ground wire could very easily contribute to an interference problem in the downstairs TV set.

While a station ground may cure some transmitter EMI problems — either by putting the transmitter chassis at a low-impedance reference point or by rearrang-ing the problem so the "hot spots" are farther away from susceptible equipment — it is not the cure-all that some literature has suggested. A ground is easy to install, and it may reduce stray fundamental or harmonic currents on your antenna lead; it is worth a try.

SPECIFIC CURES

Now that you have learned some EMI fundamentals, you can work on technical solutions. A systematic approach will identify the problem and suggest a cure. Armed with your EMI knowledge, a kit of filters and tools, your local TC and a determination to solve the problem, it is time to diagnose the problem.

Most EMI problems can be solved by the application of standard cures. If you try these cures and they work, you may not need to troubleshoot the problem at all. Perhaps if you can install a low-pass filter on your transmitter or a common-mode choke on a TV, the problem will be solved.

Here are some specific cures for different interference problems. You should also get a copy of the ARRL RFI book. It's comprehensive and picks up where this chapter leaves off. Here are several standard cures.

Transmitters

We start with transmitters not because most interference comes from transmitters, but because your station transmitter is under your direct control. Many of the troubleshooting steps in other parts of this chapter assume that your transmitter is "clean" (free of unwanted RF output).

Controlling Spurious Emissions — Start by looking for patterns in the interference. If the interference is only on frequencies that are multiples of your operating frequency, you clearly have interference from harmonics. (Although these harmonics may *not* come from your station!)

If HF-transmitter spurs are interfering with a VHF service, a low-pass filter on the transmitter will usually cure the problem. Install it after the amplifier (if used) and *before* the antenna tuner. (A second filter between the transmitter and amplifier may occasionally help as well.) Install a low-pass filter as your first step in any interference problem that involves another radio service.

Interference from nonharmonic spurious emissions is extremely rare in commercially built radios. Any such problem indicates a malfunction that should be repaired.

Television Interference (TVI)

For a TV signal to look good, it must have about a 45 to 50 dB signal-to-noise ratio. This requires a good signal at the TV antenna-input connector. This brings up an important point: to have a good signal, you must be in a good signal area. The FCC does not protect fringe-area reception.

TVI, or interference to any radio service, can be caused by one of several things:

- Spurious signals within the TV channel coming from your transmitter or station.
- The TV set may be overloaded by your transmitter's fundamental signal.
- Signals within the TV channel from some source other than your station, such as electrical noise, an overloaded mast-mounted TV preamplifier or a transmitter in another service.
- The TV set might be defective or misadjusted, making it look like there is an interference problem.

All of these potential problems are made more severe because the TV set is hooked up to *two* antenna systems: (1) the incoming antenna and its feed line and (2) the ac power lines. These two "long-wire" antennas can couple *a lot* of fundamental or harmonic energy into the TV set! The TVI Troubleshooting Flowchart in the **References** chapter is a good starting point.

Fundamental Overload

A television set can be overloaded by a strong, local RF signal. This happens because the manufacturer did not install the necessary filters and shields to protect the TV set from other signals present on the air. These design deficiencies can sometimes be corrected externally.

Start by determining if the interference is affecting the video, the sound or both. If it is present only on the sound, it is probably a case of audio rectification. (See the Stereos section of this chapter.) If it is present on the video, or both, it could be getting into the video circuitry or affecting either the tuner or IF circuitry.

The first line of defense for an antenna-connected TV is a high-pass filter. Install a high-pass filter directly on the back of the TV set. You may also have a problem with common-mode interference. The second line of defense is a common-mode choke on the antenna feed line — try this first in a cable-television installation. These two filters can probably cure most cases of TVI!

Fig 28.12 shows a "bulletproof" installation. If this doesn't cure the problem, the TV circuitry is picking up your signal directly. In that case, don't try to fix it yourself — it is a problem for the TV manufacturer.

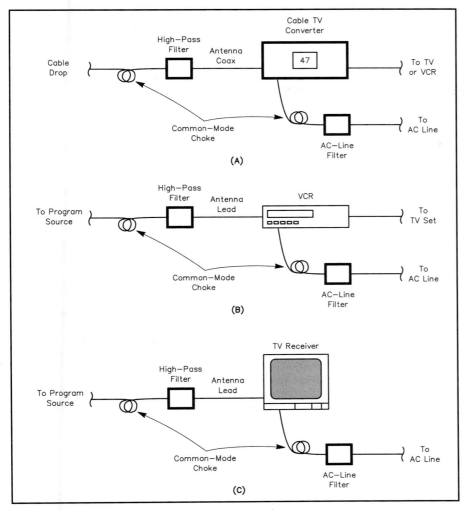

Fig 28.12 — This sort of installation should cure any kind of conducted TVI. It will not cure direct-pickup or spurious-emission problems.

VHF Transmitters — A VHF transmitter can interfere with over-the-air TV reception. Most TV tuners are not very selective and a strong VHF signal can overload the tuner easily. In this case, a VHF notch or stop-band filter at the TV can help by reducing the VHF fundamental signal that gets to the TV tuner. Star Circuits is one company that sells tunable notch filters.[3]

The Electronic Industries Association (EIA) can help you contact equipment manufacturers. Contact them directly for assistance in locating help. (Their address is in the **References** chapter Address List.)

Spurious Emissions

Start by analyzing which TV channels are affected. The TV Channel Chart in the **References** chapter shows the relation-ship of the ham allocations and their harmonics to over-the-air and cable channels. Each channel is 6-MHz wide. If the interference is only on channels that are multiples of your operating frequency, you clearly have interference from harmonics. (It is not certain that these harmonics are coming from your station, however.)

You are responsible for spurious signals produced by your station. If your station is generating any interfering spurious signals, the problem must be cured there. So, if the problem occurs only when you transmit, go back and check your station. Refer to the section on Transmitters. You must first find out if the transmitter has any spurs.

If your transmitter and station check "clean," then you must look elsewhere. The most likely cause is TV susceptibility — fundamental overload. This is usually manifest by interference to all channels, or at least all VHF channels. If the problem is fundamental overload, see that section earlier in this chapter. If not, read on.

Electrical Noise

Electrical noise is fairly easy to identify by looking at the picture or listening on an HF receiver. Electrical noise on a TV screen is shown in **Fig 28.13**. On a receiver, it usually sounds like a buzz, sometimes changing in intensity as the arc or spark sputters a bit. If you have a problem with electrical noise, go to the Electrical Noise section.

Cable TV

Cable TV has been a blessing and a curse for Amateur Radio TVI problems. On the plus side, the cable delivers a strong, consistent signal to the TV receiver. It is also (in theory) a shielded system, so an external signal can't get in and cause trouble. On the minus side, the cable forms a large, long-wire antenna that can pick up lots of external signals on its shield (in the common mode). Many TVs and VCRs and even some cable set-top converters are easily overloaded by such common-mode signals.

Leakage into a cable-TV system is called ingress. Leakage out is called egress. If the

[3]Star Circuits Model: 23H tunes 6 m, Model: 1822 tunes 2 m and Model: 46FM tunes the FM broadcast band. Their address is in the **References** chapter Address List.

PATTERN A   PATTERN B

Fig 28.13 — Two examples of TVs experiencing electrical noise.

Fig 28.14 — Several turns of coax on a ferrite core eliminate HF and VHF signals from the outside of a coaxial cable.

cable isn't leaking, there should be no external signals getting inside the cable. So, an in-line filter such as a high-pass filter is not usually necessary. For a cable-connected TV, the first line of defense is a common-mode choke. Only in rare cases is a high-pass filter necessary. It is important to remember this, because if your neighbor has several TVs connected to cable and you suggest the wrong filter (at $15 each), you may have a personal diplomacy problem of a whole new dimension. **Fig 28.14** shows a common-mode choke.

Fig 28.12 shows a bulletproof installation for cable TV. (The high-pass filter is usually not needed.) If all of the cures shown have been tried, the interference probably results from direct pickup inside the TV. In this case, contact the TV manufacturer through the EIA.

Interference to cable-TV installations from VHF transmitters is a special case. Cable TV uses frequencies allocated to over-the-air services, such as Amateur Radio. When the cable shielding is less than perfect, interference can result.

The TV Channel Chart in the **References** chapter shows which cable channels coincide with ham bands. If, for example, you have interference to cable channel 18 from amateur 2-m operation, suspect cable ingress. Contact the cable company; it may be their responsibility to locate and correct the problem. The cable company is not responsible, however, for leakage occurring in customer-owned, cable-ready equipment that is tuned to the same frequency as the over-the-air signal. If there is interference to a cable-TV installation, the cable company should be able to demonstrate interference-free reception when using a cable-company supplied set-top converter.

TV Preamplifiers

Some television owners use a preamplifier — sometimes when it's not needed.

Preamplifiers are only needed in weak-signal areas, and they often cause more trouble than they prevent. They are subject to the same overload problems as TVs, and their location on the antenna mast usually makes it difficult to install the appropriate cures. You may need to install a high-pass or notch filter at the *input* of the preamplifier, as well as a common-mode choke on the input, output and power-supply wiring (if separate) to effect a complete cure.

VCRs

A VCR usually contains a television tuner, or has a TV channel output, so it is subject to all of the interference problems of a TV receiver. It is also hooked up to an antenna or cable system and the ac-line wiring. The video baseband signal extends from 30 Hz to 3.5 MHz, with color information centered around 3.5 MHz and the FM sound subcarrier at 4.5 MHz. The entire video baseband is frequency modulated onto the tape at frequencies up to 10 MHz. It is no wonder that some VCRs are quite susceptible to EMI.

Many cases of VCR EMI can be cured. Start by proving that the VCR is the susceptible device. Temporarily disconnect the VCR from the television. If there is no interference to the TV, then the VCR is the most likely culprit.

You need to find out how the interfering signal is getting into the VCR. Temporarily disconnect the antenna or cable feed line from the VCR. If the interference goes away, then the antenna line is involved. In this case, you can probably fix the problem with a common-mode choke or high-pass filter.

Fig 28.12 shows a bulletproof VCR installation. If you have tried all of the cures shown and still have a problem, the VCR is probably subject to direct pickup. In this case, contact the manufacturer through the EIA.

Nonradio Devices

Interference to nonradio devices is not the fault of the transmitter. (A portion of the *FCC Interference Handbook,* 1990 Edition, is shown in **Fig 28.15**.[4]) In essence, the FCC views nonradio devices that pick up nearby radio signals as im-

[4]The *FCC Interference to Home Electronic Entertainment Equipment Handbook* is available at the FCC web site, http://www.fcc.gov/cib/Publications/tvibook.html, or from the US Government Printing Office. See the **References** chapter for their address and telephone number.

PART II

INTERFERENCE TO OTHER EQUIPMENT

CHAPTER 6

TELEPHONES, ELECTRONIC ORGANS, AM/FM RADIOS, STEREO AND HI-FI EQUIPMENT

Telephones, stereos, computers, electronic organs and home intercom devices can receive interference from nearby radio transmitters. When this happens, the device improperly functions as a radio receiver. Proper shielding or filtering can eliminate such interference. The device receiving interference should be modified in your home while it is being affected by interference. This will enable the service technician to determine where the interfering signal is entering your device.

The device's response will vary according to the interference source. If, for example, your equipment is picking up the signal of a nearby two-way radio transmitter, you likely will hear the radio operator's voice. Electrical interference can cause sizzling, popping or humming sounds.

Fig 28.15 — Part of page 18 from FCC *Interference Handbook* (1990 edition) explains the facts and places responsibility for interference to nonradio equipment.

properly functioning; contact the manufacturer and return the equipment. The FCC does not require that nonradio devices include EMI protection and they don't offer legal protection to users of these devices that are susceptible to interference.

Telephones

Telephones have probably become the number one interference problem of Amateur Radio. However, most cases of telephone interference can be cured by correcting any installation defects and installing telephone EMI filters where needed.

Telephones can improperly function as radio receivers. There are devices inside many telephones that act like diodes. When such a telephone is connected to the telephone wiring (a large antenna), an AM radio receiver can be formed. When a nearby transmitter goes on the air, these telephones can be affected.

Troubleshooting techniques were discussed earlier in the chapter. The suggestion to simplify the problem applies especially to telephone interference. Disconnect all telephones except one, right at the service entrance if possible, and start troubleshooting the problem there.

If any one device, or bad connection in the phone system, detects RF and puts the detected signal back onto the phone line as audio, that audio cannot be removed with filters. Once the RF has been detected and turned into audio, it cannot be filtered out because the interference is at the same frequency as the desired audio signal. To effect a cure, you must locate the detection point and correct the problem there.

The telephone company lightning arrestor may be defective. Defective arrestors can act like diodes, rectifying any nearby RF energy. Telephone-line amplifiers or other electronic equipment may also be at fault. Leave the telephone company equipment to the experts, however. There are important safety issues that are the sole responsibility of the telephone company.

Inspect the installation. Years of exposure in damp basements, walls or crawl spaces may have caused deterioration. Be suspicious of anything that is corroded or discolored. In many cases, homeowners have installed their own telephone wiring, often using substandard wiring. If you find sections of telephone wiring made from nonstandard cable, replace it with standard twisted-pair wire. Radio Shack, among others, sells several kinds of telephone wire.

Next, evaluate each of the telephone instruments. If you find a susceptible telephone, install a telephone EMI filter on that telephone. Several *QST* advertisers

sell small, attractive telephone EMI filters.

If you determine that you have interference only when you operate on one particular ham band, the telephone wiring is probably resonant on that band. If possible, install a few strategically placed inline telephone EMI filters to break up the resonance.

Telephone Accessories — Answering machines, fax machines and some alarm systems are also prone to interference problems. All of the troubleshooting techniques and cures that apply to telephones also apply to these telephone devices. In addition, many of these devices connect to the ac mains. Try a common-mode choke and/or ac-line filter on the power cord (which may be an ac cordset, a small transformer or power supply).

Cordless Telephones — A cordless telephone is an unlicensed *radio* device that is manufactured and used under Part 15 of the FCC regulations. The FCC does not intend Part 15 devices to be protected from interference. These devices usually have receivers with very wide front-end filters, which make them very susceptible to interference. A label on the telephone or a paragraph in the owner's manual should explain that the telephone must not cause interference to other services and must tolerate any interference caused to it.

It's worthwhile to try a telephone filter on the base unit and properly filter its ac line cord. (You might get lucky!) The best source of help is the manufacturer, but they may point out that the Part 15 device is not protected from interference. These kinds of problems are difficult to fix after the fact. The necessary engineering should be done when the device is designed.

Other Audio Devices

Other audio devices, such as stereos, in-tercoms and public-address systems can also pick up and detect strong nearby transmitters. The FCC considers these nonradio devices and does not protect them from licensed radio transmitters that may interfere with their operation. See Fig 28.15 for the FCC's point of view.

Use the standard troubleshooting techniques discussed earlier in this chapter to isolate problems. In a multicomponent stereo system (as in **Fig 28.16**), for example, you must determine what combination of components is involved with the problem. First, disconnect all auxiliary components to determine if there is a problem with the main receiver/amplifier. (Long speaker/interconnect cables are prime suspects.)

Stereos — If the problem remains with the main amplifier isolated, determine if the interference level is affected by the volume control. If so, the interference is getting into the circuit *before* the volume control, usually through accessory wiring. If the volume control has no effect on the level of the interfering sound, the interference is getting in *after* the control, usually through speaker wires.

Speaker wires are often resonant on the HF bands. In addition, they are often connected directly to the output transistors, where RF can be detected. Most amplifier designs use a negative *feedback* loop to improve fidelity. This loop can conduct the detected RF signal back to the high-gain stages of the amplifier. The combination of all of these factors makes the speaker leads the usual indirect cause of interference to audio amplifiers.

There is a simple test that will help determine if the interfering signal is being coupled into the amplifier by the speaker leads. Temporarily disconnect the speaker leads from the amplifier, and plug in a test set of headphones with short leads. If there is no interference with the headphones,

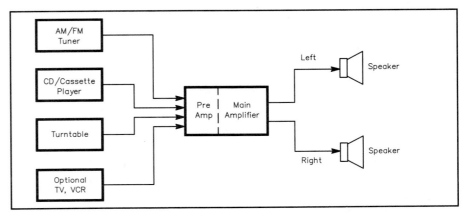

Fig 28.16 — A typical modern stereo system.

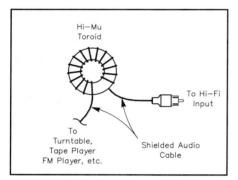

Fig 28.17 — This is how to make a speaker-lead common-mode choke. Be sure to use the correct ferrite material.

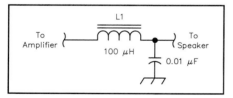

Fig 28.18 — An LC filter for speaker leads.

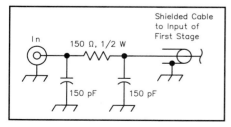

Fig 28.19 — A filter for use at the input of audio equipment. The components should be installed inside of the chassis at the connector by a qualified technician.

filtering the speaker leads will cure the problem.

The best way to eliminate RF signals from speaker leads is with common-mode chokes. **Fig 28.17** shows how to wrap speaker wires around an FT-140-43 ferrite core to cure speaker-lead EMI. Use the correct core material for the job. See the information about common-mode chokes earlier in this chapter.

Another way to cure speaker-lead interference is with an LC filter as shown in **Fig 28.18**. Sources for similar filters are listed in the ARRL Technical Information Service (TIS) "EMI/RFI Package."

Interconnect cables can couple interfering signals into an amplifier or accessories. The easiest cure here is also a common-mode choke. However, it may also be necessary to add a differential-mode filter to the input of the amplifier or accessory. **Fig 28.19** shows a home-brew version of such a filter.

Intercoms and Public-Address Systems — All of these problems also apply to intercoms, public-address (PA) systems and similar devices. These systems usually have long speaker leads or interconnect cables that can pick up a lot of RF energy from a nearby transmitter. The cures discussed above do apply to these systems, but you may also need to contact the manufacturer to see if they have any additional, specific information.

Computers and Other Unlicensed RF Sources

Computers and microprocessors can be sources, or victims, of interference. These devices contain oscillators that can, and do, radiate RF energy. In addition, the internal functions of a computer generate different frequencies, based on the various data rates as software is executed. All of these signals are digital in nature — with fast rise and fall times that are rich in harmonics.

Don't just think "computer" when thinking of computer systems. Many household appliances contain microprocessors: digital clocks, video games, calculators and more.

Computing devices are covered under Part 15 of the FCC regulations as unintentional emitters. The FCC has set up absolute radiation limits for these devices. FCC regulations state that the operator or owner of Part 15 devices must take whatever steps are necessary to reduce or eliminate any interference they cause to a licensed radio service. This means that if your neighbor's video game interferes with your radio, the neighbor is responsible for correcting the problem. (Of course, your neighbor may appreciate your

help in locating a solution!)

The FCC has set up two levels of type acceptance for computing devices. Class A is for computers used in a commercial environment. FCC Class B requirements are more stringent — for computers used in residential environments. If you buy a computer or peripheral, be sure that it is Class B certified or it will probably generate interference to your amateur station or home-electronics equipment.

If you find that your computer system is interfering with your radio (not uncommon in this digital-radio age), start by simplifying the problem. Temporarily switch off as many peripherals as possible and disconnect their cables from the back of the computer. If possible, use just the computer, keyboard and monitor. This test may indicate one or more peripherals as the source of the interference.

When seeking cures, first ensure that all interconnection cables are shielded. Replace any unshielded cables with well shielded ones; this often significantly reduces RF noise from computer systems. The second line of defense is the common-mode choke, made from a ferrite toroid. The toroids should be installed as close to the computer and/or peripheral device as practical. **Fig 28.20** shows the location of common-mode chokes in a complete computer system where both the computer and peripherals are noisy.

In some cases, a switching power supply may be a source of interference. A common-mode choke and/or ac-line filter may cure this problem. In extreme cases of computer interference you may need to improve the shielding of the computer. Refer to the ARRL RFI book for more information about how to do this. Don't forget that some peripherals (such as modems) are connected to the phone line, so you may need to treat them like telephones.

Automobiles

As automobiles have become more technologically sophisticated, questions about the compatibility of automobiles and amateur transmitters have increased in number and scope. The use of microprocessors in autos makes them computer systems on wheels, subject to all of the same problems as any other computer. Installation of ham equipment can cause problems, ranging from nuisances like a dome light coming on every time you transmit to serious ones such as damage to the vehicle electronic control module (ECM).

Only qualified service personnel should work on automotive EMC problems. Many critical safety systems on modern

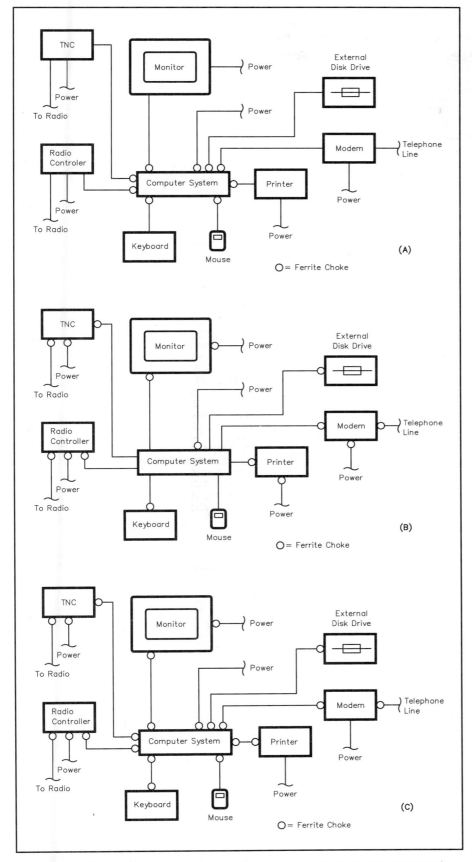

Fig 28.20 — Where to locate ferrites in a computer system. At A, the computer is noisy, but the peripherals are quiet. At B, the computer is quiet, but external devices are noisy. At C, both the computer and externals are noisy.

cars should not be handled by amateurs. Even professionals can meet with mixed results. The ARRL (TIS) contacted each of the automobile manufacturers and asked about their EMC policies, service bulletins and best contacts to resolve EMI problems. About 20% of the companies never answered, and answers from the rest ranged from good to poor. One company even said that the answers to those questions were "proprietary."

Some of the companies *do* have reasonable EMC policies, but these policies often fall apart at the dealer level. The ARRL has reports of problems with nearly every auto manufacturer. Check with your dealer before you install a transceiver in a car. The dealer can direct you to any service bulletins or information that is applicable to your model. If you are not satisfied with the dealer's response, contact the regional or factory customer service representatives.

The ARRL TIS maintains a file on each manufacturer, as well as a list of contact people that can help. The TIS can also supply a reprint of the General Motors installation guidelines.[5] For additional information about automotive EMC, refer to the Automobiles chapter in the *ARRL RFI book.*

Electrical Noise

Many electrical appliances and power lines can generate electrical noise. On a receiver, electrical noise usually sounds like a rough buzz, heard across a wide frequency range. The buzz will either have a strong 60- or 120-Hz component, or its pitch will vary with the speed of a motor that generates the noise. The appearance of electrical noise on a television set is shown in the TVI section of this chapter. This kind of noise can come from power lines, electrical motors or switches, to name just a few. Here is one quick diagnostic trick — if electrical noise seems to come and go with the weather, the source is probably outside, usually on the power lines. If electrical noise varies with the time of day, it is usually related to what people are doing, so look to your own, or your neighbors', house and lifestyle. The ARRL RFI book describes techniques for locating RFI sources.

Filters usually cure electrical noise. At its source, the noise can usually be filtered with a differential-mode filter. A differen-

[5]The ARRL "EMI/RFI — GM Installations Guidelines" is a reprint of the General Motors booklet. Send a 9×12-inch SASE with three units of First-Class postage and a specific request for the "EMI/RFI GM Installation Guidelines" to TIS at ARRL HQ.

tial-mode filter can be as simple as a 0.01-μF ac-rated capacitor, such as Panasonic part ECQ-U2A103MN, or it can be a pi-section filter like that shown in Fig 28.7.

For removing signals that arrive via power lines, a common-mode choke is usually the best defense. Wrap about 10 turns of the ac-power cord around an FT-240-43 ferrite core; do this as close as possible to the device you are trying to protect.

Electrical noise can also indicate a dangerous electrical condition that needs to be corrected. The ARRL has recorded several cases where defective or arcing doorbell transformers caused widespread neighborhood electrical interference. This subject is well covered in the *Interference Handbook* or *The ARRL RFI Book*.[6]

Power Lines — Electrical noise can also come from power lines. Diagnosis and repair of power-line problems *must* be left to professionals. It is even dangerous to tap poles with a hammer as a diagnostic tool; broken insulators can fall and possibly strike people or passing cars. If you have a problem with power-line noise, contact the utility company for help. Most electric utility companies have qualified, knowledgeable personnel to correct EMI problems.

In Conclusion

Remember that EMI problems can be cured. With the proper technical knowledge and interpersonal skills, you can deal effectively with the people and hardware that make up any EMI problem.

[6]*Interference Handbook*, ARRL Order no. 6015. *The ARRL RFI Book*, ARRL Order no. 6834. Both are available from ARRL Publication Sales or your local Amateur Radio equipment dealer.

Contents

Regulations
29

Glossary

Amateur Service—A radio communication service for the purpose of self-training, intercommunication and technical investigations carried out by amateurs, that is, duly authorized persons interested in radio technique solely with a personal aim and without pecuniary interest.

Auxiliary station—An amateur station, other than in a message-forwarding system, transmitting communications point-to-point within a system of cooperating amateur stations.

Bandwidth—The width of a frequency band outside of which the mean power of the transmitted signal is attenuated at least 26 dB below the mean power of the transmitted signal within the band.

Beacon—An amateur station transmitting communications for the purposes of observation of propagation and reception or other related experimental activities.

Broadcasting—Transmissions intended for reception by the general public, either direct or relayed.

Carrier power—The average power supplied to the antenna transmission line by a transmitter during one RF cycle taken under the condition of no modulation.

Certification—An equipment authorization granted by the FCC. It is used to ensure that equipment will function properly in the service for which it has been accepted. Most amateur equipment does not require FCC certification, although HF power amplifiers and amplifier kits do. Part 15 Rules require FCC certification for all receivers operating anywhere between 30 and 960 MHz. Amateur transmitters may not be legally used in any other service that requires FCC equipment authorization. For example, it is illegal to modify an amateur transmitter to operate on police, fire or business services.

Control operator—An amateur operator designated by the licensee of a station to be responsible for the transmissions from that station to assure compliance with the FCC Rules.

Control point—The location at which the control operator function is performed.

Covenants—Private contractual agreements between two parties. PRB-1 does not apply to such agreements

CW—International Morse code telegraphy emissions having designators with A, C, H, J or R as the first symbol; 1 as the second symbol; A or B as the third symbol; and emissions J2A and J2B. See the sidebar, "Classification of Emissions."

Data—Telemetry, telecommand and computer communications emissions having designators with A, C, D, F, G, H, J or R as the first symbol; 1 as the second symbol; D as the third symbol; and emission J2D. Only a digital code of a type specifically authorized in this Part may be transmitted. See the sidebar, "Classification of Emissions."

External RF power amplifier—A device capable of increasing power output when used in conjunction with, but not an integral part of, a transmitter.

External RF power amplifier kit—A number of electronic parts, which, when assembled, is an external RF power amplifier, even if additional parts are required to complete assembly.

FCC—Federal Communications Commission.

Frequency coordinator—An entity, recognized in a local or regional area by amateur operators whose stations are eligible to be auxiliary or repeater stations, that recommends transmit/receive channels and associated operating and technical parameters for such stations in order to avoid or minimize potential interference.

Harmful interference—Interference which endangers the functioning of a radionavigation service or of other safety services or seriously degrades, obstructs or repeatedly interrupts a radiocommunication service operating in accordance with the international Radio Regulations.

Image—Facsimile and television emissions having designators with A, C, D, F, G, H, J or R as the first symbol; 1, 2 or 3 as the second symbol; C or F as the third symbol; and emissions having B as the first symbol; 7, 8 or 9 as the second symbol; W as the third symbol. See the sidebar, "Classification of Emissions."

Information bulletin—A message directed only to amateur operators consisting solely of subject matter of direct interest to the amateur service.

International Morse code—A dot-dash code as defined in International Telegraph and Telephone Consultative Committee (CCITT) Recommendation F.1 (1984), Division B, I. Morse Code.

ITU—International Telecommunication Union.

Key clicks—Undesired switching transients beyond the necessary bandwidth of a Morse code transmission caused by improperly shaped modulation envelopes.

Mean power—The average power supplied to an antenna transmission line during an interval of time sufficiently long compared with the lowest frequency encountered in the modulation taken under normal operating conditions.

Necessary bandwidth—The width of the transmitted frequency band which is just sufficient to ensure the transmission of information at the rate and with the quality required under specified conditions.

Out-of-band emission (splatter)—An emission on a frequency immediately outside the necessary bandwidth caused by overmodulation on peaks (excluding spurious emissions).

PEP (peak envelope power)—The average power supplied to the antenna transmission line by a transmitter during one RF cycle at the crest of the modulation envelope taken under normal operating conditions.

Phone—Emissions carrying speech or other sound information having designators with A, C, D, F, G, H, J or R as the first symbol; 1, 2 or 3 as the second symbol; E as the third symbol. Also speech emissions having B as the first symbol; 7, 8 or 9 as the second symbol; E as the third symbol. See the sidebar, "Classification of Emissions."

Power—Power is expressed in three ways: (1) Peak envelope power (PEP); (2) Mean power; and (3) Carrier power.

PRB-1—The limited federal government preemption of local zoning ordinances which states that local zoning authorities must be reasonable and that regulation must represent the minimum practicable to accomplish its legitimate purpose.

Pulse—Emissions having designators with K, L, M, P, Q, V or W as the first symbol; 0, 1, 2, 3, 7, 8, 9 or X as the second symbol; A, B, C, D, E, F, N, W or X as the third symbol. See the sidebar, "Classification of Emissions."

RF—Radio frequency.

Radio Regulations—The latest ITU *Radio Regulations*.

RACES (Radio Amateur Civil Emergency Service)—A radio service that uses amateur stations for civil defense communications during periods of local, regional or national civil emergencies.

Remote control—The use of a control operator who indirectly manipulates the operating adjustments in the station through a control link to achieve compliance with the FCC Rules.

Repeater—An amateur station that simultaneously retransmits the signals of other stations on a different channel or channels.

RTTY—Narrow-band direct-printing telegraphy emissions having designators with A, C, D, F, G, H, J or R as the first symbol; 1 as the second symbol; B as the third symbol; and emission J2B. See the sidebar, "Classification of Emissions."

Space station—An amateur station located more than 50 km above the Earth's surface.

Splatter—See *Out-of-band emission*.

Spread Spectrum (SS)—Emissions using bandwidth-expansion modulation emissions having designators with A, C, D, F, G, H, J or R as the first symbol; X as the second symbol; X as the third symbol. See the sidebar, "Classification of Emissions."

Spurious emission—An emission, on frequencies outside the necessary bandwidth of a transmission, the level of which may be reduced without affecting the information being transmitted. They include harmonic emissions, intermodulation products and frequency conversion products, but exclude out-of-band emissions.

Telecommand—A one-way transmission to initiate, modify, or terminate functions of a device at a distance.

Telecommand station—An amateur station that transmits communications to initiate, modify, or terminate functions of a space station.

Telemetry—A one-way transmission of measurements at a distance from the measuring instrument.

Test—Emissions containing no information having the designators with N as the third symbol. Test does not include pulse emissions with no information or modulation unless pulse emissions are also authorized in the frequency band. See the sidebar, "Classification of Emissions."

International and national radio regulations govern the operational and technical standards of all radio stations. The International Telecommunication Union (ITU) governs telecommunications on the international level and broadly defines radio services through the international Radio Regulations. In the United States, its trust territories and possessions, the agency responsible for nongovernmental and nonmilitary stations is the Federal Communications Commission (FCC). Title 47 of the *US Code of Federal Regulations* governs telecommunications. Different rule Parts of Title 47 govern the various radio services in the US. The Amateur Radio Service is governed by Part 97. Some other Parts are described in the sidebar "Other FCC Rule 'Parts'." *The ARRL RFI Book* contains a detailed chapter on these FCC rule parts which affect Amateur Radio directly and indirectly.

Experimentation has been the backbone of Amateur Radio for almost a century and the amateur rules provide a framework

---

Other FCC Rule "Parts"

Part 97 is just a small piece of the overall regulatory picture. An up-to-date copy of Part 97 can be found on the web at **http://www.arrl.org/field/ regulations/news/part97**. The *US Code of Federal Regulations*, Title 47, consists of telecommunications rules numbered as Parts 0 through 300. These Parts contain specific rules for the many telecommunications services the FCC administers. Individuals may purchase or obtain from the Web a specific rule Part for a particular service from the Superintendent of Documents, US Government Printing Office (see the Address List in the **References** chapter). Here is a list of FCC Parts amateurs may find of interest:

Part
0   Commission organization
1   Practice and procedure
2   Frequency allocation and radio treaty matters; general rules and regulations, and type acceptance procedures
15   Low-power radio-frequency transmitting devices
17   Construction, marking and lighting of antenna structures
18   Industrial, scientific and medical equipment
73   Radio broadcast services
76   Cable Television Service
90   Private Land Mobile Radio Service
95   Personal radio services, including CB and GMRS
97   Amateur Radio Service

Classification of Emissions

§2.201 Emission, modulation and transmission characteristics.

The following system of designating emission, modulation and transmission characteristics shall be employed.

(a) Emissions are designated according to their classification and their necessary bandwidth.

(b) A minimum of three symbols are used to describe the basic characteristics of radio waves. Emissions are classified and symbolized according to the following characteristics:

(1) First symbol—type of modulation of the main carrier;

(2) Second symbol—nature of signal(s) modulating the main carrier;

(3) Third symbol—type of information to be transmitted.

Note: A fourth and fifth symbol are provided for additional information and are shown in Appendix 6, Part A of the ITU Radio Regulations. Use of the fourth and fifth symbol is optional. Therefore, the symbols may be used as described in Appendix 6, but are not required by the Commission.

(c) First symbol—types of modulation of the main carrier:

(1) Emission of an unmodulated carrier .... N
(2) Emission in which the main carrier is amplitude-modulated (including cases where subcarriers are angle-modulated):
—Double sideband .... A
—Single sideband, full carrier .... H
—Single sideband, reduced or variable level carrier ... R
—Single sideband, suppressed carrier .... J
—Independent sidebands .... B
—Vestigial sideband C
(3) Emission in which the main carrier is angle-modulated:
—Frequency modulation .... F
—Phase modulation .... G

Note: Whenever frequency modulation (F) is indicated, phase modulation (G) is also acceptable.

(4) Emission in which the main carrier is amplitude- and angle-modulated either simultaneously or in a pre-established sequence .... D
(5) Emission of pulses[1]
—Sequence of unmodulated pulses .... P
—A sequence of pulses:
—Modulated in amplitude .... K
—Modulated in width/duration L
—Modulated in position/phase .... M
—In which the carrier is angle-modulated during the period of the pulse .... Q

—Which is a combination of the foregoing or is produced by other means V
(6) Cases not covered above, in which an emission consists of the main carrier modulated, either simultaneously or in a pre-established sequence in a combination of two or more of the following modes: amplitude, angle, pulse .... W
(7) Cases not otherwise covered X

(d) Second Symbol—nature of signal(s) modulating the main carrier:

(1) No modulating signal 0
(2) A single channel containing quantized or digital information without the use of a modulating subcarrier, excluding time-division multiplex 1
(3) A single channel containing quantized or digital information with the use of a modulating subcarrier, time-division multiplex 2
(4) A single channel containing analog information 3
(5) Two or more channels containing quantized or digital information 7
(6) Two or more channels containing analog information 8
(7) Composite system with one or more channels containing quantized or digital information, together with one or more channels containing analog information 9
(8) Cases not otherwise covered X

(e) Third Symbol—type of information to be transmitted:[2]

(1) No information transmitted .... N
(2) Telegraphy, for aural reception .... A
(3) Telegraphy, for automatic reception .... B
(4) Facsimile C
(5) Data transmission, telemetry, telecommand .... D
(6) Telephony (including sound broadcasting) .... E
(7) Television (video) .... F
(8) Combination of the above .... W
(9) Cases not otherwise covered X

(f) Type *B* emission: As an exception to the above principles, damped waves are symbolized in the Commission's rules and regulations as type *B* emission. The use of type B emissions is forbidden.

(g) Whenever the full designation of an emission is necessary, the symbol for that emission, as given above, shall be preceded by the necessary bandwidth of the emission as indicated in § 2.202(b)(1).

[1]Emissions where the main carrier is directly modulated by a signal which has been coded into quantized form (e.g., pulse code modulation) should be designated under (2) or (3).

[2]In this context, the word "information" does not include information of a constant, unvarying nature such as is provided by standard frequency emissions, continuous wave and pulse radars, etc.

Note: For an electronic copy of this information, see **http://www.itu.int/radioclub/rr/aps01.htm**.

---

within which amateurs have wide latitude to experiment in accordance with the basis and purpose of the service. The rules should be viewed as vehicles to promote healthy activity and growth, not as constraints that lead to stagnation. A brief overview of Amateur Radio regulations follows with special emphasis on technical standards.

BASIS AND PURPOSE OF THE AMATEUR RADIO SERVICE

There's much more in the regulatory scheme than Part 97. The basis for the FCC regulations is found in treaties, international agreements and statutes that provide for the allocation of frequencies and place conditions on how the frequencies are to be used. For example, Article S25 of the international *Radio Regulations* limits the types of international communications amateur stations may transmit and mandates that the technical qualifications of amateur operators be verified.

It's the FCC's responsibility to see that amateurs are able to operate their stations in a manner consistent with the basis and purpose of the amateur rules. The FCC must also ensure that amateurs have the knowledge and ability to operate powerful and potentially dangerous equipment safely without causing harmful interference to others. The sidebar "The FCC's Role" discusses the Commission in a bit more detail. A review of each of the five basic purposes of the Amateur Radio Service, as they appear in Part 97, follows:

Recognition and enhancement of the value of the amateur service to the public as a voluntary noncommercial communication service, particularly with respect to providing emergency communications [97.1(a)].

Probably the best known aspect of Amateur Radio to the general public is its ability to provide emergency communications. One of the most important aspects of the service is its noncommercial nature. Amateurs are prohibited from receiving any form of payment for operating their stations.

Continuation and extension of the amateur's proven ability to contribute to the advancement of the radio art [97.1(b)].

For nearly a century, hams have carried on a tradition of learning by doing, and since the beginning have remained at the forefront of technology. Through experimenting and building, hams have pioneered advances, such as techniques for single-sideband transmissions, and are currently engaged in the development of new digital schemes which continue to improve the efficiency of such communications. Hams' practical experience has led to technical refinements and cost reductions beneficial to the commercial radio industry.

Encouragement and improvement of the Amateur Radio Service through rules which provide for advancing skills in both the communication and technical phases of the art [97.1(c)]

The FCC's Role

The Federal Communications Commission (FCC) is the US government agency charged by Congress with regulating communications involving radio, television, wire, cable and satellites. This includes Amateur Radio. The objective of the FCC is to provide for orderly development and operation of telecommunications services.

The FCC functions like no other Federal agency. It was created by Congress and it reports directly to Congress. The FCC allocates bands of frequencies to nongovernment communications services and assigns operator privileges. (The National Telecommunications and Information Administration allocates government frequencies.)

Amateurs have always been experimenters and that's what sets the Amateur Service apart from other services. The cost to the government for licensing and enforcement is minimal when compared to the benefit the public receives. Hams have contributed greatly to the development of computer communications techniques. The FCC and industry have also credited the amateur community with the development of Low-Earth-Orbit (LEO) satellite technology. The same can be said for a number of digital modes that have arrived on the scene in the 1990s, such as PacTOR, CLOVER and PSK31.

Expansion of the existing reservoir within the Amateur Radio Service of trained operators, technicians and electronic experts [97.1(d)]

Amateurs learn by doing. While all amateurs may not be able to troubleshoot and repair a transceiver, all amateurs have some degree of technical competence.

Continuation and extension of the amateur's unique ability to enhance international goodwill [97.1(e)].

Amateur Radio is one of the few truly international hobbies. It is up to amateurs to maintain high standards and to represent the US as its ambassadors, because, in a sense, all US amateurs serve that function.

A QUICK JOURNEY THROUGH PART 97

The Amateur Radio Service rules, Part 97, are organized in six major subparts: General Provisions, Station Operation Standards, Special Operations, Technical Standards, Providing Emergency Communications and Qualifying Examination Systems. A brief discussion of the highlights of each subpart follows:

General Provisions

Subpart A covers the basics that apply to all facets of Amateur Radio. The "Basis and Purpose" of Amateur Radio, discussed above, is found at the beginning of Part 97 [97.1]. Definitions of key terms used throughout Part 97 form the foundation of Part 97 [97.3].

The remainder of the subpart is devoted to Federal restrictions on amateur installations (see sidebar), which include a mention of FCC standards for RF exposure. The ARRL publication *RF Exposure and You* details these RF exposure requirements.

Station Operation Standards

Subpart B, "Station Operation Standards," concerns the basic operating practices that apply to all types of operation. Amateurs must operate their stations in accordance with good engineering and

Federal Restrictions on the Installation of Amateur Stations

The following Federal restrictions apply to amateur antennas (see *The FCC Rule Book* for complete details):

- Amateurs must take certain actions before placing an amateur station on land of environmental importance or on land significant in American history, architecture, or culture [97.13(a)].
- Amateurs must protect FCC monitoring stations from harmful interference if an amateur is operating within one mile of such a facility [97.13(b)].
- Before causing or allowing an amateur station to transmit from any place where its operation could cause RF exposure in excess of Federal guidelines, an RF exposure evaluation may be required. [97.13(c)].
- Amateurs must have the FAA's approval before installing a tower over 200 ft high and the structure must be registered with the FCC per Part 17 of the *Code of Federal Regulations* (Construction, Marking and Lighting of Antenna Structures) [97.15]. See **http://www.fcc.gov/wtb/antenna/what.html**.
- Amateurs located near airports must meet additional limitations [97.15]. Details appear in *The FCC Rule Book*.
- Amateurs may install an antenna up to 20 ft above the ground or other natural features or on top of any man-made structure except antenna towers. Such antennas are exempt from the rules described in Part 17 of FCC regulations.
- Amateurs must notify the Interference Office of the National Radio Astronomy Observatory if a beacon station is planned for the National Radio Quiet Zone, a small area in Maryland, Virginia and West Virginia [97.203(e) and 1.924]. New stations within a 10 mile radius of the Arecibo Observatory in Puerto Rico [97.203(h), 1.924] must notify the Interference Office.

amateur practice [97.101(a)]. Part 97 doesn't always tell amateurs specifically how to operate their stations, particularly concerning technical issues, but the FCC provides broad guidelines. The use of good engineering and amateur practice means, for example, that amateurs shouldn't operate a station with a distorted signal and that amateurs shouldn't operate on a busy band like 20 m just to talk to a ham across town. Also, amateurs must share the frequencies with others—no one ham or group has any special claim to any frequency [97.101(b)]. The station licensee is always responsible for the proper operation of an amateur station, except where the control operator is someone other than the station licensee, in which case both share responsibility equally [97.103(a)].

The requirements for control operators, station control and reciprocal licensing authority are also addressed in Subpart B. Each station must have a control point [97.109(a)]. A control operator must always be at the control point, except in a few cases where the transmitter is controlled automatically [97.109(b), (c), (d) and (e)]. The purpose of the Amateur Radio Service is to communicate with other amateurs [97.111]. Certain one-way transmissions are allowed. Amateurs can send a one-way transmission to:

- make adjustments to equipment for test purposes
- call CQ
- remotely control devices
- communicate information in emergencies
- send code practice and information bulletins of interest to amateurs [97.111(b)]

Broadcasting to the public is strictly prohibited [97.113(b)]. The section on prohibited transmissions states that amateurs cannot: be paid for operating a station, make transmissions on behalf of an employer, transmit music (unless otherwise allowed in the rules), transmit obscenity, use amateur stations for newsgathering purposes or transmit false signals and ciphers. The FCC has relaxed the previously restrictive business rules to encourage public service and personal communications [97.113].

Station identification is addressed in this subpart. The purpose of station identification is to make the source of its transmissions known to those receiving them, including FCC monitors. The rules cover identification requirements for the various operating modes. Section 97.119 details the station-identification requirements. Amateurs must transmit their call sign at the end of the communication and every 10 minutes during communications. CW and phone may be used to identify an

Local Zoning Ordinances, Covenants and Deed Restrictions

Amateurs may be restricted by local antenna zoning ordinances. PRB-1 is a limited preemption of local zoning ordinances. It outlines three rules for local municipalities to follow in regulating antenna structures: (1) state and local regulations that operate to preclude amateur communications are in direct conflict with federal objectives and must be preempted; (2) local regulations that involve placement, screening or height of antennas based on health, safety or aesthetic considerations must be crafted to reasonably accommodate amateur communications; and (3) such local regulations must represent the minimum practicable regulation to accomplish the local authority's legitimate purpose. Amateurs faced with unreasonable zoning restrictions should request the "PRB-1 package" from the Regulatory Information Branch at ARRL HQ. *The FCC Rule Book* gives details.

PRB-1 does not address deed, lease or rental covenants or restrictions; in fact, it specifically excludes them since such agreements are *voluntarily* entered into by the buyer or tenant and seller or landlord in their contract. Amateurs restricted by covenants or deed restrictions may contact the Regulatory Information Branch at ARRL HQ for suggested wording that may be written into such agreements.

amateur station. RTTY and data (using a specified digital code) may be used when all or part of the communications are transmitted using such an emission. Images (Amateur Television, for example) may be used to identify when all or part of the transmission is in that mode. A final section addresses restricted operation and sets forth the conditions that must exist in an interference case involving a neighbor's TV or radio before the Commission can impose "quiet hours"—hours of the day when a particular amateur may not operate an amateur transmitter [97.121(a)]. Imposition of quiet hours by the FCC is rare, however.

Special Operations

Subpart C, "Special Operations," addresses specialized activities of Amateur Radio including the various types of stations an amateur may operate. This subpart gives specific guidelines concerning repeaters, beacons, space stations, Earth stations, message forwarding systems, and telecommand (remote control) stations. These rules are of particular interest to the technically minded amateur. An amateur may send ancillary functions (user functions) of a repeater on the input of the repeater—to turn on and off an autopatch, for example. However, the primary control links used to turn the repeater on and off, for example, may be transmitted only above 222.150 MHz since such one-way transmissions are auxiliary transmissions. Every repeater trustee/licensee and user should understand the rules for repeaters and auxiliary links. An important regulatory approach to solving interference problems between repeaters is addressed

in that section: Repeater station licensees are equally responsible for resolving an interference problem, unless one of the repeaters has been approved for operation by the recognized repeater coordinator for the area and the other has not. In that case, the owner of the uncoordinated repeater has primary responsibility to resolve the problem [97.205(c)]. The control operator of a repeater that inadvertently retransmits communications in violation of the rules is not held accountable [97.205(g)]. The originator and first forwarding station of a message transmitted through a message forwarding system are held accountable for any violations of Part 97. Other forwarding stations are not held accountable [97.219]. For a detailed explanation, see the ARRL's *FCC Rule Book*.

Technical Standards

The word *standard* means consistency and order—and this is what the technical standards in Subpart D are all about. The FCC outlines the specific frequency bands available to US amateurs [97.301] as well as the sharing agreements [97.303]. (The **References** chapter of this book includes a table of frequencies allocated to the Amateur Radio Service.) The Commission made these standards a basic framework so all types of amateur operation may peacefully coexist with other radio occupants in the spectrum neighborhood. Emission standards for RTTY, data and spread spectrum are discussed, as are standards for the type acceptance of RF power amplifiers. FCC type acceptance is not needed for most amateur equipment. This gives amateurs the freedom to experiment without being bound by specific equipment standards.

Providing Emergency Communications

Subpart E, "Providing Emergency Communications," addresses disaster communications, stations in distress, communications for the safety of life and protection of property and the Radio Amateur Civil Emergency Service (RACES).

Qualifying Examination Systems

The final subpart of the rules, Subpart F, deals with the examination system and covers exam requirements and elements and standards. In 1983, the Commission delegated much of the exam administration program to amateurs themselves. The rules provide for checks and balances on volunteer examiners (VEs), who administer exams at the local and regional levels. These checks and balances protect against fraud and provide integrity for the exam process.

REGULATION, THE ITU AND FREQUENCY SHARING

The International Telecommunication Union (ITU), an agency of the United Nations, plays the vital role of dividing the spectrum for reallocation by the telecommunications authorities of individual countries. For convenience in organizing frequency allocations for the various services, the ITU divides the world into three Regions: 1, 2 and 3. See **Fig 29.1**. North and South America and the adjoining waters comprise Region 2.

Frequency allocations for Amateur Radio and other services can differ among ITU Regions. Today, there are numerous radio services vying for pieces of the spectrum, all with legitimate purposes. To accommodate them all, the international and domestic regulatory agencies may allocate the same frequency bands to more than one service. While some of our allocations are

exclusive, we share others on the basis of priority. In some cases, the Amateur Radio Service is *primary*. In other cases, amateurs are *secondary*. Stations in a secondary service must not cause harmful interference to, and are not protected from interference caused by stations in the primary service.

The FCC specifies amateur frequencies and what, if any, restrictions apply. Since the detailed list of frequencies and the sharing requirements is far too long to be included in this short chapter, they can be found in Subpart D of Part 97. *The FCC Rule Book* and *The ARRL Operating Manual* give details of the sharing requirements of each band.

Guidelines based upon the actual sharing requirements of amateur frequencies outlined in Section 97.303 follow:

- A station in a secondary service must not cause interference to and must accept interference from stations in the primary service.
- Stations in one ITU Region shall not cause harmful interference to services in another Region.
- Many other services share frequencies with amateurs on and above the 70-cm band, and amateurs must be mindful of these services.
- Amateurs near large military bases should check Section 97.303 carefully, since additional restrictions may apply.

Voluntary Band Plans

Another aspect of frequency sharing is voluntary band plans. Although the FCC Rules set aside portions of some bands for specific modes, there's still a need to further divide amateur bands among user groups by "gentlemen's agreements." These agreements usually emerge by con-

sensus of the band occupants, and are sanctioned by a national body like ARRL. These agreements allow many modes of operation to be used by amateurs in a given band. For example, amateurs avoid the domestic "DX windows" set aside on some bands, so that stations can hear and work weak-signal DX stations. Detailed band plans for the amateur HF bands can be found in the **References** chapter of this book. The complete band plans can be found in *The FCC Rule Book* and *The ARRL Repeater Directory*.

Emission Standards and Bandwidth

Like most of Part 97, the technical standards exist to promote operating techniques that make efficient use of the spectrum and minimize interference. The standards in Part 97 identify problems that must be solved. Section 97.307 spells out the standards FCC expects amateur signals to meet. It states, in part: "No amateur station transmission shall occupy more bandwidth than necessary for the information rate and emission type being transmitted, in accordance with good amateur practice" [97.307(a)]. Simply stated, don't transmit a wide signal when a narrow one will do. Specific bandwidth limits are given for RTTY and data emissions. Specific bandwidth limits are not given for other modes of operation, but amateurs must still observe good engineering and operator practice.

The rules state: "Emissions resulting from modulation must be confined to the band or segment available to the control operator" [97.307(b)]. Every modulated signal produces sidebands. Amateurs must not operate so close to the band edge that the sidebands extend out of the subband, even if the frequency readout says that the carrier is inside the band. Further: "Emissions outside the necessary bandwidth must not cause splatter or key-click interference to operations on adjacent frequencies" [97.307(b)]. The rules simply codify good operating practice. Key clicks or over-processed voice signals shouldn't cause interference up and down the band.

Spurious emissions

Spurious emissions include harmonic emissions, parasitic emissions, intermodulation products and frequency conversion products, but do not include splatter [97.307(c)]. Definitions for *necessary bandwidth* and *out-of-band emission* appear in the **Glossary** at the beginning of this chapter. Also see **Fig 29.2**.

Emission standards

The FCC is very specific concerning spurious emission standards [97.307(d)]. If an amateur transmitter or RF power am-

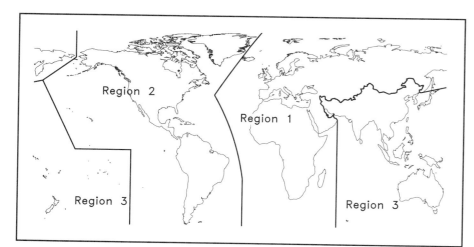

Fig 29.1—Frequency allocations for Amateur Radio and other radio services can differ, depending on location. There are three administrative ITU Regions, each with its own allocations.

plifier was built after April 14, 1977, or first marketed after December 31, 1977, and transmits on frequencies below 30 MHz, the mean power of any spurious emissions must:

• never be more than 50 mW;
• be at least 30 dB below the mean power of the fundamental emission, if the mean power output is less than 5 W; and
• be at least 40 dB below the mean power of the fundamental emission if the mean power output is 5 W or more.

The requirement that no spurious emission exceed 50 mW means that above 500 W, the suppression must be greater than 40 dB. At 1500 W, the suppression must be 44.77 dB. The requirements for transmitters operating below 30 MHz are shown graphically in **Fig 29.3**.

The following requirements apply between 30 and 225 MHz [97.307(e)]:

• In transmitters with 25 W or less mean output power, spurs must be at least 40 dB below the mean power of the fundamental emission and never greater than 25 µW (microwatt), but need not be reduced further than 10 µW. This means that the spurs from a 25-W transmitter must be at least 60 dB down to meet the 25-W restriction.
• In transmitters with more than 25 W mean output power, spurious emissions must be at least 60 dB below the mean power of the fundamental emission.

The situation for transmitters operating between 30 and 225 MHz is more complex. The combination of the requirement that spurious emissions be less than 25 µW and the stipulation that they don't need to be reduced below 10 µW makes the requirements vary significantly with power level. This ranges from 0 dB suppression required for a transmitter whose power is 10 µW to 60 dB of suppression required for power levels above 25 W. The requirements for transmitter operation between 30 and 225 MHz are shown graphically in **Fig 29.4**. There are no absolute limits for transmitters operating above 225 MHz, although the requirements for good engineering practice would still apply.

Transmitter Power Standards

Amateurs shall not use more power than necessary to carry out the desired communication [97.313(a)]. Don't use 700 W when 10 m is wide open, for example. No station may use more than 1.5 kW peak envelope power [97.313(b)] and no station may use more than 200 W in the 30-m band. Novices and Technicians are limited to 200 W in their HF segments [97.313(c)]. Amateurs may use no more than 50 W in the 70-cm band near certain

military installations [97.313(f) and (g)].

The FCC has chosen and published the following standards of measurement: (1) Read an in-line peak-reading RF wattmeter that is properly matched; and (2) calculate the power using the peak RF voltage as indicated by an oscilloscope or other peak-reading device. Multiply the peak RF voltage by 0.707, square the result and divide by the load resistance. The SWR must be 1:1.

The FCC requires that you meet the power output regulations, but does not require that you make such measurements or possess measurement equipment. The methods listed simply indicate how the Commission would measure your transmitter's output during a station inspection.

As a practical matter, most hams don't have to worry about special equipment to check their transmitter's output because they never approach the 1500-W PEP out-

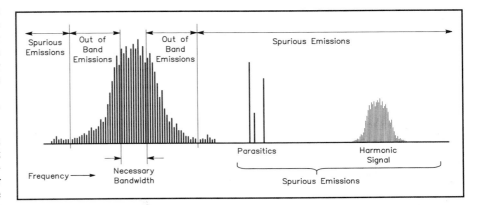

Fig 29.2—Some of the modulation products are outside the necessary bandwidth. These are out-of-band emissions, but they are not considered spurious emissions. On the other hand, these out-of-band emissions must not interfere with other stations [97.307(b)]. The harmonics and parasitics shown in this figure are spurious emissions, and they must be reduced to the levels specified in Part 97. The FCC states that all spurious emissions must be reduced "to the greatest extent practicable" [97.307(c)]. Further, if any spurious emission, including chassis or power-line radiation, causes harmful interference to the reception of another radio station, the licensee of the interfering amateur station is required to take steps to eliminate the interference.

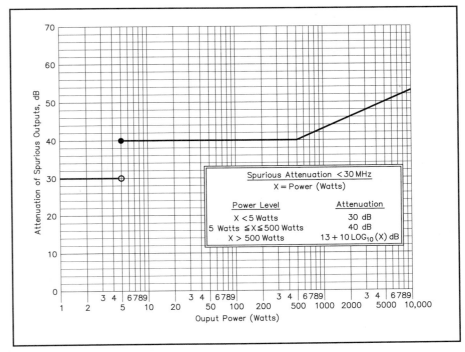

Fig 29.3—Required attenuation of spurious outputs below 30 MHz is related to output power.

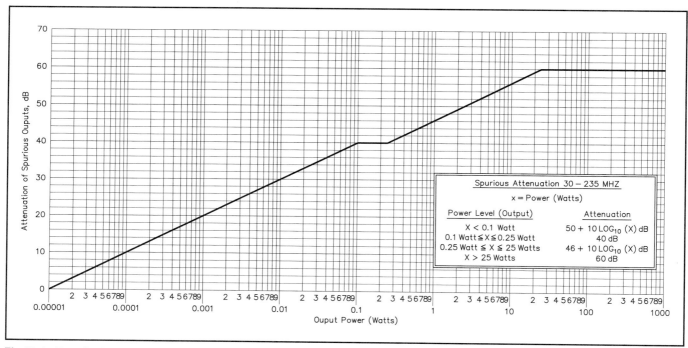

Fig 29.4—Required attenuation of spurious outputs, 30-235 MHz.

put limit. Many common amplifiers aren't capable of generating this much power. However, if you do have a capable amplifier and do operate close to the limit, you should be prepared to measure your output along the lines detailed above.

External RF Power Amplifiers: Certification and Standards

In 1978, the FCC banned the manufacture and marketing of any external RF power amplifier or amplifier kit capable of operation on any frequency below 144 MHz, unless the FCC has issued a grant of type acceptance (now called FCC certification) for that model amplifier. The FCC also banned the manufacture and marketing of HF amplifiers that were capable of operation on 10 m to stem the flow of amplifiers being distributed for illegal use in and around frequencies used by CB operators.

Amateurs may still use amplifiers capable of operation on 10 m. While the rules may make it difficult to buy a new amplifier capable of operation on 10 m, the FCC allows amateurs to modify an amplifier to restore or include 10-m capability. An amateur may modify no more than one unit of the same model amplifier in any year without FCC certification [97.315(a)].

Of course, amateurs are permitted to build amplifiers, convert equipment from any other radio service for this use or to buy used amplifiers. When converting equipment from other services, it must meet all technical standards outlined in Part 97 and it can no longer be used in the service for which is was intended since the type acceptance would have been voided. Nonamateurs are specifically prohibited from building or modifying amplifiers capable of operation below 144 MHz without FCC certification [97.315(a)]. All external amplifiers and amplifier kits capable of operation below 144 MHz must be FCC certified in order to be marketed [97.315(b)]. A number of amplifiers, manufactured prior to the April 28, 1978, cutoff were issued a waiver of the new regulations [97.315(b)(2)]. Amateurs may buy or sell an amplifier that has either been FCC certified, granted a waiver or modified so that the certification is no longer valid. There are restrictions that would be valid regardless of whether the amplifier was capable of operation below 144 MHz. Some amplifiers marketed before April 28, 1978, are covered under the waiver if they are the same model that was granted a waiver [97.315(b)(2)]. An individual amateur may sell his amplifier regardless of grants or waivers, provided that he sells it only to another amateur operator [97.315(b)(4)]. Amateurs may also sell a used amplifier to a bona fide amateur equipment dealer [97.315(b)(5)]. The dealer could sell those amplifiers only to other hams [97.315(b)(5)].

In some cases, the FCC will deny certification. Some features that may cause a denial are (1) any accessible wiring which, when altered, would permit operation in a manner contrary to FCC rules; (2) circuit boards or similar circuitry to facilitate the addition of components to change the amplifier's operation characteristics in a manner contrary to FCC rules; (3) for operation or modification of the amplifier in a manner contrary to FCC rules; (4) any internal or external controls or adjustments to facilitate operation of the amplifier in a manner contrary to FCC rules; (5) any internal RF sensing circuitry or any external switch, the purpose of which is to place the amplifier in the transmit mode; (6) the incorporation of more gain than is necessary to operate in the amateur service.

CONCLUSION

A common thread in Amateur Radio's history has been a dynamic regulatory environment that has nurtured technological growth and diversity. This thread continues to sew together the elements of Amateur Radio today and prepare it for tomorrow's challenges.

References
30

US Customary to Metric Conversion Factors

International System of Units (SI)—Metric Units

Prefix	Symbol		Multiplication Factor	
exe	E	10^{18}	=	1,000,000 000,000,000,000
peta	P	10^{15}	=	1,000,000,000,000,000
tera	T	10^{12}	=	1,000,000,000,000
giga	G	10^{9}	=	1,000,000,000
mega	M	10^{6}	=	1,000,000
kilo	k	10^{3}	=	1,000
hecto	h	10^{2}	=	100
deca	da	10^{1}	=	10
(unit)		10^{0}	=	1
deci	d	10^{-1}	=	0.1
centi	c	10^{-2}	=	0.01
milli	m	10^{-3}	=	0.001
micro	μ	10^{-6}	=	0.000001
nano	n	10^{-9}	=	0.000000001
pico	p	10^{-12}	=	0.000000000001
femto	f	10^{-15}	=	0.000000000000001
atto	a	10^{-18}	=	0.000000000000000001

Linear
1 meter (m) = 100 centimeters (cm) = 1000 millimeters (mm)

Area
$1\ m^2 = 1 \times 10^4\ cm^2 = 1 \times 10^6\ mm^2$

Volume
$1\ m^3 = 1 \times 10^6\ cm^3 = 1 \times 10^9\ mm^3$
$1\ liter\ (l) = 1000\ cm^3 = 1 \times 10^6\ mm^3$

Mass
1 kilogram (kg) = 1000 grams (g)
  (Approximately the mass of 1 liter of water)
1 metric ton (or tonne) = 1000 kg

Multiply →
Metric Unit = Conversion Factor × US Customary Unit

← Divide
Metric Unit ÷ Conversion Factor = US Customary Unit

Metric Unit =	Conversion Factor ×	US Unit		Metric Unit =	Conversion Factor ×	US Unit
(Length)				**(Volume)**		
mm	25.4	inch		mm^3	16387.064	in^3
cm	2.54	inch		cm^3	16.387	in^3
cm	30.48	foot		m^3	0.028316	ft^3
m	0.3048	foot		m^3	0.764555	yd^3
m	0.9144	yard		ml	16.387	in^3
km	1.609	mile		ml	29.57	fl oz
km	1.852	nautical mile		ml	473	pint
				ml	946.333	quart
(Area)				l	28.32	ft^3
mm^2	645.16	$inch^2$		l	0.9463	quart
cm^2	6.4516	in^2		l	3.785	gallon
cm^2	929.03	ft^2		l	1.101	dry quart
m^2	0.0929	ft^2		l	8.809	peck
cm^2	8361.3	yd^2		l	35.238	bushel
m^2	0.83613	yd^2				
m^2	4047	acre		**(Mass)**	**(Troy Weight)**	
km^2	2.59	mi^2		g	31.103	oz t
				g	373.248	lb t
(Mass)	**(Avoirdupois Weight)**					
grams	0.0648	grains		**(Mass)**	**(Apothecaries' Weight)**	
g	28.349	oz		g	3.387	dr ap
g	453.59	lb		g	31.103	oz ap
kg	0.45359	lb		g	373.248	lb ap
tonne	0.907	short ton				
tonne	1.016	long ton				

US Customary Units

Linear Units
12 inches (in) = 1 foot (ft)
36 inches = 3 feet = 1 yard (yd)
1 rod = $5^{1}/_{2}$ yards = $16^{1}/_{2}$ feet
1 statute mile = 1760 yards = 5280 feet
1 nautical mile = 6076.11549 feet

Area
$1\ ft^2 = 144\ in^2$
$1\ yd^2 = 9\ ft^2 = 1296\ in^2$
$1\ rod^2 = 30^{1}/_{4}\ yd^2$
$1\ acre = 4840\ yd^2 = 43,560\ ft^2$
$1\ acre = 160\ rod^2$
$1\ mile^2 = 640\ acres$

Volume
$1\ ft^3 = 1728\ in^3$
$1\ yd^3 = 27\ ft^3$

Liquid Volume Measure
1 fluid ounce (fl oz) = 8 fluid drams = 1.804 in
1 pint (pt) = 16 fl oz
1 quart (qt) = 2 pt = 32 fl oz = $57^3/_4\ in^3$
1 gallon (gal) = 4 qt = 231 in^3
1 barrel = $31^1/_2$ gal

Dry Volume Measure
1 quart (qt) = 2 pints (pt) = 67.2 in^3
1 peck = 8 qt
1 bushel = 4 pecks = 2150.42 in^3

Avoirdupois Weight
1 dram (dr) = 27.343 grains (gr) or (gr a)
1 ounce (oz) = 437.5 gr
1 pound (lb) = 16 oz = 7000 gr
1 short ton = 2000 lb, 1 long ton = 2240 lb

Troy Weight
1 grain troy (gr t) = 1 grain avoirdupois
1 pennyweight (dwt) or (pwt) = 24 gr t
1 ounce troy (oz t) = 480 grains
1 lb t = 12 oz t = 5760 grains

Apothecaries' Weight
1 grain apothecaries' (gr ap) = 1 gr t = 1 gr
1 dram ap (dr ap) = 60 gr
1 oz ap = 1 oz t = 8 dr ap = 480 gr
1 lb ap = 1 lb t = 12 oz ap = 5760 gr

Abbreviations List

A

a—atto (prefix for 10^{-18})
A—ampere (unit of electrical current)
ac—alternating current
ACC—Affiliated Club Coordinator
ACSSB—amplitude-compandored single sideband
A/D—analog-to-digital
ADC—analog-to-digital converter
AF—audio frequency
AFC—automatic frequency control
AFSK—audio frequency-shift keying
AGC—automatic gain control
A/h—ampere hour
ALC—automatic level control
AM—amplitude modulation
AMRAD—Amateur Radio Research and Development Corp
AMSAT—Radio Amateur Satellite Corp
AMTOR—Amateur Teleprinting Over Radio
ANT—antenna
ARA—Amateur Radio Association
ARC—Amateur Radio Club
ARES—Amateur Radio Emergency Service
ARQ—Automatic repeat request
ARRL—American Radio Relay League
ARS—Amateur Radio Society (station)
ASCII—American National Standard Code for Information Interchange
ATV—amateur television
AVC—automatic volume control
AWG—American wire gauge
az-el—azimuth-elevation

B

B—bel; blower; susceptance; flux density, (inductors)
balun—balanced to unbalanced (transformer)
BC—broadcast
BCD—binary coded decimal
BCI—broadcast interference
Bd—baud (bids in single-channel binary data transmission)
BER—bit error rate
BFO—beat-frequency oscillator
bit—binary digit
bit/s—bits per second
BM—Bulletin Manager
BPF—band-pass filter
BPL—Brass Pounders League
BT—battery
BW—bandwidth

C

c—centi (prefix for 10^{-2})

C—coulomb (quantity of electric charge); capacitor
CAC—Contest Advisory Committee
CATVI—cable television interference
CB—Citizens Band (radio)
CBBS—computer bulletin-board service
CBMS—computer-based message system
CCITT—International Telegraph and Telephone Consultative Committee
CCTV—closed-circuit television
CCW—coherent CW
ccw—counterclockwise
CD—civil defense
cm—centimeter
CMOS—complimentary-symmetry metal-oxide semiconductor
coax—coaxial cable
COR—carrier-operated relay
CP—code proficiency (award)
CPU—central processing unit
CRT—cathode ray tube
CT—center tap
CTCSS—continuous tone-coded squelch system
cw—clockwise
CW—continuous wave

D

d—deci (prefix for 10^{-1})
D—diode
da—deca (prefix for 10)
D/A—digital-to-analog
DAC—digital-to-analog converter
dB—decibel (0.1 bel)
dBi—decibels above (or below) isotropic antenna
dBm—decibels above (or below) 1 milliwatt
DBM—doubly balanced mixer
dBV—decibels above/below 1 V (in video, relative to 1 V P-P)
dBW—decibels above/below 1 W
dc—direct current
D-C—direct conversion
DDS—direct digital synthesis
DEC—District Emergency Coordinator
deg—degree
DET—detector
DF—direction finding; direction finder
DIP—dual in-line package
DMM—digital multimeter
DPDT—double-pole double-throw (switch)
DPSK—differential phase-shift keying
DPST—double-pole single-throw (switch)
DS—direct sequence (spread spectrum); display
DSB—double sideband
DSP—digital signal processing
DTMF—dual-tone multifrequency

DVM—digital voltmeter
DX—long distance; duplex
DXAC—DX Advisory Committee
DXCC—DX Century Club

E

e—base of natural logarithms (2.71828)
E—voltage
EA—ARRL Educational Advisor
EC—Emergency Coordinator
ECL—emitter-coupled logic
EHF—extremely high frequency (30-300 GHz)
EIA—Electronic Industries Assn
EIRP—effective isotropic radiated power
ELF—extremely low frequency
ELT—emergency locator transmitter
EMC—electromagnetic compatibility
EME—earth-moon-earth (moonbounce)
EMF—electromotive force
EMI—electromagnetic interference
EMP—electromagnetic pulse
EOC—emergency operations center
EPROM—erasable programmable read only memory

F

f—femto (prefix for 10^{-5}); frequency
F—farad (capacitance unit); fuse
fax—facsimile
FCC—Federal Communications Commission
FD—Field Day
FEMA—Federal Emergency Management Agency
FET—field-effect transistor
FFT—fast Fourier transform
FL—filter
FM—frequency modulation
FMTV—frequency-modulated television
FSK—frequency-shift keying
FSTV—fast-scan (real-time) television
ft—foot (unit of length)

G

g—gram (unit of mass)
G—giga (prefix for 10^9); conductance
GaAs—gallium arsenide
GDO—grid- or gate-dip oscillator
GHz—gigahertz (10^9 Hz)
GND—ground

H

h—hecto (prefix for 10^2)
H—henry (unit of inductance)
HF—high frequency (3-30 MHz)
HFO—high-frequency oscillator; heterodyne frequency oscillator

HPF—highest probable frequency; high-pass filter

Hz—hertz (unit of frequency, 1 cycle/s)

I

I—current, indicating lamp

IARU—International Amateur Radio Union

IC—integrated circuit

ID—identification; inside diameter

IEEE—Institute of Electrical and Electronics Engineers

IF—intermediate frequency

IMD—intermodulation distortion

in.—inch (unit of length)

in./s—inch per second (unit of velocity)

I/O—input/output

IRC—international reply coupon

ISB—independent sideband

ITF—Interference Task Force

ITU—International Telecommunication Union

J

j—operator for complex notation, as for reactive component of an impedance ($+j$ inductive; $-j$ capacitive)

J—joule (kg m^2/s^2) (energy or work unit); jack

JFET—junction field-effect transistor

K

k—kilo (prefix for 10^3); Boltzmann's constant (1.38x10^{-23} J/K)

K—kelvin (used without degree symbol) absolute temperature scale; relay

kBd—1000 bauds

kbit—1024 bits

kbit/s—1024 bits per second

kbyte—1024 bytes

kg—kilogram

kHz—kilohertz

km—kilometer

kV—kilovolt

kW—kilowatt

kΩ—kilohm

L

l—liter (liquid volume)

L—lambert; inductor

lb—pound (force unit)

LC—inductance-capacitance

LCD—liquid crystal display

LED—light-emitting diode

LF—low frequency (30-300 kHz)

LHC—left-hand circular (polarization)

LO—local oscillator; Leadership Official

LP—log periodic

LS—loudspeaker

lsb—least significant bit

LSB—lower sideband

LSI—large-scale integration

LUF—lowest usable frequency

M

m—meter (length); milli (prefix for 10^{-3})

M—mega (prefix for 10^6); meter (instrument)

mA—milliampere

mAh—milliampere hour

MCP—multimode communications processor

MDS—Multipoint Distribution Service; minimum discernible (or detectable) signal

MF—medium frequency (300-3000 kHz)

mH—millihenry

MHz—megahertz

mi—mile, statute (unit of length)

mi/h—mile per hour

mi/s—mile per second

mic—microphone

min—minute (time)

MIX—mixer

mm—millimeter

MOD—modulator

modem—modulator/demodulator

MOS—metal-oxide semiconductor

MOSFET—metal-oxide semiconductor field-effect transistor

MS—meteor scatter

ms—millisecond

m/s—meters per second

msb—most-significant bit

MSI—medium-scale integration

MSK—minimum-shift keying

MSO—message storage operation

MUF—maximum usable frequency

mV—millivolt

mW—milliwatt

MΩ—megohm

N

n—nano (prefix for 10^{-9}); number of turns (inductors)

NBFM—narrow-band frequency modulation

NC—no connection; normally closed

NCS—net-control station; National Communications System

nF—nanofarad

NF—noise figure

nH—nanohenry

NiCd—nickel cadmium

NM—Net Manager

NMOS—N-channel metal-oxide silicon

NO—normally open

NPN—negative-positive-negative (transistor)

NPRM—Notice of Proposed Rule Making (FCC)

ns—nanosecond

NTIA—National Telecommunications and Information Administration

NTS—National Traffic System

O

OBS—Official Bulletin Station

OD—outside diameter

OES—Official Emergency Station

OO—Official Observer

op amp—operational amplifier

ORS—Official Relay Station

OSC—oscillator

OSCAR—Orbiting Satellite Carrying Amateur Radio

OTC—Old Timer's Club

oz—ounce (force unit, $^1/_{16}$ pound)

P

p—pico (prefix for 10^{-12})

P—power; plug

PA—power amplifier

PACTOR—digital mode combining aspects of packet and AMTOR

PAM—pulse-amplitude modulation

PBS—packet bulletin-board system

PC—printed circuit

P$_D$—power dissipation

PEP—peak envelope power

PEV—peak envelope voltage

pF—picofarad

pH—picohenry

PIC—Public Information Coordinator

PIN—positive-intrinsic-negative (semiconductor)

PIO—Public Information Officer

PIV—peak inverse voltage

PLL—phase-locked loop

PM—phase modulation

PMOS—P-channel (metal-oxide semiconductor)

PNP—positive negative positive (transistor)

pot—potentiometer

P-P—peak to peak

ppd—postpaid

PROM—programmable read-only memory

PSAC—Public Service Advisory Committee

PSHR—Public Service Honor Roll

PTO—permeability-tuned oscillator

PTT—push to talk

Q-R

Q—figure of merit (tuned circuit); transistor

QRP—low power (less than 5-W output)

R—resistor

RACES—Radio Amateur Civil Emergency Service

RAM—random-access memory

RC—resistance-capacitance

R/C—radio control

RCC—Rag Chewer's Club

RDF—radio direction finding

RF—radio frequency

RFC—radio-frequency choke

RFI—radio-frequency interference
RHC—right-hand circular (polarization)
RIT—receiver incremental tuning
RLC—resistance-inductance-capacitance
RM—rule making (number assigned to petition)
r/min—revolutions per minute
RMS—root mean square
ROM—read-only memory
r/s—revolutions per second
RS—Radio Sputnik, Russian ham satellite
RST—readability-strength-tone (CW signal report)
RTTY—radioteletype
RX—receiver, receiving

S

s—second (time)
S—siemens (unit of conductance); switch
SASE—self-addressed stamped envelope
SCF—switched capacitor filter
SCR—silicon controlled rectifier
SEC—Section Emergency Coordinator
SET—Simulated Emergency Test
SGL—State Government Liaison
SHF—super-high frequency (3-30 GHz)
SM—Section Manager; silver mica (capacitor)
S/N—signal-to-noise ratio
SPDT—single pole double-throw (switch)
SPST—single-pole single-throw (switch)
SS—Sweepstakes; spread spectrum
SSB—single sideband
SSC—Special Service Club
SSI—small-scale integration
SSTV—slow-scan television
STM—Section Traffic Manager
SX—simplex
sync—synchronous, synchronizing
SWL—shortwave listener
SWR—standing-wave ratio

T

T—tera (prefix for 10^{12}); transformer
TA—ARRL Technical Advisor
TC—Technical Coordinator
TCC—Transcontinental Corps (NTS)
TCP/IP—Transmission Control Protocol/Internet Protocol
tfc—traffic
TNC—terminal node controller (packet radio)
TR—transmit/receive
TS—Technical Specialist
TTL—transistor-transistor logic
TTY—teletypewriter
TU—terminal unit
TV—television

TVI—television interference
TX—transmitter, transmitting

U

U—integrated circuit
UHF—ultra-high frequency (300 MHz to 3 GHz)
USB—upper sideband
UTC—Coordinated Universal Time (also abbreviated Z)
UV—ultraviolet

V

V—volt; vacuum tube
VCO—voltage-controlled oscillator
VCR—video cassette recorder
VDT—video-display terminal
VE—Volunteer Examiner
VEC—Volunteer Examiner Coordinator
VFO—variable-frequency oscillator
VHF—very-high frequency (30-300 MHz)
VLF—very-low frequency (3-30 kHz)
VLSI—very-large-scale integration
VMOS—V-topology metal-oxide semiconductor
VOM—volt-ohmmeter
VOX—voice-operated switch
VR—voltage regulator
VSWR—voltage standing-wave ratio
VTVM—vacuum-tube voltmeter
VUCC—VHF/UHF Century Club
VXO—variable-frequency crystal oscillator

W

W—watt (kg $m^2 s^{-3}$), unit of power
WAC—Worked All Continents
WAS—Worked All States
WBFM—wide-band frequency modulation
WEFAX—weather facsimile
Wh—watthour
WPM—words per minute
WRC—World Radio Conference
WVDC—working voltage, direct current

X

X—reactance
XCVR—transceiver
XFMR—transformer
XIT—transmitter incremental tuning
XO—crystal oscillator
XTAL—crystal
XVTR—transverter

Y-Z

Y—crystal; admittance
YIG—yttrium iron garnet
Z—impedance; also see UTC

Numbers/Symbols

5BDXCC—Five-Band DXCC
5BWAC—Five-Band WAC
5BWAS—Five-Band WAS
6BWAC—Six-Band WAC
°—degree (plane angle)
°C—degree Celsius (temperature)
°F—degree Fahrenheit (temperature)
α—(alpha) angles; coefficients, attenuation constant, absorption factor, area, common-base forward current-transfer ratio of a bipolar transistor
β—(beta) angles; coefficients, phase constant current gain of common-emitter transistor amplifiers
γ—(gamma) specific gravity, angles, electrical conductivity, propagation constant
Υ—(gamma) complex propagation constant
δ—(delta) increment or decrement; density; angles
Δ—(delta) increment or decrement determinant, permittivity
ε—(epsilon) dielectric constant; permittivity; electric intensity
ζ—(zeta) coordinates; coefficients
η—(eta) intrinsic impedance; efficiency; surface charge density; hysteresis; coordinate
θ—(theta) angular phase displacement; time constant; reluctance; angles
ι—(iota) unit vector
κ—(kappa) susceptibility; coupling coefficient
λ—(lambda) wavelength; attenuation constant
Λ—(lambda) permeance
μ—(mu) permeability; amplification factor; micro (prefix for 10^{-6})
μC—microcomputer
μF—microfarad
μH—microhenry
μP—microprocessor
ξ—(xi) coordinates
π—(pi) 3.14159
ρ—(rho) resistivity; volume charge density; coordinates; reflection coefficient
σ—(sigma) surface charge density; complex propagation constant; electrical conductivity; leakage coefficient; deviation
Σ—(sigma) summation
τ—(tau) time constant; volume resistivity; time-phase displacement; transmission factor; density
φ—(phi) magnetic flux angles
Φ—(phi) summation
χ—(chi) electric susceptibility; angles
Ψ—(psi) dielectric flux; phase difference; coordinates; angles
ω—(omega) angular velocity $2\pi f$
Ω—(omega) resistance in ohms; solid angle

ARRL Handbook Address List

These companies or individuals are cited in this edition of the *ARRL Handbook*. Please send updates to the ARRL Handbook Editor at ARRL Headquarters.

A & A Engineering
2521 West La Palma, Unit K
Anaheim, CA 92801
714-952-2114
fax 714-952-3280
e-mail w6ucm@aol.com

Advanced Receiver Research
Box 1242
Burlington, CT 06013
860-485-0310

AEA, Division of Tempo
Research Corp
1221 Liberty Way
Vista CA 92083
760-598-9677
fax 760-598-4898
[Also see Timewave]

Aero/Marine Beacon Guide
2856-G W Touhy Ave
Chicago, IL 60645

Alexander Aeroplane Co
PO Box 909
Griffin, GA 30224
800-831-2949
fax 404-229-2329

Alinco Electronics
438 Amapola Ave, #130
Torrance, CA 90501
310-618-8616
fax 310-618-8758
e-mail alinco@alinco.com
web www.alinco.com

All Electronics Corp
14928 Oxnard St
PO Box 567
Van Nuys, CA 91411
800-826-5432
818-904-0524
fax 818-781-2653
e-mail allcorp@callcorp.com
web www.allcorp.com/

Allied Electronics
7410 Pebble Dr
Fort Worth, TX 76118
800-433-5700
web www.allied.avnet.com/

Allstar Magnetics
7206 NE 37 Ave
Vancouver, WA 98665
206-693-0213
fax 206-693-0639
web www.allstarmagnetics.com/

Alpha-Delta Communications
PO Box 620
Manchester KY 40962
606-598-2029
fax 606-598-4413

Alpha/Power, Inc
14440 Mead Ct, Unit B
Longmont, CO 80504
970-535-4173
fax 970-535-0281
e-mail sales@alpha-power-inc.com
web www.alpha-power-inc.com/

Aluma Tower Company, Inc
PO Box 2806-AL
Vero Beach, FL 32961-2806
561-567-3423
fax 561-567-3432
e-mail atc@alumatower.com
web www.alumatower.com/

AM Press/Exchange
2116 Old Dover Rd
Woodlawn, TN 37191

Amateur Television Quarterly
(ATVQ)
5931 Alma Dr
Rockford, IL 61108
815-398-2683
fax 815-398-2688
e-mail atvq@aol.com
web www.cris.com/Gharlan/

AMECO Publishing Corp
224 E 2nd St
Mineola, NY 11501
516-741-5030
fax 516-741-5031
e-mail sales@amecocorp.com
web www.amecocorp.com/

American Design Components
400 County Ave
Secaucus, NJ 07094
201-601-8999
fax 201-601-8990

American National Standards
Institute (ANSI)
11 W 42nd St
New York, NY 10036
212-642-4900
web www.ansi.org/

ARRL—The national association
for Amateur Radio
225 Main St
Newington, CT 06111-1494
860-594-0200
fax 860-594-0259
e-mail tis@arrl.org
web www.arrl.org/

Ameritron
116 Willow Rd
Starkville, MS 39759
601-323-8211
fax 601-323-6551
e-mail amertron@ameritron.com
web www.ameritron.com/

Amidon, Inc
250 & 250 Briggs Ave
Costa Mesa, CA 92626
714-850-4660
fax 714-850-1163
e-mail sales@amidoncorp.com
web www.amidoncorp.com/

AMRAD
Post Box 6148
McLean, VA 22106
web www.amrad.org/

AMSAT (Radio Amateur Satellite
Corp)
PO Box 27
Washington, DC 20044
301-589-6062
fax 301-608-3410
e-mail martha@amsat.org
web www.amsat.org/

ANARC
PO Box 11201
Shawnee Mission, KS 66207-0201
913-345-1978 (BBS)
web www.anarc.org/

Anchor Electronics
2040 Walsh Ave
Santa Clara, CA 95050
408-727-3693
fax 408-727-4424

Angle Linear
PO Box 35
Lomita, CA 90717
310-539-5395
fax 310-539-8738
e-mail cangle@anglelinear.com
web www.anglelinear.com/

antenneX Magazine
web www.antennex.com/

Antique Electronic Supply
6221 South Maple Ave
PO Box 27468
Tempe AZ 85285-7468
602-820-5411
fax 602-820-4643
web www.tubesandmore.com/

Antique Radio Classified
PO Box 802-B14
Carlisle, MA 01741
978-371-0512
fax 978-371-7129
web www.antiqueradio.com/

Array Solutions
350 Gloria Rd
Sunnyvale, TX 75182
972-203-8810
fax 972-203-8811
e-mail wx0b@arraysolutions.com
web www.arraysolutions.com/

Arrow Electronics
25 Hub Dr
Melville, NY 11747
800-932-7769
516-694-6800
fax 516-585-0878

ARS Electronics
7110 De Celis Pl
PO Box 7323
Van Nuys, CA 91409
818-997-6200
800-422-4250
fax 818-997-6158

Atlantic Surplus Sales
3730 Nautilus Ave
Brooklyn, NY 11224
718-372-0349

ATV Research, Inc
1301 Broadway
PO Box 620
Dakota City, NE 68731
402-987-3771
800-392-3922 (orders)
fax 402-987-3709
e-mail atv@pionet.net
web www.atvresearch.com/

Avantek
3175 Bowers Ave
Santa Clara, CA 95054-3292
408-727-0700

Avatar Magnetics
240 Tamara Trail
Indianapolis, IN 46217

Barker and Williamson Corp (B&W)
603 Cidco Rd
Cocoa, FL 32926
407-639-1510
fax 407-639-2545
e-mail custsrvc@bwantennas.com
web bwantennas.com/

Brian Beezley, K6STI
3532 Linda Vista
San Marcos, CA 92069
760-599-8662 (product support)
e-mail k6sti@n2.net

Bencher, Inc
831 N Central Ave
Wood Dale, IL 60191
630-238-1183
fax 630-238-1186
e-mail bencher@bencher.com
web www.bencher.com/

Bird Electronics Corp
30303 Aurora Rd
Cleveland, OH 44139
440-248-1200

British Amateur Television Club
Grenehurst, Pinewood Rd
High Wycombe
Bucks HP12 4DD
United Kingdom
01494-528899
e-mail memsec@batc.org.uk

Buckmaster Publishing
6196 Jefferson Highway
Mineral, VA 23117
800-282-5628
540-894-5777
fax 540-894-9141
e-mail info@buck.com
web www.buck.com/

Caddock Electronics
1717 Chicago Ave
Riverside, CA 92507-2364
909-788-1700
fax 909-369-1151
web www.caddock.com/

Calogic
237 Whitney Pl
Fremont, CA 94539
510-656-2900

Jim Cates, WA6GER
3241 Eastwood Rd
Sacramento, CA 95821
916-487-3580

Cetron Communications
715 Hamilton St
Geneva, IL 60134
800-238-7661
708-208-3700
fax 708-208-3750

Circuit Specialists
220 S Country Club Dr, Bldg #2
Mesa, AZ 85210
602-966-0764
800-528-1417
480-464-2485
fax 480-464-5824

Coilcraft
1102 Silver Lake Rd
Cary, IL 60013
847-639-6400
fax 847-639-1469
e-mail info@coilcraft.com
web www.coilcraft.com/

Communication Concepts, Inc
    (CCI)
508 Millstone Dr
Beavercreek, OH 45434-5840
937-426-8600
fax 937-429-3811
e-mail ccidayton@pobox.com

Communications and Power
    Industries
Eimac Division
301 Industrial Way
San Carlos, CA 94070-2682
800-414-TUBE (414-8823)
650-594-4175
web www.eimac.com/

Communications Quarterly
— see CQ Communications

Communications Specialists Inc
426 West Taft Ave
Orange, CA 92865-4296
714-998-3021
800-854-0547
fax 714-974-3420 or
    800-850-0547
web www.com-spec.com/

Condenser Products Corp
2131 Broad St
Brooksville, FL 34609
904-796-3561
fax 904-799-0221

Contact East, Inc
335 Willow St
North Andover, MA 01845
508-682-2000
fax 508-688-7829

Ken Cornell
225 Baltimore Ave
Point Pleasant, NJ 08742

Courage HANDI-HAM System
3915 Golden Valley Rd
Golden Valley, MN 55422
763-520-0511
763-520-0245 (TTY)
fax 763-520-0577
e-mail handiham@mtn.org
web www.mtn.org/handiham/

CQ Communications
25 Newbridge Rd
Hicksville, NY 11801
516-681-2922 (business office)
fax 516-681-2926
e-mail cqmagazine@aol.com
web www.cq-amateur-radio.com/

Curry Communications
PO Box 1884
Burbank, CA 91507
818-846-0617

Cushcraft Corp
PO Box 4680
Manchester, NH 03103
603-627-7877
fax 603-627-1764
e-mail hamsales@cushcraft.com
web www.cushcraft.com/

Custom Computer Services, Inc
PO Box 2452
Brookfield, WI 53008
414-781-2794
web www.ccsinfo.com/picc.html

Jacques d'Avignon, VE3VIA
965 Lincoln Dr
Kingston, ON K7M 4Z3
Canada
613-634-1519
e-mail
    monitor@limestone.kosone.com

Dallas Remote Imaging Group
4209 Meadowdale Dr, Ste 3
Carrollton, TX 75010
214-394-7438 (BBS)
214-394-7325 (Voice)
fax 214-492-7747
e-mail jwallach@drig.com
web www.drig.com/

Dan's Small Parts and Kits
Box 3634
Missoula, MT 59806-3634
406-258-2782 (voice and fax)
web www.fix.net/dans.html

Davis RF Co, Div of Davis
    Associates Wire & Cable, LLC
PO Box 730
Carlisle, MA 01741
800-328-4773 (orders and general
    info)
978-369-1738 (technical info)
fax 978-369-3484
e-mail davisRFinc@aol.com
web www.davisRF.com/

DC Electronics
PO Box 3203
2200 N Scottsdale Rd
Scottsdale, AZ 85271-3203
800-467-7736
web www.dckits.com/index.htm

DCI Inc
29 Hummingbird Way
White City, SK S0G 5B0
Canada
306-781-4451
e-mail dci@dci.ca
web www.dci.ca/

Digi-Key Corp
701 Brooks Ave S
PO Box 677
Thief River Falls, MN
    56701-0677
800-344-4539 (800-DIGI-KEY)
fax 218-681-3380
web www.digikey.com/

Digital Vision, Inc
270 Bridge St
Dedham, MA 02026
617-329-5400
617-329-8387 (BBS)

Dover Research
321 W 4th St
Jordan, MN 55352-1313
612-492-3913

Down East Microwave
954 Rte 519
Frenchtown, NJ 08825
908-996-3584
fax 908-996-3072
web
    www.downeastmicrowave.com/

East Coast Amateur Radio, Inc
415 Division St
N Tonawanda, NY 14120
716-695-3929

EDI, Inc
1260 Karl Ct
Wauconda, IL 60084
708-487-3347
fax 708-487-3346

Edlie Electronics, Inc
2700 Hempstead Tpk
Levittown, NY 11756-1443
orders 800-647-4722
516-735-3330
fax 516-731-5125

Eimac (see Communications and
    Power Industries)

Electric Radio Magazine
14643 County Rd G
Cortez, CO 81321-9575
970-564-9185 (voice and **fax**)
e-mail er@frontier.net

Electro Sonic, Inc
100 Gordon Baker Rd
Willowdale, ON M2H 3B3
Canada
416-494-1666
416-494-1555

Electronic Emporium
3621-29 E Weir Ave
Phoenix, AZ 85040
602-437-8633
fax 602-437-8835

Electronic Equipment Bank (EEB)
323 Mill St, NE
Vienna, VA 22180
orders 800-368-3270
technical 703-938-3350
fax 703-938-6911
e-mail eeb@access.digex.net
web www.access.digex.net/eeb/
    eeb/html

Electronic Industries Alliance
    (EIA)
2500 Wilson Blvd
Arlington, VA 22201-3834
703-907-7500
web www.eia.org/

Electronic Precepts of Florida
11651 87th St
Largo, FL 34643-4917
800-367-4649
fax 813-544-1910

Electronic Rainbow, Inc
6227 Coffman Rd
Indianapolis, IN 46268
317-291-7262
fax 317-291-7269

Electronics Now
500-B Bi-County Blvd
Farmingdale, NY 11735
800-288-0652
    (Customer Service)
800-999-7139
    (New subscriptions)
516-293-3000 (Admin Office)
fax 516-293-3115

Elktronics
12536 T.77
Findlay, OH 45840
419-422-8206

Elna Ferrite Laboratories, Inc
234 Tinker St
PO Box 395
Woodstock, NY 12498
914-679-2497

Embedded Research
PO Box 92492
Rochester, NY 14692
e-mail embres@frontier.net
web www.frontiernet.net/~embres

EMI Filter Company
9075A 130 Ave N
Largo, FL 34643
800-323-7990
fax 813-586-9378

Encomm
1506 Capitol Ave
Plano, TX 75074

Engineering Consulting
583 Candlewood St
Brea, CA 92621
714-671-2009
fax 714-255-9984

ESF Copy Service
4011 Clearview Dr
Cedar Falls, IA 50613-6111
319-266-7040

ETO, Inc
4975 N 30th St
Colorado Springs, CO 80919
719-260-1191
719-599-3861
fax 719-260-0395

Fair Radio Sales Co, Inc
1016 E Eureka St
PO Box 1105
Lima, OH 45802-1105
419-227-6573
419-223-2196
fax 419-227-1313
e-mail fairadio@wcoil.com
web www.fairradio.com/

Fair-Rite Products Corp
PO Box J, 1 Commercial Row
Wallkill, NY 12589
914-895-2055
fax 914-895-2629
e-mail ferrites@fair-rite.com
web fair-rite.com/

Fala Electronics
PO Box 1376
Milwaukee, WI 53201-1376

FAR Circuits
18N640 Field Court
Dundee, IL 60118-9269
847-836-9148 (voice and **fax**)
web www.cl.ais.net/faircir/

Federal Emergency Management
    Agency (FEMA)
web www.fema.gov

G-O-Metric
909 Norwich Ave
Delran, NJ 08075

Gary Meyers, K9RX
1753 Elmwood Dr
Rockhill, SC 29730

Gateway Electronics
8123 Page Blvd
St. Louis, MO 63130
800-669-5810
314-427-6116
fax 314-427-3147
e-mail gateway@mo.net
web www.gatewayelex.com/

Glen Martin Engineering
Rte 3, Box 322
Boonville, MO 65233
816-882-2734
fax 816-882-7200
web www.glenmartin.com/

Grove Enterprises Inc
PO Box 98
Brasstown, NC 28902
800-438-8155 (orders)
704-837-9200 (BBS)
fax 704-837-2216
e-mail nada@grove.net
web www.grove.net/

HAL Communications Corp
1201 W Kenyon Rd
PO Box 365
Urbana, IL 61801-0365
217-367-7373 (voice and fax)
e-mail halcomm@phairienet.org
web www.halcomm.com/

Ham Equipment Buyers Guide
189 Kenilworth
Glen Ellyn, IL 60137

Hammond Mfg Co, Inc
4700 Genesee St
Cheektowaga, NY 14225-2466
716-631-5700
fax 716-631-1156

Hammond Mfg, Ltd
394 Edinburgh Rd, N
Guelph, ON N1H 1E5
Canada
519-822-8323

Hamtronics, Inc
65-Q Moul Rd
Hilton, NY 14468
716-392-9430
fax 716-392-9420
web www.hamtronics.com/

H. B. Electronics
43 Rector St
E Greenwich RI 02818-3312
e-mail
  hb.electronics@businesson.com
web www.businesson.com/
  hamparts/

Heathkit Educational Systems
PO Box 8589
Benton Harbor, MI 49023-8589
800-253-0570
fax 616-925-4876
e-mail techsupport@heathkit.com
web www.heathkit.com/

Heldref Publications
1319 Eighteenth St, NW
Washington, DC 20036
800-365-9753

Henry Radio
2050 South Bundy Dr
Los Angeles, CA 90025
310-820-1234
800-877-7979 (Orders)
fax 310-820-1234
e-mail henryradio@earthlink.net
web www.henryradio.com/

Herbach and Rademan
  (H & R Co)
16 Roland Ave
Mt Laurel, NJ 08054-1012
800-848-8001 (Orders only)
609-802-0422
fax 609-802-0465
e-mail sales@herbach.com
web www.herbach.com/

HERD Electronics
514 S Baltimore St
Dillsburg, PA 17019-9601
717-432-4519
fax 717-432-7850
e-mail herd@juno.com

Heritage Transformer Co, Inc
13483 Litchfield Rd
Eastview, KY 42732
270-862-9877
e-mail heritran@juno.com

Hi-Manuals
PO Box 802
Council Bluffs, IA 51502-0802
e-mail himan@radiks.net
web www.hi-manuals.com/

HI-Tech Software, LLC
Suite 105, 7830 Ellis Rd
Melbourne, FL 39204
800-735-5717
web www.htsoft.com/

Hosfelt Electronics
2700 Sunset Blvd
Steubenville, OH 43952
800-524-6464
fax 800-524-5414
e-mail hosfelt@clover.net
web www.hosfelt.com/

Howard W. Sams and Company
2647 Waterfront Pky E Dr
Indianapolis, IN 46214-2041
800-428-7267 (428-SAMS)
317-298-5565
fax 317-298-5604
web www.hwsams.com/

ICOM America, Inc
2380 116th Ave NE
PO Box C-90029
Bellevue, WA 98004
425-454-8155
425-450-6088 (literature)
fax 425-454-1509
e-mail
  75540.525@compuserve.com
web www.icomamerica.com/

Idiom Press
PO Box 1025
Geyserville, CA 95441-1025

IEEE
345 E 47th St
New York, NY 10017-2394
web www.ieee.org/

IEEE Operations Center
445 Hoes Ln
PO Box 1331
Piscataway, NJ 08855-1331
908-981-0060

Industrial Communications
  Engineers (ICE)
PO Box 18495
Indianapolis, IN 46218-0495
317-545-5412
800-423-2666
fax 317-545-9645

Industrial Safety Co
1390 Neubrecht Rd
Lima, OH 45801
419-227-6030
fax 419-228-5034

International Components Corp
105 Maxess Rd
Melville, NY 11747
800-645-9154
516-293-1500 (NY)
fax 516-293-4983
e-mail oemsales@icc107.com
web www.icc107.com/index/

International Crystal Mfg Co
10 North Lee
PO Box 26330
Oklahoma City, OK 73126-0330
800-725-1426
405-236-3741
fax 800-322-9426
e-mail
  customerservice@icmfg.com
web www.icmfg.com/

International Radio
13620 Tyee Rd
Umpqua, OR 97486
541-459-5623
fax 541-459-5632
e-mail inrad@rosenet.net
web www.qth.com/inrad/

International Telecommunication
  Union (ITU)
Place des Nations
CH 1211
Geneva 20, Switzerland
web www.itu.ch/

International Visual Communications Association (IVCA)
PO Box 140336
Nashville, TN 37214
web www.ivca.org/

Intuitive Circuits
2275 Brinston Ave
Troy, MI 48083
248-524-1918
web www.icircuits.com

IPS Radio and Space Services
PO Box 5606
West Chatswood NSW 2057
Australia
+61 2 414 8300
fax +61 2 414 8331
e-mail rwc@ips.gov.au
web www.ips.gov.au

Jade Products Inc
PO Box 368
East Hampstead, NH 03826
800-JADE-PRO (800-523-3776)
603-329-6995
fax 603-329-4499

Jameco Electronics
1355 Shoreway Rd
Belmont, CA 94002
800-831-4242
fax 800-237-6948
e-mail info@jameco.com
web www.jameco.com/

James Millen Electronics
PO Box 4215BV
Andover, MA 01810-4215
978-975-2711
fax 978-474-8949
e-mail info@jamesmillenco.com

JAN Crystals
2341 Crystal Dr
PO Box 06017
Fort Myers, FL 33906-6017
800-526-9825 (JAN-XTAL)
941-936-2397
fax 941-936-3750

JDR Microdevices
1850 South 10th St
San Jose, CA 95122-4108
800-538-5000 (Orders)
408-494-1400
fax 800-538-5005
e-mail sales@jdr.com
web www.jdr.com/

Jolida Tube Factory
10820 Guilford Rd
Annapolis Junction, MD 20701
800-783-2555

K-Com
PO Box 82
Randolph OH 44265
330-325-2110
fax 330-325-2525
e-mail k-com@worldnet.att.net
web www.k-comfilters.com/

Kanga US
3521 Spring Lake Dr
Findlay, OH 45840
419-423-4604
419-423-5643
e-mail kanga@bright.net
web www.bright.net/~kanga/
  kanga/

Kangaroo Tabor Software
Rte 2, Box 106
Farwell, TX 79325-9430
fax 806-225-4006
e-mail ku5s@wtrt.net
web www.taborsoft.com/

Kantronics
1202 East 23rd St
Lawrence, KS 66046-5099
785-842-7745
fax 785-842-2031
e-mail
  purchasing@kantronics.com
web www.kantronics.com/

Kenwood Communications Corp
2201 East Dominguez St
PO Box 22745
Long Beach, CA 90801-5745
310-639-5300 (customer support)
800-536-9663 - (repairs/parts)
310-761-8284 (BBS)
fax 310-631-3913
web www.kenwood.net

Kepro Circuit Systems, Inc
630 Axminister Dr
Fenton, MO 63026-2992
800-325-3878
314-343-1630
fax 314-343-0668

Kilo-Tec
PO Box 10
Oak View, CA 93022
805-646-9645 (voice and fax)

Kirby
298 West Carmel Dr
Carmel, IN 46032
317-843-2212

Kooltronic
1700 Morse Ave
Ventura, CA 93003
805-642-8521

KVG
2240 Woolbright Rd, Ste 320-A
Boynton Beach, FL 33426-6325
407-734-9007
fax 407-734-9008

K2AW's Silicon Alley
175 Friends Ln
Westbury, NY 11590
516-334-7024
fax 516-334-7024

L. L. Grace Communications
PO Box 1345
Voorhees, NJ 08043
609-751-1018
fax 609-751-9705
e-mail n2rec@amsat.org

John Langner, WB2OSZ
115 Stedman St
Chelmsford, MA 01824-1823
508-256-6907

Lashen Electronics, Inc
21 Broadway
Denville, NJ 07834
973-627-3783
fax 973-625-9501
e-mail sales@lashen.com
web www.lashen.com/

Roy Lewallen, W7EL
PO Box 6658
Beaverton, OR 97007
503-646-2885
fax 503-671-9046
e-mail w7el@teleport.com

Lodestone Pacific
4769 Wesley Dr
Anaheim, CA 92807
914-970-0900
fax 914-970-0800

The Longwave Club of America
45 Wildflower Rd
Levittown, PA 19057
215-945-0543
e-mail naswa1@aol.com
web anarc.org/lwca/

M/A-COM, Inc (an AMP Company)
1011 Pawtucket Blvd
PO Box 3295
Lowell, MA 01853-3295
508-442-4500
fax 508-442-4436
e-mail sales@macom.com
web www.amp.com/

M² Antenna Systems
7560 North Del Mar Ave
Fresno, CA 93711
559-432-8873
fax 559-432-3059
e-mail mzinc@mzinc.com
web www.mzinc.com/

MAI/Prime Parts
5736 N Michigan Rd
Indianapolis, IN 46208
317-257-6811
fax 317-257-1590
e-mail mai@iquest.net

The Manual Man
27 Walling St
Sayreville, NJ 08872-1818
908-238-8964
fax 908-238-8964

Mark V Electronics
8019 E Slauson Ave
Montebello, CA 90640
800-423-FIVE (Orders outside CA)
800-521-MARK (CA orders)
213-888-8988
fax 213-888-6868

Marlin P. Jones & Associates, Inc
PO Box 12685
Lake Park, FL 33403-0685
800-652-6733
561-848-8236 (Tech)
fax 800-432-9937
e-mail mpja@gate.net

MARS
Air Force MARS
HQ AFC4A/SYXR (MARS)
203 Losey St, Room 205
Scott AFB, IL 62225-5234
web public.afca.scott.af.mil/public/mars1.htm

Navy-Marine Corps MARS
Chief US Navy-Marine Corps
Military Affiliate Radio System (MARS)—Bldg 13
NAVCOMMU WASHINGTON
Washington, DC 20397-5161
web www.navymars.org/

Army MARS
HQ US ARMY SIGNAL COMMAND
ATTN: AFSC-OPE-MA (ARMY MARS)
Ft Huachuca, AZ 85613-5000
web www.asc.army.mil/mars/

Maxim Integrated Products
120 San Gabriel Dr
Sunnyvale, CA 94086
408-737-7600
fax 408-737-7194
web www.maxim-ic.com/

Richard Measures, AG6K
6455 LaCumbre Rd
Somis, CA 93066
805-384-3734
e-mail 2@vc.net
web www.vcnet/measures/

Mendelsohn Electronics Co, Inc
340 E First St
Dayton, OH 45402
800-344-4465
937-461-3525
fax 937-461-3391

Metal and Cable Corp, Inc
9241 Ravenna Rd, Unit C-10
PO Box 117
Twinsburg, OH 44087
216-425-8455
fax 216-425-3504
web www.metal-cable.com/

MFJ Enterprises
PO Box 494
Mississippi State, MS 39762
601-323-5869
800-647-1800
fax 601-323-6551
e-mail mfj@mfjenterprises.com
web www.mfjenterprises.com/

Micro Engineering Labs, Inc.
Box 7532
Colorado Springs, CO 80933-7532
719-520-5323
web www.melabs.com/mel/

Microchip Technology
2355 W Chandler Blvd
Chandler, AZ 85224-6199
602-786-7200
fax 602-899-9210
web www.microchip.com/

Microcraft Corp
PO Box 937Q
Thiensville, WI 53092
414-241-8144

Microwave Components of Michigan
PO Box 1697
Taylor, MI 48180
313-753-4581 (evenings)

Microwave Filter Co, Inc
6743 Kinne St
E Syracuse NY 13057
800-448-1666
315-438-4700
fax 315-463-1467
e-mail mfcsales@microwavefilter.com
web www.mwfilter.com/

Mini Circuits Labs
PO Box 350166
Brooklyn, NY 11235-0003
800-654-7949
718-934-4500
fax 718-332-4661
web www.minicircuits.com/

Mirage Communications
116 Willow Rd
Starkville, MS 39759
601-323-8287
fax 601-323-6551
web www.mirageamp.com/

Bob Mobile, WA1OUB
RFD 2, Hillsboro Cir
Hillsborough, NH 03244

Model Aviation
5151 East Memorial Dr
Muncie, IN 47302
317-287-1256
fax 317-289-4248

Modern Radio Laboratories
PO Box 14902-Q
Minneapolis, MN 55414-0902

Morse Telegraph Club, Inc
1101 Maplewood Dr
Normal, IL 61761
e-mail jramtc@mwci.net

Motorola Semiconductor Products, Inc
5005 East McDowell Rd
Phoenix, AZ 85008
512-891-2030
512-891-3773
web www.mot.com/

Mouser Electronics
2401 Hwy 287 N
Mansfield, TX 76063
800-346-6873
fax 817-483-0931
e-mail sales@mouser.com
web www.mouser.com/

National Electronics
PO Box 15417
Shawnee Mission, KS 66285
800-762-5049 (orders)
e-mail sales@national-electronics.com
web www.sound.net/~ne/

National Fire Protection Association
1 Batterymarch Park
PO Box 9101
Quincy, MA 02269-9101
800-344-3555
web www.nfpa.org/

National Semiconductor Corp
PO Box 58090
Santa Clara, CA 95052-8090
800-272-9959
408-721-5000
fax 800-432-9672

National Technical Information Service
5285 Port Royal Rd
Springfield, VA 22161
703-487-4650 (sales desk)
703-487-4639 (TDD - for the hearing impaired)
fax 703-321-8547

The New RTTY Journal
PO Box 236
Champaign, IL 61824-0236

New Sensor Corp
20 Cooper Station
New York, NY 10003
212-529-0466
800-633-5477 (orders)
fax 212-529-0486

Newark Electronics
4801 N. Ravenswood Ave
Chicago, IL 60640-4496
800-463-9275
312-784-5100
web www.newark.com/

Noble Publishing
2245 Dillard St
Tucker, GA 30084
770-908-2320
770-939-0157

NOISE/COM Co
E 49 Midland Ave
Paramus, NJ 07652
201-261-8797
fax 201-261-8339
e-mail info@noisecom.com.
web www.noisecom.com/

Northern Lights Software
2881 C.R. 21
Canton, NY 13617
315-379-0161
fax 315-379-5804
e-mail w9ip@webcom.com

Nuts & Volts Magazine
430 Princeland Court
Corona, CA 91719-9938
909-371-8497
fax 909-371-3052
e-mail subscribe@nutsvolts.com
web www.nutsvolts.com/

Oak Hills Research
2460 S Moline Way
Aurora, CO 80014
800-238-8205 (orders)
303-752-3382
fax 303-745-6792
e-mail qrp@ohr.com
web www.morsex.com/ohr/

Oakland University
ftp ftp://oak.oakland.edu/pub/hamradio/arrl/bbs/programs

Ocean State Electronics
6 Industrial Dr
Westerly, RI 02891
401-596-3080
800-866-6626
fax 401-596-3590

Old Tech—Books & Things
PO Box 803
Carlisle, MA 01741

Osborne McGraw-Hill
2600 Tenth St
Berkeley, CA 94710

PacComm Packet Radio Systems
Inc
4413 North Hesperides St
Tampa, FL 33614-7618
800-486-7388 (Orders)
813-874-2980
fax 813-872-8696
e-mail info@janrix.com
web www.paccom.com/

Palomar Engineers
PO Box 462222
Escondido, CA 92046
760-747-3343
fax 760-747-3346
e-mail
  75353.2175@compuserve.com

Pasternack Enterprises
PO Box 16759
Irvine, CA 92713
714-261-1920
fax 714-261-7451

PC Electronics
2522 Paxson Ln
Arcadia, CA 91007
626-447-4565
fax 626-447-0489
e-mail tom@hamtv.com
web www.hamtv.com/

Peter W. Dahl Co, Inc
5869 Waycross Ave
El Paso, TX 79924
915-751-2300
fax 915-751-0768
e-mail pwdco@pwdahl.com
web www.pwdahl.com/

Phillips Components
PO Box 8533
Scottsdale, AZ 85252
602-269-5974
800-880-6637 (MMDS)
602-947-7700
fax 602-947-7799

Phillips-Tech Electronics
PO Box 737
Trinidad, CA 95570
707-677-0159
fax 707-677-0934
web www.phillips-tech.com

Bob Platts, G8OZP
43 Ironwalls Ln
Tutbury,
Straffordshire DE13 9NH
United Kingdom
01 283 531443
e-mail
  bobplatts@compuserve.com

Popular Communications
— See CQ Communications

Power Supply Components
677 Palomar Ave
Sunnyvale, CA 94086
408-737-1333
fax 408-737-0502

Practical Wireless
Arrowsmith Court
Station Approach
Broadstone, Dorset BH18 8PW
United Kingdom
44-1202-659950
e-mail rob@pwpub.demon.co.uk

PRO Distributors
2811-B 74th St
Lubbock, TX 79423
800-658-2027

PSK31
web aintel.bi.ehu.es/psk31.html

The PX Shack
52 Stonewyck Dr
Belle Mead, NJ 08502

QRP Quarterly (Subscriptions)
2046 Ash Hill Rd
Carrolton, TX 75007

Quantics
PO Box 2163
Nevada City, CA 95959-1263

R & B Enterprises
20 Clipper Rd
W Conshohocken, PA 19428
610-825-1960
fax 610-825-1684
e-mail rbemc@ix.netcom.net

R & D Electronics
5363 Broadway Ave
Cleveland, OH 44127-1551
800-642-1123 (orders)
216-621-1121
fax 216-621-8628

R & L Electronics
1315 Maple Ave
Hamilton, OH 45011
800-221-7735
513-868-6399
fax 513-868-6574
e-mail sales@randl.com
web www.randl.com/

Radio Adventures Corp.
RD #4, Box 240
Summit Dr
Franklin, PA 16346
814-437-5355
fax 814-437-5432
e-mail info@radioadv.com
web www.radioadv.com/

Radio Amateur Telecommunications Society
203 Bishop Blvd
N Brunswick, NJ 08902
e-mail askrat@rats.org
web fftp://www.rats.org

Radio Book Store
PO Box 209
Rindge, NH 03461
800-457-7373
fax 603-899-6826
web www.radiobooks.com/

Radio Society of Great Britain
Lambda House
Cranborne Rd
Potters Bar
Herts EN6 3JE
United Kingdom
44—01-707-659015
web www.rsgb.org/

Radio Switch Corp.
PO Box 159
Marlboro, NJ 07746-0159
732-462-6100

Radiokit
PO Box 973
Pelham, NH 03076
603-635-2235
Telex: 887697
fax 603-635-2943

Radioware
PO Box 209
Rindge, NH 03461
800-457-7373
fax 603-899-6826
e-mail radware@
  radio-ware.com
web www.radio-ware.com/

Rainy Day Books
PO Box 775
Fitzwilliam, NH 03447
603-585-3448
fax 603-215-0046

Ramsey Electronics, Inc
793 Canning Pkwy
Victor, NY 14564
716-924-4560
fax 716-924-4555

The Raymond Sarrio Co
6147 Via Serena St
Alta Loma CA 91701
800-413-1129 (orders)
909-411-1129
fax 909-484-5125
e-mail wb6siv@cyberg8t.com
web www.sario.com/

Marius Rensen
web www.hffax.de/

RF Parts Co
435 South Pacific St
San Marcos, CA 92069
760-744-0700
800-737-2787 (orders)
fax 760-744-1943
e-mail rfp@rfparts.com
web www.rfparts.com/

RMD Technology, Inc
250 Airport Industrial Dr
Ypsilanti, MI 48198-6061
313-482-2670
fax 313-482-2671
e-mail robertrau@aol.com

Robert H. Bauman Sales
PO Box 122
Itasca, IL 60143

ROHN
PO Box 2000
Peoria, IL 61656
309-697-4400
fax 309-697-5612
e-mail rohn@rohnnet.com

Ron Boucher
PO Box 541
Goffstown, NH 03045
603-497-2988
fax 603-497-3244
e-mail
  72440.1356@compuserve.com

S & S Engineering
14102 Brown Rd
Smithsburg, MD 21783
301-416-0661
fax 301-416-0963
e-mail n3ga@aol.com
web www.xmetric.com/sseng/

Sentry Mfg Co
PO Box 250
Chickasha, OK 73023
405-224-6780
800-252-6780
fax 405-224-8808

SESCOM, Inc
2100 Ward Dr
Henderson, NV 89015-4249
702-565-3400
fax 702-565-4828

SHF Microwave Parts Company
7102 West 500 South
La Porte, IN 46350-9575
219-785-4552
e-mail prutz@shfmicro.com
web www.shfmicro.com/

Simple Design Implementations
PO Box 9303
Forestville, CT 06011-9303
860-582-8526

Sky Publishing Company
PO Box 9111
Belmont, MA 02178-9111
800-253-0245
617-864-7360
fax 617-864-6117
e-mail skytel@skypub.com
web www.skypub.com/

SkymooN
9012 Kings Dr
Manvel, TX 7578

SkyWave Technologies
17 Pine Knoll Rd
Lexington, MA 02173
617-862-6742

Small Parts, Inc
13980 NW 58th Ct
PO Box 4650
Miami Lakes, FL 33014-0650
800-220-4242 (Orders)
305-557-7955 (Customer service)
fax 800-423-9009
e-mail smlparts@smallparts.com
web www.smallparts.com/

Society of Wireless Pioneers, Inc
PO Box 86
Geyserville, CA 95441
e-mail k6dzy@netdex.com
web web.mountain.net/~carto/

Software Systems Consulting
615 South El Camino Real
San Clemente, CA 92672
714-498-5784
fax 714-498-0568

The Solar Depot
99 rear Washington St
Melrose, MA 02176
617-665-7609
e-mail rmeuse@world.std.com

Solar Energy of South Florida
1024 NE 21st St
Belle Glade, FL 33430
407-996-6290

Solder-It Co
Box 20100
Cleveland, OH 44120
216-791-4600
800-897-8989
fax 216-721-3700
e-mail fdoo6@apk.net
web www.solder-it.com/

Southern Electronics Supply
1909 Tulane Ave
New Orleans, LA 70112
504-524-2343
e-mail e-mail@southernele.com
web www.southernele.com/

Southwestern Bell
7486 Shadeland Station Way
Indianapolis, IN 46256
317-576-6847

Sparrevohn Engineering
143 Nieto Ave SW #1
Long Beach, CA 90815
562-799-1577
web www.members.aol.com/
  zsmrtfred/

SPEC-COM Journal
PO Box 1002
Dubuque, IA 52004-1002
319-557-8791
fax 319-583-6462

Spectrum International, Inc
PO Box 1084
Concord, MA 01742
508-263-2145
fax 508-263-7008

Star Circuits
PO Box 94917
Las Vegas, NV 89193
800-433-6319

Star-Tronics
PO Box 98102
Las Vegas, NV 89193-8102
702-795-7151 (voice and **fax**)

Steinmetz Electronics
7519 Maplewood Ave
Hammond, IN 46324
219-931-9316

Sunlight Energy Systems
955 Manchester Ave SW
N Lawrence, OH 44666
330-832-4161

Surplus Sales of Nebraska
1502 Jones St
Omaha, NE 68102-3112
800-244-4567 (orders)
402-346-4750
fax 402-346-2939
e-mail
   grinnell@surplussales.com
web www.surplussales.com/

Surplus Traders
PO Box 276, Winters Ln
Alburg, VT 05440
514-739-9328
fax 514-345-8303
e-mail marv@73.com
web www.73.com/a/

Svetlana Electron Devices
8200 S. Memorial Parkway
Huntsville, AL 35802
256-882-1344
800-239-6900
fax 256-880-8077
e-mail sales@svetlana.com
web www.svetlana.com/

TAB/McGraw-Hill, Inc
860 Taylor Station Rd
Blacklick, OH 43004
800-262-4729
fax 614-759-3641
e-mail
   customer.service@mcgraw-
   hill.com

Tandy National Parts
900 E Northside Dr
Ft Worth, TX 76102
800-322-3690
fax 817-870-5626

TCE Labs
2365 Waterfront Park Dr
Canyon Lake, TX 78133
830-899-4575

TE Systems
PO Box 25845
Los Angeles, CA 90025
310-478-0591
fax 310-473-4038

Tech America
PO Box 1981
Fort Worth, TX 76101-1981
web www.techam.com/
800-877-0072

Tejas RF Technology
PO Box 720331
Houston, TX 77272-0331
713-879-9300
fax 713-879-9494

Telecom Industries
1385 Akron St
Copaigue, NY 11726
516-789-5020

Teletec Corp
10101 Capital Blvd
Wake Forest, NC 27587
909-556-7800

Telex Communications, Inc
8601 East Cornhusker Highway
Lincoln, NE 68505
402-467-5321
402-465-7021 (parts and service)
fax 402-467-3279

Tempo Research Corp [see AEA]

Ten-Tec, Inc
1185 Dolly Parton Pkwy
Sevierville, TN 37862
423-453-7172
fax 423-428-4483
web www.tentec.com/

Texas Towers
1108 Summit Ave, Suite 4
Plano, TX 75074
800-272-3467
972-422-7306 (Tech)
fax 972-881-0776
e-mail sales@texastowers.com
web www.texastowers.com/

Timewave Technology, Inc
2401 Pilot Knob Rd
Suite 134
St Paul MN 55120
612-452-5939
fax 612-452-4571

Toroid Corp of Maryland
202 Northwood Dr
Salisbury, MD 21801
410-860-0300
fax 410-860-0302

Tri-Ex Tower Corp
7182 Rasmussen Ave
Visalia, CA 93291
800-328-2393 (orders)
209-651-7850 ext. 352 or 353
   (Tech Support)
fax 209-651-5157

Trinity Software
7801 Rice Dr
Rowlett, TX 75088
972-475-7132

Tucson Amateur Packet Radio
8987-309 E Tanque Verde Rd,
   #337
Tucson, AZ 85749-9399
817-383-0000 (voice mail)
fax 817-566-2544
e-mail tapr@tapr.org
web www.tapr.org

TX RX Systems, Inc
8625 Industrial Pky
Angola, NY 14006
716-549-4700
fax 716-549-4772

Typetronics
PO Box 8873
Fort Lauderdale, FL 33310-8873
954-583-1340
fax 954-583-0777

Unified Microsystems
PO Box 133
Slinger, WI 53086
414-644-9036
fax 414-644-9036
e-mail uns@nconnect.net

United Nations Bookstore
UN General Assembly Building,
   Room 32B
New York, NY 10017

Universal Manufacturing Co
43900 Groesbeck Hwy
Clinton Township, MI 48036
810-463-2560
fax 810-463-2964

US Electronics
585 North Bicycle Path
Port Jefferson Station, NY 11776
516-331-2552
e-mail info@uselec.com
web www.uselectronics.com/

US Government Printing Office
Washington, DC 20402-9371
202-783-3238
202-512-1800 (credit card orders)
web www.gpo.gov/

US Plastic Corp
1390 Neubrecht Rd
Lima, OH 45801
800-537-9724
fax 419-228-5034

US Tower Corp
1220 Marcin St
Visalia, CA 93291
209-733-2438
fax 209-733-7194

VK3UM EME Planner
Doug McArthur
30 Rolloway Rise
Chirnside Park 3116
Australia

The W5YI Group
PO Box 565101
Dallas, TX 75356
800-669-9594 (orders)
817-274-0400
e-mail w5yi@w5yi.org
web www.w5yi.org/

W6EL Software
11058 Queensland St
Los Angeles, CA 90034-3029
310-473-7322
e-mail ad363@lafn.org

W7FG Vintage Manuals
3300 Wayside Dr
Bartlesville, OK 74006
800-807-6146 (orders)
918-333-3754
e-mail w7fg@eigen.net
web www.w7fg.com/

Watkins-Johnson Company,
   Telecommunications Group
700 Quince Orchard Rd
Gaithersburg, MD 20878-1794
800-954-3577 (800-WJ-HELPS)
301-948-7550
fax 301-921-7479
e-mail wjhelps@wj.com or
   mike.cox@wj.com
web www.wj.com/

Ed Wetherhold, W3NQN
1426 Catlyn Pl
Annapolis, MD 21401-4208
410-268-0916
fax 410-268-4779

Wilderness Radio
PO Box 734
Los Altos, CA 94023-0734
650-494-3806
e-mail qrpbob@datamers.com
web www.fix.net/jparker/wild.html

Winegard
3000 Kirkwood St
Burlington, IA 52601-1007

The Wireman Inc
261 Pittman Rd
Landrum, SC 29356-9544
800-727-WIRE (800-727-9473)
   Orders only
864-895-4195 Technical
fax 864-895-5811
e-mail into@thewireman.com
web www.thewireman.com/

Worldradio
2120 28th St
Sacramento, CA 95818
916-457-3655
800-366-9192 (subscriptions)
e-mail kb6hp@ms.net
web www.wr6w.com/

Wyman Research, Inc
8339 South, 850 West
Waldron, IN 46182-9644
765-525-6452

Yaesu U.S.A.
17210 Edwards Rd
Cerritos, CA 90703
562-404-2700
fax 562-404-1210
e-mail yaesu@worldnet.att.net
web www.yaesu.com/

Zero Surge
944 State Rte 12
Frenchtown NJ 08825
908-996-7700
fax 908-996-7773

Z-TRAK
RFD 1 Box 33
Milo, ME 04463

73 Amateur Radio Today
70 Route 202N
Peterborough, NH 03458
603-924-0058
800-274-7373 (subscriptions)

The ARRL Technical Information Service (TIS)

The ARRL answers questions of a technical nature for ARRL members and nonmembers alike through the Technical Information Service. Questions may be submitted via e-mail (tis@arrl.org); phone (860-594-0214); Fax (860-594-0259); or mail (TIS, ARRL, 225 Main St, Newington, CT 06111). The TIS also maintains a home page on *ARRLWeb*: **http://www.arrl.org/tis**. This site contains links to several technical areas:

TISfind – This search engine contains over 2000 providers of products, services and information of interest to radio amateurs. Before contacting TIS for the address of someone who can repair your radio, or sells antennas, or has old manuals or schematics, look in *TISfind*. Instructions and categories are on the *TISfind* page. This file may be downloaded for use off-line.

ARRL Periodicals Index Search – This **Members Only** search engine contains the *QST* index from 1915 to the present and the *QEX* index from 1981 to the present. For *QST* issues from 1970 to the present, and some selected articles back to 1950, identifying keywords have been added to the database. By entering keywords (such as ANTENNA) or combinations of keywords (such as CONSTRUCTION ANTENNA VERTICAL HF) into the **Title words:** field, you may create dynamic bibliographies. A KEYWORD list is provided on the ARRL Periodicals Index Search page. This replaces the old bibliographies TXT files.

QST Product Reviews — There is a link to a list of Product Reviews from 1970 to the present and for **Members Only** these Product Reviews may be viewed online or downloaded to your computer for printing. There are also links to ARRL book companion software, project templates, product notes, and source code and programs mentioned in *QST* articles.

TIS Pages — The flagship of the TIS Web Page is the Technical Information Pages menu. These articles replace the old TIS Packages that were available only by mail in the past. There are now over 100 topics containing over 500 articles for viewing or download. A list of topics follows:

Access for the Blind
Antennas
• Antenna Projects
• Antenna Gain
• Antenna Ground
• Antenna Theory
• Balloon Antenna
• Grounding
• Indoor
• Receiving Wire Antennas
• Smith Charts
• Transmission Lines/SWR
• Transmatch/Antenna Tuner
Antique/Vintage Radio
Amateur Television
• ATV — Fast Scan TV
• SSTV — Slow Scan TV
Batteries
Buying a Rig
Beginner Ham
• First Steps in Radio
• Your First Antenna
Construction Projects/Techniques
• Antenna Projects
• Building Equipment
• Construction Techniques
• Projects for the Ham Shack
• QRP Projects
• Servicing Equipment
• Surface Mount Technology
• Tube Amplifiers
• VHF Projects
• Tube Transmitters/Receivers
Digital Modes
• AMTOR, APRS, Packet, RTTY, TCP/IP
• Digital Voice
• PSK31
• WEFAX
DSP — Digital Signal Processing
EME — Moonbounce

Emergency/Alternative Power
Grid Squares
Ham Radio History
Internet Ham Radio
Lab Notes, Complete
Laser Communications
Lightning Protection
Link/Remote Control
Mobile
• Marine
• VHF Antennas
Morse/CW
• Coherent CW
Propagation
QRP
Repeaters
RF Exposure
RFI/EMI
Safety
• EDS (Electrostatic Discharge Safety)
• Electrical Safety
• RF Exposure (New!)
Satellites
• General
• Phase 3D
Science Fair/Merit Badge Projects
UHF/Microwave
• UHF/Microwave Projects and Information
Check back often for new and updated TIS pages.

QST and QEX Article Reprints — Reprints are available at nominal cost from the ARRL Technical Secretary by mail (Reprints, ARRL, 225 Main St, Newington, CT 06111), by telephone (860-594-0278), or e-mail (**reprints@arrl.org**). Reprints cannot be e-mailed or attached to an e-mail message. Article copies must be prepaid. The most convenient payment method is via a major credit card.

Computer Connector Pinouts

(A)

Parallel Port (DB 25 pin)
Female

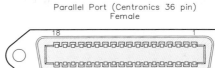

Pin	Signal	Pin	Signal
1	Strobe	10	Acknowledge
2	Data 0	11	Busy
3	Data 1	12	Paper Empty
4	Data 2	13	Select
5	Data 3	14	Auto Feed
6	Data 4	15	Error
7	Data 5	16	Initialize
8	Data 6	17	Select In
9	Data 7	18-25	GND

(B)

Parallel Port (Centronics 36 pin)
Female

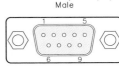

Pin	Signal	Pin	Signal
1	Strobe	13	Select
2	Data 0	14	Auto Feed
3	Data 1	15	N/C (not connected)
4	Data 2	16	Signal GND
5	Data 3	17	Frame GND
6	Data 4	18	+5 V Out
7	Data 5	19-30	GND
8	Data 6	31	Reset
9	Data 7	32	Error
10	Acknowledge	33	External GND
11	Busy	34	N/C
12	Paper Empty	35	N/C
		36	Select In

(C)

Serial Port (DB 9 pin)
Male

Pin	Signal
1	DCD (Data Carrier Detect)
2	RxD (Receive Data)
3	TxD (Transmit Data)
4	DTR (Data Terminal Ready)
5	GND (Signal Ground)
6	DSR (Data Set Ready)
7	RTS (Request To Send)
8	CTS (Clear To Send)
9	RI (Ring Indicator)

(D)

Serial Port (DB 25 pin)
Male

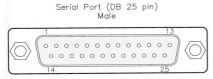

Pin	Signal	Pin	Signal
1	N/C (not connected)	20	DTR (Data Terminal Ready)
2	TxD (Transmit Data)	21	N/C
3	RxD (Receive Data)	22	RI (Ring Indicator)
4	RTS (Request To Send)	23	N/C
5	CTS (Clear To Send)	24	N/C
6	DSR (Data Set Ready)	25	N/C
7	GND (Signal Ground)		
8	DCD (Data Carrier Detect)		
9-19	N/C		

(E)

Ethernet Connector (RJ45-8 pin)
Female

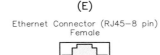

Pin	Signal
1	Output Transmit Data (+)
2	Output Transmit Data (−)
3	Input Receive Data (+)
4	N/C (not connected)
5	N/C
6	Input Receive Data (−)
7	N/C
8	N/C

(F)

Ethernet Connector (RJ45-10 pin)
Female

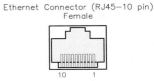

Pin	Signal
1	DCD (Data Carrier Detect)
2	DTR (Data Terminal Ready)
3	CTS (Clear To Send)
4	GND (Signal Ground)
5	RxD (Receive Data)
6	TxD (Transmit Data)
7	GND (Frame Ground)
8	RTS (Request To Send)
9	DSR (Data Set Ready)
10	RI (Ring Indicator)

(G)

Mouse Port (DB 9 pin)
Male

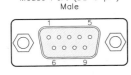
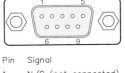

Pin	Signal
1	N/C (not connected)
2	Data
3	Clock
4	N/C
5	GND (Signal Ground)
6	N/C
7	RTS (12-9 V)
8	N/C
9	N/C

(H)

Mouse Port (mini DIN 9 pin)
Female

Pin	Signal
1	+5 V
2	X−A
3	X−B
4	Y−A
5	Y−B
6	Button 1
7	Button 2
8	Button 3
9	GND

Note: All figures not drawn to same scale.

(I)

Game/Joystick Port (DB 15 pin)
Female

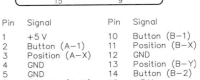

Pin	Signal	Pin	Signal
1	+5 V	10	Button (B−1)
2	Button (A−1)	11	Position (B−X)
3	Position (A−X)	12	GND
4	GND	13	Position (B−Y)
5	GND	14	Button (B−2)
6	Position (A−Y)	15	+5 V
7	Button (A−2)		
8	+5 V		
9	+5 V		

Voltage-Power Conversion Table

Based on a 50-ohm system

RMS	Peak-to-Peak	dBmV	Watts	dBm
0.01 µV	0.0283 µV	−100	2×10^{-18}	−147.0
0.02 µV	0.0566 µV	−93.98	8×10^{-18}	−141.0
0.04 µV	0.113 µV	−87.96	32×10^{-18}	−134.9
0.08 µV	0.226 µV	−81.94	128×10^{-18}	−128.9
0.1 µV	0.283 µV	−80.0	200×10^{-18}	−127.0
0.2 µV	0.566 µV	−73.98	800×10^{-18}	−121.0
0.4 µV	1.131 µV	−67.96	3.2×10^{-15}	−114.9
0.8 µV	2.236 µV	−61.94	12.8×10^{-15}	−108.9
1.0 µV	2.828 µV	−60.0	20.0×10^{15}	−107.0
2.0 µV	5.657 µV	−53.98	80.0×10^{-15}	−101.0
4.0 µV	11.31 µV	−47.96	320×10^{-15}	−94.95
8.0 µV	22.63 µV	−41.94	1.28×10^{-12}	−88.93
10.0 µV	28.28 µV	−40.00	2.0×10^{-12}	−86.99
20.0 µV	56.57 µV	−33.98	8.0×10^{-12}	−80.97
40.0 µV	113.1 µV	−27.96	32.0×10^{-12}	−74.95
80.0 µV	226.3 µV	−21.94	128.0×10^{-12}	−68.93
100.0 µV	282.8 µV	−20.0	200.0×10^{-12}	−66.99
200.0 µV	565.7 µV	−13.98	800.0×10^{-12}	−60.97
400.0 µV	1.131 mV	−7.959	3.2×10^{-9}	−54.95
800.0 µV	2.263 mV	−1.938	12.8×10^{-9}	−48.93
1.0 mV	2.828 mV	0.0	20.0×10^{-9}	−46.99
2.0 mV	5.657 mV	6.02	80.0×10^{-9}	−40.97
4.0 mV	11.31 mV	12.04	320×10^{-9}	−34.95
8.0 mV	22.63 mV	18.06	1.28 µW	−28.93
10.0 mV	28.28 mV	20.00	2.0 µW	−26.99
20.0 mV	56.57 mV	26.02	8.0 µW	−20.97
40.0 mV	113.1 mV	32.04	32.0 µW	−14.95
80.0 mV	226.3 mV	38.06	128.0 µW	−8.93
100.0 mV	282.8 mV	40.0	200.0 µW	−6.99
200.0 mV	565.7 mV	46.02	800.0 µW	−0.97
223.6 mV	632.4 mV	46.99	1.0 mW	0
400.0 mV	1.131 V	52.04	3.2 mW	5.05
800.0 mV	2.263 V	58.06	12.80 mW	11.07
1.0 V	2.828 V	60.0	20.0 mW	13.01
2.0 V	5.657 V	66.02	80.0 mW	19.03
4.0 V	11.31 V	72.04	320.0 mW	25.05
8.0 V	22.63 V	78.06	1.28 W	31.07
10.0 V	28.28 V	80.0	2.0 W	33.01
20.0 V	56.57 V	86.02	8.0 W	39.03
40.0 V	113.1 V	92.04	32.0 W	45.05
80.0 V	226.3 V	98.06	128 W	51.07
100.0 V	282.8 V	100.0	200.0 W	53.01
200.0 V	565.7 V	106.0	800.0 W	59.03
223.6 V	632.4 V	107.0	1000.0 W	60.0
400.0 V	1,131.0 V	112.0	3,200.0 W	65.05
800.0 V	2,263.0 V	118.1	12,800.0 W	71.07
1000.0 V	2,828.0 V	120.0	20,000 W	73.01
2000.0 V	5,657.0 V	126.0	80,000 W	79.03
4000.0 V	11,310.0 V	132.0	320,000 W	85.05
8000.0 V	22,630.0 V	138.1	1.28 MW	91.07
10,000.0 V	28,280.0 V	140.0	2.0 MW	93.01

Voltage, $V_{p\text{-}p} = V_{RMS} \times \sqrt{2}$

Voltage, $dBmV = 20 \times Log_{10}\left[\dfrac{V_{RMS}}{0.001\,V}\right]$

Power, Watts $=\left[\dfrac{V_{RMS}^2}{50\,\Omega}\right]$

Power, $dBm = 10 \times Log_{10}\left[\dfrac{Power\,(watts)}{0.001\,W}\right]$

Large Machine-Wound Coil Specifications

Coil Dia, Inches	Turns Per Inch	Inductance in µH
1¼	4	2.75
	6	6.3
	8	11.2
	10	17.5
	16	42.5
1½	4	3.9
	6	8.8
	8	15.6
	10	24.5
	16	63
1¾	4	5.2
	6	11.8
	8	21
	10	33
	16	85
2	4	6.6
	6	15
	8	26.5
	10	42
	16	108
2½	4	10.2
	6	23
	8	41
	10	64
3	4	14
	6	31.5
	8	56
	10	89

Inductance Factor for Large Machine-Wound Coils

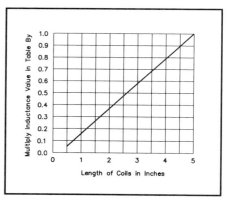

Factor to be applied to the inductance of large coils for coil lengths up to 5 inches.

Small Machine-Wound Coil Specifications

Coil Dia, Inches	Turns Per Inch	Inductance in μH
¹/₂ (A)	4	0.18
	6	0.40
	8	0.72
	10	1.12
	16	2.8
	32	12
⁵/₈ (A)	4	0.28
	6	0.62
	8	1.1
	10	1.7
	16	4.4
	32	18
³/₄ (B)	4	0.6
	6	1.35
	8	2.4
	10	3.8
	16	9.9
	32	40
1 (B)	4	1.0
	6	2.3
	8	4.2
	10	6.6
	16	16.9
	32	68

Inductance Factor for Small Machine-Wound Coils

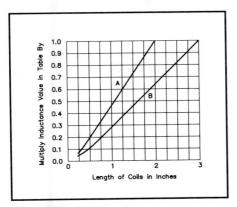

Factor to be applied to the inductance of small coils as a function of coil length. Use curve A for coils marked A, and curve B for coils marked B.

Measured inductance for #12 Wire Windings

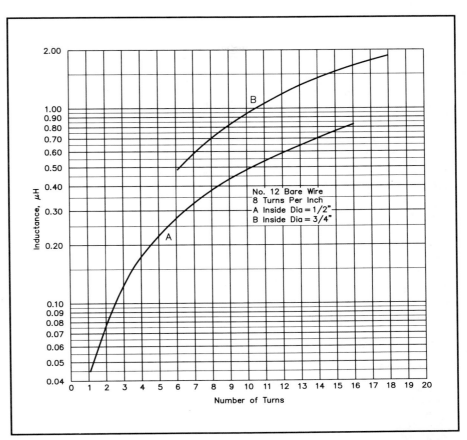

Values are for inductors with half-inch leads and wound with eight turns per inch.

How to Use the Standard Value Capacitor (SVC) Filter Tables

Detailed instructions for using these tables appear in the Filters chapter. If you are unfamiliar with filter design from tables, look there to learn the basics. This simple example is intended as a memory aid, not a tutorial.

Let's design a low-pass filter for a 20-m CW transmitter. Based on measurements of the second harmonic, insertion loss (attenuation) should be at least 20 dB at the minimum second-harmonic frequency (28 MHz). Insertion loss should be minimal at the maximum operating frequency (14.1 MHz).

When choosing a filter, look for appropriate cutoff and attenuation frequencies, but *ignore the decimal points* because the component values are easily scaled by powers of ten. A 5-element Chebyshev design looks like a good choice because designs 20 through 22 show 20-dB frequencies of 2.73 and 2.77 MHz and cutoff frequencies of 1.44 to 1.66 MHz. In fact, those numbers are *too* close to our targets (27.7 MHz is only 1.1% under 28 MHz). Using 5% components, we would be lucky to get within 5% of the design targets. It's better to move each target value 10% or so to the safe side, which yields 20 dB at 25.2 MHz and f_{co} = 15.5 MHz. rolloff. No 5-element design in the table can meet these criteria.

In the 7-element Chebyshev list (Page 30.18), however, design 25 meets the needs. It has a maximum SWR of 1.099:1, which is acceptable.

FREQUENCY (MHz)

NO.								
	F_{co}	3 dB	20 dB	40 dB	MAX SWR	C1,7 (pF)	L2,6 (µH)	C3,5 (pF)	L4 (µH)
25	1.68	1.93	2.35	3.03	1.099	1500	6.58	3300	7.72

Scaling the filter is easy. We need only divide one of the frequencies listed into the desired frequency, round to the nearest power of ten and multiply all frequencies and divide all component values by the result: 28/2.35 = 11.91, say 10; which gives:

FREQUENCY (MHz)

NO.								
	F_{co}	3 dB	20 dB	40 dB	MAX SWR	C1,7 (pF)	L2,6 (µH)	C3,5 (pF)	L4 (µH)
25	16.8	19.3	23.5	30.3	1.099	150	0.658	330	0.772

In some cases, the filter terminating impedances may not be 50 Ω. Then we need to adjust the filter values to match the required impedance. All tabulated designs are easily scaled to impedance levels other than 50 Ω, while keeping the convenience of standard-value capacitors and the "scan mode" of design selection. If the desired new impedance level differs from 50 Ω by a factor that is some power of ten, the 50-Ω design is scaled by shifting the decimal points of the component values, that is multiplying or dividing by some power of ten. The other data remain unchanged. For example, if the impedance level is increased by ten or one hundred times (to 500 or 5000 Ω), the decimal point of the capacitor is shifted to the left (dividing) one or two places and the decimal point of the inductor is shifted to the right (multiplying) one or two places. With increasing impedance, capacitor values decrease and inductor values increase. The opposite is true when impedance decreases.

When the desired impedance level differs from the standard 50-Ω value by a factor that is not a power of ten, such as 1.2, 1.5 or 1.86, the search criteria to select the design number must be adjusted by that factor:

1. Calculate the impedance scaling ratio:

$$R = \frac{Z_x}{50} \qquad (1)$$

where Z_x is the desired new impedance level, in ohms.

2. Calculate the cutoff frequency (f_{50co}) of a "trial" 50-Ω filter,

$$f_{50co} = R \times f_{xco} \qquad (2)$$

where R is the impedance scaling ratio and f_{xco} is the desired cutoff frequency of the filter at the new impedance level.

3. Select a design from the SVC tables based on the calculated f_{50co}. The capacitor values of this design are taken directly, but the frequency and inductor values must be scaled to the new impedance level.

4. Calculate the exact f_{xco} values, where

$$f_{xco} = \frac{f'50co}{R} \qquad (3)$$

and f'_{50co} is the tabulated cutoff frequency of the selected design. Calculate the other frequencies of the design in the same way.

5. Calculate the inductor values for the new filter by multiplying the tabulated inductor values of the selected design by the square of the scaling ratio, R.

For example, assume that our 20-m low-pass filter were to be used in a 1000-Ω IF stage. This requires that we apply both methods, because a change from 50 to 1000 involves factors of 10 and 2 (50 × 2 × 10 = 1000). Therefore, we must first scale the desired frequencies by from 50 Ω to 100 Ω (50 × 2 = 100):

R = 100/50 = 2

f_{50co} = 2 × 15.5 = 31 MHz

f_{20dB} = 2 × 25.2 = 50.4 MHz

Select a filter based on these two values. Design 59 from the 7-element low-pass Chebyshev list looks good.
Scale all frequencies of the final design by dividing the tabulated frequencies impedance scaling ratio, 2:

f_{co} = 3.3/2 = 1.65

f_{20dB} = 4.81/2 = 2.405

The inductor values are scaled to 100 Ω by multiplying them by the square of the impedance ratio, where R = 2 and R^2 = 4.0:

L2,6 = 4.0 × 3.24 µH = 12.96 µH

L4 = 4.0 × 3.88 µH = 15.52 µH

The 100-Ω design is now impedance scaled to 1000 Ω by shifting the decimal points of the capacitor values to the left and the decimal points of the inductor values to the right. The final scaled component values for the 1000-Ω filter are:

C1,7 = 68 pF
C3,5 = 160 pF
L2,6 = 129.6 µH
L4 = 155.2 µH.

5-Element Chebyshev Low-Pass Filter Designs
50-Ohm Impedance, C-In/Out for Standard E24 Capacitor Values

Filter No.	F_{co}	Frequency (MHz) 3 dB	20 dB	40 dB	Max SWR	C1,5 (pF)	L2,4 (µH)	C3 (pF)	Filter No.	F_{co}	Frequency (MHz) 3 dB	20 dB	40 dB	Max SWR	C1,5 (pF)	L2,4 (µH)	C3 (pF)
1	1.01	1.15	1.53	2.25	1.355	3600	10.8	6200	69	4.55	5.54	7.65	11.4	1.167	620	2.37	1200
2	1.02	1.21	1.65	2.45	1.212	3000	10.7	5600	70	5.07	5.84	7.84	11.5	1.304	680	2.16	1200
3	1.15	1.29	1.71	2.51	1.391	3300	9.49	5600	71	3.96	5.76	8.38	12.8	1.041	470	2.35	1100
4	1.10	1.32	1.81	2.69	1.196	2700	9.88	5100	72	4.39	5.84	8.31	12.6	1.079	510	2.31	1100
5	1.25	1.41	1.88	2.75	1.386	3000	8.67	5100	73	4.88	6.01	8.33	12.5	1.152	560	2.20	1100
6	1.04	1.37	1.94	2.94	1.085	2200	9.82	4700	74	5.50	6.34	8.54	12.6	1.293	620	1.99	1100
7	1.15	1.41	1.95	2.92	1.155	2400	9.37	4700	75	4.40	6.34	9.20	14.1	1.043	430	2.13	1000
8	1.32	1.50	2.01	2.96	1.332	2700	8.29	4700	76	4.91	6.45	9.13	13.8	1.087	470	2.09	1000
9	1.13	1.50	2.12	3.22	1.081	2000	9.00	4300	77	5.38	6.62	9.17	13.7	1.154	510	2.00	1000
10	1.26	1.54	2.13	3.19	1.157	2200	8.56	4300	78	6.00	6.95	9.37	13.8	1.282	560	1.83	1000
11	1.39	1.61	2.18	3.21	1.276	2400	7.88	4300	79	4.81	6.97	10.1	15.5	1.042	390	1.94	910
12	1.05	1.62	2.38	3.66	1.028	1600	8.35	3900	80	5.43	7.09	10.0	15.2	1 091	430	1.89	910
13	1.23	1.65	2.34	3.55	1.076	1800	8.19	3900	81	6.00	7.31	10.1	15.1	1.167	470	1.80	910
14	1.39	1.70	2.35	3.51	1.159	2000	7.75	3900	82	6.60	7.64	10.3	15.2	1.283	510	1.66	910
15	1.55	1.79	2.41	3.55	1.295	2200	7.05	3900	83	4.86	7.69	11.4	17.5	1.023	330	1.76	820
16	1.17	1.76	2.57	3.94	1.033	1500	7.70	3600	84	5.51	7.76	11.2	17.1	1.052	360	1.74	820
17	1.27	1.77	2.55	3.88	1.057	1600	7.64	3600	85	6.07	7.89	11.1	16.8	1.095	390	1.70	820
18	1.46	1.82	2.54	3.81	1.135	1800	7.28	3600	86	6.77	8.17	11.2	16.7	1.184	430	1.60	820
19	1.65	1.92	2.59	3.83	1.268	2000	6.64	3600	87	7.54	8.61	11.5	17.0	1.327	470	1.45	820
20	1.88	2.08	2.73	3.97	1.497	2200	5.70	3600	88	5.26	8.40	12.4	19.2	1.022	300	1.61	750
21	1.43	1.94	2.77	4.21	1.068	1500	6.96	3300	89	6.04	8.49	12.2	18.7	1.052	330	1.59	750
22	1.54	1.97	2.77	4.17	1.109	1600	6.79	3300	90	6.70	8.64	12.2	18.4	1.101	360	1.55	750
23	1.76	2.07	2.81	4.17	1.238	1800	6.21	3300	91	7.33	8.89	12.3	18.3	1.175	390	1.48	750
24	2.02	2.25	2.96	4.31	1.470	2000	5.31	3300	92	8.24	9.42	12.6	18.5	1.327	430	1.33	750
25	1.31	2.10	3.11	4.79	1.022	1200	6.43	3000	93	6.69	9.36	13.5	20.6	1.054	300	1.44	680
26	1.48	2.12	3.06	4.68	1.046	1300	6.39	3000	94	7.48	9.56	13.4	20.2	1.110	330	1.40	680
27	1.75	2.19	3.05	4.57	1.135	1500	6.07	3000	95	8.25	9.89	13.6	20.2	1.196	360	1.32	680
28	1.89	2.25	3.08	4.57	1.206	1600	5.77	3000	96	9.10	10.4	13.9	20.4	1.328	390	1.20	680
29	2.19	2.45	3.23	4.71	1.440	1800	4.92	3000	97	7.21	10.2	14.8	22.6	1.048	270	1.32	620
30	1.51	2.34	3.44	5.29	1.026	1100	5.78	2700	98	8.18	10.5	14.7	22.2	1.107	300	1.28	620
31	1.70	2.36	3.40	5.17	1.057	1200	5.73	2700	99	9.11	10.9	14.9	22.1	1.203	330	1.19	620
32	1.87	2.40	3.38	5.10	1.104	1300	5.57	2700	100	10.1	11.5	15.3	22.5	1.355	360	1.08	620
33	2.20	2.56	3.46	5.11	1.268	1500	4.98	2700	101	7.82	11.3	16.4	25.1	1.042	240	1.19	560
34	2.39	2.69	3.56	5.21	1.406	1600	4.53	2700	102	9.02	11.6	16.3	24.6	1.105	270	1.16	560
35	1.75	2.63	3.85	5.91	1.033	1000	5.14	2400	103	8.66	12.4	18.0	27.6	1.044	220	1.09	510
36	1.99	2.67	3.81	5.78	1.072	1100	5.05	2400	104	9.64	12.6	17.9	27.1	1.088	240	1.06	510
37	2.19	2.74	3.81	5.71	1.135	1200	4.85	2400	105	9.22	13.5	19.6	30.0	1.039	200	1.00	470
38	2.40	2.84	3.86	5.73	1.227	1300	4.55	2400	106	9.85	14.7	21.5	33.0	1.034	180	0.919	430
39	1.89	2.87	4.21	6.47	1.030	910	4.71	2200									
40	2.14	2.91	4.16	6.31	1.068	1000	4.64	2200									
41	2.39	2.99	4.16	6.23	1.135	1100	4.45	2200									
42	2.64	3.11	4.22	6.25	1.238	1200	4.14	2200									
43	2.93	3.29	4.36	6.39	1.398	1300	3.71	2200									
44	2.05	3.16	4.64	7.13	1.028	820	4.28	2000									
45	2.36	3.20	4.57	6.94	1.068	910	4.22	2000									
46	2.63	3.28	4.57	6.86	1.135	1000	4.05	2000									
47	2.93	3.43	4.65	6.89	1.251	1100	3.73	2000									
48	3.29	3.67	4.85	7.07	1.440	1200	3.28	2000									
49	2.34	3.51	5.14	7.88	1.033	750	3.85	1800									
50	2.63	3.56	5.08	7.71	1.069	820	3.79	1800									
51	2.96	3.66	5.09	7.62	1.145	910	3.61	1800									
52	3.30	3.84	5.19	7.67	1.268	1000	3.32	1800									
53	3.76	4.15	5.45	7.93	1.497	1100	2.85	1800									
54	2.70	3.96	5.76	8.82	1.039	680	3.42	1600									
55	3.06	4.03	5.71	8.63	1.086	750	3.34	1600									
56	3.38	4.14	5.73	8.57	1.159	820	3.18	1600									
57	3.82	4.39	5.89	8.67	1.311	910	2.86	1600									
58	2.77	4.21	6.18	9.48	1.030	620	3.21	1500									
59	3.14	4.26	6.10	9.26	1.067	680	3.17	1500									
60	3.51	4.38	6.10	9.14	1.135	750	3.03	1500									
61	3.88	4.56	6.20	9.17	1.241	820	2.82	1500									
62	4.46	4.95	6.51	9.48	1.473	910	2.41	1500									
63	3.39	4.88	7.08	10.8	1.044	560	2.77	1300									
64	3.84	4.98	7.02	10.6	1.097	620	2.70	1300									
65	4.26	5.14	7.08	10.5	1.181	680	2.55	1300									
66	4.79	5.46	7.29	10.7	1.341	750	2.28	1300									
67	3.61	5.28	7.68	11.8	1.039	510	2.56	1200									
68	4.06	5.36	7.61	11.5	1.083	560	2.51	1200									

C1 = C5 L2 = L4

(A)

(B)

The schematic for a 5-element capacitor input/output Chebyshev low-pass filter is shown at A. At B is the typical attenuation response curve.

7-Element Chebyshev Low-Pass Filter Designs
50-Ohm Impedance, C-In/Out for Standard E24 Capacitor Values

Filter No.	F_{co}	Frequency (MHz) 3 dB	20 dB	40 dB	Max SWR	C1,7 (pF)	L2,6 (µH)	C3,5 (pF)	L4 (µH)
1	1.02	1.10	1.31	1.65	1.254	3300	11.2	6200	12.6
2	1.04	1.16	1.40	1.79	1.142	2700	10.9	5600	12.6
3	1.13	1.23	1.45	1.84	1.264	3000	10.1	5600	11.3
4	1.05	1.23	1.51	1.96	1.071	2200	10.3	5100	12.3
5	1.12	1.26	1.53	1.96	1.123	2400	10.0	5100	11.7
6	1.23	1.34	1.59	2.01	1.247	2700	9.29	5100	10.4
7	1.03	1.30	1.63	2.15	1.030	1800	9.52	4700	11.9
8	1.12	1.33	1.64	2.13	1.064	2000	9.50	4700	11.4
9	1.21	1.37	1.66	2.13	1.119	2200	9.27	4700	10.8
10	1.29	1.42	1.70	2.16	1.200	2400	8.82	4700	10.0
11	1.10	1.41	1.79	2.36	1.023	1600	8.68	4300	11.0
12	1.21	1.45	1.79	2.33	1.058	1800	8.71	4300	10.5
13	1.31	1.49	1.81	2.33	1.114	2000	8.50	4300	9.91
14	1.42	1.56	1.86	2.36	1.202	2200	8.06	4300	9.14
15	1.54	1.65	1.93	2.43	1.336	2400	7.39	4300	8.18
16	1.25	1.57	1.97	2.59	1.031	1500	7.90	3900	9.85
17	1.32	1.59	1.97	2.57	1.050	1600	7.91	3900	9.62
18	1.44	1.64	1.99	2.56	1.109	1800	7.73	3900	9.04
19	1.57	1.72	2.05	2.60	1.205	2000	7.30	3900	8.27
20	1.44	1.73	2.14	2.78	1.056	1500	7.29	3600	8.82
21	1.52	1.76	2.15	2.78	1.086	1600	7.22	3600	8.54
22	1.66	1.84	2.20	2.81	1.176	1800	6.86	3600	7.83
23	1.83	1.96	2.30	2.90	1.327	2000	6.22	3600	6.90
24	1.51	1.86	2.32	3.05	1.037	1300	6.70	3300	8.27
25	1.68	1.93	2.35	3.03	1.099	1500	6.58	3300	7.72
26	1.77	1.98	2.38	3.05	1.147	1600	6.40	3300	7.37
27	1.96	2.11	2.49	3.14	1.294	1800	5.83	3300	6.50
28	1.56	2.02	2.56	3.38	1.021	1100	6.04	3000	7.68
29	1.68	2.05	2.56	3.35	1.042	1200	6.09	3000	7.47
30	1.79	2.09	2.57	3.33	1.073	1300	6.05	3000	7.21
31	1.99	2.20	2.64	3.37	1.176	1500	5.72	3000	6.52
32	2.11	2.28	2.70	3.42	1.257	1600	5.42	3000	6.08
33	1.75	2.25	2.84	3.75	1.023	1000	5.45	2700	6.89
34	1.89	2.29	2.84	3.71	1.048	1100	5.48	2700	6.68
35	2.02	2.34	2.86	3.70	1.086	1200	5.41	2700	6.40
36	2.15	2.41	2.90	3.72	1.141	1300	5.26	2700	6.06
37	2.44	2.61	3.07	3.86	1.327	1500	4.66	2700	5.18
38	2.01	2.54	3.20	4.21	1.027	910	4.86	2400	6.09
39	2.17	2.59	3.20	4.17	1.056	1000	4.86	2400	5.88
40	2.33	2.66	3.24	4.17	1.104	1100	4.77	2400	5.59
41	2.49	2.76	3.30	4.21	1.176	1200	4.57	2400	5.22
42	2.67	2.88	3.41	4.30	1.282	1300	4.27	2400	4.77
43	2.15	2.76	3.49	4.60	1.024	820	4.44	2200	5.61
44	2.35	2.82	3.49	4.55	1.053	910	4.46	2200	5.41
45	2.52	2.89	3.52	4.54	1.099	1000	4.38	2200	5.15
46	2.72	3.01	3.60	4.59	1.176	1100	4.19	2200	4.78
47	2.94	3.16	3.73	4.70	1.294	1200	3.88	2200	4.33
48	2.38	3.04	3.84	5.06	1.025	750	4.04	2000	5.09
49	2.57	3.09	3.84	5.01	1.050	820	4.06	2000	4.93
50	2.78	3.18	3.88	5.00	1.100	910	3.98	2000	4.68
51	2.99	3.31	3.96	5.05	1.176	1000	3.81	2000	4.35
52	3.26	3.50	4.12	5.19	1.308	1100	3.50	2000	3.89
53	2.67	3.38	4.26	5.61	1.027	680	3.64	1800	4.57
54	2.89	3.45	4.27	5.56	1.056	750	3.65	1800	4.41
55	3.09	3.54	4.31	5.55	1.100	820	3.59	1800	4.21
56	3.35	3.69	4.42	5.62	1.188	910	3.40	1800	3.87
57	3.65	3.92	4.60	5.80	1.327	1000	3.11	1800	3.45
58	3.07	3.82	4.80	6.30	1.033	620	3.24	1600	4.03
59	3.30	3.90	4.81	6.25	1.064	680	3.24	1600	3.88
60	3.55	4.02	4.87	6.26	1.120	750	3.15	1600	3.67
61	3.81	4.18	4.99	6.34	1.204	820	3.00	1600	3.39
62	3.16	4.05	5.12	6.75	1.024	560	3.03	1500	3.82
63	3.45	4.13	5.12	6.68	1.053	620	3.04	1500	3.69
64	3.69	4.24	5.17	6.66	1.097	680	2.99	1500	3.51
65	3.99	4.41	5.28	6.73	1.176	750	2.86	1500	3.26
66	4.31	4.64	5.48	6.91	1.297	820	2.64	1500	2.94
67	3.81	4.72	5.90	7.74	1.036	510	2.64	1300	3.26
68	4.10	4.82	5.93	7.69	1.070	560	2.62	1300	3.14
69	4.43	4.98	6.02	7.72	1.133	620	2.54	1300	2.94
70	4.78	5.21	6.19	7.85	1.230	680	2.39	1300	2.70
71	4.13	5.11	6.39	8.38	1.035	470	2.43	1200	3.01
72	4.40	5.20	6.41	8.33	1.064	510	2.43	1200	2.91
73	4.72	5.35	6.49	8.34	1.116	560	2.37	1200	2.76
74	5.12	5.60	6.67	8.48	1.214	620	2.23	1200	2.52
75	4.49	5.57	6.97	9.15	1.035	430	2.23	1100	2.76
76	4.82	5.68	7.00	9.09	1.066	470	2.22	1100	2.66
77	5.12	5.83	7.07	9.10	1.112	510	2.18	1100	2.54
78	5.52	6.07	7.24	9.21	1.196	560	2.07	1100	2.35
79	4.93	6.12	7.67	10.1	1.034	390	2.03	1000	2.51
80	5.33	6.26	7.70	10.0	1.069	430	2.02	1000	2.41
81	5.69	6.44	7.80	10.0	1.122	470	1.97	1000	2.29
82	6.08	6.68	7.97	10.1	1.198	510	1.88	1000	2.13
83	6.63	7.09	8.32	10.5	1.343	560	1.71	1000	1.89
84	5.48	6.75	8.43	11.0	1.038	360	1.85	910	2.28
85	5.84	6.87	8.46	11.0	1.068	390	1.84	910	2.20
86	6.28	7.09	8.58	11.0	1.126	430	1.79	910	2.07
87	6.75	7.39	8.80	11.2	1.213	470	1.69	910	1.91
88	5.68	7.39	9.37	12.4	1.020	300	1.65	820	2.10
89	6.17	7.52	9.36	12.2	1.043	330	1.66	820	2.04
90	6.60	7.68	9.41	12.2	1.079	360	1.65	820	1.96
91	7.01	7.89	9.53	12.2	1.131	390	1.61	820	1.86
92	7.59	8.27	9.82	12.5	1.233	430	1.51	820	1.70
93	6.72	8.21	10.2	13.4	1.042	300	1.52	750	1.87
94	7.23	8.40	10.3	13.3	1.080	330	1.51	750	1.79
95	7.72	8.66	10.4	13.4	1.138	360	1.46	750	1.69
96	8.24	9.00	10.7	13.6	1.222	390	1.39	750	1.57
97	7.36	9.04	11.3	14.8	1.039	270	1.38	680	1.70
98	7.98	9.27	11.4	14.7	1.082	300	1.37	680	1.62
99	8.58	9.59	11.6	14.8	1.148	330	1.32	680	1.52
100	9.23	10.0	11.9	15.1	1.247	360	1.24	680	1.39
101	7.91	9.86	12.4	16.2	1.032	240	1.26	620	1.56
102	8.67	10.1	12.4	16.1	1.075	270	1.25	620	1.49
103	9.39	10.5	12.7	16.2	1.145	300	1.20	620	1.39
104	8.86	11.0	13.7	18.0	1.036	220	1.14	560	1.40
105	9.49	11.2	13.8	17.8	1.068	240	1.13	560	1.35
106	9.72	12.0	15.0	19.7	1.036	200	1.03	510	1.28

The schematic for a 7-element Chebyshev low-pass filter. See page 30.17 for the attenuation response curve.

C1=C7 C3=C5 L2=L6

5-Element Chebyshev Low-Pass Filter Designs—50-Ohm Impedance, L-In/Out for Standard-Value L and C

Filter No.	F_{co}	3 dB	20 dB	40 dB	Max SWR	L1,5 (µH)	C2,4 (pF)	L3 (µH)
1	0.744	1.15	1.69	2.60	1.027	5.60	4700	13.7
2	0.901	1.26	1.81	2.76	1.055	5.60	4300	12.7
3	1.06	1.38	1.94	2.93	1.096	5.60	3900	11.8
4	1.19	1.47	2.05	3.07	1.138	5.60	3600	11.2
5	1.32	1.58	2.17	3.23	1.192	5.60	3300	10.6
6	0.911	1.39	2.03	3.12	1.030	4.70	3900	11.4
7	1.08	1.50	2.16	3.29	1.056	4.70	3600	10.6
8	1.25	1.63	2.30	3.48	1.092	4.70	3300	9.92
9	1.42	1.77	2.46	3.68	1.142	4.70	3000	9.32
10	1.61	1.92	2.63	3.90	1.209	4.70	2700	8.79
11	1.05	1.64	2.41	3.72	1.025	3.90	3300	9.63
12	1.29	1.80	2.60	3.96	1.054	3.90	3000	8.83
13	1.54	1.99	2.80	4.22	1.099	3.90	2700	8.15
14	1.80	2.19	3.03	4.53	1.164	3.90	2400	7.57
15	1.99	2.35	3.20	4.75	1.222	3.90	2200	7.23
16	1.34	2.00	2.93	4.49	1.034	3.30	2700	7.89
17	1.68	2.25	3.20	4.84	1.077	3.30	2400	7.15
18	1.92	2.43	3.40	5.11	1.118	3.30	2200	6.73
19	2.16	2.63	3.62	5.40	1.174	3.30	2000	6.35
20	1.65	2.46	3.59	5.51	1.035	2.70	2200	6.43
21	1.99	2.70	3.86	5.85	1.069	2.70	2000	5.93
22	2.34	2.97	4.15	6.24	1.118	2.70	1800	5.50
23	2.71	3.27	4.49	6.68	1.188	2.70	1600	5.13
24	2.92	3.43	4.67	6.92	1.233	2.70	1500	4.97
25	2.01	3.01	4.39	6.74	1.034	2.20	1800	5.26
26	2.52	3.37	4.80	7.27	1.077	2.20	1600	4.76
27	2.78	3.57	5.02	7.56	1.107	2.20	1500	4.55
28	3.34	4.02	5.52	8.21	1.190	2.20	1300	4.18
29	2.36	3.61	5.29	8.14	1.029	1.80	1500	4.38
30	3.12	4.14	5.89	8.92	1.080	1.80	1300	3.88
31	3.51	4.45	6.23	9.36	1.118	1.80	1200	3.67
32	3.93	4.78	6.60	9.85	1.169	1.80	1100	3.48
33	4.37	5.15	7.01	10.4	1.233	1.80	1000	3.31
34	3.10	4.51	6.56	10.0	1.041	1.50	1200	3.51
35	3.65	4.90	6.99	10.6	1.073	1.50	1100	3.27
36	4.21	5.34	7.47	11.2	1.118	1.50	1000	3.06
37	4.75	5.77	7.95	11.9	1.173	1.50	910	2.89
38	3.53	5.41	7.94	12.2	1.029	1.20	1000	2.92
39	4.30	5.94	8.53	13.0	1.060	1.20	910	2.69
40	5.09	6.53	9.18	13.8	1.106	1.20	820	2.49
41	5.73	7.04	9.75	14.6	1.155	1.20	750	2.35
42	6.42	7.61	10.4	15.4	1.219	1.20	680	2.23
43	4.40	6.60	9.65	14.8	1.033	1.00	820	2.40
44	5.27	7.20	10.3	15.7	1.064	1.00	750	2.22
45	6.15	7.87	11.1	16.7	1.108	1.00	680	2.07
46	6.95	8.51	11.8	17.6	1.160	1.00	620	1.95
47	7.80	9.22	12.6	18.6	1.227	1.00	560	1.85
48	5.23	7.96	11.7	17.9	1.030	0.82	680	1.99
49	6.33	8.72	12.5	19.0	1.061	0.82	620	1.83
50	7.45	9.56	13.4	20.3	1.106	0.82	560	1.70
51	8.44	10.3	14.3	21.4	1.158	0.82	510	1.60
52	9.28	11.0	15.1	22.4	1.211	0.82	470	1.53
53	6.41	9.66	14.1	21.7	1.032	0.68	560	1.64
54	7.75	10.6	15.2	23.1	1.064	0.68	510	1.51
55	8.83	11.4	16.1	24.3	1.100	0.68	470	1.42
56	9.97	12.3	17.1	25.6	1.148	0.68	430	1.34

7-Element Chebyshev Low-Pass Filter Designs—50-Ohm Impedance, L-In/Out for Standard-Value L and C

Filter No.	F_{co}	3 dB	20 dB	40 dB	Max SWR	L1,7 (µH)	C2,6 (pF)	L3,5 (µH)	C4 (pF)
1	1.01	1.18	1.44	1.87	1.081	5.89	4300	13.4	5100
2	1.09	1.29	1.60	2.08	1.059	5.06	3900	12.0	4700
3	1.03	1.09	1.26	1.58	1.480	10.1	4300	17.1	4700
4	1.20	1.40	1.73	2.24	1.071	4.81	3600	11.2	4300
5	1.16	1.23	1.44	1.81	1.383	8.34	3900	14.6	4300
6	1.33	1.54	1.88	2.43	1.087	4.58	3300	10.3	3900
7	1.42	1.68	2.07	2.70	1.064	3.95	3000	9.27	3600
8	1.34	1.41	1.63	2.04	1.506	7.98	3300	13.4	3600
9	1.53	1.85	2.31	3.02	1.045	3.36	2700	8.32	3300
10	1.50	1.59	1.86	2.33	1.406	6.57	3000	11.4	3300
11	1.63	2.06	2.59	3.41	1.029	2.83	2400	7.41	3000
12	1.69	1.81	2.13	2.68	1.317	5.36	2700	9.70	3000
13	1.86	2.27	2.83	3.70	1.042	2.71	2200	6.78	2700
14	1.91	2.07	2.46	3.12	1.238	4.31	2400	8.19	2700
15	2.14	2.52	3.11	4.04	1.064	2.63	2000	6.18	2400
16	2.01	2.11	2.45	3.06	1.506	5.32	2200	8.91	2400
17	2.29	2.78	3.46	4.52	1.045	2.24	1800	5.54	2200
18	2.25	2.39	2.79	3.49	1.406	4.38	2000	7.61	2200
19	2.45	3.09	3.88	5.11	1.029	1.89	1600	4.94	2000
20	2.53	2.71	3.19	4.02	1.317	3.57	1800	6.47	2000
21	2.85	3.37	4.15	5.39	1.064	1.97	1500	4.64	1800
22	2.86	3.11	3.69	4.68	1.238	2.88	1600	5.46	1800
23	3.13	3.84	4.79	6.27	1.039	1.59	1300	4.00	1600
24	3.27	4.12	5.18	6.81	1.029	1.41	1200	3.70	1500
25	3.47	3.90	4.70	6.02	1.140	2.01	1300	4.17	1500
26	3.99	4.61	5.64	7.28	1.087	1.53	1100	3.43	1300
27	4.27	5.05	6.22	8.09	1.064	1.32	1000	3.09	1200
28	4.01	4.22	4.90	6.11	1.506	2.66	1100	4.45	1200
29	4.63	5.53	6.85	8.91	1.056	1.17	910	2.81	1100
30	4.49	4.77	5.57	6.98	1.406	2.19	1000	3.81	1100
31	5.05	6.11	7.60	9.92	1.047	1.03	820	2.53	1000
32	4.93	5.23	6.10	7.64	1.416	2.02	910	3.49	1000
33	5.58	6.70	8.31	10.8	1.052	0.954	750	2.31	910
34	5.54	5.94	6.99	8.80	1.326	1.65	820	2.97	910
35	6.23	7.41	9.16	11.9	1.059	0.881	680	2.10	820
36	5.92	6.24	7.26	9.06	1.476	1.76	750	2.98	820
37	6.79	8.12	10.0	13.1	1.055	0.796	620	1.91	750
38	6.64	7.07	8.27	10.4	1.379	1.45	680	2.54	750
39	7.46	8.97	11.1	14.5	1.051	0.711	560	1.73	680
40	7.21	7.63	8.89	11.1	1.438	1.40	620	2.41	680
41	8.18	9.85	12.2	15.9	1.050	0.645	510	1.57	620
42	8.10	8.66	10.2	12.8	1.345	1.15	560	2.05	620
43	9.21	10.8	13.2	17.1	1.074	0.633	470	1.46	560
44	8.78	9.31	10.9	13.6	1.425	1.14	510	1.96	560
45	10.1	11.8	14.4	18.7	1.081	0.589	430	1.34	510

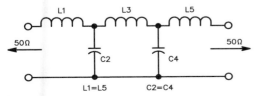

The schematic for a 5-element inductor input/output Chebyshev low-pass filter. See page 30.17 for the attenuation response curve.

The schematic for a 7-element inductor input/output Chebyshev low-pass filter. See page 30.17 for the attenuation response curve.

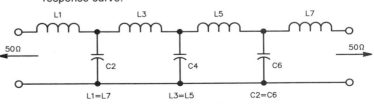

5-Branch Elliptic Low-Pass Filter Designs
50-Ohm Impedance, Standard E12 Capacitor Values for C1, C3 and C5

Filter No.	F_{co}	F_{3dB}	F_{As}	A_s	Max.	C1	C3	C5	C2	C4	L2	L4	F2	F4
		(MHz)		(dB)	SWR		(pF)				(µH)		(MHz)	
1	0.795	0.989	1.57	47.4	1.092	2700	5600	2200	324	937	12.1	10.1	2.54	1.64
2	1.06	1.20	1.77	46.2	1.234	2700	4700	2200	341	982	9.36	7.56	2.82	1.85
3	1.47	1.57	2.15	45.4	1.586	2700	3900	2200	364	1045	6.32	4.88	3.32	2.23
4	0.929	1.18	1.91	48.0	1.077	2200	4700	1800	257	743	10.2	8.59	3.11	1.99
5	1.27	1.45	2.17	46.7	1.215	2200	3900	1800	271	779	7.85	6.39	3.45	2.26
6	1.69	1.82	2.54	45.9	1.489	2200	3300	1800	287	821	5.64	4.42	3.96	2.64
7	1.12	1.44	2.41	49.8	1.071	1800	3900	1500	192	549	8.45	7.25	3.95	2.52
8	1.49	1.73	2.70	48.8	1.183	1800	3300	1500	200	570	6.75	5.62	4.33	2.81
9	2.11	2.27	3.27	47.8	1.506	1800	2700	1500	213	604	4.55	3.64	5.12	3.40
10	1.28	1.66	2.63	46.3	1.064	1500	3300	1200	192	561	7.20	6.00	4.28	2.74
11	1.79	2.06	2.99	44.8	1.195	1500	2700	1200	204	592	5.52	4.42	4.75	3.11
12	2.52	2.70	3.63	43.8	1.525	1500	2200	1200	220	636	3.71	2.82	5.58	3.76
13	1.56	2.08	3.55	50.1	1.055	1200	2700	1000	127	363	5.88	5.07	5.83	3.71
14	2.23	2.59	4.04	48.8	1.183	1200	2200	1000	133	380	4.50	3.75	6.50	4.22
15	3.17	3.41	4.90	47.8	1.506	1200	1800	1000	142	402	3.03	2.42	7.68	5.10
16	1.94	2.52	4.15	48.4	1.064	1000	2200	820	115	331	4.79	4.06	6.78	4.34
17	2.73	3.14	4.73	47.0	1.199	1000	1800	820	121	348	3.66	2.99	7.56	4.93
18	3.73	4.02	5.63	46.2	1.491	1000	1500	820	129	368	2.56	2.01	8.76	5.85
19	2.39	3.11	5.20	49.4	1.065	820	1800	680	89.3	256	3.91	3.35	8.51	5.44
20	3.26	3.79	5.85	48.2	1.185	820	1500	680	93.6	267	3.07	2.54	9.39	6.10
21	4.83	5.17	7.30	47.2	1.569	820	1200	680	100	286	1.95	1.54	11.4	7.58
22	2.85	3.71	6.15	48.8	1.063	680	1500	560	76.6	220	3.26	2.78	10.1	6.43
23	4.16	4.74	7.14	47.3	1.221	680	1200	560	81.3	233	2.40	1.97	11.4	7.44
24	5.72	6.13	8.58	46.5	1.547	680	1000	560	86.3	246	1.65	1.30	13.3	8.91
25	3.67	4.69	7.95	50.5	1.076	560	1200	470	57.6	164	2.59	2.23	13.0	8.31
26	5.02	5.77	9.01	49.4	1.212	560	1000	470	60.3	171	2.01	1.68	14.5	9.40
27	7.18	7.68	11.1	48.6	1.582	560	820	470	64.1	181	1.32	1.06	17.3	11.5
28	4.40	5.60	9.24	49.3	1.079	470	1000	390	51.4	147	2.16	1.84	15.1	9.66
29	6.17	7.01	10.6	48.0	1.236	470	820	390	54.2	155	1.63	1.34	17.0	11.1
30	8.63	9.20	12.9	47.3	1.604	470	680	390	57.6	164	1.09	0.857	20.1	13.4
31	5.47	6.91	11.8	51.3	1.086	390	820	330	38.5	109	1.76	1.52	19.3	12.3
32	7.55	8.59	13.5	50.2	1.242	390	680	330	40.4	114	1.34	1.12	21.7	14.1
33	10.9	11.5	16.8	49.5	1.659	390	560	330	42.8	120	0.862	0.695	26.2	17.4
34	6.59	8.17	13.0	47.7	1.096	330	680	270	39.0	112	1.46	1.22	21.1	13.6
35	9.10	10.2	15.0	46.5	1.267	330	560	270	41.2	118	1.09	0.881	23.7	15.6
36	12.4	13.2	18.1	45.8	1.635	330	470	270	43.9	125	0.741	0.573	27.9	18.8

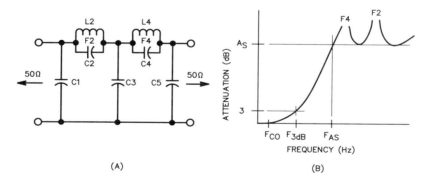

(A) (B)

The schematic for a 5-branch elliptic low-pass filter is shown at A. At B is the typical attenuation response curve.

5-Element Chebyshev High-Pass Filter Designs
50-Ohm Impedance, C-In/Out for Standard E24 Capacitor Values

Filter No.	F_{co}	3 dB	20 dB	40 dB	Max SWR	C1,5 (pF)	L2,4 (µH)	C3 (pF)
1	1.04	0.726	0.501	0.328	1.044	5100	6.45	2200
2	1.04	0.788	0.554	0.366	1.081	4300	5.97	2000
3	1.17	0.800	0.550	0.359	1.039	4700	5.85	2000
4	1.07	0.857	0.615	0.410	1.135	3600	5.56	1800
5	1.17	0.877	0.616	0.406	1.076	3900	5.36	1800
6	1.33	0.890	0.609	0.397	1.034	4300	5.26	1800
7	1.12	0.938	0.686	0.461	1.206	3000	5.20	1600
8	1.25	0.974	0.693	0.461	1.109	3300	4.86	1600
9	1.38	0.994	0.691	0.454	1.057	3600	4.71	1600
10	1.54	1.00	0.683	0.444	1.028	3900	4.67	1600
11	1.14	0.978	0.723	0.490	1.268	2700	5.09	1500
12	1.28	1.03	0.738	0.492	1.135	3000	4.64	1500
13	1.43	1.06	0.738	0.486	1.068	3300	4.44	1500
14	1.61	1.07	0.730	0.476	1.033	3600	4.38	1500
15	1.21	1.08	0.812	0.555	1.398	2200	4.82	1300
16	1.35	1.14	0.841	0.567	1.227	2400	4.29	1300
17	1.55	1.20	0.853	0.566	1.104	2700	3.94	1300
18	1.75	1.23	0.848	0.555	1.046	3000	3.81	1300
19	1.28	1.15	0.871	0.597	1.440	2000	4.57	1200
20	1.45	1.24	0.909	0.614	1.238	2200	3.99	1200
21	1.60	1.29	0.923	0.616	1.135	2400	3.71	1200
22	1.84	1.32	0.921	0.605	1.057	2700	3.54	1200
23	2.14	1.34	0.906	0.588	1.022	3000	3.50	1200
24	1.57	1.34	0.989	0.669	1.251	2000	3.69	1100
25	1.75	1.40	1.01	0.672	1.135	2200	3.40	1100
26	1.93	1.44	1.01	0.664	1.072	2400	3.27	1100
27	2.27	1.46	0.992	0.645	1.026	2700	3.21	1100
28	1.71	1.47	1.08	0.734	1.268	1800	3.39	1000
29	1.93	1.54	1.11	0.739	1.135	2000	3.09	1000
30	2.15	1.58	1.11	0.730	1.068	2200	2.96	1000
31	2.41	1.60	1.10	0.714	1.033	2400	2.92	1000
32	1.66	1.50	1.14	0.783	1.473	1500	3.54	910
33	1.82	1.59	1.18	0.803	1.311	1600	3.18	910
34	2.09	1.69	1.22	0.812	1.145	1800	2.83	910
35	2.36	1.74	1.22	0.802	1.068	2000	2.70	910
36	2.68	1.76	1.20	0.783	1.030	2200	2.66	910
37	2.12	1.81	1.33	0.898	1.241	1500	2.73	820
38	2.28	1.86	1.35	0.902	1.159	1600	2.58	820
39	2.61	1.93	1.35	0.890	1.069	1800	2.43	820
40	3.01	1.96	1.33	0.866	1.028	2000	2.39	820
41	2.17	1.90	1.42	0.970	1.341	1300	2.67	750
42	2.57	2.06	1.48	0.985	1.135	1500	2.32	750
43	2.76	2.10	1.48	0.978	1.086	1600	2.25	750
44	3.21	2.14	1.46	0.952	1.033	1800	2.19	750
45	2.45	2.13	1.58	1.08	1.304	1200	2.36	680
46	2.69	2.23	1.62	1.09	1.181	1300	2.17	680
47	3.17	2.33	1.63	1.07	1.067	1500	2.01	680
48	3.44	2.35	1.62	1.06	1.039	1600	1.99	680
49	2.70	2.34	1.74	1.18	1.293	1100	2.14	620
50	2.99	2.46	1.78	1.19	1.167	1200	1.96	620
51	3.28	2.53	1.79	1.19	1.097	1300	1.87	620
52	3.93	2.59	1.76	1.15	1.030	1500	1.81	620
53	3.02	2.60	1.93	1.31	1.282	1000	1.92	560
54	3.37	2.74	1.97	1.32	1.152	1100	1.75	560
55	3.72	2.81	1.98	1.31	1.083	1200	1.67	560
56	4.10	2.85	1.97	1.29	1.044	1300	1.64	560
57	3.31	2.86	2.12	1.44	1.283	910	1.75	510
58	3.69	3.00	2.17	1.45	1.154	1000	1.60	510
59	4.11	3.09	2.17	1.44	1.079	1100	1.52	510
60	4.59	3.14	2.15	1.41	1.039	1200	1.49	510
61	3.49	3.05	2.28	1.55	1.327	820	1.66	470
62	3.95	3.24	2.35	1.57	1.167	910	1.49	470
63	4.39	3.34	2.36	1.56	1.087	1000	1.41	470
64	4.94	3.40	2.34	1.53	1.041	1100	1.38	470
65	3.81	3.34	2.49	1.70	1.327	750	1.52	430
66	4.24	3.52	2.56	1.72	1.184	820	1.38	430
67	4.77	3.65	2.58	1.71	1.091	910	1.29	430
68	5.36	3.72	2.56	1.68	1.043	1000	1.26	430
69	4.20	3.68	2.75	1.87	1.328	680	1.38	390
70	4.72	3.89	2.83	1.90	1.175	750	1.24	390
71	5.22	4.02	2.84	1.88	1.095	820	1.17	390
72	5.93	4.10	2.82	1.85	1.042	910	1.14	390
73	4.48	3.95	2.96	2.02	1.355	620	1.30	360
74	5.01	4.18	3.05	2.05	1.196	680	1.16	360
75	5.60	4.34	3.08	2.04	1.101	750	1.09	360
76	6.23	4.42	3.07	2.01	1.052	820	1.06	360
77	4.79	4.25	3.20	2.19	1.391	560	1.22	330
78	5.44	4.55	3.33	2.24	1.203	620	1.07	330
79	6.03	4.72	3.36	2.23	1.110	680	1.00	330
80	6.77	4.82	3.35	2.20	1.052	750	0.970	330
81	7.70	4.87	3.30	2.14	1.023	820	0.962	330
82	5.28	4.68	3.53	2.41	1.386	510	1.10	300
83	5.94	4.99	3.65	2.46	1.212	560	0.978	300
84	6.66	5.20	3.70	2.46	1.107	620	0.910	300
85	7.43	5.31	3.68	2.42	1.054	680	0.882	300
86	8.56	5.36	3.62	2.35	1.022	750	0.875	300
87	6.05	5.31	3.97	2.70	1.332	470	0.956	270
88	6.69	5.58	4.07	2.74	1.196	510	0.870	270
89	7.43	5.78	4.11	2.73	1.105	560	0.817	270
90	8.39	5.91	4.08	2.68	1.048	620	0.792	270
91	7.07	6.09	4.51	3.06	1.276	430	0.818	240
92	7.84	6.38	4.61	3.08	1.155	470	0.752	240
93	8.59	6.55	4.62	3.06	1.088	510	0.719	240
94	9.64	6.66	4.58	3.00	1.042	560	0.702	240
95	7.61	6.60	4.90	3.33	1.295	390	0.760	220
96	8.53	6.95	5.02	3.36	1.157	430	0.690	220
97	9.43	7.15	5.04	3.33	1.085	470	0.658	220
98	10.4	7.26	5.01	3.28	1.044	510	0.644	220
99	7.58	6.83	5.19	3.56	1.470	330	0.776	200
100	8.53	7.33	5.42	3.67	1.268	360	0.678	200
101	9.36	7.64	5.52	3.70	1.159	390	0.628	200
102	10.4	7.88	5.54	3.66	1.081	430	0.596	200
103	8.55	7.67	5.81	3.98	1.440	300	0.685	180
104	9.69	8.24	6.06	4.09	1.238	330	0.597	180
105	10.7	8.57	6.15	4.10	1.135	360	0.556	180
106	9.80	8.73	6.58	4.50	1.406	270	0.595	160

C1 = C5   L2 = L4

(A)

(B)

The schematic for a 5-element capacitor input/output Chebyshev high-pass filter is shown at A. At B is the typical attenuation response curve.

7-Element Chebyshev High-Pass Filter Designs
50-Ohm Impedance, C-In/Out for Standard E24 Capacitor Values

Filter No.	F_{co}	3 dB	20 dB	40 dB	Max SWR	C1,7 (pF)	L2,6 (µH)	C3,5 (pF)	L4 (µH)
1	1.02	0.826	0.660	0.504	1.036	5100	6.16	2000	4.98
2	1.00	0.880	0.724	0.563	1.109	3900	5.67	1800	4.86
3	1.08	0.905	0.732	0.563	1.058	4300	5.55	1800	4.60
4	1.16	0.922	0.734	0.558	1.030	4700	5.55	1800	4.45
5	1.00	0.924	0.780	0.617	1.257	3000	5.53	1600	4.93
6	1.09	0.971	0.806	0.630	1.147	3300	5.15	1600	4.48
7	1.16	1.00	0.819	0.634	1.086	3600	4.99	1600	4.22
8	1.23	1.02	0.824	0.632	1.050	3900	4.93	1600	4.05
9	1.34	1.04	0.825	0.625	1.023	4300	4.95	1600	3.92
10	1.03	0.958	0.815	0.648	1.327	2700	5.43	1500	4.89
11	1.13	1.02	0.853	0.669	1.176	3000	4.92	1500	4.31
12	1.22	1.06	0.871	0.676	1.099	3300	4.70	1500	4.01
13	1.30	1.09	0.879	0.675	1.056	3600	4.63	1500	3.83
14	1.39	1.11	0.880	0.670	1.031	3900	4.63	1500	3.71
15	1.22	1.13	0.954	0.755	1.282	2400	4.57	1300	4.09
16	1.34	1.20	0.994	0.776	1.141	2700	4.17	1300	3.62
17	1.45	1.24	1.01	0.780	1.073	3000	4.03	1300	3.38
18	1.57	1.27	1.02	0.775	1.037	3300	4.00	1300	3.24
19	1.31	1.21	1.03	0.816	1.294	2200	4.25	1200	3.81
20	1.41	1.28	1.07	0.836	1.176	2400	3.94	1200	3.45
21	1.55	1.34	1.09	0.845	1.086	2700	3.74	1200	3.16
22	1.68	1.37	1.10	0.841	1.042	3000	3.70	1200	3.01
23	1.41	1.32	1.12	0.887	1.308	2000	3.93	1100	3.53
24	1.54	1.39	1.16	0.912	1.176	2200	3.61	1100	3.16
25	1.65	1.44	1.19	0.921	1.104	2400	3.46	1100	2.95
26	1.80	1.49	1.20	0.919	1.048	2700	3.39	1100	2.78
27	1.97	1.52	1.20	0.907	1.021	3000	3.41	1100	2.68
28	1.54	1.44	1.22	0.971	1.327	1800	3.62	1000	3.26
29	1.70	1.53	1.28	1.00	1.176	2000	3.28	1000	2.87
30	1.82	1.59	1.31	1.01	1.099	2200	3.14	1000	2.67
31	1.95	1.63	1.32	1.01	1.056	2400	3.08	1000	2.55
32	2.15	1.67	1.32	1.00	1.023	2700	3.10	1000	2.45
33	1.85	1.67	1.40	1.10	1.188	1800	3.01	910	2.64
34	2.00	1.75	1.44	1.11	1.100	2000	2.85	910	2.43
35	2.15	1.80	1.45	1.11	1.053	2200	2.81	910	2.31
36	2.31	1.83	1.45	1.10	1.027	2400	2.81	910	2.24
37	1.91	1.77	1.50	1.19	1.297	1500	2.91	820	2.61
38	2.03	1.85	1.55	1.22	1.204	1600	2.74	820	2.42
39	2.22	1.94	1.59	1.24	1.100	1800	2.57	820	2.19
40	2.41	2.00	1.61	1.23	1.050	2000	2.53	820	2.08
41	2.61	2.03	1.61	1.22	1.024	2200	2.54	820	2.01
42	2.26	2.04	1.71	1.34	1.176	1500	2.46	750	2.16
43	2.38	2.10	1.73	1.35	1.120	1600	2.38	750	2.04
44	2.60	2.17	1.76	1.35	1.056	1800	2.31	750	1.91
45	2.83	2.22	1.76	1.34	1.025	2000	2.32	750	1.84
46	2.40	2.20	1.85	1.46	1.230	1300	2.31	680	2.05
47	2.69	2.34	1.92	1.49	1.097	1500	2.13	680	1.81
48	2.82	2.39	1.94	1.49	1.064	1600	2.10	680	1.75
49	3.11	2.45	1.94	1.47	1.027	1800	2.10	680	1.67
50	2.66	2.43	2.04	1.61	1.214	1200	2.08	620	1.84
51	2.84	2.52	2.09	1.63	1.133	1300	1.98	620	1.71
52	3.16	2.64	2.13	1.63	1.053	1500	1.91	620	1.58
53	3.33	2.67	2.13	1.62	1.033	1600	1.91	620	1.54
54	2.73	2.55	2.17	1.73	1.343	1000	2.05	560	1.85
55	2.98	2.71	2.27	1.79	1.196	1100	1.86	560	1.64
56	3.19	2.82	2.32	1.81	1.116	1200	1.77	560	1.52
57	3.39	2.89	2.35	1.81	1.070	1300	1.73	560	1.45
58	3.81	2.98	2.36	1.79	1.024	1500	1.73	560	1.37
59	3.27	2.97	2.49	1.96	1.198	1000	1.70	510	1.49
60	3.53	3.10	2.55	1.99	1.112	1100	1.61	510	1.38
61	3.76	3.18	2.58	1.99	1.064	1200	1.58	510	1.31
62	4.01	3.24	2.59	1.98	1.036	1300	1.57	510	1.27
63	3.51	3.21	2.69	2.12	1.213	910	1.58	470	1.40
64	3.79	3.35	2.76	2.15	1.122	1000	1.49	470	1.28
65	4.07	3.45	2.80	2.16	1.066	1100	1.45	470	1.21
66	4.35	3.52	2.81	2.14	1.035	1200	1.45	470	1.17
67	3.79	3.47	2.93	2.31	1.233	820	1.46	430	1.30
68	4.12	3.65	3.02	2.35	1.126	910	1.37	430	1.18
69	4.42	3.76	3.06	2.36	1.069	1000	1.33	430	1.11
70	4.77	3.85	3.07	2.34	1.035	1100	1.33	430	1.07
71	4.20	3.85	3.24	2.55	1.222	750	1.32	390	1.17
72	4.52	4.02	3.32	2.59	1.131	820	1.24	390	1.07
73	4.89	4.15	3.37	2.60	1.068	910	1.21	390	1.01
74	5.27	4.24	3.39	2.58	1.034	1000	1.20	390	0.969
75	4.48	4.13	3.48	2.75	1.247	680	1.24	360	1.10
76	4.86	4.33	3.59	2.80	1.138	750	1.15	360	1.00
77	5.20	4.47	3.65	2.82	1.079	820	1.12	360	0.942
78	5.64	4.58	3.67	2.80	1.038	910	1.11	360	0.899
79	4.87	4.49	3.79	2.99	1.254	620	1.14	330	1.01
80	5.26	4.71	3.91	3.05	1.148	680	1.06	330	0.924
81	5.67	4.87	3.98	3.07	1.080	750	1.03	330	0.864
82	6.07	4.98	4.00	3.06	1.043	820	1.02	330	0.829
83	5.32	4.91	4.15	3.28	1.264	560	1.04	300	0.930
84	5.80	5.18	4.30	3.36	1.145	620	0.965	300	0.838
85	6.22	5.36	4.37	3.38	1.082	680	0.933	300	0.787
86	6.71	5.49	4.40	3.36	1.042	750	0.923	300	0.752
87	7.25	5.58	4.40	3.33	1.020	820	0.931	300	0.731
88	5.98	5.50	4.64	3.66	1.247	510	0.926	270	0.824
89	6.46	5.77	4.78	3.74	1.142	560	0.867	270	0.752
90	6.98	5.97	4.87	3.76	1.075	620	0.837	270	0.703
91	7.50	6.11	4.89	3.74	1.039	680	0.831	270	0.675
92	6.39	5.97	5.08	4.04	1.336	430	0.873	240	0.787
93	6.94	6.32	5.29	4.16	1.200	470	0.798	240	0.704
94	7.41	6.55	5.41	4.21	1.123	510	0.762	240	0.656
95	7.95	6.75	5.48	4.22	1.068	560	0.742	240	0.620
96	8.61	6.90	5.50	4.19	1.032	620	0.740	240	0.595
97	7.56	6.88	5.77	4.54	1.202	430	0.733	220	0.646
98	8.11	7.16	5.91	4.60	1.119	470	0.697	220	0.599
99	8.63	7.35	5.98	4.61	1.071	510	0.681	220	0.570
100	9.28	7.51	6.00	4.58	1.036	560	0.677	220	0.548
101	7.70	7.19	6.11	4.86	1.327	360	0.723	200	0.652
102	8.30	7.56	6.34	4.99	1.205	390	0.667	200	0.589
103	8.97	7.90	6.51	5.06	1.114	430	0.632	200	0.542
104	9.59	8.11	6.58	5.07	1.064	470	0.618	200	0.515
105	8.72	8.09	6.86	5.44	1.294	330	0.637	180	0.571
106	9.42	8.51	7.11	5.57	1.176	360	0.590	180	0.517

C1 = C7 , C3 = C5 , L2 = L6

The schematic for a 7-element capacitor input/output Chebyshev high-pass filter. See page 30.21 for the attenuation response curve.

5-Branch Elliptic High-Pass Filter Designs
50-Ohm Impedance, Standard E12 Capacitor Values for C1, C3 and C5

Filter No.	F_{co}	F_{3dB}	F_{As}	A_s	Max.	C1	C3	C5	C2	C4	L2	L4	F2	F4
		(MHz)		(dB)	SWR			(nF)			(μH)		(MHz)	
1	1.01	0.936	0.670	45.9	1.489	2.7	1.8	3.3	20.7	7.24	6.58	8.40	0.431	0.646
2	1.14	0.976	0.608	50.4	1.186	3.3	1.8	3.9	32.3	11.4	5.53	6.54	0.377	0.582
3	1.30	1.01	0.604	49.4	1.071	3.9	1.8	4.7	35.8	12.5	5.19	6.07	0.369	0.578
4	1.19	1.11	0.810	45.4	1.543	2.2	1.5	2.7	16.4	5.71	5.65	7.28	0.523	0.780
5	1.38	1.20	0.797	46.8	1.199	2.7	1.5	3.3	22.0	7.66	4.61	5.65	0.499	0.765
6	1.56	1.19	0.685	51.6	1.064	3.3	1.5	3.9	33.7	11.9	4.32	4.97	0.417	0.655
7	1.51	1.40	1.01	45.9	1.489	1.8	1.2	2.2	13.8	4.82	4.39	5.60	0.646	0.968
8	1.75	1.51	1.00	46.6	1.180	2.2	1.2	2.7	17.7	6.14	3.65	4.47	0.627	0.961
9	2.02	1.52	0.920	48.3	1.055	2.7	1.2	3.3	23.4	8.09	3.44	4.04	0.562	0.880
10	1.78	1.65	1.15	45.7	1.506	1.5	1.0	1.8	12.7	4.47	3.71	4.64	0.733	1.10
11	2.07	1.80	1.20	46.8	1.199	1.8	1.0	2.2	14.7	5.11	3.07	3.77	0.749	1.15
12	2.38	1.83	1.13	47.8	1.064	2.2	1.0	2.7	18.6	6.43	2.87	3.40	0.689	1.08
13	2.22	2.08	1.55	43.7	1.531	1.2	0.82	1.5	8.19	2.83	3.05	4.02	1.01	1.49
14	2.52	2.17	1.39	48.7	1.186	1.5	0.82	1.8	13.5	4.73	2.51	3.01	0.865	1.33
15	2.89	2.23	1.36	48.2	1.065	1.8	0.82	2.2	15.5	5.37	2.36	2.78	0.833	1.30
16	2.57	2.40	1.68	47.8	1.560	1.0	0.68	1.2	8.40	2.96	2.60	3.27	1.08	1.62
17	3.05	2.68	1.85	44.7	1.215	1.2	0.68	1.5	8.77	3.02	2.10	2.64	1.17	1.78
18	3.48	2.66	1.57	49.9	1.063	1.5	0.68	1.8	14.1	4.94	1.96	2.28	0.957	1.50
19	3.17	2.96	2.13	46.1	1.554	0.82	0.56	1.0	6.31	2.21	2.13	2.72	1.37	2.05
20	3.62	3.16	2.05	48.6	1.210	1.0	0.56	1.2	8.93	3.14	1.74	2.10	1.28	1.96
21	4.19	3.30	2.11	46.1	1.076	1.2	0.56	1.5	9.30	3.19	1.61	1.94	1.30	2.02
22	4.30	3.79	2.55	46.9	1.233	0.82	0.47	1.0	6.69	2.33	1.48	1.82	1.60	2.45
23	4.89	3.84	2.31	49.7	1.079	1.0	0.47	1.2	9.34	3.27	1.36	1.59	1.41	2.21
24	5.87	3.89	2.31	47.4	1.021	1.2	0.47	1.5	9.71	3.32	1.35	1.58	1.39	2.20
25	4.44	4.17	3.01	46.5	1.618	0.56	0.39	0.68	4.37	1.53	1.54	1.97	1.94	2.90
26	5.14	4.52	2.99	48.0	1.236	0.68	0.39	0.82	5.88	2.06	1.23	1.50	1.87	2.87
27	5.88	4.67	2.90	48.0	1.085	0.82	0.39	1.0	7.05	2.45	1.13	1.34	1.78	2.78
28	5.99	5.34	3.60	47.1	1.269	0.56	0.33	0.68	4.63	1.62	1.06	1.31	2.27	3.46
29	6.81	5.48	3.37	49.0	1.096	0.68	0.33	0.82	6.15	2.15	0.961	1.13	2.07	3.22
30	8.07	5.50	3.17	49.3	1.026	0.82	0.33	1.0	7.33	2.54	0.945	1.09	1.91	3.02
31	6.38	5.99	4.26	47.3	1.609	0.39	0.27	0.47	3.18	1.12	1.06	1.34	2.74	4.10
32	7.34	6.47	4.18	49.2	1.241	0.47	0.27	0.56	4.33	1.53	0.856	1.03	2.61	4.01
33	8.39	6.73	4.17	48.4	1.092	0.56	0.27	0.68	4.90	1.71	0.784	0.930	2.57	4.00
34	7.92	7.36	4.98	49.6	1.522	0.33	0.22	0.39	3.05	1.08	0.828	1.02	3.17	4.79
35	9.21	8.05	5.27	48.1	1.217	0.39	0.22	0.47	3.40	1.19	0.686	0.832	3.30	5.06
36	10.4	8.18	4.84	50.5	1.077	0.47	0.22	0.56	4.56	1.60	0.636	0.740	2.95	4.62

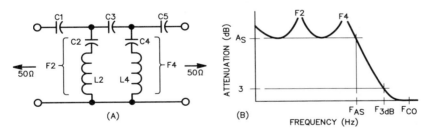

The schematic for a 5-branch elliptic high-pass filter is shown at A. At B is the typical attenuation response curve.

Relationship Between Noise Figure and Noise Temperature

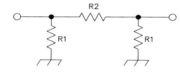

NF (dB)	K
0.1	6.75
0.2	13.67
0.3	20.74
0.4	27.98
0.5	35.39
0.6	42.96
0.7	50.72
0.8	58.66
0.9	66.78
1.0	75.09
1.1	83.59
1.2	92.29
1.3	101.20
1.4	110.31
1.5	119.64
1.6	129.18
1.7	138.94
1.8	148.93
1.9	159.16
2.0	169.62

Tower Manufacturers

Contact information appears in the Handbook Address List elsewhere in this chapter. Send updates to the Handbook Editor at ARRL Headquarters.

Aluma Tower
Glen Martin Engineering
Hy-Gain Division, Telex Communications, Inc
M² Enterprises
National Tower Co
Rohn
Texas Towers
Tri-Ex Tower Corp
Universal Manufacturing Co
US Tower Corp

Pi-Network Resistive Attenuators (50 Ω)

dB Atten.	R1 (Ohms)	R2 (Ohms)
1	870.0	5.8
2	436.0	11.6
3	292.0	17.6
4	221.0	23.8
5	178.6	30.4
6	150.5	37.3
7	130.7	44.8
8	116.0	52.8
9	105.0	61.6
10	96.2	71.2
11	89.2	81.6
12	83.5	93.2
13	78.8	106.0
14	74.9	120.3
15	71.6	136.1
16	68.8	153.8
17	66.4	173.4
18	64.4	195.4
19	62.6	220.0
20	61.0	247.5
21	59.7	278.2
22	58.6	312.7
23	57.6	351.9
24	56.7	394.6
25	56.0	443.1
30	53.2	789.7
35	51.8	1405.4
40	51.0	2500.0
45	50.5	4446.0
50	50.3	7905.6
55	50.2	14,058.0
60	50.1	25,000.0

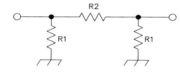

Note: A PC board kit for the Low-Power Step Attenuator (Sep 1982 *QST*) is available from FAR Circuits. Project details are in the Handbook template package STEP ATTENUATOR. See the templates section on *ARRLWeb*, **www.arrl.org/notes/1867/**.

T-Network Resistive Attenuators (50 Ω)

dB Atten.	R1 (Ohms)	R2 (Ohms)
1	2.9	433.3
2	5.7	215.2
3	8.5	141.9
4	11.3	104.8
5	14.0	82.2
6	16.6	66.9
7	19.0	55.8
8	21.5	47.3
9	23.8	40.6
10	26.0	35.0
11	28.0	30.6
12	30.0	26.8
13	31.7	23.5
14	33.3	20.8
15	35.0	18.4
16	36.3	16.2
17	37.6	14.4
18	38.8	12.8
19	40.0	11.4
20	41.0	10.0
21	41.8	9.0
22	42.6	8.0
23	43.4	7.1
24	44.0	6.3
25	44.7	5.6
30	47.0	3.2
35	48.2	1.8
40	49.0	1.0
45	49.4	0.56
50	49.7	0.32
55	49.8	0.18
60	49.9	0.10

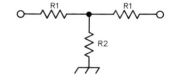

Antenna Wire Strength

American Wire Gauge	Recommended Tension[1] (pounds)		Weight (pounds per 1000 feet)	
	Copper-clad steel[2]	Hard-drawn copper	Copper-clad steel[2]	Hard-drawn copper
4	495	214	115.8	126
6	310	130	72.9	79.5
8	195	84	45.5	50
10	120	52	28.8	31.4
12	75	32	18.1	19.8
14	50	20	11.4	12.4
16	31	13	7.1	7.8
18	19	8	4.5	4.9
20	12	5	2.8	3.1

[1]Approximately one-tenth the breaking load. Might be increased 50% if end supports are firm and there is no danger of ice loading.
[2]"Copperweld," 40% copper.

Standard vs American Wire Gauge

SWG	Diam (in.)	Nearest AWG
12	0.104	10
14	0.08	12
16	0.064	14
18	0.048	16
20	0.036	19
22	0.028	21
24	0.022	23
26	0.018	25
28	0.0148	27
30	0.0124	28
32	0.0108	29
34	0.0092	31
36	0.0076	32
38	0.006	34
40	0.0048	36
42	0.004	38
44	0.0032	40
46	0.0024	—

Impedance of Various Two-Conductor Lines

Wire Size	Twists per Inch				
	2.5	5	7.5	10	12.5
no. 20	43	39	35		
no. 22	46	41	39	37	32
no. 24	60	45	44	43	41
no. 26	65	57	54	48	47
no. 28	74	53	51	49	47
no. 30			49	46	47

Measured in ohms at 14.0 MHz.

This chart illustrates the impedance of various two-conductor lines as a function of the wire size and number of twists per inch.

Attenuation per Foot for Lines

Wire Size	Twists per Inch				
	2.5	5	7.5	10	12.5
no. 20	0.11	0.11	0.12		
no. 22	0.11	0.12	0.12	0.12	0.12
no. 24	0.11	0.12	0.12	0.13	0.13
no. 26	0.11	0.13	0.13	0.13	0.13
no. 28	0.11	0.13	0.13	0.16	0.16
no. 30			0.25	0.27	0.27

Measured in decibels at 14.0 MHz.

Attenuation in dB per foot for the same lines as shown to left.

Equivalent Values of Reflection Coefficient, Attenuation, SWR and Return Loss

Reflection Coefficient (%)	Attenuation (dB)	Max SWR	Return Loss	Reflection Coefficient (%)	Attenuation (dB)	Max SWR	Return Loss	Reflection Coefficient (%)	Attenuation (dB)	Max SWR	Return Loss
1.000	0.000434	1.020	40.00	24.000	0.2577	1.632	12.40	60.749	2.0000	4.095	4.33
1.517	0.001000	1.031	36.38	25.000	0.2803	1.667	12.04	63.000	2.1961	4.405	4.01
2.000	0.001738	1.041	33.98	26.000	0.3040	1.703	11.70	66.156	2.5000	4.909	3.59
3.000	0.003910	1.062	30.46	27.000	0.3287	1.740	11.37	66.667	2.5528	5.000	3.52
4.000	0.006954	1.083	27.96	28.000	0.3546	1.778	11.06	70.627	3.0000	5.809	3.02
4.796	0.01000	1.101	26.38	30.000	0.4096	1.857	10.46	70.711	3.0103	5.829	3.01
5.000	0.01087	1.105	26.02	31.623	0.4576	1.925	10.00				
6.000	0.01566	1.128	24.44	32.977	0.5000	1.984	9.64				
7.000	0.02133	1.151	23.10	33.333	0.5115	2.000	9.54				
7.576	0.02500	1.164	22.41	34.000	0.5335	2.030	9.37				
8.000	0.02788	1.174	21.94	35.000	0.5675	2.077	9.12				
9.000	0.03532	1.198	20.92	36.000	0.6028	2.125	8.87				
10.000	0.04365	1.222	20.00	37.000	0.6394	2.175	8.64				
10.699	0.05000	1.240	19.41	38.000	0.6773	2.226	8.40				
11.000	0.05287	1.247	19.17	39.825	0.75000	2.324	8.00				
12.000	0.06299	1.273	18.42	40.000	0.7572	2.333	7.96				
13.085	0.07500	1.301	17.66	42.000	0.8428	2.448	7.54				
14.000	0.08597	1.326	17.08	42.857	0.8814	2.500	7.36				
15.000	0.09883	1.353	16.48	44.000	0.9345	2.571	7.13				
15.087	0.10000	1.355	16.43	45.351	1.0000	2.660	6.87				
16.000	0.1126	1.381	15.92	48.000	1.1374	2.846	6.38				
17.783	0.1396	1.433	15.00	50.000	1.2494	3.000	6.02				
18.000	0.1430	1.439	14.89	52.000	1.3692	3.167	5.68				
19.000	0.1597	1.469	14.42	54.042	1.5000	3.352	5.35				
20.000	0.1773	1.500	13.98	56.234	1.6509	3.570	5.00				
22.000	0.2155	1.564	13.15	58.000	1.7809	3.762	4.73				
23.652	0.2500	1.620	12.52	60.000	1.9382	4.000	4.44				

$$\rho = \frac{SWR - 1}{SWR + 1}$$

where ρ = 0.01 × (reflection coefficient in %)

$$\rho = 10^{\frac{-RL}{20}}$$

where RL = return loss (dB)

$$\rho = \sqrt{1 - \left(0.1^X\right)}$$

where X = A/10 and A = attenuation (dB)

$$SWR = \frac{1 + \rho}{1 - \rho}$$

Return loss (dB) = −8.68589 ln (ρ)
where ln is the natural log (log to the base e)

Attenuation (dB) = −4.34295 ln (1−ρ²)
where ln is the natural log (log to the base e)

Guy Wire Lengths to Avoid

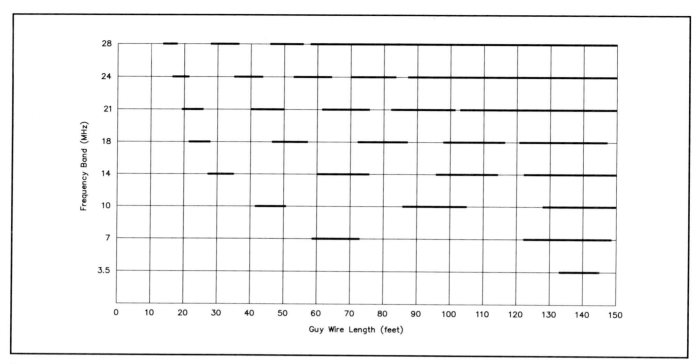

The black bars indicate ungrounded guy wire lengths to avoid for the eight HF amateur bands. This chart is based on resonance within 10% of any frequency in the band. Grounded wires will exhibit resonance at odd multiples of a quarter wavelength. *(Jerry Hall, K1TD)*

Morse Code Character Set[1]

A	didah	• —
B	dahdididit	— •••
C	dahdidahdit	— • — •
D	dahdidit	— ••
E	dit	•
F	dididahdit	•• — •
G	dahdahdit	— — •
H	didididit	••••
I	didit	••
J	didahdahdah	• — — —
K	dahdidah	— • —
L	didahdidit	• — ••
M	dahdah	— —
N	dahdit	— •
O	dahdahdah	— — —
P	didahdahdit	• — — •
Q	dahdahdidah	— — • —
R	didahdit	• — •
S	dididit	•••
T	dah	—
U	dididah	•• —
V	didididah	••• —
W	didahdah	• — —
X	dahdididah	— •• —
Y	dahdidahdah	— • — —
Z	dahdahdidit	— — ••
1	didahdahdahdah	• — — — —
2	dididahdahdah	•• — — —
3	didididahdah	••• — —
4	dididididah	•••• —
5	didididit	•••••
6	dahdidididit	— ••••
7	dahdahdididit	— — •••
8	dahdahdahdidit	— — — ••
9	dahdahdahdahdit	— — — — •
0	dahdahdahdahdah	— — — — —

Period [.]:	didahdidahdidah	• — • — • —	A̅A̅A̅
Comma [,]:	dahdahdididahdah	— — •• — —	M̅I̅M̅
Question mark or request for repetition [?]:	dididahdahdidit	•• — — ••	I̅M̅I̅
Error:	didididididididit	••••••••	H̅H̅
Hyphen or dash [−]:	dahdididididah	— •••• —	D̅U̅
Double dash [=]	dahdidididah	— ••• —	B̅T̅
Colon [:]:	dahdahdahdididit	— — — •••	O̅S̅
Semicolon [;]:	dahdidahdidahdit	— • — • — •	K̅R̅
Left parenthesis [(]:	dahdidahdahdit	— • — — •	K̅N̅
Right parenthesis [)]:	dahdidahdahddidah	— • — — • —	K̅K̅
Fraction bar [/]:	dahdididahdit	— •• — •	D̅N̅
Quotation marks ["]:	didahdididahdit	• — •• — •	A̅F̅
Dollar sign [$]:	dididahdididah	••• — •• —	S̅X̅
Apostrophe [']:	didahdahdahdahdit	• — — — — •	W̅G̅
Paragraph [¶]:	didahdidahdidit	• — • — ••	A̅L̅
Underline [_]:	dididahdahdidah	•• — — • —	I̅Q̅
Starting signal:	dahdidahdidah	— • — • —	K̅A̅
Wait:	didahdididit	• — •••	A̅S̅
End of message or cross [+]:	didahdidahdit	• — • — •	A̅R̅
Invitation to transmit [K]:	dahdidah	— • —	K
End of work:	didididahdidah	••• — • —	S̅K̅
Understood:	didididahdit	••• — •	S̅N̅

Notes:

1. Not all Morse characters shown are used in FCC code tests. License applicants are responsible for knowing, and may be tested on, the 26 letters, the numerals 0 to 9, the period, the comma, the question mark, A̅R̅, S̅K̅, B̅T̅ and fraction bar [D̅N̅].

2. The following letters are used in certain European languages which use the Latin alphabet:

Ä, Ą	didahdidah	• — • —
Á, Å, À, Â	didahdahdidah	• — — • —
Ç, Ć	dahdidahdahdit	— • — — •
É, È, Ę	didididahdit	•• — ••
È	didahdididah	• — ••—
Ê	dahdididahdit	— •• — •
Ö, Ø, Ó	dahdahdahdit	— — — •
Ñ	dahdahdidahdah	— — • — —
Ü	didididahdah	•• — —
Ź	dahdahdidit	— — ••
Z	dahdahdididah	— — •• —
CH, Ş	dahdahdahdah	— — — —

3. Special Esperanto characters:

Ĉ	dahdidahdidit	— • — ••
Ŝ	didididahdit	••• — •
Ĵ	didahdahdahdit	• — — — •
Ĥ	dahdidahdahdit	— • — — •
Ŭ	dididahdah	•• — —
Ĝ	dahdidahdidit	— — • — •

4. Signals used in other radio services:

Interrogatory	dididahdidah	•• — • —	I̅N̅T̅
Emergency silence	dididididahdah	•••• — —	H̅M̅
Executive follows	didahdididah	•• — •• —	I̅X̅
Break−in signal	dahdahdahdahdah	— — — — —	T̅T̅T̅T̅T̅
Emergency signal	dididahdahdahdididit	••• — — — •••	S̅O̅S̅
Relay of distress	dahdididahdididahdidit	— •• — •• — ••	D̅D̅D̅

Morse Abbreviated Numbers

Numeral		Long Number		Abbreviated Number	Equivalent Character
1	didahdahdahdah	• — — — —	didah	• —	A
2	dididahdahdah	•• — — —	dididah	•• —	U
3	didididahdah	••• — —	didididah	••• —	V
4	dididididah	•••• —	dididididah	•••• —	4
5	didididit	•••••	didididit	••••• or •	5 or E
6	dahdidididit	— ••••	dahdidididit	— ••••	6
7	dahdahdididit	— — •••	dahdididit	— •••	B
8	dahdahdahdidit	— — — ••	dahdidit	— ••	D
9	dahdahdahdahdit	— — — — •	dahdit	— •	N
0	dahdahdahdahdah	— — — — —	dah	—	T

Note: These abbreviated numbers are not legal for use in call signs. They should be used only where there is agreement between operators and when no confusion will result.

The following tables have been placed on the ARRL Web page **http://www.arrl.org/notes/1867/** in the file **ch30-tab.pdf**.

Tables Referred to in Chapter 12

"ITA2 (Baudot) and AMTOR Codes"

"Baudot Signaling Rates and Speeds"

"The ASCII Coded Character Set"

"Conversion from ASCII to Morse and Baudot"

"Code Conversion, ITA1 through 4 (Notes 1 and 2)"

"Data Interface Connections"

"EIA-449 37-Pin Connector Assignments"

"EIA-449 9-Pin Connector Assignments"

"RTTY Control Sequences (from CCITT Recommendation S.4)"

"ISO 3593 Pin Allocations for V.35 Interfaces"

Tables Referring to Operating and Public Service

"EME Software"

"Principles of Emergency Communication"

"Operating Aids for Public Service"

"ARES Personal Checklist"

"ARES/RACES"

"The Interaction Between the EOC/NCS and the Command Post(s) in a Local Emergency"

"Organization and Interaction of ARES and NTS"

"Typical Station Deployment for Local AREAS Net Coverage in an Emergency"

"Typical Structure of an HF Network for Emergency Communication"

The ASCII Coded Character Set

Bit Number				Hex	6: 0 / 5: 0 / 4: 0 — 1st 0	0/0/1 — 1	0/1/0 — 2	0/1/1 — 3	1/0/0 — 4	1/0/1 — 5	1/1/0 — 6	1/1/1 — 7
3	2	1	0	2nd	0	1	2	3	4	5	6	7
0	0	0	0	0	NUL	DLE	SP	0	@	P		p
0	0	0	1	1	SOH	DC1	!	1	A	Q	a	q
0	0	1	0	2	STX	DC2	"	2	B	R	b	r
0	0	1	1	3	ETX	DC3	#	3	C	S	c	s
0	1	0	0	4	EOT	DC4	$	4	D	T	d	t
0	1	0	1	5	ENQ	NAK	%	5	E	U	e	u
0	1	1	0	6	ACK	SYN	&	6	F	V	f	v
0	1	1	1	7	BEL	ETB	'	7	G	W	g	w
1	0	0	0	8	BS	CAN	(	8	H	X	h	x
1	0	0	1	9	HT	EM	)	9	I	Y	i	y
1	0	1	0	A	LF	SUB	*	:	J	Z	j	z
1	0	1	1	B	VT	ESC	+	;	K	[	k	{
1	1	0	0	C	FF	FS	,	<	L	\	l		
1	1	0	1	D	CR	GS	-	=	M	]	m	}
1	1	1	0	E	SO	RS	.	>	N	^	n	~
1	1	1	1	F	SI	US	/	?	O	—	o	DEL

ACK	acknowledge	FF	form feed
BEL	bell	FS	file separator
BS	backspace	GS	group separator
CAN	cancel	HT	horizontal tab
CR	carriage return	LF	line feed
DC1	device control 1	NAK	negative acknowledge
DC2	device control 2	NUL	null
DC3	device control 3	RS	record separator
DC4	device control 4	SI	shift in
DEL	(delete)	SO	shift out
DLE	data link escape	SOH	start of heading
ENQ	enquiry	SP	space
EM	end of medium	STX	start of text
EOT	end of transmission	SUB	substitute
ESC	escape	SYN	synchronous idle
ETB	end of block	US	unit separator
ETX	end of text	VT	vertical tab

Notes

1. "1" = mark, "0" = space.
2. Bit 6 is the most-significant bit (MSB). Bit 0 is the least-significant bit (LSB).

Voluntary HF Band Plans for Considerate US Operators

The following frequencies are generally recognized for certain modes or activities (all frequencies are in MHz).

Nothing in the rules recognizes a net's, group's or any individual's special privilege to any specific frequency. Section 97.101(b) of the Rules states that "Each station licensee and each control operator must cooperate in selecting transmitting channels and in making the most effective use of the amateur service frequencies. No frequency will be assigned for the exclusive use of any station." No one "owns" a frequency.

It's good practice—and plain old common sense—for any operator, regardless of mode, to check to see if the frequency is in use prior to engaging operation. If you are there first, other operators should make an effort to protect you from interference to the extent possible given that 100% interference-free operation is an unrealistic expectation in today's congested bands.

1.800-1.830	CW, data and other narrowband modes
1.810	QRP CW calling frequency
1.830-1.840	CW, data and other narrowband modes, intercontinental QSOs only
1.840-1.850	CW; SSB, SSTV and other wideband modes, intercontinental QSOs only
1.850-2.000	CW; phone, SSTV and other wideband modes
3.500-3.510	CW DX
3.590	RTTY DX
3.580-3.620	Data
3.620-3.635	Automatically controlled data stations
3.710	QRP Novice/Technician CW calling frequency
3.790-3.800	DX window
3.845	SSTV
3.885	AM calling frequency
3.985	QRP SSB calling frequency
7.040	RTTY DX
	QRP CW calling frequency
7.075-7.100	Phone in KH/KL/KP *only*
7.080-7.100	Data
7.100-7.105	Automatically controlled data stations
7.110	QRP Novice/Technician CW calling frequency
7.171	SSTV
7.285	QRP SSB calling frequency
7.290	AM calling frequency
10.106	QRP CW calling frequency
10.130-10.140	Data
10.140-10.150	Automatically controlled data stations
14.060	QRP CW calling frequency
14.070-14.095	Data
14.095-14.0995	Automatically controlled data stations
14.100	IBP/NCDXF beacons
14.1005-14.112	Automatically controlled data stations
14.230	SSTV
14.285	QRP SSB calling frequency
14.286	AM calling frequency
18.100-18.105	Data
18.105-18.110	Automatically controlled data stations
21.060	QRP CW calling frequency
21.070-21.090	Data
21.090-21.100	Automatically controlled data stations
21.340	SSTV
21.385	QRP SSB calling frequency
24.920-24.925	Data
24.925-24.930	Automatically controlled data stations
28.060	QRP CW calling frequency
28.070-28.120	Data
28.120-28.189	Automatically controlled data stations
28.190-28.225	Beacons
28.385	QRP SSB calling frequency
28.680	SSTV
29.000-29.200	AM
29.300-29.510	Satellite downlinks
29.520-29.580	Repeater inputs
29.600	FM simplex
29.620-29.680	Repeater outputs

Notes

ARRL band plans for frequencies above 28.300 MHz are shown in *The ARRL Repeater Directory* and *The FCC Rule Book*. For detailed packet frequencies, see *QST*, September 1987, page 54, and March 1988, page 51.

IBP/NCDXF beacons operate on 14.100, 18.110, 21.150, 24.930 and 28.200 MHz.

TVI Troubleshooting Flowchart (also see EMI chapter)

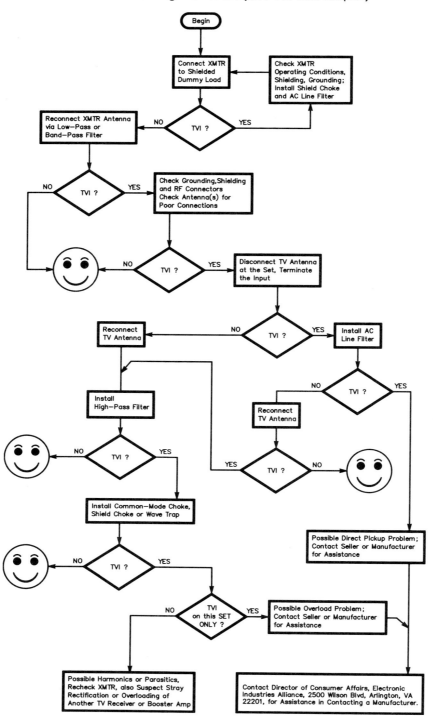

TV Channels vs Harmonics

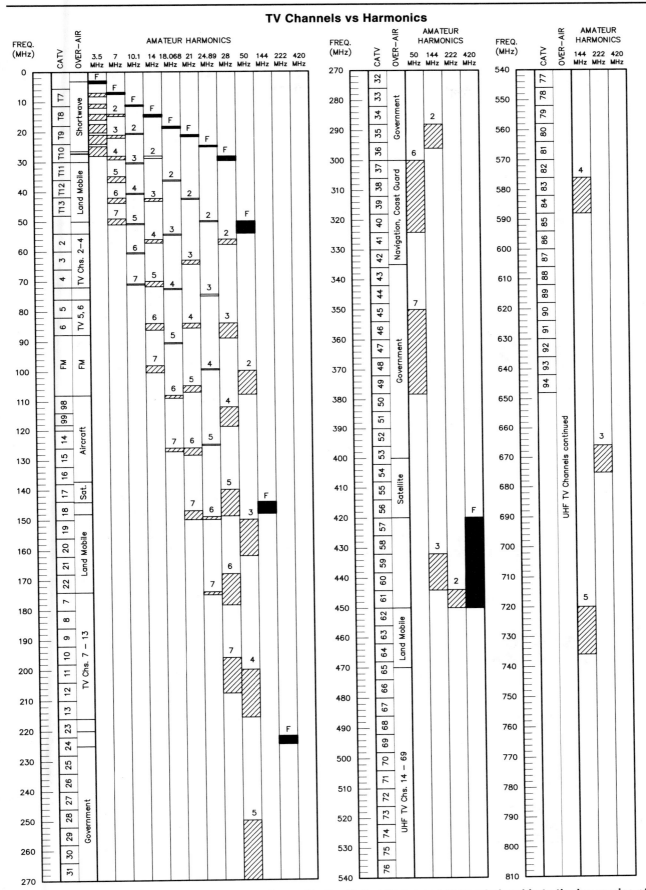

This chart shows CATV and broadcast channels used in the United States and their relationship to the harmonics of MF, HF, VHF and UHF amateur bands.

160 METERS

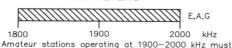

1800   1900   2000   kHz
E,A,G

Amateur stations operating at 1900–2000 kHz must not cause harmful interference to the radiolocation service and are afforded no protection from radiolocation operations.

80 METERS

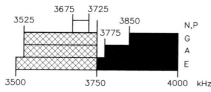

3675   3725
3525   3850
3775
N,P
G
A
E
3500   3750   4000   kHz

5167.5 kHz (SSB only): Alaska emergency use only.

40 METERS

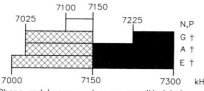

7100   7150
7025   7225
N,P
G †
A †
E †
7000   7150   7300   kHz

† Phone and Image modes are permitted between 7075 and 7100 kHz for FCC licensed stations in ITU Regions 1 and 3 and by FCC licensed stations in ITU Region 2 West of 130 degrees West longitude or South of 20 degrees North latitude. See Sections 97.305(c) and 97.307(f)(11). Novice and Technician Plus licensees outside ITU Region 2 may use CW only between 7050 and 7075 kHz. See Section 97.301(e). These exemptions do not apply to stations in the continental US.

30 METERS

10,100   10,150   kHz
E,A,G

Maximum power on 30 meters is 200 watts PEP output. Amateurs must avoid interference to the fixed service outside the US.

20 METERS

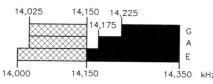

14,025   14,150   14,225
14,175
G
A
E
14,000   14,150   14,350   kHz

17 METERS

E,A,G
18,068   18,110   18,168   kHz

15 METERS

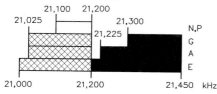

21,100   21,200
21,025   21,300
21,225
N,P
G
A
E
21,000   21,200   21,450   kHz

US Amateur Bands

April 15, 2000

Novice, Advanced and Technician Plus Allocations

New Novice, Advanced and Technician Plus licenses will not be issued after April 15, 2000. However, the FCC has allowed the frequency allocations for these license classes to remain in effect. They will continue to renew existing licenses for those classes.

12 METERS

E,A,G
24,890   24,930   24,990   kHz

10 METERS

28,100   28,500
N,P
E,A,G
28,000   28,300   29,700   kHz

Novices and Technician Plus licensees are limited to 200 watts PEP output on 10 meters.

6 METERS

50.1
E,A,G,P,T
50.0   54.0   MHz

2 METERS

144.1
E,A,G,P,T
144.0   148.0   MHz

1.25 METERS

E,A,G,P,T,N
222.0   225.0   MHz

Novices are limited to 25 watts PEP output from 222 to 225 MHz.

70 CENTIMETERS **

E,A,G,P,T
420.0   450.0   MHz

33 CENTIMETERS **

E,A,G,P,T
902.0   928.0   MHz

23 CENTIMETERS **

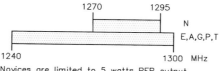

1270   1295
N
E,A,G,P,T
1240   1300   MHz

Novices are limited to 5 watts PEP output from 1270 to 1295 MHz.

US AMATEUR POWER LIMITS

At all times, transmitter power should be kept down to that necessary to carry out the desired communications. Power is rated in watts PEP output. Unless otherwise stated, the maximum power output is 1500 W. Power for all license classes is limited to 200 W in the 10,100–10,150 kHz band and in all Novice subbands below 28,100 kHz. Novices and Technicians with Morse code credit are restricted to 200 W in the 28,100–28,500 kHz subbands. In addition, Novices are restricted to 25 W in the 222–225 MHz band and 5 W in the 1270–1295 MHz subband.

Operators with Technician class licenses and above may operate on all bands above 50 MHz. For more detailed information see *The FCC Rule Book*.

KEY

= CW, RTTY and data

= CW, RTTY, data, MCW, test, phone and image

= CW, phone and image

= CW and SSB phone

= CW, RTTY, data, phone, and image

= CW only

E = EXTRA CLASS
A = ADVANCED
G = GENERAL
P = TECHNICIAN PLUS
T = TECHNICIAN
N = NOVICE

* Effective April 15, 2000, Technicians passing the Morse code exam will gain HF Novice privileges, although they still hold a Technician license.

** Geographical and power restrictions apply to these bands. See *The FCC Rule Book* for more information about your area.

Above 23 Centimeters:

All licensees except Novices are authorized all modes on the following frequencies:
2300–2310 MHz
2390–2450 MHz
3300–3500 MHz
5650–5925 MHz
10.0–10.5 GHz
24.0–24.25 GHz
47.0–47.2 GHz
75.5–81.0 GHz
119.98–120.02 GHz
142–149 GHz
241–250 GHz
All above 300 GHz

For band plans and sharing arrangements, see *The ARRL Operating Manual* or *The FCC Rule Book*.

VHF/UHF/EHF Calling Frequencies

Band (MHz)	Calling Frequency
50	50.125 SSB
	50.620 digital (packet)
	52.525 National FM simplex frequency
144	144.010 EME
	144.100, 144.110 CW
	144.200 SSB
	146.520 National FM simplex frequency
222	222.100 CW/SSB
	223.500 National FM simplex frequency
432	432.010 EME
	432.100 CW/SSB
	446.000 National FM simplex frequency
902	902.100 CW/SSB
	903.1 Alternate CW, SSB
	906.500 National FM simplex frequency
1296	1294.500 National FM simplex frequency
	1296.100 CW/SSB
2304	2304.4
	2305.2 FM simplex frequency
10000	10368.1 Narrow-band

VHF/UHF Activity Nights

Some areas do not have enough VHF/UHF activity to support contacts at all times. This schedule is intended to help VHF/UHF operators make contact. This is only a starting point; check with others in your area to see if local hams have a different schedule.

Band (MHz)	Day	Local Time
50	Sunday	6 PM
144	Monday	7 PM
222	Tuesday	8 PM
432	Wednesday	9 PM
902	Friday	9 PM
1296	Thursday	10 PM

ITU Regions

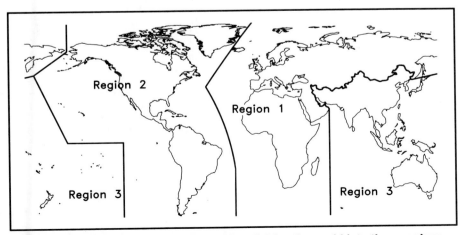

The International Telecommunication Union divides the world into three regions. Geographic details appear in *The FCC Rule Book.*

Allocation of International Call Signs

Call Sign Series	Allocated to	Call Sign Series	Allocated to	Call Sign Series	Allocated to
AAA-ALZ	United States of America	† E4A-E4Z	Palestinian Authority	OEA-OEZ	Austria
AMA-AOZ	Spain	FAA-FZZ	France	OFA-OJZ	Finland
APA-ASZ	Pakistan	GAA-GZZ	United Kingdom of Great Britain and Northern Ireland	OKA-OLZ	Czech Republic
ATA-AWZ	India			OMA-OMZ	Slovak Republic
AXA-AXZ	Australia	HAA-HAZ	Hungary	ONA-OTZ	Belgium
AYA-AZZ	Argentina	HBA-HBZ	Switzerland	OUA-OZZ	Denmark
A2A-A2Z	Botswana	HCA-HDZ	Ecuador	PAA-PIZ	Netherlands
A3A-A3Z	Tonga	HEA-HEZ	Switzerland	PJA-PJZ	Netherlands Antilles
A4A-A4Z	Oman	HFA-HFZ	Poland	PKA-POZ	Indonesia
A5A-A5Z	Bhutan	HGA-HGZ	Hungary	PPA-PYZ	Brazil
A6A-A6Z	United Arab Emirates	HHA-HHZ	Haiti	PZA-PZZ	Suriname
A7A-A7Z	Qatar	HIA-HIZ	Dominican Republic	P2A-P2Z	Papua New Guinea
A8A-A8Z	Liberia	HJA-HKZ	Colombia	P3A-P3Z	Cyprus
A9A-A9Z	Bahrain	HLA-HLZ	South Korea	P4A-P4Z	Aruba
BAA-BZZ	China	HMA-HMZ	North Korea	P5A-P9Z	North Korea
CAA-CEZ	Chile	HNA-HNZ	Iraq	RAA-RZZ	Russian Federation
CFA-CKZ	Canada	HOA-HPZ	Panama	SAA-SMZ	Sweden
CLA-CMZ	Cuba	HQA-HRZ	Honduras	SNA-SRZ	Poland
CNA-CNZ	Morocco	HSA-HSZ	Thailand	• SSA-SSM	Egypt
COA-COZ	Cuba	HTA-HTZ	Nicaragua	• SSN-SSZ	Sudan
CPA-CPZ	Bolivia	HUA-HUZ	El Salvador	STA-STZ	Sudan
CQA-CUZ	Portugal	HVA-HVZ	Vatican City	SUA-SUZ	Egypt
CVA-CXZ	Uruguay	HWA-HYZ	France	SVA-SZZ	Greece
CYA-CZZ	Canada	HZA-HZZ	Saudi Arabia	S2A-S3Z	Bangladesh
C2A-C2Z	Nauru	H2A-H2Z	Cyprus	S5A-S5Z	Slovenia
C3A-C3Z	Andorra	H3A-H3Z	Panama	S6A-S6Z	Singapore
C4A-C4Z	Cyprus	H4A-H4Z	Solomon Islands	S7A-S7Z	Seychelles
C5A-C5Z	Gambia	H6A-H7Z	Nicaragua	S8A-S8Z	South Africa
C6A-C6Z	Bahamas	H8A-H9Z	Panama	S9A-S9Z	Sao Tome and Principe
* C7A-C7Z	World Meteorological Organization	IAA-IZZ	Italy	TAA-TCZ	Turkey
		JAA-JSZ	Japan	TDA-TDZ	Guatemala
C8A-C9Z	Mozambique	JTA-JVZ	Mongolia	TEA-TEZ	Costa Rica
DAA-DRZ	Germany	JWA-JXZ	Norway	TFA-TFZ	Iceland
DSA-DTZ	South Korea	JYA-JYZ	Jordan	TGA-TGZ	Guatemala
DUA-DZZ	Philippines	JZA-JZZ	Indonesia	THA-THZ	France
D2A-D3Z	Angola	J2A-J2Z	Djibouti	TIA-TIZ	Costa Rica
D4A-D4Z	Cape Verde	J3A-J3Z	Grenada	TJA-TJZ	Cameroon
D5A-D5Z	Liberia	J4A-J4Z	Greece	TKA-TKZ	France
D6A-D6Z	Comoros	J5A-J5Z	Guinea-Bissau	TLA-TLZ	Central African Republic
D7A-D9Z	South Korea	J6A-J6Z	Saint Lucia	TMA-TMZ	France
EAA-EHZ	Spain	J7A-J7Z	Dominica	TNA-TNZ	Congo Republic
EIA-EJZ	Ireland	J8A-J8Z	St. Vincent and the Grenadines	TOA-TQZ	France
EKA-EKZ	Armenia			TRA-TRZ	Gabon
ELA-ELZ	Liberia	KAA-KZZ	United States of America	TSA-TSZ	Tunisia
EMA-EOZ	Ukraine	LAA-LNZ	Norway	TTA-TTZ	Chad
EPA-EQZ	Iran	LOA-LWZ	Argentina	TUA-TUZ	Ivory Coast
ERA-ERZ	Moldova	LXA-LXZ	Luxembourg	TVA-TXZ	France
ESA-ESZ	Estonia	LYA-LYZ	Lithuania	TYA-TYZ	Benin
ETA-ETZ	Ethiopia	LZA-LZZ	Bulgaria	TZA-TZZ	Mali
EUA-EWZ	Belarus	L2A-L9Z	Argentina	T2A-T2Z	Tuvalu
EXA-EXZ	Kyrgyzstan	MAA-MZZ	United Kingdom of Great Britain and Northern Ireland	T3A-T3Z	Kiribati
EYA-EYZ	Tajikistan			T4A-T4Z	Cuba
EZA-EZZ	Turkmenistan	NAA-NZZ	United States of America	T5A-T5Z	Somalia
E2A-E2Z	Thailand	OAA-OCZ	Peru	T6A-T6Z	Afghanistan
E3A-E3Z	Eritrea	ODA-ODZ	Lebanon	T7A-T7Z	San Marino

Call Sign Series	Allocated to	Call Sign Series	Allocated to	Call Sign Series	Allocated to
T8A-T8Z	Palau	ZKA-ZMZ	New Zealand	5XA-5XZ	Uganda
T9A-T9Z	Bosnia and Herzegovina	ZNA-ZOZ	United Kingdom of Great Britain and Northern Ireland	5YA-5ZZ	Kenya
UAA-UIZ	Russian Federation			6AA-6BZ	Egypt
UJA-UMZ	Uzbekistan	ZPA-ZPZ	Paraguay	6CA-6CZ	Syria
UNA-UQZ	Kazakhstan	ZQA-ZQZ	United Kingdom of Great Britain and Northern Ireland	6DA-6JZ	Mexico
URA-UZZ	Ukraine			6KA-6NZ	South Korea
VAA-VGZ	Canada	ZRA-ZUZ	South Africa	6OA-6OZ	Somalia
VHA-VNZ	Australia	ZVA-ZZZ	Brazil	6PA-6SZ	Pakistan
VOA-VOZ	Canada	Z2A-Z2Z	Zimbabwe	6TA-6UZ	Sudan
VPA-VQZ	United Kingdom of Great Britain and Northern Ireland	Z3A-Z3Z	Macedonia (Former Yugoslav Republic)	6VA-6WZ	Senegal
				6XA-6XZ	Madagascar
† VRA-VRZ	China—Hong Kong	2AA-2ZZ	United Kingdom of Great Britain and Northern Ireland	6YA-6YZ	Jamaica
VSA-VSZ	United Kingdom of Great Britain and Northern Ireland			6ZA-6ZZ	Liberia
		3AA-3AZ	Monaco	7AA-7IZ	Indonesia
VTA-VWZ	India	3BA-3BZ	Mauritius	7JA-7NZ	Japan
VXA-VYZ	Canada	3CA-3CZ	Equatorial Guinea	7OA-7OZ	Yemen
VZA-VZZ	Australia	• 3DA-3DM	Swaziland	7PA-7PZ	Lesotho
V2A-V2Z	Antigua and Barbuda	• 3DN-3DZ	Fiji	7QA-7QZ	Malawi
V3A-V3Z	Belize	3EA-3FZ	Panama	7RA-7RZ	Algeria
V4A-V4Z	Saint Kitts and Nevis	3GA-3GZ	Chile	7SA-7SZ	Sweden
V5A-V5Z	Namibia	3HA-3UZ	China	7TA-7YZ	Algeria
V6A-V6Z	Micronesia	3VA-3VZ	Tunesia	7ZA-7ZZ	Saudi Arabia
V7A-V7Z	Marshall Islands	3WA-3WZ	Viet Nam	8AA-8IZ	Indonesia
V8A-V8Z	Brunei	3XA-3XZ	Guinea	8JA-8NZ	Japan
WAA-WZZ	United States of America	3YA-3YZ	Norway	8OA-8OZ	Botswana
XAA-XIZ	Mexico	3ZA-3ZZ	Poland	8PA-8PZ	Barbados
XJA-XOZ	Canada	4AA-4CZ	Mexico	8QA-8QZ	Maldives
XPA-XPZ	Denmark	4DA-4IZ	Philippines	8RA-8RZ	Guyana
XQA-XRZ	Chile	4JA-4KZ	Azerbaijan	8SA-8SZ	Sweden
XSA-XSZ	China	4LA-4LZ	Georgia	8TA-8YZ	India
XTA-XTZ	Burkina Faso	4MA-4MZ	Venezuela	8ZA-8ZZ	Saudi Arabia
XUA-XUZ	Cambodia	4NA-4OZ	Yugoslavia	9AA-9AZ	Croatia
XVA-XVZ	Viet Nam	4PA-4SZ	Sri Lanka	9BA-9DZ	Iran
XWA-XWZ	Laos	4TA-4TZ	Peru	9EA-9FZ	Ethiopia
XXA-XXZ	Portugal	* 4UA-4UZ	United Nations	9GA-9GZ	Ghana
XYA-XZZ	Myanmar	4VA-4VZ	Haiti	9HA-9HZ	Malta
YAA-YAZ	Afghanistan	4XA-4XZ	Israel	9IA-9JZ	Zambia
YBA-YHZ	Indonesia	* 4YA-4YZ	International Civil Aviation Organization	9KA-9KZ	Kuwait
YIA-YIZ	Iraq			9LA-9LZ	Sierra Leone
YJA-YJZ	Vanuatu	4ZA-4ZZ	Israel	9MA-9MZ	Malaysia
YKA-YKZ	Syria	5AA-5AZ	Libya	9NA-9NZ	Nepal
YLA-YLZ	Latvia	5BA-5BZ	Cyprus	9OA-9TZ	Democratic Republic of Congo
YMA-YMZ	Turkey	5CA-5GZ	Morocco		
YNA-YNZ	Nicaragua	5HA-5IZ	Tanzania	9UA-9UZ	Burundi
YOA-YRZ	Romania	5JA-5KZ	Colombia	9VA-9VZ	Singapore
YSA-YSZ	El Salvador	5LA-5MZ	Liberia	9WA-9WZ	Malaysia
YTA-YUZ	Yugoslavia	5NA-5OZ	Nigeria	9XA-9XZ	Rwanda
YVA-YYZ	Venezuela	5PA-5QZ	Denmark	9YZ-9ZZ	Trinidad and Tobago
YZA-YZZ	Yugoslavia	5RA-5SZ	Madagascar		
Y2A-Y9Z	Germany	5TA-5TZ	Mauritania		
ZAA-ZAZ	Albania	5UA-5UZ	Niger		
ZBA-ZJZ	United Kingdom of Great Britain and Northern Ireland	5VA-5VZ	Togo		
		5WA-5WZ	Western Samoa		

• Half series
* Series allocated to an international organization
† Provisional allocation in accordance with S19.33

FCC-Allocated Prefixes for Areas Outside the Continental US

Prefix	Location
AH1, KH1, NH1, WH1	Baker, Howland Is
AH2, KH2, NH2, WH2	Guam
AH3, KH3, NH3, WH3	Johnston I
AH4, KH4, NH4, WH4	Midway I
AH5K, KH5K, NH5K, WH5K	Kingman Reef
AH5, KH5, NH5, WH5 (except K suffix)	Palmyra, Jarvis Is
AH6-7, KH6-7, NH6-7, WH6-7	Hawaii
AH7K, KH7K, NH7K, WH7K	Kure I
AH8, KH8, NH8, WH8	American Samoa
AH9, KH9, NH9, WH9	Wake, Wilkes, Peale Is
AH0, KH0, NH0, WH0	Northern Mariana Is
AL, KL, NL, WL	Alaska
KP1, NP1, WP1	Navassa
KP2, NP2, WP2	Virgin Is
KP3-4, NP3-4, WP3-4	Puerto Rico
KP5, NP5, WP5	Desecheo

DX Operating Code

For W/VE Amateurs

Some DXers have caused considerable confusion and interference in their efforts to work DX stations. The points below, if observed by all W/VE amateurs, will help make DX more enjoyable for all.

1) *Call* DX only after he calls CQ, QRZ? or signs S̅K̅, or voice equivalents thereof. Make your calls short.

2) Do not call a DX station:
   a) On the frequency of the station he is calling until you are sure the QSO is over (S̅K̅).
   b) Because you hear someone else calling him.
   c) When he signs K̅N̅, A̅R̅ or CL.
   d) Exactly on his frequency.
   e) After he calls a directional CQ, unless of course you are in the right direction or area.

3) Keep within frequency band limits. Some DX stations can get away with working outside, but you cannot.

4) Observe calling instructions given by DX stations. Example: 15U means "call 15 kHz up from my frequency." 15D means down, etc.

5) Give honest reports. Many DX stations depend on W/VE reports for adjustment of station and equipment.

6) Keep your signal clean. Key clicks, ripple, feedback or splatter gives you a bad reputation and may get you a citation from the FCC.

7) *Listen* and call the station you want. Calling CQ DX is not the best assurance that the rare DX will reply.

8) When there are several W or VE stations waiting, avoid asking DX to "listen for a friend." Also avoid engaging him in a ragchew against his wishes.

For Overseas Amateurs

To all overseas amateur stations:

In their eagerness to work you, many W and VE amateurs resort to practices that cause confusion and QRM. Most of this is good-intentioned but ill-advised; some of it is intentional and selfish. The key to the cessation of unethical DX operating practices is in your hands. We believe that your adoption of certain operating habits will increase your enjoyment of Amateur Radio and that of amateurs on this side who are eager to work you. We recommend your adoption of the following principles:

1) Do not answer calls on your own frequency.

2) Answer calls from W/VE stations only when their signals are of good quality.

3) Refuse to answer calls from other stations when you are already in contact with someone, and do not acknowledge calls from amateurs who indicate they wish to be "next."

4) Give *everybody* a break. When many W/VE amateurs are patiently and quietly waiting to work you, avoid complying with requests to "listen for a friend."

5) Tell listeners where to call you by indicating how many kilohertz up (U) or down (D) from your frequency you are listening.

6) Use the ARRL-recommended ending signals, especially K̅N̅ to indicate to impatient listeners the status of the QSO. K̅N̅ means "Go ahead (specific station); all others keep out."

7) Let it be known that you avoid working amateurs who are constant violators of these principles.

W1AW SCHEDULE

Pacific	Mtn	Cent	East	Mon	Tue	Wed	Thu	Fri
6 AM	7 AM	8 AM	9 AM		Fast Code	Slow Code	Fast Code	Slow Code
7 AM-1 PM	8 AM-2 PM	9 AM-3 PM	10 AM-4 PM	Visiting Operator Time (12 PM - 1 PM closed for lunch)				
1 PM	2 PM	3 PM	4 PM	Fast Code	Slow Code	Fast Code	Slow Code	Fast Code
2 PM	3 PM	4 PM	5 PM	Code Bulletin				
3 PM	4 PM	5 PM	6 PM	Teleprinter Bulletin				
4 PM	5 PM	6 PM	7 PM	Slow Code	Fast Code	Slow Code	Fast Code	Slow Code
5 PM	6 PM	7 PM	8 PM	Code Bulletin				
6 PM	7 PM	8 PM	9 PM	Teleprinter Bulletin				
6⁴⁵ PM	7⁴⁵ PM	8⁴⁵ PM	9⁴⁵ PM	Voice Bulletin				
7 PM	8 PM	9 PM	10 PM	Fast Code	Slow Code	Fast Code	Slow Code	Fast Code
8 PM	9 PM	10 PM	11 PM	Code Bulletin				

W1AW's schedule is at the same local time throughout the year. The schedule according to your local time will change if your local time does not have seasonal adjustments that are made at the same time as North American time changes between standard time and daylight time. From the first Sunday in April to the last Sunday in October, UTC = Eastern Time + 4 hours. For the rest of the year, UTC = Eastern Time + 5 hours.

◆ **Morse code transmissions:**

Frequencies are 1.818, 3.5815, 7.0475, 14.0475, 18.0975, 21.0675, 28.0675 and 147.555 MHz.

Slow Code = practice sent at 5, 7½, 10, 13 and 15 wpm.

Fast Code = practice sent at 35, 30, 25, 20, 15, 13 and 10 wpm.

Code practice text is from the pages of *QST*. The source is given at the beginning of each practice session and alternate speeds within each session. For example, "Text is from July 1992 *QST*, pages 9 and 81," indicates that the plain text is from the article on page 9 and mixed number/letter groups are from page 81.

Code bulletins are sent at 18 wpm.

W1AW qualifying runs are sent on the same frequencies as the Morse code transmissions. West Coast qualifying runs are transmitted on approximately 3.590 MHz by W6OWP, with K6YR as an alternate. At the beginning of each code practice session, the schedule for the next qualifying run is presented. Underline one minute of the highest speed you copied, certify that your copy was made without aid, and send it to ARRL for grading. Please include your name, call sign (if any) and complete mailing address. Send a 9×12-inch SASE for a certificate, or a business-size SASE for an endorsement.

◆ **Teleprinter transmissions:**

Frequencies are 3.625, 7.095, 14.095, 18.1025, 21.095, 28.095 and 147.555 MHz.

Bulletins are sent at 45.45-baud Baudot and 100-baud AMTOR, FEC Mode B. 110-baud ASCII will be sent only as time allows.

On Tuesdays and Fridays at 6:30 PM Eastern Time, Keplerian elements for many amateur satellites are sent on the regular teleprinter frequencies.

◆ **Voice transmissions:**

Frequencies are 1.855, 3.99, 7.29, 14.29, 18.16, 21.39, 28.59 and 147.555 MHz.

◆ **Miscellanea:**

On Fridays, UTC, a DX bulletin replaces the regular bulletins.

W1AW is open to visitors from 10 AM until noon and from 1 PM until 3:45 PM on Monday through Friday. FCC licensed amateurs may operate the station during that time. Be sure to bring your current FCC amateur license or a photocopy.

In a communication emergency, monitor W1AW for special bulletins as follows: voice on the hour, teleprinter at 15 minutes past the hour, and CW on the half hour.

Headquarters and W1AW are closed on New Year's Day, President's Day, Good Friday, Memorial Day, Independence Day, Labor Day, Thanksgiving and the following Friday, and Christmas Day.

ARRL Procedural Signals (Prosigns)

In general, the CW prosigns are used on all data modes as well, although word abbreviations may be spelled out. That is, "CLEAR" might be used rather than "CL" on radioteletype. Additional radioteletype conventions appear at the end of the table.

Situation	CW	Voice
check for a clear frequency	QRL?	Is the frequency in use?
seek contact with any station	CQ	CQ
after a call to a specific named station or to indicate the end of a message	$\overline{AR}$	over, end of message
invite any station to transmit	K	go
invite a specific named station to transmit	$\overline{KN}$	go only
invite receiving station to transmit	BK	back to you
all received correctly	R	received
please stand by	$\overline{AS}$	wait, stand by
end of contact (sent before call sign)	$\overline{SK}$	clear
going off the air	CL	closing station

Additional RTTY prosigns
SK QRZ—Ending contact, but listening on frequency.
SK KN—Ending contact, but listening for one last transmission from the other station.
SK SZ—Signing off and listening on the frequency for any other calls.

Q Signals

These Q signals most often need to be expressed with brevity and clarity in amateur work. (Q abbreviations take the form of questions only when each is sent followed by a question mark.)

QRA   What is the name of your station? The name of your station is _____.

QRG   Will you tell me my exact frequency (or that of _____)? Your exact frequency (or that of _____) is _____ kHz.

QRH   Does my frequency vary? Your frequency varies.

QRI   How is the tone of my transmission? The tone of your transmission is _____ (1. Good; 2. Variable; 3. Bad).

QRJ   Are you receiving me badly? I cannot receive you. Your signals are too weak.

QRK   What is the intelligibility of my signals (or those of _____)? The intelligibility of your signals (or those of _____) is _____ (1. Bad; 2. Poor; 3. Fair; 4. Good; 5. Excellent).

QRL   Are you busy? I am busy (or I am busy with _____). Please do not interfere.

QRM   Is my transmission being interfered with? Your transmission is being interfered with (1. Nil; 2. Slightly; 3. Moderately; 4. Severely; 5. Extremely.)

QRN   Are you troubled by static? I am troubled by static _____ (1-5 as under QRM).

QRO   Shall I increase power? Increase power.

QRP   Shall I decrease power? Decrease power.

QRQ   Shall I send faster? Send faster (_____ WPM).

QRS   Shall I send more slowly? Send more slowly (_____ WPM).

QRT   Shall I stop sending? Stop sending.

QRU   Have you anything for me? I have nothing for you.

QRV   Are you ready? I am ready.

QRW   Shall I inform _____ that you are calling on _____ kHz? Please inform _____ that I am calling on _____ kHz.

QRX   When will you call me again? I will call you again at _____ hours (on _____ kHz).

QRY   What is my turn? Your turn is numbered _____

QRZ   Who is calling me? You are being called by _____ (on _____ kHz).

QSA   What is the strength of my signals (or those of _____)? The strength of your signals (or those of _____) is _____ (1. Scarcely perceptible; 2. Weak; 3. Fairly good; 4. Good; 5. Very good).

QSB   Are my signals fading? Your signals are fading.

QSD   Is my keying defective? Your keying is defective.

QSG   Shall I send _____ messages at a time? Send _____ messages at a time.

QSK   Can you hear me between your signals and if so can I break in on your transmission? I can hear you between my signals; break in on my transmission.

QSL   Can you acknowledge receipt? I am acknowledging receipt.

QSM   Shall I repeat the last message which I sent you, or some previous message? Repeat the last message which you sent me [or message(s) number(s) _____].

QSN   Did you hear me (or _____) on _____ kHz? I did hear you (or _____) on _____ kHz.

QSO   Can you communicate with _____ direct or by relay? I can communicate with _____ direct (or by relay through _____).

QSP   Will you relay to _____? I will relay to _____

QST   General call preceding a message addressed to all amateurs and ARRL members. This is in effect "CQ ARRL."

QSU   Shall I send or reply on this frequency (or on _____ kHz)? Send or reply on this frequency (or _____ kHz).

The RST System

READABILITY
1—Unreadable.
2—Barely readable, occasional words distinguishable.
3—Readable with considerable difficulty.
4—Readable with practically no difficulty.
5—Perfectly readable.

SIGNAL STRENGTH
1—Faint signals, barely perceptible.
2—Very weak signals.
3—Weak signals.
4—Fair signals.
5—Fairly good signals.
6—Good signals.
7—Moderately strong signals.
8—Strong signals.
9—Extremely strong signals.

TONE
1—Sixty-cycle ac or less, very rough and broad.
2—Very rough ac, very harsh and broad.
3—Rough ac tone, rectified but not filtered.
4—Rough note, some trace of filtering.
5—Filtered rectified ac but strongly ripple-modulated.
6—Filtered tone, definite trace of ripple modulation.
7—Near pure tone, trace of ripple modulation.
8—Near perfect tone, slight trace of modulation.
9—Perfect tone, no trace of ripple of modulation of any kind.

If the signal has the characteristic steadiness of crystal control, add the letter X to the RST report. If there is a chirp, add the letter C. Similarly for a click, add K. (See FCC Regulations §97.307, Emissions Standards.) The above reporting system is used on both CW and voice; leave out the "tone" report on voice.

QSV Shall I send a series of Vs on this frequency (or on _____ kHz)? Send a series of Vs on this frequency (or on _____ kHz).

QSW Will you send on this frequency (or on _____ kHz)? I am going to send on this frequency (or on _____ kHz).

QSX Will you listen to _____ on _____ kHz? I am listening to _____ on _____ kHz.

QSY Shall I change to transmission on another frequency? Change to transmission on another frequency (or on _____ kHz).

QSZ Shall I send each word or group more than once? Send each word or group twice (or _____ times).

QTA Shall I cancel message number _____? Cancel message number _____

QTB Do you agree with my counting of words? I do not agree with your counting of words. I will repeat the first letter or digit of each word or group.

QTC How many messages have you to send? I have _____ messages for you (or for _____).

QTH What is your location? My location is _____

QTR What is the correct time? The correct time is _____

QTV Shall I stand guard for you? Stand guard for me.

QTX Will you keep your station open for further communication with me? Keep your station open for me.

QUA Have you news of _____? I have news of _____.

ARRL QN Signals

QNA* Answer in prearranged order.

QNB Act as relay between _____ and _____.

QNC All net stations copy. I have a message for all net stations.

QND* Net is Directed (Controlled by net control station.)

QNE* Entire net stand by.

QNF Net is Free (not controlled).

QNG Take over as net control station.

QNH Your net frequency is High.

QNI Net stations report in. I am reporting into the net. (Follow with a list of traffic or QRU.)

QNJ Can you copy me?

QNK* Transmit messages for _____ to _____.

QNL Your net frequency is Low.

QNM* You are QRMing the net. Stand by.

QNN Net control station is _____. What station has net control?

QNO Station is leaving the net.

QNP Unable to copy you. Unable to copy _____.

QNQ* Move frequency to _____ and wait for _____ to finish handling traffic. Then send him traffic for _____.

QNR* Answer _____ and Receive traffic.

QNS Following Stations are in the net.* (follow with list.) Request list of stations in the net.

QNT I request permission to leave the net for _____ minutes.

QNU* The net has traffic for *you*. Stand by.

QNV* Establish contact with _____ on this frequency. If successful, move to _____ and send him traffic for _____.

QNW How do I route messages for _____?

QNX You are excused from the net.*

QNY* Shift to another frequency (or to _____ kHz) to clear traffic with _____.

QNZ Zero beat your signal with mine.

*For use only by the Net Control Station.

Notes on Use of QN Signals

These QN signals are special ARRL signals for use in amateur CW nets *only*. They are not for use in casual amateur conversation. Other meanings that may be used in other services do not apply. Do not use QN signals on phone nets. *Say it with words.* QN signals need not be followed by a question mark, even though the meaning may be interrogatory.

CW Abbreviations

Although abbreviations help to cut down unnecessary transmission, make it a rule not to abbreviate unnecessarily when working an operator of unknown experience.

AA	All after	OM	Old man
AB	All before	OP-OPR	Operator
AB	About	OT	Old timer; old top
ADR	Address	PBL	Preamble
AGN	Again	PSE	Please
ANT	Antenna	PWR	Power
BCI	Broadcast interference	PX	Press
BCL	Broadcast listener	R	Received as transmitted; are
BK	Break; break me; break in	RCD	Received
BN	All between; been	RCVR (RX)	Receiver
BUG	Semi-automatic key	REF	Refer to; referring to; reference
B4	Before		
C	Yes	RFI	Radio Frequency Interference
CFM	Confirm; I confirm	RIG	Station equipment
CK	Check	RPT	Repeat; I repeat; report
CL	I am closing my station; call	RTTY	Radioteletype
CLD-CLG	Called; calling	RX	Receiver
CQ	Calling any station	SASE	Self-addressed, stamped envelope
CUD	Could		
CUL	See you later	SED	Said
CW	Continuous wave (i.e., radiotelegraph)	SIG	Signature; signal
		SINE	Operator's personal initials or nickname
DE	From		
DLD-DLVD	Delivered	SKED	Schedule
DR	Dear	SRI	Sorry
DX	Distance, foreign countries	SSB	Single sideband
ES	And, &	SVC	Service; prefix to service message
FB	Fine business, excellent		
FM	Frequency modulation	T	Zero
GA	Go ahead (or resume sending)	TFC	Traffic
GB	Good-by	TMW	Tomorrow
GBA	Give better address	TNX-TKS	Thanks
GE	Good evening	TT	That
GG	Going	TU	Thank you
GM	Good morning	TVI	Television interference
GN	Good night	TX	Transmitter
GND	Ground	TXT	Text
GUD	Good	UR-URS	Your; you're; yours
HI	The telegraphic laugh; high	VFO	Variable-frequency oscillator
HR	Here, hear	VY	Very
HV	Have	WA	Word after
HW	How	WB	Word before
LID	A poor operator	WD-WDS	Word; words
MA, MILS	Milliamperes	WKD-WKG	Worked; working
MSG	Message; prefix to radiogram	WL	Well; will
N	No	WUD	Would
NCS	Net control station	WX	Weather
ND	Nothing doing	XCVR	Transceiver
NIL	Nothing; I have nothing for you	XMTR (TX)	Transmitter
NM	No more	XTAL	Crystal
NR	Number	XYL (YF)	Wife
NW	Now; I resume transmission	YL	Young lady
OB	Old boy	73	Best regards
OC	Old chap	88	Love and Kisses

ITU Recommended Phonetics

A — Alfa (**AL** FAH)
B — Bravo (**BRAH** VOH)
C — Charlie (**CHAR** LEE OR **SHAR** LEE)
D — Delta (**DELL** TAH)
E — Echo (**ECK** OH)
F — Foxtrot (**FOKS** TROT)
G — Golf (GOLF)
H — Hotel (HOH **TELL**)
I — India (**IN** DEE AH)
J — Juliet (**JEW** LEE ETT)
K — Kilo (**KEY** LOH)
L — Lima (**LEE** MAH)
M— Mike (MIKE)
N — November (NO **VEM** BER)
O — Oscar (**OSS** CAH)
P — Papa (PAH **PAH**)
Q — Quebec (KEH **BECK**)
R — Romeo (**ROW** ME OH)
S — Sierra (SEE *AIR* RAH)
T — Tango (**TANG** GO)
U — Uniform (**YOU** NEE FORM or **OO** NEE FORM)
V — Victor (**VIK** TAH)
W— Whiskey (**WISS** KEY)
X — X-Ray (**ECKS** RAY)
Y — Yankee (**YANG** KEY)
Z — Zulu (**ZOO** LOO)

Note: The **Boldfaced** syllables are emphasized. The pronunciations shown in the table were designed for speakers from all international languages. The pronunciations given for "Oscar" and "Victor" may seem awkward to English-speaking people in the U.S.

ARRL Log

		See inside front cover.		Output in Watts.	UTC recommended.			RST. See back inside cover.	This column may also be used for contest-exchange info received.			

| | ----- FIXED ----- | | | | | | | ---- VARIABLE ---- | | | | |

DATE	FREQ.	MODE	POWER	TIME	STATION WORKED	REPORT SENT	REPORT REC'D	TIME OFF	QTH / NAME / QSL VIA			QSL S	QSL R
28 JUL	146.52	FM	10	0430	WA1CCR				Wallingford	Eric	New converter works!		
3 OCT	7.0	CW	150	2319	WA6VEF	001	322	Contra Cos	CALIFORNIA QSO PARTY				
				22	N6OJ	002	157	SONO					
				24	K6NA	003	331	SD					
				31	N6OP/m	004	117	Calav					
9 OCT	28.6	SSB	1 kW	0301	JA1OCA	59	57		Tokyo	Isao	Buro	✓	
	21	CW		1545	EA9GD	559	579		Melilla	Jose	Box 348	✓	✓
				56	6OØDX	599	599		Somalia		I2YAE	✓	
5 Nov	3.810.2	SSB	150	0030	W9NA	59+	59+	0117	Wausau, WI	Reno			
9 Nov	21	CW	10	1642	G4BUE	339	449	1657		1 watt!			

The ARRL Log is adaptable for all types of operating—ragchewing, contesting, DXing. References are to pages in the ARRL Log.

ARRL Operating Awards

Award	Qualification
Friendship Award	Contact 26 stations with calls ending A through Z.
Rag Chewer's Club	A single contact 1/2 hour or longer
Worked All States (WAS)	QSLs from all 50 US states
Worked All Continents (WAC)	QSLs from all six continents
DX Century Club (DXCC)	QSLs from at least 100 different countries
VHF/UHF Century Club (VUCC)	QSLs from many grid squares
A-1 Operator Club	Recommendation by two A-1 operators
Code Proficiency	One minute of perfect copy from W1AW qualifying run
Old Timers Club	Held an Amateur Radio license at least 20 years prior
ARRL Membership	ARRL membership for 25, 40, 50, 60 or 70 years

---

ARRL Membership QSL Card

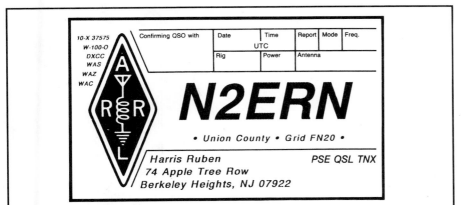

The ARRL membership QSL card. This example is from Harris Ruben, N2ERN, who designed the card. Your card would reflect your own call sign and address; awards and VUCC grid-square are optional. ARRL does not print or sell the cards. Inquire with printers who advertise in the *QST* Ham Ads.

Mode Abbreviations for QSL Cards

Abbreviation	Explanation
CW	Telegraphy
DATA	Telemetry, telecommand and computer communications (includes packet radio)
IMAGE	Facsimile and television
MCW	Tone-modulated telegraphy
PHONE	Speech and other sound
PULSE	Modulated main carrier
RTTY	Direct-printing telegraphy (includes AMTOR)
SS	Spread Spectrum
TEST	Emissions containing no information

Note: For additional information on emission types refer to latest edition of *The FCC Rule Book.*

US/Canada Map

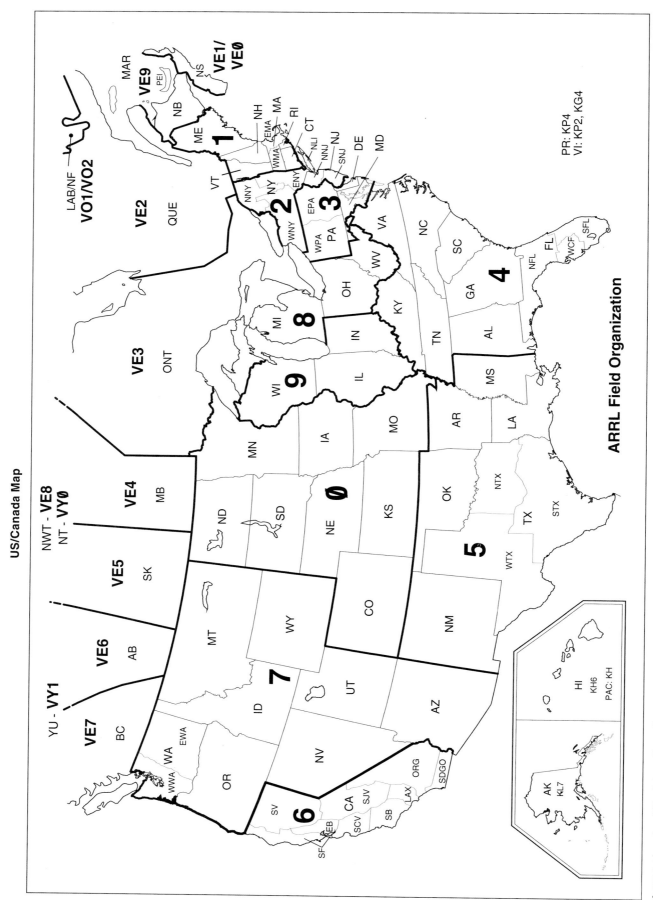

ARRL Field Organization

A map showing US states, Canadian provinces and ARRL/RAC Sections.

PR: KP4
VI: KP2, KG4

ARRL Grid Locator for North America

This ARRL Grid Locator map is based on the worldwide "Maidenhead" system. The first two characters (letters) constitute the 20° X 10° *field*. This is followed by two numbers designating the 2° X 1° *square*. To indicate location more precisely, 5th and 6th characters (letters) are used to indicate the 5' X 2.5' *subsquare*. More information on grid locations and the ARRL VHF UHF Century Club Awards (based on contacting 100 grid squares) can be obtained from the Headquarters of the American Radio Relay League, 225 Main Street, Newington, CT 06111, U.S.A.

Field boundary
Square boundary
International boundary
State and Provincial boundary

Additional Grid Square Designators

Alaska
Anchorage BP51
Fairbanks BP64
Juneau CO28

Canada
Charlottetown, PEI FN86
Edmonton, Alb DO33
Halifax, N S FN84
Saskatoon, Sask DO62
St Johns Nfl GN37

Caribbean
Havana EL83
Santo Domingo FK58
Port au Prince FK38
Guantanamo Bay FK29

Mexico
Mexico City EK09
Monterrey DL95

Puerto Rico and Virgin Islands
San Juan FK68
St Thomas and St John FK78
St Croix FK77

Cartography by
C. C. Roseman K9AKS and A Hobscheid
© 1983 The American Radio Relay League Inc

This and a World Grid Locator Map are available from ARRL.

AMATEUR MESSAGE FORM

Every formal radiogram message originated and handled should contain the following component parts in the order given.

I   PREAMBLE
- *a.* Number (begin with 1 each month or year)
- *b.* Precedence (R, W, P or EMERGENCY)
- *c.* Handling Instructions (optional, see text)
- *d.* Station of Origin (first amateur handler)
- *e.* Check (number of words/groups in text only)
- *f.* Place of Origin (not necessarily location of station of origin)
- *g.* Time Filed (optional with originating station)
- *h.* Date (must agree with date of time filed)

II   ADDRESS (as complete as possible, include zip code and telephone number)

III   TEXT (limit to 25 words or less, if possible)

IV   SIGNATURE

CW MESSAGE EXAMPLE

I   NR 1 R HXG W1AW 8 NEWINGTON CONN 1830Z JULY 1
    a   b c   d   e   f   g   h

II   DONALD SMITH $\overline{AA}$
    164 EAST SIXTH AVE $\overline{AA}$
    NORTH RIVER CITY MO 00789 $\overline{AA}$
    733 4968 $\overline{BT}$

III   HAPPY BIRTHDAY X SEE YOU SOON X LOVE $\overline{BT}$

IV   DIANA $\overline{AR}$

Note that X, when used in the text as punctuation, counts as a word.

CW: The prosign $\overline{AA}$ separates the parts of the address. $\overline{BT}$ separates the address from the text and the text from the signature. $\overline{AR}$ marks end of message; this is followed by B if there is another message to follow, by N if this is the only or last message. It is customary to copy the preamble, parts of the address, text and signature on separate lines.

RTTY: Same as cw procedure above, except (1) use extra space between parts of address, instead of $\overline{AA}$; (2) omit cw procedure sign $\overline{BT}$ to separate text from address and signature, using line spaces instead; (3) add a CFM line under the signature, consisting of all names, numerals and unusual works in the message in the order transmitted.

PACKET/AMTOR BBS: Same format as shown in the cw message example above, except that the $\overline{AA}$ and $\overline{AR}$ prosigns may be omitted. Most amtor and packet BBS software in use today allows formal message traffic to be sent with the "ST" command. Always avoid the use of spectrum-wasting multiple line feeds and indentations.

PHONE: Use *prowords* instead of prosigns, but it is not necessary to name each part of the message as you send it. For example, the above message would be sent on phone as follows: "Number one routine HX Golf W1AW eight Newington Connecticut one eight three zero zulu July one Donald Smith *Figures* one six four East Sixth Avenue North River City Missouri zero zero seven eight nine *Telephone* seven three three four nine six eight *Break* Happy birthday X-ray see you soon X-ray love *Break* Diana *End of Message* Over. "End of Message" is followed by "More" if there is another message to follow, "No More" if it is the only or last message. Speak clearly using VOX (or pause frequently on push-to-talk) so that the receiving station can get fills. Spell phonetically all difficult or unusual words—do *not* spell out common words. Do not use cw abbreviations or Q-signals in phone traffic handling.

PRECEDENCES

The precedence will follow the message number. For example, on cw 207 R or 207 EMERGENCY. On phone, "Two Zero Seven, Routine (or Emergency)."

EMERGENCY—Any message having life and death urgency to any person or group of persons, which is transmitted by Amateur Radio in the absence of regular commercial facilities. This includes official messages of welfare agencies during emergencies requesting supplies, materials or instructions vital to relief of stricken populace in emergency areas. During normal times, it will be *very rare*. On cw, RTTY and other digital modes this designation will always be spelled out. When in doubt, *do not* use it.

PRIORITY—Important messages having a specific time limit. Official messages not covered in the Emergency category. Press dispatches and other emergency-related traffic not of the utmost urgency. Notification of death or injury in a disaster area, personal or official. Use the abbreviation P on cw.

WELFARE—A message that is either a) an inquiry as to the health and welfare of an individual in the disaster area b) an advisory or reply from the disaster area that indicates all is well should carry this precedence, which is abbreviated W on cw. These messages are handled *after* Emergency and Priority traffic but before Routine.

ROUTINE—Most traffic normal times will bear this designation. In disaster situations, traffic labeled Routine (R on cw) should be handled *last*, or not at all when circuits are busy with Emergency, Priority or Welfare traffic.

Handling Instructions (Optional)

HXA—(Followed by number.) Collect landline delivery authorized by addressee within . . . . .miles. (If no number, authorization is unlimited.)

HXB—(Followed by number.) Cancel message if not delivered within . . . . .hours of filing time; service originating station.

HXC—Report date and time of delivery (TOD) to originating station.

HXD—Report to originating station the identity of station from which received, plus date and time. Report identity of station to which relayed, plus date and time, or if delivered report date, time and method of delivery.

HXE—Delivering station get reply from addressee, originate message back.

HXF—(Followed by number.) Hold delivery until . . . .(date).

HXG—Delivery by mail or landline toll call not required. If toll or other expense involved, cancel message and service originating station.

For further information on traffic handling, consult the Public Service Communications Manual or the ARRL Operating Manual, both published by ARRL.

A Simple NTS Formal Message

THIS IS A FORMAL MESSAGE. FORMAL MEANS THAT THE MESSAGE FOLLOWS A PRE-ESTABLISHED FORM OR CONVENTION. A FORMAL MESSAGE CONTAINS ALL THE NECESSARY "RECORDKEEPING" ELEMENTS THAT ARE REQUIRED TO KEEP A HISTORY OF THE MESSAGE AS IT IS SENT THROUGH THE NTS. ALL FORMAL MESSAGES CONSIST OF FOUR PARTS: THE PREAMBLE, THE ADDRESS, THE TEXT AND THE SIGNATURE.

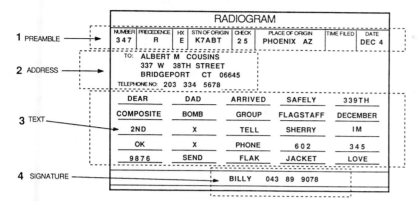

EACH OF THE ELEMENTS OF THE FORMAL MESSAGE HAS CERTAIN FORMAT REQUIREMENTS WHICH MUST BE MET IN ORDER TO AVOID CONFUSION ON THE AIR AS THE MESSAGE IS SENT, AND ALSO TO ASSURE THAT A SENDER-TO-RECEIVER TRACE CAN ALWAYS BE DONE ON THE MESSAGE.

Handling Instructions

HXA—(Followed by number.) Collect landline delivery authorized by addressee within _____ miles. (If no number, authorization is unlimited.)

HXB—(Followed by number.) Cancel messages if not delivered within _____ hours of filing time; service originating station.

HXC—Report date and time of delivery (TOD) to originating station.

HXD—Report to originating station the identity of station from which received, plus date and time. Report identity of station to which relayed, plus date and time, or if delivered report date, time and method of delivery.

HXE—Delivering station get reply from addressee, originate message back.

HXF—(Followed by number.) Hold delivery until _____ (date).

HXG—Delivery by mail or landline toll call not required. If toll or other expense involved, cancel message and service originating station.

An HX prosign (when used) will be inserted in the message preamble before the station of origin, thus: NR 207 R HXA50 W1AW 12...(etc). If more than one HX prosign is used they can be combined if no numbers are to be inserted; otherwise the HX should be repeated, thus: NR 207 R HXAC W1AW... (etc), but: NR 207 R HXA50 HXC W1AW...(etc). On phone, use phonetics for the letter or letters following the HX, to ensure accuracy.

ARL Numbered Radiograms

The letters ARL are inserted in the preamble in the check and in the text before spelled out numbers, which represent texts from this list. Note that some ARL texts include insertion of numerals. *Example:* NR 1 R W1AW ARL 5 NEWINGTON CONN DEC 25 DONALD R SMITH AA 164 EAST SIXTH AVE AA NORTH RIVER CITY MO AA PHONE 733 3968 BT ARL FIFTY ARL SIXTY ONE BT DIANA AR.

Group One—For possible "Relief Emergency" Use

ONE	Everyone safe here. Please don't worry.
TWO	Coming home as soon as possible.
THREE	Am in _____ hospital. Receiving excellent care and recovering fine.
FOUR	Only slight property damage here. Do not be concerned about disaster reports.
FIVE	Am moving to new location. Send no further mail or communication. Will inform you of new address when relocated.
SIX	Will contact you as soon as possible.
SEVEN	Please reply by Amateur Radio through the amateur delivering this message. This is a free public service.
EIGHT	Need additional _____ mobile or portable equipment for immediate emergency use.
NINE	Additional _____ radio operators needed to assist with emergency at this location.
TEN	Please contact _____. Advise to standby and provide further emergency information, instructions or assistance.
ELEVEN	Establish Amateur Radio emergency communications with _____ on _____ MHz.
TWELVE	Anxious to hear from you. No word in some time. Please contact me as soon as possible.
THIRTEEN	Medical emergency situation exists here.
FOURTEEN	Situation here becoming critical. Losses and damage from _____ increasing.
FIFTEEN	Please advise your condition and what help is needed.
SIXTEEN	Property damage very severe in this area.
SEVENTEEN	REACT communications services also available. Establish REACT communications with _____ on channel _____.
EIGHTEEN	Please contact me as soon as possible at _____.
NINETEEN	Request health and welfare report on _____. (State name, address and telephone number.)
TWENTY	Temporarily stranded. Will need some assistance. Please contact me at _____.
TWENTY ONE	Search and Rescue assistance is needed by local authorities here. Advise availability.
TWENTY TWO	Need accurate information on the extent and type of conditions now existing at your location. Please furnish this information and reply without delay.
TWENTY THREE	Report at once the accessibility and best way to reach your location.
TWENTY FOUR	Evacuation of residents from this area urgently needed. Advise plans for help.
TWENTY FIVE	Furnish as soon as possible the weather conditions at your location.
TWENTY SIX	Help and care for evacuation of sick and injured from this location needed at once.

Emergency/priority messages originating from official sources must carry the signature of the originating official.

Group Two—Routine messages

FORTY SIX	Greetings on your birthday and best wishes for many more to come.
FIFTY	Greetings by Amateur Radio.
FIFTY ONE	Greetings by Amateur Radio. This message is sent as a free public service by ham radio operators here at _____. Am having a wonderful time.
FIFTY TWO	Really enjoyed being with you. Looking forward to getting together again.
FIFTY THREE	Received your _____. It's appreciated; many thanks.
FIFTY FOUR	Many thanks for your good wishes.
FIFTY FIVE	Good news is always welcome. Very delighted to hear about yours.
FIFTY SIX	Congratulations on your _____, a most worthy and deserved achievement.
FIFTY SEVEN	Wish we could be together.
FIFTY EIGHT	Have a wonderful time. Let us know when you return.
FIFTY NINE	Congratulations on the new arrival. Hope mother and child are well.
*SIXTY	Wishing you the best of everything on _____.
SIXTY ONE	Wishing you a very merry Christmas and a happy New Year.
*SIXTY TWO	Greetings and best wishes to you for a pleasant _____ holiday season.
SIXTY THREE	Victory or defeat, our best wishes are with you. Hope you win.
SIXTY FOUR	Arrived safely at _____.
SIXTY FIVE	Arriving _____ on _____. Please arrange to meet me there.
SIXTY SIX	DX QSLs are on hand for you at the _____ QSL Bureau. Send _____ self-addressed envelopes.
SIXTY SEVEN	Your message number _____ undeliverable because of _____. Please advise.
SIXTY EIGHT	Sorry to hear you are ill. Best wishes for a speedy recovery.
SIXTY NINE	Welcome to the _____. We are glad to have you with us and hope you will enjoy the fun and fellowship of the organization.

* Can be used for all holidays.

Note: ARL numbers should be spelled out at all times.

Checking Your Message

Traffic handlers don't have to dine out to fight over the check! Even good ops find much confusion when counting up the text of a message. You can eliminate some of this confusion by remembering these basic rules:

1) Punctuation ("X-rays," "Querys") count separately as a word.
2) Mixed letter-number groups (1700Z, for instance) count as one word.
3) Initial or number groups count as one word if sent together, two if sent separately.
4) The signature does not count as part of the text, but any closing lines, such as "Love" or "Best wishes" do.

Here are some examples:
- Charles J McClain—3 words
- W B Stewart—3 words
- St Louis—2 words
- 3 PM—2 words
- SASE—1 word
- ARL FORTY SIX—3 words
- 2N1601—1 word
- Seventy-three—2 words
- 73—1 word

Telephone numbers count as 3 words (area code, prefix, number), and ZIP codes count as one, ZIP + 4 codes count as two words. Canadian postal codes count as two words (first three characters, last three characters.)

Although, it is improper to change the text of a message, you may change the check. Always do this by following the original check with a slash bar, then the corrected check. On phone, use the words "corrected to."

How to be the Kind of Net Operator the Net Control Station (NCS) Loves

As a net operator, you have a duty to be self-disciplined. A net is only as good as its worst operator. You can be an exemplary net operator by following a few easy guidelines.

1) *Zero beat the NCS.* The NCS doesn't have time to chase all over the band for you. Make sure you're on frequency, and you will never be known at the annual net picnic as "old so-and-so who's always off frequency."

2) *Don't be late.* There's no such thing as "fashionably late" on a net. Liaison stations are on a tight timetable. Don't hold them up by checking in 10 minutes late with three pieces of traffic.

3) *Speak only when spoken to by the NCS.* Unless it is a bona fide emergency situation, you don't need to "help" the NCS unless asked. If you need to contact the NCS, make it brief. Resist the urge to help clear the frequency for the NCS or to "advise" the NCS. The NCS, not you, is boss.

4) Unless otherwise instructed by the NCS, *transmit only to the NCS.* Side comments to another station in the net are out of order.

5) *Stay until you are excused.* If the NCS calls you and you don't respond because you're getting a "cold one" from the fridge, the NCS may assume you've left the net, and net business may be stymied. If you need to leave the net prematurely, contact the NCS and simply ask to be excused (QNX PSE ON CW).

6) *Be brief when transmitting to the NCS.* A simple "yes" (C) or "no" (N) will usually suffice. Shaggy dog tales only waste valuable net time.

7) *Know how the net runs.* The NCS doesn't have time to explain procedure to you. After you have been on the net for a while, you should already know these things.

Templates

Template packages are available for many *ARRL Handbook* projects and some text discussions. They may include full-size etching patterns, more detailed information on a subject, author's updates and other useful information. They are updated as new information is received at ARRL Headquarters. The templates for many *ARRL Handbook* projects are located in this section. These are full-size etching patterns, with parts-placement diagrams for some of them.

All of the templates are provided free in **Adobe Acrobat** format on *ARRLWeb:* **http://www.arrl.org/notes/1867**. In addition, the template packages are available from the Technical Secretary, ARRL HQ, 225 Main Street, Newington, CT 06111-1494, USA. The cost for template packages by mail that are *not* included in this section of the *ARRL Handbook* are given in the following Table. Please include a check (payable in US funds) with your order, made out to ARRL. A self-addressed 9 × 12 envelope will help expedite your request. You may also call between 8 AM and 4 PM Eastern time, with your credit card number. Voice 860-594-0278. Fax 860-594-0259.

Template Reference Chapter	Name	Cost Member/Nonmember
13	Wingfield Tables	$2/$4
16	Crystal Filter Evaluation	$2/$4
16	DSP-3 Filter	$2/$4
17	NorCal Sierra	$3/$5
20	Stanley SWR/Z	$2/$4
22	Taggart CW Interface	$2/$4
23	Repeater Frequency Coordination and Band Plans	$2/$4
23	S-Band Converter	$2/$4
23	L-Band Satellite Antenna/Amplifier	$2/$4
26	Six Digit Programmable Frequency Counter/Digital Dial	$2/$4
27	Dip Meter Sources	$2/$4
30	Step Attenuator	$2/$4

Sabin Power Supply

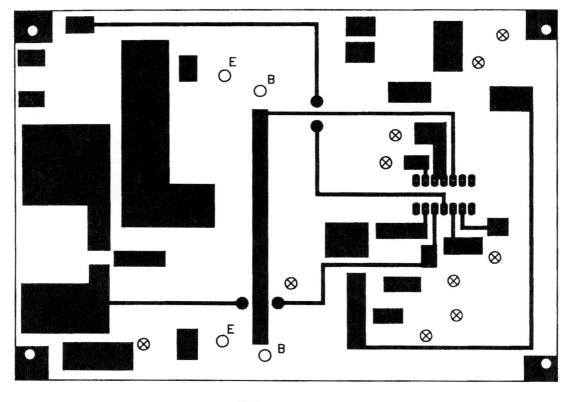

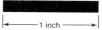
1 inch

A Series Regulated 4.5 to 25-V, 2.5-A Power Supply, Chapter 11

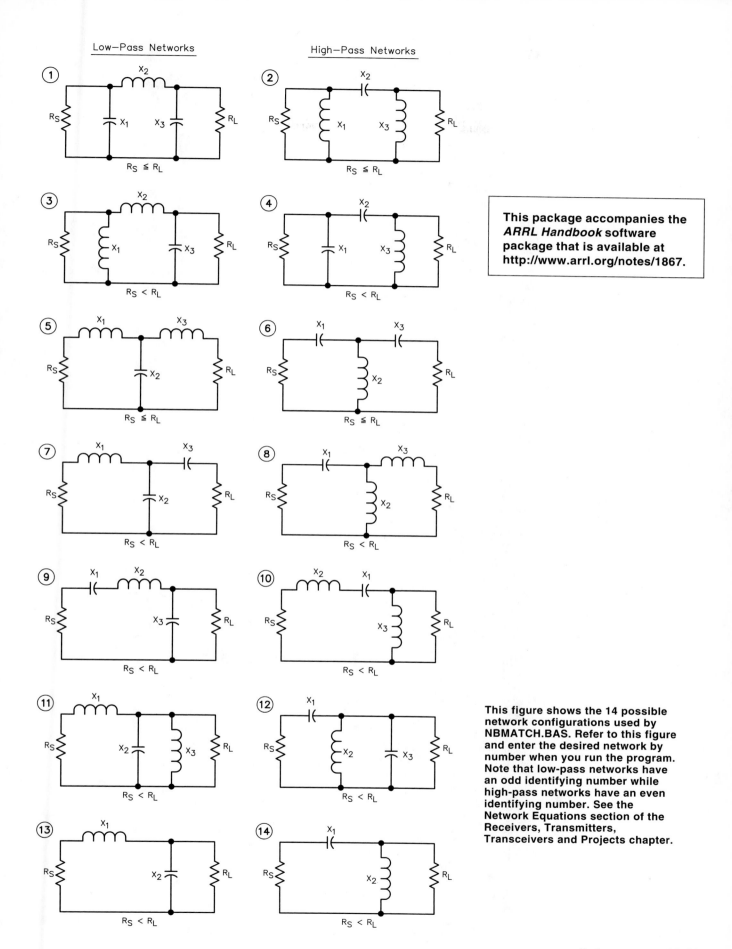

This package accompanies the *ARRL Handbook* software package that is available at http://www.arrl.org/notes/1867.

This figure shows the 14 possible network configurations used by NBMATCH.BAS. Refer to this figure and enter the desired network by number when you run the program. Note that low-pass networks have an odd identifying number while high-pass networks have an even identifying number. See the Network Equations section of the Receivers, Transmitters, Transceivers and Projects chapter.

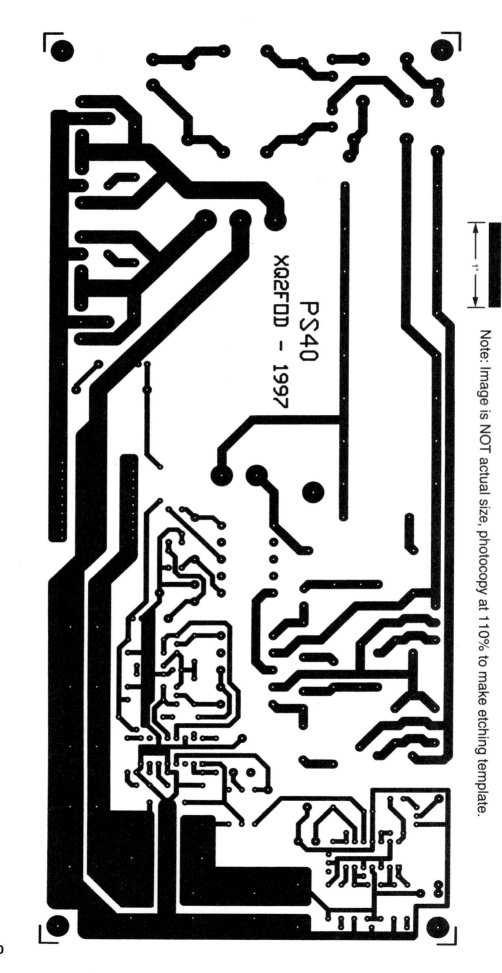

PS40
XQ2FDD - 1997

Note: Image is NOT actual size, photocopy at 110% to make etching template.

1"

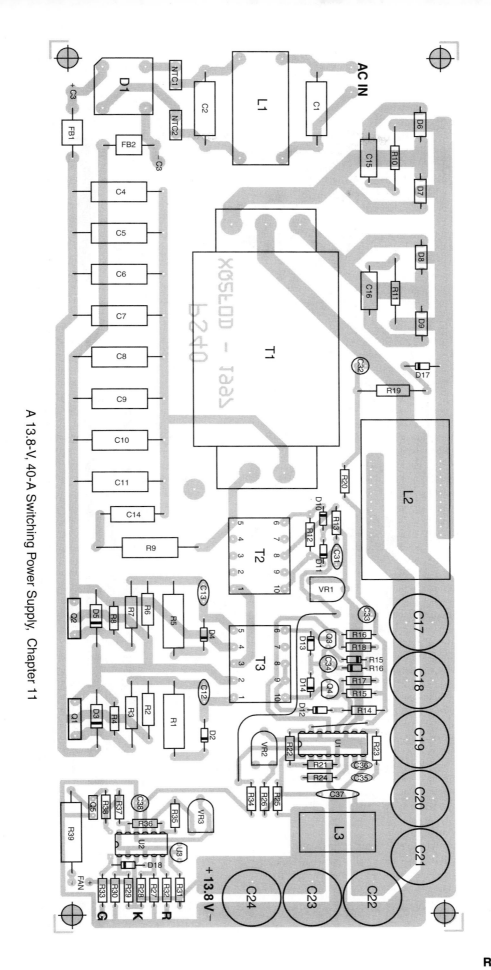

A 13.8-V, 40-A Switching Power Supply, Chapter 11

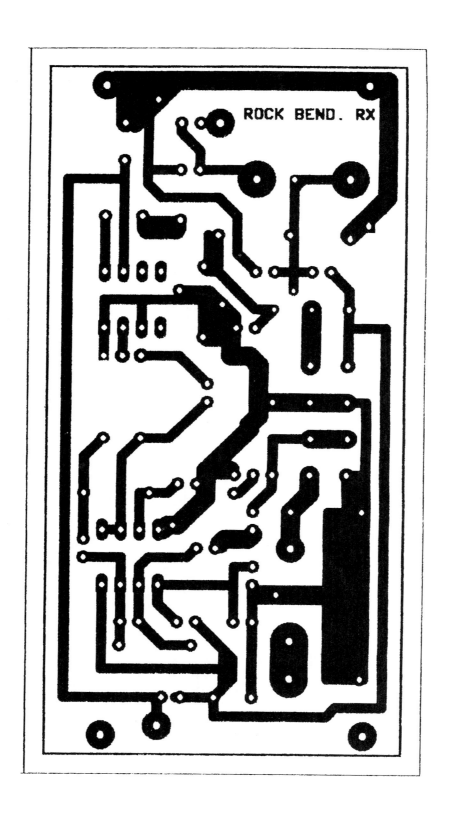

A Rock Bending Receiver for 7 MHz, Chapter 17

A Rock Bending Receiver for 7 MHz, Chapter 17

R12 (10 Ω) MUST BE ADDED IN SERIES WITH C18

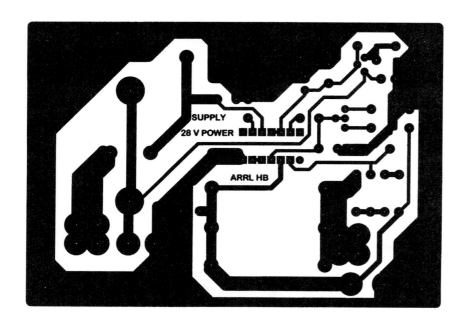

1"

PC Template for the "28-V, High-Current Supply," Chapter 11

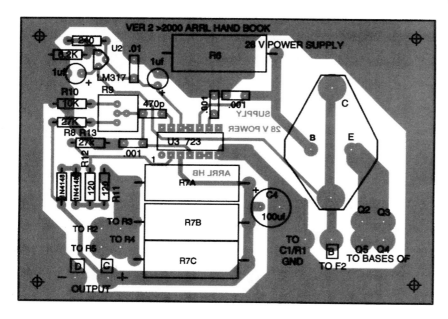

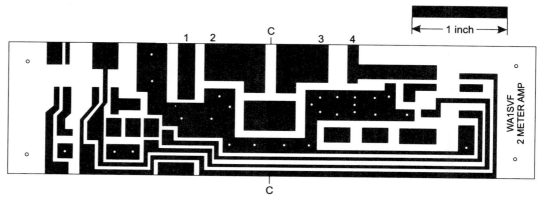

1 inch

PC Template for "A 2-M Brick Amp for Handhelds," Chapter 13

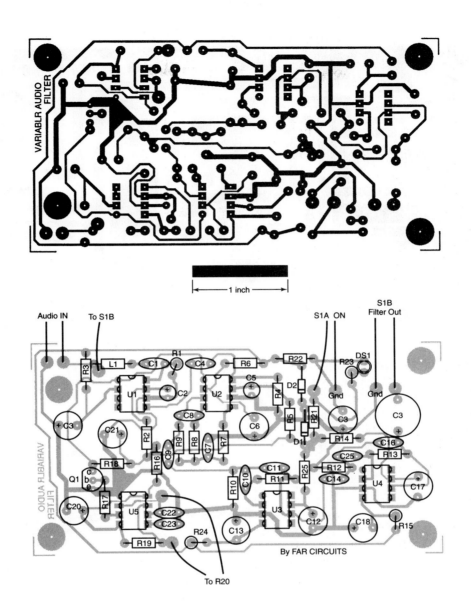

PC template and layout "A Continuously Variable Bandwidth Audio Filter," Chapter 16

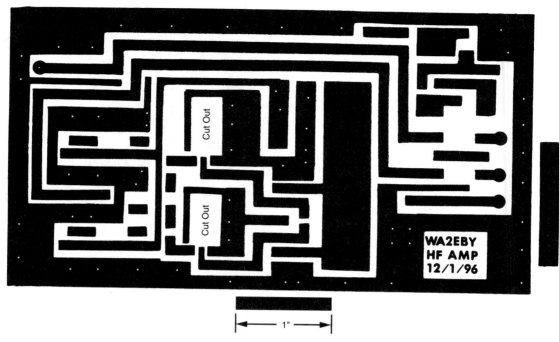

Amplifier Board Etching Pattern

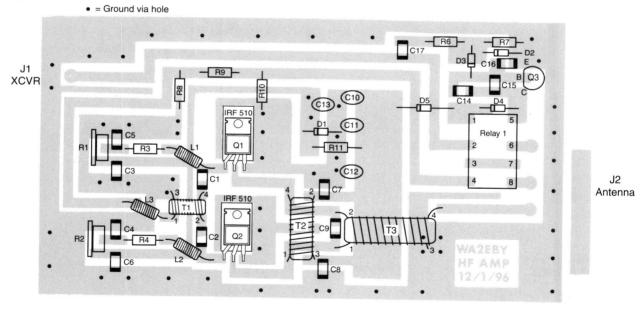

Amplifier Board Parts-Placement Diagram

This is a double-sided circuit board with the second side being a solid ground plane. All components mount to the top of the board, as shown in the parts-placement diagram. Components with wire leads are mounted by bending their leads and soldering directly to the circuit traces. Rectangular cutouts allow Q1 and Q2 to lie flat on the board. Holes are for the insertion of wire jumpers. Solder to both the top trace and the bottom ground foil.

PC template and layout "Broadband HF Amplifier," Chapter 17

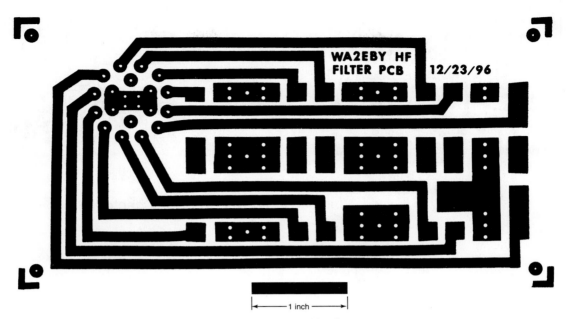

Filter Board Etching Pattern

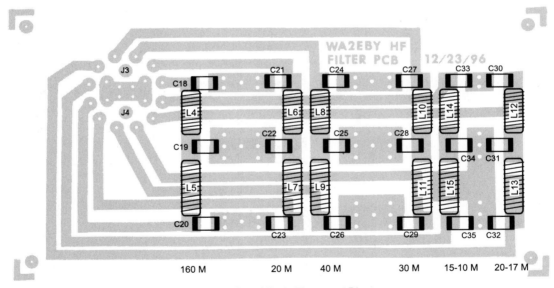

Filter Board Parts-Placement Diagram

This is a double-sided circuit board with the scond side being a solid ground plane. All components mount to the top of the board, as shown in the parts-placement diagram. Use a ¼ inch drill bit to remove the copper around each switch pin (but don't drill all the way through the circuit board with that bit!). The other holes are for the insertion of wire jumpers, soldered to both the top trace and the bottom ground foil.

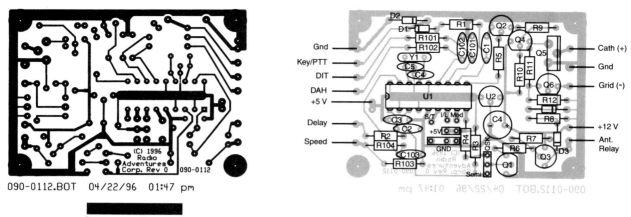

Vintage Radio T/R Adapter, Chapter 22

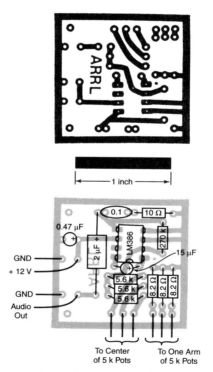

Expandable Headphone Mixer,
Chapter 22

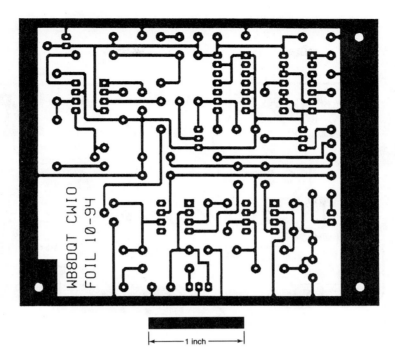

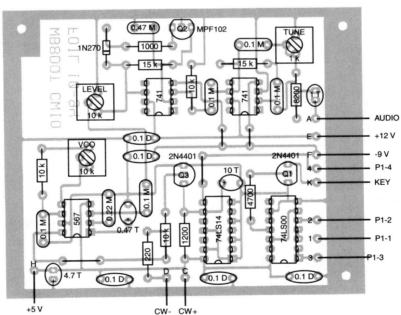

Quick and Easy CW with Your PC, Chapter 22

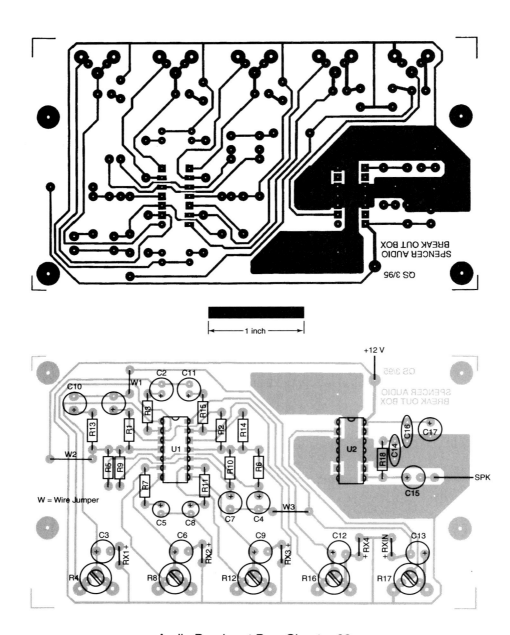

Audio Break-out Box, Chapter 22

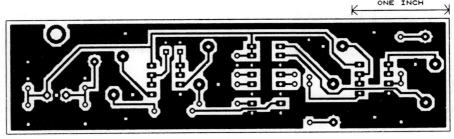

THIS READS CORRECTLY FROM SOLDER SIDE

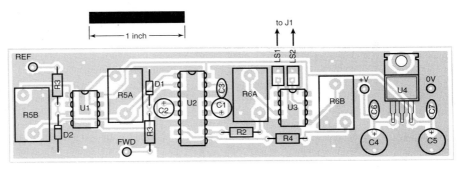

An SWR Detector Audio Adapter, Chapter 22

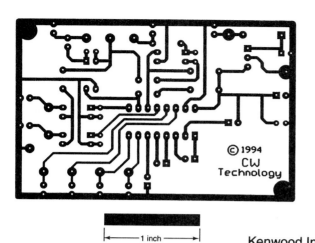

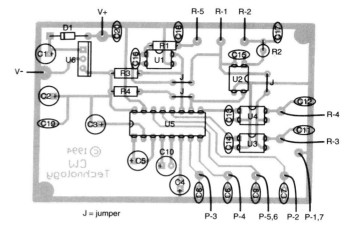

J = jumper

Kenwood Interface, Chapter 22

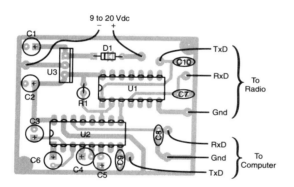

1 inch

ICOM / Ten-Tec / Yaesu Interface, Chapter 22

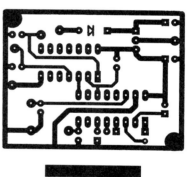

1 inch

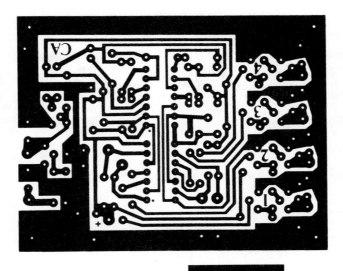

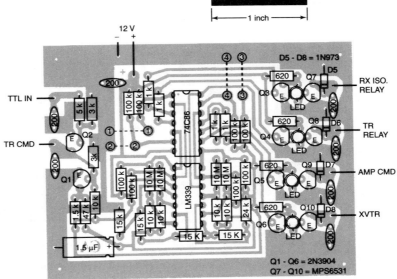

TR Time-Delay Generator, Chapter 22

**Parts placement diagram of the TR time-delay generator
and full-size etching pattern.**

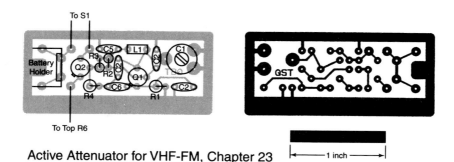

Active Attenuator for VHF-FM, Chapter 23

1 inch

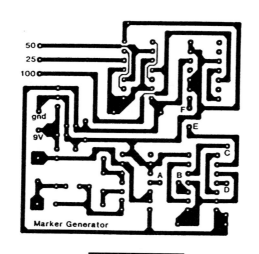

Marker Generator

1 inch

Parts-placement diagram and full-size etching pattern for the marker generator PC board. Consult Table 26.2 for information on placing the jumpers at points A, B, C, D, E and F. Extra Pads in the oscillator section of the board allow for the use of a crystal in a small HC-25 holder or a larger HC-33 or HC-6 holder. Pads are also provided for different sizes of trimmer capacitor.

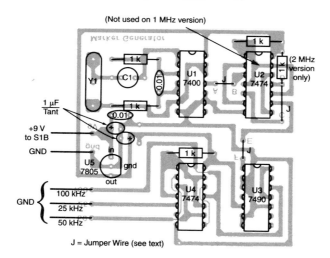

Marker Generator with Selectable Output, Chapter 26

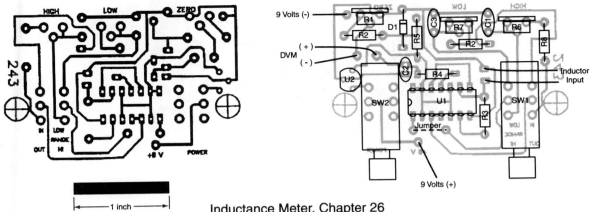

Inductance Meter, Chapter 26

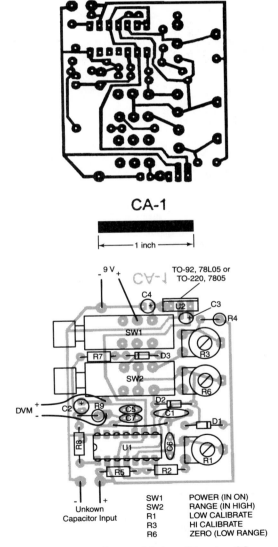

CA-1

Capacitance Meter, Chapter 26

SW1	POWER (IN ON)
SW2	RANGE (IN HIGH)
R1	LOW CALIBRATE
R3	HI CALIBRATE
R6	ZERO (LOW RANGE)

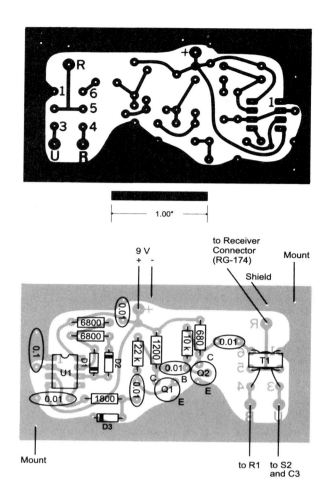

A Noise Bridge for 1.8 Through 30 MHz,
Chapter 26

**Parts-placement guide for the noise bridge as viewed
from the component or top side of the board.
Mounting holes are located in two corners of the
boards, as shown.**

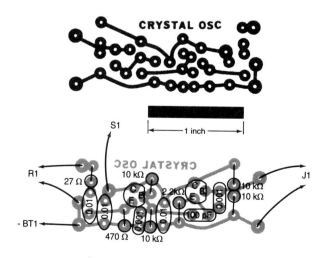

Crystal-Controlled Signal Source, Chapter 27

About the ARRL

The national association for Amateur Radio

The seed for Amateur Radio was planted in the 1890s, when Guglielmo Marconi began his experiments in wireless telegraphy. Soon he was joined by dozens, then hundreds, of others who were enthusiastic about sending and receiving messages through the air—some with a commercial interest, but others solely out of a love for this new communications medium. The United States government began licensing Amateur Radio operators in 1912.

By 1914, there were thousands of Amateur Radio operators—hams—in the United States. Hiram Percy Maxim, a leading Hartford, Connecticut, inventor and industrialist saw the need for an organization to band together this fledgling group of radio experimenters. In May 1914 he founded the American Radio Relay League (ARRL) to meet that need.

Today ARRL, with approximately 170,000 members, is the largest organization of radio amateurs in the United States. The ARRL is a not-for-profit organization that:

• promotes interest in Amateur Radio communications and experimentation
• represents US radio amateurs in legislative matters, and
• maintains fraternalism and a high standard of conduct among Amateur Radio operators.

At ARRL headquarters in the Hartford suburb of Newington, the staff helps serve the needs of members. ARRL is also International Secretariat for the International Amateur Radio Union, which is made up of similar societies in 150 countries around the world.

ARRL publishes the monthly journal *QST*, as well as newsletters and many publications covering all aspects of Amateur Radio. Its headquarters station, W1AW, transmits bulletins of interest to radio amateurs and Morse code practice sessions. The ARRL also coordinates an extensive field organization, which includes volunteers who provide technical information and other support for radio amateurs as well as communications for public-service activities. ARRL also represents US amateurs with the Federal Communications Commission and other government agencies in the US and abroad.

Membership in ARRL means much more than receiving *QST* each month. In addition to the services already described, ARRL offers membership services on a personal level, such as the ARRL Volunteer Examiner Coordinator Program and a QSL bureau.

Full ARRL membership (available only to licensed radio amateurs) gives you a voice in how the affairs of the organization are governed. ARRL policy is set by a Board of Directors (one from each of 15 Divisions). Each year, one-third of the ARRL Board of Directors stands for election by the full members they represent. The day-to-day operation of ARRL HQ is managed by an Executive Vice President and a Chief Financial Officer.

No matter what aspect of Amateur Radio attracts you, ARRL membership is relevant and important. There would be no Amateur Radio as we know it today were it not for the ARRL. We would be happy to welcome you as a member! (An Amateur Radio license is not required for Associate Membership.) For more information about ARRL and answers to any questions you may have about Amateur Radio, write or call:

ARRL—The national association for Amateur Radio
225 Main Street
Newington CT 06111-1494
(860) 594-0200

Prospective new amateurs call:
800-32-NEW HAM (800-326-3942)
You can also contact us via e-mail: **newham@arrl.org**
or check out *ARRLWeb:* **http://www.arrl.org/**

Index

FEEDBACK

Please use this form to give us your comments on this book and what you'd like to see in future editions, or e-mail us at **pubsfdbk@arrl.org** (publications feedback). If you use e-mail, please include your name, call, e-mail address and the book title, edition and printing in the body of your message. Also indicate whether or not you are an ARRL member.

Where did you purchase this book?
☐ From ARRL directly   ☐ From an ARRL dealer

Is there a dealer who carries ARRL publications within:
☐ 5 miles   ☐ 15 miles   ☐ 30 miles   of your location?   ☐ Not sure.

License class:
☐ Novice   ☐ Technician   ☐ Technician with code   ☐ General   ☐ Advanced   ☐ Amateur Extra

---

Name _____   ARRL member? ☐ Yes ☐ No

_____   Call Sign _____

Daytime Phone (   ) _____   Age _____

Address _____

City, State/Province, ZIP/Postal Code _____

If licensed, how long? _____   e-mail address: _____

Other hobbies _____

_____

Occupation _____

For ARRL use only	2001 HBK
Edition	78 79 80
Printing	1 2 3 4 5 6 7 8 9 10 11 12

From _____

_____

_____

EDITOR, ARRL HANDBOOK
ARRL—THE NATIONAL ASSOCIATION FOR AMATEUR RADIO
225 MAIN STREET
NEWINGTON CT 06111-1494

— — — — — — — — — — — — — please fold and tape — — — — — — — — — — —

THE AMERICAN RADIO RELAY LEAGUE, INC

225 MAIN STREET **NEWINGTON, CONNECTICUT 06111 USA**

Call toll free in the US to join: 1-888-277-5289   World Wide Web: http://www.arrl.org/
Outside the US call 860-594-0200   Fax 860-594-0303

A bona fide interest in Amateur Radio is the only essential requirement, but full voting membership is granted only to licensed radio amateurs of the US and possessions. Therefore, if you have a license, please be sure to indicate it below. Please Print.

❏ New member   ❏ Previous member   ❏ Renewal   ❏ Not currently licensed

Class of License	Call Sign	Date of Birth

Name

Address

City, State, ZIP

Membership Class	❏ Regular		❏ Family ❏ Blind	❏ 65 or older
	US AND POSSESSIONS	CANADA ELSEWHERE		IN THE US AND POSSESSIONS
❏ 1 year	$ 34	$ 47 $ 54	$ 5	$ 25
❏ 2 years	65	89 103	10	53
❏ 3 years	92	127 146	15	76

*These rates include the postage surcharge which partially offsets the additional cost to mail *QST* outside the US. Write for airmail rates.

IMPORTANT: Please attach your Expiration Notice to this form if you are renewing your current membership. Payment in US funds only. Checks must be drawn on a bank within the US.

A member of the immediate family of a League member, living at the same address, may become a League member without *QST* at the special rate of $5 per year. Family membership must run concurrently with that of the member receiving *QST*. Blind amateurs may join without *QST* for $5 per year.

Persons who are age 65 or older residing in the US may upon request apply for League membership at the reduced rates shown. Please be sure to provide your date of birth. If you are age 21 or younger, a special rate may apply. Please contact the Circulation Department for details.

Your membership card will be mailed to you about 2 weeks from the date we receive your application. Delivery of *QST* may take slightly longer, but future issues should reach you on a regular basis. Membership is available only to individuals. Fifty percent of dues is allocated to *QST*, and the balance for membership.

DUES ARE SUBJECT TO CHANGE WITHOUT NOTICE.

If you do not wish your name and address made available for non-ARRL related mailings, please check this box ❏.

HB/00

I am donating $ _____ ($1 minimum) to the Legal Research & Resource Fund.

JOIN ARRL
THE LARGEST ORGANIZATION OF RADIO AMATEURS IN THE US
- Receive Monthly *QST* Journal
- Access to Members Only Web Site
- Use QSL Bureau–Clearing House for Overseas QSL cards
- Equipment Insurance Available for Home and Car
- Representation in Washington for All Matters Concerning Amateur Radio
- League-Member-Only Awards

Payment enclosed ❏

Charge to
❏ VISA (13 or 16 digits)   ❏ MasterCard (16 digits)
❏ AMEX (15 digits)   ❏ Discover (16 digits)

— — — — — — — — — — — — — — — —
1 2 3 4 5 6 7 8 9 10 11 12 13 14 15 16
Card number

Expiration date _____

Signature _____
List ARRL family members:

Name _____ Call sign _____

Name _____ Call sign _____

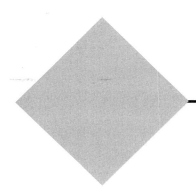

2001 ARRL Handbook
Companion Software

See page viii for software availability.

A2D.BAS runs the "PC Voltmeter and SWR Bridge" in the Station Accessories chapter. It accepts analog input and sends the digital version to your computer printer port.

ACTFILT.EXE designs some simple active filters. See "Active Filters" in the Filters chapter for details of the filters.

CTL3DV2.DLL is a Windows DLL that may be needed with TISFIND. See TISFIND below.

DSP.EXE contains *NWTNSQRT.BAS*, a program that uses Newton's method for determining square roots; *FSQRT.TXT*, a machine-language routine for doing fast square roots and *DSP.DOC*, a file that gives more information about the first two.

ELLER.EXE consists of program and support files to calculate placement and length for a coil-loaded shortened dipole using an existing pair of inductors. See "Computer-Aided Design of Loaded Short-Doublet Antennas," in the Antennas chapter.

INSTALL.EXE reads this file and installs the software on a hard disk. TISFIND requires a separate installation under Windows. See instructions under TISFIND.

GRIDLOC.BAS converts lattiude/longitude coordinates to four-character (such as FN42) grid square locations. *GRIDLOC.EXE* is a compiled version of the .*BAS* program.

MATCH.EXE is a Visual Basic program to design the 14 NBMATCH networks described in the Receivers, Transmitters, Transceivers and Projects chapter.

MSVBVM50.DLL is a Windows DLL that may be needed with Windows95 systems to run MATCH.EXE.

MORSE.EXE sends and receives CW. See "Quick and Easy CW with your PC" in the Station Accessories chapter for details.

SHADOW.EXE locates local true north in order to orient directional antennas. See "North Shadow" in the Antennas and Projects chapter for details.

PWRSPLY.BAS calculates the characteristics of various power supplies and regulators. *PWRSPLY.EXE* is a compiled version of the .*BAS* program.

PINET.EXE consists of several programs to aid design of Pi and Pi-L matching networks, as described under "Tank Output Circuits" in the Amplifiers chapter.

PRODREV.ADB is the *QST* Product Review database, a comprehensive listing of reviews from 1970 to the present. *TISFIND.EXE* (see below) is used to search this data.

README.TXT is the README file for the disk.

SAFETY.TXT consists of the complete list of RF Safety references.

SVCFILT.EXE is a program that designs passive element filters using standard value capacitors. A complete help file is included.

TIS.EXE contains TISFIND.EXE (and related files), a Windows database look-up program used to view databases distributed by ARRL. Included with TISFIND is the ARRL TIS Address Database, which contains address and contact information for over 1000 companies and organizations of interest to amateurs.

TLI.EXE contains the files TL.EXE and TL.DOC. TL.EXE computes many parameters for transmission lines, as well as for antenna tuners—including losses and stresses. See the Transmission Lines chapter.

TUNER.TXT contains tables that show losses computed under various load conditions with the new high-power antenna tuner.

UTCZONE.BAS is a handy program that gives the UTC zone and time offset for each zone. *UTCZONE.EXE* is a compiled version of the .*BAS* program.

VESTER_F.EXE contains the many files that make up the slow-scan TV system described in the Modulation Sources chapter. Be sure to read the README files.